ASTRONOMY AND ASTROPHYSICS ABSTRACTS

A Publication of the Astronomisches Rechen-Institut Heidelberg
Member of the Abstracting Board
of the International Council of Scientific Unions

Volume 20
Literature 1977, Part 2

Edited by
S. Böhme U. Esser W. Fricke I. Heinrich
D. Krahn L. D. Schmadel G. Zech

Springer-Verlag Berlin Heidelberg GmbH 1978

Astronomisches Rechen-Institut Heidelberg
Director: Professor Dr. Walter Fricke

Astronomy and Astrophysics Abstracts
Editors-in-Chief: Ute Esser, Dr. Lutz D. Schmadel

Astronomy and Astrophysics Abstracts is prepared
under the auspices of the International Astronomical Union

ISBN 978-3-662-12312-6   ISBN 978-3-662-12310-2 (eBook)
DOI 10.1007/978-3-662-12310-2

Library of Congress Catalog Card Number 72-104650.

Preface

Astronomy and Astrophysics Abstracts, which has appeared in semi-annual volumes since 1969, is devoted to the recording, summarizing and indexing of astronomical publications throughout the world. It is prepared under the auspices of the International Astronomical Union (according to a resolution adopted at the 14th General Assembly in 1970).

Astronomy and Astrophysics Abstracts aims to present a comprehensive documentation of literature in all fields of astronomy and astrophysics. Every effort will be made to ensure that the average time interval between the date of receipt of the original literature and publication of the abstracts will not exceed eight months. This time interval is near to that achieved by monthly abstracting journals, compared to which our system of accumulating abstracts for about six months offers the advantage of greater convenience for the user.

Volume 20 contains literature published in 1977 and received before February 20, 1978; some older literature which was received late and which is not recorded in earlier volumes is also included.

We acknowledge with thanks contributions to this volume by Dr. J. Bouška, Prague, who surveyed journals and publications in Czech and supplied us with abstracts in English, and by Prof. P. Brosche, Bonn, who supplied us with literature concerning some border fields of astronomy.

It is a pleasure to thank Dr. Ulrich Güntzel-Lingner for his valuable contributions. From 1969 to his retirement he was co-editor of *Astronomy and Astrophysics Abstracts.* We express our warmest thanks again to Ms. Helga Ballmann, Ms. Monika Betz, Ms. Lore Kiefert, who typed the text of this volume on IBM 72 Composers and compiled the pages from abstract slips in a perfect form for offset reproduction. We are indebted to Ms. Elisabeth Feigenbutz for punching material for the author index and for the subject index which finally were printed with a TN chain on a 1403 IBM high-speed printer. Finally we have to thank Mr. R. Jährling and Mr. W. Sanns who supported our task by careful proofreading.

Heidelberg, April 1978

Siegfried Böhme Dietlinde Krahn
Ute Esser Lutz D. Schmadel
Walter Fricke Gert Zech
Inge Heinrich

Contents

VIII   Contents

X   Contents

Introduction

Astronomical bibliographies

Astronomy and Astrophysics Abstracts begins documentation and abstracting from the year 1969. For information on astronomical literature before this date consultation of one of the following bibliographies is suggested:

(1) J. J. de Lalande, Bibliographie Astronomique, Paris 1803 (this work covers the time from 480 B. C. to the year 1803, VIII + 966 pages).

(2) J. C. Houzeau, A. Lancaster, Bibliographie générale de l'astronomie, Volume I (in two parts), Bruxelles 1882, 1887, Volume II, Bruxelles 1889. The complete title of Volume II is "Bibliographie générale de l'astronomie ou catalogue méthodique des ouvrages, des mémoires et des observations astronomiques, publiés depuis l'origine de l'imprimerie jusqu'en 1880". A new edition of these volumes was prepared by D. W. Dewhirst in 1964.

(3) Bibliography of Astronomy, 1881 - 1898. The literature of this period was recorded on standard slips by the Observatoire Royal de Belgique. From the material (some 52,000 items) a microfilm version was produced by University Microfilms Limited, Tylers Green, High Wycombe, Buckinghamshire, England, in 1970.

(4) Astronomischer Jahresbericht, 1899 gegründet von Walter Wislicenus, herausgegeben vom Astronomischen Rechen-Institut in Heidelberg (formerly in Berlin), Verlag W. de Gruyter, Berlin. For the period from 1899 to 1968 sixty-eight volumes were published, each of which, in general, covers the literature of one year.

(5) Bulletin Signalétique – Section 120: Astronomie, Physique Spatiale, Géophysique. Published by Centre de Documentation du Centre National de la Recherche Scientifique, Paris. This publication is a continuation of "Bibliographie Mensuelle de l'Astronomie" founded in 1933 by the Société Astronomique de France. The publication is continued.

(6) Referativnyj Zhurnal. Founded in 1953 and published by Vsesoyuznyj Institut Nauchnoj i Tekhnicheskoj Informatsii, Akademiya Nauk, Moskva. The publication is continued.

Concept of Astronomy and Astrophysics Abstracts

This abstracting service aims to present a comprehensive documentation of the literature in all fields of astronomy and astrophysics and their border fields. It appears in semiannual volumes. Two of these volumes cover the literature of one calendar year. The half-yearly period of issue is regarded as an optimal period for summarizing papers into subject categories and for the presentation of abstracts as quickly as possible after the publication of the original literature.

The recording, summarizing and indexing of astronomical publications of the year 1977 received from July 1977 to February 1978 are subjects of **Volume 20**. It also records a number of papers issued before 1977 but received within this period.

The main characteristics of the concept of Astronomy and Astrophysics Abstracts may be summarized as follows:
(1) The subdivision of astronomy and its border fields into subject categories is facilitated by the fact that the astronomical objects appear to be particularly well suited for the formation of categories. It may be assumed that such subdivisions can be maintained for a long period. Experience shows, however, that progress in research might imply minor changes in the classification scheme.

(2) Each paper has been classified into one of 109 numbered subject categories and given a serial number within the category. In this way each item is numbered by six figures: the first three indicate the number of the category, the following three the serial number within the category. Reference to an abstract in Volume 1 is indicated by "01" before the number of the category; for example: 01.074.028, denotes Volume 1, category 074, abstract 028.

A paper might be classified into more than one category. In this case, its abstract is placed only in one category, whereas in the other categories only cross references are given. These are listed at the end of each category.

(3) Authors' abstracts are used whenever possible. Popular articles are not abstracted.

(4) If possible, titles of papers and abstracts are given in English. A special reference is made to titles which we have not taken in the original language.

Transliteration scheme for the Russian alphabet

The transliteration of the Russian alphabet in use in Astronomy and Astrophysics Abstracts is presented here.

А	а	a	Р	р	r
Б	б	b	С	с	s
В	в	v	Т	т	t
Г	г	g	У	у	u
Д	д	d	Ф	ф	f
Е	е	e	Х	х	kh
Ё	ё	e	Ц	ц	ts
Ж	ж	zh	Ч	ч	ch
З	з	z	Ш	ш	sh
И	и	i	Щ	щ	shch
Й	й	j	Ъ	ъ	''
К	к	k	Ы	ы	y
Л	л	l	Ь	ь	'
М	м	m	Э	э	eh
Н	н	n	Ю	ю	yu
О	о	o	Я	я	ya
П	п	p			

This transliteration was recommended by the Abstracting Board of the International Council of Scientific Unions in 1969. It corresponds essentially to the transliteration proposed by the Academy of Sciences, Moscow, which is used by the Referativnyj Zhurnal. In this case the letters can be read and printed by usual data processing machines.

If the names of Russian authors in the literature are transliterated in a different scheme, we present the names as they are given in the references cited and in addition in brackets according to our transliteration table.

Sources of information

The majority of sources of information for this volume is given in section **001 Periodicals** and in section **008 Observatories, Institutes.** Section 001 records 579 periodicals indicating full titles and publishers. It may be noted that the titles of the periodicals are given in the original languages, and that Russian titles have been transliterated applying the transliteration scheme given above. Section 008 records 100 periodicals; these are publication series of observatories and astronomical institutes. Titles of the periodicals have been given following the recommendations of the "International List of Periodical Title Word Abbreviations" and its additions (see also **Abbreviations**, p. 3). In most cases they permit recognition of the full title without recourse to the key in section 001.

If other abstracting journals have been consulted in order to examine the degree of completeness of our service, we cite these papers and give reference to the abstracting service.

Author index and subject index

The subject category and the serial number have been used as a reference both in the author index and the subject index. These references are more precise than page references. They offer considerable advantages in indexing by means of data processing machines, and are more convenient for the user.

The author index of this volume contains 9310 names. A complete reference comprises six figures, three for the subject category and three for the serial number within the category. In the case of more than one reference to abstracts in one category, the number of the category is given only once and not repeated in the immediately following references. The total number of papers (some do not give names of authors) recorded in this volume amounts to 7800.

We consider the subject index as an approximation to an optimal index covering all fields of astronomy and astrophysics and their border fields. The assigning of one or more key words to a paper is, undoubtedly, a difficult task. Some jour-

nals have started giving key words together with the titles of papers. These key words are chosen by the authors themselves. Starting with Volume 18, the subject index was enlarged to a certain extent in order to provide a thesaurus of astronomical and astrophysical terms. This is done not only for the users' convenience, but also with the intention to propose the use of special key words to authors and publishers.

While each volume is scheduled to contain an author index and a subject index, the magnetic tapes containing the index information will be used to produce separate index volumes (authors and subjects) at intervals of five years.

The sorting program for the author and subject indexes is based on the IBM SORT/MERGE Program. This SORT-program sorts blank before hyphen (−) and before letters. Apostrophes are ignored by a special routine.

Examples: a) De Laeter
:
Deacon
:
De Laeter
b) A Stars
:
Absolute Magnitudes
c) Solar X Rays
:
Solar-Terrestrial Relations
d) Boehm-Vitense
Boehme

The introduction of small and capital letters in the layout caused some difficulties. Special programs had to code the capital letters into small ones. For the layout, a TN chain for a 1403 IBM high-speed printer was used. All the programs were written by Dr. H. Scholl in PL/I. The computations and printing were carried out on an IBM 360/44.

The users are requested to inform us on spelling errors within the author and subject indexes in order to assist us in eliminating mistakes in future cumulative indexes.

Abbreviations

Abbreviations used in Astronomy and Astrophysics Abstracts are primarily based on the 'International List of Periodical Title Word Abbreviations', prepared for the UNISIST/ICSU-AB Working Group on Bibliographic Descriptions (1970).

A.A.B.	Associazione Astrofili Bolognesi
Aarg.	Aargang
AAS	American Astronomical Society
AAVSO	American Association of Variable Star Observers
Abh.	Abhandlung—
Abstr.	Abstract—
Abt.	Abteilung—
Acad.	Academi—, Academy
Accad.	Accademi—
Act.	Active, Activit—
Adm.	Administr—
Adv.	Advanc—
Aehron.	Aehronomi—
Aeron.	Aeronom—
Aeronaut.	Aeronauti—
Aerosp.	Aerospace
AG	Astronomische Gesellschaft
AIAA	American Institute of Aeronautics and Astronautics
AJB	Astronomischer Jahresbericht
Akad.	Akadem—
Ala.	Alabama
Alm.	Almanac—, Almanak—
Amat.	Amateur—
An.	Anais, Anale—, Anali—, Anals
Anal.	Analis—, Analit—, Analys—, Analyt—
Angew.	Angewandt—
Ann.	Annaes, Annal—
Annu.	Annu—
Anst.	Anstalt
Anu.	Anual—, Anuar—
Anz.	Anzeiger
Appl.	Applied
Arb.	Arbeit
Arch.	Archiv—
Årg.	Årgang
Ariz.	Arizona
Ark.	Arkiv—
Arkh.	Arkhiv—
Artif.	Artifici—
ASA	Astronomical Society of Australia
Asoc.	Asocia—
ASP	Astronomical Society of the Pacific
ASSA	Astronomical Society of Southern Africa
Assem.	Assembl—
Assoc.	Associ—
Assoz.	Assozi—
Astrofis.	Astrofisic—
Astrofiz.	Astrofizi—
Astrometr.	Astrometr—
Astron.	Astronom—
Astronaut.	Astronauti—, Astronauty—
Astrophys.	Astrophys—
ASV	Astronomical Society of Victoria
ASWA	Astronomical Society of Western Australia
At.	Atom—
Atmos.	Atmosf—, Atmosph—
BAA	British Astronomical Association
Bayer.	Bayerisch—
Beitr.	Beitrag, Beiträge
Beob.	Beobacht—
Ber.	Bericht—
Bibl.	Bibliot—
Bibliogr.	Bibliograf—, Bibliograph—
BIH	Bureau International de l'Heure (Paris)
Bimest.	Bimestr—
Bl.	Blatt, Blätter
Bol.	Boletin
Boll.	Bolletino
Bul.	Buleten—, Buletin—, Bulten
Bull.	Bulletin—, Bullettino
Bur.	Bureau—
Byul.	Byuleten—, Byuletin—
Byull.	Byulleten—
C.R.	Comptes Rendus
Cah.	Cahier—
Calif.	California
Cas.	Casopis
Cent.	Center—, Central, Centrale, Centrally, Centre
Cercet.	Cercetary
Chem.	Chemi—
Chim.	Chimi—
Chron.	Chronic—, Chronik, Chronique
Chronom.	Chronometr—
Cie.	Compagnie
Cienc.	Ciencia—
Cient.	Cientific—
Circ.	Circolar—, Circolo, Circolaire—, Circular—, Circulo
Cirk.	Cirkulaer—
Cl.	Clasa, Classe—
Co.	Companies, Company
Coll.	College
Collect.	Collect—
Colloq.	Colloqui—
Colo.	Colorado
Comet.	Cometary
Commentat.	Commentat—
Commun.	Communica—
Comput.	Computation, Computer—, Computing
Comun.	Comunica—
Conf.	Conferen—
Congr.	Congres—
Conn.	Connecticut
Contract.	Contract—
Contrib.	Contribu—
Cosm.	Cosmic—
Cosmochim.	Cosmochimi—
COSPAR	Committee on Space Research
Crystallogr.	Crystallograph—
CSIRO	Commonwealth Scientific and Industrial Research Organization
Cult.	Cultur—, Cultuur
Curr.	Current
D.C.	District of Columbia
DDR	Deutsche Demokratische Republik
Del.	Delaware
Dep.	Departament, Département, Department
Dev.	Development—, Développement—
Diss.	Disserta—
Div.	Divis—
Doc.	Document—
Dok.	Dokument—
Dokl.	Doklad—
Ehksp.	Ehksperiment—
Eidg.	Eidgenössisch—
Eksp.	Eksperiment—
Electron.	Electroni—

Eng.	Engineer–	Ing.	Ingenieur
Environ.	Environment–	INIS	International Nuclear Information System
Equip.	Equipement, Equipment	Inst.	Institut–, Instytut–
Ergeb.	Ergebnis–	Instn.	Institution
ESA	European Space Agency	Instrum.	Instrument–
ESO	European Southern Observatory	Int.	Internationa–, Internazional–
ESRO	European Space Research Organization	Inter.	Intérieur–, Interior
Eval.	Evaluation–	Interplanet.	Interplanetary
Exp.	Experiment–	Intez.	Intezet–
Extraterr.	Extraterrestr–	Ionos.	Ionosfer–, Ionospher–
F. R. Germany	Federal Republic of Germany	Iskusstv.	Iskusstvenn–
Fac.	Facolt–, Faculd–, Facult–	Issled.	Issledovan–
Fak.	Fakult	Ist.	Istitut–
Fasc.	Fascicul–	Izd.	Izdatel–
Fenn.	Fenni–	Izv.	Izvesti–
Fis.	Fisic–, Fisik–	J.	Joernaal–, Jornal–, Journal–
Fiz.	Fizic–, Fizik–, Fizyk	Jaarb.	Jaarboek–
Fla.	Florida	Jahrb.	Jahrbuch, Jahrbücher
Fluid.	Fluidi–	Jahresber.	Jahresbericht–
Fond.	Fondation–, Fondazione	Jahresschr.	Jahresschrift
Fortschr.	Fortschritt–	Jahrg.	Jahrgang
Fotogr.	Fotograf–	JPL	Jet Propulsion Laboratory
Found.	Foundation–	K.	Königlich–, Koninklijk–, Kunglig–
Freq.	Frequen–	Kans.	Kansas
Fundam.	Fundamenta–	Kartogr.	Kartograf–
Fys.	Fysik–, Fysisch, Fysisk–	Kernforsch.	Kernforschung
Fyz.	Fyzik–	Kernphys.	Kernphysik–
G.	Giornale	Khem.	Khemyi–
Ga.	Georgia	Khim.	Khimi–
Gaz.	Gazeta, Gazette	Kim.	Kimija–, Kimya
Gazz.	Gazzetta	Kl.	Klass–
Gen.	General	Kolloq.	Kolloquium–
Geochem.	Geochem–	Komet.	Kometnyj
Geochim.	Geochim–	Komm.	Kommission–
Geod.	Geodaes–, Geodaet–, Geodes–, Geodet–, Geodez–	Konf.	Konfer–
Geofis.	Geofis–	Kongr.	Kongress
Geofiz.	Geofiz–	Kosm.	Kosmich–
Geofys.	Geofys–	Kosmog.	Kosmogon–
Geogr.	Geograf–, Geograph–	Kozp.	Kozponti
Geokhim.	Geokhim–	KPNO	Kitt Peak National Observatory
Geol.	Geolog–, Geolosk–	Kut.	Kutato
Geomagn.	Geomagneti–	Ky.	Kentucky
Geophys.	Geophys–	La.	Louisiana
Ges.	Gesellschaft	Lab.	Laborato–
Gl.	Glavno–	Lett.	Letter–, Lettra, Lettre
Glas.	Glasnik	Libr.	Librair–, Librar–
Gos.	Gosudarst–	Mag.	Magasin, Magazin–
Gov.	Government–	Magn.	Magneti–, Magnitn–
Grenzgeb.	Grenzgebiet–	Mass.	Massachusetts
GSFC	Goddard Space Flight Center	Mat.	Matemaat–, Matemat–
H. M.	Her Majesty's , His Majesty's	Mater.	Material–
Handb.	Handbook, Handbuch	Math.	Mathemat–
Her.	Herald–	Md.	Maryland
Hist.	History	Meas.	Measur–
Hochsch.	Hochschule	Mec.	Mecani–
Hoegsk.	Hoegskol–	Mech.	Mechani–
HR-diagram	Hertzsprung-Russell diagram	Medd.	Meddelande–, Meddelelse
Hydrogr.	Hydrograf–, Hydrograph–	Meded.	Mededeeling–, Mededeling–
IAF	International Astronautical Federation	Mekh.	Mekhani–
IAU	International Astronomical Union	Mem.	Memento–, Memoir–, Memori–, Memory–, Memuary
IBM	International Business Machines Corporation	Memo.	Memorand–
ICSU	International Council of Scientific Unions	Mens.	Mensile, Mensual–, Mensuel–
ICSU-AB	International Council of Scientific Unions– Abstracting Board	Messtech.	Messtechni–
		Meteorol.	Meteorolog–
IEEE	Institute of Electrical and Electronics Engineers	Mich.	Michigan
		Micromec.	Micromecaniq–
Ill.	Illinois	Miner.	Mineral, Minerale–, Minerali–
Inc.	Incorporated	Mineral.	Mineralog–
Ind.	Industr–	Minn.	Minnesota
Inf.	Informat–, Informaz–, Informe–	Miss.	Mississippi
		MIT	Massachusetts Institute of Technology

Mitt.	Mitteilung–
Mo.	Missouri
Mod.	Modern–
Mol.	Molecul–, Molekul–
Mon.	Monat, Monatlich–, Month–
Monogr.	Monograph–
Mont.	Montana
MPI	Max-Planck-Institut
N. C.	North Carolina
N. D.	North Dakota
N. H.	New Hampshire
N. J.	New Jersey
N. M.	New Mexico
N. Y.	New York
Nablyud.	Nablyudeni–
Nac.	Nacion–
Nachr.	Nachricht–
NASA	National Aeronautics and Space Administration
Nat.	Natur–
Natl.	National–
Naturforsch.	Naturforsch–
Naturwiss.	Naturwissenschaft–
Nauchn.	Nauchny–
Nauk.	Nauka, Naukite, Naukov–, Naukow–
Naut.	Nautic–
Nav.	Naval–
Navig.	Navigat–
Naz.	Nazion–
Nebr.	Nebraska
Nev.	Nevada
Newsl.	Newsletter–
Not.	Notationes, Notic–, Notise–, Notizi–
Nouv.	Nouveau–, Nouvell–
Nov.	Novoe
Nucl.	Nucléaire–, Nuclear–, Nucl–
Nukl.	Nukle–
Numer.	Numeri–
O-va	Obshchestva
O-vo	Obshchestvo
Obs.	Observ–
Obz.	Obzor–
Okla.	Oklahoma
Opt.	Optic–, Optik–, Optique
Oreg.	Oregon
Oss.	Osserva–
Pa.	Pennsylvania
Paleontol.	Paleontolog–
Pap.	Paper–, Papier
Part.	Particle
Perem.	Peremenn–
Period.	Periodi–
Petrol.	Petrolog–
Philos.	Philosoph–
Photogr.	Photograf–, Photograph–
Photogramm.	Photogrammetr–
Photom.	Photometr–
Phys.	Physic–, Physik–, Physique–, Physisch–
Pict.	Picture–
Planet.	Planetary
Pr.	Prac–
Prelim.	Prelimin–
Prepr.	Preprint
Prib.	Pribor–
Prikl.	Prikladnoj
Prir.	Prirodn–
Prirodoved.	Prirodoved–
Probl.	Problem–
Proc.	Proceedings
Prod.	Prodott–, Produc–, Produkt,
Prog.	Progres–
Propag.	Propagation
Prov.	Provinc–, Provints–, Provinz–
Pubbl.	Pubblicazion–
Publ.	Publicac–, Publicas–, Publicat–, Publikas–, Publikat–
Q.	Quarterly
Quant.	Quantit–
R.	Royal
R. I.	Rhode Island
Radiat.	Radiati–
Radioact.	Radioactiv–, Radioaktiv–
Radioisot.	Radioisotop–
Rap.	Raport–
Rapp.	Rapport–
RAS	Royal Astronomical Society
Rec.	Record–
Rech.	Recherche–
Ref.	Referat–, Reference–, Referieren
Relat.	Related, Relation–
Relativ.	Relativit–
Rend.	Rendicont–
Rep.	Report–
Repr.	Reprint–
Repub.	Republi–
Res.	Research–
Result.	Resultad–, Resultat–
Rev.	Review–, Revisio, Revista, Revue–
Rezul't.	Rezul'tat–
Ric.	Ricerca, Ricerche
Riv.	Rivist–
Rundsch.	Rundschau
S. C.	South Carolina
S. D.	South Dakota
SAF	Société Astronomique de France
SAI	Società Astronomica Italiana
Samml.	Sammlung–
SAO	Smithsonian Astrophysical Observatory
SAS	Société Astronomique de Suisse
Satell.	Satellite
Sb.	Sbornik–
Schr.	Schrift–
Schriftenr.	Schriftenreihe
Sci.	Scienc–, Scient–, Scienz–
Scr.	Scripta, Scritt–
Secc.	Seccion–
Sect.	Secti–
Seer.	Seeria
Sekc.	Sekci–, Sekcj–
Sekt.	Sektion–, Sektor–
Sekts.	Sektsi–
Sel.	Seleccion–, Select–, Selek–, Selezione
Selsk.	Selskab–, Selskap–
Semin.	Séminair–, Seminar–
Sep.	Separat–
Ser.	Seria–, Serie–, Seriya
Serv.	Servic–, Serviz–
Sess.	Sessi–
Signal.	Signalétique–
Simp.	Simpoz–
Sitzungsber.	Sitzungsbericht–
Skr.	Skrift–
Soc.	Sociedad–, Societ–
Sol.	Solar
Soln.	Solnechn–
Sonderdr.	Sonderdruck–
Soobshch.	Soobshchen–
South .	Southern
Spacecr.	Spacecraft
Spat.	Spatial–
Spec.	Special–
Spectrosc.	Spectroscop–
Spectrosk.	Spectroskop–
Spets.	Spetsial–

Spez.	Spezial–, Speziell–	**UK**	United Kingdom
SSR	Sovetskaya Sotsialisticheskaya	**Umsch.**	Umschau
	Respublika	**UN**	United Nations
SSSR	Soyuz Sovetskikh Sotsialisticheskikh	**Univ.**	Universidad–, Universit–, Univerzitet–
	Respublik	**US**	United States
St.	Saint–, Sankt–, Sant–	**USA**	United States of America
– St.	– Straße, Street	**USSR**	Union of Soviet Socialist Republics
Stand.	Standard–, Standart–	**Va.**	Virginia
Sternw.	Sternwarte–	**Var.**	Various
Stiint.	Stiintific–	**Ver.**	Verein–, Verenig–
Stn.	Station, Stazione	**Veränderl.**	Veränderlich–
Stud.	Studia, Studie–, Studii	**Verh.**	Verhandl–
Supl.	Suplement–, Supliment–	**Vermess.**	Vermessung–
Suppl.	Supplement–	**Vermessungswes.**	Vermessungswesen
Surv.	Survey–	**Veröff.**	Veröffentlich–
Symp.	Sympos–, Sympoz–	**Vesn.**	Vesnik
Syst.	System–	**Vestn.**	Vestnik
Sz.	Szemle	**Vetensk.**	Vetenskap–
Teach.	Teacher–, Teaching	**Vidensk.**	Videnskab–, Videnskap
Tec.	Tecni–	**Vierteljahresschr.**	Vierteljahresschrift–
Tech.	Techni–	**Vierteljahrsschr.**	Vierteljahrsschrift–
Technol.	Technolog–	**VLB**	Very Long Baseline
Tecnol.	Tecnolog–	**Volcanol.**	Volcanolog–
Teh.	Tehnic–, Tehnika, Tehnisk–	**Vopr.**	Vopros–
Tehnol.	Tehnolog–, Tehnolosk–	**Vortr.**	Vorträge
Tek.	Tekni–	**Vses.**	Vsesoyuzn–
Tekh.	Tekhni–	**Vt.**	Vermont
Tekhnol.	Tekhnolog–	**Vyp.**	Vypusk–
Teknol.	Teknolog–	**Vyssh.**	Vyssh–
Telesc.	Telescop–	**Vyzk.**	Vyzkum–
Telev.	Television–	**W. Va.**	West Virginia
Tenn.	Tennessee	**Wash.**	Washington
Teor.	Teoret–, Teori–	**West.**	Western
Terr.	Terrestr–	**Wet.**	Wetenschap–, Wetenskap–
Test.	Testing	**Wis.**	Wisconsin
Tex.	Texas	**Wiss.**	Wissenschaft–
TH	Technische Hochschule	**Wyo.**	Wyoming
Theor.	Theoret–, Theori–	**Yad.**	Yadern–
Tidschr.	Tidschrift–	**Z.**	Zeitschrift–
Tidskr.	Tidskrift–	**ZA**	Zero Age
Tidsskr.	Tidsskrift–	**ZAED**	Zentralstelle für Atomkernenergie-
Top.	Topic–		Dokumentation
Tr.	Trudy	**Zap.**	Zapisk–, Zapyisk–
Trans.	Transactions, Transazione	**Zaved.**	Zaveden–
Tsentr.	Tsentral–	**Zent.**	Zentral
Tsirk.	Tsirkulyar–	**Zentralbl.**	Zentralblatt
TU	Technical University	**Zesz.**	Zeszyt
Uch.	Uchen–	**Zh.**	Zhurnal–
Uchebn.	Uchebn–	**Zirk.**	Zirkular

Periodicals, Proceedings, Books, Activities

001 Periodicals

AAS Photo-Bull.
AAS (American Astronomical Society) Photo-Bulletin.
Published by the Working Group on Photographic Materials. Produced by Eastman Kodak Co., Rochester, N. Y.

AAVSO Bull.
Bulletin of the American Association of Variable Star Observers, 187 Concord Avenue, Cambridge, Mass., 02138, U.S.A.

Acad. R. Belgique, Bull. Cl. Sci.
Académie Royale de Belgique, Bulletin de la Classe des Sciences (Koninklijke Academie van België, Mededelingen van de Klasse der Wetenschappen). 5ᵉ Série, Palais des Académies, Bruxelles.

Acta Astron.
Acta Astronomica. An international quarterly journal.
Publisher: Polska Akademia Nauk, Komitet Astronomii (Polish Academy of Sciences, Committee of Astronomy), Warszawa – Wrocław.

Acta Astron. Sinica
Acta Astronomica Sinica. Published by Purple Mountain Observatory, Academia Sinica, Nanking, China.

Acta Astronaut.
Acta Astronautica. Journal of the International Academy of Astronautics. Publisher: Pergamon Press Inc., Elmsford, New York, U.S.A.; Pergamon Press Ltd., Oxford, England.

Acta Cienc. Indica
Acta Ciencia Indica. Pragati Prakashan, Begum Bridge, Post Box No. 62, Meerut-250001, India.

Acta Cosmologica
Acta Cosmologica. Published by Obserwatorium Astronomiczne Universytetu Jagiellońskiego, Kraków, Poland.

Acta Crystallogr. A
Acta Crystallographica, Section A: Crystal Physics, Diffraction, Theoretical and General Crystallography. Munksgaard International Booksellers and Publishers Ltd., 35 Norre Sogade, DK 1370 Kobenhavn K, Denmark.

Acta Electron.
Acta Electronica. 3 Avenue Descartes, BP 15,94 450 Limeil-Brevannes, (Val-de-Marne), France.

Acta Geophys. Sinica
Acta Geophysica Sinica. Chinese Academy of Sciences, Department of Geophysical Research. Published by Science Press, Peking, Peoples Republic of China.

Acta Phys. Acad. Sci. Hungaricae
Acta Physica Academiae Scientiarum Hungaricae. Postafiok 24, Budapest 502, Hungary.

Acta Phys. Austriaca
Acta Physica Austriaca. Springer-Verlag, A-1011 Wien, Molkerbastei 5, Postfach 367, Austria.

Acta Phys. Polonica B
Acta Physica Polonica B. ARS Polona-Ruch, Warszawa 1, P.O. Box 154, Poland.

Acta Phys. Sinica
Acta Physica Sinica. Chinese Academy of Sciences, Institute of Physics, Peking, Peoples Republic of China. [English translation in: Chinese J. Phys. (*USA*)].

Acta Phys. Slovaca
Acta Physica Slovaca. VEDA Publishing House of the Slovak Academy of Sciences, 895 30 Bratislava, Klemensova 27, Czechoslovakia.

Acta Polytech. III
Acta Polytechnica. Series III. Elektrotechnicka fakulta CVUT v Prace, Technicka ul. 2, Praha 6-Dejvice, Czechoslovakia.

Acta Tech. CSAV
Acta Technica Ceskoslovenska Akademie Ved.Academia, Publishing House of the Czechoslovak Academy of Sciences, Vodickova 40, 112 29 Praha 1, Czechoslovakia. John Benjamins N.V., Periodical Trade, Warmoesstraat 54, Amsterdam, Netherlands.

Acta Univ. Carolinae Math. Phys.
Acta Universitatis Carolinae, Mathematica et Physica. Administrace: Matematicko-fyzikálni fakulta University Karlovy, Praha.

Adv. Astron. Astrophys.
Advances in Astronomy and Astrophysics. Publisher: Academic Press, New York – London.

Adv. Phys.
Advances in Physics. Taylor & Francis Ltd., 10–14 Macklin Street London, WC2B 5NF, England.

Aerotec. Missili Spazio
L'Aerotecnica Missili e Spazio. Tamburini Editore S.P.A., Via Pascoli 55, 20133 Milano, Italy.

AFOEV Bull.
Association Française des Observateurs d'Étoiles Variables. Bulletin. Redaction and publication: M. Émile Schweitzer, 1, rue Beethoven, Strasbourg.

AIAA J.
AIAA Journal. A Publication of the American Institute of Aeronautics and Astronautics devoted to Aerospace Research and Development. Published by the American Institute of Aeronautics and Astronautics, New York, N.Y.

AIP Conf. Proc.
AIP Conference Proceedings. American Institute of Physics, 335 East 45th Street, New York, N.Y. 10017, USA.

Alta Freq.
Alta Frequenza.Ufficio Centrale AEI-CEI, Viale Monza 259, 20126 Milano, Italy.

American J. Phys.
American Journal of Physics. Published for the American Association of Physics Teachers by the American Institute of Physics, 335 East 47th Street, New York, N.Y. 10017, USA.

American Mineral.
American Mineralogist. Mineralogical Society of America, 1707 L Street, N.W., Washington, DC 20036, USA.

American Sci.
American Scientist. Society of Sigma XI, 345 Whitney Avenue, New Haven, CT 06510, USA.

An. Acad. Brasil. Cienc.
Anais da Academia Brasileira de Ciencias. Caixa Postal 229, ZC-00 Rio de Janeiro gb, Brazil.

An. Fis.
Anales de Física. Real Sociedad Española de Física y Química (Facultad de Ciencias), Ciudad Universitaria, Madrid-3, Spain.

An. Stiint. 'Al. I. Cuza' Iasi (Ser. Noua) I
Analele Stiintifice ale Universitatu 'Al. I. Cuza' din Iasi (Serie Noua), Sectiunea I Fizica. Calea 23 August, Iasi, Rumania.

Ann. Acad. Sci. Fennicae, Ser. A. VI
Annales Academiae Scientiarum Fennicae, Series A VI (Physica). Snellmaninkatu 9–11, 00170 Helsinki-17, Finland.

Ann. Françaises Chronom. Microméc.
Annales Françaises de Chronométrie et de Micromécanique, publication annuelle de l'Observatoire de Besançon, du Centre Technique de l'Industrie Horlogère et de la Société Française de Chronométrie et de Micromécanique. Rédaction et administration: Observatoire de Besançon. Publiées avec le concours du Centre National de la Recherche Scientifique et des organismes corporatifs.

Ann. Géophys.
Annales de Géophysique. Service des Publications du CNRS, 15 Quai Anatole-France, 75700 Paris, France.

Ann. Inst. Henri Poincaré A
Annales de l'Institut Henri Poincaré, Section A (Physique Theorique). 11 Rue Pierre-Curie, Paris 5, France.

Ann. Obs. Astron. Météorol. Toulouse
Annales de l'Observatoire Astronomique et Météorologique de Toulouse. Publisher: Gauthier-Villars, Paris.

Ann. Physics
Annals of Physics. Academic Press Inc., 111 Fifth Avenue, New York, NY 10003, USA.

Ann. Physik
Annalen der Physik. 7. Folge. Publisher: Johann Ambrosius Barth, Salomonstr. 18B, Leipzig 701, German Democratic Republic.

Ann. Physique
Annales de Physique. Publisher: Masson et Cie., 120 Boulevard Saint-Germain, Paris 6, France.

Ann. Sci.
Annals of Science. Taylor & Francis Ltd., 10–14 Macklin Street, London, WC2B 5NF, England.

Ann. Soc. Sci. Bruxelles I
Annales de la Société Scientifique de Bruxelles. Série I: Sciences Mathématiques, Astronomiques et Physiques. Rue de Bruxelles 61, B 5000 Namur, Belgium.

Ann. Télécommun.
Annales des Télécommunications. Centre National d'Études des Télécommunications, 38 rue du Général Leclerc, 92 Issy-les-Moulineaux, France.

Ann. Tokyo Astron. Obs.
Annals of the Tokyo Astronomical Observatory. University of Tokyo, Mitaka, Tokyo, Japan.

Ann. Univ.-Sternw. Wien
Annalen der Universitäts-Sternwarte Wien. In Kommission bei Ferd. Dümmlers Verlag, Bonn.

Annu. Rep. Astron. Inst. Greece
Annual Reports of the Astronomical Institutes of Greece. Published by the Greek National Committee for Astronomy. Academy of Athens, Research Center for Astronomy and Applied Mathematics.

Annu. Rev. Astron. Astrophys.
Annual Review of Astronomy and Astrophysics. Publisher: Annual Reviews Inc., Palo Alto, California.

Antenna
L'Antenna. Via Monte Generoso 6/a, 20155 Milano, Italy.

Anz. Österreich. Akad. Wiss. Math.-Naturwiss. Kl.
Anzeiger. Österreichische Akademie der Wissenschaften. Mathematisch-Naturwissenschaftliche Klasse. Publisher: Springer-Verlag, Wien.

APL Tech. Dig.
APL Technical Digest.Applied Physics Laboratory, The John Hopkins University, 8621 Georgia Avenue, Silver Spring, MD 20910, USA.

Appl. Opt.
Applied Optics. A monthly publication of the Optical Society of America. Published for the Optical Society of America by the American Institute of Physics, 335 East 45th Street, New York, NY 10017, USA.

Appl. Phys.
Applied Physics. Springer-Verlag, Heidelberger Platz 3, D-1000 Berlin 33, F. R. Germany.

Appl. Phys. Lett.
Applied Physics Letters. American Institute of Physics, 335 East 45th Street, New York, N.Y. 10017, USA.

Appl. Spectrosc.
Applied Spectroscopy. 428 East Preston Street, Baltimore, MD 21202, USA.

Appl. Spectrosc. Rev.
Applied Spectroscopy Reviews. Marcel Dekker Inc., 95 Madison Avenue, New York, NY 10016, USA.

Arch. Mech.
Archives of Mechanics (Archiwum Mechaniki Stosowanej). Polish Scientific Publishers, Swietokrzyska 21, Warszawa, Poland.

Arch. Ration. Mech. Anal.
Archive for Rational Mechanics and Analysis. Springer-Verlag, Heidelberger Platz 3, D 1000 Berlin 33, F. R. Germany.

Arch. Sci.
Archives des Sciences, éditées par la Société de Physique et d'Histoire Naturelle de Genève. Publisher: Imprimerie Kundig, Genève. Subscription address: Librairie Payot, Genève.

Ark. Astron.
Arkiv för Astronomi. Utgivet av Kungliga Svenska Vetenskapsakademien, Stockholm. Printed by Almqvist & Wiksell, Stockholm.

Ark. Geofys.
Arkiv för Geofysik. Kungliga Svenska Vetenskapsakademien, Stockholm. Printed by Almqvist & Wiksell, Stockholm

Artif. Satell.
Artificial Satellites. Publication of Polish Scientific Institutions. Polish Academy of Sciences, National Committee of Geophysics and Geodesy, National Committee for Space Research, Warsaw. Space Research Centre, Pałac Kultury i Nauki 2301, 00-901 Warszawa, Poland.

Asoc. Argentina Astron. Bol.
Asociación Argentina de Astronomía. Boletin. Editor: Instituto Argentino de Radioastronomía, Provincia de Buenos Aires, Argentina. Printer: Talleres Gráficos "Renovación", La Plata, República Argentina.

Assoc. Veneta Oss. Stelle Variabili Bull.
Associazione Veneta Osservatori di Stelle Variabili, Bulletin. c/o Gruppo Astrofili di Padova, Corso Garibaldi 41, 35100 Padova, Italy.

Astrofiz. Issled. Izv. Spets. Astrofiz. Obs.
Astrofizicheskie Issledovaniya. Izvestiya Spetsial'noj Astrofizicheskoj Observatorii. Akademiya Nauk SSSR. Publishers: Izdatel'stvo "Nauka", Leningradskoe Otdelenie, Leningrad.

Astrofizika
Astrofizika. Izdatel'stvo Akademii Nauk Armyanskoj SSR, Erevan. [An English translation is published in "Astrophysics".]

Astrometr. Astrofiz.
Astrometriya i Astrofizika. Respublikanskij Mezhvedomstvennyj Sbornik. Akademiya Nauk Ukrainskoj SSR, Glavnaya Astronomicheskaya Observatoriya. Naukova Dumka, Kiev.

Astron. Astrophys.
Astronomy and Astrophysics. A European Journal. Published by Springer-Verlag, Berlin – Heidelberg– New York.

Astron. Astrophys., Suppl. Ser.
Astronomy and Astrophysics. Supplement Series. A European Journal. Published by the Astronomical Insti-

tute Lausanne and Geneva Observatory, CH-1290 Sauverny, Switzerland.

Astron. Her.
Astronomical Herald. Astronomical Society of Japan, Tokyo Astronomical Observatory, Oosawa Mitaka, Tokyo, Japan.

Astron. J.
The Astronomical Journal. Published for the American Astronomical Society by the American Institute of Physics, New York, N. Y. Editorial Office: Department of Astronomy, Columbia University, New York, N. Y.

Astron. Nachr.
Astronomische Nachrichten. Publisher: Akademie-Verlag, Berlin.

Astron. Rep.
The Astronomical Reports. Polish Amateur Astronomical Society. Polskie Towarzystwo Miłośników Astronomii, Kraków, Poland.

Astron. Schule
Astronomie in der Schule. Zeitschrift für die Hand des Astronomielehrers. Herausgegeben vom Verlag Volk und Wissen, Berlin. Redaktion: Sternwarte Bautzen.

Astron. Tidsskr.
Astronomisk Tidsskrift. Edited by Astronomisk Selskab, København; Norsk Astronomisk Selskap, Oslo; Svenska Astronomiska Sällskapet, Stockholm. Printed by John Griegs Boktrykkeri, Bergen.

Astron. Tsirk.
Astronomicheskij Tsirkulyar, izdavaemyj Byuro Astronomicheskikh Soobshchenij Akademii Nauk SSSR. Tipografiya Astrosoveta AN SSSR, Moskva.

Astron. Vestn.
Astronomicheskij Vestnik. Publishers: Izdatel'stvo "Nauka", Moskva.

Astron. Zh. Akad. Nauk SSSR
Astronomicheskij Zhurnal. Akademiya Nauk SSSR. Publishers: Izdatel'stvo "Nauka", Moskva. [An English translation is published in "Soviet Astronomy AJ"].

Astronautik
Astronautik. Organ der Hermann-Oberth-Gesellschaft e.V. Astronautik-Verlag, Druckerei H. Brandt, Delmenhorst, F.R. Germany.

Astronomia
Astronomia. Periodico trimestrale dell'Unione Astrofili Italiani.

Astronomie
L'Astronomie et Bulletin de la Société Astronomique de France. Société Astronomique de France, Paris.

Astronomy
Astronomy. AstroMedia Corp., 757 North Broadway, Suite 204, Milwaukee, WI 53202, USA.

Astrophys. J.
The Astrophysical Journal. Published for the American Astronomical Society by the University of Chicago Press, Chicago, Illinois.

Astrophys. J., Lett.
The Astrophysical Journal. Letters to the Editors. Pub-

lished for the American Astronomical Society by the University of Chicago Press, Chicago, Illinois.

Astrophys. J., Suppl. Ser.
The Astrophysical Journal. Supplement Series. Published for the American Astronomical Society by the University of Chicago Press, Chicago, Illinois.

Astrophys. Lett.
Astrophysical Letters. Published by NASA–Goddard Space Flight Center. Gordon and Breach Science Publishers Ltd., New York–London–Paris.

Astrophys. Space Sci.
Astrophysics and Space Science. An International Journal of Cosmic Physics. Published by D. Reidel Publishing Company, Dordrecht, Holland.

Astrophysics
Astrophysics. A cover-to-cover translation of Astrofizika (USSR). Consultants Bureau, New York, N.Y.

Atmos. Environ.
Atmospheric Environment. Pergamon Press Ltd., Headington Hill Hall, Oxford, OX3 OBW, England.

Atomkernenergie
Atomkernenergie. Verlag Karl Thiemig, Pilgersheimer-strasse 38, 8 München 90, Postfach 900740, F.R. Germany.

Atti Accad. Naz. Lincei, Mem. Ser. Ottava
Atti della Accademia Nazionale dei Lincei. Serie Ottava. Memorie. Classe di Scienze fisiche, matematiche e naturali. Sezione I: Matematica, Meccanica, Astronomia, Geodesia e Geofisica. Published by Accademia Nazionale dei Lincei, Roma.

Atti Accad. Naz. Lincei, Rend. Ser. Ottava
Atti della Accademia Nazionale dei Lincei. Serie Ottava. Rendiconti. Classe di Scienze fisiche, matematiche e naturali. Published by Accademia Nazionale dei Lincei, Roma.

Atti Fond. G. Ronchi, Contrib. Ist. Naz. Ottica
Atti della Fondazione Giorgio Ronchi e Contributi dell' Istituto Nazionale di Ottica. Largo Enrico Fermi 6, 50125 Firenze, Italy.

Atti Soc. Astron. Italiana
Atti della Società Astronomica Italiana. Via Brera 28, Milano. Publisher: Tipografia Baccini & Chiappi, Firenze (Italy).

Australian J. Phys.
Australian Journal of Physics. Published by the Commonwealth Scientific and Industrial Research Organization, 372 Albert Street, East Melbourne, Victoria 3002, Australia.

Australian J. Phys., Astrophys. Suppl.
Australian Journal of Physics, Astrophysical Supplement. Published by Commonwealth Scientific and Industrial Research Organization, 372 Albert Street, East Melbourne, Victoria 3002, Australia.

BAV Rundbrief
BAV Rundbrief. Mitteilungsblatt der Berliner Arbeitsgemeinschaft für Veränderliche Sterne. Editor: BAV Berliner Arbeitsgemeinschaft für Veränderliche Sterne eV., Berlin.

BBSAG Bull.
Bedeckungsveränderlichen Beobachter der Schweizerischen Astronomischen Gesellschaft, [Swiss Astronomical Society's Eclipsing Variable Observers], Bulletin. To be obtained from R. Diethelm, Winterthur, Switzerland.

Bell Syst. Tech. J.
Bell System Technical Journal. American Telephone and Telegraph Co., 195 Broadway, New York, NY 10007, USA.

Bild Wiss.
Bild der Wissenschaft. Zeitschrift über die Naturwissenschaften und die Technik in unserer Zeit. Publisher: Deutsche Verlagsanstalt, Stuttgart.

Blick Weltall
Blick in das Weltall. Astronomische Veranstaltungen und Mitteilungen für Sternfreunde. Archenhold-Sternwarte Berlin-Treptow.

Bol. Acad. Cienc. Fis. Mat. Nat.
Boletin de la Academia de Ciencias Fisicas Matematicas y Naturales. Printed by Italgrafica, S.R.L. Republica de Venezuela.

Bol. Astron. Obs. Madrid
Boletín Astronómico del Observatorio de Madrid. Editor: Instituto Geografico y Catastral. General Ibáñez de Ibero, Madrid.

Bol. Inst. Mat., Astron. Fis., Univ. Nac. Córdoba
Boletin del Instituto de Matematica, Astronomia y Fisica, Universidad Nacional de Córdoba (R. A.). Dirección General de Publicaciones, Córdoba (Argentina).

Bol. Inst. Tonantzintla
Boletin del Instituto de Tonantzintla. Instituto Nacional de Astrofisica, Optica y Electronica, Apartados Postales Nos. 216 y 51, Puebla, Pue, Mexico.

Bol. Liga Latinoamericana Astron.
Boletin de la Liga Latinoamericana de Astronomia. Publicado por la Asociacion Argentina Amigos de la Astronomia, Buenos Aires, Argentina.

Bol. Obs. Ebro
Boletín del Observatorio del Ebro, Tortosa. Printed by Cooperativa Gráfica Dertosense, Tortosa.

Boll. Geod. Sci. Affini
Bolletino di Geodesia e Scienze Affini. Pubblicazione dell'Istituto Geografico Militare, Firenze.

British Astron. Assoc. Circ.
British Astronomical Association, Circular. Editorial Office: 97 Hawkswood Drive, Hailsham, Sussex.

British J. Philos. Sci.
British Journal for the Philosophy of Science. Cambridge University Press, Bentley House, 200 Euston Road, London, NW1 2DB, England.

Bul. Inst. Politeh. Iasi I
Buletinul Institutului Politehnic din Iasi. Sectia I Matematica, Mecanica, Teoretica, Fizica. Polytechnic Institute, Iasi, Rumania.

Bull. American Astron. Soc.
Bulletin of the American Astronomical Society. Published for the American Astronomical Society by the American

Institute of Physics, 335 East 45th Street, New York, N.Y. 10017, USA.

Bull. American Meteorol. Soc.
Bulletin of the American Meteorological Society. 45 Beacon Street, Boston, MA 02108, USA.

Bull. Astron. Inst. Czechoslovakia
Bulletin of the Astronomical Institutes of Czechoslovakia. Published under the auspices of the Czechoslovak Academy of Sciences by Academia, Praha. Editor: Astronomical Institutes of the Czechoslovak Academy of Sciences, Praha.

Bull. Astron., Obs. R. Belgique
Bulletin Astronomique, Observatoire Royal de Belgique. (Astronomisch Bulletin, Koninklijke Sterrenwacht van België).

Bull. Astron. Soc. India
Bulletin of the Astronomical Society of India. Edited and published by M. S. Vardya, Tata Institute of Fundamental Research, Bombay on behalf of the Astronomical Society of India, Osmania University, Hyderabad.

Bull. Géod.
Bulletin Géodésique, being the Journal of the International Association of Geodesy. Nouvelle Série. Publié par le Bureau Central de l'Association Internationale de Géodésie, 39 Rue Guy-Lussac, 75005 Paris, France.

Bull. Geogr. Surv. Inst.
Bulletin of the Geographical Survey Institute. Published by the Geographical Survey Institute, Ministry of Construction, Tokyo, Japan.

Bull. Groupe Rech. Géod. Spat.
Groupe de Recherches de Géodésie Spatiale. Bulletin. CNES/Toulouse, France.

Bull. Inst. Space Aeronaut. Sci., Univ. Tokyo, B
Bulletin of the Institute of Space and Aeronautical Science, University of Tokyo, B. Tokyo, Japan.

Bull. Obs. Astron. Belgrade
Bulletin de l'Observatoire Astronomique de Belgrade. Editor: Observatoire Astronomique de Belgrade. Printed by Naucna delo, Belgrade.

Bull. Sci. Yougoslavie
Bulletin Scientifique. Conseil des Academies des Sciences et des Arts de la RSF de Yougoslavie. Section A: Sciences Naturelles, Techniques et Médicales. Rédaction et Administration: Opatička ul. 18/II, Zagreb, Yougoslavie.

Bull. Seismol. Soc. America
Bulletin of the Seismological Society of America. Seismological Society of America, 2907 Claremont Avenue, Berkeley, CA 94705, USA.

Bull. Signal.
Bulletin Signalétique. Section 120: Astronomie, Physique spatiale, Géophysique. Centre de Documentation du Centre Nationale de la Recherche Scientifique, Paris.

Bull. Signal.
Bulletin Signalétique. Bibliographie des Sciences de la Terre. Section 220, Cahier A: Minéralogie, Géochimie, Géologie extraterrestre. Centre de Documentation du C.N.R.S., Paris; Département Documentation du B.R. G.M., Orléans.

Bull. Soc. Astron. Liège
Bulletin de la Société Astronomique de Liège. Editeur G. Mathys, 3, Avenue Laurent Gilys, 4621 Retinne.

Bull. Soc. R. Sci. Liège
Bulletin de la Société Royale des Sciences de Liège. L'Université, 15 Avenue des Tilleurs, Liège, Belgium.

Bull. Tokyo Gakugei Univ. IV
Bulletin of Tokyo Gakugei University. Series IV (Mathematics and Natural Sciences). 4-1-1 Nukui-kita-machi, Koganei, Tokyo, Japan.

Byull. Abastuman. Astrofiz. Obs.
Abastumanskaya Astrofizicheskaya Observatoriya, Gora Kanobili. Byulleten'. Akademiya Nauk Gruzinskoj SSR. Publishers: Izdatel'stvo "Metsniereba", Tbilisi.

Byull. Inst. Astrofiz.
Byulleten' Instituta Astrofiziki, Akademiya Nauk Tadzhikskoj SSR. Izdatel'stvo Donish, Dushanbe.

Byull. Inst. Teor. Astron.
Byulleten' Instituta Teoreticheskoj Astronomii. Izdatel'stvo Nauka, Leningradskoe Otdelenie, Leningrad.

C. R. Acad. Bulgare Sci.
Comptes Rendus de l'Académie Bulgare des Sciences. (Doklady Bolgarskoj Akademii Nauk). Sofiya, Bulgaria.

C. R. Acad. Sci. Paris
Comptes Rendus hebdomadaires des Séances de l'Académie des Sciences, publié avec le concours du Centre National de la Recherche Scientifique. Imprimerie: Gauthier-Villars, Paris.

Cah. Fundam. Sci.
Fundamenta Scientiae. Cahiers du Séminaire sur les Fondaments des Sciences. Fundamenta Scientiae, Groupe CBLL, Centre de Recherche Nucléaires, 67037 Strasbourg Cédex.

Canadian J. Earth Sci.
Canadian Journal of Earth Sciences. National Research Council of Canada, Ottawa KIA OR6, Canada.

Canadian J. Phys.
Canadian Journal of Physics. Published by the National Research Council of Canada, Ottawa. Printed in Canada by the University of Toronto Press, Toronto, Ont.

Carter Obs., Astron. Bull.
Carter Observatory, Astronomical Bulletin. Carter Observatory, P.O. Box 2909, Wellington 1, New Zealand.

Celestial Mech.
Celestial Mechanics. An International Journal of Space Dynamics. Publishers: D. Reidel Publishing Company, Dordrecht, Holland.

Cent. Astrophys. Prepr. Ser.
Center for Astrophysics, Preprint Series. Harvard College Observatory, Smithsonian Astrophysical Observatory. Center for Astrophysics, 60 Garden St., Cambridge, Mass. 02138.

Cent. Données Stellaires, Inf. Bull.
Centre de Données Stellaires. Information Bulletin. Compiled at Observatoire de Strasbourg, Strasbourg, France.

Ceskoslovensky Cas. Fis., A
Ceskoslovensky Casopis pro Fisiku. Sekce A. Academia

Publishing House of the Czechoslovak Academy of
Sciences, Vodickova 40, 112 29 Praha 1, Czechoslovakia.

Chem. Phys. Lett.
Chemical Physics Letters. North-Holland Publishing Co.,
P.O. Box 211, Amsterdam-C, Netherlands.

Chinese Astron.
Chinese Astronomy. A selected translation of Acta Astro-
nomica Sinica. Translated by T. Kiang. Publisher:
Pergamon Press, Headington Hill Hall, Oxford, OX3 0BW,
England – Maxwell House, Fairview Park, Elmsford,
New York 10523, USA.

Chinese J. Phys. (*Taiwan*)
Chinese Journal of Physics. Physical Society of the Re-
public of China, Physics Department, National Taiwan
University, Taipei, Taiwan, China.

Ciel Terre
Ciel et Terre. Bulletin de la Société Belge d'Astronomie,
de Météorologie et de Physique du Globe. Administration: Avenue Circulaire, 3, Bruxelles. Printed by Imprimerie R. Louis, Bruxelles.

Circ. Inf.
Circulaire d'Information. Union Astronomique Internationale. Commission des Etoiles Doubles. Address: Observatoire de Meudon, Meudon, France.

Circ. Stn. Astron. Int. Latitudine, Carloforte-Cagliari
Circolari della Stazione Astronomica Internazionale di
Latitudine, Carloforte–Cagliari. Serie A printed by Tipo-
Offset "3 T", Cagliari. Serie B printed by Multi Copy,
Milano.

Coelum
Coelum. Periodico bimestrale per la Divulgazione dell'
Astronomia. Editor: Osservatorio Astronomico Univer-
sitario di Bologna.

Commentat. Phys.-Math.
Commentationes Physico-Mathematicae. Societas Scientiarum Fennica, Helsinki–Helsingfors. Printed by Kes-
kuskirjapaino–Centraltryckeriet, Helsinki–Helsingfors.

Comments Astrophys.
Comments on Astrophysics. A Journal of Critical Dis-
cussion of the Current Literature. Comments on Modern
Physics: Part C. Publishers: Gordon and Breach, Science
Publishers Ltd., 42 William IV Street, London WC2,
England.

Comments At. Mol. Phys.
Comments on Atomic and Molecular Physics. Gordon
& Breach Science Publishers Ltd., 41 and 42 William IV
Street, London, WC2, England.

Comments Nucl. Part. Phys.
Comments on Nuclear and Particle Physics. Gordon &
Breach Science Publishers Ltd., 41 and 42 William IV
Street, London, WC2, England.

Commun. Math. Phys.
Communications in Mathematical Physics, Springer-
Verlag, Postfach 105280, 6900 Heidelberg 1, F.R.
Germany.

Comput. Math. with Appl.
Computers & Mathematics with Applications. Pergamon
Press Ltd., Headington Hill Hall, Oxford, OX3 0BW,
England.

Comput. Phys. Commun.
Computer Physics Communications. North-Holland
Publishing Co., P.O. Box 211 Amsterdam, Netherlands.

Comun. Obs. Astron. Univ. Coimbra
Comunicações do Observatório Astronomico da Universidade de Coimbra, Portugal.

Contemp. Phys.
Contemporary Physics.Taylor and Francis Ltd., 10 - 14
Macklin Street, London, WC2B 5NF, England.

Contrib. Atmos. Phys.
Contributions to Atmospheric Physics – Beiträge zur
Physik der Atmosphäre. Publisher: Friedrich Vieweg &
Sohn, Braunschweig.

Contrib. Obs. New Mexico State Univ.
Contributions of the Observatory of New Mexico State
University. Published by the Astronomy Department,
Box 4500, New Mexico State University, Las Cruces,
New Mexico 88003.

COSPAR Inform. Bull.
COSPAR. Information Bulletin. Address: COSPAR
Secretariat, Paris.

CQ Radio Amat. J.
CQ Radio Amateur's Journal. 14 Vanderventer Avenue,
Port Washington, Long Island, NY 11050, USA.

Cryogenics
Cryogenics. IPC Science and Technology Press Ltd.,
IPC House 32 High Street, Guildford, Surrey GU1 3EW,
England.

CSIO Commun.
CSIO Communications. Central Scientific Instruments
Organisation, Sector 30, Chandigarh-160020, India.

Curr. Sci.
Current Science, Current Science Association, Raman
Research Institute, Bangalore 6, India.

Czechoslovak J. Phys. B
Czechoslovak Journal of Physics, Section B. Czechoslovak
Academy of Science, Akademia, Vodickova 40, 112 29
Praha 1, Czechoslovakia.

Deutsche Geod. Komm. Bayerisch. Akad. Wiss.
Deutsche Geodätische Kommission bei der Bayerischen
Akademie der Wissenschaften. Reihe A: Höhere Geo-
däsie; Reihe B: Angewandte Geodäsie; Reihe C: Disser-
tationen; Reihe D: Tafelwerke; Reihe E: Geschichte und
Entwicklung der Geodäsie. Published by Verlag der
Bayerischen Akademie der Wissenschaften, München.

Dokl. Akad. Nauk
Doklady Akademii Nauk SSSR. Seriya Matematika,
Fizika. Publishers: Izdatel'stvo "Nauka", Moskva.

Dudley Obs. Rep.
Dudley Observatory Reports. Dudley Observatory,
Albany, N.Y., USA.

Dunsink Obs. Publ.
Dunsink Observatory Publications. The Observatory of
the School of Cosmic Physics, Dublin Institute for Advanced Studies, Dublin.

Earth Extraterr. Sci.
Earth and Extraterrestrial Sciences. Gordon & Breach

Science Publishers Ltd., 41 and 42 William IV Street, London, WC2, England.

Earth Planet. Sci. Lett.
Earth and Planetary Science Letters. A Letter Journal devoted to the Development in Time of the Earth and Planetary System. Publisher: North-Holland Publishing Company, Amsterdam, Netherlands.

El Universo
El Universo. Organo de la Sociedad Astronomica de Mexico, Mexico, D.F.

Electricidade
Electricidade. Empresa Editorial Electrotecnica Edel, Rua de Dona Estefania, 48, 3., Esq., Lisboa 1, Portugal.

Electron. Australia
Electronics Australia. 12th floor, 235–242 Jones Street, Broadway, Sydney, N.S.W., Australia.

Electron. Lett.
Electronics Letters. Institution of Electrical Engineers, Savoy Place, London, WC2R OBL, England.

Electron. Power
Electronics and Power. Journal of the Institution of Electrical Engineers. Savoy Place, London, WC2R OBL, England.

Electron. Prod. Methods Equip.
Electronic Production Methods and Equipment. Kiver-Patterson Publishing, 322 St. John Street, London, EC1 England.

Electronics
Electronics. McGraw-Hill Publishing Co., 1221 Avenue of the Americas, New York, N.Y. 10020, USA.

Elektrotech. Z. B
Elektrotechnische Zeitschrift. Ausgabe B: Der Elektro-techniker. Verband Deutscher Elektrotechniker Publication address: VDE-Verlag, Bismarckstrasse 33, 1000 Berlin 12, F. R. Germany.

Endeavour
Endeavour. A review of the progress of science, published in four languages by Imperial Chemical Industries, North Block, Thames House, Millbank, London, SW1P 4QE, England.

EOS Trans. American Geophys. Union
EOS Transactions of the American Geophysical Union. 1707 L Street, N.W., Washington, DC 20036, USA.

ESO Bull.
European Southern Observatory, Bulletin. Edited by European Southern Observatory. Office of the Director: Hamburg.

ESO Sci. Prepr.
European Southern Observatory, Scientific Preprints. Available from Preprints Service, ESO-Library c/o CERN, 1211 Geneva 23, Switzerland.

ESO Tech. Rep.
European Southern Observatory, (ESO), Technical Report. Published by the European Southern Observatory Telescope Project Division, CERN, Geneva, Switzerland.

Exp. Tech. Phys.

Experimentelle Technik der Physik, VEB Deutscher Verlag der Wissenschaften, Traubenstrasse 10, 108 Berlin 8, German Democratic Republic.

Feinwerktech. Messtech.
F & M. Feinwerktechnik und Messtechnik. Fusion of "Feinwerktechnik" and "Messtechnik" (formerly Zeit-schrift für Instrumentenkunde) beginning with Jahrgang 82, No. 5 (1974). Publishers: Karl Hanser Verlag, Kolbergerstr. 22, D-8000 München 80. F.R.Germany.

Finommech. Mikrotech.
Finommechanika-Mikrotechnika.Vadasz utca 16, Buda-pest 5, Hungary.

Fiz. Szemle
Fizikai Szemle. Kiadja a Lapkiado Vallalat, Budapest VII, Lenin korut 9–11, Hungary.

Fizika
Fizika. 'Mladost' Export-Import, Zagreb, Ilica 30, Yugoslavia.

Found. Phys.
Foundations of Physics. Plenum Publishing Co., 8 Scrubs Lane, Harlesden, London, NW10 6SE, England.

Fundam. Cosmic Phys.
Fundamentals of Cosmic Physics. Gordon and Breach Science Publishers Ltd., New York–London–Paris.

Funkschau
Funkschau. Francis-Verlag, 8 München 37, Postfach 37 01 20, Karlstrasse 37, Germany.

Fys. Tidsskr.
Fysisk Tidsskrift. Subscription address: Jul. Gjellerups Boghandel, Solvgade 87, 1307 Kobenhavn, Denmark.

G. A.A.B.
Giornale dell'A.A.B. Notiziario trimestrale delle attività culturali e scientifiche della Associazione Astrofili Bolog-nesi, Bologna, Italy.

G. Astron.
Giornale di Astronomia, Pubblicazione della Società Astronomica Italiana. Printed by Tipolitografia Lodigraf S.p.A. Lodi (MI).

Gen. Relativ. Gravitation
General Relativity and Gravitation. Published under the auspices of the International Committee on General Relativity and Gravitation GRG. Publishing Office: Plenum Publishing Corporation, 227 West 17th Street, New York, N.Y. 10011, USA.

Geochim. Cosmochim. Acta
Geochimica et Cosmochimica Acta. Journal of the Geo-chemical Society. Publishing House: Pergamon Press, Ltd., Oxford.

Geod. Geophys. Veröff., Reihe III
Geodätische und Geophysikalische Veröffentlichungen. Reihe III: Physik der festen Erde. Herausgegeben vom Nationalkomitee für Geodäsie und Geophysik bei der Akademie der Wissenschaften der Deutschen Demokra-tischen Republik.

Geod. Kartogr.
Geodezja i Kartografia. Komitet Geodezji Polskiej Aka-demii Nauk. Publisher: Państwowe Wydawnictwo Nau-kowe, Warszawa.

Geoexploration
Geoexploration. Elsevier Publishing Co., P.O. Box 211, Amsterdam, Netherlands.

Geomagn. Aehron.
Geomagnetizm i Aehronomiya. Akademiya Nauk SSSR. Izdatel'stvo "Nauka", Moskva [An English translation is published in "Geomagnetism and Aeronomy", American Geophysical Union, Washington, D.C.].

Geophys. J. R. Astron. Soc.
The Geophysical Journal of the Royal Astronomical Society. Published for the Royal Astronomical Society by Blackwell Scientific Publications, Oxford – Edinburgh.

Geophys. Res. Lett.
Geophysical Research Letters. Published monthly by the American Geophysical Union, Washington, D.C., U.S.A.

Geophysics
Geophysics. Society of Exploration Geophysicists, P.O. Box 3098, Tulsa, OK 74101, USA.

GEOS
GEOS. Department of Energy, Mines and Resources, Ottawa, Canada.

Gerlands Beitr. Geophys.
Gerlands Beiträge zur Geophysik. Publisher: Akademische Verlagsgesellschaft Geest & Portig K.-G., Leipzig.

Glasnik Mat.
Glasnik Matematicki. Published by the Society of Mathematicians and Physicists of the S. R. of Croatia. Publisher: Drustvo Matematicara i Fizicara S. R. Hrvatske, Zagreb.

GSFC Doc.
Goddard Space Flight Center, Greenbelt, Maryland. Available from Technical Information Division, Code 250, Goddard Space Flight Center, Greenbelt, Maryland 20771.

Helvetica Phys. Acta
Helvetica Physica Acta. Schweizerische Physikalische Gesellschaft. Publisher: E. Birkhäuser, Elisabethenstr. 19, CH-4000 Basel 10, Switzerland.

I.U.A.A. Bull.
I.U.A.A. Bulletin. International Union of Amateur Astronomers, Contributions. I.U.A.A. c/o Achille Leani, via Bertesi 15, 26100 Cremona, Italy.

IAU Circ.
International Astronomical Union, Circular. Central Bureau for Astronomical Telegrams, Smithsonian Astrophysical Observatory, Cambridge, Mass.

Icarus
Icarus. International Journal of Solar System Studies. Publisher: Academic Press, New York – London.

ICSU Bull.
ICSU Bulletin. International Council of Scientific Unions. Secretariat: 51, Bd de Montmorency, Paris, France.

IEE J. Microwave Opt. Acoust.
IEE Journal on Microwave, Optics and Acoustics. Institution of Electrical Engineers, Publishing Department, P.O. Box 8, Southgate House, Stevenage, Herts. SG1 1HQ, England.

IEEE Spectrum
IEEE Spectrum. Published monthly by the Institute of Electrical and Electronics Engineers, 345 East 47th Street, New York, N.Y. 10017, USA.

IEEE Trans. Aerosp. Electron. Syst.
IEEE Transactions on Aerospace and Electronic Systems. Published by the Institute of Electrical and Electronics Engineers, 345 East 47th Street, New York, N.Y. 10017, USA.

IEEE Trans. Antennas Propag.
IEEE Transactions on Antennas and Propagation. Published by the Institute of Electrical and Electronics Engineers, 345 East 47th Street, New York, N.Y. 10017, USA.

IEEE Trans. Electron Devices
IEEE Transactions on Electron Devices. Published by the Institute of Electrical and Electronics Engineers, 345 East 47th Street, New York, N.Y. 10017, USA.

IEEE Trans. Geosci. Electron.
IEEE Transactions on Geoscience Electronics. Published by the Institute of Electrical and Electronics Engineers, 345 East 47th Street, New York, N.Y. 10017, USA.

IEEE Trans. Instrum. Meas.
IEEE Transactions on Instrumentation and Measurement. Published by the Institute of Electrical and Electronics Engineers, 345 East 47th Street, New York, N.Y. 10017, USA.

IEEE Trans. Microwave Theory Tech.
IEEE Transactions on Microwave Theory and Techniques. Published by the Institute of Electrical and Electronics Engineers, 345 East 47th Street, New York, N.Y. 10017, USA.

IEEE Trans. Nucl. Sci.
IEEE Transactions on Nuclear Science. Institute of Electrical and Electronics Engineers, 345 East 47th Street, New York, NY 10017, USA.

IEEE Trans. Plasma Sci.
IEEE Transactions on Plasma Science. Institute of Electrical and Electronics Engineers, 345 East 47th Street, New York, NY 10017, USA.

Ind. Math.
Industrial Mathematics. Industrial Mathematics Society, P.O. Box 159, Roseville, MI 48066, USA.

Indian East. Eng.
Indian and Eastern Engineer. 'Piramal Mansion', 235 Dr. D. Naoroji Road, Bombay 400001, India.

Indian J. Meteorol. Hydrol. Geophys.
Indian Journal of Meteorology, Hydrology & Geophysics. Formerly: Indian J. Meteorol. Geophys. Indian Meteorological Department, Civil Lines, Delhi 110006, India.

Indian J. Phys.
Indian Journal of Physics. 2 - 3 Raja Subodhchandra, Mallik Road, Calcutta 700032, India.

Indian J. Pure Appl. Math.
Indian Journal of Pure and Applied Mathematics. National Institute of Sciences India, Bahadur Shah Zafar Marg, New Delhi 1, India.

Indian J. Pure Appl. Phys.

Indian Journal of Pure and Applied Physics. Council of Scientific and Industrial Research, Hillside Road, New Delhi 110012, India.

Indian J. Radio Space Phys.
Indian Journal of Radio & Space Physics. Council of Scientific & Industrial Research. Editorial address: Publications & Information Directorate, Hillside Road, New Delhi 110012, India.

Inf. Bull. South. Hemisphere
Information Bulletin for the Southern Hemisphere. Editorial Office: Observatorio Astronómico, La Plata, Argentina.

Inf. Bull. Variable Stars
Commission 27 of the I.A.U. Information Bulletin on Variable Stars. Konkoly Observatory, Budapest.

Infrared Phys.
An International Research Journal. Publisher: Pergamon Press Ltd., Oxford, England.

Ingenieur
De Ingenieur. Koninklijk Institut van Ingenieurs. Editorial address; 23 Prinsessegracht, Den Haag, Netherlands.

Inst. Theor. Astrophys., Blindern–Oslo, Rep.
Institute of Theoretical Astrophysics, Blindern–Oslo, Report. Universitetsforlagets trykningssentral, Oslo.

Int. At. Energy Agency Bull.
International Atomic Energy Agency Bulletin. Kärntnerring 11, P.O. Box. 590, A-1011 Wien, Austria.

Int. J. Electron.
International Journal of Electronics. Taylor and Francis Ltd., 10–14 Macklin Street, London, WC2B 5BF, England.

Int. J. Theor. Phys.
International Journal of Theoretical Physics. Plenum Publishing Co. Ltd., Davis House, 8 Scrubs Lane, London. NW10 6SE, England.

Irish Astron. J.
The Irish Astronomical Journal. A Quarterly Publication under the auspices of the Observatories of Armagh and Dunsink. Armagh Observatory, Northern Ireland.

Izv. Akad. Nauk Armyan. SSR
Izvestiya Akademii Nauk Armyanskoj SSR. Fizika. Publisher: Izdatel'stvo AN Armyanskoj SSR; Erevan.

Izv. Astron. Ehngel'gardt. Obs.
Izvestiya Astronomicheskoj Ehngel'gardtovskoj Observatorii. Izdatel'stvo Kazanskogo Universiteta, Kazan.

Izv. Glav. Astron. Obs. Pulkovo
Izvestiya Glavnoj Astronomicheskoj Observatorii v Pulkove. Akademiya Nauk SSSR. Izdanie Glavnoj astronomicheskoj observatorii v Pulkove, Leningrad.

Izv. Krymskoj Astrofiz. Obs.
Izvestiya Krymskoj Astrofizicheskoj Observatorii. Akademiya Nauk SSR. Publishers: Izdatel'stvo "Nauka", Moskva.

J. American Assoc. Variable Star Obs.
The Journal of the American Association of Variable Star Observers. Published by The American Association of

Variable Star Observers, 187 Concord Avenue, Cambridge, Mass. 02138, USA.

J. Appl. Meteorol.
Journal of Applied Meteorology. American Meteorological Society, 45 Beacon Street, Boston, MA 02108, USA.

J. Appl. Photogr. Eng.
Journal of Applied Photographic Engineering. Society of Photographic Scientists and Engineers, Suite 204, 1330 Massachusetts Avenue, N.W. Washington, D.C. 20005, USA.

J. Appl. Phys.
Journal of Applied Physics. American Institute of Physics, 335 East 45th Street, New York, NY 10017, USA.

J. Astron. Soc. Victoria
The Journal of the Astronomical Society of Victoria. McKinnon–Melbourne.

J. Astron. Soc. Western Australia
The Journal of the Astronomical Society of Western Australia. Edited by the Astronomical Society of Western Australia, Perth, W. A.

J. Astronaut. Sci.
Journal of the Astronautical Sciences. American Astronautical Society, 6060 Duke Street, Alexandria, VA 22304, USA.

J. Atmos. Sci.
Journal of the Atmospheric Sciences. American Meteorological Society, 45 Beacon Street, Boston, MA 02108, USA.

J. Atmos. Terr. Phys.
Journal of Atmospheric and Terrestrial Physics. Pergamon Press Ltd., Oxford, England.

J. British Astron. Assoc.
Journal of the British Astronomical Association. Burlington House, Piccadilly, London, W1V 0NL, England.

J. British Interplanet. Soc.
Journal of the British Interplanetary Society. British Interplanetary Society, 12 Bessborough Gardens, London. SW1V 2JJ, England.

J. Colloid Interface Sci.
Journal of Colloid and Interface Science. Academic Press Inc., 111 Fifth Avenue, New York, N.Y. 10003, USA.

J. Comput. Phys.
Journal of Computational Physics. Academic Press Inc., 111 Fifth Avenue, New York, NY 10003, USA.

J. Fluid Mech.
Journal of Fluid Mechanics. Cambridge University Press, Bentley House, 200 Euston Road, London, NW1 2DB, England.

J. Franklin Inst.
Journal of the Franklin Institute. Pergamon Press Ltd., Headington Hill Hall, Oxford, OX3 0BW, England.

J. Geomagn. Geoelectr.
Journal of Geomagnetism and Geoelectricity. Society of Terrestrial Magnetism and Electricity of Japan, Geophysical Institute, Tokyo University, Tokyo 113, Japan.

J. Geophys.
Journal of Geophysics / Zeitschrift für Geophysik.
Springer-Verlag, D-6900 Heidelberg, Postfach 105280,
F. R. Germany.

J. Geophys. Res.
Journal of Geophysical Research. Published by American
Geophysical Union, 1909 K Street, N.W. Washington D.C.
First section: Space physics; Second section: Physics and
chemistry of the solid earth, planetology, geodesy; Third
section: Oceans and atmospheres.

J. Hist. Astron.
Journal for the History of Astronomy. Published by
Science History Publications Ltd., Halfpenny Furze,
Mill Lane, Chalfont St Giles, Buckinghamshire, England,
and distributed by Neale Watson Academic Publications
Inc., 156 Fifth Avenue, New York, NY 10010, USA.

J. Indian Inst. Sci.
Journal of the Indian Institute of Science. Bangalore
560012, India.

J. Inst. Math. Appl.
Journal of the Institute of Mathematics and its Applica-
tions. Academic Press Inc. (London) Ltd., 24 - 28 Oval
Road, London NW1 7DX, England.

J. Instn. Electron. Telecommun. Eng.
Journal of the Institution of Electronics and Telecom-
munication Engineers. 2 Lodi Road, Institutional Area,
New Delhi 110003, India.

J. Math. Phys., New York
Journal of Mathematical Physics. American Institute of
Physics, 335 East 45th Street, New York, N.Y. 10017,
USA.

J. Mech. Eng. Lab.
Journal of Mechanical Engineering Laboratory. Agency
of Industrial Science and Technology, Igusa Suginami-ku,
Tokyo, Japan.

J. Mol. Spectrosc.
Journal of Molecular Spectroscopy. Academic Press Inc.,
111 Fifth Avenue, New York, NY 10003, USA.

J. Navig.
The Journal of Navigation. The Royal Institute of Naviga-
tion at the Royal Geographical Society, Kensington Gore,
London, SW7 2AT. Scottish Academic Press Ltd., 33
Montgomery Street, Edinburgh EH7 5JX.

J. Opt.
Journal of Optics. Optical Society of India, Department
of Applied Physics, University of Calcutta, 92 Acharya
Prafulla Chandra Road, Calcutta-9, India.

J. Opt. Soc. America
Journal of the Optical Society of America. American
Institute of Physics, 335 East 45th Street, New York,
N.Y. 10017, USA.

J. Phys. A
Journal of Physics A, (Mathematical, Nuclear and
General). Institute of Physics, 47 Belgrave Square,
London, SW1X 8QX, England.

J. Phys. B
Journal of Physics B, (Atomic and Molecular Physics).
Institute of Physics, 47 Belgrave Square, London,
SW1X 8QX, England.

J. Phys. Chem. Ref. Data
Journal of Physical and Chemical Reference Data.
American Chemical Society, 1155 Sixteenth Street,
N.W., Washington, DC 20036, USA.

J. Phys. E
Journal of Physics E, (Scientific Instruments). Formerly:
J. Sci. Instrum. (GB). Institute of Physics, 47 Belgrave
Square, London, SW1X 8QX, England.

J. Phys. G
Journal of Physics G, (Nuclear Physics). Institute of
Physics, 47 Belgrave Square, London, SW1X 8QX,
England.

J. Phys. Soc. Japan
Journal of the Physical Society of Japan. Room 211,
Kikai Shinko Building, Shiba Koen, Minato-ku, Tokyo
105, Japan.

J. Physique
Journal de Physique. Société Française de Physique,
87 bis Avenue du Général Leclerc, 75014 Paris, France.

J. Plasma Phys.
Journal of Plasma Physics. Cambridge University Press,
Bentley House, 200 Euston Road, London, NW1 2DB,
England.

J. Proc. R. Soc. New South Wales
Journal and Proceedings of the Royal Society of New
South Wales. Science House, Gloucester and Essex Streets,
Sydney, N.S.W. 2000, Australia.

J. Quant. Spectrosc. Radiat. Transfer
Journal of Quantitative Spectroscopy & Radiative Trans-
fer. Pergamon Press Ltd., Headington Hill Hall, Oxford,
OX3 OBW, England.

J. R. Astron. Soc. Canada
The Journal of the Royal Astronomical Society of Canada, devoted to the advancement of astronomy and allied
sciences. The Royal Astronomical Society of Canada,
124 Merten Street, Toronto, Ontario, Canada.

J. Radio Res. Lab.
Journal of the Radio Research Laboratories. Chief Planning Section, Radio Research Laboratories, Ministry of
Posts & Telecommunications, Nukui-Kitamachi, Konga-
nei-shi, Tokyo 184, Japan.

J. Res. Natl. Bur. Stand. B
Journal of Research of the National Bureau of Standards.
Section B (Mathematics and Mathematical Physics). US
Government Printing Office, Division of Public
Documents, Washington, DC 20402, USA.

J. Sci. Res. Banaras Hindu Univ.
Journal of Scientific Research of the Banaras Hindu
University. PO-Banaras Hindu University, India.

J. Spacecr. Rockets
Journal of Spacecraft and Rockets. American Institute
of Aeronautics and Astronautics, 1290 Avenue of the
Americas, New York, N.Y. 10019, USA.

J. Test. Eval.
Journal of Testing and Evaluation. American Society for
Testing and Materials, 1916 Race Street, Philadelphia,
PA 19103, USA.

Japanese J. Appl. Phys.

Japanese Journal of Applied Physics. Publication Office, 2nd Toya Kaiji Building, 24-8 Shinbashi, Minato-ku, Tokyo 105, Japan.

Jenaer Rundsch. (Jena Rev.)
Jenaer Rundschau (Jena Review). Publisher: VEB Verlag Technik, Berlin, German Democratic Republic.

JETP Lett.
JETP Letters. A translation of JETP Pis'ma v Redaktsiyu of the Academy of Sciences in the USSR. American Institute of Physics, 335 East 45th Street, New York, NY 10017, USA.

JPL Tech. Mem.
Jet Propulsion Laboratory, California Institute of Technology, Pasadena, California. National Aeronautics and Space Administration. Technical Memorandum.

JPL Tech. Rep.
Jet Propulsion Laboratory, California Institute of Technology, Pasadena, California. National Aeronautics and Space Administration. Technical Report.

K. Danske Vidensk. Selsk. Mat.-Fys. Skr.
Kongelige Danske Videnskabernes Selskab, Matematisk-fysiske Skrifter. Dantes Plads, 5, 1556 Kobenhavn V Publisher: Munksgaard Publishers, 35 Norre Sogade, DK-1370 Kobenhavn, Denmark.

Kexue Tongbao
Kexue Tongbao. Academia Sinica, Peking, People's Republic of China [English translation in: Kexue Tongbao (Scientia) (USA)].

Kometn. Tsirk. *Kiev*
Kometnyj Tsirkulyar. Gruppa po Issledovaniyu Komet Astrosoveta i Mezhduvedomstvennyj Geofizicheskij Komitet Akademii Nauk SSSR. Kievskij Universitet im. T. G. Shevchenko.

Komety i Meteory
Komety i Meteory. Akademiya Nauk Tadzhikskoj SSR. Astronomicheskij Sovet Akademii Nauk SSSR. Publishers: Izdatel'stvo "Donish", Dushanbe.

Kosm. Issled.
Kosmicheskie Issledovaniya. Akademiya Nauk SSSR. Publishers: Izdatel'stvo "Nauka", Moskva [An English translation is published as "Cosmic Research", Consultants Bureau, New York, N.Y.].

Kozmos
Kozmos. Popular Astronomical Journal of the Slovak Central Observatory in Hurbanovo. Publisher: Slovenská ústredná hvezdáren v Hurbanove.

Lab. Equip. Dig.
Laboratory Equipment Digest. Morgan-Grampian Ltd., 30 Calderwood Street, Woolwich, London, SE18 6QH, England.

Lett. Math. Phys.
Letters in Mathematical Physics. D. Reidel Publishing Co., P.O. Box 17, Dordrecht, Netherlands.

L'Universo
L'Universo. Rivista dell'Istituto Geografico Militare. Direzione, Redazione e Amministrazione: Istituto Geografico Militare, Firenze.

Magn. Polya Soln. Pyaten

Magnitnye Polya Solnechnykh Pyaten. (Supplements to Solnechnye Dannye. Byulleten' (*Solar Data*)). Publishers: Izdatel'stvo "Nauka", Leningrad.

Marconi Rev.
Marconi Review. Marconi Co., Marconi House, Chelmsford, Essex, England.

Mater. Pr.
Materiały i Prace. Publications of the Institute of Geophysics, Polish Academy of Sciences. Edited by Państwowe Wydawnictwo Naukowe, Warszawa.

Math. Proc. Cambridge Philos. Soc.
Mathematical Proceedings of the Cambridge Philosophical Society. Formerly: Proceedings of the Cambridge, Philosophical Society (Mathematical and Physical Sciences). Cambridge University Press, Bentley House, 200 Euston Road, London, NW1 2DB, England.

Mech. Res. Commun.
Mechanics Research Communications. Pergamon Press Ltd., Headington Hill Hall, Oxford, OX3 OBW, England.

Mem. Chubu Inst. Technol.
Memoirs Chubu Institute of Technology. Nagoya, Japan.

Mem. Fac. Sci. Kyoto Univ.
Memoirs of the Faculty of Science, Kyoto University. Series of Physics, Astrophysics, Geophysics, and Chemistry. Printed by Yamashiro Printing Publishing Co. Ltd., Kamigyo, Kyoto.

Mem. Japan Astron. Study Assoc.
Memoirs of the Japan Astronomical Study Association. c/o National Science Museum, Ueno Park, Taito-ku, Tokyo, Japan.

Mem. R. Astron. Soc.
Memoirs of the Royal Astronomical Society. Published for the Royal Astronomical Society by Blackwell Scientific Publications, Oxford — London — Edinburgh — Melbourne.

Mém. Sci. Rev. Metall.
Mémoires Scientifiques de la Revue de Metallurgie. 5 Rue Paul Cézanne, 75008 Paris, France.

Mem. Soc. Astron. Italiana
Memorie della Società Astronomica Italiana. Presso Laboratorio di Astrofisica Spaziale, Castella Postale 67, 00044 Frascati, Italy.

Mercury
Mercury. The Journal of the Astronomical Society of the Pacific. Published by the Astronomical Society of the Pacific, 1244 Noriega Street, San Francisco, California 94122, USA.

Messenger
The Messenger, El Mensajero. Edited by European Southern Observatory, Schleißheimer Straße 17, D-8046 Garching bei München, F. R. Germany.

Meteoritics
Meteoritics. The Journal of the Meteoritical Society. Published quarterly by The Meteoritical Society and Arizona State University Bureau of Publications. Editorial address: Center for Meteorite Studies, The Arizona State University, Tempe, Ariz. 85281, USA.

Meteoritika

Akademiya Nauk SSSR. Komitet po Meteoritam. Publishers: Izdatel'stvo "Nauka", Moskva.

Meteorol. Rundsch.
Meteorologische Rundschau. Springer-Verlag, D-1000 Berlin 33, Heidelberger Platz 3, Germany.

Metrologia
Metrologia. Springer-Verlag, Heidelberger Platz 3, D-1000 Berlin 33, Germany.

Microwave J.
Microwave Journal. To be obtained from 610 Washington Street, Dedham Plaza, Dedham, Massachusetts, U.S.A.

Microwaves
Microwaves. Hayden Publishing Co., 50 Essex Street, Rochelle Park, NJ 07662, USA.

Minor Planet Bull.
The Minor Planet Bulletin. Bulletin of the Minor Planets Section of the Association of Lunar and Planetary Observers. Editorial Office: R. G. Hodgson, Dordt College, Sioux Center, Iowa, U.S.A.

Minor Planet Circ.
Minor Planet Circulars. Published by Cincinnati Observatory. Observatory Place, Cincinnati, Ohio 4528, U.S.A.

Mitt. Astron. Ges.
Mitteilungen der Astronomischen Gesellschaft, Hamburg. Printed by G. Braun, GmbH, Karlsruhe. To be available from Astron. Inst. Univ. Bochum.

Mitt. Inst. Theor. Geod. Univ. Bonn
Mitteilungen aus dem Institut für Theoretische Geodäsie der Universität Bonn, Nußallee 17, 5300 Bonn 1, F.R. Germany.

Mitt. Veränderl. Sterne (MVS)
Mitteilungen über Veränderliche Sterne. Herausgegeben von der Sternwarte Sonneberg der Akademie der Wissenschaften der DDR, Sonneberg, German Democratic Republic.

Mod. Geol.
Modern Geology. Gordon & Breach Science Publishers Ltd., 41 and 42 William IV Street, London WC2, England.

Mon. Not. R. Astron. Soc.
Monthly Notices of the Royal Astronomical Society. Published for the Royal Astronomical Society by Blackwell Scientific Publications, Oxford – London – Edinburgh – Melbourne.

Mon. Notes Astron. Soc. South. Africa
Monthly Notes of the Astronomical Society of Southern Africa. Published by the Astronomical Society of Southern Africa, S. A. Astronomical Observatory, Cape Province, South Africa.

Mon. Notes Int. Polar Motion Serv.
Monthly Notes of the International Polar Motion Service. Published by the Central Bureau, International Latitude Observatory of Mizusawa, Mizusawa-shi, Iwate-ken, Japan.

Monitor
Monitor. Formerly: Proc. Instn. Radio Electron. Eng. Australia. Institution of Radio and Electronics Engineers Australia. Science House, 157 Gloucester Street, Sydney, N.S.W. 2000, Australia.

Moon
The Moon. An International Journal of Lunar Studies. Publisher: D. Reidel Publishing Company, Dordrecht, Holland.

Nablyud. Iskusstv. Nebesn. Tel
Nablyudeniya Iskusstvennykh Nebesnykh Tel. Published by Astronomicheskij Sovet Akademii Nauk SSSR, Moskva.

Nachr. Akad. Wiss. Göttingen II
Nachrichten der Akademie der Wissenschaften in Göttingen. II. Mathematisch-Physikalische Klasse. Vandenhoeck & Ruprecht, Göttingen.

Nachr. Elektron.
Nachrichten Elektronik. Formerly: Int. Elektron. Rundsch. Elektro-Welt-Verlag Dr. Hüthig, D - 6900 Heidelberg 1, Postfach 102869, Wilckensstr. 3 - 5, F.R. Germany.

Nachr. Karten-, Vermessungswesen
Nachrichten aus dem Karten- und Vermessungswesen. Editor: Institut für Angewandte Geodäsie (Abt. II des Deutschen Geodätischen Forschungsinstituts). Published by Verlag des Instituts für Angewandte Geodäsie, Frankfurt a. M.

Nachr. Olbers-Ges. Bremen
Nachrichten der Olbers-Gesellschaft Bremen. Werderstraße 73, Bremen.

Nachrichtentech. Elektron.
Nachrichtentechnik Elektronik. VEB Verlag Technik, DDR 102 Berlin, Oranienburger Strasse 13/14, Germany.

NASA Contract. Rep.
NASA Contractor Report. National Aeronautics and Space Administration, Washington, D.C. For sale by the National Technical Information Service, Springfield, Virginia 22161.

NASA Tech. Memo.
NASA Technical Memorandum. National Aeronautics and Space Administration, Washington, D.C. For sale by the National Technical Information Service, Springfield, Virginia 22161.

NASA Tech. Note
NASA Technical Note. National Aeronautics and Space Administration, Washington, D.C. For sale by the National Technical Information Service, Springfield, Virginia 22161.

NASA Tech. Pap.
NASA Technical Paper. National Aeronautics and Space Administration, Washington, D.C. For sale by the National Technical Information Service, Springfield, Virginia 22161.

Nature
Nature. Editorial and Publishing Offices: Macmillan Journals Limited, 4 Little Essex Street, London WC2R 3LF, 711 National Press Building, Washington, D.C. 20045.

Naturwiss. Rundsch.
Naturwissenschaftliche Rundschau. Wissenschaftliche Verlagsgesellschaft mbH. Birkenwaldstraße 44, Stuttgart N.

Naturwissenschaften

Die Naturwissenschaften. Publisher:Springer-Verlag, Berlin–Heidelberg–New York.

Nauch. Inf.
Nauchnye Informatsii. Astronomicheskij Sovet Akademii Nauk SSSR, Moskva.

Navigation (*France*)
Navigation. Institut Française de Navigation, 3 avenue Octave-Greard, Paris 7, France.

NBS Monogr.
National Bureau of Standards Monograph. U.S. Government Printing Office, Washington, D.C. 20402.

Nederlands Tijdschr. Natuurk.
Nederlands Tijdschrift voor Natuurkunde. Publisher: Martinus Nijhoff, Lange Voorhout 9, Den Haag, Netherlands.

New Phys. (Korean Phys. Soc.)
New Physics (Korean Physical Society). Seoul, Korea.

New Scientist
New Scientist. New Science Publications, 128 Long Acre, London, WC2E 9QH, England.

New Zealand Energy J.
New Zealand Energy Journal. Formerly: New Zealand Electr. J. Technical Publications Ltd., 127 Molesworth Street, P.O.Box 3047, Wellington, New Zealand.

News Lett. Astron. Soc. N.Y.
News Letter of the Astronomical Society of New York. A. G. D. Philip (Editor). Astronomical Society of New York, Dudley Observatory, 100 Fuller Road, Albany, N. Y.

Notas Fis.
Notas de Física. Centro Brasileiro de Pesquisas Físicas, Av. Wenceslau Braz 71, Rio de Janeiro, Brazil.

Nouv. Rev. Opt.
Nouvelle Revue d'Optique. Masson, 120 Boulevard Saint-Germain, F-75280 Paris Cedex 06, France.

Nucl. Instrum. Methods
Nuclear Instruments and Methods. North-Holland Publishing Co., P.O. Box 211, Amsterdam, Netherlands.

Nucl. Phys. A
Nuclear Physics, Volume A. North-Holland Publishing Co. P. O. Box 211, Amsterdam, Netherlands.

Nukleonika
Nukleonika. Polska Akademia Nauk, 00-901 Warszawa, Polac Kultury i Nauki, Poland.

Numer. Math.
Numerische Mathematik. Springer-Verlag, Berlin–Heidelberg–New York.

Nuovo Cimento
Il Nuovo Cimento. Rivista Internazionale e Organo della Società Italiana di Fisica, Series A, B. Via Degli Andalo 2, 40124 Bologna, Italy.

Nuovo Cimento, Lett.
Lettere al Nuovo Cimento, a Cura della Società Italiana di Fisica. Via Degli Andalo 2, 40124 Bologna, Italy.

Nuovo Cimento, Riv.

Rivista del Nuovo Cimento. Società Italiana di Física, Via Degli Andalo 2, Bologna 40124, Italy.

Nuovo Cimento, Suppl.
Supplemento al Nuovo Cimento. Società Italiana di Fisica. Via Degli Andalo 2, 40124 Bologna, Italy.

Obs. Artif. Earth Satell.
Observations of Artificial Satellites of the Earth (Nablyudeniya Iskusstvennykh Sputnikov Zemli). Magyar Tudományos Akadémia Csillagvizsgáló Intézete. Budapest.

Obs. Astrophys. Lab., Univ. Helsinki.Rep.
Observatory and Astrophysics Laboratory, University of Helsinki. Report. Helsinki, Finland.

Observatory
The Observatory. A Review of Astronomy. Publishers: The Editors of 'The Observatory', Royal Greenwich Observatory, Herstmonceux Castle, Hailsham, Sussex, England, BN27 1RP.

Occas. Rep. R. Obs. Edinburgh
Occasional Reports of the Royal Observatory, Edinburgh, Blackford Hill, Edinburgh EH9 3HJ, Scotland.

Occultation Newsl.
Occultation Newsletter. Published by the International Occultation Timing Association (I.O.T.A.). 6N106 White Oak Lane, St. Charles, Ill. 60174, USA.

Österreich. Z. Vermessungswes. Photogramm.
Österreichische Zeitschrift für Vermessungswesen und Photogrammetrie. Editor and Publisher: Österreichischer Verein für Vermessungswesen und Photogrammetrie, Wien, Austria.

Opt. Acta
Optica Acta. Taylor and Francis Ltd., 10 - 14 Macklin Street, London, WC2B 5NF, England.

Opt. Commun.
Optics Communications. North-Holland Publishing Co., P.O. Box 211, Amsterdam, Netherlands.

Opt. Eng.
Optical Engineering. Society of Photo-Optical Instrumentation Engineers, 337 Tejon Place, Palos Verdes Estates, CA 90274, USA.

Opt. Laser Technol.
Optics and Laser Technology. Formerly: Opt. Technol. IPC Science and Technology Press Ltd., IPC House, 32 High Street, Guildford, Surrey, England.

Opt. Lett.
Optics Letters. A publication of the Optical Society of America. American Institute of Physics, 335 East 45th Street, New York, N.Y. 10017, USA.

Opt. News
Optics News. Publication of the Optical Society, Washington, D.C.

Opt. Pura Apl.
Optica Pura y Aplicada. Serrano 121, Madrid 6, Spain.

Opt. Spectra
Optical Spectra. The Magazine of Optical/Electro-Optical/Laser Technology. Published by The Optical Publishing

Co., Inc., 59 Bartlett Ave., P.O. Box 1146, Pittsfield, Mass. 01201, USA.

Optik
Optik. Zeitschrift für das gesamte Gebiet der Licht- und Elektronenoptik. Publishers: Wissenschaftliche Verlagsgesellschaft mbH, Postfach 40, D-7000 Stuttgart, F. R. Germany.

Origins of Life
Origins of Life (Formerly Space Life Sciences). An International Journal. Publisher: D. Reidel Publishing Company, Dordrecht, Holland.

Orion
Orion. Zeitschrift der Schweizerischen Astronomischen Gesellschaft (SAG). Bulletin de la Société Astronomique de Suisse (SAS). Printed by A. Schudel & Co. AG, 4125 Riehen, Switzerland.

Oss. Astrofis. Catania, Pubbl.
Osservatorio Astrofisico di Catania, Pubblicazione. Printed by Scuola Salesiana del Libro, Catania.

Oss. Mem. Oss. Astrofis. Arcetri
Osservazioni e Memorie dell'Osservatorio Astrofísico di Arcetri. Università Degli Studi di Firenze, Firenze, Italy.

Oyo Buturi
Oyo Buturi. Japan Society of Applied Physics, Room No. 209-2, Kikai-Shinko Building, 21 Shiba-Koen Minato-ku, Tokyo, Japan.

Perem. Zvezdy, Byull.
Peremennye Zvezdy, Byulleten', izdavaemyj Astronomicheskim Sovetom Akademii Nauk SSSR. Published by Astronomicheskij Sovet Akademii Nauk SSSR, Moskva.

Perem. Zvezdy, Prilozhenie
Peremennye Zvezdy, Prilozhenie (The Variable Stars, Supplement). Astronomicheskij Sovet Akademii Nauk SSSR, Moskva.

Philos. Mag.
The Philosophical Magazine. A Journal of Theoretical, Experimental and Applied Physics. Eighth Series. Publisher: Taylor & Francis, Ltd., 10 - 14 Macklin Street, London, WC2B 5NF, England.

Philos. Trans. R. Soc. London, Ser. A
Philosophical Transactions of the Royal Society of London. Series A, Mathematical and Physical Sciences. Carlton House Terrace, London, SW1Y 5AG, England.

Photogr. Sci. Eng.
Photographic Science and Engineering. Society of Photographic Scientists and Engineers, Suite 204, 1330 Massachusetts Avenue N.W., Washington, DC 20005, USA.

Photogramm. Eng. Remote Sensing
Photogrammetric Engineering and Remote Sensing. Formerly: Photogramm. Eng. American Society of Photogrammetry, 105 North Virginia Avenue, Falls Church, VA 22046, USA.

Phys. Abstr.
Physics Abstracts. Science Abstracts, Series A. An INSPEC Publication, published by The Institution of Electrical Engineers in Association with the Institute of Electrical and Electronics Engineers Inc. Printed by Pindar & Son Ltd., Scarborough, N. Yorkshire, England.

Phys. Ber.
Physikalische Berichte. Herausgegeben von der Deutschen Physikalischen Gesellschaft e.V. und von der Deutschen Akademie der Wissenschaften zu Berlin. Physik-Verlag, Weinheim, F.R. Germany.

Phys. Bl.
Physikalische Blätter. Physik-Verlag GmbH, Pappelallee 3, Postfach 1260/1280, D-6940 Weinheim, F. R. Germany.

Phys. Bull.
Physics Bulletin. Published by the Institute of Physics, 47 Belgrave Square, London, SW1X 8QX, England.

Phys. Earth Planet. Inter.
Physics of the Earth and Planetary Interiors. A journal devoted to observational and experimental studies of the Earth and Planetary interiors and their theoretical interpretation by the physical sciences. Publisher: North-Holland Publishing Company, Amsterdam, Netherlands.

Phys. Educ.
Physics Education. Institute of Physics, 47 Belgrave Square, London, SW1X 8QX, England.

Phys. Fluids
The Physics of Fluids. Published by the American Institute of Physics, 335 East 45th Street, New York, NY 10017, USA.

Phys. Lett.
Physics Letters. Volumes A and B. Publisher: North-Holland Publishing Company, Amsterdam.

Phys. Med. Biol.
Physics in Medicine and Biology. Institute of Physics, 47 Belgrave Square, London, SW1X 8QX, England.

Phys. Norvegica
Physica Norvegica. Institute of Physics, University of Oslo, Blindern, per Oslo, Norway.

Phys. Rep. Phys. Lett. C
Physics Reports. Physics Letters Section C. North-Holland Publishing Co., P.O.Box 103, Amsterdam, Netherlands.

Phys. Rev. A
Physical Review A, General Physics. Published for the American Physical Society by the American Institute of Physics, 335 East 45th Street, New York, NY 10017, USA.

Phys. Rev. B
Physical Review B, Solid State. Published for the American Physical Society by the American Institute of Physics, 335 East 45th Street, New York, NY 10017, USA.

Phys. Rev. C
Physical Review C, Nuclear Physics. Published for the American Physical Society by the American Institute of Physics, 335 East 45th Street, New York, NY 10017, USA.

Phys. Rev. D
Physical Review D, Particles and Fields. Published for the American Physical Society by the American Institute of Physics, 335 East 45th Street, New York, NY 10017, USA.

Phys. Rev. Lett.
Physical Review Letters. Published for the American Physical Society by the American Institute of Physics, 335 East 45th Street, New York, NY 10017, USA.

Phys. Scr.
Physica Scripta. (Formerly Arkiv för Fysik). Published by the Royal Swedish Academy of Sciences, S-104 05 Stockholm 50, Sweden.

Phys. Teach.
Physics Teacher. American Institute of Physics, 335 East 45th Street, New York, NY 10017, USA.

Phys. Today
Physics Today. Published by the American Institute of Physics, 335 East 45th Street, New York, NY 10017, USA.

Physica
Physica. Publishers: North-Holland Publishing Company, Amsterdam, The Netherlands, on request of the Foundation "Physica", Utrecht.

Pis'ma v Astron. Zhurn.
Pis'ma v Astronomicheskij Zhurnal. Akademiya Nauk SSSR. Publishers: Izdatel'stvo 'Nauka', Moskva.

Planet. Space Sci.
Planetary and Space Science. Pergamon Press Ltd., Headington Hill Hall, Oxford, OX3 OBW, England.

Plasma Phys.
Plasma Physics. Pergamon Press Ltd., Headington Hill Hall. Oxford, OX3 OBW, England.

Pokroky
Pokroky matematiky, fyziky a astronomie. Editor: Jednota čs. matematiků a fyziků. Publisher: Academia, Praha.

Postepy Astron.
Postępy Astronomii. Czasopismo Poświecone Upowszechnianiu Wiedzy Astronomicznej. Polskie Towarzystwo Astronomiczne, Warszawa. Printed in Poland by Pánstwowe Wydawnictwo Naukowe, Lódź.

Pramāna
Pramāna. Indian Academy of Sciences, Bangalore 560006, India.

Priroda
Priroda. Publishers: Izdatel'stvo "Nauka", Moskva.

Probl. Kosm. Fiz.
Problemy Kosmichskoj Fiziki. Mezhvedomstvennyj Nauchnoj Sbornik. Izdatel'skoe Obedinenie Vishcha Shkola. Izdatel'stvo pri Kievskom Universitete, Kiev.

Proc. Astron. Soc. Australia
Proceedings of the Astronomical Society of Australia. Published for the Society by Sydney University Press, Sydney.

Proc. IEEE
Proceedings of the IEEE. Published by the Institute of Electrical and Electronics Engineers, 345 East 47th Street, New York, NY 10017, USA.

Proc. Indian Acad. Sci. A
Proceedings of the Indian Academy of Sciences, Section A. Bangalore 560006, India.

Proc. Instn. Electr. Eng.
Proceedings of the Institution of Electrical Engineers. Institution of Electrical Engineers, Savoy Place, London, WC2R OBL, England.

Proc. Int. Latitude Obs. Mizusawa
Proceedings of the International Latitude Observatory of Mizusawa. Published by the International Latitude Observatory of Mizusawa, Japan.

Proc. Japan Acad.
Proceedings of the Japan Academy. Ueno Park, Tokyo, Japan.

Proc. K. Nederlandse Akad. Wet. B
Koninklijke Nederlandse Akademie van Wetenschappen. Proceedings. Series B, Physical Sciences. Publisher: North-Holland Publishing Company, Amsterdam, Netherlands.

Proc. Natl. Acad. Sci. U.S.A.
Proceedings of the National Academy of Sciences of the United States of America. National Academy of Sciences, 2101 Constitution Avenue, Washington, DC 20418, USA.

Proc. R. Soc. London, Ser. A
Proceedings of the Royal Society of London. Series A: Mathematical and Physical Sciences. Published by the Royal Society, 6 Carlton House Terrace, London, SW1Y 5AG, England.

Proc. Res. Inst. Atmos. Nagoya Univ.
Proceedings of the Research Institute of Atmospherics Nagoya University. Nagoya University, 13 Honohara, 3 Chrome, Toyokawa 442, Japan.

Prog. Theor. Phys.
Progress of Theoretical Physics. Published for the Research Institute for Fundamental Physics and the Physical Society of Japan. Publication Office: Progress of Theoretical Physics, Yukawa Hall, Kyoto University, 606 Kyoto, Japan.

Prog. Theor. Phys., Suppl.
Supplement of the Progress of Theoretical Physics. Published for the Research Institute for Fundamental Physics and The Physical Society of Japan. Publication Office: Progress of Theoretical Physics. Yukawa Hall, Kyoto University, 606 Kyoto, Japan.

PTB Mitt.
PTB Mitteilungen. Forschen + Prüfen. Fachorgan für Wirtschaft und Wissenschaft. Amts- und Mitteilungsblatt der Physikalisch-Technischen Bundesanstalt. Braunschweig—Berlin. Deutscher Eichverlag, Postfach 3367, D-3300 Braunschweig, F.R. Germany.

Publ. Astron. Soc. Japan
Publications of the Astronomical Society of Japan. Published by the Astronomical Society of Japan. Office of the Society: Tokyo Astronomical Observatory, Mitaka, Tokyo. Agent: Maruzen Co. Ltd. (Export Department), Nihonbashi, Tokyo, Japan.

Publ. Astron. Soc. Pacific
Publications of the Astronomical Society of the Pacific. Published in Provo, Utah, by the Astronomical Society of the Pacific, San Francisco, California. Printed by Brigham Young University Printing Service, Provo, Utah 84602, USA.

Publ. Dominion Astrophys. Obs.
Publications of the Dominion Astrophysical Observatory, Victoria, B.C. National Research Council of Canada.

Publ. Eidg. Sternw. Zürich
Publikationen der Eidgenössischen Sternwarte Zürich.

Schulthess Polygraphischer Verlag, Zürich.

Publ. Inst. Geophys., Polish Acad. Sci., F
Publications of the Institute of Geophysics, Polish
Academy of Sciences, F: Planetary geodesy.
Państwowe Wydawnictwo Naukowe, Warszawa, Poland.

Publ. Inst. R. Meteorol. Belgique A
Publications, Institut Royal Météorologique de Belgique.
Serie A. 3 Avenue Circulaire, Uccle-Bruxelles 1180,
Belgium.

Publ. Int. Latitude Obs. Mizusawa
Publications of the International Latitude Observatory of
Mizusawa. Published by the International Latitude Ob-
servatory of Mizusawa, Japan.

Publ. R. Obs. Edinburgh
Publications of the Royal Observatory, Edinburgh.
Published by The Royal Observatory, Edinburgh,
Scotland.

Publ. Tartu Astrofiz. Obs.
W. Struve nimelise Tartu Astrofüüsika Observatooriumi,
Publikatsioonid. Eesti NSV Teaduste Akadeemia, Tartu.

Publ. United States Naval Obs.
Publications of the United States Naval Observatory.
Department of the Navy, U.S. Naval Observatory,
Washington. U.S. Government Printing Office, Washing-
ton, D.C.

Publ. Variable Star Sect. R. Astron. Soc. New Zealand
Publications of the Variable Star Section, Royal Astro-
nomical Society of New Zealand. Director: F. M.
Bateson, Greerton, Tauranga, New Zealand.

Pure Appl. Geophys.
Pure and Applied Geophysics. Birkhäuser Verlag,
CH-4010 Basel, Elizabethenstrasse 19, Switzerland.

Q. Appl. Math.
Quarterly of Applied Mathematics. American Mathematical Society, P.O. Box 1571, Providence, RI 02901, USA.

Q. Bull. Sol. Act.
International Astronomical Union, Quarterly Bulletin on
Solar Activity. Published by the Eidgenössische Stern-
warte in Zürich with financial support from UNESCO.

Q. J. R. Astron. Soc.
Quarterly Journal of the Royal Astronomical Society.
Burlington House, London, W1V ONL, England.

Q. J. R. Meteorol. Soc.
Quarterly Journal of the Royal Meteorological Society.
Cromwell House, High Street, Bracknell, Berks, England.

R. Obs. Ann.
Royal Observatory Annals. Royal Greenwich Observatory,
Herstmonceux Castle, Hailsham, East Sussex, BN 27 1RP,
England.

Radio Commun.
Radio Communication. Radio Society of Great Britain,
35 Doughty Street, London, WC1N 2AE, England.

Radio Sci.
Radio Science. American Geophysical Union, 2901
Byrdhill Road, Richmond, VA 23228, USA.

Radiochem. Radioanal. Lett.

Radiochemical and Radioanalytical Letters. Elsevier
Sequoia S.A., P.O. Box 851, CH-1001 Lausanne,
Switzerland. Akademiai Kiado, Alkotmany U 21,
Budapest 5, Hungary.

Radiotekh. Ehlektron.
Radiotekhnika i Ehlektronika. Moskva TSP-3, Pr. Karl
Marx 18, USSR.

Rech. Aerosp.
Recherche Aerospatiale. Office National d'Études et de
Recherches Aerospatiales, 29 Avenue de la Division
Leclerc, 92320-Chatillon, France.

Recherche
Recherche, 4 Place de l'Odéon, Paris 6, France.

Ref. Zh. 51. Astron.
Referativnyj Zhurnal. 51. Astronomiya. Vsesoyuznyj
Institut Nachnoj i Tekhnicheskoj Informatsii. Moskva.

Ref. Zh. 52. Geod. i Aehrosemka
Referativnyj Zhurnal. 52. Geodeziya i Aehrosemka.
Vsesoyuznyj Institut Nachnoj i Tekhnicheskoj Infor-
matsii. Moskva.

Ref. Zh. 62. Issled. kosm. prostranstva
Referativnyj Zhurnal. 62. Issledovanie Kosmicheskogo
Prostranstva. Vsesoyuznyj Institut Nauchnoj i Tekhni-
cheskoj Informatsii. Moskva.

Rep. Finnish Geod. Inst.
Reports of the Finnish Geodetic Institute. Suomen
Geodeettisen Laitoksen Tiedonantoja. Helsinki, Finland.

Rep. Inst. Phys. Chem. Res.
Reports of the Institute of Physical and Chemical Re-
search. Rikagaku Kenkyushu, Wako-shi, Saitama 351,
Japan.

Rep. Ionos. Space Res. Japan
Report of Ionosphere and Space Research in Japan.
Institute of Space and Aeronautical Science, University of
Tokyo, Komaba, Meguro-ku, Tokyo 153, Japan.

Rep. Math. Phys.
Reports on Mathematical Physics. Pergamon Press Ltd.,
Headington Hill Hall, Oxford OX3 0BW, England.
Subscription address: ARS Polona-Ruch Foreign Trade
Enterprise, Krakowskie Przedmiescie 7, 00-068
Warszawa.

Rep. NRL Prog.
Report of NRL Progress. National Technical Information
Service, U.S. Department of Commerce, Springfield,
VA 22151, USA.

Rep. Obs. Lund
Reports from the Observatory of Lund.

Rep. Prog. Phys.
Reports on Progress in Physics. Published by the Institute
of Physics, 47 Belgrave Square, London, SW1X 8QX,
England.

Rev. Acad. Cienc. Zaragoza
Revista de la Academia de Ciencias Zaragoza. Academia
de Ciencias Exactas, Fisico-Quimicas y Naturales de
Zaragoza, Plaza de Paraiso 1, Zaragoza, Spain.

Rev. Astron.

Revista Astronomica. Organo de la Asociación Argentina Amigos de la Astronomia, Avenida Patricias Argentinas 550, Buenos Aires 5, Argentina.

Rev. Brasil. Fis.
Revista Brasileira de Fisica. Sociedade Brasileira de Fisica, Cx. Postal 20553, Sao Paolo SP, Brazil.

Rev. Céthédec
Revue du Céthédec. Centre d'Études Théoriques de la Détection et des Communications, 5 bis, avenue de la Porte de Sèvres, Paris 15, France.

Rev. Fac. Sci. Univ. Istanbul C
Revue de la Faculté des Sciences de l'Université d'Istanbul (Istanbul Universitesi Fen Fakultesi Mecmuasi) Série C (Astronomie, Physique, Chimie). Beyazit, Istanbul, Turkey.

Rev. Geofis.
Revista de Geofisica, Instituto Nacional de Geofisica, Serrano 123, Madrid 2, Spain.

Rev. Geophys. Space Phys.
Reviews of Geophysics and Space Physics (formerly Reviews of Geophysics). Published by the American Geophysical Union, 1909 K Street, N.W., Washington, DC 20006, USA.

Rev. Mexicana Astron. Astrofis.
Revista Mexicana de Astronomia y Astrofisica. Dirección: Instituto de Astronomia, Universidad Nacional Autónoma de México, Apartado Postal 70-264, Mexico 20, D.F., Mexico.

Rev. Mexicana Fis.
Revista Mexicana de Fisica. Sociedad Mexicana de Fisica, Apartado Postal No. 20-364, Mexico 20, D.F, Mexico.

Rev. Mod. Phys.
Reviews of Modern Physics. Published for the American Physical Society by the American Institute of Physics, 335 East 35th Street, New York, NY 10017, USA.

Rev. Phys. Appl.
Revue de Physique Appliquée. Société Française de Physique. 87 bis, Avenue du Général-Leclerc, 75014 Paris, France.

Rev. Polytech.
Revue Polytechnique. Chemin de la Caroline 22, 1213 Petit-Lancy, Genève, Switzerland.

Rev. Radio Res. Lab.
Review of the Radio Research Laboratories.Tokyo, Japan.

Rev. Roumaine Phys.
Revue Roumaine de Physique. Academie Republicii Populare Romine, Boite Postale 134 - 135, Bucuresti, Rumania.

Rev. Sci. Instrum.
Review of Scientific Instruments.American Institute of Physics, 335 East 45th Street, New York, NY 10017, USA.

Rezul't. Nablyud. Iskusstv. Sputnikov Zemli
Rezul'taty Nablyudenij Iskusstvennykh Sputnikov Zemli. Published by Astronomicheskij Sovet Akademii Nauk SSSR, Ryazanskij Gosudarstvennyj Pedagogicheskij Institut, Ryazan'.

Rezul't. Nablyud. Sovet. Iskusstv. Sputnikov Zemli
Rezul'taty Nablyudenij Sovetskikh Iskusstvennykh Sputnikov Zemli. Published by Astronomicheskij Sovet Akademii Nauk SSSR, Moskva. Replaced after No. 140 by Rezul'taty Nablyudenij Iskusstvennykh Sputnikov Zemli.

Říše hvězd
Říše hvězd. Czechoslovak popular astronomical journal. Publisher: Orbis, Praha.

Rumanian Sci. Abstr.
Rumanian Scientific Abstracts. Natural Sciences. Publishers: The Scientific Documentation Centre of the Academy of the Socialist Republic of Romania, Bucureşti.

Sci. American
Scientific American. 415 Madison Avenue, New York, NY 10017, USA.

Sci. Cult.
Science and Culture. Indian Science News Association, 92 Acharya Prafullachandra Road, Calcutta 9, India.

Sci. Dimension
Science Dimension. National Research Council of Canada, Ottawa K1A OR6, Canada.

Sci. Pap. Coll. Gen. Educ. Univ. Tokyo
Scientific Papers of the College of General Education, University of Tokyo. Komaba, Meguro-ku, Tokyo, Japan.

Sci. Pap. Inst. Phys. Chem. Res.
Scientific Papers of the Institute of Physical and Chemical Research. Rikagaku Kenkyusho, Wako-shi, Saitama 351, Japan.

Sci. Prog.
Science Progress. Blackwell Scientific Publications, Oxford, England.

Sci. Prog. Découverte
Science Progrès Découverte (formerly Science Progrès, La Nature). Revue publiée avec la participation du Palais de la Découverte. Published by Dunod, Editeur, Paris. Imprimerie Bayeusaine, Bayeux.

Sci. Rep. Tôhoku Univ. I.
The Science Reports of the Tôhoku University. First Series (Physics, Chemistry, Astronomy). Published by the Faculty of Science, Tôhoku University, Sendai, Japan.

Sci. Sinica
Scientia Sinica. Edited by Editorial Committee of Scientia Sinica, Peking. Published by Science Press, Peking, China.

Sci. Sintering
Science of Sintering. Yugoslav Committee for ETAN, P.O. Box 356 (ETAN), 11001 Beograd, Yugoslavia.

Science
Science. American Association for the Advancement of Science, 1515 Massachusetts Avenue, N.W., Washington, D.C. 20005, USA.

Sdelovaci Tech.
Sdelovaci Technika. Publishers of Technical Literature, Spalena 51, 11302 Praha 1, Czechoslovakia.

SIAM J. Appl. Math.

SIAM Journal on Applied Mathematics. Society for
Industrial and Applied Mathematics, 33 South 17th
Street, Philadelphia, PA 19103, USA.

Siemens Rev.
Siemens Review. Siemens-Aktiengesellschaft, Postfach
325, 8520 Erlangen 2, F. R. Germany.

Simon Stevin
Simon Stevin. De Natuur - en Geneeskundige Vennoot-
schap, Rozier 44, B-9000 Gent, Belgium.

Sitzungsber. Akad. Wiss. DDR
Sitzungsberichte der Akademie der Wissenschaften
der DDR. Mathematik-Naturwissenschaften-Technik.
Akademie-Verlag, Berlin.

Sitzungsber. Bayerische Akad. Wiss.
Bayerische Akademie der Wissenschaften. Mathematisch-
Naturwissenschaftliche Klasse. Sitzungsberichte. Publisher: Verlag der Bayerischen Akademie der Wissenschaften, München.

Sitzungsber. Heidelberger Akad. Wiss.
Sitzungsberichte der Heidelberger Akademie der Wissen-
schaften. Mathematisch-Naturwissenschaftliche Klasse.
Publisher: Springer-Verlag, Heidelberg.

Sitzungsber. Österreich. Akad. Wiss.
Sitzungsberichte. Österreichische Akademie der Wissen-
schaften. Mathematisch-Naturwissenschaftliche Klasse.
Abteilung II: Mathematik, Astronomie, Meteorologie und
Technik. Publisher: Springer-Verlag, Wien.

Sky Telesc.
Sky and Telescope. Published by Sky Publishing Corpora-
tion, 49-50-51 Bay State Road, Cambridge, Mass. 02138,
USA.

Slaboproudy Obz.
Slaboproudy Obzor. Krakovska 8, Praha 1,
Czechoslovakia.

Smithsonian Astrophys. Obs., Spec. Rep.
Smithsonian Institution, Astrophysical Observatory,
Cambridge, Massachusetts 02138. Research in Space
Science. SAO Special Report.

Smithsonian Contrib. Astrophys.
Smithsonian Contributions to Astrophysics. Smithsonian
Institution Astrophysical Observatory, Cambridge, Mass.
Printed by Smithsonian Institution Press, City of Washing-
ton. For sale by the Superintendent of Documents, U. S.
Government Printing Office, Washington, D. C.

Smithsonian Year
Smithsonian Year. Annual Report of the Smithsonian
Institution, including the financial report of the Executive Committee of the Boards of Regents. Published
by the Smithsonian Institution, Washington, D.C.

Sol. Energy
Solar Energy. Pergamon Press, Maxwell House, Fairview
Park, Elmsford, NY 10523, USA

Sol. Phys.
Solar Physics. A Journal for Solar Research and the
Study of Solar Terrestrial Physics. Publisher: D. Reidel
Publishing Company, Dordrecht, Holland.

Soln. Dannye, Byull.
Solnechnye Dannye. Byulleten'. *(Solar Data).* Publishers:

Izdatel'stvo "Nauka", Leningradskoe Otdelenie, Leningrad.

Soobshch. Byurakan. Obs.
Soobshcheniya Byurakanskoj Observatorii. Akademiya
Nauk Armyanskoj SSR, Erevan.

Soobshch. Gos. Astron. Inst. Shternberg
Soobshcheniya Gosudarstvennogo Astronomicheskogo
Instituta im. P.K. Shternberga. Publishers: Izdatel'stvo
Moskovskogo Universiteta, Moskva.

Soobshch. Spets. Astrofiz. Obs.
Soobshcheniya Spetsial'noj Astrofizicheskoj Observatorii.
Izdanie Spetsial'noj Astrofizicheskoj Observatorii AN
SSSR.

South African Astron. Obs. Circ.
South African Astronomical Observatory, Circulars.
S.A. Astronomical Observatory, Observatory, Cape.

South African J. Antarct. Res.
South African Journal of Antarctic Research. Editorial
Address: P.O. Box 3718, Johannesburg, South Africa.

South. Stars
Southern Stars. The Journal of the Royal Astronomical
Society of New Zealand (Inc.). Address of the Society:
P.O. Box 3181, Wellington C1, New Zealand.

Soviet Astron.
Soviet Astronomy. A translation of Astronomicheskij
Zhurnal (Astronomical Journal). Published by the
American Institute of Physics, New York, N.Y.

Space Sci. Instrum.
Space Science Instrumentation. An International Journal
of Scientific Instruments for Aircraft, Balloons, Sounding
Rockets, and Spacecraft. Published by D. Reidel Publishing Company, Dordrecht, Holland.

Space Sci. Rev.
Space Science Reviews. Publishers: D. Reidel Publishing
Company, Dordrecht, Holland.

Spaceflight
Spaceflight. Published by the British Interplanetary
Society. Printed by Unwin Brothers Ltd., at the Gresham
Press, Old Woking, England.

Spaceworld
Spaceworld. Palmer Publications Inc., Amherst, WI
54406, USA.

Sterne
Die Sterne. Zeitschrift für alle Gebiete der Himmelskunde.
Johann Ambrosius Barth, Leipzig, German Democratic
Republic.

Sterne Weltraum
Sterne und Weltraum. Astronomische Monatsschrift.
Publisher: Verlag Sterne und Weltraum Dr. Vehrenberg,
Düsseldorf, F.R. Germany.

Sternenbote
Sternenbote. Monatsschrift für Österreichs Amateurastro-
nomen. Publisher: Astronomisches Büro, Hermann
Mucke, Wien, Austria.

Stockholms Obs. Ann.
Stockholms Observatorium Annaler. Printed by Alm-
quist & Wiksell, Stockholm, Sweden.

Stockholms Obs. Rep.
Stockholms Observatorium, Saltsjöbaden, Sweden,
Report.

Strolling Astron.
The Strolling Astronomer. The Journal of The Association of Lunar and Planetary Observers, Publication Office: The Strolling Astronomer, Box 3 AZ, University
Park, New Mexico, USA.

Stud. Appl. Math.
Studies in Applied Mathematics. American Elsevier
Publishing Co., 52 Vanderbilt Avenue, New York, NY
10017, USA.

Stud. Cercet. Astron.
Studii şi Cercetări de Astronomie. Editura Academiei Republicii Socialiste România. Editorial Office: Observato-
rul Astronomic, Bucureşti, Rumania.

Stud. Cercet. Fiz.
Studii şi Cercetări de Fizica. Academia Republicii Populare Romine. P.O. Box 134-5, Calca Victoriei 126, Bucureşti, Rumania.

Stud. Geophys. Geod.
Studia geophysica et geodaetica. Published for the Geo-
physical Institute of the Czechoslovak Academy of
Sciences by Academia, Praha.

Stud. Soc. Sci. Torunensis
Studia Societatis Scientiarum Torunensis, Toruń −
Polonia. Sectio F (Astronomia).

Stud. Univ. Babeş-Bolyai
Studia Universitatis Babeş-Bolyai. Series Mathematica-
Physica. Publishers: Intreprinderea Poligrafica, Cluj.

Tartu Astron. Obs. Teated
Tartu Astronoomia Observatoorium Teated. Eesti NSV
Teaduste Akadeemia W. Struve nim. Tartu Astrofüüsika
Observatoorium, Tartu.

Technica
Technica.E. Birkhäuser, CH-4010 Basel 10, Switzerland.

Tectonophysics
Tectonophysics. Elsevier Scientific Publishing Co., P.O.
Box 211, Amsterdam, Netherlands.

Tegnikon
Tegnikon. Die Suid-Afrikaanse Akademie wir Wetenskap
en Kuns, Engelenburghuis, Hamiltonstraat, Postbus 538,
Pretoria, S. Africa.

Tellus
Tellus, a bi-monthly Journal of Geophysics. Svenska
Geofysiska Foreningen, Arrhenius laboratoriet, Fack,
S - 104 05 Stockholm, Sweden.

Tijdschr. Ned. Elektron. Radiogenoot.
Tijdschrift van het Nederlands Elektronica- en Radio-
genootschap. Postbus 39, Leidschendam, Netherlands.

Tokyo Astron. Bull.
Tokyo Astronomical Observatory, Japan. Tokyo Astronomical Bulletin.

Tokyo Astron. Obs. Rep.
University of Tokyo, Tokyo Astronomical Observatory,
Japan. Report.

Tr. Astrofiz. Inst. Alma-Ata
Trudy Astrofizicheskogo Instituta, Alma-Ata. Akademiya
Nauk Kazakhskoj SSR. Publishers: Izdatel'stvo "Nauka"
Kazakhskoj SSR, Alma-Ata.

Tr. Astron. Obs., *Leningrad*
Uchenye Zapiski Gosudarstvennogo Universiteta im.
A. A. Zhdanova, Seriya matematicheskikh nauk = Trudy
Astronomicheskoj Observatorii. Izdatel'stvo Leningrad-
skogo Universiteta, Leningrad.

Tr. Glav. Astron. Obs. Pulkovo
Trudy Glavnoj Astronomicheskoj Observatorii v Pulkove.
Akademiya Nauk SSR. Izdanie Glavnoj astronomiches-
koj observatorii v Pulkove, Leningrad.

Tr. Inst. Teor. Astron., *Leningrad*
Trudy Instituta Teoreticheskoj Astronomii. Akademiya
Nauk SSSR. Publishers: Izdatel'stvo "Nauka", Leningrad.

Tr. Kazan. Gorod. Astron. Obs.
Trudy Kazanskoj Gorodskoj Astronomicheskoj Observatorii. Izdatel'stvo Kazanskogo Universiteta, Kazan.

Tr. Tashkent. Astron. Obs.
Trudy Tashkentskoj Astronomicheskoj Observatorii.
Akademiya Nauk Uzbekskoj SSR. Publishers: Izdatel'-
stvo "FAN" Uzbekskoj SSR, Tashkent.

Trans. Astron. Obs. Yale Univ.
Transactions of the Astronomical Observatory of Yale
University. Published by the Observatory, New Haven.

Trans. IAU
Transactions of the International Astronomical Union.
Published and distributed for the IAU (UAI) by D. Rei-
del Publishing Company, Dordrecht, Holland − Boston,
U.S.A.

Trans. Inst. Electron. Commun. Eng. Japan E
Transactions of the Institute of Electronics and Communication Engineers of Japan, Section E. Denshi
Tsushin Gakkai, Kikai-Shinko-kaikan Bldg., Shiba Park
21-1-5, Minatoku, Tokyo, Japan. English translation of
selected articles appear in Electron & Commun. Jap.
(USA) and Syst. Comput. Control (USA).

Trans. R. Soc. Canada
Transactions of the Royal Society of Canada (Mémoires
de la Société Royale du Canada). National Library,
395 Wellington Street, Ottawa 4, Canada. University of
Toronto Press, Toronto 5, Ontario.

Tsirk. Astron. Inst. Tashkent
Tsirkulyar Astronomicheskogo Instituta. Akademiya
Nauk Uzbekskoj SSR. Izdatel'stvo "FAN" Uzbekskoj
SSR, Tashkent.

Tsirk. Astron. Obs. L'vov
Tsirkulyar. Astronomicheskaya Observatoriya. L'vovskij
Ordena Lenina Gosudarstvennyj Universitet imeni Ivana
Franko. Publisher: Izdatel'stvo L'vovskogo Universiteta,
L'vov.

Umschau
Umschau in Wissenschaft und Technik. Umschau Verlag,
Stuttgarter Str. 18 - 24, D-6000 Frankfurt/M., F.R.
Germany.

United States Naval Obs., Circ.
United States Naval Observatory, Circular. U.S. Naval
Observatory, Washington, D.C. 20390.

Urania Barcelona
Urania. Revista de Astronomía y Ciencias Afines. Órgano de la Sociedad Astronómica de España y América, Barcelona; Unión Nacional de Astronomía y Ciencias Afines, Madrid, Spain.

Urania Kraków
Urania. Miesięcznik Polskiego Towarzystwa Miłośników Astronomii, Kraków. Publisher: Krakowska Drukarnia Prasowa, Kraków, Poland.

Vasiona
Vasiona. Revue d'Astronomie et d'Astronautique. Bulletin de la Société Astronomique "R. Bosković", Beograd.

Vatican Obs. Publ.
Vatican Observatory Publications, Specola Vaticana, Città del Vaticano, Castel Gandolfo, Italy.

Veröff. Astron. Rechen-Inst. Heidelberg
Veröffentlichungen des Astronomischen Rechen-Instituts Heidelberg. Verlag G. Braun, Karlsruhe, F.R. Germany.

Veröff. Sternw. Sonneberg
Akademie der Wissenschaften der DDR, Zentralinstitut für Astrophysik, Veröffentlichungen der Sternwarte in Sonneberg. Publisher: Akademie-Verlag, Berlin, German Democratic Republic.

Veröff. Zentralinst. Phys. Erde
Akademie der Wissenschaften der DDR, Forschungsbereich Geo- und Kosmowissenschaften. Veröffentlichungen des Zentralinstituts für Physik der Erde, Potsdam, German Democratic Republic.

Vesmír
Vesmír. Přírodovědecky časopis Čs. akadmie věd. Publisher: Academia, Praha.

Vestn. Khar'kov. Univ.
Vestnik Khar'kovskogo Universiteta. Seriya Astronomicheskaya. Publishers: Izdatel'stvo Khar'kovskogo Universiteta, Khar'kov.

Vestn. Kiev. Univ.
Vestnik Kievskogo Universiteta. Seriya Astronomii. Publishers: Izdatel'stvo Kievskogo Universiteta, Kiev.

Vierteljahrsschr. Naturforsch. Ges. Zürich
Vierteljahrsschrift der Naturforschenden Gesellschaft in Zürich. Printer and Publisher: Leeman AG, Zürich, Switzerland.

Vistas Astron.
Vistas in Astronomy. An international review journal. Pergamon Press, Oxford—New York—Braunschweig.

Weather
Weather. James Glaisher House, Grenville Place, Bracknell, Berks RG12 1BX, England.

Wiad. Telekomun.
Wiadomosci Telekomunikacyjne. Editorial address: Warszawa, Kazimierzowska 52, Poland.

Wireless World
Wireless World. IPC Electrical-Electronic Press Ltd., Dorset House, Stamford Street, London, SE1 9LU, England.

Wiss. Z. Friedrich-Schiller-Univ. Jena
Wissenschaftliche Zeitschrift der Friedrich-Schiller-Universität. Jena. Mathematisch-Naturwissenschaftliche Reihe; Edited by the Rektor der Friedrich-Schiller-Universität Jena, Am Anger 24, Jena, German Democratic Republic.

Wiss. Z. Humboldt-Univ. Berlin
Wissenschaftliche Zeitschrift der Humboldt-Universität zu Berlin. Mathematisch-Naturwissenschaftliche Reihe. Edited by the Rektor der Humboldt-Universität Berlin, Unter den Linden 6, 108 Berlin, German Democratic Republic.

Z. angew. Math. Mech.
Zeitschrift für angewandte Mathematik und Mechanik. Akademie-Verlag GmbH, 108 Berlin, Leipziger Strasse 3–4, German Democratic Republic.

Z. Elektr. Inf.- Energietech.
Zeitschrift für Elektrische Informations- und Energietechnik. Akademische Verlagsgesellschaft Geest & Portig K.G., DDR 701 Leipzig, Sternwartenstrasse 8, Germany. Formerly: Hochfrequenztech. & Elektroakust. (Germany) and Wiss. Z. Elektrotech. (Germany).

Z. Naturforsch.
Zeitschrift für Naturforschung. Europhysics Journal. Teil a: Astrophysik, Physik, Physikalische Chemie. Published by Verlag der Zeitschrift für Naturforschung, Tübingen, F.R. Germany.

Z. Phys. A
Zeitschrift für Physik A. Atoms and Nuclei. Springer-Verlag, Berlin—Heidelberg—New York.

Z. Phys. B
Zeitschrift für Physik B. Condensed Matter and Quanta. Springer-Verlag, Berlin—Heidelberg—New York.

Z. Vermessungswes.
Zeitschrift für Vermessungswesen. Verlag Konrad Wittwer, 7000 Stuttgart 1, Nordbahnhofstrasse 16, Postfach 147, F.R. Germany.

Zemlya i Vselennaya
Zemlya i Vselennaya. Astronomiya, Geofizika, Issledovaniya Kosmicheskogo Prostranstva. Nauchno-Populyarnyj Zhurnal Akademii Nauk SSSR. Publishers: Izdatel'stvo "Nauka", Moskva.

Zenit
Populair wetenschappelijk maandblad over sterrenkunde/weerkunde/ruimtevaart/ruimte-onderzoek/aanverwante wetenschappen en technieken. Bureau: Stichting De Koepel, Utrecht.

Zentralbl. Math. Grenzgeb. – Math. Abstr.
Zentralblatt für Mathematik und ihre Grenzgebiete – Mathematics Abstracts. Publisher: Springer-Verlag, Berlin—Heidelberg—New York.

Zvaigžņota Debess
Latvijas PSR Zinātņu Akadēmijas Radioastrofizikas Observatorijas Populārzinatnisks Gadalaiku Izdevums. Izdevnieciba "Zinātne", Riga.

002 Bibliographical Publications, Catalogues, Atlases

002.001 **Astrolabe catalogue CASF 3 of San Fernando.**
M. Sánchez.
Astron. Astrophys., Suppl. Ser., Vol. 29, 245 - 247 (1977).
In French.
This paper shows the corrections to the adopted coordinates of the stars observed with the astrolabe at San Fernando, Spain, between April 1973 and May 1976. In this third catalogue campaign 155 stars of FK4 and 63 stars of FK4 Sup have been studied. The same computation method as in the two preceding catalogues was applied.

002.002 **A uniform edition of the Stockholm Southern Milky Way Survey (magnetic tape).**
J. Andersen.
Astron. Astrophys., Suppl. Ser., Vol. 29, 257 (1977).
A magnetic-tape version in a uniform format of parts I to III of the Stockholm Southern Milky Way Survey (Sundman et al. 1974, Nordström 1975, Lodén et al. 1976) has been produced. The positions of the stars in Catalogue I have been recomputed to the same accuracy as that of the later catalogues.

002.003 **A general catalogue of *UBV* photoelectric photometry (magnetic tape).**
J.-C. Mermilliod, B. Nicolet.
Astron. Astrophys., Suppl. Ser., Vol. 29, 259 - 261 (1977).
This catalogue collects the *UBV* photoelectric photometry published since 1953 until the end of 1975, the total being more than 73,000 measurements for some 53,000 stars.

002.004 **MK spectral classifications.** Third general catalogue.
Compiled by W. Buscombe.
Northwestern University, Evanston. 263 pp. (1977). − See also 12.114.156.

002.005 **Atlas of galactic globular clusters with colour magnitude diagrams.** G. Alcaíno.
Ediciones Nueva Universidad, Universidad Católica de Chile.
Vicerrectoría de Comunicaciones. 108 pp. (1973). − Orders for purchase should be sent to: Departamento Editorial, Vicerrectoría de Comunicaciones, Universidad Católica de Chile, Alameda B. O'Higgins 340, Of. 10, Santiago-Chile.
The catalogue presents general data for 131 galactic globular clusters and the photometric results for 42 galactic globular clusters for which colour-magnitude diagrams on the UBV system are available.

002.006 **An atlas of light and colour curves of field RR Lyrae stars.** J. Lub.
Astron. Astrophys., Suppl. Ser., Vol. 29, 345 - 378 (1977).
An atlas is presented of light and colour curves in the Walraven *VBLUW* photometric system for 90 field RR Lyrae and short-period variables. The observational programme and the reduction procedure are briefly described. Tables are presented containing the colours at mean and minimum light.

002.007 **On books on physics and astrophysics published by the main editorial staff for physical-mathematical literature of the Publishing House "Nauka" in 1977.**
V. V. Vlasov.
Usp. fiz. nauk, Vol. 121, 550 - 555 (1977). In Russian.
From Ref. zh., 51. Astron., 8.51.47 (1977).

002.008 **Catalogue of magnitudes of HR stars in the uniform P_{44} and V systems.** E. Rybka.
Nakładem Uniw. Jagiellońskiego, Kraków. 74 pp. Price zł 11.00 (1977).
The present paper forms a part of a program proposed by the author in 1958 and precised afterwards by R. H. Stoy (1960) as 'General Catalogue of Magnitudes'. It is a continuation of the paper of E. Rybka (1960), where the systematic differences between the particular series of Harvard observations have been investigated. The author restricted his attention to the 'Revised Harvard Photometry' as published in H. A. 50 (1908). Harvard magnitudes (H. A. 50 and 54) may be regarded as a photometric image of the whole sky at the end of 19th century as concerns bright stars. For any comparison with other photometric systems, particularly with photoelectric magnitudes determined recently, it would be necessary to reduce the original Harvard magnitudes to a uniform system and to express them in the system V, universally adopted at present. It was done by the author in the present paper.

002.009 **Résolution des plaques de la Carte du Ciel par l'AGK 2/3 dans les zones situées entre +31° et −2°.**
A. Valbousquet.
Cent. Données Stellaires, Inf. Bull. No. 13, p. 2 - 4 (1977).

002.010 **Some thoughts on a catalog of average radial velocities.** C. Jaschek.
Cent. Données Stellaires, Inf. Bull. No. 13, p. 6 - 7 (1977).

002.011 **Summary of the bibliography on stellar radial velocities.** F. Ochsenbein.
Cent. Données Stellaires, Inf. Bull. No. 13, p. 8 - 10 (1977).

002.012 **Errata in the Catalogue of Bright Stars.**
D. Hoffleit.
Cent. Données Stellaires, Inf. Bull. No. 13, p. 14 - 23 (1977).

002.013 **Errors in the AGK2 catalogue.** J. S. Duncombe.
Cent. Données Stellaires, Inf. Bull. No. 13, p. 24 - 26 (1977).

002.014 **Errata in the "Recent Bibliography on the Galactic Polar Areas" by W. Gliese, C. Jaschek and M. McCarthy.** W. Gliese.
Cent. Données Stellaires, Inf. Bull. No. 13, p. 27 (1977).

002.015 **Enquiry on radioastronomical data.**
G. Westerhout, C. Jaschek.
Cent. Données Stellaires, Inf. Bull. No. 13, p. 28 - 53 (1977).

002.016 **The Dwingeloo 21-cm line observations on magnetic tape.** T. A. T. Spoelstra.
Cent. Données Stellaires, Inf. Bull. No. 13, p. 59 - 65 (1977).

002.017 **Linear polarization observations of galactic radio emission on magnetic tape.** T. A. T. Spoelstra.
Cent. Données Stellaires, Inf. Bull. No. 13, p. 66 - 68 (1977).

002.018 **Centre de données d'étoiles doubles de l'Observatoire de Nice.** P. Couteau, M. Fulconis.
Cent. Données Stellaires, Inf. Bull. No. 13, p. 69 - 74 (1977).

002.019 **Catalogue of stellar abundances.**
A. Strobel, J. Strobel.
Cent. Données Stellaires, Inf. Bull. No. 13, p. 75 (1977).

002.020 **Plans for a computer-readable cluster catalogue.**
G. Lyngå.
Cent. Données Stellaires, Inf. Bull. No. 13, p. 76 (1977).

002.021 **Bibliography on galactic structure in the direction of polar caps.**
W. Gliese, C. Jaschek, M. F. McCarthy.

Highlights of Astronomy, Vol. 4, Part II, (see 012.022), p. 23 - 24 (1977).

002.022   A catalogue of 0.2 Å resolution far-ultraviolet stellar spectra measured with Copernicus.
T. P. Snow, Jr., E. B. Jenkins.
Highlights of Astronomy, Vol. 4, Part II, (see 012.025), p. 353 - 354 (1977).

002.023   A study of the revised 3C catalogue. I. Confusion and resolution.   P. Véron.
Astron. Astrophys., Suppl. Ser., Vol. 30, 131 - 144 (1977).
A catalogue is given of radio sources with flux densities larger than 9.0 Jy at 178 MHz. It is believed to be nearly unbiased and complete for $\delta > -5°$ and $|b| \geqslant 10°$. It is based on the 3CR catalogue, but considerable corrections for confusion and resolution have been applied. Fifty-nine sources have been removed from the original 255 3CR sources because they are too faint after correction for confusion. Nine new sources have been added. The new catalogue contains 205 sources.

002.024   Bibliographical index of planetary nebulae for the period 1965–1976.   A. Acker, J. Marcout.
Astron. Astrophys., Suppl. Ser., Vol. 30, 217 - 221 (1977).
The list of known galactic planetary nebulae (PN) has been brought up to date; with the 148 new ones, the number of known PN is now 1184, including 199 misclassified and dubious cases. Cross-identifications between the usual names and the new denominations of PN are given, as well as the names of central stars. A bibliographic index is published both on magnetic tape and on microcard at the Centre de Données Stellaires de Strasbourg.

002.025   Astronomical catalogues 1951 - 1975.
Compiled by M. Collins.
INSPEC Bibliography Series No. 2. The Institution of Electrical Engineers, Station House, Nightingale Road, Hitchin Hertfordshire SG5 1RJ. 9 + 327 pp. Price £ 55.00, $ 110.00 (1977). ISBN 0-85296-440-4.
Presents a collection of nearly 2500 catalogues that appeared in the astronomical literature covering the years 1951 to 1975 inclusive. Some catalogues published in 1976 and 1977 are also included. The bibliography contains lists of celestial objects, phenomena and equipment as well as books and slides. Each entry contains full bibliographic details and most have an abstract and a summary of the catalogue information content. Authors and corporate authors appear in indexes and there is a designation index which is a new and unique feature that will be useful for those working with the current literature and for those starting in astronomical research.

002.026   Star catalogs and files available at the Stellar Data Center. July 1977.
Cent. Données Stellaires, Inf. Bull. No. 13, p. 77 - 114 (1977).

002.027   Ein neuer Mondatlas für Amateure.
Sternenbote, 20. Jahrg., 134 - 138 (1977).
Concerning "Mond–Mars–Venus, Taschenatlas der erdnächsten Himmelskörper", by A. Rükl. Artia Verlag, Prag; Dausien-Verlag, Hanau/Main. 256 pp. Price öS 75.00, DM 7.80 (1977).

002.028   A bibliographical catalogue of field RR Lyrae stars (magnetic tape).   A. Heck, J. M. Lakaye.
Astron. Astrophys., Suppl. Ser., Vol. 30, 397 - 398 (1977).
This catalogue, available on magnetic tape and on microfiches at the Strasbourg Stellar Data Center, gives bibliographical references for 5855 field RR Lyrae stars (6607 entries).

002.029   Katalog über 230 sichere, wahrscheinliche, mögliche und zweifelhafte Impaktstrukturen.   J. Classen.

Orion, 35. Jahrg., 198 - 206 (1977).

002.030   Worldwide growth of geophysics and astrophysics research since 1967.   C. D. Ellyett.
EOS Trans. American Geophys. Union, Vol. 58, 124 - 126 (1977). – Abstr. in Phys. Abstr., Vol. 80, Abstr. 66884 (1977).

002.031   Cosmology: list of publications.
Gen. Relativ. Gravitation, Vol. 8, 1003 - 1037 (1977).

002.032   A catalog of southern groups and clusters of galaxies.   A. Duus, B. Newell.
Astrophys. J., Suppl. Ser., Vol. 35, 209 - 219 (1977).
The authors present the results of a survey for southern clusters of galaxies. Their catalog contains all previously identified groups and clusters that have $\delta \lesssim -27°$, plus 710 newly located clusters.

002.033   Astronomy and Astrophysics Abstracts. Vol. 19: Literature 1977, Part 1.
S. Böhme, U. Esser, W. Fricke, U. Güntzel-Lingner, I. Heinrich, D. Krahn, L. D. Schmadel, G. Zech (Editors). Published for Astronomisches Rechen-Institut by Springer-Verlag, Berlin – Heidelberg – New York. 10 + 732 pp. Price DM 95.00; ca. US $ 43.70 [Subscription price DM 76.00; ca. US $ 35.00] (1977).

002.034   Pulkovo sky survey in the interstellar neutral hydrogen radio line. III. Observations for the zone $-29° \leqslant \delta \leqslant +40°$. Atlas of drift scans $T_A(\alpha)/V, \delta$.
N. V. Bystrova, I. A. Rakhimov.
Akademiya Nauk SSSR, Spetsial'naya Astrofizicheskaya Observatoriya. Izdatel'stvo "Nauka" Leningradskoe Otdelenie, Leningrad. 62 pp. Price 80 Kop. (1977). In Russian and English.

002.035   On the astronomical and mathematical manuscripts in the State A. Firdousy Library of the Tadzhik SSR.
A. Kakhkhorov, I. Khodzhiev.
Izv. AN TadzhSSR. Otd. fiz.-mat. i geol.-khim. nauk, 1977, No. 1, p. 23 - 30. In Russian. – Abstr. in Ref. zh., 51. Astron., 11.51.10 (1977).

002.036   On the edition of K. Eh. Tsiolkovskij's manuscript heritage.   S. A. Sokolova.
Istor.-astron. issled. Vyp. (No.) 3. Moskva, Nauka, 1977, p. 389. In Russian. – From Ref. zh., 62. Issled. kosm. prostranstva, 11.62.10 (1977).

002.037   A four-colour photographic atlas of the sky.
W. Schlosser, T. Schmidt-Kaler.
Vistas Astron., Vol. 21, 447 - 466 (1977).
With the spherical mirror camera the authors obtained in two campaigns UBVR-photographs of the northern and southern sky. 22 representative photographs were selected for an atlas covering the whole of the Milky Way in each of the colours UBVR. This Atlas of the Milky Way is available as a separate publication (Schlosser, Schmidt-Kaler, Hünecke; 1974). Here the authors present five of these 22 photographs. The atlas depicts the Milky Way, the zodiacal light and the Gegenschein.

002.038   Catalogue of photometric 'star boxes" in the UBV $B_1B_2V_1G$ system.   M. Golay, N. Mandwewala.
Publ. Obs. Genève, Sér. B, Fasc. 3, 4 + 106 pp. (1977).
This catalogue gives the stellar content of all the 0.01 mag photometric star boxes made using 7082 stars measured in the UBV $B_1B_2V_1G$ system.

002.039   Epitome fundamentorum astronomiae. Monographs on star catalogues.   B. M. Ševarlić, G. Teleki.

Publ. Dep. Astron., Univ. Beograd, Fac. Sci., No. 6, (see 012.040), p. 125 (1976). — This first monograph will appear in Publications of the Department of Astronomy, Belgrade University, No. 7.

002.040 Chronicle.   L. N. Bondarenko.
Astron. Tsirk., No. 946, p. 7 - 8 (1977). In Russian. Diss. at various institutions of the USSR.

002.041 Spectroscopic and photometric observations of galaxies from the ESO/Uppsala List. Second catalogue.
N. Å. S. Bergwall (*Bergvall*), T. M. Borchkhadze, J. Breysacher, A. B. G. Ekman, A. Lauberts, S. Laustsen, A. B. Muller, H.-E. Schuster, J. Surdej, R. M. West, B. E. Westerlund. ESO Sci. Prepr. No. 17, 29 pp. (1977).
Spectroscopic and photometric observations are presented for a total of about 175 southern galaxies, selected from the ESO-Uppsala Lists Nos. 1 - 5. Many of the galaxies have emission lines and several are members of multiple systems.

002.042 Catalogue of absolute right ascensions of 1023 bright and faint fundamental stars of the northern hemisphere.   A. A. Nemiro, A. I. Plyugina, K. N. Tavastsherna, V. N. Shishkina.
Tr. Glav. Astron. Obs. Pulkovo, Ser. 2, Vol. 82, 4 - 52 (1977). In Russian.
Absolute right ascensions of 505 bright (FK4) and 518 faint (FKSZ) fundamental stars of the northern hemisphere have been determined on the base of observations with the Pulkovo large transit instrument during 1954—1961. Mean epoch of observations is about 1958 for both the bright and faint stars. Mean magnitude of the bright stars is 4.3, that of the faint ones 7.8. A comparison of the Pu58 catalogue with the FK4 and KSV for bright stars, with the PFKSZ and PFKSZ' for faint stars is given in tables. The comparison indicates the presence of a magnitude equation in the FK4.

002.043 Catalogue of declinations of 710 stars for equinox and epoch 1960.0 determined from observations with the vertical circle of the Nikolaev Observatory during 1957—1964.   I. I. Bozhko, G. K. Zimmerman.
Tr. Glav. Astron. Obs. Pulkovo, Ser. 2, Vol. 82, 53 - 77 (1977). In Russian.
625 stars of the Pulkovo program and 85 zodiacal stars are contained in the catalog. Mean values of mean errors of the declinations are presented in a table. Tables give the results of a comparison of the Nik_{60} catalogue with the FK4.

002.044 Catalogue of right ascensions of 395 stars.
L. D. Voronenko, M. I. Il'kiv, N. S. Kalikhevich.
Tr. Glav. Astron. Obs. Pulkovo, Ser. 2, Vol. 82, 78 - 94 (1977). In Russian.
The catalogue NikF1 containing 395 stars with declinations −16° to +76° was compiled from 13440 observations during 1966—1969 with the photoelectric transit instrument in Nikolaev. The compiled catalogue was compared with the FK4 and KSV.

002.045 Catalogue of right ascensions of 312 stars.
V. N. Pyshnenko.
Tr. Glav. Astron. Obs. Pulkovo, Ser. 2, Vol. 82, 95 - 105 (1977). In Russian.
A catalogue of right ascensions has been compiled on the basis of 7340 photoelectric observations with the Time Service transit instrument in Nikolaev. The catalogue was compared with the FK4, KSV and NikF1.

002.046 Bibliography.
Calendar of the Tartu Astronomical Observatory for the year 1978, (see 047.026), p. 73 - 77 (1977). In Estonian.

002.047 Annotations on astrophysical papers published in

"News of Universities. Radiophysics", Vol. 20, Nos. 1, 3, 6, 7 (1977).
Astron. Zh. Akad. Nauk SSSR, Vol. 54, 1349 (1977). In Russian. English translation in Soviet Astron., Vol. 21, No. 6.

002.048 A catalogue of spectroscopically identified white dwarfs.   G. P. McCook, E. M. Sion.
Villanova Univ. Obs. Contrib. No. 2, 6 + 50 pp. (1977).

002.049 Catalogue of B, V magnitudes and spectral classes of 720 stars centered at $\alpha_{1950} = 2^h17^m9$, $\delta_{1950} = +58°59'$.   V. I. Voroshilov, L. N. Kolesnik.
Astrometriya i Astrofizika, Kiev, vyp. (No.) 33, (see 003.020), p. 21 - 29 (1977). In Russian.
A catalogue of B, V magnitudes and spectral classifications is presented for 720 stars to V = 12^m5 in a field of 5 square degrees near the galactic plane around the stellar ring SR 58 ($l = 134°2$, $b = -1°7$). The spectral classification is in the MK system. The internal accuracy of the catalogue is $± 0^m03$.

002.050 Coperniciana of the Ukrainian libraries.
D. N. Mazurenko.
Istor.-astron. issled. Vyp. (No.) 13. Moskva, Nauka, 1977, p. 399 - 422. In Russian. — Abstr. in Ref. zh., 51. Astron., 1.51.56 (1978).

002.051 Solar eclipses (a bibliography with abstracts).
Report for 1964—February 1977. D. W. Grooms.
Edited by Natl. Tech. Inf. Serv., Springfield, Va., USA. NTIS/PS—77/0250. 165 pp. (1977). — Abstr. from INIS7723 341943.

002.052 Quasars, pulsars and black holes (a bibliography with abstracts). Report for 1964—February 1977.
D. W. Grooms.
Edited by Natl. Tech. Inf. Serv., Springfield, Va., USA. NTIS/PS—77/0168. 262 pp. (1977). — From INIS7724 345331.

002.053 Differential catalogue of right ascensions of 544 bright stars from FK4 for the observational epoch and equinox 1950.0 (zone −20° to +35°).   K. N. Derkach.
Physics of the moon and planets. Fundamental astrometry, (see 003.024), Vestn. Khar'kov. Univ., No. 160, p. 42 - 55 (1977). In Russian.

002.054 Henry Norris Russell. Bibliography.
Dudley Obs. Rep. No. 13, (see 012.051), p. 159 - 170 (1977).

002.055 Bibliography.
Z. Kopal, M. Moutsoulas, F. B. Waranius (Editors). Moon, Vol. 16, 351 - 386, 465 - 492, Vol. 17, 179 - 201 (1977). — Concerning literature on lunar topics.

The atlas of Mercury.   See Abstr. 003.042.

Photographic sky atlas.   See Abstr. 003.125.

The catalogue of the history of science collections of the University of Oklahoma Libraries.
See Abstr. 003.140.

IAU Commission 45: Working Group on Spectroscopic and Photometric Data. Catalogs recently published, to be published or in preparation. List VII.
See Abstr. 010.018.

Meridian observations made in Brorfelde (Copenhagen University Observatory) 1975 - 1976.
See Abstr. 041.025.

Synthetic catalogue of right ascensions for time service.   See Abstr. 041.029.

Determination of the right ascensions of 386 stars from visual observations 1970 - 1973. See Abstr. 041.032.

On a catalogue of star positions for the polar region obtained by the method of overlapping plates. See Abstr. 041.055.

Improvement of the first Mizusawa PZT star catalog.   See Abstr. 041.047.

Solar proton flares and their prediction. See Abstr. 073.111.

Compiled system of selenodetic coordinates of 4900 points of the lunar surface (visible side of the moon). See Abstr. 094.120.

Addendum to the catalogue of meteorite collections of the Lithuanian SSR.   See Abstr. 105.064.

Improvement of stellar radial-velocity data. See Abstr. 112.007.

Proper motions of stars relative to galaxies in four areas of the Pulkovo program.   See Abstr. 112.011.

Proper motions of stars relative to galaxies in area No. 5 of the Pulkovo program.   See Abstr. 112.012.

Catalogue of proper motions of stars relative to galaxies in five Pulkovo areas.   See Abstr. 112.013.

Relative proper motions of stars in five areas of the Pulkovo program.   See Abstr. 112.014.

Proper motions of stars relative to galaxies in the areas Nos. 104, 109, 111, and 113 of the Pulkovo program. See Abstr. 112.016.

Photoelectric *uvby* and Hβ photometry of 750 A and F stars in 63 selected areas with $|b| < 30°$. See Abstr. 113.028.

A high-dispersion photometric atlas of the dM0 star HD 88230 from 3900 to 6000 Å.   See Abstr. 114.533.

The Copernicus ultraviolet spectral atlas of Tau Scorpii.   See Abstr. 114.542.

Variable stars, their discoverers and first compilers from 1006 to 1975.   See Abstr. 120.003.

On the catalogue of proper motions of 10600 stars in 41 selected areas of the sky.   See Abstr. 122.010.

A list of possible, probable, and true planetary nebulae detected since 1966.   See Abstr. 135.020.

Study of NGC 2324. I. Catalogue of stellar B magnitudes.   See Abstr. 153.029.

Basic data for galactic globular clusters. See Abstr. 154.034.

UBV color-magnitude diagrams of galactic clusters. See Abstr. 154.041.

Survey of late-type and irregular southern galaxies on plates taken with the UK 1.2-m Schmidt telescope. I. See Abstr. 158.034.

Observations of supergiant stars in the Small Magellanic Cloud.   See Abstr. 159.010.

003 Books

003.001 L'Académie Pontificale des Sciences en mémoire de son second président Georges Lemaître à l'occasion du cinquième anniversaire de sa mort.
Pontificiae Academiae Scientiarum Scripta Varia, Vol. 36, 296 pp. Ex aedibus academicis in Civitate Vaticana (1972). Included are commémorations scientifiques (see 005.001 - 005.003); L'univers en expansion (G. Lemaître); Deux conférences de Lemaître à Rome; Les nouveaux chiffres de Lemaître; Deux commémorations données par Lemaître à Rome; Essai d'une bibliographie de Georges Lemaître.

003.002 Methods in computational physics. Vol. 17: General circulation models of the atmosphere.
J. Chang (Editor).
With contributions by A. Kasahara, G. A. Corby, A. Gilchrist, P. R. Rowntree, W. M. Washington, D. L. Williamson, A. Arakawa, V. R. Lamb, W. Bourke, B. McAvaney, K. Puri, R. Thurling.
Academic Press Inc., New York—San Francisco—London. 9 + 337 pp. Price $ 35.00, DM 134.50 (1977). ISBN 0-12-460817-5.

003.003 Astrometric determinations of the coordinates of celestial bodies.
Sverdlovsk. 92 pp. Price 30 Kop. (1976). In Russian. – See abstract 041.004.

003.004 The new astronomy and space science reader.
J. C. Brandt, S. P. Maran (Editors).
W. H. Freeman and Company, San Francisco, USA—Reading, England. 9 + 371 pp. Price $ 15.00, DM 23.55 cloth, $ 7.50 paper (1977). ISBN 0-7167-0349-1 paperback, ISBN 0-7167-0350-5. – Reviews in Astron. Tidsskr., Årg. 10, 139 (1977); Sky Telesc., Vol. 54, 325 (1977).
Collected articles mainly from Natural History, Smithsonian, Scientific American, Science, Nature, and the Astrophysical Journal.

003.005 Cosmology now.
L. John (Editor), with an introduction by B. Lovell, and contributions by H. Bondi, J. Peach, M. Ryle, D. Lynden-Bell, D. Sciama, J. V. Narlikar, W. H. McCrea, R. Penrose, M. Rees, J. Taylor.
Taplinger Publishing Company, New York. 168 pp. Price $ 10.95. DM 33.80 (1976). ISBN 0-8008-1925-X. (First published 1973 – see 11.003.068). – Review in Sky Telesc., Vol. 54, 131 - 132, 1977 (J. D. Fernie).

003.006 Solar cosmic rays.
L. I. Dorman, L. I. Miroshnichenko. Translated from the Russian edition 1968.
Published for the National Aeronautics and Space Administration and the National Science Foundation, Washington, D.C. by the Indian National Scientific Documentation Centre, New Delhi. 8 + 582 pp. (1976). – See AJB 68170.

003.007 Lunar sample studies.
Preface by W. C. Phinney.
Natl. Aeronaut. Space Adm., Washington, D.C. NASA SP-418. For sale by the National Technical Information Service, Springfield, Virginia 22161. 5 + 69 pp. Price $ 4.50 (1977). The individual contributions are included in their corresponding subject categories – see abstracts 094.408 - 094.412.

003.008 Annual Review of Astronomy and Astrophysics. Volume 14.
G. R. Burbidge, D. Layzer, J. G. Phillips (Editors).
Annual Reviews Inc., 4139 El Camino Way, Palo Alto, California 94306. 9 + 500 pp. (1976). – The individual contributions are included in their corresponding subject categories – see abstracts 031.234, 031.235, 042.019, 064.022, 065.022, 102.015, 107.023, 122.056, 141.062, 141.063, 142.040, 151.029, 155.027, 156.012, 158.071, 158.505, 162.034.

003.009 Heinrich Wild 1877 - 1951.
Astronomisch-geodätische Arbeiten in der Schweiz, edited by Schweizerische Geodätische Kommission. Vol. 31, 39 pp. (1977). – See 004.025 - 004.027, 005.008, 005.009.

003.010 Highlights of Astronomy, Vol. 4, Parts I and II, as presented at the XVIth General Assembly of the I.A.U., Grenoble, 1976. E. A. Müller (Editor).
D. Reidel Publishing Company, Dordrecht, Holland – Boston, U.S.A. Part I, 7 + 370 pp., Part II, 7 + 407 pp. (1977). ISBN 90-277-0849-5, 90-277-830-4 (pbk.); ISBN 90-277-850-9, 90-277-832-0 (pbk.) – See abstracts 012.018 - 012.026, 061.016, 091.038, 133.009.

003.011 Astrometriya i Astrofizika. Vypusk 32.
A. V. Morozhenko (Editor).
Respublikanskij Mezhvedomstvennyj Sbornik. Akademiya Nauk Ukrainskoj SSR, Glavnaya Astronomicheskaya Observatoriya. Izdatel'stvo "Naukova dumka", Kiev. 100 pp. Price 1 Rbl. 24 Kop. (1977). In Russian. – The papers included are abstracted in their corresponding subject categories – see abstracts 034.014, 045.013, 071.016, 071.017, 073.049, 082.036, 102.017, 102.018, 114.045, 114.046, 122.067, 155.053, 155.054.

003.012 Annual Review of Astronomy and Astrophysics, Volume 15.
G. Burbidge, D. Layzer, J. G. Phillips (Editors).
Annual Reviews Inc., 4139 El Camino Way, Palo Alto, Calif. 94306, USA. 8 + 602 pp. Price DM 51.00 (1977). ISBN 0-8243-0915-4. – The individual contributions are included in their corresponding subject categories – see abstracts 015.010, 022.054, 072.035, 073.050, 080.040, 092.031, 099.059, 117.030, 125.017, 131.158, 131.159, 141.532, 142.081, 151.043, 155.058, 158.096, 160.053, 162.052.

003.013 Modern astronomy. Selections from the yearbook of astronomy. P. Moore (Editor).
Sidgwick & Jackson Limited, London; W. W. Norton and Company Inc., New York. 184 pp. Price $ 9.95, DM 12.40, £ 2.50 in UK (1977). ISBN 0-283-98322-1 (hard), ISBN 0-283-98369-8 (soft), ISBN 0-393-06417-4 (soft). – The individual contributions are included in their corresponding subject categories – see abstracts 004.039, 015.011, 032.022, 081.028, 082.046, 092.032, 098.043, 100.510, 104.021, 105.072, 114.048, 114.049, 115.009, 116.014, 131.160, 131.161, 158.097, 158.508.

003.014 Chemical evolution of the giant planets.
C. Ponnamperuma (Editor).
Academic Press, New York – San Francisco – London. 11 + 240 pp. Price $ 11.50, £ 8.15, DM 44.30 (1976). ISBN 0-12-561350-4. – The individual contributions are included in their corresponding subject categories – see abstracts 015.012 - 015.018, 032.527, 051.032, 091.041 - 091.044, 099.060, 099.061, 100.511. – Reviews in Nature, Vol. 269, 92; 1977 (C. T. Pillinger); Space Sci. Rev., Vol. 20, 678; 1977 (A. J. Kliore).

003.015 The new cosmos.
A. Unsöld.
2nd revised and enlarged edition. Translated by R. C. Smith based on the translation by W. H. McCrea of the 1st edition.

Heidelberg Science Library. Springer-Verlag, New York—
Heidelberg—Berlin. 12 + 451 pp. (1977). ISBN 0-387-90223-6;
ISBN 3-540-90223-6.
    Contents: I. Classical astronomy. II. Sun and stars:
Astrophysics of individual stars. III. Stellar systems: Milky
Way and galaxies; cosmogony and cosmology. Physical con-
stants and astronomical quantities.

003.016   Problems of cosmic physics. Vypusk 12.
        S. K. Vsekhsvyatskij (Editor).
Mezhvedomstvennyj Nauchnyj Sbornik. Izdatel'stvo pri
Kievskom Gosudarstvennom Universitete Izdatel'skogo
Obedineniya "Vishcha Shkola", Kiev. 126 pp. Price 1 Rbl.
23 Kop. (1977). In Russian. — The papers included are
abstracted in their corresponding subject categories — see
abstracts 011.023, 032.025, 034.023, 064.038, 071.021, 073.
051, 076.009, 077.024, 084.015, 098.045, 103.204, 103.702,
103.801, 104.022 - 104.024.

003.017   Scientific results of the Viking project.
        J. Geophys. Res., Vol. 82, No. 28, p. 3951 - 4684
(1977).
    High-quality reprints are available for any paper in this
issue. The paper number should be given when ordering. Pay-
ment must accompany orders. The first paper in an order costs
$ 3.50 ($ 2.00 for deposit accounts); each additional paper
costs $ 1.00. Send orders to American Geophysical Union,
Separates Sales, 1909 K Street, N.W., Washington, D.C. 20006.
See abstracts 031.321 - 031.323, 032.564, 032.565, 051.053,
053.009, 066.321, 074.040, 097.163 - 097.204, 097.511,
097.512.

003.018   Photometric investigations of carbon stars.
        A. Balklavs (Editor).
Radioastrofizicheskaya Observatoriya Akademii Nauk
Latvijskoj SSR. Trudy 16. Izdatel'stvo "Zinatne", Riga.
176 pp. Price 1 Rbl. 30 Kop. (1977). In Russian. — The
individual contributions are included in their corresponding
subject categories — see abstracts 113.038, 113.039, 122.123.

003.019   Investigations of the sun and red stars. 6.
        A. Balklavs (Editor).
Latvijas PSR Zinātņu akadēmija, Radioastrofizikas observa-
torija. Akademiya nauk Latvijskoj SSR, Radioastrofizicheskaya observatoriya. Izdatel'stvo "Zinatne", Riga. 80 pp.
Price 19 Kop. (1977). In Russian. — The individual contributions are included in their corresponding subject categories —
see abstracts 031.332, 064.050, 113.040, 114.059, 122.126.

003.020   Astrometriya i Astrofizika. Vypusk 33.
        I. G. Kolchinskij (Editor).
Respublikanskij Mezhvedomstvennyj Sbornik. Akademiya
Nauk Ukrainskoj SSR, Glavnaya Astronomicheskaya Observatoriya. "Naukova Dumka", Kiev. 104 pp. Price 1 Rbl. 40 Kop.
(1977). In Russian. — The papers included are abstracted in
their corresponding subject categories — see abstracts 002.049,
007.000, 031.342, 034.086, 041.054, 041.055, 045.029,
045.030, 066.343, 071.050, 071.051, 073.114, 073.115,
082.124, 082.125, 100.033, 114.561, 158.122.

**003.021   Optical investigations of the emission of the atmo-
        sphere, aurorae and noctilucent clouds aboard the
orbital scientific station Salyut 4.**
A. G. Nikolaev, V. I. Sevast'yanov, G. M. Grechko, A. I.
Lazarev, Ch. Willmann, O. A. Avaste, A. D. Povzner, L.
Riives (Editors).
Academy of Sciences of the USSR, Soviet Geophysical
Committee, Academy of Sciences of the Estonian SSR,
Institute of Astrophysics and Atmospheric Physics, Tartu.
179 pp. Price 1 Rbl. 97 Kop. (1977). In Russian. — The in-
dividual papers are abstracted in their corresponding subject
categories — see abstracts 051.071, 082.126 - 082.131, 084.

031, 094.191, 104.044.

**003.022   Illustrated glossary for solar and solar-terrestrial
        physics.**
A. Bruzek, C. J. Durrant (Editors).
Astrophysics and Space Science Library, Vol. 69. D. Reidel
Publishing Company, Dordrecht/Holland—Boston/U.S.A. 18 +
204 pp. (1977). ISBN 90-277-0825-8. — The individual contributions are included in their corresponding subject categories — see abstracts 071.056, 072.053, 073.116 - 073.118,
074.101 - 074.103, 077.053, 080.082 - 080.085, 084.301.

003.023   A geological basis for the exploration of the planets.
        R. Greeley, M. H. Carr (Editors), with an introduc-
tion by R. Greeley, M. H. Carr, C. R. Chapman.
Natl. Aeronaut. Space Adm., Washington, D.C., NASA SP-417.
For sale by the National Technical Information Service,
Springfield, Virginia 22161. 10 + 109 pp. Price $ 5.25 (1976).
The individual contributions are included in their correspond-
ing subject categories — see abstracts 091.078 - 091.082.

**003.024   Physics of the moon and planets. Fundamental
        astrometry.   V. I. Ezerskij (Editor).**
Ministerstvo Vysshego i Srednego Spetsial'nogo Obrazovaniya
USSR. Izdatel'stvo pri Khar'kovskom Gosudarstvennom
Universitete Izdatel'skogo Obedineniya "Vishcha Shkola",
Khar'kov. Vestnik Khar'kovskogo Universiteta, No. 160.
94 + 3 pp. Price 1 Rbl. (1977). In Russian. — The individual
contributions are included in their corresponding subject
categories — see abstracts 002.053, 031.348, 031.414, 031.
415, 032.055, 044.041, 063.055, 073.123, 094.612, 096.
028, 097.218, 098.091, 100.035.

**003.025   Sternzeiten, (zur 275jährigen Geschichte der
        Berliner Sternwarte, der heutigen Sternwarte**
Babelsberg). Band I, II.   G. Jackisch (Editor).
Akademie-Verlag, Berlin. 95 + 88 pp. Price M 15.00, 14.00
(1977) = Akad. Wiss. DDR, Veröff. Forschungsber. Geo-
Kosmoswiss. Heft 6, 7 = Zentralinst. Astrophys. Sternw.
Babelsberg, Mitt. Neue Folge, Nr. 178, 179. — See abstracts
004.086 - 004.098, 005.028, 066.380.

003.026   Chemistry of the moon.   L. H. Ahrens (Editor).
        Physics and chemistry of the earth, Vol. 10, No. 3.
Pergamon Press, Oxford. Price $ 78.00 (1977). — Review in
Geochim. Cosmochim. Acta, Vol. 41, 1879; 1977 (D. M. Shaw).

**003.027   Absolute photometry of Mars in 1971, 1973 and
        1975.**
Yu. V. Aleksandrov, D. F. Lupishko, T. A. Lupishko.
Vishcha shkola, Khar'kov. 128 pp. Price 1 Rbl. 40 Kop.
(1977). In Russian. — Review in Ref. zh., 51. Astron., 1.51.
39 (1978).

003.028   Native American astronomy.
        A. F. Aveni (Editor).
University of Texas Press. 286 pp. Price $ 15.95 (1977).
Review in Sky Telesc., Vol. 54, 422 (1977).

**003.029   Atomic energy levels and Grotrian diagrams. Vol. I.
        Hydrogen I — phosphorus XV.**
S. Bashkin, J. O. Stoner, Jr.
North-Holland, Amsterdam; American Elsevier, New York.
615 pp. Price $ 59.95 (1975). ISBN 0-7204-0322-7 North-
Holland; 0-444-10827-0 American Elsevier. — Review in
Astrophys. Lett., Vol. 19, 33; 1977 (L. Armstrong, Jr.).

003.030   Solar energy: the awakening science.
        D. Behrman.
Little, Brown and Co., Boston. 408 pp. Price $ 12.50 (1976).
Review in Sky Telesc., Vol. 54, 133 - 135; 1977 (C. A.
Federer, Jr.).

003.031 **Exploring the cosmos.**
L. Berman, J. C. E. Evans.
Little, Brown, Boston, 2. Ed. 10 + 436 pp. Price $ 12.95 (1977). — Review in Sky Telesc., Vol. 54, 230 (1977).

003.032 **Principles of cosmology and gravitation.**
M. Berry.
Cambridge University Press, London. 192 pp. Price £ 7.00 hardback; £ 2.50 paperback (1977). — Review in J. British Astron. Assoc., Vol. 88, 101 - 102; 1977 (*B. J. Carr*).

003.033 **Programmes for pocket calculators HP-67 and HP-97 in the field of theoretical and observational astronomy.**   F. C. Bertiau, in collaboration with E. Fierens.
Leuven University Press, Astronomisch Instituut, Katholieke Universiteit Leuven, Naamsestraat 61, 3000 Leuven. 6 + 117 pp. (1977). ISBN 90-6186-065-2.

003.034 **End of an era in space exploration. From international rivalry to international cooperation.**
J. C. D. Blaine.
American Astronautical Soc., San Diego, Calif., USA. 16 + 199 pp. (1976). ISBN 0-87703-080-4. — Abstr. in Phys. Abstr., Vol. 80, Abstr. 57149 (1977).

003.035 **Mars and its satellites: a detailed commentary on the nomenclature.**   J. Blunck.
Exposition, Hicksville, N.Y. 200 pp. Price $ 10.00 (1977). From Phys. Today, Vol. 30, No. 12, p. 62 (1977).

003.036 **Measurement of weak forces in physics experiments.**   V. B. Braginsky (*Braginskij*), A. B. Manukin.
University of Chicago Press, Chicago. 153 pp. Price $ 10.00 (1977). — Review in Sky Telesc., Vol. 54, 138 (1977).

003.037 **Soyuz and Apollo.**   K. D. Bushuev (Editor).
Politizdat, Moskva. 272 pp. (1976). In Russian.
Review in Priroda, 1977, No. 10, p. 126 - 131 (*A. A. Leonov, V. A. Denisenko*).

003.038 **The key to the universe.**   N. Calder.
British Broadcasting Corporation, London. 196 pp. Price £ 5.95 (1977). — Reviews in Astron. Tidsskr., Årg. 10, 129 - 130; 1977 (*D. Dravins*); J. British Astron. Assoc., Vol. 88, 97 - 98; 1977 (*I. Nicolson*).

003.039 **Handbook on plasma instabilities, Vol. 1.**
F. F. Cap.
Academic Press, New York — San Francisco — London. 21 + 458 pp. Price $ 19.50 (1976). — Review in Gerlands Beitr. Geophys., Band 86, 343; 1977 (*R. Treumann*).

003.040 **Introduction to applied optics for engineers.**
F. P. Carlson.
Academic Press, New York. 277 pp. Price $ 16.50 (1977). Review in Appl. Opt., Vol. 16, 3064; 1977 (*T. K. McCubbin*).

003.041 **The inner planets.** New light on the rocky worlds of Mercury, Venus, Earth, the Moon, Mars, and the Asteroids.   C. R. Chapman.
Charles Scribner's Sons, New York. 16 + 170 pp. Price $ 7.95 (1977). — Review in Sky Telesc., Vol. 54, 230 (1977).

003.042 **The atlas of Mercury.**   C. A. Cross, P. Moore.
Mitchell Beazley, London; Crown, New York. 48 pp. Price £ 5.95, $ 10.00 (1977). — Reviews in J. British Astron. Assoc., Vol. 87, 641 - 642; 1977 (*J. H. Robinson*); Nature, Vol. 270, 373; 1977 (*W. P. O'Donnell*).

003.043 **Le monde des galaxies.** Cours donné au Collège de France, Novembre 1976.   G. de Vaucouleurs.
Obs. Besançon et Lab. Astron. Fac. Sci., Collège de France,

Paris. 14 + 279 pp. Price F 35.00 (1976). — Review in Astronomie, Vol. 91, 509; 1977 (*L. Gouguenheim*).
Contents: Introduction historique; Sources d'information; Morphologie, classification et composition qualitatives; Indicateurs de distances; L'échelle des distances extragalactiques; Luminosités et dimensions des galaxies; Couleurs et spectres des galaxies; Rayonnement radio des galaxies; Rotation et dispersion des vitesses; Masses des galaxies; Formation et evolution des galaxies.

003.044 **The Edmund sky guide.**
T. Dickinson, S. Brown.
Edmund Scientific Co., Barrington, N.J. 36 pp. Price $ 3.25 (1977). — Review in J. R. Astron. Soc. Canada, Vol. 71, 473 - 474; 1977 (*B. L. Matthews*).

003.045 **À la recherche d'une vie sur Mars.**   A. Ducrocq.
Éditions Flammarion, Paris. 296 pp. (1976).
Review in Astronomie, Vol. 91, 166; 1977 (*J. Meeus*).

003.046 **The bowels of the earth.**   J. Elder.
Oxford University Press, London. 7 + 222 pp. Price £ 6.00 (1976). — Review in Nature, Vol. 263, 447; 1976 (*P. J. Smith*).

003.047 **Solar noise storms.**   E. Ø. Elgarøy.
International Series in Natural Philosophy.
Pergamon Press, Oxford—New York—Toronto—Sydney—Paris—Frankfurt. 13 + 363 pp. Price $ 18.00, £ 9.75 (1977). ISBN 0-08-021039-2.
Contents: Relations between noise storms and optically observable features on the sun. The spectrum of noise storms. Polarization of noise storm emission. Directivity. Coronal scattering of radiation from noise storm sources. Ordered behaviour of storm bursts in the time-frequency plane. Periodic and quasi-periodic phenomena. Metric noise storms and related phenomena. Discussion of some observed features of noise storms. Introduction to noise storm theories. Plasma wave theories of type I bursts. Cyclotron radiation. Some theories with possible application to storm emission.

003.048 **Methods of data reduction in scientific cosmic experiments.**   V. P. Evdokimov, V. M. Pokras.
Nauka, Moskva. 176 pp. Price 1 Rbl. 15 Kop. (1977). In Russian. — Review in Ref. zh., 62. Issled. kosm. prostranstva, 9.62.445 (1977).

003.049 **Der Stern der Weisen. Geschichte oder Legende?**   K. Ferrari d'Occhieppo.
Verlag Herold, Wien — München. 2nd revised edition. 171 pp. Price öS 128.00 (1977). — Review in Sternenbote, 20. Jahrg., 212 - 213 (1977).

003.050 **Fotografia Astronomica.**   W. Ferreri.
Il Castello, 16 via C. Ravizza, 20149 Milan, Italy. 224 pp. Price Lire 12,000 (1977). — From Sky Telesc., Vol. 54, 519 (1977).

003.051 **Surveying the Moon.**   G. Fielder.
The US Geological Professional Paper 880, US Geological Survey (1977). — Review in Nature, Vol. 270, 563 (1977).

003.052 **The solar planets.**   V. A. Firsoff.
David & Charles, Newton Abbot—London—Vancouver; Crane, Russak & Company, Inc., New York. 184 pp. Price £ 5.50, DM 26.10 (1977). ISBN 0-7153-7352-8 (UK), ISBN 0-8448-0964-0 (USA). — Reviews in J. British Astron. Assoc., Vol. 87, 531 - 532; 1977 (*S. Dunlop*); Nature, Vol. 268, 468; 1977 (*J. Feldman).

003.053 **Meteorites.**   R. V. Fodor.

Dodd, Mead. 47 pp. Price $ 4.25 (1977). — Review in Sky Telesc., Vol. 54, 230 (1977).

003.054   Analytical mechanics, 3rd edition.   G. R. Fowles.
Holt, Rinehart and Winston, New York. 334 pp.
Price $ 16.95 (1977). — From Phys. Today, Vol. 30, No. 9,
p. 62 (1977).

003.055   The moon book: exploring the mysteries of the
lunar world.   B. M. French.
Penguin, New York — Harmondsworth, U.K. 287 pp. Price
$ 4.95, £ 1.95 (1977). — Reviews in Nature, Vol. 269, 838;
1977 (*N. R. Goulty*); Sky Telesc., Vol. 54, 422 (1977).

003.056   Leben wir unter kosmischen Einflüssen?
C. Friedemann.
Urania-Verlag, Leipzig — Jena — Berlin. 128 pp. Price M 4.50
(1976). — Review in Sterne, 53. Band, 253; 1977 (*P. Ahnert*).

003.057   Apollo Soyuz.   W. Froehlich.
National Aeronautics and Space Administration.
132 pp. Price $ 3.30 paperbound (1976). — Review in Sky
Telesc., Vol. 54, 138 (1977).

003.058   Structure of the moon.
I. N. Galkin, V. V. Shvarev.
Ser. "Kosmonavtika, astronomiya", 1977, No. 2. Znanie,
Moskva. 64 pp. Price 11 Kop. (1977). In Russian. — Reviews
in Priroda, 1977, No. 8, p. 157; Ref. zh., 51. Astron., 12.51.
39; 62. Issled. kosm. prostranstva, 12.62.125 (1977).

003.059   The quiet sun.   E. Gibson.
Translated from the English edition. Mir, Moskva.
408 pp. Price 1 Rbl. 80 Kop. (1977). In Russian. — From
Ref. zh., 51. Astron., 1.51.43 (1978).

003.060   Cosmology + 1. Readings from Scientific American.
Introduction by O. Gingerich.
W. H. Freeman & Co., San Francisco — Reading, England.
10 + 114 pp. Price $ 8.50, £ 6.80 cloth; $ 4.50, £ 3.30 paper
(1977). — Reviews in Astron. Tidsskr., Årg. 10, 139 (1977);
Sky Telesc., Vol. 54, 422 (1977); Spaceflight, Vol. 19, 452;
1977 (*G. J. Day*); Strolling Astron., Vol. 27, 34; 1977
(*J. E. Westfall*).

003.061   Über Physik und Astrophysik. Ausgewählte
fundamentale Probleme.
W. L. Ginsburg (*V. L. Ginzburg*).
Wissenschaftliche Taschenbücher. Akademie-Verlag, Berlin.
151 pp. Price M 8,00 (1977). — Review in Astron. Schule,
14. Jahrg., 116 (1977).

003.062   Physique de l'ionosphère.   A. Giraud, M. Petit.
Presses Universitaires de France, Paris. 213 pp.
(1975). — Review in Space Sci. Rev., Vol. 20, 238; 1977
(*K. Rawer*).

003.063   Astronomie I. Die Sonne und ihre Planeten.
F. Gondolatsch, G. Groschopf, O. Zimmermann.
Klett-Studienbücher Physik, Ernst Klett, Stuttgart. 349 pp.
Price DM 27.80 (1977). ISBN 3-12-983830-9.
Contents: Der Forschungsbereich der Astronomie.
Bewegungsvorgänge im Planetensystem. Die großen Planeten
und ihre Monde, Planetoiden, Kometen, interplanetare Materie.
Die Sonne.

003.064   The visible scientists.   R. Goodell.
Little, Brown, Boston. 242 pp. Price $ 9.95 (1975).
Review in Phys. Today, Vol. 30, No. 11, p. 65 - 67; 1977
(*R. W. Nichols*).

003.065   Cosmic gas dynamics.   V. G. Gorbatskij.

Nauka, Moskva. 360 pp. Price 2 Rbl. 10 Kop.
(1977). In Russian. — From Ref. zh., 51. Astron., 11.51.131
(1977).

003.066   White holes: the beginning and end of space.
J. Gribbin.
Paladin, London. 200 pp. Price £ 1.50, $ 4.50 (1977).
Review in Nature, Vol. 270, 133; 1977 (*R. Znajek*).

003.067   White holes. Cosmic gushers in the universe.
J. Gribbin.
Delacorte Press, Eleanor Friede, New York. 8 + 296 pp.
Price cloth $ 8.95, paper $ 4.95 (1977). — From Science,
Vol. 198, 1148 (1977).

003.068   The large-scale structure of space-time.
S. W. Hawking, G. F. R. Ellis.
Translated from the English edition. Mir, Moskva. 431 pp.
Price 3 Rbl. 40 Kop. (1977). In Russian. — Review in Ref.
zh., 51. Astron., 11.51.139 (1977).

003.069   Klassisk astronomi.   B. E. Helt.
Akademisk Forlag, Köpenhamn. 72 pp. Price DKr.
29.50 (1975). — Review in Astron. Tidsskr., Årg. 10, 89;
1977 (*K. Särg*).

003.070   Astronomiopgaver.   B. E. Helt, H. E. Jørgensen.
Akademisk Forlag, Köpenhamn. 72 pp. Price DKr.
29.50 (1976). — Review in Astron. Tidsskr., Årg. 10, 89;
1977 (*K. Särg*).

003.071   Exposure guide for astrophotography.
R. V. Henderson.
Astronomics, Lynn, Mass. 53 pp. Price $ 7.95 (1977).
Review in Sky Telesc., Vol. 54, 230 (1977).

003.072   Kosmische Weiten. Geschichte der Entfernungs-
messung im Weltall.   D. B. Herrmann.
Johann Ambrosius Barth, Leipzig. 95 pp. Price M 9.60 (1977).
Review in Astron. Schule, 14. Jahrg., 116 (1977).

**003.073   Geschichte der Astronomie von Herschel bis
Hertzsprung.**   D. B. Herrmann.
VEB Deutscher Verlag der Wissenschaften, Berlin. 2. Edition.
284 pp. Price M 12.80; M 16.00 (1977).

003.074   Chemical kinetics.   D. R. Herschbach (Editor).
Physical Chemistry, Series two, Vol. 9, International
Review of Science, Butterworths & Co., Ltd. Price £ 13.45
(1976). — Review in Space Sci. Rev., Vol. 20, 236 - 237; 1977
(*A. Bar-Nun*).

003.075   Laser monitoring of the atmosphere.
E. D. Hinkley (Editor).
Springer-Verlag, Berlin. 15 + 380 pp. Price $ 39.80 (1976).
Review in Phys. Abstr., Vol. 80, Abstr. 25799 (1977).

003.076   Astronomy. Fundamentals and frontiers.
R. Jastrow, M. H. Thompson.
3rd edition. John Wiley and Sons, Chichester, Sussex — New
York. 16 + 532 pp. Price £ 10.30, $ 15.50 (1977). — Reviews
in Astron. Tidsskr., Årg. 10, 139 (1977); J. British Astron.
Assoc., Vol. 88, 98 - 99; 1977 (*D. W. Hughes*).

003.077   Solen — en innføring i moderne solfysikk.
E. Jensen, O. Engvold.
Universitetsforlaget, Oslo—Bergen—Tromsø. 10+256 pp. Price
N. kr. 89.50, $ 18.00 (1977). ISBN 82-00-02426-1.

003.078   Astrofysik.   H. E. Jørgensen, B. E. Helt.
Akademisk Forlag, Köpenhamn. 120 pp. Price DKr.
46.00 (1976). — Review in Astron. Tidsskr., Årg. 10, 89;

1977 (*K. Särg*).

003.079   **Optical surfaces: aspherical optical systems, X-ray optics, reflecting microscopes, reflectors, measurements.**   B. Jurek.
Elsevier Scientific Publishing Company, Amsterdam – New York. 217 pp. Price Dfl. 87.50, $ 34.95 (1976). – Reviews in Appl. Opt., Vol. 16, 2337 - 2338; 1977 (*G. Schulz*); J. Opt. Soc. America, Vol. 67, 989; 1977 (*O. N. Stavroudis*).

003.080   **Deciphering the Maya script.**   D. H. Kelley.
University of Texas Press, Austin – London. 334 pp. Price £ 22.00 (1976). – Review in J. British Astron. Assoc., Vol. 88, 104; 1977 (*D. J. Schove*).

003.081   **Automatic stations for investigation of the lunar surface.**   A. L. Kemurdzhian, V. V. Gromov, I. I. Cherkasov, V. V. Shvarev.
Izdatel'stvo "Mashinostroenie" (1976). ·In Russian. – Review in Zemlya i Vselennaya, 1977, No. 5, p. 94 - 95.

003.082   **The finest NGC objects (Set II of deep sky objects).**   G. R. Kepple.
Astro Cards, Pennsylvania. Obtainable from Bretmain Ltd., 2 Station Road, Rayne, Braintree, Essex. Price £ 4.50.
Review in J. British Astron. Assoc., Vol. 87, 425; 1977 (*W. E. Pennell*).

003.083   **Connaître les étoiles en 10 leçons.**   P. Kohler.
Librairie Hachette, 6 Avenue Pierre 1ᵉʳ de Serbie, 75116 Paris, France. 256 pp. (1977). – From Sky Telesc., Vol. 54, 519 (1977).

003.084   **Astronomers. Biographical handbook.**
I. G. Kolchinskij, A. A. Korsun', M. G. Rodriges.
Naukova Dumka, Kiev. 416 pp. Price 1 Rbl. 65 Kop. (1977). In Russian.

003.085   **This fascinating astronomy.**   V. Komarov.
Translated from the Russian edition (1972) by N. Kittell.
Mir Publishers, Moscow, U.S. distributor: Imported Publications, Chicago. 346 pp. Price $ 2.00 (1976). – Review in Sky Telesc., Vol. 54, 325 (1977).

003.086   **Thermal radiation of the planets.**
K. Ya. Kondrat'ev, N. I. Moskalenko.
Gidrometeoizdat, Leningrad. 264 pp. Price 2 Rbl. 30 Kop. (1977). In Russian. – Review in Ref. zh., 51. Astron., 9.51.75 (1977).

003.087   **Ibn aṣ-Salāḥ. Zur Kritik der Koordinatenüberlieferung im Sternkatalog des Almagest.**
P. Kunitzsch (Editor).
Arabischer Text nebst deutscher Übersetzung, Einleitung und Anhang.
Abhandlungen der Akademie der Wissenschaften in Göttingen, Vandenhoeck & Ruprecht. 160 pp. Price DM 88.00 (1975). Essay review in J. Hist. Astron., Vol. 8, 204 - 210; 1977 (*G. J. Toomer*).

003.088   **Mathematics and the universe. An interpretation based on the theory of relativity.**
E. T. Lawrence.
Vantage, New York. 6 + 106 pp. Price $ 8.95 (1977). – From Science, Vol. 198, 861 (1977).

003.089   **Apollo Soyuz mission report.**
C. M. Lee (Editor).
Advances in the Astronautical Sciences, Vol. 34. American Astronautical Society, San Diego, Calif.; Univelt Inc., P.O. Box 28130, San Diego, Calif. 14 + 322 pp. Price $ 35.00 (1977). – Review in Sky Telesc., Vol. 54, 324 - 325 (1977).

003.090   **La recherche en astrophysique.**
J. Lequeux (Editor).
Reprinted from La Recherche, Collection Points, Série Sciences. Editions du Seuil, Paris. 256 pp. Price 18.00 FF (1977). – From Science, Vol. 197, 299 (1977).

003.091   **Tensors, relativity and cosmology.**   E. A. Lord.
Tata McGraw Hill, New Delhi. 208 pp. Price Rs. 19.50 (1977). – From Phys. Today, Vol. 30, No. 10, p. 73 (1977).

003.092   **The harvest of a quiet eye. (A selection of scientific quotations).**   A. L. Mackay.
M. Ebison (Editor), with a foreword by P. Medawar.
The Institute of Physics, Bristol and London. 12 + 192 pp. Price £ 5.20 (1977). – Review in J. British Astron. Assoc., Vol. 88, 97; 1977 (*C. A. Ronan*).

003.093   **Astronomy.**   T. Maloney.
Macdonald. 96 pp. Price £ 1.00 (1977). – Review in Spaceflight, Vol. 19, 377; 1977 (*L. J. Carter*).

003.094   **Pulsars.**   R. N. Manchester, J. H. Taylor.
Freeman, San Francisco. 281 pp. Price $ 19.95 (1977). – From Phys. Today, Vol. 30, No. 12, p. 62 (1977).

003.095   **The XXXth Herzen lectures. Theoretical physics and astronomy. Scientific reports.**
Leningrad. gos. ped. inst. im. A. I. Gertsena. Leningrad. 52 pp. Price 49 Kop. (1977). In Russian. – From Ref. zh., 51. Astron., 11.51.129 (1977).

003.096   **Our fragile water planet: an introduction to the earth sciences.**   C. L. Mantell, A. M. Mantell.
Plenum, New York. 221 pp. Price $ 19.50 (1977). – From Phys. Today, Vol. 30, No. 9, p. 62 (1977).

003.097   **W poszukiwaniu Kosmitów.**   A. Marks.
Ludowa Spółdzielnia Wydawnicza, Warszawa. 252 pp. Price 40 zł. (1977). – Review in Urania Kraków, Vol. 48, 311 - 314; 1977 (*T. Z. Dworak, J. Usowicz*).

003.098   **Negative ions.**   H. Massey.
Cambridge University Press, Cambridge, 3rd edition. 741 pp. Price $ 69.50 (1976). ISBN 0-521-20775-4. – Review in Astrophys. Lett., Vol. 19, 32 - 33; 1977 (*D. W. Norcross*).

003.099   **Legal implications of remote sensing from outer space.**   N. M. Matte, H. DeSaussure (Editors).
Sijthoff, Leyden, The Netherlands. 230 pp. Price Dfl. 49.00, US $ 19.00 (1976). – Review in Space Sci. Rev., Vol. 20, 235; 1977 (*G. C. M. Reijnen*).

003.100   **Infrared, correlation, and Fourier transform spectroscopy.**
J. S. Mattson, H. B. Mark, Jr., H. C. MacDonald, Jr. (Editors).
Marcel Dekker, Inc., New York. 233 pp. Price $ 27.50 (1977). Review in Opt. Spectra, Vol. 11, No. 10, p. 52 (1977).

003.101   **Stellar evolution.**   A. J. Meadows.
Pergamon Press, Oxford. 2nd edition. 180 pp. Price $ 15.00, £ 7.50 hardcover, $ 5.00, £ 2.50 paper. ISBN 0-08-021668-4; ISBN 0-08-021669-2 f.

003.102   **The tides of the planet earth.**   P. Melchior.
Pergamon Press, Headington Hill Hall, Oxford OX3 0BW, England. Fairview Park, New York 10523, USA. 492 pp. Price $ 40.00, £ 22.50 (1977). ISBN 0-08-022047-9.

003.103 Focus on the stars.
H. Messel, S. T. Butler (Editors).
Heinemann Educational Books, London. 287 pp. Price £ 4.80
(1977). – Review in J. British Astron. Assoc., Vol. 87, 638;
1977 (*E. A. Beet*).

003.104 Storia dell'astronomia.   R. Migliavacca.
Mursia Editore. 267 pp. Price Lire 6000 (1976).
Review in Coelum, Vol. 45, 165; 1977 (*G. Romano*).

**003.105 Benjamin Martin – author, instrument maker and
    country showman.**   J. R. Millburn.
Nordhoff International Publishing, Leyden, The Netherlands.
12 + 244 pp. Price Dfl. 63.00 (1976). – Review in J. British
Astron. Assoc., Vol. 88, 100; 1977 (*C. A. Ronan*).

003.106 Gravitation. Volume I.
C. W. Misner, K. S. Thorne, J. A. Wheeler.
Translated from the English edition. Mir. Moskva. 476 pp.
Price 3 Rbl. 26 Kop. (1977). In Russian. – Review in Ref. zh.,
51. Astron., 8.51.43 (1977).

003.107 Gravitation. Volume 2.
C. W. Misner, K. S. Thorne, J. A. Wheeler.
Translated from the English edition. Mir, Moskva. 525 pp.
Price 3 Rbl. 93 Kop. (1977). In Russian. – Review in Ref.
zh., 51. Astron., 10.51.90 (1977).

003.108 Gravitation. Volume 3.
C. W. Misner, K. S. Thorne, J. A. Wheeler.
Translated from the English edition. Mir, Moskva. 512 pp.
Price 4 Rbl. 40 Kop. (1977). In Russian. – Review in Ref.
zh., 51. Astron., 11.51.140 (1977).

003.109 Relativitätstheorie.   C. Møller.
Translated from the English edition (London, 1972).
Bibliográphisches Institut, Mannheim. 316 pp. Price DM
36.00 (1976). – From Science, Vol. 198, 1078 (1977).

003.110 History of the earth.   A. S. Monin.
Nauka, Leningrad. 228 pp. Price 1 Rbl. 38 Kop.
(1977). In Russian. – Reviews in Priroda, 1977, No. 12, p.
137 - 138 (*K. K. Markov*); Ref. zh., 51. Astron., 9.51.73
(1977).

003.111 The A–Z of astronomy.
P. Moore, diagrams by C. Deakins.
Revision of The Amateur Astronomer's Glossary (1967).
Charles Scribner's Sons, New York. 192 pp. Price cloth $ 7.95,
paper $ 2.95 (1977). – Review in Sky Telesc., Vol. 54, 138
(1977).

003.112 The astronomy of Southern Africa.
P. Moore, P. Collins.
Howard Timmins, P.O. Box 94, Cape Town. 160 pp. Price
R 9.75 (1977). – Review in Mon. Notes Astron. Soc. South.
Africa, Vol. 36, 106; 1977 (*C. Turk*).

003.113 Essentials of astronomy.   L. Motz, A. Duveen.
Columbia University Press, New York. 10 + 763 pp.
Price $ 17.50 (1977). ISBN 0-231-04009-1.

003.114 Chemical petrology, with applications to the
    terrestrial planets and meteorites.
R. F. Mueller, S. K. Saxena.
Springer-Verlag, New York–Heidelberg–Berlin. 12 + 394 pp.
Price DM 72.80 (1977). – Reviews in Astron. Tidsskr., Årg.
10, 139 (1977); Phys. Earth Planet. Inter., Vol. 15, 376; 1977
(*S. K. Runcorn*).

003.115 Flight to Mercury.   B. Murray, E. Burgess.
Columbia Univ. Press, New York. 11 + 162 pp.

Price $ 12.95, DM 39.70 (1977). ISBN 0-231-03996-4.
Reviews in J. British Astron. Assoc., Vol. 88, 98; 1977
(*P. Moore*); Sky Telesc., Vol. 54, 138 (1977); Spaceflight,
Vol. 19, 450; 1977 (*C. A. Cross*).

003.116 The crime of Claudius Ptolemy.   R. R. Newton.
Johns Hopkins University Press. 411 pp. Price
$ 22.50 (1977). – From Sky Telesc., Vol. 54, 519 (1977).

003.117 Astronomy – a dictionary of space and the universe.
I. Nicolson.
Arrow Books Ltd., London. 250 pp. Price £ 1.75 (1977).
Reviews in J. British Astron. Assoc., Vol. 88, 96 - 97; 1977
(*P. Gill*); Spaceflight, Vol. 19, 452; 1977 (*T. G. Cook*).

003.118 Space chemistry.   L. Nikolaev.
Translated from the Russian edition by Y. Nadler.
Mir Publishers, Moscow, U.S. distributor: Imported Publica-
tions, Chicago. 200 pp. Price $ 2.00 (1976). – Review in Sky
Telesc., Vol. 54, 325 (1977).

**003.119 Mechanism of crater formation by impact and ex-
    plosion.**   V. N. Nikolaevskij (Editor).
Mekh. Novoe v zarubezh. nauk., No. 12, Mir, Moskva (1977).
In Russian. – From Ref. zh., 51. Astron., 12.51.43 (1977).

003.120 What's up? An hourly guide to selected astronomical
    objects.   D. H. Olson.
D. H. Olson, Minneapolis. 60 pp. Price $ 5.95 (1976).
Review in Sky Telesc., Vol. 54, 230 (1977).

003.121 Dynamics of gravitating systems of the metagalaxy.
T. B. Omarov.
Nauka, Alma-Ata. 144 pp. Price 1 Rbl. 4 Kop. (1976). In
Russian. – Review in Astron. Zh. Akad. Nauk SSSR, Vol. 54,
1143 - 1144; 1977 (*V. G. Demin*).

003.122 Introduction to astronomy.
R. A. Oriti, W. B. Starbird.
Glencoe Press, Encino, Calif. 402 pp. Price $ 12.95 (1977).
Reviews in Sky Telesc., Vol. 54, 230 (1977); Strolling Astron.,
Vol. 27, 32 - 33; 1977 (*W. S. Cameron*).

003.123 Radio galaxies.   A. G. Pacholczyk.
International Series in Natural Philosophy, Vol. 89.
Pergamon Press, Oxford, England–New York, U.S.A. 12 + 294
pp. Price $ 25.00, £ 10.00 (1976/77). ISBN 0-08-021031-7.
From Science, Vol. 198, 863 (1977).

**003.124 Einheiten und Größenarten der Naturwissen-
    schaften.**   E. Padelt, H. Laporte.
VEB Fachbuchverlag, Leipzig. 3rd revised edition. 378 pp.
Price M 10.80 (1976).

003.125 Photographic sky atlas.   C. Papadopoulos.
Pergamon Press, Oxford. Price Vol. I: $ 100.00,
Vol. II: $ 150.00 (1977). – Review in J. American Assoc.
Variable Star Obs., Vol. 6, 40 - 41; 1977 (*C. E. Scovil*).

003.126 The logical flaws of Einstein's relativity.
L. Parish.
Cortney, Luton, UK. 171 pp. Price £ 6.50 (1977). – From
Phys. Today, Vol. 30, No. 12, p. 62 (1977).

003.127 A perspective of physics. Volume 1. Selections from
    1976 Comments on Modern Physics.
Introduced and put into perspective by R. Peierls. Contribu-
tors: L. B. Okun, Ya. B. Zeldovich, B. W. Lee, C. Quigg,
H. Harari, M. Jacob, D. Cline, A. K. Mann, Q. Lewin Keller,
J. W. Negele, F. S. Stephens, A. J. Leggett, R. W. Keyes,
J. Tauc, S. R. Nagel, S. W. Lovesey, A. H. Thompson, B.
Paczynski, P. Goldreich, W. Sargent, R. B. Larson, J. E.

Grindlay, L. Spitzer, S. J. Brodsky, G. Karl, J. L. Picqué, H. H. Stroke, V. Dose, R. V. Ambartzumian (*Ambartsumyan*), V. S. Letokhov, F. M. Harris, G. Bekefi, C. Deutsch, H. P. Furth, V. N. Tsytovich, J. D. Callen.
Gordon and Breach Science Publishers, London–New York– Paris. 35 + 243 pp. Price £ 20.00 (1977). ISBN 0-677-13190-9.

003.128 **600 Jahre Astronomie in Nürnberg.** K. Pilz.
Verlag Hans Carl, Nürnberg. 376 + 32 pp. Price DM 76.00 (1977). ISBN 3-418-00447-4. – Reviews in Orion, 35. Jahrg., 177 (1977); Sterne Weltraum, Jahrg. 16, 348; 1977 (*K. Schaifers*).

003.129 **The versatile satellite.** R. W. Porter.
Oxford University Press, Oxford – New York – London. 8 + 173 pp. Price £ 4.95, $ 11.00 (1977). ISBN 0-19-885104-9. – Reviews in Astron. Tidsskr., Årg. 10, 140 (1977); Nature, Vol. 270, 374; 1977 (*L. F. Curtis*).

003.130 **Astronomia oggi.** F. Potenza.
Longanesi, Milano. 207 pp. Price Lire 6.000 (1976). Review in G. Astron., Vol. 3, 51 - 52; 1977 (*E. Proverbio*).

003.131 **Les tables astronomiques de Louvain de 1528 par Henri Baers ou Vekenstyl.**
Introduction, traduction et commentaire par E. Poulle, A. De Smet.
Edition en fac-simile. Editions Culture et Civilisation, Bruxelles. 39 + 90 pp. Price DM 33.40 (1976). – Review in J. Hist. Astron., Vol. 8, 216; 1977 (*J. Dobrzycki*).

003.132 **SI units in engineering and technology.**
S. H. Qasim.
Pergamon International Library. Pergamon, New York. 8 + 54 pp. Price $ 5.00 (1977). – From Science, Vol. 197, 979 (1977).

003.133 **Effects of solar activity in the lower atmosphere.** L. R. Rakipova (Editor).
Gidrometeoizdat, Leningrad. 143 pp. Price 68 Kop. (1977). In Russian. – From Ref. zh., 51. Astron., 10.51.80 (1977).

003.134 **Constellation and star finder.** B. V. S. Rao.
Higgin-bothams Pvt. Ltd., Mount Road, Madras-600 002. Price Rs. 22.00. – Review in Bull. Astron. Soc. India, Vol. 5, 58; 1977 (*A. Bhatnagar*).

003.135 **The lens in action.** S. F. Ray.
Hastings House, Inc., New York. 201 pp. Price $ 8.95 (1976). – Review in Sky Telesc., Vol. 54, 229 - 230; 1977 (*R. E. Cox*).

003.136 **Stellar formation.** V. C. Reddish.
Pergamon Press, Oxford, England–New York, U.S.A. ca. 225 pp. Price approx. $ 15.00, £ 7.50 (1976). ISBN 0-08-018062-0.

003.137 **Carl Friedrich Gauß, 1777/1977.** K. Reich.
Heinz Moos Verlag, München. 128 pp. Price DM 18.00 (1977). ISBN 3-7879-0099-3.

003.138 **Les astronomes et le droit de l'espace.**
G. Ringeard.
Presses Universitaires de France. 250 pp. (1977). – Review in Astronomie, Vol. 91, 445; 1977 (*J. Mauro*).

003.139 **Lectures on density wave theory.** K. Rohlfs.
Lecture Notes in Physics, Vol. 69. Springer-Verlag, Berlin – Heidelberg – New York. 6 + 184 pp. (1977). ISBN 3-540-08448-7, ISBN 0-387-08448-7.
Contents: Some elements of stellar dynamics. The density distribution perpendicular to the galactic plane. The

stability of solutions. The dynamics of a rotating disk of non-interacting particles. The theory of spiral structure of galaxies. Gas dynamical effects on density waves. Observable consequences of the density wave theory. Evolution and origin of density waves.

003.140 **The catalogue of the history of science collections of the University of Oklahoma Libraries.**
D. H. D. Roller, M. M. Goodman.
Mansell, London. 12 + 584; 608 pp. Price £ 90.00 (1976). Review in J. Hist. Astron., Vol. 8, 215; 1977 (*O. Gingerich*).

003.141 **Sunlight convergence solar burn.** C. Ross.
University of Utah Press, Salt Lake City, Utah. 38 pp. Price $ 20.00 (1976). – Reviews in Sky Telesc., Vol. 54, 138; 318 - 319; 1977 (*J. White*).

003.142 **Orbital motion.** A. E. Roy.
Adam Hilger Ltd. 300 pp. Price £ 12.50 (1977). ISBN 0-85274-322-X.

003.143 **Astronomy: principles and practice.**
A. E. Roy, D. Clarke.
Adam Hilger Ltd., Bristol, UK. 15 + 342 pp. Price hardback £ 10.00, paperback £ 7.50 (1977). ISBN 0-85274-292-4; 0-85274-346-7. – Review in Nature, Vol. 268, 570; 1977 (*B. Pagel*).

003.144 **Moon, Mars and Venus, a concise guide in colour.**
A. Rükl.
Hamlyn, London. Price £ 1.50. – Review in Nature, Vol. 270, 646; 1977 (*D. W. Hughes*).

003.145 **Lunar nomenclature. Far side of the moon, 1961 - 1973.** K. B. Shingareva, G. A. Burba.
Nauka, Moskva. 56 pp. Price 30 Kop. (1977). In Russian. Review in Ref. zh., 51. Astron., 8.51.36 (1977).

003.146 **Space, time and spacetime.** L. Sklar.
University of California Press, Berkeley. Reprint of the 1974 edition. 12 + 424 pp. Price $ 5.95 (1977). – From Science, Vol. 197, 299 (1977).

003.147 **A dual ether universe: introducing a new field theory.** L. Sokolow.
Exposition–University, Hicksville, N.Y. 12 + 157 pp. Price $ 12.50 (1977). – From Phys. Today, Vol. 30, No. 9, p. 62 (1977).

003.148 **Digital communications by satellite.** J. J. Spilker.
Englewood Cliffs, N.J., Prentice-Hall. 672 pp. Price $ 38.00 (1977). ISBN 0-13-214155-8. – Review in Proc. IEEE, Vol. 65, 1519 (1977).

003.149 **Sphärische Trigonometrie**, mit einigen Anwendungen aus Geodäsie, Astronomie und Kartographie.
K.-G. Steinert.
Kleine Naturwissenschaftliche Bibliothek, Reihe Mathematik, Band 8. Akademische Verlagsgesellschaft Geest u. Portig K.G. BSB B. G. Teubner Verlagsgesellschaft, Leipzig. 160 pp. Price M 9.50 (1977). – Reviews in Astron. Schule, 14. Jahrg., 117 (1977); Sternenbote, 20. Jahrg., 151 (1977).

003.150 **Multicolor stellar photometry.** Photometric systems and methods. V. Straižys.
Institut Fiziki Akademii Nauk Litovskoj SSR. Izdatel'stvo "Mokslas", Vil'nyus. 311 pp. Price 3 Rbl. 20 Kop. (1977). In Russian.

003.151 **Gamma-spectrometry in cosmic investigations.**
Yu. A. Surkov.
Atomizdat, Moskva. 240 pp. Price 2 Rbl. 4 Kop. (1977). In

38   Periodicals, Proceedings, Books, Activities

Russian. – Review in Ref. zh., 51. Astron., 8.51.34 (1977).

003.152 Continents in motion.   W. Sullivan.
Macmillan, London. 399 pp. Price £ 6.95 (1977).
Review in Astron. Tidsskr., Årg. 10, 140 (1977).

003.153 Lectures on selected topics in statistical mechanics.
D. ter Haar.
Pergamon Press, Oxford – New York – Toronto – Sydney –
Paris – Frankfurt. 7 + 124 pp. (1977). ISBN 0-08-017937-1.
Contents: The occupation number representation. The
Green function method in statistical mechanics. The pair
Hamiltonian model of an imperfect Bose gas. Fluctuations in
a perfect Bose gas. The equation of state of an imperfect gas.
A simple derivation of the Bloch equation. Statistical
mechanics of stellar systems. Approach to equilibrium.

003.154 From quarks to quasars. An outline of modern
physics.   E. Thomas.
Athlone Press of the University of London, London. (US
distributor, Humanities Press, Atlantic Highlands, N.J.). 10 +
294 pp. Price $ 8.00 (1977). – From Science, Vol. 197, 979
(1977).

003.155 Johannes Kepler: giant of faith and science.
J. H. Tiner.
Mott Media, P. O. Box 236, Milford, Mich. 48042. 200 pp.
Price $ 3.50 (1977). – Review in Sky Telesc., Vol. 54, 422
(1977).

003.156 Radiative properties of shock waves in gases.
M. A. Tsikulin, E. G. Popov.
Nauka, Moskva. 176 pp. Price 1 Rbl. 10 Kop. (1977). In Rus-
sian. – Review in Ref. zh., 51. Astron., 9.51.82 (1977).

003.157 Ausserhalb der Erde.
K. E. Ciolkovskij (*K. Eh. Tsiolkovskij*).
Mit einer Einführung zu Leben und Werk des berühmten russi-
schen Raketen- und Raumfahrtpioniers und erklärenden An-
merkungen zum Text von W. Petri sowie eigenhändigen
Skizzen des Autors.
Wilhelm Heyne Verlag, München. 192 pp. (1977). ISBN
3-453-30448-9 = Veröff. Forschungsinst. Deutsches Mus.
Gesch. Naturwiss. Tech., Reihe B, Nr. 9.

003.158 Explosion craters of the Ukrainian shield.
A. A. Val'ter, V. A. Ryabenko.
Naukova dumka, Kiev. 154 pp. Price 1 Rbl. (1977). In Rus-
sian. – Review in Ref. zh., 51. Astron. 1.51.42 (1978).

003.159 Astronomy for the amateur, Vol. I: Planetary
astronomy.   R. P. Van Zandt.
Published by the author, P.O. Box 3013, Peoria, Ill. 61614,
3rd edition. 243 pp. Price $ 8.75 (1977). – Reviews in Sky
Telesc., Vol. 54, 518; 1977 (*C. A. Federer, Jr.*); Strolling
Astron., Vol. 27, 33 - 34; 1977 (*M. B. Smith*).

003.160 Spectrometric techniques. Vol. 1.
G. A. Vanasse (Editor)..
Academic Press, New York–San Francisco–London. 368 pp.
Price $ 29.50, £ 20.95 (1977). ISBN 0-12-710401-1.

003.161 International cooperation in space. Legal questions.
V. S. Vereshchetin.
Nauka, Moskva. 264 pp. Price 1 Rbl. 70 Kop. (1977). In Russian. – Review in Ref. zh., 62. Issled. kosm. prostranstva,
1.62.9 (1978).

003.162 Starscapes: topics in astronomy.   G. L. Verschuur.
Little, Brown, Boston, MA. 16 + 202 pp. Price
$ 6.95 (1977). – Reviews in Sky Telesc., Vol. 54, 325 (1977);
Strolling Astron., Vol. 27, 32; 1977 (*E. F. Bailey*).

003.163 Space, time, and gravity. The theory of the Big Bang
and black holes.   R. M. Wald.
The University of Chicago Press, Chicago–London. 8 + 131 pp.
Price £ 7.70 (1977). ISBN 0-226-87030-8.
Contents: The geometry of space and time. Special
relativity. General relativity. Implications for cosmology: the
"Big Bang". The evolution of our Universe. Stellar evolution.
Gravitational collapse to black holes. Energy extraction from
black holes. The astrophysics of black holes. Quantum particle
creation near black holes.

003.164 Astronomin i vår tid.   Å. Wallenquist.
Bokförlaget Prisma, Stockholm. 248 pp. Price
S. kr. 45.00 (1977). – Review in Astron. Tidsskr., Årg. 10,
137 - 138; 1977 (*G. Lyngå*).

003.165 Handbuch der babylonischen Astronomie. Band 1:
Der babylonische Fixsternhimmel. Beiträge zur
ältesten Geschichte der Sternbilder.   E. F. Weidner.
Leipzig 1915 Reprint. 3 + 147 pp. Assyriologische Bibliothek
XXIII (1976).

003.166 Stars and planets.   K. Wicks.
Franklin Watts. 44 pp. Price $ 3.95 (1977). –
Review in Sky Telesc., Vol. 54, 422 (1977).

003.167 Progress in optics, Vol. XIV.   E. Wolf (Editor).
North-Holland Publishing Company, Amsterdam–
New York. 17 + 422 pp. Price Dfl. 140.00, $ 57.25 (1976).
Review in J. Opt. Soc. America, Vol. 67, 1132; 1977 (*M.
Young*).

003.168 Das Weltall.   B. A. Woronzow-Weljaminow
(*Vorontsov-Vel'yaminov*).
Urania-Verlag, Leipzig–Jena–Berlin. 255 pp. Price M 24.60
(1976). – Reviews in Sky Telesc., Vol. 54, 324 (1977); Sterne,
53. Band, 247 - 248; 1977 (*S. Marx*).

003.169 Principles of astronomy.   S. P. Wyatt.
Allyn and Bacon, Boston. 3rd edition. 730 pp.
Price $ 15.95 (1977). – Review in Sky Telesc., Vol. 54, 138
(1977).

003.170 Dzieje Zegara.   L. Zajdler.
Wydanie Drugie, Przerobione, Wiedza Powszechna,
"Złota Seria Literatury Popularnonaukowej", Warszawa. 397
pp. Price 85 zł (1977). – Review in Urania Kraków, Vol. 48,
314 - 316; 1977 (*T. Z. Dworak*).

003.171 Electromagnetic waves in cosmic plasma. Genera-
tion and propagation.   V. V. Zheleznyakov.
Izdatel'stvo Nauka, Glavnaya Redaktsiya Fiziko-Matematicheskoj Literatury, Moskva. 432 pp. Price 2 Rbl. 66 Kop. (1977).
In Russian. – Review in Ref. zh., 51. Astron., 9.51.83 (1977).

003.172 Commémoration du 500me anniversaire de la
naissance de Nicolas Copernic (1473 - 1543).
Institut Interuniversitaire pour l'étude de la renaissance et de
l'humanisme. Editions de l'Université de Bruxelles (1974).
Review in Ciel Terre, Vol. 93, 186 - 187; 1977 (*A.
Koeckelenbergh*).

003.173 Cosmic data.
Mesyach. obzor. No. 10. Okt. 1976. Inst. zemn.
magn., ionos. i rasprostr. radiovoln AN SSSR. Nauka, Moskva.
79 pp. (1977). In Russian. – Review in Ref. zh., 62. Issled.
kosm. prostranstva, 10.62.57 (1977).

003.174 Dynamics of cosmic plasma.
Inst. zemn. magn., ionos. i rasprostranenie radiovoln
AN SSSR. Moskva. 264 pp. Price 1 Rbl. 8 Kop. (1976). In
Russian. – From Ref. zh., 51. Astron., 9.51.84 (1977).

003.175   **McGraw-Hill encyclopedia of science and technology.**
McGraw-Hill, New York. 4. Edition. 15 volumes.Price $ 497; to institutions, $ 447 (1977). − From Science, Vol. 197, 1105 (1977).

003.176   **Meteor investigations. No. 4.**
Mezhduved. geofiz. kom. pri prezidiume AN SSSR. Sov. radio, Moskva. 120 pp. Price 1 Rbl. 20 Kop. In Russian. From Ref. zh., 51. Astron., 11.51.133 (1977).

003.177   **Publications of the 10th lectures devoted to the scientific heritage and to the development of ideas of K. Eh. Tsiolkovskij. Kaluga, 16 - 19 Sept. 1975.**
Sekts. "Issled. nauchn. tvorchestva K. Eh. Tsiolkovskogo". AN SSSR. Komis. po razrabotke nauchn. naslediya K. Eh. Tsiolkovskogo, Gos. muzej istor. kosmonavt. im K. Eh. Tsiolkovskogo. Moskva, 75 pp. Price 35 Kop. (1977). In Russian. − From Ref. zh., 62. Issled. kosm. prostranstva, 1.62.48 (1978).

003.178   **Space investigations made in the USSR in 1976.**
Nauka, Moskva (1977). 63 pp. In Russian. − Review in Ref. zh., 62. Issled. kosm. prostranstva, 12.62.23 (1977).

003.179   **Techniques and methods of radio astronomical reception.**
Trudy Fiz. inst. AN SSSR, 93. Nauka, Moskva. 148 pp. Price 1 Rbl. 29 Kop. (1977). In Russian. − From Ref. zh., 51. Astron., 11.51.127 (1977).

003.180   **The planispheric astrolabe, make-it-yourself astrolabe.**
Department of Navigation and Astronomy, National Maritime Museum, Greenwich, London, SE10 9NF, England. 56 pp. Price £ 1.00; astrolabe, 75 p. (1976). − Reviews in J. British Astron. Assoc., Vol. 87, 533 - 534; 1977 (*D. Tattersfield*); J. Navig., Vol. 30, 525; 1977 (*C. H. Cotter*); Sky Telesc., Vol. 54, 422 (1977).

Astronomical catalogues 1951 - 1975.
See Abstr. 002.025.

004 History of Astronomy, Chronology

004.001 Newton and the development of classical mechanics.
A. T. Grigor'yan, B. N. Fradlin.
Priroda, 1977, No. 7, p. 7 - 13. In Russian.

004.002 To the sources of primeval astronomy.
B. A. Frolov.
Priroda, 1977, No. 8, p. 96 - 106. In Russian.

004.003 Star map of the neolithic man.
Yu. P. Pskovskij.
Priroda, 1977, No. 9, p. 28 - 29. In Russian.

004.004 Maya dates AD 352–1296. D. J. Schove.
Nature, Vol. 268, 670 (1977).

004.005 The Earl of Rosse's experiments on reflecting
telescopes. N. S. Hetherington.
J. British Astron. Assoc., Vol. 87, 472 - 477 (1977).

004.006 Birth of the VfR (*Verein für Raumschiffahrt*): the
start of modern astronautics. F. H. Winter.
Spaceflight, Vol. 19, 243 - 256 (1977).

004.007 Carl Friedrich Gauss, Astronomer. B. G. Marsden.
J. R. Astron. Soc. Canada, Vol. 71, 309 - 323
(1977).
    Some of the many contributions of Gauss to astronomy
are described. The most notable of them, his development of a
method of orbit determination and application to the computation of the orbit of Ceres in 1801 and his computation of
the perturbations by Jupiter on Pallas, are discussed in
particular detail.

004.008 Herschel's "Large 20-foot" telescope.
J. Ashbrook.
Sky Telesc., Vol. 54, 174 - 175 (1977).

004.009 Lowell and Venus. W. G. Hoyt.
Bull. American Astron. Soc., Vol. 9, 457 (1977).
Abstract.

004.010 Archaeo-astronomy: the current surge of interest.
W. K. Hartmann.
Bull. American Astron. Soc., Vol. 9, 457 - 458 (1977).
Abstract.

004.011 Leonardo da Vinci's telescope. V. A. Gurikov.
Zemlya i Vselennaya, 1977, No. 4, p. 70 - 73. In
Russian.

004.012 Rara Astronomica in Estonia.
    Eesti NSV Teaduste Akadeemia W. Struve Nimeline
Tartu Astrofüüsika Observatoorium. Tartu Astron. Obs.,
Teated, No. 55, 47 pp. (1977). — The individual contributions
are included in their corresponding subject categories — see
abstracts 004.013 - 004.016.

004.013 Rara Astronomica in the library of the Tartu Univer-
sity: books printed before 1600. O. Nagel.
Rara Astronomica in Estonia, 1977, (see 004.012) = Tartu
Astron. Obs., Teated, No. 55, p. 6 - 20 (1977).

004.014 Rara Astronomica in the library of the Tartu astro-
nomical Observatory. H. Eelsalu, H. Silvet.
Rara Astronomica in Estonia, 1977, (see 004.012) = Tartu
Astron. Obs., Teated, No. 55, p. 21 - 26 (1977).

004.015 Rara Astronomica in the scientific library of the

Academy of Sciences: a preliminary report.
K. Robert, H. Eelsalu.
Rara Astronomica in Estonia, 1977, (see 004.012) = Tartu
Astron. Obs., Teated, No. 55, p. 27 - 35 (1977).

004.016 British astronomers' letters in the library of the
Tartu University. H. Tankler, P. Müürsepp.
Rara Astronomica in Estonia, 1977, (see 004.012) = Tartu
Astron. Obs., Teated, No. 55, p. 36 - 46 (1977).

004.017 Lambert: self-taught physicist. S. L. Jaki.
Phys. Today, Vol. 30, No. 9, p. 25 - 30, 32 (1977).
    This year marks the bicentennial of the death of Johann
Heinrich Lambert; although his contributions to photometry
are better known, his ideas in cosmology are surprisingly modern, even hinting at black holes.

004.018 The astronomy of the Hopi Indians.
S. C. McCluskey.
J. Hist. Astron., Vol. 8, 174 - 195 (1977).

004.019 Some comments by Caroline Herschel on the use
of the 40ft telescope. A. J. Turner.
J. Hist. Astron., Vol. 8, 196 - 198 (1977).

004.020 The island of Eday. A. S. Thom, T. R. Foord.
J. Hist. Astron., Vol. 8, 198 - 199 (1977).

004.021 Remarks on "Hipparchus's solar theory derived
from lunar eclipse observations" by K. P.
Moesgaard. R. R. Newton, with comments by K. P.
Moesgaard.
J. Hist. Astron., Vol. 8, 200 - 203 (1977).

004.022 Cosmographic works of E. Karnetsi.
A. G. Abramyan.
Istor. estestvozn. i tekh. v Armenii. No. 6, Erevan, 1976,
p. 58 - 99. In Armenian. — Abstr. in Ref. zh., 51. Astron.,
9.51.12 (1977).

004.023 Armenian astronomical tables — "bolorak".
B. E. Tumanyan.
Istor. estestvozn. i tekh. v Armenii. No. 6, Erevan, 1976,
p. 100 - 121. In Armenian. — Abstr. in Ref. zh., 51. Astron.,
9.51.13 (1977).

004.024 On the time of appearance of ideas of the spherical
shape of the earth. V. M. Manoyan.
Istor. estestvozn. i tekh. v Armenii. No. 6, Erevan, 1976,
p. 185 - 193. In Armenian. — Abstr. in Ref. zh., 51. Astron.,
9.51.14 (1977).

004.025 Heinrich Wild bei der schweizerischen Landestopo-
graphie (1900 - 1907). F. Kobold.
Heinrich Wild 1877 - 1951, (see 003.009), p. 15 - 20 (1977).

004.026 Heinrich Wild und Heerbrugg. G. Strasser.
Heinrich Wild 1877 - 1951, (see 003.009), p. 21 -
28 (1977).

004.027 Heinrich Wild und der Instrumentenbau bei Kern
Aarau 1935 - 1951. R. Haller.
Heinrich Wild 1877 - 1951, (see 003.009), p. 29 - 31 (1977).

004.028 La précision des techniques d'observation anciennes.
S. Arend.
O telescópio refractor e a astrometria ao serviço das estrelas
duplas, (see 012.013), p. 55 - 60 (1977).

L'auteur présente une brève synthèse historique concernant la précision des mesures d'étoiles doubles visuelles effectuées par les premiers observateurs dans ce domaine et s'intéresse également à certaines de leurs équations personnelles.

004.029 Alignements astronomiques du site mégalithique de Callanish, Lewis: une critique. J.-R. Roy.
J. R. Astron. Soc. Canada, Vol. 71, 405 (1977). – Abstract.

004.030 On sources for ideas of cosmonautics.
B. A. Starostin.
Priroda, 1977, No. 10, p. 77 - 85. In Russian.

004.031 An astronomical re-appraisal of the Star of Bethlehem – a nova in 5 BC.
D. H. Clark, J. H. Parkinson, F. R. Stephenson.
Q. J. R. Astron. Soc., Vol. 18, 443 - 449 (1977).

004.032 Megalithic astronomy – fact or fiction?
D. C. Heggie.
Q. J. R. Astron. Soc., Vol. 18, 450 - 458 (1977).

004.033 The manuscript archives of the Royal Astronomical Society. J. A. Bennett.
Q. J. R. Astron. Soc., Vol. 18, 459 - 463 (1977).

004.034 Visst hade inkafolket astronomiska kunskaper.
B. Stenholm.
Astron. Tidsskr., Årg. 10, 93 - 95 (1977).

004.035 Wann trat die Julianische Schaltregel in Kraft?
K. Ferrari d'Occhieppo.
Sternenbote, 20. Jahrg., 203 - 205 (1977).

004.036 Wie Gauß zum Astronomen wurde.
K.-R. Biermann.
Sterne, 53. Band, 146 - 150 (1977).

004.037 Tycho Brahe and the Great Comet of 1577.
O. Gingerich.
Sky Telesc., Vol. 54, 452 - 458 (1977).

004.038 Isaac Newton und die Begründung der mathematischen Prinzipien der Naturphilosophie.
H.-J. Treder.
Sitzungsber. Akad. Wiss. DDR, Math. Naturwiss. Tech., Jahrg. 1977, Nr. 12N, p. 5 - 12.

004.039 Harrison, Maskelyne and the longitude problem.
P. S. Laurie.
Modern astronomy, (see 003.013), p. 13 - 20 (1977).

004.040 Some early astronomical visual aids.
D. Gavine.
J. British Astron. Assoc., Vol. 88, 32 - 37 (1977).

004.041 Maya chronology and planetary conjunctions.
D. J. Schove.
J. British Astron. Assoc., Vol. 88, 38 - 52 (1977).

004.042 The dark night sky paradox. E. R. Harrison.
American J. Phys., Vol. 45, 119 - 124 (1977).
Abstr. in Phys. Abstr., Vol. 80, Abstr. 33758 (1977).

004.043 Chinese cosmology. J. Gribbin.
Astronomy, Vol. 5, No. 1, p. 46 - 48 (1977).
Abstr. in Phys. Abstr., Vol. 80, Abstr. 33816 (1977).

004.044 Determination of latitudes in Yuan Dynasty.
G.-q. Li, P.-r. Yi, B.-t. Li.
Acta Astron. Sinica, Vol. 18, 129 - 137 (1977).
Kuo Shou-jing, the famous astronomer in Yuan Dynasty,

and his co-workers devised and constructed new astronomical instruments and carried out many astronomical determinations for the purpose of revising the calendar system. One of the determinations was a large scale latitude measurement carried out in 1279 A. D. Latitude values were determined in twenty seven observatories or observation stations, with geographic latitudes ranging from +15° to +65°. This paper discusses the methods employed and the significance of these determinations.

004.045 La astronomía en el Antiguo Testamento.
A. Paluzíe Borrell.
Urania Barcelona, Año 61, Núm. 285, p. 163 - 168 (1976).

004.046 Delle prime lavorazioni dei cannocchialai in Italia e della loro fortuna. M. L. Righini Bonelli.
G. Astron., Vol. 3, 153 - 161 (1977).

004.047 An eighteenth-century Nova Scotia observatory.
R. L. Bishop.
J. R. Astron. Soc. Canada, Vol. 71, 425 - 442 (1977).
At Castle Frederick, in central Nova Scotia, an observatory was erected in 1765. Initially it was used as a station for the checking of surveying instruments; however, in 1767 and again in 1769 it was equipped with some of the best astronomical instruments from the Dollonds of London. Conceived by Joseph Frederick Wallet DesBarres, one of the most remarkable individuals of the eighteenth century, and eventually but reluctantly paid for by the British Admiralty, this building merits consideration as possibly the earliest, optically equipped, astronomical observatory in the Western hemisphere.

004.048 On ancient astronomical constructions on the Kazakh territory. P. I. Marikovskij.
Vestn. AN KazSSR, 1977, No. 5, p. 54 - 61. In Russian. – Abstr. in Ref. zh., 51. Astron., 11.51.7 (1977).

004.049 Astronomical determinations of points in the European part of Russia in the XIXth century.
P. P. Papkovskij.
Vopr. istor. estestvozn. i tekh. Vyp. (No.) 3 - 4 (56 - 57).
Moskva, Nauka, 1977, p. 50 - 56, 151. In Russian. – Abstr. in Ref. zh., 51. Astron., 11.51.24 (1977).

004.050 Correspondence between Kepler and Galileo.
A. Postl.
Vistas Astron., Vol. 21, 325 - 330 (1977).

004.051 The first visibility of the lunar crescent.
F. Bruin.
Vistas Astron., Vol. 21, 331 - 358 (1977).
The problem of predicting the moment when, after conjunction, the new crescent will become visible is both astronomical and physical. This paper first discusses the importance of the sighting to the peoples of Islam and then mentions the criteria which control the phenomenon. This is followed by an outline of the theoretical solution given by the early Arab astronomers. It then proceeds to give a more accurate treatment, according to modern methods, which leads to rules by which the appearance and disappearance of the crescent can be predicted to within five minutes of time. The second part of the paper presents translated extracts on the subject from the oldest sources, using modern astronomical nomenclature.

004.052 Am Himmel über Franken. H. W. Nachrodt.
Veröff. Remeis-Sternw. Bamberg, Astron. Inst.
Univ. Erlangen–Nürnberg, Band 12, Nr. 122, 17 pp. (1977).

004.053 Astronomie im frühen Buchdruck.
Veröff. Remeis-Sternw. Bamberg, Astron. Inst.
Univ. Erlangen–Nürnberg, Band 12, Nr. 128, 96 pp. (1977).
Katalog zur Ausstellung der Staatsbibliothek Bamberg

aus Anlaß des 500. Geburtstages von Johannes Schöner zum Kolloquium Nr. 42 der Internationalen Astronomischen Union.

004.054 Ein halbes Jahrhundert A. S. Eddington, "Der innere Aufbau der Sterne".   H. Lambrecht.
Sterne, 53. Band, 194 - 214 (1977).

004.055 K. F. Zöllner in seinen Beziehungen zu O. W. Struve und Rußland.   D. B. Herrmann.
Sterne, 53. Band, 226 - 236 (1977).

004.056 Zur Hypothese einer 854jährigen Planetenperiode in der babylonischen Astronomie.
K. Ferrari d'Occhieppo.
Anz. Philos.-Hist. Kl. Österreich. Akad. Wiss., Jahrg. 113, Nr. 9, p. 231 - 234 (1976).

004.057 Zur Identifizierung der Sonnenfinsternis während des Petschenegenkrieges Alexios' I. Komnenos (1084).   K. Ferrari d'Occhieppo.
Jahrb. Österreich. Byzantinistik, 23. Band, 179 - 184 (1974).

004.058 Keplers Weg zur Physik des Himmels.
K. Ferrari d'Occhieppo.
119. Jahrb. Oberösterreich. Musealver., Linz, p. 91 - 106 (1974).

004.059 Die Observatorien in Delhi und Jaipur.
W. Petri.
Mitt. Astron. Ges., Nr. 42, p. 84 (1977).

004.060 Ein astronomischer Papyrus und die ältesten Oster-tafeln.   K. Ferrari d'Occhieppo.
Mitt. Astron. Ges., Nr. 42, p. 85 - 88 (1977).

004.061 Levi Ben Gerson und die 3. Ungleichheit des Mondes.
W. Strohmeier.
Mitt. Astron. Ges., Nr. 42, p. 88 (1977).

004.062 Geistesgeschichtliche Überlegungen zur Bahnbe-stimmung bei Gauss.   F. Schmeidler.
Mitt. Astron. Ges., Nr. 42, p. 89 (1977).

004.063 Abu Nasr Ibn Irak's treatises on the astrolabe.
Kh. Tllashev, S. A. Ramazanova.
Mat. i astron. v trudakh uchenykh srednevekovogo Vost. Tashkent, "Fan", 1977, p. 89 - 107. In Russian. – Abstr. in Ref. zh., 51. Astron., 12.51.5 (1977).

004.064 Historical development of notions of sunspots.
Eh. K. Solomatina.
Istor.-astron. issled. Vyp. (No.) 13. Moskva, Nauka, 1977, p. 275 - 304. In Russian. – Abstr. in Ref. zh., 51. Astron., 1.51.4 (1978).

004.065 Development of notions of meteor phenomena from ancient time to the 19th century.   I. N. Kovshun.
Istor.-astron. issled. Vyp. (No.) 13. Moskva, Nauka, 1977, p. 305 - 318. In Russian. – Abstr. in Ref. zh., 51. Astron., 1.51.5 (1978).

004.066 Diffraction of light in the works of the 18th centu-ry's astrophysicists.   N. I. Nevskaya.
Istor.-astron. issled. Vyp. (No.) 13. Moskva, Nauka, 1977, p. 339 - 376. In Russian. – Abstr. in Ref. zh., 51. Astron., 1.51.7 (1978).

004.067 Archimedes' astronomical works.
S. V. Zhitomirskij.
Istor.-astron. issled. Vyp. (No.) 13. Moskva, Nauka, 1977, p. 319 - 337. In Russian. – Abstr. in Ref. zh., 51. Astron.,

1.51.10 (1978).

004.068 Al-Khwarizmi s astronomical treatises.
B. A. Rozenfel'd, N. D. Sergeeva.
Istor.-astron. issled. Vyp. (No.) 13. Moskva, Nauka, 1977, p. 201 - 218. In Russian. – Abstr. in Ref. zh., 51. Astron., 1.51.11 (1978).

004.069 The scientific inheritance of Abu Nasr Ibn Irak, astronomer of the 10th - 11th centuries.
G. P. Matvievskaya, Kh. Tllashev.
Istor.-astron. issled. Vyp. (No.) 13. Moskva, Nauka, 1977, p. 219 - 232. In Russian. – Abstr. in Ref. zh., 51. Astron., 1.51.12 (1978).

004.070 The double star discoveries of Fearon Fallows.
B. Warner.
Mon. Notes Astron. Soc. South. Africa, Vol. 36, 134 - 135 (1977).

004.071 Mittelalterliche astronomisch-astrologische Glossare mit arabischen Fachausdrücken.   P. Kunitzsch.
Bayer. Akad. Wiss., Philos.-Hist. Kl., Sitzungsber., Jahrg. 1977, Heft 5, 59 pp. (1977).

004.072 Werk und Wirkung von Copernicus als Gegenstand der Wissenschaftsgeschichte.   H. M. Nobis.
Sudhoffs Arch., Band 61, 118 - 143 (1977) = Veröff. For-schungsinst. Deutschen Mus. Gesch. Naturwiss. Tech., Reihe A, Kleine Mitt. Nr. 199.

004.073 Zur Deutung der Sternfarben im 18. Jahrhundert (G. C. Lichtenberg und M. Hell).   P. Brosche.
Sudhoffs Arch., Band 61, 248 - 257 (1977).
     Attention is drawn to a letter of G. C. Lichtenberg (of July 1st, 1776) which refers to a paper on the physical mean-ing of stellar colours. The available evidence points toward the authorship of M. Hell, then director of the Vienna Observatory. Presumably he explained the colours of stars in the same way as the colours of auroras, namely by the optically different nature of particles in their atmospheres.

004.074 Russell and theoretical astrophysics.
L. Spitzer, Jr.
Dudley Obs. Rep. No. 13, (see 012.051), p. 3 - 8 (1977).

004.075 Russell and stellar evolution – his "Relations be-tween the spectra and other characteristics of the stars".   R. W. Smith.
Dudley Obs. Rep. No. 13, (see 012.051), p. 9 - 13 (1977).

004.076 Russell and the composition of stellar atmospheres.
C. Payne-Gaposchkin.
Dudley Obs. Rep. No. 13, (see 012.051), p. 15 - 18 (1977).

004.077 The contributions of Henry Norris Russell to the study of close binary stars.   F. B. Wood.
Dudley Obs. Rep. No. 13, (see 012.051), p. 47 - 49 (1977).

004.078 H. N. Russell as a pioneer in trigonometric parallaxes.
D. Hoffleit.
Dudley Obs. Rep. No. 13, (see 012.051), p. 51 - 54 (1977).

004.079 Hertzsprung's contributions to the HR diagram.
K. Aa. Strand.
Dudley Obs. Rep. No. 13, (see 012.051), p. 55 - 60 (1977).

004.080 The origins of the Hertzsprung-Russell Diagram.
D. H. DeVorkin.
Dudley Obs. Rep. No. 13, (see 012.051), p. 61 - 77 (1977).

004.081 Dunkle Regenten als Vorläufer Schwarzer Löcher.

S. L. Jaki.
Nachr. Olbers-Ges. Bremen, Nr. 107, p. 1 - 10 (1977).

004.082 **Ptolemy's report on the color of Sirius and modern observations.**   N. Brosch, I. Nevo.
Bull. Israel Phys. Soc., Vol. 23, 111 (1977). — Summary. Abstr. from INIS7721 336202.

004.083 **Etīdes astronomijas vēsturē. 5. Šis Drausmīgais Ditmaršēnas Lāċazvērs ...**
I. Rabinovičs.
Zvaigžņotā debess, 1976/77. gada ziema, p. 51 - 59.

004.084 **Etīdes astronomijas vēsturē. 6. Rētiks, Kopernika māceklis.**   I. Rabinovičs.
Zvaigžņotā debess, 1977. gada vasara, p. 40 - 49.

004.085 **Āryabhaṭa.**   J. Bouška.
Vesmír, Vol. 56, 219 (1977). In Czech.

004.086 **Leonhard Eulers Programm für die Berliner Sternwarte.**   C. Kirsten.
Sternzeiten, (see 003.025), Band 1, 7 - 12 (1977).

004.087 **Leibniz und die Gründung der Berliner Sternwarte.**   H.-J. Felber, M. Faak.
Sternzeiten, (see 003.025), Band 1, 13 - 25 (1977).

004.088 **Die philosophische Position Johann Heinrich Lamberts.**   R. Wahsner.
Sternzeiten, (see 003.025), Band 1, 27 - 41 (1977).

004.089 **Die Berliner Sternwarte und Johann Heinrich Lamberts kosmologische Vorstellungen.**
G. Jackisch.
Sternzeiten, (see 003.025), Band 1, 43 - 52 (1977).

004.090 **Die Sternkarten von Johann Elert Bode.**   D. Wattenberg.
Sternzeiten, (see 003.025), Band 1, 53 - 68 (1977).

004.091 **Johann Franz Encke und Wilhelm Olbers in ihrem gegenseitigen Briefwechsel.**   D. Wattenberg.
Sternzeiten, (see 003.025), Band 1, 69 - 95 (1977).

004.092 **Carl Friedrich Gauss und Alexander von Humboldt in ihren Beziehungen zur Berliner Sternwarte.**
K.-R. Biermann.
Sternzeiten, (see 003.025), Band 2, 5 - 16 (1977).

004.093 **Arthur Auwers als Astronom der Berliner Akademie.**   J. Wempe.
Sternzeiten, (see 003.025), Band 2, 17 - 28 (1977).

004.094 **Wilhelm Foerster und die Gründung des Astrophysikalischen Observatoriums Potsdam.**
D. B. Herrmann.
Sternzeiten, (see 003.025), Band 2, 29 - 34 (1977).

004.095 **Wilhelm Foersters Verdienste um die Wissenschaftsorganisation.**   H. Oleak.
Sternzeiten, (see 003.025), Band 2, 35 - 39 (1977).

004.096 **Karl Hermann Struve und die Verlegung der Berliner Sternwarte nach Babelsberg.**   K. Fritze.
Sternzeiten, (see 003.025), Band 2, 49 - 57 (1977).

004.097 **Eugen Goldstein und die Physik an der Berliner Sternwarte.**   O. Singer.
Sternzeiten, (see 003.025), Band 2, 59 - 70 (1977).

004.098 **Dokumente zur Geschichte der Sternwarte Berlin-Babelsberg aus den Jahren 1700–1945 im zentralen Archiv der Akademie der Wissenschaften der DDR.**
I. Baumgart.
Sternzeiten, (see 003.025), Band 2, 79 - 88 (1977).

Native American astronomy.
See Abstr. 003.028.

Kosmische Weiten.   See Abstr. 003.072.

Deciphering the Maya script.
See Abstr. 003.080.

Ibn aṣ-Ṣalāḥ. Zur Kritik der Koordinatenüberlieferung im Sternkatalog des Almagest.   See Abstr. 003.087.

Storia dell'astronomia.   See Abstr. 003.104.

Benjamin Martin — author, instrument maker and country showman.   See Abstr. 003.105.

The crime of Claudius Ptolemy.
See Abstr. 003.116.

600 Jahre Astronomie in Nürnberg.
See Abstr. 003.128.

Les tables astronomiques de Louvain de 1528 par Henri Baers ou Vekenstyl.   See Abstr. 003.131.

The catalogue of the history of science collections of the University of Oklahoma Libraries.
See Abstr. 003.140.

Johannes Kepler: giant of faith and science.
See Abstr. 003.155.

Handbuch der babylonischen Astronomie. Band 1: Der babylonische Fixsternhimmel. Beiträge zur ältesten Geschichte der Sternbilder.   See Abstr. 003.165.

Teaching megalithic astronomy.
See Abstr. 014.012.

From "terrestrial" to "cosmic" astronomy.
See Abstr. 051.027.

The location of the supernova of AD 1572.
See Abstr. 125.009.

From Mädler to Kuzmin — the first step in modelling of flattened stellar systems.   See Abstr. 151.076.

Errata

004.901 **Errata: "Miscellanea from the history of celestial mechanics" [Celestial Mech., Vol. 14, 365 - 382 (1976)].**   O. Volk.
Celestial Mech., Vol. 15, 507 (1977).

005 Biography

005.001   Georges Lemaître, le savant et l'homme.
C. Manneback.
Pontificiae Acad. Sci., Scripta Varia, Vol. 36, 13 - 26 (1972).

005.002   Monseigneur Lemaître et son oeuvre.
O. Godart.
Pontificiae Acad. Sci., Scripta Varia, Vol. 36, 27 - 66 (1972).

005.003   The scientific work of Georges Lemaître.
P. A. M. Dirac.
Pontificiae Acad. Sci., Scripta Varia, Vol. 36, 67 - 83 (1972).

005.004   I. Newton, on the occasion of the 250th anniversary
of his death (1643, January 4 - 1727, March 31).
P. L. Kapitsa.
Priroda, 1977, No. 7, p. 3 - 6. In Russian.

005.005   Albert Einstein.   W. Fontana.
Coelum, Vol. 45, 147 - 152 (1977).

005.006   F. L. Whipple.
The dusty universe, (see 012.001), p. 311 - 316
(1975).

005.007   Achille Papapetrou.   J. Stachel.
Gen. Relativ. Gravitation, Vol. 8, 541 - 543, with a
bibliography, p. 545 - 548 (1977).

005.008   Leben und Wirken Heinrich Wilds.   H. Wild, Jr.
Heinrich Wild 1877 - 1951, (see 003.009), p. 9 -
14 (1977).

005.009   16 Jahre Zusammenarbeit mit Dr. h. c. Heinrich
Wild 1935 - 1951.   R. Haller.
Heinrich Wild 1877 - 1951, (see 003.009), p. 33 - 37 (1977).

005.010   The Struves of Pulkovo — a family of astronomers.
A. H. Batten.
J. R. Astron. Soc. Canada, Vol. 71, 345 - 372 (1977).

005.011   P. K. Shternberg — astronomer, partisan, hero of
the civil war.   P. G. Kulikovskij.
Priroda, 1977, No. 11, p. 20 - 27. In Russian.

005.012   Sir James Hopwood Jeans, 1877 - 1946.
R. W. Smith.
J. British Astron. Assoc., Vol. 88, 8 - 17 (1977).

005.013   Un tricentenaire: Jacques Cassini (Cassini II).
S. Grillot.
Astronomie, Vol. 91, 424 (1977).

005.014   Johannes Regiomontanus.   F. Schmeidler.
Vistas Astron., Vol. 21, 315 - 324 (1977).

005.015   Francesco Zagar.   M. G. Fracastoro.
Accad. Naz. Lincei, Celebrazioni Lincee, No. 107,
25 pp. (1977) = Pubbl. Varie Fuori Ser., Oss. Astron. Torino,
(Pino Torinese), No. 67.

005.016   P. Placidus Joseph Fixlmillner, Kremsmünsters
bedeutendster Astronom.

K. Ferrari d'Occhieppo.
Mitt. Oberösterreich. Landesarch., Band 12, 75 - 79 (1977).

005.017   William Thomson (1761–1806) a neglected
meteoriticist.   R. S. Clarke, Jr.
Meteoritics, Vol. 12, 194 - 195 (1977). — Abstract.

005.018   Great scientist of the cosmic era (on the occasion
of the 70th birthday of S. P. Korolev, member of
the Academy of Sciences).
Vestn. AN SSSR, 1977, No. 6, p. 130 - 135. In Russian.
From Ref. zh., 62. Issled. kosm. prostranstva, 12.62.4 (1977).

005.019   M. V. Lomonosov's academic biography.
L. E. Majstrov.
Istor.-astron. issled. Vyp. (No.) 13. Moskva, Nauka, 1977,
p. 167 - 180. In Russian. — Abstr. in Ref. zh., 51. Astron.,
1.51.13 (1978).

005.020   Th. Banachiewicz in Jurjew (1915 - 1918).
A. V. Shpilevskij.
Istor.-astron. issled. Vyp. (No.) 13. Moskva, Nauka, 1977,
p. 147 - 165. In Russian. — Abstr. in Ref. zh., 51. Astron.,
1.51.16 (1978).

005.021   Reminiscences of Henry Norris Russell.
R. d'E. Atkinson.
Dudley Obs. Rep. No. 13, (see 012.051), p. 19 - 25 (1977).

005.022   Collaboration with Henry Norris Russell over the
years.   C. Moore-Sitterly.
Dudley Obs. Rep. No. 13, (see 012.051), p. 27 - 41 (1977).

005.023   The man and the scholar.   J. E. Merrill.
Dudley Obs. Rep. No. 13, (see 012.051), p. 43 - 46
(1977).

005.024   Some recollections of Henry Norris Russell.
L. C. Green.
Dudley Obs. Rep. No. 13, (see 012.051), p. 79 - 81 (1977).

005.025   Kārlis Frīdrihs Gauss.   N. Cimahoviča.
Zvaigžņotā debess, 1977. gada pavasaris, p. 38 - 47.

005.026   Karl Friedrich Gauss.   H. Wussing.
Pokroky, Vol. 22, 195 - 204 (1977). In Czech.

005.027   Isaac Newton.   Z. Horák.
Pokroky, Vol. 22, 263 - 269 (1977). In Czech.

005.028   Friedrich Küstner als Observator an der Berliner
Sternwarte.   H.-U. Sandig.
Sternzeiten, (see 003.025), Band 2, 41 - 48 (1977).

Carl Friedrich Gauß, 1777/1977.
See Abstr. 003.137.

Russell and theoretical astrophysics.
See Abstr. 004.074.

Dunkle Regenten als Vorläufer Schwarzer Löcher.
See Abstr. 004.081.

006 Personal Notes

P. Ahnert, 80th birthday.
Astron. Schule, 14. Jahrg., 134 (1977).

P. Ahnert achtzig Jahre.
H. Lambrecht, K. Wiecke.
Sterne, 53. Band, 193 (1977).

S.-I. Akasofu received the Chapman Medal.
Q. J. R. Astron. Soc., Vol. 18, 410 (1977).

J. H. Black received the Robert J. Trumpler Award.
Mercury, Vol. 6, No. 4, p. 12 (1977).

B. J. Bok received the Bruce Gold Medal.
Mercury, Vol. 6, No. 4, p. 1 - 2 (1977).

B. J. Bok received the Astronomical Society of the
Pacific's Bruce gold medal.
Sky Telesc., Vol. 54, 201 (1977).

The Twelfth Annual Karl G. Jansky Lectureship has
been awarded to **E. M. Burbidge.**
Phys. Today, Vol. 30, No. 12, p. 68 (1977).

M. de Groot, director of the Armagh Observatory.
E. J. Öpik.
Irish Astron. J., Vol. 12, 239 - 240 (1976).

S. Débarbat received the prix des Dames.
Astronomie, Vol. 91, 501 (1977).

W. Dieminger, 70 Jahre alt − ein Senior der Iono-
sphärenphysik.   G. Lange-Hesse.
Phys. Bl., 33. Jahrg., 518 - 519 (1977).

C. Fehrenbach received la médaille d'or du Centre
national de la Recherche scientifique pour l'année 1977.
C. R. Acad. Sci. Paris, Vie Acad., Tome 285, 61 (1977).

W. Groubé received the prix Marius Jacquemetton.
Astronomie, Vol. 91, 502 (1977).

J.-L. Heudier received the prix Edmond Girard.
Astronomie, Vol. 91, 503 (1977).

R. G. Hohlfeld received the Prize of the Astronomical Society of New York.   A. G. D. Philip.

News Lett. Astron. Soc. N.Y., Vol. 1, No. 2, p. 1 (1977).

F. Hoyle received the Dorothea Klumpke-Roberts
Award.
Mercury, Vol. 6, No. 4, p. 2, 12 (1977).

V. I. Krasovskij, on the occasion of the 70th
anniversary of his birthday.   P. V. Shcheglov.
Zemlya i Vselennaya, 1977, No. 4, p. 49 - 51. In Russian.

Ľ. Kresák, 50th birthday.   V. Guth.
Kozmos, Vol. 8, 124 (1977). In Czech.

J. Lequeux received the prix Janssen.
Astronomie, Vol. 91, 501 (1977).

F. Link received the Prix Damoiseau.
C. R. Acad. Sci. Paris, Vie Acad., Tome 285, 122 (1977).

W. H. McCrea received the Gold Medal of the Royal
Astronomical Society.
Q. J. R. Astron. Soc., Vol. 18, 409 (1977).

E. J. Öpik, received the Catherine Wolfe Bruce Gold
Medal for 1976.   G. Burbidge.
Irish Astron. J., Vol. 12, 241 - 243 (1976).

J. A. Ratcliffe received the Gold Medal of the
Royal Astronomical Society.
Q. J. R. Astron. Soc., Vol. 18, 409 (1977).

On the occasion of the 70th birthday of L. I. Sedov,
member of the Academy.   M. A. Lavrent'ev.
Kosm. Issled., Vol. 15, 803 - 808 (1977). In Russian.

B. Šternberk, 80th birthday.   M. Kopecký.
Pokroky, Vol. 22, 109 - 110 (1977). In Czech.

H. E. Suess received the Leonard Medal of The
Meteoritical Society.   P. Pellas.
Meteoritics, Vol. 12, 161 - 164 (1977).

Victor Szebehely received the Brouwer Award.
Mercury, Vol. 6, No. 5, p. 11 (1977).

A. Terzan received the prix Henri Rey.
Astronomie, Vol. 91, 501 - 502 (1977).

007 Obituaries

C. Anderson died 1976 November 15.
C. Hurless.
J. American Assoc. Variable Star Obs., Vol. 6, 9 - 11 (1977).

I. S. Astapovich, 1908, January 11 - 1976, January 2.   V. V. Fedynskij.
Komety Meteory, No. 26, p. 43 - 48 (1977). In Russian.

H. Bouquet, 1897 - 1977.
P. de La Cotardière.
Astronomie, Vol. 91, 413 - 414 (1977).

H. Brinton: 1901 - 1977 June 1.
P. Moore.
J. British Astron. Assoc., Vol. 88, 85 (1977).

H. H. Dahlerup, 1895 August 25 - 1977 May 10.
P. Darnell.
Astron. Tidsskr., Årg. 10, 127 - 128 (1977).

L. D. de Feiter – **in memoriam.** 1927, August 17 - 1975, September 2.   C. de Jager.
Study of travelling interplanetary phenomena 1977, (see 012. 042), p. XI - XII (1977).

M. Edelberg, 8. September 1909 - 8. Februar 1977.
A. Reiz.
Astron. Tidsskr., Årg. 10, 84 - 85 (1977).

H. Haffner, 1912 November 8 - 1977 February 23.
O. Heckmann.
Mitt. Astron. Ges., Nr. 42, p. 5 - 8 (1977).

N. B. Ibragimov, 1932, December 29 - 1977, January 1.
Astron. Tsirk., No. 949, p. 7 - 8 (1977). In Russian.

A. K. Korol', 1913, March 4 - 1977, March 10.

I. G. Kolchinskij.
Astrometriya i Astrofizika, Kiev, vyp. (No.) 33, (see 003.020), p. 100 (1977). In Russian.

B. V. Kukarkin, 1909, October 30 - 1977, September 15.
Priroda, 1977, No. 10, p. 155. In Russian.

A. Oberstatter, 1905 - 1977 March 28.
A. Florsch.
Astronomie, Vol. 91, 340 (1977).

R. A. Rossiter, 1886 December 19 - 1977 January 26.   F. Holden.
Mon. Notes Astron. Soc. Southern Africa, Vol. 36, 60 - 62 (1977).

G. Ruggieri died.   L. Baldinelli.
G. Astron., Vol. 3, 63 - 64 (1977).

B. Sticker, 1906 August 2 - 1977 August 30.
F. Krafft.
Mitt. Astron. Ges., Nr. 42, p. 9 - 10 (1977).

J. van Breda Lourens, died 1977 September 3.
T. W. Russo.
Mon. Notes Astron. Soc. South Africa, Vol. 36, 79 (1977).

A. P. Vinogradov, died 1975, November 16.
Meteoritika, vyp. (No.) 36, p. 171 - 172 (1977). In Russian.

W. von Braun, 1912 March 23 - 1977 June 16.
B. Lovell.
Nature, Vol. 269, 633 - 635 (1977).

W. von Braun, 1912 March 23 - 1977 June 16.
A. V. Cleaver.
Spaceflight, Vol. 19, 320 - 322 (1977).

008 Observatories, Institutes

Reports, communications and publications of observatories and astronomical institutes are recorded in this section; included are numbered series of reprints. Whenever possible, the numbers of the abstracts referring to the publications are given. Observatories and institutes are listed in alphabetical order of their towns. In some cases observatory publications do not give the name of the town; the following list which gives names and towns of some institutions may serve as an aid in such cases.

Aarne Karjalainen Observatory	Oulu, Finland
Algonquin Radio Observatory	Lake Traverse, Ontario, Canada
Allegheny Observatory	Pittsburgh, Pennsylvania, USA
Archenhold-Sternwarte	Berlin-Treptow, German Democratic Republic
Argentine Radioastronomy Institute	Pereyra Iraola, Argentina
Arthur J. Dyer Observatory	Nashville, Tennessee, USA
Astronomical Latitude Station, Polish Academy of Sciences	Borowiec, Poland
Astronomisches Rechen-Institut	Heidelberg, F. R. Germany
Australian National University–Research School of Physical Sciences, Department of Astronomy	Mount Stromlo; Siding Spring, Australia
Bell Laboratories	Murray Hill, New Jersey, USA
Bell Telephone Laboratories	Holmdel, New Jersey, USA
Bosscha Observatory	Lembang, Indonesia
Boyden Observatory	Bloemfontein, South Africa
Bureau International de l'Heure	Paris, France
Cajigal Observatory	Caracas, Venezuela
California Institute of Technology	Pasadena, California, USA
Carter Observatory	Wellington, New Zealand
Catalina Station	Tucson, Arizona, USA
Cavendish Laboratory	Cambridge, England
Cerro Tololo Interamerican Observatory	La Serena, Chile
Ceskoslovenská Akademie Ved Astronomický Ustav	Praha, Czechoslovakia
Chamberlin Observatory, University of Denver	Denver, Colorado, USA
Columbia University, Department of Astronomy	New York, New York, USA
Commonwealth Observatory	Canberra, Australia
Cornell University, Center for Radiophysics and Space Research	Ithaca, New York, USA
Corralitos Observatory	Las Cruces, New Mexico, USA
Crawford Hill Laboratory	Holmdel, New Jersey, USA
David Dunlap Observatory, University of Toronto	Richmond Hill, Ontario, Canada
Dearborn Observatory	Evanston, Illinois, USA
Department of Astronomy and Observatory, Univ. California	Los Angeles, California, USA
Department of Astronomy, University of Texas	Austin, Texas, USA
Deutsches Hydrographisches Institut (DHI)	Hamburg, F. R. Germany
Division Radiophysics, C.S.I.R.O. University Grounds	Sydney, Australia
Dominion Astrophysical Observatory	Victoria, B.C., Canada

Dominion Observatory	Ottawa, Ontario, Canada
Dominion Radio Astrophysical Observatory	Penticton, B. C., Canada,
Dudley Observatory	Albany, New York, USA
Dunsink Observatory	Dublin, Ireland
Dyer Observatory, Vanderbilt University	Nashville, Tennessee, USA
Ege University	Izmir, Turkey
Engelhardt Observatory	Kazan, USSR
Erwin W. Fick Observatory, Iowa State University	Ames, Iowa, USA
European Southern Observatory	La Silla, Chile
Felix Aguilar Observatory	San Juan, Argentina
Fernbank Observatory	Atlanta, Georgia, USA
Five College Observatories	Amherst, Massachusetts, USA
Florida State University Radio Observatory	Tallahassee, Florida, USA
Flower and Cook Observatories, University of Pennsylvania	Philadelphia, Pennsylvania, USA
Fraunhofer Institut	Freiburg, F.R. Germany
George R. Wallace Jr. Astrophysical Observatory	Cambridge, Massachusetts, USA
Georgetown Observatory	Washington, D.C., USA
Glavnaya Astronomicheskaya Observatoriya AN SSSR	Pulkovo, USSR
Goddard Space Flight Center	Greenbelt, Maryland, USA
Goethe Link Observatory, Indiana University	Bloomington, Indiana, USA
"Guido Horn d'Arturo" Observatory	Bologna, Italy
H. M. Nautical Almanac Office, Royal Greenwich Observatory	Greenwich, England
Hale Observatories	Pasadena, California, USA
Harvard College Observatory	Cambridge, Massachusetts, USA
Harvard Radio Astronomy Station	Cambridge, Massachusetts, USA
Haute Provence Observatory	Saint Michel, France
Haystack Observatory	Westford, Massachusetts, USA
Heinrich-Hertz-Institut	Berlin-Adlershof, Germany
Herzberg Institute of Astrophysics	Victoria, B. C.; Ottawa, Canada
High Altitude Observatory, University of Colorado	Boulder, Colorado, USA
Hopkins Observatory	Williamstown, Massachusetts, USA
Horn d'Arturo Observatory	Bologna, Italy
IBM Thomas J. Watson Research Center	Yorktown Heights, New York, USA
Indian Institute of Astrophysics	Bangalore, India
Institute for Astronomy, University of Hawaii	Honolulu, Hawaii, USA
Institute for Theoretical Astronomy (Institut Teoreticheskoj Astronomii)	Leningrad, USSR
Institute of Astronomy and Space Science, University of British Columbia	Vancouver, B. C., Canada
Institute of Theoretical Astrophysics, Blindern	Oslo, Norway

Instituto Argentino de Radio-
astronomia   **Pereyra Iraola,** Argentina
Instituto de Astronomia y Fisica
del Espacio (IAFE)   **Buenos Aires,** Argentina
Instituto Venezolano de
Astronomia   **Merida,** Venezuela
Instituto y Observatorio de
Marina   **San Fernando (Cádiz),** Spain
Inter-American Observatory   **Cerro Tololo,** La Serena, Chile
International Latitude
Observatory   **Mizusawa,** Japan
Jet Propulsion Laboratory,
California Institute of
Technology   **Pasadena,** California
Joint Institute for Laboratory
Astrophysics (JILA)   **Boulder,** Colorado, USA
Kandilli Observatory   **Istanbul,** Turkey
Kansas University Observatory   **Lawrence,** Kansas, USA
Kapteyn Astronomical
Laboratory   **Groningen,** Netherlands
Karl-Schwarzschild-
Observatorium   **Tautenburg,** German
                                    Democratic Republic
Kenneth Mees Observatory   **Rochester,** New York, USA
Kitt Peak National Observatory   **Tucson,** Arizona, USA
Kwasan and Hida Observatories   **Kyoto,** Japan
Lamont-Hussey Observatory   **Bloemfontein,** South Africa
Landessternwarte Heidelberg -
Königstuhl   **Heidelberg,** F. R. Germany
Lawrence Livermore Laboratory,
University of California   **Livermore,** California, USA
Leander McCormick Observatory,
University of Virginia   **Charlottesville,** Virginia, USA
Lee Observatory   **Beirut,** Lebanon
Leopold-Figl-Observatorium   **Wien,** Austria
Leuschner Observatory   **Berkeley,** California, USA
Lick Observatory   **Santa Cruz,** (Mount Hamil-
                                    ton), California, USA
Lindheimer Astronomical
Research Center   **Evanston,** Illinois, USA
Lockheed Palo Alto Research
Laboratory   **Palo Alto,** California, USA
Lockheed Solar Observatory   **Saugus,** California, USA
Lohrmann-Observatorium der
Technischen Universität
Dresden   **Dresden,** German Democratic
                                    Republic
Louisiana State University
Observatory   **Baton Rouge,** Louisiana, USA
Lowell Observatory   **Flagstaff,** Arizona, USA
Lunar and Planetary Laboratory   **Tucson,** Arizona, USA
Max-Planck-Institut für
Astronomie   **Heidelberg,** F. R. Germany
Max-Planck-Institut für
Physik und Astrophysik   **München,** F. R. Germany
Max-Planck-Institut für
Radioastronomie   **Bonn,** F. R. Germany
McDonald Observatory   **Fort Davis,** Texas, USA
McGraw-Hill Observatory   **Kitt Peak,** Arizona, USA
McMath Hulbert Observatory   **Pontiac,** Michigan, USA
C.E.K. Mees Observatory,
University of Rochester   **Rochester,** New York, USA
Michigan State University
Observatory   **East Lansing,** Michigan, USA
Molonglo Radio Observatory,
University of Sydney   **Sydney,** Australia
Monterey Institute for Research
in Astronomy   **Carmel Valley,** California,
                                    USA
Mount Cuba Observatory   **Wilmington,** Delaware, USA
Mount John Observatory   **Lake Tekapo,** New Zealand

Mount Palomar Observatory   **Pasadena,** California, USA
Mount Wilson Observatory   **Pasadena,** California, USA
Mullard Radio Astronomy
Observatory   **Cambridge,** England
Narrabri Observatory, University
of Sydney   **Sydney,** Australia
National Bureau of Standards   **Washington,** D.C., USA
National Observatory, USA   **Kitt Peak,** Arizona, USA
National Radio Astronomy
Observatory   **Charlottesville,** Virginia, USA
                                    Green Bank, West Virginia,
                                    USA
                                    Socorro, New Mexico, USA
                                    Tucson, Arizona, USA
National Research Council
of Canada   **Ottawa,** Ontario, Canada
New Mexico State
University Observatory   **Las Cruces,** New Mexico, USA
Nicholas Copernicus Observatory
and Planetarium   **Brno,** Czechoslovakia
Nizamiah & Rangapur
Observatories   **Hyderabad,** India
Nuffield Radio Astronomy
Laboratories, Jodrell Bank
University of Manchester   **Manchester,** England
Observatoire Royal de Belgique   **Uccle,** Belgium
Observatories of the University
of Western Ontario   **London,** Canada
Observatorio Astronomico
del Vaticano   **Castel Gandolfo,** Italy
Observatório Astronômico do
Instituto de Física da Universi-
dade Federal do Rio Grande
do Sul   **Porto Alegre,** Rio Grande
                                    do Sul, Brazil
Observatorio de Cartuja   **Granada,** Spain
Observatorio del Ebro   **Tortosa,** Spain
Observatorio Fabra   **Barcelona,** Spain
Observatorio Nacional   **Rio de Janeiro,** Brazil
Observatorio Nacional de
Física Cósmica   **San Miguel,** Argentina
Observatory, University of
Michigan   **Ann Arbor,** Michigan, USA
Ohio State University
Radio Observatory   **Columbus,** Ohio, USA
Ole Roemer-Observatoriet   **Aarhus,** Denmark
Onsala Space Observatory   **Gothenburg,** Sweden
Owens Valley Radio
Observatory   **Big Pine,** California, USA
Perkins Observatory, Ohio State
and Wesleyan Universities   **Delaware,** Ohio, USA
Physical Research Laboratory   **Ahmedabad,** India
Purple Mountain Observatory   **Nanking,** China
Radcliffe Observatory   **Pretoria,** South Africa
Raman Research Institute   **Bangalore,** India
Remeis-Sternwarte   **Bamberg,** F. R. Germany
Ritter Astrophysical Research
Center of the University of
Toledo   **Toledo,** Ohio, USA
Rosemary Hill Observatory   **Gainesville,** Florida, USA
Royal Radar Establishment,
Radio Astronomy Division   **Malvern,** England
Sacramento Peak Observatory   **Sunspot,** New Mexico, USA
Sagamore Hill Radio Observatory   **Hamilton,** Massachusetts, USA
San Fernando Observatory   **El Segundo,** California, USA
Smithsonian Astrophysical
Observatory   **Cambridge,** Massachusetts,
                                    USA
Sonnenobservatorium
Kanzelhöhe   **Graz,** Austria
South African Astronomical

Observatory	**Cape Town**, South Africa
Specola Astronomica Vaticana	**Castel Gandolfo**, Italy
Specola di Padova	**Asiago**, Italy
Sproul Observatory	**Swarthmore**, Pennsylvania, USA
Stellar Data Center	**Strasbourg**, France
Sternberg Astronomical Institute	**Moskva**, USSR
Steward Observatory, University of Arizona	**Tucson**, Arizona, USA
W. Struve Tartu Astrophysical Observatory	**Tartu**, USSR
Tata Institute of Fundamental Research	**Bombay**, India
United States Naval Observatory	**Washington**, D. C., USA
University of California	**Berkeley**, California, USA
University of Florida Observatories	**Gainesville**, Florida, USA
University of Florida, Radio Observatory	**Gainesville**, Florida, USA
University of Hawaii	**Honolulu**, Hawaii, USA
University of Illinois Observatory	**Urbana**, Illinois, USA
University of Kansas Observatory	**Lawrence**, Kansas, USA
University of Maryland	**College Park**, Maryland, USA
University of Michigan Observatories	**Ann Arbor**, Michigan, USA
University of Minnesota	**Minneapolis**, Minnesota, USA
University of South Florida Observatory	**Tampa**, Florida, USA

University of Texas, Department of Astronomy	**Austin**, Texas, USA
University of Washington, Astronomy Department	**Seattle**, Washington, USA
Uttar Pradesh State Observatory	**Naini Tal**, India
Van Vleck Observatory	**Middletown**, Connecticut, USA
Vatican Observatory	**Castel Gandolfo**, Italy
Venezuelan Astronomical Institute	**Merida**, Venezuela
Wallace Astrophysical Observatory	**Cambridge**, Massachusetts, USA
Warner and Swasey Observatory	**Cleveland**, Ohio, USA
Washburn Observatory, University of Wisconsin	**Madison**, Wisconsin, USA
West Melton Observatory	**Christchurch**, New Zealand
Wilhelm-Förster Sternwarte	**Berlin**, Germany
Yale University Observatory	**New Haven**, Connecticut, USA
Yerkes Observatory	**Williams Bay**, Wisconsin, USA
Zentralinstitut für Astrophysik, Sternwarte Babelsberg, (Fachbereich Kosmische Physik)	**Potsdam-Babelsberg**, German Democratic Republic
Zentralinstitut für Astrophysik, Sternwarte in Sonneberg	**Sonneberg**, German Democratic Republic
Zentralinstitut für solar-terrestrische Physik	**Berlin-Adlershof**, German Democratic Republic

008.001 Abastumani

Abastumani Astrophysical Observatory of the Academy of Sciences of the Georgian SSR. E. K. Kharadze. Zemlya i Vselennaya, 1977, No. 6, p. 60 - 61. In Russian.

008.002 Ahmedabad

Physical Research Laboratory. R. V. Bhonsle. Bull. Astron. Soc. India, Vol. 5, 87 - 91 (1977). – Annual report for the year April 1976 to March 1977.

008.003 Albany, N.Y.

Dudley Observatory, *Albany, New York,* **Reports,** No. 13 (20.012.051).

008.004 Alger

Université d'Alger. Annales de l'Observatoire Astronomique d'Alger, Tome 5, (Fasc. 1), (20.153.029; 20.041.039).

008.005 Armagh

Armagh Observatory in 1975. E. J. Öpik. Irish Astron. J., Vol. 12, 233 - 238 (1976).

008.006 Athens

Research Center for Astronomy and Applied Mathematics, Academy of Athens, Contributions, Series I (Astronomy), Nos. 45 (20.097.210), 46 (18.155.001), 47 (20.085.036), 48 (19.097.024), 49 (19.122.020).

008.007 Austin, Tex.

Department of Astronomy and McDonald Observatory of the University of Texas, Austin, Texas, Reprints Nos. 450 (18.064.099), 451 (18.064.098), 452 (18.117.038), 453 (18.126.035), 454 (19.114.307), 455 (19.160.009), 456 (19.141.501), 457 (18.097.145), 458 (18.141.172), 459 (19.124.114), 460 (19.114.013), 461, 462 (18.114.104), 463 (19.122.001), 464 (18.160.078), 465 (17.064.037), 466 (19.141.040), 467 (18.124.103), 468 (18.131.110), 469 (20.116.018), 470 (09.062.008), 471 (17.131.530), 472 (17.131.521), 473 (19.114.028), 474 (19.131.066), 475 (19.066.014), 476 (19.093.008), 477, 478 (19.131.054), 479 (19.031.402), 480 (19.097.034), 481 (20.031.236), 482 (20.044.006), 483 (18.071.053), 484 (17.158.309), 485 (03.141.066), 486 (08.141.011), 487 (11.141.081), 488 (17.141.082), 489 (19.002.027), 490 (19.065.028), 491 (19.133.013), 492 (19.124.117), 493 (19.121.022), 494 (19.158.083), 495 (19.158.084), 496 (19.158.085), 497 (19.158.087), 498 (19.158.086), 499 (19.099.509), 500 (19.154.031).

008.008 Bamberg

Veröffentlichungen der Remeis-Sternwarte Bamberg, Astronomisches Institut der Universität Erlangen–Nürnberg, Band 10, Nr. 119 (19.102.004), Band 12, Nr. 122 (20.004.052), 123 (20.123.068), 124 (19.121.072), 125 (20.123.069),

126 (19.121.076), 127 (20.121.118), 128 (20.004.053), 129 (20.120.003), 130 (20.124.403), 131 (20.103.105).

008.009 Bangalore

Indian Institute of Astrophysics.
M. K. V. Bappu.
Q. J. R. Astron. Soc., Vol. 18, 396 - 400 (1977). — Report for the year 1975 April 1 to 1976 March 31.

008.010 Belgrade

Bulletin de l'Observatoire Astronomique de Belgrade, No. 128 (10.081.058; 20.082.114; 20.032.040; 20.041.034; 20.082.115; 20.118.030; 20.118.031; 20.118.032; 20.118. 033; 20.118.034; 20.118.035; 20.118.036; 20.118.037; 20. 098.078; 20.103.704; 20.098.079; 20.096.025; 20.045.019).

University of Beograd, Faculty of Sciences. **Publications of the Department of Astronomy** (Publications de la Chaire d'Astronomie), No. 6 (20.012.040).

008.011 Berlin

Veröffentlichungen der Wilhelm-Foerster-Sternwarte Berlin, Nos. 39 (19.120.001), 45 (19.015.009), 46 (19.124. 102), 47 (20.097.211).

008.012 Berlin-Adlershof

Heinrich-Hertz-Institut, Solare Beobachtungsergebnisse. Akademie der Wissenschaften der DDR, Zentralinstitut für Solar-Terrestrische Physik (Heinrich-Hertz-Institut), Berlin-Adlershof, HHI Solar Data, Vol. 28, 1977 April - September (20.075.008).

008.013 Berlin-Treptow

Blick in das Weltall, Archenhold-Sternwarte Berlin-Treptow. Astronomische Veranstaltungen und Mitteilungen für Sternfreunde, 25. Jahrg., Nos. 7 - 12 (1977).

008.014 Bologna

Pubblicazione dell' Osservatorio Astronomico "G. Horn d'Arturo", Nos. 22 (14.031.079), 23 (17.121.097), 24 (18.121.101), 25 (19.121.060), 26 (19.031.242).

008.015 Bombay

Radio Astronomy Centre, Ootacamund and the Radio Astronomy Group of the Tata Institute of Fundamental Research, Bombay.   G. Swarup.
Bull. Astron. Soc. India, Vol. 5, 92 - 94 (1977).

008.016 Bonn

Max-Planck-Gesellschaft zur Förderung der Wissenschaften, M.P.I. f. Radioastronomie, Bonn, Sonderdruck, Ser. A, Nos. 141 (18.131.140), 142 (20.022.095), 143 (18.157. 010), 144 (18.125.042), 145 (18.141.061), 146 (18.131. 563), 147 (18.141.176), 148 (18.158.174), 149 (18.141. 070), 150 (18.158.119), 151 (18.131.275), 152 (19.132. 008), 153 (20.022.096), 154 (18.141.343), 155 (19.131. 031), 156 (19.141.021), 157 (19.133.005), 158 (19.132. 004), 159 (19.132.023), 160 (19.141.505), 161 (19.141. 013), 162 (19.131.047), 163 (19.131.091), 164 (20.022. 097), 165 (19.131.051), 166 (19.131.048), 167 (19.125. 015), 168 (19.131.148), 169 (19.125.029), 170 (19.141. 510).

008.017 Bordeaux

Publications de l'Observatoire de l'Université de Bordeaux (Floirac), Nouvelle Série, Nos. 60 (12.131.152), 61 (13.106.017), 62 (13.106.018), 63 (14.041.013), 64 (14.106. 003), 65 (17.098.001), 66 (17.077.011), 67 (17.131.542), 68 (18.100.207), 69 (18.106.007), 70 (19.002.003).

008.018 Borowiec

Polish Academy of Sciences, Astronomical Latitude Observatory, Borowiec, Poland, Circular No. 142 (20.044.023).

008.019 Bruxelles

Astrofysisch Instituut, Vrije Universiteit Brussel, Overdruk Nos. 118 (18.114.385), 119 (19.121.010), 120 (19.113.018), 121 (18.117.048), 122 (18.074.137), 123 (19.113.053), 124 (19.113.054), 125 (18.117.049), 126 (19.117.011), 127 (19.123.024), 128 (19.113.009), 129 (19.113.036), 130 (19.113.037), 131 (19.123.029), 132 (19.122.152), 133 (20.064.030), 134 (20.117.039), 135 (20.082.017), 136 (18.142.097), 137 (20.116.010), 138 (20.117.016), 139 (20.065.021).

008.020 Budapest

A Magyar Tudományos Akadémia Csillagvizsgáló Intézetének Közleményei — Mitteilungen der Sternwarte der Ungarischen Akademie der Wissenschaften, Budapest—Szabadsághegy, No. 70 (20.122.125).

008.021 Byurakan

Byurakan Astrophysical Observatory of the Academy of Sciences of the Armenian SSR.   G. M. Tovmasyan.
Zemlya i Vselennaya, 1977, No. 6, p. 58 - 59. In Russian.

Byurakan Astrophysical Observatory, Armenia, USSR, Reprints Nos. 188 (20.158.014), 189 (20.160.011), 190 (20.158.016), 191 (20.063.001), 192 (20.158.017; 20. 141.013; 20.158.018), 193 (20.134.010), 194 (20.064.008), 195 (20.158.025), 196 (20.158.027), 197 (20.141.021), 198 (20.031.207), 199 (20.160.029), 200 (20.162.025), 201 (20. 151.025), 202 (20.122.045).

008.022   Calar Alto

Calar Alto 1977 — im Bild.
Sterne Weltraum, Jahrg. 16, 407 - 410 (1977).

008.023   Cambridge, Mass.

Smithsonian Institution. Astrophysical Observatory.
Research in Space Science. SAO Special Reports, Nos. 377
(20.081.059), 378 (20.081.060), 379 (20.052.040), 380
(20.121.091).

008.024   Cardiff

University College, Cardiff, Astronomical Communications No. 30 (20.131.130).

008.025   Catania

Report from the Catania Astrophysical Observatory
1976. List of papers and books printed during 1976.
G. Godoli, P. Maffei.
Oss. Astrofis. Catania, Pubbl. Nuova Ser. No. 158, 12 pp.
(1977).

Osservatorio Astrofisico di Catania, Pubblicazione
Nuova Serie Nos. 158 (20.008.025), 159 (20.075.007).

008.026   Crimea

Crimean Astrophysical Observatory of the USSR
Academy of Sciences.   V. M. Mozhzherin.
Zemlya i Vselennaya, 1977, No. 6, p. 57 - 58. In Russian.

Izvestiya Ordena Trudovogo Krasnogo Znameni
Krymskoj Astrofizicheskoj Observatorii, Akademiya Nauk
SSSR, Tom (Vol.) 57 (20.122.017; 20.122.018; 20.122.019;
20.117.005; 20.114.007; 20.113.007; 20.064.005; 20.157.
003; 20.072.002; 20.072.003; 20.073.008; 20.072.004;
20.077.002; 20.077.003; 20.099.006; 20.077.004; 20.033.
003; 20.031.402; 20.034.001; 20.032.502; 20.080.008;
20.080.009).

History of the Simeis Observatory.
See Abstr. 009.025.

008.027   Delaware, Ohio

Contributions from the Perkins Observatory, Ohio
State–Ohio Wesleyan Universities. Series I, No. 158 (20.131.
186), 159 (18.116.031), 160 (19.114.004), 161 (20.159.012),
162 (20.131.197).

Contributions from the Perkins Observatory, Ohio
State–Ohio Wesleyan Universities. Series II, Nos. 63 (18.131.
022), 64 (19.113.031), 65 (19.065.062), 66 (19.114.034), 67
(19.064.018), 68 (19.115.007), 69 (19.114.047), 70 (20.072.
005).

008.028   Dublin

Contributions from the Dunsink Observatory, Nos.
13 (14.155.049), 14 (17.122.066).

Dunsink Observatory Reprints Nos. 67 (07.098.051),
87 (13.008.042), 90 (14.065.011), 93 (14.142.091), 94 (14.
122.182), 95 (17.122.135), 96 (17.158.002), 97 (18.142.369),
99 (19.142.018).

008.029   Dushanbe

Byulleten' Instituta Astrofiziki, Akademiya Nauk
Tadzhikskoj SSR, No. 66 - 67 (20.151.004; 20.162.005; 20.
066.010; 20.162.006; 20.122.007; 20.113.006; 20.114.505;
20.082.004; 20.122.008; 20.122.009).

008.030   Genève

Publications de l'Observatoire de Genève, Sér. B,
Fasc. 3 (20.002.038).

008.031   Gent

Universiteit te Gent, Sterrenkundig Instituut,
Mededelingen, No. 41 (20.071.045).

008.032   Graz

Mitteilungen des Sonnenobservatoriums Kanzelhöhe,
No. 25 (18.074.135).

Mitteilungen der Universitätssternwarte Graz, Nos.
29 (18.099.210), 31 (18.042.109), 32 (19.098.021).

008.033   Green Bank, W. Va.

National Radio Astronomy Observatory, Green
Bank, Reprints, Series A, Nos. 698 (19.141.085), 699 (19.
158.091), 700 (19.133.026), 701 (19.119.015), 702 (20.033.
009), 703 (19.158.503), 704 (19.155.019), 705 (19.131.073),
706 (19.141.075), 707 (19.141.062), 708 (19.066.045), 709
(19.141.058), 710 (19.155.017), 711 (18.141.168), 712 (19.
131.074), 713 (19.158.063), 714 (19.134.020), 715 (19.
141.096), 716 (19.131.097), 717 (19.158.110), 718 (19.
133.027), 719 (19.131.098), 720 (19.158.111), 721 (19.
131.112), 722 (19.141.049), 723 (19.141.150), 724 (19.
063.016), 725 (19.131.114), 726 (19.155.041), 727, 728,
729 (19.158.117), 730 (19.131.126), 731 (19.063.022), 732
(19.141.140), 733 (19.158.141), 734 (19.141.142), 735 (19.
122.155), 736 (19.158.124), 737 (19.131.128), 738 (19.131.
166), 739 (20.133.002), 740 (20.160.013), 741 (20.141.015),
742 (20.131.037), 743 (20.155.009), 744 (20.141.016), 745
(20.141.031), 746 (20.141.032), 747 (20.141.033), 748 (20.
141.024), 749 (20.131.012), 750 (20.155.001).

National Radio Astronomy Observatory, Green
Bank, Reprints, Series B, Nos. 469 (20.155.027), 470 (20.141.
063), 473, 476 (18.141.004), 477 (19.158.147), 478.

008.034 Greenbelt, Md.

Goddard Space Flight Center, Greenbelt, Maryland,
GSFC Documents X-660-77-144 (20.022.103), X-660-77-225
(20.099.079), X-921-77-81 (20.081.062).

008.035 Greenwich

Royal Observatory Annals, [Royal Greenwich Observatory, Herstmonceux], No. 11 (20.075.009).

H. M. Nautical Almanac Office, Royal Greenwich
Observatory, Library Reprint, Nos. 336 (20.094.016), 337
(20.041.044).

008.036 Hamburg

Deutsches Hydrographisches Institut, Hamburg.
Zeit- und Breitendienst, 1976 January - December (20.044.
025).

008.037 Hyderabad

Nizamiah and Japal-Rangapur Observatories and the
Department of Astronomy, Osmania University, Hyderabad.
K. D. Abhyankar.
Bull. Astron. Soc. India, Vol. 5, 86 - 87 (1977). — Annual
report January 1976 to December 1976.

008.038 Izmir

Scientific Reports of the Faculty of Science, Ege
University, *Izmir,* No. 185 — Astron. No. 14 (20.141.547).

008.039 Kiev

Main Astronomical Observatory of the Academy
of Sciences of the Ukrainian SSR.   V. S. Kislyuk.
Zemlya i Vselennaya, 1977, No. 6, p. 61 - 62. In Russian.

Astrometriya i Astrofizika, *Kiev,* Vyp. (Nos.)
32 (20.003.011), 33 (20.003.020).

Kodaikanal see Bangalore

008.040 La Silla

ESO and the Danish telescopes in Chile.
K. Gyldenkerne.
Fys. Tidsskr., Vol. 74, No. 4, p. 145 - 166 (1976). In Danish.
Abstr. in Phys. Abstr., Vol. 80, Abstr. 53425 (1977).

The Messenger — El Mensajero. Edited by European
Southern Observatory (ESO), Garching, No. 10, 24 pp., No.
11, 24 pp. (1977).

ESO Scientific Preprint Nos. 9 (20.141.161), 10
(20.155.065), 11 (20.042.070), 12 (20.160.062), 13 (20.122.

151), 14 (20.151.072), 15 (20.160.063), 16 (20.062.077), 17
(20.002.041), 18 (20.131.213).

008.041 Leningrad

Trudy Instituta Teoreticheskoj Astronomii
Akademiya Nauk SSSR. Izdatel'stvo Nauka, Leningradskoe
Otdelenie, Leningrad. Vyp. (No.) 16 (20.151.075; 20.031.
340; 20.042.078; 20.097.214).

008.042 London, England

University College London, Mullard Space Science
Laboratory.   R. L. F. Boyd.
Q. J. R. Astron. Soc., Vol. 18, 372 - 381 (1977). — Report
for the period 1975 October 1 - 1976 September 30.

008.043 Lund

Reports from the Observatory of Lund, Nos. 12
(20.012.044), 13 (20.021.022).

008.044 L'vov

L'vovskij Ordena Lenina Gosudarstvennyj Universitet
imeni Ivana Franko, Astronomicheskaya Observatoriya,
Tsirkulyar, No. 51 (20.134.037; 20.117.037; 20.121.080;
20.123.046; 20.071.039; 20.071.040; 20.034.062).

008.045 Lyon

Observatoire de Lyon, Reprint No. 62 (17.118.004).

008.046 Manchester

Astronomical contributions from the University of
Manchester, Series III, Nos. 301 (13.141.031), 302 (12.131.
112), 303 (13.132.124), 304 (12.065.128), 305 (13.034.
014), 306 (12.125.052), 307 (13.121.028), 308 (13.131.
532), 309 (14.132.013), 310 (14.131.501), 311 (14.131.
502), 312 (13.121.067), 313 (14.132.012), 314 (13.121.
068), 315 (12.131.542), 316 (14.131.542), 317 (14.141.
104), 318 (14.125.043), 319 (14.034.020), 320 (14.125.
002), 321 (14.131.550), 322 (18.131.116), 323, 324 (14.121.
022), 325 (18.062.018), 326 (17.131.509), 327 (17.131.
520), 328 (14.131.110), 329 (17.034.007), 330 (17.121.
030), 331 (17.141.057), 332 (17.132.001), 333 (17.132.
010), 334 (14.131.536), 335 (13.133.014), 336 (13.132.
006), 337 (17.131.537), 338 (17.121.036), 339 (14.121.
078), 340 (17.121.037), 341 (17.141.100), 342 (18.141.
030), 343 (18.132.010), 344 (18.132.009), 345 (19.131.
093), 346 (18.125.001), 347 (17.131.528), 348 (18.131.
584), 349 (18.125.046), 350 (18.117.089).

008.047 Minneapolis

University of Minnesota, Minneapolis, Minnesota,
Separate print (20.112.020).

008.048 Mizusawa

Annual Report of the International Polar Motion Service for the year 1975, (20.045.031).

Bulletins, Time Service of the Mizusawa Observatory, Vol. 19/20, 1974/1975 (20.044.032).

Monthly Notes of the International Polar Motion Service, 1977 Nos. 6 - 11 (20.045.026).

Proceedings of the International Latitude Observatory of Mizusawa, No. 16 (20.082.121; 20.081.066; 20.079.104; 20.041.045; 20.045.023; 20.079.403; 20.034.080; 20.045.024).

Publications of the International Latitude Observatory of Mizusawa, Vol. 10, No. 2 (20.041.046; 20.079.105; 20.045.025; 20.041.047).

Annual report of the geophysical observations made at the International Latitude Observatory of Mizusawa for the year 1975. I. Tsubokawa.
Published by the International Latitude Observatory of Mizusawa, 45 pp. (1976).

Annual report of the meteorological observations made at the International Latitude Observatory of Mizusawa for the year 1976. I. Tsubokawa.
Published by the International Latitude Observatory of Mizusawa, 2 + 30 pp. (1977).

008.049 Moskva

State Astronomical Sternberg-Institute.
E. P. Aksenov, Yu. P. Pskovskij.
Zemlya Vselennaya, 1977, No. 6, p. 63 - 64. In Russian.

008.050 München

Lehrstuhl für Astronomische und Physikalische Geodäsie, Technische Universität München, Mitteilungen No. 144 (20.094.017).

008.051 Ottawa

National Research Council of Canada, Ottawa, Ontario, Canada, NRC Nos. 15789 (20.073.061), 15884 (19.141.132), 16110 (20.125.008).

Dominion Astrophysical Observatory Victoria, British Columbia. Herzberg Institute of Astrophysics, National Research Council of Canada. See Abstr. 008.074.

008.052 Oxford

Department of Astrophysics, University Observatory, Oxford, Publications Nos. 148 (18.032.009), 149 (18.114.021), 150 (18.071.016), 151, 152 (18.066.135), 153 (18.071.058), 154 (19.042.057), 155 (18.022.030), 156 (20.022.123), 157 (19.122.003), 158, 159 (19.162.029), 160 (18.117.075), 161 (19.071.015), 162 (19.066.053), 163 (18.122.083), 164 (19.142.052), 165 (19.162.032), 166, 167 (19.064.037), 168 (19.066.104), 169 (19.162.033), 170 (19.115.009), 171 (20.066.375), 172 (20.066.154), 173 (20.066.376), 174 (19.066.058), 175 (19.160.012), 176 (19.162.020), 177 (18.141.168), 178, 179 (19.066.157), 180 (19.122.062).

008.053 Paris

Bureau International de l'Heure, (B.I.H.), Circular A (20.044.029); D128 - D133 (20.044.030), E7 (20.044.031).

008.054 Potsdam

Akademie der Wissenschaften der DDR. Zentralinstitut für Astrophysik, Sternwarte Babelsberg, Potsdam-Babelsberg, Mitteilungen. Neue Folge, Nos. 142 (13.066.080), 143 (13.066.081), 144 (13.066.083), 145 (14.162.004), 146 (14.141.057), 147 (14.158.100), 148 (14.158.138), 151 (14.031.228), 152 (17.158.021), 153 (17.141.020), 154 (17.066.006), 155 (17.161.004), 156 (17.066.048), 157 (17.162.045), 158 (17.162.046), 159 (17.158.098), 160 (17.141.084), 161. 162, 166 (19.043.004), 167 (20.066.377), 168 (18.162.012), 169 (20.022.124), 170 (20.066.378), 171 (18.066.101), 172 (19.066.050), 173 (19.044.003), 174 (20.066.379), 175 (19.004.026), 176 (19.004.027), 177 (19.004.028), 178 (20.003.025), 179 (20.003.025).

Mitteilungen des Astrophysikalischen Observatoriums Potsdam, Nos. 175 (14.101.005), 176 (14.113.036), 177 (13.082.058), 178 (20.063.040), 179 (17.116.003), 180 (17.113.017), 181 (17.113.023), 182 (17.062.013), 183 (13.072.060), 184 (13.062.044), 185 (13.062.045), 186 (19.080.054).

Astronomische Zeit- und Breitenbestimmungen, Empfangszeiten von Zeitsignalen, Präzisionszeitvergleiche. 1976 January - December (20.044.028).

Veröffentlichungen des Zentralinstituts Physik der Erde, *Potsdam*, No. 52, Teil 1 (20.012.035).

008.055 Praha

Académie Tchécoslovaque des Sciences, Institut Astronomique, Temps et Latitude Nos. 1 - 6, Jan. - Déc. 1976 (20.044.027).

Astronomical Institute of the Technical University, Praha–ČSSR. Publikace, Nos. 43 (14.041.002), 44 (17.045.006), 45 (20.082.120), 46 (20.046.064), 47 (20.055.011).

Czechoslovak Academy of Sciences, Astronomical Institute, Publication No. 52 (20.073.111).

008.056 Pulkovo

Main Astronomical Observatory of the USSR Academy of Sciences. V. A. Krat.

Zemlya i Vselennaya, 1977, No. 6, p. 56. In Russian.

Izvestiya Glavnoj Astronomicheskoj Observatorii v Pulkove, No. 195, Astrometriya i Astrofizika (20.041.050; 20.032.048; 20.041.051; 20.031.337; 20.031.338; 20.031. 339; 20.097.213; 20.100.031; 20.041.052; 20.154.048; 20. 153.035; 20.121.099; 20.072.049; 20.080.079; 20.034.082; 20.034.083; 20.031.065; 20.032.049; 20.032.587).

Trudy Glavnoj Astronomicheskoj Observatorii v Pulkove, Seriya 2, Vol. 82 (20.002.042-20.002.045, 20.041. 048; 20.041.049).

008.057   Riga

Radioastrophysical Observatory of the Academy of Sciences of the Latvian SSR.   A. Balklavs (Editor). Latvijas PSR Zinātnu Akadēmijas. Izdevniecība "Zinātne", Riga. 32 pp. Price 48 Kop. (1977). In Latvian, Russian and English.

Radioastrofizicheskaya Observatoriya Akademii Nauk Latvijskoj SSR. Trudy (*Transactions*) Observatorii 16 (A. Balklavs, 20.003.018).

008.058   Roma

Monthly Bulletin. Osservatorio Astronomico di Roma, Nos. 225 - 232 (20.075.012).

Photographic Journal of the Sun, Osservatorio Astronomico di Roma, Nos. 121 - 124 (20.075.011).

008.059   San Fernando

La Biblioteca del Instituto y Observatorio de Marina de San Fernando.   D. Almorza. Urania Barcelona, Año 61, Núm. 285, p. 169 - 173 (1976).

Instituto y Observatorio de Marina, San Fernando (Cadiz), España, Serie C, No. 79 (20.044.026).

008.060   Santa Cruz

The University of California. **Contributions from the Lick Observatory, Santa Cruz, California,** Nos. 407 (17.154.021), 409 (18.082.045).

University of California. **Lick Observatory Bulletin,** Nos. 665 (13.115.005), 688 (18.124.104), 708 (17.132.020), 711 (18.141.006), 713 (17.064.018), 717 (17.064.017), 722 (18.131.511), 723 (18.141.010), 724 (17.141.094), 725 (18.065.009), 726 (18.158.112), 727 (18.131.514), 728 (17.141.090), 729 (18.158.029), 730 (18.082.008), 731 (18.118.004), 732 (18.117.018), 733 (18.122.064), 734 (19.065.009), 735 (19.158.008), 736 (18.131.272), 737 (19.135.003), 738 (19.114.040), 739 (19.135.009), 740 (19.141.015), 741 (18.158.034), 742 (18.158.173), 743 (18.160.059), 744 (19.141.061), 745 (19.062.037), 746 (19.158.504), 747 (19.135.017), 748 (19.141.097), 749 (19.122.115), 752 (19.124.137), 753 (19.160.035), 759 (19.141.111).

008.061   Shemakha

Shemakha Astrophysical Observatory of the Academy of Sciences of the Azerbajdzhan SSR. I. R. Salmanov, P. R. Amnuehl'. Zemlya i Vselennaya, 1977, No. 6, p. 64 - 65. In Russian.

008.062   Siding Spring

Department of Astronomy, the Australian National University Research School of Physical Sciences. O. J. Eggen. Q. J. R. Astron. Soc., Vol. 18, 486 - 506 (1977). — Report for year ending 1976 December 31.

008.063   Skalnaté Pleso

Contributions of the Astronomical Observatory Skalnaté Pleso, Vol. VI (1976) (20.012.007).

008.064   Sonneberg

Akademie der Wissenschaften der DDR, Zentralinstitut für Astrophysik. **Veröffentlichungen der Sternwarte in Sonneberg,** Band 8, Heft 7 (20.123.060).

Zentralinstitut für Astrophysik. **Mitteilungen über Veränderliche Sterne,** *Sonneberg,* Band 8, Heft 1 (20.123. 061; 20.123.062; 20.123.063; 20.113.045; 20.123.064; 20.123.065; 20.124.905; 20.121.097; 20.122.152; 20.123. 066; 20.123.067).

008.065   St. Andrews

Communications from the University Observatory, St. Andrews, Nos. 12 (14.117.007), 13 (14.113.046), 14 (18.154.028), 15 (18.113.067), 16 (19.113.001), 17 (19.158. 001), 18 (19.113.014), 19 (19.131.170).

University Observatory, St. Andrews. Reprint Nos. 63 (14.131.135), 66 (17.122.024), 67 (18.155.021), 68 (20.064.022), 69 (19.123.015), 70 (19.113.015), 71 (19.126. 026), 72 (20.131.194), 73 (20.160.005), 74 (20.113.005).

008.066   Sydney

Sydney Observatory Papers, Nos. 75 (20.112.019), 76 (20.098.083).

Division of Radiophysics, CSIRO, Sydney, Australia, **Radiophysics Publication** RPP 1845 (17.131.030), 1878 (18.077.027), 1910 (18.099.039), 1922 (19.141.001), 1925 (18.077.050), 1939 (18.077.087), 1948 (18.131.141), 1969 (18.131.130), 1985 (19.158.056), 1986 (19.131.039), 1993 (18.141.094).

008.067   Tartu

Chronicle.   Ü. Ibrus. Calendar of the Tartu Astronomical Observatory for the year

1978, (see 047.026), p. 82 - 84 (1977). In Estonian.

Eesti NSV Teaduste Akadeemia (Akademiya Nauk Ehstonskoj SSR), **W. Struve nimelise Tartu Astrofüüsika Observatooriumi, Publikatsioonid** (Publikatsii Tartuskoj Astrofizicheskoj Observatorii imeni W. Struve) Köide (Tom) 45 (20.063.044; 20.063.045; 20.063.046; 20.114.062; 20. 022.105; 20.114.559; 20.122.153; 20.103.705; 20.082.122; 20.114.063; 20.065.082; 20.065.083; 20.154.047; 20.162. 114; 20.064.056; 20.022.106; 20.063.047; 20.121.098).

Tartu Astronoomia Observatoorium, Teated, Nos. 49 (20.160.032), 55 (20.004.012).

008.068 Tashkent

A centenary of the Astronomical Institute of the Uzbek Academy of Sciences (1873 - 1973). V. P. Shcheglov. Istor.-astron. issled. Vyp. (No.) 13. Moskva, Nauka, 1977, p. 15 - 41. In Russian. − Abstr. in Ref. zh., 51. Astron., 1.51.8 (1978).

Astronomical Institute of the Academy of Sciences of the Uzbekian SSR. V. P. Shcheglov. Zemlya i Vselennaya, 1977, No. 6, p. 66. In Russian.

Trudy Ordena Trudovogo Krasnogo Znameni Astronomicheskogo Instituta Akademii Nauk UzSSR Tom 1 (20.082.108).

Tsirkulyar Ordena Trudovogo Krasnogo Znameni Astronomicheskogo Instituta, Akademiya Nauk Uzbekskoj SSR, Tashkent, Nos. 64 (20.044.020; 20.075.005; 20.032. 039), 66 (20.112.010), 68 (20.044.020; 20.075.005; 20.112. 011), 69 (20.112.012), 70 (20.044.020; 20.075.005; 20.112. 013), 71 (20.044.020; 20.075.005; 20.112.014), 72 (20.044. 020; 20.075.005; 20.044.021), 73 (20.044.022), 74 (20.112. 015), 75 (20.112.016), 76 (20.112.017), 77 (20.041.032), 79 (20.041.033).

008.069 Tokyo

Tokyo Astronomical Bulletin, Tokyo Astronomical Observatory, Second Series, Nos. 248 (20.121.101), 249 (20.134.045), 250 (20.101.044).

Contributions from the Department of Astronomy, University of Tokyo, Nos. 209 (18.064.126), 210 (18.065. 112), 211 (18.064.127), 212 (18.064.128), 213 (18.122.153), 214 (18.080.011), 215 (18.062.027), 216 (18.114.070), 217 (18.112.004), 218 (18.116.030), 219 (20.022.122), 220 (18. 080.059), 221 (18.080.121), 222 (20.080.006), 223 (20.122. 012), 224 (20.121.002), 225 (20.065.002), 226 (20.062.003), 227 (20.155.005), 229 (20.071.001), 230 (20.074.044), 231 (20.114.539), 232 (20.077.022), 233 (20.158.081), 234 (20. 062.042).

Tokyo Astronomical Observatory, Reprints Nos. 503 (17.158.123), 504 (18.034.064), 512 (20.042.004), 513 (20.114.507), 514 (20.155.006), 517 (20.033.002), 518 (19.094.155), 519 (20.045.028), 520 (20.124.901), 521 (20.158.081), 522 (20.158.082), 523 (20.044.010), 524 (20.041.019), 525 (20.074.051), 526 (20.131.215), 527 (20.114.065), 528 (20.114.560), 529 (20.132.015), 530 (20.073.090).

Time and Latitude Bulletins, Tokyo Astronomical Observatory, Vol. 51, Nos. 1 - 2 (20.044.040).

Data Report of Hydrographic Observations. Series of Astronomy and Geodesy, Maritime Safety Agency, Tokyo, Japan, No. 11 (20.096.026).

008.070 Torino

Osservatorio Astronomico di Torino, Pino Torinese. Time Service, Bulletin Nos. 15, 16 (20.044.033).

Contributi dell'Osservatorio Astronomico di Torino (Pino Torinese), Nos. 94 (19.121.006), 95 (19.042.021), 96 (19.121.016), 98 (19.098.021), 99 (19.098.010), 100 (20. 066.341), 101 (19.121.053), 103 (20.045.027).

Pubblicazioni Varie Fuori Serie dell'Osservatorio Astronomico di Torino (Pino Torinese), No. 67 (20.005.015).

008.071 Trieste

Pubblicazione Osservatorio Astronomico di Trieste, Nos. 519 (20.031.328), 525 (20.114.556), 526 (20.114.534), 527 (20.116.013), 528 (20.114.040), 529 (20.114.508), 530 (20.114.554), 531 (20.034.063), 532 (20.034.064), 533 (19.142.705), 534 (20.077.036), 535 (20.075.010), 536, 537 (20.031.329), 538, 539 (19.122.070), 540 (20.031.336), 541 (19.077.012), 542 (20.075.010), 543 (20.113.024), 544 (20. 031.410), 545 (20.031.411), 546 (20.117.026), 547 (20.122. 101), 548 (20.117.018), 549 (20.121.027), 550 (20.153.018), 551 (20.075.010), 552 (20.077.049).

008.072 Uccle

Bulletin Astronomique. (Astronomisch Bulletin). **Observatoire Royal de Belgique** (Koninklijke Sterrenwacht van België), Vol. 8, No. 6 (20.044.039; 20.098.084; 20.098.085; 20.103.401; 20.103.471; 20.075.019; 20.079.208).

Observatoire Royal de Belgique (Koninklijke Sterrenwacht van België), **Communications** (Mededelingen), Série A, Nos. 39 (18.118.014), 40 (19.031.211), 41 (19.042. 021), 42 (19.155.005).

008.073 Utrecht

Utrechtse Sterrekundige Overdrukken, Sterrewacht "Sonnenborgh", Utrecht, Nos. 381 (18.015.032), 382 (18.077. 016), 383 (18.077.017), 384 (18.032.545), 385 (18.032.558), 386 (19.114.008), 387 (19.114.318), 388 (20.080.095), 389 (18.073.133), 390 (18.077.066), 391 (18.076.026), 392 (18.080.074), 393 (19.080.011), 394 (19.022.019), 395 (19.114.303), 396 (20.062.084), 397 (19.114.319), 398 (19.074.048), 399 (19.071.016), 400 (19.142.102), 401 (19.022.094), 402 (19.162.034), 403 (20.066.004), 404 (20.077.032), 405 (20.074.054).

008.074 Victoria, B.C.

Dominion Astrophysical Observatory. J. B. Hutchings. J. R. Astron. Soc. Canada, Vol. 71, 414 - 415 (1977).

Dominion Astrophysical Observatory Victoria, British Columbia. Herzberg Institute of Astrophysics, National Research Council of Canada.   E. H. Richardson. Q. J. R. Astron. Soc., Vol. 18, 382 - 395 (1977). – Report for the year 1976 April 1 to 1977 March 31.

Contributions from the Dominion Astrophysical Observatory, Victoria, B.C., Nos. 315 (19.119.003), 316 (19.122.055), 317 (19.126.022), 318 (19.114.039), 319 (19.114.031), 320 (19.114.003), 321 (19.002.030), 322 (19.117.032), 323 (20.114.512), 324 (20.114.534), 325 (19.114.049), 326 (20.113.008), 327 (19.114.355), 328 (20.124.105), 329 (20.064.015), 330 (20.142.026).

Publications of the Dominion Astrophysical Observatory, Victoria, B.C., Vol. 15, No. 2 (20.114.067).

008.075   Villanova, Penn.

Villanova University Observatory Contributions Nos. 1 (20.121.100), 2 (20.002.048).

008.076   Vilnius

Vilniaus Astronomijos Observatorijos Biuletenis (Bulletin of the Vilnius Astronomical Observatory), Nr. 44 (20.113.021; 20.113.022; 20.113.023).

008.077   Warszawa

Warsaw University Observatory and Astronomical Institute, Polish Academy of Sciences, Reprint Nos. 380 (19.065.059), 381 (20.065.006), 382 (20.065.007).

008.078   Washington, D.C.

U.S. Naval Observatory, Washington, D.C. Time Service Publications, Series 4, Nos. 544 - 569; Series 7, Nos. 497 - 522; Series 14, No. 22 (20.044.035 - 20.044.037).

008.079   Wellington

Report of the Carter Observatory Board for the year ended 1977 March 31. Carter Obs., Astron. Bull. No. 88 (1977).

Carter Observatory, Wellington, New Zealand, Astronomical Bulletin, Nos. 87 (20.047.027), 88 (20.008.079).

008.080   Wien

Annalen der Universitäts-Sternwarte Wien, Band 31, Nr. 4 (20.031.066).

Astronomische Mitteilungen Wien, No. 19 (20.122.154).

008.081   Wrocław

Wrocław Astronomical Observatory, Reprint Nos. 101 (20.103.301), 102 (20.042.071).

008.082   Zelenchukskaya

Chronicle. Astrofiz. Issled., Izv. Spets. Astrofiz. Obs., Vol. 9, 117 - 119 (1977). In Russian.

Astrofizicheskie Issledovaniya. Izvestiya Spetsial'noj Astrofizicheskoj Observatorii, Vol. 9 (20.121.016; 20.122.046; 20.114.527; 20.031.223; 20.031.224; 20.031.225; 20.033.004; 20.033.005; 20.033.006; 20.032.011; 20.032.012; 20.032.013; 20.116.008; 20.008.082).

008.083   Zürich

Astronomische Mitteilungen der Eidgenössischen Sternwarte Zürich, Nos. 349 (20.079.207), 350 (20.075.016), 353 (20.074.099), 354 (20.075.017), 355 (20.072.051), 356 (20.079.404), 357 (20.075.018).

Publikationen der Eidgenössischen Sternwarte Zürich, Band 15, Heft 1 (20.075.015).

Quarterly Bulletin on Solar Activity (Zürich), Nos. 195 - 196 (20.075.014).

009 Notes on Observatories, Planetaria, and Exhibitions

009.001 Royal Aircraft Establishment Farnborough, Geophysical Studies in Space Department.
Q. J. R. Astron. Soc., Vol. 18, 365 - 371 (1977). — Report for the year ending 1977 March 31.

009.002 Television observations at the People's Observatory Yambol. G. Momchev.
Zemlya i Vselennaya, 1977, No. 4, p. 76 - 78. In Russian.

009.003 The achievements to date at the satellite observation station in Wettzell. K. Nottarp, M. Schneider, H. Seeger, R. Sigl, P. Wilson, E. Wolf.
Deutsche Geod. Komm. Bayerische Akad. Wiss., Reihe B: Angew. Geod., Heft Nr. 221, p. 71 - 85 (1977).

009.004 The Space Imagery Center at the University of Arizona. G. Georgenson.
NASA Tech. Mem., NASA TM X-3511, (see 012.010), p. 290 (1977). — Abstract.

009.005 Astronomical archives in Southern Africa. B. Warner.
J. Hist. Astron., Vol. 8, 217 - 222 (1977).

009.006 Investigation of variable stars at the Padua—Asiago Astrophysical Observatory. L. Rosino.
Nauk. i chelovechestvo. 1977. Moskva, Znanie, 1976, p. 234 - 245. In Russian. — Abstr. in Ref. zh., 51. Astron., 9.51.36 (1977).

009.007 A new West German planetarium. H.-U. Keller.
Sky Telesc., Vol. 54, 375 - 377 (1977).

009.008 Besuch auf der Sternwarte Violau. F. Frevert.
Sterne Weltraum, Jahrg. 16, 356 - 357 (1977).

009.009 Institute of Cosmic Investigations of the USSR Academy of Sciences. G. S. Narimanov.
Zemlya i Vselennaya, 1977, No. 6, p. 67 - 68. In Russian.

009.010 Institute of Terrestrial Magnetism, Ionosphere and Radio Wave Propagation of the USSR Academy of Sciences. V. V. Migulin.
Zemlya i Vselennaya, 1977, No. 6, p. 69 - 70. In Russian.

009.011 Institute of Physics of the Atmosphere of the USSR Academy of Sciences. A. M. Obukhov.
Zemlya i Vselennaya, 1977, No. 6, p. 71. In Russian.

009.012 Second birth of the Moscow Planetarium. K. A. Portsevskij.
Zemlya i Vselennaya, 1977, No. 6, p. 76 - 81. In Russian.

009.013 Fünfzig Jahre Wiener Planetarium. H. Mucke.
Sternenbote, 20. Jahrg., 236 - 240 (1977).

009.014 Site evaluation for the Ondřejov laser satellite-tracking station. J. Zahradník, J. Klokočník.
Bull. Astron. Inst. Czechoslovakia, Vol. 28, 371 - 373 (1977).
The Ondřejov area is reviewed in terms of its known geology and geophysics. Operational and cloud-cover data are included. The criteria used in this report are exactly the same as used by Mao and Mohr (1976).

009.015 25 Jahre Olbers-Planetarium in Bremen. W. Stein.
Jenaer Rundsch. (Jena Rev.), 22. Jahrg., 309 - 310 (1977).

009.016 Annual report for magnetic observatories — 1973. E. I. Loomer.
Geomagn. Ser. Earth Phys. Branch, No. 9, p. 1 - 8 (1976). Abstr. in Phys. Abstr., Vol. 80, Abstr. 45183 (1977).

009.017 Un Observatoire Municipal au Brésil. M. Laffineur.
Astronomie, Vol. 91, 431 - 432 (1977).

009.018 L'observatoire du «Capricornio». Station astronomique municipale de Campinas — Brésil. J. Nicolini.
Astronomie, Vol. 91, 433 - 440 (1977).

009.019 Osservatorio di Pietralacroce – Ancona. M. Veltri.
Astronomia, N. 3, p. 33 - 36 (1977).

009.020 Die Satellitenbeobachtungsstation Wettzell. H. Seeger.
Nachr. Karten- Vermessungswesen, Reihe I, No. 73, 53 - 62 (1977).

009.021 Der Aufbau des Zeitsystems der Satellitenbeobachtungsstation Wettzell in den Jahren 1972—1977. K. Nottarp, I. Zeitel.
Nachr. Karten- Vermessungswesen, Reihe I, No. 73, 87 - 91 (1977).

009.022 Activity of the astronomical observatory "Čolina Kapa" with special reference to the Sky Atlas made at Sarajevo. M. Muminović.
Publ. Dep. Astron., Univ. Beograd, Fac. Sci., No. 6, (see 012.040), p. 53 - 55 (1976).

009.023 The Texas A&M underground cosmic ray observatory. R. Benson, N. M. Duller, P. J. Green, L. Bergamasco Osborne.
Proc. Southwest Reg. Conf., Vol. 3, (see 012.043), p. 123 - 127 (1977).
Texas A&M University is currently developing a cosmic ray muon anisotropy detection system. The array will be located near Hockley, Texas. The system will be 466 meters underground inside the Hockley salt dome. The purpose of the system will be to search for anisotropy in the cosmic radiation at galactic containment energies.

009.024 A planetarium in a university department of physics and astronomy. P. R. Engle.
Proc. Southwest Reg. Conf., Vol. 3, (see 012.043), p. 155 - 159 (1977).

009.025 History of the Simeis Observatory. I. I. Neyachenko.
Istor.-astron. issled. Vyp. (No.) 13. Moskva, Nauka, 1977, p. 43 - 116. In Russian. — Abstr. in Ref. zh., 51. Astron., 1.51.9 (1978).

010 Societies, Associations, Organizations

010.001   American Association of Variable Star Observers (AAVSO)

AAVSO data processing: ten years of computerization.   R. S. Hill.
J. American Assoc. Variable Star Obs., Vol. 6, 12 - 14 (1977).

Committee reports.
J. American Assoc. Variable Star Obs., Vol. 6, 45 - 48 (1977).

Minutes of the general meeting of the A.A.V.S.O. held at Washington, D.C., May 13 - 14, 1977.   C. B. Ford.
J. American Assoc. Variable Star Obs., Vol. 6, 42 - 44 (1977).

The Journal of the American Association of Variable Star Observers, Vol. 6, No. 1 (1977).

010.002   American Astronomical Society

Abstracts of papers presented at the Dynamical Astronomy Division meeting, held 2 - 4 December 1976 at Chapel Hill, North Carolina.
Bull. American Astron. Soc., Vol. 9, 435 - 438 (1977).

Abstracts of papers presented at the Planetary Sciences Division meeting held 19 - 22 January 1977 at Honolulu, Hawaii.
Bull. American Astron. Soc., Vol. 9, 440 - 481 (1977).

Late-paper abstracts from the 150th meeting of the American Astronomical Society, held 12 - 15 June 1977 at Atlanta, Georgia.
Bull. American Astron. Soc., Vol. 9, 429 - 434 (1977).

Bulletin of the American Astronomical Society, Vol. 9, No. 3 (1977).

010.003   Association Française des Observateurs d'Étoiles Variables (AFOEV)

La vie de l'Association.   E. Schweitzer.
AFOEV Bull., Tome 11, 46 (1977).

Über die Association Française des Observateurs d'Étoiles Variables – AFOEV.   E. Schweitzer.
BAV Rundbr., 26. Jahrg., 46 - 47 (1977).

Activité de l'A.F.O.E.V. Association Française des Observateurs d'Étoiles Variables.   E. Schweitzer.
Astronomie, Vol. 91, 359 - 360 (1977).

Association Française des Observateurs d'Étoiles Variables Bulletin, Tome 11, No. 2 (1977).

010.004   Association of Lunar and Planetary Observers (ALPO)

Meetings of the association.
Strolling Astron., Vol. 27, 15 - 16 (1977).

The Strolling Astronomer. The Journal of the Association of Lunar and Planetary Observers, Vol. 26, Nos.

11, 12, Vol. 27, Nos. 1, 2 (1977).

010.005   Astronomical Society of Australia (ASA)

No publication received.

010.006   Astronomical Society of Czechoslovakia

No publication received.

010.007   Astronomical Society of the Pacific (ASP)

Abstracts of papers presented at the Claremont meeting of the Astronomical Society of the Pacific 20 - 22 June 1977.
Publ. Astron. Soc. Pacific, Vol. 89, 612 - 625 (1977).

Mercury. The Journal of the Astronomical Society of the Pacific, Vol. 6, Nos. 3, 4, 5, 6 (1977).

Publications of the Astronomical Society of the Pacific, Vol. 89, Nos. 529- 531 (1977).

010.008   Astronomical Society of Southern Africa (ASSA)

Monthly Notes of the Astronomical Society of Southern Africa, Vol. 36, Nos. 7-8, 9-10, 11-12 (1977).

Proceedings of the annual general meeting, 1977, held at the Teachers' Centre, Durban, 1977 July 29.
Mon. Notes. Astron. Soc. South Africa, Vol. 36, 80 - 85 (1977).

Centre reports, 1976 - 77. Cape Centre (*C. H. Larmuth*); Natal Centre (*J. G. Barker*); Natal Midlands Centre (*R. H. Dale*); O.F.S. Centre (*G. J. Muller*); Pretoria Centre (*C. A. Posemann*); Salisbury Centre (*J. Vincent*).
Mon. Notes Astron. Soc. South. Africa, Vol. 36, 112 - 119 (1977).

Section reports.
Mon. Notes Astron. Soc. South. Africa, Vol. 36, 76 - 78, 86 - 89, 135 (1977).

Notices.
Mon. Notes Astron. Soc. South. Africa, Vol. 36, 59, 79, 111 (1977).

010.009   Astronomical Society of Victoria (ASV)

No publication received.

010.010   Astronomical Society of Western Australia (ASWA)

Report on proceedings of the 27th annual general meeting, 11th July, 1977.

J. Astron. Soc. Western Australia, Vol. 28, July (1977).

Report on proceedings of ordinary meetings.
J. Astron. Soc. Western Australia, Vol. 28, June, October, November (1977).

010.011   Astronomische Gesellschaft (AG)

Wissenschaftliche Astronomische Tagung in Göttingen anläßlich des 200. Geburtstages von C. F. Gauss vom 1. bis 4. März 1977.   H. Mauder.
Mitt. Astron. Ges., Nr. 42, p. 11 (1977).

Zur Eröffnung der Wissenschaftlich-astronomischen Tagung Basel 1977.   See Abstr. 011.027.

Mitteilungen Astronomische Gesellschaft, Nr. 42 (1977).

010.012   Astronomisk Selskab København

No publication received.

010.013   British Astronomical Association (BAA)

Notices.
J. British Astron. Assoc., Vol. 87, 433, 560 - 561, Vol. 88, 1 - 2 (1977).

Meetings and Activities of the Association.
J. British Astron. Assoc., Vol. 87, 434 - 442, 541 - 559, 561 - 567, Vol. 88, 4 - 7 (1977).

Section reports.
J. British Astron. Assoc., Vol. 87, 493 - 512, 597 - 629, Vol. 88, 56 - 84 (1977).

BAA residential weekend course at Winchester, 1977.   R. H. Chambers.
J. British Astron. Assoc., Vol. 87, 514 - 515 (1977).

Journal of the British Astronomical Association, Vol. 87, Nos. 5, 6, Vol. 88, No. 1 (1977).

The Handbook of the British Astronomical Association 1978. (20.047.006).

010.014   British Interplanetary Society (BIS)

Society news.
Spaceflight, Vol. 19, 313 - 314 (1977).

Second BIS conference on interstellar travel and communication.   K. W. Gatland.
J. British Interplanet. Soc., Vol. 30, 443 - 444 (1977).

Journal of the British Interplanetary Society, Vol. 30, Nos. 8 - 12 (1977).

Spaceflight. A publication of the British Interplanetary Society, Vol. 19, Nos. 7 - 12 (1977).

010.015   Committee on Space Research (COSPAR)

No publication received.

010.016   European Space Agency (ESA)

Status of E.S.A. plans.   E. Høg.
Highlights of Astronomy, Vol. 4, Part I, (see 012.021), p. 349 (1977).

010.017   International Astronautical Federation (IAF)

No publication received.

010.018   International Astronomical Union (IAU)

XVIth General Assembly of the International Astronomical Union.   C. Iwaniszewska.
Postępy Astron., Tom 25, 129 - 130 (1977). In Polish.

IAU Commission 45: Working Group on Spectroscopic and Photometric Data. Catalogs recently published, to be published or in preparation. List VII.
M. Barbier, W. P. Bidelman, O. Dluzhnevskaya, B. Hauck, N. Houk, C. Jaschek, M. McCarthy, J. Mead, K. Nandy, D. Philip.
Cent. Données Stellaires, Inf. Bull. No. 13, p. 54 - 58 (1977).

IAU-Symposium 77: Struktur und Eigenschaft naher Galaxien.   See Abstr. 011.028.

Transactions of the International Astronomical Union, Volume XVI B: Proceedings of the Sixteenth General Assembly, Grenoble 1976.   See Abstr. 012.006.

010.019   Meteoritical Society

Abstracts of papers presented at the 40th annual meeting. The Meteoritical Society, July 24 - 29, 1977. University of Cambridge, Cambridge, England.
Meteoritics, Vol. 12, 167 - 395 (1977).

010.020   Nederlandse Vereniging voor Weer- en Sterrenkunde

No publication received.

010.021   Polskie Towarzystwo Astronomiczne (PTA)

No publication received.

010.022   Polskie Towarzystwo Miłośników Astronomii (PTMA)

Urania. Miesięcznik Polskiego Towarzystwa Miłośników Astronomii. Rok 48, Nr. 1 - 6, 9, 10 (1977).

010.023   Royal Astronomical Society (RAS)

Meetings of the Society.
Observatory, Vol. 97, 153 - 165, 182 - 195, 213 - 217 (1977).

Meetings and activities of the Society.
Q. J. R. Astron. Soc., Vol. 18, 404 - 427 (1977).

Star formation and the interstellar medium. RAS
discussion meeting.
Q. J. R. Astron. Soc., Vol. 18, 464 - 467 (1977). — See
065.033, 131.156, 131.157, 132.028, 132.029.

Geophysical Journal of the Royal Astronomical
Society, Vol. 50, No. 3, Vol. 51, Nos. 1, 2, 3 (1977).

Memoirs of the Royal Astronomical Society, Vol.
84, Part 2 (1977).

Monthly Notices Royal Astronomical Society,
Vol. 180, Nos. 2, 3, Vol. 181, Nos. 1, 2, 3 (1977).

The Quarterly Journal of the Royal Astronomical
Society, Vol. 18, Nos. 3, 4 (1977).

010.024   Royal Astronomical Society of Canada
          (RAS Canada)

Annual meeting of the Royal Astronomical Society
of Canada. Minutes of the annual meeting, 2 July, 1977.
J. R. Astron. Soc. Canada, Vol. 71, 329 - 333 (1977).

Eighth meeting of the Canadian Astronomical
Society at the University of Western Ontario. London, Ont.,
May 25 - 27, 1977, including abstracts of papers.
J. R. Astron. Soc. Canada, Vol. 71, 391 - 411 (1977).

The Journal of the Royal Astronomical Society of
Canada, Vol. 71, Nos. 4, 5, 6 (1977).

010.025   Royal Astronomical Society of New Zealand
          (RAS New Zealand)

Section report.   F. M. Bateson.
Publ. Variable Star Sect., R. Astron. Soc. New Zealand, No. 5
(C77), p. 49 - 52 (1977).

Publications of Variable Star Section, Royal
Astronomical Society of New Zealand, No. 5 (C77) (1977).

010.026   Schweizerische Astronomische Gesellschaft (SAG)

Orion. Zeitschrift der Schweizerischen Astronomischen Gesellschaft. Bulletin de la Société Astronomique de
Suisse, 35. Jahrgang, Nos. 161 - 163 (1977).

010.027   Sociedad Astronómica de México

No publication received.

010.028   Società Astronomica Italiana (SAI)

Notizie della Società.

G. Astron., Vol. 3, 53 - 61, 107 - 111, 201 - 202 (1977).

Giornale di Astronomia, Vol. 3, N. 1, 2, 3 (1977).

010.029   Société Astronomique de France (SAF)

Séances, commissions, activités de la Société.
Astronomie, Vol. 91, 335 - 338, 387 - 390, 397 - 403, 425 -
429, 431, 446 - 461, 493 - 500, 504 - 506 (1977).

Société Astronomique de France.
P. de La Cotardière.
Astronomie, Vol. 91, 329 - 332 (1977).

Prix et médailles décernés par la Société.
Astronomie, Vol. 91, 504 - 506 (1977).

l'Astronomie, Vol. 91, September - December (1977).

010.030   Société Astronomique "R. Boškovic"

No publication received.

010.031   Société Belge d'Astronomie, de Météorologie et de
          Physique du Globe

Bulletin de la Société Belge d'Astronomie, de la
Météorologie et de Physique du Globe, Vol. 93, Nos. 3, 4
(1977).

010.032   Société Chronométrique de France

No publication received.

010.033   Svenska Astronomiska Sällskapet

No publication received.

010.034   VAGO (Astronomical-Geodetical Society of the
          USSR)

Plenary session of the Central Council of VAGO.
V. A. Bronshtehn.
Zemlya i Vselennaya, 1977, No. 4, p. 55 - 57. In Russian.
Moscow, 1977, February 21 - 23.

010.035   Vereniging voor Sterrenkunde, België

No publication received.

010.036   **News Letter of the Astronomical Society of New**
          York, Dudley Observatory, 1202 Troy Schenectady
Rd., Latham, New York 12110. Vol. 1, No. 2 (1977).
          Including abstracts of papers presented at the Union
College and State University of New York at Oswego
meetings, 1976 November 6 - 7.

010.037   **American Geophysical Union 1976 Spring Annual Meeting. (Abstracts only).**
EOS Trans. American Geophys. Union, Vol. 57, No. 4 (1976). Abstr. in Phys. Abstr., Vol. 80, Abstr. 32969 (1977).

010.038   **American Geophysical Union Fall 1976 Annual Meeting. (Papers in summary form only received).**
EOS Trans. American Geophys. Union, Vol. 57, No. 12 (1976). Abstr. in Phys. Abstr., Vol. 80, Abstr. 32970 (1977).

010.039   **Nachrichten der Vereinigung der Sternfreunde e.V.**
Sterne Weltraum, Jahrg. 16, 304 - 307, 345, 380 - 382, 419 - 420 (1977).

010.040   **The Nantucket Maria Mitchell Association.**
Seventy-fifth annual report for the year ending December 31, 1976.

Edited by the Nantucket Maria Mitchell Assoc., Nantucket, Mass., 52 pp. (1977).

010.041   **Activities of the Astronomical and Geodetic Society (1976).**   P. Kalv, H. Raudsaar.
Calendar of the Tartu Astronomical Observatory for the year 1978, (see 047.026), p. 80 - 81 (1977). In Estonian.

010.042   **Bulletin de la Société Astronomique de Liège.**
Périodique mensuel, Septembre - Décembre 1977.

010.043   **Publications of the Astronomical Society of Japan, Vol. 29, Nos. 2, 3, 4 (1977).**

010.044   **Bulletin of the Astronomical Society of India, Vol. 5, Nos. 2, 3 (1977).**

011 Reports on Colloquia, Congresses, Meetings, Symposia, and Expeditions

011.001 **Meteorite research old and new.**
R. Hutchison.
Nature, Vol. 268, 691 - 692 (1977).
   The Fortieth Annual Meeting of the Meteoritical Society was held at Cambridge, UK on 24 - 29 July, 1977.

011.002 **Giant molecular clouds.**   M. G. Edmunds.
Nature, Vol. 269, 105 - 106 (1977).
   A workshop on giant molecular clouds in the Galaxy was held at GregynogHall, Newtown,Wales from 8–13 August 1977.

011.003 **Saturn and its satellites.**   G. E. Hunt.
Observatory, Vol. 97, 163 - 165 (1977).
   A report of the joint discussion meeting of the Royal Astronomical and Royal Meteorological Societies, held on 1976 October 8.

011.004 **Im Zeichen des 200. Geburtstages von Gauß.**
L. Engelhard.
Phys. Bl., 33. Jahrg., 370 (1977).
   37. Jahrestagung der Deutschen Geophysikalischen Gesellschaft e.V. (DGG) in Verbindung mit der Frühjahrstagung der Arbeitsgemeinschaft Extraterrestrische Physik (AEP) vom 28.–30. April 1977.

011.005 **Introduction to the workshop on polar motion.**
P. Melchior.
Satellite Doppler positioning, (see 012.002), p. 747 - 754.

011.006 **Minutes of the workshop session on polar motion.**
J. Popelar.
Satellite Doppler positioning, (see 012.002), p. 755 - 762.

011.007 **Gravitation but no levitation.**   M. MacCallum.
Nature, Vol. 269, 201 - 202 (1977).
   The Eighth International Conference on General Relativity and Gravitation was held at the University of Waterloo, Ontario, Canada on 7 - 12 August, 1977.

011.008 **Radio astronomy and cosmology.**
M. S. Longair.
Nature, Vol. 263, 372 - 374 (1976).
   Report on IAU Symposium No. 74, radio astronomy and cosmology, held at the Cavendish Laboratory, Cambridge on August 16 - 20, 1976.

011.009 **Scientific session of the Department of General Physics and Astronomy of the USSR Academy of Sciences, 29 - 30 September, 1976.**
Usp. fiz. nauk, Vol. 121, 539 - 546 (1977). In Russian.
Abstr. in Ref. zh., 51. Astron., 8.51.17 (1977).

011.010 **Scientific session of the Department of General Physics and Astronomy and of the Department of Nuclear Physics of the USSR Academy of Sciences, October 27 - 28, 1976.**
Usp. fiz. nauk, Vol. 121, 727 (1977). In Russian. − Abstr. in Ref. zh., 51. Astron. 8.51.18 (1977).

011.011 **Report on the space sciences symposium.**
V. S. Venkatavaradan.
Bull. Astron. Soc. India, Vol. 5, 57 (1977).
   The 3rd space sciences symposium was held at Vikram Sarabhai Space Centre, Trivandrum, January 18 - 21, 1977.

011.012 **Report on the symposium on Hindu astronomy.**
R. C. Gupta.

Bull. Astron. Soc. India, Vol. 5, 57 - 58 (1977).
   A symposium on Hindu astronomy was held at Lucknow on 24 - 25 October, 1976.

011.013 **Advanced course in astronomy and astrophysics.**
J. McFarland.
Irish Astron. J., Vol. 12, 244 - 247 (1976).
   Sixth Advanced Course in Astronomy and Astrophysics, March 29 - April 3, 1976.

011.014 **Report on IAU Colloquium No. 39: "The relationships of comets, minor planets and meteorites".**
J. R. Dickel.
Icarus, Vol. 32, 251 - 254 (1977). − Meeting review.
   This meeting, dedicated to Fred L. Whipple, was held in Lyon, France, in August 1976.

011.015 **Origin of cosmic rays.**
Nature, Vol. 269, 653 - 654 (1977).
   The 15th Biennial International Cosmic Ray Conference was held at Plovdiv, Bulgaria on 13 - 26 August, 1977.

011.016 **R.A.S. specialist discussion on "Chemical evolution of galaxies".**
Observatory, Vol. 97, 189 - 195 (1977).

011.017 **The fifth European symposium on cosmic rays, Leeds, England, 14 - 17 September 1976.**
G. E. Kocharov.
Vestn. AN SSSR, 1977, No. 4, p. 111 - 113. In Russian.
Abstr. in Ref. zh., 51. Astron., 9.51.47 (1977).

011.018 **All-Union conference "Systems of galaxies".**
Special Astrophysical Observatory, 1977, February 1 - 4.   K. V. Bychkov, A. I. Shapovalova.
Astron. Zh. Akad. Nauk SSSR, Vol. 54, 1147 - 1148 (1977).
In Russian. English translation in Soviet Astron., Vol. 21, No. 5.

011.019 **28ᵉ Congrès International d'Astronautique, Prague, 26 September - 1 October 1977.**   E. Brun.
C. R. Acad: Sci. Paris, Tome 285, Vie Acad., 25 - 26 (1977).

011.020 **New stage in the development of meteoritics.**
Z. L. Ponizovskij, A. N. Simonenko.
Priroda, 1977, No. 10, p. 152 - 153. In Russian.
   Report on the XVIIth All-Union conference on meteoritics, 1977, May 24 - 26.

011.021 **New view of the problem of extraterrestrial civilizations.**   B. N. Panovkin.
Priroda, 1977, No. 10, p. 153 - 154. In Russian.
   Report on the meeting of April 25 and May 5, 1977 at the Sternberg Institute.

011.022 **VIIth congress of the All-Union Society "Znanie".**
Moskva, 1977, May 25 - 27.   E. P. Levitan.
Zemlya i Vselennaya, 1977, No. 6, p. 73 - 75.   In Russian.

011.023 **VIIIth All-Union conference on physics and dynamics of comets and asteroids (Kiev, October 2 - 5, 1974).**   N. I. Il'chishina, K. I. Churyumov.
Problems of cosmic physics. Vyp. (No.) 12, (see 003.016), p. 116 - 120 (1977). In Russian.

011.024 **Scientific session of the Department of General Physics and Astronomy of the USSR Academy of Sciences, 24 - 25 November, 1976.**
Usp. fiz. nauk, Vol. 122, 5, 159 (1977). In Russian. − Abstr.

in Ref. zh., 51. Astron., 10.51.53 (1977).

011.025 **New numerical methods in mathematical physics
and problems of interaction between the solar wind
and cosmic objects.**   O. M. Belotserkovskij.
Usp. fiz. nauk, Vol. 121, 727 - 729 (1977). In Russian.
Abstr. in Ref. zh., 62. Issled. kosm. prostranstva, 10.62.194
(1977). − Report of a meeting of the Departments of Physics
and Astronomy and the Department of Nuclear Physics of the
USSR Academy of Sciences, 1976, Oct. 27 - 28.

011.026 **Report of the symposium on laboratory and astro-
physical spectroscopy of small molecular species.**
K. S. Krishna Swamy.
Bull. Astron. Soc. India, Vol. 5, 83 - 85 (1977).

011.027 **Zur Eröffnung der Wissenschaftlich-astronomischen
Tagung Basel 1977.**   U. W. Steinlin.
Sterne Weltraum, Jahrg. 16, 392 - 394 (1977).

011.028 **IAU-Symposium 77: Struktur und Eigenschaft naher
Galaxien.** W. Sieber.
Sterne Weltraum, Jahrg. 16, 399 - 400 (1977).

011.029 **96ᵉ Congrès de l'AFAS.**   S. Débarbat.
Astronomie, Vol. 91, 477 - 480 (1977).

011.030 **The eighty-eighth annual scientific and membership
meeting of the Astronomical Society of the Pacific.**
Mercury, Vol. 6, No. 5, p. 12 - 19 (1977).

011.031 **Workshop on high-resolution infrared- and sub-
millimeter spectroscopy, Bonn, 8 - 10 March 1977.**
G. V. Schultz.
Infrared Phys., Vol. 17, No. 6, p. V (1977).

011.032 **On the XVIth General Assembly of the International
Astronomical Union and on French astronomy.**
L. Luud.
Calendar of the Tartu Astronomical Observatory for the year
1978, (see 047.026), p. 41 - 49 (1977). In Estonian.

011.033 **Die Hamburger Meridiankreisexpedition nach West-
Australien 1967−1974.**   A. Behr.
Mitt. Astron. Ges., Nr. 42, p. 81 (1977).

011.034 **International symposium on electromagnetic
distance measurement and the influence of atmo-**
spheric refraction.   P. Richardus.
Bull. Géod., Vol. 51, 245 - 247 (1977).

011.035 **International symposium on satellite geodesy in
Budapest from June 28 to July 1, 1977.**
I. Joó.
Bull. Géod., Vol. 51, 249 - 252 (1977).

011.036 **A report on the Japan−U.S. seminar on rare gas
abundance and isotopic constraints on the origin
and evolution of the earth's atmosphere.**   S. Matsuo.
10th Lunar and Planetary Symposium, (see 012.050), p. 109 -
113 (1977).
The Japan−U.S. seminar was held at Hakone National
Park, Japan, from June 28 - to July 1, 1977.

011.037 **Konferences un sanāksmes.**
Zvaigžņotā debess, 1976/77. gada ziema, p. 43 - 50.
Apspriede par zvaigžņu atmosfēru modeļiem (*J.-I. Straume*),
p. 43 - 45; Vissavienības sanāksme par geodēzijas jautājumiem
celtniecībā (*U. Zuments*), p. 45 - 47; Saules pētnieku apspriede
Irkutskā (*I. Šmelds*), p. 47 - 50.

011.038 **Astronomu kongresā Grenoblē.**   A. Alksnis.
Zvaigžņotā debess, 1977. gada pavasaris, p. 1 - 7.

011.039 **Konferences un sanāksmes.**
Zvaigžņotā debess, 1977. gada pavasaris, p. 48 - 52.
III Vissavienības jauno astronomu salidojums (*J. Miezis*).

011.040 **Konferences un sanāksmes.**
Zvaigžņotā debess, 1977. gada vasara, p. 30 - 39.
Astronomijas padomes plēnums Kijevā (*A. Balklavs*), p. 30 -
38; VIII Starptautiskā kartogrāfu konference (*J. Strauhmanis*),
p. 38 - 39.

011.041 **Sixth meeting of the Slovak Astronomical Society.**
J. Svoreň.
Kozmos, Vol. 8, 98 - 99 (1977). In Slovak.

011.042 **Congress of the International Astronautical
Federation in Prague.**   J. Bouška.
Říše hvězd, Vol. 58, 225 - 227 (1977). In Czech.

011.043 **Congress of the International Astronautical
Federation in Prague.**
Kozmos, Vol. 8, 161 - 175, 178 - 179 (1977). In Slovak.

012 Proceedings of Colloquia, Congresses, Meetings and Symposia

012.001   **The dusty universe.** Proceedings of a symposium
honoring F. L. Whipple on his retirement as Director
of the Smithsonian Astrophysical Observatory. October 17 -
19, 1973.   G. B. Field, A. G. W. Cameron (Editors), with
comments by F. L. Whipple.
Published for the Smithsonian Astrophysical Observatory by
Neale Watson Academic Publications, Inc., New York. 10 +
323 pp. (1975). ISBN 0-88202-033-1. — See also 17.012.027.
The individual contributions are included in their corresponding subject categories — see abstracts 005.006, 051.003, 064.
002, 064.003, 102.001, 105.005, 105.006, 106.001, 107.005,
131.004 - 131.008.

012.002   **Satellite Doppler positioning. Volume 1, 2.**
Proceedings of the International Geodetic Sym-
posium, held at New Mexico State University, Las Cruces,
N.M. (USA), October 12 - 14, 1976.
With a keynote address by G. Veis and closing remarks by
O. W. Williams.
Hosted by: Physical Science Laboratory, New Mexico State
University, Box 3548, Las Cruces, N.M. 88003. Cosponsored
by: U. S. Defense Mapping Agency, National Ocean Survey,
NOAA. 2 Volumes. 9 + 902 pp. — The individual papers with-
in the subject scope of Astronomy and Astrophysics Abstracts
are included in their corresponding subject categories — see
abstracts 011.005, 011.006, 021.001, 045.002 - 045.004,
046.003 - 046.019, 055.001, 081.005.

012.003   **Extra-atmospheric explorations of solar active
regions.** Proceedings of the VIIIth consultative
meeting of the Academies of Sciences of the Socialist
countries on solar physics.
AN SSSR. Kom. mnogostoron. sotrudn. AN sots. stran po
kompleks. probl. "Planetar. geofiz. issled." (KAPG), Sib. otd.
AN SSSR, Sib. inst. zemn.magn., ionos. i rasprostr. radiovoln.
Nauka, Moskva, 155 pp. Price 90 Kop. (1976). In Russian.
Abstr. in Ref. zh., 51. Astron., 8.51.38 (1977).

012.004   **Recueil des séminaires, année 1975 - 1976.**
A. Hayli (Editor).
Edited by Observatoire de Besançon Laboratoire d'Astronomie de la Faculté des Sciences et des Techniques, 41 bis,
avenue de l'Observatoire, 25000 Besançon. 157 pp. (1977).
The individual contributions are included in their corresponding subject categories — see abstracts 031.221, 032.009,
042.014, 042.015, 065.014, 080.015, 096.005, 114.019,
118.009, 131.079, 131.080, 151.011, 151.017 - 151.019,
154.020, 155.016, 158.041.

012.005   **The evolution of galaxies and stellar populations.**
Conference at Yale University, May 19 - 21, 1977.
B. M. Tinsley, R. B. Larson (Editors), assisted by D. C.
Gehret, with concluding remarks by J. E. Gunn.
Published and sold by Yale University Observatory, Box 2023
Yale Station, New Haven, Conn. 06520, U.S.A. 12 + 449 pp.
(1977). — The individual contributions are included in their
corresponding subject categories — see abstracts 131.081, 131.
082, 151.020, 151.021, 155.017, 155.018, 158.042 - 158.047,
161.003, 162.024.

012.006   **Transactions of the International Astronomical
Union, Volume XVI B: Proceedings of the Six-**
teenth General Assembly, Grenoble 1976.
E. A. Müller, A. Jappel (Editors).
D. Reidel Publishing Company, Dordrecht, Holland — Boston,
U.S.A. 10 + 586 pp. (1977). ISBN 90-277-0836-3.

The present volume contains the full record of the
Inaugural Ceremony, the report of the two sessions of the
XVIth General Assembly, the resolutions it adopted, the slate
of the Executive Committee, the Presidents, Vice-Presidents
and general members of the Union's 40 Commissions, and the
alphabetical list of IAU Members as of December 1976. The
volume also contains the reports of Commission meetings
held during the assembly and the report of the Working Group
for Planetary System Nomenclature (WGPSN). This report
gives details on the topographic nomenclature of the Moon
and Planets, as formulated at the various meetings of the
WGPSN since its creation in 1973 in Sydney, and as endorsed
by the XVIth General Assembly. It also reproduces a table of
names of Apollo Landing Sites, as approved by the XVth
General Assembly, but not published in Transactions, volume
XV B.

012.007   **Solar activity and solar-terrestrial relations.**
Proceedings of the 7th Regional Consultation on
Solar Physics, Starý Smokovec, September 24 - 28, 1973.
J. Sykora (Editor).
VEDA, Publishing House of the Slovak Academy of Sciences,
Bratislava, Czechoslovakia = Contrib. Astron. Obs. Skalnaté
Pleso, Vol. VI, 402 pp. Price Kčs 65.00 (1976). — The individ-
ual contributions are included in their corresponding subject
categories — see abstracts 031.222, 031.404, 032.509, 034.
006, 062.013, 062.014, 071.008, 072.008 - 072.018, 073.018 -
073.034, 074.017 - 074.025, 076.004, 077.007 - 077.016,
079.101, 080.016 - 080.019, 083.025, 084.221, 084.222.

012.008   **Advanced stages in stellar evolution.** Seventh
Advanced Course of the Swiss Society of Astronomy and Astrophysics.   P. Bouvier, A. Maeder (Editors).
Published and sold by: Geneva Observatory, CH-1290
Sauverny, Switzerland. 5 + 363 pp. (1977). — The individual
contributions are included in their corresponding subject
categories — see abstracts 065.016 - 065.018.

012.009   **The structure and content of the Galaxy and galactic
gamma rays.** A symposium sponsored by Goddard
Space Flight Center, Greenbelt, Maryland, June 2 - 4, 1976.
C. E. Fichtel, F. W. Stecker (Editors), with opening remarks
by J. F. Clark.
National Aeronautics and Space Administration, Washington,
D.C. NASA CP-002. For sale by the National Technical Information Service, Springfield, Virginia 22161. 11 + 369 pp.
Price $ 10.75 (1977). — The individual contributions are
included in their corresponding subject categories — see
abstracts 051.018, 125.012, 141.521 - 141.523, 142.704,
143.016, 151.027, 155.020 - 155.022, 156.011, 157.007 -
157.014.

012.010   **Reports of planetary geology program, 1976 - 1977.**
Abstracts of papers presented to the annual meeting
of Planetary Geology Principal Investigators at Washington
University, St. Louis, Missouri, May 23 - 26, 1977.
Compiled by R. Arvidson, R. Wahmann.
NASA Tech. Mem., NASA TM X-3511. 11 + 294 pp. Price
$ 9.25 (1977). — The individual contributions within the
subject scope of Astronomy and Astrophysics Abstracts are
included in their corresponding categories — see abstracts
009.004, 021.008, 031.228, 034.008, 081.012 - 081.015,
091.020 - 091.032, 092.009 - 092.029, 093.026, 093.027,
094.114 - 094.119, 097.073 - 097.116, 097.505, 098.031,
102.012 - 102.014, 105.025, 107.020 - 107.022, 131.091.

012.011 **Topics in interstellar matter.** Invited reviews given for Commission 34 (Interstellar Matter) of the International Astronomical Union, at the Sixteenth General Assembly of IAU, Grenoble, August 1976.
H. van Woerden (Editor), with an introduction by E. B. Jenkins.
Astrophysics and Space Science Library, Vol. 70. Proceedings.
D. Reidel Publishing Company, Dordrecht, Holland — Boston, U.S.A. 8 + 295 pp. Price Dfl. 75.00, $ 30.00 (1977). ISBN 90-277-0835-5. — The individual contributions are included in their corresponding subject categories — see abstracts 064.021, 131.097 - 131.107, 132.014 - 132.018, 135.012, 142.039, 142.707, 155.026, 158.068 - 158.070.

012.012 **Scientific applications of lunar laser ranging.** Proceedings of a symposium, held in Austin, Tex., 8 - 10 June 1976.
J. D. Mulholland, C. A. Burk, E. C. Silverberg (Editors), with a foreword by N. A. Armstrong.
Astrophysics and Space Science Library, Vol. 62 (Proceedings).
D. Reidel Publishing Company, Dordrecht, Holland — Boston, U.S.A. 17 + 302 pp. (1977). ISBN 90-277-0790-1. — For the individual contributions within the subject scope of Astronomy and Astrophysics Abstracts see abstracts 031.236 - 031.245, 044.004 - 044.009, 045.012, 046.031 - 046.033, 066.080, 081.020, 094.005 - 094.008.

012.013 **O telescópio refractor e a astrometria ao serviço das estrelas duplas.** Comptes Rendus du Colloque Astronomique Européen, held at the Universidade de Coimbra, Observatório Astronômico, Portugal, 20 - 24 Octobre 1974.
Préface par M. G. Fracastoro, allocutions par A. Simões Da Silva et J. Dommanget et synthèse et commentaires par J. Rösch.
Publiés par l'Osservatorio Astronomico di Torino, Pino Torinese. 210 pp. (1977). — The individual contributions are included in their corresponding subject categories — see abstracts 004.028, 013.014, 031.249 - 031.251, 034.009, 117.022, 118.010 - 118.020.

012.014 **Radio astronomy and cosmology.** International Astronomical Union, Symposium No. 74, held in Cambridge, England, August 16 - 20, 1976.
D. L. Jauncey (Editor).
D. Reidel Publishing Company, Dordrecht, Holland — Boston, U.S.A. 10 + 398 pp. Price Dfl. 95.00, $ 38.00 (cloth), ISBN 90-277-0838-X; Dfl. 50.00, $ 19.50 (paper), ISBN 90-277-0839-8 (1977). — The individual contributions are included in their corresponding subject categories — see abstracts 066.082 - 066.084, 141.079 - 141.110, 161.005, 162.040 - 162.043.

012.015 **Proceedings of the Seventh Lunar Science Conference, Houston, Texas, March 15 - 19, 1976.** Compiled by the Lunar Science Institute, Houston, Texas.
R. B. Merrill (managing editor).
Vol. 1: Regolith studies (with geochemical and geophysical observations of the moon, 9 pp.); Vol. 2: Petrogenetic studies of mare and highland rocks; Vol. 3: The moon and other bodies.
Pergamon Press, New York — Oxford — Toronto — Sydney — Frankfurt. Geochim. Cosmochim. Acta, Suppl. 7, 23 + 12 + 11 + 3651 + 24 + 24 + 38 pp. Price $ 160.00, £ 90.00 (1976). ISBN 0-08-021771-0. — The individual papers are included in their corresponding subject categories — see abstracts 022.041 - 022.047, 081.023, 082.034, 094.009, 094.010, 094.126 - 094.169, 094.417 - 094.558, 094.902, 097.126 - 097.128, 098.034, 105.032 - 105.039.

012.016 **A new era in space transportation.** Proceedings of the XXVIIth International Astronautical Congress held at Anaheim, Calif., U.S.A., 10 - 16 October 1976.

L. G. Napolitano (Editor).
Pergamon Press, Oxford—New York—Toronto—Sydney—Paris—Frankfurt. 8 + 451 pp. Price $ 50.00, £ 27.50 (1977). ISBN 0-08-021710-9. — The individual contributions within the subject scope of Astronomy and Astrophysics Abstracts are included in their corresponding categories — see abstracts 032.512, 051.023, 051.024, 052.012, 053.006, 084.235, 091.036.

012.017 **Life-sciences research in space.** Proceedings of a symposium held at Cologne/Porz, Germany, 24 - 26 May, 1977. W. R. Burke, T. D. Guyenne (Editors).
European Space Agency, 8 - 10, rue Mario-Nikis, 75738 Paris, France. ESA SP-130. 21 + 353 pp. (1977).

012.018 **X-ray binaries and compact objects.** Joint discussion at the XVIth General Assembly of the I.A.U., Grenoble, 1976. E. P. J. van den Heuvel (Editor).
Highlights of Astronomy, Vol. 4, Part I, (see 003.010), p. 71 - 171 (1977). — The individual contributions are included in their corresponding subject categories — see abstracts 013.021, 142.055 - 142.067.

012.019 **Space missions to the moon and planets.** Joint discussion at the XVIth General Assembly of the I.A.U., Grenoble, 1976. S. K. Runcorn (Editor).
Highlights of Astronomy, Vol. 4, Part I, (see 003.010), p. 173 - 241 (1977). — The individual contributions are included in their corresponding subject categories — see abstracts 091.039, 092.030, 093.033, 094.170, 099.053, 099.054.

012.020 **Clusters of galaxies, cosmology and intergalactic matter.** Joint discussion at the XVIth General Assembly of the I.A.U., Grenoble, 1976.
M. S. Longair, J. M. Riley (Editors).
Highlights of Astronomy, Vol. 4, Part I, (see 003.010), p. 243 - 342 (1977). — The individual contributions are included in their corresponding subject categories — see abstracts 066.087, 141.114, 151.040, 160.044 - 160.052, 161.006.

012.021 **Prospects in space astrometry.** Joint meeting of Commissions 8 and 24, at the XVIth General Assembly of the I.A.U., Grenoble, 1976.
P. Lacroute (Editor).
Highlights of Astronomy, Vol. 4, Part I, (see 003.010), p. 345 - 370 (1977). — The individual contributions are included in their corresponding subject categories — see abstracts 010.016, 032.513, 041.020, 041.021.

012.022 **Galactic structure in the direction of the polar caps.** Joint discussion at the XVIth General Assembly of the I.A.U., Grenoble, 1976.
M. F. McCarthy, A. G. D. Philip (Editors), with concluding remarks by A. Blaauw.
Highlights of Astronomy, Vol. 4, Part II, (see 003.010), p. 3 - 98 (1977). — The individual contributions are included in their corresponding subject categories — see abstracts 002. 021, 013.022, 013.023, 111.002, 111.003, 112.008, 112. 009, 114.031, 115.007, 131.127, 155.034 - 155.052.

012.023 **Stellar atmospheres as indicator and factor of stellar evolution.** Joint discussion at the XVIth General Assembly of the I.A.U., Grenoble, 1976.
R. Cayrel (Editor), with an introductory talk by B. E. J. Pagel.
Highlights of Astronomy, Vol. 4, Part II, (see 003.010), p. 99 - 218 (1977). — The individual contributions are included in their corresponding subject categories — see abstracts 064. 028 - 064.034, 065.027 - 065.030.

012.024 **The small scale structure of solar magnetic fields.** Joint discussion at the XVIth General Assembly

of the I.A.U., Grenoble, 1976.
F. L. Deubner (Editor), with concluding remarks by
H. U. Schmidt.
Highlights of Astronomy, Vol. 4, Part II, (see 003.010),
p. 219 - 275 (1977). — The individual contributions are included in their corresponding subject categories — see abstracts
031.253, 062.022, 071.014, 072.032, 072.033, 080.032,
080.033.

012.025   **The impact of ultraviolet observations on spectral
          classification.** Joint discussion at the XVIth General
Assembly of the I.A.U., Grenoble, 1976.
L. Houziaux (Editor).
Highlights of Astronomy, Vol. 4, Part II, (see 003.010),
p. 277 - 369 (1977). — The individual contributions are included in their corresponding subject categories — see abstracts
002.022, 114.032 - 114.042, 121.025, 126.018.

012.026   **Observational evidence of the heterogeneities of the
          stellar surfaces.** Joint meeting of Commissions 25,
27, 29, 35, 36, and 42 at the XVIth General Assembly of the
I.A.U., Grenoble, 1976.
M. Hack, J. P. Swings (Editors).
Highlights of Astronomy, Vol. 4, Part II, (see 003.010),
p. 373 - 407 (1977). — The individual contributions are included in their corresponding subject categories — see abstracts
080.034, 115.008, 116.013, 121.026, 122.063, 122.064.

012.027   **Far infrared astronomy.** Proceedings of a conference
          held at Cumberland Lodge, Windsor, U.K., July
11th - 13th, 1975.   M. Rowan-Robinson (Editor), with an
introduction by J. Bastin.
Suppl. Vistas Astron. Pergamon Press, Oxford—New York—
Toronto—Sydney—Paris—Frankfurt, 14 + 335 pp. Price $ 20.00
(1976). ISBN 0-08-020513-5-Y, ISBN 0-08-020591-7-R
respectively. The individual contributions are included in their
corresponding subject categories — see abstracts 031.254 -
031.256, 032.514 - 032.520, 034.012, 034.013, 066.073,
066.088 - 066.090, 071.015, 073.048, 082.035, 099.055,
131.128 - 131.134, 132.026, 132.027, 133.010 - 133.014,
134.022, 134.023, 141.115.

012.028   **Computing in plasma physics and astrophysics.**
          Proceedings of the Second European Conference
on Computational Physics, held at Garching, Germany, 27 -
30 April 1976. Invited papers, reprinted from Comput. Phys.
Commun., Vol. 12, p. 1 - 124 (1976).
D. Biskamp (Editor).
North-Holland Publishing Co., Amsterdam—Oxford—New
York. 12 + 124 pp. Price DM 75.00, ISBN 0-7204-07133
(1976). The individual contributions are included in their
corresponding subject categories — see abstracts 021.012 -
021.016, 062.023 - 062.025, 065.031, 131.135, 141.530.

012.029   **Star formation.** International Astronomical Union,
          Symposium No. 75, held in Geneva, Switzerland,
September 6 - 10, 1976.
T. de Jong, A. Maeder (Editors), with an introduction by
L. Spitzer, Jr., and a summary of the conference by G. H.
Herbig.
D. Reidel Publishing Company, Dordrecht, Holland — Boston,
U.S.A. 14 + 296 pp. Price Dfl. 75.00 (1977). ISBN 90-277-
0796-0; ISBN 90-277-0797-9 pbk. — The individual contribu-
tions are included in their corresponding subject categories —
see abstracts 065.032, 131.140 - 131.147.

012.030   **Perte de masse des étoiles.** Compte-rendu de l'école
          de Goutelas (France), 4 - 8 avril 1977.
With an introduction by E. Schatzman.
Available from D. Ballereau, Observatoire de Meudon, F-92190
Meudon. 6 + 45 + 49 + 4 + 37 + 27 + 74 + 6 pp. Price   F 40.00
(1977). — The individual papers are included in their corre-

sponding subject categories — see abstracts 022.061, 064.039 -
064.041, 065.035, 065.036, 117.031.

012.031   **Modern utilization of infrared technology civilian
          and military.** II. San Diego, Calif., USA, 26 - 27
August 1976.   I. J. Spiro (Editor).
Proceedings of the Society of Photo-optical Instrumentation
Engineers, Vol. 95. Palos Verdes Estates, Calif., USA. 10 +
230 pp. (1976). ISBN 0-89252-122-8. — See abstracts 031.
283, 031.284, 032.538 - 032.541, 034.033, 122.084. — Abstr.
in Phys. Abstr., Vol. 80, Abstr. 45891 (1977).

012.032   **Imaging through the atmosphere.** Reston, Va., USA,
          22 - 23 March 1976.   J. G. Wyant (Editor).
Proceedings of the Society of Photo-optical Instrumentation
Engineers, Vol. 75. Palos Verdes Estates, Calif., USA. 6 + 170
pp. (1976). ISBN 0-89252-102-3. — See abstracts 031.032 -
031.041, 031.288 - 031.290, 034.037, 082.064 - 082.066.
Abstr. in Phys. Abstr., Vol. 80, Abstr. 42484 (1977).

012.033   **Methods for atmospheric radiometry.** San Diego,
          Calif., USA, 26 - 27 August 1976.
D. P. McNutt (Editor).
Proceedings of the Society of Photo-optical Instrumentation
Engineers, Vol. 91. Bellingham, Wash., USA. 4 + 124 pp.
(1976). ISBN 0-89252-118 X. — Abstr. in Phys. Abstr., Vol.
80, Abstr. 70880 (1977). — See abstracts 031.306 - 031.309,
034.049, 034.050, 082.081, 082.082, 082.084 - 082.086,
084.020.

012.034   **17 convegno internazionale scientifico sullo spazio.**
          (Proceedings of the 17th international technical
scientific meeting on space). Roma, Italy, 25 - 26 March 1977.
Rassegna Int. Elettronica Nucleare ed Aerospaziale. 628 pp.
(1977). — See abstracts 013.032, 013.033, 032.557, 032.558,
052.027, 052.028, 054.018, 054.019, 061.039, 081.047,
082.093.

012.035   **Proceedings of the 3rd International Symposium
          Geodesy and Physics of the Earth.** Part 1. Weimar,
GDR, 25 - 31 October 1976.   H. Kautzleben, A.
Massevitsch (*Masevich*), E. Tengström, E. Buschmann (Editors).
Veröff. Zentralinst. Phys. Erde, No. 52, Part 1. 232 pp. (1977).
The individual contributions within the subject scope of
Astronomy and Astrophysics Abstracts are included in their
corresponding subject categories — see abstracts 042.033,
045.015, 046.051 - 046.054, 066.320, 081.050 - 081.052.

012.036   **Dynamical and chemical coupling between the
          neutral and ionized atmosphere.** Proceedings of the
NATO Advanced Study Institute held at Spåtind, Norway,
April 12 - 22, 1977.
B. Grandal, J. A. Holtet (Editors), with an Opening Address
by E. V. Thrane.
NATO Advanced Study Institutes, Ser. C, Vol. 35. D. Reidel
Publishing Company, Dordrecht, Holland — Boston, U.S.A.
19 + 392 pp. Price Dfl. 90.00, $ 36.00 respectively (1977).
ISBN 90-277-0840-1. — The individual contributions within the subject scope of Astronomy and Astrophysics Abstracts
are included in their corresponding categories — see abstracts
051.054, 082.098 - 082.102, 083.099 - 083.105, 084.022 -
084.025.

012.037   **Attitude and orbit control systems.** Proceedings of a
          conference held in Noordwijk, The Netherlands,
3 - 6 October, 1977.
C. Rowley, B. Battrick (Editors), with an opening address by
J. C. Hawkes and an introduction by A. J. Sarnecki.
ESA Scientific & Technical Publications Branch, ESTEC,
Noordwijk, The Netherlands. ESA SP-128. 14 + 555 pp. (1977).
Available from European Space Agency, 8 - 10, rue Mario-
Nikis, 75738 Paris 15, France.

012.038 Proceedings of the meeting 'How can flares be understood? ' held during the 16th General Assembly of the IAU in Grenoble, France, on 27 August, 1976.
Z. Švestka (Editor).
Sol. Phys., Vol. 53, 213 - 301 (1977). — See abstracts 073.073 - 073.084, 077.033.

012.039 **News in the theory of relativity and gravitation.**
Mater. Soveshch. 27 - 28 okt. 1975 g. Mosk. o-vo ispyt. prirody. Nauka, Moskva. 98 pp. Price 40 Kop. (1977). In Russian. — Abstr. in Ref. zh., 51. Astron., 11.51.138 (1977).

012.040 **Astronomical proceedings of the 6th congress of Yugoslav mathematicians, physicists and astronomers held in Novi Sad 1975.** B. M. Ševarlić (Editor).
Publ. Dep. Astron., Univ. Beograd, Fac. Sci., No. 6, 140 pp. (1976). — The individual contributions are included in their corresponding subject categories — see abstracts 002.039, 009.022, 015.030, 032.041 - 032.045, 034.065, 041.035 - 041.038, 042.067, 042.068, 045.020, 046.065, 065.077, 065.078, 071.046, 072.045, 079.601, 098.080, 098.081, 118.038, 122.124.

012.041 **Early stages of stellar evolution.**
I. G. Kolesnik (Editor).
Papers presented at the All-Union conference in Kiev, 1975, October 27 - 29. Akademiya Nauk Ukrainskoj SSR. Glavnaya Astronomicheskaya Observatoriya. Naukova Dumka, Kiev. 152 pp. Price 1 Rbl. 9 Kop. (1977). In Russian. — The individual contributions are included in their corresponding subject categories — see abstracts 061.046, 064.051, 064.052, 065.079, 107.035, 114.060, 122.127, 122.128, 131.206 - 131.211, 133.015, 151.070, 151.071.

012.042 **Study of travelling interplanetary phenomena 1977.**
Proceedings of the L. D. de Feiter Memorial Symposium held in Tel Aviv, Israel, June 7 - 10, 1977.
M. A. Shea, D. F. Smart, S. T. Wu (Editors).
Astrophysics and Space Science Library, Vol. 71, Proceedings. D. Reidel Publishing Company, Dordrecht, Holland — Boston, U.S.A. 12 + 439 pp. Price Dfl. 95.00, $ 38.00 (1977). ISBN 90-277-0860-6. — The individual contributions are included in their corresponding subject categories — see abstracts 007.000, 051.069, 073.110, 074.084 - 074.091, 077.048, 078.018, 085.038, 102.033 - 102.035, 106.034 - 106.037.

012.043 **Proceedings of the southwest regional conference for astronomy and astrophysics.** Austin, Texas, May 23, 1977. Volume 3.
P. F. Gott, P. S. Riherd (Editors).
Available from Department of Physics, Texas Tech University, Lubbock, Texas, U.S.A. 196 pp. (1977). ISBN 0147-2003. The individual papers are included in their corresponding subject categories — see abstracts 009.023, 009.024, 014.020 - 014.024, 015.031, 022.098, 022.099, 031.409, 033.028, 034.066, 061.047, 064.053, 065.080, 098.082, 106.038, 115.015, 124.108, 159.013, 162.113.

012.044 **Measurements and interpretation of polarization arising in the solar chromosphere and corona.**
Proceedings of a workshop held at Lund Obs., May 9 - 13, 1977. J. O. Stenflo (Editor).
Rep. Obs. Lund, No. 12, 6 + 205 pp. (1977). — The individual contributions are included in their corresponding subject categories — see abstracts 022.104, 034.081, 063.042, 063.043, 064.054, 064.055, 071.047 - 071.049, 073.112, 073.113, 074.092 - 074.096, 076.019.

012.045 **Cepheid modeling.** The proceedings of a conference and workshop held at Goddard Space Flight Center, Greenbelt, Maryland, July 29, 1974.
D. Fischel, W. M. Sparks (Editors).

Natl. Aeronaut. Space Adm., Washington, D.C., NASA SP-383. For sale by the National Technical Information Service, Springfield, Virginia 22161. 9 + 332 pp. Price $ 9.50 (1975). The individual contributions are included in their corresponding subject categories — see abstracts 122.157 - 122.168.

012.046 **The Space Telescope.**
Natl. Aeronaut. Space Adm., Washington, D.C., NASA SP-392. For sale by the Superintendent of Documents, U.S. Government Printing Office, Washington, D.C. 20402. 11 + 231 pp. Price $ 2.30. Stock Number 033-000-00644-6 (1976).
This volume contains the authors' summaries of their papers on the Space Telescope presented at the 21st annual meeting of the American Astronautical Society at Denver, Colorado, August 26 - 28, 1975. — See abstracts 031.067 - 031.074, 031.412, 032.590 - 032.616, 051.074 - 051.078, 054.022, 054.025.

012.047 **Solar-wind interaction with the planets Mercury, Venus, and Mars.** The proceedings of a bilateral seminar of the US—USSR Joint Working Group on Near-Earth Space, the Moon, and Planets, held at the Space Research Institute of the Academy of Sciences of the USSR, Moscow, November 17—21, 1975.
N. F. Ness (Editor), with introductory remarks by O. Belotserkovskii (*Belotserkovskij*), N. F. Ness.
Natl. Aeronaut. Space Adm., Washington, D.C., NASA SP-397. For sale by the National Technical Information Service, Springfield, Virginia 22161. 11 + 170 pp. Price $ 6.25 (1976). The individual contributions are included in their corresponding subject categories — see abstracts 084.315, 091.071 - 091.077, 092.044, 093.064, 094.194, 097.148 - 097.150.

012.048 **Effect of the ionosphere on space systems and communications.** Based on ionosphere effects symposium, held in Crystal City, Arlington, Va., January 20 - 22, 1975. J. M. Goodman (Editor).
Naval Res. Lab., Washington, D.C. 20375. For sale by the Superintendent of Documents, U.S. Government Printing Office, Washington, D.C. 20402. 16 + 495 pp. Price $ 10.20 (1975). — The individual contributions within the subject scope of Astronomy and Astrophysics Abstracts are included in their corresponding categories — see abstracts 083.122 - 083.144, 084.039, 084.040.

012.049 **Comets, asteroids, meteorites — interrelations, evolution and origins.** Proceedings of the International Astronomical Union Colloquium No. 39, held at Lyon, France, 17 - 20 August, 1976.
A. H. Delsemme (Editor).
Published by The University of Toledo, Toledo, Ohio, USA. 21 + 587 pp. Price $ 36.50 (1977). — The individual contributions are included in their corresponding subject categories — see abstracts 094.199, 098.092 - 098.106, 102.038 - 102.053, 103.402, 103.602, 103.707, 104.052 - 104.056, 105.267 - 105.283, 106.049, 106.050, 107.045 - 107.058, 131.220, 131.221.

012.050 **Proceedings of the tenth ISAS Symposium on Lunar and Planetary Science,** held at Tokyo 11 - 13 July, 1977.
Published by Institute of Space and Aeronautical Science, University of Tokyo. 8 + 238 pp. (1977). — The individual contributions within the subject scope of Astronomy and Astrophysics Abstracts are included in their corresponding categories — see abstracts 011.036, 022.117 - 022.121, 053.010, 063.056, 072.058, 074.120, 080.092 - 080.094, 081.072, 082.144, 082.145, 091.085, 093.067 - 093.070, 094.200, 094.201, 094.613, 097.219, 099.085, 099.086, 102.054, 102.055, 103.106, 105.245, 105.284 - 105.286, 106.051, 107.059 - 107.063.

012.051 In memory of Henry Norris Russell. Sessions I and II of IAU Symposium No. 80, The HR diagram, held in Washington, D.C., November 2, 1977.
A. G. D. Philip, D. H. DeVorkin (Editors). Included are three papers by H. N. Russell.
Dudley Observatory Report No. 13, 12 + 181 pp. (1977). For the other contributions see abstracts 002.054, 004.074 - 004.080, 005.021 - 005.024.

012.052 Quantum field theoretical methods. Conference held at Les Houches, France, 28 July - 6 Sept. 1975.
R. Balian, J. Zinn-Justin (Editors).
North-Holland Publishing Company, Amsterdam, Netherlands. 20 + 386 pp. Price $ 39.95 (1976). ISBN 0-7204-0433-9. From Phys. Abstr., Vol. 80, Abstr. 57457 (1977).

012.053 Décalages vers le rouge et expansion de l'univers — l'évolution des galaxies et ses implications cosmo-logiques. Proceedings of two conferences held at Paris, September 1976.
C. Balkowski, B. E. Westerlund (Editors).
CNRS, Paris. 619 pp. Price FF 180.00 (1977). − From Phys. Today, Vol. 30, No. 12, p. 62 (1977).

012.054 8th International Space Rescue and Safety Sympo-sium. Lisbon, Portugal, 24 - 26 Oct. 1975.
P. H. Bolger (Editor).
American Astronautical Soc. (1976), San Diego, Calif., USA. 11 + 217 pp. (1976). − Review in Phys. Abstr., Vol. 80, Abstr. 25913 (1977).

012.055 Planetary satellites. Papers from a colloquium, Ithaca, N.Y., August 1974.
J. A. Burns (Editor).
University of Arizona Press, Tucson. 24 + 598 pp. Price $ 19.95 (1977). − Reviews in Science, Vol. 198, 1147 - 1148; 1977 (*W. M. Kaula*)ј Strolling Astron., Vol. 27, 34; 1977 (*J. W. Young*).

012.056 Space shuttle missions of the 80's. Parts 1 and 2. Proceedings of the 21st Annual Meeting of the American Astronautical Society, Denver, August 1976.
W. J. Bursnall, G. W. Morgenthaler, G. E. Simonson (Editors).
Advances in the Astronautical Sciences, Vol. 32. AAS, San Diego. Distributor: Univelt, P. O. Box 28130, San Diego, Calif. 92128. 1308 pp. Price $ 85.00 (1977). − Reviews in Phys. Today, Vol. 30, No. 9, p. 60 (1977); Sky Telesc., Vol. 54, 138 (1977).

012.057 Real-Time Devices and Novel Techniques. Conference held at San Diego, Calif., USA. 25 Aug. 1976.   D. Casasent, A. Sawchak (Editors).
Proceedings of the Society of Photo-optical Instrumentation Engineers, Vol. 83. Bellingham, Wash., USA. 4 + 156 pp. (1976). ISBN 0-89252-110-4. − Review in Phys. Abstr., Vol. 80, Abstr. 71932 (1977).

012.058 Asymptotic structure of space-time. Proceedings of a Symposium held at the University of Cincinnati, June 1976.   F. P. Esposito, L. Witten (Editors).
Plenum, London − New York. 10 + 442 pp. Price $ 42.50 (1977). − Reviews in Nature, Vol. 269, 183; 1977 (*P. Davies*); Phys. Today, Vol. 30, No. 9, p. 56 (1977).

012.059 Proceedings of the International Neutrino Con-ference. Aachen 1976.
H. Faissner, H. Reithler, P. Zerwas (Editors).
Vieweg, Wiesbaden, Germany. 12 + 748 pp. Price DM 168.00 (1977). − From Science, Vol. 198, 1077 (1977).

012.060 Low light level devices for science and technology. Conference held at Reston, Va., USA, 22 - 23

March 1976.   C. Freeman (Editor).
Proceedings of the Society of Photo-optical Instrumentation Engineers, Vol. 78. Palos Verdes Estates, Calif., USA. Soc. Photo-optical Instrumentation Engineers. 6 + 162 pp. (1976). ISBN 0-89252-105-8. − From Phys. Abstr., Vol. 80, Abstr. 39091 (1977).

012.061 Life sciences and space research XV. Papers pre-sented to the Working Group on Space Biology at the XIXth Plenary Meeting of COSPAR, held in Philadelphia, 8 - 19 June, 1976.
R. Holmquist, A. C. Stickland (Editors).
Pergamon Press, Oxford. 308 pp. (1977).

012.062 Theoretical and applied mechanics. Proceedings of the IUTAM congress on theoretical and applied mechanics. Delft, The Netherlands, Aug. 30 - Sept. 4, 1976.
W. T. Koiter (Editor).
North-Holland Publishing Company, Amsterdam, The Nether-lands. (ISBN 0-7204-0549-1). − Abstr. from INIS7720 333740.

012.063 Advances in X-ray analysis, Vol. 20. (Proc. of the 25th annual conf. on applications of X-ray analysis, Denver, August 1976).   H. F. McMurdie, C. S. Barrett, J. B. Newkirk, C. O. Ruud (Editors).
Plenum, New York. 604 pp. Price $ 42.50 (1977). − From Phys. Today, Vol. 30, No. 9, p. 63 (1977).

012.064 Bicentennial space symposium. W. C. Schneider (Editor).
Advances in the Astronautical Sciences, Vol. 35. American Astronautical Society, San Diego, Calif.; Univelt Inc., San Diego, Calif. 229 pp. Price $ 25.00 (1977). − Review in Sky Telesc., Vol. 54, 230 (1977).

012.065 Conférences d'astronomie de l'Observatoire. − 1976. Observatoire de Genève, 222 pp. Price F 40.00 (1977). − Review in Astronomie, Vol. 91, 509 (1977).

012.066 Advances in precision machining of optics. Conference held at San Diego, Calif., USA. 26 - 27 August 1976.
Proceedings of the Society of Photo-optical Instrumentation Engineers. Vol. 93. Bellingham, Wash., USA: Soc. Photo-optical Instrumentation Engineers. 8 + 159 pp. (1976). ISBN 0-89252-120-1. − Abstr. in Phys. Abstr., Vol. 80, Abstr. 42637 (1977).

012.067 Proceedings of the 11th International Symposium on Mathematical Geophysics. Conference held at Seeheim/Odenwald, Germany, 18 - 27 Aug. 1976.
J. Geophys., Vol. 43, No. 1 - 2 (1977). − Review in Phys. Abstr., Vol. 80, Abstr. 73622 (1977).

012.068 Infrared detection techniques for space research. V. Manno, J. Ring (Editors).
Translated from the English edition (see 08.012.005). Mir, Moskva. 384 pp. Price 2 Rbl. 99 Kop. (1977). In Russian. From Ref. zh., 51. Astron., 12.51.35 (1977).

012.069 Topical conference on solar and interplanetary physics. Tucson, Ariz.,Jan. 12 - 15, 1977.
Book of abstracts.
American Astronomical Society, Princeton, N.J. Solar Physics Div. American Geophys. Union, Washington, D.C. Solar Terrestrial Relationships Div., 20 pp. (1977). − From IKK 77A15004335.

012.070 Winter school on high energy astrophysics. Bombay, January 5 - 16, 1976.

Review in Bull. Astron. Soc. India, Vol. 4, 33 - 34; 1976   (*J. V. Narlikar, S. M. Chitre*).

013 Reports on Astronomy in Various Countries and Particular Fields, International Cooperation

013.001 **Progress report: Copernicus observations of solar system objects.**   E. S. Barker.
Bull. American Astron. Soc., Vol. 9, 465 (1977). – Abstract.

013.002 **Soviet neutrino astronomy.**   B. Belitsky.
Spaceflight, Vol. 19, 311 - 312 (1977).

013.003 **The New Zealand Government Time-Service. – An informal history.**   G. A. Eiby.
South. Stars, Vol. 27, 15 - 34 (1977).

013.004 **The sociology of innovation in modern astronomy.**   D. Edge.
Q. J. R. Astron. Soc., Vol. 18, 326 - 339 (1977).
   This paper describes some of the main features of the development of astronomy since 1945, stressing sociological factors, and drawing examples mainly from the history of radio astronomy.

013.005 **The Strasbourg Stellar Data Centre.**   C. Jaschek.
Vistas Astron., Vol. 21, 311 - 314 (1977).

013.006 **Recent astronomical research in China.**   T. Kiang.
Sky Telesc., Vol. 54, 260 - 263 (1977).

013.007 **International cooperation of centers for ephemerides and astrometric data.**
R. L. Duncombe, A. D. Fiala, P. K. Seidelmann.
Proc. Fifth Biennial Int. CODATA Conf., June 28 - July 1, 1976, Univ. Colorado, Boulder. Pergamon Press, p. 279 - 280 (1977).

013.008 **The experience of the Stellar Data Center as a data bank in astronomy.**   C. Jaschek, F. Ochsenbein.
Proc. Fifth Biennial Int. CODATA Conf., June 28 - July 1, 1976, Univ. Colorado, Boulder. Pergamon Press, p. 281 - 283 (1977).

013.009 **Hawaiian observatory starts lunar ranging.**
Phys. Today, Vol. 30, No. 10, p. 20 (1977).

013.010 **The effects of defence science on the advance of astronomy.**   B. Lovell.
J. Hist. Astron., Vol. 8, 151 - 173 (1977).

013.011 **About the evolution of the Soviet astronomy.**   G. S. Chromov (*Khromov*).
Astron. Schule, 14. Jahrg., 99 - 101 (1977). In German.

013.012 **Astronomy research in 1976.**
K.-H. Schmidt, K.-G. Steinert, H. Zimmermann.
Astron. Schule, 14. Jahrg., 101 - 104 (1977). In German.

013.013 **Soviet astronomy on the occasion of the 60th anniversary of the Great October.**
Astron. Zh. Akad. Nauk SSSR, Vol. 54, 929 - 931 (1977). In Russian. English translation in Soviet Astron., Vol. 21, No. 5.

013.014 **L'astronomie des étoiles doubles au Portugal.**   A. Simões Da Silva.
O telescópio refractor e a astrometria ao serviço das estrelas duplas, (see 012.013), p. 195 - 197 (1977).
   Le développement des recherches astronomiques à Coimbre est décrit d'un point de vue historique et d'entreprises récentes. Les possibilités de l'avenir sont discutées en mettant l'accent plus particulièrement sur les possibilités offertes dans le domaine des étoiles doubles visuelles.

013.015 **Main results of Soviet researches in the field of geomagnetism and aeronomy over 60 years (1917 - 1977).**
Geomagn. Aehron., Vol. 17, 777 - 783 (1977). In Russian.

013.016 **The space science coordination office.**   P. A. Forsyth.
J. R. Astron. Soc. Canada, Vol. 71, 399 (1977). – Abstract.

013.017 **The Canadian Corporation for University Space Science.**   G. F. Lyon.
J. R. Astron. Soc. Canada, Vol. 71, 400 (1977). – Abstract.

013.018 **Astronomy and astrophysics at NASA.**   R. C. Henry.
J. R. Astron. Soc. Canada, Vol. 71, 400 (1977). – Abstract.

013.019 **Long-term planning in the European Space Agency.**   G. A. H. Walker.
J. R. Astron. Soc. Canada, Vol. 71, 400 - 401 (1977). Abstract.

013.020 **Spectrophotometry at York University.**   W. G. Weller, S. Jeffers.
J. R. Astron. Soc. Canada, Vol. 71, 402 (1977). – Abstract.

013.021 **Coordinated campaign to observe X-ray binaries.**   Y. Kondo.
Highlights of Astronomy, Vol. 4, Part I, (see 012.018), p. 171 (1977).

013.022 **Some research programmes into galactic structure at the galactic caps under way at the Royal Greenwich Observatory.**   D. H. P. Jones.
Highlights of Astronomy, Vol. 4, Part II, (see 012.022), p. 27 - 28 (1977).

013.023 **Current results and suggestions for future work.**   M. McCarthy, A. G. D. Philip, I. King, U. Steinlin.
Highlights of Astronomy, Vol. 4, Part II,(see 012.022), p. 97 - 98 (1977).

013.024 **The UK 1.2-m Schmidt Telescope January—June 1977.**
Q. J. R. Astron. Soc., Vol. 18, 468 - 472 (1977).

013.025 **Progress in far-infrared and submillimeter astronomy.**   J. R. Houck.
J. Opt. Soc. America, Vol. 67, 1430 - 1431 (1977). — Abstract.

013.026 **A discussion of the magnitude errors and magnitude scales of meteor observers in Sweden and Czechoslovakia.**   B. A. Lindblad, J. Štohl.
Bull. Astron. Inst. Czechoslovakia, Vol. 28, 321 - 328 (1977).
   The present paper discusses the magnitude errors and the magnitude scales of individual observers participating in the Swedish meteor program. A comparison is also made between the magnitude scales of the Swedish and Czechoslovakian observing teams. The Swedish data were collected at the Onsala Space Observatory.

013.027 **On the activities of the Laboratory of Cosmic Geology of the Geological Faculty of the Moscow University.**   Ya. G. Kats.
Vestn. Mosk. univ., 1977, No. 1, p. 113 - 114. In Russian.
From Ref. zh., 51. Astron., 10.51.41 (1977).

013.028 **The University of Hawaii lunar ranging experiment geodetic-geophysics support programme.**
W. E. Carter, E. Berg, S. Laurila.
Philos. Trans. R. Soc. London, Ser. A, Vol. 284, 451 - 456 (1977). — Abstr. in Phys. Abstr., Vol. 80, Abstr. 56569 (1977).

013.029 **The N.A.S.A. Earth and ocean dynamics programme.**
F. O. Vonbun.
Philos. Trans. R. Soc. London, Ser. A, Vol. 284, 607 - 619 (1977). — Abstr. in Phys. Abstr., Vol. 80, Abstr. 56733 (1977).

013.030 **Astronomy and spectroscopy.**   M. G. Edmunds.
Phys. Bull., Vol. 28, No. 4, p. 168 - 170 (1977).
Abstr. in Phys. Abstr., Vol. 80, Abstr. 60623 (1977).

013.031 **The contribution of the SFB 78 in the observations of GEOS-III in the years 1975/76.**   H. Seeger.
Z. Vermessungswes., Vol. 102, No. 4, p. 165 - 173 (1977). In German. — Abstr. in Phys. Abstr., Vol. 80, Abstr. 70591 (1977).

013.032 **The contribution of SNIA Viscosa to ESA space programmes.**   D. Palladino.
17 convegno internazionale scientifico sullo spazio (see 012. 034), p. 401 - 410 (1977). In Italian. — Abstr. in Phys. Abstr., Vol. 80, Abstr. 82109 (1977).

013.033 **UTEX: a proposal from Italy for an astronomical payload of the S/L.**   G. Zappalà.
17 convegno internazionale scientifico sullo spazio (see 012. 034), p. 517 - 527 (1977). In Italian. — Abstr. in Phys. Abstr., Vol. 80, Abstr. 82112 (1977).

013.034 **Soviet-French cooperation in the field of gamma astronomy.**   S. A. Nikitin.
Priroda, 1977, No. 12, p. 32 - 39. In Russian.

013.035 **Intercosmos program. Cooperation of Socialist countries.**   B. N. Petrov.
Vestn. AN SSSR, 1977, No. 6, p. 95 - 105. In Russian.
Abstr. in Ref. zh., 62. Issled. kosm. prostranstva, 11.62.14 (1977).

013.036 **Some results of Soviet-French co-operation in satellite geodesy.**
A. G. Masevich, S. K. Tatevyan.
Nauchn. Inf., vyp. (No.) 35, p. 5 - 15 (1977). In Russian.
   Results of Soviet-French co-operative investigations based on tracking of satellites are considered. Investigations of the geopotential by means of series of laser and photographic observations of special satellites are planned. The joint experiment of laser ranging of the moon is described.

013.037 **Big astronomy in Chile: the southern observatories come of age.**   A. L. Hammond.
Science, Vol. 198, 1235 - 1239 (1977).
   They are the Cerro Tololo Interamerican Observatory (CTIO), a U.S. national observatory operated for the National Science Foundation by a group of universities; the European Southern Observatory (ESO), a collaborative venture of six European countries; and the Las Campanas Observatory, a private scientific facility financed by the Carnegie Institution of Washington, operated by the Hale Observatories and still known locally by its earlier name, the Carnegie Southern Observatory (CARSO).

013.038 **Astronomy in society: introduction to a history of contemporary astronomy.**   S. Vaghi.
Cah. Fundam. Sci., No. 73, 44 pp. (1977).

Enquiry on radioastronomical data.
See Abstr. 002.015.

Space investigations made in the USSR in 1976.
See Abstr. 003.178.

Activity of the astronomical observatory "Čolina Kapa" with special reference to the Sky Atlas made at Sarajevo.   See Abstr. 009.022.

On the XVIth General Assembly of the International Astronomical Union and on French astronomy.
See Abstr. 011.032.

Die 20 Jahre nach Sputnik 1. Der Mensch im Weltraum.   See Abstr. 051.022.

Data on the variable star studies in the USSR during 1972 - 1975.   See Abstr. 120.002.

014 Teaching in Astronomy

014.001 **Some student projects in astronomy.**
Sky Telesc., Vol. 54, 171 - 173 (1977).

014.002 **Zur pädagogischen Forschung auf dem Gebiet des Astronomieunterrichts.** K.-G. Steinert.
Astron. Schule, 14. Jahrg., 77 (1977).

014.003 **Der historische Aspekt im Astronomieunterricht.**
H. Bernhard, D. B. Herrmann.
Astron. Schule, 14. Jahrg., 84 - 86 (1977).

014.004 **Better preparation of astronomy teachers.**
A. V. Artem'ev.
Vestn. vyssh. shkoly, 1977, No. 1, p. 83 - 84. In Russian.
Abstr. in Ref. zh., 51. Astron., 8.51.20 (1977).

014.005 **Site for astronomical observations at school.**
B. Bonev, G. Momchev.
Fizika, NRB, Vol. 2, No. 1, p. 22 - 25 (1977). In Bulgarian.
Abstr. in Ref. zh., 51. Astron., 9.51.57 (1977).

014.006 **Neuere Erkenntnisse über Körper des Sonnensystems im Unterricht.** M. Schukowski.
Astron. Schule, 14. Jahrg., 104 - 107 (1977).

014.007 **Le leggi di Keplero.** F. Martino.
G. Astron., Vol. 3, 27 - 45 (1977).

014.008 **Progetto di insegnamento delle scienze integrate nella scuola elementare e media (scuola dell'obbligo).**
Gruppo di ricerca FOSDIS, Cagliari.
G. Astron., Vol. 3, 101 - 106 (1977).

014.009 **Astronomie-Lehrplan für die Gymnasien in Baden-Württemberg.**
F. Gondolatsch, O. Zimmermann.
Sterne Weltraum, Jahrg. 16, 378 (1977).

014.010 **Variable star simulator.** W. E. Hughes.
American J. Phys., Vol. 44, 1227 - 1228 (1976).
Abstr. in Phys. Abstr., Vol. 80, Abstr. 22357 (1977).

014.011 **Methodical annotations for teaching on the theme "Cosmogony and Cosmology" in the XIth class.**
G. Momchev.
Fizika (NRB), Vol. 2, No. 2, p. 16 - 20 (1977). In Bulgarian.
From Ref. zh., 51. Astron., 10.51.61 (1977).

014.012 **Teaching megalithic astronomy.** R. Manning.
American J. Phys., Vol. 45, 125 - 130 (1977).
Abstr. in Phys. Abstr., Vol. 80, Abstr. 33759 (1977).

014.013 **Method for determining the radius vector of a planet from two observations of position.** M. K. Gainer.
American J. Phys., Vol. 45, 131 - 134 (1977). – Abstr. in Phys. Abstr., Vol. 80, Abstr. 33760 (1977).

014.014 **Inverse-square gravitation from Kepler's first two laws: a Cartesian coordinate treatment.** G. Pozzi.
American J. Phys., Vol. 45, 307 - 308 (1977). – Abstr. in Phys. Abstr., Vol. 80, Abstr. 33802 (1977).

014.015 **The 1986 apparition of Halley's comet.**
T. E. Margrave, Jr.
Phys. Teach., Vol. 15, 110 - 111 (1977). – Abstr. in Phys. Abstr., Vol. 80, Abstr. 45652 (1977).

014.016 **High school teachers as astronomers (*photoelectric observations of asteroids*).** R. Sather.
Phys. Teach., Vol. 15, 86 - 89 (1977). – Abstr. in Phys. Abstr., Vol. 80, Abstr. 45347 (1977).

014.017 **Atmospheric refraction.** J. B. Johnston.
Phys. Teach., Vol. 15, 308 - 309 (1977). – Abstr. in Phys. Abstr., Vol. 80, Abstr. 57134 (1977).

014.018 **Experimental summer class in astrophotography.**
B. R. Parker.
American J. Phys., Vol. 45, 491 - 492 (1977). – Abstr. in Phys. Abstr., Vol. 80, Abstr. 64902 (1977).

014.019 **Geophysics without geology.** R. A. Rudin.
American J. Phys., Vol. 45, 572 - 573 (1977).
Abstr. in Phys. Abstr., Vol. 80, Abstr. 67155 (1977).

014.020 **Blue dwarfs and baby-pink giants. Some comments on the confusing terminology of the H-R diagram.**
U. O. Herrmann.
Proc. Southwest Reg. Conf., Vol. 3, (see 012.043), p. 141 - 147 (1977).

014.021 **Astronomy education and cognitive processes: why some students find learning about nature to be an unnatural act.** R. G. Cooper, Jr.
Proc. Southwest Reg. Conf., Vol. 3, (see 012.043), p. 149 - 154 (1977).
This article describes cognitive limitations on students' understanding of astronomy and a few strategies for ameliorating these limitations. The approach that poor students use is briefly described. The cognitive limitations which underlie this approach are examined within a cognitive-developmental framework. Specific types of concepts which are difficult are enumerated. Strategies for conveying these concepts are presented.

014.022 **A junior/senior PSI (*Personalized System of Instruction*) astrophysics course.** M. Zeilik II.
Proc. Southwest Reg. Conf., Vol. 3, (see 012.043), p. 161 - 167 (1977).

014.023 **Personalized system of instruction: an introductory astronomy course for nonscience majors.**
M. Breger.
Proc. Southwest Reg. Conf., Vol. 3, (see 012.043), p. 169 - 183 (1977).
This article describes successful applications of a PSI (selfpaced) astronomy course to relatively large sections of students. Special attention is paid to the course materials and organization, as well as the activities of the instructor, tutors and students. The grade distribution of this course relative to lecture and classical PSI courses is described. Other aspects such as student achievement rates, procrastination, days of the week are briefly analyzed.

014.024 **Observing projects for very large astronomy classes.**
R. R. Robbins.
Proc. Southwest Roy. Conf., Vol. 3, (see 012.043), p. 185 - 193 (1977).
It is argued that measurement activities are an essential ingredient of an astronomy class, and that they are effective in aiding students in developing improved reasoning patterns. Some sample activities used in a large University of Texas astronomy class are presented, and the course logistics necessary to implement such projects in large enrollment classes are discussed.

Meteorites.   See Abstr. 003.053.

Black hole physics illustrated in photon orbits.
See Abstr. 021.017.

Celestial navigation with the stereographic projection.   See Abstr. 046.042.

015 Miscellanea

015.001 **The star of Bethlehem.**
O. Edwards, S. Campbell, C. M. Botley, N. R. Flores, T. J. Palmer, E. V. Hulse, D. W. Hughes.
Nature, Vol. 268, 565 - 567 (1977). — Comments on a paper of D. W. Hughes (see 18.015.028).

015.002 **On the coincidence of the position of Mercury with the 90-day oscillation of Jupiter's Red Spot.**
E. Reese, R. Beebe.
Planet. Space Sci., Vol. 25, 890 - 892 (1977).
Recent observations have been utilized to investigate the proposed temporal connection between the 90-day oscillation of Jupiter's Red Spot and the inferior conjunction of Mercury. The oscillations appear to be synchronized with the inferior conjunction of a "mean Mercury" rather than the real Mercury implying that the period of oscillation of the Red Spot is constant. Although the probability of a synchronization due to chance is small, the failure of the oscillation to coincide with the motion of the real Mercury offers a strong argument against a physical connection between the two phenomena.

015.003 **Cosmic velocities.**   A. I. Govyadinov.
Metodika i tekh. fiz. ehksp. Saratov, 1976, p. 60 - 63. In Russian. — Abstr. in Ref. zh., 51. Astron., 8.51.22 (1977).

015.004 **Stjärnornas namn. Namnförteckning för 393 stjärnor och 71 stjärngrupper.**   P. Kalaja.
Astron. Tidsskr., Årg. 10, 49 - 61 (1977).

015.005 **Searching for extraterrestrial intelligence: the ultimate exploration.**   D. Black, J. Tarter, J. N. Cuzzi, M. Conners, T. A. Clark.
Mercury, Vol. 6, No. 4, p. 3 - 7 (1977).

015.006 **At the technological frontier: the JPL search for extraterrestrial intelligence.**   R. E. Edelson.
Mercury, Vol. 6, No. 4, p. 8 - 12 (1977).

015.007 **The earth from aboard Salyut 5.**
B. V. Volynov, V. D. Bol'shakov, V. M. Zholobov, N. P. Lavrova.
Zemlya i Vselennaya, 1977, No. 4, p. 15 - 16. In Russian.

015.008 **"Flying saucers" — a test of intellect.**
D. Ya. Martynov.
Zemlya i Vselennaya, 1977, No. 4, p. 44 - 48. In Russian.

015.009 **The prospect of astro-palaeontology.**   J. Armitage.

J. British Interplanet. Soc., Vol. 30, 466 - 469 (1977).
This paper expresses the view that a large number of life-bearing planets might well develop civilisations during the lifetime of a Galaxy, but equally points out that, because of the enormous time-span of galactic events, and the possibly limited lifetimes of many or most technological civilisations, very few such civilisations would be likely to co-exist at any given point in time.

015.010 **About dogma in science, and other recollections of an astronomer.**   E. J. Öpik.
Annu. Rev. Astron. Astrophys., Vol. 15, (see 003.012), 1 - 17 (1977).

015.011 **A run-off roof observatory.**   R. F. Spry.
Modern astronomy, (see 003.013), p. 40 - 46 (1977).

015.012 **Life on the second sun.**   R. D. MacElroy.
Chemical evolution of the giant planets, (see 003.014), p. 69 - 84 (1976).

015.013 **Microbial life at low temperatures.**   D. J. Kushner.
Chemical evolution of the giant planets, (see 003.014), p. 85 - 93 (1976).

015.014 **Possibility of growth of airborne microbes in outer planetary atmospheres.**
R. L. Dimmick, M. A. Chatigny.
Chemical evolution of the giant planets, (see 003.014), p. 95 - 106 (1976).

015.015 **Dormant and resistant stages of procaryotic cells.**   R. S. Hanson.
Chemical evolution of the giant planets, (see 003.014), p. 107 - 120 (1976).

015.016 **Life in extreme environments: biological water requirements.**   N. H. Horowitz.
Chemical evolution of the giant planets, (see 003.014), p. 121 - 128 (1976).

015.017 **Energy requirements of a biosphere.**
B. Kok, R. Radmer.
Chemical evolution of the giant planets, (see 003.014), p. 183 - 197 (1976).

015.018 **Organic synthesis in a simulated Jovian atmosphere of the planet Jupiter.**   C. Ponnamperuma.

Chemical evolution of the giant planets, (see 003.014), p. 221 - 231 (1976).

015.019 The search for extraterrestrial intelligence: tele-communications technology.
R. E. Edelson, G. S. Levy.
Communications and knowledge. I. Conference held at Dallas, Tex., USA, 29 Nov. - 1 Dec. 1976, IEEE 1976, New York, USA, p. 1.5/1 - 5. — Abstr. in Phys. Abstr., Vol. 80, Abstr. 33361 (1977).

015.020 The pyrolytic release experiment: measurement of carbon assimilation.   J. S. Hubbard.
Origins of Life, Vol. 7, 281 - 292 (1976). — Abstr. in Phys. Abstr., Vol. 80, Abstr. 41346 (1977).

015.021 Problems of radio communication with extra-terrestrial civilizations.   N. T. Petrowitsch.
Nachrichtentech. Elektron., Vol. 27, No. 4, p. 146 - 148 (1977). In German. — Abstr. in Phys. Abstr., Vol. 80, Abstr. 56845 (1977).

015.022 Probleme und Problematik der heutigen Astronomie.
H.-J. Treder.
Astron. Schule, 14. Jahrg., 131 - 134 (1977).

015.023 Der Entwicklungsgedanke in der Astronomie und das Gravitationsgesetz.   W. Spickermann.
Astron. Schule, 14. Jahrg., 135 - 138 (1977).

015.024 Die Bewohnbarkeit des Weltalls.   J. Dorschner.
Astron. Schule, 14. Jahrg., 141 - 143 (1977).

015.025 Was Noah versäumte.   G. Doebel.
Sterne Weltraum, Jahrg. 16, 423 (1977).

015.026 On the timing of an interstellar communication.
W. I. McLaughlin.
Icarus, Vol. 32, 464 - 470 (1977).
By considering prominent events that are observable from both Earth and nearby stellar systems it is possible to establish common clocks that may be useful in estimating arrival times for signals of intelligent extraterrestrial origin. The geometry and statistics of a timing strategy are developed together with quantitative estimates of its effectiveness and limits on its application. Signal opportunities for several nearby Sun-like stars are calculated using the bright Nova Cygni 1975 as a clock.

015.027 The absence of extraterrestrials on Earth and the prospects for CETI.   D. W. Schwartzman.
Icarus, Vol. 32, 473 - 475 (1977).
The absence of extraterrestrials on Earth in spite of the probable existence of a "Galactic Club" is a result of our nearly unique position on the verge of becoming a member. This supports the view that we are under surveillance by extraterrestrial intelligence, and reduces the likelihood of contact by radiotelescopes.

015.028 Scandal in the heavens: renowned astronomer accused of fraud.   N. Wade.
Science, Vol. 198, 707 - 709 (1977).

015.029 Variations in the earth's orbit: pacemaker of the ice ages?   D. L. Evans, H. J. Freeland, with a response by J. D. Hays, J. Imbrie, N. J. Shackleton.
Science, Vol. 198, 528 - 530 (1977).

015.030 On the origin of the rotation of celestial bodies.
P. Savić.
Publ. Dep. Astron., Univ. Beograd, Fac. Sci., No. 6, (see 012.040), p. 5 - 8 (1976).

015.031 The human uses of space.   H. J. Smith.
Proc. Southwest Reg. Conf., Vol. 3, (see 012.043), p. 195 - 196 (1977).

015.032 Bemerkungen zum Begriff der "beobachterischen Erfahrung".   W. W. Spangenberg.
Sterne, 53. Band, 237 - 240 (1977).

015.033 Der Einfluß der astronomischen Fernsehsendungen auf die Bevölkerung.   P. von der Osten-Sacken.
Mitt. Astron. Ges., Nr. 42, p. 89 (1977).

015.034 Tides in the universe.   M. Karovska.
Vasiona, Vol. 25, 56 - 61 (1977). In Serbo-Croatian.

015.035 Jedna grafička metoda za odredivanje azimuta izlaza i zalaza Sunca (A graphical method for deter-mination of the azimuth of sunrise and sunset).   I. Vince.
Vasiona, Vol. 25, 66 - 67 (1977).

015.036 Mining the asteroids.   M. J. Gaffey, T. B. McCord.
Mercury, Vol. 6, No. 6, p. 1 - 6, 9 (1977).

015.037 A model for a non-chemical form of life: crystalline physiology.   J. Schneider.
Origins of Life, Vol. 8, 33 - 38 (1977).

On the habitability of Mars. An approach to plane-tary ecosynthesis.   See Abstr. 097.152.

Applied Mathematics, Physics

021 Mathematics, Computing

021.001 DOPPLR – a point positioning program using integrated Doppler satellite observations.
R. W. Smith, C. R. Schwarz, W. D. Googe.
Satellite Doppler positioning, (see 012.002), p. 839 - 890.

021.002 The microcomputer: a new tool in dynamical astronomy. R. A. Broucke.
Bull. American Astron. Soc., Vol. 9, 437 (1977). – Abstract.

021.003 The nested variance power spectrum.
A. J. Owens.
J. Geophys. Res., Vol. 82, 3315 - 3318 (1977).
The nested variance technique for calculating the power spectrum is sometimes more appropriate than other power spectral estimates for data in space physics. It is especially useful for power law spectra. A description of the method, derivation of its properties, and sample computer program are given.

021.004 Kometenbahn-Bestimmung mit dem HP 67.
V. Kasten.
Sterne Weltraum, Jahrg. 16, p. 299 - 300 (1977).

021.005 A program to calculate coronal emission line strengths. P. L. Dufton.
Comput. Phys. Commun., Vol. 13, 25 - 38 (1977).
This computer program is used to calculate the intensities of emission spectral lines from the upper solar atmosphere. The optically thin approximation is used and the validity of this assumption is checked by calculating line centre optical depths.

021.006 Radio recombination lines from H^+ regions and cold interstellar clouds: computation of the b_n factors.
M. Brocklehurst, M. Salem.
Comput. Phys. Commun., Vol. 13, 39 - 48 (1977).
Emission lines produced by the recombination of hydrogen and hydrogenic ions are observed from many astronomical sources; maser amplification is frequently present. The recombination line spectrum depends upon the populations of the energy levels of the emitting species. The present program computes the ratio, b_n, of the population of energy level n to the (known) population in thermodynamic equilibrium for given values of electron temperature and density. A background radiation field may be present. The results are accurate for the range of temperatures and densities associated with cold clouds, H^+ regions, and planetary nebulae ($10-20\,000$ K, $10^{-4}-10^6$ cm^{-3}).

021.007 Multistate molecular treatment of atomic collisions in the impact parameter approximation. II – Calculation of differential cross-sections from the transition amplitudes for the straight line case.
R. D. Piacentini, A. Salin.
Comput. Phys. Commun., Vol. 13, 57 - 62 (1977).

021.008 Computer simulations of planetary accretion dynamics: sensitivity to initial conditions.
R. Isaacman, C. Sagan.
NASA Tech. Mem., NASA TM X-3511, (see 012.010), p. 6 (1977). – Abstract.

021.009 Seven-place tables of trigonometric functions with argument in time. L. S. Khrenov.
Glavnaya redaktsiya fiziko-matematicheskoj literatury, Nauka, Moskva. 305 pp. (1976). In Russian. – Review in Astron. Zh. Akad. Nauk SSSR, Vol. 54, 1144 - 1145; 1977 (*V. K. Abalakin*).

021.010 An algorithm for recurrent calculation of gravitational acceleration. A. Drożyner.
Artif. Satell., Vol. 12, No. 2, p. 33 - 39 (1977).
In this paper an algorithm for recurrent calculation of gravitational acceleration in the motion of artificial Earth satellites is obtained. For the Earth's gravitational field Standard Earth II model is adopted.

021.011 Calculs astronomiques pour amateurs.
B. Morando.
Astronomie, Vol. 91, 391 - 396 (1977).

021.012 The advance from 2D electrostatic to 3D electromagnetic particle simulation. O. Buneman.
Computing in plasma physics and astrophysics, (see 012.028), p. 21 - 31 (1976).

021.013 Numerical solution of continuity equations.
J. P. Boris.
Computing in plasma physics and astrophysics, (see 012.028), p. 67 - 79 (1976).
This paper compares and contrasts the six general methods available for solving the time dependent continuity equation numerically.

021.014 Symbolic computation of nonlinear wave interactions on MACSYMA.
A. Bers, J. L. Kulp, C. F. F. Karney.
Computing in plasma physics and astrophysics, (see 012.028), p. 81 - 98 (1976).
The authors describe the use of a large symbolic computation system – MACSYMA – in determining approximate analytic expressions for the nonlinear coupling of waves in an anisotropic plasma. MACSYMA was used to implement the solution of a fluid plasma model nonlinear partial differential equations by perturbation expansions and subsequent iterative analytic computations. By interacting with the details of the symbolic computation, the physical processes responsible for particular nonlinear wave interactions could be uncovered and appropriate approximations introduced so as to simplify the final analytic result. Details of the MACSYMA system and its use are discussed and illustrated.

021.015 **Finite difference and finite element methods.**
K. W. Morton.
Computing in plasma physics and astrophysics, (see 012.028),
p. 99 - 108 (1976).

021.016 **Computation of Tokamak transport.**
C. Mercier, J. P. Boujot, F. Werkoff.
Computing in plasma physics and astrophysics, (see 012.028),
p. 109 - 119 (1976).
The complexity of the equations describing the transport
phenomena in Tokamaks requires the use of numerical codes
for simulation. These codes assume that the plasma is a multi-
fluid medium. They involve equations for the evolution of
electrons, light ions and current density, equations for heavy
ions (impurities) and elaborate models for the evolution of the
neutrals of light atoms. These codes allow testing of the existing theories and deduction of empirical rules and coefficients.
After the authors have presented in more detail the Fontenay-
aux-Roses code, they give examples of a few applications of
this code.

021.017 **Black hole physics illustrated in photon orbits.**
H. Cohn.
American J. Phys., Vol. 45, 239 - 241 (1977). – Abstr. in
Phys. Abstr., Vol. 80, Abstr. 33783 (1977).

021.018 **Algebraic computing and the Newman-Penrose
formalism in general relativity.**
S. J. Campbell, J. Wainwright.
Gen. Relativ. Gravitation, Vol. 8, 987 - 1001 (1977).

021.019 **Equivalence of some integrals of the radiation
theory.** T. T. Chia.
Astrophys. Space Sci., Vol. 46, 239 - 246 (1977).
A definite integral which occurs in radiation theory is
shown to be equal in value to another definite integral by
evaluating the flux from a spherically symmetrical radiating
sphere in two ways. As a corollary, an alternate proof of the
invariance of the specific intensity of a ray in empty space
along its path is presented. Furthermore, this equality leads to
the conversion of members of a class of indefinite and definite
integrals involving arbitrary functions of angle into other
integrals. These transformations facilitate the calculation of
some of these integrals which arise not only in the theory of
radiation, but in other physical situations with spherical or
axial symmetry.

021.020 **On the reduction of certain integrals occurring in
Kopal's Fourier theory of eclipsing binaries.**
I. Jurkevich, W. B. Heard.
Astrophys. Space Sci., Vol. 52, 237 - 238 (1977).

021.021 **Computational techniques for solar wind flows past
terrestrial planets – theory and computer programs.**
S. S. Stahara, D. S. Chaussee, B. C. Trudinger, J. R. Spreiter.
NASA Contract. Rep., NASA CR-2924, 5 + 130 pp. (1977).
Theoretical analysis and the development of user-oriented
computer programs were carried out for the purpose of developing computational techniques for predicting the interaction
of the solar wind with terrestrial planets. The procedures are
based on a single-fluid, steady, dissipationless, magnetohydrodynamic model and are appropriate for the calculation of
axisymmetric, supersonic, super-Alfvénic solar wind flow past
both magnetic and nonmagnetic planets.

021.022 **LENAM, a standard computer program for the
calculation of multi-dimensional radiative transfer,
and LENAM-P, an extended version including polarization.**
L. G. Stenholm.
Rep. Obs. Lund, No. 13, 59 pp. (1977).

021.023 **Symbolic algebraic computer programs. Part I. The
LISP programming language.**
A. Krasiński, M. Perkowski.
Postępy Astron., Tom 25, 203 - 211 (1977). In Polish.
The paper describes the role of computers applied to
symbolic algebraic formulae manipulation in physics and
astronomy. In the present first part of the work the LISP
programming language is described on an elementary level as
an example of a symbol manipulation language. Its application
to algebraic calculation is shortly described.

**Programmes for pocket calculators HP-67 and
HP-97 in the field of theoretical and observational astronomy.**
See Abstr. 003.033.

**Rechnen mit dem Taschenrechner: Ortsbestimmun-
gen auf Sonne und Planeten.** See Abstr. 031.252.

**Refraction correction for the reduction routines
"RA".** See Abstr. 031.341.

**An interactive system for the reduction of photo-
graphic data using FORTH.** See Abstr. 031.409.

Programmsystem Geodätische Astronomie.
See Abstr. 046.001.

**Naval Surface Weapons Center reduction and
analysis of Doppler satellite receivers using the Celest computer program.** See Abstr. 046.016.

**Variations in Doppler positions resulting from
differences in computer programs and tropospheric refraction
computations.** See Abstr. 046.017.

Computation of ideal MHD equilibria.
See Abstr. 062.023.

**Recent developments in the computational aspects
of MHD stability.** See Abstr. 062.024.

**Radiative transfer calculated from a Markov chain
formalism.** See Abstr. 063.006.

Convection in stars. See Abstr. 065.031.

**Three-dimensional random Earth atmospheres for
Monte Carlo trajectory analyses.** See Abstr. 082.023.

**Lunar surface sputter erosion: a Monte Carlo ap-
proach to microcrater erosion and sputter redeposition.**
See Abstr. 094.128.

**Large scale cratering of the lunar highlands: some
Monte Carlo model considerations.** See Abstr. 094.150.

**The application of a Bessel transform to the deter-
mination of stellar rotational velocities.**
See Abstr. 116.018.

022 Physical Papers Related to Astronomy and Astrophysics

022.001 Intensity measurements in the ν_4-fundamental of methane. F. K. Ko, P. Varanasi.
J. Quant. Spectrosc. Radiat. Transfer, Vol. 18, 145 - 150 (1977).

The absolute intensities of all the J-multiplets between $R(13)$ at 1375 cm^{-1} and $P(12)$ at 1225 cm^{-1}, in the ν_4-fundamental of $^{12}CH_4$, have been measured at 300°K. The values are consistent with published band-intensity measurements and also with the theoretical line strength tabulation by Fox.

022.002 Analysis of the $\nu_3 + \nu_4$ band of ammonia. S. Sarangi.
J. Quant. Spectrosc. Radiat. Transfer, Vol. 18, 257 - 288 (1977).

The 2-micron band of ammonia has been studied in the laboratory with a spectral resolution of 0.05 cm^{-1}. Most of the strong lines, belonging to the $\nu_3 + \nu_4$ (perpendicular) band have been identified and spectroscopic constants have been derived.

022.003 Measurements of line intensities in the two-micron band of ammonia. S. Sarangi.
J. Quant. Spectrosc. Radiat. Transfer, Vol. 18, 289 - 293 (1977).

Intensities of about four hundred lines of ammonia in the 2μ region have been measured using Doppler-broadened lines. Comparison with rigid-rotor calculations gives fair agreement between theory and experiment. The total integrated intensity of the $\nu_3 + \nu_4$ (perpendicular) band has been estimated to be 17.19 cm^{-2} atm^{-1} at 296 K.

022.004 A measurement of the width and shift of the Fe I 3719.94 Å line broadened by helium.
R. D. Driver, G. Lombardi.
Astron. Astrophys., Vol. 59, 299 - 301 (1977).

The authors present results of measurements carried out on a ballistic piston compressor of the width and shift of the Fe I line at 3719.94 Å broadened by helium. At a temperature of 4000°K they find $\gamma/n = (5.2 \pm 0.8) \, 10^{-9}$ rad s^{-1} cm^3 (FWHM) and the line is blue shifted by $\beta/n = (4.9 \pm 1.1) \, 10^{-10}$ rad s^{-1} cm^3. The Fe I line at 3722.56 Å was also studied and found to have the same width and shift. Comparison is made with previous experiments and calculations of neutral line broadening of Fe I lines. The astrophysical significance of these measurements is discussed.

022.005 Laboratory wavelengths of forbidden transitions in the spectrum of Mn V. A. N. Ryabtsev.
Astron. Zh. Akad. Nauk SSSR, Vol. 54, 919 - 920 (1977). In Russian. English translation in Soviet Astron., Vol. 21, No. 4.

The wavelengths for forbidden transitions between levels of the $3d^3$ configuration of Mn V have been calculated in the region 1591–10000 Å. A comparison is made with the spectrum of the planetary nebula NGC 7027.

022.006 Distribution of molecular weight in glyceride polymerizates or aggregates of them after contact with lunar grains. S. K. Asunmaa, R. Haack.
Moon, Vol. 16, 325 - 334 (1977).

Increase in the statistical average molecular weight up to 1200–1300 in different fractions of glycerides from 250–300 starting value was determined by gel permeation chromatography for fractions kept in thin layer contact with lunar grains for two years at 25°C.

022.007 Broadening of spectral lines by electron scattering. I. Methods of calculation.
D. I. Nagirner, V. G. Vedmich.
Astrofizika, Vol. 12, 437 - 449 (1976). In Russian. – English translation in Astrophysics, Vol. 12, No. 3.

Two methods are proposed for calculating line profiles formed in a semi-infinite atmosphere under simultaneous action of resonance and electron scattering. The first method is based on a two-dimensional linear integral equation for the emergent intensity which can be solved by successive approximations when electron scattering is relatively small. For the second method the intensity is divided into three parts which correspond to the continuous spectrum, electron scattering and resonance scattering. Iterations are also necessary but in this case for a system of two coupled one-dimensional linear integral equations. Explicit expressions are given for the solutions of these equations with known free terms.

022.008 On the quantum theory of the screening effect in thermonuclear reactions. I. Relativistic electron plasma. Yu. N. Redkoborodyj.
Astrofizika, Vol. 12, 495 - 510 (1976). In Russian. – English translation in Astrophysics, Vol. 12, No. 3.

The effect of electron screening of the Coulomb field of a nucleus is considered. It causes the increase of the thermonuclear fusion rate under high densities. The quantum theory of electron screening is generalized for the relativistic case. An analytic expression for the screened barrier potential is derived. The screening factor is a function of the density and the temperature of the electron plasma and is independent of the degree of degeneracy.

022.009 On the $A \, ^1\Pi - X \, ^1\Sigma^+$ band system in CH$^+$ and CD$^+$: theoretical spectroscopic constants and lifetimes.
N. Elander, J. Oddershede, N. H. F. Beebe.
Astrophys. J., Vol. 216, 165 - 173 (1977).

Polarization propagator calculation of spectroscopic constants and radiative lifetimes for the $A \, ^1\Pi - X \, ^1\Sigma^+$ band system are presented. The spectroscopic constants agree well with experimental and other theoretical values. The radiative lifetimes calculated for CD$^+$ are between 1.3% and 3.9% larger than the corresponding CH$^+$ lifetimes.

022.010 Laboratory rest frequencies for N$_2$D$^+$.
T. G. Anderson, T. A. Dixon, N. D. Piltch, R. J. Saykally, P. G. Szanto, R. C. Woods.
Astrophys. J., Lett., Vol. 216, L85 - L86 (1977).

The authors located the $J = 1 \leftarrow 0$ transition of N$_2$D$^+$ and measured the frequencies of the two most prominent hyperfine features using the same experimental arrangement employed earlier for N$_2$H$^+$.

022.011 Absorption coefficients of ices of CH$_4$, CO$_2$, NH$_3$, H$_2$S, H$_2$O and sulfuric acid.
G. T. Sill, U. Fink, J. Ferraro.
Bull. American Astron. Soc., Vol. 9, 466 (1977). – Abstract.

022.012 An improved fit to the inversion spectrum of ammonia. L. G. Young, A. T. Young.
Bull. American Astron. Soc., Vol. 9, 470 - 471 (1977). Abstract.

022.013 Molecular analysis of organic solids produced under simulated Jovian conditions.
B. N. Khare, C. Sagan, E. L. Bandurski, B. Nagy.
Bull. American Astron. Soc., Vol. 9, 476 (1977). – Abstract.

022.014 **A critical study of discrepancies between theoretical and experimental oscillator strengths for Cr I.**
U. Becker.
Astron. Astrophys., Vol. 60, 389 - 392 (1977).

Theoretical Hartree-Fock oscillator strengths for Cr I calculated by Froese Fischer and semi-empirical values computed by Biemont and by Kurucz and Peytremann are inconsistent with recently measured values. The main purpose of this paper is to resolve the established discrepancies. Therefore the radial integrals and the signs of the mixing coefficients given by Roth (1970) for the suitable transitions are determined by a fit to the experimental lifetime data of Becker et al. (1977), which are accurate to 3%. The gf-values calculated in this way are in good agreement with the experimental data and compared with the theoretical results of other authors.

022.015 **Cosmic radiation and fundamental problems in physics.** W. Heisenberg.
Usp. fiz. nauk, Vol. 121, 669 - 677 (1977). In Russian.
Abstr. in Ref. zh., 51. Astron., 8.51.88 (1977).

022.016 **Neutron capture by ^{208}Pb at stellar temperatures.**
R. L. Macklin, J. Halperin, R. R. Winters.
Astrophys. J., Vol. 217, 222 - 226 (1977).

Neutron capture cross section data for isotopically enriched ^{208}Pb were taken at the Oak Ridge electron linear accelerator time-of-flight facility and analyzed for nuclear resonance capture parameters up to 825 keV. Two new capture resonances at energies (43.29 and 47.26 keV) near stellar interior temperatures were found. The resonance parameter data lead to improved values for ^{208}Pb neutron capture probabilities calculated for a wide range of stellar interior temperatures.

022.017 **Hydrogen atom and hydrogen molecule ion in homogeneous magnetic fields of arbitrary strength.**
R. K. Bhaduri, Y. Nogami, C. S. Warke.
Astrophys. J., Vol. 217, 324 - 329 (1977).

A simple and accurate variational method is formulated and applied to the ground and excited states of the hydrogen atom as well as the hydrogen molecule ion in a homogeneous magnetic field of arbitrary strength. Explicit calculations are carried out for the $m_l = -1$ first excited state of hydrogen and the $(\Pi_u 2p)$-state of H_2^+. Numerical comparisons are made with some recent papers in the literature.

022.018 **The absorption spectrum of CO_2 around 7740 cm^{-1}.**
F. P. J. Valero, R. W. Boese.
J. Quant. Spectrosc. Radiat. Transfer, Vol. 18, 391 - 398 (1977).

022.019 **Estimates of Stark broadening of some Si (II) lines.**
J. D. Hey.
J. Quant. Spectrosc. Radiat. Transfer, Vol. 18, 425 - 431 (1977).

Recent measurements of spectral line widths for the first five multiplets in the visible of singly-ionized silicon, are compared with the corresponding values predicted by the semi-empirical method of Griem, as well as two other Stark broadening theories. Although there is considerable disagreement between different measurements, as well as with and between the theories involved, agreement to better than 30% on average is obtained between these calculations and the computations of Sahal-Bréchot.

022.020 **Photoelectric absorption spectra of methane (CH_4), methane and hydrogen (H_2) mixtures, and ethane (C_2H_6).** K. A. Dick, U. Fink.
J. Quant. Spectrosc. Radiat. Transfer, Vol. 18, 433 - 446 (1977).

Long path absorption spectra of methane have been recorded photoelectrically from 4200 to 10,600 Å. Visual comparison with spectra of Uranus shows that pure methane explains the absorption features of Uranus quite well and that large laboratory amounts are necessary to match the weak absorptions below 6200 Å. A brief description of the complexity of the levels responsible for the methane transitions is given.

022.021 **Band model analysis of laboratory methane absorption spectra from 4500 to 10500 Å .**
U. Fink, D. C. Benner, K. A. Dick.
J. Quant. Spectrosc. Radiat. Transfer, Vol. 18, 447 - 457 (1977).

Molecular band models are used to derive absorption and pressure coefficients for the methane absorption spectrum from 4500 to 10500 Å at intervals of 10 Å. These coefficients provide a necessary basis for the interpretation of the large methane absorptions in the atmospheres of the major planets. The effects of pressure on the absorption are surprisingly small, leading to large values of the pressure coefficient quite unlike any previous application of the band-model theory. They indicate a pseudo-continuum character of the methane spectrum throughout the visible and near infrared. A spectrum synthesis calculation using the derived coefficients shows the close fit to experimental data that can be realized.

022.022 **On the oscillator strengths of Fe I lines.**
A. G. Gasanalizade.
Soln. Dannye 1977 Byull., No. 5, p. 65 - 67 (1977). In Russian.

From a least-squares solution using 265 Fe I lines for the conversion of old oscillator strength values to the new system a linear relation has been obtained.

022.023 **On the number of lines observed in the hydrogen spectrum.** L. N. Kurochka.
Soln. Dannye 1977 Byull., No. 6, p. 90 - 97 (1977). In Russian.

The electron concentration determined by the number of the ultimate observed line does not only depend on the atom velocities but upon the continuous background radiation intensity as well. The latter decreases the line image contrast and the number of lines in the spectrum. The necessary calculation including the background radiation influence on the determination of electron concentration in solar phenomena is given.

022.024 **The analysis of satellites to the H-like ion resonance lines observed in the X-ray region.**
V. A. Boiko, A. Ya. Faenov, S. A. Pikuz, U. I. Safronova.
Mon. Not. R. Astron. Soc., Vol. 181, 107 - 120 (1977).

The present paper gives a detailed analysis of the intensity of satellites to the Lyman α lines of Mg XII to S XVI, observed in X-ray spectra of a dense laser plasma, where $T_e \cong 10^6 - 10^7$ K and $N_e \cong 10^{20} - 10^{21}$ cm^{-3}. The interpretation of the intensities of the lines, which are caused by transitions of the type $1snl' - 2lnl'$, was made on the basis of relativistic calculations of radiative transition probabilities and autoionization rates. Rather good agreement between theoretical and experimental data is obtained.

022.025 **Silicon carbide and the infrared excess of carbon stars.** J. Dorschner, C. Friedemann, J. Gürtler.
Astron. Nachr., Band 298, 279 - 283 (1977) = Mitt. Univ.-Sternw. Jena, Nr. 128.

This paper reports laboratory investigations on the absorption spectrum of SiC particles in the 10 μm region. The particles had a mean diameter of 0.31 μm. The experimental results indicate that a large part of the infrared excess observed in carbon stars may be caused by SiC.

022.026 **Narrow lines from alpha-alpha reactions.**
B. Kozlovsky, R. Ramaty.
Astrophys. Lett., Vol. 19, 19 - 24 (1977).

The authors have evaluated the intensities and spectral shapes of the 0.431 and 0.478-MeV lines of ^{7}Li and ^{7}Be resulting from $\alpha\alpha$ reactions. They also discuss the observability of the lines in view of their small widths which could be caused by both kinematical effects and the delayed nature of ^{7}Be decay.

022.027   Lifetime measurements of excited Co I levels.
W. D. Klotz, U. Becker, L. H. Göbel.
Astron. Astrophys., Vol. 61, 51 - 57 (1977).

In the region of 3500 Å the lifetimes of eight excited Co I levels have been measured by means of the zero field level crossing method. The influence of the line profile of the exciting resonance lines on the lifetimes has been investigated. The results are compared with those of other authors. Furthermore absolute oscillator strengths were calculated with known branching ratios and a new absolute scale has been established.

022.028   Stimulated emission of the He$^+$ radio recombination lines.   J. Weisheit, C. M. Walmsley.
Astron. Astrophys., Vol. 61, 141 - 144 (1977).

Departure coefficients for He$^+$ can readily be obtained from available tabulations for H. The authors have used this fact to estimate the non-LTE amplification factors to be expected for He$^+$ lines in gaseous nebulae. These turn out to be similar to those calculated for H lines at neighbouring frequencies and the authors can satisfactorily explain the 2-cm observations by Chaisson and Malkan (1976) of He$^+$ 121α and H 76α lines in NGC 7027.

022.029   Analysis of the theory of high-energy ion transport.   J. W. Wilson.
NASA Tech. Note, NASA TN D-8381. 28 pp. Price $ 4.00 (1977).

The development of analytical methods to study the transport properties of high-energy ions in materials is the topic of the present report. Procedures for the approximation of the transport of high-energy ions are discussed on the basis of available data on ion nuclear reactions. A straightahead approximation appears appropriate for space applications.

022.030   Molecular synthesis in interstellar clouds: some relevant laboratory measurements.
D. Smith, N. G. Adams.
Astrophys. J., Vol. 217, 741 - 748 (1977).

Data are presented which have resulted from a systematic program of measurements of the reaction rate coefficients and product distributions of positive ion/molecule reactions which are probably important reaction channels for the synthesis of molecular species observed in interstellar gas clouds. In particular, the reactions of the CH_n^+ and $C_2H_n^+$ ions ($n = 0$ to 4) with several molecules have received special attention. The relative abundances of some hydrogen-carbon-nitrogen molecules in Ori A and Sgr B2 are explained tentatively.

022.031   Oscillator strengths of transitions between low-lying S and P states of helium-like ions.
C. D. Lin, W. R. Johnson, A. Dalgarno.
Astrophys. J., Vol. 217, 1011 - 1015 (1977).

The relativistic random phase approximation is used to calculate oscillator strengths for the transitions $m\ ^1S_0 - n\ ^1P_1$ and $m\ ^3S_1 - n\ ^3P_1$, with $2 \le m$, $n \le 5$ of the helium-like ions of Mg, Al, S, Fe, and Ni. The accuracy is checked by comparing the results of the relativistic random phase approximation for helium-like neon with elaborate variational calculations.

022.032   Theoretical and experimental investigations of multipole radiation of atoms.
L. Augustyniak, K. Dunajski.
Postępy Astron., Tom 25, 85 - 104 (1977). In Polish.

Principal theoretical ideas of multipole radiation transitions are discussed. Some experimental results of spectro-

scopic investigations are also given.

022.033   Transition in Λ-doublets of molecules induced by collisions with ions.   D. Bouloy, A. Omont.
Astron. Astrophys., Vol. 61, 405 - 410 (1977).

The different hyperfine transition rates due to ions inside any Λ-doublet with $J = 3/2$ are exactly calculated by the semi-classical method. The results are applied to $^2\Pi_{3/2}$ ground level of OH. The exact transition rates are much smaller than given by the approximation of Rogers and Barrett (1968) and of Goss and Field (1968), in the case of heavy ions at low temperature.

022.034   Intensity and transmission measurements in the ν_3-fundamental of N_2O at low temperatures.
P. Varanasi, F. K. Ko.
J. Quant. Spectrosc. Radiat. Transfer, Vol. 18, 465 - 470 (1977).

Spectral transmission of i.r. radiation through the nitrogen-broadened lines of the ν_3-fundamental of N_2O has been measured of 154°, 202° and 300°K.

022.035   Intensities and widths of H_2O lines between 1800 and 2100 cm^{-1}.   Y. S. Chang, J. H. Shaw.
J. Quant. Spectrosc. Radiat. Transfer, Vol. 18, 491 - 499 (1977).

A non-linear, least-squares program was used to obtain the line intensities and widths of 91 air-broadened lines in the ν_2 rotation-vibration band of water vapor in the region from 1800 to 2100 cm^{-1}.

022.036   Transition probabilities in Nd(II) and the solar neodymium abundance.
R. S. Maier, W. Whaling.
J. Quant. Spectrosc. Radiat. Transfer, Vol. 18, 501 - 507 (1977).

022.037   Atomic transition probabilities for Sn(I), Sn(II) and Cl(I) lines in the 5300–6850 Å wavelength range.   T. Wujec, S. Weniger.
J. Quant. Spectrosc. Radiat. Transfer, Vol. 18, 509 - 514 (1977).

022.038   Strengths of H_2O lines in the 5000–5750 cm^{-1} region.
R. A. Toth, C. Camy-Peyret, J. M. Flaud.
J. Quant. Spectrosc. Radiat. Transfer, Vol. 18, 515 - 523 (1977).

Measurements of the strengths of 311 lines of water vapor have been made with high resolution in the region 5000–5750 cm^{-1}. The strength data of lines in the (011) and (110) bands are analyzed to determine the band strengths and the coefficients of the F factors.

022.039   Nitrogen-induced absorption of oxygen in the Herzberg continuum.   Shardanand.
J. Quant. Spectrosc. Radiat. Transfer, Vol. 18, 525 - 530 (1977).

Total absorption of O_2 induced by collisions with N_2 has been measured at room temperature in the Herzberg continuum using a one meter normal incidence grating monochromator. The enhanced absorption is ascribed to the formation of $O_2 - O_2$ and $O_2 - N_2$ dimers. The interaction constants for these dimers are determined and utilized to investigate their effect on the absorption of solar radiation in the stratosphere.

022.040   On the detectability of forbidden lines.
Y. Mekler, A. Eviatar.
J. Quant. Spectrosc. Radiat. Transfer, Vol. 18, 531 - 533 (1977).

The authors show that the traditional interpretation of

the non-detection of forbidden lines in dense natural plasmas is based on a misunderstanding. The central intensity of the line, taken to be a measure of detectability, is shown to have a maximum as a function of electron density and to be a linear function of electron density in tenuous plasmas.

022.041 On the ion-bombardment reduction mechanism.
L. Yin, T. Tsang, I. Adler.
Proc. Seventh Lunar Sci. Conf., (see 012.015), 891 - 900 (1976).

022.042 Solubility of Cr, Ti, and Al in co-existing olivine, spinel, and liquid at 1 atm.
J. Akella, R. J. Williams, O. Mullins.
Proc. Seventh Lunar Sci. Conf., (see 012.015), 1179 - 1194 (1976).

022.043 Partitioning of chromium between silicate crystals and melts. J. S. Huebner, B. R. Lipin, L. B. Wiggins.
Proc. Seventh Lunar Sci. Conf., (see 012.015), 1195 - 1220 (1976).

022.044 Fe and Mg in plagioclase.
J. Longhi, D. Walker, J. F. Hays.
Proc. Seventh Lunar Sci. Conf., (see 012.015), 1281 - 1300 (1976).

022.045 Sample size and sampling errors as the source of dispersion in chemical analyses.
U. S. Clanton, C. R. Fletcher.
Proc. Seventh Lunar Sci. Conf., (see 012.015), 1413 - 1428 (1976).

022.046 Further characterization of spectral features attributable to titanium on the moon.
R. G. Burns, K. M. Parkin, B. M. Loeffler, I. S. Leung, R. M. Abu-Eid.
Proc. Seventh Lunar Sci. Conf., (see 012.015), 2561 - 2578 (1976).

022.047 Electrical conductivity of orthopyroxene to 1400°C and the resulting selenotherm.
A. Duba, H. C. Heard, R. N. Schock.
Proc. Seventh Lunar Sci. Conf., (see 012.015), 3173 - 3181 (1976).

022.048 Gapped power spectra.
G. G. Fahlman, T. J. Ulrych.
J. R. Astron. Soc. Canada, Vol. 71, 404 (1977). — Abstract.

022.049 Absolute transition-probability data for the violet band system of CN.
L. L. Danylewych-May, R. W. Nicholls.
J. R. Astron. Soc. Canada, Vol. 71, 411 (1977). — Abstract.

022.050 Effect of perturbation on the periodic solutions of the Störmer problem.
V. V. Markellos, C. Zagouras.
Astron. Astrophys., Vol. 61, 505 - 514 (1977).
The periodic motions of a charged particle in the meridian plane of a magnetic dipole are examined when a perturbing force inversely proportional to the square of the distance (Dr^{-2}) is present. It is found that for the particular case considered ($D = 0.05$) only four families of simple-periodic oscillations, symmetric with respect to the equatorial axis, exist. The stability of the periodic oscillations is determined and attention is paid to the infinitesimal periodic motions in the neighborhood of each equilibrium point.

022.051 Proton induced resonances on ^{21}Ne.
H. L. Berg, W. Hietzke, C. Rolfs, H. Winkler.
Nucl. Phys. A, Vol. A276, No. 1, p. 168 - 188 (1977).
Abstr. in Phys. Abstr., Vol. 80, Abstr. 18540 (1977).

022.052 Collisional transfer between rotational levels of OCS.
L. J. Retallack, R. M. Lees, J. van der Linde.
J. Mol. Spectrosc., Vol. 63, 527 - 536 (1976). — Abstr. in Phys. Abstr., Vol. 80, Abstr. 19016 (1977).

022.053 On the Stark broadening of isolated lines of F(II) and Cl(III) by plasmas. J. D. Hey.
J. Quant. Spectrosc. Radiat. Transfer, Vol. 18, 649 - 655 (1977).
Recent measurements of Stark widths of isolated lines emitted by singly-ionized fluorine atoms and doubly-ionized chlorine atoms are compared with the corresponding values calculated by the semi-empirical impact approximation of Griem. A discussion is given of some of the difficulties which arise particularly when this method is applied to lines from multiply charged ions. On the basis of these comparisons, some new values are proposed for the threshold Gaunt factors of the ions under consideration, and these are compared with values deduced earlier for a number of other ions. Some systematic trends are observed.

022.054 Transition probability data for molecules of astrophysical interest. R. W. Nicholls.
Annu. Rev. Astron. Astrophys., Vol. 15, (see 003.012), 197 - 234 (1977).
This article is a review of transition probability data of molecular spectra, principally of diatomic molecular spectra that are of astrophysical importance. This topic is viewed in the context of contemporary spectroscopic research, which is advancing on two major fronts: (a) wavelength studies, from which atomic and molecular structure constants may be derived from the inferred energy-level separations, and (b) intensity studies, from which transition probability data and physico-chemical conditions in the light source or absorbing layer may be inferred.

022.055 Observations of Li I and Li II absorption spectra in the grazing incidence region.
A. M. Cantù, W. H. Parkinson, G. Tondello, G. P. Tozzi.
J. Opt. Soc. America, Vol. 67, 1030 - 1033 (1977).

022.056 Absorption spectrum of Pb I between 1350 and 2041 Å. C. M. Brown, S. G. Tilford, M. L. Ginter.
J. Opt. Soc. America, Vol. 67, 1240 - 1252 (1977).

022.057 Absolute transition probability data for the CN violet band system.
L. L. Danylewych-May, R. W. Nicholls.
J. Opt. Soc. America, Vol. 67, 1432 (1977). — Abstract.

022.058 Spectral radiances calibrations between 165—300 nm: an interlaboratory comparison.
J. M. Bridges, W. R. Ott, E. Pitz, A. Schulz, D. Einfeld, D. Stuck.
Appl. Opt., Vol. 16, 1788 - 1790 (1977).

022.059 Penning discharge as a photoelectric EUV spectroscopy source. E. S. Warden, H. W. Moos.
Appl. Opt., Vol. 16, 1902 - 1904 (1977).

022.060 Vacuum ultraviolet radiation scales: an accurate comparison between plasma blackbody lines and synchrotron radiation. P. J. Key, R. C. Preston.
Appl. Opt., Vol. 16, 2477 - 2485 (1977).

022.061 Méthodes d'analyse des observations. C. Magnan.
Perte de masse des étoiles, (see 012.030), 37 pp. (1977).

022.062 **Microwave spectra of molecules of astrophysical interest. XI. Silicon sulfide.**   E. Tiemann.
J. Phys. Chem. Ref. Data, Vol. 5, 1147 - 1156 (1976).
Abstr. in Phys. Abstr., Vol. 80, Abstr. 23493 (1977).

022.063 **New isotopes of interest to astrophysics.**
C. N. Davids, E. B. Norman, R. C. Pardo, L. A. Parks.
3rd international conference on nuclei far from stability.
Corgese, Corsica, France, 19 - 26 May 1976. CERN, Geneva, Switzerland. 8 + 608 pp. (1976). p. 590 - 592. — Abstr. in Phys. Abstr., Vol. 80, Abstr. 26635 (1977).

022.064 **Some O I oscillator strengths and the interstellar abundance of oxygen.**
C. J. Zeippen, M. J. Seaton, D. C. Morton.
Mon. Not. R. Astron. Soc., Vol. 181, 527 - 540 (1977).
    The authors present the oscillator strengths for the far ultraviolet resonance lines of O I. All the values were derived from calculations with the exception of those for multiplets 1, 2 and 5 which were obtained from lifetime experiments. The value for the inter-system line is supported both by theoretical work described in this paper and by the curve-of-growth constraints resulting from all three lines observed in absorption in the interstellar gas towards ζ Oph. The first astrophysical measurement of the O I intersystem line also is reported here.

022.065 **Inverting the ground state of interstellar CH.**
M. Elitzur.
Astrophys. J., Vol. 218, 677 - 686 (1977).
    It is shown that if the collision mechanism suggested by Gwinn and Townes is correct, the observed inversion of the ground state of interstellar CH can be explained as due to collisional excitation of the first rotational level.

022.066 **Lifetime measurements of the La II $y^3F_{4,3,2}^0$ levels with the beam-laser method. (*Solar photosphere La abundance*).**
A. Arnesen, A. Bengtsson, R. Hallin, T. Noreland.
J. Phys. B, Vol. 10, 565 - 568 (1977). — Abstr. in Phys. Abstr., Vol. 80, Abstr. 30827 (1977).

022.067 **Neutron branching in the reaction $^{12}C + ^{12}C$.**
R. Dayras, Z. E. Switkowski, S. E. Woosley.
Nucl. Phys. A, Vol. A279, No. 1, p. 70 - 84 (1977). — Abstr. in Phys. Abstr., Vol. 80, Abstr. 34336 (1977).

022.068 **Di-electronic recombination.**
M. J. Seaton, P. J. Storey.
Atomic processes and applications. P. G. Burke (Editor). North-Holland Publ. Co., Amsterdam, Netherlands, 10 + 533 pp. Price $ 65.95 (1976), p. 133 - 197. — Abstr. in Phys. Abstr., Vol. 80, Abstr. 36751 (1977).

022.069 **Applications to astrophysics: absorption spectra.**
W. Whaling.
Beam-foil spectroscopy. S. Bashkin (Editor). Springer-Verlag, Berlin—Heidelberg—New York. 12 + 318 pp. Price $ 28.30 (1976). ISBN 3-540-07914-9. p. 179 - 191. — Abstr. in Phys. Abstr., Vol. 80, Abstr. 41313 (1977).

022.070 **Stellar energy-loss rates due to S, P or T neutral currents.**   D. A. Dicus, E. W. Kolb.
Phys. Rev. D, Vol. 15, 977 - 982 (1977). — Abstr. in Phys. Abstr., Vol. 80, Abstr. 53577 (1977).

022.071 **β-decay of nuclear excited states.**
K. Yokoi, M. Yamada.
Prog. Theor. Phys., Suppl., No. 60, p. 161 - 170 (1976).
Abstr. in Phys. Abstr., Vol. 80, Abstr. 41927 (1977).

022.072 **Measurement of the internal pair emission branch of the 7.654 MeV state of ^{12}C, and the rate of the** stellar triple-α reaction.
P. G. H. Robertson, R. A. Warner, S. M. Austin.
Phys. Rev. C, Vol. 15, 1072 - 1079 (1977). — Abstr. in Phys. Abstr., Vol. 80, Abstr. 54183 (1977).

022.073 **Coupled states cross sections for rotational excitation of H_2CO by He impact at interstellar temperatures.**   B. J. Garrison, W. A. Lester, Jr.
J. Chem. Phys., Vol. 66, 531 - 536 (1977). — Abstr. in Phys. Abstr., Vol. 80, Abstr. 54593 (1977).

022.074 **Helium burning of ^{40}Ca.**
E. L. Cooperman, M. H. Shapiro, H. Winkler.
Nucl. Phys. A, Vol. A284, 163 - 176 (1977). — Abstr. in Phys. Abstr., Vol. 80, Abstr. 65230 (1977).

022.075 **Proton affinities and cluster ion stabilities in CO_2 and CS_2. Applications in Martian ionospheric chemistry.**   M. Meot-Ner, F. H. Field.
J. Chem. Phys., Vol. 66, 4527 - 4531 (1977). — Abstr. in Phys. Abstr., Vol. 80, Abstr. 77760 (1977).

022.076 **Adiabatic pressure dependence of the 2.7 and 1.9 μm water vapor bands.**
C. V. Mathai, W. L. Walls, S. Broersma.
J. Opt. Soc. America, Vol. 67, 1532 - 1537 (1977).
    Acoustic excitation was used to determine the adiabatic pressure derivative of the spectral absorptance of the 2.7 and 1.9 μm water vapor bands and its dependence upon the thermodynamic parameters concentration, pressure, and temperature.

022.077 **Refractivity and dispersion of hydrogen in the visible and near infrared.**   E. R. Peck, S. Huang.
J. Opt. Soc. America, Vol. 67, 1550 - 1554 (1977).
    The refractivity of natural hydrogen gas has been measured at eighteen wavelengths from 0.4047 to 1.6945 μm by means of a corner-reflector interferometer. The resulting refractivity at 0.5462252 μm, and at standard conditions, is 139.304×10^{-6}. Sixteen points of the data are well fitted by the dispersion formula $10^6 (n-1) = 21.113 + 12723.2/(111 - \sigma^2)$, where σ is wave number in reciprocal micrometers.

022.078 **Cross sections for (p, xn) reactions, and astrophysical applications.**   R. Silberberg, C. H. Tsao.
Astrophys. J., Suppl. Ser., Vol. 35, 129 - 136 (1977).
    Nuclear reactions in which the incident proton is absorbed or suffers charge exchange and x neutrons are emitted have very large cross sections at low energies. A knowledge of these cross sections is important in an environment subject to solar flare particles or in the radiation belts; yet relatively few of these cross sections have been measured. Such reactions are also likely to be important at various astrophysical sites. Accordingly, the authors have devised semiempirical formulae for calculating the unmeasured cross sections. The calculated cross sections agree well with the available experimental ones.

022.079 **Comparison of methods for calculating cross sections at high energies in astrophysics.**
R. Silberberg, C. H. Tsao.
Astrophys. J., Suppl. Ser., Vol. 35, 137 - 144 (1977).
    The relative merits of current techniques for calculating nuclear breakup cross sections at high energies are examined. The Monte Carlo calculations and the results obtained are described and are compared with those based on semiempirical methods. The prescriptions for calculating cross sections yet unmeasured are given.

022.080 **The quadrupole vibration-rotation transition probabilities of molecular hydrogen.**
J. Turner, K. Kirby-Docken, A. Dalgarno.
Astrophys. J., Suppl. Ser., Vol. 35, 281 - 292 (1977).

Transition probabilities are calculated for all the possible spontaneous electric quadrupole rotation-vibration transitions of the ground electronic state of molecular hydrogen originating in vibrational levels up to $v = 14$ and rotational levels up to $J = 20$.

022.081 **Hyperfine structures for astrophysically interesting levels of Mn I.**   T. G. R. Beynon.
Astron. Astrophys., Vol. 61, 853 - 857 (1977).

This paper attempts to clarify and extend laboratory hyperfine structure data for Mn I. Two methods are used: (1) re-analysis of old optical dáta in the light of recent measurements; (2) cautious use of semi-empirical calculation in combination with fragmentary experimental data. Reliable hyperfine structures for many lines eminently suitable for stellar abundance analysis result.

022.082 **The effects of recombination Balmer decrements of collisional and self-absorption processes.**
S. A. Drake.
Publ. Astron. Soc. Pacific, Vol. 89, 616 (1977). – Abstract.

022.083 **Laser-induced fluorescence in TiO: lifetimes, collisions, and forbidden transitions.**   J. Feinberg.
Publ. Astron. Soc. Pacific, Vol. 89, 616 (1977). – Abstract.

022.084 **Recent advances in the calculation of oscillator strengths.**   A. Hibbert.
Phys. Scr., Vol. 16, 7 - 12 (1977).

A review of work published since the last beam-foil spectroscopy conference in Gatlinburg (1975) is presented. For allowed transitions where LS coupling is valid, the NCMET and FOTOS schemes are compared and an extension of the application of FOTOS is proposed. Where relativistic effects are important—for medium to heavy atoms and for forbidden transitions—the stability of coefficients in configuration interaction expansions, and the way in which the Breit or Breit-Pauli terms are introduced, are discussed.

022.085 **f-value measurements for $3d$-elements.**
M. C. E. Huber.
Phys. Scr., Vol. 16, 16 - 30 (1977).

The author gives a survey of data on transition probabilities of allowed lines that belong to the spectra of neutral and singly-ionized iron-group elements. The classical methods used to determine oscillator strengths of weak lines (usually on a relative scale) are reviewed and some of the difficulties arising in investigating the spectra of $3d$-elements are pointed out. The quality of experimental lifetimes and f-values of strong lines, i.e., of data which are frequently used to establish absolute scales, are discussed in the context of the methods employed. The results on weak lines are then assessed, element by element, and some applications are mentioned.

022.086 **Oscillator strengths for some Mn II lines and the solar Mn abundance.**
I. Martinson, L. J. Curtis, P. L. Smith, E. Biémont.
Phys. Scr., Vol. 16, 35 - 38 (1977).

Oscillator strengths are given for transitions belonging to the $a^5S-z^5P^0$ and $a^5D-z^5P^0$ multiplets in Mn II. Theoretical f-values are also given for these transitions, the agreement between experiment and theory being satisfactory. Using the experimental gf-values obtained in this work and empirical solar models, the solar abundance of Mn has been determined to be $\log N_{Mn} = 5.4 \pm 0.2$ (on the $\log N_H = 12.00$ scale).

022.087 **Recent advances in studies of molecular transition probabilities.**   P. Erman.
Phys. Scr., Vol. 16, 60 - 64 (1977).

022.088 **Radiative recombination in some ions of astrophysical interest.**

U. Narain, H. P. Mital, S. Chandra.
Sol. Phys., Vol. 52, 417 - 422 (1977).

Radiative recombination coefficients for some quadruply and quintuply ionized atoms, present in the Sun and its atmosphere, are investigated in the temperature range $10-10^4$ K by using the method of detailed balance. Simple expressions are given for a quick estimation.

022.089 **High n solar radio recombination lines.**   A. Greve.
Sol. Phys., Vol. 52, 423 - 427 (1977).

For a representative set of atmospheric and atomic parameters the author determines the strengths of solar dielectronic recombination lines originating in ions with $Z \leqslant 6$ at frequencies of $\gtrsim 70$ GHz. He compares the line strengths derived here with those calculated by Berger and Simon (1972).

022.090 **New measurements of the Se I resonance lines.**
B. Lindgren, H. P. Palenius.
Sol. Phys., Vol. 53, 347 - 352 (1977).

Se I resonance lines have been measured in absorption to an accuracy of a few milliångstroms by using the flash photolysis technique. As a result it was found that the $4p^3nl$ levels and the ionization limit given by Morillon and Vergès should be increased by 0.23 ± 0.02 cm^{-1}. Calculated wavelengths are given for Se I lines which may be searched for in the solar spectrum in order to find selenium in the Sun.

022.091 **Cosmic tests of Maxwell's equations. I: A photon rest mass.**   J. C. Byrne.
Astrophys. Space Sci., Vol. 46, 115 - 132 (1977).

In this paper the author shows that the upper limit on the photon rest mass, as established by laboratory and terrestrial data, does not rule out the possibility that Maxwell's equations are not applicable to large scale astrophysical phenomena.

022.092 **Rotational dynamics of a deformable medium.**
V. S. Geroyannis, J. N. Tokis.
Astrophys. Space Sci., Vol. 51, 409 - 427 (1977).

The aim of the present investigation has been to derive from the fundamental Cauchy's first law of continuum mechanics the explicit form of the Eulerian general equation which governs the three-axial generalized rotation about the centre of mass of a self-gravitating deformable finite material continuum, viscolinear (i.e., Newtonian) or not, consisting of compressible fluid of arbitrary viscosity, in an external field of force. The generalized rotation is a superposition of the so-called rigid-body (i.e., time dependent only) rotation of the continuum plus a nonrigid-body (i.e., position-time dependent) rotation of its configurations.

022.093 **Analysis of the $2p^43s$, $2p^43p$ and $2p^43d$ configurations of five-times ionized silicon (Si VI).**
M.-C. Artru, W.-Ü. L. Brillet.
Phys. Scr., Vol. 16, 93 - 98 (1977).

The Si VI spectrum has been investigated in the vacuum UV. About seventy new lines (690–1344 Å) have been identified as transitions between the $2p^43s$, $2p^43p$ and $2p^43d$ configurations, leading to the determination of energy levels belonging to these configurations. Intercombination lines between terms of different multiplicities or different parentages are observed. The ground-term transitions from the $2s2p^6$ level (246–249 Å) and from the $2p^43s$ configuration (91–101 Å) have been remeasured.

022.094 **Analysis of the P VI spectrum.**
M. Eidelsberg, M.-C. Artru.
Phys. Scr., Vol. 16, 109 - 113 (1977).

The spectrum of five-times ionized phosphorus (P VI) is observed in the 400–1500 Å wavelength range. A total of 76 new lines are measured and classified as transitions between the $2p^53s$, $3p$, $3d$ and $4f$ configurations. All energy levels in these configurations are determined. Their identification is supported by the results of parametric calculations and by an isoelectronic comparison

022.095   The other rotamer of formic acid, *cis*-HCOOH.
W. H. Hocking.
Z. Naturforsch., Band 31 a, 1113 - 1121 (1976) = Max-Planck-Inst. Radioastron. Bonn, Sonderdr. Ser. A, Nr. 142.

022.096   The molecular structures of HNC and HCN derived from the eight stable isotopic species.
E. F. Pearson, R. A. Creswell, M. Winnewisser, G. Winnewisser.
Z. Naturforsch., Band 31 a, 1394 - 1397 (1976) = Max-Planck-Inst. Radioastron. Bonn, Sonderdr. Ser. A, Nr. 153.

022.097   Inversion of the K_a = 0 and 1 rotational levels in the lowest excited vibrational state of HNCS.
K. Yamada, M. Winnewisser, G. Winnewisser, L. B. Szalanski, M. C. L. Gerry.
J. Mol. Spectrosc., Vol. 64, 401 - 414 (1977) = Max-Planck-Inst. Radioastron., Bonn, Sonderdr. Ser. A, Nr. 164:

022.098   Preliminary results of a measurement of the H^- photodetachment cross section in the continuum region.   H. Sharifian, H. C. Bryant, J. Donahue, H. Tootoonchi, P. A. M. Gram, J. C. Pratt, M. A. Yates-Williams.
Proc. Southwest Reg. Conf., Vol. 3, (see 012.043), p. 47 - 49 (1977).
The authors have measured the H^- photodetachment relative cross section in the photon energy range from 1.7 to 12 eV. This paper reports preliminary results on the cross section in the continuum.

022.099   Observation of resonances in the photodetachment cross section of the H^- ion near 11 eV.
H. Tootoonchi, H. C. Bryant, J. Donahue, H. Sharifian, P. A. M. Gram, J. C. Pratt, M. A. Yates-Williams.
Proc. Southwest Reg. Conf., Vol. 3, (see 012.043), p. 51 - 54 (1977).
Using a colliding beam method, two resonances in the photodetachment of electrons from H^- near 11 eV have been studied. The center-of-mass energy of 1.7 to 12 eV is achieved by directing a nitrogen laser beam at variable angle across a 800 MeV H^- beam.

022.100   Determination of the speed of light by absolute wavelength measurement of the $R(14)$ line of the CO_2 9.4-μm band and the known frequency of this line.
J.-P. Monchalin, M. J. Kelly, J. E. Thomas, N. A. Kurnit, A. Szöke, A. Javan, F. Zernike, P. H. Lee.
Opt. Lett., Vol. 1, 5 - 7 (1977).
A precision long-arm scanning Michelson interferometer system is described that is capable of measuring absolute laser wavelength to within several parts in 10^9 in the 10-μm spectral range and to within several parts in 10^{11} in the visible range. The $R(14)$ line of the CO_2 9.4-μm band is measured to be 9.305 385 613 (70) μm. This measured value and the known frequency of this line give a value for the speed of light: c = 299 792 457.6 (2.2) m/sec, in agreement with the recent independent measurements of c and its recommended value.

022.101   Precision measurements of NH_3 spectral lines near 11 μm using the infrared heterodyne technique.
J. J. Hillman, T. Kostiuk, D. Buhl, J. L. Faris, J. C. Novaco, M. J. Mumma.
Opt. Lett., Vol. 1, 81 - 83 (1977).

022.102   Measuring the wavelength of light with a self-calibrating grating.   T. W. Hänsch.
Opt. Lett., Vol. 1, 191 - 193 (1977).
A novel scheme for making absolute wavelength measurements with a grating spectrograph is proposed. A grating with

special multiple rulings is illuminated with a reference laser of known wavelength to project a ruler-like diffraction pattern of equidistant wavelength calibration lines directly onto the unknown spectrum.

022.103   X-ray and gamma-ray line production by nonthermal ions.   R. W. Bussard, K. Omidvar, R. Ramaty.
GSFC Doc. X-660-77-144, Prepr. 35 pp. (1977).
The authors have calculated X-ray production at ~6.8 keV by the 2p to 1s transition in fast hydrogen- and helium-like iron ions, following both electron capture to excited levels and collisional excitation. The effective X-ray line production cross-section was found to be sharply peaked in energy at about 8 to 12 MeV/amu. Since fast ions of similar energies can also excite nuclear levels, the authors have calculated the ratio of selected strong γ-ray line emissivities to the X-ray line emissivity. They use these calculations to set limits on the intensity of γ-ray line emission from the galactic center and the radio galaxy Centaurus A, and find that these limits are generally lower than those reported in the literature.

022.104   Comment on the use of the Hanle effect for magnetic field determinations.   S. I. Gopasyuk.
Rep. Obs. Lund, No. 12, (see 012.044), p. 63 (1977).

022.105   Calculation of the transition probability for C III in a single configuration approximation.
A. A. Nikitin, A. F. Kholtygin, T. Kh. Feklistova.
Publ. Tartu Astrofiz. Obs., Vol. 45, 63 - 69 (1977). In Russian.
Oscillator strengths and transition probabilities in the C III spectra have been found for astrophysically interesting lines, using the parameters of the generalized hydrogenic wave functions. Employing semi-empirical formulae, oscillator strengths states have been approximately estimated for transitions into the continuum.

022.106   Recombination spectrum of C III.   A. A. Nikitin, A. A. Sapar, T. Kh. Feklistova, A. F. Kholtygin.
Publ. Tartu Astrofiz. Obs., Vol. 45, 257 - 274 (1977). In Russian.
The recombination spectrum of C III has been calculated for 23 lower singlet and 27 lower triplet states for the temperature range 5000 - 100000°K. Some general features of ion recombination spectra are analyzed. The results are compared with the spectrum of NGC 7027.

022.107   He and Ne cross sections in natural Al and Mg targets bombarded with 18 to 72 MeV protons.
P. Pulfer, J. Beer, F. Bühler.
Meteoritics, Vol. 12, 342 (1977). – Abstract.

022.108   Production cross sections of stable and radioactive isotopes of geophysical interest.
P. Pulfer, J. Beer, F. Bühler.
Meteoritics, Vol. 12, 342 - 343 (1977). – Abstract.

022.109   Electron impact on atmospheric gases. 1. Updated cross sections.
C. H. Jackman, R. H. Garvey, A. E. S. Green.
J. Geophys. Res., Vol. 82, 5081 - 5090 (1977).
The authors update the analytic characterizations of electron impact cross sections for important atmospheric gases (namely, O_2, N_2, O, CO, CO_2, and He). With these cross sections it is simple to communicate massive quantities of experimental and theoretical results.

022.110   Electron impact on atmospheric gases. 2. Yield spectra.
A. E. S. Green, C. H. Jackman, R. H. Garvey.
J. Geophys. Res., Vol. 82, 5104 - 5111 (1977).
The authors introduce a concept 'yield spectrum' and

calculate this two-dimensional function using a modified discrete energy bin method for 50-eV to 10-keV incident electrons impacting on the gases Ar, H_2, H_2O, O_2, N_2, O, CO, CO_2, and He. The yield spectrum is amenable to physical interpretation, accurate analytic representation, and convenient application to the determination of all types of yields needed in aeronomical problems.

022.111 **Tables of spectral-line intensities. Part I, II — arranged by elements.** Second edition.
W. F. Meggers, C. H. Corliss, B. F. Scribner.
NBS Monogr. 145. For sale by the Superintendent of Documents, U.S. Government Printing Office, Washington, D.C. 20402. Part I: 15 + 387 pp. Price $ 8.55. Part II: 15 + 213 pp. Price $ 6.80 (1975).

The intensity, character, wavelength, spectrum, and energy levels of 39 000 lines between 2000 Å and 9000 Å observed in copper arcs containing 0.1 atomic percent of each of 70 elements.

022.112 **The first spectrum of hafnium (Hf I).**
W. F. Meggers, C. E. Moore.
NBS Monogr. 153, for sale by the Superintendent of Documents, U.S. Government Printing Office, Washington, D.C. 20402. 117 pp. Price $ 1.35 (1976).

The present publication terminates the work on the analysis of Hf I which was started by the late W. F. Meggers in 1928 and left unfinished in 1966. His final line list contains some 4700 lines of which about 67 percent have been classified. Observed g-values are known for 198 levels. The reliability of the Zeeman observations is indicated in tables containing sums of observed and Landé g-values for selected groups of "even" and "odd" terms. An attempt has been made to continue Meggers' analysis in LS-coupling as far as possible. This coupling is not rigorous in Hf I, and many intervals are irregular. Consequently, the levels are given also in numerical order with the even and odd levels presented in separate tables. An ionization limit of 54700 ± 600 cm^{-1}, giving an ionization potential of 6.78 ± 0.07 eV has been derived from a two-member series.

022.113 **Soft X-ray spectrum of a hot plasma.**
J. C. Raymond, B. W. Smith.
Astrophys. J., Suppl. Ser., Vol. 35, 419 - 439 (1977).

The authors have calculated the spectrum of radiation emitted by a hot, optically thin plasma with abundances and equilibrium ionization balance appropriate to interstellar conditions. They discuss the results at wavelengths shorter than 200 Å for material in the range of electron temperature $1.6 \times 10^5 \leqslant T \leqslant 1.0 \times 10^8$.

022.114 **Emission cross sections for rotational transitions of molecules of astrophysical interest.**
S. Ames, W. F. Huebner.
Los Alamos Sci. Lab., N. Mex., LA−6366.130 pp. (1977). Abstr. from INIS7720 333688.

022.115 **Lifetime measurements with the beam-laser method.**
A. Arnesen, A. Bengtsson, R. Hallin, J. Lindskog, C. Nordling, T. Noreland.
Sep. print Uppsala Univ., Sweden, Fys. Instn. UUIP−952. 13 pp. (1977). − Abstr. from INIS7724 344467.

022.116 **Evolution of matter in the universe.** I. Aničin.
Vasiona, Vol. 25, 73 - 82 (1977). In Serbo-Croatian.

022.117 **The strength of lunar analogues and its geophysical implications.** H. Mizutani, H. Spetzler.
10th Lunar and Planetary Symposium, (see 012.050), p. 26 - 30 (1977).

The present paper reports experimental and theoretical studies of the effect of an ultrahigh vacuum with attendant low water vapor pressure on the stress-strain relation of rocks. This information is found to be essential in order to understand geologic and tectonic processes of the moon and other planets.

022.118 **High-velocity impact into basaltic and metallic targets.**
A. Fujiwara, G. Kamimoto, A. Tsukamoto.
10th Lunar and Planetary Symposium, (see 012.050), p. 36 - 40 (1977). In Japanese.

022.119 **Estimation of the absolute cooling rate of rocks from the width of exsolved pyroxene and its application to eucrites.** M. Miyamoto.
10th Lunar and Planetary Symposium, (see 012.050), p. 41 - 46 (1977).

022.120 **Measurement of the absorption coefficient on silicates.** C. Koike, H. Hasegawa, N. Asada.
10th Lunar and Planetary Symposium, (see 012.050), p. 60 - 65 (1977). In Japanese.

022.121 **Rare gas occlusion into grains during their growth— an approach to study planetary formation.**
M. Honda, M. Ozima.
10th Lunar and Planetary Symposium, (see 012.050), p. 122 - 125 (1977).

The authors have made an experiment on rare gas occlusion into grains during their growth, whose mechanism has been suggested by Arrhenius and Alfvén (1971). Although argon seems to be trapped loosely in fine grains, it may be redistributed to more retentive sites by compression of grains.

022.122 **A Monte Carlo calculation of high-energy sputtering.**
T. Onaka, F. Kamijo.
Japanese J. Appl. Phys., Vol. 16, 559 - 564 (1977) = Contrib. Dep. Astron. Univ. Tokyo No. 219.

Sputtering processes of high-energy incident ions are calculated numerically by using the Monte Carlo method under the assumptions of random arrangement and no inelastic collisions. In the high-energy region, the unscreened Coulomb potential is used, while the hard-sphere type is adopted in the low-energy region. The simulation has been performed for the bombardment by deuterons with energies of 20 keV to 1 MeV on a silver target. The calculated sputtering ratios agree well with the experimental values. On the basis of these results a new formula for the sputtering ratio is derived by modifying Pease's formula.

022.123 **The absorption spectrum of europium.**
G. Smith, F. S. Tomkins.
Philos. Trans. R. Soc. London, Ser. A, Vol. 283, 345 - 365 (1976) = Univ. Oxford, Dep. Astrophys., Publ. No. 156.

022.124 **On a mechanical model of the Bopp-Podolsky potential.** H. Günther.
Ann. Physik, 7. Folge, Band 33, 448 - 454 (1976) = Zentralinst. Astrophys., Sternw. Babelsberg, Mitt. Neue Folge Nr. 169.

Atomic energy levels and Grotrian diagrams. Vol. I. Hydrogen I − phosphorus XV. See Abstr. 003.029.

Chemical kinetics. See Abstr. 003.074.

From quarks to quasars. An outline of modern physics. See Abstr. 003.154.

Multistate molecular treatment of atomic collisions in the impact parameter approximation. II −Calculation of differential cross-sections from the transition amplitudes for the straight line case. See Abstr. 021.007.

Equivalence of some integrals of the radiation

theory.   See Abstr. 021.019.

f-values and abundances of the elements in the sun and stars.   See Abstr. 071.024.

Statistical equilibrium in cometary C_2. I. See Abstr. 102.007.

Astronomical Instruments and Techniques

031 Astronomical Optics, Methods of Observation and Reduction, Data Processing, Automation

Astronomical Optics

031.001 Shorter than a Schmidt. C. G. Wynne.
Mon. Not. R. Astron. Soc., Vol. 180, 485 - 490 (1977).

This paper is concerned with spectrograph cameras. It is shown that a system consisting of an aspheric plate, a spherical mirror and a self-achromatic lens some distance in front of the focus can be corrected for spherical aberration, coma, astigmatism, field curvature and chromatic difference of focus over a wide spectral range. Compared with a Schmidt camera of the same focal length, these systems have an overall length that is considerably shorter, by an amount that can be controlled in the design. Numerical data are given for three designs, of relative aperture $f/2.5$.

031.002 Mathematische Grundlagen und genaue Formeln für die Vermessung von Parabolspiegeln.
A. Korhammer.
Orion, 35. Jahrg., 131 - 136 (1977).

031.003 A Calver mirror remounted. G. K. Moore.
J. British Astron. Assoc., Vol. 87, 478 - 481 (1977).

031.004 On the imaging properties of holographic gratings.
C. H. F. Velzel.
J. Opt. Soc. America, Vol. 67, 1021 - 1027 (1977).

The author presents a theory of the aberrations of gratings that are recorded holographically. Stigmatic imaging at one or two wavelengths is possible with such a grating. He gives a summary of the cases of stigmatism, including some newly found. The general principles of the correction of aberrations of holographic recording are described.

031.005 Apodization for maximum Strehl ratio and specified Rayleigh limit of resolution. J. E. Wilkins, Jr.
J. Opt. Soc. America, Vol. 67, 1027 - 1030 (1977).

031.006 Correlation between angle-of-arrival fluctuations on the entrance pupil of a solar telescope.
J. Borgnino, F. Martin.
J. Opt. Soc. America, Vol. 67, 1065 - 1072 (1977).

The longitudinal and transverse correlation functions for the fluctuations in the angle of arrival on the entrance pupil of a solar telescope have been estimated. The results are found to be consistent with the inertial model of turbulence. Values of the integral $\int C_N^2 \, dh$ of the structure constant over the lower atmospheric layers and of Fried's seeing parameter r_0 are deduced. They are found to be consistent with other independent estimations made with the same telescope.

031.007 Design of single element gradient-index collimator.

D. T. Moore.
J. Opt. Soc. America, Vol. 67, 1137 - 1143 (1977).

Gradient-index materials will provide new valuable degrees of freedom for the lens designer. The gradient-index single element lens is studied in terms of its third-order aberration coefficients, ray-trace analysis, and wave-front error.

031.008 Catadioptric system with a gradient-index corrector plate. D. T. Moore.
J. Opt. Soc. America, Vol. 67, 1143 - 1146 (1977).

Sands showed the equivalence of the third-order aspheric surface contribution of the aberration polynomial to the third-order inhomogeneous surface contribution. This fact is exemplified by the substitution of an axial gradient for the spheric surface of the Schmidt corrector plate. It is shown that a gradient-index corrector plate system is limited in its off-axis performance by fifth-order oblique spherical aberration just as the conventional Schmidt system is.

031.009 Cassegrain alignment using computer-derived adjustment parameters. I. M. Egdall.
J. Opt. Soc. America, Vol. 67, 1365 (1977). − Abstract.

031.010 Transfer function characterization of deformable mirrors. J. E. Harvey, G. M. Callahan.
J. Opt. Soc. America, Vol. 67, 1367 (1977). − Abstract.

031.011 Dependence of the finesse of a Fabry-Perot on the plate separation.
C. Roychoudhuri, A. Cornejo.
J. Opt. Soc. America, Vol. 67, 1385 (1977). − Abstract.

031.012 Inexpensive anastigmatic telescope.
D. R. Shafer.
J. Opt. Soc. America, Vol. 67, 1395 (1977). − Abstract.

031.013 Four-mirror unobscured anastigmatic telescope with all spherical surfaces. D. R. Shafer.
J. Opt. Soc. America, Vol. 67, 1395 (1977). − Abstract.

031.014 Three-color achromatism in the extended visible range. H. F. Bennett.
J. Opt. Soc. America, Vol. 67, 1428 (1977). − Abstract.

031.015 Lens design using spherical gradients.
S. D. Fantone, D. T. Moore.
J. Opt. Soc. America, Vol. 67, 1428 (1977). − Abstract.

031.016 Unified description of holographic grating diffraction. R. Magnusson, T. K. Gaylord.
J. Opt. Soc. America, Vol. 67, 1438 (1977). − Abstract.

031.017 The efficiencies of plane diffraction gratings meas-

ured in the conical and classical mountings.
M. Neviere, D. Maystre, W. R. Hunter.
J. Opt. Soc. America, Vol. 67, 1438 - 1439 (1977). – Abstract.

031.018 Measuring concave diffraction grating efficiencies at grazing incidence.   W. R. Hunter, D. K. Prinz.
J. Opt. Soc. America, Vol. 67, 1439 (1977). – Abstract.

031.019 Diffraction in the zonal Foucault test.
J. M. Simon.
Appl. Opt., Vol. 16, 1782 - 1783 (1977).

031.020 Correcting glass-angle of a prism.   A. S. De Vany.
Appl. Opt., Vol. 16, 2019 - 2021, with an erratum,
p. 2787 (1977).

031.021 Simultaneous optical and X-ray imaging telescopes.
R. C. Harney.
Appl. Opt., Vol. 16, 2039 - 2040 (1977).

031.022 Anastigmatic three-mirror telescope.   D. Korsch.
Appl. Opt., Vol. 16, 2074 - 2077 (1977).
A new configuration of a three-mirror telescope is introduced that combines high performance with practicality. A
geometric spot size of less than 0.1 sec of arc in an easily ac-
cessible flat field of 1.5° and excellent stray light suppression
are the outstanding features.

031.023 Primary aberrations for grazing incidence.
C. E. Winkler, D. Korsch.
Appl. Opt., Vol. 16, 2464 - 2469 (1977).
Beginning with exact relations between object and image
coordinates for a single reflective surface, a systematic analysis
of general grazing incidence systems is presented. A complete
set of primary aberrations for single-element and two-element
systems is developed. The importance of a judicious choice
for a coordinate system in showing field curvature to be
clearly the predominant aberration for a two-element system
is discussed. The validity of the theory is verified through
comparisons with the exact ray-trace results for the case of a
telescope.

031.024 Unit magnification optical system.   S. Rosin.
Appl. Opt., Vol. 16, 2568 - 2571 (1977).
A simple self-conjugate optical arrangement, originally
proposed by Dyson, consists of a spherical mirror and a plano-
convex lens. Despite the fact that all the Seidel aberrations
are corrected, the field coverage is severely limited by higher
order tangential astigmatism. This paper discusses modifications to improve the field coverage. It also treats the creation
of internal reflecting surfaces to separate the object and image
planes. Finally a few possible applications are discussed, including one to the creation of a high speed spectrograph.

031.025 Null Ronchi gratings from spot diagrams.
G. W. Hopkins, R. N. Shagam.
Appl. Opt., Vol. 16, 2602 - 2603 (1977).

031.026 Grating efficiency theory as it applies to blazed and holographic gratings.
E. G. Loewen, M. Nevière, D. Maystre.
Appl. Opt., Vol. 16, 2711 - 2721 (1977).

031.027 Infrared optical glasses for applications in 8−12 μm thermal imaging systems.
J. A. Savage, P. J. Webber, A. N. Pitt.
Appl. Opt., Vol. 16, 2938 - 2941 (1977).

031.028 Perspex lenses and mirrors.   J. Wall.
J. British Astron. Assoc., Vol. 88, 28 - 31 (1977).

031.029 Computer simulations of the image synthesis for a strip telescope.   N. Baba, K. Murata.
Opt. Acta, Vol. 24, 189 - 199 (1977). – Abstr. in Phys. Abstr.,
Vol. 80, Abstr. 29660 (1977).

031.030 Designing and testing germanium optics.   F. G. Back.
Electro-Opt. Syst. Des., Vol. 8, No. 5, p. 33 - 35
(1976). – Abstr. in Phys. Abstr., Vol. 80, Abstr. 39011 (1977).

031.031 A matrix treatment of primary aberrations.
P. B. Kosel.
Opt. Acta, Vol. 24, 757 - 772 (1977). – Abstr. in Phys. Abstr.,
Vol. 80, Abstr. 50622 (1977).

031.032 An elementary derivation of phase fluctuations of an optical wave in the atmosphere.   H. T. Yura.
Imaging through the atmosphere, (see 012.032), p. 9 - 15
(1976). – Abstr. in Phys. Abstr., Vol. 80, Abstr. 42486 (1977).

031.033 Dynamic atmospheric turbulence corrections.
R. J. Noll.
Imaging through the atmosphere, (see 012.032), p. 39 - 42
(1976). – Abstr. in Phys. Abstr., Vol. 80, Abstr. 42487 (1977).

031.034 Varieties of isoplanatism.   D. L. Fried.
Imaging through the atmosphere, (see 012.032),
p. 20 - 29 (1976). – Abstr. in Phys. Abstr., Vol. 80, Abstr.
42499 (1977).

031.035 Optical wavefront correction in real time.
V. N. Mahajan.
Imaging through the atmosphere, (see 012.032), p. 109 - 118
(1976). – Abstr. in Phys. Abstr., Vol. 80, Abstr. 42500 (1977).

031.036 The effects of atmospheric dispersion on com-pensated imaging.   E. P. Wallner.
Imaging through the atmosphere, (see 012.032), p. 119 - 125
(1976). – Abstr. in Phys. Abstr., Vol. 80, Abstr. 42501 (1977).

031.037 Post-processing of imagery from active optics – some pitfalls.   R. E. Wagner.
Imaging through the atmosphere, (see 012.032), p. 136 - 140
(1976). – Abstr. in Phys. Abstr., Vol. 80, Abstr. 42502 (1977).

031.038 Fundamental limitations in linear invariant restora-tion of atmospherically degraded images.
J. W. Goodman, J. F. Belsher.
Imaging through the atmosphere, (see 012.032), p. 141 - 154
(1976). – Abstr. in Phys. Abstr., Vol. 80, Abstr. 42503 (1977).

031.039 Using membrane mirrors in adaptive optics.
M. Yellin.
Imaging through the atmosphere, (see 012.032), p. 97 - 102
(1976). – Abstr. in Phys. Abstr., Vol. 80, Abstr. 42587 (1977).

031.040 Active image restoration with a flexible mirror.
A. Buffington, F. S. Crawford, R. A. Muller, A. J.
Schwemin, R. G. Smits.
Imaging through the atmosphere, (see 012.032), p. 90 - 96
(1976). – Abstr. in Phys. Abstr., Vol. 80, Abstr. 45319 (1977).

031.041 Wideband adaptive optics for imaging.
J. Feinleib, J. W. Hardy.
Imaging through the atmosphere, (see 012.032), p. 103 - 108
(1976). – Abstr. in Phys. Abstr., Vol. 80, Abstr. 45320 (1977).

031.042 Building a 6″ reflector. I. Grinding the mirror.
A. P. Witzgall.
Astronomy, Vol. 5, No. 4, p. 34 - 39, 54 (1977). – Abstr. in
Phys. Abstr., Vol. 80, Abstr. 56826 (1977).

031.043 Secondary spectrum correction with normal glasses.
C. G. Wynne.
Opt. Commun., Vol. 21, 419 - 424 (1977). — Abstr. in Phys. Abstr., Vol. 80, Abstr. 62032 (1977).

031.044 Fourth-order aberrations of plane grating mono-chromators. P. Miyake, K. Masutani.
J. Opt., (France), Vol. 8, No. 3, p. 175 - 180 (1977). — Abstr. in Phys. Abstr., Vol. 80, Abstr. 62056 (1977).

031.045 Mounting and application of lens systems.
N. Bardny.
Finommech.-Mikrotech., Vol. 15, 367 - 372 (1976). In Hungarian. — Abstr. in Phys. Abstr., Vol. 80, Abstr. 72050 (1977).

031.046 Electromagnetic field near the focus of wide-angular lens and mirror systems.
C. J. R. Sheppard, A. Choudhury, J. Gannaway.
IEE J. Microwave Opt. Acoust., Vol. 1, No. 4, p. 129 - 132 (1977). — Abstr. in Phys. Abstr., Vol. 80, Abstr. 72051 (1977).

031.047 Polarization effects caused by mirror overcoatings.
V. L. Williams, J. B. Goodell.
Proceedings of the IEEE 1976 National Aerospace and Electronics Conference, Dayton, Ohio, 18 - 20 May 1976. IEEE, New York, USA (1976). p. 349 - 353. — Abstr. in Phys. Abstr., Vol. 80, Abstr. 72055 (1977).

031.048 Statistical properties of the sum of partially correlated complex variables.
A. Consortini, F. Pasqualetti, L. Ronchi.
Opt. Acta, Vol. 24, 931 - 937 (1977). — Abstr. in Phys. Abstr., Vol. 80, Abstr. 79754 (1977).

031.049 Calculating the encircled energy in the point-spread function. E. C. Kintner.
Opt. Acta, Vol. 24, 1075 - 1076 (1977). — Abstr. in Phys. Abstr., Vol. 80, Abstr. 79766 (1977).

031.050 Statistical properties of the sum of two partially correlated speckle patterns.
J. Ohtsubo, T. Asakura.
Appl. Phys., Vol. 14, 183 - 187 (1977). — Abstr. in Phys. Abstr., Vol. 80, Abstr. 83813 (1977).

031.051 Holographic optical elements.
A. K. Aggarwal, P. C. Gupta.
CSIO Commun., Vol. 3, No. 3, p. 44 - 50 (1976). — Abstr. in Phys. Abstr., Vol. 80, Abstr. 83966 (1977).

031.052 Single-spindle aspheric figuring with flexible pads.
N. J. Brown.
Opt. Lett., Vol. 1, No. 1, p. 33 - 34 (1977). — Abstr. in Phys. Abstr., Vol. 80, Abstr. 84017 (1977).

031.053 About a modification of Hartmann's method.
P. V. Shcheglov.
Astron. Tsirk., No. 954, p. 2 - 5 (1977). In Russian.

031.054 Beobachtungen beim Optimieren eines Dreilinsers nach dem gedämpften Minimum-Verfahren.
H. Slevogt.
Optik, Band 47, 313 - 323 (1977).

031.055 Polarization and interference in optics. I: The transfer function — OTF. J. Ben Uri.
Optik, Band 47, 337 - 350 (1977).

031.056 Polarization and interference in optics. Part II: Interference, coherency matrix and degree of polarization. J. Ben Uri.
Optik, Band 47, 405 - 420 (1977).

031.057 Polarization and interference in optics. Part III: Reflection, refraction and diffraction. J. Ben Uri.
Optik, Band 48, 1 - 22 (1977).

031.058 Diffraction of uniform and Gaussian beams: an application of Zernike polynomials.
R. M. Sillitto.
Optik, Band 48, 271 - 277 (1977).

031.059 A simple set of formulas for tracing skew rays.
H. Determann.
Optik, Band 48, 305 - 319 (1977). In German.
Four trigonometric quantities are introduced to determine the skew ray. In the special case of the ray parallel to the meridional plane, two of these have to be replaced by others. The formulas serve to trace rotationally symmetric optical systems of plane and spherical refractive surfaces. The author also gives formulas for the heights of interaction and the length of the ray traces between the refractive faces.

031.060 Comparison of the visual and photoelectric measured imaging quality of telescopes.
J. Eggert, K.-J. Rosenbruch.
Optik, Band 48, 439 - 450 (1977). In German.
It is shown by means of the results obtained on four telescopes having equal optical data but a different correction that the MTF curves deduced by visual observation and measured by a simulated eye result in the same description of the visual imaging quality.

031.061 Formulae for tracing skew rays. H. Determann.
Optik, Band 49, 187 - 201 (1977). In German.
The formulae serve to trace rotationally symmetric optical systems composed of refracting spheres. The skew ray is described by the skew invariant and three variables. Two of these can be traced separately without knowledge of the third. Trigonometric functions are used for tracing instead of square roots. No special formulae are required for the plane surface and the ray parallel to the meridional plane. The formulae presented are particularly simple.

031.062 Polarization and interference in optics. IV: Transfer function of a glass plate, a prisma, and total reflection. J. Ben Uri.
Optik, Band 49, 263 - 275 (1977).

031.063 Proposal for phase recovery from a single intensity distribution. A. H. Greenaway.
Opt. Lett., Vol. 1, 10 - 12 (1977).
A method is proposed for the recovery of phase information from a single intensity distribution. The proposal requires the use of a specially designed spatial filter and is analyzed for a one-dimensional case using the theory of entire functions. The only, and unlikely, ambiguity is detectable a posteriori.

031.064 Imaging with obscured pupils. V. N. Mahajan.
Opt. Lett., Vol. 1, 128 - 129 (1977).
It is shown that diffraction-limited imaging systems with symmetric pupils and a noncentral obscuration yield a higher concentration of energy near the center of an image of a point object than those with central obscuration.

031.065 Estimate of the quality of the image of two-mirror telescopes and their use in cosmic astronomy.
N. V. Merman, M. A. Sosnina.
Izv. Glav. Astron. Obs. Pulkovo, No. 195, Astrofiz. Astrometr., p. 160 - 171 (1977). In Russian.
Various types of two-mirror systems used in astronomy are classified and estimated. Accurate (taking into account aberrations of the 3d order) and empirical formulas are given

for determining the useful field of view with respect to different criteria for permissible aberrations.

031.066 Optimale Dimensionierung optischer Apparaturen und Untersuchung von Ausgleichsansätzen zur Bestimmung der relativistischen Lichtablenkung.   M. G. Firneis. Ann. Univ.-Sternw. Wien, Band 31, (Nr. 4), 65 - 234 (1976).

A possible experimental verification of general relativity is the deflection of light in the gravitational field of the sun which can be observed during a solar eclipse. For this experiment investigations on the attainable systematic accuracy of stellar positions and the Einstein effect are carried out for four different adjustment models.

031.067 Optical performance control.   T. A. Facey. The Space Telescope, (see 012.046), p. 64 - 67 (1976).

031.068 Impact of focal plane dynamics on image quality. W. J. Pragluski, P. W. Abbott, J. F. Eastman. The Space Telescope, (see 012.046), p. 68 - 71 (1976).

031.069 Stray light from out of field sources.   R. J. Noll. The Space Telescope, (see 012.046), p. 72 - 75 (1976).

031.070 Design of highly stable optical support structure. M. H. Krim. The Space Telescope, (see 012.046), p. 76 - 79 (1976).

031.071 Mirror substrate material and manufacturing. W. C. Lewis. The Space Telescope, (see 012.046), p. 120 - 122 (1976).

031.072 Fabrication and test of 1.8-meter (71-inch) diameter, high-quality U.L.E.™ mirror. R. J. Wollensak, C. A. Rose. The Space Telescope, (see 012.046), p. 123 - 134 (1976).

031.073 Design and testing with a reflective null system. L. Montagnino, A. Offner. The Space Telescope, (see 012.046), p. 135 - 138 (1976).

031.074 Test results on homogeneity of expansion for a 1.8-meter (71-inch) U.L.E.™ lightweight mirror. G. Friedman, G. Gasser. The Space Telescope, (see 012.046), p. 139 - 144 (1976).

031.075 Investigation of the optical scheme of a three-lens objective.   Yu. D. Pimenov. Opt.-mekh. prom-st', 1977, No. 7, p. 18 - 21. In Russian. Abstr. in Ref. zh., 51. Astron., 1.51.86 (1978).

031.076 Quality control of large lenses.   Yu. A. Klevtsov. Issled. po geomagn., aehron. i fiz. Solntsa. Moskva, Nauka, 1977, p. 207 - 211. In Russian. — Abstr. in Ref. zh., 51. Astron., 1.51.88 (1978).

Introduction to applied optics for engineers. See Abstr. 003.040.

Optical surfaces: aspherical optical systems, X-ray optics, reflecting microscopes, reflectors, measurements. See Abstr. 003.079.

Progress in optics, Vol. XIV. See Abstr. 003.167.

Transfer functions, correlation scales, and phase retrieval in speckle interferometry.   See Abstr. 031.317.

The vulnerability of speckle photography to lens aberrations.   See Abstr. 031.318.

Interference method for control of the quality of the BTA main mirror.   See Abstr. 032.003.

Application of the isophotometric registration method to investigations and certification of the BTA main mirror.   See Abstr. 032.004.

Organization of the work for making the blank for the main mirror of the BTA (Large Azimuth Telescope). See Abstr. 032.027.

Technological control of the main mirror of the BTA (Large Azimuth Telescope) by the Hartmann method. See Abstr. 032.028.

Investigation of the main mirror of the BTA (Large Azimuth Telescope) at an observatory. See Abstr. 032.029.

3.6 m telescope. The adjustment and test on the sky of the prime focus optics with the Gascoigne plate correctors. See Abstr. 032.036.

Adaptive optics for space telescopes. See Abstr. 032.541.

Science performance considerations for the design of the Space Telescope.   See Abstr. 032.594.

Concentric spectrographs.   See Abstr. 034.056.

A review of the optical effects of the clear turbulent atmosphere.   See Abstr. 082.064.

Methods of Observation and Reduction

031.201 **Radar detectability of asteroids. A survey of opportunities for 1977 through 1987.**
R. F. Jurgens, D. F. Bender.
Icarus, Vol. 31, 483 - 497 (1977).

The capability of Earth-based radar to study asteroids is assessed with respect to determining the number of detectable objects and the number of detectable events during the next 10 years. The Goldstone radar system operating at 3.5-cm wavelength should be able to detect roughly 18 different asteroids at 34 favorable opportunities during the next 10 years. The Arecibo radar system operating at 12.5-cm wavelength may be able to detect 60 asteroids at approximately 97 favorable opportunities in the 10-year period. This sample is sufficiently large that classification of types and correlation with optical data should be possible.

031.202 **Combined position and diameter measures for lunar craters.** D. W. G. Arthur.
Icarus, Vol. 32, 127 - 129 (1977).

The note addresses the problem of simultaneously measuring positions and diameters of circular impact craters on wide-angle photographs of approximately spherical planets such as the Moon and Mercury. The method allows for situations in which the camera is not aligned on the planet's center.

031.203 **Multiple telescope measurements of particle anisotropies in space.**
T. R. Sanderson, R. J. Hynds.
Planet. Space Sci., Vol. 25, 799 - 807 (1977).

The performance of a set of particle telescopes mounted upon a spinning satellite is analysed. A simulation is used to optimize a set of 3 telescopes with a view to obtaining a full 3-dimensional particle distribution, taking into account the telescope parameters and their pointing directions.

031.204 **Considerations for the application of the lunar occultation technique.** S. T. Ridgway.
Astron. J., Vol. 82, 511 - 515 (1977).

Computations of the limiting angular resolution of the lunar occultation technique are presented for wavelengths $\lambda = 0.55$, 2.2, and 10.6 μ.

031.205 **On the measurement of the diffuse X- and γ-background as seen by non directional detectors.**
G. Cavallo, H. M. Horstman, E. Moretti-Horstman.
Astron. Astrophys., Vol. 59, 405 - 409 (1977).

The problem of the experimental determination of the diffuse X- and γ-background using intrinsically non-directional detectors (i.e. scintillators and proportional counters) is examined. It is shown that the contributions due to the internal background of the detector and to the emission from the surrounding material cannot usually be separated from the celestial background by purely experimental methods.

031.206 **Precision estimation of precession and nutation from radio interferometric observations.**
H. G. Walter.
Astron. Astrophys., Vol. 59, 433 - 440 (1977).

On employing the technique of Very Long Baseline Interferometry the precision for precession and nutation terms has been estimated as function of radio source declinations, location of baselines and time span of observations. Observations accumulated over five to ten years yield standard deviations superior to the current ones by approximately one order of magnitude for luni-solar precession and somewhat less for nutation terms.

031.207 **A method for detecting compact galaxies.**
A. T. Kalloglyan, F. Börngen.
Astrofizika, Vol. 12, 697 - 700 (1976). In Russian. – English translation in Astrophysics, Vol. 12, No. 4.

A method for detecting compact galaxies by means of comparing the ratios of two equidensity diameters of galaxies and stars located on the same plate is proposed.

031.208 **A new method for calculating stellar densities.**
J. E. Cope, B. W. Rust.
Bull. American Astron. Soc., Vol. 9, 430 (1977). – Abstract.

031.209 **Prediction of occultations of stars by solar system objects using a microdensitometer.**
P. J. Shelus, G. F. Benedict, D. S. Evans.
Bull. American Astron. Soc., Vol. 9, 436 (1977). – Abstract.

031.210 **The Martian Surface Wind Tunnel (MARSWIT).**
R. Greeley.
Bull. American Astron. Soc., Vol. 9, 448 (1977). – Abstract.

031.211 **Asteroids detectable by radar systems 1977 – 86.**
D. F. Bender, R. F. Jurgens.
Bull. American Astron. Soc., Vol. 9, 460 - 461 (1977). Abstract.

031.212 **Two types of radio occultation Fresnel zones lead to different scintillation profiles.** T. A. Croft.
Bull. American Astron. Soc., Vol. 9, 469 - 470 (1977). Abstract.

031.213 **Radio occultations by turbulent planetary atmospheres: power spectra of intensity scintillations.**
B. S. Haugstad.
Bull. American Astron. Soc., Vol. 9, 470 (1977). – Abstract.

031.214 **Analytic transform pair illustrating atmospheric occultation experiments.** V. R. Eshleman.
Bull. American Astron. Soc., Vol. 9, 470 (1977). – Abstract.

031.215 **Speckle interferometry with a linear digicon detector.**
G. D. Schmidt, J. R. P. Angel, R. Harms.
Publ. Astron. Soc. Pacific, Vol. 89, 410 - 414 (1977).

A linear digicon detector has been employed for speckle interferometry in a mode where digital on-line autocorrelation analysis was performed at the telescope. Results are presented for observations of two binary systems and Nova Cygni 1975. The extension of this method to a two-dimensional device and problems associated with speckle interferometry at very low light levels are discussed.

031.216 **Multivariate analysis of spectrophotometry.**
C. A. Christian, K. A. Janes.
Publ. Astron. Soc. Pacific, Vol. 89, 415 - 423 (1977).

Factor analysis was performed on intermediate-band photometry of K giant stars in order to determine the number of linearly independent parameters influencing the data. The preliminary analysis done on photometry available in the literature indicated that three to four factors were present in the data. These factors were not clearly identifiable as specific atmospheric parameters such as temperature, gravity, and abundance, however. In an effort to isolate some of these parameters, low-resolution scanner data in the spectral region 4000 Å–7000 Å were obtained for stars limited to spectral type K0. Factor analysis of these data indicated that at least two and possibly three factors were present.

031.217 **A method of measuring the translational velocity of a light source.**
P. V. Nikolaev, Yu. A. Sabinin, V. P. Tyukin.
Izv. vyssh. uchebn. zaved. Priborostroenie, Vol. 19, No. 12,

p. 101 - 105 (1976). In Russian. −.Abstr. in Ref. zh., 51.
Astron., 8.51.97 (1977).

**031.218 The effects of the atmospheric point-spread "seeing"
function on spatially resolved spectra of Jupiter.**
J. Gelfand, W. D. Cochran, W. H. Smith.
Astrophys. J., Vol. 217, 320 - 323 (1977).
The authors present the results of an analysis of the
effects of atmospheric seeing and of instrumental spectral and
spatial resolution on the observed variation of absorption-line
profiles across the disk of Jupiter. The technique described
may be applied equally well to the analysis of observations of
any extended astronomical source. These results show the
necessity of obtaining accurate point-spread-function information during the course of observations of this nature. The
authors also point out that in order to avoid the uncertainties
and ambiguities inherent in attempts at deconvolution of observational data, one must properly convolve the appropriate
spatial and spectral resolution functions with the models being
tested and then compare the results with the observational data.

**031.219 Speckle-Interferometrie. II. Ergebnisse und Möglich-
keiten.**
A. Lohmann, M. Reinecke, H. Ruder, G. Weigelt.
Sterne Weltraum, Jahrg. 16, 284 - 292 (1977).

031.220 Sternspektroskopie.   E. Pollmann.
Sterne Weltraum, Jahrg. 16, 296 - 298 (1977).

031.221 Observations méridiennes à Besançon.
M. Creze.
Recueil des séminaires, (see 012.004), p. 26 - 30 (1977).

**031.222 Methodical improvements of magnetographic meas-
urements.**
G. Bachmann, H. Künzel, K. Pflug, J. Staude.
Solar activity and solar-terrestrial relations, (see 012.007), p.
395 - 402 (1976).
In order to obtain correct magnetic field strengths for
both weak and strong magnetic fields by magnetographic
measurements and to avoid the well-known difficulty of calibration some methodical improvements are made.

**031.223 Restoration of radio brightness distribution in knife
diagram observations.**   B. S. Minchenko.
Astrofiz. Issled., Izv. Spets. Astrofiz. Obs., Vol. 9, 29 - 37
(1977). In Russian.
A restoration algorithm allowing to synthesize radio
images of high quality is described. Examples of restored
images are considered, their quality is evaluated.

**031.224 Correction of radio images in the program of
cartography with RATAN-600.**   B. S. Minchenko.
Astrofiz. Issled., Izv. Spets. Astrofiz. Obs., Vol. 9, 38 - 46
(1977). In Russian.
An extremely complex structure of the beamshape of the
radiotelescope RATAN-600 causes considerable distortions of
two-dimensional synthesized radio images. A possibility of
application of Högbom's method of "cleaning" for suppression
of the distortions is considered. Results of computer modelling
of the "cleaning" procedure are presented.

**031.225 Taking account of refraction and relativistic effects
in measuring of arcs using VLBI.**
A. F. Dravskikh, A. M. Finkel'shtejn.
Astrofiz. Issled., Izv. Spets. Astrofiz. Obs., Vol. 9, 47 - 52
(1977). In Russian.
The paper deals with formulas for refraction, aberration,
and relativistic effects in the VLBI measurements of mutual
angular distances between QSOs.

**031.226 On a method of measurement of solar radio fluxes in
the centimeter range.**
M. S. Durasova, O. I. Yudin.
Soln. Dannye 1977 Byull., No. 6, p. 72 - 75 (1977). In Russian.
A method is elaborated of the measurement of solar radio
fluxes in the centimeter–decimeter wavelength range for the
observations of the solar-terrestrial service. Observations of the
solar radio flux have been performed at λ = 3 cm during 2
years with a relative error < 2%.

**031.227 Angular diameter counts of galaxies: a method for
determining the selection criteria for deep Schmidt
plates.**   R. S. Ellis, R. Fong, S. Phillipps.
Mon. Not. R. Astron. Soc., Vol. 181, 163 - 182 (1977).
It is demonstrated that a sensitive method for studying
the selection effects that operate on galaxy samples is to
investigate theoretical fits to the observed angular diameter
frequency distribution. This technique is applied to a two-
colour sample taken from SRC 48-in Schmidt plates that have
been measured by the COSMOS machine. The selection effects
are discussed in terms of the selection function, i.e. the probability that an arbitrary galaxy at a certain redshift is accepted
for the sample. It is found necessary to invoke luminosity
evolution in order to account for the slope of the observed
distribution at large angular diameters.

**031.228 Adaptation of the Alpha Particle Instrument for
penetrator missions.**
A. Turkevich, T. Economou, E. Franzgrote.
NASA Tech. Mem., NASA TM X-3511, (see 012.010), p. 258 -
259 (1977). − Abstract.

**031.229 Astrometric techniques with a PDS microdensitom-
eter.**   L.-T. G. Chiu.
Astron. J., Vol. 82, 842 - 848 (1977).
Digital image centering for astrometry as done with the
Berkeley PDS machine is discussed. The performance of various
centering algorithms is tested for both bright stars and faint
stars on three common emulsion types, namely 103a, 127-02,
and IIIa-J. It is also shown that plates taken with different
telescopes can be combined astrometrically.

**031.230 Theoretical principles of determination of the
azimuth of a terrestrial object with a portable
instrument without precise level.**   N. A. Kozlov.
Izv. vyssh. uchebn. zaved. Geod. i aehrofotosemka, 1976,
No. 6, p. 63 - 68. In Russian. − Abstr. in Ref. zh., 51. Astron.,
9.51.196; 52. Geod. Aehrosemka, 9.52.124 (1977).

**031.231 Optimum methods of astronomical determinations
in the Antarctic continent.**
S. S. Uralov, L. V. Neverov.
Izv. vyssh. uchebn. zaved. Geod. i aehrofotosemka, 1976,
No. 6, p. 57 - 62. In Russian. − Abstr. in Ref. zh., 51. Astron.,
9.51.197; 52. Geod. Aehrosemka, 9.52.125 (1977).

**031.232 Influence of atmospheric turbulence on a photo-
graphic survey.**   Yu. S. Timofeev.
Tr. TsNII geod., aehrosemki i kartogr. 1977, vyp. (No.) 218,
p. 47 - 51. In Russian. − Abstr. in Ref. zh., 52. Geod. Aehrosemka, 9.52.169 (1977).

**031.233 Investigation of geometrical distortions of TV space
photographs.**   V. A. Poloznikov, A. P. Tishchenko.
Geod., kartogr. i aehrofotosemka. Resp. mezhved. nauchn.-
tekh. sb., 1977, vyp. (No.) 25, p. 92 - 95. In Russian. − Abstr.
in Ref. zh., 52. Geod. Aehrosemka, 9.52.212 (1977).

031.234 Astronomical applications of echelle spectroscopy.
F. H. Chaffee, Jr., D. J. Schroeder.
Annu. Rev. Astron. Astrophys., Vol. 14, (see 003.008), 23 - 42
(1976).
The authors review the general characteristics of echelles

and the astronomical applications for which echelle spectrographs have been used, with the emphasis on the latter. Their discussion of applications is limited to those for which an echelle spectrograph is used at the Cassegrain focus. They also consider briefly selected planned echelle instruments and intended applications.

031.235 **Radio astrometry.**   C. C. Counselman III.
          Annu. Rev. Astron. Astrophys., Vol. 14, (see 003. 008), 197 - 214 (1976).

The radio-astrometric revolution promises to benefit optical astrometry through the establishment of a more nearly inertial reference frame, defined by a system of positions of extragalactic objects, and through the more accurate determination of the Earth's motion about its center of mass. Radio astrometry has come of age, although it is still developing rapidly and is far from maturity. Its present state and directions of growth are discussed in this review.

031.236 **Mathematical modelling of lunar laser measures and their application to improvement of physical parameters.**   J. D. Mulholland.
Scientific applications of lunar laser ranging, (see 012.012), p. 9 - 18 (1977).

031.237 **Scientific expectations in the selenosciences.**
          J. Kovalevsky.
Scientific applications of lunar laser ranging, (see 012.012), p. 21 - 36 (1977).

Assuming that, in the future, a 3 cm accuracy on lunar laser ranging will be achieved, representing at least accuracies of the order 0.''01 in lunar rotation and 0.''001 in its motion in space, many dynamical or kinematic features pertaining to the Moon or the Earth—Moon system should become observable and bring new light about the physics of the Earth and of the Moon. These are reviewed in this presentation.

031.238 **Present scientific achievements from lunar laser ranging.**   J. G. Williams.
Scientific applications of lunar laser ranging, (see 012.012), p. 37 - 50 (1977).

031.239 **Whole earth dynamics and lunar laser ranging.**
          D. E. Smylie.
Scientific applications of lunar laser ranging, (see 012.012), p. 105 - 130 (1977).

031.240 **An intermediate term strategy for deployment of mobile laser stations.**
G. V. Latham, H. J. Dorman.
Scientific applications of lunar laser ranging, (see 012.012), p. 157 - 165 (1977).

031.241 **On the problems of the astrometric methods and of the lunar laser ranging in the study of the Earth's rotation.**   B. Kolaczek.
Scientific applications of lunar laser ranging, (see 012.012), p. 171 - 178 (1977).

This report is concerned with the comparison of classical astrometric methods with the lunar laser ranging (LLR) techniques in the study of the lunar rotation and of the lunar orbital motion.

031.242 **A review of perturbing parameters which affect the quality of laser distance measurements.**
P. Morgan.
Scientific applications of lunar laser ranging, (see 012.012), p. 223 - 239 (1977).

031.243 **The deformational environment of the Haleakala lunar laser ranging observatory.**
E. Berg, G. H. Sutton.

Scientific applications of lunar laser ranging, (see 012.012), p. 263 - 275 (1977).

The University of Hawaii is initiating a geodetic-geophysical program in support of the lunar laser ranging at Haleakala, Maui, Hawaii. The program is aimed at measuring and understanding possible motion of the observatory site with respect to its surroundings, so that local and regional movement can be separated from the more general plate motion, as referred to other lunar ranging or VLBI stations.

031.244 **ALSEP-quasar VLBI: complementary observable for laser ranging.**   M. A. Slade, W. S. Sinclair, A. W. Harris, R. A. Preston, J. G. Williams.
Scientific applications of lunar laser ranging, (see 012.012), p. 287 (1977). — Abstract.

031.245 **Laser ranging techniques required to test Dirac's cosmological model.**   J. L. Hughes.
Scientific applications of lunar laser ranging, (see 012.012), p. 289 - 302 (1977).

Conventional lunar laser ranging techniques will be marginally effective in any test of Dirac's cosmological model. A new approach based on the Earth-Mars system is proposed.

031.246 **Recovery of fringe visibility from recorded speckle images quantized to two levels.**   R. H. T. Bates.
Mon. Not. R. Astron. Soc., Vol. 181, 365 - 374 (1977).

Single and joint probability density functions are obtained for speckle images of a resolvable object. The formulae are valid for arbitrary statistical correlation between the intensities at arbitrarily positioned pairs of points within the seeing disc. The information distortion introduced by nonlinear recording of speckle images is examined. Explicit formulae, convenient for computation, are obtained for speckle images quantized to two levels.

031.247 **The determination of meteor stream radiants from single station observations.**
J. Jones, J. D. Morton.
Bull. Astron. Inst. Czechoslovakia, Vol. 28, 267 - 272 (1977).

An improved method for detection of faint meteor showers in the presence of a strong sporadic background and the determination of the radiants of these streams is presented. It is applicable to single station observations of both optical and radio meteors. The procedure, which has been tested using both computer simulated data and visual observations, appears to be as sensitive as predicted theoretically.

031.248 **Meteor radiant distribution using spherical harmonic analysis.**   J. Jones.
Bull. Astron. Inst. Czechoslovakia, Vol. 28, 272 - 277 (1977).

A technique based on the deconvolution of spherical harmonic transforms for the recovery of radiant activity distributions from single station observations is presented. Its use is demonstrated with some visual observations from Springhill, Ottawa and the method is seen to work well. A rough estimate of how the sensitivity of the method varies with quantity and accuracy of the individual observations is also given.

031.249 **Area scanning technique for astrometric observations of visual binaries.**   K. Rakos.
O telescópio refractor e a astrometria ao serviço das estrelas duplas, (see 012.013), p. 61 (1977). — Abstract.

031.250 **Application de l'observation des occultations d'étoiles par la Lune à la decouverte et à la mesure d'étoiles doubles.**   J. Hecquet.
O telescópio refractor e a astrometria ao serviço das estrelas duplas, (see 012.013), p. 73 - 75 (1977).

Le phénomène de diffraction observable par enregistrement photométrique ultrarapide, lors d'une occultation d'étoile par la Lune offre des possibilités d'application au cas des couples visuels en vue de leur mesure et de leur observation photométrique. Les difficultés de la méthode sont discutées mais aussi ses grandes possibilités qui se sont vu confirmées par ses premiers succès.

031.251 Application de la transformée de Fourier à la mesure des binaires.   J. Rösch, G. Coupinot.
O telescópio refractor e a astrometria ao serviço das estrelas duplas, (see 012.013), p. 77 - 88 (1977).

031.252 Rechnen mit dem Taschenrechner: Ortsbestimmungen auf Sonne und Planeten.   W. Wepner.
Sterne Weltraum, Jahrg. 16, 374 - 377 (1977).

031.253 Some comments on the measurement of small scale strong magnetic fields on the sun.   E. Wiehr.
Highlights of Astronomy, Vol. 4, Part II, (see 012.024), p. 251 - 254 (1977).

031.254 Fourier-spectroscopy from balloon platforms.   R. Hofmann, K. W. Michel, F. Naumann, J. Stocker.
Far infrared astronomy (see 012.027), p. 53 - 61 (1976).
The requirements and design of a balloon-borne Fourier spectrometer are described in view of limited pointing accuracy and suppression of atmospheric line radiation by the wobbling mirror technique.

031.255 The moon as a calibration source for submillimetre astronomy.   M. J. Pugh.
Far infrared astronomy (see 012.027), p. 63 - 69 (1976).
A limited solution to the problem of thermal emission in the presence of simple scattering is described. This is used with the temperature profiles from Apóllo 17 to produce brightness temperature curves along the thermal equator. These are proposed as standards to overcome the difficulties of absolute measurements through the terrestrial atmosphere.

031.256 Solar radiometry in the millimetre region with a local calibration source.   B. Carli.
Far infrared astronomy (see 012.027), p. 87 - 91 (1976).
A system for solar radiometry in the millimetre region, as an alternative to calibrations using the moon, is described. The effects of long wavelength diffraction and atmospheric absorption are discussed.

031.257 Using Watts angles in observing reappearances.   D. Herald.
Occultation Newsl., Vol. 1, 127 - 128 (1977).

031.258 Values of k for occultation reduction.   A. M. Sinzi.
Occultation Newsl., Vol. 1, 128 (1977).

031.259 Interferometry from 1950 to the present.   L. Genzel, K. Sakai.
J. Opt. Soc. America, Vol. 67, 871 - 879 (1977).
This review paper deals with the history, the important aspects, and the development of far-infrared Fourier transform spectroscopy and Fabry-Perot interferometry from 1950 to the present. Most of the fundamentals of Fourier spectroscopy were worked out in the period from 1951–1961 after the realization of the multiplex advantage. The following period from 1962 to the present brought new instrumental innovations and refinements and the stage of general use. Fabry-Perot interferometry became possible in the far infrared since metal mesh was found to be the ideal reflector for the interferometer. This kind of interferometry is now mainly and increasingly used in connection with far-infrared laser research.

031.260 Image sharpness, Fourier optics, and redundant-spacing interferometry.
J. P. Hamaker, J. D. O'Sullivan, J. E. Noordam.
J. Opt. Soc. America, Vol. 67, 1122 - 1123 (1977).

The authors give a simple proof of the image sharpness criterion S_1 introduced by Muller and Buffington. A close connection with interferometric techniques for diffraction-limited imaging is pointed out. The method of the proof provides indications on the limited validity of several other sharpness criteria.

031.261 Noise considerations in stellar speckle interferometry.   M. G. Miller.
J. Opt. Soc. America, Vol. 67, 1176 - 1184 (1977).
A general model for the signal-to noise ratio for stellar speckle interferometry is developed, including the effects of finite sample averaging, quantum fluctuations, and additive noise. The author finds that in most cases the controlling parameter is the ratio of the total spatial spectrum to the signal spatial spectrum. Provided the average number of photon events per frame is greater than one, the signal-to-noise ratio for an ideal quantum limited detector is insensitive to aperture size. If a signal-independent noise source is also present, either source can dominate under certain conditions. Large resolved objects must be considerably brighter than their equivalent point source if diffraction details are to be observed.

031.262 Speckle interferometry with severely aberrated telescopes.   D. P. Karo, A. M. Schneiderman.
J. Opt. Soc. America, Vol. 67, 1277 - 1278 (1977).
Quantitative measurements of the modulation transfer function for speckle interferometry in the absence and presence of severe defocus demonstrate that near-diffraction-limited resolution is possible in both cases.

031.263 Use of the Multiple-Mirror Telescope as a multibase Michelson stellar interferometer.   G. M. Sanger.
J. Opt. Soc. America, Vol. 67, 1385 (1977). – Abstract.

031.264 Energy dependence of the transfer function in stellar speckle interferometry in the poor seeing limit.
L. Sica.
J. Opt. Soc. America, Vol. 67, 1390 (1977). – Abstract.

031.265 Wavelength, bandpass, and exposure time effects in speckle interferometry.
D. P. Karo, A. M. Schneiderman.
J. Opt. Soc. America, Vol. 67, 1390 (1977). – Abstract.

031.266 Atmospheric imaging constraints measured with speckle interferometry.
A. M. Schneiderman, D. P. Karo.
J. Opt. Soc. America, Vol. 67, 1390 (1977). – Abstract.

031.267 Imaging on Mars: from concept to reality.   F. O. Huck.
J. Opt. Soc. America, Vol. 67, 1414 (1977). – Abstract.

031.268 Maximum likelihood method for obtaining near-diffraction-limited images through the turbulent atmosphere.   E. G. Hawman, R. S. Hershel.
J. Opt. Soc. America, Vol. 67, 1415 (1977). – Abstract.

031.269 Corrections for atmospheric refractivity in satellite laser ranging.   R. S. Iyer, J. L. Bufton.
Appl. Opt., Vol. 16, 1997 - 2003 (1977).
The range correction calculated from ground-level meteorological measurements is shown to consist of a series of terms. The zeroth order term of this series corresponds to the correction obtained under the assumption of a spherically symmetric refractivity profile. The higher order terms arise due to departures from the spherical-symmetry assumption. The range correction is also shown to have residual errors, and an analytic expression for these errors is derived. Pertinent surface data are analyzed to obtain the magnitudes of the higher order corrections and the residual errors.

031.270 **The effective field of view for line sources (meteors).**
M. Kresáková.
Bull. Astron. Inst. Czechoslovakia, Vol. 28, 340 - 345 (1977).
    Relations between the size of the geometric and effective field of view for the detection of line sources are derived. A novel approach makes it possible to consider a variety of field shapes, in addition to the circular case already solved. Formulae for circular and rectangular fields are given. The results are applied to meteor observations by telescopic, photographic, and television techniques.

031.271 **Radar astronomy.**   J. Oberg.
Astronomy, Vol. 4, No. 12, p. 14 - 19 (1976). −
Abstr. in Phys. Abstr., Vol. 80, Abstr. 25947 (1977).

031.272 **Ground based low light astronomy.**   B. V. Barlow.
Contemp. Phys., Vol. 18, 23 - 46 (1977). − Abstr. in Phys. Abstr., Vol. 80, Abstr. 29657 (1977).

031.273 **A simple technique to measure Doppler broadening parameters of Doppler shifts without a spectrograph.**   R. Michard.
Astron. Astrophys., Vol. 61, 729 - 736 (1977).
    There are astrophysical problems, which could be efficiently tackled with a much simpler technique than currently in use. In its simplest form, the proposed technique is equivalent to a Fourier transform spectrometer measuring only one suitably chosen component of the interferogram of the source's light. It can be useful on extended sources due to the large field of the corresponding optical system.

031.274 **Graphic determination of the centre of an astrophotograph obtained by a high-precision astronomical adjustment.**   V. M. Abragamets.
Geod., kartogr. i aehrofotosemka. Resp. mezhved. nauchn.-tekh. sb., 1977, vyp. (No.) 26, p. 3 - 4. In Russian. − Abstr. in Ref. zh., 51. Astron., 10.51.203 (1977).

031.275 **Some peculiarities in interpreting space photographs of the surfaces of the moon and planets.**
M. V. Ostrovskij.
Tr. TsNII geod., aehrosemki i kartogr., 1977, vyp. (No.) 218, p. 52 - 62. In Russian. − Abstr. in Ref. zh., 52. Geod. Aehrosemka, 10.52.151 (1977).

031.276 **A new diagnostic method for magnetospheric plasma.**
J. Etcheto, M. Petit.
C. R. Acad. Sci. Paris, Tome 285, Sér. B, 329 - 332 (1977). In French.
    A new diagnostic technique for magnetospheric plasma, based on in situ excitation of the plasma resonances, has been used for the first time on board the Geos satellite. The preliminary results are very gratifying: electron density and magnetic field intensity are derived reliably and accurately from the resonances observed; hopefully, temperature and electric field will be deduced from the data as well.

031.277 **A new possibility of measuring low fluxes of gravitational radiation.**   A. Kulak.
Acta Phys. Polonica B, Vol. B8, 153 - 158 (1977). − Abstr. in Phys. Abstr., Vol. 80, Abstr. 30075 (1977).

031.278 **Use of thin ionization calorimeters for measurements of cosmic ray energy spectra.**
W. V. Jones, J. F. Ormes, W. K. H. Schmidt.
Nucl. Instrum. Methods, Vol. 140, 557 - 568 (1977). − Abstr. in Phys. Abstr., Vol. 80, Abstr. 30713 (1977).

031.279 **High-resolution techniques in optical astronomy.**
A. Labeyrie.
Progress in optics, Vol. XIV. E. Wolf (Editor). North-Holland Publishing Company, Amsterdam, Netherlands. 17 + 422 pp.

Price $ 57.25 (1976). ISBN 0-7204-1514-4, p. 49 - 87. − Abstr. in Phys. Abstr., Vol. 80, Abstr. 33357 (1977).

031.280 **Applications of beam-foil spectroscopy to the solar ultraviolet emission spectrum.**   L. Heroux.
Beam-foil spectroscopy. S. Bashkin (Editor). Springer-Verlag, Berlin−Heidelberg−New York. 12 + 318 pp. Price $ 28.30 (1976). ISBN 3-540-07914-9. p. 193 - 208. − Abstr. in Phys. Abstr., Vol. 80, Abstr. 41314 (1977).

031.281 **Modified astronomical speckle interferometry 'speckle masking'.**   G. P. Weigelt.
Opt. Commun., Vol. 21, 55 - 59 (1977). − Abstr. in Phys. Abstr., Vol. 80, Abstr. 49308 (1977).

031.282 **Photon noise limitations on the recovery of stellar images by speckle interferometry.**
M. E. Barnett, G. Parry.
Opt. Commun., Vol. 21, 60 - 62 (1977). − Abstr. in Phys. Abstr., Vol. 80, Abstr. 49309 (1977).

031.283 **Image processing techniques applied to scanned radiometric collections of stellar sources.**
R. N. Devich, N. F. Lyon, P. W. Baker.
Modern utilization of infrared technology civilian and military. II. (see 012.031), p. 13 - 22 (1976). − Abstr. in Phys. Abstr., Vol. 80, Abstr. 49312 (1977).

031.284 **Performance evaluation of infrared sensors and associated processing.**   E. M. Winter.
Modern utilization of infrared technology civilian and military. II. (see 012.031), p. 165 - 175 (1976). − Abstr. in Phys. Abstr., Vol. 80, Abstr. 49313 (1977).

031.285 **Sensitivity analysis of short-arc station coordinate determinations from range data.**
B. E. Schutz, B. D. Tapley.
J. Astronaut. Sci., Vol. 24, 111 - 136 (1976). − Abstr. in Phys. Abstr., Vol. 80, Abstr. 44902 (1977).

031.286 **Spacecraft radio-occultation technique for the study of planetary atmospheres.**   R. Eshleman.
J. Spacecr. Rockets, Vol. 13, 768 (1976). − Abstr. in Phys. Abstr., Vol. 80, Abstr. 45316 (1977).

031.287 **A simple method of estimating the RMS phase variation due to atmospheric turbulence.**
R. J. Scaddan, J. C. Dainty.
Opt. Commun., Vol. 21, 551 - 554 (1977). − Abstr. in Phys. Abstr., Vol. 80, Abstr. 45317 (1977).

031.288 **Astronomical speckle imaging.**
P. Nisenson, D. C. Ehn, R. V. Stachnik.
Imaging through the atmosphere, (see 012.032), p. 83 - 88 (1976). − Abstr. in Phys. Abstr., Vol. 80, Abstr. 45318 (1977).

031.289 **Digital image processing of simulated turbulence and photon noise degraded images of extended objects.**   J. R. Breedlove, Jr.
Imaging through the atmosphere, (see 012.032), p. 155 - 162 (1976). − Abstr. in Phys. Abstr., Vol. 80, Abstr. 45321 (1977).

031.290 **A statistical method for post-detection compensation for atmospheric distortions of images of faint scenes.**   C. E. KenKnight.
Imaging through the atmosphere, (see 012.032), p. 163 - 167 (1976). − Abstr. in Phys. Abstr., Vol. 80, Abstr. 45322 (1977).

031.291 **Methods for determining a standard ionospheric topside profile by single measurements.**
K. B. Serafimov.
C. R. Acad. Bulgare Sci., Vol. 29, 1613 - 1615 (1976).

Abstr. in Phys. Abstr., Vol. 80, Abstr. 56739 (1977).

031.292 Applications of very-long-baseline interferometry to geodesy and geodynamics.   B. Anderson.
Philos. Trans. R. Soc. London, Ser. A, Vol. 284, 469 - 473 (1977). – Abstr. in Phys. Abstr., Vol. 80, Abstr. 56767 (1977).

031.293 Accuracy of estimating the masses of Phobos and Deimos from multiple Viking Orbiter encounters.
M. L. Mason, R. H. Tolson.
J. Astronaut. Sci., Vol. 24, No. 3, p. 221 - 241 (1976).
Abstr. in Phys. Abstr., Vol. 80, Abstr. 56831 (1977).

031.294 Photographic equidensitometry for astronomical application.   N. Owaki, T. Mizuno, M. Ono.
Bull. Tokyo Gakugei Univ., Ser. IV, Vol. 28, 237 - 257 (1976).
Abstr. in Phys. Abstr., Vol. 80, Abstr. 56833 (1977).

031.295 Decomposition of multigradation image by iso-image.   D. N. Mishev, P. V. Petrov.
C. R. Acad. Bulgare Sci., Vol. 29, 1621 - 1623 (1976).
Abstr. in Phys. Abstr., Vol. 80, Abstr. 56834 (1977).

031.296 Application of multi-channel analysis techniques to beam-foil spectroscopy of neutral cobalt in the region 3000 Å to 3950 Å.   E. H. Pinnington, P. Weinberg, W. Verfuss, H. O. Lutz.
Z. Phys. A, Vol. 281, 325 - 332 (1977). – Abstr. in Phys. Abstr., Vol. 80, Abstr. 58206 (1977).

031.297 A posteriori restoration of atmospherically degraded images using multiframe imagery.   J. W. Sherman.
Image processing. Pacific Grove, Calif., USA, 24 - 26 February 1976. Proceedings of the Society of Photo-optical Instrumentation Engineers, Vol. 74. Palos Verdes Estates, Calif., USA (1976). p. 249 - 258. – Abstr. in Phys. Abstr., Vol. 80, Abstr. 60469 (1977).

031.298 Maximum entropy restorations of Ganymede.
B. R. Frieden.
Image processing. Pacific Grove, Calif., USA, 24 - 26 February 1976. Proceedings of the Society of Photo-optical Instrumentation Engineers, Vol. 74. Palos Verdes Estates, Calif., USA (1976). p. 160 - 165. – Abstr. in Phys. Abstr., Vol. 80, Abstr. 60624 (1977).

031.299 Computer simulation studies of compensation of turbulence degraded images.   B. L. McGlamery.
Image processing. Pacific Grove, Calif., USA, 24 - 26 February 1976. Proceedings of the Society of Photo-optical Instrumentation Engineers, Vol. 74. Palos Verdes Estates, Calif., USA (1976). p. 225 - 233. – Abstr. in Phys. Abstr., Vol. 80, Abstr. 60625 (1977).

031.300 Recent developments at JPL in the application of digital image processing techniques to astronomical images.   J. J. Lorre, D. J. Lynn, W. D. Benton.
Image processing. Pacific Grove, Calif., USA, 24 - 26 February 1976. Proceedings of the Society of Photo-optical Instrumentation Engineers, Vol. 74. Palos Verdes Estates, Calif., USA (1976). p. 234 - 238. – Abstr. in Phys. Abstr., Vol. 80, Abstr. 60626 (1977).

031.301 A discussion on methods and applications of ranging to artificial satellites and the Moon. Prologue.
D. G. King-Hele.
Philos. Trans. R. Soc. London, Ser. A, Vol. 284, 421 - 430 (1977). – Abstr. in Phys. Abstr., Vol. 80, Abstr. 64183 (1977).

031.302 One-dimensional stellar and solar speckle interferometry.   C. Aime, F. Roddier.
Opt. Commun., Vol. 21, 435 - 438 (1977). – Abstr. in Phys.

Abstr., Vol. 80, Abstr. 64662 (1977).

031.303 Determination of exospheric temperature by ground measurements of ionosphere and red oxygen line.
K. B. Serafimov.
C. R. Acad. Bulgare Sci., Vol. 30, 37 - 39 (1977). – Abstr. in Phys. Abstr., Vol. 80, Abstr. 67003 (1977).

031.304 Analogue techniques for carrying out the two-dimensional Fourier transformation.   J. P. Hamaker.
Tijdschr. Ned. Elektron. Radiogenoot., Vol. 42, No. 1 - 2, p. 27 - 32 (1977). In Dutch. – Abstr. in Phys. Abstr., Vol. 80, Abstr. 67042 (1977).

031.305 Optimum coherent imaging of a limited field of view in the presence of angular and aperture noise.
G. V. Borgiotti.
J. Franklin Inst., Vol. 303, 155 - 173 (1977). – Abstr. in Phys. Abstr., Vol. 80, Abstr. 67045 (1977).

031.306 Atmospheric transmission measurements using infrared lasers and Fourier spectroscopy – techniques, results, and comparisons to computer models.   J. A. Dowling.
Methods for atmospheric radiometry, (see 012.033), p. 39 - 48 (1976). – Abstr. in Phys. Abstr., Vol. 80, Abstr. 70885 (1977).

031.307 Cryogenic spectrometry for the measurement of airglow and aurora.   A. T. Stair, Jr.
Methods for atmospheric radiometry, (see 012.033), p. 71 - 76 (1976). – Abstr. in Phys. Abstr., Vol. 80, Abstr. 70886 (1977).

031.308 Terrestrial measurement of the performance of high-rejection optical baffling systems.
J. C. Kemp, C. L. Wyatt.
Methods for atmospheric radiometry, (see 012.033), p. 85 - 92 (1976). – Abstr. in Phys. Abstr., Vol. 80, Abstr. 70888 (1977).

031.309 Radiometry and spectroscopy of the upper atmosphere from aircraft.   W. G. Mankin.
Methods for atmospheric radiometry, (see 012.033), p. 96 - 101 (1976). – Abstr. in Phys. Abstr., Vol. 80, Abstr. 70890 (1977).

031.310 The HISSA (*high-speed spectrum analysis*) method to analyze various time-varying phenomena in space physics.   T. Saito.
Rep. Ionos. Space Res. Japan, Vol. 30, No. 3 - 4, p. 69 - 80 (1976). – Abstr. in Phys. Abstr., Vol. 80, Abstr. 73896 (1977).

031.311 Extending radiative transfer models by use of Bayes' rule.   C. Whitney.
J. Atmos. Sci., Vol. 34, 766 - 772 (1977). – Abstr. in Phys. Abstr., Vol. 80, Abstr. 78443 (1977).

031.312 Atmospheric phase path corrections for very long baseline interferometry.
K. J. W. Lynn, J. S. Gubbay.
Monitor, Vol. 37, 361 - 368 (1976). – Abstr. in Phys. Abstr., Vol. 80, Abstr. 78637 (1977).

031.313 Digital processing of the Mariner 10 images of Venus and Mercury.
J. M. Soha, D. J. Lynn, J. A. Mosher, D. A. Elliott.
J. Appl. Photogr. Eng., Vol. 3, No. 2, p. 82 - 92 (1977).
Abstr. in Phys. Abstr., Vol. 80, Abstr. 78647 (1977).

031.314 High-accuracy range measurements to the Moon.   E. C. Silverberg.
High speed optical techniques, San Diego, Calif., 26 - 27 August 1976. M. A. Duguay, R. K. Petersen (Editors), Proceedings of the Society of Photo-optical Instrumentation

Engineers, Vol. 94. Bellingham, Wash. 8 + 152 pp. (1976).
ISBN 0-89252-121X. p. 83 - 88. − Abstr. in Phys. Abstr.,
Vol. 80, Abstr. 82126 (1977).

031.315 Planimetric Martian triangulations.
    D. W. G. Arthur, D. K. McMacken.
Photogramm. Eng. Remote Sensing, Vol. 43, 701 - 707 (1977).
Abstr. in Phys. Abstr., Vol. 80, Abstr. 87276 (1977).

**031.316 The application of Fourier transform spectroscopy
    to the remote identification of solids in the solar
system.**   H. P. Larson, U. Fink.
Appl. Spectrosc., Vol. 31, 386 - 402 (1977). − Abstr. in Phys.
Abstr., Vol. 80, Abstr. 87273 (1977).

**031.317 Transfer functions, correlation scales, and phase
    retrieval in speckle interferometry.**
D. P. Karo, A. M. Schneiderman.
J. Opt. Soc. America, Vol. 67, 1583 - 1587 (1977).
    Properties of speckle imaging lens-atmosphere transfer
functions that impact phase retrieval and subsequent image
reconstruction have been measured under realistic conditions
using bright stellar sources. These measurements generally
support assumptions made by those who have proposed phase
retrieval algorithms and tested them on simulated data.

**031.318 The vulnerability of speckle photography to lens
    aberrations.**   K. A. Stetson.
J. Opt. Soc. America, Vol. 67, 1587 - 1590 (1977).
    It is experimentally demonstrated that lens aberrations
can have detrimental effects in speckle photography. These
effects include not only loss of speckle correlations and subsequent halo fringes, but also distortion of the halo fringes that
can lead to false information about object motion.

031.319 Signal-to-noise ratio in Fourier spectroscopy.
    R. R. Treffers.
Appl. Opt., Vol. 16, 3103 - 3106 (1977).
    Calculations of the SNR obtainable with a Fourier trans-
form spectrometer as well as that obtainable with a photometer or scanning device are presented. It is shown that the SNR
obtained with a Fourier spectrometer is $(N/8)^{1/2}$ greater than
that obtained with a scanning device (where N is the desired
number of spectral elements scanned). Reasons why this
factor, known as the multiplex advantage, differs from other
values found in the literature are discussed.

031.320 Measurement of the Stokes parameters of light.
    H. G. Berry, G. Gabrielse, A. E. Livingston.
Appl. Opt., Vol. 16, 3200 - 3205 (1977).
    The authors describe a measuring system for determining
the state of polarization of a beam of light in terms of its
Stokes parameters. The technique which can be fully automated incorporates a monochromator and single photon counting
detection and can thus be applied over a large wavelength
range for very weak optical signals. Fourier transformation of
the data by an on-line minicomputer allows immediate calculation of the Stokes parameters. The authors discuss special
applications to light emitted from excited atomic systems with
and without cylindrical symmetry.

031.321 IPL processing of the Viking orbiter images of Mars.
    R. M. Ruiz, D. A. Elliott, G. M. Yagi, R. B.
Pomphrey, M. A. Power, K. W. Farrell, Jr., J. J. Lorre, W. D.
Benton, R. E. Dewar, L. E. Cullen.
J. Geophys. Res., Vol. 82, (see 003.017), 4189 - 4202 (1977).
Paper No. 7S0505.
    The Viking orbiter cameras returned over 9000 images of
Mars during the 6-month nominal mission. Digital image pro-
cessing was required to produce products suitable for quantita-
tive and qualitative scientific interpretation. Processing includ-
ed the production of surface elevation data using computer

stereophotogrammetric techniques, crater classification based
on geomorphological characteristics, and the generation of
color products using multiple black-and-white images recorded
through spectral filters. The Image Processing Laboratory of
the Jet Propulsion Laboratory was responsible for the design,
development, and application of the software required to
produce these 'second-order' products.

031.322 Processing the Viking lander camera data.
    E. C. Levinthal, W. Green, K. L. Jones, R. Tucker.
J. Geophys. Res., Vol. 82, (see 003.017), 4412 - 4420 (1977).
Paper No. 7S0440.
    Over 1000 camera events were returned from the two
Viking landers during the Primary Mission. A system was
devised for processing camera data as they were received, in
real time, from the Deep Space Network. A second-order
processing system was developed which allowed extensive
interactive image processing including computer-assisted photogrammetry, a variety of geometric and photometric transformations, mosaicking, and color balancing using six different
filtered images of a common scene. These results have been
completely cataloged and documented to produce an Experiment Data Record.

**031.323 Viking 1975 Mars lander interactive computerized
    video stereophotogrammetry.**
S. Liebes, Jr., A. A. Schwartz.
J. Geophys. Res., Vol. 82, (see 003.017), 4421 - 4429 (1977).
Paper No. 7S0468.
    A novel computerized interactive video stereophoto-
grammetry system has been developed for analysis of Viking
1975 lander imaging data. Prompt, accurate, and versatile
performance is achieved.

031.324 VLBI clock synchronization.
    C. C. Counselman III, I. I. Shapiro, A. E. E. Rogers,
H. F. Hinteregger, C. A. Knight, A. R. Whitney, T. A. Clark.
Proc. IEEE, Vol. 65, 1622 - 1623 (1977).
    Atomic clocks at widely separated sites can be synchro-
nized to within several nanoseconds from a few minutes of
very-long-baseline interferometry (VLBI) observations, and to
within one nanosecond from several hours of such observations.

**031.325 Observational differences between human eye and
    photoelectric visual photometry.**   H. J. Landis.
J. American Assoc. Variable Star Obs., Vol. 6, 4 - 6 (1977).

**031.326 Técnicas observacionales en espectroscopía
    astrofísica.**   M. Rego.
Urania Barcelona, Año 61, Núm. 285, p. 41 - 52 (1976).

031.327 ALSEP-Quasar differential VLBI.
    M. A. Slade, R. A. Preston, A. W. Harris, L. J.
Skjerve, D. J. Spitzmesser.
Moon, Vol. 17, 133 - 147 (1977).
    A program of ALSEP-Quasar Very Long Baseline Inter-
ferometry (VLBI) is being carried out at the Jet Propulsion
Laboratory. These observations primarily employ a '4-antenna'
technique whereby simultaneous observations with two
antennas at each end of an intercontinental baseline are used
to derive the differential interferometric phase between a compact extragalactic radio source (usually a quasar) and a number
of ALSEP transmitters on the lunar surface. A continuous
ALSEP-quasar differential phase history over a few hour
period will lead to milliarcsecond angular accuracy in measuring the lunar position against the quasar reference frame if
suitable calibration measurements are obtained. These high
accuracy observations are of value to tie the lunar ephemeris
to a nearly inertial extragalactic reference frame, to test
gravitational theories, and to measure the Earth-Moon tidal
friction interaction.

031.328 Estimate of the limiting standard deviation in the absolute timing of periodic sources.
L. Rusconi, G. Sedmak.
Astrophys. Space Sci., Vol. 46, 301 - 308 (1977).

This note demonstrates the application of a rigorous procedure for estimating the limiting standard deviation in the absolute timing, of periodic sources of known period, to the problem of the evaluation of the effects of the main source and instrumental parameters. The procedure is used to examine the case of the optical pulsar NP 0532. The results are given in diagrammatic form.

031.329 Estimate of the optimum measuring time in the synchronous photometry of optical sources.
L. Rusconi, G. Sedmak.
Astrophys. Space Sci., Vol. 48, 53 - 56 (1977).

The note shows how to estimate accurately the optimum measuring time in the measurement of a periodical optical source by means of a synchronous photometer affected by phase-tracking error. The optimum number of scans to be averaged results to be typically $N_{opt} \approx 0.5r/\epsilon$ for an optical source of Lorentzian signal pattern, where r is the duty cycle of the pulsed component and ϵ the relative truncation error of the local sync to the source period.

031.330 Recognition of objects and position recovery from microdensitometer measurements of large field plates.   G. Caprioli, D. Nanni, A. Palma, A. Vignato.
Astrophys. Space Sci., Vol. 52, 27 - 33 (1977).

The Coma region on a glass copy of the Palomar Sky Survey has been scanned by the PDS microdensitometer at the Napoli Observatory. A method is described for obtaining the photometric parameters and positions of the images: 12 316 objects have been found. The repositioning of the scanner over the computed coordinates is satisfactory. AGK3 stars were considered to evaluate the plate constants and the precision. Comparison with Dressel and Condon's positions of galaxies gives a r.m.s. scatter consistent with the internal error of the published list.

031.331 Reduction of observations of scintillations of radio sources.   V. S. Artyukh, V. M. Getmanets, L. G. Chizhova.
Fiz. inst. AN SSSR. Fiz. kosmosa. Prepr. No. 69. Moskva, 1977. 10 pp. In Russian. − Abstr. in Ref. zh., 51. Astron., 11.51.583 (1977).

031.332 Use of the dispersion relation for computation of the distribution of the radio brightness of the sun and other sources of cosmic radio emission.
A. Balklavs, V. Locāns.
Investigations of the sun and red stars. 6, (see 003.019), p. 5 - 22 (1977). In Russian.

031.333 NH₃ spectral line measurements on Earth and Jupiter using a 10 μm superheterodyne receiver.
T. Kostiuk, M. J. Mumma, J. J. Hillman, D. Buhl, L. W. Brown, J. L. Faris, D. L. Spears.
Infrared Phys., Vol. 17, 431 - 439 (1977).

Measurements of absolute line positions and shapes of eleven absorption lines of NH_3 were made using an infrared heterodyne spectrometer. Line profiles were obtained with 5 MHz resolution. Using the obtained data a search list for NH_3 spectral lines on Jupiter was generated. Observations of Jovian polar regions in search of NH_3 aR (1,1) and sQ(2,1) auroral emission revealed possible strong nonthermal features. The line widths were consistent with low local kinetic temperatures. The relatively high line intensities, determined by local pumping mechanisms, corresponded to temperatures of several hundred Kelvins. Experimental details and implications of these measurements are discussed.

031.334 Astronomical Fourier spectroscopy in the infrared.   A. F. M. Moorwood.
Infrared Phys., Vol. 17, 441 - 449 (1977).

The principle of Fourier transform spectroscopy and considerations relating to its use in infrared astronomy are briefly reviewed. A survey of existing spectrometers and current applications is presented.

031.335 Statistical properties of the sum of partially developed speckle patterns.   J. Ohtsubo, T. Asakura.
Opt. Lett., Vol. 1, 98 - 100 (1977).

The probability density function and average contrast of the sum of n uncorrelated, partially developed speckle patterns have been theoretically investigated. A new form of the probability density function is given for the sum of n uncorrelated, partially developed speckle patterns under the assumption that the individual speckle fields to be added follow the circular statistics.

031.336 Applications of the data acquisition and processing system of Trieste Astronomical Observatory − Real time spectral analysis of time-varying optical sources.
G. Sedmak, M. Pucillo, P. Santin.
Publ. Astron. Obs. Trieste, No. 540 (1977).

031.337 Correction tables for chromatic refraction in photoelectric and other astrometrical observations.
G. G. Lengauer.
Izv. Glav. Astron. Obs. Pulkovo, No. 195, Astrofiz. Astrometr., p. 26 - 36 (1977). In Russian.

031.338 Elimination of lateral refraction from absolute declinations of equatorial stars by the micrometric method.   E. I. Krejnin, S. A. Tolchel'nikova-Murri.
Izv. Glav. Astron. Obs. Pulkovo, No. 195, Astrofiz. Astrometr., p. 37 - 42 (1977). In Russian.

The micrometric method of observations near the equator (1972, 1973) can be used for a study of lateral refraction. Some recommendations are given for compiling a program of observations and their reduction in order to estimate the lateral refraction effect on the measured values of arcs between the components of stellar pairs at low altitudes and to eliminate lateral refraction from the determined absolute declinations of stars.

031.339 Some comments on phase reduction formulae for observations of planets.   G. K. Gorel'.
Izv. Glav. Astron. Obs. Pulkovo, No. 195, Astrofiz. Astrometr., p. 43 - 45 (1977). In Russian.

Phase reduction formulas used for visual and photographic observations at the Nikolaev Observatory are considered. Tables are given for corrections to the right ascensions and declinations of a planet when the thread of the measuring device is set on the planet's visible disc by the method of equal areas.

031.340 On the use of VLBI for solution of fundamental problems of astrometry, geodesy and geodynamics.
I. D. Zhongolovich, V. I. Valyaev, A. A. Malkov, T. B. Sabanina.
Tr. Inst. Teor. Astron., Leningrad, vyp. (No.) 16, p. 19 - 58 (1977). In Russian.

The paper deals with the principles of the application of VLBI to the solution of some fundamental problems of astrometry, geodesy and geodynamics. These problems are concerned with the determination of lengths and directions of terrestrial chords extending over several thousands of kilometers, the precise synchronization of atomic frequency and time standards located at the ends of these chords, the determination of the earth's polar motion, and of the variations in the angular speed of the earth's diurnal rotation. All exact formulas of the VLBI method are derived which may be necessary for these applications. Various methods of practical applications of the derived formulas are described. The accuracies of expected results have been studied on the basis of models depending upon the length and orientation of a radio-interferometer baseline, the number and location of the observed radio sources as well as upon the accuracy of directly observed quantities.

031.341 Refraction correction for the reduction routines "RA".   N. P. Erpylev.
Nauchn. Inf., vyp. (No.) 35, p. 120 - 122 (1977). In Russian.
    A modification of the refraction correction in the reduction routine "RA" is proposed. The modification allows to use this routine for large zenith distances.

031.342 On the determination of astronomical flexure.   V. V. Konin.
Astrometriya i Astrofizika, Kiev, vyp. (No.) 33, (see 003.020), p. 96 - 99 (1977). In Russian.
    An autocollimation method is suggested to measure the flexure of an astrometric instrument tube. The method uses reverse reflectors and a mirror (attached to the cube of the instrument) the normal of which is parallel to the optical axis of the instrument tube.

031.343 Recent results of VLBI observations.
    I. I. K. Pauliny-Toth, E. Preuss, A. Witzel, R. Genzel, K. I. Kellermann, D. B. Shaffer, V. A. Efanov, L. R. Kogan, V. I. Kostenko, L. I. Matveenko, I. G. Moiseev.
Mitt. Astron. Ges., Nr. 42, p. 99 (1977).

031.344 Eine statistische Methode zur Bestimmung von Radialgeschwindigkeiten unter simultaner Benützung großer Spektralbereiche mit dem PDS-1000 Mikrodensitometer.   H. Jenkner, W. W. Weiss, H. J. Wood.
Mitt. Astron. Ges., Nr. 42, p. 131 - 133 (1977).

031.345 On the accuracy of the photographic method of investigating the astronomical refraction.
A. V. Arkhangel'skij.
Geod., kartogr. i aehrofotosemka. Resp. mezhved. nauchn.-tekh. sb., 1977, vyp. (No.) 26, p. 5 - 9. In Russian. − Abstr. in Ref. zh., 51. Astron., 12.51.146 (1977).

031.346 Electronic punching device for recording moments of star transits.
V. K. Budz'ko, A. N. Otkidach.
Vrashchenie i priliv. deformatsii Zemli. Resp. mezhved. sb., 1977, vyp. (No.) 9, p. 97 - 99. In Russian. − Abstr. in Ref. zh., 51. Astron., 12.51.158 (1977).

031.347 Some peculiarities of interpretation of cosmic pictures of the lunar and planetary surfaces.
M. V. Ostrovskij.
Tr. TsNII geod., aehrosemki i kartogr. 1977, vyp. (No.) 218, p. 52 - 62. In Russian. − Abstr. in Ref. zh., 51. Astron., 12.51.198 (1977).

031.348 Integral photoelectric device.
    A. D. Egorov, L. I. Churaeva.
Physics of the moon and planets. Fundamental astrometry, (see 003.024), Vestn. Khar'kov. Univ., No. 160, p. 32 - 36 (1977). In Russian.

031.349 ZMP optiskās novērošanas metodes.
    L. Laucenieks.
Zvaigžņotā debess, 1977. gada pavasaris, p. 7 - 15.

    Some results of Soviet-French co-operation in satellite geodesy.   See Abstr. 013.036.

    Active image restoration with a flexible mirror.
See Abstr. 031.040.

    Wideband adaptive optics for imaging.
See Abstr. 031.041.

    Flight performance of the 102-cm low balloon-borne far infrared telescope.   See Abstr. 032.516.

    Viking Lander camera: performance characteristics and data reduction techniques.   See Abstr. 032.534.

    The European Space Agency study of photon counting imaging for the Space Telescope.
See Abstr. 032.598.

    Sur l'emploi d'un réseau échelle dans un spectromètre photoélectrique destiné à la mesure des vitesses radiales.
See Abstr. 034.002.

    Un équipement du type «Télévision» pour l'observation des étoiles doubles visuelles.   See Abstr. 034.009.

    Study of a photoelectric device for measurements of photographic recordings of circle reading.
See Abstr. 034.014.

    The IAS triangulation camera.
See Abstr. 034.026.

    How to build a speckle interferometer.
See Abstr. 034.037.

    Submillimeter detector calibration with a low-temperature reference for space applications.
See Abstr. 034.053.

    Image-scanning systems using tilting plane mirrors.
See Abstr. 034.071.

    Accuracy of the average of star positions obtained from overlapping plates.   See Abstr. 041.022.

    Accuracy of star positions obtained by a block adjustment.   See Abstr. 041.023.

    Nuevos coeficientes para la reducción automática de posiciones de estrellas.   See Abstr. 041.030.

    Lunar laser ranging and fundamental astrometry.
See Abstr. 041.044.

    Differential determinations of declinations from observations with a vertical circle.   See Abstr. 041.051.

    Earth rotation study using lunar laser ranging data.
See Abstr. 044.005.

    Earth rotation as inferred from McDonald Observatory lunar laser observations during October 1975.
See Abstr. 044.006.

    McDonald UTO results and implications for the EROLD campaign.   See Abstr. 044.008.

    Universal Time: lunar ranging results and comparisons with VLBI and classical techniques.
See Abstr. 044.009.

    Determinación de la latitud por distancias cenitales de la polar. Método de Littrow.   See Abstr. 045.016.

    Die Photographie von Satelliten-Laserechos mit der ZEISS BMK 75 − Erste Ergebnisse und Folgerungen.
See Abstr. 046.025.

Instrumental and atmospheric corrections for the reduction of photographic plates taken with sidereally driven satellite cameras.   See Abstr. 046.029.

Résultats de la détermination de mouvements des continents d'après les méthodes astronomiques.
See Abstr. 046.030.

The role of extremely accurate surveying techniques in existing geodetic networks.   See Abstr. 046.031.

The measurement of the positions of points on the Earth's surface using an absolute gravimeter and a multi-wavelength geodimeter as complements to extraterrestrial techniques.   See Abstr. 046.032.

Geodesy by radio interferometry.
See Abstr. 046.033.

Geodetic applications of laser ranging.
See Abstr. 046.043.

Smithsonian Astrophysical Observatory laser tracking systems.   See Abstr. 046.045.

Goddard laser systems and their accuracies.
See Abstr. 046.046.

Future developments in lunar and satellite laser ranging.   See Abstr. 046.047.

Results from lunar laser ranging.
See Abstr. 046.048.

Berechnung topozentrischer Richtungen künstlicher Erdsatelliten aufgrund photographischer Messungen mit AFU-75-Kameras.   See Abstr. 054.024.

Spatial heterodyne interferometry of VY Canis Majoris, Alpha Orionis, Alpha Scorpii, and R Leonis at 11 microns.   See Abstr. 064.016.

The results of magnetograph calibration programmes 1974.   See Abstr. 072.004.

Photographic determination of heliographic coordinates of sunspots without taking double photos.
See Abstr. 072.045.

Measurement of the solar wind using spacecraft radio scattering observations.   See Abstr. 074.086.

The analysis and interpretation of solar X-ray photographs.   See Abstr. 076.014.

Solar stereo radioastronomy.
See Abstr. 077.048.

Gravity-field determination from laser observations.
See Abstr. 081.042.

Measurements of atmospheric isoplanatism using speckle interferometry.   See Abstr. 082.045.

Scintillation statistics caused by atmospheric turbulence and speckle in satellite laser ranging.
See Abstr. 082.050.

Upper-atmosphere studies by ranging to satellites.
See Abstr. 082.077.

On the estimate of accuracy of determination of the parameters of an ellipsoid approximating the image of the horizon on a space photograph of a planet.
See Abstr. 091.070.

Free librations of the Moon from lunar laser ranging.
See Abstr. 094.006.

A numerical study of the effects of fourth degree terms in the Earth–Moon mutual potential on lunar physical librations.   See Abstr. 094.007.

An introductory review of ephemerides for lunar laser ranging.   See Abstr. 094.016.

Lunar surface chemistry: a new imaging technique.
See Abstr. 094.407.

Investigation of the figure of the physical surface of planets from space photographs of their limbs (for example of photographs obtained from space apparatus Mars 3).
See Abstr. 097.124.

Spectrophotometric and color estimates of the Viking lander sites.   See Abstr. 097.185.

Viking imaging of Phobos and Deimos: an overview.
See Abstr. 097.502.

The angular diameter of Vesta from speckle interferometry.   See Abstr. 098.050.

The spectrophotometry of meteor video data.
See Abstr. 104.046.

Methods of meteor observations.
See Abstr. 104.051.

Mg and Ca isotopic study of individual microscopic crystals from the Allende meteorite by the direct loading technique.   See Abstr. 105.021.

On the accuracy of fundamental quasi-mono-chromatic stellar photoelectric photometry.
See Abstr. 113.007.

The influence of the Balmer discontinuity in *UBV* reductions.   See Abstr. 113.009.

Comparison of results of absolute spectrophotometry and UBV photometry.   See Abstr. 113.046.

Quantitative Klassifikation von Sternspektren mittels Objektivprismenaufnahmen.   See Abstr. 114.017.

High-resolution polarization observations inside spectral lines of magnetic Ap stars. I. Instrumentation and observations of β Coronae Borealis.   See Abstr. 114.513.

The application of maximum entropy spectral analysis to the study of short-period variable stars.
See Abstr. 122.076.

A photometric study of NGC 1904.
See Abstr. 154.045.

New color photos of galaxies.
See Abstr. 158.099.

Data Processing, Automation

031.401 **Some applications of computers to optical astronomy.** R. J. Dodd.
J. British Astron. Assoc., Vol. 87, 464 - 471 (1977).

031.402 **The system of recording and processing of radioastronomical data in real time.**
S. L. Domnin, V. A. Efanov, E. S. Korsenskaya, V. A. Korsenskij, I. G. Moiseev, N. S. Nesterov, I. D. Strepka.
Izv. Krymskoj Astrofiz. Obs., Vol. 57, 205 - 208 (1977). In Russian.

A multichannel system of automatic recording and processing of observational data on the basis of the computer "Nairi-K" is described. The system records the results of observations of perforated and magnetic tape and in graphical and digital form, does processing of results in real time, and operates the receiver.

031.403 **L'informatique et les étoiles doubles.**
M. Fulconis.
Astronomie, Vol. 91, 361 - 365 (1977).

031.404 **Coronal electron temperatures as measured by X-ray photometer.**
F. Fárník, M. Macháček, B. Valníček.
Solar activity and solar-terrestrial relations, (see 012.007), p. 147 - 150 (1976).

The authors present here a rather general method for evaluating electron temperatures and emission measures from X-ray energy flux data obtained by satellites.

031.405 **Lunar height measurements made easy.**
H. D. Jamieson.
Strolling Astron., Vol. 26, 235 - 240 (1977).

031.406 **FORTRAN IV subroutines for integration of the equations of motion of artificial earth satellites.**
V. I. Prokhorenko.
Inst. kosm. issled. AN SSSR. Prepr. Pr-302. Moskva, 1977. 73 pp. In Russian. − Abstr. in Ref. zh., 62. Issled. kosm. prostranstva, 9.62.339 (1977).

031.407 **Automation of reduction of pulsar observations.**
B. V. Vyzhlov, V. V. Ivanova, V. A. Izvekova, A. D. Kuz'min, Yu. P. Kuz'min, V. M. Malofeev, Yu. M. Popov, N. S. Solomin, T. V. Shabanova, Yu. P. Shitov.
Tr. Fiz. inst. AN SSSR, Vol. 93, 69 - 77 (1977). In Russian. − Abstr. in Ref. zh., 51. Astron., 11.51.320 (1977).

031.408 **Automation of record readings of the micrometer of the Wanschaff vertical circle.**
L. A. Kukharskij, A. S. Kharin.
Astron. Tsirk., No. 940, p. 1 - 2 (1977). In Russian.

031.409 **An interactive system for the reduction of photographic data using FORTH.**
F. Schiffer, III, C. A. Harvel, T. R. Gull.
Proc. Southwest Reg. Conf., Vol. 3, (see 012.043), p. 99 - 115 (1977).

With the use of fast microdensitometers, the amount of digital data available from photographic plates has risen to the point that computer processing is a necessity. The authors have developed an interactive processor for reducing such data. To avoid the inherent difficulties of a sophisticated one-pass program, their processor is an interactive one using FORTH as its basic operating system.

031.410 **ELSPEC/1. Un software package per l'elaborazione automatizzata di spettrogrammi stellari. I. Struttura.**
L. Rusconi, G. Sedmak, L. Crivellari.
Publ. Astron. Obs. Trieste, No. 544 (1977).

031.411 **ELSPEC/1. Un software package per l'elaborazione automatizzata di spettrogrammi stellari. II. Programmi.** L. Rusconi.
Publ. Astron. Obs. Trieste, No. 545 (1977).

031.412 **Simulation of the in-orbit maintenance cycle.**
J. A. Donnelly.
The Space Telescope, (see 012.046), p. 216 - 218 (1976).

031.413 **Automation of position and photometric measurements.** V. P. Kosykh, G. P. Chejdo.
Avtometriya, 1977, No. 3, p. 88 - 98. In Russian. − Abstr. in Ref. zh., 51. Astron., 1.51.166 (1978).

031.414 **Coherent-optical computer of the Khar'kov University.** V. N. Dudinov, V. S. Tsvetkova, V. A. Krishtal', A. N. Gurenko, L. F. Shpilinskij.
Physics of the moon and planets. Fundamental astrometry, (see 003.024), Vestn. Khar'kov. Univ., No. 160, p. 65 - 76 (1977). In Russian.

031.415 **Application of coherent-optical methods for reduction of astronomical images.**
V. N. Dudinov, V. S. Tsvetkova, V. A. Krishtal', N. P. Stadnikova, L. F. Shpilinskij.
Physics of the moon and planets. Fundamental astrometry, (see 003.024), Vestn. Khar'kov. Univ., No. 160, p. 76 - 86 (1977). In Russian.

AAVSO data processing: ten years of computerization. See Abstr. 010.001.

Computer simulation studies of compensation of turbulence degraded images. See Abstr. 031.299.

Standardized modular instrumentation system for space scientific instruments. See Abstr. 032.553.

System applications of the fault tolerant memory. See Abstr. 032.615.

Space Telescope external interfaces. See Abstr. 032.616.

Using of a computer for excluding the influence of radiometer instability on measurement results. See Abstr. 033.003.

Automation of radioastronomical works on the RT-22 radio telescope with a computer. See Abstr. 033.021.

Automation of the control of the horizontal telescope for a magnetograph. See Abstr. 034.006.

A real-time spectrum analyser with on-line definition of the confidence levels. See Abstr. 034.063.

The multifunction photoelectric photometer of Torino Observatory. See Abstr. 034.064.

Computer solutions for studying correlations between solar magnetic fields and Skylab X-ray observations. See Abstr. 080.062.

Maximum-likelihood method for determination of membership in open clusters. See Abstr. 153.012.

Accurate radial velocities using cross-correlation techniques and a TV detector. I. The velocity dispersion of NGC 6397. See Abstr. 154.025.

032 Astronomical Instruments, Space Instrumentation

Astronomical Instruments

032.001 2-m-RCC-Spiegelteleskop für Bulgarien.
H. Schmieder.
Jenaer Rundsch. (Jena Rev.), 22. Jahrg., 205 (1977).

032.002 First tests of the Irénée du Pont telescope.
H. W. Babcock.
Sky Telesc., Vol. 54, 90 - 94 (1977).

032.003 Interference method for control of the quality of the BTA main mirror.
T. S. Kolomijtsova, N. V. Konstantinovskaya, N. A. Gojko.
Opt.-mekh. prom-st', 1976, No. 12, p. 3 - 6. In Russian.
Abstr. in Ref. zh., 51. Astron., 8.51.91 (1977).

032.004 Application of the isophotometric registration method to investigations and certification of the BTA main mirror.
V. A. Zverev, V. K. Kirillovskij, M. N. Sokol'skij.
Opt.-mekh. prom-st', 1976, No. 12, p. 6 - 8. In Russian.
Abstr. in Ref. zh., 51. Astron., 8.51.94 (1977).

032.005 Neuer Flug des Heidelberger Ballonteleskopes THISBE 1.   D. Lemke.
Sterne Weltraum, Jahrg. 16, p. 294 - 295 (1977).

032.006 User guide to telescope equipment. I. Isaac Newton Telescope,Herstmonceux, UK.   D. L. Harmer.
Q. J. R. Astron. Soc., Vol. 18, 351 - 364 (1977).

032.007 Le nouveau télescope de Sirahama.   A. M. Sinzi.
Astronomie, Vol. 91, 338 - 339 (1977).

032.008 ESO's 3.6 meter teleskop i brug.   S. Laustsen.
Astron. Tidsskr., Årg. 10, 62 - 78 (1977).

032.009 Le télescope de Schmidt de l'I.N.A.G.
J.-L. Heudier.
Recueil des séminaires, (see 012.004), p. 120 - 122 (1977).

032.010 The Anglo-Australian Telescope.
E. J. Wampler, D. C. Morton.
Vistas Astron., Vol. 21, 191 - 207 (1977).
The principal characteristics of the instrument are delineated, some of the problems that have been encountered are discussed, and the organizational structure is described.

032.011 Optical arrangement of a BTA-type telescope.
M. V. Lobachev, L. E. Yakukhnova.
Astrofiz. Issled., Izv. Spets. Astrofiz. Obs., Vol. 9, 99 - 107 (1977). In Russian.
The optical arrangement of a BTA-type telescope is considered. It includes the primary focus and the fixed focus system. The latter contains optical units which provide operation of the telescope at different relative apertures and different fields of view. Aberration characteristics for each unit are given. Aberrations are presented in the form of a scatter circle expressed in seconds of arc.

032.012 Calculation of information and economic efficiency of an astronomical observational instrument.
V. S. Rylov.
Astrofiz. Issled., Izv. Spets. Astrofiz. Obs., Vol. 9, 108 - 110 (1977). In Russian.

032.013 A device for remote fine correction control of a small telescope.   V. G. Shtol'.
Astrofiz. Issled., Izv. Spets. Astrofiz. Obs., Vol. 9, 111 - 114 (1977). In Russian.
A description is presented of a device developed and manufactured at the Special Astrophysical Observatory of USSR Academy of Sciences for remote fine correction control of the telescope "Zeiss-600".

032.014 First results from observations using a two-telescope interferometer.
A. Blazit, D. Bonneau, L. Koechlin.
C. R. Acad. Sci. Paris, Tome 285, Sér. B, 149 - 152 (1977). In French.
A prototype of optical aperture synthesis array — two telescopes 25 cm in diameter — begins operation in southern France. With a variable baseline from 13 to 19 m, it yields astrophysical data on star diameters and close binaries with 1 arc-millisecond resolution.

032.015 Cornell's 25-inch training telescope.
J. R. Houck, G. E. Gull.
Sky Telesc., Vol. 54, 264 - 266 (1977).

032.016 The UK 3.8 m infrared telescope.
J. British Astron. Assoc., Vol. 87, 594 - 596 (1977).

032.017 Ein auf die automatische Datenverarbeitung abgestimmtes Zeitregistriersystem für das Danjon-Astrolab.   J. Cuno.
Veröff. Bayerisch. Komm. Int. Erdmessung, Bayerisch. Akad. Wiss., Astron.-Geod. Arbeiten, Heft Nr. 36, p. 73 - 79 (1977).

032.018 The Soviet 6-meter altazimuth reflector.
B. K. Ioannisiani.
Sky Telesc., Vol. 54, 356 - 362 (1977).

032.019 First altazimuth telescope with a 6-m mirror.
B. K. Ioannisiani.
Zemlya i Vselennaya, 1977. No. 6, p. 48 - 54. In Russian.

032.020 Den sovjet-russiske 6 meter reflektor.
B. R. Pettersen.
Astron. Tidsskr., Årg 10, 96 - 100 (1977).

032.021 A new catadioptric telescope. I, II.
R. E. Cox, R. W. Sinnott.
Sky Telesc., Vol. 54, 425 - 430, 432, 521 - 527 (1977).

032.022 Cassegrain reflectors for amateurs.   H. E. Dall.
Modern astronomy, (see 003.013), p. 30 - 39 (1977).

032.023 Near infrared astronomical light collector.
J.-P. Chevillard, P. Connes, M. Cuisenier, J. Friteau, C. Marlot.
Appl. Opt., Vol. 16, 1817 - 1833 (1977).
A 4-m aperture light collector has been built using the mosaic principle specifically for high resolution stellar spectroscopy in the near infrared by the Fourier multiplex technique. The most novel feature is complete servo control of all optical elements to bring individual images to a common point. Initial goals were a spectral survey of ir objects, and the testing of techniques for a larger future collector. However, the project has been stopped by lack of interest and support; the finished collector will not be used.

032.024 Image quality and alignment tolerances of IAC (*Instituto de Astrofisica de Canarias*) 80-cm telescope.   M. J. Galán, C. S. Magro.

Appl. Opt., Vol. 16, 2040 - 2041 (1977).

032.025   Field tests of a meteor camera.   E. N. Kramer,
   L. Ya. Lakejchuk, V. F. Lemeshchenko, Yu. E.
Migach, B. A. Murnikov, Yu. D. Russo, E. A. Timchenko-
Ostroverkhova, I. S. Shestaka.
Problems of cosmic physics. Vyp. (No.) 12, (see 003.016),
p. 60 - 70 (1977). In Russian.

032.026   A linear astrolabe.   D. Tattersfield.
   J. British Astron. Assoc., Vol. 88, 53 - 55 (1977).

**032.027   Organization of the work for making the blank for
   the main mirror of the BTA (Large Azimuth Tele-
scope).**   V. A. Shestakov, S. E. Stepanov, I. M. Buzhinskij,
V. F. Sinyakov.
Opt.-mekh. prom-st', 1977, No. 3, p. 55 - 57. In Russian.
Abstr. in Ref. zh., 51. Astron., 10.51.141 (1977).

**032.028   Technological control of the main mirror of the
   BTA (Large Azimuth Telescope) by the Hartmann
method.**   V. A. Zverev, S. A. Rodionov, M. N. Sokol'skij,
V. V. Usoskin.
Opt.-mekh. prom-st', 1977, No. 3, p. 3 - 5. In Russian.
Abstr. in Ref. zh., 51. Astron., 10.51.142 (1977).

**032.029   Investigation of the main mirror of the BTA (Large
   Azimuth Telescope) at an observatory.**
V. A. Zverev, S. A. Rodionov, M. N. Sokol'skij, V. V. Usoskin.
Opt.-mekh. prom-st', 1977, No. 4, p. 3 - 5. In Russian.
Abstr. in Ref. zh., 51. Astron., 10.51.143 (1977).

**032.030   A very large optical telescope array linked with fused
   silica fibers.**
J. R. P. Angel, M. T. Adams, T. A. Boroson, R. L. Moore.
Astrophys. J., Vol. 218, 776 - 782 (1977).
   The authors propose a new approach to the problem of
building a very large optical array that makes use of single
fused silica fibers to bring together light from around 100
mirrors, each of ~2.5 m diameter. They discuss the properties
of fused silica fibers; that is, their transmission as a function of
wavelength and the effect of fiber propagation on focal ratio.
A design for a Fiber-linked Optical Array Telescope (FLOAT)
which would work well with currently available fibers is
presented. The authors contrast the properties of a fiber-linked
telescope with telescopes of more conventional design.

032.031   Building a 6 inch reflector. II. Figuring the mirror.
   A. P. Witzgall.
Astronomy, Vol. 5, No. 5, p. 42 - 47 (1977). – Abstr. in Phys.
Abstr., Vol. 80, Abstr. 64639 (1977).

**032.032   A photographic zenith tube has been constructed in
   China.**
Nanking Astronomical Instruments Factory, Academia Sinica.
Acta Astron. Sinica, Vol. 18, 32 - 38 (1977).
   In this paper, the main parts of the PZT and problems of
operating the PZT in vacuum are described and discussed.

**032.033   Thermal deformation characteristics of a six-inch
   graphite/epoxy and ultra-low-expansion mirror
telescope.**   J. M. Miller.
J. Spacecr. Rockets, Vol. 14, 315 - 316 (1977). – Abstr. in
Phys. Abstr., Vol. 80, Abstr. 78636 (1977).

032.034   A new domeless solar telescope.   C. Kuhne.
   Zeiss Inf., Vol. 22, No. 84, p. 31 (1977). – Abstr.
in Phys. Abstr., Vol. 80, Abstr. 78640 (1977).

**032.035   Building a 6″ reflector. III. Assembling your tele-
   scope.**   A. P. Witzgall.
Astronomy, Vol. 5, No. 8, p. 26 - 31 (1977). – Abstr. in Phys.

Abstr., Vol. 80, Abstr. 87266 (1977).

**032.036   3.6 m telescope. The adjustment and test on the sky
   of the prime focus optics with the Gascoigne plate
correctors.**   F. Franza, M. Le Luyer, R. N. Wilson.
ESO Tech. Rep., No. 8, 68 pp. (1977).
   This report is concerned with the prime focus station of
the 3.6 m telescope with the Gascoigne plate correctors. The
alignment procedures are described and the results of two-
dimensional Hartmann testing and polynomial analysis are
exposed in detail. Final centering was also performed from the
Hartmann analysis to 0.2 arcsec of decentering coma. The
behaviour of the Serrurier truss with inclined telescope, the
effects of top-unit exchange and removal and replacement of
the prime mirror were investigated with regard to variations of
decentering coma. The conclusion is that the optical specification, even including the "red" or "blue" Gascoigne plate
(these plates are themselves difficult technical objects), has
been met by a considerable margin. However, the potentially
excellent optical quality can only be exploited if improvements
in dome and telescope seeing are effected and a high precision
of centering maintained.

032.037   First stage of designing a large azimuth telescope.
   N. N. Mikhel'son.
Opt.-mekh. prom-st', 1977, No. 5, p. 70 - 72. In Russian. –
Abstr. in Ref. zh., 51. Astron., 11.51.187 (1977).

**032.038   Investigation of change of the position of the focal
   plane of a Danjon astrolabe with temperature.**
G. I. Balashova, V. I. Sergienko.
Vrashchenie i priliv. deformatsii Zemli. Resp. mezhved. sb.,
1977, vyp. (No.) 9, p. 93 - 97. In Russian. – Abstr. in Ref.
zh., 51. Astron., 11.51.276 (1977).

**032.039   Investigation of the levels of the Tashkent meridian
   circle.**   I. M. Boroditskij.
Tsirk. Astron. Inst., Tashkent, No. 64 (411), p. 7 - 13 (1976).
In Russian.

**032.040   A retrospect on the vacuum meridian marks of the
   Belgrade Large Transit Instrument.**
L. A. Mitić, I. Pakvor.
Bull. Obs. Astron. Belgrade, No. 128, p. 11 - 15 (1977).
   Evidence, collected so far, shows that vacuum meridian
marks meet two principal requirements: the steadiness of the
mark image through durable vacuum and the position stability
of the vacuum tubes. The precision of the vacuum marks
readings is equal to that of collimator readings in all circumstances.

**032.041   Hvar Observatory 65-cm telescope and some prelimi-
   nary results.**   Z. Ivanović, K. Pavlovski.
Publ. Dep. Astron., Univ. Beograd, Fac. Sci., No. 6, (see
012.040), p. 43 - 47 (1976).
   Description of a 65-cm telescope and its single-channel
UBV photometer is given. Determination of certain instrumental parameters which are necessary in the reduction of observations (calibration of sensitivity switch, measuring of photometer apertures, determination of extinction and color-system
coefficients) is presented. Note is given on observations carried
out so far, as well as on future research programme and improvements of the instrument.

032.042   The vacuum meridian marks of Belgrade Observatory.
   L. Mitić, I. Pakvor.
Publ. Dep. Astron., Univ. Beograd, Fac. Sci., No. 6, (see
012.040), p. 111 - 113 (1976).
   The examinations of the system of the Large Transit
Instrument – vacuum meridian marks of Belgrade Observatory
are being continued. The examinations of the motion of

vacuum tube end according to temperature variation are carried out. Obtained values show clear dependence on the temperature variations, however, they are within tolerable limits.

032.043 Observational results of the Belgrade Vertical Circle after its reconstruction.
G. Teleki, M. Mijatov.
Publ. Dep. Astron., Univ. Beograd, Fac. Sci., No. 6, (see 012.040), p. 123 - 124 (1976).

032.044 Comparison of two programs for the determination of the corrections of the micrometer screw value of the Belgrade Zenith Telescope.   M. Djokić.
Publ. Dep. Astron., Univ. Beograd, Fac. Sci., No. 6, (see 012.040), p. 137 - 138 (1976).

032.045 The constants of the transit instrument "Bamberg" 10/100 cm of Belgrade Observatory.
M. Jovanović.
Publ. Dep. Astron., Univ. Beograd, Fac. Sci., No. 6, (see 012.040), p. 139 - 140 (1976).

032.046 The United Kingdom telescope.   T. J. Lee.
Infrared Phys., Vol. 17, 485 (1977). − Abstract.

032.047 On-orbit optical control of the Space Telescope.
A. Wissinger, T. Facey.
Opt. Spectra, Vol. 11, No. 3, p. 30, 32 - 38 (1977).

032.048 On the determination of the parameters of orientation of the Pulkovo large transit instrument from observations during 1953−1961.   A. I. Plyugina.
Izv. Glav. Astron. Obs. Pulkovo, No. 195, Astrofiz. Astrometr., p. 14 - 20 (1977). In Russian.

032.049 Thermal defocusing in reflectors with zero expansion material mirrors.   N. N. Mikhel'son.
Izv. Glav. Astron. Obs. Pulkovo, No. 195, Astrofiz. Astrometr., p. 172 - 173 (1977). In Russian.
A reflector of the Cassegrain type with a steel tube and mirrors made of zero expansion material is considered. It is shown that in case temperature compensation is not provided for the disk of scattering in either of the secondary foci (Cassegrain or Coudé) will increase approximately by 0.5 sec of arc if the temperature of the air varies by 1°C.

032.050 60 cm reflector at KSC (*Kagoshima Space Center*).
M. Matsuoka, H. Tsunemi, M. Eiraku, K. Tomita.
Rep. Inst. Space Aeronaut. Sci., Univ. Tokyo, Vol. 12, 775 - 791 (1976). In Japanese.

A new optical 60-cm reflector was constructed at KSC, Kagoshima, in the summer of 1975. Photographic observations are available at the prime focus (F = 4.5) and a photon counting photometer is set up at the Gregorian focus (F = 31) to observe the fast optical variations. Since the telescope is set up at the rocket launching center, it is very convenient to carry out simultaneous X-ray and optical observation for the X-ray sources as well as normal astronomical observations.

032.051 The 1.5 m reflector of the Tartu Observatory.
A. Kipper.
Calendar of the Tartu Astronomical Observatory for the year 1978, (see 047.026), p. 30 - 35 (1977). In Estonian.

032.052 Increasing the effectivity of observations of artificial earth satellites with the AFU-75 satellite telescope.   I. D. Bregeda, M. A. Frolov.
Vopr. radiofiz. Khabarovsk, 1976, p. 75 - 79. In Russian.
Abstr. in Ref. zh., 52. Geod. Aehrosemka, 12.52.82 (1977).

032.053 Astronomische Längen- und Breitenbestimmung mit einer transportablen Zenitkamera.   G. Seeber.
Mitt. Astron. Ges., Nr. 42, p. 81 - 83 (1977).

032.054 Search-guide system of a large azimuthal telescope.
V. A. Malarev.
Opt.-mekh. prom-st', 1977, No. 7, p. 32 - 36. In Russian.
From Ref. zh., 51. Astron., 1.51.96 (1978).

032.055 On systematic errors of the Zeiss short-focus astrograph.   P. P. Pavlenko.
Physics of the moon and planets. Fundamental astrometry, (see 003.024), Vestn. Khar'kov. Univ., No. 160, p. 55 - 59 (1977). In Russian.

032.056 Riekstukalna Šmita teleskops desmit gados.
A. Alksnis.
Zvaigžņotā debess, 1976/77. gada ziema, p. 1 - 7.

The planispheric astrolabe, make-it-yourself astrolabe.   See Abstr. 003.180.

Use of the Multiple-Mirror Telescope as a multibase Michelson stellar interferometer.   See Abstr. 031.263.

Automation of record readings of the micrometer of the Wanschaff vertical circle.   See Abstr. 031.408.

Astronomical azimuths of terrestrial objects as indicators of rotational motions of continental blocks. See Abstr. 081.002.

Space Instrumentation

032.501 **Apparatus for balloon measurements of the neutron flux.** J. Dubinský, K. Kudela, Yu. E. Efimov, Yu. A. Chichikalyuk, L. Michaeli, T. Vašek.
Bull. Astron. Inst. Czechoslovakia, Vol. 28, 241 - 243 (1977).
An apparatus for measuring the integral flux of neutrons in the energy range from thermal up to 1 MeV and for estimating power spectra in the interval of 1–10 MeV is described. The apparatus is used for measuring the neutron characteristics in the atmosphere on balloons. The detector system, electronic circuits and recording system are described.

032.502 **Čerenkov source as a possible standard of radiation in the vacuum ultraviolet region.**
V. K. Prokof'ev.
Izv. Krymskoj Astrofiz. Obs., Vol. 57, 221 - 227 (1977). In Russian.
The emission in the spectral region 1200 - 4000 Å of a Čerenkov source with a plate of MgF_2 and $Sr^{90} + Y^{90}$ is calculated. The Hartmann interpolation formula for the refraction index of MgF_2 in the region 5460 - 1200 Å is given.

032.503 **Perspectives of development of instruments for space investigations.** L. S. Gorn, B. I. Khazanov.
[Tr.] Soyuz. NII priborostr. 1977, vyp. (No.) 34 - 35, tom (Vol.) 1, 134 - 148. In Russian. – Abstr. in Ref. zh., 62. Issled. kosm. prostranstva, 8.62.103 (1977).

032.504 **Suppression of the background of cosmic rays in UFS-2 and UFS-3 ultraviolet radiation spectrophotometers.** S. I. Babichenko, E. V. Dereguzov, V. A. Sklyankin, A. S. Smirnov.
[Tr.] Soyuz. NII priborostr. 1977, vyp. (No.) 34 - 35, tom (Vol.) 1, 192 - 199. In Russian. – Abstr. in Ref. zh., 62. Issled. kosm. prostranstva, 8.62.114 (1977).

032.505 **Improvement of measuring characteristics of an apparatus for investigation of X-rays of cosmic sources.** S. I. Babichenko, V. A. Sklyankin.
[Tr.] Soyuz. NII priborostr. 1977, vyp. (No.) 34 - 35, tom (Vol.) 1, 149 - 157. In Russian. – Abstr. in Ref. zh., 62. Issled. kosm. prostranstva, 8.62.115 (1977).

032.506 **Instruments for plasma measurements in space investigations.** L. S. Zhurina, D. S. Zakharov, A. A. Klimashov, V. V. Polenov, V. I. Khazanov.
[Tr.] Soyuz. NII priborostr. 1977, vyp. (No.) 34 - 35, tom (Vol.) 1, 168 - 179. In Russian. – Abstr. in Ref. zh., 62. Issled. kosm. prostranstva, 8.62.119 (1977).

032.507 **P-2 instrument for measurement of the parameters of plasma near artificial earth satellites.**
S. K. Chapkynov, M. Kh. Petrunova, T. N. Ivanova.
Nauchn. pribory. No. 11, Moskva, 1976, p. 23 - 26. In Russian. Abstr. in Ref. zh., 62. Issled. kosm. prostranstva, 8.62.120 (1977).

032.508 **Instruments for measurement of molecular streams in the circumterrestrial cosmic space.**
L. G. Ol'dekop.
[Tr.] Soyuz. NII priborostr. 1977, vyp. (No.) 34 - 35, tom (Vol.) 1, 180 - 185. In Russian. – Abstr. in Ref. zh., 62. Issled. kosm. prostranstva, 8.62.121 (1977).

032.509 **Characteristics of the proportional counter for soft X-rays, developed for Czechoslovak space experiments.** F. Fárník, B. Valníček, L. Moučka.
Solar activity and solar-terrestrial relations, (see 012.007), p. 161 - 165 (1976).
Instruments on the Intercosmos satellites, designed for solar X-ray flux measurement, were equipped with a scintillation detector, a gasfilled proportional counter and a solid-state detector. This paper reviews the development and characteristics of the proportional counters.

032.510 **Ultraviolet spectroscopy with Copernicus.**
T. P. Snow, Jr.
Sky Telesc., Vol. 54, 371 - 374 (1977).

032.511 **Determination of the orientation of the axis of a gamma-telescope aboard an artificial earth satellite. II. Method for calculation of the orientation.**
Yu. A. Gur'yan, E. P. Mazets, I. A. Sokolov.
Kosm. Issled., Vol. 15, 635 - 639 (1977). In Russian.

032.512 **Multispectral and stereo imaging on Mars.**
E. C. Levinthal, F. O. Huck.
Proc. 27th Internat. Astronaut. Congr. (see 012.016), p. 303 - 321 (1977).

032.513 **The large space telescope astrometric instrument.**
W. F. van Altena, O. G. Franz, L. W. Fredrick.
Highlights of Astronomy, Vol. 4, Part I, (see 012.021), p. 351 (1977).

032.514 **Balloon-borne transform spectroscopy.**
W. A. Traub.
Far infrared astronomy (see 012.027), p. 1 - 9 (1976).
The design and construction of a high-resolution far infrared Fourier transform spectrometer for use on the Smithsonian balloon-borne 1-metre telescope is described. The instrument will operate at a resolution of about 0.1 cm^{-1} in the region 25 to 150 μm, and will be used to obtain spectra of Jupiter, Venus, Orion and other H II and molecular cloud regions, and the terrestrial stratosphere.

032.515 **Some balloon-borne IR telescope developments at University College London.** I. Furniss, R. E. Jennings, W. A. Towlson, T. E. Venis, B. Y. Welsh.
Far infrared astronomy (see 012.027), p. 15 - 19 (1976).
The design of the balloon-borne 60-cm telescope system under construction at University College London is described, together with the modulation system to be used.

032.516 **Flight performance of the 102-cm low balloon-borne far infrared telescope.**
G. G. Fazio, E. L. Wright, F. J. Low.
Far infrared astronomy (see 012.027), p. 21 - 31 (1976).
Maps with a resolution of 1' arc FWHM have been achieved with absolute position accuracies of ±10'' arc.

032.517 **The use of a large telescope in the infrared.**
D. E. Kleinmann.
Far infrared astronomy (see 012.027), p. 33 - 45 (1976).
The impacts of the design, location, and size of a telescope upon its infrared limitations and capabilities, and therefore on its scientific use in the infrared are considered, with specific application to the Large Space Telescope (LST). The telescope, and the IR photometer proposed for it are described, and the infrared performance to be expected from their combination is evaluated. This performance is compared to that for other large telescopes. Some possible far infrared astronomy applications are indicated for the LST.

032.518 **Chopping primary for a balloon-borne IR telescope.**
D. Lemke, K. Haussecker.
Far infrared astronomy (see 012.027), p. 47 - 51 (1976).
The design of a servo controlled chopper driver for a 20-cm primary of a far infrared telescope is described. The system allows square wave modulation even at large chopper throws without causing vibrations or acoustic noise. The reliability and temperature stability of the system have been

tested during a recent balloon flight.

032.519 A lamellar grating interferometer experiment to determine the spectrum of the cosmic background radiation.   J. B. Mercer, S. Wilson, P. Chaloupka, W. K. Griffiths, P. Marchant, P. L. Marsden, C. C. Morath.
Far infrared astronomy (see 012.027), p. 103 - 114 (1976).

A balloon-borne interferometer experiment to measure the cosmic background radiation over the spectral range $3 \leqslant \tilde{\nu} \leqslant 12$ cm^{-1} is described. In addition, calculations on the expected atmospheric emission are presented.

032.520 An airborne infrared astronomy program: system description and preliminary results.   P. Turon, D. Rouan, P. Lena, J. Wijnbergen, J. W. Aalders.
Far infrared astronomy (see 012.027), p. 201 - 205 (1976).

The 32-cm airborne telescope has been mounted on the CV-990 from NASA and extensively used during and after a Spacelab operations simulation period. The instrument was equipped with a four-channels photometer covering the spectral range 30-200 μm. Extensive study of noise behaviour at the tropopause boundary was made. Preliminary results for M17, W51 and ρ Oph are given.

032.521 Verification of radiation background rates in an IR sensor system.   A. B. Holman, E. C. Smith, G. W. Autio.
IEEE Trans. Nucl. Sci., Vol. NS-23, 1775 - 1780 (1976). Abstr. in Phys. Abstr., Vol. 80, Abstr. 22011 (1977).

032.522 Space plasma chamber.   H. Mori, S. Miyazaki.
Rev. Radio Res. Lab., Vol. 22, 163 - 168 (1976). In Japanese. − Abstr. in Phys. Abstr., Vol. 80, Abstr. 22039 (1977).

032.523 Large-area multi-crystal NaI (Tl) detectors for X-ray and gamma-ray astronomy.
G. J. Fishman, R. W. Austin.
Nucl. Instrum. Methods, Vol. 140, 193 - 196 (1977). Abstr. in Phys. Abstr., Vol. 80, Abstr. 22070 (1977).

032.524 A portable He^3 cryostat for space applications.   G. Chanin, J. P. Torre.
Proceedings of the 6th International Cryogenic Engineering Conference, Grenoble, France, 11 - 14 May 1976. F. R. S. Mendelssohn (Editor). IPC Sci. & Technol. Press, Guildford, Surrey, England. Price £ 29.00 (1976), p. 96 - 98. − Abstr. in Phys. Abstr., Vol. 80, Abstr. 22077 (1977).

032.525 Optimal maneuvering and fine pointing control of large space telescope with a new magnetically suspended, single gimballed momentum storage device.
A. A. Nadkarni, S. M. Joshi.
Proceedings of the 1976 IEEE conference on decision and control including the 15th symposium on adaptive processes, Clearwater, Fla., USA, 1 - 3 Dec. 1976. IEEE, New York, USA, p. 854 - 856 (1976). − Abstr. in Phys. Abstr., Vol. 80, Abstr. 22078 (1977).

032.526 Soft X-ray performance of shallow blazed diffraction gratings.   E. G. Lowen.
J. Opt. Soc. America, Vol. 67, 1439 (1977). − Abstract.

032.527 Spin-scan imaging − application to planetary missions.   E. E. Russell, M. G. Tomasko.
Chemical evolution of the giant planets, (see 003.014), p. 147 - 164 (1976).

032.528 Effect of the shuttle contaminant environment on a sensitive infrared telescope.
J. P. Simpson, F. C. Witteborn.
Appl. Opt., Vol. 16, 2051 - 2073 (1977).

032.529 Quasi-microscope concept for planetary missions.   F. O. Huck, R. E. Arvidson, E. E. Burcher, O. Giat, S. D. Wall.
Appl. Opt., Vol. 16, 2454 - 2459 (1977).

Viking lander cameras have returned stereo and multi-spectral views of the Martian surface with a resolution that approaches 2 mm/lp in the near field. A two-orders-of-magnitude increase in resolution could be obtained for collected surface samples by augmenting these cameras with auxiliary optics that would neither impose special camera design requirements nor limit the cameras field of view of the terrain. Quasi-microscope images would provide valuable data on the physical and chemical characteristics of planetary regoliths.

032.530 Rocket calibration of the Nimbus 6 solar constant measurements.   C. H. Duncan, R. G. Harrison, J. R. Hickey, J. M. Kendall, Sr., M. P. Thekaekara, R. C. Willson.
Appl. Opt., Vol. 16, 2690 - 2697 (1977).

032.531 A focusing gas scintillation proportional counter.   D. F. Anderson, O. H. Bodine, R. Novick, R. S. Wolff.
Nucl. Instrum. Methods, Vol. 144, 485 - 491 (1977). − Abstr. in Phys. Abstr., Vol. 80, Abstr. 83461 (1977).

032.532 Construction and performance of a large-area multi-wire ionization hodoscope for use in a cosmic-ray detector.   P. L. Love, J. Tueller, J. W. Epstein, M. H. Israel, J. Klarmann.
Nucl. Instrum. Methods, Vol. 140, 469 - 576 (1977). − Abstr. in Phys. Abstr., Vol. 80, Abstr. 30714 (1977).

032.533 Viking Lander camera.   H. McCall.
Developments in semiconductor microlithography. San Jose, Calif., USA, 1 - 3 June 1976. Proceedings of the Society of Photo-optical Instrumentation Engineers, Vol. 80. D. E. Routh, E. C. Thompson, R. P. Mandel, J. W. Giffen (Editors). Palos Verdes Estates, Calif., USA (1976). p. 108. Abstr. in Phys. Abstr., Vol. 80, Abstr. 41285 (1977).

032.534 Viking Lander camera: performance characteristics and data reduction techniques.   F. O. Huck.
Developments in semiconductor microlithography. San Jose, Calif., USA, 1 - 3 June 1976. Proceedings of the Society of Photo-optical Instrumentation Engineers, Vol. 80. D. E. Routh, E. C. Thompson, R. P. Mandel, J. W. Giffen (Editors). Palos Verdes Estates, Calif., USA (1976). p. 109. − Abstr. in Phys. Abstr., Vol. 80, Abstr. 41286 (1977).

032.535 Intensified charge coupled devices for ultra low light level imaging.   S. Sobieski.
Low light level devices for science and technology, Reston, Va., USA. 22 - 23 March 1976. Proceedings of the Society of Photo-optical Instrumentation Engineers, Vol. 78. C. Freeman (Editor). Palos Verdes Estates, Calif., USA. 6 + 162 pp. (1976). p. 73 - 77. − Abstr. in Phys. Abstr., Vol. 80, Abstr. 41287 (1977).

032.536 Stable oscillator for Pioneer Venus Programme.   M. B. Bloch, M. P. Meirs, T. M. Robinson.
Proceedings of the 30th Annual Symposium on Frequency Control. Atlantic City, N.J., USA, 2 - 4 June 1976. Electronic Industries Assoc. 8 + 543 pp. Price $ 8.00 (1976). p. 279 - 283. Abstr. in Phys. Abstr., Vol. 80, Abstr. 41288 (1977).

032.537 The imaging photopolarimeter: the first extraterrestrial spin-scan imager.   W. Swindell.
Proceedings of the 22nd International Instrumentation Symposium, San Diego, Calif., USA, 25 - 27 May 1976. ISA, Pittsburgh, Pa., USA. 13 + 729 pp. (1976). ISBN 0-87664-290-3. p. 649 - 665. − Abstr. in Phys. Abstr., Vol. 80, Abstr.

41303 (1977).

032.538 Shuttle Infrared Telescope Facility.
S. G. McCarthy.
Modern utilization of infrared technology civilian and
military. II. (see 012.031), p. 2 - 7 (1976). – Abstr. in Phys.
Abstr., Vol. 80, Abstr. 49297 (1977).

**032.539 A telescope for the Infrared Astronomical Satellite
(IRAS).** R. L. Hedden.
Modern utilization of infrared technology civilian and
military. II. (see 012.031), p. 8 - 12 (1976). – Abstr. in Phys.
Abstr., Vol. 80, Abstr. 49298 (1977).

032.540 Planetary radiometers. S. C. Chase.
Modern utilization of infrared technology civilian
and military. II. (see 012.031), p. 30 - 37 (1976). – Abstr. in
Phys. Abstr., Vol. 80, Abstr. 49299 (1977).

032.541 Adaptive optics for space telescopes.
T. R. O'Meara, C. J. Swigert, W. P. Brown.
Modern utilization of infrared technology civilian and
military. II. (see 012.031), p. 126 - 133 (1976). – Abstr. in
Phys. Abstr., Vol. 80, Abstr. 49301 (1977).

**032.542 Spectrograph suitable for the mass and energy
analysis of space plasmas over the energy range 0.1–
10 keV.** T. E. Moore.
Rev. Sci. Instrum., Vol. 48, 221 - 225 (1977). – Abstr. in
Phys. Abstr., Vol. 80, Abstr. 53401 (1977).

**032.543 A simple aspect sensor for small geophysical
vehicles.** M. Ilyas, G. K. M. Thutupalli.
J. Phys. E, Vol. 10, 446 - 447 (1977). – Abstr. in Phys. Abstr.,
Vol. 80, Abstr. 45279 (1977).

032.544 The Viking orbiter visual imaging subsystem.
J. B. Wellman, F. P. Landauer, D. D. Norris,
T. E. Thorpe.
J. Spacecr. Rockets, Vol. 13, 660 - 666 (1976). – Abstr. in
Phys. Abstr., Vol. 80, Abstr. 45282 (1977).

032.545 Spacelab infrared telescope facility (SIRTF).
F. C. Witteborn, L. S. Young.
J. Spacecr. Rockets, Vol. 13, 667 - 674 (1976). – Abstr. in
Phys. Abstr., Vol. 80, Abstr. 45283 (1977).

**032.546 System for biological and soil chemical tests on a
planetary lander.**
R. J. Radmer, B. Kok, J. P. Martin.
J. Spacecr. Rockets, Vol. 13, 719 - 726 (1976). – Abstr. in
Phys. Abstr., Vol. 80, Abstr. 45284 (1977).

032.547 On a high gain Yagi antenna.
P. K. Reddy, V. K. Lakshmeesha, S. Pal, V.
Mahadevan, L. Nicholas, S. P. Kosta.
J. Inst. Electron. Telecommun. Eng., Vol. 22, 772 - 774
(1976). – Abstr. in Phys. Abstr., Vol. 80, Abstr. 45286 (1977).

032.548 Ultraviolet reflecting mirrors for space applications.
M. Blanc, A. Malherbe.
J. Opt., Vol. 8, 195 - 199 (1977). In French. – Abstr. in Phys.
Abstr., Vol. 80, Abstr. 67022 (1977).

**032.549 Compensated time-of-flight telescope for space-
borne cosmic ray measurements.** H. Rothermel.
IEEE Trans. Nucl. Sci., Vol. NS-24, 801 - 803 (1977). – Abstr.
in Phys. Abstr., Vol. 80, Abstr. 67939 (1977).

**032.550 The satellite Hermes, Canada's outer space connec-
tion.** W. Cherwinski.
Sci. Dimension, Vol. 9, No. 2, p. 10 (1977). – Abstr. in Phys.

Abstr., Vol. 80, Abstr. 70977 (1977).

**032.551 The HEAO-A scanning modulation collimator in-
strument.** A. Roy, J. Ballas, N. Jagoda,
P. McKinnon, A. Ramsey, E. Wester.
IEEE Trans. Nucl. Sci., Vol. NS-24, 804 - 809 (1977). – Abstr.
in Phys. Abstr., Vol. 80, Abstr. 71003 (1977).

**032.552 Recent progress in the development of a gas scintilla-
tion proportional counter for X-ray astronomy.**
R. D. Andresen, E.-A. Leimann, A. Peacock, B. G. Taylor,
G. Brownlie, P. Sanford.
IEEE Trans. Nucl. Sci., Vol. NS-24, 810 - 816 (1977). – Abstr.
in Phys. Abstr., Vol. 80, Abstr. 71004 (1977).

**032.553 Standardized modular instrumentation system for
space scientific instruments.**
R. C. Carden, III., D. W. Juergens, J. O. Maloy, G. R. Mohler.
IEEE Trans. Nucl. Sci., Vol. NS-24, 823 - 834 (1977). – Abstr.
in Phys. Abstr., Vol. 80, Abstr. 71005 (1977).

**032.554 Shuttle/Spacelab: platform for a cooled infrared
astronomical telescope.** S. G. McCarthy.
J. Spacecr. Rockets, Vol. 14, 345 - 350 (1977). – Abstr. in
Phys. Abstr., Vol. 80, Abstr. 73921 (1977).

**032.555 Space Experiments with Particle Accelerators:
SEPAC. Controlled active experiments in ionosphere
and magnetosphere with particle accelerators on-board the
Space Shuttle/Spacelab.** T. Obayashi.
Rep. Ionos. Space Res., Japan, Vol. 30, No. 3 - 4, p. 57 - 68
(1976). – Abstr. in Phys. Abstr., Vol. 80, Abstr. 73925 (1977).

**032.556 Optimal fine pointing control of a Large Space
Telescope using an Annular Momentum Control
Device.** A. A. Nadkarni, S. M. Joshi, N. J. Groom.
Proceedings of SOUTHEASTCON on imaginative engineering
thru education and experience, Williamsburg, Va., 4 - 6 April
1977. IEEE, New York, USA, 20 + 664 pp. (1977). p. 596 -
601. – Abstr. in Phys. Abstr., Vol. 80, Abstr. 73934 (1977).

**032.557 Technical development of scientific experiments for
spacecraft.** G. W. Cocks, J. W. Heaton.
17 convegno internazionale scientifico sullo spazio (see 012.
034), p. 483 - 494 (1977). – Abstr. in Phys. Abstr., Vol. 80,
Abstr. 82111 (1977).

**032.558 A new approach for a fine telescope pointing and
stabilization.** E. Turci.
17 convegno internazionale scientifico sullo spazio (see 012.
034), p. 411 - 419 (1977). – Abstr. in Phys. Abstr., Vol. 80,
Abstr. 82124 (1977).

032.559 A magnetometer for the Pioneer Venus Orbiter.
R. C. Snare, J. D. Means.
IEEE Trans. Magn., Vol. MAG-13, 1107 - 1109 (1977).
Abstr. in Phys. Abstr., Vol. 80, Abstr. 87229 (1977).

**032.560 Spaceborne astronomy with electro-optical image
sensors.** W. C. Bradley.
Opt. Eng., Vol. 16, 249 - 256 (1977). – Abstr. in Phys. Abstr.,
Vol. 80, Abstr. 87268 (1977).

**032.561 Balloon-borne infrared telescope for absolute sur-
face photometry of the night sky.**
W. Hofmann, D. Lemke, C. Thum.
Appl. Opt., Vol. 16, 3125 - 3130 (1977).
A dry ice cooled 15-cm ir telescope was used on board
the balloon-borne gondola THISBE for absolute surface
photometry of the Milky Way, the zodiacal light, and the airglow in the PbS wavelength region. The mechanical, optical,

electronical, and thermal design of the instrument is described. The efficiency of the baffle system for suppression of stray light from earth and balloon is discussed in detail. Recent airglow measurements are presented.

032.562 Inflight performance of the Viking visual imaging subsystem.
K. P. Klaasen, T. E. Thorpe, L. A. Morabito.
Appl. Opt., Vol. 16, 3158 - 3170 (1977).

Photography from the Viking Orbiter Visual Imaging Subsystem, taken while enroute to and in orbit about Mars, has been analyzed to determine the performance of the cameras. The cameras have remained in good focus. Random and coherent noise levels in flight were the same as measured prior to launch. A recalibration of each instrument allows photometric measurements to accuracies of less than 3% for relative measurements and 9% for absolute measurements. Geometric distortion remained close to the preflight levels of 4 pixels rms and 11 pixels maximum.

032.563 Measuring concave diffraction grating efficiencies at grazing incidence.   W. R. Hunter, D. K. Prinz.
Appl. Opt., Vol. 16, 3171 - 3175 (1977).

032.564 The Viking Lander Imaging Investigation: an introduction.   T. A. Mutch.
J. Geophys. Res., Vol. 82, (see 003.017), 4389 - 4390 (1977).
Paper No. 7S0557.

The nine articles that summarize the results of the Lander Imaging Investigation are briefly reviewed.

032.565 Calibration and performance of the Viking lander cameras.
W. R. Patterson III, F. O. Huck, S. D. Wall, M. R. Wolf.
J. Geophys. Res., Vol. 82, (see 003.017), 4391 - 4400 (1977).
Paper No. 7S0560.

The cameras on board the two Viking landers are, in effect, radiometers with an optical-mechanical scanning mechanism which provides the line-scan raster and which selects the field of view for a detector array of 12 silicon photodiodes. Since there are two cameras at known positions on each lander, their combined direction measurements make possible stereoscopic mapping of the near field, that is, high-accuracy photogrammetry.

032.566 The S-054 X-ray telescope experiment on Skylab.
G. S. Vaiana, L. van Speybroeck, M. V. Zombeck,
A. S. Krieger, J. K. Silk, A. Timothy.
Space Sci. Instrum., Vol. 3, 19 - 76 (1977).

A description of the S-054 X-ray telescope on Skylab is presented with a discussion of the experiment objectives, observing program, data reduction and analysis. Some results from the Skylab mission are given. The telescope photographically records high-resolution images of the solar corona in several broadband regions of the soft X-ray spectrum. It includes an objective grating used to study the line spectrum. The spatial resolution, sensitivity, dynamic range and time resolution of the instrument were chosen to survey a wide variety·of solar phenomena.

032.567 The gas proportional scintillation counter.
A. J. P. L. Policarpo.
Space Sci. Instrum., Vol. 3, 77 - 107 (1977).

Data published in the periodical literature concerning gas proportional scintillation counters are reviewed. The energy linearity and resolution, the effect of wavelength shifters, the process of light multiplication and the localization capabilities of the detectors are considered. Examples are given of some instruments together with their performance.

032.568 Periodic slot collimator for accurate gamma ray burst source locations.
C. Karmendy, H. Helmken.
Space Sci. Instrum., Vol. 3, 115 - 121 (1977).

The responses of a wide-angle two-dimensional position sensitive counter under a periodic slot collimator are investigated. Fast Fourier Transform analysis techniques are utilized. Computer simulations have reproduced multiple source locations to an accuracy of a few minutes of arc. Simulations involved single and dual point gamma ray sources; the latter were separated by a range of angles from 6' to several degrees.

032.569 Detectors for gamma-ray burst astronomy. (A critical comparison).
J. Carter, A. J. Dean, R. K. Manchanda, D. Ramsden.
Space Sci. Instrum., Vol. 3, 123 - 129 (1977).

In order to increase the sensitivity of detectors designed for gamma-ray burst astronomy a number of experimenters have built instruments which incorporate large blocks of plastic scintillators. The performance of this type of detector is compared with much smaller, sodium iodide, scintillators. A consideration of the burst photon spectrum, sources of background and the basic characteristics of the two detector systems suggests that a $125 \text{ cm}^2 \times 1 \text{ cm}$ NaI detector is equivalent in sensitivity to a $5000 \text{ cm}^2 \times 5 \text{ cm}$ plastic scintillator.

032.570 The LPSP (*Laboratoire de Physique Stellaire et Planétaire*) experiment on OSO-8. I: Instrumentation, description of operations, laboratory calibrations and pre-launch performances.
G. Artzner, R. M. Bonnet, P. Lemaire, J. C. Vial, A. Jouchoux,
J. Leibacher, A. Vidal-Madjar, M. Vite.
Space Sci. Instrum., Vol. 3, 131 - 161 (1977).

The LPSP instrument on OSO-8 is designed to study the solar atmosphere with high spatial, spectral and temporal resolution. The scientific objectives of the LPSP experiment concern the investigation of the dynamics of the upper atmosphere of the Sun and the fine structure of the quiet and active chromosphere. The authors describe the instrument and present its optical, electrical and photometric performances as measured on the ground before the launch.

032.571 The calibration of Bragg X-ray analyser crystals for use as polarimeters in X-ray astronomy.
K. D. Evans, R. Hall, M. Lewis.
Space Sci. Instrum., Vol. 3, 163 - 169 (1977).

The paper discusses the characteristics of X-ray analyser crystals used in the Bragg mode as polarimeters. A calibration technique is described and typical results are presented for the case of pentaerythritol.

032.572 Signal preprocessing on spacecraft.   D. Jones.
Space Sci. Instrum., Vol. 3, 171 - 185 (1977).

032.573 Digital onboard processor for directional charged particle flux measurements in space.
A. Balogh, T. Iversen.
Space Sci. Instrum., Vol. 3, 187 - 198 (1977).

Spacecraft-borne instruments for the measurement of directional particle flux distributions rely increasingly on digital onboard processors for improving their data collection efficiency. This paper describes in some detail a special purpose Data Processor from the point of view of how it complements the particle sensors for achieving optimum directional, energy and time measurements of interplanetary energetic proton fluxes.

032.574 Mariner 10 ultraviolet spectrometer: airglow experiment.
A. L. Broadfoot, S. S. Clapp, F. E. Stuart.
Space Sci. Instrum., Vol. 3, 199 - 208 (1977).

An extreme ultraviolet airglow spectrometer was flown on Mariner 10 to examine the atmosphere of Venus and Mercury. An objective grating spectrometer was used with channel electron multipliers at fixed positions in the image plane to continuously monitor the resonance-scattered emission rate of expected atomic atmospheric constituents He, H, A, Ne, O and C. A mechanical collimator placed in the entrance aperture of the spectrometer provided spectral separation of 19 Å over the wavelength range from He⁺ at 304 Å to C at 1657 Å and provided spatial separation of 0.125° consistent with the spacecraft and trajectory capabilities. The calibration techniques are discussed.

032.575 **Mariner 10 ultraviolet spectrometer: occultation experiment.**
A. L. Broadfoot, S. S. Clapp, F. E. Stuart.
Space Sci. Instrum., Vol. 3, 209 - 218 (1977).

A description is given of the spectrometer which was included in the scientific payload of the Mariner 10 spacecraft to measure the extinction properties of the Mercurian atmosphere at extreme ultraviolet wavelengths. The solar intensity was monitored continuously by a grazing incidence pinhole spectrometer in 150 Å bands centered at 485 Å, 740 Å, 810 Å and 895 Å. This instrument provided the capability of determining atmospheric surface pressures greater than 1×10^{-9} mbar.

032.576 **Plasma analyzer for the Pioneer Jupiter missions.**
D. D. McKibbin, J. H. Wolfe, H. R. Collard, H. F. Savage, R. Molari.
Space Sci. Instrum., Vol. 3, 219 - 228 (1977).

A description is given of the NASA/Ames Research Center Plasma Probe on board the Jupiter missions of the Pioneer 10 and 11 spacecraft. The instrument has two quadrispherical electrostatic analyzer units; one has high sensitivity and resolution and the other is capable of measuring large fluxes of solar wind particles. The two analyzer units measure particle energy-to-charge ratio, flux, and direction of flow for positive ions and electrons over the wide range of particle densities found in the solar wind during the Jupiter missions.

032.577 **The Voyager infrared spectroscopy and radiometry investigation.** R. Hanel, B. Conrath, D. Gautier, P. Gierasch, S. Kumar, V. Kunde, P. Lowman, W. Maguire, J. Pearl, J. Pirraglia, C. Ponnamperuma, R. Samuelson.
Space Sci. Rev., Vol. 21, 129 - 157 (1977).

The infrared investigation on Voyager uses two interferometers covering the spectral ranges 60–600 cm⁻¹ (17–170 μm) and 1000–7000 cm⁻¹ (1.4–10 μm), and a radiometer covering the range 8000–25000 cm⁻¹ (0.4–1.2 μm). Two spectral resolutions (approximately 6.5 and 2.0 cm⁻¹) are available for each of the interferometers. In the middle of the thermal channel (far infrared interferometer) the noise level is equivalent to the signal from a target at 50 K; in the middle of the reflected sunlight channel (near infrared interferometer) the noise level is equivalent to the signal from an object of albedo 0.2 at the distance of Uranus.

032.578 **The Voyager mission photopolarimeter experiment.**
C. F. Lillie, C. W. Hord, K. Pang, D. L. Coffeen, J. E. Hansen.
Space Sci. Rev., Vol. 21, 159 - 181 (1977).

The Voyager Photopolarimeter Experiment is designed to determine the physical properties of particulate matter in the atmospheres of Jupiter, Saturn, and the rings of Saturn by measuring the intensity and linear polarization of scattered sunlight at eight wavelengths in the 2350–7500 Å region of the spectrum. The experiment will also provide information on the texture and probable composition of the surfaces of the satellites of Jupiter and Saturn and the properties of the sodium cloud around Io. During the planetary encounters a search for optical evidence of electrical discharges (lightning) and auroral activity will also be conducted.

032.579 **Ultraviolet spectrometer experiment for the Voyager**

mission. A. L. Broadfoot, B. R. Sandel, D. E. Shemansky, S. K. Atreya, T. M. Donahue, H. W. Moos, J. L. Bertaux, J. E. Blamont, J. M. Ajello, D. F. Strobel, J. C. McConnell, A. Dalgarno, R. Goody, M. B. McElroy, Y. L. Yung.
Space Sci. Rev., Vol. 21, 183 - 205 (1977).

The Voyager Ultraviolet Spectrometer (UVS) is an objective grating spectrometer covering the wavelength range of 500–1700 Å with 10 Å resolution. Its primary goal is the determination of the composition and structure of the atmospheres of Jupiter, Saturn, Uranus and several of their satellites. The capability for two very different observational modes have been combined in a single instrument. Observations in the airglow mode measure radiation from the atmosphere due to resonant scattering of the solar flux or energetic particle bombardment, and the occultation mode provides measurements of the atmospheric extinction of solar or stellar radiation as the spacecraft enters the shadow zone behind the target. In addition to the primary goal of the solar system atmospheric measurements, the UVS is expected to make valuable contributions to stellar astronomy at wavelengths below 1000 Å.

032.580 **Magnetic field experiment for Voyagers 1 and 2.**
K. W. Behannon, M. H. Acuna, L. F. Burlaga, R. P. Lepping, N. F. Ness, F. M. Neubauer.
Space Sci. Rev., Vol. 21, 235 - 257 (1977).

The magnetic field experiment to be carried on the Voyager 1 and 2 missions consists of dual low field (LFM) and high field magnetometer (HFM) systems. The dual systems provide greater reliability and, in the case of the LFM's, permit the separation of spacecraft magnetic fields from the ambient fields. Additional reliability is achieved through electronics redundancy. Objectives include the study of planetary fields at Jupiter, Saturn, and possibly Uranus; satellites of these planets; solar wind and satellite interactions with the planetary fields; and the large-scale structure and microscale characteristics of the interplanetary magnetic field. The interstellar field may also be measured.

032.581 **The Low Energy Charged Particle (LECP) experiment on the Voyager spacecraft.**
S. M. Krimigis, T. P. Armstrong, W. I. Axford, C. O. Bostrom, C. Y. Fan, G. Gloeckler, L. J. Lanzerotti.
Space Sci. Rev., Vol. 21, 329 - 354 (1977).

The Low Energy Charged Particle (LECP) experiment on the Voyager spacecraft is designed to provide comprehensive measurements of energetic particles in the Jovian, Saturnian, Uranian and interplanetary environments. These measurements will be used in establishing the morphology of the magnetospheres of Saturn and Uranus, including bow shock, magnetosheath, magnetotail, trapped radiation, and satellite-energetic particle interactions. The experiment is described in detail.

032.582 **Cosmic ray investigation for the Voyager missions; energetic particle studies in the outer heliosphere— and beyond.** E. C. Stone, R. E. Vogt, F. B. McDonald, B. J. Teegarden, J. H. Trainor, J. R. Jokipii, W. R. Webber.
Space Sci. Rev., Vol. 21, 355 - 376 (1977).

A cosmic-ray detector system (CRS) has been developed for the Voyager mission which will measure the energy spectrum of electrons from ≈3–110 MeV and the energy spectra and elemental composition of all cosmic-ray nuclei from hydrogen through iron over an energy range from ≈1–500 MeV/nuc. Isotopes of hydrogen through sulfur will be resolved from ≈2–75 MeV/nuc. Studies with CRS data will provide information on the energy content, origin and acceleration process, life history, and dynamics of cosmic rays in the Galaxy, and contribute to an understanding of the nucleosynthesis of elements in the cosmic-ray sources.

032.583 **Corpuscular radiation and the problem of solar-atmospherical relations. II. Apparatus and methods** of measurements of precipitating corpuscular radiation with means of space technology.   A. P. Babaev, Yu. M. Zhuchenko, V. I. Lazarev, V. A. Lipovetskij, B. V. Mar'in, M. A. Savel'ev, M. V. Tel'tsov, V. F. Tulinov, V. M. Fejgin. Ehff. soln. aktivnosti v nizhn. atmos. Leningrad, Gidrometeoizdat, 1977, p. 106 - 119. In Russian. — Abstr. in Ref. zh., 62. Issled. kosm. prostranstva, 11.62.68 (1977).

032.584   Lamellar grating Fourier spectrometer for a balloon-borne telescope.
R. Hofmann, S. Drapatz, K. W. Michel.
Infrared Phys., Vol. 17, 451 - 456 (1977).

032.585   Cooled instrumentation for infrared astronomical investigations from space.   J. E. Beckman.
Infrared Phys., Vol. 17, 503 - 511 (1977).
    A brief review is presented of infrared astronomical instrumentation which has been proposed or accepted for observations from space platforms. Some stress is placed on cryogenic design problems which have hitherto formed barriers to progress. Technical solutions to problems are outlined, and steps in implementing some of them are described.

032.586   IR observatory slated for orbit.
Opt. Spectra, Vol. 11, No. 5, p. 19, 21 (1977).

032.587   Selection of a variant for the construction of an automatic stratospheric solar station.
V. M. Danilov.
Izv. Glav. Astron. Obs. Pulkovo, No. 195, Astrofiz. Astrometr., p. 174 - 178 (1977). In Russian.
    The method of rank correlation was used for analysing four possible variants of the construction of an automatic solar stratospheric station.

032.588   Airborne submillimeter radiometer with removable spectral filters for investigation of the atmosphere's emission in the 50—1000 μm wavelength region.
V. D. Gromov, G. B. Sholomitskij, V. A. Soglasnova, V. V. Artamonov.
Radiofiz. issled. atmos. Leningrad, Gidrometeoizdat, 1977, p. 130 - 131. In Russian. — Abstr. in Ref. zh., 62. Issled. kosm. prostranstva, 12.62.95 (1977).

032.589   Automatic radiometer "Obzor" for submillimeter investigations of the earth from board artificial earth satellites.   A. E. Solomonovich, S. V. Solomonov,
A. S. Khajkin, V. N. Bakun, V. S. Kovalev.
Radiofiz. issled. atmos. Leningrad, Gidrometeoizdat, 1977, p. 229 - 236. In Russian. — Abstr. in Ref. zh., 62. Issled. kosm. prostranstva, 12.62.96 (1977).

032.590   Program status.   J. A. Downey III.
The Space Telescope, (see 012.046), p. 21 - 28 (1976).

032.591   Scientific instruments.   G. M. Levin.
The Space Telescope, (see 012.046), p. 29 - 37 (1976).

032.592   Scientific operation plan.   D. K. West.
The Space Telescope, (see 012.046), p. 40 - 45 (1976).
    A Space Telescope ground system is described that is compatible with the operational requirements of the Space Telescope.

032.593   Automation of the Space Telescope.
W. W. Warnock, C. W. Case.
The Space Telescope, (see 012.046), p. 55 - 57 (1976).

032.594   Science performance considerations for the design

of the Space Telescope.
D. D. Ostrander, J. C. Tuttle.
The Space Telescope, (see 012.046), p. 60 - 63 (1976).

032.595   Large format Secondary Electron Conduction Orthicon integrating television sensor for the Space Telescope.   J. L. Lowrance.
The Space Telescope, (see 012.046), p. 82 - 87 (1976).

032.596   The intensified-charge-coupled device as a photon-counting imager.   J. T. Williams.
The Space Telescope, (see 012.046), p. 88 - 89 (1976).

032.597   Infrared capabilities.   R. T. Hall, T. Kelsall,
D. E. Kleinmann, G. Neugebauer.
The Space Telescope, (see 012.046), p. 90 - 96 (1976).

032.598   The European Space Agency study of photon counting imaging for the Space Telescope.
R. J. Laurance.
The Space Telescope, (see 012.046), p. 97 - 105 (1976).

032.599   Development of an infrared spectroradiometer.
W. H. Alff, J. G. Thunen.
The Space Telescope, (see 012.046), p. 106 - 108 (1976).

032.600   Faint object spectrograph.   W. P. Devereux.
The Space Telescope, (see 012.046), p. 109 - 110 (1976).

032.601   High-speed point/area photometer, conceptual design and integration.
W. Bloomquist, F. Steputis.
The Space Telescope, (see 012.046), p. 111 - 113 (1976).

032.602   High-resolution spectrograph.   K. Peacock.
The Space Telescope, (see 012.046), p. 114 - 117 (1976).

032.603   An analytical and experimental evaluation of actuator vibration on Space Telescope image distortion.   A. D. Houston, L. W. Hodge, Jr., T. J. Kertesz.
The Space Telescope, (see 012.046), p. 146 - 150 (1976).

032.604   Development of a large-inertia fine-pointing and dimensional stability simulator.
R. L. Gates, D. H. Wine, R. W. Seiferth, N. A. Osborne.
The Space Telescope, (see 012.046), p. 151 - 152 (1976).

032.605   Evaluation of communication antenna drive system design requirements to allow tracking and data relay satellite tracking during Space Telescope fine pointing.
A. J. Besonis, C. J. Chang.
The Space Telescope, (see 012.046), p. 153 - 157 (1976).

032.606   Space Telescope interferometric fine guidance sensor.   A. B. Wissinger, R. H. Carricato.
The Space Telescope, (see 012.046), p. 158 - 160 (1976).

032.607   Prismatic grating startracker.   A. H. Greenleaf.
The Space Telescope, (see 012.046), p. 161 - 165 (1976).

032.608   Thermostructural design considerations to achieve the Space Telescope line of sight requirements.
D. J. Tenerelli.
The Space Telescope, (see 012.046), p. 166 - 168 (1976).

032.609   Design of low thermal distortion Space Telescope metering structure.   J. R. Lager.
The Space Telescope, (see 012.046), p. 169 - 173 (1976).

032.610   Three-axis simulation of the pointing control sub-
system – a multidiscipline activity.
W. W. Emsley, T. D. Fehr, D. C. Fosth, D. L. Knobbs.
The Space Telescope, (see 012.046), p. 174 - 176 (1976).

032.611   Data management and mission operations concept.
R. Walker, F. Hudson, L. Murphy.
The Space Telescope, (see 012.046), p. 178 - 180 (1976).

032.612   Data management for the Space Telescope.
G. R. Hope, Jr., T. J. Rasser.
The Space Telescope, (see 012.046), p. 181 - 188 (1976).

032.613   A cost-effective data management subsystem.
J. A. Dougherty, T. D. Patterson, A. E. Cole.
The Space Telescope, (see 012.046), p. 189 - 193 (1976).

032.614   System consideration, design approach, and test of a
low-gain spherical coverage antenna for large space
vehicles.   R. E. Ferguson, T. D. Patterson, M. R. Moreno.
The Space Telescope, (see 012.046), p. 194 - 200 (1976).

032.615   System applications of the fault tolerant memory.
L. J. Murphy.
The Space Telescope, (see 012.046), p. 201 - 203 (1976).

032.616   Space Telescope external interfaces.
R. E. Collart.
The Space Telescope, (see 012.046), p. 206 - 210 (1976).

032.617   Biaxial optical guider for extra-atmospheric solar
observations.   M. Hłond.
Postępy Astron., Tom 25, 245 - 255 (1977). In Polish.
The biaxial optical guider designed for rocket investigation of solar X-ray spectra is presented. Descriptions of the
block diagram, design methods, circuit and mechanical
diagrams of the instrument are given.

Infrared detection techniques for space research.
See Abstr. 012.068.

UTEX: a proposal from Italy for an astronomical
payload of the S/L.   See Abstr. 013.033.

The efficiencies of plane diffraction gratings measured in the conical and classical mountings.
See Abstr. 031.017.

Measuring concave diffraction grating efficiencies at
grazing incidence.   See Abstr. 031.018.

Primary aberrations for grazing incidence.
See Abstr. 031.023.

Optical performance control.
See Abstr. 031.067.

Impact of focal plane dynamics on image quality.
See Abstr. 031.068.

Stray light from out of field sources.
See Abstr. 031.069.

Design of highly stable optical support structure.
See Abstr. 031.070.

Mirror substrate material and manufacturing.
See Abstr. 031.071.

Fabrication and test of 1.8-meter (71-inch) diameter,
high-quality U.L.E.™ mirror.   See Abstr. 031.072.

Design and testing with a reflective null system.
See Abstr. 031.073.

Test results on homogeneity of expansion for a 1.8-
meter (71-inch) U.L.E.™ lightweight mirror.
See Abstr. 031.074.

Multiple telescope measurements of particle
anisotropies in space.   See Abstr. 031.203.

Fourier-spectroscopy from balloon platforms.
See Abstr. 031.254.

Simulation of the in-orbit maintenance cycle.
See Abstr. 031.412.

On-orbit optical control of the Space Telescope.
See Abstr. 032.047.

High spectral resolution line observations in the
infrared.   See Abstr. 034.012.

High resolution Michelson interferometer for airborne infrared astronomical observations. 1: Concept and
performance.   See Abstr. 034.018.

High resolution Michelson interferometer for airborne infrared astronomical observations. 2: System design.
See Abstr. 034.019.

Astronomers in space. I.   See Abstr. 051.040.

The plasma experiment on the 1977 Voyager mission.
See Abstr. 051.063.

Testing the Viking Lander.
See Abstr. 053.008.

Refurbishment and support.
See Abstr. 054.022.

Space Telescope power system long-life design
techniques.   See Abstr. 054.025.

Design of gravitational antennae for use at 3 mK.
See Abstr. 066.315.

Observations of resonance-line polarization in the
solar EUV.   See Abstr. 076.019.

Low-frequency solar radio bursts observed with the
"Intercosmos-Kopernik 500" satellite.
See Abstr. 077.009.

Bistatic radar measurements of electrical properties
of the Martian surface.   See Abstr. 097.178.

The Viking X-ray fluorescence experiment: analytical
methods and early results.   See Abstr. 097.196.

Errata

032.901   Errata: "Crystals for astronomical X-ray spectros-
copy' [Space Sci. Instrum., Vol. 2, 53 - 104 (1976)].
A. J. Burek.
Space Sci. Instrum., Vol. 3, 109 (1977).

033 Radio Telescopes and Equipment

033.001 An interferometer for millimeter wavelengths.
W. J. Welch, J. R. Forster, J. Dreher, W. Hoffman,
D. D. Thornton, M. C. H. Wright.
Astron. Astrophys., Vol. 59, 379 - 385 (1977).
A description is given of the Hat Creek Millimeter Wavelength Interferometer. It is a two element, aperture synthesis instrument with the antennas movable along a T-shaped track. The system is designed to operate in the range $1-15$ mm either in the continuum or with a 128 channel spectrometer. The mapping resolution is about $1''$ at the shortest wavelength. The system is described in some detail with emphasis on the novel aspects.

033.002 A high-resolution acousto-optical radiospectrometer for millimeter-wave astronomy.
N. Kaifu, N. Ukita, Y. Chikada, T. Miyaji.
Publ. Astron. Soc. Japan, Vol. 29, 429 - 435 (1977).
A high-resolution acousto-optical radiospectrometer has been developed and tested at the Tokyo Astronomical Observatory. The results confirm that the acousto-optical spectrometer with a relatively simple optical system and a TeO_2 crystal as a light deflector provides superior performance to its filter-bank equivalents.

033.003 Using of a computer for excluding the influence of radiometer instability on measurement results.
S. L. Domnin, V. A. Efanov, V. A. Korsenskij, I. G. Moiseev, N. S. Nesterov.
Izv. Krymskoj Astrofiz. Obs., Vol. 57, 199 - 204 (1977). In Russian.
The description of a receiver that is insensitive to its gain variations is given. The principle of its realization is based on calculation of the received and reference signals ratio by computer.

033.004 An investigation of VPA [variable profile antenna] depending on aperture illumination using the method of optical simulation.   A. N. Korzhavin.
Astrofiz. Issled., Izv. Spets. Astrofiz. Obs., Vol. 9, 53 - 70 (1977). In Russian.
Using the method of optical simulation a study is made of the dependence of variable profile antenna (VPA) pattern properties on the law of aperture illumination. Different distributions of the field amplitude are imitated by means of photographic filters of variable transparency. Quantitative data are obtained by recording of diffraction patterns. The output signal is treated with a computer and the data are represented as graphs and tables.

033.005 VPA [variable profile antenna] patterns for real modes.   A. N. Korzhavin.
Astrofiz. Issled., Izv. Spets. Astrofiz. Obs., Vol. 9, 71 - 88 (1977). In Russian.
The method of optical simulation is used to find the patterns of the VPA (variable profile antenna) for several altitudes in the range 0 to 90°. These can be used to treat observations made with the radiotelescopes LPR (Large Pulkovo Radiotelescope) and RATAN-600. To imitate the real laws of illumination of the antenna aperture special filters were made photographically. A problem of correction of the measured patterns which is necessary to allow for the difference in the vertical envelopes of the patterns of the telescopes considered is discussed. The results of measurements of the simulated antenna patterns are treated by computing technique and presented in the form of graphs and tables.

033.006 On an algorithm for calculating RATAN-600-type antenna patterns taking into account aberrations and polarization effects.   G. B. Gel'frejkh.
Astrofiz. Issled., Izv. Spets. Astrofiz. Obs., Vol. 9, 89 - 98 (1977). In Russian.
A calculating algorithm of a variable profile antenna pattern is described, where the main mirror is considered as a multielement interferometer. The algorithm allows to calculate easily aberrations of the main mirror which arise at longitudinal and transverse displacements of the feed from the primary focus, as well as polarization characteristics of the pattern. Possible employments of the algorithm described are discussed. In working out the algorithm special attention has been paid to convenience of its application to observations with RATAN-600.

033.007 3-cm radiotelescope.   V. L. Rakhlin, M. M. Zubov, R. N. Rat, A. M. Starodubtsev, A. F. Dement'ev.
Izv. vyssh. uchebn. zaved. Radiofiz., Vol. 20, 156 - 158 (1977). In Russian. − Abstr. in Ref. zh., 51. Astron., 9.51.146 (1977).

033.008 RATAN-600: first observations.   Yu. N. Parijskij.
Zemlya i Vselennaya, 1977, No. 6, p. 40 - 47.
In Russian.

033.009 Multiband low-noise receivers for a very large array.
S. Weinreb, M. Balister, S. Maas, J. Napier.
IEEE Trans. Microwave Theory Tech., Vol. MTT-25, 243 - 248 (1977) = Natl. Radio Astron. Obs., Green Bank, Repr. Ser. A, No. 702.
The very large array (VLA), presently under construction by the National Radio Astronomy Observatory, is an array of 27 25-m-diam antennas. This paper describes the feed and low-noise front-end systems used on the antennas. Measured system performance is presented and some construction details are given.

033.010 On the reflector accuracy of a high precision radiotelescope.   H. Eschenauer, P. Brandt, K. Schutz.
Tech. Mitt. Krupp Forschungsber., Vol. 35, No. 1, p. 17 - 31 (1977). In German. − Abstr. in Phys. Abstr., Vol. 80, Abstr. 53427 (1977).

033.011 A refracting radio telescope.
P. Bernhardt, A. V. da Rosa.
Radio Sci., Vol. 12, 327 - 336 (1977). − Abstr. in Phys. Abstr., Vol. 80, Abstr. 45312 (1977).

033.012 Radio astronomy: radio signals from the Universe.
H. H. Klinger.
Funkschau, Vol. 49, 106 - 110 (1977). In German. − Abstr. in Phys. Abstr., Vol. 80, Abstr. 56827 (1977).

033.013 Digital correlation.   J. D. O'Sullivan.
Tijdschr. Ned. Elektron.- Radiogenoot., Vol. 42, No. 1-2, p. 5 - 13 (1977). − Abstr. in Phys. Abstr., Vol. 80, Abstr. 64650 (1977).

033.014 Basic principles of observation techniques with the WSRT (*Westerbork Synthesis Radio Telescope*).
H. C. Kahlmann.
Tijdschr. Ned. Elektron. Radiogenoot., Vol. 42, No. 1 - 2, p. 3 - 4 (1977). In Dutch. − Abstr. in Phys. Abstr., Vol. 80, Abstr. 67039 (1977).

033.015 The 5000 channel digital correlator receiver in Westerbork.   A. Bos.
Tijdschr. Ned. Elektron. Radiogenoot., Vol. 42, No. 1 - 2, p. 15 - 19 (1977). In Dutch. − Abstr. in Phys. Abstr., Vol. 80,

Abstr. 67040 (1977).

033.016 Instrumentation and atmospheric disturbances; their consequences and cure.   J. D. Bregman.
Tijdschr. Ned. Elektron. Radiogenoot., Vol. 42, No. 1 - 2, p. 21 - 25 (1977). In Dutch. — Abstr. in Phys. Abstr., Vol. 80, Abstr. 67041 (1977).

033.017 The Jodrell Bank 1024-channel digital autocorrelation spectrometer.   E. Pointon.
J. Phys. E, Vol. 10, 833 - 837 (1977). — Abstr. in Phys. Abstr., Vol. 80, Abstr. 73970 (1977).

033.018 Il radiotelescopio «Croce del Nord».
A. Ficarra, E. Gandolfi, F. Perugini.
G. Astron., Vol. 3, 115 - 151 (1977).
Il radiotelescopio «Croce del Nord» è situato in una tranquilla pianura a pochi chilometri da Medicina (BO); la sua costruzione è iniziata nel 1960 ed è terminata nel 1967 pur senza completare il preliminare progetto costruttivo. Questo articolo vuole essere una trattazione volutamente semplice dei principi fondamentali e delle tecniche che sono alla base della realizzazione e del funzionamento di questo radiotelescopio che è attualmente il più importante esistente in Italia.

033.019 Progetto e realizzazione di un radiotelescopio per la gamma 26-32 cm.   R. Paolinetti.
Astronomia, N. 3, p. 37 - 42 (1977).

033.020 The Mark II Nançay radioheliograph. Preliminary results.
J. Bonmartin, I. Jones, A. Kerdraon, A. Lacombe, M. F. Lantos, P. Lantos, C. Mercier, M. Pick, G. Trottet, M. Bruley, C. Chantelat, M. Chapuis, Y. Chapuis, B. Clavelier, C. Couteret, J. P. Drouhin, P. Gueniau, R. Tocqueville.
Sol. Phys., Vol. 55, 251 - 261 (1977).
The main characteristics of this instrument are high space and time resolutions and flexible on-line data processing. At present it produces one-dimensional images at 169 MHz. This instrument is mainly intended for solar observations. Preliminary solar results are presented.

033.021 Automation of radioastronomical works on the RT-22 radio telescope with a computer.
A. V. Kutsenko, B. A. Polos'yants, Yu. M. Polubesova, R.L. Sorochenko.
Tr. Fiz. inst. AN SSSR, Vol. 93, 3 - 19 (1977). In Russian. — Abstr. in Ref. zh., 51. Astron., 11.51.198 (1977).

033.022 On some constructions of principal assemblies of parabolic radio telescope antennas.
P. D. Kalachev, I. A. Emel'yanov, V. P. Nazarov, V. L. Shubeko, V. B. Khavaev.
Tr. Fiz. inst. AN SSSR, Vol. 93, 23 - 44 (1977). In Russian. — Abstr. in Ref. zh., 51. Astron., 11.51.199 (1977).

033.023 A precision control system of the RT-22 radio telescope of the FIAN.
V. A. Vvedenskij, P. D. Kalachev, A. D. Kuz'min, Yu. N. Semenov, R. L. Sorochenko.
Tr. Fiz. inst. AN SSSR, Vol. 93, 45 - 49 (1977). In Russian. — Abstr. in Ref. zh., 51. Astron., 11.51.200 (1977).

033.024 On a method of determining the manufacturing costs of a completely turnable radio telescope.
P. D. Kalachev, A. V. Samotsvetov. Eh. A. Tret'yakov.
Tr. Fiz. inst. AN SSSR, Vol. 93, 59 - 68 (1977). In Russian. — Abstr. in Ref. zh., 51. Astron., 11.51.201 (1977).

033.025 High-efficiency microwave reflector antennas—a review.   P. J. B. Clarricoats, G. T. Poulton.
Proc. IEEE, Vol. 65, 1470 - 1504 (1977).
The paper provides a review of current research on microwave reflector antennas with particular regard to those antennas which produce pencil-beam radiation patterns. Attention is focused on circularly symmetric antennas with axially symmetric feed systems. This class of antenna accounts for the largest number of applications which include microwave point-to-point communication, satellite communication, and radio astronomy.

033.026 Radiospectrographs for the decimeter wave range.
A. M. Gusejnov, V. P. Egorychev, Yu. D. Danilov, V. A. Tikhomirov.
Astron. Tsirk., No. 948, p. 1 - 3 (1977). In Russian.

033.027 Spectrograph for investigation of the fine structure of spectra of the S-component of solar radio emission in the 4.2—7.1 GHz region.   A. N. Vaulin, N. S. Kaverin, A. I. Korshunov, Yu. D. Panfilov, V. A. Tikhomirov, N. N. Kholodilov.
Astron. Tsirk., No. 948, p. 3 - 5 (1977). In Russian.

033.028 Low frequency orbiting radiotelescope. Structural feasibility survey.   F. Polma.
Proc. Southwest Reg. Conf., Vol. 3, (see 012.043), p. 129 - 137 (1977).
Low frequency radiation is prevented from reaching the earth by ionospheric reflection. It is desirable to build a low frequency radiotelescope in orbit. Structural feasibility of such telescopes is now enhanced by use of space shuttles, however the solar wind and solar noise may present problems. Unfurlable as well as spun and rigid structures are feasible.

033.029 Automatic meteor radar system.
B. L. Kashcheev, Yu. I. Voloshchuk, A. A. Tkachuk, B. S. Dudnik, A. A. D'yakov, V. V. Zhukov, V. A. Nechitajlenko.
Meteor. issled. No. 4. Moskva, "Sov. radio", 1977, p. 11 - 61. In Russian. — Abstr. in Ref. zh., 51. Astron., 12.51.290 (1977).

033.030 Radiotelescope RATAN-600.
M. S. Dimitrijević.
Vasiona, Vol. 25, 33 - 43 (1977). In Serbo-Croatian.

Techniques and methods of radio astronomical reception.   See Abstr. 003.179.

Experiments on harmonic generation and mixing of millimeter waves.   See Abstr. 034.034.

Antenna characteristics of whisker diodes used as submillimeter receivers.   See Abstr. 034.076.

Carbon monoxide observations at 230 GHz.
See Abstr. 131.129.

034 Astronomical Accessories (Spectrometers, Photometers, etc.)

034.001 Photon counting magnetograph and polarimeter for measurement of circular polarization in narrow and wide spectral ranges.   A. B. Bukach, L. V. Granitskij, V. N. Komissarov, V. M. Kuvshinov.
Izv. Krymskoj Astrofiz. Obs., Vol. 57, 209 - 220 (1977). In Russian.

The authors describe a photon counting system developed for measurements of circular polarization with the use of a stellar magnetograph and a polarimeter. A square-wave modulation of light and a pulse height discrimination of dark noise allow to increase the signal-to-noise ratio by a factor of two in comparison with the lock-and-amplifier method.

034.002 Sur l'emploi d'un réseau échelle dans un spectro-mètre photoélectrique destiné à la mesure des vitesses radiales.   A. Baranne, M. Mayor, J.-L. Poncet.
C. R. Acad. Sci. Paris, Tome 285, Sér. B, 117 - 120 (1977).
The advantages of using an echelle grating for photo-electric radial velocity spectrometry are discussed. The first results of measurements with such an instrument are presented.

034.003 Six-channel radiometer for spectral investigations of terrestrial and planetary atmospheres.
A. B. Burov, V. N. Voronov, A. A. Krasil'nikov, N. V. Serov.
Izv. vyssh. uchebn. zaved. Radiofiz., Vol. 19, 1795 - 1799 (1976). In Russian. – Abstr. in Ref. zh., 51. Astron., 8.51. 107 (1977).

034.004 Spectrograph for study of planets.
O. N. Gusev, N. G. Zandin, M. V. Lobachev.
Opt.-mekh. prom-st', 1976, No. 12, p. 63 - 64. In Russian. Abstr. in Ref. zh., 51. Astron., 8.51.238 (1977).

034.005 Spectrometer of energies of charged particles with improved electrostatic analyzer.
A. D. Verevkin, A. A. Klimashov, V. G. Kovalenko, L. G. Ol'dekop, B. V. Polenov, B. I. Khazanov.
[Tr.] Soyuz. NII priborostr. 1977, vyp. (No.) 34 - 35, tom (Vol.) 1, 186 - 191. In Russian. – Abstr. in Ref. zh., 62. Issled. kosm. prostranstva, 8.62.122 (1977).

034.006 Automation of the control of the horizontal tele-scope for a magnetograph.   M. Klvaňa.
Solar activity and solar-terrestrial relations, (see 012.007), p. 391 - 393 (1976).
The principle of an automatic pointing and scanning device of the horizontal telescope feeding the magnetograph at the Ondřejov Observatory is described. Further, the author describes the spectroheliograph controlled by the same auto-matic device.

034.007 A high-resolution Michelson interferometer for the Isaac Newton Telescope.   R. C. Wayte, J. Ring.
Mon. Not. R. Astron. Soc., Vol. 181, 131 - 147 (1977).
This paper describes a Michelson interferometer capable of resolving powers over one million in the visible, at the coudé focus of the Isaac Newton Telescope. The advantages of the system over conventional techniques are demonstrated, namely high resolving power plus high luminosity, an adjust-able instrumental profile and very accurate wavelength calibra-tion.

034.008 An X-ray diffractometer for Mars.
A. E. Metzger, J. B. Willett, H. W. Schnopper.
NASA Tech. Mem., NASA TM X-3511, (see 012.010), p. 255 - 257 (1977). – Abstract.

034.009 Un équipement du type «Télévision» pour l'observa-

tion des étoiles doubles visuelles.   J. Dommanget.
O telescópio refractor e a astrometria ao serviço das estrelas duplas, (see 012.013), p. 63 - 71 (1977).

034.010 A spectrographic complex of instruments for the world network of cosmic ray stations.
L. I. Dorman, Yu. Ya. Krest'yannikov, A. V. Sergeev.
Geomagn. Aehron., Vol. 17, 826 - 831 (1977). In Russian.

034.011 Vielkanal-Detektoren in der optischen Astronomie.
H. Tüg.
Sterne Weltraum, Jahrg. 16, 366 - 369 (1977).

034.012 High spectral resolution line observations in the infrared.   M. Anderegg, A. F. M. Moorwood, H. H. Hippelein, J. P. Baluteau, E. Bussoletti, N. Coron.
Far infrared astronomy (see 012.027), p. 171 - 184 (1976).

A high resolution ($\lambda/\Delta\lambda \cong 10^4$) Michelson interferometer has been built for use on NASA's AIRO (C141) telescope to study far infrared emission lines from H II regions. The instru-ment operates in the rapid scanning mode under computer control and spectra are computed, averaged and displayed on line. In preparation for the first airborne measurements the instrument has recently been used on the 1 m telescope at Observatoire du Pic du Midi to search for S IV (10.5 μm) and S III (18.7 μm) line emission from the Orion nebula and from a number of planetary nebulae. Atmospheric emission spectra were also obtained between 17.5 μm and 20 μm at a spectral resolution of 0.02 cm^{-1}.

034.013 On the sensitivity of heterodyne detectors in far infrared astronomy.   H. G. van Bueren.
Far infrared astronomy (see 012.027), p. 185 - 191 (1976).
The signal-to-noise ratio of astronomical heterodyne detection infrared spectrographs is considered, taking into account background, linewidth and seeing effects. A com-parison with incoherent detector systems is presented.

034.014 Study of a photoelectric device for measurements of photographic recordings of circle reading.
N. P. Krasnenko.
Astrometr. i Astrofiz., Kiev, vyp. (No.) 32, (see 003.011), p. 87 - 93 (1977). In Russian.

034.015 Narrow-bandpass interference filters for the far infra-red.   G. D. Holah, N. Morrison.
J. Opt. Soc. America, Vol. 67, 971 - 974 (1977).
A technique for producing narrow-band interference filters for the far infrared, $\lambda \gtrsim 50$ μm, is discussed. Using this system it is possible to produce peak transmissions of 60% in the first order of a Fabry-Perot system with half-widths less than 5%. The technique does not use spacers and has a pitching accuracy of ±1 μm. The performance of filters made using this technique at low temperatures is also considered.

034.016 Low light level detectors for astronomy.
P. B. Boyce.
Science, Vol. 198, 145 - 148 (1977).

034.017 Real-time, very-long-baseline interferometry based on the use of a communications satellite.
J. L. Yen, K. I. Kellermann, B. Rayhrer, N. W. Broten, D. N. Fort, S. H. Knowles, W. B. Waltman, G. W. Swenson, Jr.
Science, Vol. 198, 289 - 291 (1977).
The Hermes satellite, a joint Canadian-American program, has been used to provide a communication channel between radio telescopes in West Virginia and Ontario, for very-long-baseline interferometry (VLBI). This system makes possible

instantaneous correlation of the data as well as a sensitivity substantially better than that of earlier VLBI systems, by virtue of a broader observational bandwidth.

034.018 High resolution Michelson interferometer for airborne infrared astronomical observations. 1: Concept and performance. J. P. Baluteau, M. Anderegg, A. F. M. Moorwood, N. Coron, J. E. Beckman, E. Bussoletti, H. H. Hippelein.
Appl. Opt., Vol. 16, 1834 - 1840 (1977).

A Michelson interferometer has been built for use with the 91-cm telescope on NASA's Gerard P. Kuiper Airborne Observatory primarily to measure ir line emission from H II regions. Operation is in the rapid scan mode, and the achievable resolution is 0.02 cm^{-1} in the wavelength range from 10 μm to around 300 μm. A minicomputer is used to provide on-line spectrum displays and to control and monitor the instrument performance. The design and use of the instrument is discussed, and a comparison is made between the theoretical performance and that actually achieved on the first flights when measurements of line emission from the Orion nebula and from the atmosphere were made.

034.019 High resolution Michelson interferometer for airborne infrared astronomical observations. 2: System design. A. Langlet, C. Delage, D. Stefanovitch, B. Talureau, J. Tualy, J. Verveer, W. P. Fischer, J. M. Gilles, R. Scheper, J. Leblanc, G. Dambier.
Appl. Opt., Vol. 16, 1841 - 1848 (1977).

In this paper design details of the instrument are presented. These include the optics, control He−Ne laser interferometer, helium-cooled bolometer detector, and cooled passband filters. In addition, the on-line computer software which enables the operator to interact rapidly with the system to produce inflight spectra and control accordingly the observational parameters is described, as are elements of the electronics hardware developed specially for airborne observations.

034.020 Bandpass filters for use in the visible region. M. A. Res, C. J. Kok, J. Bednarik, K. Kröger.
Appl. Opt., Vol. 16, 1908 - 1913 (1977).

A series of new filter glasses showing fairly symmetrical bandpass shapes has been developed. The transmittance curves with half-widths ranging between 58 nm and 168 nm cover the whole visible region. It seems that almost any desired peak transmittance wavelength can be obtained by compositional shifting. These new filters should find useful application in optics and filter radiometry.

034.021 Single electron counting by self-scanning diode array in a Kron camera. S. B. Mende, F. H. Chaffee.
Appl. Opt., Vol. 16, 2698 - 2702 (1977).

034.022 Thallium selenide infrared detector. P. S. Nayar, W. O. Hamilton.
Appl. Opt., Vol. 16, 2942 - 2944 (1977).

The application of semiconducting thallium selenide for ir detection is described. A responsivity of 10^6 V/W and NEP of the order of 10^{-15} W/$\sqrt{}$/Hz with a response time of 3 msec can be obtained by operating the detector at 1.5 K.

034.023 Calculation of illumination in the shadow of the external occulting screen of a coronograph. I. Basic relations. A. V. Lenskij.
Problems of cosmic physics. Vyp. (No.) 12, (see 003.016), p. 41 - 48 (1977). In Russian.

034.024 Far infrared interference filters. G. D. Holah, S. D. Smith.
J. Phys. E, Vol. 10, No. 2, p. 101 - 111 (1977). − Abstr. in Phys. Abstr., Vol. 80, Abstr. 22592 (1977).

034.025 Use of absorption filters in the UV spectrophotometers UFS-2 and UFS-3. S. I. Babichenko, E. V. Dereguzov, V. G. Kurt, N. N. Romanova, V. A. Sklyankin, A. S. Smirnov.
(Tr.) Soyuz. NII priborostr., 1977, vyp. (No.) 34 - 35, tom 1, 200 - 206. In Russian. − Abstr. in Ref. zh., 51. Astron., 10.51.274 (1977).

034.026 The IAS triangulation camera. H. Debehogne, C. Lippens, E. Van Hemelrijck, E. Van Ransbeeck.
Ann. Géophys., Vol. 32, 195 - 201 (1976). In French. − Abstr. in Phys. Abstr., Vol. 80, Abstr. 33192 (1977).

034.027 The high-power X-band planetary radar at Goldstone: design, development, and early results. R. Hartop, D. A. Bathker.
IEEE Trans. Microwave Theory Tech., Vol. MTT-24, 958 - 963 (1976). − Abstr. in Phys. Abstr., Vol. 80, Abstr. 37503 (1977).

034.028 Recent advances in far infrared detectors for astronomy in the 1-1000 μm range. K. Shivanandan.
2nd international conference and winter school on submillimeter waves and their applications. San Juan, Puerto Rico, 6 - 11 December 1976. IEEE, New York, USA, 16 + 252 pp. (1976). p. 57 - 58. − Abstr. in Phys. Abstr., Vol. 80, Abstr. 41305 (1977).

034.029 A large imaging array CCD program. F. E. Vescelus, G. A. Antcliffe.
Low light level devices for science and technology, Reston, Va., USA. 22 - 23 March 1976. Proceedings of the Society of Photo-Optical Instrumentation Engineers, Vol. 78. C. Freeman (Editor). Palos Verdes Estates, Calif., USA. 6 + 162 pp. (1976). p. 60 - 64. − Abstr. in Phys. Abstr., Vol. 80, Abstr. 41316 (1977).

034.030 Astronomical applications of charge injection devices. R. S. Aikens, C. R. Lynds, R. E. Nelson.
Low light level devices for science and technology, Reston, Va., USA. 22 - 23 March 1976. Proceedings of the Society of Photo-Optical Instrumentation Engineers, Vol. 78. C. Freeman (Editor). Palos Verdes Estates, Calif., USA. 6 + 162 pp. (1976). p. 65 - 72. − Abstr. in Phys. Abstr., Vol. 80, Abstr. 41317 (1977).

034.031 Test results on intensified charge coupled devices. J. T. Williams.
Low light level devices for science and technology, Reston, Va., USA. 22 - 23 March 1976. Proceedings of the Society of Photo-Optical Instrumentation Engineers, Vol. 78. C. Freeman (Editor). Palos Verdes Estates, Calif., USA. 6 + 162 pp. (1976). p. 78 - 82. − Abstr. in Phys. Abstr., Vol. 80, Abstr. 41318 (1977).

034.032 Low light level imaging devices for the middle ultraviolet. G. Carruthers, J. Kervitsky, G. Hicks, C. Opal, E. O. Hulburt.
Low light level devices for science and technology, Reston, Va., USA. 22 - 23 March 1976. Proceedings of the Society of Photo-Optical Instrumentation Engineers, Vol. 78. C. Freeman (Editor). Palos Verdes Estades, Calif., USA. 6 + 162 pp. (1976). p. 95 - 100. − Abstr. in Phys. Abstr., Vol. 80, Abstr. 41319 (1977).

034.033 Solid Fabry-Perot etalons as high resolution infrared interferometers. A. E. Roche.
Modern utilization of infrared technology civilian and military. II. (see 012.031), p. 196 - 203 (1976). − Abstr. in Phys. Abstr., Vol. 80, Abstr. 45888 (1977).

034.034 **Experiments on harmonic generation and mixing of millimeter waves.**   P. P. Lombardini, B. Fiscella.
Atti Accad. Sci. Torino I, Vol. 109, No. 12, p. 187 - 193 (1975). In Italian. — Abstr. in Phys. Abstr., Vol. 80, Abstr. 49292 (1977).

034.035 **Astronomical proportional control autoguidance system using the quadrant photosil detector.**
P. G. Craven, D. J. Fegan.
J. Phys. E, Vol. 10, 516 - 520 (1977). — Abstr. in Phys. Abstr., Vol. 80, Abstr. 49296 (1977).

034.036 **Gaussian-spindle gravitational wave antenna and single-antenna anticoincidence experiments.**
R. H. Gowdy.
Phys. Rev. D, Vol. 15, 969 - 974 (1977). — Abstr. in Phys. Abstr., Vol. 80, Abstr. 41523 (1977).

034.037 **How to build a speckle interferometer.**
A. M. Schneiderman, D. P. Karo.
Imaging through the atmosphere, (see 012.032), p. 70 - 82 (1976). — Abstr. in Phys. Abstr., Vol. 80, Abstr. 41678 (1977).

034.038 **Aerials for ionospheric research.**
W. Skonieczny.
Wiad. Telekomun., Vol. 16, No. 10, p. 280 - 289 (1976). In Polish. — Abstr. in Phys. Abstr., Vol. 80, Abstr. 45252 (1977).

034.039 **Un système de guidage automatique pour l'astro-photographie.**   P. Campiche.
Orion, 35. Jahrg., 214 - 218 (1977).

034.040 **A high-gain narrow-band amplifier using active filters (*ionosphere measurement equipment*).**
K. K. Bahri, C. S. G. K. Setty.
J. Inst. Electron. Telecommun. Eng., Vol. 23, 29 - 31 (1977). Abstr. in Phys. Abstr., Vol. 80, Abstr. 56760 (1977).

034.041 **The electronographic camera.**
D. McMullan, R. Powell.
New Scientist, Vol. 73, 715 (1977). — Abstr. in Phys. Abstr., Vol. 80, Abstr. 56829 (1977).

034.042 **Characteristics of large area CCD imaging systems.**
G. W. Taylor, D. R. Collins, G. A. Hartsell, G. A. Antcliffe.
IEEE Trans. Nucl. Sci., Vol. NS-24, 497 - 500 (1977). — Abstr. in Phys. Abstr., Vol. 80, Abstr. 57426 (1977).

034.043 **Liquid helium-cooled MOSFET preamplifier for use with astronomical bolometer.**   J. H. Goebel.
Rev. Sci. Instrum., Vol. 48, 389 - 391 (1977). — Abstr. in Phys. Abstr., Vol. 80, Abstr. 61123 (1977).

034.044 **Low-noise preamplifier for photoconductive detectors.**   E. L. Dereniak, R. R. Joyce, R. W. Capps.
Rev. Sci. Instrum., Vol. 48, 392 - 394 (1977). — Abstr. in Phys. Abstr., Vol. 80, Abstr. 61124 (1977).

034.045 **Portable visible-infrared reflectometer.**
F. M. Melsheimer, D. M. Rank.
Rev. Sci. Instrum., Vol. 48, 482 - 483 (1977). — Abstr. in Phys. Abstr., Vol. 80, Abstr. 62033 (1977).

034.046 **Artificial polarization anomalies from holographic gratings.**   J. J. Cowan, E. T. Arakawa.
Opt. Commun., Vol. 21, 428 - 431 (1977). — Abstr. in Phys. Abstr., Vol. 80, Abstr. 62058 (1977).

034.047 **Der Zerodur-"Spiegel", optimal für die Sonnen-beobachtung.**   P. Hückel.
Sterne Weltraum, Jahrg. 16, 418 (1977).

034.048 **The Photicon.**
E. Kellogg, S. Murray, U. Briel, D. Bardas.
Rev. Sci. Instrum., Vol. 48, 550 - 553 (1977). — Abstr. in Phys. Abstr., Vol. 80, Abstr. 67499 (1977).

034.049 **Atmospheric transmissometer and radiometer for E-O sensors field evaluation and model validation.**
J. R. Moulton, F. M. Zweibaum.
Methods for atmospheric radiometry, (see 012.033), p. 30 - 38 (1976). — Abstr. in Phys. Abstr., Vol. 80, Abstr. 70884 (1977).

034.050 **A versatile radiometer for infrared emission measure-ments of the atmosphere and targets.**   R. J. Huppi.
Methods for atmospheric radiometry, (see 012.033), p. 77 - 84 (1976). — Abstr. in Phys. Abstr., Vol. 80, Abstr. 70887 (1977).

034.051 **High DQE detectors.**   C. I. Coleman.
Photogr. Sci. Eng., Vol. 21, No. 2, p. 49 - 59 (1977). Abstr. in Phys. Abstr., Vol. 80, Abstr. 71010 (1977).

034.052 **How to build a speckle interferometer.**
A. M. Schneiderman, D. P. Karo.
Opt. Eng., Vol. 16, 72 - 79 (1977). — Abstr. in Phys. Abstr., Vol. 80, Abstr. 71369 (1977).

034.053 **Submillimeter detector calibration with a low-tem-perature reference for space applications.**
I. G. Nolt, J. V. Radostitz, P. Kittel, R. J. Donnelly.
Rev. Sci. Instrum., Vol. 48, 700 - 702 (1977). — Abstr. in Phys. Abstr., Vol. 80, Abstr. 79152 (1977).

034.054 **An easy means for the proper adjustment of plane grating spectrometers.**
W. Herkt, G. Muller.
Appl. Spectrosc., Vol. 31, 466 - 467 (1977). — Abstr. in Phys. Abstr., Vol. 80, Abstr. 82611 (1977).

034.055 **Experimental studies on the preparation of near infrared filters of Fabry-Perot construction.**
D. S. Chhabra, J. Prasad.
CSIO Commun., Vol. 3, No. 4, p. 68 - 74 (1976). — Abstr. in Phys. Abstr., Vol. 80, Abstr. 84016 (1977).

034.056 **Concentric spectrographs.**   L. Mertz.
Appl. Opt., Vol. 16, 3122 - 3124 (1977).
A novel class of geometric optical configurations for dif-fraction grating spectrographs is introduced. The concentric configurations offer appreciable advantages over traditional arrangements. Excellent image quality at high numerical aperture is demonstrated experimentally.

034.057 **An improved intensified dissector scanner for Cassegrain spectroscopy.**
P. M. Rybski, A. L. Mitchell, T. Montemayor.
Publ. Astron. Soc. Pacific, Vol. 89, 621 (1977). — Abstract.

034.058 **The McDonald Observatory digital area photometer.**
P. M. Rybski, G. W. van Citters, G. F. Benedict.
Publ. Astron. Soc. Pacific, Vol. 89, 621 (1977). — Abstract.

034.059 **The DDO diode array spectrometer.**
B. Campbell.
Publ. Astron. Soc. Pacific, Vol. 89, 728 - 732 (1977).
A self-scanning linear array of 256 photodiodes to a small spectrograph has been attached. This system has a dem-onstrated capability for medium-resolution spectrophotom-etry at moderate light levels. Particular features include linear response to better than 0.25%, no perceptible lag or cross-talk, useful from 3800 Å to 9500 Å, and a dynamic range of 1500. To eliminate dark current the diode array is cooled to −90°C with liquid nitrogen boil-off. The usefulness of this type of device for accurate colorimetry is stressed.

034.060 **Fick Observatory radial-velocity spectrometer.**
W. I. Beavers, J. J. Eitter.
Publ. Astron. Soc. Pacific, Vol. 89, 733 - 738 (1977).
A recently completed photoelectric stellar radial-velocity spectrometer is described. Initial performance tests are reported and the results of a calibration of the instrument are given.

034.061 **Fotometro per la misura delle magnitudini stellari da negativi fotografici.** F. Cerchio.
Astronomia, N. 3, p. 11 - 19 (1977).

034.062 **Broadening of the dynamical region of an electro-photometer by the correction of dead time.**
E. B. Vovchik, R. F. Fedoriv.
Tsirk. Astron. Obs., L'vov, No. 51, p. 33 - 34 (1976). In Russian.

034.063 **A real-time spectrum analyser with on-line definition of the confidence levels.**
P. Santin, G. Sedmak.
Astrophys. Space Sci., Vol. 48, 57 - 63 (1977).
This paper describes a real-time spectrum analyser with on-line definition of the confidence levels built at Trieste Observatory for a research program on periodical components of unknown period in optical sources. The system is realized by a photon counting photometer on-line to a computer that elaborates in real-time the data from the photometer and displays the spectrum of the source measured. The spectrum is normalized to unity mean value and the display is calibrated by the confidence levels relative to the χ_0^2 distribution of the spectrum displayed. Some experimental data show the performance of the machine that operates in real-time at full efficiency up to 160 Hz.

034.064 **The multifunction photoelectric photometer of Torino Observatory.** S. Furlani, G. Sedmak.
Astrophys. Space Sci., Vol. 48, 65 - 78 (1977).
The paper describes the design and the realization of a multiple-function photoelectric photometer made by Trieste Observatory for Torino Observatory. The system design shows a two-beam, sequential multiband photon counting and analog photoelectric photometer configuration based on a PDP8/E computer for the control and data acquisition and elaboration.

034.065 **A D.C. photoelectric polarimeter.**
A. Kubičela, J. Arsenijević, I. Vince.
Publ. Dep. Astron., Univ. Beograd, Fac. Sci., No. 6, (see 012.040), p. 25 - 29 (1976).

034.066 **Making every photon count: new frontiers in astronomical detectors and their application to telescopes of moderate size.** R. G. Tull.
Proc. Southwest Reg. Conf., Vol. 3, (see 012.043), p. 83 - 97 (1977).
The efficiency of astronomical spectroscopic observations can be greatly increased by use of modern panoramic detectors. Examples are shown using systems developed at McDonald Observatory, and it is demonstrated that 1000-fold speed gains are possible compared with single-channel spectrum scanning.

034.067 **Filter glasses with bandpass characteristics for the green region of the visible spectrum.**
M. A. Res, J. Bednarik, F. Hengstberger, C. J. Kok.
Optik, Band 48, 83 - 94 (1977). In German.
A brief summary of the properties of bandpass filters and the present state of the art of this type of glass absorption filter is given. The goal of the present investigation was the development of bandpass filter glasses, the transmission peaks of which would cover the green region of the visible spectrum in steps of 5—10 nm.

034.068 **Filter glasses with band-pass characteristics for the yellow-orange region of the visible spectrum.**
M. A. Res, J. Bednarik, C. J. Kok, K. Kröger.
Optik, Band 48, 371 - 382 (1977).

034.069 **Filter glasses with band-pass characteristics in the violet-blue region of the visible spectrum.**
M. A. Res, J. Bednarik, C. J. Kok, K. Kröger.
Optik, Band 49, 277 - 293 (1977). In German.
The conditions for the development of band-pass filters in the violet-blue region are compared with those for the development of band-pass filters in the green and yellow-orange region. The purpose of the present investigation was the development of band-pass filter glasses whose transmission peaks would cover the violet-blue region of the visible spectrum in steps of 5—10 nm. By means of one to six colouring oxides introduced into single glasses it was possible to achieve steps of 2—9 nm. Results with newly developed filters are given both in the form of tables and figures.

034.070 **Cut-on filters containing Ce-Ti for use in the visible region.** K. Kröger, M. A. Res, J. Bednarik.
Optik, Band 49, 347 - 355 (1977).
Results are given of an investigation into a new series of glass cut-on filters containing cerium and titanium. The parameters studied included the effect of different base glasses on the transmittance characteristics of one colouring oxide and the effect of different amounts of a specific colouring ion on the cut-on effect exhibited in a specific base glass. A series of cut-on filters covering the wavelength range from 300 nm to 650 nm was developed.

034.071 **Image-scanning systems using tilting plane mirrors.**
C. J. Baddiley.
Infrared Phys., Vol. 17, 393 - 404 (1977).
A number of image-scanners using tilting plane-mirrors are described for use with optical-imaging systems. The paper compares these types and includes some novel designs particularly useful for infrared astronomy with great advantages over conventional systems. The tilting mirror-systems described either use two mirrors on a common tilting base, or for lower inertia use two mirrors independently pivoted but mechanically linked. Moving double mirror-systems may be replaced by equivalent single mirror arrangements with some considerable advantages. Two such double mirror-systems have been built and one of the equivalent single mirror systems is proposed as an optical path invariant, two-dimensional image-scanner.

034.072 **A wide field infrared photometer.**
C. J. Baddiley, J. Ring.
Infrared Phys., Vol. 17, 405 - 413 (1977).
A wide field photometer was designed, built and used for the detection of extended infrared sources. A maximum sky area of 25 mm diameter in the focal plane of a telescope is integrated on a single 4 X 4 mm PbS detector by using focal reduction optics. Maximum sky-comparison amplitudes of 25 mm are achieved at frequencies between 2 to 60 Hz using a tilting spherical mirror placed at the image of the telescope objective stop. Good square waveforms and position offsetting of the tilting mirror are achieved with a servo system using an R-C bridge position detection system. The instrument was successfully used in 1975 to detect globular clusters in the H and K bands.

034.073 **Infrared Fabry-Perot and heterodyne spectrometers.**
A. Betz.
Infrared Phys., Vol. 17, 427 - 429 (1977).
The status of infrared instrumentation for astronomical investigations at U.C. Berkeley is described with emphasis on the techniques of high spectral and spatial resolution. Present instrumentation includes three Fabry-Perot spectrometers for the 10, 20, and 100 μm wavelength regions, a submillimeter

receiver using an optically pumped laser, a 10 μm heterodyne spectrometer for studies of planetary atmospheres, and a 2-element 10 μm stellar interferometer for measuring the angular diameters of infrared stars.

034.074   Recent development of infrared detectors.
        M. F. Kimmitt.
Infrared Phys., Vol. 17, 459 - 466 (1977).
        The detectors available for 5−1000 μm are discussed with particular consideration of performance improvement in recent years. Two potentially useful detectors for the 50−1000 μm region are considered in detail: (1) Photoconductors with very shallow donor and acceptor sites. (2) Photon drag detectors as power measuring devices.

034.075   Josephson effect heterodyne receivers.
        B. T. Ulrich.
Infrared Phys., Vol. 17, 467 - 474 (1977).

034.076   Antenna characteristics of whisker diodes used as
        submillimeter receivers.
H. Kräutle, E. Sauter, G. V. Schultz.
Infrared Phys., Vol. 17, 477 - 483 (1977).
        The effectivity of whisker diodes in an open structure configuration has been improved using corner reflectors. The power patterns have been calculated. They show good agreement with experimental results. For a laser beam with a beam width of 8° and corner angles between 90° and 60° an increase in gain of 12 db has been obtained.

034.077   Infrared instrumentation of the Calar Alto telescopes.
        D. Lemke.
Infrared Phys., Vol. 17, 487 - 494 (1977).
        Photometers for the near and middle i.r., an i.r.-image tube camera and a Fabry-Perot interferometer, are in use at the 1.2 m telescope. Similar instrumentation is under development for the 2.2 m and 3.5 m-telescope. Both telescopes will be equipped with chopping secondaries.

034.078   Proposed sub-mm-photometer for the ESO 3.6-m
        telescope.   E. Kreysa, G. V. Schultz.
Infrared Phys., Vol. 17, 495 - 498 (1977).
        The design of a sub-mm-photometer for a large telescope presents several problems. Among the different telescope configurations the prime focus is shown to have certain advantages for sub-mm-photometry. Two photometer-designs for the prime focus of the ESO 3.6-m telescope are described.

034.079   Optical components of a sub-mm-photometer.
        E. M. Arnold, E. Kreysa.
Infrared Phys., Vol. 17, 499 - 502 (1977).
        The development of a photometer for ground-based astronomical observations in the 300−1000 μm wavelength range is described.

034.080   Characteristics of iodine stabilized He−Ne laser for
        the absolute determination of gravity.
T. Tsubokawa.
Proc. Int. Latitude Obs. Mizusawa, No. 16, p. 116 - 132 (1977).
In Japanese.

034.081   The Haleakala polarimeter.   D. L. Mickey.
        Rep. Obs. Lund, No. 12, (see 012.044), p. 81 - 88
(1977).

034.082   On the precision of spectral recording with a double-
        pass digital spectrophotometer.   E. K. Kokhan.
Izv. Glav. Astron. Obs. Pulkovo, No. 195, Astrofiz. Astrometr., p. 134 - 154 (1977). In Russian.
        The precision of previously published spectral data obtained at Pulkovo with the double-pass spectrophotometer with registration on punched cards is studied. The accidental

errors are analysed. The results of the study of the errors (during five-year observations) in the continuum depending on the wavelength in the center of the solar disc and at 1".6 and 0".6 from the limb are presented. The error is also studied in the profiles of spectral lines of various depths and halfwidths in the spectral region 4222−7699 Å during the period of observations 1969 - 1973.

034.083   Semi-automatic measuring machine for photograph-
        ic plates taken with a photographic vertical circle.
V. D. Shkutov.
Izv. Glav. Astron. Obs. Pulkovo, No. 195, Astrofiz. Astrometr., p. 155 - 159 (1977). In Russian.
        A visual-photoelectric measuring machine for plates taken with the vertical circle is described. The mean square error of one photoelectric measurement is equal to ± 0.7 microns, the systematic error does not exceed ± 1 micron and did not vary during one year and a half of operation.

034.084   A high speed photoelectric photometer for the 60
        cm reflector.
H. Tsunemi, M. Matsuoka, K. Tomita.
Rep. Inst. Space Aeronaut. Sci., Univ. Tokyo, Vol. 12, 793 - 813 (1976).   In Japanese.
        A high speed photoelectric photometer which consists of a photon counting system has been equipped the 60 cm telescope at KSC (Kagoshima Space Center). In this photometer, star light is divided into three colors by two dichroic mirrors. Sky background data are also obtained.

034.085   A project study for a continuum photometer.
        W. W. Weiss.
Mitt. Astron. Ges., Nr. 42, p. 126 - 131 (1977).

034.086   Investigations of half-second levels.
        Yu. S. Dobrokhotov, A. P. Buyanov.
Astrometriya i Astrofizika, Kiev, vyp. (No.) 33, (see 003.020), p. 91 - 96 (1977). In Russian.
        The Dresden automatic comparator was used to study two half-second levels made in Kiev.

034.087   Possibilities of improving the quality of the image in
        spectrographs with plane diffraction gratings.
D. A. Zhuravlev.
Redkollegiya zh. "Izv. vyssh. uchebn. zaved. Geod. i aehrofotosemka". Moskva, 1977. 11 pp. In Russian. − Abstr. in Ref. zh., 51. Astron., 12.51.186 (1977).

034.088   Stellar spectrograph with cross dispersion.
        N. G. Zandin, O. N. Gusev, I. V. Pejsakhson.
Opt.-mekh. prom-st', 1977, No. 6, p. 20 - 23. In Russian. Abstr. in Ref. zh., 51. Astron., 11.51.309 (1977).

        On the imaging properties of holographic gratings.
See Abstr. 031.004.

        Dependence of the finesse of a Fabry-Perot on the
plate separation.   See Abstr. 031.011.

        Astronomical applications of echelle spectroscopy.
See Abstr. 031.234.

        First results from observations using a two-telescope
interferometer.   See Abstr. 032.014.

        Hvar Observatory 65-cm telescope and some prelimi-
nary results.   See Abstr. 032.041.

        Calibration and performance of the Viking lander
cameras.   See Abstr. 032.565.

        Cooled instrumentation for infrared astronomical

investigations from space.   See Abstr. 032.585.

Large format Secondary Electron Conduction Orthicon integrating television sensor for the Space Telescope. See Abstr. 032.595.

The intensified-charge-coupled device as a photon-counting imager.   See Abstr. 032.596.

Infrared capabilities.   See Abstr. 032.597.

Development of an infrared spectroradiometer. See Abstr. 032.599.

Multi-channel subtractive spectrograph and filament observations.   See Abstr. 073.092.

The mean magnetic field of the Sun: observations at Stanford.   See Abstr. 080.069.

A filter-tilting type nightglow photometer using a small GaAs photocathode and aspheric lenses. See Abstr. 082.087.

Investigation of the photometric system of the AZT-8 telescope.   See Abstr. 113.006.

Polarimetric observations of variable stars. Equipment and method of observations. Observations of TU Cas in 1973.   See Abstr. 122.009.

The infrared polarization of NGC 1275, NGC 4151, Markarian 231, and 3C 273.   See Abstr. 131.017.

On the high energy proton spectrum measurements. See Abstr. 143.064.

Narrow-bandpass imagery of the Magellanic Clouds. See Abstr. 159.013.

035 Clocks and Frequency Standards, Sundials

035.001 New State standard of time and frequency of the
USSR.   V. G. Il'in, V. V. Sazhin.
Priroda, 1977, No. 8, p. 16 - 27. In Russian.

035.002 Eine zeitgemässe Sonnenuhr für alle Längen- und
Breitengrade.   H. Mendel.
Orion, 35. Jahrg., 168 - 173 (1977).

035.003 Eine achtteilige mittelalterliche Sonnenuhr an der
Stadtkirche zu Burg Stargard.   W. Hanke.
Sterne, 53. Band, 151 - 162 (1977).

035.004 Observed time-discontinuity of clock synchroniza-
tion in rotating frame of the earth.   Y. Saburi.
J. Radio Res. Lab., Vol. 23, 255 - 265 (1976). — Abstr. in
Phys. Abstr., Vol. 80, Abstr. 60333 (1977).

035.005 A proposal of accurate time dissemination using TV
signals from a broadcasting satellite.   F. Takahashi.
J. Radio Res. Lab., Vol. 23, 267 - 283 (1976). — Abstr. in
Phys. Abstr., Vol. 80, Abstr. 60334 (1977).

035.006 Réception et distribution de signaux horaires sur les
sites d'observation.   T. Flatrès.
Astronomie, Vol. 91, 473 - 477 (1977).

035.007 Problem of the steering of the International Atomic
Time Scale with primary standards.   G. Becker.
PTB Mitt., 87. Jahrg., 465 - 467 (1977).

035.008 Sundials in the Leningrad district.
L. E. Majstrov, A. K. Petrenko.
Istor.-astron. issled. Vyp. (No.) 13. Moskva, Nauka, 1977, p.
377 - 386. In Russian. — Abstr. in Ref. zh., 51. Astron.,
1.51.167 (1978).

VLBI clock synchronization.
See Abstr. 031.324.

Clocks and experimental gravitation: a null gravitational redshift experiment, laboratory tests of post-Newtonian
gravity, and gravity-wave detection by spacecraft tracking.
See Abstr. 066.296.

036 Photographic Auxiliaries

036.001 Le renouveau de la photographie astronomique. II.
J.-L. Heudier.
Astronomie, Vol. 91, 341 - 348 (1977).

036.002 Nomogram for determination of the limiting mag-
nitude obtained by direct photography.
G. G. Lengauer.
Astron. Tsirk., No. 942, p. 1 - 3 (1977). In Russian.

Positional Astronomy, Celestial Mechanics

041 Positional Astronomy, Astrometry

041.001 The possibility of using differential astrometric observations for improvement of the fundamental catalogue system. M. S. Zverev.
Astron. Zh. Akad. Nauk SSSR, Vol. 54, 875 - 883 (1977).
In Russian. English translation in Soviet Astron., Vol. 21, No. 4.

The importance of strictly absolute (independent) determinations of celestial bodies coordinates for fundamental astrometry is noted. Actually all kinds of systematic corrections to a fundamental catalogue can be determined using differential observations, provided certain conditions are adopted. Corrections to the equinox ΔA and equator $\Delta \delta_0$ are derived from differential observations of planets. The quasi-absolute method of deducing $\Delta \delta_\delta$ and $\Delta \alpha_\delta$ corrections is described.

041.002 Systematic errors of FK4. M. Sánchez.
Astron. Astrophys., Vol. 60, 61 - 65 (1977).
In French.

The systematic differences "astrolabe−FK4" and "astrolabe−FK4 Sup" derived from a previous publication of the author are investigated. The discussion comprises 194 FK4 and 84 FK4 Sup stars respectively. The differences are analysed separately as a function of right ascension and declination. Comparisons between group corrections obtained in the two separate catalogue campaigns and the systematic differences depending on α for each group are also displayed.

041.003 The phase effect in transit circle observations of planets. E. M. Standish, Jr.
Bull. American Astron. Soc., Vol. 9, 436 (1977). − Abstract.

041.004 Method of determination of the coordinates of celestial objects from observations at equal altitudes. G. S. Tyuterev.
Astrometric determinations of the coordinates of celestial bodies, 1976, (see 003.003), p. 30 - 39. In Russian. − Abstr. in Ref. zh., 51. Astron., 8.51.164 (1977).

041.005 Results of absolute determinations of the right ascensions of 433 stars from observations with the Freiberg-Kondrat'ev transit instrument in Nikolaev. R. T. Fedorova.
Nikolaev. otd. Glav. astron. obs. AN SSSR. Nikolaev, 1977. 54 pp. In Russian. − Abstr. in Ref. zh., 51. Astron., 8.51. 176 (1977).

041.006 Right ascensions of the sun, Mercury and Venus observed in 1968−1969 with the Freiberg-Kondrat'-ev transit instrument in Nikolaev. O. T. Markina, G. M. Petrov.
Nikolaev. otd. Glav. astron. obs. AN SSSR. Nikolaev, 1976. 25 pp. In Russian. − Abstr. in Ref. zh., 51. Astron., 8.51.178 (1977).

041.007 Right ascensions of the sun, Mercury and Venus observed in Nikolaev in 1970 with a transit instrument. G. M. Petrov, V. N. Pyshnenko.
Nikolaev. otd. Glav. astron. obs. AN SSSR. Nikolaev, 1976. 17 pp. In Russian. − Abstr. in Ref. zh., 51. Astron., 8.51.179 (1977).

041.008 Right ascensions of the sun, Mercury and Venus observed in Nikolaev in 1971 with a transit instrument. V. N. Pyshnenko, R. T. Fedorova.
Nikolaev. otd. Glav. astron. obs. AN SSSR. Nikolaev, 1976. 19 pp. In Russian. − Abstr. in Ref. zh., 51. Astron., 8.51.180 (1977).

041.009 Right ascensions of the sun, Mercury and Venus observed in Nikolaev in 1972 with a transit instrument. G. M. Petrov, V. N. Pyshnenko, R. T. Fedorova.
Nikolaev. otd. Glav. astron. obs. AN SSSR. Nikolaev, 1976. 17 pp. In Russian. − Abstr. in Ref. zh., 51. Astron., 8.51.181 (1977).

041.010 Right ascensions of the sun, Mercury and Venus observed in Nikolaev with a transit instrument in 1973−1975. V. N. Pyshnenko, R. T. Fedorova.
Nikolaev. otd. Glav. astron. obs. AN SSSR. Nikolaev, 1976. 32 pp. In Russian. − Abstr. in Ref. zh., 51. Astron., 8.51.182 (1977).

041.011 Right ascensions of Mars obtained during the 1971 opposition. V. N. Pyshnenko, R. T. Fedorova.
Nikolaev. otd. Glav. astron. obs. AN SSSR. Nikolaev, 1976. 8 pp. In Russian. − Abstr. in Ref. zh., 51. Astron., 8.51.183 (1977).

041.012 Declinations of the sun, Mercury and Venus obtained from observations with the vertical circle of the Nikolaev Observatory in 1960. I. I. Bozhko, G. K. Zimmerman.
Nikolaev. otd. Glav. astron. obs. AN SSSR. Nikolaev, 1975. 14 pp. In Russian. − Abstr. in Ref. zh., 51. Astron., 8.51.184 (1977).

041.013 Declinations of the sun, Mercury, Venus and Mars obtained from observations with the vertical circle of the Nikolaev Observatory in 1973−1975. O. T. Markina, V. P. Sibilev.
Nikolaev. otd. Glav. astron. obs. AN SSSR. Nikolaev, 1976, 34 pp. In Russian. − Abstr. in Ref. zh., 51. Astron., 8.51.185 (1977).

041.014 Results of photographic position observations of Venus with the zonal astrograph in Nikolaev in 1967−1972. G. K. Gorel'.
Nikolaev. otd. Glav. astron. obs. AN SSSR. Nikolaev, 1976,

12 pp. In Russian. — Abstr. in Ref. zh., 51. Astron., 8.51.186 (1977).

041.015 **Photoelectric meridian observations of Mars, Jupiter, Saturn, Uranus, Neptune and four minor planets 1968 - 1971.**   L. Lindegren, E. Høg.
Astron. Astrophys., Suppl. Ser., Vol. 30, 125 - 129 (1977).

A catalogue of positions of Mars, Jupiter, Saturn, Uranus, Neptune, Ceres, Pallas, Juno and Vesta is given for the interval 1968.0 - 1971.3. The meridian observations were obtained within the Southern References System programme by the Hamburg Observatory Expedition to Australia, using a photoelectric multislit micrometer. Accidental mean errors of single observations of all planets except Mars are very nearly the same as for stellar observations, viz. $0''.17$ and $0''.31$ in $\Delta\alpha \cos \delta$ and $\Delta\delta$, respectively. Phase corrections for defective illumination and limb darkening and (in the case of Mars) corrections for surface markings have been applied..

041.016 **A three-dimensional analysis of the kinematics of 512 FK4/FK4 Sup stars.**   B. du Mont.
Astron. Astrophys., Vol. 61, 127 - 132 (1977).

The Ogorodnikov-Milne model of a three-dimensional differential centroid velocity field has been applied to the proper motions of the 512 distant FK4/FK4 Sup stars which Fricke (1967) and Fricke and Tsioumis (1975) have analysed in using the Oort-Lindblad model. The values of the parameters of galactic rotation do not change if the two-dimensional model is extended to a three-dimensional one; but they do change if the three-dimensional model is applied to star sets restricted to one hemisphere alone. No systematic motion perpendicular to the galactic plane has been detected.

041.017 **Numerical theory of motion of the earth and Venus derived from data of radar observations and optical observations and from observations of motion of the artificial satellites Venera 9 and Venera 10.**
Eh. L. Akim, V. A. Stepan'yants.
Dokl. AN SSSR, Vol. 233, 314 - 317 (1977). In Russian. Abstr. in Ref. zh., 51. Astron., 9.51.158 (1977).

041.018 **Southern Astrographic Catalog project.**   D. Dunham, D. Herald.
Occultation Newsl., Vol. 1, 113 - 114 (1977).

041.019 **True external errors of the observations by the Tokyo meridian circle estimated from the Uranus occultation of SAO 158687.**   H. Hara, S. Isobe.
Publ. Astron. Soc. Japan, Vol. 29, 631 - 637 (1977).

Meridian circle observations of Uranus and SAO 158687 occulted by it were carried out during 1976 - 1977, and a prediction for the occultation is given on the basis of observations made before the event. This prediction shows clear discrepancies with the photoelectric observations made at both the Kuiper Airborne Observatory and the Perth Observatory. It is shown that the cause of the discrepancies is the true external error of the relative positions of the occulted star to Uranus in the Tokyo observations, which is estimated to be about $0''.1$ in both right ascension and declination.

041.020 **General principles for space astrometry.**   P. Lacroute.
Highlights of Astronomy, Vol. 4, Part I, (see 012.021), p. 353 - 359 (1977).

041.021 **Future astrometry from space and from the ground.**   E. Høg, H. J. Fogh Olsen.
Highlights of Astronomy, Vol. 4, Part I, (see 012.021), p. 361 - 367 (1977).

The space astrometry projects being studied by the European Space Agency (ESA) promise an accuracy about $\pm 0''.003$ due to photon statistics for about 100000 stars. In-herent limitations of the space astrometry projects and the future continued importance of ground based astrometry are outlined.

041.022 **Accuracy of the average of star positions obtained from overlapping plates.**   K. von der Heide.
Astron. Astrophys., Vol. 61, 545 - 552 (1977).

An analytic expression is derived that gives the accuracy of star positions computed as the average of the positions taken from overlapping plates. The variance of such a position is given by the sum of two terms, a stochastic term corresponding to the measurements on the plate and a local systematic term that is introduced by the errors of the plate constants. For some visual arrangements of plates graphs are computed for direct use.

041.023 **Accuracy of star positions obtained by a block adjustment.**   K. von der Heide.
Astron. Astrophys., Vol. 61, 553 - 562 (1977).

The accuracy of field star positions when obtained by a block adjustment is computed numerically by inversion of the normal matrices. Its dependence on all important parameters, such as density of field and reference stars, on size of the block, reduction model, etc. are discussed in detail. The accuracy obtained by this method is, in most cases, better than that achieved by single plate adjustments with added averaging. A further discussion treats the problem of rigidity of a block. The numerical calculations show that it is impossible to make a large block of several hundred plates rigid provided a usual density of reference stars at appropriate weight is used.

041.024 **Observations of Uranus made with the Danjon Astrolabe of Santiago, Chile, during 1975 and 1976.**
F. Noël, K. Contreras, H. Repetur.
Astron. Astrophys., Suppl. Ser., Vol. 30, 189 - 191 (1977).

This paper gives the results of the observations of Uranus made with the Danjon Astrolabe of Santiago, Chile, during 1975 and 1976. The residuals in zenith distance of the planet, the number of the star group where the planet was observed as well as its weight are given among other pertinent data.

041.025 **Meridian observations made in Brorfelde (Copenhagen University Observatory) 1975 - 1976.**
Positions of 712 stars from 43 areas surrounding radio sources. H. J. Fogh Olsen, L. Helmer.
Astron. Astrophys., Suppl. Ser., Vol. 30, 349 - 360 (1977).

This catalogue presents positions for selected faint stars mainly from AGK3. The stars are distributed in 43 selected areas around radio sources and are suitable as reference stars for measuring the optical counterparts to the radio sources.

041.026 **Positions of radio stars.**   H. G. Walter.
Astron. Astrophys., Suppl. Ser., Vol. 30, 381 - 386 (1977).

Places of radio stars are compiled as a preparatory step to refined observation programmes for these objects which possess the potential of intermediaries for tying the extragalactic reference frame to the fundamental system of star positions.

041.027 **Theory of the algorithm of computer compilation of ephemerides for star pairs at equal altitudes.**
L. V. Neverov, S. S. Uralov.
Izv. vyssh. uchebn. zaved. Geod. i aehrofotosemka, 1977, No. 1, p. 57 - 62. In Russian. — Abstr. in Ref. zh., 51. Astron., 10.51.199; 52. Geod. Aehrosemka, 10.52.95 (1977).

041.028 **The optimum time for determination of the azimuth from observations of the sun.**
V. N. Gan'shin, V. N. Malishevskij.
Geod. i kartogr., 1977, No. 5, p. 23 - 25. In Russian. — Abstr. in Ref. zh., 52. Geod. Aehrosemka, 10.52.93 (1977).

041.029 **Synthetic catalogue of right ascensions for time service.**
Transit Division, Shensi Observatory, Academia Sinica.
Acta Astron. Sinica, Vol. 18, 18 - 31 (1977).

The synthetic catalogue of right ascensions for time service (CTC) was compiled through the synthesis of time determination data obtained with five photoelectric transits and a visual transit in five observatories. This catalogue lists 1156 stars with magnitude interval $0^m.1 - 6^m.6$, and declination interval $-30° - +66°$. In compiling the catalogue, corrections and systematic smoothing of the right ascensions of stars were carried out on the basis of FK_4 catalogue, but no corrections of the vernal equinox were made, and no attempt was made to establish an own system of proper motions, therefore CTC is a relative catalogue. Observational data spreading over 3−5 years were utilized, the total number of star observations reached 76847, so the catalogue has a rather high precision, especially within the declination zone $-5° -+56°$, having 1043 stars, the precision of position determination in this zone is in general higher than ±4 ms. In this paper, the method of compiling the CTC catalogue is described, and the precision discussed.

041.030 **Nuevos coeficientes para la reducción automática de posiciones de estrellas.** M. J. Sevilla.
Urania Barcelona, Año 61, Núm. 285, p. 53 - 60 (1976).

041.031 **Probleme der fundamentalen Astrometrie und deren Beziehung zur Himmelsmechanik und Radioastro-**
metrie. W. Fricke.
Mitt. Astron. Ges., Nr. 42, p. 29 - 41 (1977). − Review paper.

041.032 **Determination of the right ascensions of 386 stars from visual observations 1970 - 1973.**
Eh. A. Sanakulov.
Tsirk. Astron. Inst., Tashkent, No. 77 (424), 27 pp. (1977). In Russian.

041.033 **On the improvement of the right ascensions of fundamental stars.** M. F. Bykov.
Tsirk. Astron. Inst., Tashkent, No. 79 (426), 28 pp. (1977). In Russian.

041.034 **Preliminary results of the comparison of the AGK3 catalogue with the General Catalogue of Latitude**
Stars and observational data from Zenith-Telescopes and Photographic Zenith Tubes. S. N. Sadzakov, V. A. Fomin.
Bull. Obs. Astron. Belgrade, No. 128, p. 16 - 18 (1977).

Preliminary results are presented of the analysis of systematic errors of declinations and proper motions in declination of the AGK3 catalog. The analysis was made by comparison with the General Catalogue of Latitude Stars and with the data of latitude observations with the ZT and PZT at a number of observatories during last 10-15 yr. A brief characteristic of the compared catalogues is given.

041.035 **Utilization of some technical achievements for fundamental determination of right ascension.**
J. Bem.
Publ. Dep. Astron., Univ. Beograd, Fac. Sci., No. 6, (see 012.040), p. 101 - 109 (1976).

An independent information about the irregularitis of the Earth's rotation received by VLBI techniques can be profitable utilized in the determination of the α coordinates of stars. By using the atomic time scale, the reductions of meridian observations become simplified, more objective and homogeneous. The proposed method was tested on a short series of meridian observations.

041.036 **New international refraction tables.** G. Teleki.
Publ. Dep. Astron., Univ. Beograd, Fac. Sci., No. 6, (see 012.040), p. 115 (1976). − Summary: The full text of

this paper will appear in Trudy 20. Astrometr. konf. SSSR, Leningrad.

041.037 **Some results of the current work on the catalogue of PZT programme stars.**
S. Sadžakov, D. Šaletić, M. Dačić.
Publ. Dep. Astron., Univ. Beograd, Fac. Sci., No. 6, (see 012.040), p. 117 - 118 (1976).

041.038 **Observations of the Sun, Mercury and Venus in both coordinates, made with the Large Meridian Circle of**
Belgrade Observatory. S. Sadžakov, M. Dačić, D. Šaletić.
Publ. Dep. Astron., Univ. Beograd, Fac. Sci., No. 6, (see 012.040), p. 119 - 121 (1976).

041.039 **Résultats des observations faites à Alger avec l'astro-labe impersonnel A. Danjon OPL 8. Temps et**
latitude 1975.
A. Ghezloun, M. A. Benhocine, M. Haroun, B. Haddad.
Ann. Obs. Astron. Alger, Tome 5, Fasc. 1, p. 41 - 49 (1977).

041.040 **Corrections of the oriêntation of the FK4 from photographic positional observations of Venus with**
the wide-angle astrograph of the Sternberg Institute.
B. S. Vozdvizhenskij.
Astron. Tsirk., No. 949, p. 3 - 5 (1977). In Russian.

041.041 **Absolute catalogue Nik-75 and the precision of the Time Service Compiled Catalogue (CTS).**
R. T. Fedorova.
Astron. Tsirk., No. 954, p. 1 - 2 (1977). In Russian.

041.042 **Influence of seeing on astrometric measurements. 1. Dependence of vibration of images on different**
factors. V. I. Ivanov.
Astron. Tsirk., No. 956, p. 6 - 8 (1977). In Russian.

041.043 **Influence of seeing on astrometric measurements. 2. Estimate of random errors on coordinate measure-**
ments. V. I. Ivanov.
Astron. Tsirk., No. 958, p. 5 - 7 (1977). In Russian.

041.044 **Lunar laser ranging and fundamental astrometry.**
C. A. Murray, B. D. Yallop.
Phil. Trans. R. Soc. London, Ser. A, Vol. 284, 507 - 514 (1977) = H. M. Naut. Almanac Office, Libr. Repr. No. 337.

041.045 **Place corrections of Mizusawa astrolabe stars.**
S. Sakai.
Proc. Int. Latitude Obs. Mizusawa, No. 16, p. 70 - 83 (1977). In Japanese.

Corrections to places of Mizusawa astrolabe stars are computed with data obtained from 1966 to 1976 for three cases. In the first case, groups with more than 20 "standard stars" observed are used, where the standard stars are chosen from stars observed relatively frequently. In the second and third cases, groups with more than 20 and 15, respectively, stars observed are used. Mean errors of the corrections computed for the first case turn out to be slightly larger than those obtained from other two cases because of smaller number of groups used, although the dispersion of the unit weight is significantly smaller, as expected, for the first case than for other two.

041.046 **Place corrections to FK4 stars derived from results of observations with the Mizusawa Danjon Astro-**
labe. S. Sakai, K. Horiai, T. Sasao, K. Sato, K. Iwadate.
Publ. Int. Latitude Obs. Mizusawa, Vol. 10, 109 - 127 (1976).

Corrections to places of Mizusawa astrolabe stars taken from FK4 and its supplement are calculated with the results of time and latitude observations from January 1966 to February 1976. The method of calculations is almost the same with the one adopted by Yokoyama (1968) except in the determination of group corrections where the method of least squares for correlated measurements with different accuracies is applied. Corrections to star places $\Delta\alpha$ and $\Delta\delta$ are obtained for 77 double passage stars and 69 single passage stars. The systematic corrections $\Delta\alpha_a$ and $\Delta\delta_a$ depending on right ascension are calculated.

041.047   Improvement of the first Mizusawa PZT star catalog.
S. Takagi, G. Murakami, H. Kitago, S. Sakai, K. Iwadate.
Publ. Int. Latitude Obs. Mizusawa, Vol. 10, 179 - 191 (1976).

The first Mizusawa PZT star catalog was improved with 16 years' results of observations made during 1959 - 1975. The improved catalog was compared with the new Washington PZT star catalog and other fundamental catalogs. The equator and equinox points of the improved catalog are to be shifted by $-0\overset{s}{.}1$ and -5 ms with respect to the FK4 system.

041.048   Positions of Jupiter, Saturn, Uranus and Neptune determined from observations with the Nikolaev zonal astrograph during 1967–1971.
G. K. Gorel', F. F. Kalikhevich.
Tr. Glav. Astron. Obs. Pulkovo, Ser. 2, Vol. 82, 106 - 111 (1977). In Russian.

Results of determination of 169 precise positions of major planets obtained from measurements of plates taken with the zonal astrograph of the Nikolaev Observatory are given. The coordinates and proper motions of reference stars were taken from AGK3 and SAO.

041.049   Positions of Jupiter's Galilean satellites.
G. K. Gorel'.
Tr. Glav. Astron. Obs. Pulkovo, Ser. 2, Vol. 82, 112 - 118 (1977). In Russian.

Geocentric coordinates of the Galilean satellites of Jupiter determined from observations with the Nikolaev zonal astrograph during the period 1962–1965, 1973–1974 are given.

041.050   A priori estimate of the accuracy of the absolute chain method for equalizing results of observations of right ascensions of stars with a large transit instrument.
V. S. Gubanov.
Izv. Glav. Astron. Obs. Pulkovo, No. 195, Astrofiz. Astrometr., p. 3 - 13 (1977). In Russian.

Theoretical estimates of the accuracy of the right ascension system (with respect to accidental errors) have been derived with the help of an observation model close to real conditions. The absolute chain method of equalization (Gubanov, 1975) is used. The dependence of the obtained estimate on observational accuracy, program structure and the latitude of the observatory is considered.

041.051   Differential determinations of declinations from observations with a vertical circle.   G. S. Kosin.
Izv. Glav. Astron. Obs. Pulkovo, No. 195, Astrofiz. Astrometr., p. 21 - 25 (1977). In Russian.

Some suggestions are made for the reduction of differential observations obtained with the vertical circle. They are based on excluding the zero point of the circle when determining the zenith distance of a star. Some problems connected with the determination of the systematic errors $\Delta\delta_a$ of the FK4 are considered.

041.052   Observational program of the Pulkovo Observatory's expedition to the equator.
S. A. Tolchel'nikova-Murri, M. P. Varin.
Izv. Glav. Astron. Obs. Pulkovo, No. 195, Astrofiz. Astrometr., p. 67 - 73 (1977). In Russian.

A program for determining the declinations of 91 stars with $|\delta| \leqslant 10'$ observable in Kenya ($\varphi = 0°0\overset{.}{.}0$) by the micro-

metric method is given.

041.053   Inertial coordinate system in astronomy.
K. G. Gnevysheva, D. D. Polozhentsev.
Vestn. AN SSSR, 1977, No. 7, p. 114 - 116. In Russian.
Abstr. in Ref. zh., 51. Astron., 1.51.140 (1978).

041.054   Plan for celestial photography with wide-angle astrographs.   I. G. Kolchinskij, A. B. Onegina.
Astrometriya i Astrofizika, Kiev, vyp. (No.) 33, (see 003.020), p. 11 - 16 (1977). In Russian.

A plan is presented for cooperative celestial photography with wide-angle astrographs. The aim is to compile a catalogue of positions and proper motions of stars up to $11 - 12^m$ with declinations $\delta > -20°$.

041.055   On a catalogue of star positions for the polar region obtained by the method of overlapping plates.
Yu. F. Zinchenko, V. V. Podobed.
Astrometriya i Astrofizika, Kiev, vyp. (No.) 33, (see 003.020), p. 16 - 20 (1977). In Russian.

A modification of the method of overlapping plates is described for compiling a catalogue of star positions in the polar region.

Astrolabe catalogue CASF 3 of San Fernando.
See Abstr. 002.001.

Résolution des plaques de la Carte du Ciel par l'AGK 2/3 dans les zones situées entre +31° et −2°.
See Abstr. 002.009.

Catalogue of absolute right ascensions of 1023 bright and faint fundamental stars of the northern hemisphere.
See Abstr. 002.042.

Catalogue of declinations of 710 stars for equinox and epoch 1960.0 determined from observations with the vertical circle of the Nikolaev Observatory during 1957–1964.
See Abstr. 002.043.

Catalogue of right ascensions of 395 stars.
See Abstr. 002.044.

Catalogue of right ascensions of 312 stars.
See Abstr. 002.045.

Differential catalogue of right ascensions of 544 bright stars from FK4 for the observational epoch and equinox 1950.0 (zone −20° to +35°).   See Abstr. 002.053.

Die Hamburger Meridiankreisexpedition nach West-Australien 1967–1974.   See Abstr. 011.033.

International cooperation of centers for ephemerides and astrometric data.   See Abstr. 013.007.

Observations méridiennes à Besançon.
See Abstr. 031.221.

Astrometric techniques with a PDS microdensitometer.   See Abstr. 031.229.

Radio astrometry.   See Abstr. 031.235.

ALSEP-quasar VLBI: complementary observable for laser ranging.   See Abstr. 031.244.

ALSEP-Quasar differential VLBI.
See Abstr. 031.327.

Recognition of objects and position recovery from

microdensitometer measurements of large field plates.
See Abstr. 031.330.

Elimination of lateral refraction from absolute
declinations of equatorial stars by the micrometric method.
See Abstr. 031.338.

Some comments on phase reduction formulae for
observations of planets.   See Abstr. 031.339.

Observational results of the Belgrade Vertical Circle
after its reconstruction.   See Abstr. 032.043.

Recent studies of astronomical refraction.
See Abstr. 082.114.

An introductory review of ephemerides for lunar
laser ranging.   See Abstr. 094.016.

Laser location of the moon.
See Abstr. 094.185.

Results of photographic observations of Mars with
the Pulkovo short-focus astrograph during 1973 - 1974.
See Abstr. 097.213.

Photographic positional observations of Saturn and
its satellites at Pulkovo during 1971 - 1974.
See Abstr. 100.031.

Precise positions of four emission-line stars in
Sagitta.   See Abstr. 114.558.

Precise positions of radio sources. V. Positions of 36
sources measured on a baseline of 35 km.
See Abstr. 141.059.

042 Celestial Mechanics, Figure of Celestial Bodies

042.001 On the question of determination of the Love number by satellites observations.   J. Kostelecký.
Bull. Astron. Inst. Czechoslovakia, Vol. 28, 232 - 237 (1977).
    In determining the influence of tidal forces on the motion of a satellite, the solution of Dirichlet's problem for an outer point is required. It may be possible that the spherical approximation of the Earth is not satisfactory. As the Earth resembles a rotational ellipsoid rather than a sphere, the solution of the problem for the ellipsoidal surface is indicated.

042.002 Dynamical coupling of periodic systems. Application to the n-body planetary problem. Resonance.
I. Stellmacher.
Astron. Astrophys., Vol. 59, 337 - 347 (1977). In French.
    In a recent work it was shown that Haag's theory of synchronization could be used in certain problems of celestial mechanics, such as the research of periodical solutions of the restricted three body problem. By analogy with what Haag calls the coupling of periodical systems, this theory can be extended to the study of periodical solutions of the planetary problem.

042.003 On the stability of periodic motions near the Lagrangian solutions.
A. P. Markeev, A. G. Sokol'skij.
Astron. Zh. Akad. Nauk SSSR, Vol. 54, 897 - 908 (1977). In Russian. English translation in Soviet Astron., Vol. 21, No. 4.
    The problem of the orbital stability of small periodic motions near the Lagrangian solutions for the restricted three-body problem in the plane and three-dimensional circular cases is solved in nonlinear statement.

042.004 The motion of a particle from the L_2 Lagrangian point: elliptic cases.   K. Nariai.
Publ. Astron. Soc. Japan, Vol. 29, 263 - 288 (1977).
    The author calculates the orbit of an infinitesimal particle starting from the L_2 Lagrangian point when the binary orbit is elliptic. Asymptotic or quasi-asymptotic osculating elements of the third particle are given in tables.

042.005 Some qualitative properties of n-body systems.
D. S. Saari.
Bull. American Astron. Soc., Vol. 9, 435 (1977). – Abstract.

042.006 The solar system in a binary star.
R. S. Harrington.
Bull. American Astron. Soc., Vol. 9, 435 (1977). – Abstract.

042.007 Collinear configurations may be stable.
C. Hunter, S. D. Tremaine.
Bull. American Astron. Soc., Vol. 9, 435 (1977). – Abstract.

042.008 Stability of the lunar orbit.
V. Szebehely, R. McKenzie.
Bull. American Astron. Soc., Vol. 9, 435 (1977). – Abstract.

042.009 The stability of periodic orbits in the planar general problem of three bodies.   J. B. Dunham.
Bull. American Astron. Soc., Vol. 9, 435 - 436 (1977). Abstract.

042.010 An extension of Poisson's theorem to higher orders.
P. E. Nacozy.
Bull. American Astron. Soc., Vol. 9, 436 (1977). – Abstract.

042.011 A note on the averaging method.
H. Kinoshita.
Bull. American Astron. Soc., Vol. 9, 436 (1977). – Abstract.

042.012 An improved transformation-elimination technique for the solution of perturbed Hamiltonian systems.
R. A. Howland.
Bull. American Astron. Soc., Vol. 9, 436 (1977). – Abstract.

042.013 Dispersion of particles from a point source active over an interval of time.   W. B. Heard.
Bull. American Astron. Soc., Vol. 9, 438 (1977). – Abstract.

042.014 Le problème restreint des 3 corps.   M. Hénon.
Recueil des séminaires, (see 012.004), p. 3 - 7 (1977).

042.015 Problème général des 3 corps.   M. Hénon.
Recueil des séminaires, (see 012.004), p. 8 - 10 (1977).

042.016 Orbit–orbit resonances in the solar system: varieties and similarities.   R. Greenberg.
Vistas Astron., Vol. 21, 209 - 239 (1977).
    Contents: 1. Introduction; 2. Titan-Hyperion; 3. Analysis of the resonance; 4. Small eccentricity mechanism; 5. Enceladus-Dione resonance; 6. Mimas-Tethys resonance; 7. Neptune-Pluto resonance; 8. Secular resonances; 9. Coupled librations; 10. The Laplace relation; 11. Trojan asteroid orbits; 12. Higher order commensurabilities; 13. Conclusion.

042.017 Direct computation of a planetary ephemeris.
J. Chapront.
Astron. Astrophys., Vol. 61, 7 - 11 (1977). In French.
    The author presents a program, for the non-specialist user, which produces directly a short ephemeris in the form of Chebyshev polynomials.

042.018 Bifurcations and trifurcations of asymmetric periodic orbits.   V. V. Markellos.
Astron. Astrophys., Vol. 61, 195 - 198 (1977).
    This paper reports on two as yet unreported features of the structure of the periodic orbits of the restricted three-body problem: (a) the bifurcation of a family of asymmetric periodic orbits from another family of asymmetric periodic orbits, and (b) the intersection of three different families of periodic orbits at a point of "trifurcation". A procedure of determination of the critical values of the mass parameter of the problem for which such trifurcations occur is outlined.

042.019 Orbital resonances in the solar system.
S. J. Peale.
Annu. Rev. Astron. Astrophys., Vol. 14, (see 003.008), 215 - 246 (1976).
    Known orbital resonances are listed and described. Contents: Physical description of the resonance phenomenon; Saturn's rings and the asteroids; Analytical development; Discussion.

042.020 On periodic motions of an axisymmetric satellite relative to the mass centre in an elliptic orbit.
Yu. V. Barkin, A. A. Pankratov.
Kosm. Issled., Vol. 15, 526 - 532 (1977). In Russian.

042.021 Numerical investigation of periodic orbits in the restricted simplified averaged elliptic three-body problem.   V. G. Demin, M. V. Kurchanova.
Kosm. Issled., Vol. 15, 646 - 657 (1977). In Russian.

042.022 **Sur les positions colinéaires d'équilibre relatif dans le problème des trois corps.**   B. Elmabsout.
C. R. Acad. Sci. Paris, Tome 285, Sér. A, 727 - 730 (1977).

042.023 **On a family of three-dimensional periodic orbits around the moon and planets.**   M. L. Lidov.
Dokl. AN SSSR, Vol. 233, 1068 - 1071 (1977). In Russian. Abstr. in Ref. zh., 51. Astron., 10.51.172; 62. Issled. kosm. prostranstva, 10.62.308 (1977).

042.024 **On periodic motions of an axisymmetric satellite relative to the centre of mass in a circular orbit in a non-central gravitational field.**   V. E. Ievlev.
Mosk. vyssh. tekh. uch-shche im. N. Eh. Baumana. Moskva, 1977. 12 pp. In Russian. − Abstr. in Ref. zh., 51. Astron., 10.51.173 (1977).

042.025 **Procedure for the determination of orbits of astronomical bodies.**   D. Birnbaum.
American J. Phys., Vol. 45, 135 - 137 (1977). − Abstr. in Phys. Abstr., Vol. 80, Abstr. 33761 (1977).

042.026 **Systematic formulation of perturbation theory for orbital motion.**   R. Cid, V. Camarena.
Rev. Acad. Cienc. Zaragoza, Vol. 31, No. 1 - 2, p. 17 - 23 (1976). In Spanish. − Abstr. in Phys. Abstr., Vol. 80, Abstr. 49279 (1977).

042.027 **Lagrange parenthesis and perturbation equations of a Keplerian motion.**   R. Cid, M. Palacios.
Rev. Acad. Cienc. Zaragoza, Vol. 31, No. 1 - 2, p. 25 - 34 (1976). In Spanish. − Abstr. in Phys. Abstr., Vol. 80, Abstr. 49280 (1977).

042.028 **Planetary motion calculation method.**   H. Citrynell.
Indian J. Meteorol. Hydrol. Geophys., Vol. 27, No. 2, p. 185 - 186 (1976). − Abstr. in Phys. Abstr., Vol. 80, Abstr. 53411 (1977).

042.029 **Lunar orbital theory.**   J. Kovalevsky.
Philos. Trans. R. Soc. London, Ser. A, Vol. 284, 565 - 571 (1977). − Abstr. in Phys. Abstr., Vol. 80, Abstr. 64676 (1977).

042.030 **Minimally classifying relative equilibria.**
J. I. Palmore.
Lett. Math. Phys., Vol. 1, 395 - 399 (1977). − Abstr. in Phys. Abstr., Vol. 80, Abstr. 67027 (1977).

042.031 **Two-body problems − a unified, classical, and simple treatment of spin-orbit effects.**
L.-H. Chan, R. F. O'Connell.
Phys. Rev. D, Vol. 15, 3058 - 3059 (1977). − Abstr. in Phys. Abstr., Vol. 80, Abstr. 68386 (1977).

042.032 **An interesting property of the Kepler ellipse.**
I. M. Freeman.
American J. Phys., Vol. 45, 585 - 586 (1977). − Abstr. in Phys. Abstr., Vol. 80, Abstr. 70984 (1977).

042.033 **Some problems on the dynamics of the Earth-Moon system.**   M. Burša.
Veröff. Zentralinst. Phys. Erde, No. 52, Part 1, (see 012.035), p. 83 - 118 (1977).
The force function of the Sun-Earth-Moon system has been derived keeping the main harmonics in the gravitational fields of these bodies. The corrections for the Moon's orbit as well as for the Earth's precession and rotation have been estimated due to the harmonics not considered in the fundamental three-body problem for the Sun-Earth-Moon system and in the traditional theory of the rotation of the Earth.

042.034 **Asymptotic solutions in the many-body problem. Part II: Periodic orbits in four-body systems.**
P. G. D. Barkham, V. J. Modi, A. C. Soudack.
Celestial Mech., Vol. 15, 5 - 20 (1977).
A small particle moves in the vicinity of two masses, forming a close binary, in orbit about a distant mass. Unique, uniformly valid solutions of this four-body problem are found for motion near both equilateral triangle points of the binary system in terms of a small parameter ϵ, where the primaries move in accordance with a uniformly-valid three-body solution. Accuracy is maintained within a constant error $O(\epsilon^8)$, and the solutions are uniformly valid as ϵ tends to zero for time intervals $O(\epsilon^{-3})$. Orbital position errors near L_4 and L_5 of the Earth−Moon system are found to be less than 5% when numerically-generated periodic solutions are used as a standard of comparison.

042.035 **General relativity and satellite orbits: the motion of a test particle in the Schwarzschild metric.**
D. P. Rubincam.
Celestial Mech., Vol. 15, 21 - 33 (1977).
The motion of a satellite with negligible mass in the Schwarzschild metric is treated as a problem in Newtonian physics. The disturbing function is expressed in terms of the Keplerian elements of the orbit and substituted into Lagrange's planetary equations. Integration of the equations shows that a typical Earth satellite with small orbital eccentricity is displaced by about 17 cm from its unperturbed position after a single orbit, while the periodic displacement over the orbit reaches a maximum of about 3 cm. Application of the equations to the planet Mercury gives the advance of the perihelion and a total displacement of about 85 km after one orbit, with a maximum periodic displacement of about 13 km.

042.036 **Three-dimensional branchings of plane periodic solutions.**   V. V. Markellos, P. G. Kazantzis.
Celestial Mech., Vol. 15, 35 - 40 (1977).
In the case of the restricted three-body problem with small mass parameter a family of plane symmetric periodic orbits of the direct type around the large primary is found to have branches of three-dimensional periodic orbits. One such branch has been established consisting of stable orbits for small deviations from the plane.

042.037 **On the stabilization of the collinear libration points of the restricted circular three-body problem.**
P. S. Krasilnikov, A. L. Kunitsyn.
Celestial Mech., Vol. 15, 41 - 51 (1977).
The possibility of stabilizing the collinear libration points of the circular restricted three-body problem by using an additional jet acceleration (constant in magnitude) is investigated. Three stabilization laws are considered when the jet acceleration is either directed continuously to one of the primaries m_1, m_2, or is parallel to the line joining them. The solution of the problem formulated is based on the method of the driving forces structure analysis created by W. Thomson and P. Tait. For the Earth−Moon system the numerical data of time-existence of the satellite in the vicinity of the libration point situated near the Moon are given.

042.038 **Long-term variations of the Earth's orbital elements.**   A. Berger.
Celestial Mech., Vol. 15, 53 - 74 (1977).
Critical analysis of theories of the long-term variations of the ecliptical elements of the Earth leads to the following conclusions, regarding the influence of different terms on the accuracy of the expansions used: (1) further improvement in planetary masses will not have significant influence; (2) for the (e, π) system, terms depending upon the second order as to the disturbing masses are more important than ones coming from the third degree with respect to the planetary eccentricities and inclinations; (3) for the (i, Ω) system, the latter

terms have highly significant influence, whereas additional terms in masses are negligible. The same conclusion can be drawn for (ϵ, Ψ_g). Using these results, a new solution for the long-term variations of the Earth's orbital elements is obtained. The results for e, π, i, Ω include terms depending upon the second power as to the disturbing masses and to the third degree with respect to the planetary e's and i's. For the obliquity ϵ and the annual general precession in longitude Ψ_g, a Laplace series is proposed where amplitudes, mean rates and phases are computed from those of the (i, Ω) system.

042.039   A relation in families of periodic solutions.
M. Hénon.
Celestial Mech., Vol. 15, 99 - 105 (1977).

The author shows the existence of a general relation between the parameters of periodic solutions in dynamical systems with ignorable coordinates. In particular, for time-independent systems with an axis of symmetry, the relation takes the form $\partial T/\partial A = -\partial\Phi/\partial E$, where T is the period, A is the angular momentum, Φ is the angle through which the system has rotated after one period, and E is the energy.

042.040   Analytical determination of the measure of stability of triple stellar systems.
V. Szebehely.
Celestial Mech., Vol. 15, 107 - 110 (1977).

An analytical stability criterion is given by the limiting value of the ratio of the semi-major axes of the outer star and inner binary and Harrington's (1972) numerically obtained results are slightly modified for planar systems in direct motion.

042.041   Convergence of Newton's iteration for the expansion of the planetary disturbing function.
M. S. Petrovskaya.
Celestial Mech., Vol. 15, 125 - 129 (1977).

A method is suggested for choosing the first approximation in Newton's iterations to expand the planetary disturbing function. The method ensures convergence of the process for any planetary orbits. An estimation is given for the number of iterations depending on a given accuracy of calculation.

042.042   Another regularization of the Kepler problem.
J. Stickforth.
Celestial Mech., Vol. 15, 131 (1977). – Technical note.

042.043   Preliminary orbit-determination method having no co-planar singularity.
R. M. L. Baker, Jr., N. H. Jacoby, Jr.
Celestial Mech., Vol. 15, 137 - 160 (1977).

It has long been recognized and demonstrated in the astrodynamic literature that three observations of angular position are not always sufficient to determine a preliminary orbit. One reason for this is due to the fact that as the plane of the observer's motion approaches the plane of the orbit of the observed object, the determination of the orbit of the object becomes indeterminant. It is the purpose of this paper to develop a practical and simple method of orbit determination using four observations. This method also allows one to avoid the problem of multiple orbit-determination solution roots, and provides numerical indices that are useful in assessing the degree of indeterminacy in any given observer/object geometry. The equations for the new method are developed and a numerical example is presented that demonstrates the efficiency of the method.

042.044   Stability of interplay motions.   M. Hénon.
Celestial Mech., Vol. 15, 243 - 261 (1977).

A family of rectilinear periodic solutions of the three-body problem, in which the central body collides alternately with each of the two other bodies, is investigated numerically for all values of the three masses. It is found that for every mass combination there exists just one solution of this kind. The linear stability of the orbits with respect to arbitrary three-dimensional perturbations is also investigated. Domains of stability and instability are displayed in a triangular mass diagram. Their boundaries form one-parameter families of critical orbits, which are tabulated. Limiting cases where one or two masses vanish are studied in detail. The domains of stability cover nearly one half of the total area in the mass diagram: this reinforces the conclusion that real triple stars might have motions of a kind entirely different from the usual hierarchical arrangement.

042.045   On the solution of the problem of three fixed centres.   G. T. Arazov, S. A. Habibov (Gabibov).
Celestial Mech., Vol. 15, 265 - 276 (1977).

For a special choice of the initial conditions a solution of the plane restricted problem of four bodies i. e. the problem of the motion of a passively gravitating material point P attracted according to Newton's law by three fixed point masses P_1, P_2 and P_3 has been obtained.

042.046   Theory of the rotation of the rigid Earth.
H. Kinoshita.
Celestial Mech., Vol. 15, 277 - 326 (1977).

An analytical theory is developed for planes normal to the angular-momentum axis, to the figure axis, and to the rotational axis of the triaxial rigid Earth. One of the purposes of this paper is to determine the effect on nutation and precession of Eckert et al's improvement to Brown's tables of the Moon and to check Woolard's theory from a different point of view. The present theory is characterized by the use of Andoyer variables, a moving reference plane, and Hori's averaging perturbation method.

042.047   An accelerated elimination technique for the solution of perturbed Hamiltonian systems.
R. A. Howland, Jr.
Celestial Mech., Vol. 15, 327 - 352 (1977).

A new technique is developed for the formal solution of non-degenerate or degenerate Hamiltonian systems under periodic perturbation through continually accelerated elimination of the periodic terms. Special features of the method are the ability to eliminate both short-period and long-period variables simultaneously and the attainment of [formal] 'quadratic convergence' for non-degenerate systems and nearly quadratic convergence in degenerate cases. The technique utilizes Lie transforms and is based on an approach due to Kolmogorov and Arnol'd.

042.048   Kustaanheimo-Stiefel Regularization and nonclassical canonical transformations.   I. V. Kurcheeva.
Celestial Mech., Vol. 15, 353 - 365 (1977).

This paper presents a method for constructing a transformation of coordinates in phase space yielding canonical equations. The general transformation, which does not necessarily yield a canonical system, has an arbitrary function in its right-hand members. The cases when the transformed equations may be made canonical by appropriate choice of this function are established. The method is illustrated by means of the Kustaanheimo-Stiefel Regularization and applied to the circular restricted three body problem.

042.049   An asymptotic solution for the Trojan case of the plane elliptic restricted problem of three bodies.
B. Érdi.
Celestial Mech., Vol. 15, 367 - 383 (1977).

The planar motion of a Trojan asteroid is considered within the framework of the elliptic restricted three-body problem. The solution is derived asymptotically to second order taking the square root of the Jupiter–Sun mass ratio and the orbital eccentricity of Jupiter as first order quantities. The results are given in explicit form for the coordinates as functions of the true anomaly of Jupiter including both short and long periodic terms resulting from the orbital eccentricity of Jupiter.

042.050 **Critical generating orbits for second species periodic solutions of the restricted problem.**
D. L. Hitzl, M. Hénon.
Celestial Mech., Vol. 15, 421 - 452 (1977).
The second species periodic solutions of the restricted three body problem are investigated in the limiting case of $\mu = 0$. These orbits, called consecutive collision orbits by Hénon and generating orbits by Perko, form an infinite number of continuous one-parameter families and are the true limit, for $\mu \to 0$, of second species periodic solutions for $\mu > 0$. By combining a periodicity condition with an analytic relation for criticality, isolated members of several families are obtained which possess the unique property that the stability index k jumps from $\pm \infty$ to $\mp \infty$ at that particular orbit. These orbits are of great interest since, for small $\mu > 0$, 'neighboring' orbits will then have a finite (but small) region of stability.

042.051 **The geopotential in nonsingular orbital elements.**
P. E. Nacozy, S. S. Dallas.
Celestial Mech., Vol. 15, 453 - 466 (1977).
The geopotential expansion is given entirely in terms of nonsingular orbital elements. The expansion and its derivatives are valid for zero eccentricity and inclination. The development begins with the geopotential expansion in singular, classical elements as given by Izsak (1964), Allan (1965) and Kaula (1966). The singular geopotential is then transformed into a nonsingular set of elements.

042.052 **Two-body motion under the inverse square central force and equivalent geodesic flows.**
E. A. Belbruno.
Celestial Mech., Vol. 15, 467 - 476 (1977).
In this paper the fixed energy surfaces for the two-body problem for parabolic and, in particular, hyperbolic motion are completely determined by utilizing an earlier work of J. Moser. The characterization of these fixed energy manifolds yields the explicit solutions to the above problems in an elementary way for arbitrary dimensions.

042.053 **Le problème des n corps et les invariants intégraux.**
L. Losco.
Celestial Mech., Vol. 15, 477 - 488 (1977).
This paper has two parts: mathematical tools, then an application to the n-body problem where, by the use of the 11th 'local' integral, the author found some new equations and developed them for the collinear triple collision of three bodies.

042.054 **A study of asymptotic solutions in the vicinity of the collinear libration points of the restricted three-body problem.** L. G. Lukjanov (*Luk'yanov*).
Celestial Mech., Vol. 15, 489 - 500 (1977).
Some particular solutions of the restricted three-body problem which determine outgoing or incoming orbits near libration points are considered. The solutions are obtained in the form of absolutely convergent Liapunov series. It is proved that these asymptotic solutions are plane curves situated in the orbital plane of the primaries. Each family of asymptotic solutions for every collinear point consists of four solutions which are the separatrices of a saddle point. The angles of inclination of the separatrices are determined.

042.055 **A note on expansions of functions of velocity in the two-body problem.** H. Kinoshita.
Celestial Mech., Vol. 15, 501 - 505 (1977).
Fourier expansions of functions of velocity in the two-body problem are obtained in terms of both the true anomaly and the mean anomaly.

042.056 **Statistical mechanics of Keplerian orbits. III. Perturbations.** K. A. Hämeen-Anttila.
Astrophys. Space Sci., Vol. 46, 133 - 154 (1977).
Perturbations and gravitational encounters are incorporated into the statistical theory of Keplerian orbits. The birth of planets is discussed as an application of the theory. It turns out to be a consequence of the combined action of collisions, differential rotation and gravitational interaction.

042.057 **Numerical determination of asymmetric periodic solutions.** V. V. Markellos, A. A. Halioulias.
Astrophys. Space Sci., Vol. 46, 183 - 193 (1977).
A simple predictor-corrector procedure is described for the determination of 'asymmetric' periodic solutions of dynamical systems of two degrees of freedom. An application in the case of the Störmer problem is given. The computed periodic motions of the charged particle are of the open-path type.

042.058 **Application of Lie series to investigation of nonlinear Hamiltonian systems.** V. P. Petruk.
Kosm. Issled., Vol. 15, 933 - 936 (1977). In Russian.

042.059 **The flattening of three-dimensional periodic orbits in the restricted problem.** C. Zagouras.
Astrophys. Space Sci., Vol. 50, 383 - 407 (1977).
A number of partly known families of symmetric three-dimensional periodic orbits of the restricted three-body ($\mu = 0.4$) problem are numerically continued in both ends until they terminate with orbits in the plane of motion of the primaries. The families of plane symmetric periodic orbits from which they bifurcate are identified and many orbit illustrations are given.

042.060 **On the totality of periodic motions in the meridian plane of a magnetic dipole.**
V. V. Markellos, A. A. Halioulias.
Astrophys. Space Sci., Vol. 51, 177 - 186 (1977).
The structure of the periodic solutions of the Störmer problem, representing the magnetic field of the Earth, is examined by considering the equatorial oscillations of the charged particle and their 'vertical' bifurcations with meridian periodic oscillations. An infinity of new families of simple-periodic oscillations are found to exist in the vicinity of the 'thalweg' and four such new families are actually established by numerical integration.

042.061 **Statistical mechanics of Keplerian orbits. IV. Concluding remarks.** K. A. Hämeen-Anttila.
Astrophys. Space Sci., Vol. 51, 429 - 437 (1977).
Theoretical predictions agree with computer simulations at least for those collisional systems in which the restitution coefficient is independent of impact velocity. An uncertainty principle for the orbits restricts the validity of the theory and its predictions. Discussion of the whole theory and of computer simulations shows that a velocity-dependent restitution coefficient provides the only astronomically interesting applications of the collisional processes. The Saturnian and Uranian ring systems correspond very well to theoretical expectations if the restitution coefficient is of this type.

042.062 **A class of periodic solutions of averaged equations of the restricted elliptical three-body problem.**
V. P. Evteev, B. M. Nagorev.
Dokl. AN TadzhSSR, Vol. 20, No. 3, p. 15 - 19 (1977). In Russian. – Abstr. in Ref. zh., 51. Astron., 11.51.209 (1977).

042.063 **Approximate analytical continuation in a large perturbation region of quasi-stationary solutions of equations for the perturbed motion of a point in a Newtonian gravitational field.** V. V. Laricheva.
Uch. zap. Tsentr. aehrogidrodinam. inst., Vol. 8, No. 2,

p. 69 - 79 (1977). In Russian. − Abstr. in Ref. zh., 51. Astron., 11.51.211 (1977).

042.064   On a form of differential equations for the elements of a hyperbolic orbit.   V. M. Chepurova.
Vestn. Mosk. univ., Fiz., astron., Vol. 18, 78 - 83 (1977). In Russian. − Abstr. in Ref. zh., 51. Astron., 11.51.212 (1977).

042.065   Hill's case of the averaged three-body problem and the stability of plane orbits.
S. L. Ziglin, M. L. Lidov.
Prikl. mat. i mekh., Vol. 41, 234 - 244 (1977). In Russian. − Abstr. in Ref. zh., 51. Astron., 11.51.215 (1977).

042.066   Nearly circular orbits of a planet's satellite moving near the equatorial plane.
E. I. Timoshkova, E. N. Polyakhova.
Vestn. Leningr. univ., 1977, No. 7, p. 150 - 156. In Russian. − Abstr. in Ref. zh., 51. Astron., 11.51.230 (1977).

042.067   Perturboj de planedetoj esprimitaj per korektataj suncentraj pozicio kaj rapido.   B. Popović.
Publ. Dep. Astron., Univ. Beograd, Fac. Sci., No. 6, (see 012.040), p. 69 - 82 (1976).

042.068   Numerical determination of approximate true anomalies in the proximity of quasicomplanar orbits of celestial bodies.   J. Lazović.
Publl. Dep. Astron., Univ. Beograd, Fac. Sci., No. 6, (see 012.040), p. 83 - 88 (1976).

042.069   Asymptotic distribution of particles from fragmented celestial bodies.   W. B. Heard.
Astron. J., Vol. 82, 1025 - 1035 (1977).
A general theory is developed for the asymptotic ($t \to \infty$) distribution in space of an ensemble of particles which disperses from a point. The theory is used to obtain the spatial density for, and asymptotic domain occupied by, particles in Keplerian orbits, which disperse anisotropically from an elliptical orbit. The theory is also used to determine the asymptotic domain occupied by particles orbiting an oblate primary.

042.070   Disappearance of integrals in systems of more than two degrees of freedom.   G. Contopoulos.
ESO Sci. Prepr. No. 11, 3 + 7 pp. (1977). − To appear in Celestial Mech.
The disappearance of some integrals of motion when two or more resonance conditions are approached at the same time is explained. As an example, a Hamiltonian of three degrees of freedom is considered in action-angle variables which in zero order represents three harmonic oscillators, while the perturbation contains two trigonometric terms. For any given perturbation, one integral disappears if two appropriate resonant conditions are approached sufficiently closely.

042.071   Periodic orbits in the regularized restricted three-body problem III. Some numerical results.
B. Szczodrowska-Kozar.
Acta Univ. Wratislaviensis, No. 296, Mat. Fiz. Astron. 22, p. 1 - 15 (1977) = Wrocław Astron. Obs. Repr. Nr. 102.
The present paper demonstrates the results of the numerical integration of the equations of motion of the infinitesimal mass placed in the vicinity of the libration point of the Sun−Jupiter system. The problem of suitable choice of the initial conditions permitting the continuation of the Type II Trojan's orbits revealed by Goodrich is discussed in some detail.

042.072   Groupements de planètes.   J. Meeus.
Astronomie, Vol. 91, 487 - 493 (1977).

042.073   Moderné Verfahren der Himmelsmechanik.
M. Schneider.
Mitt. Astron. Ges., Nr. 42, p. 13 - 28 (1977) = Mitt. Nr. 142 Lehrstuhl Astron. Phys. Geod.,Tech. Univ. München.
Review article.

042.074   Energetische Abschätzung zur Stabilität von Satellitenbahnen.   R. Dvorak, C. Marchal.
Mitt. Astron. Ges., Nr. 42, p. 74 - 75 (1977).

042.075   Die Jacobische Konstante im elliptischen eingeschränkten Dreikörperproblem.   R. Dvorak.
Mitt. Astron. Ges., Nr. 42, p. 75 - 77 (1977).

042.076   Pluto-artige Bewegungstypen im Bereich der Jupiterbahn.   J. Schubart.
Mitt. Astron. Ges., Nr. 42, p. 78 (1977).

042.077   Langperiodische Bewegungen der Trojaner.   R. Bien.
Mitt. Astron. Ges., Nr. 42, p. 78 - 79 (1977).

042.078   L'appréciation du terme restant des séries de Lindstedt. Partie II.   G. A. Merman.
Tr. Inst. Teor. Astron., Leningrad, vyp. (No.) 16, p. 59 - 87 (1977). In Russian.
On considère une modification des séries trigonométriques Lindstedtiennes développées suivant les puissances du petit paramètre ϵ qui satisfont formellement au système canonique des équations différentielles et qui sont également valables pour le cas de résonance ainsi que pour le cas contraire. L'appréciation du terme restant des séries montre que dans ce dernier cas les perturbations ne surpassent pas une petite quantité d'ordre ϵ sur l'intervalle ϵ^{-N}, N étant le nombre des termes tenus des séries.

042.079   On a form of equations of motion of a planetary equatorial satellite.   M. Khasanova.
Dokl. AN TadzhSSR, Vol. 20, No. 5, p. 20 - 22 (1977). In Russian. − Abstr. in Ref. zh., 51. Astron., 12.51.121 (1977).

042.080   Differential equations for nonangular elements of hyperbolic orbits.   V. M. Chepurova.
Vestn. Mosk. univ. Fiz., astron., Vol. 18, No. 2, p. 34 - 42 (1977). In Russian. − Abstr. in Ref. zh., 51. Astron., 1.51.117 (1978).

Orbital motion.   See Abstr. 003.142.

Geistesgeschichtliche Überlegungen zur Bahnbestimmung bei Gauss.   See Abstr. 004.062.

Method for determining the radius vector of a planet from two observations of position.   See Abstr. 014.013.

Inverse-square gravitation from Kepler's first two laws: a Cartesian coordinate treatment.
See Abstr. 014.014.

Effect of perturbation on the periodic solutions of the Störmer problem.   See Abstr. 022.050.

Numerical theory of motion of the earth and Venus derived from data of radar observations and optical observations and from observations of motion of the artificial satellites Venera 9 and Venera 10.   See Abstr. 041.017.

Probleme der fundamentalen Astrometrie und deren Beziehung zur Himmelsmechanik und Radioastrometrie.
See Abstr. 041.031.

A study of near-circular satellite orbits: with an application to lunisolar perturbations.   See Abstr. 052.018.

Optimal low-thrust takeoff from an orbit about an oblate planet.   See Abstr. 052.031.

Rotational distortion of polytropes by analytic approximation, II.   See Abstr. 065.061.

A boundary equation for uniformly rotating polytropes.   See Abstr. 065.062.

Perihelion precession for the charged two-body problem in general relativity.   See Abstr. 066.297.

Nach-Newtonsche Korrekturen zur Dynamik des Systems Erde—Mond und ihre Bedeutung für die relativistischen Gravitationstheorien.   See Abstr. 066.320.

Theories of lunar libration.
See Abstr. 094.015.

On the global insolubility of the Trojan asteroids problem in the light of the Brown conjecture.
See Abstr. 098.011.

Aspekte der Planetoidenforschung.
See Abstr. 098.028.

The asteroidal belt and Kirkwood gaps. I. A statistical study.   See Abstr. 098.048.

The asteroidal belt and Kirkwood gaps. II. Kinematical theory.   See Abstr. 098.049.

Comparison of Lieske's theory of the Galilean satellites with numerical integration.
See Abstr. 099.505.

Theory of Enceladus and Dione.
See Abstr. 100.504.

Long-term behavior of planetesimals and the formation of the planets.   See Abstr. 107.007.

Distribution of the mean motions of planets and satellites and the development of the solar system.
See Abstr. 107.019.

The gas drag effect on the elliptic motion of a solid body in the primordial solar nebula.
See Abstr. 107.028.

Planetary orbits in binary stars.
See Abstr. 117.014.

A new family of periodic oscillations in the Störmer problem: the principal asymmetric.   See Abstr. 143.058.

043   Astronomical Constants

043.001   Equinox — position and motion during 250 years.
      K. C. Blackwell.
Mon. Not. R. Astron. Soc., Vol. 180, 65P - 73P (1977).
      Recent meridian observations of the Sun made at Herstmonceux indicate that the right ascensions of Newcomb's equatorial fundamental stars require a correction ΔN of $+0\overset{s}{.}003$, agreeing closely with values obtained from modern occultations. An investigation has been made of the position and motion of the equinox obtained from observations of the Sun from 1755 to 1973. All observed data have been adjusted to a uniform system. The final solution from 52 equations of condition gave $\Delta N = -0\overset{s}{.}0366 \pm 0\overset{s}{.}0036 + (0\overset{s}{.}0064 \pm 0\overset{s}{.}0063)T +(0\overset{s}{.}0439 \pm 0\overset{s}{.}0135)T^2$ (s.e.) where T is measured in centuries from 1900. The implied acceleration of the equinox along the equator is 1.3 arcsec per century per century.

043.002   Das IAU (1976) – System astronomischer Konstanten.   H. Lambrecht.
Sterne, 53. Band, 243 - 244 (1977).

Precision estimation of precession and nutation from radio interferometric observations.
See Abstr. 031.206.

Probleme der fundamentalen Astrometrie und deren Beziehung zur Himmelsmechanik und Radioastrometrie.
See Abstr. 041.031.

Theory of the rotation of the rigid Earth.
See Abstr. 042.046.

Die Newtonsche Gravitationskonstante in extragalaktischen Distanzen.   See Abstr. 066.342.

The mass and moment of inertia of the earth.
See Abstr. 081.065.

Entfernungsmessungen zum Mond und die Skizzierung eines hypothesenfreien Äquatorialsystems.
See Abstr. 094.011.

New opposition of the minor planet 433 Eros.
See Abstr. 098.080.

044 Time, Rotation of the Earth

044.001 Variations in the Earth's rotation due to sectorial and tesseral harmonics.   M. Burša.
Bull. Astron. Inst. Czechoslovakia, Vol. 28, 237 - 240 (1977).
Variations in the Earth's rotation, due to the Stokes constants of the Earth (n degree, k order) $n = 2, k = 2; n = 3, k = 1; n = 3, k = 3; n = 4, k = 3$, have been derived theoretically. The zonal Stokes constants ($k = 0$) do not affect the variations.

044.002 Eine Genauigkeitsbetrachtung zur Bestimmung der momentanen lokalen Rotationsachse.
W. Schlüter.
Veröff. Bayerisch. Komm. Int. Erdmessung, Bayerisch. Akad. Wiss., Astron.-Geod. Arbeiten, Heft Nr. 36, p. 102 - 108 (1977).

044.003 Study of the system of Universal Time.
G. P. Pil'nik.
Astron. Zh. Akad. Nauk SSSR, Vol. 54, 1133 - 1139 (1977). In Russian. English translation in Soviet Astron., Vol. 21, No. 5.
Two methods of deriving the system of Universal Time TU2 are considered, and the results of a spectral analysis of the differences ΔTU2 are presented. It is shown that the amplitude and the phase of the derived sinusoidal signal depend strongly on the adopted method of filtration of the irregular nonuniformity in the earth's rotation.

044.004 Effects of oceanic tides on the rotation of the Earth.
P. Brosche, J. Sündermann.
Scientific applications of lunar laser ranging, (see 012.012), p. 133 - 141 (1977).

044.005 Earth rotation study using lunar laser ranging data.
A. W. Harris, J. G. Williams.
Scientific applications of lunar laser ranging, (see 012.012), p. 179 - 190 (1977).

044.006 Earth rotation as inferred from McDonald Observatory lunar laser observations during October 1975.
P. J. Shelus, S. W. Evans, J. D. Mulholland.
Scientific applications of lunar laser ranging, (see 012.012), p. 191 - 200 (1977).

044.007 Accuracy obtainable for universal time and polar motion during the EROLD campaign.
A. Stolz, D. Larden.
Scientific applications of lunar laser ranging, (see 012.012), p. 201 - 216 (1977).
The lunar laser range data give information on the rotations of the Earth and the Moon about their centres of mass, as well as on the lunar orbit. A vital question for the proposed EROLD campaign is how well the coordinates of the pole and changes in Earth rotation can be determined. The dependence of the results on station location, lunar declination, averaging interval and, data loss due to weather and other such factors is examined. The authors find that Universal Time and polar motion are usually obtainable to better than measuring accuracy if the averaging interval is two days.

044.008 McDonald UTO results and implications for the EROLD campaign.   P. L. Bender, A. Stolz.
Scientific applications of lunar laser ranging, (see 012.012), p. 217 - 218 (1977). – Abstract.

044.009 Universal Time: lunar ranging results and comparisons with VLBI and classical techniques.
R. W. King, T. A. Clark, C. C. Counselman III, D. S. Robertson,

I. I. Shapiro, C. A. Knight.
Scientific applications of lunar laser ranging, (see 012.012), p. 219 - 220 (1977). – Abstract.

044.010 On a relation between irregular variations of the earth's rotation and anomalous changes of the atmospheric circulation.   S. Okazaki.
Publ. Astron. Soc. Japan, Vol. 29, 619 - 629 (1977).
A relation between irregular variations of the rotational velocity of the earth and anomalous changes of the global zonal wind circulation is dealt with for periods longer than 4 months. In the residuals of rotational velocity data after eliminating the mean rotational acceleration and the mean seasonal terms, determined during 20 yr from 1955 to 1975, a quasi-periodic change with about an 8-yr period from 1957 to 1964 is clearly seen as well as short-term periodic fluctuations with periods less than 2 yr.

044.011 On the connection of irregular variations of the daily rotation of the earth and solar wind velocity.
Yu. D. Kalinin, V. M. Kiselev.
Geomagn. Aehron., Vol. 17, 964 - 965 (1977). In Russian.

044.012 Results of observations made with the astrolabe of Santiago from 1968 to 1971.   F. Noël.
Astron. Astrophys., Suppl. Ser., Vol. 30, 183 - 188 (1977).
Results in time and latitude as well as in radius obtained with the astrolabe of Santiago during the period 1968–1971 are given.

044.013 Time and latitude results of observations made at Merate Observatory with the astrolabe for the year 1976.   L. Buffoni, F. Chlistovsky, A. Manara, F. Mazzoleni.
Astron. Astrophys., Suppl. Ser., Vol. 30, 193 - 194 (1977).
Results of the observations made with the Astrolabe Danjon OPL No. 32 during 1976 are given. These results are in the FK4 system.

044.014 Earthquakes, weather and wobble.
C. R. Wilson, R. A. Haubrich.
Geophys. Res. Lett., Vol. 4, 283 - 284 (1977).
Two recently published reports propose different causes for the Chandler wobble excitation. One study suggests that earthquakes are important, while another investigation by the authors favors meteorological sources. Upon re-examination of the earthquake data, they find that the magnitude of the estimated earthquake contribution could explain about 25% of the Chandler wobble variance, but that there is no supporting evidence of correlation with observed polar motion. The authors conclude that there is presently no reason to believe the earthquake portion of the Chandler wobble excitation is any larger than 25%. On the other hand, the contribution of meteorological variation appears to be not less than 25%, and may easily be much more.

044.015 The Earth's variable rate of rotation: a discussion of some meteorological and oceanic causes and consequences.   K. Lambeck, A. Cazenave.
Philos. Trans. R. Soc. London, Ser. A, Vol. 284, 495 - 506 (1977). – Abstr. in Phys. Abstr., Vol. 80, Abstr. 56571 (1977).

044.016 Towards a theory of irregular variations in the length of the day and core-mantle coupling.
R. Hide.
Philos. Trans. R. Soc. London, Ser. A, Vol. 284, 547 - 554 (1977). – Abstr. in Phys. Abstr., Vol. 80, Abstr. 64186 (1977).

044.017 Universal time service in China in the years 1966–

1975.
The First Laboratory, Shanghai Observatory, Academia Sinica.
Acta Astron. Sinica, Vol. 18, 4 - 17 (1977).

Universal time service in the years 1966–1975 is reviewed, including astronomical time observations, the Joint System of universal time, the rapid service and the controlling of the transmission of UT_1 time signals. Comparison is made between the Chinese system and the BIH system, and their long-term stability is estimated.

044.018 **Program of observations with a prismatic astrolabe in Irkutsk.**
S. A. Sergienko, V. S. Gubanov, V. I. Sergienko.
Vrashchenie i priliv. deformatsii Zemli. Resp. mezhved. sb., 1977, vyp. (No.) 9, p. 102 - 108. In Russian. — Abstr. in Ref. zh., 51. Astron., 11.51.242 (1977).

044.019 **Influence of the observer on the results of determinations of time in automatic recording of star transits.**   M. B. Kaufman.
Tr. VNII fiz.-tekh. i radiotekh. izmerenij, 1976, vyp. (No.) 31 (61), p. 49 - 53. In Russian. — Abstr. in Ref. zh., 52. Geod. Aehrosemka, 11.52.84 (1977).

044.020 **Determination of time (TU1).**   Eh. A. Sanakulov.
Tsirk. Astron. Inst., Tashkent, Nos. 64 (411), 68 (415), 70 (417), 71 (418) (1976); 72 (419) (1977). In Russian. — 1976, January - October.

044.021 **Accuracy of determination of time from observations at transit instruments in Tashkent.**
B. V. Yasevich.
Tsirk. Astron. Inst., Tashkent, No. 72 (419), p. 10 - 16 (1977). In Russian.

044.022 **List of stars for determination of universal time with the transit instrument in Tashkent.**
Eh. A. Sanakulov.
Tsirk. Astron. Inst., Tashkent, No. 73 (420), 12 pp. (1977). In Russian.

044.023 **Time and latitude service.**
Polish Acad. Sci., Astron. Latitude Obs., Borowiec, Poland, Circ. No. 142 (1977). — 1977 April - June.

044.024 **Behaviour of atomic and mechanical oscillators during a solar eclipse.**
V. S. Kazachok, O. B. Khavroshkin, V. V. Tsyplakov.
Astron. Tsirk., No. 943, p. 4 - 6 (1977). In Russian.

044.025 **Zeit- und Breitendienst.**
Deutsches Hydrogr. Inst., Hamburg, 9 + 8 + 9 + 7 pp. (1976). — 1976 January - December.

044.026 **Rotacion de la tierra año 1976. Resultados obtenidos en San Fernando con el Astrolabio Impersonal**
Danjon OPL No. 37. I. — Tiempo y latitud. II. — Observaciones de Saturno, Marte y Jupiter.
L. Rohat-Julien, M. Sánchez, J. B. Fernández, F. Parra, M. Esparragosa, O. Gölbasi.
Published by Inst. y Obs. de Marina, San Fernando (Cadiz), Ser. C, No. 79, 24 pp. (1977).

044.027 **Temps et latitude.**   L. Webrová, V. Ptáček.
Circulaire des observatoires tchécoslovaques, Nos. 1 - 6. Acad. Tchécoslovaque Sci., Inst. Astron. (1977). Janvier - Décembre 1976.

044.028 **Astronomische Zeit- und Breitenbestimmungen, Empfangszeiten von Zeitsignalen, Präzisionszeitvergleiche.**
Akad. Wiss. DDR, Zentralinst. Phys. Erde, Abt. Geod. Astron.,

Jahrg. 1976, Nr. 1 - 6 (1977). — 1976 January - December.

044.029 **UT2–UT1 for 1978.**
Bureau International de l'Heure, (B.I.H.), Paris, Circ. A (1977).

044.030 **Universal time and coordinates of the pole; Emission time of time signals; Coordinated Universal Time; International atomic time; Information on time signals; UTC time step.**
Bureau International de l'Heure, (B.I.H.), Paris, Circ. D128 - D133 (1977). — 1977 July - December.

044.031 **UTC time step on the 1st of January 1978.**
B. Guinot.
Bureau International de l'Heure, (B.I.H.), Paris, Circ. E 7 (1977). In English and French.

044.032 **Time Service of the Mizusawa Observatory.**
Bulletins, Vol. 19, No. 1 - 12, 1973; Vol. 20, No. 1 - 12 (1974).
S. Takagi, C. Kakuta, I. Okamoto, M. Aihara, K. Yokoyama, T. Sasao, T. Hara, G. Murakami, I. Tsubokawa, H. Kitago, K. Iwadate, S. Sakai, K. Horiai, K. Sato.
Edited by the Int. Latitude Obs. Mizusawa, Mizusawa-Shi, Iwate-Ken, Japan. 2 + 94 pp. (1976).

This Bulletin contains the results of time service and astronomical observations made at the Mizusawa Observatory from 1 January, 1974 to 31 December, 1975.

044.033 **Time service.**   C. Moranzino (Editor).
Oss. Astron. Torino (Pino Torinese), Bull. Nos. 15 - 16 (1976/77). — Results of the time determinations 1976 September - 1977 April.

044.034 **The rotation of the earth between 1955.5 and 1976.5.**   J. Vondrák.
Studia, Vol. 21, 107 - 117 (1977).

044.035 **Daily phase values and time differences.**
U.S. Naval Obs., Washington, D.C. Time Service Publ., Ser. 4, Nos. 544 - 569 (1977). — 1977 July - December.

044.036 **Preliminary times and coordinates of the pole.**
U.S. Naval Obs., Washington, D.C. Time Service Publ., Ser. 7, Nos. 497 - 522 (1977). — 1977 July - December.

044.037 **Time service announcement.**   G. M. R. Winkler.
U.S. Naval Obs., Washington, D.C. Time Service Publ., Ser. 14, No. 22 (1977).

044.038 **Precise time determinations with transit instrument ASKANIA Ap 70.**   P. Jackson.
Geowiss. Mitt., Vol. 7, 151 - 175 (1975).

During 1973 and 1974 astronomical time determinations were made with a transit instrument ASKANIA Ap 70 and a digital equipment by University-Observatory Vienna (internal accuracy ± 0.006). The direct comparison with an atomic clock (National Time Service) has given the instantaneous difference between atomic time and universal time. The results (UTC-UT 1) Vienna show annual periodic deviations (amplitude 13 ms) compared with the semidefinitive values (Circular D) given by Bureau International de l'Heure in Paris. Together exists a perfect agreement in the zero point confirming the results of earlier longitude observations.

044.039 **Déterminations de l'heure et de la latitude de 1966 à 1972, effectuées à l'astrolabe de Danjon No. 7 (II).**
P. Pâquet, W. De Rop.
Bull. Astron., Obs. R. Belgique, Vol. 8, 292 - 298 (1976).

044.040 **International Time and Latitude Service at the**

Tokyo Astronomical Observatory during 1977.
Z. Suemoto.
Tokyo Astron. Obs., Time and Latitude Bulletins, Vol. 51,
Nos. 1 - 2, p. 1 - 30 (1977). — Results of time determinations
1977 January - June.

044.041   Analysis of the results of observations of the Time
           Service of the Astronomical Observatory of the
Khar'kov State University and KhGNIIM in the years 1972 -
1974.   V. I. Turenko, A. F. Vantsan, N. G. Litkevich.
Physics of the moon and planets. Fundamental astrometry,
(see 003.024), Vestn. Khar'kov. Univ., No. 160, p. 36 - 42
(1977). In Russian.

044.042   Solar eclipse of 29 April 1976 and the difference
           between ET and UT.   J. Bouška.
Acta Univ. Carolinae Math. Phys., Vol. 18, No. 2, p. 61 - 64
(1977).
     From the observation of the partial solar eclipse of 29
April 1976 the difference between the Ephemeris Time and
the Universal Time has been determined.

044.043   Greenwich time report. Royal Greenwich Observa-
           tory Time and Latitude Service 1976 April - June.
Royal Greenwich Obs. Herstmonceux Castle, 18 pp. (1977).
From Phys. Abstr., Vol. 80, Abstr. 70581 (1977).

     On the problems of the astrometric methods and of
the lunar laser ranging in the study of the Earth's rotation.
See Abstr. 031.241.

     Observed time discontinuity of clock synchroniza-
tion in rotating frame of the earth.   See Abstr. 035.004.

     Synthetic catalogue of right ascensions for time
service.   See Abstr. 041.029.

     Some problems on the dynamics of the Earth-Moon
system.   See Abstr. 042.033.

     Theory of the rotation of the rigid Earth.

See Abstr. 042.046.

     Results of observations made at Paris with the
astrolabe. Time and latitude 1976.   See Abstr. 045.001.

     Astronomische Breiten- und Längenbestimmungen
mit einem Danjon-Astrolab in Wettzell 1975.5—1976.9.
See Abstr. 045.009.

     Astronomical determinations of latitude and
longitude in 1972 - 1975.   See Abstr. 045.022.

     On the fluctuation of the length of day and its
influence on gravity.   See Abstr. 081.026.

     Inter-hemispherical thermal mechanism and changes
of the relative momentum of an atmospheric impulse.
See Abstr. 082.137.

     On a possible long-period connection between paleo-
variations of the geomagnetic field and the earth's rotation.
See Abstr. 084.239.

     Relation between solar wind characteristics and the
irregularities of the earth's rotation.   See Abstr. 085.040.

     On the accuracy of computed topocentric distances
of lunar retroreflectors in view of an application of lunar laser
ranging technique to the determination of the earth's rotation.
See Abstr. 094.017.

Errata

044.901   Erratum: "Dissipative core-mantle coupling and
           nutational motion of the earth" [Publ. Astron. Soc.
Japan, Vol. 29, 83 - 105 (1977)].
T. Sasao, I. Okamoto, S. Sakai.
Publ. Astron. Soc. Japan, Vol. 29, 437 (1977).

045 Latitude Determination, Polar Motion

045.001 **Results of observations made at Paris with the astrolabe. Time and latitude 1976.**
F. Chollet, S. Débarbat.
Astron. Astrophys., Suppl. Ser., Vol. 29, 241 - 244 (1977). In French.

Results are given for observations made with the astrolabe APP during the year 1976. No change has been made in the star catalogue since 1956. The observing technique and computation method have also been unchanged since 1972. The results are in the FK4 system.

045.002 **DMATC (*Defense Mapping Agency Topographic Center*) Doppler determination of polar motion.**
B. R. Bowman, C. F. Leroy.
Satellite Doppler positioning, (see 012.002), p. 141 - 157.

045.003 **MEDOC experiment or the French polar motion project.** B. Guinot, F. Nouel.
Satellite Doppler positioning, (see 012.002), p. 159 - 174.

045.004 **Application of Doppler satellite tracking system for polar motion studies in Canada.**
J. A. Orosz, J. Popelar.
Satellite Doppler positioning, (see 012.002), p. 783 - 792.

045.005 **Secular trend of the earth's rotation pole: consideration of motion of the latitude observatories.**
S. R. Dickman.
Geophys. J. R. Astron. Soc., Vol. 51, 229 - 244 (1977).

Secular polar motion has been recorded in ILS data over the past 75 years, an amount greater by a factor of ten than the 'true polar wandering' deduced from paleomagnetic data. In this work, the possibility that the secular trend is an observational artifact of the continental drift of the ILS stations is directly examined by consideration of several absolute plate velocity models earlier proposed by Minster et al. (1974), Kaula (1975), and Solomon, Sleep & Richardson (1975). The assumptions underlying those models are discussed.

045.006 **On the amplitude of the Chandler wobble.**
M. Bonafede, E. Boschi.
Geophys. J. R. Astron. Soc., Vol. 51, 259 - 263 (1977).

The change in the inertia tensor of the Earth, due to the mass shift following a seismic event, has been computed by several authors for non-rotating earth models. Rotation is taken into account in the present paper, and the additional change in the inertia tensor is computed for an equivalent earth model, in which the axis of geometrical symmetry becomes tilted instead of the axis of greatest inertia. Rotation is thus seen to produce an increase by a factor 1.4 in the amplitude variation of the Chandler wobble, with respect to the non-rotating case, which, when added to the 1.4 amplitude increase due to the precessional re-adjustment of the equatorial bulge, gives a factor of 2 increase of the Chandler wobble amplitude with respect to the case of a rigid earth model.

045.007 **On variations of the amplitudes of the Chandler and annual polar motion of the earth in connection with solar activity.** L. D. Kostina, V. I. Sakharov.
Soln. Dannye 1977 Byull., No. 5, p. 100 - 104 (1977). In Russian.

A comparison has been made between the variation in the amplitudes of the Chandler and the annual component of the earth's polar motion for 130 years (1846–1975) and variation of the Wolf numbers.

045.008 **How to find the earth's pole?** E. P. Fedorov.

Zemlya i Vselennaya, 1977, No. 4, p. 17 - 23. In Russian.

045.009 **Astronomische Breiten- und Längenbestimmungen mit einem Danjon-Astrolab in Wettzell 1975.5– 1976.9.** K. Kaniuth, K. Stuber, W. Wende.
Veröff. Bayerisch. Komm. Int. Erdmessung, Bayerisch. Akad. Wiss., Astron.-Geod. Arbeiten, Heft Nr. 36, p. 65 - 72 (1977).

045.010 **Das Beobachtungsprogramm zur Bestimmung der Polbewegung mit dem Zirkumzenital.**
G. Soltau.
Veröff. Bayerisch. Komm. Int. Erdmessung, Bayerisch. Akad. Wiss., Astron.-Geod. Arbeiten, Heft Nr. 36, p. 80 - 92 (1977).

045.011 **Die astronomische Bestimmung von Längen- und Breitenänderungen mit dem Zirkumzenital – Fortsetzung der 1975 begonnenen Messungen.**
W. Schlüter, G. Soltau.
Veröff. Bayerisch. Komm. Int. Erdmessung, Bayerisch. Akad. Wiss., Astron.-Geod. Arbeiten, Heft Nr. 36, p. 93 - 101 (1977).

045.012 **Dynamics of polar motion and plate tectonics.**
R. O. Vicente.
Scientific applications of lunar laser ranging, (see 012.012), p. 143 - 148 (1977).

045.013 **On some causes of the non-polar latitude variations.**
G. S. Sheptunov.
Astrometr. i Astrofiz., Kiev, vyp. (No.) 32, (see 003.011), p. 74 - 86 (1977). In Russian.

The closing error and non-polar latitude variations are shown to be due to displacements of the so-called refraction zenith, errors of some instrumental constants and errors of the adopted proper motions of stars.

045.014 **Polar motion and Earth tides from laser tracking.**
R. Kolenkiewicz, D. E. Smith, D. P. Rubincam, P. J. Dunn, M. H. Torrence.
Philos. Trans. R. Soc. London, Ser. A, Vol. 284, 485 - 494 (1977).

045.015 **Secular polar motion of the Earth and its parameters. Computation from artificial satellites observations.** Zh. S. Erzhanov.
Veröff. Zentralinst. Phys. Erde, No. 52, Part 1, (see 012.035), p. 129 - 144 (1977). In Russian with an English summary.

045.016 **Determinación de la latitud por distancias cenitales de la polar. Método de Littrow.** M. J. Sevilla.
Urania Barcelona, Año 61, Núm. 285, p. 61 - 70 (1976).

045.017 **Investigations of local microclimatic peculiarities and their influence on latitude observations.**
N. A. Popov, O. V. Chuprunova, V. K. Budz'ko.
Vrashchenie i priliv. deformatsii Zemli. Resp. mezhved. sb., 1977, vyp. (No.) 9, p. 82 - 93. In Russian. – Abstr. in Ref. zh., 51. Astron., 11.51.241 (1977).

045.018 **On photoelectric latitude observations on a zenith telescope.** V. K. Budz'ko.
Vrashchenie i priliv. deformatsii Zemli. Resp. mezhved. sb., 1977, vyp. (No.) 9, p. 99 - 102. In Russian. – Abstr. in Ref. zh., 51. Astron., 11.51.275 (1977).

045.019 **Observations à la lunette zénithale (de 110 mm) du Service de Latitude de l'Observatoire de Beograd en 1975.** R. Grujić, M. Djokić.

Bull. Obs. Astron. Belgrade, No. 128, p. 62 - 64 (1977).

045.020 Differences in the latitude values obtained by different observers at the Astronomical Observatory in Belgrade.   R. Grujić.
Publ. Dep. Astron., Univ. Beograd, Fac. Sci., No. 6, (see 012.040), p. 135 - 136 (1976).

045.021 Variation of the latitude of the Engelhardt Observatory at the night from March 4 to 5, 1977.
V. V. Lobanova, I. A. Urasina, N. N. Chudinov.
Astron. Tsirk., No. 951, p. 6 - 7 (1977). In Russian.

045.022 Astronomical determinations of latitude and longitude in 1972 - 1975.   M. Ollikainen.
Publ. Finnish Geod. Inst., No. 81, 90 pp. (1977).

045.023 Latitude observations with the visual zenith telescope attached with an electromagnetic level at Mizusawa.   S. Abe, S. Kuji.
Proc. Int. Latitude Obs. Mizusawa, No. 16, p. 84 - 89 (1977). In Japanese.
After detailed investigations, the upper Talcott level was replaced with the electromagnetic level (called TEM level) in November 1973. The reduction of the latitude observations with the visual zenith telescope is made with readings of the lower Talcott level and those of the TEM level independently. It has been found that there is discrepancy of about $0.''09$ between the monthly mean latitudes calculated with the TEM level and lower Talcott level, respectively. However, after installing the amplifier of the TEM level in a room temperature of which is kept in a range $15-23°C$, the discrepancy between them has become negligibly small.

045.024 The excitation function computed from the 5 days-mean atmospheric pressure distribution.
N. Kikuchi.
Proc. Int. Latitude Obs. Mizusawa, No. 16, p. 133 - 136 (1977). In Japanese.

045.025 The Chandler wobble and its excitation mechanism.   M. Ooe, Y. Goto.
Publ. Int. Latitude Obs. Mizusawa, Vol. 10, 163 - 177 (1976).
Every five-day value of excitation function for the Chandler wobble is estimated for both x and y components of the pole coordinates from the astronomical data. The results of statistical analysis of excitation functions inferred both from the astronomical observations and from the world air pressure data on the assumption of a rigid ocean show a good similarity in a frequency band from 1/2 to 1/6 cpy, especially, for the y component. x component of the excitation function, however, is not wholly explained by the atmospheric excitation. It is suggested that the remaining part of the excitation sources other than atmospheric one is caused from the response of the ocean to the air pressure.

045.026 Monthly Notes of the International Polar Motion Service.
IPMS Mon. Notes, Int. Latitude Obs. Mizusawa (Japan). 1977 Nos. 6 - 11, p. 53 - 118 (1977). — Announces the values of latitudes observed at the collaborating stations during 1977 June - 1977 November.

045.027 Sulla deviazione standard delle misure di latitudine al telescopio zenitale Wanschaff di Carloforte.
C. Moranzino.
Contrib. Oss. Astron. Torino (Pino Torinese), N. 103, 15 pp. (1977).

045.028 On the continuous excitation of the polar wobble.   N. Sekiguchi.
J. Geod. Soc. Japan, Vol. 23, No. 1, p. 17 - 24 (1977) =

Tokyo Astron. Obs., Repr. No. 519.
Assuming that the Chandlerian wobble is excited by a function which consists of random excitations of short durations, some relations between polar motion and the excitation function are deduced.

045.029 On the annual motion of the poles of rotation and inertia of the earth. II.
A. A. Korsun', Ya. S. Yatskiv.
Astrometriya i Astrofizika, Kiev, vyp. (No.) 33, (see 003.020), p. 3 - 7 (1977). In Russian.
Discussed are the variations in the parameters of the annual ellipses of the poles of rotation and inertia of the earth from 1846 till 1971. Periods of $\sim 10, 15, 22$ and 45 years were determined.

045.030 Relativistic variation of the latitude of Gorkij.   L. Ya. Arifov, R. K. Kadyev.
Astrometriya i Astrofizika, Kiev, vyp. (No.) 33, (see 003.020), p. 7 - 10 (1977). In Russian.
The effect of light deflection by the gravitational field of the sun was analysed and applied to the results of measuring the latitude of Gorkij. It is shown that the amplitudes of relativistic waves in the variation of the latitude reaches $\sim 0.''01$.

045.031 Annual Report of the International Polar Motion Service for the year 1975.   S. Yumi.
Published for the International Council of Scientific Unions by Central Bureau of the International Polar Motion Service, Mizusawa, Japan. 4 + 134 pp. (1977).

Determination of latitudes in Yuan Dynasty.
See Abstr. 004.044.

Whole earth dynamics and lunar laser ranging.
See Abstr. 031.239.

Ein auf die automatische Datenverarbeitung abgestimmtes Zeitregistriersystem für das Danjon-Astrolab.
See Abstr. 032.017.

Astronomische Längen- und Breitenbestimmung mit einer transportablen Zenitkamera.   See Abstr. 032.053.

Eine Genauigkeitsbetrachtung zur Bestimmung der momentanen lokalen Rotationsachse.   See Abstr. 044.002.

Accuracy obtainable for universal time and polar motion during the EROLD campaign.
See Abstr. 044.007.

Results of observations made with the astrolabe of Santiago from 1968 to 1971.   See Abstr. 044.012.

Time and latitude results of observations made at Merate Observatory with the astrolabe for the year 1976.
See Abstr. 044.013.

Time and latitude service.
See Abstr. 044.023.

Zeit- und Breitendienst.   See Abstr. 044.025.

Rotacion de la tierra año 1976.
See Abstr. 044.026.

Temps et latitude.   See Abstr. 044.027.

Astronomische Zeit- und Breitenbestimmungen, Empfangszeiten von Zeitsignalen, Präzisionszeitvergleiche.
See Abstr. 044.028.

Déterminations de l'heure et de la latitude de 1966 à 1972, effectuées à l'astrolabe de Danjon No. 7 (II). See Abstr. 044.039.

International Time and Latitude Service at the Tokyo Astronomical Observatory during 1977. See Abstr. 044.040.

Point positioning concept using precise ephemeris.

See Abstr. 046.005.

Résultats de la détermination de mouvements des continents d'après les méthodes astronomiques. See Abstr. 046.030.

Zeitabhängige Koordinatenänderungen, bestimmt durch astronomische Beobachtungen und durch Dopplerbeobachtungen. See Abstr. 046.059.

046 Astronomical Geodesy, Satellite Geodesy, Navigation

046.001 **Programmsystem Geodätische Astronomie.**
     A. Bauch, U. Schröder.
Deutsche Geod. Komm. Bayerische Akad. Wiss., Reihe B:
Angew. Geod., Heft Nr. 223, 80 pp. (1977).
     Part I contains some ALGOL-procedures and -programs
for computations in geodetic astronomy. Three of these procedures are laid out for a comfortable way to compute apparent places of stars, the other procedures offer a convenient
handling of the formulas of geodetic astronomy. An essential
part of this program-system is a program for computing the
observation ephemerides for different methods. Part II contains an ALGOL-program for the computation of apparent
places of stars at any hour angles from mean places and some
other ALGOL-programs for the evaluation of different
methods of observation in geodetic astronomy; they are
supplemented with remarks for the data input.

046.002 **Geodätisch-astronomische Feldbeobachtung hoher
     Genauigkeit − Untersuchungen an den Verfahren**
von Zinger und Pewzow.   W. Zick.
Deutsche Geod. Komm. Bayerische Akad. Wiss., Reihe C:
Diss., Heft Nr. 230, 100 pp. (1977).

046.003 **The Doppler concept and the operational Navy
     Navigation System.**   R. B. Kershner.
Satellite Doppler positioning, (see 012.002), p. 5 - 24.

046.004 **Position determination using the Transit system.**
     H. D. Black.
Satellite Doppler positioning, (see 012.002), p. 25 - 45.

046.005 **Point positioning concept using precise ephemeris.**
     R. J. Anderle.
Satellite Doppler positioning, (see 012.002), p. 47 - 75.

046.006 **Concept of satellite Doppler positioning using
     translocation techniques.**   D. E. Wells.
Satellite Doppler positioning, (see 012.002), p. 77 - 96.

046.007 **Doppler positioning by the Short Arc Method.**
     D. C. Brown.
Satellite Doppler positioning, (see 012.002), p. 97 - 140.

046.008 **The National Geodetic Survey Doppler satellite
     positioning program.**   W. E. Strange, L. D. Hothem.
Satellite Doppler positioning, (see 012.002), p. 207 - 227.

046.009 **A portable geodetic positioning system with real-
     time computation.**
W. J. Piurek, J. E. Johnson, J. Vivian.
Satellite Doppler positioning, (see 012.002), p. 245 - 248.

046.010 **Modeling of residual range error in two frequency
     corrected Doppler data.**
A. J. Tucker, J. R. Clynch, H. L. Supp.
Satellite Doppler positioning, (see 012.002), p. 357 - 376.
     In the use of NAVSAT's to obtain an accurate location,
signals are received at two frequencies (150 MHz and 400 MHz)
and the Doppler data at these frequencies combined to
eliminate the first order effects of the ionosphere. However,
there remains residual errors due to: 1) higher order terms in
the refractive index, and 2) bending of the optical path. In
this experiment, the difference between the true vacuum
phase and the measured phase along the refracted path for
each frequency has been computed using a sophisticated
model of the ionosphere. The computer model used to
simulate the ionosphere is described.

046.011 **Results from portable Doppler receivers using
     broadcast and precise ephemerides.**
C. F. Leroy.
Satellite Doppler positioning, (see 012.002), p. 399 - 416.

046.012 **Methodology and field tests of GDOP, a geodetic
     computation package for the short arc adjustment**
of satellite Doppler observations.   C. Boucher.
Satellite Doppler positioning, (see 012.002), p. 417 - 426.

046.013 **Reference orbits from range and Doppler observa-
     tions.**   J. Berbert, H. Parker.
Satellite Doppler positioning, (see 012.002), p. 465 - 477.

046.014 **A comparison of several Doppler-satellite data
     reduction methods.**   R. Brunell.
Satellite Doppler positioning, (see 012.002), p. 479 - 497.

046.015 **New positioning software from Magnavox.**
     R. Hatch.
Satellite Doppler positioning, (see 012.002), p. 499 - 517.

046.016 **Naval Surface Weapons Center reduction and
     analysis of Doppler satellite receivers using the**
Celest computer program.   J. W. O'Toole.
Satellite Doppler positioning, (see 012.002), p. 519 - 550.

046.017 **Variations in Doppler positions resulting from
     differences in computer programs and tropospheric**
refraction computations.   H. L. White.
Satellite Doppler positioning, (see 012.002), p. 551 - 576.

046.018 **Analysis of Geoceiver receiver delays.**
     F. B. Varnum.
Satellite Doppler positioning, (see 012.002), p. 577 - 586.

046.019 **Determination of geophysical parameters from long
     term orbit perturbations using navigation satellite**
Doppler derived ephemerides.   B. R. Bowman.
Satellite Doppler positioning, (see 012.002), p. 763 - 782.

046.020 **Geometric adjustment of Western European Satellite
     Triangulation. (Solution 1975).**   W. Ehrnsperger.
Deutsche Geod. Komm. Bayerische Akad. Wiss., Reihe B:
Angew. Geod., Heft Nr. 221, p. 27 - 42 (1977).
     Based on 2612 reduced plates, i.e. 72% of the total sum,
so far submitted to the Computing Centre Munich some
different solutions of the Western European Satellite Triangulation (WEST) were computed. The results were compared with
other geodetic networks.

046.021 **An investigation of the zero-point stability of the
     present Wettzell laser ranging system.**
K. Nottarp.
Deutsche Geod. Komm. Bayerische Akad. Wiss., Reihe B:
Angew. Geod., Heft Nr. 221, p. 67 - 69 (1977).

046.022 **On the accuracy of longitude determinations using
     the VÚGTK-ČSSR circumzenithal astrolabe.**
G. Soltau.
Deutsche Geod. Komm. Bayerische Akad. Wiss., Reihe B:
Angew. Geod., Heft Nr. 221, p. 99 - 106 (1977).

046.023 **Astrogeodetic geoid determination in the Western
     Harz.**   W. Torge, G. Boedecker, H.-G. Wenzel.
Deutsche Geod. Komm. Bayerische Akad. Wiss., Reihe B:
Angew. Geod., Heft Nr. 221, p. 107 - 122 (1977).

046.024   **Doppler point positioning using the N.N.S.S.** *(Navy Navigation Satellite System)* **broadcast ephemeris.**
P. Wilson.
Deutsche Geod. Komm. Bayerische Akad. Wiss., Reihe B: Angew. Geod., Heft Nr. 221, p. 131 - 136, 6 figures (1977).

046.025   **Die Photographie von Satelliten-Laserechos mit der ZEISS BMK 75 — Erste Ergebnisse und Folgerungen.**   E. Wolf.
Veröff. Bayerisch. Komm. Int. Erdmessung, Bayerisch. Akad. Wiss., Astron.-Geod. Arbeiten, Heft Nr. 36, p. 44 - 58 (1977).

046.026   **Aktuelle europäische Programme zur Doppler- und Laserentfernungsmessung nach Satelliten.**
H. Seeger, P. Wilson.
Veröff. Bayerisch. Komm. Int. Erdmessung, Bayerisch. Akad. Wiss., Astron.-Geod. Arbeiten, Heft Nr. 36, p. 150 - 160 (1977).

046.027   **Die Problematik der Bezugssysteme bei der Nutzung genauer Strecken- und Richtungsmessungen in der Satellitengeodäsie.**   E. Nagel.
Veröff. Bayerisch. Komm. Int. Erdmessung, Bayerisch. Akad. Wiss., Astron.-Geod. Arbeiten, Heft Nr. 36, p. 165 - 172 (1977).

046.028   **Zur Einbeziehung von Dopplerbeobachtungen in eine Satellitentriangulation aus photographischen Beobachtungen.**   M. Näbauer.
Veröff. Bayerisch. Komm. Int. Erdmessung, Bayerisch. Akad. Wiss., Astron.-Geod. Arbeiten, Heft Nr. 36, p. 255 - 262 (1977).

046.029   **Instrumental and atmospheric corrections for the reduction of photographic plates taken with sidereally driven satellite cameras.**   M. Scherer.
Bull. Géod., Vol. 51, 203 - 211 (1977).
To obtain the direction of a vector from the earth towards a satellite with an accuracy of 0.''2 or better, some typical instrumental and atmospheric influences can no longer be ignored or neglected when reducing photograms taken with sidereally driven satellite cameras. The wobbling error and the permanent error of the camera pole axis, the starting delay and the systematic alternations of the rotation velocity as well as the influences of changes in the amount and in the direction of the astronomical refraction are discussed.

046.030   **Résultats de la détermination de mouvements des continents d'après les méthodes astronomiques.**
N. Stoyko, A. Stoyko.
Bull. Géod., Vol. 51, 219 - 225 (1977).
La discussion des observations des latitudes faites dans des stations astronomiques situées sur les continents divers peut permettre de déterminer des mouvements des continents. L'accroissement du nombre des stations astronomiques permet de déceler non seulement le déplacement des continents mais aussi des mouvements de rotation des continents.

046.031   **The role of extremely accurate surveying techniques in existing geodetic networks.**   S. W. Henriksen.
Scientific applications of lunar laser ranging, (see 012.012), p. 149 - 156 (1977).

046.032   **The measurement of the positions of points on the Earth's surface using an absolute gravimeter and a multi-wavelength geodimeter as complements to extraterrestrial techniques.**   J. E. Faller, J. Levine.
Scientific applications of lunar laser ranging, (see 012.012), p. 277 - 283 (1977).

046.033   **Geodesy by radio interferometry.**
C. C. Counselman III.

Scientific applications of lunar laser ranging, (see 012.012), p. 285 - 286 (1977). — Abstract of a paper which appeared in Annu. Rev. Astron. Astrophys., Vol. 14, 197 - 214 (1976) — see also abstract 031.235.

046.034   **Laser rangefinders in satellite geodesy.**
N. P. Erpylev.
Zemlya i Vselennaya, 1977, No. 5, p. 34 - 41. In Russian.

046.035   **High-accuracy satellite laser ranging system.**
D. A. Byrns, S. C. Morford.
Laser and electrooptical systems, conference, San Diego, Calif., USA, 25 - 27 May 1976. IEEE,New York, USA, p. 12 (1976). — Abstract in Phys. Abstr., Vol. 80, Abstr. 21932 (1977).

046.036   **Geschichte, Probleme und Aufgaben der Geodäsie.**
P. Hörmannsdorfer.
Sternenbote, 20. Jahrg., 170 - 190 (1977).

046.037   **Ein dreidimensionales Berechnungsmodell für geodätische Punkt- und Geoidbestimmungen.**
A. Preusser.
Deutsche Geod. Komm. Bayerische Akad. Wiss., München, Reihe C: Diss., Heft Nr. 238, 102 pp. (1977).

046.038   **Determination of azimuth and latitude of an observational place from zenith distances and azimuth variations of a star.**   B. N. Sochivko.
Geod. i kartogr., 1976, No. 11, p. 8 - 13. In Russian. — From Ref. zh., 51. Astron., 10.51.200 (1977).

046.039   **On methods of combined determination of astronomical coordinates in the Antarctic.**
A. V. Butkevich, F. D. Zablotskij.
Geod., kartogr. i aehrofotosemka. Resp. mezhved. nauchn.-tekh. sb., 1977, vyp. (No.) 26, p. 9 - 18. In Russian. — Abstr. in Ref. zh., 52. Geod. Aehrosemka, 10.52.96 (1977).

046.040   **Equivalence of estimable quantities and invariants in geodetic networks.**
E. Grafarend, B. Schaffrin.
Z. Vermessungswes., Vol. 101, 485 - 492 (1976). — Abstr. in Phys. Abstr., Vol. 80, Abstr. 32980 (1977).

046.041   **Current tasks of satellite geodesy.**   R. Sigl.
Z. Vermessungswes., Vol. 102, 20 - 40 (1977). In German. — Abstr. in Phys. Abstr., Vol. 80, Abstr. 44905 (1977).

046.042   **Celestial navigation with the stereographic projection.**   D. R. Hutton.
Phys. Educ., Vol. 12, 251 - 254 (1977). — Abstr. in Phys. Abstr., Vol. 80, Abstr. 60866 (1977).

046.043   **Geodetic applications of laser ranging.**
D. E. Smith, R. Kolenkiewicz, G. H. Wyatt, P. J. Dunn, M. H. Torrence.
Philos. Trans. R. Soc. London, Ser. A, Vol. 284, 529 - 536 (1977).

046.044   **Methods and achieved results of satellite geodesy.**
K. Rinner.
Sitzungsber. Österreich. Akad. Wiss., Math.-Naturwiss. Kl., Abt. II, Vol. 185, 239 - 266 (1976). In German. — Abstr. in Phys. Abstr., Vol. 80, Abstr. 64190 (1977).

046.045   **Smithsonian Astrophysical Observatory laser tracking systems.**
M. R. Pearlman, N. W. Lanham, C. G. Lehr, J. Wohn.
Philos. Trans. R. Soc. London, Ser. A, Vol. 284, 431 - 442 (1977). — Abstr. in Phys. Abstr., Vol. 80, Abstr. 64481 (1977).

046.046 **Goddard laser systems and their accuracies.**
F. O. Vonbun.
Philos. Trans. R. Soc. London, Ser. A, Vol. 284, 443 - 450
(1977). – Abstr. in Phys. Abstr., Vol. 80, Abstr. 64482 (1977).

046.047 **Future developments in lunar and satellite laser**
ranging.   S. A. Ramsden.
Philos. Trans. R. Soc. London, Ser. A, Vol. 284, 457 - 460
(1977). – Abstr. in Phys. Abstr., Vol. 80, Abstr. 64483 (1977).

046.048 **Results from lunar laser ranging.**   J. G. Williams.
Philos. Trans. R. Soc. London, Ser. A, Vol. 284,
587 (1977). – Abstr. in Phys. Abstr., Vol. 80, Abstr. 64678
(1977).

046.049 **Ein Beitrag zur numerischen Berechnung von**
Hansen-Koeffizienten und Inklinationsfunktionen.
E.-U. Fischer.
Mitt. Inst. Theor. Geod. Univ. Bonn, No. 52, 2 + 18 pp. (1977).
Two function-subroutines are presented providing the
numerical computation of Inclination-functions and Hansen's
coefficients. The development of the Hansen's coefficients is
described in detail.

046.050 **Automated astroposition determination using sensor**
array systems.   P.-F. Chen, W. A. Allen.
IEEE Trans. Instrum. Meas., Vol. IM-26, 197 - 200 (1977).
Abstr. in Phys. Abstr., Vol. 80, Abstr. 82023 (1977).

046.051 **Results of co-operation in section 6 of the Inter-**
cosmos program.
A. G. Massevitch (*Masevich*).
Veröff. Zentralinst. Phys. Erde, No. 52, Part 1, (see 012.035),
p. 31 - 41 (1977).

046.052 **Monitoring geodetic networks by space techniques.**
B. H. Chovitz.
Veröff. Zentralinst. Phys. Erde, No. 52, Part 1, (see 012.035),
p. 43 - 58 (1977).

046.053 **Application of the method of "short arcs" to the**
determination of the coordinates of terrestrial
stations.
O. M. Bulygina, Yu. V. Surnin, S. K. Tatevyan.
Veröff. Zentralinst. Phys. Erde, No. 52, Part 1, (see 012.035),
p. 175 - 183 (1977). In Russian.

046.054 **Coordinate determination of stations of the "Large**
Chords" project by the semidynamic method.
H. Montag, G. Gendt.
Veröff. Zentralinst. Phys. Erde, No. 52, Part 1, (see 012.035),
p. 185 - 192 (1977).

046.055 **Über Vorhaben des SFB 78/IfAG zu Laserentfer-**
nungsmessung nach Satelliten.
K. Nottarp, H. Seeger, P. Wilson.
Nachr. Karten-Vermessungswesen, Reihe I, No. 73, 63 - 72
(1977).

046.056 **Die Arbeiten des SFB 78/IfAG auf dem Gebiet der**
Dopplermessungen nach Satelliten.
P. Wilson, H. Seeger.
Nachr. Karten- Vermessungswesen, Reihe I, No. 73, 73 - 80
(1977).

046.057 **Arbeiten am astronomischen Längennetz.**
G. Soltau.
Nachr. Karten- Vermessungswesen, Reihe I, No. 73, 105 - 109
(1977).

046.058 **Zur Bestimmung astronomischer Koordinaten und**
Azimute.   G. Soltau.

Nachr. Karten- Vermessungswesen, Reihe I, No. 73, 111 - 115
(1977).

046.059 **Zeitabhängige Koordinatenänderungen, bestimmt**
durch astronomische Beobachtungen und durch
Dopplerbeobachtungen.   W. Schlüter.
Nachr. Karten- Vermessungswesen, Reihe I, No. 73, 117 - 120
(1977).

046.060 **Arbeiten zur Bestimmung des Geoides in der Bundes-**
republik Deutschland.   D. Lelgemann.
Nachr. Karten- Vermessungswesen, Reihe I, No. 73, 121 - 125
(1977).

046.061 **On the analysis of accuracy of determination of the**
coordinates of points by the orbital method.
V. V. Bojkov.
Geod. i kartogr., 1977, No. 5, p. 17 - 22. In Russian. – Abstr.
in Ref. zh., 52. Geod. Aehrosemka, 11.52.71; 62. Issled. kosm.
prostranstva, 11.62.149 (1977).

046.062 **Determination of the coordinates of a point with**
the help of astrolabes according to a combined
program of observations.   V. I. Sergienko, S. A. Sergienko.
Tr. VNII fiz.-tekh. i radiotekh. izmerenij, 1976, vyp. (No.) 31
(61), p. 44 - 48. In Russian. – Abstr. in Ref. zh., 52. Geod.
Aehrosemka, 11.52.86 (1977).

046.063 **On the selection of heavenly bodies for determina-**
tion of the azimuth.   M. A. Sergeev.
Izv. vyssh. ucheb. zaved. Priborostroenie, Vol. 20, No. 4,
p. 75 - 80 (1977). In Russian. – Abstr. in Ref. zh., 52. Geod.
Aehrosemka, 11.52.87 (1977).

046.064 **Some methods of adjusting a terrestrial spatial net-**
work (Part 2).   J. Kabeláč.
Stud. Geophys. Geod., Vol. 19, 330 - 349 (1975) = Astron.
Inst. Tech. Univ. Praha, Publ. No. 46.

046.065 **Present state of geodetic astronomy.**   B. Kilar.
Publ. Dep. Astron., Univ. Beograd, Fac. Sci., No. 6,
(see 012.040), p. 127 - 134 (1976).

046.066 **Geschlossene Integration von Potentialkernen beim**
Modell der einfachen Schicht mit stückweise ebenem
Rand.   W. Bosch.
Mitt. Inst. Theor. Geod. Univ. Bonn, Nr. 49, 4 + 38 pp. (1977).
It is shown how in connection with the Simple Layer
Model integrals over potential kernels appear in boundary value
problems and in dynamic satellite geodesy.

046.067 **Les observations photographiques des satellites**
artificiels de la Terre à la station soviétique-
française à l'île Kerguelen.
N. N. Kovalenko, Yu. L. Spirin.
Nauchn. Inf., vyp. (No.) 35, p. 16 - 19 (1977). In Russian.

046.068 **Determination of the geocentric coordinates of the**
station Kerguelen by the dynamic method.
I. S. Gayazov.
Nauchn. Inf., vyp. (No.) 35, p. 20 - 22 (1977). In Russian.

046.069 **Programm zur Bestimmung von Länge und Richtung**
eines geodätischen Vektors aus synchronen photo-
graphischen und Laser-Beobachtungen künstlicher Satelliten.
T. V. Kasimenko, K.-H. Marek, N. A. Sorokin.
Nauchn. Inf., vyp. (No.) 35, p. 64 - 70 (1977). In Russian.

046.070 **Détermination des coordonnées rectangulaires des**
stations par observations des "flashes" du satellite
GEOS-B.   N. A. Sorokin, V. I. Krylov.
Nauchn. Inf., vyp. (No.) 35, p. 71 - 73 (1977). In Russian.

046.071   Détermination des longueurs et des directions
spatiales des vecteurs Haute Provence — Riga et
Haute Provence — Helsinki en utilisant des observations lasers
et optiques du satellite GEOS-B.
N. A. Sorokin, V. I. Krylov.
Nauchn. Inf., vyp. (No.) 35, p. 74 - 75 (1977). In Russian.

046.072   Einige Schätzungen der Effektivität der Benutzung
der AFU-75-Kameras bei der photographischen
Satellitenbeobachtung und Satellitengeodäsie.
K. Lapushka, L. Laucenieks, J. Badolis.
Nauchn. Inf., vyp. (No.) 35, p. 80 - 99 (1977). In Russian.

046.073   Determination of the azimuth and latitude of an
observational site from zenith distances and changes
of the azimuth of a star.   B. N. Sochivko.
Geod. i kartogr., 1976, p. 8 - 13. In Russian. — Abstr. in Ref.
zh., 52. Geod. Aehrosemka, 12.52.101 (1977).

046.074   Remarks on time and reference frames.
E. Doukakis.
Bull. Géod., Vol. 51, 295 - 300 (1977).
     The transformation of the instantaneous terrestrial co-
ordinate system to the mean or average earth-fixed one is
parameterized by the polar motion components which are continuously changing in time. Using the non-symmetricity of the
connection coefficients connecting the above frames the
errors ("misclosures") are estimated which would be present
if the instantaneous frame would be used as the geodetic
reference frame.

046.075   On the weight of the azimuth obtained from ob-
servations of stars in one vertical.
A. P. Gerasimov.
Izv. výssh. uchebn. zaved. Geod. i aehrofotosemka, 1977,
No. 3, p. 46 - 50. In Russian. — Abstr. in Ref. zh., 52. Geod.
Aehrosemka, 1.52.109 (1978).

046.076   Geodynamical research by means of artificial
earth's satellites.   J. Klokočník.
Říše hvězd, Vol. 58, 206 - 212 (1977). In Czech.

     Astronomical determinations of points in the
European part of Russia in the XIXth century.
See Abstr. 004.049.

     The achievements to date at the satellite observation
station in Wettzell.   See Abstr. 009.003.

     Die Satellitenbeobachtungsstation Wettzell.
See Abstr. 009.020.

     Der Aufbau des Zeitsystems der Satellitenbeobach-
tungsstation Wettzell in den Jahren 1972–1977.
See Abstr. 009.021.

     The University of Hawaii lunar ranging experiment
geodetic-geophysics support programme.

See Abstr. 013.028.

     Some results of Soviet-French co-operation in
satellite geodesy.   See Abstr. 013.036.

     DOPPLR — a point positioning program using in-
tegrated Doppler satellite observations.
See Abstr. 021.001.

     An intermediate term strategy for deployment of
mobile laser stations.   See Abstr. 031.240.

     The deformational environment of the Haleakala
lunar laser ranging observatory.   See Abstr. 031.243.

     Corrections for atmospheric refractivity in satellite
laser ranging.   See Abstr. 031.269.

     Sensitivity analysis of short-arc station coordinate
determinations from range data.   See Abstr. 031.285.

     Applications of very-long-baseline interferometry to
geodesy and geodynamics.   See Abstr. 031.292.

     A discussion on methods and applications of ranging
to artificial satellites and the Moon. Prologue.
See Abstr. 031.301.

     Theory of the algorithm of computer compilation of
ephemerides for star pairs at equal altitudes.
See Abstr. 041.027.

     Analysis of selected theories of motion of artificial
earth's satellites in the aspect of geodetical applications.
See Abstr. 052.011.

     Atmospheric drag analyses of low-altitude Doppler
beacon satellites.   See Abstr. 055.001.

     Improvement of the geoid in local areas by satellite
gradiometry.   See Abstr. 081.019.

     Gravity-field determination from laser observations.
See Abstr. 081.042.

     Deviation of the vertical in the South Island, New
Zealand.   See Abstr. 081.054.

Errata

046.901   Remarque sur l'article "Alignment of geodetic and
satellite coordinate systems to the average ter-
restrial system" par D. E. Wells, P. Vanicek [Bull. Géod.,
Nouvelle Sér., Année 1975, No. 117, p. 241 - 257 (1975)].
R. N. Sanchez.
Bull. Géod., Vol. 51, 239 (1977).

047 Ephemerides, Almanacs, Calendars

047.001 Astronomical yearbook for the year 1978.
Prepared by Shikin-zan Observatory, Chinese Academy of Sciences. Academic Publishing House, Peking. 365 pp. Price 5.60 Yen. (1977). In Chinese.

047.002 Nautisches Jahrbuch 1978.
Edited by Seehydrographischer Dienst der Deutschen Demokratischen Republik, Rostock. 28. Jahrg., 29 + 365 pp. (1977).

047.003 Himmelskalender 1978. Ein kleines astronomisches Jahrbuch für Österreich.
H. Mucke (Editor).
Verlag: Astronomisches Büro, H. Mucke, Sanettystr. 3, 1080 Wien, Österreich. 104 pp. (1977).

047.004 Annuaire de l'Observatoire Royal de Belgique [Jaarboek van de Koninklijke Sterrenwacht van Belgie] 1978.
Imprimerie Hayez, Bruxelles. 145ᵉ année (jaargang), 219 pp. (1977).

047.005 The Star Almanac for Land Surveyors for the Year 1978.
Prepared by H. M. Nautical Almanac Office, published by order of The Science Research Council.
Her Majesty's Stationery Office, London. 16 + 80 pp. Price £ 1.10 (1977). ISBN 0-11-881396-X.

047.006 The Handbook of the British Astronomical Association 1978.
Prepared by the Computing Section of the Association under the supervision of G. E. Taylor.
Office of the Association: Burlington House, Piccadilly, London, W1V 0NL. 100 pp. Price £ 1.50, $ 3.75 (1977).

047.007 Astronomische Grundlagen für den Kalender 1979.
Compiled by T. Lederle, edited by Astronomisches Rechen-Institut in Heidelberg.
Verlag G. Braun, Karlsruhe. 84 pp. Price DM 42.00 (1977).

047.008 Almanaque Nautico, 1978. Con suplemento para la navegacion aerea.
Published by Instituto y Observatorio de Marina, San Fernando (Cádiz). Printed in Spain by Observatorio de Marina, San Fernando (Cádiz). 419 + 30 + 5 pp. (1977). ISBN 84-500-1847-1.

047.009 Jour Julien et date du calendrier. J. Meeus.
Ciel Terre, Vol. 93, 158 - 160 (1977).

047.010 Ableitungen aus der Gaußschen Osterformel.
Zum 200. Geburtstag von Carl Friedrich Gauss (1777 - 1855). M. Oswalden.
Orion, 35. Jahrg., 148 - 150 (1977).

047.011 1978 Abridged Nautical Almanac. Pub. No. 683.
Published by Hydrographic Office of Japan, Maritime Safety Agency, Tokyo, Japan. 6 + 242 + 8 pp. (1977).

047.012 1978 Nautical Almanac. Pub. No. 681.
Published by Hydrographic Office of Japan, Maritime Safety Agency, Tokyo, Japan. 6 + 469 + 4 pp. (1977).

047.013 Ugekalenderen.
P. E. Kustaanheimo, J. Tutein.
Astron. Tidsskr., Årg. 10, 101 - 105 (1977).

047.014 Hvězdářská ročenka 1978. J. Bouška, V. Guth, B. Onderlička, J. Ruprecht (Editors).
Ročník 54. Academia, Nakladatelství Československé akademie věd, Praha. 257 pp. Price 31.00 Kčs (1977).

047.015 Astronomical Yearbook of the USSR for the year 1980. V. K. Abalakin (Editor).
Institut Teoreticheskoj Astronomii Akademii Nauk SSSR. Izdatel'stvo "Nauka", Leningradskoe Otdelenie, Leningrad. 720 pp. Price 9 Rbl. 60 Kop. (1977). In Russian.

047.016 Rocznik Astronomiczny Obserwatorium Krakowskiego 1978. International Supplement No. 49.
K. Kozieł (Editor).
Państwowe Wydawnictwo, Naukowe, Kraków. 4 + 129 pp. Price zł 72.00 (1977).

047.017 Almanaque Nautico 1977.
Estados Unidos Mexicanos, Secretaria de Marina, Direccion General de Oceanografia y Señalamiento Maritimo, Mexico, D.F. 3 + 280 + 35 pp. Price $ 70.00 (1976).

047.018 The Air Almanac 1978, January - June.
Her Majesty's Stationery Office, London; United States Naval Observatory, Washington. p. 1 - 364, A1 - A104, F1 - F4. Price £ 7.50 (1977). ISBN 0-11-772128 X.

047.019 Annuaire du Bureau des Longitudes. Éphémérides 1978. Calendriers-soleil-lune-planètes-satellites-étoiles-marées-déclinaison magnétique.
Gauthier-Villars, Paris. 12 + 253 + 10 pp. (1977). ISBN 2-04-010184-5 = Supplément à l'Astronomie de Janvier 1978.

047.020 Moon tables. Phases of the moon for times past, present, and future 601 B.C. - 2700 A.D. **Brahdes maanetabellar.** R. Brahde.
Published by Nordanger Forlag, P.O. Box 731 N-5001 Bergen, Norway. 110 pp. Price N.Kr. 74.60 (1977). ISBN 82-7051-049-1.
The tables contain the time of the two principal phases of the moon, New and Full, between the years -600 and 2700. The dates are given in the Julian calendar up to 1700, including the month of February that year. From March 1st, 1700, the Gregorian calendar is used. The time of the day is referred to the Greenwich meridian, and begins at midnight. The table may be used to estimate the tides at any time.

047.021 1978 Polaris Almanac for Azimuth Determination.
Astronomical Division, Hydrographic Departement, Tokyo, Japan. Pub. No. 685, 15 pp. (1977).

047.022 The Indian Ephemeris and Nautical Almanac for the Year 1978.
Office of preparation: Nautical Almanac Unit, Regional Meteorological Centre, Alipore, Calcutta 700027. Printed by the Manager, New Govt. of India Press (K.S.R. Unit), Santragachi, Howrah-4 and published by the Controller of Publications, Civil Lines, Delhi. 22 + 469 pp. Price Rs. 30.00, £ 3.50, $ 10.80 (1977).

047.023 Grafikoni izlaza i zalaza Sunca i Mjeseca 1978.
Edited by Hidrografski Institut Jugoslavenske Ratne Mornarice, Split, HI-N-32, 27 pp. (1977).

047.024 Nautički Godišnjak 1978.
Published by Hidrografski Institut Jugoslavenske Ratne Mornarice, Split, HI-N-31, Godina 36, 11 + 213 + 74 pp. (1977).

047.025   **Astronomical calendar of the Observatory in Sofia for the year 1978.**
D. Rajkova, Z. Krajcheva, S. Milcheva, edited by A. Bonov. Izdatelstvo na Blgarskata Akademiya na Naukite. Sofiya. 132 pp. Price 1.20 Lv. (1977). In Bulgarian.

047.026   **Calendar of the Tartu Astronomical Observatory for the year 1978.**
Academy of Sciences of the Estonian SSR, Institute of Astrophysics and Atmospheric Physics, W. Struve Tartu Astrophysical Observatory. 54th year of publication. Publishing House "Valgus", Tallinn. 29 pp. + Appendix and Published scientific papers, p. 30 - 85. Price 30 Kop. (1977). In Estonian. – See also abstracts 002.046, 008.067, 010.041, 011.032, 032.051, 063.049, 134.043, 151.076, 151.077.

047.027   **Astronomical handbook for 1977.**   B. M. Lewis. Carter Obs., Astron. Bull., No. 87, 32 pp.

047.028   **Apparent Places of Fundamental Stars 1979, containing the 1535 stars in the Fourth Fundamental Catalogue (FK 4).**
Edited by Astronomisches Rechen-Institut, Heidelberg, under the supervision of W. Fricke, T. Lederle. Published and produced by G. Braun GmbH, Karlsruhe, Germany. 44 + 510 pp. Price DM 46.00 (1977). ISBN-Nr. 37650-00787.

047.029   **Rocznik Astronomiczny na rok 1978.**
Prepared under the supervision of J. Radecki. Instytut Geodezji i Kartografii, Państwowe Przedsiębiorstwo Wydawnictw Kartograficznych, Warszawa, Vol. 33, 155 pp. Price zł 78.00 (1977).

047.030   **Anuario del Observatorio Astronómico de Madrid para 1978.**
Published by Instituto Geografico Nacional, Madrid. 422 pp. Price 250 pesetas (1977).

047.031   **Astronomske efemeride za 1978. godinu (Astronomical ephemeris for the year 1978).**
V. Ištvan, Z. Knezevic. Vasiona, Vol. 25, 93 - 112 (1977).

047.032   **Almanaque Nautico 1978.**
Estados Unidos Mexicanos Secretaria de Marina, Direccion General de Oceanografia, Mexico, D.F. 3 + 280 + 35 pp. Price $ 80.00 (1977).

047.033   **Anuário Astronômico 1978.**
Published by Instituto Astronômico e Geofísico, Universidade de São Paulo, São Paulo, Brasil. 12 + 118 + 156* pp. (1977).

047.034   **Kalender für Sternfreunde 1978. Kleines astronomisches Jahrbuch.**   P. Ahnert.
Johann Ambrosius Barth, Leipzig. 200 pp. Price M 4.80; M 7.50 (1977). – From Astron. Schule, 14. Jahrg., 116 (1977).

047.035   **Skywatcher's almanac 1978.**   R. L. Mansfield. Astronomical Data Service, 3922 Leisure Lane, Colorado Springs, Colo. 80917. 35 pp. Price $ 10.00 (1977). From Sky Telesc., Vol. 54, 519 (1977).

Wann trat die Julianische Schaltregel in Kraft? See Abstr. 004.035.

International cooperation of centers for ephemerides and astrometric data.   See Abstr. 013.007.

An introductory review of ephemerides for lunar laser ranging.   See Abstr. 094.016.

Ephemerides of minor planets for 1978. See Abstr. 098.025.

A new presentation of the ephemeris of Jupiter's Galilean satellites.   See Abstr. 099.514.

Space Research

051 Extraterrestrial Research, Spaceflight Related to Astronomy and Astrophysics

051.001 **Round-trip mission requirements for asteroids 1976 AA and 1973 EC.** J. C. Niehoff.
Icarus, Vol. 31, 430 - 438 (1977).
The recent discovery of 1976 AA has renewed interest in the possibility of modest asteroid sample-return missions. Such ventures may be logical precursors to more complex round-trip planetary missions. Both manned and unmanned mission requirements are assessed for two Apollo–Amor objects: 1976 AA and 1973 EC. It is shown that the propulsion requirements of 1-yr manned missions to either target are excessive.

051.002 **Besuch bei Jupiter und Saturn.** W. Engelhardt.
Umschau, 77. Jahrg., 505 - 509 (1977).

051.003 **Collections of cosmic dust.** C. L. Hemenway. The dusty universe, (see 012.001), p. 211 - 244 (1975).
Some recent results of collection experiments in the upper atmosphere by rocket techniques, by balloon collections in the intermediate atmosphere, and by recoverable satellites in near-earth space are described. Evidence is presented for the existence of relatively high and variable fluxes of submicron particles entering the atmosphere.

051.004 **Visiting Halley's comet.** D. W. Hughes. Nature, Vol. 268, 486 - 488 (1977).

051.005 **Voyager-Raumsonden erforschen äussere Planeten und ihre Monde.** O. Walthert.
Orion, 35. Jahrg., 108 - 112 (1977).

051.006 **Voyage to the outer planets.** S. Sharrock. Nature, Vol. 269, 98 (1977).

051.007 **Can penetrators provide useful information for planetary exploration?**
R. Reynolds, J. Westphal, M. Blanchard, V. Oberbeck, T. Bunch, T. Canning, R. Jackson, W. Quaide.
Bull. American Astron. Soc., Vol. 9, 443 (1977). – Abstract.

051.008 **Development of a penetrator emplaced seismic experiment.** W. F. Miller.
Bull. American Astron. Soc., Vol. 9, 443 (1977). – Abstract.

051.009 **Earth-approaching asteroids as raw materials for space manufacturing.** B. O'Leary.
Bull. American Astron. Soc., Vol. 9, 458 (1977). – Abstract.

051.010 **A solar sailor to Halley.** L. D. Friedman. Bull. American Astron. Soc., Vol. 9, 463 (1977). Abstract.

051.011 **A mission designed for a planetary science explosion –MJS77 (*Mariner Jupiter/Saturn 1977 mission*).** J. E. Randolph, A. L. Lane.
Bull. American Astron. Soc., Vol. 9, 469 (1977). – Abstract.

051.012 **Spaceflight, colonization and independence: a synthesis.** Part three: the consequences of colonization. M. A. G. Michaud.
J. British Interplanet. Soc., Vol. 30, 323 - 331 (1977).

051.013 **A first comet mission.** Report of the Comet Halley Science Working Group.
NASA TM-78420, 7 + 85 pp. (1977).

051.014 **Lunar research after Apollo: are we on the right track?** T. Gold.
News Lett. Astron. Soc. N.Y., Vol. 1, No. 2, p. 8 - 10 (1977). Abstract.

051.015 **20 Jahre Weltraumfahrt. Ein Rückblick – Von der ersten Erdumkreisung bis zu den Sternen.**
W. Engelhardt.
Umschau, 77. Jahrg., 633 - 636 (1977).

051.016 **Two decades of Soviet space.** V. Rich. Nature, Vol. 269, 459 - 460 (1977).

051.017 **Projekt Voyager – der Flug zu Jupiter und Saturn.** H. W. Köhler.
Sterne Weltraum, Jahrg. 16, 316 - 318 (1977).

051.018 **Preliminary results from the European Space Agency's COS-B satellite for gamma-ray astronomy.**
K. Bennett, J. J. Burger, G. G. Lichti, B. G. Taylor, R. D. Wills, G. F. Bignami, G. Boella, R. Buccheri, A. Cuccia, L. Scarsi, W. Hermsen, J. Higdon, B. N. Swanenburg, G. Kanbach, H. A. Mayer-Hasselwander, L. Koch, J. Masnou, J. A. Paul, P. G. Shukla.
The structure and content of the Galaxy and galactic gamma rays, (see 012.009), p. 27 - 64 (1977).
The COS-B experiment and mission; COS-B observations of the high-energy gamma radiation from the galactic disk; COS-B observations of localized sources of gamma-ray emission; The time structure of the gamma-ray emission from the Crab and Vela pulsars.

051.019 **Voyager to Jupiter and Saturn.**
Foreword by N. W. Hinners.
Natl. Aeronaut. Space Adm., Washington, D. C. NASA SP-420. For sale by the National Technical Information Service, Springfield, Virginia 22161. 5 + 58 pp. (1977).
This publication briefly describes the National Aeronautics and Space Administration's Voyager mission to explore the

giant planets of the outer solar system – Jupiter, Saturn, and possibly Uranus.

051.020 **Implications of outer-zone radiations on operations in the geostationary region utilizing the AE4 environmental model.** J. W. Wilson, F. M. Denn.
NASA Tech. Note, NASA TN D-8416. 70 pp. Price $ 4.50 (1977).

051.021 **Mars polar ice sample return mission – 2.**
R. L. Staehle, S. A. Fine, A. Roberts, C. R. Schulenburg, D. L. Skinner.
Spaceflight, Vol. 19, 399 - 409 (1977).

051.022 **Die 20 Jahre nach Sputnik 1. Der Mensch im Weltraum.** W. Engelhardt.
Sterne Weltraum, Jahrg. 16, 358 - 362 (1977).

051.023 **The participation of the European Space Agency in atmospheric and magnetospheric research.**
D. E. Page.
Proc. 27th Internat. Astronaut. Congr. (see 012.016), p. 51 - 61 (1977).

051.024 **Spacelab new tool for research and investigations in space.** R. Gibson, D. R. Lord.
Proc. 27th Internat. Astronaut. Congr. (see 012.016), p. 275 - 283 (1977).

051.025 **Cosmonautics: achievements and perspectives.**
R. Z. Sagdeev.
Priroda, 1977, No. 10, p. 4 - 9. In Russian.

051.026 **First twentieth anniversary of the cosmic era and astronomy.** I. S. Shklovskij.
Priroda, 1977, No. 10, p. 86 - 92. In Russian.

051.027 **From "terrestrial" to "cosmic" astronomy.**
V. G. Kurt.
Zemlya i Vselennaya, 1977, No. 5, p. 29 - 32. In Russian.

051.028 **Project Daedalus.** Daedalus Study Group.
Spaceflight, Vol. 19, 419 - 430 (1977).

051.029 **Pioneer Venus 1978.** Staff of NASA.
Spaceflight, Vol. 19, 431 - 434 (1977).

051.030 **Mars polar ice sample return mission.**
R. L. Staehle, S. A. Fine, A. Roberts, C. R.
Schulenburg, D. L. Skinner.
Spaceflight, Vol. 19, 441 - 445 (1977).

051.031 **"Intercosmos" – program of peace and progress.**
B. N. Petrov.
Zemlya i Vselennaya, 1977, No. 6, p. 23 - 29. In Russian.

051.032 **Planetary mission planning for the next decade.**
D. H. Herman, S. J. Grivas.
Chemical evolution of the giant planets, (see 003.014), p. 129 - 146 (1976).

051.033 **The tethered balloon current generator – a space shuttle-tethered subsatellite for plasma studies and power generation.** P. R. Williamson, P. M. Banks.
Plasma science. Austin, Tex., USA, 24 - 26 May 1976. IEEE, New York, USA. Price $ 10.00 (1976). p. 102. – Abstract. Abstr. in Phys. Abstr., Vol. 80, Abstr. 25912 (1977).

051.034 **Destination Mars.**
Indian East. Eng., Vol. 118, No. 9, p. 393, 395 - 397, 399 (1976). – Abstr. in Phys. Abstr., Vol. 80, Abstr. 33321 (1977).

051.035 **The planetary exploration program.** D. G. Rea.
Communications and knowledge. I. Conference held at Dallas, Tex., USA, 29 Nov. - 1 Dec. 1976, IEEE 1976, New York, USA, p. 1.2/1 - 2. – Abstr. in Phys. Abstr., Vol. 80, Abstr. 33324 (1977).

051.036 **Space travel results of USSR.** E. J. T. Kordik.
Österreichische Ing.-Z., Vol. 20, No. 2, p. 37 - 42 (1977). In German. – Abstr. in Phys. Abstr., Vol. 80, Abstr. 41283 (1977).

051.037 **Viking on Mars: a preliminary survey.**
R. S. Young.
American Sci., Vol. 64, 620 - 627 (1976). – Abstr. in Phys. Abstr., Vol. 80, Abstr. 41341 (1977).

051.038 **Outer planets atmospheric entry probes: science objectives and payloads.** H. Myers.
J. Spacecr. Rockets, Vol. 13, 712 - 718 (1976). – Abstr. in Phys. Abstr., Vol. 80, Abstr. 49267 (1977).

051.039 **The 'Raduga' experiment – exploration of the Earth from the cosmos.** A. Rot.
Nachr. Elektron., Vol. 31, 97 - 98 (1977). In German. – Abstr. in Phys. Abstr., Vol. 80, Abstr. 56765 (1977).

051.040 **Astronomers in space. I.** J. E. Oberg.
Astronomy, Vol. 5, No. 4, p. 18 - 24 (1977).
Abstr. in Phys. Abstr., Vol. 80, Abstr. 56806 (1977).

051.041 **Shuttle: the next step to the stars.**
B. T. Cummings.
Spaceworld, Vol. N-1-157, p. 4 - 13 (1977). – Abstr. in Phys. Abstr., Vol. 80, Abstr. 56811 (1977).

051.042 **Astronomers in space. II.** J. E. Oberg.
Astronomy, Vol. 5, No. 5, p. 48 - 53 (1977).
Abstr. in Phys. Abstr., Vol. 80, Abstr. 64588 (1977).

051.043 **Die 20 Jahre nach Sputnik 1. Forschen, helfen, spionieren.** W. Engelhardt.
Sterne Weltraum, Jahrg. 16, 401 - 406 (1977).

051.044 **Mariner-Jupiter-Saturn low energy charged particle experiment.**
D. P. Peletier, S. A. Gary, A. F. Hogrefe.
IEEE Trans. Nucl. Sci., Vol. NS-24, 795 - 800 (1977). – Abstr. in Phys. Abstr., Vol. 80, Abstr. 70971 (1977).

051.045 **The 'voyagers' to Jupiter, Saturn, Uranus, Neptune.**
A. Dupas.
Recherche, Vol. 8, 669 - 670 (1977). In French. – Abstr. in Phys. Abstr., Vol. 80, Abstr. 70976 (1977).

051.046 **MJS'77: a space odyssey.** C. E. Kohlhase.
Spaceworld, Vol. N-3-159, 30 - 32 (1977). – Abstr. in Phys. Abstr., Vol. 80, Abstr. 73932 (1977).

051.047 **Viking to Mars. Profile of a space expedition.**
J. S. Martin, Jr., A. T. Young.
Spaceworld, Vol. N-4-160, 4 - 27 (1977). – Abstr. in Phys. Abstr., Vol. 80, Abstr. 73933 (1977).

051.048 **HEAO-1 pointed observations.**
IAU Circ., No. 3134 (1977).

051.049 **What is a HEAO?**
Astronomy, Vol. 5, No. 7, p. 16 - 17 (1977).
Abstr. in Phys. Abstr., Vol. 80, Abstr. 78595 (1977).

051.050 **Probing the nearest star.** J. E. Oberg.
Astronomy, Vol. 5, No. 8, p. 18 - 22 (1977).

Abstr. in Phys. Abstr., Vol. 80, Abstr. 87228 (1977).

051.051 **Space projects in Japan.**   S. Saito.
Oyo Buturi, Vol. 46, 367 - 380 (1977). In Japanese.
From Phys. Abstr., Vol. 80, Abstr. 87237 (1977).

051.052 **On the international cooperation for Space Shuttle.**
S. Saito.
Oyo Buturi, Vol. 46, 432 - 434 (1977). In Japanese. − From
Phys. Abstr., Vol. 80, Abstr. 87238 (1977).

051.053 **The Viking project.**   G. A. Soffen.
J. Geophys. Res., Vol. 82, (see 003.017), 3959 -
3970 (1977). Paper No. 7S0535.
This paper is a summary of the project.

051.054 **Spacelab: a new tool for cooperative research.**
E. R. Schmerling.
Dynamical and chemical coupling between the neutral and
ionized atmosphere, (see 012.036), p. 373 - 379 (1977).

051.055 **Soft X-ray astronomy from HEAO-A.**
F. M. Walter, P. Charles, S. Bowyer.
Publ. Astron. Soc. Pacific, Vol. 89, 624 (1977). − Abstract.

051.056 **The exploration of Venus.**   L. Colin.
Space Sci. Rev., Vol. 20, 249 - 258 (1977).
The Pioneer Venus program consists of a Multiprobe and
Orbiter mission, both to be launched and to encounter Venus
in 1978. The evolution of the program is traced from its con-
ception in 1968 as the Goddard Space Flight Center Planetary
Explorer Program through its transfer to Ames Research
Center in 1971 as Pioneer Venus to the present.

051.057 **The Pioneer Venus program.**
L. Colin, C. F. Hall.
Space Sci. Rev., Vol. 20, 283 - 306 (1977).
The Pioneer Venus program encompasses two spacecraft
missions, Orbiter and Multiprobe, to be launched and to
encounter Venus during the 1978 Venus mission opportunity.
The missions are described in detail including mission and
spacecraft descriptions, scientific objective and payloads. The
ways in which the payloads address the major scientific
questions concerning Venus are treated in subsequent papers.

051.058 **Pioneer Venus experiment descriptions.**
L. Colin, D. M. Hunten (Editors), with contribu-
tions by U. von Zahn, H. Taylor, J. Hoffman, V. Oyama,
M. Tomasko, R. Boese, R. Knollenberg, B. Ragent, J. Blamont,
A. Seiff, V. Suomi, C. Counselman, A. Kliore, T. Croft,
R. Woo, H. Neimann, W. Knudsen, L. Brace, C. Russell,
J. Wolfe, F. Scarf, J. Hansen, F. Taylor, A. Stewart,
G. Pettengill, W. Evans, G. Keating, R. Phillips, I. Shapiro.
Space Sci. Rev., Vol. 20, 451 - 525 (1977).
This paper contains brief engineering descriptions of the
experiments to be integrated into the Orbiter and Multiprobe
scientific payloads.

051.059 **The Voyager missions to the outer system.**
E. C. Stone.
Space Sci. Rev., Vol. 21, 75 (1977).

051.060 **Voyager mission description.**
C. E. Kohlhase, P. A. Penzo.
Space Sci. Rev., Vol. 21, 77 - 101 (1977).
The Voyager Project, managed by the Jet Propulsion
Laboratory, involves the launching of two advanced spacecraft
to explore the Jovian and Saturnian systems, as well as inter-
planetary space. The purpose of this paper is to describe the
Voyager mission characteristics in order to establish a frame-
work upon which to better understand the objectives and goals
of scientific investigations.

051.061 **Voyager imaging experiment.**   B. A. Smith,
G. A. Briggs, G. E. Danielson, A. F. Cook, II,
M. E. Davies, G. E. Hunt, H. Masursky, L. A. Soderblom,
T. C. Owen, C. Sagan, V. E. Suomi.
Space Sci. Rev., Vol. 21, 103 - 127 (1977).
The overall objective of this experiment is exploratory
reconnaissance of Jupiter, Saturn, their satellites, and Saturn's
rings. Such reconnaissance, at resolutions and phase angles
unobtainable from Earth, can be expected to provide much
new data relevant to the atmospheric and/or surface properties
of these bodies.

051.062 **Radio science investigations with Voyager.**
V. R. Eshleman, G. L. Tyler, J. D. Anderson,
G. Fjeldbo, G. S. Levy, G. E. Wood, T. A. Croft.
Space Sci. Rev., Vol. 21, 207 - 232 (1977).
The planned radio science investigations during the
Voyager missions to the outer planets involve: (1) the use of
the radio links to and from the spacecraft for occultation
measurements of planetary and satellite atmospheres and ionospheres, the rings of Saturn, the solar corona, and the general-
relativistic time delay for radiowave propagation through the
Sun's gravity field; (2) radio link measurements of true or
apparent spacecraft motion caused by the gravity fields of the
planets, the masses of their larger satellites, and characteristics
of the interplanetary medium; and (3) related measurements
which could provide results in other areas, including the
possible detection of long-wavelength gravitational radiation
propagating through the Solar System.

051.063 **The plasma experiment on the 1977 Voyager mission.**
H. S. Bridge, J. W. Belcher, R. J. Butler, A. J.
Lazarus, A. M. Mavretic, J. D. Sullivan, G. L. Siscoe, V. M.
Vasyliunas.
Space Sci. Rev., Vol. 21, 259 - 287 (1977).
This paper contains a brief description of the plasma experiment to be flown on the 1977 Voyager Mission, its principal scientific objectives, and the expected results. The scientific
goals include studies of (1) the properties and radial evolution
of the solar wind, (2) the interaction of the solar wind with
Jupiter, (3) the sources, properties and morphology of the
Jovian magnetospheric plasma, (4) the interaction of magnetospheric plasma with the Galilean satellites with particular
emphasis on plasma properties in the vicinity of Io, (5) the
interaction of the solar wind with Saturn and the Saturnian
satellites with particular emphasis on Titan, and (6) ions of
interstellar origin.

051.064 **A plasma wave investigation for the Voyager mission.**
F. L. Scarf, D. A. Gurnett.
Space Sci. Rev., Vol. 21, 289 - 308 (1977).
The Voyager Plasma Wave System (PWS) will provide the
first direct information on wave-particle interactions and their
effects at the outer planets. The data will give answers to
fundamental questions on the dynamics of the Jupiter and
Saturn magnetospheres and the properties of the distant interplanetary medium. At Jupiter, plasma wave measurements will
also lead to understanding of the key processes known to be
involved in the decameter bursts such as the cooperative
mechanisms that yield the intense radiation, the observed
millisecond fine-structure, and the Io modulation effect.
Similar phenomena should be associated with other planetary
satellites or with Saturn's rings. Local diagnostic information
(such as plasma densities) will be obtained from wave observations, and the PWS may detect lightning whistler evidence of
atmospheric electrical discharges.

051.065 **Planetary radio astronomy experiment for Voyager
missions.**   J. W. Warwick, J. B. Pearce, R. G.
Peltzer, A. C. Riddle.
Space Sci. Rev., Vol. 21, 309 - 327 (1977).
The planetary radio astronomy experiment will measure

radio spectra of planetary emissions in the range 1.2 kHz to
40.5 MHz. These emissions result from wave-particle-plasma
interactions in the magnetospheres and ionospheres of the
planets. At Jupiter, they are strongly modulated by the Galilean
satellite Io.

051.066   **Cosmonauts investigate the earth.**
          G. A. Ivanyan, K. Ya. Kondrat'ev.
Priroda, 1977, No. 12, p. 48 - 55. In Russian.

051.067   **Numerical modelling of a satellite experiment using
          the Doppler effect at coherent frequencies in the
presence of a wave disturbance in the ionosphere.**
Yu. K. Postoev, B. V. Troitskij.
Kosm. Issled., Vol. 15, 940 - 942 (1977). In Russian.

051.068   **Measurement of radiation doses aboard AES Molniya
          1.**   I. N. Senchuro, P. I. Shavrin.
Kosm. Issled., Vol. 15, 942 - 945 (1977). In Russian.

051.069   **Overview of the Helios 1 and Helios 2 missions and
          their participation in STIP Intervals I and II.**
H. Porsche.
Study of travelling interplanetary phenomena 1977, (see 012.
042), p. 421 - 429 (1977).
          As an introduction to the contributions of Helios to STIP
investigations the orbital peculiarities of the two probes, their
relation to one another and to the earth are discussed. Addi-
tionally some characteristics of the probes show how they are
adapted to specific aspects of their research tasks.

051.070   **Mars surface Penetrator – system description.**
          L. A. Manning (Editor).
NASA Tech. Memo., NASA TM 73243, 5 + 83 pp. Price
$ 4.75 (1977).
          A Penetrator network is vital to a geophysical under-
standing of Mars and an invaluable adjunct to an Orbiter/
Rover combination in an intensive study of Mars. This report
presents a point design of a Penetrator system for a 1984 Mars
mission. The point design, including the strawman payload
and its derivation from a geophysical science rationale, is
described.

051.071   **Programme of investigations of the emission of the
          atmosphere, aurorae and noctilucent clouds aboard
the orbital station Salyut 4.**
A. I. Lazarev, V. N. Sergeevich.
Optical investigations of the emission of the atmosphere,
aurorae and noctilucent clouds aboard the orbital scientific
station Salyut 4, (see 003.021), p. 17 - 30 (1977). In Russian.
          This paper presents a short summary of the preliminary
results of investigations to be published in this collection on
the emission of the atmosphere, aurorae and noctilucent
clouds carried out aboard the orbital scientific station
Salyut 4.

051.072   **20 years cosmic era.**
          B. N. Petrov, V. Yu. Rutkovskij.
Izv. AN SSSR. Tekh. kibern., 1977, No. 4, p. 3 - 9. In Russian.
Abstr. in Ref. zh., 62. Issled. kosm. prostranstva, 12.62.3
(1977).

051.073   **Scientific photographic experiments from board the
          orbital station Soyuz 3.**   P. R. Popovich, Yu. P.
Artyukhin, V. D. Bol'shakov, N. P. Lavrova.
Izv. vyssh. uchebn. zaved. Geod. i aehrofotosemka, 1977, No.
3, p. 51 - 57. In Russian. – Abstr. in Ref. zh., 62. Issled. kosm.
prostranstva, 12.62.71 (1977).

051.074   **Potential for advancement of space astronomy.**
          A. D. Code.
The Space Telescope, (see 012.046), p. 1 - 15 (1976).

051.075   **Concepts of operation.**   C. R. O'Dell.
          The Space Telescope, (see 012.046), p. 17 - 19
(1976).

051.076   **Space Telescope operations, a typical day.**
          W. J. Pragluski, R. H. Brown.
The Space Telescope, (see 012.046), p. 46 - 50 (1976).

051.077   **Mission analysis.**   F. M. Friedlaender.
          The Space Telescope, (see 012.046), p. 51 - 54
(1976).

051.078   **Orbital crew extravehicular maintenance operations.**
          H. T. Fisher.
The Space Telescope, (see 012.046), p. 224 - 228 (1976).

051.079   **Twenty years of cosmic era.**   M. Pańków.
          Urania Kraków, Vol. 48, 290 - 294 (1977). In
Polish.

051.080   **Mensch–Erde–Weltall.**
          V. I. Sevast'janov (*Sevast'yanov*), A. D. Ursul.
Naturwiss. Rundsch., Band 28, 277 - 280 (1975) = Veröff.
Forschungsinst. Deutschen Mus. Gesch. Naturwiss. Tech.,
Reihe A, Kleine Mitt. Nr. 176.

051.081   **Astronautische Perspektiven in der Wende.**
          W. Petri.
Naturwiss. Rundsch., Band 28, 389 - 394 (1975) = Veröff.
Forschungsinst. Deutschen Mus. Gesch. Naturwiss. Tech.,
Reihe A, Kleine Mitt. Nr. 176.

051.082   **Space Report.**
          Spaceflight, Vol. 19, 256 - 265, 314 - 319, 348 -
354, 392 - 398, 435 - 440 (1977).

051.083   **Space report.**
          J. British Interplanet. Soc., Vol. 30, 316, 381, 390,
399, 400, 465 (1977).

          **End of an era in space exploration. From inter-
national rivalry to international cooperation.**
See Abstr. 003.034.

          Space investigations made in the USSR in 1976.
See Abstr. 003.178.

          **8th International Space Rescue and Safety Sympo-
sium.**   See Abstr. 012.054.

          Space shuttle missions of the 80's. Parts 1 and 2.
See Abstr. 012.056.

          **Intercosmos program. Cooperation of Socialist
countries.**   See Abstr. 013.035.

          Analysis of the theory of high-energy ion transport.
See Abstr. 022.029.

          Ultraviolet spectroscopy with Copernicus.
See Abstr. 032.510.

          **Technical development of scientific experiments for
spacecraft.**   See Abstr. 032.557.

          **The Voyager infrared spectroscopy and radiometry
investigation.**   See Abstr. 032.577.

          The Voyager mission photopolarimeter experiment.
See Abstr. 032.578.

          Ultraviolet spectrometer experiment for the Voyager

mission.   See Abstr. 032.579.

Magnetic field experiment for Voyagers 1 and 2. See Abstr. 032.580.

The Low Energy Charged Particle (LECP) experiment on the Voyager spacecraft.   See Abstr. 032.581.

Cosmic ray investigation for the Voyager missions; energetic particle studies in the outer heliosphere—and beyond. See Abstr. 032.582.

Phase-coherent dual-frequency link for high-precision Doppler tracking between spacecraft. See Abstr. 052.029.

The Viking relativity experiment. See Abstr. 066.321.

Pioneer probes Sun's magnetic field. See Abstr. 080.052.

Exploration of the planets: an invited discourse presented before the Sixteenth General Assembly of the International Astronomical Union, Grenoble, France, August, 1976.   See Abstr. 091.038.

Why explore Venus?   See Abstr. 093.046.

Composition and structure of the atmosphere of Venus.   See Abstr. 093.048.

The clouds of Venus.   See Abstr. 093.049.

Dynamics, winds, circulation and turbulence in the atmosphere of Venus.   See Abstr. 093.050.

The thermal balance of the atmosphere of Venus. See Abstr. 093.051.

The Venus ionosphere and solar wind interaction. See Abstr. 093.052.

The surface and interior of Venus. See Abstr. 093.053.

Exploration strategy for Mars and the role of the sample return mission.   See Abstr. 097.145.

052 Astrodynamics and Navigation of Space Vehicles

052.001   Space vehicles could be propelled by remote lasers. G. B. Lubkin.
Phys. Today, Vol. 30, No. 8, p. 17, 19 - 20 (1977).

052.002   Geosynchronous satellite theory in equinoctial elements.   D. L. Richardson, D. W. Dunham.
Bull. American Astron. Soc., Vol. 9, 437 - 438 (1977). Abstract.

052.003   Influence or the gradient of the geomagnetic field on the motion of a charged artificial earth satellite.
Yu. M. Manakov.
Izv. vyssh. uchebn. zaved. Geod. i aehrofotosemka, 1976, No. 6, p. 69 - 72. In Russian. — Abstr. in Ref. zh., 62. Issled. kosm. prostranstva, 8.62.370 (1977).

052.004   In the open space.
G. G. Bebenin, Yu. N. Glazkov.
Zemlya i Vselennaya, 1977, No. 4, p. 10 - 14. In Russian.

052.005   On guaranteed characteristics of accuracy of determination of the parameters of motion of space vehicles.   B. Ts. Bakhshiyan, R. R. Nazirov, P. E. Ehl'yasberg.
Kosm. Issled., Vol. 15, 546 - 553 (1977). In Russian.

052.006   Optimization of determination of an orbit with in-complete knowledge of the covariance matrix and expectation value of errors.
B. Ts. Bakhshiyan, R. R. Nazirov, P. E. Ehl'yasberg.
Kosm. Issled., Vol. 15, 658 - 667 (1977). In Russian.

052.007   Investigation of the optimum trajectory with an angular distance of 180° between coplanar elliptic orbits.   R. F. Appazov, V. I. Ogarkov.
Kosm. Issled., Vol. 15, 668 - 676 (1977). In Russian.

052.008   14th- and 28th-order lumped coefficients from the changes of the orbital inclination of the INTER-KOSMOS 9 and 10 satellites.   J. Klokočník.
Bull. Astron. Inst. Czechoslovakia, Vol. 28, 291 - 299 (1977).
The lumped coefficients of 14th- and 28th-order were computed from the changes in the inclination of orbits of the INTERKOSMOS 9 and 10 satellites at the time of the 14/1-resonance. The results obtained are compared with the lumped coefficients evaluated by Walker (1975) and those computed by means of the individual harmonic coefficients for $m = 14$ from various contemporary Earth gravity field models and the resonant Reigber and Balmino (1976) solution.

052.009   Orbit determination of the "Interkosmos-Kopernik 500" satellite.   K. Ziołkowski.
Artif. Satell., Vol. 12, No. 2, p. 15 - 23 (1977).

052.010   Determination of the "Interkosmos-Kopernik 500"
           satellite orientation on the basis of the geomagnetic
field measurements.   K. Stasiewicz.
Artif. Satell., Vol. 12, No. 2, p. 25 - 32 (1977).

052.011   Analysis of selected theories of motion of artificial
           earth's satellites in the aspect of geodetical applica-
tions.   J. Kryński.
Artif. Satell., Vol. 12, No. 3, p. 3 - 61 (1977).

052.012   Optimization of space trajectories.   C. Marchal.
           Proc. 27th Internat. Astronaut. Congr. (see 012.
016), p. 21 - 31 (1977).

052.013   Plane oscillations of a gravitational system satellite-
           stabilizer with maximal speed of response.
V. A. Sarychev, S. A. Mirer, V. V. Sazonov.
Acta Astronaut., Vol. 3, 651 - 669 (1976). — Abstr. in Phys.
Abstr., Vol. 80, Abstr. 25905 (1977).

052.014   Stability of flexible spacecrafts.   P. Santini.
           Acta Astronaut. Vol. 3, 685 - 713 (1976). — Abstr.
in Phys. Abstr., Vol. 80, Abstr. 25906 (1977).

052.015   Drag perturbations of near-circular orbits in an
           oblate diurnal atmosphere.   F. A. Santora.
AIAA J., Vol. 14, 1196 - 1200 (1976). — Abstr. in Phys.
Abstr., Vol. 80, Abstr. 25915 (1977).

052.016   Determination of the orientation of a vehicle from
           photographs of the sky.
A. F. Stetsenko, M. I. Burov.
Izv. vyssh. uchebn. zaved. Geod. i aehrofotosemka, 1977,
No. 1, p. 71 - 75. In Russian. — Abstr. in Ref. zh., 52. Geod.
Aehrosemka, 10.52.150 (1977).

052.017   Qualitative investigations of the properties of
           motion of a satellite of a spheroidal planet.
M. Kh. Khasanova.
Prikl. mat. i mekh., Vol. 41, 561 - 564 (1977). In Russian.
Abstr. in Ref. zh., 62. Issled. kosm. prostranstva, 10.62.322
(1977).

052.018   A study of near-circular satellite orbits: with an
           application to lunisolar perturbations.
S. Hughes, A. J. Meadows.
Proc. R. Soc. London Ser. A, Vol. 355, 131 - 140 (1977).
Abstr. in Phys. Abstr., Vol. 80, Abstr. 49278 (1977).

052.019   Model for solar torque effects on DSCS II.
           T. E. Suttles, R. E. Beverly.
J. Astronaut. Sci., Vol. 24, 165 - 184 (1976). — Abstr. in
Phys. Abstr., Vol. 80, Abstr. 45285 (1977).

052.020   The motion of a satellite in resonance with the
           longitude-dependent harmonics.   S. S. Dallas.
J. Astronaut. Sci., Vol. 24, 97 - 110 (1976). — Abstr. in Phys.
Abstr., Vol. 80, Abstr. 45295 (1977).

052.021   Collision avoidance for two counter-orbiting
           satellites.
D. Schaechter, J. V. Breakwell, R. A. Van Patten, F. W. Everitt.
J. Astronaut. Sci., Vol. 24, 137 - 146 (1976). — Abstr. in
Phys. Abstr., Vol. 80, Abstr. 45296 (1977).

052.022   On the nature of the radial and cross track errors
           for artificial Earth satellites.
N. L. Bonavito, R. A. Gordon, J. G. Marsh.
J. Astronaut. Sci., Vol. 24, 147 - 164 (1976). — Abstr. in
Phys. Abstr., Vol. 80, Abstr. 45297 (1977).

052.023   Probing the Earth's gravity field by means of

satellite-to-satellite tracking.   F. O. Vonbun.
Philos. Trans. R. Soc. London, Ser. A, Vol. 284, 475 - 483
(1977). — Abstr. in Phys. Abstr., Vol. 80, Abstr. 56768 (1977).

052.024   Stability of motion of a space vehicle under con-
           stantly acting disturbances.   J. R. Weiss, E. Y. Yu.
Acta Astronaut., Vol. 3, 943 - 952 (1976). — Abstr. in Phys.
Abstr., Vol. 80, Abstr. 56805 (1977).

052.025   Salvage of the 'Geos' mission.
           Telecommun. J., Vol. 44, 319 - 321 (1977). — Abstr.
in Phys. Abstr., Vol. 80, Abstr. 70979 (1977).

052.026   Analytical modelling of the motion of a drag-free
           satellite.   A. Clochet, F. Paitel.
Rech. Aerosp., No. 3, p. 139 - 146 (1977). In French. — Abstr.
in Phys. Abstr., Vol. 80, Abstr. 73936 (1977).

052.027   Dynamics and control of the SIRIO satellite in
           geostationary orbit.
A. De Agostini, G. Bergamaschi, G. Bianchini, F. Palutan.
17 convegno internazionale scientifico sullo spazio (see 012.
034), p. 375 - 388 (1977). In Italian. — Abstr. in Phys. Abstr.,
Vol. 80, Abstr. 82108 (1977).

052.028   MUPSS: Multipurpose Utility Pointing and Stabiliza-
           tion System.   G. Farnetani.
17 convegno internazionale scientifico sullo spazio (see 012.
034), p. 609 - 616 (1977). In Italian. — Abstr. in Phys. Abstr.,
Vol. 80, Abstr. 82114 (1977).

052.029   Phase-coherent dual-frequency link for high-
           precision Doppler tracking between spacecraft.
J. J. Stiffler, A. C. Berg, D. G. Young.
Space Sci. Instrum., Vol. 3, 3 - 18 (1977).
           During the Apollo-Soyuz Test Project, a phase-coherent
dual-frequency link was established between the Docking
Module (DM) and the Apollo Command Service Module
(CSM). The purpose of the radio link was to enable relative
Doppler measurements to an accuracy of the order of one
millihertz between the two modules. These measurements will
be used to detect mass concentrations (mascons) in the earth's
crust of the order of 10 milligal and greater.

052.030   Nutational dampers of spin-stabilized satellites.
           V. A. Sarychev, V. V. Sazonov.
Celestial Mech., Vol. 15, 75 - 98 (1977). In Russian.
           The stability of the stationary motions of a spin-stabi-
lized satellite is investigated using the Lyapunov second
method. The nutating motion of the satellite is damped by a
special-purpose device. The Krylov-Bogolyubov method is
used to investigate the influence of external moments on the
motion of the satellite rotation axis.

052.031   Optimal low-thrust takeoff from an orbit about an
           oblate planet.   R. A. Jacobson, W. F. Powers.
Celestial Mech., Vol. 15, 161 - 189 (1977).
           Future space missions to the outer planets may depend
upon the use of low-thrust propulsion systems. As these planets
are decidedly oblate, the question of the effect of that oblateness on a low-thrust trajectory is of some interest. In this paper
the problem of optimal energy increase is attacked under the
assumption that the coefficients for the second zonal harmonic,
i.e., J_2, and the nondimensional thrust acceleration are the same
order of magnitude. The optimal control program is found to
be oscillatory and quite similar to the optimal control for
energy increase in an inverse square gravitational field.

052.032   The equations of motion of an artificial satellite in
           nonsingular variables.   G. E. O. Giacaglia.
Celestial Mech., Vol. 15, 191 - 215 (1977).
           The equations of motion of an artificial satellite are given

in nonsingular variables. Any term in the geopotential is considered as well as luni-solar perturbations up to an arbitrary power of r/r', r' being the geocentric distance of the disturbing body. Resonances with tesseral harmonics and with the Moon or Sun are also considered. By neglecting the shadow effect, the disturbing function for solar radiation is also developed in nonsingular variables for the long periodic perturbations. Formulas are developed for implementation of the theory in actual computations.

052.033 **First-order effects of the Earth's oblateness upon coasting bodies.**   J. F. Andrus.
Celestial Mech., Vol. 15, 217 - 224 (1977).
    The Kustaanheimo-Stiefel transformation and perturbation techniques are employed in deriving approximate closed-form expressions for the trajectories of bodies coasting about an oblate Earth. The solutions are used to obtain Lambert type solutions which include oblateness effects.

052.034 **Nutational dampers of dual-spin satellites.**
V. A. Sarychev, V. V. Sazonov.
Celestial Mech., Vol. 15, 225 - 242 (1977). In Russian.
    The paper considers application of the small parameter methods (Artemjev, 1944; Volosov, 1962) in the linear theory of damping the nutating motion of dual-spin satellites. The external moments affecting the satellite are left out of consideration. The influence of the damper on the satellite motion is assumed to be small. As a special case the conditions of stability for the satellite stationary spinning and the optimal parameters ensuring a maximal rate of damping the nutating motion have been obtained.

052.035 **Perturbations of a close-Earth satellite due to sunlight diffusely reflected from the Earth. I: Uniform albedo.**   D. A. Lautman.
Celestial Mech., Vol. 15, 387 - 420 (1977).

052.036 **Periodic oscillations of a satellite in a plane elliptical orbit.**
V. A. Sarychev, V. V. Sazonov, V. A. Zlatoustov.
Kosm. Issled., Vol. 15, 809 - 834 (1977). In Russian.

052.037 **Optimization of the guaranteed reliability of the determination of orbital parameters under conditions of indeterminacy.**
B. Ts. Bakhshiyan, R. R. Nazirov, P. E. Ehl'yasberg.
Kosm. Issled., Vol. 15, 835 - 845 (1977). In Russian.

052.038 **On methods of taking into account the perturbing effect of the earth's shadow.**   E. N. Polyakhova.
Redkollegiya zh. Vestn. Leningr. univ. Mat., mekh., astron. Leningrad, 1977. 42 pp. In Russian. — Abstr. in Ref. zh., 51. Astron., 11.51.228 (1977).

052.039 **Disturbing influence of the pressure of the earth's thermal radiation on artificial earth satellites.**
Yu. M. Manakov.
Izv. vyssh. ucheb. zaved. Geod. i aehrofotosemka, 1977, No. 2, p. 73 - 78. In Russian. — Abstr. in Ref. zh., 52. Geod. Aehrosemka, 11.52.68 (1977).

052.040 **Third-order solution of an artificial-satellite theory.**   H. Kinoshita.
Smithsonian Astrophys. Obs., Special Rep. 379, 5 + 33 + 16 + 55 pp. (1977).
    A third-order solution is developed for the motions of artificial satellites moving in the gravitational field of the earth, whose potential includes the second-, third-, and fourth-order zonal harmonics. Third-order periodic perturbations with fourth-order secular perturbations are derived by Hori's perturbation method.

052.041 **Les perturbations à longue période du second ordre dans le mouvement d'un satellite artificiel de la Terre.**   N. A. Sorokin.
Nauchn. Inf., vyp. (No.) 35, p. 123 - 132 (1977). In Russian.
    Le problème des perturbations à longue période du second ordre des éléments orbitaux relativement à l'aplatissement de la Terre d'un satellite artificiel de la Terre en utilisant l'orbite intermédiaire de deux centres fixés est discuté. Le terme pour les perturbations de le demi-grand axe de l'orbite est présenté.

052.042 **Improvement of orbital elements of artificial satellites along short arcs by means of photographic and laser tracking data.**   O. M. Bulygina.
Nauchn. Inf., vyp. (No.) 35, p. 133 - 136 (1977). In Russian.

052.043 **Investigation of the stability of plane periodic motions of a satellite in a circular orbit.**
A. P. Markeev, A. G. Sokol'skij.
Izv. AN SSSR. Mekh. tverd. tela, 1977, No. 4, p. 46 - 57. In Russian. — Abstr. in Ref. zh., 62. Issled. kosm. prostranstva, 12.62.234 (1977).

052.044 **Algorithm for calculation and optimization of interplanetary swing-around trajectories with return to Earth.**   A. V. Labunskij, A. V. Leshchenko.
Tr. desyatykh chtenij, posvyashch. razrabotke nauchn. naslediya i razvitiyu idej K. Eh. Tsiolkovskogo, Kaluga, 1975. Sekts. "Mekh. kosm. poleta". Moskva, 1976, p. 148 - 159. In Russian. — Abstr. in Ref. zh., 62. Issled. kosm. prostranstva, 12.62.235 (1977).

052.045 **On spherical motions of cosmic flying apparatuses under the action of zero potential forces.**
V. V. Lunev.
Tr. desyatykh chtenij, posvyashch. razrabotke nauchn. naslediya i razvitiyu idej K. Eh. Tsiolkovskogo, Kaluga, 1975. Sekts. "Mekh. kosm. poleta". Moskva, 1976, p. 72 - 78. In Russian. — Abstr. in Ref. zh., 62. Issled. kosm. prostranstva, 12.62.237 (1977).

052.046 **On a method for calculating swing-around trajectories of the moon.**
L. I. Gusev, A. M. Nikulin.
Tr. desyatykh chtenij, posvyashch. razrabotke nauchn. naslediya i rezvitiyu idej K. Eh. Tsiolkovskogo, Kaluga, 1975. Sekts. "Mekh. kosm. poleta". Moskva, 1976, p. 85 - 94. In Russian. — Abstr. in Ref. zh., 62. Issled. kosm. prostranstva, 12.62.238 (1977).

052.047 **Advances in the interplanetary navigation.**   P. Koubský.
Říše hvězd, Vol. 58, 145 - 147 (1977). In Czech.

    An algorithm for recurrent calculation of gravitational acceleration.   See Abstr. 021.010.

    FORTRAN IV subroutines for integration of the equations of motion of artificial earth satellites. See Abstr. 031.406.

    On the question of determination of the Love number by satellites observations.   See Abstr. 042.001.

    General relativity and satellite orbits: the motion of a test particle in the Schwarzschild metric. See Abstr. 042.035.

    Preliminary orbit-determination method having no co-planar singularity.   See Abstr. 042.043.

On a form of equations of motion of a planetary equatorial satellite.   See Abstr. 042.079.

Satellite orbits perturbed by direct solar radiation pressure: general expansion of the disturbing function. See Abstr. 054.001.

053 Lunar and Planetary Probes and Satellites

053.001 **Voyaging to the outer planets.**   J. K. Beatty. Sky Telesc., Vol. 54, 93 - 99 (1977).

053.002 **Rovers: science instrumentation, mobility and control developments at JPL.**   J. D. Burke. Bull. American Astron. Soc., Vol. 9, 445 - 446 (1977). Abstract.

053.003 **Planetary science missions using an ion rocket.** K. L. Atkins. Bull. American Astron. Soc., Vol. 9, 446 (1977). – Abstract.

053.004 **The Voyager missions.**   P. H. Abelson. Science, Vol. 197, 1039 (1977).

053.005 **Das Viking-Programm in der erweiterten Missionsphase. Aktueller Stand des Unternehmens.** H. W. Köhler. Sterne Weltraum, Jahrg. 16, 369 - 370 (1977).

053.006 **The information system of planetary vehicles and motion safety problems.**   B. N. Petrov, Ye. V. Avotinsh (*E. V. Avotin'sh*), V. G. Grigoryev (*Grigor'ev*), A. L. Kemurdzhian, L. N. Lupichev, L. T. Panova, L. N. Polenov, P. S. Semenov. Proc. 27th Internat. Astronaut. Congr. (see 012.016), p. 293 - 299 (1977).

053.007 **International Sun-Earth Explorer: a three-spacecraft program.** K. W. Ogilvie, T. von Rosenvinge, A. C. Durney. Science, Vol. 198, 131 - 138 (1977). The International Sun-Earth Explorer is a three-spacecraft program of the National Aeronautics and Space Administration and the European Space Agency aimed at securing a more quantitative knowledge of the structure and stability of the magnetosphere. One spacecraft (ISEE-C) makes observations in the solar wind upstream of the earth, while the other two (ISEE-A and ISEE-B), in the same highly eccentric orbit but separated by a relatively small variable distance, observe inside the magnetosphere.

053.008 **Testing the Viking Lander.**   H. Caruso. J. Environ. Sci., Vol. 20, No. 2, p. 11 - 17 (1977). Abstr. in Phys. Abstr., Vol. 80, Abstr. 56810 (1977).

053.009 **The missions of the Viking orbiters.** C. W. Snyder. J. Geophys. Res., Vol. 82, (see 003.017), 3971 - 3983 (1977). Paper No. 7S0504. The two Viking orbiters carried the two landers into orbit around Mars, observed the planet to certify the landing sites, released the landers for the landings, and subsequently served as telemetry relays for the lander data. In addition, they conducted scientific investigations using two cameras, an infrared radiometer for temperature measurements, an infrared spectrometer for water vapor measurements, and the radio communication system.

053.010 **Feasibility study of lunar mission by Mu Launch vehicles.**   H. Matsuo, K. Uesugi. 10th Lunar and Planetary Symposium, (see 012.050), p. 31 - 35 (1977). In Japanese.

Soyuz and Apollo.   See Abstr. 003.037.

Automatic stations for investigation of the lunar surface.   See Abstr. 003.081.

Projekt Voyager – der Flug zu Jupiter und Saturn. See Abstr. 051.017.

Voyager to Jupiter and Saturn. See Abstr. 051.019.

Outer planets atmospheric entry probes: science objectives and payloads.   See Abstr. 051.038.

On a method for calculating swing-around trajectories of the moon.   See Abstr. 052.046.

Viking 1. Early results.   See Abstr. 097.151.

A summary of the Viking Project. See Abstr. 097.158.

054 Artificial Earth Satellites

054.001 **Satellite orbits perturbed by direct solar radiation pressure: general expansion of the disturbing function.**   S. Hughes.
Planet. Space Sci., Vol. 25, 809 - 815 (1977).
An expression is derived for the solar radiation pressure disturbing function on an Earth satellite orbit which takes into account the variation of the solar radiation flux with distance from the Sun's centre and the absorption of radiation by the satellite. This expression is then expanded in terms of the Keplerian elements of the satellite and solar orbits using Kaula's method. The resulting expression reduces to the form commonly used in solar radiation pressure perturbation studies, when certain terms are neglected.

054.002 **Cosmos 900 investigates the polar ionosphere.**
L. A. Vedeshin.
Priroda, 1977, No. 8, p. 132 - 133. In Russian.

054.003 **Missions to Salyut 5. II.**   G. R. Hooper.
Spaceflight, Vol. 19, 266 - 268 (1977).

054.004 **Geos. Salvaged results.**   S. Sharrock.
Nature, Vol. 269, 189 - 190 (1977).
Report on some of the preliminary results from experiments aboard Geos.

054.005 **Lunar mission Cosmos satellites.**   D. R. Woods.
Spaceflight, Vol. 19, 383 - 388 (1977).

054.006 **Data processing of the "Interkosmos-Kopernik 500" satellite.**   K. Kossacki, K. Ziołkowski.
Artif. Satell., Vol. 12, No. 2, p. 3 - 5 (1977).

054.007 **Data processing of the radiospectrograph at the "Interkosmos-Kopernik 500" satellite.**
K. Kossacki.
Artif. Satell., Vol. 12, No. 2, p. 7 - 14 (1977).

054.008 **Earth observations satellites (past, present, and future).**   M. Garbacz, H. Mannheimer, W. Stoney.
National telecommunications conference, Pt. II. Dallas, Tex., USA, 29 November - 1 December 1976. IEEE, New York, USA (1976), p. 25.4/1-8. – Abstr. in Phys. Abstr., Vol. 80, Abstr. 29631 (1977).

054.009 **The Space Shuttle.**   P. Chapman.
Electron. Power, Vol. 23, No. 1, p. 25 - 29 (1977).
Abstr. in Phys. Abstr., Vol. 80, Abstr. 33320 (1977).

054.010 **Ground stations for weather observation and earth exploration satellite systems.**   W. Hirschmann.
Nachrichtentech. Z., Vol. 30, 385 - 387 (1977). In German.
Abstr. in Phys. Abstr., Vol. 80, Abstr. 56763 (1977).

054.011 **Instrumentation requirements and provisions for Shuttle/Spacelab missions.**   C. J. Pellerin, Jr.
IEEE Trans. Nucl. Sci., Vol. NS-24, 817 - 822 (1977). – Abstr. in Phys. Abstr., Vol. 80, Abstr. 70972 (1977).

054.012 **Mission design for a halo orbiter of the Earth.**
R. W. Farquhar, D. P. Muhonen, D. L. Richardson.
J. Spacecr. Rockets, Vol. 14, 170 - 177 (1977). – Abstr. in Phys. Abstr., Vol. 80, Abstr. 73920 (1977).

054.013 **HEAO-1 launch.**
IAU Circ., No. 3099 (1977).

054.014 **The 125 MS computer in the Spacelab.**
M. G. Jacquier.
Aeronaut. Astronaut., No. 63, p. 67 - 70 (1977). In French.
Abstr. in Phys. Abstr., Vol. 80, Abstr. 78594 (1977).

054.015 **SPOT–experimental Earth observation system.**
G. Brachet, M. Courtois.
Aeronaut. Astronaut., No. 64, p. 37 - 53 (1977). In French.
Abstr. in Phys. Abstr., Vol. 80, Abstr. 82097 (1977).

054.016 **India's domestic satellite.**
Indian East. Eng., Vol. 119, No. 3, p. 131 (1977).
Abstr. in Phys. Abstr., Vol. 80, Abstr. 82099 (1977).

054.017 **GEOS satellite in stationary orbit.**
Technica, Vol. 26, 1013 (1977). In German.
Abstr. in Phys. Abstr., Vol. 80, Abstr. 82105 (1977).

054.018 **Synchronisation and Image Channel system on board the METEOSAT satellite.**   M. E. Canevari.
17 convegno internazionale scientifico sullo spazio (see 012.034), p. 463 - 473 (1977). In Italian. – Abstr. in Phys. Abstr., Vol. 80, Abstr. 82110 (1977).

054.019 **Passive atmospheric sounders for Spacelab.**
J. B. Farrow.
17 convegno internazionale scientifico sullo spazio (see 012.034), p. 547 - 557 (1977). – Abstr. in Phys. Abstr., Vol. 80, Abstr. 82113 (1977).

054.020 **Pageos: new results of photoelectric observations.**
M. V. Bratijchuk, V. P. Epishev, Ya. M. Motrunich, I. F. Najbauehr (*Neubauer*).
Astron. Tsirk., No. 941, p. 6 - 8 (1977). In Russian.

054.021 **Satellite digest.**
Spaceflight, Vol. 19, 298, 339 - 340, 362, 416, 449 - 450 (1977).

054.022 **Refurbishment and support.**   J. Henschke.
The Space Telescope, (see 012.046), p. 211 - 215 (1976).

054.023 **Photographic tracking data of satellites for the program "Atmosphere".**   M. A. Lur'e.
Nauchn. Inf., vyp. (No.) 35, p. 76 - 79 (1977). In Russian.

054.024 **Berechnung topozentrischer Richtungen künstlicher Erdsatelliten aufgrund photographischer Messungen mit AFU-75-Kameras.**   J. H. Žagars, A. J. Zariņš.
Nauchn. Inf., vyp. (No.) 35, p. 100 - 119 (1977). In Russian.

054.025 **Space Telescope power system long-life design techniques.**
O. B. Smith, R. L. Donovan, J. L. Oberg.
The Space Telescope, (see 012.046), p. 219 - 223 (1976).

Apollo Soyuz.   See Abstr. 003.057.

Apollo Soyuz mission report.
See Abstr. 003.089.

The versatile satellite.   See Abstr. 003.129.

Les observations photographiques des satellites artificiels de la Terre à la station soviétique-française a l'île Kerguelen.   See Abstr. 046.067.

Orbit determination of the "Interkosmos-Kopernik

500" satellite.   See Abstr. 052.009.

Determination of the "Interkosmos-Kopernik 500" satellite orientation on the basis of the geomagnetic field

measurements.   See Abstr. 052.010.

Tidal studies from the perturbations in satellite orbits.   See Abstr. 081.043.

055 Observations of Earth Satellites, Lunar and Planetary Probes

055.001 **Atmospheric drag analyses of low-altitude Doppler beacon satellites.** K. S. W. Champion, J. M. Forbes.
Satellite Doppler positioning, (see 012.002), p. 343 - 355.

055.002 **Explorer 19 − 1963-53-1. September - October 1976. Explorer 39 − 1968-66-1. June - September 1976. Visual observations. Equatorial coordinates (1950.0).**
Rezul't. Nablyud. Iskusstv. Sputnikov Zemli, vyp (No.). 61 (201), 63 pp. (1977). In Russian.

055.003 **Explorer 39 − 1968-66-1. October - November 1976. Oreol 1 − 1971-119-1. April - November 1976. Visual observations. Equatorial coordinates (1950.0).**
Rezul't. Nablyud. Iskusstv. Sputnikov Zemli, vyp. (No.) 62 (202), 59 pp. (1977). In Russian.

055.004 **Poljot 1 − 1963-43-1. April - November 1976. Intercosmos 10 − 1973-82-1. August - November 1976. Visual observations. Equatorial coordinates (1950.0).**
Rezul't. Nablyud. Iskusstv. Sputnikov Zemli, vyp. (No.) 63 (203), 62 pp. (1977). In Russian.

055.005 **Explorer 19 − 1963-53-1. Visual observations. Horizontal coordinates (February - October 1976).**
Rezul't. Nablyud. Iskusstv. Sputnikov Zemli, vyp. (No.) 64 (204), 66 pp. (1977). In Russian.

055.006 **Explorer 39 − 1968-66-1. May - November 1976. Intercosmos 10 − 1973-82-1. August - November 1976. Visual observations. Horizontal coordinates.**
Rezul't. Nablyud. Iskusstv. Sputnikov Zemli, vyp. (No.) 65 (205), 67 pp. (1977). In Russian.

055.007 **MFSK frequency acquisition and synchronization for the Jupiter probe-to-relay communication link.**
R. B. Fluchel, G. M. Lee, E. A. Paddon.
International telemetering conference, Los Angeles, Calif., 28 - 30 September 1976. Int. Found. Telemetering, Woodland Hills, Calif. (1976). p. 465 - 475. − Abstr. in Phys. Abstr., Vol. 80, Abstr. 64613 (1977).

055.008 **A wideband-PCM recorder for the Space Shuttle Orbiter.**   R. D. Petit.
International telemetering conference, Los Angeles, Calif., 28 - 30 September 1976. Int. Found. Telemetering, Woodland Hills, Calif. (1976). p. 562 - 576. − Abstr. in Phys. Abstr., Vol. 80, Abstr. 64614 (1977).

055.009 **An explicit method of calculating visibility of the artificial satellite.**   K. Uchida, H. Takahashi.
Rev. Radio Res. Lab., Vol. 22, 283 - 297 (1976). In Japanese.
Abstr. in Phys. Abstr., Vol. 80, Abstr. 82104 (1977).

055.010 **Visual observations of artificial earth satellites in Finland 1976.**
Prepared under the supervision of A. Tuominen, with an introduction by P. Järvi.

Observations of Satellites, No. 17, (published by the Finnish Meteorol. Inst., Helsinki, Finland), 8 + 100 pp. (1977).

055.011 **Observations of artificial earth satellites No. 14: 1974.**   J. Kabeláč.
Astron. Inst. Tech. Univ. Praha, Publ. No. 47, 21 pp. (1975). In Russian.
   The cases when the effect of parallactic refraction is strongly decreasing or equal to zero is discussed. The author suggests a method to compute it from meteorological data. The distorsion of the SBG camera has been determined assuming it is not symmetric with respect to the plate center. Estimating photoplate, Dodd's method has been used and a procedure is proposed to determine the unknown parameter.

055.012 **Expérience soviéto-françáis sur la location par laser des satellites artificiels de la Terre, effectué à Uzhgorod en 1972.**   G. Ganibé, F. Comett, G.-L. Ata, R. Piquaie, M. V. Bratijchuk, K. A. Kudak, T. I. Laslo.
Nauchn. Inf., vyp. (No.) 35, p. 23 - 33 (1977). In Russian.

055.013 **Le service de l'heure pendant la réalisation de la location par laser des satellites artificiels de la Terre à Uzhgorod.**   T. I. Laslo, K. A. Kudak, F. Comett, G. N. Palij, Yu. A. Fedorov.
Nauchn. Inf., vyp. (No.) 35, p. 34 - 39 (1977). In Russian.

055.014 **A time system for satellite laser radar.**   P. Navara.
Nauchn. Inf., vyp. (No.) 35, p. 40 - 63 (1977). In Russian.
   Timing devices, their technical parameters and a comparative estimate of their accuracies are described related to satellite tracking stations equipped with a satellite laser radar. Recommendations are given for such stations.

055.015 **Information on tracking data obtained during 1973 - 1974 by stations co-ordinated by the Astronomical Council and collaborating countries.**   N. N. Kovalenko.
Nauchn. Inf., vyp. (No.) 35, p. 137 - 142 (1977). In Russian.
   Results of photographic and laser observations of artificial satellites in 1973 - 1974 are presented. Their quality and distribution along different chords are evaluated.

Secular polar motion of the earth and its parameters. Computation from artificial satellites observations. See Abstr. 045.015.

Reference orbits from range and Doppler observations.   See Abstr. 046.013.

Determination of geophysical parameters from long term orbit perturbations using navigation satellite Doppler derived ephemerides.   See Abstr. 046.019.

Photometry of eclipses of artificial earth satellites observed in situ (part II).   See Abstr. 082.001.

Theoretical Astrophysics

061 General Theoretical Problems of Astrophysics, Neutrino Astronomy, Infrared, X-Ray, Gamma-Ray Astronomy, Origin and Abundances of Elements

061.001 A new approach to nucleocosmochronology.
B. M. P. Trivedi.
Astrophys. J., Vol. 215, 877 - 884 (1977).

A new model of nucleocosmochronology based on discrete addition of r-process elements to the presolar cloud is proposed. It is suggested that the solar system accreted its r-process (and presumably all other) elements during its passage through the galactic arms, which has been treated as a discrete event. The number of times the presolar cloud passed through the galactic arm before contraction is calculated to be 50 ± 10.

061.002 Synthesis of "by-passed" elements with neutrino sharing. Yu. S. Kopysov.
Priroda, 1977, No. 7, p. 126 - 127. In Russian.

061.003 Sound from neutrinos. M. A. Korets.
Priroda, 1977, No. 9, p. 132 - 133. In Russian.

061.004 Neutrino-induced nucleosynthesis and deuterium.
S. E. Woosley.
Nature, Vol. 269, 42 - 44 (1977).

Domogatskij and Nadezhin (1977) have suggested that the neutrino flux from a collapsing stellar core might induce nuclear transformation in the mantle and envelope of the surrounding pre-supernova star. They further suggest that such weak interactions might be responsible for producing the so called 'p-process' nuclei. This letter points out a weakness in their model as a mechanism for p-nucleosynthesis and concurrently suggests a rather novel method for the synthesis of deuterium that may have been operative during an early generation of supermassive stars ($M \sim 10^6 - 10^8 M_\odot$).

061.005 An investigation of the stability of the Bondi-Hoyle model of accretion flow. L. L. Cowie.
Mon. Not. R. Astron. Soc., Vol. 180, 491 - 494 (1977).

The author has reinvestigated the Bondi-Hoyle-Lyttleton model of gravitational accretion on to fast-moving bodies. The stability of the steady state solution has been analysed and the accretion column shown to be unstable against short-wavelength perturbations.

061.006 The exact hyperfine structure and Einstein A-coefficients of OH: consequences in simple astrophysical models.
J. L. Destombes, C. Marliere, A. Baudry, J. Brillet.
Astron. Astrophys., Vol. 60, 55 - 60 (1977), with a correction Vol. 61, 769 (1977).

A new theoretical model, which simultaneously analyzes UV, IR and microwave experimental data, is used to derive a consistent set of energy levels and Einstein A-coefficients for all electric dipole allowed transitions in $v = 0$, $^2\Pi$ states of OH. The significance of the exact hyperfine structure of OH and of the building of a new frequency/transition probability system is investigated in simple problems of astrophysical interest.

061.007 The importance of long-lived isomeric states in s-process branching. R. A. Ward.
Astrophys. J., Vol. 216, 540 - 547 (1977).

A reexamination is made of the effects of nonthermal isomeric branching for many key s-process isotopes. Attention is focused on the effect on the usual branching arguments of these isomeric states if the stellar s-process environment is incapable of quickly thermalizing their initial non-equilibrium populations. These metastable states are often significantly populated by either the gamma cascades following neutron capture on the preceding isotope or directly by the β-decay of an unstable isobar.

061.008 On s-process nucleosynthesis in thermally pulsing stars. J. W. Truran, I. Iben, Jr.
Astrophys. J., Vol. 216, 797 - 810 (1977).

The authors investigate the consequences of adopting the $^{22}Ne(\alpha, n)^{25}Mg$ reaction as a neutron source for s-process nucleosynthesis in the convective shell of a thermally pulsing star. Quantitative predictions of s-process abundance levels are presented for matter in the convective shell. The implications of the results for the interpretation of observed s-process abundance patterns in red-giant stars and for the production of s-process elements in our Galaxy are briefly explored.

061.009 Magnetic fields greater than 10^{20} gauss?
I. Lerche, D. N. Schramm.
Astrophys. J., Vol. 216, 881 - 882 (1977).

Zaumen recently showed that a uniform magnetic field will spontaneously pair-produce if greater than about 10^{20} gauss. The authors point out here that on both dynamical and quantum-mechanical grounds such fields are unlikely to occur in astrophysics. Thus, while Zaumen's calculation is of basic physical interest, it does not seem to bear on the evolution of any known astrophysical system.

061.010 Neutrino absorption cross sections for ^{37}Cl with applications. J. N. Bahcall.
Astrophys. J., Lett., Vol. 216, L115 - L118 (1977).

Neutrino absorption cross sections for ^{37}Cl are calculated Applications are made to problems involving solar neutrinos, collapsing stars, and the cosmic neutrino mass density. The present Brookhaven experiment can detect stellar collapses as far away as 5 kpc for plausible collapse parameters.

061.011 **Infrared astronomy and galactic dust. I.**
J.-C. Pecker.
Zemlya i Vselennaya, 1977, No. 4, p. 30 - 36. In Russian.

061.012 **New aspects in the theory of beta decay without neutrinos.** E. Bagge.
VIII Leningr. mezhdunar. semin. Mater. mezhdunar. semin. "Akt. protsessy na Solntse i probl. soln. nejtrino", 1976. Leningrad, 1976, p. 51 - 53. In Russian. − Abstr. in Ref. zh., 51. Astron., 9.51.214 (1977).

061.013 **More on big-bang nucleosynthesis with nonzero lepton numbers.** G. Beaudet, A. Yahil.
Astrophys. J., Vol. 218, 253 - 262 (1977).

The calculation of big-bang nucleosynthesis with nonzero lepton numbers is extended to atomic weights $A < 12$. The observed abundances of D and ^{4}He place only a lower limit on the present mean baryon density of the Universe: $\rho_b \geqslant 5 \times 10^{-31}$ g cm^{-3} for primordial D/H $= 1.4 \times 10^{-5}$ by number (assuming no astration), and $\rho_b \geqslant 3.5 \times 10^{-31}$ g cm^{-3} if the primordial abundance was twice as high, and half of it has been destroyed in stars. In the case of astration, primordial ^{7}Li can account at most for only $\sim 25\%$ of the observed abundance (15% for zero lepton numbers).

061.014 **On the possible existence of cosmological cosmic rays. III. Nuclear γ-ray production.**
T. Montmerle.
Astrophys. J., Vol. 218, 263 - 268 (1977).

In the framework of the cosmological cosmic ray hypothesis, developed in preceding papers, the flux of nuclear γ-rays arising from low-energy $p\alpha$ and $\alpha\alpha$ reactions via the decay of the excited states of ^{4}He, ^{6}Li, ^{7}Li, and ^{7}Be is calculated. It is found that this flux is at least six orders of magnitude below the observed X-ray background flux and is thus likely to be undetectable. This is the case even for the decay of ^{7}Li*(0.478 MeV) and ^{7}Be*(0.431 MeV), which dominate nuclear γ-ray production. The relevant cross sections are studied in the Appendix.

061.015 **On the negative specific heat paradox.**
D. Lynden-Bell, R. M. Lynden-Bell.
Mon. Not. R. Astron. Soc., Vol. 181, 405 - 419 (1977).

All astronomers know that when a star or a star cluster loses energy its temperature will increase in accordance with the virial theorem. However, there is a simple proof in statistical mechanics that specific heats are positive. This paradox, first resolved by Thirring, is further explored with a simple model which obeys the virial theorem.

061.016 **Astronomy and the laws of physics.**
P. Morrison.
Highlights of Astronomy, Vol. 4, Part I, (see 003.010), p. 35 (1977). − Abstract.

061.017 **The use of a-priori information in the derivation of temperature structures from X-ray spectra.**
I. J. D. Craig.
Astron. Astrophys., Vol. 61, 575 - 590 (1977).

The task of inferring the temperature structure of a plasma from its X-ray spectrum is reviewed. It is recognized, at the outset, that the ill-posed nature of the problem is reflected in the lack of stability of the basic integral equation relating the source spectrum to the emission measure function ξ. The author examines in detail the application of a well known non-classical technique to a typical spectral inversion problem. The method allows the required a-priori information to be incorporated in the form of a smoothness constraint for the solution ξ. A specific bremsstrahlung problem associated with solar flare emission is considered.

061.018 **Infrared astronomy and galactic dust. II.**

J.-C. Pecker.
Zemlya i Vselennaya, 1977, No. 5, p. 43 - 50. In Russian. Translated from English by S. N. Rodionova.

061.019 **Li, Be and B production in proton-induced reactions: implications for astrophysics and space radiation effects.** C. T. Roche, R. G. Clark, G. J. Methews, V. E. Viola, Jr.
Nuclear cross sections and technology. II. Washington, D.C., USA. 3 - 7 March 1975. R. A. Schrack, C. D. Bowman (Editors). Natl. Bur. Stand., Washington, D.C., USA (1975), p. 504 - 508. − Abstr. in Phys. Abstr., Vol. 80, Abstr. 23053 (1977).

061.020 **Weak interactions in astrophysics and cosmology.**
R. J. Tayler.
J. Phys. G, Vol. 3, 219 - 237 (1977). − Abstr. in Phys. Abstr., Vol. 80, Abstr. 25934 (1977).

061.021 **Dynamical r-process and nuclei far from the region of β-stability.** W. Hillebrandt, K. Takahashi.
3rd international conference on nuclei far from stability. Corgese, Corsica, France, 19 - 26 May 1976. CERN, Geneva, Switzerland. 8 + 608 pp. (1976). p. 580 - 583. − Abstr. in Phys. Abstr., Vol. 80, Abstr. 26004.

061.022 **X-ray astronomy.** R. Giacconi.
APL Tech. Dig., Vol. 15, No. 1, p. 16 - 32 (1976). Abstr. in Phys. Abstr., Vol. 80, Abstr. 26084 (1977).

061.023 **Heavy element nucleosynthesis.**
D. N. Schramm, E. B. Norman.
3rd international conference on nuclei far from stability. Corgese, Corsica, France, 19 - 26 May 1976. CERN, Geneva, Switzerland. 8 + 608 pp. (1976), p. 570 - 579. − Abstr. in Phys. Abstr., Vol. 80, Abstr. 29644 (1977).

061.024 **Production of O III through Auger effect.**
O. Bely, J. M. P. Serrão.
Astron. Astrophys., Vol. 61, 711 - 714 (1977).

The authors study the production of O III through the Auger effect. Initial ionization of neutral oxygen or double excitation of O II is assumed. The autoionization probabilities of O II are then calculated and rates for the population of the ground configuration of the O III ion through cascades are obtained.

061.025 **Estimate of the ^{37}Ar generation rate by cosmic ray neutrinos in an experiment of solar neutrino recording. Abstract.** G. V. Domogatskij, R. A. Ehramzhyan.
VIII Leningr. mezhdunar. semin. Mater. mezhdunar. semin. "Akt. protsessy na Solntse i probl. soln. nejtrino", 1976. Leningrad, 1976, p. 77 - 79. In Russian and English. − Abstr. in Ref. zh., 51. Astron., 10.51.226 (1977).

061.026 **Search for superheavy elements in nature.**
K. D. Tolstov.
Obedin. inst. yader. issled. Soobshch. R6−10515. Dubna, 1977. 31 pp. In Russian. − Abstr. in Ref. zh., 51. Astron., 10.51.231 (1977).

061.027 **Pion condensation I. (*Stellar interiors*). Nonrelativistic treatment for π⁻ condensation.** T. Kanai.
Prog. Theor. Phys., Vol. 56, 1802 - 1811 (1976). − Abstr. in Phys. Abstr., Vol. 80, Abstr. 30500 (1977).

061.028 **Screening effect on electron captures at high densities.** K. Yokoi, M. Yamada.
Prog. Theor. Phys., Vol. 56, 1781 - 1785 (1976). − Abstr. in Phys. Abstr., Vol. 80, Abstr. 33478 (1977).

061.029 **Submillimeter wave astronomy.** M. W. Werner.

2nd international conference and winter school on submillimeter waves and their applications. San Juan, Puerto Rico, 6 - 11 December 1976. IEEE, New York, USA. 16 + 252 pp. (1976). p. 96 - 97. – Abstr. in Phys. Abstr., Vol. 80, Abstr. 41315 (1977).

061.030 **Cosmological changes in atomic and nuclear constants.**   H. C. Ohanian (*Oganyan*).
Found. Phys., Vol. 7, 391 - 404 (1977). – Abstr. in Phys. Abstr., Vol. 80, Abstr. 49286 (1977).

061.031 **Atoms in high magnetic fields.**   R. H. Garstang.
Rep. Prog. Phys., Vol. 40, No. 2, p. 105 - 154 (1977). Abstr. in Phys. Abstr., Vol. 80, Abstr. 50383 (1977).

061.032 **Infrared astronomy.**   C. G. Wynn-Williams.
Phys. Bull., Vol. 28, 109 - 110 (1977). – Abstr. in Phys. Abstr., Vol. 80, Abstr. 53433 (1977).

061.033 **Antimatter.**   P. I. P. Kalmus.
Nature, Vol. 270, 661 - 662 (1977).

061.034 **Nuclear astrophysics – the origin of heavy elements.**   D. N. Schramm.
Nukleonika, Vol. 21, 727 - 751 (1976). – Abstr. in Phys. Abstr., Vol. 80, Abstr. 56825 (1977).

061.035 **Effect of vacuum polarization on the solar p-p reaction rate.**   G. E. Bohannon, L. Heller.
Phys. Rev. C, Vol. 15, 1221 - 1227 (1977). – Abstr. in Phys. Abstr., Vol. 80, Abstr. 57555 (1977).

061.036 **Projectile charge dependence of electron-capture cross sections.**   D. Belkic, R. McCarroll.
J. Phys. B, Vol. 10, 1933 - 1943 (1977). – Abstr. in Phys. Abstr., Vol. 80, Abstr. 68110 (1977).

061.037 **A revision of magnetic null lines of astrophysical interest.**   L. Iglesias.
Opt. Pura Apl., Vol. 9, No. 2, p. 79 - 97 (1976). – Abstr. in Phys. Abstr., Vol. 80, Abstr. 70998 (1977).

061.038 **Variation of elements in nature.**   V. S. Venkatavaradan.
Int. At. Energy Agency Bull., Vol. 19, No. 2, p. 50 - 58 (1977). Abstr. in Phys. Abstr., Vol. 80, Abstr. 71161 (1977).

061.039 **A new thermodynamic effect (heat and gravitation).**   W. Schmidt, W. Sachsze.
17 convegno internazionale scientifico sullo spazio (see 012. 034), p. 539 - 545 (1977). In German. – Abstr. in Phys. Abstr., Vol. 80, Abstr. 80049 (1977).

061.040 **The behaviour of the system of particles at high pressures.**   P. Savic.
Sci. Sintering, Vol. 9, 233 - 242 (1977). – Abstr. in Phys. Abstr., Vol. 80, Abstr. 87261 (1977).

061.041 **Neutron transport methods in neutrino transport calculations.**   S. A. Bludman, I. Lichtenshtadt, A. Ron, N. Sack, J. J. Wagschal.
Nucl. Sci. Eng., Vol. 64, 294 - 298 (1977). – Abstr. in Phys. Abstr., Vol. 80, Abstr. 87377 (1977).

061.042 **Gamma-ray astrophysics.**   C. E. Fichtel.
Space Sci. Rev., Vol. 20, 191 - 234 (1977).
The most striking feature of the celestial sphere when viewed in the frequency range of γ-rays is the emission from the galactic plane. Several point γ-ray sources have now been observed, including four radio pulsars. This last result is particularly striking since only one radio pulsar has been seen at either optical or X-ray frequencies. Nuclear γ-ray lines have

been seen from the Sun during a large solar flare and future satellite experiments are planned to search for γ-ray lines from supernovae and their remnants. A general apparently diffuse flux of γ-rays has also been seen whose energy spectrum has interesting implications; however, in view of the possible contribution of point sources and the observation of galactic features such as Gould's belt, its interpretation must await γ-ray experiments with finer spatial and energy resolution, as well as greater sensitivity.

061.043 **Solar-system abundances of nuclides and nuclear structure.**   G. S. Anagnostatos.
Astrophys. Space Sci., Vol. 50, 445 - 450 (1977).
Some new, hitherto unknown features of the abundance distribution curve for the heavy elements have been observed showing surface, symmetry of nuclear forces, and Coulomb energy effects. These features are not explainable by the theory of Burbidge, Burbidge, Fowler, Hoyle (1957) and show an intrinsic consistency in the abundance distribution, which supports the arguments for a common history of the heavy elements during nucleosynthesis, or at least during its ultimate phase.

061.044 **On the computation of optical properties of heterogeneous grains.**
C. F. Bohren, N. C. Wickramasinghe.
Astrophys. Space Sci., Vol. 50, 461 - 472 (1977).
The authors discuss a method of calculating the optical properties of heterogeneous particles with a view to astrophysical applications. Such computations are particularly important in regions of resonant absorption, where grain properties and compositions are often inferred from precise positions of absorption peaks. Shifts of resonant wavelengths, due to effects such as porosity and foreign particulate inclusions, could be crucial in affecting and altering such interpretations.

061.045 **Possibilities of experiments with high-energy cosmic neutrinos: project DUMAND (*Deep Underwater Muon Neutrino Detection*).**   V. S. Berezinskij, G. T. Zatsepin.
Usp. fiz. nauk, Vol. 122, 5, 3 - 36 (1977). In Russian. – Abstr. in Ref. zh., 51. Astron., 11.51.1010 (1977).

061.046 **Peculiarities of chemical composition of prestellar matter.**   B. V. Vajner, O. V. Dryzhakova, V. L. Zaguskin, L. S. Marochnik, L. I. Reznitskij.
Early stages of stellar evolution, (see 012.041), p. 18 - 21 (1977). In Russian.
Results of thermonuclear synthesis of light elements in a hot model of the Universe are given. The influence of entropy and temperature fluctuations on the synthesis of light elements is studied.

061.047 **Do pulsars synthesize ^{7}Li?**   E. Dwek.
Proc. Southwest Reg. Conf., Vol. 3, (see 012.043), p. 13 - 18 (1977).
The author reexamines the model presented by Clayton and Dwek proposing that ^{7}Li in the Galaxy is synthesized by αα collisions in helium-rich nebula surrounding newly-born pulsars. Using a characteristic composition for the ejected nebula and a more realistic treatment for the evolution of the energetic α-particles in the nebulae he finds that to account for the observed galactic ^{7}Li abundance, more than one-third of the pulsar's total luminosity needs to be converted into a special injection spectrum of alpha particles.

061.048 **Nuclear matter with neutrino confinement.**
Yu. L. Vartanyan, N. K. Ovakimova.
Astron. Zh. Akad. Nauk SSSR, Vol. 54, 1281 - 1284 (1977). In Russian. English translation in Soviet Astron., Vol. 21, No. 6.
The equation of state and the chemical composition of

degenerate matter in the case of confinement of neutrinos is considered under various initial ratios of the number of barions to the number of leptons. The results in the region below the nuclear density indicate a displacement of the nuclei of a given chemical composition towards the region of higher densities and difficulty of formation of free neutron gas. Under densities exceeding the nuclear one the contribution of leptons to the pressure and the density turns out to be large; the change of the sequence of birth of hyperons is observed.

061.049 **Astrophysical implications of isotopic anomalies.** D. D. Clayton.
Meteoritics, Vol. 12, 195 - 197 (1977). — Abstract.

061.050 **Neue obere Grenzen der Variation von Naturkonstanten.** W. Eichendorf, M. Reinhardt.
Mitt. Astron. Ges., Nr. 42, p. 89 - 90 (1977).
Dirac's large number hypothesis (LNH) is reanalyzed. It implies constancy of the ratios of inertial masses of elementary particles and the fine structure constant. Geophysical data on the surface temperature of the earth 3.4×10^9 a ago exclude a variation of the gravitational constant G according to Dirac's hypothesis. Also a variation of the elementary charge cannot explain the LNH. It still might be feasible, if the inertial mass of elementary particles or h or c were functions of time.

061.051 **Neutrino emission in the gravitational collapse of stars and the possibility of its detection with scintillation detectors.** G. V. Domogatskij, V. S. Imshennik, D. K. Nadezhin.
Inst. prikl. mat. AN SSSR. Prepr. No. 40. Moskva, 1977. 27 pp. Price 10 Kop. In Russian. — Abstr. in Ref. zh., 51. Astron., 1.51.473 (1978).

061.052 **Hēlijs un Visuma evolūcija.** J. Francmanis.
Zvaigžņotā debess, 1977. gada vasara, p. 4 - 10.

061.053 **X-ray astronomy.** M. Grün, P. Koubský.
Vesmír, Vol. 56, 237 - 241 (1977). In Czech.

Progress in far-infrared and submillimeter astronomy. See Abstr. 013.025.

Cross sections for (p, xn) reactions, and astrophysical applications. See Abstr. 022.078.

Comparison of methods for calculating cross sections at high energies in astrophysics. See Abstr. 022.079.

Cosmic tests of Maxwell's equations. I: A photon rest mass. See Abstr. 022.091.

An upper limit to the cosmological variation of Planck's constant derived from the spectrum of the microwave background radiation. See Abstr. 066.079.

Gamma rays from accreting black holes. See Abstr. 066.140.

Two astrophysical applications of conformal gravity. See Abstr. 066.312.

Iron: whence it came, where it went. See Abstr. 071.026.

An interpretation of galactic observations of CNO isotopes. See Abstr. 131.060.

Light-element production by cosmological cosmic rays. See Abstr. 143.020.

Galactic neutrino sources and experimental neutrino astronomy. See Abstr. 143.045.

UH cosmic rays and solar system material: the elements just beyond iron. See Abstr. 143.059.

On the possible existence of cosmological cosmic rays. I. The framework for light-element and gamma-ray production. See Abstr. 162.009.

On the possible existence of cosmological cosmic rays. II. The observational constraints set by the γ-ray background spectrum and the lithium and deuterium abundances. See Abstr. 162.011.

Limits on masses and number of neutral weakly interacting particles. See Abstr. 162.085.

Pregalactic nucleosynthesis. See Abstr. 162.109.

Evidence for the fundamental role of Planck units in cosmology. See Abstr. 162.114.

Errata

061.901 **Erratum: "Propagation of blast waves"** [Astrophys. J., Vol. 209, 424 - 428 (1976)].
A. Cavaliere, A. Messina.
Astrophys. J., Vol. 216, 972 (1977).

061.902 **Errata: "Approximation functions to the Emden and associated Emden functions near the first zero, II"** [Astrophys. Space Sci., Vol. 37, 401 - 426 (1975)].
A. P. Linnell.
Astrophys. Space Sci., Vol. 47, 253 (1977).

062 Hydrodynamics, Magnetohydrodynamics, Plasma

062.001   Thermal equilibrium in plasmas containing negative ions – II.   J. D. Hey.
J. Quant. Spectrosc. Radiat. Transfer, Vol. 18, 253 - 254 (1977).

The additional existence of positive molecular ions, such as H_2^+, in a low-temperature plasma containing appreciable numbers of negative ions, is considered. This is shown to have no effect on conclusions reached earlier with regard to the effect of negative ion formation on the achievement of LTE level populations by the neutral atoms.

062.002   First and second order approximations of the first adiabatic invariant for a charged particle interacting with a linearly polarized hydromagnetic plane wave.
R. Vanclooster.
Planet. Space Sci., Vol. 25, 765 - 771 (1977).

The effect of a sinusoidal modulation of an electromagnetic field on the invariance of the magnetic moment is studied. Such a generalized invariant plays an important role in problems concerning the motion of charged particles in the non-uniform magnetic field of the magnetosphere or the solar wind.

062.003   On the singular point of the hydromagnetic Bernoulli flow.   W. Unno, E. Ribes.
Publ. Astron. Soc. Japan, Vol. 29, 351 - 357 (1977).

The singular point of the hydromagnetic Bernoulli flow depends on the constraint imposed on the flow. If the geometry is given, the singular point appears where the Mach number is unity. If the total pressure is given as a function of position, it appears in the subsonic sub-Alfvénic region. The latter singular point is important particularly in the numerical construction of a two-dimensional hydromagnetic structure.

062.004   Microinstabilities in a moderately inhomogeneous plasma.   C. E. Singer.
J. Geophys. Res., Vol. 82, 2686 - 2692 (1977).

062.005   Determination of the transport coefficients for a plasma with quasi-collisions.
V. S. Danilova, G. F. Krymskij.
Geomagn. Aehron., Vol. 17, 727 - 734 (1977). In Russian.

062.006   Solar magnetoatmospheric waves – a simplified mathematical treatment.   J. A. Adam.
Astron. Astrophys., Vol. 60, 171 - 179 (1977).

The inhomogeneous wave equation for a special class of magnetoatmospheric waves is formally solved, and the principle of stationary phase used to provide information on the group velocity properties of such waves. General results are presented concerning the associated mechanical energy flux. The basic problem considered is relevant to waves initiated by sudden events in the solar atmosphere.

062.007   Processes of comptonization and spectra of relativistic electrons in a plasma turbulent reactor.
Yu. A. Nikolaev, V. N. Tsytovich.
Astrofizika, Vol. 12, 543 - 553 (1976). In Russian. – English translation in Astrophysics, Vol. 12, No. 3.

The processes of formation of a power-type distribution of electrons upon energy $f_e \sim \epsilon^{-\gamma}$ in a plasma turbulent reactor (PTR) have been investigated taking into account Compton scattering of reabsorbed radiation. The universality of a PTR as source of relativistic electrons with a power-type spectrum in a condition close to real cosmic conditions has been shown in the presence of magnetic fields and magnetic turbulent modes of oscillation. The dependence of the spectrum indices γ from parameters characterizing the plasma of a turbulent

reactor for various types of turbulence has been investigated. The obtained values of $\gamma \leqslant 3$ correspond to the most probable values found in an investigation of cosmic radio sources.

062.008   Accretion magnetosphere stability. II. Polar cap "drip".   F. C. Michel.
Astrophys. J., Vol. 216, 838 - 841 (1977).

The entry of plasma past the shielding magnetic field of a collapsed object is examined. It is concluded that a plausible entry mode is simply a "dripping" motion of the polar caps of the magnetopause, owing to radiation of the hot compressed plasma there. The plasma "drips" would hit the object's surface either near the magnetic poles or in a ring-shaped "auroral" zone around the poles. Insofar as this entry mode is concerned, no special role is played by finite plasma resistivity since the plasma can reach the stellar object even if the conductivity is infinite.

062.009   Évolution des chocs d'Alfvén en magnétohydrodynamique classique.   S. Giambò, A. Greco.
C. R. Acad. Sci. Paris, Tome 285, Sér. A, 301 - 304 (1977).

The law of evolution of Alfvén shocks is given, and their constance along the shock rays is shown in the case of propagation in a constant state.

062.010   Electric currents in cosmic plasmas.   H. Alfvén.
Rev. Geophys. Space Phys., Vol. 15, 271 - 284 (1977).

Since the beginning of the century, physics has been dualistic in the sense that some phenomena are described by a field concept and others by a particle concept. This dualism is essential also in the physics of cosmic plasmas. A survey is given of a number of phenomena which can be understood only from the particle aspect. These include the formation of electric double layers, the origin of 'explosive' events like magnetic substorms and solar flares, and further, the transfer of energy from one region to another. A useful method of exploring many of these phenomena is to draw the electric circuit in which the current flows and to study its properties. A number of simple circuits are analyzed in this way.

062.011   Adiabatic stability of rotating gaseous masses.   C.-H. Sung.
Astron. Astrophys., Vol. 60, 393 - 404 (1977).

The adiabatic stability of rotating gaseous masses is investigated via the initial-value problem. Formal theory is developed first and then applied to various cases with and without a toroidal magnetic field to establish sufficient conditions for stability when the Eulerian change in the gravitational potential is neglected. In the absence of rotation, some of the sufficient conditions are also necessary for stability. The effect of viscosity and magnetic diffusivity on stability is also discussed in some special cases.

062.012   Effects of gyroviscosity on self-gravitating rotating plasma of variable density.
S. L. Maheshwari, P. K. Bhatia.
Gerlands Beitr. Geophys., Band 86, 313 - 322 (1977).

The hydromagnetic stability of a self-gravitating rotating plasma of variable density is studied taking into account the effect of finiteness of the ion Larmor radius. The macroscopic equations of motion are used, where the finite ion Larmor radius effects are incorporated in the stress tensor. An explicit solution for a fluid of finite depth and with an exponential density variation is obtained by means of a variational principle characterizing the problem. The inclusion of the effects of gyroviscosity has been found to be stabilizing.

062.013   **On the damping of turbulent motions by a magnetic field in an electrically conducting medium.**
K.-H. Rädler.
Solar activity and solar-terrestrial relations, (see 012.007), p. 373 (1976). — Abstract. (See 12.062.058).

062.014   **On second order statistical moments in MHD-turbulence.**   F. Krause.
Solar activity and solar-terrestrial relations, (see 012.007), p. 381 - 382 (1976).

A brief survey concerning investigations of the magnetic field correlation tensor in the framework of the second order correlation approximation is given. A special relation connecting the mean square of the fluctuating magnetic field and the square of the mean magnetic field is applied to the fields observed at the Sun's surface.

062.015   **A method of solving the equations of a force-free field in a region with grad $\alpha \neq 0$. I.**
E. A. Rudenchik.
Soln. Dannye 1977 Byull., No. 6, p. 87 - 90 (1977). In Russian.
A method is proposed for solving the equations rot $H = \alpha H$ for any given function α. The method permits to determine the existence conditions and to solve the equations.

062.016   **Stability of shell distributions.**   D. D. Barbosa.
Planet. Space Sci., Vol. 25, 981 - 984 (1977).
Stability criteria for parallel propagating plasma waves driven unstable by a nearly isotropic bump-in-energy "half-shell" beam of electrons along a magnetic field are investigated. Comparison with the drifting Maxwellian at equal densities reveals smaller growth rates for the half-shell and a shift in the unstable wavenumbers towards smaller phase velocities. The limit of complete isotropy is stable to the waves under consideration.

062.017   **Smoothed particle hydrodynamics: theory and application to non-spherical stars.**
R. A. Gingold, J. J. Monaghan.
Mon. Not. R. Astron. Soc., Vol. 181, 375 - 389 (1977).
A new hydrodynamic code applicable to a space of an arbitrary number of dimensions is discussed and applied to a variety of polytropic stellar models. The principal feature of the method is the use of statistical techniques to recover analytical expressions for the physical variables from a known distribution of fluid elements. Starting with a non-axisymmetric distribution of approximately 80 particles in three dimensions, the method is found to reproduce the structure of uniformly rotating and magnetic polytropes to within a few per cent.

062.018   **Generation of a deeply penetrating gravity-like oscillation by a non-resonant second-order interaction of waves.**   T. Sakurai.
Publ. Astron. Soc. Japan, Vol. 29, 543 - 554 (1977).
The author has studied a non-resonant second-order interaction between an inner gravity oscillation and an unstable convection in a rotating two-layered Boussinesq fluid, the lower and the upper part of which are stably and unstably stratified, respectively. A growing gravity-like oscillation is generated by this interaction. This oscillation can penetrate deeply into the lower stable region. Related solar problems are discussed briefly.

062.019   **Model of the kinematic dynamo of three spherical vortexes. II.**
Eh. P. Kropachev, S. N. Gorshkov, P. M. Serebryanaya.
Geomagn. Aehron., Vol. 17, 927 - 929 (1977). In Russian.

062.020   **On the influence of an outer magnetic field on the structure of the electrostatic field in a trace behind a disk in a stream of rarefied plasma.**
G. P. Patalakh, V. A. Semenov, V. A. Shuvalov.

Geomagn. Aehron., Vol. 17, 936 - 939 (1977). In Russian.

062.021   **Hydrodynamic collapse calculations of cylindrical clouds.**   P. Bastien.
J. R. Astron. Soc. Canada, Vol. 71, 397 (1977). — Abstract.

062.022   **Flows in magnetic flux tubes.**   C. J. Durrant.
Highlights of Astronomy, Vol. 4, Part II, (see 012.024), p. 267 - 270 (1977).

062.023   **Computation of ideal MHD equilibria.**
K. Lackner.
Computing in plasma physics and astrophysics, (see 012.028), p. 33 - 44 (1976).
This paper reviews two- and three-dimensional ideal MHD equilibrium codes, with particular emphasis on axisymmetric toroidal calculations.

062.024   **Recent developments in the computational aspects of MHD stability.**   R. C. Grimm, J. L. Johnson.
Computing in plasma physics and astrophysics, (see 012.028), p. 45 - 52 (1976).
In recent years variational methods have been applied successfully to the numerical determination of the spectra for simple linearized ideal magnetohydrodynamic models. Currently, several extensive efforts have been undertaken to implement these approaches to determine the normal modes for general axisymmetric toroidal equilibria, especially those applicable to the Tokamak experimental program. This paper reviews the motivation, the difficulties, and the various numerical approaches employed. Several simple illustrative results are given.

062.025   **Non-linear behaviour of hydromagnetic instabilities.**   J. A. Wesson.
Computing in plasma physics and astrophysics, (see 012.028), p. 53 - 65 (1976).
There has recently been a growth of interest in the non-linear development of hydromagnetic instabilities. The theoretical attack on this problem has been largely numerical and has been carried out by obtaining the time dependent solution to the set of hydromagnetic equations appropriate to each problem. The earliest calculations used the ideal hydromagnetic equations but many of the more realistic problems require the introduction of resistive effects to allow for changes in magnetic topology and for ohmic heating. This paper gives a description of some of the recent calculations, relevant to magnetic confinement systems, concentrating particularly on Tokamak and pinch configurations.

062.026   **Enhancement of thermonuclear reaction rate due to strong screening.**
N. Itoh, H. Totsuji, S. Ichimaru.
Astrophys. J., Vol. 218, 477 - 483 (1977).
The enhancement factor for the rate of thermonuclear reactions in the strong-screening regime is calculated through explicit consideration of correlations between ions; the correlation functions are accurately estimated with the aid of the existing Monte Carlo computations. It is shown that the previous result obtained by DeWitt et al. amounts to an overestimation of the screening effect because of their neglect of spatial dependence of the screening function. Consequences of the present calculation on the explosive carbon burning in a degenerate stellar core are pointed out.

062.027   **Generalized Ohm's laws for relativistically streaming plasmas.**   R. R. Burman.
Australian J. Phys., Vol. 30, 471 - 480 (1977).
A theoretical framework is given for treating energy dissipation and the associated diffusion of magnetic field lines resulting from 'frictional' interactions between different species in plasmas in which the species have relativistic

streaming velocities. The plasmas are not necessarily neutral. One result is a form for generalized Ohm's laws. Details are given for binary plasmas, with particular attention paid to the analysis of inertial effects. Some remarks are made on the relevance of the results to pulsar magnetospheres.

062.028   Emission and absorption of Langmuir waves by anisotropic unmagnetized particles.
D. B. Melrose, J. E. Stenhouse.
Australian J. Phys., Vol. 30, 481 - 493 (1977).
    The primary purpose of this paper has been to present a theory for the emission and absorption of Langmuir waves by anisotropic unmagnetized particles. In conclusion, it is evident that some conventional ideas relating to streaming instabilities in the solar corona are oversimplified, and that other distributions such as a loss-cone anisotropy and P_1 anisotropy could be important in generating intense Langmuir turbulence and hence observable plasma emission.

062.029   Shock waves in relativistic magnetohydrodynamics under general assumptions.   A. Lichnerowicz.
J. Math. Phys.,New York, Vol. 17, 2135 - 2142 (1976).
Abstr. in Phys. Abstr., Vol. 80, Abstr. 24035 (1977).

062.030   Quantum-statistical models for multicomponent plasmas.   B. F. Rozsnyai, B. J. Alder.
Phys. Rev. A, Vol. 14, 2295 - 2300 (1976). – Abstr. in Phys. Abstr., Vol. 80, Abstr. 24053 (1977).

062.031   Photon pair production in astrophysical transrelativistic plasmas.   W. R. Stoeger.
Astron. Astrophys., Vol. 61, 659 - 669 (1977).
    After briefly reviewing pair production in a plasma with an equilibrium (Planck) spectrum, the author presents the results of pair-concentration calculations for steady-state transrelativistic (6×10^8 K to 6×10^9 K) plasmas of relatively high density ($N = 10^{14}$ cm^{-3} to $N = 10^{22}$ cm^{-3}) which are characterized by pure bremsstrahlung, comptonized bremsstrahlung, and unsaturated Compton scattering spectra. Such plasmas are often found in neutron-star and black-hole accretion environments. Finally the author indicates the qualitative relevance of the results to non-steady state regimes and to interpreting specific astrophysical (X-ray and γ-ray) phenomena.

062.032   Response of radiation of magnetohydrodynamic waves to a moving rotator.
V. P. Dokuchaev, V. Ya. Ehjdman.
Dokl. AN SSSR, Vol. 234, 1039 - 1042 (1977). In Russian.
Abstr. in Ref. zh., 51. Astron., 10.51.244 (1977).

062.033   Legendre expansion of the quasi-linear equations for anisotropic particles and Langmuir waves.
P. Hoyng, D. B. Melrose.
Astrophys. J., Vol. 218, 866 - 880 (1977).
    The quasi-linear diffusion and friction coefficients for axisymmetric electron distributions interacting with Langmuir waves are evaluated explicitly by expanding the distribution of waves in Legendre polynomials. The quasi-linear equations are then reduced to a form in which both the distributions of waves and of particles are simultaneously expanded in Legendre polynomials, and all coefficients are evaluated explicitly. It is argued that such expansions are likely to be justified in practice and that the results obtained should prove useful in discussing quasi-linear relaxation under various conditions in three dimensions rather than one dimension. New results are anticipated for the problem of the propagation of electron streams causing type III solar radio bursts.

062.034   Gravity induced magnetic instability.   M. Sinha.
Pramāṇa, Vol. 8, 214 - 216 (1977). – Abstr. in Phys. Abstr., Vol. 80, Abstr. 41295 (1977).

062.035   Collision processes involving highly-excited atoms.   K. Takayanagi.
Comments At. Mol. Phys., Vol. 6, No. 5 - 6, p. 177 - 188 (1977). – Abstr. in Phys. Abstr., Vol. 80, Abstr. 58146 (1977).

062.036   Electromagnetic pulse propagation in a weakly nonlinear plasma.   E. H. Satorius.
IEEE Trans. Antennas Propag., Vol. AP-25, 587 - 589 (1977).
Abstr. in Phys. Abstr., Vol. 80, Abstr. 58863 (1977).

062.037   Gravitational instability of a composite rotating plasma.   K. Prakash.
Arch. Mech., Vol. 29, No. 2, p. 205 - 211 (1977). – Abstr. in Phys. Abstr., Vol. 80, Abstr. 60591 (1977).

062.038   Conversion of an electromagnetic wave into longitudinal wave in inhomogeneous plasma.   S. N. Paul.
Czechoslovak J. Phys. B, Vol. B77, 873 - 879 (1977). – Abstr. in Phys. Abstr., Vol. 80, Abstr. 68800 (1977).

062.039   Plasma turbulent reactors.   C. A. Norman.
Ann. Physics, Vol. 106, 26 - 43 (1977). – Abstr. in Phys. Abstr., Vol. 80, Abstr. 70990 (1977).

062.040   The effect of finite electrical and thermal conductivities on magnetic buoyancy in a rotating gas.
P. H. Roberts, K. Stewartson.
Astron. Nachr., Band 298, 311 - 318 (1977).
    A horizontal magnetic field if increasing in strength downwards can cause a horizontal layer of electrically conducting fluid to become unstable, a phenomenon known as 'magnetic buoyancy', and sometimes thought to have relevance to magnetic A stars, and to sunspot creation. Analyses that assume infinite thermal and electrical conductivities (and zero viscosity) predict that modes of zero horizontal wave-length, in the direction perpendicular to the field, are maximally unstable but are stabilised by even small Coriolis forces. It is shown here, however, that when proper allowance is made for the finite (though large) conductivities of the fluid the layer may experience a 'conductive instability' that grows on the ohmic time-scale and is maximally unstable to a mode of non-zero horizontal extent.

062.041   Electrostatic waves in the warm magnetoplasma at the cyclotron harmonic frequencies.
A. K. Gwal, K. D. Misra.
IEEE Trans. Plasma Sci., Vol. PS-5, 146 - 150 (1977).
Abstr. in Phys. Abstr., Vol. 80, Abstr. 80308 (1977).

062.042   Dynamics of a cloud of fast electrons travelling through the plasma. II. Semi-analytical approach.
T. Takakura.
Sol. Phys., Vol. 52, 429 - 461 (1977).
    Numerical analysis of quasi-linear relaxation has been made for four models of electron beam with a finite length travelling through the plasma. A model atmosphere of the corona is adopted and also an increase in the cross-section of the electron beam is taken into account. The electron velocity distribution generally becomes a quasi-plateau form in limited velocity and time ranges. Collisional damping of plasma waves cannot be neglected. An approximate formula for the velocity distribution of the solar electrons passing through the corona has been derived analytically. A satisfactory fit between the numerical and semi-analytical results is shown.

062.043   Magnetic field transfer by two-dimensional convection and solar 'semi-dynamo'.
E. M. Drobyshevski (*Eh. M. Drobyshevskij*).
Astrophys. Space Sci., Vol. 46, 41 - 49 (1977).
    A comparison of the equations for the magnetic field transfer and for the heat transfer by two-dimensional turbulent convection of a conducting compressible medium shows the

magnetic field to be transported as a scalar admixture provided it is parallel to the convective rolls. A study of the distribution and amplification of the poloidal field in the two-dimensional convection zone of the Sun reveals the importance of considering generation mechanisms of the semi-dynamo type. An illustrative calculation of the solar poloidal field maintained by a weak Coriolis e.m.f. acting in a thin external layer of the convective envelope yields for the general near-polar field, if one somehow takes into account (1) field pumping by three-dimensional supergranulation, (2) field transfer and amplification by two-dimensional convection, and (3) ohmic diffusion of the field into a stable core, a value of the order of 10^{-1} gauss.

062.044 **Hydromagnetic break-up of bridges and jets into aligned objects.** P. S. Wesson, A. Lermann.
Astrophys. Space Sci., Vol. 46, 51 - 60 (1977).
The authors treat the stability of a galactic plasma column to hydromagnetic instability. They find the fastest-growing wave number and apply this result in a discussion of astrophysical instabilities. A conclusion is given, in which predictions are made from the hypothesis that jets with a morphology like that of M87 will fragment into ≈ 10 regularly-spaced bodies, possibly having some connection with the work of Arp (1966).

062.045 **Alfvén instabilities in streaming plasmas with anisotropic pressures and their relevance for the solar wind.** F. Verheest.
Astrophys. Space Sci., Vol. 46, 165 - 173 (1977).
Low frequency or Alfvén waves in streaming plasmas can become unstable. For these new Alfvén instabilities the streaming effects can be enhanced by a suitable pressure anisotropy. Perpendicular pressure effects are stabilizing, parallel pressure effects are destabilizing, as in the usual firehose instability. The observed velocity differences between helium and the main (hydrogen) flow in the solar wind plasma are such that the Alfvén waves are getting close to marginal instability. These new Alfvén instabilities limit the velocity differences between helium and hydrogen.

062.046 **Magneto-gravitational instability and suspended particles.** R. C. Sharma.
Astrophys. Space Sci., Vol. 46, 255 - 259 (1977).
The gravitational instability of an infinite homogeneous and infinitely conducting selfgravitating gas-particle medium in the presence of a vertical magnetic field and suspended particle is considered. It is found that in the presence of suspended particles and magnetic field, Jeans' criterion determines the gravitational instability.

062.047 **Equilibrium figures with a magnetic field.** N. P. Bondarenko, O. V. Kravtsov.
Astrophys. Space Sci., Vol. 46, 341 - 356 (1977).
A set of equations which are magnetohydrodynamic equilibrium conditions in the case of axial symmetry is solved. The possibility of splitting Maclaurin's sequence in the presence of a magnetic field is shown. The effect of a forceless magnetic field on Maclaurin's P-ellipsoid is investigated.

062.048 **Charged particle transport in turbulent magnetic fields: the perpendicular diffusion coefficient.** I. H. Urch.
Astrophys. Space Sci., Vol. 46, 389 - 406 (1977).
The diffusion of charged particles in a static turbulent magnetic field, which is superimposed on a constant magnetic field $B_0 k$, is considered. Previous calculations of the particle flux in a direction perpendicular to k have related the flux $S_\perp$ to the particle number density f by $S_\perp = -\kappa_\perp (\nabla f)_\perp$ where $\kappa_\perp$ is found from the power spectrum of the turbulent magnetic field. The author shows that this formula is incorrect for diffusion in static turbulent magnetic fields and obtains an alternative expression for $S_\perp$ when the turbulent magnetic field is composed of plane Alfvén waves propagating in the direction of the constant field B_0.

062.049 **Maximum growth rates of magneto-atmospheric instabilities.** J. A. Adam.
Astrophys. Space Sci., Vol. 47, L5 - L7 (1977).
Simple bounds on the growth rates of magneto-atmospheric instabilities are derived, both for perturbations along and perpendicular to a horizontal magnetic field. The physical significance of a singular level for stable modes is briefly discussed.

062.050 **Viscid and inviscid self-similar blast waves: spherical, isothermal flows and inferences for supernova remnants.** I. Lerche.
Astrophys. Space Sci., Vol. 47, 347 - 360 (1977).
The author investigates the spherically symmetric, self-similar flow behind a blast wave from a point explosion in a medium whose density varies with distance as $r^{-\omega}$ with the assumption that the flow is isothermal and viscid. The present paper shows that as the viscosity tends to zero, the viscid flow does not tend towards the inviscid flow pattern. Now the validity of adiabatic blast wave models has elsewhere been shown to be questionable for supernova remnants, and the inviscid blast wave models have also been shown to be inappropriate for supernova remnants. Taken together with these previous results, the results of the present calculations strongly suggest that the assumption of isothermal blast wave behavior of supernova remnants, either viscid or inviscid is not valid.

062.051 **Some questions of dynamics of magnetospheric plasma.** V. D. Pletnev, G. A. Skuridin.
Kosm. Issled., Vol. 15, 875 - 886 (1977). In Russian.

062.052 **Nonlinear magneto-acoustic dispersive waves.** S. K. Malik, M. Singh.
Astrophys. Space Sci., Vol. 48, 47 - 52 (1977).
The multiple time scale method is applied to investigate the nonlinear magneto-acoustic dispersive waves in a collisional plasma. Whereas in the absence of collisions these waves are modulationally stable, the collisions render them modulationally unstable.

062.053 **On the adiabatic pulsations of uniformly rotating bodies.** M. Tassoul.
Astrophys. Space Sci., Vol. 48, 89 - 102 (1977).
The problem of the oscillations and stability of compressible Maclaurin spheroids is reconsidered, on the basis of the third-order virial equations, in an arbitrarily rotating frame of reference. In contrast with the work of Kochhar and Trehan (1974), it is found that the frequencies evaluated in a rotating frame and those evaluated in an inertial frame are related to one another in a very simple way. Numerical calculations made for a wide range of the adiabatic exponent further clarify the effect of compressibility on the natural frequencies.

062.054 **Charged particle diffusion in a turbulent magnetic field.** I. H. Urch.
Astrophys. Space Sci., Vol. 48, 231 - 236 (1977).
The diffusion of charged particles in a turbulent magnetic field, but with no constant background electromagnetic fields, is discussed and expressions for the particle fluxes calculated.

062.055 **Electromagnetic instabilities in an anisotropic loss-cone plasma.** K. L. Vithal.
Astrophys. Space Sci., Vol. 49, 293 - 316 (1977).
A form of general dispersion relation for electromagnetic waves in a fully ionized anisotropic plasma with loss-cone that explicates the contribution of the loss-cone to the dispersion relation is developed.

062.056   New approach of studying electromagnetic mode
coupling in inhomogeneous magnetized plasma. I.
Normal incidence.   F. T. Cheng, P. C. W. Fung.
Astrophys. Space Sci., Vol. 49, 367 - 388 (1977).

The authors use a new approach to derive the system of
first-order coupled equations governing the propagation of
electromagnetic waves in an inhomogeneous magnetized plasma
for normal incidence. In this new approach they employ a step
model and use Maxwell's equations indirectly. The method
possesses simplicity in mathematical manipulation and gives a
clearer physical picture of the mechanism of mode-coupling.

062.057   The equation of polarization transfer in an inhomo-
geneous magnetized plasma. I. Formalism.
F. T. Cheng, P. C. W. Fung.
Astrophys. Space Sci., Vol. 49, 427 - 442 (1977).

In an inhomogeneous magnetized plasma, the polarization
characteristics of electromagnetic waves can vary in the course
of propagation due to emission from the medium, Faraday
rotation, differential absorption of characteristic modes, mode
coupling and inhomogeneity of the medium. The authors
formulate in the first paper of this series the polarization
transfer equation in tensorial form including terms describing
the five stated phenomena. Explicity expressions for the
relevant coefficients describing the different physical processes
are given and some properties of the transfer equation are
discussed.

062.058   Charged particle diffusion in the presence of Alfvén
waves: the perpendicular particle flux.   I. H. Urch.
Astrophys. Space Sci., Vol. 49, 443 - 472 (1977).

062.059   Conduction mode variation and inertial conductivity.   P. J. Baum, A. Bratenahl.
Astrophys. Space Sci., Vol. 49, 473 - 480 (1977).

Using separation of variables in space and time the
authors find two-dimensional solutions consistent with simple
forms of the inertial Ohm's law, Maxwell's equations, and the
equation of momentum transport. Conduction mode variation
(a temporal change in electrical conductivity due, e.g., to onset
of a plasma instability) is frequently found necessary. The
result for the effective conductivity shows that either the conduction mode variation or inertial effects may dominate.

062.060   Axially-symmetric explosion with thermal radiation.
S. Sakashita, K. Morita.
Astrophys. Space Sci., Vol. 50, 133 - 140 (1977).

A point explosion with thermal radiation in an axially
symmetric inhomogeneous medium is investigated by generalizing the method of Laumbach and Probstein to include the
effects of radiative cooling. As an example, a point explosion
in the plane stratified transparent medium with exponential
density distribution is calculated. It is shown that the focusing
effect along the symmetry axis is enhanced by radiative cooling
effect. Explosion models of extragalactic double radio sources
are briefly discussed.

062.061   Unsteady hydromagnetic thermal boundary layer
flow. II: No heat transfer.
G. C. Pande, N. Kafousias, C. L. Goudas.
Astrophys. Space Sci., Vol. 50, 141 - 152 (1977).

With viscous dissipation and Joule heating taken into
account, solution of the energy equation is obtained for unsteady hydromagnetic thermal boundary layer flow past a
porous wall (e.g., surface of a star) in presence of a transverse
magnetic field, under the condition of zero heat transfer between the fluid and the boundary – the so-called 'plate
thermometer problem' in MHD. Solution of the problem, in
the form of power series, is obtained under certain valid
simplifying assumptions.

062.062   Containment of an adiabatic plasma on magnetic

lines of force by a self-generated electrostatic field.
A. Hruška.
Astrophys. Space Sci., Vol. 50, 153 - 161 (1977).

One component of a three-fluid adiabatic plasma is under
certain conditions contained in a restricted region of space by
a large-scale electrostatic field generated within the plasma. The
containment is discussed here for plasma consisting of ions and
two populations of electrons characterized by different pitch
angle distribution functions.

062.063   Isothermal self-similar blast wave theory of super-
nova remnants driven by relativistic gas pressure.
I. Lerche.
Astrophys. Space Sci., Vol. 50, 323 - 342 (1977).

The author investigates the spherically symmetric, self-
similar flow behind a blast wave from a point explosion in a
medium whose density varies with distance as $r^{-\omega}$ with the
assumption that the flow is both isothermal and contains a
relativistic component of pressure. He concludes that isothermal self-similar blast waves do not provide a valid model
for a supernova remnant driven by a relativistic gas pressure.

062.064   New approach to studying electromagnetic mode
coupling in inhomogeneous magnetized plasma. II.
Oblique incidence.   P. C. W. Fung, F. T. Cheng.
Astrophys. Space Sci., Vol. 50, 361 - 381 (1977).

Considering mode coupling as a consequence of the
matching of boundary conditions at an infinitesimal discontinuity, a concept introduced by the authors (1977), they
derive explicit expressions for the coupling coefficients for
electromagnetic waves propagating in a rather general direction in an inhomogeneous magnetized plasma. Some special
cases of the theory are discussed.

062.065   Hydrodynamic instability of convectively unstable
atmospheres in shear flow.   J. A. Adam.
Astrophys. Space Sci., Vol. 50, 493 - 514 (1977).

Results are presented concerning the interaction between
regions of convectively unstable fluid, bounded above and below by stable fluid, with a basic horizontal flow field, sheared
in a vertical direction. The work is relevant to any astrophysical and geophysical situations in which convectively unstable
regions and shear flows are likely to be together present, but
the special motivation here is that of describing some aspects
of the interaction between supergranular flow and granular
convection.

062.066   Discontinuous transitions in a current-carrying
plasma.   H.-J. Lee, J. F. McKenzie, W. I. Axford.
Astrophys. Space Sci., Vol. 51, 3 - 32 (1977).

The properties of 'discontinuous' transitions in a current-
carrying plasma are analyzed by formulating the problem in
terms of the two-fluid (i.e., electrons and ions) equations. The
jump conditions, which connect states on either side of a
discontinuity, are derived and it is shown that these are similar
to the ordinary gas-dynamic Rankine-Hugoniot conditions
except that there are extra terms which represent an enhanced
mass flux in the momentum equation, and an additional heat
flux in the energy equation, both of which arise from the
existence of a current flow through the discontinuity.

062.067   Hydromagnetic flow near an oscillating porous
limiting surface.   G. C. Pande, N. G. Kafousias,
G. A. Georgantopoulos, C. L. Goudas.
Astrophys. Space Sci., Vol. 51, 125 - 134 (1977).

Unsteady hydromagnetic flow near a harmonically
oscillating limiting surface (e.g., of a star) is considered in
presence of a transverse magnetic field. Exact solutions, for a
periodic boundary layer without a 'mean steady flow', are obtained when the magnetic Prandtl number is unity and there is
a normal velocity of injection imposed at the wall. The results
are also presented for the case when the wall is subjected to a

normal velocity of suction instead of injection. It is observed that two distinct boundary (or hydromagnetic boundary) layers exist and tend to coalesce into a single layer when the magnetic field parameter approaches zero. The thicknesses of these boundary layers are significantly affected by the injection/suction velocity and the applied magnetic field.

062.068 **The behaviour of the magnetic lines of a rotating system.** E. Evangelidis.
Astrophys. Space Sci., Vol. 51, 319 - 327 (1977).
The behaviour of a magnetic line has been studied under continuous injection of plasma. Under increasing density the existence of three zones of possible concentration has been shown.

062.069 **On steady force-free dynamos.**
S. R. Sreenivasan.
Astrophys. Space Sci., Vol. 51, 341 - 348 (1977).
It is shown that the amplification of the magnetic energy that results in steady force-free dynamos automatically implies a depletion of the overall mechanical and thermal energies of the fluid in the region. The precise gain and loss of the field and the fluid, respectively, are demonstrated. This offers a natural and direct explanation of the relative coolness of sunspots with respect to their surroundings and also predicts lower velocities over sunspot regions resulting from the smoothing of turbulent fluctuations of velocities by viscous stresses in the magnetic region. The missing energy of the fluid in the region is shown to reside in the increased magnetic energy of the sunspots.

062.070 **The effect of a poloidal magnetic field on the linear and adiabatic oscillations of a polytrope $n = 3$.**
M. Goossens.
Astrophys. Space Sci., Vol. 52, 3 - 10 (1977).
The frequencies of the linear and adiabatic oscillations of a gaseous polytrope with a poloidal magnetic field are determined with the aid of a perturbation method. The influence of the poloidal magnetic field on the different types of spheroidal oscillation modes is discussed. The poloidal magnetic field generally strengthens the stability of the oscillation modes and this effect is the largest in the case of the non-radial p-modes.

062.071 **The equation of polarization transfer in an inhomogeneous magnetized plasma. II. Solutions.**
P. C. W. Fung, F. T. Cheng.
Astrophys. Space Sci., Vol. 52, 243 - 264 (1977).
Following the formalism on the polarization transfer equation presented recently by the same authors, solutions to this transfer equation under several special cases of interest are discussed in this paper. Analytic solutions for the Stokes parameters for several special cases of interest are given, and numerical solutions to these parameters for arbitrary propagation direction and two types of inhomogeneity of the medium are presented.

062.072 **A qualitative discussion on the possibility of gravitational instabilities at the origin of explosions in magnetospheres.** P. Kaufmann.
Astrophys. Space Sci., Vol. 52, 429 - 434 (1977).
This paper discusses the possibility of using the Chandrasekhar and Fermi (1953) model of gravitational instabilities in superconductive infinite cylinders subjected to magnetic fields, to explain explosions arising from plasma cumulations in neutral regions at certain magnetospheres, and solar explosions in particular.

062.073 **The stability of a magnetic flux element in a horizontally stratified compressible plasma.**
P. R. Wilson.
Sol. Phys., Vol. 55, 35 - 45 (1977).
The configuration of a magnetic flux element in a static,

compressible, gravitationally stratified plasma is considered. Under isothermal conditions an exact force-free solution is given for a two-dimensional cartesian flux sheath but for an axi-symmetric element, i.e. a flux tube, approximate solutions, applicable only to thin flux tubes, are obtained. Although such simple models cannot claim to approximate real sunspots, they do provide an understanding of the basic properties of flux elements in a compressible atmosphere.

062.074 **Magnetic field energy dissipation in neutral current sheets.**
A. T. Altyntsev, V. I. Krasov, V. M. Tomozov.
Sol. Phys., Vol. 55, 69 - 81 (1977).
An extended current sheet characterized by two peculiarities was formed in a configuration with opposite magnetic fields in a laboratory plasma on a 'θ-pinch' device. As a result, a stable neutral sheet has the complicated structure of a magnetic field, including closed magnetic loops elongated along the axis of the system. The experimental data, obtained over a broad range of plasma densities and magnetic field values typical for the solar atmosphere, show that the antiparallel magnetic field turbulent dissipation could play an important role in the mechanism of solar energy release. The parameters of accelerated particles agree nicely with the data of astrophysical observations.

062.075 **Forced radial oscillations of a fully ionized cylindrical plasma surrounded by a partially ionized plasma shell.** M. Bureš, E. Tennfors.
Phys. Scr., Vol. 16, 117 - 122 (1977).

062.076 **Particle aspect analysis of $E \times B$ drift instability in the presence of non-uniform electric field.**
K. D. Misra, M. S. Tiwari.
Phys. Scr., Vol. 16, 142 - 146 (1977).
A physical mechanism and the dynamical details of the drift wave instability have been studied by investigating the trajectories of the charge particles in the presence of ambient magnetic field and non-uniform electric field. The effect of finite gyro-radius corrections have also been accounted. The general dispersion relation for the electrostatic drift wave has been obtained. The expression for the growth rate has been derived from an energy balance approach. The important geophysical situations for the existence of inhomogeneous electric field have been outlined and potential application of this result has been indicated.

062.077 **Spontaneous symmetry breaking and bifurcations from the Maclaurin and Jacobi sequences.**
D. H. Constantinescu, L. Michel, L. A. Radicati.
ESO Sci. Prepr. No. 16, 2 + 35 pp. (1977).
The equilibrium of a rotating self-gravitating fluid is governed by non-linear equations. The equilibrium solutions, parametrized in terms of the angular momentum squared, exhibit the phenomenon of bifurcation, accompanied by spontaneous symmetry breaking. Under very general assumptions, a set of selection rules can be derived, which drastically restrict the patterns of symmetry breaking that are allowed to appear. Bifurcations of this kind are similar to second-order phase transitions à la Landau. The method is illustrated by the simple example of an incompressible fluid in rigid rotation. However, the selection rules are more general; they apply also to models which approximate a rotating star more realistically.

062.078 **Cord diffusion of slightly ionized plasma forming an angle with a magnetic field.**
V. A. Rozhanskij, L. D. Tsendin.
Geomagn. Aehron., Vol. 17, 1002 - 1007 (1977). In Russian.

062.079 **Amplitudenbegrenzung von Magnetfeldern durch großräumige inkompressible Strömungen bei vollkonvektiven α^2-Dynamos.** R. Hellmich.

Mitt. Astron. Ges., Nr. 42, p. 110 - 112 (1977).

062.080 Penetration of a plane monochromatic wave falling from vacuum into a semi-confined plasma.
B. S. Ryabov.
Din. kosm. plazmy. Moskva, 1976, p. 132 - 135. In Russian.
Abstr. in Ref. zh., 51. Astron., 12.51.176 (1977).

062.081 Generalization of Planck's law of radiation to anisotropic dispersive media.   K. D. Cole.
Australian J. Phys., Vol. 30, 671 - 673 (1977).
   The distribution of radiant energy in media which are not only dispersive but also anisotropic is derived. The distribution is found to depend on the phase velocity V_s, the group velocity V_{gs} and the angle α_s between them for each mode s. Planck's law is a limiting case when $V_s = V_{gs} = c$ and $\alpha_s = 0$ and applies strictly only to a vacuum.

062.082 Convective flows of viscous fluid in spherical layers. Some astrophysical applications.
I. M. Yavorskaya.
Inst. kosm. issled. AN SSSR. Pr-346. Moskva, 1977, 64 pp.
In Russian. − Abstr. in Ref. zh., 51. Astron., 1.51.185 (1978).

062.083 Electric currents in cosmic plasmas.
      H. Alfvén.
Tek. Hoegsk., Stockholm, Sweden. Instn. Plasmafys. TRITA-EPP−77-14. 75 pp. (1977). − Abstr. from INIS7723 341834.

062.084 Radiation from a source in a cold magnetoactive plasma, re-examined. Application to cyclotron and multipole radiation.   P. Hoyng, G. A. Stevens.
Phys. Fluids, Vol. 20, 520 - 532 (1977) = Utrechtse Sterrenkundige Overdrukken, No. 396.

Handbook on plasma instabilities, Vol. 1.
See Abstr. 003.039.

Electromagnetic waves in cosmic plasma.
See Abstr. 003.171.

The advance from 2D electrostatic to 3D electromagnetic particle simulation.   See Abstr. 021.012.

Numerical solution of continuity equations.
See Abstr. 021.013.

Symbolic computation of nonlinear wave interactions on MACSYMA.   See Abstr. 021.014.

Finite difference and finite element methods.
See Abstr. 021.015.

Computation of Tokamak transport.
See Abstr. 021.016.

On the detectability of forbidden lines.
See Abstr. 022.040.

Soft X-ray spectrum of a hot plasma.
See Abstr. 022.113.

Magnetic fields greater than 10^{20} gauss?

See Abstr. 061.009.

Linear waves in a radiating and scattering grey medium. II. The effect of a transverse magnetic field.
See Abstr. 063.030.

Linear waves in a radiating and scattering grey medium. III. The effect of a non-transverse magnetic field.
See Abstr. 063.035.

Non-linear scattering probability of longitudinal plasma waves in a magnetoactive plasma system.
See Abstr. 063.037.

The influence of multiple Compton scattering on the spectrum of X-radiation. Monte-Carlo computations.
See Abstr. 063.050.

Non-LTE line formation in the presence of magnetic fields.   See Abstr. 064.006.

Convection in stars.   See Abstr. 065.031.

The collapse of unstable isothermal spheres.
See Abstr. 065.042.

Mechanism of X-ray and soft γ-ray radiation from accreting neutron stars.   See Abstr. 066.014.

Comments on pulses of characteristic energy produced in solar flare detonations and its possible application to other astrophysical plasmas.   See Abstr. 073.095.

Collective plasma effects and the electron number problem in solar hard X-ray bursts.   See Abstr. 076.012.

Nonpotential waves in the auroral electrojet.
See Abstr. 084.028.

Electromagnetic wave propagation and instabilities in the magnetosphere.   See Abstr. 084.288.

On the question of the energy of the precessional dynamo.   See Abstr. 084.315.

A model of accretion disks in close binaries.
See Abstr. 117.007.

A remark on co-rotation in pulsar magnetospheres.
See Abstr. 141.542.

Propagation of magnetohydrodynamic waves from the galactic center. Origin of the 3-kpc arm and the North Polar Spur.   See Abstr. 155.013.

Errata

062.901 Damping of magnetohydrodynamic waves: a correction.   B. R. De.
Astrophys. Space Sci., Vol. 47, 457 - 458 (1977). − See 17.062.012.

063 Radiative Transfer, Scattering

063.001 **Quasi-asymptotic solutions of the radiative transfer problem in an optically finite layer. II. Nonconservative scattering.** M. A. Mnatsakanyan.
Astrofizika, Vol. 12, 451 - 473 (1976). In Russian. – English translation in Astrophysics, Vol. 12, No. 3.
Approximate analytic (quasi-asymptotic) solutions of the radiative transfer problem for an optically finite layer are obtained in the case of monochromatic isotropic nonconservative ($\lambda \leqslant 1$) scattering. Although the solutions are asymptotic, they can practically be applied to a layer of arbitrary thickness.

063.002 **Brightness coefficients for a two-layer atmosphere with anisotropic scattering. II.** A. K. Kolesov.
Astrofizika, Vol. 12, 485 - 494 (1976). In Russian. – English translation in Astrophysics, Vol. 12, No. 3.
The auxiliary functions which are a generalisation of Ambartsumyan's functions φ and ψ for the case of an anisotropically scattering two-layer atmosphere are considered. It is found that for given values of indices i and m there are three independent auxiliary functions. The linear integral equations for these functions are derived. The kernel and the free terms of these equations are expressed in terms of optical characteristics of the atmospheric layers.

063.003 **A fast invariant imbedding method for multiple scattering calculations and an application to equivalent widths of CO_2 lines on Venus.**
M. Sato, K. Kawabata, J. E. Hansen.
Astrophys. J., Vol. 216, 947 - 962 (1977).
The authors describe a fast method for solution of the invariant imbedding equation, make explicit timing comparisons with the doubling method, and make illustrative calculations for inhomogeneous model atmospheres of Venus.

063.004 **Radiative transfer in an infinite atmosphere. II.**
V. V. Ivanov.
Astrofizika, Vol. 12, 565 - 578 (1976). In Russian. – English translation in Astrophysics, Vol. 12, No. 4.
The interrelations between four typical angular distributions of intensity in plane infinite and semi-infinite media are studied. These angular distributions correspond to the cases when the source and the detector of radiation are in the same plane (for a semi-infinite medium at the boundary) and when they are infinitely separated. This enabled, in particular, to obtain from physical considerations previously unknown Fredholm equations for the boundary intensity $u(\mu)$ in the Milne problem and for Ambartsumyan's functions $\varphi_n^m(\mu)$.

063.005 **The radiation field in a plane layer illuminated by parallel rays.** Eh. Kh. Danielyan.
Astrofizika, Vol. 12, 579 - 586 (1976). In Russian. – English translation in Astrophysics, Vol. 12, No. 4.
The problem of the internal radiation regime in a slab of finite thickness for isotropic scattering has been considered. It is shown that in the case of parallel radiation the internal intensity may be found in terms of a source function without integration over the optical depth.

063.006 **Radiative transfer calculated from a Markov chain formalism.** L. W. Esposito, L. House.
Bull. American Astron. Soc., Vol. 9, 470 (1977). – Abstract.

063.007 **Non-radial radiative transfer in close binaries. Application to the bolometric reflection effect in W UMa stars.** I. Pustylnik.
Acta Astron., Vol. 27, 251 - 264 (1977).
The author considers the equation of radiative transfer

in a grey spherically symmetrical absorbing and scattering atmosphere which is subjected to irradiation by a plane parallel beam and solves this equation in the Eddington approximation with special attention given to the nonradial component of the radiation field. He uses the obtained results for a rough estimate of the influence of non-radial radiative transfer upon the light curve of a close binary system.

063.008 **Influence of the refractive index on the radiative source function of an isotropically scattering medium.**
B. F. Armaly, H. S. El-Baz.
J. Quant. Spectrosc. Radiat. Transfer, Vol. 18, 419 - 424 (1977).
The influence of refractive index on the radiative source function is presented for the case of emission from an isothermal, isotropically scattering medium. A closed-form, approximate solution is obtained and results are presented for finite and semi-infinite nonconservative cases. An increase in refractive index causes the source function to increase and that effect is more pronounced at higher scattering albedo and smaller optical depths.

063.009 **Radiative heat transfer in a magnetic field.**
G. G. Pavlov, D. G. Yakovlev.
Astrofizika, Vol. 13, 173 - 183 (1977). In Russian. English translation in Astrophysics, Vol. 13, No. 1.
The law of radiative heat transfer in a medium with a magnetic field is obtained and the expressions for the heat transfer coefficients $D\parallel$ and $D\perp$ along the magnetic lines and perpendicular to them are derived. In a white-dwarf envelope the anisotropic character of radiative heat transfer in a magnetic field may produce hydrodynamical flows similar to meridional circulations.

063.010 **Characteristic lengths in radiative transfer problems for a moving medium.** S. I. Grachev.
Astrofizika, Vol. 13, 185 - 197 (1977). In Russian. English translation in Astrophysics, Vol. 13, No. 1.
Thermalization length and diffusion length for different absorption coefficient profiles are found for an infinite one-dimensional medium moving with a constant velocity gradient $\ll 1$. The thickness of the boundary layer for a moving semi-infinite medium with uniformly distributed primary sources is also estimated.

063.011 **Radiative transfer in cylindrical media.**
J. N. Heasley.
J. Quant. Spectrosc. Radiat. Transfer, Vol. 18, 541 - 547 (1977).
A numerical method for the solution of the radiative transfer equation in a circularly symmetric, cylindrical region is developed. The transfer equation is formulated as a second-order differential equation resulting in a set of tridiagonal difference equations. This form is particularly well suited to line formation and energy balance calculations using the complete linearization method. Several numerical examples are presented.

063.012 **Non-LTE transfer – III. Asymptotic expansion for small ϵ.** U. Frisch, H. Frisch.
Mon. Not. R. Astron. Soc., Vol. 181, 273 - 280 (1977).
Radiative transfer with complete frequency redistribution in the full- or half-space with interior sources is considered in the limit of small ϵ (probability of collisional destruction). It is shown that a suitably scaled source function satisfies for $\epsilon \to 0$ a singular equation.

063.013 **Non-LTE transfer – IV. A rapidly convergent iter-**

ative method for the Wiener–Hopf integral equations.
H. Frisch, Ch. Froeschlé.
Mon. Not. R. Astron. Soc., Vol. 181, 281 - 292 (1977).

A simplified solution of the Wiener–Hopf equation of
non-coherent transfer introduced in Frisch & Frisch (1975) is
extended into a systematic approximation procedure by an
iterative under-relaxation method. Contrary to the Λ-iteration,
the number of iterations required to achieve a given accuracy
is independent of ϵ. After suitable modifications the method is
also applied to the singular integral equation obtained for the
interior solution by Frisch & Frisch (1977) when $\epsilon \to 0$ and to
Ivanov's homogeneous Wiener–Hopf equation for the surface
boundary-layer. Numerical results are given for various choices
of the thermal source and of the scattering kernel.

063.014   Theory of thin-screen scintillations for a spherical
wave.   L. C. Lee.
Astrophys. J., Vol. 218, 468 - 476 (1977).

A thin-screen scintillation theory for a spherical wave is
presented under the "quasi-optical" approximation. The
author calculates the "scattering angle", the "observed angle",
the intensity correlation function, and the temporal pulse
broadening for the random wave. It is found that as the wave
propagates outward away from the phase screen, the correlation scale of the intensity fluctuation increases linearly while
the "observed angle" decreases linearly. The calculations are
carried out for both Gaussian and power-law spectra of the
turbulent medium.

063.015   Two-frequency mutual coherence function of plane
waves in strongly turbulent media.
M. A. Plonus, S. C. H. Wang.
Antennas and propagation, International Symposium 1976,
Amherst, Mass., USA, 11 - 15 Oct. IEEE, New York, USA,
p. 592 - 593 (1976). – Abstr. in Phys. Abstr., Vol. 80, Abstr.
21907 (1977).

063.016   Radiative transfer theory for microwave remote
sensing of two-layer media.   L. Tsang, J. A. Kong.
Antennas and propagation, International Symposium 1976,
Amherst, Mass., USA, 11 - 15 Oct. IEEE, New York, USA,
p. 598 - 601 (1976). – Abstr. in Phys. Abstr., Vol. 80, Abstr.
21908 (1977).

063.017   Radiative transfer in closely packed media.
G. H. Goedecke.
J. Opt. Soc. America, Vol. 67, 1339 - 1348 (1977).

063.018   Extinction cross sections of arbitrarily shaped ran-
domly oriented nonspherical particles.   P. Chýlek.
J. Opt. Soc. America, Vol. 67, 1348 - 1350 (1977).

Extinction cross section of large arbitrarily shaped randomly oriented nonspherical particles is always larger than the
extinction cross section of spherical particles of the same
volume. This theorem is proved rigorously for particles with
the size parameter $x \gg 1$. It is conjectured that the same is true
for sufficiently wide polydispersions of particles with $x > 15$.

063.019   Non stationary multiple scattering.   A. P. Wang.
J. Math. Phys., New York, Vol. 18, 47 - 51 (1977).
Abstr. in Phys. Abstr., Vol. 80, Abstr. 29641 (1977).

063.020   Simulation of emission frequencies from angle-
dependent partial frequency redistributions.
J.-S. Lee.
Astrophys. J., Vol. 218, 857 - 865 (1977).

Simple and fast algorithms for generating random emission frequencies from angle-dependent redistribution functions
$R_I (x, n; x', n')$, $R_{II} (x, n; x', n')$, and $R_{III} (x, n; x', n')$ are
given. The motivation for applying angle-dependent redistribution functions to studies of photon diffusion by the Monte
Carlo technique is the fact that in numerous cases, particularly

those involving the dipole phase function, angle-averaged
redistributions yield a poor approximation to the angle-dependent case. Theoretical proofs of the frequency-generating
algorithms are given, and numerical values of the generated
frequency distributions are presented to validate the algorithms.

063.021   The backscattered fraction in two-stream approxima-
tions.   W. J. Wiscombe, G. W. Grams.
J. Atmos. Sci., Vol. 33, 2440 - 2451 (1976). – Abstr. in Phys.
Abstr., Vol. 80, Abstr. 45150 (1977).

063.022   The delta-Eddington approximation for radiative
flux transfer.
J. H. Joseph, W. J. Wiscombe, J. A. Weinman.
J. Atmos. Sci., Vol. 33, 2452 - 2459 (1976). – Abstr. in Phys.
Abstr., Vol. 80, Abstr. 45151 (1977).

063.023   Geometrical optics in dispersive media.
A. M. Anile, P. Pantano.
Phys. Lett. A, Vol. 61A, 215 - 218 (1977). – Abstr. in Phys.
Abstr., Vol. 80, Abstr. 61917 (1977).

063.024   Comparison of results obtained by solving the
radiative transfer equation with an iterative method
and a spherical harmonics method.
O. P. Bahethi, R. S. Fraser.
J. Atmos. Sci., Vol. 34, 553 - 556 (1977). – Abstr. in Phys.
Abstr., Vol. 80, Abstr. 66986 (1977).

063.025   Exponential-sum fitting of radiative transmission
functions.   W. J. Wiscombe.
J. Comput. Phys., Vol. 24, 416 - 444 (1977). – Abstr. in Phys.
Abstr., Vol. 80, Abstr. 70995 (1977).

063.026   A solution of the Rayleigh scattering problem for
plane-parallel atmospheres of large optical thickness.
A. B. Kahle.
Rand Corp., Santa Monica, Calif., Rep. P-5795, 146 pp. (1977).
Abstr. in Phys. Abstr., Vol. 80, Abstr. 73947 (1977).

063.027   Radiative transfer of atomic and molecular resonant
emissions in upper atmospheres. I. Basic theories in
Doppler-broadening atmospheres.
T. Tohmatsu, H. Yamamoto.
J. Geomagn. Geoelectr., Vol. 28, 437 - 460 (1976). – Abstr.
in Phys. Abstr., Vol. 80, Abstr. 78418 (1977).

063.028   A note on extinction and scattering efficiencies.
P. Chýlek.
J. Appl. Meteorol., Vol. 16, 321 - 322 (1977). – Abstr. in
Phys. Abstr., Vol. 80, Abstr. 78421 (1977).

063.029   Phase fluctuations in a turbulent medium.
A. Ishimaru.
Appl. Opt., Vol. 16, 3190 - 3192 (1977).

General relationships are shown relating the log-amplitude
and the phase fluctuations to the Rytov geometric optical
solution for the phase fluctuation under the assumption of
normal distribution for the log-amplitude and the phase fluctuations. In the weak fluctuation region, this general relationship reduces to the Rytov solution. In the strong fluctuation
region, the phase correlation function is shown to approach
the Rytov geometric optical solution.

063.030   Linear waves in a radiating and scattering grey
medium. II. The effect of a transverse magnetic field.
N. Kaneko, Y. Ōno.
Astrophys. Space Sci., Vol. 47, 147 - 150 (1977).

The equation of radiative acoustics is derived by taking
into account the effect of a transverse magnetic field, which is
quite similar to the acoustic equation derived in Paper I
(Kaneko et al., 1976). The only difference is that the adiabatic,

isothermal, and isentropic speeds of sound and the radiation-acoustic speed are replaced by the adiabatic, isothermal, and isentropic magnetoacoustic speeds and the radiation-magnetoacoustic speed, respectively. The main results shown in Paper I are valid even in the presence of a transverse magnetic field.

063.031   A novel methodology for radiative transfer in a planetary atmosphere. I. The functions a^m and b^m of anisotropic scattering.   A. L. Fymat, R. E. Kalaba.
Astrophys. Space Sci., Vol. 47, 195 - 216 (1977).

As an introduction the relevant aspects of Sobolev's solution are reviewed. The theory of anisotropic scattering for omnidirectional illumination of the atmosphere is provided. All relevant functions as algebraic relations involving the newly introduced function b^m (or h^m) are derived.

063.032   Compton effect and solar spectral lines.   P. Maltby.
Astrophys. Space Sci., Vol. 47, L21 - L23 (1977).

The conclusion reached by Missana (1975) and by Missana and Piana (1975, 1976) that the observed wavelength shift of solar spectral lines can be explained by means of a scattering theory is questioned. It is shown that both the magnitude and the direction of the wavelength shift associated with Compton scattering are incompatible with the observed shift of solar spectral lines.

063.033   A new representation of the H-functions of radiative transfer.   S. R. Das Gupta.
Astrophys. Space Sci., Vol. 50, 187 - 203 (1977).

The author obtains a new representation of Chandrasekhar's H-functions $H(z)$. This new form of $H(z)$ has proved to be very useful in solving coupled integral equations involving X-, Y-functions of transport problems.

063.034   Thomson scattering in a thick layer.   M. Missana.
Astrophys. Space Sci., Vol. 50, 409 - 419 (1977).

The equations for the transfer of light through a thick layer are studied with the aid of a computer in the presence of Thomson scattering and Compton effect. The red-shifts, top intensities and red asymmetry of Gaussian-shaped spectral lines are tabulated against the optical thickness and the normalized width. It is shown that the equivalent width of the lines is conserved in the transfer and that also the central intensity is conserved for wide enough lines. Two astrophysical applications of the resulting equations are proposed: to the spectra of white dwarfs, and to the spectra of Orion Nebula.

063.035   Linear waves in a radiating and scattering grey medium. III. The effect of a non-transverse magnetic field.   N. Kaneko, A. Habe, Y. Ōno.
Astrophys. Space Sci., Vol. 50, 451 - 460 (1977).

The equation of radiative acoustics is derived by taking into account the effect of a non-transverse magnetic field, and the solutions are schematically represented. The main results shown in Paper I (Kaneko et al., 1976) and Paper II (Kaneko, Ōno, 1977) are valid even in the presence of a non-transverse magnetic field, and the only difference is that the adiabatic, isothermal, and isentropic speeds of sound and the radiation-acoustic speed in Paper I are replaced by the sets of speeds of adiabatic, isothermal, isentropic, and radiation-acoustic fast and slow waves, respectively.

063.036   Multiple scattering in the atmosphere with a rough surface. I. Azimuth-independent case.   S. Mukai.
Astrophys. Space Sci., Vol. 51, 165 - 172 (1977).

The intensity of radiation emergent from the atmosphere bounded by a rough surface is discussed with the aid of the superposition method derived by Mukai (1973). The merit of this method is to express the laws of diffuse reflection and transmission for the planetary problem with a rough surface

in terms of a scattering and a transmission function for the standard problem. The reflection pattern of this model surface is discussed in comparison with that of a simple reflecting surface.

063.037   Non-linear scattering probability of longitudinal plasma waves in a magnetoactive plasma system.   P. C. W. Fung, K. C. C. Ko.
Astrophys. Space Sci., Vol. 52, 105 - 135 (1977).

A semi-quantum approach is used to study Compton scattering and non-linear scattering processes in a magnetoactive plasma under weak turbulence regime. Analytical expressions for these two types of scattering probabilities are derived from first principles for l-l waves. The results obtained are shown to be identical to previous derivations.

063.038   Non-stationary luminescence of an infinite homogeneous space.   D. I. Nagirner.
Vestn. Leningr. univ., 1977, No. 7, p. 138 - 143. In Russian. – Abstr. in Ref. zh., 51. Astron., 11.51.284 (1977).

063.039   An upper bound for translation kernels in scattering theory.   M. Coz.
C. R. Acad. Sci. Paris, Tome 285, Ser. A, 1121 - 1124 (1977). In French.

The paper derives a new estimate for the Riemann solution used in the representation of the translation kernel. A new estimate for the translation kernel itself, as well as a more workable expression for the fundamental equation, are deduced.

063.040   Fourier expansion of the phase matrix for Mie scattering.   H. Domke.
Z. Meteorol., Band 25, 357 - 361 (1975) = Mitt. Astrophys. Obs. Potsdam, Nr. 178.

063.041   Non-LTE radiative transfer, including multi-dimensional effects and polarization.   L. Stenholm.
Thesis Lund Obs., Sweden = Rep. Obs. Lund, 8 + 179 pp. (1977).

063.042   Multi-dimensional transfer of polarized radiation in magnetic fluxtubes.   L. G. Stenholm, J. O. Stenflo.
Rep. Obs. Lund, No. 12, (see 012.044), p. 53 - 62 (1977).

Calculations of multi-dimensional transfer of the Stokes vector in magnetic fluxtubes are used to study how the multi-dimensional effects influence the relation between the apparent field strength observed with a solar magnetograph and the true field strength. The line-ratio method used to derive true field strengths from simultaneous magnetograph recordings in two spectral lines of different Landé factor is tested.

063.043   The Hanle effect in a compact non-LTE radiative-transfer formulation.   J. O. Stenflo.
Rep. Obs. Lund, No. 12, (see 012.044), p. 65 - 79 (1977).

A non-LTE theory for the Hanle effect (magnetic-field sensitive coherence effects in resonance-line scattering) is developed. The theory allows polarization calculations for multi-level atoms in arbitrary optically thick media, and includes the effect of collisions. The main limitation is that the magnetic fields are assumed to be weak in the sense that the Zeeman splitting should be much smaller than the Doppler width of the line.

063.044   On the solution of Chandrasekhar's planetary problem.   T. Viik.
Publ. Tartu Astrofiz. Obs., Vol. 45, 3 - 12 (1977).

A solution of the integral equation for the source function of a homogeneous slab with isotropic scattering, bounded by a Lambert law reflector and illuminated by uniform parallel rays, is reduced to a Cauchy system.

063.045 Numerical results for the source function of the albedo problem in the case of a semi-infinite medium.   T. Viik.
Publ. Tartu Astrofiz. Obs., Vol. 45, 13 - 33 (1977).

The results of the solution of the Ambartsumyan equation for the source function of the albedo problem are presented. It is assumed that the half-space illuminated by uniform parallel rays consists of homogeneous matter which both absorbs and scatters radiation isotropically. Tables for computing the source function for a large variety of the albedos of single scattering are presented.

063.046 Some remarks on the non-radial component of radiative transport in close binaries.   I. Pustylnik.
Publ. Tartu Astrofiz. Obs., Vol. 45, 34 - 44 (1977).

It is shown that in the spherically symmetric atmosphere of a star illuminated by a parallel beam anisotropic scattering may give rise to the non-radial component of the radiative flux.

063.047 A fast method for finding the resolvent of the equation of radiative transfer by the use of Neumann series.   A. Sapar.
Publ. Tartu Astrofiz. Obs., Vol. 45, 275 - 283 (1977).

Using a Neumann series expansion, a fast method has been proposed for finding the resolvent of the equation of radiative transfer. The essence of the method lies in the fact that the resolvent involving scatterings up to the order of 2^{n+1} can be obtained by considering scatterings up to the order of 2^n followed by the scatterings up to the order 2^n.

063.048 The Eddington approximation generalized for radiative transfer in spherically symmetric systems. II. Nongray extended dust-shell models.   W. Unno, M. Kondo.
Publ. Astron. Soc. Japan, Vol. 29, 693 - 710 (1977).

The nongray radiative transfer in an extended dust shell surrounding a central star is solved by use of the generalized Eddington approximation. The results provide a set of model parameters if spectral photometric observations are available. The relations between these parameters and the spectral features of emergent flux are tabulated. The effects of scattering including anisotropy to the second orders are also considered.

063.049 On the solution of the equation of radiative transfer. Summary.   T. Viik.
Calendar of the Tartu Astronomical Observatory for the year 1978, (see 047.026), p. 78 - 79 (1977). In Estonian.

063.050 The influence of multiple Compton scattering on the spectrum of X-radiation. Monte-Carlo computations.   L. A. Pozdnyakov, I. M. Sobol', R. A. Syunyaev.
Astron. Zh. Akad. Nauk SSSR, Vol. 54, 1246 - 1258 (1977). In Russian. English translation in Soviet Astron., Vol. 21, No. 6.

A spherical cloud of relativistic plasma with a given optical depth from scattering and a given temperature of Maxwellian electrons is considered. There is a point source of low frequency radiation in the centre of the cloud with Planckian spectrum. Monte-Carlo computations and analytical estimates show that in the case of small optical depth the radiation escaping from the cloud has a power-law spectrum. In the case of an optically thick cloud the escaping radiation spectrum tends to a Wien spectrum. The energy loss rate of the cloud (its luminosity) is computed. The transfer of hard radiation from a central point source through a plasma cloud with $kT_e \approx 3$ keV is considered. Monte-Carlo techniques for computing such problems are described.

063.051 On the diffuse reflection of light by a semi-infinite atmosphere with strongly elongated dispersion indicatrix.   A. S. Anikonov, S. Yu. Ermolaev.
Vestn. Leningr. univ., 1977, No. 7, p. 132 - 137. In Russian. Abstr. in Ref. zh., 51. Astron., 12.51.169 (1977).

063.052 Statistical theory of nonstationary radiative transfer in a medium of variable scatterers.   Yu. N. Barabanenkov, V. D. Ozrin.
Izv. vyssh. uchebn. zaved. Radiofiz., Vol. 20, 712 - 720 (1977). In Russian. – Abstr. in Ref. zh., 51. Astron., 12.51. 170 (1977).

063.053 The problem of diffuse reflection with redistribution of radiation according to frequencies and directions.   A. G. Nikogosyan.
Dokl. AN SSSR, Vol. 235, 786 - 789 (1977). In Russian. Abstr. in Ref. zh., 51. Astron., 1.51.177 (1978).

063.054 A non-local-thermodynamic equilibrium formulation of the transport equation for polarized light in the presence of weak magnetic fields.   D. J. McNamara.
Thesis Natl. Cent. Atmos. Res., Boulder, Colo., USA. NCAR/ CT–44. 118 pp. (1977). – Abstr. from INIS7724 345377.

063.055 On the photographic method of determination of the scattering indicatrix.   Yu. V. Aleksandrov, V. P. Tishkovets.
Physics of the moon and planets. Fundamental astrometry, (see 003.024), Vestn. Khar'kov. Univ., No. 160, p. 20 - 23 (1977). In Russian.

063.056 Contrast transmittance at the top of an atmosphere bounded by a horizontally nonuniform diffuse reflector.   S. Ueno.
10th Lunar and Planetary Symposium, (see 012.050), p. 166 - 170 (1977).

Based on a two-dimensional transfer model, it is shown that contrast transmittance at top expressed in terms of the scattering and transmission functions depends not only on the background but also on the target.

Radiative properties of shock waves in gases. See Abstr. 003.156.

LENAM, a standard computer program for the calculation of multi-dimensional radiative transfer, and LENAM-P, an extended version including polarization. See Abstr. 021.022.

Extending radiative transfer models by use of Bayes' rule.   See Abstr. 031.311.

On the computation of optical properties of heterogeneous grains.   See Abstr. 061.044.

The equation of polarization transfer in an inhomogeneous magnetized plasma. I. Formalism. See Abstr. 062.057.

The equation of polarization transfer in an inhomogeneous magnetized plasma. II. Solutions. See Abstr. 062.071.

Non-LTE line formation in the presence of magnetic fields.   See Abstr. 064.006.

Contribution to spectral line formation in moving stellar envelopes. Radiation field and statistical equilibrium equations.   See Abstr. 064.010.

Stellar opacity.   See Abstr. 064.022.

Non-LTE line formation in a magnetic field. The two-level atom with a frequency independent source function.

I: Formulation.   See Abstr. 064.055.

Asymptotic behaviour near past null infinity for scattering problems.   See Abstr. 066.216.

Linear hydrodynamical equations coupled with radiative transfer in a non-isothermal atmosphere. I. Method. See Abstr. 080.067.

A method of computing the emergent radiation by the atmosphere in the region ranging from ultraviolet to infra-red.   See Abstr. 091.066.

Interstellar scattering and scintillation of radio waves.   See Abstr. 131.159.

Multiple scattering in reflection nebulae. I. A Monte Carlo approach.   See Abstr. 134.024.

Multiple scattering in reflection nebulae. II. Uniform plane-parallel nebulae with foreground stars. See Abstr. 134.025.

Multiple scattering in reflection nebulae. III. Nebulae with embedded illuminating stars. See Abstr. 134.026.

Multiple scattering in reflection nebulae. IV. The multiplicity of scattering.   See Abstr. 134.027.

064 Stellar Atmospheres, Stellar Envelopes, Mass Loss

064.001 Model stellar atmospheres – I. Numerical techniques.
D. M. Peterson.
J. Quant. Spectrosc. Radiat. Transfer, Vol. 18, 205 - 225
(1977).
The author presents a series of new numerical techniques
and novel uses of existing techniques to solve the model
stellar atmosphere problem. The techniques are applied to
the fully linearized LTE problem and procedures for generalizing to the non-LTE problem are indicated. He also shows
that the usual angle quadrature schemes, particularly the
$n = 3$ double Gauss scheme, can introduce significant errors in
the calculation of the mean intensity. Alternative schemes are
presented that substantially improve on this.

064.002 Dust in stellar atmospheres.   E. E. Salpeter.
The dusty universe, (see 012.001), p. 47 - 57 (1975).
Nucleation theory, developed for liquid-droplet growth,
is reviewed. Difficulties in the growth of crystals held together
by valence forces are pointed out. The rate of mass loss is
estimated for dust grains in cool stellar atmospheres as a function of luminosity. For intermediate values of the luminosity,
only a few high-ejection-velocity dust grains are ejected from
the atmosphere.

064.003 Circumstellar dust.   N. J. Woolf.
The dusty universe, (see 012.001), p. 59 - 87 (1975).
The infrared spectral features of circumstellar solid
matter are described. An attempt is made to identify the
particles and to predict their sizes.

**064.004 Emission and absorption in circumstellar dust shells:
an isothermal approach.**
G. Robinson, A. R. Hyland.
Mon. Not. R. Astron. Soc., Vol. 180, 495 - 524 (1977).
An isothermal approach has been used to construct an
extensive grid of circumstellar dust shell models. The effect on
the output spectrum of the star and shell temperature and the
optical depth of the shell is investigated for power opacity laws
and silicate opacity laws. It is found that, for silicate shells,
there exists a large domain in which the 10 and 20-μm features
are in absorption for all values of the star temperature and that
the principal factor influencing whether the features are in
emission or absorption is the shell temperature. The model is
compared with observations using a variety of colour–colour
diagrams and a semi-empirical silicate opacity law. It is found
that the model accounts remarkably well for observations of
cool stars with silicate emission features.

**064.005 Depths of formation of spectral lines and the
curve-of-growth method.**   L. S. Lyubimkov.
Izv. Krymskoj Astrofiz. Obs., Vol. 57, 87 - 98 (1977). In
Russian.
Using model stellar atmospheres the author estimates the
accuracy of the approximate expression $B(T) = B^{(0)} + B^{(1)}\tau$ assumed for the Planck function $B(T)$ in the calculations of the
theoretical curves of growth (τ is the optical depth in the
continuum). The ratio $B^{(0)}/B^{(1)}$ is calculated for different
values of effective temperature and surface gravity in the
atmosphere. The average optical depth of line formation τ_W
was computed as a function of the ratio of the absorption
coefficient at the center of the line to the continuous absorption coefficient. Calculations were made for $B^{(0)}/B^{(1)} = 0$,
1/3, 2/3, 4/3 and $a = 10^{-3}, 10^{-2}, 10^{-1}$, where a is the parameter connected with the damping constant. On the basis of
computations of the τ_W values and with the aid of model
atmospheres the author estimates the maximum scattering
of temperature and electron pressure obtained from different
lines.

**064.006 Non-LTE line formation in the presence of magnet-
ic fields.**
L. H. Auer, J. N. Heasley, L. L. House.
Astrophys. J., Vol. 216, 531 - 539 (1977).
The equations of radiative transfer and statistical equilibrium in the presence of a magnetic field are presented. A
general difference equation scheme for solving the vector
transfer equation in Stokes parameters, allowing for arbitrary
variations of the magnetic field and other quantities, is described. The solution of the Stokes non-LTE problem for
Ca II by the complete linearization method is described, and
numerical examples of the procedure are presented.

**064.007 Effects of CNO abundances on the Balmer jump of
late-B horizontal-branch stars.**
L. H. Auer, P. Demarque.
Astrophys. J., Vol. 216, 791 - 796 (1977).
Model stellar atmospheres have been constructed to investigate the effects of the continuous opacity of C I, N I,
and O I on the Balmer jump of late-B horizontal-branch stars.
The models provide a natural explanation for the variations
in the Balmer jump which have been observed from star to
star by Oke near 12,500 K on the horizontal branch of the
globular cluster M92. The authors conclude that some stars
on the horizontal branch of M92 have an atmospheric C
and/or N content as much as 1000 times the cluster average.
Although primordial chemical inhomogeneities within the
cluster cannot be ruled out, the most plausible interpretation
for these objects seems to be one in which surface enrichment
occurs through mixing from a helium-burning region in the
interior during the course of stellar evolution. The effects of
C I and N I opacities may also explain Newell's gap 1 in the
$(U - B, B - V)$-diagram for field blue stars in the halo.

**064.008 On the electron concentration in the photospheres
of F–G–K type stars.**   R. A. Epremyan.
Astrofizika, Vol. 12, 647 - 656 (1976). In Russian. – English
translation in Astrophysics, Vol. 12, No. 4.
An empirical relationship between the parameter
$Q[= W(2755 \text{ Fe II})/W(2967 \text{ Fe I})]$ and spectral class is derived using equivalent-width data for the absorption lines 2755
Fe II and 2967 Fe I obtained from the Orion 2 shortwave
spectral images of more than 60 F0–K2 type stars. With the
help of Q the electron concentration in the photospheres of
the F0–K2 stars is obtained. The electron concentration is
large in F0 stars and falls rapidly in stars of late types. A
comparison of the obtained results with data of other authors
is given.

**064.009 Limb darkening coefficients for cool model stellar
atmospheres.**
A. Manduca, R. A. Bell, B. Gustafsson.
Bull. American Astron. Soc., Vol. 9, 433 (1977). – Abstract.

**064.010 Contribution to spectral line formation in moving
stellar envelopes. Radiation field and statistical
equilibrium equations.**   J. Surdej.
Astron. Astrophys., Vol. 60, 303 - 311 (1977).
Using the escape probability method introduced by
Sobolev the author derives the mean intensity of the radiation
field at any point of an envelope which expands spherically
with a positive or negative radial velocity gradient. A general
discussion of the net radiative rates which populate the levels
of an atom relates the problem of spectral line formation in
a moving medium to the one in a transparent atmosphere at
rest.

064.011 C I as a source of continuous opacity in late B-type

main-sequence stars.   M. A. J. Snijders.
Astron. Astrophys., Vol. 60, 377 - 388 (1977).

The far UV continuous opacity of C I in late B-type main-sequence stars is calculated in LTE and non-LTE. In LTE C I is an important source of continuous opacity below 1240 Å for stars of spectral type B5 and later. In non-LTE the C I opacity is always much less and is important only for stars of spectral type B8 or cooler. The non-LTE predictions for the emergent flux shortward of 1240 Å are in reasonable agreement with the observations.

064.012   Thermal conductivity in stellar atmospheres II,
            without magnetic field.
T. Nowak, P. Ulmschneider.
Astron. Astrophys., Vol. 60, 413 - 416 (1977).

The authors calculate on the basis of the Chapman-Enskog-Burnett theory the coefficient of thermal conductivity for a gas mixture of stellar abundances assuming departures from local thermodynamic equilibrium. In addition, using the elementary kinetic theory, simple approximation formulas for the coefficient of thermal conductivity are given that allow evaluation under arbitrary ionization conditions and arbitrary element abundances.

064.013   Continuous spectra of Wolf-Rayet stars.
            V. I. Stebnev.
Kazan. khim.-tekhnol. inst., Kazan. inzh.-stroit. inst. Kazan', 1977. 35 pp. In Russian. — Abstr. in Ref. zh., 51. Astron., 8.51.577 (1977).

064.014   Some properties of very low temperature, pure
            helium surface layers of degenerate dwarfs.
K.-H. Böhm, T. R. Carson, G. Fontaine, H. M. Van Horn.
Astrophys. J., Vol. 217, 521 - 529 (1977).

Models of the atmospheres and envelopes of degenerate dwarfs with pure He surface layers have been constructed using a hot Thomas-Fermi calculation of the opacity and equation of state. The models have $\log g = 8$ and $T_{eff} = 3000$ (250) 6000 K, include the effects of convection even on the atmospheric structure at $\bar{\tau} \ll 1$, and extend, inward to the point where $\rho = 10^3 \mathrm{g\ cm^{-3}}$. Because of the high transparency, the densities in these atmospheres are so high that even the photospheres of the coolest models are degenerate.

064.015   Line-distortion effects in OB supergiant X-ray
            binaries.   J. B. Hutchings.
Astrophys. J., Vol. 217, 537 - 542 (1977) = Dominion Astrophys. Obs. Contrib. No. 329 = NRC No. 15611.

Calculations have been made of mean line profiles of Si IV, O II, He I, and H for OB supergiants which fill their Roche lobes in a binary system. Radial-velocity distortions and line-strength variations are presented and generalized for the T_{eff} range 25,000–40,000 K. The results are discussed in connection with the five known OB supergiant X-ray binaries.

064.016   Spatial heterodyne interferometry of VY Canis
            Majoris, Alpha Orionis, Alpha Scorpii, and R
Leonis at 11 microns.   E. C. Sutton, J. W. V. Storey,
A. L. Betz, C. H. Townes, D. L. Spears.
Astrophys. J., Lett., Vol. 217, L97 - L100 (1977).

Using the technique of heterodyne interferometry, measurements have been made of the spatial distribution of 11 micron radiation from four late-type stars. The circumstellar shells surrounding VY CMa, α Ori, and α Sco were resolved, whereas that of R Leo was only partially resolved at a fringe spacing of 0."4.

064.017   On the motion of thermal waves in polytropic en-
            velopes of stars.
V. S. Imshennik, I. A. Klimishin, I. V. Otroshchenko.
Astrofizika, Vol. 13, 103 - 115 (1977). In Russian. English translation in Astrophysics, Vol. 13, No. 1.

The one-dimensional, nonlinear problem of strong thermal wave propagation in a nonuniform medium with polytropic law density distribution is discussed. The ratio of gas energy density to radiation energy density is not limited. An approximate solution of the problem is found. The criterion of thermal wave into shock wave transformation is examined. The results are applied to the analysis of thermal wave propagation in nova envelopes.

064.018   Theoretical chromospheres of late type stars. I.
            Acoustic energy generation.   A. Renzini,
C. Cacciari, P. Ulmschneider, F. Schmitz.
Astron. Astrophys., Vol. 61, 39 - 45 (1977).

The authors discuss the acoustic energy generation in stars. For the chemical composition $X = 0.7$, $Y = 0.28$ and $Z = 0.02$ they have computed the acoustic flux and the period of the acoustic waves for a number of stellar surface gravities and effective temperatures in the range $-1 \lesssim \log g \lesssim 6$ and $3.45 \lesssim \log T_{eff} \lesssim 4.00$. The whole set of computations was repeated for three values of the ratio of the mixing length to the pressure scale height.

064.019   Non-thermal velocities in stellar atmospheres. II.
            Six giants and the Sun.   L. G. Stenholm.
Astron. Astrophys., Vol. 61, 155 - 159 (1977).

The photospheric small-scale velocity fields of six giants are determined and compared with each other and with the velocity fields of the Sun and Arcturus. The Sun is reconsidered and the previously derived disk-center velocity curve is compared with the velocity curve derived from the integrated solar disk.

064.020   On the absolute scale of mass-loss in red giants. I.
            Circumstellar absorption lines in the spectrum of
the visual companion of α^1 Her.   D. Reimers.
Astron. Astrophys., Vol. 61, 217 - 224 (1977).

Mass-loss from α^1 Her (M5 II) is rediscussed on the basis of new measurements and observations of circumstellar absorption lines in the spectrum of its visual companion α^2 Her. The rate of mass-loss is $2.7 \times 10^{-8} M_\odot/\mathrm{yr}$. Masses, luminosities and evolutionary stages of the components of α Her are discussed.

064.021   Structure and evolution of wind-driven circumstellar
            shells.   R. McCray.
Topics in interstellar matter, (see 012.011), p. 35 - 44 (1977).

The ultraviolet spectrometer on the Copernicus spacecraft has shown that almost all early-type stars with bolometric magnitude $M_v \lesssim -6$ have broad stellar resonance lines of N V λ1240 and O VI λ1035 that indicate a stellar wind. It was these stellar lines rather than the interstellar lines that led to the theory (Castor, McCray and Weaver 1975) of "bubbles" of hot gas surrounding the early type stars. The author finds that the structure of the interstellar bubble is similar to that of a supernova shell.

064.022   Stellar opacity.   T. R. Carson.
            Annu. Rev. Astron. Astrophys., Vol. 14, (see 003.008), 95 - 117 (1976).

Contents: Basic theory; Opacity calculations; Atomic absorption; Molecular absorption; Conduction; New opacities.

064.023   Linear polarization of Hα in the Be star Gamma
            Cassiopeiae.   R. Poeckert, J. M. Marlborough.
Astrophys. J., Vol. 218, 220 - 226 (1977).

The wavelength dependence of linear polarization across Hα has been measured in the Be star γ Cas. The polarization decreases toward the line center. The authors have also found a slight change in position angle of polarization across the line which is asymmetric with respect to the line center. Such position-angle changes are not due to interstellar polarization. The authors present model calculations which indicate that such position-angle changes are to be expected from a rotating disklike envelope surrounding a Be star. Because the circumstellar envelope rotates about the star, the line absorption in the two wings of the line will occur in separate parts of the envelope. In the continuum the absorptive processes are the same in both halves of the envelope, and the polarization has some mean magnitude and direction. Introducing line absorption into either half of the envelope reduces the amount of scattered flux from that half of the envelope, and the result is a rotation of the position angle of polarization.

064.024   **Transfer of angular momentum in stationary disc accretion.**   V. G. Gorbatskij.
Astron. Zh. Akad. Nauk SSSR, Vol. 54, 1036 - 1040 (1977). In Russian. English Translation in Soviet Astron., Vol. 21, No. 5.
     The conditions of stationary disc accretion are considered. It is shown that accretion may be time-independent only if there are constant inflows of matter and momentum across the outer boundary of the disc. The total input of angular momentum at the boundary is transferred through the disc to the surface of the central star.

064.025   **Improved estimates of continuous absorption from the free–free transitions of C^-.**
T. L. John, R. J. Williams.
Mon. Not. R. Astron. Soc., Vol. 181, 483 - 487 (1977).
     Improved calculations of continuous absorption from free–free transitions of the negative carbon ion are made with continuum orbitals computed with configuration interaction target atom wave-functions. The authors fit the absorption coefficient to a simple analytical expression suitable for stellar atmosphere opacity computations.

064.026   **A test of the micro-macroturbulence model of non-thermal velocities.**   D. F. Gray.
J. R. Astron. Soc. Canada, Vol. 71, 401 (1977). – Abstract.

064.027   **Mass-loss effects on stellar evolution, I: 5 $M_\odot$ and 15 $M_\odot$ population I models.**
S. R. Sreenivasan, W. J. F. Wilson.
J. R. Astron. Soc. Canada, Vol. 71, 407 (1977). – Abstract.

064.028   **Behaviour of microturbulence with evolution.**
R. Foy.
Highlights of Astronomy, Vol. 4, Part II, (see 012.023), p. 115 - 118 (1977).
     The author emphasizes three main conclusions. (1) The microturbulence in cool stars is a real phenomenon and not a spurious effect. (2) The microturbulence is a good indication of the degree of evolution of giants. It allows to distinguish between field stars on the horizontal branch and the red giant branch. (3) The microturbulence in stellar photospheres of cool stars varies as the strength of the underlying convective zone: this supports the idea that microturbulence is induced from convection, presumably by the mechanism of overshooting.

064.029   **Boundary conditions with mass-loss: general considerations.**   R. N. Thomas.
Highlights of Astronomy, Vol. 4, Part II, (see 012.023), p. 143 - 154 (1977).

064.030   **Mass loss in stars of moderate mass by stellar winds and effects on the evolution.**   C. de Loore.
Highlights of Astronomy, Vol. 4, Part II, (see 012.023), p. 155 - 173 (1977).
     Contents: Introduction; Mechanical fluxes; Heating of the corona; X-ray fluxes expected from stellar coronas; Observations of stellar coronas; Mass loss by stellar winds; Mass loss in red giants and supergiants; Evolutionary implications; Mass loss in T Tauri stars.

064.031   **Boundary condition with mass loss: the radiatively-driven wind model.**   D. Mihalas.
Highlights of Astronomy, Vol. 4, Part II, (see 012.023), p. 175 (1977). – Abstract.

064.032   **Stratification of elements in a quiet atmosphere: diffusion processes.**   G. Michaud.
Highlights of Astronomy, Vol. 4, Part II, (see 012.023), p. 177 - 191 (1977).
     Contents: Abundance anomalies: the signature of stability; Basic physics; Diffusion timescales; Diffusion and abundance anomalies in F, A and B stars; Stellar winds, line asymmetries and isotope anomalies.

064.033   **Competition between diffusion processes and hydrodynamical instabilities in stellar envelopes.**
G. Vauclair, S. Vauclair.
Highlights of Astronomy, Vol. 4, Part II, (see 012.023), p. 193 - 203 (1977).

064.034   **Decay of light elements in stellar envelopes.**
A. M. Boesgaard.
Highlights of Astronomy, Vol. 4, Part II, (see 012.023), p. 209 - 216 (1977).
     Decay of the light elements is caused by 1) nuclear destruction – including convective depletion, diffusion, convective overshoot, turbulence from rotational braking, meridional circulation and 2) dilution and 3) mass loss effects. The observations can be well understood in terms of these effects and give strong support for theoretical ideas on stellar structure and evolution.

064.035   **Accretion disk coronae and Cygnus X-1.**
E. P. T. Liang, R. H. Price.
Astrophys. J., Vol. 218, 247 - 252 (1977).
     It is plausible that an accretion disk pumps energy into an outer tenuous layer of its atmosphere to form a high-temperature corona analogous to the solar corona. General features of such a corona, not specific to any model, are discussed, and order-of-magnitude estimates are made for the physical parameters of a corona on the Cygnus X-1 disk. It is shown that a corona with a strong "wind" (mass loss flux) can produce hard ($\gtrsim 100$ keV) X-rays if it is cooled by inverse Compton scattering. An observational test for the presence of such a corona is suggested.

064.036   **Theoretical stellar chromospheres of late type stars. II. Temperature minima.**   P. Ulmschneider,
F. Schmitz, A. Renzini, C. Cacciari, W. Kalkofen, R. Kurucz.
Astron. Astrophys., Vol. 61, 515 - 521 (1977).
     The authors apply the theory of heating by short period acoustic waves to predict the height of shock formation and the acoustic flux at the base of the chromosphere for stars with T_{eff} = 4000 to 6500 K and log g = 2 to 4. These predictions are compared with heights of temperature minima and with chromospheric radiation losses computed from semi-empirical models.

064.037   **Interstellar bubbles. II. Structure and evolution.**
R. Weaver, R. McCray, J. Castor, P. Shapiro,
R. Moore.
Astrophys. J., Vol. 218, 377 - 395 (1977).
     The authors present the detailed structure of the interaction of a strong stellar wind with the interstellar medium. First they give an adiabatic similarity solution, which is applicable at early times. Second, they derive a similarity solution, including the effects of thermal conduction between the hot ($T \approx 10^6$ K) interior and the cold shell of swept-up interstellar matter. The authors then modify this solution to include the effects of radiative energy losses. They calculate the evolution of an interstellar bubble, including the radiative losses.

**064.038   On thermal waves in stars. III. Escape of a stellar
envelope.**   I. A. Klimishin, B. I. Gnatyk.
Problems of cosmic physics. Vyp. (No.) 12, (see 003.016),
p. 113 - 116 (1977). In Russian.

The paper deals with the numerical solution of the hydrodynamics equations describing the motion of strong thermal
waves in a stellar envelope. It is stated that after the thermal
wave's egression on the stellar surface there occurs an escape
of its envelope and the dynamical effects caused by a strong
shock or thermal wave are in the end approximately identical.

064.039   Pertes de masse. Modèles de vents stellaires.
N. Berruyer.
Perte de masse des étoiles, (see 012.030), 49 pp. (1977).

064.040   Perte de masse des étoiles OB.   B. Lazareff.
Perte de masse des étoiles, (see 012.030), 4 pp.
(1977).

**064.041   Émission infrarouge comme indicateur d'enveloppes
circumstellaires.**
J. L. Puget, R. Courtin, R. Gispert.
Perte de masse des étoiles, (see 012.030), 27 pp. (1977).

**064.042   LTE line blanketing of an early B star atmosphere by
ultraviolet lines using accurate damping constants.**
A. P. Phillips.
Mon. Not. R. Astron. Soc., Vol. 181, 777 - 788 (1977).

Most previous models for O and B star atmospheres including, explicitly, the bound—bound opacity of the strongest
ultraviolet lines have used a damping constant of 10 times the
classical value, which has subsequently been found to be much
too large. In order to give a more realistic estimate of the
blanketing effect of the strongest lines an LTE model atmosphere is presented for T_e = 21914 K (θ_e = 0.23), log g = 4.0,
including the strongest 97 lines using more accurate (radiative +
electron) damping constants. A comparison is made with a
model in which 10 times the classical damping constant was
used for the metallic lines.

**064.043   Meridional circulation in the surface layers of
rotating stars.**   B. L. Smith, I. W. Roxburgh.
Astron. Astrophys., Vol. 61, 747 - 754 (1977).

The influence of self-inertia on the meridional circulation in the radiative envelopes of rotating early-type stars is
examined. It is found that the inertia of the circulation becomes important in a shallow subsurface boundary layer of
relative depth $(\alpha\epsilon^2)^{1/10}$, where α is approximately the ratio
of centrifugal to gravitational force, and ϵ the ratio of the
free-fall time to the Kelvin-Helmholtz contraction time. The
structure equations appropriate to the boundary layer are
derived using stretched variable techniques, and qualitative
features of the flow discussed. There is found to be minimal
deviation from the Von Zeipel gravity darkening law through
the boundary layer for all rotation speeds.

**064.044   Molecular emission from expanding envelopes
around evolved stars. I. Nonmaser SiO emission
lines.**   M. Morris, C. Alcock.
Astrophys. J., Vol. 218, 687 - 696 (1977).

The authors consider the nonmaser emission from SiO
in oxygen-rich circumstellar envelopes. They show that the
"thermal" SiO lines are ultimately excited by the 8 μm infrared radiation field rather than by collisions, and describe how
the excitation process works. The detailed radiative transfer
calculations for molecules in a spherically symmetric, expanding, circumstellar envelope are described, the resulting line
profiles are presented, and these are compared to the existing
observations.

064.045   Grain-gas interaction in envelopes of red giants.
W. J. Maciel.

Rev. Brasil.Fis., Vol. 6, 459 - 469 (1976). — Abstr. in Phys.
Abstr., Vol. 80, Abstr. 60760 (1977).

**064.046   Limb darkening coefficients for late-type giant
model atmospheres.**
A. Manduca, R. A. Bell, B. Gustafsson.
Astron. Astrophys., Vol. 61, 809 - 813 (1977).

Limb-darkening coefficients for *UBV* and *uvby* band-
passes have been calculated for selected late-type giant and
supergiant model atmospheres from the grid given by Bell et
al. (1976). These calculations have been carried out at μ = 1.0,
0.8, 0.6, 0.4, 0.2 and 0.1 and the synthetic spectra, on which
the colors are based, were calculated at 1 Å resolution. The
degree of limb darkening decreases with increasing temperature and with increasing wavelength. The calculations are used
to study the effect of limb darkening on existing solutions
for the cool giant eclipsing binaries RZ Cnc and AR Mon
(Popper, 1976). Other possible applications are discussed.

**064.047   Frequency dependent diameters and atmospheric
structure of late-type supergiants.**   M. S. Vardya.
Astrophys. Space Sci., Vol. 47, L15 - L16 (1977).

A method is suggested for using the measured frequency
dependent diameters to test the atmospheric structure of late-
type supergiants. The efficiency of the method vis-à-vis limb-
darkening measurements is discussed.

**064.048   The interpretation of cyclical photometric variations
in certain dwarf Me-type stars.**   E. Budding.
Astrophys. Space Sci., Vol. 48, 207 - 223 (1977).

A method of determination of parameter sets characterizing models of starspots is described. The method makes use
of a systematic integral notation in the description of the
darkening due to spots and optimization procedures to evaluate appropriate parameters. The method is applied to light
curves of YY Gem and CC Eri. The physical meaning of the
derived parameter set and possibly correlated effects are considered for YY Gem.

**064.049   Equivalent width of molecular lines in stars. III:
Lines of Lyman band of H_2 and (A—X) band of
CO and SiO in stars.**   K. S. Krishna Swamy.
Astrophys. Space Sci., Vol. 49, 389 - 398 (1977).

The author has calculated the expected equivalent widths
of the individual rotational lines of the Lyman band of H_2 and
(A—X) band of CO and SiO for Main Sequence stars. The
results indicate that the lines of H_2 should be observable in
absorption up to $T_e \lesssim$ 9000 K. The lines of CO are found to
be much weaker than those of H_2 lines. A discussion of these
results is presented.

064.050   Structure of the atmospheres of late-type stars. II.
J. I. Straume.
Investigations of the sun and red stars. 6, (see 003.019),
p. 23 - 42 (1977). In Russian.

The role of Rayleigh scattering on H, H_2 and free electrons in the atmospheres of cool stars is investigated. For this
purpose model atmospheres of late-type stars for T_e = 2500,
3000 °K and log g = 0, 2, 4 and for T_e = 3500, 4000 °K and
log g = 1, 3, 5 are constructed.

**064.051   Ejection of matter from stars in early stages of
evolution.**
G. S. Bisnovatyj-Kogan, S. A. Lamzin.
Early stages of stellar evolution, (see 012.041), p. 107 - 118
(1977). In Russian.

The observational facts on the mass loss by gravitationally
contracting stars as well as the theoretical works on this
problem are given. Self-consistent models for freely escaping
hydrogen isentropic stars were constructed. Correspondence
between theory and observations as well as the evolutionary
consequence for young stars are discussed.

064.052 Detection of disc-like gas-dust stellar envelopes.
V. S. Strel'nitskij.
Early stages of stellar evolution, (see 012.041), p. 118 - 127 (1977). In Russian.

Observational evidence is discussed of the existence of cold oblate gas-dust envelopes around young stars. Probably there are such envelopes around the stars illuminating M78 in Orion, the "Egg Nebula" in Cygnus, the cometary nebulae, and also around cold supergiant "infra-red stars" with microwave OH and H_2O maser emission. One object (M78) gives evidence of a strong azimuthal magnetic field, in some others one observes an expansion of gas.

064.053 Balmer line profiles in decelerating atmospheres.
T. J. Schneeberger, H. A. Beebe.
Proc. Southwest Reg. Conf., Vol. 3, (see 012.043), p. 69 - 79 (1977).

The Sobolev approximation has been used to compute level populations and line source functions for the hydrogen Balmer lines in model envelopes with decelerating flows. T Tauri like envelope conditions are used. The resulting line profiles are examined for a range of envelope parameters.

064.054 Resonance line polarization in finite atmospheres.
D. E. Rees.
Rep. Obs. Lund, No. 12, (see 012.044), p. 25 - 34 (1977).

Some polarization effects are discussed for resonance lines formed in plane-parallel atmospheres of finite optical thickness. Magnetic fields and collisional depolarization are neglected.

064.055 Non-LTE line formation in a magnetic field. The two-level atom with a frequency independent source function. I: Formulation.   E. Landi degl'Innocenti.
Rep. Obs. Lund, No. 12, (see 012.044), p. 35 - 49 (1977).

The formation of a Zeeman-triplet in a plane parallel atmosphere with a homogeneous magnetic field is considered in the hypothesis of complete redistribution in frequency. The problem is reduced to the solution of a system of coupled integral equations for the source functions of the various Zeeman sublevels.

064.056 A rational visualized representation and analytical fit formula to emergent continuous stellar fluxes.
A. Sapar, I. Kuusik.
Publ. Tartu Astrofiz. Obs., Vol. 45, 211 - 256 (1977).

Two methods of highly visualized representation of emergent stellar fluxes are presented. For the emergent continuous stellar fluxes of different continua a rational analytical approximation formula for O and B stars has been established, using the effective temperature, surface gravity, hydrogen abundance and frequencies as arguments. The formula has been constructed by the method of least squares from a grid of model atmospheres. The results are compared with observations.

064.057 Models of outflowing envelopes of T Tau stars.
G. S. Bisnovatyj-Kogan, S. A. Lamzin.
Astron. Zh. Akad. Nauk SSSR, Vol. 54, 1268 - 1280 (1977). In Russian. English translation in Soviet Astron., Vol. 21, No. 6.

A model of the envelope of T Tauri stars which consists of hot ($T \sim 10^6 \, ^\circ K$) gas outflowing similarly to the solar wind and of cold clouds ($T \sim 2 \times 10^4 \, ^\circ K$) moving with decreasing velocities is considered. The continuum emission is determined by the corona, and hydrogen and metal emission lines are determined by cold clouds. The formation of clouds is due to thermal instability of the coronal gas. The observed delay in the excess of line emission compared to the excess of continuum emission is determined by the cooling time of coronal condensations. It follows from the model that the flux of non-thermal mechanical energy from the star may exceed the photospheric flux from such stars. A considerable flux of soft X-ray emission is predicted. The correspondence of the model proposed with observations is discussed and some observational tests are proposed.

064.058 Atmosphärische Eigenschaften der Ap-Sterne.
H. Muthsam.
Mitt. Astron. Ges., Nr. 42, p. 137 (1977).

064.059 Studies of dust shells around stars.   P. J. Bedijn.
Proefschrift Rijksuniv. Leiden, Netherlands, 91 pp. (1977). – Abstr. from INIS7712 311828.

064.060 Shells around stars.   F. M. Olnon.
Proefschrift Rijksuniv. Leiden, Netherlands, 81 pp. (1977). – Abstr. from INIS7712 311881.

064.061 Chemistry of protostar envelopes: chemical reaction systems, ionization by cosmic radiation, radiation transport by molecule transitions.   G. Hertel.
Diss. Max-Planck-Inst. Phys. Astrophys., München, F. R. Germany. MPI–PAE/Astro–107. 115 pp. (1977). In German. Abstr. from INIS7722 339255.

Numerical solution of continuity equations.
See Abstr. 021.013.

Broadening of spectral lines by electron scattering. I. Methods of calculation.   See Abstr. 022.007.

Thomson scattering in a thick layer.
See Abstr. 063.034.

The Eddington approximation generalized for radiative transfer in spherically symmetric systems. II. Nongray extended dust-shell models.   See Abstr. 063.048.

Mass loss towards the white dwarf stage.
See Abstr. 065.001.

The evolution of population II stars and mass loss and stellar evolution.   See Abstr. 065.017.

Evolution of massive stars with mass loss by stellar wind.   See Abstr. 065.021.

Observational evidence for atmospheric physical characteristics relevant to stellar evolution.
See Abstr. 065.027.

Observational evidence for atmospheric chemical composition peculiarities relevant to stellar evolution.
See Abstr. 065.028.

Sensitivity of internal structure to the surface boundary condition.   See Abstr. 065.029.

Évolution stellaire et perte de masse.
See Abstr. 065.036.

Basic concepts in the theory of stellar atmospheres.
See Abstr. 065.039.

Großräumige Plasmabewegungen auf Sternen.
See Abstr. 080.030.

Convection in a rotating deep compressible spherical shell: application to the Sun.   See Abstr. 080.068.

Carbon and nitrogen abundances in F- and G-type stars.   See Abstr. 114.020.

Neutral-ion anomaly in cool stars.
See Abstr. 114.022.

Long-term changes in ultraviolet P Cygni profiles observed with Copernicus. See Abstr. 114.024.

Diffusion, turbulence and abundance anomalies in F, B and A stars. See Abstr. 114.029.

Model atmosphere analysis of HR 8799.
See Abstr. 114.502.

Observed departures from LTE ionization equilibrium in late-type giants. See Abstr. 114.503.

Comparison of predicted and observed spectral energy distributions of A-type stars. See Abstr. 114.514.

High-resolution rocket spectra of the λ1920 and λ1720 features in the spectrum of Zeta Tauri.
See Abstr. 114.521.

A spectroscopic study of 14 Comae and other A-type shell stars. See Abstr. 114.526.

Observation of preplanetary disks around MWC 349 and LkHα 101. See Abstr. 114.535.

On the near-infrared excesses of very cool supergiants. See Abstr. 114.536.

A sensitive observation of the far-ultraviolet (1160–1700 Å) spectrum of Arcturus and implications for its outer atmosphere. See Abstr. 114.537.

High-resolution optical observations of Ca II K in Deneb and Aldebaran. See Abstr. 114.538.

Spectroscopic study of two O-type supergiants, Alpha Camelopardalis and 19 Cephei: model-atmosphere analysis. See Abstr. 114.539.

The envelope of Pleione in the new shell phase, 1972 - 1975. See Abstr. 114.540.

Radio emission from mass-outflow stars.
See Abstr. 116.002.

A magnetic field interpretation for the outburst of CH Cygni. See Abstr. 116.016.

The application of a Bessel transform to the determination of stellar rotational velocities.
See Abstr. 116.018.

The evolution of low-mass close binary systems. V. Transport processes in the envelopes of contact components.
See Abstr. 117.001.

The massive hot binary 29 Canis Majoris.
See Abstr. 117.035.

Study of the physical conditions in the envelope of the dwarf nova SS Cyg. I. See Abstr. 122.046.

Relative abundances of isotopes of Zr in R Cygni.
See Abstr. 122.113.

Chromospheres of flare stars.
See Abstr. 122.121.

Hydrodynamic effects in the atmospheres of variable stars. See Abstr. 122.161.

Radiating cosmic dust. See Abstr. 131.035.

Encounters between stars and dense interstellar clouds. See Abstr. 131.067.

The protostellar origin of interstellar grains.
See Abstr. 131.133.

Formation and destruction of dust grains.
See Abstr. 131.158.

The circumstellar envelope of IRC + 10216.
See Abstr. 133.001.

Water maser and envelope of infrared stars.
See Abstr. 133.017.

Coherent amplification and pulsar phenomena.
See Abstr. 141.533.

Large-scale winds driven by flare-star mass loss.
See Abstr. 151.046.

065 Stellar Structure and Evolution

065.001   **Mass loss towards the white dwarf stage.**
          V. Weidemann.
Astron. Astrophys., Vol. 59, 411 - 418 (1977).

It is shown that there are several cases in which mass loss towards the white dwarf stage has been larger than expected by current theory, i.e. within the framework of stellar evolution for single stars without rotation, combined with mass loss formulae. Evidence is presented for differential mass loss on a much larger scale than hitherto assumed e.g. for the explanation for the horizontal branch. It is suggested that angular momentum is correlated with mass loss: large values of both lead to white dwarfs as the final stage of evolution, even for massive stars up to about $10 \, M_\odot$, whereas only in cases of small initial angular momentum does the evolution follow the conventional picture and lead to supernova explosions for intermediate mass stars.

065.002   **On the origin of R-type carbon stars: possibility of hydrogen mixing during helium flicker.**
M. Y. Fujimoto.
Publ. Astron. Soc. Japan, Vol. 29, 331 - 349 (1977).

A possible origin of R-type carbon stars first suggested by Schwarzschild and Härm (1967) relies upon the mixing which is caused by the contact of the helium convective zone with the hydrogen-rich envelope during the thermal pulse. Thermal pulses were computed with changing conditions in the envelope for the cores of mass $1.26 \, M_\odot$ and $1.07 \, M_\odot$, and the conditions required to trigger the hydrogen mixing was sought. In the author's computation, such mixing has been realized only when the mass fraction of the hydrogen-rich envelope is as small as $10^{-5} - 10^{-6}$.

065.003   **Lane-Emden equation and Padé approximants.**
          P. Pascual.
Astron. Astrophys., Vol. 60, 161 - 163 (1977).

Using the technique of Padé approximants the author presents approximate analytical solutions of the Lane-Emden equation valid for any value of n.

065.004   **Core helium flash and the origin of CH and carbon stars.**   B. Paczyński, S. C. Tremaine.
Astrophys. J., Vol. 216, 57 - 60 (1977).

Off-center core helium flash is studied in a model of a Population II star ($M = 0.8 \, M_\odot$, $Y = 0.299$, $Z = 0.001$). The mass fraction at which helium ignites is taken as a free parameter. An evolutionary sequence with helium ignition at $M_r = 0.40 \, M_\odot$ leads to penetration of the helium core by the convective envelope soon after the peak of the helium flash. The surface carbon abundance is increased to $X_{12} = 0.013$. The authors suggest that this process may explain the carbon overabundance in CH stars and R stars.

065.005   **Evolution of helium stars.**
          R. Stothers, C.-w. Chin.
Astrophys. J., Vol. 216, 61 - 66 (1977).

The evolution of helium stars in the mass range $4-15 \, M_\odot$ has been followed from the initial helium main sequence to the end of carbon burning in the core, with the use of Carson's new radiative opacities. As compared with earlier work based on smaller opacities, the main-sequence band in the H-R diagram is now both wider and cooler than before. If neutrino losses are neglected in the stellar models, the phase of carbon burning in the core occurs in the red-supergiant region; otherwise, it occurs, as it does in the earlier models with or without neutrino emission, close to the helium main sequence. Observational data for Wolf-Rayet stars and R Coronae Borealis variables are found to lend some support to the new models.

065.006   **Light and radial velocity variations in a nonradially oscillating star.**   W. Dziembowski.
Acta Astron., Vol. 27, 202 - 211 (1977).

Formulae for light and radial velocity variations are derived in terms of solution of the linear oscillations equations. A possibility of observational identification of the oscillation mode is discussed. It is shown that nonradial oscillations cannot be distinguished from the radial ones on the basis of the amplitude ratios of radial velocity and light variations. Instead, the use of the Baade-Wesselink method is recommended. For stars with known radii, this method permits to determine the spherical harmonic order of the oscillation.

065.007   **Linear series of stellar models. VI. Supermassive stars.**   B. Paczyński, M. Rózyczka.
Acta Astron., Vol. 27, 213 - 223 (1977).

Linear series of stellar models on hydrogen burning main sequence is constructed for Population I composition ($X = 0.7$, $Y = 0.27$, $Z = 0.03$) for masses in the range $1.8 \leqslant \log M/M_\odot \leqslant 12.0$. No terminating point is found in the supermassive star region. Models with $\log M/M_\odot > 5.7$ are known to be thermally unstable even though there is no turning point of the linear series. This phenomenon is caused by the onset of dynamical instability at the same mass.

065.008   **Nonrotating superdense stars with frozen superstrong magnetic fields.**   G. A. Shul'man.
Redkollegiya zh. "Izv. vyssh. uchebn. zaved. Fiz." Tomsk, 1977. 16 pp. In Russian. – Abstr. in Ref. zh., 51. Astron., 8.51.484 (1977).

065.009   **Thermal stability of hydrogen-burning shells in white dwarfs.**   S. C. Vila.
Astrophys. J., Vol. 217, 171 - 174 (1977).

The author has constructed models for degenerate stars of $1 \, M_\odot$ in the range $-4 < \log (L/L_\odot) < 4$ and of $0.6 \, M_\odot$ in the range $-4 < \log (L/L_\odot) < 2$ with carbon-oxygen cores and hydrogen envelopes. These stars derive all their luminosity from hydrogen shell burning near their surfaces. He has investigated the thermal stability of these models and found them stable in all cases. This stability is greatest for the models with least luminosity.

065.010   **Linear radial and nonradial modes of oscillation of hot white dwarfs.**   A. J. DeGregoria.
Astrophys. J., Vol. 217, 175 - 180 (1977).

Low-order linear radial and nonradial quadrupole modes of oscillation of hot white dwarfs with hydrogen envelopes are evaluated in the quasi-adiabatic approximation. Luminosities range from 10^{36} to 10^{38} ergs s^{-1}. Masses range between 0.6 and $1.4 \, M_\odot$. Models are found with unstable radial modes as high as the third harmonic. The first nonradial quadrupole gravity mode is always unstable. The results may be relevant to pulsating X-ray sources, in general, and are applied, specifically, to Cen X-3, Her X-1, and SMC X-1.

065.011   **Neutron-capture nucleosynthesis in the helium-burning cores of massive stars.**
S. A. Lamb, W. M. Howard, J. W. Truran, I. Iben, Jr.
Astrophys. J., Vol. 217, 213 - 221 (1977).

The authors investigate the consequences for nucleosynthesis of the neutron release from the nuclear reaction $^{22}Ne(\alpha, n)^{25}Mg$ when it occurs in the convective helium-burning cores of massive stars. They follow the neutron-induced buildup from an original solar system distribution for the cores of 9, 15, and $25 \, M_\odot$ stars which have an initial composition $Y = 0.28$, $Z = 0.02$, and show that significant enhancements in nuclear abundances up to atomic mass number

$A \approx 90$ can be achieved in all but the lightest of the three models. The results may be important for considerations of the origin of galactic cosmic rays and as input to detailed supernova nucleosynthesis calculations.

065.012 **The use of collocation for the construction of stellar models.** R. L. Smith, H. Fuchs.
News Lett. Astron. Soc. N.Y., Vol. 1, No. 2, p. 15 (1977). Abstract.

065.013 **Neutrino transport in supernova models: S_N method.** W. R. Yueh, J. R. Buchler.
Astrophys. J., Vol. 217, 565 - 577 (1977).
The equations governing the transport of neutrinos in a collapsing stellar core are solved numerically. Neutrino-matter coupling is treated within the framework of Weinberg's neutral current model, with various scalings of the rates to allow for the existing experimental uncertainties. Special care is devoted to neutrino-electron Compton scattering and to the numerical conservation of neutrinos. Neutrino stresses, luminosities, and energy transfer rates to matter are computed in realistic collapsing stellar cores.

065.014 **Modèles d'évolution stellaire avec overshooting des noyaux convectifs.** A. Maeder.
Recueil des séminaires, (see 012.004), p. 14 - 19 (1977).

065.015 **On the removal of degeneracy and approximations of an ideal electron gas in a cool dense star with frozen magnetic field.** G. A. Shul'man, V. S. Sekerzhitskij.
Astrofizika, Vol. 13, 165 - 172 (1977). In Russian. English translation in Astrophysics, Vol. 13, No. 1.
The Fermi level of a cold relativistic electron gas in a superstrong magnetic field is calculated. It is shown that the presence of a superstrong magnetic field leads to removal of its degeneracy. The strength of the magnetic field capable to remove the electron gas degeneracy is determined under the assumption that the mass density corresponds to the threshold of the emergence of free neutrons.

065.016 **Stellar structure and evolution with emphasis on the evolution of intermediate mass stars.** I. Iben.
Advanced stages in stellar evolution, (see 012.008), p. 3 - 148 (1977).
Order of magnitude estimates; Solar models and neutrino fluxes; Classical stellar evolution with emphasis on composition changes; Pulsations and evolution; Thermal pulses in stars of intermediate mass; Production of s-process elements in thermally pulsing stars and in massive stars; On the contribution of low-mass and intermediate-mass stars to galactic nucleosynthesis; Urca processes in carbon-oxygen cores of asymptotic-branch stars.

065.017 **The evolution of population II stars and mass loss and stellar evolution.** A. Renzini.
Advanced stages in stellar evolution, (see 012.008), p. 151 - 283 (1977).
Evolution of pop.II stars from the main sequence to the helium flash; The horizontal branch; The post-HB evolution; Comparison of theory with observations; Mass loss in red giants and supergiants; Evolution of population II stars with mass loss and core rotation; Mass loss in advanced evolutionary stages of pop.I stars.

065.018 **Nucleosynthesis and the later stages of stellar evolution.** D. N. Schramm.
Advanced stages in stellar evolution, (see 012.008), p. 284 - 363 (1977).
Presupernova evolution of massive stars; Explosive nucleosynthesis; The origin of the elements from C to Ni; The origin of the heavy elements past the iron peak (n-processes); Alternative supernova mechanisms; Gravitational core collapse to neutron stars or black holes; Summary of supernova models and the origin of the elements.

065.019 **Eruptive phenomena in early stellar evolution.** G. H. Herbig.
Astrophys. J., Vol. 217, 693 - 715 (1977) = Lick Obs. Bull., No. 763.
The second and third members of the FU Ori class have now been discovered: V1057 Cyg and V1515 Cyg. FU Ori itself has since 1960 faded very slightly. All the spectra are much alike: F or G supergiants having wide absorption lines, P Cygni structure at Hα, displaced shell components, and strong Li I λ6707. Much detailed information is now available for V1057 Cyg, which is definitely known to have been a T Tau star before its outburst. The observed frequency of three eruptions in about 80 years, together with the known number of T Tau stars of that luminosity in the nearer associations, indicates that unless there is some fundamental misunderstanding of the situation, FU Ori-type eruptions are repetitive and recur in the average T Tau star after roughly 10^4 years. There is a speculative possibility that similar activity on a minor but more frequent level occurs in a few peculiar T Tau stars. The cause of the outbursts remains unknown.

065.020 **Thermal pulse and interpulse properties of intermediate-mass stellar models with carbon-oxygen cores of mass 0.96, 1.16, and 1.36 $M_\odot$.** I. Iben, Jr.
Astrophys. J., Vol. 217, 788 - 798 (1977).
The properties of stellar models in the asymptotic-branch phase of evolution are investigated as a function of carbon-oxygen core mass. For a model of total mass $M_s = 7 M_\odot$, and for core masses of $M_c/M_\odot = 0.96$, 1.16, and 1.36, the following properties are found: $L_s/10^4 L_\odot$ = mean luminosity during the interpulse phase = 3.2, 4.6, and 5.8; $L_H/10^4 L_\odot$ = mean hydrogen-burning luminosity during interpulse = 3.0, 4.2, and 5.0.

065.021 **Evolution of massive stars with mass loss by stellar wind.**
C. de Loore, J. P. De Grève, H. J. G. L. M. Lamers.
Astron. Astrophys., Vol. 61, 251 - 259 (1977).
The evolution of stars with initial masses of 20, 25, 30, 40 and 50 $M_\odot$ with mass loss are computed. The results are presented in evolution tracks, isochrones and lines of equal mass. They are compared with observations. The mass luminosity relation for supergiants indicates that mass loss has affected their evolution. The lack of very luminous late-B supergiants can only be explained if the mass loss rates are very high and the stars evolve from the beginning of the shell burning phase to the helium main sequence.

065.022 **Nonradial oscillations of stars: theories and observations.** J. P. Cox.
Annu. Rev. Astron. Astrophys., Vol. 14, (see 003.008), 247 - 273 (1976).
This paper is primarily concerned with a review of more general kinds of stellar oscillations than the purely radial, spherically symmetric type. Some general remarks on the nature and classification of nonradial oscillations of stars are made and the nonradial oscillations of spherical stars without rotation, magnetic fields, or other "perturbations" are considered. Some effects of certain "perturbations" on nonradial stellar oscillations are briefly discussed. Some applications of nonradial-oscillation theory to certain kinds of real stars are surveyed.

065.023 **CNO burning and the location of zero-age horizontal-branch stars.** V. Castellani, A. Tornambè.
Astron. Astrophys., Vol. 61, 427 - 431 (1977).
Zero-age horizontal-branch stars are revisited in the light of a detailed approach to the distribution of primary elements in the H-burning shell. The effect of varying the abundance of CNO elements in the original matter is studied in connection

with suggested interpretations of peculiarities in H-R diagrams of very metal-poor globular clusters.

065.024 Initiation of a collapse of degenerate carbon star cores by thermonuclear burning.
L. N. Ivanova, V. S. Imshennik, V. M. Chechetkin.
Astron. Zh. Akad. Nauk SSSR, Vol. 54, 1009 - 1026 (1977).
In Russian. English translation in Soviet Astron., Vol. 21, No. 5.

The evolution of carbon cores of stars with mass $M \approx 1.4\,M_\odot$ leads to a carbon flash under different possible values of central density. It is shown that for the central density of a star $\gtrsim 5 \times 10^9$ g/cm³ thermonuclear burning of the developing carbon flash leads to gravitational collapse with production of a neutron star. The results of calculations for two hydrostatically equilibrium models with initial central density 8.36×10^9 g/cm³ and 5.04×10^9 g/cm³ are presented. The results are compared with calculations of the collapse of an iron core of stars with masses $M \approx (1.2-2)\,M_\odot$.

065.025 On s-process nucleosynthesis during thermal pulses of helium shell-burning.   M. Y. Fujimoto.
Publ. Astron. Soc. Japan, Vol. 29, 537 - 541 (1977).

The neutron capture process during thermal pulses in intermediate-mass stars is investigated for the case of single irradiation, i.e., for the case when matter is exposed to neutrons only once. It is found that neutrons are produced through the $^{22}Ne(\alpha, n)^{25}Mg$ reaction in the case of a core mass larger than $1.26\,M_\odot$. The role of the neutron-producing reaction $^{18}O(\alpha, n)^{21}Ne$ is also discussed.

065.026 CNO cycle in convective regions.
H. Falk, R. Mitalas.
J. R. Astron. Soc. Canada, Vol. 71, 403 (1977). – Abstract.

065.027 Observational evidence for atmospheric physical characteristics relevant to stellar evolution.
R. Cayrel, G. Cayrel.
Highlights of Astronomy, Vol. 4, Part II, (see 012.023), p. 105 - 114 (1977).

The authors conclude by saying that atmospheric physical parameters are an essential tool in connecting observations with internal structure computations as they supply three of the fundamental parameters, (Z, T_{eff}, g). The ultimate goal is of course to combine these three parameters with other independent data (mass or absolute magnitude) in order to interpret or to check the theory of stellar evolution.

065.028 Observational evidence for atmospheric chemical composition peculiarities relevant to stellar evolution.   B. E. J. Pagel.
Highlights of Astronomy, Vol. 4, Part II, (see 012.023), p. 119 - 135 (1977).

Abundance peculiarities in successive stages of stellar evolution are reviewed. Main-sequence stars show anomalies in lithium and, on the upper main sequence, the Am, Ap and Bp effects, which may be largely due to separation processes, and helium and CNO anomalies to which nuclear evolution and mixing could have contributed. Red giants of both stellar Populations commonly show more or less extreme variations among the C, N, O isotopes, sometimes accompanied by s-process enhancement, due to mixing out in various evolutionary stages.

065.029 Sensitivity of internal structure to the surface boundary condition.   P. Demarque.
Highlights of Astronomy, Vol. 4, Part II, (see 012.023), p. 137 - 142 (1977).

In summary, one can say that proper surface boundary conditions for interior models of late-type stars require a detailed understanding of the structure of the stellar atmosphere.

065.030 Mixing between burned core material and surface layers.   M. Schwarzschild.
Highlights of Astronomy, Vol. 4, Part II, (see 012.023), p. 205 - 208 (1977).

065.031 Convection in stars.   E. Graham.
Computing in plasma physics and astrophysics, (see 012.028), p. 121 - 124 (1976).

This paper reviews work on stellar convection theory with a particular emphasis on numerical simulations of convection. The Boussinesq and the anelastic approximations of the fluid equations are introduced and solutions of the linearized equations for the onset of convection are described. Nonlinear numerical solutions to the Boussinesq equations are discussed. Next, the character of stellar convection is considered. Convection is found to be intimately linked with a number of processes in the stellar interior.

065.032 Early stages of stellar evolution.   S. E. Strom.
IAU Symp. No. 75, (see 012.029), p. 179 - 197, with a discussion p. 198 - 212 (1977).

The Herbig Ae and Be stars associated with nebulosity; The T Tauri stars; The Herbig-Haro objects; A scenario for pre-main-sequence evolution.

065.033 Does the initial mass function for star formation depend on epoch?   A. P. Whitworth.
Q. J. R. Astron. Soc., Vol. 18, 466 - 467 (1977). – Abstract.

065.034 Le sintesi di popolazioni stellari.
E. Milandri, F. Fusi-Pecci.
Coelum, Vol. 45, 173 - 189 (1977).

065.035 Éléments de structure interne.   A. Baglin.
Perte de masse des étoiles, (see 012.030), 45 pp. (1977).

065.036 Évolution stellaire et perte de masse.
E. Schatzman.
Perte de masse des étoiles, (see 012.030), 74 pp. (1977).

065.037 Global stability of the uniformly rotating gaseous discs.   F. Takahara.
Progr. Theor. Phys., Vol. 56, 1665 - 1667 (1976). – Abstr. in Phys. Abstr., Vol. 80, Abstr. 26000 (1977).

065.038 Red giant stars.   B. Johnson.
Astronomy, Vol. 4, No. 12, p. 26 - 31 (1976).
Abstr. in Phys. Abstr., Vol. 80, Abstr. 26010 (1977).

065.039 Basic concepts in the theory of stellar atmospheres.   E. Simonneau.
Fluid dynamics. R. Bialian, J.-L. Peube (Editors). Gordon & Breach Science Publishers Ltd., London. 14 + 677 pp. (1977). ISBN 0-677-10170-8. p. 657 - 666. In French. – Abstr. in Phys. Abstr., Vol. 80, Abstr. 87380 (1977).

065.040 The pulsational stability of models of normal and metallic-line A stars.   J. R. Percy.
Mon. Not. R. Astron. Soc., Vol. 181, 563 - 572 (1977).

The purpose of the present paper is two-fold: to compare the pulsation properties of the observed δ Scuti stars with the pulsation properties of models of normal A stars and to investigate the pulsational stability of models of Am stars, assuming these to be stars in which diffusion has taken place.

065.041 Neutrino trapping during gravitational collapse of stars.   W. D. Arnett.
Astrophys. J., Vol. 218, 815 - 833 (1977).

The beginning stage of core collapse, through initial neutronization up to core bounce, is examined. Neutral currents are treated according to the model of Weinberg (1967) and Salam (1968). Weak interaction processes are treated individually by reaction network techniques; therefore lepton number is properly conserved. Transport is treated as multigroup flux-limited diffusion; this approach is valid in both the thick and the thin limits. Scattering of neutrinos by electrons is considered to be "nonconservative" (neutrinos' energies can change). The hydrodynamic calculations begin from a stellar evolutionary model ($M_a = 8 M_\odot$).

065.042 The collapse of unstable isothermal spheres.
        C. Hunter.
Astrophys. J., Vol. 218, 834 - 845 (1977).
    Similarity solutions provide simple analytical descriptions of modes of collapse of gravitationally unstable isothermal spheres, both before and after a collapsed core has formed. A new class of similarity solutions has been found, to add to the solutions obtained earlier by Larson, Penston, and Shu. Numerical integrations of collapses were performed to study their resemblance to the similarity solutions. None of the collapses studied showed any strong tendency to a similarity solution. It was also found that perturbed gravitationally unstable spheres do not necessarily collapse; they may instead perform periodic oscillations of large amplitude.

065.043 A non-elliptic solution of the Lane-Emden equation for index n = 5.   V. D. Sharma.
Phys. Lett. A, Vol. 60A, 381 - 382 (1977). − Abstr. in Phys. Abstr., Vol. 80, Abstr. 41377 (1977).

065.044 A study on the periods of higher modes of pulsations.   M. Singh.
Indian J. Pure Appl. Math., Vol. 5, 609 - 615 (1974). − Abstr. in Phys. Abstr., Vol. 80, Abstr. 49372 (1977).

065.045 Energy loss accompanying the electron captures in highly evolved stars.
Y. Egawa, K. Yokoi.
Prog. Theor. Phys., Vol. 57, 1255 - 1261 (1977). − Abstr. in Phys. Abstr., Vol. 80, Abstr. 64752 (1977).

065.046 On deviations from linear wave motion in inhomogeneous stars.   P. J. Melvin.
Q. Appl. Math., Vol. 35, No. 1, pp. 75 - 99 (1977). − Abstr. in Phys. Abstr., Vol. 80, Abstr. 71001 (1977).

065.047 Experiment and theory relevant to explosive nucleosynthesis.   W. A. Fowler.
Proceedings of the 4th conference on scientific and industrial applications of small accelerators. Denton, Tex., USA, 27 - 29 October 1976. IEEE, New York. 3 + 611 pp. (1977). p. 11 - 14. − Abstr. in Phys. Abstr., Vol. 80, Abstr. 71110 (1977).

065.048 Static stars: some mathematical curiosities.
        C. B. Collins.
J. Math. Phys., Vol. 18, 1374 - 1377 (1977). − Abstr. in Phys. Abstr., Vol. 80, Abstr. 74045 (1977).

065.049 Terminal stages of stellar evolution in quantum chromodynamics.   M. B. Kislinger, P. D. Morley.
Phys. Lett. B, Vol. 69B, 257 - 260 (1977). − Abstr. in Phys. Abstr., Vol. 80, Abstr. 78773 (1977).

065.050 Advanced evolution of massive stars. VII. Silicon burning.   W. D. Arnett.
Astrophys. J., Suppl. Ser., Vol. 35, 145 - 159 (1977).
    The stage of stellar silicon burning has been examined by a variety of numerical and analytic techniques, with emphasis on delineating the underlying physics. In this investigation the stages of ignition, flash, and consumption of silicon are studied for helium cores of mass $M_a = 4, 6, 8, 12, 16, 24$, and $32 M_\odot$. Both hydrostatic and hydrodynamic techniques are employed. A detailed examination of physical characteristics and time

scales is made. A thermonuclear problem arises because silicon burning occurs at density so high that electron capture is important. A hydrodynamic problem appears in two aspects: (1) the nature of evolution from hydrostatic to hydrodynamic (flash) conditions (an intermediate pulsational regime may occur), and (2) convection.

065.051 Equations of state for stellar partial ionization zones.
G. Fontaine, H. C. Graboske, Jr., H. M. Van Horn.
Astrophys. J., Suppl. Ser., Vol. 35, 293 - 358 (1977).
    A composite numerical equation of state has been developed to compute thermodynamic surfaces covering the density-temperature domain of interest for stellar partial ionization zones. Emphasis has been placed on the thermodynamic constraints of stability and consistency in obtaining the ionization equilibrium conditions for nonideal, multicomponent gases. Two different theoretical models have been used. The results are presented in the form of extensive tables listing the pressure, the internal energy, the adiabatic gradient, and the pressure derivatives as functions of the temperature and the density. Data are given for three mixtures with nearly pure chemical compositions, having element abundances $X_H = 0.999$, $X_{He} = 0.999$, and $X_C = 0.999$. The results are relevant to studies of pulsational properties and outer layers of white dwarfs, and also to evolutionary calculations of low-mass main-sequence stars, red giants, and white dwarfs.

065.052 Ejection of planetary nebulae by helium shell flashes and the planetary distance scale.
V. Trimble, I.-J. Sackman.
Publ. Astron. Soc. Pacific, Vol. 89, 623 - 624 (1977). Abstract.

065.053 Late stages of stellar evolution.   G. Gilmore.
South. Stars, Vol. 27, 44 - 53 (1977).

065.054 Gravitational phase transitions and the late-type stars.   G. Magni, M. Ranieri, P. Paolicchi.
Astrophys. Space Sci., Vol. 46, 3 - 12 (1977).
    A scale transformation is obtained which allows the Hertel and Thirring gravitational phase transition theory to be extended to more generalized astrophysical conditions. It is shown that by means of this transformation the formation of core-halo structures in late-type stars can be explained. The effect of the physical conditions peculiar to this type of star is discussed.

065.055 The evolution of super-metal-rich stars and the interpretation of old galactic clusters.
V. Castellani, A. Tornambè.
Astrophys. Space Sci., Vol. 46, 195 - 203 (1977).
    The influence of initial helium content on the evolutionary characteristics of super-metal-rich stars ($Z = 0.10$) has been investigated. The evolution of models in the range $-0.05 < \log L/L_\odot < 1.0$ has been followed up to the relative luminosity maximum in the subgiant branch, and under the assumption $Y = 0.20$ or $Y = 0.40$. Comparison with previous results suggests the existence of theoretical constraints that could be adopted as metal indicators for the observed H−R diagrams of old open clusters.

065.056 Strained coordinate methods in rotating stars. II. Main-Sequence stars.   B. L. Smith.
Astrophys. Space Sci., Vol. 47, 61 - 78 (1977).
    It was shown (Smith, 1976) that the method of strained coordinates may be usefully employed in the determination of the structure of rotating polytropes. This idea is extended to Main-Sequence stars with conservative centrifugal fields. The structure variables, pressure, density and temperature are considered pure functions of an auxiliary coordinate s (the strained coordinate) and the governing equations written in a

form that closely resembles the structure equations for spherical stars but with correction factors that are functions of s. A systematic, order-by-order derivation of these factors is outlined and applied in detail to a Cowling-model star in uniform rotation.

065.057   On the age difference between the oldest Population I and the extreme Population II stars.
H. Saio, Y. Shibata, M. Simoda.
Astrophys. Space Sci., Vol. 47, 151 - 162 (1977).

The authors have constructed evolutionary model sequences for both Population I and II stars using the same computer program and input physics. After giving evidence that the open cluster NGC 188 and the extremely metal-poor globular cluster M92 are representative samples of the oldest Population I and extreme Population II stars, respectively, the authors estimate the ages of NGC 188 and M92, using isochrones constructed from model sequences. They discuss the oxygen abundance of the extremely metal-poor Population II objects, which is an important factor on their age determination.

065.058   On a type of solutions for dynamical rupture of the equilibrium of stars.   G. Deb Ray.
Astrophys. Space Sci., Vol. 47, 229 - 236 (1977).

Solutions for dynamical behaviour of unstable stellar models are obtained, in the special case where the velocity of a gaseous element varies as r/t. It is found that the only possible value for the constant of proportionality is $2/3$ and that there are only two such stellar configurations.

065.059   Stellar rotation and the thermomagnetic torque.
G. G. Spear, L. T. Wood.
Astrophys. Space Sci., Vol. 47, 341 - 345 (1977).

The thermomagnetic torque, known to exist when a gas of polyatomic molecules experiences a temperature gradient in the presence of a magnetic field, has been investigated as a possible source of stellar rotational angular momentum. The effect does not appear to be significant during pre-main-sequence evolution. To influence stellar rotation a process must be capable of generating torques on the order of 10^{40} dyn cm^{-1}, whereas the effect due to the thermomagnetic torque is only as large as 10^{17} dyn cm^{-1}.

065.060   An analysis of the linear, adiabatic oscillations of a star in terms of potential fields.
M. L. Aizenman, P. Smeyers.
Astrophys. Space Sci., Vol. 48, 123 - 136 (1977).

The linear adiabatic oscillations of a spherically symmetric star are analyzed in terms of potential fields. It is found that all displacement fields $\varrho\xi$ can be described as either spheroidal or toroidal fields.

065.061   Rotational distortion of polytropes by analytic approximation, II.   A. P. Linnell.
Astrophys. Space Sci., Vol. 48, 165 - 183 (1977).

Chandrasekhar's (1933) paper on rotational distortion of polytropes contained a perturbation term in the potential which was linear in v, the rotation parameter. The same paper, and subsequent papers by various authors, developed an analytic expression for the boundary also linear in v. The latter expression is equivalent to a two term Taylor series about the unperturbed boundary, and is in error by 12% near critical rotation, for a polytropic index 3.0. The boundary can be located directly from the functions representing density, potential, and the potential gradient. The boundary error by this procedure is 0.2% near critical rotation.

065.062   A boundary equation for uniformly rotating polytropes.   A. P. Linnell.
Astrophys. Space Sci., Vol. 48, 185 - 190 (1977).

A linear approximating equation exists for the boundary of a uniformly rotating polytrope. The equation in $\eta=(\xi_1-\xi)/\xi_1$ permits rapid calculation of the polytrope radius for any latitude, and is accurate for angular velocities of rotation nearly to critical rotation. Data in this paper apply to a polytrope index $n = 3.0$.

065.063   The main sequence location as function of chemical composition.   F. Caputo.
Astrophys. Space Sci., Vol. 49, 113 - 122 (1977).

Available data for computed zero-age lines are analyzed, and empirical formulae giving the luminosity as function of the chemical composition are derived. Mass-luminosity relations are also presented for a very wide range of the chemical composition parameters (X, Y, Z). Comparison with the observed zero-age lines are discussed.

065.064   On a uniformly rotating magnetic polytrope. I. The equilibrium structure and oscillation with a large general magnetic field.   M. K. Das, J. N. Tandon.
Astrophys. Space Sci., Vol. 49, 261 - 275 (1977).

The equilibrium general magnetic field inside a magneto-rotating star, assumed to be a polytrope, has been determined more accurately, for large general magnetic field. Furthermore the effect of such field on the structure and oscillations of a slowly rotating polytrope has been studied for $n = 1.0, 1.5, 2.0$, and 3.0.

065.065   On a uniformly rotating magnetic polytrope. II. The equilibrium structure with a large toroidal magnetic field.   M. K. Das, J. N. Tandon.
Astrophys. Space Sci., Vol. 49, 277 - 291 (1977).

The equilibrium structure of a uniformly rotating magnetic polytrope in the presence of a large toroidal field has been determined for polytropic index $n = 1.0, 1.5, 2.0$ and 3.0. It has been found that the contribution of higher-order nonspherical terms in the determination of equilibrium structure becomes significant for large magnetic field.

065.066   The collapse of iron-oxygen stars: physical and mathematical formulation of the problem and computational method.   D. K. Nadyozhin (Nadezhin).
Astrophys. Space Sci., Vol. 49, 399 - 425 (1977).

A statement of the problem of gravitational collapse and a computational method are described. The main feature of the collapse — its extremely high heterogeneity — is taken into account. The structure of a collapsing star is characterized by a dense and hot nucleon core which is opaque with respect to neutrino radiation and is embedded into an extended envelope, almost transparent to neutrinos. The envelope is gradually being accreted onto the core. The increase of the mass of the core, which is opaque with respect to neutrino radiation, is fully taken into account in the calculations of the gravitational collapse.

065.067   On the use of collocation methods for the construction of stellar models.
R. L. Smith, H. Fuchs.
Astrophys. Space Sci., Vol. 50, 63 - 73 (1977).

Collocation with piecewise continuous polynomials is studied for use in the numerical modelling of stellar evolution. Accuracy and convergence of the method are demonstrated for a $5 M_\odot$ star with a convective core. Collocation should be further studied since it is likely to lead to significant grains in computational efficiency for the construction of stellar models.

065.068   Final evolution of stars in the range $10^3-10^4 M_\odot$.
J. C. Wheeler.
Astrophys. Space Sci., Vol. 50, 125 - 131 (1977).

The final dynamical collapse of oxygen cores of 10^3 and $10^4 M_\odot$ which undergo the pair formation instability is computed. These cores are found to suffer complete collapse, presumably to form black holes. These calculations represent a first attempt to ascertain the outcome of evolution over several decades of mass previously unexplored. The outcome may have some relevance to models of X-ray sources in globular clusters.

065.069 On broadened definitions of instability for stars in thermal imbalance.   N. R. Simon.
Astrophys. Space Sci., Vol. 51, 205 - 215 (1977).

The classical theory of stability of dynamical systems is employed to demonstrate that traditional definitions of pulsational instability cannot be directly applied to stars in thermal imbalance. In particular, it is shown that, for the case of thermal imbalance, pulsational displacements and pulsational velocities have separate and distinct e-folding times. A broadened set of definitions is formulated. In accordance with the new definitions, it is argued that the development of observable pulsations requires as a necessary condition infinitesimal instability of both absolute displacement and velocity. Finally, it is shown that the stability of stars in thermal imbalance may be evaluated according to the present definitions by employing either of two existing theories — the energy approach due to Demaret (1974; 1975; 1976) or the small perturbation technique of Cox et al. (1973).

065.070 The gravitational collapse of iron—oxygen stars with masses of $2\,M_\odot$ and $10\,M_\odot$. II.
D. K. Nadyozhin (*Nadezhin*).
Astrophys. Space Sci., Vol. 51, 283 - 301 (1977).

The collapse of iron—oxygen stars with masses of $2\,M_\odot$ and $10\,M_\odot$ has been calculated. The commencement of the collapse is due to dissociation of iron-group nuclei into free nucleons. After a while, the collapse proceeds in consequence of intensive energy losses due to neutrino volume radiation. At an intermediate stage of the collapse, the core — opaque with respect to neutrino radiation (neutrino core) — is formed inside the collapsing star. Both the gradual increase of the mass of the neutrino core and the partial absorption of neutrinos radiated from the surface of the neutrino core by the stellar envelope (deposition) were taken into account.

065.071 An examination of the planetesimal impact hypothesis of the formation of CP stars.   C. R. Cowley.
Astrophys. Space Sci., Vol. 51, 349 - 362 (1977) = Contrib. Dominion Astrophys. Obs., Victoria, B.C., Canada, No. 342 = NRC No. 16101.

The impact hypothesis has been found to provide a qualitative basis for an understanding of many of the characteristics of the CP (chemically peculiar) stars. It founders primarily upon the problem of underabundances of certain elements and large variations in isotopic abundances. The author concludes, however, that the hypothesis should not be prematurely discarded, since it may be combined with other mechanisms to account for the variety of abundance peculiarities of the CP stars.

065.072 Possible non-thermal effects in stellar nuclear reactions.   G. Beaudet, G. Shaviv.
Astrophys. Space Sci., Vol. 51, 395 - 400 (1977).

The slowing down of charged ions in dense plasma is examined. It is found that under conditions prevailing in late phases of evolution, the nuclei can react while slowing down. The consequence of such an additional nuclear reaction on the product of the nuclear burning is discussed briefly.

065.073 Critical accretion flow of gas onto compact stars.   T. Okuda, S. Sakashita.
Astrophys. Space Sci., Vol. 52, 35 - 49 (1977).

The critical accretion flow of gas onto compact stars with mass of $0.6\,M_\odot$ is investigated by numerical integrations of the time-dependent hydrodynamic equations in the spherically-symmetric and optically thick case. For the compact stars surrounded by such a dense cloud of gas, the radiation pressure force decelerates the infall gas significantly and free fall regime

of the gas is not at all attained. This results in incident low velocities at the standing shock front close to the stellar surface, low temperatures of the gas around the compact stars, and no X-ray in white dwarfs but soft X-rays in neutron stars, respectively. Some applications of the results to the X-ray sources are discussed.

065.074 Vibrational stability of stars in thermal imbalance. IV. Towards a definition of vibrational stability.
J. Demaret, J. Perdang.
Astrophys. Space Sci., Vol. 52, 137 - 167 (1977).

The authors discuss a general formulation of kinematic stability coefficients σ''. They show that the signature of σ'' cannot satisfactorily be correlated with the vibrational stability of stars in thermal imbalance. The intuitive concept of vibrational stability is analyzed and this notion is identified with Lyapunov's asymptotic stability. The authors introduce a thermodynamic technique of Lyapunov function construction. The relevance and feasibility of this method is illustrated applying the procedure to supermassive stars in contraction along the Hayashi line. It is shown that these stars are vibrationally unstable.

065.075 Local stellar stability. II.   J. Perdang.
Astrophys. Space Sci., Vol. 52, 313 - 350 (1977).

The existence proof of continuous spectra of eigenvalues s developed in the framework of the function space of q-regularizations (Perdang, 1976) is extended in this paper by relaxing the severe restrictions previously imposed on the mathematical structure of the stellar stability equations. This procedure is illustrated in the case of nonradial adiabatic stability. Moreover when applied to nonadiabatic perturbations it reveals the existence of two new types of local instability which seem to prevail in the majority of stars in a thermonuclear burning phase: (a) a nonradial local secular instability; (b) a radial local nuclear instability.

065.076 An asymptotic representation of unstable gravity modes in stars.   J. Denis, P. Smeyers.
Astrophys. Space Sci., Vol. 52, 435 - 442 (1977).

An asymptotic representation is developed for high order unstable gravity modes in a star which are associated with a convective zone enclosed between two radiative layers. In the domain which contains both turning points of the differential equation, the solutions are represented by a single asymptotic expansion in terms of Weber functions

065.077 General properties of the Ap stars.   M. Hack.
Publ. Dep. Astron., Univ. Beograd, Fac. Sci., No. 6, (see 012.040), p. 31 - 34 (1976).

The general properties (surface temperature, luminosity, rotational velocity, frequency of binaries, atmospherical chemical composition, variability) are summarized, together with the main theories proposed for explaining the phenomenon of the A-type peculiar stars.

065.078 An analytical estimate of values of pressure and temperature at the boundary of a convective core.
T. Angelov.
Publ. Dep. Astron., Univ. Beograd, Fac. Sci., No. 6, (see 012.040), p. 35 - 42 (1976).

The author considers a star model which is divided in two regions: the "internal" and the "external" one. The model is spherically symmetric, in hydrostatic equilibrium, with an equation of state for a perfect gas with μ = const.

065.079 Rotational effects in the evolution of young stars.   G. S. Bisnovatyj-Kogan, S. I. Blinnikov, A. V. Fedorova.
Early stages of stellar evolution, (see 012.041), p. 40 - 46 (1977). In Russian.

A review is given of the theoretical work on the role of

rotation in the problem of formation and evolution of young stars. The authors present preliminary results of their work on the evolution of rotating stars of 0.5 and 1 $M_\odot$ on the stage of gravitational contraction.

065.080 p-nuclei processing in degenerate hydrogen burning zones and its relationship to nova outbursts.
T. G. Harrison.
Proc. Southwest Reg. Conf., Vol. 3, (see 012.043), p. 31 - 45 (1977).
The author investigated the likelihood that the proton rich (p- or bypassed nuclei) could have been synthesized in a degenerate hydrogen burning region wherein the proton-proton cycle is closed via tritium burning. The author finds that many of the details of the solar system p-nuclei abundance distribution can be reproduced, particularly the relative abundances of these elements having two or more p-isotopes. Finally, he discusses possible astrophysical sites within which this process might take place, and show where white dwarf stars accreting hydrogen rich matter at a very low rate have degenerate hydrogen burning shells suitable for supporting p-nuclei synthesis via tritium burning. Furthermore it is shown that these are precisely the stars which lead to nova outbursts and that this subsequent event provides a natural mechanism for seeding the interstellar medium with processed material.

065.081 A numerical approach to the testing of the fission hypothesis.   L. B. Lucy.
Astron. J., Vol. 82, 1013 - 1024 (1977).
A finite-size particle scheme for the numerical solution of two- and three- dimensional gas dynamical problems of astronomical interest is described and tested. The scheme is then applied to the fission problem for optically thick protostars. Results are given, showing the evolution of one such protostar from an initial state as a single, rotating star to a final state as a triple system whose components contain 60% of the original mass. The decisiveness of this numerical test of the fission hypothesis and its relevance to observed binaries are briefly discussed.

065.082 Hydrogen shell burning stellar models from red giants to hot white dwarfs.   U. Uus.
Publ. Tartu Astrofiz. Obs., Vol. 45, 173 - 181 (1977).
A set of models of stars with a degenerate helium core and a stationary hydrogen burning shell has been computed. The masses of stars range from 0.25 to 1.2 $M_\odot$ and the stellar envelope masses vary within large limits so that the models cover a region from red giants to white dwarfs. The values of the physical parameters at the surfaces of the stellar helium cores are presented.

065.083 On the interaction of stellar pulsations with convection.   U. Uus.
Publ. Tartu Astrofiz. Obs., Vol. 45, 182 - 189 (1977).
It is suggested that convective heat flux variations in pulsating stars do not vanish in the limiting case when the convection time-scale greatly exceeds the pulsation period. Corresponding general equations for the calculation of convective heat flux variations in pulsating stars are presented.

065.084 The neutron mechanism of thermonuclear carbon burning, formation of neutron stars and supernova bursts.   S. S. Gershtejn, L. N. Ivanova, V. S. Imshennik, M. Yu. Khlopov, V. M. Chechetkin.
Pis'ma v ZhEhTF, Vol. 26, 189 - 193 (1977). In Russian.
Abstr. in Ref. zh., 51. Astron., 12.51.454 (1977).

065.085 On convective instability of a gaseous sphere.
N. M. Zueva, M. S. Mikhajlova, L. S. Solov'ev.
Inst. prikl. mat. AN SSSR. Prepr. No. 65. Moskva, 1977. 24 pp. Price 7 Kop. In Russian. – Abstr. in Ref. zh., 51. Astron., 1.51.458 (1978).

Stellar evolution.   See Abstr. 003.101.

Ein halbes Jahrhundert A. S. Eddington, "Der innere Aufbau der Sterne".   See Abstr. 004.054.

Neutron capture by ^{208}Pb at stellar temperatures. See Abstr. 022.016.

Stellar energy-loss rates due to S, P or T neutral currents.   See Abstr. 022.070.

β-decay of nuclear excited states. See Abstr. 022.071.

Measurement of the internal pair emission branch of the 7.654 MeV state of ^{12}C, and the rate of the stellar triple-α reaction.   See Abstr. 022.072.

Helium burning of ^{40}Ca. See Abstr. 022.074.

The importance of long-lived isomeric states in s-process branching.   See Abstr. 061.007.

On s-process nucleosynthesis in thermally pulsing stars.   See Abstr. 061.008.

Pion condensation I. (Stellar interiors). Nonrelativistic treatment for π⁻ condensation.   See Abstr. 061.027.

Screening effect on electron captures at high densities.   See Abstr. 061.028.

Smoothed particle hydrodynamics: theory and application to non-spherical stars.   See Abstr. 062.017.

Enhancement of thermonuclear reaction rate due to strong screening.   See Abstr. 062.026.

On the adiabatic pulsations of uniformly rotating bodies.   See Abstr. 062.053.

Hydromagnetic flow near an oscillating porous limiting surface.   See Abstr. 062.067.

Stellar opacity.   See Abstr. 064.022.

Mass-loss effects on stellar evolution, I: 5 $M_\odot$ and 15 $M_\odot$ population I models.   See Abstr. 064.027.

Behaviour of microturbulence with evolution. See Abstr. 064.028.

Mass loss in stars of moderate mass by stellar winds and effects on the evolution.   See Abstr. 064.030.

Decay of light elements in stellar envelopes. See Abstr. 064.034.

Accretion by magnetic neutron stars. I. Magnetospheric structure and stability.   See Abstr. 066.005.

Admissible coordinate systems for static, spherically symmetric stellar models.   See Abstr. 066.307.

Detectable gravitational radiation from stellar collapse.   See Abstr. 066.339.

Etoiles carbonées.   See Abstr. 114.019.

The ^{12}C/^{13}C ratio in carbon stars. See Abstr. 114.064.

A luminous carbon star in Canis Major OB1. See Abstr. 115.012.

Radii of some Ap, Am and A stars. See Abstr. 115.014.

Effect of rapid mass accretion onto main-sequence stars. See Abstr. 117.002.

The evolution of massive close binaries. V: Systems containing primaries with masses between 10 $M_\odot$ and 15 $M_\odot$. See Abstr. 117.039.

Adiabatic self-similar blast waves, their radial instabilities, and their application to supernova remnants. See Abstr. 125.010.

The effects of differential rotation on the splitting of nonradial modes of stellar oscillation.

See Abstr. 126.008.

Origine e fine delle stelle. See Abstr. 131.116.

Interstellar matter and the evolution of stars. See Abstr. 131.222.

On the evolution of central stars of planetary nebulae. See Abstr. 135.029.

UH cosmic rays and solar system material: the elements just beyond iron. See Abstr. 143.059.

Chemical evolution in the solar neighborhood. III. Time scales and nucleochronology. See Abstr. 155.010.

Galactic ^{4}He and heavy elements enrichment by stellar nucleosynthesis. See Abstr. 155.063.

066 Neutron Stars, Relativistic Astrophysics, Background Radiation, Gravitation Theory

066.001 **Quadrupole test particle as a detector of gravitational waves.**   R. Tammelo.
Gen. Relativ. Gravitation, Vol. 8, 313 - 319 (1977).

066.002 **On Shirokov's "One new effect of the Einsteinian theory of gravitation".**   A. Nduka.
Gen. Relativ. Gravitation, Vol. 8, 347 - 351 (1977).

066.003 **On the propagation of photons and neutrinos in curved space-time.**   J. B. Griffiths.
Gen. Relativ. Gravitation, Vol. 8, 365 - 370 (1977).

It is shown using a classical wave theory approach that photons and neutrinos have different propagation properties in curved space-time. It is also shown that neutrinos have the anomalous property that the sign of their energy density may change as they propagate, and they may even become ghost neutrinos in some regions.

066.004 **Charged black holes and phase transitions.**
P. Hut.
Mon. Not. R. Astron. Soc., Vol. 180, 379 - 389 (1977).

The thermodynamic behaviour is discussed of a charged black hole in a box in equilibrium with neutral thermal radiation, in the thermodynamic limit (i.e. of a very massive black hole). The heat capacity at constant volume of this system exhibits several types of discontinuities which resemble phase transitions, and in an appropriate phase plane a critical point can be distinguished. The correspondence with normal thermodynamic phase changes is discussed, and various other thought experiments are briefly mentioned.

066.005 **Accretion by magnetic neutron stars. I. Magnetospheric structure and stability.**
R. F. Elsner, F. K. Lamb.
Astrophys. J., Vol. 215, 897 - 913 (1977).

The authors present the results of a study of accretion by a slowly rotating, magnetic neutron star when the accretion flow is approximately radial. They examine in detail the physical processes that occur in the neighborhood of the magnetospheric boundary and the manner in which accreting plasma enters the magnetosphere. They investigate the conditions necessary for the formation of a magnetospheric cavity and for its stability and find that although the boundary of the cavity is initially close to instability, magnetospheric models strongly suggest that it is stable until the plasma outside undergoes at least some cooling.

066.006 **On the relict recombination lines.**
I. N. Bernshtejn, D. N. Bernshtejn, V. K. Dubrovich.
Astron. Zh. Akad. Nauk SSSR, Vol. 54, 727 - 733 (1977).
In Russian. English translation in Soviet Astron., Vol. 21, No. 4.

An accurate numerical calculation of intensities and profiles of hydrogen recombination lines of cosmological origin is made. The dependence of these quantities on the parameters of matter in the Universe — the hydrogen density and the total density of all kinds of matter — is investigated.

066.007 **Upper limits for the radio pulse emission rate from exploding black holes.**   W. P. S. Meikle.
Nature, Vol. 269, 41 - 42 (1977).

It has been suggested that the explosion of a primordial black hole (pbh) might produce a coherent radio pulse detectable at a distance of ~10^4 pc with a simple antenna. The negative results obtained are used to place limits on the pbh radio pulse emission rate per unit volume. This paper presents

an analysis of limit criteria, including consideration of the effects on the pulse timescale of the interstellar medium, and examines the factors which determine the optimum search strategy for pbh radio pulses.

066.008 **Observable gravitational effects on polarised radiation coming from near a black hole.**
P. A. Connors, R. F. Stark.
Nature, Vol. 269, 128 - 129 (1977).

The effects of general relativity on the degree and plane of linear polarisation of X rays emitted by an accretion disk orbiting a black hole were described in a previous letter. Using a constant of motion along null geodesics in the Kerr metric the authors have since been able to calculate analytically polarisation rotations along any null ray. They present here more extensive results obtained by this method, for two different kinds of disk model and for different black hole angular momenta.

066.009 **Quantum uncertainty in the final state of gravitational collapse.**   J. V. Narlikar.
Nature, Vol. 269, 129 - 130 (1977).

The ratio of the action S to $\hbar$ determines whether the physical system in question is to be treated classically or quantum mechanically. The author discusses here the behaviour of a physical system which is initially in the classical domain but whose later development may well take it into the region of quantum uncertainty. He considers a specific example of this — the gravitational collapse of a spherical dust ball.

066.010 **On the analog of Larmor's theorem in general relativity.**   V. I. Stoyanov.
Byull. Inst. Astrofiz., Dushanbe, No. 66 - 67, p. 23 - 27 (1976).
In Russian.

066.011 **Sur une interprétation physique de la constante cosmologique.**   R. Kerner.
C.R. Acad. Sci. Paris, Tome 285, Sér. A, 149 - 152 (1977).

The author modifies the Einstein equation system supposing the existence of a maximal value of the scalar curvature. Then he briefly discusses the cosmological consequences of such a model. He concludes that in this case the cosmological "constant" is a function of the matter density in the Universe and that the initial radius of the Universe at the moment of the "big bang" was non-null, but of the order of 10^{-13} cm.

066.012 **Sur la cinématique de l'éjection relativiste.**
S. Kichenassamy.
C. R. Acad. Sci. Paris, Tome 285, Sér. B, 53 - 56 (1977).

066.013 **Out from under the cosmic censor: Stephen Hawking's black holes.**   D. Overbye.
Sky Telesc., Vol. 54, 84 - 89, 108 (1977).

066.014 **Mechanism of X-ray and soft γ-ray radiation from accreting neutron stars.**   A. I. Tsygan.
Astron. Astrophys., Vol. 60, 39 - 42 (1977).

It is shown that during the accretion of plasma on to a neutron star with strong magnetic field (10^{10} – 10^{13} G) a radiating accreting channel appears where the plasma velocity is of the order of the free fall velocity. The relativistic regime of plasma fall in the channel ensures direction focusing of radiation. A mechanism of pulsation of X-ray and γ-ray sources is also proposed (accumulation of plasma near the

Alfvén surface and its subsequent fast fall on to a neutron star).

066.015 Expansion and rotational momentum of large cosmic masses. H.-J. Treder.
Astrofizika, Vol. 12, 511 - 519 (1976). In Russian. – English translation in Astrophysics, Vol. 12, No. 3.

In gravidynamics leading to "absorption of gravity" (tetrad theory of gravitation) as well as in theories leading to an increase of inertial mass owing to induction of inertia by local gravitational potential, equilibrium configurations of large cosmic masses are possible. If such a large mass is brought out of the equilibrium this theory predicts its expansion. During the expansion a definite asymptotic relation between the mass and the rotational momentum is obeyed. This picture is in good correspondence with existing information on the rotational momenta of galaxies and the instability of large cosmic systems (clusters and superclusters of galaxies).

066.016 Neutrino transport in pion-condensed neutron stars. R. F. Sawyer, A. Soni.
Astrophys. J., Vol. 216, 73 - 76 (1977).

The neutrino opacity of pion-condensed neutron star matter is calculated and found to be greatly enhanced over the values for ordinary neutron matter. For neutrinos of energy $E_\nu > 0.1$ MeV the neutrino free path may be less than the radius of the neutron star. The solution to the resulting neutrino diffusion problem indicates that it may take as much as a few hours for a newly formed neutron star to cool to a temperature of $kT \sim 0.1$ MeV.

066.017 Beta decay of pion condensates as a cooling mechanism for neutron stars. O. Maxwell, G. E. Brown, D. K. Campbell, R. F. Dashen, J. T. Manassah.
Astrophys. J., Vol. 216, 77 - 85 (1977).

Using recently developed dynamical models of pion condensation, the authors reconsider the analysis of Bahcall and Wolf, who showed that the existence of free pions in neutron star interiors could stimulate neutron β-decay and thus increase dramatically the rate at which these stars cool. Although the actual pion condensate is quite different from the free pions assumed by Bahcall and Wolf and although the stimulated β-decay process is also different, the authors' final results for the luminosity arising from this mechanism are very similar to their earlier predictions.

066.018 Analytic properties of relativistic rotating bodies. M. A. Abramowicz, R. V. Wagoner.
Astrophys. J., Vol. 216, 86 - 91 (1977).

The initial-value equations of general relativity for a slowly and uniformly rotating, homogeneous, momentarily stationary body are integrated exactly. Analytic formulae for the total mass, baryon number, and angular momentum of such a body are found.

066.019 Accretion onto pregalactic black holes. J. R. Ipser, R. H. Price.
Astrophys. J., Vol. 216, 578 - 590 (1977).

If, as has recently been suggested, a large part of the mass in our Galaxy is in the form of massive ($M = 10^5 M_{(5)} \cdot M_\odot \sim 10^5 M_\odot$) black holes, a significant number of these objects should be moving supersonically through the disk of the Galaxy and accreting interstellar gas at a rate $\gtrsim 10^{17} M_{(5)}{}^2$g s^{-1}. The nearest such object should be $\sim 200 M_{(5)}{}^{1/3}$ pc from the Sun. The authors find that most of these objects would accrete more or less spherically (no disk formation) and would radiate $\sim 10^{37} M_{(5)}{}^2$ ergs s^{-1} at wavelengths $\sim M_{(5)}{}^{13/8} \mu$m. It would appear that observations constrain the local density of such objects, at least for $0.3 \lesssim M_{(5)} \lesssim 30$, to be too small by a factor $\gtrsim 100$ to provide a galactic halo that would explain the suggested linear mass-radius relation for spiral galaxies, stabilize galactic disks, or close the Universe.

066.020 Tidal radiation. B. Mashhoon.
Astrophys. J., Vol. 216, 591 - 609 (1977).

The general theory of tides is developed within the framework of Einstein's theory of gravitation. It is based on the concept of Fermi frame and the associated notion of tidal frame along an open curve in spacetime. Following the previous work of the author an approximate scheme for the evaluation of tidal gravitational radiation is presented which is valid for weak gravitational fields. The emission of gravitational radiation from a body in the field of a black hole is discussed, and for some cases of astrophysical interest estimates are given for the contributions of radiation due to center-of-mass motion, purely tidal deformation, and the interference between the center of mass and tidal motions.

066.021 Gravitational radiation from point-masses in unbound orbits: Newtonian results. M. Turner.
Astrophys. J., Vol. 216, 610 - 619 (1977).

Gravitational radiation from a system of two point-masses in unbound orbits (orbital eccentricity $e \geq 1$) is calculated to Newtonian order in the multipole formalism. Waveforms of the multipole amplitudes are presented for bound and unbound orbits. The energy emitted in gravitational waves from an unbound encounter is given by an enhancement factor, $g(e)$, times a function of the other orbital parameters (similar to the results of Peters and Mathews for bound systems). Energy spectra of the gravitational radiation emitted are also presented.

066.022 Gravitationally redshifted gamma rays and neutron star masses. R. L. Bowers.
Astrophys. J., Lett., Vol. 216, L63 - L65 (1977).

Gravitationally redshifted γ-ray lines emitted from the surface of pulsars and compact X-ray sources are discussed as constraints on neutron star masses. A recent tentatively identified 400 ± 1 keV line in the direction of the Crab Nebula, presumably due to positron annihilation on the pulsar surface, gives a mass in the range $1.3 \lesssim M/M_\odot \lesssim 1.94$ for a wide range of recent equations of state. The possibility of further restricting this range using γ-ray spectroscopy of compact X-ray sources in close binary systems is suggested, and its possible importance in eliminating some models of neutron stars is discussed.

066.023 On the Zeeman splitting of X-ray lines by neutron-star magnetic fields.
C. L. Sarazin, J. N. Bahcall.
Astrophys. J., Lett., Vol. 216, L67 - L70 (1977).

In the strong magnetic fields usually assumed to exist near neutron stars ($B \sim 10^{11}–10^{13}$ gauss), X-ray emission lines may exhibit easily detected Zeeman splitting (at least when the field in the emission region $\lesssim 5 \times 10^{11}$ gauss). Detection of this splitting would allow a direct measurement of the neutron-star magnetic field strength. The observed Zeeman pattern would determine the direction of the magnetic field; in an oblique rotator, the pattern would alternate with the pulsar period.

066.024 On tidal interactions with Kerr black holes. W. A. Hiscock.
Astrophys. J., Vol. 216, 908 - 913 (1977).

The tidal deformation of an extended test body falling with zero angular momentum into a Kerr black hole is calculated. Numerical results for infall along the symmetry axis and in the equatorial plane of the black hole are presented for a range of values of a, the specific angular momentum of the black hole. Estimates of the tidal contribution to the gravitational radiation are also given. The tidal contribution in equatorial infall into a maximally rotating Kerr black hole may be of the same order as the center-of-mass contribution to the gravitational radiation.

066.025 Tidal generation of gravitational waves from orbiting Newtonian stars. I. General formalism.
M. S. Turner.
Astrophys. J., Vol. 216, 914 - 929 (1977).

A linearized formalism is presented for the calculation of the tidally produced gravitational radiation potential from binary systems with arbitrary orbits. The stars are Newtonian, isentropic, and nonrotating. Normal-mode analysis is used to calculate the tidally generated internal motions; the resulting radiation potential and its Fourier decomposition are calculated in the Newtonian limit of the multipole formalism.

066.026 Anisotropy of the cosmic microwave background radiation.
M. V. Gorenstein, G. F. Smoot, R. A. Muller.
Bull. American Astron. Soc., Vol. 9, 431 (1977). — Abstract.

066.027 Self-similar growth of primordial black holes.
G. V. Bicknell, R. N. Henriksen.
Bull. American Astron. Soc., Vol. 9, 431 (1977). — Abstract.

066.028 The star distribution around a massive black hole in a globular cluster. II. Unequal star masses.
J. N. Bahcall, R. A. Wolf.
Astrophys. J., Vol. 216, 883 - 907 (1977).

The steady-state distribution of stars around a massive black hole in a globular cluster is determined by solving numerically the coupled time-dependent Boltzmann equations for a system containing stars of two different masses. Similar results are found for an arbitrary spectrum of masses with the aid of approximate analytic solutions of the time-independent equations. The effects of mass segregation are summarized by scaling laws that are derived both by analytic approximations and by numerical solutions. The detectability of a black hole in a globular cluster is discussed in terms of possible observations of the central star distributions.

066.029 Quantum field theory in curved space-time.
P. C. W. Davies.
Nature, Vol. 263, 377 - 380 (1976).

Recent theoretical developments indicate that the presence of gravity (curved space-time) can give rise to important new quantum effects, such as cosmological particle production and black-hole evaporation. These processes hint at intriguing new relations between quantum theory, thermodynamics and space-time structure and encourage the hope that a better understanding of a full quantum theory of gravity may emerge from this approach.

066.030 On spherically symmetric motion of a relativistic gravitating gas in the presence of a strong shock wave.   A. N. Golubyatnikov.
Dokl. AN SSSR, Vol. 233, 318 - 321 (1977). In Russian.
Abstr. in Ref. zh., 51. Astron., 8.51.913 (1977).

066.031 Possible existence of black holes in cosmic rays.
V. P. Semenov.
Izv. vyssh. uchebn. zaved. Fiz., 1976, No. 12, p. 136 - 138.
In Russian. — Abstr. in Ref. zh., 51. Astron., 8.51.914 (1977).

066.032 Gravitational waves in space and in laboratory.
L. P. Grishchuk.
Usp. fiz. nauk, Vol. 121, 629 - 656 (1977). In Russian.
Abstr. in Ref. zh., 51. Astron., 8.51.939 (1977).

066.033 The upper mass limit for neutron stars including differential rotation.   D. J. Hegyi.
Astrophys. J., Vol. 217, 244 - 247 (1977).

The author shows that any significant amount of differential rotation initially present in a neutron star will be damped out within a few days. Consequently, differential rotation need not be considered in determining the upper mass limit for neutron stars. He determines this limit with the fewest possible assumptions. Without rotation the mass may not exceed $8\,M_\odot$. Rigid body rotation may increase the mass by no more than 30% according to calculations by others. Thus he finds the upper mass limit including rotation to be $11\,M_\odot$.

066.034 Very massive neutron stars in Ni's theory of gravity.
D. R. Mikkelsen.
Astrophys. J., Vol. 217, 248 - 251 (1977).

It is shown that in Ni's theory of gravity, which is identical to general relativity in the post-Newtonian limit, neutron stars of arbitrarily large mass are possible. This result is independent, within reasonable bounds, of the equation of state of matter at supernuclear densities.

066.035 The generation of gravitational waves. III. Derivation of bremsstrahlung formulae.
S. J. Kovács, K. S. Thorne.
Astrophys. J., Vol. 217, 252 - 280 (1977).

Formulae are derived describing the gravitational waves produced by a stellar encounter of the following type: The two stars have stationary (i.e., nonpulsating), nearly Newtonian structures with arbitrary relative masses; they fly past each other with an arbitrary relative velocity; and their impact parameter is sufficiently large that they gravitationally deflect each other through an angle small compared to 90°.

066.036 The dissolution of globular clusters containing massive black holes.   S. L. Shapiro.
Astrophys. J., Vol. 217, 281 - 286 (1977).

A homological model is employed to follow the dynamical evolution of a globular cluster core containing a massive black hole at its center. Energy is supplied to the core via the inward drift and tidal disruption of stars in the cusp. This outward flow of energy inevitably halts and reverses core collapse. Re-expansion and complete dissolution is the ultimate fate of star clusters in the Galaxy whether they form tight binaries or massive black holes at their centers following catastrophic core collapse. Possible observational consequences of these results are briefly explored.

066.037 Stellar density cusp around a massive black hole.
P. J Young.
Astrophys. J., Vol. 217, 287 - 295 (1977).

This paper is concerned with the distribution of stars around a massive black hole in a dense star cluster. Spherical symmetry and isotropic distribution functions are assumed along with a power-law mass spectrum of stars to obtain a one-parameter family of equilibrium "zero-flow" solutions. The analysis is performed using a Fokker-Planck equation to describe the stellar encounters in their Coulomb potentials in an imposed gravitational potential.

066.038 A new criterion for secular instability of rapidly rotating stars.
J. M. Bardeen, J. L. Friedman, B. F. Schutz, R. Sorkin.
Astrophys. J., Lett., Vol. 217, L49 - L53 (1977).

A new variational criterion sufficient for secular instability of rapidly rotating stars to bar modes driven by gravitational radiation reaction is derived. Application to disk models indicates that differentially rotating stars can be stable beyond where they should be unstable according to the tensor virial method. The authors show why the tensor virial method is invalid.

066.039 Accretion by rotating magnetic neutron stars. I. Flow of matter inside the magnetosphere and its implications for spin-up and spin-down of the star.
P. Ghosh, F. K. Lamb, C. J. Pethick.
Astrophys. J., Vol. 217, 578 - 596 (1977).

The authors investigate the flow of accreting matter and

the configuration of the magnetic field inside the magneto-sphere of a rotating neutron star. They assume that the magnetic field of the star has a symmetry axis which is aligned with the rotation axis, that the accreting plasma becomes threaded by the stellar magnetic field near the magnetospheric boundary, and that the star is not a fast rotator.They show that for bright X-ray sources, flow within the Alfvén surface is well described by the equations of magnetohydrodynamics and that the matter moves along field lines when viewed in the frame corotating with the star. In the case of slow rotators, matter inside the Alfvén surface rotates in a sense opposite that of the net angular momentum flux toward the star. The authors also study the torque exerted on the star by the accreting matter and the flow of energy within the magneto-sphere and the transition region at the magnetospheric boundary, when the star accretes matter from a disk. Finally, they comment briefly on the implications of their results for pulsating X-ray sources.

066.040   On the integration of the relativistic equations of motion for isentropic perfect fluids.   S. Bonanos.
Astrophys. J., Vol. 217, 619 - 620 (1977).

The author points out that the integration of the relativistic equations of motion for an isentropic perfect fluid in a stationary and axisymmetric spacetime given recently by Fishbone and Moncrief (1976) can be simplified considerably.

066.041   The positive definite property of the energy density in gravitational fields.   N. Hu.
Sci. Sinica, Vol. 20, 335 - 344 (1977).

The present investigation shows that new expressions for pseudo-energy-momentum tensor which give rise to non-negative energy density can be found. For the free gravitational waves, the non-negative condition of the energy density requires that only transverse waves exist.

066.042   Post-Newtonian effects in the gravo-dynamics and Einstein effects in the theory of general relativity.
H.-J. Treder.
Astron. Nachr., Band 298, 237 - 244 (1977). In German.

Post-Newtonian effects in Hertzian gravo-dynamics are discussed and its values compared with Einstein effects in general relativity theory (GRT). Generally, gravo-dynamics gives quite the same values of the post-Newtonian effects in celestial mechanics and in the gravo-optics like in GRT. However, these values are defined by measurement operations which have quite different meanings in gravo-dynamics and in GRT. P.e., the Riemannian curvature of the three-dimensional space which is resulting in gravo-dynamics according to the Mach-Einstein doctrine becomes bigger by a factor $3/2$ than the curvature resulting by the Einstein equations. − GRT and gravo-dynamics are congruent in reference to the astronomical fundamental system. This consequence defines the fundamental system according to Mach's principle.

066.043   General relativistic incompressibility.
K. Brecher, I. Wasserman, with a reply by F. I. Cooperstock, R. S. Sarracino.
Nature, Vol. 269, 728 - 729 (1977).

066.044   Charged static fluid spheres in Einstein-Cartan theory.
A. Nduka.
Gen. Relativ. Gravitation, Vol. 8, 371 - 377 (1977).

The author gives exact interior solutions of the Einstein-Cartan equations describing charged perfect fluid distribution in general relativity. Results previously unknown for the uncharged case are deduced and it is found that the pressure is discontinuous at the boundary of the fluid sphere.

066.045   Models of the universe containing matter and gravitational radiation.   G. G. Swinerd.
Gen. Relativ. Gravitation, Vol. 8, 379 - 395 (1977).

The manner in which a field of gravitational radiation modifies the cosmological background space-time of the model universe containing it is considered. The cosmological equations for the models are solved and a number of examples of the resulting universes are presented in diagrammatic and tabular form.

066.046   Metric-torsion gravitational theories.
S. J. Aldersley.
Gen. Relativ. Gravitation, Vol. 8, 397 - 409 (1977).

It is shown that under certain assumptions the Einstein-Cartan field equations are not unique but may reasonably be modified to a degree. These modified Einstein-Cartan equations are proven to be unique under quite general conditions and are likely the most general equations in any metric-torsion gravitational theory whose field equations are derivable from a variational principle and such that their geometric part is independent of constants other than the speed of light and the gravitational constant.

066.047   Space-times carrying a quasirecurrent pairing of vector fields.   R. Rosca, S. Ianus.
Gen. Relativ. Gravitation, Vol. 8, 411 - 420 (1977).

A quasirecurrent pairing of vector fields $\{X_1, X_2\}$ is investigated on a space-time in two cases: (1) X_1 is spacelike and X_2 is timelike; (2) X_1 is null and X_2 spacelike. The physical interpretation of these vector fields is given.

066.048   On static, axially symmetric Einstein−Maxwell fields. Part I.   A. Papapetrou.
Gen. Relativ. Gravitation, Vol. 8, 421 - 427 (1977).

The Weyl solution of the problem, obtained on the assumption that g_{00} is a function of the electrostatic potential, is varied and the linearized field equations for the variation are discussed. The complete solution of the problem is determined for the special case of the Weyl solution that generalizes the Reissner-Nordström solution with $m = |e|$.

066.049   On static, axially symmetric Einstein–Maxwell fields. Part II.   A. Papapetrou.
Gen. Relativ. Gravitation, Vol. 8, 429 - 435 (1977).

Starting from the Reissner–Nordström solution with $m = e$ the author considers a variation representing a second particle situated outside the horizon. A formal dipole term in the potential of the second particle ensures equilibrium without additional stresses between the particles. The complete solution for the variation is determined and discussed in detail.

066.050   Static and time-dependent solutions of Einstein-Maxwell-Yukawa fields.
K. B. Lal, M. Q. Khan.
Gen. Relativ. Gravitation, Vol. 8, 451 - 461 (1977).

An exact solution of Einstein-Maxwell-Yukawa field equations has been obtained in a space-time with a static metric. A critical analysis reveals that the results previously obtained by Patel (1975), Singh (1974), and Taub (1951) are particular cases of this solution. The singular behavior of the solutions has also been discussed. Further, extending the technique developed by Janis et al. (1969) for static fields to the case of nonstatic fields, an exact time-dependent axially symmetric solution of EMY fields has been obtained.

066.051   Tensor-virial equations for post-Newtonian relativistic stellar dynamics.   N. Spyrou.
Gen. Relativ. Gravitation, Vol. 8, 463 - 489 (1977).

The author derives the tensor-virial theorem and the angular momentum integral for a system of bodies of finite dimensions in the post-Newtonian approximation of general relativity.

066.052   On the energy in relativistic stellar dynamics.
N. Spyrou.

Gen. Relativ. Gravitation, Vol. 8, 491 - 495 (1977).

Using the results of the preceding paper (066.051) the author evaluates in the post-Newtonian approximation the energy integral for a system of extended bodies with arbitrary internal structure and internal motions.

066.053 **A test theory of special relativity: I. Simultaneity and clock synchronization.**
R. Mansouri, R. U. Sexl.
Gen. Relativ. Gravitation, Vol. 8, 497 - 513 (1977).

The role of convention in various definitions of clock synchronization and simultaneity is investigated. The authors show that two principal methods of synchronization can be considered: system internal and system external synchronization. An ether theory is constructed that maintains absolute simultaneity and is kinematically equivalent to special relativity.

066.054 **A test theory of special relativity: II. First order tests.**   R. Mansouri, R. U. Sexl.
Gen. Relativ. Gravitation, Vol. 8, 515 - 524 (1977).

First-order tests of special relativity are based on a comparison of clocks synchronized with the help of slow clock transport with those synchronized by the Einstein procedure. This comparison enables the measurement of the one-way velocity of light and is equivalent to a measurement of the time dilatation factor. The accuracy of present measurements is of the order 10^{-7}, yielding an upper limit of 3 cm/sec for the ether drift.

066.055 **A generalization of the concept of constant mean curvature and canonical time.**   A. J. Goddard.
Gen. Relativ. Gravitation, Vol. 8, 525 - 537 (1977).

Some compact spaces of achronal hypersurfaces are constructed in various types of space-time. A variational principle is introduced on these spaces, smooth extremals of which are spacelike hypersurfaces of constant mean curvature. The family of such hypersurfaces generated by altering the value of the mean curvature is discussed and the mean curvature itself is shown to have many of the properties of a canonical time coordinate.

066.056 **Szekeres's space-times have no Killing vectors.**
W. B. Bonnor, A. H. Sulaiman, N. Tomimura.
Gen. Relativ. Gravitation, Vol. 8, 549 - 559 (1977).

Szekeres obtained some exact solutions of Einstein's equations for dust which have applications in cosmology, in the theory of collapse, and in the study of gravitational radiation. It is shown that in their most general form these space-times have no Killing vectors.

066.057 **High-frequency, self-gravitating, charged scalar fields.**
Y. Choquet-Bruhat, A. H. Taub.
Gen. Relativ. Gravitation, Vol. 8, 561 - 571 (1977).

066.058 **Stationary solutions, energy, and the Bel-Robinson tensor.**   S. Deser.
Gen. Relativ. Gravitation, Vol. 8, 573 - 579 (1977).

·The absence of stationary source-free solutions and the positive energy problem in general relativity are discussed, at the linearized level, in terms of the Bel-Robinson tensor. The possibility is raised that there may exist stationary solutions to the full Einstein equations in five dimensions.

066.059 **Multipole structure of static continuous matter distributions in general relativity.**   W. G. Dixon.
Gen. Relativ. Gravitation, Vol. 8, 595 - 601 (1977).

A mass skeleton is defined for a static extended body in a gravitational field. It is a scalar-valued distribution on a tangent space, and is equivalent to that part of the reduced multipole moment structure which describes the mass density of the body. An explicit form is given for this distribution in terms

of the mass density and the scalar potential of the field.

066.060 **Event horizon and scalar potential.**
J.-P. Duruisseau, M.-A. Tonnelat.
Gen. Relativ. Gravitation, Vol. 8, 603 - 610 (1977).

The introduction of a scalar potential with a more general schema than general relativity eliminates the "event horizon". Among possible solutions, the Schwarzschild one represents a singular case. A study of the geodesic properties of the matching with an approximated interior solution are given. A new definition of the gravitational mass and χ function is deduced.

066.061 **Extremality of mass in the bimetric theory of gravitation.**   I. Goldman, N. Rosen.
Gen. Relativ. Gravitation, Vol. 8, 617 - 621 (1977).

In the bimetric theory of gravitation, the static spherically symmetric case involving matter characterized by density and pressure is considered. It is found that the condition that the mass be stationary under small variations of the field variables (including the density) for a fixed number of baryons leads to the field equations and to the equilibrium condition.

066.062 **Gravitational law and spinning electron equation in a De Sitter symmetric space.**   L. Halpern.
Gen. Relativ. Gravitation, Vol. 8, 623 - 630 (1977).

A gravitational law is proposed on a De Sitter covariant space. Dirac's De Sitter covariant spinning electron equation is generalized to the presence of gravitational fields. The resulting equation differs from the generally covariant Dirac equation by a mass renormalization.

066.063 **On theories of gravitation with higher-order field equations.**   P. Havas.
Gen. Relativ. Gravitation, Vol. 8, 631 - 645 (1977).

Weyl and Eddington suggested three alternative general relativistic theories of gravitation with fourth-order field equations which in empty space admit the Schwarzschild metric as a solution. It is shown here that in the presence of extended sources, Weyl's and Eddington's theories (as well as all other higher-order metric theories derivable from an action principle) contradict Newton's law of gravitation in the nonrelativistic limit.

066.064 **An irradiated Schwarzschild object.**   Z. Perjés.
Gen. Relativ. Gravitation, Vol. 8, 689 - 693 (1977).

A generalization of the Schwarzschild space-time, a Petrov type-II electrovacuum solution, is obtained. The metric is stationary but the Maxwell tensor contains an arbitrary function of time. The curvature invariants and singularities are investigated using $SU(2)$ spin coefficient techniques.

066.065 **A simple proof of the generalization of Israel's theorem.**   D. C. Robinson.
Gen. Relativ. Gravitation, Vol. 8, 695 - 698 (1977).

A simple proof of the generalization of the theorem of Israel concerning the uniqueness of the Schwarzschild black hole is presented.

066.066 **A variational principle giving gravitational "super-potentials", the affine connection, Riemann tensor, and Einstein field equations.**   J. Stachel.
Gen. Relativ. Gravitation, Vol. 8, 705 - 712 (1977).

066.067 **Observation du champ stellaire au voisinage du Soleil pendant une éclipse totale.**
H. Debehogne, R. R. de Freitas Mourão.
Ciel Terre, Vol. 93, 189 - 203 (1977).

066.068 **A self-similar current of spherically symmetric accretion taking into account the pressure gradient.**
Ya. M. Kazhdan, A. E. Lutskij.
Inst. prikl. mat. AN SSSR. Prepr. No. 33. Moskva, 1977.

11 pp. Price 5 Kop. In Russian. – Abstr. in Ref. zh., 51. Astron., 9.51.849 (1977).

066.069 Mach, motion, and Minkowski. J. B. Barbour.
Acta Cosmologica, Zesz. 6, 105 - 108 (1977).
The failure of general relativity to give an entirely satisfactory explanation of the phenomenon of inertia is discussed. It is concluded that this is due to the fact that Einstein employed the Minkowskian spacetime concept, which is essentially antithetical to the Machian ideas which he was trying to implement.

066.070 Space-time structures. M. Heller.
Acta Cosmologica, Zesz. 6, 109 - 128 (1977).
This paper presents different mathematical structures (differential, topological, conformal projective, spinor, causal, chronological) which may be abstracted from the Lorentz structure of space-time. Equivalently, the Lorentz structure can be constructed out of more "elementary" structures by enriching them through a suitable system of axioms. Interstructure relations are analysed.

066.071 How do neutron stars evolve? W. Kundt.
Naturwissenschaften, 64. Jahrg., 493 - 498 (1977).
A possible evolutionary history of neutron stars is delineated, ranging from star formation through supernova explosions, X-ray and pulsar stages to an eventual pulsar turnoff via spin alignment, or magnetic field decay. New is the emphasis that there should be two populations of pulsars, the slow and the fast one, according to their successive formation in a binary system. The two pulsar populations are related to two phenotypes of supernova remnants, of which Cas A and the Crab are the best known examples.

066.072 Convective accretion disks and X-ray bursters. E. P. T. Liang.
Astrophys. J., Vol. 218, 243 - 246 (1977).
Convection driven by the turbulence of a thin accretion disk could play an important role in the disk structure and stability. A new high-surface-density state of the inner X-ray disk that is secularly stable is discovered when a parameter characterizing convective transport exceeds the viscosity parameter. A bursting mechanism based on this new state is proposed for X-ray bursters such as 3U 1820–30.

066.073 The cosmic background spectrum between 0.7 and 3.0 mm. E. I. Robson.
Far infrared astronomy (see 012.027), p. 115 - 124 (1976).
A liquid cooled twin-beam polarizing interferometer was flown by balloon to an altitude of 40 km on March 13th 1974. The instrument was sensitive to radiation in the wavelength region 3 mm to 350 μm. Interferograms were obtained which when Fourier transformed appear to show the turnover of the cosmic background radiation at wavelengths shorter than 1 mm and correspond to a blackbody temperature of around 2.9 K.

066.074 Temperature fluctuations in the microwave background due to gravitational waves.
A. G. Doroshkevich, I. D. Novikov, A. G. Polnarev.
Astron. Zh. Akad. Nauk SSSR, Vol. 54, 932 - 944 (1977). In Russian. English translation in Soviet Astron., Vol. 21, No. 5.
The influence of cosmological gravitational waves on the anisotropy of primordial electromagnetic radiation is considered. The results are given in a form suitable for interpretation of observational data. Comparison of the results of calculations with observational data gives an upper limit to the possible energy density of gravitational waves in different spectral regions. Expected anisotropy of primordial electromagnetic radiation is calculated for different assumptions on the gravitational wave spectrum.

066.075 On gravitational radiation of neutron stars and white dwarfs. Yu. L. Vartanyan, G. S. Adzhyan.
Astron. Zh. Akad. Nauk SSSR, Vol. 54, 1047 - 1050 (1977). In Russian. English translation in Soviet Astron., Vol. 21, No. 5.
The possibility of generation of gravitational radiation by pulsations of rotating oblate neutron stars and white dwarfs is considered. Angular distribution, polarization, intensity, and damping time scale of such a radiation are calculated. Estimates of these parameters for rapidly rotating neutron stars and white dwarfs are given.

066.076 The magnetopause of an accreting neutron star. M. M. Basko.
Astron. Zh. Akad. Nauk SSSR, Vol. 54, 1051 - 1061 (1977). In Russian. English translation in Soviet Astron., Vol. 21, No. 5.
Calculated are the shape of the Alfvén surface separating the magnetopause of a neutron star from the accreting plasma stream, as well as velocity and optical depth of the gas layer flowing over the Alfvén surface. Calculations are performed for the spherically symmetric falling of the gas and the disk accretion. It is shown that infall of matter from the Alfvén radius onto the surface of the neutron star is only possible if the plasma is "frozen in" in the magnetic field of the star.

066.077 Nearly collisionless spherical accretion. M. C. Begelman.
Mon. Not. R. Astron. Soc., Vol. 181, 347 - 363 (1977).
A fluid-like gas accretes much more efficiently than a collisionless gas. The ability of an accreting gas to behave like a fluid depends on the relationship of the mean free path of a gas particle at $r \to \infty$ (λ_∞), to the typical length scales associated with the star–gas system. The author examines this relationship in detail by generalizing the model of Danby & Camm to cases where λ_∞ is finite. The result may apply to the accretion of grains and the dynamics of stellar cusps around massive black holes, as well as providing a cautionary tale for those who accept the fluidity of all accreting plasmas as gospel truth.

066.078 Spectrum of a radio pulse from an exploding black hole. R. D. Blandford.
Mon. Not. R. Astron. Soc., Vol. 181, 489 - 498 (1977).
Black holes of mass $\lesssim 10^{11}$ kg may be able to radiate all of their rest mass in an essentially instantaneous explosion by the Hawking process provided that the number of available modes increases sufficiently rapidly with temperature. Rees has pointed out that should this happen, electron–positron pairs may form a significant fraction of the explosion products and that if the hole is surrounded by a typical interstellar magnetic field, the kinetic energy of the pairs may be efficiently transformed into radio waves. In this paper the spectrum of the observed radiation is calculated and conditions necessary for this to be detectable are further discussed.

066.079 An upper limit to the cosmological variation of Planck's constant derived from the spectrum of the microwave background radiation. G. M. Blake.
Mon. Not. R. Astron. Soc., Vol. 181, 47P - 50P (1977).
The blackbody spectrum of the microwave background is used to set limits on the cosmological variation of Planck's constant. This approach has accuracy and limitations similar to a method based on photoelectric observations of light from distant galaxies, but extends these observations to a much earlier epoch in the evolution of the Universe.

066.080 Verification of the principle of equivalence for massive bodies.
I. I. Shapiro, C. C. Counselman III, R. W. King.
Scientific applications of lunar laser ranging, (see 012.012), p. 89 (1977). – Abstract. Published in Phys. Rev. Lett., Vol. 36, 555 - 558 (1976). – See abstract 17.066.072.

066.081 Whistler mode radiation during gravitational

collapse of magnetized stars and gas clouds.
R. R. Burman.
Publ. Astron. Soc. Japan, Vol. 29, 639 - 642 (1977).
    When a magnetized star or gas cloud collapses towards
its event horizon, its magnetic field as seen by an external
observer decays to zero. It is pointed out that energy so released could, in special circumstances, travel through the surrounding plasma in a whistler mode.

066.082   Microwave background spectrum — survey of
          recent results.   E. I. Robson, P. E. Clegg.
IAU Symp., No. 74, (see 012.014), p. 319 - 325 (1977).

066.083   Small scale fluctuations in the microwave back-
          ground radiation associated with the formation of
galaxies.   R. A. Sunyaev.
IAU Symp., No. 74, (see 012.014), p. 327 - 334 (1977).

066.084   Fluctuations in the microwave background caused
          by anisotropy of the universe and gravitational
waves.   I. D. Novikov.
IAU Symp., No. 74, (see 012.014), p. 335 - 339 (1977).

066.085   Are white holes blue or red?   R. C. Roeder.
          J. R. Astron. Soc. Canada, Vol. 71, 406 (1977).
Abstract.

066.086   A new class of inhomogeneous white-hole solutions.
          K. Lake.
J. R. Astron. Soc. Canada, Vol. 71, 406 (1977). — Abstract.

066.087   The microwave background radiation in the direc-
          tion of clusters of galaxies.   S. F. Gull.
Highlights of Astronomy, Vol. 4, Part I, (see 012.020), p.
341 - 342 (1977).

066.088   Measurements of the polarized sky background in
          the far infrared.
A. Coletti, F. Melchiorri, V. Natale.
Far infrared astronomy (see 012.027), p. 125 - 130 (1976).
    Observations carried out with a balloon-borne polarim-
eter have shown an upper limit of about 1% for the polarization of the cosmic background radiation in the 500–1500 μm
wavelength region. Moreover, some particular features of the
atmospheric emission are discussed.

066.089   Limits on a microwave background without the big
          bang.
J. V. Narlikar, M. G. Edmunds, N. C. Wickramasinghe.
Far infrared astronomy (see 012.027), p. 131 - 142 (1976).
    The possibility of explaining the cosmic microwave back-
ground in terms of thermalisation of radiation from such
sources as galaxies by dust grains is explored further. Relevant
calculations of the optical cross-sections of graphite whiskers
are given and its is shown that a smeared out dust density of
$\sim 10^{-33}$ g cm^{-3} is required. Limits are set on the large-angle
anisotropy of the background which is to be expected on the
basis of this model. The relative merits of the conventional
explanation and the present theory are discussed and a few
discriminatory observational tests proposed. Some cosmological implications of whisker grains in the intergalactic space are
examined.

066.090   General discussion on infrared and millimetre back-
          ground.   K. Shivanandan (chairman) with
contributions by P. E. Clegg, and P. Marchant.
Far infrared astronomy (see 012.027), p. 143 - 149 (1976).

066.091   Beyond the rim of the gravitational chasm.
          I. D. Novikov.
Zemlya i Vselennaya, 1977, No. 5, p. 52 - 57. In Russian.

066.092   Debye potentials for electromagnetic waves in the
          presence of a charged black hole.
G. Stephenson.
J. Phys. A, Vol. 10, L7 - L10 (1977). — Abstr. in Phys. Abstr.,
Vol. 80, Abstr. 22261 (1977).

066.093   The Klein-Gordon equation in a Kerr-Newman
          background space.
D. J. Rowan, G. Stephenson.
J. Phys. A , Vol. 10, 15 - 23 (1977). — Abstr. in Phys. Abstr.,
Vol. 80, Abstr. 22262 (1977).

066.094   Ejection of massive black holes from galaxies.
          R. Chander Kapoor.
Pramāṇa, Vol. 7, 334 - 343 (1976). — Abstr. in Phys. Abstr.,
Vol. 80, Abstr. 22263 (1977).

066.095   Zur physikalischen Bedeutung einer Quantlung der
          Gravitation.   H.-J. Treder.
Sitzungsber. Akad. Wiss. DDR, Math. Naturwiss. Tech.,
Jahrg. 1977, Nr. 3N, p. 29 - 45.

066.096   Recurrence conditions in space-time.   G. S. Hall.
          J. Phys. A, Vol. 10, 29 - 42 (1977). — Abstr. in
Phys. Abstr., Vol. 80, Abstr. 17883 (1977).

066.097   Variational principles and gauge theories of gravita-
          tion.   G. Stephenson.
J. Phys. A, Vol. 10, 181 - 184 (1977). — Abstr. in Phys. Abstr.,
Vol. 80, Abstr. 17884 (1977).

066.098   Relativistic canonical systems: a geometric approach
          to their space-time structure and symmetries.
K. Druhl.
Group theoretical methods in physics, 4th International
Colloquium, Nijmegen, Netherlands, June 1975. A. Janner,
T. Janssen, M. Boon (Editors). Springer-Verlag, Berlin.
Price $ 20.50 (1976), p. 172 - 181. — Abstr. in Phys. Abstr.,
Vol. 80, Abstr. 17885 (1977).

066.099   The collision of plane waves in general relativity.
          J. B. Griffiths.
Ann. Physics., Vol. 102, 388 - 404 (1976). — Abstr. in Phys.
Abstr., Vol. 80, Abstr. 17886 (1977).

066.100   A note on a theory of gravity.   P. Rastall.
          Canadian J. Phys., Vol. 55, 38 - 42 (1977).
Abstr. in Phys. Abstr., Vol. 80, Abstr. 17888 (1977).

066.101   Existence and construction of neutrino fields with
          geodesic and shear-free rays and of other zero rest-
mass fields in curved space-time.   J. Audretsch.
J. Phys. A, Vol. 10, 43 - 48 (1977). — Abstr. in Phys. Abstr.,
Vol. 80, Abstr. 17890 (1977).

066.102   Static plane-symmetric solutions in Brans-Dicke and
          Sen-Dunn theories of gravitation.   D. R. K. Reddy.
J. Phys. A, Vol. 10, 55 - 58 (1977). — Abstr. in Phys. Abstr.,
Vol. 80, Abstr. 17891 (1977).

066.103   On Birkhoff's theorem for electromagnetic fields in
          a scalar-tensor theory of gravitation.
D. R. K. Reddy.
J. Phys. A, Vol. 10, 185 - 188 (1977). — Abstr. in Phys.
Abstr., Vol. 80, Abstr. 17892 (1977).

066.104   Restrictions on gauge theory of gravitation.
          K. Hayashi.
Phys. Lett. B, Vol. 65B, 437 - 440 (1976). — Abstr. in Phys.
Abstr., Vol. 80, Abstr. 17893 (1977).

066.105 Supergravity and local extended supersymmetry.
S. Ferrara, J. Scherk, B. Zumino.
Phys. Lett. B, Vol. 66B, 35 - 38 (1977). – Abstr. in Phys.
Abstr., Vol. 80, Abstr. 17894 (1977).

066.106 On the efficiency of cosmic gravitational waves
detection. A. Ceapa.
Rev. Roumaine Phys., Vol. 21, 1121 - 1123 (1976). – Abstr.
in Phys. Abstr., Vol. 80, Abstr. 17899 (1977).

066.107 The search for gravitational radiation, a large-scale
experiment requiring temperatures in the millikelvin
region. M. S. McAshan.
Proceedings of the 6th International Cryogenic Engineering
Conference, Grenoble, France, 11 - 14 May 1976. F. R. S.
Mendelssohn (Editor). IPC Sci. & Technol. Press, Guildford,
Surrey, England. Price £ 29.00 (1976), p. 3 - 10. – Abstr. in
Phys. Abstr., Vol. 80, Abstr. 17900 (1977).

066.108 Can local measurements resolve the twin paradox
in a Kerr metric? D. E. Hall.
American J. Phys., Vol. 44, 1204 - 1208 (1976). – Abstr. in
Phys. Abstr., Vol. 80, Abstr. 22449 (1977).

066.109 Gravitational field of a radiating rotating body.
M. Carmeli, M. Kaye.
Ann. Physics, Vol. 103, 97 - 120 (1977). – Abstr. in Phys.
Abstr., Vol. 80, Abstr. 22450 (1977).

066.110 Conservation laws in de Sitter space from action
principles in five-space. M. S. Drew.
Ann. Physics, Vol. 103, 469 - 495 (1977). – Abstr. in Phys.
Abstr., Vol. 80, Abstr. 22451 (1977).

066.111 On a solution of the Einstein-Maxwell equations
admitting a nonsingular field.
R. G. McLenaghan, N. Tariq.
J. Math. Phys., New York, Vol. 17, 2192 - 2197 (1976).
Abstr. in Phys. Abstr., Vol. 80, Abstr. 22452 (1977).

066.112 Spontaneous compactification of extra space
dimensions. E. Cremmer, J. Scherk.
Nucl. Phys. B, Vol. B118, 61 - 75 (1977). – Abstr. in Phys.
Abstr., Vol. 80, Abstr. 22458 (1977).

066.113 New Weyl-type vacuum space-times of Einstein's
gravitational theory. J. Meinhardt.
Nuovo Cimento B, Ser. 11, Vol. 36B, 156 - 164 (1976).
Abstr. in Phys. Abstr., Vol. 80, Abstr. 22459 (1977).

066.114 On the definition of entropy in relativistic kinetic
theory. E. Alvarez.
Phys. Lett. A, Vol. 60A, 9 - 10 (1977). – Abstr. in Phys.
Abstr., Vol. 80, Abstr. 22461 (1977).

066.115 Gauge theory of Poincaré symmetry. Y. M. Cho.
Phys. Rev. D, Vol. 14, 3335 - 3340 (1976).
Abstr. in Phys. Abstr., Vol. 80, Abstr. 22463 (1977).

066.116 Gauge theory, gravitation, and symmetry.
Y. M. Cho.
Phys. Rev. D, Vol. 14, 3341 - 3344 (1977). – Abstr. in Phys.
Abstr., Vol. 80, Abstr. 22464 (1977).

066.117 Gravitational fields with a two-dimensional group
of motion. D. Kramen.
Wiss. Z. Friedrich-Schiller Univ. Jena, Math. Naturwiss. Reihe,
Vol. 25, 447 - 454 (1976). In German. – Abstr. in Phys. Abstr.,
Vol. 80, Abstr. 22466 (1977).

066.118 Shear-free and acceleration-free fluids in Einstein's
theory of gravitation.

G. Neugebauer, M. Sust.
Wiss. Z. Friedrich-Schiller Univ. Jena, Math. Naturwiss. Reihe,
Vol. 25, 455 - 460 (1976). In German. – Abstr. in Phys. Abstr.,
Vol. 80, Abstr. 22467 (1977).

066.119 Gravitational radiation from n-body systems.
T. T. Chia.
Ann. Physics, Vol. 103, 233 - 250 (1977). – Abstr. in Phys.
Abstr., Vol. 80, Abstr. 22468 (1977).

066.120 Generalised Einstein-Cartan field equations.
R. Skinner, D. Gregorash.
Phys. Rev. D, Vol. 14, 3314 - 3321 (1976). – Abstr. in Phys.
Abstr., Vol. 80, Abstr. 22472 (1977).

066.121 Quantum gravitation and the perihelion anomaly.
J. N. Kidman.
Nuovo Cimento, Lett., Ser. 2, Vol. 18, No. 6, p. 181 - 182
(1977). – Abstr. in Phys. Abstr., Vol. 80, Abstr. 22473
(1977).

066.122 Pair production by gravitational waves in the field
of a black hole. R. A. Matzner, Y. Nutku.
Phys. Lett. A, Vol. 60A, 5 - 7 (1977). – Abstr. in Phys.
Abstr., Vol. 80, Abstr. 22474 (1977).

066.123 Quantum vacuum energy in a closed universe.
L. H. Ford.
Phys. Rev. D, Vol. 14, 3304 - 3313 (1976). – Abstr. in Phys.
Abstr., Vol. 80, Abstr. 22479 (1977).

066.124 Low-frequency-limit cross sections for charged
black holes. R. A. Matzner.
Phys. Rev. D, Vol. 14, 3274 - 3280 (1976). – Abstr. in Phys.
Abstr., Vol. 80, Abstr. 25935 (1977).

066.125 Capture of particles from plunge orbits by a black
hole. P. J. Young.
Phys. Rev. D, Vol. 14, 3281 - 3289 (1976). – Abstr. in Phys.
Abstr., Vol. 80, Abstr. 25936 (1977).

066.126 Why is a black hole hot? U. H. Gerlach.
Phys. Rev. D, Vol. 14, 3290 - 3293 (1976).
Abstr. in Phys. Abstr., Vol. 80, Abstr. 25937 (1977).

066.127 Uniqueness of perturbation of a Reissner-Nordstrøm
black hole. C. H. Lee, W. D. McGlinn.
J. Math. Phys. New York, Vol. 17, 2159 - 2165 (1976).
Abstr. in Phys. Abstr., Vol. 80, Abstr. 26022 (1977).

066.128 Absorption cross section of small black holes.
W. G. Unruh.
Phys. Rev. D, Vol. 14, 3251 - 3259 (1976). – Abstr. in Phys.
Abstr., Vol. 80, Abstr. 26023 (1977).

066.129 Particle emission rates from a black hole. II. Massive
particles from a rotating hole. D. N. Page.
Phys. Rev. D, Vol. 14, 3260 - 3273 (1976). – Abstr. in Phys.
Abstr., Vol. 80, Abstr. 26024 (1977).

066.130 On the perihelion precession as a Machian effect.
P. B. Eby.
Nuovo Cimento, Lett., Ser. 2, Vol. 18, No. 3, p. 93 - 96
(1977). – Abstr. in Phys. Abstr., Vol. 80, Abstr. 26098
(1977).

066.131 A master differential equation for perturbations of a
Schwarzschild black hole. S. Persides.
Nuovo Cimento, Lett., Ser. 2, Vol. 17, 444 - 446 (1976).
Abstr. in Phys. Abstr., Vol. 80, Abstr. 26214 (1977).

066.132 A class of exact interior solutions of the Einstein-

Maxwell equations.   J. N. Islam.
Proc. R. Soc. London, Ser. A, Vol. 353, 523 - 531 (1977).
Abstr. in Phys. Abstr., Vol. 80, Abstr. 26215 (1977).

066.133 On the bimetric theory of gravitation.   H. Yilmaz.
Gen. Relativ. Gravitation, Vol. 8, 957 - 962 (1977).
Rosen's bimetric theory is analyzed anew and is shown to
have deficiencies if the space is assumed to be Riemannian.
The problems are due mainly to the introduction of the flat
metric $\gamma^{\mu\nu}$, and the identification of the stress-energy tensor,
$T^{\mu\nu}$. It is indicated that if the Riemannian interpretation could
be avoided the theory still holds promise as a viable theory of
gravitation.

066.134 Gravitational theory with local de Sitter invariance.
Y. An, S. Chen, C.-l. Tsou, H.-y. Kuo.
Kexue Tongbao, No. 8, p. 379 - 382 (1976). In Chinese.
Abstr. in Phys. Abstr., Vol. 80, Abstr. 26219 (1977).

066.135 The necessity of quantizing the gravitational field.
K. Eppley, E. Hannah.
Found. Phys., Vol. 7, No. 1-2, p. 51 - 68 (1977). – Abstr. in
Phys. Abstr., Vol. 80, Abstr. 26220 (1977).

066.136 Superdense matter.   Y. Nagoaka.
Solid State Phys., Vol. 11, 465 - 473 (1976). In
Japanese. – Abstr. in Phys. Abstr., Vol. 80, Abstr. 29643
(1977).

066.137 Stationary electromagnetic fields around black holes.
I. General solutions and the fields of some special
sources near a Schwarzschild black hole.
J. Bicak, L. Dvorak.
Czechoslovak J. Phys. B, Vol. B27, No. 2, p. 127 - 147 (1977).
Abstr. in Phys. Abstr., Vol. 80, Abstr. 29864 (1977).

066.138 Neutron astrophysics.   R. A. Smith.
Neutron interactions with nuclei, I. International
conference, Lowell, Mass., USA, 6 - 9 July 1976. E. Sheldon
(Editor). Nat. Tech. Inf. Service, Springfield, Va., USA. Price
$ 23.75 (1976), p. 597 - 610. – Abstr. in Phys. Abstr., Vol. 80,
Abstr. 29872 (1977).

066.139 Spectral features of radiation from Nordstrøm and
Kerr-Newman white holes.   N. Dadhich.
Pramāṇa, Vol. 8, 14 - 21 (1977). – Abstr. in Phys. Abstr., Vol.
80, Abstr. 29967 (1977).

066.140 Gamma rays from accreting black holes.
L. Maraschi, A. Treves.
Astrophys. J., Lett., Vol. 218, L113 - L115 (1977).
Using the model of spherical turbulent accretion onto
black holes proposed by Mészáros, the authors estimate the
electric field produced by the motion of magnetized turbulence
cells in the accretion flow. Electrons and protons are accelerated in this field, since the runaway condition is attained. The
maximum synchrotron frequency of electrons is ~20 MeV independent of the mass of the hole and of its accretion rate.
The shape of the spectrum produced via the synchrotron and
Compton mechanisms by the runaway electrons is evaluated.

066.141 On particle detection in the de Sitter space-time.
P. Hajicek.
Nuovo Cimento, Lett., Ser. 2, Vol. 18, 251 - 254 (1977).
Abstr. in Phys. Abstr., Vol. 80, Abstr. 30040 (1977).

066.142 A classification of space-time structures.
A. Trautman.
Rep. Math. Phys., Vol. 10, 297 - 310 (1976). – Abstr. in Phys.
Abstr., Vol. 80, Abstr. 30057 (1977).

066.143 Junction conditions for the Einstein-Cartan theory.

R. Skinner, I. Webb.
Acta Phys. Polonica B, Vol. B8, 81 - 91 (1977). – Abstr. in
Phys. Abstr., Vol. 80, Abstr. 30058 (1977).

066.144 Properties of a solution of the Einstein equations
with the cosmological constant.
J. K. Kowalczynski, J. F. Plebanski.
Acta Phys. Polonica B, Vol. B8, 169 - 171 (1977). – Abstr. in
Phys. Abstr., Vol. 80, Abstr. 30059 (1977).

066.145 A new family of solutions of the Einstein field
equations.   F. J. Ernst.
J. Math. Phys., New York, Vol. 18, 233 - 234 (1977). – Abstr.
in Phys. Abstr., Vol. 80, Abstr. 30061 (1977).

066.146 On plane symmetric Einstein-Maxwell fields.
A. Banerjee, N. Chakrabarty.
J. Math. Phys., New York, Vol. 18, 265 - 266 (1977). – Abstr.
in Phys. Abstr., Vol. 80, Abstr. 30062 (1977).

066.147 Symmetry mappings concomitant to particle-number-
conservation-baryon-number-conservation.
W. R. Davis.
Nuovo Cimento, Lett., Ser. 2, Vol. 18, 319 - 323 (1977).
Abstr. in Phys. Abstr., Vol. 80, Abstr. 30063 (1977).

066.148 Conformal uniqueness of the Schwarzschild interior
metric.   M. Gurses.
Nuovo Cimento, Lett., Ser. 2, Vol. 18, 327 - 328 (1977).
Abstr. in Phys. Abstr., Vol. 80, Abstr. 30064 (1977).

066.149 On the line element of a static spherical shell of
matter in vacuum.   T. L. Chow.
Nuovo Cimento B, Ser. 11, Vol. 37B, 214 - 218 (1977).
Abstr. in Phys. Abstr., Vol. 80, Abstr. 30065 (1977).

066.150 Gravitational instantons.   S. W. Hawking.
Phys. Lett. A, Vol. 60A, 81 - 83 (1977). – Abstr. in
Phys. Abstr., Vol. 80, Abstr. 30066 (1977).

066.151 Dimensional analysis in relativistic gravitational
theories.   S. J. Aldersley.
Phys. Rev. D, Vol. 15, 370 - 376 (1977). – Abstr. in Phys.
Abstr., Vol. 80, Abstr. 30067 (1977).

066.152 The effect of gravitational waves from a spinning rod
on its rigidity.   T. T. Chia.
Int. J. Theor. Phys., Vol. 15, 283 - 291 (1976). – Abstr. in
Phys. Abstr., Vol. 80, Abstr. 30068 (1977).

066.153 Some exact solutions of charged general relativistic
fluid spheres.   A. Nduka.
Acta Phys. Polonica B, Vol. B8, 75 - 79 (1977). – Abstr. in
Phys. Abstr., Vol. 80, Abstr. 30069 (1977).

066.154 On the vacuum stress induced by uniform accelera-
tion or supporting the ether.
P. Candelas, D. Deutsch.
Proc. R. Soc. London, Ser. A, Vol. 354, 79 - 99 (1977).
Abstr. in Phys. Abstr., Vol. 80, Abstr. 30071 (1977).

066.155 Primal cosmic energy field and the theory of
relativity.   L. Covez.
Indian J. Pure Appl. Phys., Vol. 14, 1012 - 1013 (1976).
Abstr. in Phys. Abstr., Vol. 80, Abstr. 30072 (1977).

066.156 Charged particles in Einstein's unified field theory.
C. R. Johnson, J. R. Nance.
Phys. Rev. D, Vol. 15, 377 - 388 (1977). – Abstr. in Phys.
Abstr., Vol. 80, Abstr. 30073 (1977).

066.157 The Fresnel formula applied to empty space.

H. Aspden.
Int. J. Theor. Phys., Vol. 15, 263 - 264 (1976). – Abstr. in
Phys. Abstr., Vol. 80, Abstr. 30082 (1977).

066.158   Superspace formulation of supergravity.
J. Wess, B. Zumino.
Phys. Lett. B, Vol. 66B, 361 - 364 (1977). – Abstr. in Phys.
Abstr., Vol. 80, Abstr. 30084 (1977).

066.159   Scalar-particle production near the singularity in an
anisotropic universe. I. Scalar field theory.
E. Pessa.
Nuovo Cimento B, Ser. 2, Vol. 37B, 155 - 184 (1977). – Abstr.
in Phys. Abstr., Vol. 80, Abstr. 30238 (1977).

066.160   Energy per particle and effective mass of neutron
matter.   R. K. Satpathy, A. P. Mishra.
Indian J. Pure Appl. Phys., Vol. 15, No. 1, p. 32 - 35 (1977).
Abstr. in Phys. Abstr., Vol. 80, Abstr. 30499 (1977).

066.161   A relativistic model for the solid core of a neutron
star.   M. Cattani, N. C. Fernandes.
Nuovo Cimento, Lett., Ser. 2, Vol. 18, 324 - 326 (1977).
Abstr. in Phys. Abstr., Vol. 80, Abstr. 33546 (1977).

066.162   Origin of the particles in black-hole evaporation.
W. G. Unruh.
Phys. Rev. D, Vol. 15, 365 - 369 (1977). – Abstr. in Phys.
Abstr., Vol. 80, Abstr. 33550 (1977).

066.163   Tachyon emission from white-holes.
S. V. Dhurandhar.
Pramāna, Vol. 8, 133 - 143 (1977). – Abstr. in Phys. Abstr.,
Vol. 80, Abstr. 33718 (1977).

066.164   On the construction of the Taub-NUT congruence.
P. A. Hogan, T. Criss.
Int. J. Theor. Phys., Vol. 15, 207 - 212 (1976). – Abstr. in
Phys. Abstr., Vol. 80, Abstr. 33910 (1977).

066.165   Black-hole magnetostatics.   A. R. King.
Math. Proc. Cambridge Philos. Soc., Vol. 81, Pt. 1,
p. 149 - 156 (1977). – Abstr. in Phys. Abstr., Vol. 80, Abstr.
33913 (1977).

066.166   A note on the Kerr solution.   P. A. Hogan.
Phys. Lett. A, Vol. 60A, 161 - 162 (1977). – Abstr.
in Phys. Abstr., Vol. 80, Abstr. 33914 (1977).

066.167   On incompleteness of some solutions of the Einstein-
Cartan equations with motion of matter.
O. I. Bogoyavlensky (Bogoyavlenskij).
Phys. Lett. A, Vol. 60A, 163 - 164 (1977). – Abstr. in Phys.
Abstr., Vol. 80, Abstr. 33915 (1977).

066.168   Positivity of energy and stationary solutions in
general relativity.   N. O. Murchadha.
Phys. Lett. A, Vol. 60A, 177 - 178 (1977). – Abstr. in Phys.
Abstr., Vol. 80, Abstr. 33916 (1977).

066.169   The interaction of a quantum mechanical oscillator
with gravitational radiation.   N. Grafe, H. Dehnen.
Int. J. Theor. Phys., Vol. 15, 393 - 409 (1976). – Abstr. in
Phys. Abstr., Vol. 80, Abstr. 33918 (1977).

066.170   Plane wave solutions in scalar tensor theories and
solutions of source-free Einstein-Maxwell theory.
D. Ray.
J. Math. Phys., New York, Vol. 18, 245 - 249 (1977). – Abstr.
in Phys. Abstr., Vol. 80, Abstr. 33919 (1977).

066.171   Some speculations on a causal unification of

relativity, gravitation and quantum mechanics.
V. Buonomano, A. Engel.
Int. J. Theor. Phys., Vol. 15, 231 - 246 (1976). – Abstr. in
Phys. Abstr., Vol. 80, Abstr. 33923 (1977).

066.172   Axially symmetric stationary black-hole states of the
Einstein gravitational theory.   R. Meinhardt.
Nuovo Cimento, Riv., Ser. 2, Vol. 6, 405 - 447 (1976).
Abstr. in Phys. Abstr., Vol. 80, Abstr. 37614 (1977).

066.173   Laser experiments and various four-dimensional
symmetries.   J. P. Hsu.
Found. Phys., Vol. 7, 205 - 220 (1977). – Abstr. in Phys.
Abstr., Vol. 80, Abstr. 37791 (1977).

066.174   On the strengths of field equations.   R. Burman.
Czechoslovak J. Phys. B, Vol. B27, 113 - 116 (1977).
Abstr. in Phys. Abstr., Vol. 80, Abstr. 37794 (1977).

066.175   Complementary aspects of gravitation and electro-
magnetism.   P. F. Browne.
Found. Phys., Vol. 7, 165 - 183 (1977). – Abstr. in Phys.
Abstr., Vol. 80, Abstr. 37795 (1977).

066.176   Eight-dimensional unified theory.   J. Rayski.
Nuovo Cimento, Lett., Ser. 2, Vol. 18, 422 - 424
(1977). – Abstr. in Phys. Abstr., Vol. 80, Abstr. 37797 (1977).

066.177   On the renormalization of quantum gravitation with-
out matter.   P. Van Nieuwenhuizen.
Ann. Physics, Vol. 104, 197 - 217 (1977). – Abstr. in Phys.
Abstr., Vol. 80, Abstr. 37798 (1977).

066.178   Measurement at 4.2 K of the Brownian noise in a
20 kg gravitational wave antenna and upper limit for
gravitational radiation at 8580 Hz.   E. Amaldi, C. Cosmelli,
F. Bordoni, P. Bonifazi, U. Giovanardi, G. Vannaroni, G. V.
Pallottino, G. Pizella, I. Modena.
Nuovo Cimento, Lett., Ser. 2, Vol. 18, 425 - 432 (1977).
Abstr. in Phys. Abstr., Vol. 80, Abstr. 37800 (1977).

066.179   Particles in a stationary spherically symmetric gravi-
tational field.   M. Soffel, B. Muller, W. Greiner.
J. Phys. A, Vol. 10, 551 - 560 (1977). – Abstr. in Phys. Abstr.,
Vol. 80, Abstr. 41293 (1977).

066.180   Charged particle trajectories in a magnetic field on a
curved space-time.   A. R. Prasanna, R. K. Varma.
Pramāna, Vol. 8, 229 - 244 (1977). – Abstr. in Phys. Abstr.,
Vol. 80, Abstr. 41296 (1977).

066.181   Bounds on the moment of inertia of nonrotating
neutron stars.   A. G. Sabbadini, J. B. Hartle.
Ann. Physics, Vol. 104, 95 - 133 (1977). – Abstr. in Phys.
Abstr., Vol. 80, Abstr. 41387 (1977).

066.182   Miniblack holes.   B. Parker.
Astronomy, Vol. 5, No. 2, p. 26 - 31 (1977).
Abstr. in Phys. Abstr., Vol. 80, Abstr. 41388 (1977).

066.183   Cosmic background radiation.
P. L. Richards, D. P. Woody.
2nd international conference and winter school on submillimeter waves and their applications. San Juan, Puerto Rico,
6 - 11 December 1976. IEEE, New York, USA. 16 + 252 pp.
(1976). p. 93 - 95. – Abstr. in Phys. Abstr., Vol. 80, Abstr.
41422 (1977).

066.184   Gravity and inertia in a Machian framework.
J. B. Barbour, B. Bertotti.
Nuovo Cimento B, Ser. 11, Vol. 38B, 1 - 27 (1977). – Abstr.
in Phys. Abstr., Vol. 80, Abstr. 41426 (1977).

066.185 On the determination, from Cauchy data, of
         Killing fields admissible in a spacetime.   B. Coll.
Ann. Inst. Henri Poincaré Sect. A, Vol. 25, 393 - 410 (1976).
In French. — Abstr. in Phys. Abstr., Vol. 80, Abstr. 45744
(1977).

066.186 Evolution of isometries in the Bondi formalism.
         R. Berezdivin, L. Herrera.
J. Math. Phys., Vol. 18, 418 - 423 (1977). — Abstr. in Phys.
Abstr., Vol. 80, Abstr. 45758 (1977).

066.187 Meta universe (tachyon contribution to gravitational
         field).   S. K. Srivastava, M. P. Pathak.
J. Math. Phys., Vol. 18, 483 - 486 (1977). — Abstr. in Phys.
Abstr., Vol. 80, Abstr. 45760 (1977).

066.188 Diverging type-D metrics.
         G. J. Weir, R. P. Kerr.
Proc. R. Soc. London, Ser. A, Vol. 355, 31 - 52 (1977).
Abstr. in Phys. Abstr., Vol. 80, Abstr. 45761 (1977).

066.189 Effective Lagrangian for general relativity in
         presence of a quantized scalar field.
E. Streeruwitz.
Phys. Lett. B, Vol. 67B, No. 2, p. 210 - 212 (1977). — Abstr.
in Phys. Abstr., Vol. 80, Abstr. 45762 (1977).

066.190 Line sources in general relativity.   W. Israel.
         Phys. Rev. D, Vol. 15, 935 - 941 (1977). — Abstr.
in Phys. Abstr., Vol. 80, Abstr. 45763 (1977).

066.191 Possibility of a static scalar field in the Schwarz-
         schild geometry.   R. F. Sawyer.
Phys. Rev. D, Vol. 15, 1427 - 1434 (1977). — Abstr. in Phys.
Abstr., Vol. 80, Abstr. 45765 (1977).

066.192 Motion of a spinning test particle in Vaidya's
         radiating metric.
M. Carmeli, C. Charach, M. Kaye.
Phys. Rev. D, Vol. 15, 1501 - 1517 (1977). — Abstr. in Phys.
Abstr., Vol. 80, Abstr. 45766 (1977).

066.193 Symmetry breaking in superdense matter.
         C.-G. Kallman.
Phys. Lett. B, Vol. 67B, 195 - 197 (1977). — Abstr. in Phys.
Abstr., Vol. 80, Abstr. 45955 (1977).

066.194 Stability of vacuum in the curved space-time of a
         relativistic star.   H. Sato.
Prog. Theor. Phys., Vol. 57, 341 - 343 (1977). — Abstr. in
Phys. Abstr., Vol. 80, Abstr. 49290 (1977).

066.195 The Schrödinger equation and cosmic bodies.
         P. Hertel.
Acta Phys. Austriaca, Suppl. 17, p. 209 - 224 (1977). — Abstr.
in Phys. Abstr., Vol. 80, Abstr. 49418 (1977).

066.196 Specific physical consequences of Mach's principle.
         J. D. Nightingale.
American J. Phys., Vol. 45, 376 - 379 (1977). — Abstr. in Phys.
Abstr., Vol. 80, Abstr. 49524 (1977).

066.197 The back reaction effect in particle creation in
         curved spacetime.   R. M. Wald.
Commun. Math. Phys., Vol. 54, No. 1, p. 1 - 19 (1977).
Abstr. in Phys. Abstr., Vol. 80, Abstr. 49645 (1977).

066.198 Characterization of the Szekeres inhomogeneous
         cosmologies as algebraically special spacetimes.
J. Wainwright.
J. Math. Phys., Vol. 18, 672 - 675 (1977). — Abstr. in Phys.
Abstr., Vol. 80, Abstr. 49646 (1977).

066.199 Rotating steady state configurations in general
         relativity.   E. N. Glass.
J. Math. Phys., Vol. 18, 708 - 711 (1977). — Abstr. in Phys.
Abstr., Vol. 80, Abstr. 49647 (1977).

066.200 Angular momentum in general relativity. I. Defini-
         tion and asymptotic behaviour.   C. R. Prior.
Proc. R. Soc. London, Ser. A, Vol. 354, 379 - 405 (1977).
Abstr. in Phys. Abstr., Vol. 80, Abstr. 49650 (1977).

066.201 A class of solutions of the Dirac equation in the
         Kerr-Newman space.
S. Einstein, R. Finkelstein.
J. Math. Phys., Vol. 18, 664 - 671 (1977). — Abstr. in Phys.
Abstr., Vol. 80, Abstr. 49828 (1977).

066.202 π^- condensation in neutron star matter.   Y. Futami.
         Prog. Theor. Phys., Vol. 57, 457 - 469 (1977).
Abstr. in Phys. Abstr., Vol. 80, Abstr. 53420 (1977).

066.203 Dense stellar matter in asymptotically free gauge
         theories. I. The high density limit.
M. B. Kislinger, P. D. Morley.
Phys. Lett. Vol. B, Vol. 67B, 361 - 366 (1977). — Abstr. in
Phys. Abstr., Vol. 80, Abstr. 53421 (1977).

066.204 Angular momentum in general relativity. II. Perturba-
         tions of a rotating black hole.   C. R. Prior.
Proc. R. Soc. London, Ser. A, Vol. 355, 1 - 29 (1977). — Abstr.
in Phys. Abstr., Vol. 80, Abstr. 53623 (1977).

066.205 Weak electromagnetic fields around a black hole.
         Field of a point charge.   R. M. Misra.
Prog. Theor. Phys., Vol. 57, 694 - 696 (1977). — Abstr. in
Phys. Abstr., Vol. 80, Abstr. 53624 (1977).

066.206 Isotropic spherical clusters of particles in general
         relativity.   A. F. F. Teixeira, M. M. Som.
Nuovo Cimento B, Ser. 11, Vol. 38B, 28 - 36 (1977). — Abstr.
in Phys. Abstr., Vol. 80, Abstr. 41514 (1977).

066.207 Stationary axisymmetric test fields on a Kerr metric.
         B. Linet.
Phys. Lett. A, Vol. 60A, 395 - 396 (1977). — Abstr. in Phys.
Abstr., Vol. 80, Abstr. 41515 (1977).

066.208 Special exact solutions of Einstein's equations — a
         theorem and some observations.   C. B. Collins.
Phys. Lett. A, Vol. 60A, 397 - 398 (1977). — Abstr. in Phys.
Abstr., Vol. 80, Abstr. 41516 (1977).

066.209 Spacetime structure explored by elementary
         particles: microscopic origin for the Riemann-
Cartan geometry.   K. Hayashi, T. Shirafuji.
Prog. Theor. Phys., Vol. 57, 302 - 317 (1977). — Abstr. in
Phys. Abstr., Vol. 80, Abstr. 41517 (1977).

066.210 Vacuum stress tensor in an Einstein universe:
         finite-temperature effects.
J. S. Dawker, R. Critchley.
Phys. Rev. D, Vol. 15, 1484 - 1493 (1977). — Abstr. in Phys.
Abstr., Vol. 80, Abstr. 41769 (1977).

066.211 Black holes in closed universes.   F. J. Tipler.
         Nature, Vol. 270, 500 - 501 (1977).
    The author provides a firm basis for black hole physics;
he extends the concept of 'black hole' to arbitrary stably
causal spacetimes by essentially defining a black hole to be
that object which contains all the 'small' trapped surfaces. The
new concept of black hole yields a purely geometrical definition of time direction in closed universes.

066.212   **The application of SU (1,1) spin coefficients for space like symmetry.**   B. Lukacs.
Acta Phys. Acad. Sci. Hungaricae, Vol. 41, No. 2, p. 137 - 144 (1976). − Abstr. in Phys. Abstr., Vol. 80, Abstr. 53807 (1977).

066.213   **On the differentiability of space-time.**   C. J. S. Clarke.
Math. Proc. Cambridge Philos. Soc., Vol. 81, Part 2, p. 279 - 282 (1977). − Abstr. in Phys. Abstr., Vol. 80, Abstr. 53815 (1977).

066.214   **Complete covariant theory of gravitation. I. Re-examination of the curvature quantities.**
R. Burghardt.
Sitzungsber. Österreichische Akad. Wiss. Math.-Naturwiss. Kl. Abt. II, Vol. 184, No. 8 - 10, p. 427 - 437 (1975). In German. Abstr. in Phys. Abstr., Vol. 80, Abstr. 53816 (1977).

066.215   **Complete covariant theory of gravitation. II. The Einstein field equations.**   R. Burghardt.
Sitzungsber. Österreichische Akad. Wiss. Math.-Naturwiss. Kl. Abt. II, Vol. 184, No. 8 - 10, p. 439 - 449 (1975). In German. Abstr. in Phys. Abstr., Vol. 80, Abstr. 53817 (1977).

066.216   **Asymptotic behaviour near past null infinity for scattering problems.**   M. Walker.
Canadian J. Phys., Vol. 55, 855 - 860 (1976). − Abstr. in Phys. Abstr., Vol. 80, Abstr. 53819 (1977).

066.217   **SU (2,1) symmetry of the Einstein-Maxwell fields.**
Y. Tanabe.
Prog. Theor. Phys., Vol. 57, 840 - 854 (1977). − Abstr. in Phys. Abstr., Vol. 80, Abstr. 53821 (1977).

066.218   **Testing relativity and gravitational theories by radar ranging to a heliocentric satellite.**   I. W. Roxburgh.
Philos. Trans. R. Soc. London, Ser. A, Vol. 284, 589 - 593 (1977). − Abstr. in Phys. Abstr., Vol. 80, Abstr. 53823 (1977).

066.219   **Anisotropic fluid in a spherical spacetime. I. Radiation from a compact star.**   D. C. Wilkins.
Ann. Physics, Vol. 105, 187 - 211 (1977). − Abstr. in Phys. Abstr., Vol. 80, Abstr. 56821 (1977).

066.220   **Bel-Robinson superenergy tensor and the tetrad description of gravitational field.**
N. V. Mitskievic, A. I. Nesterov.
Acta Phys. Polonica B, Vol. B8, 423 - 429 (1977). − Abstr. in Phys. Abstr., Vol. 80, Abstr. 57236 (1977).

066.221   **A symplectic formulation of relativistic particle dynamics.**   W. M. Tulczyjew.
Acta Phys. Polonica B, Vol. B8, 431 - 447 (1977). − Abstr. in Phys. Abstr., Vol. 80, Abstr. 57237 (1977).

066.222   **Physical equivalence of theories.**   A. Qadir.
Int. J. Theor. Phys., Vol. 15, 635 - 641 (1976). Abstr. in Phys. Abstr., Vol. 80, Abstr. 57240 (1977).

066.223   **Comments on models of scalar-tensorial field equations in general relativity.**   C. G. Oliveira.
Int. J. Theor. Phys., Vol. 15, 643 - 656 (1976). − Abstr. in Phys. Abstr., Vol. 80, Abstr. 57241 (1977).

066.224   **Alternative vacuum states in static space-times with horizons.**   S. A. Fulling.
J. Phys. A, Vol. 10, 917 - 951 (1977). − Abstr. in Phys. Abstr., Vol. 80, Abstr. 57243 (1977).

066.225   **On a family of interior solutions for relativistic fluid spheres with possible applications to highly collapsed stellar objects.**   P. G. Whitman.
J. Math. Phys., Vol. 18, 868 - 869 (1977). − Abstr. in Phys. Abstr., Vol. 80, Abstr. 57245 (1977).

066.226   **A systematic investigation of the Petrov G_4 types.**   J. R. Ray, J. C. Zimmerman.
J. Math. Phys., Vol. 18, 881 - 884 (1977). − Abstr. in Phys. Abstr., Vol. 80, Abstr. 57246 (1977).

066.227   **Investigations of space-times with four-parameter groups of motions acting on null hypersurfaces.**
W. T. Lauten III, J. R. Ray.
J. Math. Phys., Vol. 18, 885 - 888 (1977). − Abstr. in Phys. Abstr., Vol. 80, Abstr. 57247 (1977).

066.228   **Heavens and their integral manifolds.**
C. P. Boyer, J. F. Plebanski.
J. Math. Phys., Vol. 18, 1022 - 1031 (1977). − Abstr. in Phys. Abstr., Vol. 80, Abstr. 57251 (1977).

066.229   **Some special Kerr-Schild metrics.**
M. Gurses, F. Gursey.
Nuovo Cimento B, Ser. 11, Vol. 39B, 226 - 232 (1977). Abstr. in Phys. Abstr., Vol. 80, Abstr. 57253 (1977).

066.230   **The gauge-theoretical structure of general relativity and the new conserved current.**   S. Malin.
Nuovo Cimento B, Ser. 11, Vol. 39B, 319 - 330 (1977). Abstr. in Phys. Abstr., Vol. 80, Abstr. 57254 (1977).

066.231   **The Finsler space and affinities for external electro-magnetic fields in general relativity.**   C. G. Oliveira.
Rev. Brasil. Fis., Vol. 6, 559 - 577 (1976). − Abstr. in Phys. Abstr., Vol. 80, Abstr. 57256 (1977).

066.232   **Generation of gravitational waves. Linear momentum flux to higher orders.**   D. D. Dionysiou.
J. Phys. A, Vol. 10, 969 - 973 (1977). − Abstr. in Phys. Abstr., Vol. 80, Abstr. 57258 (1977).

066.233   **Is there gravitational radiation?**   N. Rosen.
Nuovo Cimento, Lett., Ser. 2, Vol. 19, 249 - 250 (1977). − Abstr. in Phys. Abstr., Vol. 80, Abstr. 57259 (1977).

066.234   **Some information theoretical aspects of the gravitational field.**   S. Ikeda.
Nuovo Cimento, Lett., Ser. 2, Vol. 19, 141 - 144 (1977). Abstr. in Phys. Abstr., Vol. 80, Abstr. 57266 (1977).

066.235   **Differential rotation of viscous neutron matter.**
J. Nitsch, J. Pfarr, H. Heintzmann.
Rev. Brasil. Fis., Vol. 6, 531 - 545 (1976). − Abstr. in Phys. Abstr., Vol. 80, Abstr. 60788 (1977).

066.236   **Resolution of the mystery behind Chandrasekhar's black hole transformations.**   J. Heading.
J. Phys. A, Vol. 10, 885 - 897 (1977). − Abstr. in Phys. Abstr., Vol. 80, Abstr. 60790 (1977).

066.237   **Quantum theory of gravitation.**   M. J. G. Veltman.
Quantum field theoretical methods, Les Houches, France, 28 July - 6 September 1975. R. Balian, J. Zinn-Justin (Editors). North-Holland Publishing Company, Amsterdam, Netherlands. 20 + 386 pp. Price $ 39.95 (1976). ISBN 0-7204-0433-9, p. 265 - 327. − Abstr. in Phys. Abstr., Vol. 80, Abstr. 60998 (1977).

066.238   **Nuclear forces in a Kerr-Newman background space.**
D. J. Rowan.
J. Phys. A, Vol. 10, 1105 - 1109 (1977). − Abstr. in Phys. Abstr., Vol. 80, Abstr. 60999 (1977).

066.239 Four-body forces and the red shift.   T. Kimura.
Prog. Theor. Phys., Vol. 57, 1437 - 1438 (1977).
Abstr. in Phys. Abstr., Vol. 80, Abstr. 61001 (1977).

**066.240 Massive spin-1/2 wave around a Kerr-Newman black
hole.   C. H. Lee.**
Phys. Lett. B, Vol. 68B, 152 - 156 (1977). − Abstr. in Phys.
Abstr., Vol. 80, Abstr. 61002 (1977).

**066.241 Foliations of space-times by spacelike hypersurfaces
of constant mean curvature.   A. J. Goddard.**
Commun. Math. Phys., Vol. 54, 279 - 282 (1977). − Abstr. in
Phys. Abstr., Vol. 80, Abstr. 64970 (1977).

**066.242 Conditions for static balance for the post-Newtonian
two-body problem with electric charge in general
relativity.   B. M. Barker.**
Phys. Lett. A, Vol. 61A, 297 - 298 (1977). − Abstr. in Phys.
Abstr., Vol. 80, Abstr. 64973 (1977).

**066.243 An approach to one Killing-vector solutions of
Einstein's equations by rotation-coefficients.**
C. Hoenselaers.
Prog. Theor. Phys., Vol. 57, 1223 - 1238 (1977). − Abstr. in
Phys. Abstr., Vol. 80, Abstr. 64974 (1977).

066.244 Gravitational Lagrangian and internal symmetry.
E. Leibowitz.
Phys. Rev. D, Vol. 15, 2139 - 2143 (1977). − Abstr. in Phys.
Abstr., Vol. 80, Abstr. 64976 (1977).

066.245 Lorentz covariance and the Kerr-Newman geometry.
S. Einstein, R. Finkelstein.
Phys. Rev. D, Vol. 15, 2721 - 2723 (1977). − Abstr. in Phys.
Abstr., Vol. 80, Abstr. 64977 (1977).

**066.246 Spinning charged test particles in a Kerr-Newman
background.   R. Hojman, S. Hojman.**
Phys. Rev. D, Vol. 15, 2724 - 2730 (1977). − Abstr. in Phys.
Abstr., Vol. 80, Abstr. 64978 (1977).

**066.247 Gravitational radiation from distant encounters and
from head-on collisions of black holes: the zero-
frequency limit.   L. Smarr.**
Phys. Rev. D, Vol. 15, 2069 - 2077 (1977). − Abstr. in Phys.
Abstr., Vol. 80, Abstr. 64980 (1977).

**066.248 Killing inequalities for relativistically rotating fluids.
II.   R. O. Hansen, J. Winicour.**
J. Math. Phys., Vol. 18, 1206 - 1209 (1977). − Abstr. in Phys.
Abstr., Vol. 80, Abstr. 64981 (1977).

**066.249 Principles of equivalence, Eötvös experiments, and
gravitational red-shift experiments: the free fall of
electromagnetic systems to post-post-Coulombian order.**
M. P. Haugan, C. M. Will.
Phys. Rev. D, Vol. 15, 2711 - 2720 (1977). − Abstr. in Phys.
Abstr., Vol. 80, Abstr. 64982 (1977).

**066.250 Stress-tensor trace anomaly in a gravitation metric:
general theory, Maxwell field.**
L. S. Brown, J. P. Cassidy.
Phys. Rev. D, Vol. 15, 2810 - 2829 (1977). − Abstr. in Phys.
Abstr., Vol. 80, Abstr. 64983 (1977).

**066.251 Action integrals and partition functions in quantum
gravity.   G. W. Gibbons, S. W. Hawking.**
Phys. Rev. D, Vol. 15, 2752 - 2756 (1977). − Abstr. in Phys.
Abstr., Vol. 80, Abstr. 64984 (1977).

066.252 Victories and defeats in general relativity theory.
C. Moller.

Fys. Tidsskr., Vol. 75, 3 - 18 (1977). In Danish. − Abstr. in
Phys. Abstr., Vol. 80, Abstr. 64985 (1977).

**066.253 Small-scale structure of spacetime as the origin of
the gravitational constant.   P. K. Townsend.**
Phys. Rev. D, Vol. 15, 2795 - 2801 (1977). − Abstr. in Phys.
Abstr., Vol. 80, Abstr. 64990 (1977).

**066.254 'Radiation' and 'vacuum polarization' near a black
hole.   S. A. Fulling.**
Phys. Rev. D, Vol. 15, 411 - 414 (1977). − Abstr. in Phys.
Abstr., Vol. 80, Abstr. 67030 (1977).

**066.255 Comment on the Damour-Ruffini treatment of
black-hole evaporation.**
M. Martellini, A. Treves.
Phys. Rev. D, Vol. 15, 2415 - 2416 (1977). − Abstr. in Phys.
Abstr., Vol. 80, Abstr. 67031 (1977).

066.256 Neutron stars and compact X-ray sources.
N. O. Lassen.
Fys. Tidsskr., Vol. 75, 33 - 37 (1977). In Danish. − Abstr. in
Phys. Abstr., Vol. 80, Abstr. 67106 (1977).

066.257 Inside the black hole.   R. W. Brehme.
American J. Phys., Vol. 45, 423 - 428 (1977).
Abstr. in Phys. Abstr., Vol. 80, Abstr. 67107 (1977).

066.258 Trace anomalies and the Hawking effect.
S. M. Christensen, S. A. Fulling.
Phys. Rev. D, Vol. 15, 2088 - 2104 (1977). − Abstr. in Phys.
Abstr., Vol. 80, Abstr. 67110 (1977).

066.259 Einstein A and B coefficients for a black hole.
J. D. Bekenstein, A. Meisels.
Phys. Rev. D, Vol. 15, 2775 - 2781 (1977). − Abstr. in Phys.
Abstr., Vol. 80, Abstr. 67111 (1977).

066.260 Pedagogical trick for general relativity.
J. D. French.
American J. Phys., Vol. 45, 580 - 581 (1977). − Abstr. in Phys.
Abstr., Vol. 80, Abstr. 67337 (1977).

**066.261 Conservation laws for test particles with internal
structure.   H. Fuchs.**
Ann. Physik, Vol. 34, 159 - 160 (1977). − Abstr. in Phys.
Abstr., Vol. 80, Abstr. 67338 (1977).

**066.262 Finite homogeneous relativistic elastic sphere in its
own gravitational field.   B. Lukacs.**
Nuovo Cimento B, Ser. 11, Vol. 40B, 169 - 181 (1977).
Abstr. in Phys. Abstr., Vol. 80, Abstr. 67344 (1977).

066.263 Note on motion in Schwarzschild field.
G. Cavalleri, G. Spinelli.
Phys. Rev. D, Vol. 15, 3065 - 3067 (1977). − Abstr. in Phys.
Abstr., Vol. 80, Abstr. 67348 (1977).

066.264 Motion in the Schwarzschild field: a reply.
A. I. Janis.
Phys. Rev. D, Vol. 15, 3068 - 3069 (1977). − Abstr. in Phys.
Abstr., Vol. 80, Abstr. 67349 (1977).

**066.265 Lagrangian formalism up to the fifth order in the
Einstein-Infeld-Hoffman theory and gravitational
radiation.   D. D. Dionysiou.**
Nuovo Cimento Lett., Ser. 2, Vol. 19, 383 - 388 (1977).
Abstr. in Phys. Abstr., Vol. 80, Abstr. 67354 (1977).

066.266 On the inertial-gravitational field.
L. D. Raigorodski.
Acta Phys. Acad. Sci. Hungaricae, Vol. 41, No. 3, p. 153 -

158 (1976). – Abstr. in Phys. Abstr., Vol. 80, Abstr. 67362 (1977).

066.267 Radiation from moving mirrors and from black holes. P. C. W. Davies, S. A. Fulling.
Proc. R. Soc. London, Ser. A, Vol. 356, 237 - 257 (1977).
Abstr. in Phys. Abstr., Vol. 80, Abstr. 67372 (1977).

066.268 De Sitter gauge transformations of Dirac equation. P. K. Smrz.
Prog. Theor. Phys., Vol. 57, 1771 - 1780 (1977). – Abstr. in Phys. Abstr., Vol. 80, Abstr. 67563 (1977).

066.269 Absence of superradiance of a Dirac field in a Kerr background. M. Martellini, A. Treves.
Phys. Rev. D, Vol. 15, 3060 - 3061 (1977). – Abstr. in Phys. Abstr., Vol. 80, Abstr. 67578 (1977).

066.270 Theory of particle detection in curved spacetimes. P. Hajicek.
Phys. Rev. D, Vol. 15, 2757 - 2774 (1977). – Abstr. in Phys. Abstr., Vol. 80, Abstr. 67597 (1977).

066.271 Spin may prevent formation of small black holes. M. Demianski.
Acta Phys. Polonica B, Vol. B8, 591 - 592 (1977). – Abstr. in Phys. Abstr., Vol. 80, Abstr. 71125 (1977).

066.272 Stress-energy tensor near a charged, rotating evaporating black hole. W. A. Hiscock.
Phys. Rev. D, Vol. 15, 3054 - 3057 (1977). – Abstr. in Phys. Abstr., Vol. 80, Abstr. 71126 (1977).

066.273 A theory of generally invariant Lagrangians for the metric fields. II. D. Krupka.
Int. J. Theor. Phys., Vol. 15, 949 - 959 (1976). – Abstr. in Phys. Abstr., Vol. 80, Abstr. 71247 (1977).

066.274 Complete integrability conditions of the Einstein-Petrov equations, type I. C. H. Brans.
J. Math. Phys., Vol. 18, 1378 - 1381 (1977). – Abstr. in Phys. Abstr., Vol. 80, Abstr. 71249 (1977).

066.275 Bivector field theories, divergence-free vectors and the Einstein-Maxwell field equations.
D. Lovelock.
J. Math. Phys., Vol. 18, 1491 - 1498 (1977). – Abstr. in Phys. Abstr., Vol. 80, Abstr. 71251 (1977).

066.276 The electrostatic and gravitational field of spherically symmetric objects.
M. Soffel, B. Muller, W. Greiner.
Phys. Lett. A, Vol. 62A, No. 2, p. 67 - 69 (1977). – Abstr. in Phys. Abstr., Vol. 80, Abstr. 71253 (1977).

066.277 Axially symmetric two-body problem in general relativity. II. Free fall.
F. I. Cooperstock, P. H. Lim.
Phys. Rev. D, Vol. 15, 2105 - 2122 (1977). – Abstr. in Phys. Abstr., Vol. 80, Abstr. 71254 (1977).

066.278 Tree ghost in spinor gravity. M. Nouri-Moghadam.
J. Phys. A, Vol. 10, 1313 - 1317 (1977). – Abstr. in Phys. Abstr., Vol. 80, Abstr. 71260 (1977).

066.279 The Newtonian analogue of the relativistic Oppenheimer-Snyder solution. P. S. Florides.
Phys. Lett. A, Vol. 62A, 138 - 140 (1977). – Abstr. in Phys. Abstr., Vol. 80, Abstr. 73945 (1977).

066.280 The interaction between a weak magnetic field and a black hole. M. D. Pollock, W. P. Brinkmann.
Proc. R. Soc. London, Ser. A, Vol. 356, 351 - 362 (1977). Abstr. in Phys. Abstr., Vol. 80, Abstr. 74080 (1977).

066.281 Dirac equation in Kerr space-time. B. R. Iyer, A. Kumar.
Pramana, Vol. 8, 500 - 511 (1977). – Abstr. in Phys. Abstr., Vol. 80, Abstr. 74081 (1977).

066.282 The Kerr metric in cosmological background. P. C. Vaidya.
Pramāna, Vol. 8, 512 - 517 (1977). – Abstr. in Phys. Abstr., Vol. 80, Abstr. 74082 (1977).

066.283 Cosmological constant in supergravity. P. K. Townsend.
Phys. Rev. D, Vol. 15, 2802 - 2804 (1977). – Abstr. in Phys. Abstr., Vol. 80, Abstr. 74171 (1977).

066.284 A test theory of special relativity: III. Second-order tests. R. Mansouri, R. U. Sexl.
Gen. Relativ. Gravitation, Vol. 8, 809 - 814 (1977).
Various second-order optical tests of special relativity are discussed within the framework of the test theory developed previously. Owing to the low accuracy of the Kennedy-Thorndike experiment, the Lorentz contraction is known by direct experiments only to an accuracy of a few percent. To improve this accuracy several experiments are suggested.

066.285 Null strings and complex Einstein-Maxwell fields with cosmological constant.
A. García, J. F. Plebański, I. Robinson.
Gen. Relativ. Gravitation, Vol. 8, 841 - 854 (1977).

066.286 A local relativistic red-shift effect. E. Ihrig.
Gen. Relativ. Gravitation, Vol. 8, 877 - 885 (1977).
A relativistic red-shift effect is found that is distinct from the Doppler and gravitational red shifts and provides a mechanism for producing large local red shifts without raising kinetic energy problems. This effect produces red shifts in discrete amounts—a situation for which there may be some evidence in quasar observations.

066.287 Singular space-times. G. F. R. Ellis, B. G. Schmidt.
Gen. Relativ. Gravitation, Vol. 8, 915 - 953 (1977).
A classification scheme for boundary points of incomplete space-times is described. For all classes explicit examples are presented to illustrate the different behaviour of the geometry near those boundary points. – Review paper.

066.288 Connections on particular Riemann spaces. E. Vamanu.
Bul. Inst. Politeh. Iasi. Sect. I, Vol. 23, No. 1 - 2, p. 27 - 32 (1977). In Rumanian. – Abstr. in Phys. Abstr., Vol. 80, Abstr. 74274 (1977).

066.289 Stationary axially symmetric solutions of Einstein-Maxwell massless scalar field equations.
A. Eris, M. Gurses.
J. Math. Phys., Vol. 18, 1303 - 1304 (1977). – Abstr. in Phys. Abstr., Vol. 80, Abstr. 74276 (1977).

066.290 Fluid space-times including electromagnetic fields admitting symmetry mappings belonging to the family of contracted Ricci collineations.
L. K. Norris, L. H. Green, W. R. Davis.
J. Math. Phys., Vol. 18, 1305 - 1311 (1977). – Abstr. in Phys. Abstr., Vol. 80, Abstr. 74277 (1977).

066.291 Causally symmetry spacetimes. F. J. Tipler.
J. Math. Phys., Vol. 18, 1568 - 1573 (1977).

Abstr. in Phys. Abstr., Vol. 80, Abstr. 74281 (1977).

066.292 **Spinorial structures and electromagnetic hyper-
heavens.**   J. D. Finley III, J. F. Plebanski.
J. Math. Phys., Vol. 18, 1662 - 1667 (1977). – Abstr. in
Phys. Abstr., Vol. 80, Abstr. 74283 (1977).

066.293 **Space-time singularities and conformal gravity.**
J. V. Narlikar, A. K. Kembhavi.
Nuovo Cimento, Lett., Ser. 2, Vol. 19, 517 - 520 (1977).
Abstr. in Phys. Abstr., Vol. 80, Abstr. 74286 (1977).

066.294 **Scalar-metric and scalar-metric-torsion gravita-
tional theories.**   S. J. Aldersley.
Phys. Rev. D, Vol. 15, 3507 - 3512 (1977). – Abstr. in Phys.
Abstr., Vol. 80, Abstr. 74295 (1977).

066.295 **Space-time structure in a generalisation of gravita-
tion theory.**   J. W. Moffat.
Phys. Rev. D, Vol. 15, 3520 - 3529 (1977). – Abstr. in Phys.
Abstr., Vol. 80, Abstr. 74296 (1977).

066.296 **Clocks and experimental gravitation: a null gravita-
tional redshift experiment, laboratory tests of post-**
Newtonian gravity, and gravity-wave detection by spacecraft
tracking.   C. M. Will.
Metrologia, Vol. 13, No. 3, p. 95 - 98 (1977). – Abstr. in
Phys. Abstr., Vol. 80, Abstr. 74301 (1977).

066.297 **Perihelion precession for the charged two-body
problem in general relativity.**
B. M. Barker, R. F. O'Connell.
Nuovo Cimento, Lett., Ser. 2, Vol. 19, 467 - 468 (1977).
Abstr. in Phys. Abstr., Vol. 80, Abstr. 78626 (1977).

066.298 **Neutron-star mass limit in the bimetric theory of
gravitation.**   G. Caporaso, K. Brecher.
Phys. Rev. D, Vol. 15, 3536 - 3542 (1977). – Abstr. in Phys.
Abstr., Vol. 80, Abstr. 78813 (1977).

066.299 **Black holes and tachyons.**
V. De Sabbata, M. Pavsic, E. Recami.
Nuovo Cimento, Lett., Ser. 2, Vol. 19, 441 - 451 (1977).
Abstr. in Phys. Abstr., Vol. 80, Abstr. 78816 (1977).

066.300 **Black-hole dyons need not explode.**
G. W. Gibbons.
Phys. Rev. D, Vol. 15, 3530 - 3535 (1977). – Abstr. in Phys.
Abstr., Vol. 80, Abstr. 78817 (1977).

066.301 **New family of exact stationary axisymmetric gravitational fields generalising the Tomimatsu-Sato
solutions.**   C. M. Cosgrove.
J. Phys. A, Vol. 10, 1481 - 1524 (1977). – Abstr. in Phys.
Abstr., Vol. 80, Abstr. 79012 (1977).

066.302 **A generalized field theory. I. Field equations.**
F. I. Mikhail, M. I. Wanas.
Proc. R. Soc. London, Ser. A, Vol. 356, 471 - 481 (1977).
Abstr. in Phys. Abstr., Vol. 80, Abstr. 79013 (1977).

066.303 **Conservation laws and gravitational radiation.**
P. Rastall.
Canadian J. Phys., Vol. 55, 1342 - 1348 (1977). – Abstr. in
Phys. Abstr., Vol. 80, Abstr. 79014 (1977).

066.304 **A note on an interior solution for a fluid sphere of
constant gravitational mass density in general**
relativity.   J. P. Sharma.
Acta Phys. Acad. Sci. Hungaricae, Vol. 41, 241 - 244 (1976).
Abstr. in Phys. Abstr., Vol. 80, Abstr. 79015 (1977).

066.305 **The gauge theory of the translation group and under-
lying geometry.**   K. Hayashi.
Phys. Lett. B, Vol. 69B, 441 - 444 (1977). – Abstr. in Phys.
Abstr., Vol. 80, Abstr. 79020 (1977).

066.306 **Algebraic properties of extended supergravity in de
Sitter space.**   S. Ferrara.
Phys. Lett. B, Vol. 69B, 481 - 483 (1977). – Abstr. in Phys.
Abstr., Vol. 80, Abstr. 79021 (1977).

066.307 **Admissible coordinate systems for static, spherically
symmetric stellar models.**   M. A. Oliver.
Gen. Relativ. Gravitation, Vol. 8, 963 - 973 (1977).
    For static, spherically symmetric stellar models it is
shown that imposing the condition that the determinant of
the metrical coefficients takes on its flat space-time value
everywhere is sufficient to ensure that the coordinates are
admissible in the sense of Lichnerowicz. The general method
of solution is illustrated by integrating the equations for a star
of constant, uniform density.

066.308 **Variational aspects of relativistic field theories, with
application to perfect fluids.**
B. F. Schutz, R. Sorkin.
Ann. Physics, Vol. 107, 1 - 43 (1977). – Abstr. in Phys. Abstr.,
Vol. 80, Abstr. 82417 (1977).

066.309 **Relativistic spheres filled with isentropic magneto-
fluids.**   T. H. Date, L. Radhakrishna.
Acta Phys. Polonica B, Vol. B8, 713 - 722 (1977). – Abstr. in
Phys. Abstr., Vol. 80, Abstr. 82419 (1977).

066.310 **The laws of relativistic thermodynamics. I. The
macroscopic quantities.**   S. R. de Groot.
Physica A, Vol. 88A, 172 - 182 (1977). – Abstr. in Phys.
Abstr., Vol. 80, Abstr. 82423 (1977).

066.311 **The laws of relativistic thermodynamics. II. The
density, momentum and energy laws.**
S. R. de Groot.
Physica A, Vol. 88A, 183 - 189 (1977). – Abstr. in Phys.
Abstr., Vol. 80, Abstr. 82424 (1977).

066.312 **Two astrophysical applications of conformal gravity.**
J. V. Narlikar.
Ann. Physics, Vol. 107, 325 - 336 (1977). – Abstr. in Phys.
Abstr., Vol. 80, Abstr. 82428 (1977).

066.313 **Generalized Nordström-Reissner metric.**
H. Yilmaz.
Nuovo Cimento, Lett., Ser. 2, Vol. 19, 617 - 621 (1977).
Abstr. in Phys. Abstr., Vol. 80, Abstr. 82430 (1977).

066.314 **Self-gravitating cosmons.**   M. Camenzind.
Phys. Lett. A, Vol. 62A, 298 - 300 (1977). – Abstr.
in Phys. Abstr., Vol. 80, Abstr. 82431 (1977).

066.315 **Design of gravitational antennae for use at 3 mK.**
J. F. Kos, R. Barton, B. Ramadan.
J. Appl. Phys., Vol. 48, 3193 - 3200 (1977). – Abstr. in Phys.
Abstr., Vol. 80, Abstr. 82435 (1977).

066.316 **Cavity transducer for subatomic mechanical vibration.**   K. Tsubono, S. Hiramatsu, H. Hirakawa.
Japanese J. Appl. Phys., Vol. 16, 1641 - 1645 (1977). – Abstr.
in Phys. Abstr., Vol. 80, Abstr. 82436 (1977).

066.317 **Co-ordinate covariant second quantization.**
S. Blaha.
Nuovo Cimento, Lett., Ser. 2, Vol. 20, 17 - 20 (1977). – Abstr.
in Phys. Abstr., Vol. 80, Abstr. 82670 (1977).

066.318 **Gravitational effects on Yang-Mills topology.**
J. M. Charap, M. J. Duff.
Phys. Lett. B, Vol. 69B, 445 - 447 (1977). – Abstr. in Phys. Abstr., Vol. 80, Abstr. 82682 (1977).

066.319 **Electromagnetic and gravitational waves in the background of a Reissner-Nordström black hole.**
R. Fabbri.
Nuovo Cimento B, Ser. 11, Vol. 40B, 311 - 329 (1977). Abstr. in Phys. Abstr., Vol. 80, Abstr. 87470 (1977).

066.320 **Nach-Newtonsche Korrekturen zur Dynamik des Systems Erde–Mond und ihre Bedeutung für die relativistischen Gravitationstheorien.** H.-J. Treder.
Veröff. Zentralinst. Phys. Erde, No. 52, Part 1, (see 012.035), p. 119 - 127 (1977).

066.321 **The Viking relativity experiment.**
I. I. Shapiro, R. D. Reasenberg, P. E. MacNeil, R. B. Goldstein, J. P. Brenkle, D. L. Cain, T. Komarek, A. I. Zygielbaum, W. F. Cuddihy, W. H. Michael, Jr.
J. Geophys. Res., Vol. 82, (see 003.017), 4329 - 4334 (1977). Paper No. 7S0564.
   Measurements of the round-trip time of flight of radio signals transmitted from the earth to the Viking spacecraft are being analyzed to test the predictions of Einstein's theory of general relativity. According to this theory the signals will be delayed by up to $\sim 250\mu s$ owing to the direct effect of solar gravity on the propagation. A very preliminary qualitative analysis of the Viking data obtained near the 1976 superior conjunction of Mars indicates agreement with the predictions to within the estimated uncertainty of 0.5%.

066.322 **The role of mathematics in gravitational physics: from the sublime to the subliminal?**
C. B. Collins.
Gen. Relativ. Gravitation, Vol. 8, 717 - 721 (1977).
   Certain gravitational systems are described for which particular equations of state are singled out by means of purely mathematical considerations. The fact that these equations of state are realistic, and indeed the most relevant to the physical systems under study, suggests that locked into Einstein's field equations lies some information about local physical conditions, and that in order to describe gravitating systems, it is not always necessary to draw on extraneous branches of physics.

066.323 **Spatially homogeneous Lichnerowicz universes.**
I. Ozsváth.
Gen. Relativ. Gravitation, Vol. 8, 737 - 752 (1977).
   The author finds a special class of spatially homogeneous solutions of the Einstein–Lichnerowicz equations, describing the gravitational field of an electrically charged fluid with infinite conductivity.

066.324 **The matter-vacuum matching problem in general relativity. General methods and special cases.**
W. Roos.
Gen. Relativ. Gravitation, Vol. 8, 753 - 760 (1977).
   In a recent paper (Roos, 1976) methods were developed for studying the problem of matching matter and vacuum solutions of Einstein's field equations along timelike hypersurfaces in the axisymmetric stationary case. Criteria for the existence and uniqueness of perfect fluid sources for given vacuum fields were given in the form of a theorem. In this paper further investigations of this "matching problem" are made for some special cases. It is found that stationary rotating dust cannot be a source of the Kerr metric and the existence of "rotating Schwarzschild sources" is proved. These results are discussed together with further aspects of the general methods used.

066.325 **Equations of motion for rotating finite bodies in the extended PPN formalism.** S. S. Dallas.

Celestial Mech., Vol. 15, 111 - 123 (1977).
   The equations of motion for rotating finite bodies in the extended PPN formalism of Will and Nordtvedt are developed. These equations may be used in conjunction with precise radio and laser ranging data in solar system tests of viable metric gravitational theories.

066.326 **Neutrino-pair bremsstrahlung in collisions between neutrons and nuclei in neutron star matter.**
E. G. Flowers, P. G. Sutherland.
Astrophys. Space Sci., Vol. 48, 159 - 164 (1977).
   In neutron star matter over the density range $4.3 \times 10^{11} \leqslant \rho \leqslant 4 \times 10^{14}$ g cm^{-3} there are both free neutrons and neutron-rich nuclei. If there is a weak neutral current interaction between neutrinos and neutrons, as suggested by recent experiments, then when neutrons scatter off nuclei they may emit $\nu\bar{\nu}$ pairs as bremsstrahlung radiation. The authors calculate the associated emissivity for degenerate (but not superfluid) neutrons and uncorrelated (not crystallized) nuclei, and they find that, under these conditions, this emissivity can under some conditions compare with that calculated by Festa and Ruderman for $\nu\bar{\nu}$ bremsstrahlung in electron-nucleus collisions.

066.327 **The origin of gravity.** S. V. M. Clube.
Astrophys. Space Sci., Vol. 50, 425 - 443 (1977).
   The author presents the elements of a non-relativistic theory of gravity which satisfies the current tests of general relativity, and which appears to offer the prospect of successful integration with quantum theory. Whether such a theory has any basis in reality obviously depends on its verifiable consequences and, particularly, those that discriminate it from general relativity. At least two observational tests which serve to discriminate it appear to be at hand.

066.328 **Accretion onto neutron star: the X-ray spectra and luminosity.** Y. Tuchman, R. Z. Yahel.
Astrophys. Space Sci., Vol. 50, 473 - 492 (1977).
   The equations of stationary fluid motion are solved for the case of spherical accretion onto non-magnetic, non-rotating neutron star. The X-ray radiation flux is calculated in two different ways: An approximate solution of the moment equations coupled to the gas fluid equations and numerical Monte-Carlo simulation of the photon random walk. The authors show that the spectrum of the X-ray radiation in the outer parts of the inflowing envelope has a characteristic power-law behaviour.

066.329 **On the electron cap shape of a rotating neutron star with a strong magnetic field.** Yu. A. Rylov.
Astrophys. Space Sci., Vol. 51, 39 - 57, 59 - 75 (1977). In Russian and English.
   Numerical calculations of the electron cap shape of a rapidly rotating neutron star with a strong magnetic field have been provided. The total charge of the star has been calculated. It is shown that two streams of charged particles escape from the star's surface. The electron stream moves along the magnetic axis. The electron stream is enveloped by a proton-positron stream, which is generated by returning hard electrons accelerated by electromagnetic field of the star.

066.330 **On the emission of neutrinos and gravitational waves in the formation of neutron stars.**
Y. Kondo, G. E. McCluskey, Jr., S. Sofia.
Astrophys. Space Sci., Vol. 51, 187 - 196 (1977).
   The energetics involved in the formation of neutron stars in close binaries as a result of supernova explosions are considered. The gravitational binding energy of the neutron star must find proper outlets. The mass ejection and cosmic ray particles can carry away only a small fraction (up to a few per cent) of this energy. Most of the binding energy goes into rotational kinetic energy, gravitational radiation and neutrino emissions. A scenario is considered in which most of the gravi-

tational binding energy goes into rotational kinetic energy and is, ultimately, radiated away as gravitational waves.

066.331 High-frequency radiation from a particle falling into a Kerr black hole. I. G. Dymnikova.
Astrophys. Space Sci., Vol. 51, 229 - 234 (1977).

The high-frequency electromagnetic and gravitational radiation from a relativistic particle falling into a Kerr and Schwarzschild black hole is considered.

066.332 Gravitational radiation from an isolated perfect fluid in higher multipole moments. D. D. Dionysiou.
Astrophys. Space Sci., Vol. 52, 17 - 26 (1977).

The purpose of this paper is to give the unknown angular momentum loss of an isolated perfect fluid in any higher multipole moments in the linear approximation of general relativity theory. The energy and linear momentum fluxes of the given source in higher multipole moments are also discussed.

066.333 Finite Lorentzian distance and causal geodesic incompleteness. J. K. Beem, P. E. Ehrlich.
C. R. Acad. Sci. Paris, Tome 285, Ser. A, 1129 - 1131 (1977). In French.

The authors prove a singularity theorem for space-times satisfying the condition that the Lorentzian distance between any two points is finite.

066.334 Scale-covariant theory of gravitation and astrophysical applications.
V. Canuto, S. H. Hsieh, P. J. Adams.
Phys. Rev. Lett., Vol. 39, 429 - 432 (1977).

The authors present generalized Einstein equations, invariant under scale transformations, and study several astrophysical tests. It is assumed that the dynamics of atoms or clocks used as measuring apparatus is given a priori. Connection with gauge fields and broken symmetries is made through the cosmological constant.

066.335 Detection of anisotropy in the cosmic blackbody radiation.
G. F. Smoot, M. V. Gorenstein, R. A. Muller.
Phys. Rev. Lett., Vol. 39, 898 - 901 (1977).

The authors have detected anisotropy in the cosmic blackbody radiation with a 33-GHz (0.9 cm) twin-antenna Dicke radiometer flown to an altitude of 20 km. They observe an anisotropy which is well fitted by a first-order spherical harmonic with an amplitude of $(3.5 \pm 0.6) \times 10^{-3}\,°K$, and direction [11.0 ± 0.6 h right ascension and 6° ± 10° declination]. This observation is readily interpreted as due to motion of the earth relative to the radiation with a velocity of 390 ± 60 km/sec.

066.336 Is the speed of light independent of the velocity of the source? K. Brecher.
Phys. Rev. Lett., Vol. 39, 1051 - 1054, with an erratum p. 1236 (1977).

Recent observations of pulsating X-ray sources in binary star systems are analyzed in the framework of the "emission" theory of light. Assuming that light emitted by a source moving at velocity v with respect to an observer, has a speed $c' = c + kv$ in the observer's rest frame, the author finds that the arrival time of pulses from the binary X-ray sources implies $k < 2 \times 10^{-9}$.

066.337 Distinguishing between stars and galaxies composed of matter and antimatter using photon helicity detection. J. G. Cramer, W. J. Braithwaite.
Phys. Rev. Lett., Vol. 39, 1104 - 1107 (1977).

066.338 Colliding impulsive gravitational waves.
Y. Nutku, M. Halil.
Phys. Rev. Lett., Vol. 39, 1379 - 1382 (1977).

066.339 Detectable gravitational radiation from stellar collapse. A. S. Endal, S. Sofia.
Phys. Rev. Lett., Vol. 39, 1429 - 1432 (1977).

The problem of astrophysical sources of detectable gravitational radiation is considered from the stellar-evolution viewpoint. Calculations are presented which indicate that the final stages of evolution may well be dominated by rapidly rotating, collapsing cores which develop nonaxisymmetric configurations. Such events emit large amounts of gravitational radiation which should be detectable in the near future.

066.340 New test of the synchronization procedure in noninertial systems. J. M. Cohen, H. E. Moses.
Phys. Rev. Lett., Vol. 39, 1641 - 1643 (1977).

066.341 On certain transformations for black-holes energetics. A. Curir, M. Francaviglia.
Atti Accad. Naz. Lincei, Rend. Cl. Sci. Fis. Mat. Nat., Ser. VIII, Vol. 61, 448 - 454 (1976) = Contrib. Oss. Astron. Torino, (Pino Torinese), N. 100.

066.342 Die Newtonsche Gravitationskonstante in extragalaktischen Distanzen. G. Dautcourt.
Sterne, 53. Band, 222 - 225 (1977).

066.343 On the possibility of detecting black holes by the gravitational mirror effect. M. I. Fajngol'd.
Astrometriya i Astrofizika, Kiev, vyp. (No.) 33, (see 003.020), p. 35 - 40 (1977). In Russian.

The article deals with the effect of the "gravitational mirror", i. e. a specific reverberation of the signal in the black hole's gravitational field caused by strong light deflection in such a field. The intensity of the reflected "echo" is obtained as a function of the distance to the black hole and of the angular distance from the black-hole signal source line. Intensification of the back scattered signal is shown. The conclusion is made that detection of black holes by the echo effect from the gravitational mirror is possible in principle. The effect of "echo" splitting into a series of pulses ("multiple echo") is considered. Exact determination of the black hole's radius from the time interval between these pulses is shown to be possible.

066.344 On the kinetic description of perturbations of homogeneous gravitating systems.
V. B. Magalinskij, V. V. Seliverstov.
Probl. teor. gravitatsii i ehlem. chastits. Vyp. (No.) 8. Moskva, Atomizdat, 1977, p. 112 - 119. In Russian. – Abstr. in Ref. zh., 51. Astron., 12.51.727 (1977).

066.345 Orbits of test bodies in the Reissner-Nordstrøm gravitational field. V. V. Mityanok.
Probl. teor. gravitatsii i ehlem. chastits. Vyp. (No.) 8. Moskva, Atomizdat, 1977, p. 151 - 156. In Russian. – Abstr. in Ref. zh., 51. Astron., 12.51.763 (1977).

066.346 On desynchronization in the process of ultrarelativistic motion in Kerr and Reissner-Nordstrøm fields. V. V. Mityanok.
Probl. teor. gravitatsii i ehlem. chastits. Vyp. (No.) 8. Moskva, Atomizdat, 1977, p. 156 - 161. In Russian. – Abstr. in Ref. zh., 51. Astron., 12.51.764 (1977).

066.347 Relativistic corrections for orbital effects in general relativity. N. Kojshibaev, A. Meshcheryakov.
Prikl. i teor. fiz. Vyp. (No.) 8. Alma-Ata, 1976, p. 112 - 118. In Russian. – Abstr. in Ref. zh., 51. Astron., 12.51.766 (1977).

066.348 Derivation of equations of rotational motion of bodies from a variation principle in general relativity. N. Kojshibaev.
Prikl. i teor. fiz. Vyp. (No.) 8. Alma-Ata, 1976, p. 119 - 124.

In Russian. – Abstr. in Ref. zh., 51. Astron., 12.51.768 (1977).

066.349 Covariant equations of motion of a point with variable rest mass and relativistic analogues of some astrodynamical problems. U. N. Zakirov.
Inst. prikl. mat. AN SSSR. Prepr. No. 41. Moskva, 1977. 44 pp. Price 15 Kop. In Russian. – Abstr. in Ref. zh., 51. Astron., 12.51.769 (1977).

066.350 Effect of plane gravitational waves on a system of free particles. L. B. Borisova, V. D. Zakharov, N. I. Kolosnitsyn, K. P. Stanyukovich.
Probl. teor. gravitatsii i ehlem. chastits. Vyp. (No.) 8. Moskva, Atomizdat, 1977, p. 32 - 37. In Russian. – Abstr. in Ref. zh., 51. Astron., 12.51.772 (1977).

066.351 Groups of symmetries in Newtonian mechanics and general relativity. I. Groups of homogeneity and noninertial reference systems. N. V. Pavlov.
Probl. teor. gravitatsii i ehlem. chastits. Vyp. (No.) 8. Moskva, Atomizdat, 1977, p. 43 - 56. In Russian. – Abstr. in Ref. zh., 51. Astron., 12.51.778 (1977).

066.352 Groups of symmetries in Newtonian mechanics and general relativity. II. Groups of homogeneity and hydrodynamic models. N. V. Pavlov.
Probl. teor. gravitatsii i ehlem. chastits. Vyp. (No.) 8. Moskva, Atomizdat, 1977, p. 56 - 66. In Russian. – Abstr. in Ref. zh., 51. Astron., 12.51.779 (1977).

066.353 On the role of the conformity principle in general relativity.
M. E. Gertsenshtejn, K. P. Stanyukovich, V. A. Pogosyan.
Probl. teor. gravitatsii i ehlem. chastits. Vyp. (No.) 8. Moskva, Atomizdat, 1977, p. 132 - 146. In Russian. – Abstr. in Ref. zh., 51. Astron., 12.51.780 (1977).

066.354 On the velocity of particle creation in gravitational fields. Ya. B. Zel'dovich, A. A. Starobinskij.
Pis'ma v ZhEhTF, Vol. 26, 373 - 377 (1977). In Russian. Abstr. in Ref. zh., 51. Astron., 1.51.786 (1978).

066.355 Hydrodynamics of primordial black hole formation. D. K. Nadezhin, I. D. Novikov, A. G. Polnarev.
Inst. kosm. issled. AN SSSR. Pr-347. Moskva, 1977. 36 pp. In Russian. – Abstr. in Ref. zh., 51. Astron., 1.51.805 (1978).

066.356 Tetrad formulation of Kepler's problem in special relativity. L. M. Chechin.
Novoe v teor. otnositel'n. i gravitatsii. 1975 g. Moskva, Nauka, 1977, p. 25 - 31. In Russian. – Abstr. in Ref. zh., 51. Astron., 1.51.812 (1978).

066.357 On nonstationary solutions in scalar-tensor theories of the gravitational field.
N. A. Zajtsev, G. N. Shikin.
Novoe v teor. otnositel'n. i gravitatsii. 1975 g. Moskva, Nauka, 1977, p. 80 - 85. In Russian. – Abstr. in Ref. zh., 51. Astron., 1.51.816 (1978).

066.358 The problem of motion in general relativity. S. L. Galkin.
Novoe v teor. otnositel'n. i gravitatsii. 1975 g. Moskva, Nauka, 1977, p. 43 - 48. In Russian. – Abstr. in Ref. zh., 51. Astron., 1.51.818 (1978).

066.359 Singularities and causality violation. F. J. Tipler.
Ann. Physics, Vol. 108, 1 - 36 (1977). – Abstr. in Phys. Abstr., Vol. 80, Abstr. 87666 (1977).

066.360 Relativistic fluid spheres and noncomoving coordinates. I. G. C. McVittie, R. J. Wiltshire.
Int. J. Theor. Phys., Vol. 16, 121 - 140 (1977). – Abstr. in Phys. Abstr., Vol. 80, Abstr. 87673 (1977).

066.361 Paths of spinning particles in general relativity as geodesics of an Einstein connection. R. R. Burman.
Int. J. Theor. Phys., Vol. 16, 211 - 225 (1977). – Abstr. in Phys. Abstr., Vol. 80, Abstr. 87674 (1977).

066.362 A rotating dust cloud in general relativity. W. B. Bonnor.
J. Phys. A, Vol. 10, 1673 - 1677 (1977). – Abstr. in Phys. Abstr., Vol. 80, Abstr. 87675 (1977).

066.363 Exact relativistic solution of disordered radiation with planar symmetry.
A. F. da F Teixeira, I. Wolk, M. M. Som.
J. Phys. A, Vol. 10, 1679 - 1685 (1977). – Abstr. in Phys. Abstr., Vol. 80, Abstr. 87676 (1977).

066.364 On the Kerr and the Tomimatsu-Sato spinning mass solutions. M. Yamazaki.
Prog. Theor. Phys., Vol. 57, 1951 - 1957 (1977). – Abstr. in Phys. Abstr., Vol. 80, Abstr. 87677 (1977).

066.365 Gravitons in Minkowski space-time. Interactions and results of astrophysical interest.
G. Papini, S. R. Valluri.
Phys. Rep. Phys. Lett. C, Vol. 33C, No. 2, p. 51 - 125 (1977). Abstr. in Phys. Abstr., Vol. 80, Abstr. 87683 (1977).

066.366 Experimental tests for some quantum effects in gravitation. N. D. Hari Dass.
Ann. Physics, Vol. 107, 337 - 359 (1977). – Abstr. in Phys. Abstr., Vol. 80, Abstr. 87684 (1977).

066.367 An extension to the noise theory of RF SQUIDs with implications for gravitational radiation detectors. J. Hough, J. R. Pugh, W. A. Edelstein, W. Martin.
J. Phys. E, Vol. 10, 993 - 997 (1977). – Abstr. in Phys. Abstr., Vol. 80, Abstr. 87685 (1977).

066.368 On gravitational shock waves.
H.-J. Treder, W. Yourgrau.
Int. J. Theor. Phys., Vol. 16, 233 - 239 (1977).

066.369 Relativistic model of a spherical star emitting neutrinos. I. D. Soares.
Centro Brasileiro de Pesquisas Fisicas, Rio de Janeiro. CBPF–A002/77. 52 pp. (1977). – Abstr. from INIS7719 331617.

066.370 General relativistic conservation laws for axisymmetric stationary flow of magnetized plasma.
J. D. Bekenstein, E. Oron.
Bull. Israel Phys. Soc.,Vol. 23, 25 - 26 (1977). – Summary. Abstr. from INIS7721 335614.

066.371 Constraints on the variability of the gravitational constant. J. D. Bekenstein, A. Meisels.
Bull. Israel Phys. Soc., Vol. 23, 26 (1977). – Summary. Abstr. from INIS7721 336197.

066.372 Can a variable-mass Kerr metric describe a star. M. Carmeli, M. Kaye.
Bull. Israel Phys. Soc., Vol. 23, 27 (1977). – Summary. Abstr. from INIS7721 336208.

066.373 Null tetrad formulation of the Yang-Mills field equations in both curved and flat spaces.
M. Carmeli, C. Charach, M. Kaye.
Bull. Israel Phys. Soc., Vol. 23, 27 - 28 (1977). – Summary. Abstr. from INIS7721 336209.

066.374 About the mass of the neutron stars.
          Z. Urban.
Říše hvězd, Vol. 58, 149 - 150 (1977). In Czech.

066.375 Feynman propagator in curved space-time.
          P. Candelas, D. J. Raine.
Phys. Rev. D, Vol. 15, 1494 - 1500 (1977) = Univ. Oxford,
Dep. Astrophys., Publ. No. 171.

066.376 Weber antenna response in different gravitation
          theories.   A. Degasperis.
Nuovo Cimento, Lett., Ser. 2, Vol. 17, 361 - 365 (1976) =
Univ. Oxford, Dep. Astrophys., Publ. No. 173.

066.377 Fermion fields, boson fields, and gravitational field.
          U. Kasper.
Ann. Physik, 7. Folge, Band 33, 317 - 320 (1976) = Zentralinst. Astrophys., Sternw. Babelsberg, Mitt. Neue Folge Nr. 167.

066.378 Zur theoretischen Analyse einer neuen Methode zur
          Bestimmung der Gravitationskonstante. I. Mechani-
sche Größen des dynamischen Systems.
U. Bleyer, R. W. John.
Gerlands Beitr. Geophys., Band 85, 402 - 414 (1976) =
Zentralinst. Astrophys., Sternw. Babelsberg, Mitt. Neue Folge
Nr. 170.

066.379 Die Strahlungs-Temperatur bewegter Körper.
          H.-J. Treder.
Ann. Physik, 7. Folge, Band 34, 23 - 29 (1977) = Zentralinst.
Astrophys., Sternw. Babelsberg, Mitt. Neue Folge Nr. 174.

066.380 Die Einstein-Verschiebung der Spektrallinien mit
          und ohne Relativität.   H.-J. Treder.
Sternzeiten, (see 003.025), Band 2, 71 - 78 (1977).

         Quasars, pulsars and black holes (a bibliography
with abstracts). Report for 1964—February 1977.
See Abstr. 002.052.

         Principles of cosmology and gravitation.
See Abstr. 003.032.

         Measurement of weak forces in physics experiments.
See Abstr. 003.036.

         White holes: the beginning and end of space.
See Abstr. 003.066.

         White holes. Cosmic gushers in the universe.
See Abstr. 003.067.

         The large-scale structure of space-time.
See Abstr. 003.068.

         Mathematics and the universe.
See Abstr. 003.088.

         Tensors, relativity and cosmology.
See Abstr. 003.091.

         Gravitation. Volume I.   See Abstr. 003.106.

         Gravitation. Volume 2.
See Abstr. 003.107.

         Gravitation. Volume 3.
See Abstr. 003.108.

         Relativitätstheorie.   See Abstr. 003.109.

         The logical flaws of Einstein's relativity.

See Abstr. 003.126.

         Space, time and spacetime.
See Abstr. 003.146.

         Space, time and gravity. The theory of the big bang
and black holes.   See Abstr. 003.163.

         Asymptotic structure of space-time.
See Abstr. 012.058.

         Black hole physics illustrated in photon orbits.
See Abstr. 021.017.

         Algebraic computing and the Newman-Penrose
formalism in general relativity.   See Abstr. 021.018.

         Optimale Dimensionierung optischer Apparaturen
und Untersuchung von Ausgleichsansätzen zur Bestimmung
der relativistischen Lichtablenkung.   See Abstr. 031.066.

         A new possibility of measuring low fluxes of
gravitational radiation.   See Abstr. 031.277.

         A lamellar grating interferometer experiment to
determine the spectrum of the cosmic background radiation.
See Abstr. 032.519.

         Gaussian-spindle gravitational wave antenna and
single-antenna anticoincidence experiments.
See Abstr. 034.036.

         General relativity and satellite orbits: the motion
of a test particle in the Schwarzschild metric.
See Abstr. 042.035.

         Relativistic variation of the latitude of Gorkij.
See Abstr. 045.030.

         Accretion magnetosphere stability. II. Polar cap
"drip".   See Abstr. 062.008.

         Gravitational instability of a composite rotating
plasma.   See Abstr. 062.037.

         A qualitative discussion on the possibility of
gravitational instabilities at the origin of explosions in magnetospheres.   See Abstr. 062.072.

         Accretion disk coronae and Cygnus X-1.
See Abstr. 064.035.

         Neutrino trapping during gravitational collapse of
stars.   See Abstr. 065.041.

         The collapse of unstable isothermal spheres.
See Abstr. 065.042.

         Static stars: some mathematical curiosities.
See Abstr. 065.048.

         Late stages of stellar evolution.
See Abstr. 065.053.

         The gravitational collapse of iron—oxygen stars with
masses of $2\,M_\odot$ and $10\,M_\odot$. II.   See Abstr. 065.070.

         The neutron mechanism of thermonuclear carbon
burning, formation of neutron stars and supernova bursts.
See Abstr. 065.084.

         Déformation du champ stellaire au voisinage du

Soleil, durant une éclipse totale, déterminée au moyen d'une méthode de simulation et d'étoiles fictives. See Abstr. 079.001.

The wavelength dependence of the facular excess brightness.   See Abstr. 080.014.

Effect of gravitational radiation on the evolution and spins of binary systems.   See Abstr. 117.041.

Radial oscillations of zero-temperature white dwarfs and neutron stars below nuclear densities. See Abstr. 126.012.

Black hole model of quasar-like objects—infrared optical emission.   See Abstr. 141.133.

On the nature of quasars and active nuclei of galaxies.   See Abstr. 141.153.

The binary pulsar: post-Newtonian timing effects. See Abstr. 141.510.

Potential drops above pulsar polar caps: acceleration of nonneutral beams from the stellar surface. See Abstr. 141.516.

A remark on co-rotation in pulsar magnetospheres. See Abstr. 141.542.

A model for bursting X-ray sources: time-dependent accretion by magnetic neutron stars and degenerate dwarfs. See Abstr. 142.027.

X-ray bursts and neutron-star thermonuclear flashes. See Abstr. 142.078.

The X-ray source in binary stars and the spin down of neutron stars.   See Abstr. 142.096.

Cyg X-1/ a candidate of the black hole. See Abstr. 142.128.

A numerical method for integrating the stellar-dynamical Fokker-Planck equation in a fixed inhomogeneous gravitational background.   See Abstr. 151.047.

On the measurement of parallaxes of objects at cosmological distances.   See Abstr. 162.004.

The source of the X-ray background. See Abstr. 162.019.

Unborn clusters.   See Abstr. 162.020.

Polarization transport in anisotropic universes. See Abstr. 162.023.

Dirac cosmology and the microwave background. See Abstr. 162.028.

Aether drift detected at last. See Abstr. 162.038.

Black hole and galaxy formation in a cold early Universe.   See Abstr. 162.039.

Cosmological solutions of the mass integral formulation of general relativity.   See Abstr. 162.058.

Theoretical isochrones with decreasing gravitational constant – II. Dirac's multiplicative cosmology. See Abstr. 162.059.

The de Sitter universe and mechanics. See Abstr. 162.090.

A class of inhomogeneous perfect fluid cosmologies. See Abstr. 162.093.

Inhomogeneous cosmologies: new exact solutions and their evolution.   See Abstr. 162.094.

Effects of a nonvanishing cosmological constant on the spherically symmetric vacuum manifold. See Abstr. 162.095.

Covariant point-splitting regularization for a scalar quantum field in a Robertson-Walker universe with spatial curvature.   See Abstr. 162.098.

The de Sitter-Castelnuovo Universe and cosmology. See Abstr. 162.101.

On the influence of massless scalar and vector electromagnetic fields on the singularity character in anisotropic cosmology.   See Abstr. 162.106.

Does the speed of light decrease with time? See Abstr. 162.107.

Effective-potential approach to graviton production in the early universe.   See Abstr. 162.117.

Errata

066.901   Errata: 'Post-Newtonian gravitational radiation from orbiting point masses' [Astrophys. J., Vol. 210, 764 - 775 (1976)].   R. V. Wagoner, C. M. Will. Astrophys. J., Vol. 215, 984 (1977).

066.902   Errata: "Polarized radiation from a point source orbiting a Schwarzschild black hole" [Mon. Not. R. Astron. Soc., Vol. 179, 691 - 697 (1977)].   S. Pineault. Mon. Not. R. Astron. Soc., Vol. 181, 799 (1977).

Sun

071 Photosphere, Spectrum

071.001 "Hot" bands of CO and the effect of vibration-rotation interaction on the solar CO spectrum.
T. Tsuji.
J. Quant. Spectrosc. Radiat. Transfer, Vol. 18, 179 - 184 (1977).

For the first overtone bands of CO, theoretical Herman–Wallis factors and experimental values show some inconsistencies for high excitation transitions. To resolve this problem, the relative band intensities, as well as the relative line intensities, are examined with CO lines in the solar infrared spectrum.

071.002 Statistical analysis of solar Fe I lines: magnetic line broadening. J. O. Stenflo, L. Lindegren.
Astron. Astrophys., Vol. 59, 367 - 378 (1977).

A statistical analysis of 402 unblended Fe I lines in the region 400–686 nm of the solar spectrum at disk center has been made in a search for magnetic line broadening effects. Empirical relations between the line width and various other parameters like line strength, excitation potential, and wavelength have been established to sort out a possible dependence of line width on Landé factor. The results support the picture of an extremely intermittent solar magnetic field.

071.003 On the east-west asymmetry of the central residual intensities and other Fraunhofer line parameters.
A. G. Gasanalizade.
Astron. Zh. Akad. Nauk SSSR, Vol. 54, 834 - 840 (1977).
In Russian. English translation in Soviet Astron., Vol. 21, No. 4.

In going from the centre to the limb of the solar disk, observational profiles of 13 Fraunhofer lines in two spectral ranges (6880 and 6980 Å) were derived from photoelectric tracings. The tracings of some Fraunhofer lines showed systematic and real differences in profile parameters at the east and west limbs. In particular, central residual intensities are slightly higher at the west limb of the solar disk than at the east limb. It is noted that this "east-west asymmetry" of the Fraunhofer line parameters may be due to horizontal motion of matter of the large-scale field of the solar atmosphere.

071.004 Eight feet of solar spectrum. O. R. Norton.
Sky Telesc., Vol. 54, 176 - 179 (1977).

071.005 Analysis of the solar magnesium I spectrum. II. Sensitivity of λ2852 to partial redistribution effects. R. C. Canfield, L. E. Cram.
Astrophys. J., Vol. 216, 654 - 658 (1977).

The authors computed theoretical profiles of the Mg I λ2852 resonance line by using various models for the frequency redistribution of the scattered radiation. They find that throughout the line core and inner wings these profiles are highly sensitive to the assumed extent of redistribution. They conclude that partial redistribution effects in the formation of this line must be taken into account in subsequent

calculations.

071.006 A search for a turbulent-free region in the solar transition zone. U. Feldman, G. A. Doschek.
Astrophys. J., Lett., Vol. 216, L119 - L121 (1977).

A search for a turbulent free transition zone region was conducted. The data used were spectra recorded by the NRL slit spectrograph on Skylab. It was found that the nonthermal turbulent motions are smallest in certain active regions and quiescent prominences. The authors discuss the spectra of one such region, a quiescent prominence.

071.007 Scattering in the Doppler core of the solar Lyman-α line: its effect on the Lyman continuum and on the chromospheric electron number density.
E. H. Avrett, J. E. Vernazza.
Bull. American Astron. Soc., Vol. 9, 432 (1977). – Abstract.

071.008 A photospheric magnetic field structure and its evolution at the appearance of an active region.
V. M. Grigorev (*Grigor'ev*), L. V. Ermakova.
Solar activity and solar-terrestrial relations, (see 012.007), p. 169 - 173 (1976).

The paper is concerned with the analysis of simultaneous observations of the longitudinal magnetic field in lines Fe I λ 5250 and 5233 Å in a quiet photosphere near the developed sunspot and in the developing sunspot group.

071.009 Sulphur molecules on the sun.
K. Sinha, G. C. Joshi, M. C. Pande.
Bull. Astron. Soc. India, Vol. 5, 45 - 48 (1977).

The authors investigate theoretically the question of presence or absence of the molecules SH, CS, SO, NS and S_2.

071.010 Determination of the damping constant from the wings of strong Fraunhofer lines.
Eh. A. Gurtovenko.
Soln. Dannye 1977 Byull., No. 6, p. 76 - 79 (1977). In Russian.

A simple and reliable method is proposed for determination of the damping constant from the wings of strong Fraunhofer lines. Results of an analysis of four Fe I lines give a value of the damping parameter much larger than Van der Waals' parameter.

071.011 Quantitative estimates of the damping constant effect on equivalent widths of weak Fraunhofer lines. V. A. Sheminova.
Soln. Dannye 1977 Byull., No. 6, p. 80 - 86 (1977). In Russian.

Uncertain data on the damping parameter makes it difficult to carry out an accurate analysis of equivalent widths of Fraunhofer lines for abundance determination. An evaluation has been made of the effect of the uncertainty in the damping constant on equivalent widths of the lines. Curves have been calculated for selecting limiting equivalent widths with which the errors of the damping constant do not change significantly the equivalent widths of the Fe I lines.

071.012 Balloon-borne imagery of the solar granulation. I. Digital image enhancement and photometric properties. A. Wittmann, J. P. Mehltretter.
Astron. Astrophys., Vol. 61, 75 - 78 (1977).

Digital image processing and restoration techniques have been applied to high-resolution photographs of the solar granulation that were obtained with a balloon-borne telescope. The final resolution that has been achieved is 0.3″ or 220 km on the sun. From the digitized and computer-processed images the authors have determined photometric properties of the solar granulation at the disc centre.

071.013 A comparison of synthetic and observed solar spectra in the vicinity of the G band. A. W. Irwin.
J. R. Astron. Soc. Canada, Vol. 71, 402 (1977). — Abstract.

071.014 Observations of small-scale photospheric magnetic fields. J. Harvey.
Highlights of Astronomy, Vol. 4, Part II, (see 012.024), p. 223 - 239 (1977).

This paper is a survey of observational techniques and results at the small-scale end of the spectrum of sizes in the solar photosphere. This topic has been frequently reviewed so that recent work is emphasized here.

071.015 New parameters for solar spicules based on submillimetre data. J. E. Beckman, J. Ross.
Far infrared astronomy (see 012.027), p. 79 - 86 (1976).

New submillimetre observations of the quiet solar limb with angular resolution 1.5″ arc show a narrow asymmetric brightness spike centred near the optical limb, with approximate half-width 10″ arc. The authors present a semi-empirical predictive model based on the properties of an individual spicule derived from optical and UV data. From it one can derive electron densities and temperatures within a spicule, a scale height of 2000 km for groups of spicules, and a mean number density over the solar surface of $4.2 \cdot 10^{-8}$ km $^{-2}$ at 1,200 km height above the photosphere.

071.016 Comparison of the profiles of weak Fraunhofer lines with the profile of the absorption coefficient. Eh. A. Gurtovenko, V. A. Ratnikova.
Astrometr. i Astrofiz., Kiev, vyp. (No.) 32, (see 003.011), p. 3 - 6 (1977). In Russian.

The results of calculations made for concrete Fe I lines show that the absorption line profiles with central intensity $\geqslant 75\%$ correspond with sufficient accuracy to the profile of the absorption coefficient obtained at the effective depth of weak-line formation.

071.017 On the "limiting" optical depths when studying the deviation from LTE in the solar photosphere. G. L. Fedorchenko.
Astrometr. i Astrofiz., Kiev, vyp. (No.) 32, (see 003.011), p. 15 - 16 (1977). In Russian.

A method is suggested to determine the level in the solar photosphere above which LTE conditions are not valid.

071.018 Supergranulation and the dynamics of gas and magnetic field below the solar photosphere. P. Foukal.
Astrophys. J., Vol. 218, 539 - 546 (1977).

The author suggests that the scale of supergranular flow may simply reflect the depth $d \approx 1.5 \times 10^4$ km below the photosphere over which the magnetic field occupies a relatively small volume and the nonmagnetic fluid is allowed to convect and rotate freely. If magnetic buoyancy is important in bringing flux to the photosphere from a deep region of either primordial or dynamo field, then a sharp increase of magnetic flux inward below the photosphere is required to satisfy a simple continuity argument. It is remarkable that the angular velocity of the nonmagnetic fluid seems to increase with depth over this supergranular layer at the rate expected if angular momentum per unit mass is conserved in radial convection. The author demonstrates that the radial gradient of the rotation velocity in the nonmagnetic gas is capable of bending flux tubes that extend upward through this supergranular layer through an angle which agrees closely with that inferred from observations by Howard. If this inferred tilt of the photospheric fields were ascribed to the torque of the interplanetary field, it would imply a rate of loss of angular momentum from the Sun exceeding by two orders of magnitude the value estimated from the rate of mass loss in the solar wind at the present epoch.

071.019 Molecular hydrogen in the sun. A. H. Cook.
Nature, Vol. 270, 297 - 298 (1977). — Review article.

071.020 Lines of H_2 in extreme-ultraviolet solar spectra. C. Jordan, G. E. Brueckner, J.-D. F. Bartoe, G. D. Sandlin, M. E. Van Hoosier.
Nature, Vol. 270, 326 - 327 (1977).

The first detection of H_2 in the Sun, in extreme ultraviolet spectra, is reported here. The Naval Research Laboratory's High Resolution Telescope and Spectrograph (HRTS), flown in a rocket on 21 July 1975, was used for these observations. A full list of observed lines is in preparation.

071.021 Study of central depths and equivalent widths of some Fraunhofer lines of the solar photosphere. P. A. Olijnyk, S. V. Shmiger.
Problems of cosmic physics. Vyp. (No.) 12, (see 003.016), p. 16 - 19 (1977). In Russian.

A study of central depths and equivalent widths of 15 absorption lines of iron, titanium and calcium has been carried out.

071.022 The ultraviolet solar spectrum 2756 Å — 2831 Å. Identification of absorption lines. A. Greve, C. D. McKeith.
Astron. Astrophys., Suppl. Ser., Vol. 30, 387 - 395 (1977).

A list of atomic spectral lines and semi-empirical gf values published by Kurucz and Peytremann (1975) allows increased and improved identifications of solar UV absorption lines previously observed in the range 2756 Å — 2831 Å.

071.023 Faint emission features in the Mg II resonance-line wings. M. S. Allen, H. C. McAllister.
Astrophys. J., Lett., Vol. 218, L137 - L139 (1977).

Three faint emission lines have been observed in the deep wings of the Mg II resonance lines by using data obtained by the University of Hawaii rocket-borne echelle spectrograph. Tentative identifications have been made with lines of Fe II and V II, on the basis of measured wavelengths and the similarity between the behavior of the observed lines and previously analyzed counterparts in the wings of Ca II H and K. If correct, these identifications suggest that these features are potentially useful diagnostics in solar and stellar spectroscopy.

071.024 f-values and abundances of the elements in the sun and stars. E. Biémont, N. Grevesse.
Phys. Scr., Vol. 16, 39 - 47 (1977).

After a brief summary giving the reasons for which the elemental abundances are of considerable interest and the methods used to derive solar and stellar abundances, the authors review the atomic and molecular f-values needed by solar spectroscopists, with a brief account of some problems encountered in stellar spectroscopy studies.

071.025 The solar chemical composition. O. Engvold.

Phys. Scr., Vol. 16, 48 - 50 (1977).

A brief review is given of recent solar abundance results. The relative abundance of 68 chemical elements is known for the sun.

071.026 Iron: whence it came, where it went.
J. F. Kerridge.
Space Sci. Rev., Vol. 20, 3 - 68 (1977) = Publ. No. 1560, Inst. Geophys. Planet. Phys., Univ. Calif., Los Angeles.

Determinations of the abundances of iron and related elements in the photosphere, chromosphere and corona of the Sun and in solar and galactic cosmic rays are reviewed and compared with abundances derived from meteoritic data. Observed Solar System abundances are found to be in accord with predictions of nucleosynthesis under either hydrostatic or explosive conditions but cannot yet be used to define these processes uniquely.

071.027 Measurements of magnetic fluxes and field strengths in the photospheric network.
T. D. Tarbell, A. M. Title.
Sol. Phys., Vol. 52, 13 - 25 (1977).

The authors present digital pictures of an active region network cell in five quantities, measured simultaneously: continuum intensity, line-center intensity, equivalent width, magnetogram signal, and magnetic field strength. Measured Zeeman splittings show the existence of strong magnetic fields (1000–1800 G) at nearly all points. From the significant disparity between measured fluxes and field strengths, the authors conclude that large flux patches (up to 4" across) consist of closely-packed unresolved filaments. The smallest filaments must be less than 0.7" in diameter. They also observe the dark component of the photospheric network, which appears to contain sizable transverse fields.

071.028 Vertical velocity fluctuations in plage-region magnetic points.   R. G. Giovanelli, N. Brown.
Sol. Phys., Vol. 52, 27 - 34 (1977).

Observations of line-of-sight velocities of gases in magnetic fields in weak plages near disk centre confirm the systematic downward velocity of 0.5 km s^{-1}, and show fluctuations about this mean by a rather uncertain 0.2 km s^{-1}. Some of the fluctuations show a fairly regular period around 5.5 min.

071.029 Solar limb darkening in the interval 7404–24018 Å, II.
A. K. Pierce, C. D. Slaughter, D. Weinberger.
Sol. Phys., Vol. 52, 179 - 189 (1977).

The coefficients of third degree and fifth degree polynomial representations of limb darkening are tabulated for 50 wavelengths in the interval λλ7404–24018.

071.030 Some results of the photospheric large-scale velocity research from Belgrade observations.
A. Kubičela, M. Karabin.
Sol. Phys., Vol. 52, 199 - 210 (1977).

On the basis of the possibility of sight-line velocity observations by a special equatorial solar spectrograph, a research programme for detection of photospheric large-scale velocities has been initiated. The first series of observations in the Fe I 6302 Å absorption line has been limited to the central meridian. The observations started in late 1974 and some results are presented here.

071.031 Morphological properties and origin of the photospheric facular granules.   R. Muller.
Sol. Phys., Vol. 52, 249 - 262 (1977).

From time series of high resolution photographs, morphological properties of the photospheric facular granules were derived. The facular granules are cells of the common granular pattern, brighter than the normal granules when seen between cos θ = 0.6 and the limb. Their apparent diameter is smaller than that of the normal granules. From the great similitude of both morphological properties and temperature models of facular and normal granules, it appears possible that the photospheric facular granules are convective cells modified by the presence of a magnetic field of some hundreds Gauss.

071.032 The effects of partial redistribution on facular K line profiles.
J. N. Heasley, F. Kneer, G. A. Chapman.
Sol. Phys., Vol. 52, 309 - 313 (1977).

The authors present theoretical Ca II K-line profiles and filtergram contrasts for several recent models of solar faculae. The line profiles vary greatly between models and between complete and partial frequency redistribution non-LTE calculations for any given model. The filtergram contrasts are relatively insensitive to the line formation theory which greatly simplifies the calculation for comparison with observations. All of the models considered exhibit K-line contrasts smaller than the mean value observed by Mehltretter.

071.033 Gas entry into non-spot magnetic tubes.
R. G. Giovanelli.
Sol. Phys., Vol. 52, 315 - 325 (1977).

Gas penetration into twisted magnetic tubes can occur by the inward diffusion of neutral atoms in the neighbourhood of the temperature minimum between photosphere and chromosphere, where the degree of ionization is low. Again, turbulent buffeting indents tubes in the convection zone and, in particular near the photosphere, provides a larger area where the overall diffusion rate may be enhanced. These processes do not contribute rapidly to the gas content of magnetic tubes, but diffusion near the temperature minimum may well be the source of the observed downflow in magnetic points.

071.034 Comments on the low-wavenumber power of granulation brightness fluctuations.
C. Aime, J. Borgnino, F. Martin, G. Ricort.
Sol. Phys., Vol. 53, 189 - 195 (1977).

A comment is made on Edmonds' study of the low-wavenumber power of solar granulation. It is concluded that its one-dimensional power spectrum is biased, and that the corresponding two-dimensional power spectrum shows negative parts.

071.035 On the asymmetry of selected Fraunhofer lines.
R. I. Kostik (*Kostyk*), T. V. Orlova.
Sol. Phys., Vol. 53, 353 - 358 (1977).

An analysis is made of the asymmetry of Fraunhofer lines observed with the double-pass monochromator of the horizontal solar telescope ASU-5 of the Main Astronomical Observatory of the Ukrainian Academy of Sciences. The conclusion is that the character of macromotion in the radial direction varies with height; in a tangential direction the motions at different depths are homogeneous. The asymmetry of weak lines is due to convection rather than to wave motions.

071.036 A new measurement of the center-to-limb variation of the rms granular contrast.   S. L. Keil.
Sol. Phys., Vol. 53, 359 - 368 (1977).

The center-to-limb variation of the root-mean-square granular contrast at 5520 Å is deduced from a set of high-spatial-resolution filtergrams obtained with the Sacramento Peak Observatory Vacuum Tower Telescope. The rms contrast is observed to decrease monotonically between μ = 1.0 and 0.6, and then increase slightly at μ = 0.4. This result is compared with the results of Edmonds (1962) and with the results of Pravdjuk et al. (1974).

071.037 Two-dimensional spatial spectrum of the photospheric brightness field near to the solar disc center.
V. N. Karpinsky (*Karpinskij*), V. V. Mekhanikov.
Sol. Phys., Vol. 54, 25 - 30 (1977).

From high-quality direct frames taken by the Soviet Stratospheric Solar Observatory, using coherent optical methods, the two-dimensional spatial spectrum of the photospheric brightness field was obtained. This spectrum is isotropic and continuous. Spectral densities were estimated for the solar disc centre, and their statistical uncertainty calculated. The true rms value for λ4650 is equal to about 29%.

071.038 The wavelength dependence of granulation (0.38–
         2.4 µm).   F. Albregtsen, T. L. Hansen.
Sol. Phys., Vol. 54, 31 - 33 (1977).

Using 2 pinhole photometers the intensity of the undisturbed photosphere was recorded simultaneously in 6 and in 4 wavelength regions. The rms value of the intensity variation in each of the 10 wavelength regions decreases slightly with increasing value of the heliocentric angle; this result confirms recent observations by other authors. The authors report the detection of a secondary maximum in the wavelength dependence of the intensity variation at λ = 1.5 µm.

071.039 Empirical determination of the damping constant in
         weak Fraunhofer lines.
B. T. Babij, I. S. Dmytryk.
Tsirk. Astron. Obs., L'vov, No. 51, p. 28 - 29 (1976). In Russian.

071.040 On the determination of the damping constant of
         weak Fraunhofer lines.   B. T. Babij, R. E. Keryk.
Tsirk. Astron. Obs., L'vov, No. 51, p. 30 - 32 (1976). In Russian.

071.041 Improved values for the solar micro- and macro-
         turbulent filter functions.
C. de Jager, J. Vermue.
Sol. Phys., Vol. 54, 313 - 317 (1977).

The contributions of any arbitrary photospheric velocity field to (macroturbulent) line displacement, and to (microturbulent) line broadening can be expressed by the macro- and micro-turbulent filters $f_M(k)$ and $f_t(k)$, where k is the wavenumber of the energy spectrum in which the line-of-sight component of the velocity field can be decomposed. As a correction to a previous computation of f_M and f_t the authors give improved values for the filter functions for weak lines in LTE. An example of the way to use the filter functions is given.

071.042 A new determination of the granule/intergranule
         contrast.   R. J. Bray, R. E. Loughhead.
Sol. Phys., Vol. 54, 319 - 326 (1977).

A new determination of the granule/intergranule contrast is described, based on high-quality photographs taken with the Culgoora refractor operating in its photospheric mode. The measurements are corrected for scattered light and for spatial smearing due to telescope and atmosphere. The contrast calculated on the basis of the Utrecht Reference Photosphere lies just outside the range of possible values permitted by the present observations, $0.76 \leqslant C \leqslant 0.81$ at λ5500. New determinations of the mean cell size of the granulation pattern (= 1.9 of arc) and of the distribution of cell sizes are also presented.

071.043 Evolution of photospheric magnetic field patterns
         during Skylab.   R. H. Levine.
Sol. Phys., Vol. 54, 327 - 341 (1977).

The evolution of the photospheric magnetic field pattern over eleven solar rotations preceding a minimum of the activity cycle is shown to be characterized by abrupt changes of the dominant geometrical patterns of the field. These changes are associated with the onset and end of a sudden increase in the calculated total energy content of the field, which is otherwise decreasing through the period. The calculated geometrical rearrangements correspond in time to observed restructurings

of the corona, the interplanetary field, and the solar rotation pattern.

071.044 Comparison of Hα synoptic charts with the large-
         scale solar magnetic field as observed at Stanford.
T. L. Duvall, Jr., J. M. Wilcox, L. Svalgaard, P. H. Scherrer, P. S. McIntosh.
Sol. Phys., Vol. 55, 63 - 68 (1977).

Two methods of observing the neutral line of the large-scale photospheric magnetic field are compared: (1) neutral line positions inferred from Hα photographs (McIntosh, 1972, 1975; McIntosh and Nolte, 1975) and (2) observations of the photospheric magnetic field made with low spatial resolution (3') and high sensitivity using the Stanford magnetograph. The comparison is found to be very favorable.

071.045 Analysis of the solar spectrum using Voigt profiles.
N. Baeck, P. Dingens, H. Steyaert.
Bull. Acad. R. Belgique, Cl. Sci., 5e Sér., Tome 63, 499 - 505 (1977).

071.046 Limb effect along the central meridian of solar disk.
A. Kubičela, M. Karabin.
Publ. Dep. Astron.,Univ. Beograd, Fac. Sci., No. 6, (see 012.040), p. 9 - 15 (1976).

The well known Fe I 6302 Å spectral line has been measured by a special equatorial solar spectrograph designed at the Belgrade Astronomical Observatory. The first results showing the limb effect along the central meridian (for $0° < \theta < 52°$) were obtained. The proposed theories for the limb effect have been analysed. A suggestion was made for the interpretation of the obtained results in the vicinity of the solar equator.

071.047 Limb polarization in spectral lines and the con-
         tinuum.   E. Wiehr.
Rep. Obs. Lund, No. 12, (see 012.044), p. 89 - 94 (1977).

071.048 Polarization of the photospheric light near the Sun's
         limb. A review of available observations.
J. L. Leroy.
Rep. Obs. Lund, No. 12, (see 012.044), p. 161 - 170 (1977).

A review of presently available measurements of the polarization of the photospheric continuum light is given together with a first look comparison with the theory. Main results are the center to limb variation of the polarization and the wavelength variation of the polarization which are rather well known.

071.049 Measurements of the polarization of the solar
         photospheric light near the limb and near the
active regions.   A. Koeckelenbergh, A. Dollfus.
Rep. Obs. Lund, No. 12, (see 012.044), p. 171 - 183 (1977).

071.050 On the asymmetry of selected Fraunhofer lines. IV.
         R. I. Kostyk, T. V. Orlova.
Astrometriya i Astrofizika, Kiev, vyp. (No.) 33, (see 003.020), p. 51 - 55 (1977). In Russian.

The asymmetry of Fraunhofer lines was determined from observations made with the double-pass system of the horizontal solar telescope ATsU-5 for different positions on the solar disk. The value of the asymmetry decreases with an increase in the equivalent width as well as with a decrease in $\cos\theta$.

071.051 On the possibility of using weak absorption lines
         for determining the amplitude of the total photo-
spheric velocity field.   N. N. Kondrashova.
Astrometriya i Astrofizika, Kiev, vyp. (No.) 33, (see 003.020), p. 62 - 64 (1977). In Russian.

The possibility of using weak absorption lines to study the total photospheric velocity field is confirmed by calculations. Absorption lines with a depth of not more than 20 –

25% are recommended.

071.052 Über die Lebensdauer der photosphärischen Granulation.   J. P. Mehltretter.
Mitt. Astron. Ges., Nr. 42, p. 113 (1977).

071.053 Korrektur von gemessenen granularen Geschwindigkeiten aus Sonnenfinsternisbeobachtungen.
W. Mattig, A. Nesis, G. Reiss.
Mitt. Astron. Ges., Nr. 42, p. 114 (1977).

071.054 Computer-processed granulation pictures of project "spectro-stratoscope".
A. Wittmann, J. P. Mehltretter.
Mitt. Astron. Ges., Nr. 42, p. 114 - 117 (1977).

071.055 The contrast of solar filigrees.
S. Koutchmy, G. Stellmacher.
Mitt. Astron. Ges., Nr. 42, p. 142 (1977).

071.056 Quiet photosphere and chromosphere.
J. M. Beckers.
Illustrated glossary for solar and solar-terrestrial physics, (see 003.022), p. 21 - 34 (1977).

071.057 High-resolution spectra of the solar Mg II h and k lines from Skylab.   G. A. Doschek, U. Feldman.
Astrophys. J., Suppl. Ser., Vol. 35, 471 - 482 (1977).
Spectra of the Mg II h and k lines emitted by different regions in the solar atmosphere have been recorded by the NRL slit spectrograph on Skylab. The spectral resolution is 0.12 Å and the spatial resolution is $2'' \times 60''$. Several examples are presented, including spectra of a chromospheric supergranulation cell boundary and interior and of a quiet-Sun region above the limb, and selected active-region spectra on the disk and above the limb. Obvious differences among these spectra are noted and qualitatively discussed.

071.058 Fine structure of the photospheric lines Fe I $\lambda6569.2$ Å and Fe I $\lambda6575.0$ Å.
V. N. Karpinskij, L. M. Kulagina, L. M. Pravdyuk.
Soln. Dannye 1977 Byull., No. 9, p. 61 - 71 (1977). In Russian.
Brightness fluctuations were studied in various sections of two moderately weak spectral lines Fe I $\lambda6569.2$ Å and Fe I $\lambda6575.0$ Å. Mean standard photometric tracings in the continuous spectrum were obtained for each line. Mean line profiles were obtained as averaged over the whole photosphere and separately for bright regions (granules) in the continuous spectrum and dark intergranular regions.

071.059 Large-scale brightness fluctuations of the solar photosphere. II.   I. V. Yudina.
Soln. Dannye 1977 Byull., No. 10, p. 67 - 73 (1977). In Russian.
On the basis of new photoelectric recordings of brightness fluctuations of the photosphere obtained in summer 1976 a conclusion is made on the presence of brightness fluctuations of a 7000–18000 km scale on the solar disc. More exact values of the mean square contrast in the studied inhomogeneities relative to the mean photospheric background are found for the centre and limb of the solar disc.

A measurement of the width and shift of the Fe I 3719.94 Å line broadened by helium.
See Abstr. 022.004.

Transition probabilities in Nd(II) and the solar neodymium abundance.   See Abstr. 022.036.

Lifetime measurements of the La II $y^3F_{4,3,2}^{0}$ levels with the beam-laser method. (*Solar photosphere La abundance*). See Abstr. 022.066.

Oscillator strengths for some Mn II lines and the solar Mn abundance.   See Abstr. 022.086.

New measurements of the Se I resonance lines.
See Abstr. 022.090.

Lifetime measurements with the beam-laser method.
See Abstr. 022.115.

Application of multi-channel analysis techniques to beam-foil spectroscopy of neutral cobalt in the region 3000 Å to 3950 Å.   See Abstr. 031.296.

Singular points of magnetic fields in Neumann's problem.   See Abstr. 072.022.

Mg I b-lines in faculae and the photosphere.
See Abstr. 072.047.

Difference of the contours of the infrared triplet of Ca II in faculae and the photosphere.
See Abstr. 072.055.

Some characteristics of the magnetic field and photospheric structure development in the August 1972 proton-flare region.   See Abstr. 073.018.

The production of lithium in the solar chromosphere and photosphere during white light flares.
See Abstr. 073.063.

On the nature of photospheric magnetic fields beneath large coronal holes.   See Abstr. 074.073.

Heliographic maps of the photosphere for the year 1976.   See Abstr. 075.015.

The solar spectrum in the vicinity of the Si IV lines at 1122 and 1128 Å.   See Abstr. 076.006.

Open magnetic fields in active regions.
See Abstr. 076.018.

The solar CO spectrum as probe of atmospheric structure and carbon abundance.   See Abstr. 077.022.

Noise storms and particular photospheric magnetic structures.   See Abstr. 077.036.

On the influence of oscillator strengths, equivalent widths and photospheric model on the iron abundance in the solar atmosphere.   See Abstr. 080.021.

A test of the micro-macroturbulence model on the solar flux spectrum.   See Abstr. 080.036.

Infrared continuum observations of five-minute oscillations.   See Abstr. 080.059.

A comparison of solar and meteoritic abundances.
See Abstr. 105.273.

The equivalent widths of iron lines in the spectra of cool standard stars for spectral analysis.
See Abstr. 114.013.

071.901 Erratum: 'Convective velocities derived from granule contrast profiles in Fe I $\lambda6569.2$' [Sol. Phys., Vol. 49, 3 - 18 (1976)].
R. J. Bray, R. E. Loughhead, E. J. Tappere.
Sol. Phys., Vol. 55, 274 (1977).

072 Sunspots, Faculae, Activity Cycles

072.001 Statistical study of the time sequence of Ca II flocculi and series of radio fluxes during solar activity cycle 20.   M. Karlický.
Bull. Astron. Inst. Czechoslovakia, Vol. 28, 200 - 217 (1977).

This paper is an attempt to study the development of statistical characteristics of solar activity cycle 20 (power spectra, autocorrelation curves, cross-correlation, cross spectra).

072.002 Photospheric gas motions in a single sunspot. S. I. Gopasyuk.
Izv. Krymskoj Astrofiz. Obs., Vol. 57, 107 - 121 (1977). In Russian.

A method for determination of all components of the velocity vector in a cylindrical coordinate system from observations of line-of-sight velocities is suggested. Following this method gas motions in a single sunspot crossing the central meridian 12.VII.1962 were studied. On the basis of the suggested method radial, azimuthal and vertical components of velocity and their changes with the radius of the spot were determined.

072.003 Magnetic fields and proper motions of sunspots. II. Group 420, October 1968.   B. Kalman.
Izv. Krymskoj Astrofiz. Obs., Vol. 57, 122 - 132 (1977). In Russian.

In the complex group of sunspots No. 420, 20 - 27 October 1968, on the basis of maps of the longitudinal and transverse magnetic field components and a series of photoheliograms, the proper motions of the nuclei of spots were compared with the structure of the magnetic field. The proper motions were compared with the flare-activity in the group too.

072.004 The results of magnetograph calibration programmes 1974.   K. Pflug, M. Klvana, J. Suda.
Izv. Krymskoj Astrofiz. Obs., Vol. 57, 144 - 155 (1977). In Russian.

It is performed both a calibrational comparison of one pair of magnetograms observed at Potsdam (line 5253.47 Å) and Ondrejov (line 5250.22 Å) on 1974, June 17 and an estimate of the internal agreement of three consecutive magnetograms observed at Potsdam in the line 5253.47 Å on 1974, June 30.

072.005 Confirmation of the presence of iron hydride in sunspots and cool stars.
R. F. Wing, J. Cohen, J. W. Brault.
Astrophys. J., Vol. 216, 659 - 664 (1977).

A high-resolution sunspot spectrum clearly shows the presence of the 9896 Å and 8691 Å bands of iron hydride (FeH). At least 332 lines of the two bands, accounting for 80% of the lines registered in the laboratory, can be identified with certainty in the spot spectrum; most of these are unblended. No evidence for FeH is found in the spectrum of the solar disk. Image-tube spectrograms of M dwarfs and S stars have been taken in the near-infrared at 86 Å mm^{-1} to examine the feature which Nordh, Lindgren, and Wing have proposed is due to the 9896 Å band of FeH. At this dispersion enough band structure is evident to confirm the identification in both kinds of stars.

072.006 Sunspot shape, growth, and motion.
R. N. Moses.
Bull. American Astron. Soc., Vol. 9, 432 (1977). – Abstract.

072.007 Investigation of solar activity in the past by means of cosmogenic isotopes.   P. Povinets.

Izv. AN SSSR. Ser. fiz., Vol. 41, 383 - 389 (1977). In Russian. – Abstr. in Ref. zh., 51. Astron., 8.51.451 (1977).

072.008 Some results of velocity measurements in the August 1972 group.   P. Macák.
Solar activity and solar-terrestrial relations, (see 012.007), p. 31 - 33 (1976).

072.009 On the mutual influences of spatially remote sunspot groups united into flare-active complexes.
V. V. Kasinskii (*Kasinskij*).
Solar activity and solar-terrestrial relations, (see 012.007), p. 63 - 67 (1976).

The microspots activity of a large sunspot group taking place on the sun's disk from 1st to 11th August, 1972 and generating five strong flares is analysed.

072.010 On two populations of sunspot groups. G. V. Kuklin.
Solar activity and solar-terrestrial relations, (see 012.007), p. 195 - 199 (1976).

An analysis of the character of cycle to cycle variations of sunspot group distributions in observed area and in maximal area with the help of the principal component method permits to put forward the following hypothesis. Distribution variations are caused by a 80-year cycle change of a size ratio of two sunspot populations observed at the Sun. The first population corresponds to a "normal" distribution and dominates at the secular cycle maximum epoch and the second one contains a larger number of smaller sunspots and a smaller number of larger sunspots and gives an essential contribution at the minimum epoch.

072.011 The fine structure of photospheric sunspots and faculae.   R. Muller.
Solar activity and solar-terrestrial relations, (see 012.007), p. 201 - 209 (1976).

The author presents the following sketch of the penumbral fine structure: bright grains, nearly as bright as the normal photosphere, generally lined up in the form of filaments, are seen against a dark background, the intensity of which is only 60% of that of the photosphere. The fine structure of penumbras and umbras in different wavelengths and for different positions of a regular spot on the disk is shown. The fine structure of photospheric faculas near the limb has also been revealed on some high quality pictures.

072.012 The rotational motions of sunspots.   Š. Knoška.
Solar activity and solar-terrestrial relations, (see 012.007), p. 211 - 212 (1976).

072.013 Profiles of the H and K Ca II emission reversals in sunspots.
R. B. Teplitskaya, S. A. Efendeva (*Ehfendieva*).
Solar activity and solar-terrestrial relations, (see 012.007), p. 213 - 229 (1976).

The profiles of K and H Ca II lines in the umbra and the penumbra of some spots located near the solar disk centre are measured. In contrast to the results received by other authors the influence of scattered light in the atmosphere and in the instrument is taken into account. The scattered light distorts residual intensities of K- and H-emission reversals in the umbra more than in the case of usual absorption lines. The main typical feature of K- and H-lines in the umbra is an enormously high residual intensity of emission peaks.

072.014 Some parameters in forecasting of solar active region development.

A. N. Koval (*Koval'*), N. N. Stepanyan.
Solar activity and solar-terrestrial relations, (see 012.007), p. 235 - 237 (1976).

A dependence between the sunspot group flare activity and its magnetic structure (the orientation of the zero line of the field) is considered. The relation between the flare activity of plages and their brightness in H-alpha- and K-lines is studied.

072.015 **Some regularities of the occurrence of large sunspot groups.**   M. Kopecký, P. Kotrč.
Solar activity and solar-terrestrial relations, (see 012.007), p. 243 - 247 (1976).

The periodicities of sunspot groups with an average area larger than 500 millionths of the solar hemisphere surface and of sunspot groups with the maximum area larger than 1500 millionths in the years 1874–1964 are studied. Studying the occurrence of the large groups of sunspots in dependence on time and heliographic longitude, it is shown that this occurrence of the large sunspot groups provides evidence of the fact that in certain periods the outburst of solar activity occurred on the major part of the solar surface, regardless of hemispheres and heliographic longitudes.

072.016 **A cause of a very high maximum of solar activity in cycle No. 19.**   M. Kopecký.
Solar activity and solar-terrestrial relations, (see 012.007), p. 249 - 250 (1976).

072.017 **An autocorrelation method for determining the sunspot activity period.**   G. Mariş.
Solar activity and solar-terrestrial relations, (see 012.007), p. 251 - 254 (1976).

An autocorrelation method is used to estimate the solar activity period assuming an autoregressive series of the data. Discussion and suggestions about the application of this method to the forecast of solar activity and the study of north–south sunspot asymmetry are also made.

072.018 **Autocorrelation analysis of some solar indices.**   V. Letfus.
Solar activity and solar-terrestrial relations, (see 012.007), p. 255 - 262 (1976).

Two solar indices have been used for the autocorrelation analysis. The first one was the daily sunspot area for the period 1883–1959, the second one the daily flare index for the period 1936–1949. In both cases data from the whole disc, as well as those from the northern and southern hemispheres have been investigated separately. The autocorrelation function gives information about the recurrence of the solar activity centres.

072.019 **On the rotational temperature of TiO in sunspots.**   K. Sinha.
Bull. Astron. Soc. India, Vol. 5, 49 - 51 (1977).

For TiO rotational temperatures consistent with observations can be obtained through a change to a more refined and realistic model atmosphere, without invoking non-LTE.

072.020 **Large sunspot groups in solar cycle No. 19. I. Longitudinal distribution.**   Eh. P. Surkov.
Soln. Dannye 1977 Byull., No. 5, p. 47 - 54 (1977). In Russian.

In the solar cycle No. 19 four active longitudes were detected coinciding with the sector borders of a large-scale solar magnetic field.

072.021 **Some peculiarities of the velocity fields and magnetic fields in developed active regions. III. On the relationship between the imbalance of streams of matter and magnetic fluxes.**   Dzh. I. Irgashev.
Soln. Dannye 1977 Byull., No. 5, p. 55 - 58 (1977). In Russian.

It is shown that the imbalance of mass streams in an active region as a whole decreases with increase of the imbalance of magnetic fluxes in sunspots.

072.022 **Singular points of magnetic fields in Neumann's problem.**   E. A. Kornitskaya, M. M. Molodenskij.
Soln. Dannye 1977 Byull., No. 5, p. 58 - 65 (1977). In Russian.

A method is given for determining the singular points above the photospheric surface for Neumann's boundary problem. The problem is solved for four sunspots with continuously distributed photospheric field. A comparison of the solutions is made for high and low spatial resolution.

072.023 **Photoelectric photometry of pores in two regions of the continuum. I.**   I. V. Yudina.
Soln. Dannye 1977 Byull., No. 5, p. 74 - 76 (1977). In Russian.

Results of photoelectric photometry of pores are given in two regions of the continuum. Brightness ratios of a pore to the neighbouring photosphere are not corrected for light scattering and image blurring.

072.024 **Sunspot distribution over magnetic field strengths.**   A. A. Solov'ev.
Soln. Dannye 1977 Byull., No. 5, p. 77 - 80 (1977). In Russian.

Maps of sunspot magnetic fields have been used to construct the distribution function of sunspots over the magnetic field. The function obtained has a maximum near H = 1800 gauss. This fact is in good agreement with theoretical predictions based on model calculations of the binding energy of sunspots.

072.025 **Photometry of sunspots in the continuum.**   V. N. Milovanov, S. O. Obashev.
Soln. Dannye 1977 Byull., No. 5, p. 94 - 99 (1977). In Russian.

The relative intensity of a sunspot umbra is determined by photometry in continuum regions free from absorption lines. Evaluation of light scattering is made from the lines Fe II 5991.4 and 6084.1 Å.

072.026 **Photoelectric photometry of pores in two regions of the continuum. II.**   I. V. Yudina.
Soln. Dannye 1977 Byull., No. 6, p. 57 - 63 (1977). In Russian.

Using results of photoelectric photometry of pores brightness ratios of a pore to the neighbouring photospheric brightness (I_p/I_{phot}) were corrected for light scattering and image blurring. An evaluation of I_p/I_{phot} is given for several pores using results of photometric reduction of a direct photograph of the solar photosphere taken during the fourth flight of the Stratospheric Solar Observatory on 20 June, 1973.

072.027 **Physical conditions in a sunspot penumbra.**   V. N. Obridko, O. G. Badalyan.
Soln. Dannye 1977 Byull., No. 6, p. 98 - 104 (1977). In Russian.

A sunspot model is considered with the zero-point of the optical depth lower by 350 - 600 km in accordance with the Wilson effect. The penumbra is a superficial structure located above a zone of free convection. Between the umbra and penumbra there is a protective shell with radiative transfer. A two-component model of a sunspot penumbra is proposed, in good agreement with observations.

072.028 **Physique solaire: interprétation mécanique de l'activité solaire.**   A. Dauvillier.
Ciel Terre, Vol. 93, 204 - 214 (1977).

La théorie attribuant l'activité solaire aux marées planétaires permet de rendre compte des périodes de 11, 22, 78 ans, des cycles alternés, de retrouver l'activité dans le passé (1132) et de la prévoir dans l'avenir (2068), d'expliquer la formation de couples de taches de polarité magnétique suivant les lois de Hale, de prévoir le nombre de taches d'un cycle, d'élucider la loi de Spörer et le paradoxe de Maunder, la rotation différentielle et les variations de la rotation solaire.

072.029 **Formation of active regions.**   N. N. Stepanyan.
VIII Leningr. mezhdunar. semin. Mater. mezhdunar. semin. "Akt. protsessy na Solntse i probl. soln. nejtrino", 1976. Leningrad, 1976, p. 99 - 104. In Russian. − Abstr. in Ref. zh., 51. Astron., 9.51.427 (1977).

072.030 **Solar activity in the minimum of the 20th cycle.**
M. N. Nazarova, N. K. Pereyaslova, I. E. Petrenko.
VIII Leningr. mezhdunar. semin. Mater. mezhdunar. semin. "Akt. protsessy na Solntse i probl. soln. nejtrino", 1976. Leningrad, 1976, p. 133 - 135. In Russian. − Abstr. in Ref. zh., 62. Issled. kosm. prostranstva, 9.62.197 (1977).

072.031 **Photospheric faculae: the contrasts at the center of the solar disk using filigree pictures.**
S. Koutchmy.
Astron. Astrophys., Vol. 61, 397 - 404 (1977).
Fine structure of photospheric faculae embedded in intergranular lanes has been observed at the continuum level. 2-dimensional restoration of the best filigree pictures obtained by Dunn (1969) was tentatively achieved using a quasi-conventional method of deconvolution and assuming largest values of the modulation transfer function. The restored pictures show a typical filigree contrast of 100%, but this limited restoration gives only a lower limit to the actual contrast.

072.032 **Line profiles of faculae and pores.**
E. N. Frazier.
Highlights of Astronomy, Vol. 4, Part II, (see 012.024), p. 255 - 260 (1977).
The single principal conclusion of this paper is that the shape of a line profile, and therefore the $T(\tau)$ relation, of faculae changes significantly and continuously from small faculae to pores.

072.033 **Facular models, the K-line, and magnetic fields.**
G. A. Chapman.
Highlights of Astronomy, Vol. 4, Part II, (see 012.024), p. 261 - 264 (1977).

072.034 **Einige Besonderheiten im Verlauf der Sonnenfleckenminima.**   W. Schulze.
Sterne, 53. Band, 163 - 168 (1977).

072.035 **The origin of solar activity.**   E. N. Parker.
Annu. Rev. Astron. Astrophys., Vol. 15, (see 003.012), 45 - 68 (1977).
This article reviews the basic physical problems and principles of solar activity, insofar as they are presently understood, concentrating on the general question of why magnetic fields produce activity.

072.036 **A study of solar activity in the past by means of cosmogenic isotopes. Abstract.**   P. Povinets.
VIII Leningr. mezhdunar. semin. Mater. mezhdunar. semin. "Akt. protsessy na Solntse i probl. soln. nejtrino", 1976. Leningrad, 1976, p. 199. In Russian and English. − From Ref. zh., 51. Astron., 10.51.563 (1977).

072.037 **A preliminary investigation on the regularities of secular variations in solar activity.**
Y.-c. Lin, C.-c. Chang.
Kexue Tongbao, Vol. 22, No. 2, p. 59 - 67 (1977). In Chinese. From Phys. Abstr., Vol. 80, Abstr. 71080 (1977).

072.038 **Morphological analysis of a large sunspot group in October 1972.**

Y.-j. Ding, W.-b. Li, Q.-f. Hong, Z.-k. Li.
Acta Astron. Sinica, Vol. 18, 39 - 59 (1977).
(1) The main morphological feature of the proton active region is the clockwise spiral structure of the penumbral filaments of the large preceding spot "A". (2) The formation of this spiral structure may be caused by the anti-clockwise spin of umbra "A". This point is discussed in detail. (3) The anti-clockwise rotation of the N-S magnetic axis of the sunspot group was directly correlated to the spin of spot "A", and the direction of both rotations was opposite to that of the effect of the differential rotation.

072.039 **Emission measures and structure of the transition region of a sunspot from emission lines in the far ultraviolet.**   C.-C. Cheng, O. Kjeldseth Moe.
Sol. Phys., Vol. 52, 327 - 335 (1977).
Absolute intensities of emission lines in the wavelength range from 1200 Å to 1817 Å from the large sunspot in McMath region 12510 near Sun center are presented. The intensities are averaged across the umbra and penumbra of the sunspot. The observations were made with the NRL slit spectrograph on Skylab. Emission measures are derived from the measured intensities. Assuming a balance between the divergence of the conductive energy flux and the radiative energy losses, a self-consistent model of the lower transition region in the sunspot is constructed.

072.040 **The spiral configuration of sunspot magnetic fields.**
M. J. Hagyard, E. A. West, N. P. Cumings.
Sol. Phys., Vol. 53, 3 - 13 (1977).
Distributions of circularly and linearly polarized intensities are computed using an analytical magnetic field model for an isolated sunspot, and these intensity distributions are compared with observed intensities in all Stokes parameters in the λ 5250 line measured with the Marshall Space Flight Center's vector magnetograph. The qualitative agreement between measured and calculated linearly polarized intensity distributions is discussed with regard to implications as to the configuration of the transverse magnetic field of the isolated sunspot.

072.041 **Determination of stray light and sunspot intensities based on the observation during the solar eclipse of 30 June, 1973.**   B. Hadjebi.
Astrophys. Space Sci., Vol. 52, 357 - 363 (1977).
The usual method for correcting sunspot intensity measurements for stray light has been applied in the case of the partial solar eclipse of 30 June, 1973, where a sunspot was partly obscured as well as the solar disc.

072.042 **On AlF lines in sunspots.**   P. S. Murty.
Sol. Phys., Vol. 54, 377 - 378 (1977).
Intensity factors for bands of the C−A transition of the AlF molecule are calculated. A reinvestigation for the lines of (0, 0) band of the C−A system in sunspot spectra is suggested, so as to resolve the questionable existence of AlF absorption lines in sunspots.

072.043 **On the difference in darkness between sunspots.**
P. Maltby.
Sol. Phys., Vol. 55, 335 - 346 (1977).
The effects of the magnetic field as well as the velocity field on sunspot equilibrium are discussed. It appears that, at an optical depth of unity in the umbra, the density has a value close to that of the environment at the same geometric depth.

072.044 **The gross energy balance of solar active regions.**
K. D. Evans, J. P. Pye, R. J. Hutcheon, M. Gerassimenko, A. S. Krieger, J. M. Davis, J. F. Vesecky.
Sol. Phys., Vol. 55, 387 - 392 (1977).
On 26 November 1973 the active region McMath 12628 was studied. It is shown that the atmosphere did not radiate and almost certainly did not receive, more than a very small

part of the 'missing flux' of the spot group. This result is an important constraint on the plausible theories of sunspot formation.

072.045 Photographic determination of heliographic coordinates of sunspots without taking double photos.
T. Aleksandar.
Publ. Dep. Astron., Univ. Beograd, Fac. Sci., No. 6, (see 012.040), p. 65 - 67 (1976).

072.046 Geometrical and kinematic parameters of non-recurrent sunspot groups.   N. I. Kozhevnikov.
Astron. Tsirk., No. 945, p. 1 - 3 (1977). In Russian.

072.047 Mg I b-lines in faculae and the photosphere.
A. A. Galal, G. F. Sitnik.
Astron. Tsirk., No. 949, p. 1 - 3 (1977). In Russian.

072.048 Long-range forecast of solar activity.
M. E. Paupere.
Astron. Tsirk., No. 953, p. 5 - 6 (1977). In Russian.

072.049 Peculiarities of the activity centres on the descending branch of the 20th solar cycle.
Yu. I. Vitinskij.
Izv. Glav. Astron. Obs. Pulkovo, No. 195, Astrofiz. Astrometr., p. 113 - 122 (1977). In Russian.

A catalogue of activity centres for 1971 - 1973 is given. It is shown that the curve of the mean lifetime of the activity centres has many peaks during the 11-year cycle, its highest maximum taking place before the epoch of maximum for Wolf numbers. Cycle curves of the principle characteristics of activity centres manifest two peaks (or three) which are similar to those detected by Gnevyshev (1966). Their highest maximum coincides with that of the mean lifetime of the centres. Active longitudes of the number of activity centres differ significantly for the descending branches of solar cycle 19 and 20. This difference occurs only after the epoch of maximum for solar cycle 20.

072.050 Das Maunder-Minimum der Sonnenaktivität.
F. W. Jäger.
Sterne, 53. Band, 215 - 221 (1977).

072.051 The epoch of the sunspot-minimum 1976.
M. Waldmeier.
Astron. Mitt. Eidg. Sternw. Zürich, Nr. 355, 6 pp. (1977). In German.

072.052 Zum Problem der Evershed-Strömungen in Penumbra-Feinstrukturen.
E. Wiehr, G. Stellmacher, D. Soltau.
Mitt. Astron. Ges., Nr. 42, p. 117 - 119 (1977).

072.053 Spots and faculae.   A. Bruzek.
Illustrated glossary for solar and solar-terrestrial physics, (see 003.022), p. 71 - 79 (1977).

072.054 Fine structure of the amplitude-frequency spectrum of the parameters of solar activity.
N. I. Kozhevnikov.
Soln. Dannye 1977 Byull., No. 9, p. 71 - 80 (1977). In Russian.

Methods of spectral and harmonic analysis were applied to an analysis of the series of relative numbers and areas of sunspots. The periods are compatible with those of the revolution of planets about the sun and periods of heliocentric configurations of Mercury, Earth, Venus and Jupiter. It seems likely that the sun is a non-linear system with respect to external actions.

072.055 Difference of the contours of the infrared triplet of Ca II in faculae and the photosphere.   A. A. Galal.

Soln. Dannye 1977 Byull., No. 9, p. 81 - 86 (1977). In Russian.

Changes in the profiles of Ca II infrared triplet lines in faculae with respect to the profiles of these lines in quiet regions of the photosphere were recorded photoelectrically at different positions on the solar disk. The observations revealed distinctly a conspicuous weakening of the lines in faculae. Sometimes slight broadening of the intermediate parts of the line wings was registered. Differences between the profiles of the lines in faculae and the photosphere were also observed in the center of the solar disk.

072.056 Oscillations of magnetic fields of sunspots and their connection with active region events.
V. F. Chistyakov.
Soln. Dannye 1977 Byull., No. 10, p. 93 - 98 (1977). In Russian.

There are two kinds of fast oscillations of the magnetic field strength of sunspots: free oscillations, which are not connected with solar activity phenomena, and oscillations which are connected with unstable processes in active regions. Both kinds of oscillations often occur on large areas of the solar surface and have global character. The oscillations of the noise storm background appear 15 - 20 minutes later than oscillations of the sunspot magnetic field. The author believes that fast oscillations of the magnetic field strength are due to wave motions arising from the interior layers of the sun.

072.057 Magnetic history of solar active regions.
J. M. Mosher.
Thesis Calif. Inst. Technol., Pasadena, USA. 299 pp. (1977). Abstr. from INIS7724 345382.

072.058 The long-term variation of the solar activity and its effect on the earth.   K. Sakurai.
10th Lunar and Planetary Symposium, (see 012.050), p. 75 - 78 (1977). In Japanese.

072.059 Ilgi Saules aktivitātes minimumi. Cik gadu pastāv Saules vainags?   G. Ozoliņš.
Zvaigžņotā debess, 1977. gada vasara, p. 1 - 4.

072.060 About the maximum of the last 80-year period of sunspots.   M. Kopecký.
Říše hvězd, Vol. 58, 228 - 230 (1977). In Czech.

072.061 Visual observation of the sun in Czechoslovakia in the year 1976.   L. Schmied.
Říše hvězd, Vol. 58, 147 - 149 (1977). In Czech.

Historical development of notions of sunspots.
See Abstr. 004.064.

The effect of finite electrical and thermal conductivities on magnetic buoyancy in a rotating gas.
See Abstr. 062.040.

On steady force-free dynamos.
See Abstr. 062.069.

The stability of a magnetic flux element in a horizontally stratified compressible plasma.
See Abstr. 062.073.

A photospheric magnetic field structure and its evolution at the appearance of an active region.
See Abstr. 071.008.

The north—south distribution of major solar flare events, sunspot magnetic classes and sunspot areas (1955 - 1974).   See Abstr. 073.061.

Transequatorial loops interconnecting McMath

regions 12472 and 12474.   See Abstr. 073.062.

Active regions.   See Abstr. 073.116.

Coronal general magnetic field evolution as a new parameter of the solar cycle.   See Abstr. 074.044.

Solar-cycle evolution of the coronal general magnetic field of 1959–1974 and the synchronous variation of high-speed solar wind streams and galactic cosmic rays. See Abstr. 074.068.

Solar wind density and its genetic relation with solar activity on flocculi level.   See Abstr. 074.109.

L'activité solaire.   See Abstr. 075.003.

Provisional sunspot-numbers for 1977 May - November.   See Abstr. 075.006.

Photoheliographic results 1967. See Abstr. 075.009.

Solar phenomena.   See Abstr. 075.012.

Definitive Sonnenflecken-Zahlen für 1976. See Abstr. 075.013.

Sunspot relative numbers for 1976. See Abstr. 075.016.

Die Sonnenaktivität im Jahre 1976. See Abstr. 075.017.

Sunspot numbers.   See Abstr. 075.025.

EUV flux variation during end of solar Cycle 20 and beginning Cycle 21, observed from AE-C satellite. See Abstr. 076.010.

Radio emission of an intense active region on the sun in July 1974 at wavelengths 3.5, 2.5 and 1.9 cm. See Abstr. 077.002.

Investigation of fluctuations of solar radio emission and of sunspot areas.   See Abstr. 077.011.

Evolution of active regions on 21 and 9 cm. See Abstr. 077.013.

Long-term forecast of proton events in active regions.   See Abstr. 078.020.

The wavelength dependence of the facular excess brightness.   See Abstr. 080.014.

On the transformation and removal of local magnetic fields into the corona and interplanetary space. See·Abstr. 080.016.

Large-scale solar magnetic fields.

See Abstr. 080.040.

Maunder's paradox and the tidal theory of solar activity.   See Abstr. 080.051.

Solar-cycle variations in the differential rotation of solar magnetic fields.   See Abstr. 080.053.

On a possible mechanism of solar faculae heating. See Abstr. 080.061.

Computer solutions for studying correlations between solar magnetic fields and Skylab X-ray observations. See Abstr. 080.062.

Correlation of the effective depth of the convective zone with the solar activity cycle.   See Abstr. 080.074.

The magnetic field of active regions and its zero points.   See Abstr. 080.080.

Cophasal regions on the sun. See Abstr. 080.081.

Solar cycle, solar rotation and large-scale circulation. See Abstr. 080.083.

The reconciliation of an F-region irregularity model with sunspot cycle variations in spread-F occurrence. See Abstr. 083.075.

Change of the latitude dependence of night ionization of the F2 layer in the solar activity cycle. See Abstr. 083.115.

Solar cycle effects on the electric fields in the equatorial ionosphere.   See Abstr. 083.120.

Sunspot cycle influence on the geomagnetic field. See Abstr. 084.201.

Amplitude variation of 27-day period in geomagnetic activity during 19th and 20th solar cycles. See Abstr. 084.263.

An 8,000-yr palaeoclimatic record of the 'Double-Hale' 45-yr solar cycle.   See Abstr. 085.001.

Influence of solar activity on particle acceleration near Jupiter.   See Abstr. 099.038.

Influence of solar activity on the acceleration of particles near Jupiter.   See Abstr. 099.045.

Manifestation of solar active processes in the interplanetary space according to data of cosmic ray variations. Abstract.   See Abstr. 106.018.

The role of solar active regions and of the total magnetic field of the sun in the 11-year cosmic-ray cycle. See Abstr. 143.030.

073 Chromosphere, Flares, Prominences

073.001 Strong convective and shock wave behavior in solar flares.
H. W. Bloomberg, J. Davis, J. P. Boris.
J. Quant. Spectrosc. Radiat. Transfer, Vol. 18, 237 - 244 (1977).
A model has been developed to study the gasdynamics of a flare region heated by a stream of energetic electrons. It is shown that the energy deposition can introduce strong chromospheric dynamical effects. As a result of fluid motion into rarefied regions, there is considerable redistribution of mass causing a profound influence on the emitted line radiation.

073.002 Rotational and turbulent motions in the surge of September 1st, 1961.
V. Ruždjak, J. Kleczek.
Bull. Astron. Inst. Czechoslovakia, Vol. 28, 193 - 197 (1977).
The appearance of the spectrum and the line profiles of a surge which occurred on September 1st, 1961 are discussed. The spectrum can be best explained by assuming ordered rotational motions of the individual surge streamers with peripheral rotational velocities in the range between 100 km s^{-1} and 200 km s^{-1}.

073.003 Line profiles in solar spicules.   V. Ruždjak.
Bull. Astron. Inst. Czechoslovakia, Vol. 28, 198 - 200 (1977).
Line profiles which are broadened by thermal, microturbulent and rotational motions are calculated for solar spicules and compared with the observed ones. With reasonable assumptions on the temperature, microturbulence and optical thickness, a peripheral rotational velocity of about 25 km s^{-1} is obtained.

073.004 Line profiles resulting from expanding and escaping fine filaments of prominences.
T. Ciurla, B. Rompolt.
Bull. Astron. Inst. Czechoslovakia, Vol. 28, 217 - 232 (1977).
The hydrogen Hα-line profiles are given for an optically thin, expanding prominence filament, simultaneously escaping from the Sun's surface. The line profiles were computed for a filament in the shape of a cylinder or a cylindrical shell, for different kinds of radial velocity distributions, and for a number of accepted values of the rate of expansion, velocity of escape, and Doppler widths. It was assumed that the filament scatters the incident solar radiation.

073.005 Solar flare activity: evidence for large-scale changes in the past.   H. A. Zook, J. B. Hartung, D. Storzer.
Icarus, Vol. 32, 106 - 126 (1977).
An analysis of radar and photographic meteor data and of spacecraft meteoroid penetration data indicates that there probably has not been a large increase in meteoroid impact rates in the last 10^4 yr. The solar flare tracks observed in the glass linings of meteoroid impact pits on lunar rock 15205 are therefore reanalyzed assuming a meteoroid flux that is constant in time. Based on this assumption, the data suggest that the production rate of Fe-group solar flare tracks may have varied by as much as a factor of 50 on a time scale of about 10^4 yr.

073.006 On the intensity of magnetic field in quiescent prominences.   J. L. Leroy.
Astron. Astrophys., Vol. 60, 79 - 84 (1977).
Previous observations of the polarization of the He I λ 5876 Å line in prominences (Leroy et al., 1977) have been analyzed with the help of recent computations about the

Hanle effect (Sahal-Bréchot et al., 1977) which yields new data about the intensity and direction of the prominence magnetic field.

073.007 Electron densities in solar flares from line ratios of Ca XVII.   G. A. Doschek, U. Feldman, K. P. Dere.
Astron. Astrophys., Vol. 60, L11 - L13 (1977).
The authors show that the intensity ratios of certain extreme ultraviolet spectral lines of Ca XVII are sensitive to electron density in solar flares. Calculations of the line ratios as functions of density are presented. For a flare that occurred on 9 August 1973, the authors derive a density of about 5×10^{11} cm^{-3} from the Ca XVII line ratios.

073.008 On the measurement of magnetic fields in prominences and flares by the photographic method.
A. N. Koval'.
Izv. Krymskoj Astrofiz. Obs., Vol. 57, 133 - 143 (1977). In Russian.
Polarized Hα spectrograms obtained with a λ/4 plate and a Wollaston prism are investigated. The measurements show that the field strengths (250–300 gauss) in two active prominences do not exceed significantly the error limits. Field strengths up to 650–1000 gauss for some parts of the third prominence were found.

073.009 A survey of soft X-ray limb flare images: the relation between their structure in the corona and other physical parameters.
R. Pallavicini, S. Serio, G. S. Vaiana.
Astrophys. J., Vol. 216, 108 - 122 (1977).
A survey of soft X-ray limb flare images obtained by the S-054 experiment on board Skylab is presented. From a morphological point of view, limb flares have been subdivided into three groups: (A) flares characterized by compact loop structures; (B) flares with a pointlike appearance; (C) flares with large and diffuse systems of loops. The significance of this subdivision is investigated. From a comparison of the spatial structure with physical parameters such as height, volume, energy density, and characteristic times, and from the correlation with white-light coronal transients and Hα active prominences, the existence of two physically distinct classes of flares is established: class I, which consists of both morphological groups A and B, and class II, which comprises only events of group C.

073.010 An emerging flux model for the solar flare phenomenon.
J. Heyvaerts, E. R. Priest, D. M. Rust.
Astrophys. J., Vol. 216, 123 - 137 (1977).
It is suggested that many solar flares occur in three stages when loops of magnetic flux emerge from below the photosphere and interact with the overlying field. First of all, during the preflare heating phase, continuous reconnection occurs in the current sheet that forms between the new and old flux. Waves which radiate from the ends of the sheet heat the plasma that passes through them and cause an increase in soft X-ray emission. Then the impulsive and flash phases take place. The onset of a turbulent electrical resistivity in the sheet causes it to expand rapidly, leading to electric fields far stronger than the Dreicer value. Particles are accelerated to high energies and escape along field lines. Finally, in the main phase, a new state of steady magnetic field reconnection is reached, with a much larger (marginally turbulent) current sheet than before.

073.011 Morphology and physical parameters of a solar flare.
J. B. Smith, Jr., R. M. Wilson, W. Henze, Jr.

Astrophys. J., Lett., Vol. 216, L79 - L82 (1977).

The chromospheric and coronal morphology of the flare of 1973 June 15 (1B/M3) in NOAA active region 131 (McMath 12379) is discussed, and results of quantitative analysis of Skylab soft X-ray observations of the event are presented.

073.012 EUV observations of class-C X-ray flare by the LPSP (*Laboratoire de Physique Stellaire et Planétaire du Centre National de la Recherche Scientifique*) spectrometer on OSO-8. A. Jouchoux, A. Skumanich, R. M. Bonnet, P. Lemaire, G. Artzner, J. Leibacher, J. C. Vial, A. Vidal-Madjar.
Bull. American Astron. Soc., Vol. 9, 432 (1977). — Abstract.

073.013 Summary of full-disk solar fluxes between 250 and 1940 Å. L. Heroux, J. E. Higgins.
J. Geophys. Res., Vol. 82, 3307 - 3310 (1977).

A summary of full-disk solar fluxes for the wavelength region 250–1940 Å compiled from data obtained from six Air Force Geophysics Laboratory rocket spectrometers flown during the period 1969–1976 is presented. The intense spectral lines in the wavelength region from 300 to 1220 Å are emitted predominantly from the solar chromosphere and the chromosphere-corona transition region.

073.014 On the formation of a power spectrum of particles accelerated in solar flares. A. A. Korchak.
Izv. AN SSSR. Ser. fiz., Vol. 41, 288 - 292 (1977). In Russian. Abstr. in Ref. zh., 51. Astron., 8.51.433 (1977).

073.015 Gas dynamics of flare plasma.
B. V. Somov, A. R. Spektor, S. I. Syrovatskij.
Izv. AN SSSR. Ser. fiz., Vol. 41, 273 - 287 (1977). In Russian. – Abstr. in Ref. zh., 51. Astron., 8.51.438 (1977).

073.016 Un metodo per l'osservazione della cromosfera solare. G. Corbò.
Coelum, Vol. 45, 144 - 146 (1977).

073.017 Particle acceleration by strong plasma turbulence. II. Acceleration of nonrelativistic electrons in solar flares. D. F. Smith.
Astrophys. J., Vol. 217, 644 - 656 (1977).

The observations of hard X-rays and nonrelativistic electrons in solar flares are interpreted. The inferred requirements for acceleration are compared with our knowledge of acceleration processes in a plasma, and a sequence of processes is proposed.

073.018 Some characteristics of the magnetic field and photospheric structure development in the August 1972 proton-flare region.
P. Ambrož, V. Bumba, J. Suda.
Solar activity and solar-terrestrial relations, (see 012.007), p. 15 - 21 (1976).

Some of the preliminary results obtained during the investigation of the August 1972 proton-flare active region are summarized. The development of the region in the frame of the magnetic field distribution, as well as the changes of the sunspot structures in connection with the magnetic field increase are discussed. Variations in appearance of photospheric details surrounding the positions of main white-light flare knots of August 7 are demonstrated. The relation between some chromospheric and photospheric structures is presented.

073.019 On the pre-maximum phase of the solar event of August 7, 1972. L. Křivský, B. Valníček, A. Böhme, F. Fürstenberg, A. Krüger.
Solar activity and solar-terrestrial relations, (see 012.007), p. 23 - 29 (1976).

The present report deals with an analysis of the pre-maximum phase of the event of August 7, 1972 which exhibited an interesting and peculiar time structure and was very well displayed by quite different methods of observations.

073.020 Change of flare activity a few days before proton flares, especially for the event of Nov. 5, 1970. L. Křivský.
Solar activity and solar-terrestrial relations, (see 012.007), p. 59 - 61 (1976).

073.021 On the eruptive phase of proton flares.
E. I. Mogilevskii (*Eh. I. Mogilevskij*), V. N. Ishkov.
Solar activity and solar-terrestrial relations, (see 012.007), p. 69 - 84 (1976).

The time development of the flare phenomena is analysed on the basis of the detailed cinephotography of the proton flare of 4th August, 1972. Special attention is paid to the development of eruptive phenomena: arch filament flare systems, oscillating disturbances of a filament, prolonged ejections, and arch filament systems.

073.022 Plasma turbulence in the current layer of a solar flare. V. M. Tomozov.
Solar activity and solar-terrestrial relations, (see 012.007), p. 85 - 89 (1976).

A common picture of physical processes connected with an energy dissipation of a magnetic field in a current layer of the solar flare model proposed by Syrovatskij is considered. The problems of the time development of a plasma turbulence and of the quasi-stationary oscillation spectra establishment are discussed.

073.023 On the solar active zones determined by the positions of strong flares.
V. V. Kasinskii (*Kasinskij*).
Solar activity and solar-terrestrial relations, (see 012.007), p. 91 - 94 (1976).

The proper locations of strong chromospheric flares in the coordinate system connected with an active region centre are investigated. A time-latitude diagram for the vectors of flare's displacement relative to the active region centre is constructed.

073.024 Effects of magnetic and gravity fields on solar flares and prominences. E. Woyk (Chvojková).
Solar activity and solar-terrestrial relations, (see 012.007), p. 95 - 97 (1976).

The evolution and structure of solar flares and prominences are explained from the characteristic behaviour of the plasma in combined magnetic and gravity fields, if only the inner particle velocity is comparable to or exceeds the escape velocity.

073.025 Physiology of quiescent filaments.
P. Ambrož, J. Kleczek.
Solar activity and solar-terrestrial relations, (see 012.007), p. 99 - 101 (1976).

The paper represents a preliminary account of the formation and evolution of quiescent filaments. Both large-scale and small-scale points of view are considered.

073.026 Profiles of Hα-line resulting from rotating optically thin filaments of prominences. B. Rompolt.
Solar activity and solar-terrestrial relations, (see 012.007), p. 103 - 116 (1976).

073.027 Doppler-brightening effect in rotating and expanding optically thin filaments.
B. Rompolt, T. Çiurla.
Solar activity and solar-terrestrial relations, (see 012.007), p. 117 - 123 (1976).

Profiles of the Hα-line produced by an optically thin and Doppler brightened filament were computed. The main mechanism of the line broadening was assumed to be either the rotation with a simultaneous radial escape of the filament from the Sun, or the expansion of the filament. The filaments structured as full cylinders or cylindrical shells of various sizes were adopted. The computation yielded profiles of essentially different shape, depending on the velocity distribution.

073.028   The verification of Chung-chieh-Cheng ideas on the heating mechanism of solar flares.   B. Valníček.
Solar activity and solar-terrestrial relations, (see 012.007), p. 127 - 130 (1976).

As demonstrated by Cheng (1972) it is possible for some large flares to emit non-thermal, as well as thermal radiation in their initial phases. This radiation must contribute to the soft X-ray region and occurs after the hard X-ray burst above 20 keV. Using X-ray observations for a number of solar flares, made on Intercosmos satellites, this hypothesis has been checked. Soft radiation in evidence a long time before hard X-ray bursts was found, however, in many cases both types of radiation were observed practically simultaneously. This indicates the existence of a minimum of two different mechanisms during the initial stage of the flare.

073.029   Development of the proton flare and the associated hard X-ray emission of November 5, 1970.
F. Fárník, P. Kotrč, L. Křivský, B. Valníček.
Solar activity and solar-terrestrial relations, (see 012.007), p. 131 - 138 (1976).

The positions of the two basic flare ribbons in the H-alpha line were determined and their velocity development was measured. This development was compared with the X-ray emission measured on the Intercosmos 4 satellite. From elaborated experimental data it was deduced that the flare develops in three phases.

073.030   X-ray spectrum of a coronal condensation and a flare.   J. Jakimiec, V. V. Krutov, S. L. Mandel-shtam (Mandel'shtam), B. Sylwester, J. Sylwester, I. A. Zhitnik.
Solar activity and solar-terrestrial relations, (see 012.007), p. 151 - 153 (1976).

The analysed spectrum covers the wavelength range of 8–13 Å and was obtained during the descending part of a solar flare. The absolute intensities of spectral lines and the continuum have been determined. A thermal model of the emitting region was then calculated and compared with various models of coronal condensations.

073.031   Magnetic fields in solar prominences.
G. Ya. Smolkov (Smol'kov), V. S. Bashkirtsev.
Solar activity and solar-terrestrial relations, (see 012.007), p. 175 - 194 (1976).

The state of knowledge on magnetic fields in the solar corona and prominences is discussed. The most objective information on the middle corona is shown to be obtained nowadays from the direct measurements in prominences. The results of magnetographic and spectrographic measurements of the field in these solar formations are considered in detail.

073.032   Forecast of the development of active regions with the method of potential functions.
T. L. Slutskaya, N. N. Stepanyan.
Solar activity and solar-terrestrial relations, (see 012.007), p. 239 - 242 (1976). In Russian.

The probability of a plage appearing in the next rotation was calculated on the basis of 16 features of the plages that emerged on the disk. The 1968–1969 data (297 plages) were used. A BESM-6 computer was used for the solution of the problem. Some features of the plages immediately after birth, the presence of the spots and flares, and the development of the plages were taken into account.

073.033   A probable interval for the mean transit time of the flare-generated disturbances.   I. D. Niță.
Solar activity and solar-terrestrial relations, (see 012.007), p. 355 (1976). – Abstract. (See 11.073.054).

073.034   Multidimensional radiative transfer in ultraviolet resonance lines of the chromospheric flash spectrum.
I. Hubený.
Solar activity and solar-terrestrial relations, (see 012.007), p. 383 - 386 (1976).

The properties of the synthetic and analytic approaches to the study of astrophysical objects are briefly discussed. The general equation of radiative transfer and the general source function for two-level atom radiation are then utilized to describe resonance-line formation of the chromospheric flash spectrum.

073.035   Influence of the differential rotation of the sun on filament orientation.   V. S. Bashkirtsev.
Soln. Dannye 1977 Byull., No. 5, p. 81 - 82 (1977). In Russian.

Effects of solar differential rotation on filament orientation are discussed. The maximum change of filament orientation takes place at the latitude $\varphi = 45°$. It can reach $14°$ during a quarter of one rotation. This should be taken into account in the investigation of magnetic field orientation in prominences.

073.036   Spectrophotometric investigation of the solar chromosphere. I.
V. M. Sobolev, K. S. Tavastsherna.
Soln. Dannye 1977 Byull., No. 6, p. 63 - 71 (1977). In Russian.

In 1969 and 1971 some chromospheric spectra were obtained with the large grating spectrograph of the ATsU-5 horizontal solar telescope. For various areas of the chromosphere the observed values are given of equivalent widths and strengths of the lines Hα, Hβ, Hε, H_8, D_3, and λ 3888 He as well as half-widths of the lines Hε, H_8, D_3 and λ 3888 He.

073.037   Structure and dynamics of a solar flare: X-ray and XUV observations.   K. P. Dere, D. M. Horan, R. W. Kreplin.
Astrophys. J., Vol. 217, 976 - 987 (1977).

The X-ray and XUV emission from the relatively simple looplike solar flare of 1973 September 5 is analyzed. The physical parameters of the flare plasma, including temperature, density, volume, and magnetic field configuration, are derived. Electron densities have been derived and increase with temperature from 3×10^{10} cm^{-3} at 2.2×10^6 K to 2×10^{11} cm^{-3} at 5×10^6 K. The variation of electron density with temperature in the decay phase is consistent with an adiabatic expansion of the flare plasma, although energy loss by conduction and radiation may also be significant.

073.038   On the exponential decay of western hemisphere solar particle events.   M. Scholer.
Planet. Space Sci., Vol. 25, 1081 - 1084 (1977).

073.039   Magnetic field of a flare flux. Annotation.
M. I. Pudovkin, S. A. Zajtseva, I. P. Oleferenko.
VIII Leningr. mezhdunar. semin. Mater. mezhdunar. semin. "Akt. protsessy na Solntse i probl. soln. nejtrino", 1976. Leningrad, 1976, p. 123 - 124. In Russian and English. – Abstr. in Ref. zh., 51. Astron., 9.51.455 (1977).

073.040   Unusual phenomena on the sun in 1972 - 1975. Annotation.   V. F. Chistyakov.
VIII Leningr. mezhdunar. semin. Mater. mezhdunar. semin. "Akt. protsessy na Solntse i probl. soln. nejtrino", 1976. Leningrad, 1976, p. 137 - 140. In Russian. – Abstr. in Ref. zh., 51. Astron., 9.51.475 (1977).

073.041 **On the formation of the power-law spectrum of particles accelerated in solar flares. Annotation.**
A. A. Korchak.
VIII Lenigr. mezhdunar. semin. Mater mezhdunar. semin. "Akt. protsessy na Solntse i probl. soln. nejtrino", 1976. Leningrad, 1976, p. 105 - 106. In Russian. — Abstr. in Ref. zh., 62. Issled. kosm. prostranstva, 9.62.186 (1977).

073.042 **Influence of photospheric magnetic fields on peculiarities in the propagation of solar flare protons.**
N. A. Mikirova, N. K. Pereyaslova.
Dokl. AN SSSR, Vol. 234, 798 - 801 (1977). In Russian. Abstr. in Ref. zh., 62. Issled. kosm. prostranstva, 9.62.200 (1977).

073.043 **Hα macrospicules: identification with EUV macrospicules and with flares in X-ray bright points.**
R. L. Moore, F. Tang, J. D. Bohlin, L. Golub.
Astrophys. J., Vol. 218, 286 - 290 (1977).
The authors summarize their observations of Hα macrospicules, present simultaneous Hα and EUV observations of macrospicules, present observations of simultaneous cospatial Hα macrospicules and flares in X-ray bright points, and, finally, discuss the implications of these results with regard to the generation mechanism for macrospicules and spicules.

073.044 **Reverse current in solar flares.**
J. W. Knight, P. A. Sturrock.
Astrophys. J., Vol. 218, 306 - 310 (1977).
The authors examine the proposal that impulsive X-ray bursts are produced by high-energy electrons streaming from the corona to the chromosphere. It is known that the currents associated with these streams are so high that either the streams do not exist or their current is neutralized by a reverse current. Analysis of a simple model in which the reverse current is stable indicates that the primary electron stream leads to the development of an electric field in the ambient corona which (a) decelerates the primary beam and (b) produces a neutralizing reverse current. The electric field acts as an energy exchange mechanism, extracting kinetic energy from the primary beam and using it to heat the ambient plasma.

073.045 **X rays and ultraviolet radiation and physics of solar flares.** S. B. Pikel'ner, M. A. Livshits.
Astron. Zh. Akad. Nauk SSSR, Vol. 54, 1062 - 1080 (1977). In Russian. English translation in Soviet Astron., Vol. 21, No. 5.
Observations of the pulsed and slow components and of the polarization of the X radiation of solar flares are used for the determination of the contribution of the thermal and nonthermal components as well as for the derivation of the characteristics of accelerated electron flows. The energy of an electron beam and of a high-temperature cloud which is formed during a flare is estimated. The ultraviolet radiation bursts in the 300 — 1800 Å region are analysed. Heating of the lower layers of the atmosphere by heat flux and by accelerated particles is considered. The role of the magnetic field and the requirements on flare theory, which are put forward by extraterrestrial observations, are briefly discussed.

073.046 **Propagation of charged particles generated in the flare of August 7, 1972.**
E. I. Morozova, O. B. Likin, N. F. Pisarenko.
Geomagn. Aehron., Vol. 17, 811 - 819 (1977). In Russian.

073.047 **Diagnosis of protons of solar flares from accompanying electromagnetic radiation.**
T. M. Bezruchenkova, N. A. Mikryukova, N. K. Pereyaslova, S. G. Frolov.
Geomagn. Aehron., Vol. 17, 820 - 825 (1977). In Russian.

073.048 **Solar flare observations at millimetre and submilli-** metre wavelengths.   D. L. Croom.
Far infrared astronomy (see 012.027), p. 93 - 101 (1976).
Observations of solar flares at millimetre wavelengths during the sunspot cycle now ending have shown that some flares emit the bulk of their radio energy in the mm band. The problems of extending such observations to the submillimetre region are considered.

073.049 **On the electron density of quiescent prominences at locations of helium and sodium emission.**
N. N. Morozhenko.
Astrometr. i Astrofiz., Kiev, vyp. (No.) 32, (see 003.011), p. 6 - 14 (1977). In Russian.
Electron densities of quiescent prominences at the locations of helium and sodium emission were determined. The calculations were performed for 15 quiescent prominences of different brightness. The values of electron densities obtained are close to each other for different prominences.

073.050 **Mass and energy flow in the solar chromosphere and corona.**   G. L. Withbroe, R. W. Noyes.
Annu. Rev. Astron. Astrophys., Vol. 15, (see 003.012), 363 - 387 (1977).
The authors discuss some results of investigations into the mass and energy flow in the solar chromosphere and corona. The objective of these investigations is the development of a physical model that will not only account for the physical conditions in the outer atmosphere of our nearest stellar neighbor, the Sun, but also can be applied to the study of the chromospheres and coronae of other stars.

073.051 **Display of supergranulation in quiescent prominences.**   V. V. Zharkova, V. I. Ivanchuk.
Problems of cosmic physics. Vyp. (No.) 12, (see 003.016), p. 20 - 27 (1977). In Russian.
From Hα spectrograms the distribution of distances of the filament feet, which shows three nearly identical maxima belonging to 27000, 37000 and 55000 km, is obtained. A discussion is carried out on works aimed at the study of periodical arch structure in prominences and a conclusion is verified as to its connection with supergranulation in the solar atmosphere.

073.052 **The downward motions in quiescent prominences.**
C. Mercier, J. Heyvaerts.
Astron. Astrophys., Vol. 61, 685 - 693 (1977).
The authors examine two possible causes for the downward mass loss in quiescent prominences. They develop a picture of the whole electric circuit of quiescent prominences, and show that Joule dissipation effects in subphotospheric parts of the current lines is a plausible candidate to explain the downward motion of prominences as a whole. An analysis of relative motion between charged and neutral species shows that this effect is negligible in the mass balance of prominences.

073.053 **Mechanisms of solar flares (theme 2.4.2.3. KAPG). Circular No. 1.**   S. I. Syrovatskij.
Akad. Nauk SSSR Sovet "Solntse — Zemlya". Moskva, 1977. 12 pp. In Russian. — Abstr. in Ref. zh., 51. Astron., 10.51.544 (1977).

073.054 **Gas dynamics of flare plasma. Abstract.**
B. V. Somov, A. R. Spector, S. I. Syrovatskij.
VIII Leningr. mezhdunar. semin. Mater. mezhdunar. semin. "Akt. protsessy na Solntse i probl. soln. nejtrino", 1976. Leningrad, 1976, p. 113 - 116. In Russian. — Abstr. in Ref. zh., 51. Astron., 10.51.546 (1977).

073.055 **Generation of γ-rays and neutrons in solar flares. Abstract.**   I. A. Ibragimov, G. E. Kocharov.
VIII Leningr. mezhdunar. semin. Mater. mezhdunar. semin. "Akt. protsessy na Solntse i probl. soln. nejtrino", 1976.

Leningrad, 1976, p. 161 - 164. In Russian. – Abstr. in Ref. zh., 51. Astron., 10.51.547 (1977).

073.056   On spatial anisotropy of chromospheric flares.
       Short note.
G. Ya. Vasil'eva, A. A. Shpital'naya.
VIII Leningr. mezhdunar. semin. Mater. mezhdunar. semin. "Akt. protsessy na Solntse i probl. soln. nejtrino", 1976. Leningrad, 1976, p. 176 - 179. In Russian. – Abstr. in Ref. zh., 51. Astron., 10.51.569 (1977).

073.057   Iron-line X-ray emission from solar plasma: comments on ionization equilibrium and line excitation.
L. W. Acton, R. C. Catura, D. T. Roethig.
Astrophys. J., Vol. 218, 881 - 887 (1977).
It is the purpose of this paper to make a detailed comparison of observation and theory for the group of iron lines at 1.9 Å emitted by solar flare plasma at temperatures around 10^7 K. The authors find that the ionization equilibrium calculations of Jacobs et al. (1977) produce better agreement with the data than does earlier work. However, at temperatures below about 10^7 K the observed iron-line fluxes are still substantially larger than predicted by theory. Rough estimates indicate that the remaining discrepancy may be removed by including inner-shell ionization and excitation of iron in stages of ionization below Fe XXIV.

073.058   Measurements of Hβ, He D_3, and Ca^+ λ8542 line emission in quiescent prominences.
D. A. Landman, S. J. Edberg, C. D. Laney.
Astrophys. J., Vol. 218, 888 - 900 (1977).
The authors present measurements of Hβ, He D_3, and Ca^+ λ8542 line emission in a number of quiescent prominences. The wings of the profiles, especially for D_3, deviate from the predictions of a simple isothermal model with Gaussian microturbulence. A more realistic two-temperature model is proposed which is found to fit the data well and to pass several consistency tests satisfactorily. From extended observations of D_3 profiles at a given quiescent prominence position, the authors have discovered the occurrence of long-period (~25 minute), low-amplitude oscillations in line width and intensity.

073.059   Measurement of the prominences magnetic field.
       O En Den, I. S. Kim, G. M. Nikolsky (*Nikol'skij*).
Sol. Phys., Vol. 52, 35 - 36 (1977).
A new type of magnetograph has been built capable to measure weak magnetic fields in the chromosphere and corona. Measurements of magnetic field in two prominences are demonstrated as examples.

073.060   On Pikel'ner's theory of prominences.
       O. Engvold, E. Jensen.
Sol. Phys., Vol. 52, 37 - 40 (1977).
Pikel'ner computed a stationary solution for coronal gas streaming along a magnetic arch, which develops into a dense condensation similar to prominence matter. This paper discusses the choice of boundary conditions and presents additional solutions.

073.061   The north–south distribution of major solar flare events, sunspot magnetic classes and sunspot areas (1955 - 1974).   J.-R. Roy.
Sol. Phys., Vol. 52, 53 - 61 (1977) = Herzberg Inst. Astrophys., Natl. Rech. Council Canada, NRCC 15789.
The north–south incidence has been studied of 31 white-light flares observed since 1859 and of 1669 events meeting the criteria for 'major flares' of Dodson and Hedeman (1971) for the period 1955 - 1974. The asymmetry in favor of the northern hemisphere increases strikingly with the importance of the events. Similarly, magnetically complex sunspot groups (Mt. Wilson classes βγ, γ and δ) display a more pronounced asymmetry in favor of the north than non-complex groups for

1962 - 1970. Contrary to the flare asymmetry, the spottedness asymmetry is independent of the size of sunspots.

073.062   Transequatorial loops interconnecting McMath regions 12472 and 12474.
Z. Švestka, A. S. Krieger, R. C. Chase, R. Howard.
Sol. Phys., Vol. 52, 69 - 90 (1977).
The authors discuss the life-story of a transequatorial loop system which interconnected the newly born active region McMath 12474 with the old region 12472. The loop system was probably born through reconnection accomplished 1.5 to 5 days after the birth of 12474 and the loops were observed in soft X-rays for at least 1.5 days. Transient 'sharpenings' of the interconnection and a striking brightening of the whole loop system for about 6 hr appear to be caused by magnetic field variations in the region 12474. Electron temperature in the loop system, equal to 2.1×10^6 K in its quiet phase, increased to 3.1×10^6 K during the brightening. During the brightening the loops became twisted. The final decay of the loop system reflected the decay of magnetic field in the region 12474.

073.063   The production of lithium in the solar chromosphere and photosphere during white light flares.
L. Hultqvist.
Sol. Phys., Vol. 52, 101 - 106 (1977).
Recently new values of the lithium formation rate in low energy flares have been reported in the literature. These values are applied to the white light flare phenomenon on the Sun. It is found that the formation rate in the chromosphere is much larger than in the upper photosphere and that the ratio between the time integrated flare created abundance and the initial photospheric abundance is modest in the chromosphere and small in the upper photosphere.

073.064   A hard X-ray observation of a solar flare with 100 ms time resolution.   K. Hurley, G. Duprat.
Sol. Phys., Vol. 52, 107 - 116 (1977).
A solar flare which occurred on 4 July 1974 was observed in hard X-rays with a balloon-borne detector. When analyzed with a time resolution of 100 ms, four 2 s long spikes are observed, which are correlated with decimetric emission. Spectral analysis shows that the hardest X-rays were produced during the decay phase of the burst, when the microwave emission reached its peak. It is argued that the fine time structure could either be a bounce time effect, or that it could be due to the electron acceleration mechanism.

073.065   The helium 10830 Å line in the undisturbed chromosphere.   R. G. Giovanelli, D. Hall.
Sol. Phys., Vol. 52, 211 - 228 (1977).
A study of the solar spectrum near helium 10830 Å has shown that, where the line is very weak, the anomalous ratio of the two components is due almost certainly to faint blends. The centre-limb intensity variation over supergranule centres is in good agreement with an optically-thin law. The integrated absorption in 10830 Å over supergranule centres is double that at the boundaries.

073.066   Wave propagation in the quiet solar chromosphere.
N. Mein.
Sol. Phys., Vol. 52, 283 - 292 (1977).
In order to precise previous results about wave propagation in the quiet chromosphere (N. Mein and P. Mein, 1976), the author studies the behaviour of Doppler shifts and intensity fluctuations in 3 lines of Ca II. Results can be summarized as follows: (1) Phase-lag between intensity fluctuations and Doppler shifts is always near 90° in the Ca II lines, even for frequencies as high as 15 mHz, and whatever is the location in the chromospheric network. (2) Magneto-acoustic waves propagating vertically in a vertical or horizontal magnetic field could account for the observations only if they were, on one

hand reflected in the upper atmosphere, on the other hand propagating with a very high sound or Alfvén speed.

073.067   A possible edge effect in enhanced network.
H. P. Jones, D. R. Brown.
Sol. Phys., Vol. 52, 337 - 342 (1977).

K-line observations of enhanced network taken with the NASA/SPO Multichannel Spectrometer on 28 September 1975 in support of OSO-8 are discussed. The data show a correlation between core brightness and asymmetry for spatial scans which cross enhanced network boundaries. The implications of this result concerning mass flow in and near supergranule boundaries are discussed.

073.068   The fine structure of prominences. III. Small scale Doppler shifted features.
O. Engvold, J. M. Malville.
Sol. Phys., Vol. 52, 369 - 377 (1977).

Faint, Doppler shifted, emission features are detected in high resolution spectra of limb prominences. Their average line-of-sight velocity is about 3×10^6 cm s^{-1}, their average life time is 300 s, and their angular sizes are $\lesssim 10^8$ cm in the spectrograms. The emission line width of the spectral features increases with increasing line shift. Some implications on the stability of prominences by these structures are discussed briefly.

073.069   Prominence mass ejections and their effects on the corona. I: The eruptive prominence of 21 August 1973 and the surge of 4 December 1973.
J. B. Smith, Jr., D. M. Speich, R. M. Wilson, E. Tandberg-Hanssen, S. T. Wu.
Sol. Phys., Vol. 52, 379 - 391 (1977).

Skylab soft X-ray observations of two lower coronal limb events and corresponding Hα observations (Skylab and ground-based) are analyzed. The authors discuss the morphology and evolution of an eruptive prominence and of a surge. For the eruptive prominence, measured X-ray flux is used in the determination of line-of-sight temperatures, emission measures, and electron densities. A time-dependent, two-dimensional, single-fluid magnetohydrodynamic computer code has been used to simulate the coronal response to these prominences.

073.070   The formation of solar prominences by thermal instability in a current sheet.
E. A. Smith, E. R. Priest.
Sol. Phys., Vol. 53, 25 - 40 (1977).

This paper presents a simple model for the thermal equilibrium and stability of a current sheet. It is found that, when its length exceeds a certain maximum value, no equilibrium is possible and the plasma in the sheet cools. The results may be relevant for the formation of a quiescent prominence.

073.071   A model for X-ray emission from loop prominences.
K. J. H. Phillips, J. B. Zirker.
Sol. Phys., Vol. 53, 41 - 58 (1977).

A study is made of X-ray line emission observed during the developing stages of a set of post-flare loop prominences. The time behaviour of the line emission can be described by a model consisting of two flux tubes containing plasma heated impulsively at the flash phase; the plasma cools by radiation and by conduction to the chromosphere. These ideas are extended to the possible formation of Hα prominences from low-lying hot loops.

073.072   Observations of limb flares with a soft X-ray telescope.   E. G. Gibson.
Sol. Phys., Vol. 53, 123 - 138 (1977).

The structure and evolution of 26 limb flares have been observed with a soft X-ray telescope flown on Skylab. The results are discussed.

073.073   The relation of flares to 'newly emerging flux' and 'evolving magnetic features'.
M. J. Martres, I. Soru-Escaut.
Sol. Phys., Vol. 53, (see 012.038), 225 - 231 (1977). – Invited opinion.

The authors conclude that the change in the magnetic field is not a consequence of the flare but, on the contrary, the flare is the consequence of a particular evolution of the magnetic structures near an inversion line and prior to the flare start.

073.074   How flares can be understood.
A. B. Severny (*Severnyj*).
Sol. Phys., Vol. 53, (see 012.038), 233 - 234 (1977). – Invited opinion.

073.075   Basic questions in our understanding of flares.
S. I. Syrovatskii (*Syrovatskij*).
Sol. Phys., Vol. 53, (see 012.038), 247 (1977). – Invited opinion.

073.076   The thermal and non-thermal flare. A result of non-linear threshold phenomena during magnetic field line reconnection.   D. S. Spicer.
Sol. Phys., Vol. 53, (see 012.038), 249 - 254 (1977). – Invited opinion.

The topic of this talk is: Can one explain the difference between the thermal and the non-thermal flare as the result of a non-linear threshold phenomenon associated with magnetic field line reconnection?

073.077   An emerging flux model for solar flares.
J. Heyvaerts, E. Priest, D. M. Rust.
Sol. Phys., Vol. 53, (see 012.038), 255 - 258 (1977). Contributed opinion.

073.078   Location of the primary flare site and energy transfer in flares. Introductory talk.   J. C. Brown.
Sol. Phys., Vol. 53, (see 012.038), 263 - 265 (1977).

073.079   Comments regarding energy release and transfer in solar flares.   J. A. Vorpahl.
Sol. Phys., Vol. 53, (see 012.038), 271 - 275 (1977). – Invited opinion.

073.080   Multiple loop activations and continuous energy release in a solar flare.   K. G. Widing.
Sol. Phys., Vol. 53, (see 012.038), 277 - 278 (1977). – Contributed opinion.

073.081   Effects of soft X-ray flux on the lower solar atmosphere in flares.   J. C. Henoux, Y. Nakagawa.
Sol. Phys., Vol. 53, (see 012.038), 279 - 280 (1977). – Invited opinion.

Chromospheric flare as a secondary effect; Energy transfer by soft X-ray photons; Solar atmospheric response.

073.082   Comments on Salyut-4 observations of active regions on the Sun.   A. B. Severny (*Severnyj*).
Sol. Phys., Vol. 53, (see 012.038), 285 (1977). – Contributed opinion.

073.083   Photometric studies of the starting phase of flares.
R. Falciani.
Sol. Phys., Vol. 53, (see 012.038), 287 - 290 (1977). – Contributed opinion.

073.084   An overview of the energy-flow problem in flares.
P. A. Sturrock.
Sol. Phys., Vol. 53, (see 012.038), 299 - 301 (1977).

073.085   An unstable arch model of a solar flare.
D. S. Spicer.

Sol. Phys., Vol. 53, 305 - 345 (1977).

The theoretical consequences of assuming that a current flows along flaring arches consistent with a twist in the field lines of these arches are examined. It is found that a sequence of magneto-hydrodynamic (MHD) and resistive MHD instabilities driven by the assumed current can naturally explain most manifestations of a solar flare. The resulting model is in excellent agreement with present observations and has successfully predicted several flare phenomena.

073.086   A measurement of the helium D₃ profile with a birefringent filter.   B. J. LaBonte.
Sol. Phys., Vol. 53, 369 - 374 (1977).

The D_3 line profile in plages on the disk is measured using a birefringent filter. The best fit Gaussian has a $1/e$ width of 0.4 Å, with negligible instrumental contribution. The D_3 opacity is produced in regions with thermal linewidth $\simeq 0.1$ Å; the much larger observed width indicates large non-thermal motions in the chromosphere.

073.087   A statistical study of spicule inclinations.   J. M. Mosher, T. P. Pope.
Sol. Phys., Vol. 53, 375 - 384 (1977).

The apparent angles of more than 5000 quiet region spicules situated at various position angles around the Sun's limb have been measured on off-band Hα filtergrams taken during the years 1972 - 1975. The counts were made exclusively in projection on the disk, within 5–25° of the limb. The tendency of the average spicule to lean towards or away from the pole is small at most, and at no latitude exceeds 8°. The spread in angles is 30–35°.

073.088   A self-consistent model of a thermally balanced quiescent prominence in magnetostatic equilibrium in a uniform gravitational field.   I. Lerche, B. C. Low.
Sol. Phys., Vol. 53, 385 - 396 (1977).

The authors present a theoretical model of quiescent prominences in the form of an infinite vertical sheet. Self-consistent solutions are obtained by integrating simultaneously the set of nonlinear equations of magnetostatic equilibrium and thermal balance. The authors assume that the prominence plasma emits more radiation than it absorbs from the radiation fields of the photosphere, chromosphere and corona, and they interpret the hypothetical heat sink to represent the amount of radiative loss that must be balanced by a nonradiative energy input.

073.089   Chromospheric rotation during 1972 - 73, years of declining activity.
E. Antonucci, L. Azzarelli, P. Casalini, S. Cerri.
Sol. Phys., Vol. 53, 519 - 529 (1977).

The rotational behaviour of the chromosphere, observed in the Ca II K₃ line, is studied during 1972 - 1973, years of decreasing solar activity. The computed rotation rate is independent of latitude, in agreement with the results obtained for the green corona during the years before sunspot minimum. Namely both chromospheric and coronal features, with lifetime exceeding one solar rotation, are almost not affected by differential rotation before sunspot minimum.

073.090   An empirical, statistical model for the formation of the cores of chromospheric Fraunhofer lines.
Z. Suemoto.
Sol. Phys., Vol. 54, 3 - 24 (1977).

The author describes an empirical, statistical model for the formation of the cores of chromospheric Fraunhofer lines. The model has been applied to the interpretation of Ca II and Mg II resonance lines, and to Hα and Hβ. By comparing computed mean intensity profiles and profiles associated with statistical intensity fluctuations with observations, a consistent set of chromospheric parameters was determined. It is tentatively concluded that the line cores are formed by structural elements whose optical thicknesses are very large at the line centre of K, located somewhere not far above 1000 km and moving with rms velocity of about 7 km s⁻¹.

073.091   Kinematic model of loop prominences formation.   L. N. Ivanov, Yu. V. Platov.
Sol. Phys., Vol. 54, 35 - 44 (1977).

The kinematics of the material motion in a variable magnetic field in the MHD approximation of a strong field and cold plasma is investigated. The variation of magnetic moments of two dipole systems leads to the development of such phenomena as loop prominences, coronal rain and funnel prominences.

073.092   Multi-channel subtractive spectrograph and filament observations.   P. Mein.
Sol. Phys., Vol. 54, 45 - 51 (1977).

A Multi-Channel Subtractive Double Pass spectrograph (MSDP) has been achieved at the Meudon solar tower. Line profiles are obtained simultaneously in a two dimensional field. Space and time resolutions are very suitable for observation of fast chromospheric phenomena. Maps have been computed for Doppler shifts and Hα-intensities (core and wings) in a plage including a temporary activated filament. The radial velocity is zero in the core of the filament, but it increases toward the edges with opposite signs on the both sides. Velocity loops inclined at small angles on the axis of the filament are suggested.

073.093   Measurement of plasma wave electric fields in solar flares.   W. D. Davis.
Sol. Phys., Vol. 54, 139 - 149 (1977).

Measurements of electric fields in solar flares using the Stark effect exhibited by neutral helium atoms are reported. Electric field strengths as high as 700 V cm⁻¹ are observed. Measurements of electron densities indicate that the electric fields originate from nonthermal plasma waves. Analysis of the plasma wave fields coupled with plasma density and temperature measurements indicate that the Lower Hybrid Drift instability may be present in the flaring region.

073.094   Comparison between some Hα and X-ray flares.   R. Falciani, M. Giordano, M. Rigutti, G. Roberti.
Sol. Phys., Vol. 54, 169 - 178 (1977).

In the present paper, Hα-evolutive curves of chromospheric events are compared with flux evolutive curves of X-ray events observed at the same time in different spectral regions. A correspondence between the emissions $E(I_{H\alpha}/I_{chr})$ at higher and higher Hα-intensity levels, and the X-ray fluxes $F(\Delta\lambda)$ in harder and harder $\Delta\lambda$-ranges is shown. Further, the present observations seem to indicate the existence of a single triggering mechanism during the flash-phase of a flare. It is also shown that these results may be in agreement with Brown's model for chromospheric flares.

073.095   Comments on pulses of characteristic energy produced in solar flare detonations and its possible application to other astrophysical plasmas.   P. Kaufmann.
Astrophys. Space Sci., Vol. 49, 123 - 131 (1977).

A qualitative discussion of physical conditions at neutral sheets was developed in an attempt to explain the repetitive pulsed energy-production mechanism which has been suggested for solar flares.

073.096   A comparison of He II 304 Å and He I 10830 Å spectroheliograms.
J. W. Harvey, N. R. Sheeley, Jr.
Sol. Phys., Vol. 54, 343 - 351 (1977).

Spectroheliograms were obtained simultaneously in the He II 304 Å emission line and the He I 10830 Å absorption line with an angular resolution of approximately 5″. Differences between these images include the facts that: (1) Disk filaments and limb darkening are strongly visible in the 10830 Å positive image, but they are weakly visible (as lightenings) in the 304 Å negative image. (2) The contrast between the chromospheric network and the network cell centers is much greater in the 10830 Å image than in the 304 Å negative image. These results provide constraints on models of helium line formation in various types of solar features.

073.097   Medium resolution EUV observations and network structure.   N. Raghavan.
Sol. Phys., Vol. 54, 363 - 370 (1977).

Observations in recent years have shown conclusively that the complex network of emission, characteristic of the chromosphere, persists up to temperatures where Ne VIII λ770 is formed. The present work investigates the consequence of this horizontal structure on EUV observations of medium resolution. The average emitting area in O VI deduced by this method is in good agreement with the results from ATM observations. The fractional emitting areas at different values of the Mg X intensity and at different temperatures are combined to find the variation of the areas with height. This variation is in good agreement with Giovanelli's model of the fractional area of cross-section of a magnetic tube of force in the transition region.

073.098   Electrostatically unstable heat flow during solar flares and its consequences.   D. S. Spicer.
Sol. Phys., Vol. 54, 379 - 385 (1977).

The authors examine some of the consequences of an electrostatically unstable return current associated with heat conduction during a solar flare. They note that an electrostatically unstable return current will lead to strong hydrodynamic effects and more rapid magnetic field thermalization, if reconnection is the source of primary energy release during a solar flare.

073.099   A study of filament transition sheath from radio observations.   A. Pramesh Rao, M. R. Kundu.
Sol. Phys., Vol. 55, 161 - 175 (1977).

The authors have observed an Hα dark filament at 8, 15, and 22 GHz and derived the radio spectrum of the filament. They suggest that the filament has to be optically thick at radio frequencies and that the observed spectrum is due to the presence of a transition sheath surrounding the filament. They examine a model for the transition sheath in which the energy radiated away is balanced by the conduction of heat from the corona, and show that the radio observations indicate that little or no thermal energy is conducted into the main body of the filament. The model is compared with ultraviolet observations of filaments and it is discussed how the discrepancies can be removed.

073.100   Early evolution of an X-ray emitting solar active region.
C. J. Wolfson, L. W. Acton, J. W. Leibacher, D. T. Roethig.
Sol. Phys., Vol. 55, 181 - 193 (1977).

The birth and early evolution of a solar active region has been investigated from X-ray observations from the Lockheed Mapping X-ray Heliometer on board the OSO-8 spacecraft. X-ray emission is observed within three hours of the first detection of Hα plage. During the fifty hours following birth almost continuous flares or flare-like X-ray bursts are superimposed on a monotonically increasing base level of X-ray emission produced by plasma with a temperature of the order 3×10^6 K. Assuming that the X-rays result from heating due to dissipation of current systems or magnetic field reconnection, the authors conclude that flare-like X-ray emission soon after active region birth implies that the magnetic field probably emerges in a stressed or complex configuration.

073.101   Chromospheric oscillations observed in the line C II λ1336 with OSO-8.   E. G. Chipman.
Sol. Phys., Vol. 55, 277 - 285 (1977).

The data presented clearly show the existence of 300-s periodicity in velocity and intensity in the C II λ1336 line in plage and chromospheric network regions. The velocity-intensity phase relation is consistent with a propagating acoustic wave, although the rather large derived phase velocities would argue against this interpretation. The amplitudes of velocity and intensity oscillation are slightly less than was observed for lower-lying chromospheric lines.

073.102   Spectral investigation of the chromosphere. VI: Observations of Hα close to the limb.
U. Grossmann-Doerth, M. von Uexküll.
Sol. Phys., Vol. 55, 321 - 333 (1977) = Mitt. Fraunhofer Inst., Freiburg, Nr. 151.

Several hundred Hα spectrograms from areas close to the solar limb were taken with the 35 cm coudé refractor at Anacapri. The 41 spectra with the greatest spatial resolution were selected and analysed. At the supergranular boundaries a considerable fraction of the line profiles were found to correspond to Beckers' Cloud Model (BCM). Moreover, the BCM parameters of the dark mottles at the limb appear to be approximately equal to those from the center of the disk. On the other hand, the authors also obtained evidence in disagreement with the general applicability of the BCM to all features of the chromospheric fine structure. The authors attempted to present the large set of observational data in a fashion that permits their interpretation by alternative theoretical models.

073.103   Ultraviolet brightenings in active regions as observed from OSO-8.   B. W. Lites, E. R. Hansen.
Sol. Phys., Vol. 55, 347 - 358 (1977).

Repeated raster images of solar active regions taken at the line centers of the Si IV and C IV resonance lines using the University of Colorado (CU) ultraviolet spectrometer aboard OSO-8 reveal dramatic transient brightenings of up to factors of 10. These brightenings last several minutes and frequently show a repetitive character. Inspection of simultaneous Hα flare patrol records show that these transition zone events are often associated with subflare-like brightenings in the chromosphere. These observations indicate that direct excitation or heating of material already at transition zone temperatures caused by non-thermal particle streams is inadequate to explain the degree of brightening of these lines.

073.104   Current sheets as the source of heating for solar active regions.
B. V. Somov, S. I. Syrovatskii (*Syrovatskij*).
Sol. Phys., Vol. 55, 393 - 399 (1977).

Observational data and theoretical arguments suggest that the heating source for an active region is the quasi-steady dissipation of magnetic field in current sheets. Effects in the solar atmosphere which are due to the presence of current sheets are considered. The most important of them is the heating of the chromosphere by the strong ultraviolet radiation of the current sheet. This can give rise to the brightening of an active region in optical emission. The energy flux from the current sheet in different ranges of the ultraviolet spectrum and the depths (column densities) into the chromosphere where this energy is absorbed are estimated.

073.105   Evolution of the high-temperature plasma in the 15 June 1973 flare.   C.-C. Cheng.
Sol. Phys., Vol. 55, 413 - 429 (1977).

The analysis of the evolution of the high temperature plasma in the 15 June, 1973 flare is presented. The author discusses the preflare enhancement of loops in the active regions, and the determination of temperature and density as a function of time for the hot plasma. He also deduces the turbulent velocity in the high-temperature plasma from the comparison of line widths of the Fe XXIV doublet at 192 and 255 Å. The energy release processes in the flare are studied.

073.106 Multiple loop activations and continuous energy release in the solar flare of June 15, 1973.
K. G. Widing, K. P. Dere.
Sol. Phys., Vol. 55, 431 - 453 (1977).

The spatial and temporal evolution of the high temperature plasma in the flare of 1973 June 15 has been studied using the flare images photographed by the NRL XUV spectroheliograph on Skylab. The overall event involves the successive activations of a number of different loops and arches bridging the magnetic neutral line. The spatial shifts and brightenings observed in the Fe XXIII−XXIV lines are interpreted as the activation of new structures. These continued for four or five minutes after the end of the microwave burst phase, implying additional energy-release unrelated to the nonthermal phase of the flare. A shear component observed in the coronal magnetic field may be a factor in the storage and release of the flare energy.

073.107 The effect of nonlinear conduction on the cooling of flare loops.   K. R. Krall.
Sol. Phys., Vol. 55, 455 - 458 (1977).

Attention is drawn to consequences of the non-linear nature of thermal conductivity on the cooling rate of solar flares.

073.108 Coronal mass-ejections−kinematics of the 19 December 1973 event.
E. Schmahl, E. Hildner.
Sol. Phys., Vol. 55, 473 - 490 (1977).

The observations are discussed in three phases: before, during, and after the prominence eruption. Within each phase, the authors separately discuss the observations pertaining to the prominence and the corona. Finally, they summarize their analysis.

073.109 Differential rotation of short-lived solar filaments.
W. M. Adams, F. Tang.
Sol. Phys., Vol. 55, 499 - 504 (1977).

The authors have measured the rotation rate of short-lived solar filaments as a function of their latitude. The resulting rotation curve appears to be somewhat flatter than the corresponding curve for long-lived filaments.

073.110 Energetic solar flare particles and interplanetary shock waves.   R. P. Lin.
Study of travelling interplanetary phenomena 1977, (see 012. 042), p. 23 - 42 (1977).

Estimates from hard X-ray measurements show that for many flares the bulk of the flare energy is released in the form of $\sim 10-10^2$ keV energy electrons. The interaction of these electrons with the solar atmosphere can produce the optical, UV, EUV, and radio emissions observed during the flare impulsive phase. For the large solar flares which produce interplanetary shock waves, the accelerated $\sim 10-10^2$ keV electron population may produce the heating and mass motion required for mass ejection and the formation of the shock wave. The shock wave can in turn accelerate ions and electrons to higher energy as it travels through the corona and interplanetary medium.

073.111 Solar proton flares and their prediction.
L. Křivský.
Czechoslovak Acad. Sci., Astron. Inst., Publ. No. 52, 121 pp. (1977).

A review is given of papers which originated mostly in the Astronomical Institute of the Czechoslovak Academy of Sciences at the Ondřejov Observatory in the year 1946 - 1976 about study of proton flares and their effects. Catalogue of proton flares 1942 - 1976 is also included.

073.112 Quantum theory of the Hanle effect. Application to the determination of magnetic fields in quiescent

prominences.   S. Sahal-Brechot, V. Bommier.
Rep. Obs. Lund, No. 12, (see 012.044), p. 5 - 24 (1977).

The paper concerns the determination of magnetic fields in quiescent prominences by means of the interpretation of the Hanle effect in the D_3 helium line. The basic foundations of the theory together with the physical sense of the approximations and of the formulation are pointed out.

073.113 The polarization of prominence emission lines and the prominence magnetic field.   J. L. Leroy.
Rep. Obs. Lund, No. 12, (see 012.044), p. 95 - 108 (1977).

The author finds that the magnetic field is probably roughly horizontal in quiescent prominences and that the angle between the prominence long axis and the field vector is not a critical parameter for the existence of prominences.

073.114 Line profiles of inhomogeneous solar phenomena with different porosity.
L. N. Kurochka, V. V. Tel'nyuk-Adamchuk.
Astrometriya i Astrofizika, Kiev, vyp. (No.) 33, (see 003.020), p. 41 - 51 (1977). In Russian.

A formula is obtained for calculating line profiles of inhomogeneous solar phenomena taking into account their porosity. An analysis of the formula is made and simple approximations are obtained. Tables are given for constructing profiles of emission lines for inhomogeneous emission formations.

073.115 Effective time of observations of chromospheric flares.
V. I. Efimenko, E. N. Zemanek, N. N. Kondrashova.
Astrometriya i Astrofizika, Kiev, vyp. (No.) 33, (see 003.020), p. 55 - 62 (1977). In Russian.

The effective time is obtained for observations of chromospheric flares. The data may be used for statistical studies of indexes of flare activity.

073.116 Active regions.   M. J. Martres, A. Bruzek.
Illustrated glossary for solar and solar-terrestrial physics, (see 003.022), p. 53 - 70 (1977).

073.117 Flares and associated phenomena.
H. W. Dodson-Prince, A. Bruzek.
Illustrated glossary for solar and solar-terrestrial physics, (see 003.022), p. 81 - 96 (1977).

073.118 Prominences.   E. Tandberg-Hanssen.
Illustrated glossary for solar and solar-terrestrial physics, (see 003.022), p. 97 - 109 (1977).

073.119 Formation of shock waves from chromospheric flares and their evolution in the solar wind.
T. V. Stepanova, A. M. Uralov.
Simpoz. po fiz. geomagnitosfery, 1977. Tezisy dokl. Irkutsk, 1977, p. 6. In Russian. − Abstr. in Ref. zh., 51. Astron., 12.51.346 (1977).

073.120 A laboratory model of a solar flare.
L. I. Kiselevskij, G. S. Antonov.
Izv. AN BSSR. Ser. fiz.-mat. nauk, 1977, No. 3, p. 58 - 63. In Russian. − Abstr. in Ref. zh., 51. Astron., 12.51.394 (1977).

073.121 Spectrophotometric investigation of the solar chromosphere. II.
V. M. Sobolev, K. S. Tavastsherna.
Soln. Dannye 1977 Byull., No. 9, p. 87 - 98 (1977). In Russian.

In 1973 and 1975 some chromospheric spectra were obtained with the large grating spectrograph of the ATsU-5 horizontal solar telescope. For various areas of the chromosphere the observed values of equivalent widths and intensities of the lines $H\alpha$, $H\beta$, $H\epsilon$, H_8, D_3 and $\lambda 3888$ He as well as half-widths of the lines $H\epsilon$, H_8, D_3 and $\lambda 3888$ He are given.

073.122 **A two-dimensional analysis of wave processes in the lower chromosphere in the neighbourhood of an active region from filtergrams in the Ba II 4554 + 0.05 Å line.**
V. I. Polyakov, V. E. Merkulenko, V. I. Skomorovskij, N. V. Larionov.
Sol. Dannye 1977 Byull., No. 10, p. 62 - 66 (1977). In Russian.

A two-dimensional analysis of the brightness variation of chromospheric elements is made from filtergrams in the wings of the Ba II 4554 + 0.05 Å line. A region of $30'' \times 10''$ of a facular area and a $60'' \times 10''$ region of the neighbouring chromosphere is covered with 5 photometric tracings. There is a wide spectrum of possible oscillation periods from 200 to 1000 sec without any obvious maximum.

073.123 **Evolution of a large solar flare on July 4, 1974.**
I. L. Belkina, T. P. Bushueva, N. P. Dyatel.
Physics of the moon and planets. Fundamental astrometry, (see 003.024), Vestn. Khar'kov. Univ., No. 160, p. 86 - 93 (1977). In Russian.

On the number of lines observed in the hydrogen spectrum. See Abstr. 022.023.

Narrow lines from alpha-alpha reactions. See Abstr. 022.026.

The use of a-priori information in the derivation of temperature structures from X-ray spectra. See Abstr. 061.017.

Electric currents in cosmic plasmas. See Abstr. 062.010.

A search for a turbulent-free region in the solar transition zone. See Abstr. 071.006.

Scattering in the Doppler core of the solar Lyman-α line: its effect on the Lyman continuum and on the chromospheric electron number density. See Abstr. 071.007.

Vertical velocity fluctuations in plage-region magnetic points. See Abstr. 071.028.

Gas entry into non-spot magnetic tubes. See Abstr. 071.033.

Quiet photosphere and chromosphere. See Abstr. 071.056.

Statistical study of the time sequence of Ca II flocculi and series of radio fluxes during solar activity cycle 20. See Abstr. 072.001.

On the mutual influences of spatially remote sunspot groups united into flare-active complexes. See Abstr. 072.009.

Some parameters in forecasting of solar active region development. See Abstr. 072.014.

Autocorrelation analysis of some solar indices. See Abstr. 072.018.

The nonequilibrium ionization of solar flare coronal plasma and the emergent X-ray spectrum. See Abstr. 074.016.

Coronal and interplanetary shock waves generated during flares on August 2 - 13, 1972. Annotation. See Abstr. 074.028.

The calcium K-line network in coronal holes. See Abstr. 074.049.

The filament-corona transition region from OSO-VI EUV observations. See Abstr. 074.078.

Dynamic modeling of coronal and interplanetary responses to solar events. See Abstr. 074.085.

Transition region. See Abstr. 074.101.

L'activité solaire. See Abstr. 075.003.

Statistik der Sonneneruptionen 1966−1975. See Abstr. 075.018.

The exponential character of the dependence of the differential emission measure on the temperature and the structure of a coronal active region. See Abstr. 076.001.

Hard X-rays of solar flares on August 2 and 11, 1972. See Abstr. 076.015.

Dynamical implications of Si IV line profiles from OSO-8 observations. See Abstr. 076.016.

Solar EUV emission line profiles of Si II and Si III and their center to limb variations. See Abstr. 076.017.

Open magnetic fields in active regions. See Abstr. 076.018.

Radio emission of an intense active region on the sun in July 1974 at wavelengths 3.5, 2.5 and 1.9 cm. See Abstr. 077.002.

On peculiar quasi-periodic components and the possible structure of the generating region of the type IV event of August 4, 1972. See Abstr. 077.007.

Observations of the solar radio emission at IZMIRAN during the proton flare of August 4, 1972. See Abstr. 077.008.

Time of appearance of type II radio emission in solar flares. Abstract. See Abstr. 077.025.

Microwave burst spectra and solar flare magnetic fields. See Abstr. 077.038.

Simple analytical solutions for spherically symmetric production and modulation of energetic solar particles. See Abstr. 078.004.

A need for (p,n) cross sections for selected targets at lower energies. See Abstr. 078.005.

Study of coronal and interplanetary propagation of solar particles following the E45° solar flare on July 29, 1973. See Abstr. 078.008.

Coherent propagation of non-relativistic solar electrons. See Abstr. 078.010.

The influence of the chromosphere and corona on the solar atmospheric oscillations. See Abstr. 080.006.

Superadiabatic acceleration of charged particles and flares on the sun and stars. Abstract. See Abstr. 080.045.

Molecular oxygen densities and the atmospheric absorption of solar Lyman α radiation. See Abstr. 082.007.

The daytime lower ionosphere during the solar phenomena of July 27 to August 16, 1972. See Abstr. 083.025.

Solar proton event: influence on stratospheric ozone.   See Abstr. 085.006.

The enhancement of scattered Lα radiation in the geocorona during the solar flares of August 1972. See Abstr. 085.025.

Structure of a typical stream of interplanetary plasma from a powerful flare and corresponding set of types of geomagnetic (magnetospheric) disturbances. I. See Abstr. 106.002.

The modelling and calculation of some cosmic phenomena of blast type.   See Abstr. 106.022.

The structure of the solar flare stream magnetic field. See Abstr. 106.031.

Solar and stellar flares.   See Abstr. 122.115.

Solar zenith angle dependence of sudden cosmic noise absorptions.   See Abstr. 143.029.

Errata

073.901   Erratum: 'Further observations of helium and hydrogen emission in quiescent prominences' [Astron. Astrophys., Vol. 49, 277 - 283 (1976)]. D. A. Landman, R. M. E. Illing. Astron. Astrophys., Vol. 61, 299 (1977).

073.902   Erratum: 'The Compton effect in the chromosphere, II' [Astrophys. Space Sci., Vol. 43, 129 - 134 (1976)].   M. Missana, A. Piana. Astrophys. Space Sci., Vol. 50, 518 (1977).

074 Corona, Solar Wind

074.001 Étude polarimétrique de la couronne solaire observée a l'éclipse totale du 30 juin 1973 à l'aide d'un filtre neutre radial.
S. Koutchmy, J. P. Picat, M. Dantel.
Astron. Astrophys., Vol. 59, 349 - 357 (1977).

Photographs of the white light solar corona were obtained during the total solar eclipse of 30 June 1973 in two orthogonal components (I_T, I_R) of polarization. The instrumentation involved a telescope using a radial graded neutral density filter and a rotatable polarizing filter. The total intensity and fraction of polarization of the white light corona are deduced.

074.002 Determination of plasma parameters from soft X-ray images for coronal holes (open magnetic field configurations) and coronal large-scale structures (extended closed-field configurations). C. W. Maxson, G. S. Vaiana.
Astrophys. J., Vol. 215, 919 - 941 (1977).

The two distinct and general classes of "quiet", inner solar coronal features – coronal holes (CH), associated with persistent high-speed wind streams and open magnetic field configurations, and large-scale structures (LSS), associated with closed-field configurations – are examined from Skylab soft X-ray images. Using low photographic density conversion techniques, the authors derive irradiances, line-of-sight averaged temperatures, and total emission measures for CH and LSS regions.

074.003 Momentum and energy transport by waves in the solar atmosphere and solar wind.
S. A. Jacques.
Astrophys. J., Vol. 215, 942 - 951 (1977).

The fluid equations for the solar wind are presented in a form which includes the momentum and energy flux of waves in a general and consistent way. The concept of conservation of wave action is introduced and is used to derive expressions for the wave energy density as a function of heliocentric distance. The explicit form of the terms due to waves in both the momentum and energy equations are given for radially propagating acoustic, Alfvén, and fast mode waves. The effect of waves as a source of momentum is explored by examining the critical points of the momentum equation for isothermal spherically symmetric flow.

074.004 Localization of the regions of quasi-stationary high-speed solar wind streams on the sun.
V. A. Kovalenko, S. I. Molodykh.
Astron. Zh. Akad. Nauk SSSR, Vol. 54, 859 - 869 (1977).
In Russian. English translation in Soviet Astron., Vol. 21, No. 4.

An analysis of the characteristics of solar regions being the sources of quasi-stationary high-speed solar wind streams (HS) is performed. They are found to be located in coronal regions with an open, nonradially divergent magnetic field configuration. These regions are characterized by low level of corona emission, by low concentration and, as a rule, adjoin corona condensations. On this basis one succeeds in understanding the specific features of HS velocity changes within the solar activity cycle. The results obtained confirm an HS model proposed earlier by the authors.

074.005 The onset of microinstability and its consequences in the solar wind. C. E. Singer, I. W. Roxburgh.
J. Geophys. Res., Vol. 82, 2677 - 2685 (1977).

A simple and general method for applying the results of a microinstability analysis to models of the solar wind is described. Existing two-fluid models are found to become unstable at heliocentric distances varying from 3 to 11 R_s. The development of these 'heat conduction' microinstabilities

affects the energy and momentum transport, observable wave spectrum, cosmic ray diffusion, and properties of minor ions in the solar wind. A proposal which would rationally modify the energy transport is developed. It is suggested that the plasma fluctuations observed near the earth could largely be a result of these instabilities.

074.006 Multispacecraft observations of microscale fluctuations in the solar wind.
K. U. Denskat, L. F. Burlaga.
J. Geophys. Res., Vol. 82, 2693 - 2704 (1977).

Three-component magnetic field and solar wind velocity data from the Explorer 33 and 35 spacecraft have been used to study Alfvénic fluctuations in the interplanetary medium.

074.007 A dynamical model of coronal holes based on radio observations. F. Chiuderi Drago, G. Poletto.
Astron. Astrophys., Vol. 60, 227 - 231 (1977).

A dynamical, homogeneous model of coronal holes is derived under the assumptions of a constant temperature at coronal levels and of a constant ratio p^2/F_c, of the square of the pressure to the conductive flux, in the transition region. The model gives the correct values of the EUV line intensities arising from the transition region.

074.008 Hydrostatic and dynamic models of solar coronal holes. R. Rosner, G. S. Vaiana.
Astrophys. J., Vol. 216, 141 - 157 (1977).

Available X-ray, EUV, and radio observations are used to construct models of the solar atmosphere in coronal holes that take the chromospheric network structure into account. The set of equations used allows for radiative and conductive losses, heating functions of exponential form, and plasma flows, and it is used to demonstrate the implication for the atmospheric structure of the introduction of plasma motions, nonradial flux-tube expansion, direct momentum deposition, and mechanical flux dissipation scale length variation. These calculations yield what are essentially four-parameter models for the coronal-hole atmosphere: base pressure p_0, dissipation scale length s_0, nonradial expansion parameter α, and finally the base flow speed v_0.

074.009 Coronal holes and high-speed wind streams.
J. B. Zirker.
Rev. Geophys. Space Phys., Vol. 15, 257 - 269 (1977).

During a 9-month Skylab solar workshop, 50 participants established some of the basic properties of coronal holes and their associated high-speed wind streams. The author summarizes the empirical properties of coronal holes. Next he examines recent models of the holes and their associated wind streams and then turns to what is known about their origin. He discusses the implications of holes for magnetic structure of the heliosphere. In the last section he summarizes some work needed for further progress.

074.010 Coronal holes as observed in Fe XIV 5303 Å.
R. C. Altrock, S. A. Musman.
Bull. American Astron. Soc., Vol. 9, 432 (1977). – Abstract.

074.011 Fast magnetic fluctuations in the solar wind: Helios 1.
F. M. Neubauer, G. Musmann, G. Dehmel.
J. Geophys. Res., Vol. 82, 3201 - 3212 (1977).

The authors give an overview of some magnetic fluctuation phenomena observed by the search coil magnetometer experiment of the Institut für Geophysik und Meteorologie (IGM) of the Technical University of Braunschweig on board Helios 1. They concentrate on those phenomena which also show a clear signature in the dc magnetic field which is provided by the flux-gate magnetometer experiment of the IGM. This experiment has a bandwidth of 0–4 Hz. Both experiments provide a unique opportunity to study electromagnetic wave phenomena in the solar wind for frequencies up to 2.2 kHz with an excellent background noise level, particularly at frequencies below 200 Hz in the radial distance range from 0.31 to 1 AU.

074.012   A reexamination of two-fluid solar wind models.
     S. Nerney, A. Barnes.
J. Geophys. Res., Vol. 82, 3213 - 3222 (1977).

Many two-fluid solar wind models have appeared in the literature with a variety of assumptions concerning the energy equations. It is the purpose of this study to clarify the importance of the energy assumptions, to present a complete set of solutions for the variable Coulomb logarithm case, to show the nature of the numerical difficulties encountered in the solution of the two-fluid equations, to present a method of solution approximately 50 times faster than any previously developed, and to explore the nature of solutions which may correspond to coronal hole boundary conditions.

074.013   An unusual aspect of solar wind speed variations
     during solar cycle 20.
J. T. Gosling, J. R. Asbridge, S. J. Bame.
J. Geophys. Res., Vol. 82, 3311 - 3314 (1977).

Geomagnetic records from 1868 through 1975 indicate that geomagnetic activity during 1973–1975 was unusually enhanced for that phase of the sunspot cycle (5–7 years after solar maximum). Previous work indicates that long-term variations in geomagnetic activity are closely coupled to long-term variations in the bulk flow speed of the solar wind. Thus the authors infer that reported averages of the solar wind speed for the 1973–1975 era are unusually large for that phase of the sunspot cycle.

074.014   Physical conditions in the solar corona during
     flare-like events.   J. A. Vorpahl.
Izv. AN SSSR. Ser. fiz., Vol. 41, 252 - 272 (1977). In Russian. – Abstr. in Ref. zh., 51. Astron., 8.51.428 (1977).

074.015   Solar polar coronal hole – a mathematical simulation.
     S. T. Suess, A. K. Richter, C. R. Winge, S. F. Nerney.
Astrophys. J., Vol. 217, 296 - 305 (1977).

The authors use a quasi-radial approximation to the full magnetohydrodynamic equations for axisymmetric, polytropic solar wind flow to simulate a polar hole, with the benefit that they can deduce model temperature and magnetic field intensities and distributions in this polar hole. The authors conclude that from $2 R_\odot$ out to $5 R_\odot$ the temperature varies only slightly with radius, but is larger near the center of the polar hole than at the edge. They also find that the magnetic field intensity at $2 R_\odot$ could be about 1 gauss at the center of the hole, decreasing toward the edge of the hole.

074.016   The nonequilibrium ionization of solar flare coronal
     plasma and the emergent X-ray spectrum.
P. R. Shapiro, R. T. Moore.
Astrophys. J., Vol. 217, 621 - 643 (1977).

The effect of temperature changes in coronal plasma during a solar flare which are more rapid than the characteristic ionization and recombination times of the plasma ions is considered. Such temperature changes imply a lack of equilibrium between the ionization and recombination of coronal ions. A model is presented in which a magnetic loop of preflare coronal plasma in ionization equilibrium at a typical quiet coronal temperature ($\sim 2 \times 10^6$ K) and at a density of between 10^{10} and 10^{11} cm^{-3} is heated within a fraction of a second to a much higher temperature ($> 10^7$ K). The authors calculate the details of this model, including the time-dependent ionization structure of the plasma and the emergent X-ray

line and continuum spectrum from 1 to 250 Å.

074.017   Analysis of X radiation of coronal condensations
     observed with broad-band filters.   J. Sylwester.
Solar activity and solar-terrestrial relations, (see 012.007), p. 139 - 142 (1976). In Russian.

The analysis of the X radiation from the central core of a coronal condensation was carried out on the basis of the condensation model of Landini and Monsignori Fossi. The maximum temperature and emission measure in this model were treated as variable parameters, and their values were chosen on the basis of observational data. The observations made from the rocket "Vertikal 1" were used in the analysis. Moreover, from the same observations, the temperature and emission measure of the condensation core were estimated, assuming an isothermal model of this region. The comparison of the results, obtained from these two approaches, shows significant differences between the non-isothermal and isothermal interpretation of the observations.

074.018   Analysis of formation of resonance lines in the X-ray
     spectrum of coronal condensations.
B. Sylwester.
Solar activity and solar-terrestrial relations, (see 012.007), p. 143 - 146 (1976). In Russian.

The values of the X-radiation fluxes in several resonance lines, which are formed in coronal condensations have been calculated. The calculations were performed making use of the coronal condensation model worked out by Landini and Monsignori Fossi. A comparison of the obtained theoretical values of the fluxes with observational data was made in order to verify the model. The analysis indicates that this coronal condensation model does not describe the real structure of such a region adequately.

074.019   Large-scale magnetic structures and the longitudinal
     distribution of the green coronal emission.
V. Bumba, J. Sýkora.
Solar activity and solar-terrestrial relations, (see 012.007), p. 231 - 233 (1976).

A very brief review of the results concerning the close relation of the location of particle-emitting flares and a characteristic large-scale pattern in the background magnetic field distribution is given. The longitudinal distribution of the green coronal emission is again discussed in its relation to the characteristic features in the magnetic field distribution.

074.020   Investigation of the solar corona above a large active
     region on 17 - 28 February, 1969.
I. N. Garczyńska.
Solar activity and solar-terrestrial relations, (see 012.007), p. 313 - 319 (1976). In Russian.

In this paper the shock waves transitions throughout the solar corona are considered. Attention is paid to waves produced by chromospheric flares inside an active region which was seen on the solar disc during the period from 17 up to 28 February 1969. The parameters of these waves were calculated using dynamical spectra of radio-bursts of type II.

074.021   Polarization of the corona during the solar eclipse
     of June 30, 1973.   J. Sýkora.
Solar activity and solar-terrestrial relations, (see 012.007), p. 321 - 322 (1976).

074.022   Latitude distribution of the total emission and of
     the half-width of the coronal Fe X λ 6374 Å line.
N. F. Tyagun, V. E. Stepanov.
Solar activity and solar-terrestrial relations, (see 012.007), p. 323 - 332 (1976). In Russian.

A statistical treatment of the parameters of 1715 profiles was carried out on the basis of red coronal line spectra obtained with 1 Å/mm dispersion on the different heliolatitudes in 1970. Curves of total intensities of half-widths and of a line of sight velocity dispersion are constructed in relation to the heliolatitude.

074.023 Parameters of the solar wind and solar activity.
V. A. Kovalenko, V. N. Malyshkin.
Solar activity and solar-terrestrial relations, (see 012.007), p. 333 - 340 (1976). In Russian.

The change of energy flux and the solar wind velocity for the period 1962–1970 is considered. The energy balance of the solar corona is considered.

074.024 On the correlation of solar-wind energy flux and geomagnetic activity. S. Krajčovič.
Solar activity and solar-terrestrial relations, (see 012.007), p. 359 - 367 (1976).

074.025 Solar-wind velocity variations and geomagnetic disturbances. E. Ţifrea.
Solar activity and solar-terrestrial relations, (see 012.007), p. 369 - 370 (1976).

The correlation coefficient between solar wind velocities and the geomagnetic index A_p is determined for solar rotations 1543–1596 (1969–1972).

074.026 Dispersion analysis of solar wind velocity.
N. A. Lotova, I. V. Chashey (*Chashej*), W. A. Coles.
Astron. Astrophys., Vol. 61, 13 - 16 (1977).

Three site observations of interplanetary scintillations are used to estimate the dispersion in diffraction pattern velocity. The mean velocity and dispersion change considerably from day to day. For the eight source days analyzed, the standard deviation of velocity varied from 17% to 37% of the mean.

074.027 A two-spacecraft study of the preshock perturbations of the solar wind protons.
L. Diodato, G. Moreno.
J. Geophys. Res., Vol. 82, 3615 - 3622 (1977).

A two-spacecraft technique is used to study the problem of solar wind perturbations occurring in the region upstream of the earth's bow shock, where low-frequency waves are present. In contrast with some previous investigations, no significant variations are found in that region for the average values of the basic solar wind parameters (bulk speed, proton density, and temperature) relative to the unperturbed flow.

074.028 Coronal and interplanetary shock waves generated during flares on August 2 - 13, 1972. Annotation.
S. Pinter, M. Dryer.
VIII Leningr. mezhdunar. semin. Mater mezhdunar. semin. "Akt. protsessy na Solntse i probl. soln. nejtrino", 1976. Leningrad, 1976, p. 165 - 166. In Russian and English. From Ref. zh., 62. Issled. kosm. prostranstva, 9.62.191 (1977).

074.029 Polarization of the white light corona during the solar eclipse of June 30, 1973. J. Sýkora.
Bull. Astron. Inst. Czechoslovakia, Vol. 28, 312 - 316 (1977).

Photographic records of the white light corona polarization were obtained during the total solar eclipse of June 30, 1973 in El Meki (Republic of Niger). The pictures were elaborated by equidensitometric and photometric methods. The degree and direction of polarization were calculated for the interval $1.1 \, R_\odot \leqslant r \leqslant 2.8 \, R_\odot$ and are graphically presented.

074.030 Structures chromosphériques fines dans la couronne solaire. J.-R. Roy.
J. R. Astron. Soc. Canada, Vol. 71, 373 - 376 (1977).

The author draws attention to the existence of very fine structures of the order of 200 km in width seen at Hα and extending from the centre of a sunspot into the corona.

074.031 Probing the solar wind with radio measurements of the second moment field.
R. Woo, F.-C. Yang, A. Ishimaru.
Astrophys. J., Vol. 218, 557 - 568 (1977).

The analysis presented in this paper is based on results for the second moment mutual coherence function obtained using the parabolic equation method. The authors examine the dependence of spectral broadening on anisotropic electron density irregularities and wind velocity fluctuations. They find that while these effects decrease the bandwidth of the spectrum over that for isotropic irregularities and no velocity fluctuations, the shape of the spectrum remains unchanged. An analysis for interpreting and relating angular broadening to spectral broadening (and other radio scattering observations) is provided, and it is seen that the properties of the solar wind deduced from both measurements are consistent.

074.032 Mode coupling in the solar corona. III. Alfvén and magnetoacoustic waves. D. B. Melrose.
Australian J. Phys., Vol. 30, 495 - 507 (1977).

The coupling is strongest for nearly parallel (to the magnetic field lines) propagation, and the coupling ratio may be approximated by $Q = (\theta_0/\theta)^3$, where θ is the angle between the wave vector and the magnetic field lines, while $\theta_0^3 = \lambda/L$, with λ the wavelength and L the scalelength of the inhomogeneity. This result may be of significance in connection with the heating of the solar corona by the dissipation of waves generated initially as acoustic waves in the photosphere, and perhaps with the propagation of hydromagnetic waves in the interplanetary medium.

074.033 Physical conditions in the solar corona during flarelike events. Abstract. J. A. Vorpahl.
VIII Leningr. mezhdunar. semin. Mater. mezhdunar. semin. "Akt. protsessy na Solntse i probl. soln. nejtrino", 1976. Leningrad, 1976, p. 107 - 111. In Russian. – Abstr. in Ref. zh., 51. Astron., 10.51.545 (1977).

074.034 Dispersion analysis of solar wind velocity.
N. A. Lotova, I. V. Chashej.
Izv. vyssh. uchebn. zaved. Radiofiz., Vol. 20, 329 - 330 (1977). In Russian. – Abstr. in Ref. zh., 62. Issled. kosm. prostranstva, 10.62.178 (1977).

074.035 On the reverse effect of cosmic rays on solar wind.
V. Kh. Babayan.
Din. kosm. plazmy. Moskva, 1976, p. 21 - 23. In Russian. Abstr. in Ref. zh., 62. Issled. kosm. prostranstva, 10.62.180 (1977).

074.036 A comparison of solar wind streams and coronal structure near solar minimum. J. T. Nolte,
J. M. Davis, M. Gerassimenko, A. J. Lazarus, J. D. Sullivan.
Geophys. Res. Lett., Vol. 4, 291 - 294 (1977).

The authors examine the relationship between high speed streams of solar wind and coronal structure during the period September - December, 1976. The solar wind streams during this period were generally of lower amplitude, lower maximum velocity, and shorter duration than those in the preceding Skylab period. Solar wind data are compared with X-ray images of the corona taken on 16 September and 17 November, 1976. The results suggest that either there has been a change in the solar wind-coronal hole relationship or that the relationship is not as general as might be inferred from the Skylab period.

074.037 Non-spherical flow of matter in the solar wind.
M. Sroczynska.
Ann. Géophys., Vol. 32, 373 - 380 (1976). – Abstr. in Phys. Abstr., Vol. 80, Abstr. 56880 (1977).

074.038 Studies of the Sun from Skylab. A. F. Timothy.
IEEE Trans. Nucl. Sci., Vol. NS-24, No. 1, p. 20 - 28 (1977). – Abstr. in Phys. Abstr., Vol. 80, Abstr. 56891

(1977).

074.039 Observations of linear polarization of the Crab
          Nebula during an occultation by the solar corona. II.
N. Kawajiri, F. Takahashi, T. Ojima, N. Kawano, C. Miki.
J. Radio Res. Lab., Vol. 23, 305 - 317 (1976). – Abstr. in
Phys. Abstr., Vol. 80, Abstr. 60837 (1977).

074.040 The Viking solar corona experiment.
          G. L. Tyler, J. P. Brenkle, T. A. Komarek, A. I.
Zygielbaum.
J. Geophys. Res., Vol. 82, (see 003.017), 4335 - 4340 (1977).
Paper No. 7S0455.
    The 1976 Mars solar conjunction resulted in complete
occultations of the Viking spacecraft by the sun at solar minimum. During the conjunction period, coherent 3.5- and 13-cm
wavelength radio waves from the orbiters passed through the
solar corona. Data were obtained within at least 0.3 and 0.8
R_S of the photosphere. The data are used to determine the
plasma density integrated along the radio path, the velocity of
density irregularities in the coronal plasma, and the spectrum
of the density fluctuations in the plasma.

074.041 Coronal holes on the sun and their solar wind inter-
          actions.   R. M. Broussard.
Publ. Astron. Soc. Pacific, Vol. 89, 614 (1977). – Abstract.

074.042 Mapping the solar corona.   F. Espenak.
J. American Assoc. Variable Star Obs., Vol. 6, 41
(1977). – Abstract.

074.043 Plasma kinetics in the solar wind.
          M. Dobrowolny, G. Moreno.
Space Sci. Rev., Vol. 20, 577 - 620 (1977).
    An account is given of the observations and theoretical
ideas concerning the role of kinetic processes in the solar wind.
This includes, first of all, the measurements on distribution
functions of plasma electrons and protons, the relation of the
observed non-thermal electron features with the concept of an
exospheric expansion of the solar corona, and the connection
of non-thermal proton distributions with bulk flow inhomogeneities of the wind. A discussion is given of the present
understanding of the connection between observed features of
the particle distributions and anomalous values of some plasma
transport coefficients, which in turn determine the actual
values of macroscopic plasma parameters. A further topic of
the review is that of possible kinetic processes occurring within
small scale structures in the solar wind, like collisionless shocks,
various types of discontinuities and D-sheets.

074.044 Coronal general magnetic field evolution as a new
          parameter of the solar cycle.   H. Yoshimura.
Sol. Phys., Vol. 52, 41 - 52 (1977).
    This paper studies the evolution of the geometry of the
coronal magnetic field lines calculated from the surface magnetic field, the evolutionary characteristics of which have been
found to be important indices of the solar cycle.

074.045 A long-lived coronal arch system observed in X-rays.
          J. P. McGuire, E. Tandberg-Hanssen, K. R. Krall,
S. T. Wu, J. B. Smith, D. M. Speich.
Sol. Phys., Vol. 52, 91 - 100 (1977).
    A large long-lived soft X-ray emitting arch system was
observed during the last Skylab mission. This arcade stayed in
the same approximate position for several solar rotations. The
authors suggest that these long-lived arches owe their stability
to the stable coronal magnetic-field configuration. A global
constant α force-free magnetic field analysis, as developed by
Nakagawa et al. (1977), is used to describe the arches, and
results in a marked resemblance between the theoretical magnetic-field configuration and the observed X-ray emitting
feature.

074.046 Collisionless deceleration of fast electron streams in
          the solar coronal plasma.
L. L. Bazelyan, N. Yu. Goncharov, V. V. Zaitsev (*Zajtsev*), V. A.
Zinichev, V. O. Rapoport, Ya. G. Tsybko.
Sol. Phys., Vol. 52, 141 - 152 (1977).
    The collisionless deceleration of electron streams re-
sponsible for type IIIb bursts has been investigated. It is shown
that under certain assumptions the electron streams with the
initial velocities of the order of 0.4–0.8c undergo a sufficient
deceleration which is characterized by a decrease in their mean
velocity by 15–17% between plasma levels at 25 to 6.25 MHz.
The stream deceleration becomes more essential with the
growth of the initial velocity of the stream. On the other hand,
the deceleration disappears when the initial velocity of the
stream is of the order of 0.35c. This critical velocity ($\simeq 0.35c$)
is assumed to define a boundary between two different expansion regimes of fast electrons moving in the solar corona.

074.047 Regulation of solar wind heat flux by ordinary mode
          instability.   G. S. Lakhina.
Sol. Phys., Vol. 52, 153 - 162 (1977).
    The ordinary mode can frequently become unstable in
the solar wind at 1 AU provided the ratio of halo to core electrons density does not exceed the value 0.05. The growth rates
corresponding to the average conditions are typically $\sim 10\Omega_P$
(Ω_P being the proton cyclotron frequency). Because of low
threshold for onset of instability for $\beta_{\perp c} \gtrsim 1$ (where $\beta_{\perp c}$ is the
transverse beta for the core electrons), the mode is expected
to play an important role in regulating the solar wind heat flux
at 1 AU.

074.048 On the role of hydromagnetic waves in the corona
          and the base of the solar wind.   D. G. Wentzel.
Sol. Phys., Vol. 52, 163 - 177 (1977).
    Hydromagnetic waves are of interest for heating the
corona or coronal loops and for accelerating the solar wind.
This paper enumerates some of the limitations that must be
considered before hydromagnetic waves are taken seriously.

074.049 The calcium K-line network in coronal holes.
          K. A. Marsh.
Sol. Phys., Vol. 52, 343 - 348 (1977).
    Microphotometry of calcium K-line photographs in the
regions of polar coronal holes shows that the chromospheric
network exterior to a hole has a slightly broader intensity distribution than that inside the hole itself, a fact which can be
attributed to a greater number of bright network elements outside the hole. These bright elements presumably represent the
enhanced network resulting from the dispersal of magnetic
flux from old active regions, a hypothesis which is consistent
with current ideas of coronal hole formation.

074.050 Radio and EUV observations of a coronal hole.
          G. A. Dulk, K. V. Sheridan, S. F. Smerd, G. L.
Withbroe.
Sol. Phys., Vol. 52, 349 - 367 (1977).
    The authors present radio and EUV observations of a
coronal hole. They attempt to derive the density and temperature distributions in the transition region and inner corona
from the combined observations. After examining several
possible modifications of the standard models they suggest that
the discrepancy would disappear if the abundances of the
heavier elements were increased by about a factor of 10. Such
increases could result from differential diffusion in the large
temperature gradient of the transition region. They conclude
therefore that models which incorporate thermal diffusion,
as well as mass outflow and departures from ionization equilibrium, offer the greatest hope of reconciling the EUV and
radio observations of coronal holes.

074.051 Magnetic field and current sheets in the corona
          above active regions.   T. Sakurai, Y. Uchida.

Sol. Phys., Vol. 52, 397 - 416 (1977).

A new method for the calculation of coronal magnetic field is proposed and it is shown to reproduce the EUV features in the corona as observed by Skylab experiments satisfactorily well. One of the remarkable points is that it reproduces the loopy threads in the active region corona and also the large scale field lines connecting active regions. The existence of coronal current is expected wherever the present coronal-current-free model fails to represent the feature. A method of calculating the coronal sheet-current is also developed. This may be used in pinning down the possible site of the flare and in discussing the flare occurrence in terms of the energy stored there.

074.052 **Solar wind heating by heat conduction driven ion acoustic instability.**   P. Revathy, G. S. Lakhina.
Sol. Phys., Vol. 52, 471 - 475 (1977).

Nonlinear saturation of the ion acoustic instability, driven by finite heat conduction in the solar wind, via resonance broadening is considered. The calculations based on the Hartle and Sturrock model show that this process can heat ions faster than the electrons, thereby, leading to the reduction in electron to proton temperature ratio. The difficulty with this process is that it increases the electron temperature above the value predicted by the Hartle and Sturrock model.

074.053 **A pictorial comparison of interplanetary magnetic field polarity, solar wind speed, and geomagnetic disturbance index during the sunspot cycle.**
N. R. Sheeley, Jr., J. R. Asbridge, S. J. Bame, J. W. Harvey.
Sol. Phys., Vol. 52, 485 - 495 (1977).

Observations of interplanetary magnetic field polarity, solar wind speed, and geomagnetic disturbance index (C9) during the years 1962–1975 are compared in a 27-day pictorial format that emphasizes their associated variations during the sunspot cycle. This display accentuates graphically several recently reported features of solar wind streams including the fact that the streams were faster, wider, and longer-lived during 1962–1964 and 1973–1975 in the declining phase of the sunspot cycle than during intervening years. The display reveals strikingly that these high-speed streams were associated with the major, recurrent patterns of geomagnetic activity that are characteristic of the declining phase of the sunspot cycle.

074.054 **Some aspects of coronal streamer dynamics.**
W. J. Weber.
Sol. Phys., Vol. 53, 59 - 77 (1977).

Observations of coronal streamers suggest that these configurations are stationary in terms of the convective time scale, but not with respect to the diffusive processes. In this paper, the diffusive time scale is estimated and the thickness of a stationary streamer where convective and diffusive processes are balanced turns out to be very small. The author shows that the arising convective model is to be understood from evolutionary arguments; the upward flux transport leads to a time scale for the life of a streamer, related to the evolution of the underlying magnetic field. Slight differences occur between stationary models and the present model; these differences are shown not to be in conflict with radio observations.

074.055 **The polarization of the inner solar corona at the eclipse of 10 July 1972.**   G. M. Nikolsky
(*Nikol'skij*), A. A. Sazanov, A. K. Kishonkov.
Sol. Phys., Vol. 53, 79 - 96 (1977).

Photographs of the corona in the continuum spectrum (580–700 nm) have been obtained with a doublet camera ($F = 4$ m) through rotating sector polarizers with the vibrational directions oriented radially and tangentially to the solar limb. Isophotes of the total emission and its polarized component, as well as diagrams giving the degree of polarization for the ($K+F$)-corona and the electron corona proper (P_K), have been plotted up to the distance of 1 $R_\odot$ from the limb. Three-dimensional

forms of 15 different coronal rays have been ascertained. Mean values of the coronal brightness and polarization versus the distance from the Sun have been determined.

074.056 **Coronal rotation dependence on the solar cycle phase.**   E. Antonucci, M. A. Dodero.
Sol. Phys., Vol. 53, 179 - 188 (1977).

A study of the green corona rotation rate, during the period 1970–1974, confirms that the differential rotation degree varies systematically through a solar cycle and that the corona rotates in an almost rigid manner before sunspot minimum. During the first two years, 1970–1971, the differential rotation degree, characteristic of high solar activity periods is detected. While during the years of declining activity, 1972–1974, a drastic decrease of the differential rotation degree occurs and the green corona rotates almost rigidly, as the coronal holes observed in the same period. These conclusions are valid only for the rotation of coronal features with lifetime of at least one solar rotation.

074.057 **EUV analysis of polar plumes.**
I. A. Ahmad, G. L. Withbroe.
Sol. Phys., Vol. 53, 397 - 408 (1977).

Three polar plumes were studied using Skylab Mg X and O VI data. The plumes lie within the boundaries of a polar coronal hole. The authors find that the mean temperature of the plumes is about 1.1×10^6 K and that they have a small vertical temperature gradient. Densities are determined and found consistent with white light analyses. The variation of density with height in the plumes is compared with that expected for hydrostatic equilibrium. It appears that polar plumes contain about 15% of the mass in a typical polar hole and occupy about 10% of the volume.

074.058 **Coronal He$^+$ λ304 radiation.**   I. A. Ahmad.
Sol. Phys., Vol. 53, 409 - 415 (1977).

The intensity of the He$^+$ λ304 coronal line relative to the H^0 λ1216 line, including the dominant contribution due to resonance scattering, is presented. All physical processes important in the corona are included. It is found that He$^+$ λ304 is a major contributor to the XUV corona, and that the sensitivity of the He$^+$ λ304/H^0 λ1216 intensity ratio to coronal temperature is very weak, supporting the belief that this ratio is a good indicator of the coronal helium abundance.

074.059 **On coronal Fe abundances and temperatures from XUV emission lines.**
M. P. Nakada, R. D. Chapman, W. M. Neupert, R. J. Thomas.
Sol. Phys., Vol. 53, 435 - 444 (1977).

Data obtained by the OSO-7 spectroheliograph on strong XUV lines of five different Fe ions from the outer equatorial corona are presented. Interpretation of the data with a spherically symmetric model atmosphere gives average ion abundances for lines of sight at 0.3 $R_\odot$ from the limb. The deviation of measured relative abundances of Fe XII, XIV, and XVI from predictions of ionization equilibrium at one temperature seems to indicate that there are appreciable temperature variations along lines of sight.

074.060 **Coronal heating by ion acoustic waves.**
P. Revathy.
Sol. Phys., Vol. 53, 445 - 448 (1977).

A quantitative study is made based on the suggestion that coronal heating results from Landau damping of ion acoustic waves. It is shown that both ions and electrons are heated by nonlinear saturation of ion acoustic instability. Ions are heated more rapidly than the electrons.

074.061 **Electrons in the solar corona. II. Coronal streamers from K-coronameter measurements.**
A. Dollfus, M.-J. Martres.
Sol. Phys., Vol. 53, 449 - 464 (1977).

The polarimetric survey of electrons in the K-corona initiated at Pic-du-Midi and Meudon Observatories in 1964 now covers a full solar cycle of activity. The measurements are photometrically calibrated in an absolute scale. Typical fan-shaped feature: 'jets en lame', typical non-elongated feature above an active center, a feature of exceptional intensity, and higher level coronal structures are discussed.

074.062   The Si IV λ1393 line in a coronal hole compared to the quiet Sun from OSO-8 observations.
M. H. Francis, R. Roussel-Dupré.
Sol. Phys., Vol. 53, 465 - 470 (1977).

Results indicate that the line width is somewhat greater in coronal holes compared to the quiet Sun, implying a difference in the broadening mechanism. There is no evidence that the line is Doppler shifted in coronal holes relative to the quiet Sun implying there is no mass flow in holes, at the 80 000 K level, greater than 4.3 km s^{-1}. Within the uncertainty of the authors' experiment the integrated line intensities are the same in a coronal hole as in the quiet Sun.

074.063   Do surges heat the corona?
D. M. Rust, D. F. Webb, W. Maccombie.
Sol. Phys., Vol. 54, 53 - 56 (1977).

A comparison of X-ray filtergrams obtained during the Skylab mission 8 hr before and within 4 hr following 54 active region surges on the disk revealed only 6 cases of long-enduring, large-scale (> 10000 km) coronal enhancements that might have been associated with surge activity. It is concluded that there is no evidence for any substantial increase in the temperature or amount of coronal material during reported surges.

074.064   Development of a complex of activity in the solar corona.   R. Howard, Z. Švestka.
Sol. Phys., Vol. 54, 65 - 105 (1977).

Skylab observations of the Sun in soft X-rays gave the first possibility to study the development of a complex of activity in the solar corona during its whole lifetime of seven solar rotations. The basic components of the activity complex were permanently interconnected (including across the equator) through sets of magnetic field lines, which suggests similar connections also below the photosphere. However, the visibility of individual loops in these connections was greatly variable and typically shorter than one day. Each brightening of a coronal loop in X-rays seems to be related to a variation in the photospheric magnetic field near its footpoint. Only loops (rarely visible) connecting active regions with remnants of old fields can be seen in about the same shape for many days. The interconnecting X-ray loops do not connect sunspots.

074.065   The structure of coronal magnetic loops. I: Equilibrium theory.
C. Chiuderi, R. Giachetti, G. van Hoven.
Sol. Phys., Vol. 54, 107 - 122 (1977).

The authors present a model, based on observations, for the magnetic-field equilibrium of a cool coronal loop. The pressure structure, taken from the Harvard/Skylab EUV data, is used to modify the usual force-free-field form in quasi-cylindrical symmetry. The resulting field, which has the same direction but different strength, is calculated and its variation displayed. Finally, localized interchange stability is evaluated and discussed, as the first step in a subsequent complete magnetohydrodynamic-stability analysis.

074.066   Observations of the birth of a small coronal hole.
C. V. Solodyna, A. S. Krieger, J. T. Nolte.
Sol. Phys., Vol. 54, 123 - 134 (1977).

Using soft X-ray data from the S-054 X-ray spectrographic telescope aboard Skylab, the authors observed temporal changes in the emission structure of the X-ray corona associated with the birth of a small coronal hole, designated as CH6. The authors examine the X-ray structures in the immediate vicinity

of the developing hole and the global pattern of reduced X-ray emission in which CH6 is located. On the quantitative level, they compute the decrease in observed X-ray emission associated with the sudden birth of CH6 and compare the observed growth and decay rates to those predicted for the diffusion of solar magnetic fields.

074.067   The results of statistical analysis of the coronal profiles above the solar active regions.
E. I. Tetruashvili.
Sol. Phys., Vol. 54, 135 - 138 (1977).

In order to establish some regularities or variations in the distribution of widths and intensities of the coronal line profiles λ5303 and λ6374 depending upon the solar activity, a statistical analysis was made for more than 3000 profiles (the data covering the period 1966–1972). Following results obtained: (1) The distribution of coronal line profile widths changes depending upon the solar activity phase. (2) The character of the relation between the intensities and widths varies with variation of the solar activity phase.

074.068   Solar-cycle evolution of the coronal general magnetic field of 1959–1974 and the synchronous variation of high-speed solar wind streams and galactic cosmic rays.
H. Yoshimura.
Sol. Phys., Vol. 54, 229 - 258 (1977).

Contents: Scope of the study of the coronal general magnetic field and its meaning to other branches of solar-system physics; Method of analysis: potential-field approximation; Stable behavior of the general corona in the minimum phase; Solar-cycle evolution of the coronal general magnetic fields of 1959–1974; Rapid change of the general corona in the maximum phase; Comparison of the coronal field lines with coronal photographs; Summary and discussion.

074.069   The effects of diverging coronal fields on the solar wind expansion.
H. J. Fahr, H. W. Ripken, M. K. Bird.
Astrophys. Space Sci., Vol. 46, L11 - L13 (1977).

The calculations made by Fahr et al. (1976) have been subjected to a re-examination, the results of which are described in this letter.

074.070   Autoionization rate coefficients for some coronal ions.   H. P. Mital, U. Narain.
Sol. Phys., Vol. 54, 387 - 391 (1977).

Autoionization rate coefficients for some ions found in solar corona have been computed in the temperature range 10^5–10^8 K semiempirically. In each case the coefficients decrease as we go from less to more ionized ones and have a typical temperature dependence.

074.071   Soft X-ray observations of large-scale coronal active region brightenings.   D. M. Rust, D. F. Webb.
Sol. Phys., Vol. 54, 403 - 417 (1977).

156 large-scale enhancements of X-ray emission from solar active regions were studied on full-disk filterheliograms to determine characteristic morphology and expansion rates for heated coronal plasma. The X-ray photographs were compared with Hα observations of flares, sudden filament disappearances, sprays and loop prominence systems (LPS). 81% of the X-ray events were correlated with Hα filament activity, but only 44% were correlated with reported Hα flares. The X-ray enhancements took the form of loops or arcades of loops ranging in length from 60000 km to 520000 km and averaging 15000 km in width. Lifetimes, event frequency, and expansion velocities of the loops are reported.

074.072   A sheet-current approach to coronal-interplanetary modeling.   T. Yeh, G. W. Pneuman.
Sol. Phys., Vol. 54, 419 - 430 (1977).

The coronal-interplanetary space may be regarded as

being partitioned by current-sheets into several piecewise current-free regions. These current sheets overlie the photospheric neutral lines, where the vertical component of the magnetic field reverses its polarity on the solar surface. But, their locations and strengths are determined by force balance between the magnetic field and the gas pressure in the coronal-interplanetary space. Since the pressure depends on the flow velocity of the solar wind and the solar wind channels along magnetic flux tubes, there is a strong magnetohydrodynamic coupling between the magnetic field and the solar wind. The sheet-current approach presented in this paper seems to be a reasonable way to account for this complicated interaction.

074.073 **On the nature of photospheric magnetic fields beneath large coronal holes.**
S. Frankenthal, A. S. Krieger.
Sol. Phys., Vol. 55, 83 - 97 (1977).

The authors consider proposed mechanisms for the formation of coronal holes, and identify as crucial the issue whether the holes are permeated by rigidly rotating fields. It is suggested that the interaction between such a field and the differentially rotating, diffusive solar envelope will produce a fore aft asymmetry in the distribution of fields which emerge to the photosphere. An initial study is carried out in the context of an illustrative example, and the results indicate that the asymmetry may be observed for a certain range of parameters involving the properties of the solar envelope and the characteristic size of the emerging field pattern.

074.074 **Thermally conductive flows in coronal holes.**
R. S. Steinolfson, E. Tandberg-Hanssen.
Sol. Phys., Vol. 55, 99 - 109 (1977).

Polytropic solar wind flows in flow tubes whose cross-sectional area increases faster with radius than for a radial expansion have been studied by Kopp and Holzer (1976). The authors have extended their work to include thermal conduction and have compared thermally conductive and polytropic flows in the lower corona for given high-speed conditions at 1 AU. The authors also considered a modified thermal conductivity which decreases more rapidly with increasing radius than does the Spitzer (1962) value. They conclude that thermal conduction alone will not explain solar wind flows originating in coronal holes and that some other mechanism (such as wave pressure) is necessary.

074.075 **A study of the background corona near solar minimum.**   K. Saito, A. I. Poland, R. H. Munro.
Sol. Phys., Vol. 55, 121 - 134 (1977).

The white light coronagraph data from Skylab is used to investigate the equatorial and polar K and F coronal components during the declining phase of the solar cycle near solar minimum. Observational data are presented in a form suitable for the calculation of K and F coronal models. The method used for calculating density models is discussed and the results from these calculations for the equatorial background, equatorial holes, and polar holes are presented. Finally the results are discussed with respect to the reliability of the derived background model and the fractions of time streamers (50%), holes (30%), and background (20%) brightness are observed on the limb.

074.076 **Photoelectric observations of Fe XIV coronal depletion: 20 April 1976.**   R. R. Fisher.
Sol. Phys., Vol. 55, 135 - 141 (1977).

On 20 April 1976 a coronal emission line transient was observed on the west limb of the Sun using a λ5303 Fe XIV detection system and the Sacramento Peak Observatory's 40 cm coronagraph. The transient as observed at a single sample point, and the time scale for this event was found to be about 40 min. Limb scans before and after the event are used to estimate the amount of coronal mass involved in this relatively slow event. A depletion of the inner corona of 4.4×10^{38} elec-

trons $(8.8 \times 10^{14} \text{ G})$ is inferred by differencing pre- and post-event model electron density distributions.

074.077 **Temporal evolution of the equatorial K-corona.**
R. M. MacQueen, A. I. Poland.
Sol. Phys., Vol. 55, 143 - 159 (1977).

Observations of the equatorial K-coronal radiance at $2.5\,R_\odot$ from Sun center and its variation with time, on a daily basis, during the Skylab mission (May 1973—February 1974) are presented. The results are interpreted as indicating a general simplification of the coronal magnetic field through the mission and, in comparison with harmonic analysis of the surface magnetic field (Levine, 1977), as indicating a rapid response of equatorial outer coronal structures to abrupt changes in the global surface field structure.

074.078 **The filament-corona transition region from OSO-VI EUV observations.**
F. Chiuderi Drago, M. Silvi.
Sol. Phys., Vol. 55, 177 - 180 (1977).

The transition region between filament and corona is investigated measuring the intensity of six EUV lines above two filaments on the disk observed on September 1 and 2, 1969 by OSO-VI. The comparison between these intensities and those observed on quiet regions shows that there is no difference between the two transition regions.

074.079 **Subsonic flows in the coronal-interplanetary regions of closed field lines.**   T. Yeh.
Sol. Phys., Vol. 55, 241 - 250 (1977).

In the coronal-interplanetary space the plasma motion, in a reference frame corotating with the Sun, is aligned with the magnetic field. Just like the solar wind, which is the supersonically expanding flow along open field lines, the flow along closed field lines is mainly driven by the pressure gradient. The flow in the regions of closed field lines is subsonic, being determined by the conditions at the two footpoints of the magnetic flux tube.

074.080 **The structure of coronal magnetic loops. II: MHD stability theory.**
R. Giachetti, G. Van Hoven, C. Chiuderi.
Sol. Phys., Vol. 55, 371 - 386 (1977).

The authors present the second part of a complete theory for the plasma and field structure of a cool coronal arch, corresponding to those observed in the EUV from Skylab. The global magnetohydrodynamic (MHD) stability of a previously described equilibrium-loop model is evaluated, and compared with that of an unmodified ambient force-free field. The influence of the photospheric boundary condition is also evaluated, producing a specification of stability limits which depend on the relative field and plasma pressures and scale widths. The resulting restrictions on the allowable field configuration of a coronal loop are then compared with observed values. The implications of this general method for deducing small-scale coronal magnetic-field structure from the measured plasma profile of an emissive feature are also described.

074.081 **Do changes in coronal emission structure imply magnetic reconnection?**
J. T. Nolte, M. Gerassimenko, A. S. Krieger, R. D. Petrasso, Z. Švestka, D. G. Wentzel.
Sol. Phys., Vol. 55, 401 - 412 (1977).

The authors examine three major possible interpretations of observed reconfigurations of coronal X-ray and XUV emitting structures on a scale comparable to the size of the structures themselves. One possibility is that little change in the large-scale magnetic field configuration is associated with the change in emission. The other two possibilities are processes by which the magnetic field structure can change.

074.082 **A method for determining the large-scale magnetic**

field of the solar corona. A. A. Ruzmajkin, D. D. Sokolov.
Inst. prikl. mat. AN SSSR. Prepr. No. 48 za 1977 g. Moskva, 1977. 20 pp. Price 8 Kop. In Russian. — Abstr. in Ref. zh., 51. Astron., 11.51.570 (1977).

074.083 On a kinetic theory of inhomogeneities in the solar wind plasma. I. S. Veselovskij.
Simpoz. po fiz. geomagnitosfery, 1977. Tezisy dokl. Irkutsk, 1977, p. 8 - 9. In Russian. — Abstr. in Ref. zh., 62. Issled. kosm. prostranstva, 11.62.102 (1977).

074.084 Mass ejections from the solar corona into interplanetary space. E. Hildner.
Study of travelling interplanetary phenomena 1977, (see 012. 042), p. 3 - 21 (1977).

Mass ejections from the corona are common occurrences. It is suggested that the frequency of ejections varies with the solar cycle and that ejections may contribute 10 percent or more of the total solar mass efflux to the interplanetary medium at solar maximum. Since ejections are confined to relatively low latitudes, their fractional mass flux contribution is greater near the ecliptic than far from it. From the behavior of ejecta, one can estimate the magnitude of the force driving them through the corona. Ejections are associated with phenomena (flares and eruptive prominences) which occur over lines separating regions of opposite polarities.

074.085 Dynamic modeling of coronal and interplanetary responses to solar events.
S. T. Wu, Y. Nakagawa, M. Dryer.
Study of travelling interplanetary phenomena 1977, (see 012. 042), p. 43 - 62 (1977).

Recent progress in the dynamic modeling of responses of the corona and interplanetary medium to solar events (such as surges, eruptive prominences, flares, etc.) is reviewed. In particular, coronal transients and wave phenomena are discussed in some detail including pertinent mathematical requirements. Within the context of hydrodynamics and magnetohydrodynamics, a summary of both one- and two-dimensional time-dependent models is presented.

074.086 Measurements of the solar wind using spacecraft radio scattering observations. R. Woo.
Study of travelling interplanetary phenomena 1977, (see 012. 042), p. 81 - 100 (1977).

This paper reviews radio scattering measurements of the solar wind carried out with coherent, monochromatic, and point-source spacecraft signals. The observed phenomena which include spectral and angular broadening, and phase as well as intensity scintillations, have provided measurements of the solar wind previously not available from radio astronomical observations. These cover a wide range of heliocentric distances (as close as 1.7 $R_\odot$), and large- as well as small-scale electron density fluctuations.

074.087 Theoretical contributions to solar wind research — a review. S. Cuperman.
Study of travelling interplanetary phenomena 1977, (see 012. 042), p. 165 - 194 (1977).

074.088 The large scale and long term evolution of the solar wind speed distribution and high speed streams.
D. S. Intriligator.
Study of travelling interplanetary phenomena 1977, (see 012. 042), p. 195 - 225 (1977).

The spatial and temporal evolution of the solar wind speed distribution and of high speed streams in the solar wind are examined. Comparisons of the solar wind streaming speeds measured at earth, Pioneer 11, and Pioneer 10 indicate that between 1 AU and 6.4 AU the solar wind speed distributions are narrower at extended heliocentric distances. These obser-

vations are consistent with the exchange of momentum in the solar wind between high speed streams and low speed streams as they propagate outward from the sun.

074.089 Pioneer 10, 11 observations of evolving solar wind streams and shocks beyond 1 AU.
E. J. Smith, J. H. Wolfe.
Study of travelling interplanetary phenomena 1977, (see 012. 042), p. 227 - 257 (1977).

Observations between 1 and 5 AU by Pioneers 10 and 11 have led to the identification of large numbers of interplanetary shocks. Both forward and reverse shocks, which begin to develop beyond 1.5 AU and which frequently appear as shock pairs, are found to accompany solar wind streams. The number of forward shocks continues to increase out to at least 5 AU. Reverse shocks are seen less often than forward shocks and, in some instances, disappear at larger distances. There is evidence that the shocks are corotating in the solar frame.

074.090 Wave-particle interaction phenomena associated with shocks in the solar wind. F. L. Scarf.
Study of travelling interplanetary phenomena 1977, (see 012. 042), p. 259 - 275 (1977).

Microscopic wave-particle interaction phenomena must generally affect the evolution of a travelling interplanetary discontinuity such as collisionless shock, and solar wind plasma instabilities should also be associated with interplanetary acceleration, diffusion, and dissipation. Recent local measurements from diagnostics on widely separated spacecraft illustrate some examples of these interaction phenomena, and two bounding cases are considered in detail here.

074.091 Solar energetic particles below 10 MeV.
E. C. Roelof, S. M. Krimigis.
Study of travelling interplanetary phenomena 1977, (see 012. 042), p. 343 - 365 (1977).

The information on solar acceleration and coronal propagation contained in low energy solar particle observations must be extracted from the effects of propagation in a dynamic interplanetary medium and the proximity of the Earth's magnetosphere. The resulting separation reveals long-lived coronal injection and strong spatial ordering of coronal propagation.

074.092 The theory of the polarization of emission lines in the solar corona. P. Charvin.
Rep. Obs. Lund, No. 12, (see 012.044), p. 51 - 52 (1977).

074.093 Observations of Fe XIII 10747 coronal emission-line polarization. C. W. Querfeld.
Rep. Obs. Lund, No. 12, (see 012.044), p. 109 - 136 (1977).

The direction, but not the magnitude, of the solar coronal magnetic field is encoded in the linear polarization of certain coronal emission lines. Polarization brightness maps made with the 10747 A line clearly show the large scale organization of the coronal magnetic field and signatures characteristic of diverging fields over sunspots, horizontal fields above filament channels, and the Van Vleck turnover in helmet streamers.

074.094 Polarization of the Fe XIV 5303 line measured in the corona between 1.1 $R_\odot$ and 1.4 $R_\odot$.
J. Arnaud.
Rep. Obs. Lund, No. 12, (see 012.044), p. 137 - 146 (1977).

074.095 Polarization of the solar corona on the total solar eclipses of February 15, 1961 and June 30, 1973.
A. Koeckelenbergh.
Rep. Obs. Lund, No. 12, (see 012.044), p. 185 - 196 (1977).

074.096 Possible polarization effects by thermal infrared emission in the solar corona. Y. Öhman.

Rep. Obs. Lund, No. 12, (see 012.044), p. 197 - 205 (1977).

A report is given of some laboratory experiments on the thermal emission of glowing iron flakes. Clear effects of polarization are found sometimes in flakes of small size indicating polarization of a kind similar to that appearing in the thermal emission from narrow metallic filaments. It seems possible that the infrared radiation of the solar corona may contain a faintly polarized component due to thermal emission from dust particles.

074.097 **Model of shockless deceleration of the solar wind.**
E. Ya. Gidalevich.
Geomagn. Aehron., Vol. 17, 976 - 983 (1977). In Russian.

074.098 **Solar wind parameters responsible for plasma injection into the magnetospheric ring-current region.**
M. S. Bobrov.
Astron. Zh. Akad. Nauk SSSR, Vol. 54, 1335 - 1345 (1977). In Russian. English translation in Soviet Astron., Vol. 21, No. 6.

The connection of the parameter q_0 characterizing the total energy of particles injected per hour into the region of the magnetospheric ring current with parameters of the plasma and magnetic field of the solar wind in the circumterrestrial space in front of a shock wave is studied statistically and by comparison of time variations. The data on 8 sporadic (solar-flare induced) geomagnetic storms of various intensity are used. It is found that q_0 correlates not only with the magnitude and the direction of the solar wind magnetic field component normal to the ecliptic plane, but also with the variability of the total magnetic field strength vector. The solar wind bulk velocity v influences the average storm intensity but the time variations of v during any individual storm do not correlate with those of q_0.

074.099 **Shape and structure of the corona at the solar eclipse of October 23, 1976.**
M. Waldmeier, S. E. Weber.
Astron. Mitt. Eidg. Sternw. Zürich, Nr. 353, 24 pp. (1977). In German.

074.100 **Koronale Löcher und das solare Magnetfeld.**
M. Stix.
Mitt. Astron. Ges., Nr. 42, p. 119 (1977).

074.101 **Transition region.**   C. Jordan.
Illustrated glossary for solar and solar-terrestrial physics, (see 003.022), p. 35 - 38 (1977).

074.102 **Solar corona.**   S. Koutchmy.
Illustrated glossary for solar and solar-terrestrial physics, (see 003.022), p. 39 - 52 (1977).

074.103 **Solar wind and interplanetary medium.**
L. Svalgaard.
Illustrated glossary for solar and solar-terrestrial physics, (see 003.022), p. 149 - 157 (1977).

074.104 **Noncompressive density enhancements in the solar wind.**   J. T. Gosling, E. Hildner, J. R. Asbridge, S. J. Bame, W. C. Feldman.
J. Geophys. Res., Vol. 82, 5005 - 5010 (1977).

The purpose here is to focus attention on the large density signals in the solar wind, called noncompressive density enhancements, to document their principal characteristics, and to suggest how they might originate in transient processes in the solar atmosphere.

074.105 **The importance of magnetic fields in formation of solar wind parameters.**
S. I. Molodykh, V. A. Kovalenko.
Simpoz. po fiz. geomagnitosfery, 1977. Tezisy dokl. Irkutsk,

1977, p. 3 - 4. In Russian. – Abstr. in Ref. zh., 51. Astron., 12.51.342 (1977).

074.106 **Energy distribution in coronal holes and high-velocity fluxes of the solar wind.**
V. A. Kovalenko, S. I. Molodykh.
Simpoz. po fiz. geomagnitosfery, 1977. Tezisy dokl. Irkutsk, 1977, p. 4. In Russian. – Abstr. in Ref. zh., 51. Astron., 12.51.343 (1977).

074.107 **Geometry of coronal holes and high-velocity fluxes of the solar wind.**   O. S. Popov.
Simpoz. po fiz. geomagnitosfery, 1977. Tezisy dokl. Irkutsk, 1977, p. 5. In Russian. – Abstr. in Ref. zh., 51. Astron., 12.51.344 (1977).

074.108 **Coronal streamers and three-dimensional structure of the interplanetary magnetic field.**
N. P. Korzhov.
Simpoz. po fiz. geomagnitosfery, 1977. Tezisy dokl. Irkutsk, 1977, p. 5 - 6. In Russian. – Abstr. in Ref. zh., 51. Astron., 12.51.345 (1977).

074.109 **Solar wind density and its genetic relation with solar activity on flocculi level.**
V. V. Kasinskij, N. N. Lyakhov.
Simpoz. po fiz. geomagnitosfery, 1977. Tezisy dokl. Irkutsk, 1977, p. 7 - 8. In Russian. – Abstr. in Ref. zh., 51. Astron., 12.51.347 (1977).

074.110 **Non-stationary input of mass in the solar corona and the solar wind.**   V. V. Kasinskij.
Simpoz. po fiz. geomagnitosfery, 1977. Tezisy dokl. Irkutsk, 1977, p. 8. In Russian. – Abstr. in Ref. zh., 51. Astron., 12.51.348 (1977).

074.111 **Solar wind study from cometary observations.**
D. A. Andrienko.
Simpoz. po fiz. geomagnitosfery, 1977. Tezisy dokl. Irkutsk, 1977, p. 13. In Russian. – Abstr. in Ref. zh., 51. Astron., 12.51.350 (1977).

074.112 **On the effect of rapid solar fluxes and strong electromagnetic radiation on the development of hose instability of the solar wind.**
V. N. Makarenko, A. K. Yukhimuk.
Simpoz. po fiz. geomagnitosfery, 1977. Tezisy dokl. Irkutsk, 1977, p. 9 - 10. In Russian. – Abstr. in Ref. zh., 51. Astron., 12.51.351 (1977).

074.113 **Development of astronomical investigations of solar wind velocity.**   N. A. Lotova, I. V. Chashej.
Tr. Fiz. inst. AN SSSR, Vol. 93, 78 - 118 (1977). In Russian. Abstr. in Ref. zh., 51. Astron., 12.51.357 (1977).

074.114 **Evolution of the large-scale structure of the solar wind at large heliocentric distances.**
V. N. Malyshkin.
Simpoz. po fiz. geomagnitosfery, 1977. Tezisy dokl. Irkutsk, 1977, p. 5. In Russian. – Abstr. in Ref. zh., 62. Issled. kosm. prostranstva, 12.62.133 (1977).

074.115 **Mode coupling in the solar corona. IV. Magnetohydrodynamic waves.**
D. B. Melrose, M. A. Simpson.
Australian J. Phys., Vol. 30, 495 - 507 (1977).

A general theory for coupling between MHD waves obliquely indicent on a stratified medium is developed. Coupling between the Alfvén mode and the magnetoacoustic mode (the fast mode for $v_A > c_s$ and the slow mode for $v_A < c_s$) is affected little by the finiteness of c_s/v_A for $v_A \gg c_s$, and the coupling becomes weaker as c_s/v_A is increased towards unity. Coupling between the fast and slow modes for $v_A \approx c_s$ is discussed qualitatively using solutions of the MHD counterpart of the Booker quartic equation.

074.116 Mode coupling in the solar corona. V. Reduction of the coupled equations.   D. B. Melrose.
Australian J. Phys., Vol. 30, 647 - 660 (1977).
A simplified version of the mode-coupling theory of Clemmow and Heading is developed by reducing the set of coupled equations to two for the magnetoionic theory and three for the MHD theory. The simplified theory reproduces known results for coupling in the neighbourhood of coupling points. It is used to treat coupling between the MHD waves, and it is found that coupling between the fast mode and the Alfvén mode for $v_A \lesssim c_s$ is stronger than the coupling between any other pair of modes. The strongest coupling of all is between the Alfvén and slow (magnetoacoustic) modes for $v_A \ll c_s$.

074.117 Nonstationary interaction between solar wind and interstellar matter.   R. Ratkiewicz.
Postępy Astron., Tom 25, 257 - 261 (1977). In Polish.
Remote solar wind deviations from spherical symmetry induced by interaction with neutral interstellar hydrogen (photoionization and charge-exchange) were studied by a perturbation technique. Solar activity induced variations in the ionization rate of neutral galactic gas inside the heliosphere influence the deceleration rate of the solar wind and cause temperature variations of 10–20% at the orbits of Saturn – Neptune.

074.118 On the alignment of plasma anisotropies and the magnetic field direction in the solar wind.
J. R. Asbridge, S. J. Bame, W. C. Feldman, J. T. Gosling, N. F. Ness.
J. Geophys. Res., Vol. 82, 5555 - 5562 (1977).
One year's Imp 6 solar wind plasma and magnetic field data are examined to determine whether anisotropies in particle velocity distributions are aligned with the measured interplanetary magnetic field vector. By assuming cylindrical symmetry about the simultaneously measured magnetic field vector during the 1-year interval under study, three-dimensional values of selected solar wind plasma thermal parameters were constructed from the two-dimensional plasma measurements, and the statistical properties of their distributions have been tabulated.

074.119 The three dimensional solar corona. A coronal streamer.   D. C. Wilson.
Thesis Natl. Cent. Atmos. Res., Boulder, Colo., USA. NCAR-CT–40. 292 pp. (1977). – Abstr. from INIS7724 345443.

074.120 Two-component dust models of infrared emission from a circumsolar dust cloud.
T. Mukai, T. Yamamoto.
10th Lunar and Planetary Symposium, (see 012.050), p. 72 - 74 (1977).
Based on a model of a circumsolar dust cloud consisting of both graphite and silicate (obsidian) materials at 4 solar radii ($R_\odot$) from the sun, the authors can explain the observed features (energy spectrum and the flux 'bump' near $4\,R_\odot$) in the wavelength range from the near to the middle infrared in the inner F corona.

074.121 Plasma physics investigations of the solar wind.   S. Cuperman, B. Levush.
Bull. Israel Phys. Soc., Vol. 23, 109 (1977). – Summary.
Abstr. from INIS7721 336215.

074.122 Cooling of the solar wind protons. Mariner 2 data.   M. Eyni, R. Steinitz.
Bull. Israel Phys. Soc., Vol. 23, 110 - 111 (1977). – Summary.

Abstr. from INIS7721 336224.

New numerical methods in mathematical physics and problems of interaction between the solar wind and cosmic objects.   See Abstr. 011.025.

Solar-wind interaction with the planets Mercury, Venus, and Mars.   See Abstr. 012.047.

A program to calculate coronal emission line strengths.   See Abstr. 021.005.

Computational techniques for solar wind flows past terrestrial planets – theory and computer programs. See Abstr. 021.021.

Coronal electron temperatures as measured by X-ray photometer.   See Abstr. 031.404.

Plasma analyzer for the Pioneer Jupiter missions. See Abstr. 032.576.

On the connection of irregular variations of the daily rotation of the earth and solar wind velocity. See Abstr. 044.011.

Overview of the Helios 1 and Helios 2 missions and their participation in STIP Intervals I and II. See Abstr. 051.069.

First and second order approximations of the first adiabatic invariant for a charged particle interacting with a linearly polarized hydromagnetic plane wave. See Abstr. 062.002.

Dynamics of a cloud of fast electrons travelling through the plasma. II. Semi-analytical approach. See Abstr. 062.042.

Alfvén instabilities in streaming plasmas with anisotropic pressures and their relevance for the solar wind. See Abstr. 062.045.

A search for a turbulent-free region in the solar transition zone.   See Abstr. 071.006.

A survey of soft X-ray limb flare images: the relation between their structure in the corona and other physical parameters.   See Abstr. 073.009.

Morphology and physical parameters of a solar flare. See Abstr. 073.011.

X-ray spectrum of a coronal condensation and a flare.   See Abstr. 073.030.

Magnetic fields in solar prominences. See Abstr. 073.031.

Mass and energy flow in the solar chromosphere and corona.   See Abstr. 073.050.

Prominence mass ejections and their effects on the corona. I: The eruptive prominence of 21 August 1973 and the surge of 4 December 1973.   See Abstr. 073.069.

Kinematic model of loop prominences formation. See Abstr. 073.091.

Coronal mass-ejections – kinematics of the 19 December 1973 event.   See Abstr. 073.108.

Formation of shock waves from chromospheric flares and their evolution in the solar wind. See Abstr. 073.119.

Dynamical implications of Si IV line profiles from OSO-8 observations.   See Abstr. 076.016.

Second harmonic radiation and related nonlinear phenomena in type III solar radio bursts. See Abstr. 077.005.

On the nature of quasi-periodic fluctuations of solar radio flux at centimeter wavelengths. See Abstr. 077.024.

A new scattering process above solar active regions: propagation in a fibrous medium.   See Abstr. 077.027.

Type II emission mechanism.   See Abstr. 077.028.

Solar radio type III bursts and coronal density structures.   See Abstr. 077.030.

Coronal X-ray holes and the quiet radio Sun at 2800 MHz.   See Abstr. 077.040.

Radio and soft X-ray evidence for dense non-potential magnetic flux tubes in the solar corona. See Abstr. 077.042.

Solar stereo radioastronomy. See Abstr. 077.048.

Analysis of the complex solar particle event on April 29 - 30, 1973.   See Abstr. 078.013.

Solar particle propagation from 1 to 5 AU. See Abstr. 078.014.

Total brightness of the corona at the solar eclipse of October 23, 1976.   See Abstr. 079.404.

The influence of the chromosphere and corona on the solar atmospheric oscillations.   See Abstr. 080.006.

Evolution of open magnetic structures on the Sun: the Skylab period.   See Abstr. 080.026.

Aeronomic aspects of the polar D-region. See Abstr. 083.108.

Impulsive penetration of filamentary plasma elements into the magnetospheres of the Earth and Jupiter. See Abstr. 084.203.

The solar proton event of April 16, 1970. 2. Transformation of proton pitch angle distributions from the solar wind into the magnetosheath.   See Abstr. 084.220.

Long-lived active formations on the solar surface and activity of geomagnetic pulsations of Pc3-type. See Abstr. 084.221.

Evidence for the control of Pc 3, 4 magnetic pulsations by the solar wind velocity.   See Abstr. 084.244.

Nonstationary interaction between solar wind discontinuities and the system leading wave — earth's magnetosphere.   See Abstr. 084.313.

A summary of significant solar-initiated events during STIP Intervals I and II.   See Abstr. 085.038.

Production of simple molecules on the surface of Mercury.   See Abstr. 092.043.

Viscous flow in the near-Venusian plasma wake. See Abstr. 093.003.

Carrying away of ions of the Venus atmosphere by the solar wind.   See Abstr. 093.061.

The interaction of the solar wind with lunar magnetic anomalies.   See Abstr. 094.105.

Photometric studies of light scattering above the lunar terminator from Apollo solar corona photography. See Abstr. 094.131.

Some evidence of the possibility of a strong interaction between the Martian satellite Deimos and the solar wind.   See Abstr. 097.507.

Possible origins of time variability in Jupiter's outer magnetosphere. I. Variations in solar wind dynamic pressure. See Abstr. 099.062.

Possible origins of time variability in Jupiter's outer magnetosphere. 2. Variations in solar wind magnetic field. See Abstr. 099.063.

Comets in the STIP context. See Abstr. 102.033.

The comet-solar wind interaction. See Abstr. 102.034.

Solar wind interaction with type-1 comet tails. See Abstr. 102.035.

Influence of solar wind upon the drift of meteor trains.   See Abstr. 104.029.

Slow shocks in the interplanetary medium. See Abstr. 106.009.

The structure of the solar flare stream magnetic field. See Abstr. 106.031.

Large-scale three-dimensional structure of the interplanetary magnetic field.   See Abstr. 106.033.

Observations of interplanetary scintillation: solar wind velocity measurements.   See Abstr. 106.034.

Scintillation observations of the interplanetary plasma.   See Abstr. 106.035.

The three-dimensional structure of the interplanetary medium.   See Abstr. 106.036.

Influence of temporal variations of the solar wind velocity on the structure of the interplanetary magnetic field. See Abstr. 106.046.

The influence of the interplanetary magnetic field on the parameters of the flow in the transition layer. See Abstr. 106.048.

The resistive effect of the solar wind on the orbits of solid bodies in the solar system.   See Abstr. 107.003.

Diffusion of cosmic rays in the solar wind deformed by the interstellar medium.   See Abstr. 143.067.

Propagation of galactic cosmic rays in the spherically symmetric solar wind.   See Abstr. 143.068.

Connection of the relative coefficient of cosmic ray modulation with solar coronal radiation.
See Abstr. 143.073.

Errata

074.901 Erratum: "The heating of the solar corona. II. A model based on energy balance" [Astron. Astrophys., Vol. 40, 63 - 73 (1975)].
R. W. P. McWhirter, P. C. Thonemann, R. Wilson.
Astron. Astrophys., Vol. 61, 859 - 861 (1977). – Correction of many numerical values given in the original paper (see 013. 074.087).

075 Solar Patrol

075.001 Solar activity during 1976.
  V. Barocas, K. J. Medway.
J. British Astron. Assoc., Vol. 88, 56 - 63 (1977).– Report of the Solar Section.

075.002 Datos relativos a la actividad solar y geomagnética en 1975.
Urania Barcelona, Año 61, Núm. 285, p. 98 (1976).

075.003 L'activité solaire.   M.-J. Martres.
  Astronomie, Vol. 91, 370 - 371, 419, 463, 510 - 511 (1977). – Rotations 1650 - 1655.

075.004 The sun in 1976.   B. E. Stonehouse.
  South. Stars, Vol. 27, 63 - 64 (1977).

075.005 Solar activity.   F. G. Mustaeva, V. M. Tishchenko.
  Tsirk. Astron. Inst., Tashkent, Nos. 64 (411), 68 (415), 70 (417), 71 (418) (1976); 72 (419) (1977). In Russian. – 1976, January - October.

075.006 Provisional sunspot-numbers for 1977 May - November.
Yamamoto Circ., Nos. 1854, 1855, 1858, 1861, 1864, 1868, 1871 (1977). In Japanese.

075.007 Solar observations made at Catania Astrophysical Observatory during 1976.   G. Godoli.
Oss. Astrofis. Catania, Pubbl. Nuova Ser. No. 159, 70 pp. (1977).
  Contents: Sunspots; Hα and K faculae; Hα flares; Hα quiescent prominences; K quiescent prominences; Hα active prominences on disc and at limb; Hα disc and limb patrol hours.

075.008 Solare Beobachtungsergebnisse. Solar data. Solar radio emission.   C.-U. Wagner, A. Böhme, F. Fürstenberg, D. Scholz, S. Böhm.
Zentralinst. Solar-Terr. Phys. (Heinrich-Hertz-Inst.), Akad. Wiss. DDR, HHI Solar Data, Vol. 28, 25 - 73 (1977). – 1977 April - September.

075.009 Photoheliographic results 1967.
  Compiled under the supervision of P. S. Laurie, A. L. T. Powell, C. Y. Hohenkerk, C. M. Ladley, P. J. Rudd.
R. Obs. Ann., No. 11, 114 pp. Price £ 2.65 (1975).
  The following data are given: Positions and areas of sunspots for each day in the year; General catalogue of groups of sunspots; Total areas of sunspots and faculae; Mean areas of sunspots and faculae; Mean heliographic latitude of sunspots; Summary of solar activity.

075.010 Solar observations. Nos. 43 - 45, 1976 July - 1977

March.   Solar Group.
Oss. Astron. Trieste, Pubbl. Nos. 535, 542, 551 (1977).

075.011 Daily Hα chromosphere pictures, daily K_{232} chromosphere pictures, daily white light photosphere pictures.   M. Cimino (Editor).
Photographic Journal of the Sun, Oss. Astron. Roma, Nos. 121 - 124 (1976). – 1976 December 4 - 1977 March 23. Solar rotation 1649 - 1652.

075.012 Solar phenomena.   M. Cimino, M. Torelli.
  Oss. Astron. Roma, Mon. Bull. N. 225 - 232 (1977).
1977 January - August: Daily total areas of sunspot-groups; Heliographic position, classification and area of sunspot-groups; Hours of K-line cinematographic patrol; Hours of Hα cinematographic patrol; Explanations.

075.013 Definitive Sonnenflecken-Zahlen für 1976.
  M. Waldmeier.
Sterne, 53. Band, 241 (1977).

075.014 Sunspots (sunspot relative-numbers and sunspot-areas); Synoptic charts of solar magnetic fields (Mount Wilson Observatory); Eruptions chromosphériques brillantes; Intensité de la couronne solaire; Solar radio emission.   M. Waldmeier, R. Howard, P. Simon, H. Tanaka.
Q. Bull. Sol. Act., Nos. 195 - 196, p. 233 - 308 (1977).
1976 July - December.

075.015 Heliographic maps of the photosphere for the year 1976.   M. Waldmeier.
Publ. Eidg. Sternw. Zürich, Band 15, (Heft 1), 1 - 31 (1977). Rotations 1636 - 1649.

075.016 Sunspot relative numbers for 1976.
  M. Waldmeier.
Astron. Mitt. Eidg. Sternw. Zürich, Nr. 350, 10 pp. (1977).

075.017 Die Sonnenaktivität im Jahre 1976.
  M. Waldmeier.
Vierteljahrsschr. Naturforsch. Ges. Zürich, Jahrg. 122, 233 - 247 (1977) = Astron. Mitt. Eidg. Sternw. Zürich, Nr. 354.

075.018 Statistik der Sonneneruptionen 1966–1975.
  A. Pittini.
Astron. Mitt. Eidg. Sternw. Zürich, Nr. 357, 15 pp. (1977).

075.019 Observations radioélectriques solaires faites sur 600 MHz en 1975 au laboratoire de radioastronomie de Humain-Rochefort.   C. Gonze, R. Gonze.
Bull. Astron., Obs. R. Belgique, Vol. 8, 316 - 324 (1976).
  Densité de flux, variabilité; Evénements remarquables; Moyennes journalières.

075.020   **Observation of the sun.**   V. Čelebonović.
Vasiona, Vol. 25, 61 - 66 (1977). In Serbo-
Croatian.

075.021   **Daily maps of the sun and geophysical graphs.**
Soln. Dannye 1977 Byull., No. 5, p. 1 - 46; No. 6,
p. 1 - 56; No. 9, p. 1 - 60; No. 10, p. 1 - 61. In Russian.

075.022   **Magnetic fields of sunspots.**
Prilozhenie k Byulletenyu "Soln. Dannye", 1977,
No. 5, 6, 9, 10. In Russian.

075.023   **Solar and solar system activity.** Radio astronomy
section (BAA).   J. R. Smith, R. J. J. Langton.

J. British Astron. Assoc., Vol. 87, 624 - 629, Vol. 88, 70 - 72
(1977).

075.024   **Solar activity.** Solar Section (BAA).
K. J. Medway, H. Hill.
J. British Astron. Assoc., Vol. 87, 493 - 498, 597 - 602, Vol.
88, 64 - 69 (1977).

075.025   **Sunspot numbers.**
Sky Telesc., Vol. 54, 157, 252, 348, 444, 532
(1977). − 1977 May - September.

075.026   **Geomagnetic and solar data.**   J. V. Lincoln.
J. Geophys. Res., Vol. 82, 2893, 3342, 3662, 4843,
5292, 5646 (1977).

076 UV, X, Gamma Radiation

076.001   **The exponential character of the dependence of the
differential emission measure on the temperature
and the structure of a coronal active region.**
M. A. Livshits, N. N. Bulatov.
Astron. Zh. Akad. Nauk SSSR, Vol. 54, 841 - 845 (1977).
In Russian. English translation in Soviet Astron., Vol. 21,
No. 4.
The temperature distribution in a non-flare X-ray source
is deduced under the assumption of a barometric decrease of
the electron density with constant scale height. On the basis
of X-ray observations of Walker et al. (1974), a model is constructed, and physical processes leading to a considerable
increase of density in the highest-temperature filaments are
briefly discussed.

076.002   **Bright X-ray arcs and the emergence of solar magnetic flux.**   G. A. Chapman, R. M. Broussard.
Astrophys. J., Vol. 216, 940 - 946 (1977).
As part of a study of the role of solar magnetic flux
emergence and X-ray brightness, the authors present observations of McMath region 12476 during 1973 August 9–11.
They find that X-ray images of this active region consist of a
system of bright arcs embedded in a diffuse emitting region.
The arc system varies in brightness and shape over more than
a 24 hour period but maintains a fixed orientation. The
temperature in the main arc is found to be about 3×10^6 K.
The energy loss rate from the region is sufficiently great to
preclude resistive heating from a force-free current as the sole
source of energy. The main source of energy may be from
merging magnetic fields.

076.003   **Solar flux estimated from electron density and ion
composition measurements in the lower thermosphere.** P. Chakrabarty, D. K. Chakrabarty, A. K. Saha.
J. Geophys. Res., Vol. 82, 3299 - 3303 (1977).
Appropriate models of solar flux in X rays and extreme
ultraviolet bands are presented in the light of the present
status of ion chemistry in the region 90–130 km and of reliable measurements of reaction rates, electron density, and
ion composition.

076.004   **Energy-spectrum distribution of the X-radiation
under 8 Å and its relation with the solar radio emission on 2800 MHz.**   E. M. Apostolov.
Solar activity and solar-terrestrial relations, (see 012.007), p.
155 (1976). − Abstract. (See 012.076.006).

076.005   **Solar X-ray bursts in June - July 1973.**
N. V. Illarionova, N. I. Nazarova, V. M. Pankov,
I. N. Rozantsev, I. A. Savenko.
Soln. Dannye 1977 Byull., No. 5, p. 83 - 86 (1977). In
Russian.
Observations of solar X-ray bursts of 5−10 keV by
Prognoz 3 during June - July 1973 are given. These measure-
ments are discussed in connection with different phenomena
of solar activity.

076.006   **The solar spectrum in the vicinity of the Si IV lines
at 1122 and 1128 Å.**
U. Feldman, G. A. Doschek.
Astron. Astrophys., Vol. 61, 295 - 296 (1977).
The extreme ultraviolet solar spectrum in the vicinity of
the Si IV lines at 1122 and 1128 Å is presented with a wavelength resolution of 0.06 Å. The Si IV line at 1122.486 Å is
blended with an unresolved line of Fe III at 1122.526 Å. The
Si IV line at 1128.340 Å is near two faint Fe III lines.

076.007   **Etude statistique des relations entre la radiation X
solaire, les renforcements soudains d'atmosphériques
et les plages d'activité solaire.**   B. Leroy.
Ciel Terre, Vol. 93, 215 - 223 (1977).
Une étude statistique basée sur des perturbations iono-
sphériques d'origine solaire, le rayonnement X solaire et les
plages solaires actives, pendant l'année 1970, met en évidence
des relations entre différents paramètres descriptifs de ces
phénomènes directs et indirects d'activité solaire.

076.008   **Development of compound X-ray events on the sun.**
O. M. Kovrizhnykh, M. I. Kudryavtsev, A. S.
Melioranskij, I. A. Savenko, V. M. Shamolin, L. M. Chupova.
Kosm. Issled., Vol. 15, 736 - 740 (1977). In Russian.

076.009   **Study of solar radiation in the far ultraviolet spectral
region.**   Eh. E. Dubov.
Problems of cosmic physics. Vyp. (No.) 12, (see 003.016),
p. 27 - 41 (1977). In Russian.
The main results of a study of spectra and spectrograms
in the solar far ultraviolet radiation (300–3000 Å) are given.
New phenomena and physical conditions in the chromosphere,
transition layer, inner corona, in active regions and solar flares
are discussed briefly. Models of these formations are discussed.

076.010   **EUV flux variation during end of solar Cycle 20 and
beginning Cycle 21, observed from AE-C satellite.**

H. E. Hinteregger.
Geophys. Res. Lett., Vol. 4, 231 - 234 (1977).

076.011   Neutrons and gamma rays from the sun.
          P. J. Lavakare.
Bull. Astron. Soc. India, Vol. 5, 64 - 68 (1977).

076.012   Collective plasma effects and the electron number
          problem in solar hard X-ray bursts.
J. C. Brown, D. B. Melrose.
Sol. Phys., Vol. 52, 117 - 131 (1977).
    Due to the relatively high stream densities involved, col-
lective interactions with the ambient plasma are likely to be
important for the electrons producing solar hard X-ray bursts.
In thick- and thin-target bremsstrahlung models the most
relevant process is limitation of the invoked electron beams
by ion sound wave generation in the neutralizing reverse
current established in the atmosphere. In this paper the authors
attempt to evaluate three of these possible configurations —
thick- and thin-target and electron confinement (i.e., those
involving continuous primary acceleration of electrons) in a
manner more consistent with constraints from plasma collective
processes.

076.013   Magnetic properties of X-ray bright points.
          L. Golub, A. S. Krieger, J. W. Harvey, G. S. Vaiana.
Sol. Phys., Vol. 53, 111 - 121 (1977).
    Using high resolution KPNO magnetograms and sequences
of simultaneous S-054 soft X-ray solar images the authors have
compared the properties of X-ray bright points (XBP) and
ephemeral active regions (ER). All XBP appear on the magnetograms as bipolar features, except for very newly emerged or old
and decayed XBP. The authors find that the separation of the
magnetic bipoles increases with the age of the XBP, with an
average emergence growth rate of 2.2±0.4 km s^{-1}. The total
magnetic flux in a typical XBP living about 8 hr is found to be
$\approx 2 \times 10^{19}$ Mx. A proportionality is found between XBP lifetime and total magnetic flux, equivalent to $\approx 10^{20}$ Mx per day
of lifetime.

076.014   The analysis and interpretation of solar X-ray
          photographs.   J. H. Underwood, D. L. McKenzie.
Sol. Phys., Vol. 53, 417 - 433 (1977).
    This paper discusses the methods used to analyse and
interpret X-ray filtergrams obtained by solar soft X-ray telescopes such as the S-056 Skylab instrument. First, an appropriate definition of the line-of-sight emission measure $\xi_L(T)$ is
developed, and it is shown how the X-ray data may be analysed to obtain an approximation to $\xi_L(T)$. The use of a
specific model is required for the proper interpretation of the
results in terms of coronal plasma processes. Examples of such
models are provided. The 'filter ratio method' is discussed.

076.015   Hard X-rays of solar flares on August 2 and 11,
          1972.   O. M. Kovrizhnykh, M. I. Kudryavtsev,
A. S. Melioranskij, I. A. Savenko, L. M. Chupova, V. M.
Shamolin.
Kosm. Issled., Vol. 15, 895 - 900 (1977). In Russian.

076.016   Dynamical implications of Si IV line profiles from
          OSO-8 observations.
D. E. Billings, R. Roussel-Dupré, M. H. Francis.
Sol. Phys., Vol. 55, 287 - 303 (1977).
    The widths and rms variations of line centers for $\lambda 1393$
profiles from the Si IV ion in the solar transition zone, as
observed by OSO-8, are analyzed to give the amplitudes and
periods of three postulated types of disturbance: sinusoidal
acoustic waves; sinusoidal acoustic shocks; and normally
propagated MHD waves. All three assumptions lead to mean
intervals between disturbances of 40–50 s. MHD disturbances
in magnetic fields ~ 2 G are consistent with observations.
Disturbances with wave vectors at an angle to the magnetic

field are suggested.

076.017   Solar EUV emission line profiles of Si II and Si III
          and their center to limb variations.
K. R. Nicolas, G. E. Brueckner, R. Tousey, D. A. Tripp, O. R.
White, R. G. Athay.
Sol. Phys., Vol. 55, 305 - 319 (1977).
    Spectral line profiles of Si II and Si III are presented
which were observed both at solar center and near the quiet
solar limb with the Naval Research Laboratory EUV spectrograph of ATM/SKYLAB. Absolute intensities and line profiles
are derived from the photographic data. A brief discussion is
given of their center-to-limb variations and of the optical
thickness of the chromosphere in these lines. Nonthermal
broadening velocities are found for the optically thin lines
from their full width at half maximum intensity.

076.018   Open magnetic fields in active regions.
          Z. Švestka, C. V. Solodyna, R. Howard, R. H. Levine.
Sol. Phys., Vol. 55, 359 - 369 (1977).
    Soft X-ray observations confirm that some of the dark
gaps seen between interconnecting loops and inner cores of
active regions may be loci of open fields, as it has been predicted by global potential extrapolation of photospheric magnetic fields. It seems that the field lines may open only in a
later state of the active region development.

076.019   Observations of resonance-line polarization in the
          solar EUV.   J. O. Stenflo, D. Dravins, Y. Öhman,
N. Wihlborg, A. Bruns, V. K. Prokof'ev, A. Severny (Severnyj),
I. A. Zhitnik, H. Biverot, L. Stenmark.
Rep. Obs. Lund, No. 12, (see 012.044), p. 147 - 160 (1977).
    An instrument that records resonance-line polarization in
the region 120–150 nm of the solar spectrum was launched
on the satellite Intercosmos 16 on July 27, 1976. The design
of the experiment is described. The telemetry recordings are
in the process of being reduced, but some partially reduced
spectra are presented and discussed.

076.020   On solar gamma-radiation of discrete energies.
          A. B. Bajsakalova, E. V. Kolomeets.
Prikl. i teor. fiz. Vyp. (No.) 8. Alma-Ata, 1976, p. 220 - 225.
In Russian. — Abstr. in Ref. zh., 51. Astron., 12.51.391 (1977).

          Applications of beam-foil spectroscopy to the solar
ultraviolet emission spectrum.   See Abstr. 031.280.

          The S-054 X-ray telescope experiment on Skylab.
See Abstr. 032.566.

          Lines of H_2 in extreme-ultraviolet solar spectra.
See Abstr. 071.020.

          Emission measures and structure of the transition
region of a sunspot from emission lines in the far ultraviolet.
See Abstr. 072.039.

          A survey of soft X-ray limb flare images: the rela-
tion between their structure in the corona and other physical
parameters.   See Abstr. 073.009.

          EUV observations of class-C X-ray flare by the
LPSP (*Laboratoire de Physique Stellaire et Planétaire du
Centre National de la Recherche Scientifique*) spectrometer
on OSO-8.   See Abstr. 073.012.

          On the pre-maximum phase of the solar event of
August 7, 1972.   See Abstr. 073.019.

          The verification of Chung-chieh-Cheng ideas on the
heating mechanism of solar flares.   See Abstr. 073.028.

Development of the proton flare and the associated hard X-ray emission of November 5, 1970. See Abstr. 073.029.

X-ray spectrum of a coronal condensation and a flare.   See Abstr. 073.030.

Structure and dynamics of a solar flare: X-ray and XUV observations.   See Abstr. 073.037.

Hα macrospicules: identification with EUV macrospicules and with flares in X-ray bright points. See Abstr. 073.043.

Iron-line X-ray emission from solar plasma: comments on ionization equilibrium and line excitation. See Abstr. 073.057.

Transequatorial loops interconnecting McMath regions 12472 and 12474.   See Abstr. 073.062.

A hard X-ray observation of a solar flare with 100 ms time resolution.   See Abstr. 073.064.

Prominence mass ejections and their effects on the corona. I: The eruptive prominence of 21 August 1973 and the surge of 4 December 1973.   See Abstr. 073.069.

A model for X-ray emission from loop prominences. See Abstr. 073.071.

Observations of limb flares with a soft X-ray telescope.   See Abstr. 073.072.

Comments on Salyut-4 observations of active regions on the Sun.   See Abstr. 073.082.

Early evolution of an X-ray emitting solar active region.   See Abstr. 073.100.

Chromospheric oscillations observed in the line C II λ1336 with OSO-8.   See Abstr. 073.101.

Ultraviolet brightenings in active regions as observed from OSO-8.   See Abstr. 073.103.

Current sheets as the source of heating for solar active regions.   See Abstr. 073.104.

Analysis of X radiation of coronal condensations

observed with broad-band filters.   See Abstr. 074.017.

Analysis of formation of resonance lines in the X-ray spectrum of coronal condensations.   See Abstr. 074.018.

A long-lived coronal arch system observed in X-rays. See Abstr. 074.045.

Radio and EUV observations of a coronal hole. See Abstr. 074.050.

EUV analysis of polar plumes. See Abstr. 074.057.

Coronal He$^+$ λ304 radiation. See Abstr. 074.058.

On coronal Fe abundances and temperatures from XUV emission lines.   See Abstr. 074.059.

Do surges heat the corona? See Abstr. 074.063.

Observations of the birth of a small coronal hole. See Abstr. 074.066.

Soft X-ray observations of large-scale coronal active region brightenings.   See Abstr. 074.071.

Do changes in coronal emission structure imply magnetic reconnection?   See Abstr. 074.081.

A two-component model of impulsive microwave burst emission consistent with soft and hard X-rays. See Abstr. 077.032.

Coronal X-ray holes and the quiet radio Sun at 2800 MHz.   See Abstr. 077.040.

Radio and soft X-ray evidence for dense non-potential magnetic flux tubes in the solar corona. See Abstr. 077.042.

Computer solutions for studying correlations between solar magnetic fields and Skylab X-ray observations. See Abstr. 080.062.

The quantitative relationship between VLF phase deviations and 1−8 Å solar X-ray fluxes during solar flares. See Abstr. 083.031.

077 Radio, Infrared Radiation

077.001 Centre-limb variations of characteristics of type I solar radio bursts. J. L. Bougeret.
Astron. Astrophys., Vol. 60, 131 - 138 (1977).
The author reviews the concept of type I sources: the type I active region is reduced to a limited number of sources which emit homologous bursts. Using this concept, the author shows that the duration of individual type I bursts tends to increase for sources near the limbs, while no centre-limb effect is found for the diameters.

077.002 Radio emission of an intense active region on the sun in July 1974 at wavelengths 3.5, 2.5 and 1.9 cm.
A. F. Bachurin, A. S. Dvoryashin, N. N. Eryushev.
Izv. Krymskoj Astrofiz. Obs., Vol. 57, 156 - 168 (1977). In Russian.
The radio emission of the local source on the sun associated with the sunspot group N 96 (McMath region 13043) from July 1 to July 7, 1974 at wavelengths 3.5, 2.5 and 1.9 cm is described. For the high activity period of the region, besides bursts, long gradual radio emission variations are observed. These variations have a mode with mean period of 100 min superposed by constant fluctuations with significant shorter time scale. Short weak bursts for which the peak flux density increases with decreasing wavelength are found. Brief characteristics of the strong burst associated with the proton flare of July 4, 1974 are given.

077.003 Diffraction and reflection of radio waves from the limb of the moon during solar eclipses.
O. Alvares, Yu. F. Yurovskij.
Izv. Krymskoj Astrofiz. Obs., Vol. 57, 169 - 176 (1977). In Russian.
The conditions for formation of a diffraction pattern on the 10-cm wave during solar eclipses have been calculated. In the September 22, 1968 and December 24, 1973 eclipses good agreement between the records and the theoretical curve of the calculated Fresnel diffraction patterns for each case was obtained. It might be concluded that in the quoted time on the sun's disc there were sources with angular dimensions between 2 and 3 seconds of arc and effective temperatures about 5×10^8 K. It is shown that the presence of a maximum at the end of the source emersion might be satisfactorily explained by refraction of the radio waves in a plasma layer near the moon with height about 12 km over the surface and electron concentration about 10^3 cm^{-3} at the lowest level.

077.004 On a method of polarization observations of solar radio emission at the 1.9 - 3.5 cm wavelength band.
L. I. Tsvetkov.
Izv. Krymskoj Astrofiz. Obs., Vol. 57, 189 - 198 (1977). In Russian.
Some problems of methods of observations of circularly and linearly polarized components of solar radio emission at 1.9, 2.5 and 3.5 cm wavelength are considered. The results of investigation of parasitic signals are discussed. A method of partial compensation of these signals is given. A possibility of increasing the number of wavelengths of the observed spectrum in the 1.9 - 3.5 cm band is indicated.

077.005 Second harmonic radiation and related nonlinear phenomena in type III solar radio bursts.
D. F. Smith.
Astrophys. J., Lett., Vol. 216, L53 - L57 (1977).
Observations of type III radio bursts and their accompanying plasma waves in the solar wind are analyzed, and shown to provide no definitive support for coherent parametric processes in these bursts. It is shown that the incoherent weak turbulent process of induced scattering on the polariza-

tion clouds of ions is sufficient to produce a quasi-isotropic spectrum of plasma waves. This in turn leads to emission near the second harmonic of the plasma frequency.

077.006 Night-time reception of a solar radio event.
J. J. Riihimaa.
Nature, Vol. 263, 397 (1976).

077.007 On peculiar quasi-periodic components and the possible structure of the generating region of the type IV event of August 4, 1972.
S. T. Akinyan (Akin'yan), V. N. Ishkov, E. I. Mogilevskii (Eh. I. Mogilevskij), A. Böhme, F. Fürstenberg, A. Krüger.
Solar activity and solar-terrestrial relations, (see 012.007), p. 35 - 46 (1976).
Based on a detailed study of H-alpha filtergrams, photospheric magnetic field measurements, and radio flux and polarization measurements, a discussion of the physical conditions of the proton flare on 1972, August 4 is made. A possible model for the occurrence of a low frequency modulation of the radio emission especially on decameter waves is proposed.

077.008 Observations of the solar radio emission at IZMIRAN during the proton flare of August 4, 1972. S. T. Akinyan (Akin'yan), G. P. Chernov, I. M. Chertok, V. V. Fomichev, A. M. Karachun, V. A. Kovalev, A. K. Markeev.
Solar activity and solar-terrestrial relations, (see 012.007), p. 47 - 51 (1976).
The radio data obtained at IZMIRAN concerning the multicomponent type IV burst, connected with the proton flare of August 4, 1972 , are presented.

077.009 Low-frequency solar radio bursts observed with the "Intercosmos-Kopernik 500" satellite.
J. Hanasz, R. Schreiber, H. Welnowski, B. Wikierski, V. I. Aksenov.
Solar activity and solar-terrestrial relations, (see 012.007), p. 157 - 159 (1976).
The report contains a short description of the first Soviet-Polish space experiment, which has been accomplished on board of the "Intercosmos-Kopernik 500" satellite. The purpose of the experiment among others is to observe sporadic radio emissons generated in the coronal and interplanetary medium at hectometric wave-lengths range.

077.010 Low-frequency modulation spectrum of radio emission as main feature of evolution of active regions on the sun. O. G. Gontarev, Eh. I. Mogilevskij.
Solar activity and solar-terrestrial relations, (see 012.007), p. 265 - 273 (1976). In Russian.

077.011 Investigation of fluctuations of solar radio emission and of sunspot areas. M. Jakimiec.
Solar activity and solar-terrestrial relations, (see 012.007), p. 275 - 279 (1976). In Russian.
The analysis concerns the solar activity fluctuations of a duration longer than 27 days. In previous papers of the author it has been shown that the fluctuations can be represented by means of a stationary stochastic process. In the present investigation the analysis is extended to the 2800 MHz radio flux measurements.

077.012 Some properties of active regions on 9 and 21 cm.
J. Kleczek.
Solar activity and solar-terrestrial relations, (see 012.007), p. 281 - 283 (1976).

The contribution of active regions to the total radio flux of the sun at λ 9 cm is deduced from radioheliograms. The relative increase of the 9 cm flux with the size of the region is larger than for the 21 cm flux which may explain the spectral peak of the slowly varying component in the maximum of solar activity.

077.013   Evolution of active regions on 21 and 9 cm.
          J. Olmr.
Solar activity and solar-terrestrial relations, (see 012.007), p. 285 - 286 (1976).
          The author studies the evolution of large active regions in radio fluxes. The resulting mean curve of evolution is compared with the spot area and relative number of active regions. The correlation of the fluxes of an active region with its spot area and mean relative number is good.

077.014   Polarization measurements and models of solar microwave bursts.   A. Krüger.
Solar activity and solar-terrestrial relations, (see 012.007), p. 287 - 290 (1976).
          Based on results of statistical investigations of the polarization characteristics of solar microwave bursts and individual type IV bursts, interpretation facilities in terms of some most probable mechanisms for producing circularly polarized radiation are discussed. Some consequences on the properties of flare models are taken into consideration.

077.015   The spectrum and polarization of noise storms.
          A. Böhme.
Solar activity and solar-terrestrial relations, (see 012.007), p. 291 (1976). − Abstract.

077.016   Polarization structure of noise storms.
          G. P. Chernov, I. M. Chertok, V. V. Fomichev,
A. K. Markeev.
Solar activity and solar-terrestrial relations, (see 012.007), p. 293 - 298 (1976).
          The results of simultaneous, high time-resolution observations of spectra (181−224 MHz) and polarization (204 MHz) of noise storms, performed in 1972, are presented. The polarization structures of isolated bursts, chains and a group of type I bursts are investigated. A detailed comparison of the spectral and polarization features of the individual bands in the chains of type I bursts with small band-splitting is carried out.

077.017   High resolution intensity and polarization structure of the sun at 21 cm.
F. Chiuderi Drago, M. Felli, G. Tofani.
Astron. Astrophys., Vol. 61, 79 - 91 (1977).
          The sun has been observed at 21 cm with the Westerbork Synthesis Radio Telescope with a resolution of $21'' \times 61''$. The maps of total intensity and circular polarization are presented.

077.018   Submillimeter observations of solar active regions.
          G. Righini-Cohen, M. Simon.
Astrophys. J., Vol. 217, 999 - 1005 (1977).
          Observations of solar active regions in the several atmospheric windows between 350 μm and 1 mm have been obtained. All the active regions studied appear as regions of enhanced brightness temperature at all submillimeter wavelengths of observation. There is close spatial coincidence of the brightness-temperature enhancements with the regions of most intense photospheric magnetic field. Spectra suggest that the observed temperature increase at submillimeter wavelengths is caused by a general heating of the solar atmosphere at levels $\tau_{5000} < 10^{-4}$.

077.019   On the interpretation of solar radio emission bursts of types II and III.   V. V. Ivanov.
Impul's. ehlektromagn. polya bystroprotekayushchikh protsessov i izmer. ikh parametrov. Moskva, Atomizdat, 1976,

p. 177 - 185. In Russian. − Abstr. in Ref. zh., 51. Astron., 9.51.471 (1977).

077.020   Type III and type I-like decametric emission in a magnetic arch.   Y. Leblanc, M. G. Aubier.
Astron. Astrophys., Vol. 61, 353 - 362 (1977).
          Two type II−type IV events observed in the range 25−70 MHz are described. They appear with fine structures drifting positively or negatively. An interpretation in terms of local acceleration at the top of an arch after magnetic field reconnection is given to explain the formation of high-energy beams of electrons.

077.021   The morphological characteristics of solar spike radio bursts.   G. P. Chernov.
Astron. Zh. Akad. Nauk SSSR, Vol. 54, 1081 - 1101 (1977). In Russian. English translation in Soviet Astron., Vol. 21, No. 5.
          The observations of shortest time scale and narrow-band solar spike radio bursts performed at IZMIRAN from 1969 to 1975 in the 180−230 MHz range with simultaneous registration of circular polarization at 204 MHz are discussed.

077.022   The solar CO spectrum as probe of atmospheric structure and carbon abundance.   T. Tsuji.
Publ. Astron. Soc. Japan, Vol. 29, 497 - 510, with a correction p. 835 (1977).
          The first overtone bands of carbon monoxide in the solar infrared spectrum are analyzed. The excitation temperature for CO is found to be 4570 ± 40 K (p.e.) and the turbulent velocity to be 1.3 km s^{-1}. These results are best explained by the solar model M of Vernazza et al. (1976). If this model is assumed, carbon abundance from the CO analysis is $\log A_C = 8.28 \pm 0.15$. Such a low carbon abundance, however, shows serious contradiction with the high carbon abundance of $\log A_C = 8.67$ recently recommended by Lambert (1976). An inhomogeneous model that consists of a cool component like model M and a 300-K hotter component in the CO-line formation layers is proposed.

077.023   Periodic fluctuations in continuum near-IR solar intensity and CO absorption.
T. A. Clark, D. A. Burrell.
J. R. Astron. Soc. Canada, Vol. 71, 403 (1977). − Abstract.

077.024   On the nature of quasi-periodic fluctuations of solar radio flux at centimeter wavelengths.
N. P. Stasyuk.
Problems of cosmic physics. Vyp. (No.) 12, (see 003.016), p. 3 - 16 (1977). In Russian.
          A spectral correlation analysis of 14 recordings of fluctuations of the solar radiation at 4 cm wavelength and performed estimates confirm the existence of a connection of the characteristics of quasi-periodic fluctuations of the radio flux with parameters of active regions. A picture is given of the dynamics of coronal condensations confirming the instability of the inhomogeneous plasma in the magnetic field of sunspots. Noted is the possibility of using observed increases of the amplitudes and periods of the quasi-periodic fluctuations of the radio flux in the cm-range for the forecast of powerful solar flares.

077.025   Time of appearance of type II radio emission in solar flares. Abstract.
L. Křivský, S. Pinter.
VIII Leningr. mezhdunar. semin. Mater. mezhdunar. semin. "Akt. protsessy na Solntse i probl. soln. nejtrino", 1976. Leningrad, 1976, p. 167 - 168. In Russian and English. − Abstr. in Ref. zh., 51. Astron., 10.51.557 (1977).

077.026   Lower limit of intensity of solar activity on microwaves.   P. Kaufmann, J. R. Blakey, P. Iacomo,
Jr., E. H. Koppe, P. Marques Dos Santos, R. E. Schaal.

An. Acad. Brasil. Cienc., Vol. 48, 187 - 191 (1977). In
Portuguese. – Abstr. in Phys. Abstr., Vol. 80, Abstr. 78729
(1977).

077.027 A new scattering process above solar active regions:
propagation in a fibrous medium.
J. L. Bougeret, J. L. Steinberg.
Astron. Astrophys., Vol. 61, 777 - 783 (1977).

The authors show that the measurements of Type I burst
directivity deduced from the STEREO-1 experiment are not
consistent with the interpretation of many Type I burst
characteristics in terms of scattering on weak density inhomogeneities. They then suggest a new scattering model based on
the following hypotheses: (1) the radiation is generated inside
a region much smaller than the observed source; (2) the
medium above active regions, at altitudes where Type I bursts
are generated, is strongly inhomogeneous, and consists of
bundles of overdense fibers; (3) Type I bursts are generated
inside, or at the border, of such bundles; (4) the primary
radiation is very directive and is beamed more or less along
the fibers.

077.028 Type II emission mechanism.
W. D. Davis, J. Feynman.
J. Geophys. Res., Vol. 82, 4699 - 4703 (1977).

The type II emissions of August 7 - 8, 1972, are reexamined on the basis of two proposed production mechanisms,
one in which the emissions are produced in the ambient plasma
and one in which the emissions are produced in the shocked
plasma. The authors first develop a new model for the radial
dependence of the solar wind density in which the density for
large distances goes as the inverse square of distance, on the
basis of the conservation of particles and the observed lack of
radial dependence of the solar wind velocity. They note that
the density model derived from type III bursts may not apply
directly to type I bursts because of solar wind density structures due to stream interactions. The authors show that the
model in which the emissions are produced in the shocked
plasma provides the more satisfactory explanation of the data.

077.029 High resolution polarimetry of the Sun at 3.7 and
11.1 cm wavelengths.   K. R. Lang.
Sol. Phys., Vol. 52, 63 - 68 (1977).

The four Stokes parameters are presented for interferometric observations of the Sun at wavelengths of $\lambda = 3.7$ cm
and $\lambda = 11$ cm with angular resolutions between 2.7 and 36.7
seconds of arc. During a forty hour observation of sunspot
region McMath No 13926 no substantial variations in circular
polarization were observed, whereas one hour prior to the
eruption of a solar flare dramatic changes in circular polarization were observed.

077.030 Solar radio type III bursts and coronal density
structures.   Y. Leblanc, J. de la Noë.
Sol. Phys., Vol. 52, 133 - 139 (1977).

The comparison of solar radio type III bursts measured at
169 MHz with K corona observations leads to the conclusion
that about 75% of the active regions over which type III bursts
occur are associated with low density coronal structures. The
comparison with X-ray maps of the solar disk shows that all
these regions are located in low intensity regions. It is concluded that the idea generally accepted that the type III bursts
are associated with dense coronal structures and travel in these
structures is not at all proven for a large number of cases.

077.031 3.5 mm depression features associated with Hα
'disparitions brusques'.   M. R. Kundu, P. Lantos.
Sol. Phys., Vol. 52, 393 - 396 (1977).

The characteristics of 3.5 mm depression features associated with two 'disparition brusques' observed in Hα are
discussed. The millimeter depressions still exist, although
reduced in strength, after the disappearance of the Hα filament.

The two depressions correspond to temperatures of 600 and
450 K before and to 200 and 250 K after the Hα filament
disappearance.

077.032 A two-component model of impulsive microwave
burst emission consistent with soft and hard X-rays.
A. Böhme, F. Fürstenberg, G. Hildebrandt, O. Saal, A. Krüger,
P. Hoyng, G. A. Stevens.
Sol. Phys., Vol. 53, 139 - 155 (1977).

A two-component (core-halo) emission model has been
applied reconciling hard and soft X-ray burst emissions with
the microwave burst radiation. Thus probable informations
about the most appropriate magnetic field parameters as well
as about the time- and frequency-dependent source diameters
(yielding growth velocities of the core region during the impulsive phase) are deduced for the burst of 1972 May 18 as an
example. Due to the simple geometry and emission process
adopted, the model refers primarily to special impulsive bursts.
For the representation of broad band microwave bursts, e.g.
type IV μ events, a more complex source geometry and/or
other variants of the emission mechanism must be invoked.

077.033 Relationship between type III and microwave radio-
bursts and the role of magnetic configuration.
M. Pick.
Sol. Phys., Vol. 53, (see 012.038), 241 - 242 (1977). – Invited
opinion.

077.034 The effects of scattering on quiet Sun emission at
frequencies less than 200 MHz.
J. N. McMullin, H. L. Helfer.
Sol. Phys., Vol. 53, 471 - 488 (1977).

The problem of predicting the radio emission from the
quiet Sun for meter and decameter waves may be formulated
in terms of a standard radiative transfer problem with conservative scattering. One may therefore avoid the numerical complications involved in using a ray-tracing approach which incorporates a Monte-Carlo routine for representing scattering.

077.035 Observations of the quiet Sun at meter and
decameter wavelengths.
M. R. Kundu, T. E. Gergely, W. C. Erickson.
Sol. Phys., Vol. 53, 489 - 496 (1977).

The new TeePee Tee array of the Clark Lake Radio Observatory has been used to observe the quiet Sun at 121.5,
73.8 and 26.3 MHz. The equatorial brightness distributions at
all three frequencies, and the polar brightness distributions at
the two higher ones have been measured. From the observed
total fluxes and half-power diameters the authors have derived
the peak brightness temperatures of the solar disk as well as of
some sources of the slowly varying component.

077.036 Noise storms and particular photospheric magnetic
structures.   C. Zanelli, P. Zlobec.
Sol. Phys., Vol. 53, 497 - 505 (1977).

A comparison of flux and polarization of solar radio
noise storms with photospheric source position and magnetic
field configuration for six year observations is reported. Three
independent results pointing to a predominance of 'plus' magnetic structures as regards noise-storm generation are outlined.
A rather strong proof towards a cause-effect connection of
photospheric magnetic structure development and noise-storm
evolution is stressed.

077.037 Determination of the decameter wavelength spectrum
of the quiet Sun.   W. C. Erickson, T. E. Gergely,
M. R. Kundu, M. J. Mahoney.
Sol. Phys., Vol. 54, 57 - 63 (1977).

The Teepee Tee array of the Clark Lake Radio Observatory
has been used to compare the flux of the Sun with that of the
sidereal sources Tau A and Vir A at several frequencies in the
range 109.0–19.0 MHz. The observations, combined with

those available in the literature, allow to derive an accurate meter and decameter wavelength spectrum of the quiet Sun.

077.038   Microwave burst spectra and solar flare magnetic fields.   D. F. Neidig, Jr.
Sol. Phys., Vol. 54, 165 - 168 (1977).

Microwave burst spectra are compared with the position, within the active region, of their associated flares observed in Hα. The magnetic fields predicted by Takakura's burst model (1972) are found to be in reasonable agreement with the fields expected at the flare locations.

077.039   4.7s nearly periodic oscillations superimposed on the solar microwave great burst of 28 March 1976.
P. Kaufmann, L. R. Piazza, J. C. Raffaelli.
Sol. Phys., Vol. 54, 179 - 182 (1977).

An unusual fast oscillation was found superimposed on the solar great burst of 28 March 1976, as measured at 7 GHz. The period of the oscillation was 4.7 ± 0.9 s, defined over the entire duration of the event. The amplitude of the oscillation was proportional to the flux density, in the range $50 < S < 3000$ solar flux units. The degree of circular polarization has not shown any fast periodic time structures.

077.040   Coronal X-ray holes and the quiet radio Sun at 2800 MHz.   A. E. Covington.
Sol. Phys., Vol. 54, 393 - 402 (1977).

Radio cool regions observed on strip scans of the Sun made at 2800 MHz with a 1.5 min arc fan beam are associated with X-ray coronal holes and are used to derive lower envelopes which are similar to spotless Sun drift curves. The enhancement of the quiet Sun of 3.0 s.f.u. for the optically inactive hemisphere of May 20, 1974 suggests that the radio quiet Sun may vary during the sunspot cycle.

077.041   Interplanetary baseline observations of type III solar radio bursts.
R. R. Weber, R. J. Fitzenreiter, J. C. Novaco, J. Fainberg.
Sol. Phys., Vol. 54, 431 - 439 (1977).

Simultaneous observations of type III radio bursts from spacecraft separated by 0.43 AU have been made using the solar orbiters HELIOS-A and HELIOS-B. The burst beginning at 19:22 UT on March 28, 1976 has been located from the intersection of the source directions measured at each spacecraft, and from burst arrival time differences. The source positions range from 0.03 AU from the Sun at 3000 kHz to 0.08 AU at 585 kHz. The electron density along the burst trajectory, and the exciter velocity (= $0.13c$) were determined.

077.042   Radio and soft X-ray evidence for dense non-potential magnetic flux tubes in the solar corona.
R. T. Stewart, J. Vorpahl.
Sol. Phys., Vol. 55, 111 - 120 (1977).

The source positions of solar radio bursts of spectral types I, III(U) and III(J) and V observed by the Culgoora radioheliograph are found to lie almost radially above soft X-ray loops on pictures taken by the S-056 telescope aboard Skylab. The radio source positions and the X-ray loops occur near magnetic loops on computed potential field maps. However, the magnetic induction required to explain the radio observations is much greater than the computed potential field value at that height. Dense current-carrying magnetic flux tubes emanating from active regions on the Sun and extending to $\leqslant 1.5 R_\odot$ above the photosphere provide a satisfactory model for the radio bursts.

077.043   The spectrum and position of solar noise storms at decameter wavelengths.
J. de la Noë, T. E. Gergely.
Sol. Phys., Vol. 55, 195 - 209 (1977).

Decametric storm radiation during the period July—August 1970 has been observed simultaneously with a high

sensitivity spectrograph at Arecibo Observatory and with the log-periodic, swept-frequency array of the Clark Lake Radio Observatory. The observations complement each other. The authors study the relative positions of the different emissions which have been observed during the storms. The continuum emission, the type I bursts and the flare-related type III's were all emitted at different locations. The storm type III bursts, type IIIb's and drift pairs overlapped in position, but appeared at different locations than the previously mentioned sources.

077.044   Nonrelativistic electron stream propagation in the solar atmosphere and type III radio bursts.
G. R. Magelssen, D. F. Smith.
Sol. Phys., Vol. 55, 211 - 240 (1977).

Electron streams with type III burst characteristics are numerically modeled. The electron-plasma wave quasilinear interaction is assumed to be the dominant velocity diffusion process. The quasilinear equations with the addition of spontaneous emission, magnetic and collisional effects are numerically solved as an initial value and a half-space boundary value problem with time, distance and velocity as the independent variables for a solar-type background plasma and a type-III-like stream. Background density and temperature coordinate structure, spontaneous emission, magnetic fields, electron-ion collisions, stream reabsorption and wave pileup are shown to affect propagation and are incorporated into a physical description of the stream motion. The calculated electron flux-time profiles at the Earth suggest scatter-free propagation and compare well with type III stream observations.

077.045   A study of Type V solar radio bursts. I: Observations.   R. D. Robinson.
Sol. Phys., Vol. 55, 459 - 472 (1977).

Results of an observational study of Type V bursts are presented. Observations were made using the C.S.I.R.O. radioheliograph at Culgoora. Source parameters studied included flux evolution, polarization, size, shape, position, motions and brightness temperature at 160, 80 and 43 MHz. Comparisons of source characteristics observed at different frequencies are made.

077.046   Change of the sign of circular polarization in microwave components of type IV solar bursts.
V. A. Kovalev.
Din. kosm. plazmy. Moskva, 1976, p. 14 - 20. In Russian. — Abstr. in Ref. zh., 51. Astron., 11.51.616 (1977).

077.047   On the relation of the parameters of solar radio bursts and proton events.   M. N. Belovskij,
Yu. P. Ochelkov, T. S. Podstrigach, M. A. Fasakhova.
Astron. Tsirk., No. 960, p. 1 - 2 (1977). In Russian.

077.048   Solar stereo radioastronomy.   J.-L. Steinberg.
Study of travelling interplanetary phenomena 1977, (see 012.042), p. 65 - 80 (1977).

Solar radio bursts can be used to probe the corona (bursts of types I and III) and the interplanetary medium (type III). Simultaneous observations in two widely different directions can yield the angular distribution of some burst characteristics or their position in space. Type I bursts stereo observations were made in cooperation between the USSR and France, at 169 MHz (STEREO-1 experiment). They have shown that type I often radiate into a beam of less than 25° half-power width. Type III bursts when observed on meter waves during the STEREO mission are much less directive than type I bursts but more than anticipated from ray tracing in a quasi spherical corona. There is evidence of large scale dense coronal structures which are probably streamers. Although solar stereo-radioastronomy is still very young, it holds great promise as one of the few techniques which can produce a 3-dimensional picture of the corona and interplanetary medium which is essentially free of the effects of integration along the line of sight.

077.049   Interferometric observations at 408 MHz of the radio sources associated with the McMath region 14143.   A. Abrami, U. Koren.
Publ. Trieste Astron. Obs., No. 552 (1977). — Reprint from Report UAG-61.

077.050   On the influence of plasma effects on the observing conditions of radio lines in the solar spectrum.
S. A. Kaplan, E. B. Klejman, I. M. Ojringel'.
Astron. Zh. Akad. Nauk SSSR, Vol. 54, 1305 - 1308 (1977). In Russian. English translation in Soviet Astron., Vol. 21, No. 6.
The emission of plasma waves by an atom in a turbulent plasma medium is more effective than usual photon emission. This effect may be one of the reasons which prevent the observation of solar radio lines.

077.051   On the interpretation of the polarized zebra structure in the solar radio emission.   E. Ya. Zlotnik.
Astron. Zh. Akad. Nauk SSSR, Vol. 54, 1309 - 1313 (1977). In Russian. English translation in Soviet Astron., Vol. 21, No. 6.
The polarization character of the zebra structure is explained by peculiarities of the non-linear conversion of longitudinal waves excited in a magnetic trap by cyclotron instability to electromagnetic radiation. It is shown that, depending on specific conditions in the generation zone, the strips observed in the radio emission may have either weak or strong circular polarization with rotation sign corresponding to the ordinary wave.

077.052   Some features of polarization of microwave solar bursts.   A. M. Uralov, V. P. Nefed'ev.
Astron. Zh. Akad. Nauk SSSR, Vol. 54, 1319 - 1324 (1977). In Russian. English translation in Soviet Astron., Vol. 21, No. 6.
On the basis of the model of a thermal pulse microwave burst proposed earlier by the authors the following phenomena are explained: the change of the polarization sign in the frequency band, the rapid change of the polarization sign in the development of a burst at a fixed frequency, the lack of time coincidence of the moments of the burst's maxima of the polarization and those of the total flux.

077.053   Solar radio emission.   A. D. Fokker.
Illustrated glossary and solar-terrestrial physics, (see 003.022), p. 111 - 138 (1977).

077.054   On the possibility of operative determining of the coordinates of microwave burst sources.
V. A. Krylov, A. Kurbanov.
Dokl. AN TadzhSSR, Vol. 20, No. 4, p. 16 - 19 (1977). In Russian. — Abstr. in Ref. zh., 51. Astron., 12.51.395 (1977).

077.055   Investigation of the probability density of an envelope of modulated fluctuations of solar radio emission.   E. A. Aver'yanikhina.
Izv. vyssh. uchebn. zaved. Radiofiz., Vol. 20, 489 - 493 (1977). In Russian. — Abstr. in Ref. zh., 51. Astron., 1.51.425 (1978).

077.056   About the origin of the solar radio radiation.
J. Olmr.
Vesmír, Vol. 56, 274 - 276 (1977). In Czech.

077.057   Solar noise storms.   J. Olmr.
Říše hvězd, Vol. 58, 163 - 166 (1977). In Czech.

Solar noise storms.   See Abstr. 003.047.

High *n* solar radio recombination lines.

See Abstr. 022.089.

On a method of measurement of solar radio fluxes in the centimeter range.   See Abstr. 031.226.

Solar radiometry in the millimetre region with a local calibration source.   See Abstr. 031.256.

Use of the dispersion relation for computation of the distribution of the radio brightness of the sun and other sources of cosmic radio emission.   See Abstr. 031.332.

Spectrograph for investigation of the fine structure of spectra of the S-component of solar radio emission in the 4.2–7.1 GHz region.   See Abstr. 033.027.

Legendre expansion of the quasi-linear equations for anisotropic particles and Langmuir waves.
See Abstr. 062.033.

New parameters for solar spicules based on submillimetre data.   See Abstr. 071.015.

The wavelength dependence of granulation (0.38– 2.4 μm).   See Abstr. 071.038.

Statistical study of the time sequence of Ca II flocculi and series of radio fluxes during solar activity cycle 20. See Abstr. 072.001.

On the pre-maximum phase of the solar event of August 7, 1972.   See Abstr. 073.019.

Solar flare observations at millimetre and submillimetre wavelengths.   See Abstr. 073.048.

A study of filament transition sheath from radio observations.   See Abstr. 073.099.

Active regions.   See Abstr. 073.116.

A dynamical model of coronal holes based on radio observations.   See Abstr. 074.007.

Collisionless deceleration of fast electron streams in the solar coronal plasma.   See Abstr. 074.046.

Radio and EUV observations of a coronal hole.
See Abstr. 074.050.

Possible polarization effects by thermal infrared emission in the solar corona.   See Abstr. 074.096.

Solare Beobachtungsergebnisse. Solar data. Solar radio emission.   See Abstr. 075.008.

Observations radioélectriques solaires faites sur 600 MHz en 1975 au laboratoire de radioastronomie de Humain-Rochefort.   See Abstr. 075.019.

Solar and solar system activity.
See Abstr. 075.023.

Energy-spectrum distribution of the X-radiation under 8 Å and its relation with the solar radio emission on 2800 MHz.   See Abstr. 076.004.

Definition of the parameters of solar protons in the earth's vicinity from radio bursts. III. Time basic functions. See Abstr. 078.002.

Determination of the parameters of solar protons in

the vicinity of the earth from radio bursts. Time parameters.
See Abstr. 078.016.

Radio observations of interplanetary magnetic field structures out of the ecliptic.   See Abstr. 106.029.

Errata

077.901   Erratum: 'Continuous injection model for hard X-ray

correlated microwave bursts' [Sol. Phys., Vol. 49, 117 - 140 (1976)].   C. Mätzler.
Sol. Phys., Vol. 53, 197 (1977).

077.902   Erratum: "Two 'negative bursts' with moving filaments, 19 May 1969" [Sol. Phys., Vol. 51, 195 - 202 (1977)].   C. Sawyer.
Sol. Phys., Vol. 54, 516 (1977).

078 Cosmic Radiation

078.001   Solar cosmic rays and interplanetary shock waves on April 29 - 30, 1973.
N. N. Volodichev, N. L. Grigorov, G. Ya. Kolesov, O. M. Kovrizhnykh, M. I. Kudryavtsev, B. M. Kuzhevskij, V. G. Kurt, Yu. I. Logachev, N. F. Pisarenko, I. A. Savenko, A. A. Suslov, L. M. Chupova, V. F. Shesterikov, I. P. Shestopalov, T. Gombosi, J. Kóta, A. J. Somogyi.
VIII Leningr. mezhdunar. semin. Mater mezhdunar semin. "Akt. protsessy na Solntse i probl. soln. nejtrino", 1976. Leningrad, 1976, p. 181 - 183. In Russian and English.
From Ref. zh., 62. Issled. kosm. prostranstva, 9.62.187 (1977)

078.002   Definition of the parameters of solar protons in the earth's vicinity from radio bursts. III. Time basic functions.   S. T. Akin'yan, I. M. Chertok.
Geomagn. Aehron., Vol. 17, 596 - 602 (1977). In Russian.

078.003   Main characteristics of solar cosmic ray streams on September 1 - 16, 1971.
M. N. Nazarova, N. K. Pereyaslova, I. E. Petrenko.
Kosm. Issled., Vol. 15, 566 - 572 (1977). In Russian.

078.004   Simple analytical solutions for spherically symmetric production and modulation of energetic solar particles.   M. W. Gross, M. A. Lee, I. Lerche.
Astrophys. J., Vol. 218, 552 - 556 (1977).
Exact analytical solutions are presented to the standard time-independent, spherically symmetric, convection-diffusion-adiabatic deceleration equation governing the transport of cosmic rays in the interplanetary medium, for the case in which particles are produced with spherical symmetry at the Sun. It is assumed that the solar wind speed is constant and radial.

078.005   A need for (p,n) cross sections for selected targets at lower energies.   H. S. Ahluwalia.
Nuclear cross sections and technology. II. Washington, D.C., USA. 3 - 7 March 1975. R. A. Schrack, C. D. Bowman (Editors). Natl. Bur. Stand., Washington, D.C., USA (1975), p. 512 - 515. — Abstr. in Phys. Abstr., Vol. 80, Abstr. 25997 (1977).

078.006   Forbush decrease and geomagnetic storms in relation to varying interplanetary parameters.
S. P. Agrawal, R. L. Singh.
Indian J. Radio Space Phys., Vol. 5, No. 4, p. 330 - 332 (1976). Abstr. in Phys. Abstr., Vol. 80, Abstr. 70956 (1977).

078.007   Correlation of propagation characteristics of solar cosmic rays detected onboard the spatially separated space probes Mars-7 and Prognoz-3.

T. Gombosi, G. Ya. Kolesov, V. G. Kurt, B. M. Kuzhevskii (Kuzhevskij), Yu. I. Logachev, I. A. Savenko, A. J. Somogyi.
Hungarian Acad. Sci., Budapest, Rep. KFKI-77-38, 6 pp. (1977). — Abstr. in Phys. Abstr., Vol. 80, Abstr. 82089 (1977).

078.008   Study of coronal and interplanetary propagation of solar particles following the E45° solar flare on July 29, 1973.   T. Gombosi, G. Ya. Kolesov, V. G. Kurt, B. M. Kuzhevskii (Kuzhevskij), Yu. I. Logachev, I. A. Savenko, I. P. Shestopalov, A. J. Somogyi.
Hungarian Acad. Sci., Budapest, Rep. KFKI-77-39, 7 pp. (1977). — Abstr. in Phys. Abstr., Vol. 80, Abstr. 82090 (1977).

078.009   Separation of solar and interplanetary diffusion in solar cosmic ray events.
T. Ford, I. D. Palmer, R. Sanders.
J. Geophys. Res., Vol. 82, 4704 - 4710 (1977).
A study has been made of the propagation of solar cosmic rays after their release from a flare under the assumption of diffusive motion around the sun and in interplanetary space.

078.010   Coherent propagation of non-relativistic solar electrons.   V. G. Kurt, Yu. I. Logachev, N. F. Pissarenko (Pisarenko).
Sol. Phys., Vol. 53, 157 - 178 (1977).
An experimental study of the propagation of solar electrons with energy $E_e > 30$ keV was carried out. Measurements were made during the period 1972—1974 using the 'Prognoz' satellite-borne instruments. A two-component structure of electron fluxes was found. The fast component, rather well-observed after solar flares of minor importance, consists of a compact beam of electrons propagating without scattering inside a narrow cone with an opening $\leqslant 10°$ along interplanetary magnetic field lines. Characteristics of this component are given.

078.011   Influence of finite injections and of interplanetary propagation on time-intensity and time-anisotropy profiles of solar cosmic rays.
B. M. Schulze, A. K. Richter, G. Wibberenz.
Sol. Phys., Vol. 54, 207 - 228 (1977).
For an observer in space the intensities and anisotropies of solar cosmic-ray events are governed by the duration and the functional shape of the injection processes near the Sun and by the propagation along the interplanetary magnetic field from the Sun to the observer. The authors study the influence of four different types of solar injections (Gaussian, exponential, step-function and coronal diffusion), and of a purely diffusive interplanetary propagation. The model is applied to the November 18, 1968 solar event.

078.012 Influence of interplanetary shocks on solar particle events.   G. E. Morfill, M. Scholer.
Astrophys. Space Sci., Vol. 46, 73 - 86 (1977).

Energetic particles, ejected from the Sun during solar flare events, may encounter interplanetary plasma/field conditions, which deviate considerably from the quiet time values used normally to describe the particle propagation. This is due to the presence of a hydromagnetic shock, which is emitted from the Sun at the time of the explosion. In a theoretical blast wave model, which incorporates the interaction with plane polarized Alfvén waves, the authors have analysed the changes in different terms of the Fokker–Planck equation, which describes energetic particle propagation. In this treatment, the shock influence on energy changes and on the transport coefficients are discussed.

078.013 Analysis of the complex solar particle event on April 29 - 30, 1973.   T. Gombosi, J. Kóta, A. J. Somogyi, V. G. Kurt, B. M. Kuzhevskii (*Kuzhevskij*), Yu. I. Logachev.
Sol. Phys., Vol. 54, 441 - 456 (1977).

Based on the data of the high-apogee satellite Prognoz-3, the April 29 - 30, 1973 solar particle event is analysed. The event's complex energetic particle, interplanetary magnetic field and solar wind plasma properties are discussed. The unusual behaviour of solar particles up to energies ≈ 100 MeV can well be explained in terms of the interaction with an interplanetary shock wave system passing the Earth.

078.014 Solar particle propagation from 1 to 5 AU.   R. D. Zwickl, W. R. Webber.
Sol. Phys., Vol. 54, 457 - 504 (1977).

A statistical analysis of solar particle events, observed by the GSFC-UNH charged particle detector on board Pioneer 10 and Pioneer 11 from March 1972 to December 1974 (from 1 to 5 AU for each spacecraft), is carried out with the goal of experimentally determining the statistical average interplanetary propagation conditions from 3 to 30 MeV. A numerical propagation model is developed that includes diffusion, convection, adiabatic deceleration, and a variable coronal injection profile.

078.015 Enhancement of solar heavy nuclei at high energies in the 4 July 1974 event.
D. L. Bertsch, D. V. Reames.
Sol. Phys., Vol. 55, 491 - 497 (1977).

Relative abundances of energetic nuclei in the 4 July 1974 solar event are presented. The results show a marked enhancement of abundances that systematically increase with nuclear charge numbers in the range of the observation, $6 \leqslant Z \leqslant 26$ for energies above 15 MeV nucl^{-1}. The energy spectrum of oxygen is observed to be significantly steeper than most other solar events studied in this energy region. It is proposed that these observations are characteristic of particle populations at energies ~1 MeV nucl^{-1}, and that the anomalous features observed here may be the result of the high energy extension of such a population that is commonly masked by other processes or populations that might occur in larger solar events.

078.016 Determination of the parameters of solar protons in the vicinity of the earth from radio bursts. Time parameters.   S. T. Akin'yan, I. M. Chertok.
Inst. zemn. magn., ionos. i rasprostr. radiovoln AN SSSR. Prepr. No. 4 (178). Moskva, 1977, 23 pp. In Russian. – Abstr. in Ref. zh., 51. Astron., 11.51.586 (1977).

078.017 Relation of solar cosmic rays with the sectorial structure of the interplanetary magnetic field.
N. K. Pereyaslova, M. N. Nazarova, S. M. Mansurov, L. G. Mansurova.
Ehff. soln. aktivnosti v nizhn. atmos. Leningrad, Gidromete-

oizdat, 1977, p. 55 - 59. In Russian. – Abstr. in Ref. zh., 51. Astron., 11.51.591 (1977).

078.018 Signatures of solar cosmic ray events and their relation to propagation and acceleration processes.
G. Wibberenz.
Study of travelling interplanetary phenomena 1977, (see 012.042), p. 323 - 342 (1977).

Simultaneous observations from spacecraft distributed in solar longitude and radial distance from the sun can be used to separate solar and interplanetary propagation processes. Combined with statistical studies they define the average characteristics of the interplanetary propagation process. Measurements with the Pioneer 10/11 spaceprobes out to beyond 5 AU are of particular importance to reveal the average propagation characteristics of the interplanetary medium. For individual events, anisotropy measurements allow to draw conclusions on the interaction with the interplanetary medium as well as on the solar injection process.

078.019 Fast azimuthal transport of solar cosmic rays via a coronal magnetic bottle.   K. H. Schatten, D. J. Mullan.
J. Geophys. Res., Vol. 82, 5609 - 5620 (1977).

Key observations pertaining to the fast azimuthal propagation of solar cosmic ray particles are reviewed. Briefly, protons and electrons with a wide range of energies from 40° to 60° heliolongitude on either side of a flare site have access to the earth-sun interplanetary field line within an hour of flare onset. The authors propose that coronal magnetic bottles, produced by flares, serve as temporary traps for solar cosmic rays in some instances.

078.020 Long-term forecast of proton events in active regions.   P. B. Bernshtejn.
Soln. Dannye 1977 Byull., No. 10, p. 73 - 77 (1977). In Russian.

Pattern recognition methods are used to solve the problem of long-term forecasting proton events in active regions. The problem is solved with help of two algorithms. The quality of the solving rule is satisfactory. A comparative informability of parameters has been estimated.

078.021 On the exponential decay of western hemisphere solar particle events.   M. Scholer.
37. Jahrestagung Deutsche Geophys. Ges. Braunschweig, F.R. Germany, March 28 - April 1, 1977. 8 pp. (1977). – From IKK 77A15004326.

Solar cosmic rays.   See Abstr. 003.006.

Measurement of radiation doses aboard AES Molniya 1.   See Abstr. 051.068.

On the exponential decay of western hemisphere solar particle events.   See Abstr. 073.038.

Solar proton flares and their prediction. See Abstr. 073.111.

On the relation of the parameters of solar radio bursts and proton events.   See Abstr. 077.047.

The solar proton event of April 16, 1970. 1. Features in interplanetary space and in the magnetosheath. See Abstr. 084.219.

The solar proton event of April 16, 1970. 2. Transformation of proton pitch angle distributions from the solar wind into the magnetosheath.   See Abstr. 084.220.

A summary of significant solar-initiated events

during STIP Intervals I and II.   See Abstr. 085.038.

Energetic protons associated with a forward-reverse interplanetary shock pair at 1 A.U.
See Abstr. 106.008.

Solar and galactic subcosmic rays: their connection with active processes on the sun.   See Abstr. 143.012.

Solar and galactic subcosmic rays: their relation to active solar processes.   See Abstr. 143.024.

079 Solar Eclipses

079.001 Déformation du champ stellaire au voisinage du Soleil, durant une éclipse totale, déterminée au moyen d'une méthode de simulation et d'étoiles fictives.
H. Debehogne.
Bull. Acad. R. Belgique, Cl. Sci., Sér. 5, Tome 63, 165 - 178 (1977).

Observation du champ stellaire au voisinage du Soleil pendant une éclipse totale.   See Abstr. 066.067.

Diffraction and reflection of radio waves from the limb of the moon during solar eclipses.
See Abstr. 077.003.

Solar eclipse 1973 June 30

079.101 Observations of the total eclipse of the Sun of June 30, 1973.   V. Rušín.
Solar activity and solar-terrestrial relations, (see 012.007), p. 309 - 311 (1976).

079.102 Measurements of ionizing radiation in the period of the solar eclipse on June 30, 1973.
T. V. Kazachevskaya.
Geomagn. Aehron., Vol. 17, 932 - 934 (1977). In Russian.

079.103 Nota acerca de la variación de la radiación solar en Barcelona durante el eclipse solar del 30 de junio de 1973.   J. Lorente, S. Alonso.
Urania Barcelona, Año 61, Núm. 285, p. 93 - 97 (1976).
Normal incidence spectral measurements of solar irradiance taken during the solar eclipse of June 30, 1973, are presented. Their relative decreases are computed and are compared with the relative values of the covered area of the solar disc.

079.104 Observation of the contact time at the total solar eclipse of 30 June 1973 with the monochromatic Torroja camera.   K. Iwadate, C. Kakuta.
Proc. Int. Latitude Obs. Mizusawa, No. 16, p. 51 - 69 (1977). In Japanese.
A monochromatic Torroja camera was used at the total solar eclipse on 30 June 1973 for the observation of the contact time at the Atar Airport, the Islamic Republic of Mauritania. The Torroja camera is designed to run the film with a constant speed along the tangential direction of the uncovered crescent at a solar eclipse. The results of comparison between the flash spectra and the Torroja camera are: (1) precision of both observations in a determination of contact time is of the same order, (2) the Torroja camera gives more fine feature than the flash spectrum does in the Moon's profile, (3) the flash spectrum gives more precise values in the maximum logarithmic gradient of the brightness distribution at the Sun's extreme limb than the Torroja camera.

079.105 Observation of the contact time at the total solar eclipse of 30 June 1973.
C. Kakuta, K. Iwadate.
Publ. Int. Latitude Obs. Mizusawa, Vol. 10, 129 - 161 (1976).
Cinematographies of the flash spectra and monochromatic Torroja camera were performed at the total solar eclipse on 30 June 1973 for the observation of the contact time at the Atar Airport, the Islamic Republic of Mauritania. Corrections to the local prediction were obtained in consideration of the Moon's profile. The monochromatic Torroja camera was found to be of the same order of magnitude in accuracy as that of the flash spectra.

Determination of stray light and sunspot intensities based on the observation during the solar eclipse of 30 June, 1973.   See Abstr. 072.041.

Étude polarimétrique de la couronne solaire observée a l'éclipse totale du 30 juin 1973 à l'aide d'un filtre neutre radial.   See Abstr. 074.001.

Polarization of the corona during the solar eclipse of June 30, 1973.   See Abstr. 074.021.

Polarization of the solar corona on the total solar eclipses of February 15, 1961 and June 30, 1973.
See Abstr. 074.095.

Solar eclipse 1976 April 29

079.201 Coordinates of optical features on the solar disc during the solar eclipse of April 29, 1976.
P. V. Matveev, Yu. F. Yurovskij.
Soln. Dannye 1977 Byull., No. 5, p. 86 - 93 (1977). In Russian.
Data are given of optical observation in the Crimea of the partial solar eclipse on April 29, 1976 in a form convenient for a comparison with results of radio observations. Some characteristics of active features and the radio image of the sun at $\lambda = 1.35$ cm are given.

079.202 Observations of the solar eclipse on April 29, 1976.   Z. Petrova.
Geod., kartogr., zemeustr., Vol. 16, No. 6, p. 32 (1976). In Bulgarian. — Abstr. in Ref. zh., 51. Astron., 9.51.46 (1977).

079.203 Annular solar eclipse on a satellite picture.
J. P. de Jongh, G. P. Konnen.
Weather, Vol. 31, 425 - 426 (1976). — Abstr. in Phys. Abstr., Vol. 80, Abstr. 33464 (1977).

079.204 Osservazione dell'eclisse di sole del 29 aprile 1976.   G. Di Giovanni.
Astronomia, N. 3, p. 20 - 23 (1977).

079.205 **Zenith photoelectric measurements during the annular solar eclipse of April 29, 1976.**
T. Prokakis, D. Dialetis.
Sol. Phys., Vol. 53, 531 - 538 (1977).

Measurements of the light intensity in the zenith, obtained during the annular solar eclipse of April 29, 1976 are given in three wavelength bands (centered on λ's 4800, 5400 and 6100 Å). The observed differences in the three wavelength regions have been examined and compared with calculations taking into account limb darkening.

079.206 **Observation of the annular solar eclipse on 29 April, 1976.**   K. I. Churyumov.
Astron. Tsirk., No. 960, p. 6 - 7 (1977). In Russian.

079.207 **The annular solar eclipse of April 29, 1976.**   M. Waldmeier, S. E. Weber.
Astron. Mitt. Eidg. Sternw. Zürich, Nr. 349, 11 pp. (1977). In German.

079.208 **Observation photographique de l'éclipse de Soleil du 29 avril 1976 à l'équatorial Cooke-Zeiss.**
J. Dommanget, E. Van Dessel, O. Nys.
Bull. Astron., Obs. R. Belgique, Vol. 8, 325 - 326 (1976).

079.209 **Saules aptumsums 1976. gada 29. aprīlī.**   M. Dīriķis, G. Ozoliņš.
Zvaigžņotā debess, 1976/77. gada ziema, p. 7 - 11.

Behaviour of atomic and mechanical oscillators during a solar eclipse.   See Abstr. 044.024.

Solar eclipse 1977 October 12

079.301 **Partial phases of this month's solar eclipse.**
Sky Telesc., Vol. 54, 276 - 277 (1977).

079.302 **Eclipse at sea.**   D. di Cicco.
Sky Telesc., Vol. 54, 470 - 474 (1977).

079.303 **Durch Kolumbien zur Sonnenfinsternis 1977.**   S. Staub, W. Staub, C. Lüthi, W. Lüthi.
Orion, 35. Jahrg., 188 - 194 (1977).

Solar eclipse 1976 October 23

079.401 **A timing experiment performed during the total solar eclipse of 1976 October 23.**   J. E. Jones.
J. British Astron. Assoc., Vol. 88, 18 - 22 (1977).

079.402 **An in-flight observation of the antipodean eclipse, 1976 October 23.**   A. G. T. White.
J. British Astron. Assoc., Vol. 88, 23 - 27 (1977).

079.403 **Report on the observation of the Australia total eclipse on October 23, 1976.**
K. Sato, I. Okamoto, C. Kakuta.
Proc. Int. Latitude Obs. Mizusawa, No. 16, p. 90 - 97 (1977). In Japanese.

079.404 **Total brightness of the corona at the solar eclipse of October 23, 1976.**   J. Dürst.
Astron. Mitt. Eidg. Sternw. Zürich, Nr. 356, 8 pp. (1977). In German.

Shape and structure of the corona at the solar eclipse of October 23, 1976.   See Abstr. 074.099.

Solar eclipse 1972 July 10

The polarization of the inner solar corona at the eclipse of 10 July 1972.   See Abstr. 074.055.

Solar eclipse 1975 May 11

079.601 **Results of the observation of the partial solar eclipse on May 11, 1975.**   V. Ruždjak, K. Pavlovski.
Publ. Dep. Astron., Univ. Beograd, Fac. Sci., No. 6, (see 012.040), p. 61 - 63 (1976).

Solar eclipse 1961 February 15

Polarization of the solar corona on the total solar eclipses of February 15, 1961 and June 30, 1973. See Abstr. 074.095.

080 Atmosphere, Figure, Internal Constitution, Neutrinos, Magnetic Fields, Rotation, Miscellanea

080.001 Note on the rapid oscillations of the solar rotation found by Chistyakov.
R. Anttila, J. Tuominen.
Bull. Astron. Inst. Czechoslovakia, Vol. 28, 253 - 254 (1977).

It is shown that the rapid oscillations of the solar rotation found by V. F. Chistyakov (see 17.080.007) are probably an aspect effect caused by the inclination of the solar axis of rotation with respect to the ecliptic.

080.002 On the possibility of contamination of the solar surface by accretion. M. Gabriel, A. Noels.
Astron. Astrophys., Vol. 59, 427 - 431 (1977).

The authors present two arguments which seem to prevent any significant increase of the surface abundance Z of heavy elements in the sun. Firstly the observed depletion of ^7Li during the main sequence phases implies the mixing of at least the outermost 1.75% of the solar mass. Secondly any increase in Z in the outer layers produces either a dynamical or a secular non radial instability which increases the mass of the contaminated layers enough to dilute accretion effects so as to make it negligible.

080.003 The effect of sunspots and faculae on the solar constant.
P. V. Foukal, P. E. Mack, J. E. Vernazza.
Astrophys. J., Vol. 215, 952 - 959 (1977).

The authors study the available measurements of the solar constant made at ground sites and from recent space observations to determine whether sunspots or faculae produce a detectable modulation of either the solar flux or the Earth's atmospheric transmission. The data from radiometers on Mariners 6 and 7 rule out any relative change of the solar constant in space due directly to faculae or spots exceeding 0.03%. This limit is two orders of magnitude smaller than previous values obtained from ground measurements.

080.004 Dilation of force-free magnetic flux tubes.
S. Frankenthal.
Astrophys. J., Lett., Vol. 215, L131 - L134 (1977).

A general study is presented of the mapping functions which relate the magnetic field profiles across a force-free rope in segments subjected to various external pressures. The results reveal that if the external pressure falls below a certain critical level (dependent on the flux-current relation which defines the tube), the magnetic profile consists of an invariant core sheathed in a layer permeated by an azimuthal magnetic field.

080.005 Nonlinear MHD model of the solar dynamo.
T. S. Ivanova, A. A. Ruzmajkin.
Astron. Zh. Akad. Nauk SSSR, Vol. 54, 846 - 858 (1977).
In Russian. English translation in Soviet Astron., Vol. 21, No. 4.

A numerical solution of the simplest nonlinear dynamo model of generation of the large-scale magnetic field in the sun's convective zone is constructed. The nonlinearity is introduced by changing the form of the function α characterizing the helicity of the velocity field. As a result, increasing oscillations are suppressed and the solution comes to a stationary one with well-defined period and amplitude. The butterfly diagram is in qualitative accordance with the observed one. The phase relations between the meridional and azimuthal components of the magnetic field are discussed. Estimates of dynamo number, period and amplitude in the stationary regime show good agreement with observations.

080.006 The influence of the chromosphere and corona on the solar atmospheric oscillations.
H. Ando, Y. Osaki.
Publ. Astron. Soc. Japan, Vol. 29, 221 - 233 (1977).

The influence of the chromosphere and corona on the solar atmospheric oscillations is examined by solving the equations of linear nonadiabatic nonradial oscillations for a realistic solar atmosphere-envelope model including the chromosphere and corona. Some wave energy leaks to the corona by running waves for acoustic modes with small horizontal wavenumbers. It is, however, found that the energy leakage is not serious for the stability of most modes. It is shown that there exists an acoustic mode trapped in the chromosphere besides the ordinary ones trapped in the convection zone.

080.007 Photospheric oscillations. IV. An accurate ω-spectrum at low values of k.
E. Fossat, G. Grec, C. Slaughter.
Astron. Astrophys., Vol. 60, 151 - 152 (1977).

Two power spectra of the five-minute oscillation averaged respectively on $8\rlap{.}''5$ and $36''$ aperture are presented. From their comparison it is concluded that the dependence of the unresolved ω-spectrum on wavenumber is probably fairly low and that the average of these two power spectra is a good estimate of this ω-spectrum.

080.008 Interference-phase method of ray-velocity measurement in the solar atmosphere.
S. M. Gorskij, V. P. Lebedev.
Izv. Krymskoj Astrofiz. Obs., Vol. 57, 228 - 236 (1977).
In Russian.

The proposed interference-phase method of investigation of ray velocities is based on phase measurement of the interference picture of a two-beam interferometer at fixed propagation difference. The method differs in principle from the classical one and opens new possibilities in investigations of solar phenomena.

080.009 Interference-phase method of magnetic-field measurement in the solar atmosphere.
I. V. Gorbacheva, S. M. Gorskij, V. P. Lebedev, V. I. Lyubimtsev.
Izv. Krymskoj Astrofiz. Obs., Vol. 57, 237 - 241 (1977). In Russian.

An interference-phase method of magnetic-field measurement consisting in the following is described: modulation of circular light polarization within the spectral line is made; a light signal passes through an optical filter the transmission bandwidth of which is greater than the line width and is transmitted to a double-beam Michelson interferometer. An amplitude of phase oscillations of the interference pattern obtained at fixed propagation difference of the interferometer beams is measured.

080.010 The neutrino flux of inhomogeneous solar models.
S. P. Bhavsar, R. Härm.
Astrophys. J., Vol. 216, 138 - 140 (1977).

A survey has been made of initially inhomogeneous solar models to study their neutrino fluxes. In the course of this survey quantitative results have been obtained for three types of hypothetical inhomogeneities in the interior of the initial Sun: (1) a reduction of the heavy-element abundance Z causes a moderate decrease in the neutrino flux, (2) small inert cores increase this flux, and (3) an increase in the interior hydrogen content strongly decreases the neutrino flux but causes

Rayleigh-Taylor instability.

080.011 Time variations of the angular momentum of the Sun. K. H. Schatten.
Astrophys. J., Vol. 216, 650 - 653 (1977).

Time variations of density models of the Sun are investigated. This is an attempt to estimate the changing moment of inertia of the Sun in order to calculate the internal solar angular velocity based upon Newton's equation of motion. Previous estimates of dI/dt disagree with those based upon central densities in a homologously contracting model. Based upon an integration of Sear's solar model, $dI/dt = -5.5 \times 10^{34}\,\text{gm cm}^2\,\text{s}^{-1}$.

080.012 Complex investigation of the sun and isotopic ecology. G. E. Kocharov.
Izv. AN SSSR. Ser. fiz., Vol. 41, 415 - 421 (1977). In Russian. Abstr. in Ref. zh., 51. Astron., 8.51.356 (1977).

080.013 A search for solar global oscillations in the Ca II K-line. J. M. Beckers, T. R. Ayres.
Astrophys. J., Lett., Vol. 217, L69 - L72 (1977).

Recent models by Hill et al. explain the apparent $\sim 50^m$ oscillations of the solar diameter by a periodic change of the limb darkening function. The variations in the physical parameters that cause this change should also produce brightness oscillations in the inner wings of the Ca^+ H and K lines amounting to as much as 0.5 % at H_1 and K_1. The authors do not confirm the presence of oscillations of this amplitude.

080.014 The wavelength dependence of the facular excess brightness. G. A. Chapman, T. E. McGuire.
Astrophys. J., Vol. 217, 657 - 660 (1977).

It has been shown that solar faculae can produce an apparent solar oblateness. They produced some of the oblateness measured at Princeton in 1966, although how much has not been finally settled. A "classical" test to separate the effects of brightness versus a geometrical oblateness is the color dependence of the brightness. Data is presented on the color dependence of solar faculae. The implication for oblateness measurements is discussed.

080.015 Sur la dynamique de l'atmosphère solaire. P. Souffrin.
Recueil des séminaires, (see 012.004), p. 95 - 101 (1977).

080.016 On the transformation and removal of local magnetic fields into the corona and interplanetary space.
V. V. Kasinskij, G. V. Kuklin.
Solar activity and solar-terrestrial relations, (see 012.007), p. 301 - 307 (1976). In Russian.

The data on the interplanetary magnetic field (sign and value) are compared with the total magnetic flux of all sunspots on the visible solar hemisphere during 210 days (1968). The correlation and spectral methods of analysis are used.

080.017 Why Syrovatskii's mechanism of dynamic dissipation of magnetic fields does not work. U. Anzer.
Solar activity and solar-terrestrial relations, (see 012.007), p. 375 (1976). − Abstract. (See 10.080.017).

080.018 Five-minute oscillations and solar atmosphere heating. Yu. D. Zhugzhda.
Solar activity and solar-terrestrial relations, (see 012.007), p. 377 - 380 (1976).

080.019 On the spread function for solar stray light. L. Staveland.
Solar activity and solar-terrestrial relations, (see 012.007), p. 387 - 389 (1976).

Different types of mathematical expressions for the spread function are given. The normalization of the spread function is discussed. A comparison between two different analytical integrations and one numerical integration giving the observed intensity is shown with a Gaussian spread function. Typical aureoles at 3 different sites are shown.

080.020 A survey of the solar atmospheric models. K. D. Abhyankar.
Bull. Astron. Soc. India, Vol. 5, 40 - 44 (1977).

This review will be a sort of a historical account of how our ideas about the solar atmospheric model developed during the last 30 years.

080.021 On the influence of oscillator strengths, equivalent widths and photospheric model on the iron abundance in the solar atmosphere. A. G. Gasanalizade.
Soln. Dannye 1977 Byull., No. 5, p. 67 - 73 (1977). In Russian.

On the basis of Fe I lines observed photoelectrically a new value of the iron abundance in the solar atmosphere has been determined.

080.022 A theory of differential rotation based on the discussion of turbulent transport of angular momentum.
G. Rüdiger.
Astron. Nachr., Band 298, 245 - 252 (1977).

This paper deals with the theory of the solar rotational law. The author assumes the turbulence to be of the largest influence compared with the momentum flux caused by molecular viscosity and meridional circulation. Firstly he uses heuristic forms for the needed cross correlations: turbulent radial momentum flux and turbulent latitudinal momentum flux. Secondly, he determines the coefficients with a theory founded upon the hypothesis that a rotating stochastic force field maintains an anisotropic turbulence. It is suggested that horizontally directed turbulent motions with not too small radial correlation lengths and time scales of about 2 weeks could be responsible for the solar differential rotation. Finally, he shows that also short-living turbulent horizontal modes provide the observed equatorial acceleration if they occur preferably at the equatorial region.

080.023 Evolving force-free magnetic fields. II. Stability of field configurations and the accompanying motion of the medium. B. C. Low.
Astrophys. J., Vol. 217, 988 - 998 (1977).

The following two aspects of magnetic field evolution in the solar atmosphere are considered. (1) The author presents a stability analysis of the force-free magnetic field configurations given in the first paper of this series. Using the energy principle of Bernstein et al., he shows that these force-free configurations, treated as static fields, are stable when subject to infinitesimal perturbations which, like the magnetic fields, are taken to depend on two Cartesian coordinates. The dynamical basis implied by this stability property for the quasi-static evolution of solar magnetic fields is discussed. (2) As a particular illustration of plasma motions accompanying the evolution of a magnetic field, the author takes as given one of the two evolving magnetic fields treated in the first paper and derives the class of kinematic velocities which satisfy the induction equation.

080.024 The diameter of the sun. A. Wittmann.
Astron. Astrophys., Vol. 61, 225 - 227 (1977).

Measurements of the solar semidiameter made between 1836 and 1975 are discussed. By combining all measurements the mean semidiameter finally becomes $R = (960.00 \pm 0.09)''$ or $R_\odot = (696265 \pm 65)$ km.

080.025 Solar graticules. L. M. Dougherty.
J. British Astron. Assoc., Vol. 87, 582 - 588 (1977).

A concise specification of heliographic co-ordinates is

presented followed by an elementary illustration of orthographic projection applied to the Sun. The use of orthographic graticules and of Porter's rectangular graticule for determination of sunspot positions is mentioned. Commencing in 1892 the history of solar orthographic graticules is traced to the present day and the communication is concluded with a note on the present availability of graticules.

080.026 Evolution of open magnetic structures on the Sun: the Skylab period. R. H. Levine.
Astrophys. J., Vol. 218, 291 - 305 (1977).

High-resolution harmonic analysis of the measured photospheric magnetic field of the Sun is used to construct models of open magnetic structures over a period of 11 solar rotations. The models successfully reproduce the surface location and topology of all coronal holes during the Skylab period. In addition, there is persistent evidence in the models that open field lines are associated with active regions in a systematic way. Specific examples of the evolution of coronal holes and of calculated open structures are presented.

080.027 Dust from the sun? D. J. Mullan.
Astron. Astrophys., Vol. 61, 369 - 375 (1977).

The author finds that particles which are small enough to be repelled from the sun by radiation pressure are probably not of solar origin: such particles may not survive sputtering in the active solar atmosphere, and sublimation in the solar radiation field. A local origin of the dust is proposed as a possible alternative to the extraterrestrial hypothesis.

080.028 Dissipation of convective noise waves in the magnetic structure of the upper layers of the solar atmosphere. S. A. Kaplan, G. B. Rybkina.
Astron. Zh. Akad. Nauk SSSR, Vol. 54, 1102 - 1109 (1977). In Russian. English translation in Soviet Astron., Vol. 21, No. 5.

The "dissipative thickness" of Alfvén and sound waves has been found for the case of their propagation in the magnetic structure of the solar chromosphere and corona. The decay of waves and nonlinear change of phase have been taken into account. The values of the "dissipative thickness" are given in a table.

080.029 On the geoeffectivity of oscillations of the solar surface.
A. V. Gul'el'mi, B. M. Vladimirskij, V. N. Repin.
Geomagn. Aehron., Vol. 17, 930 - 932 (1977). In Russian.

080.030 Großräumige Plasmabewegungen auf Sternen. H. Wöhl.
Sterne Weltraum, Jahrg. 16, 363 - 365 (1977).

080.031 Is the sun a pulsar? P. Toth.
Nature, Vol. 270, 159 - 160 (1977).

Some observations of the solar disk seem to confirm the existence of a radial mode pulsation. External effects may not, however, have been filtered satisfactorily in all cases. Thus the author tried to find an effect of magnetospheric origin that can be related in some way to the described observations. He reports here observations of pulsations with period of 2 h 40 min.

080.032 Small scale solar magnetic fields: theory. N. O. Weiss.
Highlights of Astronomy, Vol. 4, Part II, (see 012.024), p. 241 - 250 (1977).

Intergranular magnetic fields had indeed been expected but their magnitude came as a surprise. This review is limited to the principal problems raised by these filamentary magnetic fields. The author discusses the interaction of magnetic fields with convection in the sun and attempts to answer such questions as: what is the nature of the equilibrium in a flux tube? how are the fields contained? what determines their stability? how are such strong fields formed and maintained? and what limits the maximum field strength?

080.033 Small magnetostatic flux tubes. H. C. Spruit.
Highlights of Astronomy, Vol. 4, Part II, (see 012.024), p. 265 - 266 (1977).

080.034 Heterogeneity of the solar atmosphere. D. J. Mullan.
Highlights of Astronomy, Vol. 4, Part II, (see 012.026), p. 377 - 387 (1977).

Contents: Introduction; Classification of heterogeneities; Heterogeneities due to hydrodynamic effects; Heterogeneities due to magnetic effects; Conclusion.

080.035 The sensitivity of nonradial p mode eigenfrequencies to solar envelope structure.
R. K. Ulrich, E. J. Rhodes, Jr.
Astrophys. J., Vol. 218, 521 - 529 (1977).

Eigenfrequencies are calculated for nonradial p mode oscillations in the Sun. These frequencies are shown to depend on the adiabat of the solar convective envelope. The frequency dependence is greatest for the p_2 and p_3 modes described by spherical harmonics with $l = 500$ to 1000. The solar envelope model includes a realistic representation of the chromosphere out to the base of the corona.

080.036 A test of the micro-macroturbulence model on the solar flux spectrum. D. F. Gray.
Astrophys. J., Vol. 218, 530 - 538 (1977).

This is not an investigation of solar velocity fields. Rather it is an application of the micro-macroturbulence model of velocity fields to a G2 V star; and the best data for such a star come from the spatially unresolved or flux spectrum of the Sun. The author asks what information can be obtained about photospheric turbulence from this high-quality starlike spectrum and in particular where and how the micro-macroturbulence model fails to explain the flux profile data. Because the micro-macroturbulence model does reproduce the solar flux profiles, it appears that failures of this model may not be readily detectable in analyses of stellar spectra.

080.037 The 12.2 day solar rotational period. R. H. Dicke.
Astrophys. J., Vol. 218, 547 - 551 (1977).

The periodicity of the 1966 solar oblateness residuals, representing most of the variance, has been accounted for as a signal generated by a solar distortion rotating rigidly with a 12.22 ± 0.12 day period (sidereal). It is shown that the complex noiselike character of the curve for residual variance versus frequency (including the numerous secondary resonances) is to be expected and represents the true "fingerprint" of the 12.2 day distortion signal.

080.038 Has the Sun a companion star? E. R. Harrison.
Nature, Vol. 270, 324 - 326 (1977).

Pulsars are accurate timekeepers. They are believed to be rotating neutron stars, with strong magnetic fields, and the energy they radiate is at the expense of their rotational kinetic energy. As each pulsar ages, its period P (relative to the Solar System barycentre) slowly increases, and its period derivative $\dot{P}(= dP/dt)$ slowly decreases. Certain interesting pulsars have anomalously small period derivatives, and rather surprisingly, are found grouped together in the same region of the sky. The author suggests here, as an explanation of the peculiar properties of these pulsars, that the barycentre of the Solar System is accelerated, possibly because the Sun is a member of a binary system and has a hitherto undetected companion star.

080.039 Oscillations and internal structure of the sun. A. B. Severnyj.

Zemlya i Vselennaya, 1977, No. 6, p. 36 - 39. In Russian.

080.040   Large-scale solar magnetic fields.   R. Howard.
Annu. Rev. Astron. Astrophys., Vol. 15, (see
003.012), 153 - 173 (1977).
Contents: Large-scale magnetic surface features; The
solar activity cycle and the large-scale patterns; Magnetic
fields in the corona; Concluding remarks.

080.041   Alfvén wave reflection at a density transition region.
J. A. Adam.
J. Phys. A, Vol. 9, L193 - L195 (1976). — Abstr. in Phys. Abstr.,
Vol. 80, Abstr. 29754 (1977).

080.042   Report on the Brookhaven solar neutrino experiment.
R. Davis, Jr., J. C. Evans, Jr.
VIII Leningr. mezhdunar. semin. Mater. mezhdunar. semin.
"Akt. protsessy na Solntse i probl. soln. nejtrino", 1976.
Leningrad, 1976, p. 83 - 98. — Abstr. in Ref. zh., 51. Astron.,
10.51.500 (1977).

080.043   Solar neutrino oscillations.
S. M. Bilenky, B. M. Pontecorvo.
VIII Leningr. mezhdunar. semin. Mater. mezhdunar. semin.
"Akt. protsessy na Solntse i probl. soln. nejtrino", 1976.
Leningrad, 1976, p. 59, 61 - 73, 75. — Abstr. in Ref. zh., 51.
Astron., 10.51.501 (1977).

080.044   Motions and magnetic fields on the sun. Abstract.
V. A. Krat.
VIII Leningr. mezhdunar. semin. Mater. mezhdunar. semin.
"Akt. protsessy na Solntse i probl. soln. nejtrino", 1976.
Leningrad, 1976, p. 125 - 127. In Russian and English. — From
Ref. zh., 51. Astron., 10.51.514 (1977).

080.045   Superadiabatic acceleration of charged particles and
flares on the sun and stars. Abstract.
A. A. Rumyantsev.
VIII Leningr. mezhdunar. semin. Mater. mezhdunar. semin.
"Akt. protsessy na Solntse i probl. soln. nejtrino", 1976.
Leningrad, 1976, p. 157 - 160. In Russian. — Abstr. in Ref. zh.,
51. Astron., 10.51.548 (1977).

080.046   Unsolved problems of nuclear astrophysics of the
sun. Abstract.   G. E. Kocharov.
VIII Leningr. mezhdunar. semin. Mater. mezhdunar. semin.
"Akt. protsessy na Solntse i probl. soln. nejtrino", 1976.
Leningrad, 1976, p. 43 - 49. In Russian. — Abstr. in Ref. zh.,
51. Astron., 10.51.596 (1977).

080.047   The solar neutrino experiment as a means for
testing cosmological and cosmogonic hypotheses.
Abstract.   B. Kukhovich.
VIII Leningr. mezhdunar. semin. Mater. mezhdunar. semin.
"Akt. protsessy na Solntse i probl. soln. nejtrino", 1976.
Leningrad, 1976, p. 75 - 76. In Russian and English. — Abstr.
in Ref. zh., 51. Astron., 10.51.597 (1977).

080.048   Turbulence in the interior of the sun and solar
neutrinos. Abstract.
S. S. Vasil'ev, G. E. Kocharov.
VIII Leningr. mezhdunar. semin. Mater. mezhdunar. semin.
"Akt. protsessy na Solntse i probl. soln. nejtrino", 1976.
Leningrad, 1976, p. 55 - 58. In Russian. — Abstr. in Ref. zh.,
51. Astron., 10.51.604 (1977).

080.049   Background from internal radioactivity in some
radiochemical detectors of solar neutrinos. Abstract.
Yu. I. Zakharov.
VIII Leningr. mezhdunar. semin. Mater. mezhdunar. semin.
"Akt. protsessy na Solntse i probl. soln. nejtrino", 1976.
Leningrad, 1976, p. 81 - 82. In Russian. — Abstr. in Ref. zh.,

51. Astron., 10.51.605 (1977).

080.050   Observations of nonradial p-mode oscillations on
the sun.
E. J. Rhodes, Jr., R. K. Ulrich, G. W. Simon.
Astrophys. J., Vol. 218, 901 - 919 (1977).
The solar velocity field is analyzed in space and time. A
(k_h, ω) distribution of power spectral density is derived from
these data. The observed power density is concentrated in
bands whose frequencies agree with eigenfrequencies for nonradial p-mode oscillations. Comparison of the theoretical and
observed frequencies places a lower limit on the depth of the
convective layer in the solar envelope. The authors have
determined this lower limit from a masking operation which
objectively tests the goodness of fit of the theoretical eigenfrequencies to the observed power.

080.051   Maunder's paradox and the tidal theory of solar
activity.   A. Dauvillier.
Bull. Soc. R. Sci. Liège, Vol. 45, 211 - 221 (1976). In French.
Abstr. in Phys. Abstr., Vol. 80, Abstr. 45387 (1977).

080.052   Pioneer probes Sun's magnetic field.
Indian East. Eng., Vol. 119, No. 3, p. 113, 118
(1977). — Abstr. in Phys. Abstr., Vol. 80, Abstr. 82162 (1977).

080.053   Solar-cycle variations in the differential rotation of
solar magnetic fields.   J. O. Stenflo.
Astron. Astrophys., Vol. 61, 797 - 804 (1977).
Solar-cycle variations in the rotation rate of solar magnetic fields have been studied by autocorrelation analysis of
photospheric magnetic fields recorded at the Mount Wilson
Observatory during the period August 1959 — February 1975.
The observed angular velocity and its latitudinal shear show
large fluctuations correlated in latitude-time space with the
pattern of solar activity. The regions of large latitudinal shear
and low angular velocity drift towards the equator similar to
the sunspot and prominence zones.

080.054   Solar magnetic structures.   K. P. White III.
Publ. Astron. Soc. Pacific, Vol. 89, 624 (1977).
Abstract.

080.055   Anomalous solar rotation in the early 17th century.
J. A. Eddy, P. A. Gilman, D. E. Trotter.
Science, Vol. 198, 824 - 829 (1977).
The character of solar rotation has been examined for
two periods in the early 17th century for which detailed sunspot drawings are available: A.D. 1625 through 1626 and 1642
through 1644. The second period occurred just at the commencement of the Maunder sunspot minimum, 1645 through
1715. Solar rotation in the earlier period was much like that of
today. In the later period, the equatorial velocity of the sun
was faster by 3 to 5 percent and the differential rotation was
enhanced by a factor of 3. It seems likely that the change in
rotation of the solar surface between 1625 and 1645 was
associated with the onset of the Maunder Minimum.

080.056   The mean magnetic field of the sun: method of
observation and relation to the interplanetary mag-
netic field.   P. H. Scherrer, J. M. Wilcox, V. Kotov,
A. B. Severny (Severnyj), R. Howard.
Sol. Phys., Vol. 52, 3 - 12 (1977).
The mean solar magnetic field as measured in integrated
light has been observed since 1968. Since 1970 it has been observed both at Hale Observatories and at the Crimean Astrophysical Observatory. The observing procedures at both observatories and their implications for mean field measurements
are discussed. A comparison of the two sets of daily observations shows that similar results are obtained at both observatories. A comparison of the mean field with the interplanetary
magnetic polarity shows that the IMF sector structure has the

same pattern as the mean field polarity.

080.057   **On the sun's pole-equator flux differences.**
G. Belvedere, L. Paternò.
Sol. Phys., Vol. 52, 191 - 198 (1977).

The authors study the possibility that large flux differences between the poles and the equator at the bottom of the solar convective zone are compatible with the small differences observed at the surface. The consequences of increasing the depth of the convective zone due to overshooting are explored.

080.058   **Studies of solar magnetic fields. V: The true average field strengths near the poles.**   R. Howard.
Sol. Phys., Vol. 52, 243 - 248 (1977).

An estimate of the average magnetic field strength at the poles of the Sun from Mount Wilson measurements is made by comparing low latitude magnetic measurements in the same regions made near the center of the disk and near the limb. There is still some uncertainty because the orientation angle of the field lines in the meridional plane is unknown, but the most likely possibility is that the true average field strengths are about twice the measured values (0–2 G), with an absolute upper limit on the underestimation of the field strengths of about a factor 5. The measurements refer to latitudes below about 80°.

080.059   **Infrared continuum observations of five-minute oscillations.**   C. A. Lindsey.
Sol. Phys., Vol. 52, 263 - 281 (1977).

Infrared continuum observations of the Sun at wavelengths between 10μ and 30μ show a nonisothermal response of the upper photosphere to compression waves associated with the five-minute oscillations. Observations were made with four broad-band filters with effective transmission wavelengths between 10μ and 26μ and with a $10''$ aperture.

080.060   **On the occurrence of critical levels in solar magneto-hydrodynamics.**   J. A. Adam.
Sol. Phys., Vol. 52, 293 - 307 (1977).

It is shown that the singular behaviour exhibited by a solution of the magnetoatmospheric wave equation for motion in the presence of a horizontal magnetic field is a special case of the 'valve' type critical level discussed by Acheson (1973), with the difference that the 'valve effect' does not strictly occur; waves are captured as they approach the singular level from either side and are neither reflected or transmitted, but constrained to propagate along the field line. This effect is also likely to occur for purely vertical fields. The possible importance of such critical levels to solar physics is discussed.

080.061   **On a possible mechanism of solar faculae heating.**
S. I. Vainstein (*Vajnshtejn*), G. V. Kuklin, V. P. Maksimov.
Sol. Phys., Vol. 53, 15 - 23 (1977).

A new mechanism of solar faculae heating is suggested. Interaction of the convective motion with the magnetic field results in decrease of its scale down to values providing for an ohmic dissipation and leading to heating at the photospheric level. Photospheric magnetic fields, faculae and granulation are considered as a combined problem. The heating mechanism causes the observed correlation of faculae brightness with the velocity field. Some points of observation are proposed for examining the action of the suggested mechanism.

080.062   **Computer solutions for studying correlations between solar magnetic fields and Skylab X-ray observations.**
D. Teuber, E. Tandberg-Hanssen, M. J. Hagyard.
Sol. Phys., Vol. 53, 97 - 110 (1977).

A method is described which uses the NASA-Marshall Space Flight Center (MSFC) Image Data Processing System, MSFC magnetograph data, and X-ray as well as Hα observations from the Skylab mission. Solutions of Laplace's equation in three dimensions, based on the magnetograph data, are convolved with observed X-ray and Hα regions. Matched filtering (template matching) provides a best fit of the observed X-ray regions to the computed total magnetic vector magnitude between 10000 and 15000km above the photosphere.

080.063   **Electric fields in the solar atmosphere.**
D. T. F. Möhlmann.
Sol. Phys., Vol. 54, 151 - 153 (1977).

The gross-structure of the force-free currents in the solar atmosphere and their possible dynamics have been discussed as caused by quasi short-circuited electric fields, generated by the motion of the solar magnetic features.

080.064   **Solar magnetic fields and convection. VIII. Meridional motions.**   J. H. Piddington.
Astrophys. Space Sci., Vol. 47, 237 - 252 (1977).

Solar meridional drift motions are vitally important in connection with the origin of magnetic fields, the source of differential rotation, and perhaps convection. The author attempts to interpret the wide variety of observational evidence, almost all of which concerns the proper motions of tracers or relatively small-scale, long-lived structures whose motions on the disk may be followed. Without exception these tracers represent magnetic field structures, and their various forms include sunspots, filaments and large-scale magnetic regions. The interpretation of the observed motions of these magnetic structures depends to a considerable extent on the adopted model of the submerged magnetic fields.

080.065   **Solar magnetic fields and convection. IX. A primordial magnetic field.**   J. H. Piddington.
Astrophys. Space Sci., Vol. 47, 319 - 340 (1977).

The solar magnetic fields observed in active regions and their residues are thought to be parts of toroidal field systems renewed every 11-yr cycle from a poloidal field. The latter may be either a reversing (dynamo) field or a non-reversing, primordial field. The latter view was held for some 70 yr, but the apparent reversals of the polar-cap fields in 1957−8 and the development of dynamo theory brought wide acceptance of the former. Here the author considers evidence for and against each model.

080.066   **Chemistry of the solar neutrino experiment.**
B. Banerjee, S. M. Chitre, P. P. Divakaran, K. S. V. Santhanam.
Astrophys. Space Sci., Vol. 48, 445 - 451 (1977).

The possibility of chemical 'trapping' of the Ar^+ ion in the reaction $\nu + {}^{37}Cl \rightarrow {}^{37}Ar^+ + e^-$, when it takes place in tetrachloroethylene (C_2Cl_4) liquid, is examined in detail. It is concluded that if trapping does take place, the rate is much smaller than the charge neutralization rate. Therefore, this mechanism cannot explain the observed small rate of Ar production in the Brookhaven solar neutrino experiment. A detailed examination of a number of experiments which are sensitive to possible trapping lends strong support to this conclusion.

080.067   **Linear hydrodynamical equations coupled with radiative transfer in a non-isothermal atmosphere.**
I. Method.   B. Schmieder.
Sol. Phys., Vol. 54, 269 - 288 (1977).

A method coupling the hydrodynamical equations and radiative transfer in a realistic solar model atmosphere is described. The influence of the temperature gradient of the model and the radiative dissipation is pointed out.

080.068   **Convection in a rotating deep compressible spherical shell: application to the Sun.**
G. Belvedere, L. Paternò.
Sol. Phys., Vol. 54, 289 - 312 (1977).

The authors study the interaction of rotation with convection in a deep compressible spherical shell as the Sun's convection zone. They examine how the energy transport and the large scale motions can be affected by rotation. In particular they study how a large scale meridional circulation can give rise to variations of angular velocity with latitude and depth. It is assumed that the energy transport is only due to convection and that the mixing-length theory gives an adequate representation of it. Furthermore it is assumed that rotation acts as a perturbation of the turbulent convective flux through its transport coefficient. The equations involved in the model are integrated numerically in the limit of large viscosity and slow rotation. The results show a three-cell circulation extending from the poles to the equator.

080.069　The mean magnetic field of the Sun: observations at Stanford.　P. H. Scherrer, J. M. Wilcox, L. Svalgaard, T. L. Duvall, Jr., P. H. Dittmer, E. K. Gustafson.
Sol. Phys., Vol. 54, 353 - 361 (1977).

A solar telescope has been built at Stanford University to study the organization and evolution of large-scale solar magnetic fields and velocities. The primary objective of building the new observatory is to permit dedicated synoptic observations of the large-scale structures. The main observing program to date has been sun-as-a-star integrated light observations of the mean solar magnetic field. The instrument and mean field observations are described.

080.070　Transmission of acoustic wave energy across a magnetic flux sheath.
V. Venkatakrishnan, M. H. Gokhale.
Sol. Phys., Vol. 54, 371 - 375 (1977).

The presence of finite current sheets at the boundaries of a magnetic flux sheath will lead to a somewhat reduced transmission of the energy of an incident acoustic wave.

080.071　Apparent yearly precession of the Sun.
A. Kubičela, M. Karabin.
Sol. Phys., Vol. 54, 505 - 509 (1977).

The apparent yearly precession of the solar globe has been defined and described in some detail. Line-of-sight velocities less than ±18 m s^{-1} occur at the solar disk. The importance of the line-of-sight component of the precession in general, and as a test velocity in contemporary photospheric velocity research is pointed out.

080.072　Heat flow near obstacles in the solar convection zone.　H. C. Spruit.
Sol. Phys., Vol. 55, 3 - 34 (1977).

The aim of this paper is to describe, with quantitative examples, the behaviour of inhomogeneities in the heat flow in the solar convection zone. As causes of the inhomogeneities the author considers obstacles, i.e. regions where the efficiency of heat transport is reduced. In the calculations the convection zone is treated as a horizontally homogeneous medium, characterised by a convective thermal conductivity which is a function of depth into the Sun only. The diffusion equation, and the type of boundary value problem are discussed that are used for the numerical heat flow calculations. A diffusion coefficient from mixing length principles is derived, and its directional and depth dependence discussed. A numerical approximation of the diffusion coefficient is given, convenient for analytical purposes, and the consequences of its depth dependence are discussed. The author studies by numerical examples the effect on the surface temperature due to a thermally isolating (disc-shaped) object buried in the convection zone, and the magnitude and extent of the brightening around sunspots. Quantitative estimates of the influx of heat into small flux tubes, as discussed above, are given.

080.073　The determination of vector magnetic fields from Stokes profiles.
L. H. Auer, J. N. Heasley, L. L. House.
Sol. Phys., Vol. 55, 47 - 61 (1977).

The application of Unno's (1956) solution of the transfer equation for polarized radiation to the determination of the vector magnetic field is investigated. An analysis procedure utilizing non-linear least squares techniques is developed that allows one to automate the reduction of measured spectral profiles of the Stokes parameters to determine the field angles, strength as well as other parameters. The method is applied to synthetic spectra generated using a model solar atmosphere and yields results of remarkably high accuracy. The influence of additional factors upon determination of the vector field are also considered. These factors include effects of: asymmetric profiles, magneto-optical effects, magnetic field gradients, unresolved field elements, scattered light, and instrumental noise.

080.074　Correlation of the effective depth of the convective zone with the solar activity cycle.
N. I. Kozhevnikov.
Astron. Tsirk., No. 945, p. 3 - 4 (1977). In Russian.

080.075　Model of the convective zone and its possible response to outside action.　N. I. Kozhevnikov.
Astron. Tsirk., No. 945, p. 5 - 6 (1977). In Russian.

080.076　On the problem of tides on the sun.
A. I. Khlystov.
Astron. Tsirk., No. 951, p. 1 - 2 (1977). In Russian.

080.077　The amplitudes of resonance oscillations on the sun.　A. I. Khlystov.
Astron. Tsirk., No. 951, p. 2 - 4 (1977). In Russian.

080.078　Empirical study of deviation from LTE in the solar atmosphere.　G. F. Sitnik, A. A. Galal.
Astron. Tsirk., No. 952, p. 1 - 3 (1977). In Russian.

080.079　On the role of non-stationary mechanisms in maintaining non-uniform distribution of angular velocity through the thickness of the solar convection zone.
A. S. Zherbina.
Izv. Glav. Astron. Obs. Pulkovo, No. 195, Astrofiz. Astrometr., p. 123 - 133 (1977). In Russian.

Non-stationary rotation of a viscous convective layer with friction at the lower boundary is considered. Such a layer is capable of maintaining for a long time the dependence of angular velocity on radius. It is shown that this mechanism cannot cause the equatorial acceleration of the sun.

080.080　The magnetic field of active regions and its zero points.　M. M. Molodenskij, S. I. Syrovatskij.
Astron. Zh. Akad. Nauk SSSR, Vol. 54, 1293 - 1304 (1977). In Russian. English translation in Soviet Astron., Vol. 21, No. 6.

A possibility of a potential approximation for the description of the magnetic field in the chromosphere and the corona is considered. It is shown that the number of peculiar (zero) points of the magnetic field which lie higher than the photospheric layer is determined by the number of maxima and minima of the potential on the photosphere. The arrangement of the peculiar points determines the general topologic structure of the field. Neumann's problem for the field of active regions is solved. As an example for the application of the proposed methods the results of a numerical solution of a boundary-value problem for the active region McMath 11693 are given. The structure of the magnetic field in this region is described and its peculiar points are found.

080.081　Cophasal regions on the sun.　N. I. Kozhevnikov.
Astron. Zh. Akad. Nauk SSSR, Vol. 54, 1325 - 1334 (1977). In Russian. English translation in Soviet Astron., Vol. 21, No. 6.

A subsequent development of the convective zone model

suggested by the author is given. Employing the conception of cophasal region, quantitative relations for kinematic parameters of spot groups, agreeing well with observations, are constructed. The effective depth of the convective zone is found to change with the phase of the solar activity cycle.

080.082 Solar interior.   C. J. Durrant, I. W. Roxburgh.
Illustrated glossary for solar and solar-terrestrial physics, (see 003.022), p. 1 - 6 (1977).

080.083 Solar cycle, solar rotation and large-scale circulation.   R. Howard.
Illustrated glossary for solar and solar-terrestrial physics, (see 003.022), p. 7 - 12 (1977).

080.084 Non-spot magnetic fields.   J. W. Harvey.
Illustrated glossary for solar and solar-terrestrial physics, (see 003.022), p. 13 - 20 (1977).

080.085 General theoretical terms.   C. J. Durrant.
Illustrated glossary for solar and solar-terrestrial physics, (see 003.022), p. 139 - 147 (1977).
Concerning spectral lines formed in the presence of a magnetic field.

080.086 When the sun dies (published and commented by N. K. Gavryushin).   K. Ėh. Tsiolkovskij.
Istor.-astron. issled. Vyp. (No.) 3. Moskva, Nauka, 1977, p. 391 - 397. In Russian. − Abstr. in Ref. zh., 51. Astron., 12.51.7 (1977).

080.087 Solar neutrinos and the role of exchange currents in *pp*-reaction.
Yu. M. Andreev, Eh. V. Bugaev, Yu. S. Kopysov.
Pis'ma v ZhEhTF, Vol. 25, 593 - 596 (1977). In Russian. Abstr. in Ref. zh., 51. Astron., 12.51.322 (1977).

080.088 Active processes on the sun and the problem of solar neutrinos.   G. E. Kocharov.
VIII Leningr. mezhdunar. semin. Mater. mezhdunar. semin. "Akt. protsessy na Solntse i probl. soln. nejtrino", 1976, Leningrad. 1976, p. 13 - 28. In Russian. − From Ref. zh., 51. Astron., 12.51.323 (1977).

080.089 On a necessary condition of quasistationary matter outflow from the sun.   M. V. Konyukov.
Simpoz. po fiz. geomagnitosfery, 1977. Tezisy dokl. Irkutsk, 1977, p. 3. In Russian. − Abstr. in Ref. zh., 51. Astron., 12.51.340 (1977).

080.090 On the possibility of resonance enhancement of tides in the solar convective zone.
A. I. Khlystov.
Soln. Dannye 1977 Byull., No. 10, p. 78 - 87 (1977). In Russian.
It is shown that the presence of long-period proper oscillations with a period $T \geqslant 1$ month is quite possible in the solar convective zone, a resonance with forced tidal oscillations taking place at $T = 1.3$ and 6 months. Considering the solar convective zone as a linear non-conservative oscillation system one may explain the ascending branch of the 11-year cycle as a result of resonance enhancement of oscillations while the descending branch as damping of the oscillations.

080.091 Magnetic flux tubes and transport of heat in the convection zone of the sun.   H. C. Spruit.
Proefschrift Rijksuniv. Utrecht, Netherlands, 145 pp. (1977). Abstr. from INIS 7713 314401.

080.092 Comparative magnetospherology. Part 4. Model of the neutral sheet in the inner heliomagnetosphere for the year 1976.   T. Saito.
10th Lunar and Planetary Symposium, (see 012.050), p. 206 - 214 (1977).
A warped structure of the neutral sheet in the inner helio-magnetosphere ($\lesssim 1$ AU) for the year 1976 is obtained from observed interplanetary magnetic field data. The 1976 model is found to give a synthetic explanation of various solar, interplanetary, and terrestrial phenomena.

080.093 Comparative magnetospherology. Part 5. Examination of the two-hemisphere model by means of the apparent sector structure of the Jovian magnetosphere.
T. Saito, H. Oya, K. Takahashi.
10th Lunar and Planetary Symposium, (see 012.050), p. 215 - 221 (1977).
The characteristics observed in the Jovian magnetosphere and characteristics expected to be observed in the earth's magnetosphere are compared from the viewpoint of comparative magnetospherology with the characteristics observed in the interplanetary space. It is concluded from common characteristics among these three that the two-hemisphere model does well express the actual magnetic structure of the heliosphere.

080.094 Comparative magnetospherology. Part 6. Comparison of the Japanese and American models on the snail-shell type structure of the inner heliomagnetosphere.
T. Saito.
10th Lunar and Planetary Symposium, (see 012.050), p. 222 - 228 (1977).
A snail-shell structure of the magnetically neutral sheet of the heliosphere derived from the two-hemisphere model (Saito, 1975) is compared with another snail-shell structure proposed by Svalgaard and Wilcox (1976).

080.095 Magnetic flux tubes and transport of heat in the convection zone of the sun.   H. C. Spruit.
Thesis, Utrecht University (1977) = Utrechtse Sterrenkundige Overdrukken, No. 388.

080.096 Investigation of the sun and the Intercosmos satellites.   B. Valníček.
Kozmos, Vol. 8, 146 - 150 (1977). In Czech.

The quiet sun.   See Abstr. 003.059.

Astronomie I. Die Sonne und ihre Planeten. See Abstr. 003.063.

Solen − en innføring i moderne solfysikk. See Abstr. 003.077.

Methodical improvements of magnetographic measurements.   See Abstr. 031.222.

Some comments on the measurement of small scale strong magnetic fields on the sun.   See Abstr. 031.253.

Solar radiometry in the millimetre region with a local calibration source.   See Abstr. 031.256.

Rocket calibration of the Nimbus 6 solar constant measurements.   See Abstr. 032.530.

The LPSP (*Laboratoire de Physique Stellaire et Planétaire*) experiment on OSO-8. I: Instrumentation, description of operations, laboratory calibrations and pre-launch performances.   See Abstr. 032.570.

Neutrino absorption cross sections for ^{37}Cl with applications.   See Abstr. 061.010.

Solar magnetoatmospheric waves − a simplified

mathematical treatment.   See Abstr. 062.006.

On second order statistical moments in MHD-turbulence.   See Abstr. 062.014.

Generation of a deeply penetrating gravity-like oscillation by a non-resonant second-order interaction of waves.   See Abstr. 062.018.

Flows in magnetic flux tubes.
See Abstr. 062.022.

Magnetic field transfer by two-dimensional convection and solar 'semi-dynamo'.   See Abstr. 062.043.

Hydrodynamic instability of convectively unstable atmospheres in shear flow.   See Abstr. 062.065.

Magnetic field energy dissipation in neutral current sheets.   See Abstr. 062.074.

Compton effect and solar spectral lines.
See Abstr. 063.032.

Multi-dimensional transfer of polarized radiation in magnetic fluxtubes.   See Abstr. 063.042.

A non-local-thermodynamic equilibrium formulation of the transport equation for polarized light in the presence of weak magnetic fields.   See Abstr. 063.054.

Non-thermal velocities in stellar atmospheres. II. Six giants and the Sun.   See Abstr. 064.019.

Statistical analysis of solar Fe I lines: magnetic line broadening.   See Abstr. 071.002.

The solar chemical composition.
See Abstr. 071.025.

Physique solaire: interprétation mécanique de l'activité solaire.   See Abstr. 072.028.

Facular models, the K-line, and magnetic fields.
See Abstr. 072.033.

The origin of solar activity.   See Abstr. 072.035.

Influence of photospheric magnetic fields on peculiarities in the propagation of solar flare protons.
See Abstr. 073.042.

Chromospheric rotation during 1972 - 73, years of declining activity.   See Abstr. 073.089.

Momentum and energy transport by waves in the solar atmosphere and solar wind.   See Abstr. 074.003.

Large-scale magnetic structures and the longitudinal distribution of the green coronal emission.
See Abstr. 074.019.

Coronal rotation dependence on the solar cycle phase.   See Abstr. 074.056.

Koronale Löcher und das solare Magnetfeld.
See Abstr. 074.100.

Bright X-ray arcs and the emergence of solar magnetic flux.   See Abstr. 076.002.

Nonrelativistic electron stream propagation in the solar atmosphere and type III radio bursts.
See Abstr. 077.044.

Equatorial solar rotation and its relation to climatic changes.   See Abstr. 085.010.

The role of solar active regions and of the total magnetic field of the sun in the 11-year cosmic-ray cycle.
See Abstr. 143.030.

Earth

081 Figure, Composition, and Gravity

081.001 Inversion of the attenuation data of free oscillations of the Earth (fundamental and first higher modes).
A. Deschamps.
Geophys. J. R. Astron. Soc., Vol. 50, 699 - 722 (1977).

081.002 Astronomical azimuths of terrestrial objects as indicators of rotational motions of continental blocks. V. P. Shcheglov.
Astron. Zh. Akad. Nauk SSSR, Vol. 54, 884 - 889 (1977).
In Russian. English translation in Soviet Astron., Vol. 21, No. 4.
Considerations on rotational motions of continental blocks are formulated and suggestions for a check-up of these motions with the help of modern methods are given. An analysis of azimuthal determinations of the axis direction of the meridian instrument of the Ulugh Beg Observatory in Samarkand is given.

081.003 On the stationary points of the gravitation fields of the earth, the moon and Mars. S. G. Zhuravlev.
Astron. Zh. Akad. Nauk SSSR, Vol. 54, 909 - 914 (1977).
In Russian. English translation in Soviet Astron., Vol. 21, No. 4.
The coordinates of stationary points of the gravitation fields of the earth, the moon and Mars as singular points of a surface of zero-velocity have been found.

081.004 Das "International Gravity Standardization Net 1971 (IGSN 71)" in der Bundesrepublik Deutschland (Transformation des Deutschen Schwerenetzes 1962 und Stationsbeschreibungen des IGSN 71).
W. Doergé, E. Reinhart, G. Boedecker.
Deutsche Geod. Komm. Bayerische Akad. Wiss., Reihe B: Angew. Geod., Heft Nr. 225, 61 pp. (1977).
For the transformation between the German Gravity Net 1962 (Deutsches Schwerenetz 1962, DSN 62) and the International Gravity Standardization Net 1971 (IGSN 71) parameters (differences in reference level and scale) are determined by least-squares adjustment. For the comparison gravity measurements have been made at 111 referring stations. An Appendix comprises 79 sketches of IGSN 71-stations in the Federal Republic of Germany, updated in 1976 and recommended for use.

081.005 Comparison of Doppler derived undulations with gravimetric undulations considering the zero-order undulations of the geoid. R. H. Rapp, R. Rummel.
Satellite Doppler positioning, (see 012.002), p. 389 - 397.

081.006 On the depth of the lateral inhomogeneities of the earth's density. K. Arnold.
Gerlands Beitr. Geophys., Band 86, 278 - 290 (1977).

081.007 Internal waves in the earth's core. P. Olson.
Geophys. J. R. Astron. Soc., Vol. 51, 183 - 215 (1977).
The purpose of this paper is to describe some possible modes of oscillation of the earth's outer core which arise because of its density stratification, and to develop some criteria by which observations of these waves may be used to determine the thermal state of the outer core.

081.008 A comment on 'Three-dimensional statistical gravity disturbance model' by J. G. Negi and V. P. Dimri. T. J. Ulrych.
Geophys. J. R. Astron. Soc., Vol. 51, 271 - 272 (1977).

081.009 Reducing greenhouses and the temperature history of Earth and Mars. C. Sagan.
Nature, Vol. 269, 224 - 226 (1977).
The results indicate that if primitive Mars, like primitive Earth, had even a mildly reducing atmosphere, martian global temperatures a few times 10^9 yr ago may have been in the vicinity of the freezing point of seawater and, in many latitudes, the mean temperatures may have been above the freezing point.

081.010 Origin of the Earth's ocean basins. H. Frey.
Icarus, Vol. 32, 235 - 250 (1977).
The Earth's original ocean basins are proposed to be mare-type basins produced 4 billion y.a. by the flux of asteroid-sized objects responsible for the lunar mare basins.

081.011 A method of direct gravimetrical determination of differences of geoid undulations. H. Drewes.
Deutsche Geod. Komm. Bayerische Akad. Wiss., Reihe B: Angew. Geod., Heft Nr. 221, p. 13 - 20 (1977).

081.012 Magnetic profiles diagnostic of maar craters: anomalies associated with peripheral ring dikes.
L. S. Crumpler, J. C. Aubele, W. E. Elston.
NASA Tech. Mem., NASA TM X-3511, (see 012.010), p. 122 - 125 (1977). — Abstract.

081.013 Geologic, topographic, and meteorologic influences on eolian deposition, Earth and Mars.
A. D. Howard.
NASA Tech. Mem., NASA TM X-3511, (see 012.010), p. 148 - 149 (1977). — Abstract.

081.014 Erosional and depositional features of some terrestrial 'channels'.
J. C. Boothroyd, T. J. Donlon, D. Nummedal.
NASA Tech. Mem., NASA TM X-3511, (see 012.010), p. 173 - 175 (1977). — Abstract.

081.015 Deep-sea channels: another Earth analogy with Martian channels.
P. D. Komar, C. E. Reimers, R. Dolan.
NASA Tech. Mem., NASA TM X-3511, (see 012.010), p. 176 -

177 (1977). — Abstract.

081.016 On Stokes' problem.   M. I. Yurkina.
Izv. vyssh. uchebn. zaved. Geod. i aehrofotosemka,
1977, No. 1, p. 63 - 70. In Russian. — Abstr. in Ref. zh., 52.
Geod. Aehrosemka, 9.52.69 (1977).

**081.017 Die GRIM 2-Kombinationslösung und das detail-
lierte gravimetrische GRIM 2-Geoid.**
C. Reigber, G. Balmino, B. Moynot.
Veröff. Bayerisch. Komm. Int. Erdmessung, Bayerisch. Akad.
Wiss., Astron.-Geod. Arbeiten, Heft Nr. 36, p. 230 - 240
(1977).

**081.018 Vergleich des GRIM 2- und DGG2-Geoides mit
GSFC- und SAO-Geoiden sowie Altimetrie.**
C. Reigber, G. Balmino, W. Seemüller.
Veröff. Bayerisch. Komm. Int. Erdmessung, Bayerisch. Akad.
Wiss., Astron.-Geod. Arbeiten, Heft Nr. 36, p. 241 - 254
(1977).

**081.019 Improvement of the geoid in local areas by satellite
gradiometry.**   K. P. Schwarz, J. Kryński.
Bull. Géod., Vol. 51, 163 - 176 (1977).
Satellite gradiometry is studied as a means to improve the
geoid in local areas from a limited data coverage. It is shown
that only three second-order gradients contribute significantly
to the estimation of the geoidal undulations and that it is sufficient to have gradiometer data in a 5° × 5° area around the
estimation point.

**081.020 Core-resonance effects on the Earth's angular
momentum vector and rotation axis — a generalized
model.**   P. McClure.
Scientific applications of lunar laser ranging, (see 012.012),
p. 131 - 132 (1977). — Abstract.

**081.021 Positions of the axes of the ellipsoid of inertia from
satellite observations.**   M. Burša.
Bull. Astron. Inst. Czechoslovakia, Vol. 28, 316 - 318 (1977).
An attempt was made in determining the position of the
pole of the ellipsoid of inertia of the Earth relative to a
reference pole. In the GEM 5 and GEM 6 reference systems
the position of the pole of the ellipsoid of inertia is 99.5° W
and its distance from the reference pole 1.5''. As regards the
Moon the position is 95.5° W and the pole distance 1.6°.

**081.022 Variational type finite element solution of normal
modes of simple earth models.**
W. Moon, R. A. Wiggins.
Geophys. J. R. Astron. Soc., Vol. 51, 327 - 348 (1977).

**081.023 High-explosive cratering analogs for bowl-shaped,
central uplift, and multiring impact craters.**
D. J. Roddy.
Proc. Seventh Lunar Sci. Conf., (see 012.015), 3027 - 3056
(1976).

**081.024 Terrestrial lead isotopic evolution and formation
time of the Earth's core.**   R. Vollmer.
Nature, Vol. 270, 144 - 147 (1977).
The radiogenic nature of Pb in average crust and mantle
probably requires some interaction between mantle and core
during core formation and is thus incompatible with heterogeneous accretion models for the Earth. Pb isotope evolution
models for hot and cool accretion indicate that core formation
was not quasi-simultaneous with accretion although the core
was probably 95% completed within the first 700 Myr of the
Earth's history.

081.025 Ring-shaped structure of the earth.
V. N. Bryukhanov, M. Z. Glukhovskij, A. L. Stavtsev.

Priroda, 1977, No. 10, p. 54 - 65. In Russian.

**081.026 On the fluctuation of the length of day and its
influence on gravity.**   E. Lindinger.
Österreich. Z. Vermessungswes. Photogramm., 64. Jahrg., 86 -
89 (1977). In German.

081.027 Is the Earth contracting?   G. M. Blake.
Geophys. J. R. Astron. Soc., Vol. 51, 555 - 559
(1977).
The fossil record of the variation of the solar day and the
synodic month with geological time is examined for evidence
of the steady contraction of the Earth postulated by Lyttleton
to explain a discrepancy between the apparent secular accelerations of the Sun and Moon.

081.028 The tides.   H. Brinton.
Modern astronomy, (see 003.013), p. 49 - 55 (1977).

**081.029 Some geophysical and geodetic contributions of
satellite-determined gravity results.**   M. A. Khan.
Geophys. Surv., Vol. 2, 469 - 496 (1976). — Abstr. in Phys.
Abstr., Vol. 80, Abstr. 25634 (1977).

**081.030 Models for the origin and composition of the Earth,
and the hypothesis of potassium in the Earth's core.**
K. A. Goettel.
Geophys. Surv., Vol. 2, 369 - 397 (1976). — Abstr. in Phys.
Abstr., Vol. 80, Abstr. 25663 (1977).

**081.031 On a model of magneto-mechanical interaction of
the core-mantle system.**   V. N. Plakhotnyuk.
Din. kosm. plazmy. Moskva, 1976, p. 219 - 229. In Russian.
Abstr. in Ref. zh., 51. Astron., 10.51.186 (1977).

**081.032 Variation method for solving discrete problems of
physical geodesy.**   Yu. M. Nejman.
Izv. vyssh. uchebn. zaved. Geod. i aehrofotosemka, 1977,
No. 1, p. 21 - 27. In Russian. — Abstr. in Ref. zh., 52. Geod.
Aehrosemka, 10.52.67 (1977).

081.033 Short arc reductions of GEOS-3 altimetric data.
G. Hadgigeorge, J. E. Trotter.
Geophys. Res. Lett., Vol. 4, 223 - 226 (1977).

**081.034 Expansion of the earth due to a secular decrease in
G — evidence from Mercury.**
D. J. Crossley, R. K. Stevens.
Canadian J. Earth Sci., Vol. 13, 1723 - 1725 (1976). — Abstr.
in Phys. Abstr., Vol. 80, Abstr. 32972 (1977).

081.035 On a 'constant gravity' model of the Earth's interior.
R. Chander.
American J. Phys., Vol. 45, 399 - 400 (1977). — Abstr. in
Phys. Abstr., Vol. 80, Abstr. 45645 (1977).

081.036 Modern standards for gravity surveys.
Geoexploration, Vol. 15, No. 1, p. 65 - 66 (1977).
Abstr. in Phys. Abstr., Vol. 80, Abstr. 48990 (1977).

**081.037 Gravity gradients and the interpretation of the
truncated plate.**   J. M. Stanley, R. Green.
Geophysics, Vol. 41, 1370 - 1376 (1976). — Abstr. in Phys.
Abstr., Vol. 80, Abstr. 44890 (1977).

**081.038 Observations on the validity of the meteoritic Earth
model.**   L. Forni.
Atti Accad. Sci. Torino I, Vol. 109, 109 - 117 (1975). In
Italian. — Abstr. in Phys. Abstr., Vol. 80, Abstr. 44982 (1977).

081.039 Surface temperature of early Earth.
A. Henderson-Sellers, A. J. Meadows.

Nature, Vol. 270, 589 - 591 (1977).

Changes in the surface temperature of the Earth throughout its history are important for understanding both the geological development of the Earth's surface and the origin and development of life on Earth. The authors demonstrate that the accepted boundary conditions for the early Earth can be satisfied by a model that postulates absorption by water vapour and carbon dioxide only throughout the lifetime of the Earth.

081.040 **A question of gravity.**   R. A. Gibb, M. D. Thomas.
GEOS, p. 5 - 8 (Winter 1977). – Abstr. in Phys.
Abstr., Vol. 80, Abstr. 56567 (1977).

081.041 **Inversion of gravity data by use of a method of compressed mass plane to estimate crustal structure.**
Y.-L. Liu, C.-S. Wang.
Acta Geophys. Sinica, Vol. 20, No. 1, p. 59 - 69 (1977). In Chinese. – Abstr. in Phys. Abstr., Vol. 80, Abstr. 64175 (1977).

081.042 **Gravity-field determination from laser observations.**
E. M. Gaposchkin.
Philos. Trans. R. Soc. London, Ser. A, Vol. 284, 515 - 527 (1977). – Abstr. in Phys. Abstr., Vol. 80, Abstr. 64184 (1977).

081.043 **Tidal studies from the perturbations in satellite orbits.**
A. Cazenave, S. Daillet, K. Lambeck.
Philos. Trans. R. Soc. London, Ser. A, Vol. 284, 595 - 606 (1977). – Abstr. in Phys. Abstr., Vol. 80, Abstr. 64187 (1977).

081.044 **Equation of state of liquid iron at the Earth's core conditions.**   E. Boschi, F. Mulargia.
J. Geophys., Vol. 43, 465 - 472 (1977). – Abstr. in Phys. Abstr., Vol. 80, Abstr. 73694 (1977).

081.045 **Least-squares collocation and the gravitational inverse problem.**   H. Moritz.
J. Geophys., Vol. 43, 153 - 162 (1977). – Abstr. in Phys. Abstr., Vol. 80, Abstr. 73872 (1977).

081.046 **Gravity of the Earth by means of conical shells.**
J. C. Rainwater.
American J. Phys., Vol. 45, 768 - 769 (1977). – Abstr. in Phys. Abstr., Vol. 80, Abstr. 78928 (1977).

081.047 **Studies on planetary dynamics of the Earth by means of satellite geodesy.**
H. Kautzleben, H. Montag, E. Buschmann.
17 convegno internazionale scientifico sullo spazio (see 012. 034), p. 57 - 66 (1977). – Abstr. in Phys. Abstr., Vol. 80, Abstr. 81785 (1977).

081.048 **A note on the westward drift of the Earth's magnetic field.**   W. W. Wood.
J. Fluid Mech., Vol. 82, 389 - 400 (1977). – Abstr. in Phys. Abstr., Vol. 80, Abstr. 81794 (1977).

081.049 **A model of the early evolution of the Earth.**
A. V. Vitjazev (*Vityazev*), S. V. Majeva (*Maeva*).
Tectonophysics, Vol. 41, 217 - 225 (1977). – Abstr. in Phys. Abstr., Vol. 80, Abstr. 86845 (1977).

081.050 **Some actual problems in the interpretation of geodynamic processes.**   H. Kautzleben.
Veröff. Zentralinst. Phys. Erde, No. 52, Part 1, (see 012.035), p. 17 - 29 (1977).

081.051 **Recent possibilities for the determination of the geopotential from terrestrial and satellite data.**
L. P. Pellinen.

Veröff. Zentralinst. Phys. Erde, No. 52, Part 1, (see 012.035), p. 145 - 167 (1977).

081.052 **Further evaluation of the GRIM 2 earth gravity field model.**
G. Balmino, C. Reigber, B. Moynot.
Veröff. Zentralinst. Phys. Erde, No. 52, Part 1, (see 012.035), p. 193 - 232 (1977).

In order to establish a reasonable accuracy estimate of the recently derived GRGS/SFB 78 - GRIM 2 earth model, a number of tests and comparisons with other solutions and external standards are presented. Sources of comparisons are deep space network results, other near earth satellite derived solutions, survey and global geoid informations, the new determined detailed gravimetric GRIM 2 geoid, satellite orbit computations, SKYLAB IV and GEOS 3 radar altimeter profiles.

081.053 **Crustal movements.**
South. Stars, Vol. 27, 37 (1977).

081.054 **Deviation of the vertical in the South Island, New Zealand.**   J. B. Mackie.
South. Stars, Vol. 27, 54 - 62 (1977).

The paper explains briefly what deviation of the vertical is, and goes on to show how astro-geodetic deviations, determined by precise astronomical observations in the field, may be used to define the undulations of the geoid.

081.055 **Transformation of Stokes' constants in the rotation of a coordinate system.**   A. N. Marchenko.
Geod., kartogr. i aehrofotosemka. Resp. mezhved. nauchn.-tekh. sb., 1977, vyp. (No.) 26, p. 46 - 55. In Russian. – Abstr. in Ref. zh., 52. Geod. Aehrosemka, 11.52.49 (1977).

081.056 **On the error of harmonic coefficients of the geopotential.**   V. I. Umlenski, V. G. Shkodrov.
Dokl. Bolg. AN, Vol. 30, 535 - 537 (1977). In Russian. Abstr. in Ref. zh., 52. Geod. Aehrosemka, 11.52.50 (1977).

081.057 **Method of determining the earth's gravitational potential by measurement of the topographic coordinates of the ocean's surface.**
I. P. Nedyalkoy, D. N. Mishev, T. K. Yanev.
Dokl. Bolg. AN, Vol. 30, 367 - 370 (1977). – Abstr. in Ref. zh., 52. Geod. Aehrosemka, 11.52.73; 62. Issled. kosm. prostranstva, 11.62.150 (1977).

081.058 **Neregulaĵoj da masditribuo en la terkrusto.**
B. Popovič.
Bull. Obs. Astron. Belgrade, No. 128, p. 1 - 6 (1977).

081.059 **Revised estimation of 550-km X 550-km mean gravity anomalies.**   M. R. Williamson.
Smithsonian Astrophys. Obs., Special Rep. 377, 5 + 22 pp. (1977).

The calculation of 550-km X 550-km mean gravity anomalies from $1° X 1°$ mean free-air gravimetry data is discussed. The block estimate procedure developed by Kaula is used to obtain 1504 of the 1654 possible mean block anomalies. The estimated block anomalies calculated from $1° X 1°$ mean anomalies referred to the 1967 reference ellipsoid and from $1° X 1°$ mean anomalies referred to a 24th-degree-and-order field are compared.

081.060 **Global correlation of topographic heights and gravity anomalies.**   M. C. Roufosse.
Smithsonian Astrophys. Obs., Special Rep. 378, 7 + 47 pp. (1977).

The object of this research is to study and compare, on a worldwide scale, the short-wavelength features of the earth's gravity and topographic fields to determine whether any relationship exists between them.

081.061   **Marées terrestres.**   P. Melchior (Editor).
Bull. Inf., (Obs. R. Belgique, Bruxelles), No. 76,
p. 4408 - 4503 (1977).

081.062   **Gravitational potential energy of the Earth: a
spherical harmonic approach.**   D. P. Rubincam.
GSFC Doc. X-921-77-81, Prepr., 3 + 25 + 1 + 2 pp. (1977).
A spherical harmonic equation for the gravitational
potential energy of the Earth is derived for an arbitrary density
distribution by conceptually bringing in mass-elements from
infinity and building up the Earth shell upon spherical shell.

081.063   **Chemical heterogeneity of the archaean mantle,
compositon of the earth and mantle evolution.**
S.-S. Sun, R. W. Nesbitt.
Earth Planet. Sci. Lett., Vol. 35, 429 - 448 (1977).

081.064   **A theoretical equation of state for the inner core.**
M. S. T. Bukowinski.
Phys. Earth Planet. Inter., Vol. 14, 333 - 344 (1977) = Publ.
No. 1612, Inst. Geophys. Planet. Phys., Univ. Calif., Los
Angeles, Calif., U.S.A.

081.065   **The mass and moment of inertia of the earth.**
B. Romanowicz, K. Lambeck.
Phys. Earth Planet. Inter., Vol. 15, P1 - P4 (1977).
Recent revisions of geodetic and astronomical constants
by the International Association of Geodesy and the International Astronomical Union lead to improved values for the
earth's mass and moment of inertia. Corrections to be applied
to these values before they are used as constraints in the
inversion of seismic data are discussed.

081.066   **On an instrumental phase-lag of the LaCoste &
Romberg gravimeter.**   T. Sato.
Proc. Int. Latitude Obs. Mizusawa, No. 16, p. 36 - 50 (1977).
In Japanese.

081.067   **A thermal model of the earth.**   F. D. Stacey.
Phys. Earth Planet. Inter., Vol. 15, 341 - 348 (1977).

081.068   **On the representation of a function of the inner
ellipsoid of the earth by a partial sum of the
generalized Fourier series.**
G. A. Meshcheryakov, I. N. Shopyak, Yu. P. Dejneka.
Geod. kartogr. i aehrofotosemka. Resp. mezhved. nauchn.-
tekh. sb., 1977, vyp. (No.) 26, p. 55 - 62. In Russian. — Abstr.
in Ref. zh., 52. Geod. Aehrosemka, 12.52.120 (1977).

081.069   **Determination of potential coefficients to degree 52
from 5° mean gravity anomalies.**   R. H. Rapp.
Bull. Géod., Vol. 51, 301 - 323 (1977).

081.070   **Ordinary differential equations of an elementary
oscillation of the elastic rotating earth.**
M. S. Molodenskij.
Izv. AN SSSR. Fiz. Zemli, 1977, No. 7, p. 9 - 15. In Russian.
Abstr. in Ref. zh., 52. Geod. Aehrosemka, 1.52.113 (1978).

081.071   **Frequencies of natural oscillations of the earth with
the ocean.**   M. S. Molodenskij.
Izv. AN SSSR. Fiz. Zemli, 1977, No. 9, p. 3 - 10. In Russian.
Abstr. in Ref. zh., 52. Geod. Aehrosemka, 1.52.114 (1978).

081.072   **Difference in reflecting powers of the various kinds
of terrestrial igneous rocks owing to difference in
light sources.**   T. Shimazaka.
10th Lunar and Planetary Symposium, (see 012.050), p. 1 - 6
(1977). In Japanese.

The bowels of the earth.   See Abstr. 003.046.

History of the earth.   See Abstr. 003.110.

Continents in motion.   See Abstr. 003.152.

The University of Hawaii lunar ranging experiment
geodetic-geophysics support programme.
See Abstr. 013.028.

The N.A.S.A. Earth and ocean dynamics programme.
See Abstr. 013.029.

Geophysics without geology.
See Abstr. 014.019.

Variations in the earth's orbit:pacemaker of the ice
ages?   See Abstr. 015.029.

Applications of very-long-baseline interferometry
to geodesy and geodynamics.   See Abstr. 031.292.

Characteristics of iodine stabilized He—Ne laser for
the absolute determination of gravity.
See Abstr. 034.080.

Variations in the Earth's rotation due to sectorial
and tesseral harmonics.   See Abstr. 044.001.

Dynamics of polar motion and plate tectonics.
See Abstr. 045.012.

Astrogeodetic geoid determination in the Western
Harz.   See Abstr. 046.023.

Ein dreidimensionales Berechnungsmodell für geodätische Punkt- und Geoidbestimmungen.
See Abstr. 046.037.

Arbeiten zur Bestimmung des Geoides in der Bundesrepublik Deutschland.   See Abstr. 046.060.

Cosmonauts investigate the earth.
See Abstr. 051.066.

Probing the Earth's gravity field by means of
satellite-to-satellite tracking.   See Abstr. 052.023.

Cosmological change of G and the structure of the
Earth.   See Abstr. 162.001.

082 Atmosphere Including Refraction, Scintillation, Extinction, Airglow, Site Testing

082.001 Photometry of eclipses of artificial earth satellites observed in situ (part II).
L. Neužil, I. Zacharov.
Bull. Astron. Inst. Czechoslovakia, Vol. 28, 244 - 252 (1977).

The results of measurements of the satellite eclipses carried out on board of Interkosmos 7 satellite in two spectral regions are in good agreement with the measurements of the first part of this paper (see 17.082.004). The shape of the penumbra does not agree with the theoretical hypothesis of the influence of the high absorbing layer with the generally adopted parameters.

082.002 The near infrared nightglow continuum.
P. C. Wraight.
Planet. Space Sci., Vol. 25, 787 - 794 (1977).

A review is given of observations of the nightglow continuum extending from about 7000 Å to longer wavelengths. It is suggested that the source of this emission is radiative association of atomic oxygen via the repulsive $^3\Pi_u$ state to vibrationally excited $O_2(^3\Sigma_g^-)$. Predicted spectra are compared with nightglow and laboratory observations. Other implications for the physics of the thermosphere and upper mesosphere are discussed.

082.003 Studies of the sensitivity of the components of the Earth's radiation balance to changes in cloud properties using a zonally averaged model. G. E. Hunt.
J. Quant. Spectrosc. Radiat. Transfer, Vol. 18, 295 - 307 (1977).

The accuracy of a simple radiative transfer scheme suitable for use in a general circulation model of the atmosphere is assessed by comparing the calculated radiative heat balance of the Earth/atmosphere system with the available observational data. Studies are then performed to determine the sensitivity of the radiative components to changes in the cloud data and cloud parameterisations which constitute the largest potential source of error in the model input data.

082.004 On the extinction of the atmosphere at the Gissar Observatory. I. I. Gavrilova.
Byull. Inst. Astrofiz., Dushanbe, No. 66 - 67, p. 59 (1976). In Russian.

082.005 Daily variations of temperature, pressure and density of the atmosphere in the meteor zone.
K. A. Karimov. A. A. Shnejder.
Komety Meteory, No. 24, p. 35 - 39 (1976). In Russian.

082.006 Peculiarities of circulation of the atmosphere in the meteor zone. K. A. Karimov, T. M. Mukambetov.
Komety Meteory, No. 24, p. 40 - 44 (1976). In Russian.

082.007 Molecular oxygen densities and the atmospheric absorption of solar Lyman α radiation.
J. H. Carver, B. H. Horton, M. Ilyas, B. R. Lewis.
J. Geophys. Res., Vol. 82, 2613 - 2618 (1977).

195 K cross sections have been used to analyze the data from 34 rocket measurements of the atmospheric extinction of solar Lyman α radiation. Averaging the data from the flights at all locations yields an observational mean molecular oxygen density profile which exceeds the mean Cira (1972) model by about 20% at 70 km and by about 9% at 90 km. Part of this difference may be due to residual effects on the cross section of temperature variations in the 70- to 90-km region.

082.008 Ion composition in a noctilucent cloud region.
R. A. Goldberg, G. Witt.
J. Geophys. Res., Vol. 82, 2619 - 2627 (1977).

Ion composition at mesospheric altitudes has been measured at and compared between high- and midlatitude sites under summer daytime conditions. Both rocket-borne measurements were made with pumped quadrupole ion mass spectrometers. Large quantities of hydronium cluster ions were observed through 109^+, with maximum concentrations at 55^+ and 73^+. Also cluster ions of nitric oxide were observed through 84^+. The data near the mesopause show the typical cluster ions mentioned above but also a preponderance of heavy ions between 90 and 145 amu.

082.009 Multipoint method for separation of wave components from charts of night-sky luminescence.
N. M. Gavrilov, G. M. Shved.
Geomagn. Aehron., Vol. 17, 685 - 691 (1977). In Russian.

082.010 The case of the setting Sun.
J. B. Irwin.
Sky Telesc., Vol. 54, 167 - 170 (1977).

082.011 Solar halo complexes.
Sky Telesc., Vol. 54, 185 - 187 (1977).

082.012 The 1976 Standard Atmosphere and its relationship to earlier standards. R. A. Minzner.
Rev. Geophys. Space Phys., Vol. 15, 375 - 384 (1977).

The 1976 U.S. Standard Atmosphere, representing a mid-latitude atmosphere for moderate solar activity, has recently been adopted by the Committee for the Extension of the Standard Atmosphere (COESA). This discussion examines the temperature-height profile and the corresponding density-height profile for the new standard in relation to temperature-height and density-height profiles for all previous U.S. standard atmospheres. Also discussed are the history of the development of these standard atmospheres and the relationship of these developments to the history of European standard atmospheres.

082.013 The evolution of the atmosphere of the earth.
M. H. Hart.
Bull. American Astron. Soc., Vol. 9, 452 (1977). – Abstract.

082.014 The effects of urban lighting on the brightness of the night sky. M. F. Walker.
Publ. Astron. Soc. Pacific, Vol. 89, 405 - 409 (1977) = Lick Obs. Bull., No. 760.

A new study of urban lighting and its effect on the brightness of the night sky indicates: (1) The total light output of cities of similar economic development is at least approximately proportional to their populations. (2) The artificial illumination of the night sky at 45° altitude in the direction of the illuminating city varies as $I \propto D^{-2.5}$. (3) The distance at which cities of a given population produce a brightening of the sky of 0.2 magnitude at 45° altitude toward the city varies as $P \propto D^{2.5}$.

082.015 Effect of nearby supernova explosions on atmospheric ozone.
R. C. Whitten, J. Cuzzi, W. J. Borucki, J. H. Wolfe.
Nature, Vol. 263, 398 - 400 (1976).

The authors have calculated the probable effects of a nearby supernova event on the ozone layer. They find that the effects are significant and long lasting, but are relatively rare at the location of the Earth in the Galaxy.

082.016 Extension of atomic nitrogen measurements into the lower thermosphere.
M. J. Engebretson, K. Mauersberger, W. E. Potter.
J. Geophys. Res., Vol. 82, 3291 - 3294 (1977).

082.017 Atmospheric extinction coefficients in the *uvby* system for La Silla.
C. Sterken, M. Jerzykiewicz.
Astron. Astrophys., Suppl. Ser., Vol. 29, 319 - 320 (1977).
Atmospheric extinction coefficients in the *uvby* bands obtained at La Silla on 92 nights between September 1971 and January 1976 are given.

082.018 Measurement of scattered ultraviolet radiation in the vicinity of planets and in the interplanetary medium. S. I. Babichenko, V. G. Kurt, V. A. Sklyankin, A. S. Smirnov.
[Tr.] Soyuz. NII priborostr. 1977, vyp. (No.) 34 - 35, tom (Vol.) 1, 158 - 167. In Russian. – Abstr. in Ref. zh., 62. Issled. kosm. prostranstva, 8.62.113 (1977).

082.019 The possible meteorological significance of submicron cosmic dust.
C. L. Hemenway, D. S. Hallgren.
News Lett. Astron. Soc. N.Y., Vol. 1, No. 2, p. 21 (1977). Abstract.

082.020 The meteoric night-glow. W. J. Baggaley.
Mon. Not. R. Astron. Soc., Vol. 181, 203 - 210 (1977).
There exist well-documented accounts of the observations of enhanced night-glow associated with spectacular meteor shower displays. Possible mechanisms responsible for this elusive phenomenon are examined. It is shown that the observed emission is not a direct consequence of the influx of meteors on the earth but rather has its source in scattering of solar radiation by interplanetary micrometeoroids which form the dense dust clouds ejected by the parent comets of the associated meteor streams.

082.021 Distribution of minor constituents in a thermal interface. U. Boldes, J. Colman Lerner.
Planet. Space Sci., Vol. 25, 941 - 946 (1977).
A non-steady, two-dimensional, compressible mathematical model of a fluid with constant viscosity and thermal conductivity is given in order to represent a thin atmospheric layer located at 60 km height, as a function of a known initial vertical distribution of temperature, chemical concentrations of minor components and assumed free boundary conditions. The purpose of this paper is to determine the influence of convective laminar processes in the vertical distribution of minor constituents in a thermal interface. Only three minor components are considered, atomic oxygen, molecular oxygen and ozone.

082.022 Contamination of ground-based measurements of O I (6300 Å) and N I (5200 Å) airglow by OH emissions. R. G. Burnside, J. W. Meriwether, Jr., M. R. Torr.
Planet. Space Sci., Vol. 25, 985 - 988 (1977).
High resolution spectra of the 6300 Å and 5200 Å regions of the night sky have been obtained using a 1 m spectrometer. Typical errors in measurements of $O(^1D)$ 6300 Å and $N(^2D)$ 5200 Å intensities due to contamination by overlapping OH emissions have been calculated for a fixed-filter photometer, a tilting-filter photometer and a spectrophotometer.

082.023 Three-dimensional random Earth atmospheres for Monte Carlo trajectory analyses. J. W. Campbell.
NASA Tech. Mem., NASA TM X-3529. 46 pp. Price $ 4.00 (1977).

A set of four magnetic computer tapes containing random global Earth atmospheres is available for Monte Carlo trajectory analyses below an altitude of 99 km. The atmospheres were generated by a statistical atmosphere model based on over 6000 rocket and high-altitude soundings. A readily implementable means of interfacing the tapes with an existing trajectory simulation program is described.

082.024 Atmospheric extinction of solar radiation.
K. Rawer.
J. Atmos. Terr. Phys., Vol. 39, 753 - 756 (1977).

082.025 Radiowave scattering from meteoric ionization.
E. M. Poulter, W. J. Baggaley.
J. Atmos. Terr. Phys., Vol. 39, 757 - 768 (1977).
The full wave treatment is used to calculate reflection coefficients for radiowave backscattering from a column of meteoric ionization. The validity of approximate models is discussed and inconsistencies found in previous full wave methods. Detailed comparisons show that the results of experimental work are in accord with the present theoretical solutions.

082.026 Atmospheric gamma ray angle and energy distributions from 2 to 25 MeV. J. M. Ryan, B. Dayton, S. H. Moon, R. B. Wilson, A. D. Zych, R. S. White.
J. Geophys. Res., Vol. 82, 3593 - 3601 (1977).

082.027 NAS panel is concerned over atmospheric CO_2 buildup. F. C. Bennett.
Phys. Today, Vol. 30, No. 10, p. 17 - 18 (1977).

082.028 Air density at heights near 300 km, from analysis of the orbit of China 2 rocket (1971-18B).
C. J. Brookes, F. C. E. Ryland.
Planet. Space Sci., Vol. 25, 1011 - 1020 (1977).

082.029 An investigation of the angular momentum of the atmosphere. N. S. Sidorenkov.
Izv. AN SSSR. Fiz. atmos. i okeana, Vol. 12, 579 - 587 (1976). In Russian. – Abstr. in Ref. zh., 51. Astron., 9.51.172 (1977).

082.030 The (0, 1) $b'\Sigma_g^+ - X^3\Sigma_g^-$ oxygen band in the absorption spectrum of the earth's atmosphere.
V. D. Galkin.
Opt. i spektrosk., Vol. 42, 844 - 848 (1977). In Russian. Abstr. in Ref. zh., 62. Issled. kosm. prostranstva, 9.62.285 (1977).

082.031 Observations of anomalous submillimeter atmospheric spectra. G. G. Gimmestad, R. H. Ware, R. A. Bohlander, H. A. Gebbie.
Astrophys. J., Vol. 218, 311 - 313 (1977).
Atmospheric spectra in the range $7-25$ cm^{-1} have been observed at Mount Evans, Colorado. Large discrepancies are found between the observations and a model based on known atmospheric constituents. To account for this, the authors postulate the existence in the atmosphere of vapor-phase complexes of water molecules associated with certain weather phenomena.

082.032 Infrared radiation in the energetic balance of the upper atmosphere.
B. F. Gordiets, M. N. Markov.
Kosm. Issled., Vol. 15, 725 - 735 (1977). In Russian.

082.033 Oscillatory relaxation of anharmonic N_2 molecules and concentration of nitric oxide in the disturbed thermosphere. B. F. Gordiets.
Geomagn. Aehron., Vol. 17, 871 - 878 (1977). In Russian.

082.034 Limits for the accretion time of the earth from cosmogenic ^{21}Ne produced in planetesimals.

D. Heymann, M. Dziczkaniec, R. Palma.
Proc. Seventh Lunar Sci. Conf., (see 012.015), 3411 - 3419 (1976).

082.035   Implications of atmospheric fluctuations for far infrared astronomy.   T. C. L. G. Sollner.
Far infrared astronomy (see 012.027), p. 11 - 14 (1976).
Observations of atmospheric emission and transmission fluctuations in the 350 μm window have been made under a variety of observing conditions. The amplitude of this source of noise is large compared to detector noise.

082.036   Investigation of the variations of the amplitude of image vibrations and their connection with temperature fluctuations.   Yu. Yu. Balega, I. V. Shvalagin.
Astrometr. i Astrofiz., Kiev, vyp. (No.) 32, (see 003.011), p. 67 - 73 (1977). In Russian.
The variations of the amplitude of image vibration were studied using long star trails obtained with the ABP-2 telescope. Simultaneously the temperature fluctuations near the ground layer (to 15 m) were measured. Relationships between amplitudes of image vibration and temperature fluctuations were found.

082.037   Temporal radio frequency spectra of multifrequency waves in a turbulent atmosphere characterized by a complex refractive index.   R. H. Ott.
IEEE Trans. Antennas Propag., Vol. AP-25, 254 - 260 (1977).
Abstr. in Phys. Abstr., Vol. 80, Abstr. 21834 (1977).

082.038   Seasonal variation of planetary waves in the stratosphere observed by the Nimbus 5 SCR (*Selective Chopper Radiometer*).   I. Hirota.
Q. J. R. Meteorol. Soc., Vol. 102, 757 - 770 (1976). − Abstr. in Phys. Abstr., Vol. 80, Abstr. 21868 (1977).

082.039   Atmospheric amplitude spectra in an absorption region.   R. H. Ott, M. C. Thompson, Jr.
Antennas and propagation, International Symposium 1976, Amherst, Mass., USA, 11 - 15 Oct. IEEE, New York, USA, p. 594 - 597 (1976). − Abstr. in Phys. Abstr., Vol. 80, Abstr. 21878 (1977).

082.040   Atmospheric absorption near 2400 cm^{-1}.   J. Susskind, J. E. Searl.
J. Quant. Spectrosc. Radiat. Transfer, Vol. 18, 581 - 587 (1977).
Theoretical atmospheric absorption spectra between 2385 and 2425 cm^{-1} are shown to give excellent agreement with high resolution observations. Most of the atmospheric absorption in this region arises from continuum features due to absorption of N_2 and the wings of distant CO_2 lines. The treatment of each of these factors is discussed.

082.041   Very-high-resolution far-infrared measurements of atmospheric emission from aircraft.
B. Carli, D. H. Martin, E. F. Puplett, J. E. Harries.
J. Opt. Soc. America, Vol. 67, 917 - 921 (1977).
An absolute spectrometric radiometer has been flown on a NASA CV990 aircraft to measure atmospheric emission in the spectral range 5−40 cm^{-1} with a resolution of 0.02 cm^{-1} apodized. The instrument and the results are described.

082.042   Imaginary refractive index determinations of the strong absorption in atmospheric dust in the 2.0−3.0 μm spectral region.   J. B. Gillespie, J. D. Lindberg.
J. Opt. Soc. America, Vol. 67, 1363 - 1364 (1977). − Abstract.

082.043   Transmittance of the atmosphere to infrared solar radiation.
M. P. Weinreb, W. G. Planet, G. D. Jones.
J. Opt. Soc. America, Vol. 67, 1377 (1977). − Abstract.

082.044   Telescopic "seeing" measurements during the day and night.   D. L. Walters.
J. Opt. Soc. America, Vol. 67, 1377 (1977). − Abstract.

082.045   Measurements of atmospheric isoplanatism using speckle interferometry.
P. Nisenson, R. V. Stachnik.
J. Opt. Soc. America, Vol. 67, 1391 (1977). − Abstract.

082.046   On seeing.   H. R. Hatfield.
Modern astronomy, (see 003.013), p. 21 - 29 (1977).

082.047   Clear-air turbulence and tropospheric refractivity variations observed with a new VHF-radar.
J. Röttger, P. Czechowsky.
Naturwissenschaften, 64. Jahrg., 580 - 581 (1977).
Radar experiments are described which open a possibility to investigate the structure and dynamics of atmospheric turbulence.

082.048   Atmospheric water vapor at South Pole.   W. D. Smythe, B. V. Jackson.
Appl. Opt., Vol. 16, 2041 - 2042 (1977).

082.049   Relative contribution of upper and lower atmosphere to integrated refractive-index profiles.
C. Roddier, J. Vernin.
Appl. Opt., Vol. 16, 2252 - 2256 (1977).
The authors measure the wavefront coherence and the irradiance fluctuations of stellar sources to obtain integrated refractive-turbulence profiles of two regions of the atmosphere. Comparison of experimental data during their measurement program shows an equal contribution of upper and lower layers to the limitation of the optical seeing. They also note the great variability of turbulence located above 3 km up to the stratosphere, from night to night. When simultaneously operated, these two methods are suitable for astronomical site testing.

082.050   Scintillation statistics caused by atmospheric turbulence and speckle in satellite laser ranging.
J. L. Bufton, R. S. Iyer, L. S. Taylor.
Appl. Opt., Vol. 16, 2408 - 2413 (1977).
The authors study the statistics of scintillation at the ground-based receiver for the earth-space-earth retroreflector configuration of satellite laser ranging. These statistics are governed by the joint effects of atmospheric turbulence and speckle produced by the retroreflector array. An expression for the probability density function of scintillation is obtained and evaluated numerically. Comparison of the normalized variance of scintillation calculated by using this function shows good agreement with results obtained by other methods.

082.051   Atmospheric extinction between 0.55 μm and 10.6 μm due to soil-derived aerosols.
E. M. Patterson.
Appl. Opt., Vol. 16, 2414 - 2418 (1977).

082.052   Scintillation statistics measured in an earth−space−earth retroreflector link.   J. L. Bufton.
Appl. Opt., Vol. 16, 2654 - 2668 (1977).
An experiment is described for the measurement of scintillation in a vertical path from an earth-based laser transmitter to the GEOS-III satellite and back to an earth-based receiver telescope. Measurements of the normalized variance, probability density function, and power spectral density of scintillation are presented. These results are discussed in terms of recent analytical results.

082.053   Effects of horizontal refractivity gradients on the accuracy of laser ranging to satellites.
C. S. Gardner.

Radio Sci., Vol. 11, 1037 - 1044 (1976). – Abstr. in Phys. Abstr., Vol. 80, Abstr. 25910 (1977).

082.054 Altitude distributions of O_2 atmospheric and OH Meinel emissions in the nightglow.
T. Watanabe, Y. Morioka, M. Nakamura.
Rep. Ionos. Space Res. Japan., Vol. 30, No. 1 - 2, p. 41 - 45 (1976). – Abstr. in Phys. Abstr., Vol. 80, Abstr. 29530 (1977).

082.055 On the absorption of solar radiation by the Earth's atmosphere.   G. Major.
Contrib. Atmos. Phys., Vol. 49, 212 - 215 (1976). – Abstr. in Phys. Abstr., Vol. 80, Abstr. 33183 (1977).

082.056 Comparison of exospheric temperatures at Millstone Hill and St. Santin.
J. E. Salah, J. V. Evans, D. Alcayde, P. Bauer.
Ann. Géophys., Vol. 32, 257 - 266 (1976). – Abstr. in Phys. Abstr., Vol. 80, Abstr. 33224 (1977).

082.057 Reduced hydrogen temperatures in the transition region between thermosphere and exosphere.
H. J. Fahr.
Ann. Géophys., Vol. 32, 277 - 282 (1976). – Abstr. in Phys. Abstr., Vol. 80, Abstr. 33225 (1977).

082.058 Latitudinal variation of the intensity of 5577 Å line of night airglow.   S. N. Ghosh, N. Kundu.
Indian J. Phys., Vol. 50, 668 - 677 (1976). – Abstr. in Phys. Abstr., Vol. 80, Abstr. 37474 (1977).

082.059 The upper atmosphere of the earth.
J. C. G. Walker.
Atomic processes and applications. P. G. Burke (Editor). North-Holland Publ. Co., Amsterdam, Netherlands, 10 + 533 pp. Price $ 65.95 (1976), p. 45 - 69. – Abstr. in Phys. Abstr., Vol. 80, Abstr. 37482 (1977).

082.060 An unusual lunar halo.   B. L. Cardon.
American J. Phys., Vol. 45, 331 - 335 (1977).
Abstr. in Phys. Abstr., Vol. 80, Abstr. 45634 (1977).

082.061 Analyses of atmospheric extinction data obtained by astronomers. I. A time-trend analysis of data with internal accidental errors obtained at four observatories.
B. J. Taylor, P. B. Lucke, N. S. Laulainen.
Atmos. Environ., Vol. 11, No. 1, p. 1 - 20 (1977). – Abstr. in Phys. Abstr., Vol. 80, Abstr. 49161 (1977).

082.062 Analyses of atmospheric extinction data obtained by astronomers. III. Compilation of optical depths for 31 observatory sites.   N. S. Laulainen.
Atmos. Environ., Vol. 11, No. 1, p. 29 - 33 (1977). – Abstr. in Phys. Abstr., Vol. 80, Abstr. 49162 (1977).

082.063 Analyses of atmospheric extinction data obtained by astronomers. II. Seasonal variations in astronomical extinction.   N. S. Laulainen, B. J. Taylor, P. W. Hodge.
Atmos. Environ., Vol. 11, No. 1, p. 21 - 27 (1977). – Abstr. in Phys. Abstr., Vol. 80, Abstr. 53185 (1977).

082.064 A review of the optical effects of the clear turbulent atmosphere.   R. S. Lawrence.
Imaging through the atmosphere, (see 012.032), p. 2 - 8 (1976). Abstr. in Phys. Abstr., Vol. 80, Abstr. 42485 (1977).

082.065 Characterization of atmospheric turbulence.
M. Miller, P. Zieske, D. Hanson.
Imaging through the atmosphere, (see 012.032), p. 30 - 38 (1976). – Abstr. in Phys. Abstr., Vol. 80, Abstr. 45125 (1977).

082.066 Stellar-scintillation measurement of the vertical profile of refractive-index turbulence in the atmosphere.   G. R. Ochs, R. S. Lawrence, T. Wang.
Imaging through the atmosphere, (see 012.032), p. 48 - 54 (1976). – Abstr. in Phys. Abstr., Vol. 80, Abstr. 45126 (1977).

082.067 A new type of absorption in the atmospheric infrared window due to water vapor polymers.
H. Grassl.
Contrib. Atmos. Phys., Vol. 49, 225 - 236 (1976). – Abstr. in Phys. Abstr., Vol. 80, Abstr. 45145 (1977).

082.068 Thermosphere waves – a review.   S. Kato.
J. Geomagn. Geoelectr., Vol. 28, 189 - 206 (1976). Abstr. in Phys. Abstr., Vol. 80, Abstr. 45211 (1977).

082.069 Thermospheric temperature dependence of the atomic oxygen 6300 Å emission in the twilight airglow.   Y. Kondo, T. Tohmatsu.
J. Geomagn. Geoelectr., Vol. 28, 207 - 218 (1976). – Abstr. in Phys. Abstr., Vol. 80, Abstr. 45215 (1977).

082.070 On the zenith angle dependence of the infrared oxygen emission of 1.27 μm airglow observed from high altitude.   T. Makino, H. Yamamoto, H. Sekiguchi.
J. Geomagn. Geoelectr., Vol. 28, 243 - 249 (1976). – Abstr. in Phys. Abstr., Vol. 80, Abstr. 45216 (1977).

082.071 Planetary waves in horizontal and vertical shear: the generalized Eliassen-Palm relation and the mean zonal acceleration.   D. G. Andrews, M. E. McIntyre.
J. Atmos. Sci., Vol. 33, 2031 - 2048 (1976). – Abstr. in Phys. Abstr., Vol. 80, Abstr. 49111 (1977).

082.072 Planetary waves in horizontal and vertical shear: asymptotic theory for equatorial waves in weak shear.   D. G. Andrews, M. E. McIntyre.
J. Atmos. Sci., Vol. 33, 2049 - 2053 (1976). – Abstr. in Phys. Abstr., Vol. 80, Abstr. 49112 (1977).

082.073 Similarity theory of the buoyantly interactive planetary boundary layer with entrainment.
M. I. Hoffert, Y. C. Sud.
J. Atmos. Sci., Vol. 33, 2136 - 2151 (1976). – Abstr. in Phys. Abstr., Vol. 80, Abstr. 49116 (1977).

082.074 Meridional circulation in the thermosphere. II. Solstice conditions.
R. E. Dickinson, E. C. Ridley, R. G. Roble.
J. Atmos. Sci., Vol. 34, 178 - 192 (1977). – Abstr. in Phys. Abstr., Vol. 80, Abstr. 56783 (1977).

082.075 Hydrogen escape in the terrestrial atmosphere at low oxygen levels: a photochemical model.
G. Visconti.
J. Atmos. Sci., Vol. 34, 193 - 204 (1977). – Abstr. in Phys. Abstr., Vol. 80, Abstr. 56784 (1977).

082.076 Measurements and calculations of atmospheric transmittance and radiance.
M. J. Persky, J. M. Weinberg.
Opt. Eng., Vol. 15, 521 - 524 (1976). – Abstr. in Phys. Abstr., Vol. 80, Abstr. 64431 (1977).

082.077 Upper-atmosphere studies by ranging to satellites.
D. G. King-Hele.
Philos. Trans. R. Soc. London, Ser. A, Vol. 284, 555 - 563 (1977). – Abstr. in Phys. Abstr., Vol. 80, Abstr. 64485 (1977).

082.078 Laser sounding of atmospheric sodium. Interpretation in terms of global atmospheric parameters.
G. Megie, J. E. Blamont.
Planet. Space Sci., Vol. 25, 1093 - 1109 (1977).

Measurements of the night-time atmospheric sodium layer have been performed over a period of 3 years, using a Lidar facility set up at the Haute Provence Observatory.

082.079 **Electronically excited ozone in the atmosphere.**
P. C. Wraight.
Planet. Space Sci., Vol. 25, 1177 - 1181 (1977).

082.080 **The atmosphere above Earth's surface.**
P. I. Bakhshi.
Stud. J. Inst. Electron. Telecommun. Eng., Vol. 17, No. 1, p. 32 - 38 (1976). – Abstr. in Phys. Abstr., Vol. 80, Abstr. 67002 (1977).

082.081 **Optical radiation from the atmosphere.**
D. J. Baker, W. R. Pendleton, Jr.
Methods for atmospheric radiometry, (see 012.033), p. 50 - 62 (1976). – Abstr. in Phys. Abstr., Vol. 80, Abstr. 70831 (1977).

082.082 **Observations of sky brightness spatial variations.**
M. J. Cantella.
Methods for atmospheric radiometry, (see 012.033), p. 109 - 115 (1976). – Abstr. in Phys. Abstr., Vol. 80, Abstr. 70832 (1977).

082.083 **Atmospheric transmission measurements using IR lasers, Fourier transform spectroscopy, and gas-filter correlation techniques.**   J. A. Dowling, K. M. Haught, R. F. Horton, S. T. Hanely, J. A. Curcio, D. H. Garcia, C. O. Gott.
Rep. NRL Prog., p. 1 - 4 (March 1977). – Abstr. in Phys. Abstr., Vol. 80, Abstr. 70873 (1977).

082.084 **Detecting ultraviolet radiation from and through the atmosphere.**   D. K. Prinz.
Methods for atmospheric radiometry, (see 012.033), p. 2 - 11 (1976). – Abstr. in Phys. Abstr., Vol. 80, Abstr. 70881 (1977).

082.085 **Passive microwave radiometric observations of the atmosphere.**   A. L. Cassel, D. H. Staelin.
Methods for atmospheric radiometry, (see 012.033), p. 12 - 14 (1976). – Abstr. in Phys. Abstr., Vol. 80, Abstr. 70882 (1977).

082.086 **Spectral radiometric measurement of atmospheric constituents.**   W. J. Williams, D. B. Barker, J. N. Brooks, A. Goldman, J. J. Kosters, F. H. Murcray, D. G. Murcray, D. E. Snider.
Methods for atmospheric radiometry, (see 012.033), p. 15 - 28 (1976). – Abstr. in Phys. Abstr., Vol. 80, Abstr. 70883 (1977).

082.087 **A filter-tilting type nightglow photometer using a small GaAs photocathode and aspheric lenses.**
K. Misawa, I. Takeuchi.
Rep. Ionos. Space Res. Japan, Vol. 30, No. 3 - 4, p. 103 - 107 (1976). – Abstr. in Phys. Abstr., Vol. 80, Abstr. 73889 (1977).

082.088 **Parallel intensity-variations of [O I] 5577 Å line and O_2 (0-1) atmospheric band at 8645 Å.**
K. Misawa, I. Takeuchi.
Rep. Ionos. Space Res. Japan, Vol. 30, No. 3 - 4, p. 109 - 112 (1977). – Abstr. in Phys. Abstr., Vol. 80, Abstr. 73899 (1977).

082.089 **A night sky model for satellite search systems.**
G. M. Daniels.
Opt. Eng., Vol. 16, No. 1, p. 66 - 71 (1977). – Abstr. in Phys. Abstr., Vol. 80, Abstr. 73923 (1977).

082.090 **Noctilucent clouds over western Europe and the Atlantic during 1976.**
D. H. McIntosh, M. Hallissey.
Meteorol. Mag., Vol. 106, 181 - 184 (1977). – Abstr. in Phys. Abstr., Vol. 80, Abstr. 78346 (1977).

082.091 **Measurement of nitric oxide abundance in equatorial upper atmosphere.**   T. Tohmatsu, N. Iwagami.
J. Geomagn. Geoelectr., Vol. 28, 343 - 358 (1976). – Abstr. in Phys. Abstr., Vol. 80, Abstr. 78547 (1977).

082.092 **The noctilucent cloud display of 18 - 19 June 1976.**
D. A. R. Simmons.
Weather, Vol. 32, No. 7, p. 240 - 248 (1977). – Abstr. in Phys. Abstr., Vol. 80, Abstr. 81982 (1977).

082.093 **The map of the sun for Italy.**
A. Guerrini, A. Lavagnini, F. Vivona.
17 convegno internazionale scientifico sullo spazio (see 012. 034), p. 421 - 430 (1977). In Italian. – Abstr. in Phys. Abstr., Vol. 80, Abstr. 81987 (1977).

082.094 **Balloon observation of intensity distribution of nightglow over the sky. I.**
S. Shinoki, T. Watanabe, M. Nakamura.
Bull. Inst. Space Aeronaut. Sci. Univ. Tokyo B, Vol. 12, 597 - 607 (1976). In Japanese. – From Phys. Abstr., Vol. 80, Abstr. 82075 (1977).

082.095 **Balloon observation of intensity distribution of nightglow over the sky. II.**
S. Oshima, T. Watanabe, M. Nakamura.
Bull. Inst. Space Aeronaut. Sci. Univ. Tokyo B, Vol. 12, 609 - 621 (1976). In Japanese. – From Phys. Abstr., Vol. 80, Abstr. 82076 (1977).

082.096 **Statistical analysis of wavefront random deformations produced by atmospheric turbulence near the ground. I. Description of method, first results.**
J. Borgnino, F. Martin.
J. Opt. (*France*), Vol. 8, 319 - 326 (1977). In French. – Abstr. in Phys. Abstr., Vol. 80, Abstr. 87074 (1977).

082.097 **Very high resolution Fourier spectroscopy of the atmosphere in the submillimetre region.**
J. E. Harries, B. Carli, M. J. Bangham.
European conference on precise electrical measurement, Brighton, Sussex, 5 - 9 September 1977. IEEE, London, England. 12 + 177 pp. (1977). p. 71 - 72. – Abstr. in Phys. Abstr., Vol. 80, Abstr. 87133 (1977).

082.098 **Morphology of upper atmosphere temperatures.**
C. D. Rodgers.
Dynamical and chemical coupling between the neutral and ionized atmosphere, (see 012.036), p. 3 - 16 (1977).

082.099 **The effects of changing fluxes of solar UV radiation and solar cosmic rays on the middle atmosphere temperature and ozone concentration.**
A. G. Theobald, R. G. Williams, M. J. Rycroft.
Dynamical and chemical coupling between the neutral and ionized atmosphere, (see 012.036), p. 79 - 84 (1977).

082.100 **Long term variations of OH nightglow emission – relation to stratospheric humidity?**
G. Weill, J. Christophe.
Dynamical and chemical coupling between the neutral and ionized atmosphere, (see 012.036), p. 85 - 90 (1977).

082.101 **Energy balance of the atmosphere under the influence of a disturbed sun and magnetospheric processes.**   K. D. Cole.
Dynamical and chemical coupling between the neutral and ionized atmosphere, (see 012.036), p. 203 - 216 (1977).
This paper reviews physical processes which may control the energy density of the atmosphere particularly between the altitudes of 30 and 160 km. Changes of composition and atmospheric parameters which follow from magnetospheric

processes and concommitant processes on the sun are emphasized to present a framework for discussing sun-weather relationships.

082.102 Characteristics of 1 – 500 eV electrons observed in the Earth's thermosphere from the photoelectron spectrometer experiment on the atmosphere Explorer satellites.
W. K. Peterson, J. P. Doering, T. A. Potemra, C. O. Bostrom.
Dynamical and chemical coupling between the neutral and ionized atmosphere, (see 012.036), p. 353 - 364 (1977).

082.103 Comparative water vapor measurements for infrared sites.   J. W. Warner.
Publ. Astron. Soc. Pacific, Vol. 89, 724 - 727 (1977).

Water-vapor measurements are reported for Chacaltaya north of La Paz, Bolivia, and for Mount Lemmon Infrared Observatory. Comparisons are presented for Mauna Kea, Cerro Tololo, and the Wyoming site. It is concluded that Chacaltaya could serve as a good high-altitude infrared site in the Southern Hemisphere.

082.104 Importancia del flujo de fotoelectrones procedentes del punto magnéticamente conjugado en la emisión precrepuscular de 6300 Å.
C. F. González, E. Battaner, G. Pardo.
Urania Barcelona, Año 61, Núm. 285, p. 71 - 84 (1976).

The pre-dawn enhancement of 6300 Å atmospheric atomic oxygen has been studied, in the months when the sunrise at the conjugate magnetic point precedes the local sunrise. Computed values for the photoelectron fluxes needed to produce the observed enhancements are considerably higher than experimental values.

082.105 La influencia de la desactivación de $O(^1D)$ por choques en la región F2 nocturna.
C. F. González, E. Battaner, G. Pardo.
Urania Barcelona, Año 61, Núm. 285, p. 85 - 92 (1976).

A comparison between computed and experimental values of 6300 Å atmospheric emission shows the importance of the deactivation of the atomic oxygen ^1D state, mainly by N_2 molecules. The effect is more important at the beginning of the night, negligible at midnight, and relatively important at the end of the night. Due to this fact the intensity is noticeably superestimated by using Barbier's model, especially at post-dusk. At midnight however, there is a good agreement between Barbier's model brightness and the actual emission.

082.106 Morning and evening behavior of the F region green line emission: evidence concerning the sources of $O(^1S)$.   J. P. Kopp, J. E. Frederick, D. W. Rusch, G. A. Victor.
J. Geophys. Res., Vol. 82, 4715 - 4719 (1977).

Measurements of the green line emission from atomic oxygen by a satellite-borne photometer show consistently larger volume emission rates in the evening than at comparable solar zenith angles in the morning between altitudes of 150 and 200 km. The temporal variation in the airglow is consistent with diurnal model calculations if we assume that the reaction $N + O_2^+ \rightarrow NO^+ + O(^1S)$ provides a significant source for the green line, $O(^1D - ^1S)$.

082.107 Water vapor in the lower stratosphere measured from aircraft flight.
E. Hilsenrath, B. Guenther, P. Dunn.
J. Geophys. Res., Vol. 82, 5453 - 5459 (1977).

082.108 Atmospheric refraction.
S. M. Kozik. Edited by V. P. Shcheglov, with an introduction by A. A. Mikhajlov.
Akademiya Nauk Uzbekskoj SSR. Trudy Ordena Trudovogo Krasnogo Znameni Astronomicheskogo Instituta Akademii Nauk UzSSR, Tom 1. Izdatel'stvo "FAN" Uzbekskoj SSR.

Tashkent. 132 pp. Price 1 Rbl. 20 Kop. (1977). In Russian.
A short review of the development of the theory of atmospheric refraction is given. The general theory of refraction is discussed. A detailed description of the refraction theory in a spherically layer medium is given. Some new methods of astronomical refraction calculations are substantiated.

082.109 Effect of multiple scattering on balloon observation of γ-ray bursts.
H. Horstman, L. Bassani, E. Horstman-Moretti.
Astrophys. Space Sci., Vol. 52, 265 - 269 (1977).

The search for γ-ray bursts of low intensity has been undertaken mostly from balloon-borne detectors with wide aperture. The effect of multiple Compton scattering in the atmosphere greatly increases the probability of seeing small bursts, and should be taken into account when deducing the ln N – ln S curve. Detailed calculations have been carried out for different assumed spectra in the extreme case of a completely unscreened flat horizontal detector.

082.110 On a new method of separating the diffuse cosmic X-ray component in balloon-borne measurements.
E. Horstman-Moretti, H. Horstman.
Astrophys. Space Sci., Vol. 52, 307 - 311 (1977).

The diffuse X-ray background from 20 to 200 keV has been measured many times by rocket-, satellite-, and balloon-borne detectors. As has been noted by a number of authors, the balloon measurements do not agree with those outside the atmosphere. This suggests some systematic difference in the derivation of the diffuse flux. The authors propose a method for the analysis of balloon data in which the source function (atmospherically produced and scattered primary) is derived from the experimental data and is then used to calculate the residual counting rate due only to the transmitted primary photons.

082.111 On the manifestation of a connection between variations of the terrestrial atmospherical pressure and strong geomagnetic disturbances in various seasons.
V. E. Chertoprud.
Ehff. soln. aktivnosti v nizhn. atmos. Leningrad, Gidrometeoizdat, 1977, p. 86 - 93. In Russian. – Abstr. in Ref. zh., 62. Issled. kosm. prostranstva, 11.62.143 (1977).

082.112 Tropospherical response to geomagnetic disturbances and propagation of tropospherical waves into the upper atmosphere.
V. V. Mikhnevich, V. G. Kidiyarova, B. N. Trubnikov, I. A. Shcherba.
Ehff. soln. aktivnosti v nizhn. atmos. Leningrad, Gidrometeoizdat, 1977, p. 10 - 22. In Russian. – Abstr. in Ref. zh., 62. Issled. kosm. prostranstva, 11.62.144 (1977).

082.113 Temperature feedback in a stratospheric model.
F. M. Luther, D. J. Wuebbles, J. S. Chang.
J. Geophys. Res., Vol. 82, 4935 - 4942 (1977).

Perturbing the stratospheric composition affects the stratospheric temperature profile via the radiation balance. Changes in temperature affect chemical reaction rates, which in turn feed back on stratospheric composition. The effect of temperature feedback on ozone concentration has been investigated for three types of perturbations: (1) stratospheric injection of NO_x, (2) release of fluorocarbons, and (3) doubling CO_2.

082.114 Recent studies of astronomical refraction.
G. Teleki.
Bull. Obs. Astron. Belgrade, No. 128, p. 7 - 10 (1977).

A survey is given of some of the present-day investigations of the astronomical refraction, having a broader interest. With some more details description is presented of the plan for elaboration of new international refraction tables, meant to replace Pulkovo Tables, 4th edition, currently in use.

082.115 **Problems of Earth's rotation and anomalous refraction.** G. Teleki.
Bull. Obs. Astron. Belgrade, No. 128, p. 19 - 22 (1977).

The role is studied of the refraction anomalies in the investigation of the problem of Earth's rotation, and the importance of regional refractional influences is specially emphasized. Proceeding from the analysis of the latitude variations during night at ILS stations (Teleki, 1976), the author points out the need for careful selection of the site for the astrometric instruments.

082.116 **Correlation of meteorological parameters and some optical properties of the atmosphere.**
E. F. Rizov.
Astron. Tsirk., No. 947, p. 6 - 8 (1977). In Russian.

082.117 **Seasonal variations of hydroxyl emission of the night airglow.** M. V. Shagaev.
Astron. Tsirk., No. 951, p. 4 - 5 (1977). In Russian.

082.118 **Simultaneous seeing measurements near Mt. Maidanak with a double-beam telescope and a photoelectric device.** A. I. Beslik, V. P. Goranskij, A. A. Tokovinin, P. V. Shcheglov.
Astron. Tsirk., No. 955, p. 3 - 6 (1977).

082.119 **Preliminary data concerning sky cloudiness influence on the refractive properties of the atmosphere.**
A. I. Nefed'eva.
Astron. Tsirk., No. 960, p. 2 - 4 (1977). In Russian.

082.120 **The parallactic refraction determined with the aid of meteorological balloon data.** J. Kabeláč.
Stud. Geophys. Geod., Vol. 20, 1 - 9 (1976) = Astron. Inst. Tech. Univ. Praha, Publ. No. 45.

082.121 **Effect of the headlight of automobiles on the astronomical observation.**
M. Ooe, K. Hosoyama, C. Kakuta, T. Gotō.
Proc. Int. Latitude Obs. Mizusawa, No. 16, p. 1 - 24 (1977). In Japanese.

082.122 **On the optical effects observable from a satellite of the earth at sunrise and sunset in case of refraction from the water surface.** V. Kushpil, Yu. Alekseev, E. Saar.
Publ. Tartu Astrofiz. Obs., Vol. 45, 141 - 149 (1977).

Observations of sunrise and sunset aboard a space capsule are dealt with. It is shown that alongside the refracted image of the sun one can observe another image caused by Fresnel reflection from the water surface; image distortions have been calculated.

082.123 **Determination of atmospheric absorption at 460μm.**
V. F. Zabolotnyj, E. F. Rizov, V. A. Soglasnova, B. F. Yudin.
Astron. Zh. Akad. Nauk SSSR, Vol. 54, 1314 - 1318 (1977). In Russian. English translation in Soviet Astron., Vol. 21, No. 6.

The results of determination of the complete vertical absorption in the atmospheric "window" centered at 460 μm are reported. The relation between the value of absolute humidity and the optical depth of the atmosphere at 460 μm is determined.

082.124 **Method for taking into account the absorption of stellar radiation by molecules of the earth's atmosphere.** N. S. Komarov, L. A. Zavershneva.
Astrometriya i Astrofizika, Kiev, vyp. (No.) 33, (see 003.020), p. 72 - 77 (1977). In Russian.

A method is considered for taking into account the ab-

sorption of stellar radiation by H_2O molecules of the earth's atmosphere within the region from 7000 to 10500 Å.

082.125 **On the atmospheric transparency in Uzhgorod.**
G. V. Moskaleva.
Astrometriya i Astrofizika, Kiev, vyp. (No.) 33, (see 003.020), p. 87 - 90 (1977). In Russian.

The atmospheric extinction coefficient is determined from photographic observations for 29 nights in Uzhgorod and is found to be variable. Statistical data on the number of clear nights are given.

082.126 **Visual observations and photography of noctilucent clouds aboard the orbital station Salyut 4.**
Ch. Willmann, P. I. Klimuk, I. I. Koksharov, V. I. Sevast'yanov, V. N. Sergeevich, K. Eerme.
Optical investigations of the emission of the atmosphere, aurorae and noctilucent clouds aboard the orbital scientific station Salyut 4, (see 003.021), p. 53 - 66, 167 - 169 (1977). In Russian.

27 series of observations of noctilucent clouds were carried out from space. These occurrences were recorded also by the ground-based network in the course of the second space expedition aboard Salyut 4. The morphological and multi-layered structures of the noctilucent cloud field are described. It is shown that noctilucent cloud fields cover a belt at geographic latitudes higher than 45°.

082.127 **Physical interpretation of the spectra of noctilucent clouds.**
O. A. Avaste, A. M. Alekseev, U. Veismann, Ch. Willmann, P. I. Klimuk, I. I. Koksharov, A. I. Lazarev, V. I. Sevast'yanov, V. N. Sergeevich, E. O. Fedorova, K. Eerme.
Optical investigations of the emission of the atmosphere, aurorae and noctilucent clouds aboard the orbital scientific station Salyut 4, (see 003.021), p. 67 - 78 (1977). In Russian.

The results of the investigations of noctilucent clouds in the near-infrared region aboard Salyut 4 are discussed. The four-channel near-infrared radiometer with interference filters (λ 1.35, 1.9, 2.2, 2.7 μm) is briefly described. Consideration of the $O_2(^1\Delta_g)$ and OH emissions in the near-infrared spectral region gives a good agreement between the calculated and measured values of noctilucent cloud brightnesses. The results of the investigations confirm an increase in the OH emission in the mesopause at the occurrence of noctilucent clouds established by Shefov (1968).

082.128 **Determination of the daytime and twilight profiles of $O_2(^1\Delta_g)$ at 1.27 μm from measurements aboard the orbital station Salyut 4.**
O. A. Avaste, U. Veismann, Ch. Willmann, G. M. Grechko, A. A. Gubarev, P. I. Klimuk, G. I. Lobanova, O. I. Popov, V. I. Sevast'yanov, E. O. Fedorova, K. Eerme.
Optical investigations of the emission of the atmosphere, aurorae and noctilucent clouds aboard the orbital scientific station Salyut 4, (see 003.021), p. 79 - 87 (1977). In Russian.

Data are presented on the diurnal variation of the emission of molecular oxygen at λ 1.27 μm and on the vertical profile of the volume emission when $Z_\odot = 35°$. The integral intensity of the emission $O_2(^1\Delta_g)$ in the vertical direction determined from the brightness profiles of the atmosphere for the zenith angle of the sun $Z_\odot = 35°$ equals 36.5 MR.

082.129 **Horizontal optical inhomogeneity of the upper atmosphere at mean and equatorial latitudes according to the observations aboard the orbital station Salyut 4.**
S. V. Avakyan, P. I. Klimuk, I. I. Koksharov, A. I. Lazarev, V. I. Sevast'yanov.
Optical investigations of the emission of the atmosphere, aurorae and noctilucent clouds aboard the orbital scientific station Salyut 4, (see 003.021), p. 88 - 106, 170 - 173 (1977). In Russian.

Data on visual observations of horizontal optical inhomogeneities of the upper atmosphere carried out aboard Salyut 4 in June - July 1975 are discussed in detail. Of 15 cases of observations 5 were selected, which by no means could be connected with auroral phenomena, but which occur in the twilight zone. For their explanation use was made of the hypothesis suggested earlier: link between the observed inhomogeneities and the occurrence of internal acoustic-gravity waves in the upper atmosphere. The nature of changes in rates of the main excitation mechanisms of the twilight emission, i.e. the photoelectron impact in the gap and in the loop of the acoustic-gravity waves, is discussed.

082.130   Observations of the earth's night and twilight horizons from the orbital station Salyut 4.
P. I. Klimuk, I. I. Koksharov, A. I. Lazarev, G. I. Lobanova, E. Saar, V. I. Sevast'yanov, V. N. Sergeevich.
Optical investigations of the emission of the atmosphere, aurorae and noctilucent clouds aboard the orbital scientific station Salyut 4, (see 003.021), p. 107 - 127, 174 (1977). In Russian.

Using the results of the visual observations carried out aboard Salyut 4 in the period from May 24 to July 26 at an altitude of $\simeq$ 350 km, the general picture of optical phenomena on the earth's night and twilight horizons is specified. A number of atmospheric-optical phenomena are described, part of them having been observed for the first time. Among them are the vertical ray-structure of the second night emission layer located at heights of 170–220 km, and the twilight increase of the intensity of this layer. The results of visual and photographic observations of the reflection from planetary atmospheres and comets at small sliding angles are generalized. New data on the reflection obtained aboard Salyut 4 are given. They include photos of the reflection of the moon from the earth's atmosphere, which seem to be the first of their kind. The results of visual observations of setting and rising of planets and stars are described.

082.131   Observations of the first expedition aboard the orbital station Salyut 4.
A. I. Lazarev, G. M. Grechko, A. A. Buznikov.
Optical investigations of the emission of the atmosphere, aurorae and noctilucent clouds aboard the orbital scientific station Salyut 4, (see 003.021), p. 155 - 163 (1977). In Russian.

The results of visual observations of a number of atmospheric-optical phenomena on night and twilight horizons of the earth carried out aboard Salyut 4 are presented. A comparison of the results of observations of the pronounced horizontal inhomogeneities of the earth's upper atmospheric emission accomplished on board Soyus 3, Soyus 9 and Soyus 15 with those carried out on Salyut 4 is given. The results of observations of the scintillation of planets on the earth's night horizon as well as observations of the green region of the twilight aureole of the earth are described and discussed.

082.132   Molecular oxygen concentrations and absorption cross sections in the thermosphere derived from extreme ultraviolet occultation profiles.
O. K. Garriott, R. B. Norton, J. G. Timothy.
J. Geophys. Res., Vol. 82, 4973 - 4982 (1977).
A preliminary analysis of some of the Skylab extreme ultraviolet (EUV) occultation data is presented. These EUV occultation profiles contain unique information on the neutral constituent concentrations in the terrestrial upper atmosphere at altitudes between about 90 km and the orbital height of 435 km. The procedures for the initial reduction of some of these data and the derived values of the molecular oxygen concentrations over the altitude range from 90 to 160 km are described in this paper.

082.133   Medium scale TID's and their associated internal

gravity waves as seen through height-dependent electron density power spectra.
A. L. Hearn, K. C. Yeh.
J. Geophys. Res., Vol. 82, 4983 - 4990 (1977).

082.134   Far ultraviolet atomic and molecular nitrogen emissions in the dayglow.   P. Z. Takacs, P. D. Feldman.
J. Geophys. Res., Vol. 82, 5011 - 5023 (1977).
The far ultraviolet day airglow between 1130 and 1520 Å was observed at 4.4-Å spectral resolution. Fourteen bands, or blends, of the N_2 Lyman-Birge-Hopfield (LBH) system are clearly resolved and the relative intensities are found to be in good agreement with laboratory electron excitation spectra.

082.135   Comparison of measured and calculated thermospheric molecular oxygen densities.
W. E. Potter, D. C. Kayser, H. C. Brinton, L. H. Brace, M. Oppenheimer.
J. Geophys. Res., Vol. 82, 5243 - 5248 (1977).

082.136   Departures from hydrostatic equilibrium in the global distribution of thermospheric argon.
J. M. Straus.
J. Geophys. Res., Vol. 82, 5249 - 5252 (1977).
This report describes the application of a three-dimensional two-component model to the distribution of argon in the upper atmosphere of the earth. The Cira (1972) model is used to specify the background gas density and the temperature as functions of altitude, latitude, local time, and day of the year. Emphasis is placed on examining the effects of departures from hydrostatic equilibrium on the argon density distribution. Such effects are found to be significant, producing a strong summer argon bulge. Comparisons of the model results with observational data are discussed.

082.137   Inter-hemispherical thermal mechanism and changes of the relative momentum of an atmospheric impulse.
N. S. Sidorenkov, L. G. Nemirovskaya, N. Yu. Orekhova.
Tr. Gidrometeorol. nauchn.-issled. tsentr. SSSR, 1977, vyp. (No.) 171, p. 34 - 37. In Russian. – Abstr. in Ref. zh., 51. Astron., 12.51.137 (1977).

082.138   Experience of determination of the asymmetry of the astronomical refraction.
V. V. Kirichuk, N. N. Olejnik.
Geod., kartogr. i aehrofotosemka. Resp. mezhved. nauchn.-tekh. sb., 1977, vyp. (No.) 26, p. 32 - 39. In Russian. – Abstr. in Ref. zh., 51. Astron., 12.51.147; 52. Geod. Aehrosemka, 12.52.105 (1977). .

082.139   Dynamics of the lower thermosphere from meteor observations.
B. V. Kal'chenko, B. L. Kashcheev, V. V. Fedynskij.
Meteor. issled. No. 4. Moskva, "Sov. radio", 1977, p. 62 - 74. In Russian. – Abstr. in Ref. zh., 51. Astron., 12.51.291 (1977).

082.140   Frequency spectra of scintillations of the images of the sun and moon.   V. I. Ivanov, P. G. Kovadlo, Sh. P. Darchiya.
Issled. po geomagn., aehron. i fiz. Solntsa. Moskva, Nauka, 1977, p. 104 - 110. In Russian. – Abstr. in Ref. zh., 51. Astron., 1.51.113 (1978).

082.141   Frequency spectrum of star image scintillations.
V. I. Ivanov, P. G. Kovadlo, Sh. P. Darchiya.
Issled. po geomagn., aehron. i fiz. Solntsa. Moskva, Nauka, 1977, p. 111 - 115. In Russian. – Abstr. in Ref. zh., 51. Astron., 1.51.114 (1978).

082.142   Determination of the vertical aerosol profiles in the atmosphere from results of a spectrophotometry of

the twilight horizon of the earth with spaceship Soyuz 13.
K. Ya. Kondrat'ev, A. A. Buznikov, O. M. Pokrovskij.
Dokl. AN SSSR, Vol. 235, 53 - 56 (1977). In Russian. — Abstr.
in Ref. zh., 62. Issled. kosm. prostranstva, 1.62.207 (1978).

082.143 **Ion photochemistry of the thermosphere from Atmosphere Explorer C measurements.**
M. Oppenheimer, E. R. Constantinides, K. Kirby-Docken,
G. A. Victor, A. Dalgarno, J. H. Hoffman.
J. Geophys. Res., Vol. 82, 5485 - 5492 (1977).
   A detailed description is presented of the formation and removal processes of O^+ ions in the upper atmosphere, and a model is constructed of the ion chemistry of the quiet ionosphere. The model is used in conjunction with the solar ultraviolet reference of Hinteregger and the Atmosphere Explorer C satellite data on the ion and electron temperatures and on the composition of the neutral atmosphere to predict the altitude profiles of the densities of O^+, N_2^+, NO^+, and O_2^+.

082.144 **Chemical evolution — comparative planetology.**
         M. Shimizu.
10th Lunar and Planetary Symposium, (see 012.050), p. 118 - 121 (1977).
   A brief review of evolutionary models of the terrestrial atmosphere including the recently advocated $CO + N_2$ atmosphere is presented, as well as an interpretation of the atmospheric composition of Mars and Venus observed by Viking and Venera spacecraft in terms of the catastrophic degassing theory.

082.145 **Influence of the terrestrial atmosphere on remotely sensed data.** T. Kawada.
10th Lunar and Planetary Symposium, (see 012.050), p. 162 - 165 (1977). In Japanese.

   Methods in computational physics. Vol. 17:
General circulation models of the atmosphere.
See Abstr. 003.002.

   Laser monitoring of the atmosphere.
See Abstr. 003.075.

   A report on the Japan — U.S. seminar on rare gas abundance and isotopic constraints on the origin and evolution of the earth's atmosphere. See Abstr. 011.036.

   Atmospheric refraction. See Abstr. 014.017.

   Nitrogen-induced absorption of oxygen in the Herzberg continuum. See Abstr. 022.039.

   Adiabatic pressure dependence of the 2.7 and 1.9 μm water vapor bands. See Abstr. 022.076.

   Correlation between angle-of-arrival fluctuations on the entrance pupil of a solar telescope. See Abstr. 031.006.

   An elementary derivation of phase fluctuations of an optical wave in the atmosphere. See Abstr. 031.032.

   Dynamic atmospheric turbulence corrections.
See Abstr. 031.033.

   The effects of the atmospheric point-spread "seeing" function on spatially resolved spectra of Jupiter.
See Abstr. 031.218.

   Atmospheric imaging constraints measured with speckle interferometry. See Abstr. 031.266.

   Maximum likelihood method for obtaining near-diffraction-limited images through the turbulent atmosphere.

See Abstr. 031.268.

   Corrections for atmospheric refractivity in satellite laser ranging. See Abstr. 031.269.

   Determination of exospheric temperature by ground measurements of ionosphere and red oxygen line.
See Abstr. 031.303.

   Atmospheric transmission measurements using infrared lasers and Fourier spectroscopy — techniques, results, and comparisons to computer models. See Abstr. 031.306.

   Cryogenic spectrometry for the measurement of airglow and aurora. See Abstr. 031.307.

   Terrestrial measurement of the performance of high-rejection optical baffling systems. See Abstr. 031.308.

   Radiometry and spectroscopy of the upper atmosphere from aircraft. See Abstr. 031.309.

   Extending radiative transfer models by use of Bayes' rule. See Abstr. 031.311.

   Correction tables for chromatic refraction in photoelectric and other astrometrical observations.
See Abstr. 031.337.

   Elimination of lateral refraction from absolute declinations of equatorial stars by the micrometric method.
See Abstr. 031.338.

   Refraction correction for the reduction routines "RA". See Abstr. 031.341.

   On the accuracy of the photographic method of investigating the astronomical refraction.
See Abstr. 031.345.

   Balloon-borne infrared telescope for absolute surface photometry of the night sky. See Abstr. 032.561.

   Atmospheric transmissometer and radiometer for E-O sensors field evaluation and model validation.
See Abstr. 034.049.

   A versatile radiometer for infrared emission measurements of the atmosphere and targets. See Abstr. 034.050.

   Influence of seeing on astrometric measurements. 1. Dependence of vibration of images on different factors.
See Abstr. 041.042.

   On a relation between irregular variations of the earth's rotation and anomalous changes of the atmospheric circulation. See Abstr. 044.010.

   Variations in Doppler positions resulting from differences in computer programs and tropospheric refraction computations. See Abstr. 046.017.

   The participation of the European Space Agency in atmospheric and magnetospheric research.
See Abstr. 051.023.

   Cosmonauts investigate the earth.
See Abstr. 051.066.

   Photographic tracking data of satellites for the program "Atmosphere". See Abstr. 054.023.

Observations of artificial earth satellites No. 14: 1974.   See Abstr. 055.011.

Two-frequency mutual coherence function of plane waves in strongly turbulent media.   See Abstr. 063.015.

Radiative transfer theory for microwave remote sensing of two-layer media.   See Abstr. 063.016.

The backscattered fraction in two-stream approximations.   See Abstr. 063.021.

The delta-Eddington approximation for radiative flux transfer.   See Abstr. 063.022.

Radiative transfer of atomic and molecular resonant emissions in upper atmospheres. I. Basic theories in Doppler-broadening atmospheres.   See Abstr. 063.027.

A note on extinction and scattering efficiencies. See Abstr. 063.028.

Multiple scattering in the atmosphere with a rough surface. I. Azimuth-independent case. See Abstr. 063.036.

Measurements of the polarized sky background in the far infrared.   See Abstr. 066.088.

Influence of seasonal variations of the neutral atmosphere on ionization of the ionospheric E-region. See Abstr. 083.042.

The detection and study of the visible spectrum of the aurora and airglow.   See Abstr. 084.020.

Gas evaporation from collision-determined planetary exospheres.   See Abstr. 091.001.

Sources of outgassed volatiles on the terrestrial planets.   See Abstr. 091.006.

On the refraction of radio waves in radiographic inspection of planetary atmospheres.   See Abstr. 091.034.

The formation of the atmospheres of the terrestrial planets by impact.   See Abstr. 091.063.

Measurements of the ultraviolet radiation of the moon and of the earth's upper atmosphere from the orbital station Salyut 4.   See Abstr. 094.191.

Momentum of meteoroids and its effect on the Earth's atmosphere.   See Abstr. 104.035.

083 Ionosphere

083.001 Geomagnetic storms and electric fields in the equatorial ionosphere.   R. G. Rastogi.
Nature, Vol. 268, 422 - 424 (1977).

083.002 Electron distribution function and ion concentrations in the Earth's lower ionosphere from Boltzmann–Fokker–Planck theory.   J. R. Jasperse.
Planet. Space Sci., Vol. 25, 743 - 756 (1977).

083.003 The transient response of the topside ionosphere to precipitation.   J. H. Whitteker.
Planet. Space Sci., Vol. 25, 773 - 786 (1977).
A numerical time-dependent model of the topside and F-layer ionosphere is used to describe how the density of O^+ ions and the plasma temperatures change as a result of transient electron precipitation with a soft energy spectrum (ca. 100 eV per electron). The response time for electron gas heating is about 2 min; for changes in topside scale height it is from 5 to 15 min, depending on altitude; and for changes in F-layer peak density, it is more than an hour.

083.004 Electric fields in diffuse aurora.
H. P. Mahon, M. Smiddy, R. C. Sagalyn.
Planet. Space Sci., Vol. 25, 859 - 870 (1977).

083.005 Daytime ionosphere scintillation associated with geomagnetic storms.
M. R. Deshpande, H. O. Vats, G. Sethia, B. S. Murthy.
Nature, Vol. 268, 614 - 616 (1977).

083.006 The large-scale ionospheric electric field: its variation with magnetic activity and relation to terrestrial kilometric radiation.   R. H. Holzworth, J.-J. Berthelier, D. K. Cullers, U. V. Fahleson, C.-G. Fälthammar, M. K. Hudson, L. Jalonen, M. C. Kelley, P. J. Kellogg, P. Tanskanen, M. Temerin, F. S. Mozer.
J. Geophys. Res., Vol. 82, 2735 - 2742 (1977).

083.007 Multifrequency spectra of ionospheric amplitude scintillations.
R. Umeki, C. H. Liu, K. C. Yeh.
J. Geophys. Res., Vol. 82, 2752 - 2760 (1977).

083.008 On a stationary analytical model of the night F-region of the ionosphere.
A. G. Khantadze, R. G. Gachechiladze.
Geomagn. Aehron., Vol. 17, 628 - 633 (1977). In Russian.

083.009 On disturbances in the F-region of the ionosphere caused by internal gravitational waves.
R. G. Gachechiladze, A. G. Khantadze.
Geomagn. Aehron., Vol. 17, 634 - 638 (1977). In Russian.

083.010 Determination of the profiles of effective frequency of electron collisions in the ionosphere by a Volterra equation.   V. D. Gusev, N. P. Danilkin, P. F. Denisenko, V. I. Vodolazkin.
Geomagn. Aehron., Vol. 17, 645 - 648 (1977). In Russian.

083.011 Non-stationary ion formation in the lower ionosphere under the action of gravitational waves.
M. A. Nikitin, N. M. Kashchenko.
Geomagn. Aehron., Vol. 17, 649 - 654 (1977). In Russian.

083.012 On the parametric influence of quasi-periodic disturbances of the lower ionosphere on directed hydromagnetic waves.   V. M. Davydov.
Geomagn. Aehron., Vol. 17, 655 - 662 (1977). In Russian.

083.013 Some results of an investigation of low-latitude ionospheric dynamics with the help of the Doppler method.   V. V. Belyj, V. A. Vazherkin, B. Laso, L. Lois, L. A. Lobachevskij, V. D. Novikov, O. S. Sergeenko.
Geomagn. Aehron., Vol. 17, 663 - 666 (1977). In Russian.

083.014 Spatial distribution of the electron concentration in the night-time polar ionosphere.
N. V. Isaev, N. K. Osipov.
Geomagn. Aehron., Vol. 17, 667 - 670 (1977). In Russian.

083.015 Plasma waves and instabilities in the polar cusp: a review.   N. D'Angelo.
Rev. Geophys. Space Phys., Vol. 15, 299 - 307 (1977).
A review is presented of the experimental and theoretical work performed in recent years on plasma waves and instabilities in the polar cusp (cleft).

083.016 Satellite beacon contributions to studies of the structure of the ionosphere.   J. V. Evans.
Rev. Geophys. Space Phys., Vol. 15, 325 - 350 (1977).
The author outlines two methods of measuring the total electron content. He discusses observations of the gross structure of the F layer from low-altitude satellites and geostationary satellites and briefly comments on some of the physical processes that serve to control the total content as measured by either method. He describes observations of large-scale features in the ionosphere such as the trough and the equatorial anomaly, and deals with observations of traveling ionospheric disturbances. Finally, he discusses what can be learned concerning the ionospheric irregularities.

083.017 Observations of neutral composition and related ionospheric variations during a magnetic storm in February 1974.   A. E. Hedin, P. Bauer, H. G. Mayr, G. R. Carignan, L. H. Brace, H. C. Brinton, A. D. Parks, D. T. Pelz.
J. Geophys. Res., Vol. 82, 3183 - 3189 (1977).
This paper presents observations, during a moderate storm in February 1974, of N_2, O, He, and Ar densities measured by the neutral atmosphere composition experiment (Nace) on the AE-C satellite. Data are available from perigee at 158 km ($43°$N latitude) to nearly 400 km. The time development of densities during the storm is followed from low activity on February 9 to a maximum on February 12 and back to low activity on February 15. The close relation of composition changes to ionospheric effects is again pointed out, and this fact is used in studying the storm morphology.

083.018 Distortions of the winter nighttime ionosphere at $L = 4$.
M. Mendillo, M. J. Buonsanto, J. A. Klobuchar.
J. Geophys. Res., Vol. 82, 3223 - 3232 (1977).

083.019 A comparison of ionospheric currents, magnetic variations, and electric fields at Arecibo.
R. M. Harper.
J. Geophys. Res., Vol. 82, 3233 - 3242 (1977).
Ionospheric currents calculated from incoherent scatter measurements at Arecibo are compared to ground magnetic variations at San Juan and to electric fields measured in the F region for two slightly disturbed periods and one very disturbed period, comprising a total of 7 days of data. The measurements indicate that the currents primarily responsible for the ground magnetic variations were flowing in the ionosphere on the slightly disturbed days. On the very disturbed days the variation in the horizontal intensity appears to be primarily due to an asymmetric ring current of magneto-

spheric origin.

083.020 The rate coefficient for the $O^+ + N_2$ reaction in the ionosphere.
M. R. Torr, J. P. St.-Maurice, D. G. Torr.
J. Geophys. Res., Vol. 82, 3287 - 3290 (1977).

The rate coefficient for the reaction $O^+ + N_2 \rightarrow NO^+ + N$ is determined as a function of temperature from the photochemistry of NO^+ for both day and night conditions by using a large sample (>5300) of simultaneous measurements of ion and neutral concentrations and temperatures made by the Atmosphere Explorer C satellite.

083.021 Determination of molecular oxygen density between 110 and 170 km from a 1450-Å photometer.
J. E. Higgins, L. Heroux.
J. Geophys. Res., Vol. 82, 3295 - 3298 (1977).

Measurements of molecular oxygen in the approximate altitude region 110–170 km were obtained from two rocket flights of a solar-pointed 1450-Å photometer launched from White Sands Missile Range, New Mexico.

083.022 Observation of an ionospheric acceleration mechanism producing energetic (keV) ions primarily normal to the geomagnetic field direction.
R. D. Sharp, R. G. Johnson, E. G. Shelley.
J. Geophys. Res., Vol. 82, 3324 - 3328 (1977).

O^+ ions with energies of approximately 1 keV have been observed flowing upward out of the ionosphere with a pitch angle distribution having a minimum along the magnetic field direction and maxima in about the $130°-140°$ range. The data are interpreted as resulting from a mechanism which accelerates ambient ionospheric ions in a direction perpendicular to the geomagnetic field.

083.023 Electron streams with energies $E_e > 150$ MeV from data of Cosmos 555. L. V. Kurnosova, A. T. Matachun, L. A. Razorenov, M. I. Fradkin.
Fiz. inst. AN SSSR. Fiz. vys. ehnerg. i kosm. luchej. Prepr. No. 53. Moskva, 1977. 16 pp. In Russian. − Abstr. in Ref. zh., 62. Issled. kosm. prostranstva, 8.62.275 (1977).

083.024 Investigations of streams of charged particles aboard Cosmos 555. L. V. Kurnosova, A. T. Matachun, L. A. Razorenov, M. I. Fradkin.
Fiz. inst. AN SSSR. Fiz. vys. ehnerg. i kosm. luchej. Prepr. No. 52. Moskva, 1977. 17 pp. In Russian. − Abstr. in Ref. zh., 62. Issled. kosm. prostranstva, 8.62.276 (1977).

083.025 The daytime lower ionosphere during the solar phenomena of July 27 to August 16, 1972.
G. T. Nestorov, E. M. Apostolov.
Solar activity and solar-terrestrial relations, (see 012.007), p. 53 - 56 (1976).

083.026 Seasonal variation in the asymmetry of diurnal variation of absorption in the lower ionosphere.
J. Laštovička.
J. Atmos. Terr. Phys., Vol. 39, 891 - 894 (1977).

083.027 Electric fields in the equatorial ionosphere.
B. G. Anandarao, J. N. Desai, M. Giles, G. Martelli, R. Raghavarao, P. Rothwell.
J. Atmos. Terr. Phys., Vol. 39, 927 - 931 (1977).

During a Commonwealth Collaborative Campaign between India and the U.K., artificial ion and neutral clouds were released at altitudes between 150 km and 160 km. The values obtained, $E \sim 3$ mV/m and $v_n \sim 80$ m/sec, are in reasonable agreement with those reported in the literature from earlier experiments performed near the geomagnetic dip equator.

083.028 Diffusion and heat flow equations for the mid-

latitude topside ionosphere.
J.-P. St.-Maurice, R. W. Schunk.
Planet. Space Sci., Vol. 25, 907 - 920 (1977).

In the present investigation, the general system of transport equations presented by Schunk (1975) is applied to the mid-latitude topside ionosphere. In this region of the ionosphere, the general system of transport equations reduces significantly, and it is possible to derive relatively simple diffusion and heat flow equations for a plasma composed of two major ions, electrons and a number of minor ions.

083.029 Relative flow of H^+ and O^+ ions in the topside ionosphere at mid-latitudes.
G. J. Bailey, R. J. Moffett, J. A. Murphy.
Planet. Space Sci., Vol. 25, 967 - 972 (1977).

Theoretical results on the daily variation of O^+ and H^+ field-aligned velocities in the topside ionosphere are presented. The results are for an $L = 3$ magnetic field tube under sunspot minimum conditions at equinox. They come from calculations of time-dependent O^+ and H^+ continuity and momentum balance in a magnetic field tube which extends from the lower $F2$ region to the equatorial plane (Murphy et al., 1976).

083.030 Perturbation du profil de l'ionosphère supérieure antarctique en été local au niveau du cornet polaire.
D. Roux.
C. R. Acad. Sci. Paris, Tome 285, Sér. B, 183 - 186 (1977).

L'analyse des sondages en contre-haut, enregistrés à la station antarctique Dumont-d'Urville en été local, révèle des modifications du profil de densité de l'ionosphère supérieure au niveau du cornet polaire. L'étude de l'influence de flux d'électrons, de basse énergie caractéristiques du cornet polaire permet de rendre compte des anomalies observées.

083.031 The quantitative relationship between VLF phase deviations and 1–8 Å solar X-ray fluxes during solar flares. Y. Muraoka, H. Murata, T. Sato.
J. Atmos. Terr. Phys., Vol. 39, 787 - 792 (1977).

083.032 Collision frequencies for use in the continuum momentum equations applied to the lower ionosphere. R. J. Hill, S. A. Bowhill.
J. Atmos. Terr. Phys., Vol. 39, 803 - 811 (1977).

083.033 Ionospheric dynamo calculations with semidiurnal winds. R. J. Stening.
Planet. Space Sci., Vol. 25, 1075 - 1080 (1977).

Equivalent circuit method calculations using '2,2' and '2,4' mode semidiurnal winds yield sample variations of electric currents and fields at various latitudes. The positions of the current system foci are found to vary much more with longitude than with season. Observed seasonal changes in lunar current system foci must be due to changes in winds rather than in conductivities.

083.034 The role of relative ion flows on the thermal structure of the ionosphere.
A. F. Nagy, R. G. Roble, W. E. Swartz, J. F. Vickrey.
Planet. Space Sci., Vol. 25, 1085 - 1086 (1977).

Calculations are presented which show that, contrary to previous suggestions, ion flow velocity differences do not have a significant effect on the thermal structure of the midlatitude topside ionosphere.

083.035 On some problems of dynamics of protons captured by the geomagnetic field. Annotation.
K. Kudela, Yu. Dubinskij.
VIII Leningr. mezhdunar. semin. Mater mezhdunar. semin. "Akt. protsessy na Solntse i probl. soln. nejtrino", 1976. Leningrad, 1976, p. 169. In Russian and English. − Abstr. in Ref. zh., 62. Issled. kosm. prostranstva, 9.62.220 (1977).

083.036 **Capture of photoelectrons by the geomagnetic field during their motion through the plasmosphere.** I. A. Krinberg, G. K. Matafonov.
Issled. po geomagn., aehron. i fiz. Solntsa. Vyp. (No.) 41. Moskva, Nauka, 1977, p. 27 - 35. In Russian. − Abstr. in Ref. zh., 62. Issled. kosm. prostranstva, 9.62.223 (1977).

083.037 **Energetic spectrum and latitude variations of the vertical stream of surplus electrons from measurements aboard the satellite Cosmos 490.**
R. N. Basilova, N. L. Grigorov, L. F. Kalinkin, E. I. Kogan-Laskina, G. I. Pugacheva, I. A. Savenko.
Kosm. Issled., Vol. 15, 579 - 588 (1977). In Russian.

083.038 **Measurements of electron streams at heights from 200 to 500 km.** V. V. Antonenko, V. L. Bolyshev, I. N. Kapustin, S. S. Konyakhina, L. V. Kurnosova, V. I. Logachev, L. A. Razorenov, V. G. Sinitsina, M. I. Fradkin, V. S. Chukin.
Kosm. Issled., Vol. 15, 589 - 597 (1977). In Russian.

083.039 **On the inhomogeneous structure of the low-latitude ionosphere.** A. S. Bakaj, V. I. Romanova, V. M. Sinel'nikov, G. K. Solodovnikov.
Kosm. Issled., Vol. 15, 628 - 630 (1977). In Russian.

083.040 **Role of Auger processes in formation of photoelectron spectra.** S. V. Avakyan, M. A. Koen, G. S. Kudryashev, G. V. Khazanov.
Kosm. Issled., Vol. 15, 631 - 632 (1977). In Russian.

083.041 **The earth's plasmasphere (review).**
K. I. Gringauz, V. V. Bezrukikh.
Geomagn. Aehron., Vol. 17, 784 - 803 (1977). In Russian.

083.042 **Influence of seasonal variations of the neutral atmosphere on ionization of the ionospheric E-region.**
G. S. Ivanov-Kholodnyj, L. N. Leshchenko, A. A. Nusinov, I. N. Odintsova.
Geomagn. Aehron., Vol. 17, 839 - 846 (1977). In Russian.

083.043 **Calculations of the relative ion concentrations at heights between 140 and 200 km for concrete helio-geophysical conditions. I. Height 170 km, $[O^+]/n_e$.**
L. A. Antonova, G. S. Ivanov-Kholodnyj.
Geomagn. Aehron., Vol. 17, 847 - 853 (1977). In Russian.

083.044 **International coordinate measurements of geophysical effects of solar activity in the upper atmosphere. IV. Precipitation of energetic particles during a bay-like disturbance of the mid-latitude D-region of the ionosphere.** R. Knuth, N. I. Fedorova.
Geomagn. Aehron., Vol. 17, 854 - 861 (1977). In Russian.

083.045 **Auroral electrons and conductivity of the high-latitude ionosphere.** N. V. Isaev, N. K. Osipov.
Geomagn. Aehron., Vol. 17, 862 - 866 (1977). In Russian.

083.046 **On the mid-latitude modulation of energetic electron streams and VLF radiation.**
M. S. Kovner, V. A. Kuznetsova, Ya. I. Likhter.
Geomagn. Aehron., Vol. 17, 867 - 870 (1977). In Russian.

083.047 **Connection of the altitude of the night F2-region with aeronomical parameters.** G. I. Ostrovskij.
Geomagn. Aehron., Vol. 17, 939 - 940 (1977). In Russian.

083.048 **Seasonal variations of the auroral E-layer.**
T. I. Shchuka.
Geomagn. Aehron., Vol. 17, 940 - 943 (1977). In Russian.

083.049 **Function of ion formation from meteor particles in** the ionospheric E-region.
A. V. Blokhin, G. G. Novikov, L. N. Rubtsov.
Geomagn. Aehron., Vol. 17, 943 - 944 (1977). In Russian.

083.050 **Formation of E_s layers by gravitational waves generated in the region of the solar terminator.**
S. P. Chernysheva, V. M. Sheftel', Eh. G. Shcharenskaya.
Geomagn. Aehron., Vol. 17, 945 - 946 (1977). In Russian.

083.051 **Observations of the ion composition by VLF hiss in the vicinity of the plasmapause.**
Y. Nakamura, T. Ondoh, K. Marubashi, T. Murakami.
Rev. Radio Res. Lab., Vol. 22, 123 - 134 (1976). In Japanese.
Abstr. in Phys. Abstr., Vol. 80, Abstr. 21999 (1977).

083.052 **Radiowave absorption in the ionosphere's cosmic ray layer.** P. I. Velinov.
C. R. Acad. Bulgare Sci., Vol. 29, 1137 - 1140 (1976).
Abstr. in Phys. Abstr., Vol. 80, Abstr. 22024 (1977).

083.053 **Maintenance of the middle-latitude nocturnal D-layer by energetic electron precipitation.**
W. N. Spjeldvik, R. M. Thorne.
Pure Appl. Geophys., Vol. 114, 497 - 508 (1976). − Abstr. in Phys. Abstr., Vol. 80, Abstr. 25856 (1977).

083.054 **Neutral winds and nighttime F-region.**
M. S. Narayanan, P. D. Bhavsar.
Indian J. Radio Space Phys., Vol. 5, No. 3, p. 221 - 224 (1976).
Abstr. in Phys. Abstr., Vol. 80, Abstr. 29555 (1977).

083.055 **Generation of electrostatic waves around lower hybrid resonance in the ionospheric plasma.**
A. K. Gwal, S. P. Mishra, K. D. Misra.
Indian J. Radio Space Phys., Vol. 5, No. 3, p. 240 - 244 (1976).
Abstr. in Phys. Abstr., Vol. 80, Abstr. 29557 (1977).

083.056 **Role of Kelvin-Helmholtz instability in ionospheric phenomena.** S. P. Mishra, K. D. Misra.
Indian J. Radio Space Phys., Vol. 5, No. 3, p. 250 - 253 (1976).
Abstr. in Phys. Abstr., Vol. 80, Abstr. 29558 (1977).

083.057 **Influence of ionizing agencies from outside the solar system on the world morphology of F-region ionization.** C. Prasad Chaudhary.
Indian J. Radio Space Phys., Vol. 5, No. 3, p. 257 - 258 (1976).
Abstr. in Phys. Abstr., Vol. 80, Abstr. 29559 (1977).

083.058 **Nocturnal E-region during meteor showers.**
J. V. M. Naidu, B. Ramachandra Rao.
Indian J. Radio Space Phys., Vol. 5, No. 3, p. 258 - 259 (1976).
Abstr. in Phys. Abstr., Vol. 80, Abstr. 29560 (1977).

083.059 **Threshold height and equatorial spread-F.**
V. Subrahmanyam, M. Srirama Rao.
Indian J. Radio Space Phys., Vol. 5, No. 3, p. 259 - 261 (1976).
Abstr. in Phys. Abstr., Vol. 80, Abstr. 29561 (1977).

083.060 **Latitude variation of peak E_s occurrences during a half-solar cycle.**
S. P. Manohar Rao, B. Ramachandra Rao.
Indian J. Radio Space Phys., Vol. 5, No. 3, p. 261 - 263 (1976).
Abstr. in Phys. Abstr., Vol. 80, Abstr. 29562 (1977).

083.061 **Nocturnal variation of total absorption and nighttime E-region.** S. L. Jain, R. K. Rai.
Indian J. Radio Space Phys., Vol. 5, No. 3, p. 263 - 264 (1976).
Abstr. in Phys. Abstr., Vol. 80, Abstr. 29563 (1977).

083.062 **Simultaneous 1.5- and 4-GHz ionospheric scintillation measurement.** R. R. Taur.
Radio Sci., Vol. 11, 1029 - 1036 (1976). − Abstr. in Phys.

Abstr., Vol. 80, Abstr. 29581 (1977).

083.063 On the investigation of the non-linear interaction of
         powerful radiation with ionospheric plasma in
experiments aboard AES.   G. A. Gusev.
Din. kosm. plazmy. Moskva, 1976, p. 101 - 107. In Russian.
Abstr. in Ref. zh., 62. Issled. kosm. prostranstva, 10.62.244
(1977).

083.064 Variations of the geometric parameters of N(h)-
         profiles of the F_2 region in the presence of syn-
chronous action of electric fields and auroral electron streams.
N. V. Isaev.
Din. kosm. plazmy. Moskva, 1976, p. 181 - 184. In Russian.
Abstr. in Ref. zh., 62. Issled. kosm. prostranstva, 10.62.245
(1977).

083.065 Empirical expression of the electron temperature
         for mean latitudes during the period of weak solar
activity.   G. E. Sutyrina, G. V. Shapranova.
Issled. po geomagn., aehron. i fiz. Solntsa. Vyp. (No.) 41.
Moskva, Nauka, 1977, p. 84 - 86. In Russian. — Abstr. in Ref.
zh., 62. Issled. kosm. prostranstva, 10.62.248 (1977).

083.066 VHF back scattering from ionospheric E region
         irregularities near the magnetic equator.
R. G. Rastogi.
Ann. Géophys., Vol. 32, 203 - 213 (1976). — Abstr. in Phys.
Abstr., Vol. 80, Abstr. 33248 (1977).

083.067 Critical study of the classical methods of determina-
         tion of the TEC (total electron content) from the
Faraday rotation.   I. H. Keroub.
Ann. Géophys., Vol. 32, 215 - 218 (1976). — Abstr. in Phys.
Abstr., Vol. 80, Abstr. 33249 (1977).

083.068 New methods of determination of the total electron
         content in the transverse zone.   I. H. Keroub.
Ann. Géophys., Vol. 32, 219 - 225 (1976). — Abstr. in Phys.
Abstr., Vol. 80, Abstr. 33250 (1977).

083.069 The structure of the latitudinal total electron
         content (TEC) gradients over mid-latitude stations.
I. H. Keroub.
Ann. Géophys., Vol. 32, 227 - 242 (1976). — Abstr. in Phys.
Abstr., Vol. 80, Abstr. 33251 (1977).

083.070 The F2 layer seasonal anomaly.
         H. M. Lumb, C. S. G. K. Setty.
Ann. Géophys., Vol. 32, 243 - 255 (1976). — Abstr. in Phys.
Abstr., Vol. 80, Abstr. 33252 (1977).

083.071 D-region relaxation after sudden cut-off of ioniza-
         tion source.   G. T. Nestorov.
C. R. Acad. Bulgare Sci., Vol. 29, 1433 - 1435 (1976).
Abstr. in Phys. Abstr., Vol. 80, Abstr. 33254 (1977).

083.072 Electron production by energetic particle precipita-
         tions in the equatorial ionosphere.
M. N. M. Rao, B. C. N. Rao, M. Gogoshev, I. Kutiev, K.
Serafimov, M. Karadimov.
C. R. Acad. Bulgare Sci., Vol. 29, 1439 - 1442 (1976).
Abstr. in Phys. Abstr., Vol. 80, Abstr. 33255 (1977).

083.073 Differences in seasonal variation of noon F_2-ioniza-
         tion at dip-conjugate places.
N. R. Almaula, K. M. Kotadia.
Indian J. Radio Space Phys., Vol. 5, 231 - 234 (1976).
Abstr. in Phys. Abstr., Vol. 80, Abstr. 33259 (1977).

083.074 Ionospheric scintillations induced by cloud of
         intense sporadic E layer.

R. G. Rastogi, K. N. Iyer.
Curr. Sci., Vol. 45, 685 - 686 (1976). — Abstr. in Phys. Abstr.,
Vol. 80, Abstr. 37477 (1977).

083.075 The reconciliation of an F-region irregularity model
         with sunspot cycle variations in spread-F occurrence.
D. G. Singleton.
Radio Sci., Vol. 12, 107 - 118 (1977). — Abstr. in Phys. Abstr.,
Vol. 80, Abstr. 41262 (1977).

083.076 A simplified inversion procedure for calculating
         electron density profiles from ionograms for use
with minicomputers.   A. K. Paul.
Radio Sci., Vol. 12, 119 - 122 (1977). — Abstr. in Phys. Abstr.,
Vol. 80, Abstr. 41263 (1977).

083.077 The effect of ionospheric scintillation on VHF/UHF
         satellite communications.   H. E. Whitney, S. Basu.
Radio Sci., Vol. 12, 123 - 133 (1977). — Abstr. in Phys. Abstr.,
Vol. 80, Abstr. 41264 (1977).

083.078 A method for inverting oblique sounding data in the
         ionosphere.   S. L. Chuang, K. C. Yeh.
Radio Sci., Vol. 12, 135 - 140 (1977). — Abstr. in Phys.
Abstr., Vol. 80, Abstr. 41265 (1977).

083.079 On auroral electrojet curvature and the interpreta-
         tion of azimuth-scan, radar Doppler data.
A. Brekke, R. T. Tsunoda, M. J. Baron.
Radio Sci., Vol. 12, 141 - 149 (1977). — Abstr. in Phys. Abstr.,
Vol. 80, Abstr. 41266 (1977).

083.080 The D-region ion composition.
         N. Nath, C. S. G. K. Setty.
Pure Appl. Geophys., Vol. 114, 891 - 908 (1976). — Abstr. in
Phys. Abstr., Vol. 80, Abstr. 49226 (1977).

083.081 Plasma drift in the polar ionosphere, Ny Alesund,
         6 - 10 September 1974.   E. Leer, O. Bratteng.
Phys. Norvegica, Vol. 8, 129 - 135 (1976). — Abstr. in Phys.
Abstr., Vol. 80, Abstr. 45245 (1977).

083.082 Incoherent scatter radar observations during August
         4 - 7, 1972.
T. Wedde, J. R. Doupnik, P. M. Banks, R. J. Park, J. C. Siren.
Radio Sci., Vol. 12, 285 - 306 (1977). — Abstr. in Phys. Abstr.,
Vol. 80, Abstr. 45248 (1977).

083.083 A model of the midlatitude nighttime E-region in-
         cluding metallic ions.   R. D. Harris, M. McGhan.
Radio Sci., Vol. 12, 307 - 309 (1977). — Abstr. in Phys. Abstr.,
Vol. 80, Abstr. 45249 (1977).

083.084 Multifrequency studies of ionospheric scintillations.
         R. Umeki, C. H. Liu, K. C. Yeh.
Radio Sci., Vol. 12, 311 - 317 (1977). — Abstr. in Phys. Abstr.,
Vol. 80, Abstr. 45250 (1977).

083.085 Observed temporal amplitude and phase spectra
         from ATS-6 radio beacon experiment at 40, 140,
and 360 MHz.   R. H. Ott.
Radio Sci., Vol. 12, 319 - 325 (1977). — Abstr. in Phys. Abstr.,
Vol. 80, Abstr. 45251 (1977).

083.086 The diurnal variation of the electron density of the
         mid-latitude ionospheric D-region deduced from
VLF-measurements.   J. Schafer.
J. Geophys., Vol. 42, 361 - 372 (1977). — Abstr. in Phys.
Abstr., Vol. 80, Abstr. 64520 (1977).

083.087 Modelling of the ionosphere and comparison of the
         calculated and observed cosmic radio noise absorp-

tion over Delhi. M. C. Sharma, S. B. S. S. Sarma.
J. Geophys., Vol. 42, 535 - 540 (1977). – Abstr. in Phys.
Abstr., Vol. 80, Abstr. 64521 (1977).

083.088 The winter anomaly in ionospheric absorption and the D-region ion chemistry. M. A. Hidalgo.
Planet. Space Sci., Vol. 25, 1135 - 1143 (1977).

A detailed analysis of the D-region ion composition measurements performed by Zbinden et al. (1975), during a winter day of high ionospheric absorption, has been carried out. The study examines the interactive mesosphere-D-region processes which occur in such anomalous conditions and their implication for water cluster ion chemistry.

083.089 Generalized exponential model of electron concentration profiles in low ionospheres. P. I. Velinov.
C. R. Acad. Bulgare Sci., Vol. 29, 1757 - 1760 (1976).
Abstr. in Phys. Abstr., Vol. 80, Abstr. 67005 (1977).

083.090 A possible argumentation of dependence of the green oxygen line of sporadic E_s layer.
K. Kazakov, K. B. Serafimov.
C. R. Acad. Bulgare Sci., Vol. 30, 41 - 43 (1977). – Abstr. in Phys. Abstr., Vol. 80, Abstr. 67006 (1977).

083.091 Red oxygen line 6300 Å and electron content in the night F-region.
Y. L. Trutse (*Yu. L. Truttse*), M. M. Gogoshev.
C. R. Acad. Bulgare Sci., Vol. 30, 45 - 48 (1977). – Abstr. in Phys. Abstr., Vol. 80, Abstr. 67007 (1977).

083.092 Temporary variations of critical frequencies of the F_2 layer in 9-day to 1-year periods.
E. M. Apostolov, M. M. Cohen.
C. R. Acad. Bulgare, Sci., Vol. 30, 221 - 224 (1977). – Abstr. in Phys. Abstr., Vol. 80, Abstr. 67008 (1977).

083.093 Effects of magnetic and sunspot activities on the topside ionosphere.
M. Prabhakara Rao, C. Jogulu.
Indian J. Radio Space Phys., Vol. 5, 265 - 268 (1976).
Abstr. in Phys. Abstr., Vol. 80, Abstr. 70918 (1977).

083.094 Estimation of electron heating rates in the topside ionosphere during medium solar activity.
R. Singh, B. C. N. Rao.
Indian J. Radio Space Phys., Vol. 5, 269 - 271 (1976).
Abstr. in Phys. Abstr., Vol. 80, Abstr. 70919 (1977).

083.095 Investigation of D-region plasma by modulation techniques. F. M. Ragab.
Acta Phys. Acad. Sci. Hungaricae, Vol. 41, 281 - 288 (1976).
Abstr. in Phys. Abstr., Vol. 80, Abstr. 82078 (1977).

083.096 Mean arrival time and mean pulsewidth of signals propagating through a dispersive and random medium. K. C. Yeh, C. C. Yang.
IEEE Trans. Antennas Propag., Vol. AP-25, 710 - 713 (1977).
Abstr. in Phys. Abstr., Vol. 80, Abstr. 82080 (1977).

083.097 The electron density profile in the lower ionosphere derived from a rocket measurement of VLF propagation modes. I. Nagano, M. Mambo, I. Kimura.
Trans. Inst. Electron. Commun. Eng. Japan E, Vol. E59, No. 8, p. 6 - 7 (1976). – Abstr. in Phys. Abstr., Vol. 80, Abstr. 82082 (1977).

083.098 Observations with incoherent scatter radar of ionospheric modification in the auroral zone during particle precipitation and Joule heating events.
J. D. Kelly, V. B. Wickwar.
Antennas and propagation. International Symposium, Stanford,

Calif., 20 - 22 June 1977. IEEE, New York, USA. Price $ 16.50 (1977). p. 607 - 610. – Abstr. in Phys. Abstr., Vol. 80, Abstr. 82083 (1977).

083.099 Ion composition and electron- and ion-loss processes in the Earth's atmosphere.
F. Arnold, D. Krankowsky.
Dynamical and chemical coupling between the neutral and ionized atmosphere, (see 012.036), p. 93 - 127 (1977).

083.100 Mid- and high latitude reference ionosphere.
K. Rawer.
Dynamical and chemical coupling between the neutral and ionized atmosphere, (see 012.036), p. 129 - 143 (1977).

083.101 Ionization processes. J. B. Reagan.
Dynamical and chemical coupling between the neutral and ionized atmosphere,(see 012.036), p. 145 - 160 (1977).

083.102 Influence of relativistic electron precipitation on the lower ionosphere and stratosphere.
R. M. Thorne.
Dynamical and chemical coupling between the neutral and ionized atmosphere, (see 012.036), p. 161 - 168 (1977).

083.103 Coherent wave induced particle precipitation into the upper atmosphere.
U. S. Inan, T. F. Bell, R. A. Helliwell.
Dynamical and chemical coupling between the neutral and ionized atmosphere, (see 012.036), p. 169 - 174 (1977).

083.104 Thermal electron collisions and energy transfer in atmospheric gases. M. Mentzoni.
Dynamical and chemical coupling between the neutral and ionized atmosphere, (see 012.036), p. 175 - 183 (1977).

083.105 An attempt to interprete collision frequency measurements in the D-region.
K. Torkar, M. Friedrich.
Dynamical and chemical coupling between the neutral and ionized atmosphere, (see 012.036), p. 185 - 188 (1977).

083.106 Electron density enhancements in the F region beneath the magnetospheric cusp.
C. C. Chacko, M. Mendillo.
J. Geophys. Res., Vol. 82, 4757 - 4764 (1977).

083.107 Numerical model of the convecting F_2 ionosphere at high latitudes. W. C. Knudsen, P. M. Banks,
J. D. Winningham, D. M. Klumpar.
J. Geophys. Res., Vol. 82, 4784 - 4792 (1977).

083.108 Aeronomic aspects of the polar D-region.
W. Swider.
Space Sci. Rev., Vol. 20, 69 - 114 (1977).

The polar D-region has been little studied. A major part of this review concerns observations performed during the 1969, November, 2–5, solar proton event. These extensive measurements and subsequent interpretations constitute a major source of polar D-region knowledge. The aeronomic concepts garnered from the analysis of this event are discussed and compared with results from other events and other polar mesospheric data.

083.109 On the kinetic theory of ionospheric plasma conductivity. M. Ya. Alievskij.
Geomagn. Aehron., Vol. 17, 994 - 1001 (1977). In Russian.

083.110 Boundary condition for Alfvén waves in the ionosphere. Yu. P. Mal'tsev.
Geomagn. Aehron., Vol. 17, 1008 - 1011 (1977). In Russian.

083.111 **Multi-layer model of a small-scale dynamo.**
Yu. S. Vardanyan.
Geomagn. Aehron., Vol. 17, 1012 - 1017 (1977). In Russian.

083.112 **Winter anomaly of the E layer as a consequence of seasonal variations of the upper atmosphere.**
G. S. Ivanov-Kholodnyj, A. A. Nusinov.
Geomagn. Aehron., Vol. 17, 1018 - 1023 (1977). In Russian.

083.113 **Models of the night altitudinal distribution of the $\lambda 6300$ Å emission.**
K. Serafimov, M. Gogoshev, Ts. Gogosheva.
Geomagn. Aehron., Vol. 17, 1044 - 1049 (1977). In Russian.

083.114 **Analytical description of charts of the geometrical parameters of the ionospheric F2 layer by spherical functions.**   O. V. Chernyshev, B. S. Shapiro.
Geomagn. Aehron., Vol. 17, 1111 - 1112 (1977). In Russian.

083.115 **Change of the latitude dependence of night ionization of the F2 layer in the solar activity cycle.**
A. S. Besprozvannaya, L. A. Yudovich.
Geomagn. Aehron., Vol. 17, 1113 - 1115 (1977). In Russian.

083.116 **Vertical drift of the F2 layer connected with a polar substorm.**   L. N. Makarova.
Geomagn. Aehron., Vol. 17, 1115 - 1117 (1977). In Russian.

083.117 **Observations of electron temperature gradients in mid-latitude E_s layers.**
E. P. Szuszczewicz, J. C. Holmes.
J. Geophys. Res., Vol. 82, 5073 - 5080 (1977).
In this work the authors present pulsed plasma probe measurements of electron temperature which show that substantial temperature gradients do exist within an E_s layer.

083.118 **Comparison between calculated and measured photoelectron fluxes from Atmosphere Explorer C and E.**   A. F. Nagy, J. P. Doering, W. K. Peterson, M. R. Torr, P. M. Banks.
J. Geophys. Res., Vol. 82, 5099 - 5103 (1977).
Photoelectron fluxes were calculated by using the two-stream technique and were compared with measurements obtained by the photoelectron spectrometer on Atmosphere Explorer C and E. Comparisons of theoretical and experimental results for a variety of conditions have shown that the calculated and measured photoelectron fluxes agree well at low altitudes, and even at higher altitudes, where transport effects are important, the agreement is not unreasonable considering all the uncertainties involved.

083.119 **The baselevel ionospheric trough.**
M. Mendillo, C. C. Chacko.
J. Geophys. Res., Vol. 82, 5129 - 5137 (1977).

083.120 **Solar cycle effects on the electric fields in the equatorial ionosphere.**
R. F. Woodman, R. G. Rastogi, C. Calderon.
J. Geophys. Res., Vol. 82, 5257 - 5261 (1977).
The horizontal east-west electric fields in the equatorial F region are estimated by using the vertical electron drift data over Jicamarca, Peru. The aim of this paper is to investigate the solar cycle effects in the daily variation of the east-west electric field in the equatorial ionosphere.

083.121 **Formation of small-scale inhomogeneities in the auroral F-layer during geomagnetic disturbances.**
M. G. Gel'berg.
Rasprostranenie radiovoln v polyarn. ionos. Apatity, 1977, p. 38 - 43. In Russian. – Abstr. in Ref. zh., 62. Issled. kosm. prostranstva, 1.62.193 (1978).

083.122 **High latitude morphology of ionospheric scintillations.**   J. Aarons.
Effect of the ionosphere on space systems and communications, (see 012.048), p. 1 - 7 (1975).

083.123 **The role of the magnetosphere in satellite and radio-star scintillation.**   H. G. Booker.
Effect of the ionosphere on space systems and communications, (see 012.048), p. 9 - 12 (1975).

083.124 **The development of a highly-successful worldwide empirical ionospheric model and its use in certain aspects of space communications and worldwide total electron content investigations.**   R. B. Bent, S. K. Llewellyn, G. Nesterczuk, P. E. Schmid.
Effect of the ionosphere on space systems and communications, (see 012.048), p. 13 - 28 (1975).

083.125 **A review of the recent results of in situ ionospheric irregularity measurements and their relation to electrostatic instabilities.**   M. C. Kelley, F. S. Mozer.
Effect of the ionosphere on space systems and communications, (see 012.048), p. 29 - 38 (1975).

083.126 **Recent results from theoretical and numerical modeling of E and F region irregularities.**
B. E. McDonald, S. L. Ossakow, T. P. Coffey, R. N. Sudan, A. J. Scannapieco, S. R. Goldman.
Effect of the ionosphere on space systems and communications, (see 012.048), p. 39 - 52 (1975).

083.127 **Spectra of amplitude and phase scintillation.**
R. K. Crane.
Effect of the ionosphere on space systems and communications, (see 012.048), p. 53 - 64 (1975).

083.128 **First-order signal-statistical modeling of scintillation.**   E. J. Fremouw, C. L. Rino.
Effect of the ionosphere on space systems and communications, (see 012.048), p. 65 - 75 (1975).

083.129 **Morphology of ionospheric scintillation in the auroral zone.**   R. H. Wand, J. V. Evans.
Effect of the ionosphere on space systems and communications, (see 012.048), p. 76 - 83 (1975).

083.130 **F layer scintillations and the aurora.**
E. Martin, J. Aarons.
Effect of the ionosphere on space systems and communications, (see 012.048), p. 84 - 90 (1975).

083.131 **Equatorial scintillation at 136 MHz observed over half a sunspot cycle.**   J. P. Mullen, G. S. Hawkins.
Effect of the ionosphere on space systems and communications, (see 012.048), p. 95 - 100 (1975).

083.132 **Theoretical and numerical simulation studies of ionospheric inhomogeneities produced by plasma clouds.**   S. L. Ossakow, A. J. Scannapieco, S. R. Goldman, D. L. Book, B. E. McDonald.
Effect of the ionosphere on space systems and communications, (see 012.048), p. 196 - 202 (1975).

083.133 **Prediction of ionospheric effects associated with solar wind disturbances using interplanetary scintillation observations at 34.3 MHz.**   W. M. Cronyn, F. Erskine, S. D. Shawhan, B. L. Gotwols, E. C. Roelof.
Effect of the ionosphere on space systems and communications, (see 012.048), p. 223 - 234 (1975).

083.134 **Stochastic stationarity of the scintillating equatorial ionosphere.**   H. A. Blank, G. J. Bream.

Effect of the ionosphere on space systems and communications, (see 012.048), p. 250 - 255 (1975).

083.135 **Some results of scintillation studies.**
        K. C. Yeh, C. H. Liu.
Effect of the ionosphere on space systems and communications, (see 012.048), p. 276 - 282 (1975).

083.136 **Exact analysis of ionospheric inhomogeneities by the reflection coefficient method.**
A. K. Jordan, S. Ahn.
Effect of the ionosphere on space systems and communications, (see 012.048), p. 286 - 291 (1975).

083.137 **Some problems in constructing phenomenological models of ionospheric electron density.** Y. T. Chiu.
Effect of the ionosphere on space systems and communications, (see 012.048), p. 324 - 330 (1975).

083.138 **Modeling the topside of the F region.** R. S. Allen.
        Effect of the ionosphere on space systems and communications, (see 012.048), p. 331 - 335 (1975).

083.139 **Decimeter modeling of ionospheric columnar electron content at S-band frequencies.**
K. W. Yip, F. B. Winn, M. S. Reid, C. T. Stelzried.
Effect of the ionosphere on space systems and communications, (see 012.048), p. 345 - 352 (1975).

083.140 **Solar cycle variations of the total electron content at low latitude.** Y.-N. Huang.
Effect of the ionosphere on space systems and communications, (see 012.048), p. 355 - 360 (1975).

083.141 **The construction and use of storm-time corrections for ionospheric F-region parameters.**
M. Mendillo, M. J. Buonsanto, J. A. Klobuchar.
Effect of the ionosphere on space systems and communications, (see 012.048), p. 361 - 371 (1975).

083.142 **Seasonal and diurnal variations in the total electron content of the ionosphere at invariant latitude 54 degrees.** G. S. Hawkins, J. A. Klobuchar.
Effect of the ionosphere on space systems and communications, (see 012.048), p. 372 - 396 (1975).

083.143 **Limitations of mapping techniques to predicting total electron content at a distant point.**
C. M. Rush.
Effect of the ionosphere on space systems and communications, (see 012.048), p. 397 - 401 (1975).

083.144 **In situ measurements of the structure and spectral characteristics of small scale F region ionospheric irregularities.** M. Ahmed, A. D. R. Phelps, R. C. Sagalyn.
Effect of the ionosphere on space systems and communications, (see 012.048), p. 469 - 485 (1975).

Physique de l'ionosphère.   See Abstr. 003.062.

Methods for determining a standard ionospheric topside profile by single measurements.
See Abstr. 031.291.

Aerials for ionospheric research.
See Abstr. 034.038.

A high-gain narrow-band amplifier using active filters (*ionosphere measurement equipment*).
See Abstr. 034.040.

Modeling of residual range error in two frequency corrected Doppler data.   See Abstr. 046.010.

Numerical modelling of a satellite experiment using the Doppler effect at coherent frequencies in the presence of a wave disturbance in the ionosphere. See Abstr. 051.067.

Electrostatic waves in the warm magnetoplasma at the cyclotron harmonic frequencies.
See Abstr. 062.041.

Particle aspect analysis of $E \times B$ drift instability in the presence of non-uniform electric field.
See Abstr. 062.076.

Etude statistique des relations entre la radiation X solaire, les renforcements soudains d'atmosphériques et les plages d'activité solaire. See Abstr. 076.007.

La influencia de la desactivación de $O(^1D)$ por choques en la región F2 nocturna. See Abstr. 082.105.

Medium scale TID's and their associated internal gravity waves as seen through height-dependent electron density power spectra. See Abstr. 082.133.

Ion photochemistry of the thermosphere from Atmosphere Explorer C measurements. See Abstr. 082.143.

Recent advances in the use of radar auroral backscatter to measure ionospheric electric fields.
See Abstr. 084.023.

Solar-terrestrial physics. See Abstr. 084.301.

Atmospheric waves in the ionosphere due to total solar eclipse. See Abstr. 085.008.

Simulation of the effect of a solar eclipse in the ionosphere by different representations of electron temperature.
See Abstr. 085.018.

On a connection between ionospheric and geomagnetic disturbances at low latitudes and the interplanetary magnetic field. See Abstr. 106.020.

084 Aurorae, Magnetic Field, Radiation Belts

Aurorae

084.001 *F* layer scintillations and the aurora.   E. Martin,
J. Aarons.
J. Geophys. Res., Vol. 82, 2717 - 2722 (1977).

084.002 The response of the dayside aurora to sharp northward and southward transitions of the interplanetary magnetic field and to magnetospheric substorms.
J. L. Horwitz, S.-I. Akasofu.
J. Geophys. Res., Vol. 82, 2723 - 2734 (1977).

084.003 Lunar influence on the occurrence of aurora.
K. Henriksen, C. S. Deehr, G. J. Romick.
J. Geophys. Res., Vol. 82, 2842 - 2846 (1977).
      The lunar dependence of the occurrence frequency of
bright aurora reported by Stoffregen (1968, 1970) and
thought by him to be caused by magnetospheric disturbances
at the moon's orbit may be due to a hindrance of poleward
bound plasma flow by the combined effect of the moon and
its magnetohydrodynamic wake when the moon is passing
through the plasma sheet in the geomagnetic tail at the geocentric distance of 60 R_E.

084.004 Parallactic observations of the spatially ordered
structure of a radio aurora.
E. E. Timofeev, M. V. Uspenskij.
Geomagn. Aehron., Vol. 17, 678 - 684 (1977). In Russian.

084.005 A comparison between N_2^+ 4278-Å emission and
electron flux in the auroral zone.
J. F. Kasting, P. B. Hays.
J. Geophys. Res., Vol. 82, 3319 - 3323 (1977).

084.006 Calculations of soft auroral bremsstrahlung and K_a
line emission at satellite altitude.
J. G. Luhmann, J. B. Blake.
J. Atmos. Terr. Phys., Vol. 39, 913 - 919 (1977).
      Satellite altitude (835 km) fluxes of 0.1−10 keV auroral
X-rays are calculated for some typical forms of incident electron spectra. Both bremsstrahlung and K_a line emission are
included.

084.007 Effects of primary electron transit times on power
spectra of auroral-zone X-ray microbursts.
B. S. Haugstad, T. Pytte.
J. Atmos. Terr. Phys., Vol. 39, 689 - 698 (1977).

084.008 Auroral zone E-region motions deduced from spaced
receiver observations.
G. Solvang, A. Brekke, A. Haug.
J. Atmos. Terr. Phys., Vol. 39, 823 - 831 (1977).

084.009 Measurements of 1.5- to 5.3-μm infrared enhance-
ments associated with a bright auroral breakup.
K. D. Baker, D. J. Baker, J. C. Ulwick, A. T. Stair, Jr.
J. Geophys. Res., Vol. 82, 3518 - 3528 (1977).

084.010 Time-dependent studies of the aurora: effects of
particle precipitation on the dynamic morphology
of ionospheric and atmospheric properties.
R. G. Roble, M. H. Rees.
Planet. Space Sci., Vol. 25, 991 - 1010 (1977).
      A self-consistent, time-dependent numerical model of
the aurora and high-latitude ionosphere has been developed.
It is used to study the response of ionospheric and atmospheric properties in regions subjected to electron bombard-

ment. The time history of precipitation events is arbitrarily
specified and computations are made for a variety of electron
spectral energy distributions and flux magnitudes.

084.011 Auroral recombination of N and O: a possible
source for emission in the γ and δ bands of NO.
S. C. Wofsy, M. B. McElroy.
Planet. Space Sci., Vol. 25, 1021 - 1026 (1977).
      Models are derived for the auroral ionosphere and include
estimates for the concentrations of N and NO. The concentration of NO is estimated.

084.012 Location of the region of auroral break of electrons
in the morning sector of the magnetosphere from
measurements aboard the artificial earth satellite Molniya 1.
Z. G. Zueva, V. I. Lazarev, M. V. Tel'tsov.
Geomagn. Aehron., Vol. 17, 934 - 936 (1977). In Russian.

084.013 Acceleration of auroral particles.
A. Johnstone.
Nature, Vol. 270, 101 - 102 (1977).

084.014 Radio auroral scattering anisotropy inferred from
42 MHz polarization studies.
N. Bedard, G. J. Sofko.
Canadian J. Phys., Vol. 54, 2435 - 2444 (1976). − Abstr. in
Phys. Abstr., Vol. 80, Abstr. 21968 (1977).

084.015 Radiant of auroral rays from observations in the
Tiksi Bay.   N. I. Dzyubenko, N. N. Bliznyuk.
Problems of cosmic physics. Vyp. (No.) 12, (see 003.016),
p. 49 - 55 (1977). In Russian.
      From large-scale auroral photographs obtained by Kiev
University expeditions for the period 1959 - 1966 the "instant"
and mean radiant of auroral rays are obtained. Its shifting to
the south-west direction from the magnetic zenith is noted.

084.016 Comments on the origin of infrared emissions in
the auroral regions.
K. Henriksen, C. S. Deehr, K. Ahlnas.
Phys. Norvegica, Vol. 8, 183 - 186 (1976). − Abstr. in Phys.
Abstr., Vol. 80, Abstr. 33247 (1977).

084.017 Spectral albedo corrections to ISIS 2 satellite
auroral photometer data.
A. W. Harrison, C. D. Anger.
Canadian J. Phys., Vol. 55, 663 - 670 (1977). − Abstr. in Phys.
Abstr., Vol. 80, Abstr. 49208 (1977).

084.018 Earth albedo effects in satellite auroral photometry.
A. W. Harrison, C. D. Anger.
Canadian J. Phys., Vol. 55, 929 - 936 (1976). − Abstr. in
Phys. Abstr., Vol. 80, Abstr. 56737 (1977).

084.019 The eastward movement of the structure of auroral
radio absorption events in the morning sector.
J. K. Hargreaves, M. G. Berry.
Ann. Géophys., Vol. 32, 401 - 406 (1976). − Abstr. in Phys.
Abstr., Vol. 80, Abstr. 56792 (1977).

084.020 The detection and study of the visible spectrum of
the aurora and airglow.   G. J. Romick.
Methods for atmospheric radiometry, (see 012.033), p. 63 - 70
(1976). − Abstr. in Phys. Abstr., Vol. 80, Abstr. 70903 (1977).

084.021 On the role of magnetic mirroring in the auroral
phenomena.   W. Lennartsson.
R. Inst. Technol., Stockholm, Sweden, Rep. TRITA-EPP-77-

11, 64 pp. (1976). – Abstr. in Phys. Abstr., Vol. 80, Abstr. 73901 (1977).

084.022 Auroral structure and dynamics.
G. G. Shepherd.
Dynamical and chemical coupling between the neutral and ionized atmosphere, (see 012.036), p. 275 - 290 (1977).

084.023 Recent advances in the use of radar auroral backscatter to measure ionospheric electric fields.
R. A. Greenwald.
Dynamical and chemical coupling between the neutral and ionized atmosphere, (see 012.036), p. 291 - 312 (1977).

The author presents some initial observations obtained with the new Scandinavian Twin Auroral Radar Experiment that demonstrate the spatial and temporal structure of radar auroral irregularity motions.

084.024 Auroral effects on neutral dynamics. A. Brekke.
Dynamical and chemical coupling between the neutral and ionized atmosphere, (see 012.036), p. 313 - 336 (1977).

084.025 Large-scale characteristics of field-aligned currents determined from the TRIAD magnetometer experiment. T. A. Potemra.
Dynamical and chemical coupling between the neutral and ionized atmosphere, (see 012.036), p. 337 - 352 (1977).

084.026 Pulsating aurora: local and global morphology.
O. Royrvik, T. N. Davis.
J. Geophys. Res., Vol. 82, 4720 - 4740 (1977).

Pulsating describes a low-intensity aurora that undergoes rapid alternating increases and decreases in luminosity. Extensive new data available from ground-based low-light-level television cameras and satellite scanners have allowed a detailed study of the pulsating aurora phenomenon. Pulsations occur in auroral arcs, arc segments, and patches of fixed and variable area. The temporal and spatial characteristics are highly variable over a broad and continuous spectrum; rapid changes from one set of characteristics to another frequently occur, as do reversible changes from pulsating to nonpulsating auroras.

084.027 Characteristics of polar cap sun-aligned arcs.
S. Ismail, D. D. Wallis, L. L. Cogger.
J. Geophys. Res., Vol. 82, 4741 - 4749 (1977).

Observations of polar cap sun-aligned arcs obtained with the auroral scanning photometer on Isis 2 for the period 1971 to 1975 are examined. A 2:1 asymmetry was found in the occurrence frequency between the morning and evening sectors of the polar cap. Sun-aligned arcs were observed on only 0.6% of polar cap passes and occurred most frequently during periods of low magnetic activity. Although the intensity along any single arc varied considerably, it was found that the 5577 Å/3914 Å intensity ratio remained constant. Examination of particle data and the observed intensity ratios indicate that the arcs are excited by low-energy ($\leqslant 1$ keV) electron fluxes.

084.028 Nonpotential waves in the auroral electrojet.
V. G. Korobeinikov (*Korobejnikov*), A. N. Zaitzev (*Zajtsev*).
Astrophys. Space Sci., Vol. 46, 33 - 39 (1977).

An ionospheric plasma instability in the auroral electrojet region believed to be responsible for electromagnetic waves with angular frequency smaller than ν_i is discussed. The threshold current velocity U_t and oscillation growth rate γ are found for the realistic physical situation when a nonuniform plasma density and an ionization function are considered. It is found that the gradient of plasma density may be neglected in the expressions for U_t and γ and that the spectral density of fluctuations in the density of the plasma agrees well with the experimental data.

084.029 On the role of magnetic mirroring in the auroral phenomena. W. Lennartsson.
Astrophys. Space Sci., Vol. 51, 461 - 495 (1977).

On the basis of field and particle observations, it is suggested that a bright auroral display is a part of a magnetosphere-ionosphere current system which is fed by a charge-separation process in the outer magnetosphere (or the solar wind). The upward magnetic-field-aligned current is flowing out of the display, carried mainly by downflowing electrons from the hot-particle populations in the outer magnetosphere. As a result of the magnetic mirroring of these downflowing current carriers, a large potential drop is set up along the magnetic field, increasing both the number flux and the kinetic energy of precipitating electrons. This model may be able to explain a highly diversified selection of auroral particle observations.

084.030 Abnormal 5577 Å and 6300 Å emission profiles in aurorae. V. M. Ignat'ev.
Astron. Tsirk., No. 940, p. 2 - 4 (1977). In Russian.

084.031 Observations of aurorae from space.
L. S. Evlashin, P. I. Klimuk, I. I. Koksharov, A. I. Lazarev, V. I. Sevast'yanov, V. N. Sergeevich.
Optical investigations of the emission of the atmosphere, aurorae and noctilucent clouds aboard the orbital scientific station Salyut 4, (see 003.021), p. 31 - 52, 165 - 166 (1977). In Russian.

A short summary of the literature on the spatial-temporal distribution of the optical emission of aurorae and on stable auroral red arcs is given. In consecutive order the results of visual observations of aurorae from spaceships Voskhod, Soyus 9, Soyus 15 and Salyut are presented and discussed. The visual observations aboard Salyut 4 of aurorae at the southern auroral oval situated S, SW and SE of Australia are presented. Some considerations upon the regions of the emission of aurorae observed from space are stated.

084.032 Auroral O I (989 Å) and O I (1027 Å) emissions.
A. B. Christensen, G. J. Romick, G. G. Sivjee.
J. Geophys. Res., Vol. 82, 4997 - 5003 (1977).

Rocket observations of the extreme ultraviolet spectrum of the aurora from 550 to 1250 Å are presented, and the results for the O I (989 Å) and O I (1027 Å) multiplets are examined in detail. The intensity of these emissions and the observed anisotropic radiation field are inconsistent with the relative intensity of the branching radiations. Several possible solutions to the problem are discussed, and evidence is presented that suggests the need to lower the branching ratios by several orders of magnitude.

084.033 The development of auroral and geomagnetic substorm activity after a southward turning of the interplanetary magnetic field following several hours of magnetic calm. K. Lassen, J. R. Sharber, J. D. Winningham.
J. Geophys. Res., Vol. 82, 5031 - 5050 (1977).

A comprehensive study of growth phase and substorm activity following a period of magnetic calm has been conducted through a network of all-sky camera stations, auroral zone magnetic observatories, and particle detectors aboard the Isis 1 satellite. The authors have documented the observations.

084.034 Radar observations of electric fields and currents associated with auroral arcs.
O. de la Beaujardiere, R. Vondrak, M. Baron.
J. Geophys. Res., Vol. 82, 5051 - 5062 (1977).

The incoherent scatter radar at Chatanika, Alaska, has been used to study electric fields and horizontal currents associated with auroral arcs. From several radar observations, three particular events were selected because for each of them the local time, the ambient electric field, the arc morphology, and the mean precipitating electron energy were widely different. Observations show that in spite of these differences the electric field and current vary according to a consistent pattern. A detailed analysis of the ionospheric conditions associated with auroral arcs is presented.

084.035 Auroral and equatorial two-stream irregularities: difference in nonlinear state.   T. Sato.
J. Geophys. Res., Vol. 82, 5195 - 5200 (1977).

It is shown that the electrojet velocity is reduced to the ion acoustic velocity by the two-stream irregularities if the width of the unstable layer is larger than several hundred meters. However, if it is of the order of 100 m or smaller, the electrojet velocity is not appreciably changed, but the electron density is modified, the instability thereby being stabilized. On the basis of these results it is postulated that electrojets responsible for diffuse radar auroras are occasionally highly irregular and consist of many narrow streams with widths of the order of 100 m. Then the observational fact that irregularities with velocities larger than the ion acoustic velocity exist can be explained.

084.036 Correlation of ground-based and topside photometric observations with auroral electron spectra measurements at rocket altitudes.   R. L. Arnoldy, P. B. Lewis, Jr.
J. Geophys. Res., Vol. 82, 5563 - 5572 (1977).

Spectroscopic measurements of the auroral lines 5577, 4278, and 6300 Å made at Fort Yukon, Alaska, are used in the model computations of Rees and Luckey (1974) to predict the energy influx and the characteristic energy of an assumed Maxwellian primary electron spectrum for two auroral displays. Simultaneous with the ground observations, electron detectors aboard a sounding rocket directly measured the primary electron spectrum and energy flux on the field lines which contained the auroral light in the E region observed by the ground photometers. By using the measured electron precipitation and current ionospheric models the emissions at 3914, 4278, and 5577 Å are calculated.

084.037 A statistical study of the 'instantaneous' nightside auroral oval: the equatorward boundary of electron precipitation as observed by the Isis 1 and 2 satellites.
Y. Kamide, J. D. Winningham.
J. Geophys. Res., Vol. 82, 5573 - 5588 (1977).

Electron spectrograms from 351 passes of the Isis 1 and 2 satellites were utilized to study statistically the effects of the interplanetary magnetic field, substorm activity, and the earth's dipole tilt angle on the latitude of the equatorward boundary of the nightside 'instantaneous' auroral oval. The boundary location (in invariant latitude) of the instantaneous oval at hourly local time intervals was identified in terms of the equatorward boundary of the diffuse >100-eV electron precipitation.

084.038 The spatial relationship of field-aligned currents and auroral electrojets to the distribution of nightside auroras.   Y. Kamide, G. Rostoker.
J. Geophys. Res., Vol. 82, 5589 - 5608 (1977).

Nearly simultaneous sets of ground-based magnetometer data, magnetic perturbations recorded at 800-km altitude by the polar-orbiting Triad satellite and auroral imagery and information on precipitating electrons in the energy range 200

eV $< E <$ 20 keV obtained from the DMSP satellites are analyzed for intervals of moderate magnetospheric activity during which substorms were in progress.

084.039 Amplitude and fade rate statistics for equatorial and auroral scintillations.   H. E. Whitney, C. Cantor.
Effect of the ionosphere on space systems and communications, (see 012.048), p. 91 - 94 (1975).

084.040 The role of large scale electric fields and field-aligned currents in producing field-aligned irregularities in the magnetosphere and ionosphere.
K. D. Cole.
Effect of the ionosphere on space systems and communications, (see 012.048), p. 123 - 125 (1975).

Cryogenic spectrometry for the measurement of airglow and aurora.   See Abstr. 031.307.

Stability of shell distributions.   See Abstr. 062.016.

Electric fields in diffuse aurora.   See Abstr. 083.004.

Auroral electrons and conductivity of the high-latitude ionosphere.   See Abstr. 083.045.

Variations of the geometric parameters of N(h)-profiles of the F_2 region in the presence of synchronous action of electric fields and auroral electron streams.   See Abstr. 083.064.

On auroral electrojet curvature and the interpretation of azimuth-scan, radar Doppler data.   See Abstr. 083.079.

Observations with incoherent scatter radar of ionospheric modification in the auroral zone during particle precipitation and Joule heating events.   See Abstr. 083.098.

F layer scintillations and the aurora.   See Abstr. 083.130.

Energetic solar-proton precipitation in the auroral zone associated with storm sudden commencements.   See Abstr. 084.202.

Interconnection of large-scale electric fields in the magnetosphere and energetic spectra of auroral electrons.   See Abstr. 084.241.

Minimum-effect model of geomagnetic excursions applied to auroral zone locations.   See Abstr. 084.271.

Solar-terrestrial physics.   See Abstr. 084.301.

Turbulent generation of electrostatic fields in the magnetosphere.   See Abstr. 084.305.

Dependence of substorm occurrence probability on the interplanetary magnetic field and on the size of the auroral oval.   See Abstr. 084.318.

Magnetic Field

084.201 **Sunspot cycle influence on the geomagnetic field.**
J. M. Harwood, S. R. C. Malin.
Geophys. J. R. Astron. Soc., Vol. 50, 605 - 619 (1977).
The variation of the geomagnetic field associated with the sunspot cycle is examined using a much more extensive data set than hitherto. A method which takes account of the irregular nature of the solar cycle is used to represent the magnetic variation, particular attention being paid to the confidence limits of the results. The results are used to derive a spherical harmonic model, from which the contribution that results from the sunspot cycle modulation of the daily geomagnetic variation is subsequently removed. Examination of the residuals indicates the presence of local variations, on a timescale comparable with the solar cycle, that are uncorrelated with solar activity. These variations probably result from dynamo action within the core.

084.202 **Energetic solar-proton precipitation in the auroral zone associated with storm sudden commencements.**
G. Kremser, H. Specht, E. Kirsch, K. H. Saeger, W. Riedler.
Planet. Space Sci., Vol. 25, 823 - 831 (1977).

084.203 **Impulsive penetration of filamentary plasma elements into the magnetospheres of the Earth and Jupiter.** J. Lemaire.
Planet. Space Sci., Vol. 25, 887 - 890 (1977).
Assuming that the solar wind plasma is usually non-uniform over distances of 10,000 km or less, it is shown that filamentary plasma elements stretched out from the Sun can penetrate impulsively and become engulfed into the magnetosphere.

084.204 **Maximum entropy spectral analysis of the geomagnetic activity index aa over a 107-year interval.**
V. Courtillot, J. L. Le Mouël, P. N. Mayaud.
J. Geophys. Res., Vol. 82, 2641 - 2649 (1977).
Maximum entropy spectral analysis (Mesa) of a time series consisting of 107 annual mean values of the geomagnetic activity index aa and analysis of subsets of this time series yield interesting results concerning both the maximum entropy method itself and periodicities in geomagnetic activity.

084.205 **The magnetic field of the earth.**
V. P. Golovkov.
Zemlya i Vselennaya, 1977, No. 4, p. 24 - 29. In Russian.

084.206 **Observations of magnetic merging and the formation of the plasma sheet in the Earth's magnetotail.**
R. P. Lin, K. A. Anderson, J. E. McCoy, C. T. Russell.
J. Geophys. Res., Vol. 82, 2761 - 2773 (1977).

084.207 **Excitation of an Alfvén wave in the magnetosphere's tail with the help of a proton accelerator.**
M. V. Samokhin.
Geomagn. Aehron., Vol. 17, 701 - 707 (1977). In Russian.

084.208 **On hydromagnetic pulsations in the magnetosphere and Kelvin-Helmholtz instability.**
M. S. Kovner, V. V. Mishin, E. I. Shkelev.
Geomagn. Aehron., Vol. 17, 714 - 718 (1977). In Russian.

084.209 **Bursts of day-time absorption depending on a substorm phase and on the direction of the interplanetary magnetic field.** V. E. Tsirs.
Geomagn. Aehron., Vol. 17, 764 - 767 (1977). In Russian.

084.210 **Development of a polar magnetic substorm: a two-dimensional magnetometer array study.**
J. R. Bannister, D. I. Gough.

Geophys. J. R. Astron. Soc., Vol. 51, 75 - 90 (1977).
A polar magnetic substorm on 1974 September 11 was recorded by a two-dimensional array of 25 three-component magnetometers, so located that the westward ionospheric current passed over the array. The mean perturbation fields over five-minute intervals are presented at six representative epochs of the substorm, the first just before its onset and the sixth $2^{1}/_{2}$ hr later in the coda of the event.

084.211 **Electric and magnetic fields in the high-latitude magnetosphere.** D. H. Fairfield.
Rev. Geophys. Space Phys., Vol. 15, 285 - 298 (1977).
This paper reviews the present knowledge of both electric and magnetic field configurations in the outer magnetosphere and discusses their relation to each other. The extent to which measurements can be explained by a reconnection model of the magnetosphere is evaluated.

084.212 **Anomalies in the time-averaged paleomagnetic field and their implications for the lower mantle.**
R. T. Merrill, M. W. McElhinny.
Rev. Geophys. Space Phys., Vol. 15, 309 - 323 (1977).

084.213 **Pi 1−2 magnetic field pulsations on dayside cleft field lines.** R. R. Heacock, R. D. Hunsucker.
Nature, Vol. 269, 313 - 314 (1977).
The authors show that intense Pi 1−2 activity appeared at College on two occasions when the cleft came down to the College latitude, 64.7° N, at times when College was in the midday sector. Further, it is shown that an intense flux of particle precipitation into the dayside ionosphere occurred on the second occasion, 4 August 1972.

084.214 **Nonlinear low frequency cut-off of the spectrum of the lower hybrid magnetospheric plasma turbulence.**
Yu. V. Golikov, V. F. Kovalev, A. B. Romanov, M. A. Savchenko, V. V. Pustovalov, V. P. Silin.
Planet. Space Sci., Vol. 25, 1027 - 1030 (1977).

084.215 **Hydromagnetic wave observations at large longitudinal separations.**
D. Webb, L. J. Lanzerotti, D. Orr.
J. Geophys. Res., Vol. 82, 3329 - 3335 (1977).
Global enhancements and depressions in ULF magnetic activity in the ~20- to 60-mHz frequency range (magnetic pulsations) at longitudes separated by ~5 hours in local time (geomagnetic latitudes ~45°) are apparently controlled by the radial direction of the interplanetary magnetic field on those days when the field is confined to the ecliptic plane.

084.216 **Comment on 'The theory of VLF Doppler signatures and their relation to magnetospheric density structure' by B. C. Edgar.** J. C. Cerisier.
J. Geophys. Res., Vol. 82, 3337 - 3341 (1977), with a reply by B. C. Edgar. − (See 18.084.223).

084.217 **Terrestrial kilometric radiation. 2. Emission from the magnetospheric cusp and dayside magnetosheath.**
J. K. Alexander, M. L. Kaiser.
J. Geophys. Res., Vol. 82, 98 - 104 (1977).

084.218 **Terrestrial kilometric radiation. 3. Average spectral properties.** M. L. Kaiser, J. K. Alexander.
J. Geophys. Res., Vol. 82, 3273 - 3280 (1977).
The authors present a study of the spectral properties of terrestrial kilometric radiation (TKR) derived from observations made by the Goddard radio astronomy experiments on board the Imp 6 and Radio Astronomy Explorer 2 spacecraft. They present a summary of the variations of spectral properties of TKR with source altitude, observer's local time, and substorm activity as measured by the AE index.

084.219 The solar proton event of April 16, 1970. 1. Features in interplanetary space and in the magnetosheath.   I. D. Palmer, P. R. Higbie, E. W. Hones, Jr.
J. Geophys. Res., Vol. 82, 2657 - 2664 (1977).

A description is given of the solar particle event of April 16, 1970, as recorded by two Vela satellites which were in the vicinity of the earth's bow shock at 18 R_E. Comparison with Explorer 35 in lunar orbit indicates that the early time profile of the event was due to a spatial structure convected past the earth, and the orientation of the convected front is deduced. Anisotropies and pitch angle distributions of energetic protons are obtained from a 32-point sampling of the particle flux. The net flow patterns inferred in interplanetary space reveal two reversals in the north–south component, in general alignment with the magnetic field; neither was accompanied by a magnetic field reversal.

084.220 The solar proton event of April 16, 1970. 2. Transformation of proton pitch angle distributions from the solar wind into the magnetosheath.
P. R. Higbie, I. D. Palmer.
J. Geophys. Res., Vol. 82, 2665 - 2670 (1977).

During the event of April 16, 1970, energetic (~0.7 MeV) protons were observed by Vela 6B in the solar wind and by Vela 5B in the magnetosheath. During one 4-hour period the omnidirectional flux measured by 5B was a factor of 3 larger than that measured by 6B. The pitch angle distribution observed at this time by 6B was characterized by a strong peak at 0° pitch angle. Transformation of this distribution according to the Liouville equation accounts for the enhanced omnidirectional flux and the shape of the pitch angle distributions observed at 5B.

084.221 Long-lived active formations on the solar surface and activity of geomagnetic pulsations of Pc3-type.
G. V. Kuklin, V. A. Parkhomov.
Solar activity and solar-terrestrial relations, (see 012.007), p. 341 - 354 (1976). In Russian.

084.222 Solar wind and the magnetosphere, aurorae, polar caps and magnetic storms.   E. Woyk-Chvojková.
Solar activity and solar-terrestrial relations, (see 012.007), p. 357 (1976).

084.223 Substorm related electron and proton bursts over the polar caps.
E. Kirsch, M. Scholer, L. Rossberg.
Planet. Space Sci., Vol. 25, 931 - 939 (1977).

The authors examined the energetic electron and proton data from different instruments on the dawn-dusk polar orbiting satellite AZUR during periods of high electrojet activity (AE > 500 γ) and find that there is a high probability of seeing during these periods relativistic electron bursts (≳ 0.7 MeV) and in some cases also high-energy proton bursts (≳ 250–≳ 500 keV). Fluxes, composition, energy spectra and spike forms are shown and are compared with similar burst events in the geomagnetic tail observed by other authors.

084.224 Search for the magnetic neutral line in the near-earth plasma sheet. 3. An extensive study of magnetic field observations at the lunar distance.
A. T. Y. Lui, C.-I. Meng, S.-I. Akasofu.
J. Geophys. Res., Vol. 82, 3603 - 3613 (1977).

The authors extended their search for the magnetic neutral line in the magnetotail to the lunar distance on the basis of the Explorer 35 magnetic field observations from July 1967 to December 1970. The sign of the B_z component is found to be predominantly positive during satellite crossings of the midplane (or the so-called neutral sheet) during the substorm expansive phase. Thus combining the present and the earlier results, the authors conclude that there is no supporting evidence for the formation of a neutral line within the lunar

distance during the expansive phase of most substorms.

084.225 The magnetospheric boundary layers: a geometrically explicit model.   N. U. Crooker.
J. Geophys. Res., Vol. 82, 3629 - 3633 (1977).

Recent boundary layer observations complemented with results and ideas of many workers are synthesized into a geometrically explicit model of the magnetosphere in which the plasma sheet derives from the boundary layer on closed field lines at the flanks.

084.226 Plasma dynamics in laboratory models of the magnetospheres of the earth and Uranus.
I. M. Podgrony (Podgornyj), E. M. Dubinin, P. L. Izrailevich, Yu. N. Potanin.
VIII Leningr. mezhdunar. semin. Mater. mezhdunar. semin. "Akt. protsessy na Solntse i probl. soln. nejtrino", 1976. Leningrad, 1976, p. 214 - 219, 213. – Abstr. in Ref. zh., 51. Astron., 9.51.319 (1977).

084.227 Magnetic effect of an asymmetric ring current of protons.   A. S. Kovtyukh, M. I. Panasyuk, Eh. N. Sosnovets.
Kosm. Issled., Vol. 15, 559 - 565 (1977). In Russian.

084.228 Variations of electron streams with energies W_e > 0.7 MeV near the boundary of the radiation belt during a magnetospheric substorm.
E. A. Ginzburg, A. B. Malyshev, N. F. Mal'tseva.
Kosm. Issled., Vol. 15, 573 - 578 (1977). In Russian.

084.229 Energy transmission during a magnetospheric substorm.   I. I. Alekseev.
Geomagn. Aehron., Vol. 17, 885 - 893 (1977). In Russian.

084.230 Influence of bounce resonances on excitation of Alfvén waves beyond the plasmasphere.
V. A. Pilipenko, O. A. Pokhotelov, F. Z. Fejgin.
Geomagn. Aehron., Vol. 17, 894 - 899 (1977). In Russian.

084.231 Synchronous observations of long-periodic geomagnetic pulsations aboard the ATS-6 satellite and on the earth's surface.
J. N. Barfield, N. M. Bondarenko, A. M. Buloshnikov, M. B. Gokhberg, A. L. Kalisher, R. L. McPherron, V. A. Troitskaya.
Geomagn. Aehron., Vol. 17, 900 - 906 (1977). In Russian.

084.232 60-year variations of the geomagnetic field and electroconductivity of the mantle.
S. I. Braginskij, V. M. Fishman.
Geomagn. Aehron., Vol. 17, 916 - 926 (1977). In Russian.

084.233 On velocity determination of the western drift of the non-dipole part of the geomagnetic field.
I. G. Zolotov, L. G. Kas'yanenko.
Geomagn. Aehron., Vol. 17, 957 - 961 (1977). In Russian.

084.234 On the nature of transition periods of the geomagnetic field in paleohistory.   V. P. Golovkov.
Geomagn. Aehron., Vol. 17, 961 - 962 (1977). In Russian.

084.235 The magnetosphere.   C. T. Russell.
Proc. 27th Internat. Astronaut. Congr. (see 012. 016), p. 43 - 49 (1977).

In this paper the author reviews briefly the nature of the magnetosphere, the outstanding problem areas in this field, and what space missions are needed to attack these problems.

084.236 Spectral analysis of Quaternary palaeomagnetic data from British Columbia and its bearing on geomagnetic secular variation.   C. J. Oberg, M. E. Evans.
Geophys. J. R. Astron. Soc., Vol. 51, 691 - 699 (1977).

Palaeomagnetic results obtained from a 7-m sedimentary sequence in southern British Columbia spanning approximately 9000 yr (~22000 to ~31000 yr BP) are reported and analysed. Regular oscillations in the remanence vectors are observed, and maximum entropy spectral analysis reveals peaks at periods of approximately 2000 and 5000 yr. The 2000-yr peak is associated with clockwise looping of the geomagnetic vector, and therefore most likely represents the time associated with one full cycle of the westward drift of the non-dipole field. The 5000-yr peak is associated with elliptical counterclockwise looping of the local geomagnetic vector and may be indicative of counterclockwise motion of the geomagnetic dipole axis.

084.237 **Energetic oxygen ions stream up to magnetosphere.** W. A. Flanagan.
Phys. Today, Vol. 30, No. 11, p. 17 - 19 (1977).

084.238 **Hydromagnetic wave energy in the magnetosphere.** H. B. Liemohn, C. A. Oster.
Plasma science. Austin, Tex., USA, 24 - 26 May 1976. IEEE, New York, USA. Price $ 10.00 (1976), p. 105. — Abstract. Abstr. in Phys. Abstr., Vol. 80, Abstr. 25889 (1977).

084.239 **On a possible long-period connection between paleo-variations of the geomagnetic field and the earth's rotation.** V. N. Plakhotnyuk.
Din. kosm. plazmy. Moskva, 1976, p. 230 - 236. In Russian. Abstr. in Ref. zh., 51. Astron., 10.51.187 (1977).

084.240 **Evolution of the geomagnetic field in the 20th century.** V. N. Plakhotnyuk.
Din. kosm. plazmy. Moskva, 1976, p. 212 - 218. In Russian. Abstr. in Ref. zh., 62. Issled. kosm. prostranstva, 10.62.200 (1977).

084.241 **Interconnection of large-scale electric fields in the magnetosphere and energetic spectra of auroral electrons.** N. V. Isaev, N. K. Osipov.
Din. kosm. plazmy. Moskva, 1976, p. 185 - 188. In Russian. Abstr. in Ref. zh., 62. Issled. kosm. prostranstva, 10.62.225 (1977).

084.242 **Motion of the magnetic moment in the dipole field of the earth.** Yu. I. Okulov.
Din. kosm. plazmy. Moskva, 1976, p. 45 - 54. In Russian. Abstr. in Ref. zh., 62. Issled. kosm. prostranstva, 10.62.226 (1977).

084.243 **A study of an observed and forecasted meteorological index and its relation to the interplanetary magnetic field.** M. F. Larsen, M. C. Kelley.
Geophys. Res. Lett., Vol. 4, 337 - 340 (1977).

084.244 **Evidence for the control of Pc 3, 4 magnetic pulsations by the solar wind velocity.** H. J. Singer, C. T. Russell, M. G. Kivelson, E. W. Greenstadt, J. V. Olson.
Geophys. Res. Lett., Vol. 4, 377 - 379 (1977).

084.245 **Analysis of a hundred year series of magnetic activity indices. II. Positive conservation and estimation of mean values.** P. N. Mayaud.
Ann. Géophys., Vol. 32, 283 - 300 (1976). In French. — Abstr. in Phys. Abstr., Vol. 80, Abstr. 32981 (1977).

084.246 **Electrical conductivity functions in the magnetotelluric and magnetovariation methods.** M. N. Berdichewski (*Berdichevskij*), M. S. Zhdanov, E. B. Fainberg (*Eh. B. Fajnberg*).
Ann. Géophys., Vol. 32, 301 - 318 (1976). — Abstr. in Phys. Abstr., Vol. 80, Abstr. 32982 (1977).

084.247 **Field-aligned currents observed by the Ogo 5 and Triad satellites.** M. Sugiura.
Ann. Géophys., Vol. 32, 267 - 276 (1976). — Abstr. in Phys. Abstr., Vol. 80, Abstr. 33279 (1977).

084.248 **Propagation and bouncing period of VLF waves through inhomogeneous magnetosphere.** S. P. Mishra, R. P. Singh, K. D. Misra.
Indian J. Phys., Vol. 50, 764 - 774 (1976). — Abstr. in Phys. Abstr., Vol. 80, Abstr. 53319 (1977).

084.249 **Ion and electron heating in the Earth's bow shock region.** P. Revathy, G. S. Lakhina.
J. Plasma Phys., Vol. 17, Pt. 2, p. 133 - 138 (1977). — Abstr. in Phys. Abstr., Vol. 80, Abstr. 53320 (1977).

084.250 **On the study of the Earth storms.** B.-S. Pak.
New Phys. (Korean Phys. Soc.), Vol. 16, No. 2, p. 89 - 95 (1976). In Korean. — Abstr. in Phys. Abstr., Vol. 80, Abstr. 53335 (1977).

084.251 **Theoretical modelling of the magnetic and gravitational fields of an arbitrarily shaped three-dimensional body.** C. T. Barnett.
Geophysics, Vol. 41, 1353 - 1364 (1976). — Abstr. in Phys. Abstr., Vol. 80, Abstr. 44908 (1977).

084.252 **A moment equation description of magnetic reversals in the Earth.** K. A. Robbins.
Proc. Natl. Acad. Sci. USA, Vol. 73, 4297 - 4301 (1976). Abstr. in Phys. Abstr., Vol. 80, Abstr. 44921 (1977).

084.253 **The magnetosphere.** J. A. Ratcliffe.
Contemp. Phys., Vol. 18, 165 - 182 (1977). — Abstr. in Phys. Abstr., Vol. 80, Abstr. 45253 (1977).

084.254 **Evidence of rapid changes in the Permian geomagnetic field during the Zechstein marine transgression.** P. Turner, D. J. Vaughan.
Nature, Vol. 270, 593 - 594 (1977).

084.255 **Influence of the magnetosheath on solar proton penetration into the magnetosphere.** V. Domingo, K.-P. Wenzel.
Planet. Space Sci., Vol. 25, 1111 - 1117 (1977).
The observation of solar protons (1–9 MeV) aboard HEOS-2 in the high-latitude magnetotail and magnetosheath on 9 June 1972, and their comparison with simultaneous measurements on Explorers 41 and 43, both in interplanetary space, indicate the existence of a distinct region of the inner magnetosheath (about 3 Earth radii thick) near the high-latitude magnetopause in which the solar particle flow is almost reversed with respect to the flow observed in interplanetary space. The region can also be seen by comparing magnetic field measurements on the three spacecraft. The observations in the outer layer of the magnetotail show solar protons predominantly entering the magnetosphere somewhere near the Earth, perhaps the cusp region.

084.256 **The relationship between the Harang discontinuity and the substorm injection boundary.** A. Brekke.
Planet. Space Sci., Vol. 25, 1119 - 1134 (1977).

084.257 **Magnetospheric plasma discontinuity by the geostationary version of the differential phase method during storms.** J. V. Kovalevsky (*I. V. Kovalevskij*), T. V. Kuznetsova.
Planet. Space Sci., Vol. 25, 1145 - 1150 (1977).
It is shown that the dynamics of the plasmapause, the plasmasphere plasma tails, the plasma sheet and the magnetosheath boundaries of the geomagnetosphere may be investi-

gated by means of the geostationary version of the differential phase method, by which a signal transmitted from a sounding station (a geostationary satellite) and received by a response station on the Earth may be transformed, allowing the sign of the frequency shift and of the phase lag to be changed.

084.258   **The causes of recurrent geomagnetic storms.**
        L. F. Burlaga, R. P. Lepping.
Planet. Space Sci., Vol. 25, 1151 - 1160 (1977).
        The authors studied the causes of recurrent geomagnetic activity by analyzing interplanetary magnetic field and plasma data from Earth-orbiting spacecraft in the interval from November 1973 to February 1974.

084.259   **Diurnal latitude variation of the location of the
        dayside cusp.**
V. A. Troitskaya, O. V. Bolshakova (*Bol'shakova*).
Planet. Space Sci., Vol. 25, 1167 - 1169 (1977).
        The properties of specific high-latitude pulsations reveal the existence of a significant diurnal variation in latitude of the position of the day side cusp ($\Delta\phi \sim 6°$). This systematic change of the position of the cusp during 24 hr must be taken into account when the rapid shiftings of the cusp connected with the changes of magnetic activity are studied. A method of determination of the position of the cusp, using a limited number of ground stations is suggested.

084.260   **Daily variation of the geomagnetic field in equatorial
        region and sector boundary passage.**
G. K. Rangarajan.
Planet. Space Sci., Vol. 25, 1183 - 1185 (1977).

084.261   **440 keV proton precipitation at middle latitudes
        during the recovery phase of magnetic storms.**
K. Kudela, B. Dobrovolska, A. V. Zakharov, V. A. Kuznetsova.
Planet. Space Sci., Vol. 25, 1186 - 1190 (1977).

084.262   **Problems related to macroscopic electric fields in
        the magnetosphere.**   C.-G. Fälthammar.
Rev. Geophys. Space Phys., Vol. 15, 457 - 466 (1977).
        The macroscopic electric fields in the magnetosphere originate from internal as well as external sources. The fields are intimately coupled with the dynamics of magnetospheric plasma convection. They also depend on the complicated electrical properties of the hot collisionless plasma. Macroscopic electric fields are responsible for some important kinds of energization of charged particles that take place in the magnetosphere and affect not only particles of auroral energy but also, by multistep processes, trapped high-energy particles.

084.263   **Amplitude variation of 27-day period in geomagnetic
        activity during 19th and 20th solar cycles.**
E. M. Apostolov.
C. R. Acad. Bulgare Sci., Vol. 30, 33 - 35 (1977). – Abstr. in Phys. Abstr., Vol. 80, Abstr. 66886 (1977).

084.264   **Some peculiarities of the virtual pole positions
        during reversals.**   K. S. Burakov, G. Z. Gurary, A. N. Khramov, G. N. Petrova, G. V. Rassanova, V. P. Rodionov.
J. Geomagn. Geoelectr., Vol. 28, 295 - 307 (1976). – Abstr. in Phys. Abstr., Vol. 80, Abstr. 73634 (1977).

084.265   **The solar cycle effect in the northward geomagnetic
        component at Japanese observatories.**
L. R. Alldredge.
J. Geomagn. Geoelectr., Vol. 28, 317 - 319 (1976). – Abstr. in Phys. Abstr., Vol. 80, Abstr. 73635 (1977).

084.266   **Mean-field electrodynamics and dynamo theory of
        the Earth's magnetic field.**   F. Krause.
J. Geophys., Vol. 43, 421 - 440 (1977). – Abstr. in Phys. Abstr., Vol. 80, Abstr. 73636 (1977).

084.267   **An example of nonlinear dynamo action.**
        F. H. Busse.
J. Geophys., Vol. 43, 441 - 452 (1977). – Abstr. in Phys. Abstr., Vol. 80, Abstr. 73637 (1977).

084.268   **Energetics of the Earth's core.**   D. Gubbins.
        J. Geophys., Vol. 43, 453 - 464 (1977). – Abstr. in Phys. Abstr., Vol. 80, Abstr. 73638 (1977).

084.269   **Energization of charged particles to high energies
        by an induced substorm electric field within the
magnetotail.**   R. J. Pellinen, P. O. Welling, W. J. Heikkila.
R. Inst. Technol., Stockholm, Sweden, Rep. TRITA-EPP-77-10, 30 pp. (1977). – Abstr. in Phys. Abstr., Vol. 80, Abstr. 73909 (1977).

084.270   **Annual report for magnetic observatories – 1974
        (*Canada*).**   E. I. Loomer.
Geomagn. Ser. Earth Phys. Branch, No. 11, p. 1 - 95 (1977). Abstr. in Phys. Abstr., Vol. 80, Abstr. 78094 (1977).

084.271   **Minimum-effect model of geomagnetic excursions
        applied to auroral zone locations.**   G. L. Siscoe.
J. Geomagn. Geoelectr., Vol. 28, 427 - 436 (1976). – Abstr. in Phys. Abstr., Vol. 80, Abstr. 78101 (1977).

084.272   **Quiet day variation of geomagnetic H-field at low
        latitudes.**   R. G. Rastogi, K. N. Iyer.
J. Geomagn. Geoelectr., Vol. 28, 461 - 479 (1976). – Abstr. in Phys. Abstr., Vol. 80, Abstr. 78102 (1977).

084.273   **Spatial distribution of the geomagnetic spectral
        composition for disturbed days.**   W. H. Campbell.
J. Geomagn. Geoelectr., Vol. 28, 481 - 496 (1976). – Abstr. in Phys. Abstr., Vol. 80, Abstr. 78103 (1977).

084.274   **The effect of a field of external origin on spherical
        harmonic analysis using only internal coefficients.**
F. J. Lowes.
J. Geomagn. Geoelectr., Vol. 28, 515 - 516 (1976). – Abstr. in Phys. Abstr., Vol. 80, Abstr. 78104 (1977).

084.275   **Some features of Pc 5 pulsations in the period range
        180 - 300 sec.**   J. C. Gupta.
J. Geomagn. Geoelectr., Vol. 28, 359 - 373 (1976). – Abstr. in Phys. Abstr., Vol. 80, Abstr. 78576 (1977).

084.276   **Local characteristics of the electron temperature
        profile.**   K. Hirao, K. Oyama.
J. Geomagn. Geoelectr., Vol. 28, 507 - 514 (1976). – Abstr. in Phys. Abstr., Vol. 80, Abstr. 78577 (1977).

084.277   **Spectral characteristics of field variations during
        geomagnetically quiet conditions.**   W. H. Campbell.
J. Geomagn. Geoelectr., Vol. 29, 29 - 50 (1977). – Abstr. in Phys. Abstr., Vol. 80, Abstr. 87179 (1977).

084.278   **The International Magnetospheric Study (IMS) and
        the Antarctic and Southern Hemisphere Aeronomy
Year (ASHAY).**   J. A. Gledhill.
South African J. Antarct. Res., No. 5, p. 53 - 56 (1975). Abstr. in Phys. Abstr., Vol. 80, Abstr. 87210 (1977).

084.279   **La medida del valor absoluto y variaciones del
        campo geomagnético con magnetómetros de
protones y bombeo óptico. Parte 3.ª. La medida absoluta del vector geomagnético estaciones automáticas.**
M. Catalán, F. Gómez Armario.
Urania Barcelona, Año 61, Núm. 285, p. 111 - 162 (1976).

084.280   **Systematic study of plasma flow during plasma
        sheet thinnings.**   A. T. Y. Lui, L. A. Frank,

K. L. Ackerson, C.-I. Meng, S.-I. Akasofu.
J. Geophys. Res., Vol. 82, 4815 - 4825 (1977).

Plasma flows during plasma sheet thinnings at the expansive phase of magnetospheric substorms are extensively examined. It is found that plasma flows during thinnings are directed most frequently sunward, not antisunward, in the midnight sector ($|Y_{SM}| \leqslant 10 R_E$), within $X_{SM} \simeq -30 R_E$. Thus the plasma flow observations do not support the conjecture of a magnetic neutral line formation in the near-earth plasma sheet, namely, within $X_{SM} \simeq -18 R_E$, at the substorm onset for at least most magnetospheric substorms. Several modifications of the existing neutral line model are considered. It is suggested that the plasma sheet behavior during its thinning is consistent with a deflation of the plasma sheet in which plasma sheet particles are drained earthward and the energy source for the substorm expansive phase is primarily the energy associated with the sunward plasma flow from the deflation (thinning) of the plasma sheet.

084.281   **Observations of long-period pulsations in electron precipitation at 75° magnetic latitude.**
R. R. Brown, R. H. Karas, J. R. Barcus, P. Stauning.
J. Geophys. Res., Vol. 82, 4834 - 4836 (1977).

084.282   **An overview by Pioneers observations of the distant geomagnetic tail.**   U. Villante.
Space Sci. Rev., Vol. 20, 123 - 143 (1977).

Pioneer 7 and Pioneer 8 spacecraft provided the only direct observations of the geomagnetic tail at geocentric distances as large as $1000 R_e$ and $500 R_e$ respectively. The region of tail encounters and the magnitude and direction of the observed magnetic field might be consistent with a cylindrical tail with a modestly increased cross-section. Neutral sheet observations also appear to be consistent with the most recent bidimensional tail models. Finally, as in the cislunar region, the double peaked proton energy spectra can be interpreted in terms of a partial intermingling of plasma sheet and plasma mantle populations.

084.283   **High-latitude geomagnetic variations and substorms.**
V. M. Mishin.
Space Sci. Rev., Vol. 20, 621 - 675 (1977).

The review concerns high-latitude geomagnetic variations, and in particular those caused by fluctuations of the interplanetary magnetic field. The original results are presented against a general review background, which includes a method for mathematical description of global fields of magnetic variations and substorms.

084.284   **International magnetospheric study: progress report.**
Space Sci. Rev., Vol. 21, 3 - 21 (1977).

084.285   **Global problems in magnetospheric plasma physics and prospects for their solution.**
J. G. Roederer.
Space Sci. Rev., Vol. 21, 23 - 71 (1977).

Selected problems of magnetospheric plasma physics are critically reviewed. The discussion is restricted to questions that are 'global' in nature, i.e., involve the magnetosphere as a whole, and that are beyond the stage of systematic survey or isolated study requirements. Only low-energy particle aspects are discussed. The article focuses on the following subjects: (1) Effect of the interplanetary magnetic field on topography, topology and stability of the magnetospheric boundary; (2) Solar wind plasma entry into the magnetosphere; (3) Plasma storage and release mechanisms in the magnetospheric tail; (4) Magnetic-field-aligned currents and magnetosphere-ionosphere interactions. A brief discussion of the prospects for the solution of these problems during and after the International Magnetospheric Study is given.

084.286   **Low-energy charged particle ring around equator in**

the altitude range 400–1100 km.
R. K. Kaul, H. Razdan, J. A. Lockwood.
Astrophys. Space Sci., Vol. 46, 215 - 223 (1977).

Measurements of charged particle fluxes at energies $\geqslant 13$ MeV (if protons) reveal a ring of low energy charged particles around equator in the altitude range of 400–1100 km. The ring of charged particles exists below the inner radiation belt and is restricted to ±4° of the geomagnetic equator. Distribution of the maximum flux with geomagnetic latitude and L is presented. Comparison of the observed fluxes with earlier measurements of low energy particles reveals a differential energy spectrum.

084.287   **Experimental evidence for the existence of open and closed models of the magnetosphere. I.**
Eh. M. Dubinin, I. M. Podgornyj, Yu. N. Potanin.
Kosm. Issled., Vol. 15, 866 - 874 (1977). In Russian.

084.288   **Electromagnetic wave propagation and instabilities in the magnetosphere.**   K. L. Vithal.
Astrophys. Space Sci., Vol. 49, 317 - 330 (1977).

084.289   **Computations on the interaction of an interplanetary shock with the Earth's magnetosphere.**
I. De Pater, W. J. Weber.
Astrophys. Space Sci., Vol. 51, 217 - 224 (1977).

The authors present the results of a one-dimensional computer simulation of the interaction between interplanetary shocks and the Earth's magnetosphere. The position of the bowshock as a function of solar wind velocity and interplanetary field direction is studied.

084.290   **Magneto-conjugate phenomena and meteor activity.**
Sh. N. Irkaeva.
Astron. Tsirk., No. 960, p. 4 - 6 (1977). In Russian.

084.291   **Measurements of recent geomagnetic secular variation in southeastern Australia and the question of dipole wobble.**   M. Barbetti.
Earth Planet. Sci. Lett., Vol. 36, 207 - 218 (1977).

The outline of changes in the local direction of the earth's magnetic field over the last 1500 years is given. An analysis of new data from the southern hemisphere and published data for the northern hemisphere indicates that a large change occurred in the orientation of the geomagnetic dipolar axis between about 1400 and 1650 A.D.

084.292   **Geomagnetic field variations in southeastern Europe between 6500 and 100 years B.C.**
M. Kovacheva, D. Veljovich.
Earth Planet. Sci. Lett., Vol. 37, 131 - 138 (1977).

084.293   **Magnetic extra-storm of August 4 - 5, 1972 in connection with the hydromagnetic structure of the** interplanetary plasma stream from a powerful flare.
K. G. Ivanov, N. V. Mikerina.
Geomagn. Aehron., Vol. 17, 1057 - 1063 (1977). In Russian.

084.294   **Analysis of the connection of geomagnetic bays with geomagnetic activity and the interplanetary** magnetic field in 1968.   M. Bielekova.
Geomagn. Aehron., Vol. 17, 1064 - 1069 (1977). In Russian.

084.295   **On a possible mechanism of generation of low-frequency pulsations in the polar cusp.**
M. S. Kovner, V. A. Kuznetsova.
Geomagn. Aehron., Vol. 17, 1083 - 1087 (1977). In Russian.

084.296   **Fields of currents in the magnetopause and in the radiation zone.**
Ts. D. Porchkhidze, N. M. Rudneva, Ya. I. Fel'dshtejn.
Geomagn. Aehron., Vol. 17, 1124 - 1127 (1977). In Russian.

084.297 Peculiarities of the spatial distribution of Pc3 geo-
magnetic pulsations on the earth's surface.
A. M. Buloshnikov, M. B. Gokhberg, Yu. A. Kopytenko,
Yu. P. Novikov, O. M. Raspopov, V. A. Troitskaya.
Geomagn. Aehron., Vol. 17, 1127 - 1129 (1977). In Russian.

084.298 Peculiarities of study of secular variations of the
geomagnetic field in the Carpathians.
V. G. Kuznetsova, V. E. Maksimchuk.
Geomagn. Aehron., Vol. 17, 1134 - 1136 (1977). In Russian.

084.299 On weak worldwide magnetic disturbances.
Kh. D. Kanonidi.
Geomagn. Aehron., Vol. 17, 1139 - 1141 (1977). In Russian.

084.300 Geomagnetic observations at the Kanozan Geodetic
Observatory (1973).
The Geographical Survey Institute.
Bull. Geogr. Surv. Inst., Vol. 22, 69 - 88 (1977).

084.301 Solar-terrestrial physics.   V. L. Patel.
Illustrated glossary for solar and solar-terrestrial
physics, (see 003.022), p. 159 - 193 (1977).

084.302 A comparison of ULF and VLF measurements of
magnetospheric cold plasma densities.
D. C. Webb, L. J. Lanzerotti, C. G. Park.
J. Geophys. Res., Vol. 82, 5063 - 5072 (1977).
Equatorial cold plasma density profiles determined from
VLF whistlers propagating in magnetospheric ducts are com-
pared with densities computed from the observations of ULF
hydromagnetic waves (geomagnetic pulsations).

084.303 Evidence for very weak pitch angle diffusion of
outer zone electrons.
H. Leinbach, D. J. Williams.
J. Geophys. Res., Vol. 82, 5091 - 5098 (1977).

084.304 Two types of magnetospheric ELF chorus and their
substorm dependences.
B. T. Tsurutani, E. J. Smith.
J. Geophys. Res., Vol. 82, 5112 - 5128 (1977).

084.305 Turbulent generation of electrostatic fields in the
magnetosphere.   D. W. Swift.
J. Geophys. Res., Vol. 82, 5143 - 5154 (1977).
The purpose of this paper is to identify the energy source
for electrons which is responsible for the highly structured
auroral forms.

084.306 Observations of large transient magnetospheric electric fields.   T. L. Aggson, J. P. Heppner.
J. Geophys. Res., Vol. 82, 5155 - 5164 (1977).

084.307 Electron fluxes over the polar cap. 2. Electron
trapping and energization on open field lines.
J. C. Foster, J. R. Burrows.
J. Geophys. Res., Vol. 82, 5165 - 5170 (1977).
Although solar wind and solar flare electrons can have
direct access to low altitudes on open field lines, low-energy
electron fluxes over the polar caps may at times be accelerated
and trapped by a large-scale potential barrier in the magnetotail. If the effective potential difference at the barrier fluctu-
ates randomly about an average value, a fraction of the beam
flux accelerated toward the earth from beyond the barrier can
be trapped between the barrier and the earth. Additional low-
energy electrons diffusing onto field lines closed to particle
losses by the potential barrier would also be quasi-trapped.
The effects of such a fluctuating magnetotail barrier potential
are described with a simple model.

084.308 Magnetosphere boundary observations along the

Imp 7 orbit. 1. Boundary locations and wave level
variations.   F. L. Scarf, L. A. Frank, R. P. Lepping.
J. Geophys. Res., Vol. 82, 5171 - 5180 (1977).
The authors discuss and analyze magnetosphere boundary
phenomena observed by the Imp 7 magnetic field, plasma,
and plasma wave instruments in 1972 and 1973. This report
contains a brief description of the relevant Imp 7 instrumentation, a survey of the boundary locations for a 15-month
period, a discussion of the different types of crossings (essentially, sharp and diffuse), and an analysis of the electromagnetic wave modes being detected in the broad low-frequency
channel of the wave instrument.

084.309 Magnetopause shapes: general solutions.
F. C. Michel.
J. Geophys. Res., Vol. 82, 5181 - 5186 (1977).
The problem of determining the shape of a magnetopause, namely, the tangential discontinuity separating (typically) the magnetic field of a source from external plasma, is
reduced to solving an integral equation. No symmetry assumptions whatsoever are used in the derivation, consequently,
realistic problems such as model shapes for the earth's magnetosphere (tilted dipole, flowing plasma, etc.) can in principle
be treated.

084.310 Temporal variations in slant total plasmasphere content and their relationship to the ring current intensity and the plasmapause.   D. C. Webb, L. J. Lanzerotti.
J. Geophys. Res., Vol. 82, 5201 - 5207 (1977).
The purpose of this paper is to use recently published
data from the ATS 6 beacon experiment (Fritz, 1976) in order
to study the temporal variation of the plasmasphere content
with geomagnetic activity, to deduce the filling rate of the
plasmasphere at the equator, and to determine the location of
the plasmapause after a magnetic disturbance.

084.311 Characteristics of magnetospheric particle injection
deduced from events observed on August 18, 1974.
G. K. Parks, C. S. Lin, B. Mauk, S. DeForest, C. E. McIlwain.
J. Geophys. Res., Vol. 82, 5208 - 5214 (1977).
The behavior of particles injected during three well-
defined substorm events that occurred on August 18, 1974,
has been studied in detail. Plasma characteristics from ~200
eV to 80 keV detected at the ATS 6 position have been
studied in detail, and the authors suggest that the results are
consistent with the idea that there are two particle sources for
electrons during substorms: one at small pitch angles, predominantly in the magnetic field direction, and the other at
large pitch angles. The sources of these electrons are probably
the ionosphere and the plasma sheet.

084.312 Dawn-dusk gradient of the precipitation of low-
energy electrons over the polar cap and its relation
to the interplanetary magnetic field.
C.-I. Meng, S.-I. Akasofu, K. A. Anderson.
J. Geophys. Res., Vol. 82, 5271 - 5275 (1977).
The authors report a clear indication of (1) the existence
of the dawn-dusk gradient of the intensity distribution of low-
energy electrons over the northern and southern polar caps,
(2) the reversal of the gradient over the two polar caps, and
(3) the dependence of the direction of the gradient on the
interplanetary magnetic field B_y component. These features
have a striking similarity with the distribution of the magnitude of the dawn-dusk component of the electric field in the
polar cap.

084.313 Nonstationary interaction between solar wind discontinuities and the system leading wave — earth's
magnetosphere.   S. A. Grib.
Simpoz. po fiz. geomagnitosfery, 1977. Tezisy dokl. Irkutsk,
1977, p. 12. In Russian. — Abstr. in Ref. zh., 62. Issled. kosm.
prostranstva, 12.62.154 (1977).

084.314 **The effect of screening of the field of geomagnetic pulsations by the sporadic E-layer of the ionosphere.**
L. N. Gutman, S. P. Chernysheva, V. M. Sheftel'.
Rostov. inzh.-stroit. inst. Rostov-na-Donu, 1977. 5 pp. In Russian. – Abstr. in Ref. zh., 62. Issled. kosm. prostranstva, 12.62.162 (1977).

084.315 **On the question of the energy of the precessional dynamo.** Sh. Sh. Dolginov.
Solar-wind interaction with the planets Mercury, Venus, and Mars, (see 012.047), p. 167 - 170 (1976).

084.316 **Plasma dynamics in laboratory models of the magnetospheres of the earth and Uranus.**
I. M. Podgornyj, Eh. M. Dubinin, P. L. Izrajlevich, Yu. N. Potanin.
Izv. AN SSSR. Ser. fiz., Vol. 41, 1870 - 1883 (1977). In Russian. – Abstr. in Ref. zh., 62. Issled. kosm. prostranstva, 1.62.146 (1978).

084.317 **The role of hot plasma in magnetospheric convection.** D. J. Southwood.
J. Geophys. Res., Vol. 82, 5512 - 5520 (1977).

084.318 **Dependence of substorm occurrence probability on the interplanetary magnetic field and on the size of the auroral oval.** Y. Kamide, P. D. Perreault, S.-I. Akasofu, J. D. Winningham.
J. Geophys. Res., Vol. 82, 5521 - 5528 (1977).
The dependence of substorm occurrence probability on the north-south component Bz of the interplanetary magnetic field and on the size of the auroral oval is examined on the basis of two independent data sets (Isis 1 and 2 low-energy electron data and all-sky camera data from the Alaska meridian).

084.319 **Double-peaked ion spectra in the lobe plasma: evidence for massive ions?** D. A. Hardy, J. W. Freeman, H. K. Hills.
J. Geophys. Res., Vol. 82, 5529 - 5540 (1977).
Recent work by Frank et al. [1977] has shown that there are occasional higher-energy peaks observed in the ion distribution function in the boundary layer. Similar secondary peaks have been found in the Suprathermal Ion Detector Experiment (Side) data in the lobe plasma at the lunar orbit. It is the purpose of this paper to analyze these secondary peaks and discuss their significance.

084.320 **Substorm processes in the magnetotail: comments on 'On hot tenuous plasmas, fireballs, and boundary layers in the earth's magnetotail' by L. A. Frank, K. L. Ackerson, and R. P. Lepping.** E. W. Hones, Jr., with a reply by L. A. Frank, K. L. Ackerson.
J. Geophys. Res., Vol. 82, 5633 - 5640, 5641 - 5643 (1977).

084.321 **On a connection between geomagnetic activity and the passage of sector boundaries of the interplanetary magnetic field through the earth.** G. I. Ol', A. I. Ol'.
Soln. Dannye 1977 Byull., No. 9, p. 99 - 102 (1977). In Russian.
Epochs were found when the geomagnetic activity response on the sector boundaries passing near the earth is especially great, and the epochs when it is negligible. Great effects occur during the maxima of odd 11-year cycles and declining branches of even cycles. The passing of different boundaries in different seasons may result both in an increase and a decrease of geomagnetic activity.

084.322 **Sector structure of the interplanetary magnetic field and geomagnetic activity.** V. F. Loginov.
Soln. Dannye 1977 Byull., No. 10, p. 87 - 93 (1977). In Russian.
The variation of daily values of the planetary geomag-

netic index A_p near sector boundaries of the interplanetary magnetic field for different periods during 1932-1975 is considered. A complicated character of the variation of geomagnetic activity near sector boundaries of the interplanetary magnetic field depending on sector sign, sector alternation and heliographic latitude of the earth is found.

084.323 **Effect of the NS component of the interplanetary magnetic field on geomagnetic storms.**
P. Ochabová.
Studia, Vol. 21, 360 - 365 (1977).

Annual report for magnetic observatories – 1973.
See Abstr. 009.016.

A new diagnostic method for magnetospheric plasma.
See Abstr. 031.276.

On the totality of periodic motions in the meridian plane of a magnetic dipole. See Abstr. 042.060.

The participation of the European Space Agency in atmospheric and magnetospheric research.
See Abstr. 051.023.

International Sun-Earth Explorer: a three-spacecraft program. See Abstr. 053.007.

First and second order approximations of the first adiabatic invariant for a charged particle interacting with a linearly polarized hydromagnetic plane wave.
See Abstr. 062.002.

Determination of the transport coefficients for a plasma with quasi-collisions. See Abstr. 062.005.

Some questions of dynamics of magnetospheric plasma. See Abstr. 062.051.

A probable interval for the mean transit time of the flare-generated disturbances. See Abstr. 073.033.

On the correlation of solar-wind energy flux and geomagnetic activity. See Abstr. 074.024.

Solar-wind velocity variations and geomagnetic disturbances. See Abstr. 074.025.

A pictorial comparison of interplanetary magnetic field polarity, solar wind speed, and geomagnetic disturbance index during the sunspot cycle. See Abstr. 074.053.

Solar energetic particles below 10 MeV.
See Abstr. 074.091.

Solar wind parameters responsible for plasma injection into the magnetospheric ring-current region.
See Abstr. 074.098.

Comparative magnetospherology. Part 5. Examination of the two-hemisphere model by means of the apparent sector structure of the Jovian magnetosphere.
See Abstr. 080.093.

A note on the westward drift of the Earth's magnetic field. See Abstr. 081.048.

On the manifestation of a connection between variations of the terrestrial atmospherical pressure and strong geomagnetic disturbances in various seasons.
See Abstr. 082.111.

Tropospherical response to geomagnetic disturbances and propagation of tropospherical waves into the upper atmosphere.   See Abstr. 082.112.

Geomagnetic storms and electric fields in the equatorial ionosphere.   See Abstr. 083.001.

Daytime ionosphere scintillation associated with geomagnetic storms.   See Abstr. 083.005.

A comparison of ionospheric currents, magnetic variations, and electric fields at Arecibo.
See Abstr. 083.019.

Perturbation du profil de l'ionosphère supérieure antarctique en été local au niveau du cornet polaire.
See Abstr. 083.030.

On some problems of dynamics of protons captured by the geomagnetic field. Annotation. See Abstr. 083.035.

Capture of photoelectrons by the geomagnetic field during their motion through the plasmosphere.
See Abstr. 083.036.

Electron density enhancements in the F region beneath the magnetospheric cusp.   See Abstr. 083.106.

Multi-layer model of a small-scale dynamo.
See Abstr. 083.111.

Vertical drift of the F2 layer connected with a polar substorm.   See Abstr. 083.116.

The role of the magnetosphere in satellite and radio-star scintillation.   See Abstr. 083.123.

The response of the dayside aurora to sharp northward and southward transitions of the interplanetary magnetic field and to magnetospheric substorms.
See Abstr. 084.002.

Parallactic observations of the spatially ordered structure of a radio aurora.   See Abstr. 084.004.

The eastward movement of the structure of auroral radio absorption events in the morning sector.
See Abstr. 084.019.

Large-scale characteristics of field-aligned currents determined from the TRIAD magnetometer experiment.
See Abstr. 084.025.

Characteristics of polar cap sun-aligned arcs.
See Abstr. 084.027.

The development of auroral and geomagnetic substorm activity after a southward turning of the interplanetary magnetic field following several hours of magnetic calm.
See Abstr. 084.033.

A summary of significant solar-initiated events during STIP Intervals I and II.   See Abstr. 085.038.

The magnetospheres of the earth and planets.
See Abstr. 091.046.

Photon mass and planetary magnetic fields.
See Abstr. 091.049.

Cosmic rays and ancient planetary magnetic fields.
See Abstr. 091.064.

Ionosphere and atmosphere of the moon in the geomagnetic tail.   See Abstr. 094.182.

On a mechanism of influence of the earth's magnetic field on the motion of ionized meteor trails.
See Abstr. 104.008.

Structure of a typical stream of interplanetary plasma from a powerful flare and corresponding set of types of geomagnetic (magnetospheric) disturbances. I.
See Abstr. 106.002.

Impulses in the B_z-component of the interplanetary magnetic field and disturbances in the geomagnetic field.
See Abstr. 106.003.

Interplanetary magnetic field and magnetospheric substorms.   See Abstr. 106.007.

On a connection between ionospheric and geomagnetic disturbances at low latitudes and the interplanetary magnetic field.   See Abstr. 106.020.

On low frequency oscillations in the near-Earth plasma.   See Abstr. 106.021.

The effect of interplanetary magnetic field on geomagnetic activity.   See Abstr. 106.024.

Characteristics of the association between the interplanetary magnetic field and substorms.
See Abstr. 106.028.

Probability distribution functions of microscale magnetic fluctuations during quiet conditions.
See Abstr. 106.030.

The structure of the solar flare stream magnetic field.
See Abstr. 106.031.

A contribution to ULF activity in the Pc 3-4 range correlated with IMF radial orientation.
See Abstr. 106.041.

Interplanetary magnetic field conditions associated with synchronous orbit observations of Pc 3 magnetic pulsations.   See Abstr. 106.042.

Paleoclimate, paleomagnetism and the eccentricity of the earth's orbit.   See Abstr. 107.026.

Polar coupling coefficients for cosmic ray multiple neutrons with energies up to 15 GeV.
See Abstr. 143.043.

Radiation Belts

084.401 **Equilibrium structure of equatorially mirroring radiation belt protons.**   W. N. Spjeldvik.
J. Geophys. Res., Vol. 82, 2801 - 2808 (1977).

084.402 **Mechanism for blow-off of the Earth's atmosphere: population of the Van Allen belt.**   L. M. Libby.
Indian J. Radio Space Phys., Vol. 5, No. 3, p. 199 - 204 (1976).
Abstr. in Phys. Abstr., Vol. 80, Abstr. 29524 (1977).

084.403 **On the possibility of existence of a radiation belt of the earth with energies of 100 MeV and more.**
N. L. Grigorov.
Dokl. AN SSSR, Vol. 234, 810 - 813 (1977). In Russian.
Abstr. in Ref. zh., 62. Issled. kosm. prostranstva, 10.62.227 (1977).

084.404 **Experimental results of measurements of protons and α-particles with energies > 1 MeV/nucleon in the radiation belts.**   M. I. Panasyuk, S. Ya. Rejzman,
Eh. N. Sosnovets, V. N. Filatov.
Kosm. Issled., Vol. 15, 887 - 894 (1977). In Russian.

084.405 **The time persistence of certain features of electron precipitation in the slot region.**
W. L. Imhof, E. E. Gaines, J. B. Reagan.
J. Geophys. Res., Vol. 82, 5024 - 5030 (1977).
The purpose of this paper is to present the results of a study of the temporal variability at a given local time of the fluxes of precipitating electrons and the occurrence of peaked electron energy spectra and to discuss how the persistence of these features might relate to the relative importance of long-duration hiss in the slot region.

Variations of electron streams with energies $W_e > 0.7$ MeV near the boundary of the radiation belt during a magnetospheric substorm.   See Abstr. 084.228.

Low-energy charged particle ring around equator in the altitude range 400–1100 km.   See Abstr. 084.286.

Fields of currents in the magnetopause and in the radiation zone.   See Abstr. 084.296.

085 Solar-Terrestrial Relations

085.001   An 8,000-yr palaeoclimatic record of the 'Double-Hale' 45-yr solar cycle.
R. W. Fairbridge, C. Hillaire-Marcel.
Nature, Vol. 268, 413 - 416 (1977).

085.002   Support for the astronomical theory of climatic change.   A. L. Berger.
Nature, Vol. 269, 44 - 45 (1977).
Among recent papers supporting the astronomical theory of climatic change, the work of Hays et al. (1976) is of particular importance in providing detailed quasi-periods from deep-sea cores. The author wishes to draw attention here to the fact that the periods found are very close to the periods predicted by Berger (1976, 1977) in the latest and most accurate calculation of the variations of the various 'Milankovitch' parameters.

085.003   On an interpretation of the dependence of $Pc3-4$ activity on solar wind velocity.   N. A. Barkhatov.
Geomagn. Aehron., Vol. 17, 767 - 770 (1977). In Russian.

085.004   Cosmic ray intensity in the past.   V. A. Dergachev.
Izv. AN SSSR. Ser. fiz., Vol. 41, 347 - 382 (1977). In Russian. – Abstr. in Ref. zh., 51. Astron., 8.51.452 (1977).

085.005   Active processes on the sun and the biosphere.   B. M. Vladimirskij.
Izv. AN SSSR. Ser. fiz., Vol. 41, 403 - 410 (1977). In Russian. – Abstr. in Ref. zh., 51. Astron., 8.51.467 (1977).

085.006   Solar proton event: influence on stratospheric ozone.
D. F. Heath, A. J. Krueger, P. J. Crutzen.
Science, Vol. 197, 886 - 889 (1977).
Large-scale reductions in the ozone content of the middle and upper stratosphere over the polar cap regions were associated with the major solar proton event of 4 August 1972. This reduction, which was determined from measurements with the backscattered ultraviolet experiment on the Nimbus 4 satellite, is interpreted as being due to the catalytic destruction of ozone by odd-nitrogen compounds (NO_x) produced by the event.

085.007   Solar structure, variability, and the ice ages. (Solar variability and climate).   E. J. Öpik.
Irish Astron. J., Vol. 12, 253 - 276 (1976).

085.008   Atmospheric waves in the ionosphere due to total solar eclipse.   R. N. E. Baulch, E. C. Butcher.
Nature, Vol. 269, 497 - 498 (1977).
The authors present the results of measurements made on the angle of arrival of a high frequency radio wave inside the path of totality of an eclipse, which took place on 23 October 1976.

085.009   Can sunspots influence our weather?   H. Volland.
Nature, Vol. 269, 400 - 401 (1977).
A mechanism associated with the visible spectrum of the solar radiation may be the primary source of different types of Sun-weather relationship. If short-lived meteorological phenomena such as those associated with solar flares or sector boundary crossings are generated by active longitudes on the Sun, the 11-yr solar-cycle effects may be the result of an 11-yr variation of the amplitude of the 27-d forcing function.

085.010   Equatorial solar rotation and its relation to climatic changes.   K. Sakurai.
Nature, Vol. 269, 401 - 402 (1977).

During the past 10 years, covering most of solar cycle no. 20 (1965–76), the magnitude of the solar rotation speed averaged annualy showed a good inverse correlation with the annual relative sunspot numbers. The author suggests that this variation of the equatorial solar rotation speed may be responsible for the Earth's present unusual climatic conditions.

085.011   Signal variations of LF radio-waves over the sunspot cycle.   G. T. Nestorov.
J. Atmos. Terr. Phys., Vol. 39, 741 - 743 (1977).

085.012   Time variations of concentration of stable isotopes in organic matter.
S. Kh. Akhmetkereev, V. A. Dergachev.
Izv. AN SSSR. Ser. fiz., Vol. 41, 439 - 445 (1977). In Russian. Abstr. in Ref. zh., 51. Astron., 9.51.480 (1977).

085.013   Mutual spectral analysis of data on ^{14}C concentration and geomagnetic field strength.
V. A. Dergachev, N. Tujchiev.
Izv. AN SSSR. Ser. fiz., Vol. 41, 431 - 438 (1977). In Russian. Abstr. in Ref. zh., 51. Astron., 9.51.481 (1977).

085.014   Radioactive carbon concentration in the earth's atmosphere and astrophysical and geophysical phenomena.   V. A. Dergachev, G. E. Kocharov.
Izv. AN SSSR. Ser. fiz., Vol. 41, 422 - 430 (1977). In Russian. Abstr. in Ref. zh., 51. Astron., 9.51.482 (1977).

085.015   On solar activity effects on the behaviour of the f^0F2 parameter in years of solar cycle minimum.
G. V. Egorova.
Issled. po ehlektrodin. i rasprostr. ehlektromagn. voln. Tomsk, Tomsk. univ., 1977, p. 162 - 170. In Russian. – Abstr. in Ref. zh., 51. Astron., 9.51.483 (1977).

085.016   The sun and the weather.   R. H. Olson.
Nature, Vol. 270, 11 (1977).

085.017   Observation of short-period pulsations of electron and γ-fluxes in the upper layers of the atmosphere and their connection with periodic oscillations of the solar diameter. Abstract.   A. M. Gal'per, V. M. Grachev, V. V. Dmitrenko, V. G. Kirillov-Ugryumov, A. V. Kurochkin, B. I. Luchkov, S. E. Ulin, Eh. M. Shermanzon, Yu. T. Yurkin.
VIII Leningr. mezhdunar. semin. Mater. mezhdunar. semin. "Akt. protsessy na Solntse i probl. soln. nejtrino", 1976. Leningrad, 1976, p. 117 - 121. In Russian. – Abstr. in Ref. zh., 51. Astron., 10.51.587 (1977).

085.018   Simulation of the effect of a solar eclipse in the ionosphere by different representations of electron temperature.
M. K. Ivel'skaya, G. E. Sutyrina, V. E. Sukhodol'skaya.
Issled. po geomagn., aehron. i fiz. Solntsa. Vyp. (No.) 41. Moskva, Nauka, 1977, p. 62 - 65. In Russian. – Abstr. in Ref. zh., 51. Astron., 10.51.589 (1977).

085.019   Active processes on the sun and the biosphere.   B. M. Vladimirskij.
VIII Leningr. mezhdunar. semin. Mater. mezhdunar. semin. "Akt. protsessy na Solntse i probl. soln. nejtrino", 1976. Leningrad, 1976, p. 209 - 211. In Russian and English. – From Ref. zh., 51. Astron., 10.51.594 (1977).

085.020   Response of stratospheric circulation at 10 mb to solar activity oscillations resulting from the Sun's rotation.   A. Ebel, W. Batz.

Tellus, Vol. 29, No. 1, p. 41 - 47 (1977). – Abstr. in Phys. Abstr., Vol. 80, Abstr. 45118 (1977).

085.021 **Changes in sign of the relationship between sunspots and pressure, rainfall and the monsoons.**
G. J. Bell.
Weather, Vol. 32, No. 1, p. 26 - 32 (1977). – Abstr. in Phys. Abstr., Vol. 80, Abstr. 45123 (1977).

085.022 **On the reality and nature of a certain Sun-weather correlation.**   C. O. Hines, I. Halevy.
J. Atmos. Sci., Vol. 34, 382 - 404 (1977). – Abstr. in Phys. Abstr., Vol. 80, Abstr. 66956 (1977).

085.023 **Sunspots, geomagnetic indices and the weather: a cross-spectral analysis between sunspots, geomagnetic activity and global weather data.**
E. J. Gerety, J. M. Wallace, C. S. Zerefos.
J. Atmos. Sci., Vol. 34, 673 - 678 (1977). – Abstr. in Phys. Abstr., Vol. 80, Abstr. 78313 (1977).

085.024 **A mechanism for inducing climatic variations through the stratosphere: screening of cosmic rays by solar and terrestrial magnetic fields.**   J. W. Chamberlain.
J. Atmos. Sci., Vol. 34, 737 - 743 (1977). – Abstr. in Phys. Abstr., Vol. 80, Abstr. 78398 (1977).

085.025 **The enhancement of scattered Lα radiation in the geocorona during the solar flares of August 1972.**
D. H. Morgan.
Sol. Phys., Vol. 52, 463 - 469 (1977).
The S2/68 telescope on the TD-1A satellite has observed an increase in the Lα radiation scattered in the geocorona during the major solar flares of August 1972. The history of the 7 August flare is presented and shows a maximum enhancement of about 40%.

085.026 **New aspects in the technique of studying solar-terrestrial relations.**   B. M. Rubashev,
M. G. Gubanova, K. A. Smirnova.
Ehff. soln. aktivnosti v nizhn. atmos. Leningrad, Gidrometeoizdat, 1977, p. 28 - 36. In Russian. – Abstr. in Ref. zh., 51. Astron., 11.51.630 (1977).

085.027 **Geomagnetic cycles and variability of the system ocean-atmosphere.**   R. V. Smirnov.
Ehff. soln. aktivnosti v nizhn. atmos. Leningrad, Gidrometeoizdat, 1977, p. 60 - 69. In Russian. – Abstr. in Ref. zh., 51. Astron., 11.51.634 (1977).

085.028 **Corpuscular emission and the problem of solar-atmospherical relations. I. On an experimental test of one of the possible mechanisms of solar-atmospherical relations.**   V. F. Tulinov, V. M. Fejgin.
Ehff. soln. aktivnosti v nizhn. atmos. Leningrad, Gidrometeoizdat, 1977, p. 99 - 105. In Russian. – Abstr. in Ref. zh., 51. Astron., 11.51.636; 62. Issled. kosm. prostranstva, 11.62.142 (1977).

085.029 **Corpuscular emission and the problem of solar-atmospherical relations. III. Experimental study of variations of precipitating corpuscular radiation with meteorological rockets.**   V. F. Tulinov, V. M. Fejgin,
M. A. Savel'ev, V. A. Lipovetskij, Yu. M. Zhuchenko, A. P. Babaev, V. V. Tulyakov.
Ehff. soln. aktivnosti v nizhn. atmos. Leningrad, Gidrometeoizdat, 1977, p. 120 - 132. In Russian. – Abstr. in Ref. zh., 51. Astron., 11.51.637; 62. Issled. kosm. prostranstva, 11.62.112 (1977).

085.030 **On the application of the "second-order" approach to the analysis of solar-atmospherical relations.**

G. E. Poloskin.
Ehff. soln. aktivnosti v nizhn. atmos. Leningrad, Gidrometeoizdat, 1977, p. 37 - 46. In Russian. – Abstr. in Ref. zh., 51. Astron., 11.51.638 (1977).

085.031 **Some aspects of the effect of corpuscular emission on the earth's atmosphere under quiet and disturbed conditions.**   V. G. Sochnev, V. F. Tulinov,
S. G. Yakovlev.
Ehff. soln. aktivnosti v nizhn. atmos. Leningrad, Gidrometeoizdat, 1977, p. 47 - 54. In Russian. – Abstr. in Ref. zh., 51. Astron., 11.51.639 (1977).

085.032 **Effect of solar activity on circulation and temperature of the stratosphere.**   L. R. Rakipova.
Ehff. soln. aktivnosti v nizhn. atmos. Leningrad, Gidrometeoizdat, 1977, p. 3 - 9. In Russian. – Abstr. in Ref. zh., 51. Astron., 11.51.643 (1977).

085.033 **On the effect of solar activity on temperature and height of isobaric surfaces.**   A. A. Dmitriev,
V. P. Dremina, L. P. Krasnokutskaya, N. I. Pavlyuchenko, N. M. Padalka.
Ehff. soln. aktivnosti v nizhn. atmos. Leningrad, Gidrometeoizdat, 1977, p. 94 - 98. In Russian. – Abstr. in Ref. zh., 51. Astron., 11.51.644 (1977).

085.034 **Relation of atmospheric processes with the position of planets relative to the sun and earth.**
V. D. Reshetov.
Ehff. soln. aktivnosti v nizhn. atmos. Leningrad, Giodrometeoizdat, 1977, p. 78 - 85. In Russian. – Abstr. in Ref. zh., 51. Astron., 11.51.646 (1977).

085.035 **Estimate of some possible effects of solar activity on the troposphere.**   A. A. Dmitriev.
Ehff. soln. aktivnosti v nizhn. atmos. Leningrad, Gidrometeoizdat, 1977, p. 23 - 27. In Russian. – Abstr. in Ref. zh., 51. Astron., 11.51.647 (1977).

085.036 **Analytical expression of the mean annual variation of the precipitation within various latitude zones of the earth.**   J. Xanthakis, B. Tritakis.
Praktika Akad. Athens, Tom. 51, 600 - 635 (1977) = Res. Center Astron. Appl. Math., Acad. Athens, Contrib. Ser. I (Astron.), No. 47.

085.037 **Solar activity and the energy of wind in the upper atmosphere.**
P. B. Babadzhanov, V. M. Kolmakov, R. P. Chebotarev.
Astron. Tsirk., No. 950, p. 6 - 7 (1977). In Russian.

085.038 **A summary of significant solar-initiated events during STIP Intervals I and II.**
M. A. Shea, D. F. Smart, H. E. Coffey.
Study of travelling interplanetary phenomena 1977, (see 012. 042), p. 393 - 420 (1977).
A summary of the significant solar-terrestrial events that occurred during STIP Interval I (September - October 1975) and STIP Interval II (15 March - 15 May 1976) has been prepared. The first STIP Interval was characteristic of typical conditions expected near solar minimum. In contrast, STIP Interval II was an active period.

085.039 **Influence of the rotation of the sun on free oscillations of a torsion pendulum.**
E. M. Kolesnikova, S. M. Kolesnikov.
Probl. teor. gravitatsii i ehlem. chastits. Vyp. (No.) 8. Moskva, Atomizdat, 1977, p. 201 - 214. In Russian. – Abstr. in Ref. zh., 51. Astron., 12.51.424 (1977).

085.040 **Relation between solar wind characteristics and the**

irregularities of the earth's rotation.
Yu. D. Kalinin, V. M. Kiselev.
Simpoz. po fiz. geomagnitosfery, 1977. Tezisy dokl. Irkutsk, 1977, p. 10. In Russian. − Abstr. in Ref. zh., 51. Astron., 12.51.425 (1977).

085.041   **Solar activity, rotation of the earth and total circulation of the atmosphere.**
A. A. Krechetov, N. F. Sadykina.
Simpoz. po fiz. geomagnitosfery, 1977. Tezisy dokl. Irkutsk, 1977, p. 10 - 11. In Russian. − Abstr. in Ref. zh., 51. Astron., 12.51.426 (1977).

085.042   **Height and time extent of the seasonal anomaly in the solar activity cycle.**
O. K. Gordeev, N. I. Timchenko.
Issled. po ehlektrodin. i rasprostr. ehlektromagnit. voln. Tomsk, Tomsk. univ., 1977, p. 155 - 161. In Russian. − From Ref. zh., 51. Astron., 12.51.432 (1977).

085.043   **Implications of solar evolution for the earth's early atmosphere.**   M. J. Newman, R. T. Rood.
Science, Vol. 198, 1035 - 1037 (1977).
   The roughly 25 percent increase in luminosity over the life of the sun shared by many different solar models is shown to be a very general result. Superficially, this leads to a conflict with the climatic history of the earth, and if basic concepts of stellar evolution are not fundamentally in error, compensating effects must have occurred. One possible interpretation supported by recent detailed models of the earth's atmosphere is that the greenhouse effect was substantially more important than at present even as recently as 1 billion to 2 billion years ago.

085.044   **On solar radiation variability and volcanic activity as possible sources of climatic changes.**
J. M. Mitchell, Jr.
Fiz. osnovy teor. klim. i ego modelir. Leningrad, Gidrometeo-izdat, 1977, p. 133 - 136. In Russian. − Abstr. in Ref. zh., 51. Astron., 1.51.450 (1978).

085.045   **Solar-generated quasi-biennial geomagnetic variation.**
M. Sugiura, D. J. Poros.
J. Geophys. Res., Vol. 82, 5621 - 5628 (1977).
   The existence of highly correlated quasi-biennial variations in the geomagnetic field and in solar activity is demonstrated.

085.046   **Comment on 'On the high correlation between long-term averages of solar wind speed and geomagnetic activity'** by N. U. Crooker, J. Feynman, and J. T. Gosling.
A. J. Dessler, T. W. Hill.
J. Geophys. Res., Vol. 82, 5644, (1977).

085.047   **On the distribution of the occurrence of anomalous and normal temperature seasons in Leningrad over years of the 11-year solar cycle.**
B. M. Rubashev, V. S. Fionova, M. V. Tsvetikova.
Soln. Dannye 1977 Byull., No. 10, p. 98 - 103 (1977). In Russian.

085.048   **Solar-geophysical data number 389. Part I. Prompt reports. Data for December 1976−November 1976.**
H. I. Leighton.
Edited by Natl. Geophys. Sol.−Terr. Data Cent., Boulder, Colo., USA. 115 pp. (1977). − Abstr. from INIS7724 345362.

085.049   **Solar-geophysical data number 389. Part II. Comprehensive reports. Data for July 1976− June 1976 and miscellanea.**   H. I. Leighton.
Edited by Natl. Geophys. Sol.−Terr. Data Cent., Boulder, Colo., USA. 40 pp. (1977). − Abstr. from INIS 7724 345363.

085.050   **Mechanism of solar-terrestrial relations and changes of atmospheric circulation.**   V. Bucha.
Studia, Vol. 21, 350 - 360 (1977).

Corpuscular radiation and the problem of solar-atmospherical relations. II. Apparatus and methods of measurements of precipitating corpuscular radiation with means of space technology.   See Abstr. 032.583.

On variations of the amplitudes of the Chandler and annual polar motion of the earth in connection with solar activity.   See Abstr. 045.007.

The long-term variation of the solar activity and its effect on the earth.   See Abstr. 072.058.

The effect of sunspots and faculae on the solar constant.   See Abstr. 080.003.

On the geoeffectivity of oscillations of the solar surface.   See Abstr. 080.029.

The effects of changing fluxes of solar UV radiation and solar cosmic rays on the middle atmosphere temperature and ozone concentration.   See Abstr. 082.099.

Energy balance of the atmosphere under the influence of a disturbed sun and magnetospheric processes.   See Abstr. 082.101.

Solar-terrestrial physics.   See Abstr. 084.301.

Planetary System

091 Physics of the Planetary System (Planetary Atmospheres, Figure, Interiors, Magnetic Fields, Rotation, etc.)

091.001 Gas evaporation from collision-determined planetary exospheres. H. J. Fahr, B. Weidner.
Mon. Not. R. Astron. Soc., Vol. 180, 593 - 612 (1977).
The authors intend to solve the problem of calculating exospheric particle velocity distribution functions starting with slightly disturbed equilibrium conditions at the lower boundary of the exosphere. The distribution functions of exospheric particles are then determined using the emission functions at the lower boundary and taking into account the effect of collisions in the exosphere. One of the results that can be predicted from this concept is the existence of exospheric particles moving in elliptic orbits that never penetrate the lower exospheric boundary.

091.002 Planetary magnetism. Sh. Sh. Dolginov.
Geomagn. Aehron., Vol. 17, 569 - 595 (1977).
In Russian.

091.003 Observations of giant planets at 1.4 mm and consequences on the effective temperatures.
R. Courtin, N. Coron, T. Encrenaz, R. Gispert, P. Bruston, J. Leblanc, G. Dambier, A. Vidal-Madjar.
Astron. Astrophys., Vol. 60, 115 - 123 (1977).
Results of new observations of giant planets at 1.4 mm mean wavelength are reported. More precise brightness temperatures of Jupiter, Saturn and Uranus are given relative to Mars. The first detection and temperature measurement of Neptune at 1.4 mm has been obtained ($153 \pm 30°K$). Discussion and interpretation of these data is made with the help of atmospheric models and by comparison with other infrared or radio measurements of giant planets. The effect of the H_2/He ratio on far-infrared brightness temperature is discussed. Finally, effective temperatures most compatible with observations and models are obtained for Saturn, Uranus and Neptune.

091.004 Planetary crater retention ages. D. W. Hughes.
Nature, Vol. 269, 197 - 198 (1977).

091.005 Cloud microphysics: comments on the clouds of Venus, Mars, and Jupiter. W. B. Rossow.
Bull. American Astron. Soc., Vol. 9, 452 (1977). – Abstract.

091.006 Sources of outgassed volatiles on the terrestrial planets. L. L. Wilkening, G. T. Sill.
Bull. American Astron. Soc., Vol. 9, 452 (1977). – Abstract.

091.007 Gravity gradient mapping from lunar and planetary orbiters. R. L. Forward.
Bull. American Astron. Soc., Vol. 9, 456 (1977). – Abstract.

091.008 Global remanent magnetism of the terrestrial planets. L. J. Srnka.
Bull. American Astron. Soc., Vol. 9, 456 (1977). – Abstract.

091.009 Application of mutual shadowing photometric functions to planetary surfaces. W. W. Mendell.
Bull. American Astron. Soc., Vol. 9, 457 (1977). – Abstract.

091.010 Polarization in a mineral absorption band. II: Observations. C. Pieters.
Bull. American Astron. Soc., Vol. 9, 457 (1977). – Abstract.

091.011 Identification of a new class of satellites in the outer solar system.
D. P. Cruikshank, C. B. Pilcher, D. Morrison.
Bull. American Astron. Soc., Vol. 9, 464 (1977). – Abstract.

091.012 Evolution of icy satellites. G. J. Consolmagno, J. S. Lewis.
Bull. American Astron. Soc., Vol. 9, 464 - 465 (1977). Abstract.

091.013 Gravitational experiments on space missions to the giant planets.
J. D. Anderson, G. W. Null, W. B. Hubbard.
Bull. American Astron. Soc., Vol. 9, 474 (1977). – Abstract.

091.014 Die Kleinkörper im Planetensystem und der interplanetare Raum. J. Hoppe.
Astron. Schule, 14 Jahrg., 81 - 84 (1977).

091.015 Chemistry of the planets of our solar system. A. P. Vinogradov.
"Nauka i chelovechestvo, 1977". Znanie, 1976, p. 196 - 215.
In Russian. – From Ref. zh., 51. Astron., 8.51.69 (1977).

091.016 Crater evolutionary tracks. G. Schubert, R. E. Lingenfelter, R. J. Terrile.
Icarus, Vol. 32, 131 - 146 (1977).
A description of crater morphology based on rim height/depth (h/d) and depth/diameter (d/D) ratios provides a quantitative method for assessing the relative importance of competing crater modification processes. Different classes of processes produce distinctive evolutionary tracks on an h/d vs d/D diagram. The authors have calculated such tracks for three general classes of crater modification: those processes which add material to the crater, those which redistribute material within the crater vicinity, and those which remove material from the crater vicinity. They have compared h/d and d/D ratios for craters on the Earth, Moon, and Mars with theoretical evolutionary tracks for the general classes of crater modification.

091.017 Results of analyses performed on basalt adjacent to penetrators emplaced into volcanic rock at Amboy,

California, April 1976.   M. Blanchard, T. Bunch, A. Davis, H. Shade, J. Erlichman, G. Polkowski.
NASA Tech. Pap., NASA TP-1026. 16 pp. Price $ 3.25 (1977).

During 1976, four penetrators were dropped on a test site at Amboy, California. The Amboy site was selected because it simulated penetration into basalt flows on the Martian surface. This report describes the physical and chemical modifications found in the basalt after the penetrators' impact. In summary, contaminants introduced by the penetrator occur up to 1 cm away from the penetrator's skin. Although volatile elements do migrate and new minerals are formed during the destruction of host minerals in the crushed rock, no changes were observed beyond the 1-cm distance.

091.018 Results of analyses performed on soil adjacent to penetrators emplaced into sediments at McCook, Nebraska, January 1976.   M. Blanchard, T. Bunch, A. Davis, F. Kyte, H. Shade, J. Erlichman, G. Polkowski.
NASA Tech. Note, NASA TN D-8500. 29 pp. Price $ 3.25 (1977).

During 1976 several penetrators (full and 0.58 scale) were dropped into a test site at McCook, Nebraska. The McCook site was selected because it simulated penetration into wind-deposited sediments (silts and sands) on Martian plains. This report describes the physical and chemical modifications found in the sediment after the penetrators' impact. In summary, contaminants introduced by the penetrator occur up to 2 mm away from the penetrator's skin. Although volatile elements do migrate and new minerals are formed during the destruction of host minerals in the sediment, no changes were observed beyond the 2-mm distance. The analyses indicate 0.58-scale penetrators do effectively simulate full-scale testing for soil modification effects.

091.019 Significant achievements in the planetary geology program — 1975 - 1976.   J. W. Head (Editor), with a foreword by S. E. Dwornik.
NASA Contract. Rep., NASA CR-2827, 4 + 32 pp. Price $ 4.00 (1977).

Recent developments in planetology research as reported at the 1976 NASA Planetology Program Principal Investigators meeting are summarized. Important developments are summarized in topics ranging from solar system evolution, comparative planetology, and geologic processes, to techniques and instrument development for future exploration.

091.020 Composition of the terrestrial planets.   K. A. Goettel.
NASA Tech. Mem., NASA TM X-3511, (see 012.010), p. 7 - 10 (1977). — Abstract.

091.021 Heat, stretch and erupt: the relationships among global thermal evolution, crustal tectonics and surface volcanism on the terrestrial planets.   S. C. Solomon.
NASA Tech. Mem., NASA TM X-3511, (see 012.010), p. 20 - 21 (1977). — Abstract.

091.022 The rigid hard crusts of the Moon, Mars, Mercury and Venus: implications for the role of water in crustal mobility of Earth.   G. G. Schaber, J. M. Boyce.
NASA Tech. Mem., NASA TM X-3511, (see 012.010), p. 22 - 23 (1977). — Abstract.

091.023 Present impact cratering rates on the terrestrial planets and the Moon.   E. M. Shoemaker, E. F. Helin.
NASA Tech. Mem., NASA TM X-3511, (see 012.010), p. 74 - 77 (1977). — Abstract.

091.024 Crater evolutionary tracks.   G. Schubert, R. E. Lingenfelter, R. J. Terrile.
NASA Tech. Mem., NASA TM X-3511, (see 012.010), p. 97 - 99 (1977). — Abstract. — Paper will be published in "The Moon".

091.025 Formation and obliteration of large craters on the terrestrial planets.   C. R. Chapman.
NASA Tech. Mem., NASA TM X-3511, (see 012.010), p. 103 - 104 (1977). — Abstract.

091.026 Multi-ringed basins: a model for formation in multi-layered media.   R. A. De Hon.
NASA Tech. Mem., NASA TM X-3511, (see 012.010), p. 111 - 112 (1977). — Abstract.

091.027 Global seismic effects of basin-forming impacts.   H. G. Hughes, F. N. App, T. R. McGetchin.
NASA Tech. Mem., NASA TM X-3511, (see 012.010), p. 115 (1977). — Abstract.

091.028 Gravity effects on impact crater formation.   D. E. Gault, J. A. Wedekind.
NASA Tech. Mem., NASA TM X-3511, (see 012.010), p. 116 (1977). — Abstract.

091.029 Radar geology.   R. A. Simpson, H. T. Howard, G. L. Tyler.
NASA Tech. Mem., NASA TM X-3511, (see 012.010), p. 272 - 274 (1977). — Abstract.

091.030 Photometry of planetary surfaces: studies of the validity of a Minnaert description.   J. Veverka, J. Goguen, M. Noland.
NASA Tech. Mem., NASA TM X-3511, (see 012.010), p. 275 (1977). — Abstract.

091.031 Planetary frost program.   L. A. Lebofsky, J. E. Conel.
NASA Tech. Mem., NASA TM X-3511, (see 012.010), p. 276 - 277 (1977). — Abstract.

091.032 Photometric analyses of spacecraft planetary images.   B. Hapke.
NASA Tech. Mem., NASA TM X-3511, (see 012.010), p. 278 - 280 (1977). — Abstract.

091.033 Solution of the kinetic equation for gas in the gravitational field of a planet.   V. I. Zhuk.
Dokl. AN SSSR, Vol. 233, 325 - 328 (1977). In Russian. Abstr. in Ref. zh., 51. Astron., 9.51.239 (1977).

091.034 On the refraction of radio waves in radiographic inspection of planetary atmospheres.   A. V. Plotnikov.
Kosm. Issled., Vol. 15, 603 - 606 (1977). In Russian.

091.035 Direct method of solution of the transfer equation in application to inhomogeneous planetary atmospheres of large optical depth.   E. A. Ustinov, V. M. Filimonova.
Kosm. Issled., Vol. 15, 619 - 625 (1977). In Russian.

091.036 Dynamics and thermal structure of planetary atmospheres.   R. E. Dickinson.
Proc. 27th Internat. Astronaut. Congr. (see 012.016), p. 63 - 69 (1977).

091.037 The range of validity of the Eddington approximation.   W. J. Wiscombe, J. H. Joseph.
Icarus, Vol. 32, 362 - 377 (1977).

The Eddington approximation is often assumed to be useful only for optically thick media having a single-scattering albedo near unity. The authors present detailed evidence in this paper that, for homogeneous layers illuminated by a beam of radiation, the Eddington approximation predicts albedo and absorptivity reasonably well for all values of optical depth and single-scattering albedo, for several scattering phase functions (Rayleigh, Henyey–Greenstein, and Mie) having asymmetry factors less than or equal to $^1/_2$. The Eddington approximation is further found to maintain good accuracy over almost the full range of incident beam directions and surface albedos.

091.038 Exploration of the planets: an invited discourse presented before the Sixteenth General Assembly of the International Astronomical Union, Grenoble, France, August, 1976. C. Sagan.
Highlights of Astronomy, Vol. 4, Part I, (see 003.010), p. 37 - 67 (1977).

091.039 Cratering of terrestrial planets: brief review.
W. K. Hartmann.
Highlights of Astronomy, Vol. 4, Part I, (see 012.019), p. 229 - 232 (1977).
Analysis of cratering on all terrestrial planets and satellites has produced tools to study (1) the past meteoroid and planetesimal environment, (2) the erosive environments of planetary surfaces, and (3) the relative and absolute ages of planetary surface units.

091.040 Formation of Fe-Ni-Si planetary cores.
S. K. Saxena, A. Benimoff.
Nature, Vol. 270, 333 - 334 (1977).

091.041 Chemical abundances in the atmospheres of the giant planets and their satellites. T. Owen.
Chemical evolution of the giant planets (see 003.014), p. 49 - 58 (1976).
The outer planets and their satellites offer several environments of interest to the general problem of the origin and evolution of life. The characteristics of these environments are reviewed with special emphasis on the structure and composition of planetary and satellite atmospheres. Long period comets may provide a unique opportunity for sampling the primitive organic material available at the time of solar system formation.

091.042 Ion molecule plasma chemistry in reducing planetary atmospheres. W. F. Libby.
Chemical evolution of the giant planets (see 003.014), p. 59 - 67 (1976).
Plasma (ion molecule) chemistry probably has great importance for the chemistry of the planetary atmospheres and thus for the origin of life.

091.043 Exploration of the giant planets by infrared spectroscopy. R. A. Hanel.
Chemical evolution of the giant planets, (see 003.014), p. 165 - 181 (1976).

091.044 Biology on the outer planets.
R. S. Young, R. D. MacElroy.
Chemical evolution of the giant planets, (see 003.014), p. 199 - 219 (1976).

091.045 Ionizational nonequilibrium heating during outer planetary entries. L. P. Leibowitz, T.-J. Kuo.
AIAA J., Vol. 14, 1324 - 1329 (1976). – Abstr. in Phys. Abstr., Vol. 80, Abstr. 25979 (1977).

091.046 The magnetospheres of the earth and planets.
Yu. I. Gal'perin.
Usp. fiz. nauk, Vol. 122, 5, 160 - 164 (1977). In Russian. Abstr. in Ref. zh., 62. Issled. kosm. prostranstva, 10.62.199

(1977).

091.047 Lagrange characteristics of turbulent diffusion in planetary boundary layer.
D. L. Yordanov, E. D. Syrakov.
C. R. Acad. Bulgare Sci., Vol. 29, 1429 - 1431 (1976). Abstr. in Phys. Abstr., Vol. 80, Abstr. 33131 (1977).

091.048 Asymptotic eigensolutions of Laplace's tidal equation. J. W. Miles.
Proc. R. Soc. London, Ser. A, Vol. 353, 377 - 400 (1976). Abstr. in Phys. Abstr., Vol. 80, Abstr. 33166 (1977).

091.049 Photon mass and planetary magnetic fields.
G. V. Bicknell.
J. Phys. A, Vol. 10, 407 - 411 (1977). – Abstr. in Phys. Abstr., Vol. 80, Abstr. 33477 (1977).

091.050 A scenario on polarization in a planetary atmosphere.
G. W. Kattawar.
Polarized light, instruments, devices, applications. San Diego, Calif., USA, 24 - 25 August 1976. Proceedings of the Society of Photo-optical Instrumentation Engineers, Vol. 88. Palos Verdes Estates, Calif., USA. 8 + 128 pp. (1976), p. 67 - 74. Abstr. in Phys. Abstr., Vol. 80, Abstr. 49171 (1977).

091.051 Geologic evolution of the terrestrial planets.
J. W. Head, C. A. Wood, T. A. Mutch.
American Sci., Vol. 65, No. 1, p. 21 - 29 (1977). – Abstr. in Phys. Abstr., Vol. 80, Abstr. 53442 (1977).

091.052 Phase separation for a dense fluid mixture of nuclei (Giant planet interiors).
E. L. Pollock, B. J. Alder.
Phys. Rev. A, Vol. 15, 1263 - 1268 (1977). – Abstr. in Phys. Abstr., Vol. 80, Abstr. 56849 (1977).

091.053 Propagation of hydromagnetic planetary waves on a beta-plane through magnetic and velocity shear.
I. B. Eltayeb, J. F. McKenzie.
J. Fluid Mech., Vol. 81, Part 1, p. 1 - 23 (1977). – Abstr. in Phys. Abstr., Vol. 80, Abstr. 60457 (1977).

091.054 The thermal state and evolution of the Earth and terrestrial planets. D. C. Tozer.
Sci. Prog., Vol. 64, No. 253, p. 1 - 28 (1977). – From Phys. Abstr., Vol. 80, Abstr. 64243 (1977).

091.055 Lunar volcanism and the origin of the planets.
A. Rittmann.
Arch. Sci., Vol. 30, No. 1, p. 5 - 13 (1977). In French. – Abstr. in Phys. Abstr., Vol. 80, Abstr. 64671 (1977).

091.056 Genesis of ring and radial-concentric structures of planets (from a rock mechanic's point of view).
J. J. Broul, G. N. Katterfeld.
Mod. Geol., Vol. 6, No. 2, p. 101 - 115 (1977). – Abstr. in Phys. Abstr., Vol. 80, Abstr. 64675 (1977).

091.057 Magnetic fields of planets. P. H. Stoker.
Tegnikon, Vol. 24, No. 4, p. 7 - 14 (1976). In Dutch. Abstr. in Phys. Abstr., Vol. 80, Abstr. 67054 (1977).

091.058 Present concepts of Earth-type planets.
O. Wolczek.
Fiz. Szemle, Vol. 27, 103 - 113 (1977). In Hungarian. – Abstr. in Phys. Abstr., Vol. 80, Abstr. 73987 (1977).

091.059 Calculations on the evolution of the planetary interiors.
V. P. Keondijian (Keondzhyan), A. S. Monin.
Tectonophysics, Vol. 41, No. 1 - 3, p. 227 - 242 (1977).

Abstr. in Phys. Abstr., Vol. 80, Abstr. 82137 (1977).

091.060 The phase diagram and transport properties for hydrogen-helium fluid planets.
D. J. Stevenson, E. E. Salpeter.
Astrophys. J., Suppl. Ser., Vol. 35, 221 - 237 (1977).

The present paper considers in detail the phase diagram for hydrogen-helium mixtures, and its implications for the interiors of the Jovian planets. Since these implications depend on details of the transport (including fluid-dynamical) processes, the present paper also contains a survey of the current knowledge of the microscopic transport properties of dense hydrogen-helium mixtures. The present paper concentrates on the condensed-matter physics of such mixtures, with emphasis given to the pressure-temperature domain appropriate to Jupiter and Saturn. The emphasis is on the fluid state, which is almost certainly applicable to the present interiors of Jupiter and Saturn, but there is also a discussion of melting curves for the hydrogen-helium phases. Since the Jovian planets contain constituents other than hydrogen and helium, the effects of these are considered briefly.

091.061 The dynamics and helium distribution in hydrogen-helium fluid planets.
D. J. Stevenson, E. E. Salpeter.
Astrophys. J., Suppl. Ser., Vol. 35, 239 - 261 (1977).

The results of paper I (see 091.060) are used in the present paper for a semiquantitative analysis of the thermal and compositional history of an evolving hydrogen-helium planet such as Jupiter or Saturn. First, the evolution of a homogeneous planet with no first-order phase transitions or immiscibilities is considered. Next, the effects of a first-order molecular-metallic hydrogen transition are discussed for a pure hydrogen planet: a well-defined interface between the phases persists, despite the presence of convection. Convection in the presence of composition gradient is discussed, and the importance of overstable modes and diffusive-convective equilibria established. Evolutions with helium immiscibility (but no first-order molecular-metallic hydrogen transition) are discussed. Finally, more complicated cases are discussed which include both immiscibility and the first-order character of the molecular-metallic hydrogen transition.

091.062 Planetary atmospheres and interstellar clouds.
M. J. Newman, D. M. Butler, R. J. Talbot, Jr.
Publ. Astron. Soc. Pacific, Vol. 89, 619 - 620 (1977). Abstract.

091.063 The formation of the atmospheres of the terrestrial planets by impact. A. Benlow, A. J. Meadows.
Astrophys. Space Sci., Vol. 46, 293 - 300 (1977).

It is generally supposed that the atmospheres of the terrestrial planets were formed by secondary degassing processes. The authors examine here an alternative hypothesis — that the atmospheres may, on the contrary, be primary products, derived from the impacts of the accreting material on the growing planetary embryo.

091.064 Cosmic rays and ancient planetary magnetic fields.
P. S. Wesson.
Astrophys. Space Sci., Vol. 46, 321 - 326 (1977).

The possibility is discussed of using the latitude-dependent cutoff in the intensity and flux of cosmic ray particles reaching the surface of a planet to investigate ancient magnetic fields in the Moon, Mars and the Earth. In the last case, the method could provide a validity test for conventional palaeomagnetism.

091.065 On early stages of evolution of the atmosphere and climate of the terrestrial planets.
V. I. Moroz, L. M. Mukhin.
Kosm. Issled., Vol. 15, 901 - 922 (1977). In Russian.

091.066 A method of computing the emergent radiation by the atmosphere in the region ranging from ultraviolet to infrared.
T. Takashima, C. I. Taggart, E. G. Morrissey.
Astrophys. Space Sci., Vol. 49, 331 - 337 (1977).

A method of computing the diffuse reflection and transmission radiation by an inhomogeneous, plane-parallel planetary atmosphere with internal emission source is discussed by use of the adding method. If the atmosphere is simulated by a number of homogeneous sub-layers, the radiation diffusely reflected or transmitted by the atmosphere can be expressed in terms of the reflection and transmission matrices of the radiation of sub-layers. The diffusely transmitted radiation due to the internal emission source can be also easily computed in the same manner.

091.067 Structure of the terrestrial planets.
R. A. Lyttleton.
Astrophys. Space Sci., Vol. 49, L1 - L6 (1977).

Recent reviews (cf. Runcorn, 1968; or Cook, 1972, 1975) on the structure of the planets omit reference to the phase-change hypothesis for the nature of the terrestrial core, despite that numerous prior predictions of the theory based on this hypothesis have subsequently been borne out as correct. These reviews also ignore the existence of theoretical calculations of the internal structure of Venus which can be computed with high accuracy by use of the terrestrial seismic data. Several examples of numerous mistakes committed in these reviews are pointed out.

091.068 On the variety of types of eruptive rocks in the terrestrial planets.   V. I. Shmuratko.
Izv. AN ArmSSR. Nauk. o Zemle, Vol. 30, No. 1, p. 7 - 14 (1977). In Russian. — Abstr. in Ref. zh., 51. Astron., 11.51.324 (1977).

091.069 The magnetic fields of the terrestrial planets.
D. W. Strangway.
Phys. Earth Planet. Inter., Vol. 15, 121 - 130 (1977).

A single model for the terrestrial planets based upon the structure inferred for the lunar interior is developed, which subdivides them according to size. It is probable that planetary bodies smaller than the moon have never melted but they could still carry a memory of an early intense solar-system field. The magnetic fields of Mercury, Mars and Venus can be explained in terms of a crustal remanence which is a memory of a primitive internal or external field. The earth's present field, on the other hand, is due to an active internal dynamo.

091.070 On the estimate of accuracy of determination of the parameters of an ellipsoid approximating the image of the horizon on a space photograph of a planet.
L. I. Permitina.
Geod., kartogr. i aehrofotosemka. Resp. mezhved. nauchn.-tekh. sb., 1977, vyp. (No.) 26, p. 118 - 123. In Russian.
Abstr. in Ref. zh., 52. Geod. Aehrosemka, 12.52.180 (1977).

091.071 Solar-wind control of the extent of planetary ionospheres.   S. J. Bauer.
Solar-wind interaction with the planets Mercury, Venus, and Mars, (see 012.047), p. 47 - 62 (1976).

In our solar system there are at least four magnetic planets: Earth, Jupiter, Mercury, and Mars; while at least one planet, Venus, appears to be essentially nonmagnetic. The ionospheres of the magnetic planets are imbedded in their magnetosphere and thus shielded from the solar wind, whereas the ionosphere of Venus, at least, interacts directly with the solar wind. However, the solar-wind interaction with the planetary environment, in both cases, affects the behavior of their ionospheres. In this paper, the role the solar-wind interaction plays in limiting the extent of the ionospheres of both magnetic and nonmagnetic planets is discussed.

091.072 **Effect of the frozen-in magnetic field on the formation of Venus' plasma shell boundary: experimental confirmation.** Ju. V. Andrijanov (*Yu. V. Andriyanov*), I. M. Podgorny (*Podgornyj*).
Solar-wind interaction with the planets Mercury, Venus, and Mars, (see 012.047), p. 101 - 110 (1976).

The results are given of laboratory simulation of solar-wind interaction with plasma shells of nonmagnetic planets. It is shown that the momentum transfer from the plasma flow to the shell occurs due to the presence of a frozen-in magnetic field. Without a magnetic field frozen-in, the ionosphere has no sharp boundary and a shock wave does not form in the flow.

091.073 **Solar-wind interaction with planetary ionospheres.** P. A. Cloutier.
Solar-wind interaction with the planets Mercury, Venus, and Mars, (see 012.047), p. 111 - 119 (1976).

The planetary ionosphere apparently presents a hard obstacle to the flow, with bow-shock formation required in the supersonic, super-Alfvénic flow to slow and direct most of the solar-wind plasma around the planetary ionosphere. Various aspects of the interaction are examined in the context of theoretical models in an attempt to explain observed details of the interaction regions of Venus and Mars.

091.074 **Numerical study of some solar-wind interaction models with space objects.**
O. M. Belotserkovskii (*Belotserkovskij*), V. Ya. Mitnitskii (*Mitnitskij*).
Solar-wind interaction with the planets Mercury, Venus, and Mars, (see 012.047), p. 121 - 133 (1976).

Problems in space physics are discussed whose models, in simplified form, reduce to a supersonic flow scheme with a detached shock wave, namely: (A) Solar-wind interaction with an intrinsic planetary magnetic field. (B) Solar-wind interaction with the ionized component of the atmosphere of a comet. (C) Solar-wind interaction with the ionosphere of a planet which does not possess its own magnetic field. The numerical study of the above problems is performed with the use of magnetogasdynamic equations for an ideal single-fluid model.

091.075 **Magnetohydrodynamic and gasdynamic aspects of solar-wind flow around terrestrial planets. A critical review.** J. R. Spreiter.
Solar-wind interaction with the planets Mercury, Venus, and Mars, (see 012.047), p. 135 - 149 (1976).

091.076 **Numerical simulation of the effects of magnetic field induced by plasma flow past nonmagnetic planets.** A. S. Lipatov.
Solar-wind interaction with the planets Mercury, Venus, and Mars, (see 012.047), p. 151 - 157 (1976).

The interaction of a nonstationary plasma flow with a model ionosphere is studied. On the basis of a numerical simulation, the calculation yields results of the distribution of the plasma concentration and magnetic field in the transition region.

091.077 **Planetary magnetospheres: a comparative view.** A. J. Dessler.
Solar-wind interaction with the planets Mercury, Venus, and Mars, (see 012.047), p. 159 - 166 (1976).

Why some of the bodies (sun/planets) do, and some do not, have magnetic fields? Why there is such a specialized variety of particle acceleration phenomena? Why the magnetosphere of Mars does not accelerate particles?

091.078 **Stratigraphy and structural geology.**
M. H. Carr, D. E. Wilhelms, R. Greeley, J. E. Guest, B. Murray.
A geological basis for the exploration of the planets, (see

003.023), p. 13 - 32 (1976).

091.079 **Geochemistry.** M. H. Carr, P. Toulmin III, R. Zartman, K. A. Kvenvolden, F. P. Fanale, D. M. Anderson, A. Metzger, M. A. Steggert, J. Adams, T. McCord, R. Huguenin, H. Kieffer.
A geological basis for the exploration of the planets, (see 003.023), p. 33 - 61 (1976).

091.080 **Geophysics.** M. H. Carr, P. Cassen, S. Saunders, M. Langseth, L. Tyler, T. Howard, A. W. England, J. Cuzzi.
A geological basis for the exploration of the planets, (see 003.023), p. 63 - 74 (1976).

091.081 **Geodesy and cartography.** R. Batson, D. Arthur.
A geological basis for the exploration of the planets, (see 003.023), p. 75 - 83 (1976).

091.082 **Earth-based studies.** S. Dwornik, D. U. Wise, G. E. McGill, A. Howard, M. C. Gilbert, T. Bunch, E. Helin, D. Matson, T. Johnson, G. Schaber.
A geological basis for the exploration of the planets, (see 003.023), p. 85 - 106 (1976).

091.083 **Meteorite impact ejecta: dependence of mass and energy lost on planetary escape velocity.** J. D. O'Keefe, T. J. Ahrens.
Science, Vol. 198, 1249 - 1251 (1977).

The calculated energy efficiency of mass ejection for iron and anorthosite objects striking an anorthosite planet at speeds of 5 to 45 kilometers per second decreases with increasing impact velocity at low escape velocities. The impact velocities at which ejecta losses equal meteorite mass gains are found to be approximately 20, 35, and 45 kilometers per second for anorthosite objects and approximately 25, 35, and 40 kilometers per second for iron objects striking anorthosite surfaces for the gravity fields of the moon, Mercury, and Mars.

091.084 **On figures and gravitational fields of the terrestrial planets.** D. V. Zagrebin.
Izv. vyssh. uchebn. zaved. Geod. i aehrofotosemka, 1977, No. 3, p. 40 - 45. In Russian. — Abstr. in Ref. zh., 51. Astron., 1.51.130 (1978).

091.085 **Occultation of stars by planets and by minor planets.** S. Isobe.
10th Lunar and Planetary Symposium, (see 012.050), p. 236 - 238 (1977). In Japanese.

The inner planets. See Abstr. 003.041.

The solar planets. See Abstr. 003.052.

Astronomie I. Die Sonne und ihre Planeten. See Abstr. 003.063.

Chemical kinetics. See Abstr. 003.074.

Chemical petrology, with applications to the terrestrial planets and meteorites. See Abstr. 003.114.

Planetary satellites. See Abstr. 012.055.

Progress report: Copernicus observations of solar system objects. See Abstr. 013.001.

Computational techniques for solar wind flows past terrestrial planets — theory and computer programs. See Abstr. 021.021.

Intensity measurements in the ν_4-fundamental of methane.   See Abstr. 022.001.

Absorption coefficients of ices of CH_4, CO_2, NH_3, H_2S, H_2O and sulfuric acid.   See Abstr. 022.011.

The absorption spectrum of CO_2 around 7740 cm^{-1}
See Abstr. 022.018.

Photoelectric absorption spectra of methane (CH_4), methane and hydrogen (H_2) mixtures, and ethane (C_2H_6).
See Abstr. 022.020.

Band model analysis of laboratory methane absorption spectra from 4500 to 10500 Å.   See Abstr. 022.021.

Intensity and transmission measurements in the ν_3-fundamental of N_2O at low temperatures.
See Abstr. 022.034.

Transition probability data for molecules of astrophysical interest.   See Abstr. 022.054.

Electron impact on atmospheric gases. 1. Updated cross sections.   See Abstr. 022.109.

The strength of lunar analogues and its geophysical implications.   See Abstr. 022.117.

High-velocity impact into basaltic and metallic targets.   See Abstr. 022.118.

Radio occultations by turbulent planetary atmospheres: power spectra of intensity scintillations.
See Abstr. 031.213.

Analytic transform pair illustrating atmospheric occultation experiments.   See Abstr. 031.214.

Adaptation of the Alpha Particle Instrument for penetrator missions.   See Abstr. 031.228.

Radar astronomy.   See Abstr. 031.271.

Some peculiarities in interpreting space photographs of the surfaces of the moon and planets.
See Abstr. 031.275.

Spacecraft radio-occultation technique for the study of planetary atmospheres.   See Abstr. 031.286.

Some peculiarities of interpretation of cosmic pictures of the lunar and planetary surfaces.
See Abstr. 031.347.

Photoelectric meridian observations of Mars, Jupiter, Saturn, Uranus, Neptune and four minor planets 1968 - 1971.
See Abstr. 041.015.

Orbit—orbit resonances in the solar system: varieties and similarities.   See Abstr. 042.016.

Orbital resonances in the solar system.
See Abstr. 042.019.

A fast invariant imbedding method for multiple scattering calculations and an application to equivalent widths of CO_2 lines on Venus.   See Abstr. 063.003.

A novel methodology for radiative transfer in a planetary atmosphere. I. The functions a^m and b^m of anisotropic scattering.   See Abstr. 063.031.

Multiple scattering in the atmosphere with a rough surface. I. Azimuth-independent case.
See Abstr. 063.036.

Contrast transmittance at the top of an atmosphere bounded by a horizontally nonuniform diffuse reflector.
See Abstr. 063.056.

Model of the convective zone and its possible response to outside action.   See Abstr. 080.075.

Planetary waves in horizontal and vertical shear: the generalized Eliassen-Palm relation and the mean zonal acceleration.   See Abstr. 082.071.

Planetary waves in horizontal and vertical shear: asymptotic theory for equatorial waves in weak shear.
See Abstr. 082.072.

Similarity theory of the buoyantly interactive planetary boundary layer with entrainment.
See Abstr. 082.073.

Observations of the earth's night and twilight horizons from the orbital station Salyut 4.
See Abstr. 082.130.

On the question of the energy of the precessional dynamo.   See Abstr. 084.315.

Candidate areas for in situ ancient lunar materials.
See Abstr. 094.152.

Global seismic effects of basin-forming impacts.
See Abstr. 094.190.

Equilibrium and disequilibrium. Chemistry of adiabatic, solar-composition planetary atmospheres.
See Abstr. 099.061.

The effect of gravity on crater formation: thickness of ejecta and concentric basins.   See Abstr. 105.036.

Distribution of the mean motions of planets and satellites and the development of the solar system.
See Abstr. 107.019.

Significance of a conclusive test of Dirac's Large Numbers hypothesis using precision ranging to Mars.
See Abstr. 162.105.

092 Mercury

092.001 The magnetic field of Mercury.
    D. J. Jackson, D. B. Beard.
J. Geophys. Res., Vol. 82, 2828 - 2836 (1977).
    The geomagnetic field, suitably scaled down and param-
eterized, is shown to give a very good fit to the magnetic field
measurements taken on the first and third passes of the
Mariner 10 space probe past Mercury. The excellence of the
fit to a reliable planetary magnetospheric model is good
evidence that the Mercury magnetosphere is formed by a
simple, permanent, intrinsic planetary magnetic field distorted by the effects of the solar wind.

092.002 Mercury's core: the effect of obliquity on the spin-
    orbit constraints.   S. J. Peale, A. P. Boss.
J. Geophys. Res., Vol. 82, 3423 - 3429 (1977).
    In an earlier paper by the authors (1977) rather severe
constraints were placed on the properties of a Mercurian
liquid core and certain other dynamical characteristics of the
planet for consistency with Mercury's escape from the spin-
orbit resonance with the spin angular velocity equal to twice
the orbital mean motion. One assumption under which these
constraints were established was that the obliquity $\theta \approx 0$ at
the time of resonance passage. Here the authors show that for
$\theta \neq 0$, Mercury can easily escape the spin resonance with a
large core of low viscosity, and all constraints established for
resonance passage for $\theta = 0$ would vanish. However, the same
effect would have reduced θ to negligibly small values long
before the $2n$ resonance was reached. Thus Mercury most
likely passed through $\dot{\psi}_M = 2n$ with $\theta \approx 0$.

092.003 Generalized tectonic/volcanic chronology for
    Mercury.   D. Dzurisin.
Bull. American Astron. Soc., Vol. 9, 452 - 453 (1977).
Abstract.

092.004 Photometric variations on Mercury.
    D. Dzurisin, G. E. Danielson.
Bull. American Astron. Soc., Vol. 9, 453 (1977). — Abstract.

092.005 Mercury geologic mapping program.
    H. E. Holt.
Bull. American Astron. Soc., Vol. 9, 456 (1977). — Abstract.

092.006 Mercury: evidence for an anorthositic crust from
    reflectance spectra.   J. B. Adams, T. B. McCord.
Bull. American Astron. Soc., Vol. 9, 457 (1977). — Abstract.

092.007 A spin-orbit constraint on the viscosity of a
    Mercurian liquid core.   S. J. Peale, A. P. Boss.
Bull. American Astron. Soc., Vol. 9, 457 (1977). — Abstract.

092.008 Größte Elongation und Dichotomie der inneren
    Planeten.   W. Kunz.
Sterne Weltraum, Jahrg. 16, 334 - 337 (1977).

092.009 Mercury: evidence for an anorthositic crust from
    reflectance spectra.   J. B. Adams, T. B. McCord.
NASA Tech. Mem., NASA TM X-3511, (see 012.010), p. 24
(1977). — Abstract.

092.010 Data analysis of Mariner 10 magnetic field observa-
    tions at Mercury and Venus.   N. F. Ness.
NASA Tech. Mem., NASA TM X-3511, (see 012.010), p. 25 -
26 (1977). — Abstract.

092.011 A spin-orbit constraint on the viscosity of a
    Mercurian liquid core.   S. J. Peale, A. P. Boss.
NASA Tech. Mem., NASA TM X-3511, (see 012.010), p. 27 -
28 (1977). — Abstract.

092.012 Mercury's core: the effect of obliquity on the spin-
    orbit constraints.   S. J. Peale, A. P. Boss.
NASA Tech. Mem., NASA TM X-3511, (see 012.010), p. 29 -
30 (1977). — Abstract.

092.013 Relative ages of Mercurian plains.
    L. A. Soderblom.
NASA Tech. Mem., NASA TM X-3511, (see 012.010), p. 47 -
48 (1977). — Abstract.

092.014 Tectonism and volcanism on Mercury: inferences
    from morphologic and photometric studies.
D. Dzurisin.
NASA Tech. Mem., NASA TM X-3511, (see 012.010), p. 49 -
50 (1977). — Abstract.

092.015 Global tectonics of Mercury.   B. M. Cordell.
    NASA Tech. Mem., NASA TM X-3511, (see 012.
010), p. 51 - 53 (1977). — Abstract.

092.016 Preliminary results of structural lineament pattern
    analysis of Mercury.   P. Masson, P. Thomas.
NASA Tech. Mem., NASA TM X-3511, (see 012.010), p. 54 -
55 (1977). — Abstract.

092.017 Degradation trends of Mercurian craters and correla-
    tion with the Moon.
C. A. Wood, J. W. Head, M. J. Cintala.
NASA Tech. Mem., NASA TM X-3511, (see 012.010), p. 100 -
102 (1977). — Abstract.

092.018 Morphological characterization of the Mercury large
    craters: statistical behaviour of the craters in H1, H6,
H7, H8, H11 and H15 quadrangles.
A. Carusi, M. Fulchignoni, M. Poscolieri, R. Casacchia.
NASA Tech. Mem., NASA TM X-3511, (see 012.010), p. 105 -
107 (1977). — Abstract.

092.019 Multi-planet, multi-ring basin studies.
    J. F. McCauley.
NASA Tech. Mem., NASA TM X-3511, (see 012.010), p. 113 -
114 (1977). — Abstract.

092.020 Mercury geologic mapping program.   H. E. Holt.
    NASA Tech. Mem., NASA TM X-3511, (see 012.
010), p. 233 (1977). — Abstract.

092.021 The control net of Mercury.   M. E. Davies.
    NASA Tech. Mem., NASA TM X-3511, (see 012.
010), p. 234 - 235 (1977). — Abstract.

092.022 Shaded relief map of Mercury.   R. M. Batson.
    NASA Tech. Mem., NASA TM X-3511, (see 012.
010), p. 236 (1977). — Abstract.

092.023 The geology of the Goethe (H-1) quadrangle of
    Mercury.   J. M. Boyce, M. Grolier.
NASA Tech. Mem., NASA TM X-3511, (see 012.010), p. 237
(1977). — Abstract.

092.024 Geological mapping of Mercury quadrangle H-3
    (Shakespeare) and H-4.
R. Greeley, J. E. Guest, D. E. Gault.
NASA Tech. Mem., NASA TM X-3511, (see 012.010), p. 238
(1977). — Abstract.

092.025 **Geologic mapping of the Victoria quadrangle (H-2), Mercury.**   E. A. King.
NASA Tech. Mem., NASA TM X-3511, (see 012.010), p. 239 (1977). – Abstract.

092.026 **Geologic map of the Tolstoj quadrangle of Mercury.**   G. G. Schaber, J. F. McCauley.
NASA Tech. Mem., NASA TM X-3511, (see 012.010), p. 240 - 241 (1977). – Abstract.

092.027 **Geology of the Kuiper quadrangle of Mercury.**   R. A. De Hon, J. R. Underwood, Jr., D. H. Scott.
NASA Tech. Mem., NASA TM X-3511, (see 012.010), p. 242 - 244 (1977). – Abstract.

092.028 **Geologic mapping of Michelangelo quadrangle (H12) of Mercury: structural and topographic features.**
K. R. Blasius.
NASA Tech. Mem., NASA TM X-3511, (see 012.010), p. 245 - 246 (1977). – Abstract.

092.029 **Geologic mapping of Bach (south polar) quadrangle, Mercury: a progress report.**   M. C. Malin.
NASA Tech. Mem., NASA TM X-3511, (see 012.010), p 247 (1977). – Abstract.

092.030 **The magnetic field of Mercury.**   N. F. Ness.
Highlights of Astronomy, Vol. 4, Part I, (see 012.019), p. 179 - 190 (1977).
Mercury possesses a global magnetic field, a modest magnetosphere and magnetic tail. The author briefly reviews the recent magnetic field data obtained from the Mariner 10 spacecraft, the present state of our knowledge and its implications regarding the interior of Mercury.

092.031 **Mercury.**   D. E. Gault, J. A. Burns, P. Cassen, R. G. Strom.
Annu. Rev. Astron. Astrophys., Vol. 15, (see 003.012), 97 - 126 (1977).
Contents: Size and mass; Orbit; Rotation; Atmosphere; Magnetic field and magnetosphere; Surface features; Interior; Geologic history.

092.032 **Transits of Mercury.**   J. Ashbrook.
Modern astronomy, (see 003.013), p. 56 - 67 (1977).

092.033 **Mercury: interferometry at 1.35 cm wavelength and determination of thermophysical parameters.**
L. M. Golden.
Publ. Astron. Soc. Pacific, Vol. 89, 617 - 618 (1977). Abstract.

092.034 **The escape of natural satellites from Mercury and Venus.**   S. S. Kumar.
Astrophys. Space Sci., Vol. 51, 235 - 238 (1977).
It is suggested that the slow rotations of Mercury and Venus may be connected with the absence of natural satellites around them. If Mercury and Venus possessed a satellite at the time of formation, the tidal evolution would have caused the satellite to recede. At a sufficiently large distance from the planet, the Sun's gravitational influence makes the satellite orbit unstable. The natural satellites of Mercury and Venus might have escaped as a consequence of this instability.

092.035 **Convection in Mercury.**   S. K. Runcorn.
Phys. Planet. Inter., Vol. 15, 131 - 134 (1977).
In order to explain why Mercury is trapped in a resonant state of rotation with a period two thirds of its orbital period, it has been proposed that the planet departs from hydrostatic equilibrium. It is shown that its gravity field must include a second-degree harmonic term. It is argued that this must be caused by convection in the solid interior rather than an initial distortion retained by the finite strength of the interior. The presence of an iron core in Mercury poses an interesting question as to why a second-degree harmonic convection pattern in the silicate mantle is present and suggested solutions are discussed.

092.036 **The relationship between crustal tectonics and internal evolution in the moon and Mercury.**
S. C. Solomon.
Phys. Earth Planet. Inter., Vol. 15, 135 - 145 (1977).

092.037 **Global tectonics of Mercury and the moon.**   B. M. Cordell, R. G. Strom.
Phys. Earth Planet. Inter., Vol. 15, 146 - 155 (1977).

092.038 **Origin and relative age of lunar and Mercurian intercrater plains.**   R. G. Strom.
Phys. Earth Planet. Inter., Vol. 15, 156 - 172 (1977).
The very widespread distribution of intercrater plains on Mercury compared to the moon may be related to Mercury's core formation which would have resulted in a large radius increase leading to widespread extensional fracturing to provide egress for the eruption of volcanic deposits on a global scale. This was followed by a radius decrease due to contraction of the lithosphere to produce thrust faulting represented by Mercury's lobate scarps.

092.039 **Moon–Mercury: relative preservation states of secondary craters.**   D. H. Scott.
Phys. Earth Planet. Inter., Vol. 15, 173 - 178 (1977).
Geologic mapping of the Kuiper quadrangle of Mercury and other geologic studies of the planet indicate that secondary craters are much better preserved than those on the moon around primary craters of similar size and morphology. Ejection velocities of material producing most of the secondary craters are rather low (< 1 km/s) but velocities on Mercury are about 50% greater than those on the moon for equivalent ranges. Higher velocities may produce morphologically enhanced secondary craters which may account for their better preservation with time.

092.040 **Rayed craters on the moon and Mercury.**   C. C. Allen.
Phys. Earth Planet. Inter., Vol. 15, 179 - 188 (1977).

092.041 **Endogenic modification of impact craters on Mercury.**   P. H. Schultz.
Phys. Earth Planet. Inter., Vol. 15, 202 - 219 (1977) = Contrib. Lunar Sci. Inst. No. 293.

092.042 **Interpretations of optical observations of Mercury and the moon.**   B. Hapke.
Phys. Earth Planet. Inter., Vol. 15, 264 - 274 (1977).
Optical, thermal and radar remote-sensing measurements indicate that Mercury is covered with a relatively thick layer of soil similar in texture and thickness to lunar regolith. The differential photometric functions of Mercury and the moon have a latitudinal dependence which can be completely accounted for by shadowing in craters. The lack of polar darkening on Mercury in spite of the presence of a magnetic field implies that the dominant soil-darkening process on Mercury, and by extension, on the moon is not dependent on the solar wind, but probably is deposition of material evaporated by meteorite impacts.

092.043 **Production of simple molecules on the surface of Mercury.**   E. K. Gibson Jr.
Phys. Earth Planet. Inter., Vol. 15, 303 - 312 (1977).
Irradiation experiments on lunar materials have shown that solar-wind and solar-flare irradiation of the lunar surface produces selected low-molecular-weight components. Solar-wind irradiation of Mercury's surface should also produce a

wide variety of low-molecular-weight species because of the increased solar flux, which results from Mercury being nearer the sun than the moon. The thermal regime of Mercury's surface would result in thermal evaporation of low-temperature components followed by "cold-trapping" on the night side of the planet. Such desorption–adsorption processes assist chemical weathering of Mercury's regolith.

092.044   **Interaction of solar wind with Mercury and its magnetic field.**   N. F. Ness, K. W. Behannon, R. P. Lepping, Y. C. Whang.
Solar-wind interaction with the planets Mercury, Venus, and Mars, (see 012.047), p. 87 - 99 (1976).
   It is the purpose of this report to present a brief review of the magnetic field and solar-wind electron observations and to estimate the intrinsic magnetic field of the planet Mercury and the implications of such a field for the planetary interior.

092.045   **Marial basins of the planet Mercury.**   K. Beneš.
Říše hvězd, Vol. 58, 202 - 206 (1977). In Czech.

   The atlas of Mercury.   See Abstr. 003.042.

   Flight to Mercury.   See Abstr. 003.115.

   On the coincidence of the position of Mercury with the 90-day oscillation of Jupiter's Red Spot.
See Abstr. 015.002.

   Combined position and diameter measures for lunar craters.   See Abstr. 031.202.

   Digital processing of the Mariner 10 images of Venus and Mercury.   See Abstr. 031.313.

   Mariner 10 ultraviolet spectrometer: airglow experiment.   See Abstr. 032.574.

   Mariner 10 ultraviolet spectrometer: occultation experiment.   See Abstr. 032.575.

   Right ascensions of the sun, Mercury and Venus observed in 1968–1969 with the Freiberg-Kondrat'ev transit instrument in Nikolaev.   See Abstr. 041.006.

   Right ascensions of the sun, Mercury and Venus observed in Nikolaev in 1970 with a transit instrument.
See Abstr. 041.007.

   Right ascensions of the sun, Mercury and Venus observed in Nikolaev in 1971 with a transit instrument.
See Abstr. 041.008.

   Right ascensions of the sun, Mercury and Venus observed in Nikolaev in 1972 with a transit instrument.
See Abstr. 041.009.

   Right ascensions of the sun, Mercury and Venus observed in Nikolaev with a transit instrument in 1973–1975.
See Abstr. 041.010.

   Declination of the sun, Mercury and Venus obtained from observations with the vertical circle of the Nikolaev Observatory in 1960.   See Abstr. 041.012.

   Declinations of the sun, Mercury, Venus and Mars obtained from observations with the vertical circle of the Nikolaev Observatory in 1973–1975.   See Abstr. 041.013.

   Expansion of the earth due to a secular decrease in G – evidence from Mercury.   See Abstr. 081.034.

   The magnetic fields of the terrestrial planets.
See Abstr. 091.069.

   Comparative studies of the Moon, Mercury and Mars.
See Abstr. 094.115.

   Origin and relative age of lunar and Mercurian intercrater plains.   See Abstr. 094.116.

   Rayed craters on the Moon and Mercury.
See Abstr. 094.118.

   Thermal expansion and thermal stress in the moon and terrestrial planets: clues to early thermal history.
See Abstr. 094.162.

   Comparison of impact basins on Mercury, Mars and the Moon.   See Abstr. 094.168.

   Interaction of the surfaces of the Moon and Mercury with their exospheric atmospheres.   See Abstr. 094.177.

   Lineament patterns on the Moon, Mars and Mercury.
See Abstr. 094.186.

   Planetary magnetism and the interiors of the moon and Mercury.   See Abstr. 094.188.

   Moon–Mercury: large impact structures, isostasy and average crustal viscosity.   See Abstr. 094.189.

   Ancient surfaces of the terrestrial planets.
See Abstr. 097.074.

   The effect of nongravitational factors on the shape of Martian, lunar and Mercurian craters: target effects.
See Abstr. 097.078.

   Interplanetary comparisons of fresh crater morphology: preliminary results.   See Abstr. 097.079.

   On prediction of the structure of the surface layer of the planets Mars and Mercury.   See Abstr. 097.118.

   Martian crater depth/diameter relationships: comparison with the Moon and Mercury.
See Abstr. 097.127.

   Comparison of large crater and multiringed basin populations on Mars, Mercury, and the Moon.
See Abstr. 097.128.

   Comments on: "Characteristics of fresh martian craters as a function of diameter: comparison with the Moon and Mercury" by M. J. Cintala et al.
See Abstr. 097.137.

   Characteristics of fresh martian craters as a function of diameter: comparison with the Moon and Mercury– discussion.   See Abstr. 097.138.

093 Venus

093.001   An improved Venus cloud model.   A. T. Young.
    Icarus, Vol. 32, 1 - 26 (1977).

A simple radiative-transfer theory that allows for the change in the absorptions of sulfur and carbon dioxide with depth in the atmosphere of Venus can account simultaneously for (1) the spectral reflectance of Venus; (2) the wavelength dependence of contrast in uv cloud features; (3) the CO_2 line profile; (4) the change in slope of the curve of growth from the 7820- to the 10488-Å CO_2 bands; and (5) the rotational temperature near 246°K found for all CO_2 bands. The model cloud consists of 1-μm sulfuric-acid particles, which are well mixed between about 64 km and the 49-km cloud base found by Veneras 9 and 10, plus an overlapping cloud of much larger sulfur particles that extends down to the 35-km cloud base found by Venera 8.

093.002   Venus: on the phase variation of CO_2 line profiles.
    W. Macy, Jr., L. Trafton, E. Barker.
Icarus, Vol. 32, 27 - 36 (1977).

The shapes of Venus' CO_2 profiles are found to vary with solar phase angle. High-resolution spectra of the P16 and P14 lines in the 8689- and 7820-Å bands, respectively, are presented for phase angles ranging from 6 to 158°. The scattering mean free path at 80 mbar, approximately the effective pressure, is 1.7 km. Use of the van de Hulst similarity relations with simple, parametric scattering models is inadequate to separate effects due to the scattering phase function from those due to inhomogeneities in depth when one attempts to determine the atmospheric structure by fitting a family of such models over a wide range of phase angles.

093.003   Viscous flow in the near-Venusian plasma wake.
    H. Pérez-de-Tejada, M. Dryer, O. L. Vaisberg
(*Vajsberg*).
J. Geophys. Res., Vol. 82, 2837 - 2841 (1977).

The acquisition of the Venera 9 and 10 plasma probe data has provided further support for the notion that the shocked solar wind in the flank regions of the Venusian ionosheath interacts viscously with the local ionospheric plasma and that the resulting mixing region is rapidly forced into the planetary umbra.

093.004   Spin and atmospheric tides of Venus.
    W. Kundt.
Astron. Astrophys., Vol. 60, 85 - 91 (1977).

Reviving Gold's idea of 1964, it is argued that braking by tidal torques and possibly magnetospheric friction can account for a spindown of Venus within some 10^9 years. Subsequent retrograde spinup via thermally driven atmospheric tides can explain the present slow retrograde rotation of the planet. Spin reversal has been achieved by a delicate cooperation of one-sided solar heating, longitude-dependent solar attraction, and atmospheric mass asymmetry by means of which the present 4-day atmospheric circulation was switched on just in time to reverse the rotation. The present spin period of 243.1 d cannot be understood as a stable spin-orbit resonance.

093.005   Venus: evidence of vortex circulation.
    V. E. Suomi, S. S. Limaye.
Bull. American Astron. Soc., Vol. 9, 463 - 464 (1977).
Abstract.

093.006   The surface of Venus.   R. Goldstein, R. Green,
    H. Rumsey, M. Malin, R. S. Saunders.
Bull. American Astron. Soc., Vol. 9, 466 - 467 (1977).
Abstract.

093.007   Comments on the Venus rotation pole.
    W. R. Ward, W. M. Decampli.
Bull. American Astron. Soc., Vol. 9, 467 (1977). – Abstract.

093.008   Calculations of the effects of atmospheric tides on the rotation of Venus.
A. P. Ingersoll, A. Dobrovolskis.
Bull. American Astron. Soc., Vol. 9, 467 (1977). – Abstract.

093.009   Simultaneous ultraviolet and infrared imaging of Venus.   D. J. Diner, J. A. Westphal.
Bull. American Astron. Soc., Vol. 9, 467 (1977). – Abstract.

093.010   Spectroscopic determination of the rotation period of Venus.   R. A. Schorn, A. T. Young.
Bull. American Astron. Soc., Vol. 9, 467 (1977). – Abstract.

093.011   Venus: global scale inertial oscillations.
    S. S. Limaye.
Bull. American Astron. Soc., Vol. 9, 467 (1977). – Abstract.

093.012   A semi-empirical model of the circulation of the upper atmosphere of Venus using remote temperature soundings.   L. S. Elson.
Bull. American Astron. Soc., Vol. 9, 468 (1977). – Abstract.

093.013   A self consistent model of the Venusian 4-day circulation.   S. B. Fels.
Bull. American Astron. Soc., Vol. 9, 468 (1977). – Abstract.

093.014   Further results from a Venus general circulation model.   R. E. Young, J. B. Pollack.
Bull. American Astron. Soc., Vol. 9, 468 (1977). – Abstract.

093.015   Sulfur in the clouds of Venus.   A. T. Young.
    Bull. American Astron. Soc., Vol. 9, 468 (1977).
Abstract.

093.016   Two changes in the Venus aerosol distribution revealed by analysis of CO_2 line profiles.
W. Cochran, L. Trafton, W. Macy.
Bull. American Astron. Soc., Vol. 9, 468 (1977). – Abstract.

093.017   New spectrophotometric observations of Venus from 3–4 microns.   J. V. Martonchik.
Bull. American Astron. Soc., Vol. 9, 468 - 469 (1977).
Abstract.

093.018   Microwave absorption in the atmosphere of Venus from Mariner 10 radio occultation.
A. J. Kliore, C. Elachi, I. R. Patel.
Bull. American Astron. Soc., Vol. 9, 469 (1977). – Abstract.

093.019   Microwave spectral lines in the atmospheres of Mars and Venus.
W. J. Wilson, R. K. Kakar, J. W. Waters.
Bull. American Astron. Soc., Vol. 9, 469 (1977). – Abstract.

093.020   Microwave detection of carbon monoxide on Venus and Mars.
R. K. Kakar, J. W. Waters, W. J. Wilson.
Bull. American Astron. Soc., Vol. 9, 469 (1977). – Abstract.

093.021   Excitation of the Venus night airglow.
    C. A. Barth, G. M. Lawrence.
Bull. American Astron. Soc., Vol. 9, 470 (1977). – Abstract.

093.022   Krater auf der Venus-Oberfläche.   H. W. Köhler.

Sterne Weltraum, Jahrg. 16, 282 - 283 (1977).

093.023 High spectral resolution ground-based observations of Venus in the 450- to 1250-cm^{-1} region.
V. G. Kunde, R. A. Hanel, L. W. Herath.
Icarus, Vol. 32, 210 - 224 (1977).
   The thermal emission spectrum of the central portion of the apparent disk was recorded. All statistically significant sharp line absorption features in the spectrum have been identified with gaseous CO_2. Comparison between the observed spectrum and a synthetic spectrum computed from a model atmosphere, assuming gaseous CO_2 and a sulfuric acid haze as opacity sources, indicates good agreement. A broad diffuse absorption feature associated with the sulfuric acid haze is evident in the 870- to 930-cm^{-1} region.

093.024 The 1976 - 77 eastern (evening) apparition of the planet Venus: visual and photographic investigations.
J. L. Benton, Jr.
Strolling Astron., Vol. 26, 240 - 251 (1977).
   Visual and photographic observations of the planet Venus during the 1976 - 77 eastern (evening) apparition are examined and discussed. The source of the data and the instrumentation which was utilized to gather the information are described in the report. A statistical investigation of the kinds and types of surface markings seen on Venus' apparent surface at visual wavelengths is presented.

093.025 Plasma transport in the topside Venus ionosphere.
R. W. Schunk, J.-P. St.-Maurice.
Planet. Space Sci., Vol. 25, 921 - 930 (1977).
   The authors have studied the extent to which certain transport processes affect ion composition and heat flow in the daytime, topside Venus ionosphere. Particular attention is given to the conditions that prevailed during the Mariner 5 measurements, at which time the topside Venus ionosphere appeared to be in a state of diffusive equilibrium. They have found that the ion composition is sensitive to the ion temperature, the ion temperature gradient, and to relative drifts between the ion species of a few m/sec. The electron density, on the other hand, is very insensitive to these parameters. They have also found that a relative drift between the ion species of a few m/sec induces an ion heat flow that is equivalent to a 1 K/km temperature gradient. This induced heat flow could influence the energy balance in the topside Venus ionosphere.

093.026 Venus radar imaging and geologic interpretation.
R. Saunders, M. Malin, R. Goldstein, R. Green, H. Rumsey.
NASA Tech. Mem., NASA TM X-3511, (see 012.010), p. 61 (1977). – Abstract.

093.027 Venus mapping.
H. Masursky, M. Strobell, A. L. Dial.
NASA Tech. Mem., NASA TM X-3511, (see 012.010), p. 251 (1977). – Abstract.

093.028 The Venusian clouds: calculated properties of the aerosol sulphuric acid medium in the thermal infrared. L. V. Ksanfomaliti.
Astron. Zh. Akad. Nauk SSSR, Vol. 54, 1110 - 1117 (1977). In Russian. English translation in Soviet Astron., Vol. 21, No. 5.
   The coefficients of scattering, absorption, extinction, and the single-scattering albedo in the 400–1500 cm^{-1} range are calculated for an aerosol medium consisting of spherical particles of a 75 percent solution of sulphuric acid. Characteristic bands identified in the Venus spectrum have been obtained. From comparison with data of radiometry from Venera 9 and 10 the concentration of particles in the radiating cloud layer is found to be 95 cm^{-3}.

093.029 Radiative transfer in the 2.86 - 4.16 μm transparency "window" of the Venus atmosphere.
T. G. Adiks, A. P. Gal'tsev, V. M. Osipov, V. P. Shari.
Kosm. Issled., Vol. 15, 747 - 754 (1977). In Russian.

093.030 Interpretation of the optical measurements aboard the automatic interplanetary station Venera 8.
T. A. Germogenova, N. V. Konovalov, N. L. Lukashevich, E. M. Fejgel'son.
Kosm. Issled., Vol. 15, 755 - 767 (1977). In Russian.

093.031 The inverse problem of the theory of multiple scattering and interpretation of measurements of scattered radiation in the cloud layer of Venus.
E. A. Ustinov.
Kosm. Issled., Vol. 15, 768 - 775 (1977). In Russian.

093.032 Model of the temperature dependence of the Venus atmosphere for the 52 - 90 km interval on height.
L. V. Ksanfomaliti.
Kosm. Issled., Vol. 15, 796 - 798 (1977). In Russian.

093.033 The surface of Venus as revealed by Venera 9 and 10 probes. C. P. Florensky (*K. P. Florenskij*).
Highlights of Astronomy, Vol. 4, Part I, (see 012.019), p. 225 - 227 (1977).

093.034 Venus: new evidence of tectonic activity.
L. I. Miroshnichenko, E. I. Prutenskaya.
Priroda, 1977, No. 10, p. 150. In Russian.

093.035 The atmospheres of the planets. B. J. Mason.
Observatory, Vol. 97, 217 - 234 (1977).

093.036 The elongation of Venus, 1977 January.
J. H. Robinson.
J. British Astron. Assoc., Vol. 88, 73 - 78 (1977). – Report of the Mercury and Venus Section.

093.037 Photometry of Venus from Mariner 10. B. Hapke.
J. Atmos. Sci., Vol. 33, 1803 - 1815 (1976). – Abstr. in Phys. Abstr., Vol. 80, Abstr. 29685 (1977).

093.038 Venus and Mars (from recent results of Soviet and American studies). M. Ya. Marov.
Usp. fiz. nauk, Vol. 122, 5, 159 - 160 (1977). In Russian. Abstr. in Ref. zh., 51. Astron., 10.51.298 (1977).

093.039 The new Venus.
L. V. Ksanfomaliti, K. P. Florensky (*Florenskij*), A. T. Bazilevsky (*Bazilevskij*), V. V. Zasetsky (*Zasetskij*), A. M. Trakhtman.
New Scientist, Vol. 73, 127 - 129 (1977). – Abstr. in Phys. Abstr., Vol. 80, Abstr. 45338 (1977).

093.040 A normalized view of Venus.
S. Limaye, V. Suomi.
J. Atmos. Sci., Vol. 34, 205 - 215 (1977). – Abstr. in Phys. Abstr., Vol. 80, Abstr. 56850 (1977).

093.041 Possible lava flows on Venus.
Spaceworld, Vol. N-2-158, 37 - 39 (1977). – Abstr. in Phys. Abstr., Vol. 80, Abstr. 64685 (1977).

093.042 The clouds of Venus. I. An approximate technique for treating the effects of coagulation, sedimentation and turbulent mixing on an aerosol.
W. B. Rossow, P. J. Gierasch.
J. Atmos. Sci., Vol. 34, 405 - 416 (1977). – Abstr. in Phys. Abstr., Vol. 80, Abstr. 71041 (1977).

093.043 Momentum and energy exchanges due to orographi-

cally scattered gravity waves (*Venus' atmosphere*).
S. B. Fels.
J. Atmos. Sci., Vol. 34, 499 - 514 (1977). – Abstr. in Phys.
Abstr., Vol. 80, Abstr. 67057 (1977).

093.044   Microwave absorption in Venusian atmosphere.
R. K. Johri.
Indian J. Radio Space Phys., Vol. 5, No. 4, p. 311 - 316 (1976).
Abstr. in Phys. Abstr., Vol. 80, Abstr. 71040 (1977).

**093.045   The clouds of Venus. II. An investigation of the in-
fluence of coagulation on the observed droplet size
distribution.**   W. B. Rossow.
J. Atmos. Sci., Vol. 34, 417 - 431 (1977). – Abstr. in Phys.
Abstr., Vol. 80, Abstr. 67056 (1977).

093.046   Why explore Venus?   T. M. Donahue.
Space Sci. Rev., Vol. 20, 259 - 263 (1977).
This paper develops the rationale for a program of Venus
exploration by man.

093.047   Current knowledge of Venus.
D. M. Hunten, G. E. McGill, A. F. Nagy.
Space Sci. Rev., Vol. 20, 265 - 282 (1977).
A summary is given of the current knowledge of Venus,
with emphasis on recent progress and on the contributions to
be expected from the Pioneer Venus missions. Headings are
surface and interior, clouds and lower atmosphere, dynamics
and thermal structure, neutral upper atmosphere, and ionosphere and solar-wind cavity.

**093.048   Composition and structure of the atmosphere of
Venus.**
J. H. Hoffman, G. M. Keating, H. Niemann, V. Oyama,
J. Pollack, A. Seiff, A. I. Stewart, U. von Zahn.
Space Sci. Rev., Vol. 20, 307 - 327 (1977).
The Pioneer Venus set of experiments is designed to provide information both individually and collectively to help
understand and explain first of all the present state of the
atmosphere (the composition and distribution in both the
lower and upper parts, the state property profiles, the cloud
compositions, the role of phase in the thermal structure, the
planet's surface and interior composition, the high surface
temperature, the stability of CO_2, the ionosphere – its chemistry and thermal structure, the existence of superrotation, the
response of the upper atmosphere to changes in solar EUV and
the solar wind) and secondly the origin and evolution of the
atmosphere. This paper discusses these questions and the
degree to which the Pioneer Venus instruments will respond
to them.

093.049   The clouds of Venus.   R. G. Knollenberg,
J. Hansen, B. Ragent, J. Martonchik, M. Tomasko.
Space Sci. Rev., Vol. 20, 329 - 354 (1977).
The current state of knowledge of the Venusian clouds is
reviewed. The visible clouds of Venus are shown to be quite
similar to low level terrestrial hazes of strong anthropogenic
influence. Possible nucleation and particle growth mechanisms
are presented. The Pioneer Venus experiments that emphasize
cloud measurements are described and their expected findings
are discussed in detail.

**093.050   Dynamics, winds, circulation and turbulence in the
atmosphere of Venus.**   G. Schubert,
C. C. Counselman III, J. Hansen, S. S. Limaye, G. Pettengill,
A. Seiff, I. I. Shapiro, V. E. Suomi, F. Taylor, L. Travis,
R. Woo, R. E. Young.
Space Sci. Rev., Vol. 20, 357 - 387 (1977).
With the possible exception of the lowest one or two
scale heights, the dominant mode of circulation of Venus'
atmosphere is a rapid, zonal, retrograde motion. Global albedo
variations in the ultraviolet may reflect planetary scale waves

propagating relative to the zonal winds. Other special phenomena such as cellular convection in the subsolar region and
internal gravity waves generated in the interaction of the zonal
circulation with the subsolar disturbance may also be revealed
in ultraviolet imagery of the atmosphere. The authors discuss
the contributions of experiments on the Orbiter and Entry
Probes of Pioneer Venus toward unravelling the mystery of the
planet's global circulation and the role played by waves, instabilities and convection therein.

093.051   The thermal balance of the atmosphere of Venus.
M. G. Tomasko, R. Boese, A. P. Ingersoll, A. A. Lacis,
S. S. Limaye, J. B. Pollack, A. Seiff, A. I. Stewart, V. E. Suomi,
F. W. Taylor.
Space Sci. Rev., Vol. 20, 389 - 412 (1977).
Current knowledge of the temperature structure of the
atmosphere of Venus is briefly summarized. The principal
features to be explained are the high surface temperature, the
small horizontal temperature contrasts near the cloud tops in
the presence of strong apparent motions, and the low value of
the exospheric temperature. In order to understand the role
of radiative and dynamical processes in maintaining the
thermal balance of the atmosphere, a great deal of additional
data on the global temperature structure, solar and thermal
radiation fields, structure and optical properties of the clouds,
and circulation of the atmosphere are needed. The ability of
the Pioneer Venus Orbiter and Multiprobe Missions to provide
these data is indicated.

093.052   The Venus ionosphere and solar wind interaction.
S. J. Bauer, L. H. Brace, D. M. Hunten, D. S.
Intriligator, W. C. Knudsen, A. F. Nagy, C. T. Russell, F. L.
Scarf, J. H. Wolfe.
Space Sci. Rev., Vol. 20, 413 - 430 (1977).
The current state of knowledge of the chemistry,
dynamics and energetics of the upper atmosphere and ionosphere of Venus is reviewed together with the nature of the
solar wind–Venus interaction. Because of the weak, though
perhaps not negligible, intrinsic magnetic field of Venus, the
mutual effects between these regions are probably strong and
unique in the solar system. The ability of the Pioneer Venus
Bus and Orbiter experiments to provide the required data to
answer the questions outstanding is discussed in detail.

093.053   The surface and interior of Venus.
H. Masursky, W. M. Kaula, G. E. McGill, G. H.
Pettengill, R. J. Phillips, C. T. Russell, G. Schubert, I. I. Shapiro.
Space Sci. Rev., Vol. 20, 431 - 449 (1977).
On the Pioneer Venus 1978 Orbiter mission, the radar
mapper experiment will determine surface heights, dielectric
constant values and small-scale slope values along the sub-
orbital track between 50°S and 75°N. This experiment will
also estimate the global shape and provide coarse radar images
(40–80 km identification resolution) of part of the surface.
Gravity data will be obtained by radio tracking. Maps combining radar altimetry with spacecraft and ground-based
images will be made. A fluxgate magnetometer will measure
the magnetic fields around Venus. The radar and gravity data
will provide clues to the level of crustal differentiation and
tectonic activity. The magnetometer will determine the field
variations accurately. Data from the combined experiments
may constrain the dynamo mechanism; if so, a deeper understanding of both Venus and Earth will be gained.

093.054   Ultraviolet absorbers in the Venus clouds.
M. Shimizu.
Astrophys. Space Sci., Vol. 51, 497 - 499 (1977).
Some absorption features in the ultraviolet spectrum of
Venus observed by the OAO-2 cannot be interpreted in terms
of H_2SO_4. Carbon suboxide polymer has a yellow colour and
absorption at 2000 Å. Fine graphite grains have an absorption
band at about 2175 Å as is well known in the case of the inter-

stellar extinction curves. A mixture of these substances which is inevitably formed in the Venus atmosphere by photochemical reactions is the best candidate for explaining the Venus absorption features in the ultraviolet.

093.055 **On the thermal history of Venus.**
S. V. Maeva, E. L. Ruskol.
Izv. AN SSSR. Fiz. Zemli, 1977, No. 4, p. 3 - 7. In Russian. – Abstr. in Ref. zh., 51. Astron., 11.51.328; 62. Issled. kosm. prostranstva, 11.62.86 (1977).

093.056 **Observations of Venus during 1970 - 1976 in the near infrared region (0.9–2.5 μ). 1. Phase dependences.**   A. I. Smirnov, O. G. Taranova.
Astron. Tsirk., No. 950, p. 1 - 2 (1977). In Russian.

093.057 **Observations of Venus during 1970 - 1976 in the near infrared region (0.9–2.5 μ). 2. Temporal variations of the continuum and CO_2 absorption bands.**
O. G. Taranova.
Astron. Tsirk., No. 950, p. 2 - 4 (1977). In Russian.

093.058 **Observations of Venus during 1970 - 1976 in the near infrared region (0.9–2.5 μ). 3. Geometrical albedo at $\lambda = 1.046 \mu$.**   O. G. Taranova.
Astron. Tsirk., No. 950, p. 4 (1977). In Russian.

093.059 **Observations of Venus during 1970 - 1976 in the near infrared region (0.9–2.5 μ). 4. Latitude variations of $\overline{W}$ (CO_2).**   O. G. Taranova.
Astron. Tsirk., No. 950, p. 5 - 6 (1977). In Russian.

093.060 **Geomorphological observations of the first photographs of the surface of Venus transmitted by the Soviet unmanned spacecrafts Venera 9 and Venera 10.**
P. Leonardi.
Atti Accad. Naz. Lincei, Rend. Ser. Ottava, 289 - 292 (1976).

093.061 **Carrying away of ions of the Venus atmosphere by the solar wind.**
G. M. Nedyalkova, O. E. Popov, I. E. Turchinovich.
XXX Gertsenovsk. chteniya. Teor. fiz. i astron., Leningrad, 1977, p. 44 - 48. In Russian. – Abstr. in Ref. zh., 51. Astron., 12.51.216; 62. Issled. kosm. prostranstva, 12.62.139 (1977).

093.062 **Peculiarities of observation and registration of the cloud layer of Venus.**   A. I. Lazarev.
Opt.-mekh. prom-st', 1977, No. 6, p. 7 - 10. In Russian. Abstr. in Ref. zh., 51. Astron., 12.51.218 (1977).

093.063 **Numerical modelling of the excitation of the induced magnetosphere as a result of interaction between the solar wind and the Venus ionosphere.**
A. S. Lipatov.
Inst. kosm. issled. AN SSSR. Pr-351. Moskva, 1977. 24 pp. In Russian. – Abstr. in Ref. zh., 62. Issled. kosm. prostranstva, 12.62.128 (1977).

093.064 **Interaction of the solar wind with Venus.**
H. S. Bridge, A. J. Lazarus, G. L. Siscoe, R. E. Hartle, K. W. Ogilvie, J. D. Scudder, C. M. Yeates.
Solar-wind interaction with the planets Mercury, Venus, and Mars, (see 012.047), p. 63 - 79 (1976).
Two topics related to the interaction of the solar wind with Venus are considered. First, a short review of the experimental evidence with particular attention to plasma measurements carried out on Mariner-5 and Mariner-10 is given. Secondly, the results of some recent theoretical work on the interaction of the solar wind with the ionosphere of Venus are summarized.

093.065 **Results and interpretation of light-flux measurements in the near-surface layer of the Venus atmosphere.**   Yu. M. Golovin, B. E. Moshkin, A. P. Ehkonomov.
Inst. kosm. issled. AN SSSR. Pr-338. Moskva, 1977, 40 pp. In Russian. – Abstr. in Ref. zh., 51. Astron., 1.51.262; 62. Issled. kosm. prostranstva, 1.62.132 (1978).

093.066 **Magnetosphere of the planet Venus.**
Sh. Sh. Dolginov, L. N. Zhuzgov, V. A. Sharova, V. B. Buzin, E. G. Eroshenko.
Inst. zemn. magn., ionos. i rasprostr. radiovoln AN SSSR. Prepr. No. 19 (193). Moskva, 1977. 66 pp. In Russian. Abstr. in Ref. zh., 51. Astron., 1.51.264 (1978).

093.067 **Ultraviolet absorbers in the Venus clouds.**
M. Shimizu, M. Niida.
10th Lunar and Planetary Symposium, (see 012.050), p. 174 - 177 (1977).
Some absorption features in the ultraviolet spectrum of Venus observed by the OAO-2 cannot be interpreted in terms of H_2SO_4. Carbon suboxide polymer and graphite which are inevitably formed in the Venus atmosphere by photo-chemical reactions are the best candidates for explaining the Venus absorption features in the ultraviolet.

093.068 **Infrared polarization of Venus.**
S. Sato, K. Kawara, Y. Kobayashi, K. Noguchi, T. Maihara, H. Okuda.
10th Lunar and Planetary Symposium, (see 012.050), p. 178 - 182 (1977). In Japanese.

093.069 **Annual variations of the strength of CO_2 absorption on Venus.**   K. Iwasaki.
10th Lunar and Planetary Symposium, (see 012.050), p. 183 - 187 (1977). In Japanese.

093.070 **Observation of Venus and Jupiter in the 3–4 mm band. (Part II).**   K. Akabane, S. Hata.
10th Lunar and Planetary Symposium, (see 012.050), p. 188 - 192 (1977). In Japanese.

093.071 **Venus.**   M. V. Keldyš.
Kozmos, Vol. 8, 137 - 140 (1977). In Slovak.

Moon, Mars and Venus, a concise guide in colour. See Abstr. 003.144.

Digital processing of the Mariner 10 images of Venus and Mercury.   See Abstr. 031.313.

Mariner 10 ultraviolet spectrometer: airglow experiment.   See Abstr. 032.574.

Right ascensions of the sun, Mercury and Venus observed in 1968–1969 with the Freiberg-Kondrat'ev transit instrument in Nikolaev.   See Abstr. 041.006.

Right ascensions of the sun, Mercury and Venus observed in Nikolaev in 1970 with a transit instrument. See Abstr. 041.007.

Right ascensions of the sun, Mercury and Venus observed in Nikolaev in 1971 with a transit instrument. See Abstr. 041.008.

Right ascensions of the sun, Mercury and Venus observed in Nikolaev in 1972 with a transit instrument. See Abstr. 041.009.

Right ascension of the sun, Mercury and Venus observed in Nikolaev with a transit instrument in 1973–1975. See Abstr. 041.010.

Declinations of the sun, Mercury and Venus obtained from observations with the vertical circle of the Nikolaev Observatory in 1960.   See Abstr. 041.012.

Declinations of the sun, Mercury, Venus and Mars obtained from observations with the vertical circle of the Nikolaev Observatory in 1973–1975.   See Abstr. 041.013.

Results of photographic position observations of Venus with the zonal astrograph in Nikolaev in 1967–1972. See Abstr. 041.014.

Numerical theory of motion of the earth and Venus derived from data of radar observations and optical observations and from observations of motion of the artificial satellites Venera 9 and Venera 10.   See Abstr. 041.017.

Pioneer Venus 1978.   See Abstr. 051.029.

The exploration of Venus. See Abstr. 051.056.

The Pioneer Venus program. See Abstr. 051.057.

Pioneer Venus experiment descriptions. See Abstr. 051.058.

A fast invariant imbedding method for multiple scattering calculations and an application to equivalent widths of CO_2 lines on Venus.   See Abstr. 063.003.

Measurement of scattered ultraviolet radiation in the vicinity of planets and in the interplanetary medium. See Abstr. 082.018.

Chemical evolution – comparative planetology. See Abstr. 082.144.

Cloud microphysics: comments on the clouds of Venus, Mars and Jupiter.   See Abstr. 091.005.

Sources of outgassed volatiles on the terrestrial planets.   See Abstr. 091.006.

Structure of the terrestrial planets. See Abstr. 091.067.

The magnetic fields of the terrestrial planets. See Abstr. 091.069.

Effect of the frozen-in magnetic field on the formation of Venus' plasma shell boundary: experimental confirmation.   See Abstr. 091.072.

Größte Elongation und Dichotomie der inneren Planeten.   See Abstr. 092.008.

Data analysis of Mariner 10 magnetic field observations at Mercury and Venus.   See Abstr. 092.010.

The escape of natural satellites from Mercury and Venus.   See Abstr. 092.034.

The geomagnetic dynamos of the Moon and Venus: comparisons with a recent scaling law.   See Abstr. 094.164.

CO_2 photoionization and energy distribution of photoelectrons in the atmospheres of Mars and Venus. See Abstr. 097.122.

Windblown dust on Earth, Mars and Venus. See Abstr. 097.156.

094 Moon: Dynamics, Global Properties, Local Properties

Moon, Dynamics

094.001 Topocentric aberration of the Moon.
N. G. Rizvanov.
Moon, Vol. 16, 335 - 337 (1977).

Aberrational displacement of the observed topocentric positions of the Moon differs from the aberrational effect in its apparent ephemeris geocentric coordinates. The differential aberrational corrections due to the mutual positions of the observer and the Moon may account to $0.''3$. The reduction method of astrometric observations of the Moon, which takes into account this effect, is proposed.

094.002 Determination of lunar gravitational harmonic coefficients from combined Orbiter Doppler and laser ranging data.
W. S. Sinclair, A. J. Ferrari, W. L. Sjogren, J. G. Williams.
Bull. American Astron. Soc., Vol. 9, 437 (1977). – Abstract.

094.003 On the thermal history of a Moon of fission origin.
A. B. Binder, M. Lange.
Moon, Vol. 17, 29 - 45 (1977).

Model calculations show that the thermal history of a Moon which originated by fission from the proto-Earth is the same as that for the Moon as it is currently understood. In particular, a fissioned Moon currently has a small percent of partial melt or at least near solidus temperatures below depths of 800 km in accord with the seismic data which show that the deep interior of the Moon has a very low Q. The models have moderate (20–50%) degrees of partial melting in the upper mantle (depths $<$ 300 or 200 km) in the period between 3 to 4 $\times$ 10^9 years ago and, therefore, can account for the mare filling epoch. Finally the heat flow of the models is 18 ergs cm^{-2}s^{-1} which is close to the average of 19 ergs cm^{-2}s^{-1} derived from the Apollo heat flow experiments. These findings add further support for the fission origin of the Moon.

094.004 An improved lunar moment of inertia determination: a proposed strategy. M. P. Ananda,
A. J. Ferrari, W. L. Sjogren.
Moon, Vol. 17, 101 - 120 (1977) = Contrib. Div. Geol. Planet. Sci., Calif. Inst. Technol., Pasadena, Calif. No. 2885.

This study demonstrates that an improvement in the lunar moment of inertia uncertainty of over a magnitude can be achieved by processing data from the Lunar Polar Orbiter relay satellite.

094.005 Lunar dynamics and selenodesy: results from analysis of VLBI and laser data.
R. W. King, C. C. Counselman III, I. I. Shapiro.
Scientific applications of lunar laser ranging, (see 012.012), p. 51 - 52 (1977). – Abstract. Paper is published in J. Geophys. Res., Vol. 81, 6251 - 6256 (1976) – see abstract 18.094.022.

094.006 Free librations of the Moon from lunar laser ranging. O. Calame.
Scientific applications of lunar laser ranging, (see 012.012), p. 53 - 63 (1977).

Analyses of the lunar laser ranging measures, performed at the McDonald Observatory during these last six years, has permitted the improvement of a number of parameters relative to the Earth–Moon system. Particularly because of the high level precision of these observations (on the order of 10 cm in one-way distance), it has now become possible to detect the existence of free librations in the rotational motion of the Moon; a quantitative determination of them is obtained for the three modes of these oscillations of which the amplitudes are found to be: $1.''8$ (longitude), $0.''4$ and $7.''8$ (latitude) for the terms with the respective periods 2.9 years, 27.3 days and 75 years.

094.007 A numerical study of the effects of fourth degree terms in the Earth–Moon mutual potential on lunar physical librations. W. J. Breedlove, Jr.
Scientific applications of lunar laser ranging, (see 012.012), p. 65 - 77 (1977).

An order-of-magnitude calculation shows that some torques arising from previously neglected fourth-degree "figure-figure interaction" terms of the Earth–Moon mutual potential could produce effects on the physical librations that should be modeled for range accuracies of 2 to 3 cm. An investigation of the effect of these "interaction" terms has been undertaken. It is shown that these terms should be included in a lunar physical libration model when accuracies of 0.007 to 0.01 arc seconds (2 cm to 3 cm range precision) are required.

094.008 Analytical theory for the rotation of the Moon.
A. Migus.
Scientific applications of lunar laser ranging, (see 012.012), p. 79 - 86 (1977).

094.009 The role of large bodies in the formation of the earth and moon. G. W. Wetherill.
Proc. Seventh Lunar Sci. Conf., (see 012.015), 3245 - 3257 (1976).

094.010 On the intensity of the ancient lunar magnetic field.
A. Stephenson, D. W. Collinson, S. K. Runcorn.
Proc. Seventh Lunar Sci. Conf., (see 012.015), 3373 - 3382 (1976).

094.011 Entfernungsmessungen zum Mond und die Skizzierung eines hypothesenfreien Äquatorialsystems. H. Rehse.
Sterne, 53. Band, 140 - 145 (1977).

094.012 Numerical estimates of secular effects in the moon's translatory-rotational motion. D. Z. Koenov.
Dokl. AN TadzhSSR, Vol. 20, No. 2, p. 18 - 21 (1977). In Russian. – Abstr. in Ref. zh., 51. Astron., 10.51.165 (1977).

094.013 Primaeval melting of the Moon.
S. K. Runcorn, L. M. Libby, W. F. Libby.
Nature, Vol. 270, 676 - 681 (1977).

There is evidence that the Moon melted completely 4,400 Myr ago, and between 4,000 Myr and 3,200 Myr ago had an internal magnetic field. But gravity could not have provided the heat of melting, and it must have come from short lived radioelements. Theory suggests the transuranics with atomic numbers between 114 and 126 may be relatively stable, and it is shown that these 'superheavy elements' fit the requirements of the early heat source in the moon.

094.014 Lunar ranging experiment ephemerides and the reduction of observations. J. G. Williams.
Philos. Trans. R. Soc. London, Ser. A, Vol. 284, 467 (1977). Abstr. in Phys. Abstr., Vol. 80, Abstr. 56815 (1977).

094.015 Theories of lunar libration. A. H. Cook.
Philos. Trans. R. Soc. London, Ser. A, Vol. 284, 573 - 585 (1977). – Abstr. in Phys. Abstr., Vol. 80, Abstr. 64677 (1977).

094.016   An introductory review of ephemerides for lunar
              laser ranging.   G. A. Wilkins.
Phil. Trans. R. Soc. London, Ser. A, Vol. 284, 461 - 466
(1977) = H. M. Naut. Almanac Office, Libr. Repr. No. 336.

094.017   On the accuracy of computed topocentric distances
              of lunar retroreflectors in view of an application of
lunar laser ranging technique to the determination of the
earth's rotation.   B. Kołaczek.
Lehrstuhl Astron. Phys. Geod. Tech. Univ. München, Mitt.
Nr. 144, 21 pp. (1977).
    Influences of presently known values of errors of differ-
ent parameters defining the lunar physical libration, the
moon's positions and positions of the lunar retroreflectors as
well as of terrestrial stations on measured topocentric distances
of lunar retroreflectors are discussed in view of an application
of LLR technique to the earth's rotation study.

        Bibliography.   See Abstr. 002.055.

        The first visibility of the lunar crescent.
See Abstr. 004.051.

        Mathematical modelling of lunar laser measures
and their application to improvement of physical parameters.
See Abstr. 031.236.

        Scientific expectations in the selenosciences.
See Abstr. 031.237.

        Present scientific achievements from lunar laser
ranging.   See Abstr. 031.238.

        ALSEP-Quasar differential VLBI.
See Abstr. 031.327.

        Stability of the lunar orbit.
See Abstr. 042.008.

        On a family of three-dimensional periodic orbits
around the moon and planets.   See Abstr. 042.023.

        Lunar orbital theory.   See Abstr. 042.029.

        Some problems on the dynamics of the Earth-Moon
system.   See Abstr. 042.033.

        Accuracy obtainable for universal time and polar
motion during the EROLD campaign.
See Abstr. 044.007.

        Moon tables. Phases of the moon for times past,
present, and future 601 B.C. - 2700 A.D. Brahdes maanetabellar.   See Abstr. 047.020.

        Verification of the principle of equivalence for
massive bodies.   See Abstr. 066.080.

        Nach-Newtonsche Korrekturen zur Dynamik des
Systems Erde—Mond und ihre Bedeutung für die relativistischen Gravitationstheorien.   See Abstr. 066.320.

        On the stationary points of the gravitation fields of
the earth, the moon and Mars.   See Abstr. 081.003.

        Positions of the axes of the ellipsoid of inertia from
satellite observations.   See Abstr. 081.021.

        Lunar influence on the occurrence of aurora.
See Abstr. 084.003.

        Lunar volcanism and the origin of the planets.
See Abstr. 091.055.

        A solar origin for the large lunar magnetic field at
4.0×10^9 yr ago?   See Abstr. 094.550.

        Some complexities in the determination of lunar
paleointensities.   See Abstr. 094.551.

Moon, Global Properties

094.101 **Possible effect of subsurface inhomogeneities on the lunar microwave spectrum.**
A. D. Fisher, D. H. Staelin.
Icarus, Vol. 32, 98 - 105 (1977).

Inhomogeneities beneath the lunar surface could alter the average microwave emission spectrum of the Moon in a fashion generally consistent with observations, even in the absence of an average heat flux or density gradients with depth. The lunar subsurface was modeled as an inhomogeneous lossy dielectric with three-dimensional refractive index fluctuations characterized by independent horizontal and vertical correlation lengths. The model suggests that attempts to infer the physical properties of the Moon from the lunar microwave spectrum could be significantly inaccurate if subsurface scattering were neglected.

094.102 **Luna 76: a seismic model.** I. N. Galkin.
Priroda, 1977, No. 7, p. 137 - 139. In Russian.

094.103 **A comparison of the magnetic field anomalies for the Moon and Earth.** L. L. Van'yan, E. G.
Eroshenko, V. N. Lugovenko, B. A. Okulesskij, A. G. Popov, A. L. Kharitonov.
Moon, Vol. 16, 281 - 287, 289 - 294 (1977). In Russian and English.

A comparative analysis of the anomalous magnetic field of the Moon, information about which was obtained by the Apollo 15 subsatellite, and the anomalous magnetic field of the Earth, involving data provided from surveys at various altitudes (up to 500 km) is given. As a result of spectral analysis of these fields it is shown that the main difference of the spectra is in the lower intensity of long period lunar anomalies and the increased rate of their damping with height, which is probably connected with the absence of any kind of magnetization by induction.

094.104 **The Moon's permanent magnetic field: a cratered-shell model.** H. Weiss, P. J. Coleman, Jr.
Moon, Vol. 16, 311 - 315 (1977).

Coleman and Russell (1975) have pointed out that the properties of the larger-scale lunar field may depend, in a rather straight forward manner, upon the geometrical properties of the distribution of the larger craters. In this paper, the authors consider this possibility in further detail. As part of their study of the larger-scale remanent magnetic field of the Moon, they have examined the effects of cratering in an otherwise spherically symmetrical shell magnetized by a concentric dipolar magnetic field H_0 to an intensity of magnetization cH_0, where c is a constant. Finally, they use the locations and diameters of the 10 largest craters on the Moon and the depth-to-diameter ratios of Pike's formulation to model approximately the excavation of the magnetized shell.

094.105 **The interaction of the solar wind with lunar magnetic anomalies.** L. L. Van'yan.
Moon, Vol. 16, 317 - 320, 321 - 324 (1977). In Russian and English.

A simplified model for the interaction of the cold solar wind with lunar magnetic anomalies is considered. Since on the illuminated side of the Moon the dynamic pressure of the solar wind significantly exceeds the magnetic pressure of the anomalies, upward propagation of the lunar field is possible only by means of diffusion. This process does not depend on the velocity but only on the concentration of the solar wind and the characteristic size of anomalies. Theoretical calculations are compared with the data of Apollo 12 and Explorer 35.

094.106 **Latitude effects in lunar thermal evolution.**
R. V. J. Butt, J. A. Bastin.
Moon, Vol. 16, 339 - 347 (1977).

A simple analytical model is developed from which the authors calculated the temperature throughout the lunar interior resulting from internal heat sources and the imposition of surface temperature boundary conditions. The surface temperature is determined almost entirely by the balance of solar heating and surface reradiation; as a consequence this temperature is latitude dependent, decreasing towards the lunar poles. The internal solution shows that the latitude effect exists almost undiminished to great depths within the Moon.

094.107 **Lunar gravity: a mass point model.**
M. P. Ananda.
J. Geophys. Res., Vol. 82, 3049 - 3064 (1977).

A point mass representation of a quasi-global gravity field of the moon is developed by processing Apollo data consisting of 117 point masses distributed over the region of ±30° in latitude about the lunar equator. This model resolves all the previously known 'mascons' in the nearside as broad positive gravity regions. The nearside acceleration map evaluated at 100 km above the lunar surface shows good agreement with the line of sight acceleration results. The lunar farside gravity map shows strong broad positive gravity regions for the highland areas. However, all the major ringed basins are resolved as localized negative anomalies in contrast with the nearside basins. This model does not indicate any evidence for the existence of any mascon type feature in the lunar farside.

094.108 **Lunar gravity: a harmonic analysis.**
A. J. Ferrari.
J. Geophys. Res., Vol. 82, 3065 - 3084 (1977).

A sixteenth-degree and sixteenth-order spherical harmonic lunar gravity field has been derived from the long-term Keplerian variations in the orbits of the Apollo subsatellites and Lunar Orbiter 5. This model resolves the major mascon gravity anomalies of the lunar nearside and is in very good agreement with line of sight acceleration results. The farside map shows the major ringed basins to be strong localized negative anomalies located in broad regions of positive gravity which correspond closely to the highlands.

094.109 **Lunar gravity: a long-term Keplerian rate method.**
A. J. Ferrari, M. P. Ananda.
J. Geophys. Res., Vol. 82, 3085 - 3097 (1977).

Recent reductions of Apollo subsatellite and Lunar Orbiter 5 data have determined the first plausible models for the farside lunar gravity field. This paper presents a selenodesy method which estimates gravity by fitting to the long-term variations of the Kepler element rates. Raw Doppler tracking data taken over short arcs are reduced to estimate a best set of mean orbital elements for each orbit. The element rates are determined from patched cubic spline fits to the elements. The rates are adjusted for n-body effects and along with the associated elements are used as input to a gravity estimator. Simulations performed demonstrate that farside gravity features can successfully be determined by fitting to mean elements derived from nearside tracking. Arguments are presented which conclude that a long-term gravity method of this type is the most plausible technique which can obtain realistic estimates for farside lunar gravity using the currently available data.

094.110 **Orbital gamma ray data and early lunar differentiation.** N. J. Hubbard, D. Woloszyn.
Bull. American Astron. Soc., Vol. 9, 453 (1977). — Abstract.

094.111 **Large magnetic regions in the lunar farside.**
K. A. Anderson, R. Bush, J. Costello, R. Lin.

Bull. American Astron. Soc., Vol. 9, 456 - 457 (1977). Abstract.

094.112 Basaltic magmatism and the bulk composition of the Moon. I. Major and heat-producing elements.
A. E. Ringwood.
Moon, Vol. 16, 389 - 423 (1977).
  Although not identical, the major element composition of the bulk Moon is similar to that of the Earth's mantle. Moreover, this similarity extends to the abundances of the rare earths, uranium, thorium and probably also to many other involatile, lithophile elements. A model of the internal structure of the Moon based exclusively upon petrological-geochemical considerations provides an excellent explanation of the depth-distributions of density and seismic velocities within the Moon.

094.113 Basaltic magmatism and the bulk composition of the Moon. II. Siderophile and volatile elements in Moon, earth and chondrites: implications for lunar origin.
A. E. Ringwood, S. E. Kesson.
Moon, Vol. 16, 425 - 464 (1977).
  Abundance patterns of volatile elements in the Moon differ dramatically from those in ordinary chondrites and from those to be expected from condensation of a nebula of solar composition. These differences imply that the Moon was not formed from components which themselves had condensed directly from the solar nebula. The abundances of a group of siderophile elements Ni, Co, W, Ir, Os, P, S and Se are found to be very similar in ocean-floor tholeiites and low-Ti mare basalts and this similarity is believed to extend to their respective source regions in the earth's mantle and lunar interior.

094.114 Thermal history of lunar magma ocean.
F. Herbert, M. J. Drake, C. P. Sonett, M. J. Wiskerchen.
NASA Tech. Mem., NASA TM X-3511, (see 012.010), p. 31 - 33 (1977). – Abstract.

094.115 Comparative studies of the Moon, Mercury and Mars.   V. R. Oberbeck.
NASA Tech. Mem., NASA TM X-3511, (see 012.010), p. 39 - 41 (1977). – Abstract.

094.116 Origin and relative age of lunar and Mercurian inter-crater plains.   R. G. Strom.
NASA Tech. Mem., NASA TM X-3511, (see 012.010), p. 44 - 46 (1977). – Abstract.

094.117 Lunar and Martian cratering studies, and Mars Mariner 9 geologic mapping.
G. Neukum, H. Fechtig, B. König, K. Hiller, D. U. Wise.
NASA Tech. Mem., NASA TM X-3511, (see 012.010), p. 81 - 84 (1977). – Abstract.

094.118 Rayed craters on the Moon and Mercury.
C. C. Allen.
NASA Tech. Mem., NASA TM X-3511, (see 012.010), p. 85 - 86 (1977). – Abstract.

094.119 Large impact-crater production in the inner solar system.   A. Woronow.
NASA Tech. Mem., NASA TM X-3511, (see 012.010), p. 87 - 90 (1977). – Abstract.

094.120 Compiled system of selenodetic coordinates of 4900 points of the lunar surface (visible side of the moon).
I. V. Gavrilov, V. S. Kislyuk, A. S. Duma.
Naukova dumka, Kiev. 172 pp. Price 90 Kop. (1977). In Russian. – Review in Ref. zh., 51. Astron., 9.51.78 (1977).

094.121 The effect of a heat-isolating layer on the distribution of temperatures in the lunar interior.
A. N. Tikhonov, E. A. Lyubimova, V. K. Vlasov.
Dokl. AN SSSR, Vol. 233, 320 - 322 (1977). In Russian. Abstr. in Ref. zh., 51. Astron., 9.51.323 (1977).

094.122 Geological structure and history of the lunar surface development.   A. T. Bazilevskij.
Izv. AN SSSR. Ser. geol., 1977, No. 5, p. 5 - 19. In Russian. Abstr. in Ref. zh., 51. Astron., 9.51.324 (1977).

094.123 Fortschritte der Mondforschung von 1974 bis 1975.
J. Classen.
Orion, 35. Jahrg., 150 - 152 (1977).

094.124 The lunar lithosphere from electromagnetic-sounding data.   L. L. Van'yan, I. V. Egorov.
Moon, Vol. 17, 3 - 9, 11 - 17 (1977). In English and Russian.
  Four sets of published data are used: frequency dependence of the day-side horizontal magnetic amplification, the same for the dark-side vertical decrease, the day-side transient amplification and the dark-side transient decrease. All experimental results are presented in the form of the day-side frequency response. The summarised apparent resistivity curve is obtained. It corresponds to the model with resistivity about several hundreds of Ωm to the depths of 700–800 km. It suggests the absence of significant amounts of molten material to these depths.

094.125 Optimization of the process of gravimetric surveying on the lunar surface.   V. A. Strel'tsov.
Kosm. Issled., Vol. 15, 792 - 796 (1977). In Russian.

094.126 The escape of solar-wind carbon from the moon.
R. R. Hodges, Jr.
Proc. Seventh Lunar Sci. Conf., (see 012.015), 493 - 500 (1976).

094.127 A two-gas model of the lunar terminator exosphere.
J. L. Benson, J. W. Freeman.
Proc. Seventh Lunar Sci. Conf., (see 012.015), 533 - 541 (1976).

094.128 Lunar surface sputter erosion: a Monte Carlo approach to microcrater erosion and sputter re-deposition.   W. C. Carey, J. A. M. McDonnell.
Proc. Seventh Lunar Sci. Conf., (see 012.015), 913 - 926 (1976).

094.129 Degradation of small mare surface features.
P. H. Schultz, R. Greeley, D. E. Gault.
Proc. Seventh Lunar Sci. Conf., (see 012.015), 985 - 1003 (1976).

094.130 On the evolution rate of small lunar craters.
A. T. Basilevsky (*Bazilevskij*).
Proc. Seventh Lunar Sci. Conf., (see 012.015), 1005 - 1020 (1976).

094.131 Photometric studies of light scattering above the lunar terminator from Apollo solar corona photography.   J. E. McCoy.
Proc. Seventh Lunar Sci. Conf., (see 012.015), 1087 - 1112 (1976).

094.132 Critical review of models for the evolution of high-Ti mare basalts.   M. J. Drake, G. J. Consolmagno.
Proc. Seventh Lunar Sci. Conf., (see 012.015), 1633 - 1657 (1976).

094.133 A dynamic model for mare basalt petrogenesis.
A. E. Ringwood, S. E. Kesson.

Proc. Seventh Lunar Sci. Conf., (see 012.015), 1697 - 1722 (1976).

094.134 Did mare-type volcanism commence early in lunar history?   G. Ryder, G. J. Taylor.
Proc. Seventh Lunar Sci. Conf., (see 012.015), 1741 - 1755 (1976).

094.135 Lunar farside gravity: an assessment of satellite to satellite tracking techniques and gravity gradiometry.
M. Ananda, J. Lorell, W. Flury.
Proc. Seventh Lunar Sci. Conf., (see 012.015), 2623 - 2638 (1976).

094.136 Quantitative mass distribution models for Mare Orientale.   W. L. Sjogren, J. C. Smith.
Proc. Seventh Lunar Sci. Conf., (see 012.015), 2639 - 2648 (1976).

094.137 Surface chemistry of selected lunar regions.
M. J. Bielefeld, R. C. Reedy, A. E. Metzger,
J. I. Trombka, J. R. Arnold.
Proc. Seventh Lunar Sci. Conf., (see 012.015), 2661 - 2676 (1976).

094.138 Lunar surface remanent magnetic fields detected by the electron reflection method.
R. P. Lin, K. A. Anderson, R. Bush, R. E. McGuire,
J. E. McCoy.
Proc. Seventh Lunar Sci. Conf., (see 012.015), 2691 - 2703 (1976).

094.139 Degradation of large, period II lunar craters.
W. W. Mendell.
Proc. Seventh Lunar Sci. Conf., (see 012.015), 2705 - 2716 (1976).

094.140 Ages of flow units in the lunar nearside maria based on Lunar Orbiter IV photographs.
J. M. Boyce.
Proc. Seventh Lunar Sci. Conf., (see 012.015), 2717 - 2728 (1976).

094.141 Geologic structure of the eastern mare basins.
R. A. DeHon, J. D. Waskom.
Proc. Seventh Lunar Sci. Conf., (see 012.015), 2729 - 2746 (1976).

094.142 Modes of emplacement of basalt terrains and an analysis of mare volcanism in the Orientale Basin.
R. Greeley.
Proc. Seventh Lunar Sci. Conf., (see 012.015), 2747 - 2759 (1976).

094.143 Mare ridges and related highland scarps — result of vertical tectonism?   B. K. Lucchitta.
Proc. Seventh Lunar Sci. Conf., (see 012.015), 2761 - 2782 (1976).

094.144 The scarcity of mappable flow lobes on the lunar maria: unique morphology of the Imbrium flows.
G. G. Schaber, J. M. Boyce, H. J. Moore.
Proc. Seventh Lunar Sci. Conf., (see 012.015), 2783 - 2800 (1976).

094.145 The morphological evolution of mare-highland contacts: a potential measure of relative mare surface age.   R. A. Young.
Proc. Seventh Lunar Sci. Conf., (see 012.015), 2801 - 2816 (1976).

094.146 Multiple ring structures and the problem of correlation between lunar basins.   W. J. Brennan.
Proc. Seventh Lunar Sci. Conf., (see 012.015), 2833 - 2843 (1976).

094.147 Dating of individual lunar craters.
G. Neukum, B. König.
Proc. Seventh Lunar Sci. Conf., (see 012.015), 2867 - 2881 (1976).

094.148 Secondary impact craters of lunar basins.
D. E. Wilhelms.
Proc. Seventh Lunar Sci. Conf., (see 012.015), 2883 - 2901 (1976).

094.149 The significance of substrate characteristics in determining morphology and morphometry of lunar craters.   J. W. Head.
Proc. Seventh Lunar Sci. Conf., (see 012.015), 2913 - 2929 (1976).

094.150 Large scale cratering of the lunar highlands: some Monte Carlo model considerations.
F. Hörz, R. V. Gibbons, R. E. Hill, D. E. Gault.
Proc. Seventh Lunar Sci. Conf., (see 012.015), 2931 - 2945 (1976).

094.151 On the origin of fractures radial to lunar basins.
H. J. Melosh.
Proc. Seventh Lunar Sci. Conf., (see 012.015), 2967 - 2982 (1976).

094.152 Candidate areas for in situ ancient lunar materials.
V. R. Oberbeck, R. H. Morrison.
Proc. Seventh Lunar Sci. Conf., (see 012.015), 2983 - 3005 (1976).

094.153 Impact ejecta on the moon.
J. D. O'Keefe, T. J. Ahrens.
Proc. Seventh Lunar Sci. Conf., (see 012.015), 3007 - 3025 (1976).

094.154 Seismic investigation of the lunar interior.
A. M. Dainty, M. N. Toksöz, S. Stein.
Proc. Seventh Lunar Sci. Conf., (see 012.015), 3057 - 3075 (1976).

094.155 Structure of the lunar interior from magnetic field measurements.
P. Dyal, C. W. Parkin, W. D. Daily.
Proc. Seventh Lunar Sci. Conf., (see 012.015), 3077 - 3095 (1976).

094.156 The asymmetric distribution of lunar maria and the earth's gravity.   J. B. Hartung.
Proc. Seventh Lunar Sci. Conf., (see 012.015), 3097 - 3112 (1976).

094.157 Seismic structure of the moon: a summary of current status.   Y. Nakamura, G. V. Latham,
H. J. Dorman, F. K. Duennebier.
Proc. Seventh Lunar Sci. Conf., (see 012.015), 3113 - 3121 (1976).

094.158 Filter processes applied to the scattering parts of lunar seismograms for identifying the 300 km discontinuity and the lunar grid system.
J. Voss, W. Weinrebe, F. Schildknecht, R. Meissner.
Proc. Seventh Lunar Sci. Conf., (see 012.015), 3133 - 3142 (1976).

094.159 **Revised lunar heat-flow values.**
M. G. Langseth, S. J. Keihm, K. Peters.
Proc. Seventh Lunar Sci. Conf., (see 012.015), 3143 - 3171
(1976).

094.160 **Investigation of the moon's thermal history at the**
most probable concentrations of radioactive
elements.   O. I. Ornatskaya, N. M. Tseytlin (*Tsejtlin*),
Ya. I. Al'ber, I. P. Ryazantseva.
Proc. Seventh Lunar Sci. Conf., (see 012.015), 3205 - 3219
(1976).

094.161 **Inferences concerning the early thermal history of**
the moon.   S. K. Runcorn.
Proc. Seventh Lunar Sci. Conf., (see 012.015), 3221 - 3228
(1976).

094.162 **Thermal expansion and thermal stress in the moon**
and terrestrial planets: clues to early thermal
history.   S. C. Solomon, J. Chaiken.
Proc. Seventh Lunar Sci. Conf., (see 012.015), 3229 - 3243
(1976).

094.163 **Magnetic evidence concerning a lunar core.**
B. E. Goldstein, R. J. Phillips, C. T. Russell.
Proc. Seventh Lunar Sci. Conf., (see 012.015), 3321 - 3341
(1976).

094.164 **The geomagnetic dynamos of the Moon and Venus:**
comparisons with a recent scaling law.
C. T. Russell, B. E. Goldstein.
Proc. Seventh Lunar Sci. Conf., (see 012.015), 3343 - 3355
(1976).

094.165 **On the global TRM (***thermoremanent magnetiza-***
tion) **of the lunar lithosphere.**   L. J. Srnka.
Proc. Seventh Lunar Sci. Conf., (see 012.015), 3357 - 3372
(1976).

094.166 **Petrogenesis in a modestly endowed moon.**
N. J. Hubbard, J. W. Minear.
Proc. Seventh Lunar Sci. Conf., (see 012.015), 3421 - 3435
(1976).

094.167 **Geochemical constraints on the composition of**
the moon.   S. R. Taylor.
Proc. Seventh Lunar Sci. Conf., (see 012.015), 3461 - 3477
(1976).

094.168 **Comparison of impact basins on Mercury, Mars**
and the Moon.   C. A. Wood, J. W. Head.
Proc. Seventh Lunar Sci. Conf., (see 012.015), 3629 - 3651
(1976).

094.169 **New lunar cartographic products.**
D. C. Kinsler.
Proc. Seventh Lunar Sci. Conf., (see 012.015), Vol. 3, I - X
(1976).

094.170 **Lunar magnetism.**   S. K. Runcorn.
Highlights of Astronomy, Vol. 4, Part I, (see
012.019), p. 191 - 193 (1977).

094.171 **The electrical conductivity of the moon: an applica-**
tion of inverse theory.   B. A. Hobbs.
Geophys. J. R. Astron. Soc., Vol. 51, 727 - 744 (1977).
Inverse theory of Backus & Gilbert is used to analyse the
day-side electromagnetic response of the moon to magnetic
fluctuations in the solar wind. The data consist of two transfer
functions, both tangential to the lunar surface, and in the
theoretical development the required Fréchet kernels corresponding to these transfer functions are determined. The

ensuing calculations show that the data are sufficiently good
to determine the conductivity down to a depth of about 600
km.

094.172 **Mass fractionation of the lunar surface by solar**
wind sputtering.   Z. E. Switkowski, P. K. Haff,
T. A. Tombrello, D. S. Burnett.
J. Geophys. Res., Vol. 82, 3797 - 3804 (1977).
The sputtering of the lunar surface by the solar wind is
examined as a possible mechanism of mass fractionation.
Simple arguments based on current theories of sputtering and
the ballistics of the sputtered atoms suggest that most ejected
atoms will have sufficiently high energy to escape lunar gravity. The dependence of the calculated results upon the energy
spectrum of sputtered particles is investigated. The authors
conclude that mass fractionation by solar wind sputtering is
likely to be an important phenomenon on the lunar surface
but that the complex isotopic variations observed in lunar
soils cannot be completely explained by this mechanism.

094.173 **Geochemical mapping of the Moon by orbital**
gamma-ray spectroscopy.   R. C. Reedy.
Nuclear cross sections and technology. II. Washington, D.C.,
USA. 3 - 7 March 1975. R. A. Schrack, C. D. Bowman
(Editors). Natl. Bur. Standards, Washington, D.C., USA (1975),
p. 540 - 545. — Abstr. in Phys. Abstr., Vol. 80, Abstr. 25953
(1977).

094.174 **A new source of lunar electromagnetic induction:**
forcing by the diamagnetic cavity.
C. P. Sonett, M. J. Wiskerchen.
Geophys. Res. Lett., Vol. 4, 307 - 310 (1977).

094.175 **The nature of the gravity anomalies associated with**
large young lunar craters.
J. Dvorak, R. J. Phillips.
Geophys. Res. Lett., Vol. 4, 380 - 382 (1977).
The negative Bouguer anomalies (i.e., mass deficiencies)
associated with four young lunar craters are analyzed. Model
calculations based on generalizations made from studies of
terrestrial impact structures suggest that the major contribution to the Bouguer anomaly for these lunar craters is due to a
lens of brecciated material confined within the present crater
rim crest and extending vertically to at least a depth of one-
third the crater rim diameter. Calculations also reveal a
systematic variation in the magnitude of the mass deficiencies
with the cube of the crater diameter.

094.176 **What have we learnt about the Moon?**
S. Durrani.
New Scientist, Vol. 72, 708 - 711 (1976). — Abstr. in Phys.
Abstr., Vol. 80, Abstr. 33373 (1977).

094.177 **Interaction of the surfaces of the Moon and Mercury**
with their exospheric atmospheres.
D. E. Shemansky, A. L. Broadfoot.
Rev. Geophys. Space Phys., Vol. 15, 491 - 499 (1977).
The atmospheres of the Moon and Mercury are controlled
entirely by gas atom-surface interaction. Model calculations
describing the steady state atmospheres have all been based on
the assumption that the atmospheric particle source is a
'saturated' adsorbed surface layer of gas . The authors suggest
that this is in disagreement with what is known of the physics
of gas-surface interaction. It is suggested that some peculiarities
observed in the Mercury He and H atmospheres could be explained by the nature of the gas-surface coupling.

094.178 **Two major igneous events in the evolution of the**
Moon.   G. M. Brown.
Philos. Trans. R. Soc. London, Ser. A, Vol. 286, 439 - 451
(1977). — Abstr. in Phys. Abstr., Vol. 80, Abstr. 87294 (1977).

094.179 On the moments of inertia differences of the Moon.
J. Arkani-Hamed, M. B. Kermani.
Moon, Vol. 17, 167 - 176 (1977).

It is shown that the differences of the moments of inertia of the Moon are, most likely, due to the surface irregularities, the over-all front side mare fillings and the backside topography.

094.180 Inhomogeneous structure of the lunar interior.
I. N. Galkin.
Priroda, 1977, No. 12, p. 111 - 113. In Russian.

094.181 The search for the cause of the low albedo of the
Moon.   T. Gold, E. Bilson, R. L. Baron.
J. Geophys. Res., Vol. 82, 4899 - 4908 (1977).

The effects of different weathering processes on the albedo of the lunar surface cover is discussed. The surface chemical composition of numerous lunar soil and pulverized rock samples was determined by Auger electron spectroscopy. The optical albedo of these samples was also measured.

094.182 Ionosphere and atmosphere of the moon in the geo-
magnetic tail.   W. D. Daily, W. A. Barker, M. Clark,
P. Dyal, C. W. Parkin.
J. Geophys. Res., Vol. 82, 5441 - 5451 (1977).

During the 4-day period when the moon is in the geo-magnetic tail, the principal constituents of the lunar atmosphere are neon and argon. The surface concentrations of neon and argon are calculated from a theoretical model to be 3.9×10^3 and 1.7×10^3, respectively. The lunar atmosphere is ionized by solar ultraviolet radiation, resulting in electrons at a temperature of about $1.5 \times 10^5 \,^\circ K$ and ions at about $370^\circ K$. The electromagnetic properties of the quiescent ionosphere are investigated, and it is concluded that plasma effects on lunar induction studies can be neglected for quiescent conditions in the geomagnetic tail lobes.

094.183 On transport of matter on the lunar surface.
S. V. Viktorov, G. E. Kocharov, N. A. Silin,
V. I. Chesnokov.
Fiz.-tekh. inst. AN SSSR. Prepr. 537. Leningrad, 1977. 17 pp.
In Russian. – Abstr. in Ref. zh., 51. Astron., 11.51.392; 62.
Issled. kosm. prostranstva, 11.62.80 (1977).

094.184 Composite system of selenodetic coordinates of
4900 points of the lunar surface (visible side of the
moon).   I. V. Gavrilov, V. S. Kislyuk, A. S. Duma.
Nauk. dumka, Kiev. 172 pp. Price 90 Kop. (1977). In Russian.
Review in Ref. zh., 52. Geod. Aehrosemka, 11.52.79 (1977).

094.185 Laser location of the moon.   Yu. L. Kokurin.
Tr. Fiz. inst. AN SSSR, Vol. 91, 159 - 225 (1977).
In Russian. – Abstr. in Ref. zh., 52. Geod. Aehrosemka,
11.52.93 (1977).

094.186 Lineament patterns on the Moon, Mars and Mercury.
G. Fielder, R. J. Fryer, P. J. S. Gash, J. L. Whitford-
Stark, L. Wilson.
Proceedings of the first international conference on the new basement tectonics, Salt Lake City, Utah, June 3 - 7, 1974.
Utah Geol. Assoc. Publ. No. 5, p. 379 - 388 (1974).

Topographic lineaments of the Moon, Mars and Mercury form patterns – "grid systems" – which must be interpreted in terms of the respective tectonic histories of these planets. Internal and environmental factors differ between the three planets and examination of the differences between the boundary conditions peculiar to each of the planetary surfaces will reveal the relative importance of contending theories of grid system formation, the most plausible of which theories involve tidal stressing, planetary convection and general body adjustments.

094.187 The ages of the lunar maria and the filling of the
mare basins.   P. A. Davies, A. Stephenson.
Phys. Earth Planet. Inter., Vol. 14, P13 - P16 (1977).

The age discrepancy between lunar highlands and mare is examined in terms of a hydrostatic head model incorporating an impact crater being fed continuously from depth with basaltic magma. It is shown that if the age difference is attributed to the filling time of the impact basin, the dimensions of the volcanic feeder conduits are of the order of a few metres.

094.188 Planetary magnetism and the interiors of the moon
and Mercury.   P. Cassen.
Phys. Earth Planet. Inter., Vol. 15, 113 - 120 (1977).

This paper reviews those aspects of the magnetic properties of the moon and Mercury which are most readily related to the thermal evolutions of these bodies.

094.189 Moon–Mercury: large impact structures, isostasy
and average crustal viscosity.
G. G. Schaber, J. M. Boyce, N. J. Trask.
Phys. Earth Planet. Inter., Vol. 15, 189 - 201 (1977).

094.190 Global seismic effects of basin-forming impacts.
H. G. Hughes, F. N. App, T. R. McGetchin.
Phys. Earth Planet. Inter., Vol. 15, 251 - 263 (1977).

094.191 Measurements of the ultraviolet radiation of the
moon and of the earth's upper atmosphere from
the orbital station Salyut 4.
V. B. Vasil'ev, G. M. Grechko, A. A. Gubarev, B. M. Stol'berg,
V. M. Tijt, N.-R. A. Ehl'met, R. V. Shatskina.
Optical investigations of the emission of the atmosphere, aurorae and noctilucent clouds aboard the orbital scientific station Salyut 4, (see 003.021), p. 145 - 154 (1977). In Russian.

Methods of ultraviolet radiation measurements of the moon and of the earth's horizon from space in the region $\lambda\,0.2-0.3\,\mu m$ and a treatment of experimental results are described. The experimental results are given in comparison with theoretical calculations and previous measurement data.

094.192 Model of the lunar gravitational field according to
observations of motion of its artificial satellites
Luna 10, 12, 14, 19 and 22.   Eh. L. Akim, Z. P. Vlasova.
Dokl. AN SSSR, Vol. 235, 38 - 41 (1977). In Russian. – Abstr.
in Ref. zh., 51. Astron., 12.51.254; 62. Issled. kosm.
prostranstva, 12.62.123 (1977).

094.193 Selenodetic reference network from heliometric
observations of the moon at the Engelhardt Astronomical Observatory.   A. S. Mamakov.
Astron. obs. im. V. P. Ehngel'gardta. Kazan. univ. Kazan',
1977. 43 pp. In Russian. – Abstr. in Ref. zh., 52. Geod.
Aehrosemka, 12.52.95 (1977).

094.194 Some results of cislunar plasma research.
A. S. Vyshlov, N. A. Savich, M. B. Vasilyev
(Vasil'ev), L. N. Samoznaev, A. I. Sidorenko, D. Ya. Shtern.
Solar-wind interaction with the planets Mercury, Venus, and Mars, (see 012.047), p. 81 - 85 (1976).

The main results of plasma cislunar investigations, carried out during Luna-19 and Luna-22 spacecraft flights by means of dual-frequency dispersion interferometry, are briefly outlined. It is shown that a thin layer of plasma, with a height of several tens of kilometers and a maximum concentration of the order 10^3 electrons/cm³ exists above the solar illuminated lunar surface. A physical model of the formation and existence of such a plasma in cislunar space is proposed, taking into account the influence of local magnetic areas on the Moon.

094.195 The new moon – scientific results of eighteen years

of lunar exploration.   E. Burgess.
Mercury, Vol. 6, No. 6, p. 10 - 17 (1977).

094.196 **Luna incognita for 1978.**   J. E. Westfall.
Strolling Astron., Vol. 27, 16 - 19, 20, 21 (1977).

094.197 **Luna incognita: availability and use of outline forms.**
J. E. Westfall.
Strolling Astron., Vol. 27, 19 (1977).

094.198 **Periodicity and forecast of moonquakes.**
G. P. Tamrazyan.
Izv. AN ArmSSR. Nauk.o Zemle, Vol. 30, No. 3, p. 3 - 10
(1977). In Russian. – Abstr. in Ref. zh., 51. Astron., 1.51.312
(1978).

094.199 **Mixing process of the lunar surface materials and
their cosmic-ray exposure age.**
J. Iriyama, M. Honda.
Comets, asteroids, meteorites, (see 012.049), p. 301 - 306
(1977).
     From cosmic ray exposure age data, (time scale $10^7 - 10^8$
years), of lunar surface materials, the authors discuss the
mixing process of the lunar surface layer caused by the
meteoroid impact cratering.

094.200 **To infer rocks of the lunar surface by photographs
of the Moon's surface.**
N. Kumagai, T. Shimazaka.
10th Lunar and Planetary Symposium, (see 012.050), p. 7 - 12
(1977). In Japanese.

094.201 **On the apparent brightness of lunar surface details
observed at the Kharkov Observatory with a proposal of an out-of-atmosphere measurement of the spectral reflectivity of them for identifying rocks of the lunar surface.**
N. Kumagai.
10th Lunar and Planetary Symposium, (see 012.050), p. 13 -
20 (1977).

Ein neuer Mondatlas für Amateure.
See Abstr. 002.027.

Bibliography.   See Abstr. 002.055.

The solar planets.   See Abstr. 003.052.

Structure of the moon.
See Abstr. 003.058.

Moon, Mars and Venus, a concise guide in colour.
See Abstr. 003.144.

**Lunar nomenclature. Far side of the moon, 1961 -
1973.**   See Abstr. 003.145.

**The strength of lunar analogues and its geophysical
implications.**   See Abstr. 022.117.

**Some peculiarities in interpreting space photographs
of the surfaces of the moon and planets.**
See Abstr. 031.275.

**Some peculiarities of interpretation of cosmic
pictures of the lunar and planetary surfaces.**
See Abstr. 031.347.

Lunar height measurements made easy.
See Abstr. 031.405.

**Lunar research after Apollo: are we on the right
track?**   See Abstr. 051.014.

**Diffraction and reflection of radio waves from the
limb of the moon during solar eclipses.**
See Abstr. 077.003.

**High-explosive cratering analogs for bowl-shaped,
central uplift, and multiring impact craters.**
See Abstr. 081.023.

**Observations of the earth's night and twilight
horizons from the orbital station Salyut 4.**
See Abstr. 082.130.

**Gravity gradient mapping from lunar and planetary
orbiters.**   See Abstr. 091.007.

Crater evolutionary tracks.   See Abstr. 091.016.

**The relationship between crustal tectonics and
internal evolution in the moon and Mercury.**
See Abstr. 092.036.

Global tectonics of Mercury and the moon.
See Abstr. 092.037.

**Origin and relative age of lunar and Mercurian
intercrater plains.**   See Abstr. 092.038.

**Moon–Mercury: relative preservation states of
secondary craters.**   See Abstr. 092.039.

Rayed craters on the moon and Mercury.
See Abstr. 092.040.

**A Monte-Carlo simulation of galactic cosmic ray
effects in the lunar regolith.**   See Abstr. 094.422.

Thermal movement of the regolith.
See Abstr. 094.476.

**Some trace element constraints on lunar basalt
genesis.**   See Abstr. 094.499.

**The chemistry, origin and petrogenetic implications
of lunar granite and monzonite.**   See Abstr. 094.501.

Mare basalt genesis: a cumulate-remelting model.
See Abstr. 094.502.

Chronology of the early lunar crust.
See Abstr. 094.519.

Volcanism on Mars and the Moon.
See Abstr. 097.064.

**The effect of nongravitational factors on the shape
of Martian, lunar and Mercurian craters: target effects.**
See Abstr. 097.078.

**Interplanetary comparisons of fresh crater
morphology: preliminary results.**   See Abstr. 097.079.

**Martian crater depth/diameter relationships:
comparison with the Moon and Mercury.**
See Abstr. 097.127.

**Comparison of large crater and multiringed basin
populations on Mars, Mercury, and the Moon.**
See Abstr. 097.128.

**Morphology of the Manicouagan ring-structure,
Quebec, and some comparisons with lunar basins and craters.**
See Abstr. 105.035.

Moon, Local Properties

094.401 **Distribution of 28 elements in size fractions of lunar mare and highlands soils.**
W. V. Boynton, J. T. Wasson.
Geochim. Cosmochim. Acta, Vol. 41, 1073 - 1082 (1977).

094.402 **The distribution of sulfur in lunar rocks and its relationship to carbon content.**
C. B. Moore, J. D. Cripe.
Moon, Vol. 16, 295 - 310 (1977) = Cent. Meteorite Studies, Contrib. No. 105.

A summary of total sulfur abundances representative of the Apollo Missions is presented. Rock mixing models evaluate the distribution of sulfur and define indigenous rock components and extralunar contributions of sulfur in lunar soils. Extralunar sulfur shows a positive correlation with a CC-1 like meteorite component and solar wind derived total carbon content in the Apollo 16 and 17 lunar soils.

094.403 **A multispectral mosaic of northern lunar mare regions using a silicon vidicon.**
T. V. Johnson, J. Mosher, D. L. Matson.
Bull. American Astron. Soc., Vol. 9, 450 (1977). – Abstract.

094.404 **Geologically-morphological analysis of the landing site of Luna 24.**
K. P. Florenskij, A. T. Bazilevskij, G. A. Burba.
Dokl. AN SSSR, Vol. 233, 936 - 943 (1977). In Russian.
Abstr. in Ref. zh., 62. Issled. kosm. prostranstva, 8.62.155 (1977).

094.405 **Preliminary description of a column of lunar soil supplied from Mare Crisium by space apparatus Luna 24.**   A. V. Ivanov, M. A. Nazarov, O. D. Rodeh, Yu. I. Stakheev, L. S. Tarasov, K. I. Tobelko, K. P. Florenskij.
Dokl. AN SSSR, Vol. 233, 928 - 931 (1977). In Russian.
Abstr. in Ref. zh., 62. Issled. kosm. prostranstva, 8.62.156 (1977).

094.406 **Isotopes of rare gases in regolith and its origin.**
L. K. Levskij, I. M. Morozova.
Probl. datirovaniya dokembrijsk. obrazovanij. Leningrad, Nauka, 1977, p. 198 - 210. In Russian. – Abstr. in Ref. zh., 62. Issled. kosm. prostranstva, 8.62.172 (1977).

094.407 **Lunar surface chemistry: a new imaging technique.**
C. G. Andre, M. J. Bielefeld, E. Eliason, L. A. Soderblom, I. Adler, J. A. Philpotts.
Science, Vol. 197, 986 - 989 (1977).

Detailed chemical maps of the lunar surface have been constructed by applying a new weighted-filter imaging technique to Apollo 15 and Apollo 16 X-ray fluorescence data. The data quality improvement is amply demonstrated by (1) modes in the frequency distribution, representing highland and mare soil suites, which are not evident before data filtering and (2) numerous examples of chemical variations which are correlated with small-scale (about 15 kilometer) lunar topographic features.

094.408 **The history of lunar breccia 15015.**
The European Consortium and friends.
Lunar sample studies, (see 003.007), p. 1 - 33 (1977).

094.409 **Composition of lunar basalts 10069, 10071, and 12008.**   E. Jarosewich, B. Mason.
Lunar sample studies, (see 003.007), p. 35 - 40 (1977).

094.410 **Composition of eight Apollo 17 basalts.**
B. Mason, E. Jarosewich, S. Jacobson, G. Thompson.
Lunar sample studies, (see 003.007), p. 41 - 47 (1977).

094.411 **Troctolitic and basaltic clasts from a Fra Mauro breccia.**   N. G. Ware, D. H. Green.
Lunar sample studies, (see 003.007), p. 49 - 59 (1977).

094.412 **U-Th-Pb systematics of Apollo 16 samples 60018, 60025, and 64435; and the continuing problem of terrestrial Pb contamination of lunar samples.**
P. D. Nunes, D. M. Unruh, M. Tatsumoto.
Lunar sample studies, (see 003.007), p. 61 - 69 (1977).

094.413 **Application of the method of exoelectronic emission for investigation of the physical constitution of the surface of fragments of lunar matter supplied by the Soviet automatic interplanetary stations Luna 16 and Luna 20.**
R. I. Mints, I. I. Mil'man, V. I. Kryuk, L. S. Tarasov.
Fiz. metody issled. tverd. tela. Vyp. (No.) 2, Sverdlovsk, 1977, p. 95 - 104. In Russian. – Abstr. in Ref. zh., 62. Issled. kosm. prostranstva, 9.62.122 (1977).

094.414 **A possible lunar outcrop: a study of Lunokhod-2 data.**   A. T. Basilevsky (*Bazilevskij*),C. P. Florensky (*K. P. Florenskij*), L. B. Ronca.
Moon, Vol. 17, 19 - 28 (1977).

The remotely controlled vehicle Lunokhod-2 travelled extensively around the edges of a linear depression unofficially called Fossa Recta. The edges of the Fossa are marked by elongated boulder fields. Three lines of reasoning suggest that the boulder fields are not the usual 'erratic' boulders found on a normal mare surface, but are bedrock protuberances: (1) The morphology of many boulders is reminiscent of primary lava features, (2) toward the edge of the Fossa the regolith thins out; (3) a model of lunar 'gardening' indicates that no regolith is to be expected in the upper portion of a non-impact cliff.

094.415 **Détermination de la position du cratère Mösting A à partir de plaques photographiques de la lune sur fond d'étoiles.**   M. Froeschlé.
Moon, Vol. 17, 47 - 57 (1977).

The reduction of 126 photographic plates from the Moon on stellar background, gives a new determination from the Mösting A coordinates with respect to the lunar inertial system. This determination is based on: the Eckhardt theory for the physical libration, and the ephemeride ($j = 2$) derived from Brown's theory. The S.A.O. star catalog for the star positions, is used for this reduction.

094.416 **The Aristarchus-Harbinger region of the Moon: surface geology and history from recent remote-sensing observations.**   S. H. Zisk, C. A. Hodges, H. J. Moore, R. W. Shorthill, T. W. Thompson, E. A. Whitaker, D. E. Wilhelms.
Moon, Vol. 17, 59 - 99 (1977).

This study is based principally on photographic and remote-sensing observations made from Earth and Apollo orbiting space craft. Results include (1) delineation of geologic map units and their stratigraphic relationships; (2) discussion of the complex interrelationships between materials of volcanic and impact origin, including the effects of excavation, redistribution and mixing of previously deposited materials by younger impact craters; (3) deduction of physical and chemical properties of certain of the geologic units, based on both the remote-sensing information and on extrapolation of Apollo data to this area; and (4) development of a detailed geologic history of the region, outlining the probable sequence of events that resulted in its present appearance.

094.417 **Ferromagnetic resonance and magnetic studies of cores 60009/60010 and 60003: compositional and surface-exposure stratigraphy.**
R. V. Morris, W. A. Gose.
Proc. Seventh Lunar Sci. Conf., (see 012.015), 1 - 11 (1976).

094.418 Ferromagnetic resonance studies of lunar core
          stratigraphy.
R. M. Housley, E. H. Cirlin, I. B. Goldberg, H. Crowe.
Proc. Seventh Lunar Sci. Conf., (see 012.015), 13 - 26 (1976).

094.419 Recent and long-term mixing of the lunar regolith
          based on ^{22}Na and ^{26}Al measurements in Apollo 15,
16, and 17 deep drill stems and drive tubes.
J. S. Fruchter, L. A. Rancitelli, R. W. Perkins.
Proc. Seventh Lunar Sci. Conf., (see 012.015), 27 - 39 (1976).

094.420 ^{53}Mn in the Apollo 15 and 16 drill stems: evidence
          for surface mixing.   K. Nishiizumi, M. Imamura,
M. Honda, G. P. Russ III, C. P. Kohl, J. R. Arnold.
Proc. Seventh Lunar Sci. Conf., (see 012.015), 41 - 54 (1976).

094.421 Microstratigraphy of the lunar regolith and com-
          paction ages of lunar breccias.
J. N. Goswami, D. Braddy, P. B. Price.
Proc. Seventh Lunar Sci. Conf., (see 012.015), 55 - 74 (1976).

094.422 A Monte-Carlo simulation of galactic cosmic ray
          effects in the lunar regolith.
Y. Langevin, M. Maurette.
Proc. Seventh Lunar Sci. Conf., (see 012.015), 75 - 91 (1976).

094.423 Petrographic and ferromagnetic resonance studies
          of the Apollo 15 deep drill core.
G. H. Heiken, R. V. Morris, D. S. McKay, R. M. Fruland.
Proc. Seventh Lunar Sci. Conf., (see 012.015), 93 - 111 (1976)

094.424 Radiographic enhancement analysis of sedimentary
          structures and depositional history in Apollo 15
core 15011.   N. K. Coch.
Proc. Seventh Lunar Sci. Conf., (see 012.015), 113 - 122
(1976).

094.425 Origin of the Apollo 17 deep drill coarse-grained
          layer.   G. Crozaz, A. L. Plachy.
Proc. Seventh Lunar Sci. Conf., (see 012.015), 123 - 131
(1976).

094.426 The remarkable chemical uniformity of Apollo 16
          layered deep drill core section 60002.
D. F. Nava, M. M. Lindstrom, P. J. Schuhmann, D. J. Lindstrom, J. A. Philpotts.
Proc. Seventh Lunar Sci. Conf., (see 012.015), 133 - 139
(1976).

094.427 Stratigraphy in Apollo 16 drill section 60002.
          G. E. Blanford, D. A. Morrison.
Proc. Seventh Lunar Sci. Conf., (see 012.015), 141 - 154
(1976).

094.428 The Apollo 16 drill core: statistical analysis of glass
          chemistry and the characterization of a high
alumina-silica poor (HASP) glass.
M. T. Naney, D. M. Crowl, J. J. Papike.
Proc. Seventh Lunar Sci. Conf., (see 012.015), 155 - 184
(1976).

094.429 Mineral, lithic, and glass clasts < 1 mm size in
          Apollo 16 core section 60003.
H. O. A. Meyer, R. H. McCallister.
Proc. Seventh Lunar Sci. Conf., (see 012.015), 185 - 198
(1976).

094.430 The Apollo 16 drill core: modal petrology and
          characterization of the mineral and lithic com-
ponent.
D. T. Vaniman, S. F. Lellis, J. J. Papike, K. L. Cameron.
Proc. Seventh Lunar Sci. Conf., (see 012.015), 199 - 239

(1976).

094.431 Chemical characterization of lunar core 60009.
          M. Z. Ali, W. D. Ehmann.
Proc. Seventh Lunar Sci. Conf., (see 012.015), 241 - 258
(1976).

094.432 Noble gases in 60009-60010 drive tube samples:
          trapped gases and irradiation history.
D. D. Bogard, W. C. Hirsch.
Proc. Seventh Lunar Sci. Conf., (see 012.015), 259 - 279
(1976).

094.433 Drive tube 60009: a chemical study of magnetic
          separates of size fractions from five strata.
D. P. Blanchard, J. W. Jacobs, J. C. Brannon, R. W. Brown.
Proc. Seventh Lunar Sci. Conf., (see 012.015), 281 - 294 (1976).

094.434 Comparative studies of grain size separates of
          60009.   D. S. McKay, R. V. Morris, M. A.
Dungan, R. M. Fruland, R. Fuhrman.
Proc. Seventh Lunar Sci. Conf., (see 012.015), 295 - 313
(1976).

094.435 Surface exposure indices of lunar soils: a com-
          parative FMR study.   R. V. Morris.
Proc. Seventh Lunar Sci. Conf., (see 012.015), 315 - 335
(1976).

094.436 Agglutinates and carbon accumulation in Apollo
          17 lunar soils.   A. Basu, W. G. Meinschein.
Proc. Seventh Lunar Sci. Conf., (see 012.015), 337 - 349
(1976).

094.437 Lead isotopic studies of lunar soils: their bearing
          on the time scale of agglutinate formation.
S. E. Church, G. R. Tilton, J. H. Chen.
Proc. Seventh Lunar Sci. Conf., (see 012.015), 351 - 371
(1976).

094.438 Microimpact-induced changes of textural parameters
          and modal composition of the lunar regolith.
W. v. Engelhardt, H. Hurrle, E. Luft.
Proc. Seventh Lunar Sci. Conf., (see 012.015), 373 - 392
(1976).

094.439 Chemical aspects of agglutinate formation: rela-
          tionships between agglutinate composition and the
composition of the bulk soil.   W. N. Via, L. A. Taylor.
Proc. Seventh Lunar Sci. Conf., (see 012.015), 393 - 403
(1976).

094.440 The chemistry of some individual lunar soil ag-
          glutinates.
R. V. Gibbons, F. Hörz, R. B. Schaal.
Proc. Seventh Lunar Sci. Conf., (see 012.015), 405 - 422
(1976).

094.441 Volatile element depletion and ^{39}K/^{41}K fractiona-
          tion in lunar soils.
S. E. Church, G. R. Tilton, J. E. Wright, C.-N. Lee-Hu.
Proc. Seventh Lunar Sci. Conf., (see 012.015), 423 - 439
(1976).

094.442 Characterization of lunar nitrogen components.
          R. H. Becker, R. N. Clayton, T. K. Mayeda.
Proc. Seventh Lunar Sci. Conf., (see 012.015), 441 - 458
(1976).

094.443 Sulphur isotopes in grain size fractions of lunar
          soils.   H. G. Thode, C. E. Rees.
Proc. Seventh Lunar Sci. Conf., (see 012.015), 459 - 468

(1976).

094.444 **Extralunar sulfur in Apollo 16 and 17 lunar fines.**
J. D. Cripe, C. B. Moore.
Proc. Seventh Lunar Sci. Conf., (see 012.015), 469 - 479
(1976).

094.445 **Light element geochemistry of the Apollo 15 site.**
I. R. Kaplan, J. F. Kerridge, C. Petrowski.
Proc. Seventh Lunar Sci. Conf., (see 012.015), 481 - 492
(1976).

094.446 **The long-term average of the galactic cosmic-ray**
iron group composition studied by the track
method. W. Krätschmer, W. Gentner.
Proc. Seventh Lunar Sci. Conf., (see 012.015), 501 - 511
(1976).

094.447 **Solar proton fluxes during the last million years.**
N. Bhandari, S. K. Bhattacharya, J. T. Padia.
Proc. Seventh Lunar Sci. Conf., (see 012.015), 513 - 523 (1976).

094.448 **Solar wind ^3H and ^{14}C abundances and solar surface**
processes.
E. L. Fireman, J. DeFelice, J. D'Amico.
Proc. Seventh Lunar Sci. Conf., (see 012.015), 525 - 531
(1976).

094.449 **Microcraters and solar flare tracks in crystals from**
carbonaceous chondrites and lunar breccias.
J. N. Goswami, I. D. Hutcheon, J. D. Macdougall.
Proc. Seventh Lunar Sci. Conf., (see 012.015), 543 - 562
(1976).

094.450 **Noble gases in the Apollo 16 special soils from the**
East-West split and the permanently shadowed area.
P. Eberhardt, O. Eugster, J. Geiss, N. Grögler, S. Guggisberg,
M. Mörgeli.
Proc. Seventh Lunar Sci. Conf., (see 012.015), 563 - 585
(1976).

094.451 **Atmospheric rare gases in lunar rock 60015.**
S. Niemeyer, D. A. Leich.
Proc. Seventh Lunar Sci. Conf., (see 012.015), 587 - 597
(1976).

094.452 **The excess fission xenon problem in lunar samples.**
R. J. Drozd, B. M. Kennedy, C. J. Morgan, F. A.
Podosek, G. J. Taylor.
Proc. Seventh Lunar Sci. Conf., (see 012.015), 599 - 623
(1976).

094.453 **K/Ar dating of lunar soils II.**
E. C. Alexander, Jr., A. Bates, M. R. Coscio, Jr.,
J. C. Dragon, V. R. Murthy, R. O. Pepin, T. R. Venkatesan.
Proc. Seventh Lunar Sci. Conf., (see 012.015), 625 - 648
(1976).

094.454 **Spallation deuterium in rock 70215.**
L. Merlivat, M. Lelu, G. Nief, E. Roth.
Proc. Seventh Lunar Sci. Conf., (see 012.015), 649 - 658
(1976).

094.455 **Petrography of KREEP basalt fragments from**
Apollo 15 soils. A. Basu, J. F. Bower.
Proc. Seventh Lunar Sci. Conf., (see 012.015), 659 - 678
(1976).

094.456 **Major element composition of glasses from Apollo**
11, 16, and 17 soil samples. B. P. Glass.
Proc. Seventh Lunar Sci. Conf., (see 012.015), 679 - 693
(1976).

094.457 **Geochemistry of grain-size fractions of soils from**
the Taurus-Littrow valley floor.
R. L. Korotev.
Proc. Seventh Lunar Sci. Conf., (see 012.015), 695 - 726
(1976).

094.458 **Lithophiles, siderophiles, and volatiles in Apollo 16**
soils and rocks. W. V. Boynton, C.-L. Chou,
K. L. Robinson, P. H. Warren, J. T. Wasson.
Proc. Seventh Lunar Sci. Conf., (see 012.015), 727 - 742
(1976).

094.459 **Chondrule-like particles from Luna 16 and Luna 20**
regolith samples. A. V. Ivanov, M. A. Nazarov,
O. D. Rode, I. D. Shevaleevski (*Shevaleevskij*).
Proc. Seventh Lunar Sci. Conf., (see 012.015), 743 - 757
(1976).

094.460 **Ferromagnetic-superparamagnetic granulometry of**
lunar surface materials.
F. C. Schwerer, T. Nagata.
Proc. Seventh Lunar Sci. Conf., (see 012.015), 759 - 778
(1976).

094.461 **Relationship between nickel and metallic iron**
contents of Apollo 16 and 17 soils.
C.-L. Chou, G. W. Pearce.
Proc. Seventh Lunar Sci. Conf., (see 012.015), 779 - 789
(1976).

094.462 **An experimental study of phosphate reduction and**
phosphorus-bearing lunar metal particles.
J. J. Friel, J. I. Goldstein.
Proc. Seventh Lunar Sci. Conf., (see 012.015), 791 - 806
(1976).

094.463 **Carbides in lunar soils and rocks.**
J. I. Goldstein, R. H. Hewins, A. D. Romig, Jr.
Proc. Seventh Lunar Sci. Conf., (see 012.015), 807 - 818
(1976).

094.464 **The relation of metal composition to rock type**
for clasts in Apollo 16 soils.
R. H. Hewins, J. I. Goldstein, H. J. Axon.
Proc. Seventh Lunar Sci. Conf., (see 012.015), 819 - 836
(1976).

094.465 **Subsolidus reequilibration, grain growth, and com-**
positional changes of native FeNi metal in lunar
rocks. L. A. Taylor, K. C. Misra, B. M. Walker.
Proc. Seventh Lunar Sci. Conf., (see 012.015), 837 - 856
(1976).

094.466 **Lunar metallic phase: compositional variation in**
response to disequilibrium in regolith melting
processes. H. K. Mao, P. M. Bell.
Proc. Seventh Lunar Sci. Conf., (see 012.015), 857 - 862
(1976).

094.467 **ESCA studies of the surface chemistry of lunar**
fines. R. M. Housley, R. W. Grant.
Proc. Seventh Lunar Sci. Conf., (see 012.015), 881 - 889
(1976).

094.468 **The surface chemical composition of lunar samples**
and its significance for optical properties.
T. Gold, E. Bilson, R. L. Baron.
Proc. Seventh Lunar Sci. Conf., (see 012.015), 901 - 911
(1976).

094.469 **The surface structure and composition of**
60017,43. D. A. Cadenhead, M. G. Brown.

Proc. Seventh Lunar Sci. Conf., (see 012.015), 927 - 936 (1976).

094.470· Rare gas ion probe analysis of helium profiles in individual lunar soil particles.   H. W. Müller, J. Jordan, S. Kalbitzer, J. Kiko, T. Kirsten.
Proc. Seventh Lunar Sci. Conf., (see 012.015), 937 - 951 (1976).

094.471 Ion probe analysis of artificially implanted ions in terrestrial samples and surface enhanced ions in lunar sample 76215,77.
E. Zinner, R. M. Walker, J. Chaumont, J. C. Dran.
Proc. Seventh Lunar Sci. Conf., (see 012.015), 953 - 984 (1976).

094.472 Density and chemistry of interplanetary dust particles, derived from measurements of lunar microcraters.   K. Nagel, G. Neukum, J. S. Dohnanyi, H. Fechtig, W. Gentner.
Proc. Seventh Lunar Sci. Conf., (see 012.015), 1021 - 1029 (1976).

094.473 Microcraters on lunar rocks.
J.-C. Mandeville.
Proc. Seventh Lunar Sci. Conf., (see 012.015), 1031 - 1038 (1976).

094.474 Shock metamorphic effects in lunar microcraters.
R. B. Schaal, F. Hörz, R. V. Gibbons.
Proc. Seventh Lunar Sci. Conf., (see 012.015), 1039 - 1054 (1976).

094.475 The micrometeoroid impact crater comminution distribution and accretionary populations on lunar rocks: experimental measurements.
J. A. M. McDonnell, R. P. Flavill, W. C. Carey.
Proc. Seventh Lunar Sci. Conf., (see 012.015), 1055 - 1072 (1976).

094.476 Thermal movement of the regolith.
F. Duennebier.
Proc. Seventh Lunar Sci. Conf., (see 012.015), 1073 - 1086 (1976).

094.477 The kinetics of lunar glass formation, revisited.
L. C. Klein, D. R. Uhlmann.
Proc. Seventh Lunar Sci. Conf., (see 012.015), 1113 - 1121 (1976).

094.478 Derivation of the thermal history of tektites and lunar glasses from their thermal expansion characteristics.   J. Arndt, N. Rombach.
Proc. Seventh Lunar Sci. Conf., (see 012.015), 1123 - 1141 (1976).

094.479 Shock-induced fine-grained recrystallization of olivine: evidence against subsolidus reduction of Fe^{2+}.   T. J. Ahrens, F.-D. Tsay, D. H. Live.
Proc. Seventh Lunar Sci. Conf., (see 012.015), 1143 - 1156 (1976).

094.480 Temperature and duration of some Apollo 17 boulder shadows.
S. A. Durrani, K. A. R. Khazal, A. Ali.
Proc. Seventh Lunar Sci. Conf., (see 012.015), 1157 - 1177 (1976).

094.481 Chromium in basalts: experimental determination of redox states and partitioning among synthetic silicate phases.   H. D. Schreiber, L. A. Haskin.
Proc. Seventh Lunar Sci. Conf., (see 012.015), 1221 - 1259 (1976).

094.482 Zoning in spinels as an indicator of the crystallization histories of mare basalts.
A. El Goresy, M. Prinz, P. Ramdohr.
Proc. Seventh Lunar Sci. Conf., (see 012.015), 1261 - 1279 (1976).

094.483 Composition and origin of Luna 16 aluminous mare basalts.   G. Kurat, A. Kracher, K. Keil, R. Warner, M. Prinz.
Proc. Seventh Lunar Sci. Conf., (see 012.015), 1301 - 1321 (1976).

094.484 Oxygen fugacity and other thermochemical parameters of Apollo 17 high-Ti basalts and their implications on the reduction mechanism.   M. Sato.
Proc. Seventh Lunar Sci. Conf., (see 012.015), 1323 - 1344 (1976).

094.485 The phase relations, textures, and mineral chemistries of high-titanium mare basalts as a function of oxygen fugacity and cooling rate.
T. M. Usselman, G. E. Lofgren.
Proc. Seventh Lunar Sci. Conf., (see 012.015), 1345 - 1363 (1976).

094.486 Differentiation of an Apollo 12 picrite magma.
D. Walker, J. Longhi, R. J. Kirkpatrick, J. F. Hays.
Proc. Seventh Lunar Sci. Conf., (see 012.015), 1365 - 1389 (1976).

094.487 Variations in chemical composition of Apollo 15 mare basalts.   J. C. Butler.
Proc. Seventh Lunar Sci. Conf., (see 012.015), 1429 - 1447 (1976).

094.488 Compositional interrelationships of mare basalts from bulk chemical and melt inclusion studies.
P. W. Weiblen, E. Roedder.
Proc. Seventh Lunar Sci. Conf., (see 012.015), 1449 - 1466 (1976).

094.489 Chemistry, classification, and petrogenesis of Apollo 17 mare basalts.   J. M. Rhodes, N. J. Hubbard, H. Wiesmann, K. V. Rodgers, J. C. Brannon, B. M. Bansal.
Proc. Seventh Lunar Sci. Conf., (see 012.015), 1467 - 1489 (1976).

094.490 Sulfur in the Apollo 17 basalts and their source regions.   E. K. Gibson, Jr., T. M. Usselman, R. V. Morris.
Proc. Seventh Lunar Sci. Conf., (see 012.015), 1491 - 1505 (1976).

094.491 Sr isotopic constraints on the petrogenesis of Apollo 17 mare basalts.   L. E. Nyquist, B. M. Bansal, H. Wiesmann.
Proc. Seventh Lunar Sci. Conf., (see 012.015), 1507 - 1528 (1976).

094.492 Rb—Sr ages and isotopic systematics of some Serenitatis mare basalts.
V. R. Murthy, M. R. Coscio, Jr.
Proc. Seventh Lunar Sci. Conf., (see 012.015), 1529 - 1544 (1976).

094.493 Thorium and uranium variations in Apollo 17 basalts, and K—U systematics.
J. C. Laul, J. S. Fruchter.
Proc. Seventh Lunar Sci. Conf., (see 012.015), 1545 - 1559 (1976).

094.494 Sulfur prevails in coatings on glass droplets: Apollo
15 green and brown glasses and Apollo 17 orange
and black (devitrified) glasses.   P. Butler, Jr., C. Meyer, Jr.
Proc. Seventh Lunar Sci. Conf., (see 012.015), 1561 - 1581
(1976).

094.495 Volatile compounds released during lunar lava
fountaining.   J. T. Wasson, W. V. Boynton,
G. W. Kallemeyn, L. L. Sundberg, C. M. Wai.
Proc. Seventh Lunar Sci. Conf., (see 012.015), 1583 - 1595
(1976).

094.496 Fluorine as a constituent in lunar magnetic gases.
R. H. Goldberg, T. A. Tombrello, D. S. Burnett.
Proc. Seventh Lunar Sci. Conf., (see 012.015), 1597 - 1613
(1976).

094.497 Nitrogen in lunar igneous rocks.
O. Müller, E. Grallath, G. Tölg.
Proc. Seventh Lunar Sci. Conf., (see 012.015), 1615 - 1622
(1976).

094.498 Alkali mobility in shocked basalt.
R. Jeanloz, T. J. Ahrens.
Proc. Seventh Lunar Sci. Conf., (see 012.015), 1623 - 1632
(1976).

094.499 Some trace element constraints on lunar basalt
genesis.   A. R. Duncan, A. J. Erlank, M. K. Sher,
Y. C. Abraham, J. P. Willis, L. H. Ahrens.
Proc. Seventh Lunar Sci. Conf., (see 012.015), 1659 - 1671
(1976).

094.500 Chemical constraints for mare basalt genesis.
M.-S. Ma, A. V. Murali, R. A. Schmitt.
Proc. Seventh Lunar Sci. Conf., (see 012.015), 1673 - 1695
(1976).

094.501 The chemistry, origin and petrogenetic implications
of lunar granite and monzonite.
M. J. Rutherford, P. C. Hess, F. J. Ryerson, H. W. Campbell,
P. A. Dick.
Proc. Seventh Lunar Sci. Conf., (see 012.015), 1723 - 1740
(1976).

094.502 Mare basalt genesis: a cumulate-remelting model.
C.-y. Shih, E. Schonfeld.
Proc. Seventh Lunar Sci. Conf., (see 012.015), 1757 - 1792
(1976).

094.503 Petrology of 79215: brecciation of lunar cumulate.
C. E. Bickel, J. L. Warner, W. C. Phinney.
Proc. Seventh Lunar Sci. Conf., (see 012.015), 1793 - 1819
(1976).

094.504 Apollo 17 grey breccias and crustal composition in
the Serenitatis Basin region.
M. R. Dence, R. A. F. Grieve, A. G. Plant.
Proc. Seventh Lunar Sci. Conf., (see 012.015), 1821 - 1832
(1976) = Contrib. Earth Phys. Branch, Dep. Energy, Mines,
Resources, Ottawa, Canada, No. 618.

094.505 Meteorite-free Apollo 15 crystalline KREEP.
E. Dowty, K. Keil, M. Prinz, J. Gros, H. Takahashi.
Proc. Seventh Lunar Sci. Conf., (see 012.015), 1833 - 1844
(1976).

094.506 Deformation, recovery and recrystallization of lunar
dunite 72417.   J. S. Lally, J. M. Christie,
G. L. Nord, Jr., A. H. Heuer.
Proc. Seventh Lunar Sci. Conf., (see 012.015), 1845 - 1863
(1976).

094.507 X-ray diffraction profiles and exsolution history of
pigeonite.   H. Nakazawa, S. S. Hafner.
Proc. Seventh Lunar Sci. Conf., (see 012.015), 1865 - 1873
(1976).

094.508 76535: thermal history deduced from pyroxene
precipitation in anorthite.   G. L. Nord, Jr.
Proc. Seventh Lunar Sci. Conf., (see 012.015), 1875 - 1888
(1976).

094.509 Pyroxene-spinel intergrowths in lunar and terrestrial
pyroxenes.   F. P. Okamura, I. S. McCallum,
J. M. Stroh, S. Ghose.
Proc. Seventh Lunar Sci. Conf., (see 012.015), 1889 - 1899
(1976).

094.510 Microcracks, micropores, and their petrologic interpretation for 72415 and 15418.
D. Richter, G. Simmons, R. Siegfried.
Proc. Seventh Lunar Sci. Conf., (see 012.015), 1901 - 1923
(1976).

094.511 Poikilitic KREEP impact melts in the Apollo 14
white rocks.   G. Ryder, J. F. Bower.
Proc. Seventh Lunar Sci. Conf., (see 012.015), 1925 - 1948
(1976).

094.512 Mineralogy and petrology of complex breccia
14063,14.   I. M. Steele, J. V. Smith.
Proc. Seventh Lunar Sci. Conf., (see 012.015), 1949 - 1964
(1976).

094.513 Grain size statistics, composition, and provenance of
fragmental particles in some Apollo 14 breccias.
D. Stöffler, H.-D. Knöll, W.-U. Reimold, S. Schulien.
Proc. Seventh Lunar Sci. Conf., (see 012.015), 1965 - 1985
(1976).

094.514 The realities of recoil: ^{39}Ar recoil out of small grains
and anomalous age patterns in ^{39}Ar–^{40}Ar dating.
J. C. Huneke, S. P. Smith.
Proc. Seventh Lunar Sci. Conf., (see 012.015), 1987 - 2008
(1976).

094.515 History and genesis of lunar troctolite 76535 or:
how old is old?
G. W. Lugmair, K. Marti, J. P. Kurtz, N. B. Scheinin.
Proc. Seventh Lunar Sci. Conf., (see 012.015), 2009 - 2033
(1976).

094.516 Rb–Sr age or troctolite 76535.
D. A. Papanastassiou, G. J. Wasserburg.
Proc. Seventh Lunar Sci. Conf., (see 012.015), 2035 - 2054
(1976).

094.517 Laser probe ^{39}Ar–^{40}Ar ages of individual mineral
grains in lunar basalt 15607 and lunar breccia 15465.
T. Plieninger, O. A. Schaeffer.
Proc. Seventh Lunar Sci. Conf., (see 012.015), 2055 - 2066
(1976).

094.518 Ages of highland rocks: the chronology of lunar
basin formation revisited.
O. A. Schaeffer, L. Husain, G. A. Schaeffer.
Proc. Seventh Lunar Sci. Conf., (see 012.015), 2067 - 2092
(1976).

094.519 Chronology of the early lunar crust.
E. Schonfeld.
Proc. Seventh Lunar Sci. Conf., (see 012.015), 2093 - 2105
(1976).

094.520   KREEP basalt age: grain by grain U–Th–Pb systematics study of the quartz monzodiorite clast 15405,88.   M. Tatsumoto, D. M. Unruh.
Proc. Seventh Lunar Sci. Conf., (see 012.015), 2107 - 2129 (1976).

094.521   Consortium studies of light-gray breccia 73215: introduction, subsample distribution data, and summary of results.   O. B. James, D. P. Blanchard.
Proc. Seventh Lunar Sci. Conf., (see 012.015), 2131 - 2143 (1976).

094.522   Petrology of aphanitic lithologies in consortium breccia 73215.   O. B. James.
Proc. Seventh Lunar Sci. Conf., (see 012.015), 2145 - 2178 (1976).

094.523   Major and trace element compositions of matrix and aphanitic clasts from consortium breccia 73215.
D. P. Blanchard, J. W. Jacobs, J. C. Brannon, L. A. Haskin.
Proc. Seventh Lunar Sci. Conf., (see 012.015), 2179 - 2187 (1976).

094.524   Lunar breccia 73215: siderophile and volatile trace elements.   J. W. Morgan, J. Gros, H. Takahashi, J. Hertogen.
Proc. Seventh Lunar Sci. Conf., (see 012.015), 2189 - 2199 (1976).

094.525   Argon–argon ages of consortium breccia 73215.
E. K. Jessberger, T. Kirsten, T. Staudacher.
Proc. Seventh Lunar Sci. Conf., (see 012.015), 2201 - 2215 (1976).

094.526   The magnetic characteristics of highland breccia 73215: evidence for textural control of magnetization.   A. Brecher.
Proc. Seventh Lunar Sci. Conf., (see 012.015), 2217 - 2231 (1976).

094.527   Apollo 17, Station 6 boulder sample 76255: absolute petrology of breccia matrix and igneous clasts.
J. L. Warner, C. H. Simonds, W. C. Phinney.
Proc. Seventh Lunar Sci. Conf., (see 012.015), 2233 - 2250 (1976).

094.528   Textures and compositions of metal particles in Apollo 17, Station 6 boulder samples.
K. C. Misra, B. M. Walker, L. A. Taylor.
Proc. Seventh Lunar Sci. Conf., (see 012.015), 2251 - 2266 (1976).

094.529   The chronology of the Apollo 17 Station 6 boulder.   P. H. Cadogan, G. Turner.
Proc. Seventh Lunar Sci. Conf., (see 012.015), 2267 - 2285 (1976).

094.530   The petrology of 77215, a noritic impact ejecta breccia.
E. C. T. Chao, J. A. Minkin, C. L. Thompson.
Proc. Seventh Lunar Sci. Conf., (see 012.015), 2287 - 2308 (1976).

094.531   4.4b.y.-old clast in Boulder 7, Apollo 17: a comprehensive chronological study by U–Pb, Rb–Sr and Sm–Nd methods.   N. Nakamura, M. Tatsumoto, P. D. Nunes, D. M. Unruh, A. P. Schwab, T. R. Wildeman.
Proc. Seventh Lunar Sci. Conf., (see 012.015), 2309 - 2333 (1976).

094.532   Petrology and origin of Boulders # 2 and # 3, Apollo 17 Station 2.

R. F. Dymek, A. L. Albee, A. A. Chodos.
Proc. Seventh Lunar Sci. Conf., (see 012.015), 2335 - 2378 (1976) = Div. Geol. Planet. Sci., Calif. Inst. Technol., Pasadena, Calif., Contrib. No. 2754.

094.533   Consortium investigation of breccia 67435.
R. D. Warner, H. N. Planner, K. Keil, A. V. Murali, M.-S. Ma, R. A. Schmitt, W. D. Ehmann, W. D. James, Jr., R. N. Clayton, T. K. Mayeda.
Proc. Seventh Lunar Sci. Conf., (see 012.015), 2379 - 2402 (1976).

094.534   Composition of the projectiles that bombarded the lunar highlands.   J. Gros, H. Takahashi, J. Hertogen, J. W. Morgan, E. Anders.
Proc. Seventh Lunar Sci. Conf., (see 012.015), 2403 - 2425 (1976).

094.535   Petrogenesis of KREEP.
G. A. McKay, D. F. Weill.
Proc. Seventh Lunar Sci. Conf., (see 012.015), 2427 - 2447 (1976).

094.536   Heat flow in impact melts: Apollo 17 Station 6 Boulder and some applications to other breccias and xenolith laden melts.
P. I. K. Onorato, D. R. Uhlmann, C. H. Simonds.
Proc. Seventh Lunar Sci. Conf., (see 012.015), 2449 - 2467 (1976).

094.537   Lithification of vitric – and clastic – matrix breccias: SEM petrography.   W. C. Phinney, D. S. McKay, C. H. Simonds, J. L. Warner.
Proc. Seventh Lunar Sci. Conf., (see 012.015), 2469 - 2492 (1976).

094.538   Subsolidus reduction phenomena in lunar norite 78235: observations and interpretations.
C. B. Sclar, J. F. Bauer.
Proc. Seventh Lunar Sci. Conf., (see 012.015), 2493 - 2508 (1976).

094.539   Crystallization kinetics, viscous flow, and thermal histories of lunar breccias 15286 and 15498.
D. R. Uhlmann, L. C. Klein.
Proc. Seventh Lunar Sci. Conf., (see 012.015), 2529 - 2541 (1976).

094.540   Optical spectra and electron paramagnetic resonance of lunar and synthetic glasses: a study of the effects of controlled atmosphere, composition, and temperature.   P. M. Bell, H. K. Mao, R. A. Weeks.
Proc. Seventh Lunar Sci. Conf., (see 012.015), 2543 - 2559 (1976).

094.541   Age-color relationship in the lunar highlands.
M. P. Charette, L. A. Soderblom, J. B. Adams, M. J. Gaffey, T. B. McCord.
Proc. Seventh Lunar Sci. Conf., (see 012.015), 2579 - 2592 (1976).

094.542   Electrical properties of Apollo 17 rock and soil samples and a summary of the electrical properties of lunar material at 450 MHz frequency.
T. Gold, E. Bilson, R. L. Baron.
Proc. Seventh Lunar Sci. Conf., (see 012.015), 2593 - 2603 (1976).

094.543   Ultraviolet diffuse reflectance spectroscopy for lunar, meteoritic, and terrestrial samples.
C. T. Hua, A. Dollfus, J.-C. Mandeville.
Proc. Seventh Lunar Sci. Conf., (see 012.015), 2605 - 2622

(1976).

094.544 **A geochemical anomaly contiguous with the Dorsa Geike wrinkle ridge in Mare Fecunditatis.**
C. G. Andre, I. Adler, P. E. Clark, J. R. Weidner, J. A. Philpotts.
Proc. Seventh Lunar Sci. Conf., (see 012.015), 2649 - 2660 (1976).

094.545 **Characterization of lunar mare basalt types: I. A remote sensing study using reflection spectroscopy of surface soils.** C. Pieters, T. B. McCord.
Proc. Seventh Lunar Sci. Conf., (see 012.015), 2677 - 2690 (1976).

094.546 **Cosmic ray exposure ages of Apollo 17 samples and the age of Tycho.** R. Arvidson, R. Drozd, E. Guinness, C. Hohenberg, C. Morgan, R. Morrison, V. Oberbeck.
Proc. Seventh Lunar Sci. Conf., (see 012.015), 2817 - 2832 (1976).

094.547 **Photogeology of the multi-ringed crater Haldane in Mare Smythii.** R. W. Wolfe, F. El-Baz.
Proc. Seventh Lunar Sci. Conf., (see 012.015), 2903 - 2912 (1976).

094.548 **Internal friction and velocity measurements.** B. R. Tittmann, L. Ahlberg, J. Curnow.
Proc. Seventh Lunar Sci. Conf., (see 012.015), 3123 - 3132 (1976).

094.549 **Thermal diffusivity of four Apollo 17 rock samples.** K.-i. Horai, J. L. Winkler, Jr.
Proc. Seventh Lunar Sci. Conf., (see 012.015), 3183 - 3204 (1976).

094.550 **A solar origin for the large lunar magnetic field at 4.0×10^9 yr ago?**
S. K. Banerjee, J. P. Mellema.
Proc. Seventh Lunar Sci. Conf., (see 012.015), 3259 - 3270 (1976).

094.551 **Some complexities in the determination of lunar paleointensities.** G. W. Pearce, G. S. Hoye, D. W. Strangway, B. M. Walker, L. A. Taylor.
Proc. Seventh Lunar Sci. Conf., (see 012.015), 3271 - 3297 (1976).

094.552 **Magnetic effects of shock and their implications for lunar magnetism (II).** S. M. Cisowski, J. R. Dunn, M. Fuller, Y. Wu, M. F. Rose, P. J. Wasilewski.
Proc. Seventh Lunar Sci. Conf., (see 012.015), 3299 - 3320 (1976).

094.553 **Lithium as a correlated element, its condensation behaviour, and its use to estimate the bulk composition of the moon and the eucrite parent body.**
G. Dreibus, B. Spettel, H. Wänke.
Proc. Seventh Lunar Sci. Conf., (see 012.015), 3383 - 3396 (1976).

094.554 **Zr–Hf fractionation in chemically defined lunar rock groups.** A. N. Garg, W. D. Ehmann.
Proc. Seventh Lunar Sci. Conf., (see 012.015), 3397 - 3410 (1976).

094.555 **Chemical fractionation of Ru and Os in the moon.** S. Jovanovic, G. W. Reed, Jr.
Proc. Seventh Lunar Sci. Conf., (see 012.015), 3437 - 3446 (1976).

094.556 **Convection cells in the early lunar magma ocean: trace-element evidence.**
S. Jovanovic, G. W. Reed, Jr.
Proc. Seventh Lunar Sci. Conf., (see 012.015), 3447 - 3459 (1976).

094.557 **Chemistry of lunar highland rocks: a refined evaluation of the composition of the primary matter.**
H. Wänke, H. Palme, H. Kruse, H. Baddenhausen, M. Cendales, G. Dreibus, H. Hofmeister, E. Jagoutz, C. Palme, B. Spettel, R. Thacker.
Proc. Seventh Lunar Sci. Conf., (see 012.015), 3479 - 3499 (1976).

094.558 **Characterization of crust formation on a parent body of achondrites and the moon by pyroxene crystallography and chemistry.** H. Takeda, M. Miyamoto, T. Ishii, A. M. Reid.
Proc. Seventh Lunar Sci. Conf., (see 012.015), 3535 - 3548 (1976).

094.559 **Volcanic events associated with two impact craters on the lunar farside.** C. S. Beals, R. W. Tanner.
J. R. Astron. Soc. Canada, Vol. 71, 405 (1977). — Abstract.

094.560 **Metallography of lunar iron.**
R. I. Mints, T. M. Petukhova.
Priroda, 1977, No. 10, p. 66 - 76. In Russian.

094.561 **Cell dimensions and antiferromagnetism of lunar and terrestrial ilmenite single crystals.**
A. N. Thorpe, J. A. Minkin, F. E. Senftle, C. Alexander, C. Briggs, H. T. Evans, Jr., G. L. Nord, Jr.
J. Phys. Chem. Solids, Vol. 38, 115 - 123 (1977). — Abstr. in Phys. Abstr., Vol. 80, Abstr. 21764 (1977).

094.562 **Oxygen isotope abundance measurements in fine size fractions of Luna 16 and 20 soil.**
R. D. Beckinsale, S. H. U. Bowie, J. J. Durham.
Philos. Trans. R. Soc. London, Ser. A, Vol. 284, 131 - 136 (1976). — Abstr. in Phys. Abstr., Vol. 80, Abstr. 22120 (1977).

094.563 **The separation and subdivision of two 0.5 g samples of lunar soil collected by the Luna 16 and 20 missions.** C. T. Pillinger, A. P. Gowar.
Philos. Trans. R. Soc. London, Ser. A, Vol. 284, 137 - 143 (1976). — Abstr. in Phys. Abstr., Vol. 80, Abstr. 22121 (1977).

094.564 **Carbon chemistry of the Luna 16 and 20 samples.** C. T. Pillinger, G. Eglinton, A. P. Gowar, A. J. T. Jull.
Philos. Trans. R. Soc. London, Ser. A, Vol. 284, 145 - 150 (1976). — Abstr. in Phys. Abstr., Vol. 80, Abstr. 22122 (1977).

094.565 **Magnetic characteristics of Luna 16 and 20 samples.** A. Stephenson, D. W. Collinson, S. K. Runcorn.
Philos. Trans. R. Soc. London, Ser. A, Vol. 284, 151 - 156 (1976). — Abstr. in Phys. Abstr., Vol. 80, Abstr. 22123 (1977).

094.566 **Mössbauer studies of Luna 16 and 20 lunar soils.** T. C. Gibb, R. Greatrex, N. N. Greenwood.
Philos. Trans. R. Soc. London, Ser. A, Vol. 284, 157 - 165 (1976). — Abstr. in Phys. Abstr., Vol. 80, Abstr. 22124 (1977).

094.567 **^{40}Ar-^{39}Ar dating of Luna 16 and Luna 20 samples.** P. H. Cadogan, G. Turner.
Philos. Trans. R. Soc. London, Ser. A, Vol. 284, 167 - 177 (1976). — Abstr. in Phys. Abstr., Vol. 80, Abstr. 22125 (1977).

094.568 **On the age of KREEP.** H. Palme.
Geochim. Cosmochim. Acta, Vol. 41, 1791 - 1801 (1977).

Many lunar highland rocks have been extensively meta-morphosed during the late heavy bombardment of the Moon 3.9–4.0 AE ago. Highland rocks enriched in incompatible elements in most cases are mixtures between KREEP-basalt and other highland rock types. Slight isotopic differences among KREEP-enriched rocks from different landing sites become noticeable. These differences correspond to different meteoritic groups as defined by Morgan et al. (1974). Apparently there existed slightly different KREEP basalt reservoirs, with formation ages ranging from 4.25 to 4.45 AE. These reservoirs were partly exposed through impacts of basin-forming planetesimals 3.9–4.0 AE ago.

094.569   Very low-Ti mare basalts.
        G. J. Taylor, K. Keil, R. D. Warner.
Geophys. Res. Lett., Vol. 4, 207 - 210 (1977).
        Bulk compositions, petrology, and mineralogy of lithic fragments discovered in polished sections of Apollo 17 drill core samples 70007, 70008, and 70009 and Luna 24 soil 24077, 43 are described.

094.570   Geological-morphological analysis of the landing site of Luna 24.
K. P. Florenskij, A. T. Bazilevskij, G. A. Burba.
Dokl. AN SSSR, Vol. 233, 936 - 943 (1977). In Russian.
Abstr. in Ref. zh., 51. Astron., 10.51.380 (1977).

094.571   Preliminary description of the lunar soil core from Mare Crisium (Luna 24).
A. V. Ivanov, M. A. Nazarov, O. D. Rodeh, O. I. Stakheev, L. S. Tarasov, K. I. Tobelko, K. P. Florenskij.
Dokl. AN SSSR, Vol. 233, 928 - 931 (1977). In Russian.
Abstr. in Ref. zh., 51. Astron., 10.51.382 (1977).

094.572   Inert gas isotopes in regolith and its origin.
        L. K. Levskij, I. M. Morozova.
Probl. datirovaniya dokembrijsk. obrazovanij. S., Nauka, 1977, p. 198 - 210. In Russian. – Abstr. in Ref. zh., 51. Astron., 10.51.387 (1977).

094.573   Etchable ranges of fossil and fresh heavy-ion tracks in lunar and analogous crystals.
S. A. Durrani, R. K. Bull.
Nucl. Instrum. Methods, Vol. 140, 553 - 556 (1977). – Abstr. in Phys. Abstr., Vol. 80, Abstr. 30734 (1977).

094.574   Lunar mineralogy: a heavenly detective story. II.
        J. V. Smith, I. M. Steele.
American Mineral., Vol. 61, 1059 - 1116 (1976). – Abstr. in Phys. Abstr., Vol. 80, Abstr. 45327 (1977).

094.575   Structure of lunar impact craters from gravity models.   P. Janle.
J. Geophys., Vol. 42, 407 - 417 (1977). – Abstr. in Phys. Abstr., Vol. 80, Abstr. 64674 (1977).

094.576   Adsorption of methanol and water vapor on lunar soil 15081,2.   R. Sh. Mikhail, D. A. Cadenhead.
J. Colloid Interface Sci., Vol. 61, 375 - 382 (1977). – Abstr. in Phys. Abstr., Vol. 80, Abstr. 87292 (1977).

094.577   Problems of iron gain and loss during experimentation on natural rocks: the experimental crystallization of five lunar basalts at low pressures.
M. J. O'Hara, D. J. Humphries.
Philos. Trans. R. Soc. London, Ser. A, Vol. 286, 313 - 329 (1977). – Abstr. in Phys. Abstr., Vol. 80, Abstr. 87293 (1977).

094.578   Cation distribution and exsolution in the Luna-20 pyroxenes as the chronological indicator of their thermal history.   N. R. Khisina, E. S. Makarov.
Moon, Vol. 17, 149 - 165 (1977).

094.579   Rosin's law and the lunar regolith.
        E. A. King, J. C. Butler.
Moon, Vol. 17, 177 - 178 (1977).

094.580   Barycentric selenodetic coordinates of the lunar crater Mösting A.
G. M. Stolyarov, I. G. Chugunov.
Astron. Tsirk., No. 953, p. 2 - 3 (1977). In Russian.

094.581   Barycentric selenodetic coordinates of the lunar crater Mösting A from photographic observations.
I. G. Chugunov.
Astron. Tsirk., No. 953, p. 4 (1977). In Russian.

094.582   The surface composition of lunar soil grains: a comparison of the results of Auger and X-ray photoelectron (ESCA) spectroscopy.   R. L. Baron, E. Bilson, T. Gold, R. J. Colton, B. Hapke, M. A. Steggart.
Earth Planet. Sci. Lett., Vol. 37, 263 - 272 (1977).
        ESCA spectra of lunar soil and crushed rock samples are compared to previously obtained Auger spectra of the same samples. The ESCA data confirm the increase of Fe concentration on the surface of soil samples vs. their bulk Fe content; and strongly support the existence of layers on the surfaces of lunar soil grains which are significantly enriched in Fe, some of which is in the Fe^0 state. The significance of the ESCA information on the major elemental composition of lunar soil grain surfaces for the understanding of the processes that effect the state of the regolith is discussed.

094.583   Orientale and Caloris.   J. F. McCauley.
        Phys. Earth Planet. Inter., Vol. 15, 220 - 250 (1977).
        Applications of experimental explosion-crater data to Orientale and recent geologic mapping of the basin have produced a new stratigraphy and genetic model for Orientale that are also applicable to Caloris.

094.584   Preliminary analysis of variation in Al, Mg and Si in Apollo 11, −12 and −15 basalts and regolith.
J. C. Butler.
Phys. Earth Planet. Inter., Vol. 15, 275 - 286 (1977).

094.585   Orbital gamma-ray data and large-scale lunar problems.   N. J. Hubbard, D. Woloszyn.
Phys. Earth Planet. Inter., Vol. 15, 287 - 302 (1977).

094.586   Chemistry and chronology of the Luna 24 soils and rocks.   C. J. Allègre, F. Albarède, J.-L. Birck, J.-L. Joron, G. Manhes, P. Richard, M. Treuil, A. Stettler.
Meteoritics, Vol. 12, 168 (1977). – Abstract.

094.587   Ferrobasalt and ferrogabbro from Mare Crisium, Luna 24.
A. E. Bence, T. L. Grove, J. J. Papike, D. T. Vaniman, J. Friel, J. Goldstein, S. Haggerty, E. Roedder, P. Weiblen.
Meteoritics, Vol. 12, 175 - 176 (1977). – Abstract.

094.588   High voltage electron microscope observations of micron-sized grains extracted at depth 96 cm in the Luna 24 core-tube.   J. Borg, J. C. Dran.
Meteoritics, Vol. 12, 182 (1977). – Abstract.

094.589   Exposure history and fission track ages of Apollo 15 green glass spherules.
I. W. Davie, R. K. Bull, S. A. Durrani.
Meteoritics, Vol. 12, 203 (1977). – Abstract.

094.590   Laboratory simulation of secondary lunar microcraters from micron scale hypervelocity impacts on lunar rock.   R. P. Flavill, J. A. M. McDonnell.
Meteoritics, Vol. 12, 220 - 225 (1977). – Abstract.

094.591 Analysis of carbon species in lunar samples by static mass spectrometry.
L. R. Gardiner, A. J. T. Jull, C. T. Pillinger.
Meteoritics, Vol. 12, 236 (1977). — Abstract.

094.592 Neon and xenon spallation components due to solar flare protons in lunar fines.
K. Gopalan, M. N. Rao, T. R. Venkatesan.
Meteoritics, Vol. 12, 242 (1977). — Abstract.

094.593 Correlation of Ar^{40}-Ar^{39} ages with textural subunits in lunar mare basalts.
N. Grögler, P. Eberhardt, J. Geiss, S. Guggisberg, A. Stettler, G. M. Brown, A. Peckett.
Meteoritics, Vol. 12, 245 - 246 (1977). — Abstract.

094.594 The effect of the temperature of irradiation upon the sensitivity of lunar samples.
K. A. R. Khazal, S. A. Durrani.
Meteoritics, Vol. 12, 271 - 273 (1977). — Abstract.

094.595 He and Ne depth profiles in olivine from lunar soil 71501,23.   J. Kiko, T. Kirsten, M. Warhaut.
Meteoritics, Vol. 12, 274 - 275 (1977). — Abstract.

094.596 Rare gases and modal composition of special surface sample 69004.   T. Kirsten, B. Dominik.
Meteoritics, Vol. 12, 278 - 279 (1977). — Abstract.

094.597 24170: an iron-rich basalt from Mare Crisium.
U. B. Marvin, G. Ryder, H. McSween.
Meteoritics, Vol. 12, 304 (1977). — Abstract.

094.598 Pre-cataclysmic cratering of the lunar crust.
P. Maurer, P. Eberhardt, J. Geiss, N. Grögler, A. Stettler, G. M. Brown, A. Peckett, U. Krähenbühl.
Meteoritics, Vol. 12, 306 - 307 (1977). — Abstract.

094.599 Some constraints on the origin of finely-divided iron in lunar soil.   C. T. Pillinger, L. R. Gardiner, A. J. T. Jull, M. R. Woodcock, A. Stephenson.
Meteoritics, Vol. 12, 339 (1977). — Abstract.

094.600 Impact microcraters and cosmic ray tracks in Luna 16, 20 and 24 soils.   G. Poupeau, J.-C. Mandeville.
Meteoritics, Vol. 12, 340 - 341 (1977). — Abstract.

094.601 Magnetism of lunar rocks and meteorites.
S. K. Runcorn, A. Stephenson.
Meteoritics, Vol. 12, 356 - 357 (1977). — Abstract.

094.602 Luna 24 basalts and metabasalts.
G. Ryder, H. Y. McSween, U. B. Marvin.
Meteoritics, Vol. 12, 357 - 358 (1977). — Abstract.

094.603 Admixture of fresh material, agglutination, and "reworking" as reflected in the noble gas record of lunar soil constituents.   P. Signer, H. Baur, U. Derksen, P. Etique, H. Funk, P. Horn, R. Wieler.
Meteoritics, Vol. 12, 362 - 363 (1977). — Abstract.

094.604 Composition and origin of plagioclase, pyroxene, and olivine clasts in Fra Mauro breccias.
D. Stöffler, H.-D. Knöll.
Meteoritics, Vol. 12, 366 (1977). — Abstract.

094.605 Very low-Ti mare basalts.
G. J. Taylor, K. Keil, R. D. Warner.
Meteoritics, Vol. 12, 369 - 370 (1977). — Abstract.

094.606 Noble gas record of agglutinate and bulk grain size fractions separated from soil 15601.

H. W. Weber, L. Schultz, H. Hintenberger.
Meteoritics, Vol. 12, 383 (1977). — Abstract.

094.607 Use of an exoelectronic emission method for studying the physical state of the surface of lunar soil fragments supplied by the Soviet automatic interplanetary stations Luna 16 and Luna 20.
R. I. Mints, I. I. Mil'man, V. I. Kryuk, L. S. Tarasov.
Fiz. metod. issled. tverd. tela. Vyp. (No.) 2. Sverdlovsk, 1977, p. 95 - 104. In Russian. — From Ref. zh., 51. Astron., 12.51. 262 (1977).

094.608 Infrared optical characteristics of lunar soil from Mare Crisium (Luna 24).
M. V. Akhmanova, B. V. Dement'ev, M. N. Markov, M. M. Sushchinskij.
Teor. spektrosk. Moskva, 1977, p. 246 - 247. In Russian. From Ref. zh., 51. Astron., 12.51.264; Abstr. in 62. Issled. kosm. prostranstva, 12.62.112 (1977).

094.609 Electron-probe investigations of individual particles of lunar regolith.
N. P. Il'in, L. E. Loseva, V. G. Senin.
Opt. rentgen. luchej i mikroanal. Mater. VII Mezhdunar. konf. po opt. rentgen. luchej i mikroanal., Moskva — Kiev, 1974 g. Leningrad, Mashinostroenie, 1976, p. 80 - 83. In Russian. Abstr. in Ref. zh., 51. Astron., 12.51.265; 62. Issled. kosm. prostranstva, 12.62.113 (1977).

094.610 Endogenetic craters on the floors of large lunar craters.   M. Gurnis.
Strolling Astron., Vol. 27, 7 - 14 (1977).
Lunar Orbiter photographic data have been analysed to demonstrate that there are endogenetic craters or volcanoes on the Moon. Crater counts were made on the floors and ejectas of 14 large lunar craters. It was found that on the majority of the craters studied, there are more craters on the floor than on the ejecta. The explanation given is that there are endogenetic craters on the floor. An alternate explanation is that there is more crater erosion on the ejecta than on the floor.

094.611 New Soviet investigations of lunar soil and its analogues.   I. I. Cherkasov, V. V. Shvarev.
Osnovaniya, fundam. i mekh. gruntov, 1977, No. 5, p. 20 - 23. In Russian. — Abstr. in Ref. zh., 51. Astron., 1.51.330; 62. Issled. kosm. prostranstva, 1.62.119 (1978).

094.612 Colorimetric structure of areas of the visible hemisphere of the moon.
V. I. Ezerskij, N. S. Olifer, Yu. G. Shkuratov.
Physics of the moon and planets. Fundamental astrometry, (see 003.024), Vestn. Khar'kov. Univ., No. 160, p. 14 - 19 (1977). In Russian.

094.613 Sinuous rilles on the Aristarchus and Harbinger formations.   Y. Saito.
10th Lunar and Planetary Symposium, (see 012.050), p. 21 - 25 (1977).
The distribution and characteristics of 51 sinuous rille structures on and around the Aristarchus Plateau and Harbinger Mountains region of the Moon were analysed on Lunar Orbiter 4 and Apollo 15 photographs, and a preliminary geological map is presented.

Chemistry of the moon.   See Abstr. 003.026.

Surveying the Moon.   See Abstr. 003.051.

The moon book: exploring the mysteries of the lunar world.   See Abstr. 003.055.

Automatic stations for investigation of the lunar

surface.   See Abstr. 003.081.

Distribution of molecular weight in glyceride polymerizates or aggregates of them after contact with lunar grains.   See Abstr. 022.006.

On the ion-bombardment reduction mechanism. See Abstr. 022.041.

Solubility of Cr, Ti, and Al in co-existing olivine, spinel, and liquid at 1 atm.   See Abstr. 022.042.

Partitioning of chromium between silicate crystals and melts.   See Abstr. 022.043.

Fe and Mg in plagioclase. See Abstr. 022.044.

Sample size and sampling errors as the source of dispersion in chemical analyses.   See Abstr. 022.045.

Further characterization of spectral features attributable to titanium on the moon.   See Abstr. 022.046.

Electrical conductivity of orthopyroxene to 1400°C and the resulting selenotherm. See Abstr. 022.047.

Combined position and diameter measures for lunar craters.   See Abstr. 031.202.

Solar flare activity:evidence for large-scale changes in the past.   See Abstr. 073.005.

Multi-planet, multi-ring basin studies. See Abstr. 092.019.

Interpretations of optical observations of Mercury and the moon.   See Abstr. 092.042.

Production of simple molecules on the surface of Mercury.   See Abstr. 092.043.

On the intensity of the ancient lunar magnetic field.   See Abstr. 094.010.

Critical review of models for the evolution of high-Ti mare basalts.   See Abstr. 094.132.

Quantitative mass distribution models for Mare Orientale.   See Abstr. 094.136.

Surface chemistry of selected lunar regions. See Abstr. 094.137.

Geologic structure of the eastern mare basins. See Abstr. 094.141.

Modes of emplacement of basalt terrains and an analysis of mare volcanism in the Orientale Basin. See Abstr. 094.142.

Geochemical constraints on the composition of the moon.   See Abstr. 094.167.

Comments on: "Characteristics of fresh martian craters as a function of diameter: comparison with the Moon and Mercury" by M. J. Cintala et al. See Abstr. 097.137.

Characteristics of fresh martian craters as a function of diameter: comparison with the Moon and Mercury —discussion.   See Abstr. 097.138.

Thermal model for impact breccia lithification: Manicouagan and the moon.   See Abstr. 105.034.

Trace element evidence regarding a chondritic component in howardite meteorites.   See Abstr. 105.037.

Argon, krypton and xenon in iron meteorites. See Abstr. 105.101.

Potassium isotopic determination in some meteoritic and lunar samples: evidence for irradiation effects. See Abstr. 105.119.

Errata

094.901   Errata: 'On the petrology and early development of the crust of a Moon of fission origin' [Moon, Vol. 15, 275 - 314 (1976)].   A. B. Binder. Moon, Vol. 16, 493 (1977).

094.902   Errata to Proceedings of the Apollo 11 Lunar Science Conference, and to Proceedings of the Sixth Lunar Science Conference. Proc. Seventh Lunar Sci. Conf., (see 012.015), Vol. 3, XI - XIII (1976).

095 Lunar Eclipses

095.001   Notes on the lunar eclipse of September 27th. Sky Telesc., Vol. 54, 533 - 535 (1977).

095.002   Enlargement of the earth's shadow during the lunar eclipses observed in the years 1973—1975. J. Bouška. Acta Univ. Carolinae Math. Phys., Vol. 18, No. 1, p. 51 - 63 (1977).
From the timing of crater entrances into the umbra and

their exits from the umbra obtained during the last four lunar eclipses in Czechoslovakia the enlargement of the shadow was determined. During the eclipse of 9—10 December 1973 the enlargement was found to be 1/67, during the eclipse of 4—5 June 1974—1/42, during the eclipse of 29 November 1974—1/48 and during the eclipse of 18—19 November 1975—1/47.

095.003   Little known cycles of the periodicity of eclipses. M. Dujnič. Říše hvězd, Vol. 58, 170 - 173 (1977). In Slovak.

096 Lunar Occultations

096.001 **Photoelectric observing of occultations – I.**
D. S. Evans.
Sky Telesc., Vol. 54, 164 - 166 (1977).

096.002 **Lunar occultations of Io and Ganymede.**
F. Vilas, R. L. Millis, L. H. Wasserman.
Bull. American Astron. Soc., Vol. 9, 464 (1977). – Abstract.

096.003 **Photoelectric observations of lunar occultations.IX.**
J. L. Africano, D. S. Evans, F. C. Fekel,
T. Montemayor.
Astron. J., Vol. 82, 631 - 639 (1977).
Observations of 223 occultation disappearances are
reported.

096.004 **Approach to systematic observation of occultation**
reappearances. J. L. Africano, T. Montemayor.
Astron. J., Vol. 82, 640 - 641 (1977).
A technique for photoelectric observation of dark limb
reappearances of stars is described and results for four events
reported.

096.005 **Les occultations d'étoiles par la Lune: détermina-**
tion des diamètres stellaires. P. Bartholdi.
Recueil des séminaires, (see 012.004), p. 90 - 94 (1977).

096.006 **Occultation study of the multiple star β Scorpii.**
D. S. Evans, J. L. Africano, F. C. Fekel, T. Montemayor, C. Palm, E. Silverberg, W. Van Citters, J. Wiant.
Astron. J., Vol. 82, 495 - 502 (1977).
Observations of the occultation of β Scorpii on 8 July
1976 made with four telescopes at McDonald Observatory are
discussed and reduced to a form suitable for combination
with results from elsewhere. Five components of this star are
identified and estimates made of their individual magnitudes.

096.007 **Occultation observations of β Scorpii.**
W. H. Sandmann.
Astron. J., Vol. 82, 503 - 505 (1977).
Observations of the occultation of β Scorpii on 8 July
1976 with the 0.61-m telescope at Table Mountain Observatory are discussed and the results found to be comparable
with those of the University of Texas occultation group (see
Abstr. 096.006). Three components of this multiple-star
system are identified and relative V magnitudes are estimated.
The observed timings and position angles give a tentative
separation and position angle for B relative to A of $\rho = 0.451$
arcsec, $\theta = 120°2$.

096.008 **Photoelectric observing of occultations – II.**
D. S. Evans.
Sky Telesc., Vol. 54, 289 - 292 (1977).

096.009 **Angular diameters of ψ Vir (SAO 139033) and χ^1**
Ori (SAO 077705). D. S. Evans, F. C. Fekel,
E. C. Silverberg, C. Palm, J. R. Wiant.
Astron. J., Vol. 82, 828 - 831 (1977).
Occultation observation diameters of ψ Vir (6.5 milliarcsec) and χ^1 Ori (1.3 milliarcsec) are reported. Problems connected with the determination of very small angular diameters
are discussed. Values of the surface brightness parameter F_ν are
deduced, taking observations from other sources into account.

096.010 **Analyzing HMNAO residuals – II.** C. J. Bader.
Occultation Newsl., Vol. 1, 112 - 113 (1977).

096.011 **Identification of unpredicted stars.**
W. H. Warren, Jr.

Occultation Newsl., Vol. 1, 114 (1977).

096.012 **Planetary occultations.** D. W. Dunham.
Occultation Newsl., Vol. 1, 114 - 115 (1977).

096.013 **Upcoming lunar occultations of minor planets.**
D. W. Dunham.
Occultation Newsl., Vol. 1, 115 (1977).

096.014 **Occultations of galactic-nebular objects.**
R. P. Binzel.
Occultation Newsl., Vol. 1, 117 (1977).

096.015 **Observations of cluster passages.** B. Timerson.
Occultation Newsl., Vol. 1, 117 (1977).

096.016 **Grazes reported to IOTA.** D. W. Dunham.
Occultation Newsl., Vol. 1, 117 - 119 (1977).

096.017 **Lunar occultation summary. II.**
J. J. Eitter, W. I. Beavers.
Astrophys. J., Suppl. Ser., Vol. 34, 493 - 504 (1977).
Results of a second series of 196 two-color photoelectric
lunar occultation observations made within the interval 1972
May to 1973 December are reported. Each observation has
been employed to estimate the minimum magnitude difference
between the observed object and possible undetected companions. The study contains the analyses of 143 disappearances and 53 reappearances. Eighteen of these events manifest
some degree of multiplicity. Included in the study are 18
occultation events for Pleiades members.

096.018 **Lunar occultations of the Hyades Cluster.**
D. W. Dunham.
Occultation Newsl., Vol. 1, 121 - 123 (1977).

096.019 **1978–1981 grazing occultations– Aldebaran,**
Regulus, Venus.
Occultation Newsl., Vol. 1, 124 (1977).

096.020 **Grazes reported to IOTA.** D. W. Dunham.
Occultation Newsl., Vol. 1, 128 - 129 (1977).

096.021 **Erroneous star positions and unpredicted occulta-**
tions. D. Herald.
Occultation Newsl., Vol. 1, 129 - 130 (1977).

096.022 **Strykande ockultation av Aldebaran den 26**
augusti 1978. J. Meeus.
Astron. Tidsskr., Årg. 10, 121 - 122 (1977).

096.023 **Streifende Sternbedeckungen im Januar und Februar**
1978. M. Knitsch.
Sterne Weltraum, Jahrg. 16, 425 - 426 (1977).

096.024 **Occultations rasantes en France en 1978.**
J. Meeus.
Astronomie, Vol. 91, 441 - 444 (1977).

096.025 **Occultations of stars by the moon observed at the**
Belgrade Astronomical Observatory 1974–1975.
V. M. Protitch-Benišek, M. B. Protitch.
Bull. Obs. Astron. Belgrade, No. 128, p. 60 - 61 (1977).

096.026 **Occultation observation in 1975.**
T. Mori, Y. Harada, M. Kawada, T. Kanazawa.
Data Rep. Hydrographic Observations, Ser. Astron. Geod.,
Tokyo, No. 11, p. 1 - 45 (1977).

096.027   Observations of star occultations by the moon at
          the Poltava Gravimetric Observatory in 1965–1971.
B. F. Sincheskul.
Vrashchenie i priliv. deformatsii Zemli. Resp. mezhved. sb.,
1977, vyp. (No.) 9, p. 66 - 81. In Russian. − Abstr. in Ref.
zh., 51. Astron., 1.51.157 (1978).

096.028   Observations of star occultations by the moon at
          the Astronomical Observatory of the Khar'kov
University in 1969 - 1972.
S. R. Izmajlov, V. M. Kirpatovskij.
Physics of the moon and planets. Fundamental astrometry,
(see 003.024), Vestn. Khar'kov. Univ., No. 160, p. 62 - 65
(1977). In Russian.

**Considerations for the application of the lunar
occultation technique.**   See Abstr. 031.204.

**Application de l'observation des occultations
d'étoiles par la Lune à la decouverte et à la mesure d'étoiles
doubles.**   See Abstr. 031.250.

Using Watts angles in observing reappearances.
See Abstr. 031.257.

Values of $\underline{k}$ for occultation reduction.
See Abstr. 031.258.

New double stars.   See Abstr. 118.003.

New double stars.   See Abstr. 118.021.

**On the use of lunar occultations of radio sources
for the investigation of their angular structure. IV.**
See Abstr. 141.072.

**Balloon observations on the lunar occultation of the
hard X-ray source in the Crab Nebula.**
See Abstr. 142.125.

Errata

096.901   Erratum: "The occultation of Beta Scorpii" [South.
          Stars, Vol. 26, 233 - 234 (1976)].   W. H. Allen.
South. Stars, Vol. 27, 36 (1977).

097 Mars, Mars Satellites

Mars

097.001 Possible surface reactions on Mars: implications for Viking biology results. C. Ponnamperuma, A. Shimoyama, M. Yamada, T. Hobo, R. Pal.
Science, Vol. 197, 455 - 457 (1977).

The results of two of the three biology experiments carried out on the Viking Mars landers have been simulated. The mixture of organic compounds labeled with carbon-14 used on Mars released carbon dioxide containing carbon-14 when reacted with a simulated martian surface and atmosphere exposed to ultraviolet light (labeled release experiment). Oxygen was released when metal peroxides or superoxides were treated with water (gas exchange experiment). The simulations suggest that the results of these two Viking experiments can be explained on the basis of reactions of the martian surface and atmosphere.

097.002 Internal structure and properties of Mars. D. H. Johnston, M. N. Toksöz.
Icarus, Vol. 32, 73 - 84 (1977).

Theoretical physical models of the Martian interior are presented in the light of new and revised data and constraints. These models include thermal evolution, densities, and seismic wave velocities. The interior of Mars appears to be Earth-like in many respects. Although thermal models indicate that Mars has passed its peak of evolution it may still have an asthenosphere and may be moderately active tectonically.

097.003 Martian isotopic ratios and upper limits for possible minor constituents as derived from Mariner 9 infrared spectrometer data. W. C. Maguire.
Icarus, Vol. 32, 85 - 97 (1977).

The Mariner 9 infrared spectrometer obtained data over a large part of Mars for almost a year beginning late in 1971. Mars' infrared emission spectrum was measured from 200 to 2000 cm^{-1} with an apodized resolution of 2.4 cm^{-1}. No significant deviation from terrestrial ratios of carbon ($^{12}C/^{13}C$) or oxygen ($^{16}O/^{18}O$; $^{16}O/^{17}O$) isotopes was observed. The $^{12}C/^{13}C$ isotopic ratio was found to be terrestrial with an uncertainty of 15%. Upper limits have been calculated for several minor constituents.

097.004 The global distribution of O_3 on Mars. T. Y. Kong, M. B. McElroy.
Planet. Space Sci., Vol. 25, 839 - 857 (1977).

Models are developed to describe the photochemistry of ozone on Mars. Catalytic reactions involving H, OH and HO_2 play a major role at low latitudes where they ensure a vertical column density for O_3 of less than 2×10^{-4} cm atm. The source for odd hydrogen (H + OH + HO_2) is relatively smaller at high latitudes in winter due to the small concentrations of H_2O present there at that time. Odd hydrogen is also efficiently removed from the high-latitude winter atmosphere by condensation of H_2O_2. The role of catalytic chemistry is reduced accordingly and the vertical column density of O_3 may be as large as 5.7×10^{-3} cm atm.

097.005 Analysis of the occultation of ϵ Geminorum by Mars.
L. H. Wasserman, R. L. Millis, R. M. Williamon.
Astron. J., Vol. 82, 506 - 510 (1977).

High-time-resolution photoelectric measurements of the occultation of ϵ Gem by Mars were obtained with the 36-in. telescope at Goddard Space Flight Center. Temperature profiles as a function of number density have been determined by inverting the immersion and emersion light curves on the

assumption of a pure CO_2 atmosphere. For number densities between 10^{13} and 10^{15} cm^{-3}, the temperature varies in a wave-like fashion about a mean value near 150 K.

097.006 Dust clouds and frictional generation of glow discharges on Mars. A. A. Mills.
Nature, Vol. 268, 614 (1977).

A remarkable characteristic of those samples of the martian soil which have so far been analysed is the absence of carbonaceous matter down to the parts per billion (10^9) level. The author suggests that glow discharges generated by friction within dust clouds might explain this apparent absence of carbonaceous matter. In addition glow discharges might account for some reactions noted in the Viking biological experiments.

097.007 Mars and Praesepe. J. Meeus.
J. British Astron. Assoc., Vol. 87, 482 - 484 (1977).

097.008 The aureole of Olympus Mons, Mars. S. A. Harris.
J. Geophys. Res., Vol. 82, 3099 - 3107 (1977).

A large aureole of grooved terrain surrounds Olympus Mons, a Martian shield volcano. Curvilinear scarps which face away from and gentle interior slopes which face toward the volcano characterize this terrain. It is concentrated on the downslope side of Olympus Mons and rises as an uplift above the surrounding plain. The grooved terrain is probably large-scale gravity thrust sheets. The load of the Olympus Mons shield on a thick sequence of frozen bedded volcanic and aeolian type sediments caused thrusting. Movement was probably near the base of the Martian permafrost, where internal heat melted the ice, a process creating a natural zone of yield.

097.009 Celestial mechanics results from Viking radio tracking data. A. P. Mayo, W. H. Michael, Jr., R. H. Tolson, W. T. Blackshear, J. P. Gapcynski, G. M. Kelly, D. L. Cain, J. P. Brenkle, I. I. Shapiro, R. D. Reasenberg.
Bull. American Astron. Soc., Vol. 9, 437 (1977). — Abstract.

097.010 Martian fluvial features. M. H. Carr.
Bull. American Astron. Soc., Vol. 9, 440 (1977). Abstract.

097.011 The north polar region of Mars: imaging results from Viking II. J. A. Cutts, K. R. Blasius, G. A. Briggs, M. H. Carr, R. Greeley, H. Masursky.
Bull. American Astron. Soc., Vol. 9, 440 (1977). — Abstract.

097.012 Preliminary assessment of Martian volcanic features from Viking data. R. Greeley, M. Carr, J. Guest, H. Masursky, K. Blasius.
Bull. American Astron. Soc., Vol. 9, 440 (1977). — Abstract.

097.013 Viking Orbiter observations of the Valles Marineris, the great Martian equatorial troughs.
K. R. Blasius, J. A. Cutts, H. Masursky, M. Carr, R. Greeley, J. E. Guest.
Bull. American Astron. Soc., Vol. 9, 440 (1977). — Abstract.

097.014 Viking Orbiter imaging observations: condensate phenomena. G. A. Briggs.
Bull. American Astron. Soc., Vol. 9, 440 - 441 (1977). Abstract.

097.015 The nature and distribution of suspended particles in the Martian atmosphere. W. A. Baum.
Bull. American Astron. Soc., Vol. 9, 441 (1977). — Abstract.

097.016 Thermal mapping of Mars: polar and nighttime
results.   H. H. Kieffer.
Bull. American Astron. Soc., Vol. 9, 441 (1977). − Abstract.

097.017 Thermal mapping of Mars: the Olympus-Tharsis
region.   F. D. Palluconi.
Bull. American Astron. Soc., Vol. 9, 442 (1977). − Abstract.

097.018 Thermal mapping of Mars: atmospheric temperatures.   T. Z. Martin.
Bull. American Astron. Soc., Vol. 9, 442 (1977). − Abstract.

097.019 Thermal mapping of Mars: photometric and radiometric functions.   E. D. Miner.
Bull. American Astron. Soc., Vol. 9, 442 (1977). − Abstract.

097.020 The Viking water vapor mapping experiment −
diurnal, global and seasonal variations.
C. B. Farmer, D. W. Davies, D. D. LaPorte.
Bull. American Astron. Soc., Vol. 9, 442 (1977). − Abstract.

097.021 Water vapor measurements of the polar areas of
Mars from Viking.
D. W. Davies, C. B. Farmer, D. D. LaPorte.
Bull. American Astron. Soc., Vol. 9, 442 (1977). − Abstract.

097.022 Properties of Mars and its environment as deter-
mined from Viking radio tracking data.
W. H. Michael, Jr., A. P. Mayo, W. T. Blackshear, R. H.
Tolson, J. P. Gapcynski, G. M. Kelly, J. P. Brenkle, D. L.
Cain, G. Fjeldbo, D. N. Sweetnam, C. T. Stelzried, R. B.
Goldstein, P. E. MacNeil, R. D. Reasenberg, I. I. Shapiro,
T. I. S. Boak, III, M. D. Grossi, C. H. Tang, G. L. Tyler.
Bull. American Astron. Soc., Vol. 9, 442 - 443 (1977).
Abstract.

097.023 The Viking relativity experiment.
I. I. Shapiro, R. D. Reasenberg, R. B. Goldstein,
P. E. MacNeil, J. P. Brenkle, D. L. Cain, T. Komarek, A. I.
Zygielbaum, W. H. Michael, Jr.
Bull. American Astron. Soc., Vol. 9, 443 (1977). − Abstract.

097.024 Surficial geologic mapping of Mars.
P. Spudis, R. Greeley, T. Bunch.
Bull. American Astron. Soc., Vol. 9, 443 (1977). − Abstract.

097.025 Areological setting of the Viking landing sites by
way of the Lander cameras.
A. Binder, T. Mutch, R. Arvidson, E. Morris.
Bull. American Astron. Soc., Vol. 9, 443 - 444 (1977).
Abstract.

097.026 Properties of aerosols in the Martian atmosphere:
results from the Viking Lander imaging experiment.
J. B. Pollack, D. Colburn, R. Kahn, A. Binder, R. Arvidson,
E. Carlston.
Bull. American Astron. Soc., Vol. 9, 444 (1977). − Abstract.

097.027 Fine particles on the surface of Mars.
C. Sagan.
Bull. American Astron. Soc., Vol. 9, 444 (1977). − Abstract.

097.028 Viking Lander 1: sedimentological significance of
the large near-field boulder.   R. S. Saunders.
Bull. American Astron. Soc., Vol. 9, 444 (1977). − Abstract.

097.029 Spectrophotometric properties of surface materials
by way of the Viking Lander cameras.
R. Arvidson, A. Binder, F. Huck, S. Park, W. Patterson, S. Wall.
Bull. American Astron. Soc., Vol. 9, 444 (1977). − Abstract.

097.030 The neutral composition of Mars' thermosphere.
A. O. Nier.
Bull. American Astron. Soc., Vol. 9, 445 (1977). − Abstract.

097.031 The Martian ionosphere.
W. B. Hanson, S. Sanatani, D. R. Zuccaro.
Bull. American Astron. Soc., Vol. 9, 445 (1977). − Abstract.

097.032 Entry science measurements on Viking: implications for the evolutionary history of Mars' atmosphere.   M. B. McElroy.
Bull. American Astron. Soc., Vol. 9, 445 (1977). − Abstract.

097.033 The Viking atmosphere structure experiment −
preliminary results.
A. Seiff, D. B. Kirk, R. Blanchard.
Bull. American Astron. Soc., Vol. 9, 445 (1977). − Abstract.

097.034 Results of the Viking meteorology experiments.
C. Leovy, S. Hess, J. Ryan, R. Henry, J. Tillman,
J. Mitchell.
Bull. American Astron. Soc., Vol. 9, 445 (1977). − Abstract.

097.035 Results from the molecular analysis experiment.
K. Biemann, T. Owen, J. Oro, P. Toulmin III,
L. E. Orgel, A. O. Nier, D. M. Anderson, P. G. Simmonds,
D. Flory, A. V. Diaz, D. R. Rushneck, J. E. Biller, D.
Howarth, A. LaFleur.
Bull. American Astron. Soc., Vol. 9, 446 (1977). − Abstract.

097.036 Inorganic analyses of Martian surface samples by
X-ray fluorescence spectrometry.
B. C. Clark, A. K. Baird, H. J. Rose, P. Toulmin, K. Keil,
A. Castro, W. Kelliher.
Bull. American Astron. Soc., Vol. 9, 446 (1977). − Abstract.

097.037 Mineralogic/petrologic implications of Viking geo-
chemical results from Mars: interim report.
A. K. Baird, P. Toulmin, B. C. Clark, H. J. Rose, K. Keil,
J. L. Gooding.
Bull. American Astron. Soc., Vol. 9, 446 (1977). − Abstract.

097.038 Physical properties of the surface of Mars.
R. W. Shorthill, H. J. Moore, II, R. E. Hutton,
R. F. Scott, C. R. Spitzer.
Bull. American Astron. Soc., Vol. 9, 446 (1977). − Abstract.

097.039 Rock pushing − Viking 2.   H. J. Moore,
S. Liebes Jr., D. S. Crouch, L. V. Clark.
Bull. American Astron. Soc., Vol. 9, 446 (1977). − Abstract.

097.040 Results of the Viking Lander magnetic properties
experiments.
R. B. Hargraves, D. W. Collinson, R. E. Arvidson.
Bull. American Astron. Soc., Vol. 9, 447 (1977). − Abstract.

097.041 Viking seismology experiment.
D. L. Anderson, F. K. Duennebier, G. V. Latham,
M. N. Toksöz, R. L. Kovach, T. C. D. Knight, A. R. Lazare-
wicz, W. F. Miller, Y. Nakamura, G. H. Sutton.
Bull. American Astron. Soc., Vol. 9, 447 (1977). − Abstract.

097.042 Overview of Viking.   G. Soffen.
Bull. American Astron. Soc., Vol. 9, 447 (1977).
Abstract.

097.043 The geologic evolution of Mars − a preliminary
Viking view.   H. Masursky, K. Blasius, M. Carr,
J. Cutts, R. Greeley, J. Guest.
Bull. American Astron. Soc., Vol. 9, 447 (1977). − Abstract.

097.044 Volatiles on Mars.   T. Owen.
Bull. American Astron. Soc., Vol. 9, 447 (1977).

Abstract.

097.045 Evolution of the Martian environment and the
        biological significance of the Viking missions.
C. Sagan.
Bull. American Astron. Soc., Vol. 9, 447 - 448 (1977).
Abstract.

097.046 Intercrater plains and the early history of Mars.
M. C. Malin.
Bull. American Astron. Soc., Vol. 9, 448 (1977). — Abstract.

097.047 Arecibo Viking radar studies of Mars.
R. A. Simpson, G. L. Tyler, D. B. Campbell.
Bull. American Astron. Soc., Vol. 9, 448 (1977). — Abstract.

097.048 Viking/Goldstone radar studies of Mars.
G. S. Downs, R. R. Green, P. E. Reichley.
Bull. American Astron. Soc., Vol. 9, 448 (1977). — Abstract.

097.049 Groundbased infrared spectroscopy of Mars and
        implications for surface mineralogical composition.
T. B. McCord, R. Clark, R. Huguenin.
Bull. American Astron. Soc., Vol. 9, 448 - 449 (1977).
Abstract.

097.050 Mars: surface-controlled stability of the atmosphere.
R. L. Huguenin, R. G. Prinn, M. Maderazzo.
Bull. American Astron. Soc., Vol. 9, 449 (1977). — Abstract.

097.051 Photochemical weathering and the Viking biology
        experiments on Mars.   R. L. Huguenin.
Bull. American Astron. Soc., Vol. 9, 449 (1977). — Abstract.

097.052 Photodehydration of Martian dust.
K. L. Andersen, R. L. Huguenin.
Bull. American Astron. Soc., Vol. 9, 449 (1977). — Abstract.

097.053 Mechanisms for the formation of Martian channels.
M. C. Malin.
Bull. American Astron. Soc., Vol. 9, 449 (1977). — Abstract.

097.054 Mars before Tharsis: much larger obliquity in the
        past?   J. A. Burns, W. R. Ward, O. B. Toon.
Bull. American Astron. Soc., Vol. 9, 449 (1977). — Abstract.

097.055 Climatic change on Mars: hot poles at high obliquity.
O. B. Toon, W. R. Ward, J. A. Burns.
Bull. American Astron. Soc., Vol. 9, 450 (1977). — Abstract.

097.056 Observational characteristics of the Viking landing
        sites.   C. Capen.
Bull. American Astron. Soc., Vol. 9, 450 (1977). — Abstract.

097.057 Adsorption on the Martian regolith and the Martian
        atmospheric rare gas inventory.
F. P. Fanale, T. Owen, W. A. Cannon.
Bull. American Astron. Soc., Vol. 9, 450 - 451 (1977).
Abstract.

097.058 Solar heating in the Mars dusty atmosphere.
R. W. Zurek.
Bull. American Astron. Soc., Vol. 9, 451 (1977). — Abstract.

097.059 Martian atmospheric water vapor observations:
        preViking coverage.   E. S. Barker, M. A. Perry.
Bull. American Astron. Soc., Vol. 9, 451 (1977). — Abstract.

097.060 Martian atmospheric extinction and the central
        flash.   R. G. French, J. L. Elliot, P. J. Gierasch,
E. Dunham, J. Veverka, C. Church, C. Sagan.
Bull. American Astron. Soc., Vol. 9, 451 (1977). — Abstract.

097.061 Structure of the Martian upper atmosphere from
        airborne observations of the ϵ Gem occultation.
J. L. Elliot, R. G. French, E. Dunham, P. J. Gierasch, J.
Veverka, C. Church, C. Sagan.
Bull. American Astron. Soc., Vol. 9, 451 (1977). — Abstract.

097.062 The occultation of ϵ Gem by Mars on April 8, 1976.
R. L. Millis, L. H. Wasserman, N. M. White, C. W.
McCracken, R. M. Williamon.
Bull. American Astron. Soc., Vol. 9, 451 (1977). — Abstract.

097.063 Mean Martian upper atmosphere temperature and
        temperature fluctuations from ϵ Gem occultation.
W. B. Hubbard.
Bull. American Astron. Soc., Vol. 9, 451 - 452 (1977).
Abstract.

097.064 Volcanism on Mars and the Moon.   E. Schonfeld.
        Bull. American Astron. Soc., Vol. 9, 453 (1977).
Abstract.

097.065 High resolution ultraviolet spectrophotometry of
        Mars and Saturn by the TD1A satellite.
J. Caldwell.
Bull. American Astron. Soc., Vol. 9, 473 (1977). — Abstract.

097.066 The 8 April 1976 (U.T.) occultation of ϵ Gemino-
        rum by Mars.
D. S. Hall, D. O. Hamilton, A. M. Heiser, W. C. Keel.
Acta Astron., Vol. 27, 293 - 295 (1977).

097.067 How dusty is the Martian atmosphere?
D. French.
News Lett. Astron. Soc. N.Y., Vol. 1, No. 2, p. 12 (1977).
Abstract.

097.068 Occultation of ϵ Geminorum by Mars. II. The struc-
        ture and extinction of the Martian upper atmosphere.
J. L. Elliot, R. G. French, E. Dunham, P. J. Gierasch,
J. Veverka, C. Church, C. Sagan.
Astrophys. J., Vol. 217, 661 - 679 (1977).
        The occultation of ϵ Geminorum by Mars on 1976 April 8
was observed at three wavelengths and 4 ms time resolution
with the 91 cm telescope aboard NASA's G. P. Kuiper Airborne Observatory. Temperature, pressure, and number-density
profiles of the Martian atmosphere were obtained for both the
immersion and emersion events.

097.069 Mars surface properties observed by earth-based
        radar at 70-, 12.5-, and 3.8-cm wavelengths.
R. A. Simpson, G. L. Tyler, B. J. Lipa.
Icarus, Vol. 32, 147 - 167 (1977).
        A review of Mars radar data obtained through the 1973
opposition confirms that the surface of the planet has many
diverse characteristics. Analysis of the quasi-specular echo
component shows changes in apparent reflectivity of at least
5 to 1. If attributed entirely to variations in surface material,
these correspond to dielectric constants between 1.6 and 4.0.
Values of rms surface slope on 1- to 100-m scales range from
as low as 0.5° in tablelands near Valles Marineris to more than
3.0° in certain other areas such as inside Coprates Chasma
itself.

097.070 Photochemistry of the Martian atmosphere.
        T. Y. Kong, M. B. McElroy.
Icarus, Vol. 32, 168 - 189 (1977).
        A variety of models are explored to study the photo-
chemistry of CO_2 in the Martian atmosphere with emphasis on
reactions involving compounds of carbon, hydrogen, and
oxygen. Acceptable models are constrained to account for
measured concentrations of CO and O above 90 km, with an
additional requirement that they should be in accord with ob-

servations of CO, O_2, and O_3 in the lower atmosphere.

097.071 Ultraviolet observations of Mars and Saturn by the TD1A and OAO-2 satellites.   J. Caldwell.
Icarus, Vol. 32, 190 - 209 (1977).
Ultraviolet photometric and spectrophotometric observations of Mars and Saturn obtained by two Earth-orbiting satellites are combined in this report. High-resolution data from the S59 experiment aboard TD1A reveal no definite absorption features in the spectra of either planet. Broadband photometry from OAO-2 shows that atmospheric dust on Mars during the great dust storm of 1971 - 1972 reduced the ultraviolet geometric albedo by a factor of $\simeq 3$ at the height of the storm. A real brightness variation during a clear season is observed. The combined Saturn data from the two satellites strongly suggest that NH_3 does not influence the ultraviolet spectrum of Saturn, but that some other absorber does. OAO-2 broadband photometry of Jupiter and of Saturn demonstrate that these planets have very similar albedos from 2100 to 2500 Å. This implies a common ultraviolet absorber on both planets, other than NH_3.

097.072 An interpretation of photometric parameters of bright desert regions of Mars and their dependence on wavelength.   W. R. Weaver, W. E. Meador.
NASA Tech. Note, NASA TN D-8446. 18 pp. Price $ 3.50 (1977).
The photometric function developed by Meador and Weaver (1975) has been used with photometric data from the bright desert areas of Mars to determine the dependence of the three photometric parameters of the photometric function on wavelength and to provide qualitative predictions about the physical properties of the surface.

097.073 Tharsis: static or dynamic support?   R. J. Phillips.
NASA Tech. Mem., NASA TM X-3511, (see 012.010), p. 34 - 35 (1977). – Abstract.

097.074 Ancient surfaces of the terrestrial planets.   M. C. Malin.
NASA Tech. Mem., NASA TM X-3511, (see 012.010), p. 42 - 43 (1977). – Abstract.

097.075 Correlations: Martian stratigraphy and crater density.   D. H. Scott, C. D. Condit.
NASA Tech. Mem., NASA TM X-3511, (see 012.010), p. 56 - 58 (1977). – Abstract.

097.076 Timing of deformational events in the northern Tharsis bulge of Mars.   D. U. Wise.
NASA Tech. Mem., NASA TM X-3511, (see 012.010), p. 59 - 60 (1977). – Abstract.

097.077 Fracture domains of Italy – analogue study of brittle crustal behavior in a volcanic-tectonic region.
D. Wise, R. Funiciello, M. Parotto, F. Salvini.
NASA Tech. Mem., NASA TM X-3511, (see 012.010), p. 62 (1977). – Abstract.

097.078 The effect of nongravitational factors on the shape of Martian, lunar and Mercurian craters: target effects.   E. I. Smith, J. A. Hartnell.
NASA Tech. Mem., NASA TM X-3511, (see 012.010), p. 91 - 93 (1977). – Abstract.

097.079 Interplanetary comparisons of fresh crater morphology: preliminary results.
M. J. Cintala, C. A. Wood, J. W. Head, T. A. Mutch.
NASA Tech. Mem., NASA TM X-3511, (see 012.010), p. 94 - 96 (1977). – Abstract.

097.080 Crater and block populations at the Viking 1 landing site.   R. Arvidson, E. Guinness.
NASA Tech. Mem., NASA TM X-3511, (see 012.010), p. 108 - 110 (1977). – Abstract.

097.081 Ballistic impact ejecta in a Martian atmosphere.   P. H. Schultz, D. E. Gault.
NASA Tech. Mem., NASA TM X-3511, (see 012.010), p. 117 - 118 (1977). – Abstract.

097.082 Fine particles on the surface of Mars.   C. Sagan.
NASA Tech. Mem., NASA TM X-3511, (see 012.010), p. 143 (1977). – Abstract.

097.083 Sedimentary regimes on Mars.   R. Saunders.
NASA Tech. Mem., NASA TM X-3511, (see 012. 010), p. 144 - 145 (1977). – Abstract.

097.084 Recent results from the Martian Surface Wind Tunnel (MARSWIT).
R. Greeley, J. B. Pollack, J. D. Iversen, B. R. White.
NASA Tech. Mem., NASA TM X-3511, (see 012.010), p. 146 - 147 (1977). – Abstract.

097.085 On the nature and visibility of crater-associated streaks on Mars.
J. Veverka, P. Thomas, C. Sagan.
NASA Tech. Mem., NASA TM X-3511, (see 012.010), p. 155 (1977). – Abstract.

097.086 A statistical study of ragged dark streaks in the southern hemisphere of Mars.
J. Veverka, J. Goguen, K. Cook.
NASA Tech. Mem., NASA TM X-3511, (see 012.010), p. 156 (1977). – Abstract.

097.087 Crater streaks in the Chryse Planitia region of Mars: early Viking results.
R. Greeley, R. Papson, J. Veverka.
NASA Tech. Mem., NASA TM X-3511, (see 012.010), p. 157 - 158 (1977). – Abstract.

097.088 Evaporation of ice-choked rivers: application to Martian channels.   D. Wallace, C. Sagan.
NASA Tech. Mem., NASA TM X-3511, (see 012.010), p. 161 (1977). – Abstract.

097.089 Entrainment of sediment by fluid flow on Mars.   D. Nummedal.
NASA Tech. Mem., NASA TM X-3511, (see 012.010), p. 162 - 164 (1977). – Abstract.

097.090 Martian channels – classification by morphology and time of formation.   H. Masursky, J. M. Boyce,
A. L. Dial, G. G. Schaber, M. E. Strobell.
NASA Tech. Mem., NASA TM X-3511, (see 012.010), p. 165 (1977). – Abstract.

097.091 Characterization of erosional forms on Mars by Fourier analysis in closed form.
P. J. Brown, D. Nummedal, D. T. Eppler, R. Ehrlich.
NASA Tech. Mem., NASA TM X-3511, (see 012.010), p. 166 - 167 (1977). – Abstract.

097.092 Preliminary statistical analysis of some Martian channel networks.   D. Pieri.
NASA Tech. Mem., NASA TM X-3511, (see 012.010), p. 168 (1977). – Abstract.

097.093 Viking – slashing at the Martian scabland problem.   V. R. Baker.

NASA Tech. Mem., NASA TM X-3511, (see 012.010), p. 169 - 172 (1977). – Abstract.

097.094   A large landslide on Mars.
B. K. Lucchitta.
NASA Tech. Mem., NASA TM X-3511, (see 012.010), p. 178 - 179 (1977). – Abstract.

097.095   Volatile evolution.   F. P. Fanale.
NASA Tech. Mem., NASA TM X-3511, (see 012. 010), p. 183 - 186 (1977). – Abstract.

097.096   Climatic change on Mars: inferences based on Viking and Mariner data.   J. B. Pollack.
NASA Tech. Mem., NASA TM X-3511, (see 012.010), p. 187 - 188 (1977). – Abstract.

097.097   Permafrost on Mars.   M. Coradini, R. Bianchi.
NASA Tech. Mem., NASA TM X-3511, (see 012. 010), p. 189 - 190 (1977). – Abstract.

097.098   UV radiational effect on: Martian atmospheric and regolith water.
P. H. Nadeau, R. C. Reynolds, A. R. Tice, D. M. Anderson.
NASA Tech. Mem., NASA TM X-3511, (see 012.010), p. 191 - 196 (1977). – Abstract.

097.099   Superoxides and liquid water on Mars.
R. Smoluchowski.
NASA Tech. Mem., NASA TM X-3511, (see 012.010), p. 197 - 199 (1977). – Abstract.

097.100   Carbonate formation on Mars.
M. C. Booth, H. H. Kieffer.
NASA Tech. Mem., NASA TM X-3511, (see 012.010), p. 200 (1977). – Abstract.

097.101   Mars: surface mineralogy from reflectance spectra.
R. L. Huguenin, J. B. Adams, T. B. McCord.
NASA Tech. Mem., NASA TM X-3511, (see 012.010), p. 201 - 203 (1977). – Abstract.

097.102   Spectrophotometry of Mars by way of the Viking lander cameras.   R. Arvidson, S. Bragg.
NASA Tech. Mem., NASA TM X-3511, (see 012.010), p. 204 - 206 (1977). – Abstract.

097.103   Mars geologic mapping.   D. H. Scott.
NASA Tech. Mem., NASA TM X-3511, (see 012. 010), p. 209 - 210 (1977). – Abstract.

097.104   The control net of Mars.   M. E. Davies.
NASA Tech. Mem., NASA TM X-3511, (see 012. 010), p. 211 (1977). – Abstract.

097.105   Mars 1 : 5,000,000 mapping.   R. M. Batson.
NASA Tech. Mem., NASA TM X-3511, (see 012. 010), p. 212 - 213 (1977). – Abstract.

097.106   Geologic map of the Mare Australe area of Mars (1/5m).   C. D. Condit, L. A. Soderblom.
NASA Tech. Mem., NASA TM X-3511, (see 012.010), p. 214 (1977). – Abstract.

097.107   Geology of the Ismenius Lacus quadrangle (MC-5), Mars.   B. K. Lucchitta.
NASA Tech. Mem., NASA TM X-3511, (see 012.010), p. 215 - 216 (1977). – Abstract.

097.108   Geologic map of the Amazonis quadrangle (MC-8).
E. C. Morris, S. E. Dwornik.
NASA Tech. Mem., NASA TM X-3511, (see 012.010), p. 217 -

219 (1977). – Abstract.

097.109   Geology of the Phoenicis Lacus quadrangle, Mars (MC-17).
H. Masursky, A. L. Dial, M. E. Strobell.
NASA Tech. Mem., NASA TM X-3511, (see 012.010), p. 220 (1977). – Abstract.

097.110   Geologic map of the Iapygia quadrangle of Mars.
G. G. Schaber.
NASA Tech. Mem., NASA TM X-3511, (see 012.010), p. 221 - 222 (1977). – Abstract.

097.111   Geology of the Aeolis quadrangle of Mars.
D. H. Scott, E. C. Morris.
NASA Tech. Mem., NASA TM X-3511, (see 012.010), p. 223 - 225 (1977). – Abstract.

097.112   Geology of the Phaethontis quadrangle, Mars.
J. H. Howard III.
NASA Tech. Mem., NASA TM X-3511, (see 012.010), p. 226 - 227 (1977). – Abstract.

097.113   Knob-and-mesa terrains, dissected plateaus, and knobby plains of the Atlantis-Mare Sirenum region, Phaethontis quadrangle, Mars.   J. H. Howard III.
NASA Tech. Mem., NASA TM X-3511, (see 012.010), p. 228 - 229 (1977). – Abstract.

097.114   Mars soil-water analyzer: instrument description and status.
D. M. Anderson, J. B. Stephens, F. P. Fanale, A. R. Tice.
NASA Tech. Mem., NASA TM X-3511, (see 012.010), p. 260 - 266 (1977). – Abstract.

097.115   Selection of Viking landing sites based on Viking and Mariner 9 images and ground based radar data.
H. Masursky, G. G. Schaber, C. Elachi, M. E. Strobell, A. L. Dial.
NASA Tech. Mem., NASA TM X-3511, (see 012.010), p. 270 - 271 (1977). – Abstract.

097.116   Development of a seismometer for Viking.
D. L. Anderson.
NASA Tech. Mem., NASA TM X-3511, (see 012.010), p. 281 - 283 (1977). – Abstract.

097.117   Requiem for the canals.   P. Moore.
J. British Astron. Assoc., Vol. 87, 589 - 593 (1977).

097.118   On prediction of the structure of the surface layer of the planets Mars and Mercury.   D. P. Volkov,
G. N. Dul'nev, Yu. P. Zarichnyak, B. L. Muratova.
Izv. vyssh. uchebn. zaved. Radiofiz., Vol. 20, 45 - 50 (1977).
In Russian. – Abstr. in Ref. zh., 51. Astron., 9.51.264 (1977).

097.119   Global planetary lineaments and the nature of Martian channels.   G. N. Katterfel'd.
Vopr. izuch. planetarn. treshchinovatosti. Leningrad, 1976, p. 94 - 101. In Russian. – Abstr. in Ref. zh., 51. Astron., 9.51.272; 62. Issled. kosm. prostranstva, 9.62.172 (1977).

097.120   The physical parameters of Mars' atmosphere.
C. J. Macris, B. C. Petropoulos.
C. R. Acad. Sci. Paris, Tome 285, Sér. B, 239 - 241 (1977).
In French.
    The physical parameters of Mars' atmosphere have been calculated from 0–190 km, using Viking 1 measurements.

097.121   The prime meridian of Mars and the longitudes of the Viking landers.   M. E. Davies.
Science, Vol. 197, 1277 (1977).

097.122 CO$_2$ photoionization and energy distribution of photoelectrons in the atmospheres of Mars and Venus.   A. V. Dembovskij.
Kosm. Issled., Vol. 15, 607 - 618 (1977). In Russian.

097.123 Height profile of the water vapour concentration on Mars.   V. A. Krasnopol'skij, V. A. Parshev.
Kosm. Issled., Vol. 15, 776 - 780 (1977). In Russian.

097.124 Investigation of the figure of the physical surface of planets from space photographs of their limbs (for example of photographs obtained from space apparatus Mars 3).
A. P. Tishchenko, L. I. Permitina.
Kosm. Issled., Vol. 15, 781 - 787 (1977). In Russian.

097.125 The moment of inertia of Mars and the existence of a core.   A. H. Cook.
Geophys. J. R. Astron. Soc., Vol. 51, 349 - 356 (1977).
    A relation is obtained between the true value of the moment of inertia of a planet and the value calculated from the dynamical form factor, J_2, on the assumption of hydrostatic equilibrium. The result is applied to Mars and it is shown that the difference between the true and calculated moments of inertia is probably insignificant in considering models of the interior of Mars and in particular does not affect an argument for a core based on models calculated by Lyttleton.

097.126 Shield volcanism and lithospheric structure beneath the Tharsis plateau, Mars.
K. R. Blasius, J. A. Cutts.
Proc. Seventh Lunar Sci. Conf., (see 012.015), 3561 - 3573 (1976).

097.127 Martian crater depth/diameter relationships: comparison with the Moon and Mercury.
M. J. Cintala, J. W. Head, T. A. Mutch.
Proc. Seventh Lunar Sci. Conf., (see 012.015), 3575 - 3587 (1976).

097.128 Comparison of large crater and multiringed basin populations on Mars, Mercury, and the Moon.
M. C. Malin.
Proc. Seventh Lunar Sci. Conf., (see 012.015), 3589 - 3602 (1976).

097.129 Mars:photodesorption from mineral surfaces and its effects on atmospheric stability.
R. L. Huguenin, R. G. Prinn, M. Maderazzo.
Icarus, Vol. 32, 270 - 298 (1977).
    A mechanism has been proposed for uv-accelerated desorption from Fe^{2+} sites on mineral surfaces that satisfies kinetic constraints determined in the laboratory by Huguenin. The process is an integral step of the photochemical weathering mechanism for producing dust on Mars, and it now appears that it may play primary roles in stabilizing CO$_2$ against dissociation by sunlight and in controlling the oxidation state of the atmosphere.

097.130 Search for life on Mars.   S. A. Nikitin.
Priroda, 1977, No. 10, p. 118 - 125. In Russian.

097.131 Mars – a water planet?   E. Burgess.
New Scientist, Vol. 72, 152 - 153 (1976).
Abstr. in Phys. Abstr., Vol. 80, Abstr. 22130 (1977).

097.132 Behavior of volatiles in Mars' polar areas: a model incorporating new experimental data.
D. W. Davies, C. B. Farmer, D. D. LaPorte.
J. Geophys. Res., Vol. 82, 3815 - 3822 (1977).
    A model has been developed to explain the north polar water vapor results obtained by the Viking orbiter Mars atmospheric water detector; it has also been used to compute the thickness of seasonally deposited CO$_2$ frost, the variation of the total atmospheric pressure, and wind velocities due to mass motions associated with CO$_2$ condensation.

097.133 The search for life on Mars.   C. Sagan.
Spaceworld, Vol. M-10-154, 4 - 22 (1976). – Abstr. in Phys. Abstr., Vol. 80, Abstr. 25969 (1977).

097.134 Viking-1 meteorological measurements: first impressions.   S. L. Hess, R. M. Henry, C. B. Leovy, J. A. Ryan, J. E. Tillman.
Bull. American Meteorol. Soc., Vol. 57, 1150 - 1151 (1976).
From Phys. Abstr., Vol. 80, Abstr. 29686 (1977).

097.135 Estimate of Martian soil density from data of radiophysical measurements at 3 cm.
N. N. Krupenio.
Inst. kosm. issled. AN SSSR. Prepr. Pr-328. Moskva, 1977. 31 pp. In Russian. – Abstr. in Ref. zh., 51. Astron., 10.51.306; 62. Issled. kosm. prostranstva, 10.62.159 (1977).

097.136 Cartographic coverage of the Martian surface.   K. B. Shingareva, T. G. Zargaryan, K. I. Kondratskaya.
Geod. i kartogr., 1977, No. 3, p. 71 - 76. In Russian. – Abstr. in Ref. zh., 52. Geod. Aehrosemka, 10.52.87 (1977).

097.137 Comments on : "Characteristics of fresh martian craters as a function of diameter: comparison with the Moon and Mercury" by M. J. Cintala et al.
C. P. Florensky (*K. P. Florenskij*), A. T. Basilevsky (*Bazilevskij*), V. P. Polosukhin.
Geophys. Res. Lett., Vol. 4, 243 - 244 (1977).

097.138 Characteristics of fresh martian craters as a function of diameter: comparison with the Moon and Mercury – discussion.   M. J. Cintala, J. W. Head, T. A. Mutch.
Geophys. Res. Lett., Vol. 4, 245 - 246 (1977).

097.139 Viking on Mars: exciting results.   W. K. Hartmann.
Astronomy, Vol. 5, No. 1, p. 6 - 24 (1977). – Abstr. in Phys. Abstr., Vol. 80, Abstr. 33383 (1977).

097.140 Atmospheric measurements on Mars: the Viking Meteorology Experiment.
T. E. Chamberlain, H. L. Cole, R. G. Dutton, G. C. Greene, J. E. Tillman.
Bull. American Meteorol. Soc., Vol. 57, 1094 - 1104 (1976). Abstr. in Phys. Abstr., Vol. 80, Abstr. 33384 (1977).

097.141 The search for life on Mars.   N. H. Horowitz.
Sci. American, Vol. 237, No. 5, p. 52 - 61 (1977).
    The Viking landers have completed their biological experiments. The experiments did not detect life processes, but they did reveal much of interest about the chemistry of the surface of the planet.

097.142 Life on Mars: ambiguous results.   T. A. Easton.
Astronomy, Vol. 5, No. 1, p. 26 - 33 (1977).
Abstr. in Phys. Abstr., Vol. 80, Abstr. 37546 (1977).

097.143 Labeled release – an experiment in radiorespirometry.   G. V. Levin, P. A. Straat.
Origins of Life, Vol. 7, 293 - 311 (1976). – Abstr. in Phys. Abstr., Vol. 80, Abstr. 37548 (1977).

097.144 The search for life on Mars: Viking 1976 gas changes as indicators of biological activity.   V. I. Oyama, B. J. Berdahl, G. C. Carle, M. E. Lehwalt, H. S. Ginoza.
Origins of Life, Vol. 7, 313 - 333 (1976). – Abstr. in Phys. Abstr., Vol. 80, Abstr. 37549 (1977).

097.145 **Exploration strategy for Mars and the role of the sample return mission.**   E. K. Gibson, Jr., D. D. Bogard, M. B. Duke, J. Minear, L. E. Nyquist, W. C. Phinney, J. L. Warner.
Meteoritics, Vol. 12, 236 - 238 (1977). − Abstract.

097.146 **On the structure of the Martian upper atmosphere from data of experiments aboard the Viking space vehicles.**   M. N. Izakov.
Inst. kosm. issled. AN SSSR. Pr-343. Moskva, 1977. 20 pp.
In Russian. − Abstr. in Ref. zh., 62. Issled. kosm. prostranstva, 12.62.130 (1977).

097.147 **A size-frequency study of large Martian craters.**   A. Woronow.
J. Geophys. Res., Vol. 82, 5807 - 5820 (1977).
    Martian craters in the size range 10–250 km follow a log normal size-frequency distribution law. Analysis techniques based on the log normal model yield possible evidence for the size-frequency evolution of crater-producing bodies. Some regions on Mars display excessive depletions of either large or small craters; the most likely causes of the depletions are considered.

097.148 **Magnetic field and plasma inside and outside of the Martian magnetosphere.**
Sh. Sh. Dolginov, Ye. G. Yeroshenko (*E. G. Eroshenko*), L. N. Zhuzgov, V. A. Sharova, K. I. Gringauz, V. V. Bezrukikh, T. K. Breus, M. I. Verigin, A. P. Remizov.
Solar-wind interaction with the planets Mercury, Venus, and Mars, (see 012.047), p. 1 - 20 (1977).
    This paper deals with the results of a joint consideration of the magnetic and plasma data measured with wide-angle detectors. The authors considered that the magnetosphere formed by the intrinsic magnetic field of Mars is an obstacle that creates the shock wave detected during all the near-Mars magnetic and plasma measurements.

097.149 **On the nature of the solar-wind-Mars interaction.**   O. L. Vaisberg (*Vajsberg*), A. V. Bogdanov, V. N. Smirnov, S. A. Romanov.
Solar-wind interaction with the planets Mercury, Venus, and Mars, (see 012.047), p. 21 - 40 (1976).
    The results of plasma measurements near Mars on the USSR Mars-2, -3, and -5 spacecraft are considered. The data are compared with simultaneous magnetic measurements. Strong evidence is obtained in favor of a direct interaction and mass exchange between the solar-wind plasma and the gaseous envelope of Mars.

097.150 **The nighttime ionosphere of Mars from Mars-4 and Mars-5 radio occultation dual-frequency measurements.**   N. A. Savich, V. A. Samovol, M. B. Vasilyev (*Vasil'ev*), A. S. Vyshlov, L. N. Samoznaev, A. I. Sidorenko, D. Ya. Shtern.
Solar-wind interaction with the planets Mercury, Venus, and Mars, (see 012.047), p. 41 - 46 (1976).
    Dual-frequency radio sounding of the Martian nighttime ionosphere was carried out during the exits from behind the planet of the Mars-4 spacecraft on February 2, 1974 and the Mars-5 spacecraft on February 18, 1974. In these experiments, the spacecraft transmitter emitted two coherent monochromatic signals in decimeter ($\lambda_1 \approx 32$ cm) and centimeter ($\lambda_2 \approx 8$ cm) wavelength ranges. The nighttime ionosphere of Mars measured in both cases had a peak electron density of $\sim 5 \times 10^3/cm^3$ at an altitude of 110 to 130 km.

097.151 **Viking 1. Early results.**
    Foreword by J. C. Fletcher, preface by J. E. Naugle, introduction by J. S. Martin, Jr., G. A. Soffen.
Natl. Aeronaut. Space Adm., Washington, D.C., NASA SP-408.
For sale by the National Technical Information Service,
Springfield, Va. 22151. 7 + 69 pp. Price $ 2.00 (1976).

097.152 **On the habitability of Mars. An approach to planetary ecosynthesis.**
M. M. Averner, R. D. MacElroy (Editors), with contributions by S. Berman, W. R. Kuhn, P. W. Langhoff, S. R. Rogers, J. W. Thomas, R. D. MacElroy, M. M. Averner.
Natl. Aeronaut. Space Adm., Washington, D.C., NASA SP-414.
For sale by the National Technical Information Service, Springfield, Virginia 22161. 11 + 105 pp. Price $ 5.25 (1976).

097.153 **Observing Mars VII—the 1977-78 aphelic apparition.**   C. F. Capen.
Strolling Astron., Vol. 27, 34 - 42 (1977).
    The 1977-78 Martian apparition geometric values, seasonal aspects, and observational possibilities are given. Graphs, maps, and a calendar of events are presented for the telescopic observation of Mars.

097.154 **Study of the Martian surface relief using photographs of its limb.**   I. K. Lur'e, L. I. Permitina, V. A. Poloznikov, A. P. Tishchenko.
Izv. Vses. geogr. o-va, Vol. 109, 362 - 369 (1977). In Russian.
From Ref. zh., 51. Astron., 1.51.269 (1978).

097.155 **Duststorms of Mars.**   S. B. Idso.
Astronomy, Vol. 5, No. 3, p. 34 - 39 (1977).
Abstr. in Phys. Abstr., Vol. 80, Abstr. 49324 (1977).

097.156 **Windblown dust on Earth, Mars and Venus.**   J. D. Iversen, R. Greeley, J. B. Pollack.
J. Atmos. Sci., Vol. 33, 2425 - 2429 (1976). − Abstr. in Phys. Abstr., Vol. 80, Abstr. 45083 (1977).

097.157 **Influence of planetary-scale topography on the diurnal thermal tide during the 1971 Martian dust storm.**   B. J. Conrath.
J. Atmos. Sci., Vol. 33, 2430 - 2439 (1976). − Abstr. in Phys. Abstr., Vol. 80, Abstr. 45341 (1977).

097.158 **A summary of the Viking Project.**
    Lab. Equip. Dig., Vol. 15, No. 2, p. 99, 101, 103, 105, 107, 109 (1977). − Abstr. in Phys. Abstr., Vol. 80, Abstr. 56851 (1977).

097.159 **Microwave absorption in the Martian atmosphere.**   R. K. Johri, J. C. Joshi.
Indian J. Radio Space Phys., Vol. 5, 332 - 336 (1976). − Abstr. in Phys. Abstr., Vol. 80, Abstr. 71045 (1977).

097.160 **Viking experiments on Mars.**   C. Royer.
    Mem. Sci. Rev. Metall., Vol. 74, No. 4, p. 261 - 263 (1977). In French. − Abstr. in Phys. Abstr., Vol. 80, Abstr. 71046 (1977).

097.161 **Spherical harmonic representation of the Martian gravity using a short-arc technique.**
E. F. Daniels, R. H. Tolson, J. P. Gapcynski.
J. Spacecr. Rockets, Vol. 14, 323 - 327 (1977). − Abstr. in Phys. Abstr., Vol. 80, Abstr. 73999 (1977).

097.162 **Mars as a member of the solar system.**
    Spaceworld, Vol. N-5-161, 4 - 11 (1977). − Abstr. in Phys. Abstr., Vol. 80, Abstr. 74002 (1977).

097.163 **Some Martian volcanic features as viewed from the Viking orbiters.**   M. H. Carr, R. Greeley, K. R. Blasius, J. E. Guest, J. B. Murray.
J. Geophys. Res., Vol. 82, (see 003.017), 3985 - 4015 (1977). Paper No. 7S0470.
    Contents: Arsia Mons and the south Tharsis plains; Distribution of vents; Olympus Mons and vicinity; Volcanics

of Alba Patera.

097.164 Classification and time of formation of Martian channels based on Viking data. H. Masursky, J. M. Boyce, A. L. Dial, G. G. Schaber, M. E. Strobell.
J. Geophys. Res., Vol. 82, (see 003.017), 4016 - 4038 (1977). Paper No. 7S0566.

Broad channels originate in areas of collapsed terrain that may have been formed when subsurface water-ice (permafrost) was melted by geothermal heat from deep-seated volcanic centers. Conditions are reviewed for persistence of liquid water on Mars under present and more favorable pressure and temperatures. Sinuous channels of intermediate size and other shorter, stubby channels, have multiple tributaries, climatic warming may be required to explain their formation. The final fluviatile type, dendritic channel networks, has the widest areal distribution and appears to have been formed during at least two episodes. The filamentous channels in their source areas (often the rims of craters) seem to resemble terrestrial river systems; rainfall would seem to be required to form these features. Simple and complex lava channels are common: they originate at volcanic centers and are usually morphologically distinct from the aqueous channels. Three types of lava channels are recognized.

097.165 Martian permafrost features. M. H. Carr, G. G. Schaber.
J. Geophys. Res., Vol. 82, (see 003.017), 4039 - 4054 (1977). Paper No. 7S0450.

The outgassing history of Mars and the prevailing temperature conditions suggest that ground ice may occur to depths of kilometers over large areas of the planet. The presence of permafrost is also indicated by several topographic features that resemble those found in periglacial regions of the earth. Different observations support permafrost conditions not only at present but also for much of the planet's history.

097.166 Martian impact craters and emplacement of ejecta by surface flow. M. H. Carr, L. S. Crumpler, J. A. Cutts, R. Greeley, J. E. Guest, H. Masursky.
J. Geophys. Res., Vol. 82, (see 003.017), 4055 - 4065 (1977). Paper No. 7S0488.

Several types of Martian impact craters have been recognized. The most common type, the rampart crater, is distinctively different from lunar and Mercurian craters. It is typically surrounded by several layers of ejecta, each having a low ridge or escarpment at its outer edge. The internal features of Martian craters, in general, resemble their lunar and Mercurian counterparts except that the transition from bowl shaped to flat floored takes place at about 5-km diameter, a smaller size than is true for Mercury or the moon.

097.167 Geology of the Valles Marineris: first analysis of imaging from the Viking 1 orbiter primary mission. K. R. Blasius, J. A. Cutts, J. E. Guest, H. Masursky.
J. Geophys. Res., Vol. 82, (see 003.017), 4067 - 4091 (1977). Paper No. 7S0524.

Reported here are new insights into the evolution of the canyon system and possible evidence for cyclical climate change from the equatorial region. Tectonic control appears to be the fundamental influence on canyon form and evolution, but the style or intensity of tectonism appears to be regionally variable.

097.168 Geology of Chryse Planitia. R. Greeley, E. Theilig, J. E. Guest, M. H. Carr, H. Masursky, J. A. Cutts.
J. Geophys. Res., Vol. 82, (see 003.017), 4093 - 4109 (1977). Paper No. 7S0509.

Viking orbiter pictures reveal more surface detail of the area and show the basin to be more complex than was seen on Mariner 9 images.

097.169 Geological observations in the Cydonia region of Mars from Viking. J. E. Guest, P. S. Butterworth, R. Greeley.
J. Geophys. Res., Vol. 82, (see 003.017), 4111 - 4120 (1977). Paper No. 7S0486.

The authors present a geological map of the area designated the B1 landing site in Cydonia centered on 45°N latitude, 4°W longitude.

097.170 Martian dynamical phenomena during June—November 1976: Viking orbiter imaging results. G. Briggs, K. Klaasen, T. Thorpe, J. Wellman, W. Baum.
J. Geophys. Res., Vol. 82, (see 003.017), 4121 - 4149 (1977). Paper No. 7S0480.

Contents: Clouds in the Tharsis/Amazonis regions; Memnonia brightening; South polar region: clouds, surface frost; North polar cap; Dust storm activity.

097.171 Viking orbiter observations of atmospheric opacity during July—November 1976. T. E. Thorpe.
J. Geophys. Res., Vol. 82, (see 003.017), 4151 - 4159 (1977). Paper No. 7S0457.

Viking orbiter photography during the primary mission combined with lander indications of surfac properties have permitted the estimation of atmospheric optical depth and phase functions. Highly variable time of day opacities ranging from 0.05 to 0.6 are seen to occur in three principle regions. A wavelength-dependent particulate component plus a time variable grey aerosol of higher density may explain these opacities versus time of day. These data should serve as a basis for extended mission comparisons.

097.172 Viking orbiter photometric observations of the Mars phase function July through November 1976. T. E. Thorpe.
J. Geophys. Res., Vol. 82, (see 003.017), 4161 - 4165 (1977). Paper No. 7S0456.

Over 7200 Viking orbiter pictures have provided phase function information over a large range in viewing geometry. Comparison with the earlier Mariner 9 data reveals possible significant changes. A two-component limb darkening characterization is shown to fit the data better at large phase angles than the traditional Minnaert or Lommel-Seeliger approach. The phase integral is 15% larger than the Mariner 9 observations owing in part to data obtained at larger phase angles revealing apparent condensate phenomena.

097.173 A study of variable features on Mars during the Viking primary mission. J. Veverka, P. Thomas, R. Greeley.
J. Geophys. Res., Vol. 82, (see 003.017), 4167 - 4187 (1977). Paper No. 7S0571.

Very few surface changes were seen during the Viking primary mission in 1976, a result consistent with predictions of relatively low wind velocities during northern summer. No eolian activity was detected from orbit in the vicinity of either landing site. Comparison of specific albedo boundaries in the 1972 Mariner 9 coverage and in the 1976 Viking coverage revealed that in many cases, subtle changes in outline and/or contrast have occurred during the past 4 years. In a few areas the albedo patterns in 1976 are dramatically different from those in 1972. Some new light streaks have formed since 1972 and a few old ones have disappeared. Many dark streaks have changed conspicuously in both outline and direction since 1972.

097.174 Mars: water vapor observations from the Viking orbiters. C. B. Farmer, D. W. Davies, A. L. Holland, D. D. LaPorte, P. E. Doms.
J. Geophys. Res., Vol. 82, (see 003.017), 4225 - 4248 (1977). Paper No. 7S0500.

The results of observations of the spatial and temporal

variation of water vapor during the Viking primary mission are reported. The Mars atmospheric water detector is a five-channel grating spectrometer operating in the 1.4-μm water vapor bands. The seasonal period covered here is the northern summer solstice to the following equinox.The vapor has been seen to exhibit variability with local time, elevation, and latitude, each of these in turn varying with season.

097.175 **Thermal and albedo mapping of Mars during the Viking primary mission.**
H. H. Kieffer, T. Z. Martin, A. R. Peterfreund, B. M. Jakosky, E. D. Miner, F. D. Palluconi.
J. Geophys. Res., Vol. 82, (see 003.017), 4249 - 4291 (1977). Paper No. 7S0511.

Measurements of Martian emission and reflection reveal wide variations of surface properties and indicate the presence of a larger atmospheric contribution to the observed radiances than was anticipated. Temperatures observed during the Viking primary mission range from 130 to 290 K. Surface thermal inertias from 1.6 to 11×10^{-3} cal cm^{-2} s$^{-1/2}$ K^{-1} are mapped, and they correlate with surficial geologic units. An equatorial map of bolometric albedo generally correlated with prior narrowband observations. These albedos range from 0.09 to 0.43; some regional brightenings are atmospheric in origin.

097.176 **The Viking radio science investigations.**
W. H. Michael, Jr., R. H. Tolson, J. P. Brenkle, D. L. Cain, G. Fjeldbo, C. T. Stelzried, M. D. Grossi, I. I. Shapiro, G. L. Tyler.
J. Geophys. Res., Vol. 82, (see 003.017), 4293 - 4295 (1977). Paper No. 7S0383.

The Viking radio science investigations use the highly accurate radio tracking and communications systems data from the Viking orbiters and landers to perform a number of analyses concerning properties of Mars and its environment. This paper gives a general description of the investigations and of the instrumentation used; detailed results are presented in the companion papers.

097.177 **Lander locations, Mars physical ephemeris, and solar system parameters: determination from Viking lander tracking data.**   A. P. Mayo, W. T. Blackshear, R. H. Tolson, W. H. Michael, Jr., G. M. Kelly, J. P. Brenkle, T. A. Komarek.
J. Geophys. Res., Vol. 82, (see 003.017), 4297 - 4303 (1977). Paper No. 7S0426.

The lander tracking data contain information on the physical ephemeris of Mars (rotation axis orientation, rotation rate, precession, nutation), orbits of Mars and the earth, and parameters affecting the orbital motions. Approximately 6 months of data have been analyzed to obtain the results presented in this paper. The results obtained to date are the lander locations, the Mars rotation axis orientation, and the rotation rate. The analyses indicate that the additional tracking data of the Viking extended mission are required before significant improvements in the Mars rotation axis motion and parameters affecting the orbital motion can be obtained.

097.178 **Bistatic radar measurements of electrical properties of the Martian surface.**
C. H. Tang, T. I. S. Boak III, M. D. Grossi.
J. Geophys. Res., Vol. 82, (see 003.017), 4305 - 4315 (1977). Paper No. 7S0534.

The Viking lander-to-orbiter relay links make it possible to perform measurements of the electrical properties of the Martian surface by the bistatic technique. The signal amplitude's fading patterns collected with the Lander 1 to Orbiter 1 relay link were of good quality and led to the determination of ϵ_r = 3.3 ± 0.7 in the vicinity of Lander 1. These electrical properties are similar to those of pumice and tuff. The dielectric constant of the surface near the Lander 2 site is estimated to be ϵ_r = 2.8–12.5.

097.179 **Viking radio occultation measurements of the Martian atmosphere and topography: primary mission coverage.**   G. Fjeldbo, D. Sweetnam, J. Brenkle, E. Christensen, D. Farless, J. Mehta, B. Seidel, W. Michael, Jr., A. Wallio, M. Grossi.
J. Geophys. Res., Vol. 82, (see 003.017), 4317 - 4324 (1977). Paper No. 7S0429.

Radio occultation measurements were made at approximately 50 locations on Mars with the Viking Orbiter 1 S (2.3 GHz) and X (8.4 GHz) band tracking links during October 1976. The measurements have been used to study the topography and atmosphere of Mars at latitudes ranging from about 75°S to 70°N. The mean atmospheric pressure at the areoid level was found to be 5.9 mbar during the northern midsummer season. By comparing the new electron density measurements with earlier Mariner data the authors have determined that the temperature and the plasma scale height of the upper atmosphere appear to be functions of solar activity.

097.180 **Mars gravity field: combined Viking and Mariner 9 results.**
J. P. Gapcynski, R. H. Tolson, W. H. Michael, Jr.
J. Geophys. Res., Vol. 82, (see 003.017), 4325 - 4327 (1977). Paper No. 7S0487.

A sixth-degree and sixth-order Martian gravitational field has been obtained from a short-arc analysis of Viking and Mariner 9 spacecraft tracking data. The equipotential contours obtained from this field are expected to be accurate to about 75 m within the ±65° latitude band. Analysis of the residual patterns indicates that a higher-order gravity field will be necessary to adequately represent the Tharsis and Hellas regions of Mars.

097.181 **Composition and structure of Mars' upper atmosphere: results from the neutral mass spectrometers on Viking 1 and 2.**   A. O. Nier, M. B. McElroy.
J. Geophys. Res., Vol. 82, (see 003.017), 4341 - 4349 (1977). Paper No. 7S0485.

The upper atmospheric mass spectrometers flown on Viking 1 and 2 are described, and results obtained for the composition and structure of Mars' upper atmosphere are summarized. Carbon dioxide is the major constituent of the atmosphere at all heights below 180 km. The thermal structure of the upper atmosphere is complex and variable with average temperatures below 200°K for both Viking 1 and 2. The atmosphere is mixed to heights in excess of 120 km. The isotopic composition of carbon and oxygen in the Martian atmosphere is similar to that in the terrestrial atmosphere: ^{15}N is enriched in Mars' atmosphere by a factor of 1.62 ± 0.16.

097.182 **The Martian ionosphere as observed by the Viking retarding potential analyzers.**
W. B. Hanson, S. Sanatani, D. R. Zuccaro.
J. Geophys. Res., Vol. 82, (see 003.017), 4351 - 4363 (1977). Paper No. 7S0523.

The retarding potential analyzers on the Viking landers obtained the first in situ measurements of ions from the Martian ionosphere. Ion concentration, ion temperatures, ion velocities were measured at 130 - 250 km altitude. On Viking 2, considerably more structure was observed in the height profiles of ionospheric quantities, although they were similar in shape to the Viking 1 profiles.

097.183 **Structure of the atmosphere of Mars in summer at mid-latitudes.**   A. Seiff, D. B. Kirk.
J. Geophys. Res., Vol. 82, (see 003.017), 4364 - 4378 (1977). Paper No. 7S0499.

The structure of Mars' atmosphere was measured in situ by instruments on board the two Viking landers from an altitude of 120 km to near the surface.

097.184 **Photochemistry and evolution of Mars' atmosphere:**

a Viking perspective.
M. B. McElroy, T. Y. Kong, Y. L. Yung.
J. Geophys. Res., Vol. 82, (see 003.017), 4379 - 4388 (1977).
Paper No. 7S0558.

Viking measurements of the Martian upper atmosphere indicate thermospheric temperatures below 200°K. The variability in thermospheric temperature may reflect an important dynamical coupling of upper and lower regions of the Martian atmosphere. Absorption of extreme ultraviolet solar radiation provides an important source of fast N and O atoms. It appears that the abundance of N_2 in Mars' past atmosphere may have exceeded the abundance of CO_2 in the present atmosphere and that the planet also has copious sources of H_2O.

097.185 Spectrophotometric and color estimates of the
         Viking lander sites.   F. O. Huck, D. J. Jobson,
S. K. Park. S. D. Wall, R. E. Arvidson, W. R. Patterson,
W. D. Benton.
J. Geophys. Res., Vol. 82, (see 003.017), 4401 - 4411 (1977).
Paper No. 7S0502.

The spectral radiance and color of the Martian sky and soil and the spectral reflectance of soil features are estimated from six-channel (0.4–1.0 μm) spectral data obtained with the Viking lander cameras.

097.186 Particle motion on Mars inferred from the Viking
         lander cameras.   C. Sagan, D. Pieri, P. Fox,
R. E. Arvidson, E. A. Guinness.
J. Geophys. Res., Vol. 82, (see 003.017), 4430 - 4438 (1977).
Paper No. 7S0590.

The cameras of the Viking landers have uncovered several lines of evidence for fine particle mobility on the Martian surface, including particulate drifts, rock-associated raised streaks, and probable ventifacts. Inferred peak wind directions in both Chryse and Utopia are roughly the same and are consistent with peak winds inferred by orbiter photography.

097.187 The geology of the Viking lander 1 site.
         A. B. Binder, R. E. Arvidson, E. A. Guinness,
K. L. Jones, E. C. Morris, T. A. Mutch, D. C. Pieri, C. Sagan.
J. Geophys. Res., Vol. 82, (see 003.017), 4439 - 4451 (1977).
Paper No. 7S0541.

Viking 1 landed on volcanic terrain in the plains of Chryse. Stereo pictures reveal an undulating topography.

097.188 The geology of the Viking lander 2 site.
         T. A. Mutch, R. E. Arvidson, A. B. Binder, E. A.
Guinness, E. C. Morris.
J. Geophys. Res., Vol. 82, (see 003.017), 4452 - 4467 (1977).
Paper No. 7S0537.

Viking lander 2 landed on a flat plain of fine-grained sediment overlain by dispersed, evenly distributed boulders.

097.189 Lander imaging as a detector of life on Mars.
         E. C. Levinthal, K. L. Jones, P. Fox, C. Sagan.
J. Geophys. Res., Vol. 82, (see 003.017), 4468 - 4478 (1977).
Paper No. 7S0501.

Biological goals were among the important science objectives of the Viking lander camera. The camera performance characteristics relevant to these goals are discussed. They include the ability to observe (1) morphological detail, (2) color and reflectance spectra, and (3) motion and change. The scenes obtained by the cameras were scrutinized in many ways: monoscopically, stereoscopically, in color, and by computerized differencing of camera events. At the lander sites and during the times that observations were carried out on the surface of Mars, no evidence, direct or indirect, has been obtained for macroscopic biology on Mars.

097.190 Properties of aerosols in the Martian atmosphere, as
         inferred from Viking lander imaging data.
J. B. Pollack, D. Colburn, R. Kahn, J. Hunter, W. Van Camp,

C. E. Carlston, M. R. Wolf.
J. Geophys. Res., Vol. 82, (see 003.017), 4479 - 4496 (1977).
Paper No. 7S0559.

Three types of aerosols are inferred to have been present over the landers during the summer and fall season in their hemisphere. A ground fog made of water ice particles was present throughout this period. The authors estimate that the average particle radius of the fog was about 2 μm and that the fog's depth equaled approximately 0.4 km. A higher-level ice cloud was prominent only during the fall season, when it was a sporadic source of atmospheric opacity at VL-2. The cross-section weighted mean radius of these aerosols is about 0.4 μm. They have a nonspherical but equidimensional shape and rough surfaces. The principal opaque mineral in these particles is magnetite. The authors propose that soil particles, as well as any associated water ice, are eliminated from the atmosphere, in part, by their acting as condensation sites for the growth of CO_2 ice particles in the winter polar regions. The resultant CO_2-H_2O-dust particle is much larger and therefore has a much higher fallout velocity than an uncoated dust or water ice particle.

097.191 Surface materials of the Viking landing sites.
         H. J. Moore, R. E. Hutton, R. F. Scott, C. R. Spitzer,
R. W. Shorthill.
J. Geophys. Res., Vol. 82, (see 003.017), 4497 - 4523 (1977).
Paper No. 7S0447.

Martian surface materials viewed by the two Viking landers range from fine-grained nearly cohesionless soils to rocks. The soil of Mars has both cohesion and friction.

097.192 Seismology on Mars.   D. L. Anderson, W. F. Miller,
         G. V. Latham, Y. Nakamura, M. N. Toksöz,
A. M. Dainty, F. K. Duennebier, A. R. Lazarewicz, R. L.
Kovach, T. C. D. Knight.
J. Geophys. Res., Vol. 82, (see 003.017), 4524 - 4546 (1977).
Paper No. 7S0408.

A three-axis short-period seismometer has been operating on the surface of Mars in the Utopia Planitia region since September 4, 1976. During the first 5 months of operation no large events have been seen. The seismic background correlates well with wind velocity.

097.193 The Viking magnetic properties experiment: primary
         mission results.   R. B. Hargraves, D. W. Collinson,
R. E. Arvidson, C. R. Spitzer.
J. Geophys. Res., Vol. 82, (see 003.017), 4547 - 4558 (1977).
Paper No. 7S0506.

Three permanent magnetic arrays were aboard each Viking lander. The loose Martian surface material contains from 1 to 7% highly magnetic mineral. This paper constitutes a combination and elaboration of the published preliminary results from the first and second Viking missions (Hargraves et al., 1976).

097.194 Meteorological results from the surface of Mars:
         Viking 1 and 2.   S. L. Hess, R. M. Henry,
C. B. Leovy, J. A. Ryan, J. E. Tillman.
J. Geophys. Res., Vol. 82, (see 003.017), 4559 - 4574 (1977).
Paper No. 7S0459.

The meteorology experiment aboard the Viking Mars landers was designed to measure atmospheric temperature, wind speed, wind direction, and pressure.

097.195 Report of the Viking inorganic chemical analysis
         team: introductory statement.
P. Toulmin III, A. K. Baird, B. C. Clark, K. Keil, H. J. Rose, Jr.
J. Geophys. Res., Vol. 82, (see 003.017), 4575 (1977). Paper No. 7S0465.

097.196 The Viking X-ray fluorescence experiment: analytical
         methods and early results.   B. C. Clark III,

A. K. Baird, H. J. Rose, Jr., P. Toulmin III, R. P. Christian, W. C. Kelliher, A. J. Castro, C. D. Rowe, K. Keil, G. R. Huss. J. Geophys. Res., Vol. 82, (see 003.017), 4577 - 4594 (1977). Paper No. 7S0448.

Ten samples of the Martian regolith have been analyzed by the Viking lander X-ray fluorescence spectrometers. Bulk fines at both Viking landing sites are quite similar in composition.

097.197 **The Viking X-ray fluorescence experiment: sampling strategies and laboratory simulations.**
A. K. Baird, A. J. Castro, B. C. Clark, P. Toulmin III, H. Rose, Jr., K. Keil, J. L. Gooding.
J. Geophys. Res., Vol. 82, (see 003.017), 4595 - 4624 (1977). Paper No. 7S0451.

This paper is concerned with the rationale for sample site selections, surface sampler operations, and the supportive laboratory studies needed to interpret X-ray results from Mars.

097.198 **Geochemical and mineralogical interpretation of the Viking inorganic chemical results.**
P. Toulmin III, A. K. Baird, B. C. Clark, K. Keil, H. J. Rose, Jr. R. P. Christian, P. H. Evans, W. C. Kelliher.
J. Geophys. Res., Vol. 82, (see 003.017), 4625 - 4634 (1977). Paper No. 7S0464.

The elemental analyses represent the composition of samples of Martian fines; the only undetermined major constituents thought to be present are H_2O, CO_2, Na_2O, and possibly NO_x. The samples are principally silicate particles, with some admixture of oxide and probably carbonate minerals. The overall elemental composition is dissimilar to any single known mineral or rock type and apparently represents a mixture of materials.

097.199 **The composition of the atmosphere at the surface of Mars.**   T. Owen, K. Biemann, D. R. Rushneck, J. E. Biller, D. W. Howarth, A. L. Lafleur.
J. Geophys. Res., Vol. 82, (see 003.017), 4635 - 4639 (1977). Paper No. 7S0515.

The authors have confirmed the discovery of N_2 and ^{40}Ar by the Entry Science Team, and have also detected Ne, Kr, Xe, and the primordial isotopes of Ar. The noble gases exhibit an abundance pattern similar to that found in the terrestrial atmosphere and the primordial component of meteoritic gases. Xenon appears to be underabundant in comparison to the meteoritic ratio, as it is on earth. The isotopic ratios $^{15}N/^{14}N$, $^{40}Ar/^{36}Ar$, and $^{129}Xe/^{132}Xe$ are distinctly different from the terrestrial values, implying different evolutionary histories for volatiles on the two planets. The noble gas abundances indicate that at least 10 times the present atmospheric amount of N_2 and 20 times the CO_2 abundance were released by the planet during geologic time; the outgassing of a large amount of water must also have taken place. There is thus an explanation for the high surface pressure and abundance of water required at some early epoch to cut the dendritic channels observed on the Martian surface.

097.200 **The search for organic substances and inorganic volatile compounds in the surface of Mars.**
K. Biemann, J. Oro, P. Toulmin III, L. E. Orgel, A. O. Nier, D. M. Anderson, P. G. Simmonds, D. Flory, A. V. Diaz, D. R. Rushneck, J. E. Biller, A. L. Lafleur.
J. Geophys. Res., Vol. 82, (see 003.017), 4641 - 4658 (1977). Paper No. 7S0556.

A total of four Martian samples, one surface and one subsurface sample at each of the two Viking landing sites, Chryse Planitia and Utopia Planitia, have been analyzed for organic compounds by a gas chromatograph—mass spectrometer. In none of these experiments could organic material of Martian origin be detected at detection limits generally of the order of parts per billion and for a few substances closer to parts per million. The evolution of water and carbon dioxide, but not of

other inorganic gases, was observed upon heating the sample to temperatures of up to 500°C.

097.201 **Viking on Mars: the carbon assimilation experiments.**
N. H. Horowitz, G. L. Hobby, J. S. Hubbard.
J. Geophys. Res., Vol. 82, (see 003.017), 4659 - 4662 (1977). Paper No. 7S0503.

A fixation of atmospheric carbon, presumably into organic form, occurs in Martian surface material under conditions approximating the actual Martian ones. The amount of carbon fixed is small by terrestrial standards. It is unlikely that the reaction is biological.

097.202 **Recent results from the Viking Labeled Release experiment on Mars.**   G. V. Levin, P. A. Straat.
J. Geophys. Res., Vol. 82, (see 003.017), 4663 - 4667 (1977). Paper No. 7S0469.

The current status of the Labeled Release experiment on Mars is summarized.

097.203 **The Viking gas exchange experiment results from Chryse and Utopia surface samples.**
V. I. Oyama, B. J. Berdahl.
J. Geophys. Res., Vol. 82, (see 003.017), 4669 - 4676 (1977). Paper No. 7S0508.

Immediate gas changes occurred when untreated Martian surface samples were humidified and/or wet by an aqueous nutrient medium in the Viking lander gas exchange experiment. The evolutions of N_2, CO_2, and Ar are mainly associated with soil surface desorption caused by water vapor, while O_2 evolution is primarily associated with decomposition of superoxides inferred to be present on Mars. On recharges with fresh nutrient and test gas, only CO_2 was given off, and its rate of evolution decreased with each recharge. Atmospheric analyses were also performed at both sites. The mean atmospheric composition from four analyses is N_2, 2.3%; O_2, $\leqslant 0.15\%$; Ar, 1.5%; and CO_2, 96.2%.

097.204 **The Viking biological investigation: general aspects.**
H. P. Klein.
J. Geophys. Res., Vol. 82, (see 003.017), 4677 - 4680 (1977). Paper No. 7S0439.

The Viking biological investigation has tested four different hypotheses regarding the possible nature of Martian organisms. While significant results were obtained for each of these, tests of three of the hypotheses appear to indicate the absence of biology in the samples used, while the fourth is consistent with a biological interpretation. The original assumptions for each experiment and the experimental procedures that were utilized to test these assumptions are reviewed.

097.205 **Mars and Earth: origin and abundance of volatiles.**
E. Anders, T. Owen.
Science, Vol. 198, 453 - 465 (1977).

The authors have predicted Martian abundances of 31 elements from terrestrial abundances. Comparison with the observed ^{36}Ar abundance suggests that outgassing on Mars has been about four times less complete than on Earth. A curious dichotomy seems to be emerging among differentiated planets in the inner solar system. Two large planets (Earth and Venus) are fairly rich in volatiles, whereas three small planets (Mars, the moon, and the eucrite parent body — presumably the asteroid 4 Vesta) are poorer in volatiles by at least an order of magnitude.

097.206 **Nomenclature of Martian formations.**
K. B. Shingareva.
Kosm. Issled., Vol. 15, 923 - 932 (1977). In Russian.

097.207 **Water vapour in the Martian atmosphere from data of the automatic interplanetary stations Mars 3, Mariner 9, Mars 5 and Viking 1, 2.**   A. Eh. Nadzhip.

Kosm. Issled., Vol. 15, 947 - 951 (1977). In Russian.

097.208   Mars, solar wind, and supernova — implications of the Viking data.   M. Shimizu.
Astrophys. Space Sci., Vol. 49, L21 - L23 (1977).
A scenario for the evolution of the Martian atmosphere consistent with various data of the Viking 1 and 2 and the Mariner 9 has been presented: Mars was formed from Renazzo-type meteorites polluted by the products of supernova explosion. A dense ancient Martian atmosphere has been swept away by the solar wind and the present tenuous atmosphere was supplied recently by the volcanic gas from the Tharsis region, after the occurrence of the magnetic field.

097.209   Dust storms on Mars: considerations and simulations.   R. Greeley, B. R. White, J. B. Pollack, J. D. Iversen, R. N. Leach.
NASA Tech. Memo., NASA TM 78423, 30 pp. Price $ 4.00 (1977).
Aeolian processes are important in modifying the surface of Mars at present, and appear to have been significant in the geological past. Aeolian activity includes local and global dust storms, the formation of erosional features such as yardangs and depositional features such as sand dunes, and the erosion of rock and soil. As a means of understanding aeolian processes on Mars, an investigation is in progress that includes laboratory simulations, field studies of Earth analogs, and interpretation of spacecraft data. This report describes the Martian Surface Wind Tunnel (MARSWIT) and presents some results of the general investigation.

097.210   Atmospheric pressure variations of the Martian atmosphere.   C. J. Macris, B. C. Petropoulos.
Praktika Akad. Athens, Tom. 51, 224 - 244 (1976) = Res. Center Astron. Appl. Math., Acad. Athens, Contrib. Ser. I (Astron.), No. 45.
The authors have used the computed Franck-Condon factors and the data of pressure and temperature near the surface, secured by Mariners 4, 6 and 7 flights, to calculate the pressure and the temperature function to altitude into the atmosphere of Mars.

097.211   Grain size distribution of the martian soil inside the Viking Lander 2—footpad 3.
H. Zimmer, H. J. Moore, II, R. W. Shorthill, R. E. Hutton.
Veröff. Wilhelm-Foerster-Sternw. Berlin, Nr. 47, 8 pp. (1977).
The authors have studied the grain size distribution of the clear visible particles. The calibration work was done on the Science Test Lander.

097.212   The dust storm on Mars in 1975.
Yu. V. Aleksandrov, D. F. Lupishko, T. A. Lupishko.
Astron. Tsirk., No. 941, p. 3 - 4 (1977). In Russian.

097.213   Results of photographic observations of Mars with the Pulkovo short-focus astrograph during 1973 - 1974.   N. A. Shakht, O. A. Kalinichenko.
Izv. Glav. Astron. Obs. Pulkovo, No. 195, Astrofiz. Astrometr., p. 46 - 48 (1977). In Russian.

097.214   Les oscillations de l'irradiation solaire de Mars dues aux variations séculaires de son orbite.
Sh. G. Sharaf, N. A. Budnikova.
Tr. Inst. Teor. Astron., Leningrad, vyp. (No.) 16, p. 88 - 117 (1977). In Russian.
On expose les résultats des recherches sur la théorie astronomique des oscillations du climat de Mars dues à la radiation solaire et les perturbations séculaires des éléments de l'orbite de Mars. Le train séculaire de la radiation solaire sur une unité des aires des hémisphères et les 6 latitudes de l'hémisphère du nord et du sud de Mars sur l'intervalle du temps de 2 millions années depuis 1950 est illustré par des tableaux et des graphiques. On fait la comparison des insolations de Mars et de la Terre.

097.215   Model of the composition of the Martian atmosphere.   M. N. Izakov, O. P. Krasitskij.
Inst. kosm. issled. AN SSSR. Pr-320. Moskva, 1977. 38 pp. In Russian. — Abstr. in Ref. zh., 51. Astron., 1.51.272 (1978).

097.216   On the astronomical refraction in the Martian atmosphere.   V. V. Kirichuk, A. N. Marchenko.
Izv. vyssh. uchebn. zaved. Geod. i aehrofotosemka, 1977, No. 4, p. 83 - 86. In Russian. — Abstr. in Ref. zh., 51. Astron., 1.51.274; 62. Issled. kosm. prostranstva, 1.62.135 (1978).

097.217   Analytical chemistry of the elements on Mars.   B. Kuchowicz.
Urania Kraków, Vol. 48, 258 - 265 (1977). — In Polish.

097.218   On the photometric structure of classic albedo formations on Mars.
V. I. Ezerskij, V. I. Mamnitskij, N. S. Olifer, V. A. Psarev, A. S. Selivanov, M. K. Naraeva, M. I. Bokhonov.
Physics of the moon and planets. Fundamental astrometry, (see 003.024), Vestn. Khar'kov. Univ., No. 160, p. 3 - 14 (1977). In Russian.

097.219   Mars, solar wind, and supernova — implications of the Viking data.   M. Shimizu.
10th Lunar and Planetary Symposium, (see 012.050), p. 171 - 173 (1977).
A scenario for the evolution of the Martian atmosphere consistent with various Viking 1 and 2 and Mariner 9 data is presented: Mars was formed from Renazzo type meteorites polluted by the products of supernova explosion. A dense ancient Martian atmosphere had been swept away by the solar wind and the present tenuous atmosphere was supplied recently by volcanic gas from the Tharsis region, after the occurrence of a magnetic field.

Absolute photometry of Mars in 1971, 1973 and 1975.   See Abstr. 003.027.

Mars and its satellites: a detailed commentary on the nomenclature.   See Abstr. 003.035.

Moon, Mars and Venus, a concise guide in colour.   See Abstr. 003.144.

The pyrolytic release experiment: measurement of carbon assimilation.   See Abstr. 015.020.

Proton affinities and cluster ion stabilities in CO_2 and CS_2. Applications in Martian ionospheric chemistry.   See Abstr. 022.075.

The Martian Surface Wind Tunnel (MARSWIT).   See Abstr. 031.210.

Imaging on Mars: from concept to reality.   See Abstr. 031.267.

Planimetric Martian triangulations.   See Abstr. 031.315.

IPL processing of the Viking orbiter images of Mars.   See Abstr. 031.321.

Multispectral and stereo imaging on Mars.   See Abstr. 032.512.

Inflight performance of the Viking visual imaging

subsystem.   See Abstr. 032.562.

An X-ray diffractometer for Mars.
See Abstr. 034.008.

Right ascensions of Mars obtained during the 1971
opposition.   See Abstr. 041.011.

Declinations of the sun, Mercury, Venus and Mars
obtained from observations with the vertical circle of the
Nikolaev Observatory in 1973–1975.   See Abstr. 041.013.

Mars polar ice sample return mission – 2.
See Abstr. 051.021.

Mars polar ice sample return mission.
See Abstr. 051.030.

Destination Mars.   See Abstr. 051.034.

Viking on Mars: a preliminary survey.
See Abstr. 051.037.

The Viking project.   See Abstr. 051.053.

Mars surface Penetrator – system description.
See Abstr. 051.070.

Das Viking-Programm in der erweiterten Missionsphase. Aktueller Stand des Unternehmens.
See Abstr. 053.005.

The missions of the Viking orbiters.
See Abstr. 053.009.

On the stationary points of the gravitation fields
of the earth, the moon and Mars.   See Abstr. 081.003.

Reducing greenhouses and the temperature history
of Earth and Mars.   See Abstr. 081.009.

Geologic, topographic, and meteorologic influences
on eolian deposition, Earth and Mars.   See Abstr. 081.013.

Deep-sea channels: another Earth analogy with
Martian channels.   See Abstr. 081.015.

Measurement of scattered ultraviolet radiation in
the vicinity of planets and in the interplanetary medium.
See Abstr. 082.018.

Chemical evolution – comparative planetology.
See Abstr. 082.144.

Cloud microphysics: comments on the clouds of

Venus, Mars and Jupiter.   See Abstr. 091.005.

Sources of outgassed volatiles on the terrestrial
planets.   See Abstr. 091.006.

Crater evolutionary tracks.   See Abstr. 091.016.

Results of analyses performed on basalt adjacent to
penetrators emplaced into volcanic rock at Amboy, California,
April 1976.   See Abstr. 091.017.

Results of analyses performed on soil adjacent to
penetrators emplaced into sediments at McCook, Nebraska,
January 1976.   See Abstr. 091.018.

On the refraction of radio waves in radiographic inspection of planetary atmospheres.   See Abstr. 091.034.

The magnetic fields of the terrestrial planets.
See Abstr. 091.069.

On the estimate of accuracy of determination of
the parameters of an ellipsoid approximating the image of the
horizon on a space photograph of a planet.
See Abstr. 091.070.

Microwave spectral lines in the atmospheres of
Mars and Venus.   See Abstr. 093.019.

Microwave detection of carbon monoxide on
Venus and Mars.   See Abstr. 093.020.

The atmospheres of the planets.
See Abstr. 093.035.

Venus and Mars (from recent results of Soviet and
American studies).   See Abstr. 093.038.

Comparative studies of the Moon, Mercury and
Mars.   See Abstr. 094.115.

Lunar and Martian cratering studies, and Mars
Mariner 9 geologic mapping.   See Abstr. 094.117.

Thermal expansion and thermal stress in the moon
and terrestrial planets: clues to early thermal history.
See Abstr. 094.162.

Comparison of impact basins on Mercury, Mars
and the Moon.   See Abstr. 094.168.

Lineament patterns on the Moon, Mars and Mercury.
See Abstr. 094.186.

The role of volatiles in the impact process.
See Abstr. 105.013.

Mars Satellites

097.501   **Are striations on Phobos evidence for tidal stress?**
S. Soter, A. Harris.
Nature, Vol. 268, 421 - 422 (1977).
   Viking orbiter photographs of Phobos, the inner satellite
of Mars, reveal a curious series of striations or grooves over
much of the surface. These features cross over large eroded
craters but are in turn interrupted by very small well defined
craters, which indicate an origin substantially later than the
time of formation of the satellite itself when the large scale
cratering presumably occurred. The authors suggest that some
of these features are related to readjustment of the satellite's
figure with increasing tidal stress as the orbit evolves inwards
under the action of tidal friction.

097.502   **Viking imaging of Phobos and Deimos: an overview.**
T. C. Duxbury, J. Veverka.
Bull. American Astron. Soc., Vol. 9, 441 (1977). – Abstract.

097.503   **Surfaces of Phobos and Deimos: new Viking results.**
J. Veverka, T. Duxbury.
Bull. American Astron. Soc., Vol. 9, 441 (1977). – Abstract.

097.504   **Spectral albedo of Phobos and implications on its
composition.**   K. D. Pang, J. B. Pollack, J.
Veverka, A. L. Lane, J. M. Ajello.
Bull. American Astron. Soc., Vol. 9, 464 (1977). – Abstract.

097.505   **A quantitative comparison of the surface roughness
of Phobos and Deimos.**   P. Thomas, J. Veverka.
NASA Tech. Mem., NASA TM X-3511, (see 012.010), p. 78
(1977). – Abstract.

097.506   **Mars.**
Orion, 35. Jahrg., 156 - 157 (1977).

097.507   **Some evidence of the possibility of a strong interaction between the Martian satellite Deimos and the
solar wind.**   A. V. Bogdanov.
Kosm. Issled., Vol. 15, 741 - 746 (1977). In Russian.

097.508   **Phobos.   Ø. Hauge.**
Astron. Tidsskr., Årg. 10, 106 - 107 (1977).

097.509   **Die Marsmonde Phobos und Deimos.**
N. Giesinger.
Sternenbote, 20. Jahrg., 218 - 229 (1977).

097.510   **Phobos and Deimos.   J. Oberg.**
Astronomy, Vol. 5, No. 3, p. 6 - 17 (1977).
Abstr. in Phys. Abstr., Vol. 80, Abstr. 49323 (1977).

097.511   **Viking imaging of Phobos and Deimos: an overview
of the primary mission.**
T. C. Duxbury, J. Veverka.
J. Geophys. Res., Vol. 82, (see 003.017), 4203 - 4211 (1977).
Paper No. 7S0510.
   The effective resolution of the Viking images revealed a
number of unexpected surface features such as linear chains of
irregular craters and elongated grooves and striations. Addi-
tionally, a number of images of the satellites were obtained
against star backgrounds to refine further the ephemerides.

097.512   **Viking observations of Phobos and Deimos: preliminary results.**   J. Veverka, T. C. Duxbury.
J. Geophys. Res., Vol. 82, (see 003.017), 4213 - 4223 (1977).
Paper No. 7S0567.
   The improved resolution of the Viking orbiter images has
led to the discovery of a number of unusual features on the
surface of Phobos: elongated rill-like depressions associated
with the crater Stickney; chains of irregular craters which
sometimes show a 'herringbone' pattern; and sets of almost
parallel linear striations of uncertain origin. The crater chains
are not randomly oriented but tend to lie parallel to the
orbital plane of Phobos. Similar features have not been recognized on Deimos. The Viking data demonstrate that the surfaces of both satellites are definitely saturated with craters
$\geqslant 300$ m across.

   **Mars and its satellites: a detailed commentary on the
nomenclature.**   See Abstr. 003.035.

   **Accuracy of estimating the masses of Phobos and
Deimos from multiple Viking Orbiter encounters.**
See Abstr. 031.293.

098 Minor Planets

098.001 Discovery of asteroid 1976 AA.
E. F. Helin, E. M. Shoemaker.
Icarus, Vol. 31, 415 - 419 (1977) = Contrib. Calif. Inst.
Technol., Pasadena, Calif., No. 2813.

Asteroid 1976 AA was discovered as a result of a continuing systematic search for planet-crossing asteroids. It is the first asteroid to be thoroughly investigated by means of photometry and radiometry on its discovery apparition. It is also the first asteroid found with a semimajor axis and period less than that of the Earth and the first Earth-crossing asteroid which does not cross the orbit of either Mars or Venus. The authors estimate that there might be several tens of objects to absolute magnitude 18, which are exclusively Earth crossing. Some of these objects might be exceptionally easy to reach by spacecraft.

098.002 Orbit of 1976 AA.
B. G. Marsden, J. G. Williams.
Icarus, Vol. 31, 420 - 423 (1977).

The orbit of 1976 AA is determined and its evolution studied over an interval of seven centuries into both the past and future.

098.003 Photometry of the asteroid 1976 AA at 0.56 and 2.2 μm.
G. J. Veeder, D. L. Matson, O. L. Hansen.
Icarus, Vol. 31, 424 - 426 (1977).

The authors report observations at 0.56 and 2.2 μm of the Apollo asteroid 1976 AA made during its discovery apparition. They derive a 2.2-μm relative spectral reflectance (scaled to unity at 0.56 μm) of $R(2.2\ \mu m) = 1.5 \pm 0.3$. This 2.2-μm reflectance is not compatible with a carbonaceous surface composition. However, it is compatible with a wide variety of meteoritic types including ordinary chondrites, stony irons, and mesosiderites. Thus, 1976 AA may have a silicate surface similar to other Apollo—Amor objects.

098.004 The diameter and albedo of asteroid 1976 AA.
D. P. Cruikshank, T. J. Jones.
Icarus, Vol. 31, 427 - 429 (1977).

The authors present a radiometric observation of asteroid 1976 AA, and formulate a simple model for the infrared thermal phase function so that their data can be compared with similar measurements made at different phase angles. The radiometric diameter of 1976 AA from the observation is 940^{+200}_{-100} meters and the geometric albedo is 0.18 ± 0.06, in satisfactory agreement with another published radiometric observation.

098.005 Vesta: the first pyroxene band from new spectroscopic measurements.
L. A. McFadden, T. B. McCord, C. Pieters.
Icarus, Vol. 31, 439 - 446 (1977) = RSL Publ. No. 169.

New spectral reflectance measurements of asteroid 4 Vesta were obtained using a silicon vidicon spectrometer with a resolution of 0.002–0.004 μm. The major absorption band in the near infrared has a minimum at 0.924 ± 0.004 μm with a bandwidth of 0.18 μm. The band represents a 30% absorption relative to peak reflectance at 0.75 μm. The absorption band has been interpreted to be due to electronic absorptions in ferrous iron in sixfold coordination in the pyroxene, pigeonite.

098.006 UBV photometry of small and distant asteroids.
B. Zellner, L. Andersson, J. Gradie.
Icarus, Vol. 31, 447 - 455 (1977).

Photoelectric magnitudes and colors on the UBV system are presented for 65 minor planets, including four Mars crossers, six Trojans, and main-belt objects down to 6 km in diameter. The Trojans all have very similar colors not characteristic of the main-belt population. A paucity of S-type asteroids at the smallest diameters, predicted from trends seen at larger sizes, is not observed. The newly available color data for small objects ranging from 1.0 to 5.2 AU in heliocentric distance show the main belt to be a transition zone between predominantly silicate and carbonaceous compositions.

098.007 An explication of the radiometric method for size and albedo determination. O. L. Hansen.
Icarus, Vol. 31, 456 - 482 (1977).

A new radiometric model for disk-integrated photometry of asteroids is presented. With empirical support from photometry of Mercury and the Moon, the model assumes that observed sunward beaming of the infrared emission is due to craters. In contrast to earlier theoretical studies of the lunar emission, the observable flux ratio between a cratered sphere and a smooth sphere is calculated for large ranges in wavelength, temperature, and phase angle. Revised diameters and albedos based on the crater model are given for 84 asteroids. The revised values are in good agreement with Morrison's (1977) radiometric results. It is shown that the systematic discrepancy between radiometric and polarimetric albedos (Zellner and Gradie, 1976) is probably a double-valued function of albedo.

098.008 A photometric study of the minor planet 63 Ausonia. F. Scaltriti, V. Zappalà.
Icarus, Vol. 31, 498 - 502 (1977).

Photoelectric observations of the minor planet 63 Ausonia were obtained on 12 nights during the 1976 opposition at the Astronomical Observatory of Torino. A complete lightcurve with two maxima and two minima was observed with a maximum amplitude of 0.47 mag. The synodic period of rotation, never before determined photoelectrically, was found to be $9^h17^m48^s \pm 5^s$. The absolute magnitude of the primary maximum, $V_0(1, 0) = 7.49$ mag, and the phase coefficient, $\beta_V = 0.035$ mag/deg, were deduced by the magnitude—phase relation. Comparison with other observations is briefly discussed and a mean radius is determined from a previous value of the geometric albedo.

098.009 Mining the Apollo and Amor asteroids.
B. O'Leary.
Science, Vol. 197, 363 - 366 (1977).

Earth-approaching asteroids could provide raw materials for space manufacturing. For certain asteroids the total energy per unit mass for the transfer of asteroidal resources to a manufacturing site in high Earth orbit is comparable to that for lunar materials. For logistical reasons the cost may be many times less. Optical studies suggest that these asteroids have compositions corresponding to those of carbonaceous and ordinary chondrites, with some containing large quantities of iron and nickel; others are thought to contain carbon, nitrogen, and hydrogen, elements that appear to be lacking on the moon.

098.010 Spots on Vesta. J. Degewij, J. Gradie,
T. Lebertre, W. Wisniewski, B. Zellner.
Bull. American Astron. Soc., Vol. 9, 431 (1977). – Abstract.

098.011 On the global insolubility of the Trojan asteroids problem in the light of the Brown conjecture.
B. Garfinkel.
Bull. American Astron. Soc., Vol. 9, 437 (1977). – Abstract.

098.012 Planetesimal bombardments of the asteroidal belt.

W.-H. Ip.
Bull. American Astron. Soc., Vol. 9, 455 (1977). − Abstract.

098.013   **Possible bulk composition of Vesta: evidence from eucrites.**   M. J. Drake, G. J. Consolmagno.
Bull. American Astron. Soc., Vol. 9, 459 (1977). − Abstract.

098.014   **Asteroid surface materials: a mineralogical characterization from reflectance spectra.**
M. J. Gaffey, T. B. McCord.
Bull. American Astron. Soc., Vol. 9, 459 (1977). − Abstract.

098.015   **UBV photometric survey of asteroids.**
E. Bowell.
Bull. American Astron. Soc., Vol. 9, 459 (1977). − Abstract.

098.016   **The E asteroids and the origin of enstatite achondrites.**
B. Zellner, M. Leake, D. Morrison.
Bull. American Astron. Soc., Vol. 9, 460 (1977). − Abstract.

098.017   **UBV photometric observations of the Eos and Koronis asteroid families.**   J. C. Gradie.
Bull. American Astron. Soc., Vol. 9, 460 (1977). − Abstract.

098.018   **Asteroid reflectances.**
G. J. Veeder, D. L. Matson, T. V. Johnson.
Bull. American Astron. Soc., Vol. 9, 460 (1977). − Abstract.

098.019   **The asteroid size scale.**   D. Morrison.
Bull. American Astron. Soc., Vol. 9, 460 (1977).
Abstract.

098.020   **1976 UA; second asteroid with orbit smaller than Earth's.**   E. F. Helin, E. M. Shoemaker.
Bull. American Astron. Soc., Vol. 9, 461 (1977). − Abstract.

098.021   **A collisional model for the origin of asteroid rotations.**   A. W. Harris.
Bull. American Astron. Soc., Vol. 9, 461 (1977). − Abstract.

098.022   **Further asteroid collisional evolution.**
D. R. Davis, C. R. Chapman.
Bull. American Astron. Soc., Vol. 9, 461 (1977). − Abstract.

098.023   **Asteroid families: observational evidence for common origins.**   J. Gradie, B. Zellner.
Science, Vol. 197, 254 - 255 (1977).
Colors of minor planets in the UBV system indicate compositions quite distinct from those of the field population in each of three Hirayama families. The Eos and Koronis families apparently originated from the collisional fragmentation of undifferentiated silicate bodies, and the Nysa group from a geochemically differentiated parent body.

098.024   **Positions of minor planets.**   G. K. Gorel'.
Nikolaev. otd. Glav. astron. obs. AN SSSR.
Nikolaev, 1976. 17 pp. In Russian. − Abstr. in Ref. zh., 51. Astron., 8.51.187 (1977).

098.025   **Ephemerides of minor planets for 1978.**
Editor:Institut Teoreticheskoj Astronomii Akademii Nauk SSSR, under the editorship of N. S. Yakhontova. Izdatel'stvo "Nauka", Leningradskoe Otdelenie, Leningrad. 220 pp. Price 3 Rbl. 80 Kop. (1977). In Russian and English.
Contents: Introduction, p. 3 - 8; Information on new elements, p. 9 - 12; New elements, p. 13 - 16; Elements, p. 17 - 52; Opposition dates, p. 53 - 64; Ephemerides, p. 65 - 190; Ephemerides of bright planets, p. 191 - 210; Ephemerides of some unusual planets, p. 211 - 217; Critical list, p. 218.

098.026   **Search for correlation between asteroid families and**

classes.   O. Hansen.
Icarus, Vol. 32, 229 - 232 (1977).
A correlation between membership in a dynamically defined asteroid family and membership in a given asteroid spectral class is sought. Examination of 10 families each with five or more classified members indicates a correlation for the 4 families whose existence is best established, and no correlation for the remaining 6 families. This conclusion supports the break-up hypothesis for the origin of some families, while not contradicting that hypothesis for any family.

098.027   **A photographic lightcurve of the Amor asteroid 1580 Betulia.**   C.-I. Lagerkvist.
Icarus, Vol. 32, 233 - 234 (1977).
The Amor asteroid 1580 Betulia was observed photographically with the Schmidt telescope at the Uppsala Southern Station during the opposition in May 1976. The amplitude of the lightcurve was found to be $0^{m}2$.

098.028   **Aspekte der Planetoidenforschung.**   R. Bien.
Sterne Weltraum, Jahrg. 16, 319 - 320 (1977).

098.029   **The period of rotation and the photoelectric light curve of the minor planet 471 Papagena.**
G. Lustig.
Astron. Astrophys., Suppl. Ser., Vol. 30, 117 - 119 (1977).
In German.
The minor planet 471 Papagena was observed during the 1976 opposition for five nights in November using the photoelectric photometer attached to the 60 cm telescope at the Observatoire de Haute Provence, France. A synodic period of $7^{h}06^{m}46^{s} \pm 2^{s}$ m.e. and a total amplitude of $0^{m}12$ were derived from the measurements. The composite light curve was constructed based on the period, by overlapping the observations from three nights.

098.030   **Rotation period and photoelectric light curves of asteroid 471 Papagena.**   A. Surdej, J. Surdej.
Astron. Astrophys., Suppl. Ser., Vol. 30, 121 - 124 (1977).
Asteroid 471 Papagena was observed during the 1976 opposition with a photoelectric photometer attached to the 61 cm Bochum telescope at the European Southern Observatory. The light curve shows a very unusual triple maximum and minimum. The synodic period found is $7^{h}09^{m}23^{s} \pm 19^{s}$ and the maximum amplitude of the light curve 0.13 mag.

098.031   **Asteroid fragmentation processes and collisional evolution.**
C. R. Chapman, D. R. Davis, R. Greenberg.
NASA Tech. Mem., NASA TM X-3511, (see 012.010), p. 72 - 73 (1977). − Abstract.

098.032   **Possible observation of a satellite of a minor planet.**
D. W. Dunham, P. D. Maley.
Occultation Newsl., Vol. 1, 115 - 117 (1977).

098.033   **Physical properties of asteroids. Part I.**
Z. Musielak.
Postępy Astron., Tom 25, 115 - 126 (1977). In Polish.
Recent results of investigations of physical properties of asteroids are presented. A review of photometric observation results is given.

098.034   **Asteroids and comparative planetology.**
D. L. Matson, F. P. Fanale, T. V. Johnson, G. J. Veeder.
Proc. Seventh Lunar Sci. Conf., (see 012.015), 3603 - 3627 (1976).

098.035   **On the orbital dependence of the asteroidal collision process.**   W.-H. Ip.
Icarus, Vol. 32, 378 - 381 (1977).

Collision of asteroids with the main-belt asteroid population is considered with the effect of the impact kinetic energy taken into account. It is found that objects in eccentric orbits have a larger probability of destructive collision as compared to objects in orbits with mean values of eccentricity ($e = 0.15$) and inclination ($i = 10°$); also orbits with small semimajor axes ($a \approx 2.3$ AU) are found to have peak values of the probability of destructive collision.

098.036 **Rotation period of the asteroid 52 Europa.**
F. Scaltriti, V. Zappalà.
Astron. Astrophys., Suppl. Ser., Vol. 30, 169 - 174 (1977).

The asteroid 52 Europa, which is of interest as one of the last large minor planets with an unknown period, was observed photoelectrically from November 18 to December 12, 1976 at the Astronomical Observatory of Torino. The rotation period was found to be: $P_{syn} = 11^h15^m30^s$ with an amplitude of about 0.09 mag. The present frequency distribution for the minor planets with known periods was investigated, as was the necessity of joint observations at observatories of different longitude for objects with slow spin periods.

098.037 **Positions of selected minor planets.**
S. Vaghi, V. Zappalà, G. de Sanctis, W. Ferreri, L. Bacchelli.
Astron. Astrophys., Suppl. Ser., Vol. 30, 175 - 178 (1977).

Precise positions are given for 27 minor planets observed during the period August 1975 - May 1976 at the Observatory of Torino.

098.038 **Positions of main-belt asteroids.**
C.-I. Lagerkvist, V. Zappalà.
Astron. Astrophys., Suppl. Ser., Vol. 30, 179 - 181 (1977).

Precise positions are presented for 63 main-belt asteroids observed during 1973−1975.

098.039 **Planetary occultations.** D. W. Dunham.
Occultation Newsl., Vol. 1, 125 - 126 (1977).

098.040 **Passages of minor planets across galactic clusters.**
D. Wallentine, D. W. Dunham.
Occultation Newsl., Vol. 1, 126 - 127 (1977).

098.041 **The E asteroids and the origin of the enstatite achondrites.**
B. Zellner, M. Leake, D. Morrison, J. G. Williams.
Geochim. Cosmochim. Acta, Vol. 41, 1759 - 1767 (1977).

Color polarization and albedo data are summarized for the three known minor planets of optical type E − 44 Nysa, 64 Angelina and 434 Hungaria. The surfaces of the E objects evidently consist of colorless, translucent, iron-free silicates such as plagioclase, forsterite, or enstatite. Their possible identification as the source of enstatite achondrites is consistent with new laboratory polarimetry of the Norton County aubrite. Both Nysa and Hungaria seem to be rather favorably situated for the production of meteorites.

098.042 **Evoluzione della distribuzione degli asteroidi.**
Cl. Froeschlé, H. Scholl.
Coelum, Vol. 45, 190 - 201 (1977). − Translated from French by D. Quaglia, see 19.098.003.

098.043 **Minor planet 433 Eros.** J. G. Porter.
Modern astronomy, (see 003.013), p. 68 - 73 (1977).

098.044 **Object Kowal, the most distant asteroid.**
D. W. Hughes.
Nature, Vol. 270, 385 - 386 (1977).

098.045 **Photometric observations of Eros.**
S. I. Ignatovich, S. I. Mironyuk, Ya. M. Motrunich.
Problems of cosmic physics. Vyp. (No.) 12, (see 003.016),

p. 86 - 92 (1977). In Russian.

Results of photographic and photoelectric photometry of Eros 1974 September 18, 1975 January 10, 14 - 16, 21 are given.

098.046 **Photoelectric lightcurves of minor planets 599 Luisa and 128 Nemesis during the 1976 opposition.**
H. Debehogne, A. Surdej, J. Surdej.
Astron. Astrophys., Suppl. Ser., Vol. 30, 375 - 379 (1977).

The lightcurve of 599 Luisa shows two well shaped minima and maxima together with a short time scale feature. The synodic period of rotation found for 599 Luisa is $9^h33^m58^s \pm 46^s$ and the maximum amplitude of the lightcurve 0.18 mag. Minor planet 128 Nemesis could only be observed during one night and appears to be a probable low-spin rotation asteroid.

098.047 **Why is a minor planet minor?**
T. Matsui, H. Mizutani.
Nature, Vol. 270, 506 - 507 (1977).

On the basis of the accretion model of planet formation, the authors propose a new idea in which they explain why minor planets are 'minor' and could not grow to a full-size planet. These results also substantiate Orowan's idea that terrestrial planets accreted inhomogeneously, with iron being the first to accumulate and silicates the second.

098.048 **The asteroidal belt and Kirkwood gaps. I. A statistical study.** R. Pratap.
Pramāna, Vol. 8, 438 - 446 (1977). − Abstr. in Phys. Abstr., Vol. 80, Abstr. 60711 (1977).

098.049 **The asteroidal belt and Kirkwood gaps. II. Kinematical theory.** R. Pratap.
Pramāna, Vol. 8, 447 - 456 (1977). − Abstr. in Phys. Abstr., Vol. 80, Abstr., 60712 (1977).

098.050 **The angular diameter of Vesta from speckle interferometry.**
S. P. Worden, M. K. Stein, G. D. Schmidt, J. R. P. Angel.
Icarus, Vol. 32, 450 - 457 (1977).

It is shown that the autocorrelation function of a telescope diffraction-limited image is closely approximated by a subtraction of the mean cross correlation of pairs of speckle photographs from the mean autocorrelation of the same set of data. This fact is used to derive the angular diameter of the asteroid Vesta from a series of speckle interferometry data. The resultant apparent angular diameter of $0.''40 \pm 0.''04$ corresponds to an absolute diameter of 513 ± 51 km.

098.051 **On the prograde rotation of asteroids.**
O. Hansen.
Icarus, Vol. 32, 458 - 460 (1977).

Sets of diameter determinations before and after opposition for the asteroids Ceres, Pallas, Vesta, and Fortuna have been studied statistically for indications of spin direction. All four asteroids are tentatively found to have prograde spin. For Ceres, that conclusion is virtually certain.

098.052 **(1566) Icarus.**
IAU Circ., No. 3087 (1977).

098.053 **1975 YA.**
IAU Circ., No. 3089 (1977).

098.054 **1977 HB.**
IAU Circ., Nos. 3090, 3114 (1977).

098.055 **(69) Hesperia.**
IAU Circ., No. 3098 (1977).

098.056 **1963.UA.**

IAU Circ., No. 3100 (1977).

098.057   1977 RA.
IAU Circ., Nos. 3104, 3111, 3116, 3130, 3154
(1977).

098.058   1977 HA.
IAU Circ., No. 3106 (1977).

098.059   1977 VA (object Helin-Shoemaker).
IAU Circ., Nos. 3131, 3133, 3136, 3137, 3143
(1977).

098.060   1977 VB.
IAU Circ., Nos. 3133, 3136 (1977).

098.061   1977 UB (slow-moving object Kowal).
IAU Circ., Nos. 3129, 3130, 3134, 3140, 3143,
3145, 3147, 3151, 3156 (1977).

098.062   Occultation of SAO 85009 by (2) Pallas on 1978
May 29.
IAU Circ., No. 3141 (1977).

098.063   1976 AA.
IAU Circ., No. 3142 (1977).

098.064   Occultations by (15) Eunomia on 1978 January 10
and 19.
IAU Circ., Nos. 3144, 3152 (1977).

098.065   Occultations by minor planets.
IAU Circ., No. 3149 (1977).

098.066   Minor Planet Circulars, (MPC), Nos. 4195 - 4264
(1977).
Edited by Cincinnati Observatory under the supervision of
P. Herget.
   A repository of nearly all new data for numbered and un-
numbered minor planets: Observations, elements and ephemerides, identifications, newly assigned numbers and names,
occultations.

098.067   Possible observation of a satellite of a minor planet.
D. W. Dunham.
J. American Assoc. Variable Star Obs., Vol. 6, 3 (1977).
Abstract.

098.068   Minor planet rotation report.
A. C. Porter, D. Wallentine.
Minor Planet Bull., Vol. 5, 15 - 16 (1977).

098.069   Precise positions of minor planets.   F. Pilcher,
T. Kurosaki, A. T. Son, C. McEldery.
Minor Planet Bull., Vol. 5, 17 - 19 (1977).

098.070   Minor planet news.
Minor Planet Bull., Vol. 5, 19 (1977).

098.071   Kowal object 1977 HB.
Yamamoto Circ., No. 1853 (1977). In Japanese.

098.072   Occultation of SAO 99401 by (2) Pallas on 1977
July 8.
Yamamoto Circ., No. 1854 (1977). In Japanese.

098.073   (69) Hesperia.
Yamamoto Circ., No. 1860 (1977). In Japanese.

098.074   1977 RA (Wild object).
Yamamoto Circ., Nos. 1861, 1862 (1977). In
Japanese.

098.075   1977 VA (object Helin-Shoemaker).
Yamamoto Circ., Nos. 1868, 1869 (1977). In
Japanese.

098.076   1977 VB (object Helin-Shoemaker).
Yamamoto Circ., Nos. 1868, 1869 (1977). In
Japanese.

098.077   1977 UB (Kowal slow moving object).
Yamamoto Circ., Nos. 1867, 1868, 1869, 1870,
1871 (1977). In Japanese.

098.078   Observations photographiques de la petite planète
433 Eros, faites à l'Observatoire Astronomique de
Belgrade.   M. B. Protitch, V. Protitch-Benišek.
Bull. Obs. Astron. Belgrade, No. 128, p. 54 - 55 (1977).

098.079   Observations photographiques de petites planètes
faites à l'Observatoire Astronomique de Belgrade.
M. B. Protitch, V. Protitch-Benišek.
Bull. Obs. Astron. Belgrade, No. 128, p. 58 - 59 (1977).

098.080   New opposition of the minor planet 433 Eros.
V. Protić-Benišek.
Publ. Dep. Astron., Univ. Beograd, Fac. Sci., No. 6, (see
012.040), p. 57 - 59 (1976).

098.081   A contribution to the analysis of pairs of quasicomplanar orbits of numbered asteroids.
J. Lazović, M. Kuzmanoski.
Publ. Dep. Astron., Univ. Beograd, Fac. Sci., No. 6, (see
012.040), p. 89 - 99 (1976).
   The authors have presented the distribution of quasicomplanar pairs of numbered asteroids with respect to mutual
inclination of their orbits. Mean values of these inclination are
determined for three tables. The share of numerated asteroids
in quasicomplanar pairs is tabulated when the upper limits of
the mutual inclination of orbits are $0°.500$, $0°.300$ and $0°.100$.

098.082   The 1976 apparition of Betulia.
J. D. Drummond III, E. F. Tedesco.
Proc. Southwest Reg. Conf., Vol. 3, (see 012.043), p. 67 - 68
(1977).
   The results of another photometric worldwide coopera-
tive effort, on minor planet 1580 Betulia, are reported.

098.083   Precise observations of minor planets at Sydney
Observatory during 1975.   T. L. Morgan.
J. Proc. R. Soc. New South Wales, Vol. 109, 71 - 76 (1976) =
Sydney Obs. Pap. No. 76 (1977).
   Positions of 2 Pallas, 4 Vesta, 6 Hebe, 7 Iris, 389
Industria, 433 Eros and 532 Herculina obtained with the 23
cm camera are given.

098.084   Observations photographiques de petites planètes,
effectuées à l'astrographe double de 40 cm au cours
de l'année 1975 (2e semestre).   H. Debehogne.
Bull. Astron., Obs. R. Belgique, Vol. 8, 299 - 304 (1976).

098.085   Observations photographiques de petites planètes,
effectuées en 1975 à la caméra astrographique de
25 cm de l'Observatoire National de Rio de Janeiro.
H. Debehogne, R. R. de Freitas Mouraõ.
Bull. Astron., Obs. R. Belgique, Vol. 8, 305 - 307 (1976).

098.086   The evolution of asteroids and meteorite parent-
bodies (Invited review).   C. R. Chapman.
Meteoritics, Vol. 12, 191 - 193 (1977). – Abstract.

098.087   The M type asteroids and the origin of iron
meteorites.   A. Dollfus, J.-C. Mandeville.
Meteoritics, Vol. 12, 206 - 207 (1977). – Abstract.

098.088 **Rotationslichtwechsel des Planetoiden (471)**
Papagena. G. Lustig.
Mitt. Astron. Ges., Nr. 42, p. 79 (1977).

098.089 **Photoelektrische Beobachtungen des Kleinplaneten**
(7) Iris von der Nord- und Südhemisphäre der Erde
aus. H. J. Schober.
Mitt. Astron. Ges., Nr. 42, p. 80 (1977).

098.090 **Implications of recently published diameters for**
Ceres, Pallas, and Vesta. R. G. Hodgson.
Strolling Astron., Vol. 27, 1 - 7 (1977).

098.091 **Observations of selected minor planets in 1974.**
P. P. Pavlenko.
Physics of the moon and planets. Fundamental astrometry,
(see 003.024), Vestn. Khar'kov. Univ., No. 160, p. 59 - 62
(1977). In Russian.

098.092 **Sizes and albedos of the larger asteroids.**
D. Morrison.
Comets, asteroids, meteorites, (see 012.049), p. 177 - 184
(1977).
    The purpose of the present paper is to review all asteroid
diameter measurements, current through mid-1976, and to
combine them in a consistent way to give the best available
estimates for a sample totalling 187 objects. From these
diameters it is possible to determine the size-distributions of
minor planets down to diameters of 50 km in the inner belt
and 100 km in the outer belt. The associated albedos further
indicate the distribution of objects of the C, S, and M classes
throughout the belt.

098.093 **Asteroid compositional types and their distribu-**
tions. B. Zellner, E. Bowell.
Comets, asteroids, meteorites, (see 012.049), p. 185 - 197
(1977).
    The purpose in this paper is to classify as large a sample
of asteroids as currently available observations permit and to
derive distributions of the various types over diameter and
orbital parameters with correction for observational selection
biases.

098.094 **Asteroid surface materials from reflectance spec-**
troscopy: a review. M. J. Gaffey, T. B. McCord.
Comets, asteroids, meteorites, (see 012.049), p. 199 - 218
(1977).
    Mineral assemblages have been identified on asteroid
surfaces which are comparable to most known meteorite types
or which have undergone the types of processes (e.g., melting
and differentiation) necessary to produce the meteoritic
assemblages. The ordinary chondrites, which dominate the
meteoritic flux reaching the Earth's surface, are very rare or
absent on Main Belt asteroids but appear common on the
small asteroids which approach or cross the orbit of the
Earth. The present interpretations of asteroidal spectra are
not yet quantitative enough to permit the evaluation of
specific asteroids as the sources bodies of particular meteorite
specimens in terrestrial collections.

098.095 **Asteroid surface compositions from infrared spec-**
troscopic observations: results and prospects.
H. P. Larson.
Comets, asteroids, meteorites, (see 012.049), p. 219 - 228
(1977).
    Advances in IR detector technology, the increased
availability of large aperture telescopes, and the techniques of
Fourier transform spectroscopy now permit IR ($\lambda > 1\mu$)
spectroscopic observations of asteroid surfaces. Some of the
most diagnostic features of mineral spectra are in the IR
spectral region, and for featureless spectra characterized only
by slopes the extension of the spectral reflectivity curve into

the IR provides tighter constraints on possible mineralogies.

098.096 **Asteroid infrared reflectances and compositional**
implications.
D. L. Matson, T. V. Johnson, G. J. Veeder.
Comets, asteroids, meteorites, (see 012.049), p. 229 - 241
(1977).
    This paper assesses the state of asteroid infrared
reflectance measurements and discusses what they have contributed to our understanding of asteroidal surfaces.

098.097 **The nature of asteroid surfaces, from optical**
polarimetry.
A. Dollfus, J. E. Geake, J. C. Mandeville, B. Zellner.
Comets, asteroids, meteorites, (see 012.049), p. 243 - 251
(1977).
    The purpose of this paper is to deduce the physical
nature and texture of the asteroid surfaces; this has been
achieved by means of telescopic observations of the way in
which these bodies polarize reflected sunlight.

098.098 **The physical basis of the polarimetric method for**
deriving asteroid albedos. T. Gehrels.
Comets, asteroids, meteorites, (see 012.049), p. 253 - 256
(1977).
    Physical interpretations of negative polarization and
opposition effect confirm the existence of a surface layer on
the asteroids of their own dust.

098.099 **Microwave brightnesses of 1 Ceres and 4 Vesta.**
E. K. Conklin, B. L. Ulich, J. R. Dickel, D. T. Ther.
Comets, asteroids, meteorites, (see 012.049), p. 257 - 261
(1977).
    The brightnesses of Ceres and Vesta were observed at
3 mm wavelength. For Ceres, pure rock cannot reproduce the
observed values, and a dust layer is required, much similar to
lunar material. For Vesta, its different thermal characteristics
appear to require a more compacted layer of material on its
surface.

098.100 **The evolution of asteroids as meteorite parent-**
bodies. C. R. Chapman.
Comets, asteroids, meteorites, (see 012.049), p. 265 - 275
(1977).
    The hypothesis that the asteroid belt is the source
region for nearly all meteorites remains viable and there is no
compelling reason to ascribe any meteorite to cometary
origin. The scenario for the evolution of asteroids, based on
collisional models of two distinct populations of different
physical properties, is being criticized and refined.

098.101 **Fragmentation of asteroids and delivery of frag-**
ments to earth. G. W. Wetherill.
Comets, asteroids, meteorites, (see 012.049), p. 283 - 291
(1977).
    The orbital histories of fragments of inner belt asteroids
are investigated, considering the combined effects of close
planetary encounters, secular perturbations, and secular
resonances. Particular attention is given to the low inclination
($\lesssim 15°$) objects with small semimajor axis (2.1 to 2.6 A.U.),
which can make fairly close approaches to Mars ($\lesssim 0.1$ A.U.).

098.102 **The Kirkwood gaps as an asteroidal source of**
meteorites. H. Scholl, C. Froeschlé.
Comets, asteroids, meteorites, (see 012.049), p. 293 - 295
(1977).
    In addition to the Kirkwood gap at the 2/1 commensur-
ability proposed by Zimmerman and Wetherill, the Kirkwood
gaps at the 5/2 and at the 3/1 commensurability also may be
considered as possible sources for meteorites.

098.103 **Populations of planet-crossing asteroids and the**

relation of Apollo objects to main-belt asteroids and comets.   E. M. Shoemaker, E. F. Helin.
Comets, asteroids, meteorites, (see 012.049), p. 297 - 300 (1977).

From discoveries made in several independent surveys of the sky, the number of Apollo asteroids to absolute visual magnitude 18 is estimated to be of the order of 10^3; the ratio of Mars-crossing asteroids to Apollos is estimated to lie between 10 and 60. The loss of Apollos to magnitude 18 over the last 3 billion years, by collision with the planets and ejection from the solar system, is estimated to have been several times 10^4 objects.

098.104   Asteroid versus comet discrimination from orbital data.   L'. Kresák.
Comets, asteroids, meteorites, (see 012.049), p. 313 - 321 (1977).

The orbital comet-asteroid criteria, their premises, dynamical reasoning, and ranges of applicability are reviewed. The potential evolutionary paths from different sources of active comets into short-period orbits are delineated and interfaced with the process of reducing the perihelion distances of the asteroids. The significance of resonances with Jupiter is emphasized. Statistics of observed close approaches of individual comets and asteroids to the Earth is analyzed to estimate their relative fluxes.

098.105   Some interrelations of asteroids, Trojans and satellites.   T. Gehrels.
Comets, asteroids, meteorites, (see 012.049), p. 323 - 325 (1977).

The Trojans of Jupiter have the same magnitude-frequency distribution as the asteroids, but there is a puzzling asymmetry of the densities in preceding and following Lagrangian points. The outer satellites of Jupiter have a peculiar magnitude-frequency distribution, with large and also small ones missing; they therefore cannot be captured asteroids or Trojans unless during the capture a resisting medium was present out to ~ 400 R_J for proto-Jupiter.

098.106   A statistical investigation of asteroid families: preliminary results.
A. Carusi, E. Massaro.
Comets, asteroids, meteorites, (see 012.049), p. 327 - 332 (1977).

A statistical investigation of asteroid families has been carried out, using a new clustering technique developed by A. I. Gavrishin. Proper elements for 2764 asteroids (1810 numbered and 954 Palomar-Leiden-Survey (PLS) asteroids) have been computed. Using these data, the Gavrishin method gives only ten significant classes. Five of them are coincident with the Hirayama families 1, 2, 3, 5, and the Flora group, that cannot be univocally subdivided. The PLS families are recognized.

098.107   Moving object observed at the Klet Observatory.
A. Mrkos.

Komet. Tsirk., Kiev, No. 217 (1977). In Russian.

098.108   Observations of the asteroid 1977 RA at the Kleť Observatory.   A. Mrkos.
Komet. Tsirk., Kiev, No. 218 (1977). In Russian.

Radar detectability of asteroids. A survey of opportunities for 1977 through 1987.   See Abstr. 031.201.

Asteroids detectable by radar systems 1977 - 86. See Abstr. 031.211.

Pluto-artige Bewegungstypen im Bereich der Jupiterbahn.   See Abstr. 042.076.

Langperiodische Bewegungen der Trojaner. See Abstr. 042.077.

Round-trip mission requirements for asteroids 1976 AA and 1973 EC.   See Abstr. 051.001.

Occultation of stars by planets and by minor planets. See Abstr. 091.085.

Upcoming lunar occultations of minor planets. See Abstr. 096.013.

Comets, asteroids and meteorites: unsolved problems about their relationships, evolution and origin. See Abstr. 102.053.

Comets, minor planets and meteoroids. See Abstr. 102.056.

Observation of comets and asteroids at the Kleť Observatory in the year 1975.   See Abstr. 103.008.

Characterization of earth-crossing asteroids, past and present.   See Abstr. 105.016.

History of the Pasamonte achondrite: relative susceptibility of the Sm–Nd, Rb–Sr, and U–Pb systems to metamorphic events.   See Abstr. 105.106.

Relationship between meteorites, asteroids and comets.   See Abstr. 105.268.

Errata

098.901   Erratum: "Theory of the Trojan asteroids. Part I" [Astron. J., Vol. 82, 368 - 379 (1977)].
B. Garfinkel.
Astron. J., Vol. 82, 930 (1977).

099 Jupiter, Jupiter Satellites

Jupiter

099.001 **Recovery of the mean Jovian temperature structure from inversion of spectrally resolved thermal radiance data.** G. S. Orton.
Icarus, Vol. 32, 41 - 57 (1977).
The mean thermal structure of the Jovian atmosphere near the temperature minimum (0.1 bar) is recovered by inversion of thermal radiance data. Improvements over previous studies of this type are made.

099.002 **The structure of the planets Jupiter and Saturn.** W. L. Slattery.
Icarus, Vol. 32, 58 - 72 (1977).
Planetary models for Jupiter and Saturn are computed using a fourth-order theory and a new molecular equation of state. Models for Jupiter are found that have a small amount of heavy elements either mixed with hydrogen and helium throughout the interior of the planet or concentrated in a small dense core. Saturn is modeled with a solar-composition hydrogen and helium envelope and a small dense core. The planetary models show that the enrichment of heavy elements (relative to solar composition) is approximately 3 times for Jupiter and 10 times for Saturn.

099.003 **Soft electrons as a possible heat source for Jupiter's thermosphere.** D. M. Hunten, A. J. Dessler.
Planet. Space Sci., Vol. 25, 817 - 821 (1977).
The 850 K exospheric temperature inferred for Jupiter from the radio-occultation experiments on Pioneers 10 and 11 is shown to imply a heat input of $0.25-0.5$ erg cm^{-2} s^{-1}. One possible source of this energy is precipitation of electrons from a warm plasma (temperature corresponding to energies of the order of 30-500 eV). A mechanism is suggested wherein the presence of this plasma can be accounted for by centrifugal acceleration and adiabatic compression of ionospheric electrons and protons.

099.004 **Jupiter through colour filters, 1975/1976.** A. W. Heath, J. H. Robinson.
J. British Astron. Assoc., Vol. 87, 485 - 487 (1977).

099.005 **Jovian sulfur nebula.** Y. Mekler, A. Eviatar, I. Kupo.
J. Geophys. Res., Vol. 82, 2809 - 2814 (1977).
The authors present further results of their observations of the nebula of singly ionized sulfur in the magnetosphere of Jupiter. They have calculated the occupation numbers of the five lowest energy levels of ionized sulfur and have used their results together with the observed intensities to evaluate the number density of ionized sulfur.

099.006 **Some features of Jovian decametric radio emission and its temporary variations.** L. S. Levitskij.
Izv. Krymskoj Astrofiz. Obs., Vol. 57, 177 - 188 (1977). In Russian.
The following parameters of Jovian decametric activity are considered: durations of the storms, distribution of the storms over Jovian longitudes and Io's position, distribution along frequency range, "burstiness".

099.007 **Possible Jovian methane emission at 76 GHz in coincidence with decameter activity.** K. Fox, D. E. Jennings.
Astrophys. J., Lett., Vol. 216, L83 - L84 (1977).
The authors report the tentative detection of a methane line in emission at 76.2 GHz in the atmosphere of Jupiter.

The observed feature is well correlated with the presence and absence of Jovian decameter emission activity on successive days.

099.008 **Jupiter's decameter-wave radiation and solar activity.** C. H. Barrow.
Bull. American Astron. Soc., Vol. 9, 431 - 432 (1977). Abstract.

099.009 **Study of Jupiter's corotating S II nebula at high spectral resolution.** G. Münch, J. T. Trauger, F. L. Roesler.
Bull. American Astron. Soc., Vol. 9, 465 (1977). — Abstract.

099.010 **Multiple frequency sounding of a Jovian cloud.** G. S. Orton, R. J. Terrile.
Bull. American Astron. Soc., Vol. 9, 469 (1977). — Abstract.

099.011 **Cloud motions on Jupiter.** L. H. Wasserman, S. E. Jones, W. A. Baum.
Bull. American Astron. Soc., Vol. 9, 474 (1977). — Abstract.

099.012 **High resolution imaging of Jupiter at 5, 8 - 14, and 20 microns.** R. J. Terrile, J. A. Westphal, G. S. Orton.
Bull. American Astron. Soc., Vol. 9, 474 (1977). — Abstract.

099.013 **Evaluation of Jupiter longitudes in system III (1965).** P. K. Seidelmann, N. Divine.
Bull. American Astron. Soc., Vol. 9, 474 (1977). — Abstract.

099.014 **A review and criticism of the nomenclature of atmospheric features of Jupiter and Saturn.** R. Beebe, E. Reese.
Bull. American Astron. Soc., Vol. 9, 474 (1977). — Abstract.

099.015 **Turbulent viscosity and Jupiter's tidal Q.** P. Goldreich, P. D. Nicholson.
Bull. American Astron. Soc., Vol. 9, 474 - 475 (1977). Abstract.

099.016 **Internal heat flow on Jupiter.** A. P. Ingersoll, C. Porco.
Bull. American Astron. Soc., Vol. 9, 475 (1977). — Abstract.

099.017 **On the production and interaction of planetary solitary waves: applications to the Jovian atmosphere.** T. Maxworthy, L. G. Redekopp, P. D. Weidman.
Bull. American Astron. Soc., Vol. 9, 475 (1977). — Abstract.

099.018 **Thermal convection in Jupiter.** F. M. Flasar.
Bull. American Astron. Soc., Vol. 9, 475 (1977). Abstract.

099.019 **Carbon monoxide on Jupiter and implications for atmospheric convection.** R. G. Prinn, S. S. Barshay.
Bull. American Astron. Soc., Vol. 9, 475 (1977). — Abstract.

099.020 **The distribution of ammonia, hydrazine and nitrogen on Jupiter.** S. K. Atreya, T. M. Donahue, W. R. Kuhn.
Bull. American Astron. Soc., Vol. 9, 475 (1977). — Abstract.

099.021 **Methylamine formation in the Jovian atmosphere.** W. R. Kuhn, S. K. Atreya, S. Chang.
Bull. American Astron. Soc., Vol. 9, 475 (1977). — Abstract.

099.022 Synchrotron radiation from Jupiter: comparison of earth-based data with calculations based on Pioneer 10 and 11 data.
S. Gulkis, J. L. Luthey, M. J. Klein.
Bull. American Astron. Soc., Vol. 9, 476 (1977). − Abstract.

099.023 Pioneer pictures of Jupiter.   T. Gehrels.
Bull. American Astron. Soc., Vol. 9, 476 (1977). Abstract.

099.024 Low frequency radio emission from the outer planets.   L. W. Brown.
Bull. American Astron. Soc., Vol. 9, 476 (1977). − Abstract.

099.025 A Jupiter photograph in the 886 nm methane absorption band.   L. Dunkelman, R. B. Minton.
Bull. American Astron. Soc., Vol. 9, 476 (1977). − Abstract.

099.026 The aerosol content in the upper atmosphere at several locations on Jupiter.
D. W. Smith, T. F. Greene.
Bull. American Astron. Soc., Vol. 9, 476 - 477 (1977). Abstract.

099.027 Jupiter atmospheric structure from limb darkening.
A. M. Goldberg.
Bull. American Astron. Soc., Vol. 9, 477 (1977). − Abstract.

099.028 Structure of the clouds of Jupiter by correlated 5-micron spectroscopy and areal photometry.
R. J. Terrile, F. W. Taylor, R. Beer, J. A. Westphal.
Bull. American Astron. Soc., Vol. 9, 477 (1977). − Abstract.

099.029 Temporal variations of molecular absorptions on Jupiter.   J. H. Woodman.
Bull. American Astron. Soc., Vol. 9, 477 (1977). − Abstract.

099.030 The far infrared spectra of Jupiter and Saturn: a comparative study of their potential return.
M. Combes, T. Encrenaz.
Bull. American Astron. Soc., Vol. 9, 477 (1977). − Abstract.

099.031 The N/H and ^{15}N/^{14}N Jovian ratios: a tentative determination from infrared observations.
T. Encrenaz, M. Combes.
Bull. American Astron. Soc., Vol. 9, 477 - 478 (1977). Abstract.

099.032 The ^{12}C/^{13}C ratio in Jupiter and Saturn: presumptions for a telluric value.
M. Combes, J. P. Maillard, C. de Bergh.
Bull. American Astron. Soc., Vol. 9, 478 (1977). − Abstract.

099.033 A revised value for the D/H ratio in Jupiter from the CH$_3$D phase.   R. Beer, F. W. Taylor.
Bull. American Astron. Soc., Vol. 9, 478 (1977). − Abstract.

099.034 A Jovian methane abundance based on newly identified lines.
M. E. Mickelson, J. T. Trauger, D. H. Tracy, F. L. Roesler.
Bull. American Astron. Soc., Vol. 9, 478 (1977). − Abstract.

099.035 The abundance of C$_2$H$_2$ in Jupiter.
G. S. Orton, H. H. Aumann.
Bull. American Astron. Soc., Vol. 9, 478 (1977). − Abstract.

099.036 Jupiter atmospheric structure measurements.
H. Myers.
Bull. American Astron. Soc., Vol. 9, 478 (1977). − Abstract.

099.037 Linear polarization observations of Jupiter at 6, 11 and 21 cm wavelengths.

F. F. Gardner, J. B. Whiteoak.
Astron. Astrophys., Vol. 60, 369 - 375 (1977).
    Linear polarization and total intensity observations of Jupiter were made between 1969 and 1971, when the tilt of the Jovian rotational axis was near its peak negative value of −3°. They include one set of observations at 6 cm wavelength, four at 11 cm, and two at 21 cm. As found by others, the variations with Jovian rotation of both the direction of polarization and the polarized flux density depart from the predictions for a simple inclined dipole model of the magnetic field. In addition, the apparent inclination of the dipole was found to increase with wavelength. There is a decrease of polarized flux density with decreasing wavelength, which suggests a high energy cut-off of the relativistic particles at an equivalent wavelength not much shorter than 6 cm.

099.038 Influence of solar activity on particle acceleration near Jupiter.   A. Z. Dolginov, V. A. Urpin.
Izv. AN SSSR. Ser. fiz., Vol. 41, 322 - 328 (1977). In Russian.
Abstr. in Ref. zh., 51. Astron., 8.51.395 (1977).

099.039 Die Beobachtungen von Veränderungen in der Jupiter-Atmosphäre.   J. Böing.
Sterne Weltraum, Jahrg. 16, 342 (1977).

099.040 Jupiter in 1975 - 76: rotation periods.
P. W. Budine.
Strolling Astron., Vol. 26, 217 - 231 (1977).

099.041 Measured photographic latitudes on Jupiter in 1975 - 76.   P. W. Budine.
Strolling Astron., Vol. 26, 231 - 233 (1977).

099.042 An update on South Equatorial Belt Disturbance analysis, part one.   R. Doel.
Strolling Astron., Vol. 26, 254 - 257 (1977).
    The history, characteristics, and normal development of major South Equatorial Belt Disturbances on Jupiter are summarized. The apparent periodic nature of these events and their relations to activity in other belts and zones of Jupiter are examined. Thus SEB Disturbances may be only part of a planet-wide activity pattern with a cycle of three years.

099.043 A determination of the L dependence of the radial diffusion coefficient for protons in Jupiter's inner magnetosphere.
M. F. Thomsen, C. K. Goertz, J. A. Van Allen.
J. Geophys. Res., Vol. 82, 3655 - 3658 (1977).
    The technique proposed by Thomsen et al. (1977) to determine the radial diffusion coefficient in Jupiter's inner magnetosphere from observations of energetic particle phase space density profiles across the satellite orbits is extended to permit the unique identification of the parameters D_0 and n, where the diffusion coefficient is assumed to be of the form $D = D_0 L^n$. The derived value of D_0 depends directly on assumptions regarding the nature and the efficiency of the loss mechanism operating on the particles. The value of n, however, depends only on the assumed width of the loss region, and even that dependence is not strong.

099.044 On a possible explanation of the formation mechanism of the equatorial jet on Jupiter's surface.
I. M. Yavorskaya, L. M. Simuni.
Dokl. AN SSSR, Vol. 233, 60 - 63 (1977). In Russian. − Abstr. in Ref. zh., 51. Astron., 9.51.291 (1977).

099.045 Influence of solar activity on the acceleration of particles near Jupiter.
A. Z. Dolginov, V. A. Urpin.
VIII Leningr. mezhdunar. semin. Mater. mezhdunar. semin. "Akt. protsessy na Solntse i probl. soln. nejtrino", 1976. Leningrad, 1976, p. 148 - 153. − Abstr. in Ref. zh., 62. Issled.

kosm. prostranstva, 9.62.173 (1977).

099.046 New measurements of the Stokes parameters of Jupiter's 11 cm radiation.   J. Neidhöfer, R. S. Booth, D. Morris, W. Wilson, F. Biraud, J.-C. Ribes.
Astron. Astrophys., Vol. 61, 321 - 328 (1977).

Measurements of the variation of the complete polarization properties of the Jovian 11 cm radiation during one rotation of the planet are presented. The observations were made in August 1975 with the Effelsberg 100-m telescope. Special attention was paid to the measurements of the circular polarization. The data of the present observations for the linear polarization, the position angle of the electric vector and the total intensity generally agree with observations made one orbital period earlier (1963), confirming the constancy of the general configuration of the Jovian magnetic field.

099.047 Parameters of the figure and gravitational moments of Jupiter and Saturn.
A. B. Efimov, V. N. Zharkov, V. P. Trubitsyn, A. M. Bobrov.
Astron. Zh. Akad. Nauk SSSR, Vol. 54, 1118 - 1132 (1977). In Russian. English translation in Soviet Astron., Vol. 21, No. 5.

Parameters of the figure and gravitational moments of Jupiter and Saturn are calculated to the fifth approximation.

099.048 Particle acceleration in the Jovian magnetosphere.   A. Z. Dolginov, V. A. Urpin.
Geomagn. Aehron., Vol. 17, 804 - 810 (1977). In Russian.

099.049 Méthode pour calculer les périodes de rotation de Jupiter.   M. Alecsescu.
Astronomie, Vol. 91, 387 - 390 (1977).

099.050 Silicon vidicon imaging of Jupiter. 4100- to 8300-Å absolute reflectivities and limb darkening of spatially resolved regions. D. J. Diner, J. A. Westphal.
Icarus, Vol. 32, 299 - 313 (1977) = Div. Geol. Planet. Sci., Calif. Inst. Technol., Pasadena, Calif., Contrib. No. 2838.

Jupiter was observed in six continuum wavelength channels in the region 4100–8300 Å, using a silicon vidicon imaging photometer. Spectral reflectivities and high spatial resolution limb-darkening curves for several belts and zones have been extracted from the data. Simple model fits to the data yield information regarding spectral and spatial variations in single-scattering albedos and shape of particle single-scattering phase functions. Belts appear to be more back-scattering than zones, particularly in the blue.

099.051 The abundances of ammonia in the atmospheres of Jupiter, Saturn, and Titan.
J. H. Woodman, L. Trafton, T. Owen.
Icarus, Vol. 32, 314 - 320 (1977).

An investigation of low-resolution ratio spectra of Jupiter, Saturn, and Titan in the region 5400–6500 Å has permitted new evaluations of ammonia absorption bands. The distribution of ammonia over the disk of Jupiter is very inhomogeneous. The abundance of ammonia on Saturn also shows spatial variations, but appears constant in time over a 3-yr period. Two weak, unidentified absorptions were discovered in the red region of Titan's spectrum, in the absence of any detectable ammonia.

099.052 Theoretical brightness temperature profiles of atmospheric pure H_2 rotational quadrupole lines: Jupiter and Uranus.   D. Goorvitch, C. Chackerian, Jr.
Icarus, Vol. 32, 348 - 361 (1977).

With the advent of high-resolution instruments and their use high above most of the telluric water vapor, one can expect to observe the hydrogen pure rotational quadrupole lines at 28, 17, and 12 μm from the atmospheres of the outer planets. The authors have calculated the best values for the line strengths, pressure-broadening coefficients, diffusion constants, and pressure shifts for these rotational transitions. They have used the collisionally narrowed Galatry profile to calculate brightness temperature line profiles for these H_2 transitions for the outer planets Jupiter and Uranus.

099.053 On the magnetospheres of Jupiter, Saturn, and Uranus.   J. A. Van Allen.
Highlights of Astronomy, Vol. 4, Part I, (see 012.019), p. 195 - 224 (1977).

A brief descriptive summary of Jupiter's magnetosphere is based on in situ observations with the spacecraft Pioneer 10 and Pioneer 11. Current interpretative work emphasizes particle acceleration and loss mechanisms, the determination of diffusion coefficients by satellite effects, the topology of the outer magnetosphere, the possible recirculation of energetic particles, and the controversial evidence for an extended magnetotail. Available evidence on non-thermal radio emissions of Saturn and on the solar wind flow at 10 AU is invoked to suggest that Saturn very likely has a large, well developed magnetosphere resembling that of Jupiter but with the important difference that a radiation belt cannot exist interior to the outer edge of the A ring of particulate matter. It is speculated that Uranus also has a large, well developed magnetosphere and one of unique interest during epochs when its rotational axis is approximately along the planet–sun line as in mid-1985.

099.054 Pioneer images of Jupiter.
J. W. Fountain, T. Gehrels.
Highlights of Astronomy, Vol. 4, Part I, (see 012.019), p. 233 - 241 (1977).

Pioneer 10 flew by Jupiter in December 1973 and Pioneer 11, in a more polar trajectory, one year later. The authors present some of the pictures of Jupiter made with the spin-scan technique by the imaging photopolarimeter.

099.055 The far infrared spectrum of Jupiter.
I. Furniss, R. E. Jennings, K. J. King.
Far infrared astronomy (see 012.027), p. 71 - 78 (1976).

Far infrared spectra of Jupiter in the range 60 - 220 cm^{-1} are presented and compared with theoretical models.

099.056 Evidence for a telluric value of the $^{12}C/^{13}C$ ratio in the atmospheres of Jupiter and Saturn.
M. Combes, J. P. Maillard, C. de Bergh.
Astron. Astrophys., Vol. 61, 531 - 537 (1977).

A new determination of the $^{12}C/^{13}C$ isotopic ratio in the atmospheres of Jupiter and Saturn has been derived from several manifolds of the $^{13}CH_4 - 3\nu_3$ band. It is shown that the $^{13}CH_4$ and $^{12}CH_4 - 3\nu_3$ bands cannot be significantly compared in terms of a reflecting layer model. A method of comparison of the $^{13}CH_4 - 3\nu_3$ lines with weak $^{12}CH_4$ lines in terms of a scattering atmosphere is described, which is valid whatever the scattering process may be. For both Jupiter and Saturn, the resulting $^{12}C/^{13}C$ ratio does not differ significantly from the telluric value. The results are: $^{12}C/^{13}C = 89^{-10}_{+12}$ for Jupiter and 89^{-18}_{+25} for Saturn.

099.057 Search for Jovian auroral hot spots.
S. K. Atreya, Y. L. Yung, T. M. Donahue, E. S. Barker.
Astrophys. J., Lett., Vol. 218, L83 - L87 (1977).

Auroral emission originating at the foot of the Io-associated flux tube at Jupiter has been detected with a high-resolution spectrometer/telescope on board the Orbiting Astronomical Observatory Copernicus. The emission intensity at Lα is found to be greater than 100 kR and the emission is located at zenographic latitudes greater than 65°.

099.058 Discoveries about Jupiter.
Spaceworld, Vol. M-9-153, p. 1 - 32 (1976). – From

Phys. Abstr., Vol. 80, Abstr. 22134 (1977).

099.059   Jupiter's magnetosphere.
C. F. Kennel, F. V. Coroniti.
Annu. Rev. Astron. Astrophys., Vol. 15, (see 003.012), 389 - 436 (1977).

The authors present: The Pioneer 10 and 11 magnetospheric experiments: an overview; Pioneer observations of Jupiter's middle and outer magnetosphere; 5 models of Jupiter's outer magnetosphere: the authors discuss three classes of Jovian magnetosphere models: first, earth-like reconnection models, including Ioannidis & Brice's (1971) picture of the ensuing internal convection pattern for an inertia and pressure-free plasma; second, static rotating disk models; and third, radial outflow models—either super-Alfvénic, sub-Alfvénic, or mixed. In addition they discuss the extent to which Jupiter and Io can feed the outer magnetosphere with plasma.

099.060   The two types of atmosphere of Jupiter and Saturn.
T. Gehrels.
Chemical evolution of the giant planets (see 003.014), p. 1 - 11 (1976).

Shapes and motions of Jovian clouds are discussed on the basis of spin-scan imaging during the flyby missions of Pioneers 10 and 11. There is the part of the atmosphere, between +45° and −45° latitude, that has a regime of large-scale atmospheric dynamics seen as zones and belts. The second type of atmosphere occurs at the polar regions, where these large-scale features are missing and the atmosphere has great optical depth (that is, great transparency for visible light). The situation appears to be similar on Saturn.

099.061   Equilibrium and disequilibrium. Chemistry of adiabatic, solar-composition planetary atmospheres.
J. S. Lewis.
Chemical evolution of the giant planets, (see 003.014), p. 13 - 25 (1976) = Contrib. Mass. Inst. Technol., Planet. Astron. Lab., No. 119.

The impact of atmospheric and cloud structure models on the nonequilibrium chemical behavior of the atmospheres of the Jovian planets is discussed. Quantitative constraints on photochemical, lightning, and charged-particle production of organic matter and chromophores are emphasized whenever available. These considerations imply that inorganic chromophore production is far more important than that of organic chromophores, and that lightning is probably a negligibly significant process relative to photochemistry on Jupiter.

099.062   Possible origins of time variability in Jupiter's outer magnetosphere. I. Variations in solar wind dynamic pressure.   F. V. Coroniti, C. F. Kennel.
Geophys. Res. Lett., Vol. 4, 211 - 214 (1977).

The authors qualitatively examine the effects of changes in the solar wind dynamic pressure on the structure of a centrifugally driven planetary wind from Jupiter. They suggest that dynamic pressure variations can induce a transition between a super-Alfvenic wind and sub-Alfvenic breeze on Jupiter's dayside, which could possibly account for the observed large scale changes in the structure of Jupiter's outer magnetosphere.

099.063   Possible origins of time variability in Jupiter's outer magnetosphere. 2. Variations in solar wind magnetic field.   C. F. Kennel, F. V. Coroniti.
Geophys. Res. Lett., Vol. 4, 215 - 218 (1977).

The authors attempt to merge conceptually planetary wind models of Jupiter's outer magnetosphere with reconnection models of Jupiter's outer magnetosphere. Solar wind reconnection scaling arguments predict 2—4 AU for the length of the Jovian tail, consistent with that inferred from Jovian cosmic ray propagation arguments. Two conceivable limits of solar wind reconnection driven internal magnetospheric convection emerge,

depending upon whether or not the centrifugally driven planetary wind transports magnetic flux radially outward.

099.064   Comment on low energy electron measurements in the Jovian magnetosphere.
R. J. L. Grard, S. E. DeForest, E. C. Whipple, Jr.
Geophys. Res. Lett., Vol. 4, 247 - 248 (1977).

The authors suggest that the low energy electrons reported by Intriligator and Wolfe (1974) in the outer magnetosphere of Jupiter may actually be photoelectrons and/or secondary electrons from the Pioneer spacecraft surfaces. The electron spectrum is similar to the observed photoelectron spectrum on earth-orbiting satellites where differentially charged surfaces have deflected the electrons into particle detectors.

099.065   Plasma electron measurements in the outer Jovian magnetosphere.   D. S. Intriligator, J. H. Wolfe.
Geophys. Res. Lett., Vol. 4, 249 - 250 (1977).

099.066   Jupiter's atmosphere. Meudon Observatory's planets group.
Recherche, Vol. 7, 910 - 917 (1976). In French. — Abstr. in Phys. Abstr., Vol. 80, Abstr. 33403 (1977).

099.067   Synchrotron process as the source of decametric radiation from Jupiter.
R. N. Singh, R. P. Singh, R. S. Rai.
Indian J. Radio Space Phys., Vol. 5, 285 - 288 (1976). — Abstr. in Phys. Abstr., Vol. 80, Abstr. 71056 (1977).

099.068   The abundance of acetylene in the atmosphere of Jupiter.   G. S. Orton, H. H. Aumann.
Icarus, Vol. 32, 431 - 436 (1977).

The Q and R branches of the C_2H_2 ν_5 fundamental, observed in emission in an aircraft spectrum of Jupiter near 750 cm^{-1}, have been analyzed with the help of an improved line listing for this band. The line parameters have been certified in the laboratory with the same interferometer used in the Jovian observations. The maximum mixing ratio of C_2H_2 is found to be between 5×10^{-8} and 6×10^{-9}, depending on the form of its vertical distribution and the temperature structure assumed for the lower stratosphere. Most consistent with observations of both Q and R branches are: (1) distributions of C_2H_2 with a constant mixing ratio in the stratosphere and a cutoff at a total pressure of 100 mbar or less, and (2) the assumption of a temperature at 10^{-2} bar which is near 155°K.

099.069   Love numbers of the giant planets.
S. V. Gavrilov, V. N. Zharkov.
Icarus, Vol. 32, 443 - 449 (1977).

The authors calculate the Love numbers k_n for $n = 2$ to 10, and determine the "gravitational noise" from tides. The new values k_2 for Jupiter, Saturn, and Uranus yield new estimates for the planetary dissipation functions: $Q_J \gtrsim 2.5 \times 10^4$, $Q_S \gtrsim 1.4 \times 10^4$, $Q_U \gtrsim 5 \times 10^3$.

099.070   The propagation of plasma waves in the Jovian magnetosphere.   K. U. Denskat, F. M. Neubauer.
J. Geophys., Vol. 43, 511 - 519 (1977). — Abstr. in Phys. Abstr., Vol. 80, Abstr. 87317 (1977).

099.071   Does Jupiter have a "Mini-corona"?
G. Münch, F. Roesler, J. Trauger.
Publ. Astron. Soc. Pacific, Vol. 89, 619 (1977). — Abstract.

099.072   A 1.3-cm wavelength interferometric study of Jupiter.   F. Valdes.
Publ. Astron. Soc. Pacific, Vol. 89, 624 (1977). — Abstract.

099.073   Low order effect of CRAND input in the Jovian atmosphere.   M. S. Spergel.

Moon, Vol. 17, 123 - 131 (1977).

Modeling of the Jovian atmosphere shows that cosmic ray induced albedo neutron decay is inadequate to account for Pioneer 10 and 11 projected electron levels on Jupiter. High energy solar protons must also be excluded as an important neutron decay source. Analysis of neutron flux data near the top of the Jovian atmosphere can lead to the determination of He/H_2 and $^3He/^4He$ ratios for the Jovian atmosphere.

099.074 **Analisi del movimento delle W.O.S. (*white oval spots*) di Giove dal 1941 al 1974.** P. Senigalliesi.
Astronomia, N. 3, p. 45 - 57 (1977).

099.075 **Hot interior of Jupiter.** L. I. Miroshnichenko.
Priroda, 1977, No. 12, p. 113. In Russian.

099.076 **S-bursts in Jupiter's decametric radio spectra.**
J. J. Riihimaa.
Astrophys. Space Sci., Vol. 51, 363 - 383 (1977).

The purpose of the present report is to describe a new series of spectral observations performed from 1974 to 1976. A summary is given of S-activity over 9 observation periods between 1963 and 1976.

099.077 **On the optical thickness of the upper Jovian cloud.**
K. Yu. Ibragimov.
Astron. Tsirk., No. 952, p. 3 - 5 (1977). In Russian.

099.078 **New observations of Jupiter in the 800–1000 cm^{-1} range.** T. Encrenaz, M. Combes, Y. Zeau.
Infrared Phys., Vol. 17, 551 - 555 (1977).

A new spectrum of Jupiter between 800 and 1000 cm^{-1}, obtained in January 1977 at Mauna Kea Observatory, is presented. The resolution is 1.5 cm^{-1}. In addition to the $^{14}NH_3$ absorption bands, the PH_3 molecule is identified. The identification of $^{15}NH_3$ still needs to be confirmed.

099.079 **Acceleration of protons at 32 Jovian radii in the outer magnetosphere of Jupiter.**
A. W. Schardt, F. B. McDonald, J. H. Trainor.
GSFC Doc. X-660-77-225, Prepr., 26 pp. (1977).

During the inbound pass of Pioneer 10, a rapid ten-fold increase of the 0.2 to 5 MeV proton flux was observed at 32 Jovian radii (R_J). The total event lasted for 30 minutes and was made up of a number of superimposed individual events.

099.080 **Jupiter's internal magnetic field geometry relevant to particle trapping.**
J. G. Roederer, M. H. Acuña, N. F. Ness.
J. Geophys. Res., Vol. 82, 5187 - 5194 (1977).

This article describes some field-geometric features of relevance to particle trapping in the inner magnetosphere of Jupiter and in the polar cap regions using the Goddard Space Flight Center (GSFC) internal field model of Acuña and Ness (1976).

099.081 **Carbon monoxide on Jupiter and implications for atmospheric convection.**
R. G. Prinn, S. S. Barshay.
Science, Vol. 198, 1031 - 1034 (1977).

A study of the equilibrium and disequilibrium thermochemistry of the recently discovered carbon monoxide on Jupiter suggests that the presence of this gas in the visible atmosphere is a direct result of very rapid upward mixing. As a consequence the observed carbon monoxide mixing ratio is a sensitive function of the vertical eddy mixing coefficient. The authors infer a value for this latter coefficient which is about three to four orders of magnitude greater than that in the earth's troposphere. This result directly supports existing structural and dynamical theories implying very rapid convection in the deep Jovian atmosphere, driven by an internal heat source.

099.082 **The appearance of Jupiter in 1975–76: the north hemisphere.** P. K. Mackal.
Strolling Astron., Vol. 27, 24 - 30 (1977).

099.083 **Optical parameters of the Jovian atmosphere and their latitudinal variations according to $\lambda\lambda 0.6 -$ 1.1 μm observations.** V. D. Vdovichenko.
Astrofiz. inst. AN KazSSR. Alma-Ata, 1977. 43 pp. In Russian. Abstr. in Ref. zh., 51. Astron., 1.51.298 (1978).

099.084 **On determining magnetospheric diffusion coefficients from the observed effects of Jupiter's satellite Io.** M. F. Thomsen, C. K. Goertz, J. A. Van Allen.
J. Geophys. Res., Vol. 82, 5541 - 5550 (1977).

Several previously proposed techniques for determining the radial diffusion coefficient from the observed effects of the inner Jovian satellites on the energetic particle fluxes are discussed, and important shortcomings are pointed out. A new method is proposed which avoids the most important shortcoming by dealing with data from regions somewhat removed from the actual sweeping region. The new technique is applied to data obtained at the orbit of Io by the University of Iowa proton detector on Pioneer 11 and to a published electron phase space density profile constructed from data obtained (also at Io's orbit) by the University of California at San Diego instrument on Pioneer 10.

099.085 **Thermal structure of the Jovian plasmasphere.**
A. Nishida, S. Machida.
10th Lunar and Planetary Symposium, (see 012.050), p. 193 - 198 (1977). In Japanese.

The origin and dynamics of the ambient plasma in the Jovian inner plasmasphere is investigated. The study aims at interpreting the low energy proton observations reported by Frank et al. (1976).

099.086 **Structure of the Jovian magnetosphere — magnetospheric wind within the disc.**
H. Oya, T. Aoyama.
10th Lunar and Planetary Symposium, (see 012.050), p. 199 - 205 (1977). In Japanese.

On the coincidence of the position of Mercury with the 90-day oscillation of Jupiter's Red Spot.
See Abstr. 015.002.

Organic synthesis in a simulated Jovian atmosphere of the planet Jupiter. See Abstr. 015.018.

Analysis of the $v_3 + v_4$ band of ammonia.
See Abstr. 022.002.

Measurements of line intensities in the two-micron band of ammonia. See Abstr. 022.003.

Molecular analysis of organic solids produced under simulated Jovian conditions. See Abstr. 022.013.

Refractivity and dispersion of hydrogen in the visible and near infrared. See Abstr. 022.077.

The effects of the atmospheric point-spread "seeing" function on spatially resolved spectra of Jupiter.
See Abstr. 031.218.

NH_3 spectral line measurements on Earth and Jupiter using a 10 μm superheterodyne receiver.
See Abstr. 031.333.

Plasma analyzer for the Pioneer Jupiter missions.
See Abstr. 032.576.

Positions of Jupiter, Saturn, Uranus and Neptune determined from observations with the Nikolaev zonal astrograph during 1967–1971.   See Abstr. 041.048.

Voyager imaging experiment. See Abstr. 051.061.

Radio science investigations with Voyager. See Abstr. 051.062.

The plasma experiment on the 1977 Voyager mission. See Abstr. 051.063

A plasma wave investigation for the Voyager mission. See Abstr. 051.064.

Planetary radio astronomy experiment for Voyager missions.   See Abstr. 051.065.

Comparative magnetospherology. Part 5. Examination of the two-hemisphere model by means of the apparent sector structure of the Jovian magnetosphere. See Abstr. 080.093.

Impulsive penetration of filamentary plasma elements into the magnetospheres of the Earth and Jupiter. See Abstr. 084.203.

Cloud microphysics: comments on the clouds of Venus, Mars and Jupiter.   See Abstr. 091.005.

Gravitational experiments on space missions to the giant planets.   See Abstr. 091.013.

The phase diagram and transport properties for hydrogen-helium fluid planets.   See Abstr. 091.060.

The dynamics and helium distribution in hydrogen-helium fluid planets.   See Abstr. 091.061.

The atmospheres of the planets. See Abstr. 093.035.

Observation of Venus and Jupiter in the 3–4 mm band. (Part II).   See Abstr. 093.070.

Ultraviolet observations of Mars and Saturn by the TD1A and OAO-2 satellites.   See Abstr. 097.071.

Acceleration of nucleons in the interplanetary space and modulation of electrons of Jupiter between 1 and 10 a.u. by corotating regions of solar origin. See Abstr. 106.011.

Relations between turbulent regions of the interplanetary magnetic field and Jovian decameter wave emissions from the main source.   See Abstr. 106.051.

On the origin and evolution of Jupiter and Saturn. See Abstr. 107.031.

Jupiter Satellites

099.501 **Die Monde des Jupiter.**   P. Gerber. Orion, 35. Jahrg., 112 - 117 (1977).

099.502 **Periodic variations in Io's sodium and potassium clouds.**   L. Trafton. Astrophys. J., Vol. 215, 960 - 970 (1977). Spectra of Io's sodium cloud taken $7^1/_2$ " to 45" north and south of Io at a variety of magnetic latitudes of Io obtained during 17 nights of the 1975 apparition of Jupiter confirm the weakening of the cloud in the neighborhood of Jupiter's magnetic equator. The time scale for these changes is $1^1/_2$ hours or less. The author also obtained spectra of Io's potassium cloud on eight nights during the 1975 apparition. This cloud appears to behave like the sodium cloud as Io's magnetic latitude varies. The potassium, however, may become much weaker as Io passes through the magnetic equator.

099.503 **Io's sodium emission profiles: variations due to Io's phase and magnetic latitude.** L. Trafton, W. Macy, Jr. Astrophys. J., Vol. 215, 971 - 976 (1977). The authors present recent measurements of high-resolution sodium D_2 emission line profiles of Io's atmosphere and cloud. The high-velocity skirts of these profiles are strongest when Io is on the magnetic equator where protons and heavy ions, which sputter atoms from Io's surface, are concentrated.

099.504 **Jupiter and Saturn satellite observations.** D. A. Pierce. Bull. American Astron. Soc., Vol. 9, 436 (1977). – Abstract.

099.505 **Comparison of Lieske's theory of the Galilean satellites with numerical integration.** C. F. Peters. Bull. American Astron. Soc., Vol. 9, 437 (1977). – Abstract.

099.506 **Concerning the capture of the outer satellites.** J. B. Pollack, J. A. Burns, M. Tauber. Bull. American Astron. Soc., Vol. 9, 455 - 456 (1977). Abstract.

099.507 **Radar studies of the Galilean satellites.** D. B. Campbell, S. Ostro, G. H. Pettengill. Bull. American Astron. Soc., Vol. 9, 464 (1977). – Abstract.

099.508 **Absorptions in the spectrum of Io, 3.0–4.2 microns.** D. P. Cruikshank, T. J. Jones, C. B. Pilcher. Bull. American Astron. Soc., Vol. 9, 465 (1977). – Abstract.

099.509 **Io sodium emission: 1976 patrol and multislit observations.**   J. T. Bergstralh, T. V. Johnson, D. L. Matson, J. W. Young, B. A. Goldberg, R. W. Carlson. Bull. American Astron. Soc., Vol. 9, 465 (1977). – Abstract.

099.510 **Search for color changes and brightening of Io upon eclipse reappearance.**   R. M. Nelson. Icarus, Vol. 32, 225 - 228 (1977). Medium-resolution spectra were made of Io as it emerged from two eclipses in December 1975. In the wavelength range 4000–5800 Å, no spectral changes greater than the standard deviations were observed when the spectrum of Io just after reappearance was divided by the spectrum of Io 20 min later. No substantial increase in total brightness was observed over the same time interval.

099.511 **Eine Auswertung von Galileis ersten Beobachtungen**

der Jupitermonde.   M. Becker, H. J. Becker.
Sterne Weltraum, Jahrg. 16, 331 - 333 (1977).

099.512 Identification of water frost on Callisto.
     L. A. Lebofsky.
Nature, Vol. 269, 785 - 787 (1977).

   Broadband (J, K, and L) and narrowband (3.0–3.8 μm)
observations of Ganymede and Callisto were made on the
nights of 7 November 1976 and 29 November 1976 with the
Mount Lemmon 28-inch infrared telescope. The spectra
clearly show an absorption feature centred between 3.1 and
3.4 μm. The absorption band in the spectrum of Callisto is
shallower than that of Ganymede, but is of the same general
shape. This indicates the presence of an optically significant
amount of frost on Callisto and is consistent with a surface
exposure of water frost which is somewhat less than on
Ganymede.

099.513 The sodium cloud of Io.
     Sky Telesc., Vol. 54, 479 - 480 (1977).

099.514 A new presentation of the ephemeris of Jupiter's
     Galilean satellites.   D. T. Vu.
Astron. Astrophys., Suppl. Ser., Vol. 30, 361 - 367 (1977).
In French.

   The author calculates the new ephemeris directly from
the results of Sampson's theory and not from his tables. The
approximation employed is realized by Chebyschev polynomials. He sets on tables the Chebyschev coefficients, which
give, after a short calculation, the differential equatorial coordinates of the satellites. The new tables are very simple to
use and give a better precision.

099.515 Io's surface composition: observational constraints
     and theoretical considerations.
F. P. Fanale, T. V. Johnson, D. L. Matson.
Geophys. Res. Lett., Vol. 4, 303 - 306 (1977).

   Observations of line emission from neutral and ionic
species in the Io-surrounding cloud, reflectance studies and
theoretical considerations suggest Io's surface is unlike that of
any other body in the solar system. The cloud has a peculiar
composition which the authors show is probably not due to
cloud/surface fractionation. Io's surface may be largely covered
with an endogenically produced mixture of S and dehydrated
salts, or by accretion-fractionated compounds modified by
charged particle bombardment.

099.516 Io, the strange moon of Jupiter.   A. Boischot.
     Recherche, Vol. 8, 667 - 668 (1977). In French.
Abstr. in Phys. Abstr., Vol. 80, Abstr. 74006 (1977).

099.517 Jupiter XIII (Leda).
     IAU Circ., No. 3102 (1977).

099.518 Grafico delle elongazioni dei satelliti di Giove.
     S. Rosoni.
Astronomia, N. 3, p. 58 - 61 (1977).

     Maximum entropy restorations of Ganymede.
See Abstr. 031.298.

     Positions of Jupiter's Galilean satellites.
See Abstr. 041.049.

     Gravitational experiments on space missions to the
giant planets.   See Abstr. 091.013.

     Lunar occultations of Io and Ganymede.
See Abstr. 096.002.

     Volatile evolution.   See Abstr. 097.095.

     Some interrelations of asteroids, Trojans and
satellites.   See Abstr. 098.105.

Errata

099.901 Correction: 'Comment on "Longitudinal asym-
     metry of the Jovian magnetosphere and the
periodic escape of energetic particles" by T. W. Hill and A. J.
Dessler' [J. Geophys. Res., Vol. 81, 5601 - 5602 (1976)].
C. K. Goertz.
J. Geophys. Res., Vol. 82, 2892 (1977).

100 Saturn, Saturn Satellites

Saturn

100.001   **The 2−4-μm spectrum of Saturn's rings.**
R. C. Puetter, R. W. Russell.
Icarus, Vol. 32, 37 - 40 (1977).
Observations of the rings of Saturn at 2−4 μm reveal the presence of a 3.6-μm peak in the infrared reflectivity. This peak is consistent with a particle size of ⩽ 50 μm, and a composition of pure H_2O ice. The quoted size may only be indicative of the textural scale of frost on the surface of larger particles. The presence of small amounts of CH_4 in the form of a clathrate, however, cannot be ruled out by the measurements.

100.002   **The ionospheres of Saturn, Uranus, and Neptune.**
L. A. Capone, R. C. Whitten, S. S. Prasad, J. Dubach.
Astrophys. J., Vol. 215, 977 - 983 (1977).
Models of the ionospheres of Saturn, Uranus, and Neptune are presented. It is postulated that galactic cosmic-ray ionization is an important component of these ionospheres. For example, in the case of Neptune, the level of ionization caused by cosmic rays is comparable with that due to solar extreme-ultraviolet (EUV) radiation. The existence of cosmic-ray, as well as solar EUV−produced ionization, could be a valuable diagnostic tool for investigating the atmospheric thermal structure of those planets.

100.003   **What supports Saturn's rings.**   A. I. Tsygan.
Astron. Zh. Akad. Nauk SSSR, Vol. 54, 870 - 874 (1977). In Russian. English translation in Soviet Astron., Vol. 21, No. 4.
The gravitational field induced by the quadrupole moment of Saturn stabilizes the particles of the rings in the equatorial plane of the planet. The thickness of the ring is determined by the depth of the quadrupole potential hole and by the characteristic velocity of particles across the equatorial plane (for large particles self-gravitation of the ring is essential). In other words, the rings are supported by the gravitational quadrupole moment of Saturn.

100.004   **Saturn 1975−76.**   A. W. Heath.
J. British Astron. Assoc., Vol. 87, 499 - 506 (1977). Report of the Section.

100.005   **Colour on Saturn.**   J. H. Robinson.
J. British Astron. Assoc., Vol. 87, 506 - 508 (1977).

100.006   **Azimuthal brightness variations of Saturn's rings. II. Observations at an intermediate tilt angle.**
K. Lumme, L. W. Esposito, W. M. Irvine, W. A. Baum.
Astrophys. J., Lett., Vol. 216, L123 - L126 (1977) = Contrib. Five Coll. Obs., Amherst, Mass., No. 247.
The brightness variation in Saturn's ring A with orbital phase of the ring particles increases in amplitude as the declination of the Earth $|B|$ decreases from 26° to 16°. The amplitude of this azimuthal effect also appears to diminish at opposition. There is an indication that the effect decreases with decreasing wavelength, and hence with decreasing particle albedo.

100.007   **On the azimuthal brightness variations of Saturn's ring A.**
G. Colombo, F. A. Franklin, A. W. Harris.
Bull. American Astron. Soc., Vol. 9, 436 - 437 (1977). Abstract.

100.008   **Resonances in Saturn's rings.**   A. F. Cook.
Bull. American Astron. Soc., Vol. 9, 461 (1977). Abstract.

100.009   **A dynamical model of Saturn's ring.**   A. Brahic.
Bull. American Astron. Soc., Vol. 9, 461 - 462 (1977). − Abstract.

100.010   **A model of the azimuthal variation of the brightness of Saturn's A-ring.**
R. Beebe, H. Reitsema, B. Smith.
Bull. American Astron. Soc., Vol. 9, 462 (1977). − Abstract.

100.011   **The azimuthal asymmetry in the brightness of Saturn's rings.**   C. Alcock, P. Goldreich.
Bull. American Astron. Soc., Vol. 9, 462 (1977). − Abstract.

100.012   **A dynamical explanation for the azimuthal brightness asymmetry of Saturn's A ring.**
G. Colombo, P. Goldreich, A. W. Harris.
Bull. American Astron. Soc., Vol. 9, 462 (1977). − Abstract.

100.013   **On the azimuthal brightness variations of Saturn's ring A.**   G. Colombo, F. Franklin.
Bull. American Astron. Soc., Vol. 9, 462 (1977). − Abstract.

100.014   **Saturn's rings: radial distribution of radar scatterers.**
G. H. Pettengill, S. J. Ostro, D. B. Campbell, R. M. Goldstein.
Bull. American Astron. Soc., Vol. 9, 462 (1977). − Abstract.

100.015   **Concerning the thickness of Saturn's rings.**
J. N. Cuzzi, J. A. Burns, R. Durisen, P. Hamill.
Bull. American Astron. Soc., Vol. 9, 463 (1977). − Abstract.

100.016   **Saturn's ring: physical thickness and the optical depth of Cassini's division and ring C.**
J. W. Fountain.
Bull. American Astron. Soc., Vol. 9, 463 (1977). − Abstract.

100.017   **Five color photometry of Saturn and its rings.**
K. Lumme, H. J. Reitsema.
Bull. American Astron. Soc., Vol. 9, 463 (1977). − Abstract.

100.018   **Anisotropic optical scattering within Saturn's rings.**
M. J. Price.
Bull. American Astron. Soc., Vol. 9, 463 (1977). − Abstract.

100.019   **Eclipses of Iapetus by Saturn's rings in 1977/1978.**
A. W. Harris, C. F. Peters.
Bull. American Astron. Soc., Vol. 9, 463 (1977). − Abstract.

100.020   **Evidence for a seasonal variation of Saturn's H_2 and CH_4 absorption.**   L. Trafton.
Bull. American Astron. Soc., Vol. 9, 473 (1977). − Abstract.

100.021   **A rocket observation of the far ultraviolet (1160 Å−1750 Å) spectrum of the Saturnian disk and ring system.**   H. Weiser, H. W. Moos.
Bull. American Astron. Soc., Vol. 9, 473 (1977). − Abstract.

100.022   **A model for the temperature inversion of Saturn.**
A. Tokunaga, R. Cess.
Bull. American Astron. Soc., Vol. 9, 474 (1977). − Abstract.

100.023   **Detection of Lyman α emission from the Saturnian disk and from the ring system.**
H. Weiser, R. C. Vitz, H. W. Moos.
Science, Vol. 197, 755 - 757 (1977).

A rocket-borne spectrograph detected H I Lyman α emission from the disk of Saturn and from the vicinity of the planet. The signal is consistent with an emission brightness of 700 rayleighs for the disk and 200 rayleighs for the vicinity of Saturn. The emission from the vicinity of the planet may be due to a hydrogen atmosphere associated with the Saturnian ring system.

100.024 The planet Saturn. S. J. Hynes.
Spaceflight, Vol. 19, 370 - 373 (1977).

100.025 Saturn: its thermal profile from infrared measurements. D. Gautier, A. Lacombe, I. Revah.
Astron. Astrophys., Vol. 61, 149 - 153 (1977).
Thermal profiles of the atmosphere of Saturn have been retrieved by means of an iterative method of inversion from infrared spectral measurements of the planet. The brightness temperature spectra deduced from these profiles are compared to experimental data and to the spectra calculated from the model of Wallace and the model of Caldwell. Three different thermal profiles are proposed which exhibit a similar stratosphere but differ in the convective range.

100.026 The far infrared spectrum of Saturn: observability of PH_3 and NH_3. T. Encrenaz, M. Combes.
Astron. Astrophys., Vol. 61, 387 - 390 (1977).
It is shown that NH_3 and PH_3 absorption lines are expected to be observable in the far-infrared spectrum of Saturn. From the brightness temperature measured in the continuum, the thermal profile in the convective zone and the hydrogen/helium mixing ratio can be derived. From the brightness temperatures in the absorption lines, estimates of the vertical density distributions of ammonia and phosphine can be inferred.

100.027 A model for the temperature inversion within the atmosphere of Saturn.
A. Tokunaga, R. D. Cess.
Icarus, Vol. 32, 321 - 327 (1977).
A model for the temperature inversion within the atmosphere of Saturn is proposed and is shown to be consistent with photometric data in the 17- to 25-μm region. The proposed model incorporates solar heating by some "aerosol", with the aerosol heating per unit mass of the atmosphere being uniformly distributed throughout that portion of the atmosphere overlying the upper cloud deck.

100.028 Inhomogeneous models of the atmosphere of Saturn. W. Macy.
Icarus, Vol. 32, 328 - 347 (1977).
Analyses of ultraviolet, visible, and near-infrared spectra of Saturn lead to an inhomogeneous atmospheric model, having a clear gas layer which lies above an absorbing particle layer which lies above an ammonia haze layer. The boundary between the clear layer and the absorbing particle layer is at a pressure of 0.2 atm in the equatorial region and 0.3 atm in the temperate region. The boundary between the absorbing particle layer and the haze layer is at the radiative–convective boundary. Observations of ammonia absorption lines indicate that sunlight penetrates the haze to the ammonia sublimation level at a depth of 1.1 atm. Absorbing particles cause the observed decrease in reflectivity from visible to ultraviolet wavelengths.

100.029 Far-infrared spectral observations of Saturn and its rings. D. B. Ward.
Icarus, Vol. 32, 437 - 442 (1977).
The spectrum of Saturn and its rings between 45 and 115 μm has been measured at an average resolving power of 14 from the NASA Lear Jet. The combined brightness temperature of the rings and planetary disk decreases beyond 65 μm, in disagreement with previous results. A brightness

temperature of $65 \pm 10°$K is obtained for the planetary disk in the 80–110-μm wavelength range if a large-particle, constant-emissivity model is assumed for the rings. The possible effects of small particles in the rings are briefly considered.

100.030 Photoelectric measurements of the brightness of Saturn's rings A and B. V. D. Krugov.
Astron. Tsirk., No. 940, p. 5 - 7 (1977). In Russian.

100.031 Photographic positional observations of Saturn and its satellites at Pulkovo during 1971 - 1974.
T. P. Kiseleva, G. V. Panova, O. A. Kalinichenko.
Izv. Glav. Astron. Obs. Pulkovo, No. 195, Astrofiz. Astrometr., p. 49 - 66 (1977). In Russian.

100.032 Eclipse of Iapetus by Saturn.
British Astron. Assoc. Circ., No. 581 (1977).

100.033 Physical properties of the upper layers of Saturn's atmosphere. A. V. Morozhenko.
Astrometriya i Astrofizika, Kiev, vyp. (No.) 33, (see 003.020), p. 78 - 86 (1977). In Russian.
A simple two-layer model of the vertical structure for the upper layers of Saturn's atmosphere is constructed based on an analysis of polarimetric, photometric and spectrophotometric observations. The upper layer is purely gaseous and optically thin with a pressure of 0.22 ± 0.12 atm at the clouds' level; the abundance of methane and ammonia is observed 35 ± 18 matm and $\leqslant 0.35 \pm 0.20$ matm, respectively. The lower layer is homogeneous, semi-infinite and consists of gas and aerosol.

100.034 Saturn central meridian ephemeris: 1978. J. E. Westfall.
Strolling Astron., Vol. 27, 19 - 20, 22 - 24 (1977).
Two tables give the longitude of Saturn's geocentric central meridian (C.M.) for the illuminated (apparent) disk for 0^h, U.T. for each day in 1978.

100.035 Some results of a spectrophotometry of Saturn's rings (1970 - 1972). II. A. M. Gretskij.
Physics of the moon and planets. Fundamental astrometry, (see 003.024), Vestn. Khar'kov. Univ., No. 160, p. 23 - 32 (1977). In Russian.

Saturn and its satellites. See Abstr. 011.003.

Positions of Jupiter, Saturn, Uranus and Neptune determined from observations with the Nikolaev zonal astrograph during 1967–1971. See Abstr. 041.048.

Statistical mechanics of Keplerian orbits. IV. Concluding remarks. See Abstr. 042.061.

Voyager imaging experiment.
See Abstr. 051.061.

Radio science investigations with Voyager.
See Abstr. 051.062.

The plasma experiment on the 1977 Voyager mission.
See Abstr. 051.063.

A plasma wave investigation for the Voyager mission.
See Abstr. 051.064.

Gravitational experiments on space missions to the giant planets. See Abstr. 091.013.

The phase diagram and transport properties for hydrogen-helium fluid planets. See Abstr. 091.060.

The dynamics and helium distribution in hydrogen-helium fluid planets.   See Abstr. 091.061.

High resolution ultraviolet spectrophotometry of Mars and Saturn by the TD1A satellite.
See Abstr. 097.065.

Ultraviolet observations of Mars and Saturn by the TD1A and OAO-2 satellites.   See Abstr. 097.071.

The structure of the planets Jupiter and Saturn.
See Abstr. 099.002.

A review and criticism of the nomenclature of atmospheric features of Jupiter and Saturn.
See Abstr. 099.014.

Low frequency radio emission from the outer planets.   See Abstr. 099.024.

The far infrared spectra of Jupiter and Saturn: a comparative study of their potential return.
See Abstr. 099.030.

The $^{12}C/^{13}C$ ratio in Jupiter and Saturn: presumptions for a telluric value.   See Abstr. 099.032.

Parameters of the figure and gravitational moments of Jupiter and Saturn.   See Abstr. 099.047.

The abundances of ammonia in the atmospheres of Jupiter, Saturn, and Titan.   See Abstr. 099.051.

On the magnetospheres of Jupiter, Saturn, and Uranus.   See Abstr. 099.053.

Evidence for a telluric value of the $^{12}C/^{13}C$ ratio in the atmospheres of Jupiter and Saturn.
See Abstr. 099.056.

The two types of atmosphere of Jupiter and Saturn.
See Abstr. 099.060.

Love numbers of the giant planets.
See Abstr. 099.069.

The orbits of Tethys, Dione, Rhea, Titan and Iapetus.   See Abstr. 100.501.

On the origin and evolution of Jupiter and Saturn.
See Abstr. 107.031.

Saturn Satellites

100.501 **The orbits of Tethys, Dione, Rhea, Titan and Iapetus.**   A. T. Sinclair.
Mon. Not. R. Astron. Soc., Vol. 180, 447 - 459 (1977).
Theories of the motions of Tethys, Dione, Rhea and Titan are described. These theories and a theory of the motion of Iapetus are fitted to recent photographic observations, and improved orbital elements are derived. The method of analysis used gives accurate determinations of the position of Saturn relative to the background AGK3 stars. A value of the mass of Saturn is obtained which is in good agreement with other determinations. The observational data used are given.

100.502 **Die Japetus-Verfinsterung vom 20. Oktober 1977.**   K. Locher.
Orion, 35. Jahrg., 121 - 122 (1977).

100.503 **A search for the H_2 (3, 0) $S1$ line in the spectrum of Titan.**
G. Münch, J. T. Trauger, F. L. Roesler.
Astrophys. J., Vol. 216, 963 - 966 (1977).
The spectrum of Titan, in a range 0.8 Å wide centered at the expected position of the (3, 0) $S1$ line of H_2, has been studied with a PEPSIOS spectrometer at a resolution of 0.09 Å. On the basis of the statistics of photon and dark-counted events involved in the scans, a 3 σ upper limit of 3 mÅ has been established for the equivalent width of the line. The corresponding upper limit on the normal column of H_2 is 1 kilometer-amagat, one-fifth of the currently accepted value.

100.504 **Theory of Enceladus and Dione.**
W. H. Jefferys, J. D. Mulholland, L. M. Ries.
Bull. American Astron. Soc., Vol. 9, 437 (1977). − Abstract.

100.505 **Faint inner satellites of Saturn.**
S. M. Larson.
Bull. American Astron. Soc., Vol. 9, 464 (1977). − Abstract.

100.506 **Interpretation of the red methane absorption bands in the spectrum of Titan.**
L. P. Giver, M. Podolak.
Bull. American Astron. Soc., Vol. 9, 471 (1977). − Abstract.

100.507 **3-mm observations of Titan.**
E. K. Conklin, B. L. Ulich, J. R. Dickel.
Bull. American Astron. Soc., Vol. 9, 471 (1977). − Abstract.

100.508 **Secular brightness increases of Titan, Uranus, and Neptune, 1972–76.**   G. W. Lockwood.
Bull. American Astron. Soc., Vol. 9, 471 (1977). − Abstract.

100.509 **A new satellite of Saturn?**
J. W. Fountain, S. M. Larson.
Science, Vol. 197, 915 - 917 (1977).
Analysis of all available observations of faint objects near Saturn during the 1966 passage of the earth through the plane of Saturn's rings suggests the existence of at least one previously undiscovered satellite of Saturn. The data support the previously published orbit for Janus. These satellites may be major members of an extended ring.

100.510 **Titan: a satellite with a recycled atmosphere?**
G. E. Hunt.
Modern astronomy, (see 003.013), p. 74 - 82 (1977).

100.511 **Titan's atmosphere and surface.**   D. M. Hunten.
Chemical evolution of the giant planets, (see 003. 014), p. 27 - 47 (1976).
The interior of Titan is presumably dominated by a melted $NH_3 - H_2O$ solution, and the presence of CH_4 is suggested by its abundance in the atmosphere. There is also evidence for H_2, although a very high escape rate is implied. Nitrogen, from NH_3 photolysis, could help in retarding this escape. The thermal-emission spectrum shows peaks at 8 and 12 μm, presumably due to CH_4 and C_2H_6 in a warm stratosphere, but no minimum at 17 μm due to a pressure-induced H_2 greenhouse. A surface temperature near $125°K$ is suggested

by the weight of the evidence, but a value as low as 90°K is not excluded.

100.512 **Eclipses of Saturn VIII (Iapetus).**
IAU Circ., No. 3116 (1977).

100.513 **The large abundance of methane in Titan's atmosphere.**   L. P. Giver.
Publ. Astron. Soc. Pacific, Vol. 89, 617 (1977).

100.514 **Eclipses of Saturn VIII (Iapetus).**
Yamamoto Circ., No. 1853 (1977). In Japanese.

Saturn and its satellites.   See Abstr. 011.003.

The abundances of ammonia in the atmospheres of Jupiter, Saturn, and Titan.   See Abstr. 099.051.

Jupiter and Saturn satellite observations.
See Abstr. 099.504.

Eclipses of Iapetus by Saturn's rings in 1977/1978.
See Abstr. 100.019.

Photographic positional observations of Saturn and its satellites at Pulkovo during 1971 - 1974.
See Abstr. 100.031.

Identification of a new class of satellites in the outer solar system.   See Abstr. 101.019.

Secular brightness increases of Titan, Uranus, and Neptune, 1972 - 1976.   See Abstr. 101.034.

101 Uranus, Neptune, Pluto, Transplutonian Planets

101.001 Weather on Neptune.   G. E. Hunt.
Nature, Vol. 269, 10 - 11 (1977).

101.002 Predicted occultations by the rings of Uranus,
1977—1980.   A. R. Klemola, B. G. Marsden.
Center for Astrophysics, Cambridge, Mass., Prepr. Ser. No.
788, 9 pp. (1977).
Predictions are supplied for 12 occultations by the rings
of Uranus during the three-year period beginning late 1977.
The photographic magnitudes of the occulted stars range
from 12.2 to 15.5.

101.003 Far infrared observations of Neptune.
R. F. Loewenstein, D. A. Harper, H. Moseley.
Bull. American Astron. Soc., Vol. 9, 431 (1977). — Abstract.

101.004 Evidence for weather on Neptune.
C. B. Pilcher, R. R. Joyce, D. P. Cruikshank,
D. Morrison.
Bull. American Astron. Soc., Vol. 9, 471 (1977). — Abstract.

101.005 The 6819 Å CH_4 line of Uranus and Neptune.
W. Macy, W. Smith.
Bull. American Astron. Soc., Vol. 9, 471 - 472 (1977).
Abstract.

101.006 Evidence for the depletion of ammonia in the
Uranus atmosphere.
S. Gulkis, M. Janssen, E. T. Olsen.
Bull. American Astron. Soc., Vol. 9, 472 (1977). — Abstract.

101.007 Polarization of Uranus as a function of phase angle.
J. J. Michalsky.
Bull. American Astron. Soc., Vol. 9, 472 (1977). — Abstract.

101.008 The predicted seasonal variation of the thermal
flux from Uranus.   W. M. Sinton.
Bull. American Astron. Soc., Vol. 9, 472 (1977). — Abstract.

101.009 A possible source of sudden thermal changes on
Uranus.   W. M. Sinton.
Bull. American Astron. Soc., Vol. 9, 472 (1977). — Abstract.

101.010 Narrow band imagery of Uranus.
D. H. Martins, C. A. Harvel, T. H. Morgan,
A. E. Potter.
Bull. American Astron. Soc., Vol. 9, 472 (1977). — Abstract.

101.011 Photometric determination of the rotation period
of Uranus.
C. A. Avis, H. J. Smith, J. T. Bergstralh, W. H. Sandmann.
Bull. American Astron. Soc., Vol. 9, 472 - 473 (1977).
Abstract.

101.012 The rotational period of Uranus and Neptune.
S. H. Hayes, M. J. S. Belton.
Bull. American Astron. Soc., Vol. 9, 473 (1977). — Abstract.

101.013 Uranus photography in the 890-nm absorption band
of methane.   B. A. Smith.
Bull. American Astron. Soc., Vol. 9, 473 (1977). — Abstract.

101.014 The rings of Uranus.
E. Dunham, J. Elliot, D. Mink.
News Lett. Astron. Soc. N.Y., Vol. 1, No. 2, p. 23 (1977).
Abstract.

101.015 The rings of Uranus: theory.

S. F. Dermott, T. Gold.
News Lett. Astron. Soc. N.Y., Vol. 1, No. 2, p. 24 (1977).
Abstract.

101.016 Ist der Planet Uranus von einem Satellitengürtel
umgeben?   W. Eichendorf, M. Reinhardt.
Naturwissenschaften, 64. Jahrg., 432 - 433 (1977).

101.017 The rotation of Uranus.
R. A. Brown, R. M. Goody.
Astrophys. J., Vol. 217, 680 - 687 (1977).
The authors have studied the rotation of Uranus by
measuring the tilts of reflected Fraunhofer lines. Their results
are consistent with the alignment of the rotational axis along
the pole of the satellite orbits. In northern mid-latitudes they
find the rotational period $T = 15.57 \pm 0.80$ hr (standard deviation of random errors). Their finding that the planet may not
rotate as a solid body could partially resolve the apparent conflict between their value and a shorter, disk-averaged period
reported orally by Münch, Trauger, and Roesler.

101.018 Uranus hat Ringe.
Phys. Bl., 33. Jahrg., 466 - 468 (1977).

101.019 Identification of a new class of satellites in the outer
solar system.
D. P. Cruikshank, C. B. Pilcher, D. Morrison.
Astrophys. J., Vol. 217, 1006 - 1010 (1977).
From near-infrared photometry (*JHKL*) of small bodies
in the outer solar system the authors identify a previously unrecognized class of planetary satellites which may have surfaces
mostly free of frosts of water, methane, or ammonia. These
bodies — including two satellites of Uranus (Titania and Oberon),
and one each of Saturn (Hyperion) and Neptune (Triton) —
have surfaces characterized by neutral reflectance between *J*
and *H* (1.25 and 1.60 µm), and rapidly decreasing reflectance
at *K* (2.2 µm) and *L* (3.5 µm). The authors point out that the
15 solid bodies in the outer solar system that have been satisfactorily observed (excluding the asteroids) fall into six classes
according to spectral reflectance. The presence of water frost
or ice on the surface of Enceladus is established.

101.020 Predicted occultations by the rings of Uranus, 1977 -
1980.   A. R. Klemola, B. G. Marsden.
Astron. J., Vol. 82, 849 - 851 (1977) = Lick Obs. Bull. No.
782.
Predictions are supplied for 12 occultations by the rings
of Uranus during the three-year period beginning late 1977. The
photographic magnitudes of the occulted stars range from 12.2
to 15.5.

101.021 Quantitative analysis of the Dermott—Gold theory
for Uranus's rings.   K. Aksnes.
Nature, Vol. 269, 783 (1977).
Dermott and Gold (1977) have attempted to explain the
locations of Uranus's rings in terms of resonances between ring
particles and pairs of satellites. Despite the apparent success of
the theory in predicting the main observed features of the rings,
it is necessary to investigate the strength of the resonances in
more detail before any conclusions can be drawn. This letter
summarises the outcome of such an investigation which
supplements the previous largely qualitative analysis.

101.022 Revenge of tiny Miranda.
P. Goldreich, P. Nicholson.
Nature, Vol. 269, 783 - 785 (1977).
Dermott and Gold (1977) have proposed a resonance
model for the rings of Uranus. The authors report here that,

by a wide margin, the strongest resonances are all associated with Miranda. Furthermore, they show that the hypothesis that the rings are made up of librating particles, while original and ingenious, is incorrect.

101.023 **Découverte d'un anneau autour d'Uranus.**
A. Brahic.
Astronomie, Vol. 91, 405 - 412 (1977).

101.024 **Detection of methane and ethane emission on Neptune but not on Uranus.**
W. Macy, Jr., W. Sinton.
Astrophys. J., Lett., Vol. 218, L79 - L81 (1977).
The authors have observed emission from Neptune with filters which cover the ν_4 band of methane at 7.7 μm and the ν_9 band of ethane at 12.2 μm. The brightness temperature for emission in the CH_4 band is 130 K, and it is 93 K in the C_2H_6 band. No emission was detected for Uranus in the C_2H_6 band. The new measurements indicate that Neptune has an unexpectedly warm thermal inversion layer.

101.025 **Colors and magnitudes of stars occulted by the rings of Uranus, 1977–1980.** W. Liller.
Astron. J., Vol. 82, 929 (1977).
Colors and magnitudes, measured from Harvard plates and Palomar Sky Survey prints, are listed for the 12 stars which are to be occulted by the rings of Uranus during the three-year period beginning in late 1977. Red magnitudes range from 9.3 to 14.1.

101.026 **5–20 micron observations of Uranus and Neptune.**
F. C. Gillett, G. H. Rieke.
Astrophys. J., Lett., Vol. 218, L141 - L144 (1977).
Emission features at 8 and 12 μm indicate a strong temperature inversion in the upper atmosphere of Neptune. The presence of only a weak inversion on Uranus is confirmed.

101.027 **The effective temperature of Neptune.**
R. F. Loewenstein, D. A. Harper, H. Moseley.
Astrophys. J., Lett., Vol. 218, L145 - L146 (1977).
The brightness temperature of Neptune has been measured in two broad passbands with flux-weighted mean wavelengths of 45 μm and 93 μm, permitting a direct determination of its effective temperature. The derived value of 55.5 ± 2.3 K implies that it radiates twice as much power as it receives from the Sun.

101.028 **Pluto.** R. L. Newburn, Jr.
Astronomy, Vol. 5, No. 2, p. 18 - 24 (1977).
Abstr. in Phys. Abstr., Vol. 80, Abstr. 41354 (1977).

101.029 **Uranus and Neptune.** M. J. S. Belton.
Astronomy, Vol. 5, No. 2, p. 6 - 17 (1977). – Abstr. in Phys. Abstr., Vol. 80, Abstr. 45357 (1977).

101.030 **Saturn-like ring system around Uranus.**
J. C. Bhattacharyya, M. K. V. Bappu.
Nature, Vol. 270, 503 - 506 (1977).
The authors have closely examined the Kavalur photoelectric record and describe here several new features of the satellite ring system of Uranus.

101.031 **Observing the outermost planets.** H. J. Phillips.
Astronomy, Vol. 5, No. 4, p. 42 - 44 (1977).
Abstr. in Phys. Abstr., Vol. 80, Abstr. 56874 (1977).

101.032 **The rotational periods of Uranus and Neptune.**
S. H. Hayes, M. J. S. Belton.
Icarus, Vol. 32, 383 - 401 (1977).
Observations of tilts of spectral lines in the spectrum of Uranus and Neptune yield the following rotational periods: "Uranus", 24 ± 3 hr; "Neptune", 22 ± 4 hr. Neptune is con-

firmed to rotate in a direct sense. The position angle of the pole of Uranus, projected onto the plane of the sky, is found to be 283 ± 4°. The value for Neptune is 32 ± 11°. The rotational period of Uranus is found to be consistent with modern values of its optical and dynamical oblateness and the theory of solid-body rotation with hydrostatic equilibrium. This is barely the case for the period derived for Neptune and the authors suspect that future observations made under better seeing conditions may lead to a shorter rotation period between 15 and 18 hr.

101.033 **Uranus' rotational period.** L. Trafton.
Icarus, Vol. 32, 402 - 412 (1977).
A modified spectroscopic technique yields a value for Uranus' period of 23^{+5}_{-2} hr, which disagrees with the accepted value of 10.8 hr determined spectroscopically in the 1930s. Uranus' rotational axis is probably close to the perpendicular to the orbital plane of the satellites, as often assumed. Its projection on the plane of the sky is within 5° of the projection of the normal to the orbital plane of the satellites.

101.034 **Secular brightness increases of Titan, Uranus, and Neptune, 1972 - 1976.** G. W. Lockwood.
Icarus, Vol. 32, 413 - 430 (1977).
The brightnesses of Titan, Uranus, and Neptune in b (4718 Å) and y (5508 Å) have increased linearly since 1972 at rates ranging from 0.005 to 0.025 mag yr^{-1}. The observations were made differentially on a number of nights each season with respect to a network of comparison stars whose relative magnitudes were determined by independent measurements. Solar phase coefficients were derived for each object, and all observations have been normalized to zero solar phase angle and mean heliocentric distances. No explanation for the changes has been found, but a possible influence of solar activity upon planetary albedo is suggested by the fact that all of the objects observed have brightened during the declining half of the solar cycle.

101.035 **Occultations by Uranian rings.**
IAU Circ., No. 3108 (1977).

101.036 **Appulses to Neptune.**
IAU Circ., No. 3146 (1977).

101.037 **Evolution of the Uranus rings.** J. Boynton.
Publ. Astron. Soc. Pacific, Vol. 89, 613 - 614 (1977).
Abstract.

101.038 **Uranus: the rings are black.** W. M. Sinton.
Science, Vol. 198, 503 - 504 (1977).
An upper limit of 0.05 is established for the geometric albedo of the newly discovered rings of Uranus. In view of this very low albedo, the particles of the rings cannot be ice-covered as are those of rings A and B of Saturn.

101.039 **Occultations by Uranus on 1977 August 26.**
Yamamoto Circ., No. 1854 (1977). In Japanese.

101.040 **The spectrum of Neptune in the region λλ4400–7000 Å.** A. A. Atai, N. B. Ibragimov.
Astron. Tsirk., No. 940, p. 4 - 5 (1977). In Russian.

101.041 **On the methane clouds in the atmospheres of Uranus and Neptune.**
K. Yu. Ibragimov, L. P. Sorokina.
Astron. Tsirk., No. 941, p. 4 - 6 (1977). In Russian.

101.042 **Signal-to-noise ratios for stellar occultations by the rings of Uranus, 1977 - 1980.** J. L. Elliot.
Astron. J., Vol. 82, 1036 - 1038 (1977).
Approximate signal-to-noise ratios are calculated for 12 stellar occultations by the rings of Uranus during 1977 - 1980.

Formulae for the signal-to-noise ratios are given.

101.043   Nun auch Ringsystem um Uranus.
          J. Classen.
Sterne, 53. Band, 242 (1977).

101.044   Observation of occultation of the SAO 158687 star
          by Uranus at Dodaira Station.   K. Tomita.
Tokyo Astron. Bull., Second Ser., No. 250, p. 2885 - 2888 (1977).

101.045   Rings of Uranus.   Ľ. Kresák.
          Kozmos, Vol. 8, 116 - 118 (1977). In Slovak.

Photoelectric absorption spectra of methane (CH_4), methane and hydrogen (H_2) mixtures, and ethane (C_2H_6). See Abstr. 022.020.

True external errors of the observations by the Tokyo meridian circle estimated from the Uranus occultation of SAO 158687.   See Abstr. 041.019.

Observations of Uranus made with the Danjon Astrolabe of Santiago, Chile, during 1975 and 1976. See Abstr. 041.024.

Positions of Jupiter, Saturn, Uranus and Neptune determined from observations with the Nikolaev zonal astrograph during 1967–1971.   See Abstr. 041.048.

Statistical mechanics of Keplerian orbits. IV. Concluding remarks.   See Abstr. 042.061.

Plasma dynamics in laboratory models of the magnetospheres of the earth and Uranus.   See Abstr. 084.316.

Low frequency radio emission from the outer planets.   See Abstr. 099.024.

Theoretical brightness temperature profiles of atmospheric pure H_2 rotational quadrupole lines: Jupiter and Uranus.   See Abstr. 099.052.

On the magnetospheres of Jupiter, Saturn, and Uranus.   See Abstr. 099.053.

Love numbers of the giant planets. See Abstr. 099.069.

The ionospheres of Saturn, Uranus, and Neptune. See Abstr. 100.002.

Secular brightness increases of Titan, Uranus, and Neptune, 1972–76.   See Abstr. 100.508.

Long-period comets and Transneptunian planets. See Abstr. 102.032.

102 Comets (Origin, Structure, Atmospheres, Dynamics)

102.001 **Cometary debris.** R. E. McCrosky.
The dusty universe, (see 012.001), p. 169 - 184 (1975).
Problems of ablation mechanisms of cometary nuclei are discussed in the light of some specific observations of comets and meteors. Estimates of the mass in the Geminid meteor stream are given. The outbursts of Comet P/Tuttle-Giacobini-Kresak are compared with those of the more distant comet P/Schwassmann-Wachmann 1. A formal solution of heat shock effects in comets near perihelion is given as an upper limit of the efficacy of this process for cometary disruption.

102.002 **Statistical consequences of the capture of comets on the basis of the Laplace scheme.**
V. V. Radzievskij, V. P. Tomanov.
Astron. Zh. Akad. Nauk SSSR, Vol. 54, 890 - 896 (1977). In Russian. English translation in Soviet Astron., Vol. 21, No. 4.
The authors have made calculations of the orbital elements of comets captured by Jupiter according to the Laplace scheme. Comparison of the theoretical results with real catalogue data reveals the possibility of interpreting statistical regularities in the distribution of the elements of cometary orbits on the basis of the Laplace hypothesis.

102.003 **Statistical investigation of the influence of solar activity on the development of a comet's head.**
O. V. Dobrovol'skij, R. S. Osherov, M. Z. Markovich.
Komety Meteory, No. 25, p. 31 - 43 (1976). In Russian.
A statistical investigation of the influence of solar activity on the development of a cometary head was carried out on the basis of the many-years uniform observations by Beyer.

102.004 **Polarization observations of comets.**
R. S. Osherov.
Komety Meteory, No. 25, p. 44 - 52 (1976). In Russian.
All polarization observations of comets known to the author are presented, beginning with 1819, i.e. the observations by Arago.

102.005 **On a possible application of some results of collisions to the physics of comets.**
I. V. Sushanin, S. M. Kishko.
Komety Meteory, No. 25, p. 53 - 56 (1976). In Russian.
The process of a gradual dehydrogenisation of polyatomic molecules in the field of photon and corpuscular radiation of the sun and as a result of collisions of a cometary atmosphere and charged particles with molecules in question is proposed to be considered a dominant mechanism of production of heads of comets.

102.006 **Spectroscopic research for possible parent molecules in comets.**
E. A. Kajmakov, I. N. Matveev.
Komety Meteory, No. 26, p. 9 - 17 (1977). In Russian.
Emission spectra of a series of organic compounds, aminoacids in particular, supposed to be present in cometary nuclei have been investigated. The typical comet bands of CN, C_2, CO, CH etc. were observed following the excitation of vapour by an electronic shock.

102.007 **Statistical equilibrium in cometary C_2. I.**
K. S. Krishna Swamy, C. R. O'Dell.
Astrophys. J., Vol. 216, 158 - 164 (1977).
A resonance fluorescence statistical equilibrium was calculated for C_2 molecules in comets at various heliocentric distances. The Fox-Herzberg and Ballik-Ramsay bands were used in addition to the Swan bands, which were the basis for previous calculations. Significantly lower vibrational temperatures were found.

102.008 **Sublimation lifetimes of long-period comets.**
P. R. Weissman.
Bull. American Astron. Soc., Vol. 9, 465 - 466 (1977). Abstract.

102.009 **Chemical reactions in comet comae.**
W. F. Huebner.
Bull. American Astron. Soc., Vol. 9, 466 (1977). — Abstract.

102.010 **Comets' differentiation from solar abundances.**
A. H. Delsemme.
Bull. American Astron. Soc., Vol. 9, 466 (1977). — Abstract.

102.011 **Dust grains in a cometary coma: interpretation of the infrared continuum.** T. Mukai.
Astron. Astrophys., Vol. 61, 69 - 74 (1977).
A model which reproduces the observed infrared properties of a cometary coma is proposed. In particular this model explains the enhanced color temperature, the shape of the infrared continuum spectrum and the change of color ratio as a function of solar distance. Spectra and color ratios have been computed for several models of grain material.

102.012 **The constitution of cometary nuclei.**
F. L. Whipple.
NASA Tech. Mem., NASA TM X-3511, (see 012.010), p. 65 - 67 (1977) = Cent. Astrophys., Harvard Coll. Obs. and Smithsonian Astrophys. Obs., Prepr. No. 637 (1976). Abstract.

102.013 **Orbit determination of nearly-parabolic comets and conclusions concerning the Oort comet cloud.**
B. G. Marsden.
NASA Tech. Mem., NASA TM X-3511, (see 012.010), p. 68 - 69 (1977). — Abstract.

102.014 **A new model for the split comets.** Z. Sekanina.
NASA Tech. Mem., NASA TM X-3511, (see 012.010), p. 70 - 71 (1977). — Abstract.

102.015 **Physical processes in comets.**
F. L. Whipple, W. F. Huebner.
Annu. Rev. Astron. Astrophys., Vol. 14, (see 003.008), 143 - 172 (1976).

102.016 **The cyanogen abundance of comets.**
J. B. Tatum, M. I. Gillespie.
J. R. Astron. Soc. Canada, Vol. 71, 403 (1977). — Abstract.

102.017 **Dynamics of the hydrogen atmosphere of comets. Effect of dissociative heating.** L. O. Kolokolova.
Astrometr. i Astrofiz., Kiev, vyp. (No.) 32, (see 003.011), p. 17 - 24 (1977). In Russian.
The strong $L\alpha$ radiation observed in cometary heads can be explained by resonant scattering of solar radiation on hydrogen atoms formed during photodissociation of H_2O molecules in the comets' atmospheres. A system of equations to describe the multi-component gas dynamics of a cometary atmosphere is obtained taking into account the effect of dissociative heating as well as collisions, photoionization and exchange interaction with solar wind protons.

102.018 **On the theory of the wall layer of a cometary nucleus.** L. M. Shul'man.

Astrometr. i Astrofiz., Kiev, vyp. (No.) 32, (see 003.011), p. 24 - 28 (1977). In Russian.

The solution of the kinetic equation describing the behaviour of the gas at the surface of the nucleus is obtained by the method of moments. The initial Mach number is shown to be close to unit in dense atmospheres where the free path of molecules is much less than the size of the nucleus.

102.019   The cyanogen abundance of comets.
J. B. Tatum, M. I. Gillespie.
Astrophys. J., Vol. 218, 569 - 572 (1977).

Results of calculations are presented that will enable an observer to convert absolute measurements of the radiance of the violet CN bands in a cometary spectrum to column densities of CN molecules. An intensity minimum of CN near perihelion is predicted.

102.020   On the differences between the new and old comets.
L'. Kresák.
Bull. Astron. Inst. Czechoslovakia, Vol. 28, 346 - 355 (1977).

A sample of 60 new and 60 old long-period comets with best determined orbits is subjected to a number of probability tests to assess the statistical significance of the differences between these two kinds of objects. The distinctions in support of the current evolutionary concepts (a relative prevalence of larger perihelion distances among the new comets; an increased rate of post-perihelion fading of the new comets produced by the removal of the outer frosting) are qualified as rather inconclusive. On the other hand, there are very significant irregularities in the rate of occurrence and perihelion distribution of the different types of comets. These irregularities appear on a relatively short time-scale, suggest some excess of the near-UV radiation in the new comets, and cannot be explained by observational selection alone.

102.021   Analysis of the sources of cometary material.
S. I. Gerasimenko, G. G. Novikov.
Dokl. AN TadzhSSR, Vol. 20, No. 1, p. 12 - 15 (1977). In Russian. – Abstr. in Ref. zh., 51. Astron., 10.51.416 (1977).

102.022   Ultraviolet and infrared observations of comets: recent results and prospects for the Shuttle era.
C. B. Opal, G. R. Carruthers.
Rep. NRL Prog., July 1976, p. 1 - 11. – Abstr. in Phys. Abstr., Vol. 80, Abstr. 33425 (1977).

102.023   Are comets dirty snowballs or dust swarms?
D. W. Hughes.
Nature, Vol. 270, 558 - 560 (1977).

102.024   The composition of comets.   P. D. Feldman.
American Sci., Vol. 65, 299 - 309 (1977). – Abstr. in Phys. Abstr., Vol. 80, Abstr. 78717 (1977).

102.025   The ionospheres and plasma tails of comets.
D. A. Mendis, W.-H. Ip.
Space Sci. Rev., Vol. 20, 145 - 190 (1977).

The present understanding of cometary ionospheres and plasma tails is critically evaluated. Following a brief introduction of the significance of the study of cometary ionospheres and tails, the observational statistics and spectroscopic observations are summarized. The complicated and time varying morphology of the plasma tail and the ionosphere as revealed both by photographs as well as visual drawings is discussed. The evidence for a strong comet-solar wind interaction, the possible nature of this interaction and also the use of comets as probes of the solar wind are considered. This is followed by a discussion of the various processes so far proposed for the ionization of cometary gases and their limitations. A discussion of the ion chemistry and structure of the region inside the tangential discontinuity is given.

102.026   Helical waves in type-1 comet tails.
A. I. Ershkovich, A. B. Heller.
Astrophys. Space Sci., Vol. 48, 365 - 377 (1977).

Oscillations of type-1 comet tails with plasma compressibility taken into account are studied. A comet tail is treated as a plasma cylinder separated by a tangential discontinuity surface from the solar wind. The dispersion equation obtained in the linear approximation is solved numerically with typical plasma parameters. A sufficient condition for instability of the cylindrical tangential discontinuity in the compressible fluid is obtained. The phase velocity of helical waves is shown to be approximately coincident with Alfvén speed in the tail in the reference system moving with the bulk velocity of the plasma outflow in the tail. The instability growth rate is calculated. This theory is shown to be in good agreement with observations in the tails of Comets Kohoutek, Morehouse and Arend-Roland.

102.027   Some comments on the Oort cloud.   M. E. Bailey.
Astrophys. Space Sci., Vol. 50, 3 - 22 (1977).

The arguments used by Lyttleton to prove the nonexistence of the Oort cloud are reviewed, and it is shown that Oort's hypothesis remains consistent with observation. The 1950 model of the cloud cannot be correct. An 'improved' model is described and compared with observations. It is emphasized that comparison of the predictions of theory with observations should concentrate on the a-distribution. It seems that a significant number of nearly parabolic comets must have passed through the solar system before.

102.028   Laboratory simulation of cometary erosion by space plasma.   L. Kristoferson, K. Fredga.
Astrophys. Space Sci., Vol. 50, 105 - 123 (1977).

A laboratory experiment has been made where a plasma stream collides with targets made of different materials of cosmic interest. The experiment can be viewed as a process simulation of the solar wind particle interaction with solid surfaces in space – e.g., cometary dust. Special interest is given to sputtering of OH and Na.

102.029   On the production of positive molecular ions in cometary comas.
S. P. Tarafdar, N. C. Wickramasinghe.
Astrophys. Space Sci., Vol. 50, 163 - 171 (1977).

Positively charged molecular ions, such as H_2O^+, which have been observed in cometary comas, may be efficiently produced by the evaporation of positively charged clathrate grains. Such grains may be expelled from nuclei of comets, along with gaseous molecules. Grain charging occurs via interaction with solar ultraviolet photons and/or solar wind protons. Observational data on the total quantities as well as the distributions of H_2O and H_2O^+ in cometary comas are shown to be in accord with detailed model calculations.

102.030   Volatiles in the cometary nuclei.   M. Shimizu.
Astrophys. Space Sci., Vol. 51, 241 - 243 (1977).

The argument for the similarity of the composition of cometary volatiles to that of interstellar molecules has been strengthened by the analysis of CO^+ and CO_2^+ emission of the comet West. The strong 6300 Å emission of oxygen atoms can be interpreted in terms of photo-dissociation of OH by the solar Lyman-alpha radiation, and not as being due to photodissociation of CO_2 of speculatively large amount.

102.031   Investigation of the influence of systematic errors on the determination of the elements of a cometary orbit.   I. N. Murav'eva.
Kazan. univ. Kazan', 1977. 19 pp. In Russian. – Abstr. in Ref. zh., 51. Astron., 11.51.219 (1977).

102.032   Long-period comets and Transneptunian planets.
Eh. M. Drobyshevskij.

Astron. Tsirk., No. 942, p. 3 - 5 (1977). In Russian.

102.033 Comets in the STIP context.   M. K. Wallis.
Study of travelling interplanetary phenomena 1977, (see 012.042), p. 279 - 289 (1977).

Fluid descriptions of plasma motion through a cometary coma are briefly sketched, distinguishing the bow shock and ionizing flow region mainly within it, the tail region and ray structure, and the 'ionosphere' coupled closely to the expanding cometary gas.

102.034 The comet-solar wind interaction.
D. A. Mendis.
Study of travelling interplanetary phenomena 1977, (see 012.042), p. 291 - 303 (1977).

The important roles of comets as natural probes of the solar wind, particularly at high heliographic latitudes and small heliocentric distances, unattained thus far by artificial space probes, is stressed. It is becoming clear the solar wind is not merely responsible for shaping and maintaining the cometary plasma tail, but is also indirectly responsible for the rapid ionization processes in the coma. The current views are critically reviewed and an attempt is made to identify the dominant physical mechanisms that are involved.

102.035 Solar wind interaction with type-1 comet tails.
A. I. Ershkovich.
Study of travelling interplanetary phenomena 1977, (see 012.042), p. 305 - 319 (1977).

A comet tail is considered as a plasma cylinder separated by a tangential discontinuity surface from the solar wind. Under typical conditions a comet tail boundary is shown to undergo the Kelvin-Helmholtz instability. With finite amplitude the stabilizing effect of the magnetic field increases, and waves become stable. The proposed model supplies the detailed quantitative description of helical waves observed in type-1 comet tails.

102.036 On the asymmetry in the perihelia distribution of cometary orbits.   V. P. Tomanov.
Astron. Zh. Akad. Nauk SSSR, Vol. 54, 1346 - 1348 (1977). In Russian. English translation in Soviet Astron., Vol. 21, No. 6.

An asymmetry with respect to the latitude circle of the solar peculiar motion apex in the distribution of the perihelia of nearly parabolic cometary orbits with forward as well as retrograde motions is shown to exist. Some conclusions concerning the evolution of cometary orbits are made.

102.037 Solar-cometary relations: a review.   K. S. Simmons.
Strolling Astron., Vol. 27, 14 - 15 (1977).

102.038 The pristine nature of comets.   A. H. Delsemme.
Comets, asteroids, meteorites, (see 012.049), p. 3 - 13 (1977).

The present results seem to have established the fact that comets seem much less depleted in H, C, N, O than C I chondrites, and that they probably are the most primitive objects still around in the inner solar system.

102.039 A comparison of the composition of new and evolved comets.   B. Donn.
Comets, asteroids, meteorites, (see 012.049), p. 15 - 23 (1977).

The intensity ratio of the continuum to the molecular emissions was estimated in the spectra of eighty-five comets. (1) There is no readily apparent difference in continuum to emission intensity ratio between new and more evolved comets. (2) An intrinsic distribution of this characteristic does occur. (3) Periodic comets with weak continua derived from new comets with the same property. (4) No weakening of the continuum in general occurs following perihelion passage.

102.040 The constitution of cometary nuclei.
F. L. Whipple.
Comets, asteroids, meteorites, (see 012.049), p. 25 - 35 (1977).

102.041 Surface phenomena in simulated cometary nuclei.
O. V. Dobrovolsky (*Dobrovol'skij*), E. Kajmakov.
Comets, asteroids, meteorites, (see 012.049), p. 37 - 46 (1977).

The sublimation of artificial icy cometary nuclei with different inclusions is investigated. Formation of meteoroids is discussed in connection with new polarizational observations of comet West. The possibility of chemical reactions on the sublimating surface is noted.

102.042 Differential nongravitational forces in the motions of the split comets.   Z. Sekanina.
Comets, asteroids, meteorites, (see 012.049), p. 51 - 56 (1977).

The new model, which has recently been proposed for the motions of fragments of the split comets, is critically examined through a comparison with the traditional approach. It is concluded that the new model, based on the premise that the rate of separation of any two fragments of a split comet is determined primarily by the differential nongravitational forces in their motions, is preferable in many a respect to the traditional hypothesis, built on the assumption that significant impulses are exerted on the fragments at the time of splitting.

102.043 Chemistry of the inner coma: a progress report.
W. F. Huebner.
Comets, asteroids, meteorites, (see 012.049), p. 57 - 59 (1977).

The composition of the inner coma is modeled assuming that about 30 chemical species composed of H, C, and O undergo reactions. Ionization and dissociation by solar radiation and over 100 forward and reverse reactions between atoms, molecules, and ions are considered in the kinetics.

102.044 Orbital data on the existence of Oort's cloud of comets.   B. G. Marsden.
Comets, asteroids, meteorites, (see 012.049), p. 79 - 86 (1977).

Oort's work on the cometary cloud is reviewed and extended using new data from 99 comets with high-quality orbits. These data clearly show the pronounced pile-up of the "original" reciprocals of the semi-major axes at values of less than 0.000100 AU^{-1}. This concentration is found to be even more striking for comets of large perihelion distance. Lyttleton's criticisms of the concept of the Oort cloud are reviewed and dismissed as largely irrelevant.

102.045 Initial energy and perihelion distributions of Oort-cloud comets.   P. R. Weissman.
Comets, asteroids, meteorites, (see 012.049), p. 87 - 91 (1977).

A Monte Carlo model of stellar perturbations of the Oort cloud is used to study the distributions in energy and perihelion of comets entering the planetary region for the first time. The model is run for a variety of initial states and a range of velocity perturbations. In all cases the resulting orbits are uniformly distributed in perihelion distance in the planetary region, q < 20 AU.

102.046 An alternate interpretation of the Oort cloud of comets?   Ľ. Kresák.
Comets, asteroids, meteorites, (see 012.049), p. 93 - 97 (1977).

102.047 The evolution of comet orbits as perturbed by Uranus and Neptune.   E. Everhart.
Comets, asteroids, meteorites, (see 012.049), p. 99 - 104 (1977).

When the perturbing planets are Uranus and Neptune, the perturbations on comets are so much weaker than with Jupiter and Saturn that a study of the comets' orbital evolution, using exact numerical integration, would require 200 times more revolutions. This is hardly practical with present computers. Here the author describes results with a simulation approach, the "Monte Carlo (random walk) method". The proper distribution shape for the perturbations in energy are found from a few thousand numerical integrations, then this distribution of perturbations is applied to millions of simulated orbit-revolutions.

102.048 Jovian perturbations per orbital revolution for short-period comets: statistical properties.
H. Rickman, S. Vaghi.
Comets, asteroids, meteorites, (see 012.049), p. 105 - 107 (1977).

The gravitational perturbations by Jupiter per orbital revolution are calculated for six selected groups of 1000 hypothetical random comets of low inclinations, perihelia units 6 a.u. and aphelia in the range (4 a.u., 13 a.u.). The results are organized in the form of empirical distributions and their statistical properties are investigated in detail.

102.049 The origin of comets.   A. H. Delsemme.
Comets, asteroids, meteorites, (see 012.049), p. 453 - 467 (1977).

A summary of the present discussion on the hypotheses about the origin of comets is given. The only hypothesis that seems to survive this screening is that comets were condensed some five billion years ago, in the vicinity of the giant planets or beyond, at the outer edge of the solar nebula. The mechanism of their ejection into Oort's cloud is a necessary by-product of their growth by accretion.

102.050 Comets and the cosmogony of the solar system.
S. K. Vsekhsvyatsky (*Vsekhsvyatskij*).
Comets, asteroids, meteorites, (see 012.049), p. 469 - 474 (1977).

The quantitative processes of eruptive development of planets into comets and other small bodies is studied from the physical and orbital evolution of these minor bodies.

102.051 Carbon isotope ratio in comets and interstellar matter.   V. Vanysek.
Comets, asteroids, meteorites, (see 012.049), p. 499 - 503 (1977).

The $^{12}C/^{13}C$ isotope ratio in the interstellar medium and in stellar atmospheres is discussed and compared to the value found in the solar system and especially in comets.

102.052 Low-temperature condensates in comets.
A. H. Delsemme, D. Rud.
Comets, asteroids, meteorites, (see 012.049), p. 529 - 535 (1977).

Recent observational data on the volatile fraction of comets are confronted with a model based on the fractional condensation, in the 80—100 °K range, of a higher-temperature equilibrium obtained from a solar mixture, more or less

depleted in oxygen and in hydrogen. It is possible to almost duplicate the observational data, only by assuming that the solar ratio of C/O is at least as large as 0.66 and that the hydrogen was drastically depleted by an unknown process in the primitive solar nebula.

102.053 Comets, asteroids and meteorites: unsolved problems about their relationships, evolution and origin.
A. H. Delsemme.
Comets, asteroids, meteorites, (see 012.049), p. 575 - 579 (1977).

102.054 The composition of cometary gas and the primordial solar nebula.   M. Shimizu.
10th Lunar and Planetary Symposium, (see 012.050), p. 92 - 97 (1977).

102.055 Photoelectron fluxes in a cometary atmosphere.
O. Ashihara.
10th Lunar and Planetary Symposium, (see 012.050), p. 98 - 102 (1977). In Japanese.

The photoelectron fluxes in a cometary atmosphere are calculated by the Monte Carlo method. A pure H_2O atmosphere is assumed with sublimation rate of 10^{30} molecules s^{-1} at 1 A.U. The energetics of electron gas is discussed and the elementary collisional processes for determining the fluxes are investigated.

102.056 Comets, minor planets and meteoroids.
Ľ. Kresák.
Kozmos, Vol. 8, 66 - 70 (1977). In Slovak.

Solar wind study from cometary observations. See Abstr. 074.111.

Numerical study of some solar-wind interaction models with space objects.   See Abstr. 091.074.

Asteroid versus comet discrimination from orbital data.   See Abstr. 098.104.

Relationship between meteorites, asteroids and comets.   See Abstr. 105.268.

A former major planet of the solar system. See Abstr. 107.050.

On the early scattering processes of the outer planets.   See Abstr. 107.052.

Molecules of the interstellar medium and cometary nuclei.   See Abstr. 131.032.

From dust to comets.   See Abstr. 131.221.

103 Comets: Listed Objects

103.001 Six new periodic comets in 1975.
R. J. Buckley.
J. British Astron. Assoc., Vol. 87, 444 - 456 (1977).
Concerning comets 1975a Boethin, 1975b West-Kohoutek-Ikemura, 1975c Kohoutek, 1975e Smirnova-Chernykh, 1975g Longmore, 1975o Gehrels 3.

103.002 Comet Digest.   J. E. Bortle.
Sky Telesc., Vol. 54, 28, 107, 189, 270, 383, 389, 467 (1977).

103.003 Possible comet.
IAU Circ., No. 3100 (1977).

103.004 Possible comet.
IAU Circ., No. 3131, with a correction No. 3132 (1977).

103.005 Possible comet.
IAU Circ., No. 3156 (1977).

103.006 Observations of comets.
Yamamoto Circ., No. 1856 (1977). In Japanese.

103.007 Comets in the year 1976.   J. Bouška.
Vesmír, Vol. 56, 220 (1977). In Czech.

103.008 Observation of comets and asteroids at the Klet̆ Observatory in the year 1975.   A. Mrkos.
Acta Univ. Carolinae Math. Phys., Vol. 18, No. 1, p. 65 - 74 (1977).
Precise positions of comets Boethin 1975a, P/Gunn, P/Schwassmann-Wachmann 1, Mori-Sato-Fujikawa 1975j, Suzuki-Saigusa-Mori 1975k, Kobayashi-Berger-Milon 1975h and asteroids 433 Eros, 803 Hispania, 304 Olga, 739 Mandeville, 1263 Varsavia and 1483 Mertona.

Comet 1976 VI West

103.101 Lyman-alpha observations of Comet West (1975n).
C. B. Opal, G. R. Carruthers.
Icarus, Vol. 31, 503 - 509 (1977).
Images of Comet West in atomic hydrogen (1216 Å) emission were obtained from a sounding rocket on 1976 March 5.5 ($R = 0.38$ AU). The hydrogen production rate derived from the fit of a simple radial-outflow model to the observed inner isophotes was 3.2×10^{30} atoms/sec. This production rate, taken with data on C and O obtained simultaneously by Feldman and Brune, gives $Q_H : Q_O : Q_C = 8:3.5:1$. For Comet Kohoutek the authors obtained the ratio $7:1.7:1$. The difference, if real, may be due to minor differences in composition or evolution, but in any case it appears that the two comets are similar.

103.102 Evaporation of ices from Comet West.
M. F. A'Hearn, C. H. Thurber, R. L. Millis.
Astron. J., Vol. 82, 518 - 524 (1977).
Postperihelion, narrow-band, filter photometry has been used to derive gas and dust production rates for Comet West. For heliocentric distances less than 1.25 AU, C_2 and CN vary as $r^{-2.8}$, while C_3 and OH (based on published measurements) vary as $r^{-2.1}$. C_3 production closely parallels dust production. The authors conclude that OH and C_3 parents are produced at the nucleus, while CN and C_2 parents are produced from the icy grain halo.

103.103 Evaporation rates from comet West.
M. F. A'Hearn, R. L. Millis, C. M. Thurber.
Bull. American Astron. Soc., Vol. 9, 466 (1977). — Abstract.

103.104 Microwave continuum radiation from comet West 1975n.   R. W. Hobbs, J. C. Brandt, S. P. Maran.
Astrophys. J., Vol. 218, 573 - 578 (1977).
Continuum emission at wavelength 3.71 cm was observed from the nuclear region of comet West 1975n on 1976 March 5. The flux density was 0.040 Jy, which is uncertain by 25% due to calibration. Assuming that the source was a uniformly illuminated disk, the diameter was $\lesssim 1100$ km and the brightness temperature was $\gtrsim (330 \pm 85)$ K. It appears that the microwave emission can be interpreted as thermal radiation from a temporarily enhanced icy-grain halo (IGH). If this interpretation is correct, then the actual temperature (which is assumed to be approximately equal to the nuclear surface temperature) must be in the range $200-250$ K, roughly compatible with the observations, in order to satisfy the IGH models of Delsemme.

103.105 A search for radiofrequency emission from CH in Comet West 1975n.
J. Rahe, E. Churchwell, H. U. Keller.
Astron. Astrophys., Vol. 61, 765 - 767 (1977).
A search was conducted for two hyperfine transitions ($F = 1 \rightarrow 1$) and ($F = 0 \rightarrow 1$) at 9 cm of the ground state Λ-doublet of CH in Comet West 1975n. No lines could be detected, and upper limits are given for the column densities. The results are converted into production rates and compared to Comet Kohoutek observations.

103.106 A dust model of comet West (1975n).
M. Oishi, H. Okuda.
10th Lunar and Planetary Symposium, (see 012.050), p. 103 - 108 (1977). In Japanese.
Comet West (1975n) has been observed photometrically and polarimetrically in the infrared region. On the basis of the observational results, a cometary dust model is constructed. The features of the energy spectra and the scattering angle dependences of the polarization can be reproduced by a mixture of metallic dust particles (graphite or iron) and dielectric dust particles (silicate).

103.107 Comet West 1975n.   J. Bouška.
Vesmír, Vol. 56, 204 - 207 (1977). In Czech.

Radio observations of the OH radical in comets Kobayashi-Berger-Milon (1975h) and West (1975n).
See Abstr. 103.402.

CH-Radiobeobachtungen der Kometen Kohoutek 1973 XII und West 1975n.   See Abstr. 103.706.

Comet 1969 IX Tago-Sato-Kosaka

103.121 Equidensitometry of comet Tago-Sato-Kosaka.
O. V. Dobrovol'skij, I. N. Matveev, N. N. Kiselev.
Komety Meteory, No. 25, p. 3 - 10 (1976). In Russian.
Two photographs of comet Tago-Sato-Kosaka 1969 IX were treated by the equidensit method and curves of surface brightness variation constructed. The surface brightness during the outburst of 7 February 1970 is explained as a result of free outflow of molecules from the nucleus, totally or partly liberated from the dust matrix. The number of C_2 molecules in the coma is estimated as $\simeq 8.1 \times 10^{30}$ and the mean density

of C_2 as 2 cm^{-3}.

Periodic comet Klemola

103.131 **Periodic comet Klemola (1976j).**
IAU Circ., No. 3096 (1977).

103.132 **P/Klemola (1976j).**
Yamamoto Circ., No. 1859 (1977). In Japanese.

Periodic comet Gehrels 3

103.141 **Periodic comet Gehrels 3 (1975o).**
IAU Circ., No. 3097 (1977).

Comet 1977e Helin

103.151 **Comet Helin (1977e).**
IAU Circ., Nos. 3085, 3131 (1977).

103.152 **Comet Helin (1977e).**
Yamamoto Circ., No. 1855 (1977). In Japanese.

Periodic comet Faye

103.161 **Periodic comet Faye (1976i).**
IAU Circ., No. 3097 (1977).

103.162 **P/Faye (1976i).**
Yamamoto Circ., No. 1860 (1977). In Japanese.

Periodic comet Tsuchinshan 1

103.171 **Periodic comet Tsuchinshan 1.**
IAU Circ., No. 3099 (1977).

Periodic comet Chernykh

103.181 **Periodic comet Chernykh (1977l).**
IAU Circ., Nos. 3101, 3102, 3104, 3109, 3111,
3115, 3119, 3128, 3139, 3150 (1977).

103.182 **Comet Chernykh (1977l).**
Yamamoto Circ., Nos. 1860, 1861, 1863, 1864,
1867, 1871 (1977). In Japanese.

103.183 **New comet Chernykh 1977l.**
British Astron. Assoc. Circ., No. 580 (1977).

103.184 **Short-period comet Chernykh, 1977l.**
N. S. Chernykh.
Komet. Tsirk., Kiev, No. 217 (1977). In Russian. – Observations at the Klet Observatory (*A. Mrkos*). Observations of
brightness and form of comet Chernykh.

103.185 **Comet Chernykh, 1977l.**
Komet. Tsirk., Kiev, No. 219 (1977). In Russian.

Comet 1970 II Bennett

103.201 **Monochromatic photometry of comet Bennett,**
1970 II.   I. N. Matveev, G. P. Chernova.
Komety Meteory, No. 25, p. 11 - 23 (1976). In Russian.
The results are given of slitless spectra measurements of
comet Bennett, 1970 II, taken during 30 March - 1 May 1970
in the CN (0.0) band. The CN molecule number is found and
its variation with heliocentric distance determined. An analytic expression is obtained of the curves of the surface brightness variation in the CN (0.0) band. A method is proposed of
determining the radius of the cometary nucleus.

103.202 **Photometric investigation of comet Bennett,**
1970 II.   Kh. Ibadinov, Sh. R. Daminov.
Komety Meteory, No. 25, p. 24 - 30 (1976). In Russian.
A photometric investigation of the comet is carried out
on the basis of plates taken on ORWO ZU-2 emulsion in
integrated light. The dependence of surface brightness J on the
distance R is presented as $J \sim R^{-n}$ and the parameter n is determined for different photometric sections for head and tail.

103.203 **Equidensitometry of comet Bennett.**
O. V. Dobrovol'skij, I. N. Matveev, N. N. Kiselev.
Komety Meteory, No. 26, p. 18 - 26 (1977). In Russian.
The law of surface brightness diminishing for comet
Bennett 1970 II observed during March 28 - May 14 was constructed by the equidensitometric method. It was found to be
of the type $A\rho^{-\alpha}$. It corresponds in the region of the head to a
free molecular flow of matter. The power a is shown to be
dropping with growing heliocentric distance for both the head
and the tail regions. The dimensions of the head and the
length of the tail were determined.

103.204 **Effective accelerations in the dust tail of comet**
Bennett (1970 II).   L. F. Grigor'eva, V. P.
Konopleva, V. K. Rozenbush, N. S. Chernykh.
Problems of cosmic physics. Vyp. (No.) 12, (see 003.016),
p. 92 - 101 (1977). In Russian.
Structure variations of comet Bennett in April 1970 are
described. Values of effective acceleration μ of the dust
particles were determined. The range of μ indicates essential
dispersion of dust particles by their sizes. The character of the
variation of μ indicates nonstationarity of dust tail formation.

Periodic comet Giacobini-Zinner

103.211 **Comet Giacobini-Zinner and the Draconid meteor**
stream in 1978 - 1979.   Yu. V. Evdokimov.
Komet. Tsirk., Kiev, No. 219 (1977). In Russian.

Periodic comet Kopff

103.221 **Periodic comet Kopff.**
IAU Circ., Nos. 3117, 3149 (1977).

Periodic comet Comas Solá

103.231 **Periodic comet Comas Solá (1977n).**
IAU Circ., No. 3110 (1977).

103.232 **P/Comas Solá (1977n).**
Yamamoto Circ., No. 1863 (1977). In Japanese.

103.233 **Periodic comet recovery.**
British Astron. Assoc. Circ., No. 582 (1977).
Concerning 1977n P/Comas Sola.

Comet 1977m Kohler

103.241 **Comet Kohler (1977m).**
IAU Circ., Nos. 3103, 3105, 3107, 3109, 3112,
3114, 3118, with a correction No. 3137, 3121, 3125, 3127,
3132, 3137, 3148, 3153 (1977).

103.242 **Comet Kohler (1977m).**
Yamamoto Circ., Nos. 1860, 1870 (1977). In
Japanese.

103.243 **New comet Kohler 1977m.**
British Astron. Assoc. Circ., Nos. 580 - 582 (1977).

103.244 **Comet Kohler, 1977m.**
Komet. Tsirk., Kiev, No. 217 (1977). In Russian.
Observations at the Crimea Observatory with the 40-cm
astrograph (*N. S. Chernykh*).

103.245 **Comet Kohler, 1977m.**
Komet. Tsirk., Kiev, No. 218 (1977). In Russian.
Observations at the Stara Zagora Observatory, Bulgaria. New
elements and ephemeris by Marsden.

103.246 **Photometry of comet Kohler, 1977m.**
V. M. Lyutyj, V. P. Tarashchuk.
Komet. Tsirk., Kiev, No. 218 (1977). In Russian.

103.247 **Spectrum of comet Kohler, 1977m.**
V. P. Tarashchuk.
Komet. Tsirk., Kiev, No. 218 (1977). In Russian.

103.248 **Comet Kohler, 1977m.**
Komet. Tsirk., Kiev, No. 219 (1977). In Russian.
Observations at the Main Astronomical Observatory of the
USSR Academy of Sciences (*N. M. Bronnikova*). Observations
at Tartu (*H. K. Raudsaar, G. R. Kastel'*). Observations of the
brightness of comet Kohler (*V. A. Golubev*). Observations in
Novotroitskij, Donez-Region (*A. Majdik*).

Comet 1976 IX Lovas

103.251 **Comet Lovas (1976k).**
IAU Circ., No. 3086 (1977).

103.252 **Comet Lovas (1976k).**
Yamamoto Circ., No. 1855 (1977). In Japanese.

Comet 1976 XII Lovas

103.261 **Comet Lovas (1977c).**
IAU Circ., No. 3106 (1977).

Periodic comet van Biesbroeck

103.271 **Periodic comet van Biesbroeck (1977s).**
IAU Circ., Nos. 3122, 3152 (1977).

Periodic comet Tsuchinshan 2

103.281 **Periodic comet Tsuchinshan 2.**
IAU Circ., No. 3122 (1977).

Comet 1974 III Bradfield

103.301 **Positions of comets: Bradfield (1974d) (*Bradfield**
***1974b)* and Kobayashi-Berger-Milon (1975h).**
J. Bem.
Acta Astron., Vol. 27, 291 - 292 (1977).

Periodic comet Ashbrook-Jackson

103.331 **Periodic comet Ashbrook-Jackson (1977g).**
IAU Circ., Nos. 3125, 3156 (1977).

Periodic comet Sanguin

103.341 **Comet Sanguin (1977p).**
IAU Circ., Nos. 3124, 3126, 3127, 3128, 3136,
3150 (1977).

103.342 **Comet Sanguin (1977p).**
Yamamoto Circ., Nos. 1865, 1866, 1867, 1869,
1871 (1977). In Japanese.

103.343 **New periodic comet Sanguin 1977p.**
British Astron. Assoc. Circ., No. 582 (1977).

103.344 **New short-period comet Sanguin, 1977p.**
Komet. Tsirk., Kiev, No. 218 (1977). In Russian.

Periodic comet Taylor

103.351 **Periodic comet Taylor (1977a).**
IAU Circ., No. 3090 (1977).

103.352 **P/Taylor (1977a).**
Yamamoto Circ., No. 1857 (1977). In Japanese.

Comet 1977q Tsuchinshan

103.361 **Comet Tsuchinshan (1977q).**
IAU Circ., Nos. 3132, 3135, 3146, 3155 (1977).

103.362 **Comet Tsuchinshan (1977q).**
Yamamoto Circ., Nos. 1868, 1869, 1871 (1977).
In Japanese.

103.363 **New comet Tsuchinshan, 1977q.**
Komet. Tsirk., Kiev, No. 219 (1977). In Russian.

103.364 **Orbital elements and ephemeris of new comet**
Tsuchinshan, 1977q.
Komet. Tsirk., Kiev, No. 219 (1977). In Russian.

Periodic comet Gunn

103.371 **Periodic comet Gunn.**

IAU Circ., No. 3141 (1977).

Comet 1975 IX Kobayashi-Berger-Milon

103.401   Observations photographiques de la comète
Kobayashi-Berger-Milon (1975h), effectuées en
1975 à l'astrographe double de 40 cm.   H. Debehogne.
Bull. Astron., Obs. R. Belgique, Vol. 8, 308 - 312 (1976).

103.402   **Radio observations of the OH radical in comets
Kobayashi-Berger-Milon (1975h) and West (1975n).**
E. Gerard, F. Biraud, J. Crovisier, I. Kazes, B. Milet.
Comets, asteroids, meteorites, (see 012.049), p. 65 - 68 (1977).

The radio observations of the 18 cm wavelength OH main
lines in comets Kobayashi-Berger-Milon (1975h) and West
(1975n) confirm the model of ultraviolet pumping by the sun.

103.403   **Spectrophotometry of the comet 1975 IX.**
J. Bouška, A. Mrkos.
Acta Univ. Carolinae Math. Phys.,Vol. 18, No. 2, p. 55 - 60
(1977).

From the spectrograms of the comet and of the comparison stars the monochromatic fluxes of the CN and C_2
bands the relative intensities of the continuum were obtained.
From these values the total numbers of CN and C_2 molecules
in the cometary head were computed and the relative spectrophotometric gradient was determined.

Positions of comets: Bradfield (1974d) *(Bradfield
1974b)* and Kobayashi-Berger-Milon (1975h).
See Abstr. 103.301.

Observations photographiques des comètes 1973f
(Kohoutek) et 1975h (Kobayashi-Berger-Milon) faites à
l'Observatoire Astronomique de Belgrade.
See Abstr. 103.704.

Comet 1975 VIII Lovas

103.411   **Comet Lovas (1975 VIII).**
IAU Circ., No. 3142 (1977).

Periodic comet Schwassmann-Wachmann 1

103.421   **Periodic comet Schwassmann-Wachmann 1.**
IAU Circ., No. 3147 (1977).

103.422   **P/Schwassmann-Wachmann 1.**
Yamamoto Circ., No. 1871 (1977). In Japanese.

Periodic comet Clark

103.431   **Periodic comet Clark (1973 V).**
IAU Circ., No. 3155 (1977).

Periodic comet Encke

103.451   **Periodic comet Encke.**
IAU Circ., Nos. 3091, 3092, 3098, 3115, 3118,
3139 (1977).

103.452   **P/Encke.**
Yamamoto Circ., Nos. 1853, 1856, 1857, 1858,

1860, 1864 (1977). In Japanese.

Comet 1957 V Mrkos

103.461   **On the CN (0, 0) spectrum of Comet Mrkos 1957d.**
I. R. Ferrín.
Astrophys. Space Sci., Vol. 52, 11 - 16 (1977).

The author has carried out an analysis of the (0, 0)
vibrational band of the CN molecule in Comet Mrkos 1957d,
including the effect of collisions. He found that the sum of the
squares of the residuals can be reduced by a factor of ten, if
collisions account for 46 ± 3% of the population of the lower
level. For the rotational temperature of the cometary gas was
found 410 ± 40K. For the velocity of the comet a value of
34.38 ± 0.10 km s^{-1} was found.

Comet 1975 XII Mori-Sato-Fujikawa

103.471   **Observations photographiques de la comète Mori-
Sato-Fujikawa (1975j), effectuées en 1975 à la
caméra astrographique de 25 cm de l'Observatoire National de
Rio de Janeiro.**   H. Debehogne, R. R. de Freitas Mouraǒ.
Bull. Astron., Obs. R. Belgique, Vol. 8, 313 - 315 (1976).

Periodic comet d'Arrest

103.501   **Positions de la comète P/d'Arrest.**
H. Debehogne, R. R. de Freitas Mourao.
Acta Astron., Vol. 27, 297 - 300 (1977).

Periodic comet Whipple

103.521   **P/Whipple (1977h).**
Yamamoto Circ., No. 1854 (1977). In Japanese.

103.522   **Rediscovery of comet Whipple, 1977h.**
Komet. Tsirk., Kiev, No. 219 (1977). In Russian.

Periodic comet Barnard 1

103.531   **Periodic comet Barnard 1, 1884II.**
Yamamoto Circ., No. 1856 (1977). In Japanese.

Periodic comet Wolf-Harrington

103.551   **Periodic comet Wolf-Harrington (1977j).**
IAU Circ., Nos. 3091, 3094, 3121 (1977).

103.552   **P/Wolf-Harrington (1977j).**
Yamamoto Circ., No. 1857 (1977). In Japanese.

103.553   **Recoveries of periodic comets.**
British Astron. Assoc. Circ., No. 580 (1977).

Periodic comet Halley

103.601   **Comet Halley and nongravitational forces.**
D. K. Yeomans.

Comets, asteroids, meteorites, (see 012.049), p. 61 - 64 (1977).

The motion of comet Halley is investigated over the 1607 - 1911 interval. The required nongravitational force model was found to be most consistent with a rocket-type thrust from the vaporization of water-ice in the comet's nucleus. The nongravitational effects are time-independent over the investigated interval.

The 1986 apparition of Halley's comet.
See Abstr. 014.015.

A first comet mission. Report of the Comet Halley Science Working Group. See Abstr. 051.013.

Periodic comet Tempel 1

103.651 **Periodic comet Tempel 1 (1977i).**
IAU Circ., Nos. 3092, 3153 (1977).

103.652 **P/Tempel 1 (1977i).**
Yamamoto Circ., Nos. 1855, 1858 (1977). In Japanese.

Periodic comet Tempel 2

103.661 **Periodic comet Tempel 2 (1977d).**
IAU Circ., No. 3092 (1977).

103.662 **P/Tempel 2 (1977d).**
Yamamoto Circ., No. 1858 (1977). In Japanese.

Comet 1973 XII Kohoutek

103.701 **Spectrophotometry of the comet Kohoutek 1973 XII.** J. Bouška, A. Mrkos, E. Müllerová.
Bull. Astron. Inst. Czechoslovakia, Vol. 28, 288 - 291 (1977).
The monochromatic fluxes of the CN(0, 0) and $C_2 (\Delta v = +1)$ bands were determined from the spectrograms of the comet and those of the comparison stars. These values were used to compute the total numbers of CN and C_2 molecules in the cometary head.

103.702 **Study of some physical parameters and of the activity of comet Kohoutek (1973 XII) on the basis of spectral data.** K. I. Churyumov, L. V. Yurevich.
Problems of cosmic physics. Vyp. (No.) 12, (see 003.016), p. 101 - 110 (1977). In Russian.
Seven spectrograms of comet Kohoutek (1973 XII) are studied obtained for the period January 16 - 31, 1974. 74 cometary emissions are identified. The relative energy distribution in the comet's spectrum for January 24, 25, 26, 27, and 30, 1974 is constructed. The character of variations of cometary emission intensities in connection with solar activity is studied. The absolute energy fluxes are determined; the abundances of molecules CN, C_2 and C_3 in the cometary atmosphere and their partial masses are estimated.

103.703 **Some speculations about the origin of the high speed HCN jets in Comet Kohoutek (1973f).**
W.-H. Ip, D. A. Mendis.
Astrophys. Space Sci., Vol. 46, 109 - 114 (1977).
The origin of the high speed HCN jets observed in Comet Kohoutek (1973f) is considered. Acceleration of the precursor ions of HCN by energetic electrons in the ionospheric current sheets or current arcs is suggested to play an important role. Also, a fraction of the HCN observed by electric discharges,

associated with the ionosphere current system through the cometary atmosphere. It is not possible to make a realistic estimate of this fraction at the present time, but that it may be significant cannot be entirely excluded by the existing observations.

103.704 **Observations photographiques des comètes 1973f (Kohoutek) et 1975h (Kobayashi-Berger-Milon) faites à l'Observatoire Astronomique de Belgrade.**
M. B. Protitch, V. Protitch-Benišek.
Bull. Obs. Astron. Belgrade, No. 128, p. 56 - 57 (1977).

103.705 **Photometry of comet Kohoutek (1973f).**
V. Riives.
Publ. Tartu Astrofiz. Obs., Vol. 45, 131 - 140 (1977). In Russian.
The optical and infrared radiation of the comet at $\lambda < 2 \mu m$ is solar light scattered by molecules and dust. At longer wavelengths the light of the comet is thermal radiation by dust particles. The colour temperature is near to the blackbody temperature for small rotating particles. The luminosity of dust in visual light is $1^m.5$ fainter than the observed brightness of the comet. The production rate of gas and dust is proportional to r^{-2}.

103.706 **CH-Radiobeobachtungen der Kometen Kohoutek 1973 XII und West 1975n.**
E. Churchwell, J. Rahe, H. U. Keller.
Mitt. Astron. Ges., Nr. 42, p. 143 (1977).

103.707 **Photometric parameters of Comet Kohoutek 1973 XII.** T. Kleine, L. Kohoutek.
Comets, asteroids, meteorites, (see 012.049), p. 69 - 76 (1977).
An analysis of the 2796 visual observations as well as of 282 V, B or U photoelectric observations was carried out using a two-parametric model for the light curve of the comet.

Periodic comet Grigg-Skjellerup

103.751 **Periodic comet Grigg-Skjellerup (1977b).**
IAU Circ., Nos. 3092, 3120 (1977).

103.752 **P/Grigg-Skjellerup (1977b).**
Yamamoto Circ., Nos. 1856, 1858 (1977). In Japanese.

Meteors and comet P/Grigg-Skjellerup.
See Abstr. 104.040.

Periodic comet Brorsen-Metcalf

103.801 **On the position of the axis of rotation of the cometary nucleus Brorsen–Metcalf 1847 V.**
L. M. Belous.
Problems of cosmic physics. Vyp. (No.) 12, (see 003.016), p. 110 - 112 (1977). In Russian.

Periodic comet Schuster

103.841 **Periodic comet Schuster (1977o).**
IAU Circ., Nos. 3120, 3121, 3123, 3124, 3128, 3138, 3141, 3152 (1977).

103.842 **Periodic comet Schuster (1977o).**
Yamamoto Circ., Nos. 1865, 1866, 1871 (1977). In Japanese.

103.843 **New periodic comet Schuster 1977o.**
British Astron. Assoc. Circ., No. 582 (1977).

103.844 **New short-period comet Schuster, 1977o.**
Komet. Tsirk., Kiev, No. 218 (1977). In Russian.

Comet 1975 II Schuster

103.851 **Comet Schuster (1975 II).**
IAU Circ., Nos. 3092, 3142 (1977).

103.852 **Comet Schuster (1976c = 1975 II).**
Yamamoto Circ., No. 1858 (1977). In Japanese.

Periodic comet Kowal

103.901 **Periodic comet Kowal (1977f).**
IAU Circ., Nos. 3085, 3097, 3148 (1977).

103.902 **Comet Kowal (1977f).**
Yamamoto Circ., Nos. 1853, 1854, 1855, 1860 (1977). In Japanese.

103.903 **Comet Kowal 1977f — new short-period comet of the Saturn family.**
Komet. Tsirk., Kiev, No. 218 (1977). In Russian.

Periodic comet Kojima

103.951 **Periodic comet Kojima (1970 XII) = 1977r.**
IAU Circ., Nos. 3093, 3151, 3156 (1977).

103.952 **P/Kojima (1970 XII).**
Yamamoto Circ., No. 1871 (1977). In Japanese.

103.953 **Rediscovery of comet Kojima 1970 XII − 1977r.**
Komet. Tsirk., Kiev, No. 219 (1977). In Russian.

103.954 **Prediction of the positions of comet Kojima, 1970 XII.** N. A. Belyaev.
Komet. Tsirk., Kiev, No. 219 (1977). In Russian.

Periodic comet Arend-Rigaux

103.961 **Periodic comet Arend-Rigaux (1977k).**
IAU Circ., Nos. 3095, 3103, 3116, 3142, 3155 (1977).

103.962 **P/Arend-Rigaux (1977k).**
Yamamoto Circ., No. 1859 (1977). In Japanese.

103.963 **Comet Arend-Rigaux, 1977k.**
Komet. Tsirk., Kiev, No. 219 (1977). In Russian.

Recoveries of periodic comets.
See Abstr. 103.553.

104 Meteors, Meteor Streams

104.001 **New look at meteor streams?** D. W. Hughes.
Nature, Vol. 269, 107 (1977).

104.002 **Automatic meteor radar for investigation of the circulation of the atmosphere.**
B. L. Kashcheev, G. D. Krutogolov, V. V. Lizogub,
V. A. Nechitajlenko.
Komety Meteory, No. 24, p. 3 - 18 (1976). In Russian.
The problem of automation of first processing of Doppler information in the cause of radar observations of meteor drifts is examined. The authors present data obtained from an automatic meteor wind radar. Results of laboratory and field radar tests are given. Information about changes of meteor drift obtained by means of an automatic machine and the ordinary phase-time method is compared.

104.003 **Possibilities of the bearing-time radio method for determining the radiants and velocities of individual meteors.** R. P. Chebotarev.
Komety Meteory, No. 24, p. 19 - 27 (1976). In Russian.

104.004 **The observed and theoretical height-velocity dependence of meteors up to +12^m.**
N. V. Novoselova.
Komety Meteory, No. 24, p. 28 - 34 (1976). In Russian.
Results of mean height measurements for eight velocity intervals obtained in Kharkov and their linear approximation are presented. The predicted theoretical dependence h (v) is in good agreement with that observed for $v \geq 20$ km/sec.

104.005 **On the correlation of maximum and mean hourly echo rates per day recorded by radar.**
E. I. Fialko, V. N. Donij.
Komety Meteory, No. 24, p. 45 - 48 (1976). In Russian.
Meteor radar statistics has been investigated. Some recommendations for using hourly echo rates for estimation of "meteor contribution" are given.

104.006 **Comparison of observed and computed forms of light curves of meteor flares.** I. M. Khaimov.
Komety Meteory, No. 25, p. 57 - 71 (1976). In Russian.
A theoretical calculation shows a possibility of determination of the size of particles separated from meteors during flares for different forms of light curves. Analysis of 162 meteor flares gives the initial size of particles $140\mu < 2r_0 < 280\mu$. The mean size $2r_0 = 200\mu$ is used for calculation of the form of flare light curves agreeing with observed ones. The agreement is reasonably well under specific assumptions on the law of particle separation.

104.007 **On a direction of automation of radio measurements of coordinates and drift velocities of meteor trails.** V. M. Kolmakov, E. I. Fialko.
Komety Meteory, No. 26, p. 27 - 29 (1977). In Russian.
In this paper a possible method of automatic measurements of the coordinates of reflecting points on meteor trails and of the radial component of the drift velocity is examined. The automatic measurement can be realized on the basis of a phase radio method with use of ordinary and special computers.

104.008 **On a mechanism of influence of the earth's magnetic field on the motion of ionized meteor trails.**
V. M. Kolmakov, E. I. Fialko.
Komety Meteory, No. 26, p. 30 - 33 (1977). In Russian.
A possible mechanism of transformation of a part of kinetic energy of a meteoric body into kinetic energy of the drift of a meteor trail in the magnetic field is discussed. This

mechanism can be effective only for very bright meteors.

104.009 **Change of the linear density in meteor streams due to an approach to Jupiter.** L. M. Sherbaum.
Komety Meteory, No. 26, p. 34 - 42 (1977). In Russian.
The change of the linear density in three meteor streams due to deformation under the perturbing influence of Jupiter is considered. At once after approach the linear density increases a little in the regions of meteor streams which are situated close to Jupiter. Then the increase of the linear density is replaced by a considerable decrease.

104.010 **Causes of fluctuations of the effective linear electron density of meteor traces.** I. A. Delov.
Geomagn. Aehron., Vol. 17, 762 - 764 (1977). In Russian.

104.011 **Solar radiation torque on meteoroids: complications for the Yarkovsky effect from spin axis precession.**
V. J. Slabinski.
Bull. American Astron. Soc., Vol. 9, 438 (1977). — Abstract.

104.012 **A search for ultraviolet OH emission from meteors.**
G. A. Harvey.
Astrophys. J., Vol. 217, 688 - 690 (1977).
Observations for hydroxyl (OH) emission from meteors were made during the late summers of 1975 and 1976 from altitudes of 10,600 and 14,200 feet (3.2 km and 4.45 km). Two of the meteors were Perseids, and one was an α Capricornid. The Perseid meteors produced a peak irradiance at a distance of 100 km from the meteors of about 5×10^{-5} ergs cm^{-2} s^{-1} in the OH emission region. The zero-magnitude α Capricornid meteor produced a spectral irradiance, I_{3100}, of 23×10^{-8} ergs cm^{-2} Å^{-1} s^{-1}. This may be indicative of significant amounts of H_2O in these meteors.

104.013 **Spectral analysis of a high-velocity meteor.**
G. A. Harvey.
NASA Tech. Note, NASA TN D-8505. 22 pp. Price $ 3.50 (1977).
A spectrogram of a fast optical meteor was reduced and analyzed, and 60 features were identified in the spectrum. Air and ionized elements in this meteor radiate throughout the spectrum from 3000 Å to 6800 Å. A mass of 9 mg and an effective radiation temperature of approximately 5700 K were computed for the meteor. Weight ratios of Ca:Fe, Mg:Fe, and Na:Fe were computed. A plasma particle velocity distribution for meteors was derived, and the average collision speed obtained from this distribution was compared with the relative collision speed of a Fe-N_2 gas mixture at 5700 K.

104.014 **Observations of overdense Quadrantid radio meteors and the variation of the position of stream maxima with meteor magnitude.** D. W. Hughes, I. W. Taylor.
Mon. Not. R. Astron. Soc., Vol. 181, 517 - 526 (1977).
Quadrantid across-stream activity profiles are presented for meteors of magnitudes (M) 4.3, 3.8, 3.3, 2.9 and 2.3, based on observations of overdense meteor trails. Not only does the profile become less symmetrical for brighter meteors but also the solar longitude, $\lambda_\odot$, of the stream maximum varies. Radar observations at Sheffield indicate that this variation is of the form $\lambda_\odot = (283.24 \pm 0.04) - (0.109 \pm 0.010)M$ through the range $2.3 < M < 7.2$.

104.015 **Dispersion of orbital elements within the Perseid meteor stream.** V. Porubčan.
Bull. Astron. Inst. Czechoslovakia, Vol. 28, 257 - 266 (1977).
The present study deals with the dispersion of orbital elements of the Perseids as the meteor stream with the greatest

representation among the available photographic meteor orbits.

104.016 The velocity dependence of meteoric green line emission.   W. J. Baggaley.
Bull. Astron. Inst. Czechoslovakia, Vol. 28, 277 - 280 (1977).

The atomic oxygen O I 5577 Å green line emission is a high velocity feature of meteor wakes. It is proposed that this velocity characteristic is a consequence of the dependence on meteoroid velocity of the relative composition in a meteor column of O^+ ions and meteoric neutrals. A quantitative examination shows that for low velocity meteors the combination of low initial O^+ abundance and the loss of these ions by charge exchange with Fe atoms leads to the inhibition of 5577 Å radiation.

104.017 Sunrise effect on persistent radar echoes from sporadic meteors.   B. A. McIntosh, A. Hajduk.
Bull. Astron. Inst. Czechoslovakia, Vol. 28, 280 - 285 (1977).

The authors analyze observations from the Ottawa meteor radar and show that there is a significant effect on the echo durations of meteors from the sporadic background. A possible seasonal variation is noted and the complications which these effects introduce in past and future meteor studies are discussed.

104.018 Simultaneous radar meteor observations at Ondřejov and Dushanbe.
P. B. Babadzhanov, R. P. Chebotarev, A. Hajduk.
Bull. Astron. Inst. Czechoslovakia, Vol. 28, 286 - 288 (1977).

A program of radar meteor observations of selected showers has been started in 1974 with the observations of the Orionid meteor shower. The comparison of the data obtained at the two stations (Dushanbe and Ondřejov) confirms some structural features of the stream.

104.019 On fireball model and ultrahigh energy cosmic γ ray spectrum by p-p collisions.   F.-S. Kuo.
Prog. Theor. Phys., Vol. 56, 1234 - 1244 (1976). − Abstr. in Phys. Abstr., Vol. 80, Abstr. 18364 (1977).

104.020 Dazzling Czechoslovakian fireball.
Sky Telesc., Vol. 54, 475 - 478 (1977).

104.021 Fireballs.   H. G. Miles.
Modern astronomy, (see 003.013), p. 83 - 94 (1977).

104.022 Investigation of the light curves of meteors. II. Classification of light curves.
E. N. Kramer, I. S. Shestaka, V. I. Musij, M. Toktogulov.
Problems of cosmic physics. Vyp. (No.) 12, (see 003.016), p. 55 - 60 (1977). In Russian.

From quantitative characteristics of the light curve of a meteor describing the position of the point of maximum light, broadening of the light curve, its asymmetry and steepness a classification of light curves of 360 bright meteors has been made. Several groups of light curves have been singled out for each of which the values of parameters introduced are inside a definite numerical interval. The relationship between each group of curves and physical peculiarities of meteor phenomena has been traced.

104.023 Scattering of radio waves by meteor trails.
Yu. V. Chumak, R. I. Mojsya.
Problems of cosmic physics. Vyp. (No.) 12, (see 003.016), p. 70 - 80 (1977). In Russian.

A numerical solution of the problem of radio wave scattering by meteor trails is obtained. The case of arbitrary electromagnetic vector polarization is considered.

104.024 Results of a numerical solution of the problem of destruction of a spherical meteoric body.
V. G. Kruchinenko, A. N. Shajdo.
Problems of cosmic physics. Vyp. (No.) 12, (see 003.016), p. 81 - 86 (1977). In Russian.

Method and result of numerical solution of the problem of destruction of a spherical meteoric body under different values of ablation parameters are given. The difference between the obtained solution and classical theory is shown.

104.025 Fireballs photographed in Central Europe.
Z. Ceplecha.
Bull. Astron. Inst. Czechoslovakia, Vol. 28, 328 - 340 (1977).

The paper contains data on orbits and trajectories of 42 bright fireballs computed at the Ondřejov Observatory during the past decades (mostly brighter than magnitude −8). They were photographed during three different projects, which are briefly described. Future prospects of the Ondřejov Observatory in photographic recording of meteors and fireballs are outlined.

104.026 The red afterglow in meteor wakes.
W. J. Baggaley.
Bull. Astron. Inst. Czechoslovakia, Vol. 28, 356 - 359 (1977).

An interesting meteoric feature but one that has received little attention, is the occurrence of red afterglows associated with bright fireballs. It is shown that such luminosity can be produced as a result of the excitation of atmospheric molecular oxygen by ionic processes in a meteor column.

104.027 Head echo heights.   P. Pecina.
Bull. Astron. Inst. Czechoslovakia, Vol. 28, 360 - 365 (1977).

The distributions of head echo height parameters, like the beginning and end height of echoes with respect to the K-index and the local sunrise time, as well as the differences between their daytime and night values are investigated. An association of head echo height occurrence with the concentration of the atmospheric constituents like O_2^+ and NO^+ is proposed.

104.028 Deceleration of the meteor echo drift.
P. Prikryl.
Bull. Astron. Inst. Czechoslovakia, Vol. 28, 365 - 371 (1977).

Head echoes followed by drifting body echoes on the rangetime record are analyzed. The deceleration of the radial drift of the enduring echo is found to be larger for shorter echo durations and lower drift velocities. Theoretical considerations based on the theory of the motion of the effective reflexion point along the train are introduced. Experimental values are in fairly good agreement with the theoretical results. Mean penetration depths are obtained for both shower and sporadic meteors. They are found to be somewhat smaller than those obtained by Lindblad (1967 and 1968).

104.029 Influence of solar wind upon the drift of meteor trains.
B. V. Kal'chenko, B. L. Kashcheev, V. V. Fedynskij.
Dokl. AN SSSR, Vol. 234, 1035 - 1038 (1977). In Russian. Abstr. in Ref. zh., 51. Astron., 10.51.441 (1977).

104.030 Fireball networks − a mixed blessing.
K. Hindley.
New Scientist, Vol. 72, 695 - 698 (1976). − Abstr. in Phys. Abstr., Vol. 80, Abstr. 33427 (1977).

104.031 Hollow meteor trains.   W. J. Baggaley.
Nature, Vol. 270, 588 - 589 (1977).

Enduring meteor trains that appear at a height of about 85 km as double lines of light have been reported for more than a century. The peculiar formation almost certainly results from the expanding meteor train assuming the shape of a hollow cylinder and it is shown that this behaviour is to be expected on the basis of recent models of the mechanism of long enduring train luminosity.

104.032 Meteors from near α Circini.
IAU Circ., No. 3146 (1977).

104.033 Growth and decline of a scientific specialty: the case of radar meteor research. G. N. Gilbert.
EOS Trans. American Geophys. Union, Vol. 58, 273 - 277 (1977). — Abstr. in Phys. Abstr., Vol. 80, Abstr. 78722 (1977).

104.034 The behaviour of the meteoric O I 5577 Å emission.
W. J. Baggaley, C. H. Cummack.
Canadian J. Phys., Vol. 55, 1379 - 1383 (1977). — Abstr. in Phys. Abstr., Vol. 80, Abstr. 82176 (1977).

104.035 Momentum of meteoroids and its effect on the Earth's atmosphere. B. A. McIntosh.
Astrophys. Space Sci., Vol. 47, 31 - 34 (1977).
Recent studies have attributed certain properties of the Earth's atmosphere to excess orbital angular momentum of impinging meteoroids. A realistic analysis of meteor observations does not support the existence of this excess.

104.036 Hyperbolic meteors and eruptive processes in the Galaxy. S. K. Vsekhsvyatskij.
Astron. Tsirk., No. 945, p. 6 - 8 (1977). In Russian.

104.037 Measurement of wind shifts along meteor trains.
V. V. Lizogub.
Astron. Tsirk., No. 948, p. 6 - 8 (1977). In Russian.

104.038 The bright bolide of August 17, 1976.
A. K. Terent'eva, A. A. Zhitetskij.
Astron. Tsirk., No. 949, p. 5 - 6 (1977). In Russian.

104.039 On the dilution of a meteor coma.
I. N. Kovshun.
Astron. Tsirk., No. 959, p. 6 - 7 (1977). In Russian.

104.040 Meteors and comet P/Grigg-Skjellerup.
British Astron. Assoc. Circ., No. 580 (1977).

104.041 Irish fireball of 1977 November 4.
British Astron. Assoc. Circ., No. 582 (1977).

104.042 Fireball of 1977 October 12, 17.40 hrs U.T.
British Astron. Assoc. Circ., No. 582 (1977).

104.043 Fireball of 1977 September 24, 01.12 hrs U.T.
British Astron. Assoc. Circ., No. 582 (1977).

104.044 Meteor observations and cosmic dust investigations.
Ch. Willmann, P. I. Klimuk, V. N. Lebedinets,
V. I. Sevast'yanov.
Optical investigations of the emission of the atmosphere, aurorae and noctilucent clouds aboard the orbital scientific station Salyut 4, (see 003.021), p. 128 - 144, 175 - 176 (1977). In Russian.
A short summary of up-to-date methods for the investigation of solid components of the interplanetary medium is given. A critical review of the results obtained by using different methods is presented and the most urgent tasks of future investigations are pointed out. A preliminary programme of these investigations is proposed.Some results of preliminary visual observations of meteors carried out on board Salyut 4 are discussed.

104.045 The British Fireball Network.
K. B. Hindley, M. A. Houlden.
Meteoritics, Vol. 12, 257 - 258 (1977). — Abstract.

104.046 The spectrophotometry of meteor video data.
P. M. Millman, K. S. Clifton.
Meteoritics, Vol. 12, 310 (1977). — Abstract.

104.047 Spectra of meteors. V. A. Smirnov.
Istor.-astron. issled. Vyp. (No.) 3. Moskva, Nauka, 1977, p. 235 - 274. In Russian. — Abstr. in Ref. zh., 51. Astron., 12.51.283 (1977).

104.048 State and prospects of development of radio meteor research. B. L. Kashcheev.
Meteor. issled. No. 4. Moskva, "Sov. radio", 1977, p. 5 - 10. In Russian. — Abstr. in Ref. zh., 51. Astron., 12.51.289 (1977).

104.049 Imitation modelling of meteor phenomena.
Yu. I. Voloshchuk, A. A. Tkachuk.
Meteor. issled. No. 4. Moskva, "Sov. radio", 1977, p. 103 - 114. In Russian. — Abstr. in Ref. zh., 51. Astron., 12.51.292 (1977).

104.050 Meteor research program. A. F. Cook, M. R. Flannery, H. Levy II, R. E. McCrosky, Z. Sekanina, C.-Y. Shao, R. B. Southworth, J. T. Williams.
NASA Contract. Rep., NASA CR-2109. For sale by the National Technical Information Service, Springfield, Virginia 22151. 6 + 166 pp. Price $ 3.00 (1972).
Statistics from a synoptic year sample of 20,000 radar meteor observations using an eight station radar network are presented. The synoptic year sample constitutes the largest, most accurate, and most complete body of radar meteor data in existence. Since it is such a large sample, results of individual meteors from the synoptic year sample such as orbital parameters, heights, ionization, etc., are listed on magnetic tape. Results of 29 simultaneous meteor observations are presented.

104.051 Methods of meteor observations. H. Kuźmiński.
Postępy Astron., Tom 25, 213 - 219 (1977). In Polish.
The article gives a review of modern methods of observations of meteors.

104.052 The chemical composition of cometary meteoroids.
P. M. Millman.
Comets, asteroids, meteorites, (see 012.049), p. 127 - 132 (1977).
Evidence for the chemical composition of cometary meteoroids is available from the spectra of shower meteors, from the analysis of extra-terrestrial dust particles, from a study of residues in the bottom of microcraters on plates exposed to the interplanetary environment, and from measures of the relative abundances of non-atmospheric ions in the E-region of the earth's upper atmosphere. Quantitative measures of chemical abundances in meteoroids show that in general the cometary meteoroids encountered by the earth conform to the carbonaceous chondrites type 1 in the case of the commonest metallic elements.

104.053 Chemical composition of meteoric and meteoritic matter. A. A. Yavnel'.
Comets, asteroids, meteorites, (see 012.049), p. 133 - 135 (1977).
A comparison has been made of the data of Millman's (1972) work on the relative content of Na, Mg, Ca and Fe in Draconids (Giacobinids) with corresponding data for carbonaceous chondrites of type CI, CII, CIIIV, and CIIIO and ordinary chondrites of type H, L and LL.

104.054 Meteoroid populations and orbits. Z. Ceplecha.
Comets, asteroids, meteorites, (see 012.049), p. 143 - 152 (1977).
Statistical studies of the photographic data on atmospheric trajectories and orbits of meteors (from 10^{-4}g to

hundreds of tons) point to 5 groups of bodies with different structure and composition.

104.055   Meteor streams in the making.   Z. Sekanina.
      Comets, asteroids, meteorites, (see 012.049),
p. 159 - 169 (1977).
      The well-known associations of meteor streams with periodic comets and the probable cometary origin of the zodiacal dust cloud point to the importance, in the cometary debris, of particles with masses exceeding roughly 10^{-6} gram. It is shown that these large particles dominate in the sunward-oriented anomalous tails of comets. Their study is essential for meaningful estimates of the mass of meteor streams and of the injection rate of the cometary debris that contributes to the zodiacal cloud.

**104.056   Quadrantid meteor shower: eleven years of radar
              observations.   B. A. McIntosh.**
Comets, asteroids, meteorites, (see 012.049), p. 171 - 174 (1977).

104.057   Fireball Brno.   Z. Ceplecha.
      Kozmos, Vol. 8, 175 - 176, 188 (1977). In Czech.
      Observation, geocentric and heliocentric orbit of a -17^m fireball observed on 14 September 1977 in Czechoslovakia.

**104.058   Fireball of -17^m observed on 14 September 1977
              in Czechoslovakia.**
Říše hvězd, Vol. 58, 213, 233 (1977). In Czech.
      Observation, geocentric and heliocentric orbit of the fireball "Brno".

104.059   Meteoroids.   Z. Ceplecha.
      Kozmos, Vol. 8, 74 - 77 (1977). In Czech.

      Meteorites.   See Abstr. 003.053.

      Meteor investigations. No. 4.
See Abstr. 003.176.

      Development of notions of meteor phenomena from ancient time to the 19th century.   See Abstr. 004.065.

      A discussion of the magnitude errors and magnitude scales of meteor observers in Sweden and Czechoslovakia. See Abstr. 013.026.

      The determination of meteor stream radiants from single station observations.   See Abstr. 031.247.

      Meteor radiant distribution using spherical harmonic analysis.   See Abstr. 031.248.

      The effective field of view for line sources (meteors). See Abstr. 031.270.

      Field tests of a meteor camera. See Abstr. 032.025.

      Automatic meteor radar system. See Abstr. 033.029.

      The meteoric night-glow.   See Abstr. 082.020.

      Radiowave scattering from meteoric ionization. See Abstr. 082.025.

      Dynamics of the lower thermosphere from meteor observations.   See Abstr. 082.139.

      Function of ion formation from meteor particles in the ionospheric E-region.   See Abstr. 083.049.

      Nocturnal E-region during meteor showers. See Abstr. 083.058.

      Magneto-conjugate phenomena and meteor activity. See Abstr. 084.290.

      Cometary debris.   See Abstr. 102.001.

      Surface phenomena in simulated cometary nuclei. See Abstr. 102.041.

      Comet Giacobini-Zinner and the Draconid meteor stream in 1978 - 1979.   See Abstr. 103.211.

105 Meteorites, Meteorite Craters

105.001 The metallic microstructures and thermal histories of severely reheated chondrites.
B. A. Smith, J. I. Goldstein.
Geochim. Cosmochim. Acta, Vol. 41, 1061 - 1072 (1977).

105.002 The composition of carbonaceous chondrite matrix.   H. Y. McSween, Jr., S. M. Richardson.
Geochim. Cosmochim. Acta, Vol. 41, 1145 - 1161 (1977).

105.003 Correlation between nickel and sulfur abundances in Orgueil phyllosilicates.   J. F. Kerridge.
Geochim. Cosmochim. Acta, Vol. 41, 1163 - 1164 (1977).

105.004 Molybdenite in calcium-aluminum-rich inclusions in the Allende meteorite.
L. H. Fuchs, M. Blander.
Geochim. Cosmochim. Acta, Vol. 41, 1170 - 1175 (1977).

105.005 The fine-grained structure of chondritic meteorites.
J. A. Wood.
The dusty universe, (see 012.001), p. 245 - 266 (1975).

Type I and II carbonaceous chondrites consist largely of fine-grained mixtures of types of minerals that would be thermodynamically stable in the solar nebula only at relatively low temperatures ($< 500°K$). It seems likely that the currently observable low-temperature minerals condensed directly from the vapor phase.

105.006 Dust in the solar nebula.   L. Grossman.
The dusty universe, (see 012.001), p. 267 - 292 (1975).

The chemical and mineralogical features of the white, Ca-rich inclusions in Allende and other carbonaceous chondrites are strikingly similar to those predicted from thermodynamic models for the highest temperature condensates from the solar nebula. Many of the physical and chemical properties of the chondritic minerals may thus be quite like those of the dust in interstellar regions. The oxygen isotopic composition of meteoritic condensates suggests that they contain a component of interstellar dust that survived the birth of the solar system.

105.007 Tunguska's comet and non-thermal ^{14}C production in the atmosphere.   J. C. Brown, D. W. Hughes.
Nature, Vol. 268, 512 - 514 (1977).

Hughes recently summarised the generally convincing case for the hypothesis, first put forward by Whipple and Astapovich, that a comet collided with the Earth's atmosphere. Opposition to this theory and consequent support for less conventional hypotheses involving black holes almost critical masses of extraterrestrial fissionable material, anti-matter bodies and alien spacecraft have been based substantially on the non-observance of the comet before impact, the explanation of height of the explosion above the Earth's surface, the composition of the glassy spherules found at Tunguska and finally, and of most importance, the apparent occurrence of nuclear phenomena in the explosion as indicated by the subsequent enhancement of radiocarbon in the atmosphere. The authors discuss the first and last of these points which are clearly the most important.

105.008 Cubanite: a new sulfide phase in C I meteorites.
J. D. Macdougall, J. F. Kerridge.
Science, Vol. 197, 561 - 562 (1977).

Cubanite ($CuFe_2S_3$), previously unobserved in meteorites, has been discovered in two carbonaceous chondrites, Orgueil and Alais. The association of this mineral with low-copper pyrrhotite suggests that it formed in a low-temperature environment on the meteorite parent body.

105.009 Thermal metamorphism of primitive meteorites—V. Ten trace elements in Tieschitz H3 chondrite heated at 400–1000°C.
M. Ikramuddin, S. Matza, M. E. Lipschutz.
Geochim. Cosmochim. Acta, Vol. 41, 1247 - 1256 (1977).

The authors determined ten trace elements by neutron activation analysis in Tieschitz (H3) chondrite powder heated in a low-pressure environment (initially $\sim 10^{-5}$ atm H_2) for 1 week at 100°C increments from 400–1000°C. They compared trace element abundances, patterns of statistically-significant correlations, factor analysis and two-element correlations between Tieschitz and heated Krymka (L3) and, except for factor analysis, "as-received" H3-6 chondrites. Sharp contrasts in pictures for E-, L- and H-group chondrites indicate substantial differences in genetic histories.

105.010 Composition and evolution of the eucrite parent body: evidence from rare earth elements.
G. J. Consolmagno, M. J. Drake.
Geochim. Cosmochim. Acta, Vol. 41, 1271 - 1282 (1977).

Eucrites are extraterrestrial plagioclase-pigeonite basalts. Quantitative modeling of the evolution of REE abundances in the eucrites indicates that the main group of eucrites (e.g. Juvinas) may be produced by approximately 10% equilibrium partial melting of a source region with initial REE abundances which were chondritic relative and absolute. The authors' calculations are consistent with the conclusion that the eucrites were derived from a single type of source region. The close correspondence of the age of the eucrites ($\simeq 4.6$ AE) to the age of the solar system appears to preclude the possibility of extensive chemical differentiation of the eucrite parent body prior to the event which produced the eucritic melts. Thus the calculations have yielded not only the mode of the source region but, assuming homogeneous accretion, the mode and hence the bulk composition of the eucrite parent body as well.

105.011 Origin of organic matter in the early solar system—VII. The organic polymer in carbonaceous chondrites.   R. Hayatsu, S. Matsuoka, R. G. Scott, M. H. Studier, E. Anders.
Geochim. Cosmochim. Acta, Vol. 41, 1325 - 1339 (1977).

The insoluble polymer from the Murchinson C2 chondrite was studied by a variety of degradation techniques. The studies confirm the prevailing view that the meteoritic polymer has a bridged aromatic structure with functional groups such as COOH, OH, and CO, but provides much new detail.

105.012 Volatile/mobile trace elements in Karoonda (C4) chondrite.
S. D. Matza, M. E. Lipschutz.
Geochim. Cosmochim. Acta, Vol. 41, 1398 - 1401 (1977).

Data for ten volatile/mobile trace elements and non-volatile Co in Karoonda (C4) and in heated primitive chondrites are consistent with the suggestion that Karoonda derives from low-temperature, open-system metamorphism of pristine C3-like material.

105.013 The role of volatiles in the impact process.
S. W. Kieffer.
Bull. American Astron. Soc., Vol. 9, 450 (1977). – Abstract.

105.014 Ar-40/Ar-39 dating of Plainview brecciated chondrite: evidence of regolith formation since 3.7 aeons B.P.   D. D. Bogard, L. Husain.
Bull. American Astron. Soc., Vol. 9, 458 (1977). – Abstract.

105.015 Composition of the parent body of eucritic meteorites.   J. Hertogen, J. Vizgirda, E. Anders.
Bull. American Astron. Soc., Vol. 9, 458 - 459 (1977).
Abstract.

105.016 Characterization of earth-crossing asteroids, past and present.   E. Anders.
Bull. American Astron. Soc., Vol. 9, 459 (1977). – Abstract.

105.017 Astrons – the Earth's oldest scars?
J. Norman, N. Price, M. Chukwu-Ike.
New Scientist, Vol. 73, 689 - 692 (1977). – Abstr. in Phys. Abstr., Vol. 80, Abstr. 56632 (1977).

105.018 Die Siljan-Ringstruktur – ein "Nördlinger Ries" in Mittelschweden.   H. Grabert, P. Thorslund.
Umschau, 77. Jahrg., 610 - 613 (1977).

105.019 A preliminary survey on the Kirin meteorite shower.
Joint Investigation Group of the Kirin Meteorite Shower, Academia Sinica.
Sci. Sinica, Vol. 20, 502 - 512 (1977).
Made in the present paper is a preliminary report on the phenomena of the fall of the Kirin meteorite shower, its distribution area and characteristics, the meteor's flight, and its mineral constituents, structure and chemical composition. The results of the studies have provided evidence that the Kirin meteorite does belong to ordinary H-group chondrites, or is referred to as olivine-bronzite chondrites.

105.020 Silicate inclusions in group IAB irons and a relation to the anomalous stones Winona and Mt. Morris (Wis).
R. W. Bild.
Geochim. Cosmochim. Acta, Vol. 41, 1439 - 1456 (1977).
Silicates are found in many group IAB irons. The mineralogy of the silicates is chondritic – olivine, pyroxene, albitic plagioclase – as is the bulk composition. IAB inclusions have ages of about 4.5 Gyr, $I^{129}-Xe^{129}$ formation intervals in the ranges of chondrites and contain planetary-type rare gases. Samples were examined by microprobe and bulk inclusions were analyzed by instrumental and radiochemical neutron activation analysis. The IAB silicates formed probably in a similar manner as chondrite groups but in a different region of the nebula. Two anomalous stony meteorites, Winona and Mt. Morris (Wis), are similar to IAB inclusions in mineralogy, bulk composition, FeO/(FeO + Mg) ratio of the silicates, and chromite composition and are possibly related to the IAB silicates.

105.021 Mg and Ca isotopic study of individual microscopic crystals from the Allende meteorite by the direct loading technique.
T. Lee, D. A. Papanastassiou, G. J. Wasserburg.
Geochim. Cosmochim. Acta, Vol. 41, 1473 - 1485 (1977) = Div. Geol. Planet. Sci., Calif. Inst. Technol., Pasadena, Contrib. No. 2873.
The authors have developed a direct loading technique for determining the isotopic composition of Mg in chemically un-separated samples. This technique has a sensitivity and precision comparable with those of the conventional technique of analyzing pure Mg salt and eliminates contamination intro-duced during the chemical separation of Mg. This results in a significant reduction in sample size required for an analysis. This technique was combined with other characterization techniques of microscopic samples and was applied to four single crystals of pure phases from an Allende inclusion ranging in size from 25 to 150μm. The isotopic composition of Ca was also measured along with Mg, on a directly loaded anorthite crystal from this inclusion.

105.022 Sr isotopes and trace element geochemistry of the impact melt and target rocks at the Mistastin Lake

crater, Labrador.   M. Marchand, J. H. Crocket.
Geochim. Cosmochim. Acta, Vol. 41, 1487 - 1495 (1977).
This paper examines the geochemical processes that occurred in the formation of the melt rocks at Mistastin Lake and shows that the trace element and isotopic data are con-sistent with their proposed formation by the complete melting of the local country rocks.

105.023 ^{26}Al in stony meteorites with gas losses.
G. F. Herzog.
Geochim.Cosmochim. Acta, Vol. 41, 1526 - 1529 (1977).
^{26}Al contents of 16 meteorites with ^{3}He losses have the same average values as all chondrites indicating that gas losses did not occur in a recent, prolonged exposure to unusual cosmic-ray activity.

105.024 High explosive analogue of the Tunguska event.
B. W. Augenstein, with a reply by G. H. S. Jones.
Nature, Vol. 269, 355 (1977).

105.025 Bunte breccia: continuous breccia deposits of the Ries crater, Germany.   F. Hörz, V. R. Oberbeck.
NASA Tech. Mem., NASA TM X-3511, (see 012.010), p. 119 - 121 (1977). – Abstract.

105.026 Isotopic composition of uranium in chondritic meteorites.   J. W. Arden.
Nature, Vol. 269, 788 - 789 (1977).
Measurements of the isotopic composition of U in five meteorites are reported. The data show that variations in the U isotopic composition exist in these chondrites, and indicate the presence of components of low $^{238}U/^{235}U$ ratio.

105.027 Anomalous processes in the solar system in 1971 – 1974: data on radioactivity of the just fallen meteorite Gorlovka. (Annotation).
A. K. Lavrukhina, V. D. Gorin, G. K. Ustinova.
VIII Leningr. mezhdunar. semin. Mater. mezhdunar. semin. "Akt. protsessy na Solntse i probl. soln. nejtrino", 1976. Leningrad, 1976, p. 205 - 207. In Russian and English.
Abstr. in Ref. zh., 51. Astron., 9.51.353 (1977).

105.028 Detection of a new spontaneously fissioning nuclide in some meteorites.   G. N. Flerov, G. M. Ter-Akop'yan, A. G. Popeko, B. V. Fefilov, V. G. Subbotin.
Obedin. inst. yader. issled. Prepr. R6-10581. Dubna, 1977. 11 pp. In Russian. – Abstr. in Ref. zh., 51. Astron., 9.51.357 (1977).

105.029 Trace element distribution in mineral separates of the Allende inclusions and their genetic implications.
H. Nagasawa, D. P. Blanchard, J. W. Jacobs, J. C. Brannon, J. A. Philpotts, N. Onuma.
Geochim. Cosmochim. Acta, Vol. 41, 1587 - 1600 (1977) = Lunar Sci. Inst., Houston, Tex., Contrib. No. 227.
Concentrations of the REE, Sc, Co, Fe, Zn, Ir, Na and Cr were determined by instrumental neutron activation and mass spectrometric isotope dilution analysis for mineral separates of the coarse- and fine-grained types (group I and II of Martin and Mason's classification) of the Allende inclu-sions.

105.030 Trace elements in the Allende meteorite – III. Coarse-grained inclusions revisited.
L. Grossman, R. Ganapathy, A. M. Davis.
Geochim. Cosmochim. Acta, Vol. 41, 1647 - 1664 (1977).
This paper adds new RNAA data for Ba, Sr, Zr, U, Re, Pd, Ag, Zn and Se and INAA data for Lu in coarse-grained inclusions to existing analyses by Grossman and Ganapathy (1975 (17.105.011), 1976 (18.105.006)) for 21 other ele-ments in the same suite of samples.

105.031 A ^{33}S anomaly in the Allende meteorite.
C. E. Rees, H. G. Thode.
Geochim. Cosmochim. Acta, Vol. 41, 1679 - 1682 (1977).
The isotope ratios ^{33}S/^{32}S and ^{34}S/^{32}S have been measured in sulphur fractions extracted from samples of the meteorites Allende and Eagle Station by leaching at successively greater acid concentrations and higher temperatures.

105.032 Metal spherules in Wabar, Monturaqui, and Henbury impactites.
R. V. Gibbons, F. Hörz, T. D. Thompson, D. E. Brownlee.
Proc. Seventh Lunar Sci. Conf., (see 012.015), 863 - 880 (1976).

105.033 Shocked basalt from Lonar impact crater, India, and experimental analogues.   S. W. Kieffer,
R. B. Schaal, R. Gibbons, F. Hörz, D. J. Milton, A. Dube.
Proc. Seventh Lunar Sci. Conf., (see 012.015), 1391 - 1412 (1976).

105.034 Thermal model for impact breccia lithification: Manicouagan and the moon.   C. H. Simonds,
J. L. Warner, W. C. Phinney, P. E. McGee.
Proc. Seventh Lunar Sci. Conf., (see 012.015), 2509 - 2528 (1976).

105.035 Morphology of the Manicouagan ring-structure, Quebec, and some comparisons with lunar basins and craters.   R. J. Floran, M. R. Dence.
Proc. Seventh Lunar Sci. Conf., (see 012.015), 2845 - 2865 (1976).

105.036 The effect of gravity on crater formation: thickness of ejecta and concentric basins.   B. A. Ivanov.
Proc. Seventh Lunar Sci. Conf., (see 012.015), 2947 - 2965 (1976).

105.037 Trace element evidence regarding a chondritic component in howardite meteorites.
C.-L. Chou, W. V. Boynton, R. W. Bild, J. Kimberlin, J. T. Wasson.
Proc. Seventh Lunar Sci. Conf., (see 012.015), 3501 - 3518 (1976).

105.038 Thermal metamorphism of primitive meteorites – IV. Comparison with trends for ten trace elements in terrestrial basalt BCR-1 heated at 500–1000°C.
M. Ikramuddin, C. M. Binz, M. E. Lipschutz.
Proc. Seventh Lunar Sci. Conf., (see 012.015), 3519 - 3533 (1976).

105.039 Carbonaceous chondritic xenoliths and planetary-type noble gases in gas-rich meteorites.
L. L. Wilkening.
Proc. Seventh Lunar Sci. Conf., (see 012.015), 3549 - 3559 (1976).

105.040 Some aspects of the origin of meteorites.
B. Yu. Levin, A. N. Simonenko.
Meteoritika, vyp. (No.) 36, p. 3 - 23 (1977). In Russian.

105.041 Some problems of investigation of the chemical composition of meteorites.   A. A. Yavnel'.
Meteoritika, vyp. (No.) 36, p. 24 - 30 (1977). In Russian.

105.042 On the mineralogical composition and structure of meteorites.   L. G. Kvasha.
Meteoritika, vyp. (No.) 36, p. 31 - 39, 173, 175 - 176 (1977). In Russian.

105.043 Morphological investigation of the Gorlovka stony meteorite.   E. L. Krinov.

Meteoritika, vyp. (No.) 36, p. 40 - 45, 173, 177 - 178 (1977). In Russian.

105.044 Investigation of the Gorlovka chondrite.
O. A. Kirova, M. I. D'yakonova, V. Ya. Kharitonova, L. K. Levskij.
Meteoritika, vyp. (No.) 36, p. 46 - 52, 173, 179 - 181 (1977). In Russian.

105.045 Investigation of the Egvekinot iron meteorite.
N. I. Zaslavskaya, M. I. D'yakonova, V. Ya. Kharitonova, R. L. Khotinok.
Meteoritika, vyp. (No.) 36, p. 53 - 58, 174, 182 - 183 (1977). In Russian.

105.046 On the composition and structure of the Verkhne-Chirskaia meteorite.   A. Ya. Skripnik,
M. I. D'yakonova, V. Ya. Kharitonova.
Meteoritika, vyp. (No.) 36, p. 59 - 65, 174, 184 - 185 (1977). In Russian.

105.047 Method for local determination of lithium isotopes in meteorites.   M. Sh. Kaviladze, T. A. Melashvili, L. D. Sepiashvili.
Meteoritika, vyp. (No.) 36, p. 66 - 70 (1977). In Russian.

105.048 Isotopes of rare gases in the carbonaceous chondrite Allende.   L. K. Levskij.
Meteoritika, vyp. (No.) 36, p. 71 - 74 (1977). In Russian.

105.049 Influence of cosmic ray variations upon the estimate of size and age of meteorites from cosmogenic isotopes.   V. A. Alekseev, G. K. Ustinova.
Meteoritika, vyp. (No.) 36, p. 75 - 81 (1977). In Russian.

105.050 On the determination of the pre-atmospheric size of the meteorite Marjalahti.   E. M. Kolesnikov,
O. Otgonsurehn, V. P. Perelygin, A. V. Fisenko.
Meteoritika, vyp. (No.) 36, p. 82 - 86 (1977). In Russian.

105.051 Separation of fractions of chondrules and minerals from chondrites (experience of fractionation of meteorite Saratov).   A. I. Berlinskij, L. V. Razin, G. I. Kozlova, T. Yu. Marenina, S. F. Sluzhenikin.
Meteoritika, vyp. (No.) 36, p. 87 - 90 (1977). In Russian.

105.052 On troilite of chondrites.   V. P. Semenenko,
N. S. Stetsenko, V. S. Mel'nikov.
Meteoritika, vyp. (No.) 36, p. 91 - 97 (1977). In Russian.

105.053 On the composition of nickel iron particles in chondrites.   A. A. Yavnel', V. M. Grigor'eva,
I. E. Gorbunova, G. D. Petrova.
Meteoritika, vyp. (No.) 36, p. 98 - 105 (1977). In Russian.

105.054 Investigation of porosity of the Saratov and Nikol'skoe chondrites.
V. N. Savel'ev, V. D. Kolomenskij, I. A. Yudin.
Meteoritika, vyp. (No.) 36, p. 106 - 109 (1977). In Russian.

105.055 On magnetization of olivines from pallasites.
E. G. Gus'kova, V. D. Kolomenskij.
Meteoritika, vyp. (No.) 36, p. 110 - 112 (1977). In Russian.

105.056 Peculiarities of the gravitational field of astroblemes.
A. I. Dabizha, V. V. Fedynskij.
Meteoritika, vyp. (No.) 36, p. 113 - 119 (1977). In Russian.

105.057 First find of tektites in the USSR (Zhamanshin meteorite crater, northern Aral Sea area).
P. V. Florenskij.
Meteoritika, vyp. (No.) 36, p. 120 - 122 (1977). In Russian.

105.058   On the origin of the Karskaya depression.
M. A. Maslov.
Meteoritika, vyp. (No.) 36, p. 123 - 130 (1977). In Russian.

105.059   Petrographic types of tagamites of the Popigai
astrobleme.   T. V. Selivanovskaya.
Meteoritika, vyp. (No.) 36, p. 131 - 134, 174, 186 (1977). In
Russian.

105.060   Suevites of the Nördlinger Ries crater and its ana-
logues in the Popigaian meteorite crater.
T. V. Selivanovskaya.
Meteoritika, vyp. (No.) 36, p. 135 - 139, 174, 187 (1977). In
Russian.

105.061   Petrochemical comparison of impactites of the
Popigai crater and its crystalline bed rocks.
A. I. Rajkhlin, M. S. Mashchak.
Meteoritika, vyp. (No.) 36, p. 140 - 145 (1977). In Russian.

105.062   New data on the Okhansk meteorite fall.
A. K. Stanyukovich, V. V. Chichmar'.
Meteoritika, vyp. (No.) 36, p. 146 - 147 (1977). In Russian.

105.063   Tracks of primitive irradiation in meteoritic matter
and their interpretation.
A. K. Lavrukhina, L. L. Kashkarov.
Meteoritika, vyp. (No.) 36, p. 148 - 158, 174, 188 (1977). In
Russian.

105.064   Addendum to the catalogue of meteorite collections
of the Lithuanian SSR.   V. A. Vasil'ev.
Meteoritika, vyp. (No.) 36, p. 159 - 161 (1977). In Russian.

105.065   Classification of iron meteorites Egvekinot and
Tobychan from nickel and trace element abundances.
A. A. Yavnel', G. M. Kolesov.
Meteoritika, vyp. (No.) 36, p. 162 - 164 (1977). In Russian.

105.066   Astronomical conditions at the moments of fall,
radiants and orbits of the Allende and Gorlovka
meteorites.   A. N. Simonenko.
Meteoritika, vyp. (No.) 36, p. 165 - 170 (1977). In Russian.

105.067   Fall of meteorite Innisfree photographed.
A. N. Simonenko.
Zemlya i Vselennaya, 1977, No. 5, p. 50 - 51. In Russian.

105.068   The isotopic and elemental abundance of ytterbium
in meteorites and terrestrial samples.
M. T. McCulloch, K. J. R. Rosman, J. R. De Laeter.
Geochim. Cosmochim. Acta, Vol. 41, 1703 - 1707 (1977).
    The isotopic composition of ytterbium in meteoritic and
terrestrial materials was measured using a solid source mass
spectrometer. Isotopic abundances of meteoritic and terrestrial
Yb agree within experimental errors.

105.069   Petrographic variations among carbonaceous chondrites of the Vigarano type.   H. Y. McSween, Jr.
Geochim. Cosmochim. Acta, Vol. 41, 1777 - 1790 (1977).
    The Vigarano subtype is a petrographically complex class
of meteorites. Oxidized and reduced groups can be distinguished
on the basis of metal vs magnetite abundances and Ni contents
of sulfide minerals. These meteorites also differ in the proportions of matrix and chondrules and in polymict character.
Slight bulk chemical differences correlate with the recognized
petrologic groupings. Because of metamorphic effects in the
Allende chondrite and the petrographic differences among all
meteorites of the Vigarano subtype, it is suggested that
Allende alone may not adequately reflect the wide spectrum of
properties in this important class of meteorites.

105.070   Chemical and petrographic constraints on the origin
of chondrules and inclusions in carbonaceous
chondrites.   H. Y. McSween, Jr.
Geochim. Cosmochim. Acta, Vol. 41, 1843 - 1860 (1977).
    Bulk chemical compositions of the various petrographic
types of chondrules and inclusions in Type 3 carbonaceous
chondrites (excluding those affected by metamorphism) have
been determined by microprobe defocused beam analysis. A
genetic relationship between chondrules and inclusions in
carbonaceous chondrites is suggested by the compositional
continuum between thse objects. A condensation sequence
which dips into the liquid stability field at lower temperatures
is advocated for the production of both inclusions and chondrules. Textural relationships between intergrown chondrules
and inclusions support such a sequence.

105.071   Analytical electron microscopy study of the plessite
structure in the Carlton iron meteorite.
L. S. Lin, J. I. Goldstein, D. B. Williams.
Geochim. Cosmochim. Acta, Vol. 41, 1861 - 1874 (1977).
    In order to gain a better understanding of the formation
of plessite in iron meteorites, various electron optical techniques were employed to study the range of plessite structures
observed in the Carlton fine octahedrite. The authors show that
the formation of plessite is intimately related to the formation
of martensite and the further decomposition of martensite into kamacite plus taenite during the cooling history of the
meteorite. This experimental result is in agreement with the
speculations of Massalski et al. (1966) on the importance of
plessite.

105.072   The Mundrabilla meteorite.   G. J. H. McCall.
Modern astronomy, (see 003.013), p. 95 - 105 (1977).

105.073   Analysis of elements in samples by means of laser
microprobe spectroscopy.
B. Baldanza, G. Cocco, F. Matteucci.
Atti Soc. Peloritana Sci. Fis. Mat. Nat., Vol. 20, 155 - 200
(1974). In Italian. — Abstr. in Phys. Abstr., Vol. 80, Abstr.
25323 (1977).

105.074   Taking into account atmospheric inhomogeneity in
calculations of the Tunguska meteorite blast.
V. P. Korobejnikov, P. I. Chushkin, L. V. Shurshalov.
Zh. vychisl. mat. i mat. fiz., Vol. 17, 737 - 752 (1977). In
Russian. — Abstr. in Ref. zh., 51. Astron., 10.51.495 (1977).

105.075   Correlated oxygen and magnesium isotope anomalies
in Allende inclusions. I. Oxygen.
R. N. Clayton, T. K. Mayeda.
Geophys. Res. Lett., Vol. 4, 295 - 298 (1977).
    Two Ca-Al-rich inclusions from the Allende meteorite have
been found to have undergone large mass-fractionation of
oxygen isotopes subsequent to incorporation of the nucleo-
synthetic ^{16}O-anomaly found in other Allende inclusions. The
magnitude of the oxygen isotope fractionations is in constant
ratio to the magnesium isotope fractionations found in the same
inclusions by Wasserburg, Lee and Papanastassiou. The observations support earlier theories of the addition of supernova
ejecta into the solar nebula just prior to collapse and condensation.

105.076   Correlated O and Mg isotopic anomalies in Allende
inclusions. II. Magnesium.
G. J. Wasserburg, T. Lee, D. A. Papanastassiou.
Geophys. Res. Lett., Vol. 4, 299 - 302 (1977).
    Mg in two Allende Ca-Al rich inclusions shows large iso-
topic, mass-dependent fractionation which enriched the heavier
isotopes. After normalization, Mg in these inclusions shows
negative δ ^{26}Mg which appears to require the presence of nuclear
effects in Mg distinct from ^{26}Al decay. The Mg mass fractionation is correlated with distinct but smaller fractionation effects

for O reported by Clayton and Mayeda for the same inclusions.

105.077 Dhajala meteorite.
N. Bhandari, G. M. Ballabh.
Bull. Astron. Soc. India, Vol. 5, 69 - 72 (1977).

105.078 Studies of fresh and fossil tracks in meteoritic hypersthene. R. K. Bull, S. A. Durrani.
Nucl. Track Detect., Vol. 1, No. 1, p. 75 - 80 (1977). – Abstr. in Phys. Abstr., Vol. 80, Abstr. 41276 (1977).

105.079 Application of a multielement neutron activation scheme for the determination of 38 elements in the Allende meteorite.
A. O. Brunfelt, B. Sundvoll, E. Steinnes.
Radiochem. Radioanal. Lett., Vol. 28, No. 2, p. 181 - 187 (1977). – Abstr. in Phys. Abstr., Vol. 80, Abstr. 48739 (1977).

105.080 Analysis of meteoritic samples using proton-induced X-rays. H. W. Kugel, G. F. Herzog.
Nucl. Instrum. Methods, Vol. 142, No. 1 - 2, p. 301 - 305 (1977). – Abstr. in Phys. Abstr., Vol. 80, Abstr. 49344 (1977).

105.081 Allende meteorite: isotopically anomalous xenon is accompanied by normal osmium.
H. Takahashi, H. Higuchi, J. Gros, J. W. Morgan, E. Anders.
Proc. Natl. Acad. Sci. USA, Vol. 73, 4253 - 4256 (1976). Abstr. in Phys. Abstr., Vol. 80, Abstr. 45368 (1977).

105.082 A contribution to the exploration of the structure of the Ries crater on the basis of geoelectric Schlumberger soundings. L. Engelhard, J. Hansel.
Abh. Braunschweig. Wiss. Ges., Vol. 26, 23 - 41 (1976). In German. – Abstr. in Phys. Abstr., Vol. 80, Abstr. 64207 (1977).

105.083 Acid-labile amino acid precursors in the Murchison meteorite. I. Chromatographic fractionation.
J. R. Cronin.
Origins of Life, Vol. 7, 337 - 342 (1976). – Abstr. in Phys. Abstr., Vol. 80, Abstr. 64707 (1977).

105.084 Acid-labile amino acid precursors in the Murchison meteorite. II. A search for peptides and amino acyl amides. J. R. Cronin.
Origins of Life, Vol. 7, 343 - 348 (1976). – Abstr. in Phys. Abstr., Vol. 80, Abstr. 64708 (1977).

105.085 The origin of tektites: a brief review. E. A. King.
American Sci., Vol. 65, 212 - 218 (1977). – Abstr. in Phys. Abstr., Vol. 80, Abstr. 67078 (1977).

105.086 Mössbauer spectroscopy of an ordered phase (superstructure) of FeNi in an iron meteorite.
J. F. Petersen, M. Aydin, J. M. Knudsen.
Phys. Lett. A, Vol. 62A, 192 - 194 (1977). – Abstr. in Phys. Abstr., Vol. 80, Abstr. 78727 (1977).

105.087 On the flux of low-energy particles in the solar system: the record in St. Severin meteorite.
D. Lal, K. Marti.
Nucl. Track Detect., Vol. 1, No. 2, p. 127 - 130 (1977). Abstr. in Phys. Abstr., Vol. 80, Abstr. 82096 (1977).

105.088 The Dhajala meteorite shower: atmospheric fragmentation and ablation based on cosmic ray track studies.
C. Bagolia, N. Doshi, S. K. Gupta, S. Kumar, D. Lal, J. R. Trivedi.
Nucl. Track Detect., Vol. 1, No. 2, p. 83 - 92 (1977). – Abstr. in Phys. Abstr., Vol. 80, Abstr. 82188 (1977).

105.089 Fallen stars by the tonne. K. Hindley.

New Scientist, Vol. 75, 20 - 22 (1977). – Abstr. in Phys. Abstr., Vol. 80, Abstr. 87347 (1977).

105.090 The mineralogy of iron meteorites.
V. F. Buchwald.
Philos. Trans. R. Soc. London, Ser. A, Vol. 286, 453 - 491 (1977). – Abstr. in Phys. Abstr., Vol. 80, Abstr. 87348 (1977).

105.091 Electron microscopy of some stony meteorites.
J. R. Ashworth, D. J. Barber.
Philos. Trans. R. Soc. London, Ser. A, Vol. 286, 493 - 506 (1977). – Abstr. in Phys. Abstr., Vol. 80, Abstr. 87349 (1977).

105.092 A note on the norm calculation of stony meteorites.
A. Okada, H. Yabuki, M. Shima, K. Yagi.
Sci. Pap. Inst. Phys. Chem. Res., Vol. 71, No. 2, p. 45 - 47 (1977). – Abstr. in Phys. Abstr., Vol. 80, Abstr. 87350 (1977).

105.093 Antarctica: a deep-freeze storehouse for meteorites.
W. A. Cassidy, E. Olsen, K. Yanai.
Science, Vol. 198, 727 - 731 (1977).
Meteorites that fall on the Antarctic ice cap are preserved for long periods of time under very clean conditions as they are carried toward the continental margin. If the host ice encounters a barrier it cannot flow over or around, it tends to dissipate by ablation, leaving an accumulation of meteorites on the surface.

105.094 Primordial noble gases in chondrites: the abundance pattern was established in the solar nebula.
L. Alaerts, R. S. Lewis, E. Anders.
Science, Vol. 198, 927 - 930 (1977).
Ordinary chondrites, like carbonaceous chondrites, contain primordial noble gases mainly in a minor phase comprising $\leqslant 0.05$ percent of the meteorite, probably an iron-chromium sulfide. The neon-20/argon-36 ratios decrease with increasing argon-36 concentration, as expected if the gas pattern was established by condensation from the solar nebula, and was negligibly altered by metamorphism in the meteorite parent bodies.

105.095 Extinct radioactive nucleus ^{202}Pb: possible origin of isotopic anomalies of Hg in meteorites.
S. Yanagita, R. Gensho.
Astrophys. Space Sci., Vol. 46, L1 - L6 (1977).
The abundance of extinct radioactive nucleus ^{202}Pb trapped in chondrites is calculated from the anomalous isotopic ratio ^{202}Hg/^{196}Hg, assuming the isotopic anomalies are only due to the decay of ^{202}Pb. The magnitude of nonuniformity of the distribution of ^{202}Pb inferred from the excess isotopic abundance of ^{202}Hg is the same order of magnitude suggested by anomalous isotopic abundances of the other chemical elements found in meteoritic materials.

105.096 On the origin of fission fragment tracks in whitlockite from the Bjurböle meteorite.
V. P. Perelygin. S. G. Stetsenko, N. Bhandari.
At. ehnerg., Vol. 42, 482 - 485, 514, 517 (1977). In Russian. – Abstr. in Ref. zh., 51. Astron., 11.51.480 (1977).

105.097 The nature of central and ring-shaped uplifts in meteoritic craters on the earth. G. V. Skrynnik.
Geol. zh., Vol. 37, No. 3, p. 147 - 152 (1977). In Russian. – Abstr. in Ref. zh., 51. Astron., 11.51.528 (1977).

105.098 The comet that hit the earth. I. Ridpath.
Mercury, Vol. 6, No. 5, p. 2 - 5 (1977).
The author examines the Tunguska event and sorts out the facts from the speculation.

105.099 Cosmoradiogenic ghosts and the origin of Ca–Al-rich inclusions. D. D. Clayton.

Earth Planet. Sci. Lett., Vol. 35, 398 - 410 (1977).

105.100 **Nebular condensation of moderately volatile elements and their abundances in ordinary chondrites.** C. M. Wai, J. T. Wasson.
Earth Planet. Sci. Lett., Vol. 36, 1 - 13 (1977).
 With a critique of E. Anders (p. 14 - 20) and a reply to Anders by J. T. Wasson (p. 21 - 28).

105.101 **Argon, krypton and xenon in iron meteorites.** E. W. Hennecke, O. K. Manuel.
Earth Planet. Sci. Lett., Vol. 36, 29 - 43 (1977).
 The isotopic compositions of Ar, Kr and Xe trapped in iron meteorites appear to be a complementary component to the unusual noble gas component found in carbon-rich residues of stone meteorites. The isotopic compositions of noble gases in the earth, the moon, the sun and other classes of meteorites may represent mixtures of these two planetary components.

105.102 **Cosmogenic and radiogenic noble gases in the Dhajala chondrite.** K. Gopalan, M. N. Rao, K. M. Suthar, T. R. Venkatesan.
Earth Planet. Sci. Lett., Vol. 36, 341 - 346 (1977).

105.103 **Trace element content of metals from H- and LL-group chondrites.** E. R. Rambaldi.
Earth Planet. Sci. Lett., Vol. 36, 347 - 358 (1977).

105.104 **Noble gases in the St. Mesmin chondrite: implications to the irradiation history of a brecciated meteorite.** L. Schultz, P. Signer.
Earth Planet. Sci. Lett., Vol. 36, 363 - 371 (1977).

105.105 **Tungsten in ordinary chondrites.** E. R. Rambaldi, M. Cendales.
Earth Planet. Sci. Lett., Vol. 36, 372 - 380 (1977).

105.106 **History of the Pasamonte achondrite: relative susceptibility of the Sm–Nd, Rb–Sr, and U–Pb systems to metamorphic events.** D. M. Unruh, N. Nakamura, M. Tatsumoto.
Earth Planet. Sci. Lett., Vol. 37, 1 - 12 (1977).

105.107 **Composition, mineralogy and origin of group IC iron meteorites.** E. R. D. Scott.
Earth Planet. Sci. Lett., Vol. 37, 273 - 284 (1977).

105.108 **Search for fission tracks from superheavy elements in Allende.** P. Fraundorf, G. J. Flynn, J. R. Shirck, R. M. Walker.
Earth Planet. Sci. Lett., Vol. 37, 285 - 295 (1977).

105.109 **Electron microscopy of some H-group chondrites.** J. R. Ashworth.
Meteoritics, Vol. 12, 168 - 169 (1977). – Abstract.

105.110 **The basin of Lago Tremorgio (Canton Ticino) as a possible quarternary meteorite impact crater in the Swiss Alps.** K. Bächtiger.
Meteoritics, Vol. 12, 169 - 171 (1977). – Abstract.

105.111 **Ar^{39}-Ar^{40} ages of achondrites.** A. Balacescu, H. Wänke.
Meteoritics, Vol. 12, 171 - 172 (1977). – Abstract.

105.112 **The matrix of C2 and C3 carbonaceous chondrites.** D. J. Barber.
Meteoritics, Vol. 12, 172 - 173 (1977). – Abstract.

105.113 **Some trace element retentivity studies in heated primitive chondrites.**
G. Bart, M. Ikramuddin, M. E. Lipschutz.
Meteoritics, Vol. 12, 173 - 174 (1977). – Abstract.

105.114 **Biography of an agglutinate.** A. Basu.
Meteoritics, Vol. 12, 174 - 175 (1977). – Abstract.

105.115 **High voltage electron microscope search for the sites of isotopic anomalies in meteorites.**
J. P. Bibring, Y. Langevin, M. Maurette, J. M. Uro, M. Christophe, P. Eberhardt.
Meteoritics, Vol. 12, 176 (1977). – Abstract.

105.116 **Compositions of silicate inclusions as an aid in the classification of iron meteorites and a tentative classification of Britstown.** R. W. Bild.
Meteoritics, Vol. 12, 177 (1977). – Abstract.

105.117 **Xenoliths in the chondrite Breitscheid including an unusual achondrite and a possible "meteorite within a meteorite".** R. A. Binns, F. Wlotzka.
Meteoritics, Vol. 12, 177 - 178 (1977). – Abstract.

105.118 **Mulga West, a metamorphosed carbonaceous chondrite.** R. A. Binns, W. H. Cleverley, G. J. H. McCall, S. J. B. Reed, J. H. Scoon.
Meteoritics, Vol. 12, 179 (1977). – Abstract.

105.119 **Potassium isotopic determination in some meteoritic and lunar samples: evidence for irradiation effects.** J.-L. Birck, J.-C. Lorin, C. Allègre.
Meteoritics, Vol. 12, 179 - 180 (1977). – Abstract.

105.120 **Non-equilibrium effects in the formation of chondrites.** M. Blander.
Meteoritics, Vol. 12, 181 - 182 (1977). – Abstract.

105.121 **^{40}Ar-^{39}Ar dating of Scandinavian impact craters.** R. J. Bottomley, D. York, R. A. F. Grieve.
Meteoritics, Vol. 12, 182 - 183 (1977). – Abstract.

105.122 **A brief review of the chemistry of Ca, Al-rich inclusions in the Allende meteorite.** W. V. Boynton.
Meteoritics, Vol. 12, 183 - 184 (1977). – Abstract.

105.123 **An investigation of magnetic correlates of metamorphic grade and shock level in L- and H-chondrites.**
A. Brecher, M. Fuhrman, L. Albright.
Meteoritics, Vol. 12, 185 (1977). – Abstract.

105.124 **Meteorites of igneous origin and their genetic relationships: a review.** R. Brett.
Meteoritics, Vol. 12, 185 - 187 (1977). – Abstract.

105.125 **Iron meteorites.** V. F. Buchwald.
Meteoritics, Vol. 12, 187 (1977). – Abstract.

105.126 **Observations on pallasites' pasts.** P. R. Buseck.
Meteoritics, Vol. 12, 187 - 189 (1977). – Abstract.

105.127 **Meteoritic minerals as detectors of heavy cosmic ray particles.** R. K. Bull, P. F. Green, S. A. Durrani.
Meteoritics, Vol. 12, 189 - 190 (1977). – Abstract.

105.128 **Meteorite finds near McMurdo Base, Antarctica.** W. A. Cassidy, E. Olsen, K. Yanai.
Meteoritics, Vol. 12, 190 (1977). – Abstract.

105.129 **Occurrence of chromiferous sulfides and oxides in the Allende chondrite.**
B. Cervelle, M. Christophe Michel-Lévy, C. Desnoyers.
Meteoritics, Vol. 12, 191 (1977). – Abstract.

105.130 **Lead isotopic studies of the Dhajala H3 chondrite.**
J. H. Chen, G. R. Tilton.
Meteoritics, Vol. 12, 193 - 194 (1977). − Abstract.

105.131 **SEM observations on H group chondrites.**
M. Christophe Michel-Lévy.
Meteoritics, Vol. 12, 194 (1977). − Abstract.

105.132 **Origin of Ca-Al-rich inclusions in Allende.**
D. D. Clayton.
Meteoritics, Vol. 12, 197 - 199 (1977). − Abstract.

105.133 **Oxygen isotopic compositions of separated fractions of the Leoville and Renazzo carbonaceous**
chondrites.   R. N. Clayton, T. K. Mayeda.
Meteoritics, Vol. 12, 199 (1977). − Abstract.

105.134 **Uranium microdistributions in stony meteorites and pallasites.**   G. Crozaz.
Meteoritics, Vol. 12, 200 - 201 (1977). − Abstract.

105.135 **Condensation of rare earths.**
A. M. Davis, L. Grossman.
Meteoritics, Vol. 12, 203 - 204 (1977). − Abstract.

105.136 **Terrestrial impact structures: the Canadian contribution.**
M. R. Dence, R. A. F. Grieve, P. B. Robertson, M. D. Thomas.
Meteoritics, Vol. 12, 204 - 205 (1977). − Abstract.

105.137 **The silicates in the Niger I meteorite (C2).**
C. Desnoyers, M. Christophe-Michel-Lévy.
Meteoritics, Vol. 12, 205 (1977). − Abstract.

105.138 **Elgegytgyn crater: source of Australasian tektites (and bediasites from Popigai).**   R. S. Dietz.
Meteoritics, Vol. 12, 205 - 206 (1977). − Abstract.

105.139 **Shock and thermal transformations in meteorites from the Morasko crater field.**   B. Dominik.
Meteoritics, Vol. 12, 207 - 208 (1977). − Abstract.

105.140 **Noble gases and fossil tracks in Shupiyan chondrite.**
N. Doshi, J. N. Goswami, D. Lal, M. N. Rao, T. R. Venkatesan.
Meteoritics, Vol. 12, 208 (1977). − Abstract.

105.141 **The eucrite parent body: structure and composition.**
G. Dreibus, H. Kruse, B. Spettel, H. Wänke.
Meteoritics, Vol. 12, 208 - 209 (1977). − Abstract.

105.142 **The validity of Frost's rule as applied to impact points of Přibram meteorite fragments.**
A. Drożyner, B. Lang.
Meteoritics, Vol. 12, 209 - 210 (1977). − Abstract.

105.143 **Annealing and etching studies of tracks in meteoritic and analogous crystals.**
S. A. Durrani, R. K. Bull, I. W. Davie, P. F. Green.
Meteoritics, Vol. 12, 210 - 211 (1977). − Abstract.

105.144 **Zirconium-hafnium fractionation in achondrites.**
W. D. Ehmann, A. N. Garg.
Meteoritics, Vol. 12, 211 - 212 (1977). − Abstract.

105.145 **Zhamanshin crater glasses: chemical composition and comparison with tektites.**
W. D. Ehmann, W. B. Stroube, Jr., M. Z. Ali, T. I. M. Hossain.
Meteoritics, Vol. 12, 212 - 215 (1977). − Abstract.

105.146 **Fremdlinge: potential presolar material in Ca-Al-rich inclusions of Allende.**

A. El Goresy, K. Nagel, B. Dominik, P. Ramdohr.
Meteoritics, Vol. 12, 215 - 216 (1977). − Abstract.

105.147 **Type A Ca-, Al-rich inclusions in Allende meteorite: origin of the perovskite-fassaite symplectite around rhönite and chemistry and assemblages of the refractory metals (Mo, W) and platinum metals (Ru, Os, Ir, Re, Rh, Pt).**
A. El Goresy, K. Nagel, P. Ramdohr.
Meteoritics, Vol. 12, 216 (1977). − Abstract.

105.148 **History and size of chondrite parent bodies from $^{40}Ar/^{39}Ar$ ages.**   M. C. Enright, G. Turner.
Meteoritics, Vol. 12, 217 (1977). − Abstract.

105.149 **Rare earth abundances in chondrites.**
N. M. Evensen, P. J. Hamilton, R. K. O'Nions.
Meteoritics, Vol. 12, 217 - 218 (1977). − Abstract.

105.150 **Excess ^{4}He in achondrites and irons?**
D. E. Fisher.
Meteoritics, Vol. 12, 218 - 219 (1977). − Abstract.

105.151 **Metallography of some Yamato iron meteorites.**
R. M. Fisher, J. I. Goldstein, T. Nagata.
Meteoritics, Vol. 12, 219 (1977). − Abstract.

105.152 **Chassigny revisited: a cumulate dunite with hydrous amphibole-bearing melt inclusions.**
R. J. Floran, M. Prinz, P. F. Hlava, K. Keil, C. E. Nehru, J. R. Hinthorne.
Meteoritics, Vol. 12, 225 - 226 (1977). − Abstract.

105.153 **The Johnstown orthopyroxenite (diogenite) and its relationship to meteoritic cumulates.**
R. J. Floran, M. Prinz, P. F. Hlava, K. Keil, B. Spettel, H. Wänke.
Meteoritics, Vol. 12, 226 - 227 (1977). − Abstract.

105.154 **The Zhamanshin structure: geology and petrography.**   P. V. Florensky (*Florenskij*), N. Short, S. R. Winzer, K. Fredriksson.
Meteoritics, Vol. 12, 227 - 228 (1977). − Abstract.

105.155 **The Zhamanshin structure: chemical and physical properties of selected samples.**
K. Fredriksson, A. deGasparis, W. Ehmann.
Meteoritics, Vol. 12, 229 - 231 (1977). − Abstract.

105.156 **Allende dark inclusions.**
R. M. Fruland, U. S. Clanton, W. J. A. Walton.
Meteoritics, Vol. 12, 231 - 232 (1977). − Abstract.

105.157 **Similar megascopic structures of Muong Nong-type tektites and extruded terrestrial volcanic glass.**
D. S. Futrell.
Meteoritics, Vol. 12, 232 - 235 (1977). − Abstract.

105.158 **Carriers of trapped gases in ureilites.**
R. Göbel, U. Ott, F. Begemann.
Meteoritics, Vol. 12, 238 - 239 (1977). − Abstract.

105.159 **Laboratory studies, critical inputs for the interpretation of iron meteorite structures.**
J. I. Goldstein.
Meteoritics, Vol. 12, 239 - 241 (1977). − Abstract.

105.160 **Mineralogy, petrology, and chemistry of the Itapicuru Mirim, Macau, and Santa Barbara**
chondrites.   C. B. Gomes, K. Keil, J. L. Berkley, E. Jarosewich, W. S. Curvello.
Meteoritics, Vol. 12, 241 - 242 (1977). − Abstract.

stone.   D. E. Lange, K. Keil, J. E. Welsh.
Meteoritics, Vol. 12, 286 - 287 (1977). – Abstract.

105.193 Distribution of organic compounds in carbonaceous
meteorites.   J. G. Lawless, F. M. Church, G. Yuen.
Meteoritics, Vol. 12, 287 (1977). – Abstract.

105.194 The Vigarano chondrite – a reevaluation.
G. R. Levi-Donati, J. Nelen, K. Fredriksson.
Meteoritics, Vol. 12, 287 - 290 (1977). – Abstract.

105.195 Qutrixpilco, Mexico: a new carbonaceous shower?
G. R. Levi-Donati, G. P. Sighinolfi.
Meteoritics, Vol. 12, 291 (1977). – Abstract.

105.196 The Alessandria chondrite: major components,
texture and chemistry.
G. R. Levi-Donati, G. P. Sighinolfi.
Meteoritics, Vol. 12, 291 - 292 (1977). – Abstract.

105.197 Xenon in Allende sulfides and other recent studies.
R. S. Lewis, J. Hertogen, L. Alaerts.
Meteoritics, Vol. 12, 292 - 298 (1977). – Abstract.

105.198 Microanalysis/microdiffraction of the Carlton
meteorite.
L. S. Lin, D. B. Williams, J. I. Goldstein.
Meteoritics, Vol. 12, 298 - 299 (1977). – Abstract.

105.199 Pre-irradiation stages of Djermaia chondrite.
J.-C. Lorin, P. Pellas.
Meteoritics, Vol. 12, 299 (1977). – Abstract.

105.200 The Mg isotope anomaly in carbonaceous chondrites:
an ion-probe study.
J. C. Lorin, N. Shimizu, M. Christophe-Michel-Lévy,
C. J. Allègre.
Meteoritics, Vol. 12, 299 - 300 (1977). – Abstract.

105.201 Sm-Nd systematics of the Serra de Magé eucrite.
G. W. Lugmair, N. B. Scheinin, R. W. Carlson.
Meteoritics, Vol. 12, 300 - 301 (1977). – Abstract.

105.202 Time of compaction of Orgueil.
J. D. Macdougall.
Meteoritics, Vol. 12, 301 - 302 (1977). – Abstract.

105.203 The record of extinct actinide nuclides.
K. Marti.
Meteoritics, Vol. 12, 302 - 303 (1977). – Abstract.

105.204 Physical properties of droplet chondrules.
P. M. Martin, A. A. Mills.
Meteoritics, Vol. 12, 303 - 304 (1977). – Abstract.

105.205 Extraterrestrial impact structures in the USSR.
V. L. Masaitis (Masajtis).
Meteoritics, Vol. 12, 305 (1977). – Abstract.

105.206 Mineralogy and petrology of heated Murchison: a
progress report.   S. D. Matza, M. E. Lipschutz.
Meteoritics, Vol. 12, 305 - 306 (1977). – Abstract.

105.207 Microprobe search for presolar grains in meteorites.
M. Maurette.
Meteoritics, Vol. 12, 307 (1977). – Abstract.

105.208 Thermoluminescence studies of the Estacado
meteorite.   S. W. S. McKeever, S. A. Durrani.
Meteoritics, Vol. 12, 307 - 308 (1977). – Abstract.

105.209 Thermoluminescence and meteorite orbits.

C. L. Melcher, R. M. Walker.
Meteoritics, Vol. 12, 309 - 310 (1977). – Abstract.

105.210 Ejecta at Lonar crater, India.
D. J. Milton, A. Dube.
Meteoritics, Vol. 12, 311 (1977). – Abstract.

105.211 REE and igneous differentiation of the howardite
and mesosiderite parent bodies.
D. W. Mittlefehldt.
Meteoritics, Vol. 12, 311 - 312 (1977). – Abstract.

105.212 Evaluation of a crust model of achondrites from the
width of exsolved pyroxenes and their pyroxene
crystallization trend.   M. Miyamoto, H. Takeda.
Meteoritics, Vol. 12, 312 - 313 (1977). – Abstract.

105.213 Application of pattern recognition to the classifica-
tion of metal rich meteorites.
C. B. Moore, D. D. Pratt, M. L. Parsons.
Meteoritics, Vol. 12, 314 - 318 (1977). – Abstract.

105.214 Cooling rate variations within the group IVA iron
meteorites.   A. E. Moren, J. I. Goldstein.
Meteoritics, Vol. 12, 318 - 319 (1977). – Abstract.

105.215 Ries crater: an aubritic impact?
J. W. Morgan, M.-J. Janssens, J. Hertogen, H.
Takahashi.
Meteoritics, Vol. 12, 319 (1977). – Abstract.

105.216 New lunar standards for solar flare track and
microcrater production.
D. A. Morrison, E. Zinner.
Meteoritics, Vol. 12, 320 (1977). – Abstract.

105.217 Dust from large meteoritic impacts as an agent of
climatic change.   R. J. Moyer, F. Dachille.
Meteoritics, Vol. 12, 321 (1977). – Abstract.

105.218 Crystal structure and composition of cronstedtite
from the Cochabamba carbonaceous chondrite.
W. F. Müller, G. Kurat, A. Kracher.
Meteoritics, Vol. 12, 322 (1977). – Abstract.

105.219 Rb-Sr internal isochron and the initial $^{87}Sr/^{86}Sr$ for
the Estherville mesosiderite.
V. R. Murthy, M. R. Coscio, Jr., T. Sabelin.
Meteoritics, Vol. 12, 323 (1977). – Abstract.

105.220 Yamato meteorite collected in Antarctica.
T. Nagata.
Meteoritics, Vol. 12, 323 - 324 (1977). – Abstract.

105.221 Nakhla: further evidence for a young crystallization
age.
N. Nakamura, D. M. Unruh, M. Tatsumoto, R. Hutchison.
Meteoritics, Vol. 12, 324 - 325 (1977). – Abstract.

105.222 Metal fractionation patterns in the Bencubbin
meteorite.   H. E. Newsom, M. J. Drake.
Meteoritics, Vol. 12, 326 (1977). – Abstract.

105.223 A neon-E-rich phase in Dimmitt.
F. Niederer, P. Eberhardt.
Meteoritics, Vol. 12, 327 - 331 (1977). – Abstract.

105.224 I-Xe dating of silicate inclusions from iron
meteorites.   S. Niemeyer.
Meteoritics, Vol. 12, 331 - 332 (1977). – Abstract.

105.225 Observations and comments on the chemical

behavior of an oxidized meteorite.
H. H. Nininger.
Meteoritics, Vol. 12, 332 (1977). – Abstract.

105.226 **Zr-Y oxides and high-alkali glass in an ameboid inclusion from Ornans.**
A. F. Noonan, J. Nelen, K. Fredriksson, D. Newbury.
Meteoritics, Vol. 12, 332 - 335 (1977). – Abstract.

105.227 **Searching for meteorites in Antarctica – the right way and the hard way.**   E. Olsen.
Meteoritics, Vol. 12, 335 (1977). – Abstract.

105.228 **Ten stony meteorites from the Antarctic: classification and description.**
E. Olsen, G. Moreland, E. Jarosewich, K. Fredriksson.
Meteoritics, Vol. 12, 335 (1977). – Abstract.

105.229 **Origin of isolated olivine grains in the Murchison C2 meteorite.**
E. Olsen, L. Grossman, A. Davis.
Meteoritics, Vol. 12, 336 (1977). – Abstract.

105.230 **Cosmic spherules as rounded bodies in space.**
D. W. Parkin, R. A. L. Sullivan, J. N. Andrews.
Meteoritics, Vol. 12, 336 - 337 (1977). – Abstract.

105.231 **Preatmospheric dimensions of Eagle Station pallasite.**   V. P. Perelygin, S. G. Stetsenko,
N. M. Gavrilova, G. Kurat, D. Chaillou, C. Fieni, P. Pellas.
Meteoritics, Vol. 12, 337 - 338 (1977). – Abstract.

105.232 **The Zhamanshin structure: lithophile trace element abundances and strontium isotope systematics.**
J. A. Philpotts, S. Schuhmann, S. R. Winzer, R. K. L. Lum.
Meteoritics, Vol. 12, 338 (1977). – Abstract.

105.233 **Organic compounds in carbonaceous chondrites with special reference to the Mighei meteorite.**
C. Ponnamperuma.
Meteoritics, Vol. 12, 340 (1977). – Abstract.

105.234 **Petrogenesis of the Serra de Magé cumulate eucrite.**   M. Prinz, C. E. Nehru, J. L. Berkley, K. Keil,
E. Jarosewich, C. B. Gomes.
Meteoritics, Vol. 12, 341 (1977). – Abstract.

105.235 **A new method for the determination of the isotopic composition of lithium in meteorites.**
R. S. Rajan, L. Brown, R. B. Roberts, D. J. Whitford.
Meteoritics, Vol. 12, 343 (1977). – Abstract.

105.236 **Trace elements in chondrites: whence and where.**
E. R. Rambaldi, K. Fredriksson.
Meteoritics, Vol. 12, 344 (1977). – Abstract.

105.237 **The Shaw chondrite and the chemical evolution of L-chondrite parent body.**
E. R. Rambaldi, J. W. Larimer.
Meteoritics, Vol. 12, 344 - 345 (1977). – Abstract.

105.238 **Cosmogenic radionuclide and trace element characterization of the Innisfree and Louisville meteorites.**   L. A. Rancitelli, J. C. Laul.
Meteoritics, Vol. 12, 346 - 347 (1977). – Abstract.

105.239 **Possible presence of curium-248 fission in Allende inclusions.**
M. N. Rao, K. Gopalan, T. R. Venkatesan.
Meteoritics, Vol. 12, 347 (1977). – Abstract.

105.240 **The Mössbauer effect in iron nickel meteorites.**

J. L. Remo.
Meteoritics, Vol. 12, 347 - 348 (1977). – Abstract.

105.241 **Meteorite impact and tektite and impactite formation.**   J. L. Remo, P. M. Sforza.
Meteoritics, Vol. 12, 348 - 349 (1977). – Abstract.

105.242 **Water and deuterium content in eight chondrites.**
F. Robert, L. Merlivat, M. Javoy.
Meteoritics, Vol. 12, 349 - 354 (1977). – Abstract.

105.243 **Silicate inclusions from the Mundrabilla iron.**
K. L. Robinson, R. W. Bild.
Meteoritics, Vol. 12, 354 - 355 (1977). – Abstract.

105.244 **Metallurgy of the enstatite chondrites.**
M. L. Rudee, J. M. Herndon.
Meteoritics, Vol. 12, 355 - 356 (1977). – Abstract.

105.245 **Chemistry of condensation: Allende inclusions as an indicator of the chemical conditions in the solar nebula.**   H. Nagasawa.
10th Lunar and Planetary Symposium, (see 012.050), p. 144 - 149 (1977).

105.246 **Large circles on the earth's surface.**   J. M. Saul.
Meteoritics, Vol. 12, 358 - 359 (1977). – Abstract.

105.247 **Noble gas measurements in matrix and clast samples from the Djermaia chondrite.**
L. Schultz, P. Signer.
Meteoritics, Vol. 12, 359 - 360 (1977). – Abstract.

105.248 **Origin of iron meteorite groups IC and IIE.**
E. R. D. Scott.
Meteoritics, Vol. 12, 360 - 361 (1977). – Abstract.

105.249 **The origin of meteorites 1770–1850.**   D. W. Sears.
Meteoritics, Vol. 12, 361 (1977). – Abstract.

105.250 **Condensation/accretion conditions of the major iron meteorite groups.**   D. W. Sears, H. J. Axon.
Meteoritics, Vol. 12, 362 (1977). – Abstract.

105.251 **The role of intensive parameters during the formation of chondrules in the Semarkona LL-3 meteorite.**   J. W. Snellenburg.
Meteoritics, Vol. 12, 364 - 365 (1977). – Abstract.

105.252 **Cosmochemical aspects of iron condensation.**
J. R. Stephens, B. K. Kothari, J. M. Herndon.
Meteoritics, Vol. 12, 365 (1977). – Abstract.

105.253 **Origins of cumulate eucrites.**   E. Stolper.
Meteoritics, Vol. 12, 366 - 367 (1977). – Abstract.

105.254 **Fission track dating of meteorite impacts.**
D. Storzer, G. A. Wagner.
Meteoritics, Vol. 12, 368 - 369 (1977). – Abstract.

105.255 **Gow Lake, Saskatchewan: evidence for an origin by meteorite impact.**
M. D. Thomas, M. J. S. Innes, M. R. Dence, R. A. F. Grieve, P. B. Robertson.
Meteoritics, Vol. 12, 370 - 371 (1977). – Abstract.

105.256 **Meteorite ages and $^{40}Ar/^{39}Ar$ release patterns.**
G. Turner, M. C. Enright.
Meteoritics, Vol. 12, 372 - 373 (1977). – Abstract.

105.257 **Beryllium in meteorites.**   E. Vilcsek.
Meteoritics, Vol. 12, 373 (1977). – Abstract.

105.258 **The measurement and interpretation of rare gas concentrations in iron meteorites.**
H. Voshage, H. Feldmann.
Meteoritics, Vol. 12, 373 - 374 (1977). — Abstract.

105.259 **Size of the eucrite parent body.**
D. Walker, E. M. Stolper, J. F. Hays.
Meteoritics, Vol. 12, 375 (1977). — Abstract.

105.260 **Chondrite classification and origin.** J. T. Wasson.
Meteoritics, Vol. 12, 381 - 383 (1977). — Abstract.

105.261 **Chemical and mineralogical investigation of a Mundrabilla specimen.** H. H. Weinke.
Meteoritics, Vol. 12, 384 - 387 (1977). — Abstract.

105.262 **An example of metal-silicate fraction by separation of immiscible Fe-FeS and silicate melts.**
L. L. Wilkening.
Meteoritics, Vol. 12, 387 - 388 (1977). — Abstract.

105.263 **The cooling rates of iron meteorites.**
J. Willis, J. T. Wasson.
Meteoritics, Vol. 12, 388 - 389 (1977). — Abstract.

105.264 **Petrology, petrography and geochemistry of impact melts from Tenoumer crater, Mauritania.**
S. R. Winzer, M. Meyerhoff, S. J. Stokowski, Jr., R. K. L.
Lum, S. Schuhmann, J. A. Philpotts.
Meteoritics, Vol. 12, 389 - 390 (1977). — Abstract.

105.265 **Experiments on the chemical concentration of a new spontaneously fissioning nuclide from matter**
of the Allende meteorite. I. Zvara, G. N. Flerov, B. L.
Zhujkov, T. Reetts, M. R. Shalaevskij, N. K. Skobelev.
Obedin. inst. yader. issled. Prepr. R6-10589 Dubna, 1977,
12 pp. In Russian. — Abstr. in Ref. zh., 51. Astron., 12.51.
304 (1977).

105.266 **On pre-atmospheric sizes of the Lipovskij-Khutor pallasite.** D. Lkhagvasurehn, O. Otgonsurehn,
V. P. Perelygin, S. G. Stetsenko.
Obedin. inst. yader. issled. Prepr. R14-10630. Dubna, 1977,
11 pp. In Russian. — Abstr. in Ref. zh., 51. Astron., 1.51.370
(1978).

105.267 **Thermomechanical fracturing of meteorites during atmospheric passage.** B. Lang.
Comets, asteroids, meteorites, (see 012.049), p. 153 - 157
(1977).

105.268 **Relationship between meteorites, asteroids and comets.** B. J. Levin (*B. Yu. Levin*).
Comets, asteroids, meteorites, (see 012.049), p. 307 - 312
(1977).
There seems to be no objection any more to regard the asteroid belt as the past and present source of initial and intermediate parent-bodies of meteorites.

105.269 **Interstellar material in meteorites: implications for the origin and evolution of the solar nebula.**
R. N. Clayton.
Comets, asteroids, meteorites, (see 012.049), p. 335 - 341
(1977).
This paper deals with the recent observations which show that the nebula was not completely homogenized, and that it contained relics of presolar materials which are now observable in the primitive meteorites.

105.270 **On the early thermal history of chondritic asteroids derived by 244-plutonium fission track thermo-**
metry. P. Pellas, D. Storzer.

Comets, asteroids, meteorites, (see 012.049), p. 355 - 363
(1977).
Cooling curves were determined for ordinary chondrites within the time- and temperature intervals from ~4.6 to ~4.1 X 10^9 years and ~1400 K to 300 K respectively. This was done by analyzing the 244-plutonium fission track record in whitlockite and adjacent mineral track detectors. The resulting cooling rates constrain the sizes of the parent asteroids to ~120 – 200 km in radius.

105.271 **Chondrules as condensation products.**
J. A. Wood, H. Y. McSween, Jr.
Comets, asteroids, meteorites, (see 012.049), p. 365 - 373
(1977).
The formation of meteoritic chondrules via condensation from the primordial solar nebula is discussed. Chondrule formation in regions where the gas/dust ratio was enhanced, and where transient high energy events heated the gas and temporarily vaporized the dust, is advocated. The observed diversity of chondrule types can be understood as resulting from local variations in the initial gas/dust proportions and other parameters.

105.272 **Condensation and/or metamorphism: genesis of E- and L-group chondrites from studies on**
artificially heated primitive congeners.
M. E. Lipschutz, M. Ikramuddin.
Comets, asteroids, meteorites, (see 012.049), p. 375 - 383
(1977).

105.273 **A comparison of solar and meteoritic abundances.**
H. Holweger.
Comets, asteroids, meteorites, (see 012.049), p. 385 - 388
(1977).
Recent solar abundance data confirm the relationship between the sun and carbonaceous chondrites. New results for the solar S/Ca and Na/Ca ratio discriminate between types 1, 2, and 3. These and interstellar properties pose constraints on the condensation process. Protostellar separation of gas and grains will affect stellar metal content.

105.274 **Meteorites in meteorites: evidence for mixing among the asteroids.** L. L. Wilkening.
Comets, asteroids, meteorites, (see 012.049), p. 389 - 396
(1977).
Inclusions of one type of meteorite enclosed in another have been found in several gas-rich meteorites, unequilibrated chondrites and mesosiderites. The inclusions in all but one case are chondritic; a majority are mineralogically and isotopically similar to carbonaceous chondrites. These meteorite mixtures most probably resulted from collisions among asteroids.

105.275 **Unusual primitive meteorites: chondritic inclusions from group IAB iron meteorites.** R. W. Bild.
Comets, asteroids, meteorites, (see 012.049), p. 397 - 402
(1977).
Several meteorite types, distinct from the large chondrite groups (H, L, LL, E and carbonaceous), are known to also have chondritic compositions. These meteorites preserve information on conditions at additional formation locations in the early solar nebula. Silicate inclusions from group IAB iron meteorites are one such type. Evidence for their chondritic nature is given and their formation discussed.

105.276 **The origin of iron meteorites.**
W. R. Kelly, E. R. Rambaldi, J. W. Larimer.
Comets, asteroids, meteorites, (see 012.049), p. 405 - 413
(1977).
The chemistry of iron meteorites is compared to predictions of the chemical fractionations that develop during the cosmic history of the metal phase, from condensation and accretion through melting, segregation and freezing.

105.277 **Meteoritic magnetism: implications for parent bodies of origin.**   A. Brecher.
Comets, asteroids, meteorites, (see 012.049), p. 415 - 427 (1977).

Recent progress in studying and understanding the magnetic record of meteorites is reviewed. The strong preterrestrial magnetization of iron meteorites, previously believed to have been acquired during cooling in parent-body fields of ~.6 Oe, has now been shown to be probably a spontaneous moment, directionally controlled by the octahedral Ni-Fe structure. For each class of meteorites, the magnetic record is basically in accord with conclusions based on chemical-mineralogical-petrologic characteristics.

105.278 **Parent bodies of iron meteorites.**   E. R. D. Scott.
Comets, asteroids, meteorites, (see 012.049), p. 439 - 444 (1977).

On the basis of their chemical and mineralogical composition, 420 iron meteorites have been classified into 12 different groups. Each group seems to have come from a separate parent body.

105.279 **The chemical relationship between howardites and the silicate fraction of mesosiderites.**
A. B. Simpson, L. H. Ahrens.
Comets, asteroids, meteorites, (see 012.049), p. 445 - 450 (1977).

Analyses of eleven major elements in five howardite samples and in the silicate fraction of seven mesosiderites are presented in a recalculated form and compared. The Ca/Al and Ca/Mg relationships suggest that the two meteorite groups were subject to similar genetic controls, and may therefore have had a common parent body.

105.280 **High-temperature condensates in carbonaceous chondrites.**   L. Grossman.
Comets, asteroids, meteorites, (see 012.049), p. 507 - 516 (1977).

Equilibrium thermodynamic calculations of the sequence of condensation of minerals from a cooling gas of solar composition play an important role in explaining the mineralogy and trace element content of different types of inclusions in carbonaceous chondrites.

105.281 **Cl chondrites: comparison and contrast to other meteorite types.**   J. M. Herndon.
Comets, asteroids, meteorites, (see 012.049), p. 537 - 539 (1977).

105.282 **Genetic relations among meteorites and planets.**   R. N. Clayton.
Comets, asteroids, meteorites, (see 012.049), p. 545 - 550 (1977).

On the basis of $^{18}O/^{16}O$ and $^{17}O/^{16}O$ ratios, meteorites and planets can be grouped into at least nine categories. Objects of one category cannot be derived by fractionation or differentiation from the source materials of any other category, but must represent samples of different regions of an inhomogeneous solar nebula. The isotopic classification, together with major-element abundances, provides a powerful method for recognition of interrelationships of the various meteorites and their parent bodies.

105.283 **Relationship between the composition of solid solar-system matter and distance from the sun.**
J. T. Wasson.
Comets, asteroids, meteorites, (see 012.049), p. 551 - 559 (1977).

105.284 **A parent body and genetic relationship of diogenites and eucrites.**   H. Takeda.
10th Lunar and Planetary Symposium, (see 012.050), p. 47 - 52 (1977).

105.285 **Frequency spectrum of mass distribution of the Yamato stony meteorites.**   T. Nagata.
10th Lunar and Planetary Symposium, (see 012.050), p. 53 - 55 (1977).

105.286 **Metallographic structures of two Yamato iron meteorites and a Yamato pallasite.**
T. Nagata, R. M. Fisher.
10th Lunar and Planetary Symposium, (see 012.050), p. 56 - 59 (1977).

105.287 **Meteorite Innisfree.**   M. Šolc.
Říše hvězd, Vol. 58, 187 - 189 (1977). In Czech.

105.288 **News about the Tunguska event.**   R. Rost.
Vesmír, Vol. 56, 292 (1977). In Czech.

Katalog über 230 sichere, wahrscheinliche, mögliche und zweifelhafte Impaktstrukturen.
See Abstr. 002.029.

Chemical petrology, with applications to the terrestrial planets and meteorites.   See Abstr. 003.114.

Explosion craters of the Ukrainian shield.
See Abstr. 003.158.

Partitioning of chromium between silicate crystals and melts.   See Abstr. 022.043.

Fe and Mg in plagioclase.
See Abstr. 022.044.

Astrophysical implications of isotopic anomalies.
See Abstr. 061.049.

Iron: whence it came, where it went.
See Abstr. 071.026.

Meteorite impact ejecta: dependence of mass and energy lost on planetary escape velocity.
See Abstr. 091.083.

Basaltic magmatism and the bulk composition of the Moon. II. Siderophile and volatile elements in Moon, earth and chondrites: implications for lunar origin.
See Abstr. 094.113.

Large scale cratering of the lunar highlands: some Monte Carlo model considerations.   See Abstr. 094.150.

Candidate areas for in situ ancient lunar materials.
See Abstr. 094.152.

Impact ejecta on the moon.   See Abstr. 094.153.

Isotopes of rare gases in regolith and its origin.
See Abstr. 094.406.

Microcraters and solar flare tracks in crystals from carbonaceous chondrites and lunar breccias.
See Abstr. 094.449.

Alkali mobility in shocked basalt.
See Abstr. 094.498.

Composition of the projectiles that bombarded the lunar highlands.   See Abstr. 094.534.

Ultraviolet diffuse reflectance spectroscopy for

lunar, meteoritic, and terrestrial samples.
See Abstr. 094.543.

Characterization of crust formation on a parent body of achondrites and the moon by pyroxene crystallography and chemistry.   See Abstr. 094.558.

Analysis of carbon species in lunar samples by static mass spectrometry.   See Abstr. 094.591.

Magnetism of lunar rocks and meteorites.
See Abstr. 094.601.

Possible bulk composition of Vesta: evidence from eucrites.   See Abstr. 098.013.

The E asteroids and the origin of enstatite achondrites.   See Abstr. 098.016.

The E asteroids and the origin of the enstatite achondrites.   See Abstr. 098.041.

The evolution of asteroids and meteorite parent-bodies (Invited Review).   See Abstr. 098.086.

The M type asteroids and the origin of iron meteorites.   See Abstr. 098.087.

Asteroid surface materials from reflectance spectroscopy: a review.   See Abstr. 098.094.

The evolution of asteroids as meteorite parent-bodies.   See Abstr. 098.100.

Fragmentation of asteroids and delivery of fragments to earth.   See Abstr. 098.101.

The Kirkwood gaps as an asteroidal source of meteorites.   See Abstr. 098.102.

Comets, asteroids and meteorites: unsolved problems about their relationships, evolution and origin.
See Abstr. 102.053.

Chemical composition of meteoric and meteoritic matter.   See Abstr. 104.053.

A chemical and textural comparison between carbonaceous chondrites and interplanetary dust.
See Abstr. 106.050.

Meteorites and the origins of the solar system.
See Abstr. 107.017.

Tungsten distribution between metal and silicates and its implications on the formation of the Earth-Moon system.   See Abstr. 107.038.

Evidence for ^{26}Al in the solar system.
See Abstr. 107.047.

Graphite grain surface reactions in interstellar and protostellar environments.   See Abstr. 131.015.

The 10-micron interstellar absorption band: curves of growth for terrestrial and meteoritic olivine particles.
See Abstr. 131.195.

Interstellar potassium and argon.
See Abstr. 131.214.

Anomalous processes in the solar system during 1971–1974 as derived from radioactivity of the recently fallen chondrite "Gorlovka".   See Abstr. 143.011.

106 Interplanetary Matter, Interplanetary Magnetic Field, Zodiacal Light

106.001 Dust in the solar system.   P. M. Millman.
The dusty universe, (see 012.001), p. 185 - 209 (1975).

Current knowledge of the particulate material of interplanetary space is reviewed in the mass range from 10^{-16} to 10^6 g. The total mass of the interplanetary cloud lies between 10^{19} and 10^{20} g, and over half this mass consists of particles in the mass range $10^{-6.5}$ to $10^{-3.5}$ g, with a peak at 10^{-5} g. Most of the larger particles are easily fragmented, originate in comets, and are ground to smaller sizes by collisional processes. The mean chemical composition may correspond roughly to that of carbonaceous chondrites, relatively undifferentiated material compared to that of the earth and the moon.

106.002 Structure of a typical stream of interplanetary plasma from a powerful flare and corresponding set of types of geomagnetic (magnetospheric) disturbances. I.
K. G. Ivanov, N. V. Mikerina.
Geomagn. Aehron., Vol. 17, 603 - 610 (1977). In Russian.

106.003 Impulses in the B_z-component of the interplanetary magnetic field and disturbances in the geomagnetic field.   B. V. Rezhenov, Ya. I. Fel'dshtejn.
Geomagn. Aehron., Vol. 17, 708 - 713 (1977). In Russian.

106.004 Analysis of recovered samples of interplanetary dust — implications for comets and interstellar grains.   D. E. Brownlee, D. A. Tomandl, E. W. Olszewski, P. W. Hodge.
Bull. American Astron. Soc., Vol. 9, 458 (1977). — Abstract.

106.005 Interplanetary current sheets at 1 AU.
L. F. Burlaga, J. F. Lemaire, J. M. Turner.
J. Geophys. Res., Vol. 82, 3191 - 3200 (1977).

The authors have determined the structure and nature of 'discontinuities' in the interplanetary magnetic field at 1 AU in the period March 18 to April 9, 1971, by using high-resolution magnetic field measurements from Explorer 43. Both tangential and rotational discontinuities were identified. The structure of most of the current sheets was simple and ordered; i.e., the magnetic field usually changed smoothly and monotonically from one side of the current sheet to the other.

106.006 Alignment of spherically-shaped zodiacal light particles.   E. Derringh, J. M. Greenberg.
News Lett. Astron. Soc. N.Y., Vol. 1, No. 2, p. 11 (1977). Abstract.

106.007 Interplanetary magnetic field and magnetospheric substorms.   S.-I. Akasofu, R. P. Lepping.
Planet. Space Sci., Vol. 25, 895 - 897 (1977).

The interplanetary magnetic field (IMF) changes and the associated responses of the magnetosphere on November 1, 1972, are examined. IMF B_z changes consisted of a sudden southward turning, a slow northward turning, and a subsequent steady northward sense. Magnetospheric substorms occurred throughout this period.

106.008 Energetic protons associated with a forward-reverse interplanetary shock pair at 1 A.U.   A. Balogh.
Planet. Space Sci., Vol. 25, 947 - 955 (1977).

A forward-reverse interplanetary shock was observed on 25 March 1969 by the magnetometer and plasma detector on the HEOS-1 satellite. Simultaneous observations of 1 MeV solar proton fluxes were also performed on HEOS-1. A characteristic intensity peak was observed as the forward shock

passed by the spacecraft. The evolution of the proton intensity, together with a detailed analysis of anisotropies and pitch angle distributions show a complex dynamic picture of the effect of the forward shock on the ambient proton population. Significant changes in particle fluxes are seen to be correlated with fluctuations in the magnetic field.

106.009 Slow shocks in the interplanetary medium.
P. Rosenau, S. T. Suess.
J. Geophys. Res., Vol. 82, 3649 - 3653 (1977).

The production of MHD shock ensembles in the solar wind will, in general, result in slow forward and reverse shocks in addition to fast forward and reverse shocks and a contact discontinuity. As opposed to fast shocks, which last for an extended period of time, slow shocks disappear in two stages. As they propagate outward, they are first weakened until their relative Mach number becomes close to unity. At this point they start to recede as slow magnetosonic waves and will merge asymptotically with the contact discontinuity to create a tangential discontinuity. This analytical result is demonstrated by a numerical simulation of MHD shock wave propagation in the solar wind.

106.010 Meteoroid stream widths, an interpretation of zodiacal light observations.   D. W. Hughes.
Astron. Astrophys., Vol. 61, L15 - L17 (1977).

It can be concluded from stream width data alone that the enhancements in the zodiacal light observed by Levasseur-Regourd and Dumont (1977) are produced by scattering from particles with masses between 10^{-6} and 3×10^{-4} g and sizes in the range 1×10^{-2} to 9×10^{-2} cm, the mean mass and size being 3×10^{-5} g and 4×10^{-2} cm respectively.

106.011 Acceleration of nucleons in the interplanetary space and modulation of electrons of Jupiter between 1 and 10 a.u. by corotating regions of solar origin.
C. W. Barnes, D. L. Chenette, T. F. Conlon, K. K. Pyle, J. A. Simpson.
VIII Leningr. mezhdunar. semin. Mater mezhdunar. semin. "Akt. protsessy na Solntse i probl. soln. nejtrino", 1976. Leningrad, 1976, p. 141 - 145. In Russian and English. Abstr. in Ref. zh., 62. Issled. kosm. prostranstva, 9.62.182 (1977).

106.012 On acceleration of particles to relativistic energies in the interplanetary space. Annotation.
N. P. Chirkov, A. T. Filippov.
VIII Leningr. mezhdunar. semin. Mater. mezhdunar. semin. "Akt. protsessy na Solntse i probl. soln. nejtrino", 1976. Leningrad, 1976, p. 155 - 156. In Russian and English. Abstr. in Ref. zh., 62. Issled. kosm. prostranstva, 9.62.194 (1977).

106.013 Bispectral analysis of meter wavelength interplanetary scintillation.   J. W. Armstrong.
Astron. Astrophys., Vol. 61, 313 - 320 (1977).

Bispectra of meter wavelength interplanetary scintillation are compared with model predictions based on a Rice-distributed electric field. The relevant properties of the bispectrum are reviewed. Model bispectra are derived from the intensity autocovariance for the cases of Rice and lognormally distributed electric fields. Observations of CTA 21 are reported.

106.014 Temperature measurement of interplanetary-interstellar hydrogen.   J. L. Bertaux, J. E. Blamont, E. N. Mironova, V. G. Kurt, M. C. Bourgin.
Nature, Vol. 270, 156 - 158 (1977).

Photometric observations can be used to measure the

velocity of the solar system through an interstellar medium and also provide an accurate method of measuring the temperature of that medium. The authors present the results obtained using a hydrogen absorption cell in conjunction with a Lyman-α photometer contained in the Soviet scientific spacecraft Prognoz-5. They use only the Lα results to measure the temperature of the interplanetary–interstellar hydrogen. The high result they obtained indicates that the solar system may be moving through an intercloud medium heated by cosmic- or soft X rays.

106.015 How did barium titanate particulates stick together in the nebula? H. H. Scholessin.
Nature, Vol. 270, 192 (1977).

106.016 On the photometric axis of the zodiacal light. N. Y. Misconi.
Astron. Astrophys., Vol. 61, 497 - 504 (1977).

A model of the zodiacal cloud is used to predict the position of the photometric axis (the locus of points of maximum brightness) of the zodiacal light at any elongation angle from the sun for any time of the year for various symmetry planes: the orbital planes of Venus, Mars and Jupiter, the invariable plane, and the solar equatorial plane.

106.017 Ultraviolet spectroscopy of the zodiacal light at 20° elongation. P. D. Feldman.
Astron. Astrophys., Vol. 61, 635 - 639 (1977).

The author presents the first spectrometric observations of the zodiacal light in the ultraviolet, at 15 Å resolution, obtained inadvertently on a rocket flight in January 1974.

106.018 Manifestation of solar active processes in the interplanetary space according to data of cosmic ray variations. Abstract. L. I. Dorman.
VIII Leningr. mezhdunar. semin. Mater. mezhdunar. semin. "Akt. protsessy na Solntse i probl. soln. nejtrino", 1976. Leningrad, 1976, p. 187. In Russian and English. – From. Ref. zh., 51. Astron., 10.51.529 (1977).

106.019 The influence of anisotropic shock waves on the parameters of the interplanetary medium.
S. A. Grib.
VIII Leningr. mezhdunar. semin. Mater. mezhdunar. semin. "Akt. protsessy na Solntse i probl. soln. nejtrino", 1976. Leningrad, 1976, p. 170 - 175. In Russian. – Abstr. in Ref. zh., 62. Issled. kosm. prostranstva, 10.62.184 (1977).

106.020 On a connection between ionospheric and geomagnetic disturbances at low latitudes and the interplanetary magnetic field. S. Gonsales, R. A. Zevakina.
Ionos. vozmushcheniya i metody ikh prognoza. Moskva, Nauka, 1977, p. 154 - 157. In Russian. – Abstr. in Ref. zh., 62. Issled. kosm. prostranstva, 10.62.251 (1977).

106.021 On low frequency oscillations in the near-Earth plasma. M. S. Kovner, V. Lebedev, V. A. Kusnetzova (*Kuznetsova*), T. A. Plyasova-Bakounina (*Plyasova-Bakunina*), V. A. Troitskaya.
Ann. Géophys., Vol. 32, 189 - 193 (1976). – Abstr. in Phys. Abstr., Vol. 80, Abstr. 33310 (1977).

106.022 The modelling and calculation of some cosmic phenomena of blast type.
V. P. Korobeinikov (*Korobejnikov*), P. I. Chushkin, L. V. Shidlovskaya, L. V. Shurshalov.
Proceedings of the 5th international conference on numerical methods in fluid dynamics, Enschede, Netherlands, 28 June - 3 July 1976. A. I. van der Vooren, P. J. Zandbergen (Editors). Springer-Verlag, Berlin–Heidelberg–New York. Price $ 15.20 (1976). p. 268 - 273. – Abstr. in Phys. Abstr., Vol. 80, Abstr. 60565 (1977).

106.023 Neutral hydrogen in interplanetary space. T. E. Holzer.
Rev. Geophys. Space Phys., Vol. 15, 467 - 490 (1977).

The theory and observations relevant to the problem of neutral hydrogen in interplanetary space are reviewed. Emphasis is placed on those theoretical problems whose treatment in the existing literature is not entirely satisfactory, but discussion of all significant observational and theoretical aspects of the interplanetary H problem is provided. Attention is also given to other neutral constituents (particularly He) that are relatively abundant in interplanetary space and in the local interstellar medium, which is the primary source of the interplanetary neutral gas. Some consequences of the passage of the solar system through an interstellar cloud are also briefly considered.

106.024 The effect of interplanetary magnetic field on geomagnetic activity. J. Vero.
Fiz. Szemle, Vol. 27, 98 - 103 (1977). In Hungarian. – Abstr. in Phys. Abstr., Vol. 80, Abstr. 74010 (1977).

106.025 Sector polarity of IMF and annual variation in occurrence of quiet and disturbed intervals.
B. N. Bhargava, G. K. Rangarajan.
J. Geomagn. Geoelectr., Vol. 29, 1 - 7 (1977). – Abstr. in Phys. Abstr., Vol. 80, Abstr. 86767 (1977).

106.026 Laboratory analysis of interplanetary dust. D. A. Tomandl, D. E. Brownlee, E. W. Olszewski, P. W. Hodge.
Publ. Astron. Soc. Pacific, Vol. 89, 623 (1977). – Abstract.

106.027 Landau damping and steepening of interplanetary nonlinear hydromagnetic waves.
A. Barnes, J. K. Chao.
J. Geophys. Res., Vol. 82, 4711 - 4714 (1977).

106.028 Characteristics of the association between the interplanetary magnetic field and substorms.
M. N. Caan, R. L. McPherron, C. T. Russell.
J. Geophys. Res., Vol. 82, 4837 - 4842 (1977).

The geomagnetic response to changes in the orientation of the interplanetary magnetic field (IMF) has been investigated for 18 IMF events. These events consisted of clear southward shifts of the IMF when the IMF B_Z(GSM) component had been northward for more than 2 hours. It was found that when the IMF thus shifted southward and remained southward for at least 2 hours, a magnetospheric substorm always ensued. Several properties of this subsequent geomagnetic activity were determined to be associated with IMF parameters.

106.029 Radio observations of interplanetary magnetic field structures out of the ecliptic.
R. J. Fitzenreiter, J. Fainberg, R. R. Weber, H. Alvarez, F. T. Haddock, W. H. Potter.
Sol. Phys., Vol. 52, 477 - 484 (1977).

New observations of the out-of-the ecliptic trajectories of type III solar radio bursts have been obtained from simultaneous direction finding measurements on two independent satellite experiments, IMP-6 with spin plane in the ecliptic, and RAE-2 with spin plane normal to the ecliptic. Burst exciter trajectories were observed which originated at the active region and then crossed the ecliptic plane at about 0.8 AU. The authors find a considerable large scale north–south component of the interplanetary magnetic field followed by the exciters. The apparent north–south and east–west angular source sizes observed by the two spacecraft are approximately equal, and range from 25° at 600 kHz to 110° at 80 kHz.

106.030 Probability distribution functions of microscale magnetic fluctuations during quiet conditions.
Y. C. Whang.
Sol. Phys., Vol. 53, 507 - 517 (1977).

A statistical study of microscale magnetic fluctuations in the interplanetary and magnetosheath region during quiet conditions is approached from the concept of probability distribution function. Magnetic field data from Explorer 34 were used to reconstruct the distribution functions and to calculate some of their moments. The distribution functions are found to be nearly tri-Maxwellian as the background field is relatively quiet. The direction of maximum fluctuations is found to be nearly perpendicular to that of the background magnetic field, but the fluctuations are rarely circularly polarized. Across the Earth's bow shock, the degree of fluctuation anisotropy increases, but no noticeable change in relative fluctuation intensity has been observed.

106.031   The structure of the solar flare stream magnetic field.
          M. I. Pudovkin, S. A. Zaitseva (*Zajtseva*), I. P. Oleferenko, A. D. Chertkov.
Sol. Phys., Vol. 54, 155 - 164 (1977).

The structure of the interplanetary magnetic field within the flare streams as well as associated variations of the geomagnetic disturbance are considered. It is shown that in the main body of the flare stream the magnetic field is determined by the configuration of the large scale magnetic field on the Sun at the flare region. Within the head part of the flare stream the magnetic field represents by itself the compressed field of the background solar wind and hence is determined by the distribution of the super large scale solar magnetic field outside the flare region. A certain asymmetry in the parameters of the magnetic field within the streams associated with geoeffective and non-effective flares is shown to exist.

106.032   Variations in Gegenschein polarization.
          L. W. Bandermann, R. D. Wolstencroft.
Astrophys. Space Sci., Vol. 49, 253 - 257 (1977).

The authors' observations of small scale angular structure and night-to-night variations of the polarized component of the zodiacal light near the anti-solar point have been criticized. They find that these criticisms are unsubstantiated.

106.033   Large-scale three-dimensional structure of the interplanetary magnetic field.   N. P. Korzhov.
Sol. Phys., Vol. 55, 505 - 517 (1977).

Analysis of observations of the white-light corona performed aboard OSO-7 is evidence for the existence of coronal ribbon-structures, which may be observed on the limb as coronal streamers. It is shown that prolongation of these structures into interplanetary space forms a curved surface; intersection of this surface is accompanied by a change of polarity of the interplanetary magnetic field, which existed in May - July 1973; and its connection with several phenomena in the solar atmosphere, has been found.

106.034   Observations of interplanetary scintillation: solar wind velocity measurements.   T. Kakinuma.
Study of travelling interplanetary phenomena 1977, (see 012. 042), p. 101 - 118 (1977).

Recent observations of the solar wind velocity by IPS method are reviewed. Detailed analyses show that the diffraction pattern is anisotropic; thus, one cannot make the assumption of isotropic pattern in the calculation of velocity. A latitudinal distribution of the solar wind velocity is derived from the observations; the solar wind velocity is 800 km/sec for latitudes above 45°. The observations indicate the presence of the corotating stream which extends from this high-speed region to the equator.

106.035   Scintillation observations of the interplanetary plasma.   Z. Houminer.
Study of travelling interplanetary phenomena 1977, (see 012. 042), p. 119 - 141 (1977).

Observations of interplanetary scintillation indicate the presence of corotating scintillation sectors which are aligned

along the spiral magnetic field over distances of 0.5−1.0 AU. Scintillation and velocity enhancements which occur simultaneously for radio sources located on opposite sides of the Sun are due to shock disturbances associated with solar flares. By combining IPS velocity measurements with spacecraft and geomagnetic observations, a good estimate can be obtained of the shape and extent of the shock disturbance in both heliocentric latitudes and longitudes.

106.036   The three-dimensional structure of the interplanetary medium.   W. I. Axford.
Study of travelling interplanetary phenomena 1977, (see 012. 042), p. 145 - 164 (1977).

The general structure of the interplanetary medium, especially in three-dimensions, is reviewed with special emphasis being given to the solar wind, the interplanetary magnetic field, the modulation of galactic cosmic rays and the propagation of energetic solar particles.

106.037   Acceleration and modulation of electrons and ions by propagating interplanetary shocks.
T. P. Armstrong, G. Chen, E. T. Sarris, S. M. Krimigis.
Study of travelling interplanetary phenomena 1977, (see 012. 042), p. 367 - 389 (1977).

The theory and observations of spikes and discontinuities in charged particle fluxes associated with propagating interplanetary shocks are reviewed. The physical mechanisms that produce energy changes and modification of angular distributions are examined. The current status of in situ observations is reviewed.

106.038   The Gegenschein: additional information from cosmic dust experiments on Lunar Explorer 35 and Mariner IV.   W. M. Alexander.
Proc. Southwest Reg. Conf., Vol. 3, (see 012.043), p. 19 - 30 (1977).

The initial sections of the paper give brief summaries of the data from the cosmic dust experiments on Lunar Explorer 35 (hereafter referred to as LE 35) and Mariner IV (hereafter referred to as M IV). The information is pertinent to the discussion because a possible correlation between some of the data between M IV and LE 35 has been seen and the nature of this phenomenon is explored and discussed.

106.039   On the turbulence of the interplanetary plasma.
          I. V. Chashej, V. I. Shishov.
Geomagn. Aehron., Vol. 17, 984 - 993 (1977). In Russian.

106.040   Meteoroid impact pit observations require lower lunar rock exposure ages.   H. A. Zook.
Meteoritics, Vol. 12, 390 (1977). − Abstract.

106.041   A contribution to ULF activity in the Pc 3-4 range correlated with IMF radial orientation.
E. W. Greenstadt, J. V. Olson.
J. Geophys. Res., Vol. 82, 4991 - 4996 (1977).

The data of an early report tentatively linking the radial orientation of the interplanetary magnetic field (IMF), measured by Explorer 35, with ULF activity in the Pc 3-4 range observed at Calgary in September 1969 have been supplemented by data from October and November 1969. The purpose of this investigation is to seek evidence favoring or disfavoring a model of Pc 3, 4 excitation caused by transfer of quasi-parallel bow shock oscillations of large amplitude through the subsolar magnetosheath to the midday magnetopause.

106.042   Interplanetary magnetic field conditions associated with synchronous orbit observations of Pc 3 magnetic pulsations.   C. W. Arthur, R. L. McPherron.
J. Geophys. Res., Vol. 82, 5138 - 5142 (1977).

106.043 **Large-scale non-uniform structure of the inter-**
planetary plasma according to radioastronomical
observations. V. I. Vlasov.
Simpoz. po fiz. geomagnitosfery, 1977. Tezisy dokl. Irkutsk,
1977, p. 3. In Russian. − Abstr. in Ref. zh., 51. Astron., 12.
51.341 (1977).

106.044 **Electric conductivity of the interplanetary plasma.**
A. D. Chertkov.
Simpoz. po fiz. geomagnitosfery, 1977. Tezisy dokl. Irkutsk,
1977, p. 13. In Russian. − Abstr. in Ref. zh., 51. Astron.,
12.51.349 (1977).

106.045 **Sectorial structure of the interplanetary magnetic**
field and the second harmonic of the daily cosmic
ray variation.
E. V. Kolomeets, I. P. Leongard, L. A. Mirkin.
Prikl. i teor. fiz. Vyp. (No.) 8. Alma-Ata, 1976, p. 231 - 233.
In Russian. − Abstr. in Ref. zh., 51. Astron., 12.51.361
(1977).

106.046 **Influence of temporal variations of the solar wind**
velocity on the structure of the interplanetary mag-
netic field. V. A. Kovalenko, V. M. Mordvinov.
Simpoz. po fiz. geomagnitosfery, 1977. Tezisy dokl. Irkutsk,
1977, p. 7. In Russian. − Abstr. in Ref. zh., 62. Issled. kosm.
prostranstva, 12.62.134 (1977).

106.047 **Nature of the temporal restitution of the picture of**
interplanetary scintillations.
N. A. Lotova, I. V. Chashej.
Simpoz. po fiz. geomagnitosfery, 1977. Tezisy dokl. Irkutsk,
1977, p. 6 - 7. In Russian. − Abstr. in Ref. zh., 62. Issled.
kosm. prostranstva, 12.62.137 (1977).

106.048 **The influence of the interplanetary magnetic field**
on the parameters of the flow in the transition
layer. N. V. Erkaev.
Simpoz. po fiz. geomagnitosfery, 1977. Tezisy dokl. Irkutsk,
1977, p. 11 - 12. In Russian. − Abstr. in Ref. zh., 62. Issled.
kosm. prostranstva, 12.62.155 (1977).

106.049 **Radiation pressure and Poynting-Robertson drag**
for small spherical particles.
S. Soter, J. A. Burns, P. L. Lamy.
Comets, asteroids, meteorites, (see 012.049), p. 121 - 125
(1977).

106.050 **A chemical and textural comparison between**
carbonaceous chondrites and interplanetary dust.
D. E. Brownlee, R. S. Rajan, D. A. Tomandl.
Comets, asteroids, meteorites, (see 012.049), p. 137 - 141
(1977).

106.051 **Relations between turbulent regions of the inter-**
planetary magnetic field and Jovian decameter wave
emissions from the main source.
H. Oya, A. Morioka, H. Saito.
10th Lunar and Planetary Symposium, (see 012.050), p. 229 -
235 (1977).

Collections of cosmic dust.
See Abstr. 051.003.

Overview of the Helios 1 and Helios 2 missions and
their participation in STIP Intervals I and II.
See Abstr. 051.069.

Energetic solar flare particles and interplanetary
shock waves. See Abstr. 073.110.

Coronal and interplanetary shock waves generated

during flares on August 2 - 13, 1972. Annotation.
See Abstr. 074.028.

A pictorial comparison of interplanetary magnetic
field polarity, solar wind speed, and geomagnetic disturbance
index during the sunspot cycle. See Abstr. 074.053.

A sheet-current approach to coronal-interplanetary
modeling. See Abstr. 074.072.

Subsonic flows in the coronal-interplanetary regions
of closed field lines. See Abstr. 074.079.

Mass ejections from the solar corona into inter-
planetary space. See Abstr. 074.084.

Dynamic modeling of coronal and interplanetary
responses to solar events. See Abstr. 074.085.

Measurements of the solar wind using spacecraft
radio scattering observations. See Abstr. 074.086.

The large scale and long term evolution of the solar
wind speed distribution and high speed streams.
See Abstr. 074.088.

Pioneer 10, 11 observations of evolving solar wind
streams and shocks beyond 1 AU. See Abstr. 074.089.

Wave-particle interaction phenomena associated
with shocks in the solar wind. See Abstr. 074.090.

Solar energetic particles below 10 MeV.
See Abstr. 074.091.

Solar wind and interplanetary medium.
See Abstr. 074.103.

Coronal rays and three-dimensional structure of the
interplanetary magnetic field. See Abstr. 074.108.

On the alignment of plasma anisotropies and the
magnetic field direction in the solar wind.
See Abstr. 074.118.

Two-component dust models of infrared emission
from a circumsolar dust cloud. See Abstr. 074.120.

Type II emission mechanism. See Abstr. 077.028.

Solar stereo radioastronomy.
See Abstr. 077.048.

Solar cosmic rays and interplanetary shock waves
on April 29−30, 1973. See Abstr. 078.001.

Separation of solar and interplanetary diffusion in
solar cosmic ray events. See Abstr. 078.009.

Coherent propagation of non-relativistic solar elec-
trons. See Abstr. 078.010.

Influence of finite injections and of interplanetary
propagation on time-intensity and time-anisotropy profiles of
solar cosmic rays. See Abstr. 078.011.

Influence of interplanetary shocks on solar particle
events. See Abstr. 078.012.

Analysis of the complex solar particle event on
April 29 - 30, 1973. See Abstr. 078.013.

Solar particle propagation from 1 to 5 AU.
See Abstr. 078.014.

Relation of solar cosmic rays with the sectorial structure of the interplanetary magnetic field.
See Abstr. 078.017.

Signatures of solar cosmic ray events and their relation to propagation and acceleration processes.
See Abstr. 078.018.

On the transformation and removal of local magnetic fields into the corona and interplanetary space.
See Abstr. 080.016.

The mean magnetic field of the sun: method of observation and relation to the interplanetary magnetic field.
See Abstr. 080.056.

Comparative magnetospherology. Part 4. Model of the neutral sheet in the inner heliomagnetosphere for the year 1976.   See Abstr. 080.092.

Measurement of scattered ultraviolet radiation in the vicinity of planets and in the interplanetary medium.
See Abstr. 082.018.

The possible meteorological significance of sub-micron cosmic dust.   See Abstr. 082.019.

Prediction of ionospheric effects associated with solar wind disturbances using interplanetary scintillation observations at 34.3 MHz.   See Abstr. 083.133.

The response of the dayside aurora to sharp northward and southward transitions of the interplanetary magnetic field and to magnetospheric substorms.
See Abstr. 084.002.

Characteristics of polar cap sun-aligned arcs.
See Abstr. 084.027.

The development of auroral and geomagnetic substorm activity after a southward turning of the interplanetary magnetic field following several hours of magnetic calm.
See Abstr. 084.033.

Hydromagnetic wave observations at large longitudinal separations.   See Abstr. 084.215.

The solar proton event of April 16, 1970. 1. Features in interplanetary space and in the magnetosheath.
See Abstr. 084.219.

A study of an observed and forecasted meteorological index and its relation to the interplanetary magnetic field.
See Abstr. 084.243.

High-latitude geomagnetic variations and substorms.
See Abstr. 084.283.

Global problems in magnetospheric plasma physics and prospects for their solution.   See Abstr. 084.285.

Computations on the interaction of an interplanetary shock with the Earth's magnetosphere.
See Abstr. 084.289.

Magnetic extra-storm of August 4 - 5, 1972 in connection with the hydromagnetic structure of the interplanetary plasma stream from a powerful flare.
See Abstr. 084.293.

Analysis of the connection of geomagnetic bays with geomagnetic activity and the interplanetary magnetic field in 1968.   See Abstr. 084.294.

Dawn-dusk gradient of the precipitation of low-energy electrons over the polar cap and its relation to the interplanetary magnetic field.   See Abstr. 084.312.

Dependence of substorm occurrence probability on the interplanetary magnetic field and on the size of the auroral oval.   See Abstr. 084.318.

On a connection between geomagnetic activity and the passage of sector boundaries of the interplanetary magnetic field through the earth.   See Abstr. 084.321.

Sector structure of the interplanetary magnetic field and geomagnetic activity.   See Abstr. 084.322.

Effect of the NS component of the interplanetary magnetic field on geomagnetic storms.
See Abstr. 084.323.

A new source of lunar electromagnetic induction: forcing by the diamagnetic cavity.   See Abstr. 094.174.

Density and chemistry of interplanetary dust particles, derived from measurements of lunar microcraters.
See Abstr. 094.472.

Meteor observations and cosmic dust investigations.
See Abstr. 104.044.

Meteor streams in the making.
See Abstr. 104.055.

The resistive effect of the solar wind on the orbits of solid bodies in the solar system.   See Abstr. 107.003.

Observations of galactic cosmic-ray energy spectra between 1 and 9 AU.   See Abstr. 143.010.

Equatorial modulation and north-south asymmetry of galactic cosmic rays due to the interplanetary magnetic field.   See Abstr. 143.018.

Current density and diffusion coefficient for cosmic ray particles.   See Abstr. 143.035.

Generalization of the spectrographic method for studying the extraterrestrial and magnetospheric cosmic-ray variations including penumbra.   See Abstr. 143.048.

On the interplanetary cosmic-ray scintillations.
See Abstr. 143.057.

Monoenergetic-source solutions of the steady-state cosmic-ray equation of transport.   See Abstr. 143.063.

Diffusion of cosmic rays in the solar wind deformed by the interstellar medium.   See Abstr. 143.067.

Eine Blau-Photometrie der Milchstrasse und des Zodiakallichts.   See Abstr. 155.067.

Errata

106.901   Errata: 'Effects of solar radiation on the orbits of small particles' [Astrophys. Space Sci., Vol. 44, 119 - 140 (1976)].   R. A. Lyttleton.
Astrophys. Space Sci., Vol. 48, 491 (1977).

107 Cosmogony of the Planetary System

107.001 Interactions in the early solar system.
J. R. Dormand, M. M. Woolfson.
Mon. Not. R. Astron. Soc., Vol. 180, 243 - 279 (1977).

The capture theory of the origin of the solar system predicts protoplanets formed in near coplanar elliptical orbits with fairly high eccentricities. A resisting medium, which would be a byproduct of the capture event, would serve to round-off the orbits in a time which is short compared to the age of the solar system. It is shown that such a medium would also give rise to differential rotations of the lines of apses of the early planetary orbits, leading to a high probability of close interactions or collisions between planets. The consequences of a collision between two planets are considered. It is found that the larger planet could, in some cases, be expelled from the solar system and that the fragments of the smaller planet could give rise to some of the terrestrial planets.

107.002 The effects of a resisting medium on protoplanetary
orbits.   J. R. Donnison, I. P. Williams.
Mon. Not. R. Astron. Soc., Vol. 180, 281 - 288 (1977).

The equation of motion of a body in orbit about a central star emitting a radially moving stream of gas is obtained on the assumption that all the gas hitting the body is accreted by it. Exact analytical solutions are found for the case when the body has a constant radius. The variations in the orbital parameters are discussed and the results applied to protoplanets of the type envisaged by McCrea's floccule theory. It is shown that protoplanets originally close to the Asteroidal Belt, and hence outside the Roche limit are affected in such a way that they move through this limit and have their gaseous envelopes tidally disrupted. In this way an explanation for the composition of the terrestrial planets is offered.

107.003 The resistive effect of the solar wind on the orbits
of solid bodies in the solar system.
J. R. Donnison, I. P. Williams.
Mon. Not. R. Astron. Soc., Vol. 180, 289 - 296 (1977).

The equations of motion of a body orbiting the Sun and passing through a radially moving stream of gas under the assumption that the body does not accrete any significant quantities of matter are obtained. An exact solution was found on the assumption that the velocity of the gas exceeds the orbital velocity of the body. This was applied to the solid bodies found in the solar system for the present solar wind and the enhanced wind present during the Sun's T-Tauri phase.

107.004 Computer simulations of planetary accretion
dynamics: sensitivity to initial conditions.
R. Isaacman, C. Sagan.
Icarus, Vol. 31, 510 - 533 (1977).

The authors have tested the implications and limitations of Program ACRETE, a scheme based merely on Newtonian physics and accretion with unit sticking efficiency, devised by Dole in 1970 to simulate the origin of the planets. The dependence of the results on a variety of radial and vertical density distribution laws, on the ratio of gas to dust in the solar nebula, on the total nebular mass, and on the orbital eccentricity, ϵ, of the accreting grains are explored. No terrestrial planets were generated more massive than five Earth masses. The number of planets per system is for most cases of order 10, and, roughly, inversely proportional to ϵ. All systems generated obey a relation of the Titius—Bode variety for relative planetary spacing. The ease with which planetary systems are generated using such elementary and incomplete physical assumptions supports the idea of abundant and morphologically diverse planetary systems throughout the Galaxy.

107.005 Chemistry of solar material.
S. S. Barshay, J. S. Lewis.
The dusty universe, (see 012.001), p. 33 - 46 (1975).

The calculated chemical compositions of the gaseous and condensed phases in the primitive solar nebula are presented for both equilibrium and disequilibrium condensation. The implications for the compositions of individual solar-system bodies are briefly discussed. Condensation from an otherwise solar-composition gas in which carbon is more abundant than oxygen is mentioned.

107.006 Supernovae, grains and the formation of the solar
system.
J. M. Lattimer, D. N. Schramm, L. Grossman.
Nature, Vol. 269, 116 - 118 (1977).

The observed ^{26}Mg and ^{16}O anomalies in meteorites can be consistently understood if a supernova occurred within a few million years of the condensation of the solar system. Grains condensing in the ejecta from this supernova may be an integral aspect of this process.

107.007 Long-term behavior of planetesimals and the forma-
tion of the planets.
C. Hayashi, K. Nakazawa, I. Adachi.
Publ. Astron. Soc. Japan, Vol. 29, 163 - 196 (1977).

On the origin of the solar system, the authors have investigated the long-term and long-range behaviors of an ensemble of planetesimals which are rotating around the Sun, until they are finally captured by protoplanets within their Hill spheres. In this way, the authors have evaluated the growth time of the planets. They have found that protoplanets grow to the mass of the Earth and that of Jupiter's core in periods of the order of 10^7 and 10^8 yr, respectively, and that these time scales are determined mainly by the rate of migration of planetesimals in regions far from the protoplanet. This migration is a result of the interplay of the gas drag force and the mutual encounters.

107.008 Interactions between protostars and the solar
nebula.   M. Kobrick.
Bull. American Astron. Soc., Vol. 9, 453 - 454 (1977).
Abstract.

107.009 Interaction of the solar system with dense interstellar clouds: modification of the planetary radiation environment.
R. J. Talbot, Jr., M. J. Newman, D. M. Butler.
Bull. American Astron. Soc., Vol. 9, 454 (1977). — Abstract.

107.010 Interaction of the solar system with dense interstellar clouds: effects on atmospheric accretion and
loss.   M. J. Newman, D. M. Butler, R. J. Talbot, Jr.
Bull. American Astron. Soc., Vol. 9, 454 (1977). — Abstract.

107.011 Did a supernova initiate the formation of the solar
system?   D. N. Schramm.
Bull. American Astron. Soc., Vol. 9, 454 (1977). — Abstract.

107.012 Long accretion intervals and heated proto-bodies.
G. A. Ransford, W. M. Kaula, R. J. Phillips.
Bull. American Astron. Soc., Vol. 9, 454 - 455 (1977).
Abstract.

107.013 Experiments in planet formation.
W. K. Hartmann.
Bull. American Astron. Soc., Vol. 9, 455 (1977). — Abstract.

107.014 Impact strength: a fundamental parameter of

collisional evolution.
R. Greenberg, W. K. Hartmann.
Bull. American Astron. Soc., Vol. 9, 455 (1977). — Abstract.

107.015   Planets pushing planetesimals: effects of secular
evolution in the solar system.   D. R. Davis.
Bull. American Astron. Soc., Vol. 9, 456 (1977). — Abstract.

107.016   Collisional and dynamical processes in planet
formation.   R. Greenberg, W. K. Hartmann,
D. R. Davis, C. R. Chapman, J. Wacker.
Bull. American Astron. Soc., Vol. 9, 460 (1977). — Abstract.

107.017   Meteorites and the origins of the solar system.
D. W. Sears.
J. British Interplanet. Soc., Vol. 30, 344 - 348 (1977).

107.018   Was the early solar system windswept?
D. W. Hughes.
Nature, Vol. 263, 371 (1976).

107.019   Distribution of the mean motions of planets and
satellites and the development of the solar system.
H. Jehle.
Vistas Astron., Vol. 21, 265 - 287 (1977).
    In this paper the point of view is taken that the distribu-
tion of orbital elements in the solar system should be discussed
first on a purely gravitational basis, i.e. on the basis of a set of
particles entirely under gravitational interaction, before hydromagnetic and other effects are taken into consideration. Contents: 1. Formulation of statistical dynamics of gravitational
systems; 2. Statistical dynamics of the development of the
solar system; 3. Discussion of perturbations in terms of the
Jacobi integral.

107.020   The formation of the planetary system.
H. Alfvén, D. A. Mendis.
NASA Tech. Mem., NASA TM X-3511, (see 012.010), p. 3 - 5
(1977). — Abstract.

107.021   The role of pre-solar grains in the early solar system.
C. Federico, A. Coradini, G. Magni.
NASA Tech. Mem., NASA TM X-3511, (see 012.010), p. 11 -
12 (1977). — Abstract.

107.022   Properties of giant gaseous protoplanets.
A. G. W. Cameron.
NASA Tech. Mem., NASA TM X-3511, (see 012.010), p. 17 -
19 (1977). — Abstract.

107.023   Chemistry of primitive solar material.
S. S. Barshay, J. S. Lewis.
Annu. Rev. Astron. Astrophys., Vol. 14, (see 003.008), 81 - 94
(1976).
    The chemical processes that occurred in the cooler, outer
regions of the primitive solar nebula at the time of intimate
chemical contact between preplanetary condensate and nebular
gas constitute the subject matter of this review. "Condensation"
models are described and tested against the observed properties
of the planets, their satellites, and the asteroids.

107.024   Solar system isotopic anomalies: supernova neighbor or presolar carriers?   D. D. Clayton.
Icarus, Vol. 32, 255 - 269 (1977).
    The author evaluates several nuclear and chemical
problems related both to the recent scenario suggesting that
the known isotopic anomalies in the solar system have
resulted from a supernova near the protosolar nebula and to
the model of extinct presolar carriers.

107.025   Origin of the solar system.   H. Alfvén.
APL Tech. Dig., Vol. 15, No. 1, p. 2 - 15 (1976).

Abstr. in Phys. Abstr., Vol. 80, Abstr. 25957 (1977).

107.026   Paleoclimate, paleomagnetism and the eccentricity of
the earth's orbit.   G. Wollin, W. B. F. Ryan,
D. B. Ericson, J. H. Foster.
Geophys. Res. Lett., Vol. 4, 267 - 270 (1977).
    Climatic changes, geomagnetic intensity fluctuations, and
variations of the eccentricity of the earth's orbit during the
past 0.8 m.y. are compared. In general, warm climate stages are
aligned with low magnetic field intensities and high orbital
eccentricity. The authors tentatively conclude that the eccentricity of the earth's orbit may partially modulate both the
earth's magnetic field and climate.

107.027   The planets, the Universe, and cosmic evolution.
W. E. Brunk.
Communications and knowledge. I. Conference held at Dallas,
Tex., USA, 29 Nov. - 1 Dec. 1976, IEEE 1976, New York,
USA, p. 1.1/1 - 5. — From Phys. Abstr., Vol. 80, Abstr.
32971 (1977).

107.028   The gas drag effect on the elliptic motion of a solid
body in the primordial solar nebula.
I. Adachi, C. Hayashi, K. Nakazawa.
Prog. Theor. Phys., Vol. 56, 1756 - 1771 (1976). — Abstr. in
Phys. Abstr., Vol. 80, Abstr. 33362 (1977).

107.029   Are supernovae sources of presolar grains?
S. W. Falk, J. M. Lattimer, S. H. Margolis.
Nature, Vol. 270, 700 - 701 (1977).
    Isotopic peculiarities in the meteorite Allende could be
attributable to explosive nucleosynthesis, sparking suggestions
that a nearby supernova may have contaminated the early solar
nebula. One explanation for isotopic anomalies in early nebular
condensates is the injection of supernova grains. An advantage
of this mechanism is that it could avoid excessive dilution of
the anomalies by the bulk of the nebula. The authors examine
the early expansion of supernovae to determine the likelihood
of grain nucleation and growth.

107.030   Zu einigen Hypothesen über die Entstehung des
Planetensystems.   J. Hoppe.
Astron. Schule, 14. Jahrg., 138 - 141 (1977).

107.031   On the origin and evolution of Jupiter and Saturn.
S. S. Kumar.
Astrophys. Space Sci., Vol. 49, L17 - L19 (1977).
    Arguments are presented which make it very unlikely
that Jupiter and Saturn were formed by contraction from
initially extended, gaseous states. Formation of these and
other planets (in the solar system) by the mechanism of accretion does not appear to present any difficulties.

107.032   Sedimentation of grains in a solar nebula.
M. J. Handbury, I. P. Williams.
Astrophys. Space Sci., Vol. 50, 55 - 62 (1977).
    The formation of a thin dust disc within a solar nebula is
a topic of current interest and a number of analytical examinations of the process have already been carried out. A description of a numerical investigation of the same problem is given.
It transpires that a dust disc will indeed form in an acceptable
interval of time.

107.033   The distribution of mass in the planetary system and
solar nebula.   S. J. Weidenschilling.
Astrophys. Space Sci., Vol. 51, 153 - 158 (1977).
    A model 'solar nebula' is constructed by adding the solar
complement of light elements to each planet, using recent
models of planetary compositions. Uncertainties in this
approach are estimated. The computed surface density varies
approximately as $r^{-3/2}$, Mercury, Mars and the asteroid belt are

anomalously low in mass, but processes exist which would preferentially remove matter from these regions. Planetary masses and compositions are generally consistent with a monotonic density distribution in the primordial solar nebula.

107.034   On the role of physico-chemical processes during the formation of planetary systems.
G. P. Gladyshev.
Inst. khim. fiz. AN SSSR. Prepr. Chernogolovka, 1977. 8 pp. In Russian. − Abstr. in Ref. zh., 51. Astron., 11.51.321 (1977).

107.035   General problems of planetary and stellar cosmogony.
V. S. Safronov, T. V. Ruzmajkina.
Early stages of stellar evolution, (see 012.041), p. 147 - 150 (1977). In Russian.
    Problems of planetary cosmogony are considered in the light of contemporary knowledge on star formation.

107.036   The magnetic Reynolds number of the solar nebula.
G. J. Consolmagno.
Meteoritics, Vol. 12, 200 (1977). − Abstract.

107.037   Energy and momenta of planetary systems.
F. Dachille.
Meteoritics, Vol. 12, 201 - 202 (1977). − Abstract.

107.038   Tungsten distribution between metal and silicates and its implications on the formation of the Earth-Moon system.   W. Rammensee, H. Wänke.
Meteoritics, Vol. 12, 345 - 346 (1977). − Abstract.

107.039   Supernovae, grains and the origin of the solar system.   D. N. Schramm, S. H. Margolis.
Meteoritics, Vol. 12, 359 (1977). − Abstract.

107.040   Possible controls on the bulk composition of the Earth: origin of Earth and Moon.   J. V. Smith.
Meteoritics, Vol. 12, 363 - 364 (1977). − Abstract.

107.041   Co, Ni and Fe partitioning between pallasitic phases.
K. K. Turekian, A. M. Davis, S. P. Clark, Jr.
Meteoritics, Vol. 12, 371 - 372 (1977). − Abstract.

107.042   Fractionation of the chemical elements in the solar nebula: bulk composition of the Moon and on the Moon-Earth system.   H. Wänke.
Meteoritics, Vol. 12, 375 - 377 (1977). − Abstract.

107.043 ^{26}Al in the solar system.
G. J. Wasserburg, T. Lee, D. A. Papanastassiou.
Meteoritics, Vol. 12, 377 - 381 (1977). − Abstract.

107.044   Accretion of the terrestrial planets.
G. W. Wetherill.
Meteoritics, Vol. 12, 387 (1977). − Abstract.

107.045   The satellites of the major planets: were they all captured?   S. F. Singer.
Comets, asteroids, meteorites, (see 012.049), p. 109 - 117 (1977).
    Jupiter, Saturn, and Uranus have regular satellites which are generally assumed to have been formed along with the planets and near their present orbits. The author presents evidence for their having been formed much later in the history of the solar system and in initial orbits very close to their respective planets. They then evolved to their present orbits, principally by tidal friction. Their source may be captured material, possibly of cometary origin.

107.046   Large planetesimals in the early solar system.
W. K. Hartmann.

Comets, asteroids, meteorites, (see 012.049), p. 277 - 281 (1977).
    Inventories in sizes of large planetesimals can be estimated for different periods in the late planet-forming process by different techniques. For example, tabulations of craters on Mercury, the moon, and Mars give direct evidence of asteroid-like size distribution including bodies in excess of 100 km diameter. Large bodies exceeding 1000 km diameter probably existed earlier. Consequences of interactions between planets and such bodies are considered.

107.047   Evidence for ^{26}Al in the solar system.
D. A. Papanastassiou, T. Lee, G. J. Wasserburg.
Comets, asteroids, meteorites, (see 012.049), p. 343 - 349 (1977) = Contrib. No. 2821, Div. Geol. Planet. Sci., Calif. Inst. Techn., Pasadena.
    The authors review here recent experimental data on Mg isotopic abundance anomalies and the evidence for the presence in the early solar system of now extinct ^{26}Al.

107.048   Limits to energetic-proton irradiation of the primeval nebula as the origin of isotopic anomalies.
D. D. Clayton.
Comets, asteroids, meteorites, (see 012.049), p. 351 - 353 (1977).

107.049   Pre-main sequence heating of planetoids.
C. P. Sonett, F. L. Herbert.
Comets, asteroids, meteorites, (see 012.049), p. 429 - 437 (1977).
    The problem of thermal metamorphosis of meteorites, possibly the Moon and Mercury, and perhaps other planetary objects is reviewed. Classical mechanisms of heating include fossil nuclides (especially ^{26}Al), accretional heating, tidal heating, chemical reaction heat, and electrical induction. These various mechanisms involve constraints on the early thermal profiles of the Moon or Mercury. In the case of the meteorites, the primary contenders for a viable mechanism currently are fossil nuclides and electromagnetic induction, or some combination of these.

107.050   A former major planet of the solar system.
T. C. Van Flandern.
Comets, asteroids, meteorites, (see 012.049), p. 475 - 481 (1977).
    Dynamical calculations by Ovenden, indicating the former existence of a 90-Earth-mass planet in the asteroid belt, have now been supported by a study of orbital element distributions of very-long-period comets. The indicated epoch for disintegration of the planet is just 5×10^6 years ago.

107.051   Oort's cometary cloud in the light of modern cosmogony.   V. S. Safronov.
Comets, asteroids, meteorites, (see 012.049), p. 483 - 484 (1977).

107.052   On the early scattering processes of the outer planets.   W.-H. Ip.
Comets, asteroids, meteorites, (see 012.049), p. 485 - 490 (1977).
    The gradual scattering of small bodies by the Jovian planets in the late stage of their accretion is simulated by Monte Carlo calculation. The effects of collisional interaction of the scattered planetesimals with the inner planets and asteroidal belt are estimated. The origin of the cometary Oort cloud in relation to these scattering processes is also discussed.

107.053   Catalytic reactions in the solar nebula.
E. Anders, R. Hayatsu, M. H. Studier.
Comets, asteroids, meteorites, (see 012.049), p. 517 (1977). Summary.

107.054 **Condensation and agglomeration of grains.**
          J. R. Arnold.
Comets, asteroids, meteorites, (see 012.049), p. 519 - 524 (1977).
The thermodynamics of condensation from a cooling neutral gas of solar composition, have now been explored in great detail. There seems good reason to explore the condensation process experimentally, so far as it is possible to simulate the conditions usually assumed, to compare the results with what we see in the meteorites, and with the calculations. Preliminary results of this experimental investigation are being reported here.

107.055 **The gaseous composition of the solar nebula inferred from data of comets and interstellar molecules.**
M. Shimizu.
Comets, asteroids, meteorites, (see 012.049), p. 525 - 527 (1977).
The observations of comets and other primordial objects must be used to deduce the gaseous composition of the solar nebula just before its condensation. Preliminary steps in this direction are reported here.

107.056 **The solar nebula pressure gradient and its effect on planetesimal motions.**   S. J. Weidenschilling.
Comets, asteroids, meteorites, (see 012.049), p. 541 - 544 (1977).
In a centrally condensed solar nebula, the gas is partially supported by a pressure gradient, and rotates at less than the Keplerian velocity. Solid bodies lack this support, and spiral inward due to drag. The radial velocities developed can be significant, even in low-mass nebular models. Possible effects include fractionation of particles by size or density, rapid accumulation of planetesimals, and production of regions of anomalous (non-solar) composition in the nebula.

107.057 **The role of plasma in the primeval nebula.**
          H. Alfvén, G. Arrhenius, D. A. Mendis.
Comets, asteroids, meteorites, (see 012.049), p. 561 - 568 (1977).
The role of plasma and hydromagnetic processes in the primeval solar nebula is evaluated. In the light of the present knowledge of particles and fields in space it is difficult to avoid the conclusion that they must have played a crucial role in the emplacement of the material as well as its subsequent condensation into "planetesimals". Strong evidence in support of these processes in the primeval nebula is provided by the dynamical fine structure of the asteroidal belt and the Saturnian rings. The importance of current planetary magnetospheric as well as cometary research in clarifying certain physical processes believed to have occurred in the primeval nebula is stressed.

107.058 **The role of protoplanets in the formation of the solar system.**   I. P. Williams.
Comets, asteroids, meteorites, (see 012.049), p. 569 - 571 (1977).

107.059 **Grain formation in the primordial solar nebula.**
          T. Yamamoto, H. Hasegawa, S. Nishida.
10th Lunar and Planetary Symposium, (see 012.050), p. 66 - 71 (1977). In Japanese.

107.060 **Statistical behavior of planetesimals in the primordial solar system.**   Y. Nakagawa.
10th Lunar and Planetary Symposium, (see 012.050), p. 126 - 132 (1977). In Japanese.

107.061 **Primordial solar nebula – production of plasma through thermal ionization.**   H. Oya.
10th Lunar and Planetary Symposium, (see 012.050), p. 133 - 137 (1977). In Japanese.

107.062 **Dynamics of the condensation process in the primordial solar nebula.**
H. Hasegawa, T. Yamamoto.
10th Lunar and Planetary Symposium, (see 012.050), p. 138 - 143 (1977). In Japanese.

107.063 **Accretion process in the formation of planets (I).**
          K. Nakazawa.
10th Lunar and Planetary Symposium, (see 012.050), p. 150 - 156 (1977). In Japanese.
The author has investigated the accretion process of the protoplanets and evaluated the growth time of the planets under the assumption that there is no fragmentation in direct collisions between planetesimals. He reexamines this assumption.

Computer simulations of planetary accretion dynamics: sensitivity to initial conditions.
See Abstr. 021.008.

Rare gas occlusion into grains during their growth—an approach to study planetary formation.
See Abstr. 022.121.

Statistical mechanics of Keplerian orbits. III. Perturbations.   See Abstr. 042.056.

A new approach to nucleocosmochronology.
See Abstr. 061.001.

An examination of the planetesimal impact hypothesis of the formation of CP stars.   See Abstr. 065.071.

The solar neutrino experiment as a means for testing cosmological and cosmogonic hypotheses. Abstract.
See Abstr. 080.047.

Formation of Fe-Ni-Si planetary cores.
See Abstr. 091.040.

The formation of the atmospheres of the terrestrial planets by impact.   See Abstr. 091.063.

Planetesimal bombardments of the asteroidal belt.
See Abstr. 098.012.

Concerning the capture of the outer satellites.
See Abstr. 099.506.

Some comments on the Oort cloud.
See Abstr. 102.027.

Comets and the cosmogony of the solar system.
See Abstr. 102.050.

Low-temperature condensates in comets.
See Abstr. 102.052.

The composition of cometary gas and the primordial solar nebula.   See Abstr. 102.054.

The fine-grained structure of chondritic meteorites.
See Abstr. 105.005.

Dust in the solar nebula.
See Abstr. 105.006.

Nebular condensation of moderately volatile elements and their abundances in ordinary chondrites.
See Abstr. 105.100.

History of the Pasamonte achondrite: relative

susceptibility of the Sm—Nd, Rb—Sr, and U—Pb systems to metamorphic events.   See Abstr. 105.106.

The Allende meteorite: a new titanate in condensates from the early solar nebula.   See Abstr. 105.165.

Nucleochronology and short-lived isotopes. See Abstr. 105.173.

Microprobe search for presolar grains in meteorites. See Abstr. 105.207.

Chemistry of condensation: Allende inclusions as an indicator of the chemical conditions in the solar nebula. See Abstr. 105.245.

Interstellar material in meteorites: implications for the origin and evolution of the solar nebula. See Abstr. 105.269.

High-temperature condensates in carbonaceous chondrites.   See Abstr. 105.280.

Genetic relations among meteorites and planets. See Abstr. 105.282.

Relationship between the composition of solid solar-system matter and distance from the sun. See Abstr. 105.283.

The role of dust in cosmogony. See Abstr. 131.004.

Stars

111 Parallaxes, Distances

111.001 Internal and external errors in trigonometric
parallaxes.   A. R. Upgren.
Vistas Astron., Vol. 21, 241 - 264 (1977).
     The methods for evaluating internal and external errors
in parallaxes are reviewed and the results of each method are
examined. The influence on parallax errors of variations in the
weighting of individual parallax observations and of variations
in reference-star configurations is discussed with emphasis
placed on recent investigations of these effects made at the
Van Vleck Observatory.

111.002 Parallaxes of selected stars near the North Galactic
Pole.
K. Aa. Strand, C. C. Dahn, R. S. Harrington.
Highlights of Astronomy, Vol. 4, Part II, (see 012.022),
p. 59 - 60 (1977).

111.003 Progress report on a search for parallax stars in the
region of the South Galactic Cap.   C. A. Murray.
Highlights of Astronomy, Vol. 4, Part II, (see 012.022),
p. 61 (1977).

111.004 Problems related to O-type stars. Review of errors
in distance estimation. Comments about galactic
distributions.   G. Goy.
Arch. Sci. Genève, Vol. 29, 149 - 161 (1976). In French.
Abstr. in Phys. Abstr., Vol. 80, Abstr. 22209 (1977).

          Kosmische Weiten.   See Abstr. 003.072.

          G77-61: a dwarf carbon star.
See Abstr. 112.002.

          Perspective secular changes in stellar proper motion,
radial velocity and parallax.   See Abstr. 112.004.

          Parallax, orbit, and mass of Ross 614.
See Abstr. 117.013.

          Astrometric study of the eclipsing binary VV
Cephei from plates taken with the Sproul 61-cm refractor.
See Abstr. 121.015.

          An extension of a geometric method for finding
cepheid distances.   See Abstr. 122.001.

          Distance estimate for the galactic supernova remnant
G74.9 + 1.2 using H I absorption-line measurements.
See Abstr. 125.016.

Errata

111.901 Erratum: "On the statistical use of trigonometric
parallaxes" [Astron. Astrophys., Vol. 56, 273 -
281 (1977)].   C. Turon Lacarrieu, M. Crézé.
Astron. Astrophys., Vol. 60, 437 (1977).

112 Proper Motions, Radial Velocities, Space Motions

112.001 **Proper motions of faint blue stars near the galactic anticenter. III.** K. M. Cudworth.
Astron. J., Vol. 82, 516 - 517 (1977).

Palomar Schmidt plates of the regions searched by Rubin and her coworkers for faint blue stars have been blinked for proper motions. Significant proper motions were found for a few stars which are therefore very likely to be white dwarfs. The vast majority of these faint blue stars, however, show no detectable proper motion and are thus likely to be distant stars of higher luminosity possibly useful for studies of galactic structure and kinematics.

112.002 **G77-61: a dwarf carbon star.**
C. C. Dahn, J. Liebert, R. G. Kron, H. Spinrad, P. M. Hintzen.
Astrophys. J., Vol. 216, 757 - 766 (1977).

The authors report the discovery that the cool proper-motion star G77-61, previously assumed to be an early M dwarf on the basis of the astrometry and $(B - V)$ color, is a carbon star with striking C_2 and CH features. The previous parallax determination has been reanalyzed to confirm the previous result ($M_v \approx +9.6$) but the authors show that the space motion constraints alone rule out a giant or subgiant luminosity. Scanner spectrophotometry, blue and red image-tube spectra, and new photoelectric colors are presented and compared with similar observations of normal carbon stars. Various possible evolutionary interpretations are discussed.

112.003 **Proper motions of red stars at the north galactic pole.** B. F. Jones, A. R. Klemola.
Astron. J., Vol. 82, 593 - 597 (1977) = Lick Obs. Bull., No. 761.

New absolute proper motions based on plate material from the Lick astrograph are given for 210 of the 273 faint stars of spectral type M3 and later found by Sanduleak in the region of the north galactic pole.

112.004 **Perspective secular changes in stellar proper motion, radial velocity and parallax.** P. van de Kamp.
Vistas Astron., Vol. 21, 289 - 310 (1977).

This article deals primarily with secular changes in proper motion, a subject of interest for its own sake but also for the determination of radial velocity, independent of Doppler shift. The effect of the proper motions of background stars is evaluated; the effect of acceleration on orbital motion is discussed. An illustration of a well-determined secular acceleration is that of Barnard's star; other promising stars are listed.

112.005 **Radial velocities of southern B stars determined at the Radcliffe Observatory – VIII. Stars with HD spectral types B8 and B9.** R. Wood.
Mem. R. Astron. Soc., Vol. 84, 119 - 134 (1977).

Radial velocities, spectral types and $H\beta$ indices are presented for 92 southern stars with HD spectral type B8 or B9. Previously published three colour photometry on the Cape system is also given.

112.006 **Recherche d'étoiles à grande vitesse dans la direction du pôle galactique nord.** A. Florsch.
Cent. Données Stellaires, Inf. Bull. No. 13, p. 5 (1977).

112.007 **Improvement of stellar radial-velocity data.** J. Andersen.
Cent. Données Stellaires, Inf. Bull. No. 13, p. 11 - 13 (1977).

112.008 **Search for large radial velocities in direction of NGP using the Fehrenbach techniques.** A. Florsch.

Highlights of Astronomy, Vol. 4, Part II, (see 012.022), p. 63 (1977).

112.009 **Objective-prism radial velocities at high latitudes.** J. Stock, W. Osborn, A. R. Upgren.
Highlights of Astronomy, Vol. 4, Part II, (see 012.022), p. 77 (1977).

112.010 **On the catalogue of proper motions of 10600 stars in 41 selected areas of the sky.** A. G. Rakhimov.
Tsirk. Astron. Inst., Tashkent, No. 66 (413), 31 pp. (1976). In Russian.

112.011 **Proper motions of stars relative to galaxies in four areas of the Pulkovo program.** A. G. Rakhimov.
Tsirk. Astron. Inst., Tashkent, No. 68 (415), p. 8 - 23 (1976). In Russian.

112.012 **Proper motions of stars relative to galaxies in area No. 5 of the Pulkovo program.**
A. G. Rakhimov.
Tsirk. Astron. Inst., Tashkent, No. 69 (416), 30 pp. (1976). In Russian.

112.013 **Catalogue of proper motions of stars relative to galaxies in five Pulkovo areas.** A. G. Rakhimov.
Tsirk. Astron. Inst., Tashkent, No. 70 (417), p. 10 - 31 (1976). In Russian.

112.014 **Relative proper motions of stars in five areas of the Pulkovo program.** A. G. Rakhimov.
Tsirk. Astron. Inst., Tashkent, No. 71 (418), p. 11 - 31 (1976). In Russian.

112.015 **Proper motions of stars in the region of the open cluster NGC 6709.** A. A. Latypov.
Tsirk. Astron. Inst., Tashkent, No. 74 (421), 26 pp. (1977). In Russian.

112.016 **Proper motions of stars relative to galaxies in the areas Nos. 104, 109, 111, and 113 of the Pulkovo program.** A. G. Rakhimov.
Tsirk. Astron. Inst. Tashkent, No. 75 (422), 21 pp. (1977). In Russian.

112.017 **Proper motions of stars in the open cluster NGC 6940.** A. A. Latypov.
Tsirk. Astron. Inst., Tashkent, No. 76 (423), 31 pp. (1977). In Russian.

112.018 **Galactic orbits of some near-by stars.**
I. N. Latyshev.
Astron. Tsirk., No. 953, p. 6 - 7 (1977). In Russian.

112.019 **Proper motions in the region of NGC 6025.** W. H. Robertson.
J. Proc. R. Soc. New South Wales, Vol. 109, 77 - 80 (1976) = Sydney Obs. Pap. No. 75 (1977).

Proper motions of stars in the region of the galactic cluster NGC 6025 based on plates taken with the 33 cm astrograph, are determined with the aim of identifying stars in the area of the cluster which are non-members.

112.020 **Proper motion survey with the 48-inch Schmidt telescope. L. Double stars with common proper motion.** W. J. Luyten.
Sep. print Univ. Minnesota, Minneapolis, Minnesota, 35 pp. (1977).

Data are given for 2051 double stars with common proper motion. The arrangement of the table differs from the previous lists in that no LP numbers are now given, the positions are given to one second of time in Right Ascension, and one tenth of a minute of arc in Declination, red, as well as photographic magnitudes are given, and the proper motions are given to 0.001.

Some thoughts on a catalog of average radial velocities.   See Abstr. 002.010.

Summary of the bibliography on stellar radial velocities.   See Abstr. 002.011.

Preliminary results of the comparison of the AGK3 catalogue with the General Catalogue of Latitude Stars and observational data from Zenith-Telescopes and Photographic Zenith Tubes.   See Abstr. 041.034.

The radial velocities of early-type stars within six degrees of the galactic anticenter direction. See Abstr. 114.016.

Further degenerate stars. X. See Abstr. 126.015.

The spectroscopic orbit and the masses of the components of the binary X-ray source 3U0900—40/HD 77581. See Abstr. 142.068.

Accurate radial velocities using cross-correlation techniques and a TV detector. I. The velocity dispersion of NGC 6397.   See Abstr. 154.025.

The role of proper motions in determining the luminosity and density functions.   See Abstr. 155.052.

The Local Group: the solar motion relative to its centroid.   See Abstr. 155.065.

113 Magnitudes, Colors, Photometry

113.001 The infrared fluxes of Ba II stars.
M. W. Feast, R. M. Catchpole.
Mon. Not. R. Astron. Soc., Vol. 180, 61P - 63P (1977).
J, H and K (1.2–2.2 μ) photometry is given for 15 Ba II stars. These stars have an infrared excess compared to normal giants of the same ($V-I$). This excess appears to be correlated with the known blue deficiency of Ba II stars. These results strengthen the similarity between the Ba II stars and the 'blue-deficient' stars in the globular cluster Omega Centauri.

113.002 Note on the catalogue of the photometric parameters of stars measured in the Geneva Observatory System. R. Rufener.
Astron. Astrophys., Suppl. Ser., Vol. 29, 255 (1977).
The most frequently used photometric parameters and colour indices are computed for 4670 stars measured in the photometric system of the Geneva Observatory. Concise bibliographic information as well as the V magnitude are also given in this catalogue (Rufener: Publ. Obs. Genève, Ser. B, No. 2 (1977)).

113.003 Photoelectric comparison sequences in the fields of optically active extragalactic objects.
B. Q. McGimsey, H. R. Miller.
Astron. J., Vol. 82, 453 - 455 (1977).
Photoelectric UBV comparison sequences in the fields of six known or suspected variable extragalactic sources have been calibrated to facilitate photographic photometry.

113.004 Intermediate band photometry of late-type stars. II. Some stellar groups. O. J. Eggen.
Astrophys. J., Vol. 215, 812 - 826 (1977).
Observations, on the (R, I)-system and on a modified Strömgren system, of members of six stellar groups are used to demonstrate the chemical homogeneity of some 70% or 80% of the members assigned to the groups on the basis of kinematics. The groups discussed are the Hyades, Wolf 630, Arcturus, Groombridge 1830, and Kapteyn's Star Groups and an anonymous group of a half-dozen subdwarfs with (U, V) near (-150, -320) km s^{-1}. Standards for the photometric system described by the author (see 18.113.084) are extended and additional F- and G-type standards for the (R, I)-system are presented.

113.005 *UBV* photometry and MK spectral classification of northern early-type stars at intermediate galactic latitudes. P. W. Hill, A. E. Lynas-Gray.
Mon. Not. R. Astron. Soc., Vol. 180, 691 - 702 (1977).
UBV photometry and MK spectral types are reported for 193 northern early-type stars at intermediate galactic latitudes. Spectral types determined from UBV colours and visual classification are compared for the main sequence stars.

113.006 Investigation of the photometric system of the AZT-8 telescope. N. N. Kiselev.
Byull. Inst. Astrofiz., Dushanbe, No. 66 - 67, p. 47 - 49 (1976). In Russian.

113.007 On the accuracy of fundamental quasi-mono-chromatic stellar photoelectric photometry.
V. M. Zhilin.
Izv. Krymskoj Astrofiz. Obs., Vol. 57, 82 - 86 (1977). In Russian.
An estimate of the accuracy of extraatmospherical star brightness determinations in a quasi-monochromatic instrumental system (λ_{max} = 5550 Å) is given using the methods of fundamental photoelectric photometry.

113.008 The red spot-covered star HD 224085.
S. M. Rucinski.
Publ. Astron. Soc. Pacific, Vol. 89, 280 - 284 (1977) = Contrib. Dominion Astrophys. Obs., Victoria, B.C., Canada, No. 326 = NRC No. 15608.
Variations amounting to about 0^m25 in V have been confirmed in the light of HD 224085 and they indicate a highly nonuniform longitudinal distribution of spots over the stellar surface. It is suggested that the photometric period is closer to (and probably shorter than) the already known orbital one of 6^d724. The spectroscopic and parallax data available indicate that in its spectral type (K2–3), small infrared excess, strong emissions at the H, K, and Hα lines, and location slightly above the main sequence, the star resembles the secondary components of RS CVn systems more closely than the BY Dra stars to which it was originally compared. Presence of a weak $\lambda6707$ Li I line suggests an alternative hypothesis that the star is in a late stage of pre-main-sequence evolution resembling the "post-T-Tauri" object FK Ser.

113.009 The influence of the Balmer discontinuity in *UBV* reductions. A. F. J. Moffat, N. Vogt.
Publ. Astron. Soc. Pacific, Vol. 89, 323 - 328 (1977).
A method is presented which corrects for the nonlinear influence of the Balmer discontinuity in fitting photoelectric ($U-B$) observations to the standard ($U-B$) index. An initial linear fit yields systematic deviations in ($U-B$) as a function of spectral type or color. After reduction to zero air mass it was possible to correlate these deviations unambiguously to a parameter q which is a linear combination of ($B-V$) and ($U-B$), resembling Johnson's reddening independent Q parameter. The method can be applied directly to all program stars (including reddened ones). It is emphasized that, in general, corrections for the effect of the Balmer discontinuity must always be applied even if the observer's system matches the standard system.

113.010 A photoelectric magnitude sequence for AA Aurigae. A. U. Landolt.
Publ. Astron. Soc. Pacific, Vol. 89, 403 - 404 (1977) = Contrib. Louisiana State Univ. Obs., Baton Rouge, No. 124.
A UBV photoelectric sequence has been established in the vicinity of the variable star AA Aur. A finding chart is provided.

113.011 Photoelectric photometry of the shell star 4 Herculis.
H. J. Landis, H. P. Lovell, D. S. Hall, D. G. Uckotter.
Acta Astron., Vol. 27, 265 - 271 (1977).
A total of 268 photoelectric observations obtained in 1974 and 1975 plotted modulo 46^d1943 have failed to reveal any eclipse and hence fail to provide any direct evidence in favor of the binary star hypothesis of Harmanec, Koubsky, Krpata, and Zdarsky. Nevertheless, because of the excellent phase coverage, they succeeded in showing that, within the framework of the binary hypothesis, the orbital inclination must be less than $i = 79^\circ.3$.

113.012 1974–75 UBV photometry of the radio binary UX Arietis. D. S. Hall.
Acta Astron., Vol. 27, 281 - 286 (1977).
New 1974–75 UBV photometry of the RS CVn binary UX Ari is presented. The light-curve has changed since 1972, with the amplitude of the wave having decreased to only a few hundredths of a magnitude and the overall level having dropped by over 0^m1.

113.013 CRL 3068 – a dust-enshrouded carbon star.

M. J. Lebofsky, G. H. Rieke.
Astron. J., Vol. 82, 646 - 647 (1977).

Narrow-band photometry and angular diameter measurements are combined to show that CRL 3068 is probably a carbon star with an extremely dense dust shell.

113.014   Four-colour *uvby* and H$_\beta$ photometry of field stars: double-lined spectroscopic binaries, G type dwarfs and early type FK4 stars.   E. H. Olsen.
Astron. Astrophys., Suppl. Ser., Vol. 29, 313 - 318 (1977).

Twenty-one new bright double-lined spectroscopic binaries have been searched for photometric variability. Three eclipsing systems were found (HD 20301, 38735 and 71581) and one system shows ellipsoidal variations (HD 60168). For 27 G type dwarf stars *uvby* photometry is given which has been used in an abundance study of late type stars. *uvbyβ* photometry for 61 FK4 stars of types B and early A is also given.

113.015   Photoelectric observations of the shell star *o* And.
M. Bossi, G. Guerrero, L. Mantegazza.
Astron. Astrophys., Suppl. Ser., Vol. 29, 327 - 332 (1977).

Photoelectric observations of the shell star *o* And are given for the period ranging from October 1975 through January 1976. The relative light variations indicate a pseudo-periodicity of about $0\overset{d}{.}8$ which is hence similar to that of other shell stars. The various possible hypotheses which may justify these light variations are taken into consideration and some clues of their being caused by shell activity are provided.

113.016   *VBLUW* photometry of Sk 160 (SMC X-1).
J. van Paradijs.
Astron. Astrophys., Suppl. Ser., Vol. 29, 339 - 344 (1977).

The results of photometric *VBLUW* (Walraven system) observations of the optical counterpart of the Small Magellanic Cloud X-ray source SMC X-1 are presented. The star shows brightness variations in all five passbands. A new value of the period is derived. The colour index *V–B* varies systematically with phase. It is not possible to understand these colour variations from a model, incorporating only deformation of the primary and heating by X-rays impinging on it.

113.017   *UBV* photometry of stars in a region near the south galactic pole.   J. S. Drilling.
Astron. J., Vol. 82, 714 - 717, 771 (1977) = Contrib. Louisiana State Univ. Obs., No. 127.

UBV photometry and objective prism spectral types are given for 51 stars in a region of 25 sq deg centered near $l = 215°$, $b = -86°$. Included are all stars classified as A7 or earlier, plus a representative sample of later-type stars. Little or no reddening is found to be present in the region. The colors of the early-type stars brighter than $V = 11.5$ mag are consistent with their being primarily normal A-type stars and metallic line stars. The results for the stars fainter than $V = 11.5$ mag indicate a large admixture of field horizontal-branch stars, RR Lyrae stars, and F–G subdwarfs.

113.018   Broad-band 20–33-μ photometry of young stars.
T. Simon, H. M. Dyck.
Astron. J., Vol. 82, 725 - 728 (1977).

The authors present broad-band flux measurements in the 20–33-μ spectral region for 11 young stars, including Allen's source in NGC 2264, Cohen's source near the Rosette Nebula, and V1057 Cyg. All of the spectral energy distributions peak in the near infrared and, except for the spectrum of V1057 Cyg, none show strong evidence of silicate bands at 20 μ.

113.019   Broad-band infrared colors and CO and H$_2$O absorption indices for late-type dwarf stars.
S. E. Persson, M. Aaronson, J. A. Frogel.
Astron. J., Vol. 82, 729 - 733 (1977).

Broad-band *J–H, H–K,* and *V–K* colors and intermediate-band CO (2.36 μ) and H$_2$O (2.0 μ) indices are presented for 41 dwarf stars spread over a range of effective temperature corresponding to $1.2 < V–K < 7.6$. The data are compared with colors and band strengths predicted from the models of Mould. The agreement in the details of the comparison is not entirely satisfactory.

113.020   Near infrared photometry of some stars.
P. V. Kulkarni, A. G. Ananth, S. D. Sinvhal, S. C. Joshi.
Bull. Astron. Soc. India, Vol. 5, 52 - 56 (1977).

Infrared photometric stellar measurements in the J (1.2 μ) and H (1.65 μ) bands were made at the Cassegrain focus of the 104-cm, f/13 Sampurnanand reflector at the Uttar Pradesh State Observatory during the period 29 November - 2 December 1976.

113.021   Intrinsic color indices of supergiants in the system UBVRIJHKLMN'NO.
G. Kurilienė, V. Straižys.
Bull. Vilnius Astron. Obs., Nr. 44, p. 3 - 14 (1977). In Russian.

By critical analysis of published investigations on intrinsic color indices B–V of supergiants the most reliable values of them have been determined. From these and published observations of 312 supergiants new intrinsic indices U–B, V–R, V–I, V–J, V–H, V–K, V–L, V–M, V–N', V–N, and V–O have been derived.

113.022   Intrinsic color indices in the Vilnius photometric system.   G. Kurilienė.
Bull. Vilnius Astron. Obs., Nr. 44, p. 15 - 37 (1977). In Russian.

Intrinsic color indices in the Vilnius photometric system were determined on the basis of observations of 365 main sequence stars, 465 giants, 118 subgiants and 106 supergiants.

113.023   Color excess ratios in the *uvby* system.
G. Kurilienė.
Bull. Vilnius Astron. Obs., Nr. 44, p. 38 - 43 (1977). In Russian.

It is shown how the color excess ratios in the *uvby* system change in dependence on spectral and luminosity class of the stars. Estimated are the systematic errors introduced into the parameters $[c_1]$, $[m_1]$ and $[u-b]$ when the constant color excess ratios from the work of Crawford (1975) are used. The relations of the color excesses of the *uvby* system to the $E(B-V)$ system are investigated.

113.024   Photoelectric Hα line photometry of early-type stars.
B. Cester, G. Giuricin, F. Mardirossian, M. Pucillo, F. Castelli, U. Flora.
Astron. Astrophys., Suppl. Ser., Vol. 30, 1 - 10 (1977).

293 bright stars of spectral types O, B, A, F and of luminosity classes I through V have been measured with a photoelectric photometer equipped with two interference filters of 30 Å bandwidth, one centred on Hα and the other at λ6622. A correction term has been allowed for the response of the photometric system and for the energy distribution of the continuum at the two spectral regions considered. The resulting photometric α indices of Hα line strength are compared with previous Hα, Hβ, Hγ photometric measures, Hα equivalent widths, the MK spectral type, $[u-b]$, $[c_1]$ and $b-y$ indices of the photometric system *u, v, b, y*. The results emphasize the advantage of using Hα line photometry to discriminate between emission line effects and luminosity effects in early-type stars and to detect emission line variability.

113.025   Spectrum and photometric variability of He-weak and He-strong stars.
H. Pedersen, B. Thomsen.
Astron. Astrophys., Suppl. Ser., Vol. 30, 11 - 25 (1977).

Periodic variability in the strength of the He I λ4026 Å line is shown to be a frequent phenomenon among both He-weak and He-strong stars. In a sample of 26 stars, seven were found to be spectrum variables. Three previously known He variables were also observed. The periods range from 0.9 days (or possibly 0.5 days) to 15 days. The observations reported here include He I λ4026 Å and $uvby\beta$ photometry.

113.026   High-precision differential photometry, with application to early B-type stars.
J. R. Percy.
J. R. Astron. Soc. Canada, Vol. 71, 404 (1977). — Abstract.

113.027   Photometry of faint blue stars — IV. Four-colour photometry of some northern stars.
D. Kilkenny.
Mon. Not. R. Astron. Soc., Vol. 181, 611 - 617 (1977).

Photometry on the $uvby$ system has been obtained for 33 northern faint blue stars. The stars are given photometric classification using criteria described in Paper I (Kilkenny, Hill 1975). Data from this and earlier papers provide supporting evidence for the existence of gaps in the blue horizontal branch.

113.028   Photoelectric $uvby$ and Hβ photometry of 750 A and F stars in 63 selected areas with $|b| < 30°$.
J. K. Knude.
Astron. Astrophys., Suppl. Ser., Vol. 30, 297 - 305 (1977).

Four-colour $uvby$ and Hβ photometry has been carried out for 750 stars. The catalogue gives $uvby$ indices on the Crawford-Barnes standard system and Hβ index on the Crawford-Mander standard system.

113.029   HD 193793.
IAU Circ., Nos. 3107, 3136 (1977).

113.030   A photometric peculiarity index in the Ap stars.
J. Gettys, R. E. Schild.
Publ. Astron. Soc. Pacific, Vol. 89, 519 - 523 (1977).

The authors report new intermediate-band photoelectric observations obtained to study the broad absorption features recently discovered in the continua of the Ap stars. Their observations extend the spectral range from λ3200 to λ8500 with a spectral coverage of 82% in a search for previously undetected features. No new features were found. The authors report observations of the strength of the λ5300 feature for 56 Ap stars. For several stars having multiple observations, variability of the λ5300 feature was observed.

113.031   The NGC 2818 photoelectric sequence.
L. P. Connolly.
Publ. Astron. Soc. Pacific, Vol. 89, 528 - 529 (1977).

The photoelectric sequence in NGC 2818 has been observed on one night and found to be free of any significant systematic errors.

113.032   CD −30°5135.   G. Welin.
Inf. Bull. Variable Stars, No. 1313 (1977).

113.033   On the probable variability of HR 1861.
E. H. Olsen.
Inf. Bull. Variable Stars, No. 1332, 3 pp. (1977).

113.034   CPD −55°5216: a cool Ap star showing extremely large variations at 4100 Å.   E. H. Olsen.
Inf. Bull. Variable Stars, No. 1367, 4 pp. (1977).

113.035   The strength of the 2.3 μ CO band in weak-G-band stars.
M. R. Hartoog, S. E. Persson, M. Aaronson.
Publ. Astron. Soc. Pacific, Vol. 89, 660 - 662 (1977).

Broad-band infrared magnitudes and intermediate-band CO indices for 29 weak-G-band stars in the field are presented. The $(J-K)$ color indices of the stars correspond to those of early K giants, in agreement with descriptions of their spectra. All the stars display very weak CO absorption relative to field giants. This confirms a previous conclusion that the weak-G-band stars are deficient in carbon by large factors.

113.036   Intermediate-band photometry of late-type stars III. The Geneva Observatory (GO) photometric system.   O. J. Eggen.
Publ. Astron. Soc. Pacific, Vol. 89, 706 - 715 (1977).

The Geneva Observatory (GO) photometric system for late-type stars, combined with (R, I) photometry is found to be a useful supplement to results obtained on the modified Strömgren system discussed in Papers I and II (Eggen, 1976, 1977). The purpose of this investigation was to examine the possibility that observations on the GO system could be used to supplement those on the modified Strömgren system, described in Papers I and II, in studies of the correlation between metal abundance and kinematics of stars.

113.037   Elementi di fotometria stellare ad uso dei dilettanti.   P. Tempesti.
G. Astron., Vol. 3, 163 - 200 (1977).

113.038   Photometric properties of carbon stars in the galactic latitude zone centered at $l = 90°$.
A. Alksnis, Z. Alksne.
Photometric investigations of carbon stars, (see 003.018), = Radioastrofiz. Obs., Tr. 16, p. 7 - 99 (1977). In Russian.

This is a paper of a series aimed mainly at studying statistically the differences in light variability characteristics between the samples of C-type stars from several regions of the Milky Way. 46 carbon stars are known to be in a 5° × 18° zone oriented perpendicularly and placed symmetrically to the galactic equator and centered at galactic longitude $l = 90°$.

113.039   Photographic photometry of carbon stars in the zone $l = 174°$; $−9° < b < +9°$.
I. Daube, L. Duncāns.
Photometric investigations of carbon stars, (see 003.018), = Radioastrofiz. Obs., Tr. 16, p. 100 - 156 (1977). In Russian.

27 carbon stars have been monitored in the R, V, B passbands. 17 new variables among them were found.

113.040   Middle-band photometric system for classification of carbon stars.   U. Dzērvitis.
Investigations of the sun and red stars. 6, (see 003.019), p. 43 - 54 (1977). In Russian.

A photometric system for classification of carbon stars in the visual and near-infrared spectral region is proposed.

113.041   The photometric standard +15°18.
M. S. Kazanasmas, N. A. Mis'kin.
Astron. Tsirk., No. 942, p. 5 - 7 (1977). In Russian.

113.042   Isocon TV photometry of stars.
N. D. Galinskij, V. N. Ivchenko, G. P. Milinevskij, T. V. Ovchinnikova.
Astron. Tsirk., No. 946, p. 1 - 3 (1977). In Russian.

113.043   Electrospectrophotometry of δ^2 Lyr.
N. A. Gladushina.
Astron. Tsirk., No. 954, p. 5 - 7 (1977). In Russian.

113.044   Long-term infrared monitoring of stellar sources from Earth orbit.   S. P. Maran, T. F. Heinsheimer,
T. L. Stocker, S. P. S. Anand, R. D. Chapman, R. W. Hobbs, A. G. Michalitsanos, F. H. Wright, S. L. Kipp.
Infrared Phys., Vol. 17, 565 - 574 (1977).

These are the preliminary results of the first systematic program of infrared astronomy measurements made from an artificial satellite in Earth orbit. The program consists of intensive, broad-band photometric monitoring of variable sources at wavelength 2.7 μm. The sources, red giant and supergiant stars, are in some cases associated with circumstellar molecules (OH, H_2O, SiO) that emit variable radio maser radiation that may be pumped by the stellar infrared light.

113.045   UX Bootis = BD+47°2135 sehr wahrscheinlich nicht veränderlich.   S. Rößiger.
Mitt. Veränderl. Sterne (MVS), Band 8, 6 (1977).

113.046   Comparison of results of absolute spectrophotometry and UBV photometry.   A. V. Kharitonov.
Astron. Zh. Akad. Nauk SSSR, Vol. 54, 1285 - 1292 (1977). In Russian. English translation in Soviet Astron., Vol. 21, No. 6.
    The color indices $(U-B)_{cal}$ and $(B-V)_{cal}$ and stellar magnitudes V_{cal} for the stars of the Astrophysical Institute spectrophotometric programme are obtained. The calculated color indices and stellar magnitudes are compared with the directly observed ones in the UBV system. The agreement of the spectrophotometric data is satisfactory, and the systematic error of stellar spectrophotometry does not exceed 3−4%. The usually accepted magnitude of Vega $V= +0.03$ is in error; the results give $V= -0^m04 \pm 0^m01$.

    A general catalogue of UBV photoelectric photometry (magnetic tape).   See Abstr. 002.003.

    Catalogue of magnitudes of HR stars in the uniform P_{44} and V systems.   See Abstr. 002.008.

    Errata in the Catalogue of Bright Stars.
See Abstr. 002.012.

    Catalogue of stellar abundances.
See Abstr. 002.019.

    Catalogue of photometric "star boxes" in the UBV $B_1B_2V_1G$ system.   See Abstr. 002.038.

    Catalogue of B, V magnitudes and spectral classes of 720 stars centered at $\alpha_{1950} = 2^h17^m9$, $\delta_{1950} = +58°59'$.
See Abstr. 002.049.

    Multicolor stellar photometry.
See Abstr. 003.150.

    Ptolemy's report on the color of Sirius and modern observations.   See Abstr. 004.082.

    IAU Commission 45: Working Group on Spectroscopic and Photometric Data. Catalogs recently published, to be published or in preparation. List VII.
See Abstr. 010.018.

    The interpretation of cyclical photometric variations in certain dwarf Me-type stars.   See Abstr. 064.048.

    Colors and magnitudes of stars occulted by the rings of Uranus 1977−1980.   See Abstr. 101.025.

    G77-61: a dwarf carbon star.
See Abstr. 112.002.

    Searches for faint OB stars in the southern Milky Way. II. The Vela region.   See Abstr. 114.002.

    Spectral classification of stars with the same colours in intermediate multiband photometry − the concept of photometric "star-box".   See Abstr. 114.005.

    A spectrophotometric survey of the A0535+26 field.   See Abstr. 114.010.

    Infrared stars. Review of observational data.
See Abstr. 114.011.

    Accuracy of two-dimensional spectral classes derived through DDO photometry.   See Abstr. 114.025.

    Separation of Ap stars in the Vilnius photometric system.   See Abstr. 114.026.

    Spectroscopic radial velocity and photometric observations of barium stars.   See Abstr. 114.027.

    Spectral classification using ANS photometric data.
See Abstr. 114.036.

    Spectral energy distributions of standard stars of intermediate brightness. II.   See Abstr. 114.051.

    A calibration of the uvbyβ photometric system in terms of temperature and gravity.   See Abstr. 114.055.

    An examination of Breger's catalog of spectrophotometric scans for broad, continuum features in peculiar A and mercury-manganese stars.   See Abstr. 114.057.

    Observational characteristics of young stars.
See Abstr. 114.060.

    Scanner observations of main-sequence A and F stars.
See Abstr. 114.070.

    A new model for V Pup: is the star a main sequence contact binary?   See Abstr. 117.018.

    Visual binaries among the barium stars. II. HD 105902.   See Abstr. 118.004.

    Spectrophotometry of RS Canum Venaticorum, AR Lacertae, and UX Arietis.   See Abstr. 121.004.

    Photoelectric observations of the eclipsing binary RT Ursae Minoris.   See Abstr. 121.013.

    Photometry of the dwarf cepheid EH Librae.
See Abstr. 122.036.

    A note on the reddening of Polaris B.
See Abstr. 122.092.

    Ultraviolet observations of β Canis Majoris stars with the TD-1A satellite.   See Abstr. 122.104.

    Photoelectric UBV photometry of northern cepheids.
See Abstr. 122.125.

    A study of the reddening and blanketing corrections for RR-Lyrae stars in the Walraven VBLUW photometric system.   See Abstr. 122.151.

    Photometric variations of the B6 star HR 3440.
See Abstr. 123.043.

    Photoelektrische Temperaturbestimmung bei den sehr heißen DA Zwergen.   See Abstr. 126.022.

    The reddening of K-giant stars from DDO photometry.   See Abstr. 131.162.

    Infrared photometry, extinctions and R values for

some Northern Milky Way stars.   See Abstr. 131.194.

Photometry of slow X-ray pulsars. II. The 13.9 minute period of X Persei.   See Abstr. 142.074.

Optical behaviour of HZ Her/Her X-1 in 1976. See Abstr. 142.083.

The energy distribution and the recent light history of X Persei.   See Abstr. 142.085.

Four-color and Hβ photometry of open clusters. XII. NGC 6910 and NGC 6913.   See Abstr. 153.005.

Variable-star photometry of NGC 6530. See Abstr. 153.015.

The Hyades-age cluster NGC 1817. See Abstr. 153.016.

UBV photometry of the young cluster NGC 6604. See Abstr. 153.017.

Four-color and Hβ photometry of stars in NGC 7654 and M25.   See Abstr. 153.022.

Star field surrounding NGC 6528. See Abstr. 154.004.

The distribution of horizontal-branch stars in the log T_{eff}, log g diagram.   See Abstr. 154.018.

Photoelectric UBV photometry in the anomalous globular cluster NGC 2808.   See Abstr. 154.028.

The velocity dispersion of faint red dwarf stars. See Abstr. 155.004.

Observations of blue stars near the galactic anti-center.   See Abstr. 155.014.

RGU three-colour photometry of a field in Aquila near NGC 6755.   See Abstr. 155.023.

Five-channel photometric observations of stars in the Magellanic Clouds and in the Milky Way. See Abstr. 159.007.

114 Spectra, Temperatures, Chemical Composition, Spectra of Individual Stars

Spectra, Temperatures, Chemical Composition

114.001 Recalibration and analysis of Spite's method of quantitative three-dimensional classification for the F8–K1 stars.   L. da Silva, S. Grenier.
Astron. Astrophys., Suppl. Ser., Vol. 29, 195 - 208 (1977).

The method introduced by Spite (1966) is recalibrated and analysed. This calibration renders possible new determinations of T_{eff}, [Fe/H], M_v, $(B - V)_0$, $(V - K)$ and $(G - I)$ for F8–K1 stars. The present calibration is better than Spite's because the sample of standard stars, for each parameter, covers better the extended HR diagram (T_{eff}, M_v and [Fe/H]). The authors analyse the necessity of having a complete grid of accurate standard measurements to employ this method. A discussion of the use of $(G - I)$ and $(V - K)$ for the determination of T_{eff} is given. The results obtained for M_v are compared to Oke's and Wilson-Bappu's. The good precision obtained for $(B - V)$ shows the value of the method.

114.002 Searches for faint OB stars in the southern Milky Way. II. The Vela region.
J. C. Muzzio, A. M. Orsatti.
Astron. J., Vol. 82, 474 - 479, 529 - 539 (1977).

A search for faint OB stars covering an area of 42 deg^2 in the Vela region of the Milky Way resulted in the discovery of 162 OB stars and four early-type supergiants; the blue magnitude range of the survey is about $12.0 \leqslant B \leqslant 15.0$ mag. Distance estimates for OB$^+$ and OB^0 stars which might belong to groups suggest distances of about 5 kpc for most of them.

114.003 The $^{12}C/^{13}C$ ratio in 18 cool carbon stars.
J. L. Climenhaga, B. L. Harris, J. T. Holts, J. Smolinski.
Astrophys. J., Vol. 215, 836 - 844 (1977).

The $^{12}C/^{13}C$ ratio for 18 cool carbon stars has been obtained from a comparison of observed profiles with calculated profiles in the wavelength regions of the (4, 0) CN, (2, 0) CN, and (3, 1) CN bands. The values found for the ratio are all less than 10, which is consistent with the plume method of mixing in stellar interiors.

114.004 Spectroscopic studies of very old hot stars. III. Atmospheric properties of seven planetary nuclei.
S. R. Heap.
Astrophys. J., Vol. 215, 864 - 872 (1977).

The atmospheric properties of seven planetary nuclei which have O or Of-type spectra were determined by use of the Zanstra method and by analysis of the visual line spectra. One finding is that the O-type spectral sequence may extend to effective temperatures of 90,000 kelvins or more, if the results of the Zanstra method are to be believed. However, there are some indications of internal inconsistency in the Zanstra method and some outstanding discrepancies between the effective temperatures indicated by the Zanstra method and those indicated from analysis of the visual line spectra. Two ways to resolve these discrepancies are to postulate an excess of extreme-UV stellar flux or to postulate a hot binary companion for some central stars. In either case the validity of the Harmon-Seaton evolutionary sequence for central stars becomes questionable.

114.005 Spectral classification of stars with the same colours in intermediate multiband photometry – the con-
cept of photometric "star-box".
M. Golay, N. Mandwewala, P. Bartholdi.
Astron. Astrophys., Vol. 60, 181 - 194 (1977).

The $UBVB_1B_2V_1G$ photometric system is very well suited to study the stars that have nearly the same seven colours. The authors say that such stars are contained in photometric boxes. It is found that stars in the same box have nearly the same spectral type, nearly the same values of the parameters used in β photometry, Copenhagen photometry, UV and IR photometry for 2000 Å $< \lambda <$ 11000 Å. Further, the concept of "star box" is used to discuss the classification of Am and Am: stars. Some other possible applications of "star box" are given.

114.006 Stellar Lyman α and Lyman β profiles.
J. P. Vader, S. R. Pottasch, R. C. Bohlin.
Astron. Astrophys., Vol. 60, 211 - 219 (1977).

Measurements of the Lyman α and Lyman β profiles, as obtained by the Copernicus satellite for 20 early-type stars, are reported. The authors compare the profiles to theoretical profiles computed using unblanketed non-LTE model atmospheres. They find the Ly α profiles to be systematically broader than the Ly β profiles contrary to the theoretical predictions, and ascribe this discrepancy to line-blanketing effects.

114.007 Energy distribution in the spectra of 50 stars.
V. I. Burnashev.
Izv. Krymskoj Astrofiz. Obs., Vol. 57, 57 - 81 (1977). In Russian.

The energy distribution in the spectra of 50 stars (spectral region 3200–7550 Å, resolution 30 Å) has been determined by means of fundamental photoelectric spectrophotometry. The data obtained are compared with those of other catalogues.

114.008 Ultraviolet observations of Be stars. I. Macroscopic radial motions in the atmospheres of early Be stars.
J. M. Marlborough.
Astrophys. J., Vol. 216, 446 - 456 (1977).

Ultraviolet observations of six Be stars with a range of $v \sin i$ obtained with Copernicus are described. No emission lines are present in the U2 spectra from ~950 Å to ~1450Å. For stars of large $v \sin i$, the resonance lines are violet displaced with radial velocity components up to a few hundred km s^{-1}. Such macroscopic radial motions are presumably the lower velocity, denser regions of the stellar winds previously observed by Snow and Marlborough. For stars of low $v \sin i$, the limited data available do not indicate similar behavior. N V and possibly O VI absorption lines are present in the spectra of some of the stars of large $v \sin i$. If such ions are produced primarily by collisional ionization, then some Be stars possess coronae with kinetic temperatures of ~2 × 10^5 K.

114.009 Evidence for mass loss in the mid-ultraviolet spectra of Be stars.
T. H. Morgan, Y. Kondo, J. L. Modisette.
Astrophys. J., Vol. 216, 457 - 461 (1977).

The spectral region about the Mg II doublet lines near 2800 Å of Be and shell stars has been observed. The star γ Cas shows narrow components to each resonance doublet line shifted 215 km s^{-1} toward shorter wavelengths. The resonance and subordinate lines of ζ Tau are shifted toward shorter wavelengths by 75 km s^{-1}, and the subordinate lines are greatly strengthened. The star κ Dra, which is much later in spectral type, shows a comparatively normal doublet spectrum.

114.010 **A spectrophotometric survey of the A0535+26 field.**
B. Margon, J. Nelson, G. Chanan, J. R. Thorstensen, S. Bowyer.
Astrophys. J., Vol. 216, 811 - 818 (1977).

The authors have made spectrophotometric observations in the λλ3600–7000 range, with 8 Å resolution, of 11 stars in and near the positional error box of the transient slow X-ray pulsar A0535+26. None of these stars appears unusual, and only HDE 245770, the bright Be star candidate shows emission lines. Limits have been obtained on 104 s light variations similar to the X-ray period for nine of the stars observed: for HDE 245770, this upper limit, obtained during an X-ray outburst, is 0.0002 mag. A formalism is presented to interpret these limits in terms of constraints on the system geometry; this analysis is generally applicable to other pulsing binary X-ray sources.

114.011 **Infrared stars. Review of observational data.**
G. V. Khozov.
Astrofizika, Vol. 12, 705 - 732 (1976). In Russian. – English translation in Astrophysics, Vol. 12, No. 4.

Basic data of photometric, spectral and polarization observations of cool stars are considered. The observations performed in optical, infrared and radio-wave ranges during 1965–1975 are included in the review.

114.012 **Spectra of late-type stars from 4–8μ.**
R. C. Puetter, R. W. Russell, K. Sellgren, B. T. Soifer.
Publ. Astron. Soc. Pacific, Vol. 89, 320 - 322 (1977).

The 4–8μ spectra ($\Delta\lambda/\lambda \simeq 0.015$) of three late-type stars, obtained in October 1975 from the Kuiper Airborne Observatory, are reported. These spectra exhibit mainly stellar photospheric blackbody emission in the case of the M stars and thermal circumstellar dust emission in the C star. The only significant feature seen in this region is that attributed to CO from ~4.5μ to 5.2μ. This feature is observed in both types of stars.

114.013 **The equivalent widths of iron lines in the spectra of cool standard stars for spectral analysis.**
J. B. Hearnshaw.
Publ. Astron. Soc. Pacific, Vol. 89, 356 - 365 (1977).

This paper discusses the advantages of having sets of equivalent widths for standard stars used in spectral analysis which may be considered as essentially free of systematic errors. Such a set is derived for the sun. A similar set is also derived for ε Vir. Comparison of the solar values of log (W/λ) with those given by Moore et al. (1966) is made, and a significant systematic difference is noted, probably due to errors in the earlier photographic solar measurements. The equivalent widths of Cayrel and Cayrel (1963) for ε Vir were also checked for significant systematic errors, but none was found.

114.014 **Chromium, manganese, and iron in the early peculiar stars.** M. S. Allen, C. R. Cowley.
Publ. Astron. Soc. Pacific, Vol. 89, 386 - 390 (1977).

With the results of previous atmospheric diagnoses (Allen 1977) and equivalent-width data derived mostly from plates from the Dominion Astrophysical Observatory, abundances have been computed from lines of singly ionized chromium, manganese, and iron in a sample of A- and B-type normal and metal-enriched stars. The pertinent Warner (1967) oscillator strength scales were adjusted in accordance with recent laboratory work and measurements from the solar spectrum. The stars in the sample, which includes normal, Hg-Mn, Ap, and Am stars, show a wide variation in the relative abundances of these elements; in two very pronounced Mn stars, 53 Tau and HR 2844, large manganese excesses (i.e., Mn/Fe > 1) are derived. This result indicates the operation of nonnuclear processes on the composition of the observable layers of these stars.

114.015 **Bright southern stars of astrophysical interest.**
J. Andersen, B. Nordström.
Astron. Astrophys., Suppl. Ser., Vol. 29, 309 - 312 (1977).

The paper lists a number of bright peculiar stars in the southern hemisphere, discovered on 20 Å/mm spectrograms. New information is given also for a few known peculiar objects.

114.016 **The radial velocities of early-type stars within six degrees of the galactic anticenter direction.**
J. W. Christy.
Astrophys. J., Vol. 217, 127 - 133 (1977).

Radial velocities and MK spectral types have been determined at the Flagstaff Station for 30 O and B type stars which are within 6° in latitude and in longitude of the galactic anticenter direction. The velocities and spectral types are combined with published photometry, velocities, and spectral types for these and 26 other young anticenter stars to derive distances and velocities along the galactic radius. These stars lie in the anticenter spur of the Orion spiral arm.

114.017 **Quantitative Klassifikation von Sternspektren mittels Objektivprismenaufnahmen.**
T. Schmidt-Kaler, G. Diaz-Santanilla, R. Rudolph, H. Unger.
Forschungsber. Nordrhein-Westfalen, Nr. 2595, Fachgruppe Phys./Math. Westdeutscher Verlag, Opladen. 62 pp. (1976).

A new method has been tested to calibrate accurately objective-prism spectra (90 Å/mm astrograph, 570 Å/mm Hamburg Schmidt telescope) by means of photoelectric UBV magnitudes of sequences of stars whose spectra are on the same plates. The intensity of the continuous spectrum has been determined on the basis of windows (with negligible absorptions) selected after inspection of high dispersion spectra. In the case of later type stars a pseudo-continuum has been defined and an equivalent area (= line depth × half width) to replace the equivalent width when the wings of the line are heavily disturbed by blends.

114.018 **Carbon, nitrogen, and oxygen abundances in 11 G and K giants.** D. L. Lambert, L. M. Ries.
Astrophys. J., Vol. 217, 508 - 520 (1977).

Carbon, nitrogen, and oxygen abundances have been determined for 11 G and K giants. High-resolution photoelectric scans of C_2, CH, CN, [O I], and [C I] lines have been combined with recent model atmospheres. Relative to the solar atmosphere, the C abundances are depressed, the N abundances enhanced, and the O abundances unchanged. The observed CNO abundances and the previously obtained $^{12}C/^{13}C$ ratios are in good agreement with the predictions for a giant after the convective envelope has mixed material to the surface from the zone which was partially processed during the star's main-sequence lifetime.

114.019 **Étoiles carbonées.** N. Dallaporta.
Recueil des séminaires, (see 012.004), p. 55 bis - 76 (1977).

114.020 **Carbon and nitrogen abundances in F- and G-type stars.** R. E. S. Clegg.
Mon. Not. R. Astron. Soc., Vol. 181, 1 - 30 (1977).

Carbon and nitrogen abundances have been obtained for a sample of 11 F- and G-type dwarfs covering a range in [Fe/H] from −0.8 to +0.3. Model atmospheres, which included the effects of convection and line blanketing, were used to calculate synthetic spectra of the CH, CN and NH molecular bands. Effective oscillator strengths for the bands studied were found by matching synthetic spectra calculated from a model solar atmosphere with the observed solar bands; this was done at the same resolution as was obtained from the stellar coudé spectra. A comprehensive discussion of the theoretical errors is given, and some applications to galactic evolution are noted.

114.021   Spectral classification from the ultraviolet line features of S2/68 spectra. II. Late B-type stars.
A. Cucchiaro, D. Macau-Hercot, M. Jaschek, C. Jaschek.
Astron. Astrophys., Suppl. Ser., Vol. 30, 71 - 79 (1977).

The sky survey telescope (S2/68) in the TD1 satellite has provided in the wavelength range λλ1350−2740 Å a very large number of spectra of stars classed B in the visible. This paper is the second of the series and concerns the B6 to B9 stars. As in the first one, a statistical study of the UV spectra is carried out in order to establish criteria of classification. The UV criteria have permitted a classification scheme to be made and also the singling out of non-normal UV objects.

114.022   Neutral-ion anomaly in cool stars.   V. Oinas.
Astron. Astrophys., Vol. 61, 17 - 20 (1977).

An analysis of the abundances in a group of cool stars confirms the presence of a neutral-ion anomaly found by Oinas (1974); i.e., the observed strengths of the ion lines are too strong when compared to the theoretical values calculated from model atmospheres. No satisfactory explanation has been found for the anomaly.

114.023   Image-tube spectrograms of southern young emission-line objects.   I. Appenzeller.
Astron. Astrophys., Vol. 61, 21 - 26 (1977).

31 image-tube spectrograms of 18 southern emission-line stars are described. Two of the observed objects appear to be almost normal late-type dwarfs. Three other ones were found to be peculiar early-type emission-line stars. The remaining 13 objects are T Tauri stars. At least six of these objects show spectroscopic evidence for mass accretion and seem to belong to the YY Orionis subclass of the T Tauri stars. The present data suggest that YY Orionis stars are more numerous than assumed so far.

114.024   Long-term changes in ultraviolet P Cygni profiles observed with Copernicus.   T. P. Snow, Jr.
Astrophys. J., Vol. 217, 760 - 770 (1977).

Fifteen O and B stars from the earlier survey of mass-loss effects were reobserved in 1976 to search for changes in the P Cygni profiles in times of two to four years. In most cases some variations were seen, at least in details of the profiles, and, for a few stars, in the overall structure of the features. While the time scales are not well known, there is evidence that some of the changes have occurred gradually over times of years. In at least two instances (ζ Pup and δ Ori A), it appears that the changes represent real variations in the mass-loss rates.

114.025   Accuracy of two-dimensional spectral classes derived through DDO photometry.   K. M. Yoss.
Astron. J., Vol. 82, 832 - 841 (1977).

A comparison is made of 240 spectral types and luminosity classes obtained through spectrograms and through the DDO photometric system. Agreement within two spectral subclasses is found in 89% of the sample and agreement within one-half luminosity class in 83% of the cases. CN strengths also are compared, with satisfactory agreement.

114.026   Separation of Ap stars in the Vilnius photometric system.   V. Straižys, V. Žitkevičius.
Astron. Zh. Akad. Nauk SSSR, Vol. 54, 987 - 995 (1977). In Russian. English translation in Soviet Astron., Vol. 21, No. 5.

The response curves of the magnitudes of the Vilnius photometric system are situated in optimum positions for separation and classification of Ap stars. This enables to recognize Ap stars neglecting the presence of interstellar reddening. The separation of Ap stars is based on the measurement of the Balmer jump and the continuum depressions around λ4100 and 5200 Å.

114.027   Spectroscopic radial velocity and photometric observations of barium stars.

R. M. Catchpole, B. S. C. Robertson, P. R. Warren.
Mon. Not. R. Astron. Soc., Vol. 181, 391 - 404 (1977).

The authors have obtained spectra of 74 and photometry of 122 of the 150 certain barium stars listed by MacConnell, Frye & Upgren. The spectra show that these are not a homogeneous group of stars. However, the spectra and *UBVI* photometry indicate that about two-thirds are comparable with classical barium stars. The authors use their radial velocities, combined with published values, to estimate a mean absolute magnitude of about $M_V = -1$ for the barium stars. There is a suggestion that when the barium stars are divided into two groups, according to their observed barium line strengths, they have different mean ages and luminosities.

114.028   Three-micron absorption band of carbon stars.
K. Noguchi, T. Maihara, H. Okuda, S. Sato, T. Mukai.
Publ. Astron. Soc. Japan, Vol. 29, 511 - 525 (1977).

Narrow-band spectrophotometric observations have been carried out for 22 carbon stars and 23 M-type stars between 2.85 and 4.1 μm. It is shown that all the carbon stars exhibit a broad absorption feature centered at about 3.05 μm, while no trace of absorption was found in the M-type stars. It is found that the band depth correlates with color indices $U-V$, $B-V$ as well as the abundance ratio CN/C_2. Certain kinds of dust particles and molecules are examined as absorbing agents.

114.029   Diffusion, turbulence and abundance anomalies in F, B and A stars.   G. Michaud.
J. R. Astron. Soc. Canada, Vol. 71, 401 (1977). − Abstract.

114.030   Linear polarization across Hα in the Be star γ Cassiopeiae.   R. Poeckert, J. M. Marlborough.
J. R. Astron. Soc. Canada, Vol. 71, 406 - 407 (1977). Abstract.

114.031   Mean chemical abundance of the F stars as a function of distance from the galactic plane.
A. Blaauw.
Highlights of Astronomy, Vol. 4, Part II, (see 012.022), p. 51 - 52 (1977).

114.032   Present status of spectral classification in the conventional wavelength range with emphasis upon early-type stars.   C. Jaschek.
Highlights of Astronomy, Vol. 4, Part II, (see 012.025), p. 283 - 288 (1977).

114.033   Spectral classification of early type stars from the low dispersion ultraviolet spectra.   K. Nandy.
Highlights of Astronomy, Vol. 4, Part II, (see 012.025), p. 289 - 301 (1977).

The methods of spectral classification from the low dispersion ultraviolet spectra obtained with the S2/68 experiment in the TD1 satellite have been described. The bright stars, the spectra of which are photometrically accurate, can be divided into natural groups according to the spectral appearance of the features. These features vary in strength with spectral type and luminosity, and enable separation between main sequence and luminous stars. For fainter stars the spectral data have been combined to obtain narrow band magnitudes at several wavelengths. An ultraviolet photometric system which enables determinations of spectral type and luminosity of early type stars is described and the results for about 3000 stars are presented.

114.034   Spectral classification of B and A stars from the line features of S2/68 spectra.
A. Cucchiaro, M. Jaschek, C. Jaschek.
Highlights of Astronomy, Vol. 4, Part II, (see 012.025), p. 303 (1977). − Summary.

114.035 **Behaviors of B star continua and absorption features determined from the TD1 S2/68 "Ultraviolet Bright Star Spectrophotometric Catalogue" and from Copernicus spectra.** J. M. Vreux, J. P. Swings.
Highlights of Astronomy, Vol. 4, Part II, (see 012.025), p. 305 (1977).

114.036 **Spectral classification using ANS photometric data.** R. J. van Duinen, P. R. Wesselius.
Highlights of Astronomy, Vol. 4, Part II, (see 012.025), p. 311 - 313 (1977).

114.037 **Spectral classification with objective-prism spectra from Skylab.**
K. G. Henize, S. B. Parsons, J. D. Wray, G. F. Benedict.
Highlights of Astronomy, Vol. 4, Part II, (see 012.025), p. 315 - 322 (1977).
This paper investigates the correlation of the intensities of the C IV and Si IV lines with MK spectral type, and presents a preliminary classification scheme for O4 to B2 stars based on these lines.

114.038 **An atlas of ultraviolet stellar spectra.** A. D. Code.
Highlights of Astronomy, Vol. 4, Part II, (see 012.025), p. 325 - 337 (1977).

114.039 **A new temperature scale for B stars based on OAO-2 data.** J. R. Lesh.
Highlights of Astronomy, Vol. 4, Part II, (see 012.025), p. 339 (1977).

114.040 **The near ultraviolet spectrum of Fe III as a classification criterion.**
R. Faraggiana, H. J. G. L. M. Lamers, M. Burger.
Highlights of Astronomy, Vol. 4, Part II, (see 012.025), p. 349 - 351 (1977).

114.041 **Spectral classification from Copernicus data.** W. P. Bidelman.
Highlights of Astronomy, Vol. 4, Part II, (see 012.025), p. 355 - 359 (1977).

114.042 **The Mg II features near 2800 Å and spectral classification.** Y. Kondo.
Highlights of Astronomy, Vol. 4, Part II, (see 012.025), p. 363 - 364 (1977).

114.043 **Detection of errors in spectral classification by cluster analysis.**
A. Heck, A. Albert, D. Defays, G. Mersch.
Astron. Astrophys., Vol. 61, 563 - 566 (1977).
Cluster analysis methods are applied to the photometric catalogue of $uvby\beta$ measurements by Hauck and Lindemann (1973) and point out 249 stars the spectral type of which should be reconsidered or the photometric indices of which should be redetermined.

114.044 **MK spectral classifications for southern OB stars.** R. F. Garrison, W. A. Hiltner, R. E. Schild.
Astrophys. J., Suppl. Ser., Vol. 35, 111 - 126 (1977).
New MK spectral classifications are provided for southern OB stars in the Heidelberg objective-prism survey brighter than 10th magnitude. Some fainter stars are included in the total of 1113 stars.

114.045 **Temperature sequence in M giants from the molecular spectrum of titanium monoxide.**
A. V. Shavrina.
Astrometr. i Astrofiz., Kiev, vyp. (No.) 32, (see 003.011), p. 29 - 33 (1977). In Russian.
Molecular vibrational temperatures are derived from

titanium monoxide spectra of a number of M-giant stars (M2–M8). A correlation of these temperatures with spectral type is found and compared with Johnson's scale of effective temperatures.

114.046 **On the relation between mean spectrophotometric gradients and some physical parameters of MS stars.**
R. I. Chuprina.
Astrometr. i Astrofiz., Kiev, vyp. (No.) 32, (see 003.011), p. 42 - 45 (1977). In Russian.
MS-stars are studied by spectrophotometric observations in the near-infrared spectral region. Mean gradients of the stars are compared with Jones' values obtained for the same spectral region. The variation of mean gradients with spectral subclasses, colour indices $(I-K)$ and temperature is discussed.

114.047 **Observational studies of the Herbig Ae/Be stars. I. High-resolution Hα profiles.**
L. M. Garrison, C. M. Anderson.
Astrophys. J., Vol. 218, 438 - 443 (1977).
High-resolution Hα emission-line profiles are presented for 14 of the Herbig Ae/Be stars, generally considered to be pre-main-sequence objects. Very diverse profile structures are exhibited, including simple emission lines, double-peaked lines, and variations of the P Cygni profile. Despite the small sample, some systematic velocity trends are indicated. The observations are compared with theoretical models in the literature; models including rotation seem necessary, and the infall-rotation model recently proposed for T Tauri stars by Ulrich may have applicability.

114.048 **Peculiarities among the cool stars.** B. Warner.
Modern astronomy, (see 003.013), p. 124 - 131 (1977).

114.049 **The enigmatic Wolf-Rayet stars.** D. A. Allen.
Modern astronomy, (see 003.013), p. 132 - 138 (1977).

114.050 **The chemical composition of late-type supergiants. II. Lithium abundances for 19 G and K Ib stars.**
R. E. Luck.
Astrophys. J., Vol. 218, 752 - 766 (1977).
From an analysis of high-resolution, high-signal-to-noise spectra, lithium abundances or upper limits are derived for 19 G and K Ib stars. Non-LTE effects are found to be significant, particularly in later-type stars. The relation between lithium and $^{12}C/^{13}C$ is explored and the functional form is found to be best interpreted as a step function. Various mechanisms for lowering $^{12}C/^{13}C$ ratios are constrained by the derived lithium abundances. Meridional mixing is found to be the most viable mechanism for lowering $^{12}C/^{13}C$ ratios in Ib stars.

114.051 **Spectral energy distributions of standard stars of intermediate brightness. II.** R. P. S. Stone.
Astrophys. J., Vol. 218, 767 - 769 (1977) = Lick Obs. Bull., No. 758.
Spectral energy distributions for 16 stars intended for use as spectrophotometric standards for large telescopes are presented. This list incorporates the eight stars previously described, with slightly revised magnitudes resulting from use of a more recent calibration of Vega. The monochromatic magnitudes of the stars range between 9 and 13; the eight new stars average nearly a magnitude fainter than those previously presented. The wavelength region covered extends from $\lambda 3200$ to $\lambda 8370$.

114.052 **Infrared spectral observation of stars.**
K. Komaki, K. Kodaira, W. Tanaka, Z. Suemoto.
Bull. Inst. Space Aeronaut. Sci. Univ. Tokyo B, Vol. 12, 623 - 630 (1976). In Japanese. – From Phys. Abstr., Vol. 80, Abstr. 82218 (1977).

114.053   **Spectrophotometry of cool carbon stars.**
C. E. Gow.
Publ. Astron. Soc. Pacific, Vol. 89, 510 - 518 (1977).

Observations of 75 carbon stars have been made with the Indiana rapid spectrum scanner covering a wavelength range of 5000 Å−7000 Å at 30 Å resolution. The data have been used to form molecular indices for a quantitative measurement of the strength of the C_2 Swan system and the shape of the CN red (6.1) band sequence which is sensitive to the C^{12}/C^{13} isotope ratio. Using C/O ratios from Kilston (1975) it is possible to calculate the observed CN/C_2 ratio and determine photometric C/O ratios for 61 cool carbon stars. The CN red (6,1) band photometry shows that only a few (25%) of cool carbon stars have C^{12}/C^{13} isotope ratios significantly lower than the majority of cool carbon stars.

114.054   **On the question of molecular features in cool Ap stars.**   C. R. Cowley, P. J. Etzler.
Publ. Astron. Soc. Pacific, Vol. 89, 524 - 527 (1977) = Dominion Astrophys. Obs., Victoria, B.C., Contrib. No. 336 = NRC No. 15905.

The authors conclude that lines in the violet system of CN are weakly present ($\sim 5-10$ mÅ) in the spectra of some cool Ap stars.

114.055   **A calibration of the $uvby\beta$ photometric system in terms of temperature and gravity.**
F. R. Zabriskie.
Publ. Astron. Soc. Pacific, Vol. 89, 561 - 568 (1977).

This paper presents a calibration of the $uvby\beta$ system expressed as contour lines for the indices $(b-y)$, m_1, c_1, and β on the $(\log T_{eff}, \log g)$ plane. The calibration extends to B, A, and early F supergiants. A comparison with part of the MK classification is also shown, and further applications are discussed.

114.056   **A correlation between multiplicity and metallicity for solar-type stars.**   D. C. Barry.
Publ. Astron. Soc. Pacific, Vol. 89, 612 - 613 (1977). Abstract.

114.057   **An examination of Breger's catalog of spectrophotometric scans for broad, continuum features in peculiar A and mercury-manganese stars.**
S. J. Adelman.
Publ. Astron. Soc. Pacific, Vol. 89, 650 - 657 (1977).

Published spectrophotometric scans of peculiar A and mercury-manganese stars with $(B-V) \leqslant +0.10$ are examined for broad, continuum features. Four out of 26 Ap stars show the λ4200 feature, 19 out of 33 the λ5200 feature, and 3 of 29 the λ6300 feature. The λ5200 feature may be present in 2 of 11 HgMn stars. The detection of the λ5200 features via spectrophotometry is compared with the filter photometric results of Maitzen. Problems in the detection and in the measurement of these features are discussed.

114.058   **The behavior of the Mg II lines near 2800 Å in A, B, and O stars.**
Y. Kondo, T. H. Morgan, J. L. Modisette.
Publ. Astron. Soc. Pacific, Vol. 89, 675 - 683 (1977).

The equivalent widths of the Mg II resonance doublet and their subordinate lines near 2800 Å have been determined in ten stars of spectral types A through O. Combining the current results with previously published results for four F stars, the authors find that the total equivalent widths of the Mg II lines follow a pattern and that the maximum in absorption strength occurs in late-A spectral type. Evidence of mass flow has been found in the Algol spectrum together with some other peculiarities. The general behavior of the Mg II lines in stars in this spectral range is also discussed.

114.059   **New carbon stars in the zone of galactic latitude $b = -7°$.**   A. Alksnis, Z. Alksne, V. Ozoliņa.

Investigations of the sun and red stars. 6, (see 003.019), p. 55 - 67 (1977). In Russian.

In the zone centered at galactic latitude $b = -7°$ between galactic longitudes $l = 68°$ and $l = 184°$ 32 new carbon stars have been found. Their equatorial and galactic coordinates, as well as spectral types and identification charts are given.

114.060   **Observational characteristics of young stars.**
L. V. Mirzoyan.
Early stages of stellar evolution, (see 012.041), p. 100 - 106 (1977). In Russian.

A general characteristic of photometric and spectral features of young stars is given. Possible ways for explanation of unstationarity of these objects are discussed.

114.061   **Spectral classification of some stars in the neighbourhood of γ Cygni.**   T. A. Uranova.
Astron. Tsirk., No. 946, p. 4 - 5 (1977). In Russian.

114.062   **On the structure of the C III emission triplet spectrum of WR stars.**
A. A. Nikitin, A. F. Kholtygin, T. Kh. Feklistova.
Publ. Tartu Astrofiz. Obs., Vol. 45, 45 - 62 (1977). In Russian.

Data on the C III triplet lines the laboratory sliding spark spectrum and for the WR stars are analyzed and compared. A group of dominant states for triplet recombination has been singled out, the energies of states and the parameters for the generalized hydrogenic wave functions are given in a single configuration approximation.

114.063   **Technique of a quantitative spectral classification of F5-G5 stars with determination of the metallicity parameter.**   V. Malyuto.
Publ. Tartu Astrofiz. Obs., Vol. 45, 150 - 172 (1977). In Russian.

A technique of spectral classification is described. This is based on the line-depth ratios measured in objective prism spectra ($D = 166$ Å/mm at Hγ). This technique allows the determination of the spectral classes and of the absolute magnitudes of F5-G5 stars with mean errors ±0.6 of the spectral subclass and ±0ᵐ6 of the absolute magnitude respectively. The metal-deficient F5-G5 dwarfs can be distinguished.

114.064   **The $^{12}C/^{13}C$ ratio in carbon stars.**
Y. Fujita, T. Tsuji.
Publ. Astron. Soc. Japan, Vol. 29, 711 - 730 (1977).

The carbon isotope ratio of $^{12}C/^{13}C$ is determined for nine carbon stars from an empirical pseudo curve-of-growth analysis on selected rotational lines of the CN red system. ^{13}C abundance relative to ^{12}C is decreased in normal N-type carbon stars as compared with K-M giant and supergiant stars in which $^{12}C/^{13}C$ ratios are generally lower than 20. This means that carbon enrichment in the evolution of a carbon star should have been produced by extensive mixing of ^{12}C in a majority of cool carbon stars. Such a conclusion is consistent with the recent theory of mixing during thermal pulses due to instability at the He-shell burning stage.

114.065   **Spectral classification of carbon stars by means of photoelectric photometry of line strengths.**
Y. Yamashita, S. Nishimura, M. Shimizu, T. Noguchi, E. Watanabe, K.-i. Okida.
Publ. Astron. Soc. Japan, Vol. 29 731 - 738 (1977).

Indices of line strengths for Ca I λ4227, CN λ4216, Ba II λ4554, C_2 λ5165, Na I λ5893 (D-lines), and Li I λ6708 were obtained for about 70 carbon stars. The correlations between these indices and the intensities eye-estimated on spectrograms are discussed. The C-classification was made from these correlation curves. In total, 22 carbon stars were newly classified in the C-system.

114.066 **On the TiO absorption bands in red-giant spectra.**
T. Kipper.
Astron. Zh. Akad. Nauk SSSR, Vol. 54, 1241 - 1245 (1977).
In Russian. English translation in Soviet Astron., Vol. 21,
No. 6.

The observed and computed intensities of TiO α-system
band-heads are compared. It is concluded that the oxygen-
carbon abundance ratio for the observed set of M-giants is
lower than the solar value. The effective temperature scale for
late M-giants is estimated.

114.067 **A study of the uranium stars and related objects.**
C. R. Cowley, G. C. L. Aikman, W. A. Fisher.
Publ. Dominion Astrophys. Obs., Vol. 15, 37 - 72 (1977) =
NRC No. 15613.

Most of the chemically peculiar A-stars of the magnetic
sequence show U II lines in their spectra, while those of Th II
are weak or absent altogether. A non-nuclear explanation of
these observations (diffusion or magnetic accretion) may well
be possible, but for the present, one cannot exclude the
possibility that the atmospheres of these stars contain debris
from recent r-process events. The goals of this paper are three-
fold: The first is to present theoretical calculations (in LTE)
of U II and Th II line strengths based on realistic gf-values and
partition functions. The second is to give a thorough discussion of the observational material upon which the uranium
identification is based. Finally the interpretation of this
material is briefly discussed in terms of nuclear and differentiation (non-nuclear) mechanisms.

114.068 **Neue Beobachtungsergebnisse für A$_m$ Sterne.**
E. Böhm-Vitense, P. Johnson.
Mitt. Astron. Ges., Nr. 42, p. 137 (1977).

114.069 **New red objects resembling Herbig-Haro objects.**
A. L. Gyul'budagyan, T. Yu. Magakyan.
Dokl. AN ArmSSR, Vol. 64, No. 2, p. 104 - 107 (1977). In
Russian. − Abstr. in Ref. zh., 51. Astron., 12.51.547 (1977).

114.070 **Scanner observations of main-sequence A and F
stars.** E. Böhm-Vitense, P. Johnson.
Astrophys. J., Suppl. Ser., Vol. 35, 461 - 469 (1977).

In order to understand the observed UBV colors of main-
sequence A and F stars, and especially to understand the $B-V$
gap observed for field stars and some clusters, the authors have
made scanner observations of main-sequence stars with
$0.20 < B-V < 0.45$ in different clusters and with different
$v_r \sin i$. The results are given in tables.

114.071 **Peculiarities of peculiar stars. X, XI.**
B. Kuchowicz.
Urania Kraków, Vol. 48, 270 - 275, 297 - 305 (1977). In
Polish.

**A uniform edition of the Stockholm Southern
Milky Way Survey (magnetic tape).** See Abstr. 002.002.

MK spectral classifications. Third general catalogue.
See Abstr. 002.004.

Errata in the Catalogue of Bright Stars.
See Abstr. 002.012.

Catalogue of stellar abundances.
See Abstr. 002.019.

**A catalogue of 0.2 Å resolution far-ultraviolet
stellar spectra measured with Copernicus.**
See Abstr. 002.022.

**Catalogue of B, V magnitudes and spectral classes
of 720 stars centered at** $\alpha_{1950} = 2^h17^m9$, $\delta_{1950} = +58°59'$.
See Abstr. 002.049.

**IAU Commission 45: Working Group on Spectro-
scopic and Photometric Data. Catalogs recently published, to
be published or in preparation. List VII.**
See Abstr. 010.018.

**Silicon carbide and the infrared excess of carbon
stars.** See Abstr. 022.025.

Continuous spectra of Wolf-Rayet stars.
See Abstr. 064.013.

**Limb darkening coefficients for late-type giant
model atmospheres.** See Abstr. 064.046.

**Equivalent width of molecular lines in stars. III:
Lines of Lyman band of H$_2$ and (A−X) band of CO and SiO
in stars.** See Abstr. 064.049.

**f-values and abundances of the elements in the sun
and stars.** See Abstr. 071.024.

**Confirmation of the presence of iron hydride in
sunspots and cool stars.** See Abstr. 072.005.

**Radial velocities of southern B stars determined at
the Radcliffe Observatory − VIII. Stars with HD spectral types
B8 and B9.** See Abstr. 112.005.

**UBV photometry and MK spectral classification of
northern early-type stars at intermediate galactic latitudes.**
See Abstr. 113.005.

A photometric peculiarity index in the Ap stars.
See Abstr. 113.030.

**Direct observations of the heterogeneity of super-
giant disks.** See Abstr. 115.008.

**A redetermination of the absolute magnitudes of
the barium stars.** See Abstr. 115.010.

On the nature of population I Wolf-Rayet stars.
See Abstr. 119.012.

**The carbon and nitrogen abundances of Gamma
Pavonis.** See Abstr. 126.011.

**Relation between metallicity and multiplicity for
solar type stars.** See Abstr. 131.010.

**A survey of southern dark clouds for Herbig-Haro
objects and H-alpha emission stars.** See Abstr. 131.180.

Peculiar central stars of planetary nebulae.
See Abstr. 135.002.

Spectroscopic studies of stars in Per OB2.
See Abstr. 152.001.

Spectral types in the open cluster NGC 6633.
See Abstr. 153.004.

**Scanner observations of faint dwarf M stars near the
north galactic pole.** See Abstr. 155.011.

Spectra of Individual Stars

114.501   The spectrum of the T Tauri star Lk Hα 120.
M. V. Penston, P. M. Keavey.
Mon. Not. R. Astron. Soc., Vol. 180, 407 - 413 (1977).

Spectra show a rich emission line spectrum including permitted lines of neutral and once ionized iron-peak elements. P Cygni emission from the Balmer lines and calcium K is also seen — the absence of H emission is explained by neutral iron absorption and results in fluorescent emission. The simple physical and geometrical model discussed in the conclusion seems capable of qualitatively explaining all the data.

114.502   Model atmosphere analysis of HR 8799.
T. Gehren.
Astron. Astrophys., Vol. 59, 303 - 315 (1977).

From high dispersion spectra in the wavelength region $3700 \lesssim \lambda \lesssim 6000$ Å the metal-deficient star HR 8799 = HD 218396 is analyzed by means of model atmospheres in convective equilibrium without line blanketing. Comparison of observed and computed quantities (Balmer lines, some 250 metal lines) leads to an atmospheric model with T_{eff} = 6800 ± 200 K, log g = 3.3 ± 0.5, and a mean non-thermal velocity of 3 km s^{-1}. Disagreement between spectroscopic and photometric observations leads to the conclusion that HR 8799 must be part of an optical pair or a binary system.

114.503   Observed departures from LTE ionization equilibrium in late-type giants.   L. W. Ramsey.
Astrophys. J., Vol. 215, 827 - 835 (1977).

Photoelectric scans of the Ca I line at 6572 Å and the forbidden Ca II transition at 7323 Å are studied in the K giant α Tau, the M supergiant α Ori, and the M giants β And, α Cet, μ Gem, and β Peg. The relative strengths of these lines are shown to be indicative of the ratio of the relative number densities of the neutral and ionized species in the photosphere. The analysis indicates an overionization relative to LTE in qualitative agreement with the theoretical calculations of Auman and Woodrow for the K and M giants. The M supergiant α Ori exhibits a large overionization relative to LTE.

114.504   The effects of stellar rotation on measurements from coudé spectrograms.   D. J. Stickland.
Mon. Not. R. Astron. Soc., Vol. 180, 675 - 682 (1977).

The effects of rotation on the measurement of line blocking and equivalent widths from tracings of coudé spectrograms have been studied by artificially 'spinning-up' the spectra of two sharp-lined stars. The anticipated decrease in the measured blocking coefficients with increase of ν sin i is confirmed for F stars, while abundances deduced from curve of growth analyses are increased owing to the apparent domination of line blending over depression of the continuum as factors affecting equivalent widths.

114.505   Absolute spectrophotometry of bright stars in the equatorial zone. I.   N. S. Komarov, Yu. V. Borisov.
Byull. Inst. Astrofiz., Dushanbe, No. 66 - 67, p. 50 - 58 (1976). In Russian.

The spectral energy distribution for eight bright stars situated in the neighborhood of the equatorial zone are published.

114.506   The red/infrared spectrum of CPD−56°8032.
A. D. Thackeray.
Observatory, Vol. 97, 165 - 169 (1977).

Many C II, C III lines with laboratory analysis by Edlén's group but not in the Revised Multiplet Table are identified in CPD −56°8032. The star is regarded as very probably having variable velocity.

114.507   The spectrum of the companion of Mira Ceti.

Y. Yamashita, H. Maehara.
Publ. Astron. Soc. Japan, Vol. 29, 319 - 329, with a correction p. 835 (1977).

Several spectrograms of Mira Ceti that show some traces of the spectrum of the companion have been obtained near its light minima since 1967. The spectrum observed on December 10, 1975 shows strong emission and absorption features of P Cygni type. Radial velocities are measured for both emission and absorption components of P Cygni type. The broad Balmer emission lines show a large positive velocity, while the broad Balmer absorption lines have a large negative velocity. The radial velocities and the line profiles of emission and absorption lines are interpreted in terms of an expanding and rotating disk model with the companion at the center.

114.508   Line blocking and reddening of β Orionis. A new determination of the empirical effective temperature.   R. Stalio, P. L. Selvelli, L. Crivellari.
Astron. Astrophys., Vol. 60, 109 - 114 (1977).

The ultraviolet spectral energy distribution of the B8 Ia star β Orionis, derived from OAO-2 observations, has been corrected for the line-blocking fraction measured from Copernicus scans with 0.4 Å resolution. The resulting continuum appears to be affected by a not negligible color excess; one may estimate a value of E_{B-V} = 0.04. The empirical effective temperature determined from the angular diameter and from the total absolute flux, corrected for reddening, is 12070°K.

114.509   Diffusion processes and abundance anomalies in the Hg-Mn star κ Cnc.   G. Alecian.
Astron. Astrophys., Vol. 60, 153 - 159 (1977).

Radiative forces and diffusion velocities of a number of representative elements in the atmosphere of the typical Hg-Mn star κ Cnc have been computed. A real correlation has been found between the computed diffusion velocities and the observed abundance anomalies. Abundance variations with time have been computed. The results are in good agreement with the observations. The case of yttrium has been discussed.

114.510   An analysis of the helium-rich star HD 186205.
P. Lee, A. O'Brien.
Astron. Astrophys., Vol. 60, 259 - 262 (1977) = Contrib. Louisiana State Univ. Obs., Baton Rouge, Louisiana, No. 117.

A detailed analysis of the suspected helium-rich star HD 186205 has been performed with the following results: T_{eff} = 23 500°K, log g = 3.97, He/H = 0.72 by number. Other metallic abundances are also derived. The mass, radius, luminosity and distance are found to be 12.3 $M_\odot$, 6.0 $R_\odot$, log (L/L$_\odot$) = 4.0 and 1120 pc respectively.

114.511   Line spectrum variations in the Ap star HD 215441.   A. H. Krautter.
Astrophys. J., Vol. 216, 33 - 36 (1977).

The equivalent widths of 40 lines in the spectrum of the peculiar A star HD 215441 were measured on 13 coudé spectrograms. The results show that the line strengths vary by about ±20% over the cycle of the star. The effect of the large, variable magnetic field on the line strengths is examined and found to be unable to account for the observed variations. The line-strength and photometric variations are discussed in terms of the flux redistribution theory proposed by Peterson; and it appears that this process is active in HD 215441, much as it is in α² CVn.

114.512   Element identifications in Przybylski's star.
C. R. Cowley, A. P. Cowley, G. C. L. Aikman, H. M. Crosswhite.
Astrophys. J., Vol. 216, 37 - 41 (1977) = Dominion Astrophys. Obs. Contrib. 323 = NRC No. 15605.

Line coincidence statistics confirm that the iron peak

elements are definitely, though very weakly, represented in the spectrum of HD 101065.

114.513 High-resolution polarization observations inside spectral lines of magnetic Ap stars. I. Instrumentation and observations of β Coronae Borealis.
E. F. Borra, A. H. Vaughan.
Astrophys. J., Vol. 216, 462 - 478 (1977).

The authors constructed a coudé photon-counting polarimeter capable of attaining (with a Fabry-Perot interferometer) a high resolution. A description of the instrument is given, with a discussion of various sources of systematic error in the polarimetry. Observations of linear and circular polarization in the spectrum of the Ap star β Coronae Borealis, throughout the magnetic cycle, are obtained across an Fe II and a Sm II line at a resolution of 0.086 Å. Inferences are drawn regarding the magnetic geometry of the star: the geometry appears to be devoid of any symmetry but can probably still be approximated by a decentered dipole model. The longitudinal magnetic curve of the star is derived from the available data.

114.514 Comparison of predicted and observed spectral energy distributions of A-type stars.
R. J. Panek.
Astrophys. J., Vol. 216, 747 - 756 (1977).

Emergent fluxes are determined from published low-resolution spectrum scans from 1100 to 8000 Å for seven bright A-type stars which have well-determined empirical angular diameters, effective temperatures, and parallaxes. The stellar flux is compared with the emergent flux from a line-blanketed model atmosphere which has the same effective temperature. The observational accuracy is found to be sufficient to place useful constraints on model atmosphere results.

114.515 Spectrophotometric investigation of the magnetic variable star 21 Per. II. Distribution of Fe over the star's surface and study of Hγ and Hδ hydrogen line variations during the period.
Yu. V. Glagolevskij, K. I. Kozlova, V. S. Lebedev, N. S. Polosukhina.
Astrofizika, Vol. 12, 631 - 645 (1976). In Russian. – English translation in Astrophysics, Vol. 12, No. 4.

Fe II line profiles consisting of a few components are studied in 21 Per using spectrograms with dispersions 4 and 8 Å/mm. Radial velocities and equivalent widths are obtained from the components of Fe II λ 4263.90 and λ 4351.76 Å. The data analysis shows that iron is concentrated in 4 regions of the surface of 21 Per which are evenly distributed along the equator of rotation. The coordinates of the centers of the spots and their sizes as well as relative intensities of the Fe II λ 4351.76 line in each of them are determined.

114.516 Spectroscopic study of HZ Herculis.
D. C. Koo, R. G. Kron.
Publ. Astron. Soc. Pacific, Vol. 89, 285 - 299 (1977).

Medium-dispersion spectrograms of HZ Her have been measured for radial velocities and equivalent widths of Hγ, Hδ, Ca II K, He I lines, He II λ4686 emission, and C III– N III λ4640 emission. Adopting a model for the temperature distribution across the surface of HZ Her, the authors find that the derived mass ratio depends strongly on the spectral resolution because of asymmetry in the line profiles. A value of 1.5 $M_\odot$ ($i = 90°$) for the mass of Her X-1 is consistent with the authors' data. The radial-velocity curve for the emission complex at λ4640 and He II λ4686 appears to be roughly in phase with the absorption lines.

114.517 The A2p star HD 3473.   S. A. Naftilan.
Publ. Astron. Soc. Pacific, Vol. 89, 309 - 314 (1977).

An investigation of the Ap star HD 3473 has been carried out based on three coudé spectrograms, two in the blue and one in the red. The results indicate an extreme overabundance for Si, [Si] = 10.7. Mg, Fe, and the rare earths are also over-abundant. The relatively low temperature of 9750°K, and high rotational velocity of $v_{\sin i} = 75$ km sec^{-1} are unusual for a star showing this abundance pattern. Emission fill-in is almost certainly present in the cores of the stronger Si II, and Mg II lines, but is not seen in the Ca II K line. The extreme weakness and/or absence of the spectral lines due to neutral metals is probably caused by weak emission.

114.518 Observations and interpretation of the infrared spectrum of HD 44179.   J. D. Bregman.
Publ. Astron. Soc. Pacific, Vol. 89, 335 - 338 (1977) = Lick Obs. Bull. No. 751.

Infrared spectra of HD 44179 have been obtained of the 3040 cm^{-1}, 1150 cm^{-1}, and 890 cm^{-1} emission features. The strengths of the features put constraints on the possible constituents of the materials responsible for each emission feature. The 3040 cm^{-1} feature can be due to emission either from a solid if it arises from a different region than the 890 cm^{-1} feature, or from a gas phase molecule. The 8–13 micron spectrum is fitted with a simple model employing carbonates plus a dielectric in emission, and silicates in absorption. It is suggested that amorphous carbon is the dielectric responsible for the featureless underlying continuum.

114.519 Spectral changes in the pre-main-sequence star HD 97048.   N. J. Irvine, N. Houk.
Publ. Astron. Soc. Pacific, Vol. 89, 347 - 348 (1977).

Over a six-year interval the shell of the pre-main-sequence star HD 97048 has shown remarkable variability.

114.520 A search for water-vapor absorption lines in maser stars.
G. Wallerstein, H. Harrison, R. Antonucci.
Publ. Astron. Soc. Pacific, Vol. 89, 391 - 396 (1977).

Using 8 Å mm^{-1} image-tube spectra the authors have searched for water lines in the Δν = 3 band near λ9400 in seven stars: R Leo, R Cas, U Ori, VY CMa, PZ Cas, S Per, and TW Peg. They found no lines that could be attributed to a stellar or circumstellar origin. For these stars upper limits of 3×10^{20} H$_2$O molecules cm^{-2} were set, by comparison with atmospheric H$_2$O lines. This limit is substantially below predicted column densities. Possible reasons for the discrepancy are briefly discussed.

114.521 High-resolution rocket spectra of the λ1920 and λ1720 features in the spectrum of Zeta Tauri.
S. R. Heap.
Astrophys. J., Vol. 217, 90 - 94 (1977).

High-resolution ultraviolet spectrograms of the B-shell star ζ Tau reveal two features characteristic of B supergiants, one at 1720 Å and the other at 1920 Å. The presence of these features in the spectrum of this object shows that they are indicative of an extended atmosphere – either the tenuous atmosphere of a supergiant or the envelope surrounding a rapidly rotating main-sequence star – and are therefore not purely luminosity criteria. The high spectral resolution allows an identification of the contributors to these features. The dominant contributor to the λ1920 feature is Fe^{++}, while the primary contributor to the λ1720 feature is Al$^+$.

114.522 Spectrum variations of the X-ray binary HD 153919 = 3U 1700−37.
G. G. Fahlman, R. G. Carlberg, G. A. H. Walker.
Astrophys. J., Lett., Vol. 217, L35 - L39 (1977).

The authors discuss some recent spectroscopic observations of the Of star HD 153919, the primary in the 3U 1700−37 X-ray binary system. Two distinct variable components are present in the He I λ5876 P Cygni line and in the strong Hα emission line. One component is displaced by 400–600 km s^{-1} to the red. The other component, with a radial velocity of roughly 800 km s^{-1}, appears to be a phase-dependent absorption feature. Some implications of interpreting this feature as absorption by an extended wake trailing the secondary are discussed.

114.523 Line identifications in the ultraviolet spectra of Tau Herculis (B5 IV) and Zeta Draconis (B6 III).
A. B. Underhill, S. J. Adelman.
Astrophys. J., Suppl. Ser., Vol. 34, 309 - 380 (1977).

Tables of the lines found on two tracings each of the ultraviolet spectra of τ Her (B5 IV) and ζ Dra (B6 III) made by the Copernicus satellite and possible identifications are given. The lines listed in tables should be useful as a guide for identifying features in the ultraviolet spectra of stars of types B2 to B8.

114.524 An ultraviolet spectrum of Sirius B from Copernicus.
M. P. Savedoff.
News Lett. Astron. Soc. N.Y., Vol. 1, No. 2, p. 17 (1977).
Abstract.

114.525 Content of the near-ultraviolet spectrum of Alpha Cygni (A2 Ia).   A. B. Underhill.
Astrophys. J., Vol. 217, 488 - 493 (1977).

The possible identifiers for the absorption lines in the 2000 to 3017 Å region of the spectrum of α Cyg are reviewed.

114.526 A spectroscopic study of 14 Comae and other A-type shell stars.   J. F. Dominy, M. A. Smith.
Astrophys. J., Vol. 217, 494 - 507 (1977).

The physical conditions in the circumstellar shells of the shell star 14 Comae and similar stars are investigated. A shell line list is presented, and the column densities of shell ions are obtained. It is found that the shell material lies approximately three stellar radii above the stellar photosphere. There is little evidence for rapid expansion or collapse of the shell of 14 Com, but for two other stars shell column densities have changed markedly on a time scale of ~2 years. Excitation and ionization equilibrium calculations indicate that shell hydrogen is predominantly neutral.

114.527 Spectrophotometric investigation of the Be star HD 183656.   O. Eh. Aab, N. F. Vojkhanskaya.
Astrofiz. Issled., Izv. Spets. Astrofiz. Obs., Vol. 9, 22 - 28 (1977). In Russian.

Results of investigation of the spectrum in 1974 are reported. A relation is found between brightness variation and variation in the spectrum. A classification of the star is made by spectrum and luminosity. The electron temperature and the concentration, the total number of atoms along the line of sight $N_{02}H$, the contribution of the shell to the total amount of radiation from the star are estimated. Radial velocities of the star and of the shell are measured separately from all the lines of the spectrum. On the basis of the radial velocity measurement results an assumption is made on the pulsation of the B star.

114.528 Spectral variations of the Ap star HD 216533. I. Observational results.   M. Floquet.
Astron. Astrophys., Suppl. Ser., Vol. 30, 27 - 34 (1977).

The author has studied the variations of the metallic lines and of the Balmer lines Hγ, Hδ and Hϵ. He correlates these variations with the photometric ones, found in the literature. Many elements (Sr, Mn, Ti) have an intensity maximum at the same time as the magnitude V, the metallicity index and the magnetic field. At this time, the Balmer jump index is minimum but the equivalent width and the central depth of Balmer lines are average. All these variations have the same period of P = 17.22 days. The variation of Balmer lines occurs only in the core of the line. The effective temperature and the gravity log g determined from the Hγ profile indicate that this star lies on the main sequence.

114.529 Observations of four C III lines in Zeta Orionis.
H. L. Johnson, W. Z. Wisniewski.
Astrophys. Lett., Vol. 19, 25 - 27 (1977).

The authors have used their new Michelson spectrophotometer system to obtain two spectra of Zeta Orionis covering the entire wavelength range from 4000 Å to beyond 10000 Å. They have identified and measured the C III multiplet at 9701 - 17 Å, which has not previously been measureable because of atmospheric water-vapor absorptions in this region. They do not confirm the 60-Ångstrom-wide band around the C III feature at 5696 Å.

114.530 A spectrophotometric study of ultraviolet and X-ray sources with objective prism plates.
R. Viotti, A. Altamore, G. B. Baratta.
Astron. Astrophys., Vol. 61, 133 - 136 (1977).

Several objective prism spectra of objects known to have large ultraviolet or X-ray fluxes are discussed: the ultraviolet stars BD + 37° 1977 and HZ 43 (MX 1313 + 29), the X-ray sources X Per (3U0352 + 30) and HD 226868 (Cyg X−1), and the Be star HD 200775 (in NGC 7023).

114.531 Spectra of A-type stars from 3600 to 4200 Ångstrøms.   R. J. Panek.
Astrophys. J., Vol. 217, 749 - 759 (1977).

New low-resolution spectrum scans from 3600 to 4200 Å are presented for seven bright A-type stars. Theoretical spectra which explicitly include the opacity of 30 Stark-broadened Balmer lines were computed for a variety of line-blanketed model atmospheres. These spectra accurately reproduce the observations after allowance for the instrumental resolution and for the metal-line absorptions not included in the theoretical spectrum.

114.532 The rotational velocity and barium abundance of Sirius.   R. L. Kurucz, W. A. Traub, N. P. Carleton, J. B. Lester.
Astrophys. J., Vol. 217, 771 - 774 (1977).

The authors have measured the Ba II 649.69 nm line profile in Sirius using a PEPSIOS interferometer. They find a projected rotational velocity $V \sin i$ of 16 ± 1 km s^{-1}; a heliocentric radial velocity of −8.6 ± 0.4 km s^{-1}; and a log Ba abundance of −8.18 ± 0.15 relative to all atoms by number (3.87 if log H = 12), which is greater than the solar abundance by 1.76 ± 0.18.

114.533 A high-dispersion photometric atlas of the dM0 star HD 88230 from 3900 to 6000 Å.
R. G. Tull, S. S. Vogt.
Astrophys. J., Suppl. Ser., Vol. 34, 505 - 564 (1977).

The authors present a high-dispersion spectrophotometric atlas of the dM0 star HD 88230. This spectrum was obtained with a self-scanned digicon and the coudé spectrograph of the McDonald Observatory 2.7 m reflector. The atlas extends from 3900 to 6000 Å; the original dispersion was 4.4 Å mm^{-1} and the resolution is 280 mÅ.

114.534 The ultraviolet spectrum of Beta Lyrae. III.   M. Hack, J. B. Hutchings, Y. Kondo, G. E. McCluskey.
Astrophys. J., Suppl. Ser., Vol. 34, 565 - 580 (1977).

Parts of the ultraviolet spectrum of Beta Lyrae were observed on 13 consecutive days in 1975 June, to cover the 12.9 day period thoroughly. The regions observed are λλ1036−1060, 1300−1326, 1398−1416, 2050−2098, 2580−2632, 2777−2812 at low resolution (0.2−0.4 Å), and λλ1172−1177 and 2795−2799 at high resolution (0.05−0.1 Å). Light curves are presented for all wavelength regions. The authors suggest that the secondary continuum arises from a 9000 K star or disk, above λ2100, and that the far-UV flux arises from a hot spot in the stream or disk associated with the secondary. The radial velocities of all unblended lines are presented. It is found that little variation occurs with phase for most lines.

Discussion is given of the results in terms of the energy distribution, origin of the emission lines, extended envelopes about the system, the disk surrounding the secondary, and the nature of the secondary.

114.535 Observation of preplanetary disks around MWC 349 and LkHα 101. R. I. Thompson, P. A. Strittmatter, E. F. Erickson, F. C. Witteborn, D. W. Strecker.
Astrophys. J., Vol. 218, 170 - 180 (1977).

Infrared spectra of three highly reddened emission-line objects (MWC 349, MWC 297, and LkHα 101) have been obtained from both aircraft and ground-based observatories. These spectra show bright emission lines which are consistent with recombination in an H II region. Additional optical data for MWC 349 yield a model which consists of a central O6.5 star with a circumstellar disk seen face-on, a surrounding H II region, and circumstellar dust. The parameters of the disk are discussed in terms of the viscous dissipation model of Lynden-Bell and Pringle. The observations are in excellent agreement with the predictions of the model and represent the first substantial case for a newly formed star with a preplanetary disk.

114.536 On the near-infrared excesses of very cool supergiants. W. M. Fawley.
Astrophys. J., Vol. 218, 181 - 194 (1977).

Spectroscopic and narrow-band photometric observations of ~15 G, K, and M supergiants with large infrared excesses have been made to search for line weakening and chromospheric near-infrared emission of the form proposed by Humphreys and Gilman to be present in the peculiar M stars S Per, VY CMa, and VX Sgr. The results indicate that the line weakening of S Per and VX Sgr is probably photospheric in origin and temporally variable, while that of VY CMa may be constant. None of the other supergiants show significant line weakening. There are no near-infrared excesses evident in the photometry of any of these objects. Veiling of the 4.8 μm absorption feature in late-type supergiants with extremely large infrared excesses implies that the 3.5−8 μm excesses are formed above the molecular photospheres and are probably thermal reradiation from the circumstellar shells rather than free-free emission.

114.537 A sensitive observation of the far-ultraviolet (1160−1700 Å) spectrum of Arcturus and implications for its outer atmosphere.
A. Weinstein. H. W. Moos, J. L. Linsky.
Astrophys. J., Vol. 218, 195 - 204 (1977).

A low-resolution far-ultraviolet (1160−1700 Å) spectrum of Arcturus (α Boo, K2 IIIp) has been obtained. H I λ1216, O I λ1304, and a broad unresolved emission near 1510 Å were detected. A 2 σ feature is probably O I λ1356. The ratio of O I λ1304 to O I λ1356 is similar to the solar ratio. This spectrum is very different from that of the Sun, with few emission features. A model of the chromosphere-corona transition region predicts fluxes too low to be detected at present.

114.538 High-resolution optical observations of Ca II K in Deneb and Aldebaran.
W. McClintock, R. C. Henry.
Astrophys. J., Vol. 218, 205 - 208 (1977).

High-spectral-resolution echellograms confirm the existence of an asymmetric emission core in the K line of Deneb that may be due to chromospheric emission from this A-type supergiant or that may be due to scattering from a circumstellar shell. If the core is due to chromospheric emission, its width does not agree with what is predicted using the Wilson-Bappu relation for late-type stars. This would indicate that the chromospheres of hotter stars are qualitatively different from those of cooler stars. A K-emission-core profile for Aldebaran is also presented.

114.539 Spectroscopic study of two O-type supergiants,

Alpha Camelopardalis and 19 Cephei: model-atmosphere analysis. M. Takada.
Publ. Astron. Soc. Japan, Vol. 29, 439 - 476 (1977).

The author analyzed high-dispersion spectra of α Cam and 19 Cep by using standard model-atmosphere techniques in LTE approximation in order to obtain the atmospheric parameters and elemental abundances, and in order to study how well planar LTE models can interpret the observed spectra. Comparisons of observations with calculations were also made for planar NLTE models.

114.540 The envelope of Pleione in the new shell phase, 1972 - 1975. R. Hirata, T. Kogure.
Publ. Astron. Soc. Japan, Vol. 29, 477 - 495 (1977).

The shell spectrum of Pleione has gradually developed during the new shell phase that began in 1972. A series of high-dispersion spectra was analyzed. The emission intensities of the Balmer lines were measured. The Hα emission was much weaker in 1973 as compared to the value in the pre-shell phase, and thereafter it steadily increased with strengthening of the metallic and hydrogen shell lines. The radial velocities, the central residual intensities, and the widths of the shell lines in the Balmer series were measured. No appreciable Balmer progression was found, and the mean radial velocity coincided with the stellar radial velocity within the error of measurement. A spectroscopic model was derived on the basis of these measurements. The appearance of the broad absorption feature in the K line of Ca II was interpreted as formation of a rapidly rotating layer around the star's equator.

114.541 A determination of atmospheric parameters of Arcturus. P. Martin.
Astron. Astrophys., Vol. 61, 591 - 599 (1977).

A detailed analysis of Arcturus (K 2 III p) with respect to ε Virginis (G 8 III) has been performed. The resulting atmospheric parameters are given. The causes of the differences between the results and those of Mäckle et al. (1975) are analysed. The final model is compared with that of Mäckle et al.

114.542 The Copernicus ultraviolet spectral atlas of Tau Scorpii. J. B. Rogerson, Jr., W. L. Upson II.
Astrophys. J., Suppl. Ser., Vol. 35, 37 - 110 (1977).

An ultraviolet spectral atlas is presented for the B0 V star, Tau Scorpii. It has been scanned from 949 to 1560 Å by the Princeton spectrometer aboard the Copernicus satellite. From 949 to 1420 Å the observations have a nominal resolution of 0.05 Å. At the longer wavelengths, the resolution is 0.1 Å. The atlas is presented in both tables and graphs.

114.543 The spectrum of h 4866 B. J. Sahade, O. Ferrer.
Observatory, Vol. 97, 242 - 243 (1977).

The star h 4866 B, visual companion of the eclipsing variable R Arae, has been observed spectroscopically at 42 Å/mm. On most of the plates the spectral type is B9 V but on two of them it is A3 V. The material available appears to suggest a range in radial velocity of some 80 km/s.

114.544 An abundance analysis of Tau Herculis, B5 IV. S. J. Adelman.
Mon. Not. R. Astron. Soc., Vol. 181, 667 - 675 (1977).

An abundance analysis of the sharp-lined star Tau Herculis (B5 IV) has been performed using a fully line-blanketed model atmosphere. The derived abundances are similar to those of the Sun and the normal main sequence B stars, ι Her (B3 V) and ν Cap (B9 V).

114.545 Non-anomalous diffuse interstellar absorption features in Rho Leonis.
J. C. Blades, W. B. Somerville.
Mon. Not. R. Astron. Soc., Vol. 181, 769 - 776 (1977).

Photographic spectroscopic observations are presented of ρ Leo (HD 91316), a class B1 supergiant at high galactic latitude which shows little reddening. The interstellar diffuse features λ4430, λ5797 and λ6283 are not detected. Previous reports of anomalously strong λ4430 are not confirmed. These were based on low-resolution spectroscopy and the explanation is that the appearance of a wide feature is produced by a combination of several weak atomic absorption lines in the spectrum of the star itself.

114.546 Absolute energy distributions of α Lyrae and 109 Virginis from 3295 Å to 9040 Å.
H. Tüg, N. M. White, G. W. Lockwood.
Astron. Astrophys., Vol. 61, 679 - 684 (1977).

A fundamental calibration of the fluxes of α Lyrae and 109 Virginis from 3295 Å to 9040 Å has been carried out at the Lowell Observatory using blackbody sources operating at the melting points of copper and platinum. Fluxes were measured every 50 Å with a bandpass of 10 Å from 3295 Å to 5695 Å and 20 Å from 4990 Å to 9040 Å.

114.547 o Andromedae.
IAU Circ., No. 3122 (1977).

114.548 HDE 245770.
IAU Circ., No. 3129 (1977).

114.549 Infrared spectra of the WN stars HD 50896 and HD 151932.
A. P. Bernat, T. G. Barnes, B. R. Schupler, A. E. Potter.
Publ. Astron. Soc. Pacific, Vol. 89, 541 - 545 (1977).

Observations of HD 50896 (WN5) and HD 151932 (WN7) are presented for the spectral region 6000−11,000 cm^{-1} (1.7−0.9 μm) at 35 cm^{-1} resolution. This new spectral region exhibits only emission lines of H, He I, and He II, consistent with the visual spectra of WN types.

114.550 Observation of red-displaced absorption features in the T Tauri star S Coronae Austrinae.
A. E. Rydgren.
Publ. Astron. Soc. Pacific, Vol. 89, 557 - 560 (1977).

High-dispersion blue and red spectrograms of the extreme T Tauri star S CrA show violet-displaced absorption components in the Hα, Hγ, and Hδ emission lines, as well as strong red-displaced absorption components at Hγ and Hδ (Hβ was not observed). The absence of the red-displaced absorption at Hα may be due to the greater width of this emission feature. A conventional interpretation of these line profiles implies simultaneous envelope expansion and mass infall; other possibilities are briefly considered.

114.551 High resolution profiles in A type stars: I. The Ca II K line observed with the Meudon Solar Tower.
R. Freire, J. Czarny, P. Felenbok, F. Praderie.
Astron. Astrophys., Vol. 61, 785 - 796 (1977).

The Solar Tower spectrograph at Meudon Observatory has been used to observe bright A type stars at a spectral resolution of 35−68 mÅ, in the Ca II K line; the spectra are recorded with photographic plates and Lallemand electronic camera. The authors present the line profiles for α CMa (A 1 V), γ Gem (A 0 IV), α Aql (A 7 IV-V). A non-LTE analysis of the profiles provides a new determination of the Ca abundance in these stars, which is found close to the solar photospheric value in α CMa and γ Gem (Ca/H = 2.5·10^{-6}). From the analysis of the line cores, no strong evidence for the presence of a chromosphere can be derived.

114.552 Emission-line profile studies in α Scorpii.
A. M. Boesgaard, D. M. Chesley, P. B. Kunasz.
Publ. Astron. Soc. Pacific, Vol. 89, 613 (1977). − Abstract.

114.553 The ultraviolet spectra of four binaries observed with the S59 spectrometer.
T. J. Herczeg, Y. Kondo, K. A. van der Hucht.
Astrophys. Space Sci., Vol. 46, 379 - 387 (1977).

Ultraviolet spectra of o And, α CrB, η Ori A and α Vir have been studied for the presence or the absence of the effects due to their binary nature. As may have been anticipated from their orbital and other characteristics, no indication of strong binary interactions were seen in these observations. However, there are certain spectral peculiarities suggesting the possibility of modifications of spectral classifications for some of these objects.

114.554 The high-galactic-latitude O-type star HD 93521.
M. Hack, N. Yilmaz.
Astrophys. Space Sci., Vol. 48, 483 - 489 (1977).

The spectrum of the peculiar O9 star HD 93521 is studied and compared with those of the standard stars 10 Lac and AE Aur. Several possibilities are examined which might explain the high galactic latitude of this star, corresponding to z > 750 pc, and its slight helium excess. It is suggested that HD 93521 is a 'runaway' binary system composed of the O9 star plus a neutron star left over from a supernova explosion.

114.555 Spectrographic observations of Omicron Andromedae from 1967 to 1976.
M. Fracassini, L. E. Pasinetti, L. Pastori.
Astrophys. Space Sci., Vol. 49, 145 - 167 (1977).

The results of qualitative analysis and radial velocity (RV) determinations from 1967 to 1976 are given. These analyses show sometimes the presence of a thin variable shell also in the years 1967−1974, before the appearance of the envelope. The RV do not confirm the periods suggested by the photometric observations. A periodogram analysis gives RV curves with a poor evidence of periodicity. However, the period P = 1^{d}5845 obtained from this analysis, close to that of Schmidt, seems to confirm Schmidt's hypothesis of a contact binary system.

114.556 The spectrum of ε Aurigae outside eclipse.
F. Castelli.
Astrophys. Space Sci., Vol. 49, 179 - 197 (1977).

Identification, equivalent widths, profiles and radial velocities as deduced from 18 spectra of ε Aurigae are presented and discussed.

114.557 4 new Be stars in Sagitta.
M. V. Dolidze, O. D. Dokuchaeva, G. N. Kimeridze.
Astron. Tsirk., No. 943, p. 1 - 2 (1977). In Russian.

114.558 Precise positions of four emission-line stars in Sagitta.   Yu. A. Shokin, A. Sh. Khatisashvili.
Astron. Tsirk., No. 943, p. 2 - 3 (1977). In Russian.

114.559 Study of the Wolf-Rayet star HD 192163.
T. Nugis.
Publ. Tartu Astrofiz. Obs., Vol. 45, 70 - 112 (1977). In Russian.

The spectral energy distribution in the spectral region of λλ1400 − 114000 Å, the ionization equilibrium and the energy balance of free electrons were studied and possible values of the temperature and other physical parameters of the star HD 192163 (WN6) were determined.

114.560 Wavelength dependence of the opacity in the atmosphere of the K-type component of 32 Cygni.
K. Saijo, M. Saitō.
Publ. Astron. Soc. Japan, Vol. 29, 739 - 751 (1977).

An analysis of the published light curves during the 1971 eclipse of 32 Cygni is carried out. In the photosphere of the K-type supergiant component, the opacity slightly decreases with decreasing wavelength between 4000-5500Å and increases with decreasing wavelength between 1910-4000Å. In the chromosphere, the opacity varies with wavelength in the same way as in the photosphere between 2980-5500Å while it decreases again at 2460Å and 1910Å. Molecules seem to be the main opacity source in the photosphere at the far-ultraviolet region. In the course of analysis, the colors and radii of the components are determined.

114.561 **Energy distribution in the spectra of the three variable late-type stars XY Lyr, RS Cnc, OP Her.**
R. I. Chuprina.
Astrometriya i Astrofizika, Kiev, vyp. (No.) 33, (see 003.020), p. 65 - 72 (1977). In Russian.

The energy distribution in the spectra of the variable stars XY Lyr, RS Cnc and OP Her ($\lambda\lambda$ 3220–7750 Å) is given for each moment of observation (April - June, 1974). The spectral types of the variables for each moment of observation are estimated. The mean values of the spectral type coincide with those given in the General Catalogue of Variable Stars and the Catalogue by Yamashita.

114.562 **HR 8799 – ein metallarmer Doppelstern?**
T. Gehren.
Mitt. Astron. Ges., Nr. 42, p. 139 (1977).

114.563 **The spectrum variations of Pleione from 1938 to 1975.**   A. F. Gulliver.
Astrophys. J., Suppl. Ser., Vol. 35, 441 - 459 (1977) = Contrib. Dominion Astrophys. Obs., No. 338 = NRC No. 15907.

Spectrophotometric data on the shell episode of Pleione from 1938 to 1954 have been amassed for the first time. Changes in the emission profiles, absorption-line intensities and asymmetries, and the Balmer progression are studied. The emission phase of Pleione from 1969 to 1971 and the new shell phase from 1972 to 1975 are also examined. The radial velocities of the shell and star are shown to be equal and to have no significant variations. Good agreement is found between the stellar wind model and the observed variations, with the exception of the emission intensity and the Balmer progression.

On the absolute scale of mass-loss in red giants. I. Circumstellar absorption lines in the spectrum of the visual companion of α^1 Her.   See Abstr. 064.020.

Linear polarization of Hα in the Be star Gamma Cassiopeiae.   See Abstr. 064.023.

G77-61: a dwarf carbon star.
See Abstr. 112.002.

A magnetic field interpretation for the outburst of CH Cygni.   See Abstr. 116.016.

The spectrum of HD 187399.
See Abstr. 117.027.

The ultraviolet spectrum of Beta Lyrae.
See Abstr. 121.025.

A note on the reddening of Polaris B.
See Abstr. 122.092.

A search for weak interstellar lines.
See Abstr. 131.056.

PK 6–2°1, a remarkable nitrogen-rich southern planetary nebula.   See Abstr. 135.028.

Detection of [Fe XIV] emission in HD 153919 (3U 1700–37).   See Abstr. 142.076.

A CH star in the globular cluster M22, and the nature of CH and CN anomalies.   See Abstr. 154.019.

Are there population II stars of early spectral types and low gravities?   See Abstr. 155.016.

Errata

114.901 **Erratum: "Evidence for a corona of Beta Geminorum"** [Astrophys. J., Lett., Vol. 193, L107 - L110 (1974)].   H. Gerola, J. L. Linsky, R. Shine, W. McClintock, R. C. Henry, H. W. Moos.
Astrophys. J., Lett., Vol. 218, L32 (1977).

114.902 **Erratum: "2–14 μm stellar spectrophotometry II. Stars from the 2 μm Infrared Sky Survey"** [Publ. Astron. Soc. Pacific, Vol. 88, 294 - 307 (1976)].
K. M. Merrill, W. A. Stein.
Publ. Astron. Soc. Pacific, Vol. 89, 596 (1977).

115 Luminosities, Masses, Diameters, HR-Diagrams and Others

115.001   Comparison of the S2/68 and OAO-2 stellar ultra-
violet flux measurements.   F. Beeckmans.
Astron. Astrophys., Vol. 60, 1 - 7 (1977).

In order to estimate their systematic error, the S2/68 and OAO-2 absolute calibrations are compared with each other and with other recent absolute calibrations. The influence of the UV flux differences between S2/68 and OAO-2 on the computation of effective temperatures and bolometric corrections is discussed.

115.002   The radius and mass of FG Sagittae.
C. A. Whitney.
Bull. American Astron. Soc., Vol. 9, 433 (1977). – Abstract.

115.003   Further speckle interferometric studies of α
Orionis.   M. S. Wilkerson, S. P. Worden.
Astron. J., Vol. 82, 642 - 645 (1977).

Speckle interferometry was conducted on α Orionis (Betelgeuse) to check and extend the observations of Lynds, Worden, and Harvey (1976). Digital reduction of the data indicated the diameter had not changed significantly from the value derived in the previous study. In addition it was found that α Ori is very highly limb darkened. No significant surface structure was revealed. All other results of the Lynds et al. study are confirmed.

115.004   The angular diameters of Capella A and B from
two-telescope interferometry.
A. Blazit, D. Bonneau, M. Josse, L. Koechlin, A. Labeyrie, J. L. Onéto.
Astrophys. J., Lett., Vol. 217, L55 - L57 (1977).

The authors have resolved the apparent disks of Capella A and B with the two-telescope interferometer operating at 12 to 20 meter baselines. The angular diameters are respectively 5.2 ± 1.0 and 4.0 ± 2.0 milli-arcsec. Through continued observation, further improvements are possible in the accuracy of diameter and orbital determinations.

115.005   On the luminosity function of stars.
R. A. Bartaya, E. K. Kharadze.
Astrofizika, Vol. 13, 123 - 130 (1977). In Russian. English translation in Astrophysics, Vol. 13, No. 1.

On the basis of the Abastumani Catalogue of MK classification for 10396 stars the luminosity function is constructed for the stars in 42 Selected Areas, situated almost regularly over the galactic latitudes from $-17°$ to $+72°$. The form of the luminosity function for middle and high galactic latitudes obviously differs from that in the sun's vicinity.

115.006   Radii of nearby stars: an application of the Barnes-
Evans relation.   C. H. Lacy.
Astrophys. J., Suppl. Ser., Vol. 34, 479 - 492 (1977).

A method of estimating radii of all nearby stars is presented. The method is based on the Barnes-Evans $F_v(V-R)$ relation and is free of assumptions about spectral classification, luminosity class, effective temperature, or bolometric correction. The method is applied to all nearby stars with accurate parallaxes and $V-R$ photometry, and the resulting radii are compared to theoretical models. It is found that theory and observation are in good agreement for stars of about one solar mass or greater, but theoretical models of M dwarfs have up to 25% smaller radii than real stars. This conclusion is supported by the results of three other independent studies. The author speculates that the cause of the discrepancy is an inadequate treatment of the opacity sources in the atmospheres and envelopes in current stellar evolution codes.

115.007   The luminosity function of late-type main-sequence

stars in the direction of the North Galactic Cap.
D. Weistrop.
Highlights of Astronomy, Vol. 4, Part II, (see 012.022), p. 31 (1977). – Summary.

115.008   Direct observations of the heterogeneity of super-
giant disks.
J. W. Harvey, C. R. Lynds, S. P. Worden.
Highlights of Astronomy, Vol. 4, Part II, (see 012.026), p. 405 (1977).

115.009   The Hertzsprung-Russell diagram.
R. C. Maddison.
Modern astronomy, (see 003.013), p. 109 - 123 (1977).

115.010   A redetermination of the absolute magnitudes of
the barium stars.
A. R. Upgren, P. K. Lü, D. J. MacConnell.
Publ. Astron. Soc. Pacific, Vol. 89, 552 - 553 (1977).

The absolute magnitudes of 101 barium stars and 65 marginal barium stars were redetermined from Yale proper motions.

115.011   On the position of Ap-stars toward the main
sequence.   A. S. Nikolov, I. Iliev.
Inf. Bull. Variable Stars, No. 1308, 4 pp. (1977).

115.012   A luminous carbon star in Canis Major OB1.
W. Herbst, R. Racine, H. B. Richer.
Publ. Astron. Soc. Pacific, Vol. 89, 663 - 667 (1977).

The fact that W CMa illuminates a reflection nebula is used to argue that it is spatially associated with the CMa OB1/CMa R1 complex. An apparent cluster around the carbon star is found to consist primarily of field stars, although a few probable late B-type members of CMa OB1 are identified. On the basis of its likely association with CMa OB1, a luminosity for W CMa is derived. The authors find $M_v = -4.7$ and $M_{bol} = -7.2$. It seems likely that the progenitor of W CMa was an O-type member of CMa OB1 with a mass greater than $20\,M_\odot$ and a main-sequence lifetime less than 3×10^6 years.

115.013   Chemical and physical properties of nearby Popula-
tion I stars.
D. Cardini, I. Mazzitelli, L. Rossi.
Astrophys. Space Sci., Vol. 48, 283 - 292 (1977).

An analysis of some photometric and kinematic data for nearby Population I stars has been carried out. The main results show that a good correlation exists between the dynamic state and age; but a non-negligible chemical inhomogeneity seems to be present in the disk, independently of the age.

115.014   Radii of some Ap, Am and A stars.
G. S. D. Babu.
Astrophys. Space Sci., Vol. 50, 343 - 348 (1977).

Based on the observed energy curves of nine Ap stars, three Am stars, four normal A stars and one F0 V magnetic star, their radii have been estimated. Thence, the bolometric magnitudes have been obtained and it is shown that a majority of Ap and Am stars are a little above the zero-age Main-Sequence, suggesting that they are slightly more evolved as compared to the normal A stars. The bolometric corrections derived from the above bolometric magnitudes are much closer to those computed by Mihalas than to the ones given by Davis and Webb.

115.015   Pre-main sequence masses in NGC 2264 and the

Orion Nebula cluster.   B. J. McNamara.
Proc. Southwest Reg. Conf., Vol. 3, (see 012.043), p. 81 - 82 (1977). – Abstract.

115.016   **The nature of the Antares companion α Sco B.**
R.-P. Kudritzki, D. Reimers, P. R. Wesselius.
Mitt. Astron. Ges., Nr. 42, p. 140 (1977).

The origins of the Hertzsprung-Russell Diagram.
See Abstr. 004.080.

Frequency dependent diameters and atmospheric structure of late-type supergiants.   See Abstr. 064.047.

Le sintesi di popolazioni stellari.
See Abstr. 065.034.

Éléments de structure interne.
See Abstr. 065.035.

On the age difference between the oldest Population I and the extreme Population II stars.
See Abstr. 065.057.

CRL 3068 – a dust-enshrouded carbon star.
See Abstr. 113.013.

Accuracy of two-dimensional spectral classes derived through DDO photometry.   See Abstr. 114.025.

Parallax, orbit, and mass of Ross 614.
See Abstr. 117.013.

A parametric approach to the slope of the globular clusters giant branches.   See Abstr. 154.037.

UBV color-magnitude diagrams of galactic clusters.
See Abstr. 154.041.

116 Magnetic Fields, Figure, Rotation, Radio Radiation

116.001   **Observations of radio stars at 3.3 mm.**
P. R. Schwartz, J. H. Spencer.
Mon. Not. R. Astron. Soc., Vol. 180, 297 - 303 (1977).
The authors have measured the 3.3-mm continuum flux of several stars known to be radio sources at centimetre wavelengths. Two spectral classes can be identified: spectra with a constant index from centimetre to millimetre wavelength (T Tau, P Cyg, V1016 Cyg and, possibly, α Ori), and spectra that flatten at millimetre wavelengths (LkHα 101 and MWC 349). They have also detected emission from two infrared objects (IRC + 10216 and CRL 2591), probably related to thermal dust emission.

116.002   **Radio emission from mass-outflow stars.**
C. Chiuderi, G. Torricelli Ciamponi.
Astron. Astrophys., Vol. 59, 395 - 400 (1977).
The authors consider the problem of the radio emission from the extended envelopes surrounding early type stars undergoing mass loss. The emission is interpreted as due to thermal bremsstrahlung processes. The dynamics of the expanding atmosphere is treated in the polytropic approximation For a particular value of the polytropic index, falling in the physically interesting range, the authors deduce analytical expressions for the spectral index, the radio flux and the angular size of the emitting object, as well as a relationship between these quantities and the mass loss rate.

116.003   **Magnetodynamic mechanism of magnetic stars.**
E. Woyk (Chvojková).
Astron. Astrophys., Vol. 60, L5 - L8 (1977).
Cosmic bodies with two opposite magnetic poles must possess a high, thin and very stable equatorial belt, established by the magnetically trapped high-energy particles. The belt extends along the field lines toward the poles. Only closer to the poles also the number of slower particles rapidly increases.

116.004   **A search for radio emission from late-type supergiant stars.**
J. Smoliński, P. A. Feldman, L. A. Higgs.
Astron. Astrophys., Vol. 60, 277 - 280 (1977).
Radio continuum observations at 10.5 GHz have been made of 29 high-luminosity F0–K5 supergiant stars. Two sources were detected (at the 3-σ confidence level) in this survey. One corresponds to the unusual spectrum-variable star HR 8752 = HD 217476 (G0 Ia). The other is located approximately one arcminute southwest of the star HD 18391 (G0 Ia), with a positional error box containing no obvious optical candidate(s).

116.005   **Stochastic models of rotating stars. Model of μ Cep.**   I. A. Klyus.
Odessk. univ. Odessa, 1977. 20 pp. In Russian. – Abstr. in Ref. zh., 51. Astron., 8.51.483 (1977).

116.006   **Observations with the VLA of the radio binary star AR Lacertae.**   F. N. Owen, S. R. Spangler.
Astrophys. J., Lett., Vol. 217, L41 - L43 (1977).
Radio observations of AR Lac during three optical eclipses are reported. The observations were made at 4585 MHz with the Very Large Array (VLA) telescope. No clearly defined minimum was found during any of the three eclipses observed. Thus, the radio emission of AR Lac does not originate from a very compact source located at the center of mass of the system. If the radio source is roughly spherical, uniform, and between the two stars, this result suggests that $T_b \lesssim 4 \times 10^9$ K. If the radio source is self-absorbed, as is suggested by the characteristic radio spectrum of AR Lac, then the Lorentz factor of the radiating electrons is less than 4.

116.007   **On the effect of a magnetic field on the pulsation stability of rapidly rotating supermassive stars.**
V. V. Usov.
Astrofizika, Vol. 13, 117 - 121 (1977). In Russian. English translation in Astrophysics, Vol. 13, No. 1.
The pulsational stability of rapidly rotating supermassive magnetic stars is considered. It is shown that the intensity of the poloidal magnetic field of a $10^5 M_\odot$ star required for its stabilization is of the order of the virial value. It decreases as the mass increases.

116.008   **On the photoelectric accuracy of measurements of magnetic fields of stars.**   G. A. Chuntonov.
Astrofiz. Issled., Izv. Spets. Astrofiz. Obs., Vol. 9, 115 - 116 (1977). In Russian.

116.009   **Radio emission from a normal HD 26676 star.**

R. G. Strom, D. E. Harris.
Nature, Vol. 269, 581 - 582 (1977).

The authors have discovered and report here a weak radio source coincident with the 6.2-mag B8Vn star HD26676.

116.010 On the orientation of magnetic and rotation axes in Ap stars. Theory of the decentred magnetic dipole.
H. Hensberge, W. van Rensbergen, M. Goossens, G. Deridder.
Astron. Astrophys., Vol. 61, 235 - 245 (1977).

Assuming a decentred magnetic dipole, the authors have constructed a series which gives the effective field H_e as a function of the displacement a of the dipole, the angle i between the magnetic axis and the line of sight and the limb darkening coefficient β. Analytical expressions have been obtained for the mean surface field H_s for arbitrary i and β, when a = 0 and $|a| = \sqrt{3}/2$. For general a values, H_s has been computed numerically, except for sin i = 0 and cos i = 0, where analytical expressions are also derived. These results have been used to interpret the phase diagrams of H_e and H_s in terms of an oblique rotator. A calculated set of curves can then be used to select possible models for the magnetic field geometry of those Ap stars for which the phase diagrams of H_e and H_s are available.

116.011 Apparent wavelength dependence of ν sin i for Zeta Tauri.   S. R. Heap.
Astrophys. J., Lett., Vol. 218, L17 - L19 (1977).

The projected rotational velocity derived from the spectrum of ζ Tau appears to be wavelength dependent: visual lines imply a value, ν sin $i \approx$ 300 km s^{-1}, while the ultraviolet lines imply a value, ν sin $i \lesssim$ 150 km s^{-1}.

116.012 The rotation effect in 71 Draconis.
D. P. Hube, J. Couch.
J. R. Astron. Soc. Canada, Vol. 71, 396 (1977). — Abstract.

116.013 The heterogeneity of surfaces of magnetic Ap stars.
M. Hack.
Highlights of Astronomy, Vol. 4, Part II, (see 012.026), p. 389 - 394 (1977).

The observations of spectrum-variability and light-variability of Ap stars are reviewed. It is shown that these variations are interpretable as due to the changing aspect of the spotted surface as the star rotates.

116.014 Magnetic stars.   V. Barocas.
Modern astronomy, (see 003.013), p. 149 - 157 (1977).

116.015 Models of magnetic stars – IV. The perpendicular rotator with meridional circulation.   D. Moss.
Mon. Not. R. Astron. Soc., Vol. 181, 747 - 760 (1977).

First-order perturbation theory models for uniformly rotating stars with a predominantly dipolar magnetic field with axis inclined at an angle $\pi/2$ to the rotation axis have been computed, both in strict radiative equilibrium and with the inclusion of finite resistivity and meridional circulation. In some ways the results are similar to the previous calculations with aligned rotation and magnetic axes. However, a positive correlation between rotation speed and surface flux for a given internal magnetic flux is found.

116.016 A magnetic field interpretation for the outburst of CH Cygni.   T. J. Wdowiak.
Publ. Astron. Soc. Pacific, Vol. 89, 569 - 571 (1977).

The possible appearance of kilogauss magnetic structure in and above the photosphere of a red giant during helium-shell flash is examined as a mechanism for the outburst of the apparently single star, CH Cyg. Strong magnetic fields created by dynamo action in a temporary connection zone of a red giant core, by virtue of their intrinsic buoyancy, would rise quickly to the stellar surface. It is suggested that if the field is coupled with the large-scale convective structure of the envelope, the energy contained and rate of release would be sufficient to produce the emission features of the spectrum of CH Cyg.

116.017 HR 5597 and irregular variation among the magnetic Ap stars.   W. K. Bonsack.
Publ. Astron. Soc. Pacific, Vol. 89, 613 (1977). — Abstract.

116.018 The application of a Bessel transform to the determination of stellar rotational velocities.
T. J. Deeming.
Astrophys. Space Sci., Vol. 46, 13 - 22 (1977).

A method for analysing line profiles by means of a transform using Bessel functions is described. This yields the stellar rotational velocity ν sin i, to an accuracy of about ± 1 km s^{-1} for rotational velocities greater than about 5 km s^{-1}, provided that rotation is the major source of line broadening. The theory of the method is a special case of a general theory of linear transforms in data analysis, which is outlined in an appendix.

116.019 Some considerations concerning the Zeeman effect in magnetic stars.   M. J. Stift.
Astrophys. Space Sci., Vol. 46, 465 - 469 (1977).

A detailed analysis of the multiplet structure of lines used for observations of stellar magnetic fields is presented. It is shown that LS-coupling does not hold good for many of the spectroscopic terms of Ti II and Cr II and that the magnetic fields observed in several magnetic stars are strong enough to produce transition to the Paschen-Back effect in some of these lines. It is recommended that only lines originating from Russell-Saunders terms be used for magnetic observations.

116.020 Progress report on southern magnetic star survey.
H. J. Wood, H. Jenkner, W. W. Weiss.
Mitt. Astron. Ges., Nr. 42, p. 136 (1977).

The effect of finite electrical and thermal conductivities on magnetic buoyancy in a rotating gas.
See Abstr. 062.040.

Linear polarization of Hα in the Be star Gamma Cassiopeiae.   See Abstr. 064.023.

Nonrotating superdense stars with frozen superstrong magnetic fields.   See Abstr. 065.008.

Strained coordinate methods in rotating stars. II. Main-Sequence stars.   See Abstr. 065.056.

Stellar rotation and the thermomagnetic torque.
See Abstr. 065.059.

A new criterion for secular instability of rapidly rotating stars.   See Abstr. 066.038.

On the electron cap shape of a rotating neutron star with a strong magnetic field.   See Abstr. 066.329.

Maunder's paradox and the tidal theory of solar activity.   See Abstr. 080.051.

Ultraviolet observations of Be stars. I. Macroscopic radial motions in the atmospheres of early Be stars.
See Abstr. 114.008.

The effects of stellar rotation on measurements from coudé spectrograms.   See Abstr. 114.504.

High-resolution polarization observations inside spectral lines of magnetic Ap stars. I. Instrumentation and

observations of β Coronae Borealis.. See Abstr. 114.513.

Spectrophotometric investigation of the magnetic variable star 21 Per. II. Distribution of Fe over the star's surface and study of Hγ and Hδ hydrogen line variations during the period.   See Abstr. 114.515.

The rotational velocity and barium abundance of Sirius.   See Abstr. 114.532.

Analysis of spectral variability of the Ap star 73 Dra. Part I.   See Abstr. 122.018.

Coordinated X-ray, optical, and radio observations of YZ Canis Minoris.   See Abstr. 122.024.

The magnetic cepheid W Sgr.

See Abstr. 122.026:

The magnetic field of W Sgr: new Zeeman measurements.   See Abstr. 122.054.

Spectroscopic observations of the helium silicon variable HD 124224.   See Abstr. 122.060.

Linear polarization in AM Herculis objects. See Abstr. 122.068.

On the role of a magnetic field in the evolution of T Tauri stars.   See Abstr. 122.128.

Observations of SiO masers at 43 GHz with the Parkes radio telescope.   See Abstr. 131.002.

117 Binary and Multiple Stars, Planetary Companions, Theory

117.001   The evolution of low-mass close binary systems. V.
Transport processes in the envelopes of contact
components.   R. F. Webbink.
Astrophys. J., Vol. 215, 851 - 863 (1977).

Starting from an adiabatic, isopotential, irrotational
approximation to the hydrodynamic equations, the author
discusses the nature of mass flow in the envelopes of contact
components of semi-detached and contact systems. A model of
luminosity transfer in contact binaries is proposed in which
large-scale heat transport is accomplished by the coupling of
Eddington-Sweet—type circulation with the dynamical flow
near L_1. The large-scale circulation between components
absorbs or releases energy in the envelope of each star according to whether or not it is in the same sense as the static
vertical entropy gradient in that envelope. The stability of this
configuration requires that the secondary components of
contact systems with common convective envelopes develop
higher envelope entropies (as observed in the W-type W Ursae
Majoris systems) in order to drive the circulation in the proper
sense.

117.002   Effect of rapid mass accretion onto main-
sequence stars.
S. Neo, S. Miyaji, K. Nomoto, D. Sugimoto.
Publ. Astron. Soc. Japan, Vol. 29, 249 - 262 (1977).

During the evolution of a close binary system, there is a
phase with rapid mass exchange between its component stars.
Effects of rapid mass inflow on the internal structure of the
main-sequence stars are studied. The mass-receiving star becomes overluminous and its radius becomes larger. The latter
result implies that only a part of the transferred matter is absorbed into the interior of the mass-receiving star and that the
rest is accumulated just above the original surface of the star.
When the rate of inflow is large, the radius becomes even comparable with that of a red giant star.

117.003   On the possibility of convective energy transfer
between the components of a contact binary
system.   L. N. Ivanov.
Astrofizika, Vol. 12, 475 - 484 (1976). In Russian. — English
translation in Astrophysics, Vol. 12, No. 3.

Convection in the vicinity of the Lagrangian point L_1
of a contact binary system is considered. It is found that
Schwarzschild's criterion of convective instability is independent of a variety of gravitation fields near L_1. An analysis of
the motion of the convective elements in the nonlinear
phase of perturbations indicates that convective energy transfer between the components of the system is impossible. Thus
the contact of stars cannot explain the deviation of W UMa
type systems from the standard mass-luminosity relation.

117.004   On the origin and evolutionary stage of symbiotic
stars.   A. V. Tutukov, L. R. Yungel'son.
Astrofizika, Vol. 12, 521 - 530 (1976). In Russian. — English
translation in Astrophysics, Vol. 12, No. 3.

An analysis of the parameters of symbiotic stars shows
that their hot components have to be either carbon-oxygen
dwarfs with thin hydrogen-helium envelopes or helium stars
with thin hydrogen-helium envelopes, while the cold components — red giants — loose mass at the rate of $10^{-5} - 10^{-6} M_\odot$/
year over the period of $10^5 - 10^6$ years. Such systems may be
formed from wide pairs if the envelope of the initially more
massive component is lost due to continuous mass loss or due
to rapid mass loss caused by dynamical instability on the
red-giant stage. If the initial system is not too wide, a hot star
may be formed due to mass exchange. It is shown that hot
components of symbiotic stars may accrete $10^{-6} - 10^{-9} M_\odot$/
year. Some consequences of accretion on C–O dwarfs are
discussed.

117.005   Close binary systems of early spectral type as
possible candidates of X-ray sources. I. Spectro-
scopic observations of X Persei = 2U 0352 + 30.
T. S. Galkina.
Izv. Krymskoj Astrofiz. Obs., Vol. 57, 45 - 56 (1977). In
Russian.

The spectrum of the close binary system X Per has been
analysed on the basis of 30 spectrograms with dispersions of
33 - 36 Å/mm in the regions Hα and λλ 4900 - 3650 Å obtained from November 1974 to March 1975 and in February
1976. The shapes of H I, He I, He II, Si IV lines observed in
1962/1963, 1972 and 1974 - 1975 were compared. Great
variations in the spectral lines were revealed. From 1962 to
1974 the emission lines Hα, Hβ, Hγ and He I have weakened
appreciably. The spectral type of the primary component may
be estimated as being B0 – B0.2ep. Radial velocity measurements from V and R emission components and sharp absorption of Hα and from broad Balmer absorption and He I lines
are given. The star rotation velocity has been estimated to
$v \sin i = 306 \pm 6$ km/sec.

117.006   On the structure of contact binaries. II. Zero-age
models.   S. H. Lubow, F. H. Shu.
Astrophys. J., Vol. 216, 517 - 525 (1977).

The authors construct zero-age models of contact
binaries of roughly solar composition with the contact discontinuity hypothesis discussed in an earlier communication.
With this formulation, they find it possible to construct systems with common radiative envelopes as well as systems with
common convective envelopes. They present explicitly two
models with total masses, respectively, of 1.5 $M_\odot$ and 3 $M_\odot$.
These models are compared with zero-age single stars which
have masses corresponding to the individual components.

117.007   A model of accretion disks in close binaries.
B. Paczyński.
Astrophys. J., Vol. 216, 822 - 826 (1977).

The simple periodic orbits of a test particle in the restricted three-body problem are a very good approximation
to the streamlines in the accretion disk with a very small
pressure and viscosity. The maximum size of such an accretion
disk is found as a function of mass ratio of a binary system.
The author assumes that this is identical with the largest
simple periodic orbit which does not intersect other orbits.
The maximum size of a disk is much larger than that found
by Kruszewski, by Flannery, or by Lubow and Shu, but it is
always less than the Roche lobe.

117.008   Twisted accretion disks. II. Applications to X-ray
binary systems.   J. A. Petterson.
Astrophys. J., Vol. 216, 827 - 837 (1977).

Obliquity in an X-ray binary system gives rise to forces
which tend to twist the shape of a thin accretion disk in the
system. The author derives the form of the most important
gravitational twisting forces, and calculates their effect on the
shape of the disk. He shows that pressure due to radiation
from the center of the X-ray source has influence on the structure of such a twisted disk, and will in many cases cause important changes in it.

117.009   Observational evidence for the gravity darkening of
components of close binary systems.
J. A. Eaton.
Bull. American Astron. Soc., Vol. 9, 433 (1977). — Abstract.

117.010   A search for extra-solar Jovian planets by radio

techniques.
W. F. Yantis, W. T. Sullivan III, W. C. Erickson.
Bull. American Astron. Soc., Vol. 9, 453 (1977). – Abstract.

117.011   Synthetic light curves of W Ursae Majoris binary star systems as applied to V566 Ophiuchi.
T. A. Nagy.
Publ. Astron. Soc. Pacific, Vol. 89, 366 - 373 (1977).

An efficient, highly automatic method for the computation of synthetic light curves has been developed expressly for the W UMa class of binary stars for which proximity effects are important. An intercomparison of the computer programs currently in use is given. The present method has been found to be a factor of three faster than the method of Wilson and Devinney. The results of this computer model as applied to two sets of observations of the A-type, binary system V566 Oph (Bookmyer 1969, 1976) are presented.

117.012   Motion of the W Ursae Majoris-type star CC Comae.
A. R. Klemola.
Publ. Astron. Soc. Pacific, Vol. 89, 402 (1977).

Proper motions measured for CC Com and its fifteenth magnitude companion show that neither is a member of the Coma star cluster (Mel 111). The annual motion of 0″.13 indicates that the stars are relatively nearby dwarfs.

117.013   Parallax, orbit, and mass of Ross 614.
R. G. Probst.
Astron. J., Vol. 82, 656 - 661 (1977).

An orbit for the resolved astrometric binary Ross 614 has been determined from photographs taken in the interval 1928–1975. These data, previous parallaxes, and a Sproul orbit are utilized to obtain masses of 0.13 and 0.07 $M_\odot$ for the two stars. Photoelectric photometry and estimates of Δm give absolute magnitudes of 13.1 and 16.6 in V, 10.1 and 12.2 in I.

117.014   Planetary orbits in binary stars.   R. S. Harrington.
Astron. J., Vol. 82, 753 - 756 (1977).

Numerical integrations of the general three-body problem, with one component having a planetary mass, indicate that stable planetary orbits can exist in binary stars. The limitation for stability is that the ratio of the periastron distance of the outer tertiary component to the semimajor axis of the close component be somewhere in the range 3–4, regardless of which of the components is the planet. For most known binaries, this region of stability includes the region of habitability for planets.

117.015   Mass transfer instabilities in binary systems.
P. R. Wood.
Astrophys. J., Vol. 217, 530 - 536 (1977).

The stability of mass transfer from semidetached secondaries in binary systems has been studied incorporating some improvements on the previous calculations of Bath, including a revised outer boundary condition, convective energy transport, and a line-of-centers gravitational potential. Main-sequence secondaries with $T_{eff} \gtrsim 5000$ K ($M \gtrsim 0.8 M_\odot$) are found to transfer mass in pulses, while cooler (less massive) secondaries transfer mass continuously. A lobe-filling giant with parameters similar to those of the secondary in the recurrent nova T CrB is found to transfer mass continuously.

117.016   Change of the orbital elements and run-away velocities of binaries caused by a symmetric explosion of one of the components.
J.-P. de Cuyper, C. de Loore, E. P. J. van den Heuvel.
Astron. Astrophys., Suppl. Ser., Vol. 30, 93 - 112 (1977).

Orbital parameter changes and run-away velocities of initially synchronized binary systems due to symmetric mass loss of one of the components are presented. The computations were performed for systems with the following mass ratios

(i.e. ratio of the mass of the companion and the initial mass of the exploding component) $m_2 = 0.001, 0.25, 0.5, 0.75, 1, 1.25, 1.5, 2, 3, 4, 6, 8, 15, 20, 25$ and 30. Three modes of mass decay and various mass decay velocities were taken into account. The parameters are given relative to the initial values of the system.

117.017   TU Horologii – an ellipsoidal variable.
H. W. Duerbeck.
Astron. Astrophys., Vol. 61, 161 - 163 (1977).

From UBV observations TU Hor is found to be an ellipsoidal variable ($V = 5.^m925$ to $6.^m025$) with a period of 0.935984 days. Analysis of the light variation indicates abnormal reflection behaviour in close binary systems. Spectroscopic observations yield an orbital inclination of about 35°.

117.018   A new model for V Pup: is the star a main-sequence contact binary?
B. Cester, B. Fedel, G. Giuricin, F. Mardirossian, M. Pucillo.
Astron. Astrophys., Vol. 61, 275 - 277 (1977).

New photometric elements are derived for V Pup by solving, in a homogeneous way, nine light curves taken from literature. The star seems to be a main-sequence contact binary.

117.019   Spectroscopic studies of O-type binaries. III. HDE 228766: an evolved Of system.
P. Massey, P. S. Conti.
Astrophys. J., Vol. 218, 431 - 437 (1977).

The authors have examined the binary system HDE 228766 with high-dispersion coudé spectrograms. Traditionally cited as an example of a Wolf-Rayet binary with a high mass ratio, HDE 228766 actually consists of a late O primary and an Of secondary, much as Walborn has suggested. The orbital elements indicate a mass ratio near unity, based on the ability to separate the hitherto unresolved absorption lines into two components. The estimate of the present mass-loss rate from the system from the line strength of Hα is $10^{-5} M_\odot$ yr^{-1}. This rate is significant on an evolutionary time scale. The authors discuss the possibility that a Wolf-Rayet system will be the result of this evolution.

117.020   Geburt eines Planetensystems.
Orion, 35. Jahrg., 155 (1977).

117.021   Tidal torques on accretion discs in close binary systems.   J. Papaloizou, J. E. Pringle.
Mon. Not. R. Astron. Soc., Vol. 181, 441 - 454 (1977).

The authors calculate the transfer of angular momentum between an accretion disc and orbital motion in a close binary system. If the dissipative process in the accretion disc can transport angular momentum (e.g. shear viscosity) the disc fills its Roche lobe and tidal torques always dominate the transport process in the outer parts of the Roche lobe. If not, the disc does not expand, all the transferred material is accreted and the inflow rate in the disc is controlled solely by tidal processes. The authors show that in an optically thick disc, the presence of a shear viscosity gives rise to a bulk viscosity of comparable magnitude.

117.022   The possibility for visual detection of astrometric binaries with special astrophysical interest.
S. L. Lippincott.
O telescópio refractor e a astrometria ao serviço das estrelas duplas, (see 012.013), p. 131 - 138 (1977).

Intensive astrometric studies over long intervals of time have revealed duplicity among a number of nearby stars. Only with the determination of the apparent separation of the two components and their magnitudes can their individual masses be determined. The astrometric studies often imply separations under one second of arc with large Δm rendering visual detection difficult. The importance of determining the masses of these particular systems is discussed. They include luminosity classes III, IV, VI as well as very late type V objects with masses generally not accepted as stellar. Also included are several eclipsing binaries.

117.023 **A spectroscopic study of the binary system δ Equulei.**   E. M. Hans, C. D. Scarfe, J. M. Fletcher.
J. R. Astron. Soc. Canada, Vol. 71, 408 (1977). − Abstract.

117.024 **Evolution of low-mass, close binary systems with mass exchange and gravitational-radiation losses.**
W. Y. Chau, D. Lauterborn.
J. R. Astron. Soc. Canada, Vol. 71, 408 (1977). − Abstract.

117.025 **Rectilinear trajectories of double stars. Comparison between several methods. Results for h 5094. Effect of erroneous positions.**
H. Debehogne, R. R. de Freitas Mourão.
Astron. Astrophys., Vol. 61, 453 - 457 (1977). In French.
Computation of rectilinear trajectories described by the companion B as referred to the primary A is performed by means of projections. A comparison between the direct method of least squares and the method of mean places with a differential correction was carried out. Several binaries have been studied. For all the cases investigated, the direct method of least squares seems to be the best fitted. The results for 17 observations of h 5094 are presented. The effects of the introduction of erroneous observations in the method of least squares are shown by graphs.

117.026 **Revised photometric elements of 12 semi-detached systems.**   B. Cester, B. Fedel, G. Giuricin, F. Mardirossian, M. Pucillo.
Astron. Astrophys., Vol. 61, 469 - 475 (1977).
Photometric light curves of 12 double spectrum binary systems already classified as semi-detached, have been studied with the Wood program in order to obtain homogeneous parameters. Absolute masses, radii and luminosities are given. RZ Cnc, U Cep, u Her, TU Mon, U Sge, V356 Sgr, μ^1 Sco, TX UMa, Z Vul are confirmed as sd-systems; U CrB and RS Vul might be sd-d-systems and V Pup a contact system.

117.027 **The spectrum of HD 187399.**
N. L. Ivanova, A. N. Khotyanskij.
Astrofizika, Vol. 12, 623 - 630 (1976). In Russian. − English translation in Astrophysics, Vol. 12, No. 4.
Velocities and H, He I, Mg II, Ca II line profiles are measured at different phases of the spectrum binary HD 187399. Abrupt change of line intensities was observed at phase 0.850. It may be explained by the influence of a gas stream moving from an invisible component towards the main star.

117.028 **Astrometric analyses of the unseen companions to Ci 18, 2354 and Wolf 1062 from plates taken with the 61-cm Sproul refractor.**   S. L. Lippincott.
Astron. J., Vol. 82, 925 - 928 (1977).
The analysis of the Sproul plates from 1937 to 1976 on Ci 18, 2354 yields an improved photocentric orbit. A period of 26.37 yr and $\alpha = 0\overset{''}{.}033$ yield a mass of 0.009 $M_\odot$ for the companion, assuming $m_v \gtrsim 16$. The upper limit is likely not more than 0.06 $M_\odot$. It is very likely that the companion is, therefore, substellar. The positions of Wolf 1062 from plates 1938−1976 reveal an unseen companion from a photocentric orbit with $P = 2\overset{y}{.}45$ and $\alpha = 0\overset{''}{.}026$. A likely minimum mass for the invisible component is 0.06 $M_\odot$ and the upper limit is 0.13 $M_\odot$.

117.029 **Stellar disk suggests planet formation.**
Phys. Today, Vol. 30, No. 11, p. 19 (1977).

117.030 **Consequences of mass transfer in close binary**

systems.   H.-C. Thomas.
Annu. Rev. Astron. Astrophys., Vol. 15, (see 003.012), 127 - 151 (1977).
Contents: Standard assumptions and their limits; Conservative mass transfer; Loss of mass or angular momentum from the binary system; Algol-type binary systems; W UMa systems; Evolution towards the X-ray stage; Mass transfer in the X-ray stage and beyond; Concluding remarks.

117.031 **Perte de masse des étoiles binaires serrées.**
F. Van't Veer.
Perte de masse des étoiles, (see 012.030), 6 pp. (1977).

117.032 **Wobble star, does 61 Cygni A have a planetary companion?**   W. Campbell.
Sci. Dimension, Vol. 9, No. 2, p. 18, 20 (1977). − Abstr. in Phys. Abstr., Vol. 80, Abstr. 71131 (1977).

117.033 **The fraction of solar-type stars having stellar or massive nonstellar companions.**
C. Bettis, J. Bonnell, D. Branch.
Icarus, Vol. 32, 461 - 463 (1977).
The results of the survey for multiplicity by Abt and Levy are used to estimate the fraction of solar-type stars having close companions more massive than 0.01 solar masses. Current knowledge of the multiplicity characteristics of solar-type stars does not require that the fraction be nearly unity.

117.034 **A possible eclipse in the triple system 20 Leonis.**
T. G. Barnes III, F. C. Fekel, T. J. Moffett.
Publ. Astron. Soc. Pacific, Vol. 89, 658 - 659 (1977).
Observations of a possible eclipse in the 20 Leo system are reported. A discussion of the possible implications on this interesting triple system is also presented.

117.035 **The massive hot binary 29 Canis Majoris.**
J. B. Hutchings.
Publ. Astron. Soc. Pacific, Vol. 89, 668 - 674 (1977) = Dominion Astrophys. Obs., Victoria, B.C., Canada, Contrib. No. 337 = NRC No. 15906.
The results are presented from analysis of 16 high- and medium-dispersion spectrograms of the system. Mass loss from the primary has a strong dependence on phase in an orbit of low eccentricity. No change in period is detected but the primary star amplitude shows an ~10% increase over the value derived from observations in 1957. The secondary spectrum is described and the fundamental parameters of the stars and their evolutionary status are discussed.

117.036 **On the existence of pairs of carbon stars.**
J. M. Scalo, T. Deeming, D. A. Edwards.
Publ. Astron. Soc. Pacific, Vol. 89, 688 - 692 (1977).
The data in Westerlund's (1971) catalog of carbon stars have been analyzed in order to determine whether a significant excess of carbon-star pairs exists, as has been previously claimed. The authors present a simple and consistent statistical method for comparison of observed and random separation distributions. The differential separation distribution shows no strong evidence for an excess number of pairs, except possibly at very small ($\lesssim 0\overset{''}{.}03$) separations. The frequency distribution of magnitude differences for different separations and the large implied physical separations suggest to the authors that most of these pairs are not physical.

117.037 **Method for determination of the masses of the components of close binary systems from characteristics of transfer and mass loss.**   M. Yu. Skul'skij.
Tsirk. Astron. Obs., L'vov, No. 51, p. 13 - 19 (1976). In Russian.

117.038 **Remarks on erratic period fluctuations of detached close binaries and the constancy of the orbital period**

of XY UMa. E. H. Geyer.
Astrophys. Space Sci., Vol. 48, 137 - 144 (1977).

It is shown that the erratic changes of the orbital period, which are observed in some detached eclipsing binary systems, cannot be interpreted as real if their light curves show changing anomalies with asymmetric minimum profiles. In this case the photometrically determined times of the minima do not coincide with the ideal geometric ones, which are unobservable. This fact and the applied reduction methods for deriving the photometric times of minima cause systematic period errors of cumulative character. The discussion of six photoelectrically-determined minima obtained within the last 20 yr, and earlier photographic ones of the 0^d479 period eclipsing binary XY UMa yields a constant period.

117.039 The evolution of massive close binaries. V: Systems containing primaries with masses between 10 $M_\odot$ and 15 $M_\odot$. J.-P. De Grève, C. de Loore.
Astrophys. Space Sci., Vol. 50, 75 - 85 (1977).

The final state of the primaries of binary systems with initial masses $M_{1i} = 10\,M_\odot$ to $15\,M_\odot$ is derived from the mass of their C/O-cores. The possibility of a second stage of mass transfer towards the secondary is considered. It turns out that the critical mass for the bifurcation is about 14 $M_\odot$: stars with larger masses in this range are the progenitors of neutron stars, while the lower mass stars are the ancestors of white dwarfs.

117.040 On the mass-radius relation in close binary systems. T. T. Chia.
Astrophys. Space Sci., Vol. 51, 33 - 38 (1977).

In a close binary system with circular orbits in which mass transfer occurs when material from one star fills up and possibly spills over its Roche lobe, the usual mass-radius relation cannot be used as it would lead to unreasonable results in some situations.

117.041 Effect of gravitational radiation on the evolution and spins of binary systems. T. T. Chia.
Astrophys. Space Sci., Vol. 51, 159 - 164 (1977).

By taking into account the torque arising from gravitational radiation reaction, it is shown that the spins of spherically symmetrical components of binary systems remain unchanged, while those of non-spherical components will vary. The separation between the non-spherical symmetrical components decreases faster than that of the corresponding two point or spherically symmetrical mass system. In some cases, an initially synchronized binary system will definitely not remain synchronized. In systems with negligible viscosity such that the tidal lag is always small, the spins would decrease and may possibly become retrograde.

117.042 Synthetic light curves for contact binaries. L. Binnendijk.
Vistas Astron., Vol. 21, 359 - 391 (1977).

The geometry of the Roche-Jacobi-Hill models is considered which allows us to compute the sizes of the components. A new method is discussed to compute the synthetic light curves. It was found that the mean gravity of the greater component is larger than the mean gravity of the smaller component for the innermost contact surface and the dumbbell model. This explains why the greater component is the hotter one and the dumbbell model gives only a good representation for A-type systems. An extended dumbbell model is discussed for W-type systems and the starspot model is proposed for these W-type systems which have asymmetric light curves. A comparison is made of the orbital parameters found by several astronomers. The synthetic line profiles and the synthetic radial velocity curves for contact binaries are considered.

117.043 AM Her — close binary system with a magnetic dwarf — the nearest U Gem star?
I. G. Mitrofanov, G. G. Pavlov, Yu. N. Gnedin.

Astron. Tsirk., No. 948, p. 5 - 6 (1977). In Russian.

117.044 Radio survey of close binary stars. S. R. Spangler, F. N. Owen, R. A. Hulse.
Astron. J., Vol. 82, 989 - 997 (1977).

The authors present results of a deep (5 mJy) 6-cm survey of 145 close binary stars. Eleven stars were detected (including five not previously known to be radio binaries) and seven possibly detected (six of which were not previously suspected to be radio sources). This relatively unbiased survey confirms that radio binaries later than A0 are predominantly of spectral classes G and K. In an HR diagram, the radio binaries tend to be significantly to the right of radio-quiet binaries.

117.045 Rejuvenation of helium white dwarfs by mass accretion. K. Nomoto, D. Sugimoto.
Publ. Astron. Soc. Japan, Vol. 29, 765 - 780 (1977).

In a close binary system, a white dwarf may be rejuvenated by mass accretion from a companion star. Accretion of helium onto a helium white dwarfs is studied all the way from the onset of mass accretion through the helium flash or explosion. When the inflow rate dM/dt is relatively large, a flash takes place in an outer shell. For the case of a smaller dM/dt, helium flash occurs in the central region, which grows into the formation of a detonation wave. It results in a supernova explosion and the white dwarf disrupts completely.

117.046 Erscheint bei Komponenten enger Sternpaare eine Schräglage von Rotationsachse zur Bahnebene über längere Zeiträume möglich? K. Walter.
Mitt. Astron. Ges., Nr. 42, p. 84 (1977).

117.047 Neredzamo pavadoņu meklējumi turpinās. A. Deičs (*A. Deutsch*).
Zvaigžņota debess, 1976/77. gada ziema, p. 11 - 16.

L'informatique et les étoiles doubles.
See Abstr. 031.403.

The motion of a particle from the L_2 Lagrangian point: elliptic cases. See Abstr. 042.004.

The solar system in a binary star.
See Abstr. 042.006.

Analytical determination of the measure of stability of triple stellar systems. See Abstr. 042.040.

Non-radial radiative transfer in close binaries. Application to the bolometric reflection effect in W UMa stars. See Abstr. 063.007.

Some remarks on the non-radial component of radiative transport in close binaries. See Abstr. 063.046.

Line-distortion effects in OB supergiant X-ray binaries. See Abstr. 064.015.

A numerical approach to the testing of the fission hypothesis. See Abstr. 065.081.

Tidal generation of gravitational waves from orbiting Newtonian stars. I. General formalism.
See Abstr. 066.025.

Lunar occultation summary. II.
See Abstr. 096.017.

A correlation between multiplicity and metallicity for solar-type stars. See Abstr. 114.056.

The ultraviolet spectra of four binaries observed with the S59 spectrometer. See Abstr. 114.553.

Observations with the VLA of the radio binary star AR Lacertae. See Abstr. 116.006.

DL Virginis – an eclipsing binary in a triple system. See Abstr. 121.018.

Linear polarization in AM Herculis objects. See Abstr. 122.068.

VV Puppis: the day the accretion stopped! See Abstr. 122.106.

The nova-like variable BD −7°3007. See Abstr. 122.111.

Constraints on the apsidal motion of some galactic X-ray sources. See Abstr. 142.024.

Limitations on the nature and evolution of X-ray binaries. See Abstr. 142.035.

Evidence for an accretion disk in SMC X-1. See Abstr. 142.043.

Spectrum variations of the X-ray binary HD 153919 (= 3U 1700-37). See Abstr. 142.054.

Optical observations of X-ray binaries.

See Abstr. 142.064.

The evolutionary history of X-ray binaries. See Abstr. 142.066.

Accretion flows in binary X-ray systems. See Abstr. 142.067.

The 580-day and other periodicities in X Persei. See Abstr. 142.084.

Velocity curves for broad and sharp components observed in the emission lines from AM Herculis. See Abstr. 142.093.

The eccentric-orbit binary model for the transient X-ray sources. See Abstr. 142.131.

The optical polarization of X-ray binaries. See Abstr. 142.134.

Errata

117.901 Erratum: 'The infrared variability and nature of symbiotic stars' [Mon. Not. R. Astron. Soc., Vol. 179, 499 - 508 (1977)]. M. W. Feast, B. S. C. Robertson, R. M. Catchpole.
Mon. Not. R. Astron. Soc., Vol. 181, 253 (1977).

118 Visual Binaries and Multiple Stars

118.001 New double stars (14th series) discovered at Nice. P. Couteau.
Astron. Astrophys., Suppl. Ser., Vol. 29, 249 - 254 (1977). In French.
The author gives a list of 150 double stars discovered at the 50 and 74 cm refractors.

118.002 The masses of the multiple star HD 188753 (ADS 13125). S. L. Lippincott.
Observatory, Vol. 97, 200 - 201 (1977).

118.003 New double stars. D. W. Dunham.
Occultation Newsl., Vol. 1, 119 - 120 (1977).

118.004 Visual binaries among the barium stars. II. HD 105902.
R. B. Culver, P. A. Ianna, O. G. Franz.
Publ. Astron. Soc. Pacific, Vol. 89, 397 - 399 (1977).
Photometric and spectroscopic observations are presented of both members of the binary star system HD 105902 (ADS 8448) whose primary component is a barium star of spectral class K5 Ba 2. The visual absolute magnitude for the barium star primary is found to be +2.7 and the distance to the system is 190 pc. The mass for the barium star is estimated to be 2.6 $M_\odot$.

118.005 Observational selection in binary-star eccentricities. R. S. Harrington, M. Miranian.
Publ. Astron. Soc. Pacific, Vol. 89, 400 - 401 (1977).
Observational selection effects on the distribution of eccentricities in binary-star orbits are examined by monitoring

the selection of orbits from a randomly generated sample. Selection effects fully explain the difference between predicted and observed eccentricity distributions.

118.006 Relative positions of the components of Sirius. E. van Albada-van Dien.
Astron. Astrophys., Suppl. Ser., Vol. 29, 305 - 308 (1977) = Contrib. Bosscha Obs., Lembang, No. 63.
Results are given of the photographic observations of the system of Sirius (ADS 5423, 06408S1635 (1900)).

118.007 A spectroscopic orbit for the subdwarf binary Mu Cassiopeiae. T. F. Worek, W. R. Beardsley.
Astrophys. J., Vol. 217, 134 - 139 (1977).
A spectroscopic orbit for the G5 subdwarf μ Cas is presented. Eighty-one radial velocities from the Allegheny, Kitt Peak, David Dunlap, McDonald, and Hale Observatories, along with 19 previously published ones, are used to establish the orbit. The authors find that the observations best fit a computed velocity curve represented by these final elements: $P = 23.0$ yr, $e = 0.30$, $T = 1954.0$, $K = 2.8$ km s^{-1}, and $\omega = 178°$. If the primary mass is approximated to be 0.80 $M_\odot$, it then follows that $M_B \geqslant 0.27\,M_\odot$. Thus the secondary is likely to have a spectral type no later than M3 and these magnitudes: $M_{bol} \leqslant +9.6$ and $M_v \leqslant +11.9$.

118.008 Triple system Stein 2051 (G175−34). K. Aa. Strand.
Astron. J., Vol. 82, 745 - 749 (1977).
Investigation of the orbital motion of the large proper motion binary, Stein 2051, has proven that the dwarf M

component is an astrometric binary with a period of 23 yr and a semimajor axis of 0.″070. On the basis of the determined mass ratio between the red dwarf system and the DC white dwarf component, the derived mass for the white dwarf is either 0.50 or 0.72 $M_\odot$, dependent upon whether the former is a red component with a low-mass (0.02 $M_\odot$) companion, or a close binary of nearly equal magnitude components. Present observational data favor the presence of the low-mass companion and, therefore, the smaller mass of the white dwarf.

118.009 Etude de la photométrie *uvby*β à l'aide d'étoiles doubles visuelles à grande séparation.   E. Oblak.
Recueil des séminaires, (see 012.004), p. 102 - 119 (1977).

118.010 Les étoiles doubles et leur observation dans le contexte actuel de la science astronomique.
J. Dommanget.
O telescópio refractor e a astrometria ao serviço das estrelas duplas, (see 012.013), p. 17 - 24 (1977). − Discours introductif général.

118.011 La situation présente.   P. Couteau.
O telescópio refractor e a astrometria ao serviço das estrelas duplas, (see 012.013), p. 25 - 32 (1977).

118.012 Leçons d'un survey polaire.   P. Muller.
O telescópio refractor e a astrometria ao serviço das estrelas duplas, (see 012.013), p. 33 - 44 (1977).

118.013 Premier bilan sur la découverte de 1500 étoiles doubles.   P. Couteau.
O telescópio refractor e a astrometria ao serviço das estrelas duplas, (see 012.013), p. 45 - 53 (1977).

118.014 Research of massive invisible companions in double star systems.   R. van de Wiele.
O telescópio refractor e a astrometria ao serviço das estrelas duplas, (see 012.013), p. 89 - 100 (1977).
The author has studied the conditions of a systematic search of invisible bodies, with $M \geqslant 3 M_\odot$. A criterion has been derived to select some stars in the Index-Catalogue, having the greatest chances to lead to a massive invisible disturbing body.

118.015 Compagnons invisibles dans les systèmes d'étoiles doubles visuelles.   P. Baize.
O telescópio refractor e a astrometria ao serviço das estrelas duplas, (see 012.013), p. 101 - 129 (1977).

118.016 Zeta Herculis, étoile triple.   P. Baize.
O telescópio refractor e a astrometria ao serviço das estrelas duplas, (see 012.013), p. 139 - 147 (1977).

118.017 Recherches sur les étoiles doubles à longues périodes.   P. Laques.
O telescópio refractor e a astrometria ao serviço das estrelas duplas, (see 012.013), p. 149 (1977). − Abstract. (See 11.118.013).

118.018 Radial velocities of visual binaries with an orbit derived from astrometric observations.
E. van Dessel.
O telescópio refractor e a astrometria ao serviço das estrelas duplas, (see 012.013), p. 151 - 184 (1977).
Attention has been drawn on several occasions (e.g. Dommanget, 1967: Catalogue d'Ephémérides des vitesses radiales relatives) to the importance of obtaining radial velocity observations for visual binaries for which orbital elements can be derived from astrometric measurements. This contribution is meant to give an idea about what has been accomplished so far (couples that have been extensively studied are e.g. ADS 1123, Capella, 1 Gem, ADS 10598, δ Equ) and to present a list of binaries for which it would be interesting to have more radial velocity observations.

118.019 Photometric results of the observations of visual binaries by area scanning techniques.   K. Rakos.
O telescópio refractor e a astrometria ao serviço das estrelas duplas, (see 012.013), p. 185 - 187 (1977).
A program of photoelectric photometry of close visual binaries is described and its possibilities discussed. First-hand results concerning the difference in visual brightness for 80 binaries have been collected from various papers and summarized in a table at the end of this communication.

118.020 Les étoiles doubles et les lunettes astronomiques.   P. Couteau.
O telescópio refractor e a astrometria ao serviço das estrelas duplas, (see 012.013), p. 189 - 194 (1977).

118.021 New double stars.   D. W. Dunham.
Occultation Newsl., Vol. 1, 129 (1977).

118.022 Central star of NGC 3132: a visual binary.   L. Kohoutek, S. Laustsen.
Astron. Astrophys., Vol. 61, 761 - 763 (1977).
A faint companion to HD 87892 has been discovered with the 3.6 m telescope of the European Southern Observatory at La Silla having the following parameters: separation $\rho = 1.″65$, position ange $\vartheta = 226.°3$. The companion, which can be considered as the actual planetary nucleus, is estimated to have $U = 14.^m8$, $L_*/L_\odot \cong 110$, $R_*/R_\odot \cong 0.035$. Its physical association with HD 87892 is probable.

118.023 HD 165590.
IAU Circ., No. 3088 (1977).

118.024 Study of stellar system Cor 197 = CD−48°8449/ −48°280.   R. de Freitas Mourão.
An. Acad. Brasil. Cienc., Vol. 48, 193 - 197 (1977). In French. Abstr. in Phys. Abstr., Vol. 80, Abstr. 78819 (1977).

118.025 Visual double stars measured at Las Campanas Observatory, Chile.   F. Holden.
Publ. Astron. Soc. Pacific, Vol. 89, 582 - 587 (1977).

118.026 Measures of southern visual double stars.   F. Holden.
Publ. Astron. Soc. Pacific, Vol. 89, 588 - 595 (1977).

118.027 Duplicity and spectral types of HV 10814.   G. H. Herbig.
Inf. Bull. Variable Stars, No. 1323 (1977).

118.028 A probable eclipse in HD 165590.   C. D. Scarfe.
Inf. Bull. Variable Stars, No. 1357 (1977).

118.029 HD 165590.
Yamamoto Circ., No. 1857 (1977). In Japanese.

118.030 Les orbites de deux étoiles doubles visuelles.   V. Erceg.
Bull. Obs. Astron. Belgrade, No. 128, p. 23 - 24 (1977).
L'auteur donne les éléments des orbites, les quantités astrophysiques et les parallaxes dynamiques pour les étoiles doubles visuelles ADS 1227 et ADS 1530.

118.031 The orbit of the visual double star COU 79 (BD +24°329).   D. Olević.
Bull. Obs. Astron. Belgrade, No. 128, p. 25 (1977).

118.032 Rectilinear orbit of the pair ADS 12040=Σ2454.   D. Olević.
Bull. Obs. Astron. Belgrade, No. 128, p. 26 - 27 (1977).

118.033 **Orbits of two visual binaries.**   D. J. Zulević.
           Bull. Obs. Astron. Belgrade, No. 128, p. 28 - 29
(1977).
     Orbits and dynamical parallaxes are presented for two
visual binary systems: ADS 11247, and ADS 15530.

118.034 **New double stars discovered in Belgrade with the**
           Zeiss refractor 65/1055 cm. Supplement V.
G. M. Popović.
Bull. Obs. Astron. Belgrade, No. 128, p. 30 - 32 (1977).
     Presented are positions and 60 measures of 31 double
stars discovered at Belgrade with the Zeiss refractor 65/1055
cm. The positions are related to the epochs 1900, 1950 and
2000.

118.035 **Micrometer measures of double stars. Series 25.**
           D. M. Olević.
Bull. Obs. Astron. Belgrade, No. 128, p. 33 - 36 (1977).
     This series contains 195 measures of 184 pairs carried
out by the author with the Zeiss equtorial 65 cm of Belgrade
Observatory from 21 January 1974 till 29 September, 1975.

118.036 **Micrometer measures of double stars. Series 26.**
           G. M. Popović.
Bull. Obs. Astron. Belgrade, No. 128, p. 37 - 46 (1977).
     Presented are 271 measures of 134 double stars, 29 of
which are carrying the designation GP.

118.037 **Micrometer measures of double stars. Series 27.**
           D. J. Zulević.
Bull. Obs. Astron. Belgrade, No. 128, p. 47 - 53 (1977).
     Presented here are 214 measures of 100 double stars
made with the 26-inch refractor of Belgrade Observatory.

118.038 **Relative distances of the components** C **from the**
           pairs AB **in visual triple systems.**   G. M. Popović.
Publ. Dep. Astron., Univ. Beograd, Fac. Sci., No. 6, (see
012.040), p. 49 - 52 (1976).

118.039 **Is Vega a binary star?**   I. N. Latyshev.
           Astron. Tsirk., No. 950, p. 8 (1977). In Russian.

118.040 **Orbites nouvelles.**
           Circ. Inf., No. 73 (1977).

118.041 **Étoiles doubles nouvelles: Lunette de 65 cm**
           (Belgrade); Lunette de 50 cm (Nice); KPNO 4 m
telescope (Speckle interferometry).
G. M. Popovic, P. Couteau, H. A. McAlister.
Circ. Inf., No. 73 (1977).

         Centre de données d'étoiles doubles de l'Observa-
toire de Nice.   See Abstr. 002.018.

         La précision des techniques d'observation anciennes.
See Abstr. 004.028.

         L'astronomie des étoiles doubles au Portugal.
See Abstr. 013.014.

         Area scanning technique for astrometric observations
of visual binaries.   See Abstr. 031.249.

         Application de l'observation des occultations
d'étoiles par la Lune à la decouverte et à la mesure d'étoiles
doubles.   See Abstr. 031.250.

         Application de la transformée de Fourier à la mesure
des binaires.   See Abstr. 031.251.

         Un équipement du type «Télévision» pour l'observa-
tion des étoiles doubles visuelles.   See Abstr. 034.009.

         Occultation study of the multiple star β **Scorpii.**
See Abstr. 096.006.

         Occultation observations of β **Scorpii.**
See Abstr. 096.007.

         The rotational velocity and barium abundance of
Sirius.   See Abstr. 114.532.

         The spectrum of h 4866 B.   See Abstr. 114.543.

119 Spectroscopic Binaries

119.001 **Spectroscopic binary orbits from photoelectric**
           radial velocities. Paper 14: HD 187299.
R. F. Griffin, G. A. Radford.
Observatory, Vol. 97, 169 - 173 (1977).
     HD 187299 is a binary containing a supergiant G5
primary and an unseen but massive secondary for which a
spectral type of B2−3 V is suggested on circumstantial evidence. There is a strong possibility of eclipses.

119.002 **Spectroscopic binary orbits from photoelectric**
           radial velocities. Paper 15: HR 6940.
G. A. Radford, R. F. Griffin.
Observatory, Vol. 97, 173 - 175 (1977).

119.003 **The spectroscopic binaries in NGC 6475.**
           F. Gieseking.
Astron. Astrophys., Vol. 60, 9 - 12 (1977).
     With objective prism spectra, obtained with the radial
velocity astrograph (GPO) of the European Southern Observatory, a large number of new radial velocities of 13 of the
brightest members in NGC 6475 have been determined.

Three of five previously suspected short-period spectroscopic
binaries could not be confirmed. However two of these three
stars seem to be long-period spectroscopic binaries. The remaining two suspected binaries could be confirmed and new
elements have been derived. Finally, general properties of the
GPO-technique for the determination of relative radial velocities are discussed.

119.004 **On the light variation of the spectroscopic binary**
           66 Eridani.
P. Vivekananda Rao, A. G. Kulkarni.
Bull. Astron. Soc. India, Vol. 5, 73 - 74 (1977).

119.005 **CaII H and K emissions in binaries. I. Intensity-**
           period relation.   R. Głębocki, A. Stawikowski.
Acta Astron., Vol. 27, 225 - 233 (1977).
     An analysis of intensities of Ca II emission K line in
spectroscopic binaries is made. The investigation is restricted
to giants and supergiants of spectral types G and K. A correlation between K line intensity and period is found.

119.006   Optical behaviour of HDE 226868 during a Cyg X-1 X-ray transition.   E. N. Walker, G. D. Brownlie, K. O. Mason, P. W. Sanford, A. R. Quintanilla.
Nature, Vol. 263, 393 - 395 (1976).

The blue-band optical magnitude of HDE 226868, the spectroscopic binary BO1b star which is the optical counterpart of Cyg X-1, was monitored between July and December 1975 from the Spanish Sierra Nevada. The data are shown in a figure and are compared with the X-ray behaviour of Cyg X-1 as measured by Copernicus and inferred from Ariel V and ANS data.

119.007   Revised elements for the spectroscopic binary HD 224355.   M. Imbert.
Astron. Astrophys., Suppl. Ser., Vol. 29, 407 - 409 (1977). In French.

Radial velocities were combined to yield revised period and new orbital elements for the spectroscopic binary HD 224355 (m_v 5.6, F5).

119.008   Survey of selected spectroscopic and eclipsing binaries for intrinsic linear polarization.
R. J. Pfeiffer.
Astron. J., Vol. 82, 734 - 739 (1977).

Observations of $uBVR$ linear polarization from a survey of 20 close binaries are presented. The applications of four tests to the data have indicated convincing evidence for intrinsic polarization in SX Cas, YY Gem, KS Per, o Per, and HD 47129, and corroborated the same for ϕ Per, ψ Sgr, and ζ Tau. For α And and WW Aur, the tests have yielded conflicting results. The remaining systems are either unpolarized or exhibit only interstellar polarization. If the intrinsic polarization is the result of scattering from circumstellar matter, polarimetry is in good agreement with spectroscopic and photometric techniques for the detection of such material, except for o Per and β Per. Possible explanations for this disagreement are suggested.

119.009   Spectroscopic binary orbits from photoelectric radial velocities. Paper 16: HD 13738.
R. F. Griffin, G. A. Radford.
Observatory, Vol. 97, 196 - 198 (1977).

119.010   HD 137569, a population II remnant of mass-exchange evolution.   C. T. Bolton, J. R. Thomson.
J. R. Astron. Soc. Canada, Vol. 71, 402 (1977). – Abstract.

119.011   The orbit of the Wolf-Rayet spectroscopic binary HD 190918.   D. A. Fraquelli.
J. R. Astron. Soc. Canada, Vol. 71, 407 - 408 (1977). Abstract.

119.012   On the nature of population I Wolf-Rayet stars.   A. F. J. Moffat.
J. R. Astron. Soc. Canada, Vol. 71, 408 (1977). – Abstract.

119.013   Spectroscopic binary orbits from photoelectric radial velocities. Paper 17: HR 7083.
G. A. Radford, R. F. Griffin.
Observatory, Vol. 97, 235 - 237 (1977).

119.014   V711 Tauri.
IAU Circ., No. 3089 (1977).

119.015   σ 75 Gem: a bright variable similar to HK Lac.   D. S. Hall, G. W. Henry, H. J. Landis.
Inf. Bull. Variable Stars, No. 1328, 3 pp. (1977).

119.016   HR 4665: a newer, brighter RS CVn binary.   B. W. Bopp, D. S. Hall, G. W. Henry, F. Fekel, Jr., H. J. Landis.
Inf. Bull. Variable Stars, No. 1352, 3 pp. (1977).

119.017   V 711 Tauri.
Yamamoto Circ., No. 1857 (1977). In Japanese.

The red spot-covered star HD 224085.
See Abstr. 113.008.

1974—75 UBV photometry of the radio binary UX Arietis.   See Abstr. 113.012.

Four-colour $uvby$ and H_β photometry of field stars: double-lined spectroscopic binaries, G type dwarfs and early type FK4 stars.   See Abstr. 113.014.

Bright southern stars of astrophysical interest.
See Abstr. 114.015.

Observations with the VLA of the radio binary star AR Lacertae.   See Abstr. 116.006.

HD 165590.   See Abstr. 118.023, 118.029.

Binary incidence among the BY Draconis variables.
See Abstr. 122.003.

Gaseous envelopes of BY Dra type stars.
See Abstr. 122.019.

The spectroscopic orbit and the masses of the components of the binary X-ray source 3U0900—40/HD 77581.
See Abstr. 142.068.

120   Variable Stars: Ephemerides, Miscellanea

120.001   Charts for southern variables, Series No. 9, 4 pp. + charts 351 - 400 (1977).
F. M. Bateson, M. Morel, R. Winnett.

Series 9 charts show the positions for 160 variables. Every care has been taken to accurately plot the positions of other variables that lie in the field of the charts if they are brighter than approximately 13.5 visual at maximum. Photoelectric sequences have already been published for some of the variables in this series. Other photoelectric sequences are given on two pages.

120.002   Data on the variable star studies in the USSR during 1972 - 1975.   B. V. Kukarkin.
Inf. Bull. Variable Stars, No. 1319, 5 pp. (1977).

120.003   Variable stars, their discoverers and first compilers from 1006 to 1975.   W. Strohmeier.
Veröff. Remeis-Sternw. Bamberg, Astron. Inst. Univ. Erlangen-Nürnberg, Band 12, Nr. 129, 27 pp. (1977).

120.004   Ellipsoidal variables.   T. Z. Dworak.
Urania Kraków, Vol. 48, 294 - 297 (1977). In Polish.

121 Eclipsing Binaries

121.001   Observations of β Persei at 4.8 μm.
C. Sánchez Magro, J. D. Needham, J. P. Phillips, M. J. Selby.
Mon. Not. R. Astron. Soc., Vol. 180, 461 - 464 (1977).
New measurements of β Persei at 4.8 μm are presented. The relatively small scatter of the results outside eclipse indicates the absence of a phase-dependent emission excess with amplitude $\gtrsim 0.1$ mag.

121.002   The close binary system T Leo Minoris.
A. Okazaki.
Publ. Astron. Soc. Japan, Vol. 29, 289 - 317 (1977).
Physical properties of T LMi, which is known to be a member of close binaries of the "R CMa type", are investigated on the basis of the *UBV* photoelectric light curves and intermediate dispersion spectra. The principal photometric elements are deduced. On the assumption that the secondary component fills its Roche lobe, the absolute dimensions are derived. The primary component is a normal main-sequence star which satisfies the empirical mass-luminosity relation. Thus T LMi is an ordinary semi-detached system of the Algol type.

121.003   AN Ursae Majoris — another AM Herculis?
G. S. Mumford.
Sky Telesc., Vol. 54, 194 - 196 (1977).

121.004   Spectrophotometry of RS Canum Venaticorum, AR Lacertae, and UX Arietis.
C. G. Rhombs, J. D. Fix.
Astrophys. J., Vol. 216, 503 - 507 (1977).
The wavelength dependences of the ultraviolet excesses of the RS CVn-type stars RS CVn, AR Lac, and UX Ari have been measured spectrophotometrically with a wavelength resolution of 60 Å. The authors examined a number of mechanisms to account for the excess light from these systems and conclude that free-free emission from hot circumstellar gas provides the most satisfactory explanation.

121.005   A spectroscopic study of AR Lacertae.
S. A. Naftilan, S. A. Drake.
Astrophys. J., Vol. 216, 508 - 516 (1977).
Several spectra of the double-subgiant binary system AR Lac obtained both during and outside of primary eclipse are examined. It is found that a disk (or shell) exists around the secondary star. The secondary star shows evidence of a high level of chromospheric activity, which produces strong H and K emissions. The secondary star is moderately underabundant in most metals, while the primary has solar-type abundances. These results are discussed in the context of the peculiarities seen in AR Lac itself and in the RS Canum Venaticorum binaries.

121.006   Phase-locked polarization in u Herculis: evidence for the reflection mechanism.
R. J. Rudy, J. C. Kemp.
Astrophys. J., Vol. 216, 767 - 775 (1977).
Phase-locked, linear-polarization variations of 0.06% in the blue have been found in the close binary u Herculis, as a result of intensive observations which spanned 14 months. The principal Fourier components of the variations are second harmonics, relative to the 2 day orbital period. Possible causes of the polarization are discussed. The authors favor an explanation in terms of the reflection of light from the primary by the photosphere of the secondary star. A contribution from light scattering by a gas stream is also possible.

121.007   The spotted dM4e eclipsing binary flare star CM
Draconis.   C. H. Lacy.
Bull. American Astron. Soc., Vol. 9, 433 (1977). — Abstract.

121.008   Phase dependence of emission lines in the disk of DQ Herculis.
B. Margon, P. Kieniewicz, R. P. S. Stone.
Publ. Astron. Soc. Pacific, Vol. 89, 300 - 303 (1977).
Photoelectric spectrophotometry of the He II λ4686 and C III/N III λ4640 emission in DQ Her has been obtained through several cycles of the 0^d194 binary orbit. The λ4686 emission exhibits an eclipse with phase, duration, and depth similar to the continuum light. The behavior of the λ4640 emission, however, is significantly different, with only weak evidence for a shallow, broad eclipse. The possible applicability of such observations to an understanding of X-ray binary stars is pointed out.

121.009   Radial velocities of the long-period eclipsing binaries UU Cancri and BM Cassiopeiae.
D. M. Popper.
Publ. Astron. Soc. Pacific, Vol. 89, 315 - 319 (1977).
Radial velocities based primarily on low-dispersion spectrograms obtained many years ago are presented for the long-period eclipsing binaries UU Cnc (K4 III; $P = 96^d7$) and BM Cas (A5 Ia; $P = 197^d3$). They are fitted with circular orbits. No clear evidence is found in the spectra for the second components, which are the cooler stars in both systems. The value of M_V for BM Cas is near -6. Both systems have unstable light curves. Much more work is required to obtain clear understanding of these binaries.

121.010   The variable light-curve of the eclipsing binary XY UMa.   L. Lorenzi, F. Scaltriti.
Acta Astron., Vol. 27, 273 - 279 (1977).
Blue and yellow photoelectric light-curves of the eclipsing binary XY UMa are presented. Some epochs of primary minimum are given. The continuous variability of the light-curve in the time is confirmed. The consistence of a cyclic variation in the $O-C$ plot at minima and of the spot activity on the hotter component are briefly discussed.

121.011   Analysis of yellow and blue observations of XY Boo.   L. Winkler.
Astron. J., Vol. 82, 648 - 652 (1977).
During 1976 the very short-period system XY Boo was observed photometrically simultaneously in two colors. These observations and others by Binnendijk for 1970 were analyzed separately. Since the eclipses are very shallow, the computer program of Wilson and Devinney was used for analysis. The degree of overcontact is considerable, and the mass ratio has a low value of 0.182.

121.012   On general relativistic periastron advances. II.
R. H. Koch.
Astron. J., Vol. 82, 653 - 655 (1977).
Numerous minima for five relatively long-period close binaries are listed. The periods for the primary and secondary minima are evaluated and the differences and similarities between these periods are discussed for each system. Agreement in algebraic sense between the theoretical and observed period changes is shown for two binaries but there is no agreement at the 1-σ level between the magnitudes of the theoretical and observed period changes.

121.013   Photoelectric observations of the eclipsing binary RT Ursae Minoris.
R. de Santis, P. Tempesti.
Astron. Astrophys., Suppl. Ser., Vol. 29, 333 - 337 (1977).

1415 photoelectric observations of the eclipsing binary RT UMi in V light performed from 1971 to 1973 allow to draw a light-curve showing a previously undetected secondary minimum $0\overset{m}{.}07$ deep. The orbital elements derived by a nomographic solution are given.

121.014 Period study and UBV light curves of TV Cassiopeiae. A. D. Grauer, J. McCall, L. C. Reaves, T. L. Tribble, J. S. Shaw.
Astron. J., Vol. 82, 740 - 744 (1977) = North Georgia College Obs. Contrib. No. 2.

The eclipsing variable TV Cassiopeiae was observed photoelectrically during the fall of 1976. A number (1379) of observations were made and converted to the standard UBV system. Forty-three times of minimum light were used to obtain an average period of 1.8126066 days. The residuals indicate a light time effect with a semiamplitude of 0.008 day and a period of 10500 days or two discrete period changes. Sufficient data does not yet exist to be able to differentiate between these possibilities.

121.015 Astrometric study of the eclipsing binary VV Cephei from plates taken with the Sproul 61-cm refractor. P. van de Kamp.
Astron. J., Vol. 82, 750 - 752 (1977).

Astrometric data over the interval 1938–1976 yield a relative parallax of $+0\overset{''}{.}0022 \pm 0\overset{''}{.}0011$ (p.e.), a photocentric orbit with semimajor axis $0\overset{''}{.}0120 \pm 0\overset{''}{.}0013$, inclination 74°, and node 147°. Equating the astrometric angular and the spectroscopic linear values of α, a parallax of $0\overset{''}{.}0014 \pm 0\overset{''}{.}0001_5$ is found, which yields absolute visual magnitudes of -4.0 and -2.3 for the M I and Be components.

121.016 Hydrogen and helium in the visible component of the binary system β Lyrae. V. V. Leushin, M. Yu. Nevskij, L. I. Snezhko, V. V. Sokolov.
Astrofiz. Issled., Izv. Spets. Astrofiz. Obs., Vol. 9, 3 - 15 (1977). In Russian.

The equivalent widths of the absorption lines of He I and H in the atmosphere of the brighter component of β Lyr show real variations during the period, possibly with a double wave. Quantitative determination of helium abundance with the aid of model atmospheres has yielded a value of $N(\text{He})/N(\text{H}) = 1.55$ for the brighter component, a value considerably smaller than in previous determinations.

121.017 BD$-3°5357$: an eclipsing binary with a hot subdwarf component. M. M. Dworetsky, H. H. Lanning, P. B. Etzel, D. J. Patenaude.
Mon. Not. R. Astron. Soc., Vol. 181, 13P - 18P (1977).

The authors report the discovery of an eclipsing binary (BD$-3°5357$) which contains a hot subdwarf ($T \cong 40\,100$ K). The ultraviolet data, photometric observations, and spectroscopic results are combined to produce a set of tentative models of the system. These show that the cool subgiant (G8 III:) is considerably smaller than its Roche radius. A representative model gives $M_V = +4.7$, $\log g = 6.1$ for the subdwarf. The possible evolutionary history of this object is briefly considered.

121.018 DL Virginis — an eclipsing binary in a triple system. E. Schöffel.
Astron. Astrophys., Vol. 61, 107 - 116 (1977).

Photoelectric observations in UBV have been combined with available spectroscopic and photographic data to derive absolute dimensions and masses in the triple system containing the eclipsing binary DL Vir. The results are used for a discussion of the evolution of the whole system.

121.019 V 701 Scorpii and its place among early contact binaries. R. E. Wilson, K.-C. Leung.
Astron. Astrophys., Vol. 61, 137 - 140 (1977).

A solution of light curves of the early-type eclipsing binary V 701 Scorpii shows a large degree of overcontact, with the surface of the common envelope lying about halfway between the inner and outer contact surfaces. V 701 Sco is on or close to the zero age main sequence and therefore may have been formed by fission essentially in its present state.

121.020 A note on NW Aurigae. M. Hoffmann.
Astron. Astrophys., Vol. 61, 145 (1977).

The suspected W Ursae Majoris variable NW Aurigae has been observed photoelectrically. No change in brightness could be detected.

121.021 Das bedeckungsveränderliche System WY Cancri. K. Porzel, R. Lukas.
BAV Rundbr., 26. Jahrg., 29 - 30 (1977).

121.022 TV Cassiopeiae. M. Fernandes.
BAV Rundbr., 26. Jahrg., 31 - 34 (1977).

121.023 Interpretation of light curves of eclipsing binary systems with disk-like envelopes. The system HZ Her. A. M. Cherepashchuk, A. V. Goncharskij, A. G. Yagola.
Astron. Zh. Akad. Nauk SSSR, Vol. 54, 1027 - 1035 (1977). In Russian. English translation in Soviet Astron., Vol. 21, No. 5.

A method of restoration of the structure of the accretion disk in a binary system using the light curve of the eclipse is proposed. The method is applied to the X-ray binary system HZ Her. The distribution of color and brightness temperature in the optical accretion disk associated with the neutron star is obtained. Luminosity, mean temperature and dimensions of the accretion disk correlate with the phase of the 35-day cycle of X radiation.

121.024 An observational study of the eclipsing binary RZ Ophiuchi. B. W. Baldwin.
J. R. Astron. Soc. Canada, Vol. 71, 408 - 409 (1977). Abstract.

121.025 The ultraviolet spectrum of Beta Lyrae. M. Hack, J. B. Hutchings, Y. Kondo, G. E. McCluskey.
Highlights of Astronomy, Vol. 4, Part II, (see 012.025), p. 361 - 362 (1977). — Abstract.

121.026 Star spots on AR Lac type stars. D. M. Popper.
Highlights of Astronomy, Vol. 4, Part II, (see 012. 026), p. 397 - 403 (1977).

121.027 A three-colour photoelectric investigation of the eclipsing binary DO Cas. B. Cester, G. Giuricin, F. Mardirossian, M. Pucillo.
Astron. Astrophys., Suppl. Ser., Vol. 30, 223 - 226 (1977).

CO Cas has been observed in three colours between 1975 and 1976. From new times of minimum the period is improved. The solutions of the light curves suggest that CO Cas is a contact binary.

121.028 Photoelectric photometry of β Lyr in 1971. P. Broglia, P. Conconi, G. Guerrero.
Astron. Astrophys., Suppl. Ser., Vol. 30, 231 - 234 (1977).

As part of the IAU Commission 42 second coordinated programme for the observation of β Lyr, narrow and intermediate band measurements have been obtained in the second half of 1971. The observations obtained are presented and briefly commented.

121.029 Absolute dimensions and masses of the remarkable spotted dM4e eclipsing binary flare star CM Draconis. C. H. Lacy.
Astrophys. J., Vol. 218, 444 - 460 (1977).

The physical properties of this interesting nearby system — currently the smallest, faintest, least massive main-sequence eclipsing binary known — are investigated using high-speed multicolor photometry, differential and nondifferential infrared photometry, and high-dispersion spectroscopy. With two exceptions, the system is simple, with deep, nearly equal eclipses with a short (~ 2 min) duration of totality. The exceptions: the system is known to be a flare star, but the frequency of flaring is at least a factor of 40 less than classical Population I flare stars of similar luminosity; and low-amplitude (~ 0.01 mag) sinusoidal variations attributable to starspots or nonuniform surface brightness are present.

121.030   **Interpretation of Beta Lyrae. III. A study of the disk around the secondary component.**
D. A. Brown, S.-S. Huang.
Astrophys. J., Vol. 218, 461 - 467 (1977).
    The authors have analyzed light curves of β Lyrae available in the far-ultraviolet, visual, and infrared regions of the spectrum at numerous phases in eclipse in order to investigate the physical and radiative nature of the disk surrounding the secondary component. The results of the analysis together with those of other investigators lead the authors to propose that the outer regions of the disk are dominated by free electrons. This electron-scattering envelope is most likely the source of infrared radiation, as well as the cause of the observed polarization. However, the radiation in the region from the optical to the far-ultraviolet comes mainly from submerged layers where local thermodynamic equilibrium prevails. These layers represent the photosphere of either the disk or the secondary component itself.

121.031   **Period variation and mass loss in the system AR Lacertae.**   M. B. Babaev.
Izv. AN AzSSR. Ser. fiz.-tekh. i mat. n., 1976, No. 4, p. 3 - 7. In Russian. — Abstr. in Ref. zh., 51. Astron., 10.51.857 (1977).

121.032   **The polarization of Sigma Orionis E, a curious eclipsing binary.**   J. C. Kemp, L. C. Herman.
Astrophys. J., Vol. 218, 770 - 775 (1977).
    The linear polarization of the enigmatic object σ Ori E was measured on 33 nights in 1977 January—March. Variation synchronous with the 1ᵈ19 light period was found in the B filter, with peak-to-peak amplitude of about 0.10%. Lower-accuracy data in the U filter show a different polarization structure. The polarization curves suggest a strongly inclined system with apparently $i \approx 76°$ and are compatible, for example, with a binary containing a highly nonspherical component such as a disk.

121.033   **Photoelectric photometry and computer photometric solution of the eclipsing binary MR Cygni.**
Research Group on Close Binaries, Stellar Division, Peking Astronomical Observatory, Academia Sinica.
Acta Astron. Sinica, Vol. 18, 68 - 85 (1977).
    1135 observations of MR Cyg in yellow light were obtained on 19 nights from September to November, 1975, using the 60-cm reflector at Xinglung station. The determination of the time of primary minimum by the method of Kwee and van Woerden gives the result JD☉ 2442729.09916. The primary minimum should be a partial eclipse of transit type.

121.034   **AM Herculis.**
IAU Circ., No. 3095 (1977).

121.035   **An observed eclipse of θ¹ Orionis A.**
D. B. Caton, F. W. Fallon, R. E. Wilson.
Publ. Astron. Soc. Pacific, Vol. 89, 530 - 532 (1977).
    Photoelectric observations with UBV filters were made of one descent into primary eclipse, and three runs at constant brightness, for the recently discovered eclipsing variable θ¹ Ori A. They show a smaller amplitude of variation than visual estimates reported previously for the same eclipse, but are

quite similar to Lohsen's photoelectric observations.

121.036   **UBVRI light curves of RW Tauri.**
B. B. Bookmyer.
Publ. Astron. Soc. Pacific, Vol. 89, 533 - 540 (1977).
    UBVRI observations of the Algol-type eclipsing binary system RW Tau were obtained on 19 nights. Approximately 250 observations in each wavelength region define not only the primary eclipse curve but also, it is believed for the first time, the secondary eclipse curve. The eclipse curves are symmetric and there is no marked inequality in the durations of the eclipses. Secondary minimum is slightly displaced from phase 0.5; this displacement is not in agreement with the large orbital eccentricity of 0.29 derived from spectrographic observations. The outside-eclipse intervals of the light curve show evidence of both the reflection effect and tidal interaction.

121.037   **Further light curve variations of SZ Piscium.**
J. A. Eaton.
Inf. Bull. Variable Stars, No. 1297, 3 pp. (1977).

121.038   **HD 56429: a double-lined Am eclipsing binary.**
D. M. Popper, J. Andersen.
Inf. Bull. Variable Stars, No. 1298 (1977).

121.039   **Photometry and a preliminary analysis of the Beta-Lyrae like system HD 173 198.**
J. A. de Freitas Pacheco, C. Ritté, A. D. Neto.
Inf. Bull. Variable Stars, No. 1306, 3 pp. (1977).

121.040   **A distortion wave in the light curve of MM Herculis.**
D. S. Hall, G. W. Henry, E. W. Burke, Jr., J. L. Mullins.
Inf. Bull. Variable Stars, No. 1311, 3 pp. (1977).

121.041   **Evidence of mass ejection from β Persei (Algol).**
Y. Kondo, J. L. Modisette, T. H. Morgan.
Inf. Bull. Variable Stars, No. 1312, 3 pp. (1977).

121.042   **Preliminary elements for BV 1621 CMa = HD 56429.**   U. Hopp, M. Kiehl.
Inf. Bull. Variable Stars, No. 1315 (1977).

121.043   **Photoelectric minima of W UMa.**
J. Bønes, H. K. Myrabø.
Inf. Bull. Variable Stars, No. 1316 (1977).

121.044   **HD 20301: an eclipsing, double-lined early G giant.**
E. H. Olsen.
Inf. Bull. Variable Stars, No. 1317, 4 pp. (1977).

121.045   **A reconsideration of the orbital period of AZ Cas.**
A. P. Cowley, J. B. Hutchings.
Inf. Bull. Variable Stars, No. 1318 (1977).

121.046   **On the variable star UU Sagittae.**
V. P. Tsessevich (Tsesevich).
Inf. Bull. Variable Stars, No. 1320 (1977).

121.047   **XX Cephei: new times of minimum and a study of the period.**
P. Battistini, A. Bonifazi, A. Guarnieri.
Inf. Bull. Variable Stars, No. 1325, 6 pp. (1977).

121.048   **The linear polarization of u Herculis.**
R. H. Koch, R. J. Pfeiffer.
Inf. Bull. Variable Stars, No. 1326, 5 pp. (1977).

121.049   **Minima of eclipsing variables.**   L. P. Surkova.
Inf. Bull. Variable Stars, No. 1335 (1977).

121.050   **HD 5303: a new southern RS Canum Venaticorum**

binary.   J. B. Hearnshaw, J. P. Oliver.
Inf. Bull. Variable Stars, No. 1342 (1977).

121.051   **A variable light-curve in AU Monocerotis.**
M. Cerruti-Sola, L. Lorenzi.
Inf. Bull. Variable Stars, No. 1348, with an erratum No. 1363 (1977).

121.052   **Minima of eclipsing binary stars.   C. P. Stephan.**
Inf. Bull. Variable Stars, No. 1350 (1977).

121.053   **Photoelectric observations of 44i Bootis.**
U. Hopp, S. Witzigmann, M. Kiehl.
Inf. Bull. Variable Stars, No. 1353, 3 pp. (1977) = Veröff.
Wilhelm-Foerster-Sternw., Berlin, Nr. 48.

121.054   **Photoelectric minima of eclipsing binaries.**
E. Pohl, A. Kizilirmak.
Inf. Bull. Variable Stars, No. 1358, 3 pp. (1977).

121.055   **Photoelectric minima of U Oph, AB And and**
X Tri.   Z. Tüfekçioğlu.
Inf. Bull. Variable Stars, No. 1368 (1977).

121.056   **On eclipsing binary CSV 5384 = HV 10663.**
V. Tsessevich (*Tsesevich*).
Inf. Bull. Variable Stars, No. 1369 (1977).

121.057   **The light curve of i Bootis in February 1975.**
F. Gieseking.
Inf. Bull. Variable Stars, No. 1374, 6 pp. (1977).

121.058   **BD −3°5357: a newly discovered eclipsing binary**
with a hot subluminous component.
P. B. Etzel, H. H. Lanning, D. J. Patenaude, M. M. Dworetsky.
Publ. Astron. Soc. Pacific, Vol. 89, 616 (1977). − Abstract.

121.059   **Preliminary results of photometry of VV Cephei at**
the beginning of the 1976–77 eclipse.
G. Spear, P. Avellar, M. Carolin, J. Mills, M. Ross, S. Snedden.
Publ. Astron. Soc. Pacific, Vol. 89, 621 - 622 (1977).
Abstract.

121.060   **Spectroscopic orbit of CC Comae.**
S. M. Rucinski, J. A. J. Whelan, S. P. Worden.
Publ. Astron. Soc. Pacific, Vol. 89, 684 - 687 (1977) =
Dominion Astrophys. Obs., Victoria, B.C., Canada, Contrib.
No. 333 = NRC No. 15902.
Radial velocity measurements of CC Com, the shortest
period W UMa-type system, are presented. The results are
combined with photometric results to provide an estimate of
the absolute elements. The system has total luminosity
$0.29 \pm 0.03\ L_\odot$ and masses of $0.69 \pm 0.06\ M_\odot$, and 0.36 ± 0.03
$M_\odot$. Formal errors are quoted here, and the possibility of
much larger systematic errors is discussed in the text. The
value of the mass ratio derived spectroscopically (0.52 ± 0.03)
confirms the value derived from photometry. The distance to
CC Com (83 ± 4 pc) is discussed with reference to membership of the Coma cluster (Mel 111). The importance of CC
Com for theoretical models of contact binary systems is
stressed.

121.061   **V788 Cygni − a period correction.   M. E. Baldwin.**
J. American Assoc. Variable Star Obs., Vol. 6, 1 - 3
(1977).
Recent visual observations of V788 Cygni reveal that its
photometric period must be doubled.

121.062   **A revision of earlier findings on UU Canis Majoris.**
M. E. Baldwin.
J. American Assoc. Variable Star Obs., Vol. 6, 7 - 8 (1977).

121.063   **Minima of eclipsing binary stars. V.   M. E. Baldwin.**
J. American Assoc. Variable Star Obs., Vol. 6, 24 -
32 (1977).

121.064   **Chemical composition of upper main sequence stars.**
II. Eclipsing binaries.
Y. Shibata, K. Mimura.
Sci. Rep. Tôhoku Univ., Ser. I, Vol. 60, 31 - 39 (1977).
Theoretical mass-luminosity and mass-radius relations are
derived for upper main sequence stars, taking into account
effects of stellar evolution. Basing on this expressions, chemical
compositions and ages of four eclipsing binary systems, ζ Phe,
V451 Oph, AG Per, U Oph, are obtained and discussed.

121.065   **67th, 68th list of minima of eclipsing binaries.**
Compiled by P. Albert, R. Boninsegna, J. Bourgeois,
J.-P. Clovin, P. Danthine, R. Diethelm, M. Frangeul,
R. Germann, V. Lardinois, J.-F. Le Borgne, P. Le Strat,
R. Leydon, K. Locher, A. Marot, P. Mons, M. Penna,
A. Royer, J. Remis, G. Troispoux, S. Wabniz, M. Benucci,
B. Bouzin, A. Buzzoni, J.-L. Duquesne, F. Ferrara, M. de
Francesco, Z. Hevesi, A. Livi, S. Le Jehan, R. Leyman,
T. Maniet, E. Nezry, A. del Parigi, C. Pampaloni, C. Plasmati,
H. Peter, E. Poretti, P. Ralincourt, F. Travaglino, V. Tuboly,
F. Vespe, N. Zaccaria.
BBSAG Bull., No. 34, p. 1 - 4, No. 35, p. 1 - 6 (1977).

121.066   **Provisional elements for V829 Aquilae.**
R. Diethelm.
BBSAG Bull., No. 34, p. 5 (1977).

121.067   **A reinterpretation of V868 Ophiuchi.   K. Locher.**
BBSAG Bull., No. 34, p. 5 (1977).

121.068   **A visual minimum of V1068 Cygni.**
R. Diethelm.
BBSAG Bull., No. 35, p. 6 - 7 (1977).

121.069   **Possible apsidal motion in AP Tauri.   K. Locher.**
BBSAG Bull., No. 35, p. 7 (1977).

121.070   **Fourier analysis of the light curves of eclipsing**
variables, X.   Z. Kopal.
Astrophys. Space Sci., Vol. 46, 87 - 108 (1977).
The aim of the present paper will be to develop methods
for computation of the Fourier transforms of the light curves
of eclipsing variables − due to any type of eclipses − as a function of a continuous frequency variable ν. For light curves
which are symmetrical with respect to the conjunctions (but
only then) these transforms prove to be real functions of ν,
and expressible as rapidly convergent expansions in terms of
the moments A_{2m+1} of the light curves of odd orders. The odd
moments A_{2m+1} are shown to be expressible as infinite series
in terms of the even moments A_{2m} and polynomial expressions
are developed for approximating them to any desired degree of
accuracy. The numerical efficiency of such expressions will be
tested by application to a practical case, with satisfactory
results. An appeal to the Wiener-Khinchin theorem (relating
the power spectra with autocorrelation function of the light
curves) and Parseval's theorem on Fourier series will enable
the author to extend the previous methods for a specification
of quadratic moments of the light curves in terms of the linear
ones.

121.071   **_UBV_ light variation and orbital elements of MW**
Pavonis.   E. Lapasset.
Astrophys. Space Sci., Vol. 46, 155 - 164 (1977).
Photoelectric *UBV* observations of the W UMa-system
MW Pavonis are presented; they were made at Bosque Alegre
Station of Córdoba Observatory in 1972 and 1974. The
period and linear ephemeris were obtained from nine times of
minimum observed in each color. Satisfactory orbital elements

were determined for the three light curves on the basis of the Russell model.

121.072 Determination of the elements of eclipsing variables, RW Gem and AY Cam, by Fourier analysis of their light changes.   H. M. K. Al-Naimiy.
Astrophys. Space Sci., Vol. 46, 261 - 284 (1977).

The aim of the present paper has been to present an analysis of the light curve of two eclipsing systems RW Gem and AY Cam by Fourier analysis of the light changes in the frequency domain which was developed by Kopal (1975, 1976). The subject is introduced in a general way, with the intention of laying the foundation of the light curve analysis. The evaluation of the empirical values of the theoretical moment A_{2m} is demonstrated, with the equation of the condition given. Then the equations for A_{2m} in terms of the elements of the total and the annular eclipses, including partial and annular phase of transit eclipse, follow.

121.073 Analysis of the light curves of SZ Cam using automatized Fourier techniques.   E. Budding.
Astrophys. Space Sci., Vol. 46, 407 - 428 (1977).

UBV light curves of the early type close eclipsing binary system SZ Cam have been investigated using recently developed frequency-domain techniques. The combination of both minima in the analysis results in a distinct methodological improvement over the single minimum method discussed hitherto. This improvement has two aspects: (i) increased accuracy of the determined elements, (ii) agreement of the results of the two-minimum method with those of the single-minimum method provides a criterion whereby the self-consistency of the underlying model with its representation of the light curve in the regions between minima by a cosine series and the empirically determined coefficients of such a series may be assessed.

121.074 Analysis of the V light curve of the system CW Cas.   R. Burchi, L. Milano, G. Russo.
Astrophys. Space Sci., Vol. 47, 35 - 47 (1977).

The V light curve of CW Cas, observed during 1972, has been analysed by different methods of solution. The results show great differences between rectifiable and direct methods.

121.075 An analysis of the light changes of the eclipsing system AK Herculis in the frequency domain. P. Niarchos.
Astrophys. Space Sci., Vol. 47, 79 - 97 (1977).

The aim of the present paper has been to analyse the light changes of the close eclipsing system AK Herculis in the frequency domain. This analysis is based on Kopal's new theory recently developed for the study of light variations of close eclipsing binaries whose components are distorted by axial rotation and mutual tidal action. A new method for the separation of the photometric proximity and eclipse effects directly from the observed data is also presented. New elements taking into account the photometric perturbations are also given.

121.076 The semi-detached system RT Persei and its light curve.
S. Mancuso, L. Milano, G. Russo.
Astrophys. Space Sci., Vol. 47, 277 - 298 (1977).

The observed V light curve of the eclipsing binary RT Persei has been analysed by four different methods of solution to get the geometrical and photometric elements of the system. The results show a good agreement within about 6%. The system shows a variable light curve and a tentative hypothesis is made to explain this behaviour by considering a dynamical instability that may be the cause of circulating matter through the system and of the variability of the period.

121.077 Determination of the elements of eclipsing variables

RW Tauri and U Sagittae by an analysis of the light changes in the frequency domain.   A. G. Tsouroplis.
Astrophys. Space Sci., Vol. 47, 361 - 373 (1977).

The aim of the present paper has been to present an analysis of the light changes of two eclipsing systems RW Tau and U Sge in the frequency domain, which was developed by Kopal (1975, 1976). The theoretical moments A_{2m} are evaluated. The determination of the preliminary elements and their improvement, taking into account the photometric perturbations, are given. A general discussion devoted to the whole analysis of the system is presented.

121.078 Frequency domain analysis of the light changes of the close eclipsing systems TX Ursae Majoris and S Cancri.   V. A. Caracatsanis.
Astrophys. Space Sci., Vol. 47, 375 - 384 (1977).

The aim of this paper has been to carry out an analysis of the light curves of two eclipsing systems TX UMa and S Cnc by Kopal's new method of the analysis of the light curves of eclipsing systems in the frequency domain. The elements of these systems are evaluated and the results are discussed.

121.079 Frequency domain light curve analysis of partially eclipsing binary systems: with an application to β Persei (Algol).   O. Demircan.
Astrophys. Space Sci., Vol. 47, 459 - 488 (1977).

Kopal's new iterative method for analysing partially eclipsing binary light curves in the frequency domain has been put into a form suitable for applications, and explicit forms for basic expressions developed. The automated method has been tested successfully on the light curves of β Persei (Algol). Short information on the system and the revised sets of elements including a new determination of the limb-darkening coefficient in the ultraviolet have been presented.

121.080 Improvement of the orbital elements of β Lyr.   M. Yu. Skul'skij, V. M. Smolyarchuk.
Tsirk. Astron. Obs., L'vov, No. 51, p. 20 - 23 (1976). In Russian.

121.081 Fourier analysis of the light curves of eclipsing variables, XI.   Z. Kopal.
Astrophys. Space Sci., Vol. 50, 225 - 246 (1977).

The aim of the present paper will be to introduce a new definition of the loss of light suffered by mutual eclipses of the components of close binary systems: namely, as a cross-correlation of two apertures representing the eclipsing and eclipsed discs.

121.082 Fourier analysis of the light curves of eclipsing variable stars. I. Photometric perturbations for total and transit eclipses.   H. J. Livaniou.
Astrophys. Space Sci., Vol. 51, 77 - 109 (1977).

The aim of the present paper is to establish the explicit forms of the photometric perturbations, in the frequency-domain, of close binaries, whose components are distorted by axial rotation and mutual tidal action. The light changes and the photometric perturbations within eclipses in the frequency-domain are described. The explicit forms of the perturbations for occultation eclipses terminating in totality are given; analogous results are established for transit eclipses terminating in annular phases. In this latter case the results can be expressed in terms of the photometric perturbations for total eclipses and in terms of some series. To facilitate applications to actual stars these series have been computed.

121.083 Analysis of red and infrared wide-band photometry of Algol.   H. M. Al-Naimiy, E. Budding.
Astrophys. Space Sci., Vol. 51, 265 - 282 (1977).

The basic physical picture of the Algol system is reviewed, and, using collected red and infrared observations, photometric curve fits are investigated by applying numerical quadratures to determine theoretical light curves appropriate to Roche model stars. The 'contact' nature of Algol B appears to be confirmed, and effective temperatures of the three components are given. In terms of a Lambert's law approach to the 'reflection effect', the effective heat-albedo is required to be reduced from unity to one half; and it is also found that the averaged 'gravity-darkening coefficient' is close to a value appropriate for a diffusion type of heat-transfer mechanism operating in sub-photospheric layers.

121.084 Fourier analysis of the light curves of eclipsing variables, XII. Z. Kopal.
Astrophys. Space Sci., Vol. 51, 439 - 460 (1977).

The aim of the present paper will be to make use of the expressions, established in Paper XI, for the fractional loss of light α_i^0 of arbitrarily limb-darkened stars in the form of Hankel transforms of zero order, in order to evaluate the explicit forms of the α_i^0's for different types of eclipses, as well as of the moments A_{2m} of the respective light curves — in a closed form; or in terms of expansions that converge under all circumstances envisaged. Particular attention will be directed to a connection between these expansions and other functions already available in tabular form; or to alternative forms amenable to automatic computation.

121.085 Gas stream in Algol. H. Cugier, K.-Y. Chen.
Astrophys. Space Sci., Vol. 52, 169 - 176 (1977).

Additional absorption features in the red wings of the resonance Mg II lines near 2800 Å are found in the observations of Algol made by Chen and Wood (1976) from the Copernicus satellite. The absorption features were clearly seen only during a part of the primary eclipse, in the phase interval 0.90–0.03. The observations are interpreted as produced by a stream of matter flowing from Algol B in the direction of Algol A.

121.086 Fourier analysis of the light curves of eclipsing variables. XIII. O. Demircan.
Astrophys. Space Sci., Vol. 52, 189 - 199 (1977).

New expressions for the fractional loss of light α_i^0 have been derived in the simple forms of rapidly converging expansions to the series of Chebyshev polynomials, Jacobi polynomials, and Kopal's J-integrals. In these expansions, which are a supplement to those given by Kopal (1977), variables k and h occur in different products that simplify the numerical computation. The treatment follows the new definition of α_i^0 which has been recently developed by Kopal (1977).

121.087 An analysis of the light changes of AR Lacertae. A. C. Theokas.
Astrophys. Space Sci., Vol. 52, 213 - 235 (1977).

An attempt has been made to bring photoelectric and dynamical properties of the system to a common focus. The photoelectric properties are exhibited by a series of light curves in the blue and infrared produced from a series of observations made by G. E. Kron et al. from 1938 to 1948. An analysis of these reveals a decrease of the period of the order of magnitude 10^{-5}. A Fourier analysis of the light curves is employed to estimate the elements for the period represented by each curve. These results suggest a variation of the fractional radii over an eight-month period. Dynamical effects are ruled out and the result is seen to be the outcome of some photoelectric effect. A phenomenological discussion of the nature of these photoelectric effects is presented, showing that their origin may be in gas streams rather than spot effects.

121.088 Fourier analysis of the light curves of eclipsing variable stars. II. Photometric perturbations for partial eclipses. H. Rovithis-Livaniou.
Astrophys. Space Sci., Vol. 52, 271 - 306 (1977).

The aim of the present paper is to find the eclipse perturbations, in the frequency-domain, of close eclipsing systems exhibiting partial eclipses. The results are discussed and the way in which they can be applied to practical cases.

121.089 A polarimetric investigation of the eclipsing binary XY Ursae Majoris. E. H. Geyer, K. Metz.
Astrophys. Space Sci., Vol. 52, 351 - 356 (1977).

On three nights in February 1976 the authors carried out polarimetric measurements, in V, of the short periodic eclipsing binary XY UMa, covering a complete cycle. The results are reviewed. The authors conclude that no circumstellar envelope can be made responsible for the observed long-term changes of the light curve and system brightness, supporting the earlier spectroscopic finding. The different scatter of the Stokes parameters at different phase intervals and the $P_0/2$ periodicity are in favor of the star spot model for XY UMa proposed by one of the authors (E.G.).

121.090 AK Herculis — an atypical W UMa-type system. E. J. Woodward, R. E. Wilson.
Astrophys. Space Sci., Vol. 52, 387 - 414 (1977).

AK Herculis is a contact binary of spectral class F with a number of obvious peculiarities such as a displacement of secondary eclipse from phase 0.500, unequal heights of the maxima, and a possibly sinusoidal ($\simeq$ 60 yr) period variation. The light curve is variable and shows erratic short-term behavior. The authors present new photoelectric light curves in the B and V passbands and derive several new times of minima. The new observations are compared graphically with seven earlier light curves. The new observations are analyzed.

121.091 One hundred and eight eclipsing variables in the Magellanic Clouds. S. I. Gaposhkin.
Smithsonian Astrophys. Obs., Special Rep. 380, 1 + 29 pp. (1977).

The tabulated light curves of seventy-five eclipsing variables of the Large Magellanic Cloud and thirty-three of the Small Magellanic Cloud were obtained from the photographic study of the Clouds. This set of observations of 108 eclipsing variables is of unique importance, representing the behaviour of these stars over a period of more than seventy-five years. In addition, they were used successfully by the author to determine the distances of the Clouds with a higher accuracy and, above all, completely independently. The deduced moduli are the following: $18^m38 \pm 0^m02$ for the L.M.C. $19^m15 \pm 0^m03$ for the S.M.C.

121.092 Photoelectric observations of V 822 Aql. V. Ya. Alduseva, V. M. Kovalenko.
Astron. Tsirk., No. 956, p. 1 - 3 (1977). In Russian.

121.093 Observations of EE Cephei in 1975. V. Bahýl.
Astron. Tsirk., No. 956, p. 3 - 5 (1977). In Russian.

121.094 Variable star OT Cep. N. B. Perova.
Astron. Tsirk., No. 960, p. 7 (1977). In Russian.

121.095 Preliminary solution for RT Lacertae. E. F. Milone.
Astron. J., Vol. 82, 998 - 1007 (1977) = Rothney Astrophys. Obs. Publ. No. 9.

The rectification and solution of the V light curve of the peculiar eclipsing binary RT Lacertae and previously unpublished data on which they are based are presented. Basic conclusions are that (1) the less massive star is approximately filling its inner Roche lobe; (2) the more massive component has lower luminosity, lower apparent surface brightness, and later spectral type, and is smaller than the other component; and (3) the photometric evidence points to a circumstellar envelope about this component.

121.096 New features in RW Trianguli. L. Winkler.
Astron. J., Vol. 82, 1008 - 1012 (1977).

New features in the light curves of RW Tri have been found. The circumstellar material about the small, hot component varies in effective size and brightness. Appreciable variations occur in a few revolutions and regular variations over hundreds of revolutions. This activity has persisted for more than 33000 revolutions without any appreciable change in the period.

121.097   Fotografische Beobachtungen des Bedeckungs-
          sterns AZ Vulpeculae.   D. Böhme.
Mitt. Veränderl. Sterne (MVS), Band 8, 10 (1977).

121.098   UBV photometry of ε Aurigae.   H. Albo.
          Publ. Tartu Astrofiz. Obs., Vol. 45, 284 - 292
(1977). In Russian.
    Three-colour photoelectric UBV observations of ε Aur in 1969 are presented. A dip and a fluctuation of the light curve have been observed.

121.099   Peculiarities of the Beta Lyrae system. I.
          A. N. Dadaev.
Izv. Glav. Astron. Obs. Pulkovo, No. 195, Astrofiz. Astrometr., p. 98 - 112 (1977). In Russian.
    Some previous results of the author (1974, 1975) are developed. Earlier observations and measurements of the so-called B_5-spectrum are discussed. This discussion confirms the interpretation of the absorption B_5-spectrum as the spectrum of the secondary star rather than that of an expanding nebulous ring suggested by Struve. An attempt is undertaken to bring into agreement the spectroscopic and photometric data on β Lyrae in terms of the new interpretation. The Larsson-Leander light curve (1969) with a flat bottom of the primary minimum is used for calculating the relative radii of both stars. The matter ejection from the primary star is of non-permanent character, and the ejection is probably due to a protuberance mechanism. Hence, the primary star itself is much more stable than hitherto thought. The absolute dimensions of the system using the Berkeley spectrographic data (1959) are calculated.

121.100   Analysis of TX UMa and MR Cyg on the extended
          Russell model.   E. F. Guinan.
Villanova Univ. Obs. Contrib. No. 1, 36 pp. (1975) – Paper presented at I.A.U. Colloquium No. 16 "Analytical procedures for eclipsing binary light curves".

121.101   UBV photometry of VV Cephei during the ingress
          phase of the 1976–1978 eclipse.   M. Nakagiri.
Tokyo Astron. Bull., Second Ser., No. 248, p. 2871 - 2874 (1977).

121.102   Eclipsing variables at distances up to 100 pc from
          the sun.   T. Z. Dworak.
Urania Kraków, Vol. 48, 265 - 270 (1977). In Polish.

    On the reduction of certain integrals occurring in Kopal's Fourier theory of eclipsing binaries.
See Abstr. 021.020.

    Limb darkening coefficients for late-type giant model atmospheres.   See Abstr. 064.046.

    Four-colour *uvby* and H$_\beta$ photometry of field stars: double-lined spectroscopic binaries, G type dwarfs and early type FK4 stars.   See Abstr. 113.014.

    The ultraviolet spectrum of Beta Lyrae. III.
See Abstr. 114.534.

    The spectrum of ε Aurigae outside eclipse.
See Abstr. 114.556.

    Wavelength dependence of the opacity in the atmosphere of the K-type component of 32 Cygni.
See Abstr. 114.560.

    The rotation effect in 71 Draconis.
See Abstr. 116.012.

    Synthetic light curves of W Ursae Majoris binary star systems as applied to V566 Ophiuchi.
See Abstr. 117.011.

    Motion of the W Ursae Majoris-type star CC Comae.
See Abstr. 117.012.

    The possibility for visual detection of astrometric binaries with special astrophysical interest.
See Abstr. 117.022.

    Consequences of mass transfer in close binary systems.   See Abstr. 117.030.

    A possible eclipse in the triple system 20 Leonis.
See Abstr. 117.034.

    Remarks on erratic period fluctuations of detached close binaries and the constancy of the orbital period of XY UMa.   See Abstr. 117.038.

    Synthetic light curves for contact binaries.
See Abstr. 117.042.

    Survey of selected spectroscopic and eclipsing binaries for intrinsic linear polarization.
See Abstr. 119.008.

    Variable stars in the globular cluster NGC 6235.
See Abstr. 122.041.

    Kurzer Abriss der Geschichte der Veränderlichen Sterne.   See Abstr. 122.085.

    Note on V 644 Cen = HD 306989 (B3).
See Abstr. 123.022.

    OSO-8 X-ray observations of AM Herculis.
See Abstr. 142.018.

    The spectroscopic orbit and masses of SK 160/SMC X-1.   See Abstr. 142.026.

    The 35 day cycle of the X-ray binary Hercules X-1.
See Abstr. 142.090.

    X-ray emission behaviour of Her X-1.
See Abstr. 142.132.

Errata

121.901   Errata: "Period changes of UX Ursae Majoris"
          [Mon. Not. R. Astron. Soc., Vol. 180, 5P - 10P
(1977)].   B. V. Kukarkin.
Mon. Not. R. Astron. Soc., Vol. 181, 800 (1977).

121.902   Errata: "Photoelectric minima of AB And and
          X Tri" [Inf. Bull. Variable Stars, No. 1254 (1977)].
Z. Tüfekçioğlu.
Inf. Bull. Variable Stars, No. 1300 (1977).

122 Intrinsic Variables, Spectrum Variables, Flare Stars, Pulsation Theory

122.001 An extension of a geometric method for finding cepheid distances. J. D. Fernie.
Mon. Not. R. Astron. Soc., Vol. 180, 339 - 341 (1977).

The method of Barnes et al. for finding the distance of a cepheid by a surface-brightness technique is extended by obviating the need for radial velocity data. The total linear amplitude of pulsation is shown to be given by the product of the period and light amplitude alone. The linear amplitudes so derived are equated to the angular amplitudes derived by the surface-brightness method in order to derive the cepheids' distances. These are shown to be comparable in accuracy to those found by Barnes et al. through a detailed integration of available velocity curves.

122.002 Lichtabfall von R Coronae Borealis.
Orion, 35. Jahrg., 123 (1977).

122.003 Binary incidence among the BY Draconis variables. B. W. Bopp, F. Fekel, Jr.
Astron. J., Vol. 82, 490 - 494 (1977).

Using new radial velocity data from coudé spectrograms, the authors critically examine the incidence of spectroscopic binaries among the BY Draconis variables. Of 13 such stars with sufficient velocity data, eight are double-line spectroscopic binaries, one is a single-line binary, and four show no apparent velocity variation. It appears that a close companion is a sufficient, but not a necessary, condition for the occurrence of the BY Dra phenomenon. Instead, relatively rapid rotation [v (equatorial) $\gtrsim$ 5 km sec^{-1}] is viewed as the ultimate cause of the BY Dra syndrome.

122.004 Observational characteristics of simple radial and non-radial β Cephei models.
P. A. Stamford, R. D. Watson.
Mon. Not. R. Astron. Soc., Vol. 180, 551 - 565 (1977).

Light, colour and radial velocity variations are calculated for radially oscillating models applicable to β Cephei stars. Tentative non-radial calculations are also presented. A comparison with published ΔV, $\Delta(U-B)$ and $2K$ data for these variables is made. The observed $\Delta V/\Delta(U-B)$ data suggest the presence of both radial and non-radial oscillations in the β Cephei stars. Line profile variations in the star 12 Lac are compared with the predictions of $(l, m) = (2, -2)$ and $(2, 0)$ oscillations and a reasonable agreement is obtained.

122.005 On the masses and radii of double-mode cepheids. R. S. Stobie.
Mon. Not. R. Astron. Soc., Vol. 180, 631 - 638 (1977).

The properties and importance of double-mode cepheids are reviewed. It is shown that the existence of these stars requires a substantial modification to the current understanding of pulsation theory and/or stellar evolution theory. The most plausible explanation is that the calibration of theoretical relationships based on stellar pulsation calculations is not as accurate as previously considered.

122.006 Cyanodiacetylene in W Hydrae.
R. X. McGee, L. M. Newton, J. W. Brooks.
Mon. Not. R. Astron. Soc., Vol. 180, 91P - 95P (1977).

The detection of the $J = 1 \rightarrow 0$ rotational transition of cyanodiacetylene, HC_5N, is reported in the long-period variable star W Hydrae. It is shown that the line emission is probably thermal and that its median radial velocity is the same as the stellar radial velocity.

122.007 Investigation of the light curves of cepheids with periods of 8 - 11 days. O. P. Vasil'yanovskaya.
Byull. Inst. Astrofiz., Dushanbe, No. 66 - 67, p. 32 - 46 (1976).

In Russian.

The forms of the light curves of 137 cepheids belonging to different stellar systems were studied. The light curves of 16 cepheids from the Galaxy were expanded in component oscillations. It is shown that the observed distinctions between the shapes of light curves can be attributed to changes of displacements of the phases of the maxima of the overtone pulsations with respect to the maximum of the fundamental mode.

122.008 Photoelectric observations of variable stars. IV. The long-period variable stars and their variations of light and color. Absolute bolometric magnitudes of long-period stars. T. K. Kiseleva.
Byull. Inst. Astrofiz., Dushanbe, No. 66 - 67, p. 60 - 78 (1976). In Russian.

The results obtained from observations 1966 - 1973 in the U, B, V, R system of 12 long-period stars are given. It is shown that the periodic variations of the light and colors B–V, U–B, V–R are affected by variations of the intensity of TiO bands when the effective temperature varies. The absolute bolometric magnitudes, obtained on the basis of photometric observations, characterize the long-period variables as supergiants. The long-period stars with periods smaller than 300 days belong to population II, and the long-period stars with periods greater than 300 days to population I of the Galaxy.

122.009 Polarimetric observations of variable stars. Equipment and method of observations. Observations of TU Cas in 1973. N. N. Kiselev.
Byull. Inst. Astrofiz., Dushanbe, No. 66 - 67, p. 79 - 88 (1976). In Russian.

122.010 Period-amplitude relationships for individual dwarf novae. I. D. Howarth.
J. British Astron. Assoc., Vol. 87, 488 - 492 (1977).

122.011 R Coronae Borealis, 1971–1975. C. R. Munford.
J. British Astron. Assoc., Vol. 87, 509 - 512 (1977).
Report of the Section.

122.012 Stability of Cepheid-type stars against nonradial oscillations. Y. Osaki.
Publ. Astron. Soc. Japan, Vol. 29, 235 - 248 (1977).

The vibrational stability of the Cepheid-type stars against nonradial oscillations is studied. It is shown that the damping time of waves in the deep interior of "giant stars" is much shorter than the travel time of waves, and that standing oscillations extending from the stellar center to the surface are thus impossible for such stars. However, nonradial p-modes trapped in the envelope are shown still to exist for which the envelope may be treated as an isolated pulsating unit. It is found that nonradial modes with lower-order spherical harmonics are stable because of heavy leakage of wave energy from the envelope to the core, but that those of higher l are trapped well within the envelope and some of them are unstable due to the negative dissipation in the hydrogen- and helium-ionization zones.

122.013 The bright variable stars in Messier 5. C. M. Coutts Clement, H. Sawyer Hogg.
J. R. Astron. Soc. Canada, Vol. 71, 281 - 297 (1977).

Messier 5 has been photographed on about 400 nights over 41 years with University of Toronto telescopes. These observations show that the brightest variable in M5, No. 42 has a nearly constant period at 25.738 days, but that the second brightest, No. 84 varies around 26.4 days. Amateur astronomers are encouraged to observe these variables.

122.014 Are long-period variables really pulsating?
G. Wallerstein.
J. R. Astron. Soc. Canada, Vol. 71, 298 - 308 (1977).

Using available photometry in the near-infrared the luminosity and colour changes are used to infer the temperature and radius changes of the long-period variables R Aql, S CrB and U Ori. The predicted radius changes are inconsistent with the observed radial-velocity curves and reasonable hypotheses as to the centre-of mass velocity of the stars. The author suggests that the apparent increase in radius after visual light maximum is due to an increase in opacity (probably from molecules and grains) with decreasing temperature.

122.015 Hα variability in CC Eri.
I. C. Busko, G. R. Quast, C. A. O. Torres.
Astron. Astrophys., Vol. 60, L27 - L28 (1977).

The BY Dra star CC Eri is shown to exhibit periodic variation in its Hα emission, roughly in anti-phase with its photometric variation. This may be interpreted in terms of the spot model.

122.016 Discovery of a flare star, on 18-08-1976, by means of the GPO of the ESO Observatory at La Silla, Chile. H. Debehogne.
Astron. Astrophys., Vol. 60, 281 - 283 (1977). In French.

A flare star was discovered on 18-08-1976 with the GPO of the ESO Observatory. The coordinates are $\alpha = 19^h 13^m 33\overset{s}{.}002$, $\delta = -9°32'34\overset{''}{.}87$ (1950).

122.017 Analysis of the atmospheres of semiregular variable stars.
M. E. Boyarchuk, M. A. Kipper, L. F. Hanni.
Izv. Krymskoj Astrofiz. Obs., Vol. 57, 3 - 18 (1977). In Russian.

An analysis of the atmospheres of five semiregular variable stars of spectral type M was made by using spectrograms with dispersion 12 Å/mm. Excitation temperatures, electron densities and turbulent velocities were determined. It was found that the chemical composition of the investigated stars does not differ from that of the sun. The equivalent widths of Hα in the spectra of V450 Aql and EU Del decrease in minimum light.

122.018 Analysis of spectral variability of the Ap star 73 Dra. Part I.
N. S. Polosukhina, S. N. Dodonov.
Izv. Krymskoj Astrofiz. Obs., Vol. 57, 19 - 30 (1977). In Russian.

The physical parameters of the atmosphere of the magnetic variable star 73 Dra, turbulent velocity v_t and excitation temperature T_B have been obtained by the curve-of-growth method. These parameters show remarkable inhomogeneity of the atmosphere in the region where a maximum magnetic field of negative polarity is observed. The behaviour of the hydrogen lines Hγ and Hδ agrees with changes of the physical parameters v_t, T_B, and this behaviour is similar to that of the lines of Eu II, Sr II.

122.019 Gaseous envelopes of BY Dra type stars.
P. F. Chugajnov.
Izv. Krymskoj Astrofiz. Obs., Vol. 57, 31 - 44 (1977). In Russian.

Hα emission line profiles have been studied in the spectra of the following stars: FK Com, BD +30°448, and BD +27° 4642. It is found that the width of Hα-emission lines is the larger the greater the rotation velocity of the star. In order to explain this peculiarity it is supposed that the envelopes in which the Hα-emission originates are fast rotating. The most adequate is the scheme of "rigid body" rotation. The radii of the envelopes, estimated on the supposition of "rigid body" rotation, are 5 - 10 times larger than the stellar radii. The magnetic field is supposed to be a possible cause of "rigid body" rotation.

122.020 On the rotational hypothesis for the quasi-sinusoidal light variations in BY Draconis and EQ Virginis.
C. M. Anderson, F. H. Schiffer III, B. W. Bopp.
Astrophys. J., Vol. 216, 42 - 48 (1977).

High-resolution echellograms of the supposedly rotating, spotted dK–dMe stars BY Dra and EQ Vir have been obtained. These data are directly compared with similar material for three nonemission late-type dwarfs. The absorption lines in the spectrum of EQ Vir are broader than those in the nonemission stars by an amount consistent with its radius and photometric period. Although there is a distinct visual impression, upon inspection of the BY Dra spectra, of a similar effect, it was not possible to demonstrate it with any degree of confidence.

122.021 Cepheid studies. I. Mode interaction in the beat Cepheid U Trianguli Australis.
D. J. Faulkner.
Astrophys. J., Vol. 216, 49 - 56 (1977).

Photoelectric observations of the beat Cepheid U TrA previously reported by Oosterhoff and Jansen have been reanalyzed, using Fourier techniques. The two reported periods have been confirmed, and it has been shown that no further statistically significant periodicity is present. The two observed modes are subject to a strong mode interaction, but it is not yet clear whether this will have any implications for the beat masses derived for double-mode Cepheids. If U TrA is currently "mode switching", the change in relative mode amplitudes since these observations were made should be readily detectable.

122.022 On the period and luminosity stability of Sigma Orionis E. J. E. Hesser, H. Moreno, P. Ugarte P.
Astrophys. J., Lett., Vol. 216, L31 - L33 (1977).

The highly enigmatic prototype He-rich B star σ Ori E has been observed continuously for 13 nights for a third consecutive season on the $uvby\beta$ systems, and the results have been combined with reanalyzed data from earlier years. The authors find its period to be stable over the interval 1974 December–1977 February, allowing determination of an improved value: 1.19081 ± 0.00001 days. Features of the light curve reproduce very well season-to-season, and wavelength-dependent secular variations in excess of ~ 0.015 mag seem ruled out. These findings represent important constraints for any model of the system.

122.023 Extremely high circular polarization of AN Ursae Majoris. W. Krzemiński, K. Serkowski.
Astrophys. J., Lett., Vol. 216, L45 - L48 (1977).

Circular polarization V/I of the nova-like object AN UMa changes with a 1.9 hr period from -9% to -35% in blue light. The linear polarization reaches 9% in a sharp peak near the time of minimum $|V/I|$ and maximum brightness. These variations appear to be caused by rotation of a white dwarf binary companion with a magnetic field $B \geqslant 3 \times 10^8$ gauss. The similarity of AN UMa to AM Her suggests that they both represent a very distinct type of object which, because of high polarization, the authors propose be called a "polar".

122.024 Coordinated X-ray, optical, and radio observations of YZ Canis Minoris. J. T. Karpen, C. J. Crannell, R. W. Hobbs, S. P. Maran, T. J. Moffett, D. Bardas, G. W. Clark, D. R. Hearn, F. K. Li, T. H. Markert, J. E. McClintock, F. A. Primini, J. A. Richardson, S. Cristaldi, M. Rodono, D. A. Galasso, A. Magun, G. J. Nelson, O. B. Slee, P. F. Chugainov (Chugajnov), Yu. S. Efimov, N. M. Shakhovskoy (Shakhovskoj), M. R. Viner, V. R. Venugopal, S. R. Spangler, M. R. Kundu, D. S. Evans.
Astrophys. J., Vol. 216, 479 - 490 (1977).

The authors report coordinated X-ray, optical, and radio

observations of the flare star YZ CMi. Thirty-one minor optical flares and 11 radio events were recorded. No major optical flares greater than 3 magnitudes were observed during the program. Although no flare-related X-ray emission was observed, the measured upper limits in this band enable meaningful comparisons with published flare-star models. Based on the present results, the fraction of the galactic component of the diffuse soft X-ray background contributed by UV Ceti-type flare stars is $\leqslant 9 \times 10^{-3}/n_H$, where n_H is the mean density of interstellar hydrogen within a few hundred parsecs of the Sun.

122.025 **Cluster membership of the cepheid UY Persei.**
          D. G. Turner.
Publ. Astron. Soc. Pacific, Vol. 89, 277 - 279 (1977).

The criteria by which the 5.365-day cepheid UY Per qualifies as a member of the h and χ Per association are reviewed, and disagreement in age is shown to be a major negative factor. On the other hand, the important restrictions of age and color excess are shown to be satisfied by a pair of clusters, King 4 and Czernik 8, which closely bracket the cepheid. Since the absolute magnitude of $\langle M_V \rangle = -3.52 \pm 0.06$ m.e. which results for UY Per from the assumption of membership in these clusters agrees well with the value expected from the period-luminosity relation, the importance of further observation of these clusters and the region of the cepheid is emphasized.

122.026 **The magnetic cepheid W Sgr.**
          H. J. Wood, W. W. Weiss, H. Jenkner.
Bull. American Astron. Soc., Vol. 9, 434 (1977). – Abstract.

122.027 **Bond's flare star 2329−03.**   J. L. Greenstein.
          Publ. Astron. Soc. Pacific, Vol. 89, 304 - 308
(1977).

The dM4e flare star found by Bond has $M_V = +13.0$, a distance of 17 pc, low velocity, and an ultraviolet excess. Four spectra were obtained at or near minimum light. Interpreting the ultraviolet excess as an optically thin Balmer recombination continuum in a chromosphere, T_e is near 20,000°K, the emission near 2×10^{28} ergs s^{-1}, as compared to the stellar luminosity of 2.4×10^{31} ergs s^{-1}, and an electron density 10^{11} to 10^{12} cm^{-3}. Such an optically-thin chromosphere is a poor approximation, since the Balmer decrement is slow. Emission from a small area of flare-heated hot spots could also provide the ultraviolet.

122.028 **Note on the radial velocity of AW Persei.**
          D. H. McNamara, R. Chapman.
Publ. Astron. Soc. Pacific, Vol. 89, 329 - 330 (1977).

New radial-velocity measurements of the classical cepheid AW Per indicate the center-of-mass velocity of the star has increased by ∼21 km sec^{-1} since the radial-velocity measurements secured by Miller and Preston in 1960–63. The orbital period of the binary motion must exceed 20 years.

122.029 **The nature of dwarf cepheids. III. AI Velorum.**
          M. Breger.
Publ. Astron. Soc. Pacific, Vol. 89, 339 - 344 (1977).

The variability and mass of the dwarf cepheid prototype, AI Vel, has been investigated during six nights of observations with the $uvby\beta$ filters. The two excited periods are found to have average amplitudes of 680° and 370°K, respectively. An average effective gravity of log g = 3.98 and an average effective temperature of 7620°K are found. The low-mass, high-evolution hypothesis is not confirmed for this star. The trigonometric parallax is too uncertain to determine the mass of this star. However, the metallicity index, space motion, and period-gravity relation indicate that AI Vel is in fact a δ Scuti star and in the immediate post-main-sequence (high-mass) phase of evolution.

122.030 **Comments on the paper by A. E. Rydgren "T Tauri**
          stars and the $(J-H)$, $(H-K)$ diagram".
B. Baschek, R. Wehrse.
Publ. Astron. Soc. Pacific, Vol. 89, 345 - 346 (1977).

The theoretical $(J-H)$ and $(H-K)$ color indices, calculated by Rydgren (1976) for a hydrogen envelope of about 20,000°K surrounding T Tauri stars, are noticeably changed when line radiation is taken into account in addition to bound-free and free-free continuum emission.

122.031 **UBV photometry of RR Lyrae stars in M4.**
          C. R. Sturch.
Publ. Astron. Soc. Pacific, Vol. 89, 349 - 355 (1977), with an erratum , 740 (1977).

Photoelectric UBV observations of 15 a,b RR Lyrae stars in the globular cluster M4 are presented. The following characteristics are derived for M4: (1) [Fe/H] = −0.8, (2) $\langle E_{B-V} \rangle$ = 0.^{m}34 with a range of 0.^{m}28 $\leqslant E_{B-V} \leqslant$ 0.^{m}41, interpreted as a real variation in interstellar reddening, (3) $\langle V_0 \rangle$(RR) = 12.^{m}30, with luminosity increasing with period, (4) $(m-M)_0$ = 11.^{m}7 or r = 2.2 kpc.

122.032 **The low-mass X-ray binary AM Herculis.**
          D. Crampton, A. P. Cowley.
Publ. Astron. Soc. Pacific, Vol. 89, 374 - 385 (1977) = Contrib. Dominion Astrophys. Obs., Victoria, B.C., Canada, No. 335 = NRC No. 15904.

Radial velocities and line-intensity measurements of AM Her during its bright state are presented. An intercomparison of all the recent spectroscopic data yields an improved orbital period of 0.^{d}128926 ± 0.^{d}000009, in excellent agreement with polarimetric and photometric values. It is found that at least two principal regions of line formation can be identified. One of these is occulted during the deep V-light minimum, lending further support to the suggestion that the cool companion is at inferior conjunction at that phase.

122.033 **The continuous radiation emitted by accretion**
          discs in cataclysmic binaries: the dwarf nova SS
Cyg during outburst and the old novae V 603 Aql and RR Pic.
R. Tylenda.
Acta Astron., Vol. 27, 235 - 249 (1977).

A steady-state model of the accretion disc around a white dwarf, including the boundary layer and the hot spot, has been constructed. Calculated distributions of radiation are compared with soft X-ray, ultraviolet, visual, and infrared observations of SS Cyg during outburst as well as with ultraviolet measurements of V 603 Aql and RR Pic. The results support accretion models of dwarf nova outbursts.

122.034 **Variable stars of NGC 4833.**
          S. Demers, A. Wehlau.
Astron. J., Vol. 82, 620 - 625, 663 (1977).

The authors present photographic B and V photometry of the known variable stars in NGC 4833. Preliminary periods are determined for the first time for nine variables, thus increasing the known number of RR Lyrae periods to 15. The cluster is of Oosterhoff type II with $\langle P_{ab} \rangle$ = 0.68 day and $\langle P_c \rangle$ = 0.40 day. These values represent the longest ones known for type-II systems. The color range of the RR Lyrae stars suggests that color excess of NGC 4833 is close to $E(B - V)$ = 0.4.

122.035 **HD 30466 − a double wave photometric variable**
          silicon star.   H. M. Maitzen.
Astron. Astrophys., Vol. 60, L29 - L30 (1977).

New photoelectric intermediate-band photometry of HD 30466 points to P = 2.7795 days as rotational period which is twice the previously published value. In the resulting double wave the secondary maxima are nearly half as high as the primary except in Strömgren v where the secondary maximum is almost completely suppressed. This can be produced only by different chemical composition patches on the stellar surface.

122.036   Photometry of the dwarf cepheid EH Librae.
         P. Broglia, P. Conconi.
Astron. Astrophys., Suppl. Ser., Vol. 29, 321 - 326 (1977).

Photoelectric B and V observations of the dwarf cepheid EH Lib are given. They enable us to detect small variations in the height of the maxima, whilst the minima appear to be constant. Ten new epochs of maximum are derived. An analysis of the published observations shows that the period is variable.

122.037   Infrared photometry of dwarf novae and possibly related objects.   P. Szkody.
Astrophys. J., Vol. 217, 140 - 150 (1977).

Infrared photometry for six dwarf novae, six symbiotics, two novae, one recurrent nova, and seven related objects was obtained in order to study the similarities and differences among these systems in regard to a disk component. The flux distribution as a function of phase from maximum to minimum light was studied for SS Cyg, AB Dra, AH Her, and RX And. In contrast to the smooth IR decline of these systems, the decline of Nova Cyg shows an IR increase. All the symbiotics show the IR continuum of a late-type star. AE Aqr, EZ Peg, U Cep, and RS Oph appear more like the symbiotics than the dwarf novae. UX UMa and V426 Oph are very similar to the dwarf novae at standstill, while TT Ari is more like the novae in that it shows an IR excess.

122.038   Resonance effects and the Cepheid "bump mass" anomaly.   N. R. Simon.
Astrophys. J., Vol. 217, 160 - 170 (1977).

A second-order iterative nonlinear pulsation theory is generalized to the nonadiabatic case, and employed to treat Cepheid models with linear normal-mode periods lying in the vicinity of the resonance $P_2/P_0 = 0.5$. It is argued that the results of the present investigation strengthen the case for associating the bumps on Cepheid velocity curves with the resonance $P_2/P_0 = 0.5$. A brief discussion of the Cepheid "bump mass" anomaly concludes the article.

122.039   La stella variabile T Corona Borealis.
         G. Bianco.
Astronomia, N. 9, p. 8 - 15 (1977).

122.040   Photoelectric three-color observations of the galactic cepheids CD Cyg; X, Z, RR Lac and U Vul.
G. Asteriadis, L. N. Mavridis, A. Tsioumis.
Praktika Akad. Athens, Tom. 51, 540 - 576 (1976/77).

122.041   Variable stars in the globular cluster NGC 6235.
         M. H. Liller.
Astron. J., Vol. 82, 711 - 713, 770 (1977).

B magnitudes are presented for two previously known and three new variables in the region of the globular cluster NGC 6235. Three of the variables are of RR Lyrae type and are probably cluster members. The remaining two may be eclipsing systems.

122.042   An extremely rapidly pulsating star.
         G. Duthie.
News Lett. Astron. Soc. N.Y., Vol. 1, No. 2, p. 24 (1977).
Abstract.

122.043   Flare stars: II. Physical characteristics of the flares.
         D. J. Mullan.
Irish Astron. J., Vol. 12, 277 - 315 (1976).

The author summarizes observations of stellar flares at optical, X-ray and radio wavelengths. He suggests an interpretation of these observations in terms of a "once-removed" flare model in which a volume of hot plasma appears in the atmosphere. A numerical solution for the evolution of this "once-removed" flare model accounts for the presence of precursors and stillstands in the optical light curve.

122.044   Spectral and photometric observations of fast irregular variables. II. Hα and Hβ lines in the spectrum of WW Vul, VX Cas and UX Ori.   E. A. Kolotilov.
Astrofizika, Vol. 13, 33 - 49 (1977). In Russian. English translation in Astrophysics, Vol. 13, No. 1.

In 1971 - 1975 the author has obtained 90 spectrograms of the Hα region and 13 spectrograms of the Hβ region of the variables WW Vul, VX Cas and UX Ori. The profiles of Hα emission and Hβ absorption lines in the spectra of the variables are presented. The variation of Hα emission profiles and correlations between Hα fluxes and brightness variations of the stars are discussed. The profiles of Hβ absorption lines are in good agreement with Hβ absorption lines of standard A0V– A3V type stars. The author supposes that the Hα emission is formed in the gaseous envelopes of the variables. Rotational velocities of the envelopes are estimated.

122.045   Three-colour observations of a slow flare in the Orion region.   O. S. Chavushyan, N. D. Melikyan.
Astrofizika, Vol. 13, 199 - 202 (1977). In Russian. English translation in Astrophysics, Vol. 13, No. 1.

A slow flare in the Orion region has been detected simultaneously in three colors by the 40'' and 21'' Schmidt telescopes of the Byurakan Astrophysical Observatory. Data of this flare are presented.

122.046   Study of the physical conditions in the envelope of the dwarf nova SS Cyg. I.   N. F. Vojkhanskaya.
Astrofiz. Issled., Izv. Spets. Astrofiz. Obs., Vol. 9, 16 - 21 (1977). In Russian.

The emission line spectrum (mainly hydrogen) is studied. The Balmer decrement is constructed and its variation with brightness is traced. Peculiarities of the decrement variation are considered as dependent on the kind of flare. It is shown, that during the flare the electron concentration, the degree of excitation and ionization increase in the envelope as well as its opacity. The results are considered from the viewpoint of the accretion flare model. A new criterion for forecasting flares is suggested.

122.047   A search for Beta Canis Majoris stars.
         L. A. Balona.
Mem. R. Astron. Soc., Vol. 84, 101 - 117 (1977).

The results of a search for southern β CMa stars are presented. Photoelectric photometry of thirty-one candidates shows that eight of these are previously unrecognized probable or certain β CMa variables. Four of these stars are members of the galactic clusters NGC 3293. One δ Sct variable and one eclipsing binary was discovered. A few other stars are probably ellipsoidal variables.

122.048   Visuelle Beobachtungen an RS Oph 1974 - 1976.
         U. Hopp, M. Kiehl, U. Surawski.
Sterne Weltraum, Jahrg. 16, 344 (1977).

122.049   Discovery of flare activity on the dM4e star Gliese 82.
         B. R. Pettersen.
Astron. Astrophys., Suppl. Ser., Vol. 30, 113 - 115 (1977).

Flare activity on Gliese 82 is reported. Flares 2 and 3 are interpreted as sympathetic flares.

122.050   UBVRI photometry of X Per.
         M. Ferrari-Toniolo, G. Natali, P. Persi, G. Spada.
Astron. Astrophys., Vol. 61, 47 - 50 (1977).

Photometric observations of X Per, the optical counterpart of the X-ray source 3U0352 + 30, over a period of several months in 1975–1976 indicate a slow increase in luminosity at an average rate of 0.10 mag/year. Optical variability from night to night is present at a level of 0.02–0.05 mag. No periodicities were found with amplitudes larger than 0^m01, and periods in the range 5–15 min and 10–48 h.

122.051 UBV-Photometrie des mehrfachperiodischen RR Lyrae-Sternes RV Ursae Majoris.   W. Mühle.
BAV Rundbr., 26. Jahrg., 34 - 36 (1977).

122.052 Über den Lichtwechsel von AF Cygni.
K.-P. Timm.
BAV Rundbr., 26. Jahrg., 37 - 40 (1977).

122.053 Neueres von XZ Cyg.   M. Weigele.
BAV Rundbr., 26. Jahrg., 40 - 42 (1977).

122.054 The magnetic field of W Sgr: new Zeeman measurements.   H. J. Wood, W. W. Weiss, H. Jenkner.
Astron. Astrophys., Vol. 61, 181 - 184 (1977).

Additional ESO coudé Zeeman plates and a new statistical approach to the measurement of the plates yielding higher accuracy are presented. The observations confirm that the bright southern cepheid W Sgr shows a measurable magnetic field. Two hypotheses are presented for the interpretation of the observations: a full-period irregularly-shaped curve or a sine curve with 1/10 the pulsational period. More observations are required to test the hypotheses.

122.055 SS Cygni, 1969 - 1975.   I. D. Howarth.
J. British Astron. Assoc., Vol. 87, 611 - 623 (1977).
Report of the section.

122.056 The structure of cataclysmic variables.
E. L. Robinson.
Annu. Rev. Astron. Astrophys., Vol. 14, (see 003.008), 119 - 142 (1976).

The cataclysmic variables are usually divided into the following four classes: novae; recurrent novae; dwarf novae, which are subclassified as U Geminorum stars or Z Camelopardalis stars; and novalike variables. The primary defining characteristics of each class, the amplitudes and frequencies of the eruptions, are summarized. The author discusses those observations that, in his opinion, bear most directly on the structure of cataclysmic variables at minimum light.

122.057 Hα profile in two stars of type T Tauri: AA and DL Tau.   C. Fehrenbach, Y. Andrillat.
C. R. Acad. Sci. Paris, Tome 285, Sér. B, 235 - 238 (1977).
In French.

They both present a central absorption, very strong in DL Tau, in good agreement with the chromospheric assumption. The technique used allows the authors to detect short time variations of the profile.

122.058 Cepheid studies. II. A third period in the beat Cepheid TU Cassiopeiae.   D. J. Faulkner.
Astrophys. J., Vol. 218, 209 - 219 (1977).

Photoelectric observations of the beat Cepheid TU Cas made by several investigators prior to 1960 have been reanalyzed by using Fourier techniques. The two previously reported periods (2^d13931 and 1^d51833) have been confirmed, and a small-amplitude tertiary pulsation has been discovered at 1^d25246.

122.059 Variable 2.6 mm CO emission from χ Cygni and Mira.   K. Y. Lo, K. P. Bechis.
Astrophys. J., Lett., Vol. 218, L27 - L30 (1977).

The authors have observed time-variable emission from the 2.6 mm, $v = 0$, $J = 1 \rightarrow 0$, transition of $^{12}C^{16}O$ from the Mira variables χ Cygni and Mira itself. The time variation could be due to the variable vibrational excitation of the CO molecules as a result of the changing stellar luminosity. A narrow CO emission line observed from Mira could be

enhanced by stimulated emission. CO emission was also detected in five other stars.

122.060 Spectroscopic observations of the helium silicon variable HD 124224.   J. Hardorp, C. Megessier.
Astron. Astrophys., Vol. 61, 411 - 414 (1977).

The authors covered half the period of variation with 12 Å/mm blue plates. They see two "starspots" each of Si and Mg and one of He. They rediscuss the radial velocity measurements of Abt and Snowdon (1973), the varying $\log g$ found by Weiss et al. (1976) and the oblique rotator model by Khokhlova (1970).

122.061 Spectroscopic investigation of the silicon Ap star HD 34452.
V. M. Pavlova, V. L. Khokhlova, I. A. Aslanov.
Astron. Zh. Akad. Nauk SSSR, Vol. 54, 979 - 986 (1977). In Russian. English translation in Soviet Astron., Vol. 21, No. 5.

28 spectrograms with dispersion of 4 Å/mm were used to investigate the spectral variability of HD 34452. Two maxima and two minima of W_λ of the lines of all elements can be revealed, the period being $P = 2^d466$. The variability of line profiles indicates the presence of at least seven spots with enhanced concentration of elements.

122.062 The Mira variables of the metal-rich globular cluster NGC 6356.
C. Coutts Clement, H. Sawyer Hogg.
J. R. Astron. Soc. Canada, Vol. 71, 411 (1977). – Abstract.

122.063 Starspots on BY Dra-type stars.   D. S. Evans.
Highlights of Astronomy, Vol. 4, Part II, (see 012.026), p. 395 - 396 (1977).

122.064 On the spottedness and magnetic field of T Tau-type stars.   R. E. Gershberg.
Highlights of Astronomy, Vol. 4, Part II, (see 012.026), p. 407 (1977).

122.065 The eruptive BQ[ ] star HM Sagittae.
F. Ciatti, A. Mammano, A. Vittone.
Astron. Astrophys., Vol. 61, 459 - 467 (1977).

The increase of 5 mag in the 16 mag star is followed by a protracted maximum. Forbidden lines are recorded. High density is deduced for reasonable assumptions on electron temperature. These properties and the presence of low excitation features, like O I 8446 Å and [Fe II] lines, are characteristic of BQ[ ] stars. The radial velocity is deduced, and estimates of distance are presented. An evolutionary trend is suggested, where this stellar object may evolve to the phase of a compact planetary nebula, and later on to an extended nebula.

122.066 Far-UV observations of T Tau-like stars.
K. S. de Boer.
Astron. Astrophys., Vol. 61, 605 - 608 (1977).

Out of seven T Tau-like stars of the ANS far-UV photometry program only three could be detected beyond doubt. The radiation detected from V380 Ori and from CoD–44°3318 may have a contribution of hydrogen Balmer- or 2-quantum emission.

122.067 Photometry of three variables in Aquila.
B. L. Shaganyan.
Astrometr. i Astrofiz., Kiev, vyp. (No.) 32, (see 003.011), p. 33 - 41 (1977). In Russian.

Light curves of TW, V586 and V1079 Aql are determined from photographic observations. Beyer's elements for TW Aql are confirmed; van de Voorde's elements for V586 Aql do not prove correct.

122.068 Linear polarization in AM Herculis objects.

H. S. Stockman.
Astrophys. J., Lett., Vol. 218, L57 - L60 (1977).

The three known AM Herculis-like objects, AM Her, AN UMa, and VV Pup, show a temporal pulse of linear polarization which is synchronous with the periodic variations in their circular polarization, optical flux, and spectral features. The three are thought to be close binary systems with material accreting onto a highly magnetic white dwarf. The author suggests that the observed linearly polarized light is created in the accretion column near one of the white dwarf's obliquely oriented magnetic poles.

122.069 Strong TiO-related variations in the diameters of Mira and R Leonis.   A.Labeyrie, L. Koechlin, D. Bonneau, A. Blazit, R. Foy.
Astrophys. J., Lett., Vol. 218, L75 - L78 (1977).

New speckle interferometer observations in narrow spectral bands show abrupt variations in the diameters of Mira (o Ceti) and R Leonis as a function of wavelength. In strong TiO features, the diameter is two times larger. The authors have interpreted these variations as due to the large opacity of TiO: in the strong TiO features, the optically thick region of the atmosphere extends from the continuum-formation layer to several AU outward.

122.070 Survey of the BY Draconis syndrome among dMe stars.   B. W. Bopp, F. Espenak.
Astron. J., Vol. 82, 916 - 924 (1977).

In an extension of Krzeminski's survey of variability among dM and dMe stars, the authors have obtained BVr photometry of 22 K and M dwarfs to search for slow, quasisinusoidal light variations (the BY Draconis syndrome). A high percentage of the dMe stars are variable. There is a general tendency for the stars to become redder at minimum light, but some show no detectable color changes over the photometric cycle. The high incidence of variability in the sample suggests that all dMe stars are subject to the BY Dra syndrome or a slower variant. The available data suggest that higher-than-normal rotational velocities are the necessary and sufficient condition for the BY Dra syndrome and the occurrence of dMe stars.

122.071 Optical flares on Proxima Centauri.   A. R. Walker.
Mon. Notes Astron. Soc. South Africa, Vol. 36, 97 - 102 (1977).

The flare star Proxima Centauri was monitored in a co-ordinated optical, radio and soft X-ray observing program between May 16 and 19 1977, in an attempt to ascertain the ratio of energy output at widely differing wavelengths during a flare, thus providing a powerful test of the validity of the present theoretical models of stellar flares. This paper reports the observations obtained by the author in the optical region of the spectrum.

122.072 Activité d'étoiles variables du type Mira entre 1969 et 1977.   J. Aubaud.
AFOEV Bull., Tome 11, 32 - 35 (1977).

122.073 Le comportement de Z Camelopardalis.   J. Bauer.
AFOEV Bull., Tome 11, 36 - 38 (1977). — Translated from German; see 19.122.011.

122.074 AY Lyrae, 1931 - 1975.   I. D. Howarth.
J. British Astron. Assoc., Vol. 88, 79 - 83 (1977).
Report of the Variable Star Section.

122.075 Band strengths of M stars in the Orion population.   J. R. Mould, R. E. Wallis.
Mon. Not. R. Astron. Soc., Vol. 181, 625 - 635 (1977).

Narrow band measurements of CaH and TiO band strengths are reported for T-Tauri stars of spectral type M. The observations imply a surface gravity an order of magnitude below that of the lower main sequence, consistent with the quasi-static contractional evolution of low-mass stars.

122.076 The application of maximum entropy spectral analysis to the study of short-period variable stars.   J. R. Percy.
Mon. Not. R. Astron. Soc., Vol. 181, 647 - 656 (1977).

The maximum entropy method of spectral analysis has been used to determine component periods in several variable stars of the dwarf Cepheid and δ Scuti types. The method was tested on synthetic light curves.

122.077 An observational check on the phase shifts determined by the angular diameter method of distance determination for classical Cepheids.   N. R. Evans.
Mon. Not. R. Astron. Soc., Vol. 181, 85P - 87P (1977).

Observed phase shifts between light and radial velocity curves for classical Cepheids are compared with those found by the angular diameter method of distance determination. Agreement is generally within the errors but there is an indication that the slope of the surface brightness-colour relation used by Barnes et al. is slightly too negative.

122.078 Short periodic oscillations of the dwarf nova VW Hydri.   R. Haefner, R. Schoembs, N. Vogt.
Astron. Astrophys., Vol. 61, L37 - L38 (1977) = Veröff. Sternw. München, Band 7, Nr. 24.

A coherent oscillation of approximately 88 s period and 0^m005 amplitude was detected during the decline stage at the end of the long eruption of VW Hyi in December 1975. There are indications that the amplitude depends on the phase of the orbital revolution. The new period favours models in which such oscillations are caused by the orbital motion or inhomogeneities in the disc.

122.079 A hot-plasma model of the strongest flares of EV Lacertae.   K. Kodaira.
Astron. Astrophys., Vol. 61, 625 - 634 (1977).

A quasi-static "hot-plasma" model is constructed which can represent the essential characteristics of the strongest flares of EV Lac observed in the initial-decline phase. The model consists of hot-plasma bubbles with $T \sim 10^8$ K and $n_{el} \sim 5 \times 10^{10}$ cm^{-3} and with a total volume of stellar dimension, and of relatively cool dense "foot" points of the bubbles with $T \sim 10^5$ K and $n_{el} \sim 5 \times 10^{13}$ cm^{-3}. The scaling laws for other UV Cet-type stars and the possibilities of observational tests of this model are discussed.

122.080 Spectroscopic and photometric observations of the highly variable young star DR Tauri.   C. Bertout, J. Krautter, C. Möllenhoff, B. Wolf.
Astron. Astrophys., Vol. 61, 737 - 745 (1977).

Simultaneous spectroscopic and photometric observations of the T Tauri-like star DR Tauri are reported. The observed spectral properties can be explained by assuming that DR Tauri is a protostar of about 1 $M_\odot$ which has not yet reached its hydrostatic equilibrium Hayashi track.

122.081 On the frequency-period distributions of Cepheid variables in galaxies in the Local Group.   S. A. Becker, I. Iben, Jr., R. S. Tuggle.
Astrophys. J., Vol. 218, 633 - 653 (1977).

The authors find that the evolutionary tracks of intermediate-mass stars in the H-R diagram are very sensitive to changes in chemical composition. The composition dependence is especially evident in both the theoretical and observational frequency-period distributions of Cepheid variables. The authors infer that the character of the observed distribution for any galaxy depends on the existence of a composition spread, the existence of composition gradients, the form of the birthrate function, and of course selection effects. On comparing in detail observed and theoretical distributions, they conclude that a galactic birthrate function is made up of two distinct components; both the primary and secondary components have roughly the same power dependence on mass, but the normalization of the secondary component, which applies to more massive stars is between 5 and 50 times that of the primary or background component, which applies to all stars.

122.082   The instability strip of the anomalous Cepheids in the Draco dwarf spheroidal galaxy.
R. G. Deupree, S. W. Hodson.
Astrophys. J., Vol. 218, 654 - 658 (1977).

The instability strip of the anomalous Cepheids is located, and the location verifies that the variables must be more massive than the RR Lyrae variables. It is found that the red edge of the Population II instability strip becomes bluer as the stellar mass increases, as does the blue edge. This indicates that previous RR Lyrae color-width results are effectively independent of mass. The changes in the anomalous Cepheid instability strip indicated by a new opacity table, whose prime improvement is the inclusion of a number of molecules, and by the hypothesis that the helium abundance in the Draco system is slightly less than in the Galaxy, are examined.

122.083   Further photometry and analysis of 1 Monocerotis.
A. G. Kulkarni, P. Vivekananda Rao.
Bull. Astron. Soc. India, Vol. 5, 75 - 78 (1977).

122.084   Characteristics of IR variable stars as observed from orbit.   S. P. Maran, T. F. Heinsheimer, T. L. Stocker, R. D. Chapman, R. W. Hobbs, A. G. Michalitsanos.
Modern utilization of infrared technology civilian and military. II, (see 012.031), p. 23 - 29 (1976). − Abstr. in Phys. Abstr., Vol. 80, Abstr. 49406 (1977).

122.085   Kurzer Abriss der Geschichte der Veränderlichen Sterne.   K.-P. Timm.
Orion, 35. Jahrg., 185 - 187 (1977).

122.086   S Coronae Australis − ein Stern, der noch gar kein Stern ist.   I. Appenzeller.
Sterne Weltraum, Jahrg. 16, 395 - 398 (1977).

122.087   Hot ultrashort period cepheid. A new type variable in the globular cluster M15.   Y.-h. Chu.
Kexue Tongbao, Vol. 22, No. 1, p. 26 (1977). In Chinese. From Phys. Abstr., Vol. 80, Abstr. 71121 (1977).

122.088   Observation of variables in the Ophiuchus-Scorpius region − new rapid variables and nebular variables.
Third group of Stellar Division, Purple Mountain Observatory, Academia Sinica.
Acta Astron. Sinica, Vol. 18, 60 - 67 (1977).

Preliminary results of the search for flare stars and nebular variables in Ophiuchus and Scorpius are reported. One new flare star and 10 new possible nebular variables in which some are rapid variables have been identified.

122.089   Statistical theory of the turbulent convection in pulsation variables.   D.-r. Xiong.
Acta Astron. Sinica, Vol. 18, 86 - 104 (1977).

The dynamic behavior of turbulent convection is studied. From the dynamic equations of correlation function, the author obtained steady component and pulsational-component of convections. The physical significance of these solutions are discussed.

122.090   A search for extremely red stars on the Palomar Observatory Sky Survey. II. Examination for light variability.   C. Friedemann, J. Gürtler, W. Pfau.
Astron. Nachr., Band 298, 327 - 329 (1977) = Mitt. Univ.-Sternwarte Jena, No. 129.

The extremely red stars discovered on some Palomar Sky Survey prints in an earlier paper were investigated for variability. Four certain and one probable variables were found.

122.091   The RR Lyrae stars in the Large Magellanic Cloud.
J. A. Graham.
Publ. Astron. Soc. Pacific, Vol. 89, 425 - 465 (1977).

A field covering an area $1° \times 1°3$ in the Large Magellanic Cloud (LMC) has been searched for variable stars. Sixty-eight variable stars with periods less than a day and amplitudes of several tenths of a magnitude are identified as being of the RR Lyrae type. Sixty are of Bailey class *ab* and eight of Bailey class *c*. No large amplitude *ab*-type variables with periods shorter than 0^d45 are found in the LMC. The mean periods of the *ab*- and *c*-type variables are 0^d564 and 0^d328, respectively. For all stars with periods, $\langle B \rangle$ is found to be 19^m60. Adopting a true distance modulus of 18^m5 for the LMC, a mean absolute visual magnitude of $+0^m7 \pm 0^m25$ is proposed for the RR Lyrae stars in the Large Magellanic Cloud.

122.092   A note on the reddening of Polaris B.
D. G. Turner.
Publ. Astron. Soc. Pacific, Vol. 89, 550 - 551 (1977).

Polaris B is reclassified on the MK system using new spectroscopic observations. The resulting spectral type of F3 V, when combined with existing photometric data, yields a color excess $E_{B-V} = 0.00 \pm 0.01$, in agreement with the results of McNamara (1969). It is argued that this result must also apply to Polaris A, thereby confirming the results of other recent studies indicating that this cepheid is unreddened.

122.093   A search for long period cepheids in Norma.
A. L. Cabrera, J. C. Muzzio, G. Sánchez.
Inf. Bull. Variable Stars, No. 1299, 3 pp. (1977).

122.094   Photometric variability of the nonradial pulsator 53 Persei.   J. Africano.
Inf. Bull. Variable Stars, No. 1301, 4 pp. (1977).

122.095   On the period of BS Herculis.   E. N. Makarenko.
Inf. Bull. Variable Stars, No. 1303, 4 pp. (1977).

122.096   On the variability of HR 4511.   J. D. Fernie.
Inf. Bull. Variable Stars, No. 1305, 4 pp. (1977).

122.097   Spectral changes in V1331 Cygni (LkHα 120).
G. Welin.
Inf. Bull. Variable Stars, No. 1314 (1977).

122.098   Spectroscopy of HM Sagittae, a possible embryonic planetary nebula.   B. W. Bopp.
Inf. Bull. Variable Stars, No. 1327 (1977).

122.099   Revised periods for two RR Lyrae stars: variable 24 in NGC 6171 and variable 23 in NGC 6656.
C. Coutts Clement, H. Sawyer Hogg.
Inf. Bull. Variable Stars, No. 1333, 3 pp. (1977).

122.100   RS Cha: a Delta Scuti variable.
C. J. McInally, R. D. Austin.
Inf. Bull. Variable Stars, No. 1334, 3 pp. (1977).

122.101   VZ Cancri.   B. Cester, G. Giuricin, F. Mardirossian, M. Mezzetti, M. Pucillo.
Inf. Bull. Variable Stars, No. 1338, 3 pp. (1977).

122.102   Possible relation between the pulsation constant and the period in Delta Scuti variables.
E. Antonello, M. Fracassini, L. E. Pasinetti.
Inf. Bull. Variable Stars, No. 1339 (1977).

122.103   Practical formulae to search periodicities in

variable stars.   S. Ferraz-Mello.
Inf. Bull. Variable Stars, No. 1347, 4 pp. (1977).

122.104 Ultraviolet observations of β Canis Majoris stars with the TD-1 A satellite.
F. Beeckmans, M. Burger.
Astron. Astrophys., Vol. 61, 815 - 826 (1977).

The ultraviolet observations of 21 βCMa stars obtained with the S59 and the S2/68 experiments on board the TD-1 A satellite have been combined to estimate the ranges of the light variations in the ultraviolet. The β CMa stars seem to have the same colours and spectral features as the "normal" stars in the ultraviolet (at the S2/68 resolution), except for β Cep, $ξ^1$ CMa and σ Sco. The variations of effective temperature and radius have been computed from the ultraviolet and visual light ranges, on the basis of radial pulsations, assuming that the light curves are in phase at all wavelengths used. The values of ΔR and of $(\Delta R/R)/\Delta M_{bol}$ derived from these quantities are discussed.

122.105 A method for constructing envelopes and the period-amplitude relation of Cepheids.
W. Eichendorf, M. Reinhardt.
Astron. Astrophys., Vol. 61, 827 - 832 (1977).

The authors derive a statistically sound method to construct envelopes to point diagrams. They give a procedure to define points of equal probability, to which curves can be fitted by least squares. Empirical envelopes thus defined can be compared with each other or theoretical predictions and their difference can be statistically tested. The authors apply the method to the period-amplitude relation of Cepheids in M 31 and in our Galaxy.

122.106 VV Puppis: the day the accretion stopped!
J. Liebert, N. J. Woolf, H. S. Stockman, K. Hege, J. R. P. Angel.
Publ. Astron. Soc. Pacific, Vol. 89, 618 (1977). – Abstract.

122.107 A qualitative analysis of the spectrum of FG Sagittae, 1972–76.   J. S. Tenn, M. Carolin.
Publ. Astron. Soc. Pacific, Vol. 89, 622 - 623 (1977). Abstract.

122.108 Symmetric velocity structure in the SiO maser spectrum of R Cassiopeiae.   P. R. Schwartz.
Publ. Astron. Soc. Pacific, Vol. 89, 693 - 695 (1977).

The SiO $J = 1–0$, $v = 2$ lines are exterior to the $v = 1$ in velocity, strongly suggesting shell-like structure in the emission region.

122.109 A photometric study of SW Andromedae.
D. H. McNamara, K. A. Feltz, Jr.
Publ. Astron. Soc. Pacific, Vol. 89, 699 - 703 (1977).

Photometric ($uvbyβ$) observations of the strong-line RR Lyrae variable SW And are described. Intrinsic ($b-y$) and c_1 values are used to derive the variations in T_{eff} and log g. Wesselink's method is used to derive a radius of $R/R_\odot = 4.45$. Some peculiarities in the behavior of the m_1 index are discussed. The mean m_1 value at light minimum indicates that SW And has a near solar abundance of heavy elements.

122.110 Photometry of the FU Orionis stars V1057 Cygni and V1515 Cygni.   A. U. Landolt.
Publ. Astron. Soc. Pacific, Vol. 89, 704 - 705 (1977) = Contrib. Louisiana State Univ. Obs., No. 126.

$UBVR$ photoelectric data have been obtained for V1057 Cyg in 1975, 1976, and 1977. Its rate of fading has slowed to $0.^m0005$ per day. Another FU Ori type object, V1515 Cyg, is found to have photoelectric colors essentially the same as V1057 Cyg.

122.111 The nova-like variable BD $-7°3007$.

A. P. Cowley, D. Crampton, J. E. Hesser.
Publ. Astron. Soc. Pacific, Vol. 89, 716 - 719 (1977) = Dominion Astrophys. Obs., Victoria, B.C., Canada, Contrib. No. 341 = NRC No. 16100.

Simultaneous spectroscopy and rapid photometry show that BD $-7°3007$ exhibits significant changes on the time scale of several hours. Radial velocity variations ~ 200 km s^{-1} are seen in the very broad H absorption lines and suggest an orbital period of the system near $\sim 0.^d25$. Very weak variable emission in both H and Ca II appears to arise in an accretion disk about the hot star. No spectroscopic evidence of the secondary is found. Like the other nova-like variables BD $-7°3007$ shows flickering and irregular light variation of a few tenths of a magnitude, but it is not known to have suffered any significant outbursts.

122.112 UBV observations of CD $-33°12119$.
D. J. Bord, D. E. Mook, L. Petro, J. Thomas, W. A. Hiltner.
Publ. Astron. Soc. Pacific, Vol. 89, 720 - 723 (1977).

The star CD $-33°12119$, a suggested optical counterpart of X-ray source 3U 1727–33, has been studied photometrically. Brightness and color variations are observed on a time scale of days, but no regular pattern to these fluctuations is evident. Evidence for flares in U and B is discussed.

122.113 Relative abundances of isotopes of Zr in R Cygni.
A. C. Zook.
Publ. Astron. Soc. Pacific, Vol. 89, 625 (1977). – Abstract.

122.114 BC Draconis as a population II Cepheid.
K.-i. Uji-iye, H. Saio, M. Takeuti.
Sci. Rep. Tôhoku Univ., Ser. I, Vol. 60, 27 - 30 (1977).

The period of radial pulsation of BC Dra is computed. The luminosity of the model with a mass less than 1 $M_\odot$, which fits to $P_F = 2.566$ days, is in the region of W Virginis stars. If BC Dra has one more period of 3.351 days, large mass ($\sim 4 M_\odot$) and high luminosity ($M_{bol} \sim -4$ mag) and hence a large distance above the galactic plane are required.

122.115 Solar and stellar flares.   D. J. Mullan.
Sol. Phys., Vol. 54, 183 - 206 (1977). – Invited review paper.

In this paper, after briefly reviewing optical, radio and X-ray observations of stellar flares, the author shows how a simplified model which describes conductive plus radiative cooling of the coronal flare plasma in solar flares has been modified to apply to optical and X-ray stellar flare phenomena.

122.116 A relation between ultraviolet excess and superficial gravity for Cepheids.   E. Janot-Pacheco.
Astrophys. Space Sci., Vol. 47, L1 - L3 (1977).

A linear relation is calculated between the six-colour ultraviolet excess and superficial gravity for Cepheid variables. This excess appears to be a sensitive indicator of gravity behaviour for these stars.

122.117 Study of the Delta-Scuti star HR 1170.
S. K. Gupta.
Astrophys. Space Sci., Vol. 48, 199 - 206 (1977).

The light and colour curves of the δ-Scuti star HR 1170 are presented. The absolute and bolometric magnitudes are derived and the position of the star on the colour-colour diagram is also shown. The primary and beat periods estimated from the light curves are $0.^d098299$ and $0.^d39206$, respectively.

122.118 Optimal curve fitting procedures applied to the light curves of classical cepheids.   E. Budding.
Astrophys. Space Sci., Vol. 48, 249 - 266 (1977).

Optimal curve fitting procedures based on the well known Baade/Wesselink methodology have been applied to Stebbins et al.'s 6-colour photometry of the classical cepheids δ Cep and η Aql. This fitting function requires the specification of six parameters which thus play the role of unknowns in the optimization problem, though, in fact, all six parameters cannot be independently determined. The formulation involves a simple connection between colours and brightness temperatures, and model stellar atmosphere calculations can provide such a connection. The absolute magnitudes of δ Cep and η Aql are $M_v = -3.57$ and $M_v = -3.79$, respectively; but optimal curve fits would decrease both these values by about $0.^m09$.

122.119 Flare stars and the fast electron hypothesis.
G. A. Gurzadyan.
Astrophys. Space Sci., Vol. 48, 313 - 334 (1977).

An extensive analysis is made of the theory of flare stars based on the 'fast electron hypothesis', in the light of the latest observational evidence. It is shown that an adequate agreement of theory with the observations obtains regarding the internal regular features in the flare amplitude data in UBV rays, as well as the changes of the colour characteristics of stars during the flares; in the latter case the analysis is made not only with respect to the UV Cet-type stars, but flare stars as well, forming a part of the Orion association. Problems bearing on the 'negative flare' and the screening effect are dealt with. New properties of the light curves of flares are revealed, based on the above theory. Particular emphasis is laid on the X-ray radiation from flare stars. It is shown that the observed spectrum of X-ray radiation of flare stars differs sharply from that of X-ray radiation both of the stellar corona and solar X-ray flares. It is shown that the gamma-ray bursts recorded so far have no relation whatever to flare stars.

122.120 Observational studies of 12 DD Lacertae. II. Radial velocity and the variation of line profile.   N. Sato.
Astrophys. Space Sci., Vol. 48, 453 - 470 (1977).

Analysis of the radial velocity curves of 12 DD Lacertae shows that the primary period is decreasing and the secondary is almost constant, confirming the photometric results previously reported. Wesselink's method applied to the simultaneous light and radial velocity observations can reveal physical properties that are comparable with the present situation if the temperature variation is assumed to be about 2000 K, as well as when applied to the combined normal curves of the primary period, except at the maximum contraction corresponding to the phase of line broadening which must be the splitting of the line into two components at a higher dispersion.

122.121 Chromospheres of flare stars.   G. A. Gurzadyan.
Astrophys. Space Sci., Vol. 52, 51 - 104 (1977).

The present paper contains an attempt to formulate a theory, based on fast electrons hypothesis, of the chromospheres of flare stars. The following problems are tackled: Two components of the flare light; The source of excitation energy of the emission lines; The power of ionizing radiation; Excitation of the emission lines; Luminescence duration of the star in emission lines (observations); The electron temperature in the chromosphere of the flare star; Electron concentration in the chromospheres of flare stars; Duration of luminescence in the emission lines (theory); What supports the excitation of the emission lines of the 'quiescent' star? ; Degree of ionization; Balmer decrement of emission lines; Profiles of emission lines; The dependence of the intensity of emission lines on the flare amplitude; Two types of Haro flare stars; The effect of diluting the radiation; Effect of the spectral class; On the nature of 'slow' flares; The problem of forbidden lines; The possibility of observing the forbidden line 4363 [O III]; Excitation of the emission lines of helium; Lyman-alpha line in flare stars; Emission line 2800 Mg II in the spectra of flare stars.

122.122 Period variations of RR Lyrae type variable stars in the globular cluster M53.   V. P. Goranskij.
Peremennye Zvezdy, Prilozhenie, Vol. 3, 1 - 69 (1976). In Russian.

The period variability of 31 RR Lyrae variables in the globular cluster M 53 were studied on the basis of 250 plates of the Moscow Observatory 1959–1974, on the basis of Wachmann's 1943–1955 observations, and observations being published. The total time interval of observations is 53 years.

122.123 Some photometric properties of the infrared carbon star RW LMi (CIT 6).
A. Alksnis, I. Eglitis.
Photometric investigations of carbon stars, (see 003.018), = Radioastrofiz. Obs., Tr. 16, p. 157 - 172 (1977). In Russian.

The results of photographic monitoring of the peculiar infrared carbon star RW LMi = CIT 6 during six half-year observing seasons in $R,\ V,\ B$ and U passbands are discussed.

122.124 Polarimetric analysis of a slow flare of AD Leo.
J. Arsenijević, A. Kubičela, I. Vince.
Publ. Dep. Astron., Univ. Beograd, Fac. Sci., No. 6, (see 012.040), p. 17 - 23 (1976).

122.125 Photoelectric UBV photometry of northern cepheids.
I.   L. Szabados.
Mitt. Sternw. Ungarisch. Akad. Wiss., Budapest–Szabadsághegy, No. 70, 123 pp. (1977).

New UBV photoelectric observational data on 38 northern cepheids with periods of less than 5 days are presented. The period changes of the observed cepheids are investigated. Finally, the instability of the period for different types of cepheids is discussed.

122.126 Photographic photometry of IK Tau (NML Tau).
A. Alksnis, I. Eglitis.
Investigations of the sun and red stars. 6, (see 003.019), p. 68 - 75 (1977). In Russian.

Since 1971 18 observations in V, 8 in R and 6 in B of the very cool M-type long-period variable star IK Tau (NML Tau) have been made. The values of B, V, R magnitudes and the light curve are given. The observations confirm the period of light variations to be $465-470^d$.

122.127 T Tau-type stars: contemporary observational data.
P. P. Petrov.
Early stages of stellar evolution, (see 012.041), p. 66 - 100 (1977). In Russian.

Observational data on T Tau-type stars are considered. Main attention is paid to the results of observations published after 1970. A review of characteristics of Herbig-Haro objects, FU Ori and V1057 Cyg is given.

122.128 On the role of a magnetic field in the evolution of T Tauri stars.   P. P. Petrov, A. G. Shcherbakov.
Early stages of stellar evolution, (see 012.041), p. 127 - 128 (1977). In Russian.

It is assumed that some properties of T Tauri stars can be explained by strong local surface magnetic fields.

122.129 New variable star of Mira type SVS 2190.
S. M. Bychkov.
Astron. Tsirk., No. 940, p. 7 - 8 (1977). In Russian.

122.130 On the difference of the pulsating parameter $\Delta V/ \Delta (B - V)$ of RR Lyrae stars having different metal abundances.   M. S. Frolov.
Astron. Tsirk., No. 941, p. 1 - 2 (1977). In Russian.

122.131 New RR Lyrae variable SVS 2193.
V. P. Goranskij.
Astron. Tsirk., No. 942, p. 7 - 8 (1977). In Russian.

122.132 New elements of light variation of two variable stars.   S. Yu. Shugarov.

Astron. Tsirk., No. 942, p. 8 (1977). In Russian.

122.133 New variable stars in Corvus and Hydra.
  V. P. Goranskij.
Astron. Tsirk., No. 943, p. 6 - 7 (1977). In Russian.

122.134 New variable star of Mira type SVS 2189.
  O. S. Bartunov.
Astron. Tsirk., No. 943, p. 7 - 8 (1977). In Russian.

122.135 On the cold component of the system CH Cyg.
  E. B. Gusev.
Astron. Tsirk., No. 944, p. 4 - 5 (1977). In Russian.

122.136 On the variable star GR Peg.   B. D. Pochinok.
  Astron. Tsirk., No. 944, p. 5 - 8 (1977). In Russian.

**122.137 Preliminary light curve of the new variable emission
  object SVS 2183.**   O. D. Dokuchaeva.
Astron. Tsirk., No. 946, p. 6 - 7 (1977). In Russian.

122.138 Observations of V1068 Cyg and SVS 2194.
  S. Yu. Shugarov.
Astron. Tsirk., No. 949, p. 6 - 7 (1977). In Russian.

122.139 On the period of the variable TV And.
  B. D. Pochinok.
Astron. Tsirk., No. 952, p. 6 - 7 (1977). In Russian.

122.140 New variable star SVS 2197.   V. P. Goranskij.
  Astron. Tsirk., No. 952, p. 8 (1977). In Russian.

**122.141 Photometry of V1057 Cyg in the optical and infrared
  regions in 1976−77.**   E. A. Kolotilov.
Astron. Tsirk., No. 955, p. 1 - 3 (1977). In Russian.

122.142 New eruptive variable star SVS 2198.
  V. P. Goranskij.
Astron. Tsirk., No. 955, p. 8 (1977). In Russian.

**122.143 Photoelectric observations of Pleione (BU Tau) from
  September 1975 to March 1977.**
A. S. Sharov, V. M. Lyutyj.
Astron. Tsirk., No. 956, p. 5 - 6 (1977). In Russian.

**122.144 Variations of the Hα-line emission profile in the
  spectrum of T Tauri.**
G. V. Zajtseva, E. A. Kolotilov.
Astron. Tsirk., No. 957, p. 1 - 3 (1977). In Russian.

122.145 On the period of NW Lyrae.   A. S. Gadun.
  Astron. Tsirk., No. 958, p. 8 (1977). In Russian.

122.146 The dwarf nova CU Velorum.   F. M. Bateson.
  Publ. Variable Star Sect., R. Astron. Soc. New
Zealand, No. 5 (C77), p. 1 - 5 (1977).
    Visual observations from J.D. 2,438,846 to 2,442,930 of
CU Vel are discussed. Data for the 49 outbursts observed are
tabulated. All outbursts were either narrow or wide, according
to the time the variable was brighter than magnitude 13.0. The
mean cycle is 164.7 days. Mean maximum magnitude is 10.71
for wide maxima and 11.13 for narrow maxima.

122.147 The dwarf nova V436 Centauri.   F. M. Bateson.
  Publ. Variable Star Sect., R. Astron. Soc. New
Zealand, No. 5 (C77), p. 10 - 16 (1977).
    Observations from 2,436,256 to 2,442,948 show two
main types of outbursts − supermaxima and narrow maxima.
Possibly a third type of maximum occurs with short, sharp
peaks. Lists of observed outbursts and observations are
tabulated. Light curves are illustrated.

122.148 The Mira type variable RW Puppis.
  F. M. Bateson.
Publ. Variable Star Sect., R. Astron. Soc. New Zealand, No. 5
(C77), p. 17 - 20 (1977).
    Eighteen years of observations give the elements for RW
Pup: EPOCH Max. 2,436,906 ± 340.88 days. Minimum to
maximum 160.84 days. Mean range: 9.34 to 14.13ᵥ. Mean
light curves are used to produce a standard light curve.

**122.149 Recurrent novae, symbiotic variables, and binary
  evolution.**   R. F. Webbink.
Publ. Variable Star Sect., R. Astron. Soc. New Zealand, No. 5
(C77), p. 22 - 27 (1977).
    The present evolutionary status of T CrB, as a proto-
typical recurrent nova, is discussed. It is suggested that this
system is a relatively young object of this type, having undergone perhaps ten nova-like outbursts so far. The frequency of
outbursts is expected to increase, culminating in the gradual
transformation of this binary into a symbiotic variable.

122.150 The dwarf nova − V442 Centauri.
  F. M. Bateson.
Publ. Variable Star Sect., R. Astron. Soc. New Zealand, No. 5
(C77), p. 27 - 41 (1977).
    The results of 5,359 observations of V442 Cen are dis-
cussed. These cover the interval J.D. 2,434,824 to 2,443,183.
A total of 266 outbursts were recorded. The mean cycle is
$24\overset{d}{.}48$. The light curve is reproduced.

**122.151 A study of the reddening and blanketing corrections
  for RR-Lyrae stars in the Walraven VBLUW photometric system.**   J. Lub.
ESO Sci. Prepr. No. 13, 3 + 48 pp. (1977). − To appear in
Astron. Astrophys.
    Based on complete lightcurves in the five channels of the
Walraven VBLUW photometric system, a discussion is given of
the determination of blanketing and reddening corrections for
RR-Lyrae stars. A photometric accuracy of order $0\overset{m}{.}005$ has
been reached. Blanketing derived from the bracketed colour
[B-L] is shown to be equivalent to other ways of determining
the metallicity of RR-Lyrae stars such as δ(U–B)ₛ, m_1 and ΔS.
Two independent ways of determining the interstellar redden-
ing correction are studied and shown to lead to identical
results. A table is given with high accuracy blanketing and
reddening corrections for RRab, RRc, and short period variables.

**122.152 Long-time optical behaviour of AM Her = 3U 1809+
  50. Part II.**   R. Hudec, L. Meinunger.
Mitt. Veränderl. Sterne (MVS), Band 8, 10 - 17 (1977).

122.153 Photoelectric observations of CH Cygni.
  L. Luud, M. Ruusalepp, J. Vennik.
Publ. Tartu Astrofiz. Obs., Vol. 45, 113 - 130 (1977). In Russian.
    UBV observations of the peculiar variable star CH Cygni
from 1968 up to 1974 are presented. HD 182691 served as a
comparison star. The observed brightness variations are analyzed by using autocorrelation, crosscorrelation and power-
spectral density functions. The colours and their variations are
in accord with the model for a symbiotic star.

**122.154 Die Periode-Leuchtkraft-Beziehung der kurz-
  periodischen Mira-Sterne.**
K. Ferrari d'Occhieppo.
Anz. Math.-Naturwiss. Kl. Österreich. Akad. Wiss., Jahrg.
1976, Nr. 9, p. 125 - 129 (1976) = Astron. Mitt. Wien, Nr. 19.

**122.155 Ein Enveloppentest angewandt auf die Perioden-
  Amplituden-Relation von Cepheiden in unserer
  Galaxis und M31.**   W. Eichendorf, M. Reinhardt.
Mitt. Astron. Ges., Nr. 42, p. 91 - 92 (1977).

122.156 Radialgeschwindigkeitsschwankungen des Ap-Sternes HD 224801.
W. W. Weiss, H. J. Wood, H. Jenkner.
Mitt. Astron. Ges., Nr. 42, p. 133 - 136 (1977).

122.157 Ultraviolet observations of classical Cepheids by OAO-2.   J. L. Hutchinson.
Cepheid modeling, (see 012.045), p. 5 - 25, with a discussion p. 25 - 30 (1975).

OAO-2 observations of eight bright classical Cepheids in the wavelength region 1910 Å to 4250 Å are presented. The data for RT Aurigae, α Ursa Minoris, δ Cephei, and Y Ophiuchi show excellent agreement with ground-based photometry in the wavelength region of overlap and are consistent with a simple extrapolation of light-curve properties from the visible region. However, in the ultraviolet, the light curve of β Doradus shows two small flux bumps, at phases of 0.75 and 0.85, in addition to the well-known bump at phase 0.0. All three bumps should probably be associated with the arrival of shocks at the stellar surface.

122.158 Shock waves in a Beta Doradus model.
S. J. Hill.
Cepheid modeling, (see 012.045), p. 31 - 39, with a discussion p. 39 - 41 (1975).

The accumulation of observational evidence and interpretation supports the conclusion that β Doradus has running shock waves in its atmosphere. A numerical hydrodynamical model for the atmosphere of β Doradus is constructed. The results for the β Doradus model and a comparison with the observations are presented. It is of particular interest to verify that the atmosphere contains multiple shock waves as suggested by Hutchinson for the ultraviolet observations.

122.159 Mean colors of Cepheids.   A. N. Cox, C. G. Davis.
Cepheid modeling, (see 012.045), p. 43 - 52, with a discussion p. 53 - 55 (1975).

This paper concerns the method of taking a mean of the color variations of Cepheids over their pulsational cycle. It is demonstrated that the mean color depends on the type of mean employed. Thus, color observations of Cepheids can be interpreted by a color-effective temperature relation to give different T_e values for each kind of mean color. Here, theoretical colors from numerical integrations of Cepheid pulsations are used to determine the proper method of taking the color mean in order to get, by the color-T_e relation, the correct nonpulsating T_e.

122.160 Pulsation of double-mode Cepheids and comments on the problem of Cepheid masses.
D. S. King, C. J. Hansen, R. R. Ross, R. F. Stellingwerf, J. P. Cox.
Cepheid modeling, (see 012.045), p. 57 - 66, with a discussion p. 67 - 70 (1975).

122.161 Hydrodynamic effects in the atmospheres of variable stars.   C. G. Davis,Jr., S. S. Bunker.
Cepheid modeling, (see 012.045), p. 71 - 82, with a discussion p. 82 - 83 (1975).

Numerical models of variable stars are established, using a nonlinear radiative transfer coupled hydrodynamics code. The variable Eddington method of radiative transfer is used. Comparisons are for models of W Virginis, β Doradus, and η Aquilae. From these models it appears that shocks are formed in the atmospheres of classical Cepheids as well as W Virginis stars.

122.162 The Hertzsprung progression in Cepheid calculations.   R. F. Christy.
Cepheid modeling, (see 012.045), p. 85 - 98 (1975).

The Hertzsprung bump in Cepheid models has been studied for models of a wide range of mass, radius, and effective temperature. The results are consistent with a picture that the bump results from an acoustic signal that traverses the star from the ionization zone into the center and out again to the surface. The bump and other nonlinear phenomena have been studied over a series of Cepheid models from a period of less than one day to a period of 150 days. The results show characteristic changes in amplitude and in the appearance of bumps and are in good correspondence to actual Cepheids. The masses in this series are, however, characteristically lower than evolutionary masses.

122.163 Hydrodynamic models of a Cepheid atmosphere.   A. H. Karp.
Cepheid modeling, (see 012.045), p. 99 - 114 (1975).

Instead of computing a large number of coarsely zoned models covering the entire instability strip, the author has computed one model as well as computer limitations allow.

122.164 Numerical techniques for the linear, nonadiabatic stellar pulsation problem.   T. A. Bednarek.
Cepheid modeling, (see 012.045), p. 115 - 123, with a discussion p. 123 - 127 (1975).

The linear, nonadiabatic eigenvalue problem is formulated using Castor's method. Both left and right eigenvectors are calculated. Initial eigenvalues for the linear, nonadiabatic solutions are obtained from the adiabatic eigenvalues and left and right eigenvectors. The orthogonality relation is obtained. Simple formulas for the Newton method are given. The iteration procedure is constrained to improve convergence. Application of the method to Population II Cepheids is briefly presented.

122.165 Stability of nonlinear periodic pulsation of stellar envelopes.   K. von Sengbusch.
Cepheid modeling, (see 012.045), p. 129 - 134, with a discussion p. 134 - 135 (1975).

The eigenvalue method (von Sengbusch, 1973) to calculate models of periodically pulsating stars can be used to study the stability of the resulting oscillations. This paper will discuss results for a series of RR Lyrae models.

122.166 The Cepheid mode problem.   R. F. Stellingwerf.
Cepheid modeling, (see 012.045), p. 137 - 141, with a discussion p. 142 - 156 (1975).

The modal behavior of RR Lyrae stars and low mass Cepheids has been investigated using a nonlinear relaxation technique and stability analysis. The advantages of this type of numerical approach in investigations of preferred mode of pulsation are discussed. The results obtained for both classes of variable stars are quite similar: first harmonic pulsation toward the blue, fundamental pulsation toward the red, and mixed-mode behavior at the extreme red edge of the instability strip. In addition, stars near the center of the strip can pulsate in either the fundamental or the first harmonic mode. Possible implications for observational results, including the Oosterhoff dichotomy of globular clusters and the beat Cepheids, are discussed.

122.167 Cepheid modeling: workshop.   T. A. Bednarek, R. F. Christy, A. N. Cox, C. G. Davis, D. Fischel, W. M. Sparks, A. H. Karp, D. S. King, W. H. Spangenberg, K. von Sengbusch, R. F. Stellingwerf.
Cepheid modeling, (see 012.045), p. 157 - 305 (1975).

During the Workshop proceedings, the model makers discussed the various parameters used in their models.

122.168 Appendix: standard models.   T. A. Bednarek, D. Fischel, W. M. Sparks, A. H. Karp, A. N. Cox, C. G. Davis, D. S. King, R. F. Stellingwerf.
Cepheid modeling, (see 012.045), p. 307 - 332 (1975).

122.169 HD 196517: a new δ Scuti star.   D. W. Kurtz.

Mon. Notes Astron. Soc. South. Africa, Vol. 36, 131 - 132 (1977).

122.170 **PCS: an Euler-Lagrange method for treating convection in pulsating stars using finite difference techniques in two spatial dimensions.**   R. G. Deupree.
Los Alamos Sci. Lab., N. Mex.,LA−6383.91 pp. Price $ 5.00 (1977). − Available from NTIS. − Abstr. from INIS7713 314215.

122.171 **Dynamical zoning within a Lagrangian mesh by use of DYN, a stellar pulsation code.**
J. I. Castor, C. G. Davis, D. K. Davison.
Los Alamos Sci. Lab., N. Mex., LA-6664. 33 pp. (1977). Abstr. from INIS7715 319891.

122.172 **The RR-Lyrae population of the solar neighbourhood.**   J. Lub.
Proefschrift Rijksuniv. Leiden, Netherlands (1977). − From INIS-mf−3757.
Contents: Reddening and blanketing of RR-Lyrae stars; Physical properties of RR-Lyrae stars.

An atlas of light and colour curves of field RR Lyrae stars.   See Abstr. 002.006.

A bibliographical catalogue of field RR Lyrae stars (magnetic tape).   See Abstr. 002.028.

Investigation of variable stars at the Padua−Asiago Astrophysical Observatory.   See Abstr. 009.006.

Variable star simulator.   See Abstr. 014.010.

Molecular emission from expanding envelopes around evolved stars. I. Nonmaser SiO emission lines.
See Abstr. 064.044.

Models of outflowing envelopes of T Tau stars.
See Abstr. 064.057.

Early stages of stellar evolution.
See Abstr. 065.032.

The pulsational stability of models of normal and metallic-line A stars.   See Abstr. 065.040.

On the interaction of stellar pulsations with convection.   See Abstr. 065.083.

A photoelectric magnitude sequence for AA Aurigae.   See Abstr. 113.010.

Spectrum and photometric variability of He-weak and He-strong stars.   See Abstr. 113.025.

Photometric properties of carbon stars in the galactic latitude zone centered at $l = 90°$.   See Abstr. 113.038.

Photographic photometry of carbon stars in the zone $l = 174°$; $−9° < b < +9°$.   See Abstr. 113.039.

UX Bootis = BD+47°2135 sehr wahrscheinlich nicht veränderlich.   See Abstr. 113.045.

Image-tube spectrograms of southern young emission-line objects.   See Abstr. 114.023.

The spectrum of the T Tauri star Lk Hα 120.
See Abstr. 114.501.

The spectrum of the companion of Mira Ceti.

See Abstr. 114.507.

Spectral changes in the pre-main-sequence star HD 97048.   See Abstr. 114.519.

Energy distribution in the spectra of the three variable late-type stars XY Lyr, RS Cnc, OP Her.
See Abstr. 114.561.

Stochastic models of rotating stars. Model of μ Cep.   See Abstr. 116.005.

Mass transfer instabilities in binary systems.
See Abstr. 117.015.

The spectrum of HD 187399.
See Abstr. 117.027.

σ 75 Gem: a bright variable similar to HK Lac.
See Abstr. 119.015.

The spotted dM4e eclipsing binary flare star CM Draconis.   See Abstr. 121.007.

Star spots on AR Lac type stars.
See Abstr. 121.026.

Absolute dimensions and masses of the remarkable spotted dM4e eclipsing binary flare star CM Draconis.
See Abstr. 121.029.

La page de l'observateur.   See Abstr. 124.003.

Wavelength dependence of polarization. XXXII. Narrow-band polarization effects in cool stars.
See Abstr. 131.153.

Mass loss, long-period variables, and the formation of circumnebular shells.   See Abstr. 135.005.

The detection of radio emission from the new optical emission variable HM Sagittae.
See Abstr. 135.015.

On the nature of radio sources near flare stars.
See Abstr. 141.049.

OSO-8 X-ray observations of AM Herculis.
See Abstr. 142.018.

X-ray bursters and dwarf novae: a correspondence.
See Abstr. 142.038.

Velocity curves for broad and sharp components observed in the emission lines from AM Herculis.
See Abstr. 142.093.

On the reality of the open cluster Ly 6 and the membership of the cepheid TW Nor.
See Abstr. 153.007.

Variable-star photometry of NGC 6530.
See Abstr. 153.015.

On the nature of the variables in M13.
See Abstr. 154.007.

Detection and study of variable stars in the globular cluster NGC 6638.   See Abstr. 154.031.

Mira variables in the globular cluster NGC 6356.
See Abstr. 154.046.

Errata

122.901   **Errata: 'Wesselink radii for classical cepheids'**
[Astrophys. J., Vol. 209, 135 - 140 (1976)].
N. R. Evans.
Astrophys. J., Vol. 217, 1016 (1977).

122.902   **Errata: 'Light and radial velocity observations of** **classical cepheids'** [Astrophys. J., Suppl. Ser., Vol. 32, 399 - 407 (1976)].   N. R. Evans.
Astrophys. J., Vol. 217, 1016 (1977).

122.903   **Erratum: "Light and radial velocity observations of classical cepheids"** [Astrophys. J., Suppl. Ser., Vol. 32, 399 - 407 (1976)].   N. R. Evans.
Astrophys. J., Suppl. Ser., Vol. 35, 395 (1977).

123 Variable Stars: Lists of Observations

123.001 Neue Fernrohrkarten für einige RR Lyrae-Sterne des BAV-Programms. W. Mühle.
BAV Rundbr., 26. Jahrg., 42 - 45 (1977).

123.002 SAO 115794 variabel? K.-P. Timm.
BAV Rundbr., 26. Jahrg., 50 - 51 (1977).

123.003 Maxima et minima de variables du type Mira observées par l'AFOEV.
AFOEV Bull., Tome 11, 47 (1977).

123.004 Tableaux des observations faites par l'AFOEV de mai à août 1977.
AFOEV Bull., Tome 11, 48 - 69 (1977).

123.005 3U 1908+00.
IAU Circ., No. 3088 (1977).

123.006 VV Puppis.
IAU Circ., No. 3088 (1977).

123.007 HM Sagittae.
IAU Circ., Nos. 3088, 3094, 3114 (1977).

123.008 FG Sagittae.
IAU Circ., No. 3092 (1977).

123.009 RY Sagittarii.
IAU Circ., Nos. 3098, 3107, 3115, 3124 (1977).

123.010 CH Cygni.
IAU Circ., Nos. 3101, 3102, 3105, 3113 (1977).

123.011 U Geminorum.
IAU Circ., No. 3125 (1977).

123.012 V Sagittae.
IAU Circ., No. 3134 (1977).

123.013 New variable stars in the field of γ Cygni. P. Maffei.
Inf. Bull. Variable Stars, No. 1302, 3 pp. (1977).

123.014 A suspected variable emission-line object in the direction of the Large Magellanic Cloud.
N. Sanduleak.
Inf. Bull. Variable Stars, No. 1304 (1977).

123.015 Photoelectric moments of maxima of SZ Lyn. V. G. Karetnikov, Yu. A. Medvedev.
Inf. Bull. Variable Stars, No. 1309 (1977).

123.016 BV observations of the dwarf cepheid EH Lib. V. G. Karetnikov, Yu. A. Medvedev.
Inf. Bull. Variable Stars, No. 1310 (1977).

123.017 Elements for BV 840 Cen, BV 1172 Cen, BV 1444 Ara, V 449 Cen and V 603 Cen.
U. Hopp, M. Kiehl.
Inf. Bull. Variable Stars, No. 1315 (1977).

123.018 Eclipse timings of cataclysmic variables. G. S. Mumford.
Inf. Bull. Variable Stars, No. 1321 (1977).

123.019 Spectroscopic observations of V389 Cygni. C. T. Bolton.
Inf. Bull. Variable Stars, No. 1322 (1977).

123.020 A new variable Be star: HD 218393.
P. Harmanec, J. Horn, P. Koubský, S. Kříž, Z. Ivanović, K. Pavlovski.
Inf. Bull. Variable Stars, No. 1324, 3 pp. (1977).

123.021 Photographic observations of Delta Cephei variables. M. Kiehl, U. Hopp.
Inf. Bull. Variable Stars, No. 1329 (1977).

123.022 Note on V 644 Cen = HD 306989 (B3). U. Hopp, M. Kiehl.
Inf. Bull. Variable Stars, No. 1330 (1977).

123.023 FO Virginis. E. Poretti.
Inf. Bull. Variable Stars, No. 1336 (1977).

123.024 New photoelectric observations of the Delta Scuti star HD 73576 (KW 207).
M. Bossi, G. Guerrero, L. Mantegazza.
Inf. Bull. Variable Stars, No. 1337 (1977).

123.025 o Andromedae: radial velocities and probable long period variations.
M. Fracassini, L. E. Pasinetti.
Inf. Bull. Variable Stars, No. 1341 (1977).

123.026 Notes on six long-period variable stars. U. Hopp, M. Kiehl.
Inf. Bull. Variable Stars, No. 1343 (1977).

123.027 A new flare star in Cancer. M. Lovas.
Inf. Bull. Variable Stars, No. 1345 (1977).

123.028 Gl 851.1 – another BY Dra-type star? K. Krisciunas.
Inf. Bull. Variable Stars, No. 1346 (1977).

123.029 Charts and updated results for twelve Sagittarius variables. D. Hoffleit.
Inf. Bull. Variable Stars, No. 1349, 3 pp. (1977).

123.030 Photoelectric observations of the flare star BD +13°2618 in 1973, 1974. G. Kareklidis, F. Mahmoud, L. N. Mavridis, D. Stavridis, H. Zervaki-Zoerou.
Inf. Bull. Variable Stars, No. 1354 (1977).

123.031 Photoelectric observations of the flare star BD +16°2708 in 1973, 1974. G. Kareklidis, F. Mahmoud, L. N. Mavridis, D. Stavridis, H. Zervaki-Zoerou.
Inf. Bull. Variable Stars, No. 1355, 5 pp. (1977).

123.032 Photoelectric observations of the flare star BD +55°1823 in 1974. G. Kareklidis, L. N. Mavridis, D. C. Stavridis.
Inf. Bull. Variable Stars, No. 1356, 4 pp. (1977).

123.033 B and V photometry of X Persei (= 2U 0352 +30?). P. Kalv.
Inf. Bull. Variable Stars, No. 1359, 3 pp. (1977).

123.034 An RR Lyrae star in NGC 288.
L. M. Hollingsworth, M. H. Liller.
Inf. Bull. Variable Stars, No. 1360, 3 pp. (1977).

123.035 Six new RR Lyrae stars in NGC 1261.
A. Wehlau, T. Flemming, S. Demers, C. Bartolini.
Inf. Bull. Variable Stars, No. 1361, 4 pp. (1977).

123.036 Étoiles proches suspectées de variabilité.
M. Petit.
Inf. Bull. Variable Stars, No. 1362, 3 pp. (1977).

123.037 Omicron Andromedae: another recurrent
phenomenon?   M. Fracassini, L. E. Pasinetti.
Inf. Bull. Variable Stars, No. 1363 (1977).

123.038 HR 239 and HR 8676: two δ Sct-type variables.
W. W. Weiss.
Inf. Bull. Variable Stars, No. 1364, 4 pp. (1977).

123.039 Observations of the star Gamma Bootis and of the
new variable star HR 5441.   M. Auvergne,
J. M. Le Contel, J. P. Sareyan, J. C. Valtier, J. Daguillon.
Inf. Bull. Variable Stars, No. 1365, 3 pp. (1977).

123.040 On five RR Lyrae-type stars.
V. Tsessevich (*Tsesevich*).
Inf. Bull. Variable Stars, No. 1370, 4 pp. (1977).

123.041 On two semiregular variable stars RS Sagittae and
RZ Vulpeculae.   V. Tsessevich (*Tsesevich*).
Inf. Bull. Variable Stars, No. 1371, 3 pp. (1977).

123.042 On the variable star BE Herculis.
B. Pochinok.
Inf. Bull. Variable Stars, No. 1372, 3 pp. (1977).

123.043 Photometric variations of the B6 star HR 3440.
P. Renson, C. Sterken.
Inf. Bull. Variable Stars, No. 1373, 3 pp. (1977).

123.044 Variable star notes.   J. A. Mattei.
J. American Assoc. Variable Star Obs., Vol. 6, 15 -
23 (1977).
    Peculiarities, behavior, and activity of some of the more
prominent variables for 1976 are given. Part I summarizes the
behavior of the different types of stars except U Geminorum
variables. Part II lists the dates and brightness of the outbursts
of some prominent U Gem stars in the observing program, and
Part III is a list of variables in the notes in order of constellation.

123.045 Variable star notes.   M. E. Baldwin, J. A. Mattei.
J. R. Astron. Soc. Canada, Vol. 71, 341 - 344, 475 -
480 (1977).

123.046 Four-colour photometry of AB Aur, T Tau and RY
Tau according to the cooperative program 1973.
I. V. Shpychka, V. V. Golovatyj, M. B. Girnyak.
Tsirk. Astron. Obs., L'vov, No. 51, p. 24 - 27 (1976). In Rus-
sian.

123.047 Associazione Veneta Osservatori di Stelle Variabili.
L. Rosino, G. Romano, R. Sannevigo, G. Favero.
Assoc. Veneta Oss. Stelle Variabili, Bull. N. 1, 8 pp. (1977).
    This is the first number of a series of non periodical
bulletins regarding the observations of variable stars made by
the Associazione Veneta Osservatori di Stelle Variabili. The
list of variables contains, at present, 14 objects that have some
interesting characteristics (eclipsing stars with variable periods,
suspected flare stars, nova like etc.).

123.048 HM Sagittae (emission variable).
Yamamoto Circ., Nos. 1854, 1858 (1977). In
Japanese.

123.049 FG Sagittae.
Yamamoto Circ., No. 1858 (1977). In Japanese.

123.050 CH Cyg.

Yamamoto Circ., Nos. 1860, 1862 (1977). In
Japanese.

123.051 U Geminorum.
Yamamoto Circ., No. 1867 (1977). In Japanese.

123.052 Information on photoelectric observations of
variable stars deposited at the Odessa Astronomical
Observatory.   V. P. Tsesevich. E. N. Makarenko.
Astron. Tsirk., No. 953, p. 7 - 8 (1977). In Russian.

123.053 Observations of two new variable stars SVS 2191 and
SVS 2192.   S. M. Bychkov.
Astron. Tsirk., No. 959, p. 7 - 8 (1977). In Russian.

123.054 On the two stars V 464 Aql and UU Vul.
V. P. Tsesevich.
Astron. Tsirk., No. 960, p. 8 (1977). In Russian.

123.055 Sequences for southern variables.   B. Menzies.
Publ. Variable Star Sect., R. Astron. Soc. New
Zealand, No. 5 (C77), p. 6 - 9 (1977).
    V and B–V magnitudes are published for sequences of
comparison stars for T Col; RW Pup; RR Pic; RU, SU, DI Car;
RT, RX and AQ Cen; AT Ara. Charts have already been
published.

123.056 Sequence for WX Hydri.
B. F. Marino, W. S. G. Walker.
Publ. Variable Star Sect., R. Astron. Soc. New Zealand, No. 5
(C77), p. 36 (1977).

123.057 Sequence for SY Sculptoris.   B. Menzies.
Publ. Variable Star Sect., R. Astron. Soc. New
Zealand, No. 5 (C77), p. 41 (1977).

123.058 Photoelectric UBV sequences for seventeen
southern dwarf novae.   N. Vogt.
Publ. Variable Star Sect., R. Astron. Soc. New Zealand, No. 5
(C77), p. 42 - 48 (1977).
    Photoelectric UBV magnitudes for a total of 121 stars in
17 fields of southern dwarf novae are given. The limiting magnitudes range from $13^m\!.0$ to $16^m\!.5$ in V. There are one certain
and two suspected variables among the sequence stars.

123.059 Variable stars in the globular cluster NGC 6535.
M. H. Liller, C. Coutts Clement.
Astron. J., Vol. 82, 965 - 967, 1061 (1977).
    A photometric study of one previously known and one
new variable in the region of the globular cluster NGC 6535 is
presented. One variable is type RR_a and the other type RR_c.
They are believed to be members of the cluster.

123.060 Veränderliche Sterne am Südhimmel. Teil VI.
H. Gessner.
Veröff. Sternw. Sonneberg, Band 8, (Heft 7), 341 - 412
(1977).
    This publication represents the final part of the proces-
sing of Sonneberg variables discovered by C. Hoffmeister in
the area η Arae (see 17.123.037). From JD 243 6689 to
JD 243 6838 229 variables are listed: 53 eclipsing, 93 RR Lyr,
1 δ Cep, 16 Mira, 48 SR and L, 7 RV, 3 I, 3 U Gem, 2 ? , 3
constant.

123.061 5 neue veränderliche Blaue Objekte.   G. A. Richter
Mitt. Veränderl. Sterne (MVS), Band 8, 1 (1977).

123.062 Der Lichtwechsel des Veränderlichen TZ Cancri
(= BD+21° 1966).   S. Rößiger.
Mitt. Veränderl. Sterne (MVS), Band 8, 2 - 5 (1977).

123.063 Fotografische Maxima des halbregelmäßigen Ver-

änderlichen TV Andromedae.   F. Rümmler.
Mitt. Veränderl. Sterne (MVS), Band 8, 5 (1977).

123.064   Visuelle Beobachtungen von Mira- und RV-Tauri-
          Sternen.   O. Matzek.
Mitt. Veränderl. Sterne (MVS), Band 8, 7 (1977).

123.065   Fotografische Beobachtungen von RS Cassiopeiae.
          F. Rümmler.
Mitt. Veränderl. Sterne (MVS), Band 8, 8 (1977).

123.066   Variables Blaues Objekt.   P. Lochno.
          Mitt. Veränderl. Sterne (MVS), Band 8, 17 (1977).

123.067   Lichtkurve des H_2O-Halbregelmäßigen RT Virginis.
          W. Wenzel.
Mitt. Veränderl. Sterne (MVS), Band 8, 18 (1977).

123.068   Ergebnisse der visuellen Beobachtung von 22 Ver-
          änderlichen.   M. Beyer.
Veröff. Remeis-Sternw. Bamberg, Astron. Inst. Univ. Erlangen-
Nürnberg, Band 12, Nr. 123, 47 pp. (1977).

123.069   Mitteilungen über Veränderliche der Bamberger
          Liste. Neue Veränderliche Sterne am Südhimmel.
R. Knigge, F. M. Sosna.
Veröff. Remeis-Sternw. Bamberg, Astron. Inst. Univ. Erlangen-
Nürnberg, Band 12, Nr. 125,.7 pp. (1977).

On the probable variability of HR 1861.
See Abstr. 113.033.

Photometric properties of carbon stars in the galactic latitude zone centered at l = 90°.   See Abstr. 113.038.

Photographic photometry of carbon stars in the
zone l = 174°; $-9° < b < +9°$.   See Abstr. 113.039.

Duplicity and spectral types of HV 10814.
See Abstr. 118.027.

Visuelle Beobachtungen von NQ Vul, Nova Sgr
1977 und R CrB.   See Abstr. 124.905.

Errata

123.901   Errata to AFOEV Bull., Tome 9-11.
          AFOEV Bull., Tome 11, 70 (1977).

123.902   Errata: "Périodicité d'étoiles Ap Australes" [Inf.
          Bull. Variable Stars, No. 1280 (1977)].
P. Renson.
Inf. Bull. Variable Stars, No. 1300 (1977).

123.903   Erratum: "New variable star in the Pleiades"
          [Astron. Tsirk., No. 923, p. 7 - 8 (1976). In Russian].
Eh. S. Kazaryan.
Astron. Tsirk., No. 940, p. 7 (1977).

124 Novae

124.001   The evolution of a fast nova model with a Z = .03
          envelope from pre-explosion to extinction.
D. Prialnik, M. M. Shara, G. Shaviv.
Bull. American Astron. Soc., Vol. 9, 433 (1977). — Abstract.

124.002   Fifty years of novae. Henry Norris Russell Prize
          Lecture of the American Astronomical Society.
C. H. Payne-Gaposchkin.
Astron. J., Vol. 82, 665 - 673 (1977).

124.003   La page de l'observateur.   M. Duruy.
AFOEV Bull., Tome 11, 31 (1977).

124.004   What makes novae blow up?
          D. Byrd, J. Patterson.
Astronomy, Vol. 5, No. 7, p. 50 - 54 (1977). — Abstr. in
Phys. Abstr., Vol. 80, Abstr. 78799 (1977).

124.005   A search for rapid oscillations in old novae.
          E. L. Robinson, R. E. Nather.
Publ. Astron. Soc. Pacific, Vol. 89, 572 - 573 (1977).
    The authors have tested ten old novae for the presence of
rapid oscillations similar to the 71-second periodicity in the
old nova DQ Her. None of the novae displayed coherent
periodicities, indicating that the 71-second oscillation of DQ
Her must be considered exceptional rather than normal behavior. One of the novae, T Aur, is essentially identical with
DQ Her in every respect save that it lacks a rapid oscillation.
Thus, the 71-second oscillation does not appear to have, nor
have had, any significant impact on the remaining properties
of DQ Her.

124.006   A bright nova in the surroundings of the Andromeda
          Nebula.   L. Meinunger.
Inf. Bull. Variable Stars, No. 1331 (1977).

124.007   A search for light echoes from novae.
          S. van den Bergh.
Publ. Astron. Soc. Pacific, Vol. 89, 637 - 638 (1977).
    The 1.2-m Schmidt telescope on Palomar Mountain has
been used to search for cases in which recent novae might
have illuminated interstellar clouds. No such light echoes were
detected from any of the seven galactic novae observed during
the present program.

124.008   Enveloppe de poussière circumstellaire des novae.
          I. Malakpur.
Astrophys. Space Sci., Vol. 47, 49 - 59 (1977).
    The author presents the results of a study of the circum-
stellar dust envelopes of Nova Delphini 1967, Nova Herculis
1960, Nova Serpentis 1970, and Nova Persei 1901.

124.009   The inception of novae transition zone oscillations.
          J. P. Phillips, M. J. Selby.
Astrophys. Space Sci., Vol. 49, 339 - 348 (1977).
    The nature of semi-regular oscillations in the transition
zones of certain novae is discussed. It is found that hydrostatic collapse following a reduction of radiative support for
the photospheric layers is a likely explanation of transition
zone inception.

P-nuclei processing in degenerate hydrogen burning
zones and its relationship to nova outbursts.

See Abstr. 065.080.

The continuous radiation emitted by accretion discs in cataclysmic binaries: the dwarf nova SS Cyg during outburst and the old novae V 603 Aql and RR Pic.
See Abstr. 122.033.

Infrared photometry of dwarf novae and possibly related objects.   See Abstr. 122.037.

Visuelle Beobachtungen an RS Oph 1974 - 1976.
See Abstr. 122.048.

Nova Cygni 1975 = V1500 Cygni

124.101 The interstellar reddening and distance of Nova Cygni 1975 (V1500 Cygni).   G. J. Ferland.
Astrophys. J., Vol. 215, 873 - 876 (1977).

McDonald Observatory spectrophotometric scans, combined with published photometry, are used to study the interstellar reddening of Nova Cygni 1975 (=V1500 Cygni). The early blackbody energy distribution, the later free-free and free-bound energy distribution, the hydrogen Paschen to Balmer emission-line ratios, and the He triplet spectrum imply a color excess, $E(B-V)$, of 0.50 ± 0.05 mag. This value is consistent with the interstellar line strengths and polarization measurements of other observers. The author combines his results with those of Schild to derive an independent measure of the distance, 1.95 ± 0.2 kpc.

124.102 Nova Cygni 1975: wind and variations.
A. C. Fabian, J. E. Pringle.
Mon. Not. R. Astron. Soc., Vol. 180, 749 - 754 (1977).

A simple model is proposed to explain the ~3-hr periodic variations seen in the radiation from Nova Cygni 1975. The underlying binary system creates a spiral density variation in the outflowing wind. Such a spiral may persist to large radii and is then observable where it intersects the scatter-sphere. Period changes in the observed magnitude variations can naturally occur, without any change in the underlying binary period.

124.103 The spectral development of Nova Cygni 1975.
P. A. Strittmatter, N. J. Woolf, R. I. Thompson, S. Wilkerson, J. R. P. Angel, H. S. Stockman, G. Gilbert, S. A. Grandi, H. Larson, U. Fink.
Astrophys. J., Vol. 216, 23 - 32 (1977).

Optical and near-infrared (0.3–2.5 μm) observations of Nova Cygni 1975 made at Steward Observatory during the period 1975 August 30 to 1975 December 11 are reported. The persistent strength of O I λλ8446, 11287 is shown to be due to Lβ fluorescence in clouds with high (>10^3) Hα optical depth. A simple model of the nova ejecta is presented and shown to be consistent with the observed evolution of the nova spectrum.

124.104 The spectrum of Nova Cygni 1975 around maximum light.   H. W. Duerbeck, B. Wolf.
Astron. Astrophys., Suppl. Ser., Vol. 29, 297 - 304 (1977).

42 coudé spectra (12 Å/mm) of Nova Cyg 1975 were taken around maximum light (J.D. 2442654.7, 655.6, 656.6). Radial velocities were determined for the identified lines. On the basis of interstellar lines, the distance of the nova is estimated to be 1.4 ± 0.2 kpc, and the absolute magnitude at maximum is about $M_v = -9\overset{m}{.}8$, leading to a maximum photospheric radius of about 570 $R_\odot$.

124.105 Spectroscopic studies of nova V1500 Cygni. I. The 3 hour periodicity and the nebula.
J. B. Hutchings, M. L. McCall.
Astrophys. J., Vol. 217, 775 - 780 (1977) = Contrib. Dominion

Astrophys. Obs., No. 328 = NRC No. 15610.

The 3 hour periodic variation in the nova V1500 Cyg is investigated in spectrograms obtained during the initial decline and up to 10 months after outburst. A searchlight-type model is proposed to explain the data, in contrast to the model of Campbell. A model for the nebula is proposed from line-profile considerations and discussed in terms of short-period variations. The Fe II and Hβ lines show different behavior.

124.106 Nova Cygni Rapport.   D. Kitta.
Sterne, 53. Band, 169 - 176 (1977).

124.107 V1500 Cygni.
IAU Circ., No. 3099 (1977).

124.108 The helium abundance of ejecta from V1500 Cygni (Nova Cygni 1975).   G. J. Ferland.
Proc. Southwest Reg. Conf., Vol. 3, (see 012.043), p. 139 - 140 (1977). – Abstract.

124.109 V1500 Cygni.
Yamamoto Circ., Nos. 1853, 1861 (1977). In Japanese.

A search for gamma-ray lines from Nova Cygni 1975, Nova Serpentis 1970, and the Crab Nebula.
See Abstr. 134.008.

Nova Sagittarii 1974 = V3888 Sagittarii

124.151 On the decline stage of Nova V3888 Sgr (1974).
N. Vogt, H. M. Maitzen.
Astron. Astrophys., Vol. 61, 601 - 603 (1977).

Photoelectric UBV and $uvby$ photometry of V3888 Sgr = Nova Sgr 1974 was obtained covering 25 days of its decline stage. The light curve and the colour variations are described and compared with the spectrum. V3888 Sgr seems to be a fast nova with a slightly abnormal light curve showing a rapid early decline but an extended transition stage. Its maximum brightness is estimated to be $V \approx 7\overset{m}{.}5$, its amplitude $\Delta m_v \gtrsim 13.5$ and its distance 3.5 kpc.

Nova Monocerotis 1975 = V616 Monocerotis = A0620–00

124.201 Spectroscopic observations of the X-ray nova A0620–00.   J. A. J. Whelan, M. J. Ward, D. A. Allen, I. J. Danziger, R. A. E. Fosbury, P. G. Murdin, M. V. Penston, B. A. Peterson, E. J. Wampler, B. L. Webster.
Mon. Not. R. Astron. Soc., Vol. 180, 657 - 673 (1977).

Spectroscopic observations, with the Wampler–Robinson Scanner at the Anglo-Australian Telescope, of the X-ray nova A0620–00 are reported. The data cover the interval 1975 September 8 to 1976 May 5 inclusive. The features seen in the spectrum include emission lines of Hα, Hβ, Hγ, He I, He II, N III, broad absorption underlying the Balmer emission and interstellar absorption lines. The spectrum has some similarity in its later stage to that of a dwarf nova but at no time so far has it resembled the spectrum of a common nova. Models of the outburst event in terms of a binary system, involving accretion on to a neutron star or runaway nuclear burning on a white dwarf, are briefly discussed with regard to periodicities, X-ray heating and evolutionary origin.

Photographic observations of the X-ray nova Monocerotis 1975.   See Abstr. 142.082.

Nova Sagittae 1977

124.251 Nova Sagittae 1977.
IAU Circ., Nos. 3096, 3131 (1977).

124.252 Nova Sagittae 1977.
Yamamoto Circ., Nos. 1853, 1857, 1859 (1977).
In Japanese.

124.253 Nova Sagittae 1977.
A. Sh. Khatisashvili, G. N. Kimeridze.
Astron. Tsirk., No. 954, p. 7 - 8 (1977). In Russian.

Nova LMC 1977b

124.301 Photometry of LMC Nova 1977b.
R. Canterna, R. D. Schwartz.
Astrophys. J., Lett., Vol. 216, L91 - L94 (1977).
    Photometry of Nova LMC 1977b has been obtained
from premaximum to the pretransition stage. A premaximum
rise rate of 0.035 mag per hour in V was observed. The time
and apparent visual magnitude of maximum is 2,443,216.34 ±
0.24 JD and 10.67 ± 0.04. The light curve is consistent with
a moderately fast nova. A distance modulus $(m-M)_0$ = 18.6 ±
0.2 for the LMC has been determined from the light curve
and the (absolute magnitude, rate of decline)-relation.

Nova Sagittarii 1977

124.351 The early spectrum of nova Sagittarii 1977.
T. P. Prabhu.
Bull. Astron. Soc. India, Vol. 5, 79 - 81 (1977).

124.352 Nova Sagittarii 1977.
IAU Circ., Nos. 3096, 3127 (1977).

124.353 Photoelectric observations of Nova Sagittarii 1977.
R. Sagar, H. S. Mahra, S. C. Joshi, J. B. Srivastava.
Inf. Bull. Variable Stars, No. 1307, 3 pp. (1977).

124.354 Nova Sagittarii 1977.
Yamamoto Circ., Nos. 1853, 1857, 1859 (1977).
In Japanese.

**Visuelle Beobachtungen von NQ Vul, Nova Sgr
1977 und R CrB.**   See Abstr. 124.905.

Nova V1017 Sagittarii

124.361 V1017 Sagittarii.
IAU Circ., No. 3108 (1977).

Nova Delphini 1967 = HR Delphini

124.401 The mass of the nova HR Delphini nebula.
C. M. Anderson, J. S. Gallagher.
Publ. Astron. Soc. Pacific, Vol. 89, 264 - 266 (1977).
    The Wisconsin low-resolution spectrophotometer was
used to measure the absolute Hβ flux from HR Del in August
1975. Using nebular temperatures from earlier measurements,
the mass of the nebula is about $10^{-4} M_\odot$. This is in reasonable
agreement with results from other methods of estimating the
mass of ejecta from HR Del, and is consistent with nova
models in which radiation pressure drives mass loss.

**124.402 Visuelle Beobachtungen der Nova HR Del 1967 von
1974-1976.   U. Hopp, U. Surawski.**
Sterne Weltraum, Jahrg. 16, p. 298 (1977).

124.403 Spectrophotometry of nova Delphini 1967.
H. Drechsel, J. Rahe, H. W. Duerbeck, L. Kohoutek,
W. C. Seitter.
Astron. Astrophys., Suppl. Ser., Vol. 30, 323 - 334 (1977).
    35 objective prism spectra of Nova Delphini 1967 have
been analyzed. They were obtained between 17 July 1967 and
21 August 1974. The spectra have a dispersion of 570 Å/mm
and 240 Å/mm at H$_\gamma$, respectively. The wavelength region
from λ 3400 Å to λ 6500 Å was used for reduction. From the
spectral intensity distribution in the continuum, the variations
of color temperature (at λ 4250Å), photospheric radius,
(negative) Balmer discontinuity and Balmer decrement with
time were established. The development of equivalent widths
and intensity ratios of a few lines were determined. Distance
and interstellar extinction of the nova were derived.

**124.404 Étoile centrale et condition d'ionisation de l'en-
veloppe de la nova Delphini 1967.**
I. Malakpur.
Astrophys. Space Sci., Vol. 47, 3 - 15 (1977).
    After correcting the observed flux of the forbidden lines
for the supplementary reddening (due to the circumstellar
envelope), the author has recalculated the electron density
and temperature of the envelope of the nova. He has determined the temperature and radius of the nova for 1968, 1969,
and 1970. He has determined the optical depth of the envelope in the Lyman continuum. Considering the stratification
of the envelope in different regions of ionization, he has determined the radius of the inner and outer edge and the electron
temperature of every region.

**124.405 Les raies du triplet de l'hélium de la nova Delphini
1967.   I. Malakpur.**
Astrophys. Space Sci., Vol. 47, 17 - 24 (1977).
    The aim of this paper has been to study the neutral
helium triplet emission lines identified in the spectrum of the
envelope of nova Delphini. By comparing the observed flux of
the neutral helium lines with that calculated theoretically by
Robbins, the author finds that the optical thickness in the
center of the line λ 3889 is of the order of 21.50 for summer
1969. On the other hand, he obtains the number of neutral
helium atoms in the 2 ^{3}S state $[N(2\ ^3S)]$ by considering the
equilibrium between the mechanisms that populate and de-
populate this state. He calculates the optical thickness $\tau(\lambda 3889)$
of the order of 82.37, from $N(2\ ^3S)$. The difference between
the two values is very large and it cannot be attributed to calculation errors. He concludes that this difference is due to the
heterogeneity of the envelope of the nova.

Nova Sagittarii 1968

**124.451 The belated discovery of nova Sgr 1968 and nova
Oph 1969.   D. J. MacConnell.**
Inf. Bull. Variable Stars, No. 1340 (1977).

Nova Pictoris 1925 = RR Pictoris

124.461 RR Pictoris (nova 1925).   F. M. Bateson.
Publ. Variable Star Sect., R. Astron. Soc. New
Zealand, No. 5 (C77), p. 20 - 21 (1977).
    Ten day means from 2,440,950 to 2,443,090 are given.
The means show that there has been a very slow irregular
decline during the past 2,000 days.

Nova Herculis 1934 = DQ Herculis

Phase dependence of emission lines in the disk of DQ Herculis.   See Abstr. 121.008.

Nova Ophiuchi 1969

The belated discovery of nova Sgr 1968 and nova Oph 1969.   See Abstr. 124.451.

Nova Serpentis 1970 = FH Serpentis

124.601   Thick inhomogeneous shell models for the radio emission from Nova Serpentis 1970.
E. R. Seaquist, J. Palimaka.
Astrophys. J., Vol. 217, 781 - 787 (1977).
    The authors have examined thick expanding envelope models to account for the radio-emitting properties of Nova Serpentis 1970 = FH Ser. All models are characterized by four parameters related to the density, inner and outer radii, and expansion velocity. The authors conclude that all of the radio data are best represented by a shell that is expanding at a constant rate, with the outward speed proportional to the radial coordinate in the shell ("Hubble flow model"). The radio data do not permit discrimination between spherically symmetric shells and the polar cap shell discussed by Hutchings. The authors use the best-fit model to estimate the distance, mass, and kinetic energy of the expanding nova envelope.

A search for gamma-ray lines from Nova Cygni 1975, Nova Serpentis 1970, and the Crab Nebula. See Abstr. 134.008.

Nova Persei 1974 = V400 Persei

Photoelectric observations of the novae T Pyxidis, V400 Persei, and NQ Vulpeculae.   See Abstr. 124.951.

Nova Ophiuchi 1977

124.701   Nova Ophiuchi 1977.
Sky Telesc., Vol. 54, 382 (1977).

124.702   Nova Ophiuchi 1977.
IAU Circ., Nos. 3110, 3115 (1977).

Nova Sagittarii 1975 No. 2

124.751   Nova Sagittarii 1975 No. 2.
I. Lundström, B. Stenholm.
Inf. Bull. Variable Stars, No. 1351, 3 pp. (1977).

Nova Cephei 1971

124.801   The chemical composition of the Nova Cephei 1971 ejecta.   J. A. de Freitas Pacheco.
Mon. Not. R. Astron. Soc., Vol. 181, 421 - 426 (1977).
    The chemical composition of the Nova Cephei 1971 ejecta was obtained from analysis of the emission lines. He, O and N are overabundant with respect to the cosmic values, consistent with the CNO thermonuclear runaway. The ejected shell has a mass of about $10^{-4} M_{\odot}$.

Nova Scuti 1975 = V373 Scuti

124.851   On nova Scuti 1975= V373 Sct.
A. Sh. Khatisashvili, O. M. Kurtanidze.
Astron. Tsirk., No. 959, p. 1 - 2 (1977). In Russian.

Nova Vulpeculae 1976 = NQ Vulpeculae

124.901   Light, color, and velocity curves observed for Nova Vulpeculae 1976 (NQ Vul).   Y. Yamashita, K. Ichimura, M. Nakagiri, Y. Norimoto, H. Maehara, K. Miyajima.
Publ. Astron. Soc. Japan, Vol. 29, 527 - 535 (1977).
    Photoelectric and spectroscopic observations were carried out for Nova Vulpeculae 1976 (NQ Vul). V magnitudes and $B-V$ and $U-B$ colors observed in the period from October 22 to December 26, 1976 and radial velocities observed for various components in the period from October 22 to December 11, 1976 are presented. Some features characteristic to this nova are discussed and are interpreted by Grotrian's (1937) model. An absolute visual magnitude of -7.0 mag and a distance of 1.5 kpc have been obtained.

124.902   Photoelectric spectroscopy of nova Vulpeculae 1976.   S. Jeffers, W. G. Weller.
J. R. Astron. Soc. Canada, Vol. 71, 402 - 403 (1977).
Abstract.

124.903   NQ Vulpeculae.
IAU Circ., Nos. 3096, 3135 (1977).

124.904   NQ Vulpeculae.
Yamamoto Circ., Nos. 1853, 1857, 1859 (1977).
In Japanese.

124.905   Visuelle Beobachtungen von NQ Vul, Nova Sgr 1977 und R CrB.
Mitt. Veränderl. Sterne (MVS), Band 8, 9 (1977).

Photoelectric observations of the novae T Pyxidis, V400 Persei, and NQ Vulpeculae.   See Abstr. 124.951.

Nova T Pyxidis

124.951   Photoelectric observations of the novae T Pyxidis, V400 Persei, and NQ Vulpeculae.   A. U. Landolt.
Publ. Astron. Soc. Pacific, Vol. 89, 574 - 575 (1977) = Contrib. Louisiana State Univ. Obs., Baton Rouge, No. 125.
    Multicolor photoelectric data are presented for the three novae T Pyx, V400 Per, and NQ Vul.

125 Supernovae, Supernova Remnants

125.001 On the problem of hydrogen abundance in type I supernovae and their remnants.
V. S. Bychkova, K. V. Bychkov.
Astron. Zh. Akad. Nauk SSSR, Vol. 54, 772 - 780 (1977).
In Russian. English translation in Soviet Astron., Vol. 21, No. 4.

By solving the equilibrium equations for the hydrogen atom, the conclusion on the low hydrogen abundance in type I supernovae (SN I) is confirmed. The departure of atoms from the second and the third levels is mainly due to the process of photoionization by the continuous radiation from the supernova itself. On the other hand, the filaments in the Kepler and Crab nebulae, in which hydrogen abundance is close to normal, must be formed of gas of the atmospheres of the supernovae themselves. The two facts can be reconciled in the framework of the known model of stellar wind flow before the supernova explosion, supposed the layers rich in hydrogen were torn off the star by the wind.

125.002 Supernovae and molecular clouds.
J. C. Wheeler, F. N. Bash.
Nature, Vol. 268, 706 - 707 (1977).

The authors suggest that there is a critical mass M_{crit} such that for $M > M_{crit}$ stars do not evolve to an explosive phase and hence the surrounding molecular cloud is substantially unaffected by the evolving stars. As the system ages, the most massive star with $M > M_{crit}$ finally evolves to an explosive endpoint and the resulting supernova dissipates the dense molecular cloud. From the cluster observations the authors identify $M_{crit} \sim 15-20\,M_\odot$.

125.003 The supernova of AD 1006.
F. R. Stephenson, D. H. Clark, D. F. Crawford.
Mon. Not. R. Astron. Soc., Vol. 180, 567 - 584 (1977).

The supernova of AD 1006 was the most brilliant witnessed in historical times. The authors present a reassessment of the historical records of the event, confirming the nature of the outburst, and giving greatly improved estimates of its position, peak apparent brightness, and duration of visibility. From this study they make the assertions that the only possible remnant of the event is the radio source PKS 1459–41, that the supernova was at a distance of close to 1 kpc and reached apparent magnitude about −9.5 at maximum, and was visible over a period of several years. A new high-resolution 408-MHz map of the radio-remnant of the supernova is included.

125.004 Hunting for supernovae.   J. T. Bryan, Jr.
J. British Astron. Assoc., Vol. 87, 457 - 463 (1977).

125.005 CO emission from supernova remnants.
N. Z. Scoville, W. M. Irvine, P. G. Wannier, C. R. Predmore.
Astrophys. J., Vol. 216, 320 - 328 (1977) = Contrib. Five Coll. Obs., Amherst, Mass., No. 240.

In a search for molecular gas associated with supernova remnants, the authors have observed the CO line in the three optical remnants: the Cygnus Loop, IC 443, and Cas A. In both the Cygnus Loop and IC 443, CO molecules are detected from dense clouds of mass $\sim 100\,M_\odot$, probably located at the edge of the expanding remnants. These are locations of particularly bright optical filaments and strong radio continuum emission. The foreground emission seen in Cas A exhibits changes on a small scale compared with the radio continuum, implying a low filling factor for molecular absorption lines in Cas A.

125.006 Pulsar theory of supernova light curves. I. Dynam-
ical effect and thermalization of the pulsar strong waves.   B. Gaffet.
Astrophys. J., Vol. 216, 565 - 577 (1977).

In this paper the author investigates the dynamical effect of a pulsar on a surrounding supernova envelope and the process of thermalization of the pulsar strong waves, with the view of applying the results to the calculation of the light curves of supernovae. The shock wave induced by the pulsar radiation pressure is studied by means of self-similarity methods, and the results are presented in closed form for the simple case of a homogeneous density distribution in the ejected envelope.

125.007 Pulsar theory of supernova light curves. II. The light curve and the continuous spectrum.
B. Gaffet.
Astrophys. J., Vol. 216, 852 - 864 (1977).

The author presents calculations of supernova (SN) light curves that assume the existence of a central pulsar. He has shown in Paper I that absorption of the pulsar waves can occur only after some cooling has taken place, and that the pulsar acts as a thermostat, keeping central temperature fixed around 1.3×10^4 K. The main absorption process is particle acceleration by the strong waves. Several difficulties inherent in most existing models are solved in this way, especially the rapid rise of luminosity before maximum, and the observed high values of expansion velocities. A good fit of the typical SN Type II light curves has been obtained. Furthermore, the continuous spectrum and the observed temperatures of about 1.0×10^4 K near maximum can be explained easily in this model.

125.008 The true extent of the γ Cygni supernova remnant.
L. A. Higgs, T. L. Landecker, R. S. Roger.
Astron. J., Vol. 82, 718 - 724, 772 (1977).

Aperture-synthesis observations at 1.4 GHz show that the γ Cygni supernova remnant has a shell structure of $\sim$62-arcmin diameter, centered at $l = 78°.2$, $b = +2°.1$. Radio emission from the shell is strongest in the southeast quadrant, near the star γ Cygni, where it is concentrated in several tangentially elongated structures. A second region of enhanced nonthermal emission is in the northwest quadrant. Based on a brightness-diameter relation, the observed flux density of 270 ± 40 Jy implies a diameter of $\sim$33 pc and a distance of $\sim$1.8 kpc.

125.009 The location of the supernova of AD 1572.
F. R. Stephenson, D. H. Clark.
Q. J. R. Astron. Soc., Vol. 18, 340 - 350 (1977).

The coordinates of the supernova of AD 1572 have been recomputed from the original measurements of Tycho Brahe, Thomas Digges, and Michael Maestlin, in an attempt to investigate the positional discord between the supernova and its remnant. The authors suggest that the discrepancy apparent from Tycho Brahe's measurements can be attributed to a small systematic error in his instrument and/or his observational technique.

125.010 Adiabatic self-similar blast waves, their radial instabilities, and their application to supernova
remnants.   P. A. Isenberg.
Astrophys. J., Vol. 217, 597 - 618 (1977).

For several decades, a self-similar formalism has commonly been invoked to describe the adiabatic fluid motion produced by the instantaneous release of large amounts of energy at a point – an "adiabatic spherical blast wave". In this paper, the adiabatic self-similar solution is described and the general perturbation equations are derived. Several linearized stability calculations are performed and a nonlinear solution to the equations is presented. The implications of the instabilities, and other peculiarities of self-similar motion, with particular reference to the theory of supernova remnants are discussed.

125.011 Liste de supernovae de type I dont la détermination du maximum de luminosité permet l'établissement d'échantillons homogènes.
S. Depaquit, G. Le Denmat, J.-P. Vigier.
C. R. Acad. Sci. Paris, Tome 285, Sér. B, 161 - 163 (1977).

L'analyse détaillée et la comparaison des données fournies par différents auteurs concernant des galaxies contenant des supernovae de type I (jusqu'en 1975) permet d'établir une liste homogène où le maximum de luminosité des supernovae peut être considéré comme établi avec une précision suffisante pour qu'elles soient utilisées comme indicateurs de distance secondaires.

125.012 Gamma rays and supernova explosions.
W. D. Arnett.
The structure and content of the Galaxy and galactic gamma rays, (see 012.009), p. 257 - 264 (1977).

The detection of γ-rays from supernovae will provide interesting tests of current theory. This discussion will review some current ideas on the expected γ-ray flux, as modified by recent theoretical results.

125.013 Spectra of Cassiopeia A. I. Observations.
R. P. Kirshner, R. A. Chevalier.
Astrophys. J., Vol. 218, 142 - 147 (1977).

Image-tube spectrograms of the supernova remnant Cas A in the wavelength range 4700–8000 Å have been reduced to give relative line intensities, for 19 fast moving knots and seven quasi-stationary flocculi. In the fast-moving knots, previously unobserved lines of [Ar III], [Ar IV], [Ar V], O I, [O I], [Ca II], and Fe II have been measured, while more stringent upper limits have been set on Hα, Hβ, lines of helium, and lines of nitrogen. In the quasi-stationary flocculi, new lines of [O III], [Fe II], [N I], [N II], He I, [S II], [Ar III], [Ca II], and [O II] have been measured. The results are consistent with earlier investigations, but are more detailed.

125.014 Models of type I supernovae.
D. K. Nadezhin, V. P. Utrobin.
Astron. Zh. Akad. Nauk SSSR, Vol. 54, 996 - 1008 (1977). In Russian. English translation in Soviet Astron., Vol. 21, No. 5.

The main characteristics of SN near maximum are shown to depend weakly on chemical composition and to be practically independent of the spatial distribution of the injected energy. However, they are sensitive to the total energy output and the time scale of the energy generation law, and depend strongly on the mass of the ejected envelope. A working model of an SN I outburst is suggested. It is characterized by the following phases: a) emergence of a shock wave giving rise to a comparatively low-mass atmosphere ($\sim 0.01\ M_\odot$); b) slow energy release resulting in an outburst of a massive ($\sim 0.5\ M_\odot$) envelope in the form of a thin spherical shell; c) stage of an exponential light curve tail due to the activity of the central remnant. The total outburst energy and the mass of the ejected envelope for a typical SN I are $(2-5)\cdot 10^{50}$ ergs and $(0.3-0.6)\ M_\odot$, respectively.

125.015 Structure of Kepler's SNR and the Crab Nebula at 327 MHz from occultation observations.
T. Velusamy, N. V. G. Sarma.
Mon. Not. R. Astron. Soc., Vol. 181, 455 - 464 (1977).

Strip brightness distributions across Kepler's SNR and the Crab Nebula have been obtained at 327 MHz with resolutions of 8 and 26 arcsec respectively from recent occultation observations.

125.016 Distance estimate for the galactic supernova remnant G74.9+1.2 using H I absorption-line measurements.
F. Sato.
Publ. Astron. Soc. Japan, Vol. 29, 615 - 618 (1977).

A kinematic distance of 11.5 kpc is suggested for the galactic supernova remnant G 74.9+1.2 (CTB 87) from the H I 21-cm line absorption features in equal-velocity contour diagrams derived from the Maryland–Green Bank Survey (Westerhout 1973).

125.017 The interaction of supernovae with the interstellar medium.   R. A. Chevalier.
Annu. Rev. Astron. Astrophys., Vol. 15, (see 003.012), 175 - 196 (1977).

This review concentrates on the theoretical understanding of the interaction of the energy release in the supernova explosions with the interstellar medium.

125.018 A new optical supernova remnant in Crux.
A. J. Longmore, D. H. Clark, P. Murdin.
Mon. Not. R. Astron. Soc., Vol. 181, 541 - 546 (1977).

The authors have found optical nebulosity associated with the radio supernova remnant G296.1–0.7. It has a collisionally excited spectrum which indicates electron densities as high as 2×10^3 cm^{-3}. Its distance is estimated at between 3 and 5 kpc from two independent considerations. The full width at half maximum of the internal velocity dispersion in the individual filaments is 42 km/s. This is the only new optical SNR discovered in a search at 70 radio SNR positions south of $-18°$ on SRC Schmidt Survey plates.

125.019 Photometric observations of the supernovae in NGC 7723, NGC 1325 and NGC 4402.   G. Wegner.
Mon. Not. R. Astron. Soc., Vol. 181, 677 - 684 (1977).

Photoelectric observations in UBV are reported for the recent extragalactic supernovae in NGC 7723, 1325 and 4402. The supernova in NGC 1325 was peculiar in that it faded rapidly and consequently was probably of Type II, while the remaining two were apparently of Type I. These observations are compared with the mean light and colour curves for supernovae and an improved mean $(U–B)$ colour curve for Type I supernovae is given.

125.020 On the density and energy of supernova remnants.
J. Cantó.
Astron. Astrophys., Vol. 61, 641 - 645 (1977).

The effects of an interstellar magnetic field on the gas flow behind a strong shock front are considered. The ambient density and energy of supernova remnants are estimated from the intensity ratio of sulphur lines $I(6717)/I(6731)$. It is found that, on average, the ambient density around galactic supernova remnants is 4 cm^{-3}. The total energy appears to be the same for all supernova remnants (to within a factor $\cong 5$). A mean value of 4×10^{51} erg is found.

125.021 The remnant of Kepler's supernova.
S. van den Bergh, K. W. Kamper.
Astrophys. J., Vol. 218, 617 - 632 (1977).

Proper-motion observations covering the period 1942–1976 show that the expansion time scale for the optical remnant of SN 1604 is greater than $\sim 1 \times 10^4$ years. This indicates that the remnant consists of circumstellar material that was excited by the supernova shell. The optical remnant is found to have a tangential velocity of 525 ± 117 km s^{-1}. A few emission flocculi have brightened significantly during the last 35 years. Both seventeenth-century color observations and photometry of field stars give a foreground reddening $E_{B-V} = 0.7 \pm 0.2$. This reddening value and the assumption that SN 1604 occurred in the nuclear bulge of the Galaxy yield M_V(max) = -19.3 ± 0.7.

125.022 Cosmic debris – aftermath of the supernova.
C. Devismes.
Sci. Dimension, Vol. 8, No. 6, p. 20 - 25 (1976). – Abstr. in

Phys. Abstr., Vol. 80, Abstr. 37642 (1977).

125.023 **Outburst of the supernova and neutral currents of**
         weak interactions.   M. Gmitro.
Ceskoslovensky Cas. Fis., Vol. 26, 630 - 631 (1976). In Czech.
Abstr. in Phys. Abstr., Vol. 80, Abstr. 45472 (1977).

125.024 **Supernovae.**   J. Hopkins.
         Astronomy, Vol. 5, No. 4, p. 6 - 17 (1977).
Abstr. in Phys. Abstr., Vol. 80, Abstr. 57001 (1977).

125.025 **Supernova in faint galaxy.**
         IAU Circ., No. 3122 (1977).

125.026 **Supernova in anonymous galaxy.**
         IAU Circ., No. 3140 (1977).

125.027 **Supernova remnants in the Large Magellanic Cloud.**
         A photographic study in [O III] and [S II].
B. M. Lasker.
Publ. Astron. Soc. Pacific, Vol. 89, 474 - 481 (1977).
    For three supernova remnants in the Large Magellanic
Cloud, N86, N186D, and N206, monochromatic plates are
used to compare the structure in a line of high excitation,
[O III] λ5007, with that in lines of low excitation, [S II]
λλ6716, 6731. The images in [O III] as compared to [S II]
tend to be concentrated toward the periphery of the remnants
(N86, N186D, and N206), to be sharper (N86, N186D), to lie
outside the peak of the [S II] emission (N206), and perhaps to
have less prominent filamentary structure (N86).

125.028 **A new large supernova remnant in the Southern Sky?**
         J. Meaburn, P. Rovithis.
Astrophys. Space Sci., Vol. 46, L7 - L10 (1977).
    A faint filamentary structure on a Southern Schmidt plate
is thought to be a supernova remnant.

125.029 **Lagrangian coordinate transformations in self-**
         similar supernova blast-wave models.
W. I. Newman.
Astrophys. Space Sci., Vol. 47, 99 - 108 (1977).
    The Sedov self-similar solution of the fluid equations
provides a description of a spherical supernova blast-wave
during its adiabatic phase in a stationary (Eulerian) reference
frame. The author provides an exact transformation to a
Lagrangian coordinate system, a non-inertial frame of reference
which follows each individual gas particle. The physical
properties of the gas in this comoving reference system are
evaluated as a function of time. Numerical methods to facilitate Eulerian–Lagrangian and Lagrangian–Eulerian coordinate
transformations are provided.

125.030 **Northern Polar Spur — a remnant of a supernova ex-**
         plosion?   Ya. M. Khazan.
Priroda, 1977, No. 12, p. 17 - 19. In Russian.

125.031 **On pulsar-driven isothermal blast-wave models of**
         supernova remnants.   I. Lerche.
Astrophys. Space Sci., Vol. 48, 335 - 356 (1977).
    The author investigates the 'equilibrium' and stability of
spherically-symmetric self-similar isothermal blast waves with a
continuous post-shock flow velocity expanding into a medium
whose density varies as $r^{-\omega}$ ahead of the blast wave, and which
are powered by a central source (a pulsar) whose power output varies with time as $t^{\phi-3}$. The author discusses the solutions
for all possible ω. These results strongly suggest that the evolution of supernova remnants is not according to the self-similar
form.

125.032 **Theoretical spectra of the thermal X-rays from**
         young supernova remnants.   H. Itoh.
Publ. Astron. Soc. Japan, Vol. 29, 813 - 830 (1977).

    Early evolutionary changes of the thermal X-ray spectra
of supernova remnants are investigated by using a spherically
symmetric, hydrodynamic code. The effects of time-dependent ionization and lack of temperature equilibrium between
electrons and ions at the shock front are considered. If electron and ion temperatures are not equilibrated at the shock
front, the electron temperature in the shocked circumstellar
medium is lower than the ion temperature by about an order
of magnitude. The X-ray spectrum of Cassiopeia A can be well
accounted for by two-fluid model.

125.033 **Brightness, colour, and expansion velocity curves**
         of type I supernovae as functions of the rate of
brightness decline.   Yu. P. Pskovskij.
Astron. Zh. Akad. Nauk SSSR, Vol. 54, 1188 - 1201 (1977).
In Russian. English translation in Soviet Astron., Vol. 21,
No. 6.
    A photometric classification of supernovae according to
the post-maximum photographic brightness decline rate is
proposed. A dependence of the main elements of curves of
brightness, colour, envelope expansion velocity, and absolute
magnitudes in the maximum light on the photometric class
of the supernova has been found. The mean absolute magnitude of type I supernovae has been derived. The spread of
anomalous type I supernovae found by Bertola is discussed.

125.034 **Condensation in supernova ejecta.**
         L. Grossman, D. N. Schramm, J. M. Lattimer.
Meteoritics, Vol. 12, 246 - 247 (1977). — Abstract.

125.035 **Statistische Aspekte der Supernovae-Beobachtung.**
         H.-A. Ott.
Mitt. Astron. Ges., Nr. 42, p. 140 - 141 (1977).

         Viscid and inviscid self-similar blast waves: spherical,
isothermal flows and inferences for supernova remnants.
See Abstr. 062.050.

         Isothermal self-similar blast wave theory of super-
nova remnants driven by relativistic gas pressure.
See Abstr. 062.063.

         Neutron-capture nucleosynthesis in the helium-
burning cores of massive stars.   See Abstr. 065.011.

         Neutrino transport in supernova models: S_N **method.**
See Abstr. 065.013.

         The neutron mechanism of thermonuclear carbon
burning, formation of neutron stars and supernova bursts.
See Abstr. 065.084.

         Effect of nearby supernova explosions on atmo-
spheric ozone.   See Abstr. 082.015.

         Supernovae, grains and the origin of the solar
system.   See Abstr. 107.039.

         Change of the orbital elements and run-away veloci-
ties of binaries caused by a symmetric explosion of one of the
components.   See Abstr. 117.016.

         Rejuvenation of helium white dwarfs by mass
accretion.   See Abstr. 117.045.

         The molecular cloud associated with the supernova
remnant W44.   See Abstr. 131.045.

         Nature, origin, and evolution of the hot gas: inter-
stellar?   See Abstr. 131.098.

         A theory of the interstellar medium: three compo-

nents regulated by supernova explosions in an inhomogeneous substrate.   See Abstr. 131.111.

High-velocity gas in the Monoceros loop.
See Abstr. 131.185.

Radial distribution of [Fe XIV] emission in the Cygnus Loop.   See Abstr. 134.029.

The thermal and non-thermal components of sixteen nebular complexes.   See Abstr. 134.036.

A high-resolution investigation of the shell source G55.7 + 3.4 at 610 and 1415 MHz.   See Abstr. 141.047.

A further measurement of the 38-MHz flux density of Cas A.   See Abstr. 141.076.

G127.1+0.5 – a remarkable supernova remnant centred on a very compact radio source?
See Abstr. 141.130.

Observational evidence for supernova-induced star formation: Canis Major R1.   See Abstr. 152.003.

Supernova-induced star formation in Cepheus OB3.
See Abstr. 152.006.

LMC X-1: a luminous extended X-ray source.
See Abstr. 159.008.

Supernova in NGC 5253 (1972e)

125.101   **Type I supernovae. III. The spectrum of SN 1972e in NGC 5253, 365 and 435 days after the explosion.**
C. Gordon.
Astrophys. J., Vol. 216, 67 - 72 (1977).

The empirical model deduced from the observed spectra at t = 30 or 250 days can also reproduce the spectra observed at t = 365 or 435 days. The overabundance of S and Fe is confirmed. The linear decline of the photographic magnitude can be explained by the variation with time of the intensity of the Fe III lines.

Supernova in NGC 1411

125.201   **Supernova in NGC 1411.**
IAU Circ., Nos. 3091, 3103 (1977).

Supernova in NGC 4414 (1974g)

125.301   **Type I supernova 1974g in NGC 4414.**
S. Wyckoff, P. A. Wehinger.
Astrophys. Space Sci., Vol. 48, 421 - 435 (1977).
Photographic image-tube spectra (150 Å mm^{-1}, 4500 Å – 7000 Å) of the type I supernova (1974g) in the Sc galaxy NGC 4414 obtained at phases +14 days and +40 days past maximum light have been reduced to absolute flux. An electronographic isophotal map of NGC 4414 is presented, and an accurate position of SN 1974g given. The absolute magnitude at maximum light of SN 1974g was found to be M_B(max) = −19.0. Estimates of the radius of the expanding photosphere of SN 1974g, determined by two independent methods, give $R \sim 10^{15}$ cm in the early post-maximum phases. The total (observed) luminous energy of SN 1974g was $\sim 10^{51}$ erg.

Supernova in NGC 4340

125.401   **Supernova in NGC 4340.**
Astron. Tsirk., No. 941, p. 1 (1977). In Russian.

126 Low-Luminosity Stars, Subdwarfs, White Dwarfs, Degenerate Stars

126.001   A hot magnetic DA white dwarf with a field of 2–4 × 10⁷ gauss.
D. T. Wickramasinghe, J. A. J. Whelan, M. S. Bessell.
Mon. Not. R. Astron. Soc., Vol. 180, 373 - 378 (1977).

Image tube spectrophotometry from λλ4000 to 7100 Å of the suspected magnetic white dwarf BPM 25114 shows features which correspond to Zeeman patterns for the Balmer lines Hγ, Hβ and Hα at fields of 2–4 × 10⁷ gauss. The effects of magnetic broadening are clearly apparent in the spectrum and give rise to a blue wing of Hα extending ~1000 Å shortward of λ6563. The spectrum is found to be variable.

126.002   Calcium in the helium white dwarf GD 40.
H. L. Shipman, J. L. Greenstein, A. Boksenberg.
Astron. J., Vol. 82, 480 - 486 (1977).

New detector technology has allowed to obtain spectra of the DB white dwarf GD 40, the only DB star known which shows Ca II lines in its spectrum. Ten helium lines are visible along with the H and K lines of Ca II. An analysis of the spectrum gives T_{eff} = 15 200 ± 500 K, log g = 8.2 ± 0.5, and log(n_{Ca}/n_{He}) = −7.5 ± 0.3, assuming that helium is the only donor of electrons in the atmosphere.

126.003   Hβ line profiles for white dwarfs with dipole and quadrupole magnetic fields.   G. Wegner.
Mon. Notes Astron. Soc. Southern Africa, Vol. 36, 63 - 75 (1977).

Profiles are given for the white light and circular polarization of the Hβ line, that are appropriate for comparison with observations of magnetic white dwarfs. The geometry of the magnetic fields are assumed to be dipole with H_p = 2 to 5 × 10⁷ Gauss and quadrupole with H_p = 4 and 5 × 10⁷ Gauss at orientations of $0° < i_H < 90°$.

126.004   Hydrogen in a hydrogen-poor white dwarf: Ross 640.   J. Liebert.
Astron. Astrophys., Vol. 60, 101 - 108 (1977).

Lick Observatory image tube scans of the DFp white dwarf Ross 640 (EG119) are presented which show a distinguishable Hα line, weak Hβ and possible Hγ features. Model atmospheres and line profiles are most consistent with a temperature of T_{eff} = 8500° ± 300° K and a hydrogen abundance n_H ~ 3.0 ×10⁻⁴ in a helium-dominated atmosphere. Thus, Ross 640 is the first cool helium-atmosphere degenerate star with a directly determined hydrogen abundance.

126.005   Magnetism in white dwarfs.   J. R. P. Angel.
Astrophys. J., Vol. 216, 1 - 17 (1977).

A few percent of all white dwarfs are strongly magnetic. Ten examples are now known, of which half have spectra which show Zeeman splitting in lines of H, He, or CH in magnetic fields of 5 to 25 × 10⁶ gauss. Two of these have sharply defined Zeeman subcomponents, indicative of very uniform surface fields. The remaining five have still stronger fields, such that the spectral features if present are weak and of uncertain origin. In these objects the magnetic field is identified by the elliptical polarization of the optical continuum, and is of order 10⁸ gauss. Most of the magnetic white dwarfs show no spectral or polarization variations, and may be rotating very slowly (P > 10 years). However, two are identified as oblique rotators with periods of the order of hours, in line with estimates of nonmagnetic white dwarfs.

126.006   Spectrophotometry of the extreme-ultraviolet star Feige 24.   J. Liebert, B. Margon.
Astrophys. J., Vol. 216, 18 - 22 (1977).

The authors have obtained spectrophotometry of Feige 24, the second known extrasolar source of extreme-ultra-violet (100–1000 Å) radiation, on six nights in 1975–1976. They classify the two components of the system as DAwk and M1– 2V, in agreement with previous inferences. The spectroscopic distance modulus of the system is estimated as $(m - M)$ = 4.8 ± 1; the resulting M_v = 7.9 ± 1 is consistent with the luminosities of either a composite white dwarf–red dwarf or a dwarf nova system.

126.007   White-dwarf variability and the rotation of g-modes.   C. L. Wolff.
Astrophys. J., Vol. 216, 784 - 790 (1977).

The multiperiodic light curves of many DA white dwarfs are interpreted as arising from the rotation of nonlinearly coupled g-mode oscillations. Their characteristic rotation rates and consequent beat periods can be identified in the Fourier analysis of the measured light curves of four stars: HL Tau 76, G29-38, G38-29, and G207-9. An average of six strong periodicities per star can be closely matched by periods derived from the g-modes. The observed periods matched by the model contain most of the power in the light curve. If the model is correctly applied to these variables, then they are all rotating slowly with periods in the range 250 to 500 s.

126.008   The effects of differential rotation on the splitting of nonradial modes of stellar oscillation.
C. J. Hansen, J. P. Cox, H. M. Van Horn.
Astrophys. J., Vol. 217, 151 - 159 (1977).

The problem of the splitting of adiabatic nonradial modes of stellar oscillation owing to slow differential rotation is examined. Integral expressions for the frequency splitting are derived for the case of cylindrically symmetric rotation, and, for a particular rotation "law", the results are applied to cooling white dwarfs and upper–main-sequence stars.

126.009   A spectroscopic survey of white dwarf candidates from the Luyten catalogs.
J. Liebert, P. A. Strittmatter.
Astrophys. J., Lett., Vol. 217, L59 - L63 (1977).

The first spectroscopically identified white dwarfs are reported from a new search program, based on stars discovered in the Luyten Palomar proper motion surveys. New results include (1) a star with peculiar spectral features and an energy distribution distorted by the presence of a large magnetic field (LP 790–29), (2) a "DBA" star with lines of He I and H in its blue spectrum (G200–39), (3) a bluish, high-velocity star with very strong C₂ bands (LP 93–21), and (4) several cool degenerates including a red, common-proper-motion pair (LP 543–32/33). Several blue objects listed in the Markarian catalog, and previously observed by the Willses, are confirmed as white dwarfs, but MRK 392 is not a DO star and four have other literature designations.

126.010   Subdwarfs or cool DA white dwarfs?
D. T. Wickramasinghe, M. S. Bessell, P. L. Cottrell.
Astrophys. J., Lett., Vol. 217, L65 - L68 (1977).

Spectroscopic and photometric discriminants which can be used to distinguish cool, metal-deficient DA white dwarfs from subdwarfs are discussed. It is shown that, due to the effects of convection on the atmosphere structure, a white dwarf has a Balmer-line spectrum which is quite distinct from that of a subdwarf of the same effective temperature.

126.011   The carbon and nitrogen abundances of Gamma Pavonis.   R. A. Bell, R. E. S. Clegg, A. L. T. Powell.
Mon. Not. R. Astron. Soc., Vol. 181, 31 - 35 (1977).

For the mild subdwarf γ Pavonis, the $uvby$ β colours yield [M/H] = −0.6, in agreement with the iron abundance found by

Danziger from a curve of growth analysis. The authors have determined C and N abundances using a model atmosphere and synthetic spectra of the NH, CN and CH molecular bands. They find [C/H] = −0.75, a carbon abundance higher than that found by Danziger. Nitrogen may be marginally overabundant in γ Pav.

126.012 **Radial oscillations of zero-temperature white dwarfs and neutron stars below nuclear densities.**
G. Chanmugam.
Astrophys. J., Vol. 217, 799 - 808 (1977).

The radial oscillations of zero-temperature degenerate stars with subnuclear densities have been calculated for the fundamental and first two excited modes using recent equations of state. The calculations were made under the assumptions that (a) the dynamical time scale $\tau \gg \tau_N$, where τ_N is the nuclear relaxation time, in which case the adiabatic index $\gamma = \gamma_E$ (calculated along the equation of state) and (b) $\tau \ll \tau_N$ with $\gamma = \gamma_C$ calculated at constant composition. For case a, it is shown that white dwarfs are dynamically stable for $\rho_c \lesssim 1.2 \times 10^9 \, \mathrm{g \, cm^{-3}}$ while neutron stars are dynamically stable for $\rho_c \gtrsim 10^{14} \mathrm{g \, cm^{-3}}$. For case b, white dwarfs are dynamically stable for $\rho_c \lesssim 4 \times 10^{10} \mathrm{g \, cm^{-3}}$, while neutron stars are dynamically stable for $\rho_c \gtrsim 7 \times 10^{12} \mathrm{g \, cm^{-3}}$ as had been suggested previously by the author. Case b is shown to be more appropriate for infinitesimal perturbations. It is also pointed out that arguments given earlier to indicate that high-density white dwarfs and low-density neutron stars may not exist, because of instabilities over long time scales, are lacking in rigor.

126.013 **Photoelectric observations of Sirius B.**
K. D. Rakoš, R. J. Havlen.
Astron. Astrophys., Vol. 61, 185 - 188 (1977).

New photoelectric area scanner measurements of Sirius B are presented which show the star not to be unusual with respect to other DA white dwarfs. A comparison with many photoelectric calibrations of effective temperature leads to the most probable value of 32000 °K.

126.014 **Observations of the magnetic white dwarf GD 90.**
D. N. Brown, A. Rich, W. L. Williams, G. Vauclair.
Astrophys. J., Vol. 218, 227 - 231 (1977).

The white dwarf GD 90, recently found to exhibit Zeeman structure in Hβ and Hγ, which indicates the presence of a magnetic field of ~5 $\times$ 10^6 gauss, has been observed for broad-band circularly polarized radiation in the wavelength region 3500–5500 Å. The authors have also made a spectroscopic observation of GD 90 which was simultaneous with one of the polarization runs. The results of this observation were consistent with previous work, thus eliminating any possibility that the magnetic field as detected by the Zeeman components of Hβ is variable. Such variability could have explained the lack of observed circular polarization had the polarization observations been made at the time of field minimum.

126.015 **Further degenerate stars. X.**
J. L. Greenstein, J. B. Oke, D. Richstone, W. F. van Altena, H. Steppe.
Astrophys. J., Lett., Vol. 218, L21 - L25 (1977).

This list of 51 further degenerate stars includes many faint red degenerates, as well as blue objects, selected by proper motion or color. Observations were made with the multichannel spectrophotometer and with higher-resolution linear sensors. Hydrogen lines are found down to temperatures of 7000 K. A pair containing two red degenerates has a newly measured parallax leading to discussion of some evolutionary problems of wide double white dwarfs. One remarkably carbon-rich high-velocity star near 9000 K presents some astrophysical problems.

126.016 **Elements separation and mixing processes in the envelopes of white dwarfs.**

G. Vauclair, C. Reisse.
Astron. Astrophys., Vol. 61, 415 - 425 (1977).

The authors investigate the possibility that helium rich white dwarfs could have evolved from initially hydrogen rich white dwarfs. They first estimate the time scale for helium diffusion in the envelope of a white dwarf. The properties of the variable mean molecular weight region (the μ-barrier) built by the diffusion processes are discussed. Then the authors investigate two possible sources of mixing: first they analyze the development of the differential rotation induced by the meridional circulation in rotating white dwarfs. The second mixing mechanism examined is convection.

126.017 **On the estimate of critical parameters of degenerate magnetic white dwarfs.** G. A. Shul'man.
Astron. Zh. Akad. Nauk SSSR, Vol. 54, 1041 - 1046 (1977).
In Russian. English translation in Soviet Astron., Vol. 21, No. 5.

By means of the energetic approximation method the mass limit and the central density of a degenerate magnetic white dwarf is estimated on the assumption of magnetic flux conservation in its interior.

126.018 **Extreme ultraviolet observations of white dwarfs.**
M. Lampton, B. Margon, S. Bowyer.
Highlights of Astronomy, Vol. 4, Part II, (see 012.025), p. 341 - 347 (1977).

Observations of HZ 43 and Feige 24 have been obtained with the Apollo-Soyuz extreme ultraviolet telescope; both stars are copious EUV emitters, with 4 $\times$ 10^{-9} and 3 $\times$ 10^{-9} erg/cm^2 sec in the 170–620 Å band respectively. The EUV data combined with optical spectrophotometry, allow their temperatures to be estimated as 80,000 and 60,000 K respectively. The corresponding interstellar neutral hydrogen column densities are ~ 4 $\times$ 10^{18} cm^{-2}.

126.019 **Spectra of southern white dwarfs.**
D. T. Wickramasinghe, M. S. Bessell.
Mon. Not. R. Astron. Soc., Vol. 181, 713 - 717 (1977).

Spectra of 47 white dwarf suspects south of declination −45° are discussed. Contrary to previous claims, of the six stars redder than $B-V$ = 0.5, only one is a classical white dwarf.

126.020 **Spectrum synthesis of the heavily blanketed white dwarf LP 701-29.**
P. L. Cottrell, M. S. Bessell, D. T. Wickramasinghe.
Astrophys. J., Lett., Vol. 218, L133 - L135 (1977).

Synthetic spectra are constructed for the white dwarf LP 701-29, assuming a hydrogen-rich atmospheric composition. The observed spectrum is consistent with a metal abundance, [M/H] ~ −3.0. The possibility of a helium-rich atmosphere is also discussed.

126.021 **The binary orbit and mass of the hot white dwarf Feige 24.**
J. R. Thorstensen, P. A. Charles, B. Margon, S. Bowyer.
Publ. Astron. Soc. Pacific, Vol. 89, 623 (1977). − Abstract.

126.022 **Photoelektrische Temperaturbestimmung bei den sehr heißen DA Zwergen.** D. Rakos.
Mitt. Astron. Ges., Nr. 42, p. 137 - 139 (1977).

A catalogue of spectroscopically identified white dwarfs. See Abstr. 002.048.

A measurement of the width and shift of the Fe I 3719.94 Å line broadened by helium. See Abstr. 022.004.

Atoms in high magnetic fields.
See Abstr. 061.031.

Radiative heat transfer in a magnetic field.
See Abstr. 063.009.

Some properties of very low temperature, pure helium surface layers of degenerate dwarfs. See Abstr. 064.014.

Thermal stability of hydrogen-burning shells in white dwarfs.   See Abstr. 065.009.

Linear radial and nonradial modes of oscillation of hot white dwarfs.   See Abstr. 065.010.

Equations of state for stellar partial ionization zones.   See Abstr. 065.051.

P-nuclei processing in degenerate hydrogen burning zones and its relationship to nova outbursts. See Abstr. 065.080.

Mechanism of X-ray and soft γ-ray radiation from accreting neutron stars.   See Abstr. 066.014.

On gravitational radiation of neutron stars and white dwarfs.   See Abstr. 066.075.

Rejuvenation of helium white dwarfs by mass accretion.   See Abstr. 117.045.

BD $-3°5357$: an eclipsing binary with a hot sub-dwarf component.   See Abstr. 121.017.

On the distance scale of planetary nebulae and white dwarf birth rates.   See Abstr. 135.018.

New observations and a slow rotator model of the X-ray binary AM Herculis.   See Abstr. 142.037.

Space density of low-luminosity stars using high-accuracy proper motions.   See Abstr. 155.061.

Interstellar Matter, Infrared Sources, Gaseous Nebulae, Planetary Nebulae

131 Interstellar Matter, Star Formation

131.001 **Grain temperatures and infrared emission from interstellar dust clouds.**   S. Aiello, F. Mencaraglia, A. Blanco, A. Borghesi, E. Bussoletti.
Mon. Not. R. Astron. Soc., Vol. 180, 323 - 337 (1977).
Experimentally determined refractive indices have been used to calculate optical properties of silicate grains in cosmic dust clouds of varying densities and for various silicate—graphite—iron mixtures. The transfer of scattered radiation within the cloud has been investigated. The temperature distribution in the cloud has been evaluated. In a wide range of cases of astrophysical interest the silicate grain temperatures are low enough to allow recombination and condensation reactions to take place.

131.002 **Observations of SiO masers at 43 GHz with the Parkes radio telescope.**
M. Balister, R. A. Batchelor, R. F. Haynes, S. H. Knowles, M. G. McCulloch, B. J. Robinson, K. J. Wellington, D. E. Yabsley.
Mon. Not. R. Astron. Soc., Vol. 180, 415 - 427 (1977).
A survey of 83 late-type stars for $v = 1, J = 1-0$ and $v = 2, J = 1-0$ emission from SiO at 43 GHz has yielded 10 new sources. There is a strong correlation between stars detected with SiO emission and those known to emit H_2O radiation. Observations of the new SiO sources and of some of the previously known sources have confirmed previous conclusions that source spectra vary with time, some sources having timescales of a few days.

131.003 **Three unusual wide-spectrum H_2O sources.**
W. M. Goss, R. F. Haynes, S. H. Knowles, R. A. Batchelor, K. J. Wellington.
Mon. Not. R. Astron. Soc., Vol. 180, 51P - 56P (1977).
A new type of wide-spectrum H_2O source has been detected in H_2O 291.3−0.7 (NGC 3576), H_2O 351.2 + 0.7 (NGC 6334B) and H_2O 12.2−0.1. The displaced velocity components have total widths >15 km/s and have negative velocity displacements. In two of the sources, these spectral components are stronger than the H_2O features at velocities close to that of the H II region and they show pronounced time variations.

131.004 **The role of dust in cosmogony.**
A. G. W. Cameron.
The dusty universe, (see 012.001), p. 1 - 31 (1975).
A discussion is given of the production of interstellar grain cores from stellar material, the gain and loss of grain mantles in interstellar space, chemical transformations in these grains when they become part of the primitive solar nebula, and the identification of these grains with interplanetary dust derived from comets and with the matrix material in meteorites. Thus, this paper proposes a common cosmogonic frame-

work relating studies of interstellar and interplanetary grains.

131.005 **The composition of interstellar dust.**
G. B. Field.
The dusty universe, (see 012.001), p. 89 - 112 (1975).
Direct evidence that interstellar dust is composed partly of silicates, graphite, and water ice is reviewed. Indirect evidence, from recent studies of the chemical composition of interstellar gas, is assessed in terms of two possible models for the formation of the dust: condensation under thermal-equilibrium conditions and accretion under nonequilibrium conditions. Equilibrium condensation may occur either in stellar atmospheres or in circumstellar nebulae, but arguments from stellar evolution favor the latter.

131.006 **Interaction of gas and dust in the interstellar medium.**   W. D. Watson.
The dusty universe, (see 012.001), p. 113 - 129 (1975).
Physical processes involved in the interaction of interstellar gas and dust are discussed. It is still not understood whether particles heavier than helium can be returned to the gas when they hit grain surfaces. Ejection by ultraviolet radiation seems to be the most likely process. Interpretation of observations of the H_2 molecule by the Copernicus satellite indicates that H_2 is formed in grain surfaces at a rate that is in semiquantitative agreement with theoretical predictions.

131.007 **Effects of particle shape on volume and mass estimates of interstellar grains.**
J. M. Greenberg, S. S. Hong.
The dusty universe, (see 012.001), p. 131 - 153 (1975).
Mass estimates of interstellar grain materials based on visual extinction characteristics are shown to be insensitive to shape, and so long as the wavelength dependence of extinction is defined well into the infrared, they are also insensitive to size distribution. Spheroidal particles are treated by an approximate analytical method. Spheres and cylinders (core mantle as well as homogeneous) are treated by exact methods.

131.008 **Interstellar grains as pinwheels.**
E. M. Purcell.
The dusty universe, (see 012.001), p. 155 - 167 (1975).

131.009 **Interstellar molecules.**   B. Zuckerman.
Nature, Vol. 268, 491 - 495 (1977).
Recently discovered interstellar molecules are a powerful new tool in the arsenal of the astrophysicist. Operating in a completely non-terrestrial environment, interstellar chemistry produces both everyday molecules likely to be found around the house and exotic species never before observed on the Earth. Interpretation of the molecular spectra requires an interdisciplinary approach.

131.010 Relation between metallicity and multiplicity for solar type stars.   D. C. Barry.
Nature, Vol. 268, 509 - 510 (1977).

It has been suggested that the ratio of heavy elements to hydrogen (metallicity) in a contracting interstellar cloud might affect the process of star formation. There is now statistical evidence for solar type stars that the metallicity of a forming stellar system is correlated with the multiplicity of the resultant system and may actually determine whether the system will be multiple or single.

131.011 The ionization and expansion of a globule of interstellar gas overrun by a weak R-type ionization front.
J. S. Berry.
Z. Naturforsch., Band 32a, 692 - 696 (1977).

The flow of the ionized gas behind a contracting ionization front is investigated for spherical symmetry. A similarity solution is given when the initial density distribution in the neutral hydrogen is ω_0/r^a where r is the distance from the centre of contraction.

131.012 Diffuse [O I] emission and warm interstellar gas in galaxies.   J. C. Weisheit.
Astrophys. J., Vol. 215, 755 - 758 (1977).

The author shows that measurements of the relative intensities of [O I] and Hα lines in external galaxies can be used to determine whether a significant fraction of the disk of each galaxy consists of warm ($T \sim 10^4$ K) interstellar H I regions.

131.013 Far-infrared emission of molecular clouds.
C. E. Ryter, J. L. Puget.
Astrophys. J., Vol. 215, 775 - 780 (1977).

Data available now on far-infrared (10–300 μm) thermal emission and carbon monoxide millimeter radiation are compiled in order to generate a sample of objects where the thermal radiation of the dust mixed with the molecular hydrogen can be quantitatively studied. A list of nine massive molecular clouds is obtained. The clouds are clearly heated from the interior, by newly born, still unobservable stars. But any attempt to infer the power released in the clouds by using published star-formation rates falls short by a factor $\sim$10.

131.014 High-resolution profiles of the diffuse interstellar features at 6379 and 6614 Å.
G. L. Welter, B. D. Savage.
Astrophys. J., Vol. 215, 788 - 795 (1977).

High-resolution profiles ($\Delta\lambda \approx 0.2$ Å) of the diffuse interstellar features at 6379 and 6614 Å in 13 heavily reddened stars were obtained. The feature at λ6614 is observed to be asymmetrical, with its steep side toward the blue. In contrast, the feature at λ6379 is symmetrical. Attempts to fit Fano autoionization profiles to the observed data yielded unsatisfactory results because the observed features do not exhibit the broad wings that characterize the autoionization process. The theoretical profiles for an alternate process, pure electronic impurity absorption lines in small cold grains, provide an excellent fit to the observed λ6379 profile but do not satisfactorily explain that of λ6614.

131.015 Graphite grain surface reactions in interstellar and protostellar environments.
M. J. Barlow, J. Silk.
Astrophys. J., Vol. 215, 800 - 804 (1977).

Surface reactions on warm ($T_g \gtrsim 60$ K) graphite grains between lattice C atoms and impinging H and O atoms, within or in the vicinity of H II regions, can provide a prolific source of H_2CO and other interstellar molecules. It is proposed that similar reactions in the primitive solar nebula led to the formation of the organic molecules found in carbonaceous chondrites. From this and related evidence it is argued that most of the solid material in the solar system may have originated as interstellar grains.

131.016 Grain disruption in interstellar hydromagnetic shocks.   J. M. Shull.
Astrophys. J., Vol. 215, 805 - 811 (1977).

The observed abundance variations of Ca, Fe, Si, and Ti in intermediate-velocity interstellar gas suggest that grains have been disrupted in clouds with velocities as low as 20–50 km s^{-1}. The author describes a simplified hydromagnetic shock model for such clouds, derives the dynamical equations for charged grains in the postshock region, including the collisional drag and "betatron-acceleration" effect of a magnetic field gradient, and thereby calculates the fraction of grains destroyed in evaporative collisions with other grains. He finds that for shocks of 20–50 km s^{-1} in which the fractional H-ionization remains low, 3–10% of the grain material may be destroyed – sufficient to explain in part the heavy-element depletion pattern in intermediate velocity clouds and the well-known correlation of N(Ca II)/N(Na I) with cloud velocity.

131.017 The infrared polarization of NGC 1275, NGC 4151, Markarian 231, and 3C 273.
J. C. Kemp, G. H. Rieke, M. J. Lebofsky, G. V. Coyne.
Astrophys. J., Lett., Vol. 215, L107 - L110 (1977).

The authors report initial observations obtained with a new type of infrared polarimeter, employing photoelastic modulators. Operable over the range 1–8 μm, it has vanishing instrumental polarization, $\lesssim$0.03% as verified so far. The common and surprising feature of the four objects studied here is the very small polarization at 2.2 μm. These contrast with the large infrared polarizations in for example BL Lacertae objects. At least in NGC 4151, if the radiation mechanism is non-thermal it must be of an unpolarized type.

131.018 Accurate H_2O source positions in W3.
J. R. Forster, W. J. Welch, M. C. H. Wright.
Astrophys. J., Lett., Vol. 215, L121 - L125 (1977).

The Hat Creek interferometer has been used to obtain H_2O source positions accurate to a few tenths of an arc second in the W3 region. The H_2O sources in W3 (cont) are close to IRS 5. In W3 (OH) the H_2O sources are 6″ from the main compact H II region. The authors also obtain absolute positions for the OH sources accurate to about 0.″5 and find that they lie at the edge of the compact H II region. They argue that the excitation of the H_2O masers is internal since the available pumping radiation from the nearby UV and IR sources is too small to provide the large microwave power.

131.019 A remarkable structural change in a faint cometary nebula.   M. Cohen, L. V. Kuhi, E. A. Harlan.
Astrophys. J., Lett., Vol. 215, L127 - L129 (1977), with a correction in Vol. 218, L31 (1977).

A newly discovered cometary nebula is noted as having undergone a remarkable change in appearance in the last 20 years. The major part of the nebula has disappeared, to be replaced by a much smaller fan, and the associated star has brightened by a few magnitudes in the same period.

131.020 Investigation of OH radio line absorption of galactic emission sources.
M. I. Pashchenko, V. I. Slysh.
Astron. Zh. Akad. Nauk SSSR, Vol. 54, 790 - 806 (1977).
In Russian. English translation in Soviet Astron., Vol. 21, No. 4.

The angular distribution of OH absorption was measured in the galactic sources W 3, W 12, W 41, W 43, and AMWW 52 (G 32.8 + 0.2) with the Nançay radio telescope. One-dimensional distribution of the optical depth was obtained for 4 sources, and a two-dimensional distribution of the absorption was measured for W 43. The distributions show complex kinematic and angular structure as a result of absorption by both extended and compact gas clouds. The latter seem to be physically associated with H II regions and molecular maser sources. A new compact gas cloud associated with a supernova remnant and class IIa 1720 MHz maser sources was found in W 41.

131.021 The influence of dust on the collisional pumping of an H_2O cosmic maser.
G. T. Bolgova, V. S. Strel'nitskij, I. K. Shmeld.
Astron. Zh. Akad. Nauk SSSR, Vol. 54, 828 - 833 (1977).
In Russian. English translation in Soviet Astron., Vol. 21, No. 4.

The rate equations for the populations of 48 ortho-H_2O rotational levels are solved simultaneously with the equations of radiative transfer in the rotational lines, accounting for the continuous absorption and emission of resonance photons by dust grains. The radiative transport was treated in a model of a homogeneous isothermal plane-parallel slab, approximating the region of collisional pumping behind a shock front. It is found that continuous absorption and emission may strongly influence the character of the distribution of the rotational level populations.

131.022 Radio emission of the interstellar NS molecule.
D. A. Varshalovich, V. K. Khersonskij, G. F. Chernyj.
Astron. Zh. Akad. Nauk. SSSR, Vol. 54, 915 - 918 (1977).
In Russian. English translation in Soviet Astron., Vol. 21, No. 4.

The frequencies and probabilities of radiative transitions between Λ-doublet levels and low rotational levels of the NS molecule which was discovered recently in the interstellar medium have been calculated. The optical depths and possibilities of observations of different NS lines from Sgr B2 have been estimated.

131.023 Polysaccharides and infrared spectra of galactic sources.   F. Hoyle, N. C. Wickramasinghe.
Nature, Vol. 268, 610 - 612 (1977).

Observations over the infrared waveband 2–30μm available for a number of astronomical objects are shown to be reconcilable with the transmittance properties of polysaccharides. Using an experimentally determined transmittance spectrum for cellulose the authors can readily relate astronomical data in the 2–4 μm, 8–13 μm and 15–30 μm wavebands and they obtain close fits to astronomical spectra in these several bands. From this detailed spectral agreement they consider it reasonable to infer the detection of interstellar polysaccharides.

131.024 Mechanism for formaldehyde polymer formation in interstellar space.   V. I. Goldanskii.
Nature, Vol. 268, 612 - 613 (1977).

The author argues that molecular tunnelling in condensed formaldehyde is the only viable mechanism for the formation of formaldehyde polymers in interstellar conditions. Such a polymerisation process may be initiated by the action of ionising radiation.

131.025 Origin of diffuse interstellar lines.
A. E. Douglas.
Nature, Vol. 269, 130 - 132 (1977).

The author suggests that the diffuse interstellar absorption lines are caused by the absorption of polyatomic molecules and that the line width is the result of radiationless internal conversion between stable states. Furthermore, he proposes that the absorbing species are long chain carbon molecules, C_n where n may lie in the range 5–15.

131.026 Carbonaceous compounds in interstellar dust.
R. F. Knacke.
Nature, Vol. 269, 132 - 134 (1977).

Several infrared bands found recently in astronomical sources have not been assigned to specific substances in a convincing way. The author proposes here a new substance, consisting of carbonaceous material, which may be responsible for some or all of the unidentified infrared features.

131.027 Interstellar cyanoacetylene $- J = 2 \to 1, J = 4 \to 3$ transitions.
R. X. McGee, M. Balister, L. M. Newton.
Mon. Not. R. Astron. Soc., Vol. 180, 585 - 592 (1977).

The authors report the first detection of the 36.4 GHz $J = 4 \to 3$ rotational transition of interstellar cyanoacetylene (HC_3N). The line was seen in emission from Sagittarius B2 and from the Mira variable, W Hydrae. The 18.2 GHz $J = 2 \to 1$ line from Sgr B2 was reobserved to remove some confusion about the intensity of this line. The profiles of the lower-order ($J = 1 \to 0, 2 \to 1, 4 \to 3$) lines in Sgr B2 are compared. The intensities of the transitions up to $J = 14 \to 13$ suggest a kinetic temperature of ~ 12 K for the region of origin of the Sgr B2 interstellar molecular lines.

131.028 Time variations of interstellar water masers: strong sources in H II regions.
L. T. Little, G. J. White, P. W. Riley.
Mon. Not. R. Astron. Soc., Vol. 180, 639 - 656 (1977).

The strong water-line sources W3OH, Orion A, W49N and W51S have been monitored over the period 1974 September – 1976 October. Variations on a variety of timescales from years to days have been observed. W49 and W51S exhibit more rapid variations than W3OH or Orion A. A comparison of the present results with those of Sullivan during 1969–70 has been made.

131.029 Laboratory investigation of some ion–molecule reactions related to cyanide chemistry in dense interstellar clouds.   J. P. Liddy, C. G. Freeman, M. J. McEwan.
Mon. Not. R. Astron. Soc., Vol. 180, 683 - 689 (1977).

Laboratory rate coefficients of a number of ion–molecule reactions of charged and neutral species derived from HCN are presented. A mechanism based on these ion–molecule reactions is given for the formation of HCN and HNC via H_2CN^+ in dense interstellar clouds.

131.030 Detection of interstellar DNC: difficulties of chemical equilibrium hypothesis for enrichment.
P. D. Godfrey, R. D. Brown, H. I. Gunn, G. L. Blackman, J. W. V. Storey.
Mon. Not. R. Astron. Soc., Vol. 180, 83P - 86P (1977).

The $J = 1 \to 0$ transition of DNC at 76.3058 GHz has been observed in emission in NGC 2264. Comparison with previous observations of $HN^{13}C$ indicates that deuterium is enriched in DNC similarly to the enrichment reported for DCO^+ in this source. The DNC/HNC ratio is estimated to be about $^1/_{24}$.

131.031 Observation of $J = 1 \to 0$ emission from $H^{15}NC$.
R. D. Brown, P. D. Godfrey, H. I. Gunn, G. L. Blackman, J. W. V. Storey.
Mon. Not. R. Astron. Soc., Vol. 180, 87P - 89P (1977).

Emission from the $J = 1 \to 0$ transition of $H^{15}NC$ has been detected in the direction of DR21 (OH). The transition frequency of 88865.69 MHz was measured in the laboratory by microwave absorption spectroscopy. The $^{15}N/^{13}C$ isotopic abundance ratio of 1.01 for DR21 (OH) is larger than those calculated from isotopes of HCN in other interstellar clouds, perhaps implying a localized enrichment in ^{15}N in DR21(OH).

131.032 Molecules of the interstellar medium and cometary nuclei.   O. V. Dobrovol'skij, E. A. Kajmakov, I. N. Matveev.
Komety Meteory, No. 26, p. 3 - 8 (1977). In Russian.

A comparison is made between the data on interstellar

molecules and numerous experimental results on simulating a primitive earth atmosphere. A possible conclusion is that interstellar medium molecules and cometary parent molecules would be genetically identical. It is possible that disintegration of such molecules could result in basic cometary radicals.

131.033 Fragmentation of magnetic interstellar clouds by ambipolar diffusion. II. Fragments of 1, 10, and 100 $M_\odot$ in various conditions.   T. Nakano.
Publ. Astron. Soc. Japan, Vol. 29, 197 - 205 (1977).

The contraction due to ambipolar diffusion is investigated for fragments with various masses embedded in a massive interstellar cloud with density $\bar{n}$ and in a magnetic field strength B. The contraction time is found to be a decreasing function of $\bar{n}$ and of the order of or less than 10^7 yr at $\bar{n} \gtrsim 10^4$ cm^{-3}. Therefore, the total time of star formation after the parent cloud of ordinary density begins to contract is not much longer than the free-fall time, 10^7 yr, from a state with a density 10^{-23} g cm^{-3}. The contraction time due to ambipolar diffusion is insensitive to the mass of the fragment.

131.034 On the interstellar ultraviolet radiation field.
P. Joshi, S. P. Tarafdar.
Astron. Astrophys., Vol. 60, 285 - 289 (1977).

An estimate of the interstellar ultraviolet radiation field towards o Per and ζ Oph has been obtained by studying the ionization equilibria of interstellar C, S, Fe and Ca. The radiation field has a peak at ∼1600 Å and decreases more steeply towards shorter wavelengths than indicated by earlier calculations.

131.035 Radiating cosmic dust.   C. D. Andriesse.
Vistas Astron., Vol. 21, 107 - 190 (1977).

This monograph originates from a seminar on the role of solid particles in the infrared emission from the interstellar medium. It was held during the spring of 1973 for staff and students of the Kapteyn Astronomical Institute in Groningen. The present text gives the subjects of the seminar and the subsequent study. Section1 deals with the play of electromagnetic radiation in individual grains and then with cosmic radiation fields in collections of grains. Sections 2 and 3 are devoted to the probable dielectric properties of the grains and their interaction cross-sections. In section 4 on grain temperatures the step is made towards the collective behaviour in the interstellar space. Section 5 and 6 describe radiation effects in interstellar and circumstellar dust clouds.

131.036 A survey of interstellar molecular hydrogen. I.
B. D. Savage, R. C. Bohlin, J. F. Drake, W. Budich.
Astrophys. J., Vol. 216, 291 - 307 (1977).

The Copernicus ultraviolet telescope was used to survey the column densities of interstellar H_2 in the $J = 0$ and 1 rotational levels of the $v'' = 0$ vibrational state toward 109 stars, including 26 measurements collected from previous publications. In most cases, the H_2 lines exhibit strong damping wings; and column densities are derived by fitting damping profiles to the observed spectra. The following averages are obtained for matter in the galactic plane within 500 pc of the Sun: $\langle n(H_2) \rangle = 0.143$ cm^{-3}, $\langle n(H\,I) \rangle = 0.86$ cm^{-3}, and $\langle f \rangle = 0.25$. For stars with $N(H_2)$ larger than 10^{18} atoms cm^{-2}, the $N(1)/N(0)$ population ratio provides a direct measure of cloud kinetic temperatures. The values of T_{01} range from 45 to 128 K, with average over 61 stars of 77 ± 17 (rms) K.

131.037 Excitation of OH toward interstellar dust clouds.
R. M. Crutcher.
Astrophys. J., Vol. 216, 308 - 319 (1977).

Observations of the 18 cm OH lines have been performed toward and just off the continuum sources W40 and 3C 123. Toward both clouds each of the two satellite lines is inverted in one wing of the line profile and anti-inverted in the other

wing. Pumping by infrared photons with overlapping of the infrared lines of OH probably explains the satellite-line data. Definite departures from thermal equilibrium are found for the OH main lines toward both clouds. CO observations of both clouds have also been performed. These data suggest that the W40 H II region is embedded in the far side of the molecular clouds and is heating the dust near it.

131.038 Variability of intensity of interstellar maser lines due to induced Compton scattering.   C. Montes.
Astrophys. J., Vol. 216, 329 - 345 (1977).

Induced (nonlinear) Compton interaction between spectrally narrow and incoherent radiation, corresponding to interstellar maser lines (OH and H_2O) with a high brightness temperature ($T_b \approx 10^{15}$K), and the surrounding rarefied plasma is considered in order to interpret the strong variability of intensity for unsaturated maser lines. The narrow amplified line of width $\delta \nu$ exhibits large-amplitude relaxation oscillations, generating at each oscillation a broad satellite at the Compton-Doppler distance $\langle \Delta \nu_D \rangle \gg \delta \nu$, on the longward side of the line, which behaves as a photon soliton moving downward on the frequency axis at constant speed and constant amplitude.

131.039 Walker No. 67 in NGC 2264: a candidate for strong interstellar circular polarization.
R. S. McMillan.
Astrophys. J., Lett., Vol. 216, L41 - L43 (1977).

The heavily obscured B2 V star Walker No. 67 in the young cluster NGC 2264 shows a wavelength dependence of interstellar linear polarization with a peak of 5.5% ± 0.2% at $\lambda_{max} = 0.88$ μm ± 0.03 μm. Both the infrared color excesses and the cluster distance modulus imply a ratio of total to selective extinction between 4 and 5, consistent with the large λ_{max}. These results indicate that large interstellar grains persist in a localized, dense dust cloud near the center of the cluster even after most of the dust has been dispersed by hot, luminous stars. A rotation of polarization position angle of $14°$ per μm of wavelength suggests strong circular polarization.

131.040 Detection of C_2 in the interstellar spectrum of Cygnus OB2 Number 12 (VI Cygni Number 12).
S. P. Souza, B. L. Lutz.
Astrophys. J., Lett., Vol. 216, L49 - L51 (1977), with a correction in Vol. 218, L31 (1977).

The authors have detected absorption lines of the $(1-0)$ Phillips band of C_2 in the near-infrared spectrum of Cyg OB2 No. 12. They estimate a line-of-sight column density for the $J = 2$ level of 5.4×10^{13} cm^{-2}. The relative rotational intensities are consistent with an excitation temperature of 30–40 K, which would indicate a total C_2 column density near 1×10^{14} cm^{-2}. The nature of the absorbing region is discussed.

131.041 The far-infrared spectrum of the core of Sagittarius B2.   E. F. Erickson, L. J. Caroff, J. P. Simpson, D. W. Strecker, D. Goorvitch.
Astrophys. J., Vol. 216, 404 - 407 (1977).

The measured spectrum of Sgr B2 from 40 to 200 cm^{-1} is smooth and featureless with a broad maximum at ∼85 cm^{-1}. The data can be fitted analytically with a model corresponding to thermal emission by a uniform slab of dust filling the beam, with an average temperature of ∼32 K, an optical depth at 100 μm of about 1.6, and a spectral index of the dust emissivity about 1.5. The absence of features implies either that the source is optically thick or that the emission spectrum of the individual grains is smooth in this passband. The possible physical significance of this model is discussed.

131.042 A survey of high-velocity interstellar ions.
H. Cohn, D. G. York.
Astrophys. J., Vol. 216, 408 - 413 (1977).

The authors report the results of a survey of the velocity structure in the absorption lines of a number of interstellar species for 30 lines of sight. Of six species chosen for presentation, N I and excited C II emphasize the distribution of neutral interstellar gas and circumstellar H II regions; N II, C II, C III, and Si III show, in addition, the velocity distribution of a previously unidentified type of interstellar gas, with an ionization temperature in the range 20,000 to 100,000 K. Strong Si III and C III absorption at LSR velocities as large as -80 to -120 km s^{-1} is observed in five Orion stars.

131.043 Shock models of high-velocity interstellar Si III.
J. M. Shull.
Astrophys. J., Vol. 216, 414 - 418 (1977).

Recent Copernicus observations of high-velocity interstellar Si III have been interpreted as evidence for collisionally ionized gas at 30,000 to 80,000 K. In this paper, these observations are summarized, and conductive-interface and shock-heating mechanisms are investigated as sources for the temperatures, column densities, and velocity fields.

131.044 Pulsar dispersion measures and Hα emission measures: limits on the electron density and filling factor for the ionized interstellar gas.   R. J. Reynolds.
Astrophys. J., Vol. 216, 433 - 439 (1977).

Observations of Hα emission were made in the directions of 24 high-galactic-latitude ($|b| \geqslant 5°$) pulsars with a 150 mm diameter Fabry-Perot spectrometer. If the pulsar pulse dispersions and the diffuse Hα emission are both produced by the same ionized regions, then a comparison between the dispersion measure and the emission measure in each pulsar direction can be used to set limits on the average electron density $\langle n_e \rangle$ within the ionized regions and on the fraction f of the line of sight through the galactic disk occupied by these regions.

131.045 The molecular cloud associated with the supernova remnant W44.   H. A. Wootten.
Astrophys. J., Vol. 216, 440 - 445 (1977).

Characteristics of the emission in the $J = 1 \to 0$ lines of $^{12}C^{16}O$ and $^{13}C^{16}O$ near the supernova remnant W44 are described. The author observes a broadening of the line width, an intensification of the line strength, and a maximum in the column density at positions just outside the radio continuum shell. The mean velocity gradient indicates that the dense material lies in a partial shell expanding at ~ 4 km s^{-1}. The energy requirements of the shell motion are easily met by the supernova. The observations suggest that the interstellar medium has been compressed and heated by the remnant; the density enhancement produced in the ambient medium by the remnant is estimated.

131.046 On the molecular hydrogen emission at the Orion Nebula.   J. Kwan.
Astrophys. J., Vol. 216, 713 - 723 (1977).

The author proposes that the most likely explanation for the molecular hydrogen emission is collisional excitation in a shock-heated region. He derives the conditions, namely, pre-shock densities and shock velocities, that are required to give rise to the observed emission. The emission of molecular hydrogen and the cooling of the gas behind the shock are modelled. The results from numerical calculations are presented. A possible origin for the shock is discussed, as well as future experimental tests on this model of the molecular hydrogen emission.

131.047 Copernicus studies of interstellar material in the Perseus II complex. III. The line of sight to ζ Persei.   T. P. Snow, Jr.
Astrophys. J., Vol. 216, 724 - 737 (1977).

Ultraviolet spectrophotometric data obtained with Copernicus are used to analyze the physical conditions and chemical abundances in the line of sight to the B1 Ib star ζ Persei. It is found that, in addition to the principal cloud,

there is a warm cloud which contains substantial amounts of Si III. Over the entire line of sight, about 60% of the hydrogen nuclei are in the form of H_2. The column densities are estimated from a two-component curve of growth, and show a depletion pattern. The derived silicon abundance toward ζ Per is unusually high, but this is most likely an effect of confusion between stellar and interstellar components, and is therefore probably not real. The density in the portion of the cloud containing the observed CO is of order $n_H \approx 10^3$ cm^{-3}, as estimated from the CO rotational temperature. Other molecules detected are HD and OH. The ultraviolet extinction curve shows an unusually steep rise at short wavelengths, much like o Per.

131.048 Emission from highly excited rotational states of HC$_3$N in dense clouds.
M. Morris, R. L. Snell, P. Vanden Bout.
Astrophys. J., Vol. 216, 738 - 746 (1977).

Observations of emission in the $J = 10 \to 9$, $14 \to 13$, $15 \to 14$, and $16 \to 15$ transitions of cyanoacetylene have been made toward several galactic sources. Analysis of the relative line intensities in Orion Molecular Cloud 1 (OMC-1) and Sgr B2 indicates that the emission arises from relatively dense regions which appear to be clumped on a scale much smaller than 2'. A new, very dense component of the north-south ridge in OMC-1 has been found.

131.049 Detection of interstellar DNC.
R. L. Snell, H. A. Wootten.
Astrophys. J., Lett., Vol. 216, L111 - L114 (1977).

The first detection of interstellar DNC has been obtained in the $J = 2 \to 1$ line in the sources NGC 1333, Ori A, NGC 2264, and in the dark cloud L134. An abundance ratio [HN^{13}C]/[DNC] ≤ 3.2 is found in NGC 2264, indicating substantial deuterium enhancement. An even larger enhancement is indicated in L134. Proposed fractionation chemistries can account for such enhanced ratios in regions of low temperature. The variation in the [HNC]/[DNC] abundance ratio may be a stronger function of physical conditions in clouds than of the [D]/[H] ratio.

131.050 Diffuse interstellar lines and the latest experimental laboratory data.   F. M. Johnson.
Bull. American Astron. Soc., Vol. 9, 429 (1977). — Abstract.

131.051 Polarization properties of the $v = 1$, $J = 2 \to 1$ SiO maser.
F. O. Clark, D. R. Johnson, C. E. Heiles, T. H. Troland.
Bull. American Astron. Soc., Vol. 9, 429 (1977). — Abstract.

131.052 Elemental depletions in the interstellar medium.
W. W. Duley, T. J. Millar.
Bull. American Astron. Soc., Vol. 9, 429 (1977). — Abstract.

131.053 Chemistry and observations of NH$_2$D.
D. H. Blake, P. Palmer.
Bull. American Astron. Soc., Vol. 9, 429 - 430 (1977). Abstract.

131.054 Detection of interstellar DNC.
R. L. Snell, H. A. Wootten.
Bull. American Astron. Soc., Vol. 9, 430 (1977). — Abstract.

131.055 The structure of the molecular cloud Sgr B2.
L. J. Caroff.
Bull. American Astron. Soc., Vol. 9, 430 (1977). — Abstract.

131.056 A search for weak interstellar lines.
G. J. Ferland.
Publ. Astron. Soc. Pacific, Vol. 89, 271 - 273 (1977).

The author reports the results of a search for weak interstellar lines in the spectra of three stars. Several 2.2 Å mm^{-1}

coudé spectrograms of χ^2 Ori, 55 Cyg, and 13 Cep were obtained with the 2.7-m reflector of the McDonald Observatory and were co-added to produce a synthesized spectrum for each star. The spectrograms were searched for weak new interstellar lines with negative results. Assuming a Ca depletion factor of 0.003 the author finds that Al is depleted by at least 0.03.

131.057   Survey of water vapor sources in the southern hemisphere.   P. Kaufmann, S. Zisk, E. Scalise, Jr., R. E. Schaal, R. H. Gammon.
Astron. J., Vol. 82, 577 - 586 (1977).
    An extensive, low-flux-limit survey of southern hemisphere celestial water vapor sources has been carried out. Observations were made at 115 different potential source positions, with upper limits for line detection ranging from 7 to 80 Jy. Fifteen new galactic H_2O masers were discovered, and one weak one was confirmed. Some of them were found in H II regions in which type-I OH emission has not yet been detected.

131.058   CH radio emission towards the W49 region.   A. Sume, W. M. Irvine.
Astron. Astrophys., Vol. 60, 337 - 343 (1977).
    The Onsala 25.6 m radio telescope equipped with a travelling wave maser radiometer has been used in a study of the W49 region in the 3.3 GHz radio lines of CH. Observations were made towards the two continuum peaks W49A and B, as well as in eight surrounding points. The purpose of the investigation was to see if the extended CH spiral arm gas could be traced off the continuum background source, and also to estimate the excitation temperature of this gas by the method of on/off continuum source observations both towards W49A and B.

131.059   Radio observations of CH towards various galactic objects.   A. Sume, W. M. Irvine.
Astron. Astrophys., Vol. 60, 345 - 352 (1977).
    The Onsala 25.6 m radio telescope equipped with a travelling wave maser radiometer has been used in observations of the 3.3 GHz radio lines of CH towards various galactic objects. The aim of the study was to extend the investigation of interstellar CH to small nebulae, with infrared and/or radio continuum radiation or with line radiation from an associated molecular cloud. Such regions are thought to be sites of recent or continuing star formation. The authors also observed towards two stars with known molecular envelopes, as well as in the direction of a planetary nebula, which represent objects where CH has not been found or searched for, respectively.

131.060   An interpretation of galactic observations of CNO isotopes.   J. J. Cowan, W. K. Rose.
Astrophys. J., Vol. 217, 51 - 55 (1977).
    The authors use nuclear network calculations to interpret observations of [^{15}N]/[^{14}N] and [^{17}O]/[^{18}O] in interstellar clouds. Recent observations of the [^{15}N]/[^{14}N] ratio in the galactic disk and the galactic center are consistent with the idea that ^{15}N is produced in intermediate-mass stars by means of the hot CNO tri-cycle. It is suggested that low-mass stars in the galactic center could be responsible for producing the local enhancement of ^{17}O by low-temperature ($T < 10^8$ K) CNO burning.

131.061   The detailed structure of CO in molecular cloud complexes. I. NGC 6334.
H. R. Dickel, J. R. Dickel, W. J. Wilson.
Astrophys. J., Vol. 217, 56 - 67 (1977).
    Observations of both ^{12}CO and ^{13}CO emission and of H_2CO absorption have been used in conjunction with previous OH, infrared, radio continuum, and optical information to derive a model of the molecular cloud complex associated with the diffuse H II region NGC 6334.

131.062   A search for interstellar NaH in diffuse clouds.   T. P. Snow, Jr., W. H. Smith.
Astrophys. J., Vol. 217, 68 - 70 (1977).
    High-resolution scans of the NaH $(A-X)$ 8–0 $R(0)$ line 3990.88 Å were made. Upper limits (2 σ) of 0.52 mÅ and 1.37 mÅ were placed on the equivalent width in the lines of sight to ζ Oph and σ Sco, respectively, which resulted in column density limits of 1.7×10^{11} cm^{-2} and 4.4×10^{11} cm^{-2}. These low limits imply that NaH is not formed efficiently.

131.063   Anisotropic scattering in dark clouds and formaldehyde lifetimes.   C. Bernes, A. Sandqvist.
Astrophys. J., Vol. 217, 71 - 77 (1977).
    Using the ray-tracing technique for solving radiative-transfer problems in spherical symmetry, the authors have developed a procedure for determining the radiation intensity in models of dark clouds exposed to the interstellar radiation field and/or to radiation from embedded stars. To this end, a source function that includes anisotropic scattering has been derived. By studying models of the ρ Ophiuchi cloud complex, they find that higher opacities than those implied by star counts are necessary if photodissociation of formaldehyde by interstellar radiation in the outer parts of the cloud is not to occur. Furthermore, an increased excitation temperature or a high rate of photodissociation, or a combination thereof, may be the cause of the 6 cm H_2CO line intensity decrease observed at the position of an embedded star cluster.

131.064   On the departure from translational equilibrium for interstellar molecular hydrogen.
B. Shizgal.
Astrophys. J., Vol. 217, 78 - 82 (1977).
    The effects of the excitation of the $J = 2$ rotational level on the velocity-distribution function of interstellar molecular hydrogen were examined recently by Gould and Levy. The effects could be important in estimating the cooling rate of molecular clouds. A Boltzmann equation suitably modified to include the excitation of the $J = 2$ level was considered in their study. In the present work, an exact solution of this Boltzmann equation is obtained. The recent results obtained by Gould and Levy for $X_t \sim 1$ overestimate the correction to the equilibrium rate by a factor of approximately 10^3. The results of the present work do indicate that significant departures from equilibrium may occur for much smaller values of X_t that correspond to very high temperatures. However, for high temperatures, a Boltzmann equation, other than the one employed to date, which includes excitation and de-excitation of several rotational levels, would have to be considered.

131.065   A mechanism for heating dense interstellar clouds.   T. W. Hartquist.
Astrophys. J., Lett., Vol. 217, L45 - L46 (1977).
    The rate of rotation of an interstellar cloud tends to increase as it contracts. As differential rotation occurs within a cloud, magnetic field lines are bent and twisted. The deformation of the magnetic field lines produces hydromagnetic waves and leads to a damping of the rotation. A large fraction of the rotational energy is transformed into the energy of hydromagnetic waves which heat the cloud.

131.066 ^{30}SiO in the interstellar medium.   F. O. Clark, F. J. Lovas.
Astrophys. J., Lett., Vol. 217, L47 - L48 (1977).
    The ^{30}SiO isotopic form of SiO has been detected in Orion and Sagittarius B2. The Orion isotopic abundances are decidedly terrestrial.

131.067   Encounters between stars and dense interstellar clouds.   R. J. Talbot, Jr., M. J. Newman.

Astrophys. J., Suppl. Ser., Vol. 34, 295 - 308 (1977).

From analysis of statistics of the properties of dense interstellar clouds, the authors show that a disk star of the age of the Sun has probably passed through about 135 clouds of $n(H) \gtrsim 10^2$ cm^{-3} and about 16 clouds with $n(H) \gtrsim 10^3$ cm^{-3}. They have evaluated the influence of cloud density and encounter velocity upon the possibilities for reversal of the stellar wind flows into an accretion flow, including the effects of ionization of the accreting material.

131.068 Models of interstellar clouds. I. The Zeta Ophiuchi cloud. J. H. Black, A. Dalgarno.
Astrophys. J., Suppl. Ser., Vol. 34, 405 - 423 (1977).

Techniques for constructing detailed models of diffuse interstellar clouds are described. The models are designed to reproduce the observational data on the abundance and rotational excitation of H_2, the abundances of other molecules and of atomic ions. Results are presented for the ζ Ophiuchi cloud. A model with two components accounts adequately for all the available observational data.

131.069 Infrared studies of star formation.
M. W. Werner, E. E. Becklin, G. Neugebauer.
Science, Vol. 197, 723 - 732 (1977).

131.070 Zusammenstoß im Weltall. Begegnet eine interstellare Wolke dem Sonnensystem?
A. Wittmann.
Umschau, 77. Jahrg., 608 - 609 (1977).

Recent measurements of the interstellar hydrogen and deuterium density have revealed the existence of a fairly dense interstellar cloud in the neighbourhood of the solar system. The cloud is located in the direction toward the constellations of Scorpius and Ophiuchus near the line of sight to the galactic centre. It seems to approach the solar system which it might have reached in approximately 5000 years from now. During its lifetime the solar system has undergone approximately 150 encounters with interstellar clouds, some of which may have been accompanied by considerable changes of the climatic conditions on earth.

131.071 Observations of 7-mm SiO sources.
J. H. Spencer, P. R. Schwartz, J. A. Waak, J. M. Bologna.
Astron. J., Vol. 82, 706 - 710 (1977).

Detection of 11 new stellar 7-mm SiO maser sources and possible detection of two others is reported. Two have been previously reported at 3.5 mm. A survey of H II region OH and H_2O sources and other nonstellar objects for SiO emission yielded no detections.

131.072 Temperature distribution of neutral hydrogen at high galactic latitudes.
J. Dickey, E. E. Salpeter, Y. Terzian.
News Lett. Astron. Soc. N.Y., Vol. 1, No. 2, p. 7 (1977).
Abstract.

131.073 Det tomme rum. J. K. Knude.
Astron. Tidsskr., Årg. 10, 79 - 81 (1977).

131.074 The size distribution of interstellar grains.
J. S. Mathis, W. Rumpl, K. H. Nordsieck.
Astrophys. J., Vol. 217, 425 - 433 (1977).

The observed interstellar extinction over the wavelength range 0.11 μm $< \lambda <$ 1 μm was fitted with a very general particle size distribution of uncoated graphite, enstatite, olivine, silicon carbide, iron, and magnetite. Combinations of these materials, up to three at a time, were considered. The cosmic abundances of the various constituents were taken into account as constraints on the possible distributions of particle sizes.

131.075 Observations of the SiO and H_2O masers in Orion A.
J. M. Moran, K. J. Johnston, J. H. Spencer, P. R. Schwartz.
Astrophys. J., Vol. 217, 434 - 441 (1977).

Observations of the $v = 1, J = 1 \rightarrow 0$ transition of SiO ($v = 43122.0$ MHz) were made toward Orion A in 1975 August and 1976 January with the Haystack antenna. The spectrum consisted of at least nine distinct spectral components. No detailed correspondence was found between these features and those identified in the $v = 1, J = 2 \rightarrow 1$ transition by Snyder and Buhl in 1973. A VLBI experiment was performed with three antennas on the SiO maser in Orion at 43 GHz in 1975 August. The fringe spacings varied between 0".002 and 0".006. No fringes were detected, and so the SiO features were probably larger than 0".003 or 1.5 AU.

131.076 The energetics of molecular clouds. I. Methods of analysis and application to the S255 molecular cloud.
N. J. Evans II, G. N. Blair, S. Beckwith.
Astrophys. J., Vol. 217, 448 - 463 (1977).

This paper is the first of a series that result from a program to collect complete molecular line, radio continuum, and infrared data on a group of relatively simple, nearby molecular clouds. These data are used to determine the physical properties of the molecular clouds and to analyze the energetic relationship between the gas and the dust in the clouds. The analysis includes estimates of the kinetic temperature, gas density, gas cooling rate, and dust cooling rate. The cooling rates are then compared to relevant heating rates from collapse, and from radiative heating by nearby stars and embedded infrared sources. In this paper, the methods of analysis are described and applied to the molecular cloud associated with S255.

131.077 Radio detection of nitroxyl (HNO): the first interstellar NO bond. B. L. Ulich, J. M. Hollis, L. E. Snyder.
Astrophys. J., Lett., Vol. 217, L105 - L108 (1977).

A new spectral line which the authors identify as the $J_{K^-K^+} = 1_{01} - 0_{00}$ transition of nitroxyl has been detected in emission from the directions of Sgr B2 and NGC 2024. The authors also report detection of a new unidentified line, U81. 505, which may be a weak maser, and observations of previously detected transitions of ketene (H_2CCO), dimethyl ether (CH_3OCH_3) and cyanoacetylene (HC_3N). They have reduced the identification of previously reported U81.543 to either thioformaldehyde (H_2CS) or $HC_2\,^{13}CN$.

131.078 Detection and significance of carbon recombination lines in diffuse interstellar clouds.
R. M. Crutcher.
Astrophys. J., Lett., Vol. 217, L109 - L112 (1977).

Recombination lines were searched for and detected toward ζ Oph and o Per. The strengths of the radiation fields in the two diffuse clouds are derived, and the implications for the proximity of the clouds to the stars and for the density of the clouds are discussed.

131.079 Tests observationnels sur la formation stellaire.
G. Burki.
Recueil des séminaires, (see 012.004), p. 77 - 79 (1977).

131.080 Méthodes nouvelles de détermination du taux de formation stellaire dans la Galaxie et notamment dans le voisinage solaire. L. Martinet.
Recueil des séminaires, (see 012.004), p. 80 - 89 (1977).

131.081 Rates of star formation. R. B. Larson.
The evolution of galaxies and stellar populations, (see 012.005), p. 97 - 122, with a discussion, p. 122 - 132 (1977).

The author reviews some of the theoretical problems of

star formation in galaxies, some approaches that have been considered in models of galaxy evolution, and some possible observational tests that may help to clarify which processes or models are most relevant.

131.082   Star formation and the gas content of galaxies.
K. C. Freeman.
The evolution of galaxies and stellar populations, (see 012. 005), p. 133 - 148, with a discussion, p. 148 - 156 (1977).

The author talks about two separate topics. The first is the initial mass function for star formation: this function is particularly important for understanding the chemical and dynamical evolution of stellar systems, as well as for understanding star formation itself. The second concerns the problem of gas in S0 galaxies.

131.083   Saturation effects in OH maser clouds.
R. D. Davies, R. S. Booth, J.-N. Perbet.
Mon. Not. R. Astron. Soc., Vol. 181, 83 - 105 (1977).

Recent single telescope and interferometer observations of OH maser sources are used to investigate the important question of the degree of saturation of the maser components. The parameters used include profile shapes, relative line widths and intensities, time variability and relative component positions. The data for the main-line (1665 and 1667 MHz) emission from W3 OH and W75N indicate that the components are most likely unsaturated. The maser emission from each component appears to arise along a column through the OH source rather than in a small spherical region within the source.

131.084   Limits on neutral hydrogen in the direction of PHL
957.   R. D. Davies, R. S. Booth, A. Pedlar.
Mon. Not. R. Astron. Soc., Vol. 181, 1P - 3 P (1977).

It is argued that the $z = 2.3099$ absorption system of PHL 957 may be associated with a cluster or protocluster possibly of the type envisaged by Sunyaev & Zel'dovich. A search has been made for H I emission at the corresponding redshift frequency of 429.139 MHz. Significant upper limits are placed on the H I line-integrals through any such cluster or protocluster.

131.085   Detection of the $J = 9 \rightarrow 8$ transition of interstellar cyanodiacetylene.
L. T. Little, P. W. Riley, D. N. Matheson.
Mon. Not. R. Astron. Soc., Vol. 181, 33P - 35P (1977).

The $J = 9 \rightarrow 8$ transition of cyanodiacetylene has been detected in the Heiles 2 dust cloud. The emission appears to come from a region $\geqslant 9 \times \approx 2$ arcmin in extent, aligned in a direction similar to that of the narrow dust lane in which it is probably embedded.

131.086   Enhanced metal depletions and interstellar H_2 abundances.   R. G. Tabak.
Nature, Vol. 269, 582 - 583 (1977).

The OAO 3 Copernicus ultraviolet satellite has provided important new data on the abundances of elements in the interstellar gas in the direction of many O-and B-stars along lines of sight of low optical depth. This information may make it possible to choose between the following theories of molecular hydrogen formation in interstellar clouds: (1) physical adsorption of H-atoms on to cold dielectric grains and their subsequent recombination and desorption; (2) H_2 recombination on graphite grains, and (3) hydrogen recombination by nonactivated chemisorption on transition metal grains. At present all that can be said is that neither process (1) or (3) can be dismissed without further consideration.

131.087   Interstellar grains as possible cold seeds of life.
V. I. Goldanskii.
Nature, Vol. 269, 583 - 584 (1977).

The cold pre-history of life, of prebiotic evolution in interstellar clouds postulated here is based on several main points. (1) The formation of complex molecules proceeds in the 'dirty-ice' mantles of interstellar grains in diffuse or dense clouds at very low ($T \sim 10 - 20$ K) temperatures. (2) Integration of molecules in surface regions of clouds is ultraviolet-initiated. (3) Integration of molecules in the depths of clouds can be initiated only by long-range cosmic protons. (4) The equilibrium chemical composition of cold interstellar dust can be obtained by relatively simple calculations under special assumptions. (5) When the cloud collapses and a new hot star is created in its centre, the flattened protoplanetary disk formed from the remnants of the cloud continues to be cold. This makes it possible that the formation of complex polymer molecules in cold interstellar dust can provide a real cold pre-history of life.

131.088   Possible detection of the 17-cm hyperfine emission from neutral sodium in the W51 region.
D. W. Goldsmith, E. K. Conklin.
Astron. Astrophys., Vol. 61, L1 - L3 (1977).

A search for the 17-cm hyperfine emission from neutral sodium in the W 51 region was made with the Arecibo radio telescope. A weak feature with brightness temperature 0.019 ± 0.005 K and FWHM of 3.3 ± 1.0 km s^{-1} was found at V_{LSR} = 65.2 ± 0.8 km s^{-1}. The authors believe that this feature is probably real. The brightness temperature and line width imply a neutral-sodium column density of (6.1 ± 1.8) $\times 10^{16}$ cm^{-2}. Some astrophysical implications of this possible detection are discussed.

131.089   Models for interpreting the diffuse galactic light.
P. A. Bastiaansen, H. C. van de Hulst.
Astron. Astrophys., Vol. 61, 1 - 6 (1977).

The stars which act as primary light sources have a wider distribution perpendicularly to the galactic plane than the dust which causes the diffuse galactic light. This situation is approximated by a model in which a fraction F of the starlight is distributed in proportion to the dust and a fraction $1 - F$ illuminates the dust layer from outside. Computations of the mid-layer distribution of diffuse light with latitude are made for various combinations of the single-scattering albedo a and the asymmetry factor g.

131.090 H_2O in Orion: outflow of matter in the last stages of star formation.   R. Genzel, D. Downes.
Astron. Astrophys., Vol. 61, 117 - 126 (1977).

A new map of H_2O masers in Orion shows 12 spatially distinct sources with a range of 170 km s^{-1} in radial velocity. Two types of H_2O maser are present: (1) strong, low-velocity line complexes, concentrated in two sources, each of diameter ≤2″ near the core of the Kleinmann-Low nebula; (2) weak, high-velocity features spread over a zone of diameter 60″ centered on the KL nebula.

131.091   Ice grains in space.   R. Smoluchowski.
NASA Tech. Mem., NASA TM X-3511, (see 012. 010), p. 13 - 14 (1977). – Abstract.

131.092   Contraction of the Orion Nebula cluster – molecular cloud A complex.   F. W. Fallon, H. Gerola, S. Sofia.
Astrophys. J., Vol. 217, 719 - 723 (1977).

Observational evidence is presented in support of the view that the width of the molecular lines in Orion is produced by a contraction of the cloud with a velocity directly proportional to the distance from the cloud center. The main support is provided by a close agreement between the motions of the cloud and that of the stars embedded in it. A further implication of this agreement is that star formation occurred throughout the Orion cloud, rather than only in the dense center as predicted by the conventional picture of star formation.

131.093   New infrared objects associated with OH masers.

N. J. Evans II, S. Beckwith.
Astrophys. J., Vol. 217, 729 - 740 (1977).

Infrared sources have been found at 10 μm near the position of six OH masers which were previously unidentified with any optical or 2 μm object. The OH masers selected for the search have good radio positions and emit principally at 1612 MHz in two distinct velocity ranges. The 2 to 20 μm energy distributions of the sources are extremely red, and four of the objects show silicate absorption features, in striking contrast to previously studied infrared sources associated with such OH masers. Measured OH and infrared fluxes are consistent with pumping of the OH maser by 35 μm photons, but argue strongly against pumping at 2.8 μm. The observations are consistent with a model for these objects as distant (2–10 kpc), luminous ($L \approx 10^4 L_\odot$) stars with very thick circumstellar dust shells at temperatures around 500 K.

131.094 Detection of submillimeter (.870 μm) CO emission from the Orion molecular cloud.
T. G. Phillips, P. J. Huggins, G. Neugebauer, M. W. Werner.
Astrophys. J., Lett., Vol. 217, L161 - L164 (1977).

The authors report the first measurements of the interstellar $J = 3$ to $J = 2$ carbon monoxide line at a wavelength of 870 μm (346 GHz). A major feature of the observed spectra is the great strength of the high-velocity wings of the line. The position and spatial extent of the region of high-velocity emission are determined and compared with shock wave models for CO and H_2 emission.

131.095 Observations of DCO⁺: the electron abundance in dark clouds.   M. Guélin, W. D. Langer,
R. L. Snell, H. A. Wootten.
Astrophys. J., Lett., Vol. 217, L165 - L168 (1977).

The $J = 2-1$ rotational line of DCO^+ has been definitely detected in five molecular clouds, including three dark clouds, L63, L134, and L134 N, and marginally detected in four others. The DCO^+/HCO^+ abundance ratio found at the centers of dark clouds is large and implies a fractional electron abundance of less than 10^{-8}. This low electron density sets constraints on the metals and possibly CO as well as on the hydrogen density.

131.096 Prebiotic polymers and infrared spectra of galactic sources.
N. C. Wickramasinghe, F. Hoyle, J. Brooks, G. Shaw.
Nature, Vol. 269, 674 - 676 (1977).

Infrared absorption features characteristic of molecular dust clouds in the Galaxy may be assigned to complex organic polymers or prebiotic polymers. It could be argued that such highly stable, complex polymers evolve due to radiation processing of molecular mantles on interstellar grains — essentially by a type of natural selection which operates in the interstellar medium. Such interstellar material may also account for a significant fraction of the 'insoluble organic matter' which is found in carbonaceous chondrites.

131.097 Observations of O VI.   E. B. Jenkins.
Topics in interstellar matter, (see 012.011), p. 5 - 16 (1977).

A useful spectroscopic tracer for a hot phase of interstellar gas is the O VI ion. Presently, over 70 stars have been observed for O VI absorption by the Copernicus satellite. Nearly all of the stars show broad, weak lines, but no evidence favoring a circumstellar origin for the gas can be found. The relative volume in space occupied by the hot gas regions (and hence their internal density) is uncertain, but a filling factor in the range 0.02 to 0.2 seems most plausible. Fluctuations in radial velocities and column densities suggest there are roughly 6 regions per kpc, each with $N(O\ VI) \approx 10^{13}$ cm^{-2}.

131.098 Nature, origin, and evolution of the hot gas: interstellar?   C. F. McKee.

Topics in interstellar matter, (see 012.011), p. 27 - 33 (1977).

Supernova explosions in a cloudy interstellar medium produce large volumes of hot gas with typical density. 3 × 10^{-3} cm^{-3} and temperature 5 × 10^5 K. The evolution of supernova remnants in such a medium differs significantly from the conventional picture due to evaporation of cool clouds by the hot gas inside the remnant. A steady state model of the resulting three-component medium is in reasonable agreement with observations of the interstellar pressure, the O VI density, and the diffuse soft X-ray emission.

131.099 Circumstellar masers.   L. E. Snyder.
Topics in interstellar matter, (see 012.011), p. 97 - 104 (1977).

The newest circumstellar maser, silicon monoxide, is discussed and its relationship to the other known circumstellar masers, hydroxyl and water, is briefly explored. Most silicon monoxide masers are associated with post-main sequence objects which are also water and hydroxyl masers and from this association new interpretations of maser velocities, geometries and distances have been found.

131.100 Observations of molecular clouds.   B. Zuckerman.
Topics in interstellar matter, (see 012.011), p. 107 - 112 (1977).

Observations of interstellar molecules in regions of star formation are summarized. It is concluded that kinematics in molecular clouds are still poorly understood.

131.101 Isotopic abundances in interstellar clouds.
C. H. Townes.
Topics in interstellar matter, (see 012.011), p. 113 - 123 (1977).

The observation of microwave spectra of molecules in interstellar clouds allows separation and detection of the lines of isotopes of many of the more common elements. Comparison of intensities of isotopic lines shows that the relative isotopic abundances for C, O, S, N, and Si are generally rather similar to those found on Earth. However, there are interesting and provocative differences.

131.102 Formation and excitation of molecular hydrogen.
A. Dalgarno.
Topics in interstellar matter, (see 012.011), p. 125 - 133 (1977).

131.103 Progress in interstellar molecule formation.
W. D. Watson.
Topics in interstellar matter, (see 012.011), p. 135 - 147 (1977).

Selected observational, laboratory and theoretical results from the past few years which are significant for delineating interstellar molecule reactions are presented.

131.104 The nature of dust grains.   P. G. Martin.
Topics in interstellar matter, (see 012.011), p. 149 - 154 (1977).

This review is concerned not so much with the nature of the interstellar dust particles as with the nature of the investigations into the dust properties which have been going on in the three years preceding this General Assembly.

131.105 Formation and destruction of grains.
N. C. Wickramasinghe.
Topics in interstellar matter, (see 012.011), p. 155 - 162 (1977).

Refractory grains, consisting of graphite, SiC, silicate and iron particles, may form in mass flows from cool stars, novae and possibly supernovae. Tarry polymeric mantles grow under conditions which prevail in massive molecular clouds. A large fraction of polymer-coated grains could be expelled into the general interstellar medium, and such grains could be responsible for the bulk of the observed interstellar extinction at optical wavelengths. Grain destruction occurs mainly by direct involvement in star formation.

131.106 The large-scale distribution of interstellar matter in the context of the density-wave theory.
W. W. Roberts, Jr., W. B. Burton.
Topics in interstellar matter, (see 012.011), p. 195 - 205 (1977).

The theoretically viable prospect that density waves and the associated galactic shock fronts are present in disk-shaped galaxies has received support in recent years from a variety of observational studies. Large-scale shocks in the interstellar gas may play an important role in determining the kinematics and the relative distribution of various galactic tracers. This is particularly apparent in some external spirals, because of the advantageous perspective, and for the tracers H I and CO in our own Galaxy. Simulation of CO observations according to the precepts of the density-wave theory shows that these precepts are supported by several observational results.

131.107 The helium problem.   M. Peimbert.
Topics in interstellar matter, (see 012.011), p. 249 - 254 (1977).

Helium abundance determinations based on observations of interstellar matter are reviewed. Some of the conditions that these results impose on stellar models and cosmological models are discussed.

131.108 Interstellar grains and interstellar molecules. Evidence for grain clumps?   G. Winnewisser.
Naturwissenschaften, 64. Jahrg., 526 (1977).

131.109 Comments on the origins of the diffuse interstellar bands.
W. H. Smith, T. P. Snow, Jr., D. G. York.
Astrophys. J., Vol. 218, 124 - 132 (1977).

Most work in recent years on the unidentified diffuse interstellar bands has been concentrated on the hypothesis that they are produced by some process in solid grains. The authors review that hypothesis and the arguments which support it, and find that diffuse band formation by certain classes of molecules should not yet be ruled out. Since several relevant observations have been made in the last 4 years, the observational data on the bands are reviewed. It would appear that these data are as consistent with molecular origins for the bands as with grain origins. Types of molecular transitions which give rise to "diffuse" bands are described.

131.110 The H_2CO absorption toward IC 1318 b-c in Cygnus.
H. R. Dickel, A. W. Seacord II, S. T. Gottesman.
Astrophys. J., Vol. 218, 133 - 141 (1977).

Formaldehyde absorption at 6 cm has been mapped in the prominent dust lane which separates the optical nebulae IC 1318 b and c. The apparent optical depths of H_2CO range between 0.2 and 0.8 within the dust lane and at most ~0.1 where the visual absorption falls below ~4 to 4.5 mag. However, there is not a detailed correlation between the observed H_2CO absorption and the dust. In the region of highest visual extinction, the 6 cm absorption is reduced, but here 2 mm emission from more highly excited levels of H_2CO is observed instead. It is concluded that the H II regions are physically associated with the dust and molecules. The (LSR) radial velocities of the H_2CO absorption are similar to those of the $H\alpha$ emission.

131.111 A theory of the interstellar medium: three components regulated by supernova explosions in an inhomogeneous substrate.   C. F. McKee, J. P. Ostriker.
Astrophys. J., Vol. 218, 148 - 169 (1977).

Supernova explosions in a cloudy interstellar medium produce a three-component medium in which a large fraction of the volume is filled with hot, tenuous gas. In the disk of the Galaxy the evolution of supernova remnants is altered by evaporation of cool clouds embedded in the hot medium. Radiative losses are enhanced by the resulting increase in density and by radiation from the conductive interfaces between clouds and hot gas. A self-consistent model of the interstellar medium developed accounts for the observed pressure of interstellar clouds, the galactic soft X-ray background, the O VI absorption line observations, the ionization and heating of much of the interstellar medium, and the motions of the clouds.

131.112 Absorption of X-rays in the interstellar medium.
S. K. Ride, A. B. C. Walker, Jr.
Astron. Astrophys., Vol. 61, 339 - 346 (1977).

In order to interpret soft X-ray spectra of cosmic X-ray sources, it is necessary to know the photoabsorption cross-section of the intervening interstellar material. Current models suggest that the interstellar medium contains two phases which make a substantial contribution to the X-ray opacity: cool, relatively dense clouds that exist in pressure equilibrium with hot, tenuous intercloud regions. The authors have computed the soft X-ray photoabsorption cross-section (per hydrogen atom) of each of these two phases. The calculations are based on a model of the interstellar medium which includes chemical evolution of the Galaxy, the formation of molecules and grains, and the ionization structure of each phase. These cross-sections of clouds and of intercloud regions can be combined to yield the total soft X-ray photoabsorption cross-section of the interstellar medium.

131.113 The interstellar medium in the direction of the Crab nebula: reconciling soft X-ray and radio observations.   S. K. Ride, A. B. C. Walker, Jr.
Astron. Astrophys., Vol. 61, 347 - 352 (1977).

In a previous paper (131.112) the authors computed the soft X-ray photoabsorption cross-section of both interstellar clouds and intercloud regions based on a current model of the interstellar medium. In the present paper, they use these results to calculate the cross-section of the interstellar medium in the direction of the Crab nebula. They obtain a cross-section which incorporates the evolution of interstellar abundances, the presence of molecules, the depletion of heavy elements onto interstellar grains, and the ionization structure of the interstellar medium. Equipped with this cross-section, they re-analyze the soft X-ray spectrum of the Crab observed by Charles et al. (1973).

131.114 The Oort model for interstellar clouds — a Monte Carlo simulation.
M. J. Handbury, S. Simons, I. P. Williams.
Astron. Astrophys., Vol. 61, 443 - 444 (1977).

A Monte Carlo type simulation on a system of colliding coalescing clouds is carried out to determine the steady state distribution for the situation where large clouds are destroyed and replaced by small clouds. This is a representation of the Oort model of interstellar space. Agreement with the observed mass distribution is good, but it is poor with the observed velocity distribution.

131.115 Ionizing background radiation and hydrogen at the periphery of galaxies.
N. G. Bochkarev, R. A. Syunyaev.
Astron. Zh. Akad. Nauk SSSR, Vol. 54, 957 - 966 (1977). In Russian. English translation in Soviet Astron., Vol. 21, No. 5.

Computations of the ionization of the interstellar gas at the periphery of galaxies by background radiation show the existence of a sharp cutoff in the distribution of the surface density of neutral hydrogen. Comparison with observational data gives a possibility to find an upper limit of the intensity of the ionizing background radiation in the spectral region $912 Å > \lambda > 100 Å$. Direct observations of the intergalactic radiation in this range are impossible.

131.116   **Origine e fine delle stelle.**   P. Maffei.
G. Astron., Vol. 3, 1 - 15 (1977).

131.117   **Deuterium enrichment in interstellar HCN and HNC.**   R. D. Brown.
Nature, Vol. 270, 39 - 41 (1977).

It is concluded that the explanation of molecular deuterium enrichments in interstellar HCN and HNC must lie in isotope effects that influence the formation processes rather than equilibria involving HCN or HNC.

131.118   **New OH masers in the direction of H_2O masers.**   J. L. Caswell, R. F. Haynes, W. M. Goss.
Mon. Not. R. Astron. Soc., Vol. 181, 427 - 433 (1977).

The authors discovered five new OH maser emission sources, three of which are probably associated with H_2O masers. This provides further evidence that conditions favourable to H_2O maser emission imply conditions favourable to OH main-line maser emission in the near vicinity, and vice versa.

131.119   **Polysaccharides and the infrared spectrum of OH 26.5 + 0.6.**   F. Hoyle, N. C. Wickramasinghe.
Mon. Not. R. Astron. Soc., Vol. 181, 51P - 55P (1977).

The infrared spectrum of OH 26.5 + 0.6 over the waveband 2–40 μm is explained in terms of a polysaccharide grain model. The very close agreement between theory and observation lends further support to the identification of interstellar polysaccharides.

131.120   **Local interstellar hydrogen and deuterium.**   R. C. Henry, R. Anderson, H. W. Moos, W. McClintock, J. L. Linsky.
J. R. Astron. Soc. Canada, Vol. 71, 396 (1977). – Abstract.

131.121   **Elemental abundances in the interstellar medium.**   W. W. Duley, T. J. Millar.
J. R. Astron. Soc. Canada, Vol. 71, 396 (1977). – Abstract.

131.122   **Isotope fractionation of CO in dark dust clouds.**   R. L. Dickman, W. D. Langer, W. H. McCutcheon, W. L. H. Shuter.
J. R. Astron. Soc. Canada, Vol. 71, 397 (1977). – Abstract.

131.123   **Star formation in the Cas-Per arm.**   V. A. Hughes, M. R. Viner.
J. R. Astron. Soc. Canada, Vol. 71, 397 (1977). – Abstract.

131.124   **Interstellar molecule abundances from gas-phase reactions.**   G. F. Mitchell, J. L. Ginsburg.
J. R. Astron. Soc. Canada, Vol. 71, 397 - 398 (1977). Abstract.

131.125   **Comment naissent les étoiles?**   A. Laval.
Astronomie, Vol. 91, 377 - 386 (1977).

131.126   **Cyanodiacetylene in a dark dust cloud.**   J. M. MacLeod, N. W. Broten, L. W. Avery, T. Oka.
J. R. Astron. Soc. Canada, Vol. 71, 396 (1977). – Abstract.

131.127   **Polarization measurements and extinction near the NGP.**   T. Markkanen.
Highlights of Astronomy, Vol. 4, Part II, (see 012.022), p. 57 - 58 (1977).

131.128   **Time variations and position jitter in interstellar water masers.**   L. T. Little.
Far infrared astronomy (see 012.027), p. 157 - 166 (1976).

Time variations of intensity in interstellar water masers associated with H II regions, and, in particular, the position jitter observed in components of W49 (H_2O) are discussed. It is hard to explain time variations of both intensity and position by random motions of the water clouds within the maser,

so it seems most likely that changes in excitation are responsible for the observed effects.

131.129   **Carbon monoxide observations at 230 GHz.**   A. R. Gillespie, T. G. Phillips.
Far infrared astronomy (see 012.027), p. 167 - 170 (1976).

A 230 GHz CO line receiver is described, and preliminary results of observations of M8, M17, W49, W51 and ρ Ophiuchi are presented.

131.130   **Polyformaldehyde grains.**   A. Cooke, N. C. Wickramasinghe.
Far infrared astronomy (see 012.027), p. 277 - 280 (1976).

Arguments are presented for the occurrence of polyformaldehyde mantles on interstellar grains.

131.131   **Theoretical models of dust clouds.**   M. Rowan-Robinson.
Far infrared astronomy (see 012.027), p. 285 - 297 (1976).

Models of optically thick dust clouds are fitted to the spectra of galactic (Orion and W3) and extragalactic (M82 and NGC 253) sources from 3 μm to 1 mm. If the grains are composed of ice or silicates, then radii greater than 10 μm are required. These giant grains appear to be a general feature of sources peaking in the far infrared.

131.132   **Infrared emission from grains with fluctuating temperatures.**   J. M. Greenberg.
Far infrared astronomy (see 012.027), p. 299 - 307 (1976).

The far infrared radiation by the small (≈ 0.005 μm) grains in the bimodal interstellar size distribution is shown to deviate substantially from that predicted by steady state temperature predictions. In dark clouds it is shown that the radiation corresponds to lower temperature than for the classical (≈ 0.1 μm) grains while the opposite is to be expected from hot clouds.

131.133   **The protostellar origin of interstellar grains.**   J. Silk.
Far infrared astronomy (see 012.027), p. 309 - 319 (1976).

A semi-quantitative description of the origin of interstellar grains in a protostellar environment is given. Grains grow in the protostellar collapse phase and can be expelled by radiation pressure from stars of mass $\gtrsim 5\,M_\odot$. A convective shell of grains forms in which shattering produces a power-law spectrum of fragments. A bimodal distribution of grain sizes, peaking at ~ 0.1 μm and ~ 0.01 μm results. The effects of protostellar winds and rotation are considered.

131.134   **Dust in protoplanetary systems.**   I. P. Williams.
Far infrared astronomy (see 012.027), p. 321 - 325 (1976).

It is of general interest to determine whether other planetary systems exist and by what mechanism they came to be formed. An outline is given of current thinking on planetary formation, outlining the important part played by dust grains in the process.

131.135   **The formation of protostars.**   W. Tscharnuter.
Computing in plasma physics and astrophysics, (see 012.028), p. 1 - 7 (1976).

A review is given of the physical and numerical problems which arise in the theory of star formation. The evolution of protostars based on numerical calculations is qualitatively described. From the discussion of the basic physics conclusions are drawn regarding the demands on the numerical methods. Some new ideas are outlined concerning the treatment of non-stationary radiating spherical shock fronts with the aid of an implicit difference scheme.

131.136   **Numerical calculations of the saturation behaviour of inhomogeneous cosmic masers.**

E. Bettwieser, G. Misselbeck.
Astron. Astrophys., Vol. 61, 567 - 574 (1977).

The saturation behaviour of several models for inhomogeneous, spherical cosmic masers is numerically investigated. The calculations give the values of the intensity and the center-to-limb variation of the emergent radiation, especially the dependence of these quantities on the density of the masing molecules and the pump parameter.

131.137 **H_2O in the Galaxy: sites of newly formed OB stars.**
         R. Genzel, D. Downes.
Astron. Astrophys., Suppl. Ser., Vol. 30, 145 - 168 (1977) = MPI Radioastron., Bonn, Sonderdr. Ser. A, Nr. 218.

Positions and spectra are presented for 82 water vapor sources at 22 GHz. Of this list, 32 sources are new discoveries, and nearly all of the sources probably come from expanding shells around newly formed, massive stars. Many of the H_2O sources have a remarkable symmetry in their low velocity emission, and 14 ($\sim$50%) of the strongest sources show weak, high velocity features. The luminosity function of the sources has its median at $10^{29.5}$ erg s^{-1}. The H_2O sources are near, but do not coincide with the compact H II regions mapped to date. There are two types of infrared and OH maser sources. The first type is associated with the H II regions, and the other type is directly related to the H_2O sources, on the basis of coincidence in position and velocity. The OH maser lines which are related to H_2O sources come from the red or blue shifted parts of a circumstellar shell, at the same velocities as the H_2O emission.

131.138 **Polarization observations of 77 stars within 25 pc**
         from the Sun.   V. Piirola.
Astron. Astrophys., Suppl. Ser., Vol. 30, 213 - 216 (1977).

The observations give an average value for the interstellar polarization per parsec, $\bar{p}_0/\bar{r} = 0.0004\%$ pc^{-1}, within 25 pc from the Sun. The direct mean of the standard errors of the normalized Stokes parameters for the observed stars is $\bar{\epsilon} = \pm 0.008\%$. The stars can be used as near zero polarization standards. HD 142373 = χ Her, which is one of the unpolarized standard stars observed with rotatable tube telescopes, showed variable intrinsic polarization during the period 26–31 August 1974, with a maximum value of $p = 0.082 \pm 0.012\%$ on 28–29 August 1974 UT.

131.139 **A contribution to the study of the Norma dark cloud.**
         U. Haug, K. Bredow.
Astron. Astrophys., Suppl. Ser., Vol. 30, 235 - 244 (1977).

UBV and Hβ data for 250 B type stars in the "Norma dark cloud" have been obtained at the European Southern Observatory and at the Cerro Tololo Inter-American Observatory. They are used to study the absorption with respect to distance. In the whole area the absorption is increasing strongly with distance within 1 kpc, with slight indications for two separate concentrations of absorbing material near the Sun and outward from about 0.7 kpc. Mapping the absorption for four distance ranges reveals remarkable differences in transparency which are closely correlated with the star density on photographs of this region of the Milky Way.

131.140 **Star formation and the Galaxy.**   F. J. Kerr.
         IAU Symp. No. 75, (see 012.029), p. 3 - 19, with a discussion p. 20 - 36 (1977).

This review discusses the galactic context of star formation and considers how large-scale phenomena in the Galaxy can influence the processes of star formation.

131.141 **Molecular clouds.**   P. Thaddeus.
         IAU Symp. No. 75, (see 012.029), p. 37 - 54, with a discussion p. 55 - 67 (1977).

The main subjects discussed are how molecular observations provide data on the physical state of the dense interstellar gas, and the molecular observations of H II regions, stellar associations, and dark nebulae.

131.142 **Kinematics and dynamics of dense clouds.**
         J. Lequeux.
IAU Symp. No. 75, (see 012.029), p. 69 - 94, with a discussion p. 95 - 103 (1977).

This paper reviews the observational evidences for collapse in dense clouds, and also for the factors which can play against collapse (turbulence, rotation, magnetic field). The author examines to which extent the maser sources (OH, H_2O, SiO) can be related to star formation.

131.143 **Infrared observations of star formation regions.**
         C. G. Wynn-Williams.
IAU Symp. No. 75, (see 012.029), p. 105 - 118, with a discussion p. 119 - 132 (1977).

This review is divided into three parts. The first section gives a brief introduction to the different infrared wavelength ranges and to the various kinds of infrared objects seen in regions of star formation. The second section reviews the recent progress in infrared observations, concentrating on the three years since the review by Wynn-Williams and Becklin (1974) was written. The third section describes in more detail four varied examples of star formation regions.

131.144 **Radio observations related to star formation.**
         P. G. Mezger, L. F. Smith.
IAU Symp. No. 75, (see 012.029), p. 133 - 163, with a discussion p. 164 - 177 (1977).

I. An overview: Some basic facts; Radio and sub-mm observations related to star formation; A possible evolutionary sequence of O and B stars and their protostellar shells. II. Radio observations related to the formation of single stars and star clusters: Stars less massive than O stars associated with dust clouds; The star cluster in the Ophiuchus dark cloud; O stars associated with molecular clouds; The Orion region: star formation in an interarm cloud complex; W3: formation of O stars in main spiral arms; Other H II regions: a review of reviews; Molecular masers and star formation. III. Star formation rates in the Galaxy: As a function of density; As a function of time; As a function of distance from the galactic center.

131.145 **Theoretical processes in star formation.**
         L. Mestel.
IAU Symp. No. 75, (see 012.029), p. 213 - 232, with a discussion p. 233 - 247 (1977).

Introduction: fragmentation versus dissipation; Gravitational collapse: the role of the galactic magnetic field; Magnetic fragmentation; Rotating magnetic clouds; The flux-freezing approximation; Weakly magnetic, rotating systems; The limit of fragmentation and the initial mass spectrum.

131.146 **Collapse dynamics and collapse models.**
         R. B. Larson.
IAU Symp. No. 75, (see 012.029), p. 249 - 267, with a discussion p. 268 - 281 (1977).

We still have little direct information about the crucial question of how the material in molecular clouds actually becomes condensed into stars. This report discusses briefly the current status of theoretical attempts to understand this problem, based on calculations of the dynamics of collapsing clouds and protostars.

131.147 **On why we need a good theory of star formation.**
         D. Lynden-Bell.
IAU Symp. No. 75, (see 012.029), p. 291 - 296 (1977).

131.148 **Deuterium and hydrogen in the local interstellar**
         medium.
A. K. Dupree, S. L. Baliunas, H. L. Shipman.
Astrophys. J., Vol. 218, 361 - 369 (1977).

Densities of neutral hydrogen and deuterium are found

from observation with the Copernicus satellite of the Lα line toward two nearby stars. The hydrogen density is 0.03 ± 0.01 cm^{-3} toward α Aur (Capella) and 0.20 ± 0.05 cm^{-3} in the direction of α Cen A, values indicating that the nearby (less than 14 pc) interstellar medium is inhomogeneous and can be of low density in certain directions.

131.149 **Detection of interstellar ethyl cyanide.**
D. R. Johnson, F. J. Lovas, C. A. Gottlieb, E. W. Gottlieb, M. M. Litvak, M. Guelin, P. Thaddeus.
Astrophys. J., Vol. 218, 370 - 376 (1977).

Twenty-four millimeter-wave emission lines of ethyl cyanide (CH$_3$CH$_2$CN) have been detected in the Orion Nebula (OMC-1) and seven in Sgr B2. The high abundance of ethyl cyanide in the Orion Nebula suggests that ethane and perhaps larger saturated hydrocarbons may be common constituents of molecular clouds and have escaped detection only because they are nonpolar, or only weakly so.

131.150 **Radiation transport and non-LTE analysis of interstellar molecular lines. II. Carbon monosulfide.**
H. S. Liszt, C. M. Leung.
Astrophys. J., Vol. 218, 396 - 405 (1977).

The authors present numerical solutions to the coupled equations of radiative transfer and statistical equilibrium in the rotational transitions of carbon monosulfide (CS). Static, spherical, and microturbulent models are constructed over a wide range in density, abundance, and temperature, and the results are compared with observations of massive clouds near H II regions and of dark dust clouds and globules.

131.151 **Radio detection of interstellar N$_2$D$^+$.**
L. E. Snyder, J. M. Hollis, D. Buhl, W. D. Watson.
Astrophys. J., Lett., Vol. 218, L61 - L64 (1977).

The $J = 1$-0 transition of interstellar N$_2$D$^+$ has been detected in emission from the cool dust cloud L134 N. All three of the quadrupole hyperfine components produced by the outer nitrogen nucleus are easily observed. These observations of N$_2$H$^+$ emission from L134 N reveal the quadrupole hyperfine structure produced by both the outer and inner nitrogen nuclei and thus are the most highly resolved N$_2$H$^+$ spectra yet obtained.

131.152 **The hydroxyl masers in the Orion Nebula.**
S. S. Hansen, J. M. Moran, M. J. Reid, K. J. Johnston, J. H. Spencer, R. C. Walker.
Astrophys. J., Lett., Vol. 218, L65 - L69 (1977).

The 1665 MHz hydroxyl maser emission from the Orion Nebula was observed with very long baseline interferometers sensitive to structure between 0".5 and 0".005. Most of the maser components were resolved on all baselines and had apparent angular sizes larger than 0".2.

131.153 **Wavelength dependence of polarization. XXXII. Narrow-band polarization effects in cool stars.**
G. V. Coyne, A. M. Magalhães.
Astron. J., Vol. 82, 908 - 915 (1977).

Linear polarization has been measured for a number of Mira variables and the semiregular variable V CVn in narrow- and intermediate-band filters centered on Hα, Hβ, and the adjoining continuum. The polarization was also monitored from 0.3 to 1.0 μ with wide-band filters. If in V CVn polarizing dust is present, it must be well mixed in the extended atmosphere. The simultaneous increase in the polarization near the Hβ spectral region and relative decrease near the Balmer discontinuity may be explained by a shock wave passing through the extended atmosphere.

131.154 **The formation of molecules in contracting interstellar clouds.**
H. Suzuki, S. Miki, K. Sato, M. Kiguchi, Y. Nakagawa.
Prog. Theor. Phys., Vol. 56, 1111 - 1125 (1976). — Abstr. in

Phys. Abstr., Vol. 80, Abstr. 22294 (1977).

131.155 **Identification of the λ2,200 Å interstellar absorption feature.** F. Hoyle, N. C. Wickramasinghe.
Nature, Vol. 270, 323 - 324 (1977).

A broad absorption feature centred on λ2,200 Å with a half-width of ~300 Å appears in the spectra of reddened stars. This conspicuous feature in the interstellar extinction curve might hold an important clue to the identity of a major component of interstellar matter, but it has defied identification for over a decade. Here the authors identify this band as representing the integrated effect of a set of bicyclic compounds, each with the empirical formula C$_8$H$_6$N$_2$. Such nitrogenated structures could form in stellar mass flows of the type which they have also discussed. A significant mass fraction of all interstellar material might exist in this form.

131.156 **Observations of globules.** M. E. Sim.
Q. J. R. Astron. Soc., Vol. 18, 466 (1977).
Abstract.

131.157 **Observations of the λ 18 cm OH maser emission in the Orion nebula and λ 6 cm formaldehyde absorption in the associated molecular cloud.**
R. S. Booth, R. Few, R. P. Norris.
Q. J. R. Astron. Soc., Vol. 18, 467 (1977). — Abstract.

131.158 **Formation and destruction of dust grains.**
E. E. Salpeter.
Annu. Rev. Astron. Astrophys., Vol. 15, (see 003.012), 267 - 293 (1977).

Contents: 1. An overview; 2. Element abundances; 3. Chemical physics of grain formation and destruction; 4. Optical properties of grains; 5. Circumstellar dust; 6. The interstellar medium.

131.159 **Interstellar scattering and scintillation of radio waves.** B. J. Rickett.
Annu. Rev. Astron. Astrophys., Vol. 15, (see 003.012), 479 - 504 (1977).

One of the major goals of this paper is to review the interstellar scattering and scintillations observations in the light of modern theory for extended spatially homogeneous, power law inhomogeneities. The author presents formulas, with attention to the various numerical factors, for the experimentally observable quantities and compares them with the data.

131.160 **Highly reddened stars.** D. A. Allen.
Modern astronomy, (see 003.013), p. 139 - 148 (1977).

131.161 **Organic chemistry in the Galaxy.** S. Mitton.
Modern astronomy, (see 003.013), p. 161 - 166 (1977).

131.162 **The reddening of K-giant stars from DDO photometry.** K. A. Janes.
Publ. Astron. Soc. Pacific, Vol. 89, 576 - 581 (1977).

A method is described for deriving the interstellar reddening and $(B-V)_0$ value of a single K-giant star from DDO photometry. For Population I stars, the method works well and is independent of composition. It does not work, however, for Population II stars.

131.163 **On the quadratic dependence of interstellar $N(K I)$ and $N(Na I)$ on N_H.** S. P. Tarafdar.
Astron. Astrophys., Vol. 61, 755 - 759 (1977).

The proposed explanations for the quadratic dependence of interstellar $N(K I)$ and $N(Na I)$ on N_H have been re-examined. It has been argued that the implicit assumption of a constant cloud thickness, necessary for the working of these proposed explanations, is not consistent with available observations. An alternative explanation for the approximate quadratic relations in terms of a model including effects of attenuation of the interstellar radiation field has been proposed and found to give reasonable agreement with observations.

131.164 Observations of high-frequency carbon recombination-line emission in NGC 2024 and IC 1795.
L. J. Rickard, B. Zuckerman, P. Palmer, B. E. Turner.
Astrophys. J., Vol. 218, 659 - 667 (1977).
The authors have mapped C76α emission (at 14.697 GHz) toward NGC 2024 and IC 1795. The observations indicate that at this frequency spontaneous emission is the dominant line-formation mechanism in these regions. A comparison of the observed carbon recombination-line spectra with spectra calculated for models of the two emission regions suggests that there are two different carbon-line-emitting clouds in NGC 2024 and possibly in IC 1795. The authors also searched for carbon recombination-line emission toward the dense molecular sources in NGC 2023, NGC 2264, W3(OH), K3-50, and DR 21(OH).

131.165 Observations of ammonia in selected galactic regions.
P. R. Schwartz, A. C. Cheung, J. M. Bologna, M. F. Chui, J. A. Waak, D. Matsakis.
Astrophys. J., Vol. 218, 671 - 676 (1977).
Ammonia emission has been studied from 21 galactic regions associated with interstellar molecules. Optical depths, kinetic temperatures, and a measure of the ammonia clumpiness have been determined.

131.166 The chemical evolution of molecular clouds.
E. Iglesias.
Astrophys. J., Vol. 218, 697 - 715 (1977).
The author presents a study of the non-equilibrium chemistry of dense molecular clouds. The latest published chemical data and most of the new theoretical advances are included in his model. New schemes for the synthesis of several species (e.g., NCO, HNCO, and CN) are proposed. The role played by the adsorption and condensation of molecules on the surface of the dust grains is explored and found important. He demonstrates that the chemical equilibrium time scale and the molecular concentrations are strongly dependent on these processes.

131.167 The star-formation process in molecular clouds associated with Herbig Be/Ae stars. I. LkHα 198, BD +40°4124, and NGC 7129.   R. B. Loren.
Astrophys. J., Vol. 218, 716 - 735 (1977).
Three molecular clouds associated with the young, rapidly evolving Herbig Be/Ae stars are examined in an attempt to determine the triggering mechanism of star formation. The details of the extensive CO and ^{13}CO mapping of each of the molecular clouds from which these young stars formed are presented. The velocity structure of each of these clouds is examined to determine the present conditions within the molecular cloud and to see whether there is any indication of cloud collapse or other mechanism by which star formation might occur.

131.168 Molecular clouds and star formation. I. Observations of the Cepheus OB3 molecular cloud.   A. I. Sargent.
Astrophys. J., Vol. 218, 736 - 748 (1977).
To determine the connection between newly formed stars and molecular clouds, observations were made in and around the young OB association Cepheus OB3 in the $J = 1 \rightarrow 0$ transition of ^{12}CO. An extended (20 pc × 60 pc) molecular cloud was detected and mapped, and additional observations of ^{13}CO and H_2CO were made at selected positions. Moderately enhanced temperatures and densities were noted in several portions of the cloud. It appears that star formation is still continuing in the Cepheus OB3 complex.

131.169 Dust and gas near the Pleiades.   M. Jura.
Astrophys. J., Vol. 218, 749 - 751 (1977).
Observations of molecular hydrogen show that the gas in the line of sight toward 20 Tau (Maia), a prominent star in the Pleiades, is only about 0.1 pc distant from that star. The dust which produces the observed reflection nebulosity is likely to be associated with this gas and therefore lies in front of the star. Since the foreground dust is optically thin, it should be possible to use observations of the polarization and colors to constrain models of interstellar dust, while further Copernicus observations may make it possible to determine the location of the other reflection nebulae in the Pleiades. Far-infrared observations of this region may enable the author to measure the albedo of these grains. The nearness of the interstellar dust and gas to 20 Tau may be important in explaining why the CH^+ column density is unusually high toward this star.

131.170 The interstellar molecules CH and CH^+.
A. Dalgarno.
Atomic processes and applications. P. G. Burke (Editor). North-Holland Publishing Company, Amsterdam, Netherlands. 10 + 533 pp. Price $ 65.95 (1976). ISBN 0-7204-0444-4, p. 109 - 132. – Abstr. in Phys. Abstr., Vol. 80, Abstr. 33625 (1977).

131.171 Polysaccharides as interstellar grains.
K. S. Krishna Swamy.
Bull. Astron. Soc. India, Vol. 5, 82 (1977).

131.172 Theoretical interstellar and prebiotic organic chemistry: a tentative methodology.
R. Caballol, R. Carbo, R. Gallifa, J. A. Hernandez, M. Martin, J. M. Riera.
Origins of Life, Vol. 7, 163 - 173 (1976). – Abstr. in Phys. Abstr., Vol. 80, Abstr. 37641 (1977).

131.173 The case of the missing cosmic dust.   P. Wesson.
New Scientist, Vol. 73, 207 - 209 (1977). – Abstr. in Phys. Abstr., Vol. 80, Abstr. 41403 (1977).

131.174 Origin and nature of carbonaceous material in the Galaxy.   F. Hoyle, N. C. Wickramasinghe.
Nature, Vol. 270, 701 - 703 (1977).
Astronomers generally believe that the carbonaceous material emerging from stars must be in the form of graphite, the most stable condensed form of carbon, and that such emergence must be confined to situations where the C/O ratio exceeds unity, such as in the atmospheres of carbon stars. The authors argue here that this state of affairs remains valid for mass flows from stars of sufficiently low surface temperatures, but it is not correct for low density flows from stars with colour temperatures ≳ 4,000 K.

131.175 Interstellar radio spectrum lines.
Rep. Prog. Phys., Vol. 40, 483 - 565 (1977). – Abstr. in Phys. Abstr., Vol. 80, Abstr. 71138 (1977).

131.176 Collision between the solar system and an interstellar cloud.
A. Vidal-Madjar, J. Audouze, P. Bruston, C. Laurent.
Recherche, Vol. 8, 616 - 622 (1977). In French. – Abstr. in Phys. Abstr., Vol. 80, Abstr. 74036 (1977).

131.177 Interstellar grains. The interaction of light with a small-particle system.   D. R. Huffman.
Adv. Phys., Vol. 26, No. 2, p. 129 - 230 (1977). – Abstr. in Phys. Abstr., Vol. 80, Abstr. 74111 (1977).

131.178 Interstellar methane.
IAU Circ., No. 3146 (1977).

131.179 Magnetite and the interstellar medium.

S. C. Landaberry, A. M. Magalhaes.
An. Acad. Brasil. Cienc., Vol. 48, 199 - 204 (1977). – Abstr.
in Phys. Abstr., Vol. 80, Abstr. 78838 (1977).

131.180 A survey of southern dark clouds for Herbig-Haro objects and H-alpha emission stars.
R. D. Schwartz.
Astrophys. J., Suppl. Ser., Vol. 35, 161 - 170, Plates 1 - 14 (1977).

The results of a deep red objective-prism survey of selected southern dark clouds are presented. In addition to two new Herbig-Haro objects near the Orion Nebula, 14 Herbig Haro objects have been discovered in dark clouds in Vela, Chamaeleon, Lupus, and Norma. A total of 140 Hα emission stars have been identified in or near eight dark cloud complexes in the same regions. Five of these regions were observed with the objective prism in the violet for the presence of Ca II and higher Balmer line emission in the Hα stars. Celestial coordinates (1950.0) and data pertaining to the emission line and continuum intensities of these objects are tabulated. Finder charts for each of the fields are presented.

131.181 Radio observations of interstellar CH. II.
Å. Hjalmarson, A. Sume, J. Elldér, O. E. H. Rydbeck, E. L. Moore, G. R. Huguenin, Aa. Sandqvist, P. O. Lindblad, P. Lindroos.
Astrophys. J., Suppl. Ser., Vol. 35, 263 - 280 (1977).

The Onsala Space Observatory 25.6 m telescope, equipped with a traveling wave maser preamplifier, has been used to study emission in the three hyperfine transitions of the $^2\Pi_{1/2}$, $J = 1/2$, Λ doublet state of interstellar CH. The main line ($F = 1-1$) has been detected toward more than 100 positions in optically dark nebulae. All three lines have been observed in many directions with relatively strong main line emission. A statistical analysis of the data indicates that the antenna temperature increases linearly with the cloud opacity class, or the cloud density. An increase is also found for the CH column density, but it seems that the ratio between the CH and the total hydrogen column densities decreases for more opaque clouds.

131.182 Interstellar H300α line radiation.
J. L. Casse, P. A. Shaver.
Astron. Astrophys., Vol. 61, 805 - 808 (1977).

An H300α recombination line (242 MHz) has been detected in the direction of the galactic centre, with the Dwingeloo radiotelescope. It is probably due to stimulated emission from very extended low-density H II regions, whose presence in the line of sight to the galactic centre has recently been confirmed by detection of the H200β line (Shaver, 1977). Upper limits on the line intensity towards Cas A imply for the cold H I clouds in that direction either a very low hydrogen ionization rate ($\zeta_H < 2 \cdot 10^{-17}$ s^{-1}) or a very high molecular concentration.

131.183 Interstellar neutral magnesium towards moderately reddened stars.   M. Pettini, A. Boksenberg,
B. Bates, R. F. McCaughan, C. D. McKeith.
Astron. Astrophys., Vol. 61, 839 - 851 (1977).

Column densities of interstellar Mg^0 have been derived from observations of Mg I $\lambda 2852$ in the spectra of 12 early type stars of moderate reddening ($E(B-V) < 0.30$). From a comparison with published data for Na^0, K^0 and Ca$^+$ along the same lines of sight it is concluded that the three neutral species are mutually well correlated. In a number of clouds moving with moderately high velocities in the local standard of rest and showing high $N(Ca^+)/N(Na^0)$ ratios, the ratio $N(Ca^+)/N(Mg^0)$ is found to be up to an order of magnitude greater than in lower velocity components. This finding provides additional evidence to support the suggestion that the well known enhancement of the $N(Ca^+)/N(Na^0)$ ratio with cloud velocity results from an increase in the Ca gas phase abundance

131.184 Dark nebulae, globules, and protostars.
B. J. Bok.
Publ. Astron. Soc. Pacific, Vol. 89, 597 - 611 (1977).
Contents: Star birth in our Galaxy; Large globules; Optical evidence, radio evidence, their evolutionary status; Star formation in the Star Clouds of Magellan.

131.185 High-velocity gas in the Monoceros loop.
J. G. Cohen.
Publ. Astron. Soc. Pacific, Vol. 89, 626 (1977).

The interstellar Na I lines arising from high-velocity gas in the Monoceros loop supernova remnant have been observed. When combined with previously published interstellar Ca measurements, the interstellar lines in the high-velocity component of HD 47359 show an abnormally low ratio Na I/Ca II.

131.186 The maximum temperatures of interstellar grains.
R. G. Tabak.
Astrophys. Space Sci., Vol. 46, 175 - 181 (1977).

The maximum temperature a typical interstellar grain will attain upon absorption of a photon or chemical band formation of molecules on its surface is calculated by considering the exact Debye theory of dielectrics. Other contributions to the specific heats of solids are discussed. It is shown that the use of the approximate Debye theory will lead to serious errors in the calculation of velocities of desorption of molecules from grain surfaces.

131.187 The ultraviolet absorption band at 2175 Å: correlations with other interstellar features.
J. Dorschner, C. Friedemann, J. Gürtler.
Astrophys. Space Sci., Vol. 46, 357 - 369 (1977).

The hump in the ultraviolet part of the interstellar extinction curve is interpreted as a broad diffuse absorption band. Its equivalent width is estimated for 36 stars by means of OAO-2 data. Some important correlations between the equivalent width of the λ2175 extinction hump and other parameters of interstellar matter are presented. The physical parameters half-width and oscillator strength of the band at 2175 Å are estimated.

131.188 On the ionization of interstellar magnesium.
G. A. Gurzadyan.
Astrophys. Space Sci., Vol. 46, 471 - 484, 485 - 497 (1977). In Russian and English.

It has been shown that two concentric ionization zones of interstellar magnesium must exist around each star: internal, with a radius coinciding with that of the zone of hydrogen ionization S_H; and external, with a radius greater than S_H, by one order. Unlike interstellar hydrogen, interstellar magnesium is ionized throughout the Galaxy. Ionizing radiation of ordinary hot stars cannot provide for the observed high degree of ionization of interstellar magnesium. Stars of the B5 and B0 class play the main role in the formation of ionization zones of interstellar magnesium.

131.189 Origin of the diffuse interstellar absorption bands. III. Zero phonon lines in MgO and CaO.
W. W. Duley.
Astrophys. Space Sci., Vol. 47, 185 - 193 (1977).

Evidence is presented that zero phonon lines of defect centres in MgO and CaO are responsible for the diffuse interstellar bands at 5362, 5705, 6425.7, and 6699.4 Å. Phonon sidebands of these lines are identified with the diffuse bands at 5535, 6177, 6196, 6284, and 6314 Å. These features arise in interstellar MgO and CaO particles with sizes $\gtrsim 50$ Å. Infrared spectral features due to interstellar MgO and CaO are discussed.

131.190 Organic molecules in interstellar dust: a possible spectral signature at λ2200 Å?
N. C. Wickramasinghe, F. Hoyle, K. Nandy.

Astrophys. Space Sci., Vol. 47, L9 - L13 (1977).

The λ2200 Å interstellar absorption band, generally attributed to graphite grains, could equally well arise from electronic transitions in conjugated double bonds of organic molecules. These molecules, which should comprise ~10% of the total interstellar dust mass, may be lodged within clumps of 100 Å-sized refractory grains.

131.191 **The deuterated species of large interstellar molecules.**
　　　　T. J. Millar.
Astrophys. Space Sci., Vol. 47, L17 - L19 (1977).

Deuterium enhancement in small ions and molecules is thought to occur via ion-molecule reactions. It is suggested here that large interstellar molecules formed in neutral radiative association reactions should also show deuterium enhancement. Observations of large deuterated species would provide a critical test for the radiative association mechanism.

131.192 **Grains accretion processes in a proto-planetary**
　　　　nebula. II. Accretion time and mass limit.
A. Coradini, G. Magni, C. Federico.
Astrophys. Space Sci., Vol. 48, 79 - 87 (1977).

Some mechanisms which are expected to produce the growth of dust grains in the protosolar nebula are studied during the isothermal and the adiabatic phase of the gravitational collapse. Owing to the low sticking efficiency in the grain-grain collisions and also to the impossibility of gas capture by solid particles in the physical environment considered, the main result is the production in about 10^6 yr of a set of particles similar in mass. The obtained mass limit (10^{-8}–10^{-9} g) depends on the physical properties of the grains, and seems to be independent of the turbulence model used for the gas motion.

131.193 **Large interstellar molecules.**
　　　　T. J. Millar, D. A. Williams.
Astrophys. Space Sci., Vol. 48, 243 - 248 (1977).

The authors propose that photocycloaddition reactions in molecular complexes in normal interstellar clouds will create unusually large molecules. These may be sufficiently radiation stable to be circulated with the interstellar gas, and so provide convenient nucleation centres for growth of loosely bound grains in dark regions.

131.194 **Infrared photometry, extinctions and R values for**
　　　　some Northern Milky Way stars.
I. G. van Breda, D. C. B. Whittet.
Astrophys. Space Sci., Vol. 48, 297 - 304 (1977).

Infrared wideband photometry for 21 early-type stars in the Northern Milky Way is used to determine extinction values by the colour-difference method. The mean extinction curve is similar to the van de Hulst theoretical curve No. 15, and there is no significant difference in the results for stars lying towards and away from nebulosity. None of the stars exhibit infrared excesses at 3.5 μ. A mean value of $R = 3.15 \pm 0.15$ is deduced for the ratio of total to selective extinction.

131.195 **The 10-micron interstellar absorption band: curves**
　　　　of growth for terrestrial and meteoritic olivine
particles.   J. Dorschner, C. Friedemann, J. Gürtler.
Astrophys. Space Sci., Vol. 48, 305 - 312 (1977).

Curves of growth have been determined for the interstellar absorption band at 9.75 μm using infrared spectra of small olivine particles. The mass absorption coefficient was found to depend strongly on the mean radius of the grains. A comparison was made with a sample of Allende meteorite and good agreement was found between the curves of growth for terrestrial and meteoritic olivine.

131.196 **Molecule formation and cloud collapse.**
　　　　T. J. Millar, D. A. Williams.
Astrophys. Space Sci., Vol. 48, 379 - 387 (1977).

The gravitational collapse of a uniform, nonmagnetic nonrotating, spherically symmetric interstellar gas cloud is studied taking into account the conversion of atomic to molecular hydrogen and appropriate heating and cooling mechanisms. It is shown that this conversion can result in the collapse of an initially stable gas cloud, and that in clouds with masses greater than the Jeans mass the main effect of molecule formation is to alter the temperature distribution inside the clouds.

131.197 **The temperature of interstellar iron grains.**
　　　　R. G. Tabak.
Astrophys. Space Sci., Vol. 49, 41 - 46 (1977).

In order for catalytic reactions to occur in interstellar dust clouds, it is necessary for the temperature of the grains to be about an order of magnitude hotter than usually calculated for grains of dielectric materials. However, transition metal (e.g., iron) grains should be fairly abundant, and because they absorb strongly in the visible and ultraviolet regions of the interstellar radiation field, they have equilibrium temperatures ~133 K.

131.198 **Unreddened stars and the two-phase model of the**
　　　　interstellar medium.   P. Joshi, S. P. Tarafdar.
Astrophys. Space Sci., Vol. 49, 199 - 215 (1977).

The authors have computed two phase models of the interstellar medium, with cosmic rays and X-rays assumed to be the main ionizing agents, heating due to photoelectron ejection from the interstellar grains. They show that it is possible to have a hot and tenuous intercloud medium in pressure equilibrium with the interstellar clouds for a wide range of physical conditions, possibly existing in the interstellar space. It is suggested that the intercloud medium may be predominantly neutral, with ionization rates consistent with the limits imposed by molecular observations. The mean fractional ionization of the intercloud medium is ~1%.

131.199 **Molecular hydrogen in intercloud medium.**
　　　　P. Joshi, S. P. Tarafdar.
Astrophys. Space Sci., Vol. 49, 217 - 227 (1977).

The authors discuss the formation of molecular hydrogen on the surfaces of grains in a hot intercloud medium, by the process of chemisorption of hydrogen atoms on graphite grains. It is suggested that the molecular hydrogen observed towards stars with low reddening, may be located in the intercloud medium towards these stars. A comparison of the observed population distributions of $\overline{H}_2$ with the theoretical calculations shows that the observations are in the main consistent with a gas kinetic-temperature ~8000 K and densities about 0.1 to 1 cm^{-3}, parameters which are appropriate to the intercloud phase of the two phase model of the interstellar medium.

131.200 **Electric charge of grains and interstellar radioactivity.**
　　　　S. Yanagita.
Astrophys. Space Sci., Vol. 49, L11 - L15 (1977).

The author investigates the electric potential of high speed interstellar grains trapping radioactive nuclei which are suggested to exist in interstellar medium by the study of meteorite samples and the gamma-ray line observation. These radioactive nuclei, e.g. ^{22}Na, ^{26}Al, ^{44}Ti and ^{60}Fe bring about a drastic change in grain potential. Grains with larger radius than the critical radius a_c attain a negative potential when grains are traversing a typical intercloud region.

131.201 **Polyoxymethylene co-polymers on grains.**
　　　　A. Cooke, N. C. Wickramasinghe.
Astrophys. Space Sci., Vol. 50, 43 - 53 (1977).

Conditions prevalent in dense molecular clouds are shown to favour the polymerization of H_2CO molecules and the deposition of formaldehyde co-polymer mantles on smaller refractory grains. If a significant fraction of such co-polymer coated grains are expelled with systematic gas flows into the general interstellar medium, these moderately refractory grains may be responsible for the bulk of interstellar extinction and polarization at optical wavelengths. Mie calculations for a mixture consisting of iron, graphite and polyoxymethylene particles are presented as an example. Suitably end-capped and stabilized co-polymer-coated grains, with either silicate or graphite cores, may survive at temperatures ~450 K under interstellar ambient conditions and be responsible for the 10 μ emission feature in many sources.

131.202 Interstellar graphite grain temperature and infrared optical constants.   S. Aiello, F. Mencaraglia.
Astrophys. Space Sci., Vol. 51, 111 - 116 (1977).

The temperature of graphite grains in the interstellar medium has been computed with different assumptions on the extrapolation of its optical constants into infrared. It is found that values computed up to now are generally underestimated by 10–20% in normal interstellar conditions. For extreme conditions (very dark clouds) errors by a factor of two are possible and more attention has to be paid in dealing with problems related to molecule formation.

131.203 Model for surface reactions on interstellar grains – a numerical study.
J. B. Pickles, D. A. Williams.
Astrophys. Space Sci., Vol. 52, 443 - 452 (1977).

A model of the formation of molecules by surface reactions on interstellar grains is described and assessed numerically. The model predicts that for the molecules – other than H_2 – likely to be important in the interstellar medium, the formation rates by surface reactions are insensitive to the nature of the surface. The formation rates have magnitudes which are significant when compared with other routes. The model also describes H_2 formation in high density clouds and shows it to be parameter dependent.

131.204 The chemistry of interstellar nitrogen.
J. B. Pickles, D. A. Williams.
Astrophys. Space Sci., Vol. 52, 453 - 478 (1977).

A fairly complete but limited set of gas phase reactions involving nitrogen-bearing molecules is linked to a simple model of grain surface reactions. Calculations are performed attempting to simulate the nitrogen chemistry in interstellar clouds of low and high density. While it appears probable that grain surface reactions contribute to the chemistry in both regimes, conclusive evidence awaits observational and theoretical developments.

131.205 Modulation of nuclei heavier than helium and their spectra in interstellar space.
A. A. Ajtmukhambetov, A. G. Zusmanovich, E. V. Kolomeets.
Prikl. i teor. fiz. Vyp. (No.) 8. Alma-Ata, 1976, p. 208 - 213.
In Russian. – Abstr. in Ref. zh., 51. Astron., 11.51.838 (1977).

131.206 The equation of state for a two-phase interstellar medium and the problem of condensation of interstellar gas clouds.   S. A. Kaplan.
Early stages of stellar evolution, (see 012.041), p. 5 - 10 (1977). In Russian.

131.207 Opacity effects in the evolution of interstellar clouds and star formation.   I. G. Kolesnik.
Early stages of stellar evolution, (see 012.041), p. 10 - 13 (1977). In Russian.

It is shown that before gravitational instability massive interstellar clouds become opaque for external ionizing radiation. Thermal balance changes in screened regions, and a dense core, where star formation may begin, arises.

131.208 Orientation of interstellar grains.
A. Z. Dolginov, I. G. Mitrofanov.
Early stages of stellar evolution, (see 012.041), p. 14 - 15 (1977). In Russian.

Some possible mechanisms of interstellar grain orientations are considered.

131.209 On the nature of maser sources.
V. V. Burdyuzha, D. A. Varshalovich, T. V. Ruzmajkina.
Early stages of stellar evolution, (see 012.041), p. 16 - 18 (1977). In Russian.

It is shown that behind the shock front in the shell of an infrared star thermal instability conditions are fulfilled resulting in fragmentation of the medium. In the clouds thus formed conditions for ignition of OH and H_2O masers may be fulfilled. The problem of the pumping mechanisms of OH and H_2O masers is discussed.

131.210 Hydrodynamic stages of star formation.
I. G. Kolesnik.
Early stages of stellar evolution, (see 012.041), p. 22 - 39 (1977). In Russian.

Physics of protostellar collapse is considered. Possible cooling mechanisms on the initial stages of contraction are discussed. The mathematical formulation of the protostellar collapse problem is given and the initial stage and boundary conditions are discussed. The question is considered on the mass of formed stars. It is shown that for gravitational condensation of massive stars of the galactic disk, protostars must have great density and high initial temperature, or a magnetic field frozen in protostellar matter.

131.211 Role of a magnetic field in the process of star formation.   A. E. Dudorov.
Early stages of stellar evolution, (see 012.041), p. 56 - 65 (1977). In Russian.

The magnetic fields of interstellar gas-dust clouds, geometry and the influences of a magnetic field on gravitational collapse and angular momentum of gas-dust clouds are discussed.

131.212 Search for methanol masers.   R. B. Buxton, A. H. Barrett, P. T. P. Ho, M. H. Schneps.
Astron. J., Vol. 82, 985 - 988 (1977).

Orion A is presently the only known source of microwave methanol (CH_3OH) maser emission at $\lambda \sim 1$ cm. A search has been made for emission in the $J = 2$, 4, and 6 lines of methanol in 132 sources of H_2O, SiO, and OH microwave maser emission. No new methanol sources were observed and a table of upper limits is presented. The apparent uniqueness of Orion A is probably due to its being relatively nearby (0.5 kpc).

131.213 Radio and infrared observations of the OH/H_2O source G12.2–0.1.   P. A. Shaver, A. C. Danks.
ESO Sci. Prepr. No. 18, 3 + 14 pp. (1977). – To appear in Astron. Astrophys.

High-resolution $\lambda 6$ and 21 cm radio observations and $\lambda 1 - 5$ μm infrared observations of the OH/H_2O maser source G12.2–0.1 are presented. The strongest H_2O emission originates within 4 arcsec (0.07 pc) of a compact radio source and intense near-infrared source; it is suggested that the infrared source may be the hot (> 1000 K) cocoon of a newly-formed O star. The OH and weak H_2O sources are located at the edge of a ridge of radio emission, possibly an edge-on ionization front.

131.214 Interstellar potassium and argon.   D. D. Clayton.
Earth Planet. Sci. Lett., Vol. 36, 381 - 390 (1977).

Isotopic anomalies are calculated in K and Ar that will be expected to be carried in interstellar grains. Special attention is given to the supernova condensates that precipitate during the expansion of the explosive oxygen-burning zone of the interior, because that is where ^{39,41}K and ^{36,38}Ar find their natural origins. Expectations are given for: (1) K–Ar ages in excess of the solar age in accumulations of presolar grains, and (2) K anomalies in Ca-rich minerals within Allende inclusions, because experiments searching for such effects are currently underway.

131.215 HCN emission in the Sagittarius A molecular cloud.
Y. Fukui, T. Iguchi, N. Kaifu, Y. Chikada, M. Morimoto, K. Nagane, K. Miyazawa, T. Miyaji.
Publ. Astron. Soc. Japan, Vol. 29, 643 - 667 (1977).
The molecular cloud near Sagittarius A has been mapped in the $J=1-0$ emission line of hydrogen cyanide ($H^{12}CN$). The full extent of the molecular cloud was found to be about 26'× 8' in galactic longitude and latitude, respectively, and its total mass was estimated to be $\gtrsim 3 \times 10^6$ M$_\odot$. The brightest part of the cloud defines a thin straight ridge. The molecular cloud shows an apparent anticorrelation to the distribution of radio continuum radiation, but resembles rather well the far- and near-infrared distributions. The velocity field of the cloud is complex and is not dominated by any monotonic velocity gradient.

131.216 Transfer of line radiation in collapsing molecular clouds.   S. Deguchi, Y. Fukui.
Publ. Astron. Soc. Japan, Vol. 29, 683 - 692 (1977).
The method of the escape probability is extended to the case of spherically symmetric cloud in which the collapse velocity decreases with radius. Line profiles of ^{12}CO, ^{13}CO, and CS are calculated for the molecular cloud with the velocity field $V(R) \propto R^{-0.5}$. This extended method treating the transfer of line radiation can be applied to a variety of objects, especially, stars with P Cygni profiles and maser sources associated with infrared stars.

131.217 The origin of intrinsic linear polarization of optical radiation from stars with infrared excess.
A. Z. Dolginov, I. G. Mitrofanov.
Astron. Zh. Akad. Nauk SSSR, Vol. 54, 1259 - 1267 (1977). In Russian. English translation in Soviet Astron., Vol. 21, No. 6.
The physical conditions are considered in circumstellar clouds around M giants with infrared excess. The outflow of gas and dust is shown to occur with constant relative velocity. The dust density is $(0.1-1)\%$ of the gas density. The orientation of dust grains due to paramagnetic relaxations is believed to take place in the case of strong stellar magnetic fields. The orientation due to the relative motion of dust and gas under the action of small magnetic field should occur too. The linear polarization observed may be explained in both cases, but the assumption of small magnetic fields of M giants is more preferable.

131.218 Energy spectrum of electrons in the interstellar space.
A. A. Ajtmukhambetov, A. G. Zusmanovich, E. V. Kolomeets.
Prikl. i teor. fiz. Vyp. (No.) 8. Alma-Ata, 1976, p. 214 - 219. In Russian. – Abstr. in Ref. zh., 51. Astron., 12.51.582 (1977).

131.219 Star formation in a dusty plasma cloud.
H. Alfvén.
Tek. Hoegsk., Stockholm, Sweden. Instn. Plasmafys. TRITA-EPP–77-15. 10 pp. (1977). – Abstr. from INIS 7723 341835.

131.220 Ice grains in space.   R. Smoluchowski.
Comets, asteroids, meteorites, (see 012.049), p. 47 - 49 (1977).

131.221 From dust to comets.   J. M. Greenberg.
Comets, asteroids, meteorites, (see 012.049), p. 491 - 497 (1977).
The growth and chemical evolution of a typical interstellar dust grain are followed starting from average inter-

stellar conditions to the dense cloud and contraction phase. Based on the theory of cometary accretion directly from cold interstellar dust it is shown that the bulk material of a primordial comet would consist mostly of an icy conglomerate of complex organic molecules and frozen radicals in which are imbedded approximately equal volumes (10% each) of small grains in two different sizes.

131.222 Interstellar matter and the evolution of stars.
A. Antalová.
Kozmos, Vol. 8, 106 - 107 (1977). In Slovak.

Pulkovo sky survey in the interstellar neutral hydrogen radio line. III.   See Abstr. 002.034.

Stellar formation.   See Abstr. 003.136.

Giant molecular clouds.   See Abstr. 011.002.

Radio recombination lines from H$^+$ regions and cold interstellar clouds: computation of the b_n factors. See Abstr. 021.006.

On the A $^1\Pi - X$ $^1\Sigma^+$ band system in CH$^+$ and CD$^+$: theoretical spectroscopic constants and lifetimes. See Abstr. 022.009.

Molecular synthesis in interstellar clouds: some relevant laboratory measurements.   See Abstr. 022.030.

Transitions in Λ-doublets of molecules induced by collisions with ions.   See Abstr. 022.033.

Transition probability data for molecules of astrophysical interest.   See Abstr. 022.054.

Some O I oscillator strengths and the interstellar abundance of oxygen.   See Abstr. 022.064.

Inverting the ground state of interstellar CH. See Abstr. 022.065.

Coupled states cross sections for rotational excitation of H$_2$CO by He impact at interstellar temperatures. See Abstr. 022.073.

Measurement of the Stokes parameters of light. See Abstr. 031.320.

Infrared astronomy and galactic dust. I. See Abstr. 061.011.

Infrared astronomy and galactic dust. II. See Abstr. 061.018.

On the computation of optical properties of heterogeneous grains.   See Abstr. 061.044.

Astrophysical implications of isotopic anomalies. See Abstr. 061.049.

Hydrodynamic collapse calculations of cylindrical clouds.   See Abstr. 062.021.

Circumstellar dust.   See Abstr. 064.003.

Structure and evolution of wind-driven circumstellar shells.   See Abstr. 064.021.

Interstellar bubbles. II. Structure and evolution. See Abstr. 064.037.

Molecular emission from expanding envelopes around evolved stars. I. Nonmaser SiO emission lines. See Abstr. 064.044.

Does the initial mass function for star formation depend on epoch?   See Abstr. 065.033.

The collapse of unstable isothermal spheres. See Abstr. 065.042.

Stellar rotation and the thermomagnetic torque. See Abstr. 065.059.

Nearly collisionless spherical accretion. See Abstr. 066.077.

Nonstationary interaction between solar wind and interstellar matter.   See Abstr. 074.117.

Planetary atmospheres and interstellar clouds. See Abstr. 091.062.

Carbon isotope ratio in comets and interstellar matter.   See Abstr. 102.051.

Temperature measurement of interplanetary–interstellar hydrogen.   See Abstr. 106.014.

Neutral hydrogen in interplanetary space. See Abstr. 106.023.

On the role of physico-chemical processes during the formation of planetary systems.   See Abstr. 107.034.

Non-anomalous diffuse interstellar absorption features in Rho Leonis.   See Abstr. 114.545.

S Coronae Australis – ein Stern, der noch gar kein Stern ist.   See Abstr. 122.086.

Symmetric velocity structure in the SiO maser spectrum of R Cassiopeiae.   See Abstr. 122.108.

The interstellar reddening and distance of Nova Cygni 1975 (V1500 Cygni).   See Abstr. 124.101.

Supernovae and molecular clouds. See Abstr. 125.002.

CO emission from supernova remnants. See Abstr. 125.005.

The interaction of supernovae with the interstellar medium.   See Abstr. 125.017.

Compact H II regions near type I OH maser sources: IV.   See Abstr. 132.013.

Compact H II regions.   See Abstr. 132.014.

Dust in H II regions.   See Abstr. 132.015.

H I observations of dark clouds. See Abstr. 132.024.

The 4830 MHz H_2CO absorption in the direction of NGC 6334.   See Abstr. 132.026.

Far infrared radiation from dust within H II regions. See Abstr. 132.027.

Water vapor maser "turn-on" in the H II region W3

(OH).   See Abstr. 132.039.

Evidence for optically thin CO emission from the Kleinmann-Low Nebula.   See Abstr. 133.002.

L'astronomie infrarouge et les poussières galactiques. See Abstr. 133.009.

Interpretation of far infrared emission. See Abstr. 133.010.

The distribution of ionized gas and dust in W 3(A) and W 3(OH).   See Abstr. 133.011.

Optical depths of far infrared sources. See Abstr. 133.013.

Supergrain models of far infrared sources. See Abstr. 133.014.

Microwave spectra of OH/infrared-sources. See Abstr. 133.016.

Vibrationally excited molecular hydrogen in Orion. See Abstr. 134.006.

A shock model for infrared line emission from H_2 molecules.   See Abstr. 134.017.

Search for microarcsecond structure in low-frequency variable radio sources.   See Abstr. 141.058.

Radio recombination lines towards pulsars. See Abstr. 141.501.

On the shape of a pulsar's pulse scattered in the interstellar medium.   See Abstr. 141.550.

The diffuse soft X-ray sky. Astrophysics related to cosmic soft X-rays in the energy range 0.1–2.0 keV. See Abstr. 142.129.

The age of the galactic cosmic rays derived from the abundance of ^{10}Be.   See Abstr. 143.019.

Cosmic ray propagation in a closed galaxy. See Abstr. 143.056.

Chemical evolution of galaxies. See Abstr. 151.029.

The effects of the self-gravity of the interstellar gas on galactic shock waves.   See Abstr. 151.073.

On the arm structure and the star formation in M 51 See Abstr. 151.074.

Observational evidence for supernova-induced star formation: Canis Major R1.   See Abstr. 152.003.

Supernova-induced star formation in Cepheus OB3. See Abstr. 152.006.

Extinction law in dust clouds and the young southern cluster NGC 6250: further evidence for high values of R.   See Abstr. 153.019.

Gas in globular clusters. II. Time-dependent flow models.   See Abstr. 154.029.

The galactic density wave, molecular clouds, and star formation.   See Abstr. 155.015.

The galactic distribution (in radius and z) of interstellar molecular hydrogen.   See Abstr. 155.020.

An out-of-plane galactic carbon monoxide survey.
See Abstr. 155.025.

Comparative morphology of galactic carbon monoxide and hydrogen.   See Abstr. 155.026.

Supergiants in the field of the cluster M6, and the distribution of interstellar matter in the direction of the galactic center.   See Abstr. 155.028.

Infrared astronomy and high-energy astrophysics.

See Abstr. 156.011.

Star formation in blue galaxies.
See Abstr. 158.060.

Carbon monoxide in Maffei 2.
See Abstr. 158.089.

Intragalactic factor and apparent distribution of external objects.   See Abstr. 160.064.

A constraint on the universal baryon density from the abundance of ^{7}Li.   See Abstr. 162.032.

132   H I, H II Regions

132.001   The Cygnus X region VIII: maps of calibrated surface brightness of (Hα+[NII]).
H. R. Dickel, H. J. Wendker.
Astron. Astrophys., Suppl. Ser., Vol. 29, 209 - 240 (1977).
   Photoelectric and photographic observations of the HII regions in the Cygnus X region are combined to produce contour maps of the emission of Hα + [NII]. The data, analysis, and calibration network are described. Photographs and calibrated contour maps of the absolute surface brightness (ergs cm^{-2}s^{-1}ster^{-1}) in the Hα + [NII] lines are presented for 24 plates covering the Cygnus X region. For several HII regions the ratio of $I([NII])/I(H\alpha)$ is also determined by photoelectric observations. The effectively constant ratio allows the use of the maps for the intensity of Hα alone.

132.002   Carbon monoxide observations of southern hemisphere H II regions.   A. R. Gillespie, P. J. Huggins,
T. C. L. G. Sollner, T. G. Phillips, F. F. Gardner, S. H. Knowles.
Astron. Astrophys., Vol. 60, 221 - 225 (1977).
   The first survey of carbon monoxide emission from southern hemisphere H II regions is presented. The survey contains 37 sources and 4 upper limits, and includes observations of most of the brightest southern galactic radio sources.

132.003   Aperture synthesis observations of galactic H II regions. VI. Several isolated H II regions.
F. P. Israël.
Astron. Astrophys., Vol. 60, 233 - 249 (1977).
   Radio maps at λ 21 cm of S 90, S 104, S 125, S 142 and S 184, and at λ 6 cm of S 90, DR 15 and S 228 are presented.

132.004   Abundances in 10 H II regions in the Small Magellanic Cloud.   R. J. Dufour, W. V. Harlow.
Astrophys. J., Vol. 216, 706 - 712 (1977).
   The authors present (a) photoelectric spectrophotometry of selected emission lines in 10 SMC H II regions, (b) electron temperatures in several SMC nebulae, (c) spatial variations in the chemical composition of the SMC, and (d) improve the measurement of the He/H value in the SMC.

132.005   Radio line and continuum observations of ten small Sharpless H II regions.
I. Kazès, C. M. Walmsley, E. Churchwell.
Astron. Astrophys., Vol. 60, 293 - 302 (1977).
   Ten optically visible H II regions have been observed in the H 109 α, H 137 β and H I lines. In general the H 109 α radial velocities agree with those measured in Hα. The velocities of the ionized gas are generally in good agreement with

the CO radial velocities. This correlation suggests an association of the ionized gas with the molecular cloud. Finally, distance estimates are given for the ten nebulae, and their physical parameters are derived from measured continuum and line parameters.

132.006   Common properties of H II regions in galaxies.
M. A. Smirnov, B. V. Komberg.
Inst. kosm. issled. AN SSSR. Pr-315. Moskva, 1977, 39 pp.
In Russian. – Abstr. in Ref. zh., 51. Astron., 8.51.799 (1977).

132.007   Optical, infrared, and radio observations of S88B.
J. Krassner, J. L. Pipher, S. Sharpless, M. P. Savedoff,
B. T. Soifer, M. Zeilik II, S. Varlese.
News Lett. Astron. Soc. N.Y., Vol. 1, No. 2, p. 13 (1977).
Abstract.

132.008   Infrared and radio observations of the H II region S235.   J. Krassner, J. Pipher.
News Lett. Astron. Soc. N.Y., Vol. 1, No. 2, p. 19 (1977).
Abstract.

132.009   Airborne lamellar grating observations of H II regions from 100–500 microns.
T. Herter, J. Pipher, G. Duthie, J. Krassner, M. Savedoff.
News Lett. Astron. Soc. N.Y., Vol. 1, No. 2, p. 20 (1977).
Abstract.

132.010   Observations of [S III] in NGC 604 and N/S abundance gradients.   S. A. Hawley, S. A. Grandi.
Astrophys. J., Vol. 217, 420 - 424 (1977) = Lick Obs. Bull., No. 762.
   The authors have obtained line intensities for NGC 604 – a giant H II region in M33 – in the wavelength range $\lambda\lambda$6312– 10049. Observations of [S III] $\lambda\lambda$9069, 9532 yield a S^{++} abundance and an accurate total S abundance in agreement with that for the Orion Nebula. Previous investigators have seriously overestimated N/S. The authors discuss the use of [N II]/[S II] as an abundance indicator and conclude that, while a gradient in [N II]/[S II] implies a gradient in N/S, a dependable calibration of line ratio versus abundance is unavailable.

132.011   A search for neutral hydrogen clouds in radio galaxies and in intergalactic space.
M. S. Roberts, D. G. Steigerwald.
Astrophys. J., Vol. 217, 883 - 891 (1977).
   Results of three related experiments are presented. (1) Forty-one radio galaxies have been observed at and near their

systemic velocity for H I absorption. Only one new detection, that for 3C 178, was obtained. (2) For many of these radio continuum sources, the search range extended to $z \approx 0$ in an attempt to find possible H I clouds located along the line of sight to the radio source. No detections were made, and this negative result is used to set an upper limit to clouds of intergalactic H I. (3) In a complementary analysis, observations of several hundred blank reference fields made in conjunction with H I emission studies of galaxies show no H I emission at the 0.1 Jy level corresponding to at least 5 rms.

132.012 Colliding ionization fronts. I. Ionized shells in H II regions. G. Tenorio-Tagle.
Astron. Astrophys., Vol. 61, 189 - 194 (1977).

The gas dynamical effects produced by the subsequent formation of stars within and around H II regions are studied. The evolution of ionization fronts and their interaction with shock waves and other ionization fronts are numerically followed. The appearance of dense ionized shells (or filaments) in H II regions is explained as a natural consequence of these interactions. Estimates of the total emission measure from the H II region, at different times of the evolution, and typical values of the expected velocity dispersion within the flow are also given.

132.013 Compact H II regions near type I OH maser sources: IV.
H. E. Matthews, W. M. Goss, A. Winnberg, H. J. Habing.
Astron. Astrophys., Vol. 61, 261 - 274 (1977).

Galactic fields containing type I OH maser sources have been mapped with the Westerbork Synthesis Radio Telescope in the continuum at wavelengths of 6 and 21 cm. Additional measurements were made with the 100 m telescope at Effelsberg in the continuum at 2 and 2.8 cm and in the H 109α recombination line at 6 cm. The paper is concerned with a detailed discussion of the surrounding fields. The stronger resolved sources are young H II regions which are probably more evolved than the compact H II regions associated with the OH sources.

132.014 Compact H II regions. P. A. Shaver.
Topics in interstellar matter, (see 012.011), p. 49 - 59 (1977).

This paper reviews recent developments in the study of compact H II regions, particularly from the radio observational point of view. These include the association of H II regions with molecular clouds and maser sources, H II and C II regions and infrared sources in dark clouds, and aperture synthesis observations of radio recombination lines.

132.015 Dust in H II regions. S. Isobe.
Topics in interstellar matter, (see 012.011), p. 61 - 79 (1977).

The author summarizes the observational evidence for the existence of dust grains in H II regions, and discusses globules and molecular clouds as a supply source of dust grains to H II regions.

132.016 Unusual, large-scale, motions in H II regions.
J. Meaburn.
Topics in interstellar matter, (see 012.011), p. 81 - 88 (1977).

In many H II regions, huge volumes of ionized gas emit split lines. Such splitting occurs exclusively over dark areas surrounded by bright rims; these rims, which are produced by ionization fronts eating into the adjacent neutral masses, emit single lines centred on the mean motion of the nebular complex. Detailed observations of this situation are presented for the Orion (M42), Carina, Omega (M17), and 30 Doradus Nebulae.

132.017 Recombination line observations of ionized hydrogen. L. Hart.

Topics in interstellar matter, (see 012.011), p. 187 - 193 (1977).

Recombination lines surveys of the large scale distribution of ionized hydrogen in the Galaxy indicate a concentration of material between 4 and 6 kpc from the galactic centre. This is very similar to the distribution derived for molecular hydrogen from CO observations but is unlike that found for atomic hydrogen.

132.018 H II regions in galaxies of the Local Group.
G. Courtès.
Topics in interstellar matter, (see 012.011), p. 209 - 242 (1977).

The galaxies of the Local Group showing H II regions have the great advantage of being close enough to provide, sometimes more easily than in our Galaxy: the fine morphology and distribution of those H II regions; their unambiguous positions compared with those of stars and the H II contours; the precise shape of spiral patterns and their true galactocentric distances. Recent observations have given very efficient means for understanding of galactic and extragalactic structures. Efforts have been made to obtain standard sizes of H II regions, in order to resolve extragalactic distance-scale problems. Studies of star formation at the front of spiral features rich in H II regions have been discussed in relation with the kinematics of the gas as well as with stellar distribution and evolution.

132.019 Large-beam infrared observations of compact H II regions. M. Zeilik II, P. A. Heckert.
Astron. J., Vol. 82, 824 - 827 (1977).

The authors present near-infrared observations of S 156A, S 158A, S 159A, S 162A$_1$, S 228, and G 45.5 + 0.1 (No. 2) made with a beam size $\sim 1'$ in order to obtain total-flux measurements. Correcting for line-of-sight extinction, they find that the $2-25$-μ luminosities are roughly equal to or greater than the Lyman-α luminosities inferred from radio observations.

132.020 Fine-structure line emission from selected compact H II regions. M. Zeilik II.
Astrophys. J., Vol. 218, 118 - 123 (1977).

Dusty models of the infrared line emission from G45.5+0.1 (No. 2), W3 A/IRS 1, S88 B, and G29.9−0.0 are presented. These models match the observed radio and near-infrared continua from the H II regions. For those nebulae that have been observed with infrared spectrometers, the predicted intensities fall at least a factor of 10 below the observed ones. This difference is attributed to lack of knowledge of interstellar dust's optical properties in the ultraviolet and/or the validity of stellar atmosphere models in the ultraviolet.

132.021 Aperture synthesis observations of galactic H II regions. VII. A "quick look" survey of galactic H II regions. F. P. Israel.
Astron. Astrophys., Vol. 61, 377 - 386 (1977).

A survey with limited (u, v) plane coverage of 40 galactic H II regions is presented. The observations were made with the Westerbork Synthesis Radio Telescope at wavelengths of 6 and 21 cm. Out of 23 well-detected H II regions only five show compact structure and five more can be classified as subcompact. Three planetary nebulae were also observed.

132.022 On the evolution of an H II region and the structure of its ionization fronts. J. Manfroid.
Astron. Astrophys., Vol. 61, 437 - 442 (1977).

The first 1.5 million years of the life of an H II region with an initial density of $27 \, m_H \, cm^{-3}$ have been calculated. The development of inhomogeneities is described. A detailed model of the evolved nebula and the ionization fronts is presented.

132.023 Observations of 85α recombination lines from W48.

J. M. MacLeod, L. H. Doherty, L. A. Higgs.
J. R. Astron. Soc. Canada, Vol. 71, 377 - 385 (1977).

Hydrogen 85α recombination lines have been detected from all three components of the source W48, including a component previously suggested to be non-thermal. A continuum map with a resolution of 2.7 arcmin is presented. It is suggested that W48 is part of a larger molecular cloud which reaches as far west as the neighboring supernova remnant W44.

132.024   **H I observations of dark clouds.**
        W. H. McCutcheon, W. L. H. Shuter.
J. R. Astron. Soc. Canada, Vol. 71, 396 - 397 (1977).
Abstract.

132.025   **Etude théorique de raies de recombinaison et d'émission continuum des régions H II.**
J. P. Vallée, M. R. Viner, V. A. Hughes.
J. R. Astron. Soc. Canada, Vol. 71, 398 (1977). – Abstract.

132.026   **The 4830 MHz H₂CO absorption in the direction of NGC 6334.**   F. F. Gardner, J. B. Whiteoak.
Far infrared astronomy (see 012.027), p. 151 - 156 (1976).

The distribution of 4830 MHz H_2CO absorption has been mapped over the northern radio component, G351.4 + 0.7, of NGC 6334. The absorption with an average velocity of −4 km s^{-1} has a general association with a dust lane extending across the H II region. In addition there are three H_2CO concentrations. Two coincide with OH-emission centres, one of which is an infrared source. The third, with the highest absorption and smallest angular extent, is centred on a compact continuum radio component; it is also an infrared source.

132.027   **Far infrared radiation from dust within H II regions.**
        P. A. Aannestad.
Far infrared astronomy (see 012.027), p. 257 - 275 (1976).

A spherical model of dusty H II regions containing core-mantle grains where the cores are evaporated in the innermost region and the mantles are evaporated in an outer region has been compared with the observations. Infrared spectra have been computed for the H II regions of Orion A, W3, and M17. Comparison with observations shows reasonable agreement when account is taken of the differing beam widths and chopping offsets in the observations.

132.028   **Evolution of H II regions.**   F. A. Goldsworthy.
        Q. J. R. Astron. Soc., Vol. 18, 465 (1977).
Abstract.

132.029   **Radio and infrared observations of H II regions.**
        P. F. Scott.
Q. J. R. Astron. Soc., Vol. 18, 465 - 466 (1977). – Abstract.

132.030   **Neutral hydrogen in the outer regions of M31.**
        K. Newton, D. T. Emerson.
Mon. Not. R. Astron. Soc., Vol. 181, 573 - 590 (1977).

Aperture synthesis observations of neutral hydrogen in the extreme north-east and south-west (NE and SW) regions of M31 are described. A simple model for the distribution of H I gives a velocity field close to that observed. An H I cloud with anomalous velocity has been found to the NE of M31 at an angular distance of 150 arcmin from the nucleus.

132.031   **A 1420 MHz continuum survey of the W3/W4/W5 region.**
K. Rohlfs, E. Braunsfurth, D. L. Hills.
Astron. Astrophys., Suppl. Ser., Vol. 30, 369 - 373 (1977).

The W3/W4/W5 region is mapped in continuum brightness temperature at 1420 MHz. The map is compared with the 2695 MHz survey of the same area by Wendker et al. (1977) and with the optical appearance on the Palomar Sky Survey red plates.

132.032   **The clumping factor in H II regions.**
        A. Laval, G. Monnet.
Astron. Astrophys., Vol. 61, 715 - 717 (1977).

A statistical method is developed to compute the average clumping factor inside the H II regions of M33 by comparing on the same line of sight the integrated emission measure and the integrated neutral hydrogen measure. Using 136 H II regions, this method leads to an upper limit for the clumping factor $\alpha < 2.5$ or a filling factor α^{-1} of more than 40%.

132.033   **On the effect of collisional ionization during the time evolution of H II regions.**   G. Tenorio-Tagle.
Astrophys. Space Sci., Vol. 47, 225 - 228 (1977).

The set of equations describing the time evolution of H II regions, accounting for collisional ionization, are presented. Differential forms of these equations are deduced, and it is shown that it is not necessary within this context to consider changes in the potential energy due to ionization of the gas.

132.034   **Latitude-dependent line-shift field of the local H I cloud.**   M. Moles, T. Jaakkola.
Astrophys. Space Sci., Vol. 48, L1 - L7 (1977).

Latitude-velocity contour maps given by Schober (1976) for the low-velocity neutral hydrogen show a significant correlation between these parameters. The result can be interpreted either by a contraction of the local H I cloud in the galactic z-direction, or by a non-Dopplerian red shift in a medium associated with the galactic disk.

132.035   **The forbidden line spectrum in the edges of H II regions.**   J. Manfroid.
Astrophys. Space Sci., Vol. 48, 293 - 296 (1977).

The influence of a relative motion between a hot star and the surrounding medium on the forbidden lines emission spectrum of the latter is described in an example.

132.036   **The motions of the neutral and ionized material in the supermassive Doradus Nebula.**   J. Meaburn.
Astrophys. Space Sci., Vol. 49, 241 - 251 (1977).

A close association between the large scale motions of the neutral and ionized gas in the 30 Doradus Nebular Complex is established. A series of H I/H II shells are suggested to explain this phenomenon. The evidence favours a contracting situation, but is not yet conclusive.

132.037   **On the observed infrared fluxes from extragalactic H II regions.**
K. V. K. Iyengar, K. S. Krishna Swamy.
Astrophys. Space Sci., Vol. 50, 87 - 92 (1977).

The authors have calculated for extragalactic H II regions, the expected relationship between the radio flux at 11 cm and the infrared flux at 11 and 20 μm based on the grain models and the parameters which fit the observations of galactic H II regions. It is shown that the measured infrared fluxes of extragalactic H II regions are consistent with the expected infrared fluxes for these objects.

132.038   **Two populations of high-velocity hydrogen clouds and the bending of the Galaxy.**
U. Haud, J. Einasto.
Astron. Tsirk., No. 958, p. 1 - 3 (1977). In Russian.

132.039   **Water vapor maser "turn-on" in the H II region W3 (OH).**
A. D. Haschick, B. F. Burke, J. H. Spencer.
Science, Vol. 198, 1153 - 1155 (1977).

A line in the water vapor maser spectrum of the H II region W3 (OH) was observed to increase in brightness over an 8-day period and then decline to its original intensity over the following 4 weeks. The intensity variation can be explained by a simple maser model, with an impulse of energy suddenly applied. The observed time scale and energy output are consistent with a maser on the outskirts of a dust cocoon surrounding an O5 star, with a momentary "leak", lasting a day or two, supplying the necessary energy.

Radio recombination lines from H⁺ regions and cold interstellar clouds: computation of the b_n factors. See Abstr. 021.006.

An airborne infrared astronomy program: system description and preliminary results.   See Abstr. 032.520.

High spectral resolution line observations in the infrared.   See Abstr. 034.012.

Neutral hydrogen in interplanetary space. See Abstr. 106.023.

New red objects resembling Herbig-Haro objects. See Abstr. 114.069.

Three unusual wide-spectrum H_2O sources. See Abstr. 131.003.

The ionization and expansion of a globule of interstellar gas overrun by a weak R-type ionization front. See Abstr. 131.011.

Time variations of interstellar water masers: strong sources in H II regions.   See Abstr. 131.028.

Excitation of OH toward interstellar dust clouds. See Abstr. 131.037.

The detailed structure of CO in molecular cloud complexes. I. NGC 6334.   See Abstr. 131.061.

Observations of 7-mm SiO sources. See Abstr. 131.071.

The H_2CO absorption toward IC 1318 b-c in Cygnus.   See Abstr. 131.110.

Carbon monoxide observations at 230 GHz. See Abstr. 131.129.

H_2O in the Galaxy: sites of newly formed OB stars. See Abstr. 131.137.

Molecular clouds.   See Abstr. 131.141.

Observations of high-frequency carbon recombination-line emission in NGC 2024 and IC 1795. See Abstr. 131.164.

Interstellar H300α line radiation. See Abstr. 131.182.

Radio and infrared observations of the OH/H_2O source G12.2−0.1.   See Abstr. 131.213.

Infrared sources in the compact H II region G45.5 + 0.1.   See Abstr. 133.006.

Interpretation of far infrared emission. See Abstr. 133.010.

The distribution of ionized gas and dust in W 3(A) and W 3(OH).   See Abstr. 133.011.

The distribution of ionized carbon in the Orion nebula.   See Abstr. 134.021.

Die Interpretation integrierter Spektren von Emissionsnebeln.   See Abstr. 134.028.

The thermal and non-thermal components of sixteen nebular complexes.   See Abstr. 134.036.

Radio measurements of the electron temperature distribution in Orion A at 22 GHz.   See Abstr. 141.026.

Observations of radio sources near K3−50. See Abstr. 141.129.

Identification of cosmic γ-ray sources CG135 + 1 and CG189 + 1 with H II regions.   See Abstr. 142.705.

Remarks on the overall distribution of hydrogen in the galactic disk.   See Abstr. 155.021.

The morphology of hydrogen and of other tracers in the Galaxy.   See Abstr. 155.027.

21-cm observations of nonplanar H I associated with the Perseus spiral arm.   See Abstr. 155.055.

The galactic center.   See Abstr. 155.058.

Detection of the H 200β line in the direction of the galactic centre.   See Abstr. 156.002.

Ultraviolet observations of local gas. See Abstr. 157.011.

The H I content of blue compact galaxies with emission lines.   See Abstr. 158.011.

The H I content of the elliptical galaxies NGC 3904 and 4636.   See Abstr. 158.013.

The extended H I regions around spiral galaxies: a probe for galactic structure and the intergalactic medium. See Abstr. 158.059.

An interferometer study of the neutral hydrogen associated with the optical core of the irregular galaxy NGC 6822.   See Abstr. 158.086.

Spectrophotometry of nebulae in the Magellanic Clouds suitable as standards.   See Abstr. 159.011.

Monochromatic photographs of giant H II regions in the Magellanic Clouds.   See Abstr. 159.012.

Errata

132.901   Erratum and addendum: "Two-component dust
              models of near-infrared emission from compact
H II regions" [Astrophys. J., Vol. 213, 58 - 65 (1977)].
M. Zeilik II.
Astrophys. J., Vol. 216, 972 (1977).

133 Infrared Sources

133.001   The circumstellar envelope of IRC + 10216.
   J. Kwan, F. Hill.
Astrophys. J., Vol. 215, 781 - 787 (1977).
   The circumstellar envelope of IRC + 10216, a late-type carbon star, is modeled, and theoretical calculations of CO, ^{13}CO, HCN, and H^{13}CN emission are compared with observations. A steady mass loss from the star is assumed. The gas temperature in the flow is determined from considerations of the cooling due to free expansion and molecular emission, and the heating due to gas-dust collisions. Comparing the CO with the ^{13}CO emission, a [CO]/[^{13}CO] isotope ratio of 35 ± 7 is obtained.

133.002   Evidence for optically thin CO emission from the Kleinmann-Low Nebula.
P. G. Wannier, T. G. Phillips.
Astrophys. J., Vol. 215, 796 - 799 (1977).
   A CO ($J = 2 \rightarrow J = 1$) spectrum is presented for the Kleinmann-Low source in the Orion Molecule Cloud. Comparison with an equivalent CO (1 → 0) spectrum reveals a larger "plateau" to "spike" ratio for CO(2→1), which indicates that, unusually for molecule clouds, the CO in the plateau may be optically thin. This result is consistent with the recent model of Kwan and Scoville for the central source in Orion and may allow direct measurements of the ^{12}CO/^{13}CO ratio to be made.

133.003   Infrared emission lines from IRC +10420.
   R. I. Thompson, T. A. Boroson.
Astrophys. J., Lett., Vol. 216, L75 - L77 (1977).
   Infrared spectroscopy of IRC +10420 has revealed the existence of emission lines due to neutral Mg, Na, and possibly Fe. The emission mechanism for these lines is described in terms of pumping from the photosphere of the underlying object. Absorption lines of hydrogen in the 1.8–1.4 μm regions are consistent with the optical classification of F8 Ia. Longward of 2 μm the continuum is dominated by thermal dust emission.

133.004   The brightest sources in the AFCRL survey.
   S. Harris, M. Rowan-Robinson.
Astron. Astrophys., Vol. 60, 405 - 412 (1977).
   The authors have compiled a reliable subcatalogue of 216 of the brighter AFCRL sources and used it to gain an impression of the composition of the infrared sky. The majority of the sources are normal, cool, giant stars, with no excess radiation at 4 μm. About a third of the sources in the list show strong evidence of dust emission, and may be subdivided broadly into "circumstellar dust shells" and "dust clouds".

133.005   Optical, infrared, and radio studies of AFCRL sources.   D. A. Allen, A. R. Hyland, A. J. Longmore, J. L. Caswell, W. M. Goss, R. F. Haynes.
Astrophys. J., Vol. 217, 108 - 126 (1977).
   The southern point sources in the AFCRL infrared (4, 11, and 20 μm) sky survey have been studied at infrared, optical, and radio wavelengths. The authors searched at 2.2 μm to locate the sources, secured near-infrared photometry of them, and obtained classification spectra of those with optical counterparts. OH observations have yielded 14 new type II OH/IR sources and suggest a correlation between the OH flux densities and the infrared colors. Most of the AFCRL sources are carbon or late M stars similar to the reddest objects in the IRC but extending to even redder color indices.

133.006   Infrared sources in the compact H II region G45.5 + 0.1.   P. R. Jorden, A. D. MacGregor, M. J. Selby, P. A. Whitelock, M. Sánchez Magro.
Mon. Not. R. Astron. Soc., Vol. 181, 157 - 161 (1977).
   Infrared photometric observations (1.25–10 μm) of the compact H II region G45.5 + 0.1 are reported. Three new sources, which appear point-like in nature, are discussed; two of them are possibly identified with faint stars on the Palomar Sky Survey red plate. Evidence is given which indicates that they lie behind cool dust clouds with an extinction A_v ∼ 13–30 mag. These sources appear to be M stars, possibly associated with the H II region, and at least one of them is known to be a type II OH/IR source.

133.007   A high-resolution far-infrared survey of the W31 region.   E. L. Wright, G. G. Fazio, F. J. Low.
Astrophys. J., Vol. 217, 724 - 728 (1977).
   The Center for Astrophysics-University of Arizona 1 m balloon-borne telescope was used to conduct a far-infrared survey of the W31 region at an effective wavelength of 69 μm with a 1' resolution. Within this region seven far-infrared sources were observed. Five of these sources were associated with thermal radio emission. For each of these sources the infrared luminosity is much greater than the Lα luminosity, a situation requiring either dust absorption of Lyman-continuum photons or a large nonionizing stellar luminosity. Two faint infrared sources had no radio counterparts. Far-infrared radiation was not detected from two known radio sources and from one mid-infrared source in this region.

133.008   Radio emission from the infrared source CRL 618: an extremely young planetary nebula.
C. G. Wynn-Williams.
Mon. Not. R. Astron. Soc., Vol. 181, 61P - 62P (1977).
   15-GHz thermal radio emission has been found from the infrared source CRL 618. This observation strengthens the evidence that CRL 618 is an extremely young planetary nebula.

133.009   L'astronomie infrarouge et les poussières galactiques.
   J.-C. Pecker.
Highlights of Astronomy, Vol. 4, Part I, (see 003.010), p. 3 - 33 (1977).
   After a brief review of technical developments, the reasons for studying the infrared, and the principles of diagnostics, are given. The analysis of the spectral features leads to identification of ice, graphite, and various silicates, as main constituents of the dust. Examples are given of the various sources that one meets when travelling in the Galaxy: protostars, cold and dilute, dense envelopes of young stars (cold or hot), dilute envelopes of not so young stars, and ejected clouds surrounding evolved objects.

133.010   Interpretation of far infrared emission.
   J. P. Emerson, R. E. Jennings.
Far infrared astronomy (see 012.027), p. 219 - 230 (1976).
   Far infrared sources of radiation are interpreted in terms of emission from dust clouds heated by hot stars, and the problems of the location, temperature, composition and amount of dust are discussed. The dust appears to be present both inside and outside the H II region, and grains (or grain mantles) of ice fit the data well.

133.011   The distribution of ionized gas and dust in W 3(A) and W 3(OH).   P. G. Mezger.
Far infrared astronomy (see 012.027), p. 231 - 246 (1976).
   Recent radio and infrared observations are summarized and interpreted. A model fit yields the distribution of dust and ionized gas. The models are compared to a possible evolutionary sequence of O-stars and associated shell of gas and dust. The problem of dust inside H II regions and its effect on the ionization structure is discussed.

133.012 A comment on the similarity in the spectra of far infrared sources. H. Okuda.
Far infrared astronomy (see 012.027), p. 247 - 248 (1976).

133.013 Optical depths of far infrared sources.
C. D. Andriesse.
Far infrared astronomy (see 012.027), p. 249 - 255 (1976).
The optical depths at 350 μm of far-infrared sources are discussed in the context of collapsing clouds. It is assumed that the opacity is determined by dust particles, for which the far-infrared properties can be reliably estimated. The sources are found to be optically thin at 350 μm ($\tau \approx 0.001$) and reasonable values for their mass are derived. The complete infrared spectrum of Sgr B2 is analysed, giving $\tau_{350} < 0.03$ and M $\approx 10^7 M_\odot$.

133.014 Supergrain models of far infrared sources.
M. G. Edmunds, N. C. Wickramasinghe.
Far infrared astronomy (see 012.027), p. 281 - 283 (1976).
A new model for infrared and far-infrared sources is proposed. The use of thin whisker grains of lengths up to 1 mm, perhaps multiply branched to form "snowflake"-like particles, allows much greater far-infrared emission than is possible with an equivalent mass of conventional small spherical grains. The growth of such whiskers and their application to models of galactic and extragalactic sources is discussed.

133.015 The nature of the Kleinmann-Low and Becklin-Neugebauer infrared sources.
Yu. N. Gnedin, I. G. Mitrofanov.
Early stages of stellar evolution, (see 012.041), p. 15 - 16 (1977). In Russian.
On the basis of polarimetric properties models of the BN- and KL-infrared sources were constructed.

133.016 Microwave spectra of OH/infrared-sources.
A. Winnberg.
Infrared Phys., Vol. 17, 557 - 563 (1977).
A review of OH, H_2O and SiO maser emission from Mira variables and late-type supergiants is given. The galactic distribution of the "unidentified" OH/i.r. sources is discussed and arguments are given for the stellar radial velocity being midway between the radial velocities of the two OH line components.

133.017 Water maser and envelope of infrared stars.
S. Deguchi.
Publ. Astron. Soc. Japan, Vol. 29, 669 - 681 (1977).
The populations of levels including the ν_2 vibrationally excited state for water molecules in the circumstellar envelope of maser/infrared sources are calculated. Not the near-infrared radiation, but the collisional process is responsible for the pumping. Several microwave transitions in the ν_2 state are expected to show maser emissions. The time variations of the water maser source in infrared stars are explained by the change of the density in the envelope, which suggests that the mechanism of mass loss is due to the shock wave.

133.018 IR-Quellen in W51.
D. Lemke, U. Fahrbach, H. Hefele.
Mitt: Astron. Ges., Nr. 42, p. 120 (1977).

An airborne infrared astronomy program: system description and preliminary results.   See Abstr. 032.520.

Broad-band 20–33-μ photometry of young stars.
See Abstr. 113.018.

Broad-band infrared colors and CO and H_2O absorption indices for late-type dwarf stars.
See Abstr. 113.019.

Infrared stars. Review of observational data.
See Abstr. 114.011.

Observations of radio stars at 3.3 mm.
See Abstr. 116.001.

Characteristics of IR variable stars as observed from orbit.   See Abstr. 122.084.

Polysaccharides and infrared spectra of galactic sources.   See Abstr. 131.023.

Radiating cosmic dust.   See Abstr. 131.035.

The far-infrared spectrum of the core of Sagittarius B2.   See Abstr. 131.041.

Infrared studies of star formation.
See Abstr. 131.069.

Observations of the SiO and H_2O masers in Orion A.
See Abstr. 131.075.

New infrared objects associated with OH masers.
See Abstr. 131.093.

Polysaccharides and the infrared spectrum of OH 26.5 + 0.6.   See Abstr. 131.119.

Time variations and position jitter in interstellar water masers.   See Abstr. 131.128.

Infrared emission from grains with fluctuating temperatures.   See Abstr. 131.132.

H_2O in the Galaxy: sites of newly formed OB stars.
See Abstr. 131.137.

Infrared observations of star formation regions.
See Abstr. 131.143.

The 10-micron interstellar absorption band: curves of growth for terrestrial and meteoritic olivine particles.
See Abstr. 131.195.

Polyoxymethylene co-polymers on grains.
See Abstr. 131.201.

Radio and infrared observations of the OH/H_2O source G12.2−0.1.   See Abstr. 131.213.

Compact H II regions.   See Abstr. 132.014.

Fine-structure line emission from selected compact H II regions.   See Abstr. 132.020.

The 4830 MHz H_2CO absorption in the direction of NGC 6334.   See Abstr. 132.026.

A shock model for infrared line emission from H_2 molecules.   See Abstr. 134.017.

Far infrared measurements of the Orion Nebula (M42).   See Abstr. 134.022.

Two strong infrared sources identified with double reflection nebulae.   See Abstr. 134.023.

Infrared polarization of the galactic center. II.
See Abstr. 156.006.

Airborne far-infrared observations of the galactic center region.   See Abstr. 156.007.

134 Emission Nebulae, Reflection Nebulae

134.001 Intensity ratios of Brackett to Balmer lines of hydrogen in gaseous nebulae.   K. Giles.
Mon. Not. R. Astron. Soc., Vol. 180, 57P - 59P (1977).

Brackett lines, due to transitions $n \to 4$ in hydrogen recombination spectra, have recently been observed. The intensity ratios $I(n \to 4)/I(n \to 2)$, required for the interpretation of the observations, are tabulated for a wide range of temperatures and densities.

134.002 The ionization and thermal equilibrium in the average and single filaments of the Crab Nebula.
M. Contini, B. Z. Kozlovsky, G. Shaviv.
Astron. Astrophys., Vol. 59, 387 - 393 (1977), with a correction Vol. 61, 447 (1977).

The ionization and thermal equilibrium of the filaments in the Crab Nebula is considered. The model is basically identical with the one proposed by Williams (1967) i.e. gas ionized by synchrotron radiation where the diffused radiation is taken into account. The authors find that when each filament is calculated independently, the large variations in the observed line intensities can be explained by means of the same model assuming practically identical composition but different incident flux. The incident flux so found is correlated with the projected distance from an assumed center. This correlation resembles the one discovered in X-ray observation. The best agreement with observation was obtained with He/H ~0.4–0.5.

134.003 Optical investigations of kinematics of NGC 7822 (W1).   T. A. Lozinskaya, T. G. Sitnik.
Astron. Zh. Akad. Nauk SSSR, Vol. 54, 807 - 816 (1977).
In Russian. English translation in Soviet Astron., Vol. 21, No. 4.

Internal motions in the dust-gaseous nebula W1 (NGC 7822) were investigated. The observations in Hα and [N II] 6584 Å lines were carried out by using a high-contrast Fabry-Perot etalon and a contact image converter. The radial velocity field of the greatest part of the nebula was inspected. Expansion of the faint extensive nebulosity which forms a ring centered on the bright component of W1 has been found; the expansion velocity was estimated to be about 40 km s⁻¹. A new model for the W1 complex is discussed: the faint extensive shell around the bright nebula is shown to be a supernova remnant. In terms of this model the non-thermal radio-source G 118.1 + 5.0 is not a supernova remnant but a dense H II region colliding with the expanding supernova shell. The interaction of a strong stellar wind with the surrounding gas is discussed too.

134.004 The wisps in the Crab – a cosmic laser?
W. Kundt.
Astron. Astrophys., Vol. 60, L19 - L22 (1977).

The observed pulsed γ-rays and X-rays are synchro-Compton radiation emitted by the pulsar wind inside the inner edge (= injection sphere) of the nebula. The wisps are stimulated optical emission from this inner edge.

134.005 Search for thermal X-ray line emission from the Crab Nebula.   H. L. Kestenbaum, R. S. Wolff, K. S. Long, R. Novick, M. C. Weisskopf.
Astrophys. J., Lett., Vol. 216, L27 - L29 (1977).

Observations of the Crab Nebula from 1.85 to 8 keV obtained with the large-area graphite crystal spectrometer on OSO-8 show a smooth, featureless continuum with an absence of X-ray emission lines from highly ionized Si and S. Upper limits on the line emission are used to derive an upper limit to the emission measure for a thermal source associated with the Crab Nebula. The authors discuss existing evidence for an extended thermal source and show that the evidence is not contradicted by the upper limits set with the spectrometer data.

134.006 Vibrationally excited molecular hydrogen in Orion.
D. J. Hollenbach, J. M. Shull.
Astrophys. J., Vol. 216, 419 - 426 (1977).

Physical mechanisms for producing vibrationally excited molecular hydrogen, such as has recently been detected toward the Orion Nebula, are discussed. The most likely mechanisms are: (I) collisional excitation behind a shock moving into a molecular cloud; and (II) near-ultraviolet pumping in the H_2 Lyman and Werner bands and subsequent cascade. The absolute intensities of the Orion lines require either: (I) a 10 km s⁻¹ shock moving into a 3×10^5 cm⁻³ cloud; or (II) an incident near-UV flux 10^6 times the mean interstellar value. The shock model is favored.

134.007 Observation and interpretation of temperature gradients in the Orion nebula.
S. C. Perrenod, G. A. Shields, E. J. Chaisson.
Astrophys. J., Vol. 216, 427 - 432 (1977).

Radio recombination lines of hydrogen and helium, measured at several positions in Orion A, show that the gas kinetic temperature is nearly 3000 K higher at the nebular periphery than at the core, in agreement with recent optical results. Theoretical models suggest that the effect of radiation hardening is not sufficient to account for the observed temperature gradient. The authors propose that much of the temperature gradient results from a gradient in the gas-phase metal abundance, caused by evaporation of grains near the central star; in this case an outward decrease of O/H in the gas by a factor ~4 is required.

134.008 A search for gamma-ray lines from Nova Cygni 1975, Nova Serpentis 1970, and the Crab Nebula.
M. Leventhal, C. MacCallum, A. Watts.
Astrophys. J., Vol. 216, 491 - 502 (1977).

A balloon-borne γ-ray telescope employing a large-volume lithium-drifted-germanium detector as the central element was flown in an attempt to detect γ-ray lines from two recent nova explosions and from the Crab Nebula. The energy range covered was 100 keV to 5 MeV. No lines were detected from the novae, even though a sensitivity relevant to current theories of explosive nucleosynthesis was achieved. The Crab continuum was detected between 100 keV and 1 MeV. Evidence for a line feature from the Crab Nebula at 400 ± 1 keV is presented. Such a line may be produced by positron annihilation at the surface of the Crab neutron star.

134.009 Observation of X-rays from the Crab pulsar.
A. Toor, F. D. Seward.
Astrophys. J., Vol. 216, 560 - 564 (1977).

A measurement of the pulsed X-ray emission from the Crab Nebula made during a lunar occultation is presented. Comparison is made with previous observations, and possible variations in this pulsed emission are discussed.

134.010 Spectrophotometry of NGC 2261 and R Mon. I.
J. L. Greenstein, M. A. Kazaryan, T. Yu. Magakyan, Eh. E. Khachikyan.
Astrofizika, Vol. 12, 587 - 611 (1976). In Russian. – English translation in Astrophysics, Vol. 12, No. 4.

The results of a spectrophotometry of the cometary nebula NGC 2261 and its nucleus R Mon are presented. In the spectra of both objects 111 emission and 26 absorption lines have been found and identified, 70 of which have been discovered for the first time. Equivalent widths of lines were

estimated. In the spectra of the nebula and R Mon some
variations of line intensities and their profiles have been
observed. Some lines disappear completely. The components
of the doublet λ 3727 in the spectrum of NGC 2261 have
been observed. The D_1 and D_2 interstellar lines of Na I have
also been detected in the spectrum of R Mon.

134.011 Energy L_c-spectrum of the source responsible for the T Tauri nebula emission.
V. V. Golovatyj, I. V. Shpychka.
Astrofizika, Vol. 12, 613 - 622 (1976). In Russian. – English
translation in Astrophysics, Vol. 12, No. 4.

The small emission nebula connected with T Tau is
studied. The parameters $F_0 = 6 \times 10^{-29} \, \text{W/m}^{-2} \, \text{Hz}^{-1}$ and
$\alpha = -3.8$ which characterize the L_c-spectra of the ionizing
source are found. It is shown that these quantities of F_0 and
α explain well the ionization of oxygen observed in the
nebula, and the energy balance in it. Estimates of the following values are derived: energy responsible for the heating of
gas in the nebula, optical depth of the nebula, and ionization
of hydrogen.

134.012 Observations of galactic nebulae at 5 GHz.
G. S. Rossano.
Astron. J., Vol. 82, 587 - 592 (1977).

Six-centimeter flux densities are reported for nine optically identified galactic nebulae and for two unidentified
background sources near S 36. The physical nature of the
sources is discussed.

134.013 Observations of formaldehyde and 110 α recombination lines toward bright rims in diffuse nebulae.
Nguyen-Q-Rieu, V. Pankonin.
Astron. Astrophys., Vol. 60, 313 - 320 (1977).

A search for 6-cm H_2CO and 110 α recombination lines
has been made with the 100-m Effelsberg radiotelescope in
the direction of 18 bright rims in 9 nebulae. H_2CO absorption
has been detected near the rims in IC 1848, IC 434 and NGC
2264. Hydrogen recombination lines have been detected near
the elephant trunk structures in M 16 and possibly near rim
B in IC 1396.

134.014 A model for the filamentary structure in the Pleiades reflection nebulosity.   T. Arny.
Astrophys. J., Vol. 217, 83 - 89 (1977) = Contrib. Five Coll.
Obs., Amherst, Mass., No. 234.

It is suggested that the filamentary structure in the
Pleiades reflection nebula is caused by shearing of dust clumps
in an interstellar cloud moving through the star cluster. Radiation pressure flattens a dust clump and causes it to flow
around a star, forming a shell. The anisotropy of the radiation
field shears clumps into long streamers.

134.015 Far-ultraviolet imagery of the Orion Nebula.
G. R. Carruthers, C. B. Opal.
Astrophys. J., Vol. 217, 95 - 102 (1977).

Two electrographic cameras carried on a sounding
rocket have yielded the first useful-resolution far-ultraviolet
(1000–2000 Å) imagery of the Orion Nebula. The brightness
distribution in the images is consistent with a primary source
which is due to scattering of starlight by dust grains, although
an emission-line contribution, particularly in the fainter outer
regions, is not ruled out. The results are consistent with an
albedo of the dust grains that is high in the far-ultraviolet and
is increasing toward shorter wavelengths below 1230 Å.

134.016 Spectral observations of η Carinae at 4 microns.
D. K. Aitken, B. Jones, J. D. Bregman, D. F. Lester,
D. M. Rank.
Astrophys. J., Vol. 217, 103 - 107 (1977) = Lick Obs. Bull.
No. 756.

4 μm spectral observations of the central region of η

Carinae nebula have measured the Brackett-α hydrogen recombination line intensity. The bulk of the line and continuum
radiation arises in a small central source which is slightly
narrower in the line radiation and is barely resolved by the
beamwidth of 3.″4. The continuum radiation is well fitted by
$F \propto e^{-\theta/1.3}$ to a distance of 10″ from the central source. The
line intensity requires a central source which emits 10^{49}
ionizing photons s^{-1} into a region of electron density greater
than 10^6 cm^{-3}.

134.017 A shock model for infrared line emission from H_2 molecules.
R. London, R. McCray, S.-I Chu.
Astrophys. J., Vol. 217, 442 - 447 (1977).

The authors consider a shock model for the H_2 infrared
line emission detected in the Orion Nebula by Gautier et al.
They solve the equations describing stationary gas flow and
excitation of H_2 in the radiative cooling region behind a shock.
They find that, for the high densities needed to explain the
observed line strengths, the populations of the vibration-
rotation levels are nearly in LTE. By matching the calculated
line strengths to the observations, the authors find that the
acceptable values of shock velocity and upstream molecular
density must be of order $8 \lesssim v_0 \lesssim 21$ km s^{-1} and
$10^5 \lesssim n_0 \lesssim 10^6$ cm^{-3}.

134.018 On the nitrogen and oxygen abundances in nebulae.
M. Perinotto.
Astron. Astrophys., Vol. 61, 247 - 249 (1977).

The effect of charge transfer reactions between singly
and doubly ionized oxygen and nitrogen with hydrogen is
shown to modify strongly the ionization structure of these
elements in models of gaseous nebulae. The computed fractional abundances of O^+ and N^+ are larger than in models
which do not consider these reactions. That results in a better
agreement between theory and observations in the line intensities emitted from low ionization potential ions in
planetary nebulae.

134.019 Laboratory exercises in astronomy – the Crab nebula.   O. Gingerich.
Sky Telesc., Vol. 54, 378 - 382 (1977).

134.020 The excitation of neutral-oxygen emission in the Orion nebula.   R. P. Lowe, J. M. Moorhead,
W. H. Wehlau.
J. R. Astron. Soc. Canada, Vol. 71, 398 (1977). – Abstract.

134.021 The distribution of ionized carbon in the Orion nebula.   T. B. H. Kuiper, N. J. Evans II.
J. R. Astron. Soc. Canada, Vol. 71, 398 - 399 (1977).
Abstract.

134.022 Far infrared measurements of the Orion Nebula (M42).   K. Shivanandan, D. P. McNutt,
M. Daehler, P. D. Feldman.
Far infrared astronomy (see 012.027), p. 193 - 199 (1976).

Airborne observations of Orion at 100 and 285 μm are
reported.

134.023 Two strong infrared sources identified with double reflection nebulae.   C. G. Wynn-Williams.
Far infrared astronomy (see 012.027), p. 207 - 208 (1976).
Abstract.

134.024 Multiple scattering in reflection nebulae. I. A Monte Carlo approach.   A. N. Witt.
Astrophys. J., Suppl. Ser., Vol. 35, 1 - 6 (1977).

A method is described which permits the calculation of
the surface brightness distribution on a plane-parallel reflection nebula of uniform density, illuminated by a single star
located in front of, behind, or arbitrarily inside the nebula.

The multiple scattering problem is solved by the Monte Carlo technique in a three-dimensional simulation. The models are completely parametrized by describing particle properties by their single scattering albedo and by an analytic phase function with one or three parameters.

134.025   Multiple scattering in reflection nebulae. II. Uniform plane-parallel nebulae with foreground stars.
A. N. Witt.
Astrophys. J., Suppl. Ser., Vol. 35, 7 - 19 (1977).

Parametrized Monte Carlo model calculations are presented for the surface brightness of a plane-parallel, uniformly dense nebula which is illuminated by a foreground star. The effects of relevant parameter changes are examined graphically after representing the computed surface brightness distribution by a power law. Despite substantial multiple scattering, the shape of the phase function remains dominant in affecting the predicted surface brightness distributions. Multiple scattering effects strongly increase in importance with increasing angular distance from the star, significantly lessening the steepness of the surface brightness distribution compared to single-scattering models. The observational approach of deriving relative radar cross sections by the extrapolation of observed color differences to 0° offset has been tested with the present models.

134.026   Multiple scattering in reflection nebulae. III. Nebulae with embedded illuminating stars.
A. N. Witt.
Astrophys. J., Suppl. Ser., Vol. 35, 21 - 29 (1977).

Model calculations for uniformly dense, plane-parallel reflection nebulae with immersed stars are presented and briefly discussed. The detail provided is designed primarily to distinguish this type of nebula from one with a foreground star. As in nebulae with foreground stars, multiple scattering increases in importance with angular distance from the illuminating star and therefore has a significant effect on the surface brightness distribution. Detailed comparisons between optically thick plane-parallel and spherical reflection nebulae with a central star are included. It is found that observations of nebulae with embedded stars can aid in the determination of the phase function shape at different wavelengths.

134.027   Multiple scattering in reflection nebulae. IV. The multiplicity of scattering.
A. N. Witt, E. R. Oshel.
Astrophys. J., Suppl. Ser., Vol. 35, 31 - 36 (1977).

The characteristics of higher order scattering in plane-parallel reflection nebulae of finite optical thickness have been examined. Measured both by the mean number of scatterings and the effective multiplicity, the importance of multiple scattering in such systems varies strongly with position as well as direction, and the effects cannot be taken into account by simple corrections only. It is found that the amount of higher order scattering is relatively insensitive to the shape of the phase function over a wide range of asymmetries, while the dependence of multiplicity on particle albedo, optical thickness of the nebula, and other geometrical factors is as expected.

134.028   Die Interpretation integrierter Spektren von Emissionsnebeln.   J. Köppen.
Diss. Naturwiss. Gesamtfak., Ruprecht-Karl-Univ., Heidelberg, F.R. Germany, 5 + 102 pp. (1977).

134.029   Radial distribution of [Fe XIV] emission in the Cygnus Loop.
B. E. Woodgate, R. P. Kirshner, R. J. Balon.
Astrophys. J., Lett., Vol. 218, L129 - L131 (1977).

The one-dimensional distribution of [Fe XIV] λ5303 emission has been determined along a radius of the Cygnus Loop through the use of a tilting filter photometer. The observed emission extends at least 5′ outside the optical filaments.

A simple Sedov solution model of the temperature and density distribution behind the shock agrees with the observations if the shock front is near the outer extent of the [Fe XIV] emission, the shock velocity is from 300 to 250 km s^{-1}, and the density external to the remnant is about $0.7 - 1.4$ cm^{-3}. These parameters are in reasonable agreement with X-ray maps and optical radial velocities.

134.030   New polarisation maps of nebulae.
S. Pallister, H. Perkins, S. M. Scarrott, R. Bingham.
New Scientist, Vol. 73, 712 - 714 (1977). — Abstr. in Phys. Abstr., Vol. 80, Abstr. 57081 (1977).

134.031   Polarimetry of the Crab nebula and NGC 1569.
C. White.
Thesis, Univ. Durham, England (1977). — Abstr. in Phys. Abstr., Vol. 80, Abstr. 74126 (1977).

134.032   Cometary nebula near NGC 7023.
IAU Circ., Nos. 3095, 3155 (1977).

134.033   Nebula near CL4.
IAU Circ., Nos. 3116, 3118 (1977).

134.034   Optical colours and polarization of a model reflection nebula. II. Silicate and graphite grains in a nebula with the star in the rear.   G. A. Shah.
Astron. Nachr., Band 298, 319 - 325 (1977).

The colour difference between the star and the attendant reflection nebula and polarization both caused by silicate and graphite grains have been given. The properly normalized size distribution function for each type of grains has been considered within homogeneous plane parallel slab-model of the reflection nebula with the star in the rear. Contrary to some earlier results, it has been possible to show that silicate grains can certainly play a role in the phenomena of reflection nebulae.

134.035   Large proper motions in the Orion nebula.
K. M. Cudworth, R. C. Stone.
Publ. Astron. Soc. Pacific, Vol. 89, 627 - 629 (1977).

Several nebular features, as well as one faint star, with large proper motions have been identified within the Orion nebula. The measured proper motions correspond to tangential velocities of up to ~ 70 km sec^{-1}. One new probable variable star was also found.

134.036   The thermal and non-thermal components of sixteen nebular complexes.   C. Goudis.
Astrophys. Space Sci., Vol. 47, 109 - 146 (1977).

The nature of the fine structure of sixteen galactic nebular complexes is investigated by producing their radio spectra. Various physical parameters of the thermal components are derived by adopting a homogeneous, optically thin, spherical model. The variations of the physical parameters derived from the model as a function of the electron temperature and distance are investigated. A discussion about the thermal and non-thermal spectra is included.

134.037   Investigation of the physical conditions in gaseous nebulae. I. Calculation of relative intensities of emission lines for models of a nebula transparent at frequencies of the L$_c$-continuum.
V. V. Golovatyj, V. I. Pronik, O. S. Yatsyk.
Tsirk. Astron. Obs., L'vov, No. 51, p. 3 - 12 (1976). In Russian.

134.038   Thin scattering diffusion and the size of the Crab Nebula.   G. Cavallo.
Astrophys. Space Sci., Vol. 50, 173 - 177 (1977).

The dependence of the size of the Crab Nebula on the frequency of observation is deduced from the hypothesis of thin scattering diffusion of the radiating electrons in an in-homogeneous magnetic field.

134.039   A model for a structure in the galactic nebula S206.
J. E. Dyson.
Astrophys. Space Sci., Vol. 51, 197 - 204 (1977).

The galactic nebula S206 contains a half shell of high excitation nebulosity which is centred on the associated exciting star. The suggestion has been made that this structure is caused by the interaction of stellar mass loss from the star with nebular gas. A steady state model of such an interaction is investigated quantitatively.

134.040   Cometary nebula near NGC 7023.
Yamamoto Circ., No. 1859 (1977). In Japanese.

134.041   On the distance of R Mon and NGC 2261.
J. L. Greenstein, M. A. Kazaryan, T. Yu. Magakyan, Eh. E. Khachikyan.
Astron. Tsirk., No. 947, p. 1 - 3 (1977). In Russian.

134.042   New interesting nebulous objects.
A. L. Gyul'budagyan, T. Yu. Magakyan.
Astron. Tsirk., No. 953, p. 1 - 2 (1977). In Russian.

134.043   Minkowski's footprint and the Egg nebula.
T. Tuvikene.
Calendar of the Tartu Astronomical Observatory for the year 1978, (see 047.026), p. 68 - 72 (1977). In Estonian.

134.044   Simplest models of reflection nebulae.
N. V. Voshchinnikov.
Astron. Zh. Akad. Nauk SSSR, Vol. 54, 1221 - 1231 (1977). In Russian. English translation in Soviet Astron., Vol. 21, No. 6.

$(U-B)$, $(B-V)$ and $(V-R)$ colors and U, B, V and R polarization have been calculated for a model of a reflection nebula associated with a large dust cloud. For the cases in which the illuminating star is far from the surface of the cloud, the form of the nebula has been considered to be spherical. If the star is close to the surface of the cloud, a part of the nebula boundary has been considered to be flat. A comparison of the theoretical results with the observations of the Merope nebula shows that dirty ice grains represent satisfactorily the observations if the star is embedded 0.7 pc behind the front surface of the nebula.

134.045   Observations of visual polarization of the Orion nebula.   S. Isobe.
Tokyo Astron. Bull., Second Ser., No. 249, p. 2875 - 2883 (1977).

Observations of linear polarization of the Crab Nebula during an occultation by the solar corona. II.
See Abstr. 074.039.

The mass of the nova HR Delphini nebula.
See Abstr. 124.401.

A remarkable structural change in a faint cometary nebula.   See Abstr. 131.019.

On the molecular hydrogen emission at the Orion Nebula.   See Abstr. 131.046.

Detection of submillimeter (870 μm) CO emission from the Orion molecular cloud.   See Abstr. 131.094.

The interstellar medium in the direction of the Crab nebula: reconciling soft X-ray and radio observations.
See Abstr. 131.113.

Detection of interstellar ethyl cyanide.
See Abstr. 131.149.

The hydroxyl masers in the Orion Nebula.
See Abstr. 131.152.

Observations of the λ 18 cm OH maser emission in the Orion nebula and λ 6 cm formaldehyde absorption in the associated molecular cloud.   See Abstr. 131.157.

Dust and gas near the Pleiades.
See Abstr. 131.169.

Observations of [S III] in NGC 604 and N/S abundance gradients.   See Abstr. 132.010.

Line profiles from the ionized and neutral gas in the peculiar nebula NGC 6302.   See Abstr. 135.014.

The high-energy X-ray spectrum of the Crab Nebula observed from OSO-8.   See Abstr. 142.036.

A search for the reported 400-keV γ-ray line from Crab nebula.   See Abstr. 142.708.

Extinction law in dust clouds and the young southern cluster NGC 6250: further evidence for high values of R.   See Abstr. 153.019.

The radio structure of the Cygnus Loop at 25 MHz.
See Abstr. 156.003.

135 Planetary Nebulae

135.001 The physical differences between the planetary nebulae of the common field and the planetaries in the direction of the galactic centre.   E. B. Kostyakova.
Astron. Zh. Akad. Nauk SSSR, Vol. 54, 817 - 827 (1977).
In Russian. English translation in Soviet Astron., Vol. 21, No. 4.

Mean physical parameters were estimated for the planetary nebulae of two groups.regarded as a whole: planetaries of the common field and those belonging to a large group seen in the galactic-centre direction. The mean central star temperature for the objects of each group was determined by solving the equation of the electron energy balance in the nebula. The optical depth and collisional excitation of hydrogen atoms were taken into account. The necessary values of $I(N_2)/I(H_\beta)$ were obtained from observations. For the planetaries of both groups the mean electron density, the mean electron temperature, the mean degree of ionization, the dimensions and masses of the nebular envelope, and the principal parameters of the central star were estimated.

135.002 Peculiar central stars of planetary nebulae.   J. H. Lutz.
Astron. Astrophys., Vol. 60, 93 - 100 (1977).

Some of the planetary nebulae listed in the Catalogue of Galactic Planetary Nebulae are peculiar in that they have nebular emission lines in combination with the absorption spectrum of a cool (spectral type A through K) central star. Spectral classifications and relative intensities of nebular emission lines have been obtained for sixteen peculiar central stars. The spectroscopic characteristics of peculiar central stars are compared with those of other emission line objects such as Be stars, symbiotic stars, and P Cygni stars.

135.003 Sharpless 176: a large, nearby planetary nebula.   F. Sabbadin, S. Minello, A. Bianchini.
Astron. Astrophys., Vol. 60, 147 - 149 (1977).

Spectroscopic and photometric investigations on S 176 indicate that this nebula is a planetary nebula with a radius of 0.40 pc at a distance of 140 pc. Its central star is a blue subdwarf with $M_v = +5.7$.

135.004 Statistical properties of planetary nebulae: central stars.   S. Minello, F. Sabbadin.
Astron. Astrophys., Vol. 60, L9 - L10 (1977).

Further evidences of the existence of two classes of planetary nebulae (Grieg 1971, 1972) are presented. The spectral type and the absolute magnitude of the central stars of B nebulae show different statistical characteristics when compared with the C's.

135.005 Mass loss, long-period variables, and the formation of circumnebular shells.
M. Kafatos, A. G. Michalitsanos, M. S. Vardya.
Astrophys. J., Vol. 216, 526 - 530 (1977).

The authors found that the rate of mass loss $\dot{M}$ increases with an increase in the period of pulsation for Mira-type variables. This result suggests that the rate of mass loss is accelerated with time until a maximum value is reached before the ejection of the outer envelope. The matter from the continuous mass loss during the evolution of the star produces supersonic shock waves that sweep up the interstellar gas upon encountering the interstellar medium, so that a shell is formed. This phenomenon may account for the observations of extended regions of emission that surround planetary nebulae.

135.006 Fabry-Perot interferometry of stellar planetary nebulae.   H. M. Johnson.
Astrophys. J., Vol. 216, 776 - 783 (1977).

Observations have been made of 12 planetary nebulae (PN) that are apparently small in linear scale and perhaps near the start of PN evolution. They have been photoelectrically scanned with high spectral resolution around Hα, and five of them also around [O III] λ5007. Each Fabry-Perot (FP) profile is illustrated. Tables give the heliocentric radial velocity of Hα, data of emission-line widths and shapes, and background-continuum levels. The relation of stellar PN kinematics to precursor objects is discussed in terms of ejection of matter from a red giant in terminal phase.

135.007 Far infrared observations of planetary nebulae.   H. Moseley, D. A. Harper, R. F. Loewenstein.
Bull. American Astron. Soc., Vol. 9, 433 (1977). − Abstract.

135.008 The planetary nebula 164 + 31°.1.   C. Barbieri, J. W. Sulentic.
Publ. Astron. Soc. Pacific, Vol. 89, 261 - 263 (1977).

Spectroscopic observations are presented for the high-latitude planetary nebula 164 + 31°.1. This object is frequently misidentified as NGC 2474/75. It is found to be a medium excitation object with a heliocentric radial velocity of −66 km s⁻¹.

135.009 On the density gradient in NGC 7027.   M. Perinotto.
Astron. Astrophys., Vol. 60, 433 - 435 (1977).

The density gradient in NGC 7027 found by Perinotto (1971) and by Kaler et al. (1976) with a method strictly correct only for constant density in the nebula is investigated with a model nebula which allows for density variations as well as for ionization stratification. The resulting integrated line intensity ratios well match the observed ones, so substantiating that Ne variates inwards in NGC 7027 from about 10^4 to more than 10^5 cm⁻³.

135.010 A study of the planetary nebulae Abell 30 and Abell 78.   M. Cohen, H. S. Hudson, S. L. O'Dell, W. A. Stein.
Mon. Not. R. Astron. Soc., Vol. 181, 233 - 245 (1977).

The authors have studied the central regions of the planetary nebulae A 30 and A 78 by *UBVRI* photometry, optical spectroscopy, and near-infrared photometry. The spectra contain high-excitation emission lines and strongly resemble those of Wolf−Rayet stars of the carbon sequence. Stellar temperatures >50000 K are inferred. The observed 3.5-μm flux of each nebula exceeds reasonable extrapolations of both the stellar flux and any possible free−free emission. The colour temperature of this excess between 2.28 and 3.5 μm is ∼1000 K.

135.011 The 4 to 8 micron spectrum of NGC 7027.   R. W. Russell, B. T. Soifer, S. P. Willner.
Astrophys. J., Lett., Vol. 217, L149 - L153 (1977).

Spectrophotometric observations of NGC 7027 are reported. The continuum shows a strong, broad peak near 7.7 μm, but little or no evidence for the strong peak near 7.0 μm expected from previously postulated carbonate grains. A spectrally resolved feature at 6.2 μm is found and attributed to an increase in the emissivity of some as yet unidentified dust material. Unresolved emission features are seen at 4.49 and 5.60 μm. The flux in the [Mg V] line is used to derive a lower limit on the temperature of the central star of 1.3×10^5 K.

135.012 Planetary nebulae.   G. S. Khromov.
Topics in interstellar matter, (see 012.011), p. 89 - 96 (1977). − See 18.131.131.

135.013 **Observations of the [O I], λ6300 velocity field of NGC 7293, the 'Helix' nebula.** K. Taylor.
Mon. Not. R. Astron. Soc., Vol. 181, 475 - 482 (1977).

The [O I] λ6300 velocity field of the Helix nebula NGC 7293 has been obtained. The geometry best suited to accommodate both the velocity data and the appearance of the nebula in the sky is that of a radially expanding helix. The helix, however, is not of a simply type, where the axial coordinate, z, is a linear function of the azimuthal angle θ, rather it is better approximated by an extended helix where z is proportional to θ^2. The expansion velocity in the equatorial zone is ~ 22 km/s.

135.014 **Line profiles from the ionized and neutral gas in the peculiar nebula NGC 6302.**
K. H. Elliott, J. Meaburn.
Mon. Not. R. Astron. Soc., Vol. 181, 499 - 507 (1977).

The line profiles of Hα and [N II] from the ionized material, and [O I] from the neutral gas, are presented for many positions over the curious nebula NGC 6302. Neutral, as well as ionized material, appears to be flowing from the core of the nebula. Moreover, this central region is shown to have a value of $T_e = 26700 \pm 2000$ K which indicates collisional excitation. It is suggested that an energetic stellar wind, from a hidden embedded star, could produce many of the effects observed.

135.015 **The detection of radio emission from the new optical emission variable HM Sagittae.**
P. A. Feldman.
J. R. Astron. Soc. Canada, Vol. 71, 386 - 389 (1977).

Radio emission has been detected at 10.5 GHz from the emission-line object SVS 2183 = HM Sge. By analogy with V1016 Cyg, which has exhibited similar physical characteristics, HM Sge may be a low-mass planetary nebula in an early stage of formation.

135.016 **Observations of NGC 7027 in the near-infrared.**
J. N. Scrimger.
J. R. Astron. Soc. Canada, Vol. 71, 399 (1977). – Abstract.

135.017 **Emission lines in the near infrared spectra of twelve faint planetary nebulae.**
Y. Andrillat, L. Houziaux.
C. R. Acad. Sci. Paris, Tome 285, Sér. B, 263 - 265 (1977). In French.

The authors present spectra of twelve faint planetary nebulae observed with a dispersion of 230 Å mm^{-1} over the spectral range 8,000–11,000 Å. All nebulae exhibit three strong emission lines: He I λ 10,830 and [S III] λλ 9,069 and 9,532. Generally, the high excitation nebulae show in addition a very strong He II λ 10,123 line.

135.018 **On the distance scale of planetary nebulae and white dwarf birth rates.** V. Weidemann.
Astron. Astrophys., Vol. 61, L27 - L30 (1977).

Arguments are presented which favor an increase of the distance scale of planetary nebulae by 30% compared to the Seaton-Webster scale. The consequences for evolutionary tracks, PN and white dwarf relations, and birth rates are discussed.

135.019 **Recent findings about planetary nebulae.**
Y. Terzian.
Sky Telesc., Vol. 54, 459 - 463 (1977).

135.020 **A list of possible, probable, and true planetary nebulae detected since 1966.** R. Weinberger.
Astron. Astrophys., Suppl. Ser., Vol. 30, 335 - 341 (1977).

335 objects designated as new possible, or true galactic planetary nebulae since the closing of the Perek & Kohoutek (1967) catalogue are listed with names, designations, the best available equatorial coordinates, galactic coordinates, apparent dimensions, and indications of observations in the optical, infrared and radio range. 44 new candidates which subsequently turned out to be presumably non-planetaries are also given. Six new possible planetaries detected by the author are listed for the first time.

135.021 **New planetary nebulae of low surface brightness.** R. Weinberger.
Astron. Astrophys., Suppl. Ser., Vol. 30, 343 - 348 (1977).

Twelve new planetary nebulae of low surface brightness have been detected during a search for strongly reddened galaxies on the Palomar Observatory Sky Survey prints. The maximum diameters range from 16 to 194'', the surface brightness in red from 20.0 to 23.3 mag/arcsec^2, in blue from 24.0 to > 26.5 mag/arcsec^2, the integrated magnitudes in red from 11.4 to 16.7 mag, in blue from 15.1 to $\geqslant 20.6$ mag and the estimated distances (uncorrected for extinction) from 700 to 8300 pc.

135.022 **The structure of NGC 6543.**
J. P. Phillips, N. K. Reay, S. P. Worswick.
Astron. Astrophys., Vol. 61, 695 - 703 (1977).

New monochromatic electronographic data are presented which, along with previous data, enables a re-assessment of the nature and origins of the structure of NGC 6543. It is suggested that the nebula may be represented by an intrinsically simple spheroidal structure, with three axes of differing lengths.

135.023 **Geometry effects on the formation of the hydrogen Ly α line in planetary nebulae.**
A. Peraiah, R. Wehrse.
Astron. Astrophys., Vol. 61, 719 - 722 (1977).

In order to study the effects of sphericity on the radiative transfer in the hydrogen Ly α line of planetary nebulae, the radiation field in this line is calculated for static pure hydrogen models with ratios of outer to inner radii approximately equal to 2, 4 and 8. For $r_{out}/r_{in} \approx 2$ the transfer equation is also solved in plane parallel approximation for comparison. In the spherical calculations the profiles of the emergent radiation are not found to be very different in their shapes, while there are large differences in the internal radiation fields. Substantial changes occur for most quantities, when the spherical approximation is replaced by the plane parallel one.

135.024 **IC 4997.**
IAU Circ., No. 3118 (1977).

135.025 **Identification of HV 5824 and HV 5967 with planetary nebulae.** N. Sanduleak.
Inf. Bull. Variable Stars, No. 1300 (1977).

135.026 **The spectra of NGC 7009.**
L. H. Aller, S. J. Czyzak.
Publ. Astron. Soc. Pacific, Vol. 89, 612 (1977). – Abstract.

135.027 **Planetary nebulae, models, chemical compositions, and frustrations.** C. D. Keys, L. H. Aller.
Publ. Astron. Soc. Pacific, Vol. 89, 618 (1977). – Abstract.

135.028 **PK 6–2°1, a remarkable nitrogen-rich southern planetary nebula.** M. A. Dopita.
Astrophys. Space Sci., Vol. 48, 437 - 444 (1977).

Photoelectric spectrophotometry with the IDS on the 3.9 m Anglo-Australian Telescope has shown the planetary nebula PK 6–2°1 to be unique in its overabundance of nitrogen and helium, strong reddening and broad range of excitation conditions. It is suggested that here we see part of the massive envelope of a Wolf-Rayet star of the nitrogen sequence which is still embedded in its placental interstellar cloud.

135.029 On the evolution of central stars of planetary
         nebulae.   R. Z. Yahel.
Astrophys. Space Sci., Vol. 51, 135 - 152 (1977).
     The evolution of nuclei of planetary nebulae has been
calculated from the end of the ejection stage that produces the
nebulae to the white dwarf stage. The structure of the central
star is in agreement with the general picture of Finzi (1973)
about the mass ejection from the progenitors of planetary
nebulae. The author describes theoretical evolutionary tracks
of the nuclei of a planetary nebula and compares them to the
observational results. It is shown that there is a negative
correlation between the total stellar mass and the calculated
evolutionary time scale of the central stars.

135.030 Observations of 9 planetary nebulae in the spectral
         region λ6000–7000 Å.   R. I. Noskova.
Astron. Tsirk., No. 947, p. 3 - 4 (1977). In Russian.

135.031 New absolute intensities of the emission lines of 15
         planetary nebulae seen in the direction of the
galactic centre.   E. K. Kharadze, R. A. Bartaya, B. A.
Vorontsov-Vel'yaminov, E. B. Kostyakova, O. D. Dokuchaeva,
V. P. Arkhipova.
Astron. Tsirk., No. 947, p. 4 - 6 (1977). In Russian.

     Bibliographical index of planetary nebulae for the
period 1965–1976.   See Abstr. 002.024.

     Stimulated emission of the He⁺ radio recombination
lines.   See Abstr. 022.028.

     Recombination spectrum of C III.
See Abstr. 022.106.

     Ejection of planetary nebulae by helium shell

flashes and the planetary distance scale.
See Abstr. 065.052.

     Spectroscopic studies of very old hot stars. III.
Atmospheric properties of seven planetary nuclei.
See Abstr. 114.004.

     The radius and mass of FG Sagittae.
See Abstr. 115.002.

     Central star of NGC 3132: a visual binary.
See Abstr. 118.022.

     The eruptive BQ[ ] star HM Sagittae.
See Abstr. 122.065.

     Spectroscopy of HM Sagittae, a possible embryonic
planetary nebula.   See Abstr. 122.098.

     HM Sagittae (emission variable).
See Abstr. 123.048.

     Aperture synthesis observations of galactic H II
regions. VII. A "quick look" survey of galactic H II regions.
See Abstr. 132.021.

     Radio emission from the infrared source CRL 618:
an extremely young planetary nebula.
See Abstr. 133.008.

     On the nitrogen and oxygen abundances in nebulae.
See Abstr. 134.018.

     Die Interpretation integrierter Spektren von Emis-
sionsnebeln.   See Abstr. 134.028.

Radio Sources, Quasars, Pulsars, Extreme UV, X-Ray, Gamma-Ray Sources, Cosmic Radiation

141 Radio Sources, Quasars, Pulsars

Radio Sources, Quasars

141.001 Radiation from relativistic blast waves in quasars and active galactic nuclei.
R. D. Blandford, C. F. McKee.
Mon. Not. R. Astron. Soc., Vol. 180, 343 - 371 (1977).

An analysis is presented of the synchrotron and inverse Compton radiation that would be observed from behind a strong, relativistic, spherical shock propagating outwards through an ionized, magnetized medium. It is shown that, under a wide variety of conditions, a large fraction of the total dynamical energy can be dissipated in this manner. Details of the observed spectrum and its variation with time are computed for a selection of simple assumptions about the nature of the initial explosion, the ambient external medium and the relativistic particle spectrum. Illustrative applications of the analysis are made to 3C 120, CTA 102, AO 0235 + 164 and Centaurus A.

141.002 The origin of the extragalactic background radiation between 0.5 and 400 MHz. A. J. B. Simon.
Mon. Not. R. Astron. Soc., Vol. 180, 429 - 445 (1977).

The calculations described in this paper show that the superposition of extragalactic radio sources forms a large fraction of the extragalactic radio background spectrum between 0.5 and 400 MHz derived from observations. A significant contribution at low frequencies is also expected from the low-luminosity sources. Synchrotron self-absorption in the components of the individual sources which make up the background results in a low-frequency turn-over. This result is independent of the form of cosmological evolution used. The form of the spectrum is due to the fact that for non-evolving models the dominant sources at all frequencies are the low-intensity sources which also have synchrotron turn-overs at low frequencies.

141.003 Optical identification of extragalactic radio sources from the NRAO – Bonn 5GHz survey. H. Kühr.
Astron. Astrophys., Suppl. Ser., Vol. 29, 139 - 148 (1977).

From the NRAO – Bonn strong source survey of the northern sky at 5 GHz 84 new identifications based on accurate optical and radio positions are presented. The identification procedure on Palomar Sky Survey prints is described and the number of probable misidentifications is estimated. The results show that reliable identifications can be obtained by looking for positional coincidences for sources detected by surveys made at high frequencies.

141.004 Optical positions of radio sources.
M. P. Véron, P. Véron.
Astron. Astrophys., Suppl. Ser., Vol. 29, 149 - 159 (1977).

The authors have measured 125 optical positions of radio sources on the prints of the Palomar Sky Survey. The accuracy obtained is 0''.5 if the AGK3 catalogue is used as reference catalogue ($\delta > -2°$), and 0''.8 if the SAO catalogue is used. On the basis of these optical positions and by comparison with accurate radio positions, they could confirm 85 of the identifications.

141.005 Radio sources with superluminal velocities.
M. H. Cohen, K. I. Kellermann, D. B. Shaffer, R. P. Linfield, A. T. Moffet, J. D. Romney, G. A. Seielstad, I. I. K. Pauliny-Toth, E. Preuss, A. Witzel, R. T. Schilizzi, B. J. Geldzahler.
Nature, Vol. 268, 405 - 409 (1977).

Radio data from four extragalactic sources, three quasars and one galaxy, show evidence for an apparent expansion faster than the speed of light. The data on these 'superluminal' sources are reviewed, and their implications briefly discussed.

141.006 Survey of the optical variability of compact extragalactic objects. I. The field of 3C 345.
C. Barbieri, G. Romano, S. di Serego A., M. Zambon.
Astron. Astrophys., Vol. 59, 419 - 426 (1977).

This paper is the first in a series reporting observations of the optical variability of quasi stellar objects at Asiago Observatory. Data are presented for five objects in the field of 3C 345, namely 3C 345 itself, NRAO 512, 4C 38.41, 4C 39.46 and the compact galaxy Mark 501. Analysis of all the observations available from the literature shows periodicities in 3C 345 and in NRAO 512.

141.007 Applications of statistical techniques to the angular size–flux density relation for extragalactic radio sources. J. V. Narlikar, S. M. Chitre.
Mon. Not. R. Astron. Soc., Vol. 180, 525 - 537, with a correction, Vol. 181, 799 (1977).

The data on the angular sizes and flux densities of extragalactic radio sources from the 3CR and the Ooty surveys is subjected to two independent statistical tests. The purpose of these tests is to investigate whether the data necessitates the conclusion that the properties of the radio sources evolve with the epoch of the Universe. It is shown that the present scatter in this data is such that within the usual confidence limits it is not possible to distinguish between a number of evolutionary and non-evolutionary models.

141.008 The low-frequency structure of powerful radio sources and limits to departures from equipartition.
M. A. Scott, A. C. S. Readhead.
Mon. Not. R. Astron. Soc., Vol. 180, 539 - 550 (1977).

Recent interplanetary scintillation and spectral data have been combined to derive information about source structures at frequencies below 100 MHz. Sources having low-frequency spectral turnovers are studied in detail and in several cases it is shown that the total energy is within a factor 30 of the energy of equipartition between the particles and magnetic fields. In no case is there conclusive evidence of departure from equipartition.

141.009 The variation of spectral index across the radio galaxy 3C 452.   S. F. Burch.
Mon. Not. R. Astron. Soc., Vol. 180, 623 - 629 (1977).

Recent maps of the radio galaxy 3C 452 made with the Cambridge One-Mile telescope at 408 and 1407 MHz have been compared with published maps at 2.7 and 5.0 GHz made with the same telescope. The spectral index is found to increase from 0.7 at the outer ends of the source to ≈ 1.7 at the centre. This behaviour is probably caused by synchrotron losses in the radiating electrons and the observations are consistent with a simple model based on this assumption.

141.010 The variability of flat-spectrum radio sources at 962 MHz.   D. Stannard, M. Bentley.
Mon. Not. R. Astron. Soc., Vol. 180, 703 - 708 (1977).

The variability of 50 flat-spectrum radio sources is examined over a $2^1/_2$-yr period at 962 MHz. The results are compared with measurements on the same sources over a different $2^1/_2$-yr period at 2700 MHz, and it is concluded the proportion (35—40 per cent) found variable at 962 MHz is at least as great as that at 2700 MHz. Measurements are also presented of the flux densities of the variable radio sources BL Lac, OJ 287 and 3C 454.3 during 1972—76.

141.011 The spectrum of H1—36 (=3U1746—37?) at radio wavelengths.
C. R. Purton, D. A. Allen, P. A. Feldman, A. E. Wright.
Mon. Not. R. Astron. Soc., Vol. 180, 97P - 100P (1977).

The radio spectrum of H1—36, an emission-line object which has been suggested as the optical counterpart of 3U1746—37 indicates thermal emission from a large circumstellar gaseous shell which was probably produced by pro longed mass outflow from the central stellar system.

141.012 Radio and optical observations in the fields of five unidentified X-ray sources at high latitudes.
D. E. Harris, N. A. Bahcall, R. G. Strom.
Astron. Astrophys., Vol. 60, 27 - 38 (1977).

Radio observations at 610 MHz with the Westerbork Synthesis Radio Telescope and optical plates from the 48" Palomar telescope were obtained for five fields centered on the unidentified X-ray sources 3U0042+32, 3U0917+63, 3U1443+43, 3U1555+27, and 3U1809+50. A table of positions, intensities, and suggested optical identifications is given for the 249 radio sources detected in the five fields.

141.013 Search for variability of 3C 48 and 3C 84 at 408 MHz.
V. G. Malumyan, V. A. Sanamyan, S. S. Mailyan.
Astrofizika, Vol. 12, 557 - 558 (1976). In Russian. — English translation in Astrophysics, Vol. 12, No. 3.

The QSO 3C 48 and the Seyfert galaxy 3C 84 (NGC 1275) have been observed during July 1972 and October — November 1975. These observations showed that for this period the flux densities of 3C 38 and 3C 84 at 408 MHz were practically constant.

141.014 Comments on the light curve of the quasar 3C 273.
L. M. Ozernoy (*Ozernoj*), V. E. Chertoprud, L. I. Gudzenko.
Astrophys. J., Vol. 216, 237 - 243 (1977).

The interpretation by Fahlman and Ulrych of the light curve of the quasar 3C 273 as a superposition of independent random pulses is in contradiction with essential properties of the observed variability of this object. Of the two variability

concepts—superposition of independent pulses and an essentially nonlinear system (coherent, or unique body)—the first one appears to be extremely unlikely.

141.015 Effects of nonuniform structure on the derived physical parameters of compact synchrotron sources.
A. P. Marscher.
Astrophys. J., Vol. 216, 244 - 256 (1977).

The time-independent theory of a compact incoherent synchrotron radio source which contains gradients in the magnetic field and in the relativistic electron distribution is developed and discussed. Analytic expressions for the frequency spectrum are obtained, and the author demonstrates the dependence of the spectral index below the synchrotron self-absorption turnover frequency on the gradients in the source and on the source geometry. He discusses how the important physical parameters of a nonuniform source can be directly derived from observable quantities. The results indicate that the existence of steep gradients in the magnetic field strength and/or the distribution of relativistic electrons can alleviate somewhat the energetic and inverse Compton difficulties of some compact sources, while the presence of moderate gradients only aggravates these problems relative to the homogeneous model. The theory of nonuniform components is applied to the compact radio sources 1633+38 and OQ 172.

141.016 Formation of double radio source structures and superluminal expansion.
W. A. Christiansen, J. S. Scott.
Astrophys. J., Lett., Vol. 216, L1 - L6 (1977).

If the shock wave formed by an explosion in the nucleus of an active radio galaxy fragments, the trajectories of the individual plasma fragments will be naturally collimated into two distinct directions, thus resulting in the familiar double-lobed radio source geometry. The collimation and focusing of kinetic energy is the result of anisotropic effects of ram pressure on the individual plasmoids as they traverse the disklike distribution of gas in the active galactic nucleus. Calculations of the evolution of the spatial brightness distribution from an initially isotropic ejection of relativistic plasma fragments are performed.

141.017 The absorption-line spectrum of Q0453—423.
R. F. Carswell, M. G. Smith, J. A. J. Whelan.
Astrophys. J., Vol. 216, 351 - 356 (1977).

The optically selected QSO Q0453—423 (z_{em} = 2.659) is shown to have a rich absorption-line spectrum. At least two redshift systems are shown to be present, with z_{abs} = 1.1492 and 2.2754, and there is evidence for other absorption redshifts at very high velocities relative to the QSO. The spectrum provides another example of the C IV doublet in absorption falling in the Si IV emission feature.

141.018 Ultraviolet observation of a quasar spectrum.
B. E. J. Pagel.
Nature, Vol. 269, 195 - 196 (1977).

141.019 Ultraviolet spectrum of quasi-stellar object 3C 273.
A. F. Davidsen, G. F. Hartig, W. G. Fastie.
Nature, Vol. 269, 203 - 206 (1977).

The first direct observation of the ultraviolet spectrum of a quasi-stellar object (QSO) has been made with a rocket-borne telescope. The emission line spectrum of 3C 273 is similar to the spectra of high-redshift QSOs, but no absorption is observed. The results provide important new constraints on the theoretical models of QSOs, place a severe limit on the density of neutral hydrogen in the intergalactic medium, and suggest a cosmological origin for much of the absorption seen in high-redshift QSOs. Comparision of the ultraviolet spectrophotometry of low- and high-redshift QSOs suggests that the universe is closed, with $q_o \sim 1$.

141.020 **Observations of OE323 (4C34.13), MO758+120, and OS210 (4C28.40) with the 5-km telescope.**
M. A. C. Perryman, M. Ryle.
Nature, Vol. 269, 223 - 224 (1977).

The authors have mapped three extended radio sources identified with quasars of large redshift in order to study their structures with greater angular resolution and a better beam shape. They have also re-measured the optical positions of the associated quasars from the prints of the Palomar Sky Survey.

141.021 **The radio emission of NGC 5363.**
G. M. Tovmasyań, R. A. Sramek.
Astrofizika, Vol. 12, 693 - 696 (1976).

It is shown that the radio source in the galaxy NGC 5363 is coincident with the optical nucleus of the galaxy and consists of a compact core with diameter less than 2 arc sec, and probably an extended component with a size of about 20 arc sec. The location of the radio source in NGC 5363 and its radio spectrum favour the suggestion that an explosion similar to that in M82 has taken place in this galaxy.

141.022 **Optical observations of radiogalaxy lobes.**
J. A. Tyson, P. Crane, W. C. Saslaw.
Bull. American Astron. Soc., Vol. 9, 430 - 431 (1977).
Abstract.

141.023 **Photoionization calculations for the line-emitting regions of QSOs: implications of a recent observationally-determined Lyα/Hβ ratio and of new charge exchange rates.** J. M. Shuder, G. M. MacAlpine.
Bull. American Astron. Soc., Vol. 9, 431 (1977). – Abstract.

141.024 **Variability of extragalactic sources at 2.7 GHz. III. The nature of the variations in different source classes.** M. J. L. Kesteven, A. H. Bridle, G. W. Brandie.
Astron. J., Vol. 82, 541 - 556 (1977).

The incidence of variability at 2.7 GHz of 365 sources over periods of 2–10 yr is found to depend mainly on the spectral class of the source rather than on the optical identification. Virtually all sources with Q-type spectra in the radio two-color diagram are variable. Furthermore, about one-fourth of the sources with "normal" class G spectra exhibit low-amplitude variations. Variability amplitudes at 2.7 GHz are compared with those at higher frequencies. There are significant and systematic discrepancies with the predictions of the standard expanding cloud model. An interpretation is suggested wherein no variable sources intrinsically conform to the standard model, but some sources appear to do so because of opacity in previously ejected material. The variable components which coexist with transparent nonvariable emission do not appear to be systematically different from those which exist in isolation.

141.025 **Possible fast variability of the nucleus of Cen A at 13.5 mm.** P. Kaufmann, P. Marques dos Santos, J. C. Raffaelli, E. Scalise, Jr.
Nature, Vol. 269, 311 - 313 (1977).

The authors report a study of the nucleus of Cen A (NGC 5128) at a wavelength of 13.5 mm. Possible fast variability was observed.

141.026 **Radio measurements of the electron temperature distribution in Orion A at 22 GHz.**
T. Pauls, T. L. Wilson.
Astron. Astrophys., Vol. 60, L31 - L33 (1977).

Observations of the H66α recombination line from Orion A show that (1) the electron temperature has a constant value of 8200 ± 300 K within ~2' of the peak, (2) the effects of stimulated emission are negligible at 22 GHz, and (3) the values for the electron temperatures agree with the electron temperatures derived from the recent [O III] measurements of Peimbert and Torres-Peimbert (1977).

141.027 **On the redshift–apparent diameter relation for radio sources.**
G. Grueff, P. Schiavocampo, M. Vigotti, M. Zanni.
Astron. Astrophys., Vol. 60, 321 - 325 (1977).

The redshift–apparent diameter ($\theta - z$) relation for radio galaxies is extended to $z \sim 1.2$ and compared to the same relation for quasars. The authors show that the two $\theta - z$ relations are identical and that the metric diameter distributions of quasars and radio galaxies are also identical, if the quasar redshift is cosmological. The $\theta - z$ relation for radio sources in general is shown to be conflicting with the standard curves expected in Friedmann universe models, thus implying cosmological evolution for the source metric diameters. A possible discontinuity in such evolution at $z \sim 0.5$ is briefly discussed.

141.028 **Westerbork observations of the deep 5 GHz NRAO survey.** R. Fanti, L. Padrielli.
Astron. Astrophys., Suppl. Ser., Vol. 29, 263 - 277 (1977).

Accurate positions and radio structure have been obtained for 101 sources taken from the 5 GHz survey of Davis (1971). The accuracy of the positions is of the order of one arcsec. Optical identifications have been made on the basis of position coincidence only. About half of the sources are identified with objects visible on the Palomar Sky Survey prints. An analysis of the spectral index – optical identification is also given.

141.029 **Observations of 40 low luminosity radio galaxies with the Westerbork Synthesis Radio Telescope.**
C. Fanti, R. Fanti, I. M. Gioia, C. Lari, P. Parma, M. H. Ulrich.
Astron. Astrophys., Suppl. Ser., Vol. 29, 279 - 292 (1977).

The Westerbork Synthesis Radio Telescope has been used to map 40 radio galaxies at the frequencies of 610, 1415 and 4995 MHz. The results are presented here along with various physical parameters derived for the sources.

141.030 **Accurate positions of quasars and quasar candidates south of declination −45°.** G. Adam.
Astron. Astrophys., Suppl. Ser., Vol. 29, 293 - 296 (1977).

Precise positions have been obtained for 51 quasars and quasar candidates from the $\delta < -45°$ zone of the Parkes survey. The positions were measured on ESO(B) quick blue survey plates relative to SAO catalogue reference stars, or, for the $\delta < -64°$ zone, to new Cape catalogue stars. The estimated error is 0.4" on the two coordinates.

141.031 **Interferometer observations of radio sources in clusters of galaxies. V.**
F. N. Owen, L. Rudnick, B. M. Peterson.
Astron. J., Vol. 82, 677 - 687, 761 - 765 (1977).

Interferometer observations are reported for 38 radio sources in the direction of Abell clusters of galaxies. These sources are the final set used in the construction of a statistical sample of radio sources in Abell clusters of galaxies. Contour maps and models of the visibility data at 2695 and 8085 MHz are presented. Optical identifications based on 1–2-arcsec positions measured from the Palomar Sky Survey are also presented.

141.032 **Further linear polarization measurements of extragalactic radio sources at 3.71 and 11.1 cm.**
P. P. Kronberg, J. F. C. Wardle.
Astron. J., Vol. 82, 688 - 691 (1977).

Accurate measurements are presented of the integrated linear polarization of 61 extragalactic radio sources including 42 quasars at $\lambda\lambda 11.1$ and 3.7 cm (2695 and 8085 MHz). The new data include polarization measurements at the weakest flux levels yet achieved for extragalactic sources, and make it possible to extend the investigation of polarization properties over a yet greater range of source luminosity and distance.

141.033 Optical identifications of Parkes sources with flat spectra.
J. J. Condon, P. D. Hicks, D. L. Jauncey.
Astron. J., Vol. 82, 692 - 700, 766 - 769 (1977).

Accurate radio positions of flat-spectrum sources stronger than 0.5 Jy at 2700 MHz from the Parkes $-30° < \delta < -4°$ and $+4° < \delta < +25°$ supplementary surveys have been measured. Optical identifications based on close radio-optical position coincidence were made for more than 85% of the sources. Most of the identifications are stellar and have a magnitude distribution which suggests a correlation of the radio and optical luminosities.

141.034 Optical spectra and redshifts of quasi-stellar radio sources in the NRAO 5 GHz and 4C radio catalogs.
M. Schmidt.
Astrophys. J., Vol. 217, 358 - 361 (1977).

Spectroscopic observations of quasar candidates in the S1, S2, and I catalogs of the NRAO 5 GHz survey yield 33 confirmed quasars with redshifts ranging from 0.46 to 2.43. The redshift of 0421+01 is 2.048, rather than the value 0.689 given in the literature. Also reported are spectroscopic observations of four quasi-stellar sources in the $20°-40°$ declination zone of the 4C catalog.

141.035 Faint emission-line quasi-stellar object candidates.
A. A. Hoag, M. G. Smith.
Astrophys. J., Vol. 217, 362 - 381 (1977).

The authors find, by means of transmission-grating slitless spectroscopy at the prime focus of the 4 m telescope, 75 emission-line QSO candidates which have $17 < B < 21$ and $1.6 < Z < 3.5$ in sample areas totaling 5.1 square degrees.

141.036 On the space distribution of high-luminosity quasars with $1.9 < z < 3.25$.
P. S. Osmer, M. G. Smith.
Astrophys. J., Lett., Vol. 217, L73 - L76 (1977).

The authors discuss observations with the Tololo SIT vidicon system of a uniformly chosen sample of 30 quasar candidates from the CTIO survey with $1.9 < z < 3.25$ and $b^{II} < -50°$. Nine of the quasars are sufficiently luminous to form a redshift and luminosity-limited sample. The distribution in redshift for these nine luminous quasars and the application of the V/V_m test to the whole sample rule out a cutoff in the density of luminous quasars for $2.5 < z < 3.25$. The data are consistent with either the exponential density law proposed by Schmidt or a constant density distribution over the interval $1.9 < z < 3.25$.

141.037 Extragalactic radio sources and cosmology.
G. Swarup.
Bull. Astron. Soc. India, Vol. 5, 36 - 39 (1977).

The author reviews briefly some cosmological inferences that have been derived hitherto from radio astronomical data. He then describes some results derived from angular size statistics of extragalactic radio sources, based on occultation observations made at Ooty. It is shown that angular size-flux density counts of radio sources provide an independent evidence for the big bang model.

141.038 Polarisation in quasars.   F. G. Smith.
Nature, Vol. 269, 467 (1977).

141.039 Rapid fluctuations of radio flux and polarisation in quasar 3C273.   V. A. Efanov, I. G. Moiseev, N. S. Nesterov, N. M. Shakhovskoy (*Shakhovskoj*).
Nature, Vol. 269, 493 - 494 (1977).

The data from the authors' simultaneous radio and optical observations show rapid variations of radio flux and circular polarisation at $\lambda = 1.35$ cm, and of optical linear polarisation of 3C273 within one day or several hours. These variations are possibly more rapid than those previously reported.

141.040 Observations of 104 extragalactic radio sources with the Cambridge 5-km telescope at 5 GHz.
C. J. Jenkins, G. G. Pooley, J. M. Riley.
Mem. R. Astron. Soc., Vol. 84, 61 - 99 (1977).

104 radio sources from the 3C and 4C catalogues have been mapped with a resolution of 2 arcsec in RA and 2 cosec δ arcsec in dec. The results are presented here as contour maps and in tabular form, with accurate measurements of the positions of optical objects in the fields.

141.041 Observations of the NGC 7331/Stephan's Quintet area at 1421 MHz.   A. R. Gillespie.
Mon. Not. R. Astron. Soc., Vol. 181, 149 - 155 (1977).

A survey of the NGC 7331/Stephan's Quintet area is presented which is complete down to 9 mJy at 1421 MHz. The region is shown to be normal in all its radio properties except for the distribution of the brightest sources, which are grouped into a smaller area.

141.042 Observations of 15 southern extragalactic sources with the Fleurs synthesis telescope.
W. N. Christiansen, R. H. Frater, A. Watkinson, J. D. O'Sullivan, I. A. Lockhart, W. M. Goss.
Mon. Not. R. Astron. Soc., Vol. 181, 183 - 202 (1977).

High-resolution maps of 15 southern radio galaxies have been made with the Fleurs synthesis telescope at 1415 MHz. The resolution is ~50 arcsec. Two possible new head-tail galaxies (0427−53.9 and 1610−60.5) are proposed as well as a new 'wide-angle' tail source (1610−60.8). Maps of the well-known strong sources Centaurus A and Pictor A are also presented. The source G309.7 + 1.7 (13S6A) appears to be an extragalactic double.

141.043 A study of the bridges in the radio galaxies 3C 132 and 3C 192 at metre and centimetre wavelengths.
Gopal-Krishna.
Mon. Not. R. Astron. Soc., Vol. 181, 247 - 252 (1977).

A comparison is made of the high-resolution observations of two double radio galaxies, namely 3C 132 and 3C 192, over a frequency range of more than 10 : 1. Each of these sources exhibits a wide bridge which accounts for about two-thirds of the total radio output and whose extent in the direction perpendicular to the source axis is found to remain unchanged over this large frequency range. The role of radiative and adiabatic losses during evolution of the bridges is discussed. Using the continuous flow model, physical parameters are estimated for the two sources and for the intracluster medium in which they are embedded.

141.044 Absolute magnitudes and Hubble plot for QSOs.
A. K. Kembhavi, V. K. Kulkarni.
Mon. Not. R. Astron. Soc., Vol. 181, 19P - 24P (1977).

From a sample of 570 QSOs with redshifts ranging from 0.2 to more than 3, the optically most luminous are selected and a plot of absolute magnitude versus redshift is obtained, the absolute magnitudes being calculated on the assumption that the QSOs are at the cosmological distances implied by their redshifts.

141.045 Upper limits on the Faraday rotation in variable radio sources.   J. F. C. Wardle.
Nature, Vol. 269, 563 - 566 (1977).

Data on the linear polarisation of variable radio sources are examined for evidence of a variable component of Faraday rotation. Strong upper limits are set, and these are shown to have profound implications for the physics of these sources. It is suggested that the electron energy spectrum is a relativistic Maxwellian rather than a power law. This may be consistent with the relativistic blast wave models of Blandford and McKee.

141.046 Effective collision strengths of quasar ultraviolet

emission lines.   D. E. Osterbrock, R. K. Wallace.
Astrophys. Lett., Vol. 19, 11 - 14 (1977) = Lick Obs. Bull.,
No. 768.

The best available published collision strengths for excitation of permitted and semiforbidden emission lines of
abundant ions observed or expected in quasars have been
collected and averaged over Maxwellian velocity distributions.
For a few ions for which calculations are not available, extrapolation along isoelectronic sequences or in principal quantum
number n was used to estimate values. These collision strengths
were used to correct differentially published photoionization
models of quasars, and the corrected models compared with
published observational data.

141.047 A high-resolution investigation of the shell source G55.7 + 3.4 at 610 and 1415 MHz.
W. M. Goss, U. J. Schwarz, S. G. Siddesh, K. W. Weiler.
Astron. Astrophys., Vol. 61, 93 - 98 (1977).

The authors have mapped the shell source G55.7 + 3.4
using the Westerbork synthesis radio telescope at 610 MHz
and at 1415 MHz. The spectral index distribution has been
determined by a comparison of the 49- and 21-cm Westerbork
maps and the 11-cm map by Kundu and Velusamy (1971).
A shell model has been fitted to the observations. The pulsar
PSR 1919+21 is contained within this source but the association is still unproved. It may well be that the shell source is an
old supernova remnant which has no physical connection with
the pulsar. Thirty-seven point sources in the vicinity of
G55.7 + 3.4 are catalogued, including the pulsars PSR 1919 +
21 and PSR 1920 + 21.

141.048 The absolute spectrum of Cas A; an accurate flux density scale and a set of secondary calibrators.
J. W. M. Baars, R. Genzel, I. I. K. Pauliny-Toth, A. Witzel.
Astron. Astrophys., Vol. 61, 99 - 106 (1977).

A new analysis of the absolute radio spectrum of Cassiopeia A is presented which uses the latest absolute measurements
and takes account of the frequency dependence in the secular
rate of decrease. The absolute spectra of Cygnus A and Taurus
A are also given. An accurate "semi-absolute" spectrum for
Virgo A is established from direct accurate ratios to Cas A and
Cyg A. This Virgo A spectrum is used as a basis for accurate
relative spectra of a number of sources with simple spectra.
These are proposed as secondary calibrators for the routine
calibration of flux density measurements. Their spectral data
are presented for the frequency range 0.4–15 GHz. Finally a
comparison with other flux density scales is made.

141.049 On the nature of radio sources near flare stars.
W. S. Gilmore, R. L. Brown, B. Zuckerman.
Astrophys. J., Vol. 217, 716 - 718 (1977).

The authors obtained radio interferometric observations
at 2695 MHz and 8085 MHz of the radio sources near the flare
stars AD Leo and CN Leo in an attempt to establish whether
these sources are physically associated with the flare stars. The
radio source near AD Leo is a classical double; the sizes, fluxes,
and spectral indices of the two components are essentially
identical. The radio source near CN Leo is an unresolved point
source that does not share the large proper motion of CN Leo.
The authors conclude that neither radio source is a companion
to the flare star near which it appears.

141.050 Ultraviolet observations of 3C 273 by the ANS.
C.-C. Wu.
Astrophys. J. Lett., Vol. 217, L117 - L120 (1977).

Five-band spectrophotometry between rest wavelengths
1340 Å and 2850 Å is presented. The 2200 Å absorption
feature seems to be present in 3C 273 with a depth appropriate
for an internal reddening of $E(B-V) = 0.13$. The dereddened
ultraviolet flux distribution can be described by a power law
of spectral index 1.21. This power law can account for the
observed flux in the X-rays. It provides more photons than

needed to excite the observed Hβ by photoionization and Fe II
lines by resonance fluorescence. The emission line ratios,
Lα/C IV/C III]/Hβ are 100/80/60/10.

141.051 Absorption lines in the optical spectrum of quasar AO 0827+24.   M.-H. Ulrich, F. N. Owen.
Nature, Vol. 269, 673 - 674 (1977).

141.052 Evidence bearing on the interaction of gas and high-energy particles in quasi-stellar objects.
J. Baldwin, A. Boksenberg, G. Burbidge, R. Carswell, R.
Cowsik, J. Perry, A. Wolfe.
Astron. Astrophys., Vol. 61, 165 - 170 (1977).

The authors show that by looking for evidence for spallation products in the spectra of QSO's, important conclusions
can be drawn concerning the interaction of the high-energy
particle flux with the gas. At present there is no evidence that
the emitting or absorbing gas has interacted in any significant
way with a nucleon flux to produce boron.

141.053 2.8 and 6 cm wavelength observations of NGC 7822.
P. E. Angerhofer, M. R. Kundu, R. H. Becker, T. Velusamy.
Astron. Astrophys., Vol. 61, 285 - 290 (1977).

The authors present observations of the galactic radio
source W 1 (= NGC 7822) at wavelengths of 2.8 and 6 cm. No
linear polarization has been detected over this region. The
authors conclude that the known exciting stars can account
for the observed radio flux density at 6 cm, and that W 1 is a
thermal radio source.

141.054 Physical conditions in polarized compact radio sources.   T. W. Jones, S. L. O'Dell.
Astron. Astrophys., Vol. 61, 291 - 293 (1977).

The emergence of significant linear polarization from
compact synchrotron sources places severe constraints on the
concentration of low energy electrons inside such sources. The
authors argue that such constraints favor source models, such
as relativistic blasts, which generate only relativistic matter
over more conventional models in which large amounts of
nonrelativistic matter stochastically produce relatively few
high energy particles.

141.055 Optical monitoring of quasistellar objects. II. Three-color photographic photometry of BL Lac.
P. K. Lü.
Astron. J., Vol. 82, 773 - 775 (1977).

Three-color photographic photometry of the radio source
BL Lac (VRO 42.22.01) is given. Magnitudes and light curves
are presented for 235 observations in U, B, and V colors. Very
rapid optical variations and one major outburst were detected.

141.056 Extragalactic sources with strong millimeter-wave emission.   F. N. Owen, S. L. Mufson.
Astron. J., Vol. 82, 776 - 780, 853 (1977).

The authors report observations at 90 GHz of a sample of
38 sources chosen to have flat or rising spectra between 0.318
and 5 GHz. Roughly 25% of the sources were found to have
flat or rising spectra between 5 and 90 GHz. The optical magnitude of each of the sources from the E print of the Palomar
Sky Survey is also estimated. Evidence is found for a correlation between the 90-GHz flux density and the optical magnitude, suggesting a relationship between the emission processes
at these frequencies. The properties of two of the more interesting sources are also discussed.

141.057 Structure of radio sources with remarkably flat spectra: PKS 0735+178.   A. P. Marscher.
Astron. J., Vol. 82, 781 - 784 (1977).

The author investigates three incoherent synchrotron
models which can account for the remarkably flat radio spectrum ($\alpha = 0$ from 0.4 to 90 GHz) of the BL Lac object PKS

0735+178. Multifrequency VLBI observations are proposed which will allow to determine the detailed structure (and possibly the ultimate origin) of compact radio sources.

141.058 Search for microarcsecond structure in low-frequency variable radio sources.
J. W. Armstrong, S. R. Spangler, P. E. Hardee.
Astron. J., Vol. 82, 785 - 790 (1977).

A search for interstellar scintillation (ISS) of low-frequency variable radio sources is reported. Observations of 28 confirmed or suspected low-frequency variables, 21 nonvariable sources, and two pulsars were made at 408 MHz. As expected, the pulsars showed ISS, but scintillation was not detected in any other source. A typical upper limit to the rms modulation due to ISS is 150 mJy, giving lower limits to the apparent angular diameter of $\sim 10^{-6}$ arcsec. The possibilities that a true "point" source is broadened to an apparent angular diameter $\gtrsim 10^{-6}$ arcsec by scattering local to the source or in a general intergalactic medium are discussed.

141.059 Precise positions of radio sources. V. Positions of 36 sources measured on a baseline of 35 km.
C. M. Wade, K. J. Johnston.
Astron. J., Vol. 82, 791 - 795 (1977).

Positions are given for 36 sources measured on a baseline of 35 km with the radio link interferometer at Green Bank. The results are accurate to a few hundredths of a second of arc in each coordinate.

141.060 Counts of unidentified radio sources.
J. G. Robertson.
Observatory, Vol. 97, 198 - 200 (1977).

Several authors have attached cosmological significance to the steep slope observed for the number—flux-density counts of unidentified radio sources. However, it is shown here that the steepening may be completely explained by a selection effect, arising from the tendency of weak radio sources to be also faint optically.

141.061 Radio structure of 3C 147 determined by multi-element very long baseline interferometry.
P. N. Wilkinson, A. C. S. Readhead, G. H. Purcell, B. Anderson.
Nature, Vol. 269, 764 - 768 (1977).

The authors have determined the radio structure of the quasar 3C 147 from multi-baseline VLBI data at 609 MHz. The structure is similar to the central part of M87, with a bright core and a linear 'jet' of projected length ~ 1.5 kpc which is concentrated in bright 'knots'.

141.062 The line spectra of quasi-stellar objects.
P. A. Strittmatter, R. E. Williams.
Annu. Rev. Astron. Astrophys., Vol. 14, (see 003.008), 307 - 338 (1976).

This review is limited to the observations and interpretation of the line spectra of QSOs and is consequently concerned mainly with optical data.

141.063 Extended extragalactic radio sources.
D. S. De Young.
Annu. Rev. Astron. Astrophys., Vol. 14, (see 003.008), 447 - 474 (1976).

The emphasis in this review is one of summarizing and comparing the work done to date on understanding the origins, evolution, and relevant physical processes pertaining to the extended extragalactic radio sources. Observational results are covered in a general sense to emphasize the constraints that they place upon all radio-source models. Specific observations are noted if they supply particularly pointed evidence for or against a given theoretical suggestion, but the inclusion of observational results in this review is not meant to be encyclopedic.

141.064 Flux density scale of the BP radio sources in the light of the GB observations.
R. Guła, A. Michalec.
Acta Cosmologica, Zesz. 6, 27 - 30 (1977).

The BP flux density scale at 408 MHz is discussed in the light of the observations of 4C and 4CP sources in the GB region, which permitted the determination of their spectra in the frequency range: 38 MHz–2695 MHz.

141.065 Observations of Fe II and Mg II absorption in QSOs with $z_{abs} \ll z_{em}$.
E. M. Burbidge, H. E. Smith, R. J. Weymann, R. E. Williams.
Astrophys. J., Vol. 218, 1 - 7 (1977).

Low-redshift absorption-line systems showing features of Mg II, Fe II, and Mg I have been discovered in the spectra of the QSOs 1548+114b, 4C 06.41, 4C 29.01, and PKS 0454+039 with values 0.429, 0.442, 0.846, and 0.860, respectively. The values of $R = (1 + z_{em})/(1 + z_{abs})$ for these objects, together with that of PKS 1229—02, are all large, that of 1548+114b even exceeding the well-known system in PHL 938. A second absorption system is present in 0454+039 at $z = 1.1538$; in this respect, the object resembles the BL Lacertae object AO 0235+164. Absence of the Mn II resonance lines in the spectra of PHL 938 and 0454+039 suggests that manganese may be underabundant in these objects.

141.066 The radio structure and optical field of 3C 303.
P. P. Kronberg, E. M. Burbidge, H. E. Smith, R. G. Strom.
Astrophys. J., Vol. 218, 8 - 19 (1977).

The authors present new radio and optical observations of the peculiar system 3C 303. Aperture synthesis maps obtained at 1.4, 2.7, 5.0, and 8.1 GHz show the source to be a highly asymmetric double, although it is almost certain that the various radio components are physically related. The compact VLBI component has been shown to coincide with a 17th magnitude N galaxy at $z = 0.141$, while the compact radio core of the extended western component is shown to lie within $1''$ of one member of a trio of faint UV-excess objects. Optical observations of the brightest of the UV-excess objects lying within the western radio component show it to be a QSO with a redshift $z = 1.57$. The relation between the radio and optical components is discussed.

141.067 On photoionization analyses of emission spectra of quasars.
K. Davidson.
Astrophys. J., Vol. 218, 20 - 32 (1977).

A reassessment is made of the information conveyed by emission-line intensities in the spectra of quasars, under the assumption that the line-emitting gas is photoionized by continuum radiation. The emphasis is on exploring dependences upon parameters, rather than on proposing one particular model which is "sufficient" for explaining the line intensities. An essential feature of the discussion is the inclusion of two components of gas, each characterized by its value of U_1, the radiation-to-gas ratio. Actual quasars' emission regions probably have continuous distributions of values of U_1, but the two-component approximation is appropriate to the amount of information which is actually observable at present or in the near future.

141.068 On the origin of the absorption spectra of quasi-stellar and BL Lacertae objects.
G. Burbidge, S. L. O'Dell, D. H. Roberts, H. E. Smith.
Astrophys. J., Vol. 218, 33 - 38 (1977).

Following the earlier work of Wagoner (1967), the authors have reexamined the possibility that the absorption spectra of QSOs are due to gas in the disks, coronae, or halos of intervening galaxies. Comparison of the expected number of absorption systems with the number observed in all of the QSOs and BL Lacertae objects so far cataloged shows that there is a gross discrepancy in the sense that the multiplicity of absorption systems already found is far higher than the numbers expected if the absorption is due to gas in known disks or hypothetical small coronae.

141.069 **Westerbork observations of three cluster radio galaxies.** G. K. Miley, D. E. Harris.
Astron. Astrophys., Vol. 61, L23 - L26 (1977).

Aperture synthesis measurements at 1415 MHz are presented on the tail type radio galaxies 1200 + 519, 1339 + 266 and 1709 + 397, all of which occur in Abell clusters. The observed morphologies are interpreted within the framework of current models of cluster radio galaxies.

141.070 **Observations of radio galaxies with the radio-telescope RATAN-600.** N. S. Soboleva, A. B. Berlin, V. Ya. Gol'nev, G. M. Timofeeva.
Astron. Zh. Akad. Nauk SSSR, Vol. 54, 945 - 952 (1977). In Russian. English translation in Soviet Astron., Vol. 21, No. 5.

Observations with the radiotelescope RATAN-600 of some bright radio galaxies (3C 353, 348, 327, 218, 274, 405) at three wavelengths in the centimeter range with resolution from 15 to 50 arc sec in right ascension are presented. Some new details in the radio brightness distribution of these objects are obtained.

141.071 **Scintillation observations of 3C 48, 3C 273, and 3C 295 at 25 MHz.** V. S. Artyukh, B. P. Ryabov.
Astron. Zh. Akad. Nauk SSSR, Vol. 54, 953 - 956 (1977). In Russian. English translation in Soviet Astron., Vol. 21, No. 5.

Interplanetary scintillations of 3C 48, 3C 273, and 3C 295 have been observed at 25 MHz. Models of the angular structure of these sources corresponding to the observations are proposed. 3C 48 is found to possess a halo, 3C 295 at decametric waves becomes an one-component source, and 3C 273 has the same angular structure as in the meter wavelength range.

141.072 **On the use of lunar occultations of radio sources for the investigation of their angular structure. IV.** G. L. Abramyan.
Astron. Zh. Akad. Nauk SSSR, Vol. 54, 967 - 972 (1977). In Russian. English translation in Soviet Astron., Vol. 21, No. 5.

A dependence of the angular resolution limit on the wavelength band in the problem of restoring the brightness distribution of radio sources according to observations of lunar occultations is determined for a large signal-noise ratio.

141.073 **Investigation of the spectra of H_2O sources by the autocorrelation analysis method.** L. R. Kogan, L. I. Matveenko, L. S. Chesalin.
Astron. Zh. Akad. Nauk SSSR, Vol. 54, 973 - 978 (1977). In Russian. English translation in Soviet Astron., Vol. 21, No. 5.

Method and results of measurements of spectra of radio sources emitting in the water vapour line ($\lambda = 1.35$ cm) by the method of autocorrelation analysis of tape-recorded signals are considered. Spectra of the sources W49, Ori A, W51, W3OH, W3C, RX Boo, NGC 7538 S with 18 kHz resolution in the 3.6 MHz band have been obtained.

141.074 **Radio galaxies and quasars.** L. C. Green.
Sky Telesc., Vol. 54, 384 - 389 (1977).

141.075 **A deep search for X-ray emission from radio quasars with Ariel V.** G. J. White, M. J. Ricketts.
Mon. Not. R. Astron. Soc., Vol. 181, 435 - 440 (1977).

A deep search for X-ray emission from 65 radio quasars has been made with the Ariel V Sky Survey Instrument during 1975–76 and upper limits have been put on the emission in most cases. One new X-ray quasar, PKS 0349–14, is suggested, but the authors were unable to detect the possible X-ray quasar NAB 0137–01.

141.076 **A further measurement of the 38-MHz flux density of Cas A.** P. L. Read.
Mon. Not. R. Astron. Soc., Vol. 181, 63P - 65P (1977).

Following the recent measurements of an anomalous increase in the flux density of Cas A at 38 MHz, a further measurement has been made approximately one year later. The result indicates a substantial reduction in the flux density of Cas A at this frequency, suggesting that a bright radio flare with a timescale of between 2 and 6 years has occurred at low frequencies.

141.077 **Confirmation of the highest redshift QSO candidates from the Tololo deep survey.** M. G. Smith, A. Boksenberg, R. F. Carswell, J. A. J. Whelan.
Mon. Not. R. Astron. Soc., Vol. 181, 67P - 69P (1977).

The authors report the confirmation of an optically selected QSO candidate at a redshift of 3.45. They further demonstrate the validity of a systematic procedure for discovering QSOs with redshifts up to 4.7.

141.078 **The structure of extragalactic radio sources with flat spectra.** M. Inoue.
Publ. Astron. Soc. Japan, Vol. 29, 593 - 614 (1977).

Both VLBI and linear polarization data have been analyzed for extragalactic radio sources with flat spectra. The sources are shown in general to consist of three different classes of components. Polarization properties of individual components are investigated. A simple model is proposed to explain polarization observations of the flat-spectrum radio sources, and physical conditions of individual components are discussed in detail. A magnetic field prevailing in the three components within a source is suggested, and the magnetic field direction seems to be parallel to the double or triple structure of the smallest component.

141.079 **The 6C survey.** J. E. Baldwin.
IAU Symp., No. 74, (see 012.014), p. 3 - 7 (1977).

141.080 **Survey of data for determining scales of the absolute flux densities in 10–180 MHz range and source spectra in the declination strip $10° - 20°$.** S. Ya. Braude.
IAU Symp., No. 74, (see 012.014), p. 9 - 13 (1977).

141.081 **Interim report on the Texas Survey.** J. N. Douglas, F. N. Bash.
IAU Symp., No. 74, (see 012.014), p. 15 - 24 (1977).

141.082 **The Bologna survey of radio sources at 408 MHz.** C. Fanti, C. Lari.
IAU Symp., No. 74, (see 012.014), p. 25 - 29 (1977).

141.083 **The Molonglo radio source surveys at 408 MHz.** B. Y. Mills.
IAU Symp., No. 74, (see 012.014), p. 31 - 37 (1977).

141.084 **Westerbork surveys of radio sources at 610 and 1415 MHz.** A. G. Willis, C. E. Oosterbaan, R. S. Le Poole, H. R. de Ruiter, R. G. Strom, E. A. Valentijn, P. Katgert, J. K. Katgert-Merkelijn.
IAU Symp., No. 74, (see 012.014), p. 39 - 45 (1977).

141.085 **The Green Bank surveys at 1400 MHz.** J. Maslowski.
IAU Symp., No. 74, (see 012.014), p. 47 - 54 (1977).

141.086 **The Parkes 2700 MHz survey: counts of the sources and their distribution on the sky.** J. V. Wall.
IAU Symp., No. 74, (see 012.014), p. 55 - 61 (1977).

141.087 **Surveys of radio sources at 5 GHz.**
I. I. K. Pauliny-Toth.
IAU Symp., No. 74, (see 012.014), p. 63 - 74 (1977).

141.088 **The statistical analysis of anisotropies.**
A. Webster.
IAU Symp., No. 74, (see 012.014), p. 75 - 81 (1977).

141.089 **Spectra of southern radio sources.**
J. G. Bolton.
IAU Symp., No. 74, (see 012.014), p. 85 - 97 (1977).

141.090 **Spectral index studies of extragalactic radio sources.**
P. Katgert, L. Padrielli, J. K. Katgert, A. G. Willis.
IAU Symp., No. 74, (see 012.014), p. 99 - 106 (1977).

141.091 **Radio source angular sizes and cosmology.**
R. D. Ekers, G. K. Miley.
IAU Symp., No. 74, (see 012.014), p. 109 - 117 (1977).

141.092 **The angular size − flux density relation.**
V. K. Kapahi.
IAU Symp., No. 74, (see 012.014), p. 119 - 123 (1977).

141.093 **The flux density − angular size distribution for extragalactic radio sources.**
G. Swarup, C. R. Subrahmanya.
IAU Symp., No. 74, (see 012.014), p. 125 - 132 (1977).

141.094 **The angular diameter-redshift test for large redshift quasars.**   J. M. Riley, M. S. Longair, A. Hooley.
IAU Symp., No. 74, (see 012.014), p. 133 - 138 (1977).

141.095 **The angular diameter-redshift relation for scintillating radio sources.**
A. Hewish, A. C. S. Readhead, P. J. Duffett-Smith.
IAU Symp., No. 74, (see 012.014), p. 139 - 147 (1977).

141.096 **The present status of 3CR identifications.**
J. Kristian.
IAU Symp., No. 74, (see 012.014), p. 151 - 155 (1977).

141.097 **QSO identifications from a Molonglo radio survey.**
C. Hazard.
IAU Symp., No. 74, (see 012.014), p. 157 - 163 (1977).

141.098 **Identifications from the WSRT (*Westerbork Synthesis Radio Telescope*) deep surveys.**
J. K. Katgert, H. R. de Ruiter, A. G. Willis.
IAU Symp., No. 74, (see 012.014), p. 165 - 170 (1977).

141.099 **Luminosity functions for extragalactic radio sources.**
R. Fanti, G. C. Perola.
IAU Symp., No. 74, (see 012.014), p. 171 - 182 (1977).

141.100 **Physical conditions in radio galaxies and quasars.**
D. E. Osterbrock.
IAU Symp., No. 74, (see 012.014), p. 183 - 191 (1977) =
Contrib. Lick Obs., No. 410.

141.101 **Absorption systems in high redshift QSOs.**
A. Boksenberg.
IAU Symp., No. 74, (see 012.014), p. 193 - 222 (1977).

141.102 **Relations between the optical and radio properties of extragalactic radio sources.**   E. M. Burbidge.
IAU Symp., No. 74, (see 012.014), p. 223 - 235 (1977).

141.103 **Relations between radio and optical properties of radio sources − radio astronomer's point of view.**
J. M. Riley, C. J. Jenkins.
IAU Symp., No. 74, (see 012.014), p. 237 - 243 (1977).

141.104 **Interpretation of cosmological information on radio sources.**   G. Burbidge.
IAU Symp., No. 74, (see 012.014), p. 247 - 257 (1977).

141.105 **Cosmological interpretation of redshift data on quasars through the V/V_{max} test.**   M. Schmidt.
IAU Symp., No. 74, (see 012.014), p. 259 - 268 (1977).

141.106 **The redshift-magnitude relation for radio galaxies.**
H. E. Smith.
IAU Symp., No. 74, (see 012.014), p. 279 - 293 (1977).

141.107 **The Hubble diagrams for quasars.**
J. N. Bahcall, E. L. Turner.
IAU Symp., No. 74, (see 012.014), p. 295 - 303 (1977).

141.108 **Clusters of galaxies and radio sources.**
W. Jaffe.
IAU Symp., No. 74, (see 012.014), p. 305 - 316 (1977).

141.109 **Young galaxies, quasars and the cosmological evolution of extragalactic radio sources.**
M. S. Longair, R. A. Sunyaev.
IAU Symp., No. 74, (see 012.014), p. 353 - 360 (1977).

141.110 **Polarization measurements of distant sources.**
R. G. Conway.
IAU Symp., No. 74, (see 012.014), p. 361 - 366 (1977).

141.111 **A numerical study of the continuous-beam model of extragalactic radio sources, II.**   D. R. Rayburn.
J. R. Astron. Soc. Canada, Vol. 71, 394 - 395 (1977).

141.112 **Satellite-link long-baseline interferometry.**
N. W. Broten, D. N. Fort, K. I. Kellermann,
B. Rayhrer, S. H. Knowles, W. B. Waltman, G. W. Swenson,
J. L. Yen.
J. R. Astron. Soc. Canada, Vol. 71, 395 (1977). − Abstract.

141.113 **Absorption spectra of quasars.**   Y. P. Varshni.
J. R. Astron. Soc. Canada, Vol. 71, 403 - 404
(1977). − Abstract.

141.114 **Properties of radio sources in clusters of galaxies.**
D. E. Harris.
Highlights of Astronomy, Vol. 4, Part I, (see 012.020), p.
321 - 328 (1977).

141.115 **Millimetre wave observations of galactic and extragalactic objects.**
P. E. Clegg, P. A. R. Ade, M. Rowan-Robinson.
Far infrared astronomy (see 012.027), p. 209 - 217 (1976).
   Broad-band observations of galactic and extragalactic
sources at 1 mm are described, and a discussion of the calibration procedure is given.

141.116 **A search for radio emission from a sample of optically selected quasars.**   C. Fanti, R. Fanti,
C. Lari, L. Padrielli, H. van der Laan, H. de Ruiter.
Astron. Astrophys., Vol. 61, 487 - 491 (1977).
   62 optically selected quasars have been searched for
radio emission at 1415 MHz with the Westerbork Synthesis
Radio Telescope. Eight of them have been radio detected. The
remaining ones are radio quiet at levels of about 10 mJy.
These data are combined with those of Katgert et al. (1973)
on the QSO's of Braccesi et al. (1970), obtaining a sample of
85 objects. The distribution function of the ratio of radio to
optical emission is derived. It is shown that there is a disagreement with the one obtained from radio selected samples. A
brief discussion is given on such a discrepancy.

141.117 **Analysis of "noise" in the rich absorption-line**

spectra of quasars. I. Method of scrambled standard lines.   P. C. Joss, G. J. Ruffa.
Astrophys. J., Vol. 218, 347 - 352 (1977).

The authors present a new method for estimating the number of accidental redshift systems in quasar absorption-line spectra. The method is based on the use of simulated sets of standard spectral lines, in which correlations among the actual line wavelengths of specific ions are neglected. The authors have applied their method to the observed spectra of Ton 1530, PHL 938, and PHL 957 and obtain higher estimates of the number of accidental systems than those obtained by some previous investigators.

141.118 The compact radio sources in 4C 39.25 and 3C 345.   D. B. Shaffer, K. I. Kellermann, G. H. Purcell, I. I. K. Pauliny-Toth, E. Preuss, A. Witzel, D. Graham, R. T. Schilizzi, M. H. Cohen, A. T. Moffet, J. D. Romney, A. E. Niell.
Astrophys. J., Vol. 218, 353 - 360 (1977).

Long-baseline interferometry of the quasars 4C 39.25 and 3C 345 at 10.65 and 14.77 GHz shows that the centimeter radio source in each object is double. The spectra of the individual components are derived, and shown to vary with time approximately as expected for expanding self-absorbed synchrotron sources. The magnetic fields in the components are estimated to be as high as 0.1 gauss, but the structure of the sources appears to be unrelated to the magnetic field orientation derived from low-resolution polarization measurements.

141.119 Clustering of quasars.   G. Setti, L. Woltjer.
Astrophys. J., Lett., Vol. 218, L33 - L35 (1977).

Evidence for pairing of quasars of different redshifts is discussed. The sample of optical quasars of Braccesi et al. is found not to have more pairs than expected by chance. On the assumption that a significant subset of the quasars is related to giant elliptical galaxies, it is shown that substantial clustering of quasars may be expected to set in somewhere between V = 20 mag and 23 mag.

141.120 Infrared and visible polarimetry and photometry of highly variable quasi-stellar sources.
G. H. Rieke, M. J. Lebofsky, J. C. Kemp, G. V. Coyne, S. Tapia.
Astrophys. J., Lett., Vol. 218, L37 - L41 (1977).

Infrared photometry is reported for 3C 279 and five BL Lacertae sources. There is a suggestion of an intrinsic high-frequency cutoff in the spectrum of the highly redshifted source P0735+178. Infrared and visible polarimetry of P0735+178 and OI 090.4 indicates a rotation of position angle between these two spectral regions. Infrared polarimetry is also reported for B2 1101+38.

141.121 Observations of variable radio sources at 0.95- and 1.65-cm wavelength.   J. A. Waak, R. W. Hobbs.
Astron. J., Vol. 82, 855 - 856 (1977).

Results of measurements of seven sources at 1.65-cm and six sources at 0.95-cm wavelength made with the 85-ft antenna of the Naval Research Laboratory are presented. The publication of this data makes available a data base at 0.95 cm for some sources extending from August 1966 to June 1972 and continuing into the present.

141.122 Quasars and young galaxies.   M. J. Rees.
Q. J. R. Astron. Soc., Vol. 18, 429 - 442 (1977).

Galaxies still pose unanswered questions, even at the most rudimentary level. But glimmerings of a consensus seem to be emerging, particularly on the relations between quasars and galaxies. The author tries to summarize these, and also indicates some areas where there seems a genuine hope of progress in the next few years. The timescales of galactic evolution are so long, and the distances of quasars so great, that the subject inevitably involves cosmology.

141.123 The influence of statistical fluctuations on analyses of QSO data.   A. L. Smith.
Mon. Notes Astron. Soc. South Africa, Vol. 36, 103 - 105 (1977).

A statistical analysis of data available at different periods of time on quasi-stellar object (QSO) redshifts demonstrates the difficulties encountered in attempting to draw meaningful conclusions from analyses of such data.

141.124 Extragalaktiske radiokilder og overlyshastigheder.   J. Teuber.
Astron. Tidsskr., Årg. 10, 108 - 111 (1977).

141.125 A radio study of Abell clusters.   B. Y. Mills, D. G. Hoskins.
Australian J. Phys., Vol. 30, 509 - 529 (1977).

A search for radio sources close to 247 clusters of galaxies in the Abell catalogue has been carried out at the Molonglo Radio Observatory at a frequency of 408 MHz. A list of 116 sources near 89 clusters is given, identifications have been made and criteria for cluster membership have been established. A cluster luminosity function is derived in the range 10^{23}–10^{25} W Hz^{-1} sr^{-1}, and spectra have been obtained for sources in 25 clusters utilizing published surveys made at other frequencies. It is found that there is no correlation between the richness of a cluster and its inclusion of at least one radio source, but those clusters containing multiple sources are significantly richer than average.

141.126 Quasar-galaxy pairs and surface density of quasars.   J.-L. Nieto.
Nature, Vol. 270, 411 - 412 (1977).

Discoveries of radio-quiet quasars in the vicinity of galaxies seem to contradict strongly the statistical analyses on quasar-galaxy associations. In each of these analyses the calculations used the surface densities of galaxies, but it has been claimed that the small angular separations are explicable by chance only if the surface density of quasars is much larger than believed. To test such an assumption the author studied the problem by estimating directly the probabilities involved when adopting the most probable values of the density of quasars at $B \sim 19.0$–19.5, that is, $\mu \sim 3$–5 (deg)$^{-2}$.

141.127 The 5C 9 survey of radio sources.   P. C. Waggett.
Mon. Not. R. Astron. Soc., Vol. 181, 547 - 561 (1977).

The 5C 9 survey, made with the One-Mile telescope at Cambridge, has a field centre at RA = 05^h00^m, dec = $88°30'$. It covers an area of about 4° in diameter at 408 MHz to a limiting flux density of 20 mJy, and a concentric area of diameter about 1° at 1407 MHz. The survey was undertaken to provide flux densities for a study of the distribution of spectral indices of weak radio sources near the North Celestial Pole, and the positions and flux densities of 214 sources are listed. The source counts are consistent with those derived from previous 5C surveys.

141.128 Multifrequency radio observations of 3C 31: a large radio galaxy with jets and peculiar spectra.
S. F. Burch.
Mon. Not. R. Astron. Soc., Vol. 181, 599 - 610 (1977).

The author presents here new maps at 0.4 and 1.4 GHz synthesized from 32 interferometer spacings of the Cambridge One-Mile Telescope, which reveal structure on a large scale. In addition, observations made with the 5-km Telescope at 2.7, 5 and 15 GHz show details of the structure in the central regions. Results at 10 GHz from the 25-m Chilbolton telescope of the SRC Appleton Laboratory provide further spectral information.

141.129 Observations of radio sources near K3—50.   D. Colley, P. F. Scott.

Mon. Not. R. Astron. Soc., Vol. 181, 703 - 712 (1977).

Three H II regions within 2.5 arcmin of the optical nebula K3−50 have been observed with the Cambridge 5-km telescope at 15 GHz. The radio feature designated component A, adjacent to the optical object, appears as a single resolved source having a high peak brightness temperature and a large density gradient at its eastern edge. Little or no radio emission arises from the nebula itself. The radio feature 'component C' has been resolved into two compact sources, and two additional unresolved sources have been detected. New optical data have been used to obtain a revised estimate of the extinction towards K3−50.

141.130 G127.1+0.5 − a remarkable supernova remnant centred on a very compact radio source?
J. L. Caswell.
Mon. Not. R. Astron. Soc., Vol. 181, 789 - 797 (1977).

A synthesis map obtained with the Cambridge Half-mile telescope shows that the galactic radio source G127.1+0.5 has a shell structure; in conjunction with other data the author concludes that it is probably an old supernova remnant. Near the centre of the shell the map shows a compact radio source.

141.131 Q 0002−422: a QSO with $z_{abs} \ll z_{em}$.
J. A. J. Whelan, R. F. Carswell, M. G. Smith.
Mon. Not. R. Astron. Soc., Vol. 181, 81P - 84P (1977).

The optically-selected QSO Q0002−422 ($z_{em} = 2.77$) exhibits an absorption-line spectrum which has similarities to that of Q0453−423 ($z_{em} = 2.66$). Two redshift systems with $z_{abs} = 0.8354$ and 2.2996 are present in Q0002−422. The $z_{abs} = 0.8354$ system has a velocity of $0.62 c$ with respect to the emission-line system. A difficulty in finding larger velocity differences is noted.

141.132 On the profiles of the broad lines in the spectra of QSOs and Seyfert galaxies.   H. Netzer.
Mon. Not. R. Astron. Soc., Vol. 181, 89P - 92P (1977).

Observations of QSOs and Seyfert galaxies suggest that the redshifts deduced from the midpoints of the broad permitted lines may be larger than the 'true' redshifts by up to 1000 km/s. It is argued that the above effect, and some line asymmetries, may be due to gravitational redshifts in bound systems.

141.133 Black hole model of quasar-like objects—infrared optical emission.   S. Tsuruta.
Astron. Astrophys., Vol. 61, 647 - 657 (1977).

The author presents a model in which supercritical accretion onto a massive black hole can explain various phenomena associated with quasars and related objects, most importantly their energy requirements. The model, tentatively called a "fountain model", consists of violently moving clouds of gas around a massive black hole of $10^6 - 10^9 M_\odot$ which is embedded in a very dense stellar system in the galactic nucleus with stellar density of $10^6 - 10^8 M_\odot pc^{-3}$.

141.134 The structure of quasars from the region of the 5C2 survey.   R. E. Spencer, J. F. C. Wardle.
Astrophys. J., Vol. 218, 599 - 604 (1977).

The structures of 12 suspected quasars from the region of the 5C2 survey have been mapped at wavelengths of 11.1 and 3.7 cm, with the NRAO four-element interferometer. Five of the sources are well resolved (greater than 3"). The distribution of angular sizes is found not to be significantly different from that of quasars from the 3C and 4C catalogs. The angular-size—flux-density relation for quasars is discussed briefly.

141.135 Absorption by neutral hydrogen and ionized magnesium in quasi-stellar objects and BL Lacertae objects.   B. M. Peterson, G. D. Coleman, P. A. Strittmatter, R. E. Williams.

Astrophys. J., Vol. 218, 605 - 610 (1977).

At least two QSOs, PHL 938 and 3C 286, and two BL Lacertae objects, AO 0235 + 164 and PKS 0735 + 178, show absorption lines characteristic of material in a low-ionization state. Optical and radio observations of these objects are analyzed to determine whether the absorption lines are consistent with the hypothesis that the absorption occurs in H I regions such as might be expected in an intervening galaxy. The results are consistent with this hypothesis, although other possible origins of the absorption lines cannot be excluded. The magnesium abundances and iron-to-magnesium ratios are consistent with solar values. Evidence is presented for the existence of multiple redshift systems in PHL 938.

141.136 An angular size for the compact radio source at the galactic center.
K. Y. Lo, M. H. Cohen, R. T. Schilizzi, H. N. Ross.
Astrophys. J., Vol. 218, 668 - 670 (1977).

Simultaneous very long baseline ($10^8 \lambda$) and intermediate-baseline ($4 \times 10^6 \lambda$) interferometer observations at $\lambda = 3.7$ cm in 1976.2 show that the galactic-center compact radio source has a linear size of ~ 140 AU. The observations also set an upper limit of ~ 0.1 Jy to the emission from any core of size less than ~ 10 AU at that epoch. The total flux density of the compact source was slightly higher than the value reported previously.

141.137 Spectroscopy of faint quasars and the properties of the CTIO 4 meter survey.   P. S. Osmer.
Astrophys. J., Lett., Vol. 218, L89 - L92 (1977).

Observations with the CTIO SIT vidicon system of 19 candidates from a new, deep, grating-prism survey by Hoag and Smith show that at least 13 are quasars with $1.5 < z < 3.0$ and $17 \, 1/2 < m_v (1475) < 20 \, 1/2$. These results confirm the high efficiency of the survey technique. Comparison of the Curtis Schmidt and 4 m surveys indicates that they have similarly shaped surface density—magnitude relations with respective limits of completeness of 17 1/2 and 19 1/4 mag. The two surveys suggest a slope for the luminosity function of 0.9 dex per magnitude for quasars of redshift 2 to 2.5. The importance of the luminosity function to the density evolution of quasars is discussed in terms of Schmidt's redshift-magnitude tables.

141.138 Radio polarisation of quasars with optical absorption spectra.   G. A. Seielstad.
Nature, Vol. 270, 502 - 503 (1977).

The author calculated the mean and the median linear polarisation for various types of quasars at wavelengths $\lambda\lambda 3.7, 6, 11, 31,$ and 49 cm. Results are summarised in tables. The basic division is between those quasars having, and those not having, absorption lines. In no case the author finds a difference in mean linear polarisation between absorption and non-absorption quasars as large as two standard deviations. He concludes that the linear polarisation properties of quasars having absorption lines are indistinguishable from those not having absorption lines.

141.139 An explanation of the mysterious quasars. O. Borissov.
Rev. Polytech., No. 1, p. 43, 45 (1977). In French. — Abstr. in Phys. Abstr., Vol. 80, Abstr. 57104 (1977).

141.140 Solved and unsolved puzzles in today's radio astronomy.   R. G. Strom, D. E. Harris.
Tijdschr. Ned. Elektron. Radiogenoot., Vol. 42, No. 1 - 2, p. 33 - 35 (1977). In Dutch. — Abstr. in Phys. Abstr., Vol. 80, Abstr. 67133 (1977).

141.141 A statistical analysis for quasars with radio components structure. II. Relations of absolute visual magnitude, absolute radio magnitude and colour index difference Q with the linear distance between components, and the

associated evolution feature.
Y.-y. Zhou, F.-z. Cheng, Y.-q. Chu, L.-z. Fang.
Acta Astron. Sinica, Vol. 18, 113 - 128 (1977).

Based upon the first paper and continuing the "standard candle" classification by use of the component linear distance D, the authors attain to the following statistical correlation of some type of quasars with largest angular diameter Q_{max}: the correlation between visual magnitude and redshift; the correlation between its absolute magnitude, radio-luminosity of single frequency, integral radio luminosity, colour index difference Q and D. According to these statistical laws, a series of inferences on the evolution properties have been derived.

141.142 Revised identifications in the 5C2 survey.
    H. Tiersch, P. Notni.
Astron. Nachr., Band 298, 293 - 299 (1977).

In the present paper revised distances of optical objects from radio sources of the 5C2 survey are given. A statistical investigation of the data is given for blue objects and galaxies by the statistical method of the "first neighbour". The identification rate on blue plates for both the blue objects and galaxies amounts to about 40% out of the total number of 26 identifications. For every blue object and galaxy which are proposed as an identification the statistical reliability is given.

141.143 Optical identification of 664 Ohio sources using accurate radio and optical positions measured by the Texas interferometers.   F. D. Ghigo.
Astrophys. J., Suppl. Ser., Vol. 35, 359 - 393, Plates 29 - 36 (1977).

Results of optical identification work are reported for 664 radio sources selected from the Ohio 1415 MHz survey. Radio positions were measured at 365 MHz to an accuracy of about $2''$ with the Texas broad-band synthesis interferometer. Forty previously unpublished radio positions are given. Optical positions of $0''.5$ to $0''.7$ accuracy were measured on the Palomar Sky Survey (PSS) plates with a two-coordinate laser-interferometer measuring machine. These errors are consistent with those derived by comparing the positions with other accurate optical positions. Background objects were counted on the PSS plates and were used along with counts of objects within $20''$ of the radio sources to estimate the radial distribution of the radio-optical offsets of the true identifications. The completeness and reliability of the identifications are discussed.

141.144 Observation of Lyman-α emission in 3C 273 with a rocket-borne telescope.
A. F. Davidsen, G. F. Hartig, W. G. Fastie.
Publ. Astron. Soc. Pacific, Vol. 89, 615 (1977). – Abstract.

141.145 Broad-band surface photometry of a statistically complete sample of radio faint galaxies.
P. M. Rybski, M.-H. Ulrich.
Publ. Astron. Soc. Pacific, Vol. 89, 621 (1977). – Abstract.

141.146 The optical variability of six extragalactic objects.
    T. L. Mullikin, H. R. Miller.
Publ. Astron. Soc. Pacific, Vol. 89, 639 - 642 (1977).

The history of optical variability has been investigated for six extragalactic radio sources. Three of the sources, PKS 0906 +01, B2 1340 +29, and NRAO 512, were found to exhibit significant changes in brightness. Two of the sources, B2 1208 +32A and B2 1225 +31, were found to have only marginally significant evidence suggesting optical variability. A single source, Markarian 499, was found to exhibit no evidence suggesting optical variability.

141.147 On the ln(1+z) periodicity in QSO redshifts.
    D. Wills.
Publ. Astron. Soc. Pacific, Vol. 89, 643 - 645 (1977).

The 0.203 periodicity reported by Barnothy and Barnothy (1976) in the distribution of $\ln(1+z)$ for QSOs has much lower statistical significance, by a factor ~ 1000, in a slightly different sample analyzed in the same way as theirs. Their result may be due in part to spectroscopic selection effects in the early QSO data, because a sample of QSOs for which the spectroscopic observations are complete with respect to redshift shows no evidence for the periodicity in $\ln(1+z)$. Other factors that may contribute to the high statistical significance of their result are discussed.

141.148 Quasars to date.   A. J. Baldwin.
    South. Stars, Vol. 27, 38 - 43 (1977).

This article gives an account of the discovery of the quasars and a brief review of the current knowledge about them.

141.149 A search for steep low-frequency radio spectra among quasars and clusters of galaxies.
B. N. G. Guthrie.
Astrophys. Space Sci., Vol. 46, 429 - 441 (1977).

Using published flux densities at low frequencies, radio spectra were constructed for 3C, 4C, and 4CT radio sources in Abell clusters of galaxies, radio galaxies outside Abell clusters, and quasars with known redshifts. About half the sources in rich Abell clusters have steep spectra between 38 and 178 MHz. No steep spectra were found among 170 quasars. The absence of steep spectra among quasars does not necessarily mean that quasars never occur in rich clusters of galaxies, since quasars are probably being observed only in their early high-luminosity phases. The possibility that some quasar events occur in the nuclei of the dominant cD galaxies in clusters is discussed.

141.150 O VI and He II emission lines in the spectra of quasars.   Y. P. Varshni.
Astrophys. Space Sci., Vol. 46, 443 - 464 (1977).

The plasma-laser star model for quasars, which is based on the hypothesis that there is no redshift in the spectra of quasars and that the strength of the emission lines is due to laser action, is further developed. Continuity is shown to exist between the spectra of O VI sequence planetary nuclei, Sanduleak stars, and 10 quasars. The O VI $\lambda\lambda 3811$, 3834 and He II $\lambda 4686$ emission lines in the spectra of these 10 quasars are identified. Absolute magnitudes, temperatures, and masses of these quasars are estimated. The distribution of quasars in galactic coordinates is also discussed.

141.151 On the angular size-redshift test for double radio sources.   K. Brecher, C. Kieras.
Astrophys. Space Sci., Vol. 47, 25 - 30 (1977).

The origin of discrepancy between the observed redshift dependence of the angular size of double radio sources and the relation expected for constant diameter objects in homogeneous relativistic cosmologies is reconsidered. A correlation between absolute magnitude and projected linear separation for the sources could account for this discrepancy by observational selection without requiring cosmological evolution of the entire source population. The authors conclude that it is premature to use the angular size-redshift test as support either for astrophysical models of double radio source evolution, or for particular cosmological models.

141.152 Gaps in the emission line redshift distribution of QSOs.   D. Basu.
Astrophys. Space Sci., Vol. 47, 315 - 318 (1977).

A gap in a distribution is the interval between two consecutive values. Gaps in the emission line redshift distribution of QSOs are analysed using up-to-date data comprising 371 objects. It is found that the distribution of gaps is not random, but follows a definite trend, depending on the mean value of the redshift in the region.

**141.153   On the nature of quasars and active nuclei of
          galaxies.   V. L. Ginzburg, L. M. Ozernoj.**
Astrophys. Space Sci., Vol. 48, 401 - 420, Vol. 50, 23 - 41
(1977). In Russian and English.
    The authors discuss merits and deficiencies of widely
propagated ideas on the energy source of quasars and active
nuclei of galaxies (supermassive rotating magnetoplasma body –
magnetoid, accreting black hole, compact star cluster). Considering current data (especially recent results relating to the
character of optical variability of a number of objects) the
models of a compact star cluster seem to be unlikely. The
magnetoid seems to be the most controversial model, a final
selection, however, between magnetoid and accreting black
hole will be possible only by further itemization of these
models and by new observational data.

**141.154   Clumping of quasar redshifts and the geocentric
          universe.   P. R. Owen.**
Astrophys. Space Sci., Vol. 49, L7 - L9 (1977).
    It is shown that the observed clumping of quasar redshifts
can be caused purely by chance and does not require non-cos-
mological redshifts or a geocentric universe as argued by
Varshni.

**141.155   A new method of determining the distances of some
          extragalactic radio sources.   Yu. A. Kovalev.**
Inst. kosm. issled. AN SSSR. Pr-325. Moskva, 1977. 16 pp.
In Russian. – Abstr. in Ref. zh., 51. Astron., 1.51.737 (1978).

**141.156   Abundance ratios in quasi-stellar objects.
          R. K. Thakur, P. K. Mishra.**
Astrophys. Space Sci., Vol. 51, 249 - 264 (1977).
    Relative abundances of carbon and aluminium with
respect to silicon have been calculated in the QSOs PHL957,
PKS0237-23, 1331+170, 3C191 and M132. Relative abun-
dance of Fe with respect to Mg has been also calculated in the
QSOs 1331+170 and PHL938.

**141.157   18-cm OH observations of DR21.
          M. I. Pashchenko.**
Astron. Tsirk., No. 944, p. 1 - 2 (1977). In Russian.

**141.158   On the phenomenon of rapid separation of pairs of
          components in variable radio sources.**
V. N. Kuril'chik.
Astron. Tsirk., No. 957, p. 3 - 6 (1977). In Russian.

**141.159   Identification of southern radio sources with steep
          radio spectra.   M. P. Véron.**
Astron. J., Vol. 82, 937 - 940, 1041 (1977).
    Using a sample of 14 southern radio sources having steep
radio spectra ($\alpha \geqslant 1.3$), the author has shown that ten of
these are associated with clusters of galaxies and that in eight
cases out of ten the radio source is associated with the
brightest member of the cluster. This was expected; but more
important, when a Bautz–Morgan type could be assigned to
the cluster, it is preferentially a type I, I–II, or II, rather than
II–III, or III. This confirms an earlier result by McHardy
(1974) and Roland et al. (1976).

**141.160   Radio sources near the globular clusters M13 and
          M53.   S. J. Goldstein, Jr., C. M. Wade.**
Astron. J., Vol. 82, 972 (1977).
    Observations at 6-cm wavelength with a subset of the
VLA fail to confirm the faint radio sources near the centers
of M13 and M53, which had been reported by H. M. Johnson.
Another source seen by Johnson near M53 probably is extra-
galactic.

**141.161   Optical emission in the radio lobes of radiogalaxies.
          W. C. Saslaw, J. A. Tyson, P. Crane.**
ESO Sci. Prepr. No. 9, 2 + 29 pp. (1977). – To appear in

Astrophys. J.
    The authors have found photographic and photometric
evidence for optical emission in the radio lobes of 3C285,
3C265, and 3C390.3. The optical luminosity of this emission
lies between B $\sim$ 26 and B $\sim$ 22 mag arcsec^{-2}. They discuss
possible mechanisms of optical emission and their implications
for models of radiogalaxies. If the optical radiation arises from
inverse Compton scattering of the black-body background
radiation, it provides a fairly direct method for measuring the
magnetic fields in radiogalaxies. If, on the other hand, the
optical radiation is the visible extension of the radio synchrotron, it sets strong limits on the lifetimes of the relativistic
electrons which must be regenerated in the extended source.

**141.162   Distance estimates for G35.6–0.0 and G35.6–04.
          F. Sato.**
Publ. Astron. Soc. Japan, Vol. 29, 831 - 834 (1977).
    Kinematic distances of 12.6 and 3.7 kpc are suggested
for the galactic radio sources G35.6–0.0 and G35.6–04, respectively, from their 21-cm line absorption features.

**141.163   Ausgedehnte und kompakte Radioquellen.
          K. J. Fricke.**
Mitt. Astron. Ges., Nr. 42, p. 59 - 74 (1977). – Review paper.

**141.164   Pulsar magnetic alignment. The drifting subpulses.
          P. B. Jones.**
Oxford Univ., UK, Nucl. Phys. Lab., OU-NPL–9/77. 12 pp.
(1977). – Abstr. from INIS7715 319944.

141.165   Populations of weak radio sources.   P. Katgert.
          Proefschrift Rijksuniv. Leiden, Netherlands (1977).
From INIS-mf–3770.
    Contents: A survey of the 5C2 region with the Wester-
bork synthesis radio telescope at 1415 MHz (the third Westerbork survey); Source count, spectral indices and angular sizes
of weak radio sources in the 5C2 region; Results of additional
observations at 610 MHz; Source count, angular sizes and
spectral indices (revisited).

**141.166   Color dependent variations in some rapidly varying
          extragalactic objects.   M. Rosenkrantz.**
Bull. Israel Phys. Soc., Vol. 23, 112 - 113 (1977). – Summary.
Abstr. from INIS7721 336285.

    **A study of the revised 3C catalogue. I. Confusion
and resolution.**   See Abstr. 002.023.

    **Quasars, pulsars and black holes (a bibliography
with abstracts). Report for 1964–February 1977.**
See Abstr. 002.052.

    Cosmology now.   See Abstr. 003.005.

    Radio galaxies.   See Abstr. 003.123.

    **From quarks to quasars. An outline of modern
physics.**   See Abstr. 003.154.

    Radio astrometry.   See Abstr. 031.235.

    **Reduction of observations of scintillations of radio
sources.**   See Abstr. 031.331.

    Radio astronomy: radio signals from the Universe.
See Abstr. 033.012.

    Axially-symmetric explosion with thermal radiation.
See Abstr. 062.060.

    **Upper limits for the radio pulse emission rate from
exploding black holes.**   See Abstr. 066.007.

A local relativistic red-shift effect.
See Abstr. 066.286.

1974–75 UBV photometry of the radio binary UX Arietis. See Abstr. 113.012.

Radio survey of close binary stars. See Abstr. 117.044.

V711 Tauri. See Abstr. 119.014.

V 711 Tauri. See Abstr. 119.017.

CH radio emission towards the W49 region. See Abstr. 131.058.

Radio observations of CH towards various galactic objects. See Abstr. 131.059.

Radio detection of nitroxyl (HNO): the first interstellar NO bond. See Abstr. 131.077.

Limits on neutral hydrogen in the direction of PHL 957. See Abstr. 131.084.

Interstellar scattering and scintillation of radio waves. See Abstr. 131.159.

HCN emission in the Sagittarius A molecular cloud. See Abstr. 131.215.

A search for neutral hydrogen clouds in radio galaxies and in intergalactic space. See Abstr. 132.011.

The 4830 MHz H_2CO absorption in the direction of NGC 6334. See Abstr. 132.026.

Optical investigations of kinematics of NGC 7822 (W1). See Abstr. 134.003.

Observations of galactic nebulae at 5 GHz. See Abstr. 134.012.

Radio sources in the direction of globular clusters. See Abstr. 154.005.

Radio sources in globular clusters. See Abstr. 154.017.

Spectrophotometry of Seyfert 1 galaxies. See Abstr. 158.004.

Observations of Fe II emission in Seyfert galaxies and QSOs. See Abstr. 158.005.

Spectroscopy and photometry of the distant radio galaxy 3C 343.1. See Abstr. 158.024.

Rotation axes of the optical galaxies associated with Cygnus A and 3C 33. See Abstr. 158.037.

The dressed slingshot and the symmetry of double radio galaxies. See Abstr. 158.062.

The radio continuum morphology of spiral galaxies. See Abstr. 158.071.

Distribution of different extragalactic objects in the field of the North Galactic Pole. See Abstr. 158.072.

NGC 6251, a very large radio galaxy with an exceptional jet. See Abstr. 158.079.

Radio and optical observations of the N galaxy 4C39.11. See Abstr. 158.090.

Curtis Schmidt-thin prism survey for extragalactic emission-line objects: University of Michigan List II. See Abstr. 158.109.

Curtis Schmidt-thin prism survey for extragalactic emission-line objects: University of Michigan List III. See Abstr. 158.110.

Revue des galaxies exhibant une traînée lumineuse. See Abstr. 158.112.

Radio haloes around BL Lacertae objects AO0235+164 and 4C03.59. See Abstr. 158.504.

Upper limits on nuclear radio emission from some Coma Cluster spirals. See Abstr. 160.013.

On possible associations of quasi-stellar objects and radio galaxies with rich clusters of galaxies. See Abstr. 160.014.

A statistical investigation of radio sources in the directions of Zwicky clusters of galaxies. See Abstr. 160.020.

A study of 1889 rich clusters of galaxies. See Abstr. 160.023.

Radio emission of Abell clusters in the GB region. See Abstr. 160.040.

Radio sources in clusters of galaxies. See Abstr. 160.051.

Quasar absorption lines as probes of the past intergalactic medium. See Abstr. 161.003.

Interpretation of source counts and redshift data in evolutionary universes. See Abstr. 162.040.

The physics of radio sources and cosmology. See Abstr. 162.041.

Cosmological information from new types of radio observations. See Abstr. 162.042.

Progress, problems and priorities: a personal view. See Abstr. 162.043.

More evidence for a closed Universe from QSOs. See Abstr. 162.046.

Comment on Varshni's recent paper on quasar red shifts. See Abstr. 162.111.

The red-shift hypothesis for quasars: is the Earth the center of the Universe? II. See Abstr. 162.112.

Pulsars

141.501 Radio recombination lines towards pulsars.
R. N. Manchester, U. Mebold.
Astron. Astrophys., Vol. 59, 401 - 404 (1977).

H 109α and H 110α recombination-line emission has been searched for in the directions of five pulsars. In one of these directions, towards PSR 1641–45, a line apparently originating in the diffuse interstellar medium was detected. Based on the pulsar dispersion measures, a lower limit of 130 K is put on the electron temperature in the line-emitting regions, while an upper limit of about 3300 K is derived from the observed continuum brightness temperature. The clumping factor, $\langle n_e^2 \rangle / \langle n_e \rangle^2$, for the electrons in the line-emitting region is estimated to be about 20.

141.502 Galactic distribution and evolution of pulsars.
J. H. Taylor, R. N. Manchester.
Astrophys. J., Vol. 215, 885 - 896 (1977).

The distribution of pulsars with respect to period, z-distance, luminosity, and galactocentric radius has been investigated using data from three extensive pulsar surveys — those carried out at Molonglo, Jodrell Bank, and Arecibo Observatories. It is shown that selection effects only slightly modify the observed period and z-distributions but strongly affect the observed luminosity function and galactic distribution. These latter two distributions are computed from the Jodrell Bank and Arecibo data, using an iterative procedure. The largest uncertainties in the results are the result of uncertainty in the adopted distance scale.

141.503 On the nature of gamma-ray emission from pulsars.
L. M. Ozernoj, V. V. Usov.
Astron. Zh. Akad. Nauk SSSR, Vol. 54, 753 - 765 (1977).
In Russian. English translation in Soviet Astron., Vol. 21, No. 4.

The authors discuss both generation and propagation of gamma rays stipulated by the curvature of the magnetic force lines in the vicinity of a pulsar. Accounting for creation of electron-positron pairs by gamma quanta in a strong magnetic field, the distance is evaluated beginning from which gamma rays leave freely the pulsar neighbourhood. The power of gamma-ray emission and the energy at which gamma radiation has a maximum are estimated. The values obtained are in a reasonable agreement with the observed parameters of gamma-ray emission for young pulsars, as well as for comparatively old ones. The authors note that the diagram for pulsar gamma-ray emission generally will not coincide with that of radio emission. The model developed here predicts the existence of gamma-ray pulsars without apparent radio emission.

141.504 On pulsar magnetosphere instability.
L. A. Pustil'nik.
Astron. Zh. Akad. Nauk SSSR, Vol. 54, 766 - 771 (1977).
In Russian. English translation in Soviet Astron., Vol. 21, No. 4.

The paper is devoted to an analysis of the stability of a pulsar magnetosphere in which particles ejected from the surface of the neutron star by the electric field are accumulated. It is shown that at the moment when the density of magnetospheric plasma exceeds a critical value the magnetosphere must destabilize and eject outward (onto the light cylinder) a channel of plasma with the field. The value of the critical density is found in function of the pulsar parameters. This allowed to estimate numerically observational appearances of destabilization. The values obtained agree with the observations of "jumps" in the pulsars in the Crab and Vela nebulae.

141.505 The binary pulsar observed at four radio frequencies.
A. G. Lyne, R. T. Ritchings.

Nature, Vol. 268, 606 - 607 (1977).

The authors report observations of PSR 1913+16 at four radio frequencies which show that this pulsar has properties which are similar to those of the other pulsars.

141.506 Pulsar interpulses and other off-pulse emission.
E. B. Cady, R. T. Ritchings.
Nature, Vol. 269, 126 - 127 (1977).

It would be statistically incorrect to compare the fraction of pulsars known to have interpulses with the fraction predicted by a proposed pulsar model. In order to check this possibility the authors conducted an interpulse search on the pulsars that are visible from Jodrell Bank. They report the results of this search in which one new interpulse was discovered. These observations also revealed the presence of weak emission from regions other than the main pulse and interpulse windows in the cases of three pulsars.

141.507 Proper motion of the Crab pulsar.
S. Wyckoff, C. A. Murray.
Mon. Not. R. Astron. Soc., Vol. 180, 717 - 729 (1977).

The purpose of this paper is to (1) present a new measurement of the proper motion using unique plate material spanning a time interval of 77 yr (1899–1976), (2) present a detailed discussion of the reduction of the proper motion to an inertial reference frame, and (3) re-discuss the relationship between the kinematics of the Crab nebula and the proper motion of the pulsar.

141.508 Pulsars: physics laboratories in our Galaxy.
D. J. Helfand.
Mercury, Vol. 6, No. 3, p. 2 - 7 (1977).

141.509 Modulation instability of a relativistic plasma in the vicinity of a pulsar.
M. Khakimova, F. K. Khakimov, V. N. Tsytovich.
Astrofizika, Vol. 12, 531 - 542 (1976). In Russian. — English translation in Astrophysics, Vol. 12, No. 3.

The modulation instability of a relativistic turbulent plasma has been investigated which in forming coherent coagulations may be used to explain the radio radiation of pulsars. Estimates of the time development of instability are given. The level of turbulence is determined in the case where instability can develop. The dimensions of heterogeneity have also been determined in the case when coherent radiation may take place.

141.510 The binary pulsar: post-Newtonian timing effects.
R. Epstein.
Astrophys. J., Vol. 216, 92 - 100 (1977).

Certain post-Newtonian effects may be observable in the arrival times of pulses from the binary pulsar PSR 1913 + 16. Such effects include the gravitational propagation delay and the post-Newtonian corrections to the elliptical binary orbit. Fitting a model of these effects as they are predicted by general relativity to the pulse arrival times permits the estimation of more parameters than are necessary to determine the orbit and masses of the system. This redundancy provides an important check of the assumption that the observed periastron precession rate is entirely due to the effect of general relativity on compact masses.

141.511 Requirements on pulsar models from gamma ray observation. S. Hinata.
Astrophys. J., Vol. 216, 101 - 102 (1977).

The observed γ-ray luminosity $L_\gamma \sim 10^{30}$ ergs s^{-1} around the energy $E_\gamma \sim 10^{12}$ eV for the Crab pulsar by Grindlay, Helmken, and Weekes provides the location of emission, the energy and the number of charges which radiate in the observed γ-ray band. The assumption involved in the derivation is that the curvature radiation produces the observed γ-ray luminosity. The small number of high-energy charges required from the observation indicates the plausibility of the turbulent acceleration region above the usual acceleration region where the intense static electric field plays the major role.

141.512 Pulsar velocity observations: correlations, interpretations, and discussion.
D. J. Helfand, E. Tademaru.
Astrophys. J., Vol. 216, 842 - 851 (1977).

From an examination of the current sample of 12 pulsars with measured proper motions and the z-distribution of the much larger group of over 80 sources with measured period derivatives, the authors develop a self-consistent picture of pulsar evolution. A method for calculating the unmeasurable radial velocity of a pulsar is presented; it is shown that the total space velocities thus obtained are consistent with the assumption of an extreme Population I origin for pulsars which subsequently move away from the plane with a large range of velocities. The time scale for pulsar magnetic field decay is derived from dynamical considerations. A strong correlation of the original pulsar field strength with the magnitude of pulsar velocity is discussed. This results in the division of pulsars into two classes: Class A sources characterized by low space velocities, a small scale height, and low values of $P_0 \dot{P}_0$; and Class B sources with a large range of velocities (up to 1000 km s^{-1}), a much greater scale height, and larger values of initial field strength.

141.513 A Crab pulsar model: X-ray, optical, and radio emission.   A. F. Cheng, M. A. Ruderman.
Astrophys. J., Vol. 216, 865 - 872 (1977).

The predicted electron-positron pair production in the Crab pulsar outer magnetosphere at the "outer gap", 3×10^7 cm from the neutron star, is considered as a source of pulses at all observed energies. A pulse and interpulse that are roughly symmetric with separation somewhat less than 180° are expected for arbitrary angles between the stellar magnetic moment and spin. The model gives optical and X-ray pulses from inverse Compton scattering of radio photons by e^+ and e^-. The Crab radio emissions, except for the precursor, are predicted to originate from the "outer gap".

141.514 Production and beaming of pulsar γ-ray emission.
P. E. Hardee.
Astrophys. J., Vol. 216, 873 - 880 (1977).

There are now known to be four pulsars that emit pulsed γ-rays in addition to the usual radio emission. Of these four, two are normal radio pulsars with periods on the order of 0.5 s. The other two are the Crab and Vela pulsars with very short periods. It is shown that the observed γ-ray emission can originate at or near the pulsar's surface and that radio emission should originate at some distance above the surface. The phase difference between radio and γ-ray pulses can be understood as resulting from magnetic field configuration and aberration effects. It is also shown that the high-energy γ-ray emission from the Crab pulsar cannot be produced in this fashion, but is probably due to emission from high-energy particles at the light cylinder.

141.515 Submicrosecond intensity structure in pulsar PSR 0950+08 and its interpulse.
T. H. Hankins, V. Boriakoff.
Bull. American Astron. Soc., Vol. 9, 433 (1977). − Abstract.

141.516 Potential drops above pulsar polar caps: acceleration of nonneutral beams from the stellar surface.
W. M. Fawley, J. Arons, E. T. Scharlemann.
Astrophys. J., Vol. 217, 227 - 243 (1977).

The authors develop a self-consistent, three-dimensional method for the calculation of the steady-state acceleration of nonneutral plasma at the surface of a rotating, magnetized, isolated neutron star. When the effects of surface work functions are negligible, there is a unique monotonic electrostatic

potential and associated current profile. They obtain explicit solutions in the monopole and polar-cap geometries. The resultant luminosity is sufficient to explain radio pulsars if the beam energy can be converted to radio radiation with reasonably high efficiency, but is insufficient to explain pulsed γ-ray emission.

141.517 Pulsar flux observations: long-term intensity and spectral variations.
D. J. Helfand, L. A. Fowler, J. V. Kuhlman.
Astron. J., Vol. 82, 701 - 705 (1977).

The authors present daily observations of pulsar mean pulse fluxes recorded over a four-year period at 156 MHz and, simultaneously for a shorter period, at 390 MHz. The observations yield characteristic time scales for correlated flux changes of a few tens of days, with an apparent low-frequency cutoff to the variations of several hundred days. A search for periodic modulation allows a strict limit to be set on the amplitude of neutron star precession in the period range of 2−150 days for most sources. The modulation index for the high-frequency data is generally greater than that for the low-frequency fluxes, although the time scales for variations appear to be frequency independent.

141.518 On the spindown/spinup of JP 1953.
V. N. Mansfield.
News Lett. Astron. Soc. N.Y., Vol. 1, No. 2, p. 16 (1977). Abstract.

141.519 Inertial mechanism of magnetic field generation in pulsars.
D. M. Sedrakyan, K. M. Shakhabasyan, R. Rudol'f.
Astrofizika, Vol. 13, 153 - 163 (1977). In Russian. English translation in Astrophysics, Vol. 13, No. 1.

An inertial mechanism of magnetic field generation in pulsars is considered. It is shown that, when the concentration of electrons is less than the concentration of barions, generation of a magnetic field of the order of 10^{10} gauss is possible. The time of increase of the magnetic field is $10^3 - 10^6$ years.

141.520 Angular momentum and energy loss from pulsars.
N. J. Holloway.
Mon. Not. R. Astron. Soc., Vol. 181, 9P - 12P (1977).

The connection between the angular momentum and the rotational kinetic energy of a pulsar implies constraints on the possible physical processes of energy loss. These constraints appear to suggest that the emission regions are near the light cylinders of pulsars.

141.521 Radiation mechanisms and magnetospheric structure of pulsars.   P. A. Sturrock, K. B. Baker.
The structure and content of the Galaxy and galactic gamma rays, (see 012.009), p. 99 - 108 (1977).

This article outlines the chain of thought which has led to a model of pulsars now being investigated at Stanford University. Observational data seems to support a magnetosphere model based on the Scargle-Pacini idea rather than the Goldreich-Julian (1969) model.

141.522 Gamma-ray pulsars.
H. Ögelman, S. Ayasli, A. Hacinliyan.
The structure and content of the Galaxy and galactic gamma rays, (see 012.009), p. 109 - 115 (1977).

Recent data from the high-energy γ-ray experiment have revealed the existence of four pulsars emitting photons above 35 MeV. An attempt is made to explain the γ-ray emission from these pulsars in terms of an electron-photon cascade that develops in the magnetosphere of the pulsar.

141.523 Galactic distribution of pulsars.   J. H. Seiradakis.
The structure and content of the Galaxy and galactic gamma rays, (see 012.009), p. 265 - 282 (1977).

The density distributions of pulsars in luminosity, period, Z-distance, and galactocentric distance have been derived, using a uniform sample of pulsars detected during a 408-MHz pulsar survey at Jodrell Bank. There are indications of a "fine-scale" structure in the spatial distribution and evidence that there is a general correlation with other galactic populations and the overall spiral structure. The electron layer in our Galaxy is shown to be wider than the pulsar layer and uniform on a large scale. The number of pulsars in the Galaxy has been estimated and used to derive the pulsar birthrate.

141.524 A model of the radio emission mechanism in pulsars.
K. Kawamura, I. Suzuki.
Astrophys. J., Vol. 217, 832 - 842 (1977).

It is pointed out that the simultaneous excitation of right and left circularly polarized waves by ultrarelativistic nongyrating beams of electrons and positrons provide a possible mechanism for pulsar radio emission. By using a quasi-linear approximation, it is shown that such a mechanism yields a huge brightness temperature $T_b > 10^{25}$ K, enough to explain the observed one. The radio spectrum obtained has a low-frequency cutoff or turnover and decreases approximately as ω^{-3} in the high-frequency range, a result insensitive to the assumed distribution function of the beam. Such a spectral behavior gives a good fit to some pulsar spectra. Some other features of pulsar radiation are also discussed in the model.

141.525 COS-B observations of pulsed γ-ray emission from PSR 0531+21 and PSR 0833−45.
K. Bennett, G. F. Bignami, G. Boella, R. Buccheri, W. Hermsen, G. Kanbach, G. G. Lichti, J. L. Masnou, H. A. Mayer-Hassel-wander, J. A. Paul, L. Scarsi, B. N. Swanenburg, B. G. Taylor, R. D. Wills.
Astron. Astrophys., Vol. 61, 279 - 284 (1977).

COS-B has performed detailed measurements of the high-energy γ-ray emission from the Crab and Vela pulsars. The two pulsars exhibit very similar γ-ray light curves, both with narrow peaks of roughly equal intensity separated by 0.42 ± 0.03 of the period. The differential spectrum of PSR 0531+21 can be represented by $(2.4 \pm 0.4) 10^{-7} E^{-(2.1 \pm 0.3)}$ photon cm^{-2} s^{-1} GeV^{-1} in the range 50 MeV to 4 GeV, while it is difficult to reconcile the spectrum of PSR 0833−45 with a single power law over the same energy range.

141.526 Further evidence for multiple-ring structure in the hollow-cone emission of radio pulsars.
W. Sieber L. Oster.
Astron. Astrophys., Vol. 61, 445 - 446 (1977).

PSR 2045−16 seems to be another example (besides PSRs 1237+25 and 1919+21) of pulsars showing multiple-ring structure. It is concluded that multiple-ring structure may be quite common among radio pulsars.

141.527 Hadronic photoabsorption and pair production in pulsars.   P. B. Jones.
Nature, Vol. 270, 37 - 38 (1977).

For surface magnetic flux densities $\gtrsim 10^{12}$ G, the author attempts to show that the nature and mode of acceleration of the plasma moving outwards along open magnetic flux lines are determined by the occurrence of hadronic photoabsorption reactions in the electromagnetic showers produced by ultra-relativistic electrons incident on the stellar surface.

141.528 A classification scheme for pulsars.
R. K. Kochhar.
Nature, Vol. 270, 38 - 39 (1977).

The author has already (1977) distinguished between three types of pulsars depending upon their mode of formation. Here he discusses some implications of this classification scheme.

141.529 Angular size of the Crab pulsar at 74 MHz.

W. A. Coles, J. J. Kaufman.
Mon. Not. R. Astron. Soc., Vol. 181, 57P - 59P (1977).

Observations of the compact source in the Crab Nebula indicate that its angular diameter has remained constant at 0.2 arcsec over the period 1972−1976.

141.530 Numerical modelling of pulsar magnetospheres.
M. Petravić.
Computing in plasma physics and astrophysics, (see 012.028), p. 9 - 19 (1976).

Numerical models used in the study of the pulsar magnetosphere are described: a vacuum model based only on Maxwell's equations and a more realistic model employing both Maxwell's and relativistic two-fluid equations. The general approach to solving the chosen sets of partial differential equations is outlined and the possible boundary conditions are examined. Numerical methods suitable for solving Maxwell's equations are discussed and a method is developed for solving the combined fluid plus Maxwell model. Results are presented and discussed and the possible improvements in the approach are indicated.

141.531 Pulsar polarization fluctuations at 430 MHz with microsecond time resolution.
J. M. Cordes, T. H. Hankins.
Astrophys. J., Vol. 218, 484 - 503 (1977).

This paper studies the polarization properties of pulsars 0950+08, 1133+16, 1919+21, and 2016+28 after the removal of interstellar dispersion distortion. Autocorrelation functions of the Stokes parameters were computed for narrow longitude regions and were summed to reduce statistical errors. Such autocorrelation functions quantify the mean-square properties of polarization fluctuations and reveal how the polarization decorrelates on time scales between the reciprocal receiver bandwidth (8 μs) and the maximum lag of 4 ms. The authors find that single-pulse polarization is only partially correlated with the polarization of the average profile, where the latter is measured by synchronously averaging the Stokes parameters of many single pulses. Deviations of single-pulse polarization from the average are highly correlated with intensity fluctuations such as micropulses and subpulses. The authors infer from the autocorrelation functions and from single pulses that the state of polarization is usually constant through a micropulse but may change rapidly on the extremity of a micropulse or between two micropulses. Rapid polarization changes within some micropulses cannot be ruled out, but the strength and number of such micropulses must contribute negligibly to the autocorrelation functions.

141.532 Recent observations of pulsars.
J. H. Taylor, R. N. Manchester.
Annu. Rev. Astron. Astrophys., Vol. 15, (see 003.012), 19 - 44 (1977).

One section is devoted to a summary of the existing observational material relevant to the pulse emission mechanism. Another one deals with the important problem of the origin and evolution of pulsars.

141.533 Coherent amplification and pulsar phenomena.
L. W. Casperson.
Astrophys. Space Sci., Vol. 48, 389 - 399 (1977).

A modification of the rotating-star model has been developed to interpret the periodic energy bursts from pulsars. This new configuration involves θ-directed oscillation modes in the stellar atmosphere or magnetosphere, and most aspects of the typical pulse characteristics are well accounted for. Gain is provided by resonant interactions with particles trapped in the stellar magnetic field. The most significant feature is the fact that highly directional beaming of the output energy results as a natural consequence of coherence between the radiation fields emerging from various locations about the pulsar; and a localized radiation origin is not required.

141.534 Pulsar interpulses – two poles or one?
R. N. Manchester, A. G. Lyne.
Mon. Not. R. Astron. Soc., Vol. 181, 761 - 767 (1977).

A number of observational results including the recent detection of optical pulses from the Vela pulsar appear to conflict with the usual interpretation of pulsar interpulses as radiation from the opposite pole of a basically dipolar magnetic field. An alternative model in which main pulses and interpulses originate from the same pole is proposed and compared with the observational data.

141.535 Dynamics of differential rotating sphere and pulsar.
K. U. Lu.
Int. J. Theor. Phys., Vol. 15, 411 - 415 (1976). – Abstr. in Phys. Abstr., Vol. 80, Abstr. 37609 (1977).

141.536 A generalized Ohm law for pulsar magnetospheres.
R. Burman.
Phys. Lett. A, Vol. 60A, 309 - 310 (1977). – Abstr. in Phys. Abstr., Vol. 80, Abstr. 37610 (1977).

141.537 Cherenkov radiation in a charge-separated magnetic plasma as a possible source for radio emission in pulsars. H. Kolbenstvedt.
Phys. Rev. D, Vol. 15, 975 - 976 (1977). – Abstr. in Phys. Abstr., Vol. 80, Abstr. 47163 (1977).

141.538 How pulsars pulse. D. Ferguson, R. Wielebinski.
New Scientist, Vol. 74, 586 - 588 (1977). – Abstr. in Phys. Abstr., Vol. 80, Abstr. 71124 (1977).

141.539 Rate of stellar collapses in the Galaxy.
K. Lande, W. E. Stephens.
Astrophys. Space Sci., Vol. 49, 169 - 177 (1977).

From an analysis of pulsar spatial and luminosity distributions, the number density of observed pulsars in the local region is determined to be $1.1 \pm 0.4 \times 10^{-7}$ pulsar pc^{-3}. Multiplication by the detection factor and by the ratio of Galaxy mass to local matter density and division by a mean lifetime of pulsars of 3×10^6 yr suggests a pulsar birth every 4 yr. A stellar collapse might occur even more often.

141.540 Some features of the pulsed radiation from the pulsar 1919+21 at 16.7, 20 and 25 MHz.
Yu. M. Bruck, B. Yu. Ustimenko.
Astrophys. Space Sci., Vol. 49, 349 - 366 (1977).

An intense interpulse radio frequency radiation of PSR 1919+21 has been detected in the range of 16.7, 20 and 25 MHz. The maximum number of the best pronounced interpulses is about four, their location being symmetrical relative to the centre of the main pulse. The fine frequency structure of the integrated radiation was also investigated. It has been found that the shape of the signal can differ considerably with a frequency diversity of 10 to 100 kHz and only slightly with a separation of 5 to 10 MHz, the difference diminishing with an increase of the observation bandwidth and of the averaging time. Pulse broadening was studied at 16.7 and 25 MHz and has been found to agree with the interstellar scattering mechanism. The mean intensity of the main pulse and the interpulses has been evaluated and an essential difference of their spectral indices established.

141.541 The interpulse radio emission of the pulsar PSR 1919+21 at the frequencies 16.7–38 MHz.
Yu. M. Bruck, B. Yu. Ustimenko.
Astrophys. Space Sci., Vol. 51, 225 - 227 (1977).

Results of observations of the pulsar PSR 1919+21 in the range 16.7 to 25 MHz and at 38 MHz are compared. Similarities have been disclosed in the signal wave form, the number of interpulses, and other features.

141.542 A remark on co-rotation in pulsar magnetospheres.

R. Burman.
Astrophys. Space Sci., Vol. 51, 239 - 240 (1977).

It is pointed out that a mathematical step taken by Ardavan (1976), relating the Lorentz factor of relativistically co-rotating plasma to distance from the light cylinder, needs closer investigation.

141.543 A possible mechanism for the pulsar radio emission.
S. Hinata.
Astrophys. Space Sci., Vol. 51, 303 - 318 (1977).

The possibility of radio emission is considered within a model which produces the beam-plasma system near the pulsar.

141.544 Taking into account the Doppler effect in measurement of pulsar periods. T. V. Shabanova.
Tr. Fiz. inst. AN SSSR, Vol. 93, 119 - 125 (1977). In Russian.- Abstr. in Ref. zh., 51. Astron., 11.51.779 (1977).

141.545 On the spinup/spindown of pulsar JP 1953+29.
V. N. Mansfield, J. Rankin.
Vistas Astron., Vol. 21, 393 - 405 (1977).

A new analysis of the available timing data for pulsar JP 1953+29 as well as a determination of its dispersion and proper motion is reported. The pulsar's behaviour seems quite orthodox aside from its $\dot{P}$ value which is about an order of magnitude smaller than for any other pulsar. Positive definite special relativistic effects of the object's transverse velocity, however, are comparable to the observed $\dot{P}$ and thus the intrinsic period derivative is compatible with either a spindown or spinup of scale $\gg 10^9$ years. Several mechanisms are examined that might reduce $\dot{P}$. The authors conclude that these mechanisms are quite unlikely and then discuss the physical significance of the extremely slow spindown or possible spinup of JP 1953+29.

141.546 Average characteristics of pulsars.
O. Kh. Gusejnov, I. M. Yusifov.
Astron. Tsirk., No. 951, p. 7 - 8 (1977). In Russian.

141.547 Structure and radiation mechanism of pulsars.
S. R. Kınacı.
Sci. Rep. Fac. Sci., Ege Univ., *Izmir*, No. 185 (Astron. No. 14), 48 pp. (1977). In Turkish.

A model for the structure and the radiation mechanism of pulsars is suggested. In this model, the radiation emitted at the magnetic polar region of the star sweeps across the observer, as the star rotates, like a beacon. Then the observer receives the radiations as pulses. Since the radiations are emitted continuously by a swarm of electrons in a common region of star, they are coherent, polarized and intensive.

141.548 A possible radio emission mechanism for pulsars.
Yu. A. Kovalev.
Inst. kosm. issled. AN SSSR. Pr-326. Moskva, 1977. 26 pp. In Russian. – Abstr. in Ref. zh., 51. Astron., 1.51.542 (1978).

141.549 Model of a radio pulsar. A. I. Tsygan.
Fiz.-tekh. inst. AN SSSR. Prepr. No. 547, Leningrad, 1977. 13 pp. In Russian. – Abstr. in Ref. zh., 51. Astron., 1.51.543 (1978).

141.550 On the shape of a pulsar's pulse scattered in the interstellar medium. V. E. Ostashov, V. I. Shishov.
Izv. vyssh. uchebn. zaved. Radiofiz., Vol. 20, 842 - 847 (1977). In Russian. – Abstr. in Ref. zh., 51. Astron., 1.51.550 (1978).

141.551 Magneto-hydrodynamic modes in a plasma atmosphere around a pulsar.
R. Horiuchi, N. Tajima, A. Tomimatsu.
Hiroshima Univ., Takehara, Japan, Res. Inst. Theor. Phys. RRK–77-1. 23 pp. (1977). – Abstr. from INIS7723 341966.

141.552   Possible periodicity of dips in the optical light
curve of OJ 287.   M. Rosenkrantz.
Bull. Israel Phys. Soc., Vol. 23, 112 (1977). – Summary.
Abstr. from INIS7721 336284.

Quasars, pulsars and black holes (a bibliography
with abstracts). Report for 1964–February 1977.
See Abstr. 002.052.

Pulsars.   See Abstr. 003.094.

Estimate of the limiting standard deviation in the
absolute timing of periodic sources.
See Abstr. 031.328.

Automation of reduction of pulsar observations.
See Abstr. 031.407.

Do pulsars synthesize ^{7}Li?
See Abstr. 061.047.

The behaviour of the magnetic lines of a rotating
system.   See Abstr. 062.068.

Gravitationally redshifted gamma rays and neutron
star masses.   See Abstr. 066.022.

How do neutron stars evolve?
See Abstr. 066.071.

Has the sun a companion star?
See Abstr. 080.038.

Pulsar theory of supernova light curves. I. Dynamical effect and thermalization of the pulsar strong waves.
See Abstr. 125.006.

Pulsar theory of supernova light curves. II. The
light curve and the continuous spectrum.
See Abstr. 125.007.

Pulsar dispersion measures and Hα emission measures: limits on the electron density and filling factor for the
ionized interstellar gas.   See Abstr. 131.044.

Observation of X-rays from the Crab pulsar.
See Abstr. 134.009.

A high-resolution investigation of the shell source
G55.7 + 3.4 at 610 and 1415 MHz.   See Abstr. 141.047.

The effect of vacuum birefringence on the polarization of X-ray binaries and pulsars.   See Abstr. 142.007.

The scale and strength of the galactic magnetic field
according to the pulsar data.   See Abstr. 156.018.

Scale and strength of the galactic magnetic field
from pulsar data.   See Abstr. 156.019.

On the intergalactic contribution to the rotation
measures of QSO's.   See Abstr. 161.009.

Errata

141.901   Erratum: "Photométrie de Ton 1542 par
électronographie" [Astron. Astrophys., Vol. 56,
71 - 74 (1977)].   C. Vanderriest, G. Lelièvre.
Astron. Astrophys., Vol. 60, 438 (1977).

141.902   Erratum: "The Crab Nebula pulsar: six years of
radio-frequency arrival times" [Astron. J., Vol. 82,
309 - 312 (1977)].   G. E. Gullahorn, R. Isaacman, J. M.
Rankin, R. P. Payne.
Astron. J., Vol. 82, 930 (1977).

141.903   Erratum: "Radio and optical structure of Cygnus A"
[Astron. J., Vol. 82, 315 - 318 (1977)].
P. Kronberg, S. van den Bergh, S. Button.
Astron. J., Vol. 82, 1039 (1977).

142 Extreme UV, X-Ray, Gamma-Ray Sources

Extreme UV, X-Ray Sources

142.001 Observations of the LMC X-ray sources with the Ariel V Sky Survey Instrument.
R. E. Griffiths, F. D. Seward.
Mon. Not. R. Astron. Soc., Vol. 180, 75P - 79P (1977).

Long-term X-ray light curves are presented for the X-ray sources in the Large Magellanic Cloud. The five previously known sources have all been detected. Evidence is presented for a sixth source possibly associated with the LMC. All sources are variable but no regular periodic behaviour has been observed. LMC X-1 and X-2 vary by a factor of 2 or 3, LMC X-3 and X-4 were observed to vary by a factor of at least 7, and LMC X-5 and X-6 varied from just above to below the threshold of detectability.

142.002 Cyg X-3: γ rays from a magnetic accretion disk.
P. Mészáros, F. Meyer, J. E. Pringle.
Nature, Vol. 268, 420 - 421 (1977).

The recent observation of ~100-MeV γ rays from Cygnus X-3, which have a comparable luminosity to the X-ray emission and which vary in phase with the 4.8-h period displayed by the X-ray and infrared emission, has been taken as support for the suggestion that the X-ray emission is produced by a very young pulsar in a short period binary system. In this letter, the authors show that the observed γ-ray luminosity can be produced by an accretion disk around a non-magnetic neutron star or low-mass black hole, in a magnetised corona which accelerates relativistic electrons that decay by inverse Compton losses modified through the photon-photon and pair-annihilation processes.

142.003 The puzzles of Hercules. J. Pringle.
Nature, Vol. 268, 484 - 485 (1977).

142.004 A line feature at 64 keV in the X-ray spectrum of Her X-1.
M. J. Coe, A. R. Engel, J. J. Quenby, C. S. Dyer.
Nature, Vol. 268, 508 - 509 (1977).

Trumper and his colleagues have reported observations which give evidence of a hard X-ray line feature in the spectrum of the pulsed flux of Her X-1. They suggest that this is cyclotron radiation from the poles of a neutron star. The authors discuss here some similar measurements made by the Imperial College Scintillation Telescope (ST) on Ariel 5.

142.005 0.8 day periodicity in line emission from HZ Her.
J. B. Hutchings, D. Crampton.
Astron. Astrophys., Vol. 59, 441 - 444 (1977) = Contrib. Dominion Astrophys. Obs., Victoria, No. 331 = NRC No. K 900.

Line emission at λ4686 (He II) in the spectrum of HZ Her shows significant modulation with a period of ~0.81 days. This lends considerable support to precessing disk models for the system.

142.006 Positions of X-ray sources near the galactic center measured by the Ariel-5 RMC experiment.
A. M. Wilson, G. F. Carpenter, C. J. Eyles, G. K. Skinner, A. P. Willmore.
Astrophys. J., Lett., Vol. 215, L111 - L115 (1977).

The positions of 21 X-ray sources have been measured by the Ariel-5 Rotation Modulation Collimator (RMC). The sources are all within 10° of the galactic plane and between galactic longitudes −40° and +20°.

142.007 The effect of vacuum birefringence on the polariza-tion of X-ray binaries and pulsars.
R. Novick, M. C. Weisskopf, J. R. P. Angel, P. G. Sutherland.
Astrophys. J., Lett., Vol. 215, L117 - L120 (1977).

In a strong magnetic field the vacuum becomes birefringent. This effect is especially important for pulsars at X-ray wavelengths. Any polarized X-ray emission from the surface of a magnetic neutron star becomes depolarized as it propagates through the magnetic field. The soft X-ray emission from AM Her, believed to be a magnetic white dwarf, may show about one radian of phase retardation. In this case, circular polarization of the X-ray flux would be a characteristic signature of vacuum birefringence.

142.008 X-ray bursts from dense clouds. A. C. Fabian.
Nature, Vol. 268, 607 - 608 (1977).

The author investigates the effect of the motion of a compact object through a relatively dense medium. The passage of the compact object through a denser shell, which builds up outside the X ray-heated gas, produces a short burst of X rays. This creates yet another hot region leading to the repetition of bursts on a crossing time.

142.009 Variable mid-latitude X-ray source 3U 0042+32.
S. Rappaport, G. W. Clark, R. Dower, R. Doxsey, G. Jernigan, F. Li.
Nature, Vol. 268, 705 - 706 (1977).

The authors report a precise celestial location (1′ error radius) for the X-ray source 3U 0042+32 that is situated at mid-galactic latitude ($b^{II} \sim -30°$). A study of the error box for this source indicates that it may be a distant galactic binary system similar to Her X-1.

142.010 On the optical identifications of five X-ray sources.
H. V. Bradt, K. M. V. Apparao, G. W. Clark, R. Dower, R. Doxsey, D. R. Hearn, J. G. Jernigan, P. C. Joss, W. Mayer, J. McClintock, F. Walter.
Nature, Vol. 269, 21 - 25 (1977).

The data from a recently completed survey of the galactic plane with the SAS-3 modulation collimators provide precise (20″−60″) celestial positions of galactic X-ray sources. Preliminary positions of 60″ precision are reported for five sources. One of these led to the identification of the star, γ Cas, as an X-ray source, and the others lend substantial confidence to previously proposed optical identifications: 3U0352 + 30 = X Per, 3U1145−61 = HEN 715, GX301−2 = WRA977, and GX304−1 = MMV star. These identifications seem to establish the existence of a previously suggested class of De-star X-ray emitters.

142.011 Positions of galactic X-ray sources: $20° < l^{II} < 55°$.
R. E. Doxsey, K. M. V. Apparao, H. V. Bradt, R. G. Dower, J. G. Jernigan.
Nature, Vol. 269, 112 - 116 (1977).

Precise positions, determined with data from the SAS-3 X-ray observatory, are presented for eight galactic plane X-ray sources (Aql X-1, Ser X-1, 3U 1956+11, 3U 1822−00, 3U 1915−05, A 1845−02, A 1850−08, A 1905+00). Error radii for the positions range from 20 to 50 arc s. Previously proposed optical identifications of three of the sources (Ser X-1, 3U 1956+11, A 1850−08) are supported by these results. Three (Ser X-1, A 1905+00, 3U 1915−05) have been identified as X-ray burst sources.

142.012 Variability of LMC X-4.
A. Epstein, J. Delvaille, H. Helmken, S. Murray, H. W. Schnopper, R. Doxsey, F. Primini.
Astrophys. J., Vol. 216, 103 - 107 (1977).

X-rays in the 2−11 keV band were detected from LMC X-4

using the rotating modulation collimator system on SAS-3. The derived X-ray luminosity of LMC X-4 is $\sim 10^{38}$ ergs s^{-1}. The authors observed a transition to a high-intensity state which lasted ~ 40 minutes during which time short-time-scale flares (~ 20 s) were observed. The temporal and spectral behavior of LMC X-4 is found to be similar to that of several X-ray binaries. A newly derived 90% error circle with a radius of 1' is also presented.

142.013   A1540−53, an eclipsing X-ray binary pulsator.
R. H. Becker, J. H. Swank, E. A. Boldt, S. S. Holt, S. H. Pravdo, J. R. Saba, P. J. Serlemitsos.
Astrophys. J., Lett., Vol. 216, L11 - L14 (1977).

The authors have observed an eclipsing X-ray binary pulsator consistent with the location of A1540−53. The source pulse period was 528.93 ±0.10 s. The binary nature is confirmed by a Doppler curve for the pulsation period. The eclipse angle of 30°.5 ± 3° and the 4 hour transition to and from eclipse suggest an early-type, giant or supergiant, primary star.

142.014   Discovery of a 272 second periodic variation in the X-ray source GX 304−1.
J. E. McClintock, S. A. Rappaport, J. J. Nugent, F. K. Li.
Astrophys. J., Lett., Vol. 216, L15 - L18 (1977).

Observations of GX 304−1 (3U 1258−61) performed with the SAS-3 satellite on 1977 February 15 revealed regular X-ray pulsations with a period of 272 s. Average pulse profiles, derived from 7 days of observation, are presented in four energy intervals covering the range 1−19 keV. The pulse arrival times are analyzed for effects of possible orbital motion of the X-ray star, and significant constraints are placed on the orbital parameters. For example, if it is established that the companion is an early-type star with $M \gtrsim 5\,M_\odot$, then the orbital period is likely to be greater than ~ 15 days. Three 100 s flares were observed during which the source intensity rose by a factor of 4 in ~ 10 s.

142.015   Evidence for X-ray iron line emission in Cygnus X-3 obtained with a crystal spectrometer.
H. L. Kestenbaum, K. S. Long, R. Novick, M. C. Weisskopf, R. S. Wolff.
Astrophys. J., Lett., Vol. 216, L19 - L21 (1977).

The authors report evidence of X-ray Fe line emission from Cyg X-3. The energy of the line is 6.84 ±0.17 keV and suggests Lα emission from Fe XXVI but does not rule out the resonance transition of Fe XXV or K-line emission from other highly ionized states of Fe. The results exclude K-line fluorescence from neutral Fe. The best estimate of the intrinsic width is 250 eV, although the data are also consistent with any line width less than or equal to 600 eV.

142.016   X-ray spectra of Hercules X-1. II. The pulse.
S. H. Pravdo, E. A. Boldt, S. S. Holt, P. J. Serlemitsos.
Astrophys. J., Lett., Vol. 216, L23 - L26 (1977).

The X-ray spectrum of Hercules X-1 was observed in the energy range 2−24 keV with sufficient temporal resolution to allow detailed study of spectral correlations with 1.24 s pulse phase. A well-defined segment of $\sim 1/10$ the pulse light curve is found to exhibit a marked spectral evolution. Its edges are the sharpest temporal features in the pulse, with a duration of no more than 20 ms, and this portion of the light curve may be directly associated with the intrinsic underlying beam.

142.017   Multicolor linear and circular polarization of AM Herculis. I. A preliminary geometrical model.
J. J. Michalsky, G. M. Stokes, R. A. Stokes.
Astrophys. J., Lett., Vol. 216, L35 - L39 (1977).

Observations of the linear and circular polarization of the peculiar X-ray source AM Her are presented. Assuming that the source is a binary star, the combination of polarization data

with the radial velocity, photometric, and X-ray data of many other authors severely constrains some of the geometrical properties of the system. The observations further suggest that AM Her may be the first object in which the linear polarization p is related to the circular polarization q by the expression $p \sim q^2$, as predicted by the gray-body magnetoemission model of Kemp.

142.018   OSO-8 X-ray observations of AM Herculis.
J. Swank, M. Lampton, E. Boldt, S. Holt, P. Serlemitsos.
Astrophys. J., Lett., Vol. 216, L71 - L74 (1977).

The authors report the results of X-ray observations of the binary system AM Her, which were coincident with soft X-ray and ground-based optical measurements. In the 2−60 keV band, variability was detected with an eclipse during phases 0.5 to 0.7 with respect to the 0.^{d}12892 period optical minima, synchronous with the known soft X-ray eclipse. The 2−60 keV uneclipsed flux was 9.5×10^{-10} ergs cm^{-2} s^{-1}, of which 86% lies above 10 keV. Thus AM Her contains a hard source located near the similarly eclipsed soft X-ray source. The X-ray data are interpreted in terms of thermal bremsstrahlung from accretion onto a white dwarf.

142.019   On the correlation between hardness and pulsation in galactic X-ray sources.
L. Maraschi, A. Treves, E. P. J. van den Heuvel.
Astrophys. J., Vol. 216, 819 - 821 (1977).

It is pointed out that for galactic X-ray sources there exists an observational association between hard spectra ($kT \gtrsim 20$ keV), pulsation, and early-type optical counterparts. The absence of pulsation in soft X-ray sources is attributed to the fact that photons produced at the surface of a magnetic neutron star are Compton scattered in a surrounding cloud which is heated by the Compton energy losses of the hardest photons.

142.020   Structure of the X-ray source in the Virgo cluster of galaxies.
P. Gorenstein, D. Fabricant, K. Topka, W. Tucker, F. R. Harnden, Jr.
Astrophys. J., Lett., Vol. 216, L95 - L99 (1977).

High angular resolution observations in the 0.15−1.5 keV band with an imaging X-ray telescope show the extended X-ray source in the Virgo cluster of galaxies to be a diffuse halo of about 15' core radius surrounding M87. The angular structure of the surface brightness is marginally consistent with either of the two simple models: (1) an isothermal (or adiabatic or hydrostatic) sphere plus a point source at M87 accounting for 12% of the total 0.5−1.5 keV intensity, or (2) a power-law function without a discrete point source.

142.021   Spectral characteristics of 3U 1915−05, a burst source candidate.
R. H. Becker, B. W. Smith, J. H. Swank, E. A. Boldt, S. S. Holt, S. H. Pravdo, P. J. Serlemitsos.
Astrophys. J., Lett., Vol. 216, L101 - L104 (1977).

An X-ray burst source has been discovered near the X-ray source 3U 1915−05. The continuum spectra of both the burst source and the quiescent source 3U 1915−05 are hard, with kT for thermal bremsstrahlung models above 20 keV. The spectrum of 3U 1915−05 has a feature at 9.1 keV, which, if attributed to absorption by hydrogen and helium-like iron, suggests the presence of a highly ionized cloud surrounding a central X-ray source.

142.022   Discovery of the X-ray burst source MXB1659−29.
J. Doty, J. A. Hoffman, W. H. G. Lewin.
Bull. American Astron. Soc., Vol. 9, 434 (1977). − Abstract.

142.023   A simple mathematical simulacrum of the X-ray burster phenomenon.   L. M. Celnikier.
Astron. Astrophys., Vol. 60, 421 - 422 (1977).

The author points out that the essential features of a rapid X-ray burster can be reproduced by a simple, determinate, non-linear "population dynamics" equation.

142.024 Constraints on the apsidal motion of some galactic X-ray sources.   E. G. Tanzi, A. Treves.
Astron. Astrophys., Vol. 60, 431 - 432 (1977).

The possibility that the periodicities of the order of 10 days observed in Cyg X-3, Cyg X-2 and A0620−00 are due to apsidal precession is examined. It is found that the required eccentricity of the binary systems must be $e = 0.5–0.7$.

142.025 Further spectrophotometry of the transient X-ray source A0620−00.   J. B. Oke.
Astrophys. J., Vol. 217, 181 - 185 (1977).

Observations of the transient X-ray source A0620−00 in 1976 November show that the optical object had an apparent visual magnitude $V = 18.35$. The spectrum is cool and shows that one component of the light comes from a K5 V to K7 V star. The distance is estimated to be 870 pc if the cool star is normal. Some support for the 7.8 X-ray period recently found is provided by observations made in 1975 during the outburst. In addition to the cool K star, there is a second radiation source which has a spectrum of the form $f_v = $ const. It is possible that the K star is losing mass by processes such as those that occur in T Tauri stars.

142.026 The spectroscopic orbit and masses of SK 160/SMC X-1.
J. B. Hutchings, D. Crampton, A. P. Cowley, P. S. Osmer.
Astrophys. J., Vol. 217, 186 - 196 (1977) = Dominion Astrophys. Obs. No. 330 = NRC No. 15612.

Thirty-one spectrographic observations have been obtained of SK 160, with dispersions from 12 to 52 Å mm^{-1}. Radial velocities from these lead to a spectroscopic orbit which implies masses of $16.2 \pm 1.4 M_\odot$ and $1.02 \pm 0.20 M_\odot$ for the two components, and $i = 64°–70°$ for the system. Mass exchange appears to be by Roche lobe overflow rather than a strong stellar wind.

142.027 A model for bursting X-ray sources: time-dependent accretion by magnetic neutron stars and degenerate dwarfs.   F. K. Lamb, A. C. Fabian, J. E. Pringle, D. Q. Lamb.
Astrophys. J., Vol. 217, 197 - 212 (1977).

The authors consider spherically symmetric accretion flow onto a strongly magnetic compact star. They show that under certain conditions such flow is intermittent, and that the resultant time-dependent X-ray emission from the stellar surface is akin to that observed in the bursting X-ray sources. A key feature of high-luminosity flows is the interaction of the burst radiation with the accreting matter, which determines the burst shape and the time between bursts. Four distinct types of burstlike flows are possible, depending on the luminosity and temperature of the burst radiation. In each type of flow the bursts are terminated by a different physical mechanism: compressional heating of matter at the magnetospheric boundary, Compton heating of matter at the boundary, Compton heating of matter far from the boundary, or radiation pressure.

142.028 Observations of the X-ray burst source MXB 1636−53.   J. A. Hoffman, W. H. G. Lewin, J. Doty.
Astrophys. J., Lett., Vol. 217, L23 - L28 (1977).

X-ray bursts have been observed from MXB 1636−53, almost certainly associated with the strong, steady X-ray source 2S 1636−536 (Norma X-1, 3U 1636−53). The steady source was observed in 1977 January at roughly half the intensity reported in the 3U catalog and showed ~15% variability on a time scale of hours. The spectra of the X-ray bursts are well fitted by blackbody radiation whose temperature rises rapidly to a maximum of ~28×10^6 K and then cools slowly. If the source is at a distance of 10 kpc, the radius of the projected burst emission region is ~10 km, similar to the size of a neutron star.

142.029 Discovery of a 7.68 second X-ray periodicity in 3U 1626−67.   S. Rappaport, T. Markert, F. K. Li, G. W. Clark, J. G. Jernigan, J. E. McClintock.
Astrophys. J., Lett., Vol. 217, L29 - L33 (1977).

SAS-3 observations of the X-ray source 3U 1626−67 have revealed the presence of a stable 7.68 pulse period. This source was selected for study because of its hard X-ray spectrum. The compilation of source spectra used in the selection process is also presented. Pulse arrival times are analyzed for effects of possible binary orbital motion. Upper limits to the projected orbital radius were obtained which tend to exclude orbital periods in the range $0.5 \lesssim P_{orb} \lesssim 35^d$. Binary systems with either a very long orbit ($\gtrsim 175^d$) or a very short orbit ($\lesssim 0.3$) are most probable.

142.030 Pulse profile and refined orbital elements for SMC X-1.   F. Primini, S. Rappaport, P. C. Joss.
Astrophys. J., Vol. 217, 543 - 548 (1977).

The results of a reanalysis of SAS-3 data from SMC X-1, including greatly improved orbital elements, are presented. These elements, when combined with newly available data for the optical companion Sk 160, yield improved estimates for the masses of SMC X-1 and Sk 160. A significant rate of change of intrinsic pulse period, $\dot{P}$, is detected during the course of the 4 day observation. The X-ray pulse profile, averaged over one orbital cycle and with ~15 ms time resolution, is also presented. The implications of the small observed eccentricities of SMC X-1 and other binary X-ray pulsar systems are discussed.

142.031 X-ray clusters of galaxies: correlations with optical morphology and galaxy density.   N. A. Bahcall.
Astrophys. J., Lett., Vol. 217, L77 - L82 (1977).

The entire sample of 37 X-ray clusters of galaxies detected by the Ariel 5 satellite at high latitudes is analyzed optically. The author determines morphological types (for all clusters), and central galaxy densities (for all clusters with measured redshifts). He also estimates the positions, redshifts, and richness groups of the set of southern hemisphere X-ray clusters. The X-ray luminosity is found to be correlated with both cluster morphology and central galaxy density. The observed correlations are consistent with a thermal bremsstrahlung model in which the hot intracluster gas density is proportional to the galaxy density and the temperature is proportional to the galaxy velocity dispersion.

142.032 On the true space distribution of the galactic X-ray sources.   T. Matilsky.
Astrophys. J., Lett., Vol. 217, L83 - L85 (1977).

The Uhuru catalog of X-ray sources is used to examine the latitude dependence of the galactic X-ray emitters. No preference for any plane other than the galactic equator is exhibited. However, the bright and faint sources have distinctly different dispersions about the plane. This may be related to the possible existence of two distinct classes of galactic X-ray sources. The X-ray burst sources and the strong, galactic center objects have almost identical dispersions.

142.033 Hercules X-1: the 1.24 second pulsation in hard X-rays.   E. Kendziorra, R. Staubert, W. Pietsch, C. Reppin, B. Sacco, J. Trümper.
Astrophys. J., Lett., Vol. 217, L93 - L96 (1977).

A 4 hr balloon observation of Hercules X-1 during the ON state in the 35 day cycle was performed on 1976 May 3. The 1.24 s pulsations show a pulsed fraction of 58% ± 8% in the 18−31 keV interval. A pulsed flux was discovered in the 31−88 keV range with a pulsed fraction of 51% ± 14%. In the 18−60 keV range the 1.24 s pulsations are characterized by single-peaked pulses, showing no significant change with energy of pulse shape and phase of the pulse maximum. At energies above 60 keV the authors find evidence for a double-peaked pulse profile.

142.034 Positions of galactic X-ray sources Cir X-1, TrA X-1 and 3U1626—67. H. V. Bradt, K. M. V. Apparao, R. Dower, R. E. Doxsey, J. G. Jernigan, T. H. Markert. Nature, Vol. 269, 496 - 497 (1977).

The results reported here support recently proposed optical and radio identifications of these three sources.

142.035 Limitations on the nature and evolution of X-ray binaries. D. S. P. Dearborn. Astrophys. Lett., Vol. 19, 15 - 18 (1977).

Observations of the surface CNO abundances are used to place restrictions on the evolutionary scenarios for producing X-ray binaries. The lack of ^{14}N enhancement is used to infer that the system lost large amounts of mass prior to the supernovae. It is also shown that the current secondaries have lost less than ~35 percent of their original mass.

142.036 The high-energy X-ray spectrum of the Crab Nebula observed from OSO-8. J. F. Dolan, C. J. Crannell, B. R. Dennis, K. J. Frost, G. S. Maurer, L. E. Orwig. Astrophys. J., Vol. 217, 809 - 814 (1977).

The X-ray spectrum of the Crab Nebula was measured with the scintillation spectrometer on board the OSO-8 satellite between 1976 March 9 and 23. The spectrum is well fitted from 23 to 513 keV by a power law of the form $dN/dE = C(E/E_0)^{-\alpha}$ photons $cm^{-2} s^{-1} keV^{-1}$ with $C = (4.19 \pm 0.14) \times 10^{-3}$ photons $cm^{-2} s^{-1} keV^{-1}$, $E_0 = 39.1$ keV, and $\alpha = 2.00 \pm 0.06$. No indication is found of either day-to-day or long-term variability in the observed X-ray spectrum of the source.

142.037 New observations and a slow rotator model of the X-ray binary AM Herculis. H. S. Stockman, G. D. Schmidt, J. R. P. Angel, J. Liebert, S. Tapia, E. A. Beaver. Astrophys. J., Vol. 217, 815 - 831 (1977).

The soft X-ray binary AM Her/3U 1809 + 50 is known to have a common 3.09 hour period in its soft X-ray flux and in its optical flux, polarization, and spectral lines. This paper confirms the presence of strong circular polarization which varies synchronously with the common period. Spectroscopic observations varying from moderate-resolution spectrophotometry to high-resolution photographic spectra are presented. On the assumption that AM Her is at a distance of 100 pc, the total luminosity, ~10^{33} ergs s^{-1}, is small compared to typical X-ray binaries. This fact, the unusually soft X-ray spectrum, and the unique strong circular polarization suggest that AM Her is a binary system in which material is accreting onto a magnetic white dwarf with a surface field of ~10^8 gauss. A model is developed of a system in which the magnetic white dwarf's rotation is synchronized with the 3 hour binary period.

142.038 X-ray bursters and dwarf novae: a correspondence. K. Brecher, P. Morrison, A. Sadun. Astrophys. J., Lett., Vol. 217, L139 - L141 (1977).

The authors compare directly the properties of the X-ray burster MXB 1730—335 with those of the dwarf nova SS Cygni. The striking observational parallels further support the idea that X-ray bursts are produced by mass transfer in binary star systems containing a neutron star member.

142.039 Observations of the soft X-ray background. D. P. Cox. Topics in interstellar matter, (see 012.011), p. 17 - 25 (1977).

142.040 Soft X-ray sources. P. Gorenstein, W. H. Tucker. Annu. Rev. Astron. Astrophys., Vol. 14, (see 003. 008), 373 - 416 (1976).

This review addresses itself to objects that have already been detected as discrete soft X-ray sources. They include supernova remnants, a recurrent nova, flare stars, the coronas of nearby stars, SS Cygni, a white dwarf, and a soft component of emission from binary X-ray stars. An intense, diffuse cosmic background is characteristic of the soft X-ray region.

142.041 Observational constraints on the models for Cygnus X-3. D. R. Parsignault, J. Grindlay, H. Gursky, W. Tucker. Astrophys. J., Vol. 218, 232 - 242 (1977).

Observations of Cygnus X-3 made by the ANS and Uhuru satellites have shown significant X-ray flux variability on a time scale of hours. No spectral change was observed associated with these fluctuations. The authors find significant intensity and spectral changes on a time scale of months with an inverse correlation between the source intensity and the temperature, and possibly the low-energy cutoff. The spectrum studies have confirmed the existence of an excess above the fitted continuum which can be interpreted as line emission of Fe XXIV and/or Fe XXV at ~6.7 keV. Phenomenological models for Cygnus X-3 are discussed; the observational data are best interpreted by assuming that a scattering and absorbing shell is symmetrically placed around the X-ray source.

142.042 Optical candidates for 3U 1538—52. A. P. Cowley, D. Crampton, J. B. Hutchings, W. Liller, N. Sanduleak. Astrophys. J., Lett., Vol. 218, L3 - L6 (1977) = Dominion Astrophys. Obs. Victoria, B.C., Contrib. No. 344 = NRC No. 16103.

Examination of Schmidt plates of the field surrounding 3U 1538—52 reveals several possible optical candidates. Discussion of existing data indicates that the optical counterpart is likely to be a reddened OB supergiant. The most probable stars, based on position, color, and magnitude, are pointed out.

142.043 Evidence for an accretion disk in SMC X-1. J. van Paradijs, E. Zuiderwijk. Astron. Astrophys., Vol. 61, L19 - L22 (1977).

An analysis of five-colour photometric observations of Sk 160 (SMC X-1) gives evidence for the existence of an accretion disk around the neutron star in this system. This supports the suggestion by Hutchings et al. (1977) that the mass transfer in SMC X-1 occurs through Roche-lobe overflow.

142.044 Non-ejecting novae as EUV sources. M. M. Shara, D. Prialnik, G. Shaviv. Astron. Astrophys., Vol. 61, 363 - 367 (1977).

Low mass (few $\times 10^{-5} M_\odot$) hydrogen-rich envelopes on the surface of white dwarfs can yield Eddington luminosities (few $\times 10^4 L_\odot$) radiated in the extreme ultraviolet (~100 Å) for about 100 years. About 1000 such sources may exist in our Galaxy, the nearest being ~400 pc away. These EUV objects' energy contribution to the interstellar medium should match that of the sdO stars and cooling white dwarfs.

142.045 The distance of Cir X-1. W. M. Goss, U. Mebold. Mon. Not. R. Astron. Soc., Vol. 181, 255 - 258 (1977).

During the Cir X-1 radio outbursts of 1976 December 14 - 19, H I absorption observations were made with the Parkes 64-m telescope. The results were compared with a profile obtained on 1976 December 6, when the radio source was quiescent, and it was found that H I absorption extends to the velocity associated with the tangential point. The resulting lower limit to the distance is ~8 kpc, about a factor of two larger than previous estimates.

142.046 The optical and radio counterpart of Circinus X-1 (3U 1516 - 56). J. A. J. Whelan, S. K. Mayo, D. T. Wickramasinghe, P. G. Murdin, B. A. Peterson, T. G. Hawarden, A. J. Longmore, R. F. Haynes, W. M. Goss, L. W. Simons, J. L. Caswell, A. G. Little, W. B. McAdam.

Mon. Not. R. Astron. Soc., Vol. 181, 259 - 271 (1977).

Circinus X-1 (3U1516 – 56) has a radio counterpart which, at high frequencies, shows flares with the same 16.6-day periodicity as the X-ray intensity. In each cycle the radio flare occurs shortly after the intensity drop-off which defines the X-ray modulation. The radio source is positionally coincident with a faint red star having very strong Hα and weak He I emission lines which are probably variable. The object may be an early-type emission-line star or a symbiotic star, at a distance of 10 kpc.

142.047   A spherical model for the transient X-ray source A0620–00.
C. Dilworth, L. Maraschi, G. C. Perola, E. G. Tanzi.
Mon. Not. R. Astron. Soc., Vol. 181, 339 - 345 (1977).

The continuum spectrum of the transient X-ray source A0620–00, from infrared to X-ray frequencies, is interpreted as emission from a uniform spherical cloud of hot gas in which the free–free spectrum is modified by Thomson scattering. On this basis, the radius and the density of the cloud, and the distance of the source, are derived.

142.048   X-ray stars in globular clusters.   G. W. Clark.
Sci. American, Vol. 237, No. 4, p. 42 - 55 (1977).

The dense central cores of some such clusters favor the formation of X-ray-emitting double stars in which a neutron star or black hole accretes matter from a star still consuming its nuclear fuel.

142.049   Temporal and spectral variation of Cyg X-1.
Y. Ogawara, K. Doi, M. Matsuoka, S. Miyamoto, M. Oda.
Nature, Vol. 270, 154 - 156 (1977).

Among galactic X-ray sources, Cyg X-1 is distinctive because of its marked variability. The authors report certain features of the subsecond variability concluded from a recent rocket observation.

142.050   Absence of detectable X rays from supercluster candidate 4U0134–11.   S. H. Pravdo,
R. F. Mushotzky, R. H. Becker, E. A. Boldt, S. S. Holt, P. J. Serlemitsos, J. H. Swank.
Nature, Vol. 270, 158 - 159 (1977).

The Goddard Space Flight Center Cosmic X-ray Spectroscopy experiment onboard OSO 8 scanned directly over the position of 4U0134–11. The authors did not observe X rays from this source with an upper limit to the source strength a factor of 6 below the 4U strength.

142.051   What are the bursting X-ray sources telling us?
R. N. Henriksen.
J. R. Astron. Soc. Canada, Vol. 71, 393 - 394 (1977). Abstract.

142.052   Massive X-ray binaries.   J. B. Hutchings.
J. R. Astron. Soc. Canada, Vol. 71, 406 (1977). Abstract.

142.053   On the nature of H 1-36 (3U 1746-37?).
C. R. Purton, S. Jeffers, W. Weller.
J. R. Astron. Soc. Canada, Vol. 71, 406 (1977). – Abstract.

142.054   Spectrum variations of the X-ray binary HD 153919 (= 3U 1700-37).
R. G. Carlberg, G. G. Fahlman, G. A. H. Walker.
J. R. Astron. Soc. Canada, Vol. 71, 408 (1977). – Abstract.

142.055   Periods in X-ray sources.   P. J. N. Davison.
Highlights of Astronomy, Vol. 4, Part I, (see 012.018), p. 75 - 86 (1977).

The author concentrates on recent measurements on X-ray sources showing Doppler modulated periodicities, and

on sources which may well show such changes when they are studied more closely.

142.056   Recent transient X-ray sources.   A. P. Willmore.
Highlights of Astronomy, Vol. 4, Part I, (see 012.018), p. 87 - 94 (1977).

Of 14 transients, five form a rather well defined class whose properties are rapidly becoming clearer. This class the author will call the "classical" transient. The remainder form a much more miscellaneous collection which cannot be clearly distinguished from the normal X-ray sources. There is some indication that the classical transients include two species. The author describes one of each which has been particularly well-observed by Ariel V and SAS-3.

142.057   OSO-8 X-ray polarimeter and Bragg crystal spectrometer observations.
R. Novick, H. L. Kestenbaum, K. S. Long, E. H. Silver, M. C. Weisskopf, R. S. Wolff.
Highlights of Astronomy, Vol. 4, Part I, (see 012.018), p. 95 - 97 (1977) = Columbia Astrophys. Lab. Contrib. No. 127.

142.058   OSO-8 observations of Cygnus XR-1.
J. F. Dolan, B. R. Dennis, C. J. Crannell, K. J. Frost, L. E. Orwig.
Highlights of Astronomy, Vol. 4, Part I, (see 012.018), p. 99 - 100 (1977).

142.059   X-ray bursts.   G. W. Clark.
Highlights of Astronomy, Vol. 4, Part I, (see 012.018), p. 101 - 110 (1977).

142.060   Globular cluster X-ray sources.   J. E. Grindlay.
Highlights of Astronomy, Vol. 4, Part I, (see 012.018), p. 111 - 122 (1977).

The author reviews the observational data and then the models of globular cluster X-ray sources in an effort to point out the present balance of evidence for the binary vs. black hole models as well as the most promising directions for future work. Given the possibly high incidence of X-ray bursters in globular clusters, the author's discussion will refer to both observations and recent models for bursters.

142.061   A model for bursting X-ray sources.
R. N. Henriksen.
Highlights of Astronomy, Vol. 4, Part I, (see 012.018), p. 123 (1977).

142.062   Bursting X-ray sources: a theoretical framework for accretion models.   F. K. Lamb, A. C. Fabian,
J. E. Pringle, D. Q. Lamb.
Highlights of Astronomy, Vol. 4, Part I, (see 012.018), p. 125 (1977).

142.063   X-ray bursts of nuclear origin?
L. Maraschi, A. Cavaliere.
Highlights of Astronomy, Vol. 4, Part I, (see 012.018), p. 127 - 128 (1977).

142.064   Optical observations of X-ray binaries.
J. B. Hutchings.
Highlights of Astronomy, Vol. 4, Part I, (see 012.018), p. 129 - 136 (1977).

Recent progress in identifying the optical counterparts of X-ray sources has been slow, mostly because candidates are faint and X-ray data show no periodicities by which the identification can be confirmed. This report therefore deals with the investigation of some candidates to seek confirmation of their identity with the X-ray emitters, as well as details which are new or important concerning the few known binary sources. The author also makes some general remarks on the properties of the sources as they now appear.

142.065 Masses of compact X-ray sources.   Y. Avni.
Highlights of Astronomy, Vol. 4, Part I, (see
012.018), p. 137 - 143 (1977).
The author presents the observational ingredients used
in mass determinations, discusses their uncertainties, and lists
new results obtained recently. He summarizes the derived
mass ranges for the identified X-ray binaries, and other important conclusions.

142.066 The evolutionary history of X-ray binaries.
E. P. J. van den Heuvel, G. J. Savonije.
Highlights of Astronomy, Vol. 4, Part I, (see 012.018), p.
145 - 153 (1977).
The most important recent observational discoveries in
the field of X-ray binaries are probably those of the slow
pulsars and of the winds of normal early-type main-sequence
stars. These facts yield key information on the evolutionary
history of the X-ray binaries and on the rotational slow-down
mechanism for a neutron star in a stellar wind. In the theoretical field, the X-ray binaries have triggered much fundamental work, notably on the detailed processes of mass
transfer and on tidal evolution.

142.067 Accretion flows in binary X-ray systems.
R. McCray.
Highlights of Astronomy, Vol. 4, Part I, (see 012.018), p.
155 - 170 (1977).
The subject of accretion flows in binary X-ray systems
has been reviewed recently by Lightman et al. (1977), and
the subject of accretion disks by Shu (1976). The author
emphasizes developments that have occurred since those
reviews were written, and concentrates on issues concerning
the gas flows rather than the radiation mechanisms.

**142.068 The spectroscopic orbit and the masses of the com-
ponents of the binary X-ray source 3U0900—40/HD
77581.**   J. van Paradijs, E. J. Zuiderwijk, R. J. Takens,
G. Hammerschlag-Hensberge, E. P. J. van den Heuvel, C. de
Loore.
Astron. Astrophys., Suppl. Ser., Vol. 30, 195 - 211 (1977).
Radial velocity measurements are presented of the B0.5 Ib
supergiant HD 77581, the optical counterpart of the binary
X-ray system 3U0900—40 (Vela X-1). A description is given of
a new computer programme for the determination of the
orbital parameters of spectroscopic binaries from both radial
velocity and pulse-time delay data. New orbital parameters for
the Vela X-1 system were determined from spectral lines of
different ions. The differences between the orbital elements
obtained from lines of different ions are in some cases not in-
significant. The small discrepancy between the orbital parameters of HD 77581 and those derived from X-ray data is still
within the range expected from the distortion of the radial
velocity curve of the primary due to its tidal and rotational
deformation. Clear evidence is found for correlated night-to-
night variations of the radial velocity. The implications of these
non-orbital effects on the mass determination are discussed.

142.069 X-ray burst sources.   W. H. G. Lewin, P. C. Joss.
Nature, Vol. 270, 211 - 216 (1977).
More than 30 X-ray burst sources have been discovered in
the past 2 years, and information on the observational properties of the burst emission is accumulating rapidly. The physical
mechanism responsible for the bursts is so far undetermined,
but the most promising models are based on the accretion of
matter onto a compact object of stellar mass.

142.070 Evidence for a 39-d period in Cyg X-1.
J. C. Kemp, L. C. Herman, R. J. Rudy, M. S. Barbour.
Nature, Vol. 270, 227 - 228 (1977).
The detection of a 39-d period in Cyg X-1 (HDE226868),
as seen in the ultraviolet (ultraviolet-filter) polarisation would
have at least one interesting interpretation, that of a third-body

orbital period. The authors report here the result of long-term
monitoring of the optical polarisation of Cyg X-1. Power
spectra of the ultraviolet-filter polarisation suggested a peak at
the approximate period 39 d.

142.071 Possible 39-d polarisation period in Cyg X-1.
M. Milgrom, J. Shaham.
Nature, Vol. 270, 228 - 229 (1977).
Kemp has reported that the power spectra of the Q and U
Stokes polarisation parameters, for the U band radiation from
Cyg X-1, show a $\sim 3\sigma$ peak at 39.26 ± 0.10 d. The peak was
found to represent a modulation in which Q and U vary, in
phase, at 0.27% peak-to-peak total polarisation. While this was
reported to be the strongest U band periodicity in the frequency range considered, it was not found in the B and V
bands. Further observations are necessary to set limits on this
modulation for the various radiation bands. This letter suggests
that such variable polarisation component may arise from
those optical photons which, after being radiated by the disk,
are Thomson scattered into our line of sight by the flared
material at the outer parts of the disk and in the incoming
flow.

142.072 The 39-d period in Cyg X-1.
E. N. Walker, M. G. Watson, S. S. Holt.
Nature, Vol. 270, 229 - 230 (1977).
Kemp et al. recently claimed to have discovered a 39-d
period in the linear polarisation of the U band light for
HDE 226868, the optical counterpart of Cygnus X-1. Within
the errors their period is exactly seven times the 5.60-d orbital
period of this binary system. Based on this result Milgrom and
Shaham produced a model for Cyg X-1 involving a three-body
system in which the third component is in a 39-d orbit about
the 5.6-d X-ray binary system. This note draws attention to
the absence of a 39-d period in other data on this star.

**142.073 On Compton and thermal models for X-ray emission
from clusters of galaxies.**   Y. Rephaeli.
Astrophys. J., Vol. 218, 323 - 332 (1977).
Aspects of Compton and thermal models for cluster X-ray
emission are discussed. Some restrictions on the origin and
propagation of the relativistic electrons, whose existence in
large numbers is postulated in the Compton model, are imposed; and implications of the model are discussed. Within the
context of a thermal interpretation of the X-ray emission, the
author investigates the possible role of heat conduction in determining the state of the intracluster gas. A hydrostatic-con-
ductive model is treated under different assumptions on the
nature of possible heating mechanisms.

**142.074 Photometry of slow X-ray pulsars. II. The 13.9
minute period of X Persei.**
B. Margon, J. R. Thorstensen, S. Bowyer, K. O. Mason, N. E.
White, P. W. Sanford, G. Parkes, R. P. S. Stone, J. Bailey.
Astrophys. J., Vol. 218, 504 - 510 (1977).
Using spectrophotometry and narrow-band photometry,
the authors have searched unsuccessfully for an optical analog
of the 13.9 minute X-ray periodicity of 3U 0352+30 in the
bright Be star X Persei. Simultaneous X-ray observations
utilizing OAO Copernicus prove that the strong X-ray periodicity is present during the photometry; thus variability of the
X-ray pulse cannot be responsible for this discrepancy. A
simple energetic argument is used to demonstrate that the
previously reported $\lambda4686$ variations are of such large amplitude that they are probably irreconcilable with the canonical
model of X-ray binaries, where the X-ray source is the unseen
spectroscopic secondary and where synchronous optical
variations are due to the reflection effect.

**142.075 Spatial structure in the soft X-ray background as
observed from OSO-8, and the North Polar Spur as
a reheated supernova remnant.**

R. J. Borken, DeA. C. Iwan.
Astrophys. J., Vol. 218, 511 - 520 (1977).

Soft X-ray sky maps are presented in three channels spanning 0.17 to 1.8 keV, covering parts of the area $l \approx 300°$ to 60° and $b \approx 15°$ to 60°. Taken with an angular resolution of $\sim 3°$ and high exposure, these maps show detailed structure in several emission features in the soft X-ray background, including the North Polar Spur and "hot spots" in Hercules, Centaurus, and Ophiuchus. The structure of the X-ray emission in association with the neutral hydrogen and radio continuum emission from the North Polar Spur is discussed. Finally, evidence is presented showing that the North Polar Spur could be an old, but reheated, supernova remnant.

142.076   Detection of [Fe XIV] emission in HD 153919 (3U 1700–37).
A. K. Dupree, S. L. Baliunas, J. B. Lester.
Astrophys. J., Lett., Vol. 218, L71 - L74 (1977).

Spectra of HD 153919 show a previously unreported broad emission line near 5293 Å that is present at phase $\phi = 0.2$ and $\phi = 0.8$ and is absent or weaker at other phases, including the time of eclipse of the X-ray source. Approximate calculations suggest identification of this feature with the [Fe XIV] transition at 5303 Å. This line may arise in the extended atmosphere of the primary ionized by the compact X-ray source or may occur in a high-temperature corona.

142.077   X-rays in astronomy.   H. W. Schnopper.
Inner shell ionization phenomena, Part 1, 2nd International Conference, Freiburg, F. R. Germany, 29 March - 2 April 1976. W. Mehlorn, R. Breun (Editors). Published by Univ. Freiburg, F. R. Germany. Price DM 25.00 (1976), p. 14 - 20. – Abstr. in Phys. Abstr., Vol. 80, Abstr. 22061 (1977).

142.078   X-ray bursts and neutron-star thermonuclear flashes.
P. C. Joss.
Nature, Vol. 270, 310 - 314 (1977).

Some of the properties of thermonuclear flashes that should occur in the surface layers of accreting neutron stars are investigated. Such flashes may account for many of the observed properties of X-ray burst sources. Helium seems to be the most promising type of nuclear fuel for producing flashes that result in X-ray bursts.

142.079   Optical candidates for two X-ray bursters and an X-ray pulsar.   J. E. McClintock, C. R. Canizares, H. V. Bradt, R. E. Doxsey, J. G. Jernigan, W. A. Hiltner.
Nature, Vol. 270, 320 - 321 (1977).

The authors suggest faint ($B \sim 18$), blue stars as the optical counterparts of two X-ray bursters, 4U 1636–53 (MXB1636–53) and 4U1735–44 (MXB1735–44; KGX345–6?), and for the 7-s X-ray pulsar, 4U1626–67. The candidate stars have large ultraviolet excesses and were discovered well within the 20″ and 30″ X-ray error radii determined using the rotating modulation collimator experiment aboard the SAS-3 X-ray Observatory.

142.080   Positions of three X-ray burst sources.
J. G. Jernigan, K. M. V. Apparao, H. V. Bradt, R. E. Doxsey, J. E. McClintock.
Nature, Vol. 270, 321 - 323 (1977).

Precise (20–30″) positions of three steady X-ray sources which have been identified recently as X-ray burst sources 4U1636–53 = MXB1636–53, 4U1728–33 = MXB1728–34, and 4U1735–44 = MXB1735–44 are reported.

142.081   Extragalactic X-ray sources.
H. Gursky, D. A. Schwartz.
Annu. Rev. Astron. Astrophys., Vol. 15, (see 003.012), 541 - 568 (1977).
Contents: Active galaxies; The unidentified high-latitude sources; X-ray emitting clusters of galaxies; The diffuse X-ray background; Prospects for future observations.

142.082   Photographic observations of the X-ray nova Monocerotis 1975.   R. Hudec.
Bull. Astron. Inst. Czechoslovakia, Vol. 28, 374 - 375 (1977).

The B magnitudes of X-ray nova Monocerotis 1975 were measured on fourteen photographic plates taken with the astrograph and Tessar camera at the Sonneberg Observatory.

142.083   Optical behaviour of HZ Her/Her X-1 in 1976.   R. Hudec.
Bull. Astron. Inst. Czechoslovakia, Vol. 28, 376 (1977).

The photographic magnitudes of the HZ Her = Her X-1 system were measured on 30 astrograph plates in 1976. The system remains in its active state.

142.084   The 580-day and other periodicities in X Persei.   J. B. Hutchings.
Mon. Not. R. Astron. Soc., Vol. 181, 619 - 624 (1977).

Spectroscopic observations for 1975–76 are presented, together with some photometric measurements. The 580-day periodic variation in velocity is much less marked, but still significant in the 1920–76 data sample. Searches were made for 22.4-hr periodicity and upper limits for velocity variations are derived. Proposed binary models for the star are discussed.

142.085   The energy distribution and the recent light history of X Persei.
P. Persi, R. Viotti, M. Ferrari-Toniolo.
Mon. Not. R. Astron. Soc., Vol. 181, 685 - 693 (1977).

The aim of this paper is to study the energy distribution of X Per during two different epochs with particular regard to the infrared spectrum and the nature of the luminosity variations. For this purpose the authors discuss the colour excess of the star and the Balmer discontinuity, and derive the physical parameters of the Be star. The infrared energy distribution of X Per, and the long-term variability are discussed. The proposed model for X Per is described and the origin of the X-ray emission is discussed.

142.086   The binary X-ray pulsar 3U 1538–52.
P. J. N. Davison, M. G. Watson, J. P. Pye.
Mon. Not. R. Astron. Soc., Vol. 181, 73P - 79P (1977).

Ariel 5 SSI and Experiment C observations of 3U 1538–52 are presented which confirm the binary nature of this source and resolve its positional uncertainty. Combining Ariel 5 data with that from OSO-8 leads to improve estimates of the orbital parameters which are discussed in terms of the implications for the mass and radius of the as yet unidentified companion star.

142.087   Ariel V Sky Survey: X-ray variability of NGC 5128.   A. Lawrence, J. P. Pye, M. Elvis.
Mon. Not. R. Astron. Soc., Vol. 181, 93P - 99P (1977).

Long-term monitoring of the X-ray source associated with the nucleus of NGC 5128 has produced the most complete light curve in any frequency band for this peculiar active galaxy. The X-ray data indicate activity on at least two timescales: (1) a slow outburst with a timescale of 4–7 yr, (2) superimposed flares with a timescale of 20–50 day. This pattern of behaviour is related to other X-ray active galaxies.

142.088   Discovery of optical pulsations in HD 77581 = Vela X-1.   J. E. Steiner.
Astron. Astrophys., Vol. 61, L35 - L36 (1977).

Pulsed Hβ flux from HD 77581, the optical counterpart of Vela X-1, was detected. The observed period is P = 282.8 sec., coincident with the X-ray period. The modulation in flux is up to 2%.

142.089   Cyclotron lines in the Her X-1 spectrum: structure

and higher harmonics.
J. K. Daugherty, J. Ventura.
Astron. Astrophys., Vol. 61, 723 - 727 (1977).

Following the suggestion by Trümper that the recently observed hard X-ray peak in the Her X-1 spectrum represents cyclotron emission in a magnetic field $\sim 5 \times 10^{12}$ Gauss, the authors examine the observability of additional spectral features as predicted by the quantum theory of cyclotron radiation.

142.090 **The 35 day cycle of the X-ray binary Hercules X-1.**
        J. A. Petterson.
Astrophys. J., Vol. 218, 783 - 791 (1977).

The assumption that the clock of the 35 day period is a precession in HZ Her is tested. A model for the accretion region is constructed, based on this assumption, which consists of four distinguishable components: a tilted, twisted, optically thick accretion disk; an Alfvén shell whose optical depth varies gradually with polar angle; a hot, optically thin corona; and a time-varying accretion stream between HZ Her and the disk. The twisted shape of the disk is calculated in detail. The assembled configuration of disk, shell, corona, and stream is shown to provide an adequate qualitative explanation for practically all the 35 day related phenomena of the system. Arguments based on tidal evolution theory imply that a precession in HZ Her can only be present if a (so far unknown) mechanism has opposed the expected gradual decrease of obliquity in the system.

142.091 **Observations of galactic X-ray sources by OSO-7.**
        T. H. Markert, C. R. Canizares, G. W. Clark, D. R. Hearn, F. K. Li, G. F. Sprott, P. F. Winkler.
Astrophys. J., Vol. 218, 801 - 814 (1977).

The authors present the MIT data from the OSO-7 satellite for observations of the galactic plane between 1971 and 1974. A number of sources discovered in the MIT all-sky survey are described in detail. General results describing these observations of galactic sources are presented.

142.092 **Long-term behavior of MXB 1730–335.**
        J. E. Grindlay, H. Gursky.
Astrophys. J., Lett., Vol. 218, L117 - L120 (1977).

The rapid burster MXB 1730–335 was detected on at least two occasions in 1971 and 1972 by Uhuru, and coverage is available from late 1970 to early 1973. Combined with ANS coverage in 1975 and 1976, as well as published SAS-3 and Ariel V observations (1976 - 1977), a unique record of long-term burst activity is now available for this source. There appear to be burst-active states of $\sim$ one to two months' duration that occur with a $\sim 15\%$ duty cycle. The burst activity appears to be recurrent with a $\sim 0.5-1$ yr time scale. Implications for burster models are discussed.

142.093 **Velocity curves for broad and sharp components observed in the emission lines from AM Herculis.**
J. L. Greenstein, W. L. W. Sargent, T. A. Boroson, A. Boksenberg.
Astrophys. J., Lett., Vol. 218, L121 - L127 (1977).

The authors have obtained high-resolution observations of the profiles of the emission lines He I $\lambda 4471$ and He II $\lambda 4686$ around a complete cycle of the X-ray binary AM Her. The profiles are resolved into broad and sharp components: velocity curves are given for each component for both the He I and He II lines. Implications for models of AM Her are discussed.

142.094 **Binary star model for recurrent X-ray bursts.**
        R. Hoshi.
Prog. Theor. Phys., Vol. 56, 1772 - 1780 (1976). – Abstr. in Phys. Abstr., Vol. 80, Abstr. 33701 (1977).

142.095 **Positions of galactic X-ray sources: $0° < l^{II} < 20°$.**
        R. E. Doxsey, K. M. V. Apparao, H. V. Bradt,

R. G. Dower, J. G. Jernigan.
Nature, Vol. 270, 586 - 588 (1977).

Precise (20–25") positions of six X-ray sources located in the galactic bulge, GX1+4, GX9+9, GX3+1, GX9+1, GX13+1 and GX17+2 are reported.

142.096 **The X-ray source in binary stars and the spin down of neutron stars.**
Q.-y. Qu, Z.-r. Wang, T. Lu, L.-f. Luo.
Acta Astron. Sinica, Vol. 18, 138 - 142 (1977).

142.097 **3U 1735-44 = 2S 1735–444.**
        IAU Circ., No. 3085 (1977).

142.098 **MXB1916–05.**
        IAU Circ., No. 3087 (1977).

142.099 **3U 1626–67.**
        IAU Circ., No. 3088 (1977).

142.100 **MXB1637–53.**
        IAU Circ., No. 3088 (1977).

142.101 **Novalike X-ray source in Norma.**
        IAU Circ., Nos. 3090, 3094, 3101 (1977).

142.102 **MXB0512–40.**
        IAU Circ., No. 3092 (1977).

142.103 **X-ray bursts.**
        IAU Circ., Nos. 3095, 3117 (1977).

142.104 **LMC X-4.**
        IAU Circ., No. 3095 (1977).

142.105 **Circinus X-1.**
        IAU Circ., Nos. 3095, 3106, 3108 (1977).

142.106 **Optical counterparts of X-ray sources.**
        IAU Circ., No. 3096 (1977).

142.107 **Variable X-ray sources.**
        IAU Circ., No. 3099 (1977).

142.108 **A0538–66.**
        IAU Circ., No. 3100 (1977).

142.109 **X-ray flare.**
        IAU Circ., Nos. 3104, 3144, 3154 (1977).

142.110 **4U 1630–47.**
        IAU Circ., No. 3104 (1977).

142.111 **X-ray flare near 4U 1755–33.**
        IAU Circ., No. 3106 (1977).

142.112 **HD 77581 (Vela X-1).**
        IAU Circ., No. 3107 (1977).

142.113 **4U 1608–52.**
        IAU Circ., Nos. 3108, 3129 (1977).

142.114 **MXB1730–335.**
        IAU Circ., No. 3108 (1977).

142.115 **H 1743–32.**
        IAU Circ., No. 3113 (1977).

142.116 **Hercules X-1.**
        IAU Circ., No. 3116 (1977).

142.117 **4U 1145–61.**

IAU Circ., No. 3118 (1977).

142.118 SMC X-2 and SMC X-3.
IAU Circ., Nos. 3125, 3127, 3134, 3143, 3154 (1977).

142.119 MXB1706-43.
IAU Circ., No. 3134 (1977).

142.120 2S 0114+650.
IAU Circ., Nos. 3144, 3146 (1977).

142.121 WRA 977.
IAU Circ., No. 3146 with a correction No. 3152 (1977).

142.122 Cygnus X-1 (HDE 226868).
IAU Circ., No. 3149 (1977).

142.123 Extragalactic identification of 4U 0241+62.
IAU Circ., No. 3150 (1977).

142.124 The X-ray universe. T. Markert, T. L. Pencek.
Astronomy, Vol. 5, No. 7, p. 6 - 15, 18 - 22 (1977).
Abstr. in Phys. Abstr., Vol. 80, Abstr. 78892 (1977).

142.125 Balloon observations on the lunar occultation of the hard X-ray source in the Crab Nebula.
Y. Fukada. S. Hayakawa, I. Kasahara, F. Makino, H. Akiyama, M. Matsuoka, J. Nishimura, M. Oda, Y. Tanaka, M. Nakagawa, T. Sakurai, V. S. Iyengar, R. K. Manchanda, P. K. Kunte.
Bull. Inst. Space Aeronaut. Sci. Univ. Tokyo B, Vol. 12, 647 - 655 (1976). In Japanese. – From Phys. Abstr., Vol. 80, Abstr. 82319 (1977).

142.126 Was the bright transient X-ray source Centaurus XR-4 I a globular cluster?
F. D. Seward, W. Liller.
Publ. Astron. Soc. Pacific, Vol. 89, 696 - 698 (1977).

The globular cluster NGC 5824 lies within the error box of the transient X-ray source Cen XR-4. The characteristics of the cluster make it a likely a candidate for a globular cluster X-ray source. Observations of the transient X-ray source and the characteristics of the globular cluster are briefly reviewed. If the source originated in the globular cluster, the maximum X-ray luminosity was 3×10^{40} erg sec^{-1}, two orders of magnitude more luminous than any other observed galactic or globular cluster X-ray source. The implication of this result is that a massive black hole might exist within the cluster.

142.127 Sco X-1. S. Miyamoto, M. Matsuoka.
Space Sci. Rev., Vol. 20, 687 - 755 (1977).

The physical properties of X-ray, optical and radio emissions from Sco X-1 are reviewed. Sco X-1 is a typical X-ray source which has an optically thick hot plasma. The observational spectra of X-ray and optical emissions are consistent with theoretical ones from the hot plasma, but the radio emission shows a non-thermal feature. The restrictive conditions for the model of Sco X-1 are discussed from the observational facts. In spite of numerous observational facts on Sco X-1 further detailed and elaborate studies are necessary to understand this object and general compact X-ray sources comprehensively.

142.128 Cyg X-1/ a candidate of the black hole. M. Oda.
Space Sci. Rev., Vol. 20, 757 - 813 (1977).

Among discrete galactic X-ray sources, Cyg X-1 has been noted for its peculiar features in several respects. The most remarkable incident was that its optical identification with a spectroscopic binary HDE226868 has led to a presumption that it is a black hole. This possibility has induced continuous interests in the physical character of this source in conjunction with the nature of the black hole. The purpose of this paper is to summarize presently available pieces of knowledge on this source.

142.129 The diffuse soft X-ray sky. Astrophysics related to cosmic soft X-rays in the energy range 0.1–2.0 keV.
Y. Tanaka, J. A. M. Bleeker.
Space Sci. Rev., Vol. 20, 815 - 888 (1977).

The current status of the investigation of the soft X-ray diffuse background in the energy range 0.1–2.0 keV is reviewed. A consistent model, based on the soft X-ray brightness distribution and the energy spectrum over the sky, is derived. The observed diffuse background is predominantly of galactic origin and considered as thermal emission for the most part from a local hot region of temperature $\sim 10^6$ K which includes the solar system. Several pronounced features of enhanced emission are interpreted in terms of hot regions with temperatures up to 3×10^6 K, some of which are probably old supernova remnants. The properties of the soft X-ray emitting regions are discussed in relation to the observational results on O VI absorption.

142.130 On the lifetime of bright X-ray sources.
P. R. Amnuel (*Amnuehl'*), O. H. Guseinov (*O. Kh. Gusejnov*).
Astrophys. Space Sci., Vol. 46, L19 - L21 (1977).

The lifetime of massive X-ray binaries is $\sim(2-5) \times 10^5$ yr, this time close to the nuclear one. The lifetime of nonmassive X-ray binaries close to thermal one, $\sim(0.5-1) \times 10^7$ yr. Massive systems may be conserved at supernova explosion, the probability of the conservation of nonmassive system is $\sim(1-3) \times 10^{-3}$.

142.131 The eccentric-orbit binary model for the transient X-ray sources. T. Okuda, S. Sakashita.
Astrophys. Space Sci., Vol. 47, 385 - 395 (1977).

Eccentric-orbit binary models for transient X-ray sources are investigated. In these models, a compact star is in an eccentric orbit around a more massive star. As the compact star accretes mass from the stellar wind of the massive star, the accretion rate becomes time-dependent. The accretion rate is determined by Bondi's accretion radius, which depends on both the relative velocity of the stellar wind to the compact star and the sound velocity through the stellar wind. The authors obtain the variations of the light curves compatible with observations for the transient X-ray sources. It is likely that many transient X-ray sources are explainable by eccentric-orbit binary models.

142.132 X-ray emission behaviour of Her X-1.
R. K. Manchanda.
Astrophys. Space Sci., Vol. 50, 179 - 185 (1977).

The paper presents evidence for a steady-state flux of high-energy photons from Her X-1. The results are obtained from the analysis of the existing spectral measurements of the X-ray source. The continuum emission has three times more energy than that emitted during the pulse mode. The possible circumstances in which the source can simultaneously emit in two modes are presented within the framework of standard model generally accepted for this binary source.

142.133 On the nature of the outbursts from Aql X-1.
R. K. Manchanda.
Astrophys. Space Sci., Vol. 51, 501 - 506 (1977).

The first observations of hard X-ray emissions from Aql X-1 were reported by Kunte et al. (1973). The data were obtained during a balloon flight from Hyderabad, South India on March 9, 1973. In this note the author presents a second survey of the data as well as those reported by Brini et al. (1967) from this region of the sky, in the context of a recent analysis of the outbursts from Aql X-1 (Kaluzienski et al., 1977). It will be seen that the observations of hard X-rays impose new model constraints and suggest very interesting possibilities about the type of this source.

142.134 The optical polarization of X-ray binaries.
J. F. Dolan.
Astrophys. Space Sci., Vol. 52, 201 - 211 (1977).
   Polarimetric observations of close binaries may reveal the presence of a black-hole secondary. The Einstein photometric effect will introduce a characteristic, time-varying signature upon the interstellar polarization. For several reasons, it is concluded that the short time-scale variability in the polarization in HDE 226868 is caused by Rayleigh scattering from gas streams known to exist in the system. X Persei may have a variable polarization consistent with the predicted effect.

142.135 New bursts in astronomy.   J. Grindlay.
Mercury, Vol. 6, No. 5, p. 6 - 11 (1977).

142.136 Evidence for strong cyclotron emission in the hard X-ray spectrum of Her X-1.
J. Trümper, W. Pietsch, C. Reppin, B. Sacco, E. Kendziorra, R. Staubert.
Mitt. Astron. Ges., Nr. 42, p. 120 - 126 (1977).
   The authors have measured the energy spectrum of the 1.24 second pulses of Her X-1 in the energy range $15-125$ keV during a four hour balloon observation on May 3, 1976 from Palestine, Texas. The spectrum of the pulsed flux can be represented by an exponential with $kT = 7.9$ keV up to 50 keV. At about 53 keV a strong and rather narrow line feature occurs which the authors interpret as electron cyclotron emission from the polar cap plasma of the rotating neutron star. The corresponding magnetic field strength is 4.6×10^{12} Gauss.

142.137 X-ray sources in old stellar systems.   A. Finzi.
Bull. Israel Phys. Soc., Vol. 23, 113 (1977).
Summary. − Abstr. from INIS7721 336228.

142.138 Rentgenstaru uzliesmojumi lodveida kopās.
U. Dzērvītis.
Zvaigžņotā debess, 1977. gada pavasaris, p. 15 - 19.

Coordinated campaign to observe X-ray binaries.
See Abstr. 013.021.

Soft X-ray spectrum of a hot plasma.
See Abstr. 022.113.

On the measurement of the diffuse X- and γ-background as seen by non directional detectors.
See Abstr. 031.205.

The calibration of Bragg X-ray analyser crystals for use as polarimeters in X-ray astronomy.
See Abstr. 032.571.

X-ray astronomy.   See Abstr. 061.022.

Accretion magnetosphere stability. II. Polar cap "drip":   See Abstr. 062.008.

Line-distortion effects in OB supergiant X-ray binaries.   See Abstr. 064.015.

Accretion disk coronae and Cygnus X-1.
See Abstr. 064.035.

Linear radial and nonradial modes of oscillation of hot white dwarfs.   See Abstr. 065.010.

Critical accretion flow of gas onto compact stars.
See Abstr. 065.073.

Accretion by magnetic neutron stars. I. Magnetospheric structure and stability.   See Abstr. 066.005.

On the Zeeman splitting of X-ray lines by neutron-star magnetic fields.   See Abstr. 066.023.

Accretion by rotating magnetic neutron stars. I. Flow of matter inside the magnetosphere and its implications for spin-up and spin-down of the star.
See Abstr. 066.039.

How do neutron stars evolve?
See Abstr. 066.071.

Convective accretion disks and X-ray bursters.
See Abstr. 066.072.

Neutron stars and compact X-ray sources.
See Abstr. 066.256.

Accretion onto neutron star: the X-ray spectra and luminosity.   See Abstr. 066.328.

On a new method of separating the diffuse cosmic X-ray component in balloon-borne measurements.
See Abstr. 082.110.

***VBLUW* photometry of Sk 160 (SMC X-1).**
See Abstr. 113.016.

A spectrophotometric survey of the A0535+26 field.   See Abstr. 114.010.

Spectroscopic study of HZ Herculis.
See Abstr. 114.516.

Spectrum variations of the X-ray binary HD 153919 = 3U 1700−37.   See Abstr. 114.522.

A spectrophotometric study of ultraviolet and X-ray sources with objective prism plates.   See Abstr. 114.530.

Close binary systems of early spectral type as possible candidates of X-ray sources. I. Spectroscopic observations of X Persei = 2U0352 + 30.
See Abstr. 117.005.

Twisted accretion disks. II. Applications to X-ray binary systems.   See Abstr. 117.008.

Consequences of mass transfer in close binary systems.   See Abstr. 117.030.

Optical behaviour of HDE 226868 during a Cyg X-1 X-ray transition.   See Abstr. 119.006.

AN Ursae Majoris − another AM Herculis?
See Abstr. 121.003.

Extremely high circular polarization of AN Ursae Majoris.   See Abstr. 122.023.

Coordinated X-ray, optical, and radio observations of YZ Canis Minoris.   See Abstr. 122.024.

The low-mass X-ray binary AM Herculis.
See Abstr. 122.032.

UBVRI photometry of X Per.
See Abstr. 122.050.

***UBV* observations of CD −33°12119.**

See Abstr. 122.112.

3U 1908+00.   See Abstr. 123.005.

U Geminorum.   See Abstr. 123.011.

Spectroscopic observations of the X-ray nova A0620–00.   See Abstr. 124.201.

Theoretical spectra of the thermal X-rays from young supernova remnants.   See Abstr. 125.032.

Search for thermal X-ray line emission from the Crab Nebula.   See Abstr. 134.005.

The spectrum of H1–36 (=3U1746–37?) at radio wavelengths.   See Abstr. 141.011.

Radio and optical observations in the fields of five unidentified X-ray sources at high latitudes. See Abstr. 141.012.

A deep search for X-ray emission from radio quasars with Ariel V.   See Abstr. 141.075.

Solved and unsolved puzzles in today's radio astronomy.   See Abstr. 141.140.

A Crab pulsar model: X-ray, optical, and radio emission.   See Abstr. 141.513.

Hard X-ray spectra of cosmic gamma-ray bursts. See Abstr. 142.702.

Nature of gamma-ray bursts. See Abstr. 142.711.

Evidence for ionized hydrogen in the cores of globular clusters.   See Abstr. 154.010.

X-rays from globular clusters. See Abstr. 154.016.

Sources X dans les amas globulaires. See Abstr. 154.020.

UBV photometry of globular clusters containing X-ray sources.   See Abstr. 154.023.

Soft diffuse X-rays in the southern galactic hemisphere.   See Abstr. 157.005.

Electron scattering in X ray-emitting galaxies. See Abstr. 158.064.

Observations of NGC 4151 from Uhuru. See Abstr. 158.077.

An increase in the X-ray absorption of NGC 4151. See Abstr. 158.080.

Low energy γ-ray observation of NGC 4151. See Abstr. 158.093.

LMC X-1: a luminous extended X-ray source. See Abstr. 159.008.

The velocity dispersion of the X-ray cluster of galaxies in Centaurus (3U 1247–41).   See Abstr. 160.001.

X-ray, optical and radio data for the cluster of galaxies Klemola 44.   See Abstr. 160.003.

Subsonic accretion of cooling gas in clusters of galaxies.   See Abstr. 160.004.

Radiative regulation of gas flow within clusters of galaxies: a model for cluster X-ray sources. See Abstr. 160.008.

X-ray and radio observations of the structure of Abell 478.   See Abstr. 160.021.

The redshift and optical structure of the X-ray galaxy cluster A478.   See Abstr. 160.022.

On the spatial distribution of heavy elements in X-ray emitting clusters of galaxies.   See Abstr. 160.030.

The X-ray temperatures of eight clusters of galaxies and their relationship to other cluster properties. See Abstr. 160.031.

X-ray line emission for clusters of galaxies. II. Numerical models.   See Abstr. 160.041.

X-ray clusters of galaxies: What to plot against X-ray luminosities?   See Abstr. 160.049.

X-rays from clusters of galaxies. See Abstr. 160.050.

X-ray clusters of galaxies: correlation of X-ray luminosity with galactic content.   See Abstr. 160.055.

Constraints on a dense hot intergalactic medium. See Abstr. 161.001.

The source of the X-ray background. See Abstr. 162.019.

Primeval gas clouds and the low-energy X-ray background.   See Abstr. 162.065.

Gamma-Ray Sources

142.701 **Diffuse cosmic and atmospheric MeV gamma radiation from balloon observations.**
V. Schönfelder, U. Graser, J. Daugherty.
Astrophys. J., Vol. 217, 306 - 319 (1977).
Final results on diffuse cosmic and atmospheric γ-radiation between 1.5 and 10 MeV from two balloon flights with a double Compton telescope flown from Palestine, Texas, in 1973 and 1974 are presented. From these flights, the energy spectrum of the diffuse cosmic γ-ray component, its dependence on galactic latitude, the energy spectrum of the vertical atmospheric γ-ray component at various atmospheric depths, and the zenith angle distribution of atmospheric γ-radiation at 2.5 g cm^{-2} were determined.

142.702 **Hard X-ray spectra of cosmic gamma-ray bursts.**
S. R. Kane, G. H. Share.
Astrophys. J., Vol. 217, 549 - 564 (1977).
Hard X-ray measurements of six γ-ray bursts observed during the period 1969 October − 1971 April are presented. The measurements were made with the University of California (Berkeley) detector on the OGO-5 satellite and the NRL detector on the OSO-6 satellite. Spectra for five of the six bursts have been determined using measurements from both satellites in order to reduce ambiguities due to uncertain source locations. The time-integrated spectra have been fitted by power-law, exponential, and thermal bremsstrahlung functions. Evidence for spectral variability from event to event in the hard X-ray region is presented.

142.703 **New high energy γ-ray sources observed by COS B.**
W. Hermsen, B. N. Swanenburg, G. F. Bignami, G. Boella, R. Buccheri, L. Scarsi, G. Kanbach, H. A. Mayer-Hasselwander, J. L. Masnou, J. A. Paul, K. Bennett, J. C. Higdon, G. G. Lichti, B. G. Taylor, R. D. Wills.
Nature, Vol. 269, 494 - 495 (1977).
The authors describe a search for γ-ray sources using data from the ESA γ-ray satellite COS B which revealed 10 new unidentified sources. These sources seem to be galactic with typical γ-ray luminosities above 100 MeV in excess of 10^{35} erg s^{-1}.

142.704 **SAS-2 galactic gamma-ray results − II. Localized sources.**   R. C. Hartman, C. E. Fichtel, D. A. Kniffen, R. C. Lamb, D. J. Thompson, G. F. Bignami, H. Ögelman, M. Özel, T. Tümer.
The structure and content of the Galaxy and galactic gamma rays, (see 012.009), p. 15 - 25 (1977).
Gamma-ray emission has been detected from the radio pulsars PSR 1818-04 and PSR 1747-46, in addition to the previously reported γ-ray emission from the Crab and Vela pulsars. For five of the closest pulsars, upper limits for γ-ray luminosity are found to be at least three orders of magnitude lower than that of the Crab pulsar.

142.705 **Identification of cosmic γ-ray sources CG135 + 1 and CG189 + 1 with H II regions.**   A. W. Strong.
Nature, Vol. 269, 394 (1977).
Eleven unidentified γ-ray sources have been reported by the COS B group. Here, the author points out two likely associations of the new sources with giant H II regions.

142.706 **On the gamma-ray source CG 195 + 4 (Geminga).**
L. Maraschi, A. Treves.
Astron. Astrophys., Vol. 61, L11 - L13 (1977).
Two different pictures of the pulsating γ-ray source CG 195 + 4 observed by the SAS II and COS B satellites are discussed.

142.707 **Gamma-ray astrophysics and galactic structure.**
J. L. Puget.
Topics in interstellar matter, (see 012.011), p. 179 - 185 (1977).

142.708 **A search for the reported 400-keV γ-ray line from Crab nebula.**
J. C. Ling, W. A. Mahoney, J. B. Willett, A. S. Jacobson.
Nature, Vol. 270, 36 - 37 (1977).
Leventhal et al. (1977) have reported a possible γ-ray line at 400 ± 1 keV from the Crab nebula on 10 - 11 May 1976. The authors of the present paper did not observe the 400-keV γ-ray line on 10 June 1974. Their result is in contradiction with Leventhal's measurement if a constant source intensity is assumed. The two observations were separated by 2 yr, however, and one cannot rule out the possibility that the feature may vary with time.

142.709 **Energetic neutrons and gamma rays measured on the Aryabhata satellite.**
S. V. Damle, R. R. Daniel, P. J. Lavakare.
Pramāṇa, Vol. 7, 355 - 368 (1976). − Abstr. in Phys. Abstr., Vol. 80, Abstr. 22023 (1977).

142.710 **Cosmic diffuse gamma rays from 2 to 25 MeV.**
R. S. White, B. Dayton, S. H. Moon, J. M. Ryan, R. B. Wilson, A. D. Zych.
Astrophys. J., Vol. 218, 920 - 927 (1977).
The flux of cosmic diffuse γ-rays with energies from 2 to 3, 3 to 5, 5 to 7, and 7.5 to 10 MeV are reported and upper limits are given for energies of 10−15 and 15−25 MeV. The observed fluxes are compatible with the energy distribution of $2.65 \times 10^{-2} E^{-2.08}$ photons (cm^2 s sr MeV)$^{-1}$ proposed by Dennis et al. at lower energies, steepened at higher energies to meet the slope of $E^{-2.4}$ of Fichtel et al. for energies above 35 MeV. Several regions of the celestial sphere were observed. No statistically significant deviation from isotropy in direction of the cosmic diffuse γ-rays was observed. These observations give upper limits for a number of possible sources.

142.711 **Nature of gamma-ray bursts.**   A. Dauvillier.
C. R. Acad. Sci. Paris, Tome 285, Sér. B, 341 - 344 (1977). In French.
The paper suggests that the so-called gamma-ray bursts of unknown origin are probably powerful X-ray bursts occurring in binaries containing an eruptive star associated with a pulsar.

142.712 **Gamma rays and the origin of cosmic radiation.**
T. Weekes.
Astronomy, Vol. 5, No. 6, p. 6 - 13, 16 - 17 (1977). − Abstr. in Phys. Abstr., Vol. 80, Abstr. 73912 (1977).

142.713 **Origin of cosmic gamma rays.**   D. M. Worrall.
Thesis Univ. Durham, England (1977). − Abstr. in Phys. Abstr., Vol. 80, Abstr. 87213 (1977).

142.714 **Observations of periodic γ-ray emission from the discrete source Cyg X-3.**   A. M. Gal'per, V. G. Kirillov-Ugryumov, A. V. Kurochkin, N. G. Lejkov, B. I. Luchkov, Yu. T. Yurkin.
Pis'ma v ZhEhTF, Vol. 26, 381 - 385 (1977). In Russian. Abstr. in Ref. zh., 51. Astron., 1.51.564 (1978).

X-ray and gamma-ray line production by nonthermal ions.   See Abstr. 022.103.

On the measurement of the diffuse X- and γ-background as seen by non directional detectors. See Abstr. 031.205.

Periodic slot collimator for accurate gamma ray burst source locations.   See Abstr. 032.568.

Detectors for gamma-ray burst astronomy. (A critical comparison).   See Abstr. 032.569.

Preliminary results from the European Space Agency's COS-B satellite for gamma-ray astronomy. See Abstr. 051.018.

Gamma-ray astrophysics.   See Abstr. 061.042.

Gravitationally redshifted gamma rays and neutron star masses.   See Abstr. 066.022.

Effect of multiple scattering on balloon observation of γ-ray bursts.   See Abstr. 082.109.

On fireball model and ultrahigh energy cosmic γ ray spectrum by *p-p* collisions.   See Abstr. 104.019.

Gamma rays and supernova explosions. See Abstr. 125.012.

On the nature of gamma-ray emission from pulsars. See Abstr. 141.503.

Requirements on pulsar models from gamma ray observation.   See Abstr. 141.511.

Production and beaming of pulsar γ-ray emission. See Abstr. 141.514.

Gamma-ray pulsars.   See Abstr. 141.522.

COS-B observations of pulsed γ-ray emission from PSR 0531+21 and PSR 0833−45.   See Abstr. 141.525.

SAS-2 observations of the diffuse gamma radiation in the galactic latitude interval $10° < |b| \leqslant 90°$. See Abstr. 157.004.

SAS-2 galactic gamma-ray results − I. Diffuse emission.   See Abstr. 157.007.

Very high-energy gamma-ray astronomy. See Abstr. 157.009.

Picture of the sky in gamma radiation. See Abstr. 157.017.

Low energy γ-ray observation of NGC 4151. See Abstr. 158.093.

On the possible existence of cosmological cosmic rays. II. The observational constraints set by the γ-ray background spectrum and the lithium and deuterium abundances. See Abstr. 162.011.

Errata

142.901 Errata: "Further observations of the burst source MXB 1728−34" [Mon. Not. R. Astron. Soc., Vol. 179, 57P - 64P (1977)].
J. A. Hoffman, W. H. G. Lewin, J. Doty.
Mon. Not. R. Astron. Soc., Vol. 181, 799 (1977).

142.902 Erratum: "Structure of the iron fluorescence line in X-ray binaries" [Astrophys. J., Vol. 215, 285 - 290 (1977)].   S. Hatchett, R. Weaver.
Astrophys. J., Vol. 218, 931 (1977).

142.903 Erratum: "Effects of line blending on the black hole candidacy of Cygnus X-1" [Astron. Astrophys., Vol. 58, 393 - 402 (1977)].   M. Moulding.
Astron. Astrophys., Vol. 61, 863 (1977).

143 Cosmic Radiation

143.001 **Cosmic ray composition at high energies.**
Nature, Vol. 268, 584 - 585 (1977).

143.002 **27-day modulation of cosmic rays taking into account radial fading of the azimuthal asymmetry**
of the solar wind.   L. Kh. Shatashvili.
Geomagn. Aehron., Vol. 17, 611 - 614 (1977). In Russian.

143.003 **Green's function of the transport equation for the simplest models of cosmic ray propagation.**
L. I. Dorman, M. E. Kats.
Geomagn. Aehron., Vol. 17, 615 - 621 (1977). In Russian.

143.004 **Mean mass temperature and spectrographic method of investigation of cosmic ray variations.**
L. I. Dorman, Yu. Ya. Krest'yannikov.
Geomagn. Aehron., Vol. 17, 622 - 627 (1977). In Russian.

143.005 **Interactions between cosmic ray protons and background radiation photons.**   K. D. Barker.
Astron. Astrophys., Vol. 60, 291 (1977).
The mean free path for cosmic ray protons between Compton interactions with microwave background photons has been calculated to be 1.4×10^{10} lt-yr. This distance is close to the accepted value of the Hubble radius and this coincidence is structurally distinct from those discussed by Dirac.

143.006 **Does electromagnetic radiation accelerate galactic cosmic rays?**   D. Eichler.
Astrophys. J., Vol. 216, 174 - 176 (1977).
The "reactor" theories of Tsytovich and collaborators of cosmic ray acceleration by electromagnetic radiation are examined in the context of galactic cosmic rays. It is shown that any isotropic synchrotron or Compton reactors with reasonable astrophysical parameters can yield particles with maximum relativistic factor γ_{max} of only $\sim 10^3$. If they are to produce particles with higher γ, the losses due to inverse Compton scattering of the electromagnetic radiation in them outweigh the acceleration, and this violates the assumptions of the theory. This is a critical restriction in the context of galactic cosmic rays, which have a power-law spectrum extending up to $\gamma = 10^6$.

143.007 **Simple analytic solutions appropriate for galactic cosmic-ray modulation.**   R. Cowsik, M. A. Lee.
Astrophys. J., Vol. 216, 635 - 645 (1977).
Exact analytic solutions for the galactic cosmic-ray density to the standard time-independent spherically symmetric convection-diffusion-adiabatic deceleration equation governing the transport of cosmic rays in the interplanetary medium are presented. It is assumed that the solar wind speed is constant and that the spatial diffusion coefficient is of the form $\kappa_0 p^\alpha$ (p = momentum), independent of distance from the Sun in accordance with Pioneer 10 and Pioneer 11 measurements and recent calculations.

143.008 **Mean mass of cosmic-ray Ne, Mg, Si at 1.2 GeV amu^{-1}.**   R. Dwyer, P. Meyer.
Astrophys. J., Vol. 216, 646 - 649 (1977).
The authors measured the mean mass of cosmic-ray Ne, Mg, and Si in the energy range of 800–1800 MeV amu^{-1} using a technique employing the effect of the geomagnetic field on the cosmic-ray fluxes. The values for the neutron excess, $\langle A \rangle - 2Z$, at the top of the atmosphere, are 0.45 ± 0.10, 0.32 ± 0.11, and 0.26 ± 0.16 for Ne, Mg, and Si respectively. These results, when account is taken of the effects of galactic propagation, imply values for the neutron excess

in the cosmic-ray source which are in agreement with solar system values.

143.009 **Isotropy of cosmic rays caused by magnetic discontinuities.**   D. G. Wentzel.
Astrophys. J., Lett., Vol. 216, L59 - L62 (1977).
When cosmic rays stream past a bend in the magnetic field, they acquire a correlation in their phase of gyration about the magnetic field. The resulting plasma instability generates Alfvén waves and simultaneously reduces the cosmic rays' mean streaming velocity along the field. The growth rate of the instability is proportional to a fractional power of the cosmic-ray density and is rapid even for cosmic rays above 10^{11} eV. A few interstellar shocks or interplanetary rotational discontinuities suffice to reduce the streaming anisotropy to the local Alfvén velocity. Cosmic rays of 10^{11} eV may be made nearly isotropic even within the solar system.

143.010 **Observations of galactic cosmic-ray energy spectra between 1 and 9 AU.**   F. B. McDonald, N. Lal,
J. H. Trainor, M. A. I. Van Hollebeke, W. R. Webber.
Astrophys. J., Vol. 216, 930 - 939 (1977).
The major objective of these studies is to understand the properties of low- and medium-energy galactic cosmic rays. At 9 AU, new spectral features are found for helium nuclei below 60 MeV per nucleon, while at energies above 100 MeV per nucleon the radial gradient is much smaller than had been theoretically predicted. It is shown that conventional modulation theory can be made consistent with these observations by several very different interpretations of the data.

143.011 **Anomalous processes in the solar system during 1971–1974 as derived from radioactivity of the recently fallen chondrite "Gorlovka".**
A. K. Lavrukhina, V. D. Gorin, G. K. Ustinova.
Izv. AN SSSR. Ser. fiz., Vol. 41, 395 - 402 (1977). In Russian.
Abstr. in Ref. zh., 51. Astron., 8.51.403 (1977).

143.012 **Solar and galactic subcosmic rays: their connection with active processes on the sun.**   L. I. Dorman.
Izv. AN SSSR. Ser. fiz., Vol. 41, 293 - 299 (1977). In Russian.
Abstr. in Ref. zh., 51. Astron., 8.51.404 (1977).

143.013 **The effect of increase of the cross-section of inelastic interaction on the shape of the energy spectrum of cosmic rays.**   N. L. Grigorov.
Yader. fiz., Vol. 25, 788 - 801 (1977). In Russian. – Abstr. in Ref. zh., 51. Astron., 8.51.686 (1977).

143.014 **Astronomia con particelle di alta energia: i raggi cosmici.**   N. Mandolesi, G. G. C. Palumbo.
Coelum, Vol. 45, 133 - 143 (1977).

143.015 **An experimental test for the charge state of the "anomalous" helium component.**
R. B. McKibben.
Astrophys. J., Lett., Vol. 217, L113 - L116 (1977).
Observations of phase lags between intensity variations for various particle species and energy ranges in the low-energy galactic cosmic radiation during the general intensity decrease observed in 1974–1975 show that, for particles whose charge state is known (i.e., "normal" cosmic ray components), particles with higher rigidities respond more quickly to changes in modulation conditions than do those with lower rigidities. When compared to particles of known energy and charge, the behavior of the "anomalous" low-energy helium component is consistent with these observations only if the helium is singly rather than doubly charged.

143.016 **Cosmic-ray propagation and containment.**
         E. N. Parker.
The structure and content of the Galaxy and galactic gamma rays, (see 012.009), p. 283 - 299 (1977).
The cosmic rays are an active gaseous component of the disk of the Galaxy, and their propagation and containment is a part of the general dynamics of the disk. The purpose of this review is to outline the problem as it faces us at the present time.

143.017 **A measurement of the isotopic composition of cosmic-ray Fe and other nuclei with $Z = 20-25$.**
G. A. Simpson, J. Kish, J. A. Lezniak, W. R. Webber.
Astrophys. Lett., Vol. 19, 3 - 10 (1977).
The authors present new results on the isotopic composition of Fe group nuclei using an improved version of their original Cerenkov $\times$ total energy telescope (Webber et al., 1973). From this new data they infer a mass resolution $\sigma \lesssim 0.4$ AMU for Fe over a limited range of energies − a factor > 2 better than earlier measurements.

143.018 **Equatorial modulation and north-south asymmetry of galactic cosmic rays due to the interplanetary magnetic field.**   J. T. A. Ely.
J. Geophys. Res., Vol. 82, 3643 - 3648 (1977).
Measurements of galactic cosmic rays made in 1967 on a low-altitude polar satellite exhibited two kinds of quasi-periodic variations which were synchronous with the changes in sector polarity of the interplanetary magnetic field (IMF). A modulation of 30% in the equatorial flux and a north-south asymmetry of roughly the same magnitude were observed. Both effects have been found in the records of surface neutron monitors. Relevance of the equatorial modulation to IMF structure is discussed briefly.

143.019 **The age of the galactic cosmic rays derived from the abundance of ^{10}Be.**
M. Garcia-Munoz, G. M. Mason, J. A. Simpson.
Astrophys. J., Vol. 217, 859 - 877 (1977).
The isotopic composition of galactic cosmic-ray beryllium has been measured in the energy interval 30−150 MeV per nucleon with the use of cosmic-ray telescopes carried on the IMP-7 and IMP-8 Earth satellites during 1973−1975. The measured cosmic-ray isotopic abundances of ^{7}Be and ^{9}Be are found to be in agreement with calculations based on steady-state models for the interstellar propagation of cosmic-ray nuclei. However, the measured abundance of the radioactive ^{10}Be component (half-life = 1.5×10^6 years) is substantially less than that calculated under the assumption that the cosmic rays propagate in regions of the galactic disk that have the traditionally accepted average interstellar density of 1 atom cm^{-3}. From the observed ratio ^{10}Be/Be = 0.028 ± 0.014, the authors deduce an average interstellar density of about 0.2 atoms cm^{-3}, and a cosmic-ray lifetime for escape of 1.7×10^7 years. The cosmic rays may be spending the major part of their life in the galactic halo or in specific regions of the disk with low matter density. The authors discuss the implications of this low interstellar density for models of the confinement and propagation of the cosmic rays in galactic magnetic fields.

143.020 **Light-element production by cosmological cosmic rays.**   T. Montmerle.
Astrophys. J., Vol. 217, 878 - 882 (1977).
This paper examines the possibility of the existence of cosmological cosmic rays (CCR), assumed to be responsible for the 1−100 MeV γ-ray background spectrum. The CCR are supposed to appear in a burst at some (high) redshift z_s. The light elements D, ^{3}He, ^{6}Li, ^{7}Li, and ^{7}Be are produced together with γ-rays, and their abundances (relative to H, as well as relative to one another) are considered as observational constraints on the existence of CCR. Results obtained previously on the abundance of ^{6}Li and ^{7}Li (by solving a system of cou-

pled time-dependent transport equations) are summarized; new results on D and ^{3}He are presented. In short, it is possible to account simultaneously for the γ-ray background spectrum and the ^{7}Li abundance.

143.021 **Secondary antiprotons: a valuable cosmic-ray probe.**
         G. Steigman.
Astrophys. J., Lett., Vol. 217, L131 - L133 (1977).
Even in the absence of antiprotons in the primary cosmic rays, a flux of secondary antiprotons will be produced in collisions between cosmic rays and interstellar gas. The predicted antiproton fraction increases with increasing cosmic-ray confinement, so that observations of antiprotons will provide a probe of models of cosmic-ray confinement. It is shown that the expected antiproton fraction [for $E(\bar{p}) \gtrsim 10$ GeV] ranges between 2.3×10^{-4} for the "leaky box" model and 18×10^{-4} for the "closed box" model. In addition, attention is called to the fact that a detection of cosmic-ray antiprotons at, or above, a level of 2×10^{-4} will provide a valuable lower limit to the antiproton lifetime.

143.022 **The cosmic-ray antiproton flux: an upper limit near that predicted for secondary production.**
G. D. Badhwar, R. R. Daniel, T. Cleghorn, R. L. Golden,
J. L. Lacy, S. A. Stephens, J. E. Zipse.
Astrophys. J., Lett., Vol. 217, L135 - L138 (1977).
Data gathered from the 1976 September 16 balloon flight of the Johnson Space Center superconducting magnet spectrometer have been examined for the presence of cosmic-ray antiprotons. The 95% confidence level upper limit for the ratio of antiprotons to protons is 6.6×10^{-4}. This upper limit is in strong contradiction to the prediction of the closed-galaxy model of Rasmussen and Peters, but is not inconsistent with the prediction of the modified closed-galaxy model of Peters and Westergaard. It is nearly equal to the predictions of conventional propagation models. This result provides an independent confirmation of the absence of primary antimatter in the cosmic rays at a level of approximately a few times 10^{-4}.

143.023 **Intensity of cosmic rays in the past.**
         V. A. Dergachev.
VIII Leningr. mezhdunar. semin. Mater. mezhdunar. semin. "Akt. protsessy na Solntse i probl. soln. nejtrino", 1976. Leningrad, 1976, p. 193 - 197. In Russian. − Abstr. in Ref. zh., 51. Astron., 9.51.664 (1977).

143.024 **Solar and galactic subcosmic rays: their relation to active solar processes.**   L. I. Dorman.
VIII Leningr. mezhdunar. semin. Mater. mezhdunar. semin. "Akt. protsessy na Solntse i probl. soln. nejtrino", 1976. Leningrad, 1976, p. 185 - 186. In Russian and English. − Abstr. in Ref. zh., 51. Astron., 9.51.665 (1977).

143.025 **Peculiarities of the inhomogeneous anisotropic diffusion of cosmic rays in the circumsolar space.**
Annotation.   M. V. Alania, L. I. Dorman.
VIII Leningr. mezhdunar. semin. Mater mezhdunar. semin. "Akt. protsessy na Solntse i probl. soln. nejtrino", 1976. Leningrad, 1976, p. 189 - 191. In Russian. − Abstr. in Ref. zh., 62. Issled. kosm. prostranstva, 9.62.195 (1977).

143.026 **Charge composition and energy spectra of cosmic-ray nuclei at energies above 5 GeV per nucleon.**
J. H. Caldwell.
Astrophys. J., Vol. 218, 269 - 285 (1977).
A scintillation-Cerenkov counter telescope, with three gas Cerenkov counters for energy determination between 5 and 90 GeV per nucleon, has been exposed for a net total of 4.5 m^2 sr hr in two balloon flights in 1974. The measurement yields the chemical composition and energy spectra of cosmic-ray nuclei in the charge range $5 \leqslant Z \leqslant 28$. The differential spectral indices of oxygen and of the iron group are measured

to be 2.67 ± 0.04 and 2.50 ± 0.08 above 5.4 GeV per nucleon, respectively. The results are interpreted in the context of the "leaky-box" model of cosmic-ray confinement and propagation.

143.027   Variations of galactic cosmic rays during the decrease of solar activity in 1971 - 1974.
P. P. Ignat'ev, T. E. Shvidkovskaya.
Kosm. Issled., Vol. 15, 633 - 635 (1977). In Russian.

143.028   Phenomenological model of nuclear primary air showers.   D. R. Tompkins, Jr., S. F. Saterlie.
Phys. Rev. D, Vol. 14, 1245 - 1250 (1976). — Abstr. in Phys. Abstr., Vol. 80, Abstr. 22028 (1977).

143.029   Solar zenith angle dependence of sudden cosmic noise absorptions.
S. B. S. S. Sarma, M. C. Sharma.
Australian J. Phys., Vol. 30, 531 - 532 (1977).
The zenith angle dependence of the flare-time absorption of sudden cosmic noise absorptions is investigated experimentally using riometer data at five widely spaced stations. A $\cos^n \chi$ dependence is found with $n = 1.4 \pm 0.04$.

143.030   The role of solar active regions and of the total magnetic field of the sun in the 11-year cosmic-ray cycle. Abstract.   T. N. Charakhch'yan.
VIII Leningr. mezhdunar. semin. Mater. mezhdunar. semin. "Akt. protsessy na Solntse i probl. soln. nejtrino", 1976. Leningrad, 1976, p. 129 - 132. In Russian. — Abstr. in Ref. zh., 51. Astron., 10.51.571 (1977).

143.031   On measuring the impulse of particles of the hadron component of cosmic rays.   D. T. Vardumyan,
G. A. Marikyan, K. A. Matevosyan, A. P. Oganesyan.
Erevan. fiz. inst. Nauch. soobshch. EFI-210-(2)-77. Erevan, Arus, 1977. 18 pp. Price 7 Kop. In Russian. — From Ref. zh., 51. Astron., 10.51.823 (1977).

143.032   Studies of the angular distribution and of the altitude dependence of the cosmic-ray muon and general components using scintillation telescopes.
Ya. L. Blokh, L. I. Dorman, I. Ya. Libin.
Nuovo Cimento B, Ser. 11, Vol. 37B, 198 - 202 (1977).
Abstr. in Phys. Abstr., Vol. 80, Abstr. 30469 (1977).

143.033   A proposed direct measurement of 10^{14} eV iron primaries.   M. P. Gough.
J. Phys. G, Vol. 2, 965 - 969 (1976). — Abstr. in Phys. Abstr., Vol. 80, Abstr. 37490 (1977).

143.034   Plastic nuclear track detector measurements of high-LET particle radiation on Apollo Skylab and ASTP space missions.   E. V. Benton, R. P. Henke, D. D. Peterson.
Nucl. Track Detect., Vol. 1, No. 1, p. 27 - 32 (1977). — Abstr. in Phys. Abstr., Vol. 80, Abstr. 41272 (1977).

143.035   Current density and diffusion coefficient for cosmic ray particles.   M. Stehlik, J. Dubinsky.
Acta Phys. Slovaca, Vol. 26, 257 - 266 (1976). — Abstr. in Phys. Abstr., Vol. 80, Abstr. 49247 (1977).

143.036   The energy spectrum of cosmic ray positrons.
M. Giler, J. Wdowczyk, A. W. Wolfendale.
J. Phys. A, Vol. 10, 843 - 859 (1977). — Abstr. in Phys. Abstr., Vol. 80, Abstr. 49253 (1977).

143.037   Anisotropy of galactic cosmic radiation.
T. Gombosi, J. Kota, A. Somogyi, A. Varga.
Fiz. Sz., Vol. 26, 407 - 410 (1976). In Hungarian. — Abstr. in Phys. Abstr., Vol. 80, Abstr. 53343 (1977).

143.038   Have we seen a heavy antinucleus?
R. Hagstrom.
Phys. Rev. Lett., Vol. 38, 729 - 732 (1977). — Abstr. in Phys. Abstr., Vol. 80, Abstr. 45272 (1977).

143.039   Whence cosmic rays?   A. Watson.
New Scientist, Vol. 73, 408 - 410 (1977). — Abstr. in Phys. Abstr., Vol. 80, Abstr. 56799 (1977).

143.040   Isotope resolution of the iron peak.
R. P. Henke, E. V. Benton.
Nucl. Instrum. Methods, Vol. 142, 521 - 523 (1977). — Abstr. in Phys. Abstr., Vol. 80, Abstr. 56801 (1977).

143.041   Energy spectrum and interaction characteristics of cosmic rays in the energy range 10^{13}–10^{16} eV.
J. Olejniczak, J. Wdowczyk, A. W. Wolfendale.
J. Phys. G, Vol. 3, 847 - 864 (1977). — Abstr. in Phys. Abstr., Vol. 80, Abstr. 56803 (1977).

143.042   Is the cosmic ray source composition energy dependent at high energy?   F. Le Guet.
J. Phys. G, Vol. 3, 1005 - 1018 (1977). — Abstr. in Phys. Abstr., Vol. 80, Abstr. 67013 (1977).

143.043   Polar coupling coefficients for cosmic ray multiple neutrons with energies up to 15 GeV.
D. Juraj, F. Andrej, I. Jozef.
Acta Phys. Slovaca, Vol. 27, No. 2, p. 142 - 148 (1977). — Abstr. in Phys. Abstr., Vol. 80, Abstr. 70954 (1977).

143.044   Study of sudden increases and sharp decreases in cosmic ray intensity.   R. Prasad, R. S. Yadav.
Indian J. Radio Space Phys., Vol. 5, 289 - 292 (1976). — Abstr. in Phys. Abstr., Vol. 80, Abstr. 70955 (1977).

143.045   Galactic neutrino sources and experimental neutrino astronomy.   R. Gallino, A. Masani.
Nuovo Cimento, Riv., Ser. 2, Vol. 6, 495 - 528 (1976). — Abstr. in Phys. Abstr., Vol. 80, Abstr. 70964 (1977).

143.046   Density spectrum of the electromagnetic component associated with high-energy cosmic ray muons.
B. A. Khrenov, J. Olejniczak.
J. Phys. G. Vol. 3, L229 - L231 (1977). — Abstr. in Phys. Abstr., Vol. 80, Abstr. 82092 (1977).

143.047   Experiment on very high energy muons and neutrinos in cosmic rays.   M. Koshiba.
J. Phys. Soc. Japan, Vol. 43, 701 - 702 (1977). — Abstr. in Phys. Abstr., Vol. 80, Abstr. 82095 (1977).

143.048   Generalization of the spectrographic method for studying the extraterrestrial and magnetospheric cosmic-ray variations including penumbra.
L. I. Dorman, G. Sh. Shkhalakhov.
Nuovo Cimento B, Ser. 11, Vol. 40B, 371 - 380 (1977). Abstr. in Phys. Abstr., Vol. 80, Abstr. 87217 (1977).

143.049   Radioemission from extensive air showers and the composition of primary cosmic rays.
R. Baggio, N. Mandolesi, G. Morigi, G. G. C. Palumbo.
Nuovo Cimento B, Ser. 11, Vol. 40B, 289 - 302 (1977). Abstr. in Phys. Abstr., Vol. 80, Abstr. 87219 (1977).

143.050   Cosmic ray propagation assuming two different confinement regions in the Galaxy.   M. Simon.
Astron. Astrophys., Vol. 61, 833 - 838 (1977).
In dealing with the propagation of cosmic ray particles in the Galaxy the nuclear collisions with the interstellar matter have to be considered. A Monte Carlo propagation program was developed to be able to follow the individual particles

through space by varying the boundary conditions between a confinement region around the source with higher density and the interstellar space. Assuming an exponential path length distribution in both regions, this two-zone model leads to a general enhancement of heavy primary source nuclei in order to explain the measured abundance of the cosmic ray nuclei. This model offers an explanation for the observed energy dependent abundance of different cosmic ray nuclei.

143.051 **Cosmic ray kinetics in space.**
L. I. Dorman, M. E. Katz (*Kats*).
Space Sci. Rev., Vol. 20, 529 - 575 (1977).
This work gives a systematic description of the statistical theory of the propagation of cosmic ray charged particles through random electromagnetic fields in space. A kinetic equation is derived for the cosmic ray distribution function averaged over the statistical ensemble corresponding to a random field. Transition to the diffusion approximation is considered, and the problems of the scattering and acceleration of charged particles are analyzed. The theory of fluctuation effects in cosmic rays is briefly discussed.

143.052 **The electron-proton abundance ratio in cosmic rays.**
K. M. V. Apparao, R. R. Daniel.
Astrophys. Space Sci., Vol. 46, 225 - 237 (1977).
An attempt has been made to understand the electron-proton abundance ratio in cosmic rays observed near the Earth. After correction for interplanetary and interstellar effects, the ratio has been obtained near the 'source' boundary. A leaky source model which can describe consistently all components of the cosmic radiation was then used to obtain the abundance inside the 'source'. Possible effects of injection and acceleration processes on the ratio are examined. In this model, electrons and protons are accelerated in a leaky source region which modulates and injects them into interstellar space.

143.053 **Energy spectra of cosmic ray nuclei: $4 \leqslant Z \leqslant 26$ and $0.3 \leqslant E \leqslant 2$ GeV amu^{-1}.**
R. C. Maehl, J. F. Ormes, A. J. Fisher, F. A. Hagen.
Astrophys. Space Sci., Vol. 47, 163 - 184 (1977).
Energy spectra of cosmic ray nuclei in the charge range $5 \leqslant Z \leqslant 26$ have been derived from the response of an acrylic plastic Čerenkov detector. Data were obtained using a balloonborne detector and cover the energy range $320 \lesssim E \lesssim 2200$ MeV amu^{-1}. Spectra are derived from a formal deconvolution using the method of Lezniak (1975). Relative spectra of different elements are compared by observing the charge ratios.

143.054 **Spatial distribution of high energy cosmic ray electrons perpendicular to the galactic plane.**
R. Schlickeiser, K. O. Thielheim.
Astrophys. Space Sci., Vol. 47, 415 - 421 (1977).
With the help of empirical data concerning the latitudinal distribution of galactic gamma rays the contribution of inverse Compton scattered gamma rays is calculated using various models concerning the distribution of high energy cosmic ray electrons perpendicular to the galactic plane. It is shown that gamma ray astronomy from regions with vanishing stellar and interstellar matter densities at energies greater than 100 MeV provides instructive information on the cosmic ray electron density.

143.055 **Ionization states and the origin of low energy cosmic ray nuclei.** N. Durgaprasad.
Astrophys. Space Sci., Vol. 47, 435 - 445 (1977).
The origin of the new component of cosmic ray nuclei in $1-30$ MeV amu^{-1} is investigated in detail. The author examines the possibility that these particles may be coming from nearby sources with energies $E \gtrsim 10^5$ eV amu^{-1} and are accelerated to energies $E \gtrsim 10^7$ eV amu^{-1} within the heliosphere. He calculates their charged states in interstellar space and interplanetary media, the modulations they suffer in interplanetary space and

thus finally their relative abundances near the Earth and source regions.

143.056 **Cosmic ray propagation in a closed galaxy.**
B. Peters, N. J. Westergaard.
Astrophys. Space Sci., Vol. 48, 21 - 46 (1977).
A simple model of cosmic ray propagation is proposed from which the major experimental results can be derived: The model reproduces the observed nuclear abundances and accounts for the observed changes of nuclear composition with energy, the high degree of isotropy of cosmic ray flux at all energies, and the high degree of its constancy throughout the history of the Solar System. The model is characterized by the two basic assumptions: (1) that cosmic rays have been injected at an unchanging rate by sources located in the galactic spiral arms and (2) that a large-scale magnetic field retains all particles in our galaxy, where they interact with interstellar gas, so that all complex nuclei are finally fragmented and their energy dissipated in meson production and electro-magnetic interactions.

143.057 **On the interplanetary cosmic-ray scintillations.**
I. N. Toptygin, V. N. Vasilijev (*Vasil'ev*).
Astrophys. Space Sci., Vol. 48, 267 - 281 (1977).
The equation for the two-particles cosmic-ray distribution function is derived by means of the Boltzmann kinetic equation averaging. This equation is valid for arbitrary ratio of regular and random parts of the magnetic field. On the basis of the derived equation the dependence between power spectra of cosmic-ray intensity and random magnetic field is obtained.

143.058 **A new family of periodic oscillations in the Störmer problem: the principal asymmetric.**
V. V. Markellos, S. Klimopoulos.
Astrophys. Space Sci., Vol. 48, 471 - 482 (1977).
A complete family of asymmetric periodic oscillations of a charged particle in the meridian plane of a magnetic dipole is presented. It begins at a bifurcation with the principal family of symmetric periodic oscillations and terminates in a flat equatorial oscillation. It consists of many stable and unstable segments.

143.059 **UH cosmic rays and solar system material: the elements just beyond iron.**
J. P. Wefel, D. N. Schramm, J. B. Blake.
Astrophys. Space Sci., Vol. 49, 47 - 81 (1977).
The origin of the elements from Cu to As in the UH (ultra-heavy) cosmic rays is investigated and related to current concepts of the nucleosynthesis of solar system material. The charge spectrum of the UH cosmic rays in the interval $29 \leqslant Z \leqslant 60$ is studied via a fully developed propagation calculation for source abundances given by solar system material, the r-process, the massive-star core helium-burning s-process, and explosive carbon burning. None of these sources considered individually can explain the cosmic ray observations. However a combination of material produced in the r-process, the core helium-burning s-process and in explosive carbon burning provides a good representation of the experimental data.

143.060 **The velocity correlation function in cosmic-ray diffusion theory.** M. A. Forman.
Astrophys. Space Sci., Vol. 49, 83 - 97 (1977).
The concept of velocity correlation functions is introduced and applied to the calculation of cosmic ray spatial diffusion coefficients. It is assumed that the pitch angle scattering coefficient is already known from some other theory and is reasonably well-behaved. It is found that the parallel diffusion coefficient is reduced in proportion to the amplitude of the field fluctuations, and that the ratio of the perpendicular to parallel diffusion coefficients cannot be greater than $\langle \delta B_x^2 \rangle / B_0^2$.

143.061 **Radio-emitting electrons and cosmic ray confinement.**
G. D. Badhwar, R. R. Daniel, S. A. Stephens.
Astrophys. Space Sci., Vol. 49, 133 - 143 (1977).
    The propagation of cosmic ray electrons in the framework of the disk-halo diffusion model in which the diffusion coefficient $D \propto z^6 E^\mu$ (where z is the distance from the galactic plane and E is the energy), and the magnetic field $H \propto z^{-\xi}$ has been examined by making use of the recently available radio data up to 8 GHz toward the anticenter and halo minimum.

143.062 **Green's Theorem and Green's Functions for the steady-state cosmic-ray equation of transport.**
G. M. Webb, L. J. Gleeson.
Astrophys. Space Sci., Vol. 50, 205 - 223 (1977).
    The authors consider solutions of the spherically-symmetric steady-state cosmic-ray equation of propagation in interplanetary space; in particular they develop a Green's Theorem by means of which they can use the solution for the case of mono-energetic release at fixed heliocentric distance to obtain general solutions for cases in which a distribution of sources is specified and in which the spectrum is specified on the boundaries.

143.063 **Monoenergetic-source solutions of the steady-state cosmic-ray equation of transport.**   G. M. Webb.
Astrophys. Space Sci., Vol. 50, 349 - 360 (1977).
    A non spherically-symmetric monoenergetic-point-source solution of the steady-state equation of transport for cosmic rays in the interplanetary region, in which monoenergetic particles are released isotropically and continuously from a fixed heliocentric position is derived by a Laplace transform method. The solution is for a spherically-symmetric model of the propagating region incorporating anisotropic diffusion, with a diffusion tensor symmetric about the radial direction, and the solar wind velocity is radial and of constant speed.

143.064 **On the high energy proton spectrum measurements.**
    R. W. Ellsworth, A. Ito, J. MacFall, F. Siohan,
R. E. Streitmatter, S. C. Tonwar, P. R. Vishwanath, G. B. Yodh, V. K. Balasubrahmanyan.
Astrophys. Space Sci., Vol. 52, 415 - 427 (1977).
    The steepening of the proton spectrum beyond 1000 GeV and the rise in inelastic cross sections between 20 and 600 GeV observed by the PROTON 1−2−3 satellite experiments may be explained by systematic effects of energy dependent albedo (back-scatter) from the calorimeter.

143.065 **27-day modulation of cosmic rays by the solar wind with marked asymmetry.**   A. V. Belov.
Din. kosm. plazmy. Moskva, 1976, p. 24 - 31. In Russian. − Abstr. in Ref. zh., 51. Astron., 11.51.589 (1977).

143.066 **On propagation of cosmic rays in the Galaxy.**
V. S. Ptuskin.
Din. kosm. plazmy. Moskva, 1976, p. 7 - 9. In Russian. − Abstr. in Ref. zh., 51. Astron., 11.51.858 (1977).

143.067 **Diffusion of cosmic rays in the solar wind deformed by the interstellar medium.**   S. F. Nosov.
Geomagn. Aehron., Vol. 17, 969 - 975 (1977). In Russian.

143.068 **Propagation of galactic cosmic rays in the spherically symmetric solar wind.**
B. D. Naskidashvili, L. Kh. Shatashvili.
Geomagn. Aehron., Vol. 17, 1109 - 1111 (1977). In Russian.

143.069 **Determination of the neutron-proton ratio in primary cosmic rays.**   R. K. Adair, H. Kasha,
R. G. Kellogg, L. B. Leipuner, R. C. Larsen.
Phys. Rev. Lett., Vol. 39, 112 - 115 (1977).

143.070 **The nuclear particles of the cosmic radiation.**
    C. J. Waddington.
Fundam. Cosmic Phys., Vol. 3, 1 - 85 (1977).
    Contents: Introduction; The detection of energetic cosmic nuclei; Observed chemical composition of cosmic rays; The propagation of cosmic rays; The lifetime of cosmic rays; Source abundances; Astrophysical significance of the charge spectrum; Stable isotopes.

143.071 **The anomalous component of low-energy cosmic rays: a comparison of observed spectra with model calculations.**   B. Klecker.
J. Geophys. Res., Vol. 82, 5287 - 5291 (1977).
    Using current modulation theory for the transport of low-energy cosmic ray particles and the model of Fisk (1976) for their acceleration by transit time damping large-scale field variations in the outer solar system, the author constructs a set of parameters which reproduce the quiet time spectra of He, O, N, and Ne as observed at 1 AU during the time period 1973−1975. With an analytic approximation for the acceleration and a numerical solution for the steady state spherical symmetric transport equation, both the observed spectral shapes and the relative intensities for He, O, N, and Ne can be fitted simultaneously remarkably well.

143.072 **Determination of the modulation function for galactic cosmic rays.**
A. A. Ajtmukhambetov, A. G. Zusmanovich, E. V. Kolomeets.
Prikl. i teor. fiz. Vyp. (No.) 8. Alma-Ata, 1976, p. 202 - 207. In Russian. − Abstr. in Ref. zh., 51. Astron., 12.51.360 (1977).

143.073 **Connection of the relative coefficient of cosmic ray modulation with solar coronal radiation.**
R. R. Ashirov, A. G. Zusmanovich, E. V. Kolomeets.
Prikl. i teor. fiz. Vyp. (No.) 8. Alma-Ata, 1976, p. 234 - 238. In Russian. − Abstr. in Ref. zh., 51. Astron., 12.51.362 (1977).

143.074 **The nature of cosmic ray modulation.**
    Kh. Z. Aldagarova, E. V. Kolomeets, V. T. Pivneva.
Prikl. i teor. fiz. Vyp. (No.) 8. Alma-Ata, 1976, p. 239 - 244. In Russian. − Abstr. in Ref. zh., 51. Astron., 12.51.363 (1977).

143.075 **Recurrent fluxes of low-energy nuclei (1−8 MeV/nucleon) from measurements aboard the automatic station Prognoz 4.**   A. A. Kolchin, V. V. Lebedev, V. F. Levchenko, A. I. Repin, G. P. Skrebtsov, V. L. Shubin.
Izv. AN SSSR. Ser. fiz., Vol. 41, 1819 - 1826 (1977). In Russian. − Abstr. in Ref. zh., 62. Issled. kosm. prostranstva, 1.62.163 (1978).

143.076 **The effects of fluctuations and noise on the neutron monitor diurnal anisotropy. 2. Non-field-aligned diffusion.**   A. J. Owens.
J. Geophys. Res., Vol. 82, 5551 - 5554 (1977).

143.077 **A measurement in nuclear emulsion of the isotopic abundance of cosmic ray nitrogen and oxygen.**
L. Jacobsson.
Lund Univ. Sweden, Fys. Instn., LUIP−7702. 49 pp. (1977). Abstr. from INIS7720 335367.

143.078 **Cosmic radiation interactions with extraterrestrial matter.**   R. Michel, H. Weigel.
37. Jahrestagung Deutsche Geophys. Ges. Braunschweig, F. R. Germany. March 28 - April 1, 1977. 9 pp. (1977). − From IKK 77A15004324.

    The Texas A&M underground cosmic ray observatory.
See Abstr. 009.023.

    Use of thin ionization calorimeters for measurements

of cosmic ray energy spectra.   See Abstr. 031.278.

Cosmic ray investigations for the Voyager missions; energetic particle studies in the outer heliosphere—and beyond. See Abstr. 032.582.

Measurement of radiation doses aboard AES Molniya 1.   See Abstr. 051.068.

On the possible existence of cosmological cosmic rays. III. Nuclear γ-ray production. See Abstr. 061.014.

Projectile charge dependence of electron-capture cross sections.   See Abstr. 061.036.

Neutron-capture nucleosynthesis in the helium-burning cores of massive stars.   See Abstr. 065.011.

Possible existence of black holes in cosmic rays. See Abstr. 066.031.

On the reverse effect of cosmic rays on solar wind. See Abstr. 074.035.

Solar-cycle evolution of the coronal general magnetic field of 1959–1974 and the synchronous variation of high-speed solar wind streams and galactic cosmic rays. See Abstr. 074.068.

Cosmic rays and ancient planetary magnetic fields. See Abstr. 091.064.

A Monte-Carlo simulation of galactic cosmic ray effects in the lunar regolith.   See Abstr. 094.422.

Etchable ranges of fossil and fresh heavy-ion tracks in lunar and analogous crystals.   See Abstr. 094.573.

Influence of cosmic ray variations upon the estimate of size and age of meteorites from cosmogenic isotopes. See Abstr. 105.049.

Meteoritic minerals as detectors of heavy cosmic ray particles.   See Abstr. 105.127.

Manifestation of solar active processes in the interplanetary space according to data of cosmic ray variations. Abstract.   See Abstr. 106.018.

Energetic neutrons and gamma rays measured on the Aryabhata satellite.   See Abstr. 142.709.

Cosmic ray age, soft X-rays and the galactic wind. See Abstr. 156.001.

The galactic dynamo and its relation to the propagation of ultra-high-energy cosmic rays. See Abstr. 156.015.

The relationship between the galactic matter distribution, cosmic-ray dynamics, and gamma-ray production. See Abstr. 157.001.

Galactic centre γ-rays from trapped cosmic rays. See Abstr. 157.002.

Point gamma-ray sources and the distribution of cosmic rays over the Galaxy.   See Abstr. 157.003.

Gamma rays, cosmic rays, and galactic structure. See Abstr. 157.014.

Galactic gamma rays and the origin of cosmic rays. See Abstr. 157.015.

On the possible existence of cosmological cosmic rays.I. The framework for light-element and gamma-ray production.   See Abstr. 162.009.

On the possible existence of cosmological cosmic rays. II. The observational constraints set by the γ-ray background spectrum and the lithium and deuterium abundances. See Abstr. 162.011.

Stellar Systems

151 Kinematics and Dynamics of Stellar Systems, Evolution of Galaxies

151.001 Enhancement of the gravothermal catastrophe in two-component isothermal spheres.
A. P. Lightman.
Astrophys. J., Vol. 215, 914 - 918 (1977).

The author extends the work of Saito and Yoshizawa on the equilibrium configurations available to a confined, self-gravitating, isothermal gas containing particles of mass m_1 and m_2. He finds that mass apportionment has a destabilizing effect, which represents an enhancement of the "gravothermal catastrophe" of Lynden-Bell and Wood. Over a wide range of $m_2/m_1 \gtrsim 3$, the enhancement is maximal when the fraction of total mass in heavy particles $\equiv M_2/M \sim 0.40$. This destabilizing effect is in qualitative agreement with the semirealistic, numerical time-dependent calculations of several investigators.

151.002 The stability of non-linear non-radial oscillations of two-dimensional models of rotating stellar systems. V. A. Antonov, S. N. Nuritdinov.
Astron. Zh. Akad. Nauk SSSR, Vol. 54, 745 - 752 (1977).
In Russian. English translation in Soviet Astron., Vol. 21, No. 4.

The problem of non-linear stability of a circular cylinder and Maclaurin disk with respect to non-radial oscillations is discussed. Time-independent phase invariants are defined. Under the condition of their existence, the total energy of a two-dimensional model is minimized. For non-linear oscillations stability conditions in form of a limitation from above of the centroid velocity are found.

151.003 Paths of formation of elliptic and spiral galaxies.
B. V. Komberg.
Priroda, 1977, No. 9, p. 46 - 52. In Russian.

151.004 On the kinetics of excitation of drift density waves by the Doppler effect. M. N. Maksumov.
Byull. Inst. Astrofiz., Dushanbe, No. 66 - 67, p. 3 - 13 (1976). In Russian.

Quasilinear effects accompanying the drift density waves excited by the Doppler effect are considered. The adequacy of the quasilinear model of the excitation of such waves is shown.

151.005 Thermal instability in a pre-galactic gas cloud.
Y. Sabano, Y. Yoshii.
Publ. Astron. Soc. Japan, Vol. 29, 207 - 219 (1977).

The thermal stability of a contracting pre-galactic gas cloud, in which a low fraction of molecular hydrogen works as a coolant, is investigated with due regard to the change of fractional abundance of molecular hydrogen. An instability criterion is derived for the condensation mode by the linear perturbation analysis.

151.006 Amplification of galactic density waves by non-adiabatic processes in gas. II. S. Kato.
Publ. Astron. Soc. Japan, Vol. 29, 359 - 368 (1977).

For tightly wrapped density waves, a general criterion for their amplification due to non-adiabatic processes in gas is derived. The criterion is expressed in terms of the work integral known as the Carnot cycle, and is similar to that used in stellar pulsation. It is shown that the non-adiabatic acceleration of clouds' random motion, which occurs behind the galactic shock, seems to be able to amplify the wave against the damping due to the shock.

151.007 Massive compact objects in an elliptical galaxy and their dynamical relation to the halo formation.
T. Saito.
Publ. Astron. Soc. Japan, Vol. 29, 421 - 428 (1977).

A large number of massive compact objects are assumed to be distributed throughout an elliptical galaxy at its formation. The strong scattering of field stars is considered to be a result of their binary encounter with these objects. The author evaluated the contribution of the scattered stars to the formation of the halo around the elliptical galaxy. In the elliptical galaxy M87, about one tenth of the observed halo mass is explained by the present model.

151.008 A computational study of the fission of self-gravitating, rotating, and elongated gaseous disks.
M. Fujimoto, S.-A. Sørensen.
Astron. Astrophys., Vol. 60, 251 - 257 (1977).

The non-linear motions of a self-gravitating gaseous disk are followed numerically by the fluid-in-cell method. For an elongated rotating disk a shock wave occurs along the major axis when the circulating gases cross the equipotential surface of the disk. The resultant non-axisymmetric condensation leads to the growth of a "pear-shaped" deformation which divides into two independent objects. For highly elongated disks and at large gas densities multiple fragmentation takes place. A short-lived ring-like condensation is sometimes observed.

151.009 Stellar dynamics in thin disk galaxies. I. A unified approach to hydrodynamic and orbit theories.
R. H. Berman, J. W-K. Mark.
Astrophys. J., Vol. 216, 257 - 270 (1977).

The authors outline a procedure in stellar dynamics whereby one can evaluate closed moment equations as well as nonlinear stellar orbits. This formalism allows density wave disturbances which may be of finite amplitude. Furthermore, it facilitates close comparison between the collective hydrodynamic behavior and the individual orbits of the stars. The general analysis is presented within the context of nonlinear axisymmetric disturbances which are slowly varying in time and space.

151.010 Orbit segregation in evolving galaxies and clusters of galaxies. W. C. Saslaw.
Astrophys. J., Vol. 216, 690 - 693 (1977).

As a galaxy or a cluster of galaxies evolves, part of its

mass may slowly become more centrally concentrated. The author shows that such a nonhomologously changing non-linear gravitational field tends to segregate orbits according to their eccentricity. It also tends to produce an extended halo distribution of orbits. Some astronomical implications of this effect are discussed briefly.

151.011   A theory of galaxy formation and clustering.
J. R. Gott III.
Recueil des séminaires, (see 012.004), p. 1 - 2 (1977).

151.012   Growth of spiral waves in disk galaxies.
Y. Y. Lau, G. Bertin.
Bull. American Astron. Soc., Vol. 9, 430 (1977). − Abstract.

151.013   The equilibrium of a spiral galaxy in the region of
the inner Lindblad resonance.
P. O. Vandervoort, D. G. Monet.
Bull. American Astron. Soc., Vol. 9, 435 (1977). − Abstract.

151.014   Exchange collisions between binary and single stars.
J. G. Hills.
Astron. J., Vol. 82, 626 - 630 (1977).
The author calculates the probability that existing binaries in various stellar systems have suffered exchange collisions with single stars. After such a collision the former single star is one of the new binary components while the star it replaces is ejected from the binary. In the solar neighborhood, about one visual binary in 10^3 has a nonprimordial stellar component which it captured from the general field. It is likely that many of the X-ray binaries in the galactic disk and bulge have captured their relativistic components by exchange collisions. In open clusters most binaries with semi-major axes $a \gtrsim 10^3$ AU have suffered one or more exchange collisions. In the cores of dense globular clusters, this is true for binaries with $a \gtrsim 1$ AU.

151.015   Drift und Verbreiterung von Spiralarmen im Rahmen
der Dichtewellentheorie.   H. Schwerdtfeger.
Diss. Naturwiss. Gesamtfakultät, Ruprecht-Karl-Univ.,
Heidelberg, 117 pp. (1977).

151.016   Negative mass instability of flat galaxies.
R. V. E. Lovelace, R. G. Hohlfeld.
News Lett. Astron. Soc. N.Y., Vol. 1, No. 2, p. 6 (1977).
Abstract.

151.017   The ageing of spiral arms.   R. Wielen.
Recueil des séminaires, (see 012.004), p. 20 - 25
(1977).

151.018   Structure galactique à la corotation.   J. Colin.
Recueil des séminaires, (see 012.004), p. 31 - 35
(1977).

151.019   Numerical experiments on galaxies in three dimensions.   R. H. Miller.
Recueil des séminaires, (see 012.004), p. 136 - 157 (1977).

151.020   On the dynamical evolution of galaxies in clusters.
J. P. Ostriker.
The evolution of galaxies and stellar populations, (see 012.005), p. 369 - 393, with a discussion, p. 393 - 400 (1977).

151.021   Mergers and some consequences.   A. Toomre.
The evolution of galaxies and stellar populations, (see 012.005), p. 401 - 416, with a discussion, p. 417 - 426 (1977).

151.022   The influence of co-rotation singularity on density
waves in disk galaxies.   W.-r. Hu.
Sci. Sinica, Vol. 20, 325 - 334 (1977).

In this paper, the influence of co-rotation singularity on the neutral density wave of marginal stability is studied, using the gas-dynamic approach in the galactic disk model. The result shows that the whole dispersion relationship of the wave is affected by the co-rotation singularity. Calculation is made for the galactic model similar to our Galaxy.

151.023   On the origin of galaxy rotation. II.
A. D. Chernin.
Astrofizika, Vol. 13, 69 - 78 (1977). In Russian. English translation in Astrophysics, Vol. 13, No. 1.
The picture of pregalactic turbulence in the metagalactic medium indicates the existence of large condensations of matter that move with supersonic velocities. They are generated by shock waves induced by gravitational instability. Such clouds undergo inelastic collisions leading to their coalescence. A considerable part of the energy of the initial relative motion of the clouds is converted into heat, and the corresponding initial orbital momentum causes rotation of the new cloud resulting from coalescence. Estimates show that the parameters of the clouds allow a final angular momentum that can be near the typical angular momentum of giant spiral galaxies.

151.024   Magneto-gas dynamic mechanism of rotation of
galaxies with spiral structure.
V. A. Krol', S. A. Silich, P. I. Fomin.
Astrofizika, Vol. 13, 79 - 94 (1977). In Russian. English translation in Astrophysics, Vol. 13, No. 1.
A magneto-gas dynamic approach to the problem of rotation of spiral galaxies is developed. This approach is based on two observational facts: longitudinal magnetization of the spiral arms and the existence of recurrent outputs of supersonic gas fluxes from the central regions. The mechanism of rotation (Segner's mechanism) consists in the transfer of a rotational moment to the arm at the passage of shock waves along the magnetized spiral with decreasing density and subsequent shock output of gas masses from the end of the arm into space. It is shown that this mechanism can ensure the accumulation of the observed rate of rotation of the spiral during the period of about 10^9 years and results in a stationary regime of the rotation. The period of the steady rotation is proportional to the radius of the galaxy.

151.025   On the origin of star clusters.
G. M. Tovmasyan.
Astrofizika, Vol. 13, 131 - 138 (1977). In Russian. English translation in Astrophysics, Vol. 13, No. 1.
The ratios of gas masses of young star clusters in relation to their stellar masses are considered. The results contradict the hypothesis concerning the origination of the cluster stars from the existing gas cloud by means of its condensation and are rather in favour of Ambartsumyan's hypothesis on the common origin of stars and gas clouds from superdense protostellar matter.

151.026   Simulation of the Magellanic Stream to estimate
the total mass of the Milky Way.
D. N. C. Lin, D. Lynden-Bell.
Mon. Not. R. Astron. Soc., Vol. 181, 59 - 81 (1977).
Computer simulation of the Magellanic Stream as the tidally torn debris from the last passage of the Magellanic Clouds past the Galaxy can give good fits in both position and velocity. However, to explain the high radial velocity observed at the tip of the stream either the local circular velocity of the Galaxy must be large so that much of the observed velocity reflects the observer's motion, or the mass of the Milky Way must be large so that infall produces the speed. Possible pairs of values are given. The pair V_c = 290 km s^{-1}, M_G = 4.3 × $10^{11} M_\odot$ fits the determinations of V_c and M_G from the dynamics of the Local Group of galaxies (see 155.019). This paper summarizes results from some 200 trial initial conditions with the Magellanic Clouds modelled as a flat disk of test particles

circulating around a single spherical mass.

151.027 **Density wave theory.**   W. W. Roberts, Jr.
The structure and content of the Galaxy and galactic gamma rays, (see 012.009), p. 119 - 150 (1977).

The prospect that density waves and galactic shock waves are present on the large scale in disk-shaped galaxies has received support in recent years from both theoretical and observational studies. Large-scale galactic shock waves in the interstellar gas are suggested to play an important governing role in star formation, molecule formation, and the degree of development of spiral structure. Through the dynamics of the interstellar gas and the galactic shock-wave phenomenon, a new insight into the physical basis underlying the morphological classification system of galaxies is suggested.

151.028 **The effects of dissipation on the gas response to oval distortions of disk galaxies.**   R. H. Sanders.
Astrophys. J., Vol. 217, 916 - 927 (1977).

The time-dependent response of an ensemble of noninteracting test particles to a rotating oval distortion of a realistic axisymmetric galactic disk is numerically calculated for several assumed values of the distortion angular velocity.

151.029 **Chemical evolution of galaxies.**
J. Audouze, B. M. Tinsley.
Annu. Rev. Astron. Astrophys., Vol. 14, (see 003.008), 43 - 79 (1976).

Studies of chemical evolution aim primarily to account for the distribution and abundances of the chemical elements in the stars and interstellar matter of galaxies. This review begins with a summary of the observational and theoretical material that constitute the rather disjoint set of patchwork contributions available for compiling a picture of galactic evolution. It describes attempts to tie these pieces together into coherent evolutionary models. The authors discuss models, constraints, and unsolved problems for chemical evolution, respectively, in the solar neighborhood and in galaxies (including other regions of our own Galaxy).

151.030 **Purely discontinuous random processes in a field of irregular forces.**
I. W. Pietrowskaja (*I. V. Petrovskaya*).
Postępy Astron., Tom 25, 59 - 83 (1977). In Polish.

The scheme of a purely discontinuous random process is applied to the investigation of irregular forces effects in stellar systems. Kolmogorov-Feller's equations are reduced to a form convenient for the description of variations of star motion parameters under encounters with field stars. The evolution of a velocity distribution function for groups of stars with given mass and initial velocity distribution is investigated for the case of stars relaxing among the field stars. The escape rate of stars of different masses and the amount of energy taken away by the dissipated stars are evaluated.

151.031 **Escape of stars from isolated clusters. Part II. Statistical theory.**   T. Kwast.
Postępy Astron., Tom 25, 105 - 113 (1977). In Polish.

The effect of interaction of stars in a stellar system may be treated as diffusion of stars in the velocity space. Chandrasekhar showed that there also exists so-called dynamical friction in the motion of stars. The function of velocity distribution of stars is then governed by the Fokker-Planck equation. One can find the escape rate of stars from the system from the solution of this equation. The result is very close to results of classical theories which, as well as the statistical theory described here, assume that escapes are a result of accumulation of many small changes in the velocity of stars.

151.032 **Scattering of trapped star orbits in a flat spiral galaxy.**   G. Bertin, B. Coppi, A. Taroni.
Astrophys. J., Vol. 218, 92 - 95 (1977).

A combined analytical numerical study of the scattering of trapped star orbits ('bananas") in flat galaxies with a spiral wave structure is described. Standing modes localized about the corotation region and with frequencies about the star libration frequency are shown to have a relatively strong detrapping effect. It is suggested that these scattering processes affect the growth and saturation of the spiral wave if it is assumed that these depend on wave-particle resonance processes. The relevant modes can be excited either by interaction with a companion galaxy or, possibly, by an internal instability.

151.033 **The validity of statistical results from *N*-body calculations.**   H. Smith, Jr.
Astron. Astrophys., Vol. 61, 305 - 312 (1977).

The effects of errors in numerical integration of the *N*-body equations of motion by a conventional technique have been studied differentially for systems containing 16 equal masses initially in virial equilibrium, with the calculations made from the same 5 sets of initial conditions with different accuracy. The initial conditions were uniform density inside a sphere of unit radius and a step-function velocity distribution. The sets of initial conditions differ in the random numbers chosen for positions and velocities of the individual particles.

151.034 **Analytical models for spherical stellar systems.**
E. Davoust.
Astron. Astrophys., Vol. 61, 391 - 396 (1977).

Two series of analytical models, defined by two-parameter phase density distributions, are presented for the structure of spherical stellar systems. Two of the models can be adjusted to the results of numerical simulations by the Monte-Carlo method. The time evolution of the model and scaling parameters obtained in the fit follows a power law. The evolution of the core does not seem to be homologous.

151.035 **On the non-existence of the additional integral of motion of a star in a cluster.**
A. S. Baranov, M. S. Volkov.
Astron. Zh. Akad. Nauk SSSR, Vol. 54, 1140 - 1141 (1977). In Russian. English translation in Soviet Astron., Vol. 21, No. 5.

It is shown that the conditions of Poincaré's theorem on the non-existence of the additional integral are valid for the case of the motion of a star inside an axisymmetrical heterogeneous ellipsoidal cluster.

151.036 **Survival and disruption of galactic substructure.**
S. M. Fall, M. J. Rees.
Mon. Not. R. Astron. Soc., Vol. 181, 37P - 42P (1977).

The authors consider the tidal and evaporative disruption of an initial spectrum of galactic substructure which is here assumed to span a wide range of masses and sizes. The survival of subclusters having masses comparable with those of known globular clusters is best explained if the initial distribution was one of roughly constant surface brightness. Some recent observations on the relationship between the properties of globular clusters and their parent galaxies have a natural explanation within this context.

151.037 **Potentials satisfying resonance assumptions.**
P. Andrle.
Bull. Astron. Inst. Czechoslovakia, Vol. 28, 307 - 311 (1977).

The properties of the third integral of motion (in a system possessing a potential of the fourth degree) were studied. The problem of the existence of such a formal integral (Andrle, 1966) and two special resonances (Andrle, 1972 and 1974) were discussed in greater details.

151.038 **The stability of Freeman's elliptical stellar cylinder.**
M. T. Nishida, T. Ishizawa.

Publ. Astron. Soc. Japan, Vol. 29, 555 - 566 (1977).

The overall stability of Freeman's collisionless elliptical stellar cylinder system is studied by integrating the linearized Vlasov equation along an unperturbed stellar orbit. The dispersion relations are obtained for second- and third-order harmonic perturbations. The cylinder is stable for second-order harmonic perturbations, but it is unstable for third-order harmonic perturbations for some values of parameters.

151.039 Marginal mode to the second order for spherical stellar systems.
D. Gillon, J.-P. Doremus, G. Baumann.
C. R. Acad. Sci. Paris, Tome 285, Sér. A, 689 - 692 (1977). In French.

The authors write explicitly to the second order a time-independent displacement mode which is a solution of the Vlasov-Poisson equations. The authors show that this aspherical mode corresponds to a perturbation on the energy ϵ and on the square of the angular momentum J. They establish that for this mode, contrary to the case limited to the first order, the second variation of energy δW vanishes.

151.040 The dynamical evolution of clusters of galaxies.
S. D. M. White.
Highlights of Astronomy, Vol. 4, Part I, (see 012.020), p. 265 - 270 (1977).

Contents: 1. N-body models for clusters; 2. Relaxation and dynamical friction; 3. A model for the missing mass; 4. The intergalactic background.

151.041 Inner Lindblad resonance in galaxies. Nonlinear theory. II. Bars. G. Contopoulos, C. Mertzanides.
Astron. Astrophys., Vol. 61, 477 - 485 (1977).

The properties of the orbits in a galaxy composed of an axisymmetric background and a weak bar are derived theoretically by means of a new integral besides the Hamiltonian. Inside the inner Lindblad resonance there is only one periodic orbit, while outside the resonance there are one unstable and two stable periodic orbits. The non periodic orbits surround one of the stable periodic orbits, or both of them. A comparison with numerical calculations of orbits in various models shows very good agreement.

151.042 Stochastic tidal disruption of clusters.
E. Knobloch.
Astrophys. J., Vol. 218, 406 - 414 (1977).

The author supposes that the tidal disruption of star clusters can be described by a stochastic process, and formulates the problem in terms of the theory of stochastic processes. This method has several advantages. As an example he considers a spherically symmetric cluster with uniform mean density, and calculates the rate of increase in energy of a test star to second order in the interaction strength in terms of the autocorrelation tensor of the stochastic process.

151.043 Theories of spiral structure. A. Toomre.
Annu. Rev. Astron. Astrophys., Vol. 15, (see 003.012), 437 - 478 (1977).

Contents: Some of Lindblad's ideas; Nearly kinematic waves; Lin-Shu density waves; Shocks and other consequences; A quick synopsis; Global instabilities; Shearing bits and pieces.

151.044 Non-linear effect on the gravitational instability for the adiabatically deformed case compared with the isothermal one. S. Aoki.
Astron. Astrophys., Vol. 61, 609 - 624 (1977).

The non-linear gravitational wave theory of a collisionless system in one dimension is developed under the assumption that the frequency function in phase space is a gaussian in the action variable. Numerical comparisons of this and the isothermal case are presented. From these comparisons, it is concluded that compressibility is not higher than that of the

isothermal case. This is a general characteristic of adiabatic change, but the details are quite different from those of a polytropic gas.

151.045 Massive galactic halos. I. Formation and evolution.
J. E. Gunn.
Astrophys. J., Vol. 218, 592 - 598 (1977).

It is shown that massive halos form naturally around galaxies if there is a supply of material left over from galaxy formation which is of the appropriate character; viz., condensed bodies or noninteracting neutral particles. The evidence is strong that halos are made of such stuff. A fairly detailed model for the Local Group fits the total masses, time scales, and dynamics very well with essentially all the mass in halos. Predictions for galaxies in great clusters like Coma are quite different, and calculations indicate that only about 10% of the total mass should be in halos, the rest distributed throughout the cluster. It is argued that the ratio of halo stuff to "visible" matter is the same everywhere; the arguments for a low-density universe are strengthened thereby.

151.046 Large-scale winds driven by flare-star mass loss.
G. D. Coleman, S. P. Worden.
Astrophys. J., Vol. 218, 792 - 800 (1977).

The effect of injecting substantial quantities of high-temperature material into the interstellar medium from flare-star activity is examined. Using models like those developed by Mathews and Baker for calculating supernovae-driven elliptical galaxy winds, the authors consider the effects of flare-star mass loss in elliptical galaxies and globular clusters. It is found that steady outflowing winds will develope in these objects. Such winds may explain the observed absence of substantial quantities of interstellar material in globular clusters and elliptical galaxies. Assuming the presence, of elliptical galaxy winds in clusters of galaxies, the authors consider the effects of such winds on intergalactic medium dynamics.

151.047 A numerical method for integrating the stellar-dynamical Fokker-Planck equation in a fixed inhomogeneous gravitational background. J. R. Ipser.
Astrophys. J., Vol. 218, 846 - 856 (1977).

A stable numerical method is presented for solving the stellar-dynamical Fokker-Planck equation in time and phase space, for certain situations involving inhomogeneous gravitational fields. For the situations of interest, which involve three phase-space dimensions plus time, a test distribution of stars is scattered by a fixed, spherical background distributions of stars. In this paper attention is focused on background distributions that compose an equilibrium spherical star cluster of the Michie-King type, and an efficient method for constructing such clusters is presented. Simple examples of results obtained by integrating the Fokker-Planck equation are exhibited.

151.048 Unstable spiral modes in disk-shaped galaxies: enhancement of growth rate.
Y. Y. Lau, J. W-K. Mark.
Proc. Natl. Acad. Sci. USA, Vol. 73, 3785 - 3787 (1976). Abstr. in Phys. Abstr., Vol. 80, Abstr. 41414 (1977).

151.049 Evolution of dynamical systems with time-varying gravity. A. Marchant, V. Mansfield.
Nature, Vol. 270, 699 - 700 (1977).

Various attempts have been made to place limits on the time-variability of gravity through observations of astrophysical objects. The authors demonstrate that contrary to previous studies, no useful limits can be obtained from observations of N-body systems such as galaxies and star-clusters.

151.050 The origin of galaxies. R. B. Larson.
American Sci., Vol. 65, 188 - 196 (1977). – Abstr. in Phys. Abstr., Vol. 80, Abstr. 67125 (1977).

151.051 Resonant three-wave interactions of density waves in a self-gravitating disk.   S. Ikeuchi.
Prog. Theor. Phys., Vol. 57, 1239 - 1254 (1977). – Abstr. in Phys. Abstr., Vol. 80, Abstr. 67130 (1977).

151.052 The properties of galactic shock waves for the theory of density wave.   W.-r. Hu.
Kexue Tongbao, Vol. 22, No. 2, p. 79 - 83 (1977). In Chinese. From Phys. Abstr., Vol. 80, Abstr. 71145 (1977).

151.053 Theory of spiral structure.   C. C. Lin.
Proceedings of the 14th IUTAM congress on theoretical and applied mechanics. Delft, Netherlands, 30 August - 4 September 1976. W. T. Koiter (Editor). North-Holland Publishing Company, Amsterdam, Netherlands. 10 + 491 pp. Price $ 48.95 (1977). ISBN 0-7204-0549-1. p. 57 - 69. Abstr. in Phys. Abstr., Vol. 80, Abstr. 71146 (1977).

151.054 Evolution of galactic density waves at the unstable stage.   Z.-y. Yue.
Acta Astron. Sinica, Vol. 18, 105 - 112 (1977).
    The evolution law of galactic density waves is given from the equation of stellar dynamics, and compared with the results obtained from the theory of fluid mechanics by the author, showing that the results at the unstable stage of these two theories are close to each other.

151.055 Modelo de potencial para galaxias espirales del tipo Sc.   A. López García, C. Zaragoza Ruvira.
Urania Barcelona, Año 61, Núm. 285, p. 17 - 40 (1976).
    The gravitational potential of a spiral galaxy model of the Sc type is studied, in the plane of symmetry, decomposed in two terms: potential due to the galactic mass with axial symmetric distribution, and potential due to the mass condensed in two symmetrical spiral arms.

151.056 Test computations on the dynamical evolution of star clusters.   L. Angeletti, P. Giannone.
Astrophys. Space Sci., Vol. 46, 205 - 214 (1977).
    Test calculations have been carried out on the evolution of star clusters using the fluid-dynamical method devised by Larson (1970). Large systems of stars have been considered with specific concern with globular clusters. The influence of varying in turn various free parameters has been studied for the results. The partial release of some simplifying assumptions with regard to the relaxation time and distribution of the 'target' stars has been considered. The change of the structural properties is discussed, and the variation of the evolutionary time scale is outlined. An indicative agreement of the results obtained here with structural properties of globular clusters as deduced from previous theoretical models is pointed out.

151.057 A classification of galactic collisions.   K. S. Sastry, S. M. Alladin.
Astrophys. Space Sci., Vol. 46, 285 - 291 (1977).
    The effects of collisions between two galaxies on the test galaxy considered are classified as follows – Type A: The changes in the size and mass of the test galaxy are both negligible; Type B: There is significant increase in the size or decrease in the mass of the test galaxy or in both; Type C: The test galaxy becomes a component of a double galaxy by tidal capture; Type D: The test galaxy is disrupted by the tidal forces of the field galaxy. The type of collision is given as a function of the distance and speed at closest approach and also as a function of the initial impact parameter and speed at infinite separation of the two galaxies for two density models of the galaxies.

151.058 Density wave – star interaction in the theory of spiral galaxies.   E. Evangelidis.
Astrophys. Space Sci., Vol. 46, 309 - 319 (1977).
    The dispersion relation has been derived for density waves

propagating at an arbitrary angle. The analysis has shown the existence of a resonance which for a two-arm galaxy can be stable, neutral or unstable as $\omega/2\Omega \lessgtr 2$, respectively.

151.059 Substructure in clusters of galaxies: a clue to the 'missing' mass?   P. S. Wesson, A. Lermann.
Astrophys. Space Sci., Vol. 46, 327 - 334 (1977).
    The influence of subclustering in rich clusters of galaxies is examined using results from numerical n-body experiments. It is found that, under some conditions, the standard virial theorem is satisfied. No physical missing mass is needed because its role is replaced by the gravitational energy of the subclustering. The authors find that, in the Coma cluster, this effect masquerades as a 'missing' mass about 7 times that of the physical mass, so that the apparent extant virial discrepancy ($M_{VT}/M \cong 8$) in this cluster is explained.

151.060 A hypothesis to explain the virial discrepancy in clusters of galaxies.   P. S. Wesson, A. Lermann.
Astrophys. Space Sci., Vol. 46, 335 - 339 (1977).
    On the basis of numerical experiments on n-body binding energies the authors tentatively consider the following hypothesis: If the distance between two galaxies forming a binary system is a_g, and a cluster of galaxies that is substructured in a hierarchical fashion on all scales from a_g upwards has a total mass M, then the total gravitational binding energy of the cluster is $\Omega_{TH} = -GM^2/2a_g$. As an explanation for 'missing' masses up to order 100 the authors test this hypothesis in three different ways, finding remarkable agreement with observation, with no need for physical missing mass.

151.061 Stochastic simulation of fields of galaxies.   R. J. Dodd.
Astrophys. Space Sci., Vol. 46, 499 - 506 (1977).
    A method for generating model fields of galaxies stochastically is described. The output from several such models is compared with observational data derived from a region on a UK 48-in. Schmidt plate measured by the COSMOS machine. The results yield, for model matching observations, a maximum probability of 9.5% for a model with Gaussian distribution inpute parameters of 5 ± 5 galaxies per cluster and cluster radius 4 ± 4 Mpc.

151.062 A theory for the gravitational potentials of spheroidal stellar systems and its application to the Galaxy.
M. Clutton-Brock, K. A. Innanen, K. A. Papp.
Astrophys. Space Sci., Vol. 47, 299 - 314 (1977).
    A new formula for the gravitational potential for spheroidal systems is proposed, and is applied to the galactic system. The applied model consists of a disk, a nucleus and a massive halo. Although this model is not quite so accurate as those produced by the superposition of large numbers of simple spheroids, it has the important advantage of retaining analytic simplicity for its acceleration formulae, thereby producing significant economies for orbit computations.

151.063 Numerical experiments on the effect of massive central systems on the virial binding energy of clusters.   P. S. Wesson, A. Lermann, R. E. Goodson.
Astrophys. Space Sci., Vol. 48, 357 - 363 (1977).
    A program has been developed to evaluate the gravitational binding energy of a clumpy cluster composed of N particles with a given mass dispersion, a given abundance of binaries and a given massive binary system at the centre. In application to clusters of galaxies nucleated by massive cD binaries, it is found that the gravitational binding energy is typically greater than the equivalent smooth distribution by a factor in the range 2–12. This suggests that the 'missing' mass problem in clusters of galaxies can be reduced by an order of magnitude if the correct binding energy is used in the virial theorem.

151.064   *BiL(E)* **and the Hubble sequence: a new theory of
        galaxies.**   R. N. Henriksen, M. Reinhardt.
Astrophys. Space Sci., Vol. 49, 3 - 39 (1977).
    The authors show that magnetic fields can be important in
the formation and evolution of galaxies and that they might be
indeed the missing parameters to explain the Hubble sequence.
They use the self-consistent theory of spiral magneto-hydrodynamic flow developed by Henriksen and co-workers over the
last few years. The authors envisage the following scenario: The
first objects to form after recombination in a canonical hot big-
bang universe with turbulence and magnetic fields have masses
of order $10^9 M_\odot$. In a violent burst of activity – possible
mechanisms are discussed – they ionize the surrounding medium, raising the Jeans mass to a galactic scale, and becoming the
condensation seeds of galaxies. The subsequence evolution of
these nuclei, including recurrent activity, is discussed in some
detail. Some of the observationally testable predictions of the
theory concern: the energetics, duration and frequency of
nuclear activity, the absence of dwarf spiral galaxies, rigidly
rotating nuclear regions in galaxies, the mass and structure of
galactic halos, leading and trailing spiral arms and their pitch
angle, the bulge-to-disc ratio, the frequency distribution of
morphological types, and the warping of galactic discs.

151.065   **The dispersion relation of a gravitating spiral system.**
        E. Evangelidis.
Astrophys. Space Sci., Vol. 49, 481 - 495 (1977).
    The dispersion relation has been found for a galaxy,
without the assumption that the centrifugal force is balanced
by the gravitational force. It has been shown that such a system
(1) can be gravitationally unstable under appropriate condi-
tions, and (2) that there is no resonance at $\omega = 2\Omega$ (Ω=angular
velocity of the Galaxy).

151.066   **A dynamical interpretation of the Hubble sequence
        of galaxies.**   N. Dallaporta, L. Secco.
Astrophys. Space Sci., Vol. 50, 253 - 279 (1977).
    Brosche (1970) has outlined a theory of early galaxy
formation, based on a model of a protogalaxy formed by a
cluster of N gas clouds, whose evolutionary drag is entirely
due to the energy loss occurring in collisions between the
clouds. The main changes in the formalism of Brosche's scheme
are indicated when one assumes the rotating polytropes as
models, before the centrifugal condition is reached at the
equator. The centrifugal shedding of matter at the equator is
discussed and the new models following this point are presented. Further changes to the original Brosche's approach are
indicated and the whole set of equations necessary to integrate
the problem is completed. Numerical calculations seem to
indicate that one can obtain in this way, by varying the angular
momentum and the initial number of clouds, different galaxy
types (elliptical, lenticular, spiral) resembling those of the
Hubble sequence.

151.067   **Dynamical evolution of clusters with two stellar
        groups.**   L. Angeletti, P. Giannone.
Astrophys. Space Sci., Vol. 50, 311 - 322 (1977).
    The generalization of the fluid-dynamical approach from
one-component star clusters to clusters with several stellar
groups (as far as the star masses are concerned) has been applied
to the study of two-component clusters. Rather extreme values
of stellar masses and masses of groups were chosen in order to
emphasize the different dynamical evolutions and asymptotic
behaviours. Escape of stars from clusters and the problem
of equipartition of kinetic energy among the two star groups
are discussed. Some characteristic properties have been discussed in relation with some observed features of galactic
globular clusters.

151.068   **The distribution of the angular momenta of galaxies.**
        P. Brosche.
Astrophys. Space Sci., Vol. 51, 401 - 408 (1977).

    Since the average relation between the angular momenta
P and the masses M of galaxies can be represented by a power
law $P \sim M^\alpha$, the author can define a relative angular momentum
$\rho = P/M^\alpha$ (or a constant time P/M^α). For a random motion
picture within protogalaxies, ρ should follow a Maxwellian
distribution and consequently the dispersion σ of log ρ should
be 0.210. For the reasonable range of α (7/4 to 2), the limited
sample of galaxies with known dynamical parameters gives σ
between 1/2 and 1 times the Maxwellian value. For the
plausible special case $\alpha = 2$ the reciprocal of the maximum
rotational velocity v_m is already a measure of ρ and the larger
sample of v_m-values not only yields the Maxwellian σ but,
moreover, shows the shape of the distribution.

151.069   **Constructing the formal third integral of motion in
        stellar systems for resonance cases. III. Comparison
of Kruskal's and Contopoulos' methods for the case** $m/n = 2$.
L. P. Osipkov.
Vestn. Leningr. univ., 1977, No. 7, p. 144 - 149. In Russian. –
Abstr. in Ref. zh., 51. Astron., 11.51.920 (1977).

151.070   **Energy variation during fragmentation of a collapsing
        cloud. The phenomenon of gravitational explosion.**
B. E. Zhilyaev, L. M. Shul'man.
Early stages of stellar evolution, (see 012.041), p. 47 - 56
(1977). In Russian.
    A hypothesis is proposed to explain the origin of expand-
ing stellar systems. It is shown that a collapsing rotating cloud
can acquire a toroidal form if it possesses sufficient rotational
momentum. The fragmentation of the toroidal cloud generates
a set of gravitation-linked subsystems.

151.071   **Origin and evolution of open star clusters.**
        A. V. Tutukov.
Early stages of stellar evolution, (see 012.041), p. 128 - 146
(1977). In Russian.
    Parker's instability on the leading side of a spiral shock
wave leads to the birth of massive gravitationally bound clouds-
protoclusters. Young stellar clusters closely connected with
gas clouds arise as a result of gravitational collapse and fragmentation of those protoclusters. Formation of bright massive
stars leads to rapid dissipation of gas and this way leads to full
disruption of the young stellar cluster and to the birth of O–B
associations. Clusters that remain gravitationally bound after
the dissipation of gas disrupt in 108 years due to dynamical
dissipation of their members and to their collisions with interstellar gas clouds.

151.072   **Periodic orbits near the particle resonance in galaxies.**
        G. Contopoulos.
ESO Sci. Prepr. No. 14, 2 + 27 pp. (1977).
    Near the particle resonance of a spiral galaxy the almost
circular periodic orbits that exist inside the resonance (direct)
or outside it (retrograde) are replaced by elongated trapped
orbits around the maxima of the potential L_4 and L_5. These
are the long-period trapped periodic orbits. The long-period
orbits shrink to the points L_4, L_5 for a critical value of the
Hamiltonian h. The evolution of the periodic orbits with h is
followed, theoretically and numerically, from the untrapped
orbits to the long-periodic orbits and then to the short-periodic
orbits, mainly in the case of a bar. In a tight spiral case an
explanation of the asymmetric periodic and banana orbits is
given, and an example of short-period orbits not surrounding
L_4 or L_5 is provided.

151.073   **The effects of the self-gravity of the interstellar gas
        on galactic shock waves.**   T. Sawa.
Publ. Astron. Soc. Japan, Vol. 29, 781 - 794 (1977).
    Galactic shock waves in a galaxy modeled as an infinite
homogeneous cylinder in which the self-gravity of the interstellar gas is taken into account are considered. A large phase
shift between the gaseous density wave and the stellar density

wave occurs when the self-gravity of the interstellar gas becomes dominant compared with the gravity of the stellar density wave. It is shown that a purely gaseous density wave which contains galactic shock waves cannot exist in the present approximation if the stellar density wave is absent.

151.074 On the arm structure and the star formation in M 51
T. Mizuno.
Publ. Astron. Soc. Japan, Vol. 29, 795 - 811 (1977).

On the basis of the density-wave and shock-wave theories of spiral arms, the author made a detailed analysis of dark lanes and luminous regions of the spiral arms of M 51. If the rate of star formation is proportional to the n-th power of interstellar gas density, it is found from comparison with the observed brightness distribution across the arms that n is 2 to 3, closer to 2. The case of $n = 1$ is ruled out.

151.075 Perturbations of the first order in the motion of a star inside an axisymmetric heterogeneous ellipsoidal cluster.   A. S. Baranov.
Tr. Inst. Teor. Astron., Leningrad, vyp. (No.) 16, p. 3 - 18 (1977). In Russian.

The secular perturbations of the first order of a star orbit moving inside an ellipsoid of rotation with homothetical strata are investigated. The ellipsoid departs only slightly from a sphere. The stellar density is assumed to be subject to a parabolic law. As an undisturbed problem the motion of the star inside the sphere with parabolic density law is researched. It is shown that in the perturbations of the first order the dimensions and the shape of the closed restricted undisturbed orbit and also its inclination do not vary in secular manner. The longitude of the ascending node, the distance of the pericenter from the node and the instant of passage through the pericenter in the perturbations of the first order vary linearly with time. A method to calculate the secular perturbations of the first order for a periodic undisturbed star orbit is outlined.

151.076 From Mädler to Kuzmin − the first step in modelling of flattened stellar systems.   H. Eelsalu.
Calendar of the Tartu Astronomical Observatory for the year 1978, (see 047.026), p. 36 - 40 (1977). In Estonian.

151.077 The accretion spiral.   J. Jaaniste.
Calendar of the Tartu Astronomical Observatory for the year 1978, (see 047.026), p. 50 - 67 (1977). In Estonian.

151.078 Das Altern von Spiralarmen.
H. Schwerdtfeger, R. Wielen.
Mitt. Astron. Ges., Nr. 42, p. 90 - 91 (1977).

151.079 Berechnung des Energieinhalts von aus mehreren Populationen zusammengesetzten Galaxien.
W. Neutsch.
Mitt. Astron. Ges., Nr. 42, p. 102 - 103 (1977).

151.080 Modulation instability of a non-linear travelling wave of star density.   S. N. Nuritdinov.
Dokl. AN UzSSR, 1977, No. 6, p. 36 - 38. In Russian. − Abstr. in Ref. zh., 51. Astron., 12.51.668 (1977).

151.081 Spiral density waves in plane galaxies − moving solitons.   A. B. Mikhajlovskij, V. I. Petviashvili, A. M. Fridman.
Pis'ma v ZhEhTF, Vol. 26, 129 - 133 (1977). In Russian. Abstr. in Ref. zh., 51. Astron., 1.51.722 (1978).

151.082 Some discussion and speculation on the development of galaxies.   A. P. Fairall.
Mon. Notes Astron. Soc. South. Africa, Vol. 36, 120 - 131 (1977).

151.083 Galaxies and entropy from nonlinear fluctuations: a simple wave analysis.   E. P. T. Liang.
Astrophys. J., Vol. 216, 206 - 211 (1977).

Shock damping of nonlinear adiabatic fluctuations and entropy production are studied in the context of simple waves in a radiation-dominated Friedmann model. Assuming a power-law spectrum of clumpiness at recombination, the author derives the primeval spectrum when fluctuations first enter the Jeans length. It is found that weakly nonlinear ($\delta_0 \lesssim 3$) fluctuations with masses below a critical mass M_{crit} ($\sim 10^{11} - 10^{12} M_\odot$ for $\Omega = 1 - 0.1$) cannot survive till recombination with sufficient amplitude. But strongly nonlinear fluctuations ($\delta_0 \gg 1$, $\Delta v_0 \sim 1$) could survive even if they have very short wavelengths.

Lectures on density wave theory.
See Abstr. 003.139.

Theoretical and applied mechanics.
See Abstr. 012.062.

A new method for calculating stellar densities.
See Abstr. 031.208.

Dispersion of particles from a point source active over an interval of time.   See Abstr. 042.013.

A relation in families of periodic solutions.
See Abstr. 042.039.

The star distribution around a massive black hole in a globular cluster. II. Unequal star masses.
See Abstr. 066.028.

On the kinetic description of perturbations of homogeneous gravitating systems.   See Abstr. 066.344.

Chemical and physical properties of nearby Population I stars.   See Abstr. 115.013.

Encounters between stars and dense interstellar clouds.   See Abstr. 131.067.

Rates of star formation.   See Abstr. 131.081.

Quasars and young galaxies.   See Abstr. 141.122.

Dynamical models for M15 without a black hole.
See Abstr. 154.032.

Shock-wave model of the expanding ring in the galactic center region.   See Abstr. 155.005.

Shock-wave model of the kinematic features of neutral hydrogen and molecules CO, OH, and H_2CO near the galactic center.   See Abstr. 155.006.

The diffusion of stellar orbits derived from the observed age-dependence of the velocity dispersion.
See Abstr. 155.008.

On the motions of the Sun, the Galaxy and the Andromeda nebula.   See Abstr. 155.019.

H I sheets ejected by M32 from M31 and multiple disk radial velocities.   See Abstr. 158.076.

Structure and evolution of galaxies.
See Abstr. 158.105.

Interferometric study of NGC 1313.
See Abstr. 158.113.

Dynamical friction in aspherical clusters.
See Abstr. 160.054.

Cosmology and galaxy formation.

See Abstr. 162.024.

Recent theories of galaxy formation.
See Abstr. 162.052.

152 Stellar Associations

152.001   **Spectroscopic studies of stars in Per OB2.**
H. H. Guetter.
Astron. J., Vol. 82, 598 - 605 (1977).
Spectral types on the Morgan-Keenan system are presented for 111 B- and A-type stars in the association Per OB2. By the variable extinction method and the color difference method, a value of 3.2 ± 0.2 is computed for the total-to-selective absorption ratio R across the region of the association. A mean intrinsic distance modulus of 8.08 ± 0.07 mag is derived from the spectroscopic parallaxes. A band of A-type giants and subgiants is observed above the main sequence in the HR diagram. Six Am and five Ap stars are found.

152.002   **Photoelectric and spectroscopic observations of Bochum 15.**
P. D. Jackson, M. P. FitzGerald, A. F. J. Moffat.
Astron. Astrophys., Vol. 60, 417 - 420 (1977) = Contrib. Univ. Waterloo Obs., No. 54.
Photoelectric UBV photometry of 33 stars in the stellar aggregate Bochum 15 is presented. Twenty-one of these stars have been classified on the MK system using slit spectra. Bochum 15 is found to consist of two superimposed groups of stars lying at 3.0 ± 0.3 kpc, $\langle E_{B-V} \rangle = 0.^{m}54 \pm 0.^{m}16$ (s.d.) (group A) and 5.2 kpc, $\langle E_{B-V} \rangle = 0.^{m}56 \pm 0.^{m}08$ (s.d.) (group B). Group A has 20 members, two of which are early type supergiants, and group B has 7 members. One star is a faint white dwarf with distance about 70 pc.

152.003   **Observational evidence for supernova-induced star formation: Canis Major R1.**
W. Herbst, G. E. Assousa.
Astrophys. J., Vol. 217, 473 - 487 (1977).
The R association CMa R1 is found to lie at the edge of a large-scale ring of emission nebulosity. The form of the ring and the absence of luminous stellar objects at its center suggest that it may be a relatively old supernova remnant (SNR). This suggestion is greatly strengthened by the discovery of an expanding H I shell coincident with the optical feature and the discovery of a runaway star, HD 54662, in CMa OB1. An age of order 5×10^5 years is derived for the SNR. The close agreement between the likely ages of the stars and the age of the SNR, as well as the location of the recently formed objects with respect to the supernova shell, strongly support the hypothesis that a supernova event triggered star formation in CMa R1. Several other cases where evidence exists for supernova-induced star formation are briefly discussed.

152.004   **On helium abundance variations in associations and clusters of OB stars.**   L. S. Lyubimkov.
Astrofizika, Vol. 13, 139 - 151 (1977). In Russian. English translation in Astrophysics, Vol. 13, No. 1.
The method proposed previously (1974, 1976) for helium abundance determination in atmospheres of O6-B2 stars is extended to later B stars. As a result more accurate N(He)/N(H) values are obtained for O-associations and young clusters considered earlier, and the helium abundance is determined for some new groups of hot stars. 138 stars of spectral types O6.5-

B8 and luminosity classes V—III are investigated. The conclusion (1975, 1976) that the helium abundance tends to increase with the age of stellar groups is confirmed.

152.005   **Study of the NGC 2439 association.**   D. G. Turner.
Astron. J., Vol. 82, 805 - 809 (1977).
The reality of the association of OB stars surrounding the Puppis cluster NGC 2439 is examined with the aid of photometric and spectroscopic observations of member stars. The association stars are found to be very closely related in terms of age, color excess, and distance to stars in NGC 2439, provided that the distance derived by White (1975) for this cluster is revised to 3400 pc. The data for four luminous supergiant members of the association are used to test the calibration of the β index for luminous stars.

152.006   **Supernova-induced star formation in Cepheus OB3.**
G. E. Assousa, W. Herbst, K. C. Turner.
Astrophys. J., Lett., Vol. 218, L13 - L15 (1977).
The authors have reinvestigated the neutral hydrogen distribution in the direction of Cep OB3 and confirm the existence of an expanding shell. The expansion velocity, 35 km s^{-1}, and radius, 53 pc, imply that the feature is a supernova remnant (SNR) with an age of $\sim 4.3 \times 10^5$ years. The younger subgroup of Cep OB3 has a comparable age, and the authors suggest that this is another example of supernova-induced star formation. The kinematics of the association can also be explained by this hypothesis. The pulsar PSR 2223+65 may be a remnant of the SN event.

152.007   **A study of the NGC 2439 association.**
D. G. Turner.
J. R. Astron. Soc. Canada, Vol. 71, 409 (1977). — Abstract.

152.008   **The Lambda Orionis association.**
P. Murdin, M. V. Penston.
Mon. Not. R. Astron. Soc., Vol. 181, 657 - 665 (1977).
The λ Orionis association has the photometric properties of a typical young cluster with an age of about 4×10^6 yr. Its distance is 400 ± 40 pc. Attention is drawn to the lack of a dense molecular cloud and associated infrared sources in this young grouping.

152.009   **Ara OB1, NGC 6193 and Ara R1: an optical study of a very young southern complex.**
W. Herbst, R. J. Havlen.
Astron. Astrophys., Suppl. Ser., Vol. 30, 279 - 295 (1977).
The authors present photoelectric and photographic $UBVRI$ photometry for over 700 stars in the field of Ara OB1, including many in the central cluster NGC 6193. Association members have been identified on the basis of their reddenings and color-magnitude diagrams are presented. Stars in Ara OB1 are found to exhibit a near infrared anomaly identical to that seen in other young clusters and associations. The authors show that this anomaly cannot be due to thermal radiation from hypothesized circumstellar dust shells. Evidence of very recent star formation activity is present in the (Ara R1) dust cloud adjacent to NGC 6193. The presence of a clear ionization

front at the edge of this cloud should make it an ideal case for testing the model of ionization-front-shock-induced star formation once molecular line mapping of the cloud is accomplished.

152.010   Photoelectric photometry in the CrA T1 association.

V. I. Kardopolov, G. I. Filip'ev.
Astron. Tsirk., No. 959, p. 2 - 3 (1977). In Russian.

A luminous carbon star in Canis Major OB1.
See Abstr. 115.012.

153 Galactic Clusters

153.001   On the distance of the Hyades.
        H. A. McAlister.
Astron. J., Vol. 82, 487 - 489 (1977).

Hanson's (1975) convergent-point solution for the Hyades has been analyzed for possible systematic magnitude effects. The test originally applied for such effects does not rule out their existence, and a well-defined magnitude equation is found in the right ascension components of proper motion. Consideration for this effect is given by eliminating stars with $m_v \leqslant 9.5$ from the X solution of the proper motion gradient, and the distance modulus is found to decrease from 3.42 ± 0.20 to 3.18 ± 0.16.

153.002   The distance to the Hyades Cluster and the extra-galactic distance scale.   S. van den Bergh.
Astrophys. J., Lett., Vol. 215, L103 - L105 (1977).

It is shown that only about half of any change in the adopted distance to the Hyades Cluster will be reflected in the extragalactic distance scale. This is so because the calibration of some types of distance indicators (supernovae, globular clusters, RR Lyrae stars) involves techniques that bypass the Hyades. It is pointed out that the Cepheid distance scale is quite sensitive to the assumption that typical young clusters have the same (high) metallicity as do the Hyades. If $<\delta(U-B)> = +0.03$ for typical young stars, then the Cepheid distance scale has to be decreased by 7%.

153.003   The young cluster NGC 5367 and A1353 - 40.
        P. M. Williams, P. W. J. L. Brand, A. J. Longmore,
T. G. Hawarden.
Mon. Not. R. Astron. Soc., Vol. 180, 709 - 715 (1977).

The sparse young cluster embedded in and illuminating NGC 5367 in the head of Cometary Globule 12 is shown to suffer about 1 mag of visual extinction in the cloud and to be about 630 pc away. Infrared photometry and scanner spectroscopy of the two most luminous members, comprising the double h 4636, indicates that one component is surrounded by a circumstellar shell producing Balmer line emission and thermal re-radiation by grains. The authors do not favour the proposed identification of NGC 5367 with the transient X-ray source A1353−40. They suggest that NGC 5367 lies on an H I loop and that star formation in it was induced by a supernova explosion near $l = 320°$, $b = 30°$ about 10^7 yr ago.

153.004   Spectral types in the open cluster NGC 6633.
        H. Levato, H. A. Abt.
Publ. Astron. Soc. Pacific, Vol. 89, 274 - 276 (1977).

Spectral types are given for the 26 brightest stars in the region of NGC 6633. Only two of the giants are definitely cluster members. Two apparent blue stragglers are present, as well as one extreme Ap star, four marginal Am stars, and two stars with weak shell lines.

153.005   Four-color and Hβ photometry of open clusters.
        XII. NGC 6910 and NGC 6913.
D. L. Crawford, J. V. Barnes, G. Hill.
Astron. J., Vol. 82, 606 - 611 (1977) = Dominion Astrophys.

Obs., Victoria, B.C., Canada, Contrib. No. 334 = Natl. Res. Council No. 15903.

Intermediate-band photoelectric photometry was obtained for 23 stars in NGC 6910 and 25 in NGC 6913 (M29). By analysis of the data, the authors conclude that 16 stars are members of the cluster NGC 6910 and 17 of NGC 6913. The average reddening of cluster members is high. The authors find a distance modulus of 10.5 mag for NGC 6910 and 10.2 mag for NGC 6913. Neither cluster is rich in members, but both certainly appear real. For NGC 6913, the distance modulus from the photometry does not agree well with that found by use of MK spectra; the agreement is good for NGC 6910.

153.006   The Hyades-age cluster NGC 1817.
        G. L. H. Harris, W. E. Harris.
Astron. J., Vol. 82, 612 - 619 (1977) = Contrib. Univ. Waterloo Obs., Waterloo, Ontario, Canada, No. 63.

The authors describe the results from a newly completed UBV photometric study of NGC 1817, which is shown to be a populous open cluster similar to the Hyades in age. The foreground reddening is calculated to be $E_{B-V} = 0.28 \pm 0.03$ and is apparently uniform, while the distance modulus is $(m-M)_0 = 11.3 \pm 0.3$. The color-magnitude diagram reveals a well-defined main sequence, turnoff region, and populous red giant clump. The latter is $\cong 0.15$ mag bluer than the Hyades/Praesepe giants, suggesting that NGC 1817 may have a slightly lower heavy-element abundance.

153.007   On the reality of the open cluster Ly 6 and the
        membership of the cepheid TW Nor.   P. S. Thé.
Astron. Astrophys., Vol. 60, 423 - 424 (1977).

The results of an analysis of photoelectric observations of the open cluster Ly 6 on the Walraven $VBLUW$ photometric system are reported. Seventeen stars in the surrounding of TW Nor were observed. The conclusion by several astronomers that the cluster Ly 6 is real, and that the cepheid TW Nor is a member of this cluster is generally supported.

153.008   *RGU* photometry of the galactic star cluster
        NGC 3247.   C. Grubissich.
Astron. Astrophys., Suppl. Ser., Vol. 29, 379 - 382 (1977). In German.

The discussion of RGU photometry data leads to a true distance modulus of 10.83 mag. (1420 pc). Age and probable membership, as well as other cluster parameters are discussed. NGC 3247 is too old to be a spiral indicator. In the surroundings of the cluster there seems to be a local group of late-type giants at a distance of 1.7 kpc as it is the case with IC 2581 which is only 0.3° away from NGC 3247.

153.009   Membership of M39.
        B. J. McNamara, W. L. Sanders.
Astron. Astrophys., Suppl. Ser., Vol. 30, 45 - 62 (1977).

Proper motions are presented for 1710 stars in the vicinity of M39. Only 30 of these stars are suggested as cluster members. The cluster is unusual both for its sparseness and its lack of a faint stellar population. The cluster distance is re-determined and found to be 265 pc and its age is estimated as 7×10^8 years.

153.010 A study of the diameter and luminosity function of open clusters based on star counts.
P. Pişmiş, Ş. Bozkurt.
Astron. Astrophys., Suppl. Ser., Vol. 30, 81 - 87 (1977).

Star counts in and around six open clusters NGC 1647, 2099, 2439, 2482, T9 and NGC 2539, performed on Schmidt plates have yielded the diameters of the clusters and the luminosity function for three of them. In NGC 2099 the bright stars give a smaller diameter than the fainter ones. This indicates that the bright stars are located close to the center. The luminosity function in three clusters NGC 1647, 2099 and T9 shows that the range in absolute magnitude is around 7.

153.011 Places of formation of 24 open clusters.
J. Palouš, J. Ruprecht, O. B. Dluzhnevskaya, T. Piskunov.
Astron. Astrophys., Vol. 61, 27 - 37 (1977).

The authors have derived the places of formation of 24 open clusters. With the help of given kinematical data and estimates of the ages they have derived best fit angular rotation speeds for the spiral structure: $\Omega_p = 13.5$ km s^{-1}kpc^{-1} and $\Omega_p = 20.0$ km s^{-1}kpc^{-1}. Adopting these values, the places of formation closely correlate with the spiral arms according to the prediction of the density-wave theory.

153.012 Maximum-likelihood method for determination of membership in open clusters. M. H. Slovak.
Astron. J., Vol. 82, 818 - 823 (1977).

The method of determining cluster membership by modeling the distribution of proper motions in the field of a cluster is examined in detail. Using a Monte Carlo acception–rejection technique, a simulated set of proper motions is generated and used to test the stability and uniqueness of the solution for the distribution parameters as given by a maximum-likelihood method. Such a treatment of the proper motion data defines an independent criterion which, in conjunction with photometric and radial velocity data, provides a useful discriminant for distinguishing actual cluster members from surrounding field stars.

153.013 NGC 6383: a very young open cluster with pre-main-sequence evolution. M. P. Fitzgerald,
P. D. Jackson, M. Luiken, E. J. Grayzeck, A. F. J. Moffat.
J. R. Astron. Soc. Canada, Vol. 71, 409 (1977). – Abstract.

153.014 Trumpler 27: a heavily-reddened young cluster with blue and red supergiants.
A. F. J. Moffat, M. P. Fitzgerald, P. D. Jackson.
J. R. Astron. Soc. Canada, Vol. 71, 409 - 410 (1977). Abstract.

153.015 Variable-star photometry of NGC 6530.
G. C. Kilambi.
J. R. Astron. Soc. Canada, Vol. 71, 410 (1977). – Abstract.

153.016 The Hyades-age cluster NGC 1817.
G. L. H. Harris, W. E. Harris.
J. R. Astron. Soc. Canada, Vol. 71, 410 (1977). – Abstract.

153.017 UBV photometry of the young cluster NGC 6604.
D. Forbes, D. L. DuPuy.
J. R. Astron. Soc. Canada, Vol. 71, 410 (1977). – Abstract.

153.018 Photoelectric photometry of the open cluster M34.
B. Cester, G. Giuricin, F. Mardirossian, M. Pucillo.
Astron. Astrophys., Suppl. Ser., Vol. 30, 227 - 229 (1977).

UBV photometry of 43 stars, mostly members of the open cluster M34, is reported. Using cluster members of spec-tral type earlier than A0, together with other stars observed by Johnson, a new determination of the true distance modulus was made. Several stars not located on the main sequence are discussed. The age of the cluster is estimated to be about 1×10^8 years.

153.019 Extinction law in dust clouds and the young southern cluster NGC 6250: further evidence for high values of R. W. Herbst.
Astron. J., Vol. 82, 902 - 907, 934 (1977).

UBVRI photometry and spectral types have been obtained for stars in the young southern cluster NGC 6250. These include three stars which illuminate reflection nebulae. The properties of this cluster are $E(B-V) = 0.37$ mag, $V_0 - M_v = 10.05$, earliest spectral type = B1, and age = 1.4×10^7yr. Two of the stars in reflection nebulae are highly reddened and permit accurate determinations of their reddening laws. One has an unusually steep law characterized by a ratio of total to selective extinction, $R \geq 4.9$. The other appears to be normal. These results not only confirm the existence of abnormal reddening laws in high-density interstellar regions, but also show that location in a dense cloud is not sufficient reason to expect a steeper-than-normal law.

153.020 The galactic cluster NGC 2527.
R. J. Dodd, H. T. MacGillivray, R. W. Hilditch.
Mon. Not. R. Astron. Soc., Vol. 181, 729 - 734 (1977).

UBV photographic photometry of 368 stars to a limiting V magnitude of 16.6 is used to investigate the cluster status of NGC 2527. The distance to NGC 2527 is estimated to be 590 pc. Star counts to $B = 17.5$ mag using the COSMOS measuring machine confirm the lack of central condensation of this cluster which, combined with the fainter extension of the main sequence and the increased membership found from the three-colour photometry, leads to a revised Trumpler classiciation of III–2m for NGC 2527. The cluster age is estimated to be 10^9 yr.

153.021 The determination of distance, absorption, probable physical members and age for the open clusters Stock 16, Basel 18, Basel 19, Cr 272 and NGC 5168.
R. P. Fenkart, B. Binggeli, D. Good, D. Jenni, L. Labhardt, A. Tschumi.
Astron. Astrophys., Suppl. Ser., Vol. 30, 307 - 313 (1977).

With Becker's three-colour photometric method, using both (RGU-) colour-magnitude diagrams, the authors determined distance, absorption, probable physical members and age for the five open clusters Stock 16, Basel 18, Basel 19, Cr 272 and NGC 5168. The distance of the clusters varies between 600 and 2900 pc, and their absorption in G covers the range between 0.5 and 2.7 mag. Only upper limits for earliest spectral type and age could be determined for these clusters. Two of them, Stock 16 and Cr 272, prove to be sufficiently young to be considered as reliable spiral arm indicators, confirming essentially the established distribution pattern in the galactic disk.

153.022 Four-color and Hβ photometry of stars in NGC 7654 and M25. E. G. Schmidt.
Publ. Astron. Soc. Pacific, Vol. 89, 546 - 549 (1977).

Four-color and Hβ photometry has been obtained of stars in the galactic clusters NGC 7654 and M25. An analysis of these data shows variable extinction in both clusters. The distance modulus of NGC 7654 is found to be $10^{m}.99 \pm 0.2$ and that of M25 is $8^{m}.68 \pm 0.1$.

153.023 A comparison of observations with main-sequence evolutionary models.
A. G. D. Philip, P. Demarque, A. V. Sweigart, R. B. Ciardullo.
Publ. Astron. Soc. Pacific, Vol. 89, 554 - 556 (1977).

A comparison of four-color observations, transformed to

the ($\log g - \log T_{\text{eff}}$) plane, with new theoretical isochrones for the composition ($Y, Z = 0.2, 0.04$) shows good agreement between theory and observation in the range from 0.5 to 1×10^9 years. Ages are derived for seven open clusters which are in accord with previous estimates.

153.024 Spectral types in the open cluster NGC 2169. H. A. Abt.
Publ. Astron. Soc. Pacific, Vol. 89, 646 - 647 (1977).

Spectral classification of the eight brightest stars in the field shows that six are cluster members and the earliest type is B2 III. The cluster distance is 1500 ± 200 pc and the mean reddening is $0^m.17$.

153.025 Spectral types in the open cluster M34. H. A. Abt, H. Levato.
Publ. Astron. Soc. Pacific, Vol. 89, 648 - 649 (1977).

Spectral types for the 18 brightest members of M34 indicate the following: one Hg-Mn Ap star, one pronounced Si-Cr-Sr Ap star, one marginal Am star, a star with a weak shell spectrum, and five subgiants or giants.

153.026 Der offene Sternhaufen NGC 6067. W. Lohmann.
Astrophys. Space Sci., Vol. 47, 447 - 455 (1977) = Astron. Rechen-Inst., Heidelberg, Mitt. Ser. A.

The open star cluster NGC 6067 was investigated by the strip method. Because of the large distance (1820 pc) of the cluster its luminosity function is known up to date only between $M_v = -4.5$ and $M_v = +1.5$. In this paper it is continued to $M_v = 4.4$ and further extrapolated by means of 2 variants. The cluster contains 419 respectively 476 stars with total masses of 1453 respectively 1483 $M_\odot$. On account of its radius of 5.9 pc the cluster is a rather extended object, which can be described by the generalized density law of Schuster with $n = 4.42$ and the central star density 8.9 stars pc^{-3}. The mean velocity of the stars amounts to 1.03 km s^{-1}, the mass-brightness relation is 0.031 in solar units.

153.027 Photoelectric photometry of the open cluster Tr-1. U. C. Joshi, R. Sagar.
Astrophys. Space Sci., Vol. 48, 225 - 230 (1977).

The H-R diagram of the open cluster Tr-1 based on Oja membership and U, B, V photoelectric photometry is presented. The colour excess $E(B-V)$ is $0^m.52$. The distance modulus to the cluster and its age are, respectively, estimated at $11^m.6$ and 2.6×10^7 yr.

153.028 Der offene Sternhaufen NGC 2362. W. Lohmann.
Astrophys. Space Sci., Vol. 51, 173 - 176 (1977) = Astron. Rechen-Inst., Heidelberg, Mitt. Ser. A.

The very young open star cluster NGC 2362 was investigated by the strip method on charts of two photographs taken with the 1-m Schmidt telescope of the European Southern Observatory. Up to the limiting magnitude $M_v = 5^m.8$ the cluster contains 100 stars and can be described by the Gaussian density law. Further results are: Mass $M = 246 M_\odot$, central mass density $\rho_0 = 43.1 M_\odot$ pc^{-3}, radius $R \approx 2.6$ pc, mean velocity of the stars $\overline{v} = 0.64$ km s^{-1}.

153.029 Study of NGC 2324. I. Catalogue of stellar B magnitudes. A. Ghezloun, M. Svetchnikov, A. Aîad, B. Haddad, K. A. Barkhatova, L. P. Chachkina, O. P. Pylskaia (Pyl'skaya).
Ann. Obs. Astron. Alger, Tome 5, Fasc. 1, p. 1 - 40 (1977). In French.

Plates taken with the astrograph of the Alger observatory in the years 1905, 1909 and 1973 were measured. Blue magnitudes are tabulated and intercompared. The cluster contains about 640 stars in an approximately elliptic form whose center is near the star N°270. The great diameter lies in the direction 60° West of North. The nucleus and the corona extend to about $5'.5$ and $8'.0$ respectively. 22 stars are found to be probable variables.

153.030 UBV photometry and proper motions of four open star clusters in Cassiopeia. V. N. Frolov.
Astron. Tsirk., No. 957, p. 6 - 8 (1977). In Russian.

153.031 On the spatial distribution of open star clusters. K. A. Barkhatova, O. P. Pyl'skaya.
Astron. Tsirk., No. 958, p. 3 - 5 (1977). In Russian.

153.032 The (U−B)−(B−V) two-colour diagram for bright stars of the cluster NGC 6913. R. M. Raznik.
Astron. Tsirk., No. 959, p. 3 - 6 (1977). In Russian.

153.033 uvbyβ photometry of NGC 2244. A. M. Heiser.
Astron. J., Vol. 82, 973 - 977 (1977).

uvbyβ photometry is presented for 33 stars of NGC 2244, the open cluster associated with the Rosette Nebula. A color excess, $E(b-y) = 0.34$ mag, and a distance modulus, $V_0-M_V = 10.96$ mag, have been determined using all the available photometric data. An age of about 3×10^6 yr is found from the cluster's five O-type main-sequence stars. None of the stars in this study appear to be pre-main-sequence objects.

153.034 Photometry of new possible members of the Hyades cluster. A. R. Upgren, E. W. Weis.
Astron. J., Vol. 82, 978 - 984 (1977).

New photoelectric photometry in B, V, R, and I colors has been obtained for 91 of the 119 stars identified from their proper motions as new members of the Hyades cluster by Pels, Oort, and Pels−Kluyver (1975). On the basis of the photometric data, some of these stars can be rejected as members of the cluster.

153.035 Photometry and proper motions of stars of four galactic clusters in Cassiopeia. V. N. Frolov.
Izv. Glav. Astron. Obs. Pulkovo, No. 195, Astrofiz. Astrometr., p. 80 - 97 (1977). In Russian.

Relative proper motions were determined for stars within an area of $60' \times 60'$ with the center at $\alpha = 23^h54^m2$, $\delta = +60°57'$ (1950.0) using three plate pairs with a mean epoch difference of 49 years. Negatives taken with the normal astrograph of the Tashkent Observatory during 1923 - 1925 were used as first epoch plates. The second epoch plates were taken with the normal astrograph of the Pulkovo Observatory. Proper motions (μ_x and μ_y) of 2169 stars with the limiting magnitude B = $16^m.5$ were determined relative to reference stars with magnitudes $13^m.5 < B \leqslant 14^m.5$. B and V magnitudes of 2165 and 1930 stars respectively were determined using Sandage's photoelectric standards (1958). Members of the clusters were selected according to two criteria: proper motions and position on the colour-magnitude diagram. Altogether four clusters were investigated: NGC 7790, NGC 7788, Berkeley 58 (Be 58) and the anonymous galactic cluster found by the author with coordinates at $\alpha = 23^h54^m9$, $\delta = +61°21'$ (1950.0). V vs B−V diagrams for cluster members of the separate clusters are given.

The evolution of super-metal-rich stars and the interpretation of old galactic clusters. See Abstr. 065.055.

On the age difference between the oldest Population I and the extreme Population II stars. See Abstr. 065.057.

Observations of cluster passages. See Abstr. 096.015.

Lunar occultation summary. II. See Abstr. 096.017.

Lunar occultations of the Hyades Cluster.
See Abstr. 096.018.

Passages of minor planets across galactic clusters.
See Abstr. 098.040.

Proper motions of stars in the region of the open
cluster NGC 6709.   See Abstr. 112.015.

Proper motions of stars in the open cluster NGC
6940.   See Abstr. 112.017.

Intermediate band photometry of late-type stars. II.
Some stellar groups.   See Abstr. 113.004.

The NGC 2818 photoelectric sequence.
See Abstr. 113.031.

Scanner observations of main-sequence A and F stars.
See Abstr. 114.070.

The spectroscopic binaries in NGC 6475.
See Abstr. 119.003.

Cluster membership of the cepheid UY Persei.
See Abstr. 122.025.

Walker No. 67 in NGC 2264: a candidate for strong
interstellar circular polarization.   See Abstr. 131.039.

Tests observationnels sur la formation stellaire.
See Abstr. 131.079.

A model for the filamentary structure in the
Pleiades reflection nebulosity.   See Abstr. 134.014:

Stochastic tidal disruption of clusters.
See Abstr. 151.042.

Origin and evolution of open star clusters.
See Abstr. 151.071.

On helium abundance variations in associations and
clusters of OB stars.   See Abstr. 152.004.

The galactic density wave, molecular clouds, and
star formation.   See Abstr. 155.015.

Stellar populations of the disk and halo of the
Galaxy.   See Abstr. 155.017.

154 Globular Clusters

**154.001 Magellanic Cloud investigations – V. The LMC
blue cluster NGC 1854.** L. P. Connolly, W. G. Tifft.
Mon. Not. R. Astron. Soc., Vol. 180, 401 - 405 (1977).

Photographic photometry of the LMC blue globular
cluster NGC 1854 is presented. The main sequence appears
0.1 redder than the surrounding field population. The cluster
contains both blue and red supergiants and stars which fall
in the instability strip. The instability strip stars do not appear
variable and may be binaries containing a main sequence star
and a red supergiant. Difficulties are discussed concerning the
determination of the ratio of the number of blue supergiants
to the number of red supergiants. The age spread is found to
be quite low and not incompatible with a value of zero.

154.002 On the space distribution of globular clusters.
G. de Vaucouleurs.
Astron. J., Vol. 82, 456 - 458 (1977).

The recent data of Harris on the spatial structure of the
globular cluster system are analyzed to derive an improved
expression of their distribution law. It is found that the two-
dimensional distribution $\sigma(R)$ projected on the Y, Z plane
obeys closely the $R^{1/4}$ law, characteristic of the spheroidal
component of galaxies and of globular clusters of galaxies.
The effective radius of the cluster system is $R_e = 2.8$ kpc, if
$R_0 = 8.5$ kpc. Upper and lower limits of the total mass of the
Galaxy derived from the virial theorem applied to the cluster
system with and without allowance for rotational energy are
in the range $2.77 \times 10^{10} < M_T^* / M_\odot < 2.3 \times 10^{11}$.

154.003 Remote halo globular cluster Palomar 5.
A. Sandage, F. D. A. Hartwick.
Astron. J., Vol. 82, 459 - 464, 525 (1977).

The color-magnitude diagram for the sparse, remote
globular cluster Palomar 5 is obtained from a UBV photoelectric sequence of 29 stars. The horizontal branch occurs at
$V = 17.35$ which, with $E(B-V) = 0.03$ and $M_v(RR) = +0.6$,
gives $(m-M)_0 = 16.66$, or $D = 21.5$ kpc. The cluster is on the
other side of the Galactic center, at $X \simeq 5$ kpc, and is 15 kpc
above the plane. The intermediate metallicity of this distant
halo cluster, compared with the lower metal abundance of
less remote clusters, suggests that the chemical enrichment of
the halo was spotty, caused by random enrichment events
with inefficient stirring on the free-fall time scale of $< 10^9$
yrs, so that the halo was not mixed with much spatial
uniformity during the Galactic collapse.

154.004 Star field surrounding NGC 6528.
S. van den Bergh, J. deRoux.
Astron. J., Vol. 82, 465 - 467, 527 (1977).

The field surrounding NGC 6528 is found to contain stars
similar to those that occur in the region near NGC 6522. The
cluster NGC 6528 itself also appears to contain quite metal-
rich stars. An error in previously published photometry is noted.

154.005 Radio sources in the direction of globular clusters.
Y. Terzian, E. K. Conklin.
Astron. J., Vol. 82, 468 - 470 (1977).

Observations at 2380 MHz were made in the direction of
five globular clusters. The authors detected six weak radio
sources in the direction of four of the clusters. The nature of
these sources is discussed and it is suggested that some of
these sources may be due to planetary nebulae in the clusters.
One source near the center of NGC 7089 (M2) has an unusually steep nonthermal spectrum.

**154.006 Relation between $(V-K)$ color and metal abundance
for globular clusters.** C. Pritchet.
Astron. J., Vol. 82, 471 - 473 (1977).

The integrated $(V-K)$ colors of globular clusters observed
by Grasdalen, when corrected for reddening, display the expected correlation with metal abundance in the sense that
redder globular clusters are more metal rich. Several globulars
are observed to possess unusually early spectral types for
their infrared colors. It is suggested that the spectral types of
these globular clusters are affected by the presence of a
stronger than usual blue horizontal branch.

154.007 On the nature of the variables in M13.
C. D. Pike, C. J. Meston.
Mon. Not. R. Astron. Soc., Vol. 180, 613 - 621 (1977).

Improved two-colour photographic photometry of the
variable stars in the globular cluster M13 (NGC 6205) is
presented. Contrary to previous suggestions it is shown that
variables 1, 2 and 8 have magnitudes and colours consistent
with their classification as normal population II variables. A
pulsational mass of $0.54\,M_\odot$ is derived for variable no. 1. One
of the red giants observed is suspected of variability.

154.008 A survey of element abundances in globular clusters.
E. A. Mallia.
Astron. Astrophys., Vol. 60, 195 - 203 (1977).

Medium dispersion spectra of stars in four globular
clusters (NGC 5139, NGC 6397, NGC 6656 and NGC 6752)
are used to derive element abundances by means of spectrum
synthesis.

**154.009 A comment on the metal abundance of the LMC
cluster NGC 2209.**
B. Gustafsson, R. A. Bell, P. M. Hejlesen.
Astrophys. J., Lett., Vol. 216, L7 - L9, with a correction in
Vol. 218, L147 (1977).

The metal abundance of the LMC cluster NGC 2209 is
discussed, on the basis of the observational data recently
presented by Gascoigne et al. It is found that a metal abundance of [Fe/H] = −0.5, in reasonable agreement with that
found for the LMC interstellar medium, seems more probable
than the value of [Fe/H] = −1.2 obtained by Gascoigne et al.

**154.010 Evidence for ionized hydrogen in the cores of
globular clusters.** J. E. Grindlay, W. Liller.
Astrophys. J., Lett., Vol. 216, L105 - L109 (1977).

Photometric and spectrographic observations of 27
globular clusters reveal evidence for the existence of an emission component of Hα in the cores of several globular clusters,
notably NGC 5824 and the X-ray clusters NGC 1851, 6624,
and 7078.

**154.011 Radial distribution and total number of globular
clusters in M104.** K.-i. Wakamatsu.
Publ. Astron. Soc. Pacific, Vol. 89, 267 - 270 (1977).

Stellar images around M104 fainter than a threshold
magnitude are concentrated toward the nucleus, thus they are
certainly globular clusters associated with M104. By counting
their images, the radial distribution of their projected number
density is found to be similar to the luminosity profile of the
spheroidal central bulge and can be represented by de
Vaucouleurs' "1/4-law". The total number of globular clusters
in M104 is estimated to be $(2.0^{+3.5}_{-1.0}) \times 10^3$.

**154.012 Suspected globular clusters in the Fornax I cluster
of galaxies.** J. A. Dawe, R. J. Dickens.
Nature, Vol. 263, 395 - 396 (1976).

The authors report the existence of a significant clustering of faint objects around three galaxies in the Fornax I
cluster of galaxies. The nature of these objects is uncertain,
although on the scant evidence available, the authors believe

at least some of them to be unresolved globular clusters.

154.013 The metal-poor globular cluster NGC 6656.
G. Alcaino.
Astron. Astrophys., Suppl. Ser., Vol. 29, 383 - 395 (1977).

A colour-magnitude diagram (CMD) for the nearby southern globular cluster NGC 6656 (M22) was obtained from 374 stars measured photographically. The main features of the CMD are: a well-defined rather steep giant branch sequence, a large sample of blue horizontal branch stars and the absence of red horizontal branch stars. There is a deficiency of stars in the sub-giant branch. The deduced values for the metallicity parameters $\Delta V = 2^{m}.9$ and $S = 5.3$ corroborate the tendency of metal-poorness suggested by its integrated spectral type and colours of F5 and $Q = -0.41$, respectively.

154.014 The globular cluster NGC 6397. G. Alcaino.
Astron. Astrophys., Suppl. Ser., Vol. 29, 397 - 405 (1977).

A BV photometric investigation of the nearby southern globular cluster NGC 6397 was carried out. The main features of the colour-magnitude diagram (CMD) are: a well-defined rather steep giant branch sequence and a large sample of blue horizontal branch stars. The deduced values for the metallicity parameters $\Delta V = 3^{m}.0$ and $S = 6.4$ corroborate the metal deficiency tendency suggested by its integrated spectral type and colours of F5 and $Q = -0.41$, respectively.

154.015 Why globular clusters are so spherical.
A. B. Marchant.
News Lett. Astron. Soc. N.Y., Vol. 1, No. 2, p. 14 (1977). Abstract.

154.016 X-rays from globular clusters. W. Liller.
News Lett. Astron. Soc. N.Y., Vol. 1, No. 2, p. 22 (1977). – Abstract.

154.017 Radio sources in globular clusters.
Y. Terzian, E. K. Conklin.
News Lett. Astron. Soc. N.Y., Vol. 1, No. 2, p. 23 (1977). Abstract.

154.018 The distribution of horizontal-branch stars in the log T_{eff}, log g diagram. A. G. D. Philip.
News Lett. Astron. Soc. N.Y., Vol. 1, No. 2, p. 26 (1977). Abstract.

154.019 A CH star in the globular cluster M22, and the nature of CH and CN anomalies.
R. D. McClure, J. Norris.
Astrophys. J., Lett., Vol. 217, L101 - L104 (1977).

Spectroscopic evidence is presented that the star III-106 in the globular cluster M22 is a CH star. The authors also discuss the CN anomalies in globular cluster giants, and suggest that while both triple-α and CNO mixing may be necessary to explain some CN-strong stars, one would expect CN variations in the metal-rich clusters from CNO mixing alone.

154.020 Sources X dans les amas globulaires.
C. Chevalier.
Recueils des séminaires, (see 012.004), p. 123 - 130 (1977).

154.021 The luminosity distribution of globular clusters in the Virgo cluster of galaxies. D. A. Hanes.
Mem. R. Astron. Soc., Vol. 84, 45 - 60 (1977).

Luminosity distributions are presented for globular clusters in 16 fields centred on 20 bright galaxies in the Virgo cluster. The luminosities of the brightest clusters associated with a galaxy depend strongly upon the sample size, a manifestation of the Scott effect (Scott 1957); and it is suggested that the number of clusters is directly proportional to the mass of the parent galaxy. Photographic sequences in the G (=103aJ+GG13) system are presented for the fields.

154.022 An optical search for ionized hydrogen in globular clusters. II. J. E. Hesser, S. J. Shawl.
Astrophys. J., Lett., Vol. 217, L143 - L147 (1977).

Photoelectric Hα line profiles are used to set probable upper limits on the presence of ionized hydrogen in 30 galactic and one extragalactic globular clusters. No ionized gas is detected in any of the clusters. Among the most stringent limits ($M_{\text{H II}} < 1\ M_{\odot}$) on accumulated gas are those for four clusters (NGC 1851, 6441, 6624, and 7078) commonly associated with X-ray sources. Radial velocities, also crucial to 21 cm emission studies where cluster emission must be disentangled from galactic H I emission, are discussed for several clusters.

154.023 UBV photometry of globular clusters containing X-ray sources. S. van den Bergh.
Astron. J., Vol. 82, 796 - 797 (1977).

New UBV observations are presented for the X-ray globular clusters NGC 1851, NGC 6440, NGC 6441, NGC 6624, and NGC 6712. Within the rather limited accuracy of the data, the nuclear colors of these clusters do not differ from their integrated colors.

154.024 Integrated colors of globular clusters in the Galaxy, Fornax, and M31. H. C. Harris, R. Canterna.
Astron. J., Vol. 82, 798 - 804 (1977).

Integrated colors in the University of Washington C, M, T_1, T_2 photometric system are presented for 51 globular clusters in the Galaxy, four in Fornax, and six in M31. Several methods of evaluating the reddening relations are discussed and intrinsic colors are determined. A metallicity index based on color indices is defined which makes a significant improvement over the $UBV\ Q$ index. The Fornax clusters are very metal poor, with a range suggesting a possible enrichment process; those observed in M31 are metal rich.

154.025 Accurate radial velocities using cross-correlation techniques and a TV detector. I. The velocity dispersion of NGC 6397. G. S. Da Costa, K. C. Freeman, A. J. Kalnajs, A. W. Rodgers, T. E. Stapinski.
Astron. J., Vol. 82, 810 - 817 (1977).

A digital TV detector system for the measurement of accurate radial velocities of stars of intermediate to late spectral types is described. The radial velocities are derived by cross correlating the spectrum of the unknown star with that of a radial velocity standard of similar spectral type. The radial velocities of 11 stars in the globular cluster NGC 6397 have been measured to a mean accuracy of 1.5 km sec^{-1}. The velocity dispersion is 3.1 ± 0.7 km sec^{-1} and the ratio of central mass to blue luminosity is 1.3 ± 0.6 in solar units.

154.026 A spectroscopic survey of the giant branch of M5.
R. Zinn.
Astrophys. J., Vol. 218, 96 - 104 (1977).

Spectrograms of 50 stars in the globular cluster M5 have been examined for spectral peculiarities. The stars in this sample brighter than $V = 14$ ($M_v = -0.5$) represent an unbiased sample of the asymptotic and red-giant branches. Seven, or 78% of the asymptotic branch stars are weak-G-band stars. Weak-G-band stars are not found near the tip of the giant branch. This result and the color-magnitude diagram of Simoda and Tanikawa (1970) suggest that the asymptotic branch in M5 does not merge with the giant branch, but stops at $M_v \approx -1.7$. None of the 50 stars is a CH star, but eight, or 21%, of the giant branch stars have unusually strong $\lambda4216$ or $\lambda3883$ CN bands. These stars uniformly populate the giant branch over the range $-0.5 \gtrsim M_v \gtrsim -2$. A comparison of the survey of M5 with the surveys of ω Cen by other observers reveals that the CN stars in M5 are only mildly peculiar relative to the CN and CH stars in ω Cen. The peculiar giants in M5 have a negligible effect on the integrated light of the cluster. The origins of the peculiar giants in M5 are briefly discussed in terms of the theory of stellar evolution.

154.027 *J, H, K* photometry in the globular cluster Omega Centauri and the spread of the giant branch.
I. S. Glass, M. W. Feast.
Mon. Not. R. Astron. Soc., Vol. 181, 509 - 516 (1977).

J, H, K observations are presented and discussed for a number of the brighter stars in ω Cen. The large spread of the giant branch in the V, $(B-V)$ diagram, which is greatly reduced in an R, $(R-I)$ plot, reappears again in a K, $(J-K)$ plot. At a given R and $(R-I)$, the giants with redder $(B-V)$'s are fainter in the blue and brighter in the infrared.

154.028 Photoelectric UBV photometry in the anomalous globular cluster NGC 2808.   W. E. Harris.
J. R. Astron. Soc. Canada, Vol. 71, 410 - 411 (1977).
Abstract.

154.029 Gas in globular clusters. II. Time-dependent flow models.   D. A. VandenBerg, D. J. Faulkner.
Astrophys. J., Vol. 218, 415 - 430 (1977).

A numerical method is presented for implicit solution of the hydrodynamic equations appropriate to gas flow studies in globular clusters. Assuming spherical symmetry, the system of equations is solved in Eulerian coordinates by a suitable modification of the Henyey method. Shocks are automatically treated by including artificial viscous pressure terms in the difference scheme. Time-dependent flow models have been computed for a $10^6 M_\odot$ cluster and for three selected values of β, corresponding to $(2\beta)^{1/2} = 150$ km s^{-1}, 100 km s^{-1}, and 50 km s^{-1}, where β is the specific energy of injection of gas from stars into the flow. In each case, the ionization of hydrogen and helium and the radiative cooling of the gas have been accurately treated throughout the flow, on the assumption that ionization by electron collisions is in equilibrium with radiative and dielectronic recombination.

154.030 Near infrared photometry of globular clusters – VIII. The anomalous giant branch of ω Centauri.
T. Lloyd Evans.
Mon. Not. R. Astron. Soc., Vol. 181, 591 - 598 (1977).

Photometry and radial velocities confirming membership are given for additional stars in ω Cen. The anomalous giant stars have strong CN bands as a rule. Their line spectra are generally similar to those of 47 Tuc stars of similar $V-I_K$ and differ from those on the main giant branch in that the strongest lines are stronger. This is consistent with the interpretation by Lloyd Evans (1977, Paper V), that they are of higher metal content than the cluster as a whole, but the interpretation is not unique. The λ4554 line of Ba II appears possibly enhanced in the anomalous stars, suggesting a connection with the stars on the main giant branch whose strong CN bands are attributed to mixing of reaction products to the surface.

154.031 Detection and study of variable stars in the globular cluster NGC 6638.   B. Rutily, A. Terzan.
Astron. Astrophys., Suppl. Ser., Vol. 30, 315 - 322 (1977).
In French.

The photometric study of the variable stars detected in the globular cluster NGC 6638 or its close vicinity is carried out with the help of new m_{pg} and m_r sequences. The periods of five Mira Ceti stars and 15 RR Lyrae stars are computed. It results from this computation a first estimation of the distance of the cluster to the Sun: $r = 6.6$ kpc ± 2.3 kpc.

154.032 Dynamical models for M15 without a black hole.   G. D. Illingworth, I. R. King.
Astrophys. J., Lett., Vol. 218, L109 - L112 (1977).

The authors have fitted the observed velocity dispersion and brightness profile of M15 with dynamical models that have no need for the central black hole suggested by Newell, Da Costa, and Norris (1976). The stellar mixture is the same as in other globular clusters, but M15 has a higher central concentration. In the authors' models the central brightness peak is caused by the gravitational effect of neutron stars; their number is in accord with a reasonable initial mass function and retention probability. Some modes of binary formation are discussed.

154.033 Photometric calibration in the metal-rich globular cluster M69.   W. E. Harris.
Publ. Astron. Soc. Pacific, Vol. 89, 482 - 484 (1977).

Photoelectric BV photometry is reported for sequence stars in the field of NGC 6637, a globular cluster of relatively high heavy-element abundance. The data are used to recalibrate the photographic color-magnitude (CM) diagram by Hartwick and Sandage (1968) for the cluster. Considerable systematic corrections to the previous photographic photometry are found to be needed in both V and $(B-V)$, in the sense that the giant branch becomes both bluer and markedly steeper. The recalibrated photometric parameters describing the morphology of the CM diagram become $\Delta V = 2.4$, $(B-V)_{0,g} = 0.95$ and $S = 5$, and the distance modulus is estimated as $(m-M)_V = 15.6$.

154.034 Basic data for galactic globular clusters.   G. Alcaino.
Publ. Astron. Soc. Pacific, Vol. 89, 491 - 503 (1977).

Forty-seven basic parameters are tabulated for 132 globular clusters in the Galaxy. The material is summarized in four tables: (1) coordinates, distances, and velocities; (2) richness and metallicity; (3) magnitudes, color indices, reddenings, luminosities, diameters, and densities; (4) variables.

154.035 Correlation between the number of globular clusters and the luminosity of the spheroidal system for external galaxies.   K.-I. Wakamatsu.
Publ. Astron. Soc. Pacific, Vol. 89, 504 - 509 (1977).

A compilation was made of the total number of globular clusters associated with M104, M87, and all the members of the Local Group except for the Magellanic Clouds. The author found that the total number of clusters in a galaxy increases approximately linearly as the luminosity of the spheroidal system of a parent galaxy increases, where the spheroidal system means the spheroidal central bulge for the spiral galaxies, or the whole system for elliptical galaxies.

154.036 The structure of the Reticulum cluster.   C. J. Peterson, W. E. Kunkel.
Publ. Astron. Soc. Pacific, Vol. 89, 634 - 636 (1977).

Star counts on CTIO 4-meter plates of the Reticulum globular cluster have been used to define the density profile and to determine structural parameters of the cluster.

154.037 A parametric approach to the slope of the globular clusters giant branches.
V. Castellani, D. Palma.
Astrophys. Space Sci., Vol. 47, 217 - 223 (1977).

A new parameter is introduced to measure the slope of the red giant branches. It is shown that an adequate description of the shape of the observed branches requires a bi-parametric approach; this procedure allows the authors to keep track of the slope variation along the branches.

154.038 Limits on the orbital eccentricities of globular clusters.   F. House, R. Wiegandt.
Astrophys. Space Sci., Vol. 48, 191 - 197 (1977).

By use of an inverse-square mass-model for the Galaxy, the range in eccentricities for the orbits of 57 globular clusters is computed. On the assumption that all clusters have the same apogalacticon distance, various values of this distance are considered. It is found that low eccentricities are possible for small apogalacticon distances.

154.039 **Ages of globular clusters. II.** H. Saio.
Astrophys. Space Sci., Vol. 50, 93 - 103 (1977).
Evolutionary model sequences for $(X, Z) = (0.7, 4 \times 10^{-3})$ are constructed by using the same input physics and programming code as those of Saio et al. (1977). From these results the ages of globular clusters are estimated under the assumption of constant helium abundance $(Y = 0.3)$. The results suggest that there is a correlation between age and metal abundance for the globular clusters and that metal enrichment in the Galaxy slowly proceeded in several billion years from the value of the extreme Population II stars to that of the Population I stars.

154.040 **The diameters of globular clusters.** H. Wilkens.
Peremennye Zvezdy, Prilozhenie, Vol. 2, 393 - 462 (1976). In Russian.
This investigation consists in new estimates of the diameters of globular clusters, apparent and absolute.

154.041 **UBV color-magnitude diagrams of galactic clusters.** A. G. D. Philip.
Vistas Astron., Vol. 21, 407 - 445 (1977).
A catalogue of over 37,000 UBV observations is now available in the form of 165 color-magnitude and color-color plots of galactic globular clusters. A bibliography of references is given and a table of data for 115 clusters summarizing much of the current information concerning globular clusters. The UBV observations are available on tape from the Strasbourg Stellar Data Center.

154.042 **Coordinates of the center of the globular cluster M53 (NGC 5024).** G. V. Panova.
Astron. Tsirk., No. 946, p. 3 (1977). In Russian.

154.043 **V, B – V diagram of stars in the central region of M13.** E. N. Pastukhova.
Astron. Tsirk., No. 955, p. 6 - 7 (1977). In Russian.

154.044 **Search for globular clusters in M31. I. The disk and the minor axis.** W. L. W. Sargent, C. T. Kowal, F. D. A. Hartwick, S. van den Bergh.
Astron. J., Vol. 82, 947 - 953, 1045 - 1057 (1977).
A search has been made for globular clusters in M31 on plates of 29 fields obtained with the KPNO 4-m telescope. The fields cover the disk and minor axis of M31. Preliminary inspection of these plates yielded a total of 307 clusters. One hundred objects that had previously been noted as possible clusters by other observers, but that were not found during the initial survey, were subsequently reexamined on the 4-m plates. On the basis of this second examination, 48 objects were accepted as compact clusters and 52 were rejected. A preliminary extrapolation from observations in the inner halo of M31 yields a total halo population of ~ 200 globulars with $r > 1.^{\circ}0$ (11 kpc).

154.045 **A photometric study of NGC 1904.** P. B. Stetson, W. E. Harris.
Astron. J., Vol. 82, 954 - 964, 1059 - 1060 (1977).
New photoelectric and photographic UBV photometry is presented for the halo globular cluster NGC 1904 (M79), which lies near the Galactic anticenter at intermediate latitude. From five different methods, the authors estimate the cluster's foreground reddening to be $E_{B-V} = 0.01 \pm 0.01$. The cluster lies at 13 kpc from the Sun and 21 kpc from the Galactic center. Several metallicity indicators as well as the CM diagram morphology, suggest that NGC 1904 is of intermediate metallicity and similar in nature to M13. In the Appendix they describe a general technique for reduction of photographic photometry which they have used here.

154.046 **Mira variables in the globular cluster NGC 6356.** C. Coutts Clement, H. Sawyer Hogg.
Astron. J., Vol. 82, 968 - 971, 1063 (1977).

Observations in 35 of the dark Moon periods from 1946 through 1977 have been used to study six variables in the globular cluster NGC 6356. Five of the variables have Mira periods. The other appears to be an irregular variable. No RR Lyrae stars have been found in this cluster.

154.047 **Photoelectric observations on the brightness distribution in globular clusters.** G. G. Kuzmin, Ü.-I. K. Veltmann, J. A. Vennik, O. A. Tõeleid, E. V. Tago.
Publ. Tartu Astrofiz. Obs., Vol. 45, 190 - 203 (1977).
In Tartu (at Tõravere) photoelectric profiles of the brightness of the globular clusters M3, M5, M92, M15 and M2 have been measured. For that purpose a long narrow slit $(13.^{\prime\prime}6 \times 447^{\prime\prime})$ moving in the α and δ axes perpendicular to its length was used. The original and other observations are compared with generalized isochronic models. The agreement is good.

154.048 **Results of measurements of equidensity curves of several globular clusters.** Z. I. Kadla, N. Richter, W. Högner, A. A. Strugatskaya.
Izv. Glav. Astron. Obs. Pulkovo, No. 195, Astrofiz. Astrometr., p. 74 - 79 (1977). In Russian.
The method of measurement and reduction of equidensity curves of globular clusters is described. The derived results for M 3, M 5, M 13 and M 15 are given in a table.

154.049 **UBV photometry of the cluster NGC 5053.** L. V. Zhukov, V. V. Salitra.
XXX Gertsenovsk. chteniya. Teor. fiz. i astron. Leningrad, 1977, p. 48 - 49. In Russian. – Abstr. in Ref. zh., 51. Astron., 12.51.654 (1977).

154.050 **Globular cluster age determination.** M. Różyczka.
Postępy Astron., Tom 25, 221 - 231 (1977). In Polish.
Problems of chemical composition and age determination of globular clusters are reviewed.

Atlas of galactic globular clusters with colour magnitude diagrams. See Abstr. 002.005.

Plans for a computer-readable cluster catalogue. See Abstr. 002.020.

CNO burning and the location of zero-age horizontal-branch stars. See Abstr. 065.023.

On the age difference between the oldest Population I and the extreme Population II stars. See Abstr. 065.057.

The star distribution around a massive black hole in a globular cluster. II. Unequal star masses. See Abstr. 066.028.

The dissolution of globular clusters containing massive black holes. See Abstr. 066.036.

Stellar density cusp around a massive black hole. See Abstr. 066.037.

The bright variable stars in Messier 5. See Abstr. 122.013.

UBV photometry of RR Lyrae stars in M4. See Abstr. 122.031.

Variable stars of NGC 4833. See Abstr. 122.034.

Variable stars in the globular cluster NGC 6235.

See Abstr. 122.041.

The Mira variables of the metal-rich globular cluster NGC 6356.   See Abstr. 122.062.

Hot ultrashort period cepheid. A new type variable in the globular cluster M15.   See Abstr. 122.087.

Revised periods for two RR Lyrae stars: variable 24 in NGC 6171 and variable 23 in NGC 6656. See Abstr. 122.099.

Period variations of RR Lyrae type variable stars in the globular cluster M53.   See Abstr. 122.122.

An RR Lyrae star in NGC 288. See Abstr. 123.034.

Six new RR Lyrae stars in NGC 1261. See Abstr. 123.035.

Variable stars in the globular cluster NGC 6535. See Abstr. 123.059.

Star formation and the gas content of galaxies. See Abstr. 131.082.

Radio sources near the globular clusters M13 and M53.   See Abstr. 141.160.

X-ray stars in globular clusters. See Abstr. 142.048.

Globular cluster X-ray sources. See Abstr. 142.060.

Was the bright transient X-ray source Centaurus XR-4 I a globular cluster?   See Abstr. 142.126.

Enhancement of the gravothermal catastrophe in two-component isothermal spheres.   See Abstr. 151.001.

Survival and disruption of galactic substructure. See Abstr. 151.036.

Stochastic tidal disruption of clusters. See Abstr. 151.042.

Test computations on the dynamical evolution of star clusters.   See Abstr. 151.056.

Stellar populations of the disk and halo of the Galaxy.   See Abstr. 155.017.

The evolution of stellar populations. See Abstr. 155.018.

The expected number density of globular clusters near the galactic center.   See Abstr. 155.059.

The chemical composition of old stellar populations. See Abstr. 158.045.

Two new faint stellar systems discovered on ESO Schmidt plates.   See Abstr. 158.084.

Globular clusters and the Virgo cluster distance modulus.   See Abstr. 160.002.

Theoretical isochrones with decreasing gravitational constant – II. Dirac's multiplicative cosmology. See Abstr. 162.059.

Primeval entropy fluctuations and the present-day pattern of gravitational clustering.   See Abstr. 162.064.

Errata

154.901   Addendum: 'Integrated *UBV* photometry for seven globular clusters' [Astron. J., Vol. 82, 193 - 194 (1977)]   H. G. Corwin, Jr. Astron. J., Vol. 82, 757 (1977).

155 Structure and Evolution of the Galaxy

155.001 **Broadband observations of H I within 10° of the direction of the galactic center.**
W. B. Burton, J. S. Gallagher, M. A. McGrath.
Astron. Astrophys., Suppl. Ser., Vol. 29, 123 - 138 (1977).

The 140-foot (43 m) telescope of the National Radio Astronomy Observatory has been used to measure profiles of 21 cm HI emission from the region $349° \leqslant l \leqslant 12°$, $-10° \leqslant b \leqslant 10°$, sampled at 1° intervals. Each profile covers the velocity range $|v| < 500$ km s^{-1} with 5.5 km s^{-1} velocity resolution to an rms sensitivity limit of 0.08 K. The reduced observations are given as maps of antenna temperature contours in velocity, latitude coordinates.

155.002 **Ist unsere Milchstraße eine explodierte Radiogalaxie?**
Y. Sofue.
Umschau, 77. Jahrg., 542 - 543 (1977).

Characteristic features in the brightness distribution of nonthermal radio emission of our galaxy, in particular galactic spurs, are mostly explained in terms of the spiral structure. A new interpretation of the north polar spur is presented that it may be a front of magnetohydrodynamic shock wave originating from the galactic center explosion and associated with the 3 kpc expanding ring.

155.003 **Star density curves and the mass density in the neighborhood of the sun.** R. Whitley.
Astron. Astrophys., Vol. 59, 329 - 335 (1977).

The mass density and potential in the solar neighborhood are studied by means of a perturbation of the equations which, unperturbed, are satisfied by Camm's isothermal solution. Estimates for the mass density in the solar neighborhood of 0.08–0.09 solar masses per cubic parsec are obtained. The perturbation studied is that due to a small (truncated) quadratic term in a function which, in Camm's solution, is linear. Using the perturbed curves an estimate for the local density of 0.10–0.11 solar masses per cubic parsec is obtained.

155.004 **The velocity dispersion of faint red dwarf stars.**
D. Weistrop.
Astrophys. J., Vol. 215, 845 - 850 (1977).

Broad-band photometry is presented for 21 late-type stars near the north galactic pole. Many of the stars can be identified as dwarfs from their location in the $(B-V)-(V-I)$ diagram. Photometric parallaxes are combined with proper motions to yield velocity dispersions in the galactic plane that are consistent with those for nearby M dwarf stars. With these velocity dispersions, the late-type dwarfs present no problem for the stability of the galactic disk.

155.005 **Shock-wave model of the expanding ring in the galactic center region.** T. Kato.
Publ. Astron. Soc. Japan, Vol. 29, 369 - 385 (1977).

A shock-wave model is presented for the expanding ring found at 270 pc from the galactic center. A shock wave which is considered to originate from an explosion at the galactic center will propagate through the rotating medium of the galactic disk. Effects of radiative cooling, gravity, and magnetic field on the shock-wave propagation are taken into account. In the initial stage, the shock is seen to propagate adiabatically and then the effect of cooling becomes dominant. The shock front is decelerated by the effect of radiative cooling, and the shock turns into an isothermal shock. The expanding ring at 270 pc from the galactic center seems to correspond to this stage.

155.006 **Shock-wave model of the kinematic features of neutral hydrogen and molecules CO, OH, and H$_2$CO near the galactic center.** M. Saitō, Y. Saito.

Publ. Astron. Soc. Japan, Vol. 29, 387 - 414 (1977).

A method of calculating the propagation and structure of a shock wave is proposed for the shock wave originating impulsively at the galactic nucleus and propagating outward through the interstellar medium. The negative high-velocity features observed at the longitude range $0.3° \gtrsim l \gtrsim -1.7°$ in molecules CO, OH, and H$_2$CO and in neutral hydrogen are consistently interpreted to be due to the expanding shock wave whose front is propagating at ~ 290 pc from the center on the near side of the center; the hydrogen with the highest velocity corresponds to the gas just behind the shock front.

155.007 **Cosmic rays, spiral structure and molecular clouds in the Galaxy.**
C. J. Cesarsky, M. Cassé, J. A. Paul.
Astron. Astrophys., Vol. 60, 139 - 145 (1977).

In the light of recent work concerning the structure and content of the galactic disc, the authors re-examine the interpretation of the high-energy galactic gamma-ray emission observed by the SAS-II and the COS-B satellites. In particular, they show that the contribution of point sources is probably nowhere dominant; that the degree of penetration of cosmic rays in molecular clouds cannot yet be known; that all of the data available at present point to the existence of a gradient of the cosmic ray density in the Galaxy, and of a correlation between gamma-ray emission and spiral structure. The authors present a renewed version of the type of model discussed in Paul et al., 1976.

155.008 **The diffusion of stellar orbits derived from the observed age-dependence of the velocity dispersion.**
R. Wielen.
Astron. Astrophys., Vol. 60, 263 - 275 (1977).

It is the purpose of the paper (1) to point out that the observed increase in the velocity dispersion of stars with age does strongly suggest the existence of a significant component of the galactic gravitational field with a rather stochastic behaviour, and (2) to investigate the implications of such an irregular field for stellar orbits.

155.009 **Carbon monoxide in the inner Galaxy.**
T. M. Bania.
Astrophys. J., Vol. 216, 381 - 403 (1977).

Emission from the $J = 1 \rightarrow 0$ rotational transition of $^{12}C^{16}O$ has been surveyed at $b = 0°$ in the region $10° \geqslant l \geqslant 352°$ with high angular and velocity resolution. Although CO emission is more clumped in (l, v)-space than is the 21 cm radiation, a detailed comparison of the observations in the inner Galaxy reveals that there is no large-scale CO feature which is not also seen in H I. The longitude dependence of the CO centroid velocity suggests that the CO in the inner Galaxy rotates at velocities consistent with the observed 21 cm rotation curve. The major emission features seen in both CO and H I are the 3 kpc arm, which is found over the entire l-range of the survey, and the nuclear disk.

155.010 **Chemical evolution in the solar neighborhood. III. Time scales and nucleochronology.**
B. M. Tinsley.
Astrophys. J., Vol. 216, 548 - 559 (1977).

Numerical models for chemical evolution are used to supplement previous analytical approximations, to generalize results from published special models, and to assess long-lived radionuclides as chronometers and as probes of chemical evolution. Because of the stringency of constraints related to metallicity, gas content, star-formation time scales, etc., fairly general conclusions can be drawn from detailed models for a few limiting cases. Three such models are considered here, with (1) infall controlling the star-formation rate at all times, (2) infall at early times only, and (3) prompt initial enrichment by massive stars.

155.011   Scanner observations of faint dwarf M stars near the north galactic pole.   R. G. Kron.
Publ. Astron. Soc. Pacific, Vol. 89, 331 - 334 (1977).

Image-tube scans have been obtained for ten of the faintest and latest M dwarf stars on the proper-motion list of Murray and Sanduleak. The scans have been measured for magnesium hydride strength, which is shown to be correlated with the motion component $(U^2 + V^2)^{1/2}$. Only two stars have emission at $H\alpha$. From all appearances, the stars are similar to M dwarfs found in the solar neighborhood, and in particular there is no evidence for a dominant young population.

155.012   On the solar motion with respect to external galaxies.   P. L. Schechter.
Astron. J., Vol. 82, 569 - 576 (1977).

The ScI galaxy data presented by Rubin, Ford, Thonnard, Roberts, and Graham have been examined to determine whether the accuracy of the solar motion derived from anisotropy in the redshift-magnitude diagram can be substantially improved by the application of the "diameter correction" employed by Rubin et al. The author finds that it cannot. Analysis of a sample of nearby bright galaxies gives a solution for the solar motion with three times the formal accuracy obtained with the ScI sample, but with a possible systematic error arising from the motion of the sample galaxies toward the Virgo cluster.

155.013   Propagation of magnetohydrodynamic waves from the galactic center. Origin of the 3-kpc arm and the North Polar Spur.   Y. Sofue.
Astron. Astrophys., Vol. 60, 327 - 336 (1977).

Magnetohydrodynamic waves isotropically radiated from the galactic center are reflected in the galactic halo, and more than 80% of the wave front focuses on a ring in the galactic disk. Due to the focusing effect, a compression wave of small amplitude develops into a strong shock as it approaches the ring, which may explain the 3-kpc arm. A small portion of the wave front escapes into the galactic corona, where the front forms a large shell structure. A tangential view of this shell from the sun could be responsible for the North Polar Spur.

155.014   Observations of blue stars near the galactic anticenter.   F. R. Chromey.
News Lett. Astron. Soc. N.Y., Vol. 1, No. 2, p. 25 (1977). Abstract.

155.015   The galactic density wave, molecular clouds, and star formation.
F. N. Bash, E. Green, W. L. Peters III.
Astrophys. J., Vol. 217, 464 - 472 (1977).

The results of this paper are: (1) a suggestion based on the authors' dynamical model that CO is associated with a specific age range for stars which is confirmed by recombination line observations and by an observational test , (2) the use of this association to predict the relative distribution of young stars in the Galaxy, (3) the predictions of the dynamical model concerning their birthplaces and relative birthrates, (4) the suggestion that these observations imply an upper mass limit for stars that become supernovae, and (5) the comparison of the authors' independent dynamical time scale with conventional stellar evolution time scales.

155.016   Are there population II stars of early spectral types and low gravities?   L. Carrasco.
Recueil des séminaires, (see 012.004), p. 131 - 133 (1977).

155.017   Stellar populations of the disk and halo of the

Galaxy.   P. Demarque, R. D. McClure.
The evolution of galaxies and stellar populations, (see 012.005), p. 199 - 215, with a discussion, p. 216 - 217 (1977).

New ages for halo and disk star clusters have been obtained by fitting color-magnitude diagrams to a new homogeneous set of theoretical isochrones. Ages of 14 - 16 billion years were obtained for the extreme metal-poor halo globular clusters M15 and M92. Metal-rich clusters appear significantly younger with 47 Tucanae having an age of 11 - 12 billion years, possibly an indication of a slow collapse of the Galaxy. The oldest known disk cluster, NGC 188 appears much younger, 5 - 6 billion years. Evidence is also discussed concerning helium or CNO abundance differences between the halo and disk populations.

155.018   The evolution of stellar populations.   L. Searle.
The evolution of galaxies and stellar populations, (see 012.005), p. 219 - 230, with a discussion, p. 231 - 237 (1977).

155.019   On the motions of the Sun, the Galaxy and the Andromeda nebula.   D. Lynden-Bell, D. N. C. Lin.
Mon. Not. R. Astron. Soc., Vol. 181, 37 - 57 (1977).

The sum of the velocity of the Galaxy, **G**, and the velocity of the Sun may be determined from their reflection in the radial velocities of the members of the Local Group of galaxies excluding Andromeda. The knowledge of the observed radial velocity of Andromeda and the mass ratio allows the authors to determine both the velocity of the Galaxy and the circular velocity of its rotation at the Sun, V_c. The method gives $V_c = 294 \pm 42$ km/s, $G = (-34, +7, -16) \pm 23$ km/s, where **G** is with respect to the centre of mass of the Local Group. The authors make further assumptions and obtain $V_c = 290^{+10}_{-15}$ km/s.

155.020   The galactic distribution (in radius and z) of interstellar molecular hydrogen.
N. Z. Scoville, P. M. Solomon, D. B. Sanders.
The structure and content of the Galaxy and galactic gamma rays, (see 012.009), p. 151 - 161 (1977) = Contrib. Five Coll. Obs., Amherst, Mass., No. 232.

New observations of the galactic longitude and latitude distributions of $\lambda = 2.6$ mm CO emission are presented. Analysis of these spectral-line data yields the large-scale distribution of molecular clouds in the galactic disk and their z-distribution out of the disk. Strong maxima in the number of molecular clouds occur in the galactic nucleus and at galactic radii 4 to 8 kpc. The width of the cloud layer perpendicular to the galactic plane between half-density points is 105 ± 15 pc near the 5.5-kpc peak. The total mass of molecular gas in the interior of the Galaxy exceeds that of atomic hydrogen and is 3×10^9 M$_\odot$ based on these observations.

155.021   Remarks on the overall distribution of hydrogen in the galactic disk.   W. B. Burton.
The structure and content of the Galaxy and galactic gamma rays, (see 012.009), p. 163 - 187 (1977).

Several current problems concerning the overall distribution of hydrogen in the Galaxy are discussed in general terms.

155.022   Gamma rays and large-scale galactic structure.   D. A. Kniffen, C. E. Fichtel, D. J. Thompson.
The structure and content of the Galaxy and galactic gamma rays, (see 012.009), p. 301 - 313 (1977).

Gamma-ray astronomy is now beginning to provide a new look at the galactic structure and the distribution of cosmic rays, both electrons and nucleons, within the Galaxy. The observations are consistent with a galactic spiral-arm model in which the cosmic rays are linearly coupled to the interstellar gas on the scale of the spiral arms.

155.023   *RGU* three-colour photometry of a field in Aquila near NGC 6755.   A. M. Spaenhauer.

Astron. Astrophys., Suppl. Ser., Vol. 30, 63 - 70 (1977).

A field of 0.097 sq. degrees containing 1530 stars, in the direction of Aquila, has been measured on eighteen 48″ Palomar Schmidt plates in the *RGU* system down to a limiting magnitude of 17.6 mag in *G*. The major finding of this paper is the density maximum for the stars absolutely brighter than $M(G) = 3^m$ at a distance of about 2 kpc (more pronounced for the giants), which the author considers to be caused by touching the next inner spiral arm (−I). The stars fainter than $M(G) = 3^m$ cannot be seen beyond 2 kpc, because of the large absorption amounting to $A(G) = 3$ mag at about 1800 pc. The densities of these absolutely fainter stars in the near interarm region amount to about five times the value of the solar neighborhood, a fact also observed in fields directed to the galactic centre.

155.024 **The galactic halo question: new size constraints from galactic γ-ray data.** F. W. Stecker, F. C. Jones.
Astrophys. J., Vol. 217, 843 - 858 (1977).

The recent acquisition of data on the distribution of galactic γ-ray emission has made possible a reconsideration of the long-standing controversy concerning the existence and extent of the galactic halo. Analysis of the implications of the SAS-2 γ-ray data, making use of recent CO line emission and other data for determining the large-scale distribution of galactic gas, implies that there is a nonuniform distribution of cosmic rays in the Galaxy. This fact rules out large trapping halo models and particularly the recently proposed closed halo models in the same way that it rules out extragalactic origin models. Assuming the sources to be supernova remnants or pulsars, cosmic-ray-nucleon halo models with scale heights greater than 3 kpc are found to provide a poor fit to the γ-ray longitude data (probability of 6% or less). Thin halo models, or source-dominated diffusion models, are found to provide a good fit to the γ-ray data, with an upper limit scale height of ~3 kpc.

155.025 **An out-of-plane galactic carbon monoxide survey.** R. S. Cohen, P. Thaddeus.
Astrophys. J., Lett., Vol. 217, L155 - L159 (1977).

Galactic CO line emission at 115 GHz has been surveyed in the region $15° < l < 60°$ and $-1°.5 \lesssim b \lesssim 1°.5$. In addition to confirming the findings of previous in-plane surveys that galactic CO emission is concentrated in a ring 6 kpc in radius, a fit of a cylindrically symmetric galactic model to the authors' observational data has provided the first determination of the thickness of this molecular ring and its displacement from the conventional galactic plane, both as functions of galactocentric distance. The average half-thickness at half-maximum of the molecular ring is 59 pc, and the average displacement of the ring with respect to the $b = 0°$ plane is −40 pc.

155.026 **Comparative morphology of galactic carbon monoxide and hydrogen.** W. B. Burton, M. A. Gordon.
Topics in interstellar matter, (see 012.011), p. 165 - 177 (1977).

The authors compare overall galactic characteristics derived from observations of the λ2.6-mm J = 1→0 rotational transition of ^{12}CO with those derived from observations of the λ21-cm hyperfine transition of atomic hydrogen.

155.027 **The morphology of hydrogen and of other tracers in the Galaxy.** W. B. Burton.
Annu. Rev. Astron. Astrophys., Vol. 14, (see 003.008), 275 - 306 (1976).

Contents: The galactic distribution of neutral atomic hydrogen; The galactic distribution of CO and H_2; The galactic distribution of ionized hydrogen; The galactic distribution of γ-radiation and synchrotron radiation; The galactic distribution of supernova remnants and pulsars; Summary remarks on the relative distributions.

155.028 **Supergiants in the field of the cluster M6, and the distribution of interstellar matter in the direction of the galactic center.** J. W. Warner, R. F. Wing.
Astrophys. J., Vol. 218, 105 - 117 (1977).

Near-infrared eight-color classification photometry is presented for IRC stars in a field in Scorpius containing the open cluster M6. Distances and extinction values are computed from the photometric spectral types, luminosity classes, and colors. Of the six M stars observed, five are found to be supergiants lying at considerably greater distances than that of the cluster. A comparison of their distances with recent determinations of galactic structure in the direction of the galactic center indicates that the stars are likely to be associated with spiral-arm features approximately 2 and 4 kpc from the sun. This study shows that the dark clouds which produce the rift in the nuclear bulge lie predominantly at a distance of about 1.5 kpc and cause a visual extinction of about 8 mag. This screen of absorbing material is thus seen to be a major component of the Sagittarius arm.

155.029 **Neutral hydrogen in the galactic center region: a preliminary survey with the 100 m telescope.**
R. H. Sanders, G. T. Wrixon, U. Mebold.
Astron. Astrophys., Vol. 61, 329 - 337 (1977).

Results of a neutral hydrogen 21-cm line survey of the galactic center region are presented. The observations, made with the 100-m telescope of the Max-Planck-Institut für Radio-astronomie, cover the region $-3° < l < 3°$, $b = 0°$, and -300 km s$^{-1} < V < 300$ km s^{-1} with 9′ spatial resolution and 2.75 km s^{-1} velocity resolution. The neutral hydrogen emission in the negative longitude, negative velocity quadrant of the nuclear disk is reanalyzed using the modelling technique of Sanders and Wrixon (1973).

155.030 **The effect of crossing times through the solar neighborhood on the observed stellar age and metallicity distributions.**
M. Mayor, C. Turon Lacarrieu, L. Martinet.
Astron. Astrophys., Vol. 61, 433 - 436 (1977).

The effect of crossing times through the solar neighborhood on the observed stellar age and metallicity distributions is estimated. The bias due to motions parallel to the galactic plane is less than 10% in the age distribution. The modification of the cumulative metallicity distribution, taking into account motions in the plane, is negligible as compared with the discrepancy between the observations and the predictions of the simple model of chemical evolution.

155.031 **Density wave propagation through the galactic disk.** S. N. Paul, M. R. Khan.
Bull. Astron. Inst. Czechoslovakia, Vol. 28, 300 - 306 (1977).

Three-dimensional dynamical behaviour of gas particles of the galactic disk has been studied by the density wave theory. Theoretically, it is found that the thermal pressure of the gas plays a minor role in the formation and maintenance of the spiral arms at the outer region of the Galaxy. It is also observed that the amplitudes of the perturbed velocities of the gas are decreased by the pressure. It is suggested that the pressure of the gas may partially save the galactic disk from gravitational collapse. The wave propagation through the galactic disk has also been studied under certain special conditions.

155.032 **Evoluzione chimica della Galassia.** M. Hack.
G. Astron., Vol. 3, 85 - 99 (1977).

155.033 **The sun's galactic orbit, the spiral density wave, and geological history.**
A. Scarpelli, K. A. Innanen, W. W. Duley.
J. R. Astron. Soc. Canada, Vol. 71, 395 (1977). − Abstract.

155.034 **M dwarfs at lower z distances.** W. Gliese.

Highlights of Astronomy, Vol. 4, Part II, (see 012.022), p. 11 - 19 (1977).

Report on the recent discussions about the luminosity function of red dwarfs and about the problem of the missing mass in the solar neighbourhood. How far do new photometric measurements near the South Galactic Pole combined with proper motion data contribute to an answer to these questions?

155.035   Available stellar data in galactic polar areas.
            C. Jaschek, B. Hauck.
Highlights of Astronomy, Vol. 4, Part II, (see 012.022), p. 21 - 22 (1977).

155.036   M dwarfs and the missing mass.   P. S. Thé.
            Highlights of Astronomy, Vol. 4, Part II, (see 012.022), p. 25 (1977).

155.037   The space density of M dwarfs – an observational program.   P. Pesch.
Highlights of Astronomy, Vol. 4, Part II, (see 012.022), p. 29 - 30 (1977).

155.038   Galactic mass density in the vicinity of the sun.
            M. Jöeveer, J. Einasto.
Highlights of Astronomy, Vol. 4, Part II, (see 012.022), p. 33 (1977). – Abstract. (See 18.155.028).

155.039   The frequency of faint M giant stars at high galactic latitudes.   N. Sanduleak.
Highlights of Astronomy, Vol. 4, Part II, (see 012.022), p. 35 - 36 (1977).

155.040   The stellar distribution above the galactic plane: an introduction.   I. R. King.
Highlights of Astronomy, Vol. 4, Part II, (see 012.022), p. 41 - 48 (1977).

155.041   Some results of classification of stars in Kapteyn Areas applied to galactic structure in NGP.
E. K. Kharadze, R. A. Bartaya.
Highlights of Astronomy, Vol. 4, Part II, (see 012.022), p. 49 - 50 (1977).

155.042   RGU three-colour-photometry towards the north galactic pole: halo-to-disk mass ratio.
R. P. Fenkart, U. W. Steinlin.
Highlights of Astronomy, Vol. 4, Part II, (see 012.022), p. 53 - 54 (1977).

155.043   Density law, vertical distribution and vertical gradient of metal abundances for G and K giants.
M. Grenon.
Highlights of Astronomy, Vol. 4, Part II, (see 012.022), p. 55 (1977).

155.044   The gravitational field of the Galaxy in the z direction.   R. F. Griffin.
Highlights of Astronomy, Vol. 4, Part II, (see 012.022), p. 65 - 66 (1977).

155.045   Berkeley studies of faint stars at high latitudes.
            R. G. Kron, L.-T. G. Chiu, K. O. Brooks.
Highlights of Astronomy, Vol. 4, Part II, (see 012.022), p. 67 (1977).

155.046   Swedish programmes concerning the z distribution of stars.   T. Oja.
Highlights of Astronomy, Vol. 4, Part II, (see 012.022), p. 69 (1977).

155.047   Studies of A and F stars in the region of the north galactic pole.   R. W. Hilditch, G. Hill.
Highlights of Astronomy, Vol. 4, Part II, (see 012.022), p. 71 (1977).

155.048   The distribution of field horizontal-branch stars in the galactic halo.   A. G. D. Philip.
Highlights of Astronomy, Vol. 4, Part II, (see 012.022), p. 73 - 74 (1977).

155.049   A new program to determine K(z) from main sequence stars.   A. R. Upgren.
Highlights of Astronomy, Vol. 4, Part II, (see 012.022), p. 75 - 76 (1977).

155.050   Discussion of the calculation of the density law (Dz).
            C. Turon Lacarrieu.
Highlights of Astronomy, Vol. 4, Part II, (see 012.022), p. 79 - 82 (1977).

A matrix method involving eigenvector expansion is used to solve the "fundamental equation" of stellar statistics. This method is applied to M dwarfs and K giants.

155.051   Motions of near-polar K giants along the z co-ordinate.   A. N. Balakirev.
Highlights of Astronomy, Vol. 4, Part II, (see 012.022), p. 83 - 86 (1977). (See 17.155.009).

155.052   The role of proper motions in determining the luminosity and density functions.   W. J. Luyten.
Highlights of Astronomy, Vol. 4, Part II, (see 012.022), p. 89 - 93 (1977).

155.053   Study of the spatial distribution of stars and dust matter in the direction of the Cygnus–Vulpecula constellations ($l = 65°7$, $b = +1°2$).
V. I. Voroshilov, N. B. Kalandadze.
Astrometr. i Astrofiz., Kiev, vyp. (No.) 32, (see 003.011), p. 46 - 56 (1977). In Russian.

The article deals with results of studying the region with an area of 18.3 square degrees and with coordinates $l = 65°7$, $b = +1°2$. The accuracy of determination of interstellar absorption, distances of dust clouds and individual stellar distances was estimated. The position of the local spiral arm was determined. Comparison of O-B2 and B3-B5 star distributions shows that the latter form a wider and less pronounced arm. The peculiarities of distributions of stars of other spectral intervals, namely, of late giants are studied, as well as their connection with absorbing matter. The luminosity function was constructed, which shows the abundance of F1-K8 dwarfs in the studied direction.

155.054   Spatial distribution of stars and interstellar matter in the direction $l = 76°9$, $b = +0°6$.
N. B. Kalandadze, L. N. Kolesnik.
Astrometr. i Astrofiz., Kiev, vyp. (No.) 32, (see 003.011), p. 57 - 66 (1977). In Russian.

Photographic photometry and spectral classification of about 3213 stars ($V_{lim} = 13^m$) in a field of about 18 square degrees near the galactic plane centered at $l = 76°9$, $b = +0°6$ were used to study the space distribution of stars and dust. Figures show the run of colour excesses E_{B-V} with distance modulus $V-M_V$ and interstellar absorption A_V with distance from the sun in 9 regions of the field. Distances and space densities were calculated for all stars. The luminosity function was determined for $-5^m5 \leqslant M_V \leqslant +7^m5$.

155.055   21-cm observations of nonplanar H I associated with the Perseus spiral arm.   E. J. Grayzeck.
Astron. J., Vol. 82, 886 - 889 (1977).

An intermediate-velocity cloud (-55 km sec^{-1}) centered at $l = 122°$, $b = -15°$ has been mapped at 21 cm to determine its extent and possible connection to gas in the plane. The extent of the gas is presented in the form of contour maps of the H I column density over the velocity range of −45 to −70 km sec^{-1}. This material appears to be a continuation of the H I seen in the plane at velocities between −40 and −60 km sec^{-1}. Additional information on young OB stars located near $l = 122°$, $b = -15°$, particularly HD 4460, which has a distance of 2 kpc and $V_{LSR} = -55$ km sec^{-1}, further substantiate the association of this intermediate-velocity gas with the Perseus spiral arm.

155.056 The local complex of O and B stars. II. Kinematics.
    J. A. Frogel, R. Stothers.
Astron. J., Vol. 82, 890 - 901 (1977).
    The space velocities of O-B5 stars in the solar neighborhood are analyzed in the present paper. The stars are divided into members of the Gould and galactic belts on the basis of their positions in space. This permits a homogeneous kinematical comparison to be made between the two belts. The local galactic velocity field is found to be perturbed by the presence of the Gould belt, as reflected in the derived values of the Oort constant B and of the K term for the youngest stars. It is suggested that the Gould belt had a violent origin close to the galactic plane and that each passage of this belt through the galactic plane initiated a major period of star formation. Inferred kinematical ages for most of the stars concerned are 2×10^7 or 6×10^7 yr, which agree well with the observed frequency distribution of their nuclear ages.

155.057 Die chemische Entwicklung der Galaxis.
    K.-H. Schmidt.
Sterne, 53. Band, 129 - 139 (1977).

155.058 The galactic center.   J. H. Oort.
    Annu. Rev. Astron. Astrophys., Vol. 15, (see 003.012), 295 - 362 (1977).
    Contents: 1. Position and distance of the nucleus; 2. The gravitational field and the nuclear disk; 3. Expanding H I features; 4. Origin of the radial motions and the deviations from the galactic plane; 5. Molecular clouds and total gas density; 6. Emission at radio frequencies and in the 12.8-μ neon II line; 7. The infrared core.

155.059 The expected number density of globular clusters near the galactic center.   J. H. Oort.
Astrophys. J., Lett., Vol. 218, L97 - L101 (1977).
    Estimates are given of the distribution of globular clusters in the central region of the Galaxy. Within 1 or 2 kpc from the center, where direct observations are invalidated by absorption, the run of the density is computed from the gravitational field derived from the rotation of the H I nuclear disk combined with a Maxwellian cluster-velocity distribution. The resulting initial density of the clusters as a function of distance R from the center is shown. For $R < 1$ kpc the present density has been greatly reduced by dynamical friction. The mass accumulated around the nucleus by the dynamical capturing is found to be about $2 \times 10^7 M_\odot$, with an estimated uncertainty of a factor of 2.

155.060 Galactic structure and gamma radiation.
    M. Casse, C. Cesarsky, J. Paul.
Recherche, Vol. 8, 112 - 121 (1977). In French. − Abstr. in Phys. Abstr., Vol. 80, Abstr. 60829 (1977).

155.061 Space density of low-luminosity stars using high-accuracy proper motions.   L.-T. G. Chiu.
Publ. Astron. Soc. Pacific, Vol. 89, 614 - 615 (1977).
Abstract.

155.062 The spiral structure in the inner parts of the Galaxy.
    R. J. Quiroga.
Astrophys. Space Sci., Vol. 50, 281 - 310 (1977).
    The spiral structure of the inner parts of the Galaxy is studied using 21 cm line data and stellar data. To study the neutral hydrogen distribution in the galactic layer a parameter proportional to the mean densities is calculated using a first approximation for the velocity gradients due to differential rotation. The obtained distribution shows spiral features completely consistent with the early star distribution and with the H II regions. The corrugation effect of the galactic layer is observed in all the studied zones in neutral hydrogen and in the distribution of the OB stars in the Carina zone. The pattern obtained indicates four spiral arms for the inner parts of the Galaxy. The observed wave form distribution of the hydrogen cloud layer is completely consistent with the theoretical predictions of Nelson (1976) but there are no indications of such an effect in the intercloud hydrogen. The spiral features in our Galaxy show characteristics quite similar to the features in the Andromeda nebula, not only in the component materials but also in their kinematics.

155.063 Galactic ^{4}He and heavy elements enrichment by stellar nucleosynthesis.
M. Busso, R. Gallino, A. Masani.
Astrophys. Space Sci., Vol. 52, 479 - 495 (1977).
    The contribution to the galactic abundance of He and heavy elements by stellar nucleosynthesis is calculated as a function of time, keeping account of present knowledge about stellar and galactic evolution. A model is used which distinguishes the phase of the contracting halo from the subsequent history of the disc. Various uncertainties involved both in stellar and in galactic evolutionary theory are discussed. The authors find that stellar activity provides a significant contribution to the cosmic ^{4}He, though not sufficient to explain the observed abundance.

155.064 On the luminosity function of stars in the direction of the Galactic North Pole.   E. I. Zajtseva.
Soobshch. AN GruzSSR, Vol. 85, 593 - 595 (1977). In Russian. − Abstr. in Ref. zh., 51. Astron., 11.51.864 (1977).

155.065 The Local Group: the solar motion relative to its centroid.
A. Yahil, G. A. Tammann, A. Sandage.
ESO Sci. Prepr. No. 10, 44 pp. (1977). − To appear in Astrophys. J.
    A new solution for the motion of the local standard of rest (LSR) relative to the centroid of the Local Group (LG) of galaxies, based on 21 cm redshifts for a number of candidates, gives v (LSR) = 200 km s^{-1} toward $l = 107°$, and $b = -8°$. Three other solutions are given using different precepts for membership within the LG. This motion of the LSR corresponds to a best fit solar motion relative to the LG centroid of v (⊙) = 308 km s^{-1} toward $l = 105°$ and $b = -7°$.

155.066 Distribution of metals among galactic populations and the evolution of the Galaxy.
V. A. Marsakov, A. A. Suchkov.
Astron. Zh. Akad. Nauk SSSR, Vol. 54, 1232 - 1240 (1977). In Russian. English translation in Soviet Astron., Vol. 21, No. 6.
    It is found that frequency distributions of different types of stars and globular clusters in metal abundance have two gaps which separate the galactic population in three groups. One of them is the disk population and the two others belong to the conventional halo population. It is pointed out that similar discontinuities appear in kinematical parameters while passing from old populations to young ones. The results lead to conclusions on the evolution of the Galaxy, on metal and heavy element enrichment and star formation in it.

155.067 Eine Blau-Photometrie der Milchstrasse und des Zodiakallichts.   C. Classen.
Diss. Math.-Naturwiss. Fak. Rheinische Friedrich-Wilhelms-Univ., Bonn, 141 pp. (1976).

A photoelectric surface photometry of the sky is presented. The resolution is $1-2°$, the color is very close to the Johnson B. Isophote maps of the Milky Way and tables with entries for every square degree covering nearly the whole southern sky ($\delta < +45$) are given, and compared to other photometries. Color indices U–B and B–V are plotted. The zodiacal light outside the brightest parts of the main cone derived under the assumption of symmetry to the ecliptic plane and to the solar longitude, is tabulated in B and U, averaged over $5° \times 5°$ regions.

155.068   Differential rotation of the Galaxy.
    S. Ninković.
Vasiona, Vol. 25, 43 - 56 (1977). In Serbo-Croatian.

155.069   Old and new epicycles.
    J. Milogradov-Turin, S. Ninković.
Vasiona, Vol. 25, 82 - 89 (1977). In Serbo-Croatian.

Bibliography on galactic structure in the direction of polar caps.   See Abstr. 002.021.

A four-colour photographic atlas of the sky. See Abstr. 002.037.

Stellar formation.   See Abstr. 003.136.

Some research programmes into galactic structure at the galactic caps under way at the Royal Greenwich Observatory.   See Abstr. 013.022.

A three-dimensional analysis of the kinematics of 512 FK4/FK4 Sup stars.   See Abstr. 041.016.

Accretion onto pregalactic black holes. See Abstr. 066.019.

Hyperbolic meteors and eruptive processes in the Galaxy.   See Abstr. 104.036.

Progress report on a search for parallax stars in the region of the South Galactic Cap.   See Abstr. 111.003.

Proper motions of faint blue stars near the galactic anticenter. III.   See Abstr. 112.001.

Search for large radial velocities in direction of NGP using the Fehrenbach techniques. See Abstr. 112.008.

Objective-prism radial velocities at high latitudes. See Abstr. 112.009.

UBV photometry of stars in a region near the south galactic pole.   See Abstr. 113.017.

Photometry of faint blue stars – IV. Four-colour photometry of some northern stars. See Abstr. 113.027.

Searches for faint OB stars in the southern Milky Way. II. The Vela region.   See Abstr. 114.002.

The radial velocities of early-type stars within six degrees of the galactic anticenter direction. See Abstr. 114.016.

Mean chemical abundance of the F stars as a function of distance from the galactic plane. See Abstr. 114.031.

On the luminosity function of stars.

See Abstr. 115.005.

The luminosity function of late-type main-sequence stars in the direction of the North Galactic Cap. See Abstr. 115.007.

Chemical and physical properties of nearby Population I stars.   See Abstr. 115.013.

Northern Polar Spur – a remnant of a supernova explosion?   See Abstr. 125.030.

Survey of water vapor sources in the southern hemisphere.   See Abstr. 131.057.

An interpretation of galactic observations of CNO isotopes.   See Abstr. 131.060.

Models for interpreting the diffuse galactic light. See Abstr. 131.089.

The large-scale distribution of interstellar matter in the context of the density-wave theory. See Abstr. 131.106.

The Cygnus X region VIII: maps of calibrated surface brightness of ($H\alpha + [NII]$).   See Abstr. 132.001.

Recombination line observations of ionized hydrogen.   See Abstr. 132.017.

H II regions in galaxies of the Local Group. See Abstr. 132.018.

Latitude-dependent line-shift field of the local H I cloud.   See Abstr. 132.034.

Two populations of high-velocity hydrogen clouds and the bending of the Galaxy.   See Abstr. 132.038.

An angular size for the compact radio source at the galactic center.   See Abstr. 141.136.

Galactic distribution and evolution of pulsars. See Abstr. 141.502.

Galactic distribution of pulsars. See Abstr. 141.523.

Rate of stellar collapses in the Galaxy. See Abstr. 141.539.

On propagation of cosmic rays in the Galaxy. See Abstr. 143.066.

Simulation of the Magellanic Stream to estimate the total mass of the Milky Way.   See Abstr. 151.026.

Density wave theory.   See Abstr. 151.027.

Chemical evolution of galaxies. See Abstr. 151.029.

Density wave – star interaction in the theory of spiral galaxies.   See Abstr. 151.058.

A theory for the gravitational potentials of spheroidal stellar systems and its application to the Galaxy. See Abstr. 151.062.

The dispersion relation of a gravitating spiral system. See Abstr. 151.065.

Places of formation of 24 open clusters.
See Abstr. 153.011.

On the space distribution of globular clusters.
See Abstr. 154.002.

Remote halo globular cluster Palomar 5.
See Abstr. 154.003.

Diffusion models for the low-frequency radio emission from the Galactic halo. See Abstr. 156.013.

Faraday rotation and the turbulent structure of the Galaxy. See Abstr. 156.017.

Gamma rays, cosmic rays, and galactic structure.
See Abstr. 157.014.

The Magellanic Stream: the turbulent wake of the Magellanic Clouds in the halo of the Galaxy.
See Abstr. 159.003.

The Large and Small Magellanic Clouds in a binary state, the bending of the galactic disk and the Magellanic Stream. See Abstr. 159.006.

Five-channel photometric observations of stars in the Magellanic Clouds and in the Milky Way.
See Abstr. 159.007.

The Local Group: the solar motion relative to its centroid. See Abstr. 160.033.

Quasar absorption lines as probes of the past intergalactic medium. See Abstr. 161.003.

156 Galactic Magnetic Field, Galactic Radio, Infrared Radiation

156.001 Cosmic ray age, soft X-rays and the galactic wind.
J. J. Quenby.
Nature, Vol. 268, 401 - 402 (1977).

156.002 Detection of the H 200β line in the direction of the galactic centre. P. A. Shaver.
Astron. Astrophys., Vol. 59, L31 - L32 (1977).
The H 200β line in the direction of Sgr A has been detected, with approximately the LTE intensity. This contradicts two published upper limits which had implied extreme physical conditions in the line-emitting gas. The observations can now be interpreted simply in terms of line-of-sight, low-density H II regions, similar to those seen in other directions.

156.003 The radio structure of the Cygnus Loop at 25 MHz.
Eh. P. Abranin, L. L. Bazelyan, N. Yu. Goncharov.
Astron. Zh. Akad. Nauk SSSR, Vol. 54, 781 - 789 (1977).
In Russian. English translation in Soviet Astron., Vol. 21, No. 4.
Total flux density and brightness distribution of the Cygnus Loop observed at 25 MHz with the UTR-2 radio telescope are reported. Radio spectra of both the whole source and the part identified with the filamentary nebulae NGC 6992 and NGC 6995 are presented. In a logarithmic scale the spectrum of this part in the Cygnus Loop appears to be rather linear than flat.

156.004 Hydrogen recombination lines at decimetric wavelengths from the direction of the galactic centre.
M. J. Kesteven, A. Pedlar.
Mon. Not. R. Astron. Soc., Vol. 180, 731 - 747 (1977).
H166α lines have been measured at 21 positions within 1° of the galactic centre. H192α and 241β spectra from the direction of Sgr A are also presented. The observations usually show narrow low-velocity emission lines, probably from gas in the line of sight, and broad lines which are presumably from the central part of the Galaxy. The latter show evidence of an expanding or contracting disk of ionized gas at positive longitudes within 70 parsecs of Sgr A. Some H166α emission may be associated with the H I rotating nuclear disk.

156.005 Infrared polarization of the galactic center.
T. Maihara, K. Noguchi, H. Okuda, S. Sato, M. Oishi.
Publ. Astron. Soc. Japan, Vol. 29, 415 - 419 (1977).
The infrared linear polarization of the galactic center

has been reexamined. The polarizations of 5.2 ± 0.3% and 16.7 ± 5.7% have been detected at the K- and H-band respectively, and the planes of the polarization are nearly parallel to the galactic plane. These polarizations are probably caused by the interstellar extinction in the inner region of the Galaxy.

156.006 Infrared polarization of the galactic center. II.
R. F. Knacke, R. W. Capps.
Astrophys. J., Vol. 216, 271 - 276 (1977).
Linear polarization observations of the galactic center (Sgr A) at 2.2, 3.5, and 11.5 μm are reported. At 2.2 and 3.5 μm the polarization orientation is nearly parallel to the galactic plane. At 11.5 μm the polarization is approximately perpendicular to the plane, but there are spatial variations across the infrared sources. The observations suggest that the near-infrared radiation is polarized by grains in the spiral arms and the 11.5 μm flux is polarized by grains in or near the infrared sources in Sgr A.

156.007 Airborne far-infrared observations of the galactic center region.
I. Gatley, E. E. Becklin, M. W. Werner, C. G. Wynn-Williams.
Astrophys. J., Vol. 216, 277 - 290 (1977).
Maps of a region 10′ in diameter around the galactic center made simultaneously in three wavelength bands at 30, 50, and 100 μm with ~1′ resolution are presented, and the distribution of far-infrared luminosity and color temperature across this region is derived. The position of highest far-infrared surface brightness coincides with the peak of the late-type stellar distribution and with the H II regions Sgr A West. The high spatial and temperature resolution of the data is used to identify features of the far-infrared maps with known sources of near-infrared, radio continuum, and molecular emission. It is found that the visual extinction across the central 10 pc of the Galaxy is only about 3 mag, and that the dust density is fairly uniform in this region. An upper limit of 10^7 $L_\odot$ is set on the luminosity of any still unidentified source of 0.1 to 1 μm radiation at the galactic center.

156.008 Extended infrared emission from the galactic plane.
S. D. Price.
Bull. American Astron. Soc., Vol. 9, 430 (1977). – Abstract.

156.009 **The galactic magnetic field.**   T. A. Spoelstra.
Usp. fiz. nauk, Vol. 121, 679 - 694 (1977). In
Russian. – Abstr. in Ref. zh., 51. Astron., 8.51.89 (1977).

156.010 **Determination of the parameters of the galactic
magnetic field by Faraday rotations of radio
sources.**   A. A. Ruzmajkin, D. D. Sokolov, A. V. Kovalenko.
Inst. prikl. mat. AN SSSR. Prepr. No. 20. Moskva, 1977.
33 pp. Price 12 Kop. In Russian. – Abstr. in Ref. zh., 51.
Astron., 8.51.789 (1977).

156.011 **Infrared astronomy and high-energy astrophysics.**
G. G. Fazio, F. W. Stecker.
The structure and content of the Galaxy and galactic gamma
rays, (see 012.009), p. 203 - 214 (1977).

Observations of the diffuse far-infrared flux from the
galactic plane, as well as far-infrared measurements of the
properties of dense molecular clouds, when combined with
recent high-energy γ-ray measurements and radio observations
of carbon monoxide, can yield new information about the
total mass of molecular clouds, the large-scale structure of the
inner Galaxy, and the density of cosmic rays.

156.012 **The interstellar magnetic field.**   C. Heiles.
Annu. Rev. Astron. Astrophys., Vol. 14, (see 003.
008), 1 - 22 (1976).
Contents: The local field; The field on the galactic scale;
The field in clouds; The field in very dense clouds.

156.013 **Diffusion models for the low-frequency radio emission from the Galactic halo.**   A. W. Strong.
Mon. Not. R. Astron. Soc., Vol. 181, 311 - 322 (1977).
The low-frequency drift-scan observations of the Galactic
radio emission at high latitudes are discussed in the context of
simple diffusion-plus-energy-loss models for the propagation of
electrons away from the Galactic plane. It is shown that for
certain parameters such models can reproduce the observations
quite well. The halo emissivity in the region just outside the
disc at the solar radius is ~7000 K/kpc at 17.5 MHz. The
average magnetic field in the halo must be ~0.2 of that in the
disc, and the diffusion mean free path about 1 pc. The full
width to half-maximum of the 17.5 MHz emission is about
6 kpc.

156.014 **Ne II 12.8 micron emission from the galactic center.
II.**   E. R. Wollman, T. R. Geballe, J. H. Lacy,
C. H. Townes, D. M. Rank.
Astrophys. J., Lett., Vol. 218, L103 - L107 (1977).
Recent observations of the 12.8 μm Ne II emission from
the galactic center have revealed a region of primarily blueshifted emission in addition to the previously detected redshifted emission. It appears most likely that the blueshifted
and redshifted emission come from separate clouds, and that
the dynamics of the ionized gas is dominated by the gravitational potential of the massive core at the galactic center. On
the basis of the velocities and velocity dispersions, the total
mass within a radius of 1 pc about the galactic center is estimated to be ~ $4 \times 10^6 M_\odot$.

156.015 **The galactic dynamo and its relation to the propa-
gation of ultra-high-energy cosmic rays.**
M. P. White.
Thesis, Univ. Durham, England (1977). – Abstr. in Phys.
Abstr., Vol. 80, Abstr. 74146 (1977).

156.016 **Synchrotron radiation and galactic field configurations.**
C. E. Jäkel, K. O. Thielheim, H. Wiese.
Astrophys. Space Sci., Vol. 51, 329 - 340 (1977).
Models of the galactic magnetic field are discussed with
respect to their influence on calculated contour maps as well
as longitudinal distributions of the synchrotron brightness

temperature. A comparison is made with Landecker-Wiele-
binski (1970) data.

156.017 **Faraday rotation and the turbulent structure of the
Galaxy.**   J. C. Carvalho, D. ter Haar.
Astrophys. Space Sci., Vol. 51, 385 - 393 (1977).
The authors extend Jokipii and Lerche's analysis of the
turbulent structure of our Galaxy by means of a study of the
rotation measure of extragalactic sources. Like them they use
a simple, statistically homogeneous and isotropic disc model
of the Galaxy and assume that the magnetic field has both an
average component and a fluctuating one. There is a correlation between the fluctuations in the electronic density and
those in the magnetic field strength and they consider different
forms for the two-point correlation function. They discuss the
observational data and present the basic equations for the
model used to interpret the observational data on the Faraday
rotation.

156.018 **The scale and strength of the galactic magnetic field
according to the pulsar data.**
A. A. Ruzmajkin, D. D. Sokolov.
Astrophys. Space Sci., Vol. 52, 365 - 374, 375 - 385 (1977).
In English and Russian.
In accordance with the data on the Faraday rotation,
angular coordinates, and dispersion measurements and dis-
tances of 38 pulsars, the strength $\mathbf{B} = 2.1 \pm 1.1$ μG and direction $l = 99° \pm 24°$, $b \simeq 0°$ of the large-scale galactic magnetic
field and the mean electron density in the galactic disc $N_e =$
0.03 ± 0.01 cm^{-3} are determined. A comparison with the
results of a study of the measures of rotation of extragalactic
radio sources enabled the authors to estimate the characteristic half-width of the distribution of the electron density on
the Z-coordinate ($h \simeq 400$ pc). The characteristic size of
galactic magnetic field fluctuations is shown to be $L = 100$–
150 pc.

156.019 **Scale and strength of the galactic magnetic field from
pulsar data.**   A. A. Ruzmajkin, D. D. Sokolov.
Inst. prikl. mat. AN SSSR. Prepr. No. 39. Moskva, 1977. 20 pp.
Price 8 Kop. In Russian. – Abstr. in Ref. zh., 51. Astron., 11.
51.939 (1977).

156.020 **Balloon observations of infrared surface brightness
of our Galaxy.**
S. Hayakawa, K. Ito, T. Matsumoto, Uyama.
Bull. Inst. Space Aeronaut. Sci. Univ. Tokyo B, Vol. 12, 631 -
646 (1976). In Japanese. – From Phys. Abstr., Vol. 80, Abstr.
82308 (1977).

156.021 **Radio radiation of the corona of the Galaxy.**
N. M. Lipovka.
Astron. Zh. Akad. Nauk SSSR, Vol. 54, 1211 - 1220 (1977).
In Russian. English translation in Soviet Astron., Vol. 21,
No. 6.
The results of observations of the spherical component
at λ = 4 cm are presented. The interpretation of the observed
character of radioemission and the spatial distribution of the
spherical component is given within the framework of the
model of synchrotron radiation of relativistic electrons with
curved energetic spectrum moving in the magnetic field with
intensity decreasing to the poles.

**Linear polarization observations of galactic radio
emission on magnetic tape.**   See Abstr. 002.017.

Infrared astronomy and galactic dust. I.
See Abstr. 061.011.

Infrared astronomy and galactic dust. II.
See Abstr. 061.018.

Radiating cosmic dust.   See Abstr. 131.035.

Prebiotic polymers and infrared spectra of galactic sources.   See Abstr. 131.096.

Radio observations related to star formation.

See Abstr. 131.144.

Interstellar H300α line radiation.
See Abstr. 131.182.

Broadband observations of HI within 10° of the direction of the galactic center.   See Abstr. 155.001.

157 Galactic Extreme UV, X, Gamma Radiation

157.001 The relationship between the galactic matter distribution, cosmic-ray dynamics, and gamma-ray production.   D. A. Kniffen, C. E. Fichtel, D. J. Thompson.
Astrophys. J., Vol. 215, 765 - 774 (1977).

Theoretical considerations and analysis of the results of γ-ray astronomy suggest that the galactic cosmic rays are dynamically coupled to the interstellar matter through the magnetic fields; hence the cosmic-ray density should be enhanced where the matter density is greatest on the scale of galactic arms. This concept has been explored in a galactic model using recent 21 cm radio observations of neutral hydrogen and 2.6 mm observations of carbon monoxide, which is considered to be a tracer of molecular hydrogen. A constant cosmic-ray density, as might be expected in the simplest concept of a universal cosmic-ray model, gives too small a ratio between the γ-ray intensity in the central region and the general anticenter region and does not give rise to significant peaks along galactic spiral arm features in the galactic γ-ray data. With regard to the galactic center itself, whereas there may be a small additional component, such as Compton scattering from a high photon density, there is quite reasonable agreement between the observed γ-ray distribution and that predicted on the basis of cosmic-ray nucleons and electrons interacting with the current best estimate of the interstellar matter.

157.002 Galactic centre γ-rays from trapped cosmic rays.   A. W. Wolfendale, D. M. Worrall.
Astron. Astrophys., Vol. 60, 165 - 170 (1977).

A calculation has been made of the emissivity of γ-rays of energy above 100 MeV from the decay of neutral pions produced in cosmic ray-interstellar gas nucleus collisions in dense regions. The analysis is applied to the dense region within a few hundred parsecs of the galactic centre and the total γ-ray yield from all production processes is calculated.

157.003 Point gamma-ray sources and the distribution of cosmic rays over the Galaxy.
A. A. Stepanyan.
Izv. Krymskoj Astrofiz. Obs., Vol. 57, 99 - 106 (1977). In Russian.

The supposition is made that some part (~40%) of gamma-ray emission of energy $> 10^8$ eV from the galactic center region ($330° < l^{II} < 40°$) is produced by point sources. This supposition makes it easier to explain the diffuse galactic disc emission because it decreases the amount of gas necessary for the explanation. In the framework of this supposition the distribution of the electron component of cosmic rays in the Galaxy can be coordinated with the distribution of the nuclear component derived from data of gamma-ray emission of the Galaxy.

157.004 SAS-2 observations of the diffuse gamma radiation in the galactic latitude interval $10° < |b| < 90°$.
C. E. Fichtel, R. C. Hartman, D. A. Kniffen, D. J. Thompson,

H. B. Ögelman, M. E. Özel, T. Tümer.
Astrophys. J., Lett., Vol. 217, L9 - L13 (1977).

An analysis of all the second Small Astronomy Satellite (SAS-2) γ-ray data for galactic latitudes with $|b| > 10°$ has shown that the intensity varies with galactic latitude, being larger near 10° than 90°. For energies above 100 MeV the γ-ray data are consistent with a latitude distribution of the form $I(b) = C_1 + C_2/\sin b$, with the second term being dominant. This result suggests that the radiation above 100 MeV is coming largely from local regions of the galactic disk. Between 35 and 100 MeV, a similar equation is also a good representation of the data.

157.005 Soft diffuse X-rays in the southern galactic hemisphere.   W. T. Sanders, W. L. Kraushaar, J. A. Nousek, P. M. Fried.
Astrophys. J., Lett., Vol. 217, L87 - L91 (1977).

A map is presented of the soft X-ray diffuse background flux in the C band (~0.13−0.28 keV) covering almost all of the southern galactic hemisphere. A comparison at constant galactic latitude of both C band and B band (~0.1−0.18 keV) soft X-ray data with neutral hydrogen maps shows that the intensity does decrease with increasing neutral hydrogen column density, but in a manner that is inconsistent with photoelectric absorption. The authors suggest that the inverse correlation is a displacement effect. X-ray emission regions appear to be where the cool gas is not. Further, the evidence against photoelectric absorption implies that the bulk of the cool gas is beyond the X-ray emitting regions. The Sun appears to be surrounded by a soft X-ray emission region of ~million-degree gas.

157.006 Are galactic γ rays point sources or diffuse emission?   J. J. Quenby.
Nature, Vol. 269, 466 - 467 (1977).

157.007 SAS-2 galactic gamma-ray results – I. Diffuse emission.   D. J. Thompson, C. E. Fichtel, R. C. Hartman, D. A. Kniffen, G. F. Bignami, R. C. Lamb, H. Ögelman, M. E. Özel, T. Tümer.
The structure and content of the Galaxy and galactic gamma rays, (see 012.009), p. 3 - 13 (1977).

Continuing analysis of the data from the SAS-2 high energy γ-ray experiment has produced an improved picture of the sky at photon energies above 35 MeV. On a large scale, the diffuse emission from the galactic plane is the dominant feature observed by SAS-2. This galactic plane emission is most intense between galactic longitudes 310° and 45°, corresponding to a region within 7 kpc of the galactic center. Within the high-intensity region, SAS-2 observes peaks around galactic longitudes 315°, 330°, 345°, 0°, and 35°. These peaks appear to be correlated with galactic features and components such as molecular hydrogen, atomic hydrogen, magnetic fields, cosmic-ray concentrations, and photon fields.

157.008   Low- and medium-energy galactic gamma-ray observations.   G. H. Share.
The structure and content of the Galaxy and galactic gamma rays, (see 012.009), p. 65 - 80 (1977).

Current detection techniques for 0.1- to 100-MeV γ-ray measurements are summarized, and their capabilities for measuring the diffuse galactic emission are evaluated.

157.009   Very high-energy gamma-ray astronomy.
J. E. Grindlay.
The structure and content of the Galaxy and galactic gamma rays, (see 012.009), p. 81 - 98 (1977).

Recent results in ground-based very high-energy ($>10^{11}$ eV) γ-ray astronomy are reviewed. The positive detections (at $\gtrsim 10^{12}$ eV) of the Crab pulsar that suggest a very flat spectrum and time-variable pulse phase are discussed. Observations of other pulsars (particularly Vela) suggest that these features may be general. The southern sky observations are reviewed, and the significance of the detection of an active galaxy (NGC 5128) is considered for source models and future observations.

157.010   The nonthermal radiation in the Galaxy.
J. E. Baldwin.
The structure and content of the Galaxy and galactic gamma rays, (see 012.009), p. 189 - 201 (1977).

This paper does not attempt to review all aspects of the nonthermal continuum radiation in the Galaxy, but concentrates on two topics of particular interest for γ-ray studies: 1. The distribution of nonthermal emissivity with height z above the galactic plane. 2. The relationship between the nonthermal emissivity and the neutral gas.

157.011   Ultraviolet observations of local gas.
E. B. Jenkins.
The structure and content of the Galaxy and galactic gamma rays, (see 012.009), p. 215 - 227 (1977).

From satellite measurements of ultraviolet spectra of stars, an average density of approximately 1.1 cm^{-3} for hydrogen atoms, in both atomic and molecular form, is estimated for regions of space along the galactic plane within about 1 kpc of the Sun.

157.012   Small-scale local gamma-ray features.
J. L. Puget, C. Ryter, G. Serra.
The structure and content of the Galaxy and galactic gamma rays, (see 012.009), p. 229 - 236 (1977).

In order to draw implications from nearby γ-ray emission, the different ways that can be used to obtain an estimate of the amount of matter on each line of sight are investigated. Then, it is shown that, within present uncertainties, the cosmic-ray intensity inside molecular clouds within 1 kpc from the Sun is the same as the cosmic-ray intensity measured at the Sun. In the last part, what can be learned from a comparison of far infrared and γ-ray data is discussed.

157.013   Diffuse galactic gamma-ray lines.
R. E. Lingenfelter, R. Ramaty.
The structure and content of the Galaxy and galactic gamma rays, (see 012.009), p. 237 - 252 (1977).

The authors studied the origin and observability of diffuse γ-ray line emission from our galaxy. They find that such lines could be formed by nuclear excitation interactions of low-energy cosmic rays with both interstellar gas and dust grains. They present here a detailed evaluation of the production rate of the 4.44-MeV line for a variety of assumed cosmic-ray spectra, compare these results with reported galactic γ-ray line intensities, and conclude that the measurements are consistent with a low-energy cosmic-ray density which increases toward the galactic center in proportion to the molecular gas density.

157.014   Gamma rays, cosmic rays, and galactic structure.
F. W. Stecker.
The structure and content of the Galaxy and galactic gamma rays, (see 012.009), p. 315 - 345 (1977).

The relation of the recent SAS-2 observations of galactic γ-rays to the large-scale distribution of cosmic rays and interstellar gas in the Galaxy is reviewed and reexamined.

157.015   Galactic gamma rays and the origin of cosmic rays.
D. Dodds.
Thesis, Univ. Durham, England (1977).

157.016   High energy gamma-ray emission from the region of the galactic center.   B. Agrinier, M. Forichon,
B. Parlier, G. Boella, G. Gerardi, B. Sacco, R. Palmeira, M. Niel.
Astrophys. Space Sci., Vol. 47, 401 - 413 (1977).

In two balloon flights carried out in the Southern Hemisphere, a region of the sky near the galactic center has been explored with a spark chamber telescope with the aim of investigating the gamma-ray emission at energies above 20 MeV from possible celestial sources.

157.017   Picture of the sky in gamma radiation.
A. M. Gal'per, V. G. Kirillov-Ugryumov, B. I. Luchkov.
Priroda, 1977, No. 12, p. 20 - 31. In Russian.

X-ray and gamma-ray line production by nonthermal ions.   See Abstr. 022.103.

Preliminary results from the European Space Agency's COS-B satellite for gamma-ray astronomy.   See Abstr. 051.018.

On the interstellar ultraviolet radiation field.   See Abstr. 131.034.

Observations of the soft X-ray background.   See Abstr. 142.039.

The diffuse soft X-ray sky. Astrophysics related to cosmic soft X-rays in the energy range 0.1−2.0 keV.   See Abstr. 142.129.

SAS-2 galactic gamma-ray results − II. Localized sources.   See Abstr. 142.704.

Gamma-ray astrophysics and galactic structure.   See Abstr. 142.707.

The age of the galactic cosmic rays derived from the abundance of ^{10}Be.   See Abstr. 143.019.

Spatial distribution of high energy cosmic ray electrons perpendicular to the galactic plane.   See Abstr. 143.054.

Gamma rays and large-scale galactic structure.   See Abstr. 155.022.

The galactic halo question: new size constraints from galactic γ-ray data.   See Abstr. 155.024.

Galactic structure and gamma radiation.   See Abstr. 155.060.

158 Single and Multiple Galaxies, Peculiar Objects

Single and Multiple Galaxies

158.001 150 southern compact and bright-nucleus galaxies.
A. P. Fairall.
Mon. Not. R. Astron. Soc., Vol. 180, 391 - 400 (1977).

Galaxies having regions of exceptionally high surface brightness have been selected from the ESO Quick Blue Survey and investigated by 'grating photography' – direct photography plus low-dispersion slitless spectroscopy. Two new Seyfert galaxies and a peculiar multiple system have been discovered. Differences in red continua are also noted.

158.002 Absolute magnitude-color relation for early type spirals. N. Visvanathan, D. Griersmith.
Astron. Astrophys., Vol. 59, 317 - 328 (1977).

New narrow band observations of $(u-V)$ color and V magnitude have been made for 28 early type spirals in the Virgo cluster complex and for another 13 galaxies in 11 other groups. The colors $(u-V)_c$, corrected for galactic reddening, standard aperture, K term, and tilt are negatively correlated with the standard magnitudes $(V_{26})_c$ of the galaxies in the Virgo I cluster. An absolute calibration of the CM relation between absolute magnitude $(M_V)_S$ of early type spirals and the color $(u-V)_c$ yields: $(M_V)_S = -6.07 - 7.73 \ (u-V)_c$.

158.003 The radio continuum emission of M51.
P. C. van der Kruit.
Astron. Astrophys., Vol. 59, 359 - 366 (1977).

Westerbork radio continuum observations of M51 at 610, 1415 and 4995 MHz are analysed and discussed. It is shown that at 4995 MHz the thermal emission from the brightest H II-regions has been detected and that this accounts for the changes of spectral index between 1415 and 4995 MHz across the spiral arms. Hα-measurements and determinations of Balmer decrements, existing in the literature, also predict that thermal emission must have a serious effect on the radial variation (steepening with radius) of the spectral index between 610 and 1415 MHz, and can account for the observed effect.

158.004 Spectrophotometry of Seyfert 1 galaxies.
D. E. Osterbrock.
Astrophys. J., Vol. 215, 733 - 745 (1977) = Lick Obs. Bull., No. 757.

Relative emission-line intensities are given for 36 Seyfert 1 galaxies. Equivalent widths for Hβ were measured to link the emission-line strengths to the continuum. Information is also given on the broad emission-line profiles, which cover a wide range in velocity and often appear somewhat asymmetric. Nearly every Seyfert 1 galaxy has Fe II emission in its spectrum, but there is a wide range in its strength. Broad-line radio galaxies in general have much weaker (if any) Fe II emission and steeper Balmer decrements than Seyfert 1 galaxies. The Fe II emission strengths are not well correlated with ultraviolet excess or broad emission-line width, in apparent disagreement with the resonance-fluorescence excitation mechanism, though specific models will be needed to test this conclusion.

158.005 Observations of Fe II emission in Seyfert galaxies and QSOs. M. M. Phillips.
Astrophys. J., Vol. 215, 746 - 754 (1977) = Lick Obs. Bull., No. 755.

Spectrophotometric observations of Seyfert galaxies and QSOs with strong Fe II emission are presented. The Seyfert galaxy Markarian 478 is shown to have a spectrum which is very similar to I Zw 1. The Fe II lines of these "narrow-line" objects appear to have the same widths and profiles as the hydrogen Balmer lines. Observations of several Seyfert galaxies with broader emission lines confirm this finding. This result is consistent with models where the Fe II and H I emission arise from rapidly moving filaments, but is not with the supra-thermal particle model.

158.006 On the nature of M 87 jet. I. S. Shklovskij.
Astron. Zh. Akad. Nauk SSSR, Vol. 54, 713 - 721 (1977). In Russian. English translation in Soviet Astron., Vol. 21, No. 4.

The conclusion is founded that the M 87 jet is formed by clouds of magnetized plasma with $M \sim 1 \ M_\odot$ outflowing from the nucleus of this galaxy with relativistic velocities. The "one-side-directionality" of the jet is explained by the relativistic Doppler effect. From energetic and dynamical considerations, angular dimensions of single jet condensations and the parameter $\gamma = (1-\beta^2)^{-1/2}$ are estimated. The synchrotron spectrum of the jet is discussed. The deceleration of the jet "knots" by the surrounding plasma is estimated; it can be the heat source for the intergalactic gas. In the light of the results obtained some general problems of the activity of galactic nuclei are discussed.

158.007 Large-scale distribution and motion of galaxies.
A. G. Doroshkevich, S. F. Shandarin.
Astron. Zh. Akad. Nauk SSSR, Vol. 54, 734 - 744 (1977). In Russian. English translation in Soviet Astron., Vol. 21, No. 4.

In the framework of the adiabatic theory of galaxy formation the problems of large-scale distribution and motion of galaxies and clusters of galaxies are discussed. It is shown that the mean peculiar velocity of an object increases rapidly with growth of the distance to the object and reaches the limit value $\approx 300-1000$ km/sec only at a distance of about 300 Mpc. An anisotropy in the distribution and in the motion of galaxies is possible at a scale of about 100 Mpc. The results are compared with data on the motion of the Galaxy and with Sandage and Tamman's data on the motion of galaxies.

158.008 A new Sculptor-type dwarf elliptical galaxy in Carina. R. D. Cannon, T. G. Hawarden, S. B. Tritton.
Mon. Not. R. Astron. Soc., Vol. 180, 81P - 82P (1977).

A star system discovered on a plate taken with the UK 1.2-m Schmidt telescope appears to be a dwarf elliptical galaxy and a new member of the Local Group of galaxies.

158.009 Detection of the H 102α recombination line in NGC 253. E. R. Seaquist, M. B. Bell.
Astron. Astrophys., Vol. 60, L1 - L4 (1977).

The authors have detected a broad H102α emission line in the radio spectrum of the galaxy NGC 253. The feature is centered at a velocity of 132 km s^{-1} (LSR) and is in good agreement with the velocity of the ionized gas in the central region of the galaxy.

158.010 NGC 1510: a young elliptical galaxy?
M. J. Disney, S. R. Pottasch.
Astron. Astrophys., Vol. 60, 43 - 54 (1977).

NGC 1510 is a southern E0 galaxy with an A-type spectrum, strong emission lines and extremely blue colours. It has the relaxed light distribution of a normal elliptical and there is no sign of interaction with its brighter SB0 companion NGC 1512. The authors suggest that NGC 1510 may be a young galaxy that has formed from the large amount of hydrogen near NGC 1512.

158.011   **The H I content of blue compact galaxies with emission lines.**   P. Chamaraux.
Astron. Astrophys., Vol. 60, 67 - 78 (1977).

H I 21-cm line measurements carried out with the Nançay radiotelescope for 27 blue compact galaxies with optical emission lines and partly compact ones are presented and discussed. The total properties of these objects are compared to those of Hubble sequence galaxies, through relations between various parameters. It is shown that partly compact galaxies are normal in this respect, whereas blue compact ones are characterized by a high and nearly constant mean H I surface density, exceeding that of normal galaxies by a factor of 2 to 4.

158.012   **A dust model for M82: constraints on the nature of intergalactic dust.**   H. I. Abadi, C. F. Bohren.
Astron. Astrophys., Vol. 60, 125 - 130 (1977).

A dust model for light scattering in the halo of the irregular II galaxy M82 is constructed based on the assumption that the galaxy is moving through an intergalactic dust cloud. Completely general expressions for the Stokes parameters of the light received by an observer are derived on the assumption that the galaxy is a point source. For a uniform density distribution of dust, good agreement with polarization and intensity observations is possible only if the dust particles are small, needle-like, and metallic. Graphite and iron are considered as plausible candidates for the composition of the dust.

158.013   **The H I content of the elliptical galaxies NGC 3904 and 4636.**   L. Bottinelli, L. Gouguenheim.
Astron. Astrophys., Vol. 60, L23 - L25 (1977).

The elliptical galaxies NGC 3904 and 4636 have been measured in the 21-cm line of neutral hydrogen with the Nançay radiotelescope. Their neutral hydrogen masses are 7.5×10^8 and 2.6×10^8 solar masses respectively, leading to H I mass to luminosity ratios equal to 0.060 and 0.014.

158.014   **Galaxies with ultraviolet continuum. VIII.**   B. E. Markarian, V. A. Lipovetskij.
Astrofizika, Vol. 12, 389 - 396 (1976). In Russian. – English translation in Astrophysics, Vol. 12, No. 3.

The eighth list of galaxies having intense ultraviolet continuum is presented. The list contains data for 97 objects. The presence of emission lines is either established or suspected among 64 of them. The presence of Seyfert characteristics can be certainly expected on the objects No. 704, 705 and 771. Seyfert characteristics may be suspected among the objects No. 716 and 734.

158.015   **Spectral observations of the galaxy NGC 1275.**   V. T. Doroshenko, V. Yu. Terebizh, K. K. Chuvaev.
Astrofizika, Vol. 12, 417 - 429 (1976). In Russian. – English translation in Astrophysics, Vol. 12, No. 3.

The galaxy NGC 1275 has been observed from December 1973 to January 1975. The observational material consists of about 40 spectrograms covering the wavelength range $\lambda\lambda$ 4400–6800 Å with a dispersion of 58 and 110 Å/mm. A number of rather weak lines have been observed which mainly belong to the forbidden lines of Fe. The existence of the double structure of N_1, N_2 and Hβ lines discovered by Dibaj and Esipov (1968) is confirmed. The equivalent widths and contours of the emission lines of [O III] and Hβ in the spectra taken during 1972–1975 have been compared. No variations have been found beyond observational errors. The intensities of hydrogen lines indicate almost full absorption of the ultraviolet "tail" of the nonthermal nuclear continuum in the surrounding gas envelope. Permitted lines of Fe II are weaker than forbidden ones. This can be accounted for the frame of Sobolev's theory of moving star envelopes.

158.016   **Measurements of brightness variability of Markarian**

509.   O. V. Magnitskaya, K. A. Saakyan.
Astrofizika, Vol. 12, 431 - 435 (1976). In Russian. – English translation in Astrophysics, Vol. 12, No. 3.

Results of three-colour photometry of the galaxy Markarian 509 during its minimum light are given. It has had a maximum brightness during four separate remote epochs, and minimum brightness only in a period of four months was maintained, i.e. a fall of luminosity took place of the order of one magnitude.

158.017   **Infrared emission and the Byurakan classification of galaxies.**   G. M. Tovmasyan.
Astrofizika, Vol. 12, 555 - 557 (1976). In Russian. – English translation in Astrophysics, Vol. 12, No. 3.

It is shown that galaxies with starlike or split nuclei have larger colour indices in the infrared than galaxies of other Byurakan classes, and thus the probability of detection of infrared radiation should be higher in the case of observations of galaxies with separate nuclei.

158.018   **The dependence of emission line intensity of Markarian galaxies upon the colour index.**
M. A. Arakelyan.
Astrofizika, Vol. 12, 559 - 562 (1976). In Russian. – English translation in Astrophysics, Vol. 12, No. 3.

It is shown that there exists a dependence of equivalent widths of [O III] λ 5007 and Hβ of non-Seyfert-type Markarian galaxies upon ultraviolet excess. Thereby the correlation between the intensity of [O III] and the colour index is noticeably stronger than for Hβ.

158.019   **The angular momentum of galaxies.**   T. X. Thuan, J. R. Gott III.
Astrophys. J., Vol. 216, 194 - 205 (1977).

The tidal interaction picture is applied to elliptical and spiral galaxies. It gives the correct amount of angular momentum for our Galaxy. Monte Carlo calculations are used to predict the distribution of total angular momenta for galaxies of a given mass (or luminosity). The data are consistent with a dissipationless collapse for elliptical galaxies. The tidal theory predicts a distribution of relative sizes for elliptical galaxies at a given absolute magnitude in agreement with Fish's observations, but for spiral galaxies predicts too many high-surface-brightness small-scale-length disks. For tidal interactions to produce the correct amount of angular momentum, most of the mass in the universe must be contributed by galaxies and their halos and not by some smooth diffuse component.

158.020   **The main body of NGC 1275 in the visible ultraviolet.**   T. F. Adams.
Publ. Astron. Soc. Pacific, Vol. 89, 488 - 490 (1977).

New image-tube direct plates of NGC 1275 in the visible ultraviolet and the near infrared are presented. The respective filters were chosen to reduce the effect of line emission and to emphasize the hot and cool stellar populations. These plates are compared with other published plates in order to study the distribution of hot stars in the main body of NGC 1275. Hot stars are found to be associated with the line-emitting knots in the high-redshift system, but not the low-redshift system. The presence of two important outlying structures that emit in the ultraviolet continuum is also discussed.

158.021   **The color-absolute magnitude relation for E and S0 galaxies. I. Calibration and tests for universality using Virgo and eight other nearby clusters.**
N. Visvanathan, A. Sandage.
Astrophys. J., Vol. 216, 214 - 226 (1977).

Spectrum scanner observations of E and S0 galaxies in the Virgo cluster confirm the previously known correlation between color and absolute magnitude, and show it to be strongly wavelength-dependent. Galaxies in eight other groups and clusters show a color-magnitude effect that has the same wavelength dependence, slope, absolute magnitude calibration, and cosmic scatter as the Virgo cluster. If one adopts the Virgo cluster modulus to be $m - M = 31.70$, the absolute calibration of the C-M effect for E and S0 galaxies becomes $M_{V_{26}}c = -10.327 (u - V)_c + 2.19$, with an intrinsic dispersion of $\sigma(M) = \pm0.5$ mag for a single galaxy. Hence photometric distances can be determined for single E and S0 galaxies by the color-magnitude method within errors that are distributed as $\sigma(\delta r/r) = \pm0.22$.

158.022 Isolated galaxies. J. Huchra, T. X. Thuan.
Astrophys. J., Vol. 216, 694 - 697 (1977).

A search is made for isolated galaxies by using the statistical sample of "field" galaxies of Turner and Gott. Only 39 of the 350 Turner-Gott field galaxies do not have companions between 14 and 15.7 mag within a 45' radius. These 39 galaxies are, on the average, nearer than the remainder of the sample; two-thirds of them appear to be associated with nearby de Vaucouleurs groups of galaxies, which are clustered on angular scales much larger than that searched by Turner and Gott. There is little observational evidence at present for a significant population of isolated galaxies.

158.023 Submillimeter photometry of extragalactic objects.
R. H. Hildebrand, S. E. Whitcomb, R. Winston, R. F. Stiening, D. A. Harper, S. H. Moseley.
Astrophys. J., Vol. 216, 698 - 705 (1977).

The authors present results for eight extragalactic objects observed at wavelengths between 390 and 1100 μm. The best-determined spectrum at these wavelengths and in the far-infrared is that for the Seyfert galaxy NGC 1068. The spectrum can be fitted well, assuming thermal emission from an optically thin dust cloud. The galaxies NGC 253 and NGC 5236 also appear to be thermal sources with large masses of interstellar material. The submillimeter measurements for 3C 273, BL Lac, NGC 1275, and NGC 5128 give no evidence for a far-infrared peak.

158.024 Spectroscopy and photometry of the distant radio galaxy 3C 343.1.
H. Spinrad, J. Westphal, J. Kristian, A. Sandage.
Astrophys. J., Lett., Vol. 216, L87 - L89 (1977).

Spectrophotometry of the faint radio galaxy 3C 343.1 shows a strong emission line in the red. If identified with $\lambda3727$ of [O II], it yields $z = 0.750$. The identification is aided by noting that $\lambda3869$ of [Ne III] is marginally visible. The optical continuum is relatively blue in color; it is likely to be caused by a population of stars younger than often found in nearby E galaxies. By comparison with other distant galaxies of established z, the isophotal diameter of about 7", measured for 3C 343.1, is consistent with the proposed redshift, and corresponds to a linear size of 70 kpc to a surface brightness of $\mu_r \approx 26$ mag arcsec^{-2}.

158.025 Galaxies with ultraviolet continuum. IX.
B. E. Markarian, V. A. Lipovetskij.
Astrofizika, Vol. 12, 657 - 664 (1976). In Russian. – English translation in Astrophysics, Vol. 12, No. 4.

The ninth list of galaxies with ultraviolet continuum is given. The present list contains data for 98 objects. The presence of emission lines is either established or suspected among 64 of them. The presence of Seyfert characteristics can be certainly expected in the objects No. 817, 841, 849 and 876. Seyfert characteristics can be suspected among objects No. 830, 845, 854 and 871. The QSO nature may be certainly predicted for the objects No. 813 and 877.

158.026 Spectral observations of Markarian galaxies. II.
Eh. K. Denisyuk, V. A. Lipovetskij, V. L. Afanas'ev.
Astrofizika, Vol. 12, 665 - 681 (1976). In Russian. – English translation in Astrophysics, Vol. 12, No. 4.

Spectroscopic observations of 75 galaxies from Markarian's lists IV–VII of galaxies are presented. Emission lines have been found in the spectra of 64 galaxies for which redshifts and relative intensities of lines have been determined. 10 objects have no lines in the red part of the spectrum, Markarian 396 has only hydrogen absorption lines (white dwarf). The objects Markarian 595, 609, 622, 668, 699 and 700 possess the characteristics of Seyfert galaxies. Broadened emission lines are also observable or may be suspected in Markarian 414, 584, 612, 617, 646, 670, 684 and 693. Markarian 586 is a possible QSO. The accuracy of radial velocity determinations is considered. It is shown that the real error is 60–70 km/sec.

158.027 Spectral observations of galaxies of high surface brightness. IV.
M. A. Arakelyan, Eh. A. Dibaj, V. F. Esipov.
Astrofizika, Vol. 12, 683 - 687 (1976). In Russian. – English translation in Astrophysics, Vol. 12, No. 4.

The results of spectral observations of 44 galaxies of high surface brightness compiled by Arakelyan (1975) are presented. Emission lines are detected and redshifts are measured in the spectra of 22 objects. The galaxy No. 564 has pronounced spectral properties of nuclei of Seyfert galaxies.

158.028 Spectra of galaxies of high surface brightness.
Eh. A. Dibaj, V. T. Doroshenko, V. Yu. Terebizh.
Astrofizika, Vol. 12, 689 - 691 (1976). In Russian. – English translation in Astrophysics, Vol. 12, No. 4.

The spectra of galaxies of high surface brightness from the list compiled by Arakelyan (1975) have been obtained. Thirty-five of the eighty galaxies have emission-line spectra. Redshifts and absolute magnitudes of galaxies with emission lines are determined.

158.029 Shape, normalization, and local enhancement of the luminosity function for field galaxies.
J. E. Felten.
Bull. American Astron. Soc., Vol. 9, 431 (1977). – Abstract.

158.030 Colors as indicators of the presence of spiral and elliptical components in N galaxies.
B. M. Tinsley.
Publ. Astron. Soc. Pacific, Vol. 89, 245 - 250 (1977).

N galaxies have properties in many respects intermediate between those of quasars and normal galaxies, so it is assumed that they are composite systems with a quasar in the nucleus of an otherwise ordinary galaxy. The question of whether the underlying galaxies are spirals (as in most nearby Seyfert galaxies) or ellipticals (as often suggested for quasars) is approached here via the positions of N galaxies in the $(U - B, B - V)$ diagram. Because of the differences between color indices of various types of normal galaxies at the redshifts of the N systems, some conclusions can be drawn. It is found that some N galaxies are best interpreted as having an elliptical galaxy component, while others more plausibly have underlying spirals. Three of the 17 N galaxies considered have UBV colors that cannot be explained using combinations of normal galaxies and quasars.

158.031 Emission-line spectra of seven Arakelian galaxies.
D. E. Osterbrock, M. M. Phillips.
Publ. Astron. Soc. Pacific, Vol. 89, 251 - 254 (1977) = Lick Obs. Bull., No. 764.

Descriptions of the emission-line spectra of seven Arakelian galaxies are given. Two of the galaxies have very low-excitation, weak emission lines, and are definitely not Seyfert galaxies. Two have emission-line spectra which are similar to M51 and M81, and a third is closer to these objects than to Seyfert 2's. One galaxy, Akn 79, is a typical Seyfert 2 object, while another, Akn 120, has the spectrum of a typical Seyfert 1. Relative emission-line intensities are given for Akn 120, and the line profiles of Hβ and Hα are shown to differ significantly in form and width. The equivalent width of Hβ in Akn 120 is very large (190 Å), and is difficult to account for by photoionization alone.

158.032 A study of the optical variability of the Seyfert galaxy X Comae.
R. F. Green, J. P. Huchra, H. E. Bond.
Publ. Astron. Soc. Pacific, Vol. 89, 255 - 260 (1977) = Contrib. Louisiana State Univ. Obs., Baton Rouge, No. 123.

Light curves in photographic and photographic visual magnitudes have been derived for the variable galaxy X Com from plates taken during the years 1907–74. Three results were obtained from the investigation. A photographic and photoelectric sequence of stars has been established around the galaxy, with positions near the center of the Coma cluster of galaxies. A technique has been developed for removing the effects of seeing from iris photometry of a nonstellar object with respect to a stellar reference system. The light curves show some evidence for both short outbursts and slow, long-term variations in brightness, consistent with the optical activity of a Seyfert galaxy nucleus.

158.033 Blue objects near M 31.   G. Romano.
Acta Astron., Vol. 27, 287 - 289 (1977).

Many blue objects and some optical counterparts of 5C3 radio sources in the field of M 31 have been studied. Only one blue object (Richter no. 792) shows irregular light variations between 16.8 and 17.9 mag.

158.034 Survey of late-type and irregular southern galaxies on plates taken with the UK 1.2-m Schmidt telescope. I.   H. G. Corwin, Jr., A. de Vaucouleurs, G. de Vaucouleurs.
Astron. J., Vol. 82, 557 - 568 (1977).

A survey of southern galaxies generally larger than ∼1 arcmin was made. The present paper gives classifications, luminosity classes, and inner and outer diameters for 267 galaxies of types Scd and later at $\delta < -22°$ in 101 fields. Among them are over 120 new, mostly low-surface-brightness dwarf objects.

158.035 Radio continuum observations of Markarian galaxies at 1410 MHz and 2700 MHz.   J. H. Bieging, P. Biermann, K. Fricke, I. I. K. Pauliny-Toth, A. Witzel.
Astron. Astrophys., Vol. 60, 353 - 360 (1977).

The authors have measured the continuum flux densities of a selection of Markarian and Zwicky galaxies. They find that all Markarians which are neither Seyfert galaxies nor BL Lac objects have radio fluxes which are consistent with recent theoretical models by Biermann and Fricke (1977). They also discuss Seyfert galaxies, most of which give radio data also consistent with the theory.

158.036 Radio continuum and H I observations of S0 galaxies.   J. H. Bieging, P. Biermann.
Astron. Astrophys., Vol. 60, 361 - 368 (1977).

The authors have observed a selection of S0 galaxies with the Arecibo 305-m telescope in both the 21 cm H I line and the 21 cm continuum. The data are consistent with the ideas that S0 galaxies undergo repeated weak bursts of star formation (Biermann, 1977), and may have a significant amount of ionized hydrogen. The authors suggest that S0s need not suffer from gas loss to account for the low H I content, but rather may use up the gas in bursts. A comparison between S0s, normal spirals and Markarian galaxies is made using some additional H I observations.

158.037 Rotation axes of the optical galaxies associated with Cygnus A and 3C 33.   S. M. Simkin.
Astrophys. J., Vol. 217, 45 - 50 (1977).

Emission-line velocity curves for the galaxies associated with Cyg A and 3C 33 imply a rotation axis for Cyg A which lies in the plane of the sky and is nearly aligned with that object's radio axes. The emission-line velocities for 3C 33 appear to reflect expansion but may also include a rotational component whose axis is nearly aligned with this object's radio axis. Absorption-line velocities for 3C 33 show that the gas which gives rise to the emission lines is blueshifted with respect to the absorption-line system by 500 km s⁻¹. The prominent absorption lines near 5100 Å in the spectrum of 3C 33 are those of a normal E galaxy. In the ultraviolet, however, the absorption lines appear to arise from early-type stars.

158.038 Extended rotation curves of high-luminosity spiral galaxies. I. The angle between the rotation axis of the nucleus and the outer disk of NGC 3672.
V. C. Rubin, N. Thonnard, W. K. Ford, Jr.
Astrophys. J., Lett., Vol. 217, L1 - L4 (1977).

A large velocity gradient is observed in the excited nuclear gas of NGC 3672 when the slit is aligned along the minor axis, and a nuclear velocity gradient smaller than that implied by the rotation curve of the outer disk is observed with the slit along the major axis. The most direct interpretation of this observation is that the rotation axis of the nuclear gas disk makes a large angle with the rotation axis of the outer galaxy as a whole. An alternative explanation is that the nuclear gas is not rotating but is contracting (∼50 km s⁻¹) in the plane of the outer disk. Both models imply that we are observing a transient phenomenon.

158.039 Hva skjer når galakser kolliderer?
E. Jensen.
Astron. Tidsskr., Årg. 10, 82 - 83 (1977).

158.040 Brightness distributions in compact and normal galaxies. III. Decomposition of observed profiles into spheroid and disk components.   J. Kormendy.
Astrophys. J., Vol. 217, 406 - 419 (1977).

The author has examined the problem of decomposing galaxy brightness profiles into their underlying spheroid and disk components. One major result is that disks in the present compacts must have inner brightness cutoffs where the spheroids begin to dominate. These disks seem to be nearly uniform-brightness plateaus, with sharp inner and outer edges. This is very different from the classic exponential disk.

158.041 The galaxy M82: an everlasting puzzle.
L. Carrasco.
Recueil des séminaires, (see 012.004), p. 134 - 135 (1977).

158.042 Galaxies and their populations – the view on a cloudy day.   I. R. King.
The evolution of galaxies and stellar populations, (see 012. 005), p. 1 - 17, with a discussion, p. 17 (1977).

158.043 Musings on galaxy classification.
S. van den Bergh.
The evolution of galaxies and stellar populations, (see 012. 005), p. 19 - 37, with a discussion, p. 37 - 41 (1977).

158.044 Qualitative and quantitative classifications of galaxies.   G. de Vaucouleurs.
The evolution of galaxies and stellar populations, (see 012. 005), p. 43 - 94, with a discussion, p. 94 - 96 (1977).

1. Qualitative classifications: morphological type and luminosity class: luminosity index; morphological type and Yerkes color class; morphological type and Byurakan nuclear type; morphological type and BGC nuclear class; morphological type and revised DDO system. 2. Correlations with measurable parameters: concentration indices; color indices; hydrogen index; radio continuum indices; ellipticities; angular

velocities; maximum rotational velocities; velocity dispersion.
3. Quantitative classifications: luminosity distribution laws:
photometric parameters; color distribution laws: colorimetric
parameters; kinematical parameters; dynamical parameters;
physical parameters.

158.045 The chemical composition of old stellar populations.
         S. M. Faber.
The evolution of galaxies and stellar populations, (see 012.
005), p. 157 - 191, with a discussion, p. 191 - 197 (1977).

**158.046 Surface brightness and color distributions of
         elliptical and S0 galaxies in the Coma cluster.**
K. M. Strom, S. E. Strom.
The evolution of galaxies and stellar populations, (see 012.
005), p. 239 - 291, with a discussion, p. 291 - 300 (1977).
    This contribution represents a summary of the surface
brightness and color distributions derived for a large sample of
elliptical and S0 galaxies in the Coma cluster. These distributions provide, respectively, estimates of the density and chemical composition of the dominant stellar constituents in galaxies
of these types. The authors believe that these estimates will
prove fundamental to synthesizing plausible models describing
the star-forming and element-producing histories in E and S0
galaxies.

**158.047 The ultraviolet spectra and color evolution of
         galaxies at large redshifts.   H. Spinrad.**
The evolution of galaxies and stellar populations, (see 012.
005), p. 301 - 333, with a discussion, p. 333 - 338 (1977).
    Galaxies at large redshift give us information on their past
evolutionary status. All present data and the most reasonable
theoretical models suggest giant E galaxies as having been much
more luminous, and somewhat bluer in the past. This is noticeable in $U-B$ rest-frame colors and also in the vacuum ultraviolet, where comparison is made of distant galaxy spectra with
the energy distribution of local galaxies observed by satellites.
The author discusses colors of galaxies in distant rich clusters;
Butcher and Oemler have shown evidence for evolution of
cluster galaxy content. Galaxy (field) counts are also indication
of evolution; here the author comments on future prospects.

158.048 Optical outbursts in the Seyfert galaxy NGC 1275.
         Z. I. Tsvetanov, I. M. Yankulova.
Astrofizika, Vol. 13, 21 - 31 (1977). In Russian. English translation in Astrophysics, Vol. 13, No. 1.
    Using results of radio observations of NGC 1275 made in
1960 - 1973 available from the literature and UBV observations
made by Lyutyj the spectrum of the source in the interval
$10^7 - 10^{15}$ Hz is given. Considering the synchrotron nature of
emission in the radio and optical regions, the physical parameters of the relativistic electrons are found. Their energy distribution is also obtained. It is concluded that the optical outbursts differ from one another by the number of the injected
electrons. The energy density per unit energy interval at the
minimum of the outbursts varies more than at maximum.

158.049 Gaseous envelopes of Seyfert nuclei.
         V. I. Pronik.
Astrofizika, Vol. 13, 51 - 61 (1977). In Russian. English translation in Astrophysics, Vol. 13, No. 1.
    It is pointed out that the gas densities derived from the
[O III] lines for the majority of Seyfert nuclei are of the same
order ($10^6 - 5 \times 10^6$ cm^{-3}), whereas there is a rather large interval of densities (10^4 to $\gtrsim 10^7$ cm^{-3}) obtained by using other
lines. An attempt to interpret this fact leads to a spherically
symmetric model of the envelope of the nucleus in which the
gas density changes as n(r) ~ r^{-2}. Because of this and of the
existence of a critical density for a forbidden line, the maximal
contribution to the radiation in an emission line is given by a
spherical layer with density nearer the critical one for this line.

**158.050 Seyfert galaxies with large z: an electronographic
         survey.   P. A. Wehinger, S. Wyckoff.**
Mon. Not. R. Astron. Soc., Vol. 181, 211 - 231 (1977).
    The purpose of this paper is to investigate the nature and
structure of the faint outlying galactic envelopes of a sample
of Seyfert galaxies, using electronographic observations.

**158.051 Correlation analysis of the space and surface distribu-
         tion of galaxies.   G. Dautcourt.**
Astron. Nachr., Band 298, 253 - 274 (1977).
    The paper presents a number of calculations which are
useful for the interpretation of catalogues of galaxies. It is
suggested that a correlation function approach based on a
Mayer cluster expansion for the N-point probability function
of galaxies leads to a complete description of galaxy clustering.
For the classical cell count method smoothing effect estimates
are given and the Zwicky dispersion curve technique is studied.
The relation between the space distribution and the surface
distribution of galaxies is discussed in cosmological models of
Friedmann type. It is shown that the clustering amplitude
decreases within increasing redshift.

**158.052 The equidensitometric determination of angular
         diameters of compact galaxies.**
N. Richter, W. Högner.
Astron. Nachr., Band 298, 275 - 277 (1977).
    On Tautenburg Schmidt plates in system V taken at
seeings of about 2" the determination of real angular diameters of compact galaxies down to 4" is possible. Diameter-
magnitude diagrams of objects in four fields around the
globular cluster M3 show the possibility for getting homogeneous material concerning both the uniformity of the
equidensitometric copying process and the measuring process
with the iris photometer in order to determine the integral
brightness of single galaxies from their limiting equidensities.

158.053 Does the "exploding" galaxy explode?
         B. P. Artamonov.
Zemlya i Vselennaya, 1977, No. 4, p. 38 - 43. In Russian.

**158.054 Chemische Entwicklung von Populationen in
         Galaxien.   C. Trefzger.**
Sterne Weltraum, Jahrg. 16, 321 - 325 (1977).

**158.055 Sixteen southern interacting galaxy systems with
         emission lines.   T. M. Borchkhadze, J. Breysacher,**
S. Laustsen, H.-E. Schuster, R. M. West.
Astron. Astrophys., Suppl. Ser., Vol. 30, 35 - 43 (1977).
    Spectroscopic and photometric observations are reported
for sixteen southern interacting galaxy systems in which at
least one component shows emission lines. Some of the
galaxies have fairly broad lines, and with the exception of one
doubtful case (ESO 234–IG49), the measured radial velocities
indicate that the systems are physical entities.

158.056 Tidal interactions and the massive halo hypothesis.
         S. D. M. White, N. A. Sharp.
Nature, Vol. 269, 395 - 396 (1977).
    The idea that galaxies may be surrounded by large
amounts of unseen material has become very popular. It is
disquieting that there seems to be little chance of direct observation of this dark material, and this letter points out that
dynamical constraints can be put on either the extent or the
universality of galactic haloes.

**158.057 A first step toward the radial distribution of carbon
         monoxide in the galaxy M 31.**
F. Combes, P. J. Encrenaz, R. Lucas, L. Weliachew.
Astron. Astrophys., Vol. 61, L7 - L9 (1977).
    A partial radial distribution of carbon monoxide emission
in M31 is presented. The CO emission is correlated with dark

areas, presumably dusty, on the inner side of H I spiral arms. The CO and H I velocities are in good agreement.

158.058   The spectral index distribution of M51.
A. Segalovitz.
Astron. Astrophys., Vol. 61, 59 - 67 (1977).

The radial variation in M51 of the spectral index of the non-thermal radio spectrum between the two wavelengths 21 cm and 49 cm is discussed. A diffusion model for the relativistic electrons movement in the galaxian magnetic field is constructed so as to generally fit the observations.

158.059   The extended H I regions around spiral galaxies: a probe for galactic structure and the intergalactic medium.   J. Bergeron, J. E. Gunn.
Astrophys. J., Vol. 217, 892 - 902 (1977).

The H I disks observed at large radii around nearby spiral galaxies provide sensitive probes for the mass distributions in these galaxies and of their environments. The authors show, for a few well-observed systems, that there is an unseen component which dominates the mass at large radii. This additional matter cannot be gas, either neutral or ionized. The authors investigate the thermal interaction between a hot intergalactic medium near the closure density and these extended H I regions in the assumption of magnetic field lines extended outward into the intergalactic medium (IGM). They show that, with plausible initial conditions, the intergalactic temperature at present cannot exceed 1×10^7 K if the H I is to have survived until now. Consideration of conditions in the past places even more stringent limits on the temperature and density of the IGM. Survival of the H I disk also implies that these galaxies cannot have persistent hot, dense halos.

158.060   Star formation in blue galaxies.   J. P. Huchra.
Astrophys. J., Vol. 217, 928 - 939 (1977).

Photometry of blue galaxies in the lists of Haro, Zwicky, and Markarian have extended the observed galaxy two-color distribution significantly to the blue. Existing evolutionary models of galaxy colors do not match the observed colors of these blue galaxies. Models of galaxies are constructed with a variety of star-formation histories and initial mass functions. The models include the effects of emission by gas ionized by the hot stars in the galaxy. Three general classes of models were studied. "Old" galaxies are systems that are more than 10^{10} years old. "Young" galaxies are systems that are less than one-tenth that age. "Composite" galaxies are "old" galaxies, with properties similar to ordinary spirals and ellipticals, plus a burst of recent star formation. Power-law initial mass functions were used. Comparison of model predictions with the observed metal abundance, color-magnitude distribution, and internal reddening leads one to conclude that "composite" models best fit the observed properties.

158.061   On dust as the source of the infrared luminosity of type 1 Seyfert galaxies.   R. Ptak, R. Stoner.
Astrophys. J., Vol. 217, 940 - 943 (1977).

Dust, heated by trapped Lα line radiation, is proposed as the source of IR radiation from type 1 Seyfert galaxies. The Lα derives from the interaction of fast particles with relatively small, dense clouds that contain the dust, rather than from an ionizing UV continuum. No correlation between reddening (Balmer decrement) and IR luminosity is necessarily expected in this picture. Conditions of energy balance are used to estimate the physical properties of the radiating dust, and these properties seem consistent with observed IR spectra of type 1 Seyferts.

158.062   The dressed slingshot and the symmetry of double radio galaxies.   D. N. C. Lin, W. C. Saslaw.
Astrophys. J., Vol. 217, 958 - 963 (1977).

The authors calculate the conditions in which accretion disks around massive objects can survive tidal disruption during ejection by the gravitational slingshot. The probability that the disks around all the massive objects survive is greatest when the ejection is symmetric. This may produce a tendency for powerful, long-lived double radio galaxies to be nearly symmetric. It also leads to a new observational test.

158.063   2 to 8 micron spectrophotometry of M82.
S. P. Willner, B. T. Soifer, R. W. Russell, R. R. Joyce, F. C. Gillett.
Astrophys. J., Lett., Vol. 217, L121 - L124 (1977).

The authors report new spectrophotometric observations of M82 from 2 to 8 μm. These observations show further similarities with infrared sources in our Galaxy. The observed hydrogen recombination lines are used to derive properties of the ionized gas, extinction to the H II region, and an improved Ne abundance estimate, while a measured [Ar II] line gives the argon abundance.

158.064   Electron scattering in X ray-emitting galaxies.
A. C. Fabian.
Nature, Vol. 269, 672 - 673 (1977).

158.065   The velocity field in the central region of the peculiar, hot-spot nucleus galaxy NGC 2782.
P. C. van der Kruit.
Astron. Astrophys., Vol. 61, 171 - 176 (1977).

This paper reports the results of a spectroscopic study of the central region of the peculiar, "hot-spot nucleus" galaxy NGC 2782. The velocity field in the region containing the hot spots is measured and shown to be consistent with that of pure rotation. The rotation curve over the region indicates a strong increase in mass-density in the region of the hot spots. There is no evidence for non-circular motions. The nucleus of NGC 2782 resembles in its photometric properties that of a weak type 1 Seyfert nucleus.

158.066   A new determination of the gas-to-dust ratio in M 31.   E. Bajaja, T. E. Gergely.
Astron. Astrophys., Vol. 61, 229 - 234 (1977).

The mean gas-to-dust ratio in M 31 is derived from observations of the H I column density and the color excess in the direction of 121 globular clusters. The mean gas-to-dust ratio turns out to be $\sim 33 \times 10^{20}$ atoms cm^{-2} mag^{-1}, the value depends on radial distance from the center of the galaxy. Some effects that may influence this determination are discussed.

158.067   Neutral hydrogen in NGC 4248.
G. D. van Albada.
Astron. Astrophys., Vol. 61, 297 - 298 (1977).

NGC 4248 is the small optical companion of the large S(B)bc galaxy NGC 4258. In this paper the author presents neutral hydrogen data on this galaxy. Results are listed in a table. The radial velocity data strongly suggest that NGC 4248 is an actual satellite of NGC 4258.

158.068   Interstellar abundances in external galaxies.
B. L. Webster.
Topics in interstellar matter, (see 012.011), p. 243 - 247 (1977).

This field has been excellently reviewed recently by Peimbert (1975), Burbidge and Burbidge (1975) and Searle (1975). The present discussion will cover work since these and in particular the evidence for composition gradients across galaxies and the systematics with galactic type.

158.069   Galactic warps: observations.   R. Sancisi.
Topics in interstellar matter, (see 012.011), p. 255 - 259 (1977).

Recent observations of neutral hydrogen in nearby galaxies have revealed a significant bending of their outer gas layers. The most striking warp is found in NGC 5907. Most of these galaxies have no bright close companions.

158.070 The gas content of early-type galaxies.
H. van Woerden.
Topics in interstellar matter, (see 012.011), p. 261 - 283 (1977).

This paper reviews recent observations of neutral hydrogen in elliptical and lenticular galaxies. In most early-type galaxies, the amount of gas observed is much smaller than expected from mass loss by cooling stars; galactic winds, intergalactic ram pressure, or formation of new stars may be responsible for this.

158.071 The radio continuum morphology of spiral galaxies.
P. C. van der Kruit, R. J. Allen.
Annu. Rev. Astron. Astrophys., Vol. 14, (see 003.008), 417 - 445 (1976).

Contents: The disk radio emission in spiral galaxies; Spiral structure; H II regions; Supernova remnants; Central radio sources; z extent of the radio emission and halos.

158.072 Distribution of different extragalactic objects in the field of the North Galactic Pole. S. Zięba.
Acta Cosmologica, Zesz. 6, 75 - 101 (1977).

Using a certain modification of the method of statistical reduction, the distribution of galaxies, clusters of galaxies and 4C radio sources contained in an area of 2237 square degrees centered at the North Galactic Pole is investigated.

158.073 Contributions to galaxy photometry. VI. Revised standard total magnitudes and colors of 228 multiply observed galaxies. G. de Vaucouleurs, G. Bollinger.
Astrophys. J., Suppl. Ser., Vol. 34, 469 - 477 (1977).

A catalog of 228 multiply observed galaxies of both hemispheres having standard total magnitudes B_T with mean errors less than 0.10 mag and a range of determinations less than 0.30 mag is derived from weighted means of the fully corrected data of Papers I to V. The range of magnitudes is $4.3 < B_T < 14.6$ and of mean errors, 0.03 to 0.09 mag. Mean total color indices $\langle B-V \rangle_T$ from the data in Papers II and III are also given for 191 galaxies with mean errors 0.01 to 0.05 mag.

158.074 Electronographic study of NGC 4151, NGC 1265, and IC 310. H. Netzer, L. Formiggini.
Astrophys. J., Vol. 218, 58 - 69 (1977).

The authors present new electronographic observations of three active galaxies, NGC 1265 and IC 310 (in the Perseus cluster) and the Seyfert galaxy NGC 4151. They present surface-brightness maps and profiles and compare them with previous photographic results.

158.075 Neighborhoods of galaxies. II. NGC 4151.
H. Arp.
Astrophys. J., Vol. 218, 70 - 85 (1977).

Deep exposure and superposition photographs of NGC 4151 show outer, low-surface-brightness spiral arms. The largest nearby galaxy to NGC 4151 is shown to be an active galaxy in the sense that a luminous filament and other material emerge from it. This apparent companion, NGC 4156, is projected on the end of the outermost extension of an NGC 4151 spiral arm. Another companion galaxy appears to be associated with the end of the opposite spiral arm. These companion galaxies have redshifts of $z = 6400$ km s^{-1} and $z = 6700$ km s^{-1}. Despite the discordance of these redshifts with the redshift of NGC 4151, $z = 967$ km s^{-1}, evidence is discussed which indicates that these two galaxies are physically associated with the low-redshift NGC 4151. Farther away from NGC 4151, but still in its immediate vicinity, two new examples of discordant galaxy redshifts are found.

158.076 H I sheets ejected by M32 from M31 and multiple disk radial velocities. G. G. Byrd.
Astrophys. J., Vol. 218, 86 - 91 (1977).

Earlier work has shown by computer simulation that the gravity of M32, the companion galaxy which orbits M31, can produce asymmetries in the radial-velocity pattern of the projected disk of M31 similar to those observed at 21 cm wavelengths. The same orbit also ejects a sheet of material far enough from the plane to produce multiple-valued H I radial velocities over areas where both sheet and disk are superposed. The present investigation is a detailed study of the sheet generated by computer simulation. Reasonable agreement can be obtained between the simulated sheet radial-velocity pattern and 21 cm observations over part of the disk of multiple-valued velocities by Roberts and Whitehurst. The model is used to predict the multiple velocity patterns over the rest of M31's disk.

158.077 Observations of NGC 4151 from Uhuru.
M. P. Ulmer.
Astrophys. J., Lett., Vol. 218, L1 - L2 (1977).

The author reports X-ray observations of NGC 4151 with the Uhuru satellite. He sets limits to the variability of NGC 4151 of a factor of 2 for five observations made during 1970, 1971, and 1972. The average flux was 8×10^{-11} ergs cm^{-2} s^{-1} (2—10 keV). The results are compared with Ariel 5 observations, and models of NGC 4151 are briefly discussed.

158.078 Detection of an optical halo surrounding the spiral galaxy NGC 4565. D. J. Hegyi, G. L. Gerber.
Astrophys. J., Lett., Vol. 218, L7 - L11 (1977).

An optically luminous halo has been detected out to a distance of 34 kpc from the plane of the giant spiral galaxy NGC 4565 by using a new type of photometer which scans along an annular path. At this distance the halo was observed to have a surface brightness equal to approximately 1 part in 1000 of the night sky in the spectral band of the instrument. Preliminary color data indicate that the observed light is emitted by stars redder than K7 and is suggestive of a massive halo of cosmological significance.

158.079 NGC 6251, a very large radio galaxy with an exceptional jet.
P. C. Waggett, P. J. Warner, J. E. Baldwin.
Mon. Not. R. Astron. Soc., Vol. 181, 465 - 474 (1977).

A newly discovered radio galaxy, NGC 6251, has an angular size of $1°.2$, the largest of any in the northern sky. It has a projected physical size of 3.0 Mpc and a remarkable jet which is straight for nearly all of its 200-kpc length. The radio spectral index of the jet is 0.54 and the nucleus has a flat spectrum with a cut-off at low frequencies. Although the radio luminosity of the source is low, the total stored energy is comparable with the highest values known.

158.080 An increase in the X-ray absorption of NGC 4151.
P. Barr, N. E. White, P. W. Sanford, J. C. Ives.
Mon. Not. R. Astron. Soc., Vol. 181, 43P - 46P (1977).

The Ariel V Experiment C has been used to observe NGC 4151 in 1976 December. A hydrogen column density of 1.8×10^{23} atom cm^{-2} was measured, four a factor larger than that reported by Ives, Sanford & Penston (1976). This column density change is sufficient to explain the previously reported variability in the 2—10 keV energy range and indicates that the X-ray emission originates in the central area of the nucleus. An iron absorption edge is seen and an iron abundance relative to other heavy elements of 2.1 ± 1.4 times the solar value measured.

158.081 Surface photometry of the Sc III galaxy NGC 2403.
S. Okamura, B. Takase, K. Kodaira.
Publ. Astron. Soc. Japan, Vol. 29, 567 - 581 (1977).

Surface photometry of NGC 2403 is carried out by a

computerized digital method. The results are presented in various kinds of maps, figures, and tables, including photometric parameters. These data are compared with those of M 101, a typical Sc I galaxy (Okamura et al. 1976), and of some Sc galaxies in different luminosity classes.

158.082 Near-infrared surface photometry of the central region of M31.
T. Matsumoto, H. Murakami, K. Hamajima.
Publ. Astron. Soc. Japan, Vol. 29, 583 - 591 (1977).

Isophotes of the surface brightness of the central region of M31 were obtained at 7000–9000 Å and 2.2 μm by means of photographic and photoelectric methods, respectively. The position angle and axial ratio in 7000–9000 Å show almost similar features to the previously observed ones in the visible region. On the other hand both the position angle and axial ratio at 2.2 μm are found to have larger values than those in 7000–9000 Å in the inner region of.M31. This result suggests the presence of a weak bar-like structure in the central bulge, for which giant stars are responsible.

158.083 Near-infrared velocity dispersions for the nuclear bulges of M 31 and M 81.   C. Pritchet.
J. R. Astron. Soc. Canada, Vol. 71, 395 (1977). – Abstract.

158.084 Two new faint stellar systems discovered on ESO Schmidt plates.   D. A. Cesarsky, S. Laustsen, J. Lequeux, H.-E. Schuster, R. M. West.
Astron. Astrophys., Vol. 61, L31 - L33 (1977).

Two new faint stellar systems have been discovered on photographic plates obtained with the ESO Schmidt telescope. The first system is a dwarf irregular galaxy near the local group member NGC 6822, and presumably at about the same distance, ~600 kpc. The second object, near the star SAO 169422, is a globular cluster at a large distance.

158.085 Photometry of interacting galaxies with compact components.   A. Ardeberg, N. Bergvall.
Astron. Astrophys., Vol. 61, 493 - 496 (1977).

UBV photometry is presented for 13 interacting galaxies, 11 of which are compact or nearly compact. Half of the compact or nearly compact objects occupy a region in the two-colour diagram, where the compact galaxies normally exhibit emission-line spectra. The colour relations between interacting components and their possible interpretations are discussed.

158.086 An interferometer study of the neutral hydrogen associated with the optical core of the irregular galaxy NGC 6822.   S. T. Gottesman, L. Weliachew.
Astron. Astrophys., Vol. 61, 523 - 530 (1977).

A 2.3 (330 pc) synthesis of the neutral hydrogen in NGC 6822 has been made. Owing to resolution effects, the discussion is confined to the optical core of the galaxy. An analysis of the bright star distribution yields the star formation rate proportional to the H I surface density raised to the 1.5 power. The large-scale H I distribution and kinematics of the optical core are discussed. It is found that non-circular motions are important in the structure of the system.

158.087 Brightness distributions in compact and normal galaxies. II. Structure parameters of the spheroidal component.   J. Kormendy.
Astrophys. J., Vol. 218, 333 - 346 (1977).

Systematic properties of galaxy spheroids are studied by using simple fitting functions to derive the brightness, size, and shape parameters of their brightness profiles. Fitting the de Vaucouleurs $r^{1/4}$-law to the profiles of 16 red compact and 19 normal galaxies yields the following results. 1. Ellipticals with close, massive neighbors have bright outer halos which are not possessed by more isolated objects. 2. The brightness and radius parameters are related by $B_{0V} = 3.02 \log r_0 + 19.74 B$ mag arcsec^{-2}. 3. Most red compacts are not abnormally com-

pact. 4. The $B_{0V}(\log r_0)$ relation can be used to measure relative distances.

158.088 Rotation (?) in 13 elliptical galaxies.
G. Illingworth.
Astrophys. J., Lett., Vol. 218, L43 - L47 (1977).

Rotation data are presented for 13 elliptical galaxies, and compared with the oblate models of Gott, Larson, and Wilson. It is found that, on average, most elliptical galaxies have one-third the peak rotational velocity required by these models. Oblate models in which the flattening is a result of rotation appear to be precluded by these data. Plausible alternatives appear to be that ellipticals are (a) triaxial ellipsoids (or oblate spheroids) in which the flattening is due to remnant anisotropy in the velocity distribution, as suggested by Binney, or (b) rotating prolate ellipsoids, as suggested by Miller. Curiously, one giant elliptical rotates as expected.

158.089 Carbon monoxide in Maffei 2.
L. J. Rickard, B. E. Turner, P. Palmer.
Astrophys. J., Lett., Vol. 218, L51 - L55 (1977).

The authors have detected CO toward the galaxy Maffei 2, permitting an unobscured study of its inner structure. Although principally concentrated toward the nucleus, the CO emission is observed along the galactic disk at least as far as 6 kpc from the center. The mass in molecular clouds within 750 pc of the nucleus is approximately twice that of the same region in our Galaxy, and the ratio of masses in molecular and atomic gas decreases rapidly away from the nucleus. Using a partial rotation curve, the authors discuss the mass and fractional gas content of the galaxy.

158.090 Radio and optical observations of the N galaxy 4C 39.11.   M. T. Adams, L. Rudnick.
Astron. J., Vol. 82, 857 - 860, 931 (1977).

4C39.11 has been mapped at 2695 and 8085 MHz with the NRAO interferometer and identified with a high-surface-brightness optical object. Steward Observatory 2.3-m image-tube plates and spectra lead to its classification as an N galaxy with a redshift $z = 0.161$.

158.091 Study of the luminosity function for field galaxies.
J. E. Felten.
Astron. J., Vol. 82, 861 - 878 (1977).

Nine determinations of the luminosity function (LF) for field galaxies are analyzed and compared. Corrections for differences in Hubble constants, magnitude systems, galactic absorption functions, and definitions of the LF are necessary prior to comparison. The LF data suggest that there is little, if any, distinction between "field" galaxies and those in small groups. The large-scale mean LF of galaxies (mostly field galaxies) is about a factor of 2.3 below the "local" LF derived by Schechter. The nominal mean luminosity density arising within the $B(0)$ isophotes of galaxies at the blue band is $\mathcal{L} \approx 8.6 \times 10^7 \, (H/50) \, L_\odot Mpc^{-3}$ for $\alpha_B = 0.25$ mag; dependence of $\mathcal{L}$ on input parameters is shown in a table. The true value of $\mathcal{L}$ is probably within a factor of 1.6 of this.

158.092 Compact blueshifted galaxy RMB 56 (1216+141).
T. D. Kinman, V. C. Rubin, N. Thonnard, W. K. Ford, Jr., C. J. Peterson.
Astron. J., Vol. 82, 879 - 885, 932 - 933 (1977).

RMB 56 is an unusually compact dwarf galaxy with an emission line spectrum like that of an H II region in the outer arm of a spiral galaxy; it also has a strong continuum with Balmer absorption lines. UBV observations have been used to construct a model consisting of a central source with the colors of an isolated H II region which is concentric with an exponential disk similar to an Sm or Im galaxy seen face-on.

158.093 Low energy γ-ray observation of NGC 4151.
G. Di Cocco, G. Boella, F. Perotti, R. Stiglitz,

G. Villa, R. E. Baker, R. C. Butler, A. J. Dean, S. J. Martin, D. Ramsden.
Nature, Vol. 270, 319 - 320 (1977).

Observations of the Seyfert galaxy NGC 4151 by Ariel V, OSOI VII and UCSD have demonstrated that it has a relatively flat X-ray spectrum up to photon energies greater than 100 keV. The X-ray luminosity of this object is of the order of 4×10^{43} erg s^{-1} if a distance of 20 Mpc is assumed, and exceeds the integrated luminosity at all the other observed greater wavelengths. Long-term observations by Ariel V indicate that the X-ray emission may originate from a compact object with dimensions less than 8×10^{15} cm, presumably the nucleus of the Galaxy. Here the authors present the preliminary results of a positive measurement in the low energy γ-ray region of the spectrum, which are in good agreement with the X-ray data and confirm the intense high energy luminosity of the object.

158.094 **Photoelectric photometry of 45 bright galaxies.**
        J. G. Godwin, M. J. Bucknell, K. L. Dixon, M. R. Green, J. V. Peach, R. E. Wallis.
Observatory, Vol. 97, 238 - 241 (1977).

The authors present UBVRI observations of 45 bright galaxies made through a number of circular apertures.

158.095 **Some data on little-known southern galaxies.**
        E. Agüero, G. J. Carranza.
Observatory, Vol. 97, 241 - 242 (1977).

158.096 **Seyfert galaxies.**   D. W. Weedman.
        Annu. Rev. Astron. Astrophys., Vol. 15, (see 003.012), 69 - 95 (1977).

Contents: 1. Seyfert galaxies currently known; 2. Surveys and morphology; 3. Emission-line spectra; 4. Balmer decrement and line profiles; 5. Continuum radiation and luminosities; 6. Relation to QSOs; 7. Conclusions.

158.097 **Seyfert galaxies.**   S. Mitton.
        Modern astronomy, (see 003.013), p. 167 - 175 (1977).

158.098 **Galaxy correlation: the inversion of Limber's formula.**   W. E. Parry.
Phys. Lett. A, Vol. 60A, 265 - 266 (1977). — Abstr. in Phys. Abstr., Vol. 80, Abstr. 33646 (1977).

158.099 **New color photos of galaxies.**   L. A. Thompson.
        Astronomy, Vol. 5, No. 3, p. 18 - 23 (1977).
Abstr. in Phys. Abstr., Vol. 80, Abstr. 49303 (1977).

158.100 **Detection of H$_2$O emission from galaxy NGC 253.**
        J. R. D. Lépine, P. M. Dos Santos.
Nature, Vol. 270, 501 (1977).

The authors report the second detection of H$_2$O emission from an external galaxy, NGC 253, a large edge-on spiral galaxy situated at 3.4 Mpc, about five times farther away than M33.

158.101 **Can life evolve in elliptical galaxies?**   J. Gribbin.
        Astronomy, Vol. 5, No. 5, p. 18 - 24 (1977).
Abstr. in Phys. Abstr., Vol. 80, Abstr. 64845 (1977).

158.102 **ESO 113-IG 45.**
        IAU Circ., No. 3134 with a correction No. 3143 (1977).

158.103 **III Zw 2.**
        IAU Circ., Nos. 3145, 3154 (1977).

158.104 **The NGC 1275 enigma.**   S. van den Bergh.
        Astron. Nachr., Band 298, 285 - 287 (1977) = Lick Obs. Bull., No. 765.

The following arguments suggest that NGC 1275 does not consist of a giant elliptical (E) galaxy that is colliding with (or is superimposed on) a late-type spiral (L): (1) The total diameter of the region containing young associations is 33 $(100/H)$ kpc. This size is characteristic of ScI galaxies. Neither the morphology nor the integrated luminosity of the L component of NGC 1275 supports such a classification. (2) The chaotic appearance of the L component of NGC 1275 is unlikely to be due to tidal damage. This is so because: (a) the E and L components are still approaching each other, (b) their relative velocity is ≈ 3000 km s^{-1}, (c) no stripped galaxy core (which would survive a catastrophic tidal encounter) is seen near NGC 1275. (3) The core of the Perseus cluster contains only one (anemic) spiral. The a priori probability that NGC 1275 represents a chance superimposition (or collision) of a spiral and an elliptical galaxy is therefore low. (4) The assumption that the L component of NGC 1275 is superimposed on, but not interacting with, the E component does not account for (a) the presence of an active Seyfert nucleus, (b) the peculiar filamentary H II shell, discovered by Lynds, (c) the presence of recently-formed stars, (d) the X-ray emission and the radio emission of NGC 1275.

158.105 **Structure and evolution of galaxies.**   W.-S. Tai.
        Kexue Tongbao, Vol. 22, No. 6, p. 231 - 244 (1977). In Chinese. — Abstr. in Phys. Abstr., Vol. 80, Abstr. 87551 (1977).

158.106 **Galaxies: outstanding problems and instrumental prospects for the coming decade. (A review).**
J. P. Ostriker.
Proc. Natl. Acad. Sci. USA, Vol. 74, 1767 - 1774 (1977). Abstr. in Phys. Abstr., Vol. 80, Abstr. 91626 (1977).

158.107 **Light variation of the nucleus of Markarian Galaxy 358.**   M. Lovas.
Inf. Bull. Variable Stars, No. 1344 (1977).

158.108 **The nature of Markarian galaxies.**   J. P. Huchra.
        Astrophys. J., Suppl. Ser., Vol. 35, 171 - 195 (1977).

The author presents new and summarizes existing *UBV* photometry, morphological classifications, and spectroscopic data for Markarian galaxies. A Hubble constant of 50 km s^{-1} Mpc^{-1} is used throughout. These and other available data are compared with data for field galaxies. Other properties of the Markarian galaxies are discussed. The results of these comparisons are discussed and conclusions as to the nature of Markarian galaxies are presented.

158.109 **Curtis Schmidt-thin prism survey for extragalactic emission-line objects: University of Michigan List II.**
G. M. MacAlpine, S. B. Smith, D. W. Lewis.
Astrophys. J., Suppl. Ser., Vol. 35, 197 - 201, Plates 15 - 21 (1977).

Descriptions, positions, and finding charts are presented for 100 extragalactic emission-line objects and possible QSOs. They were detected optically with the 61 cm aperture Curtis Schmidt telescope in combination with a thin objective prism and a IIIa-J photographic emulsion at the Cerro Tololo Inter-American Observatory. This list, the second in a series being prepared at the University of Michigan, contains sources to about continuum magnitude 18.5 which are located in the regions of sky with equatorial coordinates $1^h3 \lesssim \alpha \lesssim 2^h0$, $+2° \lesssim \delta \lesssim +7°$ and $23^h2 \lesssim \alpha \lesssim 0^h1$, $-2°5 \lesssim \delta \lesssim +2°5$. Included are 59 apparent emission-line galaxies, 18 "probable" QSOs with estimated redshifts, and an additional 23 "possible" QSOs.

158.110 **Curtis Schmidt-thin prism survey for extragalactic emission-line objects: University of Michigan List III.**   G. M. MacAlpine, D. W. Lewis, S. B. Smith.
Astrophys. J., Suppl. Ser., Vol. 35, 203 - 207, Plates 22 - 28

(1977).

Descriptions, positions, and finding charts are presented for 100 extragalactic emission-line objects and possible QSOs. Also included is one apparent galactic nova. They were detected optically with the 61 cm aperture Curtis Schmidt telescope in combination with a thin objective prism and a Kodak IIIa-J photographic emulsion at the Cerro Tololo Inter-American Observatory. This list, the third in a series being prepared at the University of Michigan, contains sources to about continuum magnitude 18.5–19 which are located in the region with 1950 equatorial coordinates $0^h0 < \alpha < 1^h1$ and $-2°5 \lesssim \delta \lesssim +2°5$. There are 48 apparent emission-line galaxies, 27 "probable" QSOs, and an additional 24 "possible" QSOs.

158.111   Spectra of additional Arakelian galaxies.
D. E. Osterbrock.
Publ. Astron. Soc. Pacific, Vol. 89, 620 (1977). – Abstract.

158.112   Revue des galaxies exhibant une traînée lumineuse.
J. P. Vallée.
J. R. Astron. Soc. Canada, Vol. 71, 443 - 468 (1977).

This article is a review of "head-tail" galaxies (defined as an optical galaxy near one end of a highly elongated radio structure).

158.113   Interferometric study of NGC 1313.
G. J. Carranza, E. L. Agüero.
Astrophys. Space Sci., Vol. 46, 23 - 31 (1977).

The kinematics and structure of NGC 1313 are discussed on the basis of interferometric observations. Several uniformly rotating components, a total mass of 2×10^{10} solar masses, and deviations from pure circular movement of an amplitude of almost 20 km s^{-1} are found.

158.114   Interferometric study of NGC 7793.
G. J. Carranza, E. L. Agüero.
Astrophys. Space Sci., Vol. 47, 397 - 400 (1977).

Results of Hα interferometric observations of NGC 7793 are reported. This galaxy contains about 93 H II regions and a general emission background. Its radial velocity is 215 km s^{-1}. The total mass is low, $2-3 \times 10^9$ solar masses, as well as the average density; the mean M/L is 0.4 suggesting a high proportion of young objects.

158.115   Étude cinématique de la galaxie barrée NGC 5383.
M. F. Duval.
Astrophys. Space Sci., Vol. 48, 103 - 122 (1977).

Radial velocities were measured at different points in the nucleus and in the bar. In the bar the kinematics is quite complex. Solid body rotation cannot be accepted. Spectra along the edge of the nucleus provides evidence for transverse motions in the bar of 100 km s^{-1} at 4 kpc from the center. The rotation curve is drawn; in the hypothesis of a radial motion in the bar the author has calculated the distribution of mass according to the method of Burbidge and Prendergast inside a 14 kpc radius; the mass is $10^{10} M_\odot$. The ratio Hα/[N II] between 2 and 5 at several spots indicates that H II regions are highly excited in the nucleus and at the extreme end of the bar. The region of the bar where the ratio is less than 1 suggests high excitation by collision of energetic particle perhaps coming from the nucleus.

158.116   Redshifts of compact galaxies in systems of galaxies.
T. Jaakkola.
Astrophys. Space Sci., Vol. 49, 99 - 111 (1977).

82 compact galaxies with measured redshift present in systems containing partly normal galaxies have been found, using Zwicky's Catalogues and Morphological Catalogue. For them the mean residual redshift $\Delta V = +163 \pm 76$ km s^{-1} has been obtained. The chance probability for this result is 0.015. Redshift is correlated with magnitude in groups and pairs of compact galaxies, this depending in pairs on the linear separa-

tion and colours of the components. If a colour difference is present, the bluer member has the larger redshift in general. In contrast with the large luminosity deduced from the redshift for many field compacts, compact galaxies in systems are faint. The majority of the proposed associations between quasars and systems of normal galaxies are shown to be probably physical.

158.117   Optical observations of the nucleus of NGC 4151.
G. Romano, S. Minello.
Astrophys. Space Sci., Vol. 50, 421 - 423 (1977).

Photographic observations of the nucleus of the Seyfert galaxy NGC 4151, carried out during the last seven years, are reported. The object shows irregular variations between photographic magnitudes 11.2 and 13.0.

158.118   Photometry of the nucleus of the Seyfert galaxy NGC 4151.   V. L. Oknyanskij.
Astron. Tsirk., No. 944, p. 2 - 4 (1977). In Russian.

158.119   Radio and infrared measurements of S0 galaxies.
W. A. Sherwood.
Infrared Phys., Vol. 17, 575 - 578 (1977).

S0 and other early type galaxies selected from neutral hydrogen (HI) search programmes have been observed in the near i.r.(JKL). There may be a correlation between the presence of either HI or forbidden emission and the absence of the other. In the infrared data there is a slight trend for galaxies with HI and without forbidden emission to have bluer K-L indices than galaxies without HI and with forbidden emission. Selection effects may be important.

158.120   Spectrum of the halo of the cD galaxy in Abell 401.
S. M. Faber, D. Burstein, A. Dressler.
Astron. J., Vol. 82, 941 - 946, 1043 (1977) = Lick Obs. Bull. No. 772.

The spectrum of the cD galaxy in Abell 401 has been obtained both in the nucleus and at a radius of 23 arcsec in the halo. The stellar velocity dispersions of the nucleus and the halo have been derived. The value in the nucleus is 480 ± 120 km sec^{-1} and the value in the halo is 470 ± 250 km sec^{-1}. Although this value for the halo is extremely uncertain, it does rule out the hypothesis that the cD galaxy consists exclusively of tidal debris stripped from other cluster members, moving with speeds typical of ordinary cluster members. At the very least, the cD must have formed around an already-existing, massive elliptical galaxy. A comparison of the absorption line strengths of the halo with those of normal elliptical galaxies suggests that the mean metal abundance of the halo population is approximately solar at this distance from the nucleus.

158.121   Optical variability of the nuclei of Seyfert galaxies. II. UBV- and Hα-photometry.   V. M. Lyutyj.
Astron. Zh. Akad. Nauk SSSR, Vol. 54, 1153 - 1167 (1977). In Russian. English translation in Soviet Astron., Vol. 21, No. 6.

The results of further (1972–76) photoelectric UBV observations of the nuclei of Seyfert galaxies NGC 1068, 1275, 3227, 3516, 4051, 4151, 5548, 7469 and Markarian 10 are presented. The observations of the variability of Hα + [N II] emission line intensity in the nuclei of the galaxies NGC 1068, 3516, 4151, and 7469 are also given. There are two components in the light curves of all Seyfert galaxies: a fast (flare) component having the characteristic time of variability of the order of tens of days, and a slow one with time scale of variability of a few years.

158.122   Investigation of position angle distribution of galaxies from the Uppsala catalogue.
A. P. Kobushkin, A. V. Mandzhos, V. V. Tel'nyuk-Adamchuk.
Astrometriya i Astrofizika, Kiev, vyp. (No.) 33, (see 003.020),

p. 30 - 34 (1977). In Russian.

The orientation of brighter and fainter spiral galaxies of the Uppsala Catalogue were investigated. The position angle distribution differs essentially from a uniformly random distribution.

158.123   Stellar population synthesis of galactic nuclei.
C. Pritchet.
Astrophys. J., Suppl. Ser., Vol. 35, 397 - 418 (1977).

A Fourier spectrometer has been used to obtain continuous spectrophotometric data for the nuclear bulges of M31, M32, M51, M81, M86, M87, M94, NGC 3115, and NGC 5195. These galaxies represent a variety of morphological types. The data have been used to construct stellar population models of galactic nuclei with potentially higher accuracy than has been attained previously. There appears to be strong evidence that stars in the nuclei of intrinsically luminous galaxies are not spectroscopically similar to stars in the solar neighborhood. A number of uncertainties preclude an accurate determination of the visual mass-to-light ratio of galaxies from the present work.

Spectroscopic and photometric observations of galaxies from the ESO/Uppsala List. Second catalogue. See Abstr. 002.041.

Le monde des galaxies.   See Abstr. 003.043.

Radio galaxies.   See Abstr. 003.123.

R.A.S. specialist discussion on "Chemical evolution of galaxies".   See Abstr. 011.016.

A method for detecting compact galaxies. See Abstr. 031.207.

Angular diameter counts of galaxies: a method for determining the selection criteria for deep Schmidt plates. See Abstr. 031.227.

Neuer Flug des Heidelberger Ballonteleskopes THISBE 1.   See Abstr. 032.005.

Hydromagnetic break-up of bridges and jets into aligned objects.   See Abstr. 062.044.

Expansion and rotational momentum of large cosmic masses.   See Abstr. 066.015.

On the frequency-period distributions of Cepheid variables in galaxies in the Local Group. See Abstr. 122.081.

The instability strip of the anomalous Cepheids in the Draco dwarf spheroidal galaxy.   See Abstr. 122.082.

Diffuse [O I] emission and warm interstellar gas in galaxies.   See Abstr. 131.012.

Rates of star formation.   See Abstr. 131.081.

Star formation and the gas content of galaxies. See Abstr. 131.082.

Ionizing background radiation and hydrogen at the periphery of galaxies.   See Abstr. 131.115.

Comment naissent les étoiles? See Abstr. 131.125.

Theoretical models of dust clouds. See Abstr. 131.131.

Common properties of H II regions in galaxies. See Abstr. 132.006.

Observations of [S III] in NGC 604 and N/S abundance gradients.   See Abstr. 132.010.

H II regions in galaxies of the Local Group. See Abstr. 132.018.

Neutral hydrogen in the outer regions of M31. See Abstr. 132.030.

Radiation from relativistic blast waves in quasars and active galactic nuclei.   See Abstr. 141.001.

The variation of spectral index across the radio galaxy 3C 452.   See Abstr. 141.009.

Search for variability of 3C 48 and 3C 84 at 408 MHz.   See Abstr. 141.013.

The radio emission of NGC 5363. See Abstr. 141.021.

Optical observations of radiogalaxy lobes. See Abstr. 141.022.

Possible fast variability of the nucleus of Cen A at 13.5 mm.   See Abstr. 141.025.

Observations of 40 low luminosity radio galaxies with the Westerbork Synthesis Radio Telescope. See Abstr. 141.029.

Observations of 15 southern extragalactic sources with the Fleurs synthesis telescope.   See Abstr. 141.042.

A study of the bridges in the radio galaxies 3C 132 and 3C 192 at metre and centimetre wavelengths. See Abstr. 141.043.

The radio structure and optical field of 3C 303. See Abstr. 141.066.

Radio galaxies and quasars. See Abstr. 141.074.

Physical conditions in radio galaxies and quasars. See Abstr. 141.100.

Relations between radio and optical properties of radio sources — radio astronomer's point of view. See Abstr. 141.103.

The redshift-magnitude relation for radio galaxies. See Abstr. 141.106.

Clusters of galaxies and radio sources. See Abstr. 141.108.

Quasar-galaxy pairs and surface density of quasars. See Abstr. 141.126.

Multifrequency radio observations of 3C 31: a large radio galaxy with jets and peculiar spectra. See Abstr. 141.128.

On the profiles of the broad lines in the spectra of QSOs and Seyfert galaxies.   See Abstr. 141.132.

Broad-band surface photometry of a statistically complete sample of radio faint galaxies.

See Abstr. 141.145.

The optical variability of six extragalactic objects.
See Abstr. 141.146.

On the nature of quasars and active nuclei of
galaxies.   See Abstr. 141.153.

Optical emission in the radio lobes of radiogalaxies.
See Abstr. 141.161.

Ariel V Sky Survey: X-ray variability of NGC 5128.
See Abstr. 142.087.

Paths of formation of elliptic and spiral galaxies.
See Abstr. 151.003.

Massive compact objects in an elliptical galaxy and
their dynamical relation to the halo formation.
See Abstr. 151.007.

Stellar dynamics in thin disk galaxies. I. A unified
approach to hydrodynamic and orbit theories.
See Abstr. 151.009.

On the dynamical evolution of galaxies in clusters.
See Abstr. 151.020.

Mergers and some consequences.
See Abstr. 151.021.

The influence of co-rotation singularity on density
waves in disk galaxies.   See Abstr. 151.022.

On the origin of galaxy rotation. II.
See Abstr. 151.023.

The effects of dissipation on the gas response to oval
distortions of disk galaxies.   See Abstr. 151.028.

Survival and disruption of galactic substructure.
See Abstr. 151.036.

Unstable spiral modes in disk-shaped galaxies: enhancement of growth rate.   See Abstr. 151.048.

Evolution of galactic density waves at the unstable
stage.   See Abstr. 151.054.

$BiL(E)$ and the Hubble sequence: a new theory of
galaxies.   See Abstr. 151.064.

A dynamical interpretation of the Hubble sequence

of galaxies.   See Abstr. 151.066.

On the arm structure and the star formation in M 51.
See Abstr. 151.074.

Radial distribution and total number of globular
clusters in M104.   See Abstr. 154.011.

Correlation between the number of globular clusters
and the luminosity of the spheroidal system for external
galaxies.   See Abstr. 154.035.

Search for globular clusters in M31. I. The disk and
the minor axis.   See Abstr. 154.044.

The evolution of stellar populations.
See Abstr. 155.018.

Statistical analysis of catalogs of extragalactic
objects. VIII. Cross-correlation of the Abell and the 10' Shane-
Wirtanen catalogs.   See Abstr. 160.007.

Upper limits on nuclear radio emission from some
Coma Cluster spirals.   See Abstr. 160.013.

On possible associations of quasi-stellar objects and
radio galaxies with rich clusters of galaxies.
See Abstr. 160.014.

A study of 1889 rich clusters of galaxies.
See Abstr. 160.023.

Statistical analysis of catalogs of extragalactic objects.
VII. Two- and three-point correlation functions for the high-
resolution Shane-Wirtanen catalog of galaxies.
See Abstr. 160.028.

A list of hypergalaxies.   See Abstr. 160.032.

Radio galaxies in the Coma cluster. II.
See Abstr. 160.034.

Cannibalism among the galaxies: dynamically produced evolution of cluster luminosity functions.
See Abstr. 160.036.

The luminosity function of galaxies in cluster
A2670.   See Abstr. 160.042.

Faint dwarf galaxies in the Virgo cluster.
See Abstr. 160.060.

Peculiar Objects

158.501 **BL Lacertae objects.**
M. J. Disney, P. Véron.
Sci. American, Vol. 237, No. 2, p. 32 - 39 (1977).

158.502 **The extended source in AP Librae.**
N. Visvanathan, D. Griersmith.
Astrophys. J., Vol. 215, 759 - 764 (1977).

Photoelectric photometry of the extended source surrounding AP Librae in the wavelength range 3466–6738 Å, using an annular aperture, is presented. The extended source is shown to have a continuum energy distribution similar to that of a redshifted giant elliptical galaxy in the blue-red region. This result is combined with the observation of central source plus extended component to derive the energy distribution of the central source.

158.503 **A recent photometric investigation of the BL Lacertae object, B2 1101 + 38.**
H. R. Miller, B. Q. McGimsey, R. M. Williamson.
Astrophys. J., Vol. 217, 382 - 384 (1977).

High-time-resolution photometry of the BL Lacertae object, B2 1101+38, has been made in an attempt to detect small-amplitude fluctuations in brightness such as those observed by Zhukov. No evidence is found for any small-amplitude short-term optical variability for this object. However, significant variations are observed on a time scale of months, with the object exhibiting a general brightening trend from 1976 January to 1976 December.

158.504 **Radio haloes around BL Lacertae objects AO0235+164 and 4C03.59.**   Gopal-Krishna.
Nature, Vol. 269, 780 - 781 (1977).

The author reports the lunar occultation and interplanetary scintillation observations of two BL Lac sources, AO0235+164 and 4C03.59 at 327 MHz. These observations indicate that at 327 MHz, part of the emission of both these sources originates in components larger than 1 arc s.

158.505 **The BL Lacertae objects.**
W. A. Stein, S. L. O'Dell, P. A. Strittmatter.
Annu. Rev. Astron. Astrophys., Vol. 14, (see 003.008), 173 - 195 (1976).

Contents: Summary of known sources; Description of some individual sources of particular interest; Observed properties of BL Lac objects and comparison with QSOs and nuclei of Seyfert galaxies; Theoretical questions raised by the BL Lac objects; Areas for future research.

158.506 **A study of the optical and radio absorption-line systems in AO 0235+164.**
A. M. Wolfe, B. J. Wills.
Astrophys. J., Vol. 218, 39 - 52 (1977).

The authors investigate physical conditions in the two absorption systems present in AO 0235+164. A scan of the optical spectrum is presented. They confirm the redshift system at $z = 0.852$, and give accurate equivalent widths for lines in the system at $z = 0.524$. A study of the optical and 21 cm absorption spectra of the lower redshift gas shows that both types of absorption do not arise in the same line-producing region. Whereas the 21 cm features are formed in a few clouds with velocity widths $\Delta v \sim 6$ km s^{-1}, the optical lines are produced in a "turbulent" region with $\Delta v \sim 100$ km s^{-1}. An analysis of the optical spectrum of the gas at $z = 0.852$ shows that the absorbing region is characterized by $\Delta v < 20$ km s^{-1}. Excitation conditions in the $z = 0.524$ system are deduced from the absence of UV lines arising from excited fine-structure energy levels in Fe II. The low excitation level implies that the absorber must be farther than 0.45 kpc from the continuum source.

158.507 **A search for redshifted hydrogen absorption in the BL Lacertae object PKS 0735 + 178.**   J. A. Galt.
J. R. Astron. Soc. Canada, Vol. 71, 394 (1977). – Abstract.

158.508 **The BL Lacertae objects.**   D. A. Allen.
Modern astronomy, (see 003.013), p. 176 - 184 (1977).

158.509 **The nature of the nebulosity associated with the BL Lacertae object AO 0235 + 164.**
H. E. Smith, E. M. Burbidge, V. T. Junkkarinen.
Astrophys. J., Vol. 218, 611 - 616 (1977).

The authors present photographic and spectrophotometric observations of the faint nebulosity associated with the active BL Lacertae object AO 0235 + 164. The nebula is centrally concentrated and shows relatively strong emission features which they identify with [O II], Hβ, and [O III] at $z = 0.525$, coincident with the lower of the two absorption systems in AO 0235 + 164. The luminosity in the lines is an order of magnitude greater than that expected for even the most luminous intervening late-type galaxy. Although no definite conclusion can be reached, this and other arguments suggest that the absorption and emission may be produced by material which is physically associated with the BL Lacertae object.

158.510 **B2 1308+32 = CSVS 6997 = OP 313.**
IAU Circ., No. 3140 (1977).

158.511 **The peculiar object Anon 0050+72.8.**
M. Cohen, L. V. Kuhi, H. Spinrad.
Publ. Astron. Soc. Pacific, Vol. 89, 485 - 487 (1977).

The system Anon 0050+72.8 presents the appearance of a faint, galactic, bipolar nebula. However, optical spectra reveal it to be an extragalactic object, at velocity 4550 ± 50 km s^{-1}, including an extremely active nucleus marked by strong emission lines of H, He I, [N II], and [O III].

158.512 **Optical and radio properties of a newly discovered BL Lac object.**   A. J. Pica.
Astron. J., Vol. 82, 935 - 936 (1977).

A study of the optical history of the newly discovered BL Lac object, 0109+22, reveals a total amplitude of variability in excess of 3 mag. A comparison sequence, summary of optical data, and discussion of optical and radio properties are presented. The source appears to have the same relationship between 5 GHz and optical flux densities as other comparable BL Lac objects.

The variability of flat-spectrum radio sources at 962 MHz.   See Abstr. 141.010.

Optical monitoring of quasistellar objects. II. Three-color photographic photometry of BL Lac.
See Abstr. 141.055.

Structure of radio sources with remarkably flat spectra: PKS 0735 + 178.   See Abstr. 141.057.

On the origin of the absorption spectra of quasi-stellar and BL Lacertae objects.   See Abstr. 141.068.

Relations between the optical and radio properties of extragalactic radio sources.   See Abstr. 141.102.

Infrared and visible polarimetry and photometry of highly variable quasi-stellar sources.   See Abstr. 141.120.

Black hole model of quasar-like objects—infrared optical emission.   See Abstr. 141.133.

Absorption by neutral hydrogen and ionized mag-

nesium in quasi-stellar objects and BL Lacertae objects.
See Abstr. 141.135.

Submillimeter photometry of extragalactic objects.
See Abstr. 158.023.

158.901   Errata: "UBV photometry of bright southern
          galaxies" [Observatory, Vol. 96, 61 - 64 (1976)].
M. J. Bucknell, J. V. Peach.
Observatory, Vol. 97, 212 (1977).

159 Magellanic Clouds

159.001   MK classification in the Small Magellanic Cloud.
          P. Dubois, M. Jaschek, C. Jaschek.
Astron. Astrophys., Vol. 60, 205 - 210 (1977).
    This paper compares in detail the spectra of SMC super-
giants with those of galactic supergiants. The helium and
metallic lines are weaker in the former, but not uniformly
so for all metals. Also the Balmer jump is small compared
to standards, and some stars have quite large hydrogen
profiles. Other peculiar features are described.

159.002   A main-sequence luminosity function for the
          Large Magellanic Cloud.   H. Butcher.
Astrophys. J., Vol. 216, 372 - 380 (1977).
    A main-sequence luminosity function between $M_v = 0$
and $M_v = +4$ is derived for a field region in the Large Magellanic Cloud. Several techniques developed to permit photometry in very crowded fields are described, and sources of
systematic error near the plate limit are discussed. The final
luminosity function qualitatively resembles the solar neighborhood main-sequence luminosity function but changes slope
at a point close to a magnitude brighter. A natural interpretation of this result is that the bulk of star formation began
$3-5 \times 10^9$ years ago in the LMC, rather than 10×10^9 years
ago as in the Galaxy. This interpretation permits recovery
from the LMC data of Salpeter's initial luminosity function
for star formation in the solar neighborhood.

159.003   The Magellanic Stream: the turbulent wake of the
          Magellanic Clouds in the halo of the Galaxy.
D. S. Mathewson, M. P. Schwarz, J. D. Murray.
Astrophys. J., Lett., Vol. 217, L5 - L8 (1977).
    New observations of the Magellanic Stream show that it
is composed of six discrete gas clouds with looped or horse-
shoe-shaped structures which appear to have predominantly
radial velocities. To explain this, it is proposed that cold,
dense clouds are formed as a result of thermal instabilities in
the wake of the Magellanic Clouds during their passage through
the hot halo of the Galaxy and that they are now sinking
toward the galactic center.

159.004   Der Magellansche Strom.   F. House.
          Sterne Weltraum, Jahrg. 16, 328 - 330 (1977).

159.005   uvbyR surface photometry of the 30 Doradus
          region.   A. C. Danks.
Astron. Astrophys., Suppl. Ser., Vol. 30, 89 - 91 (1977).
    Surface photometry data in the uvbyR system for 177
positions in the 30 Dor region are presented. Comparison of
these observations with those of the Astronomical Netherlands
Satellite are also made.

159.006   The Large and Small Magellanic Clouds in a binary
          state, the bending of the galactic disk and the
Magellanic Stream.   M. Fujimoto, Y. Sofue.
Astron. Astrophys., Vol. 61, 199 - 215 (1977).
    The authors obtain several series of orbits for the LMC
and SMC round the Galaxy, along which the two clouds were

in a binary state for the last 5 to 10×10^9 years. Approximately eight hundred test particles are distributed so as to simulate
their continuous media within the Galaxy, LMC and SMC.
The dynamical behavior of these particles is followed numerically and compared with the observed bending of the galactic
disk and the Magellanic Stream. Trials to reproduce the
Magellanic Stream, high-velocity clouds and other gas complexes fail however to reproduce the high negative radial
velocity of the Magellanic Stream. The authors examine
dynamically the primordial-gas model for the Stream proposed by Mathewson, Cleary and Murray (1974) and
Mathewson (1976) to explain this velocity discrepancy and
the positional coincidence of the Stream and most members of
the Local Group.

159.007   Five-channel photometric observations of stars in the
          Magellanic Clouds and in the Milky Way.
T. Walraven, J. H. Walraven.
Astron. Astrophys., Suppl. Ser., Vol. 30, 245 - 260 (1977).
    Photoelectric observations in the VBLUW five-colour
system are given for stars in the Magellanic Clouds, galactic
supergiants, bright field stars of mostly early spectral type, and
stars in the Orion association, together with notes on the
principal characteristic properties in this photometric system.

159.008   LMC X-1: a luminous extended X-ray source.
          A. Epstein.
Astrophys. J., Lett., Vol. 218, L49 - L50 (1977).
    LMC X-1 was observed by the rotating modulation col-
limator experiment on the SAS-3 satellite. Results of the
observation include a refined position within a $30'' \times 40''$
error region. In addition, LMC X-1 is determined to have
complex structure. The suggestion that LMC X-1 is a supernova remnant yields SNR parameters: $t = 5200$ yr, $D = 20$ pc,
$W_0 = 6 \times 10^{51}$ ergs.

159.009   The Magellanic Clouds.   G. L. Verschuur.
          Astronomy, Vol. 4, No. 12, p. 6 - 13 (1976).
Abstr. in Phys. Abstr., Vol. 80, Abstr. 26060 (1977).

159.010   Observations of supergiant stars in the Small
          Magellanic Cloud.   A. Ardeberg, E. Maurice.
Astron. Astrophys., Suppl. Ser., Vol. 30, 261 - 278 (1977).
    From observations at the European Southern Observatory,
spectrographic and photometric data are presented for 91
supergiant stars belonging to the Small Magellanic Cloud and
its Wing. The data include MK classes for 52 stars, radial
velocity data for 51 stars and UBV photometry for 90 stars.
Whenever possible, radial velocities for interstellar Ca II and
[O II] have been measured and listed. In Remarks to the
Catalogue notes are given for the individual stars concerning
the obtained data and comparisons with results of previous
investigations.

159.011   Spectrophotometry of nebulae in the Magellanic
          Clouds suitable as standards.
R. J. Dufour, W. V. Harlow.

Publ. Astron. Soc. Pacific, Vol. 89, 630 - 633 (1977).

Accurate photoelectric line strengths of selected emission lines in the λλ3727–7330 spectral region for three small H II regions in the Magellanic Clouds are presented. The nebulae are Henize N81 in the Small Cloud and Henize N4A and N8 in the Large Cloud. The observations should be useful in calibrating spectra of nebulae obtained in the Southern Hemisphere using modern spectrophotometric devices, particularly image-tube spectrographs.

159.012 **Monochromatic photographs of giant H II regions in the Magellanic Clouds.**
S. J. Czyzak, L. H. Aller.
Astrophys. Space Sci., Vol. 46, 371 - 378 (1977).

Narrow band-pass direct photographs have been secured of the central region of 30 Doradus and of several other nebulosities in the Magellanic Clouds. Interference filters whose band passes were selected to allow for the radial velocities of the Magellanic Clouds have been utilized to record monochromatic images of [N II] λ6584, Hα, He I λ5876, and [O III] λ5007. The implications of the observations for chemical compositions of nebulosities in the Magellanic Clouds would appear to be that spectroscopic and scanner studies should be supplemented by monochromatic photographs.

159.013 **Narrow-bandpass imagery of the Magellanic Clouds.**
T. R. Gull.
Proc. Southwest Reg. Conf., Vol. 3, (see 012.043), p. 55 - 65 (1977).

A wide field camera was used with narrow bandpass filters to study the Magellanic Clouds. Further analysis will yield a rough estimate of excitation, structure and type of emission nebula for nebulae larger than one arc minute.

159.014 **A morphological investigation of the Large Magellanic Cloud.** E. H. Geyer.
Mitt. Astron. Ges., Nr. 42, p. 93 - 96 (1977).

VBLUW photometry of Sk 160 (SMC X-1).
See Abstr. 113.016.

One hundred and eight eclipsing variables in the Magellanic Clouds. See Abstr. 121.091.

The RR Lyrae stars in the Large Magellanic Cloud.
See Abstr. 122.091.

Photometry of LMC Nova 1977b.
See Abstr. 124.301.

Supernova remnants in the Large Magellanic Cloud.
A photographic study in [O III] and [S II].
See Abstr. 125.027.

Abundances in 10 H II regions in the Small Magellanic Cloud. See Abstr. 132.004.

Identification of HV 5824 and HV 5967 with planetary nebulae. See Abstr. 135.025.

Observations of the LMC X-ray sources with the Ariel V Sky Survey Instrument. See Abstr. 142.001.

Variability of LMC X-4. See Abstr. 142.012.

The spectroscopic orbit and masses of SK 160/SMC X-1. See Abstr. 142.026.

Pulse profile and refined orbital elements for SMC X-1. See Abstr. 142.030.

Evidence for an accretion disk in SMC X-1.
See Abstr. 142.043.

LMC X-4. See Abstr. 142.104.

SMC X-2 and SMC X-3.
See Abstr. 142.118.

Simulation of the Magellanic Stream to estimate the total mass of the Milky Way. See Abstr. 151.026.

Magellanic Cloud investigations – V. The LMC blue cluster NGC 1854. See Abstr. 154.001.

A comment on the metal abundance of the LMC cluster NGC 2209. See Abstr. 154.009.

160 Groups, Clusters of Galaxies, Superclusters

160.001   The velocity dispersion of the X-ray cluster of galaxies in Centaurus (3U 1247−41).
N. V. Vidal, D. T. Wickramasinghe.
Mon. Not. R. Astron. Soc., Vol. 180, 305 - 308 (1977).
Redshifts of 25 galaxies in the Centaurus cluster have been measured. The mean velocity of the cluster is 3742 km/s with a velocity dispersion of 945 km/s. The X-ray luminosity is 5.5×10^{43} erg. The reality of the Centaurus cluster and its location in the X-ray luminosity versus velocity dispersion diagram are discussed.

160.002   Globular clusters and the Virgo cluster distance modulus.   D. A. Hanes.
Mon. Not. R. Astron. Soc., Vol. 180, 309 - 321 (1977).
Photometry presented by Hanes for the globular clusters associated with 20 bright galaxies in the Virgo cluster is used in an evaluation of a pure Population II distance modulus for the Virgo cluster of galaxies. The assumption of a normally distributed universal globular cluster luminosity function, calibrated in the Galaxy sample of globulars and based upon an absolute magnitude of $M_V = 0.6$ mag for RR Lyrae stars in globular clusters, leads to Virgo cluster distance modulus estimates near $(m-M)_{app} = 30.4$ mag, with an uncertainty of ±0.5 mag. Confirmation of a model-independent nature is presented.

160.003   X-ray, optical and radio data for the cluster of galaxies Klemola 44.
T. Maccacaro, B. A. Cooke, M. J. Ward, M. V. Penston, R. F. Haynes.
Mon. Not. R. Astron. Soc., Vol. 180, 465 - 469 (1977).
A newly discovered Ariel V X-ray source is identified with the cluster of galaxies Klemola 44 and a low-frequency radio source is consistent with other correlations between X-ray intensity, steep radio spectral index and the presence of supergiant D galaxies in the cluster.

160.004   Subsonic accretion of cooling gas in clusters of galaxies.   A. C. Fabian, P. E. J. Nulsen.
Mon. Not. R. Astron. Soc., Vol. 180, 479 - 484 (1977).
Slow-moving galaxies in the cores of X-ray emitting clusters can accrete large quantities of cooling gas. The accretion flow is most likely to occur in a subsonic fashion, stagnating at some finite radius. The authors apply these ideas to NGC 1275 in the Perseus cluster, and suggest that it explains the soft X-ray enhancement in that region, as well as the stationary optical filaments.

160.005   Photographic surface photometry of galaxies in the Virgo cluster.   C. W. Fraser.
Astron. Astrophys., Suppl. Ser., Vol. 29, 161 - 194 (1977).
Standard photometric parameters are obtained for 48 galaxies in the Virgo cluster, using homogeneous plate material obtained at the University Observatory, St. Andrews. The data from this work more than doubles the information of this type now available in the literature. Investigations of the dependence of the parameters on morphological type reveal systematic relations for both brightness and effective diameters, and also for concentration indices. In particular, the central surface brightness of exponential disk components is found to vary with type. Suggestions are put forward for the possible reclassification of a few objects, based upon the values of some of their photometric parameters.

160.006   Note on the supergalactic redshift anisotropy.
P. Teerikorpi, T. Jaakkola.
Astron. Astrophys., Vol. 59, L33 - L36 (1977).
Luminosity selection has been thus far neglected in the studies of the supergalactic redshift anisotropy. It probably causes, in cooperation with the inhomogeneous distribution of galaxies, the suggested center-anticenter anisotropy.

160.007   Statistical analysis of catalogs of extragalactic objects. VIII. Cross-correlation of the Abell and the 10′ Shane-Wirtanen catalogs.   M. Seldner, P. J. E. Peebles.
Astrophys. J., Vol. 215, 703 - 716 (1977).
The authors describe an analysis of the cross-correlation between the angular positions of the Abell clusters of galaxies and the Shane-Wirtanen galaxy counts in the original 10′ × 10′ cells. The data are well approximated by a simple model in which the mean surface density of galaxies around each Abell cluster varies with angular distance from the cluster center.

160.008   Radiative regulation of gas flow within clusters of galaxies: a model for cluster X-ray sources.
L. L. Cowie, J. Binney.
Astrophys. J., Vol. 215, 723 - 732 (1977).
The authors argue that in X-ray clusters a steady inflow of material takes place, in which the gravitational energy of the accreting material powers the X-ray luminosity of the source and accounts for the observed spatial extent of the gas distribution. They consider self-contained gas flows, in which mass ejected by galaxies in the outer regions of the cluster flows toward the core of the cluster and ultimately accretes onto the central galaxies. This model, which depends only on a single parameter measuring the ratio of the mass-injection rate to the cooling in the central regions, accounts for the observed X-ray luminosities of the clusters, the surface-brightness profiles, and energy spectra in terms of thermal bremsstrahlung emission from the gas.

160.009   Continuous increase of Hubble modulus behind clusters of galaxies.   L. Nottale, J. P. Vigier.
Nature, Vol. 268, 608 - 610 (1977).
Karoji and Nottale (1976) have examined the possibility of a variation of the Hubble constant H associated with the passage of incoming light through rich clusters of galaxies. The authors confirm their positive answer by a more detailed and precise statistical analysis. They show that there is an effective continuous effect depending on the relative distance of the line of sight of any given source to the centre of the intervening clusters. They thus conclude that they are dealing with a real physical effect, tied to the presence of these clusters.

160.010   Constraints on dynamical properties of clusters of galaxies.   A. Cavaliere, L. Danese, G. De Zotti.
Astron. Astrophys., Vol. 60, L15 - L17 (1977).
The authors derive bounds to dynamical properties of clusters and groups of galaxies from a non-linear theory of gravitational instability of density enhancements developing since the recombination epoch. Catalogue data are shown to conform remarkably to the theoretical bounds.

160.011   A study of the cluster of galaxies A 193.
F. Börngen, A. T. Kalloglyan.
Astrofizika, Vol. 12, 397 - 408 (1976). In Russian. − English translation in Astrophysics, Vol. 12, No. 3.
V magnitudes for 146 and B−V color indices for 125 galaxies in the cluster Abell 193 are determined on Schmidt plates of the Tautenburg 2-m telescope. There exist local peaks in the luminosity functions in B and V at B = 18.4 and V = 17.2 mag respectively. Logarithmic integral functions in B and V are given. There is a tendency for bright galaxies to be more concentrated to the cluster center.

160.012 Compact groups of compact galaxies. VII.
F. W. Baier, H. Tiersch.
Astrofizika, Vol. 12, 409 - 415 (1976). In Russian. — English translation in Astrophysics, Vol. 12, No. 3.

The seventh list of compact groups of compact galaxies is presented. It contains 45 new objects of this class. Identification charts for all 45 groups of the list are given.

160.013 Upper limits on nuclear radio emission from some Coma Cluster spirals.   W. J. Jaffe.
Astrophys. J., Vol. 216, 212 - 213 (1977).

High-resolution 2.7 GHz observations of seven previously detected, possibly overluminous spiral galaxies in the Coma Cluster were made in order to determine nuclear to total luminosity ratios. Two of the seven were detected at 2.7 GHz. These fluxes plus the upper limits on the rest indicate normal nuclear to total ratios.

160.014 On possible associations of quasi-stellar objects and radio galaxies with rich clusters of galaxies.
D. H. Roberts, S. L. O'Dell, G. R. Burbidge.
Astrophys. J., Vol. 216, 227 - 236 (1977).

A comparison of the cataloged coordinates of QSOs and of 3CR galaxies with those of Abell's rich clusters of galaxies yields the following results: (1) There is no statistically significant evidence that high-redshift QSOs lie preferentially close to Abell clusters. (2) There is no statistically significant evidence that low-redshift QSOs lie preferentially close to Abell clusters. (3) There is (as is well known) highly significant evidence that 3CR galaxies lie preferentially close to Abell clusters. (4) The distributions of angular separations between QSOs and clusters and between 3CR galaxies and clusters differ at statistically significant (but not highly significant) levels. In view of these results, a generic relationship between low-redshift QSOs and radio galaxies seems questionable. This result for the QSOs is also entirely consistent with the idea that they are "local" objects with redshifts of noncosmological origin.

160.015 Groups of galaxies. IV. The multiplicity function.
J. R. Gott III, E. L. Turner.
Astrophys. J., Vol. 216, 357 - 371 (1977).

The spectrum of galaxy cluster sizes is a valuable cosmological datum. The problem is formalized by defining the multiplicity function as the luminosity function of groups of galaxies which satisfy a surface density enhancement criterion, $\sigma \geqslant \sigma_g$.

160.016 Mean density and the correlation function of rich clusters of galaxies: theory and observations.
A. G. Doroshkevich, S. F. Shandarin.
Inst. prikl.mat. AN SSSR. Prepr. No. 8. Moskva, 1977. 22 pp. Price 8 Kop. In Russian. — Abstr. in Ref. zh., 51. Astron., 8.51.848 (1977).

160.017 The angular size-redshift relation. I. Sizes and shapes of nearby clusters of galaxies.   P. Hickson.
Astrophys. J., Vol. 217, 16 - 23 (1977).

Sixty-four cosmologically near clusters are examined on the basis of the projected separations between their brightest galaxies. Cluster angular sizes determined from these separations are found to be inversely proportional to the cluster redshifts, with a relatively small dispersion. The size and structure parameters correlate with Rood and Sastry type. New type classifications are presented for 23 clusters. Mean-surface-density profiles are determined for the various cluster types. It is found that many clusters are deficient in separations of 1 to 2 Mpc.

160.018 The mass-to-light ratio of late-type binary galaxies: luminosity- versus number-weighted averages.
E. L. Turner, J. P. Ostriker.
Astrophys. J., Vol. 217, 24 - 26 (1977).

The dynamical analysis of a relatively large and precisely selected group of binary galaxies by Turner is repeated, weighting each galaxy by its luminosity rather than, as before, weighting giant and dwarf galaxies equally. The new estimate of $\langle M/L_{pg} \rangle_L = 100 (M_\odot / L_\odot)$ within a characteristic radius of 370 kpc (for $H_0 = 50$ km s^{-1} Mpc^{-1}) is more suitable than the number-weighted $\langle M/L_{pg} \rangle_N = 65 (M_\odot / L_\odot)$ within 270 kpc for use in cosmological discussions. Using the Gott and Turner estimate of the cosmological luminosity density, the authors find Ω(galaxies) ~ 0.08 compared with 0.05 found from $\langle M/L_{pg} \rangle_N$. The data, which strongly indicate a relative paucity of luminous close pairs, may be interpreted in the context of dynamical friction and mildly strengthen the evidence for the massive halo hypothesis.

160.019 On the dynamics of binary galaxies.   A. Yahil.
Astrophys. J., Vol. 217, 27 - 33 (1977).

Recent data on widely separated binaries are reanalyzed. It is shown that the radial velocity difference Δv is uncorrelated with projected separation r_p over the entire range of separations sampled, in agreement with the hypothesis that galaxies contain halos, whose mass grows with radius $M(r) \propto r$. It is further shown that this mass, which may be characterized by the observable virial variable $r_p \Delta v^2$, is not correlated with the luminosity of the galaxies. It is concluded that the large-scale distribution of matter, which governs the dynamics of binaries, is not strongly coupled to the gas processes at small radii, which determine the rate of star formation, and hence the luminosity. The concept of mass-to-light ratio is therefore not useful on a distance scale much greater than 10 kpc, except as an average measure of the ratio of mass to luminosity in the universe.

160.020 A statistical investigation of radio sources in the directions of Zwicky clusters of galaxies.
J. O. Burns, F. N. Owen.
Astrophys. J., Vol. 217, 34 - 44 (1977).

In order to extend the available information on radio sources in clusters of galaxies, the authors have made a statistical investigation of the positional coincidences between 3CR radio sources and Zwicky clusters of galaxies. Although their analysis of Zwicky's catalog suggests that the catalog is incomplete for less compact clusters at large distances, several general conclusions are possible. The number of coincidences between 0.5 and 2 cluster radii is consistent with the number expected by chance, whereas an excess of 14 ± 4 3CR sources was found within 0.5 cluster radii. This statistical excess agrees well with the number of identified 3CR radio galaxies associated with Zwicky clusters (16). Thus few if any of the other sources in the directions of Zwicky clusters (blank fields, QSOs, etc.) are likely to be physically associated with clusters.

160.021 X-ray and radio observations of the structure of Abell 478.
H. W. Schnopper, J. P. Delvaille, A. Epstein, H. Helmken, D. E. Harris, R. G. Strom, G. W. Clark, J. G. Jernigan.
Astrophys. J., Lett., Vol. 217, L15 - L18 (1977).

The complex X-ray and radio structures of the cluster of galaxies Abell 478 have been observed in the 2−11 keV energy band by the rotating modulation collimators on the SAS-3 X-ray observatory and at 1415 MHz by the Westerbork synthesis radio telescope. The X-ray data imply an extended source ($L_x \sim 10^{45}$ ergs s^{-1}) and are consistent with the presence of a coincident pointlike source ($L_x \sim 3 \times 10^{44}$ ergs s^{-1}). The X-ray source position is also coincident with a cD radio galaxy at the core of the cluster.

160.022 The redshift and optical structure of the X-ray galaxy cluster A478.
N. A. Bahcall, W. L. W. Sargent.
Astrophys. J., Lett., Vol. 217, L19 - L21 (1977).

The redshift and optical structure of the X-ray cluster of galaxies A478 are determined. The redshift of the cluster is 0.09. The cluster center is $\alpha = 4^h10^m40^s$; $\delta = +10°20'.3$ (1950), and the classification type is cD. The core radius of the density profile is 0.20 ± 0.08 Mpc. The observed galaxy distribution is somewhat asymmetric.

160.023 A study of 1889 rich clusters of galaxies.
A. A. Leir, S. van den Bergh.
Astrophys. J., Suppl. Ser., Vol. 34, 381 - 403 (1977).

A sample of 1889 clusters drawn from Abell's Catalogue of Rich Clusters of Galaxies have been classified on the Bautz-Morgan system. For the brightest galaxy in each cluster the authors have also determined the magnitude, diameter, apparent flattening, and distance from the cluster center. The cluster diameter and the magnitudes of the brightest and of the tenth-brightest cluster galaxy were combined to obtain improved estimates for the redshifts of all clusters in the sample. Owen's survey of radio radiation from 503 Abell clusters has been used to show that rich clusters do not have a significantly higher probability of being radio sources than do poor clusters. It follows that individual bright galaxies in poor clusters have a higher probability of becoming radio sources than do those in rich clusters.

160.024 Nearby cluster of galaxies at low galactic latitude.
J. Huchra, J. Hoessel, J. Elias.
Astron. J., Vol. 82, 674 - 676, 759 - 760 (1977).

Observations of two galaxies at low galactic latitude discovered by Weinberger et al. (1976) show that they are the brightest members of a nearby cluster. The redshift of the cluster is 4200 ± 100 km/sec. Visual and near-infrared photometry of the two galaxies indicates that the visual extinction in this direction is 1.4 ± 0.1 mag.

160.025 The clustering of galaxies.   G. Efstathiou.
Q. J. R. Astron. Soc., Vol. 18, 321 - 325 (1977).

160.026 The Coma/Abell 1367 supercluster of galaxies.
S. A. Gregory.
News Lett. Astron. Soc. N.Y., Vol. 1, No. 2, p. 5 (1977).
Abstract.

160.027 Intrinsic properties of 15 galaxy clusters.
S. A. Gregory.
News Lett. Astron. Soc. N.Y., Vol. 1, No. 2, p. 27 (1977).
Abstract.

160.028 Statistical analysis of catalogs of extragalactic objects. VII. Two- and three-point correlation functions for the high-resolution Shane-Wirtanen catalog of galaxies.
E. J. Groth, P. J. E. Peebles.
Astrophys. J., Vol. 217, 385 - 405 (1977).

The authors present estimates of the two- and three-point angular correlation functions for the high-resolution Shane-Wirtanen catalog of galaxies. The two-point function is well approximated by a power law for $\theta \lesssim 2°.5$, corresponding to a projected distance of $\sim 9h^{-1}$ Mpc ($H = 100h$ km s^{-1} Mpc^{-1}), but breaks sharply below the power law at larger angles. Several arguments indicate that the break is an intrinsic feature of the galaxy distribution, not an artifact of the analysis. New scaling relations taking account of redshift and curvature are derived and used to compare the correlation functions estimated for the Zwicky, Shane-Wirtanen, and Jagellonian samples. The two-point spatial function is estimated to be $\xi(r) = (r_0/r)^{1.77}$, with $r_0 = 4.7h^{-1}$ Mpc, 0.05 Mpc $\lesssim hr \lesssim 9$ Mpc. The three-point function at $\theta \lesssim 3°$ is well represented by the model $\zeta(1, 2, 3) = Q[\xi(1)\xi(2) + \xi(2)\xi(3) + \xi(3)\xi(1)]$ for the spatial function, with $Q = 1.29 \pm 0.21$.

160.029 A study of the cluster of galaxies Zw Cl 1710.4+6401.
F. Börngen, A. T. Kalloglyan.

Astrofizika, Vol. 13, 5 - 20 (1977). In Russian. English translation in Astrophysics, Vol. 13, No. 1.

The results of a BV photometry of galaxies in the Zw Cl 1710.4+6401 within a radius of 12'.6 are presented. Equidensity diameters are measured and mean surface brightnesses of galaxies are calculated. The B−V colour indices decrease with decrease of galaxy luminosity. There exists no local maximum in the luminosity function in B or V. The integral luminosity functions in B and V are given. There are many compact galaxies in the cluster.

160.030 On the spatial distribution of heavy elements in X-ray emitting clusters of galaxies.
A. C. Fabian, J. E. Pringle.
Mon. Not. R. Astron. Soc., Vol. 181, 5P - 7P (1977).

X-ray observations of gas in clusters of galaxies indicate that the Fe/H ratio of the gas in the central regions is approximately solar. If this ratio were constant throughout mostly primordial cluster gas, this would have far-reaching implications for processes occurring in the early stages of galaxy and cluster formation. The authors discuss a number of mechanisms by which abundance gradients can arise in the cluster gas.

160.031 The X-ray temperatures of eight clusters of galaxies and their relationship to other cluster properties.
R. J. Mitchell, J. C. Ives, J. L. Culhane.
Mon. Not. R. Astron. Soc., Vol. 181, 25P - 32P (1977).

The authors present new results on the X-ray spectra of eight clusters of galaxies that have recently been obtained with the Mullard Space Science Laboratory proportional counter spectrometer on the Ariel V spacecraft. The relationship which is established between X-ray temperature and velocity dispersion further strengthens the case for the presence of hot gas in the cluster potential wells.

160.032 A list of hypergalaxies.   J. Einasto, M. Jõeveer, A. Kaasik, P. Kalamees, J. Vennik.
Tartu Astron. Obs., Teated, No. 49, 73 pp. (1977).

The authors call small aggregates of galaxies and intergalactic matter with one concentration center hypergalaxies. The present paper contains the first list of hypergalaxies. The procedure of selecting galaxies for membership is outlined. This list of hypergalaxies is compared with lists of groups of galaxies and double galaxies compiled by other investigators.

160.033 The Local Group: the solar motion relative to its centroid.
A. Yahil, G. A. Tammann, A. Sandage.
Astrophys. J., Vol. 217, 903 - 915 (1977).

A new solution for the motion of the local standard of rest (LSR) relative to the centroid of the Local Group (LG) of galaxies, based on 21 cm redshifts for a number of candidates, gives v(LSR) $= 300$ km s^{-1} toward $l = 107°$, $b = -8°$. Three other solutions are given using different precepts for membership within the LG. This motion of the LSR corresponds to a best-fit solar motion relative to the LG centroid of $v(\odot) = 308$ km s^{-1} toward $l = 105°$, $b = -7°$.

160.034 Radio galaxies in the Coma cluster. II.
W. G. Tifft, M. Tarenghi.
Astrophys. J., Vol. 217, 944 - 950 (1977).

Optical data are presented for 50 galaxies in the Coma cluster, including all the known radio sources brighter than $m_p \approx 17$. The redshift distribution of the best radio source sample deviates markedly from that of the cluster as a whole. The correlations of redshift with magnitude and radio flux at 610 MHz are given.

160.035 Observations of a distant cluster of galaxies.
R. G. Kron, H. Spinrad, I. R. King.
Astrophys. J., Vol. 217, 951 - 957 (1977).

Faint clusters of galaxies discovered on deep Kitt Peak 4

m plates are much bluer than expected. The authors have used image-dissector systems to obtain scans of one such cluster; the indicated redshift is $z = 0.947$.

160.036   Cannibalism among the galaxies: dynamically produced evolution of cluster luminosity functions.
J. P. Ostriker, M. A. Hausman.
Astrophys. J., Lett., Vol. 217, L125 - L129 (1977).
   The authors show how the merging of galaxies in clusters will tend to produce supergiant systems of low surface brightness like the known cD systems and also the apparently nonstatistical features seen at the bright end of the cluster luminosity function. Tests of the theory are proposed and methods of deriving a luminosity correction (for dynamical evolution) are suggested.

160.037   Microwave emission from dust and hidden matter in the Coma cluster.   P. S. Wesson.
Astron. Astrophys., Vol. 61, 177 - 180 (1977).
   It is found that a dust medium composed of particles of glass or material of greater conductivity and of density up to 10^{-27} g cm^{-3} could be present in Coma. This may provide enough hidden matter to bind this and similar clusters.

160.038   Investigation on the structure of the Jagellonian Field.   H. Bereś.
Acta Cosmologica, Zesz. 6, 7 - 17 (1977).
   In the present paper the method of "shuffling of cells" is applied to investigate the clustering effect of galaxies in the Jagellonian Field. The structure of this clustering is irregular and extremely complex.

160.039   The clustering of galaxies in the Jagellonian Field.   P. Flin.
Acta Cosmologica, Zesz. 6, 19 - 26 (1977).
   The statistical reduction method has been used to study the distribution of galaxies in the Jagellonian Field. An examination of the concentration, grouping, structure and anisotropies indices shows that structures of different sizes exist there. Individual structures cannot be in general separated from each other. Therefore, the clustering effect in the Jagellonian Field is apparently continuous.

160.040   Radio emission of Abell clusters in the GB region.   A. Michalec.
Acta Cosmologica, Zesz. 6, 47 - 54 (1977).
   In the GB survey region (Masłowski 1972) there are 102 Abell clusters (Abell 1958). 31 of them coincide with the positions of GB radio sources. The number of random coincidences was estimated from a Poisson distribution. For 19 clusters from this group, the observations at 2695 MHz were made with the same instrument. The mean spectral index ($S \sim \nu^{-\alpha}$) for these sources is equal to $\alpha_{1400}^{2695} = +0.81$ with $\sigma = 0.51$. The clusters' redshifts were estimated. An analysis of the luminosity function for these clusters was carried out.

160.041   X-ray line emission for clusters of galaxies. II. Numerical models.   C. L. Sarazin, J. N. Bahcall.
Astrophys. J., Suppl. Ser., Vol. 34, 451 - 467 (1977).
   The predicted X-ray spectra of clusters of galaxies have been calculated assuming that the X-ray emission arises from hot intracluster gas in collisional ionization equilibrium. An extensive grid of models is presented for isothermal models, polytropic hydrostatic models, and wind models. The integrated X-ray spectrum is given for each model, including line and recombination radiation.

160.042   The luminosity function of galaxies in cluster A2670.   J. Mottmann, G. O. Abell.
Astrophys. J., Vol. 218, 53 - 57 (1977).
   Photovisual magnitudes are obtained for galaxies in the rich, compact cluster A2670, and also for galaxies in the surrounding field, on the system of extrafocal photographic photometry defined by Abell and Mihalas. The luminosity function, corrected for the field, is in good agreement with that found by Oemler with a different technique and in a different spectral bandpass, and matches that found by Abell for galaxies in the Coma cluster. Comparison of the luminosity functions of Coma and of A2670 suggests relative distances for those clusters that are in the same ratio as their redshifts, within the uncertainties of the corrections for galactic extinction. The total luminosity of cluster A2670 is estimated to be $7 \times 10^{12} L_\odot$.

160.043   Studies of rich clusters of galaxies – IV. Photometry of the Coma Cluster.
J. G. Godwin, J. V. Peach.
Mon. Not. R. Astron. Soc., Vol. 181, 323 - 337 (1977).
   The authors have measured V_{25} magnitudes for 923 galaxies in a 1.22 deg square field centred on the Coma Cluster. They form a complete sample of 497 galaxies to a limit of $V_{25} = 17.5$ with 426 galaxies at fainter magnitudes. The luminosity function has been constructed. Galaxies of intermediate brightness ($14 < V_{25} \leqslant 16$) are more strongly concentrated to the cluster centre than either brighter or fainter members.

160.044   Density profiles of clusters of galaxies.   N. A. Bahcall.
Highlights of Astronomy, Vol. 4, Part I, (see 012.020), p. 247 - 251 (1977).

160.045   The galaxy content of clusters.   A. Oemler, Jr.
Highlights of Astronomy, Vol. 4, Part I, (see 012.020), p. 253 - 260 (1977).

160.046   Velocity dispersions in clusters of galaxies.   R. J. Dickens, C. Moss, J. A. Dawe, B. Peterson.
Highlights of Astronomy, Vol. 4, Part I, (see 012.020), p. 261 (1977).

160.047   Groups of galaxies.   J. R. Gott III.
Highlights of Astronomy, Vol. 4, Part I, (see 012.020), p. 271 - 277 (1977).
   The author reports on some work that Ed Turner and he have done on groups of galaxies and how groups may be used to obtain useful cosmological information.

160.048   The existence of high-order clusters of galaxies.   M. Kalinkov.
Highlights of Astronomy, Vol. 4, Part I, (see 012.020), p. 279 - 289 (1977).

160.049   X-ray clusters of galaxies: What to plot against X-ray luminosities?   N. V. Vidal.
Highlights of Astronomy, Vol. 4, Part I, (see 012.020), p. 291 (1977).

160.050   X-rays from clusters of galaxies.   J. L. Culhane.
Highlights of Astronomy, Vol. 4, Part I, (see 012.020), p. 293 - 309 (1977).
   Progress in understanding the extended X-ray sources and in determining the emission mechanism has come from observations of the X-ray structure and spectra of the cluster sources and it is the purpose of this review to present the current status of these observations. The available information on cluster spectra and structure is reviewed and a list of proposed identifications presented.

160.051   Radio sources in clusters of galaxies.   I. McHardy.
Highlights of Astronomy, Vol. 4, Part I, (see 012.020), p. 311 - 319 (1977).

160.052   Hot gas in clusters of galaxies.   S. M. Lea.

Highlights of Astronomy, Vol. 4, Part I, (see 012.020), p. 329 - 339 (1977).

The observations seem to favor the presence of a hot gas in rich clusters of galaxies, whose origin is partially in primordial matter which has collapsed into the cluster, and partially in gas lost from the galaxies. The gas is in a static (or almost static) equilibrium in the cluster, and is either tightly bound or has an equation of state corresponding to $\gamma < 5/3$.

160.053   **Clusters of galaxies.**   N. A. Bahcall.
Annu. Rev. Astron. Astrophys., Vol. 15, (see 003.012), 505 - 540 (1977).

The author reviews the observational properties of clusters of galaxies and mentions some of the theoretical interpretations. The subjects reviewed are: catalogs, static properties, dynamics, X-ray emission, and radio emission.

160.054   **Dynamical friction in aspherical clusters.**
J. Binney.
Mon. Not. R. Astron. Soc., Vol. 181, 735 - 746 (1977).

This paper investigates the way the morphologies of rich clusters of galaxies will be affected by the action of dynamical friction on the brighter galaxies. The author reviews the evidence that the distributions of bright galaxies in several clusters are more elongated on the sky than the distributions of less bright galaxies. Application to the formation of cD galaxies and the capture of globular clusters by galactic nuclei is discussed.

160.055   **X-ray clusters of galaxies: correlation of X-ray luminosity with galactic content.**
N. A. Bahcall.
Astrophys. J., Lett., Vol. 218, L93 - L95 (1977).

All 14 X-ray clusters of galaxies detected by the Ariel 5 satellite at high galactic latitudes with redshifts $z < 0.05$ are analyzed optically for galaxy content. A strong correlation is found between cluster X-ray luminosity and galaxy type: the fraction of spiral galaxies in a cluster decreases rapidly with increasing X-ray luminosity. The observed correlation is consistent with a combined model of thermal bremsstrahlung emission from a hot intracluster gas and stripping of the spiral galaxies caused by the ram pressure of the same intracluster gas.

160.056   **The clustering of galaxies.**   E. J. Groth,
P. J. E. Peebles, M. Seldner, R. M. Soneira.
Sci. American, Vol. 237, No. 5, p. 76 - 78, 84, 87 - 90, 95 - 98 (1977).

Galaxies tend to form small groups, which in turn form larger clusters, and so on. Such a hierarchical organization has long been suspected, but only recently has it been clearly perceived.

160.057   **Superclusters: fact or fancy?**   J. Darius.
New Scientist, Vol. 74, 383 - 385 (1977). – Abstr. in Phys. Abstr., Vol. 80, Abstr. 71151 (1977).

160.058   **Further investigations of clusters of galaxies.**
F. W. Baier, W. Mai.
Astron. Nachr., Band 298, 301 - 310 (1977).

The number-density distribution in ten clusters of galaxies was derived by counting galaxies on the red Palomar Sky Survey prints. For seven isolated clusters the radial number-density distribution and the radial cumulative galaxy distribution were calculated.

160.059   **Photographic measurements of the diffuse light in the Coma cluster.**   T. X. Thuan, J. Kormendy.
Publ. Astron. Soc. Pacific, Vol. 89, 466 - 473 (1977).

The diffuse background light in the Coma cluster is measured using isodensity tracings of B, G, V, and R photographic plates. Between 4 and 14 arc min from the center, the surface brightness of the diffuse light decreases from ~ 26 to ~ 28 G magnitudes arc sec^{-2}. The total magnitude in this annulus is $G = 11.22$, which is $\sim 45\%$ of the light in galaxies alone, or $\sim 30\%$ of the total. This does little to alleviate the "missing mass" problem. The isodensity contours and the equivalent profile of the diffuse light closely parallel the distribution of light in galaxies, implying no strong mass segregation. However, the background light appears to be bluer than the galaxies. This is consistent with the hypothesis that the background consists of stars tidally stripped from galaxies, which generally become bluer at larger radii.

160.060   **Faint dwarf galaxies in the Virgo cluster.**
G. Reaves.
Publ. Astron. Soc. Pacific, Vol. 89, 620 - 621 (1977). Abstract.

160.061   **Dynamical evolution in clusters of galaxies with low-frequency radio emission.**   B. N. G. Guthrie.
Astrophys. Space Sci., Vol. 52, 177 - 188 (1977).

Clusters of galaxies in which radio emission at low frequencies ($\leqslant 178$ MHz) has been detected were classified on the Bautz-Morgan (BM) system according to the dominance of the brightest galaxy. Radio sources with steep low-frequency spectra occur in clusters of all BM types but more often in rich clusters; the distributions of BM types for clusters with high and low spectral indices between 38 and 178 MHz are similar. Some clusters were found to have central cores of bright galaxies which may reflect mass segregation of galaxies due to dynamical friction.

160.062   **The structure of nearby clusters of galaxies. I. Hierarchical clustering and an application to the Leo region.**   J. Materne.
ESO Sci. Prepr. No. 12, 3 + 28 pp. (1977). – To appear in Astron. Astrophys.

A new method of classifying groups of galaxies, called hierarchical clustering, is presented as a tool for the investigation of nearby groups of galaxies. The method is free from model assumptions about the groups. The scaling of the different coordinates is necessary, and the level from which one accepts the groups as real has to be determined. Hierarchical clustering is applied to an unbiased sample of galaxies in the Leo region.

160.063   **The galactic neighbourhood.**
G. A. Tammann, R. Kraan.
ESO Sci. Prepr. No. 15, 1 + 17 pp. (1977). – To appear in I.A.U. Symposium No. 79.

Several properties of the 131 galaxies known within 9.1 Mpc are investigated. 88 of these galaxies are concentrated into eight groups, leaving 33 percent of true field galaxies. The groups have small velocity dispersion which limits the mean mass-to-light ratio for the different types of group galaxies to $M/L < 20$. The sample galaxies are strongly concentrated toward the supergalactic plane; at a distance of 4 Mpc of the plane the luminosity density drops to half its value. There is also a pronounced luminosity density decrease with increasing distance from the Virgo cluster centre; at a distance of 30 Mpc the density has decreased by more than a factor of 10^4. The best estimate of the mean luminosity density within a sphere of 30 Mpc radius centered on the Virgo cluster is $1.5 \times 10^8 L_\odot$ Mpc^{-3}.

160.064   **Intragalactic factor and apparent distribution of external objects.**   B. I. Fesenko.
Astron. Zh. Akad. Nauk SSSR, Vol. 54, 1202 - 1210 (1977). In Russian. English translation in Soviet Astron., Vol. 21, No. 6.

Holmberg's data on the cosecant law for a cluster of galaxies on one hand and the cloud model for the interstellar matter on the other hand indicate large fluctuations in cluster number due to visibility condition variations. This conclusion is confirmed by the existence of a correlation between the apparent distributions of distant clusters and of the interstellar hydrogen density. All that testifies against the reality of superclusters with angular diameters of several degrees discovered earlier. It is possible that false superclusters emerge from the existence of windows in the interstellar matter and from the variations in the observational conditions for clusters of galaxies.

160.065   The radio luminosity function of cluster galaxies
            at 5 GHz.   I. Khan, J. Pfleiderer.
Mitt. Astron. Ges., Nr. 42, p. 96 - 98 (1977).

160.066   Probleme der dreidimensionalen Cluster-Analyse
            von Gruppen von Galaxien.   J. Materne.
Mitt. Astron. Ges., Nr. 42, p. 104 - 107 (1977).

160.067   Über die Ursachen der Strukturen in Galaxienhau-
            fen.   P. v. d. Osten-Sacken.
Mitt. Astron. Ges., Nr. 42, p. 107 - 109 (1977).

160.068   Hot gas in clusters of galaxies.
            A. G. Doroshkevich, A. A. Klypin.
Inst. prikl. mat. AN SSSR. Prepr. No. 77. Moskva, 1977.
14 pp. Price 5 Kop. In Russian. – Abstr. in Ref. zh., 51.
Astron., 1.51.755 (1978).

160.069   Giant clusters of galaxies.   O. Obůrka.
            Říše hvězd, Vol. 58, 161 - 163 (1977). In Czech.

            A catalog of southern groups and clusters of galaxies.
See Abstr. 002.032.

            Expansion and rotational momentum of large
cosmic masses.   See Abstr. 066.015.

            The microwave background radiation in the direc-
tion of clusters of galaxies.   See Abstr. 066.087.

            Interferometer observations of radio sources in
clusters of galaxies. V.   See Abstr. 141.031.

            Observations of the NGC 7331/Stephan's Quintet
area at 1421 MHz.   See Abstr. 141.041.

            Westerbork observations of three cluster radio
galaxies.   See Abstr. 141.069.

            Clusters of galaxies and radio sources.
See Abstr. 141.108.

            Properties of radio sources in clusters of galaxies.
See Abstr. 141.114.

            A radio study of Abell clusters.
See Abstr. 141.125.

            A search for steep low-frequency radio spectra
among quasars and clusters of galaxies.
See Abstr. 141.149.

            Identification of southern radio sources with steep
radio spectra.   See Abstr. 141.159.

            Structure of the X-ray source in the Virgo cluster
of galaxies.   See Abstr. 142.020.

            X-ray clusters of galaxies: correlations with optical
morphology and galaxy density.   See Abstr. 142.031.

            Absence of detectable X rays from supercluster
candidate 4U0134−11.   See Abstr. 142.050.

            On Compton and thermal models for X-ray emission
from clusters of galaxies.   See Abstr. 142.073.

            Extragalactic X-ray sources.   See Abstr. 142.081.

            Orbit segregation in evolving galaxies and clusters
of galaxies.   See Abstr. 151.010.

            A theory of galaxy formation and clustering.
See Abstr. 151.011.

            On the dynamical evolution of galaxies in clusters.
See Abstr. 151.020.

            The dynamical evolution of clusters of galaxies.
See Abstr. 151.040.

            Massive galactic halos. I. Formation and evolution.
See Abstr. 151.045.

            Substructure in clusters of galaxies: a clue to the
'missing' mass?   See Abstr. 151.059.

            A hypothesis to explain the virial discrepancy in
clusters of galaxies.   See Abstr. 151.060.

            Stochastic simulation of fields of galaxies.
See Abstr. 151.061.

            Suspected globular clusters in the Fornax I cluster
of galaxies.   See Abstr. 154.012.

            The luminosity distribution of globular clusters in
the Virgo cluster of galaxies.   See Abstr. 154.021.

            Correlation between the number of globular clusters
and the luminosity of the spheroidal system for external
galaxies.   See Abstr. 154.035.

            On the motions of the Sun, the Galaxy and the
Andromeda nebula.   See Abstr. 155.019.

            Absolute magnitude-color relation for early type
spirals.   See Abstr. 158.002.

            Large-scale distribution and motion of galaxies.
See Abstr. 158.007.

            The color-absolute magnitude relation for E and S0
galaxies. I. Calibration and tests for universality using Virgo
and eight other nearby clusters.   See Abstr. 158.021.

            Isolated galaxies.   See Abstr. 158.022.

            Musings on galaxy classification.
See Abstr. 158.043.

            Surface brightness and color distributions of
elliptical and S0 galaxies in the Coma cluster.
See Abstr. 158.046.

            The ultraviolet spectra and color evolution of
galaxies at large redshifts.   See Abstr. 158.047.

            Correlation analysis of the space and surface distribu-
tion of galaxies.   See Abstr. 158.051.

            Distribution of different extragalactic objects in
the field of the North Galactic Pole.   See Abstr. 158.072.

            Redshifts of compact galaxies in systems of galaxies.
See Abstr. 158.116.

Photometry of the intergalactic optical surface brightness in the Coma cluster.   See Abstr. 161.002.

Photometry of the intergalactic background light in the Coma cluster.   See Abstr. 161.006.

Dynamical collapse of the intracluster gas and the formation of a hot plasma in clusters of galaxies. See Abstr. 161.007.

Detection of intergalactic gas in distant, rich clusters. See Abstr. 161.008.

Hot and cold intracluster gas. See Abstr. 161.010.

On estimating correlations in the spatial distribution of galaxies.   See Abstr. 162.013.

The source of the X-ray background. See Abstr. 162.019.

Unborn clusters.   See Abstr. 162.020.

The correlation function for density perturbations in an expanding universe. II. Nonlinear theory. See Abstr. 162.022.

The angular-size-redshift relation. II. A test for the deceleration parameter.   See Abstr. 162.029.

Local supercluster and anomalous Hubble expansion. See Abstr. 162.031.

On the integration of the BBGKY equations for the development of strongly nonlinear clustering in an expanding universe.   See Abstr. 162.037.

Primeval gas clouds and the low-energy X-ray background.   See Abstr. 162.065.

Galaxy clusters as relativistic spherically-symmetrical inhomogeneities.   See Abstr. 162.110.

161 Intergalactic Matter

161.001 Constraints on a dense hot intergalactic medium.
G. B. Field, S. C. Perrenod.
Astrophys. J., Vol. 215, 717 - 722 (1977).

The authors propose a model in which exploding galaxies heat the intergalactic gas (IGG) to 10^8-10^9 K. The thermal bremsstrahlung from the model agrees with spectral measurements of the X-ray background (XRB). They show that recent submillimeter measurements of the cosmic microwave background (CMB) are consistent with a spectrum distorted from blackbody by Compton scattering on the same IGG. The authors also show that the isotropy and intensity of the XRB rule out its origin from discrete gas clouds. Because of the large energy requirement to heat the IGG and other considerations, one must regard the existence of a cosmologically significant amount of hot IGG as uncertain.

161.002 Photometry of the intergalactic optical surface brightness in the Coma cluster.   K. Mattila.
Astron. Astrophys., Vol. 60, 425 - 430 (1977).

Photoelectric surface brightness observations have been made of the Coma cluster of galaxies in order to measure the intensity and colour of the intergalactic background light (IBL). As an explanation of the IBL four possible mechanisms are discussed: (1) light from extended envelopes of galaxies, (2) light from dwarf galaxies, intergalactic globular clusters or individual stars, (3) thermal bremsstrahlung from a hot intergalactic gas, (4) scattering of the light of galaxies by intergalactic dust grains. The first two mechanisms seem to be the most probable ones, but the fourth mechanism may also contribute a significant fraction of the observed IBL intensity.

161.003 Quasar absorption lines as probes of the past intergalactic medium.   W. L. W. Sargent.
The evolution of galaxies and stellar populations, (see 012. 005), p. 427 - 442, with a discussion, p. 443 - 444 (1977).

The author discusses the information that quasar absorption lines give about the intergalactic medium, and possibly about the outer parts of galaxies. He assumes that the absorption lines are due to intervening material in the line of sight, although this point is considered by some to be controversial.

161.004 Search for intergalactic matter basing on the Catalogue of the Jagellonian Field.   I. Tarraro.
Acta Cosmologica, Zesz. 6, 55 - 64 (1977).

Basing on the data regarding the visibility of galaxies in the "Jagellonian Field Catalogue" the lower limit of the coefficient of the selective intergalactic absorption was estimated. It amounts to $0.^m0020 \pm 0.^m0004$ per Mpc in the blue-yellow colour index and to $0.^m0005 \pm 0.^m0003$ in the yellow-red. A sub-region of the Jagellonian Field is supposed to have larger intergalactic extinction than the whole field in average.

161.005 Rotation measures and cosmology.
P. P. Kronberg.
IAU Symp., No. 74, (see 012.014), p. 367 - 377 (1977).

161.006 Photometry of the intergalactic background light in the Coma cluster.   K. Mattila.
Highlights of Astronomy, Vol. 4, Part I, (see 012.020), p. 263 (1977).

161.007 Dynamical collapse of the intracluster gas and the formation of a hot plasma in clusters of galaxies.
F. Takahara, S. Ikeuchi, N. Shibazaki, R. Hoshi.
Prog. Theor. Phys., Vol. 56, 1093 - 1103 (1976). − Abstr. in Phys. Abstr., Vol. 80, Abstr. 22322 (1977).

161.008 Detection of intergalactic gas in distant, rich clusters.
G. Lake, R. B. Partridge.
Nature, Vol. 270, 502 (1977).

The authors have detected a reduction in intensity or 'cooling' of the cosmic microwave background as this radiation passes through several rich clusters of galaxies. These observations imply the presence of substantial quantities of hot intergalactic gas in the clusters observed.

161.009 On the intergalactic contribution to the rotation measures of QSO's.
P. P. Kronberg, M. Reinhardt, M. Simard-Normandin.
Astron. Astrophys., Vol. 61, 771 - 776 (1977).

The authors have developed some simple models of a random intergalactic (i.g.) magneto-ionic medium, and calculated their effect on the rotation measures (*RM*) of radio sources with substantial redshift (i.e. QSO's). For ionized gas clouds which co-expand and conserve magnetic flux, the contribution to *RM* increases sharply for $z > 1$. Using new rotation measure data on QSO's the authors have searched for a possible increase with redshift of the variance ($V(z)$) of the *RM*'s for sources at high galactic latitudes, where the galactic rotation measure is small. They find no difference in V when comparing 31 QSO's at $1.0 < z \lesssim 2.5$ with 34 which have $z < 1.0$. This result establishes a new, more stringent upper limit on any ionized i.g. medium. This rotation measure test for an i.g.m. is an important complement to other tests for intergalactic material. It has the advantage of being simple, relatively free of redshift-dependent selection effects, and, due to the large redshifts of QSO's ($0.1 < z \lesssim 3$), probes the i.g.m. into the interesting epoch in which galaxies and clusters of galaxies are being formed.

161.010   Hot and cold intracluster gas.
       P. R. Preussner, M. Grewing, P. Biermann.
Mitt. Astron. Ges., Nr. 42, p. 99 - 102 (1977).
       Recent X-ray observations strongly suggest the presence of hot intergalactic gas of almost normal chemical composition within clusters of galaxies. Very likely, the high temperatures are caused by the galaxy motions in the clusters which generate shock waves. The authors describe here some aspects of the time-dependent behavior of the shocked gas and its effects on the ambient medium.

161.011   New upper limits to an intergalactic magneto-ionic medium.
P. Kronberg, M. Reinhardt, M. Simard-Normandin.
Mitt. Astron. Ges., Nr. 42, p. 103 - 104 (1977).

       A search for neutral hydrogen clouds in radio galaxies and in intergalactic space.   See Abstr. 132.011.

       Large-scale winds driven by flare-star mass loss.
See Abstr. 151.046.

       The extended H I regions around spiral galaxies: a probe for galactic structure and the intergalactic medium.
See Abstr. 158.059.

       Radiative regulation of gas flow within clusters of galaxies: a model for cluster X-ray sources.
See Abstr. 160.008.

       Hot gas in clusters of galaxies.   See Abstr. 160.052.

       Photographic measurements of the diffuse light in the Coma cluster.   See Abstr. 160.059.

       Hot gas in clusters of galaxies.
See Abstr. 160.068.

162 Structure and Evolution of the Universe, Cosmology

162.001 **Cosmological change of G and the structure of the Earth.**   R. A. Lyttleton, J. P. Fitch.
Mon. Not. R. Astron. Soc., Vol. 180, 471 - 477 (1977).

The application by Hoyle & Narlikar of their cosmological theory with a decreasing G to determine changes of radius of a two-zone Earth is shown to be invalid by their homology-treatment. The initial model is not a polytropic distribution and the condition of pressure-continuity is breached in their scaled models. In the present paper the accurate solution of the problem has been found by numerical means and gives a rate of change of radius only about one-quarter that of Hoyle & Narlikar. The established existence of an accelerative component of the angular velocity of the Earth shows that the moment of inertia is diminishing, and discussions of changes of internal distribution and radius cannot adequately be made merely by taking account of a possibly changing G.

162.002 **Large number hypothesis and continuous creation cosmologies.**   I. W. Roxburgh.
Nature, Vol. 268, 504 - 507 (1977).

The large number hypothesis and the condition that general relativity is satisfied in Einstein units, allows a family of cosmological models, two of which are the Dirac model without creation and the more recent Dirac model with multiplicative creation. The new models have multiplicative creation, a cosmological scale factor $S(t) \propto t^m$, and are spatially flat; a multiplicative steady state model also satisfies the hypotheses. How these models affect the temperature of the Earth and the cosmological deceleration parameter is important.

162.003 **Le problème cosmologique et ses hypothèses. V.**   J. Dubois.
Orion, 35. Jahrg., 117 - 121 (1977).

162.004 **On the measurement of parallaxes of objects at cosmological distances.**   I. D. Novikov.
Astron. Zh. Akad. Nauk SSSR, Vol. 54, 722 - 726 (1977). In Russian. English translation in Soviet Astron., Vol. 21, No. 4.

The parallaxes of objects at cosmological distances for a universe which is homogeneous only on the average are calculated. The results obtained are shown to differ from those obtained for idealized completely homogeneous models. The disturbances originating during the passage of light through the gravitational fields of celestial bodies and the interstellar medium are analyzed.

162.005 **On thermal instability in an expanding hot universe.**   S. G. Pomagaev.
Byull. Inst. Astrofiz., Dushanbe, No. 66 - 67, p. 14 - 22 (1976). In Russian.

The thermal instability in an expanding hot universe is considered. It is supposed that the universe is filled up by radiation and hydrogen plasma, and energy losses in the presence of perturbations are due to free-free and bound-free transitions. It is found that in consequence of the thermal instability density perturbations increase with time t according to the law $t^{2-\beta}$, (where $\beta < 1$). The minimum mass being contained in these fluctuations is equal to about $10^{15} M_{\odot}$.

162.006 **Inhomogeneous cosmological models and Hubble's law.**   V. I. Stoyanov.
Byull. Inst. Astrofiz., Dushanbe, No. 66 - 67, p. 28 - 31 (1976). In Russian.

The parabolic, hyperbolic and elliptic models which accomodate to variable cosmic density claimed by de Vaucouleurs are considered. It is shown that in general only an inhomogeneous hyperbolic model admits to retain Hubble's law and besides results in an open Friedmann model under certain conditions. However, this specified model differs from the corresponding Friedmann model in the next (post-Hubble) approximation and will never evolve to it.

162.007 **Primordial star formation in a cold early Universe.**   B. J. Carr.
Astron. Astrophys., Vol. 60, 13 - 26 (1977).

Possible models for a cold early Universe are discussed. It is shown how initial density fluctuations in such a Universe can become bound after 10^{-4} s and, in some circumstances, produce regions of nuclear activity. It is described how these regions of nuclear activity may evolve into main-sequence stars. The expected mass spectrum of primordial stars is derived. The nucleosynthesis effects of the stars are discussed.

162.008 **Cosmology today.**   L. C. Green.
Sky Telesc., Vol. 54, 180 - 184 (1977).

162.009 **On the possible existence of cosmological cosmic rays. I. The framework for light-element and gamma-ray production.**   T. Montmerle.
Astrophys. J., Vol. 216, 177 - 191 (1977).

This paper examines the possibility of the existence of cosmological cosmic rays (CCR), in the framework of big-bang cosmology. Following Stecker, the CCR are assumed to be born in a burst at some (high) redshift z_s. Gamma rays originate from π^0-decay resulting from interactions of the high-energy part of the CCR, and light elements are produced via $(p\alpha) + (p\alpha)$ reactions by the low-energy part, both of them by collisions with the ambient matter (of density corresponding to a deceleration parameter q_0). The $1-100$ MeV γ-ray background spectrum and the lithium abundance are considered as observational constraints on the possible CCR flux intensity. To this end, a theoretical framework is set for simultaneous γ-ray and light-element production by solving a system of coupled time-dependent transport equations, taking ionization and expansion losses into account.

162.010 **Conductivity in type VI_0 cosmologies with electromagnetic field.**   B. O. J. Tupper.
Astrophys. J., Vol. 216, 192 - 193 (1977).

The case of spacelike 4-current arising in the class of type VI_0 cosmological models with electromagnetic field discussed recently by Dunn and Tupper is considered in more detail. It is shown that the requirement that the conductivity of the plasma be positive reduces the number of possible models. When the magnetic field is nonzero, the universe is shown to have a finite time span.

162.011 **On the possible existence of cosmological cosmic rays. II. The observational constraints set by the γ-ray background spectrum and the lithium and deuterium abundances.**   T. Montmerle.
Astrophys. J., Vol. 216, 620 - 634 (1977).

The context of the cosmological cosmic ray (CCR) hypothesis was extensively discussed in an earlier paper (Montmerle 1977). In this paper the corresponding numerical results are presented and discussed by taking the $1-100$ MeV γ-ray background spectrum and the lithium abundance as observational constraints on the CCR flux. The parameters are the deceleration parameter q_0 and the CCR burst redshift z_s.

162.012 **The correlation function for density perturbations in an expanding universe. I. Linear theory.**
J. McClelland, J. Silk.
Astrophys. J., Vol. 216, 665 - 681 (1977).

In this paper the authors find analytic solutions for the evolution of linearized spherically symmetric adiabatic density perturbations and the two-point correlation function for these perturbations in the radiation-dominated portion of the early universe. These results are extended to the regime after decoupling. In an appendix the authors describe a technique which enables them to calculate the evolution of perturbations and correlation functions with very general radial dependences.

162.013   On estimating correlations in the spatial distribution of galaxies.   S. M. Fall, S. Tremaine.
Astrophys. J., Vol. 216, 682 - 689 (1977).

The authors derive an analytic inversion of Limber's equation, the basic relation between the spatial pair-correlation function ξ and its angular analog w for a magnitude-limited sample of galaxies. The inversion is very unstable owing to the large distance spread in such a sample and the deprojection of random orientation angles; the authors illustrate this with a simple model for w. They suggest a simple method for smoothing correlation data in order to estimate ξ directly and then apply the method to some recent data. Finally, they discuss what conclusions may be reasonably inferred from present angular correlation data and how they may be improved.

162.014   On the dynamics of anisotropic homogeneous generalizations of cosmological Friedmann models.
V. A. Ruban.
Zh. ehksp. i teor. fiz., Vol. 72, 1201 - 1216 (1977). In Russian. − Abstr. in Ref. zh., 51. Astron., 8.51.904 (1977).

162.015   Influence of density fluctuations upon the expansion of the universe far from the singularity.
V. V. Petrov.
Izv. vyssh. uchebn. zaved. Fiz., 1977, No. 1, p. 7 - 11. In Russian. − Abstr. in Ref. zh., 51. Astron., 8.51.905 (1977).

162.016   Viscosity effects in isotropic cosmology.
V. A. Belinskij, I. M. Khalatnikov.
Zh. ehksp. i teor. fiz., Vol. 72, 3 - 17 (1977). In Russian. Abstr. in Ref. zh., 51. Astron., 8.51.906 (1977).

162.017   Continuous parallelism and infinity of the universe.
A. P. Trofimenko.
Filos. nauk. Vyp. (No.) 8. Alma-Ata, 1976, p. 116 - 131. In Russian. − Abstr. in Ref. zh., 51. Astron., 8.51.910 (1977).

162.018   Cosmological de Sitter model in the tetrad theory of gravitation.   V. N. Tunyak.
Izv. vyssh. uchebn. zaved. Fiz., 1976, No. 6, p. 118 - 120. In Russian. − Abstr. in Ref. zh., 51. Astron., 8.51.911 (1977).

162.019   The source of the X-ray background.
T. W. Hartquist.
Astrophys. J., Vol. 217, 3 - 5 (1977).

Under the hypothesis that protoclusters of galaxies were the first objects to condense in the early universe, a sufficient mass of the universe was heated to temperatures $\gtrsim 10^7$ K to produce the observed soft X-ray background. A further observational test to distinguish between this hypothesis and the hypothesis that galaxies formed before protoclusters is suggested.

162.020   Unborn clusters.
A. Cavaliere, L. Danese, G. de Zotti.
Astrophys. J., Vol. 217, 6 - 15 (1977).

If primordial perturbations of large masses ($\gtrsim 10^{13} M_\odot$) contained a substantial component of gaseous matter when they collapsed to a virialized stage, they dissipated energy in the X-ray band while they increased their binding. This exploratory study uses the simplest structure and cosmological scenario to suggest observations in the microwave band ($\Delta T/T$ for individual large sources) and in the X-ray band (type and counts of sources), and to stress correlations with optical counts: these data can directly elicit the epoch, total mass, and time scale of the diffuse matter in the forming protoclusters.

162.021   A review of cosmologies with varying gravity.
V. Mansfield, M. Bocko.
News Lett. Astron. Soc. N.Y., Vol. 1, No. 2, p. 18 (1977). Abstract.

162.022   The correlation function for density perturbations in an expanding universe. II. Nonlinear theory.
J. McClelland, J. Silk.
Astrophys. J., Vol. 217, 331 - 352 (1977).

The authors develop a formalism to find the two-point and higher-order correlation functions for a given distribution of sizes and shapes of perturbations which are randomly placed in three-dimensional space. The perturbations are described by two parameters such as central density and size, and the two-point correlation function is explicitly related to the luminosity function of groups and clusters of galaxies.

162.023   Polarization transport in anisotropic universes.
A. M. Anile, R. A. Breuer.
Astrophys. J., Vol. 217, 353 - 357 (1977).

The authors derive and integrate the transport equations for the intensity and polarization of the microwave background radiation, including first-order corrections to geometrical optics. In the case of an axisymmetric Bianchi type I universe they show that the first-order corrections are always negligibly small. Hence polarization is essentially unchanged and the intensity is, in fact, exactly conserved (except for redshift effects) by propagation over cosmological distances under the assumption that no Thomson scattering occurs. The implications of this result for more general situations are discussed.

162.024   Cosmology and galaxy formation.
M. J. Rees.
The evolution of galaxies and stellar populations, (see 012.005), p. 339 - 360, with a discussion, p. 360 - 368 (1977).

162.025   Cosmic numbers and the rotation of the Metagalaxy.
R. M. Muradyan.
Astrofizika, Vol. 13, 63 - 67 (1977). In Russian. English translation in Astrophysics, Vol. 13, No. 1.

Stewart's and Dirac's cosmological relations are derived and a new expression for the possible angular momentum of the Metagalaxy is obtained and expressed as a combination of fundamental constants.

162.026   Nonhomogeneous large-scale magnetic field and the global structure of the universe.
A. A. Ruzmajkin, D. D. Sokolov.
Astrofizika, Vol. 13, 95 - 102 (1977). In Russian. English translation in Astrophysics, Vol. 13, No. 1.

It is proposed to connect the large-scale metagalactic magnetic field with the topological structure of the universe. It is suggested that the anisotropy of the topological joining and the field direction have a common origin; the characteristic scale of the field determines the scale of joining.

162.027   Anisotropy in the Hubble parameter and large-scale cosmological inhomogeneity.   A. J. Fennelly.
Mon. Not. R. Astron. Soc., Vol. 181, 121 - 130 (1977).

The author fits the data to a tilted self-similar cosmology with a Gaussian density distribution superposed on a homogeneous Universe. The particular model sets the inhomogeneous region's diameter at 540 Mpc, with our Galaxy 120 Mpc from the centre which is toward $\alpha = 12^h42^m$, $\delta = +20°$. This eliminates the conflict between the anomaly and the isotropy of the microwave background.

162.028 Dirac cosmology and the microwave background.
V. Canuto, S. H. Hsieh.
Astron. Astrophys., Vol. 61, L5 - L6 (1977).

It is shown that contrary to recent claims by Mansfield (1976), Dirac cosmology is not in disagreement with observations concerning QSOs.

162.029 The angular-size-redshift relation. II. A test for the deceleration parameter.   P. Hickson.
Astrophys. J., Vol. 217, 964 - 975 (1977).

An angular-size-redshift cosmological test is presented based on the projected separations between bright cluster galaxies as the size statistic. The method is applied to 95 clusters of galaxies with redshifts ranging from 0.02 to 0.46, yielding a value of the deceleration parameter before evolution of $q_0 = -0.9$. After correcting for known evolutionary effects, the author obtains $q_0 = -0.8$. The formal standard deviation in q_0 at these values is 0.2, which corresponds to an uncertainty of about 0.4 at $q_0 = 0$.

162.030 Some considerations on the thermodynamics of the universe.   H. Hönl.
Gen. Relativ. Gravitation, Vol. 8, 647 - 654 (1977).

162.031 Local supercluster and anomalous Hubble expansion.
S. Mavrides, A. Tarantola.
Gen. Relativ. Gravitation, Vol. 8, 665 - 672 (1977).

The local supercluster of galaxies (LSG) is considered as an expanding region inside a vacuole which is itself embedded in a Friedmann model. It is shown that de Vaucouleurs' data for the LSG can be accounted for by this inhomogeneous cosmological model. On the contrary, the interpretation of these observations would meet some difficulty with a theory of tired light.

162.032 A constraint on the universal baryon density from the abundance of ^{7}Li.   S. M. Austin, C. H. King.
Nature, Vol. 269, 782 (1977).

The authors point out that ^{7}Li can be used to place an upper limit on the universal baryon density, even if other production mechanisms are important, and that this limit also strongly favours an open universe.

162.033 Effect of viscosity on the evolution of the universe: Bianchi type II.   S. L. Parnovskij.
Zh. ehksp. i teor. fiz., Vol. 72, 809 - 819 (1977). In Russian.
Abstr. in Ref. zh., 51. Astron., 9.51.833 (1977).

162.034 Observational tests of antimatter cosmologies.
G. Steigman.
Annu. Rev. Astron. Astrophys., Vol. 14, (see 003.008), 339 - 372 (1976).

In approaching the problem of the amount and astrophysical role of antimatter in the Universe, it is valuable to distinguish between two separate questions: (1) Must the Universe be symmetric? (2) Is the Universe symmetric? The extent to which these questions can be, and have been, answered is the subject of this review.

162.035 Uniform relativistic models of the universe with pressure. II. Observational tests.
J. Krempeć, B. Krygier.
Acta Cosmologica, Zesz. 6, 31 - 46 (1977).

The magnitude-redshift and angular diameter-redshift relations are discussed for uniform (homogeneous and isotropic) relativistic models of the universe with pressure. The inclusion of pressure into the energy-momentum tensor has given larger values of the deceleration parameter q. An increase of the deceleration parameter has led to the brightening of objects as well as to a little larger angular diameters.

162.036 On uniform world models with matter and radiation.
E. Wojciulewitsch.
Acta Cosmologica, Zesz. 6, 65 - 74 (1977).

Some properties of a universe containing matter with density ρ_m and radiation with density ρ_r have been investigated. The use of a density parameter for matter strongly suggests the use of an analogous parameter for radiation. Both parameters are associated with deceleration and their evolution in time can be calculated. The definition of a radiation density parameter allows for a generalization of the Stabell-Refsdal classification of uniform matter universes to universes containing both matter and radiation. In this paper no interaction between matter and radiation has been assumed.

162.037 On the integration of the BBGKY equations for the development of strongly nonlinear clustering in an expanding universe.   M. Davis, P. J. E. Peebles.
Astrophys. J., Suppl. Ser., Vol. 34, 425 - 450 (1977).

This paper deals with the question of whether the observed galaxy correlation functions could have evolved out of "reasonable" initial conditions in the early universe. The evolution of density correlations in an expanding universe can be described by the BBGKY equations. The equations admit a similarity transformation if (1) the effects of the discreteness of particles can be ignored, (2) the initial spectrum of density perturbations assumes a power law shape, and (3) the universe is described by an Einstein–de Sitter model ($\Omega \approx 1$). The numerical results presented here are based on this similarity solution. The main results are the shape of the galaxy two-point correlation function $\xi(r)$ and the value of the dimensionless coupling parameter Q in the three-point function ζ. The computed Q is in good agreement with the observations. These results suggest that the velocity dispersion within a protocluster grows as it is developing as a density perturbation, so that when the cluster fragments out of the general expansion it is already "virialized".

162.038 Aether drift detected at last.
M. Rowan-Robinson.
Nature, Vol. 270, 9 - 10 (1977).

162.039 Black hole and galaxy formation in a cold early Universe.   B. J. Carr.
Mon. Not. R. Astron. Soc., Vol. 181, 293 - 309 (1977).

The only hot models of the early Universe capable of producing primordial black holes would tend to produce them too prolifically to be consistent with observation. However, if the early Universe was cold (photonless), one could expect black holes to form prolifically without contravening observation. This is a self-consistent situation in the sense that the most plausible way to heat the Universe after the hadron era is through black hole accretion.

162.040 Interpretation of source counts and redshift data in evolutionary universes.
J. V. Wall, T. J. Pearson, M. S. Longair.
IAU Symp., No. 74, (see 012.014), p. 269 - 277 (1977).

162.041 The physics of radio sources and cosmology.
P. A. G. Scheuer.
IAU Symp., No. 74, (see 012.014), p. 343 - 352 (1977).

162.042 Cosmological information from new types of radio observations.   W. C. Saslaw.
IAU Symp., No. 74, (see 012.014), p. 379 - 387 (1977).

162.043 Progress, problems and priorities: a personal view.
H. van der Laan.
IAU Symp., No. 74, (see 012.014), p. 389 - 395 (1977).
Paper concerning radio astronomy and cosmology.

162.044 Observation in a hierarchical universe.

C. C. Dyer.
J. R. Astron. Soc. Canada, Vol. 71, 405 (1977). – Abstract.

162.045 The magnitude-redshift and angular diameter-redshift relations for a matter and radiation filled Universe. B. Krygier, J. Krempeć.
Astron. Astrophys., Vol. 61, 539 - 543 (1977).

The magnitude-redshift and angular diameter-redshift relations are discussed for matter and radiation filled Universe models with non-vanishing cosmological constant. The inclusion of homogeneous pressure into the energy-momentum tensor gives larger deceleration parameters in comparison with dustlike cosmological models. It changes the slope of the magnitude-redshift and angular diameter-redshift relations and leads to brightening of distant sources and to an increase of their diameters at the same value of the density parameter.

162.046 More evidence for a closed Universe from QSOs. T. Kiang.
Nature, Vol. 270, 205 - 206 (1977). – Review article.

162.047 An evaluation of parallax in Friedmann universes. P. D. Noerdlinger.
Astrophys. J., Vol. 218, 317 - 322 (1977).

The dependence of parallax on redshift is evaluated for homogeneous Friedmann universes, following a suggestion by Weinberg that space curvature might be evaluated thereby. Although the method is valid in principle, the only interesting variations in parallax are near the antipode in closed models. The prospects for seeking secular parallax for distant sources even in such models are poor unless our velocity through the microwave background exceeds Conklin's limit, or unless it proves possible to use source structure resolved at less than 10^{-4} arcsec, over a long time base.

162.048 Evolutive mechanics. N. Ionescu-Pallas.
Rev. Roumaine Phys., Vol. 21, 1065 - 1086 (1976). Abstr. in Phys. Abstr., Vol. 80, Abstr. 17800 (1977).

162.049 A plane symmetric universe filled with disordered radiation. S. R. Roy, P. N. Singh.
J. Phys. A, Vol. 10, 49 - 54 (1977). – Abstr. in Phys. Abstr., Vol. 80, Abstr. 22335 (1977).

162.050 Linear density perturbation in relativistic and Brans-Dicke cosmologies. N. Bandyopadhyay.
J. Phys. A, Vol. 10, 189 - 195 (1977). – Abstr. in Phys. Abstr., Vol. 80, Abstr. 22336 (1977).

162.051 Non-singular cosmologies in the conformally invariant gravitation theory. A. K. Kembhavi.
Pramāna, Vol. 7, 344 - 354 (1976). – Abstr. in Phys. Abstr., Vol. 80, Abstr. 22337 (1977).

162.052 Recent theories of galaxy formation. J. R. Gott III.
Annu. Rev. Astron. Astrophys., Vol. 15, (see 003.012), 235 - 266 (1977).

Galaxy formation is an essential part of any cosmological theory. The author reviews the observational data, describes how galaxies may originate in a standard big-bang model, and considers several detailed theories of formation.

162.053 Hubble's constant determined from super-luminal radio sources. D. Lynden-Bell.
Nature, Vol. 270, 396 - 399 (1977).

Recent data on the so-called super-light expansion velocities observed in the radio galaxy 3C120 show a good fit to the light echo theory, provided the Hubble constant is 110 ± 10 km s^{-1}Mpc^{-1}. Statistics on three super-luminal sources interpreted with that theory give the same Hubble constant.

A method of determining q_0 from each source with data that may already exist, looks very promising.

162.054 Viscous phenomena and entropy production in the early universe. N. Caderni, R. Fabbri.
Phys. Lett. B, Vol. 69B, 508 - 511 (1977). – Abstr. in Phys. Abstr., Vol. 80, Abstr. 82337 (1977).

162.055 High-density and high-temperature symmetry behaviour in gauge theories. A. D. Linde.
Phys. Rev. D, Vol. 14, 3345 - 3349 (1976). – Abstr. in Phys. Abstr., Vol. 80, Abstr. 22664 (1977).

162.056 A generalized de Sitter solution. G. L. Murphy.
Phys. Lett. A, Vol. 60A, No. 1, p. 8 (1977).
Abstr. in Phys. Abstr., Vol. 80, Abstr. 26100 (1977).

162.057 Red-shifting of light passing through clusters of galaxies: a new photon property?
Z. Maric, M. Moles, J. P. Vigier.
Nuovo Cimento, Lett., Ser. 2, Vol. 18, 269 - 276 (1977).
Abstr. in Phys. Abstr., Vol. 80, Abstr. 29642 (1977).

162.058 Cosmological solutions of the mass integral formulation of general relativity. I. W. Roxburgh.
Mon. Not. R. Astron. Soc., Vol. 181, 637 - 645 (1977).

The cosmological solutions of general relativity give three isotropic homogeneous cosmological models determined by the curvature of three space ($k = 0, +1, -1$). In the mass integral formulation of Hoyle & Narlikar (1964, 1972), the differential form of the theory is identical to general relativity but because of the integral form of the mass field, these solutions must satisfy a self-consistency condition. By mapping the $k = -1$ model into the uniformly expanding Milne model the mass integral is evaluated and shown to be self-consistent. Thus this formulation of general relativity does not uniquely determine the cosmological solution.

162.059 Theoretical isochrones with decreasing gravitational constant – II. Dirac's multiplicative cosmology.
D. A. Vanden Berg.
Mon. Not. R. Astron. Soc., Vol. 181, 695 - 701 (1977).

Theoretical isochrones and luminosity functions appropriate to old stellar systems have been computed assuming Dirac's multiplicative theory, whereby $G \propto t^{-1}$ and $M \propto t^2$. Results are shown to be indistinguishable from normal (constant G, constant M) stellar evolution calculations in so far as the appearance of the colour–magnitude diagram is concerned. Comparison of theoretical results with published observations of globular clusters and old galactic clusters is presented.

162.060 The homogeneity and isotropy of the Universe. J. D. Barrow, R. A. Matzner.
Mon. Not. R. Astron. Soc., Vol. 181, 719 - 727 (1977).

The authors analyse those models of the Universe consistent with the observed isotropy, entropy, element abundances and with the existence of galaxies. The finiteness of the entropy per baryon in the Universe today, $S_b \sim 10^8$, limits the amount of dissipation that could have taken place in the past and hence the degree of irregularity allowed in the singularity structure. This observation essentially rules out the chaotic cosmology in its full generality and appears to constrain the singularity to be of simultaneous Robertson–Walker character containing only small curvature fluctuations.

162.061 On the origin of matter in the universe. II. G. Schäfer, H. Dehnen.
Astron. Astrophys., Vol. 61, 671 - 677 (1977).

Starting from the particle creation rate per volume element, which is determined by the square of the Hubble-parameter and of the rest mass of the created particles only, it is possible to extend the considerations of a preceding paper (1977) to the open universes with flat and hyperbolic 3-dimensional space. The solutions of the cosmological differential equations of Einstein's theory of gravitation are discussed, in which the created particles represent the main matter of the universe and determine its expansion.

162.062   **A tepid model for the early Universe.**
B. J. Carr, M. J. Rees.
Astron. Astrophys., Vol. 61, 705 - 709 (1977).
If the Universe started off with a photon-to-baryon ratio much less than that presently observed, massive black holes would have formed at early times even if the initial density fluctuations were very small. These holes could have generated the rest of the background radiation through accretion; in this way, such a Universe might automatically evolve to have the photon-to-baryon ratio observed today. This scenario could explain why the times of decoupling and matter-radiation equilibrium are comparable and might provide a critical density of primordial black holes; it could also produce galaxies with black "halos".

162.063   **Quantum gravitation, elements of vacuum physics and the problem of vacuum energy. II. Quantization of the Friedmann model, quantum cosmology and quantum astrophysics.**   V. G. Lapchinskij, V. A. Rubakov.
Inst. yader. issled. AN SSSR. P-0052. Moskva, 1977. 31 pp. In Russian. — From Ref. zh., 51. Astron., 10.51.954 (1977).

162.064   **Primeval entropy fluctuations and the present-day pattern of gravitational clustering.**
D. Eichler.
Astrophys. J., Vol. 218, 579 - 581 (1977).
It is argued that chaotic conditions in the universe during and prior to the epoch $(1 + z) \sim 10^4$, such as proposed by Rees, give rise to a white noise spectrum of entropy fluctuations down to a mass scale of $\sim 10^6 M_\odot$. At this scale, the baryon number density contrast is of order unity. This would account for the densities of globular clusters (if they are of primeval origin) as well as all larger systems.

162.065   **Primeval gas clouds and the low-energy X-ray background.**   E. M. Kellogg.
Astrophys. J., Vol. 218, 582 - 591 (1977).
In this paper the author considers a possible origin of the low-energy ($E \lesssim 3$ keV) background as due to fragmenting clouds of hot gas — the collapsing protoclusters forming in the early universe. Under a maximal heating model in which some of these clouds are heated up to a temperature such that their thermal pressure stabilizes them against gravitational collapse, he finds a range of temperatures, X-ray luminosities, and lifetimes against cooling. He predicts the appearance of the X-ray background on a fine angular scale for two cases: formation of the clouds at $z = 2$ and at $z = 4$. Such an ensemble of protocluster hot gas clouds will create features in the X-ray background on a small angular scale, observable with X-ray telescopes now under construction.

162.066   **Radiation damping and the expansion of the universe.**   P. C. Aichelburg, R. Beig.
Phys. Rev. D, Vol. 15, 389 - 401 (1977). — Abstr. in Phys. Abstr., Vol. 80, Abstr. 33717 (1977).

162.067   **Closure in anisotropic cosmological models.**
G. J. Galloway.
J. Math. Phys., New York, Vol. 18, 250 - 252 (1977). — Abstr. in Phys. Abstr., Vol. 80, Abstr. 37676 (1977).

162.068   **Propagators for a scalar field in some Bianchi-type I universe.**   H. Nariai.
Prog. Theor. Phys., Vol. 57, 67 - 81 (1977). — Abstr. in Phys. Abstr., Vol. 80, Abstr. 41427 (1977).

162.069   **Hydrodynamics of the Universe.**
Ya. B. Zel'dovich.
Annual Review of Fluid Mechanics, Vol. 9. M. Van Dyke, J. V. Wehausen (Editors). Annual Reviews, Palo Alto, Calif., USA. 8 + 509 pp. (1977). ISBN 0-8243-0709-7, p. 215 - 228. Abstr. in Phys. Abstr., Vol. 80, Abstr. 49516 (1977).

162.070   **Cosmology: man's place in the universe.**
V. Trimble.
American Sci., Vol. 65, No. 1, p. 76 - 86 (1977). — Abstr. in Phys. Abstr., Vol. 80, Abstr. 53715 (1977).

162.071   **Trouser-type universe metric.**   H. Ishikawa.
Prog. Theor. Phys., Vol. 57, 339 - 341 (1977).
Abstr. in Phys. Abstr., Vol. 80, Abstr. 45623 (1977).

162.072   **Le problème cosmologique et ses hypothèses. V.**
J. Dubois.
Orion, 35. Jahrg., 194 - 198 (1977).

162.073   **Toward a geometrodynamical theory of subatomic particles as singularities and a classification by pseudogroups.**   R. Grisell.
Multidisciplinary Res., Vol. 4, 60 - 96 (1976). — Abstr. in Phys. Abstr., Vol. 80, Abstr. 57111 (1977).

162.074   **Cosmologies with varying gravity.**
V. N. Mansfield.
Phys. Teach., Vol. 15, 263 - 267 (1977). — Abstr. in Phys. Abstr., Vol. 80, Abstr. 60858 (1977).

162.075   **The large numbers: a key to the Universe?**
M. Lachieze-Rey, L. Vigroux.
Recherche, Vol. 8, 166 - 167 (1977). In French. — Abstr. in Phys. Abstr., Vol. 80, Abstr. 60859 (1977).

162.076   **The end of time.**   B. Parker.
Astronomy, Vol. 5, No. 5, p. 6 - 17 (1977). — Abstr. in Phys. Abstr., Vol. 80, Abstr. 64890 (1977).

162.077   **Polytropic universe.**   R. Chattopadhyay.
Indian J. Phys., Vol. 50, 959 - 960 (1976). — Abstr. in Phys. Abstr., Vol. 80, Abstr. 64891 (1977).

162.078   **Thermodynamics and cosmology.**
J. Abellan, A. Navarro, E. Alvarez.
J. Phys. A, Vol. 10, L129 - L130 (1977). — Abstr. in Phys. Abstr., Vol. 80, Abstr. 64892 (1977).

162.079   **Static inhomogeneous cosmological model.**
M. Novello.
Phys. Lett. A, Vol. 61A, 293 - 294 (1977). — Abstr. in Phys. Abstr., Vol. 80, Abstr. 67147 (1977).

162.080   **Singularity avoidance and quantum conformal anomalies.**   P. C. W. Davies.
Phys. Lett. B, Vol. 68B, 402 - 404 (1977). — Abstr. in Phys. Abstr., Vol. 80, Abstr. 67148 (1977).

162.081   **Torsion singularities.**   J. M. Nester, J. Isenberg.
Phys. Rev. D, Vol. 15, 2078 - 2087 (1977). — Abstr. in Phys. Abstr., Vol. 80, Abstr. 67149 (1977).

162.082   **Magnetohydrodynamic type-I cosmologies.**
B. O. J. Tupper.
Phys. Rev. D, Vol. 15, 2123 - 2124 (1977). — Abstr. in Phys. Abstr., Vol. 80, Abstr. 67150 (1977).

162.083   **Exact Bianchi IV cosmological model.**
A. Harvey, D. Tsoubelis.
Phys. Rev. D, Vol. 15, 2734 - 2737 (1977). — Abstr. in Phys. Abstr., Vol. 80, Abstr. 67151 (1977).

162.084 **Cosmological event horizons, thermodynamics, and particle creation.** G. W. Gibbons, S. W. Hawking.
Phys. Rev. D, Vol. 15, 2738 - 2751 (1977). – Abstr. in Phys. Abstr., Vol. 80, Abstr. 67152 (1977).

162.085 **Limits on masses and number of neutral weakly interacting particles.** P. Hut.
Phys. Lett. B, Vol. 69B, 85 - 88 (1977). – Abstr. in Phys. Abstr., Vol. 80, Abstr. 71000 (1977).

162.086 **The universe as Bólyai-Lobachevsky velocity space.** S. J. Prokhovnik.
Acta Phys. Acad. Sci. Hungaricae, Vol. 41, No. 3, p. 201 - 209 (1976). – Abstr. in Phys. Abstr., Vol. 80, Abstr. 71156 (1977).

162.087 **A conservative explanation of a Machian argument.** M. Reinhardt.
Phys. Lett. A, Vol. 62A, 62 (1977). – Abstr. in Phys. Abstr., Vol. 80, Abstr. 74168 (1977).

162.088 **Graviton viscosity in the early Universe.** G. L. Murphy.
Phys. Lett. A, Vol. 62A, 75 - 77 (1977). – Abstr. in Phys. Abstr., Vol. 80, Abstr. 74169 (1977).

162.089 **The cosmological term and a modified Brans-Dicke cosmology.** M. Endō, T. Fukui.
Gen. Relativ. Gravitation, Vol. 8, 833 - 839 (1977).
Adding the cosmological term Λ, which is assumed to be variable in this paper, to the Brans-Dicke Lagrangian, the authors try to understand the meaning of the term and to relate it to the mass of the universe. They also touch upon the Dirac large-number hypothesis, applying the results obtained from the application of their theory to a uniform cosmological model.

162.090 **The de Sitter universe and mechanics.** G. Arcidiacono.
Gen. Relativ. Gravitation, Vol. 8, 865 - 870 (1977).
If one studies the de Sitter universe with the methods of projective geometry, one obtains a new mechanics valid on a cosmic scale and for hyperdense matter. In this projective mechanics the mass varies with the space-time distance, and the linear and angular momentum are reunited in a single projective tensor.

162.091 **Remarks on the impact of photon-scalar boson scattering on Planck's radiation law and Hubble effect.** M. Moles, J. P. Vigier.
Astron. Nachr., Band 298, 289 - 290 (1977).
The paper contains a reply to Treder's (see 19.162.002) argument that the existence of a third scalar mode of photons contradicts the laws of heat radiation.

162.092 **A remark on the paper "Remarks on the impact of photon-scalar boson scattering on Planck's radiation law and Hubble effect".** H.-J. Treder.
Astron. Nachr., Band 298, 291 (1977).

162.093 **A class of inhomogeneous perfect fluid cosmologies.** D. A. Szafron, J. Wainwright.
J. Math. Phys., Vol. 18, 1668 - 1672 (1977). – Abstr. in Phys. Abstr., Vol. 80, Abstr. 74284 (1977).

162.094 **Inhomogeneous cosmologies: new exact solutions and their evolution.** D. A. Szafron.
J. Math. Phys., Vol. 18, 1673 - 1677 (1977). – Abstr. in Phys. Abstr., Vol. 80, Abstr. 74285 (1977).

162.095 **Effects of a nonvanishing cosmological constant on the spherically symmetric vacuum manifold.** K. Lake, R. C. Roeder.
Phys. Rev. D, Vol. 15, 3513 - 3519 (1977). – Abstr. in Phys. Abstr., Vol. 80, Abstr. 74288 (1977).

162.096 **Generalized Doppler formula in a nonstatic universe.** P. G. Gross.
American J. Phys., Vol. 45, 642 - 644 (1977). – Abstr. in Phys. Abstr., Vol. 80, Abstr. 78900 (1977).

162.097 **Stress tensor and conformal anomalies for massless fields in a Robertson-Walker universe.** T. S. Bunch, P. C. W. Davies.
Proc. R. Soc. London, Ser. A, Vol. 356, 569 - 574 (1977). Abstr. in Phys. Abstr., Vol. 80, Abstr. 78903 (1977).

162.098 **Covariant point-splitting regularization for a scalar quantum field in a Robertson-Walker universe with spatial curvature.** T. S. Bunch, P. C. W. Davies.
Proc. R. Soc. London, Ser. A, Vol. 357, 381 - 394 (1977). Abstr. in Phys. Abstr., Vol. 80, Abstr. 82434 (1977).

162.099 **Does the universe oscillate?** J. Gribbin.
Astronomy, Vol. 5, No. 8, p. 50 - 55 (1977). – Abstr. in Phys. Abstr., Vol. 80, Abstr. 87579 (1977).

162.100 **Vacuum instability and the inhomogeneous distribution of celestial objects.** L.-Z. Fang.
Kexue Tongbao, Vol. 22, No. 6, p. 258 - 262 (1977). In Chinese. – Abstr. in Phys. Abstr., Vol. 80, Abstr. 87581 (1977).

162.101 **The de Sitter-Castelnuovo universe and cosmology.** G. Arcidiacono.
Urania Barcelona, Año 61, Núm. 285, p. 3 - 16 (1976).

162.102 **The Robertson-Walker metric and the symmetries belong to the family of contracted Ricci collineations.** L. H. Green, L. K. Norris, D. R. Oliver, Jr., W. R. Davis.
Gen. Relativ. Gravitation, Vol. 8, 731 - 736 (1977).
It is shown that the general form of the Robertson-Walker cosmological metric admits symmetry properties that are members of the symmetry family of contracted Ricci collineations.

162.103 **On the cosmical constant.** R. Chandra.
Gen. Relativ. Gravitation, Vol. 8, 787 - 793 (1977).
On the grounds of the two correspondence limits, the Newtonian limit and the special theory limit of Einstein field equations, a modification of the cosmical constant has been proposed which gives realistic results in the case of a homogeneous universe. Also, according to this modification an explanation for the negative pressure in the steady-state model of the universe has been given.

162.104 **Problems of matter-antimatter boundary layers.** B. Lehnert.
Astrophys. Space Sci., Vol. 46, 61 - 71 (1977).
This paper outlines the problems of the quasi-steady matter-antimatter boundary layers discussed in Klein-Alfvén's cosmological theory, and a crude model of the corresponding ambiplasma balance is presented.

162.105 **Significance of a conclusive test of Dirac's Large Numbers hypothesis using precision ranging to Mars.** J. L. Hughes.
Astrophys. Space Sci., Vol. 46, L15 - L18 (1977).
It is now possible to test Dirac's Large Numbers hypothesis in the Earth–Mars system using microwave ranging techniques such as those associated with the Viking lander. The consequences of such a test are discussed.

162.106 **On the influence of massless scalar and vector elec-**

tromagnetic fields on the singularity character in anisotropic cosmology.   V. A. Ruban.
Astrophys. Space Sci., Vol. 46, L23 - L28 (1977).

It is shown that, for the scalar-tensor cosmology by Jordan-Brans-Dicke, in general anisotropic solution the oscillatory 'mixmaster' regime near the singularity will be destroyed by the scalar source-free field and replaced by monotonous V_3 − collapse into the 'point' or into the 'line' and 'plane' (only in case $G \to 0$) even in the presence of the primordial source-free electromagnetic field.

162.107 Does the speed of light decrease with time?
S. Bellert.
Astrophys. Space Sci., Vol. 47, 263 - 276 (1977).

Rust (1974) stated that the 'classical' (e.g., Doppler) explanations of the cosmological redshift contradict the results of astronomical observations of the period of changes in the brightness of supernovae. This paper is an attempt at explaining this discrepancy between observations and the theoretical predictions on the grounds of a hypothesis published by the author (Bellert, 1969). That hypothesis explains the cosmological redshift by the geometry of the space of events, which is a static space.

162.108 Entropy per baryon in a 'many-worlds' cosmology.
M. Clutton-Brock.
Astrophys. Space Sci., Vol. 47, 423 - 433 (1977).

One imagines the universe split into infinitely many branches, or 'worlds', only one of which one can observe. Our world has an entropy per baryon $\xi \sim 10^9$: other worlds can have all possible values of entropy per baryon. High-entropy worlds do not form galaxies, but only giant black holes. Low entropy worlds do form galaxies, but only metal-poor dwarf galaxies with no planets.

162.109 Pregalactic nucleosynthesis.
T. W. Hartquist, A. G. W. Cameron.
Astrophys. Space Sci., Vol. 48, 145 - 158 (1977).

The authors discuss the behavior of density fluctuations in an expanding universe and show that these should lead to the early formation of pregalactic hydrogen-helium stars of several hundred to several thousand solar masses. These stars flood the universe with radiation having a color temperature $\gtrsim 10^5$ K; this terminates star formation but permits galaxy formation to continue. About 10^{-2} of the mass of the galaxies is converted into heavy elements by pregalactic nucleosynthesis, with an error factor of a few.

162.110 Galaxy clusters as relativistic spherically-symmetrical inhomogeneities.   D. Trevese, A.Vignato.
Astrophys. Space Sci., Vol. 49, 229 - 240 (1977).

Spherically symmetric exact solutions of the Einstein equations in the absence of pressure are used to construct a dynamical model for the clusters of galaxies, embedded in a parabolic Universe. Radial motions only are responsible for velocity dispersion. The density profile, which shows a secondary maximum, and the variation of the velocity dispersion versus the distance from the cluster centre, are in qualitative agreement with the observations. The problem of the 'missing mass' is also partially overcome.

162.111 Comment on Varshni's recent paper on quasar red shifts.   C. B. Stephenson.
Astrophys. Space Sci., Vol. 51, 117 - 119 (1977).

A recent paper (Varshni, 1976) analyses the distribution of quasar red shifts for randomness, in an incorrect manner. A correct analysis shows that this distribution is in agreement with random expectation. Were the distribution highly non-random, the original conclusion was that, for the red shifts to be cosmological, the Earth would have a strongly privileged position in the Universe. A simple alternative model, in which this would not be so, is pointed out.

162.112 The red-shift hypothesis for quasars: is the Earth the center of the Universe? II.   Y. P. Varshni.
Astrophys. Space Sci., Vol. 51, 121 - 124 (1977).

It is pointed out that Stephenson (1977) has used incorrect Δz, and has also made an arithmetical error, which invalidate his claims. Tests for randomness of quasar red-shifts clusters, using correct Δz, have been carried out and it is shown that at least for clusters having three red shifts or more, the distribution is highly non-random. The model of the Universe proposed by Stephenson does not in any way explain these red-shift clusters; it merely substitutes one paradox by another.

162.113 A variable-G cosmological model.
C. W. Tittle.
Proc. Southwest Reg. Conf., Vol. 3, (see 012.043), p. 117 - 121 (1977).

A new cosmological model is based on the Einstein-Thirring inertial interaction of general relativity, but the gravitational constant G is allowed to vary in time so that integration over the effective universe produces the correct inertial mass of a particle. Inertial and gravitational mass are assumed to be equal. If the deceleration parameter $q_0 = 0.04$, the predicted d $(G/G_0)/dt = -1.4 \times 10^{-11}$ yr^{-1}. This compares with the average of the two recent astronomical determinations by Van Flandern and by Muller, viz., $(-2.6 \pm 1.2) \times 10^{-11}$ yr^{-1}.

162.114 Evidence for the fundamental role of Planck units in cosmology.   A. Sapar.
Publ. Tartu Astrofiz. Obs., Vol. 45, 204 - 210 (1977).

It has been demonstrated that the Planck units obtained from the fundamental constants c, G, $\hbar$ and k apparently have a fundamental meaning for cosmology, because they coincide with physical parameters of the universe.

162.115 Cosmological lower bound on heavy-neutrino masses.   B. W. Lee, S. Weinberg.
Phys. Rev. Lett., Vol. 39, 165 - 168 (1977).

162.116 Cosmological upper bound on heavy-neutrino lifetimes.   D. A. Dicus, E. W. Kolb, V. L. Teplitz.
Phys. Rev. Lett., Vol. 39, 168 - 171, with an erratum p. 973 (1977).

162.117 Effective-potential approach to graviton production in the early universe.   J. B. Hartle.
Phys. Rev. Lett., Vol. 39, 1373 - 1376 (1977).

162.118 Anisotropic uniform model universes with cosmical constant. I. Flat space.   K. Hara.
Publ. Astron. Soc. Japan, Vol. 29, 753 - 763 (1977).

Solutions of Einstein's field equations with cosmical constant are derived for a flat axially-symmetrical uniform model universe containing only dustlike matter. They are classified by means of the transverse deceleration parameter q_0 and the transverse density parameter σ_0. The age of the universe is computed and illustrated by contours on the (q_0, σ_0) plane.

162.119 On the relativity of spatial and temporal finiteness and infiniteness of a universe filled with matter.
A. L. Zel'manov.
Astron. Zh. Akad. Nauk SSSR, Vol. 54, 1168 - 1181 (1977). In Russian. English translation in Soviet Astron., Vol. 21, No. 6.

One of Friedmann's models (the model with hyperbolic co-moving space filled with dust-like matter, the cosmological constant being zero) is considered in two non co-moving frames of reference. A comparison of the properties and behaviour of the model in three frames of reference (including the co-moving one) shows that such features of an expanding universe as infiniteness or finiteness of the volume of space, of the amount of mass, of the number of particles are not invariant: these quantities are finite in some frames of reference and infinite in others. The same is true for the time elapsed from the beginning of the expansion epoch.

162.120   On the relativity of spatial finiteness and infiniteness of some Friedmann cosmological models.
L. I. Kharbediya.
Astron. Zh. Akad. Nauk SSSR, Vol. 54, 1182 - 1187 (1977). In Russian. English translation in Soviet Astron., Vol. 21, No. 6.

In the present paper, in which Zelmanov's work has been used and continued, new examples of the relativity of spatial finiteness and infiniteness for the case of Friedmann cosmological models are given.

162.121   Hubble-Konstante und Hubble-Fluß.
G. A. Tammann.
Mitt. Astron. Ges., Nr. 42, p. 42 - 58 (1977).

The extragalactic distance scale is reviewed. The distance indicators within the Local Group are discussed as well as the calibration and application of other distance indicators with wider range (brightest stars, largest H II regions, luminosity classes of spiral galaxies, 21 cm-profiles). They lead to a value of $H_0 = 55 \pm 7$ km s^{-1} Mpc^{-1}. This value is strongly and independently supported by supernovae, globular clusters, the size of the Galaxy and M 31, and brightest cluster galaxies.

162.122   On the role of dissipative processes in the lepton stage of the expansion of the universe.
V. E. Yakimov.
Probl. teor. gravitatsii i ehlem. chastits. Vyp. (No.) 8. Moskva, Atomizdat, 1977, p. 146 - 150. In Russian. — Abstr. in Ref. zh., 51. Astron., 12.51.723 (1977).

162.123   On the dynamics of anisotropic homogeneous cosmological models. I.   V. A. Ruban.
Leningr. inst. yader. fiz. AN SSSR. Prepr. No. 327, maj 1977. Leningrad, 1977. 42 pp. In Russian. — Abstr. in Ref. zh., 51. Astron., 12.51.729 (1977).

162.124   On problems of quantum cosmology.
Yu. N. Barabanenkov, V. N. Mel'nikov.
Probl. teor. gravitatsii i ehlem. chastits. Vyp. (No.) 8. Moskva, Atomizdat, 1977, p. 85 - 103. In Russian. — Abstr. in Ref. zh., 51. Astron., 12.51.735 (1977).

162.125   Cosmological restrictions of the masses of neutral leptons.   M. I. Vysotskij, A. D. Dolgov, Ya. B. Zel'dovich.
Pis'ma v ZhEhTF, Vol. 26, 200 - 202 (1977). In Russian. Abstr. in Ref. zh., 51. Astron., 1.51.780 (1978).

162.126   Quantum effects and the Friedmann model.
V. Ts. Gurovich.
Zh. ehksp. i teor. fiz., Vol. 73, 369 - 376 (1977). In Russian. Abstr. in Ref. zh., 51. Astron., 1.51.785 (1978).

162.127   Cosmology in the Einstein-Cartan theory.
V. G. Krechet, V. N. Ponomarev.
Novoe v teor. otnositel'n. i gravitatsii. 1975 g. Moskva, Nauka, 1977, p. 66 - 70. In Russian. — Abstr. in Ref. zh., 51. Astron., 1.51.796 (1978).

162.128   The problem of the space-time infinity of the universe.   E. Skarżyński.
Postępy Astron., Tom 25, 233 - 244 (1977). In Polish.

Removing the singularity in Cartan-Einstein's cosmology is a serious argument for the "straight-line" time model. Global isotropy of the background radiation speaks in favour of the simplest models with Robertson-Walker metric. In this review the author discusses various problems concerning infinity of time and space.

162.129   About the origin of galaxies.   B. Onderlička.
Kozmos, Vol. 8, 100 - 105 (1977). In Czech.

162.130   Expansion of the universe and the Hubble constant.
O. Obůrka.
Říše hvězd, Vol. 58, 230 - 233 (1977). In Czech.

Cosmology: list of publications.
See Abstr. 002.031.

Cosmology now.   See Abstr. 003.005.

Principles of cosmology and gravitation.
See Abstr. 003.032.

Cosmology + 1.   See Abstr. 003.060.

Mathematics and the universe.
See Abstr. 003.088.

Tensors, relativity and cosmology.
See Abstr. 003.091.

Space, time, and gravity. The theory of the big bang and black holes.   See Abstr. 003.163.

The dark night sky paradox.
See Abstr. 004.042.

Dunkle Regenten als Vorläufer Schwarzer Löcher.
See Abstr. 004.081.

Décalages vers le rouge et expansion de l'univers — l'évolution des galaxies et ses implications cosmologiques.
See Abstr. 012.053.

Laser ranging techniques required to test Dirac's cosmological model.   See Abstr. 031.245.

More on big-bang nucleosynthesis with nonzero lepton numbers.   See Abstr. 061.013.

On the possible existence of cosmological cosmic rays. III. Nuclear γ-ray production.
See Abstr. 061.014.

Antimatter.   See Abstr. 061.033.

Variation of elements in nature.
See Abstr. 061.038.

Possibilities of experiments with high-energy cosmic neutrinos: project DUMAND (*Deep Underwater Muon Neutrino Detection*).   See Abstr. 061.045.

Peculiarities of chemical composition of prestellar matter.   See Abstr. 061.046.

Sur une interprétation physique de la constante cosmologique.   See Abstr. 066.011.

Models of the universe containing matter and gravitational radiation.   See Abstr. 066.045.

An upper limit to the cosmological variation of Planck's constant derived from the spectrum of the microwave background radiation.   See Abstr. 066.079.

Microwave background spectrum — survey of recent results.   See Abstr. 066.082.

Small scale fluctuations in the microwave back-

ground radiation associated with the formation of galaxies.
See Abstr. 066.083.

Fluctuations in the microwave background caused by anisotropy of the universe and gravitational waves. See Abstr. 066.084.

Limits on a microwave background without the big bang.   See Abstr. 066.089.

Conservation laws in de Sitter space from action principles in five-space.   See Abstr. 066.110.

Quantum vacuum energy in a closed universe. See Abstr. 066.123.

On particle detection in the de Sitter space-time. See Abstr. 066.141.

Properties of a solution of the Einstein equations with the cosmological constant.   See Abstr. 066.144.

Scalar-particle production near the singularity in an anisotropic universe. I. Scalar field theory. See Abstr. 066.159.

Symmetry breaking in superdense matter. See Abstr. 066.193.

Characterization of the Szekeres inhomogeneous cosmologies as algebraically special spacetimes. See Abstr. 066.198.

Special exact solutions of Einstein's equations – a theorem and some observations.   See Abstr. 066.208.

Vacuum stress tensor in an Einstein universe: finite-temperature effects.   See Abstr. 066.210.

Black holes in closed universes. See Abstr. 066.211.

Pedagogical trick for general relativity. See Abstr. 066.260.

The Kerr metric in cosmological background. See Abstr. 066.282.

Cosmological constant in supergravity. See Abstr. 066.283.

Singular space-times.   See Abstr. 066.287.

Causally symmetry spacetimes. See Abstr. 066.291.

Spinorial structures and electromagnetic hyper-heavens.   See Abstr. 066.292.

Space-time structure in a generalisation of gravitation theory.   See Abstr. 066.295.

Spatially homogeneous Lichnerowicz universes. See Abstr. 066.323.

Detection of anisotropy in the cosmic blackbody radiation.   See Abstr. 066.335.

The solar neutrino experiment as a means for testing cosmological and cosmogonic hypotheses. Abstract. See Abstr. 080.047.

The planets, the Universe, and cosmic evolution. See Abstr. 107.027.

The helium problem.   See Abstr. 131.107.

Applications of statistical techniques to the angular size–flux density relation for extragalactic radio sources. See Abstr. 141.007.

Ultraviolet spectrum of quasi-stellar object 3C 273. See Abstr. 141.019.

On the redshift–apparent diameter relation for radio sources.   See Abstr. 141.027.

Extragalactic radio sources and cosmology. See Abstr. 141.037.

The statistical analysis of anisotropies. See Abstr. 141.088.

Radio source angular sizes and cosmology. See Abstr. 141.091.

The angular size – flux density relation. See Abstr. 141.092.

The flux density – angular size distribution for extragalactic radio sources.   See Abstr. 141.093.

The angular diameter-redshift relation for scintillating radio sources.   See Abstr. 141.095.

Interpretation of cosmological information on radio sources.   See Abstr. 141.104.

Cosmological interpretation of redshift data on quasars through the V/V_{max} test.   See Abstr. 141.105.

The redshift-magnitude relation for radio galaxies. See Abstr. 141.106.

The Hubble diagrams for quasars. See Abstr. 141.107.

Young galaxies, quasars and the cosmological evolution of extragalactic radio sources. See Abstr. 141.109.

On the $\ln(1+z)$ periodicity in QSO redshifts. See Abstr. 141.147.

Quasars to date.   See Abstr. 141.148.

On the angular size-redshift test for double radio sources.   See Abstr. 141.151.

Clumping of quasar redshifts and the geocentric universe.   See Abstr. 141.154.

On the spinup/spindown of pulsar JP 1953+29. See Abstr. 141.545.

Massive galactic halos. I. Formation and evolution. See Abstr. 151.045.

Numerical experiments on the effect of massive central systems on the virial binding energy of clusters. See Abstr. 151.063.

$BiL(E)$ and the Hubble sequence: a new theory of galaxies.   See Abstr. 151.064.

Galaxies and entropy from nonlinear fluctuations: a simple wave analysis.   See Abstr. 151.083.

Continuous increase of Hubble modulus behind clusters of galaxies.   See Abstr. 160.009.

The angular size-redshift relation. I. Sizes and shapes of nearby clusters of galaxies.   See Abstr. 160.017.

The mass-to-light ratio of late-type binary galaxies: luminosity- versus number-weighted averages. See Abstr. 160.018.

Rotation measures and cosmology. See Abstr. 161.005.

On the intergalactic contribution to the rotation measures of QSO's.   See Abstr. 161.009.

Errata

162.901   Erratum: 'Galaxy counts, color-redshift relations, and related quantities as probes of cosmology and galactic evolution' [Astrophys. J., Vol. 211, 621 - 637 (1977)]. B. M. Tinsley. Astrophys. J., Vol. 216, 349 - 350 (1977).

162.902   Erratum: 'Toward the resolution of the local missing mass problem' [Astron. J., Vol. 82, 195 - 197 (1977)].   K. Krisciunas. Astron. J., Vol.82, 662 (1977).

Author Index

The authors are listed in alphabetical order

according to the initial letter following the first names.

Subject Index

Starting with Volume 18 of Astronomy and Astrophysics Abstracts, some alterations concerning formation, arrangement, and versatility of the key words have been made. In order to provide an adequate description of a paper, specific key words are used as frequently as possible. References to a whole subject category are suppressed now. The user, therefore, has to refer to the contents at the beginning of each volume.

Whenever possible, the key words are formed in such a way that there are two different supplementary terms, e.g. the pair

interstellar matter

molecules.

An effort is made to choose preferably terms which can be inverted in order to increase the usefulness of this index. In the example given there are the two entries

interstellar matter

molecules

and

molecules

interstellar matter.

Exceptions to the rule of inversion of terms are given in all cases where the second key word is either a very specific one (e.g. Urca processes) or a general one (e.g. history). The use of substantives is preferred. In order to obtain the possibility to extend a one-term key word in a two-term one, combinations as

Mars   or   sun

atmosphere   active regions

are changed into

Mars atmosphere   and   solar active regions,

respectively. The number of cross references indicating such slightly different entries is reduced drastically. In previous volumes combinations like

close binaries   and   binaries

close binaries,

peculiar A stars   and   A stars

peculiar,

groups of galaxies   and   galaxies

groups

have been used. Now only the specific key words

close binaries

peculiar A stars

groups of galaxies

have to be considered as a substitute.

The user is requested to look for more specialized entries, as further references to this topic might exist elsewhere in the index under another current astronomical term.

ASTRONOMY AND ASTROPHYSICS ABSTRACTS

A Publication of the Astronomisches Rechen-Institut Heidelberg

Member of the Abstracting Board
of the International Council of Scientific Unions

Editors:
S. Böhme, U. Esser, W. Fricke, I. Heinrich, D. Krahn,
L. D. Schmadel, G. Zech

Published for the Astronomisches Rechen-Institut by
Springer-Verlag Berlin Heidelberg New York

Beam-Foil Spectroscopy

Editor: S. Bashkin

1976. 91 figures. XIII, 318 pages
(Topics in Current Physics, Volume 1)
ISBN 3-540-07914-9

Contents:
S. Bashkin: Experimental Methods. – *I. Martinson:*
Studies of Atomic Spectra by the Beam-Foil
Method. – *L.J. Curtis:* Lifetime Measurements. –
O. Sinanoglu: Theoretical Oscillator Strengths of
Neutral, Singly-Ionized, and Multiply-Ionized
Atoms: The Theory, Comparisons with Experiment,
and Critically-Evaluated Tables with New Results. –
W. Wiese: Regularities of Atomic Oscillator Strengths
in Isoelectronic Sequences. – *W. Whaling:* Appli-
cations to Astrophysics: Absorption Spectra. –
L. Heroux: Applications of Beam-Foil Spectroscopy to
the Solar Ultraviolet Emission Spectrum. – *R. Marrus:*
Studies of Hydrogen-Like and Helium-Like Ions of
High Z. – *J. Macek, D. Burns:* Coherence, Aligment,
and Orientation Phenomena in the Beam-Foil Light
Source. – *I. A. Sellin:* The Measurement of
Autoionizing Ion Levels and Lifetimes by Fast
Projectile Electron Spectroscopy.

High-Resolution
Laser Spectroscopy

Editor: *K. Shimoda*

1976. 132 figures. XIII, 378 pages
(Topics in Applied Physics, Volume 13)
ISBN 3-540-07719-7

Contents:
K. Shimoda: Introduction. – *K. Shimoda:* Line
Broadening and Narrowing Effects. – *P. Jacquinot:*
Atomic Beam Spectroscopy. – *V.S. Letokhov:* Satu-
ration Spectroscopy. – *J.L. Hall, J.A. Magyar:* High
Resolution Saturated Absorption Studies of Methane
and Some Methyl-Halides. – *V.D. Chebotayev:* Three-
Level Laser Spectroscopy. – *S. Haroche:* Quantum
Beats and Time-Resolved Fluorescence Spectros-
copy. – *N. Bloembergen, M.D. Levenson:* Doppler-
Free Two-Photon Absorption Spectroscopy.

Laser Monitoring
of the Atmosphere

Editor: *E.D. Hinkley*

1976. 84 figures. XV, 380 pages
(Topics in Applied Physics, Volume 14)
ISBN 3-540-07743-X

Contents:
E.D. Hinkley: Introduction. – *S.H. Melfi:* Remote
Sensing for Air Quality Management. – *V.E. Zuev:*
Laser-Light Transmission through the Atmosphere. –
R.T.H. Collis, P.B. Russell: Lidar Measurement of
Particles and Gases by Elastic Backscattering and
Differential Absorption. – *H. Inaba:* Detection of
Atoms and Molecules by Raman Scattering and
Resonance Fluorescence. – *E.D. Hinkley, R.T. Ku,*
P. L. Kelley: Techniques for Detection of Molecular
Pollutants by Absorption of Laser Radiation. –
R.T. Menzies: Laser Heterodyne Detection
Techniques.

Laser Speckle and
Related Phenomena

Editor: *J.C. Dainty*

1975. 133 figures. XIII, 286 pages
(Topics in Applied Physics, Volume 9)
ISBN 3-540-07498-8

Contents:
J.C. Dainty: Introduction. – *J.W. Goodman:* Statisti-
cal Properties of Laser Speckle Patterns. – *G. Parry:*
Speckle Patterns in Partially Coherent Light. –
T.S. McKechnie: Speckle Reduction. – *M. Françon:*
Information Processing Using Speckle Patterns. –
A.E. Ennos: Speckle Interferometry. – *J.C. Dainty:*
Stellar Speckle Interferometry.

Springer-Verlag
Berlin
Heidelberg
New York